COMPUTER METHODS AND ADVANCES IN GEOMECHANICS
VOLUME 2

PROCEEDINGS OF THE NINTH INTERNATIONAL CONFERENCE
ON COMPUTER METHODS AND ADVANCES IN GEOMECHANICS
WUHAN / CHINA / 2-7 NOVEMBER 1997

Computer Methods and Advances in Geomechanics

Edited by

JIAN-XIN YUAN

Institute of Rock and Soil Mechanics, The Chinese Academy of Sciences, Wuhan, China

VOLUME 2

6 *Determination of parameters and data analysis;* 7 *Constitutive modeling;* 8 *Flow and consolidation;* 9 *Coupled thermo-hydro-mechanical analysis;* 10 *Disaster and geo-environmental engineering;* 11 *Tunnel engineering;* 12 *Underground opening and mining engineering*

A.A. BALKEMA / ROTTERDAM / BROOKFIELD / 1997

The texts of the various papers in this volume were set individually by typists under the supervision of each of the authors concerned.

Published by
A. A. Balkema, P.O. Box 1675, 3000 BR Rotterdam, Netherlands (Fax: +31.10.413.5947)
A. A. Balkema Publishers, Old Post Road, Brookfield, VT 05036-9704, USA (Fax: 802.276.3837)

For the complete set of four volumes, ISBN 90 5410 904 1
For Volume 1, ISBN 90 5410 905 X
For Volume 2, ISBN 90 5410 906 8
For Volume 3, ISBN 90 5410 907 6
For Volume 4, ISBN 90 5410 908 4

Printed in the Netherlands

Computer Methods and Advances in Geomechanics, Yuan (ed.) © *1997 Balkema, Rotterdam, ISBN 90 5410 904 1*

Scheme of the work

VOLUME 1

Keynote papers;
Invited papers;
1. Numerical method, analytical and semi-analytical methods;
2. Discontinuous deformation analysis and discrete element method;
3. Reliability, fuzzy and random problems;
4. Artificial neural networks;
5. Mechanical properties.

VOLUME 2

6. Determination of parameters and data analysis;
7. Constitutive modeling;
8. Flow and consolidation;
9. Coupled thermo-hydro-mechanical analysis;
10. Disaster and geo-environmental engineering;
11. Tunnel engineering;
12. Underground opening and mining engineering.

VOLUME 3

13. Slope stability analysis;
14. Dam and embankment;
15. Damage and progressive failure;
16. Joints and excavation;
17. Shallow foundation and soil-structure interaction;
18. Pile and pile foundation;
19. Dynamics and cyclic loading;
20. Laboratory and field testing;
21. Ground improvement.

Computer Methods and Advances in Geomechanics, Yuan (ed.) © *1997 Balkema, Rotterdam, ISBN 90 5410 904 1*

Table of contents

6 *Determination of parameters and data analysis*

7 *Constitutive modeling*

8 *Flow and consolidation*

9 *Coupled thermo-hydro-mechanical analysis*

11 *Tunnel engineering*

12 *Underground opening and mining engineering*

6 Determination of parameters and data analysis

Computer Methods and Advances in Geomechanics, Yuan (ed.) © 1997 Balkema, Rotterdam, ISBN 90 5410 904 1

A statistical approach on geotechnical parameter estimation for underground structures

I.M.Lee
Department of Civil Engineering, Korea University, Seoul, Korea

D.H.Kim
Geospace E.C. Team, Kolon Construction Co., Ltd, Korea

K.Y.Lo
Department of Civil Engineering, The University of Western Ontario, Ont., Canada

ABSTRACT: Developed in the present study is a new feedback system for underground structures in which the field measurements and the prior information are systematically combined together. The Extended Bayesian Method (EBM) was adopted for the back analysis and the finite element analysis was used to predict the ground response. The best mathematical model was chosen using the EBM along with the Akaike Information Criterion (AIC) which had been theoretically developed by Akaike (1973). The proposed method was applied to an actual tunnel site in Korea and shown to be highly effective in actual field problems.

1. INTRODUCTION

One of the most important and difficult tasks in designing underground structure is the accurate estimation of geotechnical parameters. In order to minimize the uncertainties in the parameter estimation, field measurements obtained during construction stage are compared with initially estimated ground properties. This process, known as the feedback analysis, can be used to obtain the optimized geotechnical parameters.

Bayesian approach usually has an advantage of dealing with both the prior information and the measurement data (Cividini et al., 1983). In the conventional Bayesian approach, the objective function is composed of two components,

$$J(\theta) = J_o(\theta) + J_p(\theta) \tag{1}$$

where, $J_o(\theta)$ and $J_p(\theta)$ are the observed and predicted objective functions, respectively. A significant drawback of the Bayesian method is the incommensurate matching between the two components.

To overcome this, Neuman and Yakowitz (1979) introduced the adjusting positive scalar β term, which adjusts the weights of $J_o(\theta)$ and $J_p(\theta)$:

$$J(\theta) = J_o(\theta) + \beta J_p(\theta) \tag{2}$$

This concept is called 'Extended Bayesian Method'. The main advantage of the EBM is the introduction of the hyperparameter, β to overcome the difficulty of incommensurate matching between the prior informations and the observations (Honjo et al., 1994a, Honjo et al., 1994b). The β is influenced by uncertainties from both informations and by sensitivity of each parameter. The small β means that the observation data are gaining more weight than initially estimated information, and vice versa. Therefore, by observing the β, we can easily see relative importance of information in feedback analysis.

2. FORMATION OF FEEDBACK SYSTEM

2.1 *Model identification*

Most of the techniques for parameter estimation could give only the estimated model parameters for a given model, and they do not provide any information on the selection of the most appropriate model among alternative models. To make possible the selection of the best model in Bayesian approach, the Akaike Information Criterion was proposed (Akaike, 1973). The AIC for the alternative model is expressed as,

$$AIC(\beta) = (-2)\ln\{p(\boldsymbol{x} \mid \beta)\} + 2\dim\beta \qquad (3)$$

where, $p(\boldsymbol{x} \mid \beta) = \int \Pi(\boldsymbol{\theta} \mid \beta) f(\boldsymbol{x} \mid \boldsymbol{\theta}) d\boldsymbol{\theta}$. (3a)

$\Pi(\boldsymbol{\theta} \mid \beta)$ is a family of prior distribution

The first term in equation (3) means the goodness of fit of a model to the observed data, and the dim β of the second term represents the number of model parameters. Equation (3) is called the Bayesian version of the AIC.

2.2 Theory of EBM

The objective function of the EBM could be presented in equation (4) by using normal distribution of measured and predicted values.

$$J(\boldsymbol{\theta} \mid \beta) = \sum_{k=1}^{K}\{\boldsymbol{u}^{*k} - \boldsymbol{u}^{k}(\boldsymbol{x} \mid \boldsymbol{\theta})\}^{T} V_u^{-1} \{\boldsymbol{u}^{*k} - \boldsymbol{u}^{k}(\boldsymbol{x} \mid \boldsymbol{\theta})\} + \beta(\boldsymbol{\theta} - \boldsymbol{p})^{T} V_p^{-1}(\boldsymbol{\theta} - \boldsymbol{p}) \qquad (4)$$

where, $\boldsymbol{u}^{*k}$ is field observation data vector at step k, $\boldsymbol{u}^{k}$ is calculated results vector at step k by an employed physical model with a chosen parameter vector, K is total number of measured steps, $\boldsymbol{x}$ is known input data vector, $\boldsymbol{\theta}$ is model parameter vector to be estimated, V_u is a covariance matrix of measurements, $\boldsymbol{p}$ is prior (initially estimated) mean vector of the model parameter vector $\boldsymbol{\theta}$, and V_p is a covariance matrix of the prior estimation.

The main difference of the EBM from the conventional Bayesian analysis is the introduction of the scalar, β. The β parameter can be estimated again by the Bayesian theorem maximizing the following log-likelihood function (Honjo et al., 1994a, Honjo et al., 1994b):

$$l(\beta \mid \boldsymbol{u}^{*}, \boldsymbol{p}) = \ln\{L(\beta \mid \boldsymbol{u}^{*}, \boldsymbol{p})\}$$

$$\cong -\frac{1}{2} NK \ln\{J_0(\widehat{\boldsymbol{\theta}}) + \beta J_p(\widehat{\boldsymbol{\theta}})\} + \frac{1}{2}\ln\left\{\frac{\beta^{M} \mid V_p^{-1} \mid}{\mid \sum_{k=1}^{K} S^{(k)T} V_u^{-1} S^{(k)} + \beta V_p^{-1} \mid}\right\} + const \qquad (5)$$

The AIC value can be expressed, for the present study, as equation (6), and the β parameter can be also obtained by minimizing the AIC value.

$$AIC = -2\ln\{L(\beta \mid \boldsymbol{u}^{*}, \boldsymbol{p})\} + 2\dim\beta \qquad (6)$$

Once we obtain the β from equation (5), the parameter $\boldsymbol{\theta}$ is to be estimated. The Gauss-Newton method and the modified Box-Kanemasu iteration method is used to estimate the θ by minimizing the equation (4) (Beck and Arnold, 1977).

2.3 Numerical Method for Back Analysis

A finite element method is developed for the purpose of back analysis of underground excavation by modifying the program originally developed by Owen and Hinton (1980). The primary support system in underground structures is modelled as the strut element, since axial force might be a major source of force in arch shape structures. To simulate the excavation, initial stress is applied to the ground; and then, the stress obtained in the initial stress state is applied in the opposite direction.

The code developed in this study is applied to the example problem and the results are compared with the benchmark analytical solution (Kirsch equation) and the commercial finite difference code, FLAC-2D. The maximum relative error between the results of analytical solution and the modified FEM analysis is less than 8 % in the check points (Lee et al., 1996)

3. APPLICATION

3.1 Ground condition

The proposed method is applied to a subway tunnel in Pusan, Korea. The ground condition of this site is shown in Figure 1 and Table 1. The tunnel was constructed through the highly weathered rock layer, which is classified as 'Poor Rock' by RMR description. The typical cross section of the tunnel is shown in Figure 2.

The elastic modulus and the strength parameters, can be empirically estimated from correlations

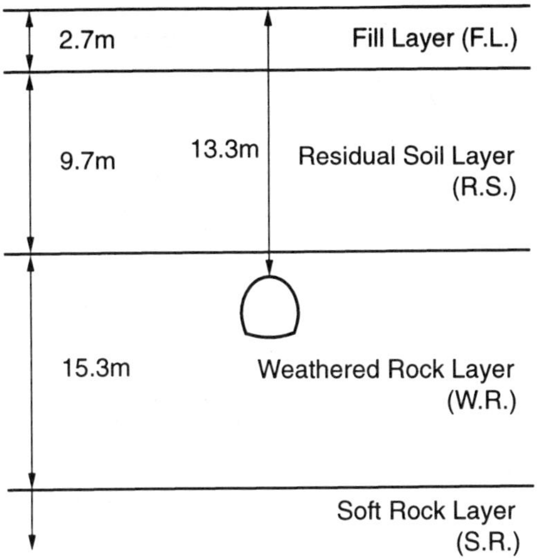

Figure 1. Idealized geological condition

with SPT N-value. The coefficient of variation of SPT N-value is about to be 0.26 (Harr, 1987). The coefficient of variation of the E and K_0 are selected to be 0.3 including uncertainty of SPT N-value.

The plastic zone was assessed by comparing the stresses obtained from the elastic finite element calculation and the strength obtained from the Hoek-Brown

Table 1. Typical values of geotechnical parameters

	F. L.	R. S.	W. R.	S. R.
Elastic Modulus (t/m^2)	2,000	3,000	20,000	100,000
Earth Pressure Coefficient	0.5	0.5	0.5	1.0
Poisson' Ratio	0.35	0.33	0.23	0.20
Unit Weight (t/m^3)	1.9	2.0	2.2	2.4

model. Since the ground motion is almost elastic except at the invert corner zones (Lee et al., 1994), the elastic modulus (E) and horizontal earth pressure coefficient at rest (K_o) are adopted as the model parameters to be estimated.

3.2 Measurement

The absolute displacements corrected from thc oserved data by adding up the missed deformations are presented in Figure 3. Two stations located closely are chosen for the tunnel convergence measurements; however, for the application of the EBM, two stations are combined by endowing each section with time step 1 and 2.

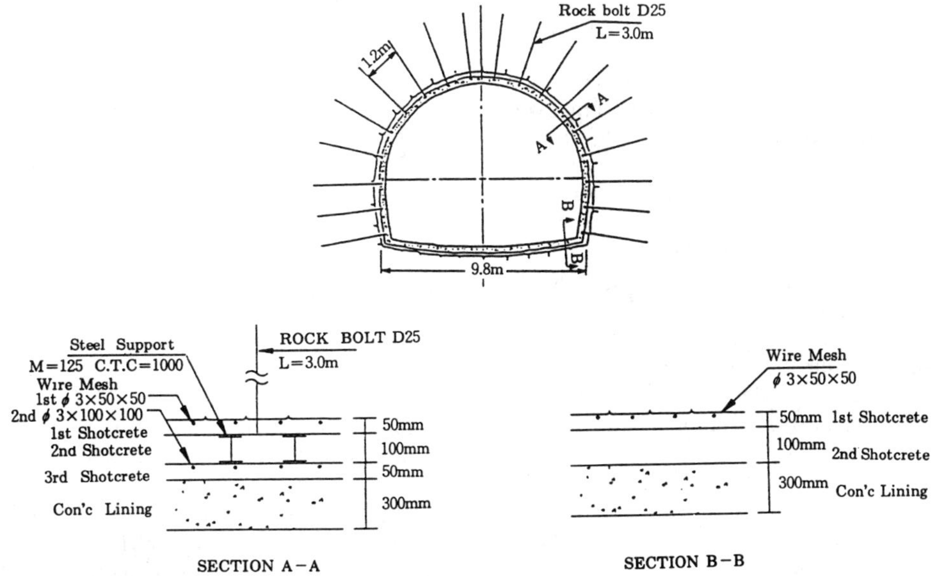

Figure 2. Typical tunnel cross section

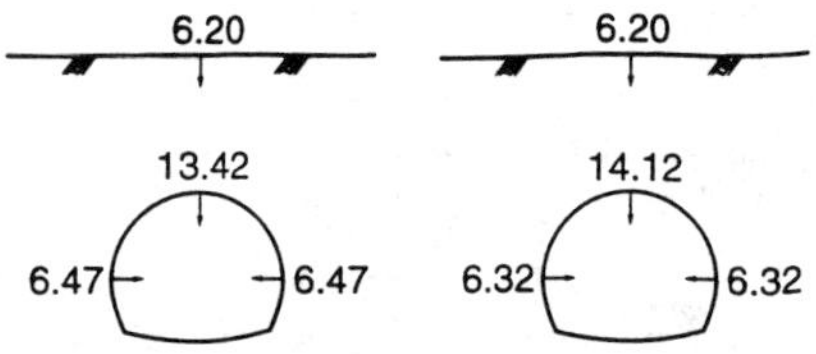

(a) Station I (b) Station II

Figure 3. Absolute displacements

3.3 Model identification

3.3.1 Description of alternative models

In order to assess the best model, three models are introduced. The model I -1 is adopted directly from the site investigation (see Figure 1) with suggested typical values of elastic moduli in Korea. Since the elastic modulus of residual soil selected in the model I -1 seems too small, the model I -2 adopted the larger elastic modulus for residual soil layer. The estimated elastic modulus of granitic soil for model I -2 was assumed to be about 10,000 ton/m^2 obtained from triaxial test results performed by Kim(1994). In the model II, residual soils and weathered rocks are considered as a single 'weathered zone'.

3.3.2 Selection of the best model for back analysis

The measurement error, V_u, might then be significant in field conditions; however, since the relative magnitude of V_u to V_p can be adjusted by applying the β, V_u was assumed to be unit matrix for the comparison under the same condition. The comparison of each of the model I -1, model I -2 and model II by the AIC is shown in Figure 4. It is found from this figure that the model I -2 gives the minimum AIC among the three models. The AIC value of the model II looks also small, even though slightly larger than model I -2.

3.4 Parameter estimation

The analyses were performed to obtain the optimized parameters for the three models. Table 2 shows the results of analysis. As shown in Table 2, the objective functions of the model I -2 and model II are quite small compared with the function of the model I -1.

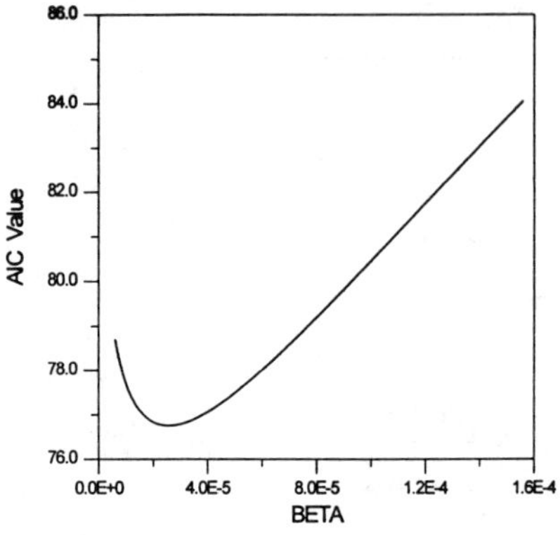

(a) Model I -1

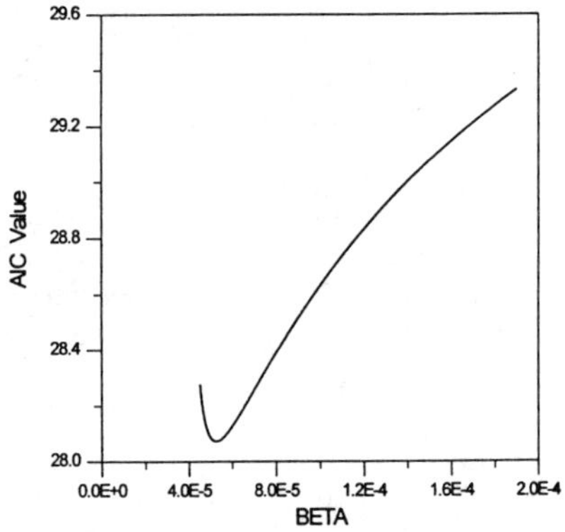

(b) Model I -2

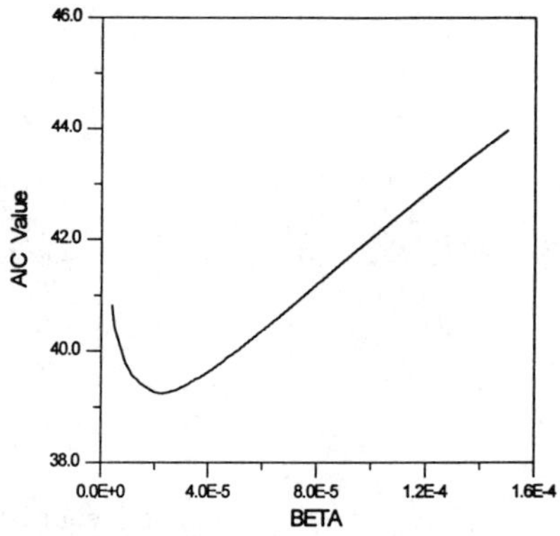

(c) Model II

Figure 4. The hyperparameter β versus AIC value in each model

Table 2. Results of parameter estimation

	Ground Condition	Elastic Modulus (t/m^2)		Earth Pressure Coefficient		Objective Function
		Prior Information	Optimized Value	Prior Information	Optimized Value	
Model I -1	R.S.	3,000	4,610	0.50	0.49	28.60
	W.R.	20,000	22,630	0.50	0.50	
Model I -2	R.S.	10,000	11,990	0.50	0.50	0.26
	W.R.	20,000	19,620	0.50	0.51	
Model II	W.Z.	20,000	18,990	0.50	0.49	0.76

note : (1) R.S. : Residual soil layer
(2) W.R. : Weathered rock layer
(3) W.Z. : Weathered zone

3.5 Uncertainty evaluation

It is our concern to know how to reduce the uncertainties by adopting the proposed technique; the comparison between the uncertainty of initially estimated parameters (the prior estimation) and the uncertainty of estimated values utilizing the EBM (the posterior estimation).

The posterior covariance matrix based on the conventional Bayesian theory is:

$$\Sigma_p=\left\{ \sum_{k=1}^{K} (S^k)^T V_u^{-1} S^k + \beta V_p^{-1} \right\}^{-1} \qquad (7)$$

where, Σ_p is the covariance matrix of the posterior estimation.

The measurement errors of 4%, 20%, and 30% are considered in this study to evaluate the influence of the measurement accuracy on the magnitude of uncertainties. Table 3 shows the reduction of uncertainties from the prior to the posterior with the variation of measurement errors in model I -2. The elastic modulus of weathered rock zone is more sensitive by changing of measurement accuracy.

Table 3. C.O.V. of geotechnical parameters

		Ω (E)		Ω (K_0)	
		R.S.	W.R.	R.S.	W.R.
Prior error		0.3	0.3	0.3	0.3
Posterior error	Measurement error : Ω=0.3	0.261	0.237	0.295	0.292
	Measurement error : Ω=0.2	0.215	0.189	0.252	0.248
	Measurement error : Ω=0.04	0.052	0.045	0.064	0.062

※ Note: Ω (·)=Coefficient of variation

4. CONCLUSIONS

1. The model identification is carried out between the initial models based on the site investigation and the simplified model. From the model identification, the model I -2 was selected as the optimized model in this site. It is again confirmed by the smallest value of objective function. And the model II was also found to be a good choice showing low AIC value and small value of objective function (even though larger than model I -2). This result shows that the selection of initially estimated parameter is the most important work during the feedback system, and shows that sophisticated model which has more uncertain parameters does not always mean to give better engineering predictions than simpler model.

2. The uncertainty of the posterior is always less than that of the prior information, and the uncertainties are proportionally reduced to the accuracy of field measurements.

ACKNOWLEDGEMENT

This paper was supported by the Korea Research Foundation, Ministry of Education, Republic of Korea, and was initiated when the first author was on sabbatical leave in the Geotechnical Research Centre, the University of Western Ontario, Canada.

REFERENCES

Akaike, H. 1973, *Information Theory and an Extension of the Maximum Likelihood Principle*, 2nd Int. Symp. on Information Theory, Akademiai Kiado, 267-281.

Beck, J. V. and Arnold, K. J. 1977, *Parameter Estimation in Engineering and Science*, John Wiely & Sons, Inc.

Cividini A., Maier G., and Nappi A. 1983, *Parameter Estimation of a Static Geotechnical Model Using a Bayes's Approach*, Int. J. Rock Mech. Min. Sci. & Geomech. Abstr., 20(5), 215-226.

Harr, M. E. 1987, *Reliability-Based Design in Civil Engineering*, McGraw-Hill Book Company.

Honjo, Y. Wen-Tsung, L. and Sakajo,S. 1994a, *Application of Akaike Information Criterion Statistics to Geotechnical Inverse Analysis : the Extended Bayesian Method*, Structural Safety, 14, 5-29

Honjo, Y. Wen-Tsung, L. and Guha, S. 1994b, *Inverse Analysis of an Embankment on Soft Clay by Extended Bayesian Method*, Int. J. Numer. Anal. Methods Geomech., 18, 709-734.

Kim, Y. J. 1994, *Constitutive Characteristics of Decomposed Korean Granites*, Ph. D. Thesis, Dept. of Civil Eng., Korea University

Lee, I. M., Kim, D. H., Park, Y. J., Baik D. H., and Choi, S.I. 1994, *A Study on Feedback System for the Geotechnical Parameter Estimation in Underground Construction,* Proceedings of the KGS Fall '94 National Conference, Seoul, 191-198

Lee, I. M., Kim, D. H. 1996, *A Systematic Approach on Geotechnical Parameter Estimation for Underground Structures*, Proceedings of the Korean Society of Civil Engineers, 16 (III-5), 423-431

Neuman, S. P. and Yakowitz, S. 1979, *A Statistical Approach to the Inverse Problem of Aaquifer Hydrology-1. Theory*, Water Resources Research, 15(4), 845-860.

Owen, D.R.J. and Hinton, E. 1980, *Finite Elements in Plasticity : Theory and Practice,* Pineridge Press, Swansea.

Computer Methods and Advances in Geomechanics, Yuan (ed.) © 1997 Balkema, Rotterdam, ISBN 90 5410 904 1

Cavity expansion modeling of cone penetration mechanism in granular soils

Anand J. Puppala
University of Texas at Arlington, Tex., USA
Yalçın B. Acar*
Louisiana State University, Baton Rouge, La., USA
Mehmet T. Tumay
Louisiana Transportation Research Center, Louisiana State University, Baton Rouge, La., USA

ABSTRACT: Cavity expansion theoretical models have been used to analyze in situ cone penetration test results. This paper evaluates two of the existing model theories by comparing their cone test interpretations with the measured test results. The cavity models considered in this paper have adapted same soil behavior but different analysis types such as small strain and large strain analysis. The model limiting pressure are first computed and then correlated with measured cone tip resistances from calibration chamber tests. Both theories predicted similar correlations with tip resistances. Factors that influence the cavity model interpretations in granular soils are discussed. Future research directions to improve the model interpretations are noted.

1 INTRODUCTION

Cavity expansion theoretical models have been used to analyze in situ cone penetration test results. The penetration of a cone is equivalent to the expansion of a cavity in a soil medium from zero radius to a cone tip radius. Several cavity models have analyzed this cavity expansion with different theoretical soil models and then reported the limiting pressures in the form of analytical expressions. These expressions are derived as functions of soil strength and deformation properties. The limiting pressure is normally correlated with the tip resistance in cone penetration testing. The measure or success of the interpretations by the cavity models generally depends on type of soil model used in the original model formulations. Chadwick (1959) developed a pressure-expansion relationship for an elastic-perfectly plastic material. Vesic (1972) also developed limiting pressure relationships which are functions of the rigidity index values. These earlier models have certain limitations with respect to soil deformation behavior particularly in the dilation zone. Recent spherical expansion models developed by Carter et al. (1986) and Yu and Houlsby (1991) have modified those limitations by incorporating a more realistic soil deformation behavior in their model formulations.

The cavity expansion models developed by Carter et al. (1986) and Yu and Houlsby (1991) have adopted similar yield criterion, the Mohr-Coulomb yield criterion. However, the major variation between these models is the type of strain analysis used in each model formulation. Yu and Houlsby (1991) considered a large strain analysis whereas Carter et al. (1986) used a small strain analysis in the derivation of limit pressure and expansion relationships. It is important to understand the variations in the strain analysis of cavity expansion theories and their influence on the model interpretations of cone test results. Therefore, an attempt is made in this paper to evaluate these two existing model theories by comparing their cone test interpretations with the measured test results. This comparison analysis will provide an insight on the influence of soil properties and the analysis type used in each model on the cone test interpretations.

Calibration chamber and triaxial tests conducted on dilatant uncemented Monterey No. 0/30 sand provided the measured test data for the interpretations (Puppala, 1993; Puppala et al. 1995). The limiting pressures are first computed from the two spherical cavity expansion models and then they are correlated with the measured tip resistances through relative densities and confining pressures. The influence of strain analysis, confining pressure, and relative density on the correlations are explained. Other factors that influence the interpretations are mentioned. Future improvements necessary for improving the accuracy of the cavity model interpretations are stated.

**deceased*

2 EXPERIMENTAL PROGRAM

Cone penetration test results used in the evaluation were obtained by testing a sub-rounded to sub-angular Monterey No. 0/30 sand in a calibration chamber under controlled conditions (Puppala et al. 1995). Dry pluviation technique was adopted in the specimen preparation. Cone penetration tests were conducted on these specimens by using a miniature cone penetrometer under zero lateral strain conditions traditionally known as Boundary Condition 3 or BC3. The miniature cone which has a diameter of 1.27 cm renders a diameter ratio of 42 when tested in the LSU calibration chamber (de Lima and Tumay, 1992). This diameter ratio is perceived to reduce boundary condition effects on cone test results (Puppala, 1993).

Specimens prepared at three ranges of relative densities (D_r) (45 - 55%, 65 - 75%, 85 - 90%) were tested at three levels of confining pressures (100, 200, and 300 kPa). A typical cone penetration test conducted in a calibration chamber is presented in Figure 1. Drained triaxial tests were also conducted on identical sand specimens. Figure 2 presents typical triaxial test results on dense sand. These tests provided strength parameters (effective friction angles) and deformation parameters (dilation angles) (Puppala and Acar, 1994). These parameters are used in the cavity expansion models to determine the limiting pressures.

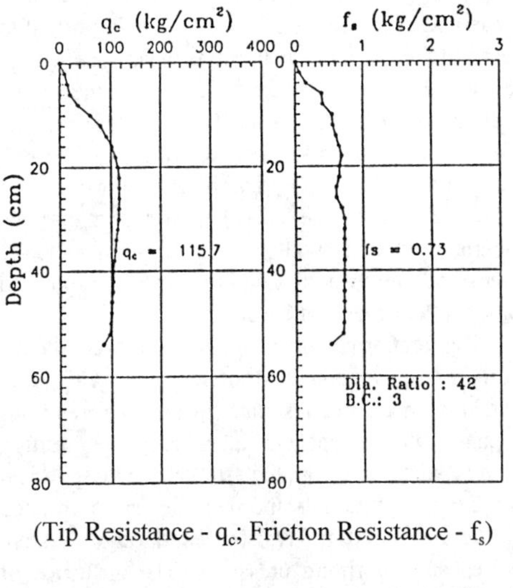

(Tip Resistance - q_c; Friction Resistance - f_s)

Figure 1: Cone Penetration Test in the Calibration Chamber (D_r = 54.8%; Confining Pressure = 300 kPa)

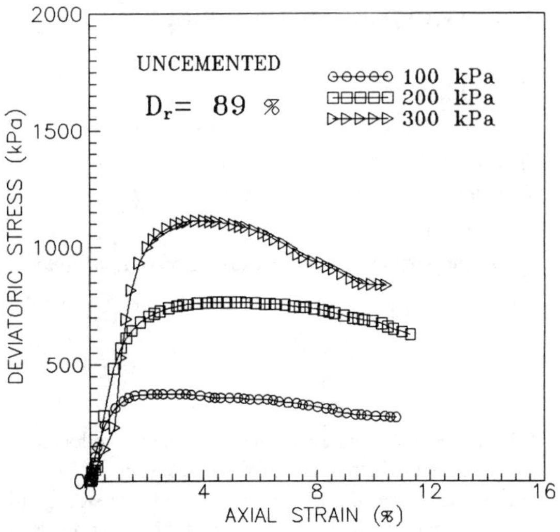

Figure 2: Drained Triaxial Test Results (Arslan, 1993)

3 CAVITY EXPANSION MODELS

3.1 Carter et al. (1986)

This closed form solution for cohesive frictional soil is developed by considering small strain analysis. The final form of this solution is the derivation of limiting pressure expression as a function of soil properties. Detailed description of this solution can be found in the respective reference. The limiting pressure can be correlated with the measured tip resistance of cone penetration tests.

3.2 Yu and Houlsby (1991)

This closed form solution is developed by considering a large strain analysis. The derivation of limiting pressure relationship and other details on yield function can be found elsewhere (Yu and Houlsby, 1996).

4 ANALYSIS OF RESULTS

Both model solutions are first used to estimate the limiting pressures. This step uses both strength and deformation properties obtained from triaxial test results. The ratios between cone tip resistance and limiting pressures are then computed. Figure 3 (Carter et al. 1986) and 4 (Yu and Houlsby, 1991) present these ratios and their variations with respect to relative density and confining pressure.

Both figures indicate that the ratios decrease with relative density. They, however, appear to show no consistent trend with respect to confining pressure. The decrease in ratios at higher relative densities is attributed to higher strength and deformation (dilation) properties at those densities. Higher properties yielded significantly larger limiting pressures and lower tip resistance to limiting pressure ratios.

An attempt is also made to compare the limiting pressures computed by both model solutions depicted in figures 3 and 4. This comparison indicates that there is no significant variation between model predictions at lower to medium relative densities. At higher relative densities, the Yu and Houlsby solution predicted higher limiting pressures than the Carter et al. Solution. This means that large strain terms may have certain influence on the final limiting pressures, particularly at higher densities.

The sand, at higher relative densities, undergoes significantly larger deformation under a cone tip penetrometer. Therefore, the solution that has considered large strain terms has provided better interpretations for sands at higher relative densities.

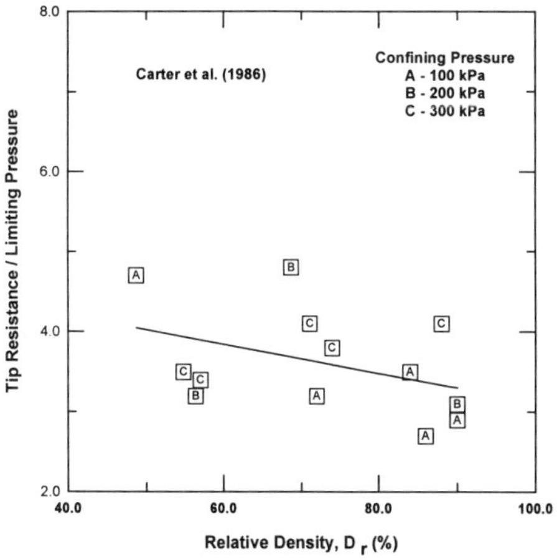

Figure 3: Ratios Versus Relative Density (Carter et al. 1986)

Another study conducted on a uniform quartz sand, Ooosterschelde sand, reported tip resistance to limiting pressure ratios which ranged between 3.5 and 4 (Greeuw et al. 1988). This study also reported a decrease in ratios with increase in relative density. The variations in the ratio magnitudes between this and present studies are attributed to the differences in sand

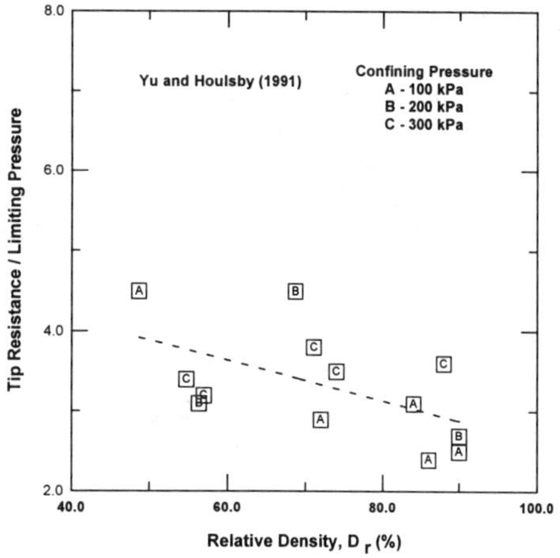

Figure 4: Ratios Versus Relative Density (Yu and Houlsby, 1991)

properties, equipments used (different cone sizes), testing procedures, and the approaches used in determining the dilation angles (Puppala, 1993).

5 SUMMARY AND CONCLUSIONS

This paper presents an evaluation on two cavity expansion theories in their interpretation of cone penetration test results. Spherical cavity expansion results are only considered since the expansion around a cone tip is considered to be close to spherical in shape. Both models provided similar predictions. The only variation is reported at higher relative densities, which may explain the importance of large strain terms and their contributions to limiting pressures at those densities. Further studies are still necessary to understand this aspect. The ratios between tip resistance and limiting pressures are higher than one. This indicates that there is a variation between idealized cavity expansion and actual expansion near the cone penetrometer.

The ratios between tip resistance and limiting pressure decrease with an increase in relative densities and also they do not follow any consistent trend with respect to confining pressure. This is attributed to the incorporation of dilational behavior of soils in the model theories. Though this dilational behavior incorporation is a significant improvement, the cavity model formulations are still limited and influenced by the unrealistic representation of a constant dilation angle during entire shearing. Future cavity expansion models need to incorporate a more realistic dilation angle assumption in their formulations.

ACKNOWLEDGMENTS

This study was supported by the National Science Foundation under the grant MSS-9020368. The authors wish to express their appreciation for this support. The authors also wish to thank Semih Arslan for conducting drained triaxial tests, Barbara Wallace and Padmaja Puppala for reviewing the paper.

REFERENCES

Arslan, S. (1993) "Cone Penetration in Cemented Deposits: A Field Study," *Masters Thesis*, Louisiana State University, Civil Engineering Department, August 1993.

Carter, J. P., Booker, J.R., Crooks, A., Rothenburg, L. (1986) "Cavity Expansion in Cohesive Frictional Soil," *Geotechnique*, London, England, 36, 349-358, 1986.

Chadwick, P. (1959) "The Quasi-Static Expansion of a Spherical Cavity in Metals and Ideal Soils," *Q. J. Mech.,, Applied Math.*, Part 1, XII, pp 52-71, 1959.

de Lima, D.C., and Tumay, M.T. (1992) "Calibration and Implementation of Miniature Cone Penetrometer and Development, Fabrication and Verification of the In Situ Testing Calibration Chamber (LSU/CALCHAS)." Res. Rep. No. GE-92/08, Louisiana Transportation Research Center, Baton Rouge, LA.

Greeuw, G., Smits, F. P., van Driel, P., "Cone Penetration Tests in Dry Oosterschelde Sand and the Relation with Cavity Expansion Model," *Proceedings of the First International Symposium on Penetration Testing*, Volume. 2, ISOPT-1, Orlando, Florida, pp. 771-776, March, 1988.

Puppala A. J. (1993) "Effect of Cementation on Cone Resistance in Sands: A Calibration Chamber Study", *Ph.D. Thesis*, Louisiana State University, May, 362 p.

Puppala, A.J., Acar, Y.B. and Senneset, K. (1993) "Cone Penetration in Cemented Sands: Interpretation by Two Bearing Capacity Theories," ASCE, *Journal of Geotechnical Engineering*, Vol 119, No. 12, December 1993, pp. 1990-2001.

Puppala, A.J. and Acar, Y.B. (1994) "Constitutive Modeling of Cemented Sands: Drained Behavior," *Proccedings of the Eigth International Conference on Computer Methods and Advances in Geomechanics*, Balkema Publishers, Rotterdam, Morgantown, West Virginia, May, pp. 653-658.

Puppala, A.J., Acar, Y.B., and Tumay, M.T. (1995) "Cone Penetration in Very Weakly Cemented Sand," *Journal of Geotechnical Engineering*, ASCE, Vol 121, No.8, August, 1995.

Vesic, A.S., "Expansion of Cavities in Infinite Soil Mass," *Journal of Soil Mechanics and Foundation Div.*, ASCE, Vol. 98, pp. 265-290, 1972.

Yu, H.S. and Houlsby, G.T., "Finite Cavity Expansion in Dilatant Soils: Loading Analysis," *Geotechnique 41*, No. 2, pp. 173-183, 1991.

Computer Methods and Advances in Geomechanics, Yuan (ed.) © *1997 Balkema, Rotterdam, ISBN 90 5410 904 1*

Test study of soft clay preloading rheology

X.R.Zhu, Q.Y.Pan & G.X.Zeng
Institute of Geotechnical Engineering, Zhejiang University, Hangzhou, China

ABSTRACT: There are many modern large scale airports and express highways constructed and to be constructed in the coastal soft clay area of China. The allowable unequal settlement of these types of structures is quite serious. Therefore it seems necessary to study the deformation of the soft clay ground even in the case of preloading in the light of the modern knowledge of rheology. In this paper the principal and the method of rheology test for soft clay preloaded are introduction. The test results are analyzed. And a rheology model for Ningbo soft clay preloaded is recommended. Several valuable conclusions have been obtained through laboratory tests.

1 INTRODUCTION

Since the eighties, with the social advance and economic development a large scale of engineering have been constructed and to will be constructed in China. Owing to the lack of good ground, especially in the coastal area of China, a great number of projects have to be constructed on the soft clay, such as many modern large scale airports, express highways, oil tanks, etc. It is common knowledge that the compressibility of the soft clay is very high and the settlement of soft clay ground is very large, and the allowable unequal settlement of these types of structures is quite serious, so this is a critical contradictory. In order to meet the needs of usual usage the soft clay ground must be improved. It is proved in practice that the preloading is a kind of economical method that the soft clay grounds for airport runways and express highways are improved. Because the soft clay preloaded has rheological properties still and the grounds may still produce some settlements after preloading and on the other hand the allowable settlement of the projects is small after all, it is valuable that the soft clay preloading rheology is studied through tests.

Since born in 1928, the rheology gains a large development and there are wide applications in every field of live and production. The sentence that everything is in flow shows that the rheology is an important course. Long ago it was known that the soft clay has rheological phenomenon, but the rheological study was not underway until 1948. After 1948, some workers have studied the rheological properties of the soft clay in the theory, test and application(Tan,T.K., 1957; Qian, J.H., 1958; Xu,Z.Y,1964; Adams, J.I., 1965; Aboshi,H., 1973; Keedwell, M.J., 1984; Zao, W.B.,1987; Imai,G., 1989; Zheng, R.M., etc., 1996). Some researchers have developed specialized rheological instrument(Li, Z.Q.,1989). In general speaking, firstly on the base of test study or theoretical study these studies assume some rheological models according to different combination of basic elements and secondly analysis linear or some time nonlinear rheological behavior of normal consolidation soft clay. The studies do not discuss the rheological behavior of preloading soft clays which there are in much practical engineering, such as large scale airports and express highways. To meet the need of practical engineering in this paper through tests the rheological behavior of preloading soft clays is studied.

2 RHEOLOGICAL TEST METHOD

2.1 *Specimen Condition*

The clay tested is Ningbo airport clay which the physical properties indexes are shown in Table 1. In order to reduce disturb, a kind of special specimen trunk which there is the spring is used.

Table 1. Physical propertice of soft clay

ω (%)	ρ (g/cm^3)	e_0	I_P	I_L
45.4	1.74	1.285	17.1	1.36

2.2 *Test methods*

To imitate the surcharge precompression condition in practical engineering the test is in progress in follow sequence: preloading, unloading, loading or not, rheological test. The test pressures are decided according to Ningbo airport runway load. The overconsolidation ratios are shown in Table 2. In order to compare the test results to those of no preloading specimen, that is normal consolidation clay, the rheological tests of clay for OCR=1 are in progress.

The test is underway in the triaxial compression instrument and the direct shear instrument. The creep test and the relaxation test are gone on the stress instrument and the strain instrument respectively.

It is assumed that the creep can linearly add, that is the creep produced by changing load is equal to sum of the creep produced by the load increment. In this way the test uses step load method to increase load.

Table 2. Overconsolidation ratios of specimen

Triaxial compression test		Direct shear test	
Creep	Relaxation	Creep	Relaxation
1.43	1.43	1.152	1.152
1.0	1.0	1.0	1.0

3 TEST RESULTS AND ANALYSIS

3.1 *Test results*

The triaxal creep test results are shown in Figure 1 to Figure 3, the triaxial relaxation test results are shown in Figure 4 to Figure 5.

3.2 *Analysis of test results*

Figure 1 and Figure 3 show the relationships of clay strain and time under different vertical load for preloading, OCR=1.43, and Figure 2 for normal consolidation clay, OCR=1.0. Follow results can be obtain from the figures.

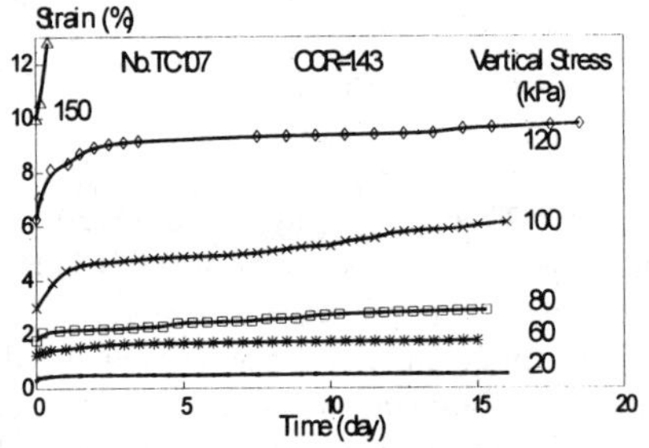

Figure 1 Creep curves of clay preloaded for p_c=100 kPa

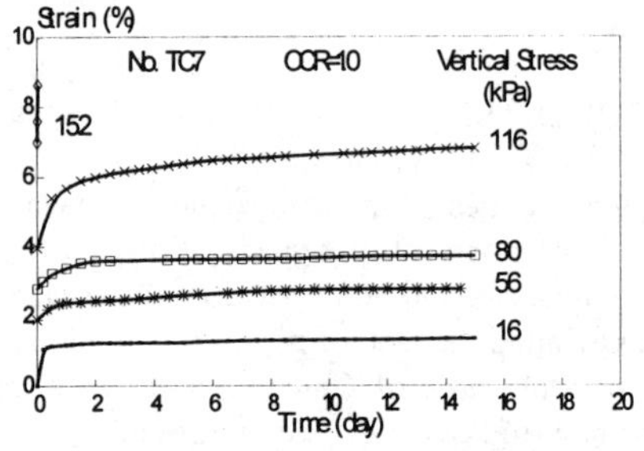

Figure 2 Creep curves of soft clay for p_c=70 kPa

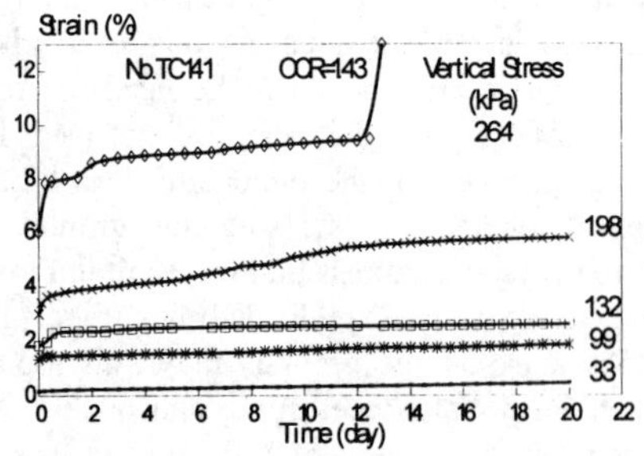

Figure 3 Creep curves of clay preloaded for p_c=143 kPa

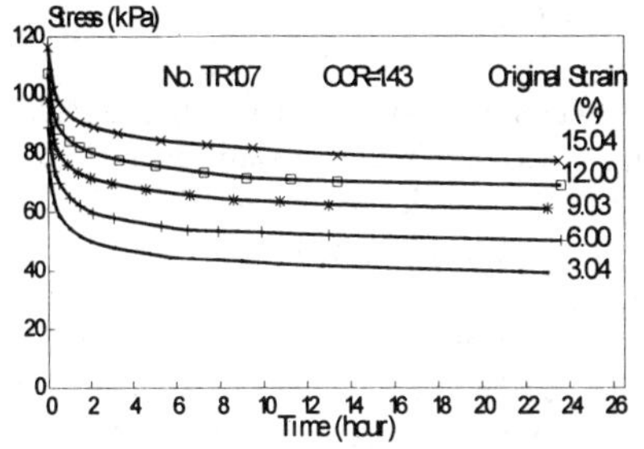

Figure 4 Relaxation curves preloaded clay for p_c=100 kPa

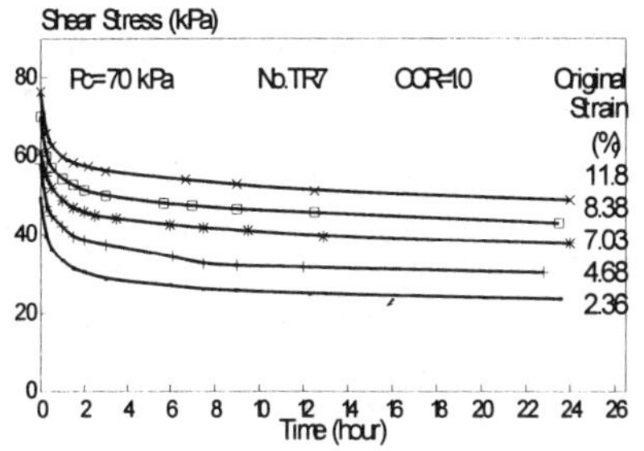

Figure 5 Relaxation curves of clay for p_c=70 kPa

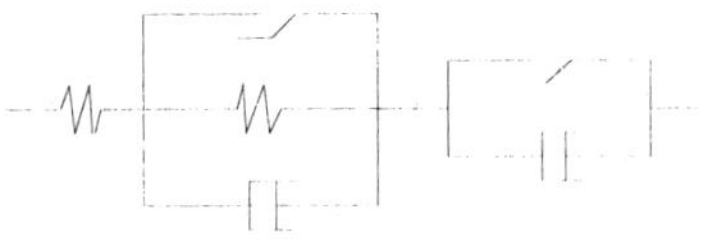

Figure 6. Rheological model for preloading clay

1.The deformation will gradually increase with time after loading and may stop in a short time when the vertical load is small.

2.With load increasing the time in which creep produces will increase and the creep velocity increase.

3.There is a threshold stress after which the deformation will quickly increase to soil failure.

4.The strain of clay after is smaller than that of no precompression soil under same load.

5. The deformation of specimen after preloading can maintain unchanged in larger stress than the normal consolidation clay.

6. The clay after preloading may destroy brittlenessly in high stress.

Figure 4 and figure 5 show the relationship of relaxation stress and time for different strain. From the figures we can gain follow results.

1. For a original strain the stress will quickly decrease within a short time, and gradually decrease to a steady amount with time.

2. The specimen preloaded has a larger steady amount than that of the normal consolidation clay in the case of same original strain.

3. The original stress and residual steady stress increase with the increasing of over pressure.

From the test results we can still know that the timeliness of soft clay behavior is very evident and the preload can clearly change the rheological behavior of soft clay. Therefore it is necessary that the principal and the method are used to research the behavior of soft clay.

4 RHEOLOGICAL MODEL AND CONSTITUTIVE RELATION

4.1 *Rheological model*

According to the basic principal of rheology and the test results above we know that while the stress is small, $\sigma < \sigma_{s1}$, the soil deformation almost immediately happens as load while and the behavior is similar to a pure elasticity body. Therefore the model must include a spring system.

While the stress level is relatively high, that is between two threshold σ_{s2} and σ_{s1}, $\sigma_{s2} > \sigma \geqslant \sigma_{s1}$, the soft clay will produce the rheology which decreases with time and finally tend to a ready value. In other word, in this stage most of the deformation can not recover. In accordance with analyzing we can find that independent the Kelvin system or the Bingham system can not completely describe the rheological behavior and combining the Kelvin system or the Bingham system, the visco-elastic-plasticity system strengthen, has the property the same as the test result. Therefore the rheological model of preloading clay must include the strengthening system.

It is obtained from the test results that while the stress is large, $\sigma \geqslant \sigma_{s2}$, the clay produces the unstable rheological deformation and the Bingham system may expresses the behavior. So the rheological model may include the Bingham system. On the base of analysis above the rheological model of preloading soft clay advanced in this paper is shown as Figure 6.

4.2 *Constitute equation*

As shown in Figure 6, while $\sigma < \sigma_{s1}$ the rheological model is only a spring, so the constitute equation is described in equation (1).

$$\sigma < \sigma_{s1}, \quad \sigma = E_H \varepsilon \tag{1}$$

$\sigma_{s2} > \sigma \geqslant \sigma_{s1}$,

$$E_K E_H \varepsilon + \eta_1 E_H \dot{\varepsilon} = -\sigma_{S1} E_H + (E_K + E_H)\sigma + \eta_1 \dot{\sigma} \tag{2}$$

The creep equation is

$$\varepsilon(t) = [\frac{1}{E_H} + \frac{\sigma_c - \sigma_{s1}}{E_K \sigma_c}(1 - e^{-\frac{E_K}{\eta_1}t})]\sigma_c \tag{3}$$

$$J(t) = \frac{1}{E_H} + \frac{\sigma_c - \sigma_{s1}}{E_K \sigma_c}(1 - e^{-\frac{E_K}{\eta_1}t}) \tag{4}$$

$$E(t) = \frac{(E_H \varepsilon_c - \sigma_s)E_H}{(E_H + E_K)\varepsilon_c}(e^{-\frac{E_K + E_H}{\eta_1}} - 1) + E_H \varepsilon_c \tag{5}$$

where $J(t)$ and $E(t)$ represent the creep modular and the relaxation modular correspondingly.

$\sigma_u > \sigma \geqslant \sigma_{s2}$,

$$\dot{\varepsilon} = \frac{\dot{\sigma}}{E_H} + \frac{1}{\eta_2}(\sigma - \sigma_{s2}) + \frac{\sigma - \sigma_{s1}}{\eta_1} - \frac{E_K}{\eta_1} e^{-\frac{E_K}{\eta_1}t} \int_0 \frac{\sigma - \sigma_{s1}}{\eta_1} e^{\frac{E_K}{\eta_1}} dt \tag{6}$$

$$\varepsilon(t) = \frac{\sigma_c}{E_H} + \frac{\sigma_c - \sigma_{s2}}{\eta_2}t + \frac{\sigma_c - \sigma_{s1}}{E_K}(1 - e^{-\frac{E_K}{\eta_1}t}) \tag{7}$$

In the equations above, E_H and E_k are the elastic modular of the springs, η_1 and η_2 are the coefficients of viscosity, and σ_{s1} and σ_{s2} are the threshold stresses of the plastic elements as shown in Figure 6.

The rheological model and constitute equation can be used to analysis the rheological behavior of preloading soft clay which is discussed in follow-up papers.

5 CONCLUSIONS

1.The clay exhibits remarkable rheological features, namely the stress-strain relationship, shear resistance and modules of the soil are all time depending. The stress-strain curve as the time keeps constant is close to be a hyperbola.

2.The rheological features will get improved by preloading, i.e. the modules, coefficient of viscosity and the threshold stress below which the strain is not time depending will increase after preloading.

3.The rheological behavior is nonlinear, i.e., the modules decreases as the strain increases, and the coefficients of viscosity is a function of stress and time.

4.There are two threshold stresses, σ_{s1} and σ_{s2}. When $\sigma < \sigma_{s1}$ the deformation is elastic, when $\sigma_{s1} \leqslant \sigma < \sigma_{s2}$ decay creep occurs and when $\sigma_{s2} \leqslant \sigma < \sigma_u$, where u is the failure stress, the rate of creep will gradually increase and finally the clay will fail.

5.The ratio of the long-term strength to the instantaneous strength of the clay is 0.6 to 0.75 and long-term modules is only 0.4 to 0.7 of the instantaneous one.

6.The paper gives out the rheological model and the constitute equations for Ningbo airport soft clay preloaded which is consist of a spring system, a visco-elastic-plasticity system strengthen and the Bingham system.

ACKNOWLEDGMENT

The financial supported from China National Natural Science Foundation are gratefully acknowledge.

REFERENCE

Imai, G. 1989. A unified theory of one dimensional consolidation with creep. XII ICSMFE. Vol.1,57-60.

Keedweel, M. J. 1984. Rheology and soil mechanics. New York: Elsevier Applied Science Publishes LTD.

Tan, T. K.1957. Three dimensional theory on the cosolidation and flow of the clay-layers. Academic Sinica.

Computer Methods and Advances in Geomechanics, Yuan (ed.) © 1997 Balkema, Rotterdam, ISBN 90 5410 904 1

Flow pump permeability test on Hong Kong marine clay

P.K.K.Lee & W.H.Wong
The University of Hong Kong, China

Y.M.Lam
Carleton University, Ottawa, Ont., Canada

ABSTRACT: The demand of accurate measurement for soil permeability, especially for fine grained soils, is gradually increasing in the recent years. In order to overcome the shortcomings of conventional constant head and falling head permeability tests, a rapid and accurate technique for the direct measurement of the coefficient of permeability is worthwhile to be considered. Experimental investigation of the coefficient of permeability of Hong Kong marine clay was carried out using the triaxial system. With the use of the flow pump technique for permeability measurement, a lower hydraulic gradient and flow rate can be simulated. The correlation between the void ratio and the coefficient of permeability is reported in this paper and it has also been found that the relationship between the change of the coefficient of permeability and the change in effective stress is in good agreement with the theoretical viewpoint.

1 INTRODUCTION

Soil permeability is always an influential but at times, a decisive parameter in the design and construction of civil engineering projects, in particular, in those involved with soft clay. It is one of the fundamental properties of soil governing groundwater regime in excavated slopes or stratified deposits and even the migration of pollutants from waste disposal facilities..

The value of the coefficient of permeability of a soil depends largely on the size and the nature of void spaces which in turn depends on the size, shape, and state of packing of the soil grains. In general, a clayey soil will have much lower permeability coefficient than sand even though it has the same void ratio.

Conventional laboratory methods for measuring permeability such as the constant head permeability test and the falling head permeability test have been found to be not suitable for clayey samples. The flow pump method provides a faster and more accurate alternative and its superiority over conventional methods has been recognized ([1],[2],[3]). Tavenas ([4]) had investigated systematically the permeability of natural undisturbed clay samples by using triaxial assemblies. He had also compared the direct and indirect methods of measurement of permeability.

In the reports ([1],[2],[3]) on the investigations carried out using the flow pump method, Olsen derived theoretical formulation for the constant flow rate test and gave a brief presentation on the relationship between permeability and stress state. Apart from these, very few measurements with flow pump method have been reported especially on undisturbed samples. Tests on some undisturbed Hong Kong marine clay specimens have been carried out and the test results are presented in this paper.

2 LOCATION AND DESCRIPTION OF SOIL SAMPLES

Soil samples tested were collected from Penny's Bay of Lantau Island situated at the north-west of Hong Kong island(Figure 1). Undisturbed piston tube samples from six boreholes were available for testing. Physical property indices determined in accordance with British Standard BS1377 were listed in Table 1. Mean specific gravity and plasticity index are 2.72 and 32.8% respectively.

According to the Unified Soil Classification System (Figure 2), the soil samples can be classified as soft dark grey inorganic clays of high plasticity with occasional shell fragments.

Particle size distribution curves for samples from individual borehole (Figure 3) indicate uniformity of

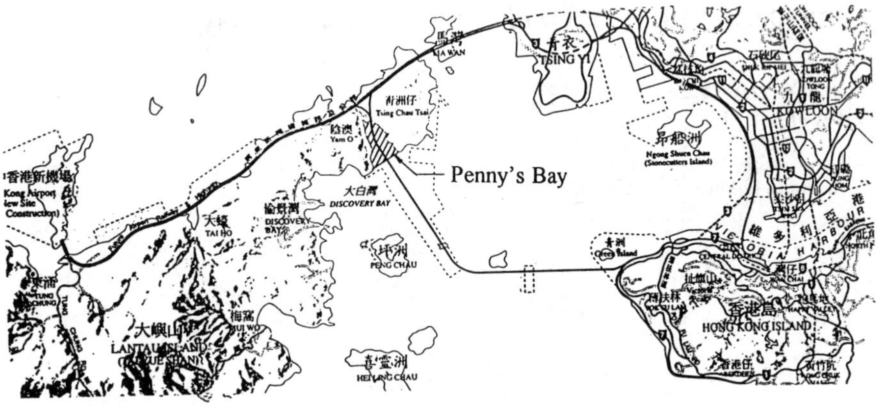

Figure 1 Location Plan

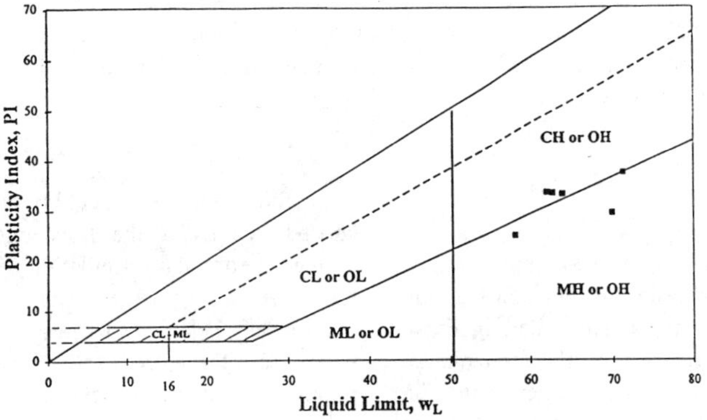

Figure 2 The plasticity chart by USCS

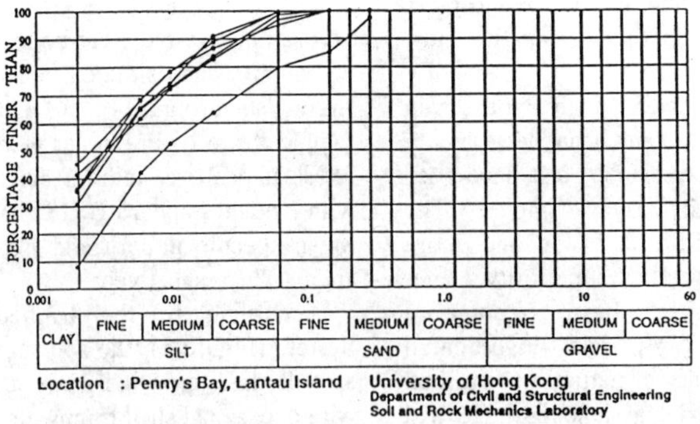

Figure 3 Particle size distribution curves

soil sample except borehole 1 which has exceptionally high sand content compared to others. This discrepancy is also reflected in the plasticity index result shown in Table 1.

Table 1 Index properties of clay samples

Borehole No.	Specific Gravity	w_L (%)	w_P (%)	P.I. (%)
1	2.70	58.1	33.3	24.8
2	2.70	69.9	32.6	29.3
3	2.73	63.8	30.8	33.0
4	2.73	71.3	38.0	37.3
5	2.72	62.0	29.7	33.3
6	2.73	62.6	29.4	33.2
Mean	2.72	64.6	32.3	32.8

3 EXPERIMENTAL SETUP AND PROCEDURE

Undisturbed soil specimens were tested in a triaxial cell. Block diagram of the system is shown in Figure 4. The triaxial cell was equipped with two independent constant pressure supplies, two electronic pressure transducers, an automatic volume change unit and a digital flow pump unit. Automatic datalogging system was used and the test was controlled by a computer.

Undisturbed soil specimens extracted from piston tubes were trimmed to 38mm diameter by 76mm high. All specimens were isotropically consolidated to at least twice of the over-burden pressure. This, therefore, brought all specimens to normally consolidated state.

After completion of consolidation, water was infused into specimens at constant rate from bottom upwards by a digital flow pump unit. As pointed out by Oslen ([2]), the method of infusion would induce less disturbance to soil specimen than extraction of water especially for normally consolidated samples.

4 EXPERIMENTAL RESULTS

A total of sixteen specimens were tested by the flow pump method. The specimens were tested under four consolidation pressures, namely, 50kPa, 100kPa, 200kPa and 400kPa. Coefficient of permeability and normal consolidation pressure can therefore be correlated.

Applied flow rate ranged from 0.05 to 0.07 mm^3/s. Typical curve of induced pressure head (normalized by σ_c) during the flow pump test was plotted in Figure 5. It can be observed that steady state was reached after about 10 hours. Permeability of clay specimen can then be readily obtained by Darcy's Law of fluid motion

$$q = k\,i\,A \tag{1}$$

where i is the hydraulic gradient across specimen;
A is the cross-sectional area of specimen perpendicular to flow direction.

Alternatively, coefficient of permeability can also be obtained indirectly from results available during consolidation. Consolidation theory can be used in the evaluation of permeability coefficient through the following relationship

$$\frac{\partial u}{\partial t} = C_v \frac{\partial^2 u}{\partial z^2} \tag{2}$$

The coefficient of consolidation, C_v can be expressed as

$$C_v = \frac{k}{\gamma_w M_v} \tag{3}$$

where k is the coefficient of permeability;
M_v is the coefficient of volume compressibility;
γ_w is the unit weight of water

Hence, k can be obtained from consolidation test results.

Summary of results is shown in Table 2. Note that three values of void ratio have been included in this table. e_1 and e_2 are different due to the change of effective stress during permeability test. However, the difference between them has been small. The tabulated results are also plotted in Figures 6 to 8.

Calculated coefficient of permeability from indirect method is shown in Figure 6. As the data points were too scatter to show a consistent trend, no rigid conclusion can be drawn. Actually, applicability of Terzaghi's consolidation theory on prediction of k value is questionable ([4]) and this will be discussed in the next section.

Coefficient permeability obtained from direct measurement, however, shows a clear trend in Figure 7. Samarasinghe [5] has presented in his article an expression to correlate the measured k value with the void ratio as:

$$k = C\frac{e^n}{1+e} \tag{4}$$

where C is a constant in the same unit as k;
n is a constant depending on the type of soil.

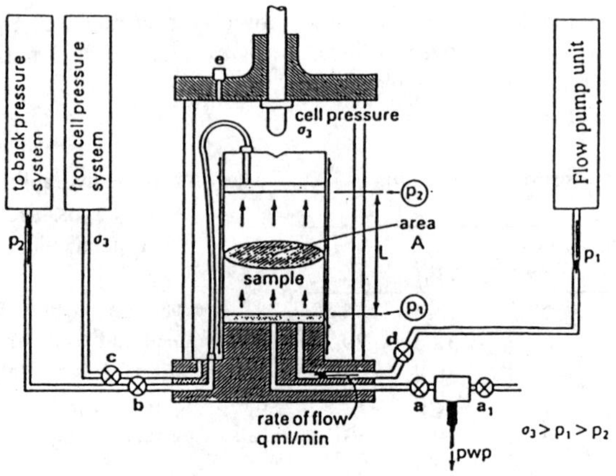

Figure 4 Block diagram of the flow pump test

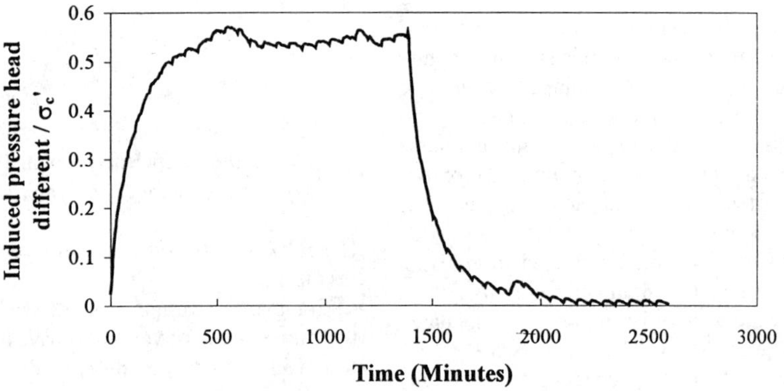

Figure 5 Excess pore pressure development during infusion of water and dissipation of pore pressure after the stoppage of flow pump.

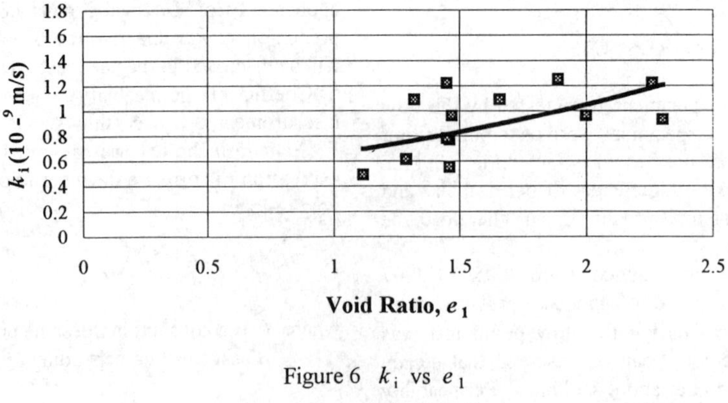

Figure 6 k_i vs e_1

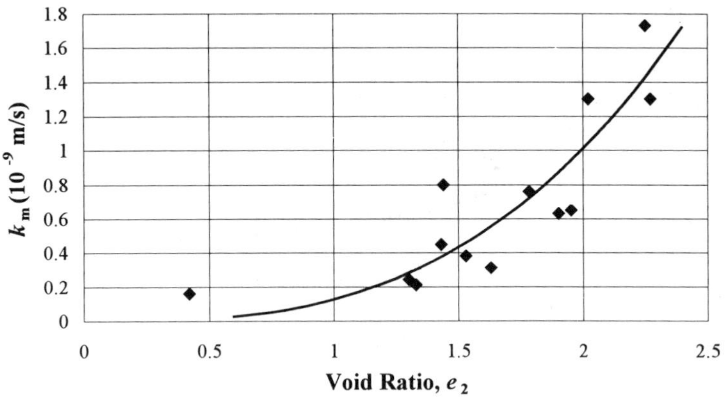

Figure 7 k_m vs e_2

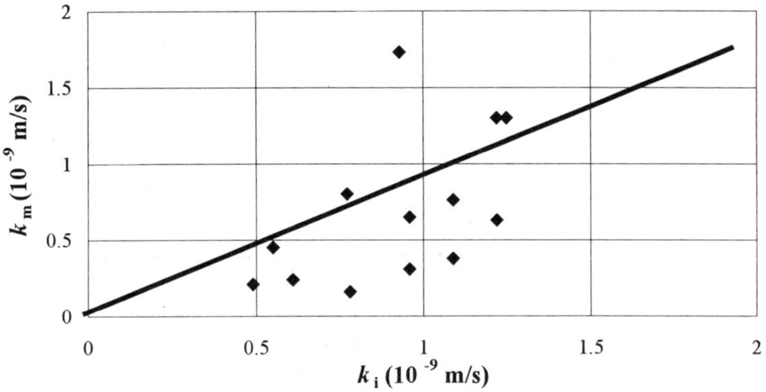

Figure 8 k_m vs k_i

It can be seen from Figure 7 that the experimental results can be matched very well by this expression. The corresponding constants C and n are 2.5m/s and 3.3 respectively.

By comparing k_i with k_m (Figure 8), it is evident that coefficient of permeability obtained from the indirect method is generally over-estimated. This phenomenon is largely due to the assumptions used in Terzaghi's theory.

Besides verifying the relationship between coefficient of permeability and void ratio established by Samarasinghe's hypothesis, relationship between k and mean pressure (σ_c') can be similarly correlated. A k_m vs σ_c' plot is shown in Figure 9 and a clear trend can be traced. This phenomenon can be explained by incorporating the well known correlation between void ratio and mean pressure, i.e.

$$e = e_o - C_c \ln \sigma_c' \tag{5}$$

where e_o is the initial void ratio;
C_c is the compression index.

Substituting (5) into (4),

$$k = C\frac{(e_o - C_c \ln \sigma_c')^n}{1 + e_o - C_c \ln \sigma_c'} \tag{6}$$

The best fit curve from Eq.(6) is superimposed in Figure 9 and it can be seen that the curve agrees well with the data points.

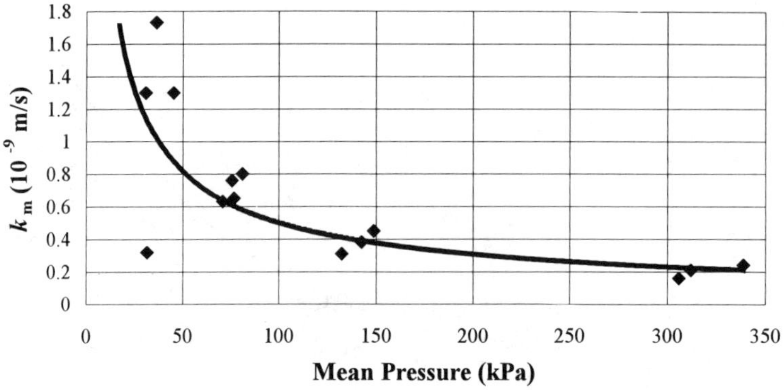

Figure 9 k_m vs σ_c'

Table 2 Summary of results

Specimen No.	e_o	e_1	e_2	k_i 10^{-9}m/s	k_m 10^{-9}m/s
S1	2.52	2.00	2.05	2.11	0.32
S2	1.61	1.46	1.44	0.77	0.80
S3	2.26	1.32	1.43	1.09	0.38
S4	2.41	1.36	1.34	0.45	0.20
S5	2.56	2.30	2.25	0.93	1.73
S6	2.90	2.00	1.95	0.96	0.65
S7	3.04	1.82	1.85	1.14	0.30
S8	2.86	1.45	1.42	0.78	0.16
S9	2.70	2.26	2.27	1.22	1.30
S10	2.65	1.45	1.49	1.22	0.63
S11	2.62	1.47	1.63	0.96	0.31
S12	1.93	1.29	1.30	0.61	0.24
S13	2.07	1.89	2.02	1.25	1.90
S14	1.97	1.66	1.78	1.09	0.76
S15	1.89	1.46	1.43	0.55	0.45
S16	2.1	1.12	1.23	0.49	0.21

Note : e_o is the initial void ratio; e_1 is void ratio after consolidation; e_2 is void ratio after compression; k_i is permeability obtained from indirect method; k_m is permeability obtained from direct measurement.

DISCUSSION & CONCLUSION

Applicability of indirect method of measurement for coefficient of permeability has been examined by Tavenas [4]. He pointed out that the most serious weakness lies in the assumption of the constants k, M_v and c_v in Terzaghi's consolidation theory. Actually, it has been shown in most experimental results that these parameters exhibit significant variations during consolidation.

The k_i value estimated from the indirect method is generally higher than the direct measured value of k_m. This is true because both c_v and M_v used in the calculations are the mean values only. They are, in general, higher than the corresponding values at the end of consolidation. The k value evaluated is, therefore, the mean coefficient of permeability of the specimen during the entire period of testing .

Coefficient of permeability of normally consolidated Hong Kong marine clay was found to range from 0.16 to 1.90 x 10^{-9} m/s in this study. It varies with void ratio and the relationship can be approximated satisfactorily by Samarasinghe's hypothesis. The constants C and n in Samarasinghe's expression for Hong Kong marine clay was found to be 2.5m/s and 3.3 respectively.

The flow pump method is a fast and reliable method when compared with traditional methods. It is also flexible and can be carried out together with testing involving oedometer, rowe cell, triaxial cell, etc. to enable the determination of permeability under various stress states.

Reference :

[1] *Pane V., Croce V., Znidarcic D., Ko H.Y., Olsen H.W. "Effects of consolidation on permeability measurements for soft clay," Geotechnique 33, 1983, pp67-71*

[2] *Olsen H.W., Nichols R.W. and Rice T.L. "Low Gradient Permeability Measurements in a Triaxial System," Geotechnique 35, 1985, pp145-147*

[3] *Morin R.H. and Olsen H.W. "Theoretical Analysis of the Transient Pressure Response From a Constant Flow Rate Hydraulic Conductivity Test" Water Resources Research Vol. 23, pp1461-1470*

[4] *Tavenas F., Jean P., Lebloud P. and Leroueil S., Canadian Geotechnical Journal Vol. 20, 1983, pp628-659*

[5] *Samarasinghe A.M., Yang H.H. and Vincent P., "Permeability and Consolidation of Normally Consolidated Soils," Proceedings of the American Society of Civil Engineers Vol. 108, 1982, pp628-659*

Computer Methods and Advances in Geomechanics, Yuan (ed.) © 1997 Balkema, Rotterdam, ISBN 90 5410 904 1

Influence of testing procedure on resilient modulus of aggregate materials

Ping Tian, Musharraf M. Zaman & Joakim G. Laguros
School of Civil Engineering and Environmental Science, University of Oklahoma, Norman, Okla., USA

ABSTRACT: Resilient Modulus (RM) is an important material property in the design of a flexible pavement. The influence of different testing procedures on RM is investigated for one limestone aggregate and the resulting RM values vary within a range of 25 to 55 percent. Also, there are some discernible changes in the static material properties, such as cohesion, friction angle, and unconfined compressive strength, which are measured after different RM testing. The RM values obtained in this research are in the range reported in the literature.

1. INTRODUCTION

As a result of more than 10 years of testing, the testing procedure for the determination of resilient modulus (RM) was finally standardized in 1994. Historically, AASHTO proposed several test methods for RM test, namely, AASHTO T274-82 (1986), T292-91I (1991), T294-92I (1992), and T294-94 (1994). In these testing procedures, the basic differences are particularly related to: (i) sample conditioning prior to testing, (ii) applied stress sequence, (iii) number of loading cycles, (iv) waveform of cyclic loading, and (v) location of LVDTs.

Since its introduction, the testing procedure T274-82 has been the target of widespread criticisms (Pezo et al. 1992), the main one being that the required loading conditions are so severe that a specimen may fail in the conditioning stage. For example, in documenting unsatisfactory experience with AASHTO T274-82, Vinson (1989) stated that the required heavy conditioning of the sample may cause different levels and types of stresses for both cohesive and cohesionless soils. Also, Ho (1989) observed that the conditioning stage was very severe for many soils. Consequently, researchers questioned the validity of and the need for such an extensive process. In 1991, AASHTO modified the testing procedure T274-82 and proposed T292-91I, and then in 1992 AASHTO adopted T294-92I in accordance with the Strategic Highway Research Program recommendations. Later, in 1994, AASHTO proposed the standard testing procedure T294-94 which is same as the interim procedure T294-92I except for the units used.

The testing procedures T292-91I and T294-94 by AASHTO were intended to overcome the deficiencies in its earlier procedure T274-82. However, there are many differences between the two testing procedures. Table 1 shows a comparison between the salient features of the two test methods.

A number of investigations have been conducted in the past on the resilient response of granular materials (e.g., Rada et al. 1981, Raad et al. 1992, and Zaman et al. 1994). These studies have contributed significantly to the understanding of the resilient properties of granular materials. However, most of the tests were conducted by using the interim testing procedure. It has been reported that different testing procedures will result in different RM values, and hence differences in the design of a pavement. Thus, it becomes important to investigate the influence of testing procedure on RM values for aggregate materials, which is the subject of this paper.

2. MATERIAL ORIGIN, ENGINEERING INDEX PROPERTIES AND RM TESTING

The Richard Spur (R. S.) limestone aggregate, which is commonly used in Oklahoma as the base material for pavements, was selected for this study. The median gradation (Figure 1) selected from the Standard Specifications for Highway Construction

Table 1. Salient features of resilient modulus testing procedures for granular materials.

	T292-91I	T294-94
Sample conditioning	Confining Pressure: 138 kPa Deviatoric Stress: 103 kPa	Confining Pressure: 103 kPa Deviatoric Stress: 103 kPa
Stress Sequence	Starts with a higher confining pressure and deviatoric stress Ends with a lower confining pressure and deviatoric stress	Opposite to T292-91I
Number of loading cycles	Conditioning: 1000 RM testing: 50	Conditioning: 1000 RM testing: 100
Wave Form	Haversine, Triangular, Rectangular	Haversine
LVDT Location	Internal, at 1/3 to 1/4 of the specimen; or external, at the top of the specimen	External, at the top of the specimen
Compaction Method	Vibration	Vibration
Bulk Stress	From 97 to 690 kPa	From 83 to 690 kPa

(Oklahoma DOT 1988) and the corresponding optimum moisture content were used for RM sample preparation. The engineering properties including liquid limit (LL), plasticity index (PI), maximum dry density (MDD), optimum moisture content (OMC), Los Angeles abrasion (Lp), cohesion (C), friction angle (ϕ), and unconfined compressive strength (U_c) for this aggregate were evaluated. A summary of the test results is presented in Table 2. It may be noted that following the repeated triaxial testing, static triaxial compression or unconfined compressive tests were performed to obtain the cohesion (C), friction angle (ϕ) or the unconfined compressive strength (U_c). The repeated triaxial tests served as a "conditioning" of the sample for triaxial compression tests that could be imposed by moving vehicles.

Two sets of RM tests were conducted using the T292-91I and T294-94 procedures, respectively. For each case, six duplicate specimens, which are 15.24 cm (6 in.) in diameter and 30.48 cm (12 in.) high, were prepared by employing a vibratory method and tested under drained conditions. The variability of the test results was investigated within a statistical framework. The mean RM values were calculated from the six individual test results, and are given in Table 3 as related to bulk stress ($\theta = \sigma_1 + \sigma_2 + \sigma_3$). The data indicate that the RM values obtained in the present study ranged from 78 to 368 MPa (11 to 53 ksi). These values are in the range reported by the Asphalt Institute (1981) for the design of a flexible pavement.

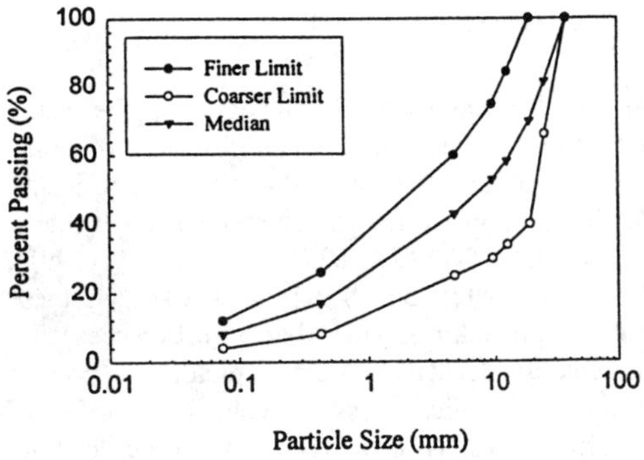

Figure 1. The median gradation used in this study.

3. DISCUSSION OF RESULTS

3.1 *Sample conditioning*

In order to minimize the effects of initially imperfect contact between the end platens and the test specimen, the sample conditioning stage is applied before RM testing in both testing procedures. The sample conditioning stage can also simulate the real situation of the pavement base in service. The sample conditioning stages for T292-91I and T294-94 differ only in the magnitude of the confining pressure σ_c applied. This difference is not expected to have any significant effect on the RM test results.

Table 2. Summary of index properties.

	LL (%)	PI	MDD (g/cm3)	OMC (%)	C (kPa)	ϕ (degree)	Uc (kPa)	Lp (%)
T 292-91I	13	3.6	2.38	4.6	68.9	58.2	347.9	24
T 294-94	13	3.6	2.38	4.6	120.6	50.1	299.0	24

Table 3. Average RM values from different testing procedures.

Bulk Stress (kPa)	T292-91I (MPa)	Bulk Stress (kPa)	T294-94 (Haversine) (MPa)	T294-94 (Triangular) (MPa)	T294-94 (Rectangular) (MPa)
483	228.9	84	118.2	65.3	71.0
551	253.6	104	141.3	87.5	87.7
621	241.0	125	149.2	104.6	101.1
689	234.6	136	158.2	108.0	88.0
378	143.8	171	172.4	116.5	116.8
447	174.8	205	182.5	118.2	132.8
516	195.1	276	249.3	127.3	147.4
585	202.0	345	247.5	161.0	170.5
241	93.0	414	240.9	188.3	174.2
276	112.1	378	252.6	142.0	152.0
345	147.8	412	274.0	167.0	163.2
414	168.9	516	302.6	210.7	199.3
136	77.7	517	311.4	168.3	179.0
171	103.8	552	334.7	175.7	201.3
205	122.6	690	367.6	224.8	246.8
97	80.8				
111	92.6				
125	102.8				

3.2 *Stress sequence*

The RM values of aggregate materials can be influenced by various factors among which the applied confining pressure is a very important factor (Rada et al. 1981). Thus, in order to better characterize such materials, it is desirable to evaluate RM tests under a wide range of confining pressures expected within the base and subbase. The AASHTO T292-91I and T294-94 methods use a variety of constant confining pressures and dynamic deviatoric stresses. However, the sequence of the applied pressures and stresses is different. The T292-91I starts with a higher confining pressure and deviatoric stress and ends with a lower confining pressure and deviatoric stress. On the other hand, the T294-94 uses a reverse sequence which starts with a lower confining pressure and deviatoric stress and ends with a higher confining pressure and deviatoric stress. Zaman et al. (1994) investigated these two stress sequences, in which two sets of RM tests were conducted under identical conditions, except for the stress application sequence. Their test results indicate that the stress sequence used in T294-94 procedure yields higher RM than the stress sequence used in T292-91I. This was attributed to the cyclic stress having a stiffening effect on the specimen structure because the stress application sequence goes from lower to higher in T294-94. The influence on the RM values due to the two different stress sequences was found to be approximately 15-34 percent according to their test results.

3.3 *Number of loading cycles*

To determine the number of repetitions necessary to reach a stable permanent deformation, the T292-91I method suggests comparing the recoverable axial deformation at the twentieth and fiftieth repetition. If the difference is greater than 5 percent, an additional 50 repetitions are necessary at that stress state. It has been reported by Khedr (1985) that the response of granular materials is fairly steady and stable after approximately 100 cycles of constant dynamic loading because the rate of permanent strain accumulation decreases logarithmically with the number of load repetitions. The number of loading cycles required by the T292-91I and T294-94 methods in the conditioning stage is the same (1000); however, it is different in the RM testing stages (50 and 100, respectively). Test results from the present study indicate that the RM values obtained from the T291-

91I method are very stable for each loading cycle and the difference of the recoverable axial deformation at the twentieth and fiftieth repetition is less than 5 percent. However, in T294-94 procedure, the loading frequency is 1.0 Hz (0.56 Hz for the T292-91I method as employed in this study), which is higher than that of T292-91I. Therefore, T294-94 needs a little longer time and a higher number of load cycles to reach the stable permanent deformation. Test results from the present study indicate that the RM values obtained from T294-94 procedure are stable after the seventieth loading cycle and the difference of the recoverable axial deformation at the seventieth and hundredth repetition is less than 5 percent. Hence, it is concluded that 50 and 100 loading cycles are adequate for testing procedure T292-91I and T294-94, respectively, to reach the stable permanent deformation.

3.4 *Loading waveform*

The AASHTO T292-91I method suggests that the triangular and rectangular wave forms are applicable to RM testing of subgrade soils and base/subbase materials in simulating traffic loading. However, T294-94 recommends that a haversine shaped load pulse with 0.1 second loading followed by a 0.9 second rest period be used for both soil and granular materials. A fixed loading duration of between 0.1 to 1.0 seconds and a fixed cycle duration of between 1.0 and 3.0 seconds are specified by T292-91I. Further, for a granular specimen, a minimum of 0.9 second relaxation between the end and the beginning of consecutive load repetitions is required.

Seed et al. (1967) stated that the loading duration, although having an influence on RM values, is not of major importance. RM values increase slightly when the time of loading is reduced. Loading frequency, on the other hand, may influence RM results significantly; RM values increase with increased frequency of load application.

Terrel et al. (1974) studied the influence of the wave pulse shape on the total and resilient strains induced in an asphalt treated base material. It was found that the triangular and the sinusoidal stress pulses produce similar effects on the resilience characteristics of the materials and that an equivalent square pulse, which is a reasonable approximation of the actual conditions, can be obtained.

In order to compare the effect of different loading wave forms, two sets of RM tests with triangular and rectangular waveforms were conducted by using the T294-94 procedure. The three wave forms used in the present study are shown schematically in Figure 2. In order to make the test results comparable, the areas under the triangular and rectangular loading forms are kept nearly the same. These tests are comparable because only wave forms vary while all other factors are kept the same. The average of the test results are given in Table 3 and also plotted in Figure 3, wherein it is observed that the haversine waveform produced substantially higher RM values (nearly 28-48 percent higher) than the triangular and rectangular wave forms. However, the RM values are nearly equal for the triangular and rectangular wave forms. The reason for this disparity could be that the higher loading frequency and the shorter loading duration used in the haversine waveform produced smaller elastic strains compared with the strains produced by the triangular and rectangular waveforms, hence, it gives higher RM values. It can be concluded that RM values increase with increased loading frequency and decreased loading duration.

3.5 *Location of LVDTs*

Compared with T292-91I, T294-94 requires two external LVDTs mounted at the top of the testing chamber to measure the amount of deformation in the specimen. However, T292-91I requires two LVDTs within the test chamber. In general, the externally mounted LVDT, which measures the entire length of the specimen, yields higher deformation and hence, gives lower RM values than the internally mounted LVDT. It was reported by Burczyk (1994) that RM measurements made with LVDTs on the ring inside the testing chamber consistently gave higher values than the LVDTs located on the loading piston. It should be noted that it is difficult to mount the internal LVDTs around the specimen and the internal LVDTs also tend to slightly disturb the specimen. Thus, in the standard testing procedure T294-94, only outside LVDTs are specified. Any internal LVDTs were not attempted in the present study.

3.6 *Comprehensive comparison*

In order to generally compare the effect of testing procedure on RM values, the mean RM values obtained from T292-91I and T294-94 are grouped

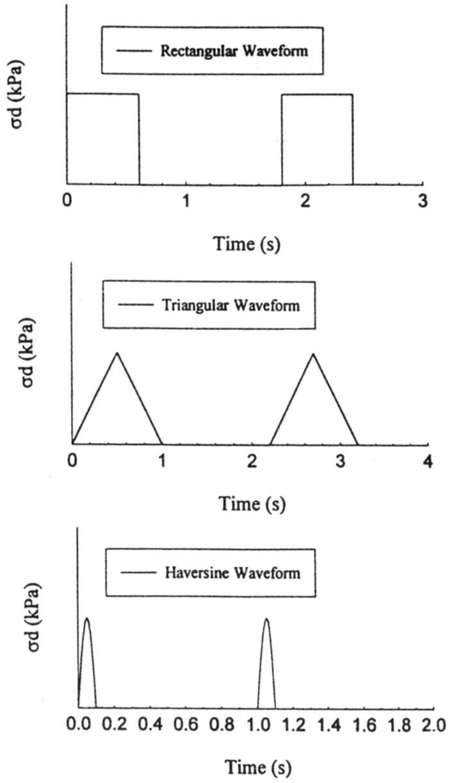

Figure 2. Three waveforms used in this study.

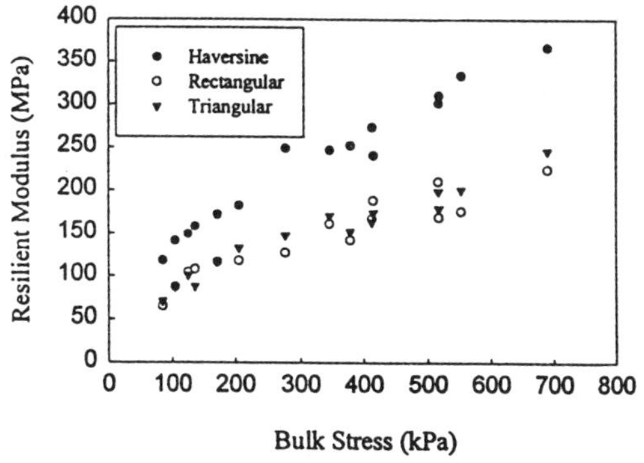

Figure 3. Mean RM values from different loading waveforms.

together (Figure 4). It can be observed that the RM values from T294-94 are nearly 25-55 percent higher than the values from T292-91I. The main reasons, as explained above, could be (i) the stress sequence used in T294-94 procedure has a stiffening effect on the specimen structure, (ii) the haversine waveform used in T294-94 has a higher loading frequency and a shorter loading duration, thus producing smaller elastic strains compared to the strains produced by the triangular and rectangular waveforms.

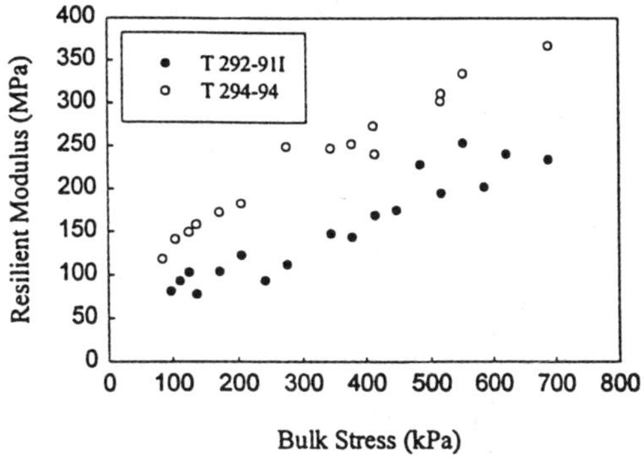

Figure 4. Mean RM values from T292-91I and T294-94 procedures.

From Table 2, it can be observed that there are some discernible changes in the static material properties which were measured after RM testing. Except for cohesion (C), other properties such as the friction angle (ϕ) and the unconfined compressive strength (U_c) present lower values for the specimens which were subjected to the T294-94 RM testing. The reason could be that the T294-94 procedure applies more loading cycles than T292-91I, and hence, it affects the structural integrity of a specimen. However, the RM tests are designed as nondestructive tests for both the testing procedures and the specimens are at a similar state after the RM testing. Hence, the changes in the static material properties are not significant.

4. CONCLUSIONS

The AASHTO T292-91I and T294-94 testing procedures were used in this study to investigate the influence of testing procedure on RM values for a limestone aggregate. On the basis of the data obtained the following are concluded: (1) The RM values obtained from the T294-94 method are nearly 25-55 percent higher than the values from the T292-91I method due to the different stress sequence and loading waveforms used in the two testing procedures. (2) There are some discernible changes in the static material properties of the aggregate which are measured after RM testing. However, the changes are not significant because the RM tests are designed as nondestructive tests. (3) The stress sequence used in the T294-94 procedure has a stiffening effect on the specimen structure because the stress application sequence goes from a lower to a higher level. (4) The haversine waveform used in the T294-94 method has a higher loading frequency and a shorter loading duration, and it produces smaller elastic strains compared to the strains produced by the triangular and rectangular waveforms. RM values increase with increased loading frequency and decreased loading duration. (5) The numbers of 50 and 100 loading cycles are adequate for the testing procedure T292-91I and T294-94, respectively, to reach a stable permanent deformation.

5. ACKNOWLEDGMENTS

Financial support from the Oklahoma Department of Transportation and the Federal Highway Administration is duly acknowledged.

6. REFERENCES

American Association of State Highway and Transportation Officials (AASHTO) 1986. AASHTO T274-82. Suggested Method of Test for Resilient Modulus of Subgrade Soils, Washington, DC.

AASHTO T292-91I, 1991. Interim Method of Test for Resilient Modulus of Subgrade Soils and Untreated Base/Subbase Materials, Washington, DC.

AASHTO T294-92I, 1992. Interim Method of Test for Resilient Modulus of Unbound Granular Base/Subbase Materials and Subgrade Soils-SHRP Protocol P46, Washington, DC.

AASHTO T294-94, 1994. Standard Method of Test for Resilient Modulus of Unbound Granular Base/ Subbase Materials and Subgrade Soils-SHRP Protocol P46, Washington, DC.

Burczyk, J.M., Ksaibati, K., Sprecher, R.A. and Farrar, M.J. 1994. Factors Influencing Determination of a Subgrade Resilient Modulus Value. Transportation Research Record 1462, TRB, Washington, DC, pp. 72-78.

Ho, R.K.H. 1989. Repeated Load Tests on Untreated Soils, a Florida Experience. Rep. No. FHWA-TS-90-031, Natl.Tech.Info. Service, Springfield, VA.

Khedr, S. 1985. Deformation Characteristics of Granular Base Course in Flexible Pavements. Trans. Res. Record 1043, TRB, Washington, DC, pp. 131-138.

Pezo, R.F., Claros, G., Hudson, W.R. and Stoke, K.H. 1992. Development of a Reliable Resilient Modulus Test for Subgrade and Non-Granular Subbase Materials for Use in Routine Pavement Design. Res. Rep. 1177f, University of Texas, Austin, TX.

Raad, L., Minassian, G.H. and Gartin, S. 1992. Characterization of Saturated Granular Bases Under Repeated Loads. Trans. Res. Record No.1369, TRB, Washington, DC, pp. 73-82.

Rada, C. and Witczak, W.M. 1981. Comprehensive Evaluation of Laboratory Resilient Moduli Results for Granular Material. Trans. Res. Record 810, TRB, Washington, DC, pp. 23-33.

Seed, H.B. 1967. Factors Influencing the Resilient Deformations of Untreated Aggregate Base in Two Layer Pavements Subject to Repeated Loading. Highway Res. Record No. 190, pp.19-57.

Standard Specifications for Highway Construction. 1988. Oklahoma Dept. of Transportation.

Terrel, R.L., Awad, I.S. and Foss, L.R. 1974. Techniques for Characterizing Bituminous Materials Using a Versatile Triaxial Testing System. STP 561, ASTM, Philadelphia, PA, pp. 47-66.

Thickness Design-Asphalt Pavements for Highways and Streets. 1981. Manual Series No. 1 (MS-1). Asphalt Institute, Lexington, KY.

Vinson, T.S. 1989. Fundamentals of Resilient Modulus Testing. Rep. No. FHWA-TS-90-031, National Tech. Info. Service, Springfield, VA.

Zaman, M.M., Chen, D.H. and Laguros, J.G. 1994. Resilient Moduli of Granular Materials. J. of Trans. Eng., ASCE, Vol. 120, No. 6, pp. 967- 988.

Computer Methods and Advances in Geomechanics, Yuan (ed.) © 1997 Balkema, Rotterdam, ISBN 90 5410 904 1

Back analysis method for determination of ground surface subsidence parameters due to subway construction*

Yang Junsheng & Liu Baochen
Changsha Research Institute of Mining and Metallurgy, China

Abstract: The back analysis method for determination of the basic two parameters such as the main influence angle β and convergence ΔA in the surface subsidence prediction method due to subway construction is introduced in this paper. This method is developed on the basis of field observations and of the stochastic medium theory for subsidence prediction of ground surface movements and deformations. Three history cases including Beijing subway in China, tunnel Thunder Bay in Canade and Healthrow Express Trial tunnel in London are analyzed in this paper.

1 INTRODUCTION

The method of Stochastic Medium Theory for prediction of ground surface subsidence due to subway construction is a new developed method in evaluation of environmental and ecological aspects. This method can be used in different conditions of soil with different method of excavation. The key problem of this method is that how can determination of basic parameters for the prediction. Those parameters are the main influence angle depended on the properties of the overburdens and the convergence of the excavation depended on the method of excavation. The determination method of those parameters above mentioned are two: determine the parameters by engineers experience and determine the parameters by using the back analysis method. It is clear that the last method is more comfortable for the practical use. In many cases there are displacement observations on the ground surface for the evaluation of influence of subway excavation on environmental conditions on ground surface. Then engineer can use the results of those observations for back analysis of determine the paramaters of ground surface subsidence. Using the parameters obtained from back analysis tanβ and ΔA for prediction the ground surface subsidence gives the best result.

2 BACK ANALYSIS METHOD

According to the optimization principle the following definitions are accepted:

W_i^0 is the observed ground surface subsidence in mm,

U_i^0 is the observed ground surface horizontal displacement in mm,

W_i is the predicted ground surface subsidence in mm,

U_i is the predicted ground surface horizontal displacement in mm,

and the object function F(X), then

$$F(X) = \sum_{i=1}^{m}(W_i - W_i^0) + \sum_{j=1}^{n}(U_j - U_j^0) \qquad (1)$$

* Project supported by National Natural Science Fundation of China.

where m is the number of the subsidence observation points on ground surface,

n is the number of the horizontal displacement observation points on ground surface,

and $X = (tan\beta, \Delta A)$.

From[1] in the theory of stochastic medium the ground surface subsidence W(X) and ground surface horizontal displacement U(X) are:

$$W(X) = \iint_{\Omega - \omega} \frac{tan\beta}{\eta} exp[-\frac{\pi tan^2\beta}{\eta^2}(X-\xi)^2] d\xi d\eta \quad (2)$$

$$U(X) = \iint_{\Omega - \omega} \frac{(X-\xi)tan\beta}{\eta^2} exp[-\frac{\pi tan^2\beta}{\eta^2} (X-\xi)^2] d\xi d\eta \quad (3)$$

where: Ω is the original area of cross-section before convergence of subway,

ω is the area after the cross-section convergence of subway.

It is supposed that the cross-section of the subway has a circular shape or like circular shape and the ΔA is convergence or average convergence. Then the following relationship is obtained:

$$\Delta A = A - \sqrt{A^2 - \frac{1}{\pi}(\Omega - \omega)} \quad (4)$$

where A is the oraginal radius of subway cross-section.

The object funtion is the function of parameters (tanβ, ΔA). The purpose of back analysis is to fined a group of parameters which gives the minimum velum of the object function.

Because of the complex of function (2) and (3), the Powell method is used as the optimization method[2].

According to the optimization design method and formulas (2) and (3) a microcomputer program SUBOPT is finished for determination of ground surface subsidence parameters tanβ and ΔA as show in figure 1.

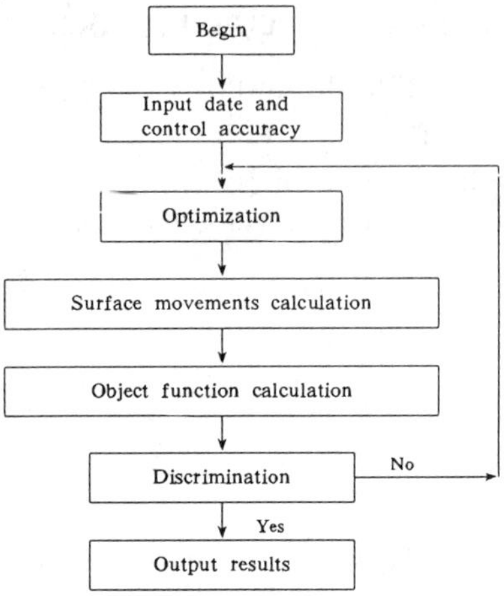

Fig. 1 Diagram of SUBOPT

3 HISTORY CASES ANALYSIS

3.1 *Fuxinmen section of Beijing subway*[3]

The subway with cross-section of 8 meters wide and 6.4 meters hight is excavated in soft soil at 11 meters below the ground surface. The observation results are showing in Tab. 1.

Tab. 1 Observation results

Distance from the centre (m)	−12.6	−8.4	−4.2	0	3.0	8.0
Surface subsidence (mm)	0.94	5.38	18.35	29.70	22.74	9.84

The object function is

$$F = [W(-12.6) - 0.94]^2 + [W(-8.4) - 5.83]^2 + [W(-4.2) - 18.35]^2 + [W(0) - 29.70]^2 + [W(3.0) - 22.74]^2 + [W(8.0) - 9.84]^2$$

Using the program SUBOPT the parameters can be obtained as tanβ = 1.521, ΔA = 15.60mm. Input those parameters into program SUBWAY1 it can be obtain the ground surface movements W(X), U(X) and deformations T(X), K(X) and

E(X). The comparison between the subsidence prediction (curve) and the observation (points) is showing in Fig.2 for the cross-section B(S) 161+56. The predicted maximum subsidence is 29.77mm, and the measured maximum subsidence is 29.70mm. Then the accuracy is 0.07mm.

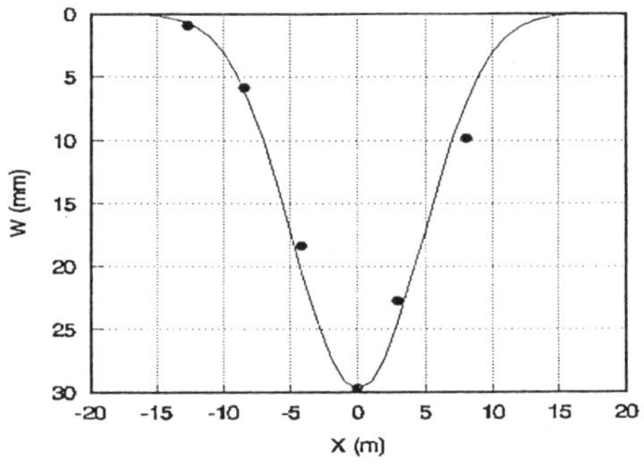

Fig.2 Comparison between prediction and observation

3.2 *The tunnel Thunder Bay in Canade*[4]

The tunnel with a diameter of 2.47 meters is excavated in soft soil at 9.27 meters below the ground surface. The observation results are showing in Tab.2.

Tab.2 Observation results

Distance from the centre (m)	-8.5	-6.0	-3.8	-2	0
Surface subsidence (mm)	0	6	15.5	40.8	50
Distance from the centre (m)	2	3.8	6.0	8.5	
Surface subsidence (mm)	39.8	22	6	0	

Using the progranm SUBOPT the parameters can be obtained as $\tan\beta = 1.487$, $\Delta A = 47.81$mm. The comparison between the predicted subsidence and observed results is showing in Fig.3.

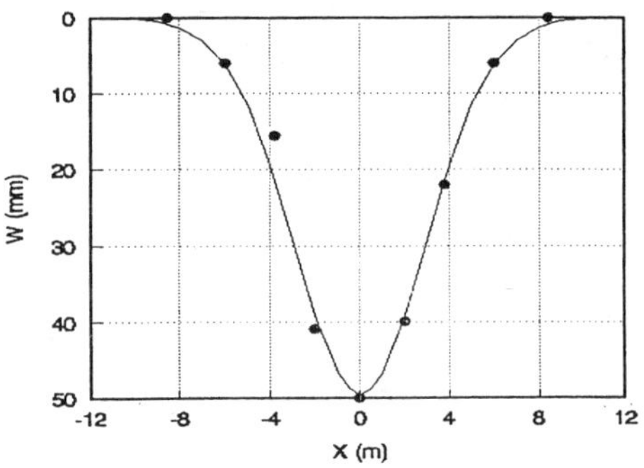

Fig.3 Comparison between prediction and observation

3.3 *Heathrow Express Trail tunnel in London*[5]

The tunnel with radius of 4.334 meters is excavated in clay layers at 16.67 meters below the ground surface. The observation results are showing in Tab.3.

Tab.3 Observation results

Distance from the centre (m)	-22	-8	-3	0	4
Surface subsidence (mm)	9	18.8	35.4	40.3	37.7
Distance from the centre (m)	7	10	16	20	32
Surface subsidence (mm)	27.7	18.1	6.5	3.8	1.0

Using the back analysis program SUBOPT the parameters can be obtained as $\tan\beta = 1.106$, $\Delta A = 30.06$mm. The comparison between the predicted subsidence and observed results in showing in Fig. 4.

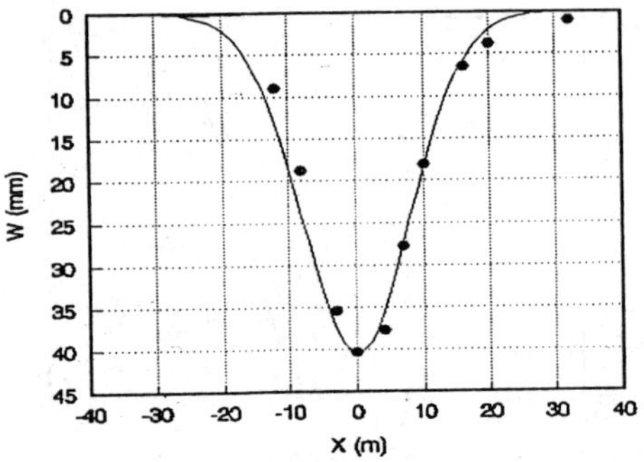

Fig.4 Comparison between prediction and observation

4 CONCLUSION

The history cases show that the back analysis method developed in this paper can applied for determination of the ground surface subsidence parameters in the subway engineering. The prediction accuracy in quite satisfactory for the engineering purpose.

REFERENCES

[1]Liu Baochen Ground surface movements due to underground excavation in P.R. China. Chap. 29 Vol. 4 "Comprehensive Rock Engineering" Pergamon Press. 1993.

[2] Zhang Kezhung Engineering optimization method and analysis. Xian Jaoton University Press, 1989(in Chinese).

[3]Wang Monshu, Rou Zhung Underground excavation of Fuxinmen section of Beijing Subway. "Tunneling and Underground Engineering" 1989.2.(in Chinese).

[4]B.M. lee, R.K. Rowe Analysis of three-dimensional ground movements: the Thunder Bay tunnel. Canadian Geotechnical Journal. Vol. 28. №1. 1991

[5]B. M. New, K. H. Bowes Ground movement model validation at the Healthrow Express trial tunnel. Tunneling '94

Computer Methods and Advances in Geomechanics, Yuan (ed.) © 1997 Balkema, Rotterdam, ISBN 90 5410 904 1

Stability reliability analysis and optimization design of slope

Baotian Wang
Hohai University, Nanjing, China

Yuwen Luo
Hydroelectric Bureau of Longqian, Zhejang, China

ABSTRACT: According to the maximum principle which was put forward by academician Jiazheng Pan,the limit state equations have been derived for the method of ultimate equilibrium analysis of slope stability to general sliding surfaces.The calculating steps of reliability analysis method have been given out from JC method. A model of optimzation design of slope has built up. A example makes clear the advantage of slope design by the method of reliability optimizatoin analysis.

1 RELIABILITY ANALYSIS METHOD OF SLOPE STABILITY

Traditionaly,the method of ultimate equilibruim analysis of slope stability is as follows. Make the physical and mechanical parameters of soil and rock of slope and the force which put on the boundary of the slope or in the slope be constants. Calculate the safety factor F_s of the slope. Compare the safety factor F_s with the value of acceptable safety factor $[F_s]$. The slope is considered safety when $F_s > [F_s]$. This is called safety factor methed. Different $[F_s]$ is needed for different slopes because the importance of slopes are not same and the consequences are different if slope slides. The value of $[F_s]$ are different in different departments and are only experience values. This is not reasonable.

Reliability theory puts into use in slope stability .It uses stability reliability Pr(or failure probability $P_f = 1 - P_r$) to express the safety of slope. We use formula (1) to difine the expecting total engineering cost.

$$C = C_0 + P_f \cdot C_f \tag{1}$$

In which: C---expecting total engineering cost;

C_0---construction cost;

C_f---lost cost if slope slipped;

P_f---failure probability of slope

Formula (1) gives a direct relationship between the importance(expressed by C_f) and failure probability. The more important the slope, the higher the lost if it slipped.

1.1 Limited state equation of stability analysis of slope for Bishop's slice method

There is a slope as in Fig.1. Its probable sliding surface is a general curved surface. If the slice method is used, the forces on every slice are as follows.

a. W_i: the force applied on vertical direction, it includes gravity and inertia force of earthquake etc.;

b. Q_i: the force acted on horizontal direction,it includes seepage force and inertia force of earthquake etc.;

c. E_i, E_{i+1}: the normal force applied by next slices on both sides of slice i;

d. N_i':the effective normal force on sliding surface;

e. U_i: the force of pore water pressure applied on sliding surface;

f. S_i: the resistance applied on sliding surface.

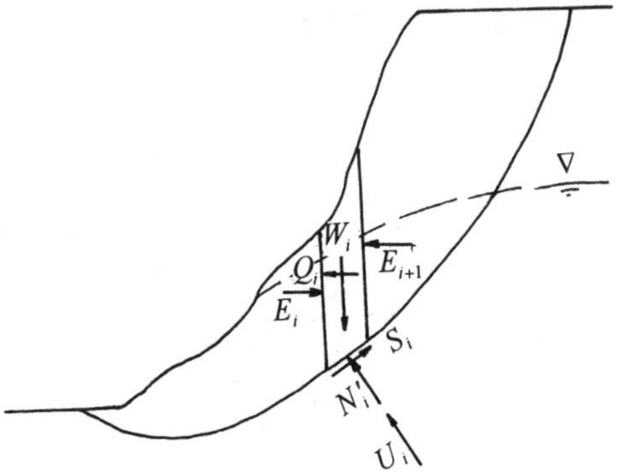

Fig.1 Analysis of force applied on slice

According to the maximum principle, all points on sliding surface are at ultimate state when the slope is at ultimate state. It means that the sliding surface of any slice i satisfies the ultimate state equation as follows.

$$T_{fi} - S_i = 0 \tag{2}$$

In which: i=1,2,...,N, N is slice number;

T_{fi} ---ultimate resistance of slice i on slice surface

S_i --- sliding force of slice i.

From Mohr-Coulomb criterion it has a equation as follows.

$$T_{fi} = c'_i \times l_i + N'_i \times tg\varphi'_i \tag{3}$$

In which: c'_i, φ'_i ---effective strength indexes; the others have the same meanings as before.

There is a equation from the equilibrium condition on the moving direction of slice i as:

$$S_i = \Delta E_i \cos\alpha_i + Q_i \cos\alpha_i + W_i \sin\alpha_i \tag{4}$$

In which: $\Delta E_i = E_{i+1} - E_i$;

the others have the same meanings as before.

The equilibrium condition on vertical direction is that:

$$W_i = N'_i \cos\alpha_i + U_i \cos\alpha_i + S_i \sin\alpha_i \tag{5}$$

From formulae (2)、(3) and (4) get the formula:

$$\Delta E_i = \sec\alpha_i (c'_i l_i + N'_i tg\varphi'_i) - Q_i - W_i tg\alpha_i \tag{6}$$

Equation (7) can be obtained according to equilibrium of landslide as:

$$\sum_{i=1}^{N} \Delta E_i = 0 \tag{7}$$

Put formula (6) to formula (7) :

$$\sum_{i=1}^{N} \left[\sec\alpha_i (c'_i l_i + N'_i tg\varphi'_i) - Q_i - W_i tg\alpha_i \right] = 0 \tag{8}$$

Formula (9) is led out from formulae (2) (3) and (5)

$$N'_i = \frac{W_i - U_i \cos\alpha_i - c'_i l_i \sin\alpha_i}{\cos\alpha_i + \sin\alpha_i tg\varphi'_i} \tag{9}$$

Substitute N'_i in formula (8) with that in formula (9) to get the limit state equation as follows.

$$\sum_{i=1}^{N} \left[\sec\alpha_i \left(c'_i l_i + \frac{W_i - U_i \cos\alpha_i - c'_i l_i \sin\alpha_i}{\cos\alpha_i + \sin\alpha_i tg\varphi'_i} tg\varphi'_i \right) - Q_i - W_i tg\alpha_i \right] = 0 \tag{10}$$

When the static stability problem of slopes is considered, the parameters in limit state equation (10) that must be considered as random variables are c'_i 、 φ'_i and ρ etc. But usually only c'_i and φ'_i are sensible random variables.

1.2 Reliability analysis method of slope stability

The steps to calculate slope stability with reliability analysis method are as follows.

(1) Divide soil and rock space of slope into several layers according to the engineering property of soil and rock. Obtain the statistics of physical and mechanical property indexes of every soil and rock layers. Give reasonable density functions (or distribution functions) and characteristic values (means, standard deviations,etc.) to the indexes which need to be random variables.

(2) Give out the interested and probable range of sliding surface according to the shape of slope section and the engineering property of the soil and rock of slope.

(3) Research all probable sliding surface in the range giving above to obtain the minimum P_r of

them. The minimum P_r is called the reliability of the slope[4].

Some people profer reliable index β to reliability P_r. Reliable index β is defined as:

$$\beta = \phi^{-1}(P_r) \qquad (11)$$

2 THE OPTIMIZATION DESIGN PRINCIPLE OF SLOPE

2.1 Optimizatiom design foundation of slope

We have obtained the failure probability $P_f(=1-P_r)$ of slope above. All engineerings allow a small failure probability [P_f]. It is easy to understand that different engineerings allow different [P_f]. The value of [P_f] of a slope depends upon the importance of it and the amount of loss if landslide takes place. This means if the failure probability of a slope designed is no more than [P_f], the design of the slope is stable. But we do not know whether this slope economical. If we stop here, the reliability design is very similar to the safety factor design.

The expecting total engineering cost C has been defined as formula (1). The optimization design of slope is to make C become minimum. If a slope is dug or filled at steep gradient, the C_0 in formula (1) must be low,but $P_f \times C_f$ in formula (1) must be high because of the high value of P_f; If the slope is at gentle gradient, the value of $P_f \times C_f$ becomes low because of the decrease of P_f, but C_0 becomes high. There is a reasonable gradient that makes the C be the least. This gradient of a slope which the expecting total cost is the least is the optimization design of the slope.

2.2 Example of the optimization of a fill slope

Fill slopes are widely used in engineering , embakment 、 earth dam 、 tailing dam, etc. The cost of construction C_0 and the lost cost C_f if landslide takes place are very different for the different engineerings. The analysis of cost combination is the most important to use optimization method to design a slope. there is a example as follows.

A embankment of highway(Fig.2) is 2.4m in height, 28m in road surface wide and 2m in road shoulder wide. The foundation is clay. Earthfilled cost is $CK_1 = 6.0$ RuanRMB / m^3.Land acquisition cost is $CK_2 = 80$ RuanRMB / m^2. Road surface cost is $CK_3 = 1000$ RuanRMB / m. If landslide has taken place, the cost of decreasing traffic capacity and the cost of probable traffic accident failure is $C_{f1} = 10^5 \cdot L^{1.2}$ YuanRMB(L:the length which includes in landslide from boundry of road surface to road center, unit: m),the repaired cost is $C_{f2} = 10^3 \cdot V_f^{1.2}$ (V_f: volume of landslide at a meter length,unit: m^3). The model of landslide(Fig.2) is that: sliping surface is in foundation, vertical tensile fractures take place in embankment. The parameters are that: q=18 kN/m^2 is traffic load; the coherence of clay in foundations c is a random variable, the distribution of c is lognormal, mean $\bar{c} = 30kP_a$, coefficient variation $V_c = 0.18$; internal friction angle of clay in foundation φ is the other random variable, the distrbution of φ is normal, mean $\bar{\varphi} = 10°$, coefficient variation $V_\varphi = 0.10$; $r_{c\varphi} = -0.3$ is correlation coefficient of c and φ; $\gamma_1 = 18.5 kN/m^3$ is unit weight of embankment; $\gamma_2 = 19.0 kN/m^3$ is unit weight of foundation soil.

The cost of construction of a meter highway C_0 can be expressed as:

$$C_0 = CK_1 \cdot V_0 + CK_2 \cdot S + CK_3 \qquad (12)$$

In which: $CK_1 = 6.0, CK_2 = 80.0, CK_3 = 1000$;

V_0: the embankment volume of a meter highway ,unit: m^3;

S: the land acquisition of a meter highway,unit m^2.

The lost cost if landslide takes place is as follows.

$$C_f = 10^5 \cdot L^{1.2} + 10^3 \cdot V_f^{1.15} \qquad (13)$$

The meanings of L and V_f are as before

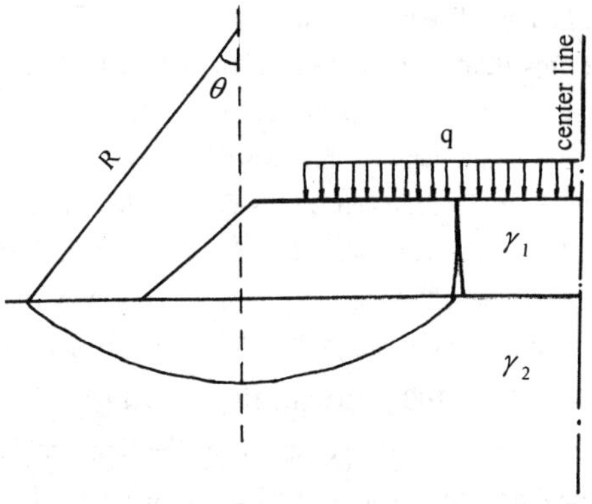

Fig.2 Failure model of highway embankment

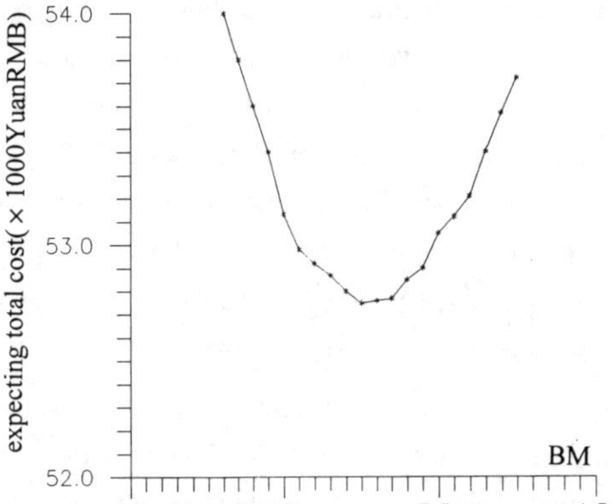

Fig.3 Relationship of C and slope gradients

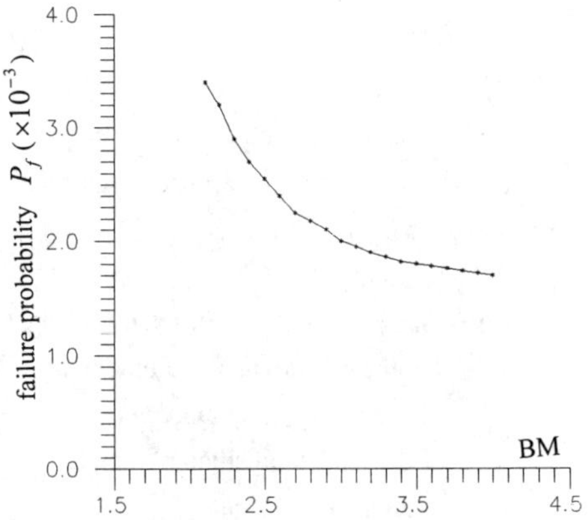

Fig.4 Relationship of P_f and slope gradients

In the example, The steps to get min C and the optimized slope gradient corresponding to min C is that:

(1) Assume a gradient 1:BM, calculate its P_f according to the steps in 1.2; Obtain L and V_f corresponding to the sliding surface which failur probabilty is P_f the maximum.

(2) Calculate C_0 and C_f using formulae (12) and (13) according to gradient 1:BM, then calculate expecting total cost C using formula (1) of the slope which gradient is 1:BM.

(3) Assume other gradients $1:BM_i$ and repeat steps (1) and (2) to get a series of C_i and $1:BM_i$. Use C_i and BM_i drawing C ~ BM curve in Fig.3. We know that min C is 5.275×10^4 YuanRMB / m, the optimized design slope gradient is 1:3. The failure probability of the slope at gradient 1:3 is 0.2%, the reliability index is 2.87. The expecting construction total cost C (Fig.4) is almost constant when the gradient varies from 1:3 to 1:3.2. We could using any gradients from 1:3 to 1:3.2, but the cost C_0 increases as the gradient decreases.The failure probability P_f and lost cost $P_f \cdot C_f$ decrease as the gradient decreases(Fig.4).

REFERENCES

1.P.thoft-Christensen, Reliability theory and its application in structural and soil mechanics. Applied Sciences No.70

2.Shiwei Wu, Structural Reliability Analysis, People's Communication Publishing House,1992

3.Vanmarke E.E., Reliability of earth slopes. ASCE, Vol.103

4. Baotian Wang, Reliability analysis of complex earth slopes, Joural of Hohai University, Vol.22, No.4, 1994

Computer Methods and Advances in Geomechanics, Yuan (ed.) © 1997 Balkema, Rotterdam, ISBN 90 5410 904 1

Analysis of the performance parameters of the hydraulic rock driller with wavelet theory

Li Mingfa, Zhang Yafang & Huang Li
Wuhan Automotive Polytechnic University, China

Abstract: This article deals with the analytical process of the performance parameters of the hydraulic rock driller with wavelet theory. In such analysis, different wavelet analytical methods are applied according to the characteristics of the testing parameters. The working conditions of the equipment can be known more clearly through the wavelet analysis of the hydraulic rock driller,and the corresponding performance supervision and failure diagnosis can be conducted accordingly.

1. Preface

The hydraulic rock driller is a mining machine powered by high-pressure liquid and with highly technical difficulties. With high-speed reciprocating motion and output of large moment of torsion asits basic characteristics, it is smaller in power consumption, good in nechanical performence, and high in drilling speed. Its appearence has remarkably increased the efficiency of mine tunneling, which is welcomed by rock-boring field workers.

However, it remains the crucial task to probe the technical responsibility of hydraulic rock drillers and apply the research results into practice. Therefore, the testing and checking of their performance parameters are very important and new methods are needed in the parameter analysis.

In this article, the wavelet analysis is introduced as a way to study different performance parameters of hydraulic rock drillers,and analyse their dynamics, so as to facilitate the performences upervision and failure diagnosis. It is a significant method for the product improvement of hydrualic rock drillers in the drilling quality and reliability.

2. Testing table

The versatile testing table for hydraulic rock drillers is a automatic system with a computer as the principal part,which can check and measure all the basic parameters of the machine. The flow chart of the system is shown in Fig 1.

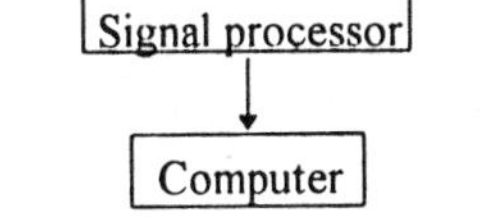

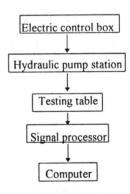

Fig 1. Testing system flow chart

The electric control box logically controls the sequential actions of the pump station, which provides power to the loading system on the testing table. Different parameter signals measured of the hydraulic rock driller are transmitted to the signal processor through corresponding sensors for sampling and A/D transformation, and then transmitted to the computer for analysis and process. The results will be displayed and printed by the computer.

3. The binary wavelet transform of the digital signals

Each performance parameter of the hydraulic rock driller obtained by sampling is a digital signal. We may assume it to be $\{d_n\}_{n\in z}$. With aproper unit of time selected, the sampling density can always be thought of as 1. $\{d_n\}$ can be analogized in the way as follows: $d_n = f*\varphi(n)$. where $f \in L^2$; It can be proved mathematically that f does exist. Though there can be different ways of analogue for a digital signal, the ununiqueness of $\{d_n\}_{n\in z}$ does not influence the uniqueness of f transform in discussing only the binary wavelet transform of digital signals. This is the basic concept. Following is the method.

Examine a low flux filter $\phi(t)$. Assume $\phi(t)$ is a window function,satisfying $\phi(0)=1$, and $\hat{\phi}(+\infty)=0$. Let $\phi_s(t)=\left(\frac{1}{s}\right)\phi\left(\frac{t}{s}\right)$, then it is easy to see that the window of $\hat{\phi}_{2^{j-1}}(\omega)$ is contained in the window of $\hat{\phi}_{2^j}(\omega)$, thus $f*\phi_{2^{j-1}}(t)$ can be derived from $f*\phi_{2^j}(t)$ .Sppose $\phi(t)$ also satisfies the condition $\hat{\phi}(2\omega)=H(\omega)\hat{\psi}(\omega)$, where

$H(\omega)=\sum_n h_n e^{in\omega}$, and $H(\omega)\le 1$

then,

$$\phi_{2^j}=\sum_n k_n\phi(t-n) \tag{1}$$

Equation (1) is called the double-measure equation of $\phi(t)$. The function which satisfies such double-measure equations is referred to as "measure function". What we mean here is that the function $\phi_{2^j}(t)$ in each measure derived from the measure function can be represented by the linear combination of the parallel translation of all the integers of $\phi_{2^{j-1}}(t)$, the fuction in the preceding measure, It is thus guaranteed that there is a recurrence formula for the roll product of $f(t)$ and $\phi_{2^j}(t)$ From (1) it is easy to derive $\phi_{2^{j-1}}=\sum_n h_n\phi_{2^{j-1}}\left(t-2^{j-1}n\right)$,then, we have the recurrence formula

$f*\phi_{2^j}(t)=\sum_n h_n f*\phi_{2^{j-1}}\left(t-2^{j-1}n\right)$.

Assume $\hat{\psi}(2\omega)=G(\omega)\hat{\phi}(\omega)$. Since $\psi(t)$ is a band filter, naturally it requires $G(0)=0$.And we may also assume $G(\omega)=\sum_n g_n e^{in\omega}$, then we have

$\psi_{2^j}(t)=\sum_n g_n\phi_{2^{j-1}}\left(t-2^{j-1}n\right)$,thus

$f*\psi_{2^j}(t)=\sum_n g_n f*\phi_{2^{j-1}}\left(t-2^{j-1}n\right)$. Given that the sample value of $f*\psi(t)$ at integer k is $C_k^0=f*\psi(k)$ then from $\{h_n\}$ and $\{g_n\}$ the integal value of binary wavelet transform in each measure $d_k^j=f*\psi_{2^j}(t)$ can be derived. The formula is as follows: $C_k^0=f*\phi(t)$; $d_k^j=\sum_n g_n C^{j-1}_{k-2^{j-1}n}$;

$$C_k^j=\sum_n k_n C^{j-1}_{k-2^{j-1}n} \tag{2}$$

In the equation, $k=0,\pm1,\pm2,\cdots,\quad j=1,2,3,\cdots,M$. $\psi(t)$ is a binary wavelet function, and there is the following recursion formula

$f*\psi_{2^{j-1}}(t)=\sum_n\left[k_{-n}f*\phi_{2^j}\left(t-2^{j-1}n\right)+g_{-n}f*\psi_{2^j}\left(t-2^{j-1}n\right)\right]$

Thus there should be the recursion formula corresponding to Equation (2):

$$C_k^{j-1}=\sum_n\left[k_{-n}c^j_{k-2^{j-1}n}+g_{-n}d^j_{k-2^{j-1}n}\right] \tag{3}$$

$k=0,\pm1,\pm2,\cdots,\quad j=M,M-1,\cdots,1$

In practical use only with $\{h_n\}$ and $\{g_n\}$ known, wavelet transform and recursive computation can be conducted with Equations (2) and (3).When different binary wavelet functions are selected, only different coefficients $\{h_n\}$ and $\{g_n\}$ should be selected.

4. Computation methods

Each performence parameter signal $f(x)$ of the hydraulic rock driller is given in the form of a group of discrete sequences $D=(d_n)_{n\in z}$. It can be proved that any discrete signals with limited energy can be regarded as the homogeneous sample values after smoothing the original signals with a smoothing operator whose measure is 1. Therefore we have $S_1 f(n)=d_n$, which can be substituted as initial condition into there currence analytic formula to compute the wavelet transform of each order by recursion. If there are N sample values of $D=(S_1 f(n))_{n\in z}$, then there will also be N sample values of the smooth signals $S_{2^j}^d = S_{2^j} f(n)$, the wavelet transform $\omega_{2^j} f(n)$. In analysing all the performane parameters of the hydraulic rock driller,the number of sample values for each parameter is 512.

Let $\omega_{2^j}^d = (\omega_{2^j} f(n))_{n\in z}$ and $S_{2^j} f(n)_{n\in z}$, from the recurrence analytic formula we can write the computing program of the wavelet transform:

$j=0$

$$\textit{while}\ (j<J)$$
$$\omega_{2^j}^d f = S_{2^j}^d f * G_j$$
$$S_{2^{j+1}}^d f = S_{2^j}^d f * H_j$$
$$\textit{END of while}$$

and the inverse transform program is

$j=J$

$$\textit{while}\ (j>0)$$
$$S_{2^{j-1}}^d = \omega_{2^j}^d f * \overline{G}_{j-1} + S_{2^j}^d f * \overline{H}_{j-1}$$
$$\textit{END of while}$$

Since only 512 sample values are taken for each performance parameter, when encountering the boundary problem in computation, we adopt the symmetric extension with the first and last points of each signal channel ascenters.

5. Wavelet analysis of the performance parameters of the hydraulic rock driller

5.1 Wavelet analysis of the moment of torsion

The wavelet analysis of the moment of torsion is conducted by utilising the sigularity of the testing signals of the binary wavelet transform. Fig 2 shows the wave patterns of the signals of the moment of torsion measured on the testing table of the hydraulic rock driller.the number of samples is 512. The signals have been analysed to 2^4 in resolving power with the equations $c_k^j = \sum_n h(n-2k)c_n^{j-1}$ and $d_k^j = \sum_n g(n-2k)c_n^{j-1}$ Fig 2(b)-(e) shows the binary wavelet transforms of the signal pattern in 5 measures. Patterns of the measure function and the wavelet function in the field of time and frequency are shown in Fig 3 and Fig 4. The filter coefficients are shown in Table 1.

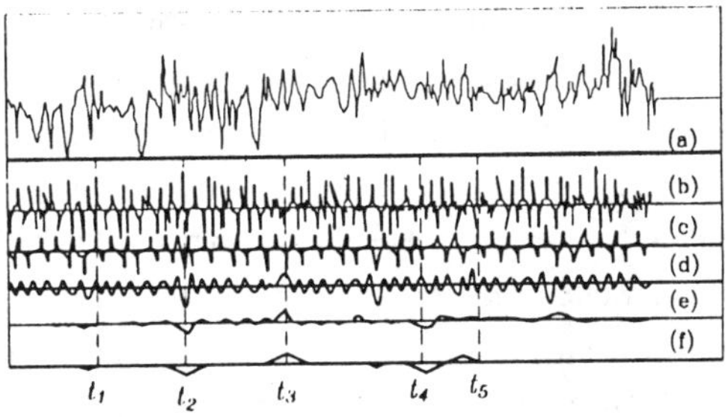

Fig 2 Signal wave patterns and wavelet transforms of the moment of torsion

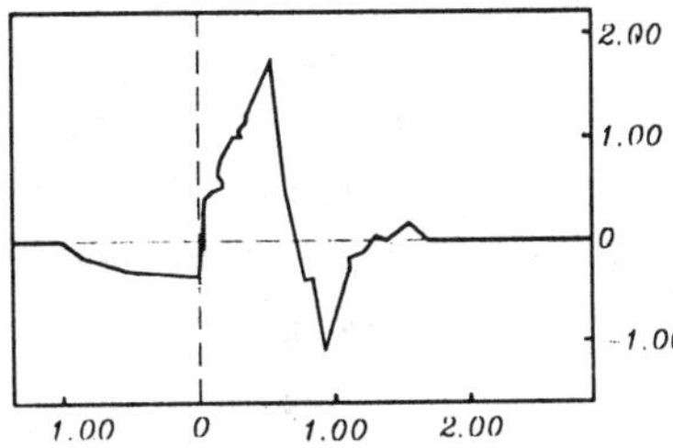

Fig 3 Pattern of the measure function of the moment of torsion

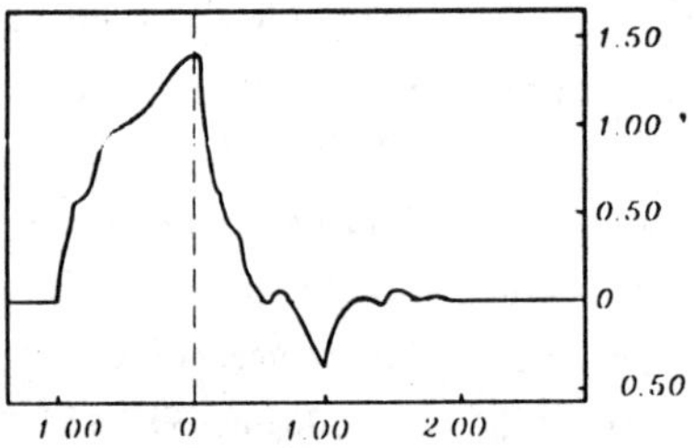

Fig 4 Pattern of the wavelet transform of the moment of torsion

Table 1 Filter coefficients

n	0	1	2	3	4
h(n)	0.483	0.837	0.224	-0.129	-0.086

Apart from clearly showing the singular point and the singular degree of the signals of moment of torsion, the results of the wavelet transform shown in the figure can also separate the frequency. In the first measure, large jumps disappear, and signals vibrate with the horizontal axis as the balance position. In the second measure, the high-frequency component disappears, with only the low-frequency component left. And from the third measure, low-frequency component disappears, with only the singular signal sign left. The singular degree of the signals of the moment of torsion can be determined according to the position and the value of the largest point in each measure after wavelet transformation. In Fig2 we can see clearly that the wavelet transforms of the signals of the moment of torsion near t_2、t_3、t_4、t_5 approach the extreme value. Therefore, sudden changes takeplace in t_2、t_3、t_4 and t_5. These sudden changes reflect the instability of the function of the signals of moment of torsion in these points.The moments are either too large or too small. Too large a moment of torsion will cause damage to the rear sleeve of the drill, while too small a moment of tortion will cause insufficient pressure and flow inthe system.

5.2 Wavelet analysis of the work flow

The signal in Fig 5(a) is the work flow wave-pattern measured on the testing table when the hydraulic rock driller works normally. The number of sample points is 512. Fig 5(b)-(j) show the binary wavelet transforms in the 9 measures of the pattern. The wavelet function and the measure function are respectively shown in Fig 6 and Fig 7. Table 2 gives the filter coefficients.

Table 2 Filter coefficients

n	0	1	2	3	4
h(n)	0.160	0.603	0.724	0.138	-2.242
n	5	6	7	8	
h(n)	-0.322	0.077	-0.006	0.003	

Fig 5 shows the result of the work flow signal wave-pattern and the binary wavelet transform of the hydraulic rock driller in a testing cycle $\left(d^1 - d^9\right)$. The driller's work flow sees the cyclic process of "normal flow → the emergence of flow instability → flow instability (higher or lower) → steadily higher or lower flow → eakening of the higher or lower flow phenomenon → normal flow". This process is clearly illustrated by the results of the wavelet transforms in Fig 5, thus providing simple and clear evidence for the determination of the starting and ending momemts of each stage.

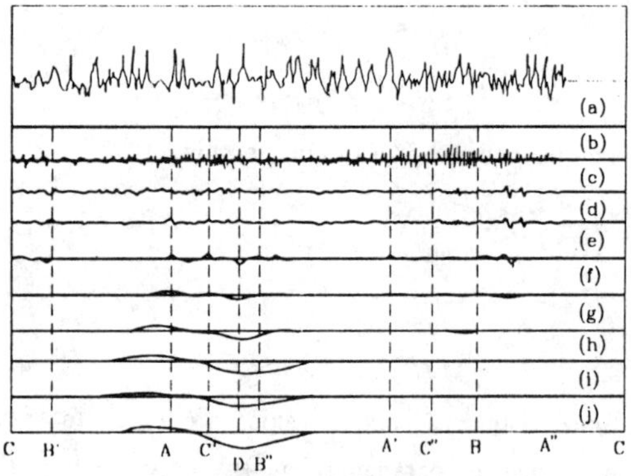

Fig 5 Flow signal wave-pattern and the pattern of wavelet analysis

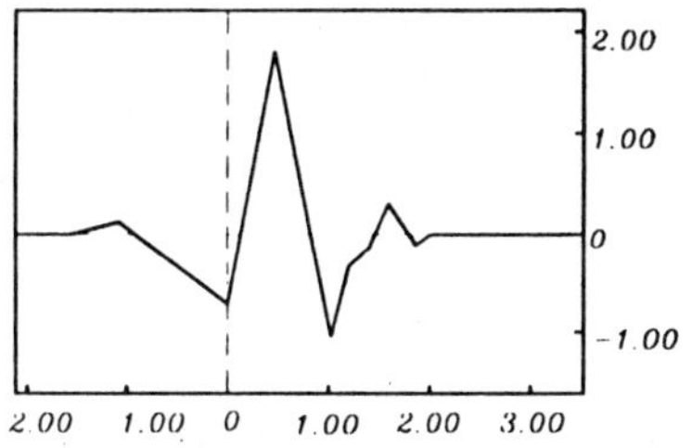

Fig 6 Pattern of the wavelet function of the flow

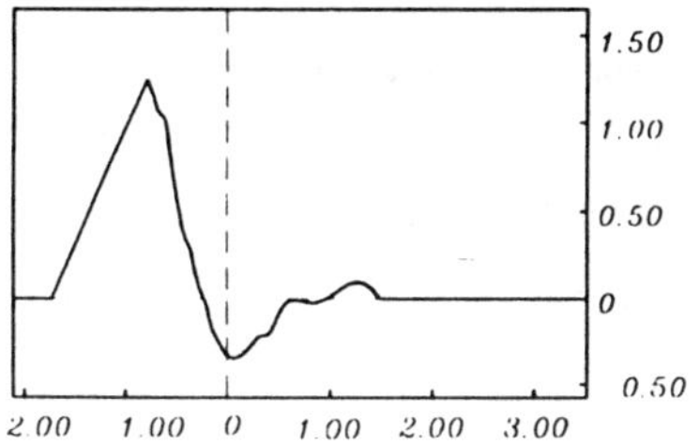

Fig 7 Pattern of the measure function of the flow

5.3 Wavelet analysis of the impulsive frequency

The technique of data compression and feature extraction is adopted in the wavelet analysis of the impulsive frequency. The signal in Fig 8 is the wave-pattern of the impulsive frequency when the hydraulic rock driller works normally. The number of sample points is 512. Fig 8(b)-(i) show the binary wavelet transforms of the pattern in 7 measures. The pattern of the wavelet function is shown in Fig 8. Table 3 gives thefilter coefficients.

The theoretical deduction of the returning from the largest localpoint to the original signal will not be introduced here because it involves a lot of deeper concepts of mathometics. With these largest points the impulsive frequency signals can be returned. The returned wave-pattern naturally bears the information of the original signal. It is a feature extracted from the signal with wavelet transform and can be utilized as an evidence to judge the working condition of the hydraulic rock driller.

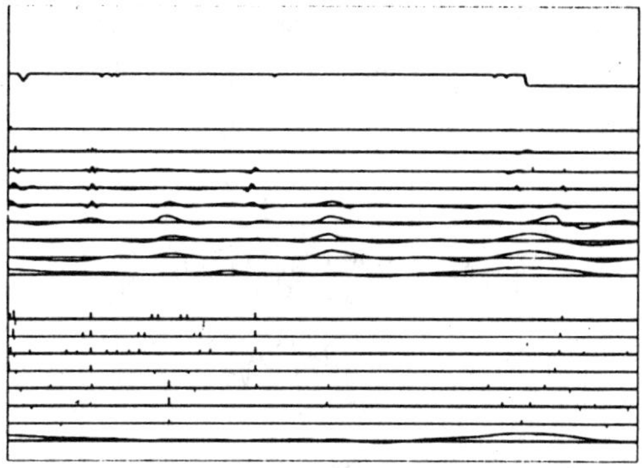

Fig 8 Pattern of the wavelet analysis of the impulsive frequency

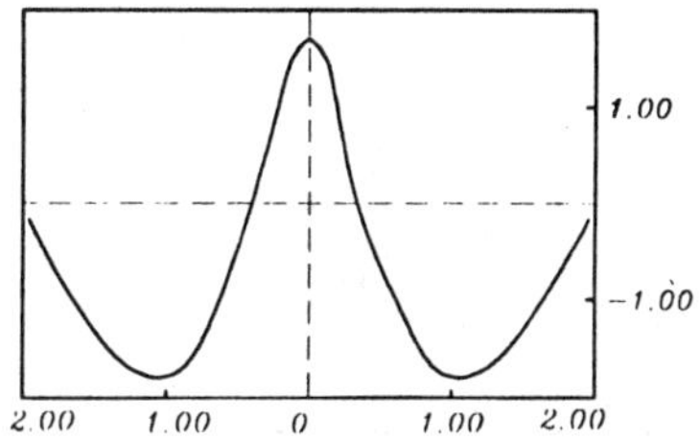

Fig 9 Pattern of the wavelet function of the impulsive frequency

Table 3 Filter coefficients

n	0	1	2	3	4	5	6
h(n)	0.026	0.188	0.527	0.688	0.281	-0.249	-0.177

5.4 Wavelet analysis of the pressure

In the wavelet analysis of the pressure, information-noise separation should be realised based on the multi-differentiation wavelet analysis. Fig 10(a) shows the sample values of the pressure wave-pattern during the operation of the hydraulic rock driller. The number of sample points is 512. Fig 2 shows the wave-pattern obtained by eliminating a one-gauss white noise from the pressure signals. The wavelet function is shown in Fig 11, and Table 4 gives the filter coefficients.

Since the effects of noice concentrates in the first measure of the wavelet transform, it is not quite significant for the singularity of the analytical signals.

Thus only the wavelet analysis should be conducted of the wave-pattern after the noice elimination, and it more clearly reflects the signal features.

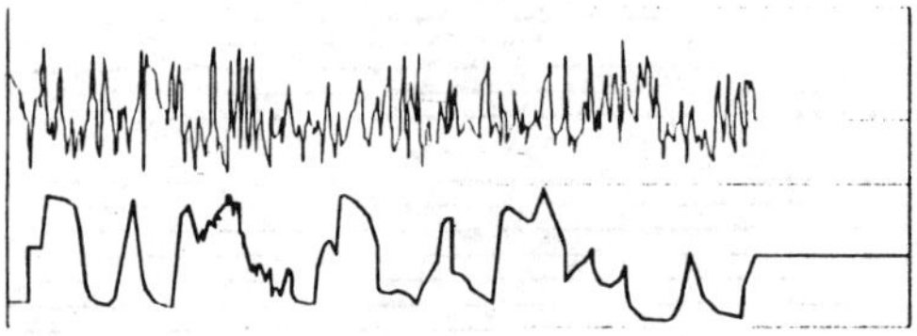

Fig 10 The wave-pattern of the pressure signals and the pressure wave-pattern after the noice elimination

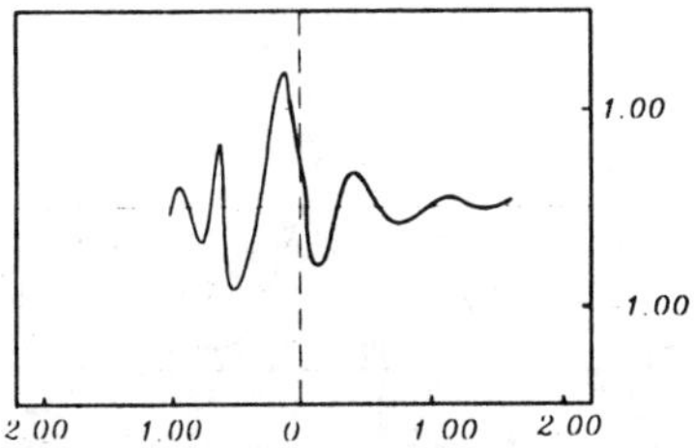

Fig 11 Wavelet function pattern of the pressure

Table 4 Filter coefficients

n	0	1	2	3	4	5
h(n)	0.038	0.243	0.604	0.657	-0.242	-0.032

6. Conclusion

Wavelet analysis as a new method is quite promosing in its application in information analysis and processing. The process of wavelet analysis has been illustrated through the analysis of different performance parameters of the hydraulic rock driller. Meanwhile, the working condition of the driller can be known clearly through the wavelet analysis of its different performance parameters, based on which the performance supervision and failure diagnosis of the machine can be conducted.

REFERENCES

1. Charles K. Chui: An Introduction to Wavelets Academic Press, Inc. 1992

2. Daubechies I: The Wavelet Transform, Time-Frequency, Localization and Signal Analysis. IEEE Trans on Information Theory, Vol 36 1990, pp961-1005

3. Flanarin P: Wavelet Analysis and Synthesis of Fraction Brownian motion, IEEE Trans, on Information Theory, Vol 38, 1992. pp910-916.

4. Qin Qian-qing, Yang Zong-kai: Practical Wavelet Analysis, Xian University of Electronic Science and Technology Press, 1994.1._0-916.

4. Qin Qian-qing, Yang Zong-kai: Practical Wavelet Analysis, Xian University of Electronic Science and Technology Press, 1994.1

Computer Methods and Advances in Geomechanics, Yuan (ed.) © 1997 Balkema, Rotterdam, ISBN 90 5410 904 1

A study on the interaction between rock specimens and loading platens

Takafumi Seiki & Yasuaki Ichikawa
Department of Geotechnical and Environmental Engineering, Nagoya University, Japan

Gyo-Cheol Jeong
Department of Geology, Andong National University, Korea

Naohiko Tokashiki
Department of Civil Engineering and Architecture, University of the Ryukyus, Nishihara, Japan

Ömer Aydan
Department of Marine Civil Engineering, Tokai University, Shimizu, Japan

ABSTRACT: Generally, to determine the mechanical properties for rock mass, laboratory experiments such as uniaxial or tri-axial compression test as element test are carried out, although the contact condition between the edges of rock specimen and loading platens may enormously influence the strain and stress. In a previous work, the authors carried out the numerical analyses under the elastic state. Based on these results, scale effect caused by the platens could not be recognized. In this paper, the authors consider the strength and the failure process of the specimen. For this purpose, we carried out axi-symmetric elasto-plastic FEM analyses to investigate stress distribution in the specimen; (1) Contact condition of platens, (2) Stiffness ratio between specimen and platens, (3) Ratio of height to diameter and (4) Scale effect. The results indicate that there is difference between the failure modes of specimen as a result of restraining effect of loading platens. It is also clear that the uniaxial strength of specimens may strongly depend on this interaction and stress-strain relation is affected by the contact condition.

1 INTRODUCTION

To determine the material parameters of rocks such as Young's modulus and Poisson ratio, laboratory tests such as uniaxial and/or triaxial compression test are carried out. In these tests, specimens are generally regarded as a uniformly strained element. Its Young's modulus is determined from the stress-strain relation computed from the relative displacement between two platens. However, the strain in samples is not distributed uniformly, because of the influence of contact condition between the specimen and the loading platens. Although the experimental data is disturbed mainly by the inhomogeneity of its internal structure, it seems that the boundary condition also influences data extremely. The experimental data of rocks is influenced by (Lama et al. 1992)

(a) The friction between the ends of rock specimen and the loading platen of testing machine,

(b) The shape of specimen, i.e. (i) its shape, (ii) the ratio between the diameter or width of a specimen and its height, and (iii) the scale,

(c) The loading rate, and

(d) The experimental environment such as the temperature, the humidity, e.t.c.

Although many researchers investigated and discussed about the scale effect of specimens (Bieniawski 1968; Hudson et al. 1971; Lundborg 1967), these researches are focused on the behavior of rock in post failure state, and did not pay attention on the elastic state. We simulated this situation for elastic state of uniaxial compression tests by using an axi-symmetrical FEM code (Seiki et al. 1996). However, the results indicated that there is no significant scale effect for the elastic state. Whereas, the uniaxial compression tests carried out on prismatic specimens with joints (e.g. Natau et al. 1995). Referring these results, internal defects, however, has more influence than the contact condition between the specimen and the loading platens. In this paper, we are concerned with items (a) and (b), and investigate the mechanical behavior under various contact conditions between the edges of a specimen and the loading platens by considering the elasto-plastic behaviour. These conditions are 1) smooth contact and 2) rough contact. The results of the numerical analyses on the stress-strain relations are used for clarifying the effect of contact conditions.

2 AXI-SYMMETRIC FEM ANALYSIS

The material parameters and deformation obtained in an experiment are affected by

(1) The inhomogeneous distribution of internal microscopic structure,

(2) The initial irregularity of specimen geometry, and

(3) The contact condition between the specimen edges and the loading platens.

Items (1) and (2) are often discussed, and their influence on the data is greater than the item (3). However, the failure and strength characteristics are influenced by the contact conditions and the contact condition appear to be a boundary value problem. To be distinguish the difference due to the contact condition, we assume the uniaxial compression of homogeneous and isotropic specimens, which has a cylindrical shape, compressed by loading platens having the same cross section. The following four conditions are considered, and the cylindrical specimens are analyzed by an axi-symmetrical elasto-perfect plastic FEM code under the uniaxially forced displacement state imposed on the loading platens (Fig.1). The macroscopic stress-strain curves are also presented for comparing the global behaviour under the uniaxial compression state. Here, to consider the perfect-plastic behaviour, it is assumed that the yielding criterion is of Drucker-Prager type which is the most simple and popular yield criterion for solid materials together with associate flow rule. We also emphasize that the loading platens have the same section as that of the specimen and the constant thickness.

2.1 *The restraint caused by the interaction of the specimen end and the loading platen*

We consider following two cases to analyze the contact condition between the specimen edges and the loading platens.

(a) Rough contact (The specimen edge and the loading platen are completely adhered), and,

(b) Smooth contact.

Even though these two cases are somewhat extreme cases, all contact conditions are actually between two cases. To represent these contacts among the loading platens and the specimen, we introduced an axi-symmetric joint element proposed by Yuan & Chua (1992) and varied its shear stiffness while keeping the normal stiffness constant. Poisson ratio of the loading platen ν_S and the rock-like specimens ν_R are chosen 0.30 and 0.15, respectively. The parameters of yield function are determined from the uniaxial strength and the internal friction angle.

2.2 *Influence of stiffness ratio on stress-strain relation*

Under the specimen edges and loading platen which have a constant contact condition and variable stiffness ratio, the degree of its influence on the stress distribution may vary. Thus, we varied the ratio of the Young's modulus of loading platen E_S to that of the specimen E_R from 10 : 1 (Hard rock), 100 : 1 (Soft rock) and 1000 : 1 (Sand or Clay) while keeping E_S constant. To consider the same situation, we varied the uniaxial strength of these three kind of specimens by the same ratio while keeping the internal friction angle be constant.

2.3 *Influence on a ratio between the height and the diameter of specimen*

We changed the ratio between the height (H) of the cylindrical specimen and its diameter (D) as follows: H : D = 1 : 1, 2 : 1 and 3 : 1. Usually, H : D = 2 : 1 is used in experiments. As there is no theoretical background for such a selection, these cases are carried out for verifying the this experimental practice.

2.4 *Influence of a scale effect on the stress distribution*

To clarify whether the stress distribution and the stress-strain relation are influenced by the scale effect or not, we predict the stress-strain curve. In these analyses, the height (H) and the diameter (D) of specimens were increase while keeping the ratio H/D be constant.

3 NUMERICAL RESULTS

According to the procedure given in the previous section, we investigated the stress distribution on the cross section of the cylindrical specimen. Especially the distribution of the yielding zone in models are compared. During the calculation, the

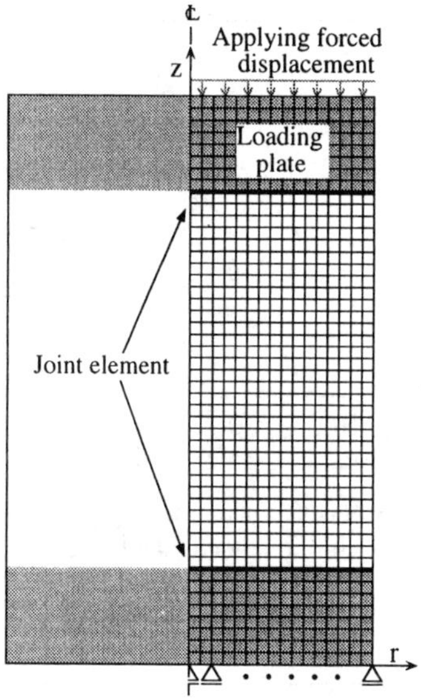

Figure 1: Mesh for axi-symmetric FEM (H : D = 1 : 1, Element number 800, Node number 867)

stress state at each loading increment is obtained, and the forced displacement and reaction force at the bottom of the specimen are stored to obtain the macroscopic stress-strain relation.

3.1 *Influence of restraint caused by the loading platen on the specimens*

We consider that the contact surface is either smooth or rough (complete contact). Under these conditions, we check the effect of varying the stiffness ratio E_S/E_R, the ratio H / D and the scale of H and D in both elastic and plastic state. The stress-strain curve on the cross section of a specimen are shown in Fig.2 and Fig.3. Fig.2 indicates that the stress-strain relation for perfectly smooth contact condition is similar to the typical perfect plastic behaviour, whereas the behaviour of rough contact condition looks like the strain-hardening plasticity. It should be noted that H : D is 1 : 1. In elastic state, the difference can not be recognized between two curves. It indicates that the apparent stiffness is not influenced by contact condition. Fig.3 shows the stress-strain relation of H : D = 2 : 1. This figure indicates that there is a hardly difference between two contact conditions. It proves that if H/D ratio is 2.0, the end effects on the uniaxial compression test are negligible.

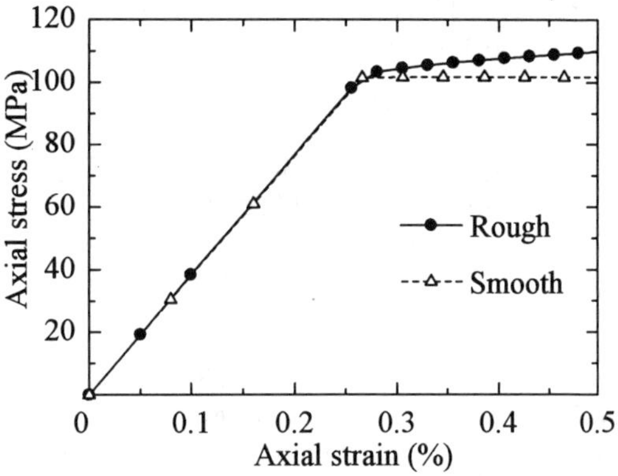

Figure 2: Stress strain curve between the rough and smooth contact condition (H : D = 1 : 1, E_S : E_R = 10 : 1)

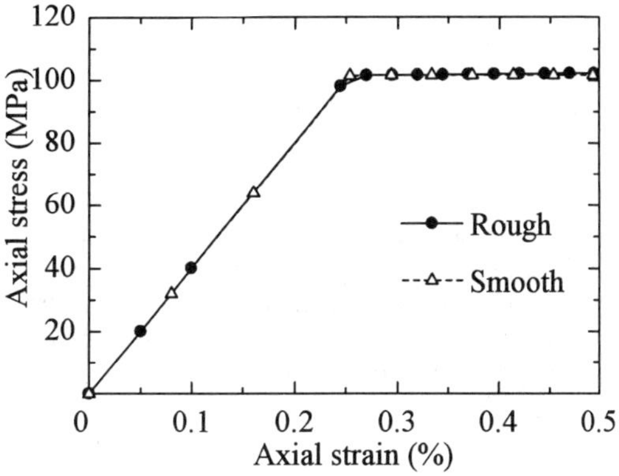

Figure 3: Stress strain curve between the rough and smooth contact condition (H : D = 2 : 1, E_S : E_R = 1000 : 1)

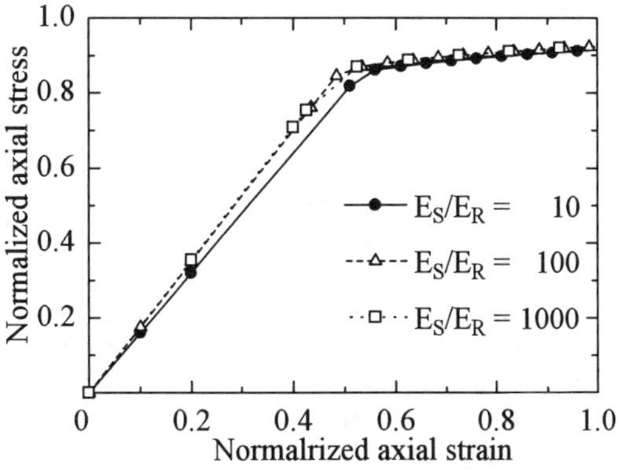

Figure 4: Normalized stress-strain curve (H/D = 1, Rough contact condition)

3.2 *Influence of the stiffness ratio between the specimen and loading platen*

In this section, to consider the influence of the stiffness ratio between the specimen and loading platen, the numerical results are presented as normalized stress-strain curves. Fig.4 shows the stress-strain curve of which are normalized with the maximum value of the stress and strain for each stiffness ratio. This relation is evaluated for smooth contact condition. This figure indicated that there is a little difference between $E_S : E_R = 1000 : 1$ and, $E_S : E_R = 10 : 1$ in elastic state. The case of $E_S : E_R = 100 : 1$ almost behave as a same $E_S : E_R = 100 : 1$. There is hardly difference in plastic state. These results indicate that the harder specimen is, the easier the stress gets the effect of the contact condition. Once the specimen becomes plastic, yield zone is no longer affected by the contact condition.

3.3 *Influence of the ratio of the specimen height to its diameter*

For various ratio of a specimen height (H) to its diameter (D), the stress-strain curve are shown in Fig.5 for stiffness ratio $E_S / E_R = 10$. This figure shows the some differences among the stress-strain curves for H/D = 1 and others. Stress-strain relations for specimen of H/D = 1 is slightly different. From this figure, it is clear that the influence of the specimen shape, namely, the ratio H/D is important to obtain the reliable data. Whereas, the curves on $E_S / E_R = 1000$ (Fig.6) have no difference each others. It means that the harder the specimen becomes, the more sensitive H/D ratio is. In other words, the stress-strain relation is easier influenced by the high stiffness specimen.

3.4 *Influence of scale effect on the stress-strain relation*

We carried out a series of numerical analyses to investigate the relation between the scale effect and the restraining caused by the loading platen. Stress-strain relation is specially influenced by the scale effect in elastic state. Whereas, there is no difference in plastic state. It is clear that the contact condition has no influence after initial yielding. Fig.7 and Fig.8 can be compared to see the difference between stress-strain curves for the effect of the scale effect with changing stiffness ratio. There is scale effect for stiffness ratio $E_S : E_R =$

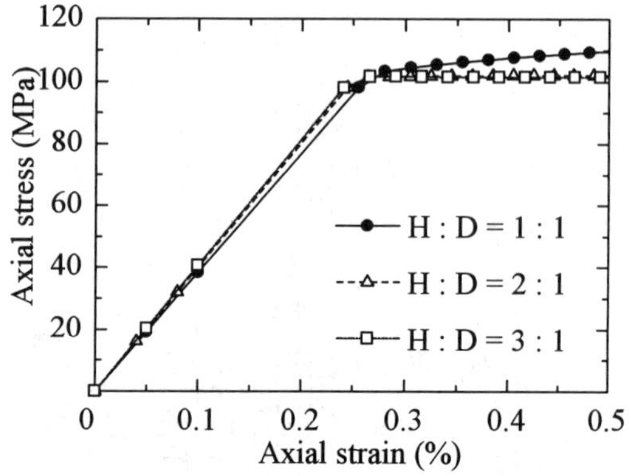

Figure 5: Stress strain curve for the investigating the ratio of the specimen height to its diameter ($E_S : E_R = 10 : 1$, Rough condition)

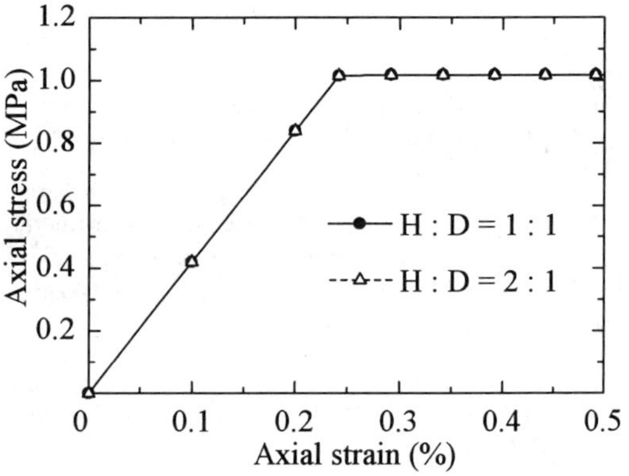

Figure 6: Stress strain curve for the investigating the ratio of the specimen height to its diameter ($E_S : E_R = 1000 : 1$, Rough contact)

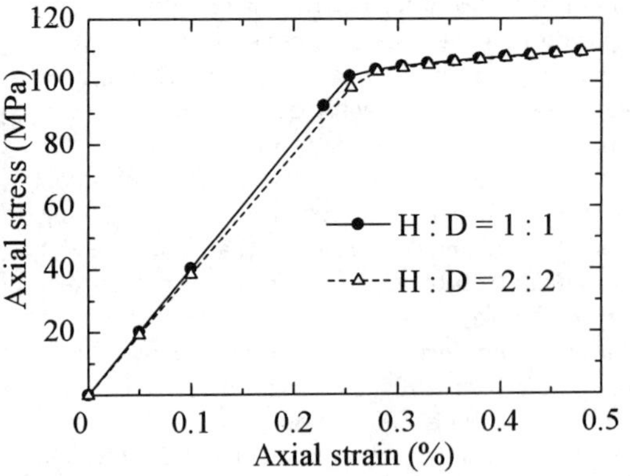

Figure 7: Stress-strain curve for scale effect ($E_S : E_R = 10 : 1$, Rough contact)

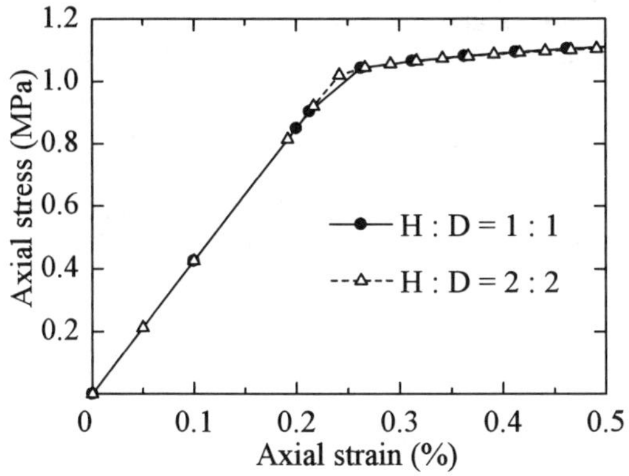

Figure 8: Stress-strain curve for scale effect (E_S : E_R = 10 : 1, Smooth contact)

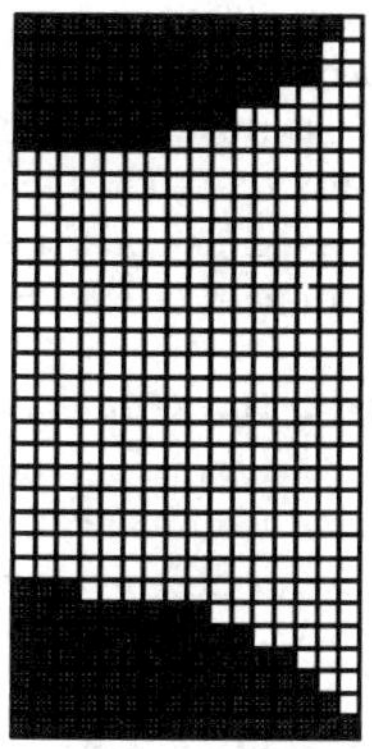

Figure 9: Yielding zone in the specimen (E_S : E_R = 10 : 1, Rough contact, Stress 103.3 MPa, Strain 0.280 %) Shadowed area indicates the elastic region. The left side of mesh is a center axis

10 : 1 (Fig.7). On the other hand, the soft specimen does not show any scale effect due to the contact condition (Fig.8).

3.5 *Comparison of the yielding state and the strength for two contact conditions*

During compression under constant strain rate, the failure propagation for two conditions are different. For the smooth contact condition, after initial yielding, whole specimen suddenly yields. Whereas, after initial yielding, the yielding spreads gradually in specimens with rough contact condition. Finally, a wedge shaped zone just beneath the loading plate remains elastic (Fig.9). This is probably the reason for a strain-hardening behaviour. From these results, we can conclude that the failure should be initiated at edge of the specimen with rough contact condition, and the splitting failure may appear for smooth contact condition. If the strength of the specimen is determined from the macroscopic stress-strain curve, it is clear that the uniaxial strength for rough contact condition is about 10 MPa larger than that of smooth contact condition (Fig.2), However, the initial yield point has almost the same value. In other words, the edge restraining has an influence on the uniaxial strength of specimens.

4 CONCLUSION

In this study, we modelled the cylindrical specimens for numerical analyses and carried out a series of axisymmetric elasto-plastic FEM analyses. From the results, the stress-strain curves of the specimens are calculated, and it is considered how the loading platens affect to the stress-strain and yielding. We have drawn the following conclusions:

1) **Influence of restraint caused by the loading platen on the specimens**: On the rough contact condition, it indicates that the stress-strain relation shows a typical perfect plastic behaviour, whereas the behaviour of rough contact condition looks like the strain-hardening plastic. In elastic state, the difference can not be recognized between two curves. It indicates that the apparent stiffness is not influenced by contact condition. It proves that the conventional specimen with H/D ratio of 2 is suitable for the uniaxal compression test.

2) **Influence of the stiffness ratio between the specimen and loading platen**: To consider the influence of the stiffness ratio between the specimen and loading platen, the numerical results are presented as normalized stress-strain curves. These results indicate that the harder specimen is, the easier the stress gets the effect of the contact condition. Once the specimen becomes plastic, yielding zone is no longer influenced by the contact condition.

3) **Influence of the ratio of specimen height to its diameter**: In the case of H/D = 1 and E_S / E_R = 10, stress-strain curve shows the different behaviour from others. It means that as the specimen becomes harder, it is more sensitive to H/D ratio. In other words, the stress-strain relation is easily influenced by the high stiffness spec-

imen. Furthermore, although the strain distribution is also influenced, it is clear that the larger H/D ratio becomes, the smaller the influence on the strain distribution is. Thus, reliable data can be obtained, if a specimen has a certain ratio of the height to the diameter, e.g. H : D = 2 : 1 ~ 2.5 : 1.

4) **Influence of scale effect on the stress-strain relation** : From the computed results, stress-strain relation is influenced by the scale effect. But there is no difference in plastic state. It is clear that the contact condition has no influence after initial yielding. But, in the elastic state, stress-strain relation is especially influenced by the contact condition.

5) **Comparison of the yielding state and the strength for two contact conditions**: During compression under constant strain rate, the failure propagation for two conditions are different. For the smooth contact condition, after initial yielding, the whole specimen yield suddenly. Whereas, after initial yielding, the yield spreads gradually in specimens with rough contact condition. Finally, the wedge shaped zone remains elastic. This is probably the reason for observing a strain-hardening behaviour restraining the edge of the specimen affects to the uniaxial strength, even though the initial yield point has almost the same value.

REFERENCES

Lama, R. D., V. S. Vutukuri & S. S. Saluja 1992. *Handbook on Mechanical Properties of Rocks I*, (Trans Tech Publications).

Bieniawski, Z. T. 1968. The effect of specimen size on compressive strength of coal, *Int. J. Rock Mech. Min. Sci.*, 5: 325-335.

Hudson, J. A., E. T. Brown & C. Fairhurst 1971. Shape of the complete stress-strain curve for rock, *Proc. 13th Symp. on Rock Mechanics*: 773-795.

Lundborg, N. 1967. The strength-size relation of granite, *Int. J. Rock Mech. Min. Sci.*, 4: 269-272.

Yuan, Z. & K. M. Chua 1992. Exact formulation of axi-symmetric interface-element stiffens matrix, *J. Geotech. Eng. ASCE*, 118: 1264-1271.

Natau, O., Fliege, O., Mutschler TH. & Stech H.-J. 1995. The triaxial tests of prismatic large scale samples of jointed rock masses in laboratory, *Int. Cong. on Rock Mech. ISRM*, 353-358.

Seiki T., Ichikawa, Y., Jeong G.-C., Tokashiki, N. & Aydan Ö. 1996. A study on the influence of contact condition between loading platens and rock specimens, *Korean-Japan Joint Symp. on Rock Eng.*, 125-129.

Computer Methods and Advances in Geomechanics, Yuan (ed.) © 1997 Balkema, Rotterdam, ISBN 90 5410 904 1

Mechanical characterization of an underground laboratory site in granite

K. Su
Groupement pour l'Etude des Structures Souterraines de Stockage (G.3S), LMS (URA), Ecole Polytechnique, Palaiseau, France

N. Hoteit
Agence Nationale pour la Gestion des Déchets Radioactifs (ANDRA), France

ABSTRACT: In the framework of a feasibility study of radioactive waste disposal in France, one of the three underground laboratories planned to be constructed is located in a granite formation. This paper is devoted to the geomechanical tests undertaken on the cores taken from this site. More than one hundred mechanical tests have been performed: axisymmetric deviatoric compression, isotropic compression, relaxation, compression under high temperature, Brazilian test, Luong's shear test. Damage processes of granite in a compression test is discussed. A particular attention is paid to the influence of discontinuities on the mechanical behavior. In fact, intact granite and granite with discontinuities (disturbed granite) exhibit very different mechanical behavior. A visual observation on the cores allows us to qualify a few parameters (discontinuity density, number of discontinuity planes). We establish a correlation between those observations and the obtained mechanical properties.

1. INTRODUCTION

In relation to the feasibility studies of a deep radioactive waste disposal in France, three possible sites are under evaluation to implement an underground laboratory. One of the potential site is located in a granite formation. In this project, geomechanical characterisation of this site has an important role for its construction. We have to know the changes of thermo-hydro-mechanical properties of surrounding rock mass as function of depth. Geomechanical data will lead us to chose the best level in regard to the objectives of the laboratory. The mean involved geomechanical parameters are:

mechanical behavior:
elastic properties, rupture criterion, damage criterion, mechanical properties of natural fractures.

thermal properties :
heat capacity, heat conductivity, thermal expansion coefficient.

hydraulic properties:
intrinsic permeability of rock mass, hydraulic properties of natural fractures.

This paper presents the results of a set of geomechanical tests undertaken on the cores coming from the Vienne province of France. These cores were extracted from depth of -260 meter to -539 meter, that is to say 275 meter of thickness. In the first part of the paper, a thermo-hydro-mechanical triaxial cell is presented. Using the theory of damage mechanic, we have studied the mechanical behavior of Vienne granite under deviatoric compression tests. Special attention is paied on the influence of discontinuities and heterogeneities. The mechanical results of both intact rocks and disturbed rocks are then compared.

2. EXPERIMENTAL DEVICES

Laboratory tests are undertaken in a cell developed by G.3S on a MTS servo-controlled machine with a high stiffness (10^9 N/m). A hydraulic system applying the confining pressure has been also conceived. It allows to impose a proportional loading or an unloading path. Confining pressure may change during a test as a function of an external variable (axial force, strains, ...). It can reach 100 MPa with imposed temperature up to 200°C. Figure 1 shows a simplified scheme of the cell. Rock samples of diameter up to 50 mm and 220 mm of height can be used. The cell is also equipped with four hydraulic circuits for which volume or pressure is controlled. Evolution of permeability can be measured during a test with respect to the damage state of sample.

Axial and lateral deformations are measured by means of two MTS extensometers located in the cell, on a jacket, the purpose of which is to avoid percolation of confining fluid into the sample. This can modify significantly the mechanical behavior of the sample even if its porosity and permeability are small. MTS extensometers have a resolution of 10^{-6}. A force transducer is also installed inside the triaxial cell to give directly the deviatoric force on the sample without important friction between joints and the cell piston.

To check the strain measurement, the strains measured by MTS extensometers and those measured by gauges are compared during a uniaxial compression test. The results show that the deformations given by the two methods are almost the same.

The results given in this paper correspond to the first set of tests performed by using the recent experimental set up. New results related to coupled thermo-hydro-mechanical aspects are being interpreted and not presented here.

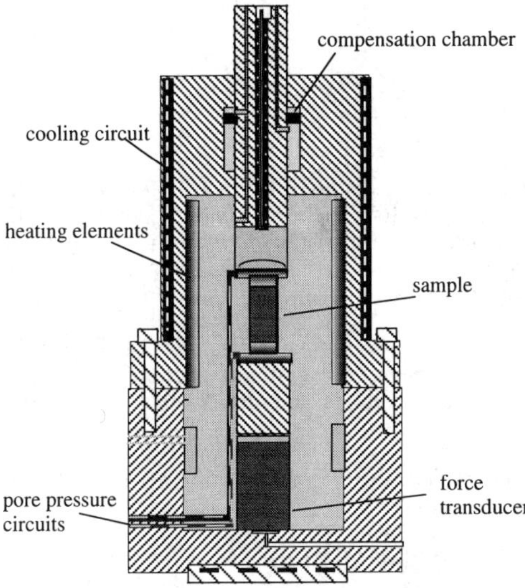

Fig. 1. Simplified scheme of thermo-hydro-mechanical cell.

3. PHYSICAL PROPERTIES OF CORES

The density of the Vienne granite is about 2.72. Its average porosity is less then 1%, measured by the classical method of saturation in water.

Sound velocity is a parameter sensitive to the pre-existing fissures. Fourmaintraux in 1976 has proposed an index describing the degree of fissuring, which is the ratio of measured velocity over theoretical velocity. In our case, cores come from different levels. Their mineral composition may be quite different with each other, so the theoretical velocity is not unique. We have prefered to compare the relative change of sound velocity with depth. Measurement of longitudinal and transversal velocities are undertaken in all specimens made from cores for mechanical tests. Longitudinal wave velocity of Vienne granite is between 5300 m s^{-1} to 6500 m s^{-1}, and 2600 m s^{-1} to 4000 m s^{-1} for transversal waves. In figure 2a. the dispersion of sound velocity is shown. One can find that the cores extracted at depths of more than 400 meters have higher velocity than those of other zones. Pre-existing fissuring at this depth in this region is probably less important compared to that of the cores extracted between 200 to 400 meter depth. The distribution of density of Vienne granite (figure 2b) exhibits the same tendency.

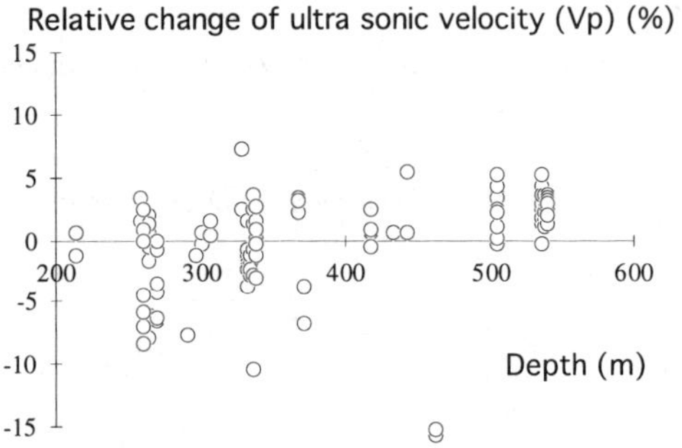

Fig. 2a. Dispersion of untra sonic velocity

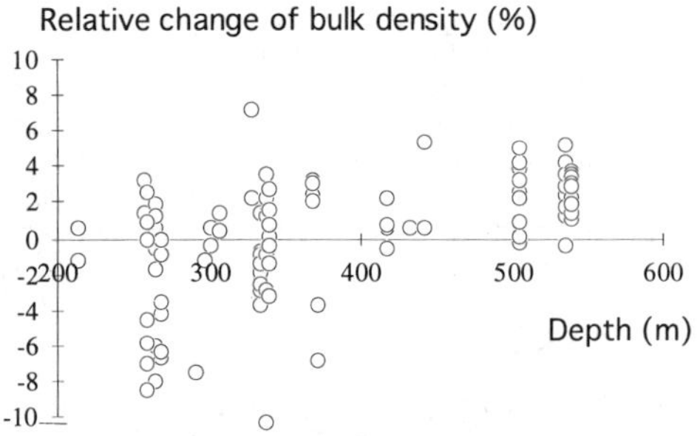

Fig. 2b. Dispersion of bulk density

Fig. 2. Variation of ultrasonic velocity and bulk mass with depth

4. MECHANICAL BEHAVIOR OF VIENNE GRANITE UNDER AXISYMETRIC TRIAXIAL COMPRESSION TEST

4.1 Damage of granite

Studying the mechanical behavior of crystalline rocks considered for underground nuclear waste repositories requires damage consideration. This is one of the key parameters controlling the stability and the permeability of the host medium. Extension of the damaged zone around a gallery depends on the mechanical behavior of the rock mass and the technique of excavation. Rocks are discontinuous media. Their aggregates are never perfectly compacted and can contain voids (pores) and intercrystalline and intracrystalline defects (cracks). The cracks are very flat, with ratio of thickness over the length of the order of 10^{-3} to 10^{-4}. The most significant mechanical effect is the initiation of cracks which consequently affects deformability and strength of rocks [Habib 1973]. Rock collapse is essentially produced by development of natural cracks under deviatoric loading. Coalescence of cracks leads to reduction of cohesion. Only friction forces may be active.

Damage theory in mechanics traduces the transition between undisturbed to disturbed material. It has been studied during recent years, in laboratory [Hunsche, et al.1990, Li et Nordlund 1995, Thorel et al. 1996], and from a theoretical point of view [Aubertin et al. 1995, Cristescu 1989, Lemaitre and Chaboche 1988, Dragon et al. 1979]. Evolution of damage can be characterised by number of parameters sensitive to cracking such as, acoustic velocities, permeability, irreversible volumetric strain, etc. In this paper irreversible strain is used to characterise the damage of granite since this parameter is directly linked to the number of active cracks.

Figures 3, 4 and 5 show stress-strain plots obtained by a deviatoric compression test at confining pressure of 20MPa. Granite sample has 60 mm diameter and 120 mm height. Young's modulus in axial direction is determined during the unloading path since loading path is associated with hardening.

There is negligible axial irreversible deformation after unloading. Therefor from this point of view, damage behavior is elastic and reversible. This is always the case with or without confining pressure. Young's modulus obtained in axial direction decreases as damage progresses.

The plots of deviatoric stress versus lateral strain show irreversible lateral strain after unloading (figure 4). The former increases as the damage of granite progresses. In lateral direction, damage is irreversible and can be considered as elasto-plastic [Dragon et al. 1979]. A remarkable hysteresis appears in loading-unloading cycles. This phenomena is certainly associated with non-linear behavior due to opening or closure of cracks. The hysteresis becomes significant under confining pressure, in comparison with uniaxial compression.

Axial Young's modulus decreases as the damage state advances. Evolution of Young's modulus as a function of irreversible volumetric strain (ε_v^{irr}) is shown in figure 6. The diminution is significant whatever the confining pressure is and reaches to 50% while ε_v^{irr} is less then 1%. This confirms that mechanical properties of granite change with respect to damage. Without confining pressure, elastic modulus is more sensitive to damage than with confining pressure.

As a matter of fact, irreversible volumetric strain presented in following figures can be replaced by lateral strain as there isn't any axial irreversible strain. For an intact material, unloading is always associated with dilatancy. For example, in figure 5, the first loading-unloading cycles show increase of volumetric strain at the end of unloading, compared to the strain at the beginning of unloading. But for the next cycles, volumetric strains decrease (contractancy) while unloading the deviatoric stress. The change of volumetric strain denotes the development of cracks. At a certain stage of damage, volumetric strain corresponding to the cracks closure due to unloading of deviatoric stress is greater than elastic dilatancy. This leads to a contractancy during unloading. On the contrary, dilatancy is observed during a loading path where the cracks tend to be open. Therefore, the kinetic behavior of cracks induces non linear behavior in damaged material.

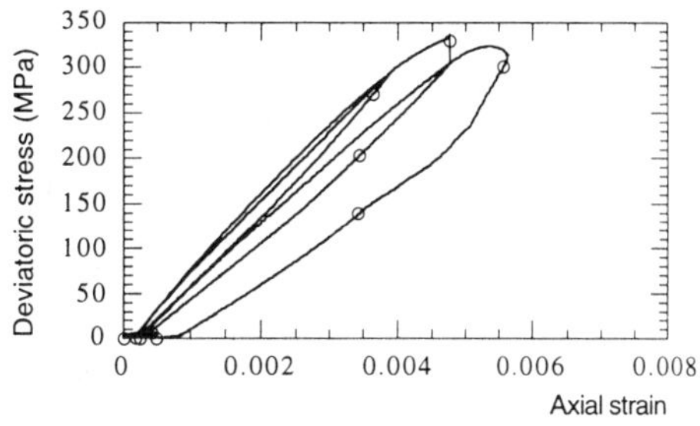

Fig. 3. Axial strain versus deviatoric stress

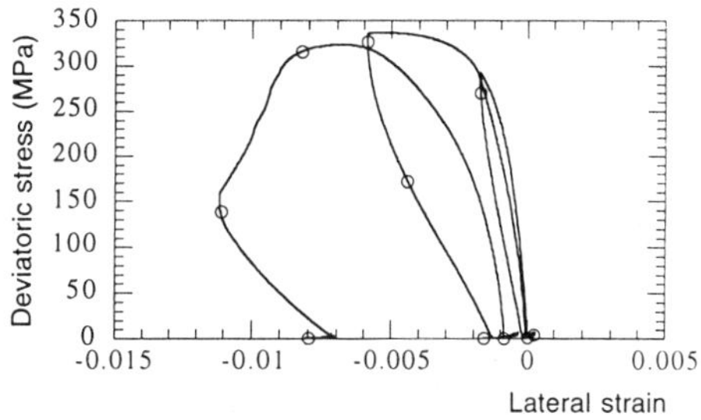

Fig. 4 : Lateral strain versus deviatoric stress

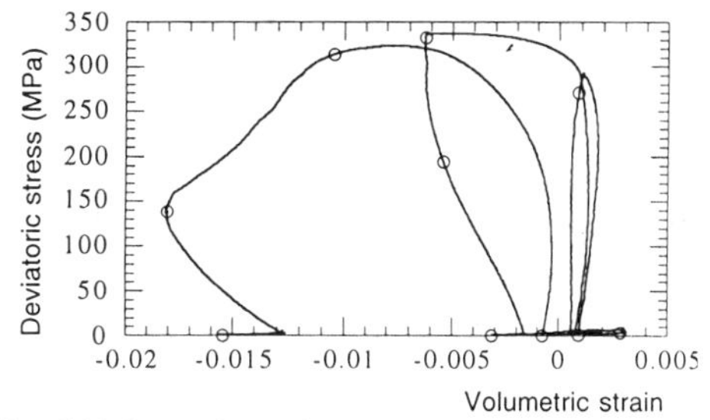

Fig. 5.Volumetric strain versus deviatoric stress

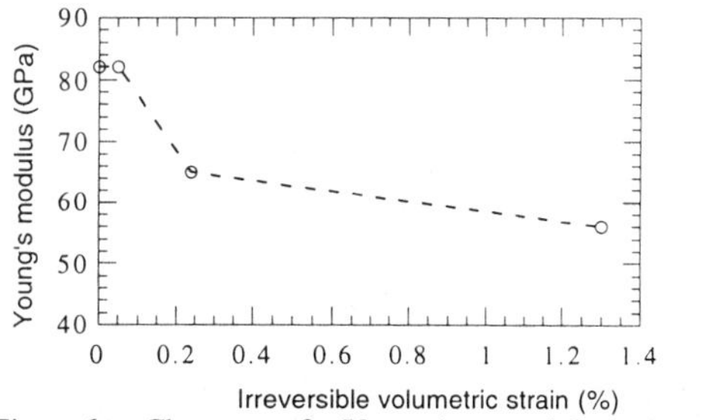

Fig. 6. Change of Young's modulus versus irreversible volumetric strain

Figures 7 shows axial and lateral strain of a damaged granite under isotropic compression. Lateral deformation is much greater, and more non-linear than axial strain. That is to say that damaged granite has anisotropic mechanical properties. In fact, induced anisotropy reflects characteristic of damage of a crystalline rock. This may not be the case for plastic rocks. For example, the experimental results carried out on rocksalt by Thorel [1995] show isotropic damage of this material.

Experimental data indicate that the ratio of lateral strain over axial strain is about 3, when irreversible volumetric strain reaches 0.25 %. Diminution of compressibility modulus is due to softening of granite with respect to lateral strain, since cracks are developed mainly in a plan parallel to the sample axis. The effect of closure or recovery of cracks during isotropic compression is visible essentially on lateral deformation (figure 7).

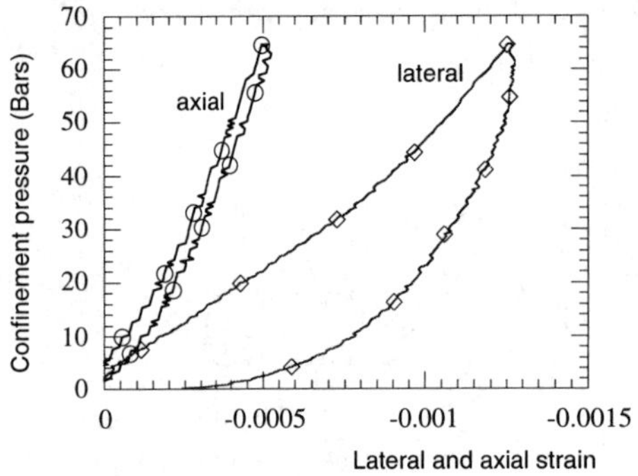

Fig. 7. Lateral and axial strain versus isotropic confinement pressure

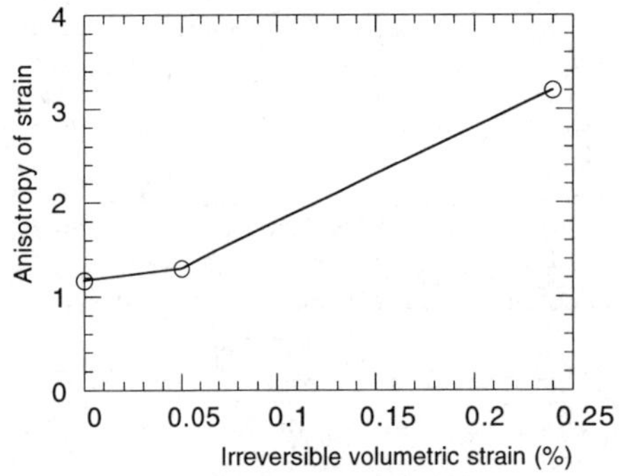

Fig. 8. Anisotropy of strain (ratio of lateral strain over axial strain) observed during isotropic compression test

4.2 Strength of intact granite

Figure 9 is a classical diagram of mean stress-deviatoric stress (P-Q), in which strengths of granite obtained in deviatoric compression test, brazilian test and shear test by Luong's method are plotted. Linear fitting of these points gives a line with a slope of 2.2 and intersecting axe Q at 29 MPa.

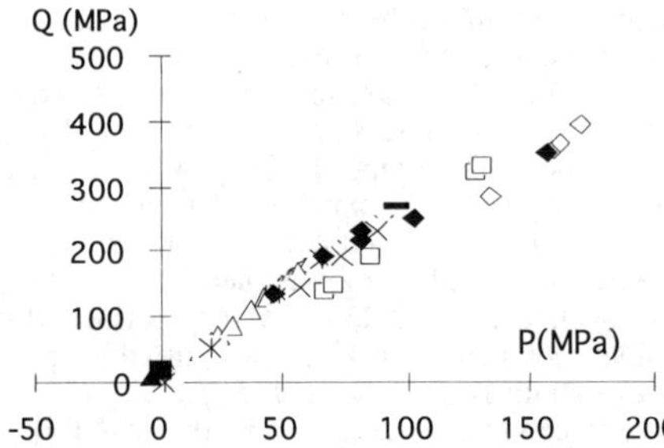

Fig. 9. Strength of granite obtained at various test conditions

Using Mohr-Coulomb criterion, one obtains the parameters of the failure criterion of the Vienne granite :

intact granite:
angle of internal friction: 53°
internal cohesion 18 MPa

4.3 Strength of the disturbed granite

Several deviatoric tests are also undertaken on samples having one or several initial marked discontinuity, considered as disturbed granite. These samples exhibit very different behavior compared to intact granite. Strength of granite is significantly affected by pre-existing marked discontinuity. Sample with marked discontinuity should be considered as a structure. Mechanical behavior observed in testing can lead us to study discontinuity behavior. From the results of deviatoric compression tests at different level of confining pressure, the shear strength of discontinuity is calculated. In the $\tau - \sigma$ diagram, figure 10 shows the shear strength of granite with discontinuity. Taking Mohr-Coulomb criterion, we obtain :

disturbed granite:
angle of internal friction: 27°
internal cohesion 4 MPa

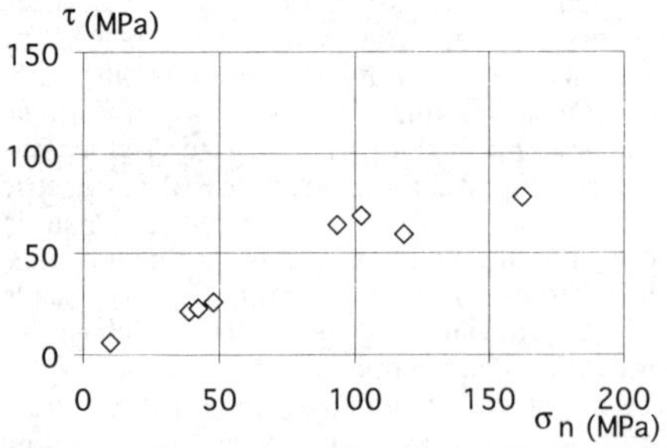

Fig. 10. Shear strength of natural discontinuity in granite

Friction angle of a natural discontinuity in granite is only one half of that of intact granite, and a quarter for its internal cohesion. So for an underground construction in granite, it is natural discontinuities which control the stability of the structure.

4. 4 Correlation between mechanical results and pre-existing fissures

Heterogeneity and discontinuity of granite affect the strength of granite. We have chosen three visual parameters to describe initial state of cores:
-discontinuity density (number of discontinuities per meter)
-number of discontinuity plans having different orientations per meter
-index of heterogeneity, defined as the ratio of largest grain size over mean grain size.

They are only determined by visual observation over the cores; no specific apparatus has been employed for this purpose at this stage of the study.

We plot in figure 11, 12 and 13 relative changes of these three parameters with respect to their mean value. Strong dispersion with depth is observed. Nevertheless, the tendency to improvement of the granite quality with depth is clearly manifested. But there isn't any zone more than 100 meter thickness having a quality absolutely superior to the other zones.

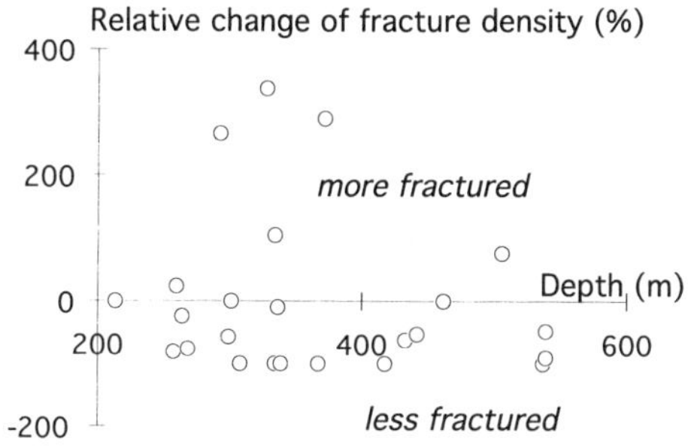

Fig. 11. Relative change of fracture density versus depth

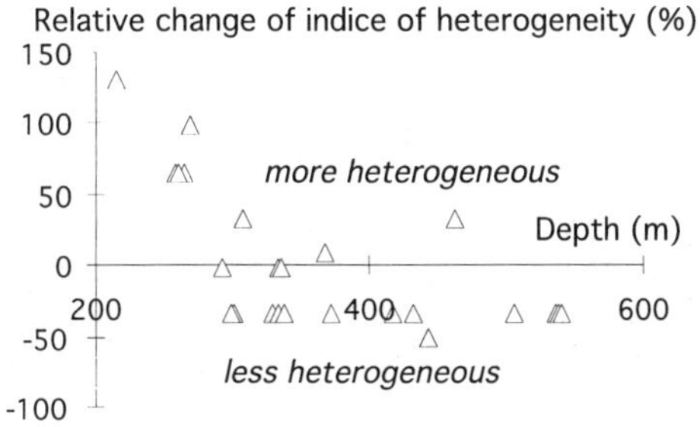

Fig. 12. Relative change of heterogeneity versus depth

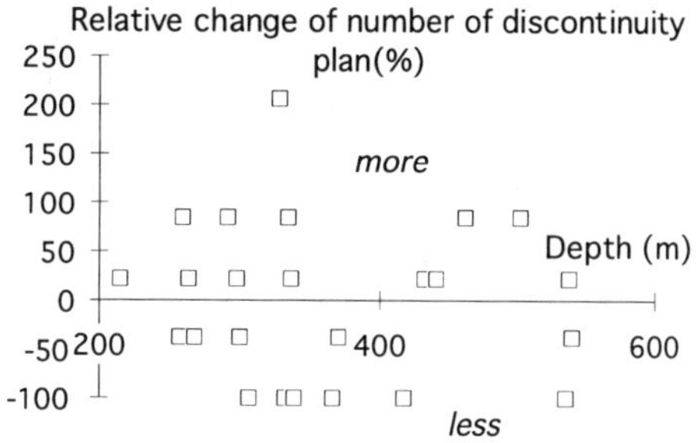

Fig. 13. Relative change of number of plan of discontinuity versus depth

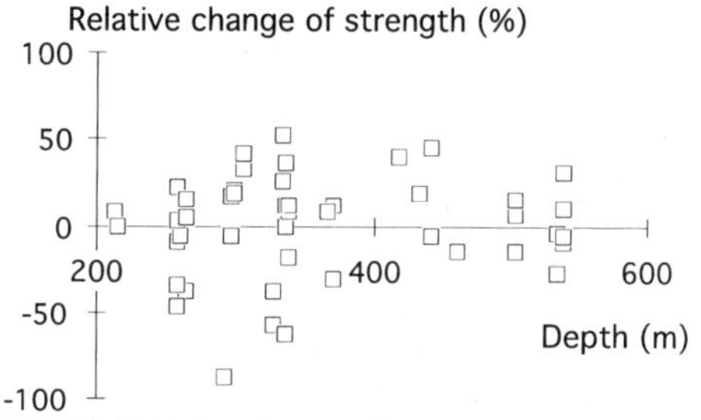

Fig. 14. Relative change of strength versus depth

It appears that at the depth between 400 to 550 meters of depth, relative changes of these parameters respect to mean value is less then changes observed between 200 and 400 meters. Variation of strength obtained at various test conditions versus depth is also plotted in figure 14. Strength improvement of granite is also observed in the region deeper than 400 meters. From the density and sound velocity evolution (figure 3), the same conclusion is obtained. Therefor physical parameters (bulk mass, sound velocity), three visual parameters related to discontinuity and heterogeneity, and the strength all confirm that the best quality of granite is located at a depth near to 500 meters.

5. CONCLUSIONS

Thanks to the new experimental set up, local deformation under high pressure offers the possibility to access to the strain tensor. During axisymetric triaxial deviatoric compression tests, axial Young's modulus of granite decreases as damage progresses. Regarding axial deformation, damage behavior is reversible (elastic). But lateral deformation shows a typical elasto-plastic behavior. Irreversible volumetric strain is mainly due to the lateral deformation. Because of the preferential direction of the cracks, anisotropic mechanical properties appear clearly during isotropic compression loading paths. The ratio of lateral deformation over axial deformation can reach 2.5. Comparison of mechanical strength of both intact granite and disturbed granite shows that natural discontinuities greatly influence the strength of granite. Angle of internal friction of disturbed granite is half of the one of intact granite. Changes of mechanical properties, physical properties and discontinuity parameters with depth give a global view of the quality of the Vienne granite.

6. AKNOWLEDGEMENTS

The authors are grateful to Dr. M. Ghoreychi, Director of G.3S, for his suggestions and comments. They also appreciate the help of S. Chanchole and T. Le Grasse for performing the tests.

REFERENCES

Abou-Ezzi N.L. 1989. Modélisation du comportement non-linéaire du béton par la mécanique de l'endommagement. PhD. Thesis, *Ecole Nationale des Ponts et Chaussées* , Paris.

Aubertin M., Gill D.E., Ladanyi B. 1991. Laboratory validation of unified viscoplastic model for rock salt. *7th Int. Cong. of ISRM*. Aachen, German. pp. 183-186.

Cristescu N. 1989. *Rock Rheology*. Kluwer Academic Publ.

Dragon A., Mroz Z. 1979. A model for plastic creep of rock-like materials accounting the kinetics of fracture. *Int. J. Rock Mech. Min. Sci. & Geomech. Abstr*. 16, 253-259.

Fourmaintraux D. 1976. *La mécanique des roches appliquée aux ouvrages du génie civil* (avec Marc Panet). édition ENPC, Paris.

Habib P. 1973. La fissuration des massifs rocheux. *Annales de l'Institut technique du bâtiment et des travaux publics.* 306, pp 63-69.

Hunsche U., Albrech H. 1990. Results of true triaxial strength tests on rock salt. *Eng. Fract. Mech.*, Vol.35, N°4/5, pp867-877.

Lemaitre J., Chaboche J.L. 1988. *Mécanique des matériaux solides*. Dunod édition.

Li C., Nordlund E. 1995. Micromechanial study of deformation behavior. *8th Int. Cong. of ISRM*. Tokyo, Japan. pp. 237-240

Thorel L., 1995. *Plasticité et endommagement des roches ductiles*, PhD. Thesis, *Ecole Nationale des Ponts et Chaussées* , Paris.

Thorel L. Ghoreychi M., Su K. 1996. Instantaneous plasticity and damage of rocksalt applied to underground structures, *Eurock'96*, September 1996, Torino, Italy, pp. 1119-1125.

Computer Methods and Advances in Geomechanics, Yuan (ed.) © 1997 Balkema, Rotterdam, ISBN 90 5410 904 1

Application of signal processing in dynamic torsional shear test

Zhaozhen Lu & Hanlong Liu
Geotechnical Engineering Research Institute, Hohai University, Nanjing, China

ABSTRACT: In the paper, the discrete dynamic torsional test data is simulated as a signal accompanying with many different frequency noises, by employing the Butterworth low-pass filter scheme of signal processing, an analytical method of test results is established. It is verified by performing a recovering analysis from a pure theoretical signal mixed with a random noise. A case study is introduced for application.

1 INTRODUCTION

Shear modulus is the most important dynamic property for use in studies of dynamic soil response. Torsional testing method provide one of approach for measuring this property in the laboratory. For solid circular soil samples, an integration equation is derived by Taylor(1975) shown as below:

$$G(\gamma) = \frac{2l}{\pi R^4}\left(\frac{3}{4}\frac{M}{\theta} + \frac{1}{4}\frac{dM}{d\theta}\right) \tag{1}$$

in which, $G(\gamma)$ is the value of the shear modulus which corresponds to the shear strain at the periphery of the sample; l and R are the length and radius of sample respectively; θ is the angular displacement caused by the application of a torque M.

From Eq.(1), it can be seen that, if the torque-angular displacement data getting from experiment fits good for the curve which can be simulated by an equation, the shear modulus and shear strain relation may be easily obtained. As a mater of fact, however, the data sampled from experiment is always in discrete distribution, and the derivation $dM/d\theta$ can not be calculated exactly. In the paper, the discrete irregular data is regarded as a signal accompanying with many different frequency noises, by using of the Butterworth low-pass filter scheme of signal processing, and employing the difference scheme, a dynamic torsional test analytical method is established. It is verified by performing a recovering analysis from a pure theoretical signal mixed with a random noise, and then, a case study is presented in the paper too.

2 ANALYSIS METHOD

2.1 *The constitutive model*

Past experience in geotechnical engineering indicates that the normalized stress-strain curves for a large variety of soils fall within a narrow band(Hardin and Drnevich,1972b). Two types of equations are used to describe these normalized stress-strain curves. The first is Hardin-Drench model:

$$\frac{G}{G_{\max}} = \frac{1}{1+\gamma/\gamma_R} \tag{2}$$

in which, $G_{\max}$ is the maximum shear modulus; $\gamma_R = \tau_{\max}/G_{\max}$ is the reference strain; $\tau_{\max}$ is the maximum dynamic shear stress.

The second is Ramberg-Osgood model, which is originally proposed by Ramberg and Osgood(1943) and modified for soils to take the following form(Streeter et.al. 1974):

$$\frac{G}{G_{\max}} = \frac{1}{1+\alpha\left|\frac{\tau}{c_1\tau_{\max}}\right|^{R-1}} \tag{3}$$

where, α, R, and c_1 are model parameters used to obtain the best fit with the laboratory data.

In the paper, the stress-strain properties of San Francisco Bay mud from Hamilton Air Force Base is used. For this material, G_{max} equals 50,00kpa, and τ_{max} equals 100KPa at a confining pressure of 207 Kpa, so the reference shear strain $\gamma_R = 0.02$, for the Ramberg and Osgood model, α=0.7, R=5.5, and c_1=0.6. Based on the above parameters, the stress-strain relation of both Hardin-Drnevich model and Ramberg-Osgood model are shown in Fig.1.

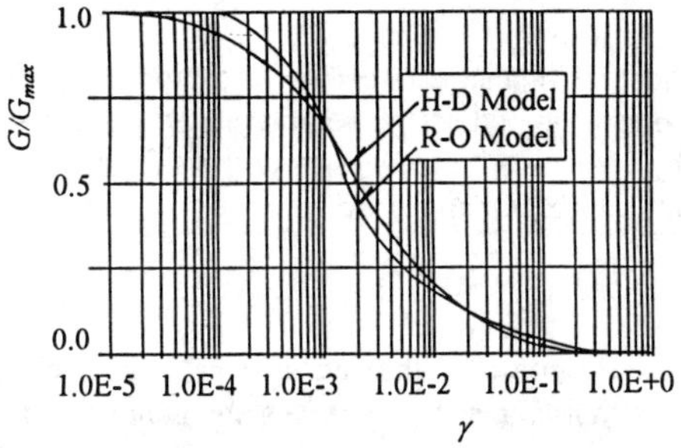

Fig.1 The relation of Shear modulus Vs shear strain

2.2 *Torque-angular displacement relation*

For torsional loading of solid circular shafts, The unit length angular displacement θ and the applied torque M, are related to the cross-sectional distribution of shear stress τ and shear strain γ, it has

$$\theta = \frac{\gamma}{\rho} \tag{4}$$

where, ρ is radial length. According to the principle of force equilibrium, it has

$$M = \int_A \tau \rho dA \tag{5}$$

where, $dA = 2\pi\rho d\rho$, $\tau = G\gamma$, put equation (4) into (5), it is given by

$$M = 2\pi\vartheta\int_0^{R_0} G(\gamma)\rho^3 d\rho \tag{6}$$

where, R_0 is the radius of sample, here R_0=6.35 cm; Since G(γ) is a non-linear function of shear strain γ, substituting equation (2) and (3) into (6), and by using of numerical integration scheme, the relation between torque M and angular displacement named as pure signal in the paper, may be got shown in Fig. 2. By sampling for the curves in Fig.2, the sampling interval is 0.025 second, based on Fast Fourier Transformation(FFT), the response spectrum of torque to angular displacement is transformed shown in Fig.3. It can be seen that from Fig.3, the torque energy distribution is mainly in the range of lower frequency.

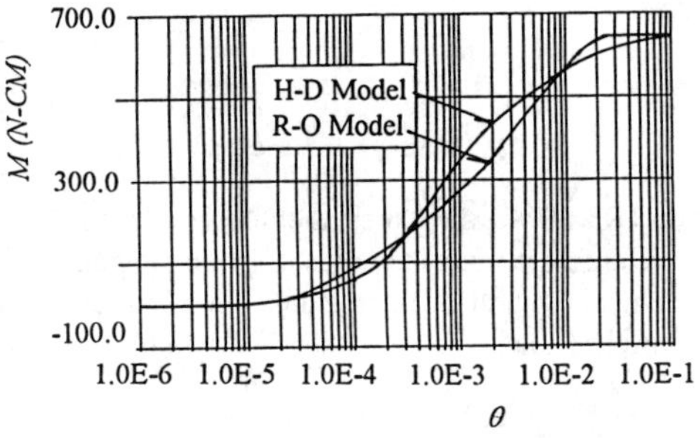

Fig. 2 The relation of torque Vs angular displacement

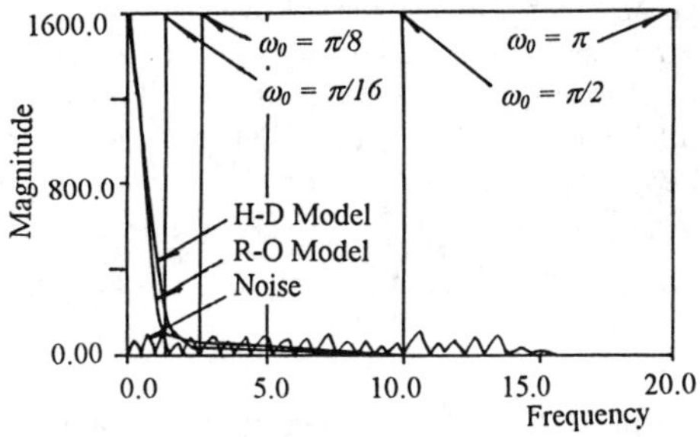

Fig. 3 The response spectral of torque to angular displacement

2.3 *The Butterworth low-pass filter*

Butterworth filter is defined by the property that magnitude response is maximally flat in the passband(Brown et.al, 1983). For an Nth-order lowpass filter, this means that the first 2N-1 derivatives of the squared magnitude function are zero at $\omega = 0$. Another property is that the approximation is monotonic in the passband and the stopband. The squared magnitude function for an analog Butterworth filter is of the form:

$$|H(j\omega)|^2 = \frac{1}{1+(\omega/\omega_0)^{2N}} \tag{7}$$

where, ω_0 is the cut-off frequency. As the parameter N in Eq.(7) increases, the filter characteristics become sharper, that is, they remain closer to unity over more of the passband and become close to zero more rapidly in the stopband, although the magnitude function at the cut-off frequency ω_0 will be $1/\sqrt{2}$ because of the nature of Eq.(7).

There are two ways to filter a function in time domain by Butterworth low-pass filter. The first way is that, by using the Fast Fourier Transformation, transform the time domain function into frequency domain function, and then, multiply the frequency response function of Butterworth filter of Equation (7), finally, make the back-transformation of FFT, the updated function in time domain can be reached. Secondly, taking the convolution directly in time domain to the time domain function of Butterworth filter and the function which to be filtered, it can also get the new filtered function. In the study, the second way is employed, and N=4.

2.4 *Filter analysis*

By using of the above method, the filter for two theoretical curves in Fig.2 are performed. When the cut-off frequency ω_0 equals π, π/8, π/16, respectively, the filtered curves are consistent with original theoretical curves very well. In fact, it can be seen from Fig.3 that, though α = π/16, there is little energy of torque lost. By using the difference scheme for Eq.(1), the recovered shear modulus to shear strain relation from the filtered signals are also consistent well with the curves in Fig.1.

In order to simulate the irregular discrete data measured in laboratory, a random noise signal which has zero average value is created by the random number generator in FORTRAN program library.

$$x(i) = k_0\left[x_0(i) - 0.5\right] \quad (8)$$

where, $x_0(i)$ is random number, k_0 is a coefficient which adjust the amplitude of noise, in the paper, k_0=60. The spectrum of noise is also shown in Fig.3. Putting the noise signals into the pure signals in Fig.2, the signals mixed with noise are induced shown in Fig.4.

By using of the Butterworth filter method, the computed results show that, when the cut-off frequency ω_0 equals π/16, the recovered shear modulus and shear strain relation agree far well in the original theoretical curves during the range of shear strain exceeding 0.0004. From Fig.3 it is seen that, even through the cut-off frequency ω_0 equals π/16, the noise energy exists still in the range of lower frequency, and it is impossible to get rid of it completely. Thus, the whole recovered curve can not be reached. However, the optimum output may be able to get from the Optimum Filter scheme of signal processing, for instance Wiener filter scheme(Oppenheim et.al 1975) etc, it will be studied further in future. On the other hand, as a matter of fact, it is known from both theory and experience that the shear modulus changing with shear strain is very slowly when shear strain is in small range. Therefore when shear strain is lower 0.0004, the experience formula is suggested as follow:

$$\frac{G}{G_{max}} = a_1 + b_1 \log\gamma + d_1(\log\gamma)^2 \quad (9)$$

where, a_1, b_1 and d_1 are parameters which can be calibrated from the boundaries and derivative consistent conditions. Therefore, the whole recovered shear modulus and shear strain relations is gotten shown in Fig.5.

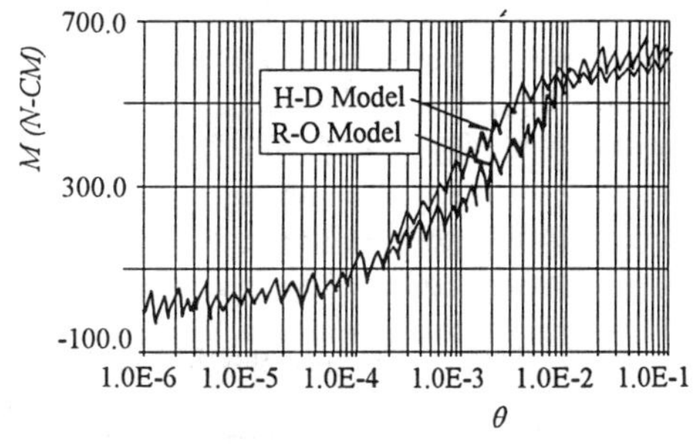

Fig. 4 The relation of Torque Vs angular displacement with noise

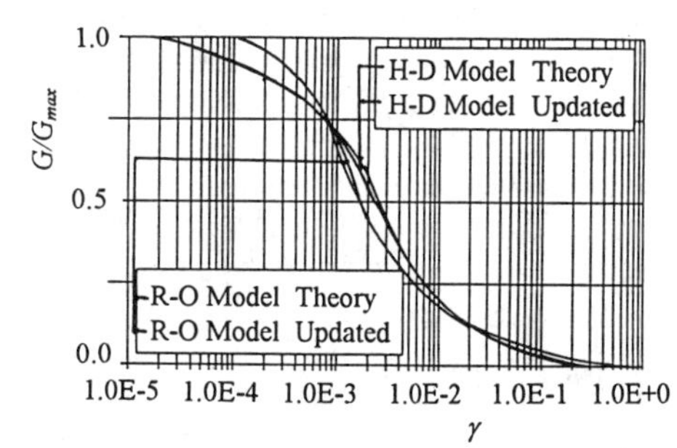

Fig. 5 The recovered shear modulus Vs shear strain

It is seen from Fig.5 that, good relations exist between the updated and the theory response. Because the noise signal employed is of stochastic, it may be explained that the method established in the paper is reasonable.

3 CASE ANALYSIS

Fig.6 is a laboratory torsional shear test results of Nanjing sand, the maximum shear modulus is 5250kpa which determined by in-situ wave velocity scheme. For there are many type of errors existing during the test, so the big discrete data induced, that means the signal is mixed with noise signal. Based on the present method, this test signal is filtered by the Butterworth filter method, the strain part lower in 0.0004 is simulated by equation(9), the computed shear modulus and shear strain is shown in Fig.7. If the Ramberg-Osgood model is fitted, the parameters will be got as α=0.9, R=5.1, c_1=0.5.

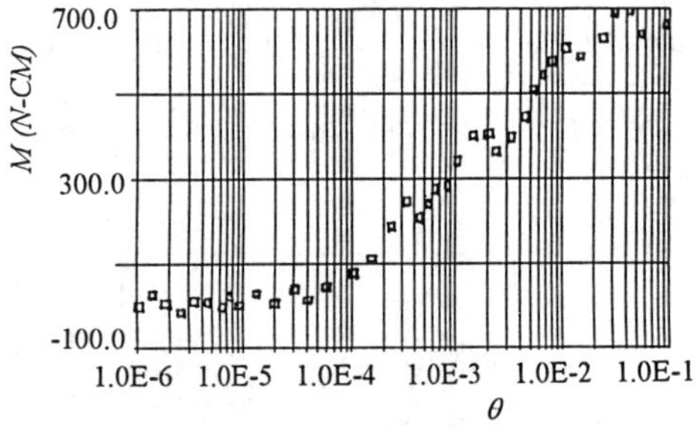

Fig. 6 The relation of Torque Vs angular displacement of Nanjing sand

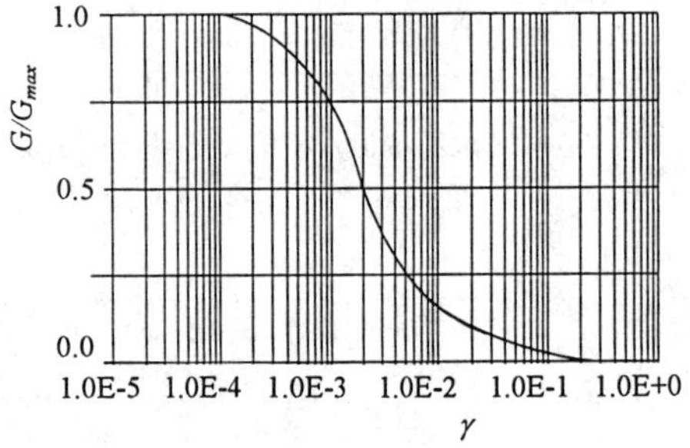

Fig. 7 The relation of shear modulus and shear strain of Nanjing sand

4 CONCLUSION

The dynamic torisional shear test is a general method to measure the shear modulus. The discrete irregular test data is simulates the as signal accompanying with noise, by employing the Butterworth low-pass filter scheme of signal processing, a test analytical method is presented. By performing a recovering analysis from a pure theoretical signal mixed with a random noise the reasonable is confirmed. The present method in the paper can also be referred in the same type test results analysis.

REFERENCES

Brown, R.G., and Hwang, P.Y.C. 1983, Introduction to Random Dignals and Applied Kalman Filtering, Published by John Wiley & Sons Inc., pp.171-210

Hardin, B.O., and Drnevich, V.P.1972, Shear Modulus and Damping in Soil: Measurement and Parameter Effects, Journal of Soil Mechanics and Foundations Division, ASCE, Vol.98, No.SM6, Proc. Paper 8977, June, pp.603-624

Oppenheim, A.V. and Schafer, R.W. 1975, Digital Signal Processing, Published by Prentice-Hall Inc., Englewbod Cliffs, New Jersey, 07632, pp.26-30

Ramberg, W., and Osgood, W.R. 1943, Description of Stress-Strain Curves by Three Parameters, Technical Note 902, National Advisory Committee for Aeronautics, Washington, D.C.

Streeter, V.L., Wylie, E.B., and Richart, F.E.,Jr 1974, Soil Motion Computations by Characteristics Method, Journal of the Geotechnical Engineering Division, ASCE, Vol.100, No.GT3, Proc. Paper 10410, March, pp.247-263

Taylor, P.W. 1975, Interpretation of Dynamic Torsion Tests on Soils, Report No.120, School of Engineering, University of Auckland, Auckland, New Zealand

Computer Methods and Advances in Geomechanics, Yuan (ed.) © 1997 Balkema, Rotterdam, ISBN 90 5410 904 1

Experimental methods for determining constitutive parameters for non-linear rock modelling

P.A. Nawrocki & M. B. Dusseault
Department of Earth Sciences, University of Waterloo, Ont., Canada

N. D. Cristescu
University of Florida, Gainesville, Fla., USA

R. K. Bratli
Saga Petroleum ASA, Sandvika, Norway

ABSTRACT: A general method of dealing semi-analytically with the effect of material non-linearities on borehole stresses is presented in this paper. To introduce material non-linearities into the analysis we assume that the hydrostatic deformation is governed by a mean stress-dependent bulk modulus $K(\sigma)$, whereas deviatoric deformation is governed by a simultaneously mean stress-dependent and shear stress-dependent shear modulus, $G(\sigma,\tau)$. Experimental methods of collecting model parameters are discussed and a special method of interpreting triaxial test data to obtain the $G(\sigma,\tau)$ function is recommended.

1 INTRODUCTION

Knowledge of material constitutive behaviour is required both for model formulation and to apply existing theories in opening design. Different predictive models have been developed which link rock stresses to rock deformation through experimentally determined elasticity constants. The Kirsch equations for linear elastic (LE) stresses induced around a plane strain circular opening are usually used to analyze wellbore stability. However, there is now general agreement that LE analyses invariably underpredict opening stability, and that models which are more realistic (and less conservative) in their predictions should be utilized. These are non-linear (NL), elasto-plastic (EP) and damage models. At the same time, NL rock properties have not yet attracted adequate attention. One limited but widely used function for simulation of σ-ε curves in finite element (FE) analysis was formalized by Duncan and Chang (1970), using Kondner's finding (1963) that a plot of stress vs strain in triaxial compression is nearly a hyperbola. Except for this hyperbolic model, incorporation of NL and inelastic material behaviour has been avoided except in models suitable only for FE analysis (e.g. Vaziri 1986). We will present an alternative NL model, hoping that it can fill the gap between simple, linear analytical solutions and FE approaches.

Constitutive models for soils and rocks (Cristescu, 1989; Christian and Desai, 1977) link stresses with strains through a set of phenomenological or experimentally determined mathematical relationships. Load paths in geomechanics are mainly compressive, therefore uniaxial, hydrostatic, and triaxial compression data are most often used for constitutive models. Material NL behavior results from inherent flaws of all scales. It is also related to internal damage which can arise during drilling, thermal loading, etc. (Dusseault & Gray, 1993). Damage reduces the stiffness modulus, which leads to stress redistribution. Because damage is related to strain (or stress level, as there is a direct coupling), realistic analyses should address issues of damage and NL behaviour where the stress response is modified by the variation of stiffness. The PDM model (Santarelli et al., 1986) and the RDM model (Nawrocki & Dusseault, 1995) provide examples of models where stiffness is not constant. For example, the PDM model introduced a σ_3-dependent Young's modulus, $E=f(\sigma_3)$, and the RDM model used an assumption of stiffness related to damage or radial distance measured from the opening wall, $E=f(r)$. Stresses are similar in these approaches, but the radial formulation allows more rapid calculations because the inverse σ_3-dependent function $1/E(\sigma_3)$ which arises in PDM methods cannot be easily integrated. Both approaches provide radial distributions of σ_Θ which correspond well to experimental results on hollow cylinders, (c.f.Daemen and Fairhurst 1971). Thus, it is now believed that for realistic simulation of rock stresses, elasticity "constants" have to be considered as functions rather then constants, and different NL models postulate different mathematical representations for those functions.

The basic goal of this research is to provide a more general representation for the functions mentioned above. That goal has been achieved by conducting a testing program designed to collect data for borehole stability analysis based on stress distributions calculated using NL elastic relationships (Nawrocki et al. 1996), and by careful interpretation of the test results. Those relationships constitute an extension of the constitutive equations of linear elasticity to NL materials which consists in introducing stress- (or strain-) dependent Lamé parameters into the constitutive model instead of fixed values of these parameters, as in linear elasticity. This permits the incorporation of several typical types of pre-peak material non-linearities with considerable accuracy. Thus, we assume that mechanical behaviour of geomaterials can be presented as a superposition of hydrostatic and deviatoric states of deformation (Nawrocki et al., 1996). To directly introduce material non-linearities, we assume that hydrostatic deformation is governed by a mean stress-dependent bulk modulus, $K=K(\sigma)$, whereas deviatoric deformation is governed by both mean stress- and shear stress-dependent shear modulus, $G=G(\sigma,\tau)$. Such an assumption seems to be a natural one and it may represent a better method of borehole stability analysis, as current models are based almost exclusively on linear elasticity or σ_3-dependent elasticity models. In practical terms, our approach seems more general then previous models, as it introduces shear modulus dependency on σ and on τ. It is also more general than PDM or RDM models because both σ and τ depend on σ_3. Also, we propose a somewhat different method of collecting model parameters and interpreting triaxial test results. To simplify the preliminary experimental program, we assume that the effects of factors such as temperature and geochemistry are of second-order and can be neglected.

2 EXPERIMENTAL PROCEDURES

The material used for testing was shale, and the experimental program used to determine its compressional behaviour consisted of hydrostatic (HT) and triaxial (TT) compression tests. The HT is intended to measure the volumetric deformation of rock specimens and to provide data necessary to establish the $K(\sigma)$ dependency, and TT provide information on shale behaviour in shear and shear modulus variation with both σ and τ.

2.1 *Sample preparation*

Tests have been conducted on shale samples collected from the Queenston Formation near Windsor, Ontario. Queenston shale is a non-fissile compaction shale, and occurs as a red to green silty calcareous mudstone of relatively low durability. The shale is calcareous, with harder and more durable bands parallel with the bedding and occasionally at right angles. The colour of the formation is brick red with occasional green bands (calcium or durable material), it is homogeneous and isotropic in texture with approximately 60% clay minerals.

The shale specimens were collected from six boreholes drilled close to Windsor. Upon collection, the cores were wrapped, marked, and stored in a refrigerator at a temperature of about -10 to 0 degrees Celsius. The cores were cut into right circular cylinders with a height to diameter ratio of 2.0 to 3.0 and a diameter of 45 mm. The ends of the specimen are cut flat and perpendicular to the longitudinal axis, and the sides are smooth. Specimens are taped, wrapped in Saran Wrap, marked and then stored in a cool place. Once the specimen is properly prepared, four strain gauges are glued onto the specimen in the middle of the sample and 90 degrees apart. The specimen is sealed using a rubber membrane, and then attached to the lid of the hydrostatic cell. The cell is completely filled with hydraulic fluid and is connected to a hydraulic line and pressure intensifier.

2.2 *Hydrostatic compression tests*

To accurately determine the $K(\sigma)$ function, HT are performed using unloading-reloading cycles, with K modulus determined during unloading (Cristescu, 1989). Load is increased in small increments, $\Delta\sigma$=5MPa, and with each stress increment $\Delta\sigma$, an instantaneous volumetric strain $\Delta\varepsilon_v$ takes place. In a first-order approximation, this can be obtained from a relationship $\Delta\sigma=K(\sigma)\Delta\varepsilon_v$, where K depends on σ. To allow deformation to stabilize, after each load increase the sample is kept under constant load for 10 minutes before partial unloading. During unloading the pressure is reduced by 5% of its current value each time at one minute intervals. Unloading continues until 75% of the initial pressure is reached, and then the reloading sequence is initiated until the axial stress increases by 5 MPa (the second loading step). This loading-unloading sequence is repeated until high stresses are achieved, where virtually no deformation increase is observed in 10 minute periods of constant load.

2.3 *Triaxial compression tests*

Triaxial tests in geomechanics have in general concentrated on the peak and the residual shearing resistances, in order to delineate peak strength criteria. Often, pre-peak non-linearity is ignored, and the relationships are linearized to obtain "constants" for analysis. In order to obtain information on $G(\sigma,\tau)$ behaviour, TT are performed for many values of σ_3, so as to delineate the deviatoric response over the range of stresses expected in situ. We have used σ_3's ranging from 2.5 to 35 MPa, differing by 2.5 MPa from test to test. About 30 TT were performed using ISRM Suggested Methods for Rock Testing and the best results were used in the analysis.

Tests are carried out using successive loadings

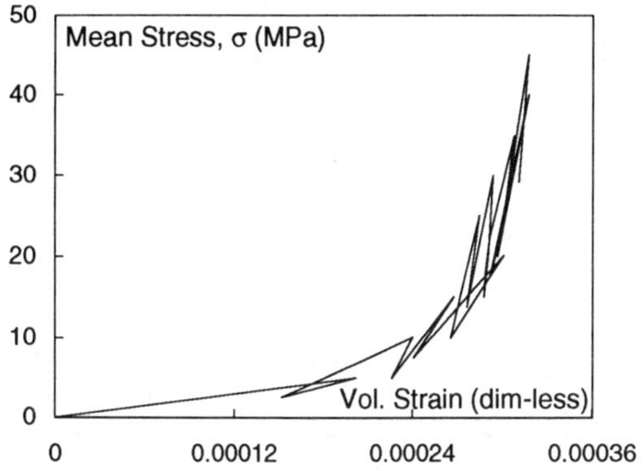

Fig. 1: Hydrostatic test results.

similar to the procedure for the isotropic tests. After attaining the proper confining stress, deviatoric stress is gradually increased to the desired level, held constant for 20 minutes, and partial unloading begun. Elastic moduli (Young's modulus and Poisson's ratio) are determined by taking the slope of the unloading curve. The load is then increased to a higher level, and the procedure is repeated until failure. Then, the shear modulus is calculated using the relationship $G=E/2(1+\nu)$.

3 TEST RESULTS AND INTERPETATION

In the case of isotropic tests, laboratory results are presented in a σ vs ε_v format, and the bulk modulus, K, corresponding to the current stress level σ, is determined by finding the slope of the curve during the unloading sequence of the HT. We believe this to be more representative of unloading moduli which clearly govern the deformation response when a borehole or a tunnel is advanced.

Results of one of the hydrostatic compression tests are shown in Fig. 1. The maximum mean stress attained during that test was 45.08 MPa. Loading-unloading steps were analyzed separately, and the K-modulus determined at unloading. Fig. 2 presents summary results for that test. In Fig. 2, K-modulus is presented as a function of σ, and the data are approximated by functions. Actually, two functions are proposed:

$$f_1\!: K(\sigma_n) = K_1 + \alpha[1 - e^{-\beta(\sigma_n - 1)}] \qquad f_2\!: K(\sigma) = -0.78\sigma^2 + 65.32\sigma - 369 \tag{1}$$

Both functions are shown in Fig.2. f_1 is an exponential function which has a horizontal asymptote $K_1+\alpha$. In Eq.(1) σ_n is a normalized mean stress, $\sigma_n=\sigma/\sigma_{(1)}$; α and β are constants; K_1 is the bulk modulus corresponding to the first data point, and normalization is with respect to the mean stress $\sigma_{(1)}$ of that point. Thus, $K_1=K(\sigma_{n(1)})$, and α can be determined from the boundary condition at the last data point $(\sigma_{(2)};K_2)$ giving $\alpha=(K_2-K_1)/[1-1/\exp(\beta(\sigma_{n(2)}-1)]$, where $\sigma_{n(2)}=\sigma_{(2)}/\sigma_{(1)}$. By changing β it is possible to obtain a good approximation of data with f_1. In the case of the curve shown in Fig.2, $\beta=0.4$.

Note, that K increases with σ and tends towards a constant value $\sigma_o=1000$ GPa, when all pores and microcracks are closed. Thus, $K(\sigma)$ is a monotonic function of σ in the interval $0<\sigma<\sigma_o$ only. For $\sigma>\sigma_o$, volumetric creep is no longer observed and the relationship σ-ε_v of Fig.1 becomes nearly linear for both loading and unloading during the cyclic process. Thus, σ_o is the smallest mean stress for which all the cracks are already closed, and that is why for $\sigma>\sigma_o$ the behaviour of the rock is nearly linearly elastic and reversible. The ultimate value of $K(\sigma)$ corresponds to the bulk modulus for the rock with all microcracks closed. This value can be a reference value for $K(\sigma)$ used to generate dimensionless variables. Therefore, f_1 increases and tends towards a horizontal plateau at high stresses. Although theoretically correct, such a function can be difficult to incorporate in NL stress analysis. That is why the polynomial expression f_2 has been also proposed. Thus, we think that f_1 is more appropriate from a physical point of view, but f_2 seams to be better for practical purposes, as it can be more easily handled by a program for NL stress analysis. It should be

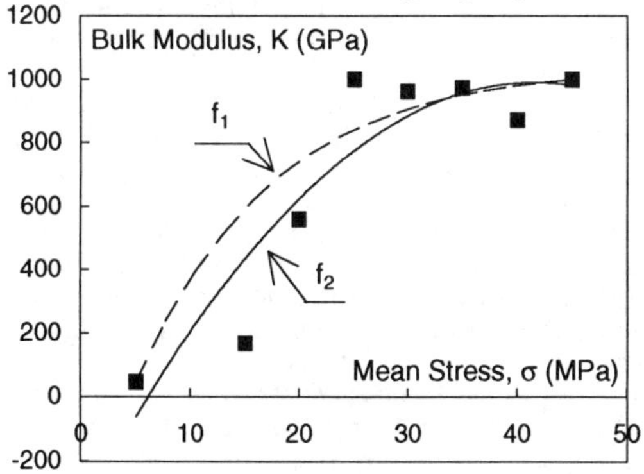

Fig. 2: Hydrostatic test results.

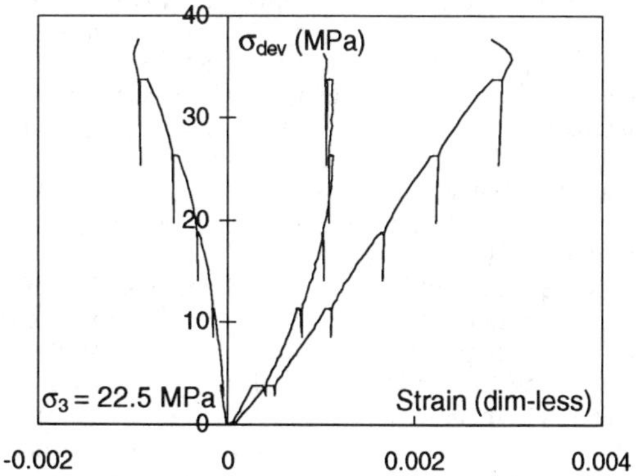

Fig. 3: Triaxial test results.

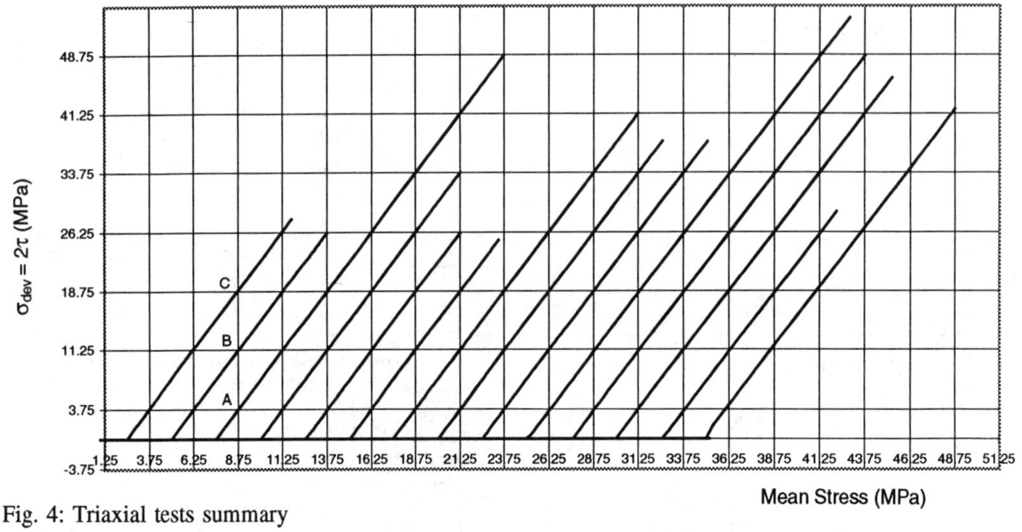

Fig. 4: Triaxial tests summary

understood, however, that f_2 is not valid for high stresses.

In order to obtain stable values of the bulk modulus, hydrostatic tests were conducted up to high stresses, as in Fig.1 . However, for porous materials such as shales, it is not possible to "close" all the pores, and a pseudo-asymptote can be chosen, beyond which the non-linearity becomes unimportant for the application considered.

Typical results of one of the triaxial tests (σ_3=22.5 MPa), following the procedure discussed previously, are shown in Fig. 3. After attaining the confining stress level prescribed for the test, σ_3 was kept constant and axial stress increased up to the unloading stress level. Unloading began at deviatoric stresses σ_{dev}=3.75, 11.25, 18.75 MPa, and so on, with 7.5 MPa increments. These values of σ_{dev} were selected because of the way we intended to read test results (see below). After attaining the deviatoric stress level prescribed for unloading, the specimen sat at constant load for 20 minutes to allow creep deformation to stop and not influence the measured elastic moduli. When creep deformation was negligible, the specimen was partially unloaded to allow the measurement of elastic constants corresponding to the current level of σ_{dev}. The rock specimen is reloaded, deviatoric stress increased to its next unloading value, and the same unloading-reloading sequence repeated to obtain elastic constants corresponding to next value of σ_{dev}, or τ.

The elastic moduli (Young's modulus, E, and Poisson's ratio, ν) were calculated directly from unloading slopes of the compression curve. With E and ν known, the shear modulus G and bulk modulus K can be calculated. Thus, the set of four elastic moduli is available for each point of unloading. Elastic moduli corresponding to the compression curve shown in Fig. 3 are given in Table 1.

Table 1: Elastic constants from test run at 22.5MPa.

σ_{dev} @unload. (MPa)	Young's Modulus (GPa)	Poisson's ratio ν	Bulk Modulus (GPa)	Shear Modulus (GPa)
3.75	99	0.13	44	44
11.25	247	0.16	119	107
18.75	290	0.20	161	121
26.25	263	0.26	184	104
33.75	223	0.32	208	85

The experimental data on elastic moduli collected in this way are now used to obtain the $G(\sigma,\tau)$ function. Collected data were fitted with approximating functions to be used in NL stress analysis. For given σ, G-modulus dependency on τ was determined first, then G dependency on σ can be introduced by comparing coefficients of functions approximating the $G(\tau)$-dependency obtained for different mean stresses.

Fig. 4 summarizes all the triaxial tests conducted. It is also used to read test results. Actually, triaxial tests, and especially the deviatoric stress levels at unloading, were designed based on that figure. Fig. 4 is plotted in a σ_{dev} - σ format, and each inclined line in Fig. 4 corresponds to one test. During the isotropic part of testing, axial stress is approximately equal to confining stress, so σ_{dev} is zero, thus the isotropic line coincides with the horizontal axis, whereas inclined lines correspond to deviatoric stresses. Moreover, crossing points of test lines and vertical lines where we read test results define the deviatoric stress of unloading; that is, the elastic moduli E, ν, K, and G are determined at each crossing point. In other words, if we use the data along the inclined lines of equation σ-σ'=const or $\sigma-\tau_{oct}/\sqrt{2}$=const (here σ' is the equivalent stress and τ_{oct} the octahedral shear stress) we will determine the dependency of the elastic parameters on mean

Table 2: Elastic parameters for several tests.

Test; σ_3= (MPa)	τ (MPa)	E (GPa)	ν	K (GPa)	G (GPa)
σ=18.75 MPa					
17.50	1.88	16.64	0.24	10.83	6.69
15.00	5.63	163.12	0.40	283.19	58.09
12.50	9.38	269.87	0.39	401.59	97.21
10.00	13.13	242.89	0.91	x	x
σ=26.25 MPa					
25.00	1.88	-52.63	0.46	x	x
22.50	5.63	246.76	0.16	119.21	106.82
20.00	9.38	236.91	0.21	137.10	97.73
17.50	13.13	338.83	0.31	295.66	129.42
σ=38.75 MPa					
35.00	1.88	-153.38	0.29	x	x
32.50	5.63	31.20	0.35	34.90	11.55
30.00	9.38	354.20	0.28	269.56	138.25
27.50	13.13	165.21	0.29	128.07	64.28
25.00	16.88	313.72	0.89	x	x

stress and on the combination of the invariants, say σ-$\tau_{oct}/\sqrt{2}$. In order to directly determine the dependency on the two invariants σ and τ_{oct}, one reads the data along the vertical line first to find the moduli dependency on τ (as if the tests would be done in "true" triaxial conditions), and afterwards one reads the data along the horizontal line to find the dependency of the coefficients on mean stress.

Once elastic moduli have been determined for all the TT, results can be interpreted by graphing the elastic modulus G as a function of τ. Thus, to obtain the G(σ,τ) function, we look at G dependency on τ first, reading TT results on vertical lines, that is, for σ=const. For example, the vertical line corresponding to σ=8.75 MPa intersects TT lines at points A (test run @ σ_3=7.5 MPa), B (test run @ σ_3=5.0 MPa), and C (test run @σ_3=2.5 MPa). Note that different shear (or deviatoric, as $\sigma_{dev}=2\tau$) stress values correspond to these points. Also note, that at this time elastic constants are already known for A, B, and C, as they have been previously calculated. Thus, for given value of mean stress, G modulus can be plotted as a function of shear stress, τ, by picking moduli values from the unloading points of TT crossed by the vertical line being used. In our case, 13 such graphs can be plotted (the case of σ=3.75 MPa is disregarded because only one inclined line is crossed by the vertical line plotted at σ=3.75).

Table 2 presents the results of such analysis for several values of mean stress (σ=18.75, 26.25, and 38.75 MPa), and Fig. 5 provides the graphical representation of the data from Table 2. In Fig 5, the elastic modulus G has been presented as a function of shear stress τ, and the G dependency on τ has been approximated using a simple function. It has been found that the experimental data obtained can be best modelled by logarithmic functions. Such a G(τ) function can be found for any value of σ for which valid TT results are available. Some of the functions found are listed below:

$$\begin{aligned}
G(\tau) &= 54.7\ln(\tau) - 29.8 && (for\ \sigma = 18.75)\\
G(\tau) &= 23.8\ln(\tau) + 9.2 && (for\ \sigma = 21.25)\\
G(\tau) &= 23.1\ln(\tau) + 61.1 && (for\ \sigma = 26.25) \quad (2)\\
G(\tau) &= 21.7\ln(\tau) + 12.8 && (for\ \sigma = 28.75)\\
G(\tau) &= 16.7\ln(\tau) + 87.1 && (for\ \sigma = 31.25)
\end{aligned}$$

The final step involved is to examine the graphs (functions) corresponding to different σ's to find the coefficients of a function approximating how the G-dependency on τ varies as a function of σ. In order to do that, we plot coefficients of our logarithmic

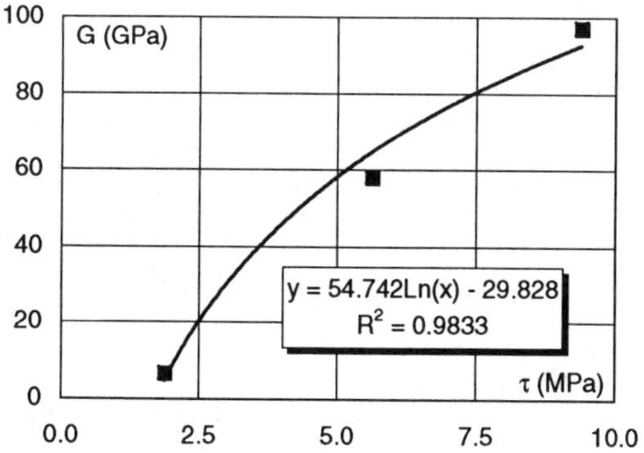

Fig. 5: Determination of the G(τ)-function (σ=-18.75MPa)

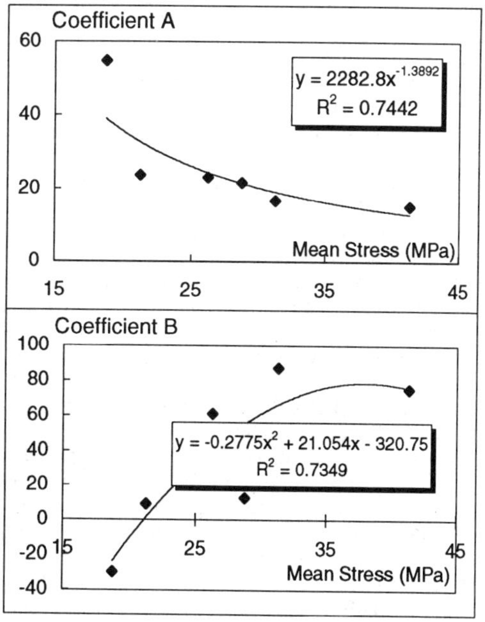

Fig.6: Finding coefficients dependency on σ

functions against σ to find their dependency on mean stress. If the general form of the logarithmic function is $G(\tau)=A\ln(\tau)+B$, then using Eqs.(2) one can find how A and B depend on σ. $A(\sigma)$ and $B(\sigma)$ functions corresponding to $G(\tau)$ functions (2) are shown in Fig. 6. Thus, A dependency on σ is now approximated by a power law:

$$A(\sigma) = 2282\sigma^{-1.3892} \tag{3}$$

and B dependency on σ is now approximated by a second-order polynomial:

$$B(\sigma) = -0.27\sigma^2 + 21.05\sigma - 320 \tag{4}$$

Knowing $A(\sigma)$ and $B(\sigma)$ functions, one can finally write the full $G(\sigma,\tau)$ function:

$$G(\sigma,\tau) = 2282\sigma^{-1.3892}\ln(\tau) - 0.27\sigma^2 + 21.05\sigma - 320 \tag{5}$$

Together with the previously determined $K(\sigma)$ functions (1), such a function can be used in NL analysis of borehole stresses using a method presented by Nawrocki et al. (1996).

4 SUMMARY AND CONCLUSIONS

Our data show several important aspects of shale compressive behaviour. First, shale deformation is markedly non-linear. Second, there is a significant effect of the mean stress on bulk modulus, as well as shear modulus dependency on σ and τ. However, the effects of both hydrostatic and the deviatoric components can be measured, analyzed, and included in modelling. Furthermore, results allow us to study the Young's modulus dependency on σ_3, provide failure envelopes, and provide information on shale behaviour in unloading. However, this does not fit within the scope of this paper; we will make it a subject of separate study.

It is believed that the experimental procedures and methods of analysis presented in this paper should be effective in better characterizing NL properties of rocks to aid NL analyses; they can be used for any rock showing NL compressional behaviour. However, because our model requires extensive triaxial testing, it is perceived as a tool for advanced investigations which can be performed when enough core is available for testing, and when 'fine tuning' in the analysis is required. Although we have not yet done so, one may also simultaneously measure acoustic emissions and permeability changes associated with damage, which be of value in evaluating the consequences of a microfissured and damaged zone around a repository structure in shales, or in carrying out coupled stress-flow-damage transient analyses.

Methods of analysis presented in this paper provide means for estimating constitutive parameters both for linear and NL models. Generally, such models are used first before introducing more complex viscoelastic or viscoplastic models, which may nonetheless be warranted by experimental and field performance data.

Realistic NL models can play a vital role in petroleum engineering (borehole stability, reservoir engineering, natural gas storage projects), allowing parametric analyses varying rock strength and stiffness, in situ stresses, geometry, and so on. As such parametric analysis tools seem to be lacking, development of these into a comprehensive analysis package for such applications seems warranted. Perhaps these approaches can aid in this task.

ACKNOWLEDGEMENTS

Financial asistance of NSERC, Saga Petroleum, and several other companies involved in the Waterloo Shale Project is gratefully acknowledged. Special thanks go to our colleagues and technical staff at the University of Waterloo who carried out the testing: Mr. B. Davidson, Ms. M. Kim, Mr. D. Hirst, and Ms. J. Fooks.

REFERENCES

Christian, J.T. & C.S. Desai 1977. Constitutive laws for geologic media. *Num. Meth. in Geotechnical Engineering*. New York: McGraw Hill, 65-115.

Cristescu, N. 1989. *Rock Reology*. Kluver, 330 pp.

Daemen, J.J.K. & C. Fairhurst 1971. Influence of failed rock properties on tunnel stability. Dynamic Rock Mech. *Proc.12th U.S.Symp. on Rock Mech.* (G.B.Clark, Ed.). AIME, New York: 855-875.

Duncan, J.M. & C.Y. Chang 1970. Non-linear analysis of stress and strain in soils. *J. Soil Mech. Found. Div. ASCE.* 96, No. SM5 Sept. 1970: 1629-1653.

Dusseault, M. & K.Gray 1992.Mechanisms of stress-induced wellbore damage. *Proc. Conf. on Wellbore Damage*, Lafayette, SPE #23825: 511-521

Kondner, R. 1963. Hyperbolic stress-strain response: cohesive soils. *J. Soil Mech. Found. Div.* Febr. 1963, ASCE 89 (SM1): 115-143.

Nawrocki,P.A. & M. Dusseault 1995. Modelling of damaged zones around openings using radius-dependent Young's modulus. *Rock Mech. and Rock Engng.* 28 (4): 227-239.

Nawrocki, P.A. Dusseault M.B. & R.K. Bratli 1996. Addressing the effects of material non-linearities on wellbore stresses using stress- and strain-dependent elastic moduli. *Proc. 35th U.S.Symp. Rock Mech.*, Lake Tahoe, Nevada, June 4-7, 1996.

Santarelli F., Brown E.T. & V.Maury 1986. Analysis of borehole stresses using pressure-dependent linear elasticity, *Int. J. Rock Mech. Min. Sci & Geomech. Abstr.* 23: 445-449.

Vaziri, H.H. 1986. Non-linear temperature and consolidation analysis of gassy soils, PhD Thesis, University of British Columbia.

Computer Methods and Advances in Geomechanics, Yuan (ed.) © 1997 Balkema, Rotterdam, ISBN 90 5410 904 1

Theoretical analysis of the undrained instability of very loose sand in triaxial compression and extension using an elastoplastic model

Ph. Dubujet
LTDS-UMR, Ecole Centrale de Lyon, Ecully, France

T. Doanh
Ecole Nationale des Travaux Publics de l'Etat, Laboratoire Géomatériaux - URA, Vaulx-en-Velin, France

ABSTRACT: In this paper an elastoplastic model is used to analyse the loss of the stability associated with the initiation of the static liquefaction phenomenon. For this theoretical study a laboratory programme was performed on very loose Hostun RF sand. The samples are consolidated isotropically or anisotropically along constant effective stress ratio paths. The possibilities and the limits of the model to describe the behaviour of very loose sand are examined. In particular the study focus on the ability to characterize the instability domain in compression and extension triaxial tests in undrained condition. It is shown that the elastic components, and in particular the influence of its non linearity is highlighted.

1 INTRODUCTION

Static liquefaction which refers to a complete and sudden loss of effective mean pressure p' under monotonic undrained condition, occurs only in very loose satured sand. Experimentally in q-p' plane (q deviator stress) the different deviator stress peaks obtained for different initial consolidation pressure can be connected by a single line for the same void ratio. To describe this phenomenon Lade (Lade 1988), Pradel (Pradel 1990) proposed the instability concept adapted to loose sand under undrained condition. Moreover factors such as anisotropic consolidation, overconsolidation or drained axial strains prior to undrained compression influenced this concept. The effects of anisotropic consolidation is studied by Matiotti (Mattioti 1996) and Canou (Canou 1991) in the compression side of the triaxial plane. It is shown that the positive initial shear produced by an initial anisotropic consolidation is an important factor. Matiotti also examines the Lade' instability concept in the case of extension tests.

Different approaches have been proposed to model these experimental observations. For example Lade (Lade 1988) Molenkamp (Molenkamp 1991) formulated the mathematical conditions for the undrained instability line in a generalized elastoplasticity non associated framework. In particular Mattioti (Mattioti 1996) showed that elastoplasticity models can describe correctly the observed behaviour of very loose sands.

This paper examines the possibilities offered by an elastoplastic model to forecast the static liquefaction. In particular we focus on the ability to characterize the instability domain boundary. The stability line in compression and extension triaxial tests given by the model are examined. The role of the non linearity of the model elastic component is examined.

For this theoretical study a laboratory programme was performed (Mattioti 1996) on very loose Hostun RF sand in order to validate the model response for the static liquefaction aspects. All samples have the same initial void ratio at fabrication tests (around 1) and Dr is less than 15 %. All tests were sheared axially in a strain controlled mode.

2 AN ELASTOPLASTIC MODEL

2.1 *The general framework of the model:*

The model used in this study is the CJS model .A detail of this model can be found in (Cambou 1988) and in this paper only essential aspects are recalled.

This model is elastoplastic and non associated. The strain rate is then defined by:

$$\dot{\varepsilon}_{ij} = \left[D^{e}_{ijkl} + \frac{1}{H} \frac{\partial f}{\partial \sigma_{kl}} \frac{\partial g}{\partial \sigma_{kl}} \right] \dot{\sigma}_{kl} \qquad (1)$$

where H is the hardening modulus, f the yield criterion, g the plastic potential surface and D^e the elastic tensor.

The elasticity is isotropic and non linear. The bulk and shear elastic moduli are expressed as

$$G^e = G^e_o \left(\frac{I_1}{3P_a} \right)^n \qquad K^e = K^e_o \left(\frac{I_1}{3P_a} \right)^n \qquad (2)$$

where n is a parameter which quantifies the non linearity, G^e_o and K^e_o two constant parameters, $I_1 = \sigma_{kk}$ and P_a a reference pressure.

Concerning plasticity the most important model features are the following:

- the model includes two stress hardening mechanisms: a kinematic and an isotropic hardenings.
- The yield criterion is non symmetric in compression and extension triaxial. In octahedral plane the criterion shape is about like the Lade model one,
- A kinematic condition imposes to the strain to verify the characteristic state concept described by Luong (Luong 1980). This state materializes in the stress space the surface where the material changes from a contractant state to a dilatant state. This surface is a non symmetric conical surface in the stress space and for the CJS model the characteristic state is supposed intrinsic.

2.2 *Identification of parameters*:

The parameters of the model are identified using different isotropically consolidated undrained compression tests in order to describe correctly the experimental observations. The elastic parameters are identified with the results of an isotropic drained compression test which is not provided in this paper. The non symmetry of the criterion in compression and extension triaxial is adjusted to the Mohr Coulomb one.

The identification procedure is conducted with a purely contractant soil, the very loose Hostun sand (initial void ratio = 1). Therefore the characteristic state is assimilated to the failure surface. Figures 1 and 2 show the realistic predictions of the model in compression and extension triaxial tests. Nevertheless it can be noted that the entirely liquefaction is always obtained in the simulations.

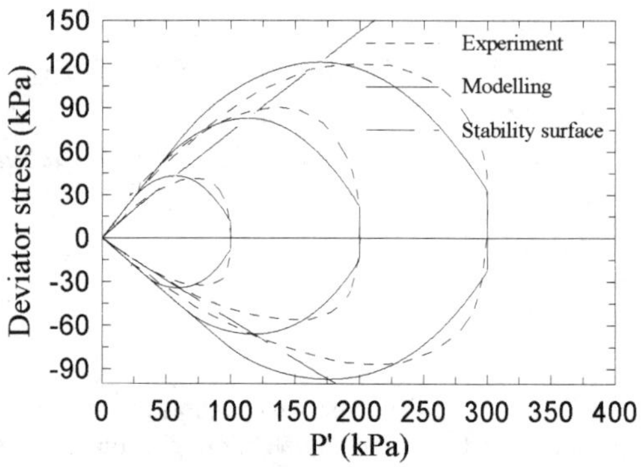

Figure 1. Stress path in extension and compression tests on isotropically consolidated samples - Very loose Hostun RF sand.

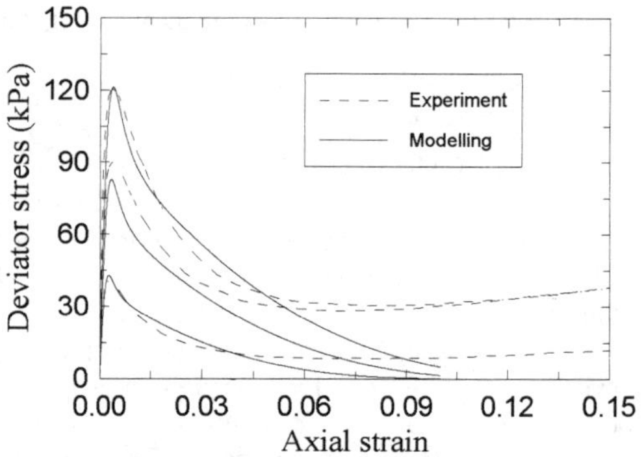

Figure 2. Undrained triaxial compression - Very loose Hostun RF sand - Axial strain - deviator stress

3 CHARACTERIZATION OF THE STABILITY SURFACE

The concept of instability line which was introduced by Lade (Lade 1988) consists in defining a zone in the stress space and where the Hill criterion (Hill 1958) is not satisfied, i.e.

$$d^2W = d\sigma_{ij} d\varepsilon_{ij} > 0 \qquad (3)$$

where d^2W is the second order incremental work.

In triaxial variables the strain rate versus the stress rate can be written as:

$$\begin{Bmatrix} \dot{\varepsilon}_v \\ \dot{\varepsilon}_q \end{Bmatrix} = \begin{bmatrix} \frac{3}{H}\frac{\partial g}{\partial I_1}\frac{\partial f}{\partial I_1} + \frac{1}{3K^e} & \frac{3}{H}\frac{\partial g}{\partial I_1}\frac{\partial f}{\partial s_{II}} \\ \frac{1}{H}\frac{\partial f}{\partial I_1}\frac{\partial g}{\partial s_{II}} & \frac{1}{H}\frac{\partial g}{\partial s_{II}}\frac{\partial f}{\partial s_{II}} + \frac{1}{2G^e} \end{bmatrix} \begin{Bmatrix} \dot{I}_1 \\ \dot{s}_{II} \end{Bmatrix} \tag{4}$$

with

$$\varepsilon_v = \varepsilon_{kk}$$
$$\varepsilon_q = \sqrt{\varepsilon_{ij}\varepsilon_{ij}} \tag{5}$$
$$s_{II} = \sqrt{(\sigma_{ij} - I_1\delta_{ij}/3)(\sigma_{ij} - I_1\delta_{ij}/3)}$$

Taking into account the incompressibility condition (undrained condition)

$$\dot{\varepsilon}_v = 0 \tag{6}$$

it can be obtained

$$\dot{s}_{II} = 6G^eK^eH \frac{\frac{3}{H}\frac{\partial g}{\partial I_1}\frac{\partial f}{\partial I_1} + \frac{1}{3K^e}}{9K^e\frac{\partial g}{\partial I_1}\frac{\partial f}{\partial I_1} + 2G^e\frac{\partial g}{\partial s_{II}}\frac{\partial f}{\partial s_{II}} + H}\dot{\varepsilon}_q \tag{7}$$

i.e.

$$\dot{s}_{II} = 6GKH\frac{f^{inst}}{f^{unic}}\dot{\varepsilon}_q \tag{8}$$

where appear f^{inst} the instability criterion associated to the loose of stability, and f^{unic} the criterion associated to the loose of uniqueness. In fact these two criteria give the boundaries of respectively the instability domain and the non uniqueness domain. In particular if

$$f^{inst}<0 \tag{9}$$

the stress state does not satisfy the Hill criterion and the static liquefaction occurs. The set of the stress states where $f^{inst}=0$ corresponds to the deviator stress peaks in the p'-q plane for compression tests. It can be noted that with this kind of model (stress hardening) and for an undrained axisymetrical triaxial state the surface defined by the annulment of f^{inst} is intrinsic.

In figure 3 is shown the instability domain boundary provided by the used model in the mean effective pressure - deviator stress plane for a loose sand for different values of the n parameter (constant defined in relation (2)). In the particular case of the very loose Hostun sand n is about 0.6, and this boundary nearly becomes a straight line.

For a dense sand it can also be noted the high influence of the non linearity of the elasticity on the instability domain shape. For example in the figure 4 the elasticity is linear (n=0) and in the figure 5 the non linearity corresponds to the Hostun sand (n=0.6). In the two cases temporary instabilities during undrained shearing tests can be observed. These temporary instabilities were already shown by Lade in experimental tests (Lade 1988).

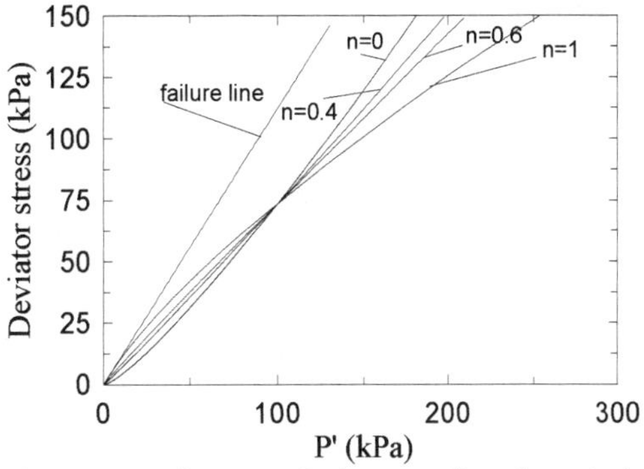

Figure 3. Influence of the non linearity of the elasticity on the stability curve - Very loose Hostun RF sand.

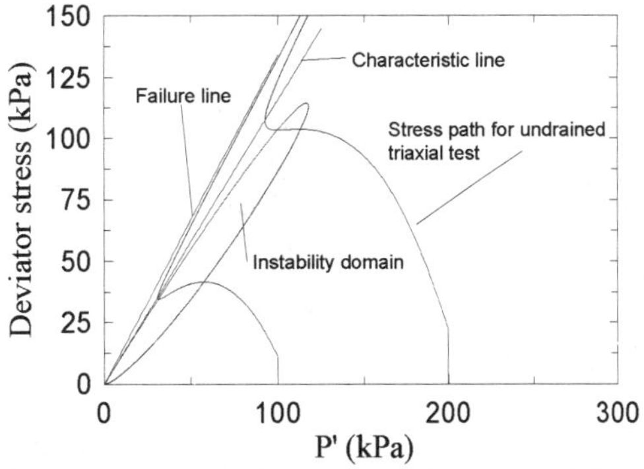

Figure 4. Instability domain for a dense sand - Linear elasticity (n=0).

With the CJS model the loss of uniqueness appears only if the Poisson's ratio is about 0.5, i.e. if the elastic component is nearly incompressible. But this value of ν does not correspond to the identified parameters for the loose Hostun sand. Nevertheless if this condition is verified it can be observed in figure 6 the loss of uniqueness domain which appears for a

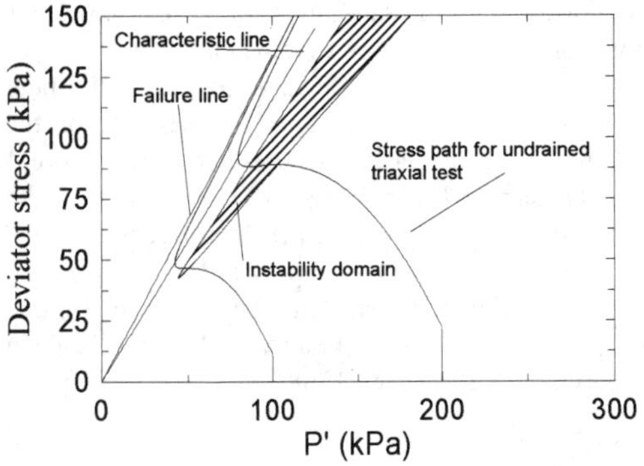

Figure 5. Instability domain for a dense sand - Non linear elasticity (n=0.6).

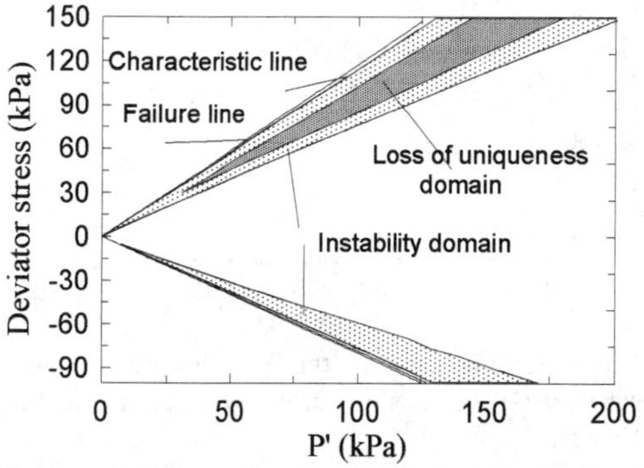

Figure 6. Loss of uniqueness domain - Small dilatancy and n=0.6 (non linear elasticity).

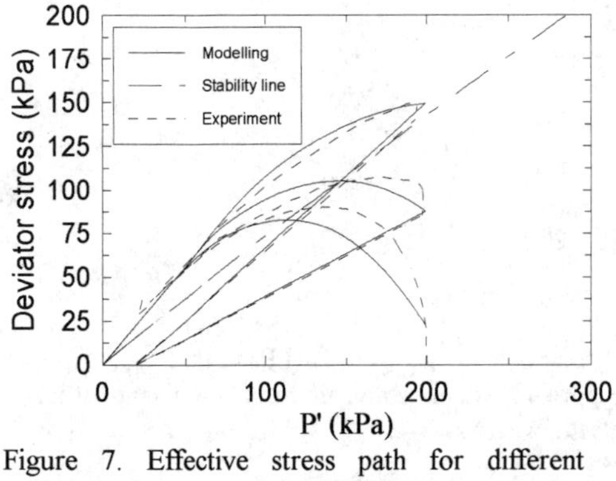

Figure 7. Effective stress path for different anisotropic consolidation- Very loose Hostun sand.

sand with a small dilatancy. It can easily be shown that the loss of uniqueness domain is always strictly located inside the instability domain. It can be noted in this figure that the size of the stability and uniqueness domains of the extension side are much more smaller than that of compression.

4 EFFECTS OF AN ANISOTROPIC CONSOLIDATION

Many authors showed that initial anisotropic consolidation or axial strains attained before undrained shearing modify the instability domain. In the case of a model with stress hardening which is the elastoplastic framework adopted in this paper, this domain is independent of the stress history. Nevertheless it seems worth to check the validity of the modelling in this case. At this aim a set of tests and modelling are performed to show the influence of the initial drained deviator stress on the instability concept of very loose sand. From an initial confining pressure of 20 kPa an anisotropic consolidation was achieved along a constant effective stress ratio path up to 200 kPa. Then a compression triaxial test under undrained condition is performed until failure. Figure 7 shows the results for three consolidation levels ($K=\sigma 1/\sigma 3=1$, 0.66, and 0.5).

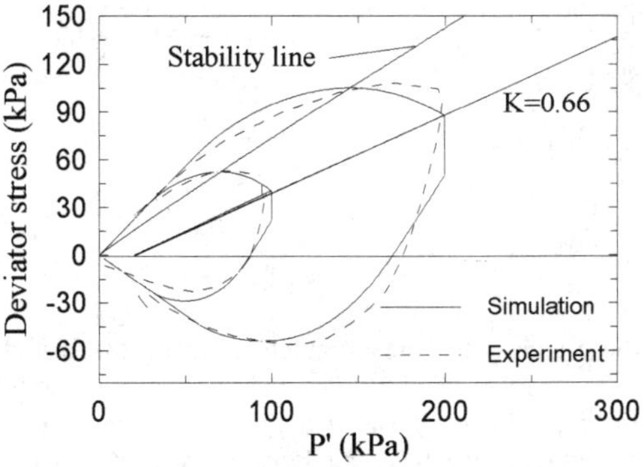

Figure 8. Anisotropic consolidation (K=0.66) Effective stress paths in extension and compression.

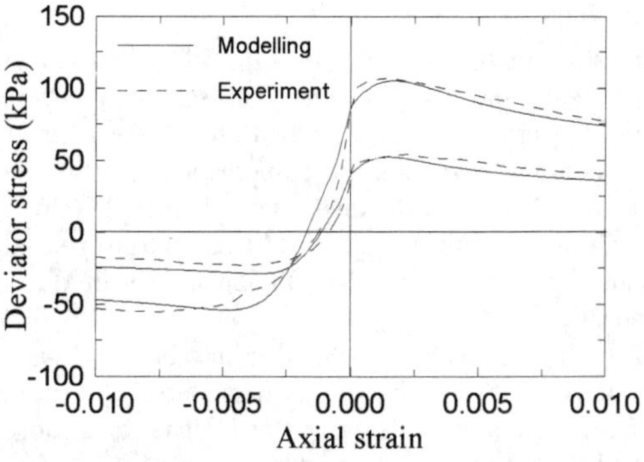

Figure 9. Anisotropic consolidation (K=0.66) - Stress strain behaviours in extension and in compression at small strains - Very loose Hostun sand.

As expected and contrary to the experience, the simulated stress peaks are always located on the instability domain boundary.

Undrained triaxial extension tests after a positive initial drained consolidation are presented in figures 8 (stress paths) and 9 (strain - stress). In particular it is shown the interest of a kinematic hardening in the used elastoplastic framework. This one permits to provide a realistic stress path in extension.

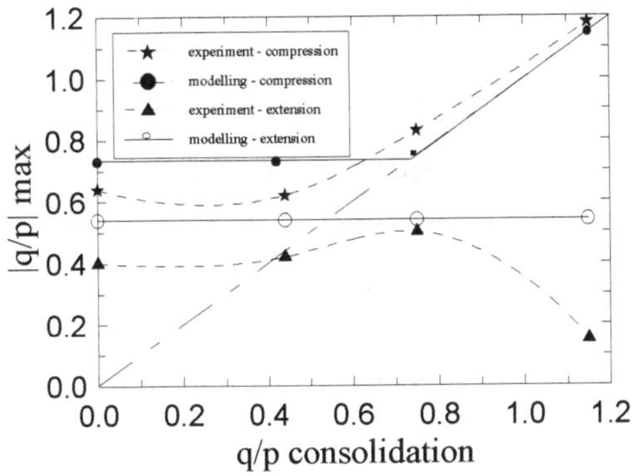

Figure 10. Effects of anisotropic consolidation on the stress deviator peak.

The effective stress ratio at peak corresponding to the maximum of the stress deviator at peak against the consolidation level for different consolidation levels is shown in figure 10. Globally experimental results suggest a little dependence of small anisotropic consolidation level. In contrast to the previous experimental evidences the theoretical analysis gives only two distinct lines in compression: one line associated to the stress deviator at peak and one line simply associated to the stress deviator at the end of the initial consolidation.

Qualitatively the theoretical simulations give an overall good agreement and demonstrates the interest of the model.

5 CONCLUSIONS

1. The elastoplastic modelling using the characteristic state concept is efficient to describe the undrained behaviour of very loose sand. A realistic shape for the instability surface is given.

2. In particular the role of the non linearity of the elasticity on the shape of the instability surface is given with a great emphasis.

3 The validity of the model concerning the influence of an anisotropic consolidation is checked.

4- The uniqueness domain is characterized for the CJS model and its high dependence on the elastic compressibility is shown

REFERENCES

Cambou, C., Jafari, K., 1988. Modèle de comportement des sols non cohérents. *Revue Française de Géotechnique*, Vol. 44, pp 53-55.

Canou, J., Thorel, L., De Laure, E. 1991. Influence d'un déviateur de contrainte initial sur les caractéristiques de liquéfaction statique du sable. *In Proc. ECSMFE*, Vol. 1, pp 49-52.

Di Prisco, C., Matiotti, R., & Nova, R., 1995. Theoretical investigation of the undrained stability of shallow submerged slopes. *Géotechnique*, 45, pp 479-496.

Hill, R., 1958.A general theory of uliqueness and stability in elastic plastic solids. *J. Mec. Phys. Solids*, (6), pp 236-249.

Lade, P.V., Nelson, Y.M., 1988. Instability of granular materials with non associated flow. *Journal Geotech. Eng., ASCE*, 114(12), pp 2173-2191.

Luong, M.P., Loret, B. 1980. Phénomènes cycliques dans les sols. *Revue Française de Géotechnique*. Vol. 10, pp 1-20.

Mattioti, R., Ibraim, E. & Doanh, T., 1996. Comportement non drainé du sable Hostun RF très lâche en consolidation anisotrope. *Revue Française de Géotech.*, 75, pp 1-12.

Molemkamp, F., 1991. Materiel instability for drained and undrained behaviour. Part 2: Combined uniform deformation and shear band generation. *Int. J. for Num. and Anal. Meth. in Geom.*, 15, pp 169-180.

Pradel, D., Lade, P. 1990. Instability and plastic flow of soils. ii: Analytical investigation. *Journal of engineering Mechanics*, Vol. 116(11), pp. 2551-2566.

7 Constitutive modeling

Computer Methods and Advances in Geomechanics, Yuan (ed.) © 1997 Balkema, Rotterdam, ISBN 90 5410 904 1

A constitutive model of sand with inherent transverse isotropy

Kiichi Suzuki
Department of Civil and Environmental Engineering, Saitama University, Urawa, Japan
Eiji Yanagisawa
Tohoku University, Sendai, Japan

ABSTRACT: The purpose of this paper is to propose the constitutive model of sand with inherent transverse isotropy and to verify the validity in the finite element analysis. In order to apply the proposed constitutive model to the general boundary condition problem, it was implemented to a 3-dimensional finite element program code. The proposed constitutive model can be classified into the category of multi-surface model based on the theory of plasticity. The validity was verified by comparing of the experimental results with the calculated results for some examples.

1 INTRODUCTION

Sand ground has inherent transverse isotropy due to a horizontal sedimentation of long axis of the sand. Therefore, it is very important to consider the effects of the inherent transverse isotropy of sand under the complicated stress condition such as soil-structure interaction behavior. And, it is necessary to establish the constitutive model in such as finite element method, in order to apply it to various boundary condition problem.

There have been very few in the field of the constitutive model of sand with inherent anisotropy or inherent transverse isotropy, though many researchers have conducted the study of the constitutive model. Miura et al.(1986) and Gutierrez et al.(1991) have proposed the constitutive model on the basis of the experimental results used inherent transverse isotropic sand. On the other hand, Ghaboussi and Momen(1984) and Toki et al.(1989) have obtained the constitutive model of sand with inherent transverse isotropy, adding an anisotropic tensor to that of initially isotopic sand.

So far, in order to reveal the fundamental characteristics of sand, the authors(Yanagisawa and Suzuki (1995) and Suzuki et al.(1995a)) have carried out monotonic loading tests and cyclic tests with fixed principal stress axes and loading tests with rotation of principal stress axes under drained condition, using a large hollow cylindrical apparatus. Through the analysis of the experiment results, the characteristics regarding the hardening law, the dependency of the principal deviatoric strain increment ratio on the principal deviatoric stress increment ratio and the dependency of the maximum principal strain increment direction on the stress increment direction etc., have been clarified by taking the effect of b-value ($= (\sigma_2 - \sigma_3)/(\sigma_1 - \sigma_3)$) and non-coaxiality into consideration.

The objective of this paper is to propose a constitutive model considering the effects of inherent transverse isotropy of sand on the basis of the experimental results, and to verify the fundamental validity through the numerical simulation. Toward this end, the proposed constitutive model was implemented to a 3-dimensional finite element analysis code using loading function, plastic potential and hardening rules based on the theory of plasticity. And, the validity of the proposed constitutive model was verified by comparing of the experimental results with the results obtained by finite element analysis for some examples.

2 CHARACTERISTICS OF SAND

Through the analysis of the experimental data under drained condition, the principal characteristics of sand with inherent transverse isotropy have been summarized by Yanagisawa and Suzuki(1995) and Suzuki et al.(1995a), by taking the effects of b-value and non-coaxiality into consideration as follows:

1) An unique failure surface is defined on the deviatoric stress plane normalized by the mean stress, using the angle β between the perpendicular direction to the bedding plane of sand and the direction of maximum principal stress.

2) Stress-strain relationship is obtained uniquely using a modified stress ratio q^* / p, which means the radius of the loading surface.

3) Moreover, dilatancy relationship is uniquely obtained by the relationship between normalized plastic work W^p/p and the equivalent strain e^p.

4) Principal deviatoric strain increment ratio depends on the increment ratio of the principal deviatoric stress.

5) The direction of principal strain increment depends on the direction of principal stress increment and the normal direction to the loading surface, as a function of the modified stress ratio.

3 PROPOSED CONSTITUTIVE MODEL

The proposed constitutive model belongs to the category of multi-surface model, as the concept of a field of hardening moduli (Mroz 1967) is used on the deviatoric stress plane normalized by mean stress. Park et al. (1990) showed that even sand with inherent transverse isotropy behaves as elastic below 0.01% strains, from plane strain tests measuring local strains in the specimen. Therefore, it is reasonable to assume that the sand in elastic region is isotropy. The elasto-plastic matrix, loading function, plastic potential and hardening rules used in the proposed model are described as followings.

3.1 *Elasto-plastic matrix*

Elasto-plastic matrix $\left[D^{ep}\right]$ used in finite element method is presented by

$$\left[D^{ep}\right]=\left[D^{e}\right]-\frac{\left[D^{e}\right]\left\{\frac{\partial g}{\partial \sigma}\right\}\left\{\frac{\partial f}{\partial \sigma}\right\}^{T}\left[D^{e}\right]}{H+\left\{\frac{\partial f}{\partial \sigma}\right\}^{T}\left[D^{e}\right]\left\{\frac{\partial g}{\partial \sigma}\right\}} \tag{1}$$

where $\left[D^{e}\right]$ is the elastic modulus, f is the loading function. g is the plastic potential, and H is the plastic modulus associated with loading surface.

3.2 *Loading function*

The failure function is presented as

$$F=\left\{(q/p)_f\cos 2\beta-\alpha_f^*\right\}^2+\left\{c_1(q/p)_f\sin 2\beta\right\}^2-\left(q^*/p\right)_f^2 \tag{2}$$

where $(q/p)_f$ is the failure strength for an arbitrary angle β, and α_f^*, $(q^*/p)_f$ represent respectively horizontal-coordinate of the center and the radius of the surface. c_1 is the shape parameter of the ellipse.

The loading function is presented by the following equation using the relationship that the loading function is assumed to have similar shape as the failure surface.

$$f=\left\{(q/p)\cos 2\beta-\overline{\alpha}_x^*\right\}^2+\left\{c_1(q/p)\sin 2\beta-\overline{\alpha}_y^*\right\}^2-\left(q^*/p\right)^2 \tag{3}$$

where $\overline{\alpha}_x^*, \overline{\alpha}_y^*, (q^*/p)$ indicate the horizontal and vertical coordinate of the center and the radius of each loading surface on the deviatoric stress plane, respectively. Also, $\overline{\alpha}_x^*, \overline{\alpha}_y^*$ imply the kinematic hardening parameters, (q^*/p) is called as the modified stress ratio considering the inherent anisotropy.

3.3 *Plastic potential*

The increments of plastic strains $d\varepsilon_{ij}^p$ are given by

$$d\varepsilon_{ij}^p=\lambda\frac{\partial g}{\partial \sigma_{ij}} \tag{4}$$

$$or \quad d\varepsilon_{ij}^p=\frac{1}{H}\left\{\frac{\partial f}{\partial \sigma}\right\}^T\{d\sigma\}\left\{\frac{\partial g}{\partial \sigma}\right\} \tag{5}$$

where λ is a proportionality constant. The increments of plastic strains are divided into the deviatoric and volumetric components as

$$d\varepsilon_{ij}^p=de_{ij}^p+\frac{1}{3}d\varepsilon_v\delta_{ij} \tag{6}$$

where de_{ij}^p are the components of plastic deviatoric strains, $d\varepsilon_v^p$ is the plastic volumetric strain, and δ_{ij} is Kronecker's delta. Also, the partial derivative of the plastic potential are divided into the deviatoric and the mean stress components as

$$\frac{\partial g}{\partial \sigma_{ij}}=\frac{\partial g}{\partial s_{ij}}+\frac{1}{3}\frac{\partial g}{\partial p}\delta_{ij} \tag{7}$$

Therefore, the following equations are obtained.

$$de_{ij}^p=\lambda\frac{\partial g}{\partial s_{ij}} \qquad d\varepsilon_v^p=\lambda\frac{\partial g}{\partial p} \tag{8}$$

Namely, only the partial derivation of the plastic potential are needed.

The principal deviatoric strain increment ratio depends on the increment ratio of the principal deviatoric stress. This is illustrated in Fig. 1. Here, $\theta_{d\sigma}$ represents the angle of the principal deviatoric stress increment from the direction of $d\sigma_1$, and α indicates the discrepancy of the principal deviatoric strain increment from the principal deviatoric stress increment. We can write

$$\tan\theta_{d\sigma}=\frac{\sqrt{3}(d\sigma_2-d\sigma_3)}{(2d\sigma_1-d\sigma_2-d\sigma_3)} \qquad or \quad \theta_{d\sigma}=\tan^{-1}\left\{\frac{\sqrt{3}(d\sigma_2-d\sigma_3)}{(2d\sigma_1-d\sigma_2-d\sigma_3)}\right\} \tag{9}$$

Also, these length in Fig. 1 are defined by the

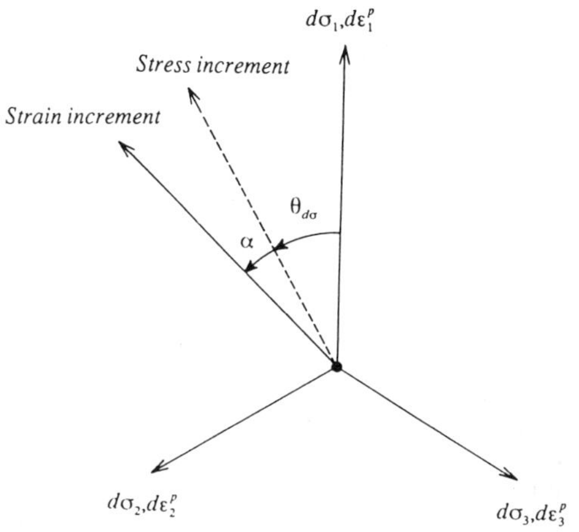

Fig. 1 Dependency of deviatoric strain increment on deviatoric stress increment

following equations.

$$\sqrt{ds_1^2 + ds_2^2 + ds_3^2} = \sqrt{\frac{2}{3}}dq$$
$$\sqrt{de_1^{p2} + de_2^{p2} + de_3^{p2}} = \sqrt{\frac{3}{2}}de^p \qquad (10)$$

Accordingly, the following equation is geometrically formed from Fig.1 as

$$de_1^p = \cos(\theta_{d\sigma} + \alpha)de^p$$
$$de_2^p = \frac{1}{2}\left\{\sqrt{3}\sin(\theta_{d\sigma} + \alpha) - \cos(\theta_{d\sigma} + \alpha)\right\}de^p \qquad (11)$$
$$de_2^p = -\frac{1}{2}\left\{\sqrt{3}\sin(\theta_{d\sigma} + \alpha) + \cos(\theta_{d\sigma} + \alpha)\right\}de^p$$

As de^p becomes λ, we can rewrite Eq.(11) as

$$\frac{\partial g}{\partial s_1} = \cos(\theta_{d\sigma} + \alpha)$$
$$\frac{\partial g}{\partial s_2} = \frac{1}{2}\left\{\sqrt{3}\sin(\theta_{d\sigma} + \alpha) - \cos(\theta_{d\sigma} + \alpha)\right\} \qquad (12)$$
$$\frac{\partial g}{\partial s_3} = -\frac{1}{2}\left\{\sqrt{3}\sin(\theta_{d\sigma} + \alpha) + \cos(\theta_{d\sigma} + \alpha)\right\}$$

On the other hand, the following equation is obtained from Fig.2 as

$$de_x^p = \frac{1}{2}\left(de_1^p + de_3^p\right) - \frac{1}{2}\left(de_1^p - de_3^p\right)\cos 2\beta_{d\varepsilon^p}$$
$$de_y^p = \frac{1}{2}\left(de_1^p + de_3^p\right) + \frac{1}{2}\left(de_1^p - de_3^p\right)\cos 2\beta_{d\varepsilon^p} \qquad (13)$$
$$de_z^p = -\left(de_1^p + de_3^p\right)$$
$$\frac{1}{2}d\gamma_{xy}^p = \frac{1}{2}\left(de_1^p - de_3^p\right)\sin 2\beta_{d\varepsilon^p}$$

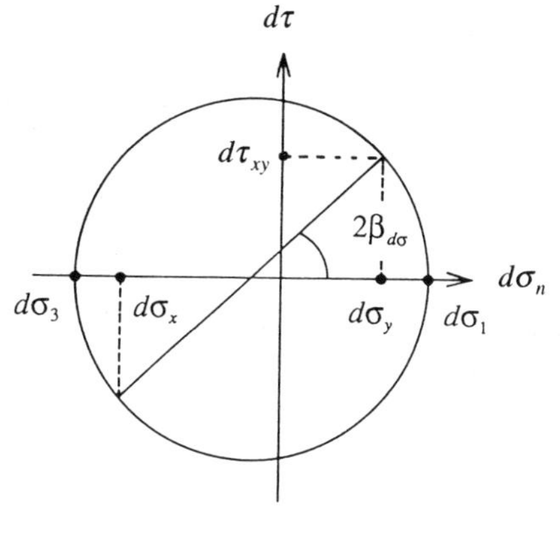

(a) stress increment

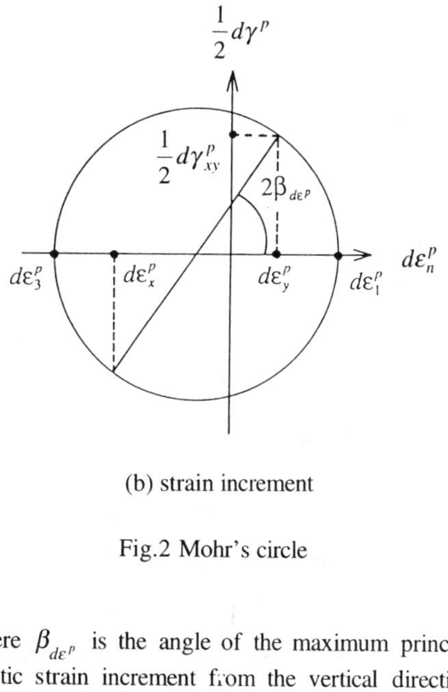

(b) strain increment

Fig.2 Mohr's circle

where $\beta_{d\varepsilon^p}$ is the angle of the maximum principal plastic strain increment from the vertical direction, and it is assumed as

$$\beta_{d\varepsilon^p} = \left(\beta_f - \beta_{d\sigma}\right)\frac{\left(q^*/p\right) - \left(q^*/p\right)_0}{\left(q^*/p\right)_f - \left(q^*/p\right)_0} + \beta_{d\sigma} \qquad (14)$$

where

$$\tan 2\beta_{d\sigma} = \frac{2d\tau_{xy}}{d\sigma_y - d\sigma_x}$$
$$\tan 2\beta_f = \frac{\sqrt{b^2 - b + 1}\,\frac{2\tau_{xy}}{p} - c_1\overline{\alpha}_y^*}{\sqrt{b^2 - b + 1}\,\frac{\sigma_y - \sigma_x}{p} - \overline{\alpha}_x^*} \qquad (15)$$

$$\tan 2\beta_{d\varepsilon^p} = \frac{d\gamma_{xy}^p}{d\varepsilon_y^p - d\varepsilon_x^p}$$

in which $\beta_{d\sigma}$, β_f represent the angle of the maximum principal stress increment and that of the normal to the loading surface, from the vertical direction at the arbitrary modified stress ratio, respectively. $(q^*/p)_0$ is the maximum modified stress ratio at the elastic region. Therefore, the following equations are obtained using Eq.(11) and Eq.(13).

$$\frac{\partial g}{\partial s_x} = \left\{ \frac{1}{2}\left(\frac{\partial g}{\partial s_1} + \frac{\partial g}{\partial s_3}\right) - \frac{1}{2}\left(\frac{\partial g}{\partial s_1} - \frac{\partial g}{\partial s_3}\right)\cos 2\beta_{d\varepsilon^p} \right\}$$

$$\frac{\partial g}{\partial s_y} = \left\{ \frac{1}{2}\left(\frac{\partial g}{\partial s_1} + \frac{\partial g}{\partial s_3}\right) + \frac{1}{2}\left(\frac{\partial g}{\partial s_1} - \frac{\partial g}{\partial s_3}\right)\cos 2\beta_{d\varepsilon^p} \right\} \tag{16}$$

$$\frac{\partial g}{\partial s_z} = \left\{ -\left(\frac{\partial g}{\partial s_1} + \frac{\partial g}{\partial s_3}\right) \right\}$$

$$\frac{\partial g}{\partial \tau_{xy}} = \left\{ \frac{1}{2}\left(\frac{\partial g}{\partial s_1} - \frac{\partial g}{\partial s_3}\right)\sin 2\beta_{d\varepsilon^p} \right\}$$

Next, the dilatancy relation is written as

$$d\varepsilon_v^p = \mu de^p - \frac{s_{ij} de_{ij}^p}{p} \tag{17}$$

where μ indicates the stress ratio at transformation line, and

$$\frac{\partial g}{\partial p} = \mu - \frac{s_{ij} de_{ij}^p}{p de^p} \tag{18}$$

Finally, the partial derivative form Eq.(7) of the plastic potential is obtained from Eqs.(16) and (18).

3.4 *Hardening rules*

The stress-strain relationship for monotonic loading is uniquely presented regardless b-value and β as

$$e^p = \frac{a^* b^* \left\{ (q^*/p) - (q^*/p)_0 \right\}^2}{1 - b^* \left\{ (q^*/p) - (q^*/p)_0 \right\}} \tag{19}$$

where a^*, b^* are experimental data. The gradient of the stress-strain h^p could be obtained as the reciprocal of the differential equation of Eq.(19) by (q^*/p).

The gradient of the stress-strain h_{cyc}^p at cyclic loading could be almost presented as a function of normalized plastic work as

$$\frac{h_{cyc}^p}{h^p} = \frac{1}{1 - 1.6(W^p/p)} \qquad \left(\frac{W^p}{p} \le 0.5\right)$$

$$\frac{h_{cyc}^p}{h^p} = 5 + 40(W^p/p - 0.5) \qquad \left(\frac{W^p}{p} \ge 0.5\right) \tag{20}$$

The plastic modulus H is obtained as the following equation using the above relations and Eq.(5).

$$H = \frac{\{\partial f/\partial(q/p)\}}{(dq/p)\left|\cos(2\beta_{d\sigma} - 2\beta_f)\right|} h^p \tag{21}$$

4 NUMERICAL SIMULATION

The calculations for the drained and undrained condition are presented in order to verify the validity of the proposed constitutive model. However, the experimental data were obtained only for the drained condition using large hollow cylindrical apparatus. For the undrained condition, only qualitative study was conducted. The calculations by finite element method were done using an eight-nodes isoparametric element as an element test.

4.1 *Drained condition*

The experimental results have been already shown by Suzuki et al.(1995b). The material used in the tests was Toyoura sand ($G_s = 2.65, e_{max} = 0.969, e_{min} = 0.614$), and the specimens were made using the multiple sieve pluviation method giving a strong inherent transverse isotropy. The relative density D_r is within the range $73 \pm 2.5\%$ after consolidation.

The experiments were carried out using a large hollow cylindrical apparatus under a constant mean stress $98\,kPa$ and a constant b-value 0.5. The value of β was set for 0° and 45° at the beginning of two kind of loading, respectively. The loading were executed until failure condition after 1 cyclic loading with principal stress axes change of 90° discontinuously.

The values for the material constants are

$E = 160MPa, \nu = 0.2, c_1 = 1.0, (q^*/p) = 1.11$

$(q^*/p)_0 = 0.15, \alpha_f^* = 0.25, \mu = 0.7, \alpha = 16.1^\circ$

$a^* = 0.002\ (1/MPa),\ b^* = 1.04(1/MPa)$

Fig.3 and Fig.5 show the comparison regarding the deviatoric strain and the volumetric strain between calculated results and the experimental ones, for respectively $\beta = 0^\circ$ and $\beta = 45^\circ$ at the beginning of the loading. Fig.4 shows the comparison of measured and calculated different strains for $\beta = 0^\circ$. These results agree well with each other both.

4.2 *Undrained condition*

The calculations were done using the same material parameters and similar loading type as that for the drained condition. In order to consider the dependency of the stiffness on the mean stress, the following equation is used

$$E = E_0 \left(\frac{p}{p_0}\right)^{0.5} \tag{22}$$

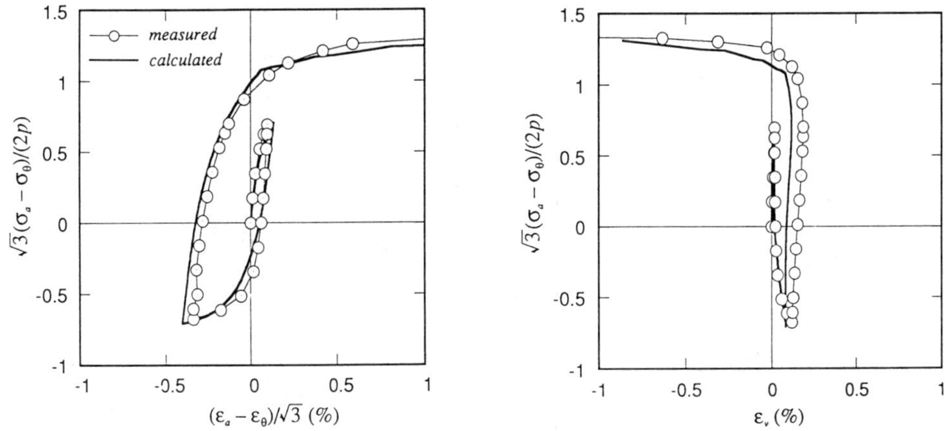

Fig. 3 Comparison of measured deviatoric, volumetric strain and calculated ones under drained condition ($\beta = 0^\circ$)

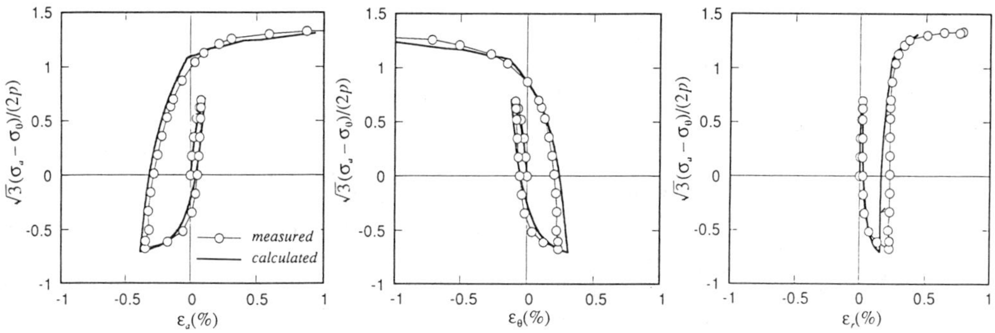

Fig. 4 Comparison of measured and calculated each strain components under drained condition ($\beta = 0^\circ$)

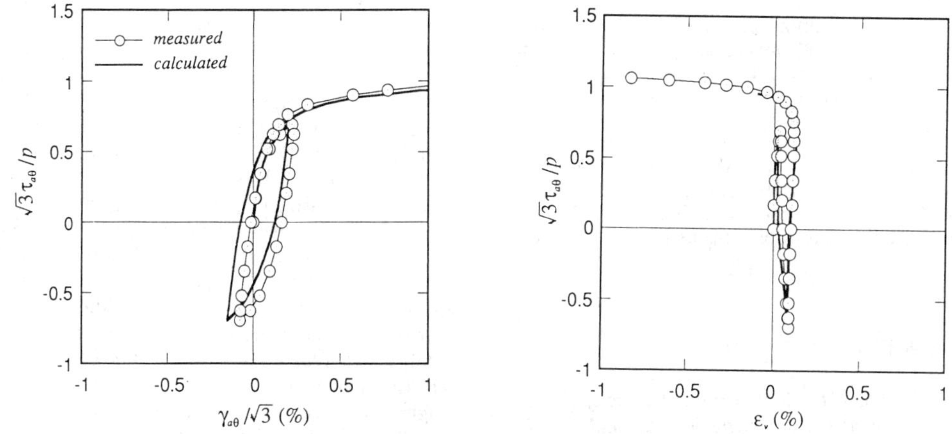

Fig. 5 Comparison of measured deviatoric, volumetric strain and calculated under drained condition ($\beta = 45^\circ$)

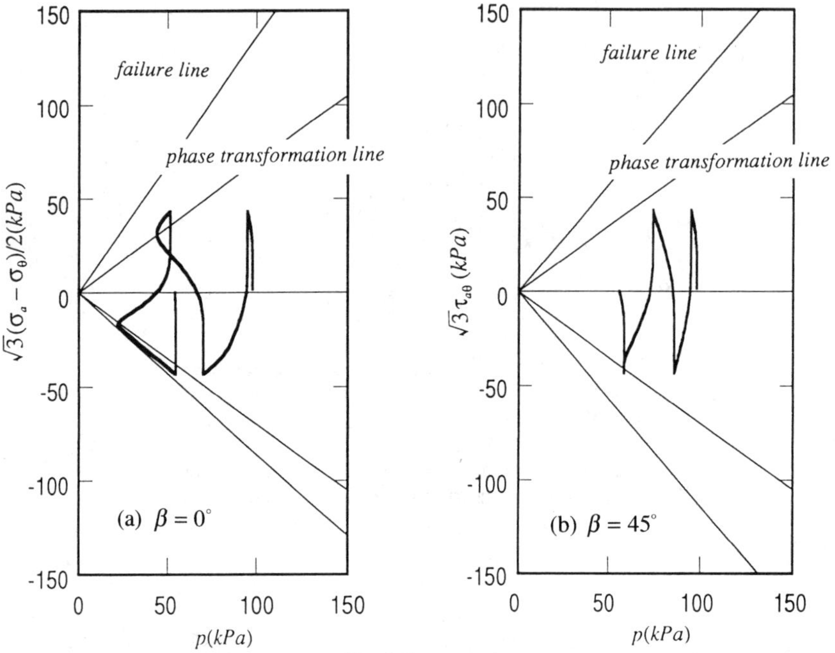

Fig.6 Stress path

where E_0, p_0 are the initial values.

Only stress path is shown in Fig.6. It shows the difference of the behaviors about cyclic mobility depending on the value of β, and seems to be qualitatively satisfied. In this case, the effects due to cyclic loading, i.e. Eq.(20) was not considered, as the behavior are too stiff due to the effects. Therefore, future subjects lie in this region.

5 CONCLUSION

In this paper, the authors proposed the constitutive model of the sand with inherent transverse isotropy for finite element analysis. The proposed constitutive model was developed on the basis of the experimental data using a large hollow cylindrical apparatus.

Through the numerical simulation of an element test, the validity of the proposed constitutive model is clarified at least for both drained condition and undrained condition as shown in this paper, though the future subjects are still remained in the detailed parts.

REFERENCES

Ghaboussi, J. and H. Momen 1984. Plasticity model for inherently anisotropic behaviour of sand, *Int. J. Numer. Analty. Methods Geomech.* 8: 1-17.

Gutierrez, M., K. Ishihara and I. Towhata 1993. Model for the deformation of sand during rotation of principal stress directions, *Soils and Foundations* 33(3): 105-117.

Park, C. S., S. Sinmei, F. Tatsuoka and S. Shibuya 1990 Anisotropy in deformation and strength of Silver Leighton Buzzard sand during plane strain compressive test, *Proc. 25th Japan Natl. Conf. of Soil Mech. Found. Eng.* : 439-442 (in Japanese).

Mroz, Z. 1967. On the description of anisotropic workhardening, *J. Mech. Phys. Solids* 15: 163-175.

Miura, K., S. Toki and S. Miura 1986. Deformation prediction for anisotropic sand during the rotation of principal stress axes, *Soils and Foundations* 26(3): 42-56.

Suzuki, K., T. Sugano and E. Yanagisawa 1995a. Drained cyclic shear behavior of sand with inherent transverse isotropy, *Proc. 5th Int. Symp. Numer. Models Geomech.*, Davos: 63-68.

Suzuki, K., T. Sugano and E. Yanagisawa 1995b. Drained cyclic shear behaviour of sand with inherent anisotropy, *J. Geotech. Eng. Div. JSCE.* 481(25): 117-124 (in Japanese).

Toki, H., T. Nakai and G. Matsumoto 1989. A constitutive elasto-plastic model for anisotropic sand, *Proc. 24th Annual Conf. of JSCE* Ⅲ: 510-511 (in Japanese).

Yanagisawa, E. and K. Suzuki 1995. Drained shear behavior of sand with inherent anisotropy, *Proc. 2nd Int. Conf. Seismology Earth. Eng.*, Tehran: 1257-1268.

Computer Methods and Advances in Geomechanics, Yuan (ed.) © 1997 Balkema, Rotterdam, ISBN 90 5410 904 1

An approximate constitutive model of non-linear viscoelastic materials

Jong Ryeol Kim
ERES Consultants, Inc., Champaign, Ill., USA
Byong Kee Park
Chonnam National University, Kwang-Ju, Korea

ABSTRACT: An approximate constitutive model of a non-linear viscoelastic material is presented in order to predict the time-dependent surface settlement in asphalt concrete pavement. The model is based upon the concept of regions of linear and non-linear behavior, and superposition of linear and non-linear strain components. The model parameters are extracted from a series of uniaxial compression creep/recovery tests. It is shown that the elastic and plastic strains of tested material are linearly proportional to the stress level, but the viscoelastic strain is non-linear with respect to the stress level and the stress history. The reliability of this model was evaluated with verification tests. It was shown that there is a good agreement between the tested and predicted data.

1. INTRODUCTION

In many geo-materials plastic and viscous properties are usually coupled with elastic properties and asphalt concrete is not exceptional. The term visco-elasto-plasticity is used to emphasize the coupling. Since the separation of these effects in laboratory tests is difficult, often either plastic properties or viscous properties have been neglected. Consequently, simplified elastoplastic and/or linear viscoelastic theories have been employed to predict the permanent deformation in asphalt concrete (Carpenter and Freeman, 1986; Moavenzadeh and Soussou, 1968). Therefore, it is not surprising that poor correlations between laboratory results and actual field performance were noticed since permanent deformation in asphalt concrete takes place because of not only its plasticity but also its viscosity (Perl, Uzan and Sides, 1983; Abdulshafi and Majidzadeh, 1984; Drescher, Kim and Newcomb, 1993).

This paper shows that elastic, plastic, viscoelastic and viscoplastic strain components are present simultaneously in the total creep strain during the loading process and all are incorporated in a model to predict permanent deformation in asphalt concrete. These components are extracted from a series of creep/recovery tests conducted under a constant compressive stress. The total creep strain has recoverable and irrecoverable parts, of which some are time-dependent and some are time-independent.

Emphasis is placed on the methodology of data analysis and developing a constitutive model with rheological elements to predict the permanent deformation (strain) based upon experimental data.

2. STRAINS IN CREEP TESTS

The simplest creep test loading/unloading program is given in Fig.1a with a typical strain response shown in Fig.1b. The applied stress is denoted as σ_0 and the time to unloading is denoted as t_1. The strain during creep (creep strain) is denoted as $\varepsilon_c(t)$ and the strain during recovery (recovery strain) is denoted as $\varepsilon'_r(s)$, where $s=t-t_1$. The recoverable strain is denoted as $\varepsilon_r(s)$, and it is measured from the extension of the creep curve $\varepsilon_c(t)$ rather than from the strain $\varepsilon_c(t_1)$.

The creep/recovery curve shown in Fig.1b, obtained from a test with one value of σ_0 and of t_1, is insufficient to determine whether or not all types of strain are present. This curve merely indicates that creep strains $\varepsilon_c(t)$ consist of an instantaneous portion and a time-dependent portion, and that creep strains $\varepsilon_c(t)$ are partially recovered after unloading. Only if recoverable strains $\varepsilon_r(s)$ equal creep strains $\varepsilon_c(t)$ at equal time intervals (both instantaneous and time-dependent strains), i.e.:

$$\varepsilon_r(s,\sigma_0) = \varepsilon_c(t,\sigma_0) \qquad for \qquad t=s \tag{1}$$

is the material response linearly viscoelastic. In this case instantaneous strains are elastic, the time-dependent strains are viscous or viscoelastic, and no plastic strains are present. This can be shown by the Boltzmann's Superposition Principle, valid for a linear viscoelastic material. Mathematically, the Boltzmann Superposition Principle can be expressed as:

$$\varepsilon(t,\sigma) = \int_{-\infty}^{t} J(t-\tau)\frac{d\sigma(\tau)}{d\tau}d\tau \tag{2}$$

where J is the compliance expressed as a function of time only. The equivalent differential form of eq.(2) is:

$$p_0\sigma + p_1\dot{\sigma} + p_2\ddot{\sigma} + ... = q_o\varepsilon + q_1\dot{\varepsilon} + q_2\ddot{\varepsilon} + ... \tag{3}$$

where dots denote the derivative of the variable with respect to time, and p_0, p_1, p_2,... and q_0, q_1, q_2,... are material constants.

If recoverable and creep strains do not satisfy eq.(1), the only conclusion which can be drawn is that the material is not linearly viscoelastic. It is possible, for instance, to regard it as nonlinearly viscoelastic or as visco-elasto-plastic. For the former case, a generalized viscoelastic superposition principle by Green and Rivlin (1957) can be used to described the different recoverable and creep strains, both instantaneous and time-dependent. However, this is not attempted, because many tests at different stress levels σ_0 and times to unloading t_1 are necessary to determine the material properties. Instead, the later case is focused in this paper. Other non-linear superposition principles also have been presented in the literature (Onat, 1966; Pipkin and Rogers, 1968; Stafford, 1969).

3. EXPERIMENT

3.1 *Material and Specimens*

All tests were performed on one asphalt concrete mixture. Three types of aggregate (95% crushed gravel, 100% screened gravel, and 100% crushed limestone) combined in a ratio of 1:2:1 were mixed with an asphalt binder with the penetration grade 120/150.

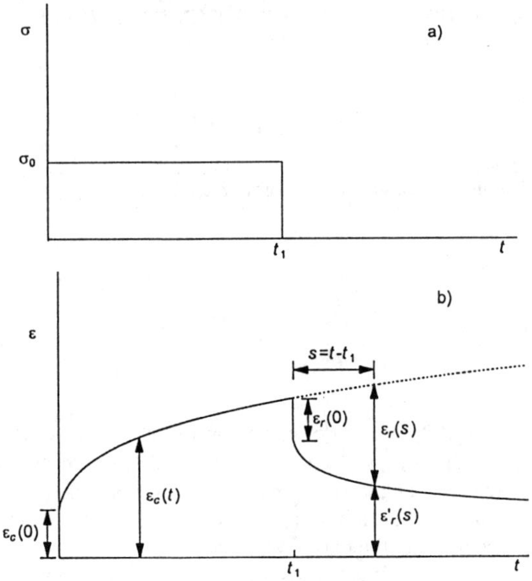

Fig.1 Creep Test

Cylindrical specimens with 100-mm diameter and 200-mm height were manually compacted in three layers (50 mm, 100 mm, and 50 mm final height) in a 250-mm high mold by means of a vibratory hammer with a 100-mm diameter tamping foot. The compaction temperature was 135±5°C.

3.2 *Experimental Set-up*

A series of uniaxial compression creep tests with a single loading/unloading cycle was performed at temperature of 25°C. All tests were performed in a MTS 810 universal loading frame. A servo-controller connected to the load cell attached to the hydraulic actuator maintained the required stress history. The specimen was placed in between two enlarged and silicone-lubricated loading platens to avoid loading platens punching and to minimize the end-friction. For measuring axial displacements, three LVDTs equally spaced around the specimen were mounted to two clamps located on the specimen 100-mm apart. Stresses and strains were recorded automatically at 10-second intervals.

4. TEST RESULTS AND DATA ANALYSIS

Figure 2 shows the dependence of averaged axial strains on time for three levels of axial stress. The scatter in axial creep strains, due primarily to the

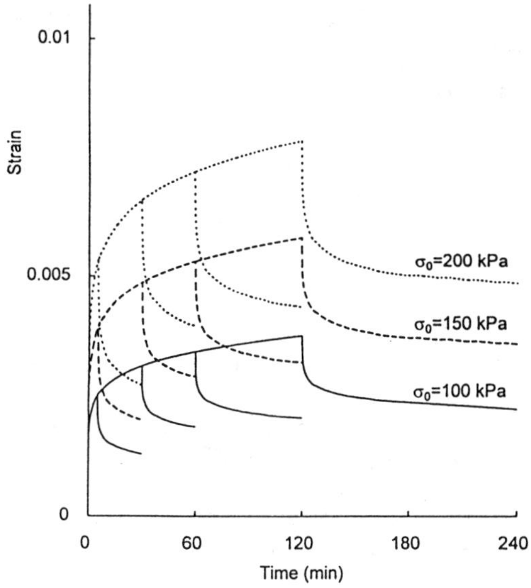

Fig.2 Test Results

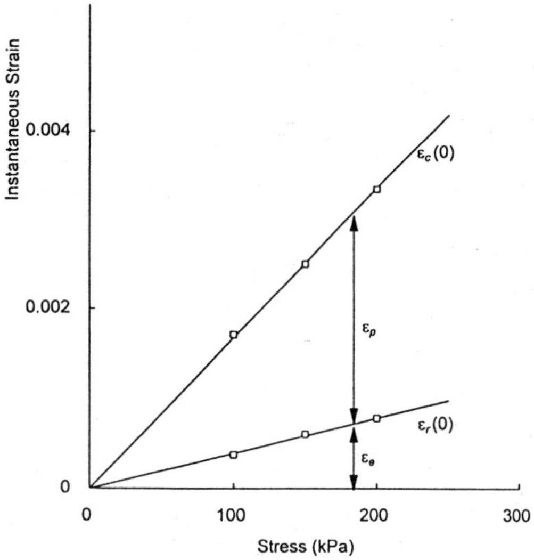

Fig.3 Instantaneous Strain *vs*. Stress

variation in the air void content, ranged from ±5% at $t \approx 0$ to ±15% at $t \approx 120$ minute. The scatter in the recoverable strains was smaller.

It is seen from Fig.2 that recoverable strains $\varepsilon_r(s)$ at all stress levels σ_0 and times to unloading t_l are significantly smaller than creep strains $\varepsilon_c(t)$. Also, the magnitude of recoverable strains seems to vary little with time t_1. Clearly, at all stress levels applied, the instantaneous recoverable strains $\varepsilon_r(0)$ do not depend on t_1 and, by definition, they are elastic strains ε_e. Consequently, the difference between the instantaneous creep strain $\varepsilon_c(0)$ and the instantaneous recoverable strains $\varepsilon_r(0)$ are the plastic strain ε_p.

Figure 3 presents the relationship between the instantaneous creep $\varepsilon_c(0)$ and recoverable strain $\varepsilon_r(0)$, extrapolated to $t_1=0$ using Fig.2, and stress. It is evident that both strains differ significantly which, according to the previous discussion, implies the existence of plastic strains ε_p and a yield stress σ_y close to 0. Since both the instantaneous creep $\varepsilon_c(0)$ and recoverable strain $\varepsilon_r(0)$ are linear functions of stress, the plastic strain is also a linear function of stress work hardening. A similar result, obtained from repeated loading/unloading tests where the instantaneous recoverable strain $\varepsilon_r(0)$ was unaffected by the number of cycles, was reported by Perl, et. al. (1983) and by Kim (1995).

By subtracting the plastic strain ε_p from creep strains $\varepsilon_c(t)$, it is possible to evaluate whether time-dependent strains are a linear function of the stress level and stress history with viscoelasticity. This can

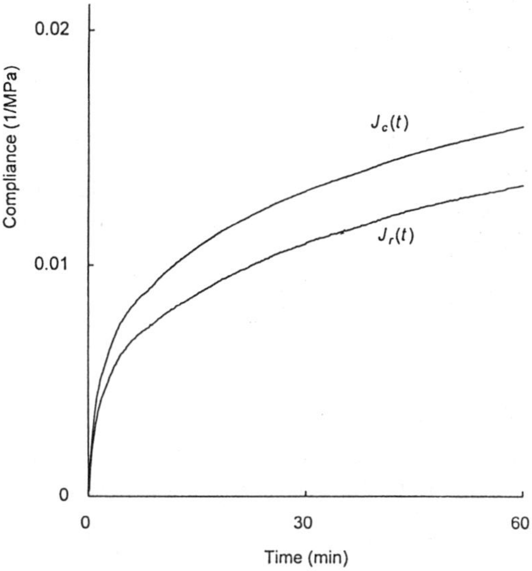

Fig.4 Compliance *vs*. Time

be done by comparing the following creep and recovery compliances:

$$J_c(t,\sigma_0) = \frac{\varepsilon_c(t) - \varepsilon_p}{\sigma_0}, \qquad J_r(s,\sigma_0) = \frac{\varepsilon_r(s)}{\sigma_0} \tag{4}$$

where $J_c(t,\sigma_o)$ and $J_r(s,\sigma_o)$ are creep and recovery compliances, respectively. It is assumed that the elastic strain ε_e is a part of viscoelastic strain $\varepsilon_{ve}(t)$

since the instantaneous viscoelastic strain $\varepsilon_{ve}(0)$ is same as the elastic strain. For a linearly viscoelastic material the creep compliance $J_c(t)$ is identical to the recovery compliance $J_r(s)$ at equal time intervals ($J_c(t)=J_r(s)$ for $t=s$). Figure 4 shows the creep and recovery compliance for σ_0=100 kPa, t_1=60 min. Since compliances differ, time-dependent responses upon loading cannot be regarded as linear viscoelastic.

It is noticed that creep $\varepsilon_c(t)$ and recovery strains $\varepsilon'_r(s)$ are proportional to stress. This implies that recoverable strains $\varepsilon_r(s)$ are also proportional to stress and, by definition, the recoverable strains $\varepsilon_r(s)$ are linear viscoelastic. In other words, the recoverable strains $\varepsilon_r(t)$ can be decomposed into two components; elastic strain ε_e and viscoelastic strains $\varepsilon_{ve}(t)$. Also, the creep strains $\varepsilon_c(t)$ can be decomposed into four components; elastic strain ε_e, plastic strain ε_p, viscoelastic strains $\varepsilon_{ve}(t)$ and other strains which are time-dependent and irrecoverable (viscoplastic).

5. DEVELOPMENT OF MODEL

Based upon test results a constitutive model of asphalt concrete was developed with rheological elements. A spring, dashpot, and sliding block represent the elastic, viscous, and plastic (linear work-hardening)response, respectively. Since the creep behavior can be expressed as:

$$\varepsilon_c(t) = \varepsilon_e + \varepsilon_p + \varepsilon_{ve}(t) + \varepsilon_{vp}(t) \qquad (5)$$

$$\varepsilon_r(t) = \varepsilon_e + \varepsilon_{ve}(t) \qquad (6)$$

the permanent deformation can be predicted with a constitutive model which contains a Burgers model, a Bingham model and a sliding block in series (Fig.5). The Burgers model can represent both linear elastic and linear viscoelastic response since the linear elastic response is a special form of viscoelastic response (instantaneous viscoelastic response). Since the behavior of both sliding blocks (linear work-hardening) are similar to that of spring when $\sigma_0 \geq \sigma_{y,i}$, the constitutive equation for this model, which may be changed due to the relation between yield stress $\sigma_{y,i}$ and applied stress σ_0, can be written in differential forms:

1. $\sigma_0 \geq \sigma_{y,i}$ (both sliding blocks are active)

$$\sigma + p_1\dot{\sigma} + p_2\ddot{\sigma} + p_3\dddot{\sigma} = q_1\dot{\varepsilon} + q_2\ddot{\varepsilon} + q_3\dddot{\varepsilon} \qquad (7a)$$

2. $\sigma_{y,1} \leq \sigma_0 < \sigma_{y,3}$ (only the sliding block for plasticity is active)

$$\sigma + p_1\dot{\sigma} + p_2\ddot{\sigma} = q_1\dot{\varepsilon} + q_2\ddot{\varepsilon} \qquad (7b)$$

3. $\sigma_{y,3} \leq \sigma_0 < \sigma_{y,1}$ (only the sliding block for viscoplasticity is active)

$$\sigma + p_1\dot{\sigma} + p_2\ddot{\sigma} + p_3\dddot{\sigma} = q_1\dot{\varepsilon} + q_2\ddot{\varepsilon} + q_3\dddot{\varepsilon} \qquad (7a)$$

4. $\sigma_0 < \sigma_{y,i}$ (neither sliding block is active)

$$\sigma + p_1\dot{\sigma} + p_2\ddot{\sigma} = q_1\dot{\varepsilon} + q_2\ddot{\varepsilon} \qquad (7b)$$

It is important to note that eqs.(7a) and eq.(7b) are identical when p_3 and q_3 in eq.(7a) are vanished. This implies that eq.(7a) can be a governing constitutive equation for any case. Solving eq.(7a) for a creep test and using the superposition principle gives:

$$\varepsilon(t) = <1>\{\frac{\sigma-\sigma_{y,1}}{H_1} + \frac{\sigma-\sigma_{y,3}}{H_3}[1-\exp(-\frac{H_3 t}{\eta_3})]\} + \frac{\sigma}{E_1} + \frac{\sigma}{\eta_1}t + \frac{\sigma}{E_2}[1-\exp(-\frac{E_2 t}{\eta_2})] \qquad (8)$$

where E is an elastic modulus, η is a viscosity, and H is a plastic (work-hardening) modulus. The

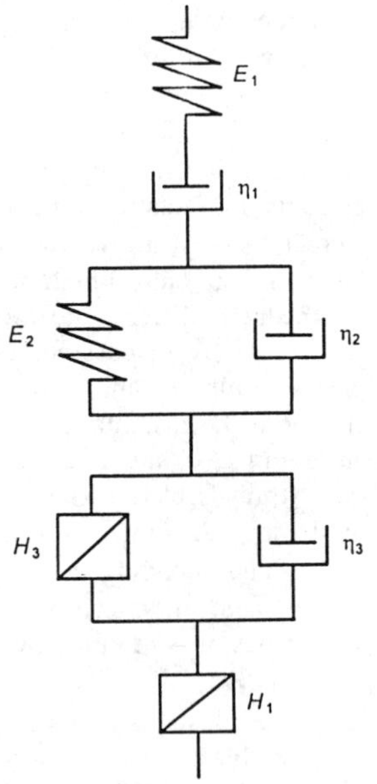

Fig.5 Visco-Elasto-Plastic Model

subscript numbers represent the corresponding elements. The symbol <1> is a switch function which is 1 for $\sigma-\sigma_{y,i}\geq 0$ and 0 for $\sigma-\sigma_{y,i}<0$. If the stress σ is removed at time t_1, the recovery strain $\varepsilon'_r(s)$, which is sometimes represented as the permanent strain, can be obtained from eq.(8):

$$\varepsilon'_r(s) = \frac{\sigma-\sigma_{y,1}}{H_1} + \frac{\sigma-\sigma_{y,3}}{H_3}[1-\exp(-\frac{H_3 t_1}{\eta_3})] + \frac{\sigma}{\eta_1}t_1 + \frac{\sigma}{E_2}[\exp(-\frac{E_2 s}{\eta_2}) - \exp(-\frac{E_2 t}{\eta_2})] \quad (9)$$

where $s=t-t_1$ and $t>t_1$. If the loading period is finite and the unloading period is infinite ($t=s=\infty$) eq.(9) gives:

$$\varepsilon'_r(s) = \frac{\sigma-\sigma_{y,1}}{H_1} + \frac{\sigma-\sigma_{y,3}}{H_3}[1-\exp(-\frac{H_3 t_1}{\eta_3})] + \frac{\sigma}{\eta_1}t_1 \quad (10)$$

While the first and last terms of eq.(10) represent the permanent strains due to the plasticity (H_1) and viscosity (η_1), respectively, the second term represents the permanent strain due to the combination of the plasticity and viscosity (H_3, η_3).

6. MATERIAL CONSTANTS DETERMINATION

With the separation of the total creep strain into three component strains (linear work-hardening plastic, linear viscoelastic, and linear work-hardening viscoplastic strains) and the use of constitutive equations for corresponding models (sliding block with friction, a Burgers model, and a Bingham model with linear work-hardening) the material constants can be determined from experimental data.

6.1 *Linear Work-Hardening Plastic Strain*

The relationship between the linear work-hardening plastic strain ε_p and the stress σ can be expressed as:

$$\varepsilon_p = \frac{\sigma-\sigma_{y,1}}{H_1} \quad (11)$$

For the tested material $\sigma_{y,1}=0$ and $H_1=77.85$ MPa.

6.2 *Linear Elastic and Viscoelastic Strains*

The Burgers model represents the linear elastic and viscoelastic behaviors. Its constitutive equation for the creep test is:

$$\varepsilon_{ve}(t) = \frac{\sigma}{E_1} + \frac{\sigma}{\eta_1}t + \frac{\sigma}{E_2}[1-\exp(-\frac{E_2 t}{\eta_2})] \quad (12)$$

For the tested material $E_1=256.41$ MPa, $E_2=131.52$ MPa, $\eta_1=10{,}351$ MPa-min, and $\eta_2= 606.53$ MPa-min.

6.3 *Linear Work-Hardening Viscoplastic Strain*

The Bingham model with linear work-hardening represents the viscoplastic behavior, and its constitutive equation for the creep test is:

$$\varepsilon_{vp}(t) = \frac{\sigma-\sigma_{y,3}}{H_3}[1-\exp(-\frac{H_3 t}{\eta_3})] \quad (13)$$

For the tested material $\sigma_{y,3}=0$, $H_3=200$ MPa, and $\eta_3=1{,}651.99$ MPa-min.

7. MODEL VERIFICATION

The result of the verification tests (repeated creep tests with 20 min loading and 20 min rest period for three cycles) and the predictions from the model were compared to evaluate the accuracy and the

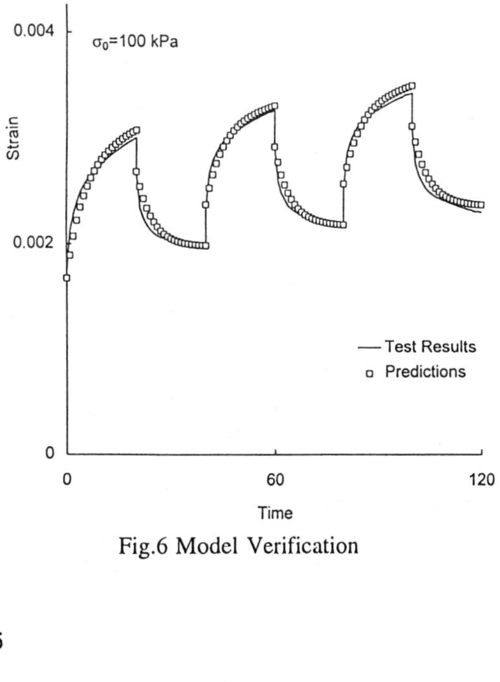

Fig.6 Model Verification

reliability of the model in predicting the material behavior. The material constants and the stress level were substituted into eq.(13) to obtain the prediction curve. The superposition principle was used for all decomposed strains. In addition, for the viscoplastic strain, $\varepsilon_{vp}(t)$, the strain-hardening concept was used (Kim, 1995): Given a test specimen, a constant stress is applied at time t=0, and a viscoplastic strain curve is obtained. During the first rest period (time t_1 to t_2), the magnitude of the viscoplastic strain is constant. When the stress is reapplied at time t_2, the continued viscoplastic strain curve is obtained, which is a translation of the viscoplastic strain curve (for constant loading) from time t_1 parallel to the time axis to time t_2 (for repeated loading), and so on.

Figure 6 is a comparison of the test results and the predictions for σ_0=100 kPa. There are no significant differences between them. The maximum deviation between the test results and the predicted values is less than 7% for both the load and rest periods.

8. CONCLUSIONS

Based upon the study presented in this paper, the following conclusions are made for tested material within a framework:

1. In a creep response the total creep strain can be separated into four components (elastic, plastic, viscoelastic, and viscoplastic strains), and the recoverable strain can be separated into two components (elastic and viscoelastic strains).
2. A set of creep tests with various times to unloading are required to separate these components.
3. Elastic and plastic strains are proportional to stress. The yield stress of plasticity for the tested material seems to be zero.
4. The viscoelastic response can be expressed with a Burgers model.
5. The viscoplastic response can be expressed with a Bingham model. The yield stress of viscoplasticity for the tested material seems to be zero.
6. The model presented in this paper gives satisfactory agreement with the test results. It can be used to predict permanent deformation in asphalt concrete and to evaluate the material properties.
7. Permanent strain in the presented model are due to plasticity (H_1), viscosity (η_1), and a combination of plasticity and viscosity (H_3, η_3).

REFERENCES

Abdulshafi,A.H. and K.Majidzadeh. (1984). "Combo Viscoelastic-Plastic Modeling and Rutting of Asphalt Mixtures." *Transp.Res.Rec.*, **968**, 19-31.

Carpenter, S.H., and T.J. Freeman. (1986). "Characterizing Permanent Deformation in Asphalt Concrete Placed over Portland Cement Concrete Pavement." *Transp.Res.Rec.*, **1070**, 30-41.

Drescher, A., J.R. Kim, and D.E. Newcomb. (1993). "Permanent Deformation in Asphalt Concrete." *J.Mat.Civi.Eng.*, ASCE, **5**, 112-128.

Green, A.E., and Rivlin, R.S.. (1957). "The Mechanics of Non-linear Materials with Memory." *Arch. Rat. Mech. Anal.*, **1**, 1-21.

Kim, J.R. (1995). "Viscoelastic Analysis of Triaxial Tests on Asphalt Concrete." Ph.D. Thesis, Univ. of Minnesota.

Moavenzadeh, F. and Soussou, J. (1968). "Viscoelastic Constitutive Equation for Sand-Asphalt Mixtures." *High. Res. Rec.*, **256**, 36-52.

Onat, E.T. (1966). "Description of Mechanical Behavior of Inelastic Solids." *Proc. 5th U.S .Nat. Congr. Appl. Mech.*, 421-434.

Perl, M., Uzan, J., and Sides, A. (1983). "Visco-Elastic-Plastic Constitutive Law for A Bituminous Mixture under Repeated Loading." *Transp. Res. Rec.*, **911**, 20-27.

Pipkin, A.C., and Rogers, T.G. (1968). "A Non-linear Integral Representation for Viscoelastic Behavior." *J. Mech. Phys. Solids*, **16**, 59-72.

Stafford, R.O. (1969). "On Mathematical Forms for The Material Functions in Nonlinear Viscoelasticity." *J. Mech. Phys. Solids*, **17**, 339-358.

Computer Methods and Advances in Geomechanics, Yuan (ed.) © 1997 Balkema, Rotterdam, ISBN 90 5410 904 1

A double hardening plasticity model for sands using a modified stress dilatancy approach

R.G.Wan & P.J.Guo
Department of Civil Engineering, University of Calgary, Alb., Canada

ABSTRACT: Due to their particulate nature, sands in general exhibit very distinctive behavioural features such as void ratio and shear induced volumetric strains when loaded. The paper presents a simple constitutive model based on the classical theory of plasticity enriched with a modified stress dilatancy equation. The original Rowe's stress dilatancy equation is modified in order to account for initial density and pressure level sensitivities by using a void ratio dependent factor. The constitutive model predictions for sands under various stress and density conditions are in good agreement with classical experimental data available for Sacramento River Sand.

1 INTRODUCTION

The plastic yield behaviour of sand is strongly influenced by the phenomenon of dilatancy which was first revealed by Reynolds (1885), and later formalized by Rowe (1962) in his famous stress-dilatancy theory. Rowe's stress dilatancy theory provides a relationship between stress ratio and dilatancy factor which can be used as a flow rule within the framework of plasticity theory for the determination of plastic strains. Although the concept is powerful, obstacles are met towards capturing density and pressure sensitivities in sand response. For example, it is not possible to describe the complete evolution of shear dilation and contraction during deformation history for sands at different initial void ratios and stress levels. To the authors' knowledge, the above issue has been addressed to a very limited extent although other aspects such as stress induced anisotropy have been examined, (Ueng et. al. 1990 & Oda 1978). Only in recent literatures (Bauer and Wu 1993 & Gudehus 1996) that attempts have been made to include both density and pressure sensitivities in the description of soil behaviour using a hypoplasticity approach. The hypoplasticity approach (Kolymbas 1977) deviates from the classical elasto-plasticity theory in that the former relates the rate of Cauchy granular stress to deformation rates through functionals and thus does not require the definition of yield surfaces and flow rules.

The main thrust of this paper is to develop a simple elasto-plasticity constitutive model for sand which can capture both density and stress level dependencies in a consistent manner. The salient features of the model include a failure surface governed by Mohr-Coulomb failure criterion, a family of shear yield surfaces which assume the same form as the failure surface, a vertical cut-off cap for compaction or consolidation, and a special flow rule. Based on experimental evidence and physical arguments, Rowe's stress dilatancy relationship is modified in order to accommodate for the effects of void ratio and stress level.

2 CONSTITUTIVE MODEL

The total strain increment in the material when loaded is divisible into elastic and plastic strain components so that $d\boldsymbol{\varepsilon} = d\boldsymbol{\varepsilon}^e + d\boldsymbol{\varepsilon}^p$, with $d\boldsymbol{\varepsilon}^e$ as the elastic strain increment calculated by generalized Hooke's law, and $d\boldsymbol{\varepsilon}^p$ as the total plastic strain increment determined from the plastic constitutive law. Two plastic strain mechanisms are assumed to contribute to the total plastic strain increment, one due to compaction ($d\boldsymbol{\varepsilon}^p_c$) and the other one due to dilatant or compressive shearing ($d\boldsymbol{\varepsilon}^p_s$) such that $d\boldsymbol{\varepsilon}^p = d\boldsymbol{\varepsilon}^p_c + d\boldsymbol{\varepsilon}^p_s$. Thus, the incremental form of the constitutive law is typically written as

$$d\boldsymbol{\sigma} = \mathbf{C}^e : (d\boldsymbol{\varepsilon} - d\boldsymbol{\varepsilon}^p) \tag{1}$$

where $\boldsymbol{\sigma}$ = Cauchy effective stress tensor and $\mathbf{C}^e$ = rank four elasticity tensor.

The following sections briefly summarize the major components of the formulation in which all stresses referred to are effective stresses.

2.1 *Elastic deformations*

The elastic strain increments which are recoverable upon unloading are calculated from a generalized Hooke's law with a non linear variation of the shear modulus G based on void ratio e and mean stress p as follows

$$G = G_0 \frac{(2.17 - e)^2}{1 + e} \sqrt{p} \quad (2)$$

where G_0 is a material constant. It is further assumed that the Poisson's ratio v remains constant so that the Young's modulus E varies non-linearly as implied by the empirical relationship (2).

2.2 *Failure criteria*

For describing the shear deformation mechanism, the failure condition at both peak and residual states are assumed to be path independent and governed by a Mohr-Coulomb criterion written in terms of σ_1 and σ_3 which are the major and minor principal stresses respectively. The expression of the failure surface is simply given as

$$F = \frac{1}{2}(\sigma_1 - \sigma_3) - \frac{1}{2}(\sigma_1 + \sigma_3) \sin \varphi = 0 \quad (3)$$

in which the frictional angle φ can assume values of φ_f or φ_{cv} depending on whether the peak or residual (constant volume) states are described respectively.

2.3 *Yield surfaces*

Two distinct yield surfaces are used for describing shear and hydrostatic compaction mechanisms.

The plastic shearing mechanism is described by yield surfaces that are assumed to take the same form as the failure surface given in Eq. (3). Also, plastic shear yielding is regarded as a continuous mobilization of the frictional angle resulting into an isotropic expansion of the yield surface in the stress space. At a given yielding state, the expression of the shear yield surface is

$$F_s = \frac{1}{2}(\sigma_1 - \sigma_3) - \frac{1}{2}(\sigma_1 + \sigma_3) \sin \varphi_m = 0 \quad (4)$$

in which φ_m is the mobilized frictional angle. Here positive stresses correspond to compression and negative stresses to tension.

For describing the hydrostatic compaction mechanism, a vertical cut-off cap surface which also moves in the stress space is used for simplicity, i.e.

$$F_c = p - p_0 = 0 \quad (5)$$

with p_0 referring to the consolidation pressure as a function of compaction. The cap surface can only harden due to physical restrictions but its movement is intimately coupled with the shear yield surface by virtue of the evolution rules used.

2.4 *Flow rules*

An associated plasticity flow rule is used for the compaction process. However, for the shearing process, the plastic strain increments are calculated from a non-associated flow rule, i.e.

$$d\varepsilon_s^p = d\lambda \frac{\partial Q_s}{\partial \boldsymbol{\sigma}} \quad (6)$$

in which $d\lambda$ is the plastic multiplier and Q_s is a conveniently chosen plastic potential function. Normally, the plastic potential function can be mathematically derived from some plastic energy dissipation equation or a stress-dilatancy law. A more expedient way of achieving the same result is to assume that the plastic potential takes the same form as the failure surface, i.e.

$$Q_s = \frac{1}{2}(\sigma_1 - \sigma_3) - \frac{1}{2}(\sigma_1 + \sigma_3) \sin \psi_m \quad (7)$$

in which the parameter ψ_m refers to the mobilized dilation angle at a given stress level. The dilation angle relates to the negative sign of the ratio of plastic volumetric (ϵ_v^p) to deviatoric (γ_s^p) strain increments. Thus, $\psi_m > 0$ corresponds to a shear induced increase in volume and $\psi_m < 0$ to shear induced contraction. The dilatancy angle evolves with deformation history following a modified Rowe's stress dilatancy law. Figure 1 shows the evolution of the yield and plastic potential surfaces together with the direction of the plastic strain increment vector.

2.5 *Modified stress dilatancy equation*

The original Rowe's stress dilatancy equation couples the dilatancy factor, $D = 1 - d\epsilon_v^p / d\varepsilon_1^p$, to effective stress ratio $R = \sigma_1/\sigma_3$, so as $R = K_{cv} D$. The factor $K_{cv}=\tan^2(\frac{\pi}{4} + \frac{\varphi_{cv}}{2})$ is a material constant derived from energy dissipation considerations, and φ_{cv} represents the frictional angle at constant volume deformations. When written in terms of mobilized dilation and frictional angles, Rowe's stress dilatancy equation takes the classical form

$$\sin \psi_m = \frac{\sin \varphi_m - \sin \varphi_{cv}}{1 - \sin \varphi_m \sin \varphi_{cv}} \quad (8)$$

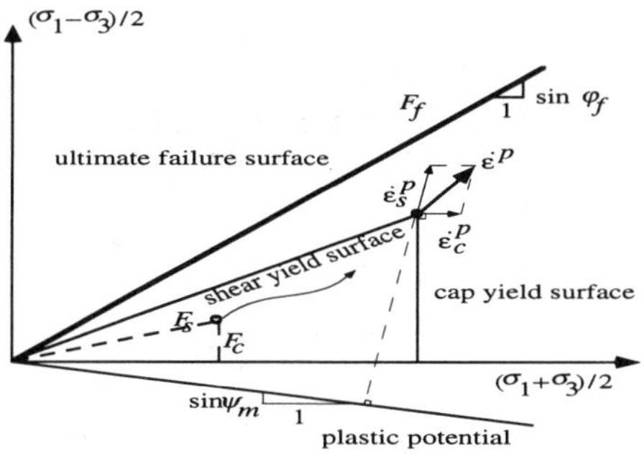

Figure 1. Plastic strain directions together with yield, failure and plastic potential surfaces

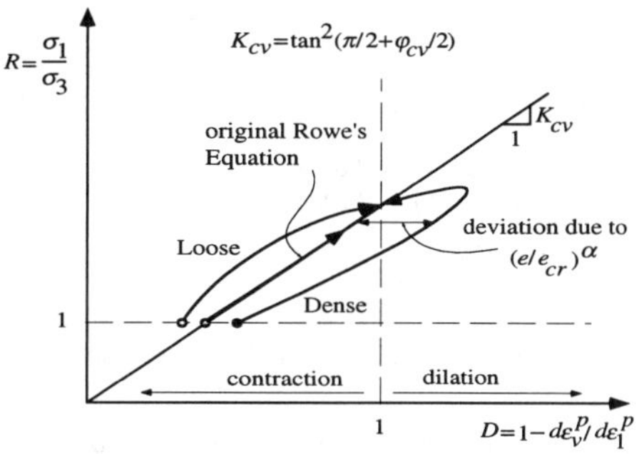

Figure 2. Modification of Rowe's theory to account for initial void ratio dependency

Since the factor K_{cv} is a constant set by the value of φ_{cv}, Rowe's theory seems to indicate that the stress ratio R varies linearly with dilatancy factor D. This linear variation cannot describe the density and void ratio dependencies during deformational history, neither the difference between pre-peak and post-peak responses for dense sands. In order to address the above deficiencies, a modification of the original Rowe's equation is proposed based on the value of the current void ratio e relative to the critical void ratio e_{cr}, i.e.

$$\sin\psi_m = \frac{\sin\varphi_m - (e/e_{cr})^\alpha \sin\varphi_{cv}}{1-(e/e_{cr})^\alpha \sin\varphi_m \sin\varphi_{cv}} \tag{9}$$

in which α is a material parameter to be determined. The validation of Eq. (9) hinges on a modified energy dissipation equation which embeds a so-called state parameter related to the current void ratio and the critical one. In fact, the energy based factor K_{cv} in the original Rowe's equation is made void ratio sensitive so that the modified stress dilatancy relationship becomes

$$\left(\frac{\sigma_1}{\sigma_3}\right)_m = K^*_{cv}\, D\ ;\ K^*_{cv} = \frac{1+(e/e_{cr})^\alpha \sin\varphi_{cv}}{1-(e/e_{cr})^\alpha \sin\varphi_{cv}} \tag{10}$$

It is obvious that void ratio dependent K^*_{cv} reverts to the original K_{cv} when the void ratio is fixed to the critical void ratio, i.e. $(e/e_{cr})^\alpha = 1$. Figure 2 illustrates how the original Rowe's equation is modified in order to account for the initial void ratio dependency. Details of the model and its calibration for classical sands can be found in Wan and Guo (1996).

2.6 *Hardening-softening rules*

The evolution laws for both compaction and shear mechanisms are governed by internal plastic variables, ϵ^p_v the plastic volumetric strain and γ^p the plastic deviatoric strain defined as follows

$$\epsilon^p_v = \varepsilon^p_1 + \varepsilon^p_2 + \varepsilon^p_3\ ;\ \ \gamma^p = \int_0^t (\frac{3}{2}\dot{\mathbf{e}}^p : \dot{\mathbf{e}}^p)^{1/2} dt \tag{11}$$

in which $\dot{\mathbf{e}}^p$ is the deviatoric part of the plastic strain rate tensor $\dot{\boldsymbol{\varepsilon}}^p$ such that $\dot{\mathbf{e}}^p = \dot{\boldsymbol{\varepsilon}}^p - \dot{\epsilon}^p_v \mathbf{1}$; $\mathbf{1}_{ij} = \delta_{ij}$ = Kronecker delta.

Shear mechanism

During deviatoric loading history, it is postulated that both volumetric strains (void ratio) and plastic deviatoric strains control hardening and softening according to a simple function given as

$$\sin\varphi_m = \frac{\gamma^p f_d(e)}{a+\gamma^p}\sin\varphi_{cv}\ ; \qquad f_d(e) = (e/e_{cr})^{-\beta} \tag{12}$$

with a, β as constants. Equation (12) basically represents a hyperbolic variation of mobilized frictional angle with plastic shear strains, as well as volumetric strains by virtue of the void ratio function f_d. Depending upon whether the current void ratio e is denser or looser of critical, the factor f_d will adjust the functional representation of $\sin\varphi_m$ so as to make it evolve into either a softening or hardening trend.

Hydrostatic compaction mechanism

For describing the hydrostatic compaction process, the cap surface expands isotropically in the stress space with cumulative irrecoverable volumetric plastic strains. The evolution of total void ratio (elastic and plastic components) with mean stress p is governed by an exponential law, i.e.

$$e = e_0 \exp\left[-(p/h_l)^m\right] \quad (13)$$

in which h_l is a modulus and m an exponent number. In the unloading path, the rebound curve is considered to be also an exponential with the following incremental form

$$\frac{de}{e} = -\frac{n}{h_u}\,(p/h_u)^{n-1}\,dp \quad (14)$$

where h_u and n are unloading parameters.

2.7 *Critical void ratio dependence on stress level*

The achievement of a critical void ratio by a random arrangement of grains largely depends on the applied confining pressure and the geometrical shape of the grains. At a macroscopic level, an appropriate functional description of the critical void ratio dependency on stress level is given as

$$e_{cr} = e_{cr0}\exp\left[-\left(\frac{p}{h_{cr}}\right)^{n_{cr}}\right] \quad (15)$$

where e_{cr0} is the critical void ratio at very small confining stress, n_{cr} is an exponent number and h_{cr} some parameter. For practical purposes, the value of e_{cr0} can be usually regarded as the maximum void ratio of a sand.

3 MODEL CONSTANTS

The model requires thirteen material constants. These can be categorized into elastic, shear dilatancy and compaction parameters which can all be derived from standard drained triaxial tests. Material constants for Sacramento River sand are listed in Table 1.

Table 1. Constants for Sacramento River sand

$G_0 = 3900$ KPa	$v = 0.25$	
$a = 0.011$	$\alpha = 1.2828$	$\beta = 1.25$
$m = 0.8324$	$h_u = 111.78$ MPa	$n = 0.7898$
$h_{cr} = 22.139$ MPa	$n_{cr} = 0.7075$	$e_{cr0} = 1.03$
$\varphi_{cv} = 34^0$	$h_l = 63.90$ MPa	

4 SIMULATION OF STRESS DILATANCY

Using the material parameters given in Table 1, the model performance was verified with well known experimental data published by Lee and Seed (1967) for dense and loose Sacramento River sands. Basically, the stress strain relationship in Eq. (1) had to be numerically integrated. Test results for the drained triaxial compression of both dense ($e_0 = 0.61$) and loose ($e_0 = 0.87$) Sacramento River sand over a large range of confining pressure were compared to numerical simulations.

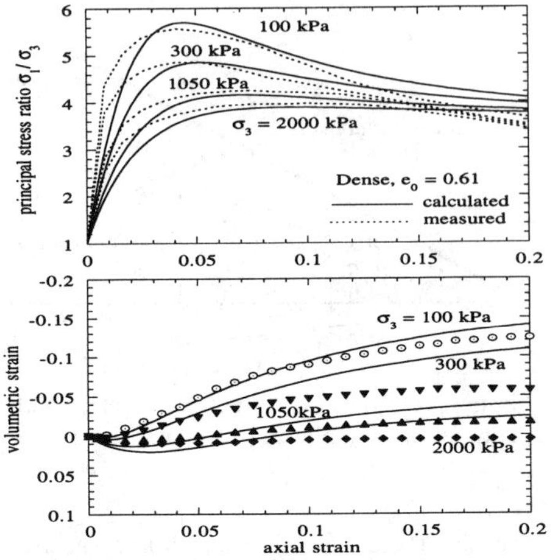

Figure 3. Comparison between calculated and measured behaviour of dense Sacramento River sand in drained triaxial compression tests

4.1 *Effect of initial void ratio and stress level*

The numerical simulations for both principal stress ratio and volumetric responses are in very good agreement with experimental results. Figure 3 shows that both strain softening and volumetric dilation are correctly captured at the low range of confining pressures for dense Sacramento River sand. Also, at a high confining pressure of 2 MPa, strain softening is almost suppressed and the volumetric response is virtually compressive even though the sand was initially at a dense state, i.e. $e_0 = 0.61$. Figure 4 shows the case of loose Sacramento River sand where strain softening is less prominent and the volumetric strain mainly contractant. However, at low confining pressure of 100 kPa, some dilation with strain softening is captured even though the sand is initially loose at $e_0 = 0.87$.

A closer look at the numerical simulations in Figs. 3 and 4 reveal that the model predicts consistently more compaction at small strains and more dilation at large strains than the experimental results. These discrepancies are mainly due to the fact that in the experiments, deformations are not uniform, especially in the large strain range where strain localization takes place. Hence, due to non-uniformities of deformation fields, the measured volumetric strains are average ones. On the other hand, the numerical simulations assume uniformity of deformations.

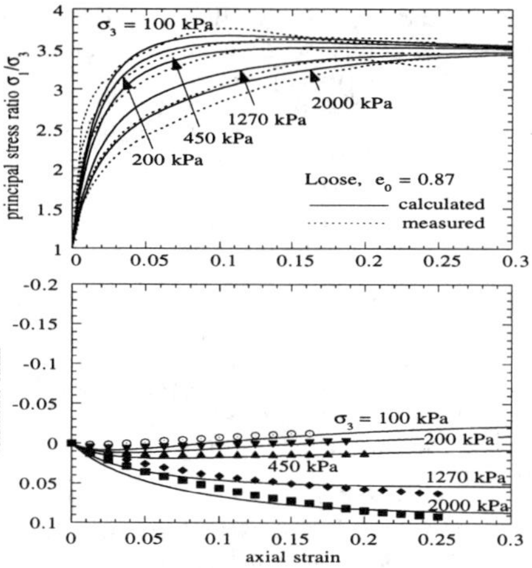

Figure 4. Comparison between calculated and measured behaviour of loose Sacramento River sand in drained triaxial compression tests

Still referring to Figs. 3 and 4, all calculated stress ratio versus axial strain curves tend consistently to a unique value equal to $\sigma_1/\sigma_3 = 3.54$ corresponding to $\varphi_{cv} = 34^0$. Here also there are small deviations from experimental curves in the regions of pre and post peak due to strain localization. The emergence of a shear band normally tends to give a higher rate of strain softening than the model which assumes homogeneous deformations.

4.2 *Maximum mobilized friction*

It is also of interest to look at the maximum (peak for strain softening cases) frictional angle achieved during deformation history. The model naturally gives the evolution of maximum mobilized frictional angle with both density and stress level according to the hardening-softening rule given in Eq. (12). Figure 5 shows the variation of calculated maximum mobilized frictional angle with confining pressure for the two initial void ratios as compared with experimental data. It is found that there is generally good agreement between calculated and experimental data for confining pressures less than 2 MPa, especially for dense sand. However, for confining pressures above 2 MPa, the experimental data seems to indicate that there is a slight drop in mobilized maximum frictional angle until approximately 5 MPa, after which the frictional angle increases again achieving a constant value at very large stresses. The drop in frictional angle may be due to grain crushing which was not included in the model. This explains the discrepancy between the calculated and measured frictional angles at high stress levels above 2 MPa.

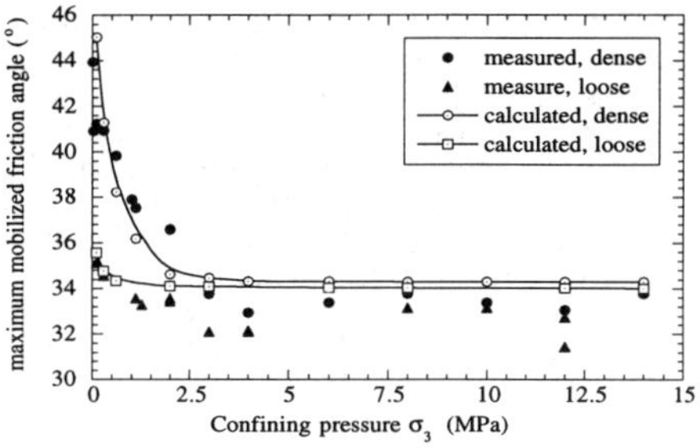

Figure 5. Influence of confining pressure on maximum mobilized friction angle

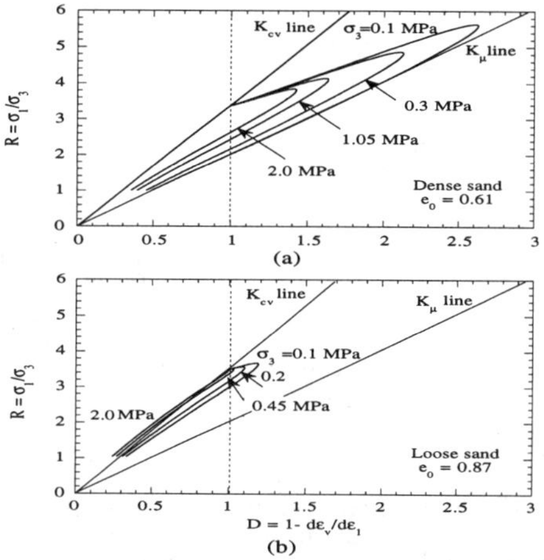

Figure 6. Computed stress-dilatancy plots for Sacramento River sand: (a) dense (b) loose

4.3 *Stress dilatancy plot*

The evolution of stress-dilatancy with deformation history is plotted in Fig. 6 for the two initial void ratios discussed earlier. Here, only numerical predictions are shown because no experimental data was available for comparison. Fig. 6a, referring to the initially dense sand, shows consistent increase in mobilized stress ratio with dilatancy factor D. The rate of dilatancy is maximum for the lowest confining pressure, i.e. $\sigma_3 = 100$ KPa, while for σ_3 larger than 2 MPa, the volumetric strains tend to be mainly contractant ($D < 1$) with less dilation. If the confining pressure is further increased, say above 4 MPa, the tendency for the sand to dilate will be completely suppressed even though the sample was initially very dense ($e_0 = 0.61$). The stress-dilatancy plots for initially loose sand ($e_0 = 0.87$) are shown in Fig.6b. They reveal mainly contractant behaviour except for confining pressures greater than 450 KPa. Furthermore, it is interesting to note that under the same confining pressure, for example $\sigma_3 = 100$ KPa, the initially dense sand has a much higher maximum dilatancy rate and stress ratio than the initially loose sand. In general, the maximum ratio R is influenced by the dilatancy factor. In order to sustain an increasing dilatancy rate, the stress ratio has to increase. After reaching a peak dilatancy rate, the stress ratio will decrease gradually to approach Critical State with zero dilation rate ($D = 1$).

In conclusion, we emphasize that the dilatancy plots obtained from the model indeed describe the complete dilatancy history of sand, an aspect which the original Rowe's stress-dilatancy relationship ($R = K_{cv}D$) could not capture adequately.

5 CONCLUSIONS

In closing, this paper demonstrates that a constitutive model based on simple concepts can be used to describe successfully the mechanical behaviour of sands for different densities and stress levels. All parameters of the model have clear physical meaning and can be easily obtained from some standard tests. By introducing a void ratio dependent term into the original Rowe's stress-dilatancy equation, and a special hardening rule, we can replicate various aspects of sand behaviour. In particular, the stress and volumetric responses exhibited by sand over a large range of density and stress levels can be rationally modelled by using a consistent set of material parameters. As such, it was found that numerical results are in good agreement with the experimental results of Lee and Seed (1967) on Sacramento River sand. Further developments are currently being done on the model with regard to stress induced anisotropy and its impact on stress-dilatancy.

6 ACKNOWLEDGMENTS

The authors are grateful to funding received from the Natural Science and Engineering Research Council of Canada and the National Energy Board of Canada.

REFERENCES

Bauer, E. and Wu, W. 1993. A hypoplastic model for granular soils under cyclic loading. *Proceedings of the International Workshop on Modelling Approaches to Plasticity, Elsevier*, 247-258.

Gudehus, G. A. 1996. Comprehensive constitutive equation for granular materials. *Soils and Foundations*, **36** (1), 1-12.

Kolymbas, D. 1977. A rate dependent constitutive equation for soils. *Mech. Res. Comm.*, **4**, 367-372.

Lee, K. L. and Seed, H. B. 1967. Drained characteristics of sands. *J. Soil Mech. Found. Div., ASCE,* **93, SM6**, 117-141.

Oda, M. 1978. Significance of fabric in granular mechanics. *Proc. U.S./Japan Seminar on Continuum Mechanical and Statistical Approaches in the Mechanics on Granular Materials,* Cowin, S. T. and Satake, M. Eds., National Science Foundation and Japan Society for the Promotion of Science, 7-26.

Reynolds, O. 1885. On the dilatancy of media composed of rigid particles in contact. *Phil. Mag.* (5 Series) **20**, 469-481.

Rowe, P. W. 1962. The Stress-dilatancy relation for static equilibrium of an assembly of particles in contact, *Proc. of Royal. Soc. A*, **269**, 500-527.

Ueng, T. S. and Lee. C. J. 1990. Deformation of sand under shear-particulate approach, *J. Geotech. Eng.*, **116**, 1625-1640.

Wan, R. G. and Guo, P. J. 1996. A pressure and density dependent dilatancy model for granular materials. Submitted to *Soils and Foundations.*

Computer Methods and Advances in Geomechanics, Yuan (ed.) © 1997 Balkema, Rotterdam, ISBN 90 5410 904 1

A constitutive theory for progressively deteriorating geomaterials

H. Harry Cheng
Ardaman and Associates, Inc., Orlando, Fla., USA

Maurice B. Dusseault
Department of Earth Sciences, University of Waterloo, Ont., Canada

ABSTRACT: Geomaterials such as rock, concrete, and dense sand exhibit some characteristic mechanical properties: softening, dilatancy, and the reduction of strength and stiffness, *etc*. These behaviours imply that the microcracks and microvoids in the materials grow, nucleate and coalesce when loaded, which is exactly the same of those normally associated with continuum damage mechanics. In this paper, a continuum damage theory is developed within the general framework of the internal variable theory of thermodynamics. The initiation, growth, nucleation and coalescence of microcracks and microvoids are described in term of a damage evolution equation involving a damage variable. Furthermore, softening, dilatancy, and the reduction of strength and stiffness of geomaterials are obtained as a consequence of damage. Comparisons of numerical predictions to available experimental data for rocks are given herein. Also, numerical simulations considering the direction effect of damage field on geomaterial behaviour are presented to illustrate the predictive capability of the model.

1 INTRODUCTION

Nearly all geomaterials such as rock and concrete and even dense, crystalline varieties such as granite or quartzite, contain tiny cavities called micro-fissures, microcracks, micropores or microcavities which may amount to 1 % or more by volume. Microcavities in the form of open microcracks may profoundly influence the properties of geomaterials. In addition to these flaws which have a discrete volume, geomaterials contain many closed flaws or incipient flaws such as closed microcracks, grain boundaries, crystal defects and other boundaries which, when strained, also act as microscopic stress concentrators. The role played by open and closed microflaws is related not only to elastic, but to electrical, transport, thermal and strength characteristics as well.

In the past two decades, *the initiation, growth, nucleation and coalescence of microcracks* have been established as dominant mechanisms in the short-time deformation and failure of brittle geomaterials such as rock and concrete under compressive loading (Wawersik and Brace, 1971; Hallbauer *et al.*, 1973; Tapponier and Brace, 1976; Paterson, 1978). This deformation mechanism causes characteristic properties of the geomaterials such as strain-softening; a dilatancy; a decrease of stiffness and strength; and increase of the peak strength with increasing confining pressure. In the past, plasticity theory and fracture mechanics have been employed to model the mechanical behaviour. The former attempts to replicate the *dislocation* or *slip* of the material at the microscale, while the latter deals with the influence of one or several macroscopic cracks, usually assumed to be embedded in an intact, non-deteriorating material, on the strength and deformation of material.

The concept of continuum damage, initially introduced by Kachanov (1958) and Rabotnov (1963), provides a viable procedure for deriving a continuum description of the effect of microcracks on the deformation and strength of geomaterials.

In this paper, a continuum damage constitutive model based on irreversible thermodynamics with internal state variables and macroscopic material properties such as strength and stiffness is presented to model the continuous, progressive process of brittle geomaterial failure. The model proposed herein incorporates damage in the material response functions as a set of internal vector variables representing irreversible microstructural rearrangements and regarded as a continuous measure of the material degradation. A specific Helmholtz free energy and a novel damage evolution law for the internal state variables in constitutive equations based on both microscopic and macroscopic experimental observations on rock and concrete are proposed. Comparisons of numerical predictions to available experimental data for various rocks in compression show good agreement. Also, numerical simulations considering the influence of direction of damage field on geomaterial behaviour are presented to illustrate the predictive capability of the model.

2 GENERAL FORMULATION

2.1 *Thermodynamical Restrictions*

Continuum thermodynamics plays a central role in establishing constitutive relationships of engineering materials. In continuum thermodynamics, there are two fundamental laws: the energy conservation law and the entropy production inequality. The energy conservation law states that, in an infinitesimal quasi-static process, the energy change of a system is equal to the sum of the mechanical work and heat put into the system. If r denotes the specific heat source and $\boldsymbol{q}$ the heat flux, the energy conservation law can be mathematically expressed as

$$\rho\dot{e} = \sigma_{ij}v_{i,j} + \rho r - q_{i,i} \tag{1}$$

in which σ_{ij}= Cauchy stress tensor; ρ= mass density of the material; e=specific internal energy of the material per unit mass; v_i=velocity vector of the solid particles; and $(\cdot)$=a total time derivative.

The entropy production inequality states that the change of entropy in a system and its surroundings is nonnegtive. Conventional arguments and the assumption that temperature T is always positive result in the Clausius-Duhem inequality

$$\sigma_{ij}d_{ij} - \rho(\dot{\psi} + s\dot{T}) - q_i\frac{gradT}{T} \geq 0 \tag{2}$$

in which ψ= the Helmholtz free energy; s=specific entropy of the material per unit mass; and d_{ij}= the deformation rate tensor.

2.2 *Thermodynamical Implications*

The Helmholtz free energy ψ in continuum thermodynamics is taken as a state function which can be expressed by a set of basic state variables. In this study, we select the strain tensor ϵ_{ij}, the temperature T and a set of damage variables $D_i^{(\alpha)}$ (α =1,2,...m),which represent irreversible microstructural material rearrangements, as basic state variables. Thus, ψ is expressed as

$$\psi = \psi(\epsilon_{ij}, T, D_i^{(\alpha)}) \tag{3}$$

By substituting equation (3) into inequality (2), the following expression for specific entropy production rate in a thermodynamically admissible process can be written:

$$(\sigma_{ij}-\rho\frac{\partial\psi}{\partial\epsilon_{ij}})\Delta\epsilon_{ij}-\rho(s+\frac{\partial\psi}{\partial T})\Delta T-\sum_\alpha\rho\frac{\partial\psi}{\partial D_i^{(\alpha)}}\Delta D_i^{(\alpha)} -q_i\frac{gradT}{T}\geq 0 \tag{4}$$

It follows from equation (4) that, in order to have the specific rate of entropy production be more than zero at X and t for every choice of ϵ_{ij}, we must have

$$\sigma_{ij}(\epsilon_{ij}, T, D_m^{(\alpha)}) = \rho\frac{\partial\psi(\epsilon_{ij}, T, D_m^{(\alpha)})}{\partial\epsilon_{ij}} \tag{5}$$

$$s(\epsilon_{ij}, T, D_m^{(\alpha)}) = -\frac{\partial\psi(\epsilon_{ij}, T, D_m^{(\alpha)})}{\partial T} \tag{6}$$

and the Clausius-Duhem inequality reduces to

$$\sum_\alpha R_i^{(\alpha)}\Delta D_i^{(\alpha)} - q_i\frac{gradT}{T}\geq 0 \tag{7}$$

$R_i^{(\alpha)}$ is the thermodynamic force associated with the internal variable $D_i^{(\alpha)}$, which is related to the fracture mechanics energy release rate, or the crack resistance force.

2.3 *Helmholtz Free Energy* ψ

Equation (5) gives the general relationship between the stress σ_{ij} and the strain ϵ_{ij} through the Helmholtz free energy $\psi(\epsilon_{ij}, T, D_m^{(\alpha)})$. Once the form of ψ is given, the relationship between σ_{ij} and ϵ_{ij} will be determined. From a purely mathematical viewpoint, the determination of ψ as a polynomial scalar-valued invariant of two or more tensors of any rank (such as ϵ_{ij}, D_m) is a relatively straightforward task involving the theory of invariants (Wang, 1970; Spencer, 1971). The irreducible integrity basis for the Helmholtz free energy is determined providing that the mathematical representation for the *internal damage state variable* D_m and the level of approximation (i.e., the highest orders of involved variables) are decided upon. The most general expression of $\psi(\epsilon_{ij}, D_m)$, not imposed by the restrictions of thermodynamics, can therefore be given as a function of these basic invariants of ϵ_{ij} and D_m and their combinations in any powers.

For specific engineering materials, however, an expression for ψ must be subjected to more restrictions. For example, for geomaterials, a particular form of ψ can be obtained by considering restrictions from the second law of thermodynamics, the principle of material frame indifference, and the assumptions of small deformation, initial isotropy and homogeneity of brittle materials, etc. A detailed discussion on a particular form of ψ is given by Cheng and Dusseault (1996).

Following Davison and Stevens (1973), Krajcinovic and Fonseka (1981), Cheng and Dusseault (1993, 1996), and Cheng *el at.* (1994a, b), we can write

$\psi(\epsilon_{ij},D_i^{(\alpha)})$ for geomaterials in a form quadratic with respect to strain:

$$\rho\psi = \frac{1}{2}\lambda\epsilon_{kk}\epsilon_{nn} + \mu\epsilon_{kl}\epsilon_{lk} + \sum_{\alpha}[\tilde{C}_1 D_k^{(\alpha)}\epsilon_{kl}D_l^{(\alpha)}\epsilon_{mm} + \tilde{C}_2 D_k^{(\alpha)}\epsilon_{kl}\epsilon_{lm}D_m^{(\alpha)}] \quad (8)$$

in which λ, μ=Lamé's parameters; α=independent damage fields, (α = *1, 2, ... n*); C_1, C_2=material constants, and $D_i^{(\alpha)}$=the damage vector, defined by the local void density ω in a cross-section with normal $\boldsymbol{n}$

$$D_i^{(\alpha)} = \omega^{(\alpha)}n_i \quad (9)$$

The physical meaning of each term and the range of parameters C_1 and C_2 in equation (8) are discussed in detailed by Cheng (1994) and Cheng and Dusseault (1996).

2.4 *Damage Evolution Equation: Phenomenological Considerations*

A material whose internal energy *e* (hence also ψ) is independent of the internal state variables $D_i^{(\alpha)}$ is elastic. For such a material the Coleman relations (5) are sufficient so that, with the balance equations, any well-posed boundary value problem may be solved, and inequality (7) is trivially satisfied. For a general case, however, constitutive equations must be specified for the internal state variables; the aforementioned inequality then acts as a restriction on such constitutive equations. These internal state variables in constitutive laws require the establishment of additional rate equations and criteria to describe their evolution and the mechanical behaviour of damaged materials.

Two basic approaches have been used in the past to establish the damage kinetic equations describing the evolution of damage and relating the flux rate $\dot{D}_i^{(\alpha)}$ and their affinities. The first is to assume that a dissipation potential *g(R,D)* exists in the space of stress and affinities. Following arguments similar to those used in the theory of plasticity, the kinetic equation can be derived from the dissipation potential. As pointed out by Krajcinovic (1984, 1989), it would be advantageous to formulate a phenomenological CDM model in the form of the dissipation potential. This leads to a formulation identical to the theory of plasticity, enabling use of most existing algorithms and computer codes after relatively minor modifications. However, the determination of the exact form of the dissipation potential *g(R,D)* is at the present time largely a matter of speculation; the experimental evidence is still limited.

Another approach to the kinetic equation is damage evolution laws based mainly on experimental observations of the material, which will be used in the present study. Using skilfully chosen evolution laws, this approach has been successfully used to model fairly complex response modes for metal (Murakami and Ohno, 1981; Lemaitre, 1985a,b, 1986; Chaboche, 1988) and for rock and concrete (Dragon and Mroz, 1979; Cheng and Dusseault, 1993, 1996; Cheng *et al.*, 1994a,b).

Based on both microscopic and macroscopic experimental observations of geomaterials such as rock and concrete, we propose the following function to define local void density ω:

$$\Delta\omega = c\delta\frac{\gamma\xi_D}{1+\gamma\xi_D}(1-\frac{\omega}{\omega_c})\Delta\xi_D \quad (10)$$

in which ω_c=critical damage level ($0 < \omega_c \leq 1.0$); γ,δ=material constants; c=loading/unloading controller; and ξ_D=an equivalent strain.

The influence of parameters used in equation (10) on the form of damage evolution is discussed by Cheng and Dusseault (1996).

2.5 *A Summary of Constitutive Damage Theory Derived*

For an isothermal process ψ is only a function of strain ϵ_{ij} and the damage internal variable $D_m^{(\alpha)}$. Hence, equation (5) becomes

$$\sigma_{ij}(\epsilon_{ij},D_m^{(\alpha)}) = \rho\frac{\partial\psi(\epsilon_{ij},D_m^{(\alpha)})}{\partial\epsilon_{ij}} \quad (11)$$

The incremental stress-strain law is expressed as:

$$\Delta\sigma_{ij} = K_{ijkl}\Delta\epsilon_{kl} + \tilde{K}_{ijm}\Delta D_m \quad (12)$$

where

$$K_{ijkl} = \lambda\delta_{ij}\delta_{kl} + 2\mu\delta_{ik}\delta_{jl} + \sum_{\alpha}[\tilde{C}_1(\delta_{ij}D_k^{(\alpha)}D_l^{(\alpha)} + \delta_{kl}D_i^{(\alpha)}D_j^{(\alpha)}) + \tilde{C}_2(\delta_{jk}D_i^{(\alpha)}D_l^{(\alpha)} + \delta_{il}D_j^{(\alpha)}D_k^{(\alpha)})] \quad (13)$$

and

$$\begin{aligned}\tilde{K}_{ijm} = \sum_{\alpha}\{&\tilde{C}_1\epsilon_{kl}[\delta_{ij}(\delta_{km}D_l^{(\alpha)} + \delta_{lm}D_k^{(\alpha)}) + \delta_{kl}(\delta_{im}D_j^{(\alpha)} + \delta_{jm}D_i^{(\alpha)})] + \tilde{C}_2\epsilon_{kl}[\delta_{jk}(\delta_{im}D_l^{(\alpha)} + \delta_{lm}D_i^{(\alpha)}) \\ &+ \delta_{il}(\delta_{jm}D_k^{(\alpha)} + \delta_{km}D_j^{(\alpha)})] \\ &- \frac{1}{2}[\tilde{C}_1(\delta_{ij}D_k^{(\alpha)}D_l^{(\alpha)} + \delta_{kl}D_i^{(\alpha)}D_j^{(\alpha)}) \\ &+ \tilde{C}_2(\delta_{kj}D_i^{(\alpha)}D_l^{(\alpha)} + \delta_{il}D_k^{(\alpha)}D_j^{(\alpha)})]\frac{D_m^{(\alpha)} + D_m^{(\alpha)}}{D_p^{(\alpha)}D_p^{(\alpha)}}\epsilon_{kl}\}\end{aligned} \quad (14)$$

From equation (9), we can express the incremental damage variable ΔD_i as follows:

$$\{\Delta \boldsymbol{D}\} = \{\Theta\}\Delta\omega \quad (15)$$

where $\{\Theta\}$ is a direction cosine matrix, defined as

$$\{\Theta\} = \{\cos\varphi\cos\theta \quad \cos\varphi\sin\theta \quad \sin\varphi\} \quad (16)$$

Equation (10) is rewritten as

$$\Delta\omega = c\tilde{H}\Delta\xi_D \tag{17}$$

where

$$\tilde{H} = \delta\frac{\gamma\xi_D}{1+\gamma\xi_D}(1-\frac{\omega}{\omega_c}) \tag{18}$$

From the equivalent strain ξ_D, we have

$$\Delta\xi_D = \{H\}\{\Delta\epsilon\} \tag{19}$$

where $\{H\}$ is a 1x6 transfer matrix for equivalent strain ξ_D and strain tensor ϵ_{ij}.

The relationship between damage variable D and strain ϵ_{ij} can be established through substituting for $\Delta\omega$ from equation (17) into (15) while using (19) as follows:

$$\{\Delta D\} = c\tilde{H}\{\Theta\}\{H\}\{\Delta\epsilon\} \tag{20}$$

Hence, from equations (12) and (20), an incremental form for the stress-strain relations written in matrix notation is derived as follows:

$$\{\Delta\sigma\} = [C]\{\Delta\epsilon\} \tag{21}$$

where

$$[C] = [K] + c\tilde{H}[\tilde{K}]\{\Theta\}\{H\} \tag{22}$$

Equation (22) represents the incremental form of the damage constitutive model. $[C]$ is an unsymmetrical tangential stiffness matrix and its value depends on the current state of strain ϵ_{ij} and damage D_i.

3 NUMERICAL SIMULATIONS: THEORY APPLICATION AND VERIFICATION

Analytical results are compared with experimental data from three representative rocks, including stress-strain curves and dilatancy in uniaxial compression. The three representative rocks are limestone, marble, and norite.

The comparisons between the numerical results and the experimental data for stress-strain curves and volumetric strain, reported by Wawersik and Fairhurst (1970), Rummel and Fairhurst (1970), and Bienawski, *et al.* are shown in Figures 1 and 2.

Good predictions for axial stress-strain and dilatancy behaviour of these materials are obtained by the proposed damage model. Based on these figures, the model replicates with a reasonable accuracy the stress-strain relationship of rocks in compression, and emulates the process of volumetric strain variation in compression, in which the material initially compacts and then dilates. These figures also indicate that the *strain-softening* and *residual strength* behaviour for different types of rocks are fully captured by the model.

Figure 2 illustrates continuous variation in volumetric strain in compression. With respect to the peak

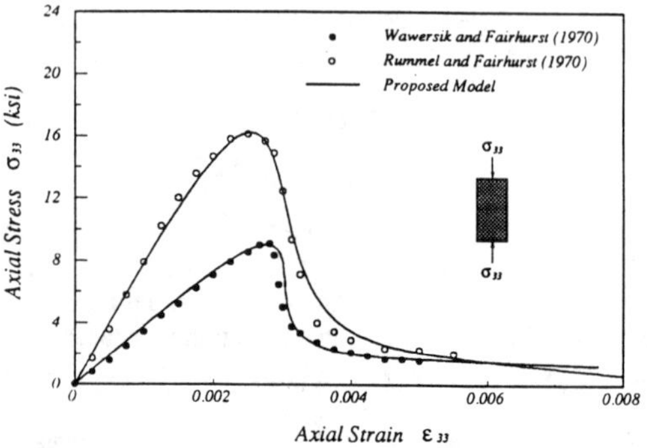

Figure 1 Comparison of analytical with experimental complete stress-strain curves in uniaxial compression tests

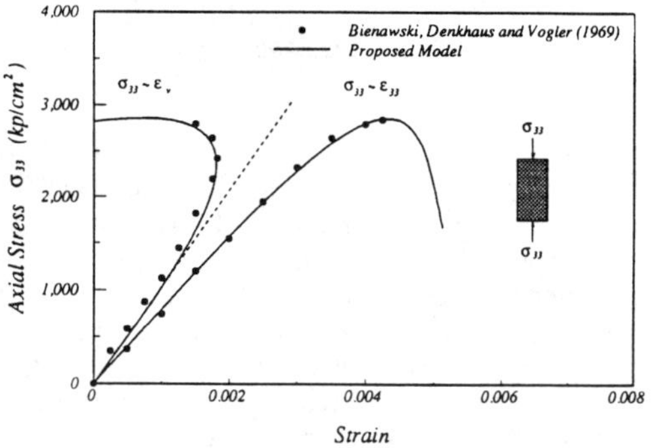

Figure 2 Comparison of analytical with experimental stress-strain curves for norite in uniaxial compression test

stress, the volumetric strain ϵ_v reaches its minimum at 80% for norite. This reversal of the volume change behaviour is typically observed in experiments, and is identified as the point at which irreversible microcracking and dilation processes become dominant over elastic strains. The damage model has adequately captured this behaviour.

Figures 3 - 5 illustrate continuous variations in volumetric strain, apparent Poisson's ratio and damage when the damage model is computed to simulate the stress-strain curves of Indiana limestone and Tennessee marble under uniaxial compression, as illustrated in Figure 1. The volumetric strain ϵ_v versus axial strain ϵ_{33} is plotted in Figure 3, and shows that dilatancy begins prior to the peak stress. Volumetric strain ϵ_v sharply increases just after peak stress, then further increases very little as the residual strength is reached, largely due to the dramatic increase in the apparent Poisson's ratio $(-\epsilon_{11}/\epsilon_{33})$, as shown in Figure 4. The apparent Poisson's ratio $(-\epsilon_{11}/\epsilon_{33})$ versus ϵ_{33}, as shown in Figure 4, indicates a rather dramatic decrease in lateral *stiffness*, especially after peak strength. The apparent Poisson's ratio exceeds unity, which is

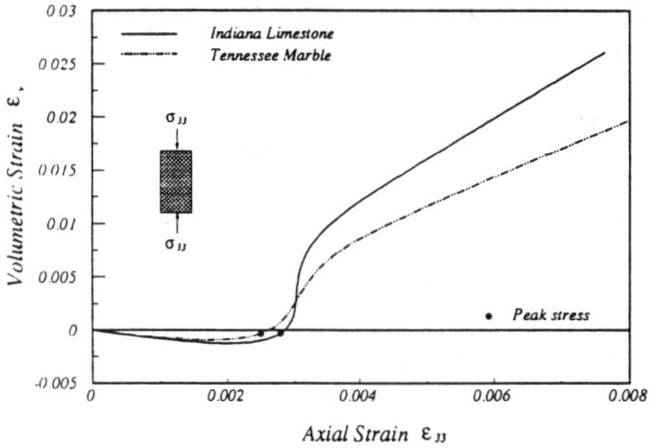

Figure 3 Variation of axial and volumetric strain for limestone and marble in uniaxial compression tests

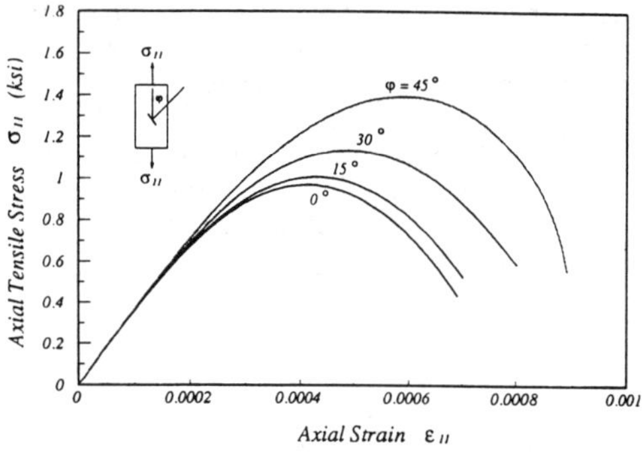

Figure 6 Effect of the damage direction on stress-strain curves in uniaxial tension

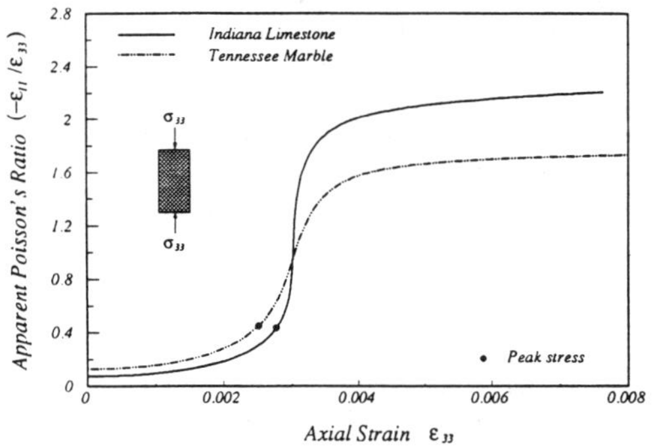

Figure 4 Apparent Poisson's ratio for limestone and marble in uniaxial compression tests

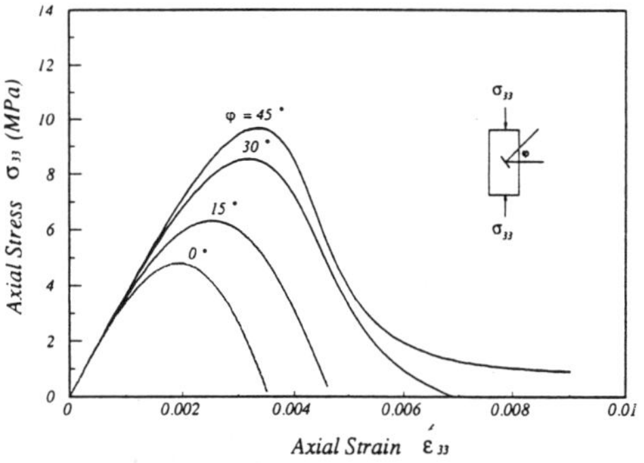

Figure 7 Effect of the damage direction on stress-strain curves in uniaxial compression

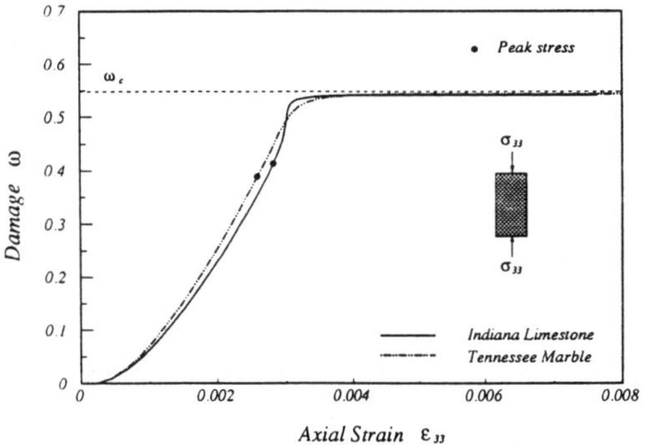

Figure 5 Damage evolution for limestone and marble

significantly more than the theoretical maximum of 0.5 for an incompressible solid body; this is because of the dilational effects arising from microcracking.

Experimental results for geomaterials reveal that microcracking begins at loads well before the peak stress is reached. It is now widely recognized that the process of geomaterial failure is a continuous, progressive one. Figure 5 shows damage evolution as a function of axial strain ϵ_{33}. It has already been noted that the present model simulates quite well the continuous, progressive process of failure. Figure 5 indicate that geomaterials steadily deteriorate, rather than suddenly beginning to exhibit damage at the so-called stress threshold σ_a. For limestone and marble, the damage growth quickly reaches approximately 80% of the critical damage level ω_c prior to peak stress, and just beyond peak stress, the damage rapidly reaches ω_c. This rapid change corresponds to the steeper slope of the axial stress-strain curve in the postfailure region.

In a general loading program, the direction of the principal strain is not always perpendicular to the plane of microdefects generated by a previous system of external loads. Thus, it is often necessary to consider a case in which there is an angle between the plane of the initial damage field and the direction of the principal strain.

The present theory, based on a vectorial damage variable, incorporating both the density and the direction of microcracks, is ideally suited for the analysis of this class of problems.

The computed stress-strain curves for geomaterial with four different orientations of the previously incurred damage under both uniaxial tension and compression are plotted in Figures 6 and 7. The difference in the peak stress and the shape of stress-strain curves is quite apparent. The results indicate the effect of strongly directional damage on geomaterial behaviour.

4 FINAL REMARKS

A damage constitutive theory for geomaterials has been systematically established based on the internal variable theory of thermodynamics. It is shown that present model is a suitable theory for describing the continuous, progressive process of geomaterial failure, and can replicate the salient aspects of the mechanical behaviour of geomaterials. Particularly, strain-softening, dilatancy, decrease of stiffness and strength observed in laboratory are well captured.

REFERENCES

Bieniawski, Z. T., Denkhaus, H. G. and Vogler, U. W. (1969). Failure of fractured rock. *Int. J. Rock Mech. Min. Sci.* **6**, 323-341.

Chaboche, J. L. (1988). Continuum damage mechanics: parts I & II. *J. Appl. Mech.* **55**, 59-72.

Cheng, H. and Dusseault, M. B. (1993). Deformation and diffusion behaviour in a solid experiencing damage: a continuous damage model and its numerical implementation. *Int. J. Rock Mech. Min. Sci. & Geomech. Abstr.* **30**, No. 7, 1323-1331.

Cheng, H., Dusseault, M. B., and Rothenburg, L. (1994a). A coupled constitutive model of strain with damage for geomaterials. *8th Int. Conf. for Computer Methods and Advances in Geomechanics* **I**, 549-554.

Cheng, H., Dusseault, M. B., and Rothenburg, L. (1994b). A continuum damage mechanics model for geomaterials. Accepted by *J. Appl. Mech.*

Cheng, H. (1994). *Continuum damage theory for geomaterials and its application to coupling in transport processes*. Ph.D. thesis, University of Waterloo, Canada.

Cheng, H. and Dusseault, M. B. (1996). Progressive failure modelling of brittle solids based on continuum damage mechanics. Accepted by *Int. J. Numer. Anal. Meth. Geomech.*

Davison, L. and Stevens, A. L. (1973). Thermomechanical constitution of spalling elastic bodies. *J. Appl. Phys.* **44**, 668-674.

Dragon, A. and Mroz, Z. (1979). A continuum model for plastic-brittle behaviour of rock and concrete. *Int. J. Engrg. Sci.* **17**, 121-137.

Hallbauer, D. K., Wagner, H., and Cook, N. G. W. (1973): Some observations concerning the microscopic and mechanical behaviour of quartzite specimens in stiff, triaxial compression tests. *Int. J. Rock Mech. Min. Sci. & Geomech. Abstr., **10**, 713-726.*

Kachanov, L. M. (1958). Time of the rupture process under creep conditions. *IVZ Akad. Nauk., SSSR, Otd. Tech. Nauk.* No. 8, 26-31.

Krajcinovic, D. and Fonseka, G. U. (1981). The continuous damage theory of brittle materials: Part 1 & II. *J. Appl. Mech.* **48**, 809-815.

Krajcinovic, D. (1984). Continuum damage mechanics. *Appl. Mech. Rev.* **37**, 1-6.

Krajcinovic, D. (1989). Damage mechanics. *Mech. Mater.* **8**, 117-197.

Lemaitre, J. (1985a). Coupled elasto-plasticity and damage constitutive equations. *Comput. Meths. Appl. Mech. Eng.* **51**, 31-49.

Lemaitre, J. (1985b). A continuous damage mechanics model for ductile fracture. *J. Eng. Mater. Technol.* **107**, 83-89.

Lemaitre, J. (1986). Local approach of fracture. *Engrg. Fract. Mech.* **25**, No. 5/6, 523-537.

Murakami, S. and Ohno, N. (1981). A continuum theory of creep and creep damage. in *Creep of Structures*, IUTAM Symposium, (edited by Ponter, A. and Hayhurst, D.), Springer, Berlin, 422-444.

Paterson, M. S. (1978). *Experimental Rock Deformation-The Brittle Field.* Springer, Berlin.

Rabotnov, Y. N. (1963). On the equations of state for creep. *Progress in Applied Mechanics, Prager Anniversary Volume*, MacMillan, New York, 307-315.

Rummel, F. and Fairhurst, C. (1970). Determination of the post-failure behaviour of brittle rock using a servo-controlled testing machine. *Rock Mech.* **2**, 189-204.

Spencer, A. J. M. (1971). Theory of invariants. in *Continuum Physics* (edited by Eringen, A. C.), Mathematics, Academic Press, New York, **I**, 239-353.

Tapponnier, P. and Brace, W. F. (1976): Development of stress-induced microcracks in Westerly Granite. *Int. J. Rock Mech. Min. Sci. & Geomech. Abstr.,**13**, 103-112.*

Wang, C.-C. (1970). A new representation theorem for isotropic functions. *Arch. Rational Mech. Anal.* **36**, 166-197.

Wawersik, W. R. and Fairhurst, C. (1970). A study of brittle rock fracture in laboratory compression experiments. *Int. J. Rock Mech. Min. Sci.* **7**, 561-575.

Wawersik, W. R. and Brace, W. F. (1971). Post-failure behaviour of a granite and diabase. *Rock Mech.* **3**, 61-85.

Computer Methods and Advances in Geomechanics, Yuan (ed.) © 1997 Balkema, Rotterdam, ISBN 90 5410 904 1

An elastoplastic constitutive model for rocks and its verification

G. Xu, M. B. Dusseault & P.A. Nawrocki
Geomechanics Group, PMRI, University of Waterloo, Ont., Canada

ABSTRACT: An elastoplastic constitutive model for rocks with mild anisotropy is proposed in this paper. Model verification through numerical simulations of triaxial and cylindrical tests is also provided. The proposed formulation invokes a path-independent, generalized 3-D Hoek-Brown criterion, the evolution of yield surfaces with damage, and a non-associated plastic flow rule. Material anisotropy is accounted for through incorporation of the anisotropic invariants of the second-order stress and fabric tensors.The material characteristics predicted by the model are strongly influenced by confining stress and display a brittle-ductile transition and a compaction-dilatancy transition. Numerical simulations of hole deformation characteristics and the distribution of stress, damage and failure around cylindrical openings are carried out.

1. INTRODUCTION

To investigate the deformation and failure characteristics of underground openings, both experimental and numerical approaches have been employed. In experiments, thick-walled hollow cylinders of rocks or rectangular rock blocks with pre-drilled holes have been subjected to equal and unequal external stresses or different stress paths. Experimental observations not only aid in identifying deformation and failure mechanisms of underground openings, but also can be applied to verify numerical models. Numerical predictions depend, to a large extent, on the choice of an appropriate constitutive law. Usually, non-linear elasticity and elasto-plasticity with perfectly plastic or perfectly brittle-plastic behaviour are applied in numerical analyses. The objective of this paper is: (i). to propose an elastoplastic model which can adequately describe typical mechanical behaviour of rocks; (ii). to verify the proposed formulation through numerical simulations of the experimental results obtained in both triaxial and cylindrical opening tests.

In Section 2, basic assumptions incorporated in the formulation are outlined. A generalized 3-D Hoek-Brown criterion and its validation are provided, followed by the description of yield, plastic potential loci, and strain hardening/softening functions. For the illustration of the proposed formulation, numerical simulations of brittle-ductile and compaction-dilatancy transitions are also presented. To account for material anisotropy, the anisotropic invariants of the second-order stress and fabric tensor are introduced. Dependence of material response on the loading orientation relative to the bedding plane is simulated. To verify the proposed framework, numerical simulations of experimental results of hole deformation characteristics, fracture, and failure around cylindrical openings are carried out in Section 3.

2 MATHEMATICAL FORMULATION

2.1 *A generalized 3-D Hoek-Brown Criterion*

Based on a comprehensive review of existing rock failure criteria, Pan and Hudson (1988) concluded that the Hoek-Brown failure criterion is the most suitable and realistic one for the prediction of the strength of rocks and rock masses. The Hoek-Brown criterion was developed by assuming that the strength of rock is independent of the intermediate principal stress. Recent studies by Addis and Wu (1993) and Ewy (1991) demonstrated the significance of the incorporation of all the principal stresses in wellbore failure analysis. Thus, in this study, a generalized 3-D Hoek-Brown criterion is derived from its original form, which, in terms of principal stresses, is given by

$$\sigma_1=\sigma_3+\sqrt{m\sigma_c\sigma_3+\sigma_c^2} \tag{1}$$

where σ_1 and σ_3 are the major and minor principal stresses at failure. m and σ_c are material parameters, and the parameter σ_c is related to the uniaxial compressive strength. This form can be expanded under isotropic conditions and formulated in terms of stress invariants. Thus, after some algebra, the failure surface in the meridional plane for this formulation can be obtained as

$$F_{HB}=\frac{9}{m}\left(\frac{\sqrt{J_2}}{\sigma_c}\right)^2+\sqrt{3}\left(\frac{\sqrt{J_2}}{\sigma_c}\right)-\left(\frac{3}{m}+\frac{I}{\sigma_c}\right)=0 \tag{2}$$

In this equation, $I=\sigma_{ii}$, $\theta=1/3\sin^{-1}(3\sqrt{3}\,J_3/2\,J_2^{3/2})$, $J_2=(1/2s_{ij}s_{ij})^{1/2}$ and $J_3=1/3s_{ij}s_{jk}s_{ki}$ are stress invariants. To overcome the dificiencies, such as the over prediction of tensile strength and non-smoothness in the deviatoric plane, a generalized 3-D Hoek-Brown criterion is proposed

$$F=c_1\left(\frac{\sqrt{J_2}}{g(\theta)\sigma_c}\right)^2+c_2\left(\frac{\sqrt{J_2}}{g(\theta)\sigma_c}\right)-\left(c_3+\frac{I}{\sigma_c}\right)=0 \tag{3}$$

or equivalently

$$F=\sqrt{J_2}-g(\theta)\sqrt{J_{2c}};$$

$$\sqrt{J_{2c}}=\frac{-c_2+\sqrt{(c_2^2+4c_1(c_3+I/\sigma_c))}}{2c_1}\sigma_c \tag{4}$$

It is obvious that $c_1=9/m$, $c_2=\sqrt{3}$ and $c_3=3/m$ corresponds to the original Hoek-Brown criterion when $g(\theta)=1$. The phenomenological function $g(\theta)$ provided by Jiang and Pietruszczak (1988) is adopted here to define the stress-dependent π-sections.

In order to validate the proposed failure criterion, experimental data for Indiana limestone obtained by Amadei and Robison (1986) are analysed. The following material parameters were chosen: $\sigma_c=42$MPa, $\sigma_t=4.0$MPa, $\sigma_{bc}=0.9\sigma_c$ and $c_2=1.0$. where σ_t is uniaxial tensile strength and σ_{bc} is the equi-biaxial compressive strength. The numerical results obtained by both the Hoek-Brown criterion and the proposed formulation, together with experimental data, are presented in Fig.1. Fig.1a shows the failure surface in the meridional plane and Fig.1b presents the π-plane sections. The failure surface in the biaxial plane with different values of

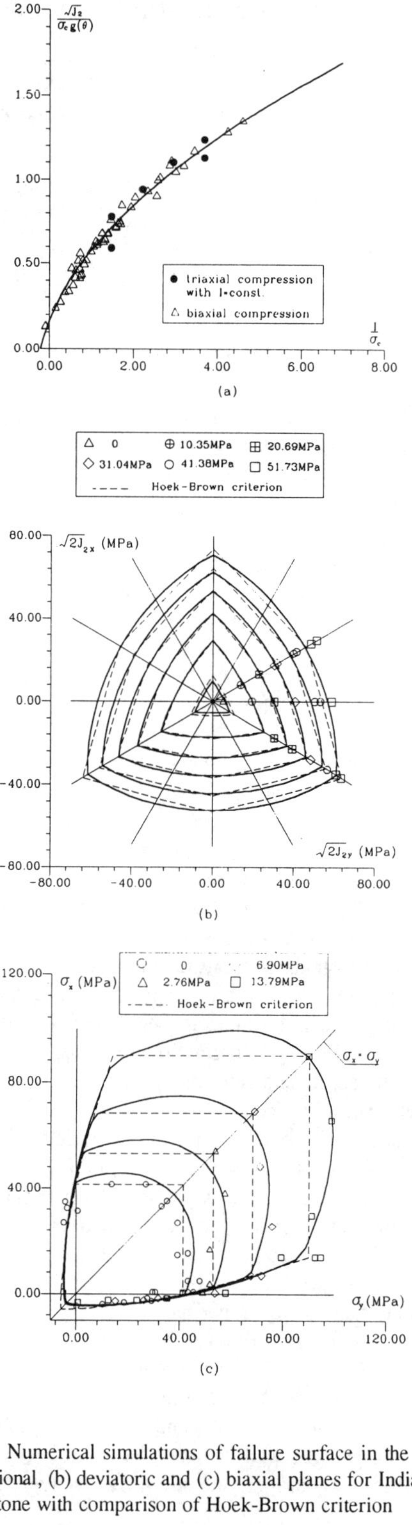

Fig.1 Numerical simulations of failure surface in the (a) meridional, (b) deviatoric and (c) biaxial planes for Indiana limestone with comparison of Hoek-Brown criterion

stress in the out-of-plane direction is provided in Fig.1c. It is evident that the proposed criterion gives a good fit to experimental data. It should be emphasized that, for triaxial compression, the proposed criterion can give exactly the same results as the Hoek-Brown criterion, as discussed above. Moreover, experimental data under complex stress paths can also be accommodated by the proposed formulation.

2.2 *Yield loci and plastic potential surface*

According to Pietruszczak et al. (1988), the family of yield loci may be defined to have a functional form similar to the failure surface

$$f=\sqrt{J_2}-\beta(\xi)g(\theta)\sqrt{J_{2c}} \tag{5}$$

where $\beta(\xi)$ represents a hardening/softening function and ξ is a suitably chosen damage parameter. It should be noted that the function $\beta(\xi)$ will satisfy the following conditions: (i). $\beta(\xi)=0$ for $\xi=0$, which indicates no material damage; (ii). $\beta(\xi)\to 1$ when the yield surface approaches the failure surface. The damage parameter ξ is defined by the history of the accumulated plastic distortions.

The direction of plastic flow is governed by a non-associated flow rule and the corresponding plastic potential is selected in a form similar to that proposed in Pietruszczak et al (1988)

$$\psi=\sqrt{\bar{J}_2}+m_c g(\theta)\bar{I}\ln(\bar{I}/\bar{I}_0) \tag{6}$$

where $\bar{I}=c_0\sigma_c+I$, and c_0 define the location of the apex of the current potential surface in the tensile domain. In eq.(6), m_c represents the value of $m=\sqrt{J_2}/(g(\theta)\bar{I})$ at which a transition from compaction to dilatancy takes place ($d\epsilon_{ii}^p=0$). It is assumed that such a transition occurs along the locus defined by

$$\bar{f}=\sqrt{\bar{J}_2}-\alpha g(\theta)\sqrt{J_{2c}}=0 \tag{7}$$

in which α is a material constant. In terms of eqs.(6) and (7), c_0 and m_c can be identified, and the compaction and dilatancy transition is controlled by the parameter α.

2.3 *A generalized hardening function and a ductile-brittle transition*

Experimental evidence shows that the mechanical response of many rocks is largely influenced by the value of confining stress. At relatively large confining stress, a pattern of numerous microcracks develops during straining, and material deformation occurs in a stable ductile mode. As the confining stress decreases, a gradual transition from ductile to brittle behaviour takes place. In the brittle regime, distinct macrocracks form, generating an unstable response. To model this transition from a stable(ductile) to an unstable(brittle) response, a suitable generalized hardening function is chosen:

$$\beta(\xi)=\frac{\xi}{A+B\xi}[1-\phi_r(1-e^{-C<\xi-\xi_f>^{\gamma}})]$$

$$C=H\left[\frac{(I/\sigma_c)_T-(I/\sigma_c)_f}{c_3+(I/\sigma_c)_T}\right]^{\mu} \tag{8}$$

where the function $f(x)=<x>$ represents $f(x)=0$ if $x<0$ and $f(x)=x$ if $x>0$, and ξ_f represents the value of ξ corresponding to the maximum value of $\beta=\beta_f$, $(I/\sigma_c)_f$ is evaluated at $\beta=\beta_f$, and $(I/\sigma_c)_T$ denotes a normalized value of confining stress at which the transition from ductile to brittle behaviour takes place. Moreover, A, B, γ, H and μ represent material constants and ϕ_r defines the residual strength of material. The constant C controls the rate of strain softening in the unstable regime.

2.4 *Numerical simulations of triaxial compression tests*

The proposed formulation is now applied to simulate a series of triaxial compression tests with different initial confining stresses. It should be noted that some parameters are assumed because of the lack of appropriate experimental data. The main objective here is to demonstrate the performance of the proposed formulation, particularly, the simulation of a ductile-brittle transition and a compaction-dilatancy transition, and to obtain a set of material parameters for further numerical simulations of boundary-valued problems.

The constitutive relation for rocks can be formulated using standard plasticity procedures as described by Pietruszczak et al. (1988). To obtain a complete response including strain softening, the strain-controlled procedure has be used instead of the stress-controlled one. The integration of the constitutive relation is accomplished by using the forward-Euler explicit algorithm and imposing a mixed boundary condition, i.e. $\dot{\epsilon}_{11}>0$ ($\dot{\epsilon}_{ij}=0$, if i ≠

j) under $\dot{\sigma}_{22}=\dot{\sigma}_{33}=0$ for the triaxial case. The results of numerical simulations with different initial values of confining stress are presented in Fig.2. Fig.2a shows the deviatoric stress characteristics whereas Fig.2b presents the volumetric strain characteristics for Indiana limestone. It is evident that the response is unstable and that a compaction-dilatancy transition occurs for a low value of confining stress; however, for a high value of confining stress, a stable response along with compaction takes place.

2.5 *Anisotropic invariants*

In this study, a simple representation of anisotropic invariants similar to that provided in Pastor (1990) is employed. It has been widely used in plasticity theory to account for anisotropic behaviour of soils.

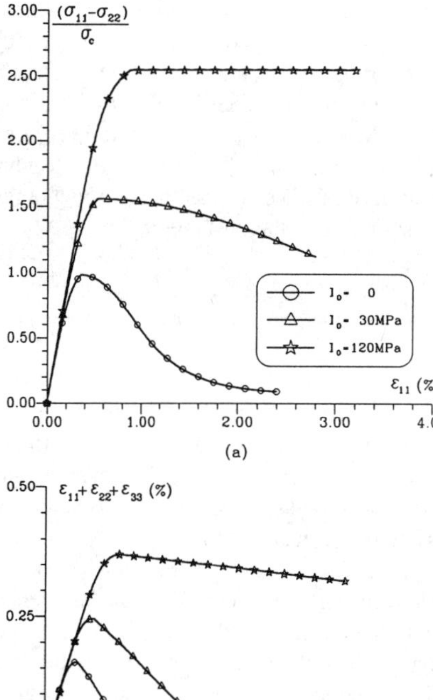

(a)

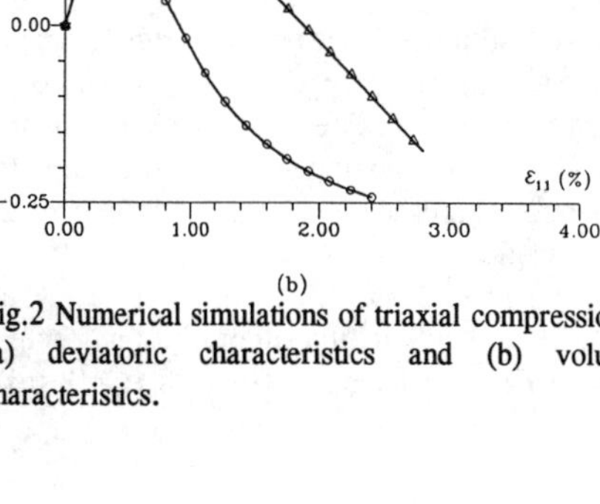

(b)

Fig.2 Numerical simulations of triaxial compression tests (a) deviatoric characteristics and (b) volumetric characteristics.

Based on the direct and joint invariants of the stress tensor σ_{ij} and the fabric tensor α_{ij}, the anisotropic invariants are defined as

$$\bar{I}=\sigma_{ij}\alpha_{ij},\ \bar{J}_2=(\frac{1}{2}\bar{s}_{ij}\bar{s}_{ij})^{1/2},\ \bar{\theta}=\frac{1}{3}\arcsin\frac{3\sqrt{3}}{2}(\frac{\bar{\Sigma}_3}{\bar{\Sigma}_2})^3 \quad (9)$$

For further details on the complete formulation may be found in Pastor (1990), as space here is limited. Thus, the proposed constitutive relation derived for isotropic rocks in terms of I, J_2 and θ can be generalized to the anisotropic case by substituting the anisotropic invariants $\bar{I}$, $\bar{J}_2$ and $\bar{\theta}$. It should be noted that the case $\alpha_{ij}=\delta_{ij}$ represents an isotropic response. Transverse anisotropy can be described by a fabric tensor α_{ij} with two distinct eigenvalues, and orthotropy corresponds to the case when all the eigenvalues of α_{ij} are distinct.

To illustrate the above approach, the following numerical simulations for anisotropic behaviour of Berea sandstone and Indiana limestone were carried out. Experimental evidence indicates some degree of transverse isotropy for both rocks. For Berea sandstone, the compressive strength perpendicular to the bedding plane is 1.3 to 1.35 times that parallel to the bedding plane with a smooth variation from 0° to 90° orientations. For Indiana limestone, a difference of only 1.07 to 1.12 has been found. To simulate anisotropic behaviour of both rocks, the following fabric tensor has been assumed: $a_{11}/a_{22}=1.15$ and $a_{22}=a_{33}$ for Berea sandstone, and $a_{11}/a_{22}=1.05$ and $a_{22}=a_{33}$ for Indiana limestone. The numerical simulations of the dependence of compressive strength on the orientation of loading relative to the bedding plane are presented in Fig.3. It is clear that

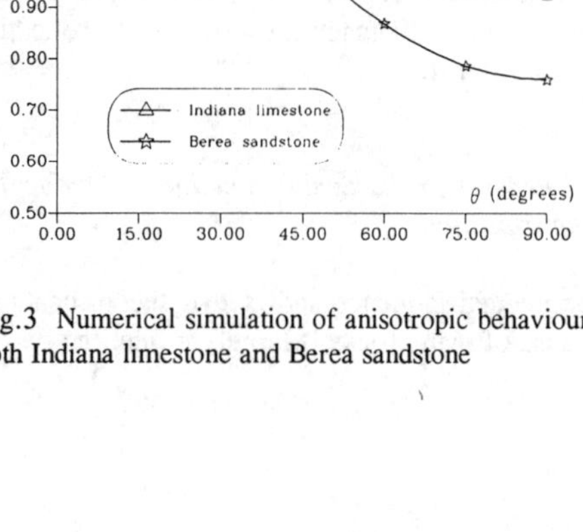

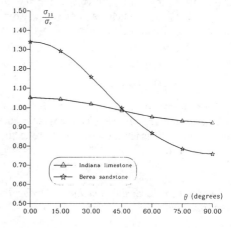

Fig.3 Numerical simulation of anisotropic behaviour for both Indiana limestone and Berea sandstone

different degrees of anisotropy can be accommodated through the use of fabric tensors.

3. NUMERICAL SIMULATIONS

Ewy and Cook (1990) investigated deformation, fracture and failure around cylindrical openings through experiments on thick-walled hollow cylinders of Berea sandstone and Indiana limestone with a hole diameter of 25.4mm and an external diameter of 89mm. Their studies incorporated plane strain conditions and the application of different stress paths. In our simulations, the case of an increase of external stress with zero internal stress is considered for both rocks. Because of the plane strain condition and symmetry, a finite element model with 375 four-noded isoparametric elements and 2x2 Gauss quadrature is employed, and the orientation of the bedding plane is assumed to be along the x direction (refer to Fig.4b). To present the numerical results, for brevity, only the distribution of two parameters will be provided:

(a). *stress intensity factor* β defined through eq.(7), which is a measure of material damage and failure. Before reaching failure, β varies from 0 to 1. After that, two failure mechanisms will occur: a stable response for high confining stress and an unstable response for low confining stress. The latter corresponds to β decreasing from 1 to a residual value. To distinguish β in the different stages, the value of $(2-\beta)$ is drawn to highlight this type of failure in the plots. As discussed before, extensive fracture occurs along with this type of failure.

(b) *tangential stress distribution*. The ratio (σ_θ/σ_c) is plotted, where σ_c is the uniaxial compressive strength.

The numerical results are presented in Figs.4 and 5. Fig.4a and Fig.5a show the hole deformation characteristics (twice diametral closure) together with experimental data for both rocks. It is clear that the prediction shows similar trends to the experimental results. Because of material anisotropy in Berea sandstone, the deformation along the internal wall is not uniform, as indicated in Fig.5a, and a preferred orientation of failure has been observed; this is evident from Fig.5c. The Indiana limestone has less pronounced strength anisotropy and failure is not as strongly directional as in Berea sandstone, as shown in Fig.4c. The distribution of tangential stress in Fig.4b and Fig.5b is also influenced by the material anisotropy. A maximum tangential stress zone with a value of about four times the uniaxial compressive strength has been predicted, shifted away from the internal wall. In general, the predicted hole deformation

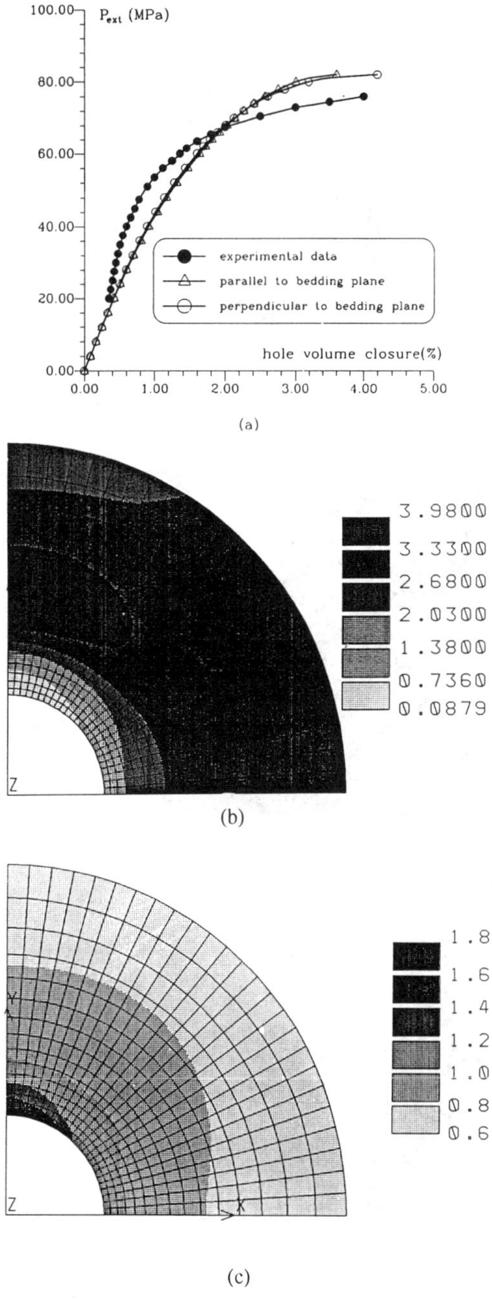

Fig.4 Hole deformation characteristics (a), tangential stress distribution (σ_θ/σ_c) (b) and damage and failure distribution β (c) for Indiana limestone

characteristics and the failure zone with extensive fractures are similar to the experimental observations for both rocks.

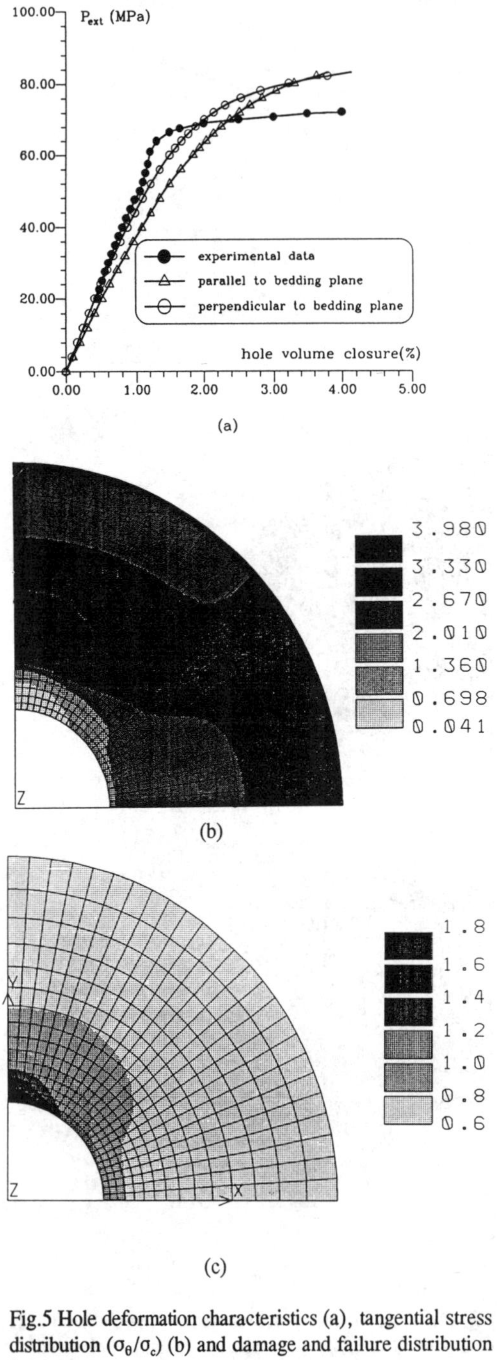

Fig.5 Hole deformation characteristics (a), tangential stress distribution (σ_θ/σ_c) (b) and damage and failure distribution β (c) for Indiana limestone

4. FINAL REMARKS

The proposed model captures the most important behaviour trends: initial anisotropy, damage evolution, sensitivity of material characteristics to confining stress, a compaction-dilatancy transition and a brittle-ductile transition. Numerical simulations of cylindrical openings for Indiana limestone and Berea sandstone were carried out. Even under an isotropic stress state, the proposed model can predict the preferred orientation of a failure zone perpendicular to the bedding plane. The maximum tangential stress is progressively displaced away from the internal wall and its value is about three to four times the uniaxial compressive strength. The influence of pore pressure on deformation, damage and failure around underground openings is under further investigation.

REFERENCES

Addis, M.A. & Wu, B. 1993. The role of the intermediate principal stress in wellbore stability studies: evidence from hollow cylinder tests. *Int. J. Rock Mech. Min. Sci. & Geomech. Abstr.*. **30**(7): 1027-1030.

Amadei, B. & Robison, M.J. 1986. Strength of rock in multiaxial loading conditions. *Proc. 27th U.S. Symposium on Rock Mechanics: Key to Energy Production.* 47-55.

Ewy, R.T. 1991. 3-D stress effects in elastoplastic wellbore failure models. *Proc. 32nd U.S. Symp.: Rock Mechanics as a Multidisciplinary Science.* 951-960.

Ewy, R.T. & Cook, N.G.W. 1990. Deformation and fracture around cylindrical openings in rock - I. Observations and analysis of deformations, II. Initiation, growth and interaction of fractures. *Int. J. Rock Mech. Min. Sci. & Geomech. Abstr.* **27**(5): 387-427.

Pan, X.D.& Hudson, J.A. 1988. A simplified three dimensional Hoek and Brown yield criterion. in: *Rock Mechanics and Power Plants*, ISRM, Spain. 95-103.

Pastor, M. 1990. Modelling of anisotropic sand behaviour.*Computers and Geotechnics***11**:173-208.

Pietruszczak, S. & Jiang, J. 1988. Convexity of yield loci for pressure sensitive materials. *Computers and Geotechnics*. **5**:51-63.

Pietruszczak, S., Jiang, J. & Mirza, F.A. 1988. An elastoplastic constitutive model for concrete. *Int. J. Solids Structures*. **24**(7):705-722.

Computer Methods and Advances in Geomechanics, Yuan (ed.) © 1997 Balkema, Rotterdam, ISBN 90 5410 904 1

DSC-based constitutive model for sand-geosynthetic interfaces

Surajit Pal & G.Wije Wathugala
Department of Civil and Environmental Engineering, Louisiana State University, Baton Rouge, La., USA

ABSTRACT: An elasto-plastic constitutive model based on the disturbed state concept (DSC) for geosynthetic-soil interfaces have been described. The proposed model is capable of capturing most of the important charteristics of interface response, such as dilation, hardening and softening. The model has been used to predict the behavior of geogrid-sand interfaces and the predictions are found to be very satisfactory. The proposed model can be used for the stress-deformation study of geosynthetic reinforced embankments through numerical simulation.

1 INTRODUCTION

Interfaces between two dissimilar materials are very common in many geotechnical engineering problems. One such problem is reinforced earth embankments where interfaces exist between reinforcement materials and adjoining soils. Geosynthetic materials, such as geogrids and geotextiles, are used extensively as reinforcements in earth embankments. To study the stress-deformation behavior of such structures through numerical simulation, it is very important to model the interfaces accurately. The present trend of modelling the interfaces between geosynthetic reinforcement and soil is to use linear or nonlinear elastic models where the onset of sliding is defined using a limiting shear stress criterion such as Mohr-Coulomb. The shortcomings of these models are that they describe the shear behavior of the interface in an approximate sense, and do not consider the coupling of normal and shear behavior. The models are incapable of capturing restrained dilation which is one of the governing factors for pull-out load capacity of an reinforcement. In this paper, an elasto-plastic constitutive model based on the disturbed state concept (DSC) for the geosynthetic-sand interface has been described. The main feature of this model is its ability to capture strain softening in addition to strain hardening and dilation.

2 DISTURBED STATE CONCEPT

The disturbed state concept is based on the idea that the response of a material can be related to and expressed as the responses of the reference states (Desai 1987). As a result, the observed behavior of a material is thus treated as a disturbance to the behavior of the reference states. This disturbance takes place in the process of certain physical changes: alteration in density and structural changes. The latter is analogous to the concept of damage caused by micro-cracking and fracture. Nonassociativeness, anisotropy and strain softening can be expressed as disturbances with respect to the reference states. In the DSC approach the observed material behavior is considered to be composed of two reference states called *intact state* and *disturbed state.* The materials in the intact state behave as a continuously hardening materials without disturbance. The disturbed state may be assumed to be an invariant state during the stress deformation process such as the ultimate, failure, and zero shear stress and/or volume change condition. The proposed model can be considered as a specialized version of the *damage* based model for granular materials proposed by Wathugala and Desai (1989). The constitutive model based on the DSC can allow proper modeling of shear transfer, volumetric behavior and localized slip in the interface zone.

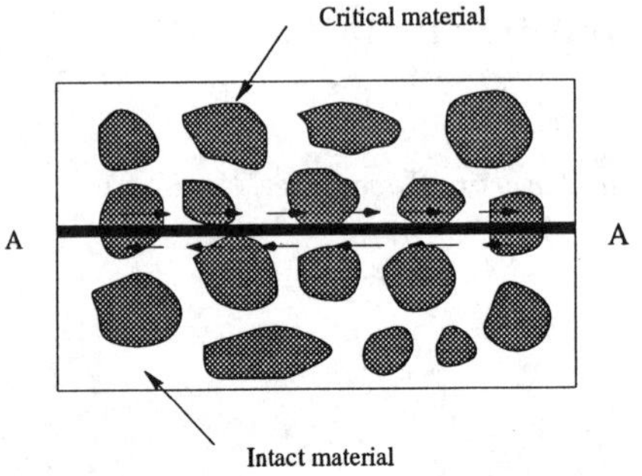

Figure 1: Disturbed state as a mixture of intact and critical states

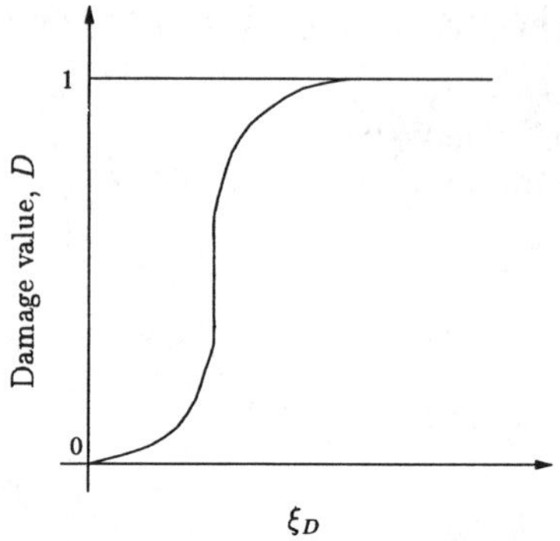

Figure 2: Evolution of damage function

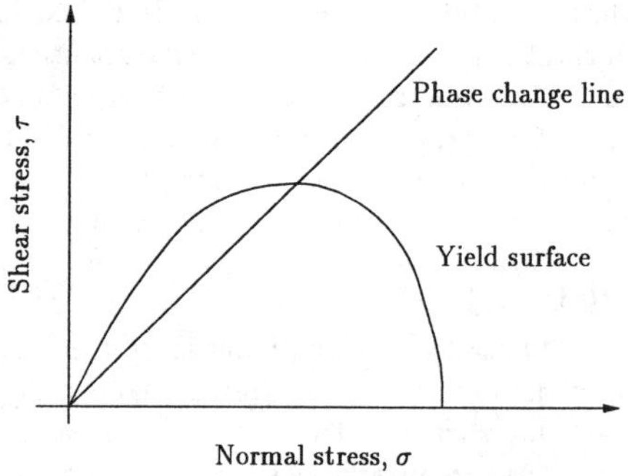

Figure 3: Yield surface for intact material

With the onset of the deformation process in a soil mass, the affected soil particles change their structural arrangement which initiates the disturbed state. Soil particles affected in such a manner are assumed to be in a critical state. The remaining part of the material is assumed to be in the intact state, as shown in Figure 1. The disturbance in the soil mass is defined as

$$D = \frac{M^c}{M} \tag{1}$$

where M is the mass of solids in the material and M^c is the mass of solids in the critical state, and D represents the extent of disturbance in the material. Initially, $D = 0$ and it can attain a maximum value of 1, which represents a fully disturbed material. Desai et al. (1986) have proposed the following expression to describe the disturbance

$$D = D_u[1 - exp(-A\xi_D^k)] \tag{2}$$

where D_u is the ultimate or critical value of D, A and k are material parameters and ξ_D is the trajectory of the deviatoric plastic strain. A schematic diagram of the disturbance function is shown in Figure 2.

3 PROPOSED MODEL

In the proposed model, the intact behavior is defined by an elasto- plastic volumetric hardening associative constitutive model. The yield surface of this model is a specialized two dimensional form (Desai 1991) of the *HiSS* δ_0^* model (Wathugala & Desai 1993) for clays which is defined as:

$$F \equiv \left(\frac{\tau}{P_a}\right)^2 + \alpha\left(\frac{\sigma}{P_a}\right)^{n_1} - \gamma\left(\frac{\sigma}{P_a}\right)^{n_2} = 0 \tag{3}$$

where, τ = shear stress
σ = normal stress
P_a = atmospheric pressure
α = hardening function
n_1, n_2, γ = material parameters

The hardening function, α is definrd as:

$$\alpha = \frac{a_1}{\xi_v{}^{\eta_1}} \tag{4}$$

where, ξ_v = trajectory of volumetric plastic strain
a_1, η_1 = material parameters

In the *HiSS* δ_0^* model, n_2 is taken as 2. In this model, the material parameter n_1 is called the phase change parameter which governs the point of transition from compressive to dilative volume change behavior. Since this is a volumetric hardening model, the material fails when it reaches the phase change line, i.e., the shear stiffness of the

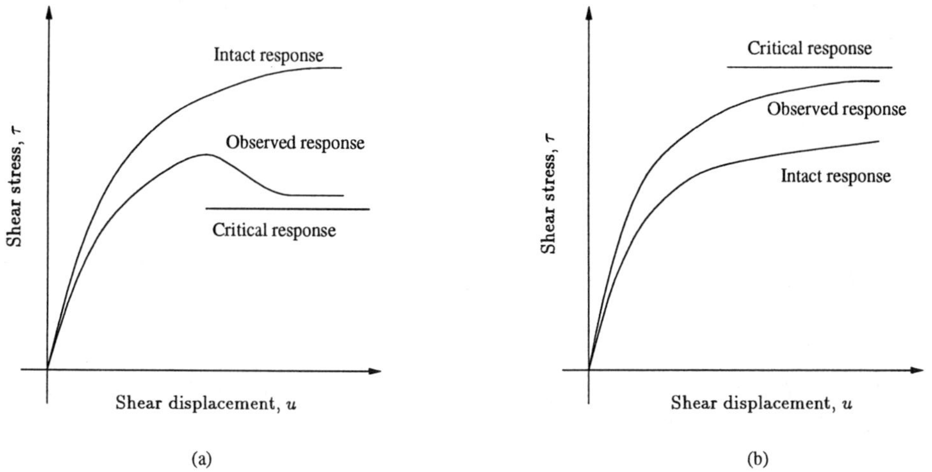

Figure 4: Schematic representation of observed response: (a) strain softening, (b) strain hardening

material becomes zero. A schematic figure of the yield surface is shown in Figure 3. The materials at the disturbed state in the proposed model are assumed to behave similar to those in a critical state at which no change in shear stress or volume occurs for a given normal stress. The shear stress and normal displacement at the critical state are related with the normal stress by Equations 5 and 6.

$$\tau^c = M\sigma \tag{5}$$

where, M = slope of the critical state line

$$v^c = a + b\ ln\left(\frac{\sigma}{P_a}\right) \tag{6}$$

where, a, b = material parameters

To describe the average or observed behavior of interfaces, it is assumed that the normal stress and shear displacement of the three phases are same. The observed or average shear stress and normal displacement are related to the behavior of the disturbed and the intact states through the damge function, D. Now, the average or observed behavior of an interface can be obtained as:

$$\sigma^{average} = \sigma^{intact} = \sigma^{disturbed} \tag{7}$$

$$u^{average} = u^{intact} = u^{disturbed} \tag{8}$$

$$\tau^{average} = (1 - D_s)\tau^{intact} + D_s\tau^{disturbed} \tag{9}$$

Table 1: Material constants for interface

Elastic constants	K_n	550 kPa/mm
	K_s	88.83 kPa/mm
Intact state	γ	95.7
	n_1	1.4
	n_2	1.395
	a_1	96.9324
	η_1	0.01965
Critcal state	M	0.456
	a	-0.2635
	b	0.0444
Disturbed state	k_s	2.2935
	A_s	0.0896
	k_v	1.17
	A_v	1.1158

$$v^{average} = (1 - D_v)v^{intact} + D_v v^{disturbed} \tag{10}$$

Where D_s and D_v are disturbances for stresses and displacements respectively and can be obtained from the following relationships:

$$D_s = D_u[1 - exp(-A_s\xi_D^{k_s})] \tag{11}$$

$$D_v = D_u[1 - exp(-A_v\xi_D^{k_v})] \tag{12}$$

where A_s, k_s, A_v and k_v are material parameters. A schematic representation of the stress-strain responses of intact, disturbed and average states are shown in Figure 4

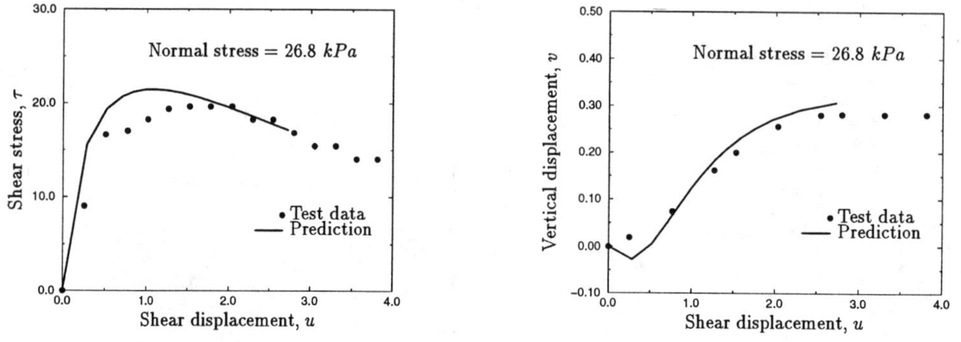

(a) (b)

Figure 5: Test data and prediction of direct shear test at normal stress of 26.8 kPa: (a) Shear stress vs. shear displacement, (b) Normal displacement vs. shear displacement

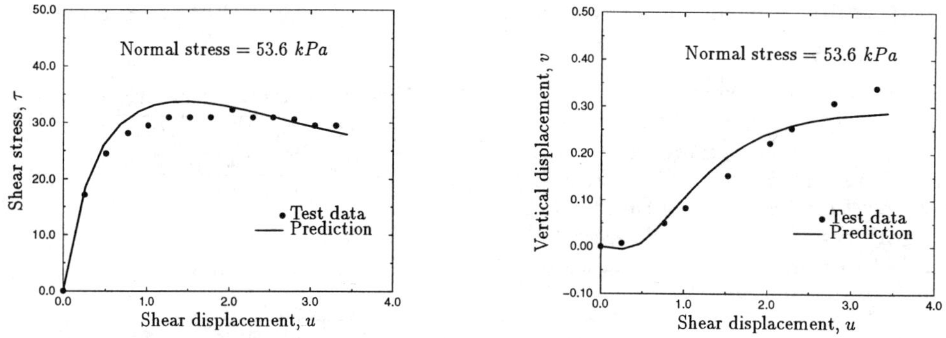

(a) (b)

Figure 6: Test data and prediction of direct shear test at normal stress of 53.6 kPa: (a) Shear stress vs. shear displacement, (b) Normal displacement vs. shear displacement

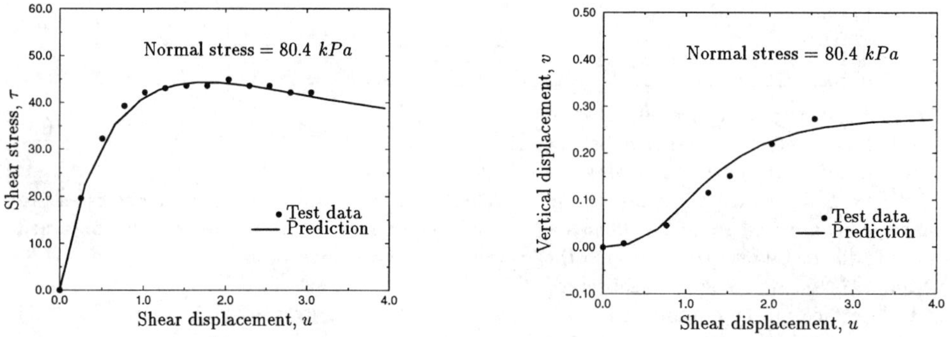

(a) (b)

Figure 7: Test data and prediction of direct shear test at normal stress of 80.4 kPa: (a) Shear stress vs. shear displacement, (b) Normal displacement vs. shear displacement

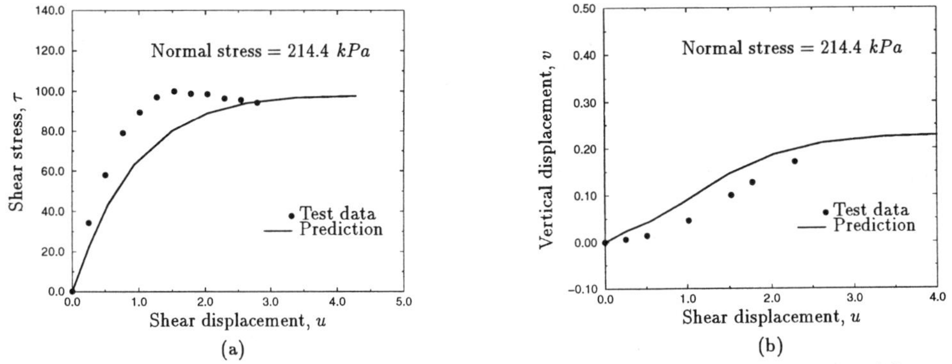

Figure 8: Test data and prediction of direct shear test at normal stress of 214.4 *kPa*: (a) Shear stress vs. shear displacement, (b) Normal displacement vs. shear displacement

4 MODEL VERIFICATION

The proposed model has been verified by predicting the behavior of the interface between sand and geogrid (geosynthetic reinforcement). In the pull-out condition it is found that the slip occurs in a plane between sand particles in the vicinity of geogrids instead of along the physical interface between geogrid and sand (Johnston & Romstad 1989, Karpurapu & Bathurst 1995). In view of this observation, the direct shear tests have been performed on sand only instead of on sand-geogrid interface. Though it has been observed that the peak shear strength of interfaces between well compacted sand and geogrids is 0–10% higher than that of the soil alone (Farrag 1990), they are assumed to be same in the present study. Four direct shear tests have been performed at normal stresses of 26.8 *kPa*, 53.6 *kPa*, 80.4 *kPa* and 214.4 *kPa*, and the results obtained from these tests have been used to calibrate the proposed model. The material parameters obtained from these tests are shown in Table 1.

Figures 5, 6, 7 and 8 show the plots of shear stress vs. shear displacement and normal displacement vs. shear displacement for test data as well as for predictions. It can be observed in these figures that the predictions for the tests at normal stresses of 26.8 *kPa*, 53.6 *kPa* and 80.4 *kPa* agree very well with the test data. However, the predictions for the test at normal stress of 214.4 *kPa* are slightly off from the test data. Considering the fact that the same material parameters are used to predict the tests at normal stresses in the wide range of 26.8 *kPa* to 214.4 *kPa*, the performance of the proposed model can be considered satisfactory.

5 CONCLUSIONS

An elastoplastic constitutive model based on the disturbed state concept is presented for sand-geosynthetic interfaces. The model is capable of characterizing dilation, hardening and softening responses of interfaces. The proposed model is verified with a series of direct shear test data and the model predictions are found to be highly satisfactory. The proposed model can be used in the numerical simulation of large-scale reinforced soil walls. Such simulation is verv useful in understanding stress-deformation behavior of reinforced soil walls.

6 ACKNOWLEDGMENT

Partial financial support from Louisiana Transportation Research Center and Louisiana Board of Regents (Grant Nos. LEQSF(1994-97)-RD-A-08 and LEQSF(1994-96) -ENH-TR-28) are greatfully acknowledged.

REFERENCES

Desai, C.S. 1987. 'Further on unified hierarchical models based on alternative correction or damage approach', *Report*, Dept. of Civil Eng. and Eng. Mech., The University of Arizona, Tucson, Arizona.

Desai, C.S. & Fishman, K.L. 1991. 'Plasticity-based constitutive model with associated testing for joints', *International Journal of Rock Mechanics and Mining Sciences*, Vol. 28, No. 1, 15-26.

Desai, C.S., Somasundaram S. & Frantziskonis, G. 1986. 'A hierarchical approach for constitutive modelling of geologic materials', *Int. J. Num. Analyt. Meth. Geomechanics*, Vol. 10, 225-257.

Farrag, K.A. 1990. 'Interaction properties of geogrids in reinforced-soil walls, testing and analysis', *Ph.D. Dissertation*, Louisiana State University, Louisiana.

Johnston, R.S. & Romstad, K.M. 1989. 'Dilation and boundary effects in large scale pull-out tests', *Proc. XIII Int. Conf. Soil Mech. and Found. Engrg*, Rio de Janeiro, 1263-1266.

Karpurapu, R. & Bathurst, R.J. 1995. 'Behavior of geosynthetic reinforced soil retaining walls using the finite element method', *Computers and Geotechnics*, Vol. 17, 279-299.

Wathugala, G.W. & Desai, C.S. 1987. 'Damage based constitutive model for soils', *Proc. 12th Canadian Congress of Appl. Mech.*, Ottawa, Canada.

Wathugala, G.W. & Desai, C.S. 1994. 'Constitutive model for cyclic behavior of clay I: Theory', *Journal of Geotechnical Engineering, ASCE*, Vol. 119, No. 4, pp. 714-729.

Computer Methods and Advances in Geomechanics, Yuan (ed.) © 1997 Balkema, Rotterdam, ISBN 90 5410 904 1

Decomposition of stress increment and the general stress-strain relation of soils

Liu Yuanxue & Zheng Yingren
Department of Architectural Engineering, Logistical University, Chongqing, China

ABSTRACT: In the light of matrix theory, the character of stress increment which cause the rotation of principal stress axes is analysed and the general stress increment is decomposed into two part: coaxial part and rotational part. Based on these, the complex three dimensions (3—D) problem involving the rotation of principal stress axes is simplified to the combination of the 3—D coaxial model and the theory about pure rotation of principal stress axes that is only around one principal stress axe. The computative method in detail of general 3—D problem is provided and other applications are presented also.

1 INTRODUCTION

The study of the constitutive relation of soils is mostly limited in such stress state that the rotation of principal stress axes is able to be ignored. So only the value of principal stress need to be considered and the direction of principal stress is regarded to be unchangeable. On such condition, the principal axes of stress increment, strain increment, stress and strain are the same. Some experiments and engineerings show that the rotation of principal stress axes will generate significant plastic deformation and the noncoaxality of stress and strain.

In the first, it is necessary to analysis why the rotation of soils principal stress axes will cause the noncoaxality of stress and strain. There are problems of the rotation of principal stress axes in elasticity, but it never bring about such result. It is because elastic strain increment is proportion to the stress increment and their directions are the same. So the rotation of stress and strain are synchronous. But the deformation of soils is influenced by stress histry. The value and direction of strain increment are determined not only by the stress increment, but also by stress. So the hysteresis exist during the strain increment rotates with the stress increment and the noncoaxality of stress and strain appear.

The study about the coaxial stress strain relation of soils is abundant. The study of the pure principal axe rotation is relatively less and it mostly concentrated on the two dimensional (2—D) problem. It is necessary to synthesize the 2—D study to the general state and popularize it to 3—D problem.

2 DECOMPOSITION OF STRESS INCREMENT

The variation of principal stress value and the rotation of the principal stress axes are the results of actions of two kinds increment. So it is necessary to study the characters of these stress increments.

2.1 *The decomposition of 2—D stress increment*

Let the principal stress values are σ_1, σ_2, the corresponding principal directions are N_1, N_2, then:

$$\sigma=(N_1 \quad N_2)\begin{pmatrix}\sigma_1 & 0\\ 0 & \sigma_2\end{pmatrix}\begin{pmatrix}N_1\\ N_2\end{pmatrix}=T_1 \Lambda T_1^T \tag{1}$$

Without Loss the general case, let the angle between N_1 and X-axe are θ.

We yield:

$$T_1=\begin{pmatrix}\cos\theta & -\sin\theta\\ \sin\theta & \cos\theta\end{pmatrix} \qquad T_1^T=\begin{pmatrix}\cos\theta & \sin\theta\\ -\sin\theta & \cos\theta\end{pmatrix} \tag{2}$$

1. The character of coaxial stress increment $d\sigma_c$ (Its principal directions are unchangeable, but the principal values vary)

It means that T_1, T_1^T are constant matrixes, but

σ_1, σ_2 change in equation(1).

$$d\sigma_c = d(T_1 \Lambda T_1^T) = T_1 (d\Lambda) T_1^T = T_1 \begin{pmatrix} d\sigma_1 & 0 \\ 0 & d\sigma_2 \end{pmatrix} T_1^T \tag{3}$$

It shows that the properties of $d\sigma_c$ are: at least one diagonal element is not zero, but that of subdiagonal are zero.

2. The character of rotational stress increment $d\sigma_r$

It means that the diagonal matix Λ in Eq. (1) is a constant matix, but T_1, T_1^T are changeable.

$$d\sigma_r = d(T_1 \Lambda T_1^T) = dT_1 \Lambda T_1^T + T_1 \Lambda dT_1^T \tag{4}$$

Differentiating equation (2):

$$dT_1 = \begin{pmatrix} -\sin\theta & -\cos\theta \\ \cos\theta & -\sin\theta \end{pmatrix} d\theta$$

$$dT_1^T = (dT_1)^T = \begin{bmatrix} -\sin\theta & \cos\theta \\ -\cos\theta & -\sin\theta \end{bmatrix} d\theta \tag{5}$$

Combining Eq. (2), (3), (5), we have:

$$T_1^T dT_1 = \begin{pmatrix} 0 & -1 \\ 1 & 0 \end{pmatrix} d\theta \quad dT_1^T T_1 = \begin{pmatrix} 0 & 1 \\ -1 & 0 \end{pmatrix} d\theta \tag{6}$$

$$\begin{aligned} d\sigma_r^T &= (dT_1 \Lambda T_1^T) + (T_1 \Lambda dT_1^T)^T \\ &= T_1 \Lambda dT_1^T + dT_1 \Lambda T_1^T = d\sigma_r \end{aligned} \tag{7}$$

So $d\sigma_r$ is a symmetrical tensor.

$d\sigma_r$ is transformed to the principal stress space:

$$\begin{aligned} T_1^T d\sigma_r T_1 &= T_1^T (dT_1 \Lambda T_1^T + T_1 \Lambda dT_1^T) T_1 \\ &= (T_1^T dT_1) \Lambda (T_1^T T_1) + (T_1^T T_1) \Lambda (dT_1^T T_1) \\ &= \begin{pmatrix} 0 & -1 \\ 1 & 0 \end{pmatrix} d\theta \begin{pmatrix} \sigma_1 & 0 \\ 0 & \sigma_2 \end{pmatrix} I \\ &\quad + I \begin{pmatrix} \sigma_1 & 0 \\ 0 & \sigma_2 \end{pmatrix} \begin{pmatrix} 0 & 1 \\ -1 & 0 \end{pmatrix} d\theta \\ &= d\theta \left[\begin{pmatrix} 0 & -\sigma_2 \\ \sigma_1 & 0 \end{pmatrix} + \begin{pmatrix} 0 & \sigma_1 \\ -\sigma_2 & 0 \end{pmatrix} \right] \\ &= \begin{pmatrix} 0 & d\theta(\sigma_1 - \sigma_2) \\ d\theta(\sigma_1 - \sigma_2) & 0 \end{pmatrix} \end{aligned} \tag{8}$$

Where T_1 is an orthogonal matrix:

$$T_1 T_1^T = T_1^T T_1 = I \tag{9}$$

$$d\sigma_r = T_1 \begin{pmatrix} 0 & d\theta(\sigma_1 - \sigma_2) \\ d\theta(\sigma_1 - \sigma_2) & 0 \end{pmatrix} T_1^T \tag{10}$$

From above equations, we can induce that the characters of rotational stress increment: the diagonal elements are zero, but the subdiagonal elements are not zero and equal; the rotational angle is the subdiagonal element divided by the difference of the two principal stress value when $d\sigma_r$ is expresed in the principal space.

3. Decomposition of stress increment

Arbitrary stress increment $d\sigma$ can be decomposited according to the character of $d\sigma_c$ and $d\sigma_r$.

First of all, arbitrary stress increment $d\sigma$ is transformed to the principal space.

$$\begin{aligned} B &= T_1^T d\sigma T_1 \\ B^T &= (T_1^T d\sigma T_1)^T \\ &= T_1^T d\sigma^T (T_1^T)^T = T_1^T d\sigma T_1 = B \end{aligned} \tag{11}$$

So B is a symmetric tensor and assumed that it has the following form.

$$B = \begin{pmatrix} K_1 & K_2 \\ K_2 & K_3 \end{pmatrix} = \begin{pmatrix} K_1 & 0 \\ 0 & K_3 \end{pmatrix} + \begin{pmatrix} 0 & K_2 \\ K_2 & 0 \end{pmatrix} \tag{12}$$

$$\begin{aligned} d\sigma &= T_1 B T_1^T \\ &= T_1 \begin{pmatrix} K_1 & 0 \\ 0 & K_3 \end{pmatrix} T_1^T + T_1 \begin{pmatrix} 0 & K_2 \\ K_2 & 0 \end{pmatrix} T_1^T \\ &= d\sigma_c + d\sigma_r \end{aligned} \tag{13}$$

According to equation (3) and (10), obtain:

$$d\sigma_1 = K_1 \quad d\sigma_2 = K_3 \quad d\theta = K_2/(\sigma_1 - \sigma_2) \tag{14}$$

So arbitrary stress increment can be decomposed to two parts (coaxial part $d\sigma_c$ and rotational part $d\sigma_r$). This is feasible and unique.

2.2 *Decomposition of 3—D stress increment*

The 3—D stress increment can be decomposed with the same principle as that of 2—D.

Let the value of three principal stresses are σ_1、σ_2、σ_3 respectively and the corresponding principal direction are N_1, N_2 *and* N_3.

$$\begin{aligned} \sigma &= (N_1 \quad N_2 \quad N_3) \begin{bmatrix} \sigma_1 & 0 & 0 \\ 0 & \sigma_2 & 0 \\ 0 & 0 & \sigma_3 \end{bmatrix} \begin{bmatrix} N_1 \\ N_2 \\ N_3 \end{bmatrix} \\ &= T \Lambda_1 T^T \end{aligned} \tag{15}$$

Arbitrary 3—D stress increment $d\sigma$ should have the following form in the principal stress space:

$$T^T d\sigma T=\begin{bmatrix} M_1 & A_1 & C_1 \\ A_1 & M_2 & B_1 \\ C_1 & B_1 & M_3 \end{bmatrix} \quad (16)$$

From the equations above, the coaxial part and the rotational parts can be obtained:

$$d\sigma_c=T\begin{bmatrix} M_1 & 0 & 0 \\ 0 & M_2 & 0 \\ 0 & 0 & M_3 \end{bmatrix}T^T \quad (17)$$

$$d\sigma_1=M_1 \quad d\sigma_2=M_2 \quad d\sigma_3=M_3 \quad (18)$$

$$d\sigma_{r1}=T\begin{bmatrix} 0 & A_1 & 0 \\ A_1 & 0 & 0 \\ 0 & 0 & 0 \end{bmatrix}T^T \quad (19)$$

Where $d\sigma_{r1}$ is the rotation stress increment around the third principal stress axe with the rotational angle $d\theta_1$:

$$d\theta_1=A_1/(\sigma_1-\sigma_2) \quad (20)$$

$$d\sigma_{r2}=T\begin{bmatrix} 0 & 0 & 0 \\ 0 & 0 & B_1 \\ 0 & B_1 & 0 \end{bmatrix}T^T \quad (21)$$

Where $d\sigma_{r2}$ is the rotation stress increment around the first principal stress axe with the rotational angle $d\theta_2$:

$$d\theta_2=B_1/(\sigma_2-\sigma_3) \quad (22)$$

$$d\sigma_{r3}=T\begin{bmatrix} 0 & 0 & C_1 \\ 0 & 0 & 0 \\ C_1 & 0 & 0 \end{bmatrix}T^T \quad (23)$$

Where $d\sigma_{r3}$ expresses the rotation stress increment around the second principal axe with the rotational angle $d\theta_3$:

$$d\theta_3=C_1/(\sigma_1-\sigma_3) \quad (24)$$

From equations (17) to (24), arbitrary 3—D stress increment can be decomposed to the coaxial part and rotational parts that each one only involves the rotation around one principal axe. The general 3—D problem is simplified to the combination of the 3—D coaxial model and the pure princial axe rotation theory around one principal axe. The further analysis is reduced significantly.

3 THE GENERAL MODEL OF THE STRESS—STRAIN RELATION OF SOILS

In the above section, arbitrary 3—D stress increment has been decomposed into the coaxial part and the rotational parts. According to elastoplastic theory, the strain increment can be decomposed into elastic part and plastic part and the elastoplastic coupling be ignored. The elastic strain increment can be calculated by the generalized Hook's law. The plastic part is the total of coaxial part and rotational parts which can be calculated by the proper model.

3.1 *3—D General case*

According to the decomposed results of equations (24) to (32), there are three kinds of strain increment (elastic part, coaxial part and rotational parts of plastic deformation) need to be computed in the 3—D general case.

3.1.1 *The elastic strain increment*

It is assumed that elastic deformation obeys the generalized Hook's Law. Elastic module and Poisson's ratio of soils are E, μ. Thus:

$$d\varepsilon^e=[C_e]d\sigma=\frac{1}{E}\begin{bmatrix} 1 & -\mu & -\mu \\ -\mu & 1 & -\mu \\ -\mu & -\mu & 1 \end{bmatrix}d\sigma \quad (25)$$

3.1.2 *The coaxial part of plastic strain increment*

The 3—D models which reflect the basic mechanics character of soils better are the multi-yielding model of soils. A three multi—yielding model whose parameters can be determined through experiments of simple stress paths is presented in reference [6]. Its results of simulation coincide with the experimental results of many stress paths. This model can be used conveniently because it is expressed in the principal stress space.

$$\varepsilon_i^p=A_i\sigma_1^2+B_i\sigma_2^2+C_i\sigma_3^2+D_i\sigma_1\sigma_2+E_i\sigma_2\sigma_3+F_i\sigma_1\sigma_3 +R_i\sigma_1+S_i\sigma_2+T_i\sigma_3+P_i \quad (26)$$

Where ε_i^p is the principal value of plastic strain.

These parameters (A_i, B_i, C_i, D_i, E_i, F_i, R_i, S_i, T_i, P_i) can be fitted from the results of the following simple experiments:

1. unaxial compresion test ($\sigma_2=\sigma_3=0$, $\sigma_1\uparrow$)
2. extension test ($\sigma_3=0$, $\sigma_1=\sigma_2\uparrow$)
3. hydrostatic compression test ($\sigma_1=\sigma_2=\sigma_3\uparrow$)

The differentiating form of equation (26) can be as:

$$d\varepsilon_c^p=[C_p]d\sigma_c \tag{27}$$

$$\begin{Bmatrix} d\varepsilon_{c1}^p \\ d\varepsilon_{c2}^p \\ d\varepsilon_{c3}^p \end{Bmatrix} = \begin{bmatrix} 2A_1\sigma_1+D_1\sigma_2+F_1\sigma_3+R_1 & 2B_1\sigma_2+D_1\sigma_1+E_1\sigma_3+S_1 & 2C_1\sigma_3+E_1\sigma_2+F_1\sigma_1+R_1 \\ 2A_2\sigma_1+D_2\sigma_2+F_2\sigma_3+R_2 & 2B_2\sigma_2+D_2\sigma_1+E_2\sigma_3+S_2 & 2C_2\sigma_3+E_2\sigma_2+F_2\sigma_1+R_2 \\ 2A_3\sigma_1+D_3\sigma_2+F_3\sigma_3+R_3 & 2B_3\sigma_2+D_3\sigma_1+E_3\sigma_3+S_3 & 2C_3\sigma_3+E_3\sigma_2+F_3\sigma_1+R_3 \end{bmatrix} \begin{Bmatrix} M_1 \\ M_2 \\ M_3 \end{Bmatrix} \tag{28}$$

In the general space, the strain increment will be:

$$d\varepsilon_c^p=T\Lambda_c T^T=T\begin{bmatrix} d\varepsilon_{c1}^p & 0 & 0 \\ 0 & d\varepsilon_{c2}^p & 0 \\ 0 & 0 & d\varepsilon_{c3}^p \end{bmatrix}T^T \tag{29}$$

3.1.3 *The plastic strain increment caused by the rotational stress increment*

According to the conclusion of decomposition, only soils deformation properties of pure principal stress axe rotation around one principal axe is need to be studied, then it can be extended to the general case. A model in reference [7], based on the cross influence of hydrostatic stress and deviator stress, can simulate the deformation of pure principal axe rotation of soils better.

This model is given in the following form:

$$d\varepsilon^p=(\frac{1}{3}AI+B\bar{n})F(\sigma_m+\bar{\sigma}_m)d\sigma_m + (\frac{1}{3}CI+Dn)\sigma_m n^a dr \tag{30}$$

In the condition of pure principal stress axe rotation, we yield:

$$d\sigma_m=0 \quad dr=d\sigma/\sigma_m \tag{31}$$

$$d\varepsilon^p=(\frac{1}{3}CI+Dn)\sigma_m n^a dr \tag{32}$$

$$\begin{cases} \sigma_m=(\sigma_1+\sigma_2+\sigma_3)/3 \quad r=\sigma/\sigma_m-1 \\ \bar{n}=r/(rr)^{0.5} \qquad \hat{n}=dr/(drdr)^{0.5} \\ n=\bar{n}/3+2\hat{n}/3 \\ a=\cos^{-1}\left(\dfrac{\Sigma(r_{ij}^{n+1}-r_{ij}^{n})/(r_{ij}^{n}-r_{ij}^{n-1})}{[\Sigma(r_{ij}^{n+1}-r_{ij}^{n})^2\Sigma(r_{ij}^{n}-r_{ij}^{n-1})^2]^{0.5}}\right) \\ n_{ij}^a=\cos(a/3) \end{cases} \tag{33}$$

Where C、D are the plastic coefficient.

The plastic deformation caused by the three rotational stress increments is computed seperately:

1. The plastic deformation caused by $d\sigma_{r1}$

$$dr_1=d\sigma_{r1}/\sigma_m = \frac{A_1}{\sigma_m}T\begin{bmatrix} 0 & 1 & 0 \\ 1 & 0 & 0 \\ 0 & 0 & 0 \end{bmatrix}T^T = \frac{d\theta_1(\sigma_1-\sigma_2)}{\sigma_m}TE_1T^T \tag{34}$$

Substituting Eq. (34) into Eq. (32), we have:

$$d\varepsilon_{r1}^p=(CI/3+Dn)\sigma_m n^a d\theta_1(\sigma_1-\sigma_2)(TE_1T^T)/\sigma_m = d\theta_1(\sigma_1-\sigma_2)(CI/3+Dn)n^a(TE_1T^T) \tag{35}$$

2. The plastic deformation caused by the rotational stress increments $d\sigma_{r2}$ and $d\sigma_{r3}$ are calculated in the same way as that of $d\varepsilon_{r1}^p$

$$dr_2=d\sigma_{r2}/\sigma_m=\frac{d\theta_2(\sigma_2-\sigma_3)}{\sigma_m}TE_2T^T \tag{36}$$

$$d\varepsilon_{r2}^p=d\theta_2(\sigma_2-\sigma_3)(CI/3+Dn)n^a(TE_2T^T) \tag{37}$$

$$dr_3=d\sigma_{r3}/\sigma_m=\frac{d\theta_3(\sigma_1-\sigma_3)}{\sigma_m}TE_3T^T \tag{38}$$

$$d\varepsilon_{r3}^p=d\theta_3(\sigma_1-\sigma_3)(CI/3+Dn)n^a(TE_3T^T) \tag{39}$$

$$E_2=\begin{bmatrix} 0 & 0 & 0 \\ 0 & 0 & 1 \\ 0 & 1 & 0 \end{bmatrix} \qquad E_3=\begin{bmatrix} 0 & 0 & 1 \\ 0 & 0 & 0 \\ 1 & 0 & 0 \end{bmatrix} \tag{40}$$

3.1.4 *The total strain increment*

According to the decomposed results, the total strain increment will be the sum of elastic part, coaxial plastic part and rotational parts of plastic deformation.

$$d\varepsilon=d\varepsilon^e+d\varepsilon_c^p+d\varepsilon_{r1}^p+d\varepsilon_{r2}^p+d\varepsilon_{r3}^p \tag{41}$$

3.1.5 *Noncoaxality caused by principal stress axe rotation*

The principal axe direction of stress incremrnt, strain increment, strain as well as stress will be different if the rotation of principal stress axes occurs. Our interesting mostly focus on the noncoaxality of stress and strain.

The three generalized principal directions of

strain (N_{1e}, N_{2e} *and* N_{3e}) can be obtained by the same means of Eq. (15). The angle θ_1 between the first principal stress axe and that of strain is:

$$N_{1e} \cdot N_1 = |N_{1e}| \cdot |N_1| \cdot \cos\theta_1 = \cos\theta_1$$
$$\theta_1 = \arccos(N_{1e} \cdot N_1) \quad (42)$$

The angles between ε_2 and $\sigma_2(\theta_2)$, ε_3 and $\sigma_3(\theta_3)$ are:

$$\theta_2 = \arccos(N_{2e} \cdot N_2) \quad (43)$$

$$\theta_3 = \arccos(N_{3e} \cdot N_3) \quad (44)$$

3.2 *The 3—D special problem*

Some problems in the engineering are not so complex as Eq. (41). Only elastic part and coaxial plastic part need to be computed in the problem which does not involve the rotation of principal stress axe. The rotational parts are calculated according to the actual situation in the case of pure principal stress axe rotation.

3.3 *The plane problem*

In certain condition, a 3—D problem can be simplified to a plane problem. The decomposition of 2—D stress increment is expressed in Eq. (13) and Eq. (14). The strain increment can be computed in the same way as that of 3—D. The coaxial model and rotational model are provided in reference [5]、[3].

4 CONCLUSION

Based on matrix theory, the characters of coaxial and rotational stress increments are analysed and the general stress strain relation of soils are studied in a new point of view in this paper. In this paper, a new path to solve the complex 3—D problem involving the rotation of principal stress axes is provided.

1. The characters of the coaxial and ratational stress increment are provided. The general 2—D and 3—D stress increments are decomposed into two parts: coaxial part and rotational part.

2. On the basis of stress increment decomposition, a complex 3—D problem involving the principal stress rotation are simplified to the combination of 3—D coaxial model and the theory about pure rotation of principal stress axes that is only around one principal stress axe. The diffculty of analysis is reduced significantly.

3. In order to let this model to be more general, the study and test of coaxial model and rotational model are necessary.

REFERENCES

Zheng Yingren, Gong Xiaolan 1990. *The Basis of Plastic Mechanics of rock and soils.* Beijing: Press of Chinese Architectural Industry.

Ying Zhongzhe 1988. A Stress-Strain Model of Double Yield Surface of Soils. *Chinese Jounal of Geotechnical Engineering* 10(4): 64—71.

H. Matasuoka et al. 1987. A Constitutive Model for Sands and Clay Evaluating principal stress rotation. *Soils and Foundations*, 27(4): 73—88.

F. Miaru, S. Toki et al. 1986. Deformation Prodiction for Anisotropic sand during the Rotation of principal stress axes. *Soils and Foundations*, 26(3): 42—56.

Sheng Zhujiang 1987. A New Stress—Strain Model of Soils. In: edited by Luo Zhao juan et al. . *The 5th Chinese conference of Soils Mechanics and Foundation Engineering*: 101—105. Beijing: Press of Chinese Architectural Industry.

Zheng Yingren, Yan Dejun 1994. Muti—yield Surface Model for Soils on the Basis of Test Fitting. In: edited by Zheng Yingren et al. . *The 5th Chinese Conference of Numerical and Analysis Method in Geomechanics*: 9—14. Wanhan: Press of Wanhan Science and Technical University of Surveying.

Chen Shengshui, Shen Zhujiang, et al. 1995. A Elastoplastic Model for Cohesionless Soils under Complex Stress Paths. *Chinese Jounal of Geotechnical Engineering*, 17(2): 20—28.

Computer Methods and Advances in Geomechanics, Yuan (ed.) © *1997 Balkema, Rotterdam, ISBN 90 5410 904 1*

An elasto-plastic model for c-ø materials under complex loading

De'an Sun & Hajime Matsuoka
Department of Civil Engineering, Nagoya Institute of Technology, Japan

ABSTRACT: Described herein is an elasto–plastic model with an isotropic and rotational hardening rule for frictional and cohesive materials (briefly known as c–ϕ materials). The model is capable of predicting the stress–strain behavior of c–ϕ materials under complex loading in three–dimensional stresses. It can also simulate in principle the deformation characteristics of c–ϕ materials under the change in cohesion like collapse of unsaturated soil. The triaxial and true triaxial drained tests on cemented soil, which is considered as a kind of c–ϕ materials, under complex loading condition have been carried out. Comparisons of the model prediction with test data are made to demonstrate its capacity. Results of comparisons show the model provides satisfactory prediction.

1 INTRODUCTION

In the study of constitutive model for geological materials, most of current available constitutive models were developed either for frictional materials or for cohesive materials. However, geological materials encountered in geotechnical practice indicate characteristics of both frictional material, such as granular materials without bond between grains, and cohesive material, such as metals with strong bond due to existence of crystalline structure. Therefore, there is the requirement to develop a constitutive model for this "intermediate material", being referred to as c–ϕ material. On the other hand, the c–ϕ materials such as soils and rocks in situ are always subjected to complex loading in three–dimensional stresses. The simple stress conditions such as triaxial compression are seldom encountered in engineering practice. In strict sense of the word, most available constitutive models for soils such as Cam–clay model are only effective for these simple stress conditions. It is necessary to develop the constitutive model which can be used to predict the stress–strain behavior of c–ϕ materials under complex loading. In this paper, an isotropic and rotational hardening elasto–plastic model for c–ϕ material is introduced from an elasto–plastic model for frictional material(Nakai et al. 1989) and the concept of the " Extended Spatially Mobilized Plane(Extended SMP)"(Matsuoka et al. 1995), which is a modification of the original concept of the Spatially Mobilized Plane. The propósed model can simulate in principle what c–ϕ materials behave under constant and variable cohesion in three–dimensional stresses.

The triaxial and true triaxial tests on cemented soils, which are the mixture of cement, sand and clay powder with water, under the complex loading conditions were performed by using ordinary triaxial apparatus and the true triaxial apparatus with three pairs of rigid loading plates(Matsuoka & Sun 1995). The cemented soil is considered as a kind of c–ϕ material, because it has the property of both frictional material and cohesive material. The results of the triaxial and true triaxial tests on cemented soil are analyzed by this model and the analytical results are compared with the measured ones. The overall good agreement between predictions and measurements indicates that the proposed model is able to simulate the deformation characteristics of cemented soils, under various kinds of three–dimensional stress paths. Especially the model is able to simulate the stress paths dependency of the strain increment directions.

2 CONSTITUTIVE MODELLING

An isotropic hardening elasto–plastic model for c–ϕ materials has been presented which can predict the behavior of c−ϕ material under the simple stress paths such as constant θ (=$\tan^{-1}[\sqrt{3}(\sigma_2-\sigma_3)/\{(\sigma_1-\sigma_2)+(\sigma_1-\sigma_3)\}]$)(see Fig.4) in three-dimensional stresses by using the concept of the Extended SMP(Matsuoka & Sun 1994, Sun & Matsuoka 1994). In order to predict the behavior of c−ϕ materials under complex loading, it is necessary to develop an elasto-plastic model with an isotropic and rotational hardening rule. A kinematic hardening model (named as kinematic t_{ij} sand model) has been proposed which can describe the behavior of sands under complex loading(Nakai et al. 1989). The presented model in this paper was developed for c−ϕ materials by following kinematic t_{ij} sand model.

2.1 A modified stress tensor for c−ϕ materials

The modified stress tensor for c−ϕ materials is defined as follows(Sun & Matsuoka 1994):

$$\bar{\sigma}^*_{ij}=\sigma_{ik}\hat{b}_{kj} \tag{1}$$

where $\hat{b}_{ij}$ is a symmetric tensor whose principal values are $\sqrt{3}\hat{a}_1$, $\sqrt{3}\hat{a}_2$ and $\sqrt{3}\hat{a}_3$. $\hat{a}_i$ is the direction cosines of the normal to "Extended Spatially Mobilized Plane(Extended SMP)" which can be expressed as follows:

$$\hat{a}_i=\sqrt{\hat{J}_3/(\hat{\sigma}_i\hat{J}_2)} \quad (i=1,2,3) \tag{2}$$

in which $\hat{J}_2$ and $\hat{J}_3$ are the second and third invariants of a translated stress tensor respectively, and $\hat{\sigma}_i$ is the principal value of the translated stress tensor $\hat{\sigma}_{ij}$ which is defined by

$$\hat{\sigma}_{ij}=\sigma_{ij}+\sigma_0\delta_{ij} \tag{3}$$

where δ_{ij} is Kronecker's delta and σ_0 is a cohesion parameter defined by

$$\sigma_0=c\cdot\cot\phi \tag{4}$$

where c is cohesion and ϕ is internal friction angle.

In order to describe the stress-strain behavior of c−ϕ materials under three-dimensional stress states uniquely, it is used as stress parameters that the components, $\bar{\sigma}_{SMP}^*$ and $\bar{\tau}_{SMP}^*$, of the principal value vector of $\bar{\sigma}_{ij}^*$ are normal and parallel to the Extended SMP.

$$\bar{\sigma}^*_{SMP}=\bar{\sigma}^*_1\hat{a}_1+\bar{\sigma}^*_2\hat{a}_2+\bar{\sigma}^*_3\hat{a}_3=\bar{\sigma}^*_{ij}\hat{a}_{ij} \tag{5}$$

$$\bar{\tau}^*_{SMP}=\sqrt{(\bar{\sigma}^*_1\hat{a}_2-\bar{\sigma}^*_2\hat{a}_1)^2+(\bar{\sigma}^*_2\hat{a}_3-\bar{\sigma}^*_3\hat{a}_2)^2+(\bar{\sigma}^*_3\hat{a}_1-\bar{\sigma}^*_1\hat{a}_3)^2}$$
$$=\sqrt{\bar{\sigma}^*_{ij}\bar{\sigma}^*_{ij}-(\bar{\sigma}^*_{ij}\hat{a}_{ij})^2} \tag{6}$$

where $\hat{a}_{ij}=\hat{b}_{ij}/\sqrt{3}$, and $\bar{\sigma}_i^*$ is the principal value of $\bar{\sigma}_{ij}^*$.

2.2 The division of total strain

The strain increment $d\varepsilon_{ij}$, under general three-dimensional stress condition, is expressed by

$$d\varepsilon_{ij}=d\varepsilon^e_{ij}+d\varepsilon^p_{ij}=d\varepsilon^e_{ij}+d\varepsilon^{p(IC)}_{ij}+d\varepsilon^{p(AF)}_{ij} \tag{7}$$

where $d\varepsilon_{ij}^e$ is the elastic strain increment, $d\varepsilon_{ij}^{p(IC)}$ is the plastic strain increment caused by the increment of mean effective principal stress, and $d\varepsilon_{ij}^{p(AF)}$ is the plastic strain increment which obeys isotropic hardening rule or rotational hardening rule. The elastic strain component is calculated from Hooke's law, the two plastic strain components will be given in the following.

2.3 The isotropic plastic strain component $d\varepsilon_{ij}^{p(IC)}$

It is well known that the behavior of c−ϕ materials is isotropic hardening during the change in mean effective principal stress under the isotropic compression. The plastic volumetric strain increment under isotropic compression is expressed as

$$\varepsilon^p_V=(C_t-C_e)\{(\hat{\sigma}_m/p_a)^m-(\hat{\sigma}_{mo}/p_a)^m\} \tag{8}$$

where $\hat{\sigma}_{mo}$ is the value of $\hat{\sigma}_m$ for $\varepsilon_v^p=0$, $\hat{\sigma}_m$ is the translated mean effective principal stress, C_t, C_e and m are isotropic compression and swelling parameters, and p_a is the atmospheric pressure. From the differentiation of Eq.(8), the plastic strain component $d\varepsilon_{ij}^{p(IC)}$ is given by

$$d\varepsilon^{p(IC)}_{ij}=\frac{d\varepsilon^p_V}{3}\delta_{ij}=K_1(\hat{\sigma}^{m-1}_m\langle d\hat{\sigma}_m\rangle-\hat{\sigma}^{m-1}_{mo}\langle d\hat{\sigma}_{mo}\rangle)\delta_{ij} \tag{9}$$

where

$$K_1=\frac{m(C_t-C_e)}{3p_a^m} \tag{10}$$

and < > is the Macauley symbol, which applies to any scalar variable x

$$\langle x\rangle=x,\ if\ x\geq 0 \quad \langle x\rangle=0,\ if\ x<0 \tag{11}$$

2.4 The approach for determining $d\varepsilon_{ij}^{p(AF)}$

The stress tensor $\hat{X}_{ij}$ and the stress ratio $\hat{X}$ for c−ϕ materials are defined as follows:

$$\hat{X}_{ij}=\frac{\bar{\sigma}^{*\prime}_{ij}}{\bar{\sigma}^{*}_{SMP}+\sqrt{3}\sigma_0} \tag{12}$$

$$\hat{X}=\frac{\bar{\tau}^{*}_{SMP}}{\bar{\sigma}^{*}_{SMP}+\sqrt{3}\sigma_0}=\sqrt{\hat{X}_{ij}\hat{X}_{ij}} \tag{13}$$

where

$$\bar{\sigma}^{*\prime}_{ij}=\bar{\sigma}^{*}_{ij}-\bar{\sigma}^{*}_{SMP}\hat{a}_{ij} \tag{14}$$

In case of metals , it is reasonable to assume that the shear behavior obeys kinematic hardening rule in deviatoric stress space. On the other hand, in case of granular materials, it is considered that the shear behavior obeys kinematic hardening rule in the stress ratio space or rotational hardening rule in the stress space. Considering the hardening rules of metals and granular materials , we assume the following hardening rule for c−ϕ materials.

$$\hat{X}^{*}=\sqrt{(\hat{X}_{ij}-\hat{n}_{ij})(\hat{X}_{ij}-\hat{n}_{ij})}=\xi=const. \tag{15}$$

where ξ is the material parameter, and $\hat{n}_{ij}$ the center of yield surface in the stress ratio space for c−ϕ materials, its increment is determined by

$$d\hat{n}_{ij}=\mu_1(\hat{X}_{ij}-\hat{n}_{ij}) \tag{16}$$

Combining Eq.(15) and Eq.(16) gives

$$d\hat{n}_{ij}=\frac{(\hat{X}_{kl}-\hat{n}_{kl})d\hat{X}_{kl}}{\hat{X}^{*2}}(\hat{X}_{ij}-\hat{n}_{ij}) \tag{17}$$

In accordance with the isotropic hardening model for c−ϕ materials(Matsuoka & Sun 1994), the yield function for c−ϕ materials is assumed as follows:

$$\left.\begin{aligned} &f=\ln\hat{\sigma}^{*}_{SMP}-\frac{\hat{\alpha}}{1-\hat{\alpha}}\ln\left[1-(1-\hat{\alpha})\frac{\hat{X}^{*}+\hat{n}}{\hat{M}}\right]-c=0 && (\hat{\alpha}\neq 1)\\ &f=\ln\hat{\sigma}^{*}_{SMP}+\frac{\hat{X}^{*}+\hat{n}}{\hat{M}}-c=0 && (\hat{\alpha}=1)\end{aligned}\right\} \tag{18}$$

where $\hat{\sigma}_{SMP}^{*}=\bar{\sigma}_{SMP}^{*}+\sqrt{3}\sigma_0$, $\hat{\alpha}$ and $\hat{M}$ are the material parameters determined by the dilatancy characteristics of c−ϕ materials, and

$$\hat{n}=\sqrt{\hat{n}_{ij}\hat{n}_{ij}} \tag{19}$$

The plastic strain increment $d\varepsilon_{ij}^{p(AF)}$, which obeys the associated flow rule in $\bar{\sigma}_{ij}^{*}$−space, is expressed by

$$d\varepsilon^{p(AF)}_{ij}=\Lambda\frac{\partial f}{\partial\bar{\sigma}^{*}_{ij}} \tag{20}$$

where Λ is the plastic multiplier, and can be obtained as follows:

$$\Lambda=\frac{d\bar{W}^{*p}-K_1(\hat{\sigma}_m^{m-1}\langle d\hat{\sigma}_m\rangle-\hat{\sigma}_{m0}^{m-1}\langle d\hat{\sigma}_{m0}\rangle)\bar{\sigma}^{*}_{ij}\delta_{ij}}{\bar{\sigma}^{*}_{ij}\dfrac{\partial f}{\partial\bar{\sigma}^{*}_{ij}}} \tag{21}$$

where $d\bar{W}^{*p}$ is given by the following experimentally derived expression:

$$\left.\begin{aligned} &d\bar{W}^{*p}=K_2\left\{\hat{\sigma}^{*}_{SMP}\left[1-(1-\hat{\alpha})\frac{\bar{X}}{\hat{M}}\right]^{\frac{\hat{\alpha}}{\hat{\alpha}-1}}-\sqrt{3}\sigma_0\right\}^m\left\{\left[1-(1-\hat{\alpha})\frac{\bar{X}}{\hat{M}}\right]^{\frac{\hat{\alpha}}{\hat{\alpha}-1}}\left[d\hat{\sigma}^{*}_{SMP}+\frac{\hat{\alpha}\hat{\sigma}^{*}_{SMP}\overline{dX}}{\hat{M}-(1-\hat{\alpha})\bar{X}}\right]-\sqrt{3}d\sigma_0\right\} && (\hat{\alpha}\neq 1)\\ &d\bar{W}^{*p}=K_2\left[\hat{\sigma}^{*}_{SMP}e^{\frac{\bar{X}}{\hat{M}}}-\sqrt{3}\sigma_0\right]^m\left\{e^{\frac{\bar{X}}{\hat{M}}}\left[d\hat{\sigma}^{*}_{SMP}+\frac{\hat{\sigma}^{*}_{SMP}}{\hat{M}}\overline{dX}\right]-\sqrt{3}d\sigma_0\right\} && (\hat{\alpha}=1)\end{aligned}\right\} \tag{22}$$

in which

$$K_2=\frac{m(C_t-C_e)}{3^{(m+1)/2}p_a^{\,m}} \tag{23}$$

$$\bar{X}=\frac{(\hat{X}_{kl}-\hat{n}_{kl})\hat{X}_{kl}}{\hat{X}^{*}} \tag{24}$$

$$\overline{dX}=\frac{(\hat{X}_{kl}-\hat{n}_{kl})d\hat{X}_{kl}}{\hat{X}^{*}} \tag{25}$$

3 MODEL PARAMETERS DETERMINATION

Presented herein are predictions of triaxial and true triaxial tests conducted on specimens of two kinds of cemented soil, referred to as A–I and C–I. The component ratio for cemented soil A–I by weight is Toyoura sand : Portland Cement : water : clay powder=11.7 : 1 : 4.8 : 1.5 and for cemented soil C–I is Toyoura sand : Portland cement : water=15 : 1 : 3. The experimental data is after Matsuoka and Sun(1995). The material parameters are determined from the results of isotropic consolidation test with loading–unloading–reloading process and subsequent triaxial compression test with constant σ_m. The three material constants related to consolidation(C_t, C_e and m) are determined from the results of isotropic consolidation test. Other four material parameters related to shear($\hat{\alpha}$, $\hat{M}$, σ_0, ξ) are determined from the results of unconfined compression test and triaxial test under constant σ_m. σ_0 is assumed to be constant for cemented soils during shear, and can be determined from Mohr's stress circle diagram at failure(Matsuoka and Sun, 1991). $\hat{\alpha}$ and $\hat{M}$ are the

Table 1 Model parameters for cemented soils

	Consolidation			Shear			
	C_t(%)	C_e(%)	m	$\hat{M}$	$\hat{a}$	σ_0(kPa)	ξ
A-I	0.090	0.024	1.0	0.48	0.50	200	0.3
C-I	0.100	0.040	0.8	0.50	0.60	350	0.3

slope and intercept of the so-called stress-dilatancy equation based on the Extended SMP. The parameter ξ implies the size of yield surface during its rotation. The values of the relevant soil parameters used in the predictions are listed in Table 1.

4 COMPARISONS WITH EXPERIMENTAL RESULTS

4.1 Modelling Triaxial Behavior

Fig.1 shows the experimental and predicted results of triaxial compression tests on cemented soil A-I. In this and following figures, $\hat{\sigma}_1/\hat{\sigma}_3$ denotes the translated principal stress ratio($=(\sigma_1+\sigma_0)/(\sigma_3+\sigma_0)$), and the sign ○ represents the measured behavior of triaxial compression test, and the solid line represents the model predictions. It can be seen that the influence of the stress paths and mean effective principal stress on the stress-strain-volume change behavior of cemented soil is well reflected by the model.

4.2 Modelling True Triaxial Behavior

The results of true triaxial tests on the cemented soil C-I, with radial stress paths(constant θ) and bending stress path are shown in Figs.2, 3, 4 and 5, together with the predictions of the model. The isotropic consolidation pressure is 800kPa for all cases shown. Figs.2 and 3 show the mutual relationships among the principal strains (ε_1, ε_2 and ε_3), the volumetric strain (ε_v) and the translated principal stress ratio ($\hat{\sigma}_1/\hat{\sigma}_3$) along the radial and bending stress paths. It can be seen that the model is capable of predicting the deformation characteristics of cemented soil under general three-dimensional stress condition. Fig.4(a) shows the projection of strain increment vectors in the π plane along radial stress paths (θ=15°, 30° and 45°), measured by true triaxial tests on the cemented soil. It is seen from Fig.4(a) that the directions of strain increment vectors deviate from the directions of stress increment vectors along radial stress paths with the

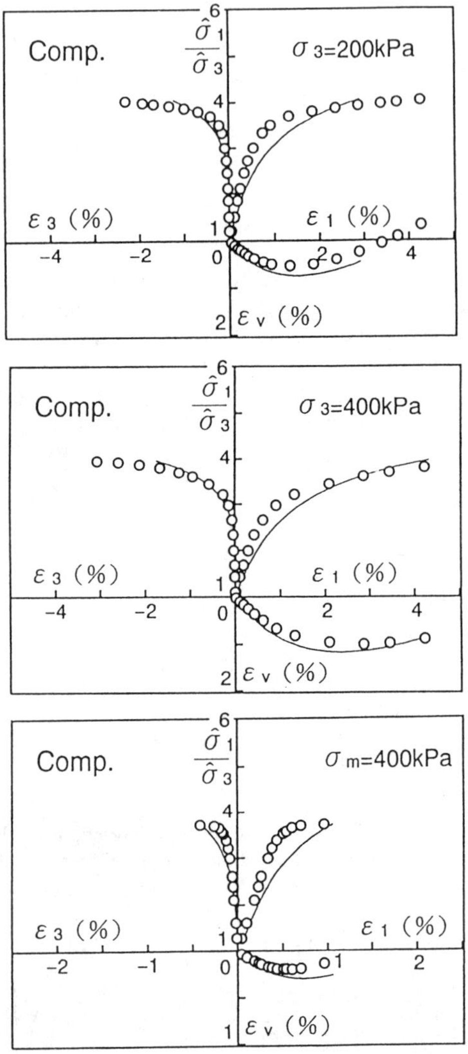

Fig.1 Predicted and experimental results of triaxial compression tests on Cemented Soil A-I

same trend. This deviation can be considered to be caused by induced anisotropy of the cemented soil(Matsuoka and Sun 1995). Fig.4(b) shows that the model can predict the induced anisotropy of the cemented soil. Fig.5 shows the experimental and predicted projection of strain increment vectors in the π plane along bending stress paths(OAC, OAD, OAF and OBE). From Fig.5(a), it can be seen that the direction of the strain incremental vector at point G along stress path OAD is different from the direction along stress path OBE, and the model explains well this stress paths dependency of strain incremental vectors from Fig.5(b).

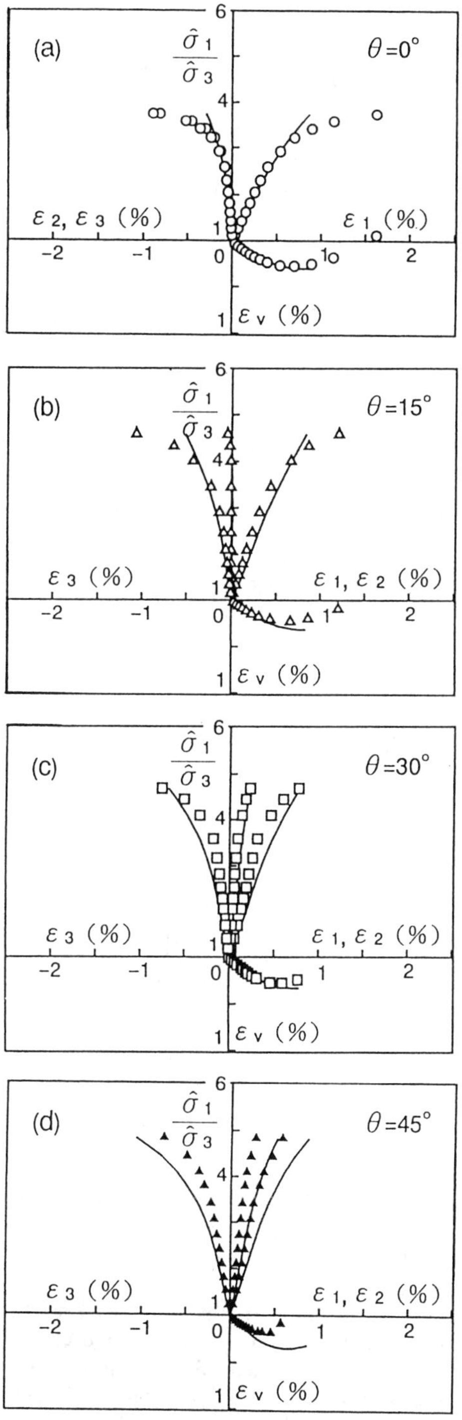

Fig.2 Predicted and experimental results of true triaxial tests on Cemented Soil C–I along radial stress path

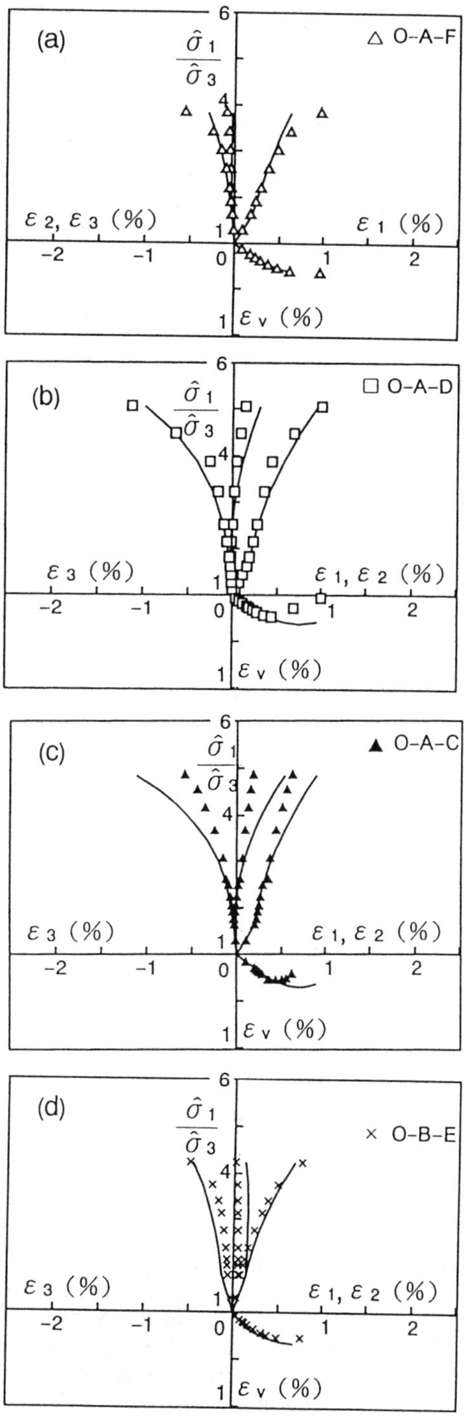

Fig.3 Predicted and experimental results of true triaxial tests on Cemented Soil C–I along bending stress path

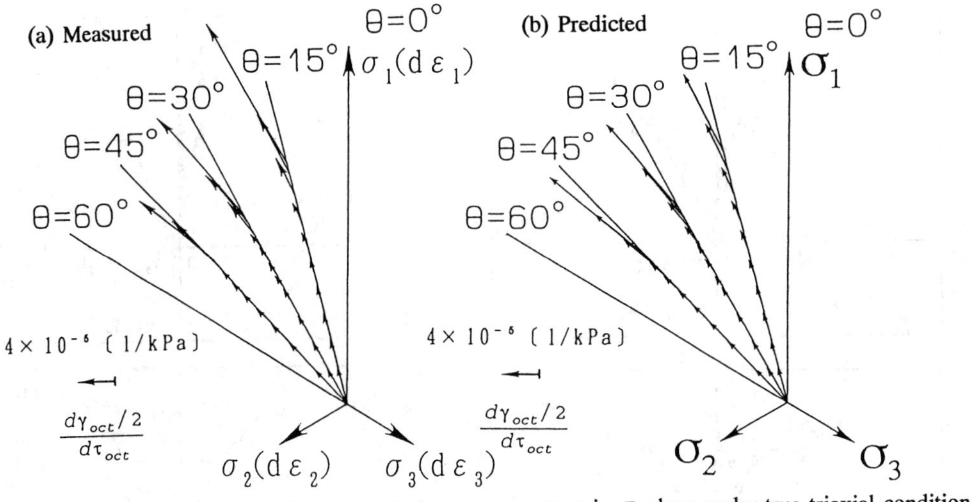

Fig.4 Predicted and measured directions of strain increment vectors in π plane under true triaxial condition along radial stress paths

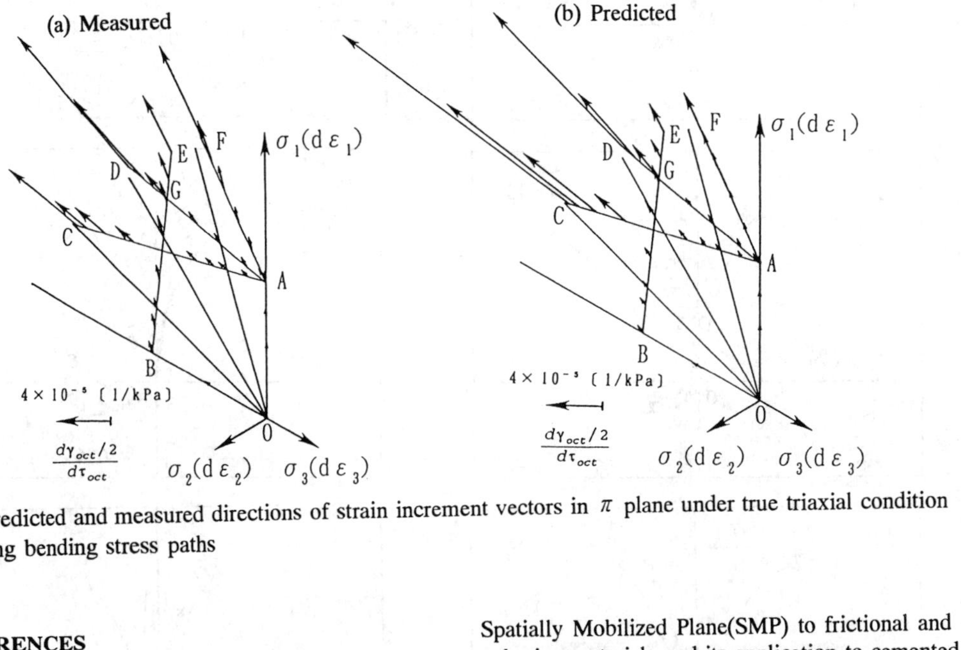

Fig.5 Predicted and measured directions of strain increment vectors in π plane under true triaxial condition along bending stress paths

REFERENCES

Matsuoka, H. & D. A. Sun 1991. A constitutive law covering granular materials to metals, Proc. of 9th Asian Regional Conference on Soil Mechanics and Foundation Engineering, Vol.1, Bangkok,pp.55–58.

Matsuoka, H. & D. A. Sun 1994. An elasto–plastic model for frictional and cohesive materials including granular materials and metals, Proc. Japan Society of Civil Engineers(JSCE), No.505, pp.201–210 (in Japanese).

Matsuoka, H. & D. A. Sun 1995. Extension of Spatially Mobilized Plane(SMP) to frictional and cohesive materials and its application to cemented sand, Soils and Foundations,Vol.35, No.4, pp.63–72.

Nakai, T., J. Fuji & H. Taki 1989. Kinematic extension of an isotropic hardening model for sand, Proc. of 3rd International Symposium on Numerical Methods in Geomechanics, Niagra, pp.36–45.

Sun, D. A. & H. Matsuoka 1994. An elasto–plastic model for frictional and cohesive materials, Proc. on Pre–failure deformation of geomaterials, Sapporo, Vol.1, pp.433–438.

Computer Methods and Advances in Geomechanics, Yuan (ed.) © 1997 Balkema, Rotterdam, ISBN 90 5410 904 1

Formulation of an anisotropic model based on experiments

M.S.A.Siddiquee
Bangladesh University of Engineering and Technology, Dhaka, Bangladesh
F.Tatsuoka
University of Tokyo, Japan

ABSTRACT: Most of the natural deposits show definite formation of stratification. Modelling of this kind of stratified materials needs particular attention to the different behaviour of the material in two particular direction, specially in the direction parallel and perpendicular to the stratification. A series of tests were carried out to find out the behaviour of granular material in the hydrostatic direction of stress. Isotropic compression tests were performed using two cylindrical and four rectangular prismatic triaxial specimens. It was seen that for a hydrostatic loading the response was not hydrostatic. It showed definite trend of anisotropy in the deformation pattern. This anisotropy is modelled here, in this paper by treating soil-dilatancy as a variable quantity. Equation of the plastic potential surface (non-associated flow rule) varying with three main variables of this model, confining pressure, σ_3, void ratio, e and the angle of bedding plane orientation with respect to σ_1-direction was derived.

1 INTRODUCTION

The present study attempted to obtain both very accurate deformation characteristics of Toyoura sand during loading and unloading of isotropic compression. A series of six isotropic compression tests were performed on two types of specimens, square prismatic and cylindrical specimen. All the tests were carried out under high compressive stress. Specimens were prepared by pluviating sand in air. Both loose and dense and loose specimen were prepared and ends were lubricated to reduce the effect of end-friction. A pair of Local Deformation Transducers (LDT) were used to measure axial deformations free from the effect of bedding error. Three sets of Radial Deformation Transducers (RDT) (for cylindrical specimen) and four sets of lateral LDTs (square prismatic specimen) were used to measure the change of the specimen diameter (side). The confining pressure was controlled by vacuum (up to 76.40 kN/m^2) and subsequently by air pressure (up to 686 kN/m^2).

The anisotropic behaviour of dense sand is modelled here. Equation of the yield or plastic potential surface varying with three controlling factors, confining pressure, σ_3, void ratio, e and the angle of bedding plane orientation (δ) with respect to σ_1-direction was derived. This model formed a variable cap in the isotropic stress direction, which was supplemented with a shearing surface of Mohr-Coulomb type (Tatsuoka and Siddiquee et al., 1994) to form a complete model.

2 EXPERIMENTAL SETUP

In order to establish the yield or plastic potential surface, which describes the development of the anisotropic plastic strains, isotropic and K_0 compression tests were performed. Strain anisotropy in the isotropic loading was proved by performing several isotropic compression tests at high pressure

Table 1 Test cases of isotropic and K_0-compression test using in a rectangular prismatic TC specimen.

Test	Type of material	Type of specimen	Void ratio (*)	Load type	Specimen size (mm)
1	T. sand	Round	0.785	K_{is}	600X300
2	T. sand	Round	0.673	K_{is}	600X300
3	T. sand	Square	0.687	K_{is}	230X570
4	T. sand	Square	0.696	K_0	230X570
5	SLB sand	Square	0.504	K_{is}	230X570
6	SLB sand	Square	0.524	K_0	230X570

(*) Initial value at σ_c=20 kN/m^2.

(up to 686 kN/m^2). This section is devoted for the description of the details of those tests. Following table describes the testing specifications.

Two isotropic compression tests were performed using two cylindrical triaxial specimens (Test-1 and Test-2). Specimens were prepared by pluviating either of Toyoura sand and SLB sand in air. The density was controlled by changing the size of the nozzle through which sand was flowed out into a set of criss-crossed sieves. One specimen was loose (e=0.785) and the other was dense (e=0.673). In both cases, lubricated ends were used to reduce the effect of end-friction. The thickness of grease layer was 0.06 μm The thickness of the end rubber disks was 0.3 mm and the lateral rubber membrane was 0.8 mm. A pair of Local Deformation Transducers (LDT) were used to measure the deformations free from the effect of bedding error. Three sets of Radial Deformation Transducers (RDT) were used to measure the change of the specimen diameter (Fig. 1). Since the diameter change as measured involved the effect of membrane penetration (MP), it was corrected by assuming: (1) the effect of MP is a function of σ_c, but the same during loading and unloading of σ_c, and (2) the behavior during unloading is isotropic. The confining pressure was controlled by vacuum (up to 76.40 kN/m^2) and subsequently by air pressure (up to 686 kN/m^2), while the axial load was applied by using a set of Belloframe cylinders so as to maintain the isotropic stress condition at the top of specimen. Loading was controlled manually, while all data was monitored and stored in a computer.

Another four cases of experiments were performed (Test-3,4,5,6) using rectangular prismatic samples with a square horizontal cross-section. The cross-section was designed to be square in order to attach lateral LDTs (A, B, C, D) as shown in Fig. 2. The probable corner stress-concentration due to inhomogeneity did not have much effect, as seen from the comparison of the test results with those of cylindrical specimen. All other procedures of testing were same as those for of cylindrical ones.

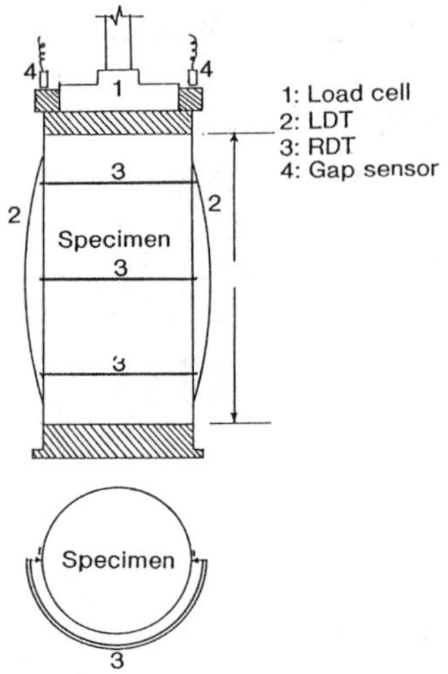

Figure 1 Experimental setup for cylindrical specimen.

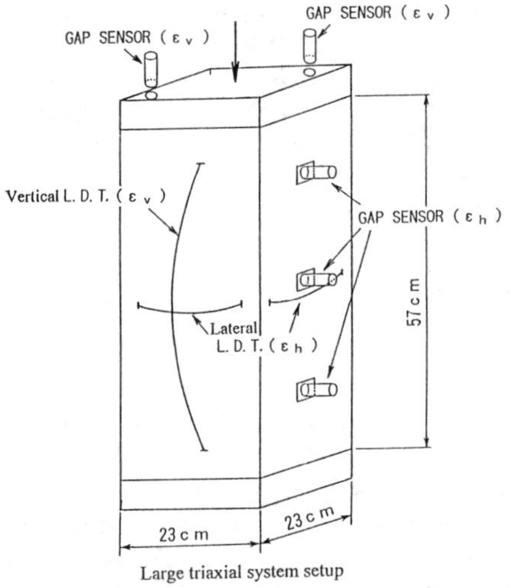

Figure 2 Experimental setup for prismatic specimen with square cross-section.

3 EXPERIMENTAL RESULTS

Fig. 3 shows clearly the anisotropy during loading for both loose and dense specimens of Toyoura sand.

Tests 3 and 5 followed isotropic stress path while tests 4 and 6 followed K_0 stress path using rectangular prismatic specimen. It has been observed that some of the stress-paths were little ragged due to the manual control of the stress-paths. There was isotropy in the horizontal layer, which has been confirmed by measuring lateral strains in two perpendicular direction in the horizontal plane (Siddiquee, 1994). The results show that the external measurement shows larger strain due to the bedding error, and LDT measured the uniform strain in the specimen. The lateral strains measured with all the lateral LDTs for four cases Tests-3,4,5,6 were almost unique. The uniquely designed lateral LDTs really worked to measure the lateral strain, ε_r in separately four sides of a rectangular prism specimen (Siddiquee, 1994). The lateral strain measured with gap sensors at three levels showed erroneous results

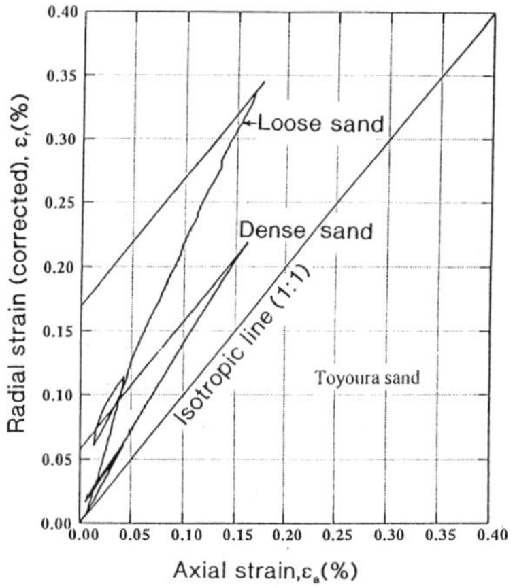

Figure 3 Strain-anisotropy of granular material (cylindrical sample)

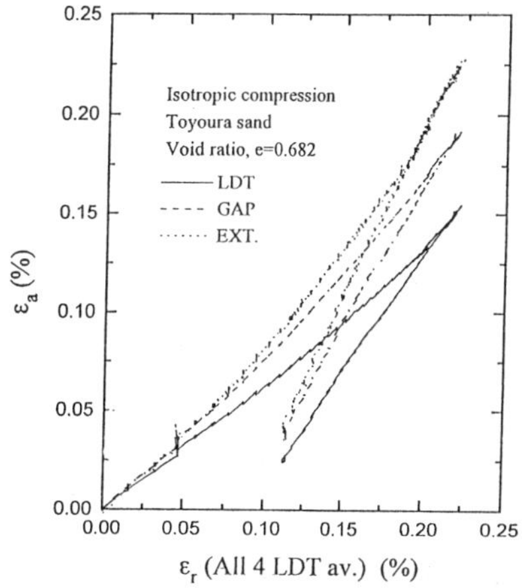

Figure 4 Averaged vertical strain, ε_a vs. lateral strain, ε_r (strain-anisotropy) for Test-3 measured with LDT, gap sensors and external devices.

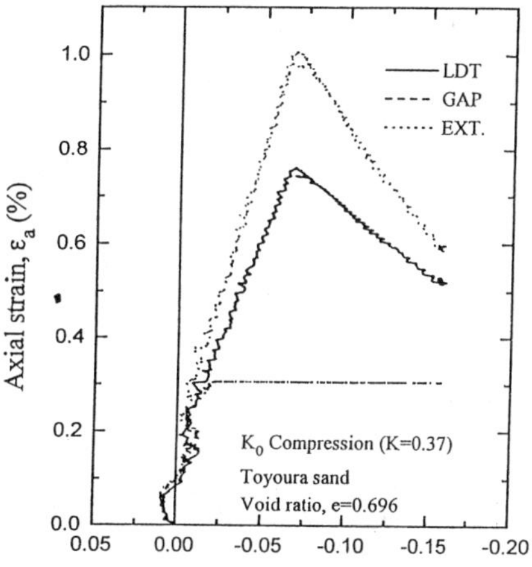

Figure 5 Averaged vertical strain, ε_a vs. lateral strain, ε_r (strain-anisotropy) for Test-4 measured with LDT, gap sensors and external devices.

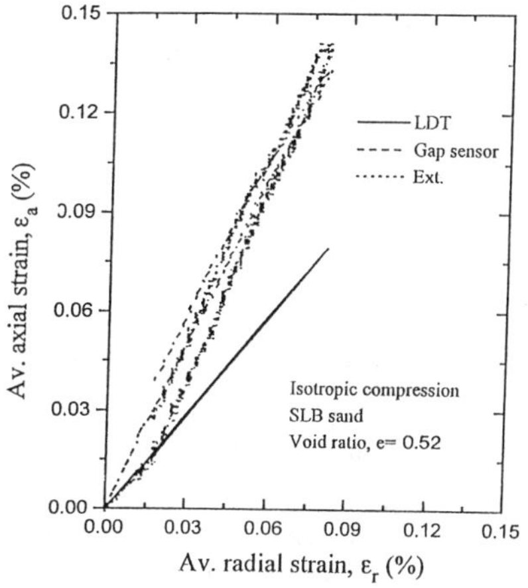

Figure 6 Averaged vertical strain, ε_a vs. lateral strain, ε_r (strain-anisotropy) for Test-5 measured with LDT, gap sensors and external devices.

due to bending of the bottom plate of the triaxial apparatus caused by the change in the cell pressure. The results have been corrected for errors due to rotation of the holding rod of the gap sensors standing on both sides of the specimen (Siddiquee, 1994). Fig. 4 through 7 show the plot of strain-anisotropy for isotropic and K_0 loading conditions for four cases of rectangular prismatic samples. Although the plot shows data from all the measuring devices, only LDT data are counted reliable as no bedding or membrane penetration error are included.

Fig. 11 shows the variation of initial strain-anisotropy with stress ratio. This figure is very similar to the stress-dilatancy curve obtained from a TC test at a constant pressure but shifted to the left due to the inherent anisotropic fabric of the granular material. This inherent anisotropy is a result of the specimen preparation technique used. Based on this curve, it is assumed that it is a straight line and a cap plastic potential surface is formulated as shown below.

4 FORMULATION OF THE MODEL

A single surface elasto-plastic model of Mohr-

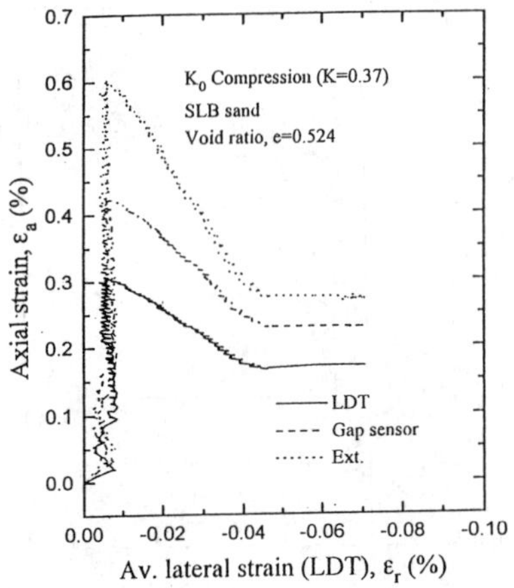

Figure 7 Averaged vertical strain, ε_a vs. lateral strain, ε_r (strain-anisotropy) for Test-6 measured with LDT, gap sensors and external devices.

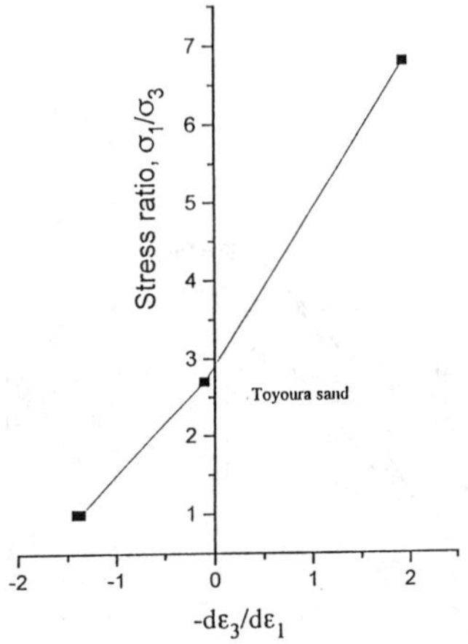

Figure 8 Stress ratio versus initial strain anisotropy for Toyoura sand.

Coulomb type cannot capture the anisotropic volumetric plastic strain in the isotropic loading direction as it is open in that direction. Herein described is a cap surface, which has been developed to combine with Mohr-Coulomb type model in such a manner that plastic volumetric strain can be captured by both shearing with dilatancy and compression.

The assumptions used are:

1. For Toyoura sands the parameter K for the stress-dilatancy relation ($R=KD$) is constant i.,e., independent of δ, and σ_3 (for PSC at a constant p)
2. For anisotropic and isotropic compression at a constant q/p, the principal strain increment ratio at isotropic stress condition varies with the change in δ linearly from PSC (δ=90°) to PSE (δ=0°).

From Fig. 9, the equation of a straight line at δ=90° can be written as a generic equation as follows;

$$\frac{\sigma_1}{\sigma_3} = -m\frac{d\varepsilon_3}{d\varepsilon_1} + c \tag{1}$$

using, $s = \frac{1}{2}(\sigma_1 + \sigma_3), t = \frac{1}{2}(\sigma_1 - \sigma_3)$

$$\frac{s+t}{s-t} = -m\frac{d\varepsilon_3}{d\varepsilon_1} + c \tag{2}$$

Hence, we get:

$$\frac{d\varepsilon_3}{d\varepsilon_1} = \frac{c(s-t)-(s+t)}{m(s-t)} \tag{3}$$

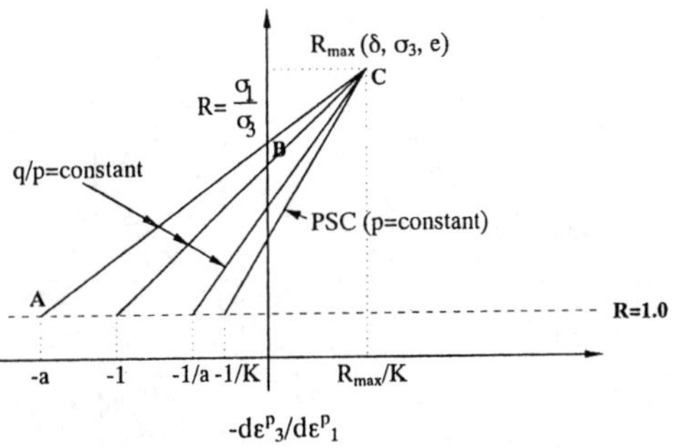

Figure 9 Variable stress-dilatancy relation

From Eq. (3), we get:

$$\frac{d\varepsilon_v}{d\gamma} = \frac{m(s-t)+c(s-t)-(s+t)}{m(s-t)-c(s-t)+(s+t)} \tag{4}$$

From the above Fig. 10, we can write:

$$\frac{dt}{ds} = -\frac{tc_1 + sc_2}{tc_3 + sc_4} \tag{5}$$

Where,

$$c_1 = -1 - c - m$$
$$c_2 = -1 + c + m$$
$$c_3 = 1 + c - m$$
$$c_4 = 1 - c + m$$
$$c_5 = c_1 + c_4$$

$$(tc_1 + sc_2)ds + (tc_3 + sc_4)dt = 0 \tag{6}$$

Substituting $t = vs$ and $dt = vds + sdv$ into Eq.

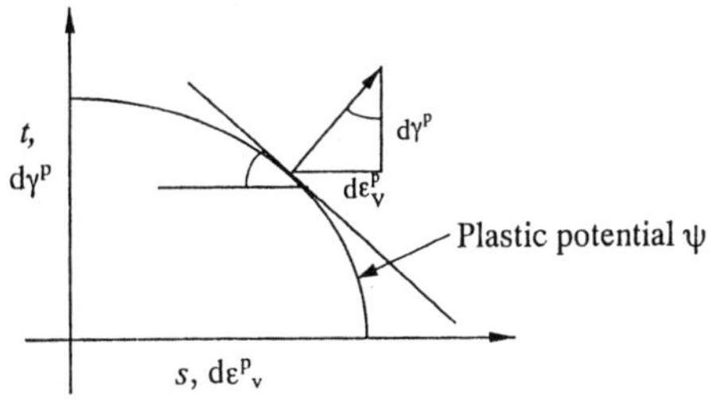

Figure 10 Shows a segment of the plastic potential.

(6) and integrating, the following equation is obtained;

$$\int \frac{1}{s} ds + \int \frac{c_3 v + c_4}{c_3 v^2 + (c_1 + c_4)v + c_2} dv = 0 \quad (7)$$

After completing the integration,

$$(c_3 t^2 + c_5 st + c_2 s^2)\left(\frac{t + b_1 s}{t + b_2 s}\right)^{b_4} = X \quad (8)$$

When $t = 0$, $s = s_0$, then,

$$X = c_2 s_0^2 \left(\frac{b_1}{b_2}\right)^{b_4} \quad (9)$$

Hence the required equation for the plastic potential is;

$$(c_3 t^2 + c_5 st + c_2 s^2) = c_2 s_0^2 \left(\frac{b_1 t + \frac{c_2}{c_3} s}{b_2 t + \frac{c_2}{c_3} s}\right)^{b_4} \quad (10)$$

Where,

$$b_1 = \frac{c_5}{2c_3} - \sqrt{\frac{c_5^2}{4c_3^2} - \frac{c_2}{c_3}} \quad (11)$$

$$b_2 = \frac{c_5}{2c_3} + \sqrt{\frac{c_5^2}{4c_3^2} - \frac{c_2}{c_3}}$$

$$b_4 = \frac{c_4 - c_1}{2c_3 \sqrt{\frac{c_5^2}{4c_3^2} - \frac{c_2}{c_3}}}$$

Determination of 'm' and 'c' depending on different angle of bedding plane orientation (δ) with respect to major principal stress direction.

Fig. 12 shows an example of the variable plastic potential surface varying with three main variables of

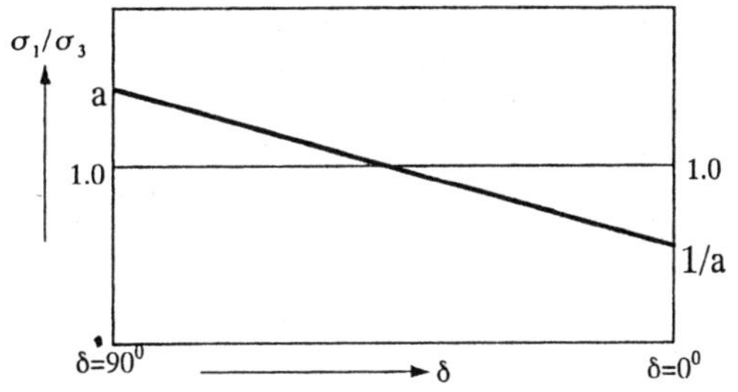

Figure 11 Variation of parameter 'a' with the angle of bedding plane

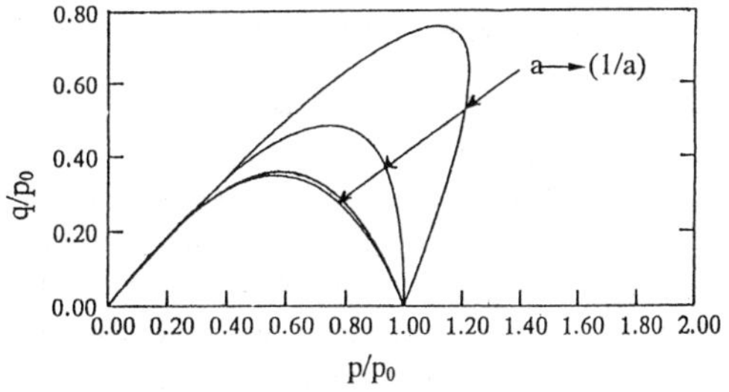

Figure 12 Example of variable plastic potential curves for anisotropic model

$$m = \frac{K(R_{max} - 1)}{\varpi K + R_{max}} \quad (12)$$

$$c = \frac{R_{max}(\varpi K + 1)}{\varpi K + R_{max}}$$

where,

$$\varpi = \left(\frac{1 - a^2}{90a}\right)\delta + a, \text{ where } a = \frac{\varepsilon_r}{\varepsilon_a} \quad (13a)$$

$$K = \frac{1 + \sin\varphi_r}{1 - \sin\varphi_r} \quad (13b)$$

$$R_{max} = R_{max}(\sigma_3, e, \delta)$$

this model, confining pressure, σ_3, void ratio, e and the angle of bedding plane orientation (δ) with respect to σ_1-direction. At any stage of loading, mean pressure will determine its main starting position as shown in Fig. 12 in the x-axes. Other positions are determined by the inherent anisotropy factor, *a*. Formulation of R_{max} is discussed in Tatsuoka and Siddiquee et al. (1994).

Eq. 13 describes the need for the parameter *a* which shows that loading in isotropic stress direction is also pronounced with anisotropic strains (Fig. 11).

The main reason is the unequal plastic strain development as a result of inherent anisotropy due to particles' deposition in a preferred direction during bed preparation. Figs. 3 through 7 show the curves from which the parameter is determined to be ***a=1.42***. Fig. 8 shows the experimental stress-ratio vs. initial strain increment-ratio, which asserted the assumption of a straight line variation of stress-ratio vs. initial strain increment-ratio.

5 CONCLUSIONS

The following conclusions can be drawn from the above analysis:

1. The results of experiments showed definite trend of anisotropy in the deformation pattern.
2. The anisotropy is modelled here. Equation of the plastic potential surface varying with three main affecting factors, confining pressure, σ_3, void ratio, e and the angle of bedding plane orientation with respect to σ_1-direction was derived.

REFERENCES

Siddiquee, M . S. A. 1994. FEM Simulation of deformation and failure of stiff geomaterials based on element test results. Doctor Thesis, The University of Tokyo, Japan.

Siddiquee, M . S. A. and Tanaka, T. and Tatsuoka, F. 1991. An FEM Simulation of Model Footing Tests on Sands. *JSSMFE Proceedings of 26th National Conference,* 1309--1312.

Tatsuoka, F. and Nakamura, S. and Huang, C. C. and Tani, K. 1989a. Strength anisotropy and shear band direction in plane strain tests of sand. *Soils and Foundations,* 30 (1) : 35--54.

Tatsuoka, F., Siddiquee, M. S. A., Park, C.-S, et al. 1994. Modelling of stress-strain relations of sand. *Soils and Foundations*, Vol. 33, No. 2, 60--81.

Computer Methods and Advances in Geomechanics, Yuan (ed.) © 1997 Balkema, Rotterdam, ISBN 90 5410 904 1

Constitutive modelling of unsaturated sandy silt

F. Geiser, L. Laloui & L. Vulliet
Soil Mechanics Laboratory, EPFL, Lausanne, Switzerland

ABSTRACT: The mechanical behaviour of a sandy silt is characterised experimentally in saturated and unsaturated states. First, these test data are used to examine the performance of a constitutive model developed by Alonso *et al.* (1990) for partially saturated soils. The corresponding parameters are determined and predictions based on this model are compared with the experimental data. Then a new approach based on the Disturb State Concept (DSC) is proposed to incorporate some typical features of unsaturated soils to obtain better predictions.

1. INTRODUCTION

Until 1990, strength and volumetric behaviours of unsaturated soils had been modelled separately (Fredlund *et al.*, 1978, Lloret & Alonso, 1985). The first works taking into account both aspects were based on an extension of the theories developed for saturated soils. In 1990 Alonso *et al.* proposed a constitutive model using the critical state concept to couple strength and volumetric response in unsaturated soils. This model is used in this paper to simulate experimental data of an unsaturated sandy silt.

In the first part, a brief presentation of the experimental characterisation of the behaviour of the material is given. Then the determination of the constitutive parameters for the model is explained. Using the model, predictions are made for some stress paths and compared with experimental values. The last section proposes a new approach based on the Disturb State Concept to predict the unsaturated soil behaviour.

2. EXPERIMENTAL CHARACTERISATION

The authors have undertaken a research on an unsaturated sandy silt, which involves rheological characterisation performing both extensive testing procedures under different stress paths and numerical simulations. Part of the experimental results are used in this paper.

2.1 *Material and apparatus*

The studied soil is a washing sandy silt from the region of Sion (Switzerland). Its characteristics are given in Table 1.

Table 1: Characteristics of the Sion silt

w_L(%)	w_P(%)	I_P	%<2μm	%>60μm	γ_s(kN/m^3)
25	17	8	8	20	27.4

The sample preparation is explained in Laloui *et al.*, (1995).

Standard experimental apparatus were adapted to run unsaturated tests with the air-pressure technique: triaxial cells, oedometers and Richards cells (pressure plate extractors). This technique consists in imposing air-pressure at the top of the sample, while the pore water-pressure is measured or imposed at the bottom, where a special high air entry value ceramic is located (for details see Laloui *et al.*, 1997). The suction s is then obtained as the excess of pore air-pressure u_a to water-pressure u_w:

$$s = u_a - u_w \tag{1}$$

2.2 *Solicitation paths*

The different types of experimented stress paths are summarised by Figure 1. σ_1 and σ_3 are the principal vertical and lateral stresses. Samples are submitted to mechanical loading at different initial suctions and to hydric loading (variation of suction) at different initial mechanical stresses.

In the triaxial apparatus, samples are first consolidated to a given confining pressure (A-B) in saturated conditions. This should erase all the mechanical history of the sample as the confining pressure is always higher than the preconsolidation pressure of the fabricated soil. Then, in an unsaturated test, samples are submitted to an hydric path (B-C): the suction increases with the application of u_a, while keeping u_w equal to the atmospheric

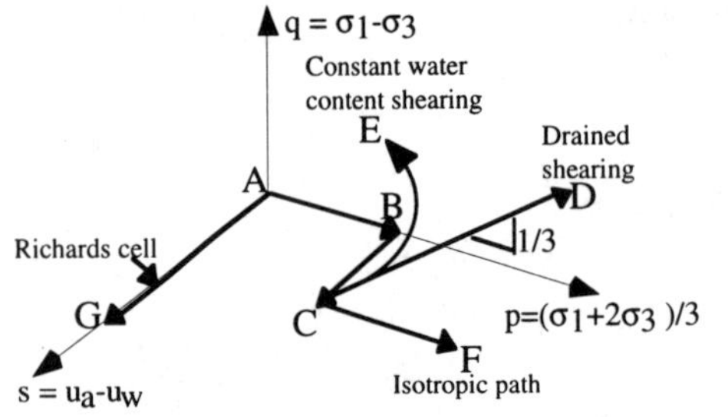

Fig. 1: Hydro-mechanical stress paths

pressure. Finally, when the equilibrium is reached (point C), several stress paths are analysed separately:

1- C-F : *"isotropic"* condition. This path is obtained by increasing the confining pressure p while keeping both u_a and u_w constant (thus allowing free flow of air and water).

2- C-E : *"constant water content shearing"* condition. This path is driven in water undrained conditions while maintaining a constant air-pressure u_a. The water-pressure u_w is measured during the shearing, so that the suction is known.

3- C-D: *"drained shearing test"*. It consists in increasing the deviatoric stress while keeping both u_a and u_w constant.

In the Richards cell, the sample is submitted to an hydric path (A-G) without any "mechanical" stress.

3. ALONSO *et al.* CONSTITUTIVE MODEL

3.1 *Introduction to the model*

The constitutive law proposed by Alonso *et al.* is based on the modified Cam-Clay model. It incorporates many features of unsaturated soils. This model is formulated within the framework of hardening plasticity using two independent sets of stress variables: the excess of total mean pressure over air pressure $p^* = p-u_a$ (defined as the net mean pressure) and the suction s. The detailed description of the model is given in Alonso *et al.* (1990).

3.1.1 *Parameters associated with isotropic behaviour (loading-collapse yield curve)*

– $\lambda(0)$: compressibility coefficient for the saturated state along virgin loading

– κ: compressibility coefficient along elastic (unloading-reloading) stress paths

When the samples are unsaturated, two more parameters are needed to describe the evolution of the compressibility λ with the suction:

$$\lambda(s)=\lambda(0)\ [(1-r)\exp(-\beta s)+r] \qquad (2)$$

- r: establishes the minimum value of the compressibility coefficient for high values of suction
- β: controls the rate of increase in stiffness with suction
- p^c is the reference stress

3.1.2 *Parameters associated with changes in suction*

- λ_s: compressibility coefficient for increments of suction across virgin states
- κ_s: compressibility coefficient for changes in suction within the elastic region

3.1.3 *Parameters associated with changes in shear stress and shear strength*

- ν: Poisson's ratio
- M: the slope of the critical state line
- k: parameter that controls the increase in cohesion with the suction.

3.2 *Determination of the parameters*

The determination of the parameters for saturated and unsaturated states is explained in the following sections.

3.2.1 *Parameters associated with isotropic behaviour*

$\lambda(0)$ and κ are deduced from Figure 2 showing an isotropic consolidation test of the saturated silt. This silt shows an increase in mechanical compressibility $\lambda(0)$ with the mean pressure for the experimented stress range. As a result, two compressibilities (slopes in the e-lnp diagram) are chosen for two different ranges of stresses: $\lambda_1(0)=0.032$ for a mean pressure p between 100 and 400 kPa and $\lambda_2(0)=0.047$ for p higher than 400 kPa. The unloading slope κ is around $\kappa=0.007$.

The parameters r and β are deduced from unsaturated oedometric tests at different constant suctions (Charlier *et al.*, 1997). Figure 3 shows the calibration of parameters for the compressibility decrease with the suction as the model suggests. The best fit was find with r=0.65 and $\beta=0.005$ kPa^{-1}. The experimental data also show that the unloading slope κ can be considered as independent of the suction, as proposed in the model.

The last term associated with the isotropic stress conditions is the reference stress p^c. It is chosen as equal to 20 kPa.

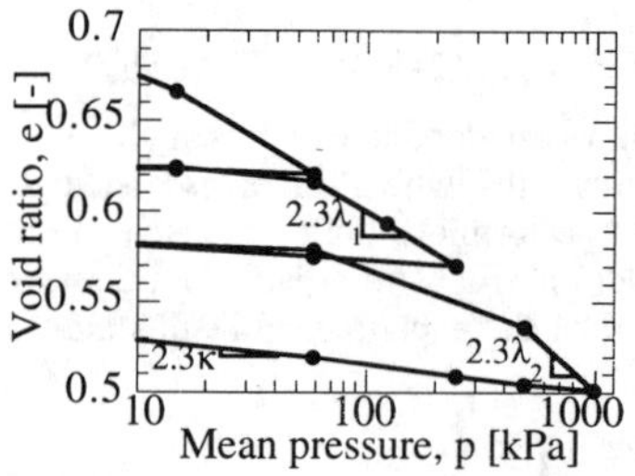

Fig. 2: Consolidation in saturated conditions

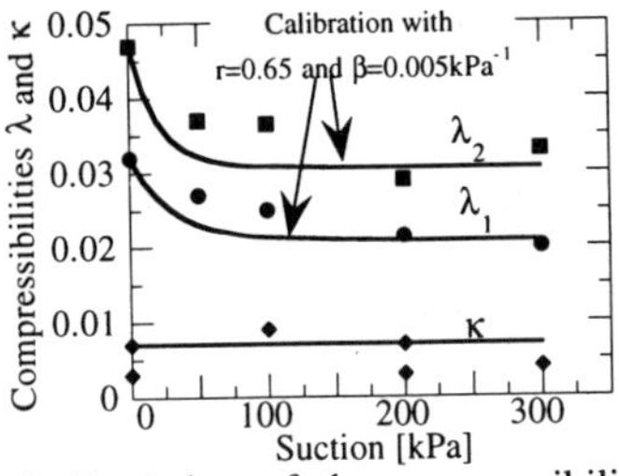

Fig. 3: Evolution of the compressibilities with the suction

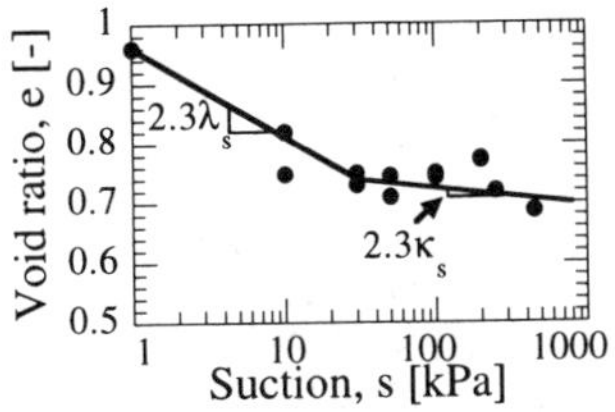

Fig. 4: Hydric compressibilities

3.2.2 *Parameters associated with changes in suction*

λ_s and κ_s are deduced from the Richards cell tests on the drying paths (A-G). Figure 4 shows the evolution of the void ratio vs. suction.

We can notice that the suction has the same type of influence as a mechanical stress, but with different compressibility slopes: λ_s=0.06 and κ_s=0.009. As long as the sample is almost saturated (s<30kPa, see Laloui *et al.*, 1997) the void ratio decreases highly with suction.

3.2.3 *Parameters associated with changes in shear stress and shear strength*

The Poisson's ratio is deduced from the unloading slopes of the deviatoric saturated triaxial tests: ν is taken as equal to 0.4.

To find the parameters M and k we use the critical state points from the shearing tests (paths C-D and C-E) in the p* vs. q (deviatoric stress) plane. Figure 5 shows a decrease of the slope of the critical state line M with the suction. Simultaneously the cohesion c (value of the deviatoric stress at p*=0) increases with the suction. Alonso *et al.* assume that the slope of the critical state line is independent of the suction s. Only the cohesion increases with the suction.

The slope M is equal to 1.33 in saturated conditions. As the model does not take in account an increase of M, the cohesion c is calculated for each critical point with the slope of the critical state line in saturated conditions:

$$c=q-1.33*(p-u_a) \tag{3}$$

Figure 6 shows the points resulting from this assumption for the drained tests. The parameter k is

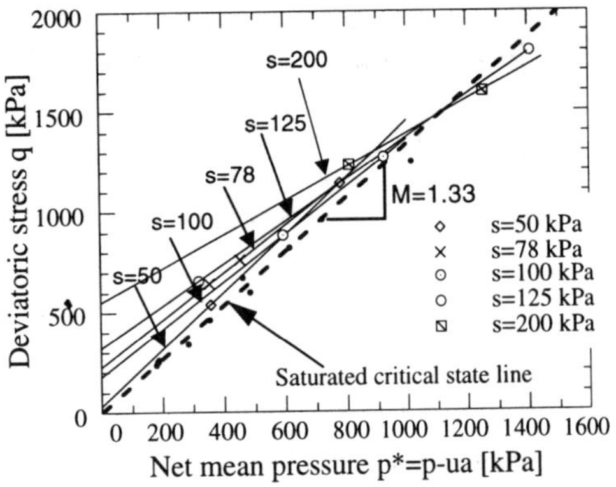

Fig. 5: Evolution of the critical state line with the suction

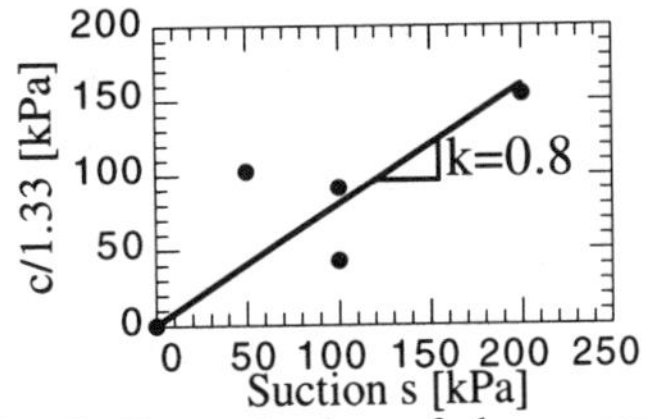

Fig. 6: Determination of the parameter k from the experimental results.

chosen to be equal to 0.8.

The application of the model requires of course, in addition to the constitutive parameters, information on the initial state of stress, suction and void ratio of each test.

3.3 *Comparison of model predictions with experimental data*

Two types of stress paths were predicted with the Alonso *et al* model: isotropic path (C-F) and drained shearing path (C-D).

3.3.1 *Isotropic path*

Figure 7 shows the predictions of the model for the isotropic behaviour of the Sion sandy silt at different saturation states. Note that on the figure the volumetric strain represents the water volume change divided by the total volume.

At small suctions (100 kPa) the observed and predicted behaviours do not match very well. The more the suction increases, the more the predictions approach the experiment. At a suction of 100 kPa, the model predicts an elastic domain up to a mean pressure of about 500 kPa, what is not observed in the experiment. However the observed and calculated slopes are not very different. Nevertheless the predicted volume variations are underestimated because of this elastic domain.

For higher suctions (s=200 and 280 kPa), the

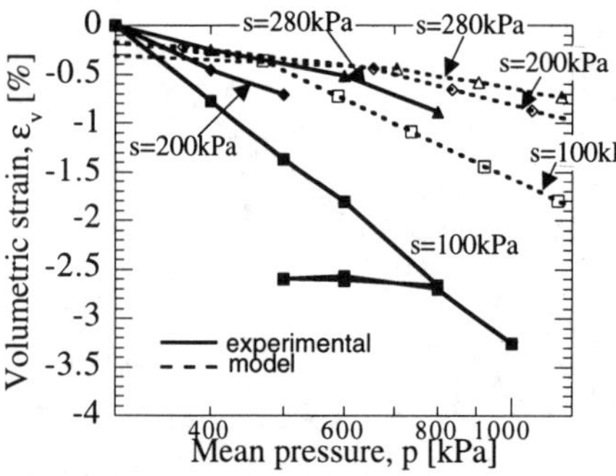

Fig. 7: comparison of the numerical predictions (Alonso *et al.* model) and the experimental data.

elastic domain of the model increases. The experience also shows a pseudoelastic phase up to mean pressures of about 600 kPa. In these conditions, with close slopes, the volumetric predictions are still underestimated but qualitatively better.

3.3.2 *Drained shearing path*

Figures 8 to 13 compare the predictions and the experimental data for different confining pressures and suctions (path C-D). The volumetric strain still represents the water volume change divided by the total volume.

Figure 8 and 11 show the saturated state. The model reproduces the experimental data very well in the deviatoric plane - axial strain versus deviatoric stress. The predictions of the model in the volumetric plane - axial strain versus volumetric strain - are quantitatively good.

Figure 9 shows an unsaturated test at a low suction (s=50 kPa) and quite high degree of saturation ($Sr \approx 0.90$). The model predicts well the observed behaviour in this partially saturated case.

Figures 10, 12 and 13 show the results of drained shearing tests at higher suctions (100 and 200 kPa). As before, the predicted ultimate resistance is in good accordance with the test results in the deviatoric plane. In the volumetric plane, the experimental data show a typical feature of unsaturated silts (see Maâtouk, 1993). The water content of the sample continuously decreases without a clear stabilisation level. Due to the fact that the constitutive model does not include this phenomenon, the predicted volume change is much less than the experimental value and does not show any continuous increase of the water volumetric strain.

Note that for the stress-strain behaviour, the experimental results show first an increase of strength and then a loss of resistance like a brittle failure (see Figure 13). This behaviour was already observed, with no significant consequences, for smaller suctions (Figures 10 & 12). The model of Alonso *et al.* is not able to predict such phenomenon which

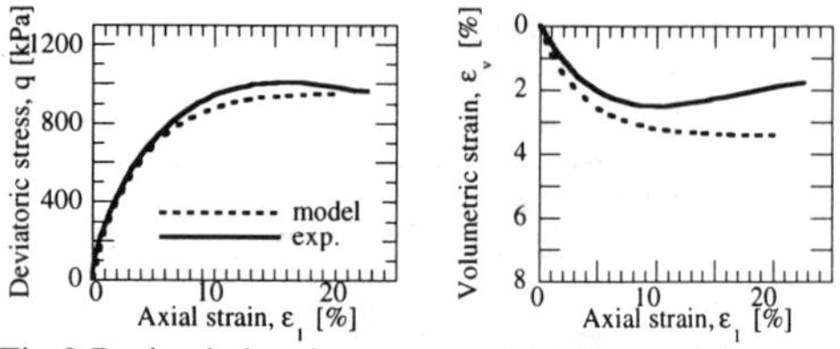

Fig.8:Drained shearing test, σ_3=400 kPa, s=0 kPa

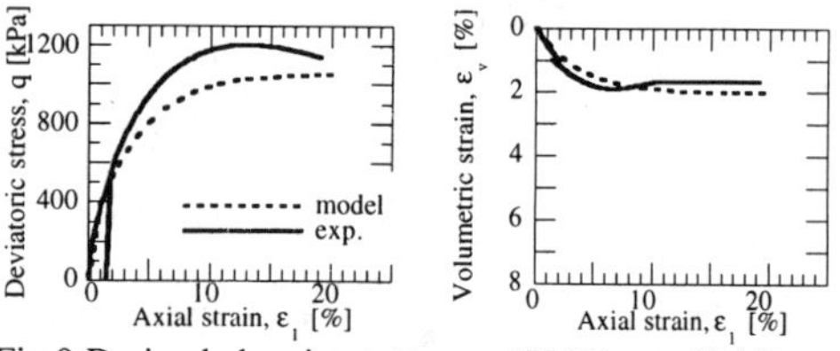

Fig.9:Drained shearing test, σ_3=450 kPa, s=50 kPa

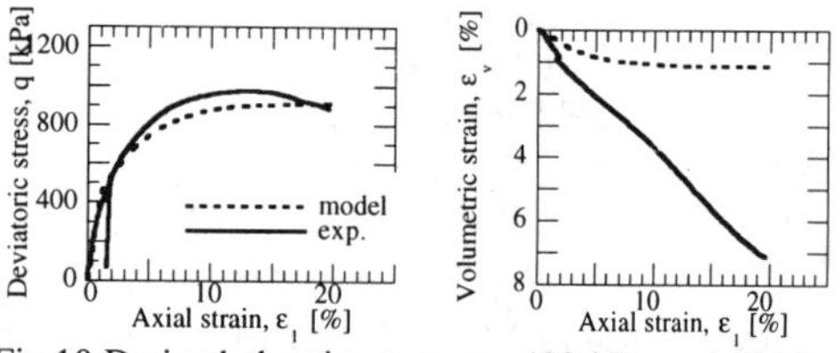

Fig.10:Drained shearing test, σ_3=400 kPa, s=100 kPa

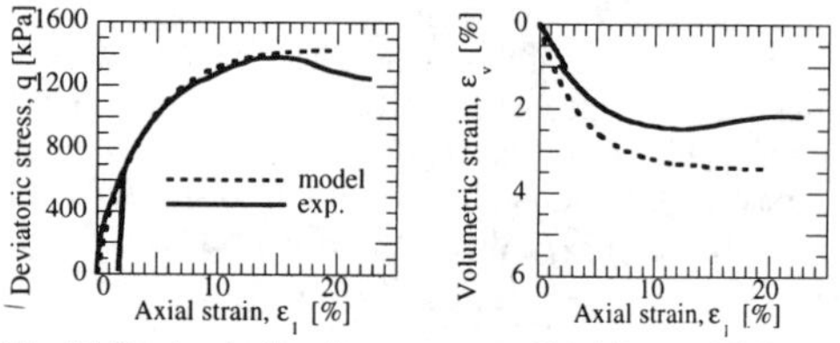

Fig.11:Drained shearing test, σ_3=600 kPa, s=0 kPa

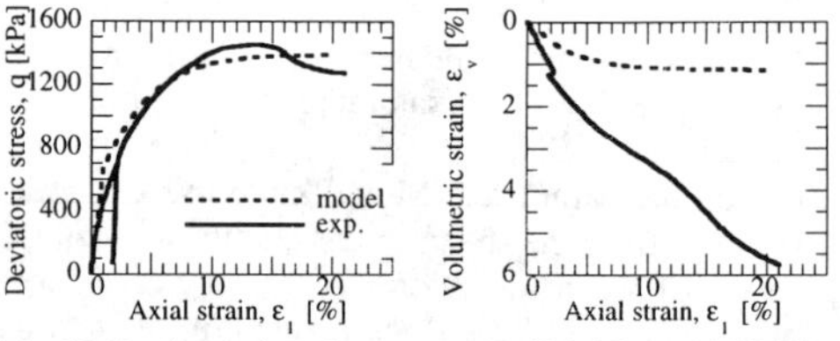

Fig.12:Drained shearing test, σ_3=600 kPa, s=100 kPa

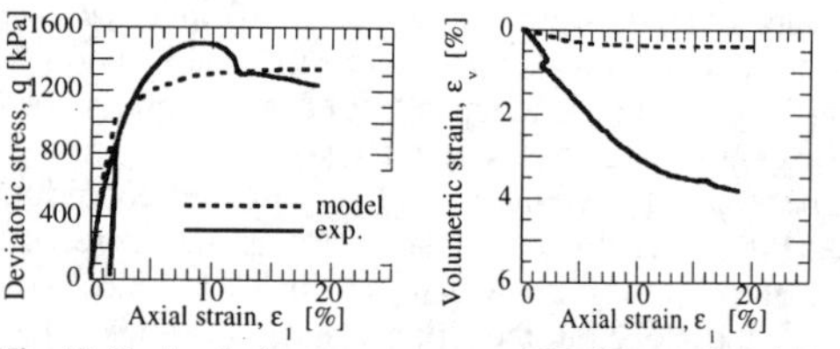

Fig.13:Drained shearing test, σ_3=600 kPa, s=200 kPa

tends to be more pronounced for higher suctions. However this model is applicable to the principal paths (for qualitative responses in isotropic conditions and for resistance and volume predictions in deviatoric paths) when the soil is partially saturated ($0.9 \leq Sr \leq 1$ as a first approximation).

4. NEW APPROACH BASED ON THE DSC

The use of a general concept in the framework of the Disturb State Concept (DSC) (Desai, 1994) is proposed to model the particular volumetric behaviour of the unsaturated soils and the loss of strength in the stress-strain relationship due to suction. The disturb state concept was first presented by Desai and co-workers. We present here first steps of the use of this approach in the modelling of unsaturated soils (Desai *et al.*, 1996).

The DSC is based on the idea that a deforming material element can be treated as a mixture of two constituent parts in the relative intact (RI) and fully adjusted (FA) states, referred to as reference state. During external loading, the material experiences internal changes in its microstructure due to a self-adjustment process, and as a consequence, the initial RI state transforms continuously to the FA state. The observed stress σ^a is defined as:

$$\sigma^a = (1-D)\sigma^i + D\,\sigma^c \tag{4}$$

where σ^i is the RI stress, σ^c the FA stress and D the disturbance function ($0 \leq D \leq 1$).

The FA state is considered as the stress state with D=1. It corresponds to the saturated state.

Figure 14 shows schematically the first choices made for the modelling of the unsaturated behaviour of the silt.

4.1 *Modified δ_l-model*

To represent the relative intact soil behaviour, we modified the nonassociative Hierarchical Single Surface Model δ_1-HISS (Desai, 1994). The expanding yield surface F is defined by:

$$F \equiv J_{2D}^* - \left[-\alpha(J_1^*)^n + \gamma(J_1^*)^2\right](1 - \beta\, S_r)^{-0.5} \tag{5}$$

where the material parameters are explicit functions of the suction.

Expressions in Eq. (5) are the following:

$J_{2D}^* = J_{2D}/p_a^2$, J_{2D} is the second invariant of the deviatoric stress tensor, t_{ij};

$J_1^* = J_1/p_a$, J_1 is the first invariant of the net stress tensor $J_1=3(p-u_a)$;

p_a = atmospheric pressure constant;

γ and β are the ultimate parameters;

S_r is the stress ratio $= \sqrt{27/2}\; J_{3D} \cdot J_{2D}^{-3/2}$, J_{3D} is the third invariant of the deviatoric stress tensor t_{ij};

α is the hardening function defined as:

$$\alpha = \frac{a_1}{\xi^{\eta_1}} \tag{6}$$

where a_1 and η_1 are the hardening parameters and ξ is the trajectory of total plastic strains given by

$$\xi = \int (d\varepsilon_{ij}^p d\varepsilon_{ij}^p)^{1/2} \tag{7}$$

n is the phase change parameter related to the state of stress at which transition from compaction to dilation occurs or at which the change in the volume vanishes. The plastic potential function is defined in terms of F and a correction function introduced through α. It requires a nonassociative parameter κ.

For the elasticity (here linear) two more parameters are needed: the Young's modulus E and the Poisson's ratio ν.

4.2 *RI and FA states*

The RI-state is represented by the modified δ_1-model. As a first approach, we only take into account the evolution of the three parameters γ, a_1 and n with the suction. Further, we do not change the flow rules and we keep D=0. The parameters β, η_1, κ, E and ν are assumed to be independent of the suction.

The FA-state is the saturated case. It can be seen as a particular case of the RI-state when s = 0.

The parameter evolution with the suction are determined experimentally (see Figure 15). With increasing suction, the parameter γ (which defines the slope of the ultimate envelope) increases and the hardening parameter a_1 decreases. This denotes a stiffening of the soil with increasing suction. The phase change parameter n seems also to be affected by the suction.

Three simulations are done for the RI-state (see Figure 16) with the modified δ_1-model for the same initial net mean pressure p*=400 kPa:

- saturated case: σ_3=400 kPa, s=0kPa
- small suction: σ_3=450 kPa, s=50kPa
- high suction: σ_3=600 kPa, s=200kPa

The saturated simulation is in good accordance with the experimental data in both deviatoric and volumetric plane. The test at a small suction (s=50 kPa) gives also quantitatively good results. However as for the former model, similar observations can be done (especially for the case of a high suction) for the strength and the volume variation.

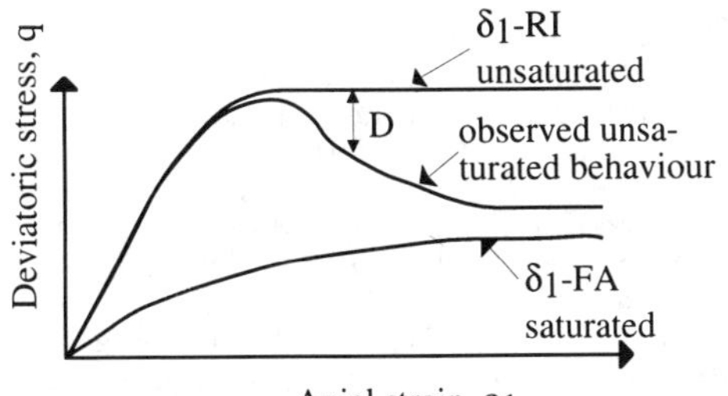

Fig.14: DSC concept for unsaturated soil

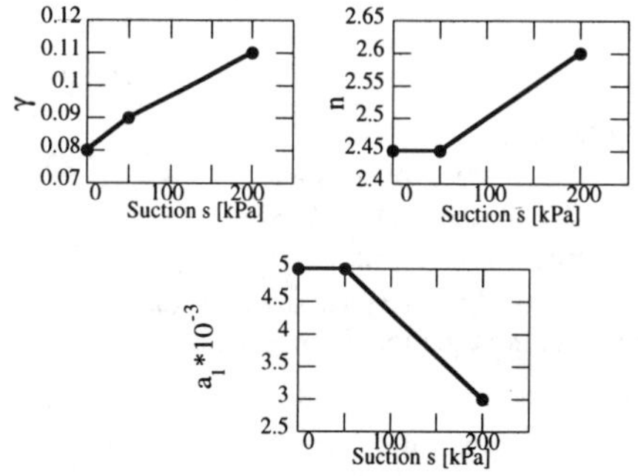

Figure 15: Evolution of the parameters γ, n and a_1 with the suction

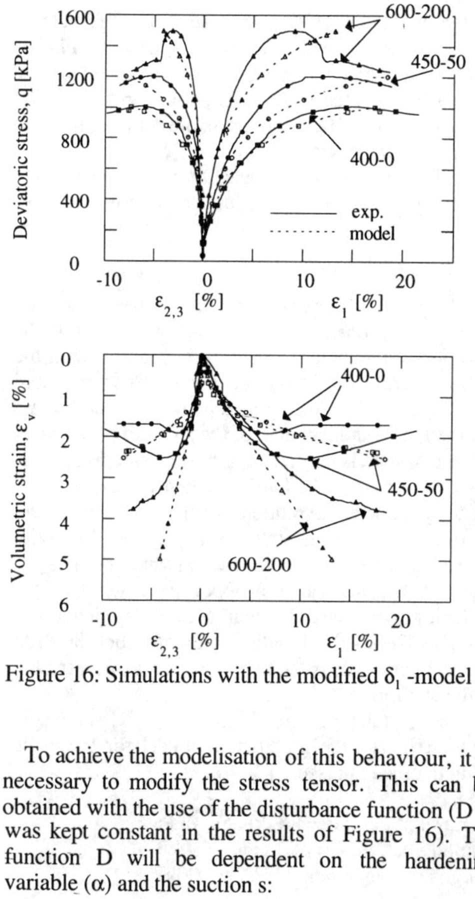

Figure 16: Simulations with the modified δ_1 -model

To achieve the modelisation of this behaviour, it is necessary to modify the stress tensor. This can be obtained with the use of the disturbance function (D was kept constant in the results of Figure 16). The function D will be dependent on the hardening variable (α) and the suction s:

$$D = D(\alpha, s) \qquad (8)$$

This approach is actually developed.

5. CONCLUSION

The test data obtained on a sandy silt are used to examine the performance of the constitutive model proposed in 1990 by Alonso *et al.* The model reproduces relatively well the principal characteristics in the partially saturated state ($0.9 \leq Sr \leq 1$). Under deviatoric solicitation, both predicted resistance and volumetric behaviour match the experimental results. The predictions are only qualitatively good along isotropic path. However this model is not able to predict the particular volumetric behaviour (continuous volume contraction) and the loss of strength for small degrees of saturation ($0.5 \leq Sr \leq 0.9$). To improve the constitutive approach, we present a first step towards the use of the DSC to take into account these particular features of unsaturated soils.

6. ACKNOWLEDGEMENTS

The authors thank Prof. Alonso and J. Vaunat from Uni. Barcelona for their help to use their numerical model inside the ALERT Geomaterial Network. They also thank Prof. Desai for working with them on the DSC for unsaturated soils during his stay at the EPFL in Summer 1996. This work was supported by the Swiss NSF, grant 21-42360.94.

7. REFERENCES

Alonso, E.E., A. Gens & A. Josa 1990. A constitutive model for partially saturated soils. *Géotechnique* 40(3): 405-430.

Charlier, R., X.L. Li, A. Bolle, F. Geiser, L. Laloui & L. Vulliet 1997. Mechanical behaviour modelling of an unsaturated sandy silt. *Proc. XIV Int. Conf. on Soil Mech. & Found. Eng.*: Hamburg: Balkema.

Desai, C.S. 1994. *Hierarchical single surface and the disturb state constitutive models with emphasis on geotechnical applications* in Geotechnical Engineering: Emerging trends in design and practice (ch. 5) New Delhi: Oxford and IBH.

Desai, C.S., L. Vulliet, L. Laloui & F. Geiser 1996. *Disturb state concept for constitutive modeling of partially saturated porous materials* Internal report EPFL.

Fredlund, D.G., N.R. Morgenstern & R.A. Widger 1978. The shear strength of unsaturated soils. *Can. Geotech. J.* 15: 313-321.

Laloui, L., F. Geiser, L. Vulliet, X.L. Li, A. Bolle & R. Charlier 1997. Characterisation of the mechanical behaviour of an unsaturated sandy silt. *Proc. XIV Int. Conf. on Soil Mech. & Found. Eng.*: Hamburg: Balkema.

Laloui, L., L. Vulliet & G. Gruaz 1995. Influence de la succion sur le comportement mécanique d'un limon sableux. *Proc. First Int. Conf. on Unsaturated Soils*: 133-138. Paris: Balkema.

Lloret, A. & E.E. Alonso 1985. State surfaces for partially saturated soils. *Proc. 11th Int. Conf. Soil Mech. Fdn. Eng.*: 557-562. San Francisco.

Maâtouk, A. 1993. *Application des concepts d'état limite et d'état critique à un sol partiellement saturé effondrable*. Doctoral thesis, University of Laval.

Computer Methods and Advances in Geomechanics, Yuan (ed.) © 1997 Balkema, Rotterdam, ISBN 90 5410 904 1

A strain softening constitutive model for rock interface

A.Yashima & F.Oka
Department of Civil Engineering, Gifu University, Japan

T.Sumi
Teikoku Kensetsu Consultant Co., Ltd, Japan (Formerly: Gifu University), Japan

ABSTRACT: An elasto-plastic constitutive model for rock interface is proposed based on a stress history vector. The stress history vector is used in order to express both strain hardening and strain softening characteristics of rock interface. The applicability of the proposed constitutive model is examined by means of a finite element analysis. A comparison between the numerical results and the experimental results on mortar interface is carried out. It is found that the proposed constitutive model can simulate both strain-hardening and strain-softening behaviors of rock interface.

1 INTRODUCTION

It is well known that the shear stress of rock interface with a certain roughness increases in the early stages of shear loading, reaches a peak strength, then decreases, and finally reaches its residual shear strength. In designing civil engineering structures on/in rock masses with discontinuities, it is very important to take into account this rather complicated behavior of rock interface.

In the present study, an elasto-plastic constitutive model for rock interface is proposed based on a concept of the history of the stress vector acting on the interface. The stress history vector is used in order to express both strain-hardening and strain-softening characteristics. A history of stress vector is defined as a functional of the stress vector history with respect to a new displacement measure. The proposed displacement measure is similar to the intrinsic time used in the endochronic theory proposed by Valanis (1971). Although the observed shear stress of rock interface increases to a peak strength and then decreases to a residual strength, the stress history vector monotonously increases during the entire deformation process and leads to a residual strength state. At a large displacement, the difference between stress and the stress history vector devreases. Therefore, it can be said that the difference between stress and the stress history vector corresponds to the strength due to the interface roughness (Fig.1).

In the first half of this paper, the stress history vector, the plastic potential function, and the hardening rule in the proposed constitutive model are explained. The procedure for determining the values of the material parameters in the constitutive model is discussed based on the experimental results on mortar interface. The correlation between the values of the material parameters, normal stress, and interface roughness are examined.

In the second half of this paper, the applicability of the proposed constitutive model is examined by means of a finite element analysis. A comparison between the numerical results and the experimental results on mortar interface is conducted. It is found that the proposed constitutive model can simulate both strain-hardening and strain-softening behaviors of interface. The procedure for determining material parameters is also applied to another experimental case on andesite interface. The proposed procedure is found to be effective in the simulation of the behavior of a real rock interface.

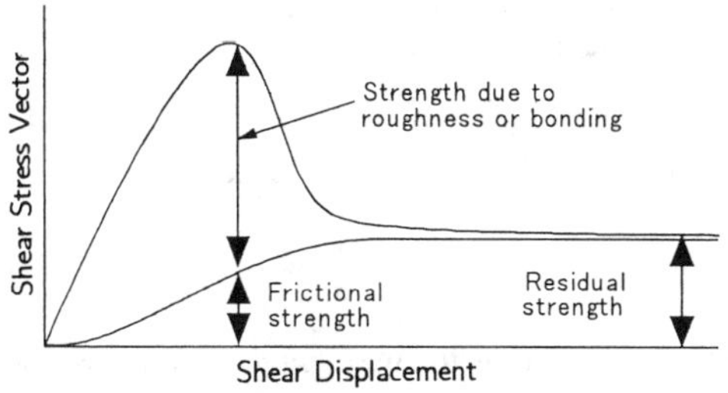

Fig.1 Stress-strain relation for rock interface.

2 CONSTITUTIVE MODEL FOR ROCK INTERFACE

Adachi and Oka (1992) extended an elasto-plastic model for overconsolidated clay by use of the stress history tensor in order to express both strain hardening and strain softening characteristics. They proposed that the stress history tensor is a functional of the stress history with respect to a new strain measure. In this study, we propose stress history vector $\vec{\sigma}^*$ using relative displacement measure z to model the strain-haardening and strain-softening behaviors of the rock joint, namely,

$$\vec{\sigma}^* = \frac{1}{\beta}\int_0^z \exp(-\frac{z-z'}{\beta})\vec{\sigma}(z')dz' , \tag{1}$$

$$dz = (du \cdot du + dv \cdot dv)^{1/2} , \tag{2}$$

$$\vec{\sigma} = \begin{pmatrix} \tau \\ \sigma_n \end{pmatrix}, \quad \vec{\sigma}^* = \begin{pmatrix} \tau^* \\ \sigma_n^* \end{pmatrix} \tag{3}$$

where β is the material parameter which controls the strain-hardening and strain-softening phenomena, du is an increment of the relative shear displacement of the joint, dv is an increment of the relative normal displacement of the joint, τ and τ^* are shear component of the stress vector and the stress history vector, and σ_n and σ_n^* are the normal components of the stress vector and the stress history vector, respectively.

Stress increment vector $d\vec{\sigma}$ is calculated by

$$d\vec{\sigma} = Dd\vec{\varepsilon}^e = D(d\vec{\varepsilon} - d\vec{\varepsilon}^p), \tag{4}$$

$$D = \begin{pmatrix} k_s & 0 \\ 0 & k_n \end{pmatrix}, \quad d\vec{\varepsilon} = \begin{pmatrix} du \\ dv \end{pmatrix}, \quad d\vec{\varepsilon}^p = \begin{pmatrix} du^p \\ dv^p \end{pmatrix} \tag{5}$$

where k_s is the shear stiffness of the joint, k_n is the normal stiffness of the joint, du^p is an increment of the plastic relative shear displacement of the joint, and dv^p is an increment of the plastic relative normal displacement of the joint. We summarize the constitutive model as follows:

Yield function and hardening parameter:

Yield function f is not dependent on the real stress vector $\vec{\sigma}$, but is dependent on stress history vector $\vec{\sigma}^*$ and hardening parameter κ, namely,

$$f = \eta^* - \kappa = 0 \tag{6}$$

in which η^* is the stress ratio defined by

$$\eta^* = (\frac{\tau^*}{\sigma_n^*} \cdot \frac{\tau^*}{\sigma_n^*})^{\frac{1}{2}} . \tag{7}$$

The evolution rule for hardening parameter κ is defined by

$$d\kappa = \frac{M_f^2 \cdot G' \cdot d\gamma^p}{(M_f + G'\gamma^p)^2} \tag{8}$$

where γ^p is the accumulated plastic relative shear displacement of the joint

$$d\gamma^p = (du^p \cdot du^p)^{\frac{1}{2}} , \quad \gamma^p = \int d\gamma^p \tag{9}$$

in which M_f is the value of η^* at the residual strength state and G' is the initial tangential gradient of the hardening function.

Plastic potential and boundary surface:

The flow rule concept for the plastic strain in continuum mechanics is used for the calculation of the plastic relative displacement of the joint, namely,

$$d\vec{\varepsilon}^p = \Lambda\frac{\partial g}{\partial \vec{\sigma}} \tag{10}$$

where g is the plastic potential. This plastic potential is defined with parameters σ_{nb} and b, on shown in Fig.2.

$$g = \eta + Mln\{(\sigma_n + b)/(\sigma_{nb} + b)\} = 0 \tag{11}$$

where the stress ratio η is defined by

$$\eta = (\frac{\tau}{\sigma_n + b} \cdot \frac{\tau}{\sigma_n + b})^{\frac{1}{2}} . \tag{12}$$

M is the parameter that controls the development of the relative normal displacement of the joint. In this study, boundary surface f_b is introduced to determine the value of M.

$$f_b = \eta + M_m ln\{(\sigma_n + b)/(\sigma_{nb} + b)\} = 0 \tag{13}$$

where M_m is the value of η when the maximum contractive normal displacement of the joint takes place. In the overconsolidated state ($f_b < 0$), M is defined as

$$M = -\eta/ln\{(\sigma_n + b)/(\sigma_{nb} + b)\} \tag{14}$$

On the other hand, M is kept constant in the normally consolidated region ($f_b \geq 0$) as

$$M = M_m . \tag{15}$$

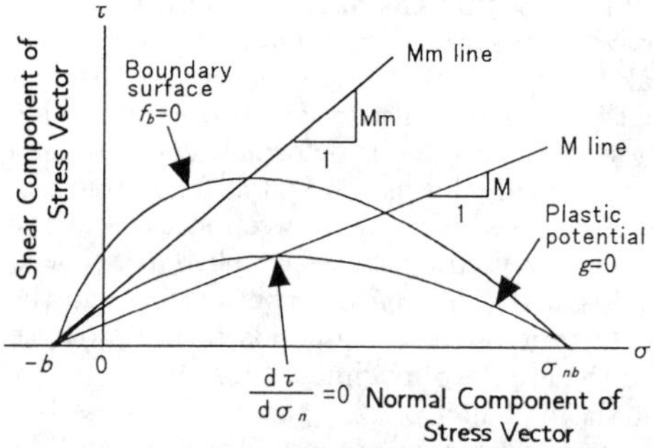

Fig.2 Plastic potential surface and boundary surface.

3 SIMULATION OF JOINT SHEAR TEST ON MORTAR

Nishikawa (1991) carried out direct shear tests on a mortar joint. The joint had a roughness coefficient (RSRI) of 0.61931~0.86982 and a slope of 2° from the horizontal plane. The direct shear tests were carried out for the joint with positive or negative slopes. The normal pressure was kept constant at 0.5, 1.0, 2.0 or 4.0 MPa. The rate of horizontal displacement was 1 mm/min. The direct shearing with the same normal pressure was repeated three times and the effect of the breakage in joint roughness for the mechanical behavior of the joint was investigated. After a horizontal displacement of 11 mm was applied, the upper block of mortar was reset at the initial position and then second or third shear tests were conducted.

Material parameters k_s and M_f can be determined from the experimental results. Fig.3 summarizes the relationship between initial joint shear stiffness k_s and the normal component of stress vector, σ_n. For the first shear, the linear relation is obtained in a double logarithmic diagram for the joint with both positive and negative slopes. The initial shear stiffness for first, second and third shear tests is shown with respect to the accumulated horizontal displacement in Fig.4. It is found from this figure that during the first direct shear test, the roughness of the joint was broken and the shear stiffness decreased.

The variation in the residual shear strength, with respect to the normal component of stress vector, is summarized in Fig.5. The unique linear relationship between the residual shear strength and the normal stress is obtained. The slope of this straight line is regarded as the value of M_f. From this figure, M_f=0.73 is obtained.

To determine the values of other material parameters, finite element analyses are carried out with the finite element mesh drawn in Fig.6. In this figure, the case for a joint with a negative slope is shown. To reproduce the experimental conditions, the horizontal displacements are applied at nodes 5 and 7.

The simulated shear stress-horizontal displacement relation is very sensitive to the values of parameters β and G'. The relationship between β and G' in Fig.7 which gives reasonable numerical results is plotted in the double logarithmic diagram. From this figure, it is found that there is a unique linear relationship between β and G' in the double logarithmic diagram. If we optimize either of these two parameters, the other parameter can easily be determined based on the unique relationship shown in Fig.7. A reduction in the value of β, with respect to the accumulation of horizontal displacement, is shown in Fig.8. It is easily understood that the value of β decreases due to a breakage in joint roughness.

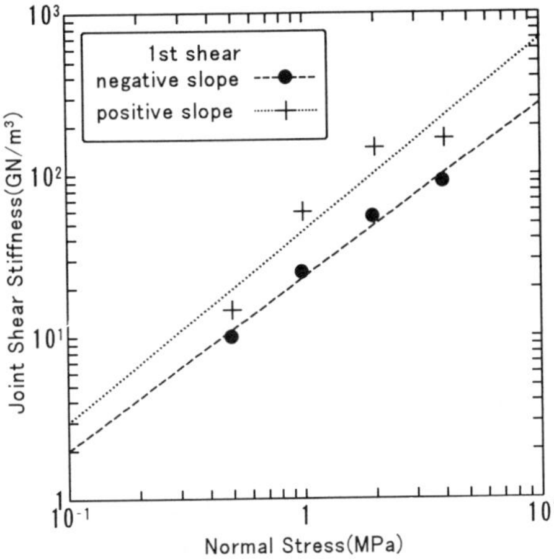

Fig.3 Relationship between initial joint shear stiffness and the normal component of stress vector (after Nishikawa, 1991),

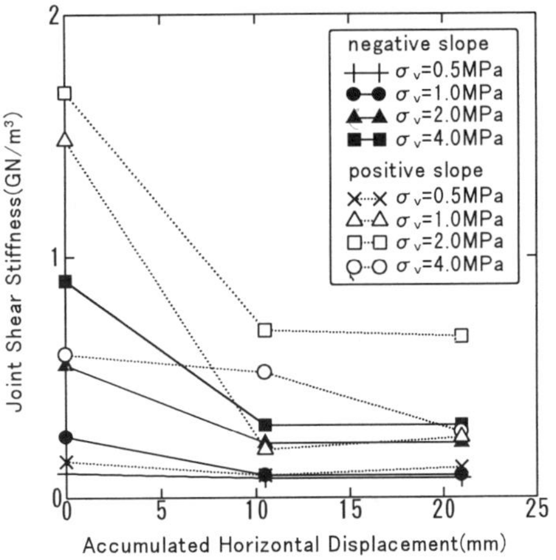

Fig.4 Initial shear stiffness for first, second and third shear tests with respect to the accumulated horizontal displacement (after Nishikawa, 1991).

The simulated results are not sensitive to other material parameters such as k_n, σ_{nb} and b in the range of the normal stress considered. In this study, therefore, the value of k_n is considered to be equal to the value of k_s. σ_{nb}=15 MPa and b=4 MPa, respectively.

Figs.9 and 10 show the shear stress-horizontal displacement relations for the mortar joint with negative and positive slopes. It is seen that the simulated results agree well with the results ob-

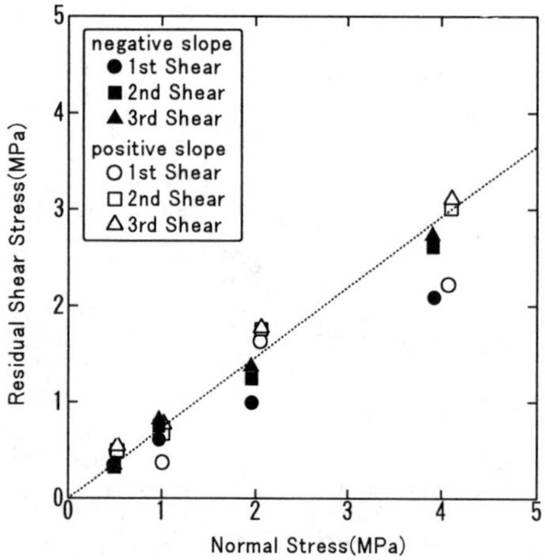

Fig.5 Variation in the residual shear strength with respect to the normal component of stress vector (after Nishikawa, 1991).

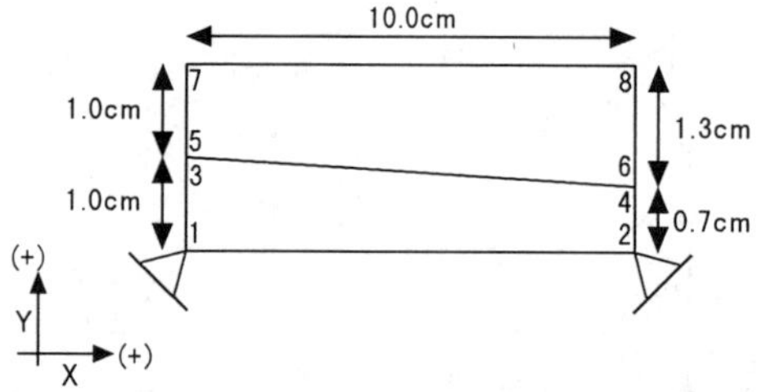

Fig.6 Finite element mesh for the mortar joint with a negative slope.

tained from direct shear tests. The effect of the normal stress on the behavior of the joint may be evaluated by comparing the results of different normal stresses. The joint subjected to a higher normal stress level exhibits greater peak and residual shear strengths. The effect of the breakage in joint roughness can be understood by comparing the results of the first, second and third shear tests. It is found that the strain-softening behavior is no longer observed once the joint roughness has been broken and the shear stress monotonously increases to the residual strength.

4 APPLICATION TO ROCK INTERFACE

The proposed constitutive model is applied to analyze the behavior of an andesite joint at several normal pressure levels carried out by the technical committee of the Japanese Geotechnical Society. First, the material parameters for the normal stresses of 0.20 MPa and 3.73 MPa are adapted. Then, β and G' for the normal stress of 0.79 MPa

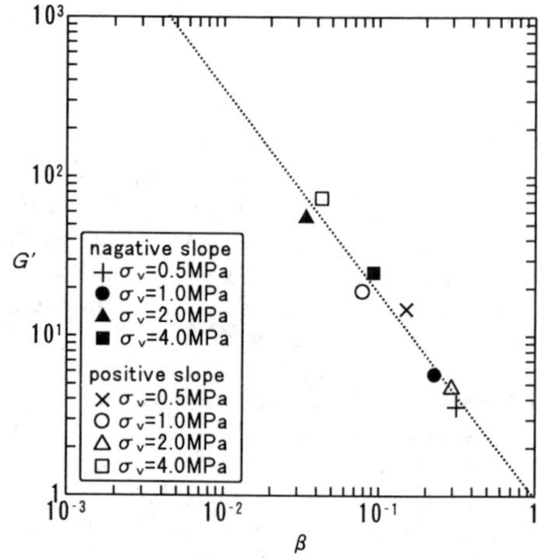

Fig.7 Relationship between G' and β.

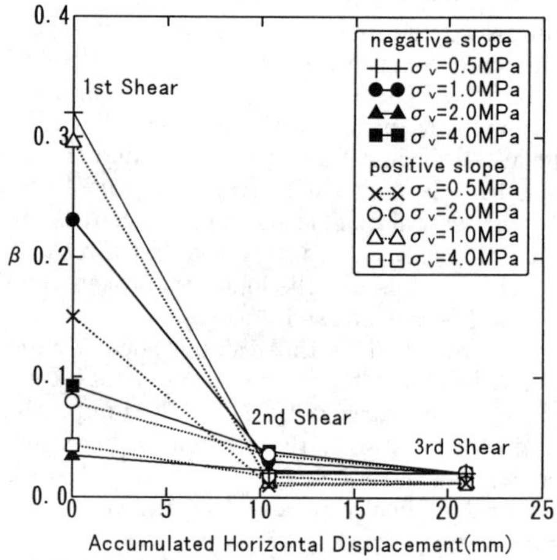

Fig.8 Reduction in the value of β with respect to the accumulated horizontal displacement.

are determined based on the linear relationship between β and G' in the double logarithmic diagram. For the andesite joint, the following relation is obtained.

$$logG' = 0.275 - 2.88log\beta \tag{16}$$

Fig.11 shows the shear stress-horizontal displacement relation for the andesite joint at the normal stress of 0.79 MPa. The results of the prediction agree well with the experimental observations.

CONCLUSIONS

An elasto-plastic constitutive model for rock interface is proposed and applied to an analysis of direct shear tests on mortar and andesite joints. The procedure for determining the values of the

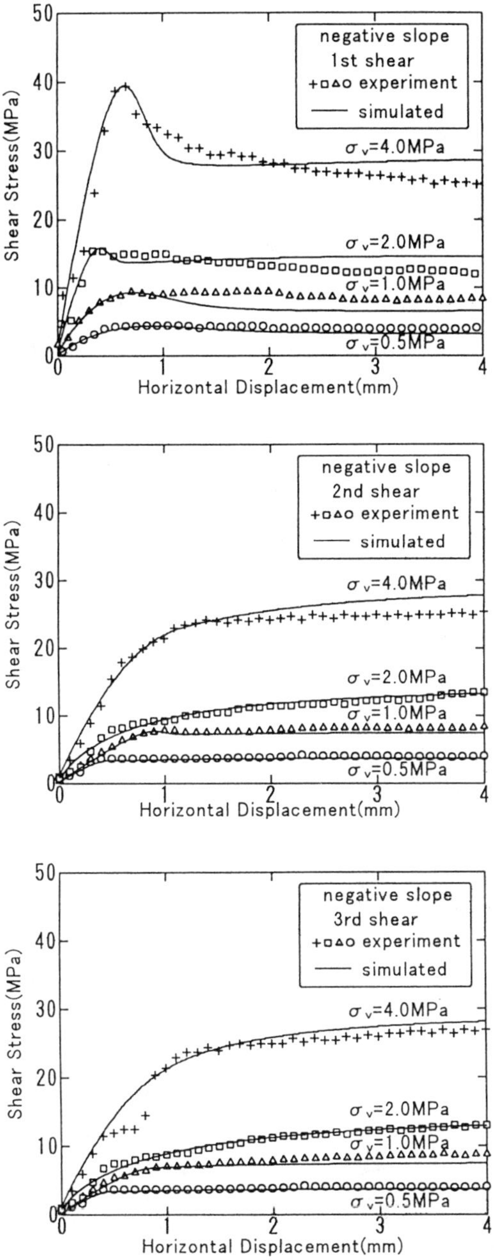

Fig.9 Shear stress-horizontal displacement relations for the mortar joint with a negative slope.

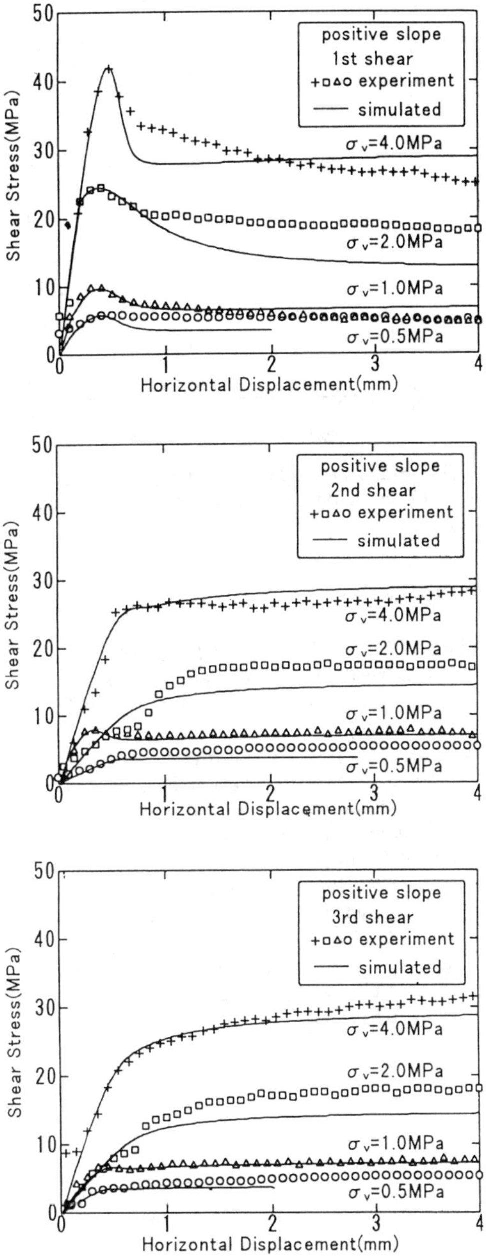

Fig.10 Shear stress-horizontal displacement relations for the mortar joint with a positive slope.

material parameters in the constitutive model is also proposed. The results of the analysis indicate that the constitutive model can well predict the behavior of the rock interface, which exhibits strain-hardening and strain-softening characteristics.

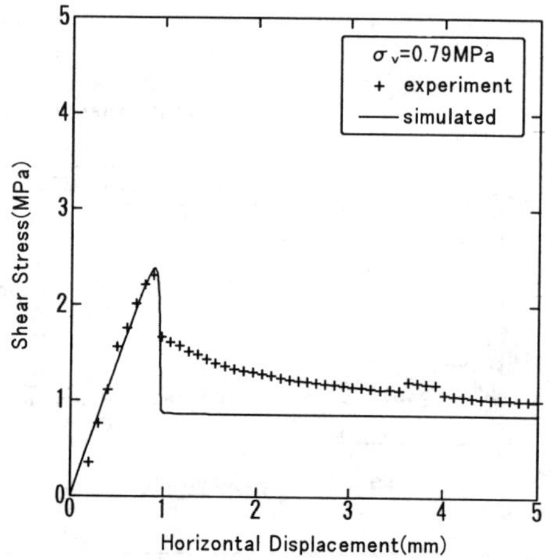

Fig.11 Shear stress-horizontal displacement relations for the andesite joint at the normal stress of 0.79 MPa.

REFERENCES

Adachi, T. and Oka, F. 1992. An elasto-plastic constitutive model for soft rock with strain softening. *Proc. JSCE.* 445/3-18: 9-16. (in Japanese).

Nishikawa, N. 1991. Experimental study on the strength and deformation characteristics of rock discontinuity. *Bachelor Thesis of Kyoto University* (in Japanese).

Technical Committee of Rock Mechanics. 1995. Research Report on discontinuous rock mass and structure. *The Japanese Geotechnical Society.* 79.

Valanis, K.C. 1971. A theory of viscoplasticity without a yield surface. *Arch. Mech. Stos.* 23(4): 517-533

Computer Methods and Advances in Geomechanics, Yuan (ed.) © 1997 Balkema, Rotterdam, ISBN 90 5410 904 1

Numerical formulation of dynamic behavior within saturated soil characterized by elasto-viscous behavior with an application to Las Vegas Valley

J. Li & D.C. Helm
School of Engineering, Morgan State University, Md., USA

ABSTRACT: In the present paper, a unified governing equation is employed for a saturated soil matrix with nonlinear behavior under dynamic conditions. This governing equation for aquifer movement was developed by Li and Helm (1995) from a new constitutive law and from first principles such as mass balance and momentum balance. The governing equation is discussed and is further simplified for numerical analysis. Based on the simplified governing equation, numerical formulations with the finite element method (FEM) in space and with the finite difference method (FDM) in time step are described. In the present paper, the soil skeleton is assumed to be a very viscous non-Newtonian fluid. Nonlinear behavior of the particulate structure allows the soil to behave as an "elasto-viscous" material. The total viscous matrix in numerical formulation shows two viscous components. One is a skeletal viscous matrix that comes directly from structural viscous damping. The other is the fluid viscous matrix due to the drag of viscous pore flow through porous material. These two damping components comprise the total *damping effect* that is usually based entirely on observed empirical results. An example of numerical modeling is conducted for a simplified case in Las Vegas Valley with an input of an actual earthquake record (Mexico, 1976).

1. INTRODUCTION

In earthquake engineering, the dynamic response of saturated sedimentary material such as soil and rock is of primary concern. A new governing equation developed by Li and Helm (1995) with the assumption of soil skeletal nonlinear behavior is employed in the present paper. Consideration of viscous pore fluid and nonlinear soil skeleton behavior allows this governing equation to describe a unified physical processes (e.g., wave, diffusion and creep). This governing equation also overcomes one common shortcoming in previous investigations (Biot, 1962; Fukuo, 1969; Zienkiewicz and Bettess, 1982) in that the significance of relative acceleration $\mathbf{a}^r$ between solid and fluid phases is now accounted for. Based on this governing equation, the previous numerical formulations established by earlier investigators can be generalized by the new numerical formulation developed in this paper. In a simplified case in Las Vegas Valley, dynamic response to an earthquake for saturated soil is portrayed to occur with a combination of wave and diffusion behavior. The conceptual and numerical modeling is based on numerical formulations developed in this paper.

2 GOVERNING EQUATION FOR SKELETAL MOVEMENT

From mass and momentum conservation as well as a constitutive law for skeletal movement characterized with a non-Newtonian behavior, the governing equation describing aquifer movement in terms of the displacement of the solid skeleton is given by the following expression (Li and Helm, 1995):

$$c_1 \ddot{\mathbf{u}} - c_2 \dot{\mathbf{u}} - [\mathbf{D}\nabla(\mathbf{L}\dot{\mathbf{u}}) + \dot{\mathbf{D}}\nabla(\mathbf{L}\mathbf{u})] = \mathbf{R} \qquad (1)$$

where the right-hand side vector $\mathbf{R}$ is given by:

$$\mathbf{R} = c_3\mathbf{b} - c_4 \dot{\mathbf{q}}^b - c_2\mathbf{q}^b - c_5 - \mathbf{R}^\alpha \qquad (2)$$

where coefficients $\mathbf{c}_1$ through $\mathbf{c}_5$ are given by:

$$\mathbf{c}_1 = \rho^s[\delta(1 + 1/eG) - \mathbf{m}^{-1}/2n^2G] \quad (3a)$$

$$\mathbf{c}_2 = \rho^s (\ln \dot{n}) (\delta/Gn - \mathbf{m}^{-1}/2n^2G) + \gamma_w \mathbf{K}^{-1} \quad (3b)$$

$$\mathbf{c}_3 = \rho^s n(1/G - 1)\delta \quad (3c)$$

$$\mathbf{c}_4 = \rho^s(\mathbf{m}^{-1}/2n^2G - \delta/nG) \quad (3e)$$

$$\mathbf{c}_5 = 3\rho_w \sqrt{ng/2eK\pi}\, \mathbf{m}^{-1}\mathbf{f}(t) \quad (3f)$$

where G ($=\rho_s/\rho_w$) is the specific weight, e and n are void ratio and porosity, **m** (=**K**/K, **K** and K is conductivity in forms of a matrix and a scalar) is a second order directional tensor that accounts for anisotropic material (Li, 1995). For isotropic material, **m** reduces to the Kronecker delta δ. g is gravitational acceleration. Superscripts "s" and "w" denote soil and water phases, ρ^s and ρ^w represent the weighted phase densities that equal the product of phase density (ρ_s and ρ_w) and volume fraction (1-n and n) for the volume ratio occupied by each phase, γ_w is the unit weight of water. **b** denotes the body force vector. **L** is an operational matrix for definition of strain and **u** represents the displacement field of the solid phase. **D** is a fourth order tensor of viscosity for the constitutive law and is usually defined as a function of stress, strain, rate of stress and strain, time, space and temperature. $\mathbf{q}^b$ [$= n\, \mathbf{v}^w + (1-n)\, \mathbf{v}^s$] is defined as bulk flux. $\mathbf{R}^\alpha$ is a function of shear dilatancy. **f**(t) is a function of time t and is given as:

$$\mathbf{f}(t) = \int_{-\infty}^{t} [(d\mathbf{v}^r/d\tau)/\sqrt{t-\tau}\,]d\tau \quad (4)$$

where $\mathbf{v}^r$ ($= \mathbf{v}^w - \mathbf{v}^s$) is relative velocity.

Expression (1) is a new governing equation developed by Li and Helm (1995) to describe saturated solid matrix movement under dynamic and viscous conditions. One of the advantages of (2) written in terms of $\mathbf{q}^b$ and its rate is that it is convenient for one to reduce governing equation (1) by invoking the incompressibility condition (i.e., $\nabla\cdot\mathbf{q}^b = 0$ and $\nabla\cdot\dot{\mathbf{q}}^b = 0$).

It is worthwhile to discuss the characteristics of the employed governing equation (1).

The first term on the left-hand side of (1) is the inertial force, the second is the viscous force on the skeleton caused by viscous drag of pore water flow, the third represents a viscous force on the particulate skeleton caused by structural resistance of the matrix itself and the last is an apparent structural elastic resistant force on the skeletal matrix. The third and the fourth terms together behave as a "Kelvin Model" due to non-Newtonian behavior. The first term and the fourth dominate wave propagation in terms of displacement **u**; the first and third dominate diffusion in termsof velocity $\mathbf{v}^s$ ($\equiv\dot{\mathbf{u}}$); the second and the fourth control diffusion in terms of displacement **u**. The right-hand vector **R** has several components which include any change in body force, drag force and other forces caused by the bulk flux $\mathbf{q}^b$ and $\dot{\mathbf{q}}^b$ whose divergence can be reduced by invoking constituent material incompressibility conditions.

3. NUMERICAL FORMULATION FOR THE SIMPLIFIED GOVERNING EQUATION

A simplified form of equation (1) is formulated numerically with the finite element method (FEM). The following development is to discretize the simplified governing equation in both space and time.

3.1 Discretization of the simplified governing equation in space with the finite element method (FEM)

By applying the assumption of small relative acceleration, one can simplify coefficients $\mathbf{c}_1$ through $\mathbf{c}_5$ greatly and can write a reduced form of (1) as:

$$\rho_s^* \ddot{\mathbf{u}} - \gamma_w \mathbf{K}^{-1} \dot{\mathbf{u}} - \nabla(\mathbf{D}\mathbf{L}\,\dot{\mathbf{u}}) = \rho_s^* \mathbf{b} - \gamma_w \mathbf{K}^{-1}\mathbf{q}^b \quad (5)$$

where ρ_s^* denotes the effective density that is defined by $\rho_s^* = (1-n)(\rho_s - \rho_w)$. By assuming negligibly small effect of shear dilatance, namely $\mathbf{R}^\alpha = 0$ in (2), the principle of virtual work and the technology of discretization of finite elements in space can be employed. One can write (1) in terms of displacement **u** by the expression:

$$\int_V \delta\mathbf{u}\{\rho_s^* \ddot{\mathbf{u}} - \gamma_w \mathbf{K}^{-1} \dot{\mathbf{u}} - \nabla[\mathbf{D}(\mathbf{L}\,\dot{\mathbf{u}})]$$

$$-\rho_s^* \mathbf{b} + \gamma_w \mathbf{K}^{-1}\mathbf{q}^b\}dV = 0 \quad (6)$$

where $\delta\mathbf{u}$ is the virtual displacement that is based on the principle of virtual work and V is the volume of interest in space. From the theory of discretization for the finite element method (FEM) (Zienkiewicz, 1977), (6) can be expressed in a compact form and written in terms of the discrete displacement **d** at nodes of an element:

$$\mathbf{M}_n \ddot{\mathbf{d}} + (\mathbf{D}_n - \mathbf{C}_d)\dot{\mathbf{d}} + \dot{\mathbf{D}}_n \mathbf{d} = \mathbf{R} \quad (7)$$

where the mass matrix $\mathbf{M}_n$ is defined by:

$$\mathbf{M}_n = \int_V (\mathbf{N}^T \rho_s^* \mathbf{N}) dV \quad (8)$$

in which **N** is the shape function that depends on the geometric shape of an element. In the second term of (7), $\mathbf{D}_n$ and $\mathbf{C}_d$ are together defined as an apparent viscous matrix. They are caused individually by viscous behavior of skeletal matrix and pore water flow and are given by:

$$\mathbf{D}_n = \int_V (\mathbf{B}^T \mathbf{D} \mathbf{B}) dV \quad (9)$$

$$\mathbf{C}_d = \int_V \mathbf{N}^T (\gamma_w \mathbf{K}^{-1}) \mathbf{N} dV \quad (10)$$

where **B** is called the transform matrix of strain used to associate strain with displacement on element node **d**. For a linear function of **N** (say a triangle element in Cartesian coordinates), **B** will be a constant. The second coefficient ($\mathbf{D}_n$ - $\mathbf{C}_d$) in (7) is defined as a total viscous matrix and includes two components. The term $\mathbf{D}_n$ is related to the third term in (1) caused by skeletal viscous forces and $\mathbf{C}_d$ is related to the second term in (1) caused by a drag force due to viscous fluid flow through the particulate structure. In the third coefficient in (7), $\dot{\mathbf{D}}_n$ denotes the rate of change of nonlinear viscous matrix $\mathbf{D}_n$. It serves the same function as a rigidity matrix in an elastic case. It is given by the expression:

$$\dot{\mathbf{D}}_n = \int_V (\mathbf{B}^T \dot{\mathbf{D}} \mathbf{B}) dV. \quad (11)$$

The right-hand vector **R** in (7) is given by

$$\mathbf{R} = \int_S (\mathbf{N}^T \mathbf{t}) d\mathbf{A} + \int_V \mathbf{N}^T (\mathbf{f} + \rho_s^* \mathbf{b} + \frac{\gamma_w}{\mathbf{K}} \mathbf{q}^b) dV \quad (12)$$

where **A** denotes area, **t** is traction due to the surface force $\mathbf{F}^s$ on the surface S that bounds the volume V and is defined by the following expression:

$$\mathbf{t} = \lim_{\Delta S \to 0} \frac{\Delta \mathbf{F}^s}{\Delta S} = \frac{d\mathbf{F}^s}{dS}. \quad (13)$$

$\mathbf{f}[=D(\nabla \mathbf{D}\varepsilon)/Dt]$ denotes a body force per unit volume.

The first term in (12) stands for a surface force and the second and third denote the skeletal and submerged body force of solids. The fourth is also associated with a body force caused by a bulk driving force in terms of bulk flux $\mathbf{q}^b$.

If one recalls the discrete governing equation for the traditional visco-elastic case when applied in earthquake engineering or geotechnical engineering and compares it to (7), one finds that traditional *mass, damping and rigid matrixes* (Zienkiewicz, 1977) correspond individually to the matrixes $\mathbf{M}_n$, ($\mathbf{D}_n$-$\mathbf{C}_d$) and $\dot{\mathbf{D}}_n$ in (7). However, the significant difference is that in the visco-elastic case, the second term on the left hand-side of (7) is usually based entirely on observed empirical results ascribed ad hoc to a *damping effect* (called $\mathbf{C}_n$) in the soil dynamic response. The traditional form of $\mathbf{C}_n$ is empirically determined by a linear combination of the mass matrix $\mathbf{M}_n$ and a rigidity matrix $\mathbf{G}_n$:

$$\mathbf{C}_n = \alpha \mathbf{M}_n + \beta \mathbf{G}_n \quad (14)$$

where α and β are constants and are related to two parameters called the damping ratio λ and the natural angle frequency ω of an object. Relation (14) is also known as the *Reyleigh Damp* (Zienkiewicz, 1977). In contrast to the traditional empirical ad hoc development of (14), equation (7) is based on a new non-Newtonian constitutive relationship of skeletal deformation and results from rigorous analytical derivation. According to the present theory, the empirical damping matrix $\mathbf{C}_n$ consists of two viscous resistances in the second coefficient of (7), namely $\mathbf{D}_n$-$\mathbf{C}_d$, which gives a clear physical interpretation. The first term is defined as a viscous matrix and is generated from grain-to-grain viscosity and another parameter such as density. The second term is caused by drag due to pore water flow through porous sedimentary material. The actual drag caused by viscous flow is associated with both viscosity of pore fluid and the particulate structure (e.g., porosity), and contributes to viscous damping. Both structural and viscous damping comprise the total damping effect that is apparently observed in experiments due to energy that is dissipated under dynamic loads.

3.2 Discretization in time with the finite difference method (FDM)

The finite difference method is employed to discretize a time step for equation (7). In the present paper,

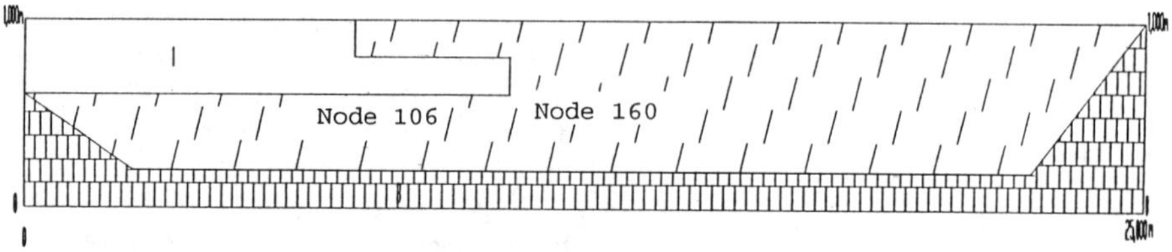

Fig. 1. Heterogenous model for a Las Vegas cross section

Newmark's approach (Zienkiewicz, 1977) is applied, which is an implicit and self-starting method. It starts with step one that is related to the initial conditions. The discretized form for a time step for Equation (7) is given by the following expression:

$$\mathbf{H}_i \mathbf{d}_i = \mathbf{\Phi}_{i-1} \qquad (15)$$

in which:

$$\mathbf{H}_i = \frac{1}{\beta\delta t^2}\mathbf{M}_n + \frac{\gamma}{\beta\delta t}\mathbf{C}_n + \dot{\mathbf{D}}_n \qquad (16)$$

$$\mathbf{\Phi}_{i-1} = \mathbf{R}^n_{i-1} + (\frac{1}{\beta\delta t^2}\mathbf{M}_n + \frac{\gamma}{\beta\delta t}\mathbf{C}_n)\mathbf{d}_{i-1}$$

$$+[\frac{1}{\beta\delta t}\mathbf{M}_n + (\frac{\gamma}{\beta} - 1)\mathbf{C}_n]\,\dot{\mathbf{d}}_{i-1} + \{(\frac{1}{2\beta} - 1)\mathbf{M}_n$$

$$+[(\frac{\gamma}{2\beta} - 1)\delta t\,\mathbf{C}_n]\}\,\ddot{\mathbf{d}}_{i-1} \qquad (17)$$

where t is time, δ not bold is incremental change in a variable, $\mathbf{R}$ is the right-hand vector in (7), $\dot{\mathbf{d}}$ is the rate of nodal displacement, $\mathbf{C}_n$ has been redefined to equal $\mathbf{D}_n$-$\mathbf{C}_d$. γ and β are constants. When γ >or =1/2 and β >or = 0.25 (γ+0.5) (γ+0.5), (14) is unconditionally stable. For a case of γ=1/2 and β=1/4, (14) is also unconditionally stable. When γ=1/2 and β=1/6, (15) is conditionally stable and reduces to Wilson's method that is also a self-starting method.

3.3 An example of numerical modeling for a simplified case in Las Vegas Valley

Based on (15), a computer program has been coded using the finite element method. A preliminary numerical calculation for Las Vegas Valley has been carried out and demonstrates the combined wave and diffusion behavior. According to a fence diagram of lithology in Las Vegas Valley (Plume, 1981), a heterogeneous cross-section is drawn in Fig. 1 in which, three different kinds of sedimentary materials are identified (i.e., sand in zone 1, clay in zone 2 and bed rock in zone 3). From Fig. 1 a mesh 50 x 5 with each element 200 m x 500 m is applied in this case. Actual two dimensional earthquake records (Mexico, 1978) are used to represent dynamic random excitement input along the left hand boundary. Two dimensionality means that an earthquake source at a distance is assumed. On the boundary, components of body forces such as submerged body and bulk driving body forces in (11) are assumed to be negligibly small. Outputs of two components of displacement (horizontal and vertical) from node 160 and one component (vertical) of calculated displacement from node 106 are illustrated in Fig. 2 a, b and c. Node 106 is located at the mouth of material interfingering and 160 is well within clayey material. Node 106 and 160 are respectively located 8.5 km and 13 km away from the input of seismicity along the left hand of boundary (see Fig. 1). For comparison, dynamic responses for homogeneous (dot line) and heterogeneous (solid line) cases are calculated and plotted in the same diagram (Fig. 2 a, b and c). The homogeneous case means that only zone 1 material occupies the entire region. Based on results in Fig. 2, the following conclusions are drawn: 1) Dynamic response with seismic wave behavior is successfully demonstrated. Velocities of seismic waves for horizontal and vertical components at node 160 are about 110m/s and 85 m/s respectively. 2) Attenuation in magnitude of node displacement is significant in this case (e.g., from Fig. 2, magnitude decay of displacement is about 22% per kilometer between nodes 106 and 160). 3) Although frequency scattering is not very significant for both cases, it can still be observed by measuring the peak-to-peak time period on the displacement-time curves of nodes 106 and 160 (e.g., for the heterogeneous case, the difference in period T is about 10 seconds over the first cycle of horizontal components between this two nodes). 4) Homogeneity and heterogeneity as shown in Fig. 1 play interesting roles. The horizontal component at node 106 and the vertical component at node 160, for example, show differences in velocities of propagation, attenuation of magnitude, scattering in frequency and changes in phase due to

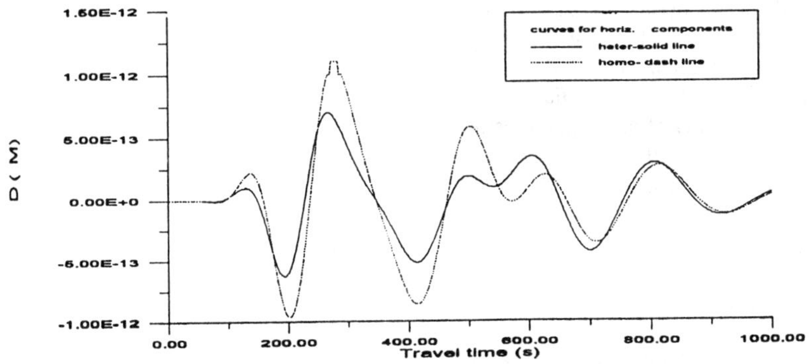

Fig. 2a. Comparison of homogeneous and heterogeneous cases at node #106 (8.5 km) for horizontal components

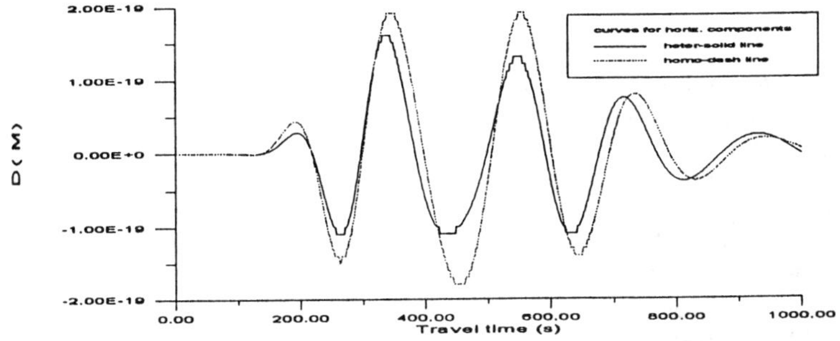

Fig. 2b. Comparison of homogeneous and heterogeneous cases at node #160 (13 km) for horizontal components

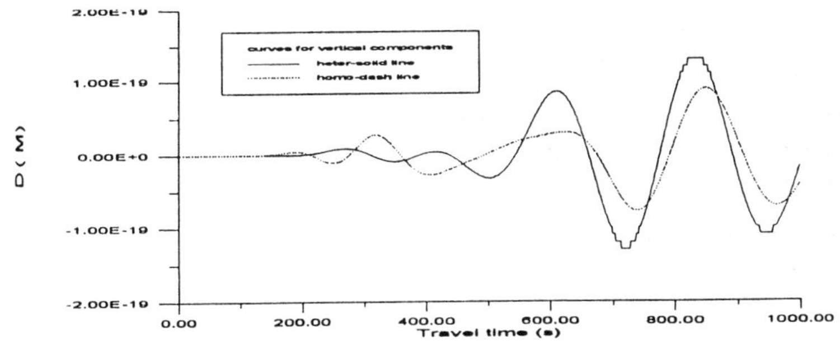

Fig. 2c. Comparison of homogeneous and heterogeneous cases at node #160 (13 km) for vertical components

material properties and geometry. 5) Modification of input for magnitude of seismicity and for soil parameters may be needed so that the magnitude of the displacement-time curve becomes large enough for realistic analysis of velocity and acceleration at a node.

In brief, the simplified governing equation (1) has been successfully employed for a numerical formulation. Numerical modeling and calculation for Las Vegas Valley demonstrate satisfactory results for this unified governing equation that describes the physical process of combined wave and diffusion behavior. A full sensitivity analysis of physical parameters in the numerical calculation is beyond the scope of the present paper and will be investigated in further studies. Physical parameters for both the pore fluid and the skeleton (such as coefficients of propagating velocity and diffusion) are related in this paper directly to phase densities ρ_w and ρ_s, porosity n, hydraulic conductivity **K** and skeletal viscosity **D**.

ACKNOWLEDGMENT

This research was supported by the Nevada Bureau of Mines and Geology. The writing of this paper is supported by Morgan State University.

REFERENCES

Biot, M. A. (1962). Mechanics of deformation and acoustic propagation in porous media. *J. Appl. Phys.* **33**, No. 4, p. 1483-1498.

Fukuo, Y. (1969). Visco-elastic theory of the deformation of a confined aquifer, in Tison, ed. *Land Subsidence*: Internat. Assoc. of Sci. Hydrology, Pub. 88, Vol. **2**, p. 547-562, Tokyo, 1969.

Helm, D.C. (1987). Three-dimensional consolidation theory in terms of the velocity of solids. *Geotechnique* **37**, No. 3, p. 369-392.

Li, J. and Helm D.C. (1995). A General Formulation for Saturated Aquifer Deformation under Dynamic and Viscous Condition. In F.B.J. Barends, F.J.J. Brower and F.H. Schroder (editors), *Land Subsidence*: Internat. Assoc. of Sci. Hydrology Pub 234, p 323-332, Netherlands, 1995.

Plume, R.W. (1981). Ground water condition in Las Vegas Valley, Clark County, Nevada: Part 1, Hydrogeologic framework. U.S.G.S. Water Supply Paper 2320-A.

Zienkiewicz, O.C. (1977). *The Finite Element Method* (third edition), McGraw-Hill.

Zienkiewicz, O.C. and Bettess, P. (1982). Soil and other saturated media under transient, dynamic conditions; General formulation and the validity of various simplifying assumptions, in *Soil Mechanics-Transient and Cyclic Loads*, (eds, G.N. Pande and O.C. Zienkiewicz), John Wiley & Sons Ltd., p. 1-8.

Computer Methods and Advances in Geomechanics, Yuan (ed.) © 1997 Balkema, Rotterdam, ISBN 90 5410 904 1

The dissipated energy equation of lightly cemented clay in relation to the critical state model

Noriyuki Yasufuku, Hidetoshi Ochiai & Kiyonobu Kasama
Department of Civil Engineering, Kyushu University, Fukuoka, Japan

ABSTRACT : In order to evaluate the stress-strain behaviour of clay with an artificially light cementation, a dissipated energy equation is proposed in relation to the critical state model based on the newly assumed critical state concept. The characteristics of the proposed energy equation are first discussed, and then compared with those of Cam-clay and Modified Cam-clay models. The applicability of the critical state model derived from the energy equation was examined on the artificially light cemented clay, using the results of undrained triaxial compression and K_o consolidation tests.

1 INTRODUCTION

Artificially cemented clays are widely utilized as geomaterial to improve the soft ground. It is considered that the mechanical behaviours are remarkably different from those of the reconstitutive clays. Considering the recent progress of numerical analysis and a need of introducing the deformable properties into design, it is essential to develop the constitutive model incorporating the effects of cementation, taking care of the compression and shear properties of cemented clay.

The aim of this paper is originally to present the dissipated energy equation and critical state concept which reflects the cementation effect of clay, based on the critical state framework. This will be made by extending the Cam-clay type energy equation. Further, using the newly presented energy equation and critical state concept, an extended critical state model will be derived for lightly cemented clay, which are formulated by introducing a key internal variable which controls the effect of cementation..

The verification of the constitutive modeling presented is made by an experimental evidence through a series of isotropic consolidations and undrained triaxial compression tests of lightly cemented clay. In addition, the predicted behaviours for undrained and K_o triaxial compression tests are compared with the available experimental data.

2 GENERALIZATION OF CAM-CLAY TYPE ENERGY EQUATION

It is well known that the difference between the Cam-clay model and the Modified Cam-clay one lies in the assumption of the dissipated energy equation, that is, each model assumes the following dissipated energy equation respectively (Roscoe and Burland, 1968; Schofield and Wroth, 1968),

$$dW_{in} = pMd\varepsilon^p \qquad (1)$$

$$dW_{in} = p\sqrt{\left(dv^p\right)^2 + \left(Md\varepsilon^p\right)^2} \qquad (2)$$

where, $d\varepsilon^p$ and dv^p mean plastic shear and volumetric strain increments, respectively, and M, q and p are well-known Cam-clay parameters which mean stress ratio at critical state, deviatoric stress and mean effective principal stress, respectively. In relation to the theoretical consideration of implication of both energy equations, a generalized dissipated energy equation have been found as follows (Yasufuku and Ochiai, 1991; Yasufuku et al., 1993):

$$dW_{in} = p\sqrt{\left(dv^p\right)^2 - Xdv^p d\varepsilon^p + \left(Md\varepsilon^p\right)^2} \qquad (3)$$

Here, the term "$Xdv^p d\varepsilon^p$" can be considered as a soil dilatancy dependent coupling term, in which X is a factor which characterizes the dilatancy property. Using Eq.(3), the following stress-dilatancy relationship is obtained:

$$\frac{dv^p}{d\varepsilon^p} = \frac{M^2 - \eta^2}{X + 2\eta} \qquad (4)$$

Here, it is important to emphasize that, when X=0, Eq.(4) reduces to the stress-dilatancy relationship of the Modified Cam-clay model derived from Eq.(2), and also that, when X=(M-η), the above equation reduces to that of the Cam-clay model derived from Eq.(1), such that:

Table 1 Typical stress-dilatancy relationships

$$dW_{in} = p\sqrt{(dv^p)^2 - X dv^p d\varepsilon^p + (Md\varepsilon^p)^2}$$

$$X = M-\eta \rightarrow \frac{dv^p}{d\varepsilon^p} = M-\eta \quad : \quad X=0 \rightarrow \frac{dv^p}{d\varepsilon^p} = \frac{M^2-\eta^2}{2\eta}$$

$$X = (c-2)\eta \rightarrow \frac{dv^p}{d\varepsilon^p} = \frac{M^2-\eta^2}{c\eta} \quad : \quad X=-2\alpha\eta \rightarrow \frac{dv^p}{d\varepsilon^p} = \frac{M^2-\eta^2}{2(\eta-\alpha)}$$

$$X = (M\eta)^2 \rightarrow \frac{dv^p}{d\varepsilon^p} = \frac{M^2-\eta^2}{(2+M^2\eta)\eta}$$ (Miura et al., 1984)

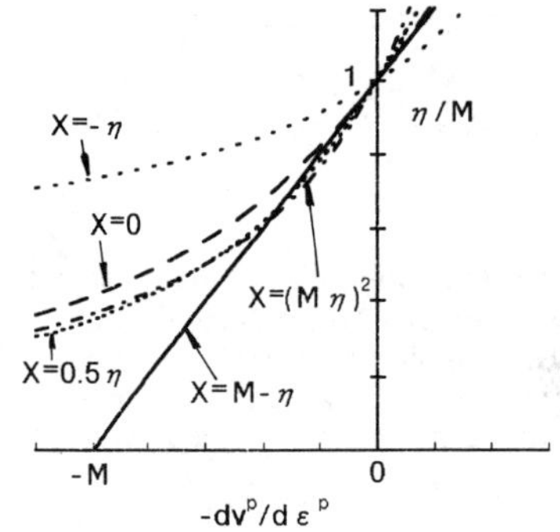

Fig.1 Examples of stress-dilatancy relationships

$$X = 0 \rightarrow \frac{dv^p}{d\varepsilon^p} = \frac{M^2-\eta^2}{2\eta} \tag{5a}$$

$$X = M-\eta \rightarrow \frac{dv^p}{d\varepsilon^p} = M-\eta \tag{5b}$$

Thus, based on the above considerations, it is important to point out that:

1. The dissipated energy equations of the Cam-clay and Modified Cam-clay models give the special cases of Eq.(3), that is, the difference between two models lies only in the assumed value of parameter X in Eq.(3),
2. Eq.(3), which is a generalized energy equation, supports that the Cam-clay equation has an implicit volumetric strain term, which has already been asserted by Ohta (1975), because Eq.(3) becomes equivalent to the Cam-clay equation of Eq.(1) when $X=M-\eta$,
3. Therefore considering the energy equation through Eq.(3), both energy equations for two models can be treated as the level equations with a volumetric strain term.

Fig.1 shows some examples of the stress-dilatancy relationships, which are depicted by assumed stress-dilatancy relationships in Table 1 derived from Eq.(4) with various values of parameter X. In addition, when $X=(c-2)\eta$, for a example in Table 1, Eq.(3) derives a yield function as follows:

$$c = 1: \quad f = q^2 + 2(Mp)^2 \ln\left(\frac{p}{p_o}\right) = 0 \tag{6a}$$

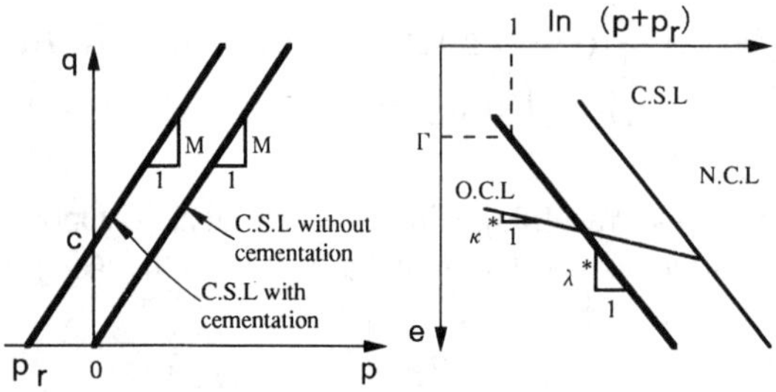

Fig.2 Schematic diagram of critical state line introducing cementation effect

$$c \neq 1: \quad f = p^2 - p^2\left(\frac{p_o}{p}\right)^{\frac{2(c-1)}{c}} + \frac{(c-1)}{M^2}q^2 = 0 \tag{6b}$$

It should be emphasized that an appropriate selection of the value for parameter X through a consideration of stress dilatancy will lead the generalized energy equation into the correct form.

3 INTRODUCTION OF CEMENTATION EFFECT INTO ENERGY EQUATION

The clay considered in this study is assumed to be characterized as a frictional geomaterial with some cohesion due to cementation effects. Therefore, the clay holds the usual compressible and dilative properties as clayey soil.

3.1 *Critical state condition with cementation*

The conditions of critical state in q-p and e-ln p stress space are generally given by :

$$q = Mp \quad (7): \quad e = \Gamma - \lambda \ln p \tag{8}$$

where, Γ: void ratio e at which p is unit pressure, λ: slope of critical sate and/or normally consolidated lines (see Fig.2). In order to newly introduce a cementation effect into the above critical state condition, Eqs.(7) and (8) are extended as:

$$q = M(p+p_r) \quad (9): \quad e = \Gamma - \lambda^* \ln(p+p_r) \tag{10}$$

where, as shown in Fig.2, p_r is a parameter defined by p at q=0, which means that the cementation effect inherently exists at critical state. In general, p_r seems to be an internal variable depending on the deformation characteristics of clay, however, in order to clear the essence of this study, p_r is treated as constant, which is approximately true in a limited stress range. M is given by modified stress ratio η^* at critical state, in which η^* is defined by:

$$\eta^* = \frac{q}{(p+p_r)} \tag{11}$$

Here, when related to the strength parameters c_{cs} and

ϕ_{cs} at critical state, p_r is presented by:

$$p_r = c_{cs} \tan \phi_{cs} \tag{12}$$

It is clear that when p_r=0, Eqs.(9) and (10) reduce to the critical state condition in Eqs.(7) and (8).

3.2 *Dissipated energy equation*

In order to introduce the cohesion component p_r to the generalized energy equation as shown in Eq.(3), the following internal dissipated work dW_{in} is proposed as:

$$dW_{in} = p^* \sqrt{(dv^p)^2 - X dv^p d\varepsilon^p + (M d\varepsilon^p)^2} - p_r dv^p \tag{13}$$

where, p^* is defined as follows:

$$p^* = p + p_r \tag{14}$$

It is important to note that when p_r=0, Eq.(13) reduces to Eq.(3). This expression may be combined with the outer work done (dW_{out} =pdv^p+$qd\varepsilon^p$) to derive the stress-dilatancy relationship considering the cohesion component p_r such that:

$$\frac{dv^p}{d\varepsilon^p} = \frac{M^2 - \eta^{*2}}{X + 2\eta^*} \tag{15}$$

It is obvious that when p_r=0 in this equation, from the definition of η^* in Eq.(11), Eq.(15) coincides with Eq.(3). It is therefore concluded that Eq.(13) is a more generalized form in Eq.(3), which can represent the cementation effect.

4. FORMULATION OF MODEL

4.1 *General stress strain increment*

To formulate a model in general space, Cam-clay stress-strain parameters are generalized as;

$p=\sigma_{ij}\delta_{ij}/3$, $q = \sqrt{(3/2)s_{ij}s_{ij}}$, $\eta=q/p$, $dv=d\varepsilon_{ij}\delta_{ij}$, $d\varepsilon = \sqrt{(2/3)de_{ij}de_{ij}}$, in which σ_{ij} and s_{ij} are stress and deviatoric stress tensor, which are related to s_{ij}= σ_{ij}-$p\delta_{ij}$. and also $d\varepsilon_{ij}$ and de_{ij}, which are defined as de_{ij}=$d\varepsilon_{ij}$-$(dv/3)\delta_{ij}$, are incremental strain and deviatoric strain tensor, respectively, and δ_{ij} is Kronecker delta.

Based on the linear incremental approach in plasticity, it is assumed that the total strain increment $d\varepsilon_{ij}$ can be divided into elastic and plastic parts as follows;

$$d\varepsilon_{ij} = d\varepsilon_{ij}^e + d\varepsilon_{ij}^p \tag{16}$$

For persisting isotropy, the elastic strain increment can be expressed by

$$d\varepsilon_{ij}^e = \frac{1+\nu}{E} d\sigma_{ij} - \frac{\nu}{E} d\sigma_{kk}\delta_{ij} \tag{17}$$

where E and ν are Young's modulus and Poisson's ratio, respectively, which are also associated with the bulk modulus and shear modulus G such that;

$$K = \frac{E}{3(1-2\nu)} \quad (18); \qquad G = \frac{E}{2(1+\nu)} \quad (19)$$

In addition, in the case of critical state approach, based on the $e - \ln p^*$ straight line plot during swelling, K in Eq.(18) is also given by

$$K = \frac{(1+e_o)}{\kappa^*} p^* \tag{20}$$

with e_o the initial value of e and κ^* the slope of $e - \ln p^*$ swelling line.

The plastic strain increment $d\varepsilon_{ij}^p$ with associated flow rule can be expressed by:

$$d\varepsilon_{ij}^p = \langle\Lambda\rangle \frac{\partial f}{\partial \sigma_{ij}} \tag{21}$$

where Λ is a proportional factor, $\langle\ \rangle$ is the Macauley bracket. When the expressions of f and Λ are precisely formulated by the function of state of stress, the plastic strain increments are then clearly computed through Eq.(21).

4.2 *Critical state model with cementation effect*

According to the original critical state model, the isotropic elastic relationship in Eqs.(17)-(21) is assumed to be valid. For the specification of the yield surface with cementation effect, the plastic dissipated energy in Eq.(13) is introduced, which is believed to be a new feature in this study. When assuming $X=(c-2)\eta^*$ as an isotropic hardening, Eq.(13) can be rewritten as

$$dW_{in} = p^* \sqrt{(dv^p)^2 + (2-c)\eta^* dv^p d\varepsilon^p + (M d\varepsilon^p)^2} - p_r dv^p \tag{22}$$

where c is a soil constant which characterizes the shape of yield surface. Based on Eq.(22), the following $\eta^* - dv^p/d\varepsilon^p$ relationship can be derived as follows:

$$\frac{dv^p}{d\varepsilon^p} = \frac{M^2 - \eta^{*2}}{c\eta^*} \tag{23}$$

Applying the normality rule to Eq.(23), the yield function is easily derived, which is dependent on the constant c, such that:

$$c = 1: \; f = q^2 + 2(Mp^*)^2 \, ln\left(\frac{p^*}{p_o^*}\right) = 0 \qquad (24a)$$

$$\frac{\partial f}{\partial p^*} = 2p^*\left(M^2 - (q/p^*)^2\right); \quad \frac{\partial f}{\partial q} = 2q; \qquad (24b)$$

$$\frac{\partial f}{\partial p_o^*} = -2M^2 p^{*2}\left(1/p_o^*\right) \qquad (24c)$$

$c \neq 1$:

$$f = p^{*2} - p^{*2}\left(\frac{p_o^*}{p^*}\right)^{\frac{2(c-1)}{c}} + \frac{(c-1)}{M^2}q^2 = 0 \qquad (25a)$$

$$\frac{\partial f}{\partial p^*} = \frac{2(c-1)p^*}{c}\left(1 - \left(\frac{\eta^*}{M}\right)^2\right); \quad \frac{\partial f}{\partial q} = \frac{2(c-1)q}{M^2} \qquad (25b)$$

$$\frac{\partial f}{\partial p_o^*} = \frac{2(c-1)p^*}{c}\left(\frac{p_o^*}{p^*}\right)^{\frac{2-c}{c}} \qquad (25c)$$

where, p_o^* is defined by $p_o^* = p_o + p_r$ in which p_o is the value of p at which yield surface intersects the p axis. It is important to point out that when p_r=0, Eq.(25) reduces to Eq.(6), in addition, in the case of c=2, Eq.(25) reduces to that of Modified Cam-clay model. The effects of constant c on the shape of yield curves with cementation effect p_r are shown in Fig.3.

Now, based on the associated flow rule, Eq.(21) can be rewritten as follows:

$$d\varepsilon_{ij}^p = \frac{1}{H_p}\left(\frac{\partial f}{\partial \sigma_{ij}}\right)\frac{\partial f}{\partial \sigma_{ij}} d\sigma_{ij} \qquad (26)$$

where,

$$H_p = -\left(\frac{\partial f}{\partial p_o^*}\frac{\partial p_o^*}{\partial e^p}\frac{\partial e^p}{\partial v^p}\right) = -\frac{\partial f}{\partial p_o^*}\bar{p}_o^* \qquad (27)$$

Here, p_o^* is an internal variable defining the hardening characteristics, which represents a change in size of yield surface, and $\bar{p}_o^*$ is a scalar quantity as will be mentioned later. Therefore, in order to calculate the plastic strain increment using Eqs.(26) and (27), the flow vector $(\partial f/\partial \sigma_{ij})$ and hardening modulus H_p must be determined as functions of stress state.

First, for simplicity, both critical state condition and yield function are assumed to be independent on the third deviatoric stress invariant. Then, the yield function has the following general form:

$$f = f(\sigma_{ij}, \; p_o^*) = f(p^*, \; q, \; p_o^*) = 0 \qquad (29)$$

Using this equation, the flow vector can be derived as follows:

$$\frac{\partial f}{\partial \sigma_{ij}} = A_1\frac{\partial p^*}{\partial \sigma_{ij}} + A_2\frac{\partial q}{\partial \sigma_{ij}} \qquad (30)$$

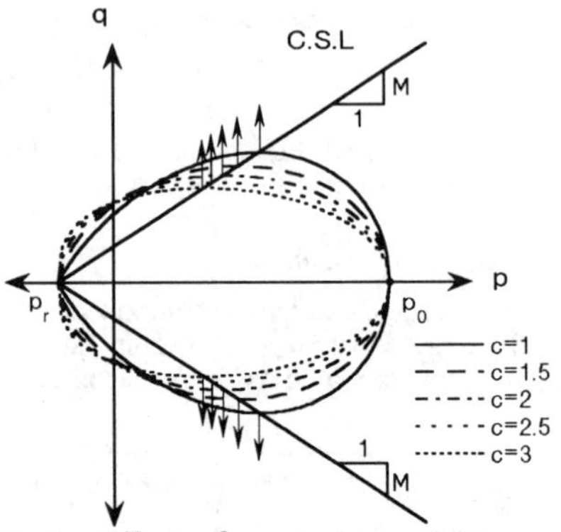

Fig.3 Effects of constant c on yield curve

where, $A_1 = \partial f/\partial p^*$ and $A_2 = \partial f/\partial q$, which are given by Eq.(24-b) or Eq.(25-b), depending on the value of constant c, and also two partially derivatives in Eq.(30) can be defined as $\partial p^*/\partial \sigma_{ij} = (1/3)\delta_{ij}$, $\partial q/\partial \sigma_{ij} = (3/2q)s_{ij}$.

The remaining work to complete the model is to present the evolution equation of p_o^* in Eq.(27), namely, $dp_o^* = \langle\Lambda\rangle\bar{p}_o^*$, which will give H_p according to Eq.(27). Based on the observed $e - ln\ p^*$ straight line during the proportional loading and the $de^p = -(1+e)dv^p$ relationship, the following equations are obtained:

$$\frac{dp_o^*}{de^p} = -\frac{p_o^*}{\lambda^* - \kappa^*}; \quad \bar{p}_o^* = \frac{1+e}{\lambda^* - \kappa^*}\frac{\partial f}{\partial p}p_o^* \qquad (31a)$$

$$H_p = -\frac{\partial f}{\partial p_o^*}\bar{p}_o^* = -\left(\frac{1+e}{\lambda^* - k^*}p_o^*\right)\frac{\partial f}{\partial p_o^*}\frac{\partial f}{\partial p} \qquad (31b)$$

Here, the terms $\partial f/\partial p_o^*$ and $\partial f/\partial p$ are already shown in Eqs.(24) or (25) in relation to constant c. It is noted that when η^*=M, H_p becomes zero, and critical failure occurs.

Based on this critical state model derived by assuming a new dissipated energy equation, which can evaluate the cementation effect of lightly cemented clay, some useful equation can be shown. The evaluation of undrained strength and K_o value in closed form seems to be useful in practical point of view. The undrained strength $q_f = Mp_f^*$ can be calculated by setting $\eta^* = M$ in the following equation:

$$c = 1: \; \frac{p^*}{p_o^*} = exp\left\{-\left(1 - \kappa^*/\lambda^*\right)\frac{\eta^{*2}}{2M^*}\right\} \qquad (32a)$$

$$c \neq 1: \; \frac{p^*}{p_o^*} = \left\{\frac{M^2}{(c-1)\eta^{*2} + M^2}\right\}^{\frac{c(1-\kappa^*/\lambda^*)}{2(c-1)}} \qquad (32b)$$

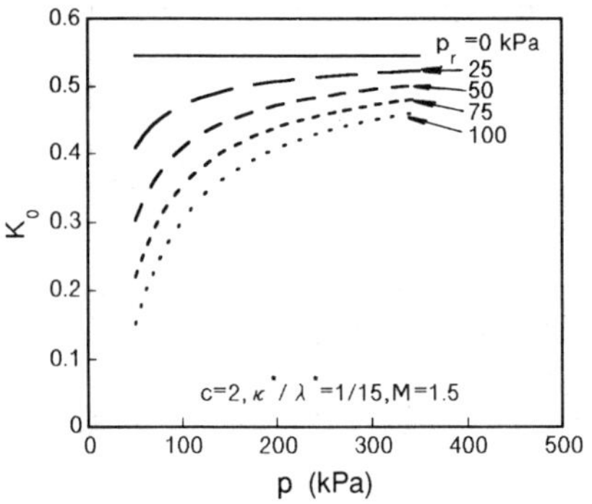

Fig.4 K_o value's dependence on cementation effect

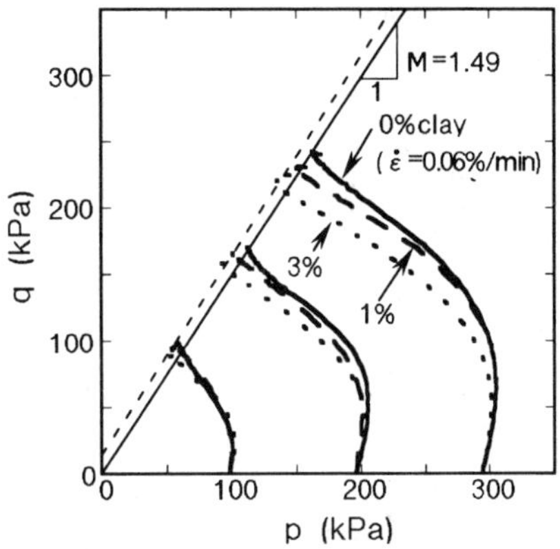

Fig.5 Undrained stress paths for each artificially cemented clay

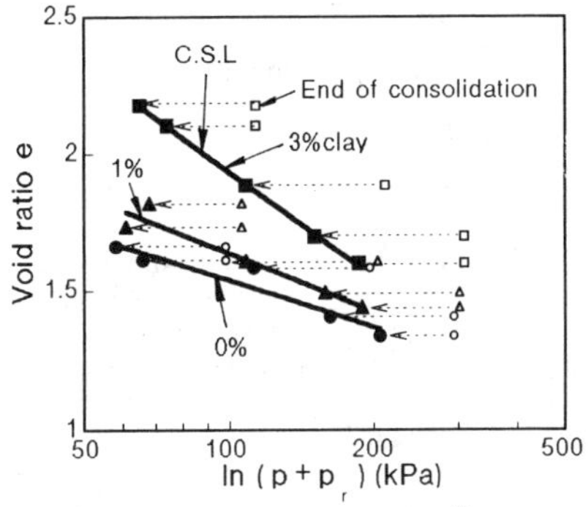

Fig.6 Critical state line in $e - ln\ p^*$ space for each cemented clay

and also when considering K_o consolidation condition, namely, $(d\varepsilon/dv)_{K_o} = 2/3$, the evaluation of K_o value is finally given by :

$$K_o = \frac{3-(1+p_r/p)\eta^*_{K_o}}{2(1+p_r/p)\eta^*_{K_o}+3} \qquad (33a)$$

where,

$$\eta^*_{K_o} = \frac{-\frac{3}{2}c\left(1-\frac{\kappa^*}{\lambda^*}\right)+\sqrt{\left\{\frac{3}{2}c\left(1-\frac{\kappa^*}{\lambda^*}\right)\right\}^2+4M^2}}{2} \qquad (33b)$$

with $\eta^*_{K_o}$ the value of η^* at K_o condition. The cementation effect on the K_o value can be evaluated using this equation.

Fig.4 shows the effect of parameter p_r on the predicted K_o value in the case of c=2, $\kappa^*/\lambda^* = 1/15$ and M=1.5. It can be seen that the predicted K_o value tends to decrease with the increasing p_r, which reflects the cementation effect, and it is also important to note that K_o value tends to close with K_o value at p_r=0 with increasing confining pressure.

5. APPLICATION TO LIGHTLY CEMENTED CLAY

5.1 *Samples and testing method*

First, Ariake clay (w_L=86.5%, I_p=51.3, ρ_s=2.609 g/cm^3) in a slurry of 2w_L water content was mixed adequately with portland cement. In order to set up the Ariake clay sample with light cementation, at which the sample can hold the soil properties as a geomaterial, the cement contents of 1% and 3% per dried sample weights were selected.

Each reconstituted clay with 1% and 3% cements, so called 1%-Clay and 3%-Clay, was one-dimensionally preconsolidated up to 49 kPa confining pressure. The preconsolidated samples were wrapped in moisture proof bags and were cured for 35 days in a humid room at the temperature of 20±3℃. The specimen size tested was 3.5 cm diameter and 7 cm height. According to the standard of JGS(Japanese Geotechnical Society), a series of isotropic and K_o consolidation and then undrained compression tests were carried out using 0%-, 1%- and 3%-clay.

5.2 *Verification of critical state concept*

The applicability of the critical state concept extended here is examined in this section.

Fig.5 shows the undrained stress path in q-p stress space for each cemented clay. Assuming that M value is constant and identical for each specimen, p_r values can easily be obtained from this figure. The determined values are summarized in table 2. We can say from this figure that the introduction of p_r is an effective measure to evaluate the cementation effect of clay.

Fig.6 shows the critical state line in $e - ln\ p^*$ space for each specimen. It can be seen that the slope of critical state lines shown by black marks are remarkably different for each specimen and they can be approximated by straight lines in $e - ln\ p^*$ space.

Table 2 Soil constants used

	e_{o100}	λ^*	κ^*	M	p_r (kPa)	c
0%-clay	1.663	0.243	0.004	1.50	0	2.0
1%-clay	1.881	0.308	0.037	1.50	8	2.0
3%-clay	2.170	0.554	0.021	1.50	15	2.0

e_{o100} : the values of e at p=100 kPa

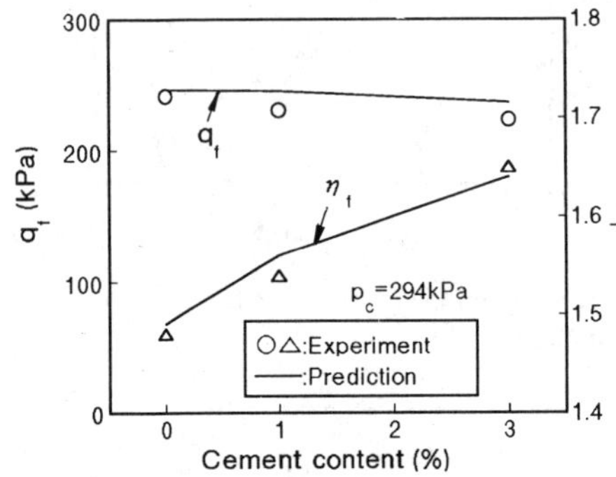

Fig.7 Comparison of observed q_f and η_f with predicted ones

It is interesting to point out that the initial void ratio for 3%-clay is much higher than that of 0%-clay, even if the preconsolidation period (totally 11 days) for each sample is completely the same, and also that the slope of critical state line increases with the increasing cementation effect, namely, values of p_r.

5.3 *Comparison with experimental data*

We will now demonstrate the capability of the presented critical state model with undrained triaxial compression and K_o consolidation tests. Six constants are needed to predict the stress-strain curves, in which four constants are Cam-clay parameters and the other constants c and p_r are newly introduced. Here, for simplicity, it is assumed that c=2. The soil constants are summarized in table 2.

Fig.7 shows the predicted and observed strength q_f and the corresponding stress ratio η_f for each specimen. It is found that the model reasonably represents the tendency of observed results, in which q_f decreases with and also η_f increases with the increasing values of p_r.

Fig.8 shows the applicability of the model to K_o consolidation test results for 0% and 3% cemented Ariake clay, comparing the predicted K_o values with observed ones. It is found that the predicted results give somewhat large K_o values against p, compared with the observed ones. However, the model can reasonably represent the observed tendency that K_o value decreases with the increasing cementation effects and also K_o value of 3%-clay approaches that of 0%-clay with the increasing values of p.

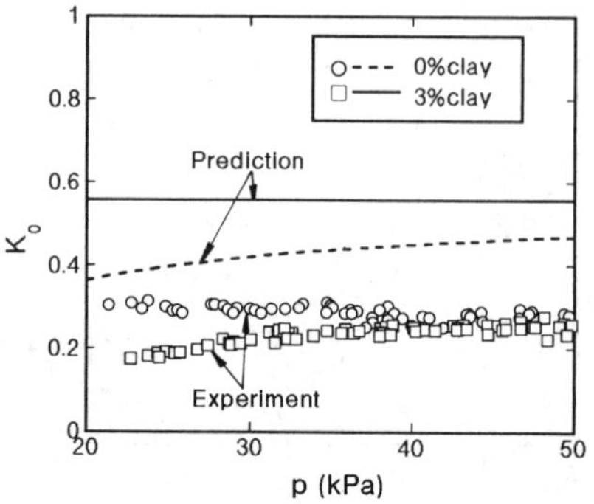

Fig.8 Evaluation of K_o value's depencence on cementation effect

6. CONCLUSIONS

Based on experimental and theoretical considerations, an extended energy equation and critical state concept have been proposed to evaluate the cementation effect of clay. The key characteristic is the introduction of an internal variable which controls the effects of cementation, in which the energy equation is believed to be a generalization of Cam-clay type energy equations. Using the proposed energy equation and critical state concept, a generalized critical state model has been derived for inherently light cemented clay.

It has been shown that when four original Cam-clay parameters and two other additional parameters can practically be determined, the model can reasonably represent the changes in undrained strength and K_o values of artificially light cemented clay with the increasing cementation effects.

REFERENCES

Burland, J. B. 1965. The yielding and dilation of clay, Correspondence, Geotechnique, Vol.15, 211-214.

Miura, N., Murata, H. & Yasufuku, N. 1984. Stress-strain characteristics of sand in a particle-crushing region, Soils and Foundations, Vol.24, No.1, 77-89.

Ohta, H., Yoshitani, S. & Hata, S. 1975. Anisotropic stress-strain relationship of clay and its application to finite element analysis, Soils and Foundations, Vol.15, No.4, 61-79.

Roscoe, K. H. & Burland, J. B. 1968. On the generalized stress-strain behaviour of wet clay, Engineering Plasticity, Cambridge University Press., 535-609.

Schofield, A. N. & Wroth, C. P. 1968. Critical State Soil Mechanics, McGraw-Hill Book Company, New York.

Yasufuku, N & Ochiai, H. 1991. Anisotropic hardening model for sandy soils over a wide stress region, the Memoirs of the Faculty of Engineering. Kyushu University, Vol.51, No.2, 81-118.

Yasufuku, N., Sugiyama, M., Hyodo, M. and Murata, H. 1993. Extended critical state model for cohesive clay with initial induced anisotropy, Technology Reports of the Yamaguchi University, Vol.5, No.2, 103-116.

Computer Methods and Advances in Geomechanics, Yuan (ed.) © *1997 Balkema, Rotterdam, ISBN 90 5410 904 1*

Nonlinear anisotropic modeling of dilative granular material behavior

E.Tutumluer

Department of Civil Engineering, University of Illinois, Urbana-Champaign, Ill., USA

ABSTRACT: A new cross-anisotropic model is proposed to predict the nonlinear dilative granular material behavior. A cross-anisotropic representation has different material properties (i.e., elastic modulus and Poisson's ratio) assigned in the horizontal and vertical directions. Repeated load triaxial test results with vertical and lateral deformation measurements performed on a variety of aggregate types were used to characterize the nonlinear stress dependent granular material behavior for horizontal, vertical, and the shear stiffnesses. Anisotropic elastic properties employed in the models satisfied thermodynamic considerations. The ratio of vertical to horizontal moduli, expressed as a measure of stiffness anisotropy, was generally larger than 1 with ratios rapidly increasing at higher principal stress ratios (σ_1 / σ_3). Dilation was observed and accurately modeled using the anisotropic approach. Depending on the material type, varying degrees of dilation - as defined by the principal strain ratios ($\varepsilon_1 / \varepsilon_3$) - were predicted at high principal stress ratios.

1 INTRODUCTION

The behavior of a granular medium at any point depends on the arrangement of particles which is usually determined by particle size and characteristics, construction methods, and the loading conditions. Granular materials loaded under anisotropic conditions typically exhibit a directional dependency of material stiffnesses. The load transfer is achieved through compression and shear forces between the particles oriented in a chainlike manner (Dobry et al., 1989). As the material stiffens, the deformability decreases with a higher modulus attained in the primary loading direction. Both the important effect of load induced directional stiffening and the dilative behavior of granular materials can be modeled with relative ease if an anisotropic approach is followed.

Elastic behavior of granular particles is generally determined in the small strain range during static unloading and/or repeated load triaxial tests for resilient response after shakedown is reached. The law of elasticity has been shown to be nonlinear, with a dependency of modulus upon the stress state (Boyce, 1980; Lade and Nelson, 1987; Uzan et al., 1992). In triaxial conditions, the elastic modulus E has been best obtained from the experimental data when both the mean pressure $p = (\sigma_1 + 2\sigma_3)/3$ and the deviator stress $\sigma_d = (\sigma_1 - \sigma_3)$ are included in the material characterization models. Any irreversible axial strain accumulated due to anisotropic loading will undoubtedly create an anisotropic structure leading to an anisotropic elastic domain and an anisotropic elastic law (Biarez and Hicher, 1994).

A new nonlinear cross-anisotropic constitutive model is proposed in this paper to predict the dilative response of granular materials as observed from laboratory triaxial test results. A cross-anisotropic representation has different material properties (i.e., elastic modulus and Poisson's ratio) assigned in the horizontal and vertical directions. For a cylindrical triaxial sample, a realistic assignment of in-plane and out-of plane stiffnesses is, therefore, achieved under axial symmetry with the axial stiffness increasing relative to the radial stiffness.

2 DILATION IN GRANULAR MATERIALS

Perhaps the most significant characteristic of a particulate medium is that dilation (increase in volume) takes place under shear loading in a dense state of packing. Many isotropic elastic models, however, fail to predict this kind of dilative behavior which often results in considerable tensile stresses computed in unbound granular media. This is in contrast with the very limited tensile stress taking ability of these materials. Furthermore, when modeling dilative behavior, Poisson's ratios often exceeding 0.5 need to be assigned in the analysis, which are generally inadmissible under the assumption of a homogeneous, isotropic material.

Several researchers have addressed the dilatancy behavior of granular materials to take place under high major to minor principal stress loading conditions. Frydman (1968) found that dilation of dense sands starts to occur at a constant principal stress ratio of 3.5. Later, Billam (1971) observed that the tendency to dilate results in the material experiencing confining stresses in addition to those induced by the load. Similarly, Witczak and Uzan (1988) concluded that under increased shear loading, the material dilates and develops confining pressures in excess of the pressure induced by compaction.

Based on an extensive review of the laboratory test data, Uzan et al. (1992) presented a comprehensive granular material characterization procedure. In addition to the nonlinear stress dependent modulus, a nonlinear stress dependent Poisson's ratio model was also developed. Experimental results from Allen (1973) and other sources which included both axial and radial measured specimen deformations were used in the model development. Predictions of the moduli and Poisson's ratios were in good agreement with the experimental values. The Poisson's ratios often reached values in excess of 0.5 even with the assumption of homogeneous, isotropic material behavior. This was attributed to the dilation effects which have been measured at high major to minor stress ratios.

3 CROSS-ANISOTROPY

The effects of anisotropic behavior of cohesionless soils have been reported by several researchers to influence the computed stress-strain response. Borowicka (1943) indicated an increase in the calculated vertical stresses near the load when overburden stresses were considered to cause an initial anisotropic material behavior. Similar results were obtained by Barden (1963), and Gerrard and Mulholland (1966) when anisotropy was taken into account.

Using a highly sophisticated, true triaxial testing device, Desai et al. (1983) performed extensive tests on three uniform sized aggregates utilized as ballast in track support structures. In each test, the material was spooned into the cubical mold (10 x 10x 10 cm) and then compacted by vibration. An apparent deviation from isotropy exhibited by the test specimens was attributed to both material anisotropy and specimen preparation with the lowest strains measured in the vertical direction of compaction. As the aggregate size decreased, similar strain responses were observed in the two horizontal directions. This kind of behavior can be modeled by using cross-anisotropy under axial symmetry.

An isotropic model has the same elastic material properties in all directions. A cross-anisotropic representation, however, has different elastic properties (i.e., modulus and Poisson's ratio) in the horizontal and vertical directions as shown in Figure 1. A general formulation of a cross-anisotropic layered system in terms of the in plane and normal to the strata modulus (*E*) and Poisson's ratios (ν) has been given by Zienkiewicz and Taylor (1989). The variables *n* and *m*, which are commonly substituted for horizontal modulus and shear modulus (*G* in z direction) in the formulation, represent the ratios of horizontal modulus to vertical modulus and shear modulus to vertical modulus, respectively.

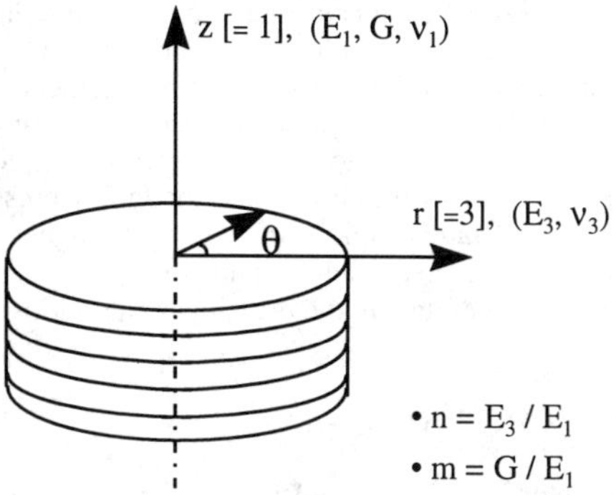

Figure 1. Stratified anisotropic material under axial symmetry.

Pickering (1970) studied the bounds of the elastic parameters in a cross-anisotropic material. Using the quadratic form for the matrix of elastic cross-anisotropic constants, he deduced the bounding values of the Poisson's ratios necessary to ensure that the resulting strain energy was positive. In addition to the three moduli being greater than zero, the following are the requirements for a positive definite quadratic form:

$$-1 < \nu_3 < 1 \tag{1}$$

$$\frac{E_3}{E_1} > \frac{2\nu_1^2}{(1-\nu_3)} \tag{2}$$

The constants ν_1 and ν_3 are defined as the Poisson's ratios for strain in any horizontal direction due to a horizontal stress at right angles, and for strain in the vertical direction due to a horizontal direct stress, respectively. Selection of the appropriate Poisson's ratios to be used in the analysis must be from these equations such that for any value of the modular ratio, the thermodynamic considerations are satisfied.

Lo and Lee (1990) studied experimentally the response of a granular soil tested along constant stress increment ratio paths under axisymmetric conditions. Deformation behavior for unloading was found to be essentially elastic in nature. The elastic response was anisotropic and the degree of anisotropy increased with the principal stress ratio. Based on these findings, a cross-anisotropic model, with the degree of anisotropy increasing with the principal stress ratio, was proposed to model elastic granular material behavior.

4 ANISOTROPIC PROPERTIES FROM TRIAXIAL TESTS

The repeated load triaxial compression test is currently the most commonly used method to measure the resilient (elastic) deformation characteristics of unbound aggregates for use in pavement design. The resilient modulus test is performed on a cylindrical specimen of granular materials subjected to repeated axial compressive (deviator) stresses. To simulate the lateral stresses caused by the initial in situ pressure and that from applied wheel loadings, the specimen is subjected to a constant all-around confining pressure. An advantage of the triaxial test is that the axial and radial (or volumetric and shear) strains can be determined relatively easily. Determination of lateral strains in a triaxial specimen is essential for characterizing the anisotropic elastic properties of granular bases.

Anisotropic resilient response - elastic response obtained from the repeated load triaxial tests due to the pulse deviator stress - can be defined from triaxial test data with measured vertical and lateral deformations as follows:

Vertical Modulus: $$E_1 = \sigma_d / \varepsilon_1 \quad (3)$$

Horizontal Modulus: $$E_3 = |\sigma_3 / \varepsilon_3| \quad (4)$$

Shear Modulus: $$G = \sigma_d / 2(\varepsilon_1 - \varepsilon_3) \quad (5)$$

where the horizontal modulus (E_3) is newly defined for anisotropic elasticity. Since for a cylindrical triaxial sample there is co-axiality between the material and principal stress axes, the horizontal and vertical directions, as referred to in the above definitions, are used in the same context with the radial (r) and vertical (z) directions under axial symmetry.

5 MODEL DEVELOPMENT

A new improved way of modeling granular materials using cross-anisotropic nonlinear elasticity is proposed here to predict the dilative granular material behavior as observed from laboratory triaxial test results. Previous studies have shown that granular material response can reasonably be characterized by using stress dependent models which express the modulus as nonlinear power functions of stress states (Boyce, 1980; Lade and Nelson, 1987; Uzan et al., 1992). Such a characterization model must include in the formulation the two triaxial stress conditions, i.e., the mean effective pressure p and the deviator stress σ_d, to account for the effects of both confinement and shear loading, respectively (Uzan et al., 1992).

Repeated load triaxial test results which included vertical and lateral deformation measurements performed on a variety of aggregate types have been recovered for this study from the work of Allen (1973). Based on these results, nonlinear stress dependent material characterization models were developed for the horizontal, vertical, and shear stiffnesses to define mainly the vertical load-induced anisotropic behavior under axisymmetric conditions. Several combinations of the stress variables were investigated for modeling the stiffnesses as functions of stress states. Uzan type models (Uzan et al., 1992), which relate the moduli to both mean effective and deviator stresses, provided a good statistical fit with high correlation coefficients (R^2 = 0.95 to 0.99) when the model parameters were obtained from the multiple regression analyses of the individual test data.

The vertical and horizontal moduli are, therefore, modeled from the triaxial stress states using the equation

$$E_i = K_{1i}(3p/p_a)^{K_{2i}}(\sigma_d/p_a)^{K_{3i}} \quad (6)$$

where i = "1" or "3" for "vertical" or "horizontal (radial)" directions, respectively; K_{1i} to K_{3i} are model parameters, and p_a = unit pressure (1 kPa).

Similarly, the shear modulus can also be given by

$$G = K_4(3p/p_a)^{K_5}(\sigma_d/p_a)^{K_6} \quad (7)$$

where K_4 to K_6 are model parameters. Note that these model parameters are obtained from the multiple regression analyses of the triaxial test results.

Alternatively, anisotropy of the stiffnesses expressed as the horizontal and shear stiffness ratios [vertical to horizontal modular ratio ($1/n$), and vertical to shear modular ratio ($1/m$)] can be obtained as functions of principal stress ratio R and confining stress σ_3 as follows:

$$\frac{E_1}{E_3} = (1/A)\left(\frac{\sigma_3}{p_a}\right)^{1-B-C}[R+2]^{-B}[R-1]^{1-C} \quad (8)$$

$$\frac{E_1}{G} = (1/D)\left(\frac{\sigma_3}{p_a}\right)^{1-E-F}[R+2]^{-E}[R-1]^{1-F} \quad (9)$$

where $R = \sigma_1 / \sigma_3$, p_a = unit pressure (1 kPa), and

$A = K_{13}/K_{11}$, $D = K_4/K_{11}$,

$B = K_{23}-K_{21}$, $E = K_5-K_{21}$,

$C = K_{33}-K_{31}+1$, $F = K_6-K_{31}+1$.

Substituting $E_1 = \sigma_d / \varepsilon_1$ and $E_3 = -\sigma_3 / \varepsilon_3$ into the above stiffness ratio models, the principal strain ratio can then be given by

$$\frac{\varepsilon_1}{\varepsilon_3} = A\left(\frac{\sigma_3}{p_a}\right)^{B+C-1}[R+2]^{B}[R-1]^{C} \quad (10)$$

where p_a = unit pressure (1 kPa).

It is interesting to note that principal strain ratios derived from the nonlinear anisotropic model (Equation 11) are independent of the assumed Poisson's ratio values of ν_1 and ν_3. On the other hand, using isotropic linear elasticity, Poisson's ratios (ν) must be provided to compute the principal strain ratios as given by the following equation

$$\frac{\varepsilon_1}{\varepsilon_3} = \frac{(R-2\nu)}{[1-\nu(R+1)]}. \quad (11)$$

6 VALIDATION OF THE MODEL

The nonlinear anisotropic model developed in the previous section was based on laboratory tests conducted at the University of Illinois at Urbana-Champaign (Allen, 1973). A total of 9 individual repeated load triaxial tests were performed on gravels, partially crushed, and crushed aggregates at varying densities and saturation levels. In each test, increasing levels of deviator stresses were applied on 15 cm diameter by 30 cm height samples with confining pressures held constant at 14, 35, 55, 76, and 103 kPa. An average of 19 sets of vertical and lateral deformation measurements corresponding to various stress states were taken from each specimen.

Typical variations of the vertical and horizontal moduli (Equations 3 and 4) with strain levels are presented in Figure 2 for a high density, high quality partially saturated crushed stone (HD1) tested by Allen (1973). As the vertical modulus increases with increasing strains, which is similar to the generally observed trends for the shear modulus at different confining pressures, the horizontal modulus tends to decrease with increasing radial strains. This variation with strain levels can also be perceived for the moduli when plotted with increasing bulk and deviator stresses. The observed horizontal modulus reduction in the large strain zone is in accordance with the dilation phenomenon often encountered in the granular bases under the wheel load (Witczak and Uzan, 1988; Uzan et al., 1992).

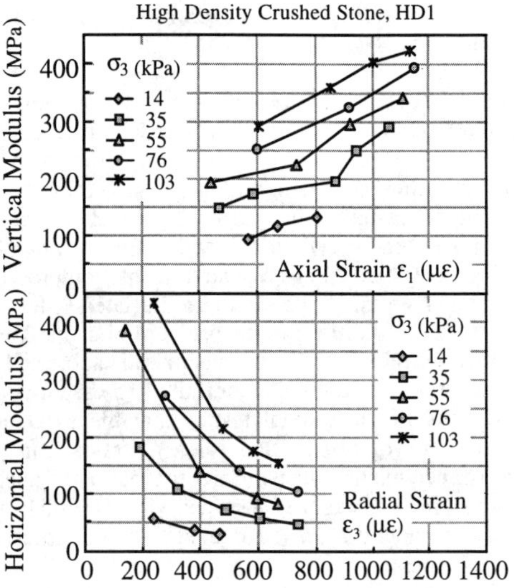

Figure 2. Variation of measured stiffnesses with strain levels for the HD1 material.

The experimental response of the HD1 granular material shown in Figure 2 was modeled by using the nonlinear anisotropic model. The following are the model parameters obtained for each modulus from multiple regression analyses of the test results:

E_3: K_{13} = 40.76 MPa, K_{23} = 3.416, K_{33} = -2.808, ($n = 19$, $R^2 = 0.992$)

E_1: K_{11} = 533.92 MPa, K_{21} = 0.640, K_{31} = 0.065, ($n = 19$, $R^2 = 0.995$)

G: K_4 = 143.75 MPa, K_5 = 0.834, K_6 = -0.167, ($n = 19$, $R^2 = 0.996$)

Figure 3 shows for the HD1 material the variation of principal strain ratios with increasing major to minor principal stress ratios. Any value of strain ratio less than 2 corresponds to the case of elastic dilation. Strain ratios predicted using the nonlinear

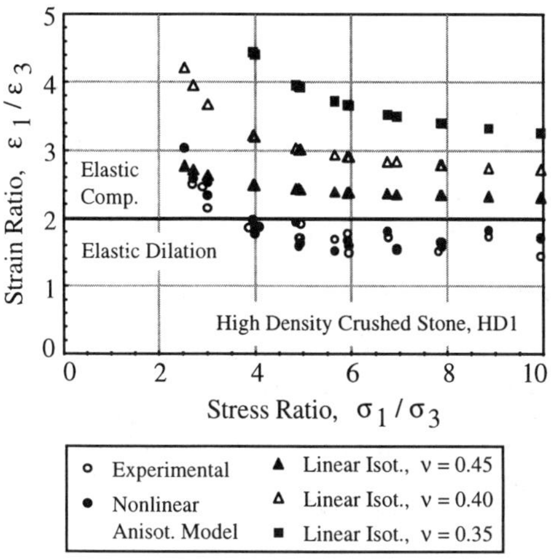

Figure 3. Variation of principal strain ratio with principal stress ratio for the HD1 material.

anisotropic model (Equation 11) match very closely with the experimental ones. Dilation begins to take place at a stress ratio of 3.5, as found for sands by Frydman (1968), and adequately modeled by the nonlinear anisotropic model. An almost constant degree of dilation was predicted for the HD1 material even at high principal stress ratios. Also shown on the compression side in Figure 3 are the linear isotropic model predictions obtained using Equation 12 for v = 0.35, 0.4, and 0.45. Under the assumption of linear isotropy, modeling of the dilative behavior could only be achieved with the improper Poisson's ratio assignments exceeding 0.5.

The anisotropy of the stiffnesses in the cylindrical triaxial sample is next analyzed for the HD1 material using Equation 8. Predictions obtained using the nonlinear anisotropic model for confining pressures σ_3 = 14 to 103 kPa are indicated in Figure 4 with the dashed and solid lines. The model is capable of matching all the experimental points, and also captures the trend of increasing degree of stiffness anisotropy with increasing principal stress ratios (Lo and Lee, 1990). This kind of behavior is in good agreement with the previous findings of Konishi (1978) who concluded that an increase in axial stiffness can occur at the expense of a reduction in radial stiffness, thus leading to a compensating effect on the volumetric stiffness.

To investigate the effects of large principal stress ratios on the degree of dilation, graphs similar to Figure 3 were plotted for various granular materials tested by Allen (1973). It is interesting to note that for the lowest quality (lowest shear strength properties) material, referred to here as the low density partially saturated gravel (LD2), dilation starts to occur even at a lower stress ratio of 2.2 (see Figure 5). Moreover, the LD2 material exhibits a continually decreasing strain ratio response (down to 0.9) with increasing principal stress ratios as shown in Figure 5. This type of highly dilative behavior has been once again adequately modeled by the nonlinear anisotropic model.

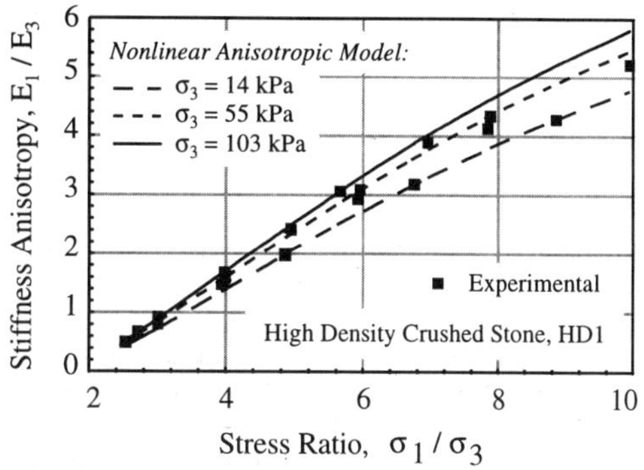

Figure 4. Variation of the stiffness anisotropy with principal stress ratio for the HD1 material.

6.1 Modeling of granular bases in pavements

The cross-anisotropic model was incorporated into a nonlinear finite element program, named GT-PAVE, (Tutumluer, 1995) to predict the performance of a granular base layer in a conventional flexible pavement. The elastic response, referred to as the resilient response under many repeated applications of the wheel load of the layered pavement system, was solved using the GT-PAVE program. The nonlinear anisotropic approach was shown to effectively account for the dilative behavior observed under the wheel load (Tutumluer and Thompson, 1997). The predicted horizontal and shear stiffnesses were typically less than the vertical. The main advantage of using a cross-anisotropic model in the unbound aggregate base is the drastic reduction or elimination of the high inadmissible horizontal tensile stresses generally predicted by isotropic linear elastic layered programs.

7 CONCLUSIONS

1. A new cross-anisotropic elastic model is developed to predict the nonlinear dilative granular material behavior under axisymmetric conditions.

2. Repeated load triaxial test results with vertical and lateral deformation measurements suffice to characterize the nonlinear stress dependent granular material behavior for horizontal, vertical, and the shear stiffnesses.

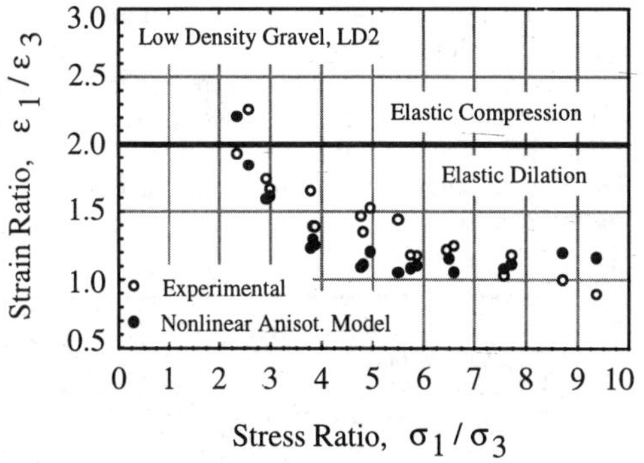

Figure 5. Variation of principal strain ratio with principal stress ratio for the LD2 material.

3. Dilation observed in the materials tested was accurately modeled using the anisotropic approach.

4. The ratio of vertical to horizontal moduli, expressed as a measure of stiffness anisotropy, was generally larger than 1 with ratios rapidly increasing at higher principal stress ratios (σ_1/σ_3).

5. An almost constant degree of dilation - as defined by the principal strain ratios ($\varepsilon_1/\varepsilon_3$) - was predicted for a high density, high shear strength crushed stone even at high principal stress ratios. Gravels with lower strength properties may result in lower principal strain ratios.

REFERENCES

Allen, J.J. 1973. *The effects of non-constant lateral pressures on the resilient response of granular materials.* Ph.D. dissertation, Department of Civil Engineering, University of Illinois, Urbana, IL, May.

Barden, L. 1963. Stresses and displacements in a cross-anisotropic soil. *Geotechnique*, September, p. 198.

Biarez, J. and Hicher, P.Y. 1994. *Elementary mechanics of soil behavior.* Balkema, Rotterdam, The Netherlands.

Billam, J. 1971. Aspects of the behavior of granular materials at high pressures. Stress-strain behavior of soils. *Proceedings of the Roscoe Memorial Symposium*, Cambridge University, UK, pp. 69-80.

Borowicka, H. 1943. Pressure distribution in a halfspace with a linearly varying modulus of elasticity. *Ingenieur-Archiv*, 14 (2), pp. 75.

Boyce, H.R. 1980. A nonlinear model for the elastic behavior of granular materials under repeated loading. *International Symposium on Soils under Cyclic and Transient Loading.* Balkema, Rotterdam, The Netherlands, pp. 285-294.

Desai, C.S., Siriwardane, H.J., and Janardhanam, R. 1983. Interaction and load transfer in track support structures, part 2: "Testing and constitutive modeling of materials and interfaces", Final report, Contract No. DOT-US-05-80013, US Department of Transportation, Office of University Research, Washington D.C., January.

Dobry, R, Ng, T.T., and Petrakis, E. 1989 Deformation characteristics of granular soil in the light of particulate mechanics. *Proceedings, 14th Conference on Geotechnics*, Italian Geotechnical Association, Torino, November 28-30.

Frydman S. 1968. *The effect of stress history on the stress deformation behavior of sand.* M.Sc. thesis. Technion Israel Institute of Technology.

Gerrard, C.M. and Mulholland, P. 1966. Stress-strain and displacement distributions in cross-anisotropic and two-layer isotropic elastic systems. In *Proceedings of the 3rd Conference of Australian Road Research Board, Part 2*, p. 1123.

Konishi, J. 1978. Microscopic model studies on the mechanical behavior of granular materials. *US-Japan Seminar on Continuum Mechanics and Statistical Approaches in Mechanics of Granular Materials*, June, Sendai, Japan, pp. 27-45.

Lade, P.V. and Nelson, R.B. 1987. Modeling the elastic behavior of granular materials. *International Journal for Numerical and Analytical Methods in Geomechanics*, Volume 11, pp. 521-542.

Lo, S-C.R., and Lee, I.K. 1990. Response of granular soil along constant stress increment ratio path. *Journal of Geotechnical Engineering*, Vol. 116, No. 3, March, pp. 355-376.

Pickering, D.J. 1970. Anisotropic elastic parameters for soil. *Geotechnique* 20, No. 3, pp. 271-276.

Tutumluer, E. 1995. *Predicting behavior of flexible pavements with granular bases.* Ph.D. dissertation, School of Civil and Environmental Engineering, Georgia Institute of Technology, Atlanta, GA, Sept.

Tutumluer, E. and Thompson, M.R. 1997. Anisotropic modeling of granular bases in flexible pavements. Paper No. 970539. *76th Annual Meeting of Transportation Research Board*, Washington, D.C., January.

Uzan, J., Witczak, M.W., Scullion, T., and Lytton, R.L. 1992. Development and validation of realistic pavement response models. In *Proceedings, 7th International Conference on Asphalt Pavements, Nottingham, U.K.*, Vol. 1, pp. 334-350.

Witczak, M.W. and Uzan, J. 1988. The universal airport pavement design system, Report I of V: Granular material characterization. University of Maryland, Department of Civil Engineering, MD.

Zienkiewicz, O.C. and Taylor, R.L. 1989. *The finite element method.* Volume 1, Basic formulation and linear problems, 4th Edition, McGraw Hill Book Co. (UK) Limited.

Computer Methods and Advances in Geomechanics, Yuan (ed.) © 1997 Balkema, Rotterdam, ISBN 90 5410 904 1

Updated hypoplastic analysis of the behaviors of soils under cyclic loading

Limin Zhang
Sichuan Union University, Chengdu, China (Presently: University of Florida, Gainesville, Fla., USA)

ABSTRACT: The hypoplasticity theories have been able to depict successfully the main characteristics of granular materials under monotonic loading. However, their validity under cyclic loading is not satisfactory as respect to excess pore water pressure generation and plastic strain development. In this paper, a comprehensive hypoplasticity theory was modified to take into account the effects of strain reversals under repeated loading. Element tests for stress controlled triaxial tests showed a good agreement between the measured and predicted excess pore water pressure. The cyclic mobility process following initial liquefaction was also successfully simulated by this model.

1 INTRODUCTION

Among the advances in the constitutive laws for granular materials in recent years, a significant one is the hypoplasticity theory which is characterized by its superior capacity to depict the hardening and softening of granular materials up to critical state. It does not use the concepts of yielding surfaces and therefore does not require decomposing the strain into a plastic and an elastic part. This theory was originally put forward by Kolymbas (1988, 1993), and was finally extended and generalized into a comprehensive hypoplastic constitutive equation for granular materials by Gudehus (1995), namely,

$$\dot{T}_s = f_b f_e [L(\hat{T}_s, D_s) + f_d N(\hat{T}_s)\|D_s\|] \quad (1)$$

$$N(\hat{T}_s, D_s) = a_1(\hat{T}_s + \hat{T}_s^*) \quad (2)$$

$$L(\hat{T}_s, D_s) = a_1^2 + \hat{T}_s tr(\hat{T}_s D_s) \quad (3)$$

The factors f_d, f_b, and f_e are defined as,

$$f_d = (\frac{e - e_d}{e_c - e_d})^\alpha \quad (4)$$

$$f_b = \frac{h_s}{n h_i} \frac{1 + e_i}{e_i} (\frac{3p_s}{h_s})^{1-n} \quad (5)$$

$$h_i = \frac{1}{c_1^2} + \frac{1}{3} - (\frac{e_{i0} - e_{d0}}{e_{c0} - e_{d0}})^\alpha \frac{1}{c_1\sqrt{3}} \quad (6)$$

$$f_e = (\frac{e_c}{e})^\beta \quad (7)$$

$$a_1^{-1} = c_1 + c_2\|\hat{T}_s^*\|[1 + \cos(3\theta)] \quad (8)$$

where $\mathbf{T}_s$=$\mathbf{T}_s$/tr$\mathbf{T}_s$, $\mathbf{T}_s$ is the effective stress tensor; $\mathbf{D}_s$ is the granulate strain rate tensor; c_1, c_2, n, α, h_s, and β are material constants; f_e and f_d are factors controlling the pyknotropy and dilatancy of materials, and f_b controls the barotropy of matcrials. The material behavior under monotonic loadings of different stress paths, such as drained oedometer tests, triaxial tests, strain controlled simple shear tests,were modeled using this model with satisfaction (Wu,1992,1993, Bauer, 1995). This theory was also used to analyze the bearing capacity of shallow foundations, granular flow of bulk materials, stability of loose sand slopes, deformation of sheet pile wall system, and spontaneous liquefaction of loose sand(Gudehus, 1993, 1994). Later, it was incorporated into a finite element code to analyze the localizations and bifurcations in granular materials (Tejchman, 1994). Preliminary studies on generalizing this theory for cohesive soils and unsaturated soils have also been carried out (Wu et al, 1993, Gudehus,1995).

While great successes been reported in monotonic loading problems, the hypoplastic theory demonstrates shortcomings in handling situations involving cyclic loading. Under drained condition, the model is able to describe the characteristics of volumetric changes of samples toward shake-down in strain controlled simple shear tests. However it overestimates the dilation or contraction, and in the later case the model will predict the same stable void ratio after certain cycles of strain reversals even though the strain may be infinitesimally small. In case of undrained tests, the model predicts a much faster excess pore water pressure increase than the experiments do, and a tiny cyclic stress or strain action will always lead to complete liquefaction which in reality could remain in elastic state.

To cope with the above problems, Bauer and Wu (1993) suggested a structural tensor **S** which relates to the so-called memory function f_m and the transition function f_t.

$$T_t = T + S = T(1+f_m)(1+f_t) \qquad \textbf{(9)}$$

The final expression of the hypoplastic equation was,

$$\hat{T} = L(T_t, D) + N(T_t)\|D\| I_e \qquad \textbf{(10)}$$

$$I_e = (1-a)\frac{e-e_m}{e_c-e_m} + a \qquad \textbf{(11)}$$

However, this modification was not able to solve the basic shortcomings (Niemunis, 1993). The structural tensor **S** could not prevent ratcheting under infinitesimally small strain cycles, and is not continuously differentiable for σ: ϵ=0. The transition function I_e increases the stiffness in both loading and unloading due to densification. Wu (1992) considered the inertial effect associated with the later process of cyclic response, and assumed a jump in stress to eliminate ratcheting.

Niemunis and Herle (1996), on the other hand, revised the model by borrowing the concepts in elstoplastic theories. The plastic deformation was assumed not to occur during unloading process, and occur during reloading only after the stress state has reached the historical value. This revise is effective to limit the deformation and pore pressure accumulation at small strains. However, it implies yielding surface concepts that we try to avoid in the hypoplastic theory.

In this paper, characterization of unloading and reloading and the energy dissipation during strain reversals are considered by updating the dilatancy factor f_d in the original model.

2 BEHAVIOR OF SOILS UNDER CYCLIC LOADS

2.1 Hypoplastic characterization of unloading and reloading behavior

The hypoplastic equation is able to describe the unloading behavior of granular materials by the switch function $\|\mathbf{D}_s\|$ in the nonlinear part of the equation. However, it fails to describe the energy dissipation occurred in repeated strain reversals, and treats a reloading process as a virgin loading.

During reloading, the pyknotropy factor f_d', which represents the dilatancy effect and the degree of nonlinearity, differs from the corresponding one for virginal loading. In ideally elastic condition, f_d' =0, a very small disturbance or infinitesimally small stress cycles will not lead to densification of soils to shakedown under drained conditions, neither to complete liquefaction under undrained conditions. Whereas when the cyclic strain amplitude is very large, additional plastic volume change will occur as the results of the contract of yielding surfaces caused by strain reversals, as described in the elastoplastic models (Carter, 1982, Zhang, 1995). Therefore, the pyknotropy factor f_d' shall change with strain amplitude and take a value between f_d'=0 for perfectly linear elastic loading and f_d' = f_d as defined by Eq.(4) for virgin loading. As shown in Fig.1, the curve for f_d'=0 corresponds to elastic loading ($e=e_d$), and the curve f_d'=f_d* corresponds to the reloading at certain stress increment. The ratio of the virgin loading strain increment to the total strain increment $\epsilon_{AD}/\epsilon_{OD}$ can be roughly expressed as,

$$\frac{\epsilon_{AD}}{\epsilon_{OD}} = 1 - \frac{E}{E_{\max}} \qquad \textbf{(12)}$$

Accordingly, the average pyknotropy factor during a reloading or an unloading process can be approximately expressed as,

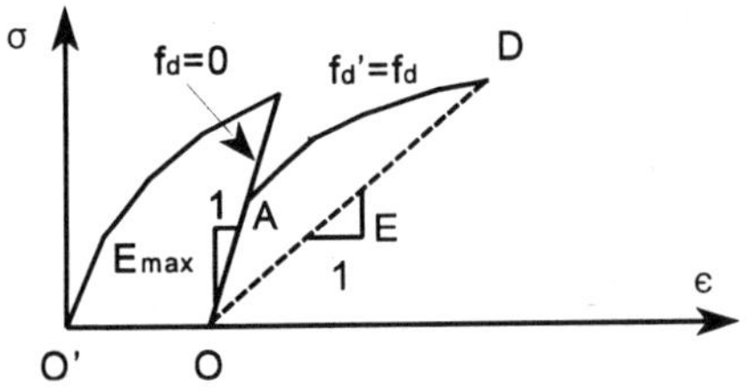

Fig.1 Dilatancy factor f_d in hypoplastic model

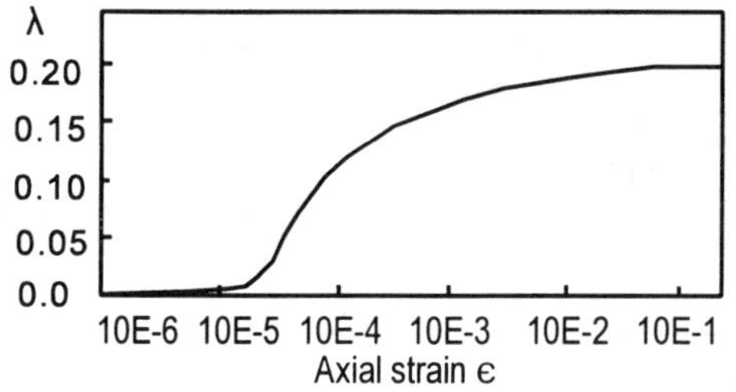

Fig.2 Typical variation of λ with strain amplitude

$$f_d^* \propto \left(1 - \frac{E}{E_{max}}\right) f_d \tag{13}$$

where, E and E_{max} are the secant modulus for elastic loading and reloading respectively. This equation can be simplified as,

$$f_d^* = k \frac{\lambda}{\lambda_{max}} f_d \tag{14}$$

where λ is the damping ratio at certain strain, λ_{max} is the maximum damping ratio at large strain, k is an experimental coefficient. This expression relates the energy dissipation with the dilatancy factor f_d*.

The term λ is the well-known dynamic damping ratio which changes with shear strains. It can be measured in a resonant column or a dynamic triaxial apparatus, and is always a prerequisite for dynamic response analysis. As shown in Fig.2 is a typical curve measured from triaxial tests on a sand (He and Zhang, 1994). From the above figure, the dilatancy factor $f_d^*=0$ when the amplitude of cyclic strains is extremely small, generally $\epsilon<10^{-5}$. It is always smaller than f_d even when ϵ is large because a cyclic loading cycle is not a completely virgin loading process.

The updated constitutive law now reads,

$$\dot{T}_s = f_b f_e [L(\hat{T}_s, D_s) + f_d^* N(\hat{T}_s)\|D_s\|] \tag{15}$$

It can be further proved that this equation is in accordance with the requirements of strains and stresses at the critical state and at the peak state.

2.2 Undrained behavior of granular materials

The solution of drained tests is straight forward with Eq.(15). For undrained tests, the generation of pore pressure is one of the key factors. Unlike other nonlinear elastic or elastoplastic models which always need a pore water pressure generation model if effective stress analyses are required, the present comprehensive constitutive law hides in itself the pore pressure generation mechanism. For undrained tests in general three-dimensional stress conditions where the total stress rate tensor $\mathbf{T}^t$ is known, as in a number of construction and excavation problems, the excess pore water pressure can be determined from the following equations,

$$D = E^{-1}\dot{T}_s \tag{16}$$

$$\dot{T}_s = \dot{T} - \dot{u} \tag{17}$$

$$tr[E^{-1}(\dot{T}^t - \dot{u})] = 0 \tag{18}$$

where **E** is the matrix of stiffness along the principal stress directions as defined by Eq.(15), $\hat{\mathbf{u}}$ is pore water pressure rate. For strain controlled tests where strain increments are given, the changes in the effective stress can be calculated by Eq.(15).Under triaxial test conditions with constant confining pressure, the total radial stress rate tensor T_r^t is zero, therefore,

$$\dot{u} = -\dot{T}_r \tag{19}$$

For stress controlled triaxial tests,

$$\dot{u} = \frac{E_2\dot{T}_1^t + 2E_1\dot{T}_2^t}{E_2 + 2E_1} \tag{20}$$

where $\mathbf{T}_1^t$ and $\mathbf{T}_2^t$ are the total axial stress rate and radial stress rate respectively, E_1 and E_2 are the axial and radial stiffness respectively. The situations with a given stress path or a strain path can be viewed as special cases described by Eq.(16) through Eq.(18).

3 PERFORMANCE OF THE MODEL

In the following, comparisons between the test results

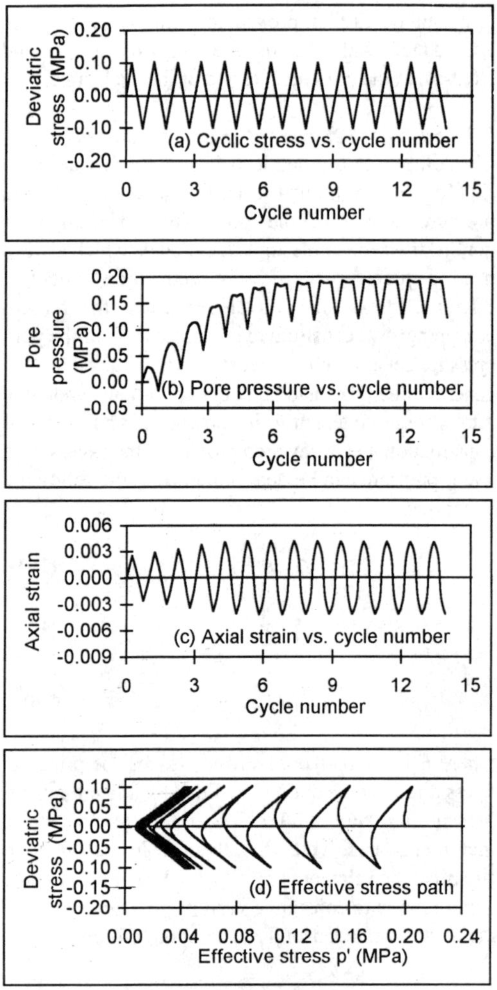

Fig.3 Predicted undrained response of a sample under triaxial condition, σ_{33}=200kPa, σ_{11}=200kPa, uniform cyclic stress amplitude q_{max}=100kPa

and the predicted results are made only for undrained triaxial test conditions.

The medium dense sand in the sand deposits beneath a rockfill dam was intensively investigated (He and Zhang, 1994). The particle gradation parameters of this sand are d_{60}=0.26mm, d_{50}= 0.22mm, d_{30}= 0.11mm, C_u=14.4, C_c=2.6. The specific weight γ_s/γ_w=2.68, the maximum and minimum void ratios are e_{max}=1.02, e_{min}=0.43. The site dry unit weight is γ_d=16.2 kN/m^3 and the corresponding void ratio is 0.624. The peak internal friction angle and the mobilized friction angle at critical state are ϕ=39.8^0 and ϕ_c=32.0^0 respectively.

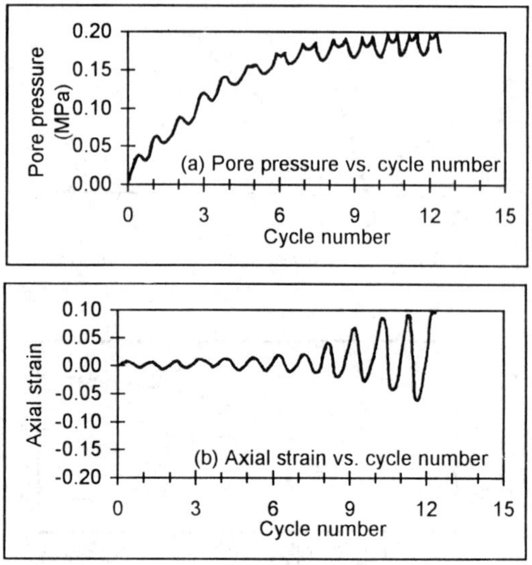

Fig.4 Measured pore water pressure and axial strain from dynamic triaxial tests, σ_{33}=200kPa, σ_{11}=200kPa, cyclic stress amplitude q_{max}=100kPa

The model parameters are $c_1=(3-\sin\phi_c)(3/8)^{0.5}/\sin\phi_c$ =2.8545, $c_2=3(3+\sin\phi_c)/(8\sin\phi_c)$=2.498, h_s= 215MPa, α =0.15, β=1.05, λ=1.46, e_{c0} =0.82, e_{d0}=0.51. It was shown by trial-and-error method that the parameter k can take the value of the maximum damping ratio λ_{max} which is readily available from Fig.2.

Fig.3(a)~(d) show the predicted pore water pressure, axial strain, and effective stress path in the p'-q space ($p'=(\sigma_1'+\sigma_2'+\sigma_3')/3$, $q=\sigma_1'-\sigma_3'$) of a saturated sample under uniform cyclic loading in undrained triaxial stress conditions. This sample was isotropically consolidated at $k_c= \sigma_{33}/\sigma_{11}$ =1.0, the amplitude of the uniform cyclic stress was 100 kPa. The total ambient stress kept constant during the test. Fig.4 shows the measured pore water pressure and axial strain curves from dynamic triaxial tests (He and Zhang, 1994). It can be seen that the predicted pore pressure is quite similar with the measured one, the reduction of pore water pressure due to dilatancy of soil was clearly simulated. Furthermore, the cyclic mobility process following initial liquefaction is modeled by both the pore pressure curve and the typical dilation loops in the effective stress path.

Illustrated in Fig.5 are the predicted results with the original comprehensive hypoplastic equation. As can be seen, the pore pressure reaches the ambient pressure in only one stress cycle, and the axial tensile strain is as large as 10% at the end of the first cycle, while the compresive strain hardly develops.

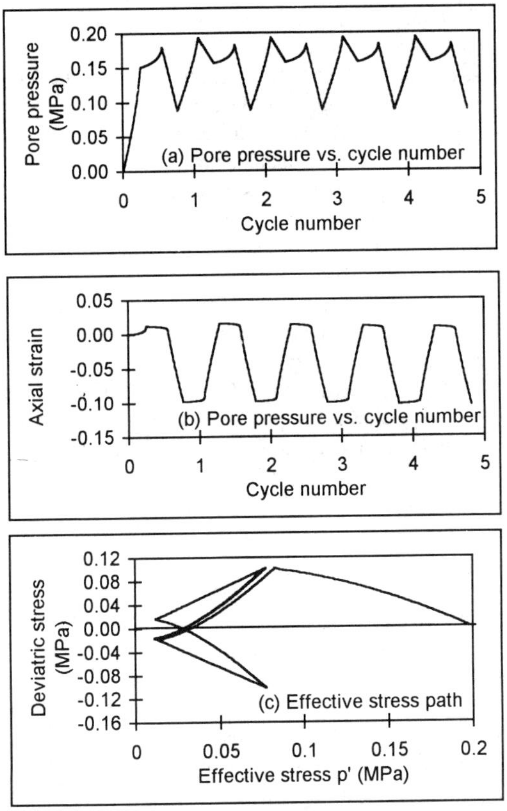

Fig.5 Predicted pore water pressure and strain with the original hypoplastic equation, σ_{33}=200kPa, σ_{11}=200kPa, cyclic stress amplitude q_{max}=100kPa

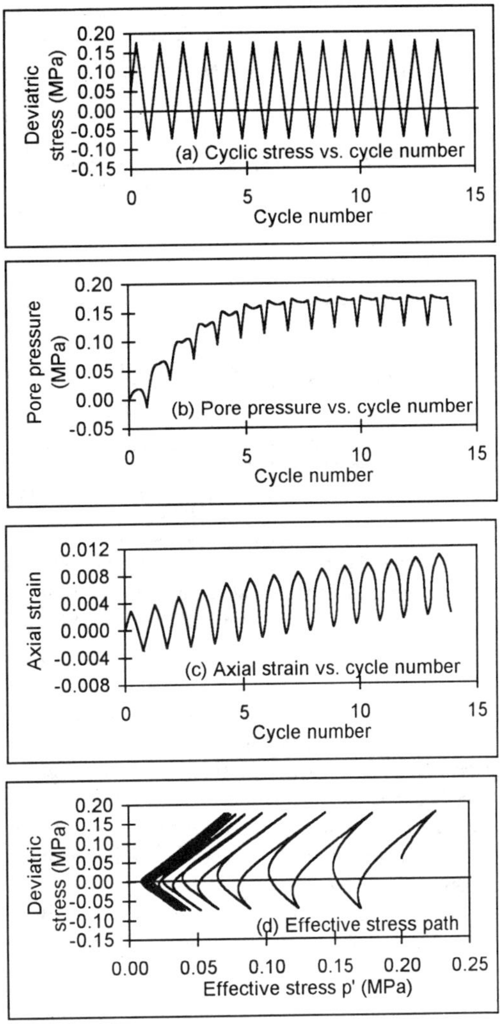

Fig.6 Predicted undrained response of an anisotropically consolidated sample, σ_{33}=183kPa, σ_{11}=234kPa, uniform cyclic stress amplitude q_{max}=123kPa

Fig.6 and Fig.7 show the predicted and measured results of an anisotropically consolidated sample (k_c= 1.38). The predicted pore pressure and stress path are in accordance with the measured results. As the sample approaches complete liquefaction, the effective stress path goes toward the origin and loses shear strength.

In both of the predicted cases as shown in Fig.3(c) and Fig.6(c) respectively, however, the rate of the axial strain does not satisfactorily reflect the measured data. In the experiments, the axial strains developed adruptly after the build-up of excess pore water put the samples to limit equilibrium states. Whereas in the results predicted by the present hypoplastic model, significant strain developed during the first stress cycle, but the deformation was not able to fully develop after the pore pressure built up. This can be explained by the fact that the model retains only the first power parts of stress and train tensors. Inclusion of the second power parts of the tensors will improve the model performance but at the same time in the sacrifice of its simplicity.

4 CONCLUSIONS

The above updating made to the comprehensive hypoplastic constitutive equation takes into account the effects of hysteresis energy dissipation on the performance of soils under cyclic loading. It shows strong capacity to describe the dilatancy of soils and

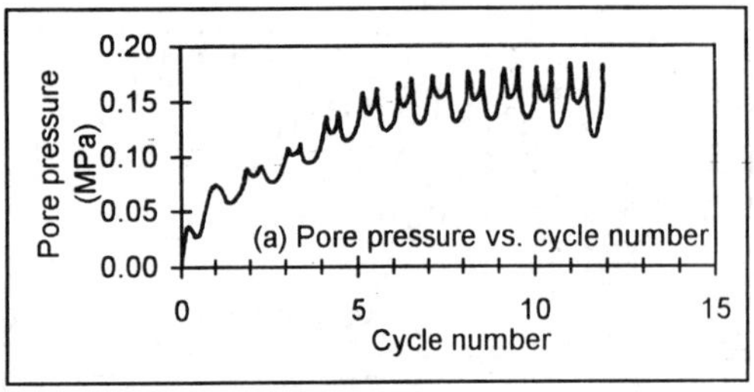

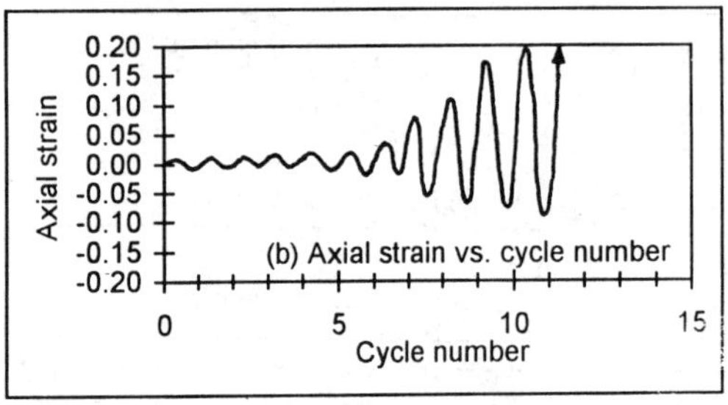

Fig.7 Measured pore water pressure and axial strain on an anisotropically consolidated sample, σ_{33}=183kPa, σ_{11}=234kPa, cyclic stress amplitude q_{max}=123kPa

therefore the cyclic mobility process of sandy soils. However, the typical strain development patterns in undrained stress-controlled tests are not satisfactorily described because the model retains only the first power parts of the stress and train tensors. Inclusion of the second power parts of the tensors will improve the capacity but in the sacrifice of simplicity. What is important in the paper is the ability to qualitatively depict the complicated processes of soils under cyclic loading.

ACKNOWLEDGEMENTS

The author wishes to express sincere thanks to Prof. G.Gudehus for introducing the hypoplasticity theories during his stay at Karlsruhe University, Germany from July 1995 to January 1996. Financial support from the Natural Science Foundation of China under grant No. 59479017 is gratefully acknowledged.

REFERENCES

Bauer, E. and Wu, W.(1993): A hypoplastic model for granular soils under cyclic loading. Proc. Int.Workshop on Modern Approaches to Plasticity for Granular Materials, Horton, Greece, Ed. Kolymbas, Elsevier:247-258.

Carter, J.P. et al. (1982): A critical state soil model for cyclic loading. Soil Mechanics-- Transient and Cyclic Loads, Ed. Pande and Zienkiewicz, Balkema.

Gudehus,G.(1993): Spontaneous liquefaction of saturated granular bodies. Proc. Int. Workshop on Modern Approaches to Plasticity for Granular Materials, Horton, Greece, Ed. Kolymbas, Elsevier:691-714.

Gudehus, G. (1995): A comprehensive constitutive equation for granular materials. Soils and Foundations, accepted.

He, Changrong and Zhang, L. (1994): Studies on the ynamic deformation of coarse granular soils under high ambient stresses, Report 85-208-02-04-01-11, Sichuan Union University.

Kolymbas, D. (1988): Eine konstitutive theorie fur boden und andere körnige stoffe. Habilitation, No.109, Institute for Soil Mechanics and Rock Mechanics, Karlsruhe University.

Kolymbas, D. and Wu, W. (1993):Introduction to hypoplasticity. Proc. Int. Workshop on Modern Approaches to Plasticity for Granular Materials, Horton, Greece, Ed. Kolymbas, Elsevier. 214-223.

Niemunis, A.(1993): Some aspects of the hypoplasticity. Proc. Int. Workshop on Modern Approaches to Plasticity for Granular Materials, Horton, Greece, Ed.Kolymbas, Elsevier.

Niemunis A. And Herle I. (1996): Extended hypo-plastic model for cohesionless soil with elastic strain range. Private communication.

Tejchman, J. (1994): Numerical study on localized deformation in a Cosserat continuum. Localization and Bifurcation Theory for Soils and Rocks, Balkema, 1994.

Wu, W. and Bauer, E. (1993): A hypoplastic model for barotropy and pyknotropy of granular soils. Proc. Int. Workshop on Modern Approaches to Plasticity, Ed. Kolymbas, Elsevier: 225-245.

Zhang, L. (1995): Elastoplastic modeling of soil behavior under cyclic loading, Proc. First Int. Conf. on Earthquake Geotechnical Engineering, Tokyo, Ed.Kokusho and Ishihara, Balkema.

Computer Methods and Advances in Geomechanics, Yuan (ed.) © 1997 Balkema, Rotterdam, ISBN 90 5410 904 1

A new nonlinear uncoupled K-G model and its applications to concrete faced rock-fill dams

Lian-shi Gao, Zhao-hua Wang & Xian-pin Luo
Department of Hydraulic Engineering, Tsinghua University, Beijing, China

ABSTRACT:In this paper, a new non–linear uncoupled K–G model and its increment regression method of model parameters are presented. This model is applicable to different stress paths of dams and foundations. The model parameters can be regressed from conventional triaxial tests. Adopting this model, the three dimensional nonlinear finite element analyses are applied in two concrete faced rock–fill dams of Guan–Men–Shan and Tian–Sheng–Qiao. The calculated results are in accord with the prototype measurements.

1 INTRODUCTION

The concrete faced rock–fill dam is a new type of dams, which is mainly designed by experience and the computed deformation prediction is not satisfactory to yet. The basic cause for this condition is no a reasonable model which can be applied in different stress paths. Duncan's model and other elastic–plastic models also can not satisfied the changes in various stress paths or can not easily regress the model parameters. Pointing at this situation, we conducted large sized triaxial tests (φ=300 mm, $\sigma_{3,max}$=2.5 Mpa) in cooperation with Kuan–Ming Hydro–electric Power Institute. The testing materials include GMS andesite and TSQ limestone, thier specific densities are 2.65 and 2.70, and compressive strength are 80 and 115 Mpa. These testings are conducted in different stress paths, including proportional loading, conventional tri–axial shearing and tipical complex stress paths. Based on these tests, a new non–linear uncoupled K–G model (abbr. Tsinghua model) is developed[1]. The simulated results for various complex stress paths tests show that the model is unified to different stress paths[2].

2 BRIEF OF THE UNCOUPLED K–G MODEL

In the non–linear uncoupled K–G model, the mean stress p,($p\sim\varepsilon_v$), the generalized shear stress q, ($q\sim\varepsilon_s$), and stress ratio $\eta=q/p$, are adopted.

2.1 *Strength criterion*

For reflecting the relationship of strength and stress–strain, denote the failure stress ratio η_f, ($\eta_f=(q/p)_f$), as the peak strength in model, then the strength failure criteron is

$$\eta-\eta_f\geq 0 \tag{1}$$

$$\eta_f = \eta_o\cdot(p/p_\alpha)^{-\alpha} \tag{2}$$

here, η_0 and α are the test parameters.

Acoording to triaxial shear tests under different stress paths, the strength parameters have uniqueness, so they can be determined by simple conventional triaxial shear tests ($\sigma_3=c$).

Otherwise, the tests of materials show that the dilatation occurrs only as the strength approaches to peak value due to the damage of material structure. Before and after occurrence of dilatation, the properties of materials are very different. So in deformation computation, the dilatation may be not considered before damage stage. At the beginning of dilatation occurrence, the stress ratio can be termed as the critical dilatation strength η_1. Then

$$\eta-\eta_1=0 \tag{3}$$

The critical dilatation strength η_1 is a little less than the peak strength η_f. According to the testing experience, the critical dilatation strength is about 95 percent of peak strength.

2.2 *Loading criterion*

In the uncoupled K–G model, the flow rule is not used to establish the increment relationship of stress and strain. The loading function is only used in loading and unloading judgement. The testing results under the complex stress path with turning show that the materials have the deformation behavior of partial loading and unloading. The double loading criteria for volumetric strain and shear strain can be expressed respectively as following:

1. The loading criterion for volumetric strain

$$p \geq p_{max}, \quad (4)$$

$$\eta_c \geq \eta_{c.max} \quad \text{and} \quad f \geq f_{max} \quad (5)$$

here

$$f = p^2 + (1+\eta_c)^2 \cdot q \quad (6)$$

2. The loading criterion for shear strain

$$q \geq q_{max} \quad (7)$$

$$\eta \geq \eta_{max} \quad (8)$$

In formulas, $\eta_c = q/p_{max}$, p_{max}, q_{max} and η_{max} all are the historical maximun values of element stress. In computation, adopting above separately judging principles we can discriminate loading or unloading condition of p and q. In addition, through the stress ratio η and the loading function f, the effects of stress history and the direction of stress increment on the judgement of loading and unloading characteristics can be reflected.

2.3 *Increment relatonship of stress and strain during plastic loading*

In the non–linear uncoupled K–G model, the relationship of stress and strain can be obtained under proportional loading tests. Then increment relationship can express in the complete differential as following:

$$d\varepsilon_v = (1/K_v)\cdot(1+\eta_c^2)^m \cdot (H/p_a)\cdot(p/p_a)^{H-1}\cdot dp + (1/K_v)\cdot(p/p_a)^H \cdot m\cdot(1+\eta_c^2)^{m-1}\cdot 2\eta_c \cdot d\eta_c \quad (\eta_c<\eta_1) \quad (9)$$

$$d\varepsilon_s = (1/G_s)\cdot(p/p_a)^{-d}\cdot F_s \cdot (B/p_a)\cdot(q/p_a)^{B-1}\cdot dq + (1/G_s)\cdot(p/p_a)^{-d}\cdot(q/p_a)^B\cdot F_s\cdot(s/(\eta_u-\eta))\cdot d\eta \quad (\eta<\eta_1) \quad (10)$$

$$F_s = (1/(1-\eta/\eta_u))^s \quad (\eta<\eta_1) \quad (11)$$

In formulas (9) and (11), p_a is the pressure of atmosphere, its unit is as same as p and q; $\eta_c = q/p_{max}$; K_v, H, m, G_s, B, d and s are the parameters without dimension, which can be determined by a set of conventional triaxial shear tests. All the parameters have their physical meanings. F_s is the factor of strength mobilization. η_u is the limit of stress ratio, which can be gained by conventional triaxial shear tests(σ_3=c).

Based on the tests (σ_3=c), we can assume the relationship of η and ε_s as a hyperbolic function, $\eta = \varepsilon_s/(a+b\varepsilon_s)$, or written in $(\varepsilon_s/\eta)=(a+b\varepsilon_s)$, thus b can be determined, $(1/b)= \eta_u$. The testing data show that η_u is determined with p increasing, i.e. $\eta_u = \eta_{uo}(p/p_a)^{-\beta}$.

In above formulas, the effects of stress state, stress history and the direction of stress increment are included. Therefore, they can reflect the comprssive hardenability, shear shrinkability and the aniso-tropy caused by stress, as well as the effects of strength mobilization on deformation.

2.4 *Tangential modulus*

In nonlinear FEM computation, the tangential modulus in certain stress increment segment can be computed according to strain increment of formulas (9) and (10).

Formula (9) can be re–written as:

$$d\varepsilon_v = (1/(K_1\cdot p_a))\cdot dp + (1/K_2)\cdot d\eta \quad (12)$$

in which,

$$K_1 = (K_v/H)\cdot(1+\eta_c^2)^{-m}\cdot(p/p_a)^{1-H} \quad (13)$$

$$K_2 = (K_v/m)\cdot(1+\eta_c^2)^{1-m}\cdot(1/2\eta_c)\cdot(p/p_a)^{-H} \quad (14)$$

here, K_1 is the tangential compression modulus; K_2 is the tangential shear shrinkage modulus which is the action of shear stress to volumetric strain.

For some engineering problem such as in concrete faced rock–fill dams, their loading condition has some particularity, where the stress path of p in unloading and q in loading can not occur at same time. In this case, we can compute the coupled tangential volumetric modulus K_t by using formulas (9). That makes the coefficient matrix in FEM become symmetric and the computation will be simplify.

$$K_t = dp / d\varepsilon_v \qquad (15)$$

Similarly, using formula (10) we can gain the coupled tangential shear modulus G_t:

$$G_t = dq / 3d\varepsilon_s \qquad (16)$$

In formulas (9), (10), (15) and (16), because of taking total differential, the interaction between p and q has been included in K_t and G_t. Therefore, using the coupled tangential modulus can compute the shear shrinkability of materials. But for some engineering problems which are in the condition of q in loading as p in unloading, the computation model with three parameters of K_1, K_2 and G_t will be adopted.

2.5 *Unloading and reloading Modulus*

In unloading and reloading conditions, the volumetric modulus K_{ur} and shear modulus G_{ur} can be written as:

$$K_{ur} = K_{uo} \cdot p_a \cdot (\sigma_3/p_a)^n \qquad (17)$$

$$G_{ur} = G_{uo} \cdot p_a \cdot (\sigma_3/p_a)^n \qquad (18)$$

here, K_{uo} and G_{uo} is respectively the volumetric modulus and shear modulus in unloading and reloaing; n is the exponent of modulus which is gained by unloading tests.

3 INCREMENT REGRESSION METHOD FOR PARAMETERS

For the regression of geotechnical deformation parameters, the total regression method is a conventional method used previously. This kind of method can not reflect the charateristic of plastic deformation and different stress path. In our model, the difference is that the parameter regression is an increment method. This method can use the results of any monotone loading triaxial tests, including conventional triaxial shear tests, to regress parameters which have uniqueness and can fit various stress paths.

3.1 *Increment Regression Analysis for Parameters of Volumetric Strain*

Based on formula (9), in monotone loading case, if the loading increment dp is enough small, then the normalized volumetric strain ε_{v1} can be written in

$$\varepsilon_{v1} = (1/K_v) \cdot (p/p_a)^H \qquad (19)$$

and the normalized volumetric strain of i-th loading (i from 1 to n) can yet be accumulated:

$$\varepsilon_{v1} = \sum_{i=1}^{n} [\Delta(\varepsilon_v)_i \cdot (N_v)_i] \qquad (20)$$

$$(N_v)_i = 1/[(1+\eta_i^2)^m + p_i^H \cdot m \cdot (1+\eta_i^2)^{m-1} \cdot 2\eta_i \cdot \Delta\eta_i / \Delta(p_i^H)] \qquad (21)$$

if $\eta=0$, $\varepsilon_{v1} = \varepsilon_v$, then

$$\varepsilon_v = (1/K_v) \cdot (p/p_a)^H \qquad (22)$$

In formula (19) –(22) $(N_v)_i$ is the normalization factor of volumetric strain; the deformation parameters K_v, m and H can be defined by the method of increment regression with the results of conventional triaxial shear tests. Fig.1 and 2 shows that the regressed paraments of volumetric strain are uniqueness for different stress paths(σ_3=c and η=c).

3.2 *Increment Regression Analysis for Shear Strain Parameters*

As above, based on the formula (10), in monotone loading case as dq is enough small, then:

$$\varepsilon_{s1} = (1/G_s) \cdot (q/p_a)^B \qquad (23)$$

and ε_{s1} can yet be written in

$$\varepsilon_{s1} = \sum_{i=1}^{n} [\Delta(\varepsilon_s)_i \cdot (N_s)_i] \qquad (24)$$

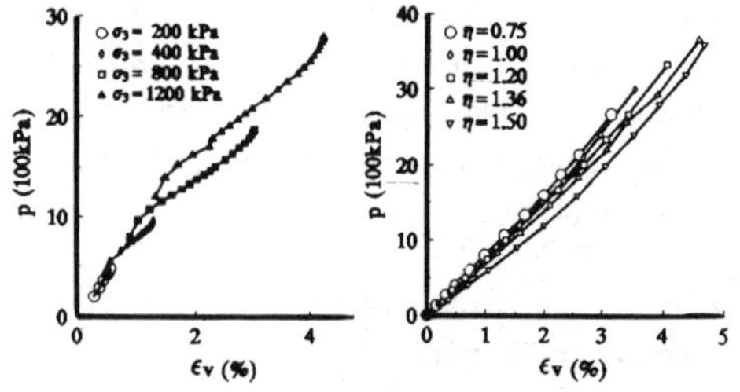

(a) σ_3=c (b) η=c

Fig.1 $p\sim\varepsilon_v$ curves of rockfill material (GMS)

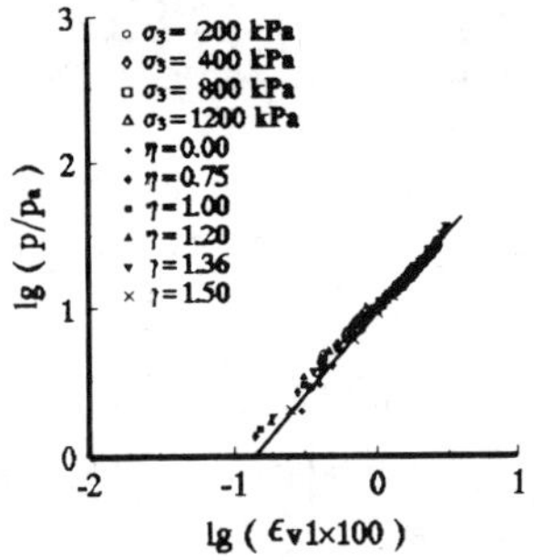

Fig.2 Normalisation of volumetric strain(σ_3=c, η=c)

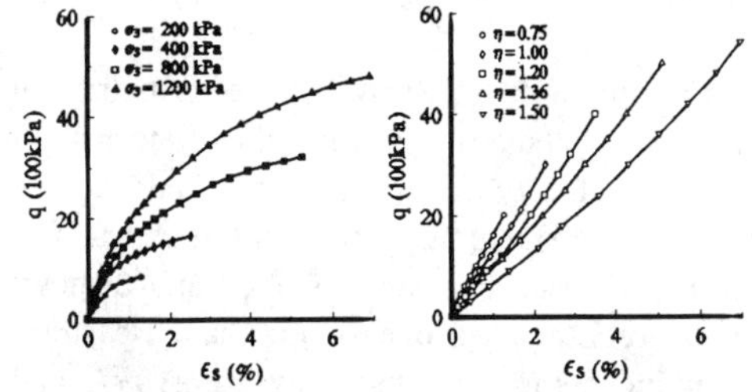

(a) σ_3=c (b) η=c

Fig.3 $q\sim\varepsilon_s$ curves of rockfill material (GMS)

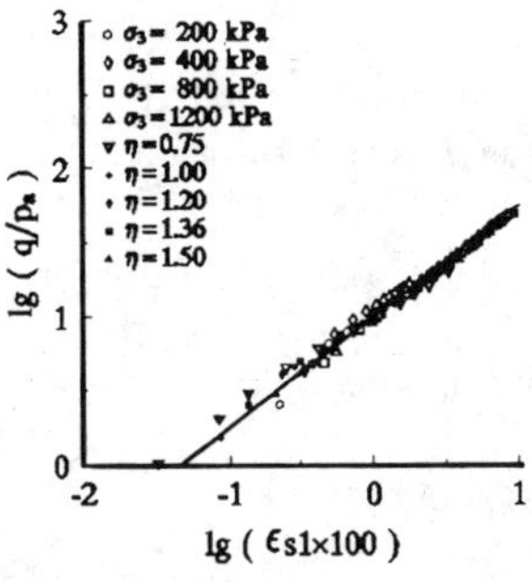

Fig.4 Normalization of shear strain (σ_3=c, η=c)

$$(N_s)_i=(p/p_a)^d/(F_s)_i/\{1+q_i{}^B\cdot[s/(\eta_{ui}-\eta_i)]\cdot\Delta\eta_i/\Delta(q_i{}^B)\} \quad (25)$$

In formula (23)–(25),the deformation parameters of G_s, d, s and B can be defined by the method of increment regression with the results of conventional triaxial shear tests, seen in Fig.3 and 4. It shows that the regressed parameters of shear strain are uniqueness for different stress paths (σ_3 =c and η=c).

3.3 *Strength and deformation parameters*

According to the testing results of conventional triaxial tests, the strength and deformation parameters of GMS and TSQ are obtained, which are shown in table 1 and table 2 respectively.

Table 1: Strength parameter

Material	γ_d(g/m³)	η_o	α	η_{uo}	β	σ_3(MPa)
GMS andesite	2.00	1.92	0.03	2.04	0.02	0.2–1.6
TSQlimestone	2.10	2.08	0.07	2.11	0.05	0.1–1.2
TSQsandy marl	2.15	1.78	0.04	1.85	0.02	0.2–1.6

Table 2: Deformation parameters

Material	K_V	m	H	G_S	d	s	B
GMS andesite	710	0.35	0.88	2000	0.50	0.60	1.35
TSQlimestone	1500	0.80	0.96	3300	0.50	0.70	1.42
TSQsandy marl	400	0.60	0.85	820	0.46	0.60	1.30

4 FINITE ELEMENT ANALYSES

4.1 *CFRD of Guan-Men-Shan*

This dam is 58.5 m in height and 183.6 m in length. Its rock–fill material is andesite. The river valley is near symetric. Using three dimensional finite element mathod with uncoupled K–G model calculates the right half part of dam. Its loadings of selfweight and water pressure are divided into 9 loading stages. The calculations of uncoupled K–G model are compared with prototype observations. The main results are shows in Fig.5 and Fig.6. The settlement values of measured points a, b, c and d are closed to calculated ones (measured: 6.7, 5.7, 5.5, 5.3, 5.8 cm; calculated: 6.8, 6.6, 7.8, 7.7 cm).

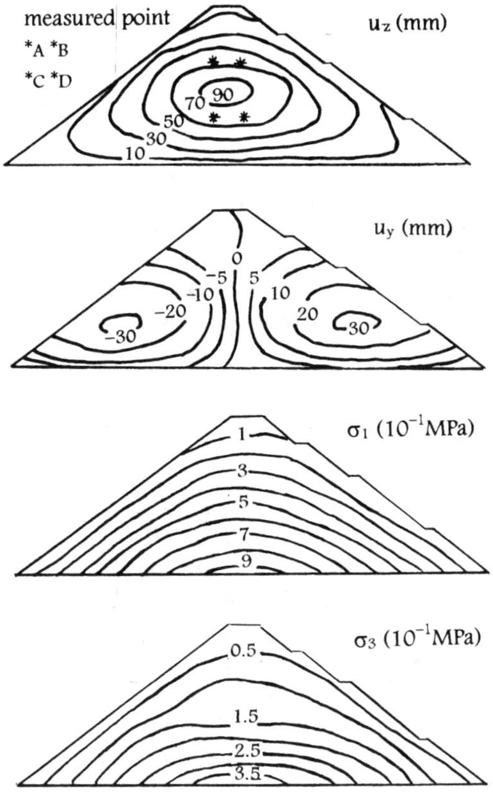

Fig.5 Isograph of displacement and stress
(GMS dam, construction period)

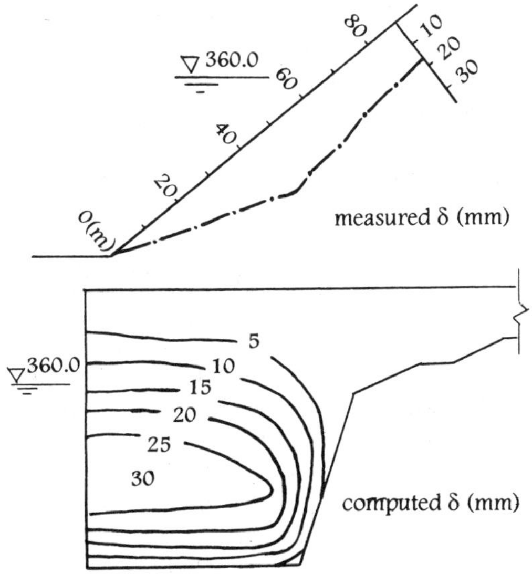

Fig.6 Normal displacement (δ) of face slab
(GMS dam, water filling period)

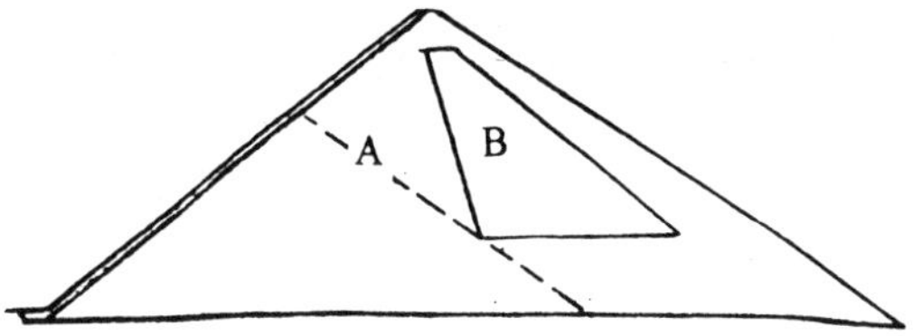

A: Limestone B: Sandy marl
Fig.7 Max cross section of TSQ dam

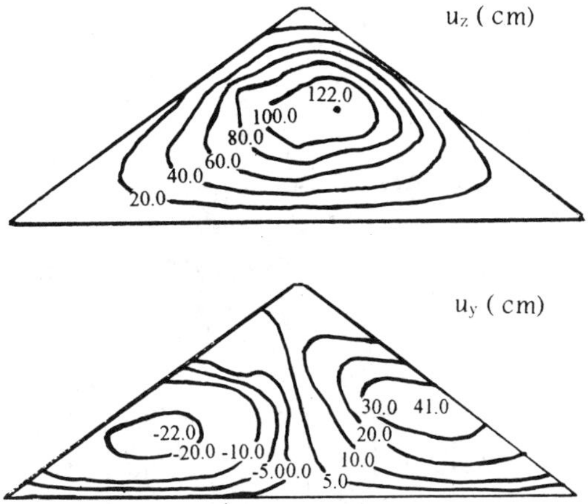

Fig.8 Displacements of TSQ dam
(construction period)

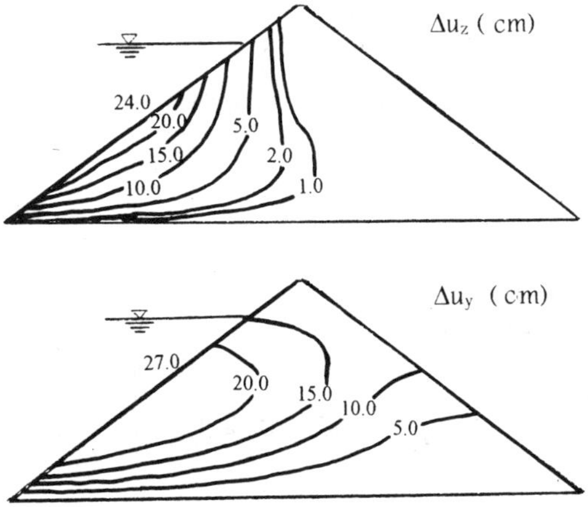

Fig.9 Displacement increments of TSQ dam
(water filling period)

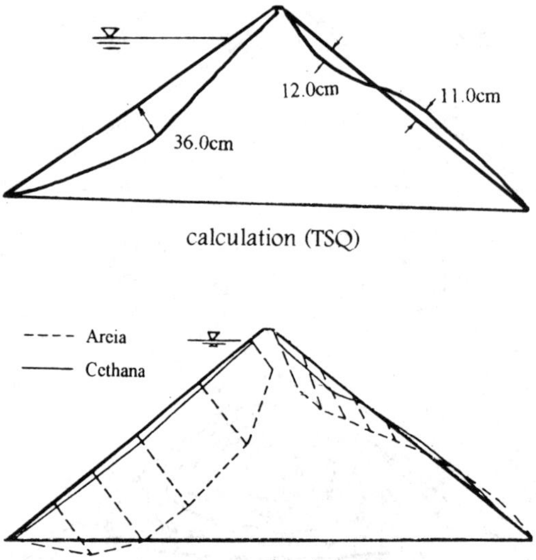

Fig.10 Displacement comparison of TSQ and others

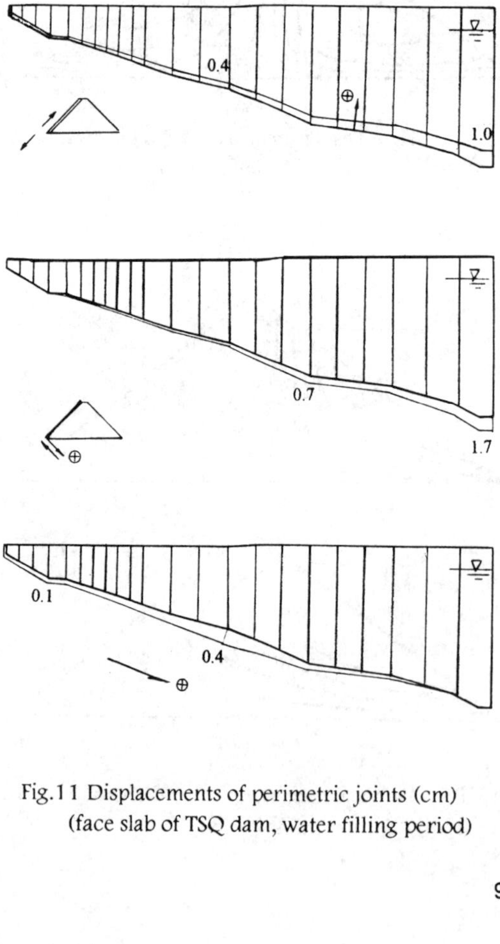

Fig.11 Displacements of perimetric joints (cm)
(face slab of TSQ dam, water filling period)

4.2 *CFRD of Tian-Shen-Qiao*

This is a 180 m high dam with a crest of 1178 m. A temporary section is filled first, then heightened to a full section. The rock–fill material zones of dam include limestone and sandy marl, shown in Fig.7. Due to the symetric valley, the left half dam is caculated. The loading is divided into 23 stages. The calculated results are shown in Fig.8 to Fig.11. In Fig.10, calculated slope deformation is compared with measured results of other dams[3], the tendency is in accord.

5 CONCLUSION

The non–linear uncoupled K–G model is a unified model, which suits to different stress paths. Its deformation parameters can be regressed and normalized from monotone loading tests. Due to the uniqueness of parameters the model suits to predict the deformation of dams and foundations, which are loaded under complex stress paths. The calculations of FEM show that the calculated results of concrete faced rock–fill dams basicly tally with the prototype observations.

REFFERENCES

1. Gao Lian–Shi, *The non–linear uncoupled K–G model for rock–fill materials and its verification.* 14th International Conference on Soil Mechannics and Foundation Engineering, Hamburg, Sept, 1997.

2. Gao Lian–Shi, H.Q. Zhao,C.Y Yi, etc., *Large sized triaxial tests of rock–fill materials and the verification of non–linear uncoupled K–G model,* Earth and Rock–fill Dam Engineerings, vol 1& 2, Beijing,1996.

3.Imaizumi, H. and A.E. Sardinha, *A study of deformation in concrete faced rockfill dams,* in “CFRD, Design, Construction and Performance”, ASCE, 1985.

Computer Methods and Advances in Geomechanics, Yuan (ed.) © 1997 Balkema, Rotterdam, ISBN 90 5410 904 1

Development of plastic potential theory and its application in constitutive models of geomaterials

Zheng Yingren & Liu Yuanxue
Department of Civil Engineering, Logistical Engineering University, Chongqing, China

ABSTRACT: Traditional plastic potential theory has been successfully used in metals, but for geomaterials, it is not quite applicable. In this paper, the traditional plastic potential theory is developed so that it is suitable for both metals and geomaterials. To distinguish the traditional and the developed plastic potential theory, the later is called as the generalized plastic potential theory to which the traditional plastic potential theory is just a special case. So, the argument of views about whether plastic potential theory should be used in the present research of geotechnical constitutive models can be united. The generalized plastic potential theory can be looked upon as the theoretical basis of various constitutive models of geomaterials.

1 INTRODUCTION

Traditional plastic potential theory, which is suitable to metals, will meet a number of challenges in its application in geomaterials. The following deformation mechanisms, which are based on a large number of geotechnical tests, are being realized by people.

1. Test results of Poorshasb, Frydman, Lode etc. show that geomaterials don' t obey associated flow rule and Drucker' s plastic postulate.
2. According to traditional plastic potential theory, the direction of plastic strain increment is determined by stress states only and is not related to the stress increment. However, tests developed by Balashablamaniam, Sheng Zhujiang etc. show that the direction of plastic strain increment is influenced by stress increment, that is, geomaterials don' t satisfy the uniqueness assumption between the direction of plastic strain increment and stress. It means that geomaterials don' t obey the traditional plastic potential theory.
3. Tests developed by Matsuoka etc. show that the rotation of stresses (the change of the direction of principal stress axes) will cause plastic strain even though the values of principal stresses are keep unchanged, but this kind of plastic strain can not be considered in the traditional plastic potential theory.
4. If shear yield surfaces like Mohr-Coulumb yield surface and the associated flow rule are used, the single yield surface model based on the traditional plastic potential theory will lead unreasonable huge dilative deformation, this situation can not be changed even if the non-associated flow rule is used. Otherwise, yield surfaces based on Cambridge model can not reflect shear yield of geomaterials as good as possible, these surfaces can reflect the volume contract of soil, but can not reflect the dilatancy of it. Both volume contraction and dilatancy of geomaterials can not be reflected successfully even if the single yield surface model with a closed form is used.

To overcome the above disadvantages, a lot of works have been done by scholars in the world. Some constitutive models which disobey plastic potential theory, such as double yield surfaces models, multi-yield surfaces models as well as models involving the rotation of principal stress axes were presented. At the same time, the non-associated flow rule was used to remedy these disadvantages. Many outstanding results were given by Chinese scholars Sheng Zhujiang, Yang Guanghua, Ying Zhongzhe, the authors and Japaneses Matsuoka. In this paper, the mentioned results were further developed and a generalized plastic potential theory which is suitable to both metals and geomaterials was estabilished. It is obvious that the generalized plastic potential theory is the theoretical basis of estabilishing constitutive models of geomaterials and there are some weakness in geotechnical constitutive models based on the traditional plastic potential theory.

2 ANALYSIS OF THE TRANDITIONAL PLASTIC POTENTIAL THEORY

In the development of geotechnical plastic mechanics, plastic potential theory was widely used. On the other hand, some scholars expressed a clear objection of it. To remove this puzzle, it is necessary to analyze the traditional plastic potential theory.

According to the traditional theory, if a function $Q(\sigma_{ij}, H_a)$ exists in the stress space, the principal directions of stress and plastic strain increment are assumed identical and the following relation is satisfied.

$$d\varepsilon_{ij}^p = d\lambda \frac{\partial Q}{\partial \sigma_{ij}} \tag{1}$$

Then, the function $Q(\sigma_{ij}, H_a)$ can be called plastic potential function. According to Eq. (1), we yield

$$d\varepsilon_1^p : d\varepsilon_2^p : d\varepsilon_3^p = \frac{\partial Q}{\partial \sigma_1} : \frac{\partial Q}{\partial \sigma_2} : \frac{\partial Q}{\partial \sigma_3} \tag{2}$$

Eq. (2) is the basic character of the traditional plastic potential theory from which the following relation between principal stress increment and principal plastic strain increment can be deduced.

$$d\varepsilon_i^p = [A_p] d\sigma_i \tag{3}$$

Elements in each line of the matrix $[A_p]$(a_{1i}, a_{2i}, a_{3i}, $i=1,2,3$) must keep the following relation

$$a_{1i} : a_{2i} : a_{3i} = \frac{\partial Q}{\partial \sigma_1} : \frac{\partial Q}{\partial \sigma_2} : \frac{\partial Q}{\partial \sigma_3} \tag{4}$$

Eq. (2) and Eq. (4) show that elements in each line of matrix $[A_p]$ are in propotion to each other and a plastic potential function exists. It can be further proved that the order of $[A_p]$ is one and only one unit vector exists. It means that the direction of plastic strain increment is controlled only by stress states and has no relation with the stress increment, this is also the physical meaning of Eq. (1). From what mentioned above, it is clear that the uniqueness assumption of the relation between the direction of plastic strain increment and stress has been made in the traditional plastic potential theory.

The assumptions of traditional plastic potential theory in which the yield surface is a function of invariants of stress rather than stress components, neglect the plastic deformation caused by the rotational parts of stress increment. That is why the traditional plastic potential theory cann't reflect the plastic deformation caused by rotation of principal stress axes.

In the traditional plastic mechanics, materials are assumed to obey Drucker's postulate, that is, to follow the associated flow rule and the plastic potential function is the same as the yield surface. In mathematics, it implies that if the gradient vector of plastic potential function is in propotion to that of yield surface, the elasto-plastic matrix must be symmetrical, otherwise, it is unsymmetrical.

In general, the following three assumptions are involed in the traditional plastic potential theory.

1. There is a unique plastic potential function in stress space, the order of matrix $[A_p]$ is one, elements of each line in $[A_p]$ is in proportion to each other, plastic deformation follows the uniqueness assumption between the direction of plastic strain increment and stress.
2. Principal axes of plastic strain increment are the same as that of stress and the plastic deformation caused by rotation of principal stress is ignored.
3. The plastic deformation obeies the associated flow rule.

As mentioned above, all of the three assumptions are contrary to the deformation mechanism of geomaterials. To applicate the plastic potential theory in geomaterials, the three assumptions should be abandoned. Based on this idea, a generlized plastic potential theory is put forward.

3 THE GENERALIZED PLASTIC POTENTIAL THEORY

If the first and the third of the mentioned three assumptions are called off and the second is remained, Eq. (1) can be written as

$$d\varepsilon_{ij}^p = \sum_{k=1}^{3} d\lambda_k \frac{\partial Q_k}{\partial \sigma_{ij}} \tag{5}$$

Eq. (5) was deduced by use of tensor law without any assumption in reference [8]. Stress and strain here should be a tensor with two order, according to tensor law, we have

$$d\varepsilon_{ij}^p = \sum_{k=1}^{3} d\varepsilon_k^p \frac{\partial \sigma_k}{\partial \sigma_{ij}} \tag{6}$$

where σ_k and ε_k are three principal stresses and principal strains resprestively.

By the difinition of gradient, we have

$$d\varepsilon_i^p = \sum_{k=1}^{3} d\lambda_k \frac{\partial Q_k}{\partial \sigma_i} \quad (7)$$

where Q_k ($k=1,2,3$) are three arbitrary potential functions which are linear irrelevant. Substituting Eq. (7) into Eq. (6), Eq. (5) can be obtained.

Eq. (5) should be a generialized plastic potential theory without considering the rotation of principal stress axes or a generalized plastic flow rule. It shows that three rather than one plastic potential functions are needed to determine the direction of plastic strain increment in a more general case, in this case, plastic potential function should not be defined as Eq. (1) and has no characters of Eq. (2) and Eq. (4). Metal is a special material which satisfies the characters of Eq. (2) and Eq. (4), so, only one plastic potential function is needed and Eq. (5) is changed into Eq. (1). It means that the traditional plastic potential theory obeys the uniqueness assumption between the directions of plastic strain increment and stress. However, the generalized plastic potential theory doesn' t obey the assumption. From Eq. (5), it can be known that when the number of plastic potential function are more than one, the direction of plastic strain increment is related to $d\lambda_k$ which is related to the stress increment. Because plastic potential functions are three arbitrary functions which are linear irrelevant, the best one which satisfy the condition are three coordinate axes in the stress space (such as $\sigma_1, \sigma_2, \sigma_3$, or p, q, θ).

Let $\sigma_1, \sigma_2, \sigma_3$ to be plastic potential functions[6][11], then

$$d\varepsilon_{ij}^p = d\lambda_1 \frac{\partial \sigma_1}{\partial \sigma_{ij}} + d\lambda_2 \frac{\partial \sigma_2}{\partial \sigma_{ij}} + d\lambda_3 \frac{\partial \sigma_3}{\partial \sigma_{ij}} \quad (8)$$

From Eq. (8), we have

$$d\lambda_1 = d\varepsilon_1^p,\ d\lambda_2 = d\varepsilon_2^p,\ d\lambda_3 = d\varepsilon_3^p, \quad (9)$$

Eq. (9) shows that $d\lambda_k$ ($k=1,2,3$) are the values of plastic strain increment in the direction of three principal strain respectively and it is in keeping with the physical meaning of $d\lambda_k$ in Eq. (8).

let p, q, θ to be platic potential function, then

$$d\varepsilon_{ij}^p = d\lambda_1 \frac{\partial p}{\partial \sigma_{ij}} + d\lambda_2 \frac{\partial q}{\partial \sigma_{ij}} + d\lambda_3 \frac{\partial \theta}{\partial \sigma_{ij}} \quad (10)$$

in which

$$d\lambda_1 = d\varepsilon_v^p,\ d\lambda_2 = d\gamma_q^p,\ d\lambda_3 = d\gamma_\theta^p$$

where $d\varepsilon_v^p$, $d\lambda_\theta^p$, $d\lambda_\theta^p$ are plastic volume strain, plastic shear strains in the directions of q and θ seperately.

Hardening law is used in the traditional plastic mechanics to determine the proportion coefficient $d\lambda_k$, but in the generalized plastic mechanics, three mothods can be introduced to get $d\lambda_k$. 1) using hardening law, 2) fitting from experiment data, 3) introducing the uniqueness assumption of the relation between the total plastic strain and the stresses, that is, let yield surface to be

$$\varepsilon_k^p = f_k(\sigma_{ij}),\ (k = 1,2,3,\ or\ v,\ \gamma_q,\ \gamma_\theta) \quad (11)$$

The third method is the same as the first one, it uses ε_k^p as hardening parameter. Eq. (11) expresses an isoplethic surface of three principal plastic strains in stress space or an isoplethic surface of plastic volume strain and an isoplethic surface of plastic shear strains in the directions of q and θ. There is a corresponding relationship between yield surface and plastic potential surface, it shows they are associated but doesn' t mean they are the same. Because the condition $d\lambda_k \geqslant 0$ can not be ensured all the time, plastic volume strain $d\lambda_1$ may be both positive (volume constraction) and negative (volume expanding).

Differentiating Eq. (11), then

$$d\lambda_k = d\varepsilon_k^p = \frac{\partial f_k}{\partial \sigma_1} d\sigma_1 + \frac{\partial f_k}{\partial \sigma_2} d\sigma_2 + \frac{\partial f_k}{\partial \sigma_3} d\sigma_3 \quad (k = 1,2,3) \quad (12)$$

$$\text{or} \quad d\varepsilon_k^p = \frac{\partial f_k}{\partial p} dp + \frac{\partial f_k}{\partial q} dq + \frac{1}{q}\frac{\partial f_k}{\partial \theta} d\theta \quad (k = p, \gamma_q, \gamma_\theta) \quad (13)$$

Substituting Eq. (12) into (7), then

$$d\varepsilon_i^p = [A_p] d\sigma_i = \begin{bmatrix} \frac{\partial f_1}{\partial \sigma_1}\frac{\partial Q_1}{\partial \sigma_1} & \frac{\partial f_1}{\partial \sigma_2}\frac{\partial Q_1}{\partial \sigma_1} & \frac{\partial f_1}{\partial \sigma_3}\frac{\partial Q_1}{\partial \sigma_1} \\ \frac{\partial f_2}{\partial \sigma_1}\frac{\partial Q_2}{\partial \sigma_2} & \frac{\partial f_2}{\partial \sigma_2}\frac{\partial Q_2}{\partial \sigma_2} & \frac{\partial f_2}{\partial \sigma_3}\frac{\partial Q_2}{\partial \sigma_2} \\ \frac{\partial f_3}{\partial \sigma_1}\frac{\partial Q_3}{\partial \sigma_3} & \frac{\partial f_3}{\partial \sigma_2}\frac{\partial Q_3}{\partial \sigma_3} & \frac{\partial f_3}{\partial \sigma_3}\frac{\partial Q_3}{\partial \sigma_3} \end{bmatrix} d\sigma_i$$

$$= \begin{bmatrix} \frac{\partial f_1}{\partial \sigma_1} & \frac{\partial f_1}{\partial \sigma_2} & \frac{\partial f_1}{\partial \sigma_3} \\ \frac{\partial f_2}{\partial \sigma_1} & \frac{\partial f_2}{\partial \sigma_2} & \frac{\partial f_2}{\partial \sigma_3} \\ \frac{\partial f_3}{\partial \sigma_1} & \frac{\partial f_3}{\partial \sigma_2} & \frac{\partial f_3}{\partial \sigma_3} \end{bmatrix} d\sigma_3 \quad (14)$$

where f_1, f_2, f_3 are three yield surfaces of principal strain.

If Q_i ($i=1,2,3$) in Eq. (14) can be expressed in a potential function Q and f_i ($i=1,2,3$) can be expressed in a yield function f, it will be a single yield surface situation. If f=Q, it will be the traditional plastic potential theory. In such a condition, elements in each line of $[A_p]$ will have

$$a_{1i} : a_{2i} : a_{3i} = \frac{\partial Q}{\partial \sigma_1} : \frac{\partial Q}{\partial \sigma_2} : \frac{\partial Q}{\partial \sigma_3} \qquad (15)$$

The plastic flexibility matrix $[C_p]$ in stress-strain relation can be obtained by coordinate transformation.

$$\{d\varepsilon\} = [C_p]\{d\sigma\} = [T_\varepsilon]_{6\times3}[A_p]_{3\times3}[T_\sigma]_{3\times6}\{d\sigma\} \qquad (16)$$

in which $[T_\varepsilon]$, $[T_\sigma]$ are transformation matrixes Eq. (13) can be written in matrix form and the influence of θ is neglected, then

$$\begin{Bmatrix} d\varepsilon_v^p \\ d\gamma^p \end{Bmatrix} = \begin{bmatrix} \frac{\partial f_v}{\partial p} & \frac{\partial f_v}{\partial q} \\ \frac{\partial f_\gamma}{\partial p} & \frac{\partial f_\gamma}{\partial q} \end{bmatrix} \begin{Bmatrix} dp \\ dq \end{Bmatrix} = \begin{bmatrix} A & B \\ C & D \end{bmatrix} \begin{Bmatrix} dp \\ dq \end{Bmatrix} \qquad (17)$$

Substituting Eq. (17) into Eq. (10), we obtain

$$\begin{aligned} d\varepsilon_{ij}^p &= \left(\frac{\partial f_v}{\partial p}dp + \frac{\partial f_v}{\partial q}dq\right)\frac{\partial p}{\partial \sigma_{ij}} \\ &\quad + \left(\frac{\partial f_\gamma}{\partial p}dp + \frac{\partial f_\gamma}{\partial q}dq\right)\frac{\partial q}{\partial \sigma_{ij}} \\ &= (Adp + Bdq)\frac{\partial p}{\partial \sigma_{ij}} \\ &\quad + (Cdp + Ddq)\frac{\partial q}{\partial \sigma_{ij}} \end{aligned} \qquad (18)$$

in which f_v, f_γ are volume yield surface and shear yield surface respectively. A、B、C、D are plastic coefficients.

Eq. (18) is the relation between plastic strain increment and stress increment of double yield surfaces model[1] developed by Sheng Zhujiang. If the coefficients A, B, C, D are fitted by test data, Eq. (18) will be turned into the multi-potential surfaces model in reference [10]. The same results can be obtaind from double yield surfaces models which are often used in the world when non-associated flow rule is used and p, q are adopted as plastic potential functions. From what mentioned above, the generalized plastic potential theory can be looked upon as the basis of double yield surfaces model, multi-yield surfaces model as well as multi-potential surfaces model.

If the second assumption of the traditional plastic potential theory is abandoned, the plastic deformation caused by the rotation of principal stress axes must be considered. At present, the research on this subject isn' t good enough. A plane model in which the rotation of principal stress axes considered was presented and verified by Matsuoka in 1986. A new approch of considering the rotation of principal stress axes is provided in this paper, it starts from writting the stress into the following form.

$$d\sigma_{ij} = d\sigma_{cij} + d\sigma_{r1ij} + d\sigma_{r2ij} + d\sigma_{r3ij} \qquad (19)$$

where $d\sigma_c$ is the coaxial part of stress increment $d\sigma_{r1}$, $d\sigma_{r2}$, $d\sigma_{r3}$ are the rotational parts of stress increment which rotate around the three principal axes seperately.

Then, the plastic deformation caused by three rotational parts of stress increment $d\varepsilon_{r1}^p$, $d\varepsilon_{r2}^p$, $d\varepsilon_{r3}^p$ can be calculated, that is

$$d\varepsilon_{ij}^p = d\varepsilon_{cij}^p + d\varepsilon_{r1ij}^p + d\varepsilon_{r2ij}^p + d\varepsilon_{r3ij}^p \qquad (20)$$

Eq. (20) is the complete formulation of the generalized plastic potential theory in which $d\varepsilon_c^p$ is the coaxial part of plastic strain. A model is needed to calculate the plastic deformation caused by the rotational part of stress increment, or the model developed by Matsuoka etc. can be used.

4 APPLICATION OF THE GENERALIZED PLASTIC POTENTIAL THEORY IN CONSTITUTIVE MODELS OF GEOMATERALS

The single yield surface models based on the traditional plastic potential is the most popular kind of models in the world (such as Cambridge model, Desai hierarchical model and new Lode model). These models express disadvantages of single yield surface mentioned althrough they are widely used. Some models are modified partly, such as the adoption of non-associated flow rule. Even though the modification can reflect the dilatance more effictively, sometime it will bring about more serious mistake because the disadvantages have not been thoroughly overcome. Four single yield surface models are verified[10] (two of them are single yield surface model with different hardening parameters, the others are Desai hierarchical models based on the associated and the non-assiciated flow rule). As a result, none of these models can reflect all the deformation characters of Pueblo sands.

In view of the disadvantages of single yield surface models, the double yield surfaces models are accepted expansively in the world. They have two yield surfaces, one is isoplethic surface of plastic volume strain called volume yield surface, the other is isoplethic surface of plastic shear strain called shear yield surface. In this kind of models, both volume and shear strain can be reflected better and the uniqueness assumption between the direction of plastic strain increment and stress is not be followed, but the disadvantages of using associated flow rule and neglecting the influence of Lode angle still exist. Using of the associated flow rule implies that shear deformation will be caused by volume yield surface, and volume deformation will be caused by shear yield surface. This will run counter to the original physical meaning of the two yield surfaces. If the non-associated flow rule is used, the double yield surfaces models obey the generalized plastic potential theory.

The multi-yield surfaces model has a few yield surfaces, system formulas about the model have been deduced by authors in [3],[5],According to the generalized plastic potential theory,the number of yield surface shoudn't be more than three,and the non-associated flow rule should be used. By test fitting,three isoplethic surfaces of principal plastic strain are obtained and looked upon as yield surfaces,after that,the plastic matrix can be calculated by Eq. (14). This model is in keeping with the generalized plastic potential theory completely.

Multi-plastic petential surface madel is based on the generalized plastic potential theory too, but it don't need yield surface. The proportion coefficient $d\lambda_k$ is obtained by test fitting rather than introducing yield surfaces.

5 CONCLUSION

1. There are some assumptions in the traditional plastic potentlial theory,so it can not reflect the deformation mechanism of geomaterials as good as possible.
2. Plastic potential theory is suitable to both metals and geomaterials,the traditional plastic potential is just a special case of it. Most models which are popular at present are based on the plastic potential theory,so,plastic potential theory shouldn't be abandoned but developed in the field of geomechanics.
3. The complete generialized plastic potential theory should include the non-coaxality situation between stress and stress increment as well as plastic strain increment and stress, it will be discussed in other papers.

REFERANCE

Sheng Zhujiang, The Reasonable Form of Elasto-plastic Stress Strain Relation of Soils, Chinese Journal of Geotechnical Engineering, No. 2, 1980

Zheng Yingren, Yan Dejun, Multi-yield Surface Model for Soils on the Basis of Test Fitting, Computer Methods and Advances in Geomechanics, A · ABALKEMA/ROTTERDAM/BROOKFIELD/1994

Formulation of the Elastoplsatic Theory and Its Finite Element Implementation, Computer and Geotechnics, 2(1986)

Matsuoka H, Sakakibara K, A Constitutive Model for Sands and Clays Evaluating Principal Stress Rotation, Soil and Foundations, Vol. 27, No 14, PP73-88,1987

Zheng Yingren, Gong Xiaonan, Basis of Geotechnical Plastic Mechonics, China Building Industry Press, 1989

Zheng Yingren, Yan Dejun, Theory of Multiple Yield Surfaces for Soil Material, Advances in Constitutive Laws for Engineering Materials (Ⅱ), International Academic Publishers, 1989

Zheng Yingren, Multi-yield Theory for Soils, Computer Methods and Advances in Geomechanics, 715-720, A · ABALKEMA/ROTTERDAM/BROOKFIELD/1991

Yang Guanghua, Elasto-plastic Constitutive Model with Multi-potential Surfaces for Geomaterials, Chineses Journal of Geotechnical Engineering, No. 5, 1991

Zheng Yingren, Theory and Model of Multi-yield Surfaces, Proceedings of Plastic Mechanics and Semomechanics,Beijing University Press,1992

Yang Guanghua, A new Elastoplastic Constitutive Model for Soil, Proceedings of International Conference on Soft Engineering, Science Press, 1993

Najjar Y M, Zamman M M, Tabbaa, Constitutive Modeling of Sand, A Comparative Study, Computer Methods and Advances in Geomechanics, Vol. 1,1994

Computer Methods and Advances in Geomechanics, Yuan (ed.) © 1997 Balkema, Rotterdam, ISBN 90 5410 904 1

Prediction of extension test results of soft cemented clays with a simple constitutive model

B.R.Srinivasa Murthy, A.Vatsala & T.G.Sitharam
Indian Institute of Science, Bangalore, India

K.Nagendra Prasad
Sri Venkateswara University, Tirupati, India

In the present paper, the constitutive model is proposed for cemented soils, in which the cementation component and frictional component are treated separately and then added together to get overall response. The modified Cam clay is used to predict the frictional resistance and an elasto-plastic strain softening model is proposed for the cementation component. The rectangular isotropic yield curve proposed by Vatsala (1995) for the bond component has been modified in order to account for the anisotropy generally observed in the case of natural soft cemented soils. In this paper, the model proposed is used to predict the experimental results of extension tests on the soft cemented soils whereas compression test results are presented elsewhere. The model predictions compare quite satisfactorily with the observed response. A few input parameters are required which are well defined and easily determinable and the model uses associated flow rule.

INDRODUCTION

During the last four decades, many constitutive models for clay have been proposed and studied. The most important ones are the Cam clay models proposed by the Cambridge group. The Cam clay model has been used for clays in one form or another by various authors such as Adachi and Okamo (1974) and Nova and Wood (1979). All these models were initially concerned with ideal states of saturated uncemented soils. Attempts have been made to extend these models to more complex states such as cemented soils (Gens and Nova, 1993; Adachi and Oka, 1995) only recently.

Owing to complex mechanisms involved, understanding the behaviour of cemented soils has lagged behind. Recently, important contributions have been made (Leroueil and Vaughan, 1990) highlighting the facts that the cemented soils also exhibit behaviour similar to other elasto-plastic materials i.e., they show well defined yield locus beyond which significant plastic strains occur while within the yield locus, the strains are relatively smaller and mostly recoverable. Conlon (1966), Feda (1982), and Oka et al. (1989) indicated that the behaviour of cemented soils can be bifurcated into those of frictional and cementation components. From the observed patterns of the behaviour of this class of materials, there has been a new thinking (Vatsala, 1989; Nagaraj et al., 1991; Gens and Nova, 1993; Adachi and Oka, 1995) that for modelling the behaviour of cemented soils it is necessary to consider the behaviour of an equivalent unbonded state. This has opened up new direction in modelling the behaviour of these soils.

A few constitutive models have been tried recently (Gens and Nova, 1993; Adachi and Oka, 1995) to characterise the behaviour of cemented soils, especially strain softening which is a typical feature of these materials. These models adopt non-associated flow rule and require further clarification regarding certain aspects. From the examination of mechanisms of stress transfer for cemented soils, it has been shown (Nagaraj et al., 1991) that the deformation is due essentially due to changes in increments on equivalent unbonded soil skeleton, while the cement bonds exhibit rigid response. During debonding with progress of shearing, the stress increments corresponding to cement bonds are transmitted either to the pore pressure or to the soil skeleton depending on the drainage conditions that exist.

Based on these facts a simple constitutive model with associated flow rule based on superposition of cementation component resistance over resistance of equivalent unbonded component to describe the behaviour of soft cemented soil such as sensitive clay has been proposed (Vatsala, 1995, Nagendra Prasad, 1996). As the detailed account of the proposed model is being published elsewhere only salient features are briefly given in the present paper.

CONSTITUTIVE MODELLING

Constitutive laws are described separately for the two components and then added to get the over all response. While the modified Cam clay is used to predict the frictional resistance, an elasto-plastic, strain softening model is proposed for the cemented component as given below.

The yield surface for the bond componet is defined by the general expression represented by:

$$f_B = f(q_B, p_B) - N = 0 \tag{1}$$

where p_B = bond component of effective mean principal stress, q_B = effective deviatoric bond stress and N = the hardening parameter reflecting the degree of bonding. The initial yield curve can be obtained by detailed experimental programme. To circumvent the difficulty associated with elaborate experimental procedure, yield curve is obtained based on the microstrucure analysis. From the criteria proposed (Appendix), the yield curve can be arrived which looks like a rectangle (Figure 1) oriented along K_o compression path. This yield curve is approximated as an ellipse in order to avoid the corner.

$$\left\{\frac{\left(\frac{p_B}{\sigma_y}-k_1\right)\cos\theta+\left(\frac{q_B}{\sigma_y}\right)\sin\theta-k_2}{a_1}\right\}^2 + \left\{\frac{\left(\frac{q_B}{\sigma_y}\right)\cos\theta+\left(\frac{p_B}{\sigma_y}-k_1\right)\sin\theta}{b_1}\right\}^2 = 1 \tag{2}$$

The elliptic parameters namely a_1, b_1, k_1, k_2 and θ could be readily obtained from curve fitting. The equation is a normalized equation with respect to σ_y, and hence can be used for any soil.

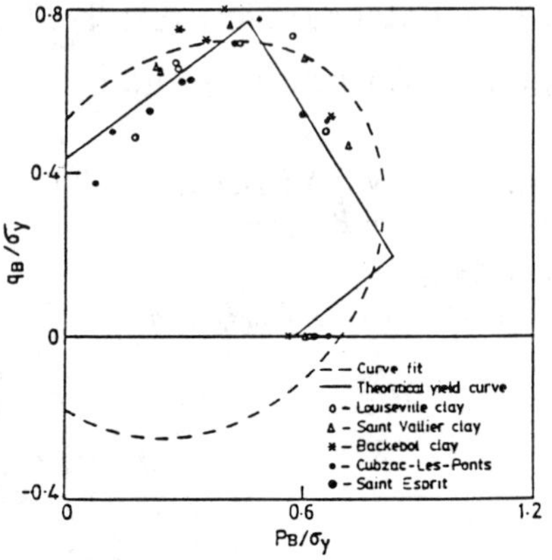

Fig.1 Yield curve for the bond component

The yield curve is assumed to harden with plastic volumetric compression and to soften with plastic shear strains. In the present investigation, just a linear hardening modulus $K_B{}^p$ and an exponential law of softening with shear strains are used. The expression for hardening or softening can be written as :

$$dN = \frac{K_B^p}{p_y} d\varepsilon_v^p - \alpha N d\varepsilon_s^p \tag{3}$$

where α = softening parameter, p_y = yield stress corresponding to isotropic compression.

The plastic strain rates are obtained by the usual methods of hardening plasticity. For the present case of associated flow rule, the plastic strain rate takes the form:

$$d\varepsilon_{rs}^p = \frac{1}{H}\frac{\partial g_B}{\partial \sigma_{Brs}}\frac{\partial f_B}{\partial \sigma_{Bij}} d\sigma_{Bij} \tag{4}$$

where $H = -\dfrac{\partial f_B}{\partial N}\dfrac{\partial N}{\partial \varepsilon_{rs}^p}\dfrac{\partial g_B}{\partial \sigma_{Brs}}$ is called hardening modulus.

The elastic strain rates are given by:

$$d\varepsilon_v^e = \frac{dp_B}{K_B} \tag{5}$$

$$de_{rs}^e = \frac{d(s_B)_{rs}}{2G_B} \tag{6}$$

where e_{rs} and s_{Brs} are the deviatoric strain and deviatoric stress components respectively.

With these defined features, the stiffness matrices D_B is obtained for the bond component. The stiffness matrix for the uncemented component, D_R is obtained from adopting the modified Cam clay model. Hence overall stiffness matrix is obtained by adding the stiffnesses of the two components as:

$$D = D_R + D_B \qquad (7)$$

The stress-strain relations for the soil are worked out incrementally, updating the stresses, strains and hardening separately for the two components at each stage.

PREDICTION OF EXTENSION TEST RESULTS

Prediction of drained and undrained stress strain curves for compression paths has been taken up for the soil tested in the investigation (Nagendra Prasad, 1996) and for other published data. In the present paper the predictive capability of the model for the extension tests is shown.

The extension test results of Osaka clay (Adachi et al., 1995) and Marl clay (Magnan and Serratrice, 1994) have been used for the present purpose. The index properties of the clays are presented in Table 1.

Table 1 Index properties

Property	Osaka clay	Marl clay
Natural water content	65-72	24.8
Liquid limit	69.2-75.2	40
Plastic limit	24.5-27.3	20
Plasticity index	41.9-50.6	18
Sensitivity	14.5	-

The material parameters for the clays under consideration used in the computation are given in Table 2. The ususal Cam clay parameters for Marl clay are selected with some judgment.

Figures 2 and 3 show the prediction for isotropically consolidated undrained triaxial extension test results at confining pressures of 40 kPa, 80 kPa and 120 kPa on Osaka clay. It may be noticed that the stress-strain behaviour and the mean principal stress versus deviatoric stress behaviour agree reasonably well with the model prediction.

Table 2 Material parameters

Property\Clay	Osaka clay	Marl clay
λ	0.194	0.056
κ	0.02	0.012
M	1.41	0.83
p_y (kPa)	83	1800
K_B (kPa)	1200	33000
G_B (kPa)	2300	60000
K_B^p (kPa)	240	25000
σ_y (kPa)	94.1	2400
α	5	5

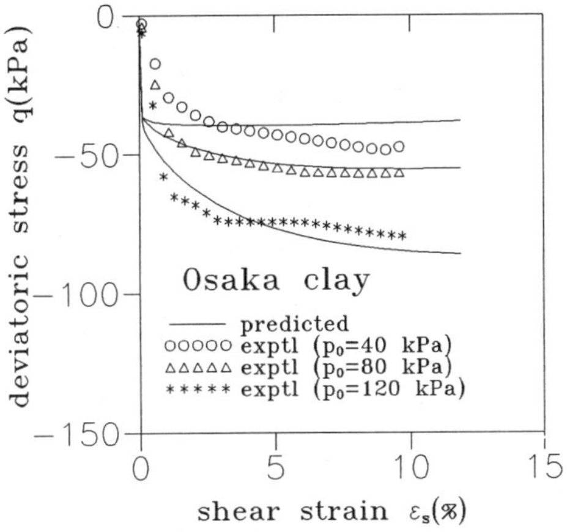

Fig.2 Predicted and experimental stress-strain curves for Osaka clay

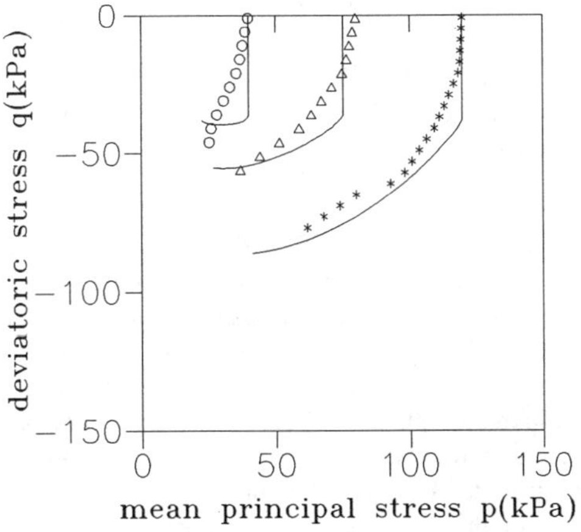

Fig.3 Predicted and experimental stress paths for Osaka clay

Figures 4 and 5 present the prediction for undrained triaxial extension test results of Marl clay at

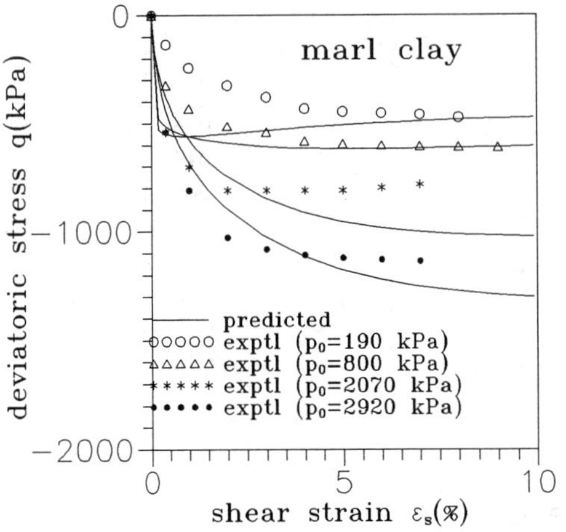

Fig.4 Predicted and experimental stress-strain curves for Marl clay

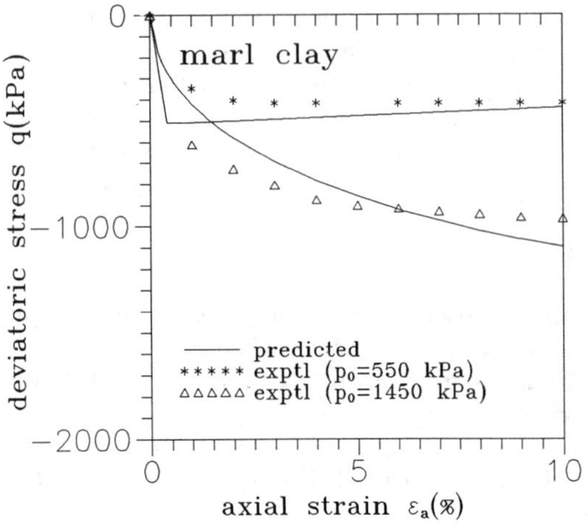

Fig.6 Predicted and experimental stress-strain curves for Marl clay

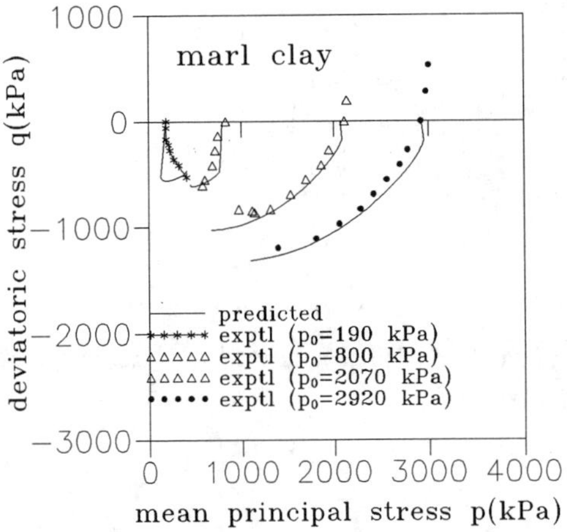

Fig.5 Predicted and experimental stress paths for Marl clay

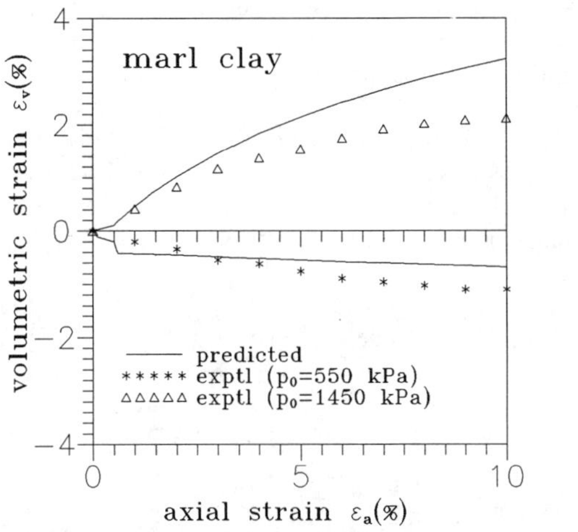

Fig.7 Predicted and experimental axial strain - volumetric strain curves for Marl clay

confining pressures of 190 kPa, 800 kPa, 2070 kPa and 2920 kPa. The experimental test results for 2920 kPa and 2070 kPa correspond to initially anisotropically compressed and then sheared by triaxial extension. The predictions presented correspond to initially isotropic state of stress before shearing by triaxial extension. This has been done as the initial states of equivalent unbonded soil skeleton corresponding to anisotropic compression are not known. Nevertheless, it is believed that the difference in initial states between experimental and predicted would have affected the behaviour only marginally as is evident from the predicted behaviour. An attempt has also been made to predict the drained extension behaviour of Marl clay as presented in figures 6 and 7. It may be noticed that the simulated behaviour is in close agreement with stress-strain - volumetric strain behaviour. It may be pointed out that the observed deviations could be attributed to the fact that the yield curve used for bond component is obtained from microstructure analysis for a K_o value of 0.5. It may be quite possible that the values of K_o for the soils considered in the present paper could be different. Further the material properties used are considered to be constant for all the confining pressures for the sake of simplicity. Also, assumed values for reconstitute states may be different from the real values. The additional complexity such as variation of parameters with stress could always be incorporated for further improving the predictive capabilities of the model.

CONCLUDING REMARKS

The model presented in the present formulation is based on superposition of the behaviour of bond component over frictional component to get the overall response. This process may be visualized as two stiffness acting in parallel for the same strain response. In the validation of the model the extension test results on two different clays for both drained and undrained conditions have been used satisfactorily.

REFERENCES

Adachi, T. and Okamo, M. (1974). A constitutive equation for normally consolidated clay, Soils and Foundations, 14(4):55-73.

Adachi, A., Oka, F., Hirata, T., Hashimoto, T., Nagaya, J., Mimura, M. and Tej Pradhan, B.S. (1995). Stress-strain behaviour and yielding characteristics of eastern Osaka clay, Soils and Foundations, 35(3):1-30.

Adachi, A. and Oka, F. (1995). An elasto-plastic constitutive model for soft rock with strain softening, Int. J. Numerical and Analytical Methods in Geomechanics, 19:233-247.

Conlon, R. J. (1966). Landslide on the Toulnustouc River, Quebec, Can. Geotech. J., 3(3):113-144.

Feda, J. (1982). Mechanics of particulate materials - the principles, Elsevier, Amsterdam, pp:437.

Gens, A. and Nova, R. (1993). Conceptual bases for a constitutive model for bonded soils and weak rocks, geotechnical engineering of Hard Soils-Soft Rocks Anagnostopoulos et al., (eds) Balkena, Rotterdam, ISBN 90:54103442.

Leroueil, S. and Vaughan, P.R. (1990). The general and congruent effects of structure in natural soils and weak rocks, Geotechnique, 40(3): 467-488.

Magnan, J.P. and Serratrice, J.F. (1995). Determination de la courbe d'etat limite d'une marne, Proc. XI ECSFMFE, Cophenhague (a paraitre).

Nagaraj, T.S., Srinivasa Murthy, B.R. and Vastala, A. (1991). Prediction of soil behaviour part III-cemented saturated soils, Indian Geotech. J., 21(2): 169-186.

Nagaraj, T.S., Srinivasa Murthy, B.R. and Vatsala, A.(1994). Analysis and prediction of soil behaviour, Wiley Eastern Limited, New Delhi.

Nagendra Prasad, K. (1996). Constitutive modelling of soft cemented soils, Doctoral Thesis submitted to Indian Institute of Science, Bangalore.

Nagendra Prasad, K., Srinivasa Murthy, B.R., Vatsala, A. and Sitharam, T.G. (1996). Yielding of sensitive clays - Micromechanical considerations, submitted to Canadian Geotechnical Journal, Canada.

Nova, R. and Wood, D.M. (1979). A constitutive model for sand in triaxial compression. Int. Journal for Num. and Anal. methods in Geomechanics, Vol.3.

Oka, F., Leroueil, S. and Tavenas, F. (1989). A constitutive model for natural soft clay with strain softening, Soils and Foundations, Vol. 29(3):54-66.

Roscoe, K.H., Schofield, A.N. and Wroth, C.P. (1958). On the yielding of soils, Geotechnique, 8(1):22-53.

Vatsala, A. (1995). Mechanical behaviour of weakly cemented soils,, Report submitted to European commission under Marie-Curie fellowship under supervision of Robert Nova.

Vatsala, A. (1989). Development of Cam-clay models for overconsolidated and sensitive soils, Ph D thesis, Indian Institute of Science, Bangalore, India.

APPENDIX

YIELD CURVE FOR THE BOND COMPONENT

The existance of intra aggregate pores, intra aggregate small pores and inter aggregate large pores has been discussed (Nagaraj et al., 1990 and Nagaraj et al., 1994) very clearly based on earlier experimental findings by various researchers on microstructure of natural clays. In the microstructure analysis proposed (Nagendra Prasad et al., 1996) the shape of the large pore under K_o compression is assumed to be elliptical, with particle domains are clusters aligned along its pheriphery. The size of the semi axes of the ellipse are such that they correspond to field K_o value with its long axis in the direction of major principal axis. The analysis of forces along the elliptic microstructure in relation to the

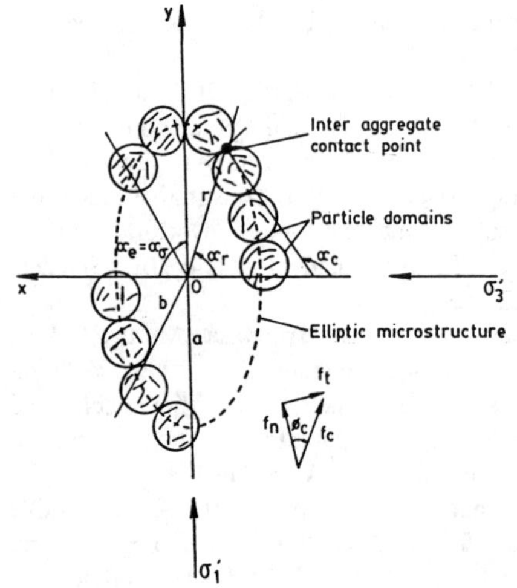

Fig.8 Geometry of elliptic microstructure

experimental yield stress values indicate that the yielding of microstructure may be initiated when the maximum contact normal force or the maximum contact shear force reaches a value which the cementation bonds are not capable to resist. The following is the criteria proposed:

Criterion based on limiting contact normal force

The maximum contact normal force can be found out for a given yield state along K_o compression by using the expression given by:

$$f_{nl} = \frac{r}{\sigma_y}\left\{\frac{\sigma_1+\sigma_3}{2} - \frac{\sigma_1-\sigma_3}{2}\cos 2(\alpha_\sigma-\alpha_r)\right\}\sin(\alpha_c-\alpha_r)$$

$$+\frac{r}{\sigma_y}\left\{\frac{\sigma_1-\sigma_3}{2}\sin 2(\alpha_\sigma-\alpha_r)\right\}\cos(\alpha_c-\alpha_r)$$

where $\alpha_r \approx 90°$, $\sigma_1 = \sigma_y$, $\sigma_3 = K_o\,\sigma_y$, f_{nl} = limiting contact normal force. α_σ, α_r, α_c correspond to geometry of elliptic mircostructure (Fig.8).

Criterion based on limiting contact shear force

The maximum contact shear force can be found out for a given yield state from the typical shear test results by using the expression given by:

$$f_{tl} = -\frac{r}{\sigma_y}\left\{\frac{\sigma_1+\sigma_3}{2} - \frac{\sigma_1-\sigma_3}{2}\cos 2(\alpha_\sigma-\alpha_r)\right\}\cos(\alpha_c-\alpha_r)$$

$$+\frac{r}{\sigma_y}\left\{\frac{\sigma_1-\sigma_3}{2}\sin 2(\alpha_\sigma-\alpha_r)\right\}\sin(\alpha_c-\alpha_r)$$

where $\alpha_r = 60°$, f_{tl} = limiting contact shear force, σ_1 and σ_3 correspond to the yield point which is determined experimentally from a typical shear test result.

Once the limiting contact normal and shear forces are determined the yield stress in terms of major principal stress for any given stress ratio (K) is obtained as the minimum of the two values from the following:

From limiting normal force criterion:

$$\frac{\sigma_1}{\sigma_y} = \frac{f_{nl}}{r[\{\frac{1+K}{2} - \frac{1-K}{2}\cos 2(\alpha_\sigma-\alpha_r)\}\sin(\alpha_c-\alpha_r) + \{\frac{1-K}{2}\sin 2(\alpha_\sigma-\alpha_r)\}\cos(\alpha_c-\alpha_r)]}$$

From limiting shear force criterion:

$$\frac{\sigma_1}{\sigma_y} = \frac{f_{tl}}{r[-\{\frac{1+K}{2} - \frac{1-K}{2}\cos 2(\alpha_\sigma-\alpha_r)\}\cos(\alpha_c-\alpha_r) + \{\frac{1-K}{2}\sin 2(\alpha_\sigma-\alpha_r)\}\sin(\alpha_c-\alpha_r)]}$$

Deviatoric and mean principal stress values corresponding to yield can now be obtained as:

$$\frac{q}{\sigma_y} = \frac{\sigma_1(1-K)}{\sigma_y}$$

$$\frac{p}{\sigma_y} = \frac{\sigma_1(1+2K)}{\sigma_y}$$

The limiting contact forces and hence the complete yield curve can be obtained from only two test results. The theory based on the microstructural analysis predicts the yield values with reasonable degree of accuracy. This can reduce the need for elaborate experimental programme required to determine the yield curve.

Computer Methods and Advances in Geomechanics, Yuan (ed.) © 1997 Balkema, Rotterdam, ISBN 90 5410 904 1

Verification of a soil model with predicted behavior of a sheet pile wall

T. Schanz
Institute of Geotechnical Engineering, University of Stuttgart, Germany

P.G. Bonnier
Plaxis BV, Netherlands

ABSTRACT: This paper introduces a constitutive model which is formulated within the framework of the classical theory of plasticity. The model is first classified with reference to existing classes of models and it is then demonstrated that it applies well to a restricted class of soils including dense to loose sand and overconsolidated clays which under deviatoric loading chiefly display plastic shear strains.
The rate of elastic strain is defined by using a hyperelastic formulation introducing a complementary strain energy function which best fits experimental data. The plastic part of the model can be described as isotropic shear strain-hardening with a non-associated flow rule which matches ROWE's theory of dilatancy by means of introducing a plastic potential as a function of a mobilized angle of dilatancy. The model is first developed for triaxial states of stress. Thereafter simplifications are introduced when implementing the model in a commercial FE-code for general 3D states of stress.
In the next section of the paper attention is paid to determining soil parameters for the model from standard laboratory tests. Empirical relationships are presented in order to relate model parameters to well known parameters from engineering practice.
The following section of the paper presents a verification of the model by performing a back-calculation of a sheet pile wall field test.

1 INTRODUCTION

In many cases of daily geotechnical engineering, it is no help to employ complex stress-strain models for calculating boundary value problems. Instead of using Hooke's single-stiffness model with linear elasticity in combination with an ideal plasticity according to Mohr-Coulomb a new constitutive formulation using a double-stiffness model for elasticity in combination with isotropic shear-strain hardening is presented.
Summarizing the existing double-stiffness models the most dominant type of model is the Cam-Clay model (HASHIGUSCHI, 1985). To describe the non-linear stress-strain behaviour of hard soils, beside the Cam-Clay model the pseudo-elastic (hypo-elastic) type of model has been developed. There a Hookean relationship is assumed between increments of stress and strain and non-linearity is achieved by means of varying Young's modulus. By far the best known model of this category ist the Duncan-Chang model, also known as the hyperbolic model (DUNCAN & CHANG, 1970). This model captures hard soil behaviour in a very tractable manner on the basis of only two stiffness parameters and is very much appreciated among consulting geotechnical enginners.
The major inconsistency of the Duncan-Chang model which is the reason why it is not accepted by scientists is that, in contrast to the elastoplastic type of model, a hypo-elastic model cannot consistently distinguish between loading and unloading. In addition, the model is not suitable for collapse load computations in the fully plastic range.
These restrictions will be overcome by reformulating the model in an elastoplastic framework in this paper. The essential features of the original model will be fully preserved so that existing experience on input parameters is also retained.

2 RATE FORMULATION

Within the framework of the classical theory of

elasto-plasticity the rate of total strain $\dot{\varepsilon}_{ij}$ consists of an elastic, $\dot{\varepsilon}^e_{ij}$, and a plastic component, $\dot{\varepsilon}^p_{ij}$, respectively

$$\dot{\varepsilon}_{ij} = \dot{\varepsilon}^e_{ij} + \dot{\varepsilon}^p_{ij} \tag{1}$$

In the following sections we present formulations for both these components. The model is described for triaxial stress conditions in p/q-space. An extension for general 3D stress conditions is given in SCHANZ & BONNIER (1997).

2.1 *Non-linear elasticity*
The elastic part of the model is based on the experimental work by EL-SOHBY & ANDRAWES (1973) suggesting a coaxiality between the ratio of stresses and the ratio of strain in combination with a secant Poisson ratio of $\nu \approx 0$ for proportional triaxial unloading. These findings are modeled by using an elastic stress strain relation of the form

$$\varepsilon^e_{ij} = \frac{1}{E}\sigma_{kl} \tag{2}$$

where E is a stress dependent secant Youngs modulus (OHDE, 1939) of the form

$$E = E^{pa}\left(\frac{\sigma}{p_a}\right)^m \tag{3}$$

Here $\sigma = \frac{1}{3}\sigma_{ii} = p$ is an arbitrary stress measure, p_a a reference pressure, E^{pa} a reference Young's modulus and m an additional material parameter. The formulation for the elastic strain rate used in this paper is based on a hyperelastic model by VERMEER (1980, 1982), including the whole stress tensor. With the assumption $\nu = 0$ we define the elastic potential W_c as

$$W_c = W_0\sigma^{2-m} = \frac{3}{E^{pa}}\frac{p_a^{2-m}}{2-m}\sigma^{2-m} \tag{4}$$

The elastic strain and the elastic strain rate are given as the first and the second derivative of the elastic potential W_c.

$$\varepsilon^e_{ij} = \frac{\partial W_c}{\partial \sigma_{ij}} = \frac{\sigma_{ij}}{E^{pa}}\left(\frac{p_a}{\sigma}\right)^m = \frac{1}{E}\sigma_{ij} \tag{5}$$

$$\begin{aligned}\dot{\varepsilon}^e_{ij} &= \frac{\partial^2 W_c}{\partial \sigma^2_{ij}}\dot{\sigma}_{ij} \\ &= \frac{1}{E}\left(\delta_{ij} - \frac{2-m}{\sigma_{mn}\sigma_{mn}}\sigma_{ij}\sigma_{kl}\right)\dot{\sigma}_{kl}\end{aligned} \tag{6}$$

We obtain the elastic stiffness matrix D^e_{ijkl} as

$$D^e_{ijkl} = E\left[\delta_{ij} + \frac{1}{1-m}\left(\frac{m}{\sigma_{mn}\sigma_{mn}}\right)\sigma_{ij}\sigma_{kl}\right] \tag{7}$$

2.2 *Plasticity*
In many standard cases of engineering practice with quasi-static loading the use of an isotropic strain-hardening theory leads to results that are as good as those obtained by the use of more complicated constitutive models based on kinematic (or mixed) strain-hardening theory or bounding surface models. From an engineering perspective a simpler model is preferred in which model parameters are easily obtained from standard soil tests. The basic elements of such a model are described in the following.

2.2.1 Yield surface and hardening law
To describe the shape of the yield surface in the p/q-plane we take into account the experimental results by TATSUOKA (1974b) and TATSUOKA & ISHIHARA (1974a) on sand.
For both biaxial and triaxial stress states these results indicate so-called equi-γ lines in p/q-space showing decreasing curvature for increasing mobilisation of shear strength resulting in an ultimate linear (Mohr-Coulomb-) condition. In the model instead of total shear strain γ the plastic component γ^p is chosen as the hardening parameter. For hard soils, plastic volume changes tend to be relatively small and this leads to the following approximation for the rate of plastic shear strain, $\dot{\gamma}^p$, defined in the case of triaxial stress states as

$$\dot{\gamma}^p = \dot{\varepsilon}^p_1 - \dot{\varepsilon}^p_2 - \dot{\varepsilon}^p_3 \;\rightarrow\; \dot{\gamma}^p = 2\dot{\varepsilon}^p_1 - \dot{\varepsilon}^p_v \approx 2\dot{\varepsilon}^p_1 \tag{8}$$

Therefore the yield surface used in this model is written as ($q = \sigma_1 - \sigma_3$)

$$F = 2\varepsilon^p_1 - \gamma^p = \underbrace{2\frac{q}{E_i\left(1-\frac{q}{q_a}\right)}}_{\gamma} - \underbrace{2\frac{q}{E_{ur}}}_{\gamma^e} - \gamma^p \tag{9}$$

Here $R_f \approx 0.9$ is a reduction factor and

$$q_a = \frac{q_f}{R_f} = M\,(p + c\,\cot\varphi)\,R_f^{-1} \qquad (10)$$

where

$$M = \frac{6\sin\varphi}{3 - \sin\varphi} \qquad (11)$$

the ultimate deviatoric stress regarding triaxial compression. The model includes two different moduli, one for initial loading, E_i, and another one for un- and reloading, E_{ur}, to take into account usually much stiffer behaviour for soils for the latter. E_{ur} is equivalent to the purely elastic Young's modulus E. In contrast E_i contains both elastic and plastic strain components. Considering q_a, E_i and E_{ur}, Eq. 9 implicitly includes the stress measure p. With the relationship $E_{ur} = \alpha\,E_i$ Eq. 9 can be rewritten as

$$\begin{aligned} q &= \frac{1}{2}q_a(1-\alpha) - \frac{1}{4}\gamma^p\alpha E_i \\ &+\sqrt{(2q_a(\alpha-1)+\gamma^p\alpha E_i)^2 + 8q_a\gamma^p\alpha E_i} \end{aligned} \qquad (12)$$

The formulation of the yield surface according to Eq. 9 gives a good match with the well-known hyperbolic stress strain relation (KONDNER, 1963) for primary triaxial loading conditions ($q < q_f$).
The former was formulated for triaxial compression states of stress where $\sigma_2 = \sigma_3$. For general states of stress, $\sigma_1 \geq \sigma_2 \geq \sigma_3$, we use a different definition of q such that Eq. 13 describes the angular Coulomb failure surface:

$$\tilde{q} = \sigma_1 - \sigma_2 + \frac{3+\sin\varphi}{3-\sin\varphi}(\sigma_2 - \sigma_3) \qquad (13)$$

2.2.2 Plastic potential and flow rule
In order to derive the plastic strain rate $\dot{\varepsilon}^p_{ij}$ according to

$$\dot{\varepsilon}^p_{ij} = \dot{\lambda}\frac{\partial g}{\partial \sigma_{ij}} \qquad (14)$$

we introduce a plastic potential function g

$$g = \tilde{q}' - M'\,(p + c\,\cot\psi_m) \qquad (15)$$

with

$$M' = \frac{6\sin\psi_m}{3 - \sin\psi_m} \qquad (16)$$

$\tilde{q}'$ uses the same formulation as for $\tilde{q}$ but the angle of mobilised friction, φ_m, is replaced with the angle of mobilised dilatancy, ψ_m.
For triaxial conditions the flow rule resulting from the above definition of the plastic potential provides the relationship between the plastic volumetric strain rate $\dot{\varepsilon}^p_v$ and the plastic shear strain rate $\dot{\gamma}^p$.

$$\dot{\varepsilon}^p_v = \sin\psi_m\dot{\gamma}^p \qquad (17)$$

Therefore the mobilised dilatancy angle ψ_m is defined using the expression

$$\sin\psi_m = \frac{\sin\varphi_m - \sin\varphi_{cv}}{1 - \sin\varphi_m\sin\varphi_{cv}} \qquad (18)$$

where φ_{cv} is the critical state friction angle, which is a material constant independent of the density, and φ_m is the mobilised friction angle. Instead of φ_{cv} the maximum value of the angle of dilatancy ψ_p is used with the approximation

$$\varphi_{cv} \approx \varphi_p - \psi_p \qquad (19)$$

The above equations correspond to the well-known stress-dilatancy theory by ROWE (1962), as explained by SCHANZ & VERMEER (1996).

3 INCREMENTAL IMPLEMENTATION

The model as described above has been implemented in a finite element code. To do this, the model equations have to be written in incremental form to arrive at the following equations:

$$\Delta\sigma = \mathbf{D}\left(\Delta\varepsilon - \lambda\frac{\partial g}{\partial\sigma}\right) \quad , \quad \gamma^p = \gamma_0^p + \Delta\gamma^p \tag{20}$$

where $\mathbf{D}$ is Hooke's elasticity matrix, based on the unloading-reloading stiffness, the strain increment $\Delta\varepsilon$ follows from the global equilibrium iteration process, g is the plastic potential function and λ is a non-negative multiplier. Although the stiffness and the plastic potential function (mobilised dilation angle) may change during an increment, we use the values at the beginning of the increment. The change in the elastic stiffness is usually not important as the deformations are dominated by plasticity. In case of unloading/reloading we iterate on the average stiffness during the increment. The change in the mobilised dilation angle is usually so small that the influence of updating the value during the increment is negligible.
This multiplier λ has to be determined from the condition that the function $F(\sigma, \gamma^p)$ has to be zero for the new stress- and deformation state.
The plastic potential function depends on the stresses and the mobilised dilation angle. The dilation angle for these derivatives is taken at the beginning of the step. The implementation uses an implicit scheme for the derivatives of the plastic potential function g. The derivatives are taken at a predictor stress, following from elasticity and the plastic deformation in the previous iteration.

The calculation of the stress increment can be performed in principal stresses. Therefore initially the principal stresses and principal directions have to be calculated, based on the elastic prediction. Principal plastic strain increments are now calculated and finally the Cartesian stresses have to be back calculated from the resulting principal constitutive stresses. The calculation of the constitutive stresses can be written as

$$\begin{aligned} \sigma &= \sigma^0 + \mathbf{D}\Delta\varepsilon - \lambda\mathbf{D}\frac{\partial g}{\partial\sigma} \\ &= \sigma^e - \lambda\mathbf{D}\frac{\partial g}{\partial\sigma} = \sigma^e - \lambda\,\mathbf{b} \end{aligned} \tag{21}$$

From this the deviatoric stress q ($\sigma_1 - \sigma_3$) and the asymptotic deviatoric stress q_a can be expressed in the elastic predictor stresses and the multiplier λ,

$$q = \sigma_1^e - \sigma_3^e - \lambda(b_1 - b_3) = q^e - \lambda(b_1 - b_3) \tag{22}$$

$$\begin{aligned} q_a &= \frac{2\sin\varphi}{1-\sin\varphi}\frac{1}{R_f}\sigma_3^\star \\ &= \frac{2\sin\varphi}{1-\sin\varphi}\frac{1}{R_f}(\sigma_3^{\star,e} - \lambda b_3) \end{aligned} \tag{23}$$

with $\sigma_3^\star = \sigma_3^e + c\cot\varphi$. For these stresses the function $F(\sigma)$ should be zero.

$$F(\sigma) = f(\sigma) - f^0 - \Delta\gamma^p = 0 \tag{24}$$

$$f(\sigma) = 2\frac{q}{E_i}\frac{q_a}{q_a - q} - 2\frac{q}{E_{ur}} \tag{25}$$

As the increment of the plastic shear strain also depends linearly on the multiplier λ, the above formulars result in a (complicated) quadratic equation for the multiplier λ which can be solved easily. Using the resulting value of λ, one can calculate (incremental) stresses and the (increment of the) plastic shear strain.
In the above formulation it is assumed that there is a single yield function. In case of triaxial compression or triaxial extension states of stress there are two yield functions and two plastic potential functions. Following KOITER (1960) one can write

$$\sigma = \sigma^e - \lambda_{13}\mathbf{D}\frac{\partial g_{13}}{\partial\sigma} - \lambda_{12}\mathbf{D}\frac{\partial g_{12}}{\partial\sigma} - \lambda_{23}\mathbf{D}\frac{\partial g_{23}}{\partial\sigma} \tag{26}$$

where the subsripts indicate the principal stresses used for the yield and potential functions. At most two of the multipliers are positive. In case of triaxial compression we have $\sigma_2 = \sigma_3$, $\lambda_{23} = 0$ and we should have $F_{12}(\sigma) = F_{13}(\sigma) = 0$ at the end of the step. To simplify these calculations, we can use a different combination of the two active functions.

$$g_a = \frac{1}{2}(g_{12} + g_{13}) \quad , \quad \lambda_a = \lambda_{12} + \lambda_{13} \tag{27}$$

$$g_b = \frac{1}{2}(g_{12} - g_{13}) \quad , \quad \lambda_b = \lambda_{12} - \lambda_{13} \tag{28}$$

The second function (b) is only used to assure $\sigma_2 = \sigma_3$. From this condition, we can calculate the value of λ_b and use this so an equation like Eq. 21 follows. The resulting equation can be solved in the same way to give the value of λ_a.
When the stresses are calculated one still has to check if the stress state violates the failure criterion ($q > q_a$). When this happens the stresses have to be returned to the Mohr-Coulomb failure surface.

4 VERIFICATION

4.1 *Determination of model parameters*

Model parameters used in the Hard-Soil model have either an explicit geotechnical meaning, so that they can be deduced from standard laboratory tests, or they can be estimated by fitting experimental data. Table 1 shows all the model parameters used in the Hard-Soil model compared to those of a common Mohr-Coulomb-type model using a non-associated flow rule.
The un- / reloading (reference) stiffness E_{ur}^{pa} is related to the oedometric stiffness E_{oed} according to

$$E_{ur} = E = \frac{(1+\nu_{ur})(1-2\nu_{ur})}{(1-\nu_{ur})} E_{oed} \tag{29}$$

To get realistic values for E_{oed} test data from one-dimensional compression tests are plotted according to Fig. 1. From the linear regression coefficients α and β in

$$\ln \varepsilon_z = \alpha \ln \frac{\sigma_z}{p_a} + \beta \tag{30}$$

Tab. 1: Model parameters of the Hard Soil model

	Mohr-Coulomb	Hard-Soil model
shear strength	c	c
	φ_p	φ_p
dilatancy	$\psi_p \approx \varphi_p - \varphi_{cv}$	$\psi_m (\rightarrow$ Eq. 18)
elasticity	E$\leftrightarrow E_s$	$E_{ur}^{pa} \leftrightarrow E_s$
	ν	$\nu_{ur} \approx 0.15 - 0.2$
add. parameters	-	$E_i^{pa} \neq E_s$
	-	m$\approx 0.4 - 0.7$
	-	$R_f \approx 0.9$

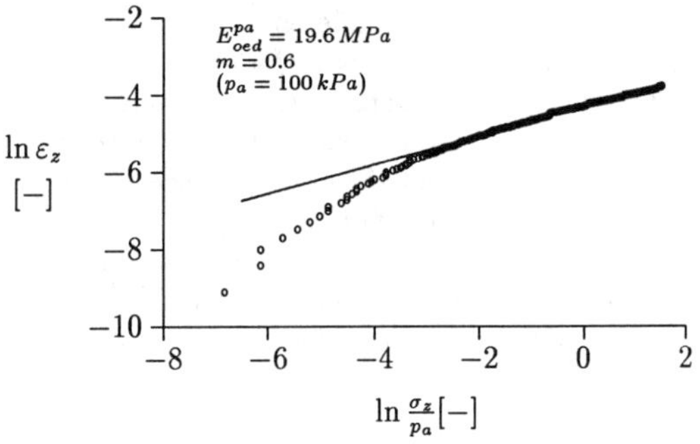

Fig. 1: Interpretation of 1D compression tests

m and E_{oed}^{pa} can be calculated:

$$m = 1 - \alpha \tag{31}$$

$$E_{oed}^{pa} = \frac{1}{\alpha} \frac{p_a}{e^\beta} \tag{32}$$

The Poisson ratio for un- and reloading, ν_{ur}, can be determined by back-calculation of one-dimensional compression tests: by varying ν_{ur} we obtain different ratios of axial to radial stress. These ratios are compared to values from JAKY's equation $K_0 = 1 - \sin\varphi$ or the solution from the theory of elasticity $K_0 = \frac{\nu_{ur}}{1-\nu_{ur}}$.
Instead of determining the initial tangent stiffness, E_i, a secant modulus from a drained triaxial test, E_{50}, is used. This modulus is not directly related to the Youngs modulus because strain includes both an elastic and a plastic component for virgin loading. For standard engineering problems we advise the use of a value of E_{50} for 50% mobilisation of the maximum shear strength.
As a rough estimation the initial value E_i may then be derived from E_{50}

$$E_{50} \approx \left(\frac{1}{2} \rightarrow \frac{1}{4}\right) E_i \tag{33}$$

Detailed relationships between modulus from triaxial tests and one-dimensional compression tests are given in SCHANZ & VERMEER (1997). Based on this study one finds an empirical accordance of the oedometric with the triaxial stiffness modulus,

$$E_{oed}^{pa} \approx E_{50}^{pa} \tag{34}$$

taking into account different reference stresses $p_a = \sigma_1$ for E_{oed} and $p_a = \sigma_3$ for E_{50}.
The shear strength parameters c, φ_p and φ_{cv} (ψ_p) are determined in the usual way. Attention is given to the fact that the maximum shear strength under plane strain conditions, φ_p^{ps}, is equal to or larger than the strength under triaxial conditions, φ_p^{tr}, depending on the residual angle of friction φ_{cv} (SCHANZ & VERMEER, 1996).

$$\varphi_p^{tr} \approx \frac{1}{5}(3\,\varphi_p^{ps} + 2\,\varphi_{cv}) \qquad (35)$$

4.2 *Backcalculation of field test*

The model was implemented in the commercial PLAXIS code (VERMEER ET. AL., 1995). To verifiy its capabilities we present the backcalculation of a sheet pile wall field test which is well documented in V. WOLFFERSDORFF (1994).
For the calculation of the different stages of construction we used the mesh shown in Fig. 2. Here stage 7 is shown with the deepest level of excavation, one supporting strut A ($EA = 2.45 * 10^5\,kN/m$) and a traction load of $q = 10kN/m^2$.
The subsoil is modelled in 3 layers with soil parameters according to Tab. 2 and the interface friction (I) $\delta = 0.7\,\varphi$, $m = 0.75$, $\gamma_{wet} = 19\,kN/m^3$, $\gamma_{dry} = 16.5\,kN/m^3$, $R_f = 0.9$ for all layers. E_{50} was estimated from CPT-resistance q_c using $E_{50} = 5\,q_c$. For E_{ur} we used the empirical relation $E_{ur} = 5\,E_{50}$. The friction and dilatancy angles result from recent triaxial tests (V. WOLFFERSDORFF, 1997) and the amount of (capilar-) cohesion is empirical, reaching down to the water table.
For convinience we display only some of the results of the calculations. For this stage of construction

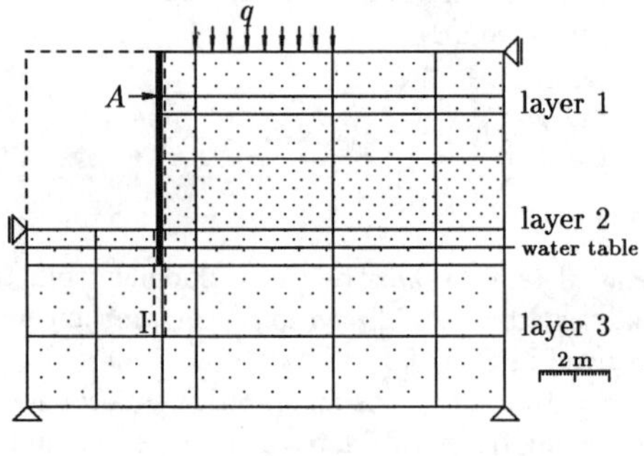

Fig. 2: FEM simulation of the sheet pile wall ($EI = 2.03 * 10^3\,kNm^2/m, EA = 2.2 * 10^6\,kN/m$) field test: stage 7

Tab. 2: Soil parameters used in the back-calculation

	ν [-]	E_{ur} [MPa]	E_{50} [MPa]	c [kPa]	φ_p [°]	ψ_p [°]
1	0.2	500	100	8	41	11
2	"	"	"	6	"	"
3	"	"	"	0	"	"

we found a value of $A = 32.9\,kN/m$ compared to a measured value of $A = 33.7\,kN/m$. Fig. 3 gives a further comparison of the measured and calculated values of earth pressure and bending moments in the sheet pile wall.

We find a horizontal deformation of the top of the wall of about $\approx 0.2\,mm$ compared to a measured value of $\approx 5\,mm$. This (general) difference in the results for the wall deformations can also be observed in the summary of the numerous other predictions in V. WOLFFERSDORFF (1997).
The final stage of the field test was the relief of the strut force, which was also simulated with the model. We found an ultimate value of A=4.03 kN/m compared to a measured value of A=4.22 kN/m.
From the results of the backcalculation it can be concluded, that the model is capable of simulating a rather complex boundary value problem as an excavation. By using input parameters from standard lab tests and some common empirical relations we manage to get relatively close to the observed behavior without any fitting procedure at all.

5 CONCLUSIONS

In this paper a new consitutive model was introduced which is able to describe the essential characteristics of hard soil behaviour: stress-dependent elasticity, isotropic shear hardening and non-asso-

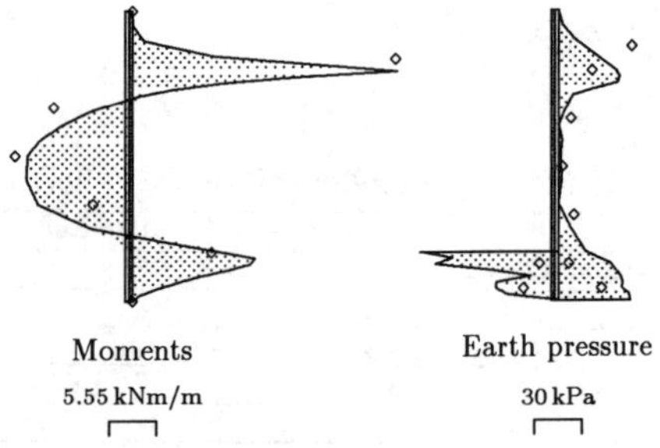

Fig. 3: Comparison of calculated and measured (◇) results

ciated plastic flow. The model was presented in its differential and incremental formulations. The main additional improvement compared with a classical Mohr-Coulomb-type model is the implicit consideration of the varying loading, un- and reloading behaviour. For all model parameters required, straightforward rules for their evaluation are given. Finally the model was implemented in a commercial FE code and the simulation of sheet pile wall field test is presented. Future work will concentrate on the implementation of softening and the introduction of a cap in order to model K_0 stress path more realistically.

REFERENCES

DUNCAN, J. M., CHANG, C. (1970). Nonlinear analysis of stress and strain in soil. Journal of the Soil Mech. and Found. Div. (ASCE, SM 5), Proc. of the Am. Society of Civil Engineers. pp. 1629-1651.

EL-SOHBY, M.A., ANDRAWES, K.Z. (1973). Experimental examination of sand isotropy, Proc. 8th Int. Conf. Soil Mech. Found. Eng., Moscow, 1.1, pp. 103-109.

HASHIGUSCHI, K (1985). Two- and three-surface models for plasticity, 5th Int. Conf. on Num. Methods in Geomech., Nagoya, pp. 285-292.

KOITER, W. T.(1960). General Theorems for Elastic-Plastic Solids, In: Progress in Soild Mechanics, Vol. 1 (Eds. Sneddon and Hill), North Holland Publishing Co., Amsterdam, pp. 165-221.

KONDNER, Z. (1963). A hyperbolic stress strain formulation for sands. 2nd Pan. Am. ICOSFE Brazil, Vol. 1, pp. 289-324.

OHDE, J. (1939). Zur Theorie der Druckverteilung im Baugrund, Der Bauingenieur, 20, Heft 33/34.

ROWE, P. W. (1962). The stress-dilatancy relation for static equilibrium of an assembly of particles in contact, *Proc. Roy. Soc. A.*, **269**, pp. 500-527.

SCHANZ, T. , BONNIER, P.G. (1997). An adequate Model for Hard Soil Behaviour, Submitted for publication in: *International Journal for Numerical Methods in Geomechanics.*

SCHANZ, T., VERMEER, P. A. (1996). Angles of friction and dilatancy of sand, *Géotechnique* 46, No. 1, pp. 145-152.

SCHANZ, T., VERMEER, P.A. (1997). On the stiffness of sands, accepted for publication: *Géotechnique* Symposium in Print: Pre-failure Deformation Behaviour of Geomaterials.

TATSUOKA, F., ISHIHARA, K. (1974a). Yielding of sand in triaxial compression, Soils and Foundations, Vol. 14, No. 2, pp. 63-76.

TATSUOKA, F., ISHIHARA, K. (1974b). Drained Deformation Of Sand Under Cyclic Stresses Reversing Direction, Soils and Foundations, Vol. 14, No. 3, pp. 51-65.

VERMEER, P.A. (1980). Formulation and Analysis of Sand Deformation Problems, Ph.D. thesis, TU Delft.

VERMEER, P.A. (1982). A five-constant model unifying well-established concepts, in: Results of the International Workshop on Constitutive Relations for Soils, eds.: Gudehus, Darve, Vardoulakis, Grenoble, Balkema, pp. 175-197.

VERMEER, P.A., BRINKGREVE, R.B.J. ET. AL. (1995). PLAXIS: Finite Element Code for Soil and Rock Analyses, Version 6, Balkema, Rotterdam.

V. WOLFFERSDORFF, P.-A. (1994). Feldversuch an einer Spundwand in Sandboden: Versuchsergebnisse und Prognosen, *Geotechnik*, 17(2), pp. 73-83.

V. WOLFFERSDORFF, P.-A. (1997). Verformungsprognosen von Stützkonstruktionen, Habilitationsschrift, Veröffentlichung des Instituts für Bodenmechanik und Felsmechanik der Universität Karlsruhe, In preperation.

Computer Methods and Advances in Geomechanics, Yuan (ed.) © 1997 Balkema, Rotterdam, ISBN 90 5410 904 1

Modification for 3-D failure stress state on the basis of stress path*

Zhan Meili & Su Baoyu
Department of Hydropower Engineering, Hohai University, Nanjing, China

ABSTRACT: In this paper, based on the analysis of stress path during the loading and unloading , a modification procedure is developed for the 3-D failure stress state. In the procedure , the effect of medium principle stress is considered , and the effect of the stress path of loading and unloading is also considered , thus , the modified stress state is more reasonable. After analyzing the known results, this paper proposes the describing method of stress level considering the medium principle stress under complex stress state . So, a more reasonable criterion is provided for the 3-D stability analysis .

1. INTRODUCTION

In geotechnical engineering, the numerical analysis , which based on the test of representative sample, is a effective procedure for solving the actual problem . Here, the finite element method is the most representative method. It is well-known that the geotechnical media is a kind of nonlinear elastic or elastoplastic media. In the numerical analysis, the load increment is finite. So, after one stage loading, some elements may be under failure stress state, i.e. the calculation stress may be larger than the limit strength of the media. In fact, the stress in soil can not be larger than its limit stress. To solve the aforesaid problem, some researchers perform a lot of research and propose some modification methods (Yin 1980 & Allen et al.). In their procedures, there are two deficiencies. First, when modifying the calculation stress state, the effect of medium principle stress is not considered. Second, the modified stress state has nothing to do with the stress path of loading and unloading. Therefore, their procedures can not represent the real actual condition, On the other hand, their modification methods are not for 3-D stress state, and should be improved. This paper emphasizes the research on that problem.

* Supported by National Natural Science Foundation .

2. STRESS LEVEL ON P-Q PLANE

Firstly, the problem that how to describe the stress level under complex stress state considering the medium principle stress σ_2 must be solved. Under general condition, the complex stress state is described as the point on p-g plane. However, the stress level, unconsidering the medium principle stress σ_2, is usually used in FEM analysis of geotechnical engineering (Yin 1980)

$$SL = \frac{(1-\sin\varphi)(\sigma_1-\sigma_2)}{2C\cos\varphi + 2\sigma_3\sin\varphi} \tag{1}$$

Where c is the cohesion; φ is the internal friction angle; σ_1 and σ_3 are the maximum principle stress and minimum principle stress, shown as in Fig. 1. Eq.(1) represents the ratio between the radius of the stress

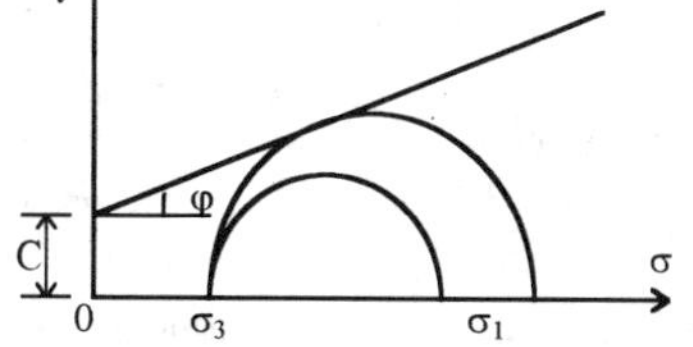

Fig. 1 Stress circle of Mohr's Strength

circle of current stress state and the radius of the stress circle of limit stress state, where the effect of the medium principle stress is not considered.

However, the parameters of shear strength are obtained by the normal triaxial tests, where $\sigma_2=\sigma_3$ is always satisfied. Therefore, strictly speaking, eq.(1) is only tenable when $\sigma_2=\sigma_3$. Under actual condition, σ_2 usually does not equal toσ_3, so the calculation result is approximate in a degree, and it is necessary to modify the eq.(1).

In geotechnical engineering, it is well-known that p and q are defined as

$$p=\frac{1}{3}(\sigma_1+\sigma_2+\sigma_3) \quad (2)$$

$$q=\frac{1}{\sqrt{2}}\sqrt{(\sigma_1-\sigma_2)^2+(\sigma_2-\sigma_3)^2+(\sigma_3-\sigma_1)^2} \quad (3)$$

Where p is the volume stress and q is generalized shear stress. In normal triaxial test, it can be obtained

$$p=\frac{1}{3}(\sigma_1+2\sigma_3) \quad (4)$$

$$q=\sigma_1-\sigma_3 \quad (5)$$

Substituting eq.(5) into eq.(4), it can be got

$$\sigma_3=\mathrm{p}-\frac{\mathrm{q}}{3} \quad (6)$$

Substituting eq.(5) and eq.(6) into eq.(1), the formula of stress level considering the medium principle stress is obtained

$$sl'=\frac{(1-\sin\varphi)q}{2C\cos\varphi+2(p-\frac{q}{3})\sin\varphi} \quad (7)$$

Where p and q can be given by eq.(2) and eq.(3).

To prove that it is reasonable for the aforesaid modification, it is assumed that the stress level is equal to 1, i.e. sl′=1, so the generalized shear stress can be solved as following

$$q=M(p+p_r) \quad (8)$$

Where

$$\left.\begin{aligned} M&=\frac{6\sin\varphi}{3-\sin\varphi} \\ p_r&=C\cdot ctg\varphi \end{aligned}\right\} \quad (9)$$

The p-q strength equation given by eq.(8) and (9), shown as in Fig.2, is consistent with the result in the reference (Yin 1980). Thus, it is unnecessary to calculate the stress level by use of eq.(1) under complex stress state.

Generally speaking, many yielding model are established in p-q coordinate system, shown as in Fig.2, when performing the elastoplastic numerical analysis in geotechnical engineering, and there are two kinds of failure equation given by following

$$q=Mp \quad (10)$$

and

$$q=M(p+p_r) \quad (8)$$

But when analyzing, the stress level calculated by eq.(1) is adopted, and if the stress level is larger than 1.0, the element is failure. Obviously, it is contrary in a degree. therefore, it is recommended that the elastoplastic constitutive equation be established by use of the eq.(8), and when analyzing, the stress level is calculated by eq.(7).

3. STRESS STATE MODIFICATION OF FAILURE ELEMENT BASED ON STRESS PATH

When stage loading method is used in the numerical analysis , the calculation stress of some element may larger than the limit strength of geotechnical media because the increment of load can not be infinite small. But the failure stress state do not exist. To solving that problem, a modification method based on the concept of stress circle has been proposed (Yin 1980). However, there are two differences in that method. First, the method can not deal with the complex stress state problem; second, the modification do not consider the effect of stress path. Therefore, this pare will develop a new stress state modification method in p-q coordinate system to improve the aforesaid shortcomings.

For one element, the stress state before loading is as follows

$$\{\sigma\}_0=[\sigma_x^0,\ \sigma_y^0,\ \sigma_z^0,\ \tau_{xy}^0,\ \tau_{yz}^0,\ \tau_{zx}^0]^T \quad (11)$$

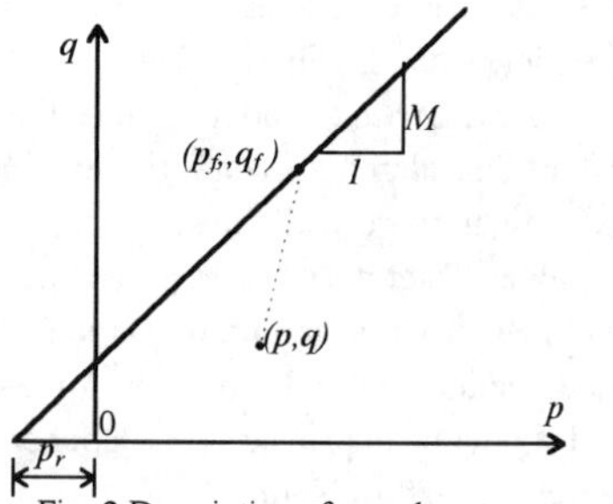

Fig. 2 Description of complex stress state

and after loading one stage increment, the stress state of the element is

$$\{\sigma\}=[\sigma_x, \sigma_y, \sigma_z, \tau_{xy}, \tau_{yz}, \tau_{zx}]^T \qquad (12)$$

In p-q coordinate system, the stress state before loading is corresponding to point A(p_0, q_0), shown as in Fig.3, and after loading, the point A moves to point B (p, q), where

$$p=(\sigma_x+\sigma_y+\sigma_z)/3 \qquad (13)$$

$$q=\frac{1}{\sqrt{2}}\sqrt{(\sigma_x-\sigma_y)^2+(\sigma_y-\sigma_z)^2+(\sigma_z-\sigma_x)^2+6(\tau_{xy}^2+\tau_{yz}^2+\tau_{zx}^2)} \qquad (14)$$

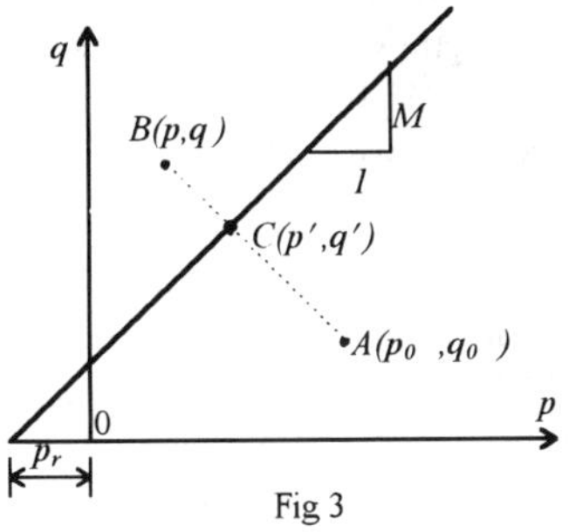

Fig 3

Here, assuming that the stress after loading is larger than the limit strength of soil, thus, the stress level given by eq. (7) is larger than 1.0, and the stress state of point B should be modified. In light of physical meaning, the point of modified stress state can only be on the failure line. But which point is on the failure line? We consider that the position of the point is given by the stress state before loading. Here the assumption that the stress path is a straight line is adopted. Obviously, it is reasonable for the increment method. According to the geometry relation ship in Fig.2, the modified stress invariant can be obtained as following:

$$p'=\frac{p_0[q-M(p+p_r)]-p[q_0-M(p_0+p_r)]}{[q-M(p+p_r)]-[q_0-M(p_0+p_r)]} \qquad (15)$$

$$q'=\frac{q_0[q-M(p+p_r)]-q[q_0-M(p_0+p_r)]}{[q-M(p+p_r)]-[q_0-M(p_0+p_r)]} \qquad (16)$$

For application, the stress components after modifying must be given. therefore, it is assumed that the modified stress can be described as

$$\{\sigma'\}=[\sigma'_x, \sigma'_y, \sigma'_z, \tau'_{xy}, \tau'_{yz}, \tau'_{zx}]^T \qquad (17)$$

but there are 6 unknown stress components in eq.(17), and we only have two equations, i.e. eq.(15) and (16), and some other equations must be replenished

Considering eq.(14) and (15), the deviatoric derivatives of generalized shear stress to the stress components can be got as following:

$$\begin{cases} \dfrac{\partial q}{\partial \sigma_x}=\dfrac{3(\sigma_x-p)}{2q} \\ \dfrac{\partial q}{\partial \sigma_y}=\dfrac{3(\sigma_y-p)}{2q} \\ \dfrac{\partial q}{\partial \sigma_z}=\dfrac{3(\sigma_z-p)}{2q} \\ \dfrac{\partial q}{\partial \tau_{xy}}=\dfrac{3\tau_{xy}}{q} \\ \dfrac{\partial q}{\partial \tau_{yz}}=\dfrac{3\tau_{yz}}{q} \\ \dfrac{\partial q}{\partial \tau_{zx}}=\dfrac{3\tau_{zx}}{q} \end{cases} \qquad (18)$$

Thus, for modified stress state, the following relationship is derived:

$$\begin{cases} \sigma'_x=p'+\dfrac{2q'}{3}\dfrac{\partial q'}{\partial \sigma'_x} \\ \sigma'_y=p'+\dfrac{2q'}{3}\dfrac{\partial q'}{\partial \sigma'_y} \\ \sigma'_z=p'+\dfrac{2q'}{3}\dfrac{\partial q'}{\partial \sigma'_x} \\ \tau'_{xy}=\dfrac{q'}{3}\dfrac{\partial q'}{\partial \tau'_{xy}} \\ \tau'_{yz}=\dfrac{q'}{3}\dfrac{\partial q'}{\partial \tau'_{yz}} \\ \tau'_{zx}=\dfrac{q'}{3}\dfrac{\partial q'}{\partial \tau'_{zx}} \end{cases} \qquad (19)$$

Where p′ and q′ are stress invariant for modified stress state given by eq.(15) and (16). Thus, eq. (19) can be used to modify the stress components. But, the change rate of generalized shear stress to stress components can not be defined. Here the assumption that the change rate of generalized shear stress after modifying equals to the one before modifying is adopted, thus the following relationship can be obtained:.

$$\begin{cases} \sigma'_x = p' + \frac{q'}{q}(\sigma_x - p) \\ \sigma'_y = p' + \frac{q'}{q}(\sigma_y - p) \\ \sigma'_z = p' + \frac{q'}{q}(\sigma_z - p) \\ \tau'_{xy} = \frac{q'}{q}\tau_{xy} \\ \tau'_{yz} = \frac{q'}{q}\tau_{yz} \\ \tau'_{zx} = \frac{q'}{q}\tau_{zx} \end{cases} \tag{20}$$

and the modified stress state is determined uniquely.

4. DISCUSSION

To prove the reasonability of eq.(20), the eq.(20) is degenerated to 2-D stress state:

$$\begin{cases} \sigma'_x = p' + \frac{q'}{q}(\sigma_x - p) \\ \sigma'_z = p' + \frac{q'}{q}(\sigma_z - p) \\ \tau'_{zx} = \frac{q'}{q}\tau_{zx} \end{cases} \tag{21}$$

and

$$\begin{cases} p = \frac{\sigma_1 + \sigma_3}{2} = \frac{\sigma_x + \sigma_z}{2} \\ q = \frac{\sigma_1 - \sigma_3}{2} = \sqrt{\left(\frac{\sigma_x - \sigma_z}{2}\right)^2 + \tau_{xz}^2} \end{cases} \tag{22}$$

Substituting eq.(22) into eq.(21), it can be got

$$\begin{cases} \sigma'_x = p' + \frac{q'}{\sqrt{\left(\frac{\sigma_x - \sigma_z}{2}\right)^2 + \tau_{xz}^2}}(\sigma_x - p) \\ \sigma'_z = p' + \frac{q'}{\sqrt{\left(\frac{\sigma_x - \sigma_z}{2}\right)^2 + \tau_{xz}^2}}(\sigma_z - p) \\ \tau'_{zx} = \frac{q'}{\sqrt{\left(\frac{\sigma_x - \sigma_z}{2}\right)^2 + \tau_{xz}^2}}\tau_{zx} \end{cases} \tag{23}$$

where p′ and q′ are determined by stress path. If the effect of stress path is neglect, and the assumption that the volume stress after modifying equals to the one before modifying is adopted, eq.(23) is consistent with the result in reference (Yin 1980).

Therefore , the formula of stress state modification is reasonable, and it includes the result summed by our predecessors. In order to show the effect of stress path on stress state modification, here an example is given. Before loading, the stresses of element I and II are assumed as

$$\{\sigma\}_{\mathrm{I}} = [30,\ 35,\ 55,\ 5,\ 8,\ 10]^T$$

$$\{\sigma\}_{\mathrm{II}} = [8,\ 9,\ 13,\ 1,\ 1,\ 2]^T$$

Where the unit of stress is kp_a.

The strength of the two elements are equal, i.e. c=3kp_a, φ=24^0, and after loading, the calculation stresses of the two elements are equal also as following:

$$\{\sigma\} = [15,\ 18,\ 27,\ 9,\ 10,\ 13]^T$$

Obviously, although the stress of the two elements are equal, their stress paths are different, and the corresponding volume stress and generalized shear stress are obtained from their stress components.

$$p_{\mathrm{I}} = 40, \quad q_{\mathrm{I}} = 33.0$$

$$p_{\mathrm{II}} = 10, \quad q_{\mathrm{II}} = 6.2$$

$$p = 20, \quad q = 34.2$$

According to eq.(9), the strength parameters in p-q coordinate system are

$$M = 0.94, \quad p_r = 6.7kp_a$$

The corresponding points of the two elements are shown as in Fig.4.

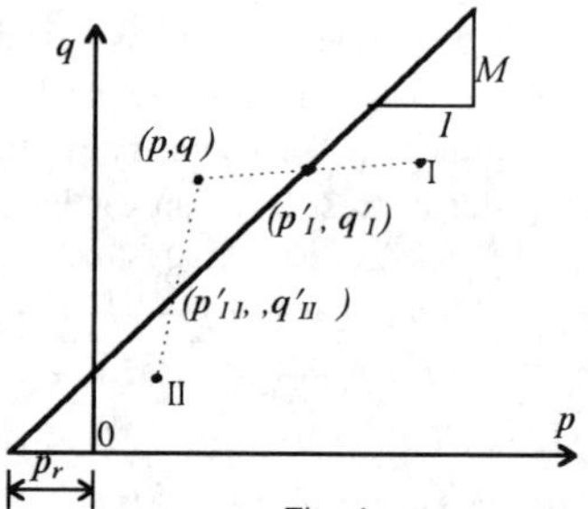

Fig. 4

From eq.(7), it can easily be obtained that the stress level after loading is 1.63. Obviously, the stress is larger than the limit strength of soil, so the stress of the two elements must be modified. The volume stress and generalized shear stress can be obtained from eq.(15) and (16)

$$p'_{\mathrm{I}} = 29.10kp_a, \quad q'_{\mathrm{I}} = 33.65kp_a$$

$$p'_{\mathrm{II}} = 15.10kp_a, \quad q'_{\mathrm{II}} = 20.5kp_a$$

and the corresponding stress components are

$$\sigma'_{\mathrm{I}x} = 24.1,\quad \sigma'_{\mathrm{I}y} = 27.1,\quad \sigma'_{\mathrm{I}z} = 36.0$$

$$\tau'_{\mathrm{I}xy} = 8.9,\quad \tau'_{\mathrm{I}yz} = 9.8,\quad \tau'_{\mathrm{I}zx} = 12.8$$

$$\sigma'_{\mathrm{II}x} = 12.1,\quad \sigma'_{\mathrm{II}y} = 13.9,\quad \sigma'_{\mathrm{II}z} = 19.3$$

$$\tau'_{\mathrm{II}xy} = 5.4,\quad \tau'_{\mathrm{II}yz} = 6.0,\quad \tau'_{\mathrm{II}zx} = 7.8$$

Obviously, if the stress path are different, the modified value of stress are different also for failure elements, and the difference is obvious. However, if the effect of stress path is not considered, the modified values of the two elements are equal, and the result is unreasonable.

5. CONCLUSION

The proposed method in this paper can improve the numerical analysis in geotechnical engineering., it has not only the theoretical significance, but also the application value in actual engineering.

REFERENCES

1. Allen, M. J. Marr and J.C.John. 1973. Finite element analysis of elastoplastic solis. M.I.J. Contract No. Nasa-9990, Solis Publication No.301, Research Report, 7
2. Yin, z.z. etal 1980. geotechnical theory and computation. China Waterpower Press, 299–301.
3. Zhan, M. L. & J. h. qian, 1993. Application of viscoelastic plastic finite element method in tunnel analysis, Jour, of Civil Eng, 26(3), P13–2

Computer Methods and Advances in Geomechanics, Yuan (ed.) © 1997 Balkema, Rotterdam, ISBN 90 5410 904 1

A boundary element approach to study the microstructure of soft soils

P. Delage
Centre d'Enseignement et Recherche en Mécanique des Sols (CERMES), Ecole Nationale des Ponts et Chaussées, Marne la Vallée, France

M. Cerrolaza
Instituto de Materiales y Modelos Estructurales, Facultad de Ingenieria, Universidad Central de Venezuela, Caracas, Venezuela

ABSTRACT
This paper presents and discusses a numerical approach to analyze the microstructure of soft soils, based on the Boundary Element Method (BEM). According to microstructure experimental results, the soil was modelled as a bidimensional linear elastic matrix containing circular pores. The matrix obeys a Tresca failure criterion and the pore size distribution a Gaussian normal law. The pores are randomly generated. The pores collapse is considered by numerically renumbering the BEM model in each step analysis, thus leading to a non-linear analysis procedure. Isotropic loading and unloading processes are studied and discussed, showing that the model is able to reproduce the experimental observations related to volume change of soft soils and other porous geomaterials. The numerical results indicate that the macroscopic non-linear behavior can be obtained from the microstructure characteristics.

INTRODUCTION

The microstructure of soils is often represented as an assembly of grains with punctual contacts. This representation is valid for sands, but approximate for fine-grained soils, which are made of clay platelets, generally aggregated together. Another big difference is related to the physico-chemical bonds acting between the clay minerals and the water molecules in fine-grained soils, which do not exist in granular soils. These clay-water interactions are quantified at a macroscopic level by the Atterberg limits. However, fine-grained soils are more often considered in soil mechanics as granular media, the grain being either the platelet, or the aggregate of platelets, and their macroscopic behaviour is often interpreted in terms of relative movements and spatial reorganisation of grains. For sands, various micromechanical analysis have been developed for understanding the macroscopic behaviour. Fewer attempts have been made in fine-grained soils.

The mechanism of pore collapse was analytically studied in a very simple two dimension model made of a linear isotropic elastic matrix containing a unique circular pore (Morlier[1]). Taking a Tresca failure criterion for the matrix, the failure criterion of the porous solid was determined analytically. In this work, a similar porous medium containing more pores of different sizes was considered in order to extend this approach to soils and other porous media. A specific numerical algorithm based on the Boundary Element Method (BEM) was developed to tackle this problem, and a macroscopic model for fine grained soils based on mechanisms occurring at a microscopic level is proposed.

SOFT SOIL MICROSTRUCTURE

Models of the microstructure of fine-grained soils have been considered very early on by geotechnical engineers in order to help in understanding some aspects of their behaviour, such as sensitivity (Casagrande[2]). In all early conceptual models, the elementary particle type considered was the single clay platelet. This is the case of Lambe's model[3] for compacted soils, which was based on the double-layer theory. Since Lambe developed his theory, there have been significant improvements in the techniques of observing microstructures, leading to a more and more complete description of the microstructure of fine grained soils in relation to their mechanical behaviour. This evolution is described in various papers concerning sensitive clays (Gillott[4]), various natural soils (Collins & McGown[5]), and in special conferences dedicated to soil microstructure (Gothenburg[6], Kingston[7]). Recent conferences on this subject provide a complete description of many related problems, including the latest technical developments, physico-chemical phenomena affecting the formation and evolution of microstructures in relation with environmental conditions, and applications related to environmental engineering (Bennett et al.[8]). As compared to earlier simple conceptual fabric models, it has been shown that the basic element of the microstructure of natural soils is not the single clay platelet, but domains constituted of various platelets aggregated together (Aylmore & Quirk[9]). Indirect evidence of the existence of aggregates was also obtained from hydraulic conductivity tests by Olsen[10]. In some cases, these aggregates can be resistant to some macroscopic mechanical actions. For a medium-sensitive Canadian clay, Delage & Lefebvre[11] showed that aggregates were still existing in the remoulded state.

An extensive study of the pore-size distributions of various sensitive clays of Eastern Canada (Delage[12]) was performed using mercury intrusion porosimetry. Mercury porosimetry is based on the principle that a non-wetting fluid, such as mercury, does not enter a porous medium unless a pressure is applied. Comparing pores to circular capillaries, the pressure P is related to an entrance pore radius r by Laplace's law :

$$P = \frac{2\sigma\cos\theta}{r} \qquad (1)$$

where σ is the surface tension of the intruding liquid and θ the contact angle (for mercury, θ=141° and σ =0.484 N/m). As the pressure P is increased, pores of smaller and smaller radius are intruded. The results are plotted in a cumulative curve, giving the pore size distribution of the porous medium.

More generally, various similarities between weak rocks and structured soils have been evidenced by Leroueil & Vaughan[17]. Another typical feature of their common brittle behaviour is observed in triaxial shear tests at low confining stresses. The stress strain curves present a fairly steep initial linear segment, and failure occurs at small strains and corresponds to a sharp peak. Before the peak, the structure remains apparently intact, with a fairly reversible response to cyclic loading. Yield is abrupt and coincides with the formation of shear surfaces. In the following, based on the observation of the microstructure of sensitive clays, the previous approach will be extended to soils and other porous media. A two dimension model made of a linear isotropic elastic matrix containing circular pores of various diameters will be considered. Due to the large complexity of the problem, a model based on the BEM will be used.

BOUNDARY ELEMENT MODEL OF THE SOIL MICROSTRUCTURE

The Boundary Element Method (BEM) has made a tremendous progress for the simulation of complex phenomena which usually arise in engineering problems and applied physics and mathematics situations. Therefore, the interest in the BEM has reached the industrial environment, thus stimulating several new developments as well as specialised research into the academic world. Much effort with the BEM has been done recently in the area of geomechanics, both at macroscopic and microscopic scales.

At the macroscopic level, the technical literature displays several works doing numerical simulation with boundary elements in soil mechanics and flow in porous media. Dominguez[19] presented a BEM direct formulation for evaluating the dynamic response of poroelastic bidimensional media, while Rajapakse & Senjuntichai[20] have presented an indirect formulation for the same topic. Ohkami et. al.[21] worked with orthotropic geomaterials by applying BEM. The simulation of consolidation processes in elastic soils is discussed by Chiou & Chi[22], who have analysed the Biot's consolidation in three dimensional layered domains. Chopra & Dargush[23] also applied the BEM to the time domain analysis of consolidation process in an axisymmetric model. In the field of flow in unsaturated media, a great amount of scientific research has also been done. Martinez & McTigue[24] presented a BEM formulation for the analysis of a partially saturated medium using a hydraulic conductivity that varies with the capillary-pressure head.

At the microscopic scale, many authors have investigated multi-phase Navier-Stokes flow in disordered microporous media by using lattice-Boltzman models. Gunstensen & Rothman[25] presented the ideas to analyse a two-phase flow porous media. Soll et. al.[26] analysed the multi-fluid flow in porous media, also using lattice-Boltzman methods, while Ginzsbourg & Adler[27] studied the boundary conditions in three dimensional lattice-Boltzman models. These works, among other well-known works, show the interest of the scientific community in the study of the microstructure of soils. Other authors have presented and discussed models for soil-pore analysis based on pseudo-fractal sets (see McBratney & Moran[28]). Also, Moran & Kirby[29] presented a soil-pore model based on random generation of bidimensional continua by using a set of linear porosity elements. These elements can represent the changes in porosity by a porosity-value, which takes the value of zero in fully-solid elements and the value of one in fully-pored elements.

The boundary element method

A detailed formulation of the BEM can be found in the technical literature as for instance, Brebbia et. al.[30], Lachat[31], Crouch & Starfield[32], Bannerjee & Butterfield[33], among others. This work uses the BEM direct formulation, based on the Somigliana's identity, which provides the unknown variable fields over the boundary. Likewise, a two dimensional model is used, since the problem herein is well described by a plane strain model, and by considering an *interior problem*: a domain (soil microstructure) enclosed by a finite boundary. The classical expression for the Somigliana's identity can be written as

$$C_{ij}(P)u_j(P) = \int_S [u_{ij}^*(P,Q)t_j(Q) - t_{ij}^*(P,Q)u_j(Q)]dS(Q) + \int_\Omega u_{ij}^*(P,Q)F_j(Q)d\Omega(Q) \quad (2)$$

When the integral expression (2) is written for the N nodes of the boundary (points P^k, k=1 to N) the following matrix expression is obtained

$$A_{ij}u_j - B_{ij}t_j = 0 \quad (3)$$

$$A_{ij}^{kl}(P^k) = C_{ij}(P^k) + \int_{S_k} t_{ij}^*(P^k,Q)F^l(Q)dS(Q) \quad (4)$$

$$B_{ij}^{kl}(P^k) = \int_{S_k} u_{ij}^*(P^k,Q)F^l(Q)dS(Q) \quad (5)$$

The introduction of the boundary conditions to expression (3) leads to a system of equations of order 2N, which is written as follows

$$K_{ij}u_j = P_i \quad (6)$$

u_i being the vector containing all the unknowns (displacements and tractions at the boundary) and K_{ij} the stiffness matrix. P_i is the vector containing the known quantities at the boundary.

A simplified 2D model: analysis of a randomly generated microstructure

The bidimensional model proposed in this work is derived from PSD measurements on soft clays. It is based on the random generation of circular pores into a bidimensional continuous medium simulating the soil skeleton. The distribution of the radii of the pores is gaussian normal, as shown in Figure 1b.

The algorithm randomly generates the required number of pores until the previously specified void ratio $e=V_v/V_s$ is reached (V_v being the void volume and V_s the volume of solids). The following gaussian-distribution parameters have to be supplied:

e = desired void ratio
Φ_{min} = minimum radius of pores

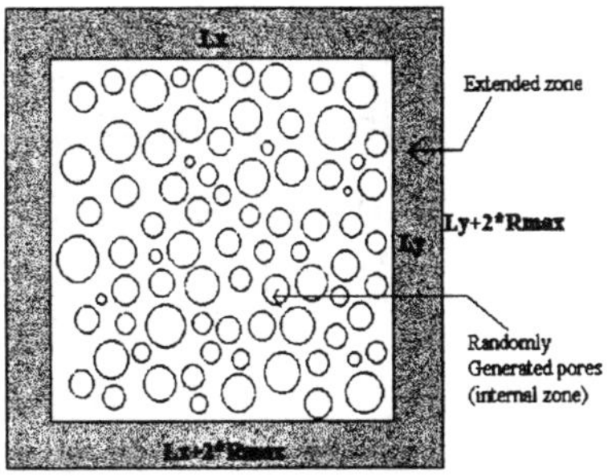

Figure 1.a Random model of soil microstructure
Rmax= maximum diameter of pores

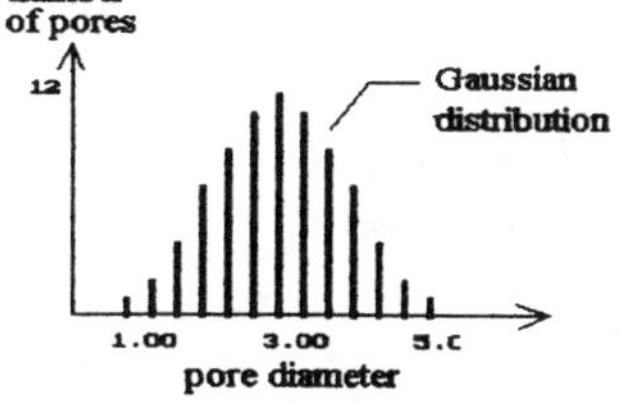

Figure 1.b Typical Gaussian distribution of pores

Φ_{max} = maximum radius of pores
Φ_{med} = mean radius of pores
S_{dev} = Standard deviation of Gaussian distribution.

As shown in Figure 1.a, the bidimensional model displays two zones: the *internal* zone (L_x*L_y dimensions) which contains the pores generated, while the *extended* zone of larger is necessary to properly apply the external pressures to the boundary element model. Numerical experiments have shown that these extensions in both directions are enough for practical purposes. The algorithm for generating the soil microstructure requires also the definition of:

dm = minimum distance between two pores
E = Young's modulus of soil
υ = Poisson's ratio of soil

The cartesian coordinates of the central point P_i of each pore are, therefore, randomly generated inside the central zone L_x*L_y. Once the algorithm has verified that the pore being generated is valid, i.e., the pore fits into the central zone and it does not intersect any other previously generated pore, the pore is stored. This process is repeated until the desired void ratio is reached.

When the generation of the model is completed, the pores are discretized in the same way as the exterior boundary of the extended zone by using straight boundary elements, as illustrated in Figure 2. Six boundary elements were used to discretize the boundary of each pore.

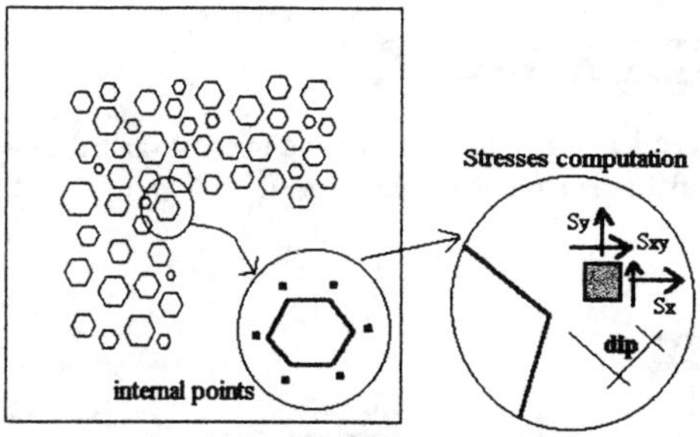

Figure 2. Boundary element discretization and internal points to compute stresses.

The boundary of the extended zone is also discretized by using six boundary elements per side, in order to apply the external pressures. Some sensitivity tests have shown that six boundary elements are accurate enough to properly represent the pore. The algorithm can create a refined model by using, for instance, eight, twelve or sixteen boundary elements per pore, but the CPU time will substantially increase and it is not necessary from an engineering and practical point of view. The boundary elements used in this work are two-noded and linear interpolation of the variable fields are used. Of course, a refined interpolation could be employed, say higher order elements of three or more nodal points (Lachat[31], Cerrolaza[36]). However, as pointed out before, our sensitivity analyses have shown that no relevant improvements is obtained by using higher interpolations. The algorithm also generates the required internal points where the stress tensor must be computed. These internal points are generated very close to the pore boundary, displaying a maximum distance called «Dip», which is determined using the following criterion: if dm<0.1*R_i then Dip = dm/2 else Dip = 0.05*R_i, «dm» being the minimum distance allowed between two adjacent pores and R_i the radius of the pore.

Once the boundary discretization is done, the user is requested to provide the external pressures, which are to be applied in the boundary element analysis. It is possible to carry out non-isotropic simulations by simply applying different pressures at the external boundary. Both normal and shear stresses can be defined in the model, thus allowing the possibility to carry out many different numerical simulations.

VALIDATION AND DISCUSSION OF THE NUMERICAL RESULTS

The validation of the previous numerical simulation is presented and discussed through two simulations: an incremental loading process, and a loading-unloading process.

Continuously incremental loading

The model is defined by the following parameters :

Number of pores = 58 (randomly generated)
ϕ_{MIN} = 0.1 μm
ϕ_{MAX} = 0.46 μm
ϕ_{MED} = 0.18 μm
dm = 0.1 μm
Sdev = 0.2 μm

$e = V_v/V_s = 1.40$

Figure 3 displays the BEM model, which has been discretized by using six boundary elements per pore. The choice of the Young modulus of the solid matrix is not an evident and easy task. In fact, it may be derived from a compression test performed on a soil at a high stress, where most of the pores have been collapsed, and where the compressibility of the clay matrix is involved. A value of 50 MPa has been considered as being realistic. Of course, more information on this value can be obtained by running a specific experimental program. A standard value of 0.3 has been taken for the Poisson's ratio.
This also could be experimentally confirmed by running tests with measurements of lateral stresses.

The initial applied pressure was 100 kPa. The test simulated a isotropic compression test where the pressure was increased by 20 kPa increments up to a stress of 280 kPa. Once the simulation process is completed, the algorithm creates a file containing all relevant information as: number of iterations of the non-linear analysis, external pressures and void ratio at each step, value of the highest stress of each pore on its boundary.

Figure 4 shows the compression curve calculated with the numerical model proposed

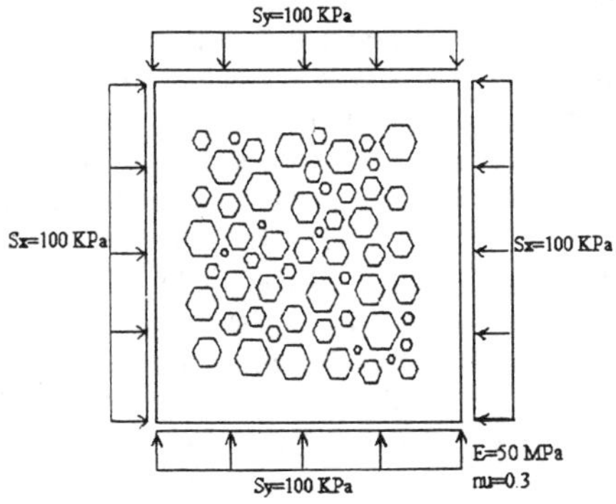

Figure 3. BEM model for the case example

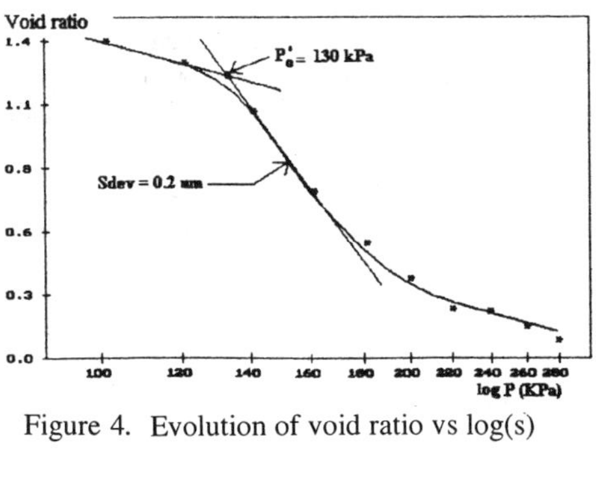

Figure 4. Evolution of void ratio vs log(s)

herein, giving the void ratio as a function of the logarithm of the total stress.

The shape of the curve is in a satisfactory agreement with that of structured soft soils: a negligible strain is observed below a given stress value, followed by an increasing slope with a maximum between 140 and 160 kPa. Finally, the slope decreases as the stresses increases. As for natural soils, this trend can be interpreted in terms of pore collapse. Below 100 kPa, all pores are able to sustain the load, leaving intact the microstructure. The pore collapse starts above 100 kPa, and a large pore volume is destroyed between 140 and 160 kPa, defining a compression index C_c of 5.24. This value is typical of a very compressible soil. Also, as shown in Figure 4, a preconsolidation pressure of 130 kPa can be determined. This pressure corresponds to the collapse of the larger pores, which start destroying the initial microstructure of the soil. At higher stresses, fewer pores remain intact, and the slope of the curve in the semi-log diagram decreases. This is also in a good agreement with the behaviour of soft soils at higher stresses.

The smallest void ratio reached by our soil model at 300 kPa was 0.1, which is very dense and not physical indeed. In soft soils, small pores are located within the aggregates, between the clay platelets. Experiments show that very high stresses are necessary to affect these pores. It is likely that our mechanism is more adapted to describe the collapse of inter-aggregate pores. So the void ratio to be considered here would rather be the inter-aggregates void ratio. By considering the Canadian sensitive clays, PSD measurements showed that the order of magnitude of this intra-aggregate pore ratio was approximately 0.6. So it is suggested to add this value to the value of pore ratio represented in the figure. Ideally, the compression of the soil on the whole range of stress should be modelled by using a second population of intra-aggregates pores, which was not made in this preliminary study.

Figure 5 shows the deformed shape of the generated pore-model compared to the undeformed one, at a loading step of 120 KPa . The decrease of the volume of the pores and their displacement towards the centre of the sample are clearly illustrated.

The overall deformation of the model is reasonably isotropic, showing that the choice of the pore number is adapted, and that their random disposition does not affect significantly the macroscopic response. More compression is observed in the middle of the sides of the sample,

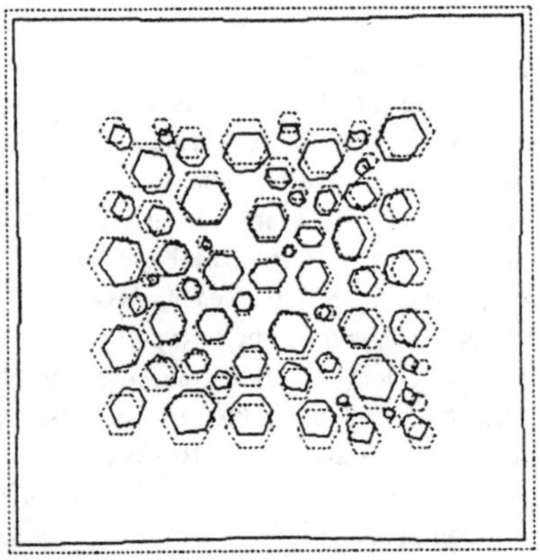

Figure 5. Deformed shape of generated pore-model at 120 kPa.

showing that ideally, the ratio between the length of the external zone and the one of the internal zone containing the pores should be minimised. There is however a limit, below which numerical problems occur.

CONCLUDING REMARKS

Most often, relationships between the microstructure of soils and their macroscopic mechanical behaviour are derived from studying the equilibrium of an assembly of grains in contact together, considering the elementary forces existing at the contacts. In this work, a different model was considered for soft clays in order to take into account the effects of the microstructure on the compressibility. The approach is derived from pore-size distribution-curves measurements on intact and compressed samples. The soil is modelled as a bidimensional porous medium made up of a linear elastic matrix containing circular pores of various sizes. The matrix obeys a Tresca failure criterion.

The calculations were carried out by using a boundary element model, which was able to contain a number of pores which keeps the CPU times within reasonable limits on a standard PC. Typically, the average CPU time was of 30 minutes on a Pentium-based microcomputer. Details of the implementation of the algorithm, as well as of the global procedure were presented. The pores of the soil are generated in a random manner, according to a previous specified Gaussian normal distribution compatible with experimental observation. When the model is subjected to an increased isotropic stress, the Tresca criterion is reached at the boundary of some pores, which collapse, resulting in a volume decrease. A remeshing procedure is then invoked in order to adjust the new microstructure. The boundary element discretization employed here has proven to be able to reproduce the complex elastic relations that exist inside the assembly of pores.

The algorithm proposed herein is also able to simulate both elastic and plastic behaviours, by simply taking into consideration the failure (collapse) of the pores as the external pressure increases. It appears from the analysis of the examples presented that the model is robust and it also allows the modelling of loading-unloading process. The responses of the model during an isotropic compression tests are in a fairly good agreement with the volume change behavior of fine-grained soils. The linear shape of the compression curve in a standard semi-logarithmic oedometric diagram is well reproduced, as well as the loading-unloading curves. This shows that hardening elastoplastic behaviour can be obtained at a macroscopic level by a microscopic model made of an elastic matrix with a Tresca failure criterion. The irreversible features are obtained by a mechanism of pore collapse. Experimental observations relating the compressibility of the soil to the pore size distribution curve have also been reproduced.

Some further extensions and improvements to the proposed model appear to be straightforward, as for instance, the possibility of simulating the microstructure of porous media by using a full three dimensional boundary element model, thus allowing the analysis of some specific 3D characteristics of the soil behaviour. The extension to the analysis of partially saturated soils is also being undertaken, as well as the exciting possibility to apply the damage mechanics theory (see Lemaitre & Chaboche[41]) to evaluate the soil stresses and deformation.

ACKNOWLEDGEMENTS

The authors wish to express their acknowledgements to the CERMES of the Ecole Nationale des Ponts et Chaussées for the support provided for this research. Also, the grant given by the Consejo de Desarrollo Científico y Humanístico and CONICIT (Caracas) is greatly appreciated.

REFERENCES

1. P. Morlier, 'Comportement mécanique des solides poreux. Domaine élastique des corps poreux. Rôle de la pression de pore.' (in French), *Rev. Industrie Minérale*, N° spécial, 1-17 (1970)
2. A. Casagrande, 'The structure of clay and its importance in foundation engineering. Contributions to Soil Mechanics', *J. of Boston Soc. of Civ. Eng.*, 15, (1932).
3. T.W. Lambe, 'The structure of compacted clay', *ASCE J. Soil Mech. and Found. Div*, 84(SM2), 1-34 (1958).
4. J.E. Gillott, 'Fabric of Leda clay investigated by optical, electron-optical, and X-ray diffraction methods', *Eng. Geology*, 4(2), 133-153 (1970)
5. K. Collins and A. McGown, 'The form and function of microfabric features in a variety of natural soils', *Géotechnique*, 24, 233-254 (1974).
6. *Proceedings of the International Symposium on Soil Structure*, Gothenburg, Sweden (1973).
7. *Proceedings of the 4th Int. Working-Meeting on Soil Micromorphology*, Kingston, Ont., Rutherford Ed. (1973).
8. H.R. Bennett, W.R. Bryant and M.H. Hulbert, *Microstructure of Fine Grained Sediments. From Mud to Shale*, Springer-Verlag, Berlin, 582 p. (1991).
9. L.G.A. Aylmore and J.P. Quirk, 'Domain or turbostratic structure of clays', *Nature*, 187, 1046-1056 (1960).
10. H.W. Olsen, 'Hydraulic flow through saturated clays', *Clays clay min*, 9(2), 131-161 (1962).
11. P. Delage and G. Lefebvre, 'Study of the structure of a sensitive Champlain clay and of its evolution during consolidation', *Canadian Geotech. J.*, 21(1), 21-35 (1984).
12. P. Delage, 'Microstructure and compressibility of some eastern canadian sensitive soft clays', *Int. Symp. on Geotech. Eng. of Soft Soils*, Mexico, 33-38 (1987).
13. P. Delage and J.P. Le Bihan, 'Microstructure et compressibilité d'argiles molles sensibles de l'Est canadien' (in French), *C.R. Acad. Sciences Paris*, t. 303, II(19), 1697-1702 (1986).
14. B.J. Botter, 'Pore collapse measurements on chalk cores', *Proc. Chalk Research Symp.*, Stavanger, Norway (1985).
15. I. Ruddy et al., 'Rock compressibility, compaction and subsidence in a high porosity chalk reservoir: a case study of Valhall field', *J. Petroleum Tech.*, 12, 741-746 (1989).
16. Ch. Schroeder, 'Le "pore collapse" : aspect particulier de l'interaction fluide-squelette dans les craies?' (in French), *C.R. Coll. Int. Groupement Belge de Mécanique des Roches*, Bruxelles, Belgium (1995).
17. S. Leroueil and P. Vaughan, 'The general and congruent effects of structure in natural soils and weak rocks', *Géotechnique*, 40(3), 467-488 (1990).
18. J. Goodier, 'Concentrations of stress around spherical and cylindrical inclusions and flaws', *Trans. Amer. Soc. Mech. Eng.*, 55, 39-44 (1933).
19. J. Dominguez, 'Boundary element approach for dynamic poroelastic problems', *Int. J. Num. Meth. in Eng.*, 35, 307-324 (1992).
20. R.K. Rajapakse and T. Senjunctichai, 'An indirect boundary integral equation method for poroelasticity', *Int. J. of Num. and Anal. Meth. in Geomech.*, 19(9), 587-614 (1995).
21. T. Ohkami, Y. Ichikawa and T. Kawamoto, 'A boundary element method for identifying orthotropic material parameters', *Int. J. of Num. and Anal. Meth. in Geomech.*, 15(9), 609-626 (1991).
22. Y.J. Chiou and S.Y. Chi, 'Boundary element analysis of Biot consolidation in layered elastic soils', *Int. J. of Num. and Anal. Meth. in Geomech.*, 18(6), 377-396 (1994).
23. G.F. Chopra and G.F. Dargush, 'Boundary element analysis of stresses in axisymmetric soil mass undergoing consolidation', *Int. J. of Num. and Anal. Meth. in Geomech.*, 19(3), 195-218 (1995).
24. M.J. Martinez and D.F. McTigue, 'A boundary integral method for steady flow in unsaturated porous media', *Int. J. of Num. and Anal. Meth. in Geomech.*, 16(8), 581-602 (1992).
25. A.K. Gunstensen and D.H. Rothman, 'Lattice-Boltzman studies of two-phase flow through porous media', *J. Geophys. Res.*, 98, 6431-6441 (1992).
26. W. Soll, S. Chen, K. Eggert, D. Grunau and D. Janecky, 'Application of the lattice-Boltzman /lattice-gas technique to multi-fluid flow in porous media', in *Computational Methods in Water Resources X* (Ed. A. Peters), Kluwer Academic Pub., 991-999 (1994).
27. I. Ginzsbourg and P.M. Adler, 'Boundary flow conditions analysis for the three dimensional lattice-Boltzman model', *J. of Physique II*, 4, 191-214 (1994).
28. A.B. McBratney and C.J. Moran, 'Soil-pore structure modelling using fuzzy random

pseudofractal sets', *Proc. 9th Int. Working Meeting on Soil Micromorphology,* Townsville, Australia (1994).
29. C.I. Moran and J.M. Kirby, 'Numerical simulation of deformation of soil-pore structure'. *Proc. 1st Int. Conf. on Unsaturated Soils, Vol. 2* (Eds. Alonso & Delage), Balkema Publ., 1123-1128 (1995).
30. C.A. Brebbia, J.F. Telles and L. Wrobel, *Boundary element techniques: theory and applications in engineering,* Ed. Springer-Verlag, Berlin (1984).
31. J.C. Lachat, 'A further development of the boundary integral technique in elastostatics', Ph.D. Thesis, University of Southampton, UK (1975).
32. S.L. Crouch and A.M. Starfield, *Boundary element methods in solid mechanics,* 1st ed, George Allen & Unwin (Publishers), London, UK (1983).
33. P.K. Bannerjee and R. Butterfield, *Boundary element methods in engineering science*, Ed. Mc Graw Hill, Chichester, UK (1981).
34. A.E.H. Love, *A treatise on the mathematical theory of elasticity,* Ed. Dover, New York (1927).
35. M. Cerrolaza and E. Alarcón, 'The p-adaptive BIEM approach for two-dimensional elasticity, *J. of Microcomputers in Civil Eng.*, 4(3), (1989).
36. M. Cerrolaza, 'The p-adaptive boundary integral equation method', *J. Adv. in Eng. Soft.*, 15(3/4), (1992).
37. M. Guiggiani, G. Krishnasamy, T.J. Rudolphi and F.J. Rizzo, 'A general algorithm for the numerical solution of hypersingular boundary integral equations', *ASME J. Appl. Mech.*, 59, 604-614 (1992).
38. L.J. Gray, L.F. Martha and A.R.Ingraffea, 'Hypersingular integrals in boundary element fracture analysis', *I. J. Num. Meth. in Eng.*, 29, 1135-1158 (1990).
39. M. Cerrolaza and E. Alarcon, 'A bicubic transformation for the numerical integration of Cauchy Principal Value integrals in boundary methods', *I. J. Num. Meth. in Eng.*, 29, (1990).
40. J. Hildebrand and G. Kuhn, 'Numerical computation of hypersingular integrals and applications to the boundary integral equation for the stress tensor', *J. of Eng. Anal. with Bound. Elem.*, 10, 209-217 (1992).
41. J. Lemaitre and J.L. Chaboche, *Mécanique des matériaux solides,* (in french), Ed. Dunod, Paris (1988).

Computer Methods and Advances in Geomechanics, Yuan (ed.) © 1997 Balkema, Rotterdam, ISBN 90 5410 904 1

Numerical modeling for cementitious granular materials

Xiaoxiong Zhong & Ching S.Chang
University of Massachusetts, Amherst, Mass., USA

ABSTRACT: Crack damage is commonly observed in cementitious granular materials such as concrete, calcareous sand, and ceramic materials. Previous analytical models based on continuum mechanics have limitations in analyzing localized damages at a micro-scale level. In this paper, a micromechanics approach is utilized. A contact law is derived for the inter-particle behavior of two particles connected by a cement binder. The binder initially contains micro-cracks which propagate and grow under external loading. As a result the binder is weakened with lower strength in shear and tension. Theory of fracture mechanics is employed to model the propagation and growth of these micro-cracks. The contact law is then incorporated in the analysis for the overall damage behavior of material using a statistical micromechanics approach. The results show that the analytical model is capable of modeling the behavior of a cementitious material in tension.

1 INTRODUCTION

Most constitutive models for cementitious granular materials are based on continuum mechanics. However, theories of continuum mechanics do not account for the microstructures in granular assemblies. In addition, these models do not explicitly account for the properties of binder and particles. Therefore, their applicability is limited to a narrow range of problems.

To explicitly consider the microstructures and the binder properties, Trent (1987) and Bazant et al. (1990) employed discrete element models to analyze the strength of cementitious granular materials. Trent and Margolin (1992, 1994) extended their work to study the fracture behavior of binder material based on Griffith crack theory. These models do not account for micro-crack growth and fracture toughness of the binder.

Fracture toughness has been often studied for cementitious granular materials (Dharmarajan and Vipulanandan 1988; Jenq and Shah 1985) and for rocks or concrete (Atkinson 1987; Bazant 1992; Wecharatna and Shah 1982; Ziegeldolf 1983). However, very little attention has been paid to the fracture toughness of binder and its influence on the granular assembly. Therefore, it is desirable to include binder fracture toughness in the modeling of cementitious material. The present approach is first to establish a binder model considering its fracture toughness. Then based on the binder model, we analyze the overall damage behavior of bonded aggregates using a statistical micromechanics approach. Stress-strain behavior of a randomly packed cementitious granulates is predicted to demonstrate the applicability of the present model.

2 STRESS-STRAIN MODEL FOR CEMENT

Cement binder is considered as an elastic material with preexisting micro-cracks. When the applied stress intensity exceeds the fracture toughness, the micro-cracks start to grow in size. As a consequence, elastic moduli of the cement material decreases.

2.1 *Moduli for binder material with micro-cracks*

The intact cementitious material without micro-cracks is isotropic and has a Poisson's ratio ν and Young's elastic modulus E. Considering a two-dimensional condition, the cementitious material containing a dilute random distribution of micro-cracks, all parallel to the x_1-direction. The stress-strain relationship is given by Nemat-Nasser (1993):

$$\begin{bmatrix} \varepsilon_{11} \\ \varepsilon_{22} \\ \varepsilon_{12} \end{bmatrix} = \begin{bmatrix} \frac{1}{E} & -\frac{\nu}{E_2} & 0 \\ -\frac{\nu}{E_2} & \frac{1}{E_2} & 0 \\ 0 & 0 & \frac{1}{\mu_{12}} \end{bmatrix} \begin{pmatrix} \sigma_{11} \\ \sigma_{22} \\ \tau_{12} \end{pmatrix} \quad (1)$$

where

$$E_2 = \frac{E}{(1-\nu^2)(1+2\pi\varphi)}$$

$$\mu_{12} = \frac{E}{2(1+\nu)\left(1+\frac{\pi(1-\upsilon)}{4}\varphi\right)} \quad (2)$$

In Eq. (2), the micro-crack density φ is a unitless quantity. The micro-crack density, in a two-dimensional case, is defined as

$$\varphi = \frac{N}{A}a_c^2, \quad (3)$$

where a_c is the length of micro-cracks and N is the total number of micro-cracks in the area A.

2.2 *Micro-crack growth criterion for tensile and shear modes*

The fracture criterion for a mixed mode micro-crack growth can be expressed in terms of the first mode and second mode stress intensity factors K_I and K_{II}, defined as:

$$K_I = \sqrt{\pi a_c}\,\sigma_n \quad ; \quad K_{II} = \sqrt{\pi a_c}\,\sigma_s \quad (4)$$

where a_c is the length of micro-crack, σ_n is the normal stress and σ_s is the shear stress with respect to the micro-crack orientation. The criterion for micro-crack growth used here represents a closed elliptical surface in the coordinate system made of K_I and K_{II}, given by

$$\frac{K_I^2}{K_{IC}^2} + \frac{K_{II}^2}{K_{IIC}^2} = 1 \quad (5)$$

where the first and second mode fracture toughness, K_{IC} and K_{IIC}, are material properties. During loading, when the stress density factors (K_I and K_{II}) satisfy Eq. (5), micro-cracks start to grow.

After the micro-crack growth, based on Eq. (4), the size of micro-cracks, a_c, is a function of stress and toughness, given by

$$a_c = \frac{1}{\pi}\left(\frac{K_{IC}^2}{\sigma_n^2} + \frac{K_{IIC}^2}{\sigma_s^2}\right) \quad (6)$$

2.3 *Behavior after crack growth*

The stress-strain relationship after crack growth-can be obtained by substituting Eq. (6) and Eq. (3) into the stress-strain relationship in Eqs. (1) and (2). Thus the stress-strain curves after crack growth are derived as fourth order equations, given by

$$\varepsilon_n = \frac{1-\nu^2}{E}\sigma_n\left[1+\frac{1}{\pi A/N}\left(\frac{K_{IC}^2}{\sigma_n^2}+\frac{K_{IIC}^2}{\sigma_s^2}\right)^2\right] \quad (7)$$

for the normal direction, and

$$\varepsilon_s = \frac{2(1+\nu)}{E}\sigma_s\left[1+\frac{1-\nu}{4\pi A/N}\left(\frac{K_{IC}^2}{\sigma_n^2}+\frac{K_{IIC}^2}{\sigma_s^2}\right)^2\right] \quad (8)$$

for the shear direction.

Eq. (7) is plotted in Fig. 1 for different values of shear stress. Eq. (8) is plotted in Fig. 2 for

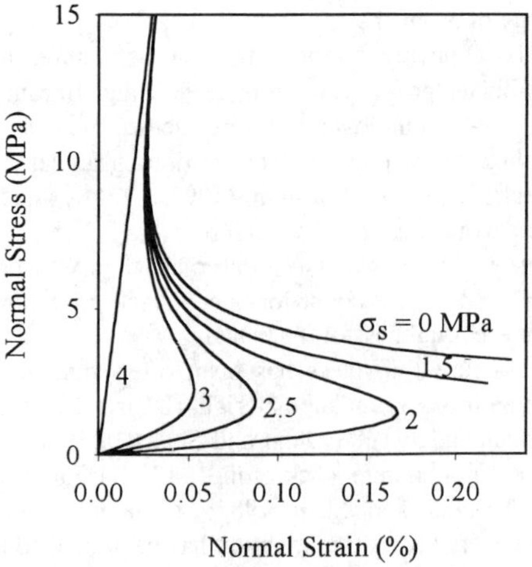

FIG. 1 Graphic representation of Eq. (7)

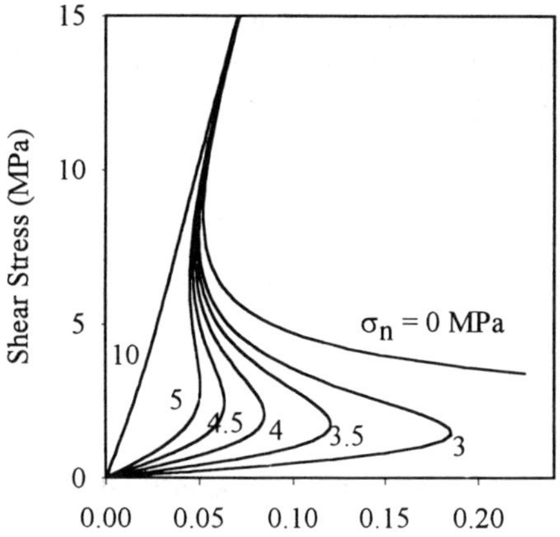

FIG. 2 Graphic representation of Eq. (8)

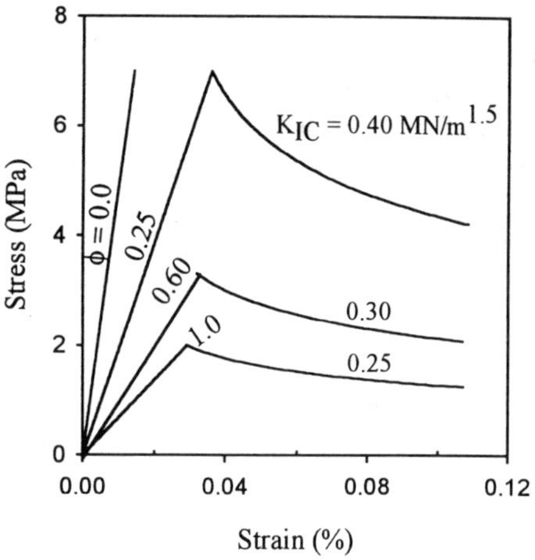

FIG. 3 Typical stress-strain curves for binder

different values of normal stress. The results in Fig. 1 and Fig. 2 are calculated based on the following material constants: E = 50 GPa ; ν = 0.15; K_{IC} = 0.45 MN/m$^{1.5}$; K_{IIC} = $0.82\ K_{IC}$; and N/A = 23.12 cm^{-2}.

Fig. 1 and Fig. 2 are similar in behavior trend but different in values of stress and strain. These curves, similar to Griffith loci (1921), represent the after-peak stress-strain behavior. The loci tangent to the elastic line of no micro-crack condition (i.e., φ=0, E=50Gpa). On each curve, there exist two points corresponding to the slope $d\sigma/d\varepsilon$ = 0. They represent the limits of unstable conditions.

In Fig. 3, the predicted stress-strain curves are plotted for cement under a simple uniaxial tension condition. These curves are predicted using different values of K_{IC}. and micro-crack density. The value of micro-crack density depends significantly on the water content at the time of mixing. For water content ranging from 0.35 to 1.0, the estimated micro-crack density ranges from 0.24 to 1.0. The fracture toughness for cement paste after curing typically ranges from 0.25 MN/m$^{1.5}$ to 0.45 MN/m$^{1.5}$ (Wittmann 1987). The N/A ranges from 23.12 cm^{-2} for K_{IC} = 0.4 MN/m$^{1.5}$to 3.17 cm^{-2} for K_{IC} = 0.2 MN/m$^{1.5}$.

Using equations (1), (2) and (7), the complete stress-strain curves can be obtained. In Fig. 3, stress-strain curves of cement are shown for three typical cases: (1) high strength binder, K_{IC}=0.4 MN/m$^{1.5}$ and φ =0.25, (2) medium strength binder, K_{IC}=0.3 MN/m$^{1.5}$ and φ =0.6, and (3) low strength binder, K_{IC}=0.25 MN/m$^{1.5}$ and φ =1.0.

The predicted range of tensile strength is comparable to that measured from experiments on cement by Welch (1966).

3 CONTACT LAW FOR TWO PARTICLES

Here we derive the mechanical behavior of two particles connected by a cement binder. Let the relative movement of the two particles be denoted as δ_n in the normal direction and δ_s in the tangential direction. The relative angular rotation is denoted as θ. Due to the relative movement of the two particles, the particle-binder interface is subjected to a stress distribution which result in a normal force f_n, a shear force f_s, and a moment M.

Now we assume that the particles are relatively rigid compared to the binder such that the stress distribution at the particle-binder interface can be assumed linear (Zhu et al. 1996). Thus for the two particles, the contact compliance relationship between contact forces and relative displacements can be described as:

$$\begin{pmatrix} f_n \\ f_s \\ M \end{pmatrix} = \begin{pmatrix} D_n & 0 & 0 \\ 0 & D_s & 0 \\ 0 & 0 & D_\theta \end{pmatrix} \begin{pmatrix} \delta_n \\ \delta_s \\ \theta \end{pmatrix} \tag{9}$$

where D_n, D_s, and D_θ are the contact stiffness respectively for tension, sliding and rolling modes. Using the stress-strain relationship in Eqs. (1) and (2) for cement binder, the contact stiffness can be derived from the binder property as follows:

$$D_n = \frac{bE}{(1-\nu^2)h(1+2\pi\varphi)}$$
$$D_s = \frac{bE}{2(1+\nu)h\left(1+\frac{\pi(1-\nu)\varphi}{4}\right)} \qquad (10)$$
$$D_\theta = \frac{bhE}{3(1+\nu)(1-\nu)(1+2\pi\varphi)}$$

where b is the width and h is the thickness of the cement binder.

The fracture criterion in Eq. (5) is used for a mixed mode cracking of the binder. The stress intensity factors for the binder can be expressed by the contact forces, given by:

$$K_I = \sqrt{\pi a_c}\frac{f_n}{b} \quad ; \quad K_{II} = \sqrt{\pi a_c}\frac{f_s}{b} \qquad (11)$$

Because of the effect of moment, the stress concentrations at two ends of a micro-crack are of different magnitudes. Therefore the first mode stress intensity factor at the two ends needs to be evaluated separately for micro-crack growth. Based on the solution of stress concentration at ends of a crack derived by Cherepanov (1979) under a linearly distributed far field normal stress, we derive the stress intensity factor K_I in terms of contact force and contact moment, given by

$$K_I = \sqrt{\pi a_c}\left(\frac{f_n}{b} \pm \frac{15 a_c M}{16 b^3}\right) \qquad (12)$$

in Eq. (12), the plus sign is for the end of a micro-crack with higher tension and minus sign is for the other end of the micro-crack.

4 CONSTITUTIVE RELATIONSHIP FOR CEMENTITIOUS MATERIAL

In this section, the contact law for inter-particle behavior is incorporated in the analysis for the overall damage behavior of cementitious material using a statistical micromechanics approach similar to that by Chang (1996). Neglecting the effect of particle rolling, the tensor expression of contact law is given by:

$$f_i^m = D_{ij}^m \delta_j^m \qquad (13)$$

The stiffness tensor in Eq. (13) can be expressed as follows:

$$D_{ij} = D_n n_i n_j + D_s s_i s_j \qquad (14)$$

where n_i and s_i are unit vectors normal and tangential to the contact area, respectively. D_n and D_s are respectively the normal and tangential contact stiffness given in Eq. (10). Based on the principle of visual work and the concept of average stress (Chang, 1993, 1994), the global stress-strain relationship becomes:

$$\sigma_{ij} = C_{ijkl}\varepsilon_{kl} \qquad (15)$$

where

$$C_{ijkl} = \frac{1}{V}\sum_{c=1}^{M_c} l_i^c D_{jk}^c l_l^c \qquad (16)$$

where V is the volume of the sample and l_i is the branch vector which connects the centers of two particles in contact. Noting that the stress and the strain tensors are symmetric, the stiffness tensor must have the symmetries. Further, since the local stiffness tensor is symmetric (i.e., $D_{ij} = D_{ji}$), the overall constitutive tensor C_{ijkl} is symmetric.

5 EXAMPLE PREDICTION

A sample of randomly packed aggregates bonded by cement is generated by computer as shown in Fig.4. The size of sample is 24 cm × 40 cm consisting of 320 particles. The sample is composed of five different particle sizes: radii are 5 mm, 6 mm, 7 mm, 8 mm, and 9 mm, with equal number of particles for each particle size. As a result of the random generation of particles, the binder thickness varies from contact to contact. The average binder thickness is about 2 mm with an average width of 6 mm. For this sample, the ratio of binder volume to the aggregate volume is 0.29. Thus the void ratio is 0.15 corresponding to a porosity of 0.3.

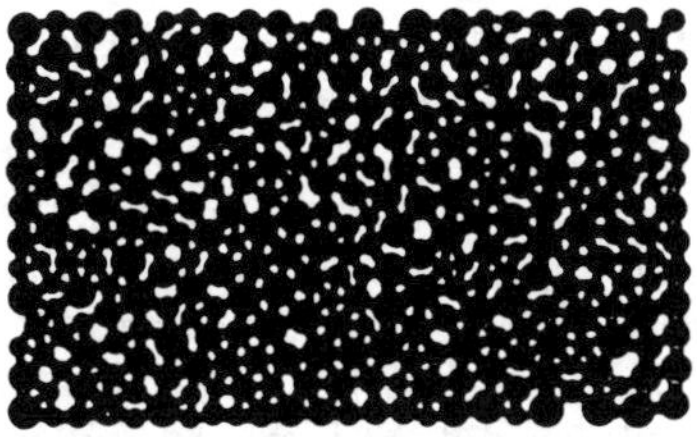

FIG. 4 A randomly generated sample of bonded aggregates

The properties of cement binder used in the analysis are as follows: K_{IC}= 0.25, 0.3, 0.35, 0.4 MN/m$^{1.5}$, $K_{IIC} = 0.82\ K_{IC}$, and φ= 0.25.

For the cementitious aggregates, four stress-strain curves are plotted in Fig. 5 and Fig. 6, corresponding to the four sets of cement fracture toughness. In Fig. 5, the calculation is performed by using a very small value of shear modulus for the binder to simulate binders that are soft against shear. In Fig. 6 the shear modulus is calculated by Eq. (2) to simulate stiff binders. The computed stress-strain curves for the two cases are different in strength and stiffness.

Due to different values of binder toughness K_{IC}, the computed range of tensile strength for the cementitious aggregates varies from 2 - 6 MPa. The results fall in the range of measured tensile strength from experiments on concrete by Welch (1966). Compared to the range of binder strength 3 - 7 MPa, the tensile strength of cementitious aggregates is less than that of cement binder, although in the same order. This result is also in agreement with the experimental observation. However, the computed modulus of cimentitious aggregates is over-predicted; approximately five times stiffer than the modulus of binder material. The over-prediction of modulus for cementitious material is caused by the assumption of rigid aggregates used in the analysis.

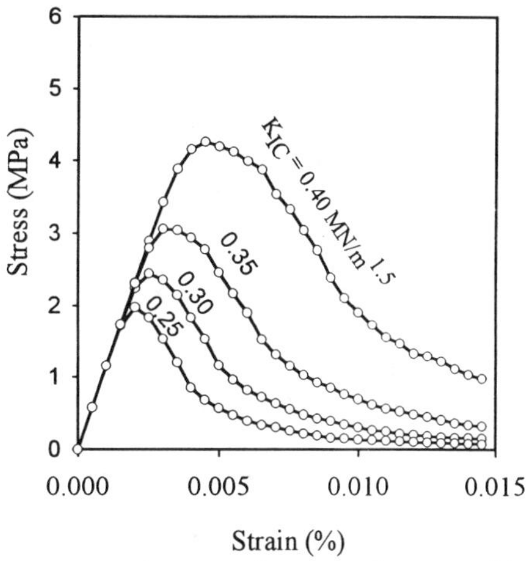

FIG. 5 Stress-strain curves for cementitious material with soft shear stiffness

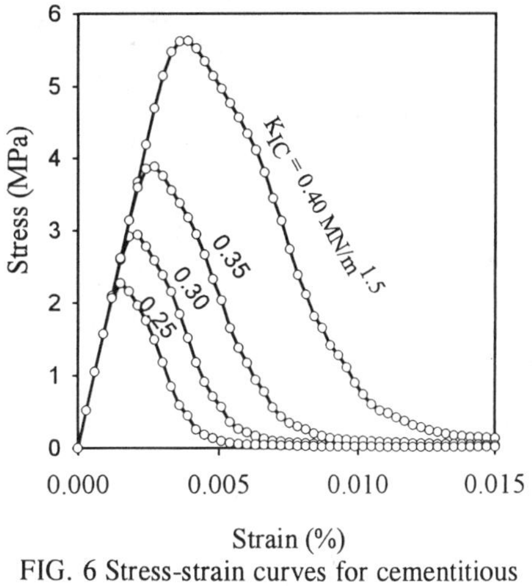

FIG. 6 Stress-strain curves for cementitious material with stiff shear stiffness

6 SUMMARY

In this paper, a stress-strain model is proposed that accounts for the fracture toughness of cement binde. The stress-strain model for cement binder is then used to derive the contact law which describe the inter-particle behavior of two round aggregates connected by a cement binder. Based on the contact law, analysis for the overall damage behavior of cementitious material can be performed using a statistical micromechanics approach. The results show that the analytical model is capable of modeling the behavior of a cementitious granular material in tension.

ACKNOWLEDGMENTS

This work was sponsored in part by the Air Force Office of Scientific Research, USAF, under grant number F49620-95-1-0117.

REFERENCES

Atkinson, B. L., ed. (1987). *Fracture Mechanics of Rock*. Academic Press Inc., London.

Bazant Z. P., ed. (1992). *Fracture Mechanics of Concrete Structures*. Elsevier Applied Science, New York.

Bazant Z. P., Tabbara M. R., Kazemi, M. T., and Pijaudier-Cabot, Gilles (1990). "Random Particle Simulation of Damage and Fracture in Particulate or Fiber-reinforced Composites," Damage Mechanics in Engineering Materials, ed. Ju. J. W., Krajcinovic, D., and Schreyer, H. L., ASME, AMD-Vol.109, pp.41-55.

Chang, C. S. (1993). "Micromechanics Modeling for Deformation and Failure of Granulates with Frictional Contacts," *Mechanics of Material*, Elsevier Science Publishers, Amsterdam, Vol. 16, No.1-2, pp. 13-24.

Chang, C. S., and Gao, J. (1996). "Kinematic and Static Hypotheses for Constitutive Modelling of Granulates Considering Particle Rotation," *Acta Mechanica*, Springer-Verlag, Vol. 115, No. 1-4, pp.213-229.

Chang, C. S., and Liao, C. L. (1994). "Estimates of Elastic Modulus for Media of Randomly Packed Granules," *Applied Mechanics Reviews*, ASME, Vol. 47, No. 1, Part 2, pp. 197-206.

Cherepanov, G. P. (1979). *Mechanics of Brittle Fracture*, McGraw Hill, Inc., New York.

Dharmarajan, N. and Vipulanandan, C. (1988). "Critical Stress Intensity Factor for Epoxy Mortar." *Polymer Engineering and Science*. Vol. 28, pp. 1182-1191.

Griffith, A. A. (1921). "The Phenomenon of Rupture and Flow in Solids." *Phil. Trans. R. Soc. London*, Series A, Vol. 221, pp. 163-198.

Jenq, Y. S. and Shah, S. P. (1985). "A Fracture Toughness Criterion for Concrete." *Engineering Fracture Mechanics*, Vol. 21, No. 5, pp. 1055-1069.

Nemat-Nasser, S. and Hori, M. (1993). *Micromechanics: Overall Properties of Heterogeneous Materials*. North-Holland, Amsterdam.

Trent, B. C., (1987). "The Effect of Microstructure on the Macroscopic Behavior of a Cemented Granular Material." Ph.D Thesis, University of Minnesota.

Trent, B. C. and Margolin, L. G., (1992). "Numerical Validation of Constitutive Theory for an Arbitrary Fractured Solid." *Proceedings of Second International Conference on Discrete Element Method*, MIT, Cambridge, Massachusetts.

Trent, B. C. and Margolin, L. G. (1994). "Modeling Fracture in Cemented Granular Materials." *Fracture Mechanics Applied to Geotechnical Engineering*, ASCE, Edited by L. E. Vallejo and R. Y. Liang, pp. 54-69.

Wecharatna, M., and Shah, S. P. (1982). "Slow Crack Growth in Cement Composites." *Journal of Structural Division*, ASCE, Vol. 108, No. SR6, pp. 1400-1413.

Welch, G. B. (1966). "Tensile Strain in Unreinforced Concrete Beams." *Magazine of Concrete Research*, Vol.18, No. 54, pp. 9-17.

Wittmann, F. H. (1987). "Structure of Concrete and Crack Formation.", in *Fracture of Nonmetallic Material*, ed. Herrmann, K. P. and Larsson, L. H., D. Reidal Publishing Co., Boston, pp. 309-340.

Zhu, H., Chang, C. S. and Rish, J. W. (1996). "Tangential and Normal Compliance for Conforming Binder Contact. I: Elastic Binder," *International Journal of Solids and Structures*, Pergamon Press, Vol. 33 , No. 29, pp. 4337-4349.

Ziegeldorf, Z. (1983). "Fracture Mechanics of Hardened Cement Paste, Aggregates and Interfaces." *Fracture Mechanics of Concrete*, Edited by F. H. Wittman, Elsevier Science Publishers, pp. 371-409.

Computer Methods and Advances in Geomechanics, Yuan (ed.) © 1997 Balkema, Rotterdam, ISBN 90 5410 904 1

Constitutive modelling of jointed rock masses based on equivalent joint compliance model

T. Nagai & J. S. Sun
Aoki Corporation, Technical Research Institute, Tsukuba, Japan

S. Sakurai
Department of Civil Engineering, Kobe University, Japan

ABSTRACT: It is well known that rock bolts and rock anchors are quite effective support measures for stabilizing jointed rock masses. However, it is extremely difficult for engineers to make an accurate prediction of the mechanical behaviour of jointed rock masses. Therefore, in engineering practice, the design of rock reinforcement is carried out on the basis of the experience and judgement of well-trained engineers. In this paper, the authors propose a mechanical model to represent the deformational behaviour of a jointed rock mass, and formulate a constitutive equation for characterizing the behaviour of a rock mass. And its validation for engineering applications is demonstrated by the comparison of numerical results with laboratory data using physical models of a blocky medium.

1 INTRODUCTION

In order to simulate the behaviour of rock masses, numerical analysis methods such as the Finite Element Method are quite often used. In general, there are two approaches available in modelling a rock mass; the discontinuum approach and the continuum approach. In the discontinuum approach, rock matrix and discontinuities are directly modelled. It is inapplicable when the rock mass contains many discontinuities, because it is very difficult to obtain detailed data regarding the position, direction, size, shape and mechanical characteristics of each of these discontinuities. In the continuum approach, on the other hand, modelling of a rock mass is done by assuming that the mechanical parameters of a continuum body are the same as that of a jointed rock mass. A numerical analysis method based on continuum mechanics is then used on this model for simulating the behaviour of a rock mass. In the continuum approach, there is no need to investigate the position, direction, size, shape or mechanical parameters of each joint. Hence, the continuum approach is preferable as far as design works are concerned.

However, in the continuum approach, the main problem lies in how to model a jointed rock mass. There are two broad categories of modelling it regarding joint geometry; the regular distribution and the random distribution. In the former case, a regularly jointed rock mass such as a stratified rock mass is considered and characterized as a pseudo-elastic material if the joints are tight, i.e., they don't slip and separate each other (Wardle & Gerrard, 1972; Chappell, 1986). If, however, there is some movement between the joints then they are described by shear and normal stiffnesses and the jointed rock mass is considered an anisotropic composite material (Kulhawy, 1978; Amadei & Goodman, 1981). While in the latter case, a randomly jointed rock mass is considered, and represented by a cracked elastic solid (Oda et al., 1984; Wei & Hudson, 1986).

In this paper, a constitutive equation is derived for a jointed rock mass which has several joint sets. The authors propose an equivalent joint compliance model in order to represent the reinforcement effects of rock bolts and/or rock anchors on the jointed rock mass, and take it into a constitutive equation. In this model, a joint is considered to be a discontinuum described by shear and normal stiffnesses. The former one is assumed to be a stress-dependent material.

2 CONSTITUTIVE MODELLING

2.1 *Stress-strain relation for a jointed rock mass*

The first step in the derivation of a constitutive equation is to characterize the stress-strain relations for a regularly jointed rock mass as illustrated in Figure 1. The intact rock has isotropic elastic properties denoted by Young's modulus E and Poisson's ratio ν. A joint is considered to be a one-

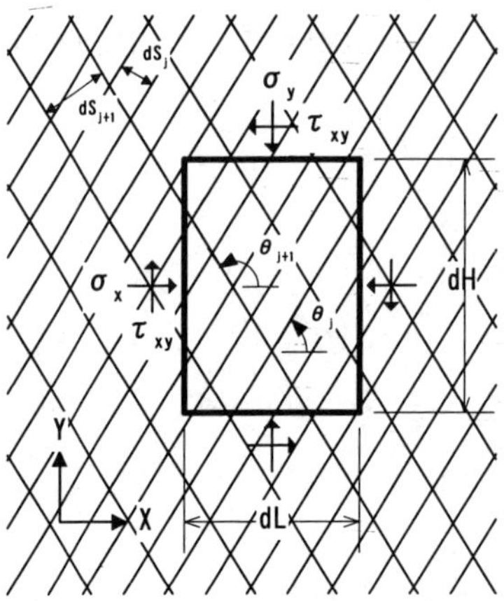

Figure 1. Model of a regularly jointed rock mass

dimensional material, therefore it is described by a shear and normal stiffness K_s and K_n, respectively. Total deformation of a regularly jointed rock mass is assumed to be the sum of the total deformation of intact and joint material. Then strain components of the rock mass are defined by using the height and width of the rock mass being considered dH and dL, respectively. Finally, the stress-strain relations for the rock mass under plain strain conditions can be obtained as follows:

$$\varepsilon_x = \frac{(1+\nu)\{(1-\nu)\sigma_x - \nu\sigma_y\}}{E} + \sum_{j=1}^{M}\left(\frac{N_{xj}\,\sigma_{nj}\sin\theta_j}{dL\,K_{nj}} - \frac{N_{xj}\,\tau_j\cos\theta_j}{dL\,K_{sj}}\right) \tag{1}$$

$$\varepsilon_x = \frac{(1+\nu)\{-\nu\sigma_x + (1-\nu)\sigma_y\}}{E} + \sum_{j=1}^{M}\left(\frac{N_{yj}\,\sigma_{nj}\cos\theta_j}{dH\,K_{nj}} + \frac{N_{yj}\,\tau_j\sin\theta_j}{dH\,K_{sj}}\right) \tag{2}$$

$$\gamma_{xy} = \frac{2(1+\nu)\tau_{xy}}{E} + \sum_{j=1}^{M}\left(\frac{-N_{yj}\,\sigma_{nj}\sin\theta_j}{dH\,K_{nj}} + \frac{N_{yj}\,\tau_j\cos\theta_j}{dH\,K_{sj}} - \frac{N_{xj}\,\sigma_{nj}\cos\theta_j}{dL\,K_{nj}} - \frac{N_{xj}\,\tau_j\sin\theta_j}{dL\,K_{sj}}\right) \tag{3}$$

where N_{xj} = the number of joints which intersect the part concerning dL in the jth joint set; N_{yj} = the number of joints which intersect the part concerning dH in the jth joint set; θ_j = the angle of the jth joint set; M = the total number of joint set; σ_{nj} = the normal stress of the jth joint set; and τ_j = the shear stress of the jth joint set. σ_{nj} and τ_j are the following:

$$\sigma_{nj} = \sigma_x \sin^2\theta_j + \sigma_y \cos^2\theta_j - \tau_{xy}\sin 2\theta_j$$

$$\tau_j = \frac{-\sigma_x + \sigma_y}{2}\sin 2\theta_j + \tau_{xy}\cos 2\theta_j \tag{4}$$

In order to verify the applicability of equations (1), (2) and (3), one of the authors compared them with the triaxial tests results using jointed models (Nagai, 1992). As a result, it was pointed out that the normal stiffness K_{nj} could be considered constant and the shear stiffness K_{sj} should be assumed to have stress-dependent properties:

$$K_{sj} = K_j\,\gamma_w\,(\sigma_{nj}/P_a)^{nj} \tag{5}$$

where K_j = the shear stiffness number; n_j = the shear stiffness index; γ_w = the unit weight of water; and P_a = the atmospheric pressure.

2.2 *Constitutive equation for a jointed rock mass*

In a constitutive equation, when once its material parameters are determined in one coordinate system, the parameters in the other coordinate system must be obtained by using the coordinates transformation. From such point of view, in order to treat the stress-strain relations for a regularly jointed rock mass (equations (1)-(3)) as the constitutive equation for it, the numbers of joints N_{xj} and N_{yj} need to be considered not scalars but vectors as follows:

$$\begin{Bmatrix} N_{xj}' \\ N_{yj}' \end{Bmatrix} = \begin{bmatrix} \cos(\theta_j - \theta_j') & \dfrac{-dL\sin(\theta_j - \theta_j')}{dH} \\ \dfrac{dH\sin(\theta_j - \theta_j')}{dL} & \cos(\theta_j - \theta_j') \end{bmatrix} \begin{Bmatrix} N_{xj} \\ N_{yj} \end{Bmatrix} \tag{6}$$

where N_{xj}' and N_{yj}' = the numbers of joints in the coordinate system *X'-Y'* corresponding N_{xj} and N_{yj}, respectively; and θ_j' = the angle of the jth joint set in the coordinate system *X'-Y'*.

In Figure 2, if $\theta_j' = 0°$ then $N_{xj}' = 0$ and $N_{yj}' = dH/dS_j$. When these relations are substituted into equation (6), the following is obtained:

$$\{N_{xj}\ \ N_{yj}\}^{\mathrm{T}} = \left[\frac{dL}{dS_j}\sin\theta_j \quad \frac{dH}{dS_j}\cos\theta_j\right]^{\mathrm{T}} \tag{7}$$

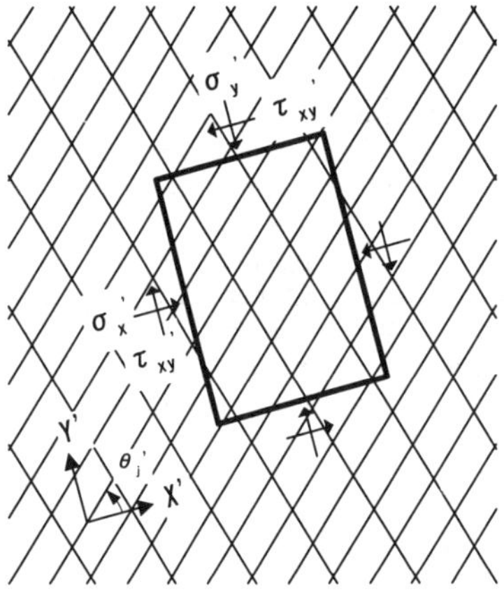

Figure 2. Model in another coordinate system

Equations (1)-(3) together with equations (4) and (7) give the constitutive equation for a regularly jointed rock mass in the form of incremental amounts as follows:

$$\{\Delta\varepsilon_x \ \ \Delta\varepsilon_y \ \ \Delta\gamma_{xy}\}^{\mathrm{T}} = [\,C\,]\{\Delta\sigma_x \ \ \Delta\sigma_y \ \ \Delta\tau_{xy}\}^{\mathrm{T}} \tag{8}$$

where $\Delta\varepsilon_x$, $\Delta\varepsilon_y$ and $\Delta\gamma_{xy}$ = the incremental strains in coordinates *X-Y*; and $\Delta\sigma_x$, $\Delta\sigma_y$ and $\Delta\tau_{xy}$ = the incremental stresses in coordinates X-Y. The compliance matrix $[\,C\,]$ has the following components C_{ij} (i, j = 1, 2, 3), and six components of them are independent each other:

$$C_{11} = \frac{(1-\nu^2)}{E} + \sum_{j=1}^{M}\left(\frac{\sin^2\theta_j}{K_{nj}} + \frac{\cos^2\theta_j}{K_{sj}}\right)\frac{\sin^2\theta_j}{dS_j}$$

$$C_{12} = C_{21} = \frac{-\nu(1+\nu)}{E} + \sum_{j=1}^{M}\left(\frac{1}{K_{nj}} - \frac{1}{K_{sj}}\right)\frac{\sin^2\theta_j\cos^2\theta_j}{dS_j}$$

$$C_{13} = C_{31} = \sum_{j=1}^{M}\left(\frac{-2\sin^2\theta_j}{K_{nj}} - \frac{\cos^2\theta_j - \sin^2\theta_j}{K_{sj}}\right)\frac{\sin\theta_j\cos\theta_j}{dS_j}$$

$$C_{22} = \frac{(1-\nu^2)}{E} + \sum_{j=1}^{M}\left(\frac{\cos^2\theta_j}{K_{nj}} + \frac{\sin^2\theta_j}{K_{sj}}\right)\frac{\cos^2\theta_j}{dS_j}$$

$$C_{23} = C_{32} = \sum_{j=1}^{M}\left(\frac{-2\cos^2\theta_j}{K_{nj}} + \frac{\cos^2\theta_j - \sin^2\theta_j}{K_{sj}}\right)\frac{\sin\theta_j\cos\theta_j}{dS_j}$$

$$C_{33} = \frac{2(1+\nu)}{E} + \sum_{j=1}^{M}\left(\frac{4\sin^2\theta_j\cos^2\theta_j}{K_{nj}} + \frac{(\cos^2\theta_j - \sin^2\theta_j)^2}{K_{sj}}\right)\frac{1}{dS_j} \tag{9}$$

One of the authors demonstrated the applicability of equations (8) and (9) by means of comparison with the triaxial tests results using jointed models (Nagai, 1992).

3 PHYSICAL MODEL TESTS

3.1 *Model material for a jointed rock mass*

The tests were carried out on 180×140×15cm samples built up from blocks of model material arranged such that two orthogonal sets of joints were produced. There were two different sizes of blocks, that is, 8×4×15cm and 4×2×15cm, which make it possible to simulate joint sets with different spacings. The material used as a substitute rock material in this study was a mix of plaster, sand and water in the proportion 3:5:2 by weight. The mechanical characteristics of the model materials are listed in Table 1 (Nagai et al., 1996).

Table 1. Properties of model material

Properties	Value
Density	14.2 kN / m^3
Uniaxial compressive strength	7.74 MPa
Young's modulus	1.55 GPa
Poisson's ratio	0.15
Peak joint friction angle	42.8°

3.2 *Model material for rock anchor*

Rock anchor was modelled by a brass rod, two brass plates and nuts (see Figure 3). The diameter of a brass rod was 1mm and 3mm. While blocks were building up, the holes for anchor models were made in certain blocks. The anchor models were tensioned immediately after the opening were excavated step by step. In this study, three anchors were used for the arch of opening and eight anchors were for both sidewalls (as indicated in Figure 4).

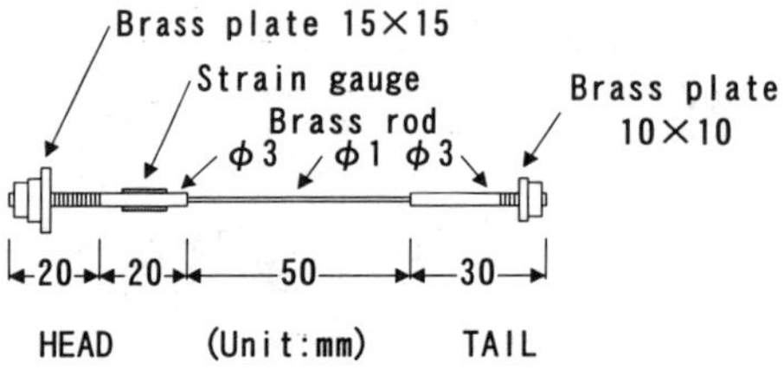

Figure 3. Model of a rock anchor

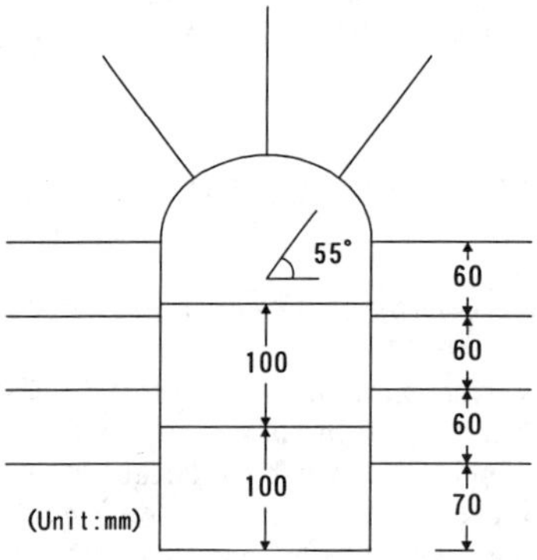

Figure 4. Pattern of anchor models

3.3 *Testing equipment*

The biaxial loading system for large model was designed to provide independent control of the applied pressure in each of the two principle loading directions (see Figure 5). The main components of it were as follows:

1. The rigid steel frame of the Institute of Rock and Soil Mechanics, Chinese Academy of Sciences, which served as the reaction structure.
2. Six oil jacks in the vertical direction and sixteen oil jacks in the horizontal direction (capacity of each jack is 49kN).
3. Top and side steel loading plates whose friction against a rock mass model was reduced by inserting Teflon sheets.

3.4 *Testing procedure*

Firstly, in order to produce a certain in situ state of stress, vertical and horizontal pressures were applied with oil jacks in five steps. Secondary, in order to simulate the excavation effect on a rock masses, opening was excavated in three steps keeping the initial stresses constant.

The displacements adjacent to the opening were measured by means of two different measuring systems in front and back of the rock mass model. The former is the non-contact measuring system based on image processing technique with CCD cameras or high-accuracy optical cameras. The latter is the contact measuring technique with dial indicators.

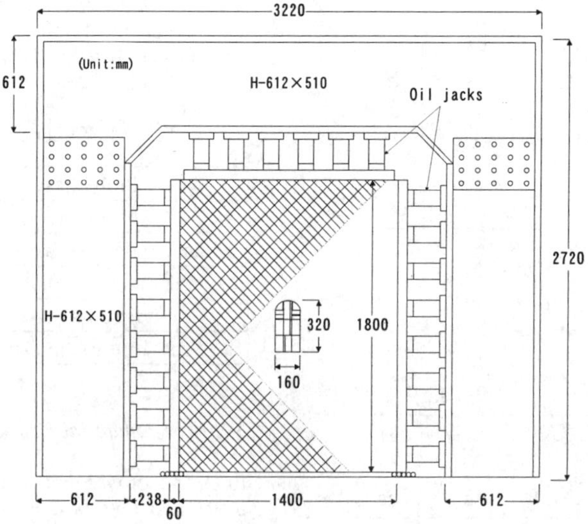

Figure 5. Biaxial testing apparatus for large scale model

3.5 *Test results*

In this study, the following parameters were investigated: joint orientation, joint spacing, in situ stresses and state of anchor models. Therefore the number of tests conducted here were sixteen.

According to the test results, the influence of joint configuration and tension force by anchoring on the behaviour of opening excavated in a jointed rock mass model was distinct. Test results and interpretation had already been described in detail (Nagai et al., 1996).

4 NUMERICAL SIMULATION

4.1 *Description of the model*

The finite element mesh, shown in Figure 6, was composed of 340 elements and 361 nodal points. The system was calculated as a plane strain problem by means of the nonlinear finite element code using Equivalent Joint Compliance Model with Stress Dependent Properties (EJOCS, see equations (5),(8) and (9)).

Table 2. Parameters of joint stiffnesses

Parameters	Value
Normal stiffness	110 GN / m^3
Shear stiffness number	2.00×10^5
Shear stiffness index	0.98

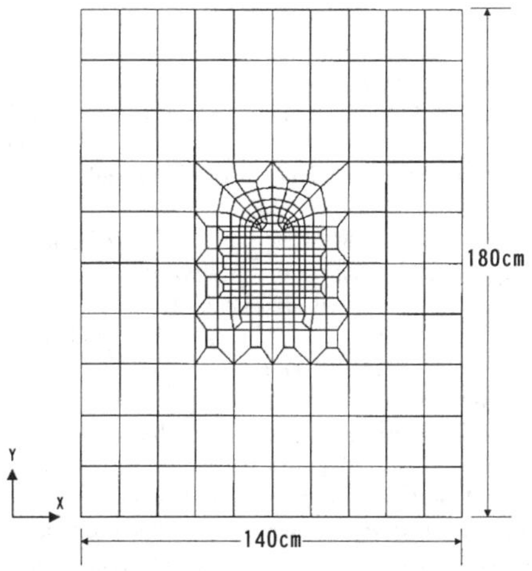

Figure 6. Finite element mesh for simulation

The same parameters as shown in Table 1 were used for a substitute rock material. The parameters for normal and shear stiffnesses of a substitute joint were obtained as indicated in Table 2 by means of a joint shear test using model blocks.

4.2 *Modelling of reinforcement effect of anchors*

In order to represent the reinforcement effect of rock bolts and/or rock anchors on jointed rock masses, the authors proposed EJOCS (Nagai, 1992). In the case of rock anchors, their effects were considered as the improvement of joint stiffness by pretensioned and additional stresses of rock anchors. The latter were calculated by the deformation occurred along rock anchors. Then improved joint stiffnesses were taken into EJOCS.

4.3 *Numerical results*

One of the typical test and numerical results in which anchor models were used is shown in Figure 7. The angle and spacing of the first joint set were 45° and 4cm, respectively. Those of the second joint set were 135° and 4cm, respectively. The vertical and horizontal pressures applied in this case were 0.1 and 0.5MPa, respectively. The anchor models were tensioned so that the equivalent supporting pressure equaled to be 0.01MPa. The displacement vectors in the numerical model as indicated in Figure 7 (b) have a good agreement with those in the physical model as indicated in Figure 7 (a).

The maximum displacements without any anchor models using elastic and proposed nonlinear analysis were 0.113mm and 1.11mm. The maximum displacements with anchor models were 0.108mm and 0.345mm, respectively. It is clear that the maximum displacement was greatly confined by means of anchor models. The same phenomenon was observed in the physical model test.

5 CONCLUSIONS

In this study, the authors proposed a mechanical model to represent the deformational behaviour of a jointed rock mass, and formulated a constitutive equation which is called EJOCS. In order to verify its applicability, the physical model tests were conducted and those results were simulated by means of the proposed method.

Some simulation analyses with the results of field measurement in underground opening are carrying out for engineering applications now using proposed method.

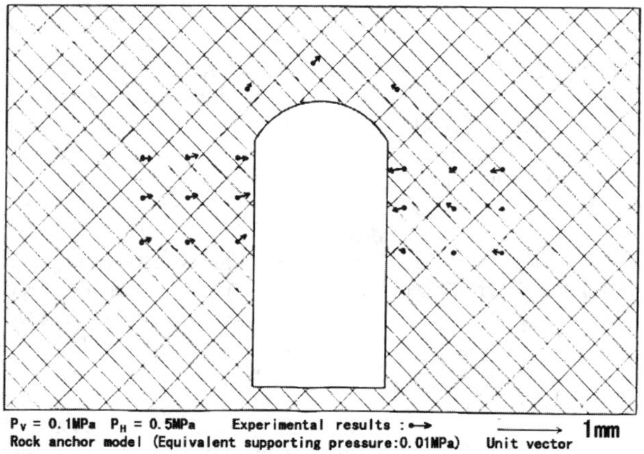

(a) test results with anchor models

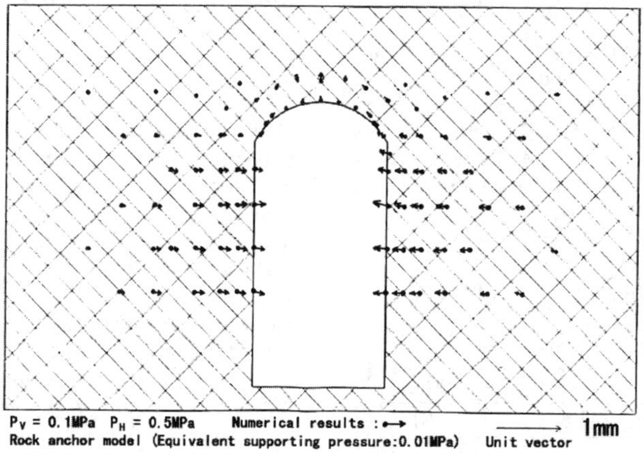

(b) numerical results with anchor models

Figure 7. Displacement vectors in physical and numerical models

6 ACKNOWLEDGEMENTS

The authors wish to thank the staff of the Institute of Rock and Soil Mechanics, Chinese Academy of Sciences. In particular, the enthusiastic cooperation of Prof. W. Zhu, Mr. W. Ren and Y. Zhang during tests is greatly appreciated.

7 REFERENCES

Wardle, L.J. & Gerrard, C.M. 1972. The "Equivalent" anisotropic properties of layered rock and soil masses. *Rock Mechanics*. 4: 155-175.

Chappell, B.A. 1986. Stress distribution in anisotropic compliance of jointed rock, *J. Geotech. Engrg*. ASCE. 112 (7): 682-700.

Kulhawy, F.H. 1978. Geomechanical model for rock foundation settlement, *J. Geotech. Engrg*. ASCE. 104 (GT2): 211-227.

Amadei, B. & Goodman, R.E. 1981. A 3-D constitutive relation for fractured rock masses. *Proc. Int. Sympo. on the Mechanical Behaviour of Structured Media*. Ottawa. Canada. Part.B: 249-268.

Oda, M., Suzuki, K. & Maeshibu, T. 1984. Elastic compliance for rocklike materials with random cracks. *Soils and Foundations*. 24 (3): 27-40.

Wei, Z.Q. & Hudson, J.A. 1986. Moduli of jointed rock masses. *Proc. Int. Sympo. Large Rock Caverns*. Helsinki: 1073-1086.

Nagai, T. 1992. An investigation into the mechanical behaviour of a jointed rock mass reinforced with rock bolts. Dissertation of Ph.D. Kobe University.

Nagai, T. et al. 1996. Behaviour of jointed rock masses around an underground opening under excavation using large-scale physical model tests. *Proc. Japanese 27th Sympo. on Rock Mechanics*: 116-120.

Computer Methods and Advances in Geomechanics, Yuan (ed.) © 1997 Balkema, Rotterdam, ISBN 90 5410 904 1

An inquiry of the mechanism about the stress-strain relationship of sand depending on the stress path

Li Boqiao
Zhuhai Shanzhao District Construction Commission, China

ABSTRACT: Some experimental information of sand follow a number of stress paths is presented at first. It indicates the stress–strain relationship of sand is influenced significantly by the stress path. In addition, some further information which is obtained from specially designed stress path tests is presented. It shows that the effects of stress path on the stress–strain relationship of sand can be divided into two parts. One is connected with the applied sequence of the different stress components .Another is connected with the fabric change of sand caused by the stress path. Finally, the microfabric of sand is analysed by dividing the contacts of any two adjacent particles into four different possible types according to the effect of the contacts on the very begining movement of the particles. The mechanism about the stress–strain relationship of sand depending on the stress path is inquiried further by analysing the effect of the constituent of the contacts on the behaviour of sand and how the stress path influence the constituent of the contacts.

1 INTRODUCTION

Since sand is an assemblage of particles, the forces applied to sand produce relative displacement between particles, the fabric of the sand is changed, so does the stiffness of the sand. The deformation of the sand is affected in turn. Assuming an element of sand is changed from one state of stress to another. If the stress paths are different, the displacement between particles will be different, and so are the fabric and the deformation of the sand. Some experimental information of sand follow a number of different stress paths is presented at first in this paper. The result shows that the strain is quite different under two equal stress conditions if the stress paths are different, and the directions of the plastic strain increment are different also. It indicates the stress path significantly influence the plastic potential function. A number of profitable studies about the stress–path dependent behavior of cohesionless soil have been made. (1)(2)(3)(4)

The stress paths in engineering practice are complex. Therefore, it is necessary for a constitutive model to account for the effect of the stress path. In addition, some further information which is obtained from specially designed stress path tests is presented. It indicates that the effect of stress path on the stress–strain relationship of sand can be divided into two parts. One is connected with the applied sequence of the different stress components. This is the direct effect of the stress path on the stress–strain relationship. The another is connected with the fabric change of sand caused by the stress path. This is the effect of the anisotropic behaviour of sand caused by the histories of stress path. Finally,the microfabric of sand is analysed by dividing the contacts of any two adjacent particles into four different possible types according to the effect of the contacts on the very begining movement of the particles. The mechanism about the stress–strain relationship of sand depending on the stress path is inquiried further by analysing the effect of the constituent of the contacts on the behaviour of sand and how the stress path influence the constituent of the contacts. The purpose of this paper not only is the inquiry of the mechanism about the stress–strain relationship of sand depending on the stress path but also is to do some fundamental work to develop an universal constitutive model for soil which is available for considering the effect of variable complex stress paths.

2 THE EFFECT OF THE STRESS PATH ON THE STRESS–STRAIN OF SAND

2.1 The experiments of the stress paths

2.1.1 Test materials

Table 1. Materials and physical index

sand	G	d_{10}	C_u	e_{max}	e_{min}
Xiao Langdi	2.68	0.16	3.56	0.68	0.36
Pingtan	2.65	0.33	1.61	0.88	0.61

in Table 1
G—Specific gravity
d_{10}—Effective diameter
c_u—Coefficient of uniformity
e_{max}—Maximum void ratio
e_{min}—Minimum void ratio
The particle size distribution curves are shown in Fig.1.

2.1.2 Test equipment

Conventional triaxial apparatus

2.1.3 Test method

Drained shear tests of saturated sand were carried out. The height and the diameter of the specimen were 15cm and 6.18cm respectively. The stress paths in test are shown in Fig.2 and Fig.3. The relative density of sand is 0.70.

2.2 Test results

2.2.1 The effect of stress path on the strain.

Two different strains are shown in Table 2 under the same stress conditions from different stress paths (Fig.2).

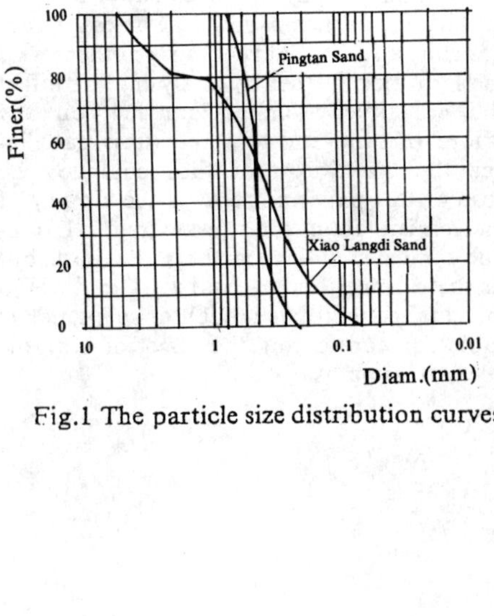

Fig.1 The particle size distribution curves

Table 2

Point	Stress(MP_a)		Stress Path	ε_{oct}(%)	γ_{oct}(%)
	P	$\sigma_1-\sigma_3$			
A	0.166	0.200	$\sigma_1=C, \sigma_3\downarrow$	0.68	1.36
			$\sigma_1\uparrow, \sigma_3=C$	0.52	1.65
B	0.300	0.300	$\sigma_1=C, \sigma_3\downarrow$	0.99	1.12
			$P=C$	0.77	1.69
C	0.367	0.200	$\sigma_1=C, \sigma_3\downarrow$	0.99	0.48
			$\sigma_1\uparrow, \sigma_3=C$	0.96	0.80
D	0.500	0.600	$P=C$	1.30	3.13
			$\sigma_1\uparrow, \sigma_3=C$	1.03	3.30

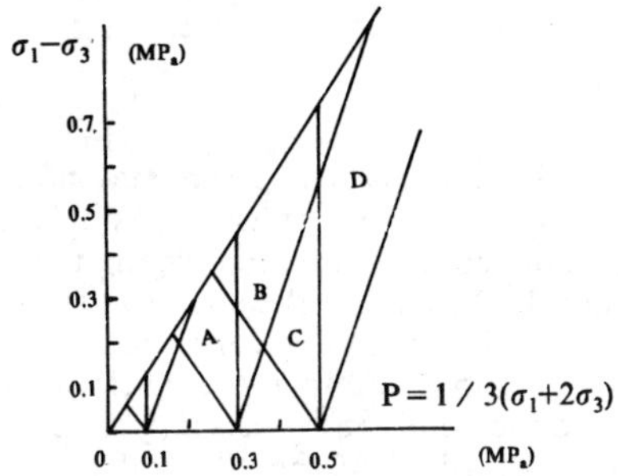

Fig.2 The stress paths in test for Xiao Langdi sand

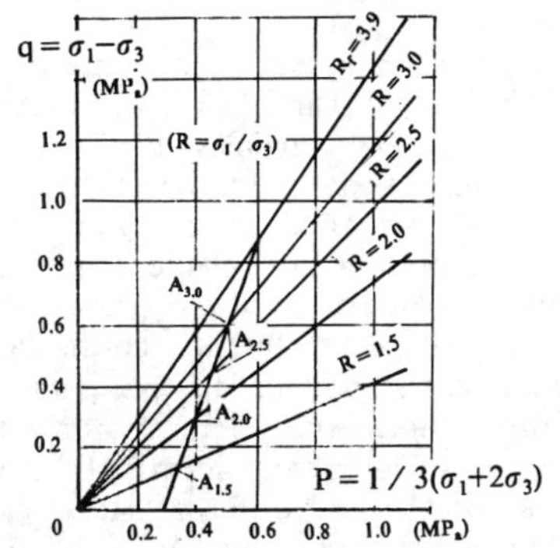

Fig.3 The stress paths in test for Pingtan sand

2.2.2 The effect of stress path on the plastic potential

Two different directional angles of plastic strain increment are shown in Table 3 under the same stress conditions from different stress paths (Fig.3).

Table 3

Point	Stress path	$\triangle\gamma^p$(%)	$\triangle\varepsilon_v$(%)	$\triangle\varepsilon_v^p$(%)	α
$A_{1.5}$	q/p=C	0.037	0.210	0.300	7°
	σ_3=c	0.308	0.270	0.350	41°20′
$A_{2.0}$	q/p=c	0.081	0.180	0.270	16°40′
	σ_3=c	0.429	0.330	0.394	47°28′
$A_{2.5}$	q/p=c	0.132	0.190	0.280	25°12′
	σ_3=c	0.404	0.200	0.241	49°26′
$A_{3.0}$	q/p=c	0.249	0.190	0.280	41°40′
	σ_3=c	0.944	0.230	0.281	73°24′

in Table 3

$\triangle\gamma^p$—Increments of plastic shear strain

$\triangle\varepsilon_v$—Increments of volumetric strain

$\triangle\varepsilon_v^p$—Increments of plastic volumentric strain

α—Directtional angles of plastic strain increment

$\alpha = tg^{-1}(\triangle\gamma^p / \triangle\varepsilon_v^p)$

3 THE COMPONENTS OF THE EFFECT OF STRESS PATH ON THE STRESS-STRAIN RELATIONSHIP

Three special stress paths designed for investigating the components of the effect of the stress path on the stress-strain relationship are shown in Fig.4. The different strains fron different stress path are shown in Fig.5 and Table 4.

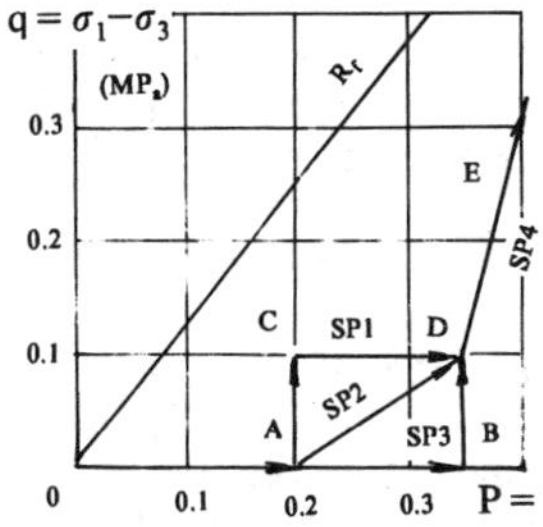

Fig.4 Three special stress paths (Pingtan sand)

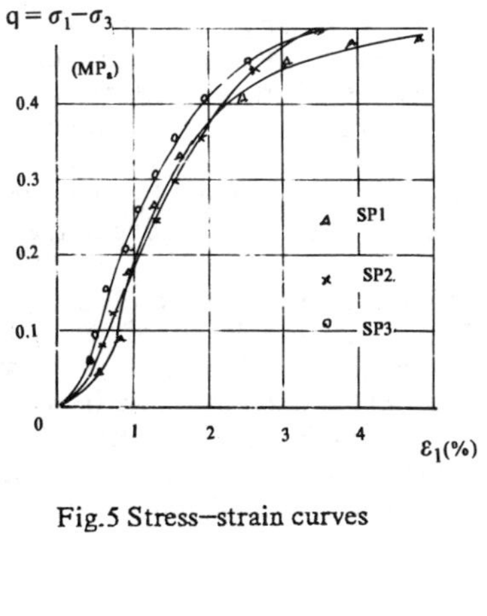

Fig.5 Stress-strain curves

Table 4

Point	stress (MP$_a$) p	stress (MP$_a$) q	stress path		ε_1(%)	$\triangle\varepsilon_1$(%)
A	0.200	0.000	SP1		0	0
			SP2		0	0
			SP3		0	0
D	0.350	0.100	SP1		0.41	0.41
			SP2		0.35	0.35
			SP3		0.25	0.25
E	0.370	0.175	SP4	from sp1	0.49	0.08
				from sp2	0.49	0.14
				from sp3	0.44	0.09

3.1 The effect of the applied sequence of the different stress components

If the applied sequence of the different stress component ($\triangle$q and $\triangle$p) is variant (such as in sp1, sp2, and sp3), the strain ε_1 will be variant (Fig.4, Fig. 5, and Tab.4). The stress component $\triangle$q(AC)is applied before $\triangle$p(CD)in sp1. But the stress component $\triangle$q(BD)is applied after $\triangle$p(AB)in sp3. Since the average stress level in sp1 is higher than that in sp3, the strain ε_1 in sp1 is larger than that in sp3. The average stress level in sp2 lies between sp1 and sp3, so the strain ε_1 in sp2 lies between sp1 and sp3.

The average sfress level is a important factor of the effect of the stress path on the stress-strain relationship of sand.

3.2 The effect of the fabric

Though the stress condition and stress paths are the same in sp4(DE), the strain increments $\triangle\varepsilon_1$ are different if their histories of stress paths are not equal (Tab.4, Fig.4, and Fig.5). The reason is the fabric of sand were changed by the effect of the histories of the stress path. The term fabric refers to the microstracture of the granular mass, namely the relative arrangment of the particles within the overall assembly. [9]

4 INQUIRY OF THE MECHANISM ABOUT THE STRESS-STRAIN RELATIONSHIP OF SAND DEPENDING ON THE STRESS PATH

Sand is an assemblage of discrete particles. Its mechanical behaviour is depend on the interraction between particles. Terzaghi proposed the following statement in 1920, "The way out of the difficulty lies in dropping the old fundamental principles and starting again from the elementary fact that the sand consists of individual grains." [5] A lot of useful researches have been made based on the consideration of the forces between particles. [6][7][8] Stress in sand is transmited through the contacts between adjacent particles, and its deformation is produced by the movement of the contacts also. So the properties of the contacts take important effect on the stress–strain relationship of sand. The contacts can be divided into four different possible types for any two adjacent particles according to the effect of the contacts on the very begining movement of particles(Fig.6)

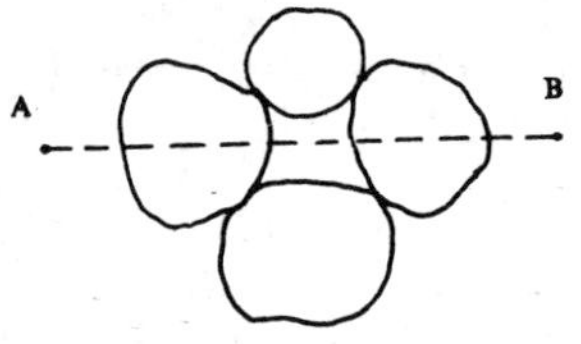

a. The noncontact

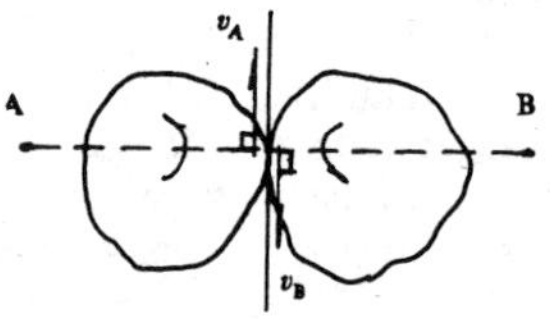

b. The normal contact

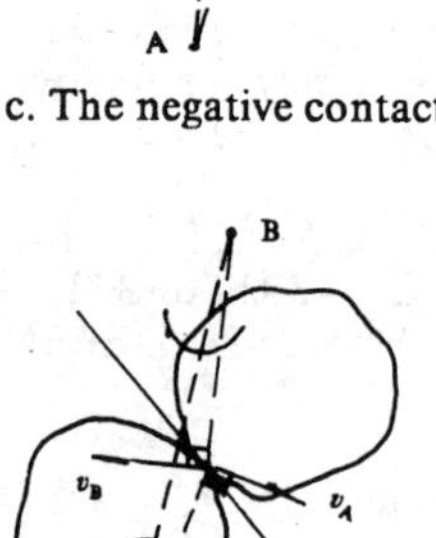

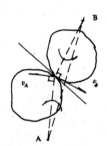

c. The negative contact

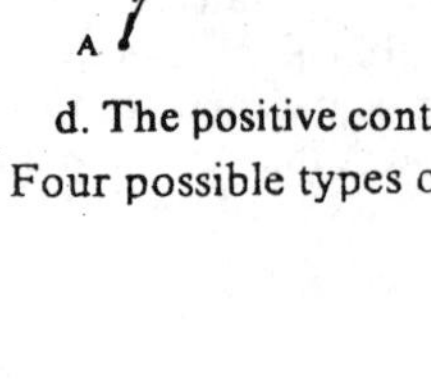

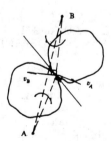

d. The positive contact

Fig.6 Four possible types of contacts

4.1 The types of contacts

4.1.1 The noncontact

Two adjacent particles under investigation are separated by other around particles. There is not direct effect to each other when they move(Fig.6, a).

4.1.2 The normal contact

The linking line of the instantaneous turning centres A,B of two adjacent particles passes through the contact point and is normal to the tangent line which passes through the contact point. There is a certain frictional resistance, but it does not increase the normal pressure between this two particles when they move(Fig.6, b).

4.1.3 The negative contact

The linking line of the instantaneous turning centres A,B of the two adjacent particles does not pass through the contact point or is not normal to the tangent line at that point. The two particles tend to departe from each other and the contact pressure will be decreased if the moving directions of the two particles are opposite at the contact point and the velocity vector of one particle does not direct to the interior of the another when they move(Fig.6, c).

4.1.4 The positive contact

The linking line of the instantaneous turning centres A,B of the two adjacent particles also does not pass through the contact or is not normal to the tangent line at the contact point. The normal pressure and the frictional resistance between the two particles tend to getting more and more increase so that there will be a tendency to resist the further move of the two particles if the turning directions of the two particles are opposite at the contact point and the velocity vector of one particle directs to the interior of the another when they move(Fig.6, d).

4.2 The different actions of the normal stress and the shear stress to the move of the particles

Sample will be contracted in all directions, and the relative slide and the nondirectional turn between particles will be taken place when the sample is only applied by the isotropical normal stress. This will make the average pressure at the contact points increase, the porosity of the sample decrease, and the number of the contact points between particles increase. But the proportion of the four types of contact will not be changed obviously.

The shear stress will force most of the particles to slide and turn along the shear strain direction when the shear stress is applied on the sample. As the positive contacts can produce a larger frictional resistance against the particle move, they will become more steady contacts. The sliding and turning for the other three types of the contacts are relatively easy, so they tend to become the positive contacts, then the proportion of the positive contacts will be increased. Furthermore, the turning of the particle at the positive contacts caused by shear stress will make the pressure on the positive contacts increased further. This will make the resistance against the shear strain increase until they could balance the external forces.

The main effective factors on the sfress–strain relationship of sand are the summation, the proportion of the four types of contacts and the average value of the pressure at the contacts though the types of contacts may be varying during the deformaton of the sampls. These effective factors are affected by the stress paths. The different stress paths can be taken as a combination of a series of different normal stress and shear stress increments. Different combination of stress will correspond to different summation and proportion of the four types of contacts and the value of the pressure at the contacts so that they will affect the stress–strain relationship of sand.

4.3 The dilatation and the contraction

4.3.1 The dilatation

When a dense sand is sheared, the summation and proportion of the positive contacts will increased, and the dilatation will appear. As shear strain increases, the relative sliding and turning of particles will make the pressure on the positive contacts increase. And most particles have a tendency to depart their neighbor particles. If the external force can not resist the tendency, the void ratio of dense sand will be increased, and the dilatation will take place. If the external force is large enorgh to resist the tendency, a number of the corners of particles at the positive contacts will be crushed during sliding and turning of particles, the dilatation will not occur.

4.3.2 The contraction

When a loose sand is sheared, most of the normal contacts will be turned into the negative contacts, the summation and proportion of the negative contacts will increase. As shear strain increases, the relative sliding and turning of particles will make the pressure on the negative contacts decrease. And most particles have a tendency to near their neighbor particles, the void ratio of the loose sand will decrease, and the contraction will appear.

4.4 The peak strength and the residual strength

4.4.1 The peak strength

When the shear of the dense sand is continued, the summation and proportion of the positive contacts will increase continuously. The positive contacts will be changed into the normal contacts, the average pressure on the contacts will increase gradually. When the summation and proportion of the positive contacts reach to maximum, the shear stress is equal to the peak strength.

4.4.2 The residual strength

After the peak strength, the normal contacts will change into negative contacts step by step, and the average pressure on the contacts reduces gradually. Finally, the summation and proportion of the four types of contacts will maintain a dynamic equilibrium, and the void ratio will not change, the shear stress is equal to the residual strength.

5 CONCLUSIONS

1.The experimental information in this paper shows that the stress–strain relationship of the saturated sand under drained triaxial shear tests is affected by the stress path obviously.

2.The effect of stress path on the stress–strain relationship of sand can be divided into two parts. One is connected with the applied sequence of the different stress component. Another is connected with the fabric change of sand caused by the stress path.

3.The contacts of any two adjacent sand particles may be divided into four different possible types according to the effect of the contacts on the very begining movement of the particles.

4.The summation and the proportion of the four types of contacts can reflect the behaviour of the fabric of sand.

5.The assumption of the four types of contacts can explain the effect of the stress path on the stress–strain relationship, the dilatation, the contraction,the peak strength and the residual strength reasonably.

REFERENCES

〔1〕 Lade, P. V. and Duncan, J. M. Stress–path dependent behavior of cohesionless soil. ASCE, Vol. 102, No. GT1, 1976.

〔2〕 Lambrechts, J. R. and Leonards, G. A. Effects of stress history on deformation of sand. ASCE, Vol. 104, No. GT11, 1978.

〔3〕 Boqiao, L. The influence of stress path on the stress–strain relationship of sand and its application to the calculation of earth dam design. Journal of the Wuhan Institute of Hydraulic and Electric Engineering, Vol. 44, NO. 3, 1982.

〔4〕 Varadarajan, A., Sharma, K. G. Mishra, S. S. and Kuberan, R. Effect of stress–path on stress–strain–volume change behaviour of Jamuna sand. Proc. Inter. Conf. on Constitutive Laws for Engineering Materials. Edited by C. S. Desai, 1983.

〔5〕 Terzaghi, K. Old earth–pressure theories and new test results. Engineering News–Record, 85, No. 14, 1920. (1960 Reprinted in From theory to practice in soil mechanics. New York, J. Wiley and Sons.)

〔6〕 Rowe, P. W. The stress–dilatancy relation for static equilibrium of an assembly of particles in contact. Proc. Roy. Soc. A269, 500–527, 1962.

〔7〕 Matsuoka, H. A microscopic study on shear mechanism of granular materials. Soils and Foundations, Vol. 14, No. 1, 1974.

〔8〕 Oda, M., Konishi, J., and Nemat–Nasser, S. Experimental micromechanical evaluation of the strength of granular materials: effects of particle rolling. Mechanics of Granular Materials: New Models and Constitutive Relations, Edited by J. T. Jenkins and M. Satake, 1983.

〔9〕 Nemat–Nasser, S. and Mehrabadi, M. Stress and fabric in granular masses.Mechanics of Granular Materials: New Models and Constitutive Relations, Edited by J. T. Jenkins and M. Satake, 1983.

Computer Methods and Advances in Geomechanics, Yuan (ed.) © 1997 Balkema, Rotterdam, ISBN 90 5410 904 1

Thermodynamic foundation of endochronic theory with application to description of creep behavior

Chunhe Yang & J.J.K. Daemen
Mining Engineering Department, University of Nevada, Reno, Nev., USA

ABSTRACT: Similar to the endochronic plastic theory proposed and modified by Valanis (1971) a time dependent endochronic constitutive model incorporating temperature effects is suggested . A new definition of inner variable is proposed in order to include a rational representation of temperature effects on creep behavior.

1. INTRODUCTION

The endochronic theory of viscoplasticity first proposed and modified by Valanis (1971) has interested many material constitutive theory researchers (e. g. Wu and Yip (1980), Lee (1996)). The theory expresses the thermomechanical history of materials through a time scale—the intrinsic time scale z. The intrinsic time scale is material dependent through a material function , which is defined by:

$$dz = d\xi / f(\xi) \quad \text{and} \quad d\xi^2 = \alpha^2 d\varsigma^2 + \beta^2 dt^2 \qquad (1)$$

where dz is intrinsic time scale; $d\xi$ is the intrinsic time measure; f(ξ) is a material function which represents the softening and hardening behavior of the material; $d\varsigma^2 = p_{ijkl}\varepsilon_{ij}\varepsilon_{kl}$; p_{ijkl}, α and β are material parameters.

Much work concentrated on finding the definition of dz for modeling different material properties. Wu and Yip (1980) proposed a definition of dz in order to consider dynamic properties under high strain rate. Lee (1996) suggested a simple endochronic transient creep model using :

$$d\varsigma^2 = P_{ijkl} d\varepsilon^c_{ij} d\varepsilon^c_{kl} \qquad (2)$$

where P_{ijkl} is a fourth order material constant tensor, ε^c_{ij} is creep strain. Detailed introduction to and discussion of the definition of the intrinsic time measurement can be seen in Lee (1996) and Yang and Li (1986).

A new viscoelastic constitutive formulation incorporating time dependent and temperature effects is introduced based on the principle of internal variable (Lee,1996).

2. THE VISCOELASTIC CONSTITUTIVE THEORETICAL FOUNDATION AND ANALYSIS

According to the principle of irreversible thermodynamics, the local thermodynamic state can be determined by strain tensor (ε_{ij}), temperature

(T) and entropy per unit volume (η), and a set of internal variables (q_i). The internal variable may be a component of second-rank tensors (for example, dislocation loop density) or scalars (grain size, hardening and softening).

Since the internal energy is a function of the thermodynamic state, the Helmholtz free energy function can be written as :

$$\varphi = \varphi(\varepsilon_{ij}, \dot{\varepsilon}_{ij}, T, T_{\bullet j}, q^n_{ij}, D) \qquad (3)$$

where $\dot{\varepsilon}_{ij}$ is strain rate tensor, $T_{\bullet j}$ is temperature gradient in space, D is damage of material.

According to the second law of thermodynamics (Lubliner, 1971)

$$\dot{T}\eta - \dot{\varphi} + \sigma_{ij}\dot{\varepsilon}_{ij} - hT_{\bullet i} \geq 0 \qquad (4)$$

where h is the heat flow vector.

Expanding equation (3), we have:

$$\dot{\varphi} = \frac{\partial\varphi}{\partial\varepsilon_{ij}}\dot{\varepsilon}_{ij} + \frac{\partial\varphi}{\partial\dot{\varepsilon}_{ij}}\ddot{\varepsilon}_{ij} + \frac{\partial\varphi}{\partial T}\dot{T} + \frac{\partial\varphi}{\partial T_{\bullet i}}\dot{T}_{\bullet i} + \frac{\partial\varphi}{\partial q^n_{ij}}\dot{q}^n_{ij} \qquad (5)$$

The evolution equation of internal variable given by many authors, e. g. (Lubliner(1971) and Valanis (1971)) is:

$$\dot{q}_{ij} = f(q_{ij}, \varepsilon_{ij}, T, T_{\bullet j}) \qquad (6)$$

That means that the internal variable depends on the current thermodynamic state and is independent of the rate of that internal variable.

Combining equations (4) and (5) yields

$$(-\eta - \frac{\partial\varphi}{\partial T})\dot{T} - \frac{\partial\varphi}{\partial\dot{T}}\ddot{T} + (\sigma_{ij} - \frac{\partial\varphi}{\partial\varepsilon_{ij}})\dot{\varepsilon}_{ij} - \frac{\partial\varphi}{\partial\dot{\varepsilon}_{ij}}\ddot{\varepsilon}_{ij} - \frac{\partial\varphi}{\partial q_{ij}}\dot{q}_{ij} - \frac{\partial\varphi}{\partial q_{\bullet i}}\dot{q}_{\bullet i} - hT_{\bullet j} \geq 0 \qquad (7)$$

Since, however, $\dot{T}$, $\dot{\varepsilon}_{ij}$, $\ddot{\varepsilon}_{ij}$ and $\ddot{T}$ can be specified arbitrarily thermodynamic state, their coefficients must vanish. The Colemen relationships can thus be obtained from equation (7):

$$\eta = -\frac{\partial\varphi}{\partial T}, \quad \sigma_{ij} = \frac{\partial\varphi}{\partial\varepsilon_{ij}}, \quad \frac{\partial\varphi}{\partial\dot{T}} = 0 \qquad (8)$$

$$\frac{\partial\varphi}{\partial\dot{\varepsilon}_{ij}} = 0, \qquad -h * T_{\bullet j} - \frac{\partial\varphi}{\partial q^n_{ij}}\dot{q}_{ij} \geq 0$$

From the above equation, we know that the free energy is independent of strain rate tensor $\dot{\varepsilon}_{ij}$, temperature rate $\dot{T}$ and gradient $T_{\bullet j}$. Based on rational mechanics principle, the form of stress tensor is same as free energy. So, we reason that stress tensor the function is independent of strain rate and temperature rate if the assumption of equation (6) is correct. However, it is a fact that the stress tensor depends strongly on strain rate for most materials. It is reasonable that we assume $\dot{q}_{ij}$ also to depend on $\dot{\varepsilon}_{ij}$ and $\dot{T}$. In this condition, the Coleman relationship does not follow (Lubliner, 1971). In general, we suppose that:

$$\dot{q}_{ij} = f_{ij} + x_{ijkl} * \dot{\varepsilon}_{kl} + \beta\dot{T}\delta_{ij} \qquad (9)$$

From equation (7) and (9), we obtain a group of modified Coleman relationships similar to those suggested by Lubliner (1971):

$$\eta = -\frac{\partial\varphi}{\partial T} - \beta\frac{\partial\varphi}{\partial q^n_{ij}} \qquad (10)$$

$$\sigma_{ij} = \frac{\partial\varphi}{\partial\varepsilon_{ij}} + x_{ijkl}\frac{\partial\varphi}{\partial q_{kl}} \qquad (11)$$

where: x_{ijkl} is a material constant tensor which represents time(rate) – dependent properties.

According to the evolution equation of internal variable proposed and theoretically verified by Valanis (1971):

$$\frac{\partial \varphi}{\partial q_{ij}^n} + b_{ijkl}\frac{\partial q_{kl}^n}{\partial z} = 0 \quad \text{(n not summed)} \tag{12}$$

Under small deformation and low temperature change, the Helmholtz free energy (equation (5)) can be expanded as follows:

$$\varphi_D = \frac{1}{2}E_1 e_{ij} e_{ij} + \sum_{n=1}^{n} F_1 e_{ij} q_{ij}^n + \frac{1}{2} H_1 q_{ij} q_{ij} \tag{13}$$

$$\varphi_H = \frac{1}{2}E_2 \varepsilon_{kk}\varepsilon_{kk} + \sum_{n=1}^{n} F_2 \varepsilon_{kk} q_{kk} + \sum_{n=1}^{n} H_2 q_{kk} q_{kk} + D_2 T\varepsilon_{kk} + E_2 T q_{kk} + \frac{1}{2} GTT \tag{14}$$

where: $e_{ij} = \varepsilon_{ij} - \frac{1}{3}\varepsilon_{kk}$, $E_1, F_1, H_1, E_2, F_2, H_2 D_{2,} E_2$, and G are material constant parameters. Here, we assume that temperature affects only the spherical state stress tensor and does not act on the deviator tensor.

The evolution equation of internal variable is given by (Yang and Li (1986)):

$$\begin{aligned} &\frac{\partial \varphi_D}{\partial q_{ij}^n} + b_1' \frac{d}{dz}(q_{ij}^n + \mu_1(t) e_{ij}) = 0 \\ &\frac{\partial \varphi_H}{\partial q_{kk}} + b_2' \frac{d}{dz}(q_{kk}^n + \mu_2(t)\varepsilon_{kk}) = 0 \end{aligned} \tag{15}$$

where b_1' and b_2' are material parameters. q_{ij}^n and q_{kk}^n are the value of the internal variables under reference state (e.g. static state). $\mu_1(t)$ and $\mu_2(t)$ are the material parameters which represent the time effects.

Based on the equations (10), (13) and (15), we obtain:

$$\begin{aligned} s_{ij} &= (E_1 + x_1(t)) e_{ij} + (F_1 + a\mu_1(t)) q_{ij} \\ \sigma_{kk} &= (E_2 + x_2(t))\varepsilon_{kk} + (F_2 + b\mu_2(t)) q_{kk} \\ &+ (D_2 + a_2 x_2(t)) T \end{aligned} \tag{16}$$

where: $s_{ij} = \sigma_{ij} - \frac{1}{3}\sigma_{kk}$; a and b are material parameters.

Solving equation (15) and (16) and transforming through Laplace integral equation, the explicit creep constitutive equation is obtained:

$$s_{ij} = \int_0^{Z_D} ((a_1 + b_1\mu_1(t)) H(Z_D - Z_D') + (c_1 + d_1\mu_1(t)) e^{-E_1(Z_d - Z_D')}) \frac{de_{ij}}{dZ_D} dZ_D' \tag{17}$$

and:

$$\sigma_{KK} = \int_0^{Z_H} (a_2 + b_2\mu_2(t)) H(Z_H - Z_H') + (c_2 + d_2\mu_2(t)) e^{-e_2(Z_H - Z_H') dZ_D'} + \int_0^{Z_H} (a_2' + b_2'\mu_1(t)) H(Z_H - Z_H') + (c_2' + d_2'\mu_2(t)) e^{-e_2(Z_d - Z_H')} \frac{dT}{dZ_H} dZ_H' \tag{18}$$

In which $dz = \alpha d\xi + \lambda dt$; $d\xi = g_{ijkl} d\varepsilon_{ij}^c d\varepsilon_{kl}^c$;

$a_1, b_1, c_1, d_1, E_1, a_2, b_2, c_2, d_2, a_2', b_2', c_2', d_2'$ and e_2' are material parameters.

So far, the three dimensional creep constitutive equations incorporating time and temperature effects are given.

In one dimension, combining the equations (17) and (18), the constitutive equation can be simplified to:

$$\sigma = \int_0^Z (((a + b\mu(t)) H(Z - Z') + (c + d\mu(t)) e^{-f(Z - Z')dZ'}) \frac{\partial \varepsilon}{\partial Z} dZ' + \int_0^Z (((a_1 + b_{1(}\mu(t)) H(Z - Z') + (c_1 + d_1\mu(t)) e^{-f_1(Z - Z')}) \frac{\partial T}{\partial Z} dZ' \tag{19}$$

Equation (19) can be inverted by Laplace transformation to obtain:

$$\varepsilon^c = \int_0^t (c_2 + d_2 x_2(t)) e^{-f_2(Z - Z')dZ} (\frac{\partial \sigma}{\partial Z} + \beta \frac{\partial T}{\partial Z}) dZ' \tag{20}$$

where: ε^c is creep strain ($\varepsilon^c = \varepsilon - \varepsilon^e$) and ε^e is initial strain of material that represents elastic strain.

This term (ε^c) represents time—dependent strain e.g creep strain and time independent strain e. g. plastic strain. In fact, it is very difficult to distinguish the creep strain and plastic strain when we carry out high pressure creep tests. However, it is not necessary to distinguish between them. We can neglect plastic strain when brittle materials are considered.

So far, a one-dimensional creep constitutive equation is obtained. Equation (20) is very similar to the traditional constitutive equation obtained by Flugge (1975) and Cristensen (1982) only when we select dZ=dt.

3. CONCLUSION

A simple endochronic model of creep incorporating temperature effects is proposed. A new group of Colemen relations is given with a rational definition of a new internal variable evolution equation. Because of the limit of length of paper required by the conference, the detailed theoretical analysis and verification of the creep experiment will be given in further published paper by authors.

REFERENCE

Christensen, R. M (1982). Theory of Viscoelasticity, New York: Academic Press.

Flugge W.(1975). Viscoelasticity. 2nd Edition . New York: Springer-Verlag.

Lee. C. F.(1996). A Simple Endochronic Transient Creep Model of Metal with Applications to Variable Temperature Creep. *International Journal of Plasticity*, 12:229-253.

Lubliner (1971). On the Thermodynamic Foundations of Non-linear Solid Mechanics. *Int. J. Non-linear Mechanics.* 7: 237-254.

Valanis, K. C (1971). A Theory of Viscoelasticity Without a Yield Surface. Part I, General Theory, *Arch. Mech.* 23: 517-533.

Valanis, K. C.(1980). Fundamental Consequences of a New Intrinsic Time Measure, Plasticity as a Limit of the Endochronic Theory. *Arch. Mech.,*32: 171-176.

Wu, H.C and Yip. M.C (1980). Strain Rate and Strain Rate History Effects on the Dynamic Behavior of Metallic Materials. *Int. J. Solids Structure*, 16: 515-519.

Yang. C. and T. Li. (1986). A Study of Rate-dependent Inner Variable constitutive Theory of Geologic Materials. *Soil and Rock Mechanics (In Chinese)*,13: 74-80.

8 Flow and consolidation

Computer Methods and Advances in Geomechanics, Yuan (ed.) © *1997 Balkema, Rotterdam, ISBN 90 5410 904 1*

Mound profiles for a semi-coupled analysis of a lightly loaded foundation on expansive soil

Stephen Fityus & David W. Smith
Department of Civil, Surveying and Environmental Engineering, The University of Newcastle, Callaghan, N.S.W., Australia

ABSTRACT In this paper, a simplified method for estimating the final volumetric moisture content beneath a covered area is presented, and by integration of the change in moisture content in the soil beneath the covered area, a mound shape due to the cover is estimated. In summary it may be said that provided the geotechnical engineer can estimate a set of fundamental parameters such as the site geometry, appropriate simplified boundary conditions, initial conditions and soil parameters, then is possible that more realistic moisture distributions may be estimated based on flow net theory. These distributions can be used in the calculation of surface movements using relatively simple techniques. The geotechnical engineer can then systematically go about quantifying the effects of increased edge beam downturn, covered areas adjacent to the foundation and initial moisture content.

1. INTRODUCTION

State of the art modelling of moisture and stress changes beneath a slab-on-ground dwelling situated on an expansive soil is very complicated. Rigorous modelling involves a set of fully coupled field equations describing moisture and temperature distributions, non–linear deformations in non–homogeneous soils, together with complex boundary conditions, which may be time dependent. For accurate representation of specific field behaviours, it may also be necessary to take into account various types of vegetation and possibly leakages from plumbing. Clearly, the implementation of such a complete model is well beyond the scope of routine engineering design. An important requirement of any design tool is that the parameters employed in the model should be fairly easily and quickly obtained from routine geotechnical tests at modest cost.

Indeed, it is because of the difficulty in fulfilling this last requirement that the design of lightly loaded foundations often proceeds largely on the basis of past experience, by-passing engineering calculations altogether. The Australian Standard for the design of lightly loaded foundations, AS2870 (1996), incorporates several alternative design approaches, including tabulated designs partially derived from engineering analysis and modified by experience. These are widely employed throughout Australia. Less frequently employed in Australia is design by 'engineering calculation' (also accommodated by AS 2870) despite its the potential for more innovative engineering solutions. This paper is concerned with improving design by engineering calculation.

2. SIMPLIFIED DESIGN BY ENGINEERING CALCULATION.

Semicoupled analysis

The design method by engineering calculation suggested in AS2870 is a simplified two dimensional semi-coupled analysis. It proceeds by first choosing a 'mound profile', this being an estimate of the deflected shape of a weightless, flexible, impermeable membrane with the same geometric attributes as a foundation slab (including edge beam downturns). The mound profile forms as the initial soil moisture profile changes after slab construction (referred to as 'time zero' in the model) to a new (final) moisture distribution, under the influence of extreme climatic conditions (i e. prolonged wet or dry conditions). The calculation then proceeds using a coupled spring model and usually requires a computer for analysis.

The main focus of this paper is to improve the estimation of mound profiles for use in the semicoupled analysis, using easily applied, simply understood techniques. Current practice assumes a profile based only on an estimate of the maximum free surface movement of an uncovered area, with little regard for initial moisture conditions. This can be greatly improved by specifically estimating the changes in the spatial distribution of moisture due to the construction of a cover and using the estimates in the calculation of the resulting differential surface movements.

Estimation of Moisture Changes

Pullan (1990) presents a comprehensive overview of the substantial body of unsaturated flow theory which has been developed to date. Only the final equations are presented here.

Unsaturated moisture flow can be described by

$$D(\vartheta).\nabla^2\theta - \alpha.D(\vartheta).\frac{\partial\theta}{\partial z} = \frac{\partial\theta}{\partial t} \qquad (1)$$

where

- $D(\vartheta)$ is the unsaturated moisture diffusivity second order tensor $D(\vartheta) = K(\vartheta)\ d\psi/d\vartheta$
- ϑ is the volumetric moisture content
- $K(\vartheta)$ is the hydraulic conductivity function
- ψ is the matric suction
- θ is the transformed volumetric moisture content, given by $\theta = \int_0^{\vartheta} D(\vartheta)\ d\vartheta$

Equation (1) is referred to as the "*theta*-based" form of the Richards equation, being expressed only in terms of the volumetric moisture content, θ, and the soil diffusivity. Analytical solutions to the Richards equation are not known and are usually only achieved by semi–analytical (Fityus and Smith, 1994) or numerical techniques.

In the present work, a simple method of estimating changes in volumetric moisture content is proposed, based on steady state solutions of a linearised form of Richards equation.

$$D_x\frac{\partial^2\vartheta}{\partial x^2} + D_z\frac{\partial^2\vartheta}{\partial z^2} = 0 \qquad (2)$$

Equation (2) is obtained from equation (1) by assuming the soil to have a constant moisture diffusivity, and by neglecting the first order differential, which is insignificant in heavier clay soils. It is shown later that by employing typical moisture diffusivities backfigured from monitoring sites in expansive soils, and assuming reasonable durations of extreme climatic conditions (eg. prolonged drought), the final moisture condition beneath a flexible covered area may be reasonably approximated by solution of the steady-state unsaturated moisture equation.

Equation (2) is now a simple elliptic equation, analogous to that of multidimensional saturated flow. Thus, assuming an homogeneous soil profile, the steady-state moisture condition can now be estimated by the use of flow net theory. This approach eliminates the need for elaborate computer software, as more familiar graphical solution techniques are now appropriate. These can easily accommodate specific site features such as anisotropic moisture diffusivity, edge beam downturns, adjacent paving and varying soil depth beneath the slab.

Estimation of mound profiles

At the present time in Australian practice, one of several simple mound profile shapes may be assumed. The mound profile of Lytton (1970) describes a mound of the form

$$z_o(x) = cx^m \qquad (3)$$

where

- c is a fitted constant to give the calculated maximum displacement, z_{max}, at the appropriate position beneath the slab
- m is the mound index, taken as an integer exponent.

Mitchell (1980), proposed a similar expression based on integrated soil suction solutions obtained analytically. The Mitchell mound profile accommodates non integer mound exponents and is given explicitly in terms of the geometrical parameters L, the slab width; H, the depth of the soil layer and z_{max}, the maximum differential soil displacement across the covered areaIn earlier work (1980) the Mitchell mound exponent, m, is given by $m = 0.75L/H$, while in later work (AS2870–1996) it was modified to take account of experience with observed slab moments, giving $m = 1.5\frac{L}{(H/7 + z_{max}/25)}$ (for a slab without downturns).

We propose to estimate the shape of the mound profile by evaluating the vertical strains which result from spatially varying moisture changes estimated from the solution of equation (2) and the initial condition. The proposed simplified analyses estimate surface movements by integration of the moisture changes over the depth of the layer at various surface locations.

There are many methods for estimating surface movements due to soil expansion. Here we will compare two, each of which is ideally suited to spreadsheet calculations. The first is the calculation suggested by Richards (1967) ;

$$z_o(x) \approx \frac{1}{3}\sum_{i=1}^{N}\frac{G_s\Delta\omega_i}{1 + e_{oi}}\Delta h_i \qquad (4)$$

where,

- $\Delta\omega_i$ is the (average) change in gravimetric moisture content in layer i,

e_{oi} is the (average) initial void ratio in layer i
G_s is the specific gravity of the soil particles
Δh_i is the thickness of the ith layer in a total of N layers.

The factor of one third in equation (4) accounts for no lateral restraint.

The second approach is a more conceptually rigorous method, and is an adaptation of the soil volume change theories of Fredlund and Rahardjo (1993), where one dimensional volume changes are related to changes in the stress state of the unsaturated soil. The proposed adaptation is to confine the analysis of each sublayer to a plane of constant net normal stress. The parameters for such an analysis can be obtained from a simplified oedometer test in which an unsaturated sample is inundated and allowed to swell under a constant load. The proposed volume change expression is

$$z_o = \frac{1}{3} \sum_{i=1}^{N} \frac{C_{sh,i}}{1+e_{oi}} \log \frac{P_{fi}}{P_{oi}} \Delta h_i \qquad (5)$$

where

$C_{sh,i}$ is the slope of a void ratio–log(matrix suction) plot for layer i, from a simple oedometer test under a constant load of σ_{vi}
e_{oi} is the initial void ratio of layer i
P_o, P_f are the initial and final stress states in the soil (at a point), equal to the sum of the net normal stress and the matrix suctions,viz, $= (\sigma_{vi}-u_a) + (u_a-u_{wo}), (\sigma_{vi}-u_a) + (u_a-u_{wf})$
u_a is the pore air pressure
u_w is the total soil suction

To use this method requires volumetric moisture contents to be converted to suctions by means of the soil moisture characteristic. The mound profiles predicted by the various methods are compared in a following section.

3. APPLICATION OF THE APPROACH.

The approach outlined above will be illustrated by considering edge shrinkage mounds beneath a 10m wide slab on a 2m deep layer of Maryland clay. The typical attributes of Maryland clay are shown in Figure 1.

The 'active zone' of 2m, approximates the depth at which volumetric moisture contents are invariant, with a value of 0.37 being typical at Maryland. For the results presented, initial surface moistures of 0.37 (dry), 0.44 (typical) and 0.49(wet) were considered. A boundary change to 0.31 at uncovered surfaces was applied in all cases from time zero and a no flow condition was maintained at the slab. A diffusivity of 1×10^{-2}

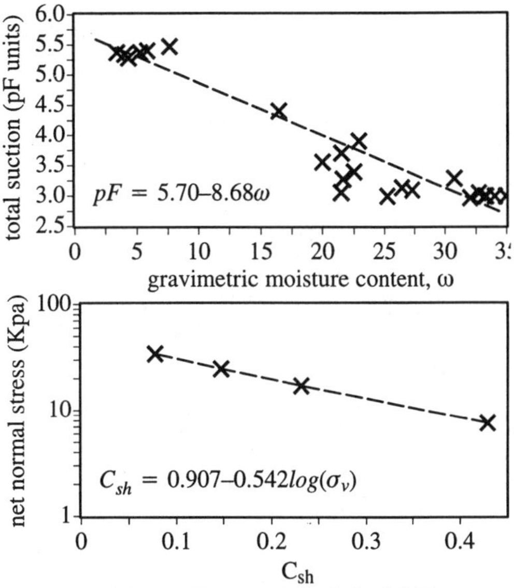

Figure 1 Moisture Characteristic (after McPherson e al., 1994) and Volume Change Index relationships.

m^2/day was based upon fitting numerical solutions to observations of time varying wetting and drying moisture front profiles in the field.

4. RESULTS.

Figures 2 and 3 compare moisture distributions from semi-analytical analyses (Fityus, Smith and Kleeman, 1996) with hand drawn flow net solutions; that is, Figs 2a, 3a, 3b with Figs 2b, 3c. It is obvious that if the rules of flow net construction are followed, good approximations to numerical solutions are possible. The distributions in Figure 3 illustrate the ease with which soil anisotropy is accommodated, at least for a single layer.

Figures 2c and 3d compare computed mound profiles after the methods of Richards and Fredlund. The method of Fredlund predicts consistently greater surface deflections both beneath and adjacent to the slab. Interesting though, the shapes of the mounds in each case are similar, and so are the differential movements. For both methods the initial condition affected the total, but not differential displacements. In the case of the Richards method, this is due to the assumed linear initial moisture profiles, while in the case of the Fredlund method it is due to the assumed linear moisture characteristic. In general, these conditions would not be fulfilled, and the estimated differential displacements would vary with the initial moisture content.

The Australian Standard 2870 (1996) recommends that the maximum differential movement across the

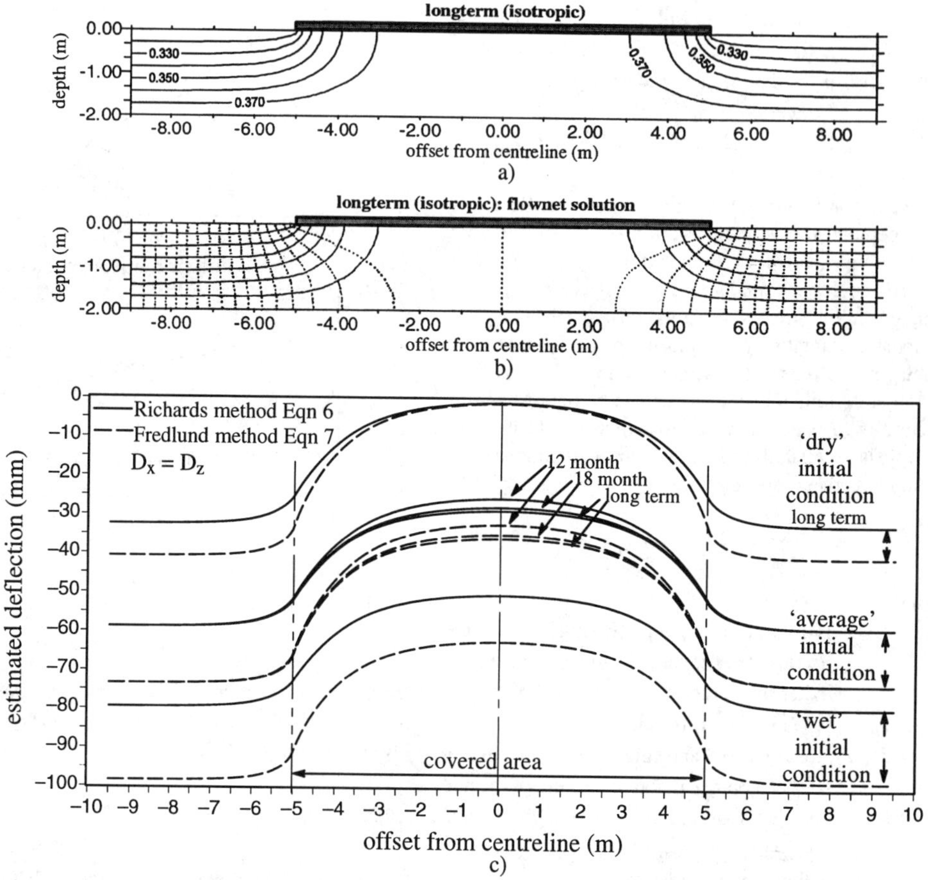

Figure 2 Moisture Content Distributions and Mound Profiles assuming Isotropic Moisture Diffusivity.

covered area be taken as 0.7 of the design potential surface movement. This value is largely based on experience. Figure 2c shows that there is a discrepancy between this recommendation and the calculated values: the recommendation being best for the 'dry' initial conditions but an overestimation for the 'average' and 'wet' initial conditions. Work is continuing to explore the effects of different boundary conditions and transient states on this important parameter.

Figures 2c and 3d show the mound profiles for 12 month, 18 month and long term semi-analytical analyses. It validates the assumption that extreme climatic events of 12 to 18 months (appropriate to the region under consideration) can produce near steady state soil moisture distributions.

Figure 4 compares the computed Richards mound profiles estimated from the flow net with the standard shapes of Lytton and Mitchell. Under isotropic conditions, the Mitchell(1980) profile is a reasonable approximation to the estimated shape, while the Lytton profile gives a fair approximation for an integer exponent of 3. Under anisotropic conditions (Figure 4b) both Mitchell profiles give a much poorer approximation to the profile estimated from the flow net. The Lytton profile though, offers a good approximation, but in this case with an exponent of 2.

Of course, the ultimate test of the adequacy of an estimated profile is how well bending moments in the slab are estimated. Based on recent research, Mitchell (AS 2870, 1996) has modified his mound profile to one which is much flatter in the middle with steep sides near the edge of the covered area (see Figure 4a). This conforms much more closely to the mound profile advocated by Walsh (AS 2870, 1996).

Assuming that the modified Mitchell profile (AS 2870, 1996) does give a better estimate of the bending moments in the slab, we look for reasons why the mound profile estimate from flow net theory is not predicting a shape closer to the modified Mitchell profile. One idea is that the most adverse mound profile is not

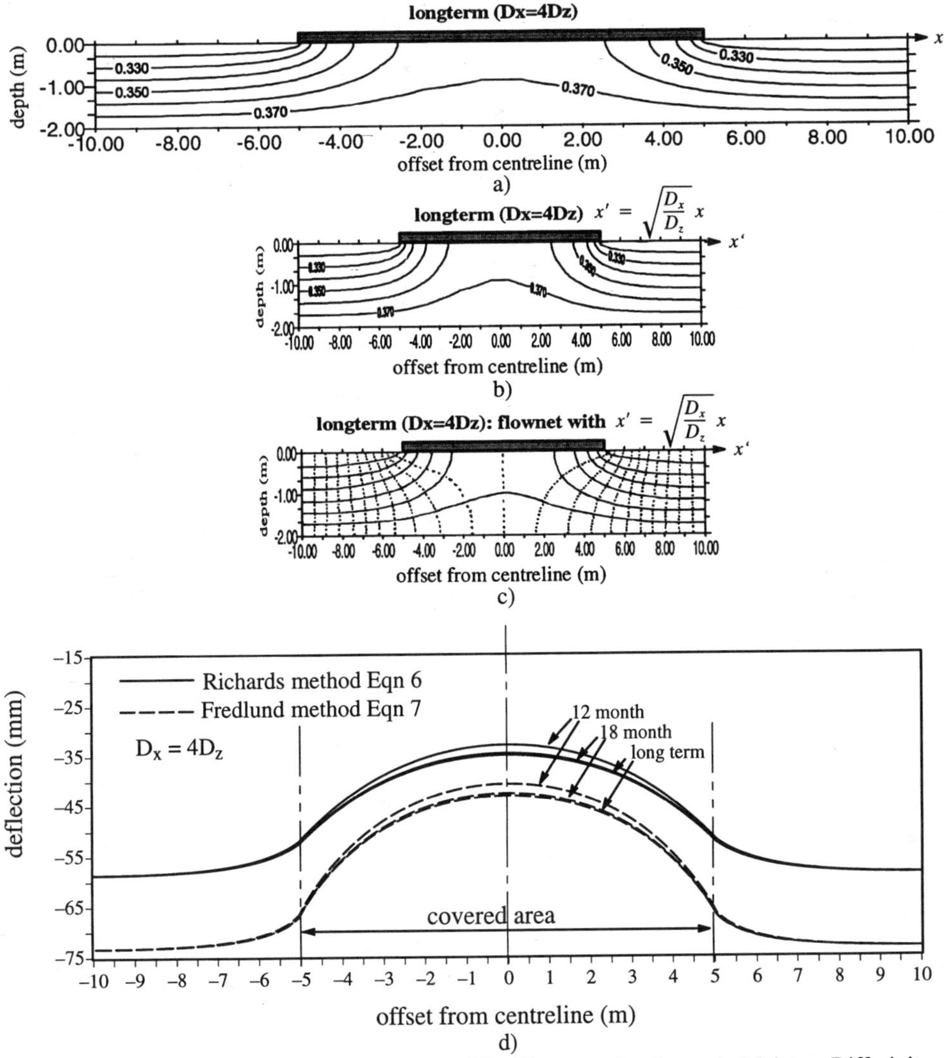

Figure 3 Moisture Content Distributions and Mound Profiles assuming Isotropic Moisture Diffusivity..

that estimated from the long term solution to equation (1), rather it should be estimated at some intermediate time. In this case the flow net solution for the moisture distribution is inappropriate. This is being investigated further by the authors. An alternative suggestion is that the flow net is an appropriate method for estimating the moisture distribution, but that account needs to be taken of edge turndown to represent the edge beam of the slab. This is also the subject of further work.

Finally we note that both the Mitchell and Lytton mound shapes require prior specification of the mound height, while the methods proposed here enables mound heights and shapes to be estimated simultaneously. Indeed, the method offers a fairly simple way of investigating the effect on mound shape of a tapering active zone, of varying the edge downturn on the slab and varying the width of impervious edging next to the slab .

5. CONCLUSIONS.

In conclusion, it may be said that provided the geotechnical engineer knows the site geometry, the depth of the active zone, appropriate simplified boundary conditions, the horizontal and vertical moisture diffusivity, the initial moisture content and possibly the soil moisture characteristic, then is possible that more real-

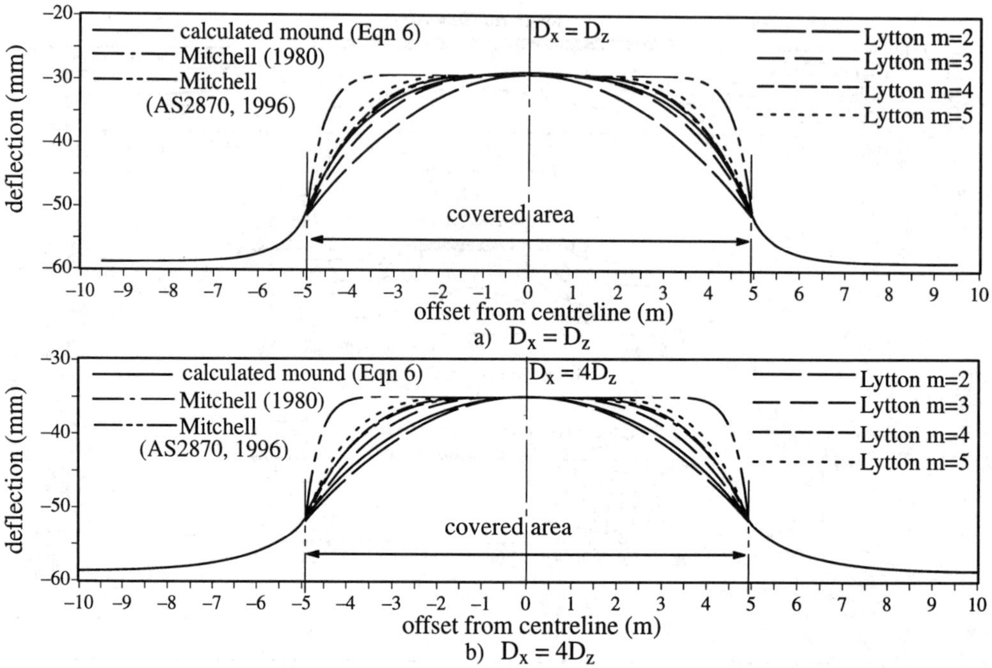

Figure 4 Comparison of Typically Adopted Mound Profile Shapes.

istic moisture profiles may be estimated. The engineer can then systematically go about quantifying the effects of differing horizontal and vertical diffusivity, tapering 'active zone', increased edge beam downturn, and covered areas adjacent to the foundation and initial moisture content. Of course, with resort to a finite element program, variable moisture diffusivity, moisture sinks due to vegetation and moisture sources due to watering and leaking pipes may be investigated in more detail.

ACKNOWLEDGEMENTS.

Financial support for this research from the Mine Subsidence Board of New South Wales, Robert Carr and Associates and the Australian Research Council is appreciated.

Fityus, S., Smith, D., (1994) "Moisture Distribution in a Partially Saturated Homogeneous Soil Layer.", *Proc. 8th. Int. Conf. on Computer Methods and Advances in Geomechanics.*, 2:1055–1061.

Fityus, S. G., Smith; D. W., Kleeman, P. W., (1996) "Two Dimensional Simulation of Soil Moisture Around a Leaking Water Pipe Adjacent to a Concrete Slab.", *Proc. 7th Aust. N.Z. Conf. on Geomech.,* Institution of Engineers. Aust., Canberra.

Fredlund, D .G., Rahardjo, H., (1993) "Soil Mechanics for Unsaturated Soils.", *John Wylie & Sons Inc.*, New York.

Lytton, R. L. (1970) "Analysis for Design of Foundations on Expansive Clay", *Proc. Symp. Soils and Earth Struct. in Arid Climates*, Aust. Geomech. Soc., Adelaide, 29–37.

McPherson, B. J., Swarbrick. G. E., (1995) "Use of Water Retention Characteristic of Newcastle Soils to Predict Ground Surface Movements.", *Proc. Buildings and Structures Subject to Ground Movement, 3rd Triennial Conf. of the Mine Subsidence Technological Society*, 53–61.

Mitchell, P. W., (1980) "The Structural Analysis of Footings on Expansive Soils.", *Research Report No. 1., K. W. Smith Consulting Engineers*, Newtown, South Australia.

Pullan, A. J., (1990) "The Quasilinear Approximation for Unsaturated Porous Media Flow.", *Water Resour. Res.*, 26:1219–1234.

Richards, B. G., (1967) "Moisture Flow and Equilibrium in Unsaturated Soils for Shallow Foundations.", *Permeability and Capillarity of Soils, A.S.T.M. S.T.P.417* Am. Soc. Testing Mats. P4–34.

Standards Association of Australia (1996) AS 2870. Residential Slabs and Footings.

Computer Methods and Advances in Geomechanics, Yuan (ed.) © 1997 Balkema, Rotterdam, ISBN 90 5410 904 1

Ground settlement and water inflows due to deep shaft excavation

Jian-Hua Yin
Department of Civil and Structural Engineering, Hong Kong Polytechnic University, China
Caizhao Zhan
Binnie Consultants Limited, Hong Kong, China
Zhong-Qi Yue
Halcrow Asia Partnership Limited, Hong Kong, China

ABSTRACT: This paper presents a practical application of finite element modelling and back-analysis to the interpretation and prediction of ground surface settlements and water inflows due to deep shaft excavation. The factors affecting the settlements and water inflows are discussed. Back-calculated parameters are useful for the prediction of the future performance of the excavation. This paper points out that the engineering judgment in the estimation of modelling parameters is necessary. Based on the prediction, the most effective and feasible measure is suggested for controlling the settlements and water inflows for further excavation.

1 INTRODUCTION

A 134.5 m deep shaft (approximately 10m in diameter) was constructed at a site in Hong Kong. When the excavation of the shaft was down from the ground surface (+4.5mPD) to approximately a half way (-64mPD) of the designed depth (-130mPD), ground surface settlements and groundwater inflows into the shaft were larger than the values expected. The settlements of the ground surface caused the cracks of strip footings supporting temporary structures on the ground surface. The rate of water inflows was beyond the capacity of the water pumps. Further excavation might cause more settlement and water inflows.

This paper presents results of applying a back-analysis technique to the prediction of the ground surface settlements and water inflows due to further shaft excavation. Both finite element (FE) seepage analysis and FE deformation analysis were carried out to back-calculate soil parameters and the thickness of a fractured layer by matching the computed settlement and water inflows to the measured values. The goodness of the back-computed parameters are not only based on the FE modelling results but also the limited site investigation (SI) data, field observations and engineering judgment. The back-calculated soil parameters and the likely thickness of a fractured layer were then used in the FE models for the prediction of settlements and water inflows due to further excavation of the shaft.

2 SUB-SURFACE CONDITIONS AND UNCERTAINTIES

The site of the shaft excavation was a piece of reclaimed land. The northern east boundary of the site was sea (at approximately 230m from the shaft centre). The site investigation data were limited especially regarding the fractures of the rock mass. Only one borehole was sunk to -140mPD along the central line of the shaft excavation. The strata of the subsoil/rock is summarised in Table 1.

The measured groundwater table was approximately at +1.5mPD at the site. The following uncertainties were presented regarding the site geology: (a) horizontal extension of the layers, (b) the thickness of the fractured tuff layer, (c) permeability and stress-strain behaviour of the soils and rock mass, and (d) the hydraulic boundary approximately 230m from the shaft. The field inspection showed apparent fractures from -50mPD to -64mPD.

Table 1. Soil strata and SPT-N values

Stratum	Elevation (mPD) (average)	SPT N
Sandfill	+4.5 to -5.5	38
Marine deposit (Clay)	-5.5 to -8.5	7
Marine deposit (Sand)	-8.5 to -10	15
Alluvium (Sand)	-10 to -11.5	?
C.D.Tuff (Sand)	-11.5 to -16	17
H./M.D. Tuff (IV/III)	-16 to -20	>200
Tuff (III/II)	-20 to -50	
Fractured Tuff layer	-50 to -75 (?)*	
Tuff (III/II)	-75 (?)* to -140	

Note for : The position of the bottom of the fractured tuff layer is to be estimated based on back-analysis, SI data, observations, and judgement.

3 MEASURED SETTLEMENTS AND WATER INFLOWS

The shaft excavation in soils had a diameter of 10m and was supported by diaphragm wall. The diaphragm wall was constructed before excavating. The shaft had a diameter of 8m in the rock section without any support. Fig.1(a) presents the measured ground surface settlements at 8 points around the shaft when the excavation was down to -60mPD. The location of the 8 points ranged approximately from 0.9m to 15m from the circumference of the shaft. The measured settlements varied from 13mm to 76mm. The variation of the measured settlement values indicated that the subsoil was non-homogeneous. The average of the settlements measured at the 8 points was 33.8mm. The surface settlements had caused cracks and differential settlement of concrete strip footings supporting temporary structures on the ground surface.

The measured average total water inflows into the shaft at excavation depth between -55mPD to -64mPD varied from 5.0×10^{-3} m^3/sec to 8.33×10^{-3} m^3/sec. The rate of inflows was larger than that expected. A highly fractured tuff layer was found from -50mPD to -64mPD and probably down further. This fractured zone was considered a main factor contributing to the unexpected high inflow rate.

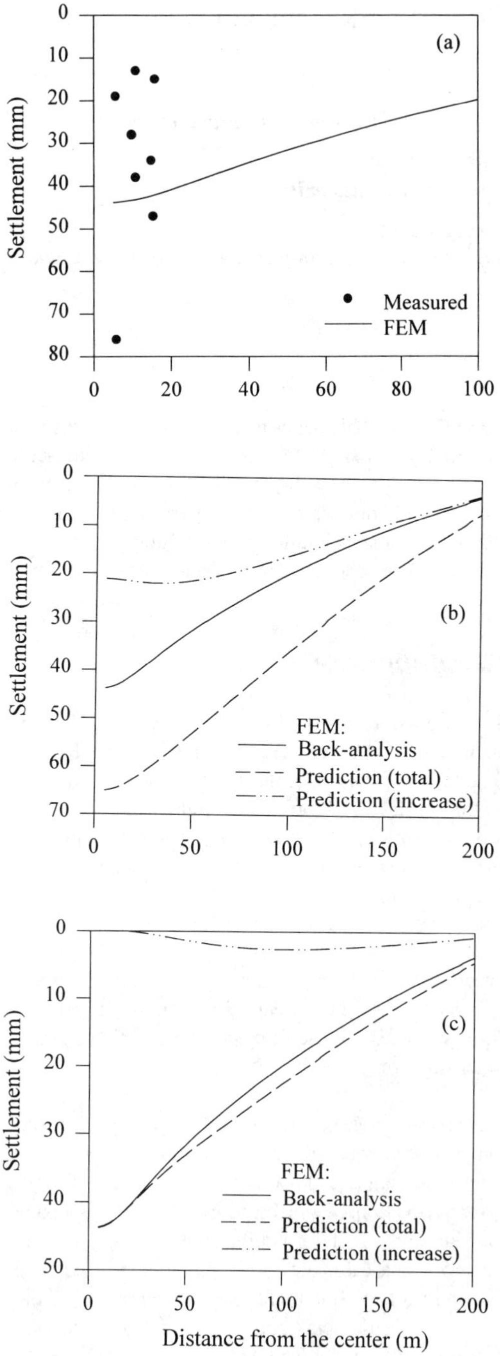

Fig.1 (a) Measured and FEM back-computed settlements, (b) FEM back-computed and predicted settlement (no sealing), and (c) FEM back-computed and predicted settlement (sealing)

The water inflows into the excavated shaft caused drawdown of the groundwater table. The settlements at the ground surface were mainly caused by the groundwater table drawdown in the subsoil. The surface settlements would increase due to the further increase in hydraulic gradient and further water table drawdown as the shaft excavation proceeded deeper. The likely maximum rate of water inflows and the increase of the settlements with further excavation were the concern of the design engineers and contractor.

4 FE SEEPAGE MODELLING - BACK-ANALYSIS AND PREDICTION

As discussed above, the water table drawdown was the main factor causing the settlement of the ground surface. In order to estimate the ground settlements at the final level of the shaft excavation, the final water table drawdown must be known. The purpose of the FE (finite element) seepage modelling was to predict the drawdown and provide porewater data for the deformation/settlement analysis using a FE deformation model.

There were two major uncertainties which would affect the seepage prediction: one was the variation in permeability and the other was the thickness of a highly fractured and highly permeable layers in the rock mass. A back-analysis using the FE seepage model was first carried out to estimate the permeability values of the soils and rock mass and the most likely location of the highly fractured tuff layer by matching the computed inflow rates to the measured inflow rates.

The FE seepage model was based on a finite element program SEEP/W (Geo-Slope, 1995). The permeability values for the soils and rock mass were estimated and listed in Table 2

The most uncertainty was the thickness of the highly fractured tuff layer. A parametric study was conducted to estimate the thickness. The computed inflow rates for different thickness are listed in Columns (1) to (4) in Table 3 for excavation down to -60mPD to -65mPD.

Fig.2 shows the FE mesh, contours of total heads in meter, flux sections and water table drawdown. Two flux sections are shown with a value of 1.12×10^{-3} m^3/sec for flux Section 1 and $9.13\times10^{-3}m^3$/sec for flux Section 2. Since this is an axisymmetrical problem, the flow rate is equal to the computed flux value multiplied by 2π in SEEP/W (Geo-Slope, 1995). For example for flux Section 1, the flow rate, $q=2\pi\times1.12\times10^{-3}=7.03\times10^{-3}m^3$/sec. The total calculated inflow rates are $7.03\times10^{-3}m^3$/sec at the excavation level of -60mPD (see Fig.2) and $8.14\times10^{-3}m^3$/sec at -65mPD. The two values are within the measured range of 5.0×10^{-3} m^3/sec to 8.33×10^{-3} m^3/sec. The highly fractures zones are assumed to vary from -50mPD to -75mPD, considering both the borehole data and the back-analysis results. The porewater data in Fig.2 at the level of -60mPD are used for FE deformation/settlement modelling.

Table 2. Parameters of soils and rock mass from back-analysis and for prediction

Stratum	k m/s	E (kPa)	υ
Sandfill	10^{-5}	4×10^4	0.3
M.D. Clay	5×10^{-7}	1.2×10^4	0.4
M.D. Sand	10^{-6}	1.6×10^4	0.3
Alluvial Sand	10^{-6}	3.4×10^4	0.3
C.D.T. Sand	10^{-6}	5×10^4	0.3
Tuff (IV)	10^{-6}	4×10^6	0.2
Tuff (III/II)	5×10^{-7}	2.5×10^7	0.2
Fractured Tuff	2×10^{-6}	1×10^7	0.2
Tuff (III/II)	5×10^{-7}	2.5×10^7	0.2

Table 3. Influence of thickness of fractured tuff layer

	(1)	(2)	(3)	(4)	(5)
Fractured Tuff Zone (mPD)	-50 to -60	-50 to -65	-50 to -70	-50 to -75	-50 to -75
Shaft Excavation (mPD)	-60	-60	-60	-65	-130
Total Flux (m^3/s)	5.3 $\times10^{-3}$	6.2 $\times10^{-3}$	7.03 $\times10^{-3}$	8.14 $\times10^{-3}$	12.2 $\times10^{-3}$

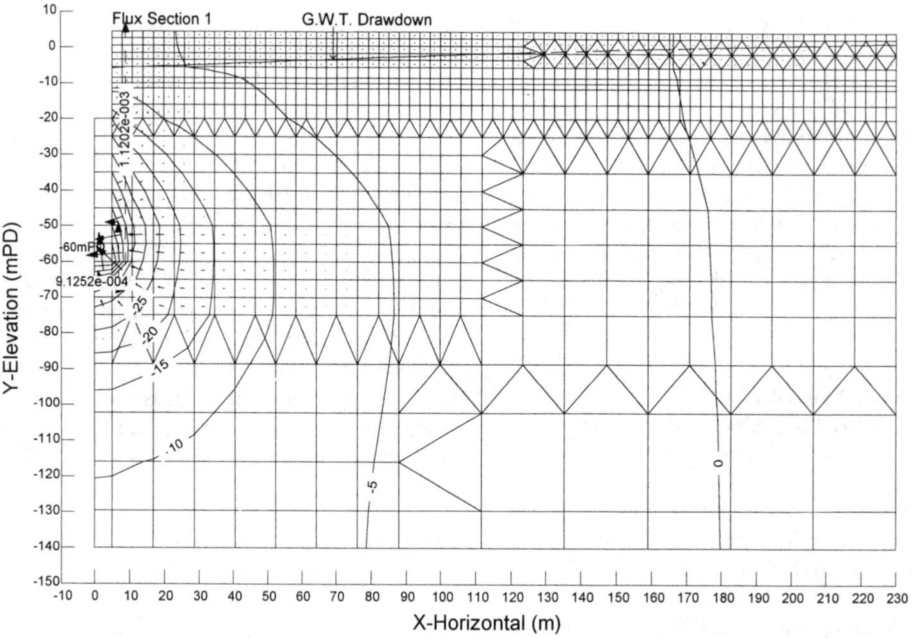

Fig.2 FEM seepage back-analysis: FE Mesh, boundary conditions, water table drawdown and flux sections

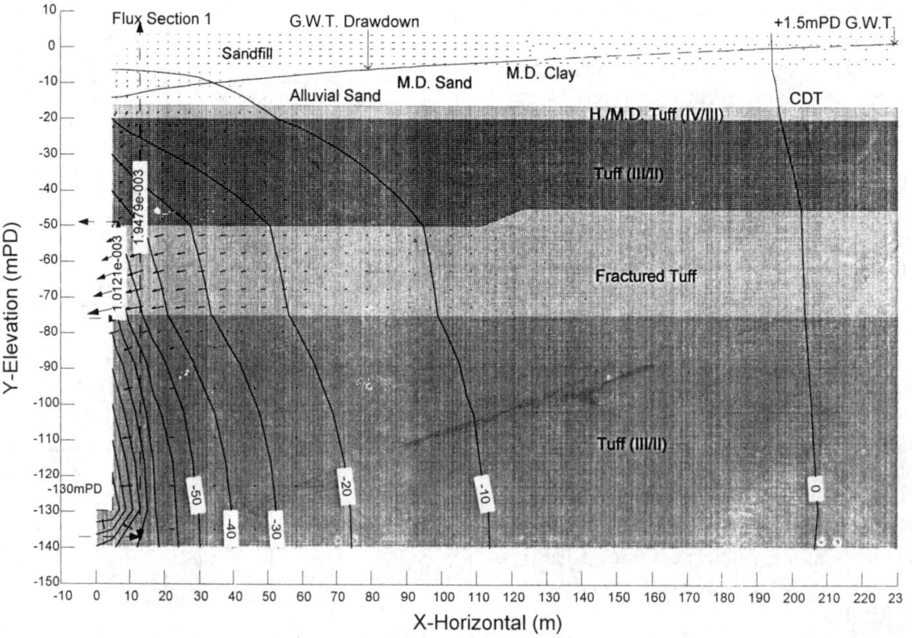

Fig.3 FEM seepage prediction: soil layers, velocity vectors , water table drawdown and flux sections

No sealing for the rock face of the excavation: Using the parameters in Table 1 and assuming the fractured tuff layer from -50mPD to -75mPD, the FE seepage model is then used to predict flux (flow rate), water table drawdown, and porewater pressures for excavation down to -130mPD (see Fig.3). The predicted total inflow rate is $12.2 \times 10^{-3} m^3/sec$, 1.73 times the inflow rate at the level of -60mPD. The predicted porewater pressure data are used in the FE deformation analysis.

Sealing for the rock face of the excavation: If the rock face of the excavated shaft from -89mPD to -45mPD is be sealed by concrete ring liners or cement grouting, the water inflows are smaller than that in Fig.2. The computed flow rate is only 6.28×10^{-3} m^3/sec. The porewater pressure data computed are to be used in the FE deformation analysis for excavation and dewatering down to -130mPD. This shows that sealing the fractured layer is a very effective measure for controlling the inflow rate.

5 FE DEFORMATION MODELLING - BACK-ANALYSIS AND PREDICTION

The fractured rock zone is assumed to exist from -50mPD to -75mPD as discussed in Section 4. A number of trial runs by changing the Young's modulus of the soils and rock mass have been carried out to match the computed settlement to the measured settlement in order to determine the appropriate deformation parameters for the soils and the rock mass. Those parameters unchanged are listed in Table 1. The computed maximum settlements due to excavation and dewatering to -60mPD are listed in Columns (1) to (4) in Table 4. Fig.4 shows the deformed FE mesh, soil layers and boundary conditions for the case in Column (4). The computed settlements are compared to measured values in Fig.1(a) and are found to be in a reasonable agreement considering the scattering of measured data and other uncertain factors. The computed maximum settlement is 43.8mm, which is in the measured range of 13 to 76mm. The deformation parameters from the case in Column (4) are then used for the settlement prediction for excavation and dewatering to -130mPD.

No sealing for the rock face of the excavation: Assuming that the rock face of in the excavated shaft is not sealed by concrete ring liners or cement grouting, the water drawdown is that as shown in Fig.2. The porewater pressure data are then used in the FE deformation analysis for excavation and dewatering down to -130mPD. The computer ground surface settlements are shown in Fig. 1(b) and Fig.5. The total maximum settlement is 65.0mm. The increase in settlement is 22.2mm compared to the case in Column (4) in Table 4 from the back-analysis.

Sealing for the rock face of the excavation: Assuming that the rock face of the excavated shaft from -89mPD to -45mPD is sealed by concrete ring liners or cement grouting, the water drawdown is smaller than that in Fig.2. The porewater pressure data has been computed and are then used in the FE deformation analysis for excavation and dewatering down to -130mPD. The computed ground surface settlements are shown in Fig. 1(c). The total maximum settlement is 43.8mm. There is no increase in the maximum settlement except for small increase at certain distance from the excavation. This demonstrates that the effective measure for controlling the ground settlement is by sealing the fractured zone.

6 DISCUSSION AND REMARKS

Many uncertainties exist for the prediction of the water inflows and settlements due to the shaft excavation. The back-calculated parameters are based

Table 4. Summary of back-analysis and prediction

Stratum	(1)	(2)	(3)	(4)	(5)
Sandfill: E (kPa)	1 $\times 10^4$	1.5 $\times 10^4$	2 $\times 10^4$	4 $\times 10^4$	4 $\times 10^4$
M.D. Clay: E (kPa)	6 $\times 10^3$	9 $\times 10^3$	1.2 $\times 10^4$	1.2 $\times 10^4$	1.2 $\times 10^4$
M.D. Sand: E (kPa)	8 $\times 10^3$	1.2 $\times 10^4$	1.6 $\times 10^4$	1.6 $\times 10^4$	1.6 $\times 10^4$
Allu. Sand: E (kPa)	1.7 $\times 10^4$	2.55 $\times 10^4$	3.4 $\times 10^4$	3.4 $\times 10^4$	3.4 $\times 10^4$
C.D.T.: E (kPa)	2 $\times 10^4$	3 $\times 10^4$	5 $\times 10^4$	5 $\times 10^4$	5 $\times 10^4$
Tuff (IV): E (kPa)	2 $\times 10^6$	2 $\times 10^6$	4 $\times 10^6$	4 $\times 10^6$	4 $\times 10^6$
Exca. & Dew. to (mPD)	-60	-60	-60	-60	-130
Max. Settl. (mm)	116	77.4	56.5	43.8	65.0

on limited SI data, field observations, FE modelling results, and engineering judgement. Using FE modelling alone is inadequate. In other words, the back-analysis can not determine uniquely the uncertain parameters. All available SI data including laboratory testing results are helpful in the elimination of those uncertain factors. Other information such as the correlation of SPT-N values to the Young's modulus of soils is also important. Sound engineering judgement in using those information and data is one of the most important elements in the prediction.

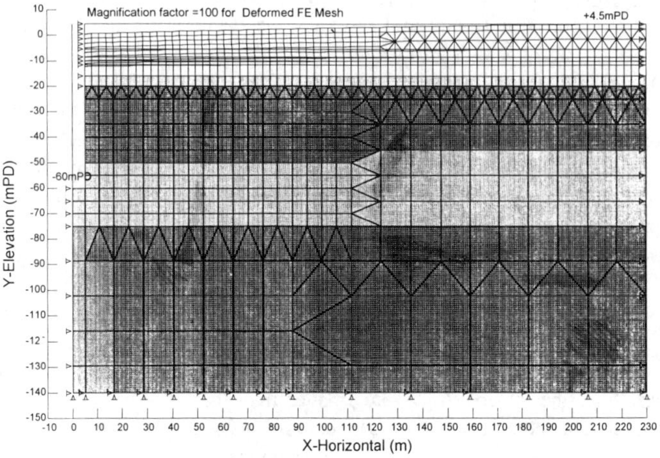

Fig.4 FEM deformation back-analysis: deformed mesh and soil layers

7 CONCLUSIONS

From the analysis, the settlement is mainly the compression of the soils especially the sandfill due to the water table drawdown (the effective stress increase). The water inflows are mainly from the highly fractured tuff layer. Based on the analysis, the most effective measure is to use concrete liners or cement grouting to seal the fractures in the fractured tuff layer near the shaft excavation. The FE modelling shows that both the water flow rate and the settlements are under control using the proposed measure.

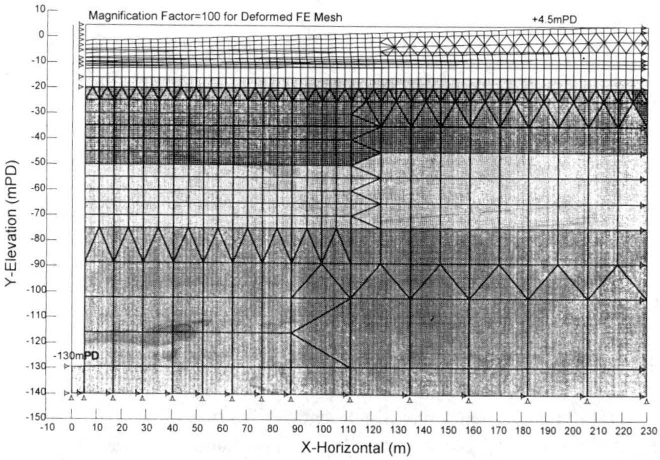

Fig.5 FEM deformation prediction: deformed mesh and soil layers

It should be pointed out that the back-analysis itself alone is inadequate for the determination of all those uncertain parameters. All available SI data, experience and sound engineering judgement are important for reasonable prediction of settlements and water inflows for real excavation projects.

ACKNOWLEDGEMENT

Financial support from the Hong Kong Polytechnic University and RGC is acknowledged.

REFERENCE

SEEP/W Manual and SIGMA/W Manual, Geo-Slope International Limited, Calgary, Canada, 1995.

Computer Methods and Advances in Geomechanics, Yuan (ed.) © *1997 Balkema, Rotterdam, ISBN 90 5410 904 1*

A geostatistical approach to estimate the groundwater flow path in fractured rocks

Shinji Nakaya
CRC Research Institute, Inc., Osaka, Japan

Akihisa Koike & Masato Horie
Kansai Electric Power Co., Japan

Tetuhiro Hirayama & Toshiaki Yoden
NEWJEC Inc., Japan

ABSTRACT: Groundwater flow paths in complex rock formations was computationally estimated by an application of geostatistics to field data obtained by the conventional measurements. Single-hole permeability measurements (Lugeon water test) were carried out for 33 bore-holesdrilled into the basement rocks of the studied area, and permeabilities of the rocks were obtained as Lugeon water values. Using the in situ Lugeon water values (Lu), the spatial distribution of Lu was three-dimensionally estimated by a geostatistical method, kriging. The results demonstrated that the estimated high and low permeable zones and the fault fracture lines, drawn based on the field observation, overlapped each other on a horizontal plane. The preliminary estimation of major groundwater flow paths from our computational procedure is skilful to determine the tactics of detailed field hydro-geological surveys and measurements.

1 INTRODUCTION

It is important to know the groundwater flow paths, which are highly permeable and fracture-concentrated zone, in rock mass before underground construction. Scince it has been difficult to estimate the groundwater flow paths from the limited data obtained by the conventional field measurements, geostatistical estimation from scattered *in situ* permeability data is an advantage to determine the spatial distribution of permeable zones in complex rock formations. In this paper, three dimensional estimation of spatial distribution of permeable zones was tried for complex rock formations in the study area by an application of the kriging, which was one of the most authorized method for geostatistical estimation in the hydroscience (Delhomme 1978, Delhome 1979, Clifton & Neuman 1982, Ahmed & Marsily 1987, Issaks & Srivatava 1989).

The studied area is located in the central Japan, and a power plant of a dam has been planned to construct in underground space. Prior to the construction, the area was geologically, geophysically and hydrogeologically surveyed. Single-hole permeability measurements (Lugeon water test) were carried out for 33 bore-holes drilled in the basement rocks within 300m x 200m rectangular area. The obtained permeability of rocks was expressed by Lugeon water values. From optically observed fractures in a survey tunnel and the fracture image of a bore-hole television system (BTV) introduced into the selected four bore-holes among the 33 bore-holes, the basement rocks were considerably fractured, from five to ten fractures in every one meter, and orientation of the fractures looked randomly distributed. Thus, basement rocks are highly permeable, and the permeability ranges from about $1x10^{-7}$ to $2x10^{-3}$ cm/s.

Several major groundwater flow paths may be present in the complex and fractured rock formations. The result was that the three dimensional estimation of spatial distribution of Lugeon water values by kriging can distinguish the major groundwater flow paths from the other permeable zones. The estimated groundwater flow path and the fault fracture planes estimated by field survey overlapped each other on horizontal plane.

2 HYDROGEOLOGICAL CHARACTERISTICS

Fig.1 shows the contour lines of groundwater level and fault lines on the horizontal plane at 180 m elevation. The basement rocks of the area comprise the Mesozoic sedimentary formations, overlying rhyolitic welded tuff and intruded granitic porphyry into the above rocks. The fracture zones of faults have been identified in the bore-holes shown by black circles in Fig. 1 and survey tunnel, although all faults in the studied area are hiden by terrace and talus deposits. The fault lines in the area were drawn using the obtained orientation data of the fracture zones. General strike of the faults are in south-north direction. On the contour map of groundwater level(Fig.1), the groundwater flow from the upper to lower from the

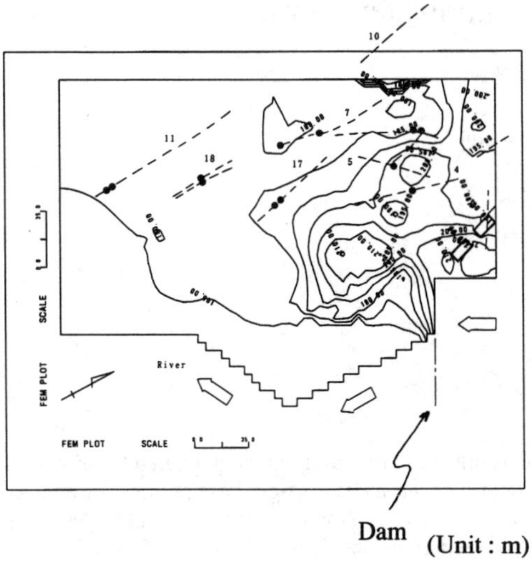

Fig.1 Contour lines of groundwater level and fault lines on the horizontal plane at 180 m elevation.

dam site remarkably appears. The zone of groundwater, which level is lower than 185 m contour, extends to mountain side. The distribution of groundwater level may reflect the hydrogeological structure, espe cially fault system.

3 GEOSTATISTICAL ESTIMATION OF PERMEABILTY

One of the geostatistical method, block kriging is applied to *in situ* Lugeon water values, here. In blockkriging, the following equation is solved for the unknown weights, b of each sample points, i, j around a target block, B, using sample values, w_i.

$$\left.\begin{aligned}&\sum_{j=1}^{n} b_i\gamma(h_{ij}) + \mu = \gamma(h_{iB})\\&\sum_{j=1}^{n} b_i = 1\,,\end{aligned}\right\}\quad (i=1, 2, 3, \ldots\ldots, n) \tag{1}$$

where : $\gamma(h_{ij})$ is semi-variogram,

$$\gamma(h_{ij}) = \frac{1}{2n(h)} \sum_{i=1}^{n(h)} (w_i - w_j)^2 \tag{2}$$

$$w = \sum_{i=1}^{n} w_i b_i \tag{3}$$

3.1 Spatial continuity of Lugeon water values

Fig.2 shows the histogram of measured Lugeon water values (Lu) in the field where one Lugeon is theo

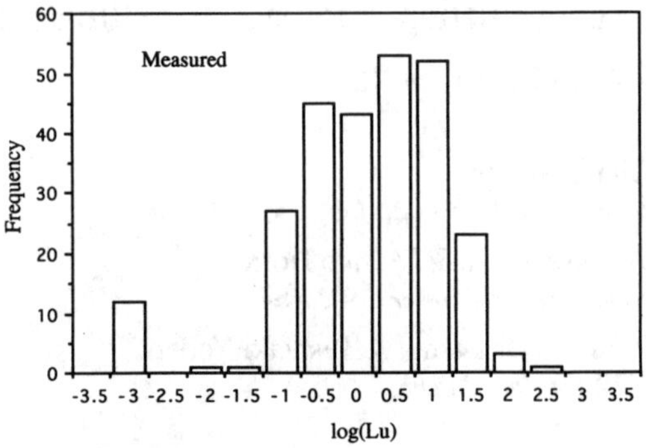

Fig.2 Histogram of measured Lugeon water values (Lu) in the field.

retically calculated to about 10^{-5} cm/s assumed to be in homogeneously isotropic porous media with laminar flow. Since the histogram shows a log-normal distributions, the Lu data can be statistically dealt with. The semi-variograms of ln(Lu) is calculated for the distance (h) of data pairs scattered in all directions, in order to check the spatial continuity. Fig.3 shows the semi-variogram of Lu, where the hole effect appears. The pattern of semi-variogram can be sufficiently approximated to a spherical model where its range has 70 m. Using the spherical model for the semi-variogram, kriging equation of Eq.(1) is solved for unknown weights of each sampling points within 70 m distance around each target blocks, of which Lu would be determined.

3.2 Estimation of spatial distribution of Lu

Fig. 4 shows the three-dimensional block Lugeon map that is estimated by the block kriging described above. Studied area is divided into 5 m cube because the injection interval of Lugeon water test is 5 m in length along a bore-hole. The distribution of Lu is concordant with that of groundwater level shown inFig. 1. A high permeable zone appears from the upper to the lower direction from the dam site. Another high permeable zone overlaps the zone of groundwater level lower than 185 m contour.

4 ESTIMATION OF GROUNDWATER FLOW PATH

Faults often behave as groundwater flow paths. In the studied area, all of the faults have *gouge* in the fracture zone. Because of low permeability of fault *gouge*, fault fracture zones would be detected as low Lu blocks as well as high Lu blocks. Fig. 5 shows blocks having the Lugeon values lower than 0.1 and higher than 50 on the horizontal slices of three-di

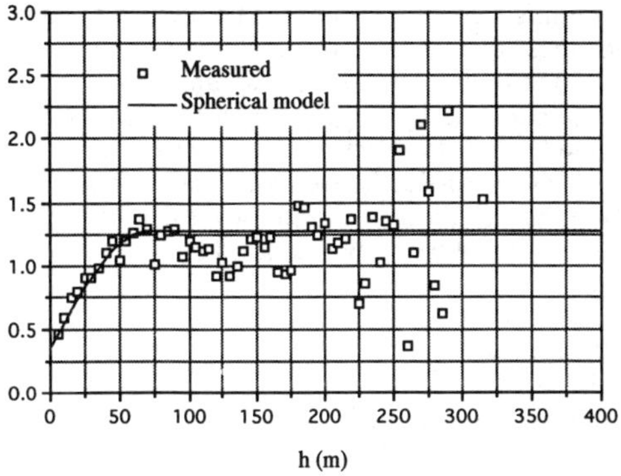

Fig. 3 Semi-variogram of ln(Lu)

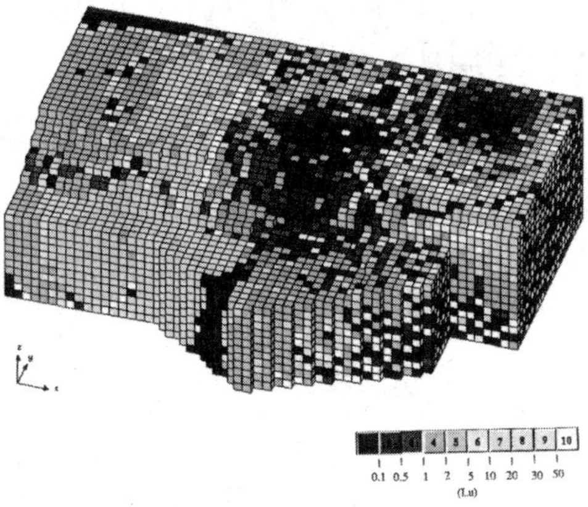

Fig. 4 Three-dimensional block Lugeon map estimated by the block kriging.

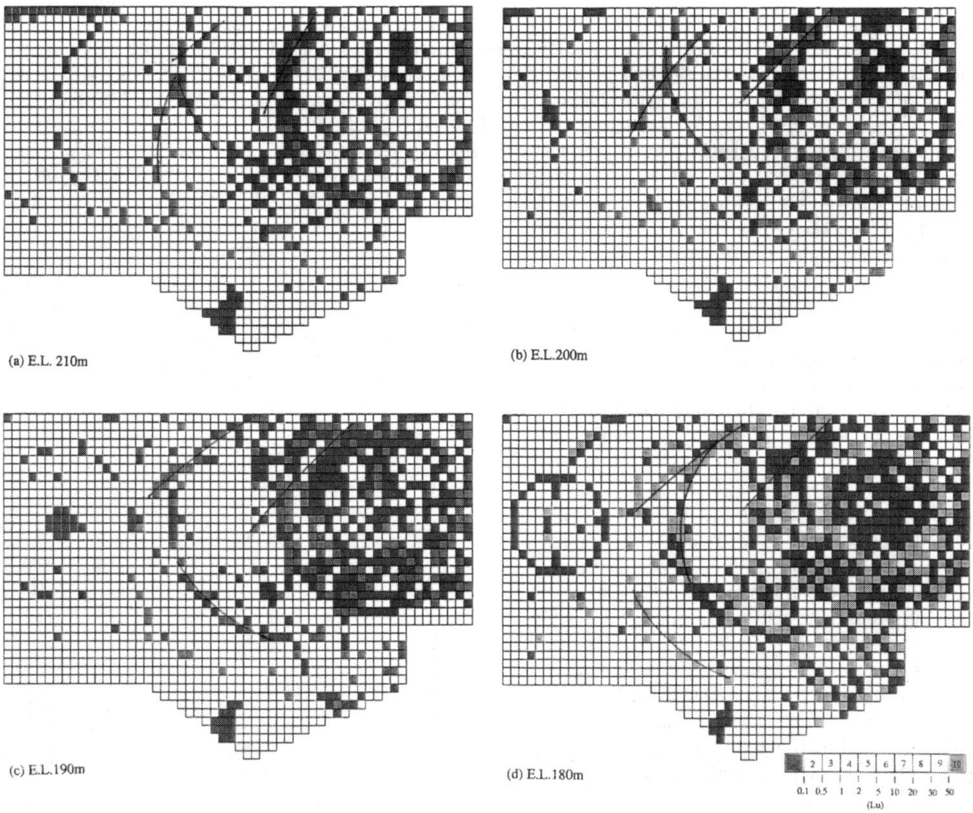

Fig. 5 Blocks having the Lugeon values lower than 0.1 and higher than 50 on the horizontal slices of three-dimensional block model.

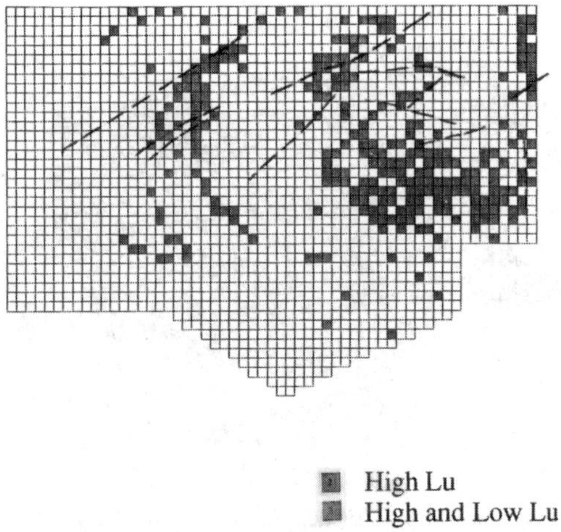

Fig.6 Major groundwater flow paths geostatistically estimated.

mensional block model. The high and low Lu blocks form several lineaments. Since the low Lu blocks mean both low permeable rocks and the low permeable parts of fault fracture zone, it is not easy to distinguish the fault fracture zones in Fig.5. In order to detect the major groundwater flow paths, the following procedures are performed.

1) Picking out the block areas where two or more blocks have the lower Lu values than 0.1 among the vertically connecting eight blocks on the horizontal planes from 220 to 180 m elevation.

2) Picking out the block areas where two or more blocks have the higher Lu values than 50 among the vertically connecting eight blocks on the horizontal planes from 220 to 180 m elevation.

3) Picking out the areas where the block areas searched in the procedures 1) and 2) overlap each other.

At the block areas selected in the procedure 1) or 3), the column comprising vertically connected eight blocks satisfies a necessary condition of permeability for a fault fracture zone which show the high and low Lugeon water values in the same column. Fig.6 shows the major groundwater flow paths, estimated from the block areas picked out by the above procedure on a horizontal plane. The geostatistically estimated major flow paths and nos.7, 11 and 18 fault fracture planes of Fig.1 which were drawn based on the field survey, overlapped each other on horizontal planes.

5 CONCLUSIONS

Applying a block kriging to measured Lugeon water values of basement rocks, the three-dimensionally hydrological sub-structure can be estimated. The followings are concluded in this study.

1. The three-dimensionally hydrological sub-structure estimated by kriging are concordant with observed sub-surface flow in complex rock formations.

2. Geostatistically estimated major groundwater flow paths coinsides with several fault fracture zoneswhich were drawn based on the field survey.

REFERENCES

Ahmed, S. and G. de Marsily : Comparison of geostatistical method for estimating transmissivity using data on transmissivity and specific capacity, *Water Resour. Res.*, 23, 1717,1737, 1987.

Clifton, P. M. and S. P. Neuman : Effect of kriging and inverse modeling on conditional simulation of the Avra Valley Aquifer in Southern Arizona, *Wa ter Resour. Res.*, 18(4), 1215-1234, 1982.

Delhomme, J. P. : Kriging in the Hydrocience, *Adv. Water Resour.*, 1(5), 251-266, 1978.

Delhomme, J. P. : Spacial Variability and Uncertainty in Groundwater Flow Parameters - A Geostatistical Approach, *Water Resour. Res.*, 15(2), 269-280, 1979.

Edward H. Isaaks & R. Mohan Srivatava : *Applied Geostatistics*, Oxford University Press, p.561, 1989.

Computer Methods and Advances in Geomechanics, Yuan (ed.) © 1997 Balkema, Rotterdam, ISBN 90 5410 904 1

Finite element drainage substructure technique for solution to free surface seepage problems with numerous draining holes

Yueming Zhu
College of Hydroelectric Engineering, Hohai University, Nanjing, China

ABSTRACT: This paper exhaustedly presents the drainage substructure technique for simulating the flow behavior of draining holes and for solution to seepage problems with numerous draining holes. The new advance makes the procedure of nodal virtual flow rate much more powerfully and adaptively be able to solve the practical seepage problems in geoengineering. A case study pertaining to an underground hydropower station is given to demonstrate the advantages of the technique proposed.

1 INTRODUCTION

In practice, draining holes used to be the most effective and important seepage control measures. Due to the facts that the holes usually have the characteristics that they are thin (internal diameter is about 0.10m), numerous (spacing at 3~5m centers to form drainage curtain), and long (from few meter to decades meter, and even over 100m) and that the tri-dimensional behavior of the draining is pretty complicated, even as yet there had been no proper numerical procedure for solution to problems with numerous draining holes. As shown in Fig.1, a draining hole may or may not intersect the free surface, *bc*, whose exact location is unknown prior to the solution. Thus, the internal peripheral surface of the hole is a possible seepage surface. Section, *ac*, of the hole is above the free surface and located out of the real saturated seepage domain and has no influence on the seepage field. The internal surface of section, *cd*, is a seepage surface. That one of section, *de*, below the elevation of the interior water level, H_0, is an isohead surface. An appropriate and high accurate solution method or technique for the problem has been being striven for many years. Based on the finite element procedures for solution to free surface seepage problems with a fixed mesh[1~4] and the concept of drainage substructure[5], the present author has give the correct definition of drainage substructures and develop a complete and powerful drainage substructure technique for the solution. The development and implementation of the technique can considerably improve the adaptability, effectiveness, and accuracy of finite element method for solving practical geotechnic seepage problems[6~7].

2 PROCEDURE OF NODAL VIRTUAL FLOW RATE WITH A FIXED MESH

The subdomains, Ω_1 and Ω_2, which respectively represent the areas upper and below the free surface, *AE*(Fig.2), are here refered to as the real and virtual seepage areas. According to Darcy's law, the governing equation and physical geometric conditions of the steady-state saturated seepage problem are Eqs.(1) and (2).

$$(k_{ij}H,_i),_j=0,\ x_i\in\Omega_1\ . \quad (1)$$

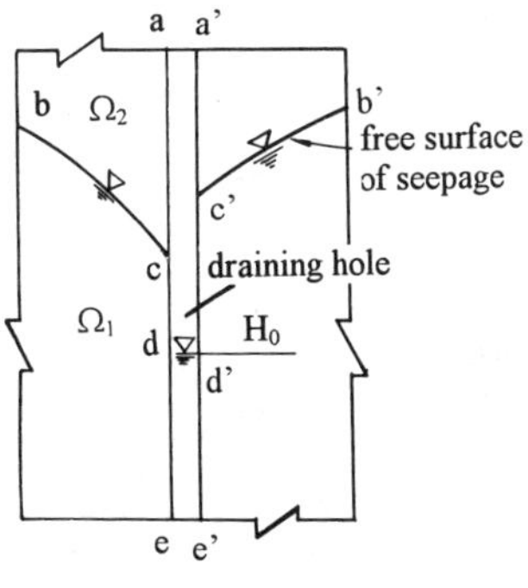

Fig.1 Seepage behavior of a draining hole

$H \geq x_3, \quad x_i \in \Omega_1$; and $H < x_3, \quad x_i \in \Omega_2$. (2)

The relevant boundary conditions are as follows:

$$H = H_1, \; x_i \in AB \qquad (3)$$

$$H = H_2, \; x_i \in CD \qquad (4)$$

$$\begin{cases} k_{ij}H,_j n_i = 0, \; x_i \in BC; \\ k_{ij}H,_j n_i = 0 \quad and \quad H = x_3, \quad x_i \in AE; \\ k_{ij}H,_j n_i \leq 0 \quad and \quad H = x_3, \quad x_i \in BC. \end{cases} \qquad (5)$$

Where k_{ij} is the permeability tensor of the medium, $H = x_3 + p/\gamma_w$ indicates the total hydraulic head, and n implies the unit outward normal on boundary surface.

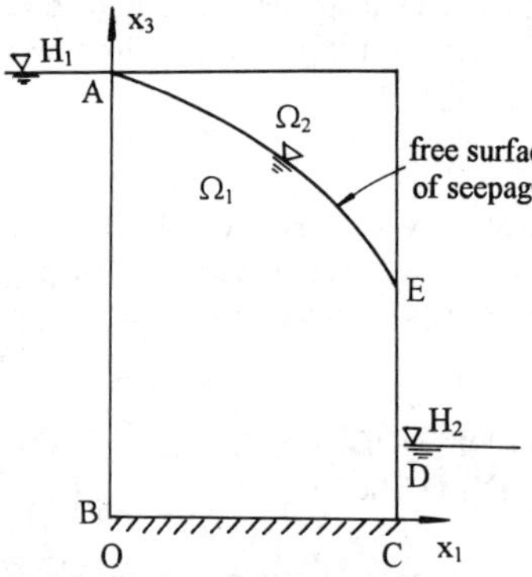

Fig.2 Steady-state saturated flow with a free surface of seepage

It is well known that the functional, and the related finite element method equation for solving the foregoing problem are Eqs.(6) and (7).

$$\prod(H) = \frac{1}{2}\int_{\Omega_1} k_{ij}H,_j H,_j d\Omega, \qquad (6)$$

$$[K_1]\{H_1\} = \{Q_1\}. \qquad (7)$$

Where $[K_1]$, $\{H_1\}$, and $\{Q_1\}$ stand for the global permeability matrix, and vectors of unknown nodal point heads and nodal equivalent flow rates, respectively.

The exact location of the free surface, and thus the size of the real seepage domain are unknown at the beginning of the solution. The simple problem has to be solved through numerical iterations. In the finite element procedure of nodal virtual flow rate the calculation domain, Ω, normally includes both of the real and virtual subdomains, i.e., $\Omega = \Omega_1 \cup \Omega_2$. Likewise, we may also establish the respective equations for domains Ω_2 and Ω , as follows:

$$[K_2]\{H_2\} = \{Q_2\}. \qquad (8)$$

$$[K]\{H\} = \{Q\}. \qquad (9)$$

Theoretically, we can appropriately so put some zero entries into $[K_1]$,$[K_2]$; $\{H_1\}$, $\{H_2\}$; and $\{Q_1\}$, $\{Q_2\}$ that there are

$$[K]=[K_1]+[K_2], \qquad (10)$$

or $$[K_1]=[K]-[K_2], \qquad (11)$$

and $$\{Q_1\}=\{Q\}\{Q_2\}. \qquad (12)$$

Substituting Eqs.(11) and (12) into Eq.(7), we have

$$[K]\{H\}=\{Q\}-\{Q_2\}+\{\Delta Q\} \qquad (13)$$

Where $$\{\Delta Q\}=[K_2]\,\{H\} \qquad (14)$$

is refered to as the vector of nodal virtual equivalent flow rate and is only contributed by the virtual elements located in domain Ω_2. In fact, $\{\Delta Q\}$ has the function to counterweigh the virtual flow rates on the left-hand-side of Eq.(13).

Since Eq.(13) is defined in the global domain, Ω, which is known prior to the solution, the problem may be solved without mesh iteration. It is apparent that both of the vectors, $\{Q_2\}$ and $\{\Delta Q\}$, are dependent on the solution,$\{H\}$. For the fact that the entries of the vector $\{Q_2\}$ are normally small and they have, in general, a small influence on the whole solution of Eq.(13). Thus, during the course of the iterations Eq(13) is updated dominantly by the revision of vector, $\{\Delta Q\}$, based on the last updated approximate midsolution and finally has the same solution of Eq.(7). Therefore, the numerical method is refered to as the procedure of nodal virtual flow rate by the author.

3 SUBSTRUCTURE TECHNIQUE FOR SOLVING SEEPAGE PROBLEMS WITH DRAINING HOLES

3.1 Drainage substructure

Fig.3 is the definition of a typical drainage substructure including one draining hole extracted from a hexahedral eight-node brick element mesh. In order to refinedly simulate the seepage behavior of the hole, the substructure is discretized into (m-1)layers elements(m>2) in elevation and into 3 layers in horizontal direction. The substructure has 12(m-1) elements. Any interaction between the substructure and the adjacent domain is dependent on and implemented by the substructure's external nodes n^1_1, n^1_2, n^1_3, n^1_4,..., n^i_1, n^i_2, n^i_3, n^i_4,..., n^m_1, n^m_2,n^m_3, n^m_4, and those located on the top and bottom faces. The total number of the external nodes is 4m+24.

In order to establish the equations used in static condensation of substructure, it is assumed that the permeability matrix and corresponding nodal point head and flow rate vectors of the substructure under consideration are partitioned into the form

$$\begin{bmatrix} k_{11} & k_{12} \\ k_{21} & k_{22} \end{bmatrix} \begin{Bmatrix} h_1 \\ {}_2 \end{Bmatrix} = \begin{Bmatrix} q_1 \\ q_2 \end{Bmatrix} \tag{15}$$

$$h_1 = k_{11}^{-1}(q_1 - k_{12}h_2) \tag{16}$$

$$k_{22}^{\cdot}h_2 = q_2^{\cdot} \tag{17}$$

Where h_1 and h_2 stand for the vectors of nodal point hydraulic heads to be condensed out and retained; The submatrices, k_{11}, k_{12}, k_{21}, and k_{22}, and vectors, q_1 and q_2, correspond to the vectors, h_1 and h_2; $k_{22}^{\cdot} = k_{22} - k_{21}k_{11}^{-1}k_{12}$ and $q_2^{\cdot} = q_2 - k_{21}k_{11}^{-1}q_1$ are the permeability matrix and right-hand-side vector of the substructure after condensation.

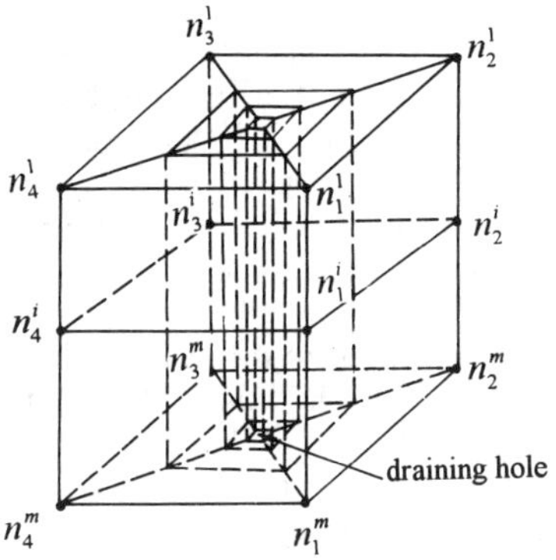

Fig.3 Definition of a drainage substructure including one draining hole

It is noteworth that there may be more than one drainage hole (even a few of rows and columns of draining holes) involved in one substructure when solving practical engineering problems. There are different types of substructure. Because many sub-structures are identical, it is effective in a computer program to establish a library of substructures from which a condensed global permeability matrix can be formed.

3.2 *Substructure technique for solution to seepage problems with draining holes*

3.2.1 *Condition one: draining hole not intersecting the free surface*

If draining holes do not intersect the free surface and are located in the real seepage domain. All of the elements involved in substructures are real elements. Those substructures can be treated as real superelements. According to the given boundary conditions of the respective draining holes and the partitions of the internal and external nodes of the substructures, the related permeability matrices and the right-hand-side vectors can directly be formed and participate in the assemblage of Eq.(13). If a whole drainage substructure is located upper the free surface, all of the elements involved in the substructure are virtual elements and the substructure totally has no seepage influence and need not to be assembled into Eq.(13).

3.2.2 *Condition two: draining holes intersecting the free surface*

In practice, draining holes often intersect the free surface. Thus, there are some substructures whose parts are located in the real domain and parts in the virtual domain. The substructures are partly composed of real elements, partly of virtual elements, and partly of transition elements which do intersect the free surface. Because the location of the free surface in substructures are also unknown at the beginning, the solution under this condition is relatively more complicated. We have to plant the concept and theory of the foregoing procedure of nodal virtual flow rate into the drainage substructure technique so that the permeability matrix and right-hand-side vector of such a substructure can correctly be revised and the influence of the virtual elements can be eliminated. After the revision the permeability matrix, $k_{22}^{\cdot\cdot}$, and the right-hand-side vector, $q_2^{\cdot\cdot}$, of the substructure which takes part in the assemblage of Eq.(13) are as follows:

$$k_{22}^{\cdot\cdot} = k_{22} - k_{21}k_{11}^{-1}k_{12} - \Delta k_{22}^{\cdot} \tag{18}$$

and

$$q_2^{\cdot\cdot} = q_2 - k_{21}k_{11}^{-1}q_1 - \Delta q_2^{\cdot} \,. \tag{19}$$

Where $\Delta k_{22}^{\cdot}$ and $\Delta q_2^{\cdot}$ are the virtual permeability matrix and the virtual equivalent nodal point flow rates contributed by the transition and virtual elements involved in the substructure.

It is worthy to be noted that the drainage substructure techniques for solution to problems under such two conditions can numerically be implemented together in a program and even under the later more complicated situation we also only need, according to Eqs.(18) and (19), to revise the permeability matrices and right hand side vectors of those substructures which the free surface goes through during iterations4. It is also the fact that the revision can be realized at the same time of the calculating of the vector of nodal point virtual flow rates, $\{\Delta Q\}$, in Eq.(13). The additional computation effort for the treatment of substructures is, thus, small. Our numerical experiments have proven that the iterations can converge quickly even for solution to pretty complicated problems.

4 SOLUTION OF THE SEEPAGE FIELD OF DACHAOSHAN UNDERGROUND HYDRO-POWER STATION

4.1 Generals and the design scheme of seepage control

The hydropower station is located on Renchan River in Yunnan Province, China. On the right bank where the underground station is laid out there are a few of major openings: the powerhouse cavern, baseline tunnel, transfer chamber, tailwater tank and tunnels, and six intake and outlet conduits, etc. The fractured rock mass surrounding the openings may, hydrogeologically speaking, be treated as imhomogeneous and anisotropic porous media and partitioned into 16 subdomains with different tensors of permeability which are back analysed and calibrated[9]. A few of major faults in the area have separately been taken into account and incorporated into the numerical model. The main measures of the seepage control scheme are as follows:

---In the foundation beneath the intake dam blocks there is a vertical grout curtain which is about 130m deep and 3m thick and just behind the relative impervious curtain a 40m-deep drainage curtain composed of draining holes at 3~4m centers is also installed.

---There are 13 horizontal drainage galleries which are laid out surrounding the powerhouse, transfer chamber, and tailwater tank either longitudinally or transversely and vertically elevated at 3 layers. On horizontal plan the galleries are intersected each other to form a complete rectagular tri-dimensional drainage curtain system with the help of the following-said draining holes.

---In the midarea between the transfer chamber and tailwater tank there also are two draining galleries located longitudinally parallel to the openings and at different elevations.

---In the bottom and/or top areas of all of the above-mentioned 15 galleries more than 1000 draining holes are laid out at 4~6m centers with an internal diameter of 0.07m and depths from 25 to 50m. The holes are installed either vertically or declinedly respective of the local hydrogeologies.

---A lot of short and irregular auxiliary draining holes are also arranged in the peripheral surficial rocks of all major openings.

Due to the complicated characters of the permeabilities of the rock mass and the intricacies of the influence of the draining measures, in particular the effectiveness of the numerous holes, the behavior of the seepage field in the area under consideration is very complicated. Without the drainage substructure technique mentioned above the seepage field couldn't be solved properly.

4.2 Solution of the seepage field

Based on the local geography, hydrogeology, chronical instrumentations of the underground water, and results of the back analysis[9], the numerical computation domain is considered to be properly selected. The internal vertical face of the domain on the bank side and the top surface below the elevation of the reservior water level are taken as the inflow boundaries. The contribution of the hydraulic heads on the vertical face is also determined through the back analysis. The top surface of the domain above the elevation of the reservior water level is assumed as possible exit surface of seepage.

The domain is preponderately discretized by hexahedral eight-node isoparametric elements. In order to obtain a high accurate solution of the seepage field and on the condition we have developed the procedure of nodal virtual flow rate, the seepage behaviors of the internal peripheral surfaces of all the major and draining openings are simulated mathematically strictly and all of the numerous draining holes except the auxiliary ones are also strictly modelled with the drainage substructure technique. The finite element mesh comprises 8855 elements and 10968 nodes excluding the internal elements and nodes of all of the substructures. If the internal elements and nodes are also counted there are 28903 elements and 47320 nodes involved in the mesh. Therefore, the way adopted here for the solution is unprecedent. The iteractions and results of the solution have been realized and obtained with a PC computer. Fig.4 shows a multilayer drainage substructure including 5 draining holes for treating the draining holes located in the area having

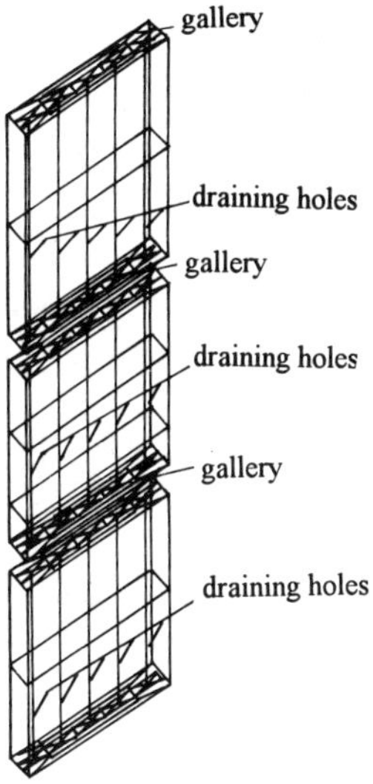

Fig. 4 Multilayer drainage substructure including 5 draining holes

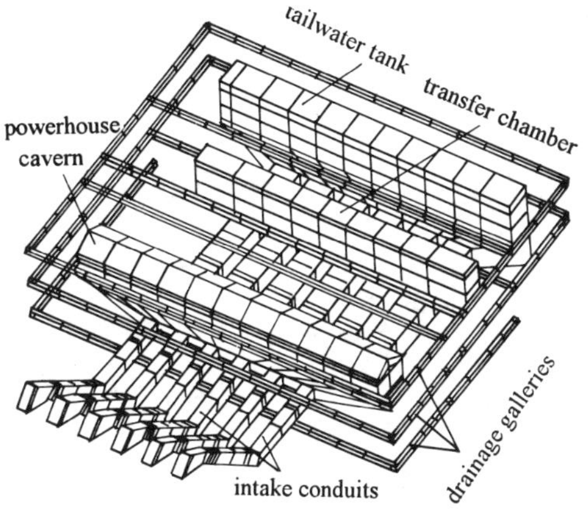

Fig. 5 Tri-D. apparent picture of structure openings involved in the mesh

multilayers of drainage galleries upstream of the powerhouse. A tri-dimensional apparent drawing of the structure openings involved in the mesh is depicted (Fig. 5). A contribution of isohead lines on a typical vertical section in the powerhouse area is given in Fig. 6.

5 CONCLUSIONS

The foregoing advance that the drainage substructure

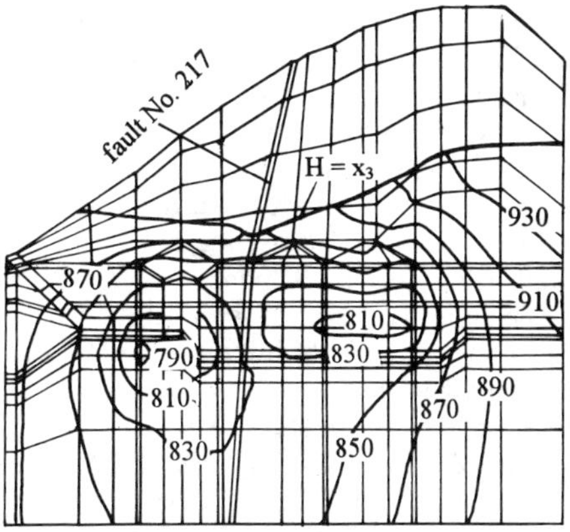

Fig. 6 Contribution of isohead lines on a typical vertical section in the powerhouse area

technique have been incorporated into the procedure of nodal virtual flow rate has gigantically improved the adaptability and capability of finite element method for solution to complicated seepage problems. With the help of the technique the seepage field of the underground hydropower station of Dachaoshan with a lot of structure openings, 15 draining galleries, and numerous (more than 1000) draining holes has been solved refinedly and unprecedently. It is supposed that the foregoing procedure and technique deserve their high application worth to practical flow problems in geoengineering. We have also successfully applied them to the analysis of the seepage behavior and control design of the 192.5m-high RCC gravity dam of Longtan, China. The concept and technique of the drainage substructure may directly be cited to strictly model the seepage behaviors of major discontinuities in rock masses. Recently, we have been working with this idea for a dam project.

REFERENCES

[1] K. J. Bathe and M. R. Khoshgoftaar: Finite element free surface analysis without mesh iteration, Int. J. for Numer. Methods in Geomech., Vol.3, 1979.

[2] W. Wittke: Felsmechanik---Grundlagen fuer Wirtschaftliches Bauen im Fels, Springer-Verlag, 1984.

[3] Baoyu Su and Yueming Zhu: Locating the free surface of seepage with a fixed mesh(in Chinese), J. of Hohai University, Vol.19, No.6,1991.

[4] Yueming Zhu et al.: Some adaptive techniques for solution of free surface seepage flow through arch dam abutments, In: Procs. of the Intern. Symp. on Arch Dams, Nanjing, Oct.,1992.
[5] Lei Wang, Zhong Liu, and Youtian Zhang: Analysis of seepage field with drainage curtain(in Chinese), J. of Hydraulic Engineering, No.4, 1992.
[6] Yueming Zhu et al.: Solution of complicated seepage field in the area of structure openings complex of underground hydropower station with improved drainage substructure method(in Chinese), J. of Hydraulic Engineering,No.9,1996.
[7] Yueming Zhu and Liaojing Zhang: Drainage substructures for solution of seepage fields including draining holes intersecting free surface of seepage (in Chinese), J. of Geotechnic Engineering(to be published).
[8] Yueming Zhu: On calculation of Darcy's seepage rate with higher accuracy based on a finite element method hydraulic head solution(in Chinese), J. of Hohai University (to be published).
[9] Yueming Zhu and Liaojing Zhang: Back analysis of permeability tensors of fractured anisotropic rock masses(in Chinese), J. of Geomechanics(to be published).
[10] Yueming Zhu et al.: Analysis of the seepage control design of the high roller-compacted-concrete gravity dam of Longtan(in Chinese), J. of Hydraulic Engineering(to be published).

Computer Methods and Advances in Geomechanics, Yuan (ed.) © 1997 Balkema, Rotterdam, ISBN 90 5410 904 1

High advective transport in mesh oblique flow using a symmetrical streamline stabilization

G. Schmid
Department of Civil Engineering, Ruhr-Universität Bochum, Germany

E. Wendland
Institute for Applied Geology, Ruhr-Universität Bochum, Germany

ABSTRACT: In this paper a high advective transport problem in mesh oblique flow is simulated using the Symmetrical Streamline Stabilization (S^3) method. The basis of the technique is to treat the transport equation in two steps. In the first step the dispersion part is approximated by a standard Ritz–Galerkin approximation, while in the second one the advection is approximated by the least squares method. The two parts are again assembled, resulting in one system of equations. The coefficient matrix results to be symmetric. Only half part of the sparse matrix has to be stored. Fast iterative algorithms for symmetrical systems of equations like preconditioned conjugate gradient method (PCG) can successfully be used. The new method leads to the introduction of an artificial diffusion term. Solute transport with high Peclet and Courant numbers does not lead to oscillations due to an inherent upwind damping. The upwind effect acts only in flow direction. The efficiency of the new formulation in terms of accuracy and computation time is shown in comparison with the standard nonsymmetrical approach.

1 INTRODUCTION

The reliability analysis of underground repositories involves in many cases the simulation of fluid flow and contaminant spreading in fractured porous media. Due to the high flow velocity in the fractures, the transport process is dominated by the advective component. The solution of this problem by the traditional finite element approach becomes numerically inefficient. In order to avoid spurious oscillations, a fine discretization is needed, leading to a huge effort concerning storage and computation time.

A new method combining computational savings, stability and accuracy is proposed in this paper. According to the operator splitting technique described by Marchuk (1995), the governing equation will first be decomposed into two parts. Extending the method of König (1994) the advective and dispersive terms will be approximated by different finite element techniques. After some mathematical operations, the two parts can be coupled again, leading to an unique system of equations. Although the advective term is considered in the implicit part of the equation, the resulting coefficient matrix remains symmetric. Under this condition, a fast and robust preconditioned conjugate gradient method (PCG), e.g. Schmid and Braess (1988), can be used. Another characteristic is the presence of an 'upwinded' advection, which assures the stability of the numerical solution.

2 GOVERNING EQUATIONS

We consider a single phase (water) flow with a single dissolved solute moving through a homogeneous, saturated porous medium. The flow field is independent of the solute concentration and is at steady–state. Chemical reactions and decay will not be considered, though they can easily be incorporated to the mathematical and numerical models.

In this case, the transient transport of dissolved solutes is governed by the well-known advection–dispersion equation

$$\frac{\partial c}{\partial t} + \mathbf{v}\nabla c - \nabla(\mathbf{D}\nabla c) = Q\,(c^* - c) \qquad (1)$$

where c is the solute concentration, t is the time, $\mathbf{v}$ is the velocity vector, $\mathbf{D}$ is the hydrodynamic dispersion tensor, defined by the mechanical dispersion and molecular diffusion and Q is the source term with concentration c^*.

3 NUMERICAL APPROACH

3.1 *Splitting up*

The differential equation (1) can be discretized in time considering a first–order finite difference approach with the weighting factor θ

$$\frac{(c^+ - c^-)}{\Delta t} + \theta(L_1 + L_2)c^+ + (1-\theta)(L_1 + L_2)c^- = f \quad , \qquad (2)$$

where

c^+ = solute concentration at time level $(t + \Delta t)$;

c^- = solute concentration at time level (t);

Δt = time step ;

$L_1 = -\nabla(\mathbf{D}\nabla) + Q$;

$L_2 = \mathbf{v}\nabla$;

$f = Qc^*$.

We introduce the dispersive component c^d of the concentration as a temporary unknown. According to Marchuk (1995), equation (2) can be decomposed in a dispersive part, which depends only on the linear operator L_1

$$\frac{c^d}{\Delta t} + \theta L_1 c^d = \frac{c^-}{\Delta t} - (1-\theta)L_1 c^- + f \qquad (3)$$

and an advective part, which depends on the linear operator L_2

$$\frac{c^+}{\Delta t} + \theta L_2 c^+ = \frac{c^d}{\Delta t} - (1-\theta)L_2 c^d \quad . \qquad (4)$$

The spatial derivatives will be discretized following the techniques proposed by König (1994). The exact solution for each step will be approximated by means of the finite element method using the standard interpolation scheme

$$c(x,y,z,t) \simeq \sum_{j=1}^{N} \varphi_j(x,y,z)\, c_j(t) \qquad (5)$$

in which φ_j is a linear base function.

3.2 *Dispersive term*

For the dispersive part given in equation (3), the solution is obtained by using the standard Ritz–Galerkin approach. The minimum is obtained by weighting the residuum with the test function φ_i

$$\int_\Omega \varphi_i \left[\frac{\Sigma\varphi_j c_j^d}{\Delta t} + \theta L_1\, \Sigma\varphi_j c_j^d\right] d\Omega = \int_\Omega \varphi_i \left[\frac{\Sigma\varphi_j c_j^-}{\Delta t} - (1-\theta)L_1\, \Sigma\varphi_j c_j^- + f\right] d\Omega \qquad (6)$$

which can be rewritten in matrix form as

$$\left(\frac{\mathbf{M}}{\Delta t} + \theta\mathbf{B}\right)\underline{c}^d = \left[\frac{\mathbf{M}}{\Delta t} - (1-\theta)\mathbf{B}\right]\underline{c}^- + \mathbf{F} \qquad (7)$$

with

$$\mathbf{M} = \int_\Omega \varphi_i\varphi_j\, d\Omega \;;$$

$$\mathbf{B} = \int_\Omega \mathbf{D}\,\nabla\varphi_i\nabla\varphi_j\, d\Omega + \int_\Omega Q\,\varphi_i\varphi_j\, d\Omega \;;$$

$$\mathbf{F} = \int_\Omega Qc^*\,\varphi_i\, d\Omega \;.$$

3.3 *Advective term*

The advective part given in equation (4) can be solved by means of the least squares method. The residuum of the approximation will be minimized, weighting it with the test function $\frac{\varphi_i}{\Delta t} + \theta\mathbf{v}\nabla\varphi_i$

$$\int_\Omega \left(\frac{\varphi_i}{\Delta t} + \theta\mathbf{v}\nabla\varphi_i\right)\left[\frac{\Sigma\varphi_j c_j^+}{\Delta t} + \theta L_2 \Sigma\varphi_j c_j^+\right] d\Omega = \int_\Omega \left(\frac{\varphi_i}{\Delta t} + \theta\mathbf{v}\nabla\varphi_i\right)\left[\frac{\Sigma\varphi_j c_j^d}{\Delta t} + (1-\theta)L_2 \Sigma\varphi_j c_j^d\right] d\Omega \;. \qquad (8)$$

It follows in matrix form

$$\left[\frac{\mathbf{M}}{\Delta t} + \theta\left(\mathbf{V} + \mathbf{V}^T\right) + \mathbf{U}^*\right]\underline{c}^+ = \left[\frac{\mathbf{M}}{\Delta t} - (1-\theta)\mathbf{V} + \theta\mathbf{V}^T - \frac{(1-\theta)}{\theta}\mathbf{U}^*\right]\underline{c}^d \qquad (9)$$

where

$$\mathbf{V} = \int_\Omega \mathbf{v}\,\varphi_i\nabla\varphi_j\, d\Omega \;;$$

$$\mathbf{U}^* = \theta^2\Delta t \int_\Omega \mathbf{v}^T\mathbf{v}\;\nabla\varphi_i\nabla\varphi_j\, d\Omega \;.$$

3.4 *Symmetrical Streamline Stabilization (S^3)*

The traditional procedure of the splitting technique is to solve the equation (7) and consequently substitute the temporary concentration c^d into the equation (9) getting the desired solution at the new time level. This substitution can be done analytically (for a detailed discussion see Wendland 1995).

The vector $\frac{\mathbf{M}}{\Delta t}\underline{c}^d$ appears in both expressions (7) and (9) through which they can be coupled to one equation. After extrapolation for elimination of the temporary concentration c^d it results in the following assembled form:

$$\left[\frac{\mathbf{M}}{\Delta t} + \theta(\mathbf{B} + \mathbf{V}) + \theta\mathbf{V}^T + \mathbf{U}^*\right]\underline{c}^+ =$$

$$\left[\frac{\mathbf{M}}{\Delta t} - (1-\theta)(\mathbf{B} + \mathbf{V}) + \theta\mathbf{V}^T - \frac{(1-\theta)}{\theta}\mathbf{U}^*\right]\underline{c}^-$$

$$+ \mathbf{F} \tag{10}$$

The analysis of the resulting equation shows the advantages of the new method. Using the least squares method, the resulting coefficient matrix is always symmetric, even when the advective term appears in the implicit side of the equation. The symmetry is due to the presence of the term $\theta\mathbf{V}^T\underline{c}^+$ on the left side of the equation.

Another characteristic is the upwind effect on the advective term, which is responsible for the stabilization of the numerical solution. It results from the combination of the terms $\mathbf{V}$ and $\mathbf{U}^*$, as shown in Fig. 1.

This upwind term $\mathbf{U}^* = \theta^2\Delta t\mathbf{v}^2$ can be analyzed in terms of the stability criteria for the one dimensional case. Considering the Courant number $C_o = \frac{v\Delta t}{\Delta l}$ it can be transformed in

$$\mathbf{U}^* = \theta^2 C_o \Delta l v \ . \tag{11}$$

After introduction of the Peclet number $P_e = \frac{v\Delta l}{D}$ one achieves:

$$\mathbf{U}^* = \theta^2 C_o P_e D \ . \tag{12}$$

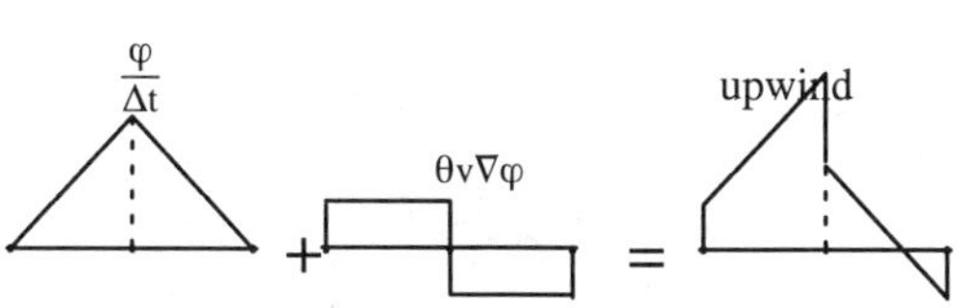

Fig. 1 : Weighting function for the advective term (modified from König 1994).

The upwind term appears to be an upscaling of the natural dispersion of the problem, through which the numerical computation is stabilized avoiding spurious oscillation. In the S^3–procedure the artificial diffusion introduced by the numerical method is controlled by the time (Θ, C_o) and space (P_e) discretization. Due to the vector product $\mathbf{v}^T\mathbf{v}$ the stabilization acts only in the flow direction. This is also true in two or three dimensional problems.

A comparison of the developed scheme (eq.10) with a traditional method using an unsymmetrical weighting function leads to the observation that some higher–order derivatives are missing in the approximation equation. In a general case (using quadratic or cubic interpolation function) this property can be seen as an inconsistency. Such problems can be avoided choosing linear interpolation functions for which the higher–order derivatives naturally disappear reducing the inconsistency.

An important characteristic is the behavior of the method for divergence free flow fields. In this case the symmetrization leads to an explicit approximation of the advective term and oscillation can be expected for high Courant numbers. As it will be shown in the application the oscillation does not occur due to the stabilization generated by the upwind effect.

4 APPLICATION

The efficiency of the new method is demonstrated in a two–dimensional test case. The system consists of a rectangular domain uniformly discretized. The flow field is diagonal oriented. A continuous contaminant source is placed on the upstream boundary as shown in Fig.2. The remaining conditions are given as no flux over the boundary. All dimensions are in [m]. Considering a conductivity of K=100.0 m/s the flow equation leads to an average velocity of 1.5 m/s, which was used to compute the Courant and Peclet numbers.

The results obtained with the proposed method (S^3) are compared with the standard Ritz–Galerkin approach (G). For this transport problem there is no analytical solution and the convergence of the methods to

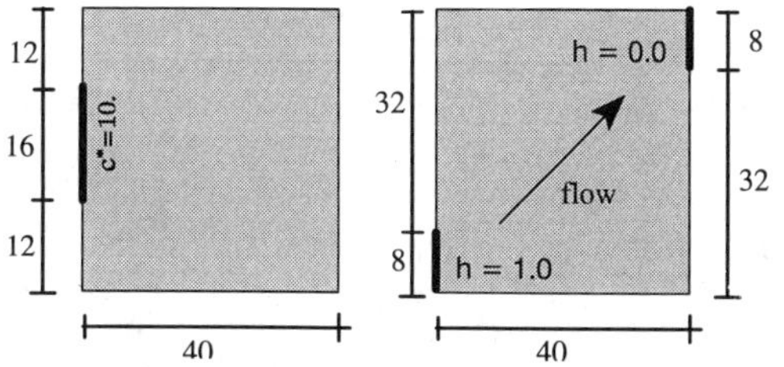

Fig. 2 : Test case (dimensions in m). Remaining conditions are given as no flux over the boundary

the correct solution is achieved by refining the mesh. The problem is solved with initial condition c=0 over all the domain using a regular mesh for different values of Peclet and Courant numbers up to the time t=40s. The time discretization was weighted with θ=1.0 (implicit Euler method). The parameters used for the different cases are shown in Table 1.

The triangles in Figs. 3–6 are spatial references in order to compare the numerical concentration distributions.

In the first case, the transport is diffusion dominated and of parabolic character. The stability criteria (P_e<2.0 and C_o<1.0) are observed. The numerical solution of this problem with the S^3–scheme is well behaved (Fig. 3) and approximates the standard results although one can observe the increased numerical diffusion due to upwind. After mesh refinement for case 2 the results with the S^3–scheme converge to the solution with the standard method (Fig. 4).

Table 1: Discretization parameters for the numerical simulation.

Case	Δt (s)	D_L (m^2/s)	D_T (m^2/s)	Δx (m)	P_e	C_o
1	2.0	2.0	0.2	2.0	1.0	1.5
2	1.0	2.0	0.2	0.5	0.25	3.0
3	1.0	0.02	0.002	2.0	100.	0.75
4	1.0	0.02	0.002	0.5	25.0	3.0

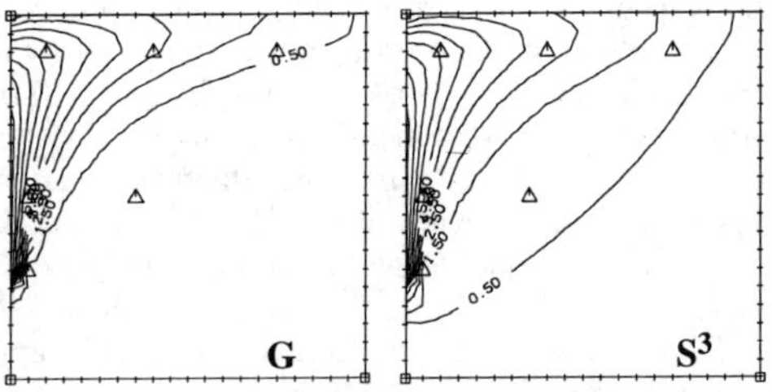

Fig. 3 : Concentration distribution for case 1.

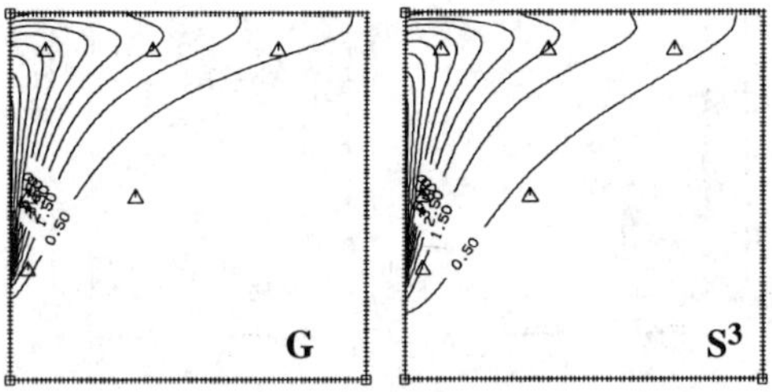

Fig. 4 : Concentration distribution for case 2.

In the third case, the transport is advection dominated and of hyperbolic character. Because of the discretization chosen, the stability criteria are not respected (P_e>>2.0). In this situation difficulties due to numerical dispersion can be expected. The computed concentration distributions after 40s are shown in Fig. 5. The standard method presents serious numerical problems due to the high Peclet number. The results oscillate strongly resulting in an unusable solution. For the S^3–scheme the problem does not occur due to the oscillation damping created by upwind.

For case 4 the mesh would refined and the results for the standard method (G) agree with the S^3–scheme (Fig. 6) converging to the correct solution. There remain isolines with negative concentration (–0.50) which appear due to the difficult numerical conditions of this case.

In Table 2 a comparison of storage requirements and computation time after 40 time steps for a mesh with 6400 nodes is shown.

The standard method leads to a nonsymmetrical matrix, for which a direct solver is chosen. The full matrix has to be stored using a *band technique* (with

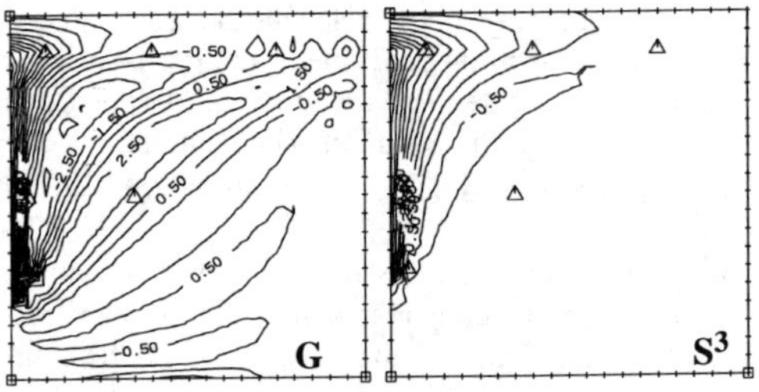

Fig. 5 : Concentration distribution for case 3.

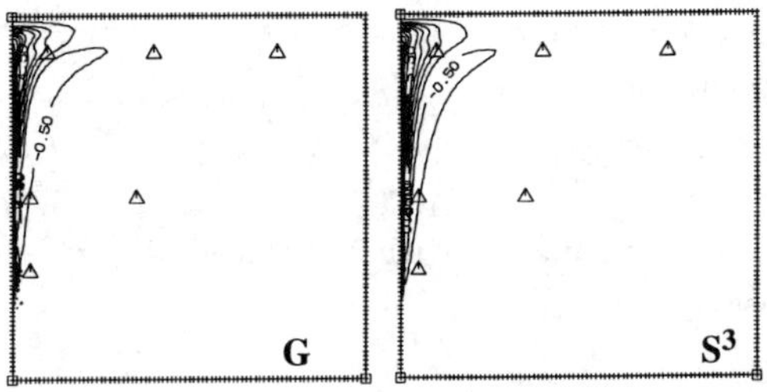

Fig. 6 : Concentration distribution for case 4.

Table 2: Storage and CPU requirements for the test case.

	Standard	S^3
Storage	MxN	15xN
CPU(s)	3296.0	110.0

M as bandwith). For the S^3–scheme a robust preconditioned conjugate gradient solver (PCG) with *sparse storage* can be used. The savings on computer memory for large models (M>>15) is evident.

The computation times (CPU) given in Table 2 relate to a workstation RISC6000/550 (24,8 MFLOPS). The time consumed by the new method is considerable smaller then for the standard case. Although conjugate gradient solvers exist for nonsymmetrical matrices, they are not so robust and efficient as a symmetric one. Therefore the nonsymmetrical solvers have not been considered.

5 CONCLUSION

The results presented here allow the following conclusion: for high advective transport problems in mesh oblique flow the standard finite element method leads to numerical problems with spurious oscillation. By refining the mesh, this problem can be overcome, but the numerical effort increases.

The Symmetrical Streamline Stabilization (S^3) scheme presents a good alternative. It proves to be a robust algorithm for solving the advection–dispersion equation, as well as for advection as for dispersion dominated problems. Starting with an operator split, the method leads to a single symmetric coefficient matrix with the advective term being considered in the implicit part of the equation. As a consequence, a symmetric preconditioned conjugate gradient solver can be used. Furthermore, the advective term is upwinded, assuring a stable solution without numerical oscillation, even for high advective transport.

In test cases for advection dominated transport the new method demonstrates to be more accurate than the standard approach. The computational effort in terms of time and storage is clearly reduced.

REFERENCES

König, C. : *Operator Split for Three Dimensional Mass Transport Equation,* Proceedings of Computational Methods in Water Resources X, 1, 309–316, Heidelberg, 1994.

Marchuk, G. I. : *Adjoint equations and analysis of complex systems*, Dordrecht: Kluwer, 1995.

Schmid, G. and Braess, D. : *Comparison of fast equation solvers for groundwater flow problems,* Proceedings of Groundwater Flow and Quality Modelling*: 173–188*, Dordrecht: Reidel Publ. Comp., 1988.

Wendland, E. : *Numerical simulation of flow and high advective solute transport in fractured porous medium*, (Dissertation in German), Department of Civil Engineering, Ruhr–University Bochum, 1995.

Computer Methods and Advances in Geomechanics, Yuan (ed.) © 1997 Balkema, Rotterdam, ISBN 90 5410 904 1

Effects of loading direction on localized flow in fractured rocks

Xing Zhang & David J. Sanderson
Geomechanics Research Group, Department of Geology, University of Southampton, UK

ABSTRACT: Understanding of fluid flow in fractured rocks remains an important issue in geomechanics. The localization of flow is of fundamental importance to all aspects of ground-water flow, which is characterized with a sudden transition from diffuse flow through fracture networks to highly localized flow generated by the openings at fracture intersections. The magnitude and orientation of horizontal stress is one of the most important parameters controlling the fluid flow and transport in fractured rocks. In this paper, numerical modelling of the coupled mechanical and hydraulic behaviour of fractured rocks using UDEC (Universal Distinct Element Code) was carried out for an improved understanding and prediction of such localized flow which likely occurred associated with unstable deformation of rocks where a highly differential stress existed.

1 INTRODUCTION

Localized behaviour is a common phenomenon in the upper crust. Localized flow may occur in different scales. For a scale of engineering site, any engineering activities, such as mine excavating, water flooding of oil reservoirs and waste discharge, which change the state of stress and/or pore pressure in a fractured rock mass, may lead to a sudden change of fluid flow in direction and magnitude (Montazer *et al.* 1982; Harper & Last 1990; Stormont 1990; Zhang 1993; Heffer *et al.* 1995).

In this paper, numerical modelling using UDEC has been carried out for a better understanding and prediction of such localized flow in three naturally fractured rocks. Combining with a new technique (Zhang *et al.* 1996; Zhang & Sanderson 1996), the permeability of three naturally fractured rocks has been evaluated, and the influence of *in situ* stress direction on the critical phase between diffuse plate flow and localized pipe flow has been investigated.

2 FLOW MODEL

Commonly, fractures in sedimentary rocks are normal to bedding plane surface (Lorenz *et al.* 1991; Engelder & Gross 1993; Gross 1993). Where the vertical or sub-vertical fractures dominantly control the permeability of fracture systems, it is reasonable to assume that the direction of one of principal permeability components is parallel to the vertical direction. When a fractured rock masses is loaded, some fractures may close, but some openings may be created by dilation. Hence, it is expected that the permeability will change in both the horizontal and vertical directions when loading. In this modelling, the calculation of horizontal flow-rates, q_h, is based on the cubic law of flow in fractures. The vertical flow-rates, q_v, in the third dimension is calculated with cubic law or/and pipe formula, depending on the shape of openings. Where a fracture is much longer than its aperture, the cubic law is used and q_v through a fracture within a unit cube is given by equation (1).

$$q_v = \frac{a^3 \Delta p\, \ell}{12\rho}, \qquad \frac{\ell}{a} \geq R \tag{1}$$

where a is the hydraulic aperture of the fracture; Δp is the hydraulic pressure differential between two ends of the fractures; ρ is the dynamic viscosity of the fluid, and ℓ is the length of the fracture. Where $\ell/a > R_c$, equation (1) is used to calculated flow-rate,

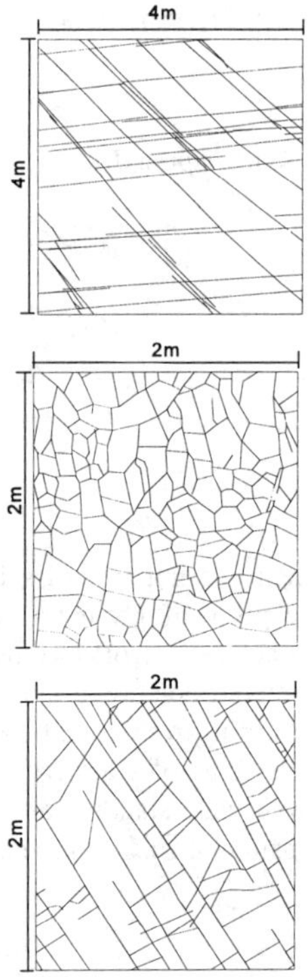

Fig.1 Natural fracture networks

Table 1 Material properties used for rocks and fractures for modelling

	Value	Units
Rock property		
Density	2500	kg m^{-3}
Shear modulus	13.8	GPa
Bulk modulus	21.8	GPa
Tensile strength	3.9	MPa
Cohesion	7.9	MPa
Friction angle	36	degree
Fracture property		
Shear stiffness	20	GPa m^{-1}
Normal stiffness	50	GPa m^{-1}
Tensile strength	0	MPa
Cohesion	0	MPa
Friction angle	25	degree
Dilation angle	5	degree
Residual aperture	0.001	m
Zero stress aperture	0.0001	m
Fluid property		
Density	1000	kg m^{-3}
Viscosity	0.00035	Pa s

q_v. Here, R_c is selected at a value of 3 because where R_c is larger than 3, the friction factor (Reynolds number) increases more quickly for an equivalent pipe, which is the function of average velocity in the pipe, the equivalent radius and the dynamic viscosity of the fluid. Otherwise, an opening is treated as an equivalent pipe (Sabersky *et al.* 1989) and the flow-rate, q_v, is calculated with equation (2)

$$q_v = \frac{\pi r^4 \Delta p}{8\rho} \qquad (2)$$

where r is the radius of an equivalent pipe:

$$r = \sqrt{\frac{(a\,\ell)}{2}} \qquad (3)$$

The hydraulic aperture of fractures, a, is given by equation (4) in which a_o is the aperture at zero effective normal stress, and u_n is the contact normal displacement of the fracture, controlled by the normal stress and the rock properties.

$$a = a_o + u_n \qquad (4)$$

3 GEOMETRIC MODEL AND LOADING SCHEME

Three natural fracture networks in sandstone (Table 1. and Fig.1) have been selected to examine the localized behaviour of fluid flow under loading, representing the range of fracture patterns typical for the Dounreay area, Scotland, and being commonly encountered in fractured sedimentary sequences.

It has been assumed that the rock mass is at a depth of 600m with a rock mass density of 2500kg/m^3. The overburden is 15MPa as a principal stress in the vertical direction, and the maximum and minimum horizontal stresses are 18 and 12MPa, respectively. Hydrostatic pore pressure is used (6MPa), so the effective principal stresses are 9, 12 and 6 MPa respectively in the three orthogonal directions. The maximum horizontal stress, S_H, is parallel to the y-direction and S_h parallel to the x-direction. Under the initial stress conditions, S_h decreases with an increment of 0.25MPa while S_H increases with the same increment, so only the differential stress increases and the mean stress

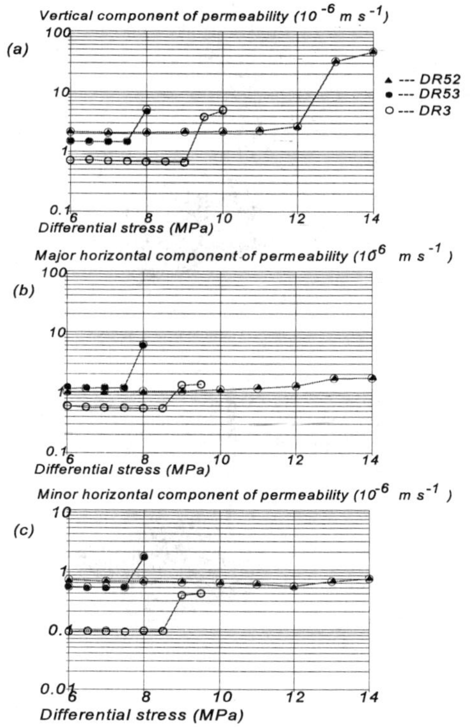

Fig. 2 Relationship between permeability components and applied differential stress

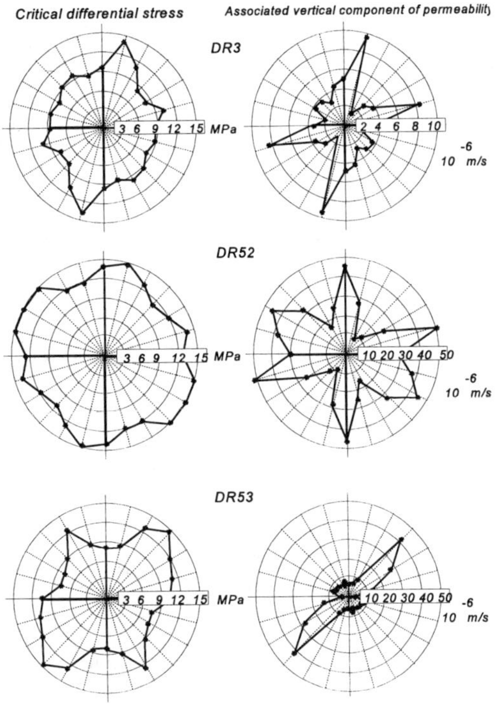

Fig. 3 Critical differential stress and vertical component at different directions

remains constant. In this way, it is expected that the deformation and the associated flow of these fractured rocks can be investigated in relation to the differential stress applied. Where a sudden transition of flow occurs, the applied differential stress is considered as the critical differential stress.

4 RESULTS

Figure 2 shows the permeability variation of the three fractured rocks with differential stress. obviously, there is a critical differential stress at which the permeability of a fractured rock has a sudden increase, indicating the transition between diffuse flow to localized flow.

For fracture networks DR3 and DR53, much lower differential stress can promote sudden transition from a low diffuse flow to a high permeability. This might be caused by the strong anisotropic features of the network geometry and a preferred loading direction. In order to examine the effects of loading direction, the three natural fracture networks are loaded at various directions with an interval of 15°. For the numerical experiments hereafter, only the vertical flow-rates are calculated, and the vertical component of permeability and its deformation is compared.

Figure 3 shows the critical differential stress at which the permeability transition occurs for the three networks at different directions and the associated vertical component of permeability.

Fracture network DR3 consists of two sets of systematic fractures, and as expected the loading direction (75^0) at which the peak of the critical differential stress (14MPa) is required to cause the permeability transition approximately bisects the obtuse angle of the two fracture sets. Where the loading direction has a preferred angle of about 25^0 with one of those continuous fractures, much lower differential stress of 8MPa can promote the permeability transition, such as at 30^0, 45^0 and 150^0. Also, the localized vertical flow is characterized in different ways at different directions. At 30^0 where the lowest differential stress is required, the localized flow is characterized by the slip dilation of a single

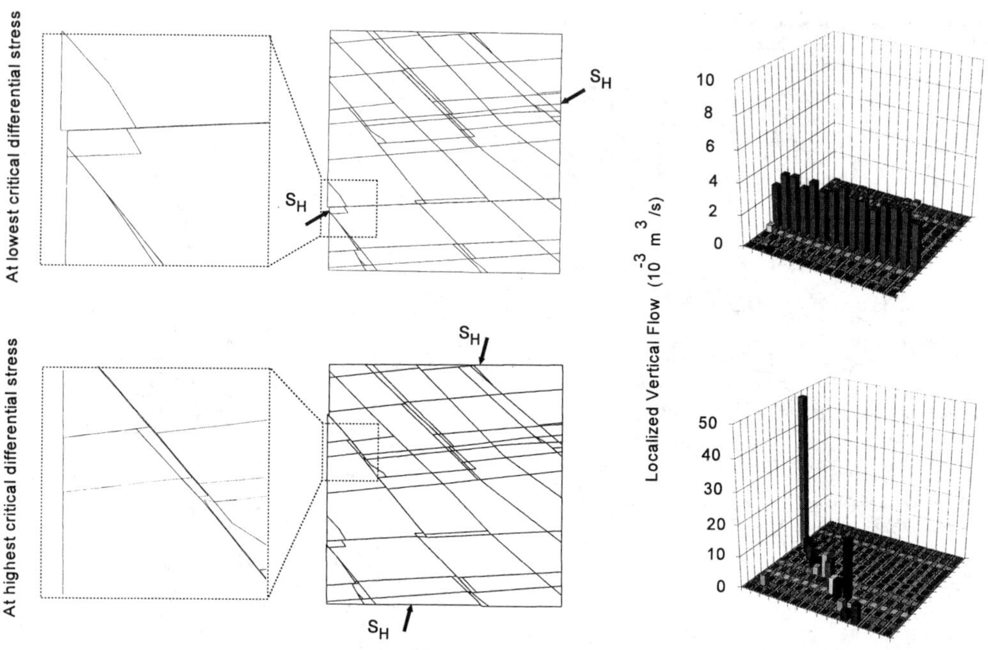

Fig.4a Geometry and vertical flow of DR3 after loading at two particular directions

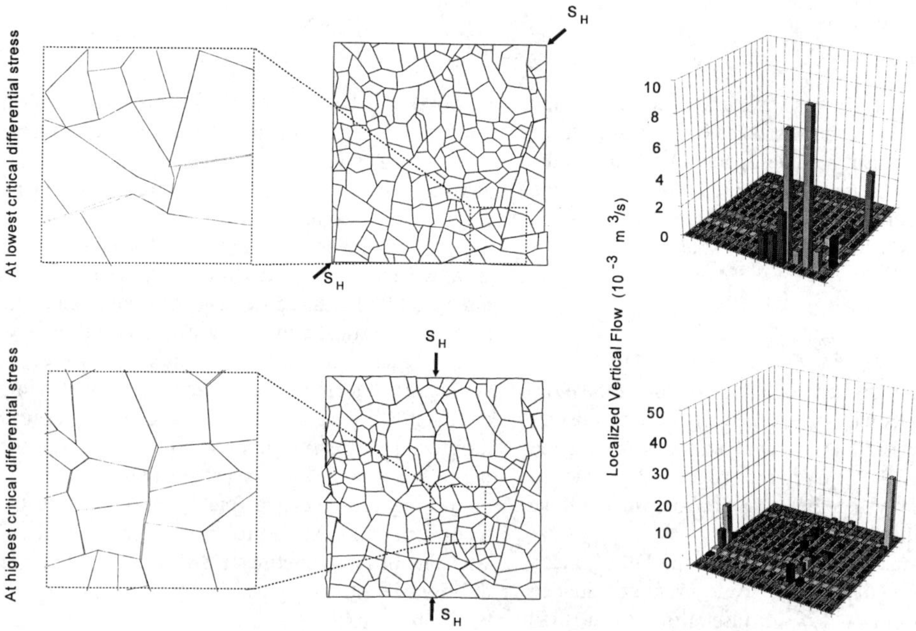

Fig.4b Geometry and vertical flow of DR52 after loading at two particular directions

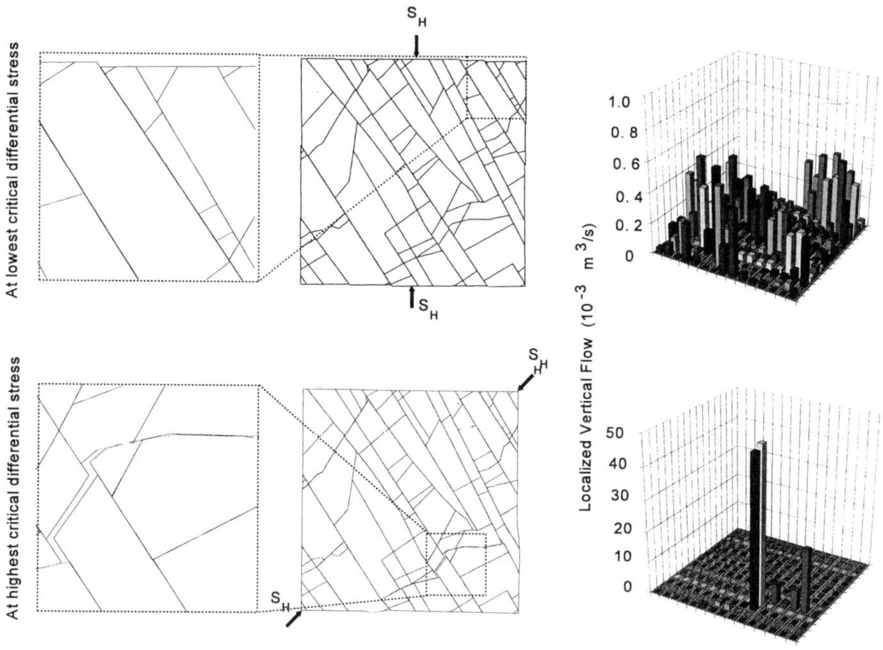

Fig.4c Geometry and vertical flow of DR53 after loading at two particular directions

fracture, and the vertical flow has a relatively low value. However, at 75^0 where the highest differential stress is required, the localized flow is characterized by the 'pipes' caused by the openings of some fracture intersections , and the vertical flow has a relatively high value (Fig.4a).

Non-systematic fracture network of DR52 behaves rather differently. There is no significant difference of the critical differential stress at various directions, but the associated vertical component of permeability has a dramatic variation with different loading directions (Fig.3). Where the intersects of fractures likely creates 'pipes' locally under loading, large vertical flow is observed. For example, the vertical flow at loading direction of 90^0 has a value of 4 times as high as that at loading direction of 45^0 because more 'pipes' are created in the former (Fig.4b).

Where there is a set of systematic fractures (DR53), the critical differential stress and associated vertical flow has very strong anisotropic features. At the direction normal or parallel to the systematic fracture set (45^0 or $120^{0)}$, higher differential stress is required, particularly normal to it. However, only at the direction normal to the systematic fracture set, the associated vertical flow has a much higher value than the others, including at the direction parallel to the fracture set (Fig.3). Where the loading direction is normal to the systematic fracture set, some of those cross fractures make large 'pipes' between the systematic fractures while at the other directions the instability is characterized by the dilational sliding of the systematic fractures. The vertical flow-rates at 45^0 and 90^0 are shown in Fig.4c.

5 CONCLUSIONS

The permeability of naturally fractured rock masses is predicted using a numerical approach (UDEC) for an enhanced understanding of localization of flow in these fractured rock masses. The effects of loading direction have been studied. The numerical modelling suggests that:

(1) At a given loading direction for a given fractured rock mass, there is a certain critical differential stress under which the localization of flow can occur.

(2) For a given fractured rock mass, there are two particular directions at one of which the lowest critical differential stress can promote the localization of flow, and at the other the highest critical differential stress is required.

(3) Those fracture networks consisting of systematic fractures likely have a noticeable anisotropic feature in the magnitude of the critical differential stress and the associated, localized flow.

REFERENCES

Engelder, T. & M. R. Gross 1993. Curving cross joints and the lithospheric stress field in eastern North America. *Geology*, **21:** 817-8.

Gross, M. R. 1993. The origin and spacing of cross joints: examples from the Monterey Formation, Santa Barbara Coastline, California. *J. Struc. Geol.*, **15:** 737-751.

Heffer, K. J., R. J. Fox, C. A. McGill & N. C. Koutsabeloulis 1995. Novel techniques show links between reservoir flow directionality, earth stress, fault structure and geomechanical changes in mature waterfloods. **SPE 30711.**

Harper, T. R. & N. C. Last 1990. Response of fractured rock subject to fluid injection. Part I: Characteristic behaviour. *Tectonophysics*, **172:** 33-51.

Lorenz, J. C., W. Teufel & N. R. Warpinski 1991. Regional fractures I: A mechanism for the formation of regional fractures at depth in flat-lying reservoirs. *American Association of Petroleum Geologists Bulletin*. **75:** 1714-1736.

Montazer, P., G. Chitomobo, R. King & W. Ubber 1982. Spatial distribution of permeability around CSM.ONWI Room, Edger Mine, Idaho Springs, Colorado, pp. 47-56, *Proc. 23rd Symp. on Rock Mechanics,* University of California, Berkely.

Sabersky, R. H., A. J. Acosta & E. G. Hauptmann 1989. *Fluid Flow*, Macmilan Publishing Company, New York, Third Edition.

Stormont, J. C. 1990. Discontinuous behaviour near excavation in a bedded salt formation. *Int. J. Mining and Geol. Eng.*, **8:** 35-5.

Zhang, X. 1993. Shear resistance of jointed rock masses and stability calculation of rock slopes. *Geotech. Geol. Eng.*, **11:** 107-124.

Zhang, X., D. J. Sanderson, R. M. Harkness & N. C. Last 1996. Evaluation of the 2-D permeability tensor for fractured rock masses. *Int. J. Rock Mech. Min. Sci. & Geomech. Abstr.*, **33:** 17-37.

Zhang, X. & D. J. Sanderson 1996. Effects of stress on the 2-D permeability tensor of natural fracture networks. *J. Geophysics Int.* **125:** 912-924.

Computer Methods and Advances in Geomechanics, Yuan (ed.) © 1997 Balkema, Rotterdam, ISBN 90 5410 904 1

Analyses of the seepage field in a mountain by high slope with rain infiltrating

Jiafa Zhang & Sishen Li
Yangtze River Scientific Research Institute, Wuhan, China

Y.Ohnishi & M.Tanaka
Kyoto University, Japan

ABSTRACT: Saturated-unsaturated flow model and FEM are adopted to simulate the seepage field in a mountain by high slope under rain infiltration condition in this paper. The results show that the infiltrating water under heavy rain would be detained over the top of less permeable zone and lower the efficacy of the drainage system designed on the bases of saturated model simulation results.

1 INTRODUCTION

Unsaturated water movement has been the object of studies in agricultural management and mass transportation estimation for a long time. It is concerned for the estimation of ground water pressure in slopes recently.

It is well known that heavy rain is often an important factor of slope movement. Because of weathering, generally rock permeability decreases as depth increases from ground surface. As it rains heavily, water infiltrates. With moving down into mountain, water encounters impedance greater and greater. Especially at an interface where permeability decreases sharply downward, it would become more easily for water to move laterally than downward, and at some condition water would accumulate on the interface to form perched water table. So water movement in the unsaturated area above ground water surface is complex, effecting water content and pressure distribution in rock mass greatly.

Water movement in a mountain by high slope under rain infiltration is simulated numerically by FEM in this paper. With the results, water pressure distribution and the efficacy of designed drainage system are analyzed.

2 MODEL OF THE SEEPAGE FIELD

Mathematical models of saturated-unsaturated water movement and numerical simulation methods have been reported by many authors. This paper adopted the model reported by Akai et al.(1977). The computer code is a revised one of theirs. Dealing with transient rain infiltration boundary condition is one of the new functions of the code (J. Zhang, 1997)

2.1 Simulated domain

The studied seepage field is in a granite rock

mountain, which is cut through by canals with high slopes formed. Just the seepage field in the mountain by the stretch of the highest slope is simulated The top of the mountain is cut off. The height of the slope is about 160 m, with the lower 56 m section being vertical. The farthest domain boundary is about 530 m away from the vertical slope. According to the simulated results of saturated seepage flow model, seven drainage tunnels are designed at different levels, and drainage holes with 90 mm diameter and 5m interval are drilled upward from the top of tunnels. Thanks to symmetry, it is reasonable just to simulate a stretch between two adjacent drainage wells for simplification.

2.2 Parameters

Canals cut through the completely-strongly and weakly weathered zone of the granite rock mass, and cut into the fresh rock. Hydrogeological survey and hydraulic tests have been done to investigate the fracture distribution and saturated permeability of the rock. On the bases of these results, it is considered that continuum model is appropriate for the simulation. Samples are taken from completely-strongly weathered zone to test unsaturated hydraulic properties, and a sample of fractured fresh rock was taken to test those properties recently. The results are not yet enough for the numerical simulation. So that the used parameters are based on assumption to some extent.

2.3 Boundary condition

This mountain locates in an area of seasonal torrential rain weather. The average annual rainfall is 1147 mm, and mainly concentrated on the duration from May to August. The largest daily rainfall on record is 386 mm. The initial water content distribution is not known. So in this work, a rain condition close to average annual rainfall is taken as a steady flow boundary condition. With these results as initial condition, unsteady flows are simulated under heavy rain conditions. From hydrological estimation and back analysis of saturated seepage field the average rain infiltration ratio(RIR) is obtained as 0.145 - 0.165. In this work RIR is taken as 0.165. Since the terrain rises and falls, and some drainage system is laid out on ground surface, it is assumed not to pond even under heavy rain. Therefore rain infiltration condition will be removed where and while it begins to pond.

The cut slope will be protected with concrete. Holes are drilled from the slope surface at slow angles up into rock mass. Thus the cut slope surface is taken as the boundary of drainage, but not of rain infiltration.

Canal bottom will be water exit surface at some time. The vertical surface opposite the high slope may be taken as a boundary of constant water head.

3 ANALYSIS OF RESULTS

Fig.1 shows FEM simulated results of the seepage field. The ground water surface at t=0 corresponds to the steady seepage field under rain condition similar to average annual rainfall, and is the initial state of following unsteady seepage field under heavy rain condition. It can be seen that ground water surface at t=0 is low, with only the lower part of the drainage system locating in saturated zone. This does not

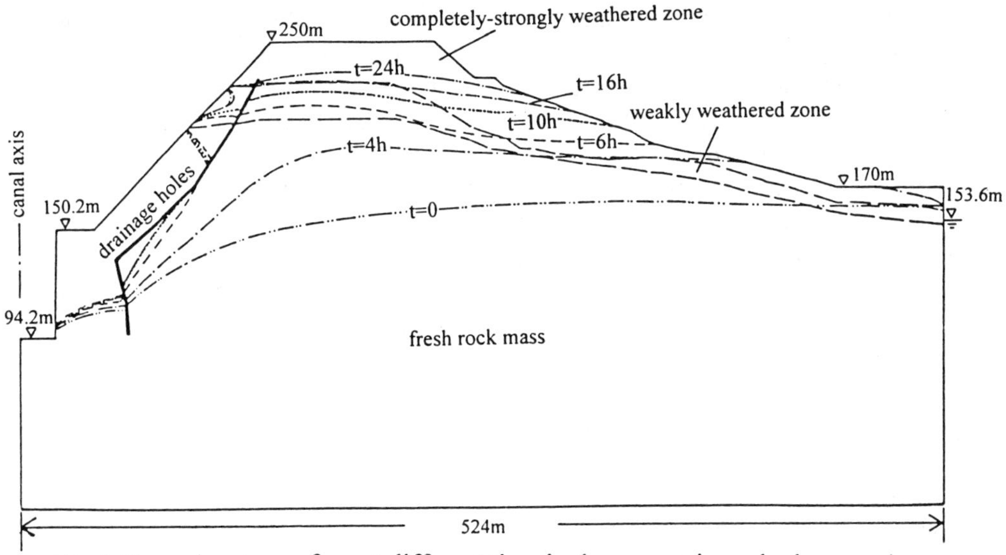

Fig.1 Ground water surface at different time in the mountain under heavy rain.

seem reasonable as real yearly situation, but might be taken as a situation at certain time before a heavy rain process.

As it rains heavily, ground water surface rises quickly. At 6 hour, the saturated zone has expanded beyond drainage holes near the interface between the weakly weathered zone and fresh rock mass, and has arrived on slope surface at the interface. As time goes on,saturated zone in the weakly weathered zone enlarges. At 16 hour, the whole weakly weathered zone has become saturated except where near the slope surface, and ground water surface has risen above the interface between the weakly weathered and the completely-strongly weathered zone.

It can be seen from the results that the drainage system in the fresh rock mass almost intercepts saturated water flow and let the rock mass between the drainage holes and the slope surface remain at unsaturated state. This implies that the drainage system would work well in uniform continuum media. It is after the ground water surface having risen above the top of the fresh rock and into the weakly weathered zone that saturated water flows through the interval between drainage holes toward the slope surface. This is due to the great permeability difference across the interface. The saturated permeability of the weakly weathered zone is about 180 times that of the fresh rock. So that, while ground water surface has risen above the interface, saturated water flows rather easily in the weakly weathered zone and moves toward the slope surface. The drainage system in this zone lowers water pressure but can not intercept saturated water flow, which exudes at the slope surface near the bottom of this zone. Because water is drained partly by the drainage holes and at the slope surface, the upper part of the slope surface of this zone remains unsaturated. Similarly, after ground water surface having risen above the top surface of the weakly weathered zone into completely-strongly weathered zone with saturated permeability increasing about 40 times, saturated water flows toward the slope surface and exudes.

These results indicate that less permeable zone in the rock mass would impede infiltrating water flow, and make it move easily through the interval between drainage holes toward slope surface. Thus water exudes out off the slope surface and the efficacy of the designed drainage system is lowered.

4 CONCLUSION

The seepage field in a mountain by high slope under rain infiltration condition was simulated by saturated-unsaturated flow model and FEM. Form the analyses of the results we can conclude:

1. The simulated results under a condition close to average yearly rain fall does not seem reasonable to the practical situation.
2. The drainage system designed on the bases of simulated results of saturated seepage model might intercept well the saturated water flow toward slope surface.
3. The permeability nonuniformity of the granite rock mass at vertical direction slope surface and exude. Thus the effic effects infiltrating water movement. After ground water table having risen up into a more permeable zone, saturated water can move rather easily in the zone. It would flow through the intervals between adjacent drainage holes toward acy of the drainage system might be lowered.

REFERENCES

Akai, K., Y. Ohnishi & M. Nishigaki, Finite Element Analysis of Three-dimensional Flows in Saturated-Unsaturated Soils, in Proceeding of the 3rd International Conference on Numerical Methods in Geomechanics, 1977.

Zhang, J., Simulation of Three Dimensional Saturated-unsaturated Steady-unsteady Seepage Field by FEM, Journal of Yangtze River Scientific Research Institute (in press).

Computer Methods and Advances in Geomechanics, Yuan (ed.) © 1997 Balkema, Rotterdam, ISBN 90 5410 904 1

Analytical solutions of pore pressure for finite permeable stone column with equal strain

Cai Yuanqiang, Wu Shiming, Xu Changjie & Chen Yunmin
Department of Civil Engineering, Zhejiang University, Hangzhou, China

ABSTRACT: The authors propose an analytical solutions for pore pressure in sand layer considering permeability of stone column in the case of equal strain. Based on the solution, the influences of the permeability of stone column on the liquefaction of sand layers during earthquake are discussed and some diagrams are presented for practical use.

1 INTRODUCTION

Xu(1985) proposed an analytical solution for pore water pressure in stone column of composite foundation in the case of free strain not taking resistance through wall into account. In fact the permeability of stone column is finite. By introducing the concept of radial average pore pressure, radial seepage and equal strain, the authors deduced an analytical solution in this paper, taking the finite permeability of stone column into account.

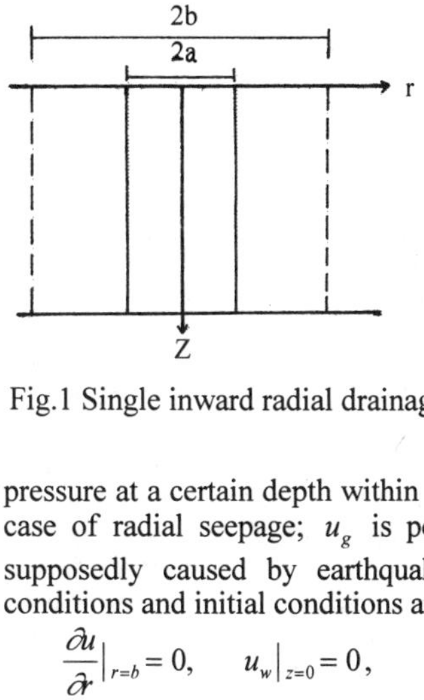

Fig.1 Single inward radial drainage stone column

2 EQUATIONS AND SOLUTIONS

For inward radial seepage in stone column with equal strain drainage problem showen in Fig.1, the Eq.s satisfying the continuity condition are:

$$-\frac{k_r}{\rho_w g}\frac{1}{r}\frac{\partial}{\partial r}\left(r\frac{\partial}{\partial r}\right)=\frac{\partial \varepsilon_v}{\partial t} \tag{2.1}$$

$$\frac{\partial \varepsilon_v}{\partial t}=-m_v\left(\frac{\partial \bar{u}}{\partial t}-\frac{\partial u_g}{\partial t}\right) \tag{2.2}$$

The continuity condition is

$$2\pi a\frac{k_r}{\rho_w g}\frac{\partial u}{\partial r}\Big|_{r=a}=-\pi a^2\frac{k_w}{\rho_w g}\frac{\partial^2 u}{\partial z^2}\Big|_{r=a} \tag{2.3}$$

Let $u|_{r=a}=u_w$, the continuity condition can be expressed as

$$\frac{\partial^2 u_w}{\partial z^2}=-\frac{2k_r}{k_w}\frac{\partial u}{\partial r}\Big|_{r=a}$$

where ε_v is volume strain at any point in effective zone, taking radial drainage into account only; u is pore water pressure at any point in effective zone in the case of radial seepage; $\bar{u}$ is radial average pore pressure at a certain depth within sand deposit in the case of radial seepage; u_g is pore water pressure supposedly caused by earthquake. The boundary conditions and initial conditions are as following

$$\frac{\partial u}{\partial r}\Big|_{r=b}=0,\qquad u_w\big|_{z=0}=0,$$

$$\frac{\partial u}{\partial z}\Big|_{z=h}=0,\qquad u\big|_{t=0}=0$$

where h is the length of stone column.
integrating both sides of equation (2.1) and letting

$$\frac{\partial u}{\partial r}\Big|_{r=b}=0$$

then obtain

$$\frac{\partial u}{\partial r}=\frac{\rho_w g}{2k_r}\left(\frac{b^2}{r}-r\right)\frac{\partial \varepsilon_v}{\partial t} \tag{2.4}$$

Similarly, integrating both sides of equation (2.4), then it gives

$$u=\frac{\rho_w g}{2k_r}\left[b^2\ln\left(\frac{r}{a}\right)-\frac{r^2-a^2}{2}\right]\frac{\partial \varepsilon_v}{\partial t}+u_w \tag{2.5}$$

According to the works by Xie and Zen(1989), radial

average pore water pressure at any depth in sand deposit can be expressed

$$\bar{u} = \frac{1}{\pi(b^2 - a^2)}\int_a^b 2\pi r u dr \tag{2.6}$$

Substituting Eq.(2.5) into Eq. (2.6), it yields

$$\bar{u} = \frac{\rho_w g F}{2k_r}\frac{\partial \varepsilon_v}{\partial t} + u_w \tag{2.7}$$

where:

$$F = (\ln n - \frac{3}{4})\frac{n^2}{n^2 - 1} + \frac{1}{n^2 - 1}(1 - \frac{1}{4n^2}), (n = \frac{b}{a})$$

During earthquake there exits

$$\frac{\partial u_g}{\partial t} = \frac{N_{eq}}{N_l}\frac{\sigma'_0}{t_d}$$

Finally, one can get

$$u = \frac{M}{\alpha}\left[1 - \sum_{n=0}^{\infty}\frac{4}{(2n+1)\pi}e^{-At}\sin\frac{(2n+1)\pi z}{2h}\right]$$

$$(\ln\frac{r}{a} - \frac{r^2 - a^2}{2b^2}) + \sum_{n=0}^{\infty}\frac{16MI^2h^2}{\alpha[(2n+1)\pi]^3}[1 - e^{-At}]$$

$$\sin\frac{(2n+1)\pi z}{2h}$$

where:

$$\alpha = \frac{2C_r}{bF}, M = \frac{N_{eq}}{N_l}\frac{\sigma'_0}{t_d},$$

$$A = \frac{\alpha[(2n+1)\pi]^2}{[(2n+1)\pi]^2 + 4I^2h^2}, I^2 = \frac{2k_r(n^2 - 1)}{b^2 F k_w}$$

C_r is the horizontal consolidation coefficient of soil. Letting $k_w \to \infty$, then Eq.(2.8) becomes

$$u = \frac{N_{eq}}{N_l}\frac{\sigma'_0}{\alpha t_d}(\ln\frac{r}{a} - \frac{r^2 - a^2}{2b^2})[1 - \exp(-\alpha t)] \tag{2.9}$$

Eq.(2.9) is the pore water pressure corresponding to no drainage resistance in stone column, which is the same as the work by Xu(1985).

From Eq.(2.8), it can be seen that during earthquake pore water pressure increases with time and reach to the maximum at the end of earthquake ($t=t_d$). After that time pore water pressure starts to dissipate.

3 CALCULATION AND ANALYSIS

From the viewpoint of engineering practice, what an engineering concerns is how build up of pore water pressure can be eliminated not to reach its initial effective stress σ'_0. From Eq.(2.8) the results of pore water pressure can be obtained.

Fig.2 shows the relations of maximum ratio of pore water pressure over the initial effective stress at z=8m versus dimensionless time t/t_d, for the case of stone column length, 8m, with parameters: a=0.5m, b=2.5m, $N_{eq}/N_l = 2$, k_w/k_r =100, where $T_{bd} = C_r t_d/b^2$.

Fig.3 gives out the relationship of maximum ratio of pore water pressure at z=4m versus dimensionless time, whereas the value of other parameters are same as in Fig.2.

According to Fig.2 and Fig.3, it can be seen that the increase of the value of T_{bd} leads to decease of pore water pressure. As the values of t_d and b in Fig.2 are equal to those in Fig.3, the increase of the value of T_{bd} is caused by the increase of the value of C_r certainly favors the dissipation of pore water pressure. Compared with Xu's solutions(1985) ignoring drainage resistance, the presented results in Fig.2 and Fig.3 are little bit greater because of

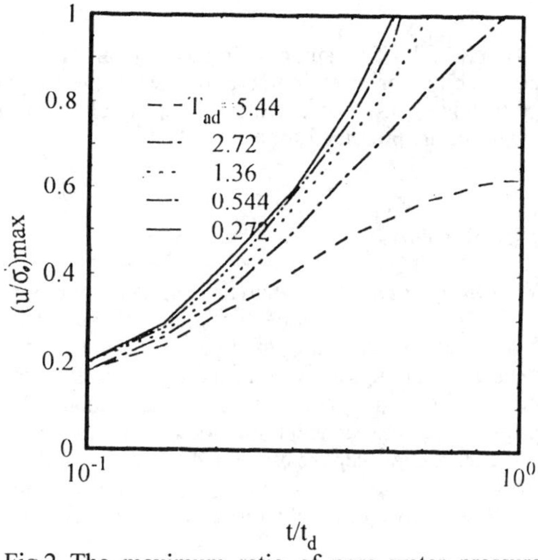

Fig.2 The maximum ratio of pore water pressure versus time at z=8m

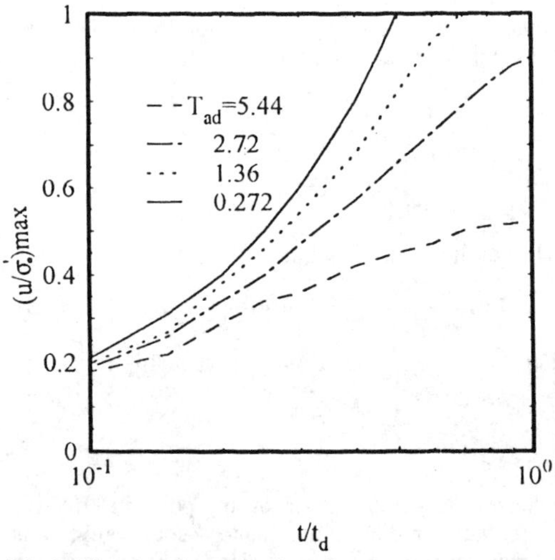

Fig.3 The maximum ratio of pore water pressure versus time at z=4m

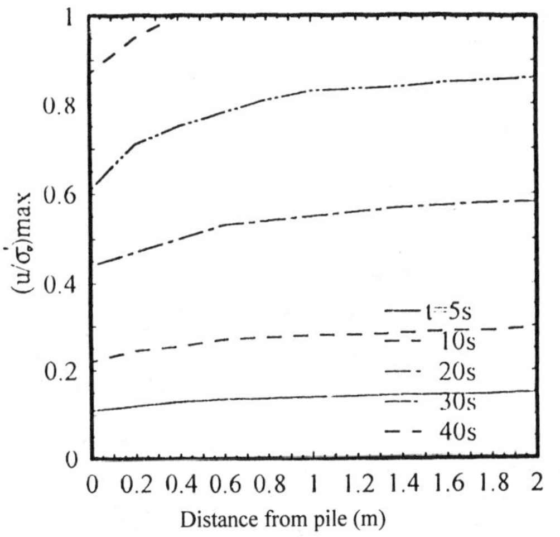

Fig.4 Pore water pressure versus the distance to pile with time, $k_w/k_r=100$

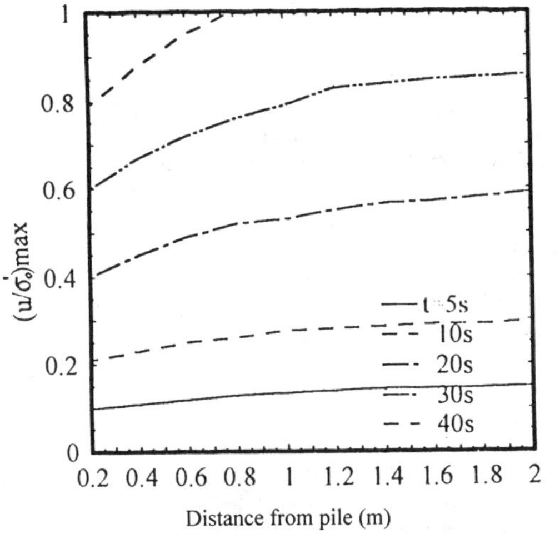

Fig.5 Pore water pressure versus the distance to pile with time, $k_w/k_r=200$

$k_w/k_r = 100$. While the value of T_{bd} is small enough, as time goes, pore water pressure will keep increasing and sooner closes too the value of σ_0'. Then, sand layer reaches its initial liquefaction condition during earthquake. After the earthquake, the pore water pressure tends to dissipate. It is clear that liquefaction easily occurs in sand layer if the value of T_{bd} is small or the coefficient of permeability is small at the case of large pile spacing. For the case both having greater value of T_{bd} and C_r, and small spacing of piles. The maximum ratio of pore water pressure won't get to 1, that is, initial liquefaction can't occur in this sand layer.

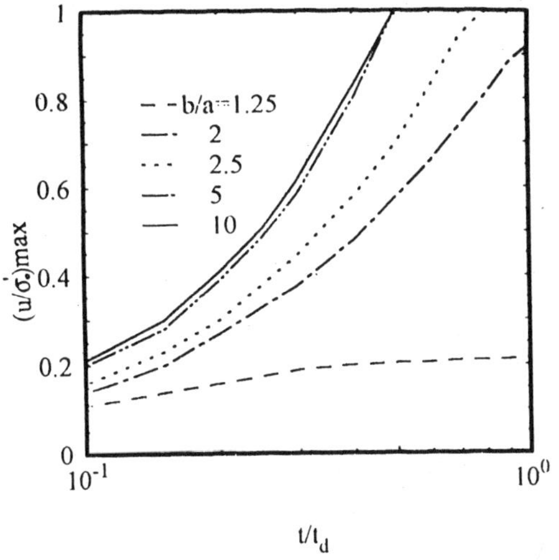

Fig.6 The maximum ratio of pore water pressure versus the diameter of pile

Considering Fig.2 and Fig.3, one can find out that different pore pressure build up for different depths, for example, pore water pressure at z=4m increases slower than that at z=8m because of their different drainage conditions.

Fig.4 and Fig.5 show the cases for $N_{eq}/N_l = 2$, a=0.5m, b=2.5m, t_d=70s, C_r=0.0243m^2/s. The values of k_w/k_r are 100 and 200 for Fig.4 and Fig.5, respectively. Both Figs. show pore water pressure ratio at the bottom of pile, it also can be seen that when $t<0.5t_d$, the farther from the edge of the pile, the greater the pore water pressure builds up. However, it generally can't reach its initial liquefaction. But for $t>0.5t_d$, the initial liquefaction sooner occurs.

Comparing Fig.4 and Fig.5, one can see the influence of horizontal coefficient of permeability on resistance to liquefaction of sand layer. As k_w/k_r are increases, pore water pressure reduces correspondingly. In fact, since the value of k_r remain unchanged for a certain case, the increase of drainage ability of stone column, so that pore water pressure dissipates quickly. For the case of $k_w/k_r=\infty$, pore water pressure is lower than those mentioned above, which basically agrees Xu's solution(1985).

Fig.6 shows the effect of b/a on pore water pressure, as $N_{eq}/N_l=2$, $k_w=100k_r$, $T_{bd}=0.272$, z=8m, with length of pile, 8m, from which the influence of the diameter of drainage of pile on resistance to liquefaction can be seen. For b/a>5, the pore water pressure increases rather fast with time, initial liquefaction occurs at sometime between $1/2t_d$ and t_d in sand layer. For small value of b/a, pore water pressure ratio can't small to 1. Therefore, increasing the diameter of pile can eliminate build up of water pressure dramatically.

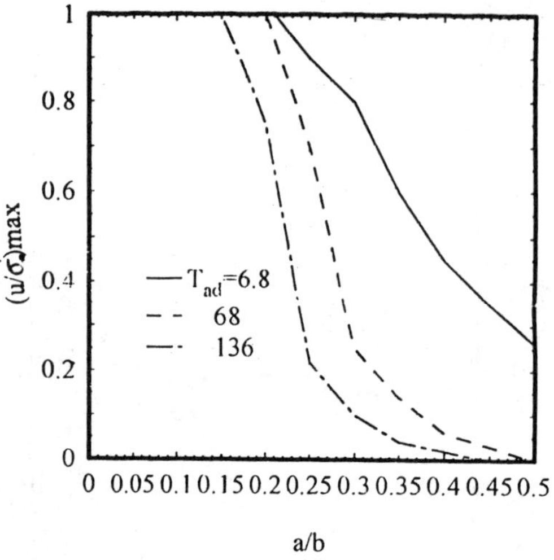

Fig.7 The maximum ratio of pore water pressure versus time and distance between piles

Other analysis for different length of pile indicates that the effect of drainage resistance of stone column to liquefaction in shallow layer is greater than in deep layer.

For $N_{eq}/N_l=1$, $k_w/k_l=\infty$, as n>4, it will lead to initial liquefaction in the range of T_{ad} from 6.8 to 136. Increase of T_{ad} results in deduction of pore water pressure correspondingly if the value of n remains constant, which means that the smaller diameter of drainage is better than that of greater one. Following is an example to demonstrate the results shown in Fig.7, where $T_{ad}=C_r t_d/a^2$.

Assuming the coefficient of permeability in sand layer, $k_r=10^{-5}$m/s, coefficient of volume change $m_v=4.2\times10^{-5}$m^2/kN, the sand layer is supposed to bear an earthquake with 24 cycles of duration 70 seconds. The sand reaches its initial liquefaction after bearing such 24 cycles of shocks by means of undrained cyclic triaxial tests. Assuming diameter of stone column, 0.8m, determine the piles spacing to control the maximum pore water pressure ratio not excess 0.6.

From the known parameters, the value of T_{ad} can be calculated, that is, T_{ad}=10.7. As $(u/\sigma'_0)_{max}<0.6$, from Fig.7 the value of a/b can be determined approximately, a/b<0.32 for diameter of drainage pile, 0.4m, and b=1.92m for a/b<0.25.

It can be concluded that it will be an efficient and ecomomic design with "finer and denser" stone column to resist the occurrence of liquefaction. In this example, keeping the value of $(u/\sigma'_0)_{max}$ equal to 0.6, for pile diameter of 0.8m the replacement factor of stone column is 11.4%; while for pile diameter of 0.4m, the corresponding replacement factor is equal to 6.7%, it is obvious that finer piles with diameter 0.4m costs less for the same effectiveness.

4 SUMMARY

The pore water pressure analytic equation reflects that the excessive pore water pressure ratio is proportional to N_{eq}/t_d(equivalent cycles in unit time),inverse proportional to N_l(liquefaction cycle time), roughly direct proportional to consolidation coefficient, meaning that direct proportional to coefficient of volume change and roughly inverse proportional to coefficient of permeability, during earthquake.

According to previous engineering experience, setting up drainage pile in sand layer is an economical and effective method to resist potential liquefaction in many cases. Eq.(2.8) provides the theoretical basis for designing stone column to prevent the sand layer from liquefaction. For various engineering conditions, it is feasible to directly calculate Eq.(2.8) or directly find the value of a/b corresponding to control value of $(u/\sigma'_0)_{max}$ from Fig.7.

It should be pointed that adapting the principal of "finer and denser" stone column in design can make the system more effectively and economically, in which construction should be taken into account and the cost of large number of stone columns as well. The piles with diameters from 400mm to 800mm are mostly recommended.

Further research should be carried out in future on the influence of vertical drainage on pore water pressure caused by earthquake.

REFERENCE

Astuo. O. 1988. In-situ experiment and analysis on wall resistance of gravel drains. Soil and Foundation, Vol.27,No.2.

Xu, Z. Y. 1985. The analysis and calculation of stabilization of potentially liquefiable sand deposits using gravel drain system. Technique of Exploration Science(in Chinese),No.1:1 ~ 6.

Xie. G. H.1987. Sand drainage consolidation theory with equal strain. Dissertation for Ph.D. in Zhejiang University.

Computer Methods and Advances in Geomechanics, Yuan (ed.) © 1997 Balkema, Rotterdam, ISBN 90 5410 904 1

Smoothed particle hydrodynamics model for flow through porous media

Yi Zhu, Patrick J. Fox & Joseph P. Morris
School of Civil Engineering, Purdue University, West Lafayette, Ind., USA

ABSTRACT: This paper presents the first application of Smoothed Particle Hydrodynamics (SPH) to problems in geomechanics. Originally developed for astrophysics applications, SPH is extended to model incompressible flows of moderate to low Reynolds numbers as encountered in groundwater flow systems. For such flows, modification of the standard SPH formalism is required to minimize errors associated with the use of a quasi-incompressible equation of state. Treatment of viscosity, state equation, kernel interpolation, and boundary conditions are described. Simulations using the method show close agreement with series solutions for Couette and Poiseuille flows. The model is demonstrated for two-dimensional flow through an idealized porous medium. The relationship between discharge velocity and hydraulic gradient is presented, and the corresponding value of hydraulic conductivity is computed.

1 INTRODUCTION

Over the past few decades, research on problems involving fluid flow through porous media has increased dramatically. This is especially true in the fields of civil engineering and hydrogeology, where there has also been a concurrent shift of emphasis from seepage and water supply engineering to mass transport of contaminants through the subsurface and environmental site remediation. For such problems, most research circumvents the details at the pore-scale level and begins from the empirical laws for calculating fluid and chemical fluxes.

The objective of our research is to simulate the microscopic flow process through porous media, and thereby gain insight into the macroscopic flow behavior. A porous media flow model has been developed for this purpose using Smoothed Particle Hydrodynamics (SPH). SPH is a fully Lagrangian technique in which the numerical solution is achieved without a grid. Using this approach, fluid velocity and pressure distributions, energy dissipation mechanisms, and volumetric flow rate can be computed. In addition, the flow paths of individual fluid masses can be monitored as they travel through the void system of the medium. Another advantage of SPH is the relative ease with which new physics may be incorporated into the formulation. Thus, the geometry of the flow domain can be specified, and detailed information about the flow process can be obtained that would be difficult or impossible to observe experimentally or with many other numerical techniques.

2 OVERVIEW OF STANDARD SPH

Originally developed for astrophysics applications (Lucy 1977, Gingold and Monaghan 1977), SPH is a truly Lagrangian method for modeling fluid dynamics. The fluid is represented by particles, which may also be regarded as interpolation points. The particles move with the local fluid velocity, advecting contact discontinuities, preserving Galilean invariance, and reducing the computational diffusion of various fluid properties (including momentum). Each particle carries mass m, velocity $\mathbf{v}$, and other fluid quantities specific to a given problem.

The standard approach to SPH is reviewed by Monaghan (1988,1992) and Benz (1990). SPH allows any function to be expressed in terms of values over a set of disordered particles. To illustrate, consider a field quantity $A(\mathbf{r})$ expressed by,

$$A(\mathbf{r}) = \int A(\mathbf{r}')\delta(\mathbf{r}-\mathbf{r}')d\mathbf{r}', \tag{1}$$

where $\mathbf{r}$ and $\mathbf{r}'$ are position vectors, and δ is the delta function. If we replace δ with an interpolation kernel, $W(\mathbf{r},h)$, we obtain an integral interpolant of the function,

$$A_i(\mathbf{r}) = \int A(\mathbf{r}')W(\mathbf{r}-\mathbf{r}',h)d\mathbf{r}'. \tag{2}$$

The kernel typically takes the form,

$$W(\mathbf{r},h) = \frac{1}{h^\sigma} f\left(\frac{|\mathbf{r}|}{h}\right), \tag{3}$$

where σ is the number of dimensions, h is the smoothing length, and the function f is discussed in §3.6. The kernel has the following properties,

$$\int W(\mathbf{r}-\mathbf{r}',h)d\mathbf{r}' = 1, \tag{4}$$

and,

$$\lim_{h\to 0} W(\mathbf{r}-\mathbf{r}',h) = \delta(\mathbf{r}-\mathbf{r}'). \tag{5}$$

For numerical work, the integral interpolant is approximated by a summation interpolant over a collection of discrete points (particles) in space,

$$A_s(\mathbf{r}) = \sum_a \frac{m_a}{\rho_a} A_a W(|\mathbf{r}-\mathbf{r}_a|,h), \tag{6}$$

where ρ is density, and field quantities at particle a are denoted by subscript a.

Using the above concepts, the SPH equations governing fluid motion can be obtained. For example, if a fluid is represented by a set of particles, ρ_a may be evaluated using,

$$\rho_a = \sum_b m_b W_{ab}, \tag{7}$$

where W_{ab} denotes,

$$W_{ab} = W(\mathbf{r}_{ab},h), \tag{8}$$

and,

$$\mathbf{r}_{ab} = \mathbf{r}_a - \mathbf{r}_b. \tag{9}$$

Other expressions for derived quantities at the particles are obtained by summation involving the kernel and its derivatives. As the derivatives can be obtained by ordinary differentiation, there is no need for a grid. For example, the most common SPH expression for the pressure gradient acceleration is,

$$-\left(\frac{1}{\rho}\nabla p\right)_a = -\sum_b m_b \left(\frac{p_a}{\rho_a^2} + \frac{p_b}{\rho_b^2}\right) \nabla_a W_{ab}, \tag{10}$$

where p is pressure and ∇_a denotes the gradient with respect to the coordinates of particle a. Provided the kernel is an even function of $\mathbf{r}$, Equation (10) conserves momentum exactly since the forces acting between individual particles are antisymmetric.

3. SPH FOR INCOMPRESSIBLE FLOW

SPH is well suited to model compressible flows because the fluid is driven by local density fluctuations at the particles. Monaghan (1994) extended SPH to incompressible flow problems involving free surface flows for high Reynolds numbers and free-slip boundary conditions. The extension of SPH to slow viscous flows, such as groundwater flow, calls for further modifications which are outlined in the following sections, and are discussed in detail by Morris et al. (1997).

3.1 *Equation of state*

SPH cannot model a truly incompressible fluid. Rather, water is treated as a quasi-incompressible fluid using an artificial equation of state. Theoretically, the actual state equation for water could be used, but this would result in a prohibitively small time step. The state equation we use is,

$$p = c^2 \rho, \tag{11}$$

where c is the speed of sound. The chosen value for c is low enough to be practical, yet high enough to limit density fluctuations to about three percent.

3.2 *Evolution of density*

To simulate incompressible flows with SPH, Monaghan (1994) evolved particle densities according to the following equation for continuity,

$$\frac{d\rho_a}{dt} = \sum_b m_b \mathbf{v}_{ab} \cdot \nabla_a W_{ab}, \tag{12}$$

where $\mathbf{v}_{ab} = \mathbf{v}_a - \mathbf{v}_b$. Equation (12) allows ρ to be evolved concurrently with particle velocities and other field quantities, thus significantly reducing the computational effort. While Equation (12) does not conserve mass exactly (Equation (7) does, provided the total number and mass of particles are constant), this is usually only important for faster flows involving shocks. The simulations presented here use Equation (12).

3.3 *Dynamic pressure*

For our simulations, fluid particles move in response to an imbalance of forces due to gravity and a large-scale static pressure field. For slow viscous flows, the effect of local variations in pressure on fluid motion can be very small compared with that of the static pressure gradient. Consequently, we use SPH to model the dynamic pressure, p_d, defined as,

$$p_d = p_t - p_s, \tag{13}$$

where p_t is the total pressure and p_s is the static pressure, respectively. Thus, it is p_d which is modeled using Equation (11). Since pressure appears only as a gradient in the Navier-Stokes equations, the effect of p_s is that of a body force. The flow is driven by a net body force per unit mass, $\mathbf{F}$, defined as,

$$\mathbf{F} = \mathbf{g} - \frac{\nabla p_s}{\rho}, \tag{14}$$

where $\mathbf{g}$ is the gravitational acceleration. This approach has advantages for the simulation of flows with periodic boundary conditions.

3.4 *Boundary conditions*

In our work, real SPH particles are used to create a no-slip boundary condition for all solid boundary surfaces. These particles contribute to the usual SPH expressions for density and pressure gradients, and their own densities (but not their positions) are evolved. Boundary particles are assigned artificial velocities such that antisymmetry in the velocity field is created across the boundary surface. Obstacles within the flow field and solid boundary walls are created by placing all the particles on an initial regular lattice throughout the computational domain and designating those particles which fall within solid objects as boundary particles. Numerical simulations have shown that the results obtained using this approach are stable and accurate.

3.5 *Viscosity*

Most implementations of SPH employ an artificial viscosity that was first introduced to permit the modeling of strong shocks (Benz 1990, Monaghan 1992). This formulation has been used to model real viscosity, however, we found that it produced inaccurate velocity profiles for our applications. Flebbe et al. (1994) proposed a more realistic form of viscosity, but it involves nested summations over the particles, and hence twice the computational effort. Our method employs an estimation of viscous diffusion which is similar to an expression used by Monaghan (1995) to model heat conduction,

$$\left\{\left(\frac{1}{\rho}\nabla \cdot \mu \nabla\right)\mathbf{v}\right\}_a = \sum_b \frac{m_b(\mu_a + \mu_b)\mathbf{r}_{ab} \cdot \nabla_a W_{ab}}{\rho_a \rho_b(\mathbf{r}_{ab}^2 + 0.01h^2)}\mathbf{v}_{ab}, \tag{15}$$

where μ is the dynamic viscosity. This expression may be considered as a hybrid of a standard SPH first derivative applied to a finite difference approximation of a first derivative. It conserves linear momentum exactly, while angular momentum is only approximately conserved. If the kernel takes the form of Equation (3), then,

$$\nabla_a W_{ab} = \frac{\mathbf{r}_{ab}}{|\mathbf{r}_{ab}|}\frac{\partial W_{ab}}{\partial r_a}, \tag{16}$$

and we can simplify Equation (15) to,

$$\left\{\left(\frac{1}{\rho}\nabla \cdot \mu \nabla\right)\mathbf{v}\right\}_a = \sum_b \frac{m_b(\mu_a + \mu_b)\mathbf{v}_{ab}}{\rho_a \rho_b}\left(\frac{1}{|\mathbf{r}_{ab}|}\frac{\partial W_{ab}}{\partial r_a}\right). \tag{17}$$

Thus, the SPH form of the momentum equation is,

$$\frac{d\mathbf{v}_a}{dt} = -\sum_b m_b\left(\frac{p_a}{\rho_a^2} + \frac{p_b}{\rho_b^2}\right)\nabla_a W_{ab} + \sum_b \frac{m_b(\mu_a + \mu_b)\mathbf{v}_{ab}}{\rho_a \rho_b}\left(\frac{1}{|\mathbf{r}_{ab}|}\frac{\partial W_{ab}}{\partial r_a}\right) + \mathbf{F}. \tag{18}$$

3.6 *Interpolation kernel*

Most SPH simulations employ a cubic spline kernel,

$$f(s) = \frac{10}{7\pi}\begin{cases} 1 - 3s^2/2 + 3s^3/4, & \text{if } 0 \le s \le 1; \\ (2-s)^3/4, & \text{if } 1 \le s \le 2; \\ 0, & \text{if } s \ge 2, \end{cases} \tag{19}$$

(here normalized for two dimensions) since it resembles a Gaussian while having compact support. For low Reynolds number simulations, it was found that a cubic spline resulted in significant noise in the pressure and velocity fields. Thus, we use the following quintic spline (normalized for two dimensions),

$$f(s) = \frac{7}{478\pi}\begin{cases} (3-s)^5 - 6(2-s)^5 + 15(1-s)^5, & \text{if } 0 \le s < 1; \\ (3-s)^5 - 6(2-s)^5, & \text{if } 1 \le s < 2; \\ (3-s)^5, & \text{if } 2 \le s < 3; \\ 0, & \text{if } s \ge 3. \end{cases} \tag{20}$$

Although the quintic spline is computationally more intensive (by approximately a factor of 2), it was found to be the most reliable for the simulations in this study.

3.7 *Time integration*

For our SPH model, explicit time integration is performed using the modified Euler method in which the time step is limited by stability constraints. The CFL condition (Courant et al. 1928) requires,

$$\Delta t \le 0.25\frac{h}{c}. \tag{21}$$

Additional constraints arise from the magnitude of particle accelerations f_a (Monaghan 1992),

$$\Delta t \le 0.25 \min_a \sqrt{\left(\frac{h}{f_a}\right)}, \tag{22}$$

and viscous diffusion,

$$\Delta t \le 0.125 \frac{\rho h^2}{\mu}. \tag{23}$$

Equation (23) is based upon the usual condition for an explicit finite difference method simulating diffusion. At sufficiently high resolution (small h) or large viscosity, Equation (23) is typically the dominant time constraint. In addition, the choice of kernel and the initial arrangement of particles may allow these conditions to be somewhat relaxed.

4. MODEL TESTING AND VERIFICATION

4.1 *Couette flow*

The first test case is Couette flow between infinite plates located at y=0 and y=L. The fluid between the plates is initially at rest. At time t=0, the upper plate moves at a constant velocity V_0 parallel to the x-axis. This flow was simulated for $\mu=10^{-3}$kgm^{-1}s^{-1}, $L=10^{-3}$ m, $\rho=10^3$kgm^{-3}, and $V_0=1.25\times10^{-5}$ms^{-1}. This corresponds to a Reynolds number of 1.25×10^{-2} using,

$$Re = \frac{\rho V_0 L}{\mu}. \tag{24}$$

Figure 1a shows a comparison between the x-velocity (v_x) obtained using a series solution and SPH at several times (Morris et al. 1997). The steady state solution is denoted by $t=\infty$. The results are in close agreement, confirming the accuracy of the approach used to evaluate viscous and boundary forces with SPH.

4.2 *Poiseuille flow*

The second test case is Poiseuille flow between stationary infinite plates at y=0 and y=L. The fluid is initially at rest, and is driven by an applied net body force **F** (of magnitude F) acting parallel to the x-axis for $t\geq0$. The flow was simulated for $\mu=10^{-3}$kgm^{-1}s^{-1}, $L=10^{-3}$m, $\rho=10^3$kgm^{-3}, and $F=10^{-4}$ms^{-2}. This corresponds to a peak fluid velocity $V_0=1.25\times10^{-5}$ ms^{-1}, and a Reynolds number of 1.25×10^{-2} using Equation (24). A comparison between v_x obtained using a series solution and SPH appears in Figure 1b (Morris et al. 1997). The results are again in close agreement, with the largest discrepancy being about 0.7% for the steady state case.

5. FLOW THROUGH A PERIODIC POROUS MEDIUM

The model was used to simulate the flow of water ($\mu=10^{-3}$kgm^{-1}s^{-1}, $\rho=10^3$kgm^{-3}) through an idealized two-dimensional porous medium (Figure 2). The unit cell is square with a side length of L=1.6mm.

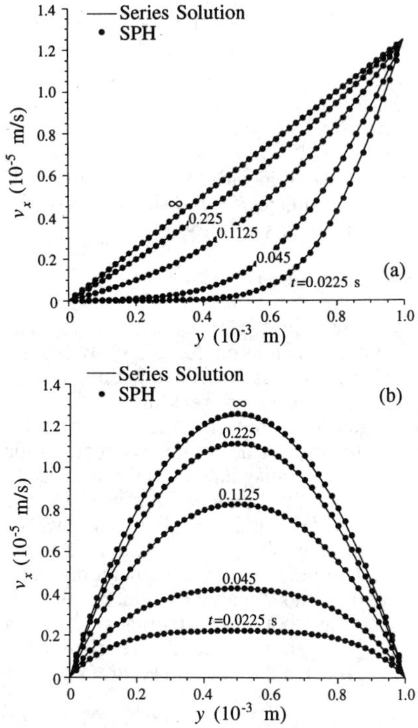

Figure 1. Comparison of SPH and series solutions for: (a) Couette flow, and (b) Poiseuille flow.

To model an infinite array, we applied periodic boundary conditions to this cell. The resulting porous medium is composed of circular soil grains with radii equal to 0.5mm and 0.4mm. The porosity of the array is 0.495. The flow is driven by **F** acting parallel to the x-axis, which is related to the hydraulic gradient, i, by,

$$i = \frac{F}{g}, \tag{25}$$

where g is the gravitational constant (9.81ms^{-2}).

Five simulations were conducted using different values of F. At steady state, the volumetric flow rate was calculated by summing particle volumes passing out of the unit cell for a given time period. The discharge velocity, v, was obtained by dividing the volumetric flow rate by the gross cross-sectional area of the cell perpendicular to the direction of flow. A total of 5574 particles was used for the simulations. Table 1 gives the values of body force, hydraulic gradient, sound speed, discharge velocity, Reynolds number, and Mach number ($Ma=v/c$) for each simulation at steady state. The Reynolds numbers were computed using L'=0.8mm (Figure 2). The small values of Ma confirm that each flow is nearly incompressible.

Figure 3 shows discharge velocity as a function of hydraulic gradient for the five simulations. In

Table 1. Summary of Darcy's law simulations.

Simulation	1	2	3	4	5
Body force, F (ms^{-2})	0.980	0.490	0.098	0.049	0.0098
Speed of sound, c (ms^{-1})	0.131	0.093	0.0416	0.0294	0.0131
Hydraulic gradient, i	0.100	0.050	0.010	0.005	0.001
Discharge velocity, v (ms^{-1})	2.49×10^{-3}	1.25×10^{-3}	2.49×10^{-4}	1.27×10^{-4}	2.41×10^{-5}
Reynolds number, Re	1.995	0.997	0.199	0.102	0.019
Mach number, Ma	1.90×10^{-2}	1.34×10^{-2}	5.98×10^{-3}	4.33×10^{-3}	1.84×10^{-3}

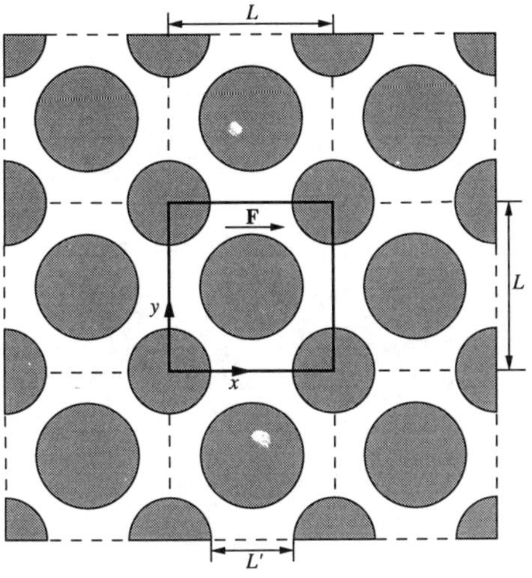

Figure 2. Periodic porous medium.

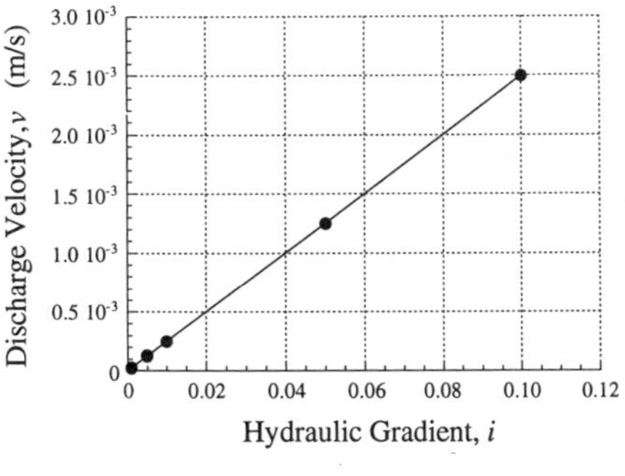

Figure 3. Discharge velocity versus hydraulic gradient for periodic porous medium.

accordance with Darcy's law, v varies linearly with i in the range of Reynolds numbers simulated. A hydraulic conductivity of $0.0249ms^{-1}$ is obtained from the slope of the linear relationship. Figure 4 shows the configuration of fluid particles at steady state for the simulation i=0.1.

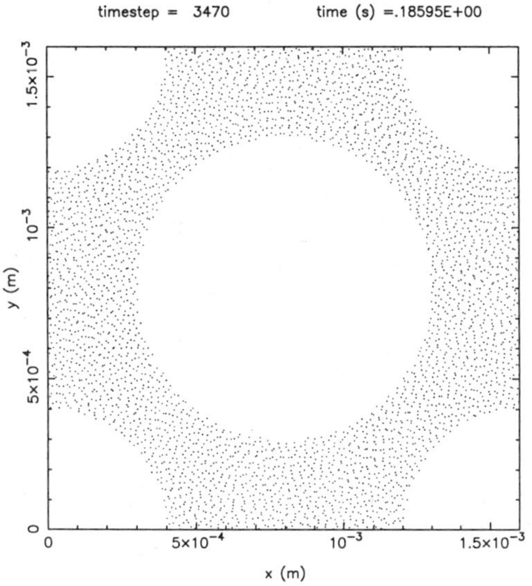

Figure 4. Fluid particles at steady state for i=0.1 (boundary particles not shown).

6. CONCLUSIONS AND FUTURE WORK

Many of the necessary extensions have been implemented and tested which allow Smoothed Particle Hydrodynamics (SPH) to model incompressible flow through porous media. Test results confirm that the proposed modifications to the equation of state, viscosity formulation, boundary conditions, and interpolation kernel result in a method which is stable and accurate. Good agreement with analytical solutions has been achieved for Couette and Poiseuille flows, confirming the method's ability to model viscous forces. Application to flow through a periodic porous medium shows discharge velocity proportional to hydraulic gradient, as required by Darcy's law. Future work will include a study of the hydraulic conductivity, tortuosity, and dispersion coefficients for a wide range of idealized porous media. Surface tension is planned to be incorporated into the model, thus permitting investigation of the migration behavior of light and dense non-aqueous phase liquids (LNAPLs and DNAPLs). Extension of the method to three dimensions is straightforward in theory, but will likely require parallelism of the code.

ACKNOWLEDGMENT/DISCLAIMER

This work was sponsored by the Air Force Office of Scientific Research, USAF, under grant number F49620-96-1-0020. The views and conclusions contained herein are those of the authors and should not be interpreted as necessarily representing the official policies or endorsements, either expressed or implied, of the Air Force Office of Scientific Research or the U.S. Government.

REFERENCES

Benz, W. (1990). Smooth particle hydrodynamics: a review, J.R. Buchler (ed.), *The Numerical Modelling of Nonlinear Stellar Pulsations*, 269-288.

Courant, R., Friedrichs, K., and Lewy, H. (1928). *Mathematische Annalen*, 100, 32-74.

Flebbe, O., Munzel, S., Herold, H., Riffert, H., and Ruder, H. (1994). Smoothed particle hydrodynamics: physical viscosity and the simulation of accretion disks, *The Astrophys. J.*, 431, 754-760.

Gingold, R.A., and Monaghan, J.J. (1977). Smoothed particle hydrodynamics: theory and application to non-spherical stars, *Mon. Not. R. Astron. Soc.*, 181, 375-389.

Lucy, L.B. (1977). A numerical approach to the testing of the fission hypothesis, *The Astron. J.*, 82, 12, 1013-1024.

Monaghan, J.J. (1988). An introduction to SPH, *Comput. Phys. Comm.*, 48, 89-96.

Monaghan, J.J. (1992). Smoothed particle hydrodynamics, *Annu. Rev. Astron. Astrophys.*, 30, 543-574.

Monaghan, J.J. (1994). Simulating free surface flows with SPH, *J. Comput. Phys.*, 110, 399-406.

Monaghan, J.J. (1995). Heat conduction with discontinuous conductivity, Applied Mathematics Reports and Preprints, Monash University, Australia.

Morris, J.P., Fox, P.J., and Zhu, Y. (1997). Modeling moderate and low Reynolds number flows using SPH, *J. Comput. Phys.*, in review.

Computer Methods and Advances in Geomechanics, Yuan (ed.) © 1997 Balkema, Rotterdam, ISBN 90 5410 904 1

Study of the management of groundwater reservoir by numerical three dimensional flow model

Bang-Woong Shin & Woo-Seok Bae
Department of Civil Engineering, Chungbuk University, Cheongju, South Korea

H. Immamoto
Kyoto University, Japan

Hee-Sung Kim
Rural Development Corporation Seoul, South Korea

ABSTRACT: At the initial stage of the underground reservoir design one should thoroughly consider surface and subsurface hydrology, hydrogeologic characteristics of aquifer system and the function of cut-off wall because it is linked to the effective management.

In the study, three dimensional finite difference model was applied to analyse the function of Ian underground reservoir at Kyungbuk Province in the Republic of Korea.

The steady and unsteady state conditions after construction of the underground dam were simulated through the model and from these results the groundwater budget and the safe yield were determined.

The model simulation indicates the infiltration of irrigation water to be one of the major factors of seasonal fluctuation of groundwater level. The recharge rates of irrigation water were estimated as 4.3mm/d during May and June, and 1.7mm/d during Jury and August.

Groundwater recharge from the watershed area estimated to about $0.04m^3/s$, almost consistent through the year. In 1984, groundwater discharge through the transverse section of the dam was $0.002m^3/s$ and the optimum yield for two months (July and August) was $254{,}000m^3$, however, the discharge became $0.013m^3/s$ in 1993, implying the failure of cut-off function. Without appropriate supplement of the cut-off wall, optimum yield during the irrigation period would be $93{,}000m^3$.

1. INTRODUCTION

Underground reservoir is constructed to rise constant level of groundwater table at the outlet of groundwater catchment area where unconfined aquifer layers are distributed. According to the ascension of groundwater level, increasing recharge rate depend on groundwater flow type and permeability of cut-off wall. Therefore, to satisfy water requirement in fixed groundwater catchment area, the analysis for characteristics of aquifer and cut-off wall must be done. Not only underground condition and groundwater flow are not observed directly but also quantitative analysis has various difficulty. So due to the stress pumping, destruction of aquifer or due to the carelessness of cut-off wall management, lowering state of cut-off function maybe occured. To remove the above trouble factor of groundwater, soil characteristics and permeability of aquifer, water budget analysis of groundwater catchment area, permeability of cut-off wall are discussed synthetically. But the case that point source is in-situ test and laboratory test, various values are not considered as model value of complex in-situ soil, therefore it is necessary that reappearance of actual conditions are made. In this study the role of aquifer for lowering state of underground reservoir function, the problem for optimum management and the cut-off effects for constructed underground dam are analysed quantitatively. In a certain way of in-situ test for borehole such as values of permeability coefficient(k), storage coefficient(S) and porosity(n), soil charact -eristics are understood until actual situation is reappeared by constructing the 3 dimensional groundwater flow analytical model. The result of simulation define water budget before and after for the cut-off wall establishment, suggest safety yield to prevent aquifer destruction, analyse the function of cut-off wall that turn out groundwater outflux through the cross section of underground cut-off wall. By the above mentioned analytical methods, the function of many underground reservoirs maybe appreciated and permeability testing method using constructed simulation for in-situ test data has confidence to determine that the upper limit for permeability test of basic foundation of the structure such as examination that after the underground cut-off wall established and test of dam foundation establishing the grouting method.

2. THEORETICAL DEVELOPMENT

As discussed above, the numerical model used in this study is based upon a groundwater flow governing equation. A mathematical model of a groundwater flow system at steady state consists of so called governing equation, such as Laplace equation or Poisson's equation, and boundary conditions(Wang and Anderson,1982). A steady state is defined as the condition where the magnitude and direction of the flow velocity are constant with time at any point in a flow field(Freeze and Cherry,1979). For steady state conditions, the amount of water flowing into a unite volume of saturated porous material equals to the amount flowing out. The groundwater flow through an isotropic, homogenious aquifer under steady state conditions is expressed as the following (Jacob, 1950 : Wang and Anderson,1982) :

$$\nabla^2 h = \frac{w(x,y,z)}{k} \tag{2.1}$$

h:hydraulic head(L) k:hydraulic conductivity (L/T)

where x, y, and z are Cartesian coordinates. $w(x,y,z)$ is the volume of water added(sources) or removed(sinks) per unit time per unit aquifer volume to the infinitesimal volume around the point(x,y,z) and has units of time.

For transient conditions, the governing equation is modified such that the volume outflow rate equals the volume inflow rate plus the rate of release of water from storage. The mathematical equation is described as the following(Wang and Anderson,1982) :

$$\frac{\partial h}{\partial x^2} + \frac{\partial^2 h}{\partial y^2} + \frac{\partial^2 h}{\partial z^2} = \frac{Ss}{k}\frac{\partial h}{\partial t} + \frac{w(x,y,z)}{k} \tag{2.2}$$

Ss : specific storage (L^{-1}) t: time (T)

In a heterogenious and anisotropic aquifer, if the cartesian coordinate axes, x, y, and z, are aligned with the principal component of the hydraulic conductivity tensor, the above equation may be rewritten as the equation(2) (Trescott et al., 1976) :

$$\frac{\partial}{\partial x}\left(Kxx\frac{\partial h}{\partial x}\right) + \frac{\partial}{\partial y}\left(Kyy\frac{\partial h}{\partial y}\right) + \frac{\partial}{\partial z}\left(Kzz\frac{\partial h}{\partial x}\right) = Ss\frac{\partial h}{\partial t} + w(x,y,z,t) \tag{2.3}$$

where, Kxx, Kyy, Kzz are the components of the hydraulic conductivity tensor(L/T). This equation constitutes a mathematical model of groundwater flow together with the specified flow and head conditions at the boundary of an aquifer system.

The solution of analytical model such as an equation(2) are hardly obtainable in a complex system such as heterogeneous and anisotropic aquifer, so approximate solutions using numerical method are desirable. The finite difference method is one of the numerical approximation method of the analytical model. In the finite difference method, the continuous system is subdivided into a finite number of discrete points in terms of space and time, and the partial dirivatives are replaced by differences between functional values at these points. This process results systems of simultanious linear algebraic difference equations and the solution obtained from this approximation is the head values at specific points and time(Mcdonald and Harbaugh, 1984). The finite difference equation from the partial differential equation(2) was derived by Pinder and Bradefoeft (1968).

For a heterogenious aquifer with irregular boundaries, the aquifer system is discrete into rectangular blocks, termed cells, which is divided by a grid system. The numerical model used in this study is using a block-centered finite difference grid configuration, in which the nodes are located at the center of the cells. A solution obtained from this method for each nodal point represents over the extent of a cell. The advantage of this grid formulation is the variable grid spacing(Trescott et al.1976).

3. GEOLOGY AND GROUNDWATER

The study area, Ian-Myon Sangju-gun Gyungbuk Province, is located in the southeastern part of the Republic of Korea, at lattitude of 128 ° 06′ 00″ to 128 ° 09′ 40″, and longitude 36 ° 34′ 00″ and 36 ° 38′ 00″. The area, which is located at the southern foothills of the Taebaek Mountains, is surrounded by a series of low-level mountains with elevations ranging from 650 to 100 meters above sea level (Fig. 1.). This area consists geologically of a Quaternary alluvial deposit overlying Cretaceous banded gneiss and undated granitic gneiss and schist.

4. UNDERGROUND DAM

Underground dam is located at the outlet of catchment area. The dam, which is 230 meters long and 5 meters high, was built using a grouting techniques: a special designed chemical material (SiO_2, cement, S.G.R.) was used as a grout. Based on the pilot hole test data(table 1), the application of Maag's formula, calculations showed equilateral spacing of holes should be planned. This was achieved by a primary grid 2.0 meters(parallel to dam axis)×1.0meters (normal to dam axis), followed by secondaries at intermediate points on the main lines.

4.1 *Surficial hydrology*

To understand water budget in the studied area, it is cited the climatological parameters at Moon -Kyung observatory, and made soil moisture accounting from potential evaportranspilation by Thorthwaite method and runoff amount by SCS metod. According to Table 2, when the anual rainfall has 953.1㎜ runoff amounts are 733㎜ and infiltration rate is 220.1㎜. In the irrigation season since April, percoration rate is not calculated because of potential evapotranspilation larger than infiltration rate of rainfall, however, because the studied

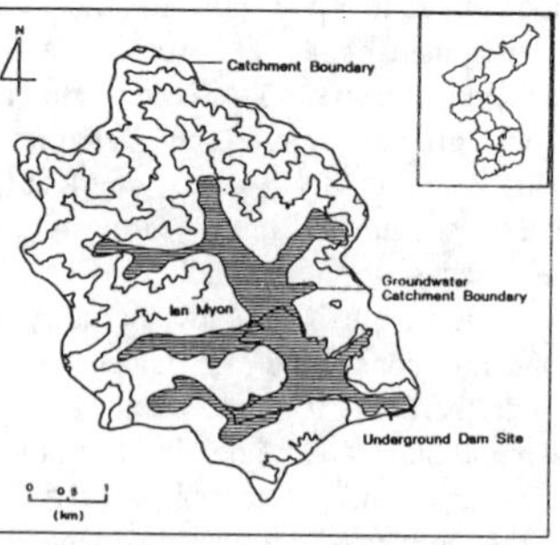

Fig. 1. Location map of Ian underground reservoir.

Table 1. Permeability test data

Pilot hole					Check hole				
hole No.	El. (m)	water table(m)	depth(m)	*k*(m/sec)	hole No.	No;	water table(m)	depth(m)	*k*(m/sec)
BH -12	80.19	4.58	0-8.0	2.45×10^{-4}	CH-1	1+16	0.9	3.5-6.0	4.6×10^{-7}
			12.3-16.0	6.64×10^{-7}	CH-2	3+4	1.3	0.3-2.5	5.56×10^{-7}
BH -13	78.64	3.14	0-8.0	6.87×10^{-5}	CH-3	6	1.3	2.5-5.0	5.5×10^{-7}
			9.0-13.0	4.34×10^{-7}	CH-4	9+17	0.9	2.0-4.5	4.39×10^{-7}
BH -14	76.86	0.91	0-5.0	4.96×10^{-4}	CH-5	4+11	1.7	0.3-2.0	2.95×10^{-7}
			10.0-13.0	4.60×10^{-7}				2.0-5.5	1.28×10^{-7}
BH -15	79.70	4.46	0-6.3	9.49×10^{-4}	CH-6	8+13	1.3	2.0-6.3	3.2×10^{-7}
			9.0-12.0	4.34×10^{-7}					

area mostly consist of farmland, irrigation water influx to the aquifer continuously. Also, in the observation, data of long term water level at the observation well in studied well, groundwater table is constant of pre-irrigation(Jan.-Apr.) and since May it has increasing trend although groundwater consumption is high at that season. For such reasons, infiltration rate of irrigation season since April has 4.8㎜/day as minimum infiltration rate and flow into the alluvium aquifer due to soil survey data.

5. GROUNDWATER MODELING

5.1 *Conceptual model*

Conceptual model is pictorial form of groundwater flow system and flow characteristics. It is simplified in defined range of small difference for flow system feature, and used as preliminary state model to apply analytical or numerical model among mathmatical models. Groundwater flow province in studied area has alluvium aquifer system of main aquifer and 3-dimensional structure geology. The outer boundary of studied area is mountaineous district of isolated ellipsodal catchment area. Such a groundwater flow system is basic and important feature in this study. Therefore within the range of possibility, to simplify this model it is assumed that vertical leakage has homogenious 2-dimensional flow verticallyt through alluvium weathered zone.

5.2 Before-dam construction condition simulation Steady state simulation

The first step in steady state simulation is the designing of the grid network and the allocation of all important parameters to every block(or cell) delineated by the grid network. The grid system used in this study is shown in Fig. 3.

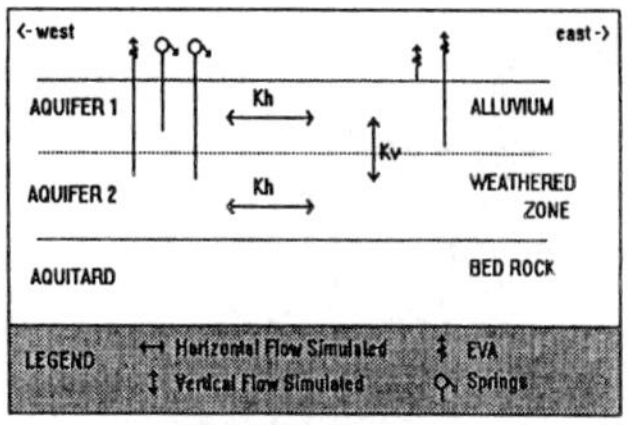

Fig.2. Conceptual model of Ian aquifer system.

The grid consists of 21×32 nodes with variable grid spacings ranging from 50 to 200 meters. The grid size near to the proposed groundwater dam site is designed to have a smaller size than the upstream cells to see more detailed water variation after the dam construction in the area. The parameters required in the steady state simulation are the boundary conditions, hydraulic heads, and the initial estimation of hydraulic conductivities. With these parameters, information on the hydrological setting of the study area used for the model design is discussed below.

5.3 *Determination of boundary conditions*

For the physical boundaries existing between alluvium and the mountain foothill in the model area, a constant head boundary is chosen.

The average depth of the weathered surface is about 4 meters in the mountain area and this layer has great potential to supply water. This 4-meter layer enables the alluviul aquifer to receive a certain amount of water from the surrounding mountain area throughout the year. This supply constitutes the main source to the alluvial aquifer because there is no other source in the study area except the low rate of infiltration from the rainfall, particularly before the irrigation season.

Boundary cells located near the outlet of the water basin, which was the proposed groundwater dam location, are defined as impermeable boundaries. This is because the geological and hydrological survey conducted on the area found outcrops of crystalline rocks exposed in these cells.

Table 2. Soil moisture accounting table

Descr.	JAN	FEB	MAR	APR	MAY	JUN	JUL	AUG	SEP	OCT	NOV	DEC	Annu
PE	0.1	0.0	18.7	57.1	93.5	129.5	164.2	134.9	84.8	43.9	9.0	0.1	735.8
P	28.9	15.8	46.6	26.6	21.8	85.4	267.6	255.0	150.3	14.1	30.4	10.6	953.1
P-PE	28.9	15.8	27.9	-30.5	-71.7	-44.1	103.4	120.1	65.5	-29.8	21.5	10.5	217.5
R/O	10.0	3.9	21.8	8.7	6.4	69.8	244.9	210.7	137.5	1.8	15.8	1.7	733.0
I	18.9	11.9	24.8	17.9	15.4	15.6	22.7	44.3	12.8	12.3	14.6	8.9	220.1
I-PE	18.9	11.9	6.0	-39.2	-78.1	-113.9	-141.5	-90.6	-72.0	-31.7	5.6	8.8	-515.8
NEG (I-PE)	0.0	0.0	0.0	-39.2	-117.3	-231.2	-372.7	-463.3	-535.3	-566.9	0.0	0.0	
ST	200	200	200	164	110	62	38	19	13	11	16.7	25.5	
ST	0.0	0.0	0.0	-36.0	-54.0	-48.0	-24.0	-19.0	-6.0	-2.0	5.6	8.8	-174.6
AET	0.0	0.0	18.7	53.9	69.4	63.6	46.7	63.3	18.8	14.3	9.0	0.1	357.8
PERC	18.9	11.9	6.0	0.0	0.0	0.0	0.0	0.0	0.0	0.0	0.0	0.0	

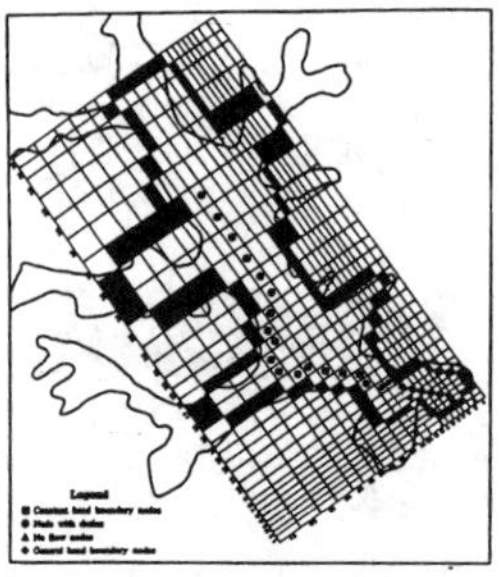

Fig. 3. Grid Design (before dam)

Another reason for this impermeable boundary setting is that the water divide area, from which these cells may receive a water supply, is relatively small in comparision with others(Fig. 3).

5.4 *Areal recharge*

Infiltration of rainfall into the ground constitutes the source of water to the alluvial aquifer system together with the recharge from the mountain area during the non-irrigation season. Since the water level was measured before the irrigation season, it is assumed that the rate of infiltration of the area is approximately the storage of the groundwater at that time.

5.5 *Calibration and Verification*

The steady state calibration is concentrated on the estimation of the hydraulic conductivity distribution in the alluvial aquifer(first layer) over the modeled area. The transmissivity value for the weathered zone (second layer) is assumed to be uniform over the study area because the basement of the area is mostly composed of single intrusive granite body. Initial guess of the hydraulic conductivity values were given by the pump test analysis results. In the process of calibration of the hydraulic conductivity, two main objectives and calibration criteria were employed as the following :

1) Reproduction of the observed piezometric head distribution as close as possible

2) Matching of groundwater outflow rate through the cross-sectional area at the outlet of the water basin estimated from a number of permeability tests performed at this location

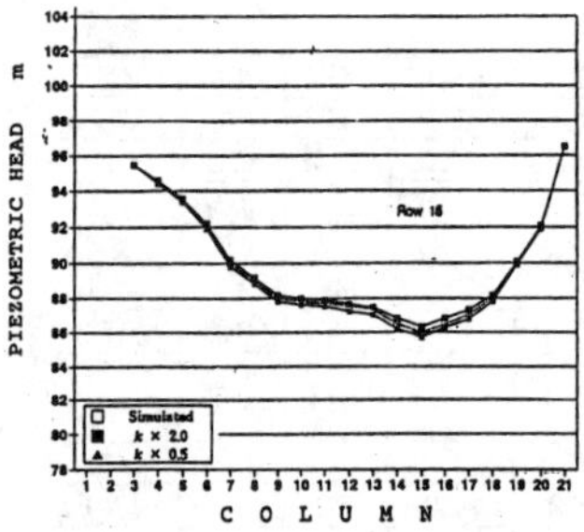

Fig. 4. Permeability sensitivity analyses

To verify the simulated steady state, the following are performed :

1) History matching - matching between the observed and simulated water level

2) Sensitivity analysis of hydraulic conductivity.

The resulted hydraulic head distribution at the end of the steady state simulation is compared with the observed one and shows very close match between them. The sensitivity analysis of the hydraulic conductivity is performed by multiplying and dividing the simulated hydraulic conductivity values by a factor of 2 (Fig. 4).

5.6 *Discussion of the Simulation Results*

Higher hydraulic conductivities accompony those cells the stream flows through. Since the stream bed is composed mainly of boulders and gravels it has a higher permeability than the alluvium composing the neighboring blocks. Not all the stream-containing blocks have higher permeabilities. The stream containing cells located between row 15 and row 18 (about 400meters in length) show lower permeability values than adjacent cells(Fig. 5). This is because the river path was changed artificially in 1960's to expand the area of the rice paddy fields. But the stream was not simulated in the steady state simulation, because the water level is assumed to exit under the stream bed.

The water flux rate flowing through the constant head boundary, is simulated as 0.04 cubic meters per second and shows close approximation to the flux rate of 0.043 calculated from the permeability and pumping tests data.

The mass balance at the end of the steady state simulation is presented together with the mass balance of subsequent simulations in Table 3.

5.7 *Transient condition simulation*

The transient model was constructed based on the simulated steady state to identify the factors that affect variation of groundwater quantitatively. The transient condition simulation for the before dam situation is performed to produce more reliable result in the after dam situation. The parameters used in the transient condition simulation is summarized in Table 4.

5.8 *Calibration and Verification*

As same as in the steady state simulation, the simulated hydraulic head is compared with the observed data to verify the simulation results. Also a series of sensitivity analysis on the storage coefficients was performed in the verification step.

Table 3. Steady state water balance(before dam)

unit : CMS

Desc		Stor-age	Const-antherd	General head	Rech-arge	EVT	Drain	Total
steady state	Inflow		0.042		0.002			0.044
	Outflow		0.044					0.044

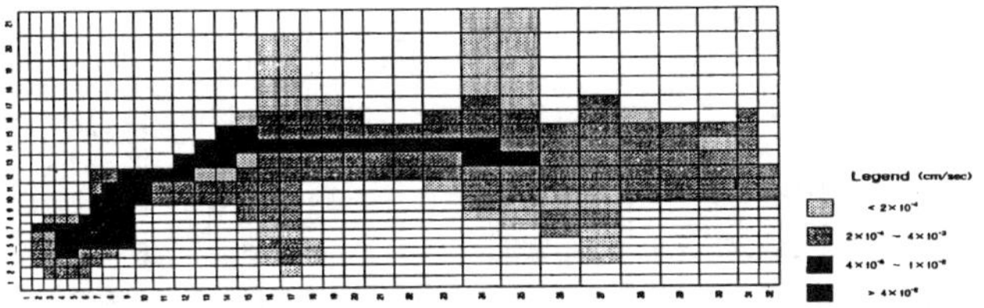

Fig. 5 Simulated permeability distribution

Table 4. Summary of input parameters

Seasons	Pre-irrigation season	Earlier irrigation season	Late irrigation season	Post irrigation season
Months	Jan.-Apr.	May-Jun	Jul.-Aug.	Sep.-Dec
Inflow	-Recharge from mountain area -Recharge from rainfall	-Recharge from mountain area -Recharge from irrigation water	-Recharge from mountain area -Recharge from irrigation water	-Recharge from mountine area -Recharge from rainfall
Outflow	-Subsurface outflow	-Subsurface outflow	-Subsurface outflow -Groundwater pumping	-Subsurface outflow

5.9 *Discussion of simulation Results*

From the transient condition simulation, the hydrologic parameters were estimated Quantitatively. The major factors which affect the seasonal variation of the groundwater table are identified as the recharge from the irrigation water and groundwater pumping and areal recharge from rainfall. This transient model will be used in the after-dam condition simulation to see the change of seasonal variation of groundwater table after the dam construction.

5.10 *After-dam condition simulation Steady state simulation*

In this study, the new steady state is developed for the after-dam construction situation using the transient head simulated with the before-dam steady state. In this simulation, cells comprising the dam site were assigned a lower permeability than before.

The dam was not designed as a impermeable boundary because of the following reasons:

1) An impermeable boundary is not able to show the possible leakage through the dam body.

2) An impermeable boundary is not correct in simulating overflow over dam.

3) The elevation of the dam crest was designed as 77 meters above sea level, therefore the water table is approximately constant with time.

5.11 *Calibration and Verification*

The water level increase was noticed during the simulation and the necessity of the adjustment of the boundary condition was found through the simulation. The constant head boundary, in which head remains constant with time, was used for the simulation of before-dam situation, but was not able to operate with the changed situation after the dam construction. Despite the constant boundary cells should keep supplying water even after the dam construction, the cells were releasing water through the boundary cells due to its fixed head, which is lower than the raised water level in the adjacent cells, rather than storing the water in the aquifer. Therefore, the adjustment of the boundary condition was necessary for the steady state simulation to prevent the increased storage from flowing out through the fixed head cells and changed to a variable head boundary.

At a general head boundary, the flux across the boundary changes in response to changes in head within the aquifer. The general head boundary allows the heads in boundary cells to vary with time and fluctuate with the heads in the adjustment cells. The change to a general head boundary was made under the assumption that there exists a point, where the head in the aquifer does not change, beyond each boundary cell(Wilson and Gerhart, 1982). The point, of which its head elevation is 120 meters, was selected and considered as the mid point of hillside in the associateed water divide.

5.12 *Transient Condition simulation*

The transient condition was simulated using the same hydrologic parameters with the before-dam transient condition simulation to examine the seasonal variation of groundwater table. Therefore this simulation is based on the assumption that such parameters are not changed even after the dam construction. For this phase of the simulation, the caliblation and verification steps does not have significant meaning because most of the parameters were calibrated in the previous transient simulation.

5.13 *Discussion of Simulation Results*

During the simulation of the after-dam construction situation, the followings are observed :

1) The dam installation resulted the rise of the groundwater level.

2) The seasonal water level fluctuation showed almost same pattern in both simulation(Fig. 6).

3) Most of the water interrupted to flow by the installation of the dam was flowing out from the study area through the stream in the form of surface water.

5.14 *Prediction model*

The constructed model is simulated to estimate the storage capacity and permeability of cut-off wall. According to the height of underground dam, the results of simulation have constant outflux through the unit area(Table 5). In the simulation for cut-off wall effect, the results of simulations at 7 steps of

Table 5. Water balance (Predict model)

Desc.	PREDICT																	
	Permeability														DAM CREST			
	2.4×10^{-3} m/s		5×10^{-4} m/s		5×10^{-5} m/s		5×10^{-6} m/s		4×10^{-7} m/s		5×10^{-8} m/s		5×10^{-9} m/s		EL 78		EL 79	
	In flow	Out flow	In flow	Out flow	In flow	Out flow	In flow	Out flow	In flow	Out flow	In flow	Out flow	In flow	Out flow	In flow	Out flow	In flow	Out flow
Storage																		
Constant head	0.041	0.025	0.041	0.013	0.041	0.003	0.041	0.001	0.041	0.001	0.041	0.001	0.041	0.001	0.041	0.001	0.041	0.001
General head																		
Recharge	0.002		0.002		0.002		0.002		0.002		0.002		0.002		0.002		0.002	
EVT.																		
Drain		0.018		0.030		0.040		0.042		0.042		0.042		0.042		0.042		0.042
Total	0.043	0.043	0.043	0.043	0.043	0.043	0.043	0.043	0.043	0.043	0.043	0.043	0.043	0.043	0.043	0.043	0.043	0.043

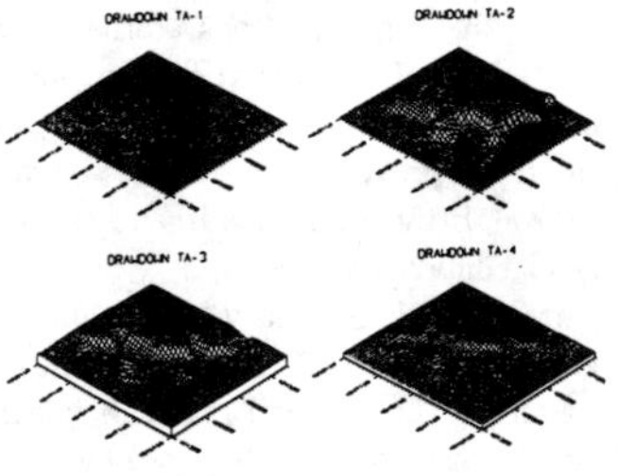

Fig. 6 Drawdown map (after dam)

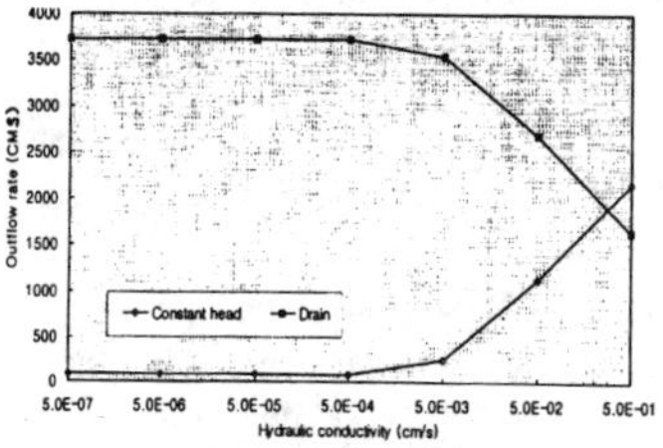

Fig. 7 Outflow rate in predict model.

hydraulics are as follows. In case of above $k = \alpha \times 10^{-4}$m/s, outflux amount increased through unit area as conductivity increases, but in case of under $k=\alpha \times 10^{-6}$m/s, outflux amount kept steady even conductivity decreases(Fig. 7).

6. CONCLUSIONS

To identify the hydrological environment of the aquifer system of the study area, parameters of surface and subsurface hydrology were evaluated in terms of a water budget. Based upon the evaluated parameters, a three dimensional numerical groundwater flow model was constructed and the effect of a underground dam was examined.

Several conclusions derived from the model simulation can be summarized as follows :

1. The aquifer system, largely consisting of Quarternary Alluvial deposits and a thin weathered zone of the Cretaceous granite, is recieving an almost constant rate of recharge throughout the year from the surrounding mountain area. The amount of annual recharge was determined as 0.04 CMS. This recharge rate did not change even after the dam construction.

2. The groundwater discharge rates through the cross-sectional area at the dam site were computed as 0.044 and 0.0009 CMS before and after the dam construction, respectively. After the dam was constructed the water levels in the aquifer rose to a new equiblium. This new water level was not far reaching in the up- stream direction because of the steep hydraulic gradient of the area(1/90).

3. In the simualtion for cut-off effect, the results of simulations at 7 steps of hydraulice are as follows. In the case over $k=\alpha \times 10^{-4}$m/s, outflux amount increased through unit area as conductivity increases, but in the case under $k = \alpha \times 10^{-6}$m/s, outflux amount kept steady even conductivity decreases. The results can help determine limited conductivities of soil of dam in this area.

REFERENCES

Bell, F. G., " Foundation of Engineering Geology", Butterworth, 1983, pp. 258-264.

Blaney, H. F. and Criddle, W. D., "Determining Water Requirements in Irrigation Areas from Climatological and Irrigation Data", U. S. Department of Agriculture, Soil Conservation Service Technical Paper 96.

Chow, "Handbook of Applied Hydrology", 1964, Chap. 13.9-13.32. Prentice Hall, 1986, pp.63-67.

Freeze, R. A. and Cherry, J. A.,"Ground Water",

McDonald, M. G. and Harbaugh, A. Q., "A Modular Three dimensional Finite Difference Ground-Water Flow Model", U. S. Geological Survey, Open File Report 83-875, 1984, pp. 528.

Saito, T., " Variation of physical properties of igneous rocks in weathering", International symposium on weak rocks, Tokyo, 1981, pp. 191 - 204.

Thornthwaite, C. W., "An Approach Toward a Rational Classification of Climate", Geographic Review, Vol. 38, 1946, pp. 55-94.

Trescott, P.C., Pinder, G.F. and Larson,S. P.,"Finite difference model for aquifer simulation in two dimensions with results of numerical experiments", U.S.goverment printing office, 1976, pp. 1 - 63.

Computer Methods and Advances in Geomechanics, Yuan (ed.) © 1997 Balkema, Rotterdam, ISBN 90 5410 904 1

A theory of consolidation for soils exhibiting rheological characteristics under cyclic loading

K.H.Xie, S.Guo, B.H.Li & G.X.Zeng
Institute of Geotechnical Engineering, Zhejiang University, Hangzhou, China

ABSTRACT: This paper presents analytical solutions to one-dimensional consolidation problem taking consideration of rheological property of soft clay under different cyclic loadings, in which a three-unit rheological model is introduced and the loading type can be triangular, rectangular, or isosceles-trapezoidal cyclic loading . Based on the solutions, the influence of the rheological parameters and loading conditions on the consolidation process are investigated. It has been shown that the consolidation behavior is mainly governed by the dimensionless parameters a_1, b and T_{V0}. Load shape has great influence on the rate of consolidation. For isosceles-trapezoidal cyclic loading, the consolidation rate in each cycle reaches the maximum at the end of constant loading phase and the minimum at the end of this cycle.

1. INTRODUCTION

Cyclic loading is one of the time-dependent loading types often encountered in geotechnical engineering. For example, some structures, such as storage oil tanks and silos usually apply cyclic loading on soft clay ground just beneath them since such structrues are periodically filled and emptied during their life time. Schiffman (1958) presented a general treatment of cosolidation under time-dependent loading. Wilson and Elgohary (1974), Balaigh and Lavadox (1978) have considered a rectangular cyclic loading of the clay layer. More recently, Favaratti and Soranzo (1995) presented a simple consolidation theory under triangular cyclic loading, while Chen et.al (1996) gave the one corresponding to trapezoidal cyclic loading. However, in all these studies ,soft clay is treated as linear elastic material although rheological model may be more sutiable.

In other hand , many consolidation theories taking consideration of rheological property of soft clay were developed during past decades (Gibson and Lo,1961; Xie and Liu,1995; Xie et.al ,1996). But when concerning cyclic loading, only a few are available.

In this paper, a three-unit rheological model is used to represent the rheological property of soft clay and theoretical solutions for one dimensional consolidation of soft clay layer under different cyclic loadings are obtained. Accordingly parameter study and the consolidation behavior are discussed.

2. DESCRIPTION OF THE PROBLEM STUDIED

The consolidation problem studied herein is a clay layer of thickness H with a pervious upper boundary and an impervious lower boundary under vertical uniform trapezoidal cyclic loading shown in Fig.1(a) in which α and β are the coefficients of proportion; t_0 is the time of one loading cycle. when $\alpha = 0.5$, the load becomes triangular one as shown in Fig.1(b); when $\alpha = 0$ it changes into rectangular cyclic loading(Fig.1(c)). Fig.2 shows the rheological model, which includes a spring of modulus E_0 and a Kelvin body with a spring of modules E_1 and a dashpot of adhesion coefficient K_1. When E_1 approaches infinite, the model turns into linear elastic one.

The constitutive equation in an integral form for the model shown in Fig.2 is

$$\varepsilon=\frac{\sigma}{E_0}+\int_0^t\frac{\sigma(\tau)}{K_1}e^{-\frac{E_1}{K_1}(t-\tau)}d\tau \tag{1}$$

where σ is effective stress acting on the model

The principle of effective stress for the load shown in Fig1(a) can be expressed as

$$\sigma+u=q(t) \tag{2}$$

where

$$q(t)=\begin{cases}\frac{q_0}{\alpha t_0}[t-(N-1)\beta t_0] & (N-1)\beta t_0\le t\le[(N-1)\beta+\alpha]t_0\\ q_0 & [(N-1)\beta+\alpha]t_0\le t\le[(N-1)\beta+(1-\alpha)]t_0\\ q_0\left\{1-\frac{1}{\alpha t_0}[t-(N-1)\beta t_0-(1-\alpha)t_0]\right\} & [(N-1)\beta+(1-\alpha)]t_0\le t\le[(N-1)\beta+1]t_0\\ 0 & [(N-1)\beta+1]t_0\le t\le N\beta t_0\end{cases}$$

Based on the above equations and all the basic assumptions made in Terzaghi's one dimensional consolidation theory except for loading condition and constitute model, the differential equation governing the consolidation process of the clay layer can be written as follows:

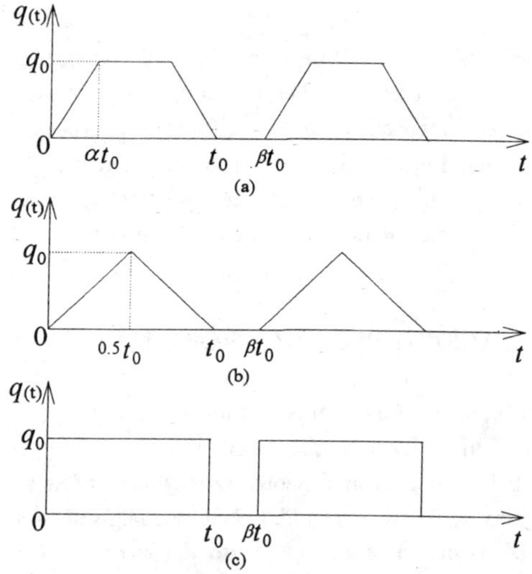

Fig.1 Loading types

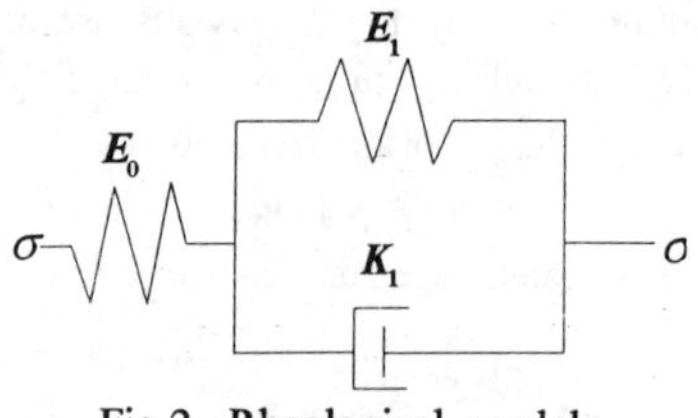

Fig.2 Rheological model

$$C_V\frac{\partial^2u}{\partial z^2}=\frac{\partial u}{\partial t}+\frac{E_0}{K_1}\int_0^t\frac{\partial u}{\partial\tau}e^{-\frac{E_1}{K_1}(t-\tau)}d\tau+f(t) \tag{3}$$

where

$$f(t)=\frac{dq}{dt}\left(\frac{E_0}{E_1}e^{-\frac{E_1}{K_1}t}-1-\frac{E_0}{E_1}\right) \tag{4}$$

$C_V=\frac{K_VE_0}{\gamma_W}$, the vertical coefficient of consolidation.

K_V is the vertical coefficient of permeability of clay; γ_W is the unit weight of water.

The solution conditions for Eqn.(3) are:

(1) $z=0$; $u=0$

(2) $z=H$; $\partial u/\partial z=0$

(3) $t=0$; $u=q(0)=0$

3. SOLUTION

The solution for Eqn.(3) can be expressed as

$$u=\sum_{m=0}^{\infty}T_m(t)\sin\frac{M}{H}z \tag{5}$$

in which $M=\frac{1}{2}(2m+1)\pi$, $m=0,1,2,\cdots$;

$T_m(t)$ is the function of time t.

Eqn.(5) satisfies the solution conditions (1) and (2). Substituting Eqn.(4) into Eqn.(3), the solution is satisfactory if, for each function $T_m(t)$, the following differential equation is satisfied:

$$C_V\frac{M^2}{H^2}T_m(t)+T_m'(t)+\frac{E_0}{K_1}\int_0^tT_m'(\tau)e^{-\frac{E_1}{K_1}(t-\tau)}d\tau$$
$$+\frac{2}{M}f(t)=0 \tag{6}$$

Applying Laplace transformation to Eqn.(6) with the solution condition (3), and then inverting it, the following expression can be obtained:

$$T_m(t)=\frac{2q_0}{MT_{V0}K}Y(t)$$

where

$$Y(t)=1+\frac{K+X_2}{X_1-X_2}e^{\frac{X_1T_V}{ba_1}}-\frac{K+X_1}{X_1-X_2}e^{\frac{X_2T_V}{ba_1}}$$

$K=\frac{M^2a_1b}{1+a_1}$, $T_{V0}=\frac{C_Vt_0}{H^2}$, $T_{V0}=\frac{C_Vt}{H^2}$

$a_1=\frac{E_0}{E_1}$, $b=\frac{K_VK_1}{H^2\gamma_W}$

$$X_{1,2}=-\frac{1}{2}\Big[1+a_1+M^2a_1b\mp\sqrt{\left(1+a_1+M^2a_1b\right)^2-4M^2a_1b}\Big]$$

Once excess pore water pressure u is found, the average degree of consolidation can be conveniently obtained from the following equation:

$$U=\frac{Hq-\int_0^t u\,dz}{Hq_0}$$

where $q=q(t)$, the load applied at the time of analysis.

3.1 *General solution*

For convenience, t is defined as: $t=(N-1)\beta t_0+t'$, $N=1,2,3,\cdots$, $0\le t'\le \beta t_0$.

The general solution for the general load shown in Fig.1(a) can be written as follows:

For $(N-1)\beta t_0\le t\le[(N-1)\beta+\alpha]t_0$, that is $0\le t'\le \alpha t_0$ (loading phase)

$$u=\sum_{m=0}^{\infty}\frac{2q_0a_1b}{\alpha MT_{V0}K}\sin(\frac{M}{H}z)\left[1+\overline{X_1}e^{\frac{X_1T'}{a_1b}}\left(B_1+\frac{1}{C_1}e^{\frac{X_1(N-1)\beta T_{V0}}{a_1b}}\right)-\overline{X_2}e^{\frac{X_2T'}{a_1b}}\left(B_2+\frac{1}{C_2}e^{\frac{X_2\beta(N-1)T_{V0}}{a_1b}}\right)\right]$$

$$U=\frac{T'}{\alpha T_{V0}}-\sum_{m=0}^{\infty}\frac{2a_1b}{\alpha M^2T_{V0}}\left[1+\overline{X_1}e^{\frac{X_1T'}{a_1b}}\left(B_1+\frac{1}{C_1}e^{\frac{X_1\beta(N-1)T_{V0}}{a_1b}}\right)-\overline{X_2}e^{\frac{X_2T'}{a_1b}}\left(B_2+\frac{1}{C_2}e^{\frac{X_2(N-1)\beta T_{V0}}{a_1b}}\right)\right]$$

where

$$\overline{X_1}=\frac{K+X_2}{X_1-X_2}C_1,\quad \overline{X_2}=\frac{K+X_1}{X_1-X_2}C_2$$

$$C_1=\frac{1-e^{\frac{X_1(N-1)\beta T_{V0}}{a_1b}}}{1-e^{\frac{X_1\beta T_{V0}}{a_1b}}},\quad C_2=\frac{1-e^{\frac{X_2(N-1)\beta T_{V0}}{a_1b}}}{1-e^{\frac{X_2\beta T_{V0}}{a_1b}}}$$

$$B_1=1+e^{\frac{X_1(\beta-1)T_{V0}}{a_1b}}-e^{\frac{X_1(\beta-\alpha)T_{V0}}{a_1b}}-e^{\frac{X_1(\beta-1+\alpha)T_{V0}}{a_1b}}$$

$$B_2=1+e^{\frac{X_2(\beta-1)T_{V0}}{a_1b}}-e^{\frac{X_2(\beta-\alpha)T_{V0}}{a_1b}}-e^{\frac{X_2(\beta-1+\alpha)T_{V0}}{a_1b}}$$

$$T'=\frac{C_Vt'}{H^2}$$

For $[(N-1)\beta+\alpha]t_0\le t\le[(N-1)\beta+(1-\alpha)]t_0$, that is, $\alpha t_0\le t'\le(1-\alpha)t_0$ (constant loading phase)

$$u=\sum_{m=0}^{\infty}\frac{2q_0a_1b}{\alpha MT_{V0}K}\sin(\frac{M}{H}z)\left[\overline{X_1}e^{\frac{X_1T'}{a_1b}}\left(B_1+\frac{D_1}{C_1}\right)-\overline{X_2}e^{\frac{X_2T'}{a_1b}}\left(B_2+\frac{D_2}{C_2}\right)\right]$$

where

$$D_1=e^{\frac{X_1(N-1)\beta T_{V0}}{a_1b}}-e^{-\frac{X_1\alpha T_{V0}}{a_1b}},\quad D_2=e^{\frac{X_2(N-1)\beta T_{V0}}{a_1b}}-e^{-\frac{X_2\alpha T_{V0}}{a_1b}}$$

$$U=1-\sum_{m=0}^{\infty}\frac{2a_1b}{\alpha M^2T_{V0}K}\left[\overline{X_1}e^{\frac{X_1T'}{a_1b}}\left(B_1+\frac{D_1}{C_1}\right)-\overline{X_2}e^{\frac{X_2T'}{a_1b}}\left(B_2+\frac{D_2}{C_2}\right)\right]$$

For $[(N-1)\beta+(1-\alpha)]t_0\le t\le[(N-1)\beta+1]t_0$, that is, $(1-\alpha)t_0\le t'\le t_0$ (unloading phase)

$$u=\sum_{m=0}^{\infty}\frac{2q_0a_1b}{\alpha MT_{V0}K}\sin\frac{Mz}{H}\left[\overline{X_1}e^{\frac{X_1T'}{a_1b}}\left(B_1+\frac{1}{C_1}(D_1-e^{-\frac{X_1(1-\alpha)T_{V0}}{a_1b}})\right)-\overline{X_2}e^{\frac{X_2T'}{a_1b}}\left(B_2+\frac{1}{C_2}(D_2-e^{-\frac{X_2(1-\alpha)T_{V0}}{a_1b}})\right)-1\right]$$

$$U=\frac{T_{V0}-T'}{\alpha T_{V0}}-\sum_{m=0}^{\infty}\frac{2a_1b}{\alpha M^2T_{V0}K}\left[\overline{X_1}e^{\frac{X_1T'}{a_1b}}\left(B_1+\frac{1}{C_1}(D_1-e^{-\frac{X_1(1-\alpha)T_{V0}}{a_1b}})\right)-\overline{X_2}e^{\frac{X_2T'}{a_1b}}\left(B_2+\frac{1}{C_2}(D_2-e^{-\frac{X_2(1-\alpha)T_{V0}}{a_1b}})\right)-1\right]$$

For $[(N-1)\beta+1]t_0\le t\le N\beta t_0$, that is, $t_0\le t'\le\beta t_0$ (no-loading phase)

$$u=\sum_{m=0}^{\infty}\frac{2q_0a_1b}{\alpha MT_{V0}K}\sin\frac{Mz}{H}\left[\overline{X_1}e^{\frac{X_1T'}{a_1b}}\left(B_1+\frac{1}{C_1}(D_1-E_1)\right)-\overline{X_2}e^{\frac{X_2T'}{a_1b}}\left(B_2+\frac{1}{C_2}(D_2-E_2)\right)\right]$$

where

$$E_1=e^{-\frac{X_1(1-\alpha)T_{V0}}{a_1b}}-e^{-\frac{X_1T_{V0}}{a_1b}},\quad E_2=e^{-\frac{X_2(1-\alpha)T_{V0}}{a_1b}}-e^{-\frac{X_2T_{V0}}{a_1b}}$$

$$U=-\sum_{m=0}^{\infty}\frac{2a_1b}{\alpha M^2T_{V0}K}\left[\overline{X_1}e^{\frac{X_1T'}{a_1b}}\left(B_1+\frac{1}{C_1}(D_1-E_1)\right)-\overline{X_2}e^{\frac{X_2T'}{a_1b}}\left(B_2+\frac{1}{C_2}(D_2-E_2)\right)\right]$$

3.2 *Triangular loading*

If $\alpha=0.5$, the loading type becomes triangular one

(Fig. 1(b)), the corresponding solution is the following :

For $(N-1)\beta t_0 \le t \le [(N-1)\beta + 0.5]t_0$, that is, $0 \le t' \le 0.5t_0$ (loading phase)

$$u=\sum_{m=0}^{\infty}\frac{4q_0a_1b}{MT_{V0}K}\sin(\frac{M}{H}z)\left[1+\overline{X_1}e^{\frac{X_1T'}{a_1b}}\left(\overline{B_1}+\frac{1}{C_1}e^{\frac{X_1(N-1)\beta T_{V0}}{a_1b}}\right)\right.$$
$$\left.\overline{X_2}e^{\frac{X_2T'}{a_1b}}\left(\overline{B_2}+\frac{1}{C_2}e^{\frac{X_2(N-1)\beta T_{V0}}{a_1b}}\right)\right]$$

$$U=\frac{2T'}{T_{V0}}-\sum_{m=0}^{\infty}\frac{4a_1b}{M^2T_{V0}K}\left[1+\overline{X_1}e^{\frac{X_1T'}{a_1b}}\left(\overline{B_1}+\frac{1}{C_1}e^{\frac{X_1(N-1)\beta T_{V0}}{a_1b}}\right)\right.$$
$$\left.-\overline{X_2}e^{\frac{X_2T'}{a_1b}}\left(\overline{B_2}+\frac{1}{C_2}e^{\frac{X_2(N-1)\beta T_{V0}}{a_1b}}\right)\right]$$

where

$$\overline{B_1}=1+e^{\frac{X_1(\beta-1)T_{V0}}{a_1b}}-2e^{\frac{X_1(\beta-0.5)T_{V0}}{a_1b}}$$
$$\overline{B_2}=1+e^{\frac{X_2(\beta-1)T_{V0}}{a_1b}}-2e^{\frac{X_2(\beta-0.5)T_{V0}}{a_1b}}$$

For $[(N-1)\beta+0.5]t_0 \le t \le [(N-1)\beta+1]t_0$,that is, $0.5t_0 \le t' \le t_0$ (unloading phase)

$$u=\sum_{m=0}^{\infty}\frac{4q_0a_1b}{MT_{V0}K}\sin\frac{Mz}{H}\left[1+\overline{X_1}e^{\frac{X_1T'}{a_1b}}\left(\overline{B_1}+\frac{\overline{D_1}}{C_1}\right)\right.$$
$$\left.-\overline{X_2}e^{\frac{X_2T'}{a_1b}}\left(\overline{B_2}+\frac{\overline{D_2}}{C_2}\right)-1\right]$$

where

$$\overline{D_1}=e^{\frac{X_1(N-1)\beta T_{V0}}{a_1b}}-2e^{-\frac{X_1T_{V0}}{2a_1b}},\ \overline{D_2}=e^{\frac{X_2(N-1)\beta T_{V0}}{a_1b}}-2e^{-\frac{X_2T_{V0}}{2a_1b}}$$

$$U=\frac{2(T_{V0}-T')}{T_{V0}}-\sum_{m=0}^{\infty}\frac{4a_1b}{M^2T_{V0}K}\left[\overline{X_1}e^{\frac{X_1T'}{a_1b}}\left(\overline{B_1}+\frac{\overline{D_1}}{C_1}\right)\right.$$
$$\left.-\overline{X_2}e^{\frac{X_2T'}{a_1b}}\left(\overline{B_2}+\frac{\overline{D_2}}{C_2}\right)-1\right]$$

For $[(N-1)\beta+1]t_0 \le t \le N\beta t_0$, that is, $t_0 \le t' \le \beta t_0$ (no-loading phase)

$$u=\sum_{m=0}^{\infty}\frac{4q_0a_1b}{MT_{V0}K}\sin\frac{Mz}{H}\left[1+\overline{X_1}e^{\frac{X_1T'}{a_1b}}\left(\overline{B_1}+\frac{1}{C_1}(\overline{D_1}+e^{-\frac{X_1T_{V0}}{a_1b}})\right)\right.$$
$$\left.-\overline{X_2}e^{\frac{X_2T'}{a_1b}}\left(\overline{B_2}+\frac{1}{C_2}(\overline{D_2}+e^{-\frac{X_2T_{V0}}{a_1b}})\right)\right]$$

$$U=-\sum_{m=0}^{\infty}\frac{4a_1b}{M^2T_{V0}K}\left[\overline{X_1}e^{\frac{X_1T'}{a_1b}}\left(\overline{B_1}+\frac{1}{C_1}(\overline{D_1}+e^{\frac{X_1T_{V0}}{a_1b}})\right)\right.$$
$$\left.-\overline{X_2}e^{\frac{X_2T'}{a_1b}}\left(\overline{B_2}+\frac{1}{C_2}(\overline{D_2}+e^{-\frac{X_2T_{V0}}{a_1b}})\right)\right]$$

3.3 *Rectangular loading*

When α=0, the loading type turns into rectangular one (Fig. 1(c)). The corresponding solutions is given by:

For $(N-1)\beta t_0 \le t \le (N-1)\beta t_0 + t_0$, that is, $0 \le t' \le t_0$ (loading phase)

$$u=\sum_{m=0}^{\infty}\frac{2q_0}{MK}\sin\frac{Mz}{H}\left[1+\overline{X_1}X_1e^{\frac{X_1T'}{a_1b}}\left(1+\frac{1}{C_1}e^{\frac{X_1(N-1)\beta T_{V0}}{a_1b}}-e^{\frac{X_1(\beta-1)T_{V0}}{a_1b}})\right)\right.$$
$$\left.-\overline{X_2}X_2e^{\frac{X_2T'}{a_1b}}\left(1+\frac{1}{C_2}e^{\frac{X_2\beta(N-1)T_{V0}}{a_1b}}-e^{\frac{X_2(\beta-1)T_{V0}}{a_1b}}\right)\right]$$

$$U=1-\sum_{m=0}^{\infty}\frac{2}{M^2K}\left[1+\overline{X_1}X_1e^{\frac{X_1T'}{a_1b}}\left(1+\frac{1}{C_1}e^{\frac{X_1(N-1)\beta T_{V0}}{a_1b}}-e^{\frac{X_1(\beta-1)T_{V0}}{a_1b}}\right)\right.$$
$$\left.-\overline{X_2}X_2e^{\frac{X_2T'}{a_1b}}\left(1+\frac{1}{C_2}e^{\frac{X_2\beta(N-1)T_{V0}}{a_1b}}-e^{\frac{X_2(\beta-1)T_{V0}}{a_1b}}\right)\right]$$

For $(N-1)\beta t_0 + t_0 \le t \le N\beta t_0$,that is, $t_0 \le t' \le \beta t_0$ (no-loading phase)

$$u=\sum_{m=0}^{\infty}\frac{2q_0}{MK}\sin\frac{Mz}{H}\left[\overline{X_1}X_1e^{\frac{X_1T'}{a_1b}}\left(1+\frac{D_3}{C_1}-e^{\frac{X_1(\beta-1)T_{V0}}{a_1b}}\right)\right.$$
$$\left.-\overline{X_2}X_2e^{\frac{X_2T'}{a_1b}}\left(1+\frac{D_4}{C_2}-e^{\frac{X_2(\beta-1)T_{V0}}{a_1b}}\right)\right]$$

in which

$$D_3=e^{\frac{X_1(N-1)\beta T_{V0}}{a_1b}}-e^{-\frac{X_1T_{V0}}{a_1b}},\ D_4=e^{\frac{X_2(N-1)\beta T_{V0}}{a_1b}}-e^{-\frac{X_2T_{V0}}{a_1b}}$$

$$U=-\sum_{m=0}^{\infty}\frac{2}{M^2K}\left[\overline{X_1}X_1e^{\frac{X_1T'}{a_1b}}\left(1+\frac{D_3}{C_1}-e^{\frac{X_1(\beta-1)T_{V0}}{a_1b}}\right)\right.$$
$$\left.-\overline{X_2}X_2e^{\frac{X_2T'}{a_1b}}\left(1+\frac{D_4}{C_2}-e^{\frac{X_2(\beta-1)T_{V0}}{a_1b}}\right)\right]$$

3.4 *Constant loading*

When α=0 and β=1.0 , the general load (Fig. 1(a)) turns to be constant loading , that is $q = q_0$. The corresponding average degree of consolidation can be simply obtained as follows :

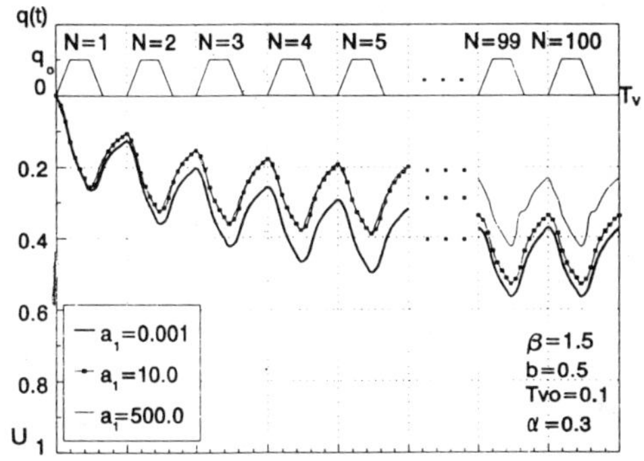

Fig. 3 The influence of a_1 on the consolidation rate

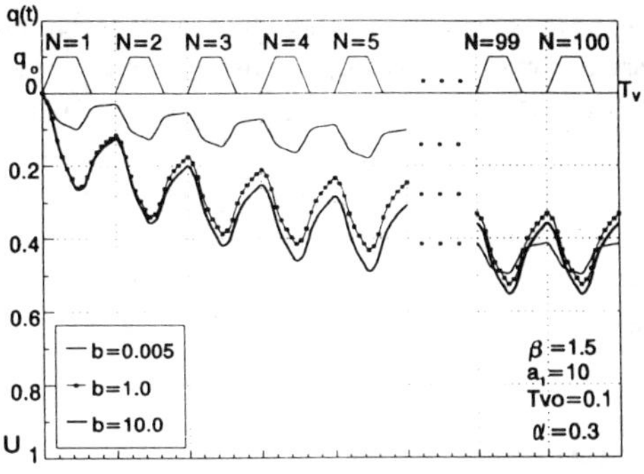

Fig. 4 The influence of b on the consolidation rate

$$U = 1 - \sum_{m=0}^{\infty} \frac{2}{M^2} \left[\frac{1 + a_1 + X_1}{X_1 - X_2} e^{\frac{X_1 T_V}{a_1 b}} - \frac{1 + a_1 + X_2}{X_1 - X_2} e^{\frac{X_2 T_V}{a_1 b}} \right]$$

Furthermore, as E_1 approaches to infinite (i.e. $a_1 \to 0$), the rheological model becomes linear elastic one, and the consolidation rate turns into

$$U = 1 - \sum_{m=0}^{\infty} \frac{2}{M^2} e^{-M^2 T_V}$$

This is just the Terzaghi's one dimensional consolidation theory.

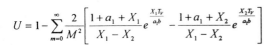

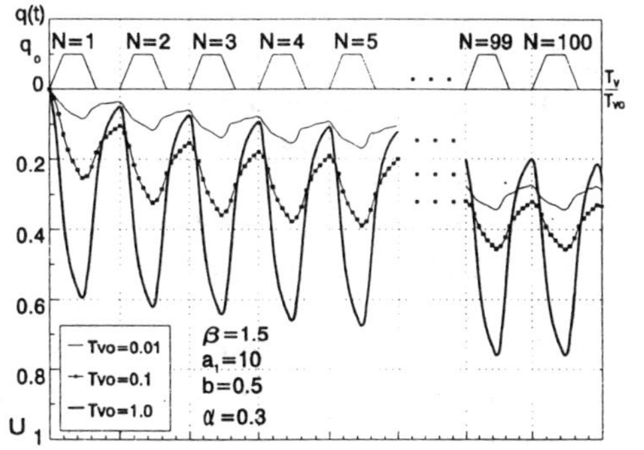

Fig. 5 The influence of T_{V0} on the consolidation rate

4. DISCUSSIONS

It can be seen from above solutions that there are great differences among solutions corresponding to different loading types. This shows that the shape of load has great influence on the process of consolidation.

To investigate the consolidation behavior, Figs.3-6 are prepared based on the general solution presented above.

Fig.3 shows the influence of $a_1 (= E_0/E_1)$ on the consolidation rate, from which it can be seen that the influence is more and more significant as the number of cycles increases, and the rate of consolidation decreases as E_1 decreases (i.e. the increase of a_1).

Fig.4 is plotted to investigate the influence of K_1. As $b (= K_V K_1/\gamma_W H^2)$ increases (i.e. K_1 increases), the consolidation is accelerated in the early stage, but in the later stage (e.g. $N = 100$) the influence of b becomes insignificant .

In Fig.5 T_{V0} varies from 0.01 to 1. It shows that the amount of relative variation of consolidation rate within each cycle becomes larger for higher T_{V0}. The consolidation rate curves tend to steady state after a certain number of loading cycles.

It can also be seen from these figures that the consolidation rate in each cycle reaches the maximum at the end of constant loading phase and the minimum at the beginning of the next cycle.

5. CONCLUSION

The following conclusions may be drawn from this study :

(1) An analytical solution has been developed for one-dimensional consolidation of soft clay exhibiting rheological characteristics under cyclic isosceles-trapezoidal loading. It shows that the consolidation behavior is mainly governed by three dimensionless parameters , that is ,a_1 ,b and T_{V0} .

(2) The solutions corresponding to triangular and rectangular cyclic loading are also presented, which shows that load shape has great influence on the rate of consolidation .

(3) For isoscess-trapezoidal cyclic loading, the consolidation rate in each cycle reaches the maximum at the end of constant loading phase and the minimum at the end of this cycle.

REFERENCES

Baligh, M.M. and Levadoux, J.N. 1978. Consolidation theory of cyclic loading. *J. Geotech. Eng. Div., ASCE* 104: GT4, 415-431..

Chen J.Z. et.al 1996. One dimensional consolidation of soft clay under trapezoidal cyclic loading. *Proc. of the Second International Conference on Soft Soil Engineering,* 211-216 .

Favaretti,M. and Soranzo,M.1995. A simplified consolidation theory in cyclic loading conditions. *Proc. of Int. Symposium on Compression and Consolidation of Clayey Soils,* 1, 405-409 .

Gibson, R.E. and Lo, K.Y.1961. A theory of consolidation of soils exhibiting secondary compression. *Acta Polytechnica Scandinavia* 296, Chapter 10.

Lo, K.Y.1961. Secondary compression of clays. *ASCE, JSMFD* 87: SM4.

Shifftman, R.L.1958. Consolidation of soils under time dependent loading and varying permeability. *Proceedings Hihgway Research* 37,584-617 .

Wilson, N.E. and Elgohary, M.M. 1974. Consolidation of soils under cyclic loading. *Canadian Geotechnical Journal* 11:3, 420-423.

Xie, K.H. and Liu, X.W. 1995. A study on one dimensional consolidation of soils exhibiting rheological characteristics. *Proc. of Int. Symposium on Compression and Consolidation of Clayey Soils,* 1, 385-388

Xie, K. H., Guo,S., and Zeng, G. X. 1996. On the consolidation of the soft clay ground beneath large steel tank . *Proc. of Int. Conf. on Advances in Steel Structures ,* 1199-1204

Computer Methods and Advances in Geomechanics, Yuan (ed.) © *1997 Balkema, Rotterdam, ISBN 90 5410 904 1*

Effects of the properties of sandy layers on the consolidation behaviour of clay

Guofu Zhu & Jian-Hua Yin
Department of Civil and Structural Engineering, Hong Kong Polytechnic University, China

ABSTRACT: This paper presents the results of a parametric study on the influence of the permeability and compressibility of two sandy soil layers on the consolidation behaviour of a clay layer. A finite element model is developed and used for the consolidation modelling of the three layer system. In the finite element model, Terzaghi's model is used for the sandy soils and Yin and Graham's elastic visco-plastic model is used for the clay. Main results from the FE modelling are presented and discussed regarding the influence of the permeability and compressibility of the sandy soils on the compression, strain and excess porewater pressure of the clay layer.

1 INTRODUCTION

The consolidation of clayey soils has practical interests to geotechnical engineers in the calculation and control of the settlement of foundations and reclamation. The stress-strain behaviour of clay is nonlinear and time-dependent (Yin and Graham 1989, 1994). The time-dependent settlement of a clay layer is due to (a) porewater pressure dissipation and (b) the creep nature of the soil skeleton. The conventional Terzaghi's method can't consider the creep nature of clayey soils and the multi-layers of a soil profile. A number of methods have been proposed to deal with the consolidation problem, including finite element method, finite difference method and semi-analytical method (Small and Booker 1988). Those methods may consider multi-layered soils. However soil models commonly used can not account for the creep behaviour of clayey soils.

Another issue in the consolidation analysis is that the sandy soil layers overlying and underlying a clay layer are normally simplified as free drainage boundaries. The permeability and deformation behaviour of the sandy layers are ignored. When the permeability of the sandy soil is very large, the above simplification is reasonable. However when the permeability of the sandy soils is not very large, the above simplification may lead to large errors in the prediction of the compression of the clay layer.

This paper presents the results of a parametric study on the influence of the permeability and compressibility of the sandy soil layers on the porewater pressure dissipation and the deformation of a clay layer under one-dimensional (1-D) vertical loading.

2 FINITE ELEMENT MODEL

A finite element (FE) model is developed for simulating the consolidation of a clay layer overlain by a sandy soil layer and underlain by a sandy soil layer. The basic equations used in the FE model are:

(a) The continuity equation

$$\frac{\partial q}{\partial z} = \frac{\partial \varepsilon}{\partial t} \tag{1}$$

where q = flow rate and ε =vertical strain, z=vertical co-ordinate and t=time.

(b) The constitutive equation

$$\frac{\partial \varepsilon}{\partial t} = \frac{\partial f(\sigma')}{\partial t} + g(\sigma', \varepsilon, t) \tag{2}$$

In Eqn.(2a), for Terzaghi's model

$$f = m_v \sigma' \qquad g = 0 \tag{3}$$

where m_v=soil compressibility, σ'=vertical effective stress.

For nonlinear elastic model

$$f = \frac{\lambda}{V}\ln\sigma' \qquad g = 0 \tag{4}$$

where λ/V is a constant.

For the elastic visco-plastic model proposed by Yin and Graham (1989, 1994)

$$f = \frac{\kappa}{V}\ln\sigma' \qquad g = \frac{\psi}{Vt_o}\exp(-\varepsilon\frac{V}{\psi})(\frac{\sigma'}{\sigma'_o})^{\lambda/\psi} \tag{5}$$

where κ/V=constant, λ=constant, ψ/V=constant, t_o=constant in unit of time, σ'_o =constant.

(c) The Darcy's law

$$q = -k\frac{\partial u}{\partial z} \tag{6}$$

where k=permeability, u=excess porewater pressure.

Using the above set of equations and standard discretization method, finite element equations can be established for consolidation modelling of a multi-layer system. Due to the nonlinear behaviour of soils, the stiffness matrix is not symmetric in general. When using a linear interpolation function, the FE stiffness matrix can be simplified to a symmetric matrix.

Table 1. Model parameters

Soil Layers	Model Parameters
Sandy soil layers (top and bottom) (Terzaghi's model)	for reference case: $m_{v,o}$=1.25×10^{-4} /kPa k_o=1.08×10^{-5} m/hour
Clay layer (Yin and Graham's model)	k_o=1.08×10^{-5} m/hour κ/V=0.01, λ/V=0.184, ψ/V=0.00911, σ'_o=67kPa t_o=0.00805 hour

3 SOIL PROFILE AND MODEL PARAMETERS

The soil profile considered in the FE modelling is

Top Sandy Soil Layer:
from elevation +6.44m to +3.22m
Middle Clay Layer:
from elevation +3.22m to 0m
Bottom Sandy Soil Layer:
from elevation 0m to -3.22m

The model parameters used the FE modelling are summarised in Table 1.

Two kinds of parametric study have been carried out. The first one is by increasing the permeability of the sandy soil layers as $1k_o$, $5k_o$, $10k_o$, $50k_o$, $100k_o$, and $1000k_o$. The second one is by increasing the permeability and the compressibility of the sandy soils using the same factor α (αk_o and $\alpha m_{v,o}$ where α=1, 5, 10, 20, and 50).

4 FE RESULTS AND DISCUSSIONS

4.1 *The Influence of k of the Sandy Soil Layers*

For simplicity, only main data are selected from the computation records of the FE modelling. Figs.1-3 illustrate the isochrones of excess porewater pressure, effective stress, and strain of the soil layers for the case of $10k_o$ for the two sandy soil layers in the first kind of parametric study. It can be seen that the water pressure in the middle of the clay layer firstly increase from its initial value 43 kPa to a maximum 47kPa, then decrease with time. The excess porewater pressure at the interface between the sandy soil layer and the clay layer is not zero as normally assumed as a free drainage boundary. As shown in Fig.2, the effective stress in the middle of the clay layer decreases to a certain extent at the initial stage of the loading corresponding to the excess porewater pressure increases as shown in Fig.1. Fig.3 shows that the vertical strain remains almost unchanged at the same initial stage in which the porewater pressure

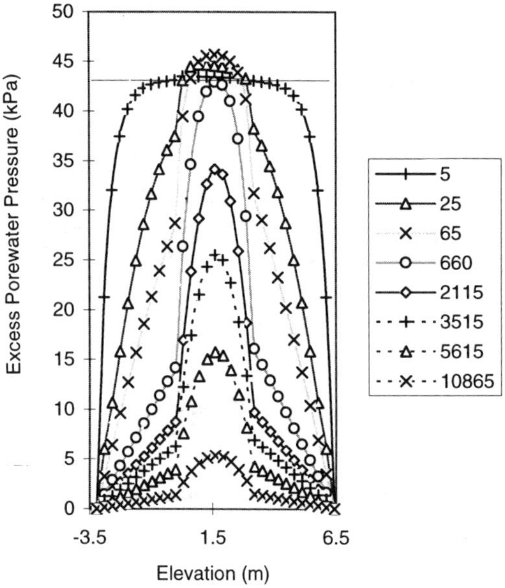

Fig.1 Isochrones of excess porewater pressure

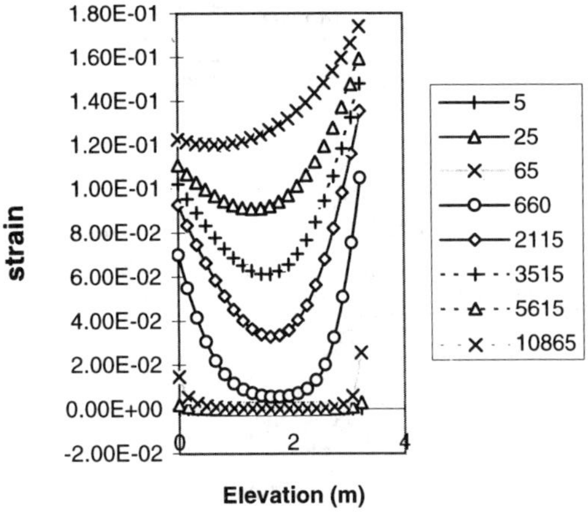

Fig.3 Isochrones of strain

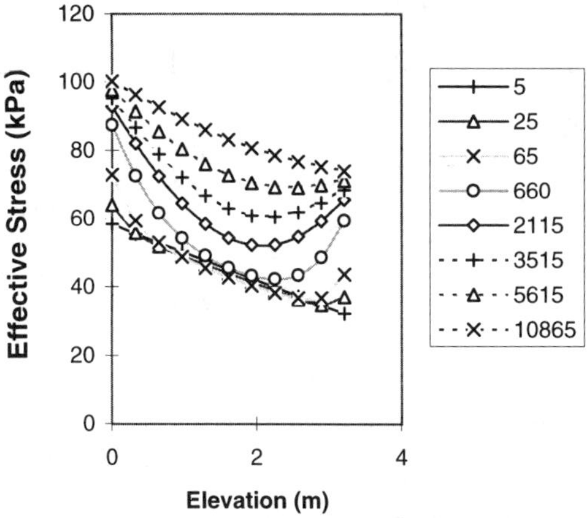

Fig.2 Isochrones of effective stress

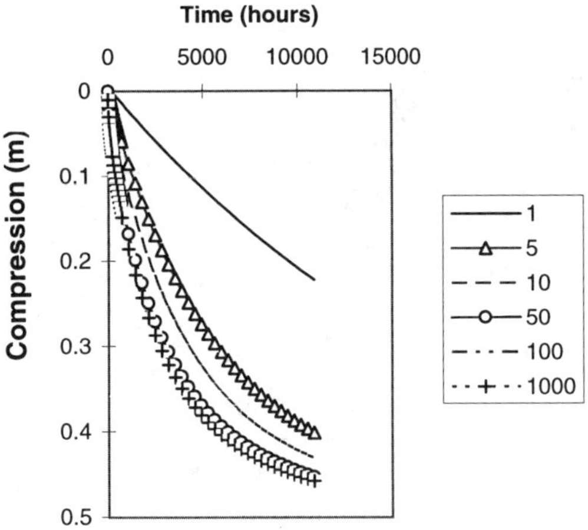

Fig.4 Compression vs. time for different k

increases and the effective stress decreases as shown in Fig.1 and Fig.2. With the increase in time, the strain accumulates as porewater pressure decreases.

Fig.4 shows that the effect of the increase in the permeability of the two sandy soil layers on the compression of the clay layer. It is seen that the compression increases with the increase in permeability. When the permeability k is more than $50k_o$, the effect is negligible.

Figs.5 and 6 shows the strain and the excess porewater pressure in the middle of the clay layer for different values of k ($k=1k_o$, $5k_o$, $10k_o$, $50k_o$, $100k_o$, $1000k_o$) of the two sandy soil layers respectively. Similarly, when the permeability k is more than $50k_o$, the influence on the strain and the excess porewater pressure is negligible.

4.2 *The Influence of factor α of the Sandy Soil Layers*

Fig.7 shows the compression of the clay layer vs. time by increasing the permeability and compressibility of the sandy layers at the same factor, α, (αk_o and $\alpha m_{v.o}$

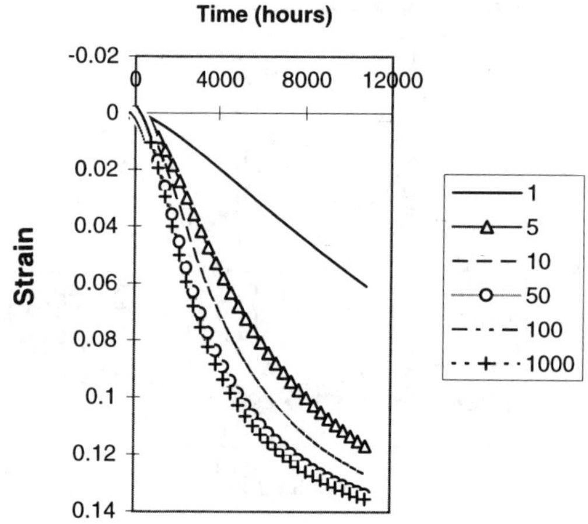

Fig.5 Strain vs. time for different k

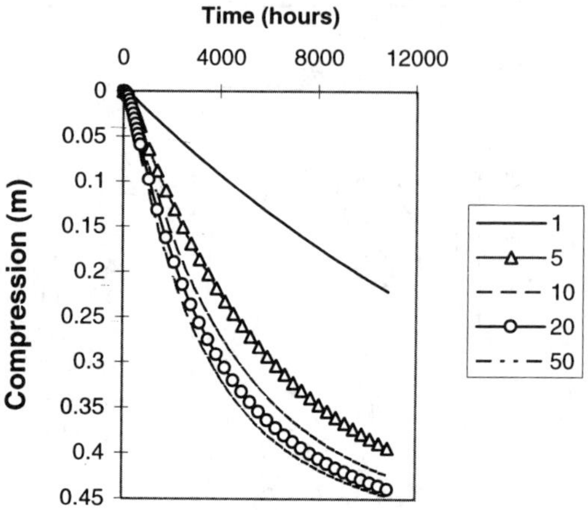

Fig.7 Compression vs. time for different α

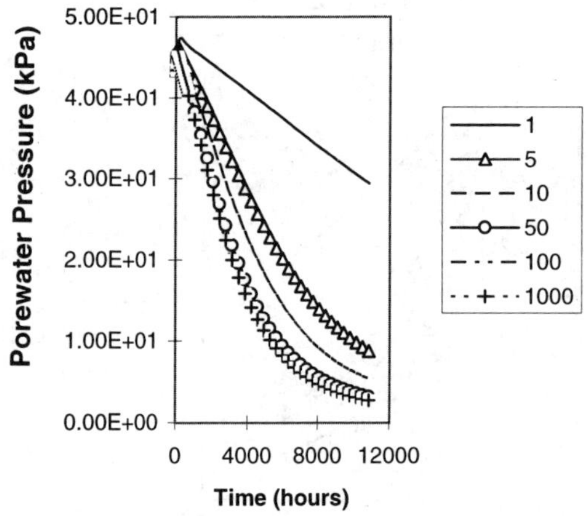

Fig.6 Excess water pressure vs. time for different k

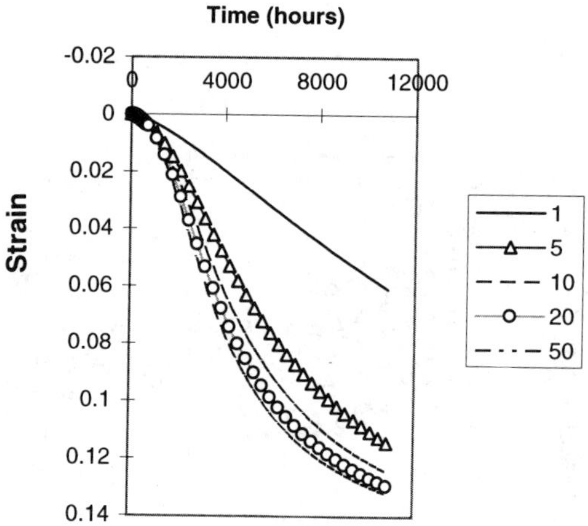

Fig.8 Strain vs. time for different α

where α=1, 5, 10, 20 and 50). Figs.8 and 9 show the strain and excess porewater pressure at the middle of the clay layer for the same ratio, α. It is seen from Figs.7,8,9 that when the ratio α is more than 20, the influence on the compression, strain and excess porewater pressure may be considered negligible. It is believed that the increase in the compressibility has little influence. But the permeability has significant influence when k is less than approximately $30k_o$. According to Terzaghi's theory, the constant ratio of k/m_v means a constant coefficient of consolidation. This means that the time factor is constant and the degree of consolidation (excess porewater) is only a function of time not the factor α. But the factor α has significant influence on the excess porewater pressure dissipation of the clay layer.

CONCLUSIONS

From the results and discussion presented in the preceding sections, the following conclusions can be made:

1. The permeability of the sandy soil layers overlaying and underlying a clay layer has significant influence on the compression, strain, and excess porewater pressure dissipation of the clay layer when

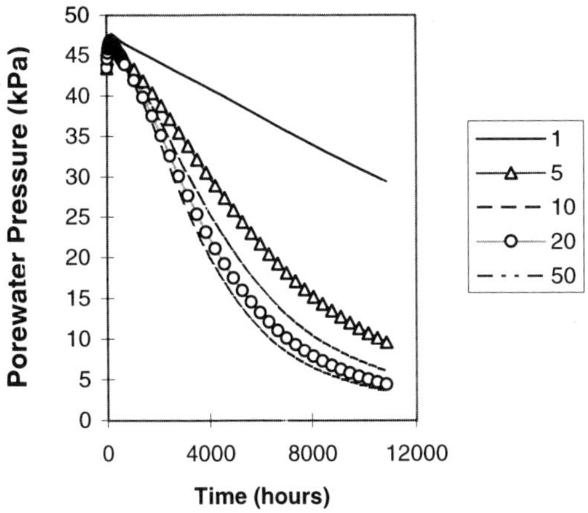

Fig.9 Excess water pressure vs. time for different α

the permeability k of the sandy soils is less than 30 times of the permeability of the clay.

2. The factor α has significant influence on the compression, strain and excess porewater pressure dissipation of the clay layer when α is less than 30. The influence of the factor α is mainly due to the influence of the permeability k of the sandy soils not the compressibility m_v.

3. Since the permeability of the sandy soils affects the porewater pressure dissipation in the clay layer, the correct assumption of the drainage conditions in the sandy soil layers is important for the accurate prediction of the compression of the clay layer.

4. It is found that the excess porewater pressure in the middle of the clay layer becomes larger than its initial values at the initial stage of loading in the three layer system. This phenomenon is due to the creep nature of the clay as explained before by Yin et al. (1994). This phenomenon is first simulated in this paper for a multi-layer soil profile.

ACKNOWLEDGEMENT

Financial support from the Hong Kong Polytechnic University and RGC are acknowledged.

REFERENCE

Small, J.C. and Booker, J.R., 1988. Consolidation of layered soils under time-dependent loading. Num. Methods in Geomechanics, Innsbruck 1988, pp.593-597.

Yin, J.-H. and Graham, J., 1989. Viscous elastic plastic modelling of one-dimensional time dependent behaviour of clays. Can. Geotech. J., Vol.26, 199-209.

Yin, J.-H. and Graham, J. (1994). Equivalent times and elastic visco-plastic modelling of time-dependent stress-strain behaviour of clays. Canadian Geotechnical Journal, Vol. 31. 42-52.

Yin, J.-H., Graham, J., Clark, J.I., and Gao, L. (1994). Modelling unanticipated porewater pressures in soft clays. Canadian Geotechnical Journal, Vol. 31. 773-778.

Computer Methods and Advances in Geomechanics, Yuan (ed.) © 1997 Balkema, Rotterdam, ISBN 90 5410 904 1

An elasto-viscoplastic analysis of self weight consolidation during continuous sedimentation

Goro Imai & Bipul Chandra Hawlader
Department of Civil Engineering, Yokohama National University, Japan

ABSTRACT: This paper puts forward an idea to model and analyze one dimensional self weight consolidation during continuous sedimentation of cohesive soil. The sedimentation process is assumed to occur step by step and the mechanism of consolidation is considered to be elasto-viscoplastic with varying permeability. An implicit finite difference technique is used in the numerical analysis with the Crank-Nicolsoln's theorem to assure computation stability. The numerical solution obtained is compared with a non linear elastic solution, and the effect of viscosity on sedimentation-consolidation process is discussed.

1 INTRODUCTION

Soil particles are generally carried as suspension in moving water. When its flow velocity reduces sufficiently, these soil particles flock together and then settle on the bed. At a certain moment these flocks join together and start to get sufficient stiffness to form fresh soil, which becomes a part of bed soil and consolidates under its self weight. When making a theoretical model for this process of sedimentation with consolidation, it is required to consider this process under very low stress level, i. e. large void ratio and very high compressibility with largely changing permeability. Few solutions available for this type of problem are elastic or plastic type with linear or nonlinear compressibility. But in this sedimentation-consolidation process viscous effect should be taken into account. Imai (1995) developed an elasto-viscoplastic consolidation theory, in which viscosity inherent to soil skeleton is considered to act during the consolidation process after viscoplastic compression yielding. In this paper, this theory is used with some restriction.

2 MODELING OF SEDIMENTATION-CONSOLIDATION PROCESS

The process of soil formation consists of three different phenomena ; settling, sediment formation and consolidation. Here the settling is the in-water sinking of suspended soil particles. The sediment formation means the process through which the particles settle down on the bed touching each other then compress and get an enough strength to sustain the load of the overlying material. And the consolidation means the one caused by the self weight of both the soil previously formed and the additional new overlying sediment. This process of sedimentation with consolidation may be explained by Fig. 1.

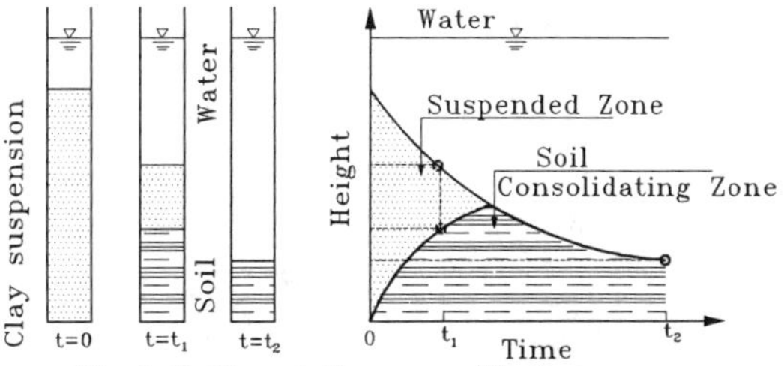

Fig. 1 Sedimentation -consolidation process

When particles and/or flocks, being separated from the other one in the suspension, settle down on the bed soil, they become a part of the bed soil. Therefore, it is reasonable to consider that there exists one boundary which distinguishes the sedimented soil from the suspension. This boundary shown in Fig. 1 was once defined as "sediment formation line" (Imai, 1981), but it should not be a line because soil formation is not an instant process. This boundary should be expressed as a zone. However, we cannot know the precise mechanism of soil formation in this zone. In this study, therefore, the authors consider this fresh sedimented zone to be initially uniform in terms of void ratio and effective stress. With the progress of further sedimentation, the top of this zone rises up and the lower soil layer consolidates due to its self weight. This process is continuous, but for the purpose of numerical analysis it is considered as a process of step by step accumulation of fresh sediment. It instantly covers the top surface of the previously sedimented soil and starts to consolidate just after this formation. It is to be noted here that this paper concerns about the consolidation process only, not the behavior of suspension.

3 CONSOLIDATION MECHANISM

Based on his experimental evidence, Imai (1992) showed that the e-logσ state path followed by each clay element in a consolidating layer is different as shown in Fig. 2. The change in void ratio with increasing effective stress is initially very small but becomes large over a certain effective stress level termed yield stress, of which definition will be given in section 8. This yield stress is higher for the element near the drainage boundary. He interprets this difference in state paths as a result of the viscosity inherent to soil skeleton. Although the behavior before yielding is not yet clarified, the behavior after yielding is formulated by Imai (1995) as follows.

$$\Gamma = e + C_c \log \sigma' = a \log(-\dot{e}) + b \qquad (1)$$

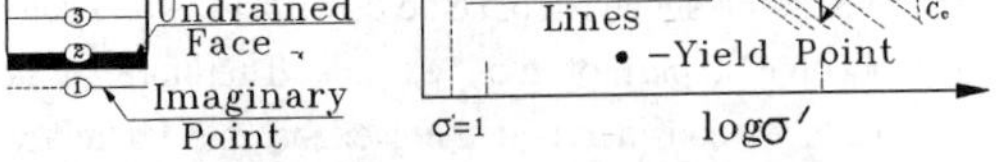

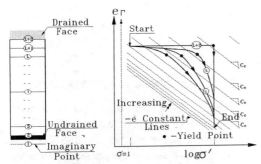

Fig.2 State paths for different soil elements

Γ is a specific parameter which defines the relative position of a state point (e,σ') measured from the basic compression line which passes through the starting state point (e_0, σ_0') and having the slope of C_c. The a and b are constants and can be obtained from simple oedometer testing. The constant a expresses the effect of viscosity and is here assumed to be equal to the coefficient of secondary compression, C_α, which is obtained from the secondary compression stage in oedometer test. The equation (1) means that the consolidation state of any element can be expressed by a state point in three dimensional space $(e,\dot{e},\sigma')$, and further the point moves, after yielding, on a single state surface $f(e,\dot{e},\sigma') = 0$.

4 PARAMETER SELECTION

In this study, very important is the selection of soil parameters about compressibility and permeability under very low stress levels.

Compressibility

Been et. al. (1981) assumed in their simulation a linear e-σ' relationship under very low stress level. But their test result shows unsatisfactory agreement with their calculation results. Yano et. al. (1988) obtained some useful data for general cohesive soils in Japan resulting in the linear relationship $\log(1+e) = A \log\sigma' + D$. In this paper this relationship is used, thereby the compression index C_c is $(1+e)A/0.434$.

Permeability

For a highly compressible range corresponding to very low effective stress levels, Been et. al. (1981) used, in their analysis, a linear relationship between k and e. But, their results show that this assumption is not well satisfactory. Alternatively, we use the relationship $\log c_v = B \log\sigma' + \log c_{v0}$ proposed by Yano (1988) and Yamauchi (1991), then we can get the following nonlinear relationship between permeability and effective stress using the definitions $k = \gamma_w m_v c_v$ and $m_v = (1/\sigma').[-A(1+e)/(1+e_0)]$.

$$\log k = (A+B-1)\log \sigma' + \log(-A\gamma_w c_{v0})$$

Initial Effective Stress (σ_0')

Soil particles suspended in water are in the stress free state. But in the transitional zone between the suspension and the existing bed soil, the soil particles just become closer to each other with increasing effective stress to form a certain structure as soil. The value of this critical effective stress, over which a consolidation analysis can be applied, is difficult to determine. Yamauchi et. al. (1991) suggested an acceptable method to select this initial effective stress and showed that it is very reasonable to consider the initial effective stress to be 0.1 gf/cm^2. In this study this value is used; that is, the additionally sedimented soil starts to consolidate from the initial uniform void ratio e_0 which corresponds to the initial effective stress of 0.1 gf/cm^2. From that state to the next time steps the consolidation analysis stated below is applied to the total layer of soil including the fresh one.

5 FORMULATION OF THE NUMERICAL MODEL

Finite difference technique is used in the present analysis. In the following description, superscript 'k' means the time grid number and subscript 'n' means the space grid number.

The change in void ratio at a consolidating soil element, Δe, generally consists of elastic, plastic and viscous components. According to the suggestion by Imai (1995), however, the total change may be separated into the two components, recoverable one (Δe^r) and irrecoverable one (Δe^{ir}). The recoverable component, which occurs whenever effective stress increases, can be defined simply by the following equation.

$$-\Delta e^r = C_s \Delta \log \sigma' = 0.434 C_s \frac{\Delta \sigma'}{\sigma'} \qquad (2)$$

The irrecoverable component, (Δe^{ir}), takes place after yielding and consists of plastic and viscous part of Δe. Since the equation (1) should be applied to the process of after-yield consolidation, we should replace e in Eq. (1) with e^{ir} in this case. By the use of the center approximation ($\Delta e^{ir} = [(\dot{e}^{ir,k} + \dot{e}^{ir,k+1})/2]\Delta t$, the following relationship between Δe^{ir}, $\Delta\sigma'$ and Δt can be obtained.

$$-\Delta e^{ir} = \frac{f_{\sigma'} \Delta t}{2 + f_{e^{ir}} \Delta t} \Delta \sigma' - \frac{2 \dot{e}^{ir,k} \Delta t}{2 + f_{e^{ir}} \Delta t} \qquad (3)$$

Where,

$$f_{e^{ir}} = \frac{\partial(-\dot{e}^{ir})}{\partial e^{ir}} = \frac{-\dot{e}^{ir}}{0.434 a} \text{ and } f_{\sigma'} = \frac{\partial(-\dot{e}^{ir})}{\partial \sigma'} = \frac{C_c}{a} \frac{-\dot{e}^{ir}}{\sigma'}$$

When combining the equation (2) with (3), it is possible to get a realistic constitutive equation to determine the void ratio change in the after-yield consolidation process.

$$-\Delta e = -(\Delta e^r + \Delta e^{ir}) = A_1 \Delta \sigma' + B_1 \Delta t \qquad (4)$$

Where,

$$A_1 = 0.434 \frac{C_s}{\sigma'} + \frac{f_{\sigma'} \Delta t}{2 + f_{e^{ir}} \Delta t} \text{ and } B_1 = \frac{-2\dot{e}^{ir,k}}{2 + f_{e^{ir}} \Delta t}$$

A problem still remaining in our consolidation analysis is the treatment of the before-yield compression process. It is very difficult to get its exact behavior experimentally. Although the authors are aware of not small effects of viscosity in this range, they disregard its effect and consider the process to be elastic.

$$-\Delta e = A_1 \Delta \sigma' \qquad (4.b)$$

In this equation $A_1 = \frac{0.434 C_t}{\sigma'}$ and C_t is the compression index in the region of before yield. The way to determine the value of C_t will be described later, and A_1 is assumed to be constant for one time step of calculation.

6 SOLUTION FOR THE NON-LINEAR DIFFERENTIAL EQUATION BY FINITE DIFFERENCE METHOD

Four fundamental equations, (1) mass conservation, (2) momentum balance, (3) Darcy's law and (4) constitutive equations are required to solve a consolidation process. In this calculation the reduced coordinate system z is used, in which dz = dz_0/(1+e_0), and z_0 is the real initial thickness of the soil layer over which suspension exists.

The exact law of momentum balance is expressed as

$$i = \frac{-\partial h}{\partial \xi} = \left[\frac{1+e_0}{1+e}\right] \frac{-\partial h}{\partial z_0} = \frac{1}{\gamma_w}\left[\frac{1}{1+e}\frac{\partial \sigma'}{\partial z} + \frac{\gamma_s - \gamma_w}{1+e}\right]$$

By using the Darcy's law, $v = ki$, throughout the whole process of consolidation and the following exact law of mass conservation

$$-\frac{\partial e}{\partial t} = \frac{\partial v}{\partial z}$$

we can get the following combined equation using submerged unit weight γ'.

$$-\frac{\partial e}{\partial t} = \alpha\left[\frac{\partial^2 \sigma'}{\partial z^2} + \beta\left\{\frac{\partial \sigma'}{\partial z} + \gamma'\right\}\right]$$

Where $\alpha = \frac{1}{1+e}\frac{k}{\gamma_w}$ and $\beta = \frac{1}{k}\frac{\partial k}{\partial z} - \frac{1}{1+e}\frac{\partial e}{\partial z}$

The values of α, β and γ' are assumed to be constant for one step of calculation. Under this assumption we apply the principle of finite difference method to the above equation, then we are left with

$$-\frac{\Delta e}{\Delta t} = \alpha\left[\frac{\partial^2 \sigma'}{\partial z^2} + \beta\left\{\frac{\partial \sigma'}{\partial z} + \gamma'\right\}\right] \tag{5}$$

Now, using the value of void ratio change Δe obtained from the constitutive equation (4), for the nth element and the kth time level, we have

$$\sigma_n'^{k+1} = \sigma_n'^{k} + \begin{bmatrix} \frac{\alpha}{A_1}\frac{\partial^2 \sigma_n'^{k}}{\partial z^2} + \frac{\alpha\beta}{A_1}\frac{\partial \sigma_n'^{k}}{\partial z} \\ +(\frac{\alpha\beta\gamma'}{A_1} - B_1) \end{bmatrix}\Delta t$$

That is,

$$\sigma_n'^{k+1} = \sigma_n'^{k} + \left[P\sigma_{n+1}'^{k} + Q\sigma_n'^{k} + R\sigma_{n-1}'^{k} + S\right]\Delta t$$

Where P, Q, R and S are calculation constants.

Now the Crank-Nicolson's scheme is applied to this problem. Because P, Q, R and S are constants between the time interval k and k+1, then we get

$$\sigma_n'^{k+1} - \begin{bmatrix} P\sigma_{n+1}'^{k+1} + Q\sigma_n'^{k+1} \\ +R\sigma_{n-1}'^{k+1} + S \end{bmatrix}\Delta t = \sigma_n'^{k} + \begin{bmatrix} P\sigma_{n+1}'^{k} + Q\sigma_n'^{k} \\ +R\sigma_{n-1}'^{k} + S \end{bmatrix}\Delta t$$

and this can be rewritten as follows,

$$-P\Delta t\sigma_{n+1}'^{k+1} + (1-Q\Delta t)\sigma_n'^{k+1} - R\Delta t\sigma_{n-1}'^{k+1}$$
$$= \sigma_n'^{k} + \left[P\sigma_{n+1}'^{k} + Q\sigma_n'^{k} + R\sigma_{n-1}'^{k} + 2S\right]\Delta t$$

Since all information at the kth time level is already known, the above equation can be simply expressed as follows.

$$C_{1,n}\sigma_{n-1}'^{k+1} + C_{2,n}\sigma_n'^{k+1} + C_{3,n}\sigma_{n+1}'^{k+1} = C_{4,n}$$

Where, $C_{4,n}$ is a function of σ' at the kth time level.

The way described above gives the constants $C_{i,n}$ at the mesh points 3 to L. But for the mesh points 2 and L+1 the additive conditions described below should be incorporated .

7 BOUNDARY CONDITION

Drained Boundary

It is assumed here that a newly sedimented soil element having the void ratio e_0 and some thickness Δz instantly overlies on the existing soil layer. The top surface of that fresh element is considered as a drained boundary for the next time step until another fresh sediment layer overlies on that element. During this time interval, it is further assumed that effective stress of this element does not change from 0.1 gf/cm^2 but its void ratio irrecoverably decreases due to viscosity from the first time, therefore, its total void ratio change is equal to the irrecoverable one. Based on these assumptions we can get the following relationship by the use of Eq. (1).

$$\Delta e = \frac{\dot{e}^k + \dot{e}^{k+1}}{2}\Delta t = a\log\left(\frac{\dot{e}^{k+1}}{\dot{e}^k}\right) = 0.434a\frac{\dot{e}^{k+1} - \dot{e}^k}{\dot{e}^k}$$

Rewriting this equation, we get

$$\dot{e}^{k+1} = \dot{e}^k + \frac{2\dot{e}^{k^2}\Delta t}{0.868a - \dot{e}^k\Delta t}$$

Applying the Crank-Nicolson's scheme to this one step process, we are left with the following finite difference scheme

$$\dot{e}^{k+1} - \frac{2\dot{e}^{k+1^2}\Delta t}{0.868a - \dot{e}^{k+1}\Delta t} = \dot{e}^k + \frac{2\dot{e}^{k^2}\Delta t}{0.868a - \dot{e}^k\Delta t}$$

The above equation is a quadratic equation of $\dot{e}^{k+1}$ and can be easily solved for $\dot{e}^{k+1}$. Once the value of $\dot{e}^{k+1}$, is calculated, then Δe can be easily calculated from the following relationship.

$$\Delta e = \frac{\dot{e}^k + \dot{e}^{k+1}}{2}\Delta t$$

But, unfortunately, authors have no idea about the initial void ratio rate for this top element. Its value is, therefore, considered to be zero in this calculation.

Undrained Boundary

At the undrained side i.e. at the mesh point 2, flow velocity v is zero throughout consolidation, therefore

$$\left.\frac{\partial \sigma'}{\partial z}\right|_2 = -(\gamma_s - \gamma_w)$$

Since the central finite difference method is used throughout the calculation, the following relationship holds good for the imaginary point 1 at a distance Δz from the undrained face.

$$\sigma_1' = \sigma_3' - 2\Delta z(\gamma_s - \gamma_w)$$

8 YIELDING CONDITION

In this study, the constitutive relationship for before- and after-yield process is different to each other. In order to define the effective stress at yielding, the value

$\Delta f = [e + C_c \log\sigma'] - [a\log(-\dot{e}) + b]$ is calculated for each step by using a set of current values (e, σ', $\dot{e}$). Since the first [ ] is the current Γ-value before yield and the second [ ] is the expected after-yield Γ-value against the current $\dot{e}$-value, the condition of $\Delta f = 0$ means that the element under consideration just yields in terms of compression. By the use of this Δf, the compression index C_t for the before-yield process is here defined as $C_t = C_s + \frac{C_c - C_s}{1 + \mu\Delta f}$. The value of $\mu = 100$ is used in this calculation, according to the suggestion of Imai and Tang (1992).

9 COMPUTATION RESULTS

We are still left with a set of equations to be solved, of which form is $A\sigma' = \mathbf{B}$. The coefficient A is a tridiagonal matrix with a special structure as described above, σ' is the stress for the interior grid points at the time step $(k+1)^{st.}$ and **B** is a function of σ', e and $\dot{e}$ values at the kth time level. Then the above equations are solved by using the Gaussian elimination.
Once the value of σ' for the $(k+1)^{st}$ level is determined, the value of Δe for this time interval can be easily calculated by the use of Eq. (4). This value of e^{k+1} and σ'^{k+1} is used to determine $\dot{e}$-value for the $(k+1)^{st}$ level by Eq. (5).

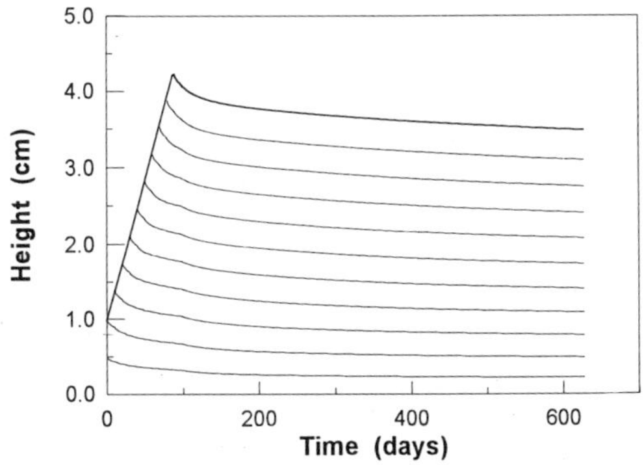

Fig.3 Analytical results on sedimentation and consolidation process

Whole consolidation process has been solved by a computer program coded in FORTRAN 90. The computational parameters used in this analysis are listed in Table 1. Consolidation settlement for different positions in an accumulating soil layer is shown in Fig. 3. From this figure it is clear that the rate of settlement is initially very high and gradually decreases with time. Figure 4 shows effective stress increase with time for the first three steps of sedimentation-consolidation process, and Fig. 5 shows the corresponding time isochrones of the void ratio distribution in the soil. At any time, the void ratio change with depth is very high for the elements near the boundary of sediment formation. But its change once drastically decreases with depth then gradually increases toward the bottom. This analytical result is justified by the experimental results obtained by Been et. al. (1981), whose analysis with the use of their simple elastic model, however, shows a quite different solution.

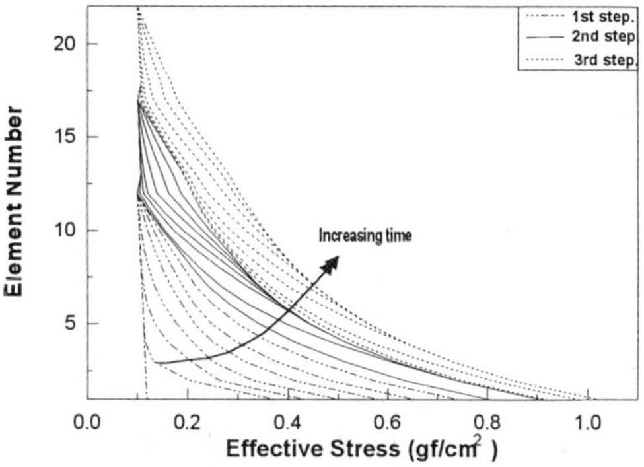

Fig. 4 Analytical result of effective stress distribution

Table 1. Parameters used in computation

Parameter	Value	Parameter	Value
A	-0.127	C_s	0.16
B	0.465	a	0.05
C_{v0}	6.17 cm²/d	γ_s	2.7 g/cm³
σ'_0	0.1 gf/cm²	γ_w	1.0 g/cm³
e_0	7.25	Δz	0.1 cm

Figure 6 shows the difference between the calculation results based on a non linear elastic model (Yamauchi, 1991) and the present elasto-viscoplastic model. From this figure it is clear that the non linear elastic solution shows a definite value of final settlement but elasto-viscoplastic solution shows endless settlement. This difference can be easily understood to be caused by viscosity's effect alone. On the other hand, in the process before the end of sedimentation, elasto-

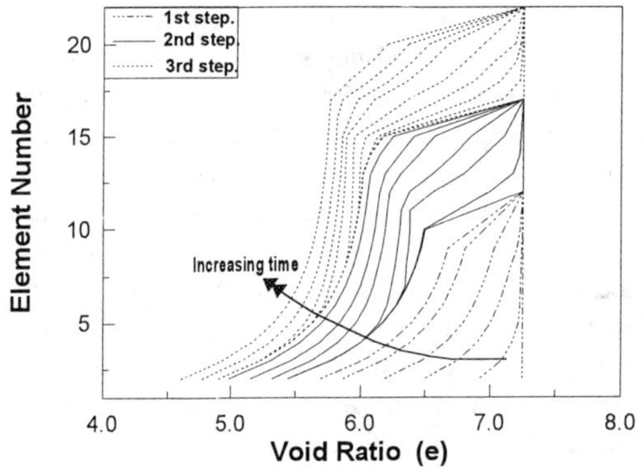

Fig. 5 Analytical result on void ratio profile

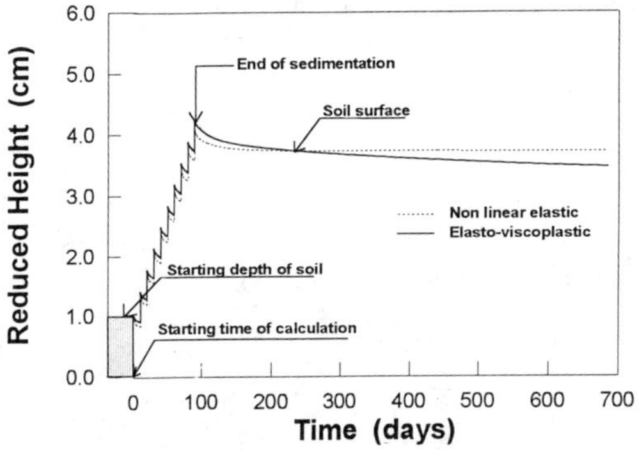

Fig 6 Difference between non linear elastic and elasto-viscoplastic model

viscoplastic solution gives thicker sedimented soil. This is mainly caused by two reasons; one is the viscosity action, which retards the compression of skeleton and another is the existance of yielding in compression, which also retards the consolidation progress.

10 CONCLUDING REMARKS

A new model to simulate consolidation process during continuous sedimentation is proposed with a finite difference solution method with the use of the Crank-Nicolson's scheme. Although the results obtained show the more realistic behavior than the one obtained by available models, the authors still have the following problems to be further studied;

(1) the way to find out the values of e_0 and $\dot{e}_0$; i.e. the definition of soil formation,

(2) the change in C_α -value during soil formation followed by consolidation.

REFERENCES

Been, K.& Sills, G.S. 1981.Self weight consolidation of soft soil: an experimental and theoretical study. *Géotechnique*, 31 No. 4, 519-535.

Imai, G. 1995. Analytical examination of the foundation to formulate consolidation phenomena with inherent time dependence. *Key Note Lecture, compression and consolidation of clayey soil, Balkema, IS-Hiroshima, Japan,* vol.2, 891-935.

Imai, G. 1981 Experimental studies on sedimentation mechanism and sediment formation of clay materials.. *Soils and Foundations*, 21(1): 7-20.

Imai, G. and Tang, Y. X. 1992 A constitutive equation of one-dimensional consolidation derived from inter-connected tests. *Soils and Foundations*, 32(2): 83-96.

Yamauchi, H., Imai, G., Watnabe, K. & Ogata, K. 1991 Sedimentation-consolidation analysis of pump-dredged cohesive soils. *Geo-coast'91,* Yokohama, 129-134.

Yano , K., Makimoto, Y., & Suzuki, Y. 1988. On multi -stage sedimentation test of cylindrical clay sample for pump dredging. *Proc. Of 23rd annual meeting of JSSMFE.* 223-226. (in Japanese).

——— x ———

Computer Methods and Advances in Geomechanics, Yuan (ed.) © 1997 Balkema, Rotterdam, ISBN 90 5410 904 1

An application of anchored discontinuous jointed rock mass fracture-damage model to the stability analysis of TGPFL high slopes*

Shucai Li, Weishen Zhu & Shiwei Bai
Institute of Rock and Soil Mechanics, The Chinese Academy of Sciences, Wuhan, China

Longqing Shi
Shandong Institute of Mining and Technology, Taian, China

ABSTRACT: Based on the hypothesis of equivalent strain enery, this paper establishes the constitutive relation of anchored discontinuous jointed rock mass under the state of complex stress, and applies it to the stability analysis of high slopes of Three Gorges Flight Lock. It puts the emphasis on the analysis of the possibility and the evolution length of subcracks developed in the joint tips of the slopes during the excavation process of the flight lock, and also discusses the effects to reduce the damage evolution zones of joints and deformation quantity with the aid of anchorage.

1 INTRODUCTION

The rock mass cut by quantities of discontinuous joints is often met in enginering. Various bolts are usually used to reinforce rock mass in roder to limit its deformation failure. The reinforcement function of bolts is obvious in jointed rock mass, but the effetive calculating method is still absent at present. When there are too many joints and bolts, we can nerther stimulate the joints and bolts one by one with joint elements and bolt elements, nor neglect anistropic and weakened characteristics of rock mass because of the existence of joints and the reinforcement function of bolts. Therefore, it is necessary to search for a calculating model suitable for the anchored jointed rock mass.

Adopting the method of damage mechanics, this paper establishes the constitutive relation and the damage evolution equation of anchored jointed rock mass, which are used to evaluate the deformation behaviours of rock mass. There are high slopes on both sides of Three Gorges Permanent Flight Lock and a 60m wide pier in the middle of the two lock channels. The total height of the high slopes is between 70m and 120m. The average height of vertical walls on the sides of lock channels is 50m, and 70m in maximum. The safety of passing ships and the normal work of the flight will be directly influenced by the stability of the high slopes and the deformation quantity of the side walls of lock channels during the excavation and operation. In addition, most surrounding rock of Three Gorges Flight Lock is granite in which there are many discontinuous joints so that the anchorage amount and the expenses are gigantic. Hence, it is an important research subject to calculate accurately the damage evolution zones and judge the stability of the high slopes. In this paper rock mass is regarded as jointed one and the fracture-damage model is adopted to analyse the stability of the high slopes.

2 THE CONSTITUTIVE RELATION OF ANCHORED JOINTED ROCK MASS

We put E as the stiffness matrix of the volume element of anchored rock mass whose volume is V and which contains n sets of joints and k_0 sets of bolts. $\underline{N}^j$ as the two order tensor of the j set of joints representing its direction, ρ^j as the joint desity, and c^j as the average characteristic size. Suppose that α_j^k is the intersecting angle of the k set of bolts and the planes of the j set of joints, and every set of bolts are even and paralled to each other.

As shown in Figure 1, an anchored jointed constitutive element is equivalent to the combination of the four parts on the right. Here σ and τ respectively refer to the projections of the apparent stress tensor $\underline{\sigma}$ in analysed constitutive element in the normal direction and in the tangent

• Financially supported by the National Natural Science Fundation of China.

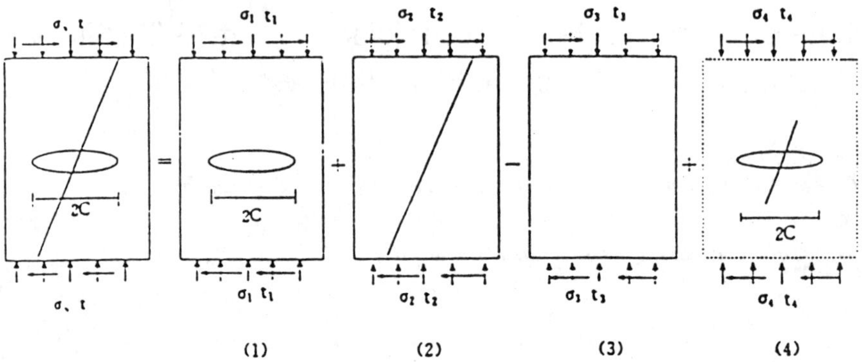

Figure 1. Equivalent model of anchored jointed constitutive element

direction to the plane of the joint. The formulas are

$$\left.\begin{aligned} \sigma &= \underline{N} : \underline{\sigma} \\ \tau &= [\underline{N} : (\underline{\sigma} \cdot \underline{\sigma} - (\underline{N} : \underline{\sigma})^2]^{1} \end{aligned}\right\} \quad (1)$$

where $\underline{N} = N_k N_l \underline{e_k}\, \underline{e_l}$ denotes the tensor of order 2 representing the direction of the j set of joints (upper mark j is omitted), $\underline{e_k}$ or $\underline{e_l}$ denotes the unit base vector in the direction of coordinate axis. The values of the lower mark k, l are 1, 2 in two-dimensional space and 1,2,3 in three-dimensional space.

As shown in Figure 1, the left state is equivalent to the combination of four parts on the right. Hence, the constitutive element strain energy is also the combination of four parts on the right in consideration of the equivalent strain energy. That is

$$U = U_1 + U_2 - U_3 + U_4 \quad (2)$$

in addition, under the condition of strain equivalence we get the formula

$$\frac{\partial U}{\partial \underline{\varepsilon}} = \frac{\partial U_1}{\partial \underline{\varepsilon}} + \frac{\partial U_2}{\partial \underline{\varepsilon}} - \frac{\partial U_3}{\partial \underline{\varepsilon}} + \frac{\partial U_4}{\partial \underline{\varepsilon}} \quad (3)$$

where:

$$U = \frac{1}{2}\underline{\varepsilon} : \underline{E} : \underline{\varepsilon}$$

$$U_1 = \frac{1}{2}\underline{\varepsilon} : \underline{E_1} : \underline{\varepsilon}$$

$$U_2 = \frac{1}{2}\underline{\varepsilon} : \underline{E_2} : \underline{\varepsilon}$$

$$U_3 = \frac{1}{2}\underline{\varepsilon} : \underline{E_3} : \underline{\varepsilon}$$

$$U_4 = \frac{1}{2}\underline{\varepsilon} : \underline{E_4} : \underline{\varepsilon}$$

where $E_i (i = 1,2,3,4)$ is the equivalent stiffness of corresponding part on the right, so we can deduce

$$E = E_1 + E_2 - E_3 + E_4 \quad (4)$$

In references [1] and [2], we know that the softness matrix of the first part on the right in Figure 1 is

$$[C] = [C_0] + \sum_{i=1}^{n} [G_i]^T [\Delta C_i][G_i] \quad (5)$$

In two-dimensional space

$$[C_0] = \begin{bmatrix} \frac{1}{E} & -\frac{\mu}{E} & 0 \\ -\frac{\mu}{E} & \frac{1}{E} & 0 \\ 0 & 0 & \frac{1}{G} \end{bmatrix} \quad (6)$$

where E is elastic modulus, μ is poisson's ratio, G is shear modulus.

$$[G_i] = \begin{bmatrix} \cos^2 B_i & \sin^2 B_i & \sin 2B_i \\ \sin^2 B_i & \cos^2 B_i & -\sin 2B_i \\ -\frac{1}{2}\sin 2B_i & \frac{1}{2}\sin 2B_i & \cos^2 B_i \end{bmatrix} \quad (7)$$

where B_i denotes the angle between the joint orientation of the i set and x coordinate axis (clockwise rotate B_i from the joint orientation to x coordinate axis direction). In addition.

$$[\Delta C_i] = \begin{bmatrix} 0 & 0 & 0 \\ 0 & \frac{c_n^i}{k_n^i}\frac{c_i}{2b_i d_i} & 0 \\ 0 & 0 & \frac{c_s^i c_i}{2k_s^i b_i d_i} \end{bmatrix} \quad (8)$$

where k_n^i and k_s^i are the normal stiffness and the shear stiffness of the i set of joints, c_n^i and c_s^i are the transmission factor of pressure and the transmission factor of shear respectively, b_i and d_i are half the joint spacing of the i set and half the center distance of adjoining joints in line respectively.

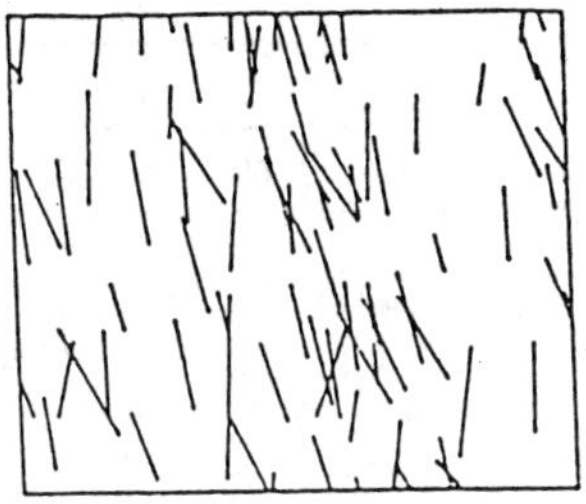

Figure 2. Joint network

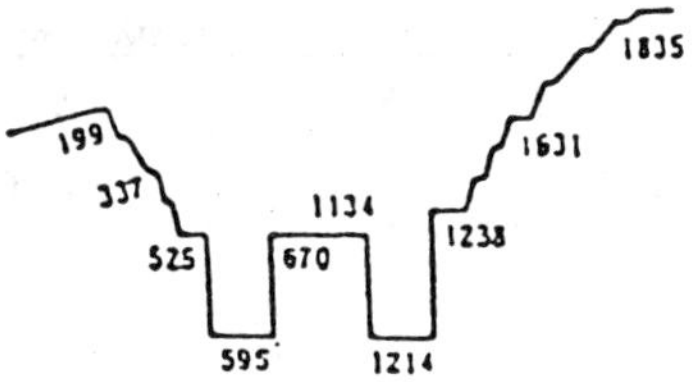

Figure 3. Location of some nodes on excavation boundary

In Figure 1 the equivalent stiffness of the second part and the third part on the right is equal to the equivalent stiffness developed in the direction of the axis of bolts. In two-dimensional space

$$E^{2\sim 3} = \sum_{k=1}^{k_0}[A_k][D_k][A_k]^T \tag{9}$$

where:

$$[D_k] = \begin{bmatrix} P_0^k E_b^k & 0 & 0 \\ 0 & 0 & 0 \\ 0 & 0 & 0 \end{bmatrix} \tag{10}$$

$$[A_k] = \begin{bmatrix} \cos^2 B_k & \sin^2 B_k & -\sin 2B_k \\ \sin^2 B_k & \cos^2 B_k & \sin 2B_k \\ \frac{1}{2}\sin 2B_k & -\frac{1}{2}\sin 2B_k & \cos 2B_k \end{bmatrix} \tag{11}$$

where P_0^k is the sectional steel ratio of the k set of anchoring component, E_b^k is the elastic modulus of the k set of anchoring component. The stiffness E^4 of the fourth part in Figure 1 can be derived from bolting nail function strain energy of bolts and residual strain energy of cracks. The former strain energy can be obtained by the deformation characteristics of bolts at the joint plane [2].

$$E_{pqkl}^4 = \sum_{j=1}^{n_1}\rho_b^j[f_1^j(E_{pqkl}^{4(1)})^{(j)} + f_2^j(E_{pqkl}^{4(3)})^{(j)} + f_3^j(E_{pqkl}^{4(2)})^{(j)}] - \sum_{j=1}^{n_1}R^j\left[\frac{c_v^{j^2}}{k_n^j}(E_{pqkl}^{4(4)})^{(j)} + \frac{c_s^{j^2}}{k_s^j}(E_{pqkl}^{4(5)})^{(j)}\right] + \sum_{j=1}^{n_1}\rho_b^j q_j^k[(b_1^k)^{(j)}(E_{pqkl}^{4(1)})^{(j)} - (b_2^k)^j(E_{pqkl}^{4(3)})^{(j)} + (b_3^k))^j(E_{pqkl}^{4(2)})^{(j)}] \tag{12}$$

The relevant marks are shown in reference [2], the constitutive relation under the state of tension-shear stress is shown in reference [3], and the damage evolution equation is shown in reference [2].

3 DISTRIBUTION CHARACTERISTICS AND RELEVANT PARAMETERS OF JOINTS IN FLIGHT LOCK ZONES

3.1 *Joint distribution and its parameters*

We mainly consider the influence caused by two sets of joints, NE—NEE and NWW, intersecting the axis of the lock channels at a small angle because other sets of joints intersecting the axis of the lock channels at a big angle have little influence on the stability of the high slope. According to the statistical results, we can get the probabilistc figure (Figure 2. Joint network) of the two sets of joints by means of Monte-Carlo method. Statistically the linkage rate of cracks is about 10 percent in the two sets of joints. Mechanical parameters of structure plane are shown in Table 1.

Table 1. Strength indexes of structure planes in flight lock zones

rock	types of structure plane		shear strength		tangent stiffness	normal stiffness
			f	C (MPa)	(MPa/cm)	(MPa/cm)
granite	hard structure plane		0.70	0.10	30	75
	weak structure plane	weak weathering	0.60	0.15	10	25
		heavy weathering	0.40	0.10	10	25

Table 2. Displacement values of some charatteristic nodes on excavation boundary

node number	199	337	525	595	670	1134	1214	1238	1631	1835
displacement (cm)	5.73	1.81	1.94	1.06	3.16	3.39	2.85	3.25	1.69	1.38

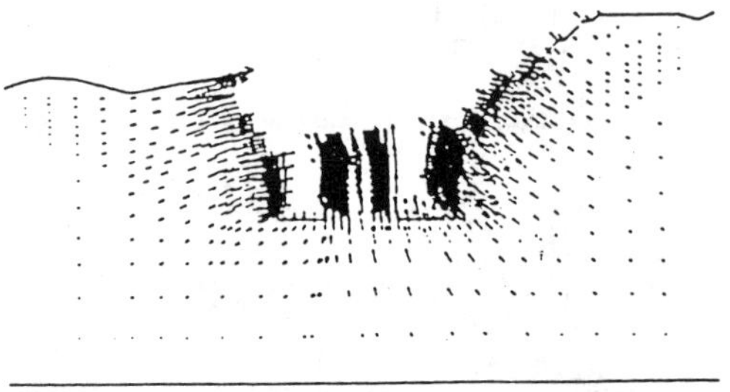

Figure 4. Displacement distribution ont the whole area

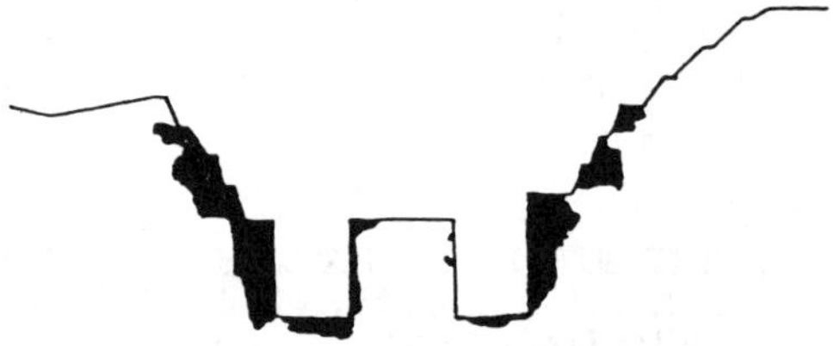

Figure 5. Anchored damage evolution zones

3.2 *Mechanical parameters of rock mass and original ground stress*

Rock in flight lock zones belongs to Pre-Sinian plagiogranite and can be divided into complete weathering layer, heavy weathering layer, slight weathering layer and slightly new layer from the surface down. In reference [4] we can know the mechanical parameters of every layer, original ground stress and bolt parameters.

4 CALCULATING METHOD AND RESULTS

Applying the model proposed by this paper and programming for calculation, we can stimulate and calculate the excavation process of the twentieth cross section of the fligh lock. The reference [4] can provide us with the calculating area of the cross section and the nets of the finite elements. In calculating area there are 1959 nodes and 2117 elements which mainly consist of squdrilateral equal parameter elements, and partly of triangular elements and joint elements. The excavation process is composed of seven steps. The last calculating results are shown in Figure 3 (Location of some nodes on excavation boundary), Table 2 (Displacement values of these nodes), Figure 4 (Displacement distribution in the whole area), Figure 5 (Anchored damage evolution zones). In Figure 5 the damage evolution zones are caused by 459 damage evolution elements and 226 tension elements.

In order to contrast the differences between the anchored condition and the unanchored condition, we have calculated the results under unanchored condition and found 670 damage evolution elements and 256 tension elements. It is calculated that unanchored damage evolution zones are 89.4 percent bigger than anchored damage evolution zones.

5 CONCLUSIONS

(1) In view of reasonableness it is more appropriate to adopt the fracture-damage model in order to analysze the stability of high slopes with sets of joints of Three Gorges Flight Lock, and the calculated results are more crditable.

(2) This paper mainly considers the influence of two sets of joints intersecting the axis of the flight lock at a small angle. Hence, whether the two sets of joints can cause damage evolution determines the stability of high slope. From the calculated results of this paper, although the damage evolution of joints has taken place in some ranges, large area fissure linking has not been found. Therefore, the flight lock are basically stable.

(3) We can conclude that the bolts have obvious effects on the reinforcement of jointed rock mass, and can effectively decrease the length and the area of damage evolution of joints.

REFERENCES

Xu J. N., Zhu W. S. & Bai S. W. 1993. Mechical characteristics-constitutive model for jointed rock mass under the state of compressive-shear stress. *Rock and Soil Mechanics*. 14(4):1—14.

Li S. C. 1996. Anchored discontinuous jointed rock mass fracture-damage model and its application. *Dissertation*, *Inst. of Rock and Soil Mech.*, CAS.

Yang Y. Y. & Wang S. Y. 1995. A study of the model of toughness increase and fracture prevention for anchored jointed rock mass. *Chinese Journal of Geotechnical Engineering*. 17(1):9—17.

Zhu W. S. & Zhang Y. J. 1996. Research on stability analysis and reinforcement schemes optimization for jointed rock mass of high slope of *Three Gorges Flight Lock*. *Chinese Journal of Rock Mechanics and Engineering*. 15(4):305—311.

Computer Methods and Advances in Geomechanics, Yuan (ed.) © 1997 Balkema, Rotterdam, ISBN 90 5410 904 1

Consolidation analysis of lumpy fills using a homogenization method

Jian-Guo Wang, C.F. Leung & Y.K. Chow
Department of Civil Engineering, National University of Singapore, Singapore

ABSTRACT: A homogenization method is developed for the Terzaghi-Rendulic consolidation theory to study the heterogeneous consolidation behavior of lumpy fill made of dredged clay lumps. A global problem and a local problem are defined. The global problem is of the same form as that of Terzaghi's consolidation theory, but the equivalent consolidation coefficients involve the heterogeneity of dredged materials. The predicted consolidation settlement is compared with those obtained from a centrifuge model study on lumpy fill.

1 INTRODUCTION

Recent coastal development works in Singapore lead to high demand of dumping grounds for soils dredged from the seabed. On the other hand, land reclamation works require large quantities of fill material. A research study is being carried out at the National University of Singapore to evaluate the feasibility of using dredged soils as reclamation fill. Along the West Coast of Singapore, the seabed soils generally consist of residual soils and weathered rocks of sedimentary origin with in-situ standard penetration resistance of at least 20 blows/30 cm. During dredging operation, soils are typically removed from the seabed using a grab resulting in lumps with sizes ranging from 0.5 m to 1.5 m. The exterior of these lumps would be significantly softened while their inner cores remain relatively stiff for a long period. When these lumps are placed as reclamation fill, the profile of the fill can be expected to be highly variable due to the presence of inter-lump voids.

The consolidation characteristics of reclamation fill made up of such clay lumps are rather complex because when the clay lumps consolidate, the inter-lump voids also close up simultaneously. Hence conventional consolidation theories are not applicable to analyze the consolidation behavior of lumpy fill. In this paper, a homogenization method for the Terzaghi-Rendulic consolidation theory is put forward to study the heterogeneous consolidation problem of lumpy fill.

2 GOVERNING EQUATIONS OF THE CONSOLIDATION PROBLEM

Wang et al.(1997) developed the following continuity equation for a soil-water mixture by means of micromechanics:

$$\frac{\partial \widetilde{H_i^0}}{\partial x_i} + \frac{\partial \varepsilon_v}{\partial t} = 0 \tag{1}$$

where the Darcy's law for specific discharge, $\widetilde{H_i^0}$, is given by

$$\widetilde{H_i^0} = -K_{ij}\frac{\partial u^\varepsilon}{\partial x_j} \tag{2}$$

where ε_v is volumetric strain, x_i is the ith component of $\boldsymbol{x}$-coordinates, t the real time, u^ε is pore water pressure, and K_{ij} the permeability of the lumpy fill.

Thus, the equation of continuity is

$$\frac{\partial}{\partial x_i}\left(K_{ij}\frac{\partial u^\varepsilon}{\partial x_j}\right) = \frac{\partial \varepsilon_v}{\partial t} \tag{3}$$

Now the key is how to introduce the constitutive law of soil skeleton into the continuity equation. For elastoplastic materials, the general constitutive relation can be expressed as

$$d\varepsilon_{ij} = C_{ijkl}d\sigma'_{kl} \tag{4}$$

or

$$\frac{d\varepsilon_v}{dt} = C_{iijj}\frac{d\bar{\sigma}}{dt} + C_{iikl}\frac{dS_{kl}}{dt} - C_{iijj}\frac{du^\varepsilon}{dt} \tag{5}$$

where ε_{ij} is strain components, C_{ijkl} the compliance coefficients, $\bar{\sigma}$ is the mean total stress, S_{ij} is the derviatoric stress, and subscripts i, j, k, l refer to the indices which take 1, 2, 3 for three-dimension coordinates. Terzaghi's consolidation theory assumes that the mean total stress $\bar{\sigma}$ keeps constant during consolidation and the dilatancy of soil skeleton is ignorable, that is, the effect of dilantancy on excess pressure

$$C_{iikl}\frac{dS_{kl}}{dt} = 0$$

At this time, the continuity equation becomes

$$\frac{\partial u^{\varepsilon}}{\partial t} = \frac{\partial}{\partial x_i}\left(M_{ij}^{\varepsilon}\frac{\partial u^{\varepsilon}}{\partial x_j}\right) \quad \text{on} \quad S_s \tag{6}$$

where M_{ij}^{ε} is consolidation coefficient and S_s is the domain for the soil-water mixture.

The boundary condition is given by

- Given excess pore pressure

$$u^{\varepsilon} = \tilde{u}^{\varepsilon} \quad \text{on} \quad S_T \tag{7}$$

where $\tilde{u}^{\varepsilon}$ is the pore water pressure on the pore water pressure boundary S_T.

- Flux

a) Flux boundary

$$-\boldsymbol{n} \bullet K^{\varepsilon} \bigtriangledown u^{\varepsilon} = \tilde{q}^{\varepsilon} \quad \text{on } S_m \tag{8}$$

Where K^{ε} is a tensor of permeability, its components are K_{ij}^{ε}. $\boldsymbol{n}$ is the outer unit vector normal to the boundary surface. $\bigtriangledown$ is the gradient operator, $\tilde{q}^{\varepsilon}$ is the specific discharge on discharge boundary S_m.

b) Mixed boundary

$$-\boldsymbol{n} \bullet K^{\varepsilon} \bigtriangledown u^{\varepsilon} = \tilde{q}^{\varepsilon} + \gamma(u^{\varepsilon} - \bar{u}) \quad \text{on } S_n \tag{9}$$

where γ is a parameter, and $\bar{u}$ is the pore water pressure outside the mixed boundary S_n.
The initial condition is

$$u^{\varepsilon} = u_0^{\varepsilon} \quad \text{when} \quad t = t_0 \tag{10}$$

3 LAPLACE TRANSFORMATION SPACE

The Laplace transformation and numerical inverse Laplace transformation are employed to obtain the general solutions. The methodology is shown in Fig.1.The forms in Laplace space are as follows (p is Laplace parameter). The consolidation coefficients $\hat{M}_{ij}^{\varepsilon}$ are assumed of time independence.

$$(p\hat{u}^{\varepsilon} - u_0^{\varepsilon}) = \frac{\partial}{\partial x_i}\left(\hat{M}_{ij}^{\varepsilon}\frac{\partial \hat{u}^{\varepsilon}}{\partial x_j}\right) \tag{11}$$

where the $\wedge$ denotes the variables in Laplace space. The Boundary condition is given by

- Given excess pore pressure

$$\hat{u}^{\varepsilon} = \hat{\tilde{u}}^{\varepsilon} \quad \text{on} \quad S_T \tag{12}$$

- Flux

a) Flux boundary

$$-\boldsymbol{n} \bullet \hat{K}^{\varepsilon} \bigtriangledown \hat{u}^{\varepsilon} = \hat{\tilde{q}}^{\varepsilon} \quad \text{on } S_m \tag{13}$$

b) Mixed boundary

$$-\boldsymbol{n} \bullet \hat{K}^{\varepsilon} \bigtriangledown \hat{u}^{\varepsilon} = \hat{\tilde{q}}^{\varepsilon} + \gamma(\hat{u}^{\varepsilon} - \hat{\bar{u}}) \quad \text{on } S_n \tag{14}$$

The problem above is only associated with spatial co-ordinates with fast oscillatory coefficients. For such a complicated structure, direct calculation will take a long time and makes the solution complicated. It is necessary to find an effective medium that has the same macro-response and that can involve microstructural effect. Homogenization theory(Sanchez-Palencia 1980; Wang 1996) is a good alternative.

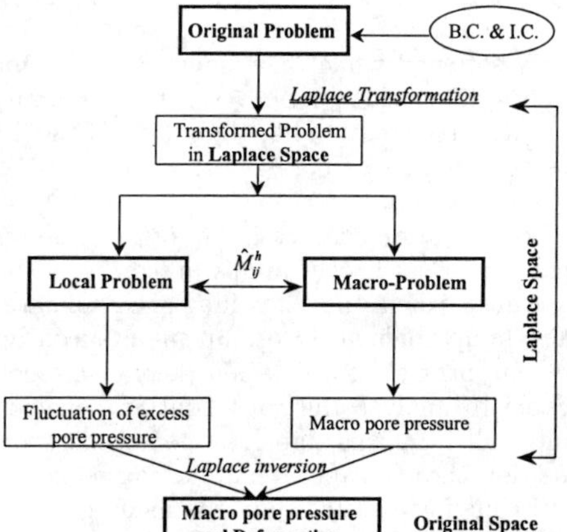

Fig. 1 Flowchart of analytical method

4 ASYMPTOTIC SPATIAL EXPANSION

The fundamental function, excess pore pressure u^ε, is assumed to be expanded as the series of scaling parameter ε.

$$\hat{u}^\varepsilon(\boldsymbol{x},\boldsymbol{y};p) = \hat{u}^0(\boldsymbol{x},\boldsymbol{y};p)+\varepsilon\hat{u}^1(\boldsymbol{x},\boldsymbol{y};p)+\cdots \quad (15)$$

and each term satisfies Y-periodicity, that is

$$\hat{u}^\alpha(\boldsymbol{x},\boldsymbol{y};p) = \hat{u}^\alpha(\boldsymbol{x},\boldsymbol{y}+\boldsymbol{Y};p) \quad (16)$$

$$\text{and} \quad \boldsymbol{y} = \frac{\boldsymbol{x}}{\varepsilon} \qquad (\alpha = 0,1,2,\cdots) \quad (17)$$

Eq. (16) means that $\hat{u}^\alpha(\boldsymbol{x},\boldsymbol{y};p)$ takes values that are almost the same in neighbouring, but very different in long distance. Such a function is called 'locally periodic'. The differential chain is

$$\frac{d}{dx_i} \Longrightarrow \frac{\partial}{\partial x_i} + \frac{1}{\varepsilon}\frac{\partial}{\partial y_i} \quad (18)$$

and

$$\hat{M}^\varepsilon \bigtriangledown \hat{u}^\varepsilon = \hat{M}^\varepsilon_{ij}\left(\frac{\partial \hat{u}^\varepsilon}{\partial x_j}\boldsymbol{i} + \frac{1}{\varepsilon}\frac{\partial \hat{u}^\varepsilon}{\partial y_j}\boldsymbol{i}\right) \quad (19)$$

The two-rank differential should be

$$\begin{aligned}
&\bigtriangledown \bullet \left(\hat{M}^\varepsilon \bigtriangledown \hat{u}^\varepsilon\right) \\
&= \frac{\partial}{\partial x_i}\left(\hat{M}^\varepsilon_{ij}\frac{\partial \hat{u}^\varepsilon}{\partial x_j}\right) + \frac{1}{\varepsilon^2}\frac{\partial}{\partial y_i}\left(\hat{M}^\varepsilon_{ij}\frac{\partial \hat{u}^\varepsilon}{\partial y_j}\right) \\
&+\frac{1}{\varepsilon}\left[\frac{\partial}{\partial y_i}\left(\hat{M}^\varepsilon_{ij}\frac{\partial \hat{u}^\varepsilon}{\partial x_j}\right) + \frac{\partial}{\partial x_i}\left(\hat{M}^\varepsilon_{ij}\frac{\partial \hat{u}^\varepsilon}{\partial y_j}\right)\right]
\end{aligned} \quad (20)$$

Therefore, Eq. (11) is expanded as

$$\sum(\,\bullet\,)\varepsilon^\beta = 0 \quad (\beta = -2,-1,0,1,2,\cdots) \quad (21)$$

because Eq. (21) holds for any ε. This infers that

a) ε^{-2}-term

$$\frac{\partial}{\partial y_i}\left(\hat{M}^\varepsilon_{ij}\frac{\partial \hat{u}^0}{\partial y_j}\right) = 0 \quad (22)$$

That means

$$\hat{u}^0(\boldsymbol{x},\boldsymbol{y};p) = \hat{u}^0(\boldsymbol{x};p) \quad (23)$$

Eq. (23), being the leading term, is of special meanings. Its existence means that the homogenization method is applicable.

b) ε^{-1}-term

$$\begin{aligned}
&\left[\frac{\partial}{\partial x_i}\left(\hat{M}^\varepsilon_{ij}\frac{\partial \hat{u}^0}{\partial y_j}\right) + \frac{\partial}{\partial y_i}\left(\hat{M}^\varepsilon_{ij}\frac{\partial \hat{u}^0}{\partial x_j}\right)\right] \\
&+\frac{\partial}{\partial y_i}\left(\hat{M}^\varepsilon_{ij}\frac{\partial \hat{u}^1}{\partial y_j}\right) = 0
\end{aligned} \quad (24)$$

$\hat{u}^1$ is determined by Eq. (24) if $\hat{u}^0$ is given. This is called as a **local problem**:

$$\frac{\partial}{\partial y_i}\left[\hat{M}^\varepsilon_{ij}\left(\frac{\partial \hat{u}^1}{\partial y_j} + \frac{\partial \hat{u}^0}{\partial x_j}\right)\right] = 0 \quad (25)$$

Under the periodicity condition $\hat{u}^1(\boldsymbol{x},\boldsymbol{y};p) = \hat{u}^1(\boldsymbol{x},\boldsymbol{y}+\boldsymbol{Y};p)$.

c) ε^0-term

$$\begin{aligned}
&\frac{\partial}{\partial y_i}\left(\hat{M}^\varepsilon_{ij}\frac{\partial \hat{u}^2}{\partial y_j}\right) \\
&+\left[\frac{\partial}{\partial x_i}\left(\hat{M}^\varepsilon_{ij}\frac{\partial \hat{u}^1}{\partial y_j}\right) + \frac{\partial}{\partial y_i}\left(\hat{M}^\varepsilon_{ij}\frac{\partial \hat{u}^1}{\partial x_j}\right)\right] \\
&+\frac{\partial}{\partial x_i}\left(\hat{M}^\varepsilon_{ij}\frac{\partial \hat{u}^0}{\partial x_j}\right) - (p\hat{u}^0 - u^0_0) = 0
\end{aligned} \quad (26)$$

A volume average operator is defined as

$$< \bullet >= \frac{1}{|Y|}\int_Y \bullet dy \quad (27)$$

in a unit cell, and let

$$\hat{u}^1 = W^r(\boldsymbol{y})\frac{\partial \hat{u}^0}{\partial x_r} + C(\boldsymbol{x}) \quad (28)$$

where $W^r(\boldsymbol{y})$ is the characteristic function, being the function of local variable $\boldsymbol{y}$. Then, the **global problem** is obtained as

$$\frac{\partial}{\partial x_i}\left(M^h_{ir}\frac{\partial \hat{u}^0}{\partial x_r}\right) - (p\hat{u}^0 - u^0_0) = 0 \quad (29)$$

where

$$M^h_{ir} = \frac{1}{|Y|}\int_Y \hat{M}^\varepsilon_{ij}\left(\delta_{jr} + \frac{\partial W^r}{\partial y_j}\right)dy \quad (30)$$

M^h_{ir} is the equivalent or effective consolidation coefficient, and $\delta_{jr} = 1$ when $j = r$, and $\delta_{jr} = 0$ when $j \neq r$. Eq. (29) is of the same form as the conventional Terzaghi-Rendulic consolidation theory. The only difference is that the present consolidation theory involves the microstructural effect.

Macro-consolidation coefficients M_{ij}^h may be heterogeneous. Again, for the **local problem**, Eq. (25), is expressed in characteristic function W^r

$$\frac{\partial}{\partial y_i}\left(\hat{M}_{ij}^{\varepsilon}\frac{\partial W^r}{\partial y_j}\right)+\frac{\partial \hat{M}_{ir}^{\varepsilon}}{\partial y_i}=0 \tag{31}$$

with $W^r(\boldsymbol{y};p)=W^r(\boldsymbol{y}+\boldsymbol{Y};p)$.

5 DISCUSSION

5.1 *Weak form for a local problem*

Assume that weighting function V is of Y-periodicity. The weak form of the local problem denoted by Eq. (31) is

$$\int_Y \frac{\partial}{\partial y_i}\left(\hat{M}_{ij}^{\varepsilon}\frac{\partial W^r}{\partial y_j}\right)V dy=-\int_Y \frac{\partial \hat{M}_{ir}^{\varepsilon}}{\partial y_i}V dy \tag{32}$$

By using Y-periodicity, one has

$$\int_Y \hat{M}_{ij}^{\varepsilon}\frac{\partial W^r}{\partial y_j}\frac{\partial V}{\partial y_i}dy=-\int_Y \hat{M}_{ir}^{\varepsilon}\frac{\partial V}{\partial y_i}dy \tag{33}$$

From Lax-Milgram Lemma, W^r is uniquely determined if its mean value is assumed to be zero. Therefore, the weak form of a local problem is

$$\begin{cases} \text{Find } W^r\in V_y; \quad <W^k>=0 \\ \int_Y \hat{M}_{ij}^{\varepsilon}\frac{\partial W^r}{\partial y_j}\frac{\partial V}{\partial y_i}dy=-\int_Y \hat{M}_{ir}^{\varepsilon}\frac{\partial V}{\partial y_i}dy \end{cases} \tag{34}$$

The mechanical property of effective consolidation coefficient M_{ir}^h is symmetric and positive definite.

5.2 *Excess pore pressure*

Fluctuation of excess pore pressure is given by

$$\hat{u}^1=W^r(\boldsymbol{y})\frac{\partial \hat{u}^0}{\partial x_r}+C(\boldsymbol{x})$$

because $C(\boldsymbol{x})$ is a function of $\boldsymbol{x}$ co-ordinate. It should be classified as $\hat{u}^0(\boldsymbol{x})$. Thus, it is reasonable to assume that $C(\boldsymbol{x})\equiv 0$. At this time, the total excess pore pressure is

$$\hat{u}\approx\hat{u}^0+\varepsilon\hat{u}^1$$

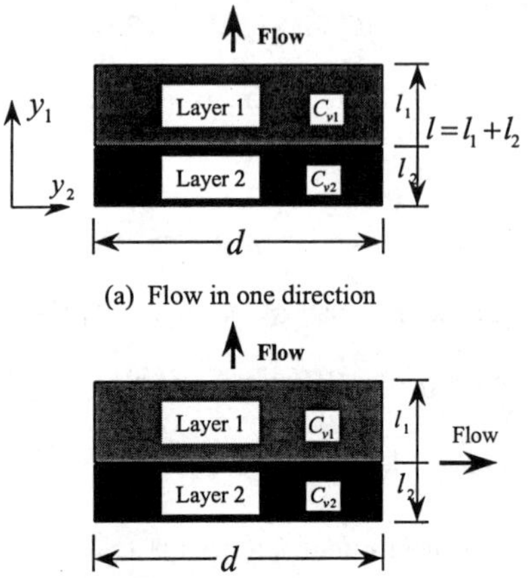

(a) Flow in one direction

(b) Flow in two directions

Fig. 2 Two-layer unit cell

6 APPLICATIONS

6.1 *Layer soil consolidation*

There is one macro-variable x and one micro-variable y. The period $\boldsymbol{Y}$ is a segment of length l, for example $(0,l)$ as the unit cell shown in Fig. 2. Thus

$$\frac{\partial \hat{u}^0}{\partial x_i}\Longrightarrow\frac{\partial \hat{u}^0}{\partial x} \qquad \frac{\partial W^r}{\partial y_j}\Longrightarrow\frac{\partial W}{\partial y} \tag{35}$$

The local problem becomes

$$\frac{\partial}{\partial y}(\hat{M}^{\varepsilon}\frac{\partial W}{\partial y})+\frac{\partial \hat{M}^{\varepsilon}}{\partial y}=0 \tag{36}$$

Its solution is

$$W=\int_0^y \frac{b}{\hat{M}^{\varepsilon}}da-y+C \tag{37}$$

$W(0)=W(l)$ determines the coefficient b. The homogenized consolidation coefficient $\hat{M}^h$ is

$$\frac{1}{\hat{M}^h}=\frac{1}{l}\int_0^l \frac{1}{\hat{M}^{\varepsilon}}da \tag{38}$$

Eq. (38) is suitable for a multi-layer medium.

The medium is assumed to be two-dimensionally periodic: That is, $\hat{M}_{ij}^{\varepsilon}=\hat{M}_{ij}^{\varepsilon}(y_1)$ and only $\hat{M}_{11}^{\varepsilon}\neq 0$ and $\hat{M}_{22}^{\varepsilon}\neq 0$. This is an axisymmetric problem.

$$\frac{\partial u^0}{\partial x_1}\neq 0 \qquad \frac{\partial u^0}{\partial x_2}=\frac{\partial u^0}{\partial x_3}\neq 0$$

The fluctuation is expanded as

$$u^1=W^1\frac{\partial u^0}{\partial x_1}+W^2\frac{\partial u^0}{\partial x_2} \tag{39}$$

From Eq. (31), the local problem is expressed as

$$\frac{\partial}{\partial y_1}(\hat{M}^{\varepsilon}_{11}\frac{\partial W^r}{\partial y_1}) + \frac{\partial}{\partial y_2}(\hat{M}^{\varepsilon}_{22}\frac{\partial W^r}{\partial y_2})$$

$$+\frac{\partial \hat{M}^{\varepsilon}_{1r}}{\partial y_1} + \frac{\partial \hat{M}^{\varepsilon}_{2r}}{\partial y_2} = 0 \quad (40)$$

When $r = 1$, its solution is

$$W^1(y_1) = \frac{\int_0^{y_1} \frac{1}{\hat{M}^{\varepsilon}} da}{\frac{1}{l}\int_0^l \frac{1}{\hat{M}^{\varepsilon}} da} - y_1 + C \quad (41)$$

When $r = 2$, the solution is that W^2=Constant because $\hat{M}^{\varepsilon}_{22}$ is just function of y_1, $\frac{\partial \hat{M}^{\varepsilon}_{22}}{\partial y_2} = 0$.. Therefore, the homogenized consolidation coefficient in the first direction is of the same form as the previous case. In the second direction, the homogenized consolidation coefficient $\hat{M}^h_2$ is

$$\hat{M}^h_2 = \frac{1}{l}\int_0^l \hat{M}^{\varepsilon}_{22} dy \quad (42)$$

For multi-layer materials

$$\hat{M}^h_2 = \frac{1}{\sum_1^n l_n}(\sum_1^n l_n M^{n\varepsilon}_{22}) \quad (43)$$

6.2 *General case*

Numerical inverse Laplace transformation is used here. Schapery (1962) proposed an analytical method such that

$$f(t) \approx \left[p\hat{f}(p)\right]|_{p=\frac{0.5}{t}} \quad f(t) \approx (2t)^{-1}\hat{f}(\frac{1}{2t}) \quad (44)$$

where $f(t)$ is the original function, and $\hat{f}(p)$ is the function in Laplace space. This method will produce very large error when $t \to 0$ because of the singularity at $t = 0$. For our computation, the singularity at $t = 0$ is avoided by use of $p \times \bullet$ instead of only $\bullet$.

The above proposed method is applied to back analyze the centrifuge model tests on dredged lumpy materials (Leung et al., 1996). This is a one-dimensional consolidation problem in macro-scale. But the microstructure of lumpy fill is complicated. A simplified homogenized method is used to determine the homogenized consolidation coefficient $M^h_1(= C_v)$. The lumpy fill pores are divided into inter-lump voids and in-lump pores. The consolidation coefficient M_h reaches its lowest value $(C_v)_{min}$ at the complete closing of inter-lump void, while it reaches its peak value $(C_v)_{max}$ at the beginning of the loading. The $C_v - e$ curve is plotted in Fig. 3 which leads to the following implications. (a) At the initial stage, consolidation process is very fast. This makes deformation almost instant. Dissipation of excess pore water pressure is difficult to be measured. The deformation is steep at the beginning of loading. On the other hand, the stress-strain curve for lumpy fills shows that the initial deformation at low loading is huge and plastic (Leung et al. 1996). The lumpy balls change its shape completely or collapse. If this deformation is mixed with the conventional consolidation one, Terzaghi consolidation theory even the large deformation one (Gibson 1967) is not suitable. (b) For a long term, the consolidation follows the usual Terzaghi's consolidation theory. Usually, deformation is small and consolidation coefficient is stable. Terzaghi consolidation theory is applicable. (c) For the whole stage, if the initial stage is treated separately, the conventional Terzaghi's theory is applicable. Thus, for the whole process to be taken one stage, the deformation consolidation degree and dissipation consolidation degree is not identical.

(d) The lump softening with water is an important factor to affect the deformation and consolidation. The direct function of this softening is reducing the strength of a clay lumpy. The lumpy may collapse and transfer its excess pore water pressure into inter-lump voids. Therefore, the consolidation process for lumpy fills is not a complete dissipation process of excess pore water pressure. Because of its multi-scale property, the dissipation and gener-

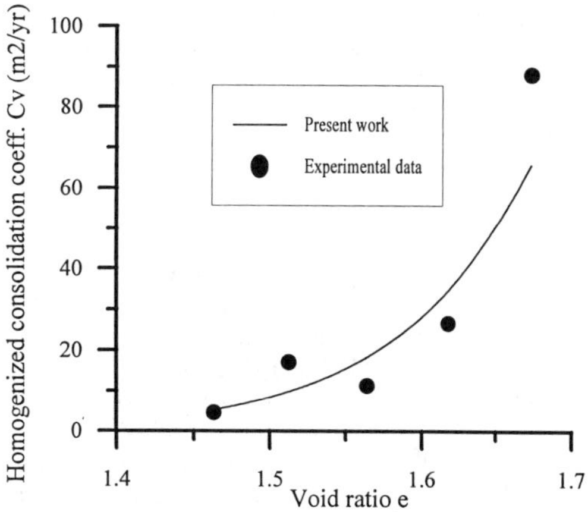

Fig. 3 Prediction of the consolidation coefficient

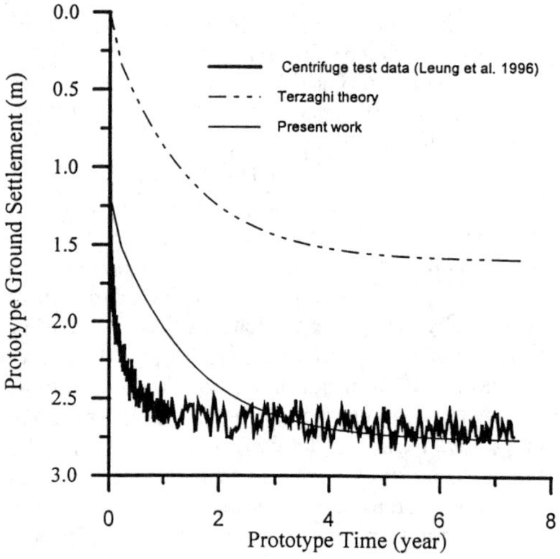

Fig.4 Comparison of prototype ground settlement

ation of excess pore water pressure in lumpy fills may exist at the same time, although the excess pore water in a lumpy ball dissipates. This coexistence depends on external loading, strength of a lumpy ball and other mechanical properties of clay.

In the centrifuge tests, the prototype height of the filling is 15 m and its final measured settlement is 2.6m, but its initial settlement is about 1.20m, as shown in Fig. 4. The prediction values from the present work are in fair agreement with the experimental data except for the initial settlement. At the beginning of loading, the microstructures of lumpy fill deform a lot. The inter-voids reduce significantly, thus the permeability reduces quickly and the consolidation coefficient changes drastically. This change makes the excess pore water pressure dissipate quickly. After that initial period, the microstructures becomes more and more stable. The consolidation is getting to the normal consolidation process. This is why the present work does still give poorer prediction at initial consolidation stage.

7 CONCLUSION

The heterogeneity of lumpy fills is analyzed by a homogenization method through a Terzaghi consolidation problem. The global and local problems are obtained. The global problem is of the same form as the conventional Terzaghi theory except for the values of consolidation coefficient. The heterogeneity of microstructures of lumpy fills is taken into consideration through the homogenized consolidation coefficient. The homogenized consolidation coefficient is analytically obtained for a two layer material. For a general lumpy filling, the homogenized coefficient is the function of inter-void ratio. The computation for a centrifuge test reveals some interesting mechanisms: (a) Coexistence is found for dissipation and generation of excess inter-void water pressure. This depends on loading and microstructures. The deformation is almost instant at the beginning of loading. The plastic deformation is the major source. This coexistence makes the consolidation degree for settlement and dissipation different. (b) Pore water pressure is heterogeneous in microstructure. The excess water pore pressure tends to become homogeneous with the breakage of lumpy balls. (c) For the long term, the behavior is dependent on external loading, mechanical property of clay balls. The conventional consolidation theory can be applied if the consolidation coefficient is revised according to the proposed method in this paper.

ACKNOWLEDGEMENT

The present study is supported by The National University of Singapore – National Science and Technology Board grant RP3940674.

REFERENCES

- Leung, C.F., Lau, A.H., Wong, J.C. and Karunaratne, G.P. 1996, Centrifuge model tests of dredged material, *Proc. of the 2nd Int. Conf. on Soft Soil Engineering*, Nanjing, May 27-30, 1:401-406
- Gibson, R.E., G.L. England and M.J.L. Hussey 1967, The theory of one-dimensional consolidation of saturated clays, *Geotechnique*, 17:261-273
- Sanchez-Palencia, E. 1980, *Non-Homogeneous Media and Vibration Theory*, Lecture Notes in Physics, Vol. 127, Springer, Berlin
- Schapery, R.A. 1962, Approximation methods of transform inversion for viscoelastic stress analysis, *Proceedings of the Fourth U.S. National Congress of Applied Mechanics*, ASME, 2:1075-1085
- Wang, J.G. 1996, *A Homogenization Theory for Geomechanics: Nonlinear Effect and Water Flow*, Dr.Engrg. Dissertation, Nagoya University, Japan
- Wang, J.G., Leung, C.F., and Chow, Y.K. 1997, Permeability model of porous media, Proc. of 3rd YGEC, Singapore, May 14-16

Computer Methods and Advances in Geomechanics, Yuan (ed.) © *1997 Balkema, Rotterdam, ISBN 90 5410 904 1*

Adaptive time integration in finite element analysis of elastic consolidation

S.W. Sloan & A.J. Abbo
Department of Civil, Surveying and Environmental Engineering, University of Newcastle, N.S.W., Australia

ABSTRACT: When Biot consolidation problems are solved using finite elements, the accuracy of the computed displacements and pore pressures depends on the temporal discretisation used in the analysis. Choosing an appropriate time step regime by hand is not a trivial task, even for an experienced analyst, since the optimal size of the time steps may grow by several orders of magnitude as consolidation takes place. This paper presents a scheme which automatically selects the temporal discretisation so that, for a given mesh, the time stepping error in the displacements does not exceed a specified tolerance.

1 INTRODUCTION

The analysis of consolidation is usually based on the theory of Biot (1941), for which Sandhu and Wilson (1969) were the first to present a solution using finite elements. Finite element techniques for modelling consolidation usually employ the θ-method. In order to solve elastic coupled consolidation problems efficiently with this algorithm, it is generally necessary to use an implicit time integration scheme with $\theta \geq 0.5$. With this choice of integration parameter, Booker and Small (1975) proved that the solution process is unconditionally stable so that large time increments may be used with safety. Explicit integration methods, which employ $\theta = 0$, are only conditionally stable and may require the use of very small time steps.

Somewhat surprisingly, very little work has been done on the development of automatic time stepping algorithms for finite element analysis of consolidation. The algorithm presented here uses an estimate of the local truncation error to regulate the step size and is based on the backward Euler and Thomas and Gladwell (1988) integration formulas. A key advantage of the new method is that it operates in single step mode and, hence, does not need to use values generated in previous time increments. Moreover, its local error estimator is embedded and can be computed cheaply with no extra matrix factorisations.

The new procedure treats the governing consolidation relations as a system of first order differential equations and automatically subincrements a prescribed series of time increments. The prescribed time increments, which are termed coarse time steps, serve to start the procedure and are chosen by the user. The new algorithm attempts to choose the time subincrements so that, for a given mesh, the time-stepping (or temporal discretisation) error in the displacements lies close to specified tolerance.

2 BASIC SOLUTION SCHEMES

The governing finite element equations for Biot consolidation can be expressed as a system of coupled differential equations of the form

$$\begin{bmatrix} \mathbf{K}_e & \mathbf{L} \\ \mathbf{L}^T & \mathbf{0} \end{bmatrix} \begin{Bmatrix} \dot{\mathbf{U}} \\ \dot{\mathbf{P}} \end{Bmatrix} + \begin{bmatrix} \mathbf{0} & \mathbf{0} \\ \mathbf{0} & \mathbf{H} \end{bmatrix} \begin{Bmatrix} \mathbf{U} \\ \mathbf{P} \end{Bmatrix} = \begin{Bmatrix} \dot{\mathbf{F}}^{ext} \\ \mathbf{Q} \end{Bmatrix} \quad (1)$$

where $\mathbf{K}_e$ is the elastic stiffness matrix, $\mathbf{L}$ is a coupling matrix, $\mathbf{H}$ is a flow matrix, $\mathbf{F}^{ext}$ is a vector of applied external forces, $\mathbf{Q}$ is a fluid supply vector, $\mathbf{U}$ are the unknown displacements, $\mathbf{P}$ are the unknown pore water pressures, and the superior dot denotes a derivative with respect to time. Precise definitions of these matrices can be found, for example, in Lewis and Schrefler (1987).

For the analysis of elastic solids with constant permeabilities, the relations (1) constitute a system of linear first order differential equations of the form

$$\mathbf{C}\dot{\mathbf{X}} + \mathbf{K}\mathbf{X} = \mathbf{F}(t) \tag{2}$$

where

$$\mathbf{C} = \begin{bmatrix} \mathbf{K}_e & \mathbf{L} \\ \mathbf{L}^T & \mathbf{0} \end{bmatrix}; \mathbf{K} = \begin{bmatrix} \mathbf{0} & \mathbf{0} \\ \mathbf{0} & \mathbf{H} \end{bmatrix}; \ \mathbf{F}(t) = \begin{Bmatrix} \dot{\mathbf{F}}^{ext} \\ \mathbf{Q} \end{Bmatrix} \tag{3}$$

and $\mathbf{X} = \{\mathbf{U}, \mathbf{P}\}^T$ with $\dot{\mathbf{X}} = \{\dot{\mathbf{U}}, \dot{\mathbf{P}}\}^T$. This type of system occurs in many areas of engineering and science and has been widely studied. A very comprehensive summary of the stability and accuracy of various solution strategies for solving (2) may be found in Wood (1990).

2.1 *The θ-Method*

The simplest strategy for solving (2) is commonly known as the θ-method. For a time step ranging from t_{n-1} to $t_n = t_{n-1} + h$, this algorithm may be expressed in the form

$$[\mathbf{C} + \theta h\mathbf{K}]\mathbf{X}_n = [\mathbf{C} - (1-\theta)h\mathbf{K}]\mathbf{X}_{n-1} + h\{(1-\theta)\,\mathbf{F}_n + \theta\mathbf{F}_{n-1}\} \tag{4}$$

where θ is an integration parameter in the interval $0 \le \theta \le 1$, the subscripts n and $n-1$ denote, respectively, quantities evaluated at the start and end of the step, and all values except $\mathbf{X}_n$ are known. The process assumes that $\mathbf{X}_0$ is known at time t_0. The θ-method is at least first order accurate and, provided $\theta \ge 0.5$, is unconditionally stable. Unconditionally stability is an essential characteristic for an efficient consolidation scheme since it is often necessary to integrate over very long time periods using large time steps. For the special case of $\theta = 0.5$, the θ-method is second order accurate and corresponds to the ubiquitous Crank-Nicolson scheme. Although appealing because of its high accuracy, the Crank-Nicolson method may generate spurious oscillations in the solution, especially if there are abrupt changes in the forcing function, and often requires special smoothing procedures. Choosing a value of $\theta = 1$ gives the well known backward Euler scheme which is first order order accurate, unconditionally stable, and oscillation free (Wood, 1990). The last of these characteristics has led to the Backward Euler method being widely used in finite element consolidation studies, even though it is less accurate than the Crank-Nicolson scheme.

As an alternative to (4), the θ-method can be expressed in the more compact form

$$[\mathbf{C} + \theta h\mathbf{K}]\mathbf{V} = (1-\theta)F_{n-1} + \theta F_n - \mathbf{K}X_{n-1} \tag{5}$$

where

$$\mathbf{V} = (\mathbf{X}_n - \mathbf{X}_{n-1})/h$$

is an average estimate of $\dot{\mathbf{X}}$ over the time step h and $\mathbf{X}$ is updated according to

$$\mathbf{X}_n = \mathbf{X}_{n-1} + h\mathbf{V} \tag{6}$$

In the finite element literature, this form of the θ-method is often referred to as the *SS11* procedure, where the terminology *SSpj* stands for a Single-Step algorithm which uses an approximation of degree p to solve a differential equation of order j.

2.2 *The Thomas and Gladwell Method*

A new method for solving systems of first and second order differential equations has recently been proposed by Thomas and Gladwell (1988). Their procedure uses three integration parameters and, when applied to (2), may be written as

$$[\varphi_2 h\mathbf{C} + \varphi_3 h^2\mathbf{K}]\mathbf{A} = \mathbf{F}(t_{n-1} + \varphi_1 h) - \mathbf{C}\dot{\mathbf{X}}_{n-1} - \mathbf{K}\{\mathbf{X}_{n-1} + \varphi_1 h\dot{\mathbf{X}}_{n-1}\} \tag{7}$$

where

$$\mathbf{A} = (\dot{\mathbf{X}}_n - \dot{\mathbf{X}}_{n-1})/h$$

is an average estimate of $\ddot{\mathbf{X}}$ over the time step h. After solving (7) for $\mathbf{A}$, the updates for $\mathbf{X}_n$ and $\dot{\mathbf{X}}_n$ are found from

$$\mathbf{X}_n = \mathbf{X}_{n-1} + h\dot{\mathbf{X}}_{n-1} + \tfrac{1}{2}h^2\mathbf{A}$$
$$\dot{\mathbf{X}}_n = \dot{\mathbf{X}}_{n-1} + h\mathbf{A}$$

This scheme, which is commonly known as a two-stage single step algorithm, is second order accurate and unconditionally stable provided

$$2\varphi_3 > \varphi_1 \ge 0.5 \tag{8}$$
$$\varphi_2 \ge 0.5 \tag{9}$$

Unlike the single-step θ-method, the procedure advances the solution for both $\mathbf{X}$ and $\dot{\mathbf{X}}$ and only uses values from the current time step. These characteristics explain why the algorithm is termed a two-stage single-step method. The chief advantage of this type of scheme is that the step size may be adjusted easily as the integration proceeds. The price of this flexibility is the need to compute and store $\dot{\mathbf{X}}$ for each time step.

3 AUTOMATIC SOLUTION SCHEME

The automatic algorithm assumes that a number of (coarse) time increments are defined and subdivides these into a number of smaller subincrements if necessary. The coarse time step is assumed to start at t_0 and end at $t_0 + \Delta t$, and is thus of size Δt. The nth time subincrement is assumed to be of size h, and range from t_{n-1} to $t_n = t_{n-1} + h$.

The adaptive procedure uses the *SS11* version of the θ-method and the Thomas and Gladwell method to provide an estimate of the local truncation error in the displacements and pore pressures for each subincrement. The integration parameters for these two schemes are specially chosen so that only one matrix assembly and factorisation is required for each substep, thus making the error indicator cheap to compute.

3.1 *Theory*

To derive an efficient solution scheme with an embedded error estimator, the integration parameters for the *SS11* and Thomas and Gladwell algorithms are selected so that only one matrix factorisation is needed for each time step. The constraints that this imposes on the integration parameters may be seen by rewriting equations (5) and (7) as

$$\mathbf{CV} + \mathbf{K}\left\{\tilde{\mathbf{X}}_{n-1} + \theta h\mathbf{V}\right\} = (1-\theta)\mathbf{F}_{n-1} + \theta\mathbf{F}_n \qquad (10)$$

$$\mathbf{C}\left\{\dot{\mathbf{X}}_{n-1} + \varphi_2 h\mathbf{A}\right\} + \mathbf{K}\left\{\mathbf{X}_{n-1} + \varphi_1 h\dot{\mathbf{X}}_{n-1} + \varphi_3 h^2\mathbf{A}\right\} = \mathbf{F}\left(t_{n-1} + \varphi_1 h\right) \qquad (11)$$

where $\tilde{\mathbf{X}}_{n-1}$ now denotes the estimate of $\mathbf{X}$ computed from the first order scheme. Comparing (10) and (11), the *SS11* and Thomas and Gladwell schemes give rise to an identical system of equations if

$$\mathbf{V} = \dot{\mathbf{X}}_{n-1} + \varphi_2 h\mathbf{A} \qquad (12)$$

$$\tilde{\mathbf{X}}_{n-1} + \theta h\mathbf{V} = \mathbf{X}_{n-1} + \varphi_1 h\dot{\mathbf{X}}_{n-1} + \varphi_3 h^2\mathbf{A} \qquad (13)$$

and

$$(1-\theta)\mathbf{F}_{n-1} + \theta\mathbf{F}_n = \mathbf{F}\left(t + \varphi_1 h\right) \qquad (14)$$

For the purposes of controlling the local error, it is assumed that the $\mathbf{X}$ values for both methods are identical at the start of each time subincrement so that

$$\tilde{\mathbf{X}}_{n-1} = \mathbf{X}_{n-1} \qquad (15)$$

in (10). This implies that the second order (rather than the first order) solution is the one that is propagated throughout the analysis and compensates for the fact that the error measure is only local. Neglecting, for the moment, the forcing function terms, the required constraints on the integration parameters are obtained by substituting (12) and (15) in (13) to give

$$\theta = \varphi_1 = \varphi_3/\varphi_2 > 0.5$$

where the Crank-Nicolson special case for the θ-method is excluded. Combining these with the unconditional stability requirements of (8) and (9) furnishes the final set of constraints

$$2\varphi_3 > \theta = \varphi_1 = \varphi_3/\varphi_2 > 0.5 \qquad (16)$$

Inspection of (14) indicates that the forcing functions for the two schemes are identical if

$$\theta = \varphi_1 = 1 \qquad (17)$$

Provided the constraints (16) and (17) are satisfied, first and second order accurate estimates for $\mathbf{X}_n$ may now be found using the *SS11* method of equation (5). This equation is solved with a second order accurate starting value $\mathbf{X}_{n-1}$ to give

$$\mathbf{V} = [\mathbf{C} + \theta h\mathbf{K}]^{-1}\left\{(1-\theta)\mathbf{F}_{n-1} + \theta\mathbf{F}_n - \mathbf{K}\mathbf{X}_{n-1}\right\} \qquad (18)$$

with the update of (6) being modified to

$$\tilde{\mathbf{X}}_n = \mathbf{X}_{n-1} + h\mathbf{V} \qquad (19)$$

$\mathbf{A}$ is then found from equation (12)

$$\mathbf{A} = \frac{\mathbf{V} - \dot{\mathbf{X}}_{n-1}}{\varphi_2 h} \qquad (20)$$

where $\dot{\mathbf{X}}_{n-1}$ is assumed known and the second order updates are

$$\mathbf{X}_n = \mathbf{X}_{n-1} + h\dot{\mathbf{X}}_{n-1} + \tfrac{1}{2}h^2\mathbf{A} \qquad (21)$$

$$\dot{\mathbf{X}}_n = \dot{\mathbf{X}}_{n-1} + h\mathbf{A} \qquad (22)$$

Since the local truncation errors in the updates (21) and (19) are, respectively, $O(h^3)$ and $O(h^2)$, the lower order estimate may be subtracted from the higher order estimate to give the local truncation error measure

$$\mathbf{E}_n = \mathbf{X}_n - \tilde{\mathbf{X}}_n = h\dot{\mathbf{X}}_{n-1} + \tfrac{1}{2}h^2\mathbf{A} - h\mathbf{V}$$

Substituting equations (20) and (22), this estimator may be expressed in the alternative form

$$\mathbf{E}_n = h\left\{\left(\varphi_2 - \tfrac{1}{2}\right)\dot{\mathbf{X}}_{n-1} + \left(\tfrac{1}{2} - \varphi_2\right)\dot{\mathbf{X}}_n\right\}$$

Note that the undesirable special case of $\varphi_2 = 0.5$, which gives a zero estimate of the local error regardless of h, is automatically excluded by the constraints (16). For the purposes of error control, $\mathbf{E}_n$ may be replaced by the more useful dimensionless relative error measure

$$R_n^u = \|\mathbf{E}_n^u\| / \|\mathbf{U}_n\| \qquad (23)$$

where

$$\mathbf{E}_n^u = h\left\{\left(\varphi_2 - \frac{1}{2}\right)\dot{\mathbf{U}}_{n-1} + \left(\frac{1}{2} - \varphi_2\right)\dot{\mathbf{U}}_n\right\}$$

and $\mathbf{U}_n$ holds the displacement components of $\mathbf{X}_n$ and $\dot{\mathbf{U}}_n$ holds the velocity components of $\dot{\mathbf{X}}_n$. Note that (23) measures the relative error for the displacements only. This is sufficiently accurate for practical step size control. A relative error estimate for the pore pressures can also be computed, but special care is then needed to avoid problems with division by (near) zero quantities in the later stages of consolidation.

In order to control the step size, the error measure R_n^u is computed for each time substep. The current time subincrement is accepted if R_n^u is less than some specified tolerance on the local truncation error, *DTOL*, and rejected otherwise. In either case, the size of the next time step h_{n+1} is found from

$$h_{n+1} = q h_n$$

where q is a factor which is chosen to limit the predicted truncation error. Since the truncation error for the next time subincrement, R_{n+1}^u, is approximately related to R_n^u by

$$R_{n+1}^u \approx q^2 R_n^u \qquad (24)$$

the required factor q is found by insisting that $R_{n+1}^u \leq DTOL$ to give

$$q \leq \sqrt{DTOL/R_n^u}$$

In practice, the factor q is chosen conservatively to minimise the number of rejected time subincrements according to

$$q = 0.9\sqrt{DTOL/R_n^u} \qquad (25)$$

with the additional constraint that

$$0.1 \leq q \leq 2 \qquad (26)$$

The coefficient of 0.9 in (25) acts as safety factor and prevents steps from failing because the local truncation error is slightly too large. Equation (26) controls the minimum and maximum ratios of consecutive time substeps, and is necessary to ensure that the extrapolation implicit in (24) is not carried too far. The factors in (25) and (26) were determined by numerical experiments on a wide variety of examples and ensure that most of the substeps are successful without making the step selection mechanism too conservative. In order to negotiate highly nonlinear segments of the response, it is also prudent to prohibit the step size from growing immediately after a failed time subincrement.

Assuming initial values for h and $\mathbf{X}_0$, with the latter typically zero, the integration scheme is started by solving (18) for **V**. For the first coarse time increment, h is typically set to Δt, but in subsequent coarse time increments it may be initialised to the value that gave the last successful subincrement. In order to compute the second order update for $\mathbf{X}$ using equation (21), a starting value for $\dot{\mathbf{X}}$ at $t = 0$ is needed. Assuming that the matrix $\mathbf{C}$ has an inverse, $\dot{\mathbf{X}}_0$ may be found by solving the governing differential equation (2) at $t = 0$ according to

$$\dot{\mathbf{X}}_0 = [\mathbf{C}]^{-1}\{\mathbf{F}_0 - \mathbf{K}\mathbf{X}_0\}$$

This type of procedure is valid for elements with a pore pressure expansion which is one order lower than the displacement expansion. For elements where the expansions are the same, $\mathbf{C}$ does not have an inverse and another starting procedure must be used.

The discussion so far has assumed that it is convenient to evaluate the external force rate, $\dot{\mathbf{F}}^{\text{ext}} = d\mathbf{F}^{\text{ext}}/dt$, analytically in the overall forcing function defined by equation (3). For cases where this is not so, this derivative can be approximated using discrete values of the external force vector. For example, a forward Euler approximation gives

$$\dot{\mathbf{F}}_n^{\text{ext}} = (\mathbf{F}_n^{\text{ext}} - \mathbf{F}_{n-1}^{\text{ext}})/h + 0(h)$$

where h is the current time step and the subscripts $n-1$ and n denote values computed at the times t_{n-1} and $t_n = t_{n-1} + h$. For problems where the external loading is piecewise linear with time, which covers most practical situations, this approximation is exact.

3.2 *Implementation*

In deciding upon an implementation of the automatic integration scheme described in Section 3.1, it is necessary to choose specific values of the integration parameters θ, φ_1, φ_2 and φ_3 which satisfy the constraints (16). After a series of numerical experiments covering a wide range of problems, these were set to

$$\theta = \varphi_1 = \varphi_2 = \varphi_3 = 1$$

The advantages of this choice are as follows:

i) Setting $\theta = 1$ implies that the first order method corresponds to the backward Euler scheme. This procedure is known to damp out unwanted oscillations quickly and thus provides a reliable first order solution for estimating the local truncation error.

ii) Selecting $\varphi_1 = \theta = 1$ means that the forcing functions for the first order and second order schemes are identical. Moreover, there is no need to evaluate the forcing function outside of the current time step.

iii) Setting $\varphi_2 = 1$ means that the vector **V**, computed from equation (18), corresponds automatically to $\dot{\mathbf{X}}_n$, the value of $\dot{\mathbf{X}}$ at the end of the current step for the second order scheme. This feature, which can be seen by comparing equations (22) and (12), results in a compact algorithm.

As mentioned previously, the implementation of the adaptive integration algorithm assumes that a series of coarse time increments have been specified. These coarse increments are, if necessary, subincremented automatically to satisfy a tolerance on the local truncation error. The automatic time stepping algorithm for elastic consolidation may be summarised as follows.

1. Enter with the time at the start of the coarse increment t_0, the current displacements and pore pressures $\mathbf{X}_{t_0}$, their corresponding derivatives $\dot{\mathbf{X}}_{t_0}$, the coarse time increment Δt, the last successful time substep h_{last}, the current effective stresses at each integration point $\boldsymbol{\sigma}'_{t_0}$, and the specified displacement error tolerance $DTOL$. For the first coarse time step, set $h_{\text{last}} = \Delta t$.

2. Set $t = t_0$ and $h = \min\{h_{\text{last}}, \Delta t\}$

3. While $t < t_0 + \Delta t$ do steps 4 to 9.

4. Compute $\dot{\mathbf{X}}_{t+h}$ according to

$$\dot{\mathbf{X}}_{t+h} = [\mathbf{C} + h\mathbf{K}]^{-1}\{\mathbf{F}_{t+h} - \mathbf{K}\mathbf{X}_t\}$$

where

$$\mathbf{F}_{t+h} = \begin{Bmatrix} \dot{\mathbf{F}}^{\text{ext}}_{t+h} \\ \mathbf{Q}_{t+h} \end{Bmatrix} = \begin{Bmatrix} (\mathbf{F}^{\text{ext}}_{t+h} - \mathbf{F}^{\text{ext}}_t)/h \\ \mathbf{Q}_{t+h} \end{Bmatrix}$$

5. Estimate the local truncation error in the displacements for the current subincrement using

$$E^u_{t+h} = \tfrac{1}{2}h\,\|\dot{\mathbf{U}}_t - \dot{\mathbf{U}}_{t+h}\|_\infty$$

where $\dot{\mathbf{U}}$ denotes the velocity component of $\dot{\mathbf{X}}$.

6. Update the displacements and pore water pressures and hold them in temporary storage according to

$$\overline{\mathbf{X}}_{t+h} = \mathbf{X}_t + \frac{h}{2}\left(\dot{\mathbf{X}}_t + \dot{\mathbf{X}}_{t+h}\right)$$

7. Estimate the relative error for current subincrement using

$$R^u_{t+h} = \max\left\{EPS, E^u_{t+h} / \|\overline{\mathbf{U}}_{t+h}\|_\infty\right\}$$

where $\overline{\mathbf{U}}_{t+h}$ is the displacement component of $\overline{\mathbf{X}}_{t+h}$ and EPS is a machine constant.

8. If $R^u_{t+h} > DTOL$ then go to step 9. Else this step is successful so update displacements, pore pressures and integration point effective stresses according to

$$\mathbf{X}_{t+h} = \overline{\mathbf{X}}_{t+h}$$
$$\boldsymbol{\sigma}'_{t+h} = \mathbf{D}\mathbf{B}\mathbf{u}_{t+h}$$

where **D** is the elastic stress-strain matrix, **B** is the element strain-displacement matrix and $\boldsymbol{u}_{t+h}$ is the element displacement vector. If the previous subincrement was successful, estimate a new subincrement size by computing

$$q = \min\left\{0.9\sqrt{DTOL/R^u_{t+h}}, 2\right\}$$

and setting

$$h \leftarrow qh$$

Store successful subincrement size $h_{\text{last}} = h$ and, before returning to step 4, update time and check that integration does not proceed beyond $t_0 + \Delta t$ by setting

$$t \leftarrow t + h$$
$$h \leftarrow \min\left\{h, t_0 + \Delta t - t\right\}$$

9. This subincrement has failed, so estimate smaller time substep by computing

$$q = \max\left\{0.9\sqrt{DTOL/R^u_{t+h}}, 0.1\right\}$$

and then setting

$$h \leftarrow qh$$

before returning to step 4.

10. Exit with displacements and pore pressures, $\mathbf{X}_{t_0+\Delta t}$, their corresponding rates, $\dot{\mathbf{X}}_{t_0+\Delta t}$, and integration point effective stresses, $\boldsymbol{\sigma}'_{t_0+\Delta t}$, at end of coarse time increment.

In step 7, *EPS* again represents the smallest relative error that can be computed on the host machine, and is typically set to around 10^{-16} for double precision arithmetic on a 32-bit architecture. Typical values for the tolerance on the time stepping error in the displacements, *DTOL*, are in the range 10^{-2} to 10^{-4}, with a value of 10^{-3} being adequate for most practical computations.

4 APPLICATION

In this section, we consider the consolidation of an elastic plane strain layer compressed between two rigid plates. The boundary conditions, material properties and finite element mesh for this problem, which has been solved analytically by Mandel (1953), are shown in Figure 1. The grid comprises six-noded triangles and uses a quadratic expansion for the displacements and a linear expansion for the pore pressures. Load is applied to the plates in the form of prescribed pressures, and the rigid boundary is modelled by constraining the nodal displacements along the plate interface to be equal. As shown in Figure 2, the finite element analysis as-

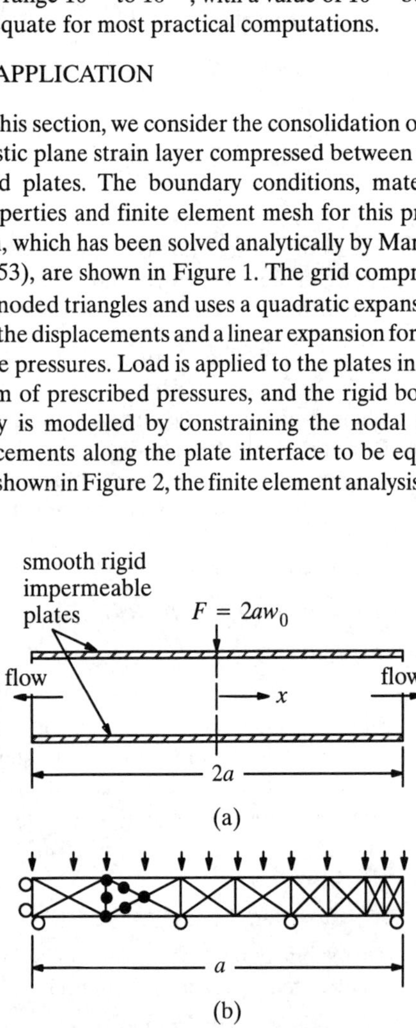

Figure 1 Consolidation between two rigid plates
(a) Geometry and boundary conditions
(b) Finite element mesh

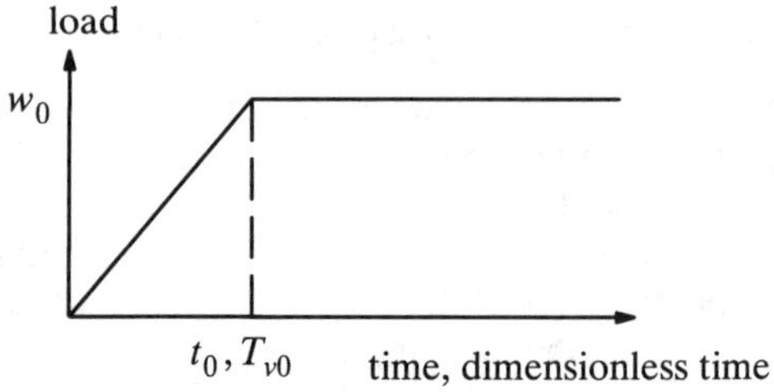

Figure 2 Load vs time.

sumes that a ramp load is imposed over the dimensionless time period $T_{v0} = 0.001$, where

$$T_{v0} = c_v \frac{t_0}{3a^2} \tag{27}$$

and c_v, the one dimensional coefficient of consolidation, is defined by

$$c_v = \frac{kE'(1 - \nu')}{\gamma_w(1 + \nu')(1 - 2\nu')}$$

In the above equation, k denotes the isotropic permeability and γ_w is the unit weight of water. After the total pressure, w_0, has been applied to the plates, consolidation is analysed over a dimensionless time factor increment of $\Delta T_v = 1.200$. Thus, at the end of the analysis, the total dimensionless time is equal to $T_v = 0.001 + 1.200 = 1.201$.

To gauge the performance of the various time stepping algorithms, the global temporal errors in the transient displacements $\mathbf{U}_t$ are estimated using

$$u_{\text{error}} = \frac{\|\mathbf{U}_{\text{ref}} - \mathbf{U}_t\|_\infty}{\|\mathbf{U}_{\text{ref}}\|_\infty} \tag{28}$$

where $\mathbf{U}_{\text{ref}}$ are a set of reference displacements calculated at the corresponding time. The reference displacements, which have a very small temporal error, were calculated using the second order scheme of Gladwell and Thomas (1988) described in section 2.2. This analysis used 1000 equal size increments to apply the load and 10,000 equal size time increments for the subsequent consolidation. All runs were performed on a SUN Ultra 170 workstation with the SUN FORTRAN 77 compiler.

To assess the performance of a typical solution method, the problem was first analysed using the backward Euler scheme with various numbers of equal-size time increments. Since the material is elastic, the backward Euler scheme requires only two formations and factorisations of the global equations to complete each analysis. These assemblies and factorisations occur at the start of the loading and consolidation phases. The CPU times

Table 1 Results for consolidation between two rigid plates using backward Euler scheme.

No. time increments		CPU time
Loading	Consolidation	(s)
1	10	0.8
10	100	1.0
100	1000	3.6
1000	10000	29.9

and temporal displacement errors for the various backward Euler runs are shown, respectively, in Table 1 and Figure 3.

The results in Table 1 suggest that, for the analyses with large to moderate time steps, the bulk of the computational work occurs in the assembly and factorisation stages and the CPU time is not proportional to the number of time steps used. For the runs with small time steps, the assembly and factorisation times are less dominant and the overall

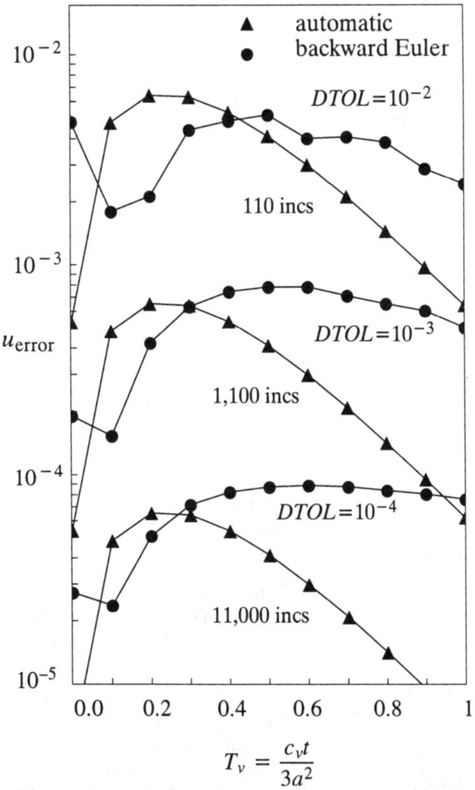

Figure 3 Variation of displacement temporal discretisation error with time for consolidation between rigid plates

CPU time grows in the expected manner. Figure 3 indicates that the displacement time stepping errors decrease as the number of time increments is increased. For all the backward Euler analyses, the time stepping error in the displacements is greatest at $T_v \approx 0.02$ and drops off significantly in later stages of consolidation. The runs with 110, 1100 and 11,000 time steps give maximum temporal discretisation displacement errors of roughly 7×10^{-2} 7×10^{-3} and 7×10^{-4}. These results clearly exhibit the first order accuracy of the backward Euler scheme.

Results for analyses using the automatic time stepping scheme are shown in Table 2 and Figure 3. Data are presented for $DTOL$ values of 10^{-2}, 10^{-3} and 10^{-4}. To test the sensitivity of the automatic scheme to the starting conditions, three runs were performed for each tolerance using 2, 6 and 11 coarse time steps. In each case, the load was applied in the first time step which had a time factor increment of $\Delta T_{v0} = 0.001$. The remaining increments were of uniform size and applied a total time factor increment of $\Delta T_v = 1.2$. Note that entries in the Table of the form $i+j=k$ indicate that i steps occurred in the loading phase, j steps occurred in the consolidation phase, and k steps occurred overall.

The results shown in Table 2 indicate that, for each value of $DTOL$, the automatic scheme always chooses a similar number of subincrements, regardless of the number of coarse increments that are specified initially. With $DTOL=10^{-3}$, for example, the automatic scheme selects 69, 70 and 73 time substeps when 2, 6, and 11 coarse time steps are specified. Each of these analyses automatically

Table 2 Results for consolidation between two rigid plates using automatic time incrementation scheme

$DTOL$	No. coarse time incs.	No. time subincrements		CPU time
		Successful	Failed	(s)
10^{-2}	1+1=2	1+23=24	0+1=1	1.2
	1+5=6	1+26=27	0+1=1	1.5
	1+10=11	1+29=30	0+1=1	1.9
10^{-3}	1+1=2	9+60=69	4+2=6	3.0
	1+5=6	9+61=70	4+2=6	3.3
	1+10=11	9+64=73	4+2=6	3.6
10^{-4}	1+1=2	30+173=203	5+3=8	7.9
	1+5=6	30+174=204	5+3=8	8.2
	1+10=11	30+178=208	5+3=8	8.6

chose 9 substeps during the loading phase. For all values of $DTOL$, the number of failed substeps is a small proportion of the total number of successful substeps. This suggests that the adaptive substepping strategy is correctly tuned and does not suffer from spurious oscillations.

The variation of the temporal discretisation error during the automatic analyses with 11 coarse time increments is shown in Figure 3. In each case, the maximum time stepping errors are just below the specified tolerance $DTOL$. With $DTOL = 10^{-3}$, for example, the maximum temporal discretisation error occurs at $T_v \approx 0.5$ and is approximately equal to 8×10^{-4}. These results suggest that the automatic scheme is able to constrain the global temporal discretisation error to lie near the specified tolerance $DTOL$. Because the automatic scheme increases the time step size as consolidation takes place, the temporal discretisation error in the displacements is roughly constant over most of the total time interval. For the run with 11 coarse time steps and $DTOL = 10^{-2}$, a plot of the excess pore pressure (at the centre of the layer) versus time is shown in Figure 4 . The numerical predictions are close to the analytic values derived by Mandel (1953) and the time steps are smallest in regions of greatest change. A more detailed plot of the time steps chosen by the automatic scheme is presented in Figure 5. This indicates that the time steps grow from a minimum value of $T_v \approx 0.001$ to a maximum of $T_v \approx 0.17$.

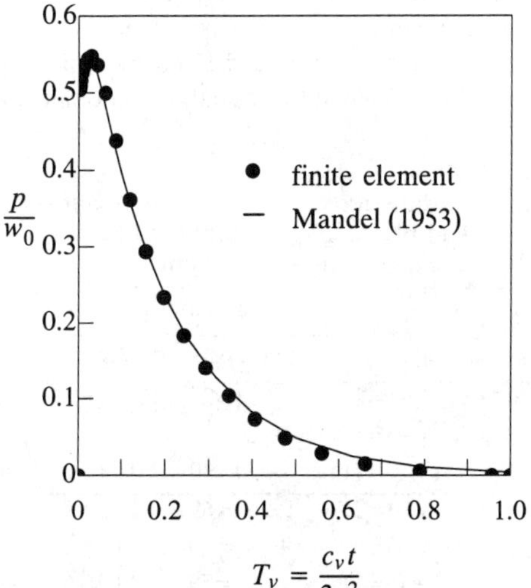

Figure 4 Pore pressure (at centre of layer) vs time factor for consolidation between two rigid plates with $DTOL = 10^{-2}$

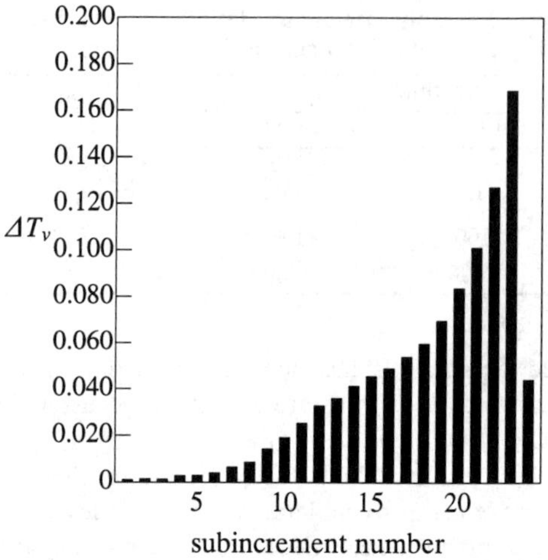

Figure 5 Subincrement selection for consolidation between two rigid plates with two coarse time steps and $DTOL = 10^{-2}$

5 CONCLUSION

A new automatic time stepping scheme for integrating elastic consolidation equations has been described. The method is simple, efficient and controls the time stepping error to lie near a specified tolerance.

REFERENCES

Biot, M.A. (1941) 'General theory of three dimensional consolidation', *Journal of Applied Physics*, 12, 155-164.

Booker, J.R. and Small, J.C. (1975). An investigation of the stability of numerical solutions of Biot's equations of consolidation, *International Journal of Solids and Structures,* 11, 907-917.

Lewis, R.W. and Schrefler, B.A. (1987). *The Finite-Element Method in the Deformation and Consolidation of Porous Media*, Wiley, Chichester.

Mandel, J. (1953). 'Consolidation des sols', *Geotechnique*, III, 287-299.

Sandhu, R.S. and Wilson, E.L. (1969). 'Finite element analysis of seepage in elastic media', *Journal of Engineering Mechanics.* ASCE, 95, 641-652.

Thomas, R.M. and Gladwell, I. (1988). Variable-order variable step algorithms for second-order systems. Part 1: The methods, *International Journal for Numerical Methods in Engineering*, 26, 39-53.

Wood, W.L. (1990). *Practical Time Stepping Schemes*, Clarendon Press, Oxford.

Computer Methods and Advances in Geomechanics, Yuan (ed.) © 1997 Balkema, Rotterdam, ISBN 90 5410 904 1

Nonlinear consolidation in finite element modelling

J.M. Dłużewski
Department of Environmental Engineering, Warsaw University of Technology, Poland

ABSTRACT: Consolidation problems in terms of the finite element approximation are analysed in the paper. Both material and geometrical non-linear behaviours are involved in the numerical analysis. The changes of the permeability are defined in relation to the current void ratio. The linear relation (in the semi-logarithmic scale) is applied to describe the current permeability for the orthotropic properties of the soil. The pseudo-viscoplastic iteration procedure is used to modelled the elasto-plastic behaviour of the soil skeleton. There large stain formulation based on the updated Lagrange procedure for two phase medium is applied to solve the boundary value problems. As the example, the staged construction of the embankment on soft organic subsoil is analysed. The embankment of the reservoir to meet the sugar factory wet wastes is numerically modelled.

1. INTRODUCTION

Numerical modelling of the embankment rising on the soft organic subsoil as peat or gytia, requires definition and description of a few coupled nonlinear problems. Four nonlinear effects, the most important from the engineering applications point of view, are listed bellow: First, the material nonlinearity, due to elasto-plastic behaviour of the soil skeleton should be modelled. Second, the large strain description is required. Next, the saturation of the soil and the position of the phreatic line is often changed during the consolidation process. The numerical algorithm should allow for the changes of the soil weight due to the movement of the phreatic line. Rising of the embankment on the soft organic subsoil most often causes large changes in the permeability due to closing of the pores. The test results after Chaciński, Dłużewski (1995) show that the changes of the permeability can reach three orders in magnitude, in realistic scope of the peat deformation. This changes influence time of consolidation.

The four nonlinear effects listed above are coupled. The equations describing consolidation of the multiphase medium constitute the complicated systems. The finite element method is used to solve the boundary value problems. To solve the nonlinear coupled system the incremental iterative algorithms are used. During the iteration the nonlinear effects are superimposed on each other and they are solved simultaneously. If the strains are very large the mesh refreshment may be needed to avoid numerical difficulties. Decoupling of the above nonlinear effects during consolidation is sometimes done. Soil skeleton and fluid are considered sequentially in the time stepping process, but the control of the error generated during calculations is lost.

In this paper the pseudo visco-plastic iterative algorithm based on the Perzyna theory (1966) and applied by Zienkiewicz and Cormoau (1974) to model the elasto-plastic behaviour is used. The method has been applied for the large deformation by Kanchi M. B., Zienkiewicz O. C., Owen P.R.J. (1978). Herein the updated Lagrange procedure, Bathe (1982), is developed for the consolidation problems. The position of the phreatic line is considered to be known. Permeability is defined in relation to the current void ratio for the orthotropic properties of the soil. The equations for soil and for the fluid are considered to be fully coupled.

2. BASIC EQUATIONS

In this section the fundamental equations which are used in the numerical modelling are given very briefly. In the present study the elasto-plastic soil models are used. The pseudo visco-plastic iteration algorithm is applied. The rate of the visco-plastic strains are defined by the flow rule

$$\dot{\varepsilon}_{ij}^{vp} = \gamma < F > \frac{\partial Q}{\partial \sigma_{ij}} \quad (1)$$

in which $\dot{\varepsilon}_{ij}^{vp}$ is the rate of the viscoplastic strains, γ is the fluidity parameter, F is the yield function, Q is the potential, brackets $< >$ are used to indicate the value of the term $< F >$. If $F > 0$, then $< F > = F$ otherwise $< F > = 0$. In the calculations the Coulomb-Mohr yield function is assumed. The nonassociated flow rule for the dilatation angle equal zero is applied.

The medium is composed of soil skeleton and water. The total stresses are the sum of the effective ones and the pore pressure

$$\sigma_{ij} = \sigma'_{ij} + \delta_{ij} p \quad (2)$$

where σ'_{ij} is the effective stress tensor, δ_{ij} is the Kronecker delta, p is the pore pressure.

The updated Lagrange procedure is adopted for large strains analysis during consolidation process. The description of the method can be found in the book by Bathe (1982). The starting point is the virtual work principle defined at the beginning of the incremental process:

$$\int_V S_{ij}\, \delta E_{ij}\, dV = \int_V F_i\, \delta u_i\, dV + \int_S T_i\, \delta u_i\, dS \quad (3)$$

where S_{ij} is the II Piola-Kirchhoff stress tensor, E_{ij} is the Green strain tensor, F_i is the vector of the volumetric forces, T_i is the vector of the surface traction, u_i is the displacement.

The continuity equation for the mass balance is the fundamental principle for the flow problem in porous medium. Assuming the pores to be fully saturated, the fluid accumulation in the unit volume is equal to the sum of its outflow and inflow what can be written as follows

$$\frac{\partial(\rho_w n)}{\partial t} + \frac{\partial v_i \rho_w}{\partial x_i} = 0 \quad (4)$$

where ρ_w is the water density, n is the porosity, v_i is the seepage velocity. The Darcy flow rule is adopted but permeability is not constant during deformation process.

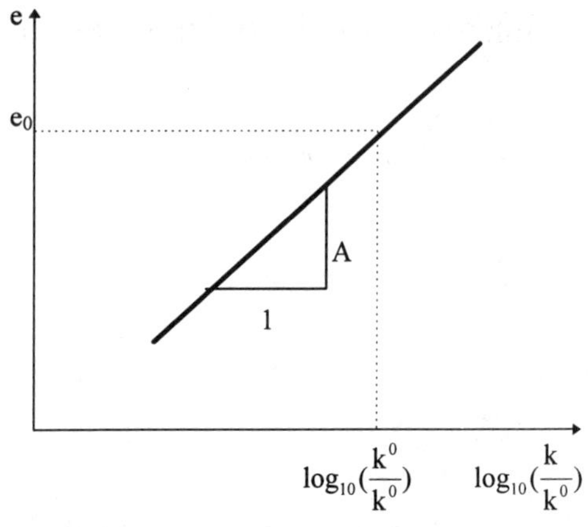

Fig.1 Equation (5) in the semi-logarithmic scale

To describe the permeability changes the following relation is introduced

$$\log_{10}\left(\frac{k_{ij}}{k_{ij}^0}\right) = \frac{e - e_0}{A} \quad (5)$$

The graphical interpretation of the material parameter A is shown in the Fig.1.

The finite element method is applied to solve the problem defined above. The coupled set of equations can be written in the following way

$$\begin{bmatrix} \mathbf{K} & \mathbf{L} \\ \mathbf{L}^T & -\mathbf{S} \end{bmatrix} \begin{bmatrix} \frac{d\mathbf{u}}{dt} \\ \frac{d\mathbf{p}}{dt} \end{bmatrix} = \begin{bmatrix} \mathbf{0} & \mathbf{0} \\ \mathbf{0} & -\mathbf{H} \end{bmatrix} + \begin{bmatrix} \frac{d\mathbf{R}}{dt} \\ \mathbf{Q} \end{bmatrix} \quad (6)$$

All arrays in the above equations depend on the unknown configuration of the body. The first equation stems form the equilibrium equation, the second one takes its roots from the continuity of the fluid. The stiffness array depends on three parameters $\mathbf{K} = \mathbf{K}(\mathbf{u}, \sigma, \varepsilon^{vp})$, the water compressibility array $\mathbf{S} = \mathbf{S}(\varepsilon_v)$, the flow array depends on the current permeability $\mathbf{H} = \mathbf{H}(\mathbf{k}(e))$, $\mathbf{L}$ is the coupling array, $\mathbf{R}$ and $\mathbf{Q}$ are the nodal forces and flux respectively. Applying the time stepping procedure (Lewis R. W., Schrefler B. A. (1987)) the above system can be rewritten

$$\begin{bmatrix} \mathbf{K} & \mathbf{L} \\ \mathbf{L}^T & -(\mathbf{S} + \alpha\Delta t\mathbf{H}) \end{bmatrix} \begin{bmatrix} \Delta\mathbf{u} \\ \Delta\mathbf{p} \end{bmatrix} = \begin{bmatrix} \mathbf{0} & \mathbf{0} \\ \mathbf{0} & -\Delta t\mathbf{H} \end{bmatrix} \begin{bmatrix} \mathbf{u} \\ \mathbf{p} \end{bmatrix} + \begin{bmatrix} \Delta\mathbf{R} \\ \Delta t\mathbf{Q} \end{bmatrix} \quad (7)$$

Incremental iterative procedures inside each time step are applied to engage the pseudo-viscoplastic algorithm and the updated Lagrange procedure for large strains.

3 . STAGED EMBANKMENT

The embankments of the reservoir for the sugar factory wet wastes are analysed. The embankments has been risen on the multilayer foundation. The peat and next the mud are founded on the deep sand layers. The analysis of the foundation is limited to two following layers: peat layer 3.1m thick and mud layer 1.2m thick. The sand below in not taken into account. The fully permeable bottom boundary is assumed. The values of the material parameters are listed in the Table 1.

The elasto-plastic models based on the Coulomb-Mohr yield surface are used. The isotropic hardening rule is assumed for the peat and mud in the range from initial cohesion c to the final cohesion c_r. H is the parameter for the strain hardening. The permeability in horizontal and vertical directions are defined in relation to the current porosity n. A is the material parameter with graphical interpretation shown in the Fig.1.

Table 1 Material parameters

soil	No	E [kPa]	ν	γ [kN/m³]	c_o [kPa]	φ [°]	H [kPa]	c_r [kPa]	k_H [m/d]	k_V [m/d]	n	A
wastes 1	1, 2	1500.	0.35	15.	15.	5.	-	-	1000.	1000.	-	-
wastes 2	3, 4	3500.	0.35	19.	27.	15.	-	-	1000.	1000.	-	-
wastes 3	5, 6	2000.	0.35	18.	27.	9.	-	-	1000.	1000.	-	-
sand + wastes 4	7, 8	2500.	0.25	17.5	1.	35.	-	-	1000.	1000.	-	-
peat	9, 10	90.	0.30	11.	8.	12.	50.	20.	0.7	0.36	0.75	1.
mud	11	120.	0.35	10.5	10.	13.	50.	20.	0.086	0.086	0.75	1.

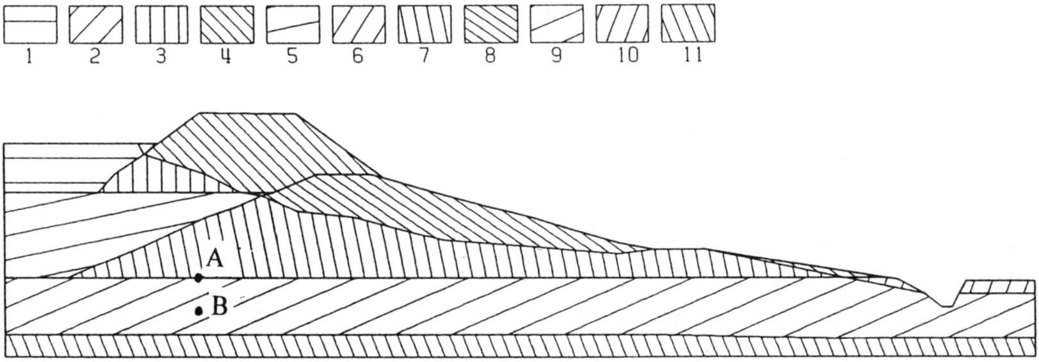

Fig. 2 Materials of the subsoil and embankment. Phreatic line divides the embankment into saturated and not fully saturated zones

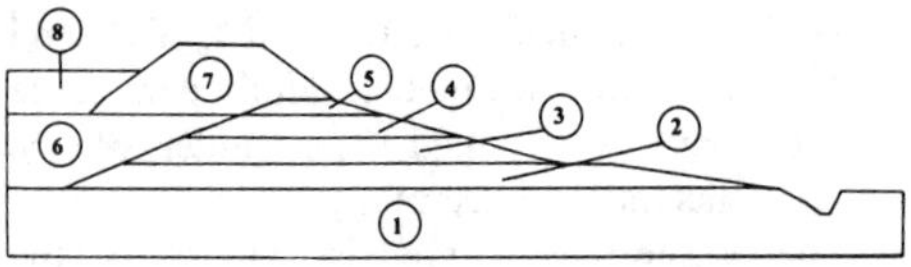

Fig.3 Stages of the embankment construction

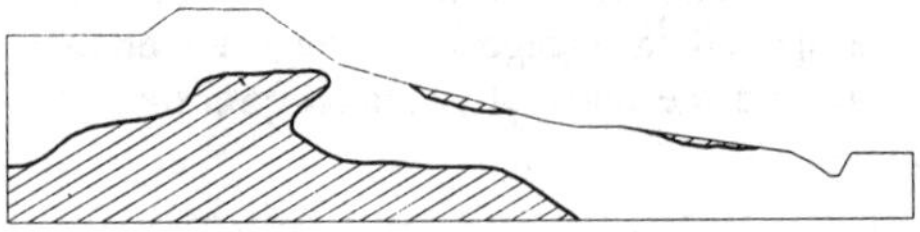

Fig.4 Plastic zones

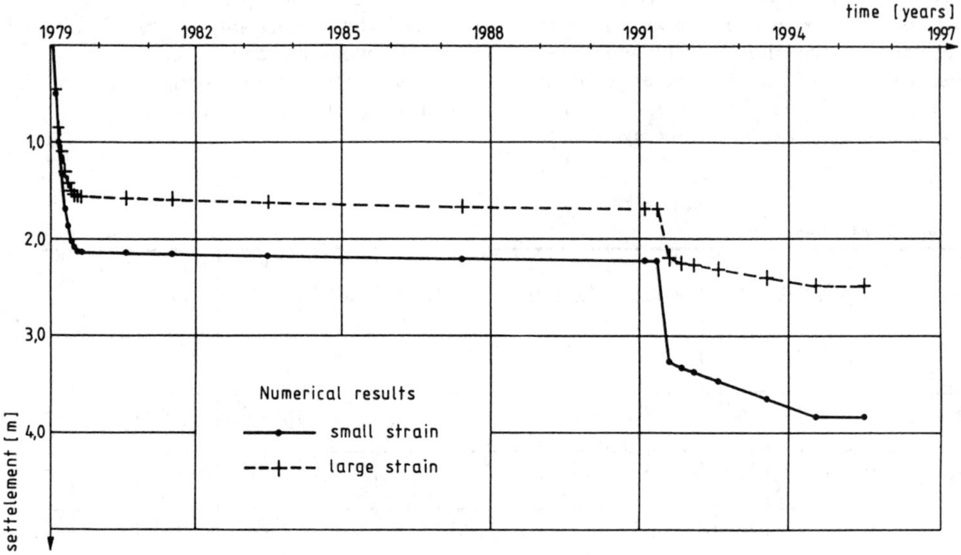

Fig.5 Settlement at the point A

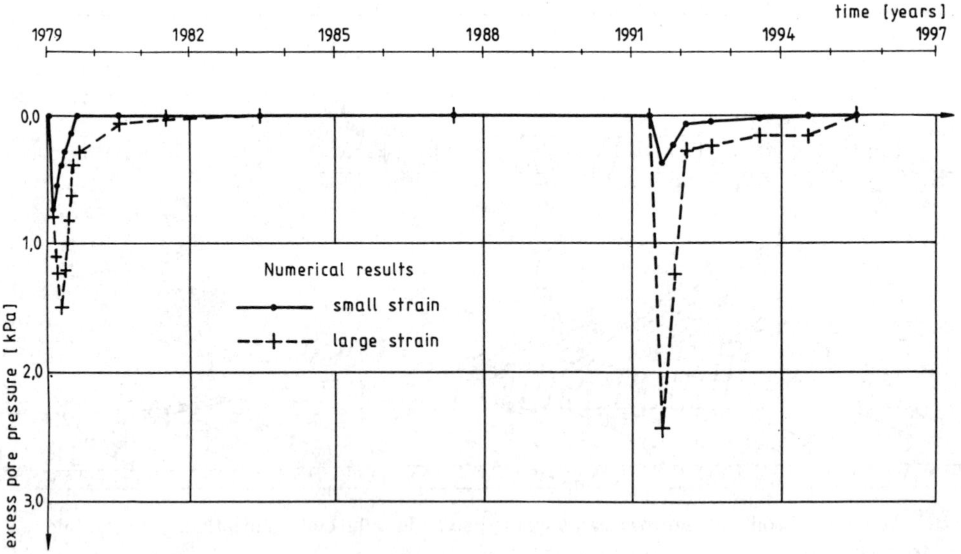

Fig. 6 Excess of the pore pressure at point B

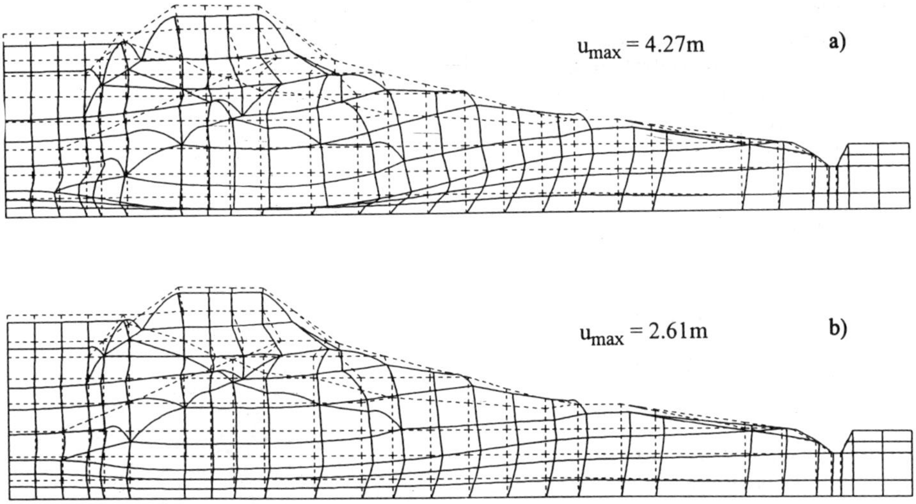

Fig.7. The final deformation, a) small strains analysis, b) large strains analysis

The embankment has been constructed in stages starting from June 1979. The reservoir has been filled with the wet wastes next risen up and fulfilled again. The stages of the embankment rising with timetable of filling the reservoir are listed in the Table 2.

Numerical modelling of the embankment rising is done by adding the elements according to the Table 2. The stages are shown in the Fig.3. The calculations are performed for small and large strains. For the small strains the permeability is kept constant during the whole analysis.

The plasic zones are depicted in the Fig.4. The comparison of the displacements for the point initially situated at the ground level are shown in the Fig.5. The displacements for the point initially located at the ground level reached 2.5m for large strains and 3.8m for small strains. The redistribution of the excess pore pressure is shown in the Fig 6. The changes of the permeability due to closing of the pores influence the consolidation process. The final deformation for large strains calculation is depicted in the Fig.7.

4. CONCLUSIONS

The updated Lagrange procedure for the large strains is successfully applied for the consolidation analysis. The permeability can be defined in relation to the current void ratio in the semi-logarithmic scale.

Table 2 Stages of the embankment rising

State	Stage number	Material no.	Time increment [days]
introduction of initial stresses	1	9, 10, 11	-
rising of embankment	2	8	22.5
			22.5
	3		22.5
			22.5
	4		22.5
			22.5
	5		22.5
			22.5
fulfilment of reservoir	6	5, 6, 7	30
			330
			360
			720
			1440
			1350
			90
rising of embankment	7	4	90
fulfilment of reservoir	8	1, 2, 3	90
			90
			180
			360
			360
no activity	9		360

The pseudo visco-plastic iterative algorithm can be used to model the elasto-plastic behaviour in such coupled situations. The numerical results obtained for the staged embankment show the differences in displacements and pore pressures for small and large strains analysis.

5. ACKNOWLEDGEMENT

The help of Mr. Paweł Popielski in preparing data for the calculations is gratefully acknowledged

6. REFERENCES

Bathe K. J. (1982) "Finite element procedures in engineering analysis" Prentice-Hall, Inc. New

Biot M. A. (1941) "General theory of three-dimensional consolidation", J. Appl. Phys., vol. 12, 155-164,

Chaciński Z., Dłużewski J.M. (1995)."Permeability of peat - test results" (in Polish), II Seminar of Institute of Water Supply and Hydrotechnical Structures, Warsaw University of Technology, 69-75

Dłużewski J. M. (1993) "Numerical Modelling of Soil Structure Interactions in Consolidation problems", Warsaw University of Technology Publications, Civil Engineering, Vol. 123, pp 1-116

Kanchi M. B., Zienkiewicz O. C., Owen P.R.J. (1978) "The visco-plastic approach to problems of plasticity and creep involving geometric nonlinear effects", Int. J. Num. Meth. Eng., Vol.12, pp 169-181

Lewis R. W., Schrefler B. A. (1987) "A finite element method in the deformation and consolidation of porous media", John Wiley, New York

Meroi E.A., Schrefler B.A., Zienkiewicz O.C. (1995) „Large strain static and dynamic semisaturated soil behaviour" ,Int. J. Numer. Analytic Meth. Geomech., Vol. 19, 81-106

Monte J. L., Kritzen R. J. (1976) "One dimensional mathematical model for large strain consolidation", Geotechnique 26(3)

Perzyna P. (1966) "Fundamental problems in visco - plasticity" Advan. in Appl. Mech., Vol. 9

Zienkiewicz O.C., Cormeau I. C. (1974). "Visco-plasticity-plasticity and creep in elastic solids - a unified numerical solution approach", Int. J. Num. Meth. Eng., Vol. 8, 821-845

9 Coupled thermo-hydro-mechanical analysis

Computer Methods and Advances in Geomechanics, Yuan (ed.) © *1997 Balkema, Rotterdam, ISBN 90 5410 904 1*

A joint element for coupled hydro-thermo-mechanical analysis of porous media

A.J.García-Molina, A.Gens, S.Olivella & E.Alonso
Geotechnical Engineering Department, School of Civil Engineering, Technical University of Catalonia, Barcelona, Spain

ABSTRACT

A general formulation for a joint element intended for the modelling of discontinuities in coupled thermo-hydro-mechanical analysis is presented. The formulation incorporates all the significant interactions between the various aspects of the problem. Specific constitutive laws are given for the behaviour of the joint concerning its thermal, hydraulic and mechanical components. The possibility that the joints may be in an unsaturated state is explicitly considered. The performance of the joint element is demonstrated in two examples of application carried out in connection with studies of disposal of high level radioactive waste.

1. INTRODUCTION

Recent work on isolation of waste in deep repositories has led to the development of models for thermo-hydro-mechanical (THM) analysis of compacted expansive materials (engineered barrier) and a variety of host rocks: clays, granite and salt (Thomas et al. 1995, Gens et al. 1997).

Continuum models are however limited in handling discontinuities that may appear either in the host rock (fractures, cracks) or in the engineering barrier. In fact, when blocks of compacted clay are used as appropriate components for the engineering barrier, joints of varying thickness are naturally created between blocks. Even when granular fills are used in bulk, a gap between the fill and the tunnel vault usually develops. Natural expansive soils also develop cracking patterns, due to swelling-shrinkage cycles, that affect lateral swelling pressures (Wallace and Lytton 1992).

Joint elements to simulate mechanical interaction in geomechanics have been developed by a number of authors (Goodman et al. 1968, Gens et al. 1988). In parallel, joints for modelling preferential water flow along fractures have also been developed including or not mechanical coupling (Gringarten et al. 1974). Thermal coupling has also been included recently in connection with the evaluation of nuclear repositories in fractured rock (Jing et al. 1995; Millard et al. 1995).

The work presented in this paper has, however, some interesting new aspects. It presents a fully coupled THM formulation of joints in granular media and allows two species within the joint: air and water, that may be in liquid or gas phases. Mechanical interaction requires the definition of an "effective stress" in the joint which takes into account its unsaturated nature. In addition, mass and heat transfer are allowed both along and across the joint and this feature enables the representation of insulation effects.

2. PHYSICAL PHENOMENA AND GOVERNING EQUATIONS

A joint between two continuum solid elements has been plotted in Figure 1a. The relevant transport and mechanical interactions within the joint have been schematically summarized in Figure 1b. Changes in joint opening due to effective stress changes imply changes in air and water conductivity and storage. In turn, air and water pressures control effective stress within the joint and they will affect its mechanical response. Temperature distribution along the joint may imply phase changes. In addition, it modifies air and water density and viscosity which in turn changes head and permeability distributions. Finally, water and air flows, in gas or liquid phase, transport energy and modify the temperature distribution. Since the medium is unsaturated, changes in degree of saturation also affects the current value of the coefficient of thermal conductivity introducing a further coupling of the mass balance equations.

Governing equations are conveniently formulated in the local coordinate system (x' , y', z'), where the plane (x' , y') is tangent to the "mean" surface, S^m , of the joint. This surface is at an equal distance of the borders (+) and (-) of the joint (Figure 1a). Joint thickness will be denoted by e . Vector or tensor quantities defined in the local coordinate system will be named F^l . Also, the notation for thermodynamic variables (i.e. flow rates, mass fractions) indicates the species through superscript, β (air = a ; water = w) and phase through subscript α (liquid = l ; gas = g).

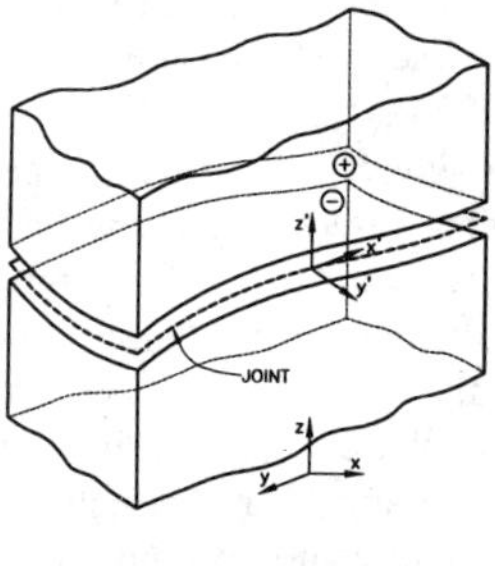

Figure 1a. Joint in a continuum medium

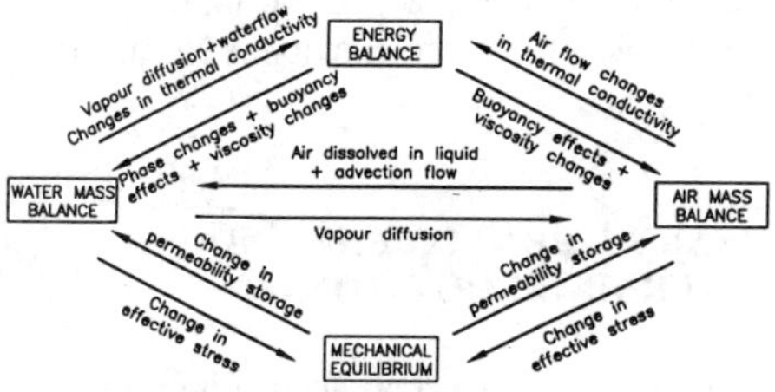

Figure 1b. Coupled phenomena occurring in a THM joint.

2.1 Mechanical equilibrium

The incremental mechanical work done in the (mean) joint surface S^m is defined through the virtual work expression

$$[w]=\int_{S^m}\delta\,[\boldsymbol{u}]^l\,\boldsymbol{t}^l d\,S^m \qquad (1)$$

where $[\boldsymbol{u}]=\boldsymbol{u}^+-\boldsymbol{u}^-$ is the relative displacement across the discontinuity and $\boldsymbol{t}^l$ is the stress vector on the joint "mean" surface.

2.2 Mass transport

State variables may vary from boundary (+) to boundary (-) at a given point of the joint surface S^m. Mass balance formulation for a species β is therefore expressed in a general three dimensional way (Olivella et al. 1994) as:

$$\frac{\partial}{\partial t}\left(\theta_l^\beta S_l+\theta_g^\beta S_g\right)+\nabla\cdot\left(\boldsymbol{j}_l^\beta+\boldsymbol{j}_g^\beta\right)=f^\beta \qquad (2)$$

where $\theta_\alpha^\beta=\rho_\alpha w_\alpha^\beta$ is the density of species β in phase α , ρ_α the density of phase α and w_α^β the mass fraction of species β in phase α . S_l and S_g are the volumetric fractions of liquid and air in the joint space $\left(S_l+S_g=1\right)$. Mass flux vectors $\boldsymbol{j}_\alpha^\beta$ may be split into its advection $\left(\boldsymbol{j}_{a\alpha}^\beta\right)$ and dispersion components $\left(\boldsymbol{i}_\alpha^\beta\right)$. The advection components may be expressed in terms of Darcy's velocities, $\boldsymbol{q}_\alpha\left(\boldsymbol{j}_{a\alpha}^\beta=\theta_\alpha^\beta\boldsymbol{q}_\alpha\right)$. As an example, when dealing with the flux of water in gas phase, the dispersion flux, $\boldsymbol{i}_g^w$, models the vapour migration along the joint void space.

Taking into account the preceding comments and relationships, the mass balance equations for water and air may be written:

$$\frac{\partial}{\partial t}\left(\theta_l^\beta S_l+\theta_g^\beta S_g\right)$$
$$+\nabla\cdot\left(\theta_l^\beta\boldsymbol{q}_l+\theta_g^\beta\boldsymbol{q}_g+\boldsymbol{i}_l^\beta+\boldsymbol{i}_g^\beta\right)=f^\beta \qquad (3)$$

where β refers either to water $(\beta\cdot w)$ or air $(\beta=a)$.

2.3 Energy Balance

If thermal equilibrium between phases is assumed the energy balance may be applied to the total energy of species (gas and water) present within the joint. Total energy is computed by adding contributions from each species. Accordingly, the balance equation becomes:

$$\frac{\partial}{\partial t}\left(E_l S_l + E_g S_g\right) + \nabla\cdot\left(\boldsymbol{i}_c + \boldsymbol{j}_{el} + \boldsymbol{j}_{eg}\right) = f^l \qquad (4)$$

where E_l, E_g are the internal energies of phases per unit volume, i_c is the thermal conduction flux, $\boldsymbol{j}_{e\alpha}$ is the convection component of phase α and f^e the internal/external energy supply.

3. CONSTITUTIVE ASSUMPTIONS

The following set of primary variables have been selected to define constitutive models: the relative displacement vector between opposite joint surfaces $[\boldsymbol{u}]$, the phase pressures P_g and P_l and the temperature T.

3.1 Mechanical

Since the joint is, in general, in an unsaturated state, an appropriate effective stress should be defined. The local expression:

$$\hat{\boldsymbol{t}}^l = \boldsymbol{t}^l + S_l P_l \boldsymbol{m} + \left(1 - S_l\right) P_g \boldsymbol{m} \qquad (5)$$

where $\boldsymbol{m}^t = (1,0,0)$, and $\boldsymbol{t}^l = \left(\sigma_{nz'}, \sigma_{tx'}, \sigma_{ty'}\right)$ is consistent with the concept of effective stress for both $S_l = 1$ and $S_l = 0$. For nonzero degrees of saturation S_l and $\left(1 - S_l\right)$ may be regarded as weighting factors for each of the phase pressures. An incremental relationship between local effective stress and (local) relative displacement may be written:

$$\dot{\hat{\boldsymbol{t}}}^l = \boldsymbol{D}[\dot{\boldsymbol{u}}]^l \qquad (6)$$

A simple uncoupled model requires only three stiffness coefficients: $k_{nz'}$ (normal) and $k_{tx'}$, $k_{ty'}$ (shear):

$$\boldsymbol{D} = \begin{bmatrix} k_{nz'} & 0 & 0 \\ 0 & k_{tx'} & 0 \\ 0 & 0 & k_{ty'} \end{bmatrix} \qquad (7)$$

3.2 Mass transport

Volumetric phase velocities are defined by a generalized Darcy's law in the local coordinate system:

$$q_\alpha^l = \frac{\boldsymbol{k}^l k_{r\alpha}}{\mu_\alpha}\left(\nabla^l p_\alpha + \rho_\alpha \boldsymbol{g}^l\right) \qquad (8)$$

where $\boldsymbol{k}^l$ is the local intrinsic permeability tensor, k_α is the relative permeability of phase α, μ_α the dynamic viscosity of phase α and $\boldsymbol{g}^l$ the gravity vector. In order to take into account joint roughness and turbulent flow some authors (Barton et al. 1985, Chan et al. 1985) use an equivalent hydraulic opening, e^h, smaller than e which depends on e and on some roughness index k^*:

$$e^h = e^h\left(l, k^*\right) \qquad (9)$$

Then, the intrinsic permeability tensor is computed as:

$$\boldsymbol{k} = \frac{12}{e_o^{h^2}} \frac{e^{h^2}}{12} \boldsymbol{ko} \qquad (10)$$

where $\boldsymbol{ko}$ is a constant tensor. Very limited information is available concerning fracture flow in unsaturated conditions. The preceding relationships may be generalized to an unsaturated state by multiplying the intrinsic permeability by a relative permeability scalar which should be a function of the degree of saturation. Non-advective fluxes are defined by means of Fick's law:

$$\boldsymbol{i}_\alpha^\beta = -\boldsymbol{D}_\alpha^\beta \nabla w_a^\beta \qquad (11)$$

where the dispersion tensor incorporates molecular as well as hydrodynamic dispersion

$$\boldsymbol{D}_\alpha^\beta = \rho_\alpha D_\alpha^\beta \boldsymbol{I} + d_t q_\alpha I + \left(d_l - d_t\right)\frac{\boldsymbol{q}_\alpha \boldsymbol{q}_\alpha^t}{q_\alpha} \qquad (12)$$

where D_α^β is the molecular diffusion content which depends on temperature and d_t, d_l the transversal and longitudinal dispersivities.

Densities of species in gas phase are computed in accordance with the law of perfect gases. In addition liquid density is made dependent on liquid pressure and temperature:

$$\rho_l = \rho_{lo}\, exp\left(\beta\left(p_l - p_{lo}\right) - \alpha\left(T - T_o\right)\right) \qquad (13)$$

where ρ is the volumetric compressibility and α the thermal dilation coefficient. In this way buoyancy effects due to temperature density changes may be accounted for.

3.3. Heat conduction

Fourier's law:

$$\boldsymbol{i}_l = -\boldsymbol{\lambda}\ \nabla T \qquad (14)$$

is used to model heat conduction. In an isotropic case the thermal conductivity tensor λ may be expressed as a geometric mean of phase conductivities λ_l and λ_g:

$$\lambda = \lambda_l^{S_l} \lambda_g^{(1-S_l)} \boldsymbol{I} \qquad (15)$$

In general, discontinuities are not a preferential path for heat conduction since the adjacent solid walls have higher conductivity values. Internal energies of species are defined in the usual way through the concept of specific heat. In unsaturated joints heat transfer by vapour convection may be significant.

3.4 Equilibrium restrictions

Henry's law:

$$w_l^a = \frac{P_g^c}{H} \frac{M_c}{M_w} \qquad (16)$$

where P_g^a is the partial air pressure in gas, H is Henry's constant, M_a and M_w the molecular weights of air and water control the dissolution of air in liquid phase. Finally, the psychrometric law provides the vapour density in equilibrium with liquid subjected to suction effects (curved interphase):

$$\theta_g^w = \left(\theta_g^w\right)^o exp\left(-\frac{\left(P_g - P_l\right)M_w}{RT\rho_l}\right) \qquad (17)$$

where $\left(\theta_g^w\right)^o$ is the vapour density for a planar interphase (no suction effects) which depends only on temperature.

4. EXAMPLES OF APPLICATION

A finite element finite difference discretization procedure has been established concerning the preceding set of equations. It has been incorporated into a general finite element code

Table 1. Boundary conditions for mock-up test

Boundary coordinates	Hydraulic condition	Thermal condition
$r = 0.806$ m	$P_l = 0.1$ MPa	—
$r = 0.881$ m	—	Mixed condition: $\gamma = 0.114$; $T_o = 20°C$ (1)
$z = 0$ m	Impervious boundary	Non transmitting boundary
$z = 0.0625$	Impervious boundary	Non transmitting boundary
$r = 0.15$ m	—	Heat flux : 20.65 W/m

(1) A mixed boundary condition may be expressed as follows:
$j = \gamma(T\text{-}T_o)$ where j is the energy flow

for THM analysis (CODE BRIGHT, Olivella et al. 1994). The solved cases presented below include solid and joint elements but intend to highlight the effect of discontinuities.

4.1 Non isothermal multiphase flow in a partially saturated jointed medium

A layout considered for deep underground repository of nuclear waste involves the horizontal storage of canisters placed in tunnels surrounded by a buffer of compacted bentonite blocks. This layout is being tested, at an scale close to prototype geometry in a laboratory experiment ("mock-up" test) being conducted by CIEMAT in Madrid. Figure 2a offers a schematic cross section of the test and Figure 2b a longitudinal section. The test is conducted inside a steel cylinder which simulates the repository tunnel with two central heaters simulating the canisters. The buffer is made of compacted calcium bentonite blocks (as compacted conditions: $\gamma_d = 1.65 \mathrm{g/cm^3}$; $w_o = 14\%$). Bentonite blocks are unsaturated and maintain a strong initial suction (about 100 MPa). Blocks are prisms whose weight is around 25kg. Construction joints indicated in Figure 2b have in practice a variable opening ranging from 0 to 10mm.

Hydration water is forced from the outer boundary of the blocks by means of a permeable mat connected to an external water supply under pressure.

If the effect of joints located in planes through the cylinder axis is disregarded, a "slice" limited by two parallel planes normal to the longitudinal axis may be used to analyze the combined effect of internal heating and external hydration on a set of two blocks limited by joints. This geometry can be further simplified, due to symmetries, as indicated in Figure 3. Only four joints and two blocks are then considered. The set of boundary conditions used in the analysis are given, with respect to the coordinates indicated in Figure 3, in Table 2.

The heat flux at the inner boundary is kept constant until a temperature of 100°C is reached. Thereafter this temperature is kept constant. The set of constitutive parameters used in the analysis is summarized in Table 2. Also indicated in the table are some relationships which complete the description of constitutive laws made before.

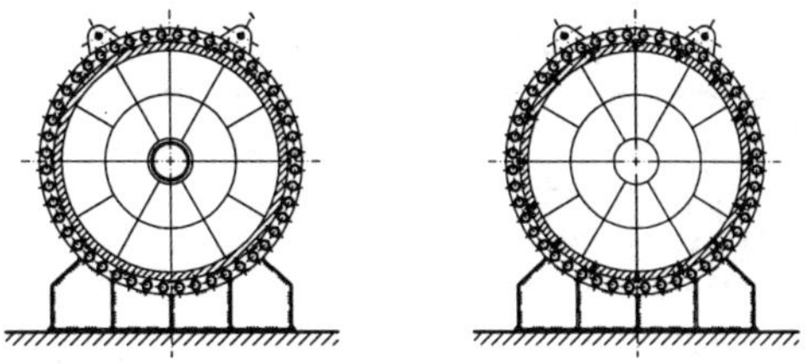

Figure 2a. Mock-up test. Cross section

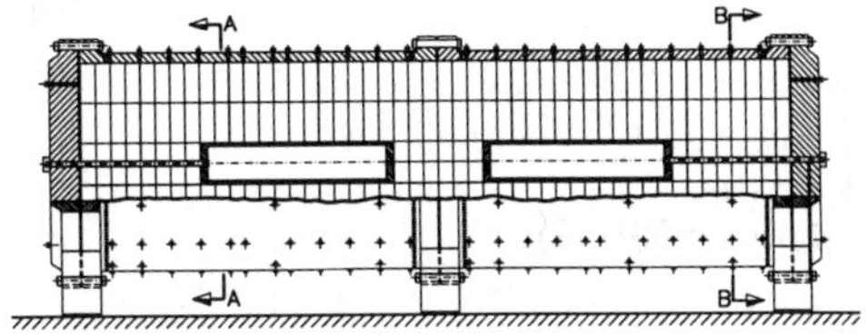

Figure 2b. Mock-up test. Longitudinal section

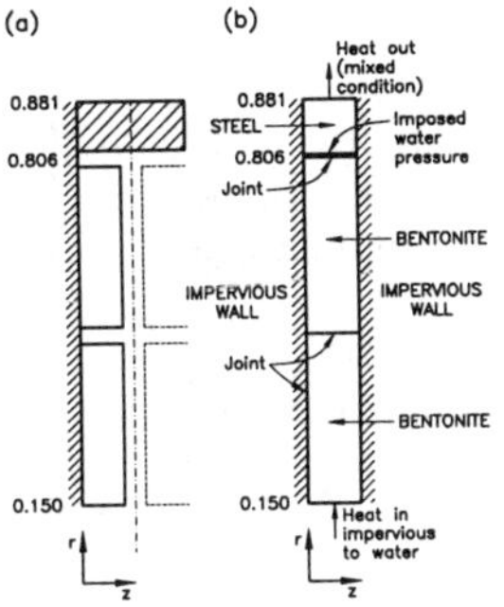

Figure 3. Mock-up geometry used in the analysis.

These parameters are based on a comprehensive set of laboratory tests whose details are outside the scope of this paper.

Contours of water pressure, 1, 4.5 and 7 days after the beginning of the experiment are shown in Figure 4. Joints become immediately saturated due to their high hydraulic conductivity. Water coming from the (open) joints hydrates the blocks in a progressive manner. Note also the different behaviour of the block closer to the heater if compared with the outer one. Vapour originating close to the heater diffuses away but condensates as temperature drops, increasing the water content of the bentonite and therefore its hydraulic conductivity. This, in turn, facilitates the infiltration from the nearly saturated joint.

Table 2. Parameters for mock-up test

Parameter	Bentonite block	Joint
Water retention (Van Genuchten law) (1)		
P_o	50 MPa	10^{-2} MPa
σ_o	0.0072 N/m	0.072 N/m
λ	0.75	0.75
S_{rl}	0.01	0.01
S_{sl}	1	1
μ (Permeability factor) ?		
Permeability (2), (3)		1 (0.25)
A	2 x 1 x 10^{-12} MPa s	
B	1808.5 K^{-1}	
C	1	1
m	3	1
k_o	2 x 10^{-20} m^2	
Heat conduction (4)		
λ_d	0.4181 W/mK	0.026 W/mK
λ_w	1.507 W/mK	0.426 W/mK
c_s (specific heat, solid)	1091.44 J/kg	
Vapour diffusion (5)		
D_o^w	5.9 x 10^{-6} m^2/s	5.9 x 10^{-6} m^2/s
τ	1	1
n	2.3	2.3
Liquid density (6)		
ρ_{wo}	1002.6 kg/m^3	1002.6 kg/m^3
β	4.5 x 10^{-4} MPa^{-1}	4.5 x 10^{-4} MPa^{-1}
α	3.4 x 10^{-4} °C^{-1}	3.4 x 10^{-4} °C^{-1}
Porosity		
ϕ	0.4064	
Solid density		
ρ	2650 kg/m^3	

(1) Water retention law

$$S_{lw} = \left(\frac{1}{1 + \left(\frac{P_g - P_l}{P} \right)^{\frac{1}{1-\lambda}}} \right)^{\lambda}$$

$$S_{lw} = \frac{S_l - S_{rl}}{S_{sl} - S_{rl}}$$

$$P = P_o \, \sigma / \sigma_o$$

(2) Liquid viscosity

$$\mu_l = A\,exp\left(\frac{B}{T + 273} \right)$$

(3) Relative permeability

$$k_r = C S_l^m$$

(4) Thermal conductivity

$$\lambda = \lambda_w^{S_l} \lambda_d^{1-S_l}$$

(5) Coefficient of molecular diffusion of water vapour

$$D_m^w = \tau D_o^w \frac{T^n}{P_g}$$

(6) State equation for liquid density

$$\rho_l = \rho_{lo}\,exp\left(\beta (P_l - P_{lo}) - \alpha (T - T_o) \right)$$

The computed distribution of temperature at the same times is indicated in Figure 5. These patterns indicate a one dimensional heat transport phenomena within the blocks since joints are not a preferential path in this case due to their higher thermal conductivity.

4.2 Hydromechanical behaviour of a joint in swelling medium.

This case is related to the mock-up experiment described above but the geometry is simplified. Consider in Figure 6 a single joint between two bentonite blocks. Water is acting on one of the edges of the joint at a given pressure. Due to symmetries only one block needs to be discretized. This problem has been solved under plain strain conditions. Hydraulic properties of the bentonite and joint have already been given in Table 2. Boundary conditions are given in Table 3 with respect to the coordinate system indicated in Figure 6.

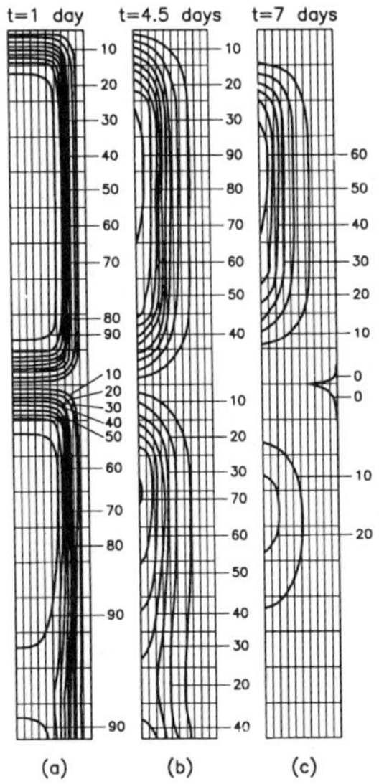

Figure 4. Contours of liquid pressure at different times. Mock-up test analysis.

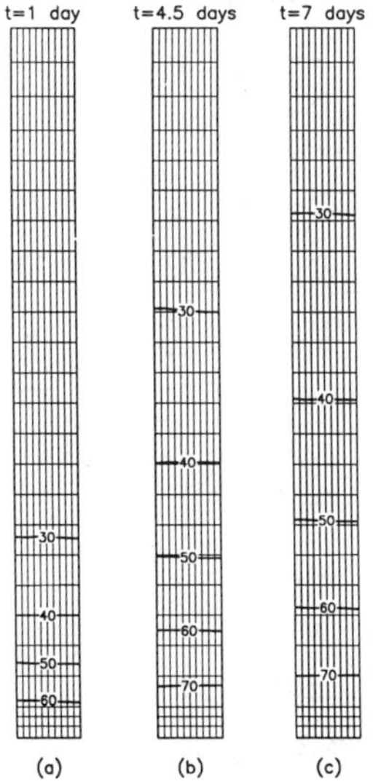

Figure 5. Temperature contours at different times. Mock-up test. analysis.

Mechanical properties of the block and joint are given in Table 4. Note in particular that the swelling behaviour of the bentonite has been modelled by a state surface approach as described in detail in (Lloret and Alonso 1985).

For a given constitutive behaviour of the bentonite the response of this case depends critically of the geometrical dimensions of blocks and joint and the pressure of the injected water. In the case presented here the joint thickness is 10^{-4} m (0.1mm) and the applied water pressure is 0.1 MPa.

Water entering the joint infiltrates into the block (Figure 7), which swells and closes the initial joint aperture. The process of joint closure can be observed in Figure 8.

As water infiltrates in the block the transverse joint displacement increases until the initial gap disappears and movements at a point (A) in the boundary cease. It is interesting to note that away from the boundary (point B) the joint opens again after closure due to the block contraction

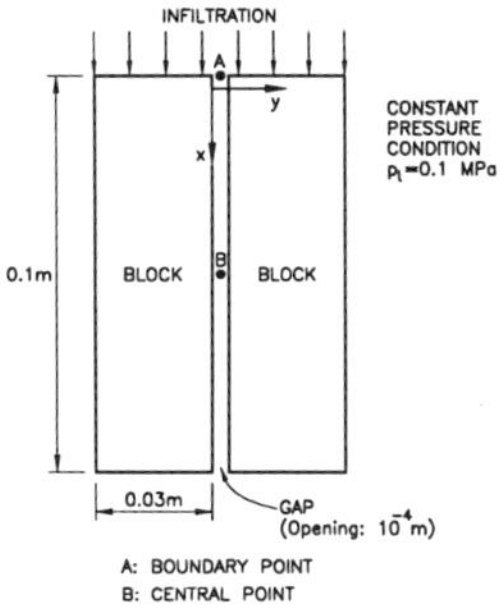

Figure 6.Geometry used in hydromechanical analysis.

Table 3. B.C. for the hydromechanical problem

Hydraulic B.C.	
x = 0.0 m	P_l = 0.1 MPa
x = 0.1 m	Impervious
y = 0.0 m	Impervious
y = 0.01 m	Impervious
Mechanical B.C.	
x = 0.0 m	$u_x = 0$
x = 0.1 m	$u_x = 0$
y = 0.0 m	$u_y = 0$
y = 0.01 m	$u_y = 0$

Table 4. Parameters for the hydromechanical problem

Parameter	Value
Joint stiffness	
Normal k_n	10^6 MPa/m
Shear k_t	10^3 MPa/m
α_1	0.999
α_2	1.0
β	10^5
Bentonite volume change behaviour (1)	
a_p	-0.12
a_s	-0.04
a_{ps}	0.006
Bentonite shear modulus	
G	10 MPa

associated with the redistribution of water content.

(1) State surface for volume change

$$\Delta\varepsilon_v = a_p \Delta\, ln\, p^* + a_s \Delta\, ln\left(\frac{s+0.1}{0.1}\right) +$$

$$+ a_{ps} \Delta\, ln\left(\frac{s+0.1}{0.1}\right) \Delta\, ln\, p^*$$

$$P^* = p - P_g \quad ; \quad s = P_g - P_l$$

Joint closure also affects the history of water pressures in the gap (Figure 9). When the joint is open all points have the same pore water pressure, equal to the applied boundary conditions. After the gap disappears, the central point of the joint (B) is no longer connected to the boundary water pressure. As the water is absorbed by the unsaturated block, the water pressure in the central part of the joint decreases.

5. CONCLUSIONS

It is possible to perform coupled thermo-hydro-mechanical analyses using joint elements especifically developed for this purpose. The comprehensive formulation presented in this paper allows one to consider all the important phenomena involved in a general THM coupled problem.

By incorporating the formulation into a numerical code it is possible to perform computations required in problems of engineering significance, notably those associated with the dispposal of high level radioactive waste.

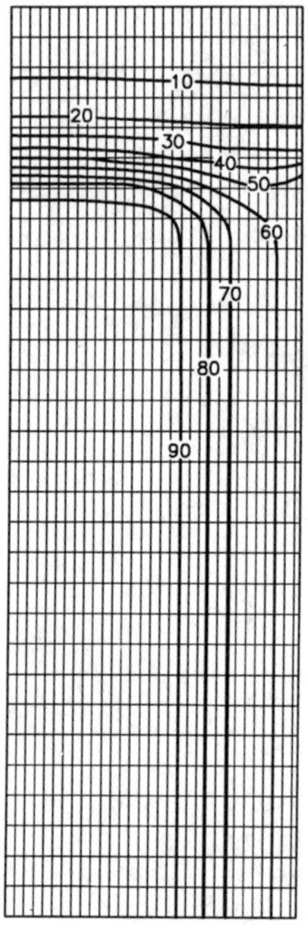

Figure 7. Liquid pressure contours. Hydromechanical analysis.

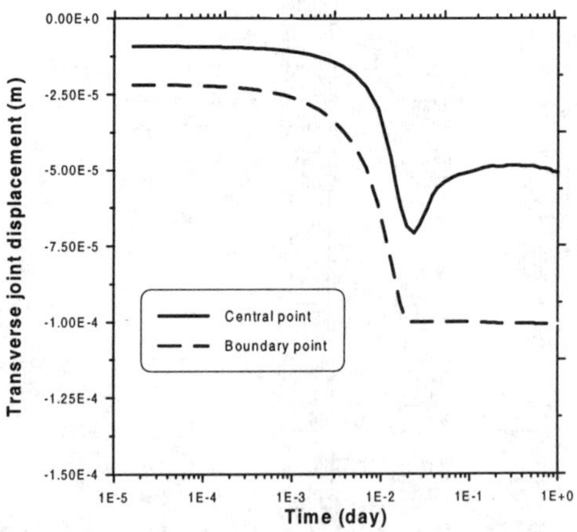

Figure 8. Time evolution of transverse joint displacement at different times.

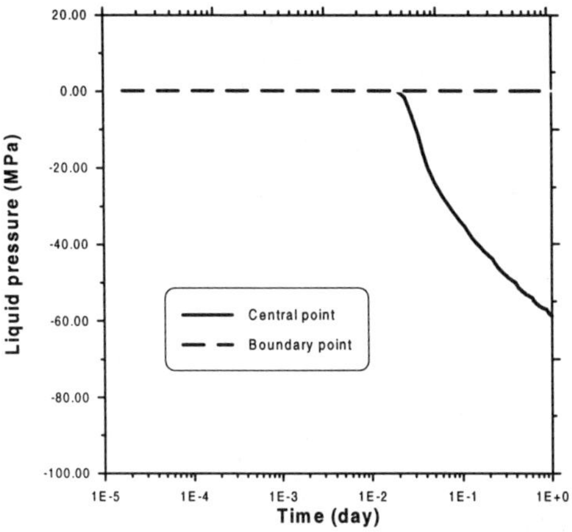

Figure 9. The evolution of liquid pressure at two different points.

ACKNOWLEDGEMENTS

The authors are grateful for the financial and technical assistance of ENRESA (Spanish Agency for Radioactive Waste) and the EEC. The contribution of the Dirección General de Investigación Científica y Técnica of Spain (P895-0771; P893-0964) is also thankfully acknowledged.

REFERENCES

Barton, N., M. Bandis and K. Bakhtar 1985. Strength, deformation and conductivity coupling of rock joints. Int. Jnl. of Rock Mechanics Geomechanical Abstracts, 22, 121-140.

Chan, T., K. Khair, L. Jing, M. Ahola, J. Noorishad, E. Vuillod 1995. International comparison of coupled thermo-hydro-mechanical models of a multifracture benchmark problem: DECOVALEX phase I, benchmark test 2, Int. Jnl. of Rock Mechanics & Geomechanical Abstracts, 32, 5, 435-452.

Gens, A., I. Carol and E.E. Alonso 1989. An interface element formulation for the analysis of soil-reinforcement interaction. Comp. Geothech. 7 (1,2), 133-151.

Gens, A., A.J. García-Molina, S. Olivella, E.E: Alonso and F. Huertas 1997. Analysis of a full scale in situ heating test simulating repository conditions. Int. Jnl. Num. and Anal. Meth. in Geomech. (submitted for publication).

Goodman, R.E., R.L. Taylor and T.L. Brekke 1968. A model for the mechanics of jointed rocks. Jnl. of Soil Mechanics and Foundations Division, ASCE, 94 (SM3), 637-659.

Gringarten, A.C., H.J. Ramey and R. Raghavan 1974. Unsteady-state pressure distributions created by a well with a single infinite-conductivity vertical fracture. Soc. Pet. Eng. Jnl., 14 (4), 347-360.

Jing, L., C.F. Tsang, O. Stephansson 1995. DECOVALEX - An international cooperative research project on mathematical models of coupled THM processes for safety analysis of radioactive waste repositories. Int. Jnl. of Rock Mechanics and Mining Sciences Geomechanical Abstracts, 32, 5, 389-398.

Lloret, A. and E.E. Alonso 1985. State surfaces for partially saturated soils. Proc. 11th ICSMFE, San Francisco, 2, 557-562.

Millard, A., M. Durin, A. Stietel, A. Thoraval, E. Vuillod, H. Baraoudi, F. Plas, A. Bougnoux, G. Vouille, A. Kobayashi, K. Hara, T. Fujita and Y. Ohnishi 1995. Discrete and continuum approaches to simulate the thermo-hydro-mechanical couplings in a large, fracturated rock mass. Int. Jnl. of Rock Mechanics and Mining Sciences Geomechanical Abstracts, 32, 5, 409-434.

Olivella, S., J. Carrera, A. Gens and E.E. Alonso 1994. Nonisothermal multiphase flow of brine and gas through saline media. Transport in porous media, 15, 271-293.

Thomas, H.R., E.E. Alonso and A. Gens 1995. Modelling thermo-hydraulic-mechanical processes in the containment of nuclear waste unsaturated soils. E.E. Alonso and P. Delage eds. Balkema, Rotterdam, 3. 1135-1141.

Wallace, K.B and R.L. Lytton 1992. Lateral pressures and swelling in a cracked expansive clay profile. Proc. 7th I.C. Exp. Soils. Dallas. Vol 1. 245-250.

Computer Methods and Advances in Geomechanics, Yuan (ed.) © 1997 Balkema, Rotterdam, ISBN 90 5410 904 1

Coupled thermo-hydro-mechanical analysis of unsaturated soil-implementation in a parallel computing environment

H. R.Thomas, Haitian Yang & Yang He
Cardiff School of Engineering, University of Wales, UK

Abstract: A parallelized sub-structuring based solution is given in this paper for the analysis of the thermo-hydro-mechanical behaviour of unsaturated soil. The temperature, displacement, pore water pressure and pore air pressure are treated as primary variables. A parallel code for the fully coupled analysis has been developed on a Paramid parallel computer at UWC. A laboratory experiment based numerical simulation has been implemented. The numerical results obtained indicate the efficiency of the proposed algorithm. This encourages the extension of the work.

1 INTRODUCTION

Coupled heat and moisture transfer in a deformable partly saturated porous medium plays an important role in a number of engineering applications. Of particular relevance at the present time is the issue of the safe disposal of high level nuclear waste(Chapman and Mckinley, 1987), where the performance of the engineered barriers is of paramount importance.

The phenomena involved are complex, consisting of the interrelated effects of heat, liquid water, water vapour and air transfer in a deforming soil. A theoretical formulation has been developed by the authors(Thomas and King, 1991; Thomas and Sansom, 1995; Thomas and He, 1995) which combines the above features in a mechanistic manner.

Due to the complexities of the problem, computer based numerical techniques are invariably required in order to produce solutions of practical problems. Furthermore, given i) the size of the domain of interest, ii) the length of the time scale involved viz-a-viz the life time of nuclear waste repositories, iii) the complexity of the formulation as described above and iv) the nature of the resulting governing differential equations, it can be seen that the problem is potentially computationally demanding.

It is therefore of considerable interest to explore the development of parallel computing solutions. Two general approaches are now being developed at UWC, an "iterative solution" method and a substructuring methodology.

In this paper a substructuring based parallel algorithm for the fully coupled thermo-hydro-mechanical behaviour of unsaturated soil is presented. A sub-structuring based algorithm is amenable to parallelization since it is able to consider each sub-structure independently. Previous examples of its use in structural analysis have been presented in the literature(Yagawa and Shioya 1993, Saxena and Perucchio 1993,).

The work presented was programmed in FORTRAN. All computations were performed on a Paramid parallel computer at UWC. The MIMD Paramid parallel computer, with distributed memory architecture, is based on the Transtech parallel system. The system has 48 processors, each consisting of a 100 MFLOP Intel i860-XP, running at 50 MHz.

A number of parallel numerical tests, including analyses of deformation, heat transfer, coupled thermo-moisture, displacement-moisture and heat-moisture-displacement effects, were implemented to verify the correctness and efficiency of the model proposed. A laboratory experiment based numerical simulation was then performed and the results achieved, both in terms of comparisons with experimental results and computational efficiency are presented.

2 BASIC EQUATIONS AND FINITE ELEMENT FORMULATION

The behaviour of unsaturated soil consisting of solid, liquid and gas phases, can be described mathematically as a set of non-linear differential equations. The formulation employed here has been presented elsewhere in considerable detail, (Thomas and King 1991, Thomas and Sansom 1995, Thomas and He 1995), and therefore only a brief description of the salient points will be given here.

2.1 *MoistureTransfer*

The moisture transfer equation , accommodating both the liquid and vapour phases is expressed in the form

$$(n\rho_l S_l)_{,t} + (n\rho_v S_a)_{,t} + (\rho_l (V_l)_j)_{,j} + (\rho_l (V_v)_j)_{,j} + (\rho_v (V_a)_j)_{,j} = 0 \qquad (1)$$

where n denotes the porosity, ρ the density, S the degree of saturation, t the time and V the velocity. The subscripts *l, a* and *v* refer to liquid, air and water vapour respectively. The subscript j, ranging from 1 to 3, is tensor index.

2.2 *Dry air mass transfer*

Using Henry's law to take account of dissolved air in the pore water, conservation of dry air flow dictates that

$$\left[n\rho_{da}\left(S_a + H_c S_l\right)\right]_t + \left[\rho_{da}\left((V_a)_j + H_c (V_l)_j\right)\right]_j = 0 \qquad (2)$$

where H_c is Henry's coefficient of solubility, and, ρ_{da} is the density of the dry air.

2.3 *Temperature*

From consideration of conservation of energy, the governing equation for heat transfer can be written as

$$\Phi_{,t} = Q_{i,i} \qquad (3)$$

where Q, the global heat flux per unit volume and Φ, the heat capacity of the soil per unit volume are defined as

$$\Phi = n\left(\rho_l S_l C_{pl} + \rho_v S_a C_{pv} + \rho_a S_a C_{pa}\right)\left(T - T_r\right) + (1-n)\rho_s C_{ps}\left(T - T_r\right) + n\rho_v S_a L$$

$$Q_j = -\lambda_T T_{,j} + \left(\rho_l (V_v)_j + \rho_v (V_a)_j\right)L + \left(C_{pl}\rho_l (V_l)_j + C_{pv}\rho_l (V_v)_j\right)\left(T - T_r\right) + \left(C_{pv}\rho_v (V_a)_j + C_{pa}\rho_a (V_a)_j\right)\left(T - T_r\right)$$

In the above λ_T is the intrinsic thermal conductivity of the soil, C_{pl}, C_{pv}, C_{pa} and C_{ps} are the specific heat capacities of pore liquid, water vapour, pore air and the solid particles respectively, T and T_r are the temperature and reference temperature respectively, ρ_s is the density of the solid particles, and L is the latent heat of vaporisation.

2.4 *Stress equilibrium equation and constitutive relationship.*

The stress equilibrium equation can be specified in an incremental form

$$d(\sigma_{ij} - \delta_{ij}u_a)_{,j} + (du_a)_{,i} + db_i = 0 \qquad (4)$$

where σ_{ij} denotes the total stress tensor, δ_{ij} the Kronecker index, b_i the body force, and u_a the air pressure.

The displacement-strain relation can be written in an incremental form

$$d\varepsilon_{ij} = (du_{i,j} + du_{j,i})/2 \qquad (5)$$

where ε_{ij} denotes the strain tensor of the solid skeleton of soil, and u_i refers to the displacement.

For the solid skeleton, the total strain increment can be decomposed into 3 parts

$$d\varepsilon_{ij} = d(\varepsilon_{ij}^{\sigma}) + d(\varepsilon_{ij}^{s}) + d(\varepsilon_{ij}^{T}) \qquad (6)$$

where the subscript σ, *s*, *T* refer to net mean stress, suction and temperature respectively.

The first and second terms on the right hand side of equation (6) can be decomposed in the form

$$d\varepsilon_{ij}^{k} = d(\varepsilon_{k}^{e})_{ij} + d(\varepsilon_{k}^{p})_{ij} \quad (k = \sigma, s) \qquad (7)$$

where the superscripts e and p refer to elasticity and plasticity respectively.

The elasto-plastic constitutive relation can be specified in the form

$$d(\varepsilon^e_\sigma)_{ij} = D_{ijlm} d\sigma_{lm} \quad (8)$$

$$d(\varepsilon^e_s)_{ij} = \delta_{ij} D^s dS \quad (9)$$

where D_{ijlm} and D^s are the elastic tensors.

$$d(\varepsilon^p_\sigma)_{ij} = \chi_1 (\frac{\partial Q_1}{\partial p}\frac{\partial p}{\partial \sigma_{ij}} + \frac{\partial Q_1}{\partial q}\frac{\partial q}{\partial \sigma_{ij}}) \quad (10)$$

$$d(\varepsilon^p_s)_{ij} = \delta_{ij}\chi_2 \frac{\partial Q_2}{\partial S} \quad (11)$$

where χ_1 and χ_2 denote the plastic flow factors, and Q_1 and Q_2 the plastic potential functions. p and q are defined as

$$p = \sigma_{ii}/3 - u_a \quad (12)$$

$$q^2 = 3/2\sigma'_{ij}\sigma'_{ij} \quad (13)$$

σ'_{ij} represents the deviatoric net mean stress.
The plastic potentials are specified as

$$Q_1 = \alpha_q q^2 - M^2(p + p_s)(p_0 - p) \quad (14)$$
$$Q_2 = S - S_0 \quad (15)$$

The yield criterion can be modelled as

$$F_1 = q^2 - M^2(p + p_s)(p_0 - p) = 0 \quad (16)$$
$$F_2 = S - S_0 = 0 \quad (17)$$

where M is the slope of the critical state line, p_s relates the effect of suction on the cohesion of the soil, p_0 is the preconsolidation stress, S_0 is the maximum value of suction previously attained and α_q is a plastic parameter related to M.
The third term in the right hand side of equation (6) can be written as

$$d\varepsilon^T_{ij} = \delta_{ij}\alpha_T dT \quad (18)$$

where α_T is the thermal expansion coefficient.

2.5 Finite element formulation

By exploiting the internal relations between variables, and employing the Galerkin method, equations (1), (2) (3) and (4) can be expressed in terms of the primary variables, $u_l, T, u_a, \underline{u}$, according to

$$\begin{bmatrix} \mathbf{K}_{ll} & \mathbf{K}_{lT} & \mathbf{K}_{la} & \mathbf{0} \\ \mathbf{K}_{Tl} & \mathbf{K}_{TT} & \mathbf{K}_{Ta} & \mathbf{0} \\ \mathbf{K}_{al} & \mathbf{0} & \mathbf{K}_{aa} & \mathbf{0} \\ \mathbf{0} & \mathbf{0} & \mathbf{0} & \mathbf{0} \end{bmatrix} \begin{Bmatrix} u_l \\ T \\ u_a \\ \underline{u} \end{Bmatrix} + \begin{bmatrix} \mathbf{C}_{ll} & \mathbf{C}_{lT} & \mathbf{C}_{la} & \mathbf{C}_{Tu} \\ \mathbf{C}_{Tl} & \mathbf{C}_{TT} & \mathbf{C}_{Ta} & \mathbf{C}_{Tu} \\ \mathbf{C}_{al} & \mathbf{C}_{Tl} & \mathbf{C}_{aa} & \mathbf{C}_{Tu} \\ \mathbf{C}_{ul} & \mathbf{C}_{uT} & \mathbf{C}_{ua} & \mathbf{C}_{uul} \end{bmatrix} \begin{Bmatrix} \dot{u}_l \\ \dot{T} \\ \dot{u}_a \\ \dot{\underline{u}} \end{Bmatrix} = \begin{Bmatrix} f_l \\ f_T \\ f_a \\ f_u \end{Bmatrix} \quad (19)$$

where u_l represents pore water pressure, u_a the pore air pressure, and $\underline{u}$ the displacement vector, $\mathbf{K}_{ij}$ and $\mathbf{C}_{ij}$ represent the corresponding matrices of the governing equations, (i,j=l, T, a, u), $\langle\cdot\rangle$ refers to the derivative with respect to time.
Equation (19) can be rewritten more simply as

$$K\{\varphi\} + C\left\{\frac{\partial\varphi}{\partial t}\right\} = \{R\} \quad (20)$$

where φ refers to the global unknowns, that is $\{u_l \quad T \quad u_a \quad \underline{u}\}^T$

Equation (20) can be discretised temporally by employing a two level difference technique(Cook, 1981), having the form

$$\chi K\{\varphi^{n+1}\} + C\{\varphi^{n+1}\}/\Delta t = (1-\chi)\{R^n\} + \chi\{R^{n+1}\} - (1-\chi)K^n\{\varphi^n\} + C\{\varphi^n\}/\Delta t \quad (21)$$

where χ is an integration factor, which defines the required time interval ($\chi \in [0,1]$).
Rewriting equation (21) in an alternative notation yields

$$A\{\varphi^{n+1}\} = \{F^{n+1}\} \quad (22)$$
where $A = \chi K + C/\Delta t$ (23)

$$F^{n+1} = (1-\chi)\{R^n\} + \chi\{R^{n+1}\} - (1-\chi)K^n\{\varphi^n\} + C\{\varphi^n\}/\Delta t \quad (24)$$

3 SUBSTRUCTURING BASED PARALLEL APPROACH FOR THESE FULLY COUPLED PROBLEMS

The technique of sub-structuring generally comprises the following steps:-

i) Assembly of the element matrices of a substructure to give the global 'stiffness' matrices of the substructures.

ii) Condensation of the terms associated with the internal nodal points(i.e. nodes not on the interface between substructures)

iii) Solution of the system of equations obtained by assembling all the substructures.

iv) Recovery of the solution of internal terms.

One of the advantages of a sub-structuring based parallel algorithm is that it does not have any of the convergence problems often associated with iterative solvers. This is especially true when there is no guarantee that the system equations are symmetric positive-definite, as is the case for the above highly non-linear fully coupled problem. Processors will work for a large proportion of the parallel computing process without communicating with each other. In that sense an efficient computing performance can be expected.

On the basis of the sub-structure principle, equation (22) can be written in another form

$$\Sigma A_j\{\varphi_j^{n+1}\} = \Sigma\ \{F_j^{n+1}\} \tag{25}$$

where

$$A_j = \chi\begin{bmatrix} K_{11}^j & K_{12}^j \\ K_{21}^j & K_{22}^j \end{bmatrix} + \begin{bmatrix} C_{11}^j & C_{12}^j \\ C_{21}^j & C_{22}^j \end{bmatrix} / \Delta t \tag{26}$$

$$\{F_j^{n+1}\} = (1-\chi)\begin{Bmatrix} R_{j1}^n \\ R_{j2}^n \end{Bmatrix} + \chi\begin{Bmatrix} R_{j1}^{n+1} \\ R_{j2}^{n+1} \end{Bmatrix} -$$

$$(1-\chi)\begin{bmatrix} K_{11}^j & K_{12}^j \\ K_{21}^j & K_{22}^j \end{bmatrix}^n \begin{Bmatrix} \varphi_{j1}^n \\ \varphi_{j2}^n \end{Bmatrix} + \tag{27}$$

$$\chi\begin{bmatrix} C_{11}^j & C_{12}^j \\ C_{21}^j & C_{22}^j \end{bmatrix}^n \begin{Bmatrix} \varphi_{j1}^n \\ \varphi_{j2}^n \end{Bmatrix} / \Delta t$$

In the above superscript and subscript j refers to the number of the substructure; subscripts 1 and 2 refer to the terms on the interface and inside domains

By condensing the internal terms, equation (25) can be written, in terms of unknowns on the interface, as follows

$$\Sigma A_j^0\{\varphi_{j1}^{n+1}\} = \{F_{j1}^{n+1}\} \tag{28}$$

Solving the system equation (28), and recovering internal terms on the individual substructures, the solution of φ^{n+1} is achieved.

A parallel strategy for solving this non-linear problem can be described as follows

Parallel strategy for the fully coupled problem
A. Loop for time step(for Lth time step(L=1 2....))
B. Loop for non-linear iteration(for ith step of non-linear iteration(i=1, 2...))
1. For P_j(j=1,.....N)
Generate A_j^0 and (F_{j1}^{n+1})
2. Send A_j^0 and $(F_{j1}^{n+1})$to a specific processor
C. For the specific processor.
1. Receive A_j^0 and (F_{j1}^{n+1}) from each processor P_j(j=1,.....N)
2. Solve $\Sigma A_j^0\{\varphi_{j1}^{n+1}\} = \Sigma\{F_{j1}^{n+1}\}$
3. Send (φ_{j1}^{n+1}) to P_j((j=1, 2......)
D. For P_j(j=1,2....N)
1. Receive (φ_{j1}^{n+1}) from a specific processor
2. Recover the internal solution (φ_{j2}^{n+1}).
E. Monitor convergence for ith iteration
F. Start a new round of non-linear iteration or enter next time step
G. End of looping for non-linear iteration
H. End of looping for time step

In order to obtain a satisfactory computing efficiency, the structure needs to be carefully decomposed so that i) All sub-structures contain approximately the same amount of computation. ii) Grid points on the interface between sub-structures are reduced to a minimum.

4. VERIFICATION EXERCISES

The proposed approach was verified numerically via a number of tests, including a deformation analysis, a heat transfer problem, coupled thermo-moisture analysis, coupled displacement-moisture analysis, and a fully coupled analysis.

5. APPLICATION

A laboratory experiment was performed on montmorillonite clay uniaxially compacted at an initial water content of 12.4% to a dry density of 1.62 g/cm³ inside a stainless steel cell which was 14.6 cm high and 15 cm in inner diameter (EX4 in Villar et al, 1993). The overall configuration is shown in Figure 1. In the upper part of the cell, a heating device of 1.5cm in height and 1.0cm in diameter was placed along the axis of the cylinder and heated up to 100°C. The temperature distribution within the sample was measured by 9 thermocouples at different levels. The thermocouples which are of chromel-alumel type with a stainless steel capsule of 0.1 cm, have an accuracy of 0.1°C.

The temperature was fixed at 100°C at the nodes of the heater. The temperature along the outside boundaries was fixed at 28°C due to the thermo-shower. Initial conditions for the simulation were estimated from measured experimental data (Villar and Martín, 1993) as follows:

Degree of saturation $S_l = 0.5$, Void ratio $e = 0.72$, Temperature $T = 20.0$ °C

Initial deformation was regarded as being zero

Due to symmetry, only half of the sample needed to be modelled. A mesh, 146 mm high and 75 mm wide, with 168 8-node isoparametric elements, was chosen to represent the test sample. In the parallel substructuring analysis, the computing domain was decomposed into 4 and 6 sub-structures, each of which had 42 and 28 isoparametric elements respectively.

The numerical results, in the form of contour plots of the distributions of the temperature, degree of saturation and void ratio of the soil sample at steady state are given in Figures 2-4, together with the experimental values obtained.

The Speedup and Efficiency are shown in Table 1, where Speedup (S_a) and Efficiency (E_a) are defined by

$$S_a = \frac{t_1}{t_p} \qquad (29)$$

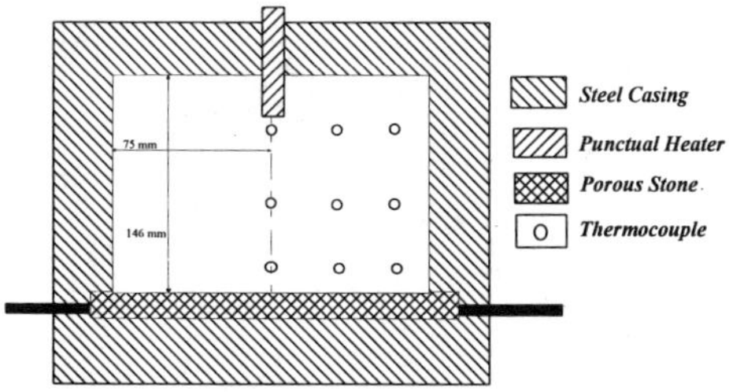

Figure 1 Configuration of Cell for Thermal Test

$$E_a = \frac{t_1}{p t_p} \times 100\% \qquad (30)$$

where subscript p refers to the number of processors and t_1 and t_p are the computational times on a single processor and on p processors respectively

Table 1 Speedup and Efficiency

Number. of processor	Speedup	Efficiency
4	3.62	90
6	4.472	74.5

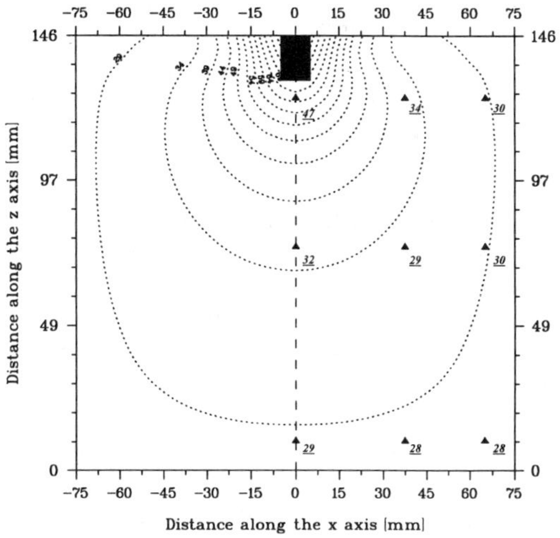

Figure 2 Plot of temperature

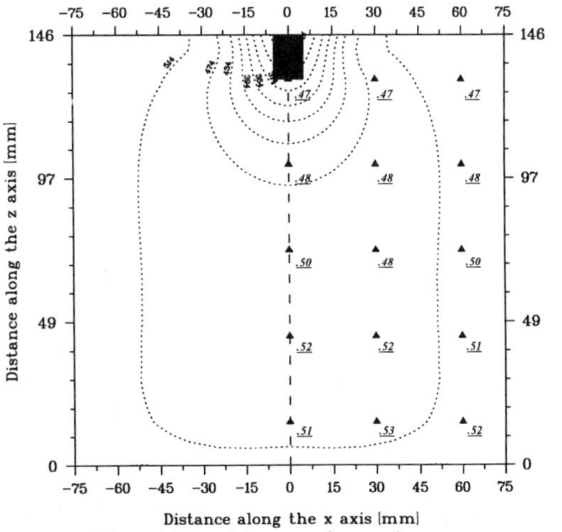

Figure 3 Plot of degreee of saturation

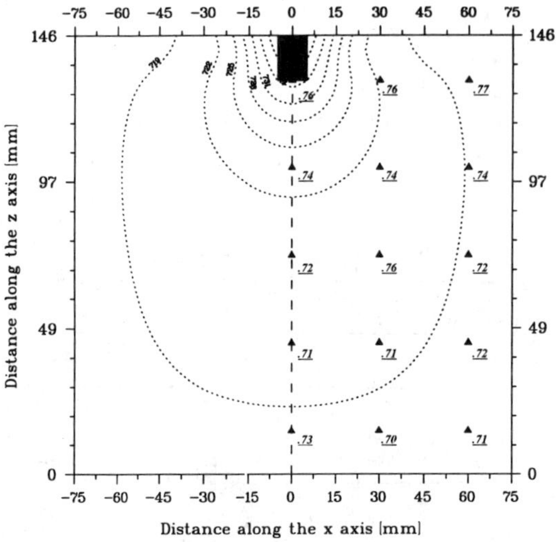

Figure 4 Plot of void ratio

The overall comparison that has been obtained between the experimental results and the sub-structuring solution is encouraging, indicating that the approach being pursued is capable of describing, in a qualitative sense, the relevant features of interest.

6 CONCLUSION

The principal objective of this work was to present a substructuring based parallel algorithm for the analysis of thermo-hydro-mechanical behaviour of porous media. A corresponding code has been developed on the Paramid parallel computer in UWC to make computing more effficient. Encouraging results have been obtained.
Various numerical examples have been computed to verify the correctness and efficiency of the algorithm. Comparison of numerical results with analytical solutions, nonsubstructuring solutions and experimental results yields good correlation.
The overall approach presented has proved to be capable of i) simulating the physical problem under consideration and ii) improving computing efficiency. It is therefore proposed as a useful tool for further calculations.

REFERENCES

Chapman, N. A. and I. G. Mckinley (1987). *The Geological disposal of nuclear waste,* Chichester: John Wiley & Sons.

Cook, R. D. (1981). *Concepts and applications of finite element analysis*, John Wiley and Sons, Inc, New York.

Saxena M. and R. Perucchio (1993). "Parallel FEM algorithm based on recusive spatial decomposition-II. Automatic analysis via hierarchical substructuring." *Computers and Structures*, 47(1), 143-154.

Thomas, H. R and He Y.(1995) "Analysis of coupled heat, moisture and air transfer in a deformable unsaturated soil." *Geotechnique*, 45(4),677-689

Thomas, H. R. and M. R. Sansom (1995). "Coupled heat, moisture and air transfer in unsaturated soil." *ASCE, Journal of Eng. Mech.*, 121(3), 392-405.

Thomas, H. R. and S. D. King (1991). "Coupled temperature/capillary variations in unsaturated soil." J. *Engrg. Mech., ASCE*, 117(11), 2475-2491.

Villar, M. V. and P.L.Martin (1993). "Suction controlled oedometric tests in montmorillonite clay." *Proc. 29th Annual Conference of the Engineering Group of the Geological Society of London*, 337-342.

Villar, M.V., J. Cuevas, A.M. Femàndez & P.L.Martin (1993). "Effect of the interaction of heat and water flow in compact bentonite." *International Workshop on Thermomechanics of clays and clay barriers*, Bergamo.

Yagawa, G. and Shioya (1993). "Parallel finite element on a massively parallel computer with domain decomposition." Computing Systems in Engineering, 4, 495-503.

Computer Methods and Advances in Geomechanics, Yuan (ed.) © 1997 Balkema, Rotterdam, ISBN 90 5410 904 1

Three-dimensional computer modeling of coupled geomechanics and multiphase flow

L.Y.Chin
Phillips Petroleum Company, Bartlesville, Okla., USA

J.H.Prevost
Princeton University, N.J., USA

ABSTRACT

For hydrocarbon reservoirs with weak formations, the coupled interaction between the stresses in the porous geomaterial and the pressures in the fluid phases during reservoir fluid production becomes very significant and the coupling effect can not be ignored when analyzing the reservoir production problems. In general, the full system of governing equations resulting from coupled geomechanics and multiphase flow is complicated to solve numerically. In this paper, a computer method for numerically solving the coupled equations of geomechanics and multiphase flow is developed and applied to finite-element analysis of practical 3-D field problems. The method is iterative, based on a multistagger solution strategy in which the full system of coupled equations defined in the problem domain is partitioned into smaller subsystems of equations. Each subsystem is then solved separately assuming that the unknown variables of the other subsystems are temporarily frozen. The subsystems are solved sequentially and repeatedly in a predetermined sequence until all subsystems have converged to a self consistent set of solution variables. The results from this modeling study show that the proposed numerical procedure is very stable, accurate, efficient and economical for analyzing highly coupled field problems. Application examples are presented in this paper.

1 INTRODUCTION

When analyzing production problems for hydrocarbon reservoirs with weak formations, it is often desirable to conduct the computer modeling of the complex processes in a coupled fashion that integrates geomechanics with reservoir fluid flow in porous media, in order to capture the coupling effect (e.g., Tortike and Farouq Ali (1993); Settari and Mourits (1994); Gutierrez and Hansteen(1994); Lewis and Sukirman (1993a, 1993b)). In general, the coupled geomechanics and multiphase flow problems are complicated and computational intensive to solve numerically. However, the continued trend of advancement in high power, low cost computing technologies makes it economically feasible to numerically analyzing complicated field problems that involve coupled geomechanics and multiphase flow in a deforming reservoir environment.

The objective of this study was to investigate how the coupled geomechanics and multiphase flow problems could be numerically solved at a reasonable computational cost while still providing accurate solutions. We achieved this objective by using a multistagger, iterative solution procedure instead of the conventional simultaneous solution procedure. For the cases studied, the proposed multistagger procedure successfully reduced the computational time by a factor of 12 compared with the conventional method and was more stable than the conventional method. The details of the governing equations formulated for the coupled field problems and the proposed computing strategy were presented in this paper.

2 FIELD EQUATIONS

The equations governing isothermal two-phase fluid flow in a deformable porous skeleton are derived in this section. The equations are developed along the lines of Biot's self consistent theory (e.g., Biot (1941, 1955, 1956)). The solid phase is assumed to consist of a solid porous skeleton whose voids are filled by one or two fluids (typically: water and oil). The shear stresses in the fluid phases are assumed small, their effects neglected and an all-around pressure state is assumed to prevail in each fluid phase. Fluid flow through the moving porous matrix is assumed to be governed by Darcy's law (Biot (1955); Verruijt (1969)).

2.1 *Basic Relations*

The three basic principles used to derive the governing equations are: balance of linear momentum, mass conservation and Darcy's law. In the following, the superscript/subscript s, w and o refer to the solid, water and oil phases, respectively. The equations are stated as follows:

- Balance of Linear Momentum:

$$\underset{\sim}{\nabla}\cdot(\underset{\sim}{\sigma}'^{s} - \bar{p}\underset{\sim}{\delta}) + \rho\underset{\sim}{b} = \underset{\sim}{0} \tag{1}$$

- Balance of Mass:

$$\text{solid:}\quad \underset{\sim}{\nabla}\cdot(\rho_s(1-\phi)\underset{\sim}{v}^{s}) + \frac{\partial(\rho_s(1-\phi))}{\partial t} = 0$$

$$\text{water:}\quad \underset{\sim}{\nabla}\cdot(\rho_w\phi S_w\underset{\sim}{v}^{w}) + \frac{\partial(\rho_w\phi S_w)}{\partial t} = 0 \tag{2}$$

$$\text{oil:}\quad \underset{\sim}{\nabla}\cdot(\rho_o\phi S_o\underset{\sim}{v}^{o}) + \frac{\partial(\rho_o\phi S_o)}{\partial t} = 0$$

- Darcy's Law:

$$\text{water:}\quad S_w\phi(\underset{\sim}{v}^{w} - \underset{\sim}{v}^{s}) = -\frac{\underset{\sim}{k}_w}{\mu_w}\cdot(\underset{\sim}{\nabla}\, p_w - \rho_w\underset{\sim}{b})$$

$$\text{oil:}\quad S_o\phi(\underset{\sim}{v}^{o} - \underset{\sim}{v}^{s}) = -\frac{\underset{\sim}{k}_o}{\mu_o}\cdot(\underset{\sim}{\nabla}\, p_o - \rho_o\underset{\sim}{b}) \tag{3}$$

where: $\underset{\sim}{\sigma}'^{s}$ = solid effective stress (see e.g., Terzaghi (1943)); $\underset{\sim}{b}$ = body force; p_w = water pressure; p_o = oil pressure; ϕ = effective porosity; ρ_s = solid mass density (grains); ρ_w = water mass density; ρ_o = oil mass density; S_w = water saturation; S_o = oil saturation (note: $S_w + S_o = 1$); ρ = total mean density, viz.

$$\rho = (1-\phi)\rho_s + \phi(S_w\rho_w + S_o\rho_o) \tag{4}$$

μ_w = water viscosity; μ_o = oil viscosity; $\underset{\sim}{k}_w$ = water effective permeability; and $\underset{\sim}{k}_o$ = oil effective permeability. The effective permeability of phase i, $\underset{\sim}{k}_i$ is related to the absolute permeability $\underset{\sim}{k}$ as:

$$\underset{\sim}{k}_i = k_{ri}(S_w)\underset{\sim}{k} \qquad (i=o,w)$$

where the relative permeability of phase i, k_{ri} is a nonlinear function of the degree of water saturation S_w. For a two-phase porous system, the degree of water saturation S_w is related to the capillary pressure p_c, viz., $S_w = S_w(p_c)$ with

$$p_c = p_o - p_w \tag{5}$$

In turn, p_c is determined from p_w and p_o which are primary unknown variables in Eqs 3. It has long been recognized that the inclusion of the effects of capillary pressure and relative permeability variations are important in reservoir simulations of multiphase flow (see e.g., Aziz and Settari (1979); Norman (1990)).

In Eq. 1, $\bar{p}$ is the effective average pore fluid pressure. It has been defined in different ways by various researchers (see e.g., Ertekin and Farouq Ali (1979); Tortike and Farouq Ali (1979); Lewis and Schrefler (1987); Lewis and Sukirman (1993a, 1993b)). In the following, the effective average pore fluid pressure is defined as:

$$\bar{p} = S_w p_w + S_o p_o \tag{6}$$

In Eqs. 2 and 3, both fluid ($\underset{\sim}{v}^{w}$ and $\underset{\sim}{v}^{o}$) and solid ($\underset{\sim}{v}^{s}$) velocities are local volume averaged values with respect to a stationary reference frame. Note that Darcy's law (Eqs 3) is expressed in terms of the fluid velocity relative to the moving (deformable) solid matrix. In order to derive the full set of governing equations, Eqs 1-3 must be supplemented with appropriate constitutive equations for the solid porous matrix and equations of states for the various constituents.

2.1.1 *Equations of State*

The fluid densities are assumed to be function of the fluid pressure only, and in the following C_i (i=o,w) denotes the fluid compressibility

$$C_i = \frac{1}{\rho_i}\frac{\partial\rho_i}{\partial p_i} \qquad (i = o,w) \tag{7}$$

The solid grains density is assumed to be function of the effective average fluid pressure and in the following C_s denotes the compressibility of the solid:

$$C_s = \frac{1}{\rho_s}\frac{\partial\rho_s}{\partial\bar{p}} \tag{8}$$

as the pore fluid pressures place all of the solid phase in a purely hydrostatic pressure state.

2.1.2 *Constitutive Equations*

The solid effective stress is related to the solid skeleton strain via a nonlinear constitutive equation as:

$$\underset{\sim}{\sigma}'^{s} = \underset{\sim}{f}^{s}(\underset{\sim}{\epsilon}^{s}, K) \tag{9}$$

where $\underset{\sim}{\epsilon}^{s} = \int \dot{\underset{\sim}{\epsilon}}^{s}\,dt$ = solid strain; $\dot{\underset{\sim}{\epsilon}}^{s}$ = solid strain rate; K denotes a set of history dependent tensorial quantities. The solid strain rate is related to solid velocities and the rate of effective average fluid pressure change as:

$$\dot{\underset{\sim}{\epsilon}}^{s} = \underset{\sim}{\nabla}\, \underset{\sim}{v}^{s}_{()} + \frac{1}{3}C_s\frac{d\bar{p}}{dt}\underset{\sim}{\delta} \tag{10}$$

where $\nabla \underset{\sim}{v}^s_{()}$ = symmetric part of the solid velocity gradient; $\underset{\sim}{\delta}$ = Kronecker delta. The correction term in Eq. 10 represents the volumetric strain caused by the uniform compression of the solid phase (as opposed to the skeleton) by the effective pore fluid pressure. These nonlinear constitutive equations (Eqs 9 and 10) are usually appropriate to describe the nonlinear behavior of the solid skeleton such as very weak reservoir rocks.

For a linear isotropic poroelastic solid skeleton, Eq. 9 simply writes as:

$$\underset{\sim}{\sigma}'^s = \lambda_s \mathrm{tr}\,\underset{\sim}{\epsilon}^s \underset{\sim}{\delta} + 2\mu_s \underset{\sim}{\epsilon}^s \tag{11}$$

where λ_s, μ_s = Lame's constant.

2.2 *Governing Equations*

Introducing Darcy's law (Eqs 3) into both fluid mass conservation equations (Eqs 2) give the following:

$$\underset{\sim}{\nabla}\cdot[\frac{\underset{\sim}{k}_i}{\mu_i}\cdot(\underset{\sim}{\nabla}p_i - \rho_i \underset{\sim}{b})] + C_i \underset{\sim}{\nabla}p_i \cdot \frac{\underset{\sim}{k}_i}{\mu_i}\cdot(\underset{\sim}{\nabla}p_i - \rho_i \underset{\sim}{b}) =$$

$$S_i \underset{\sim}{\nabla}\cdot \underset{\sim}{v}^s + S_i(1-\phi)C_s \frac{d\bar{p}}{dt} + \phi[S_i C_i - \frac{\partial S_w}{\partial p_c}]\frac{dp_i}{dt} +$$

$$\phi\frac{\partial S_w}{\partial p_c}\frac{dp_j}{dt} \quad (i,j = w,o;\ j \neq i) \tag{12}$$

where d(.)/dt is the material derivative with respect to the moving solid defined as:

$$\frac{d(\cdot)}{dt} = \frac{\partial(\cdot)}{\partial t} + \underset{\sim}{v}^s\cdot\underset{\sim}{\nabla}(\cdot) \tag{13}$$

and

$$\frac{d\bar{p}}{dt} = (S_w + p_c\frac{\partial S_w}{\partial p_c})\frac{dp_w}{dt} + (S_o - p_c\frac{\partial S_w}{\partial p_c})\frac{dp_o}{dt} \tag{14}$$

3 COMPUTING STRATEGY

3.1 *Finite Element Discretization*

The finite element discretization of the solid balance of linear momentum (Eq. 1) and fluid flow equations (Eqs 12) is expressed in terms of the nodal displacements $\bar{\underset{\sim}{u}}^s$, and nodal fluid pressures $\bar{p}_w, \bar{p}_o$ by using the Galerkin method. The unknowns are related to their nodal values by :

$$\underset{\sim}{u}^s = \underset{\sim}{N}\,\bar{\underset{\sim}{u}}^s \qquad p_i = \underset{\sim}{N}\,\bar{\underset{\sim}{p}}_i \qquad (i=o,w) \tag{15}$$

where $\underset{\sim}{N}$ is the shape function, and the discretization proceeds along standard lines (see e.g., Zienkiewicz and Taylor (1991)), to obtain the following discretized form of Eqs 1 and 12 as follows:

$$\underset{\sim}{n}_s(\underset{\sim}{u}^s) - S_w \underset{\sim}{L}\,\bar{\underset{\sim}{p}}_w - S_o \underset{\sim}{L}\,\bar{\underset{\sim}{p}}_o = \underset{\sim}{f}_s$$

$$\underset{\sim}{M}_w \dot{\bar{\underset{\sim}{p}}}_w + \underset{\sim}{C}\,\dot{\bar{\underset{\sim}{p}}}_o + S_w \underset{\sim}{L}^T \bar{\underset{\sim}{v}}^s + \underset{\sim}{n}_w(\underset{\sim}{p}_w, \underset{\sim}{p}_o, S_w) = \underset{\sim}{f}_w \tag{16}$$

$$\underset{\sim}{M}_o \dot{\bar{\underset{\sim}{p}}}_o + \underset{\sim}{C}\,\dot{\bar{\underset{\sim}{p}}}_w + S_o \underset{\sim}{L}^T \bar{\underset{\sim}{v}}^s + \underset{\sim}{n}_o(\underset{\sim}{p}_o, \underset{\sim}{p}_w, S_w) = \underset{\sim}{f}_o$$

where $\underset{\sim}{L}$ is a coupling matrix arising from the pressure gradient term in Eq. 1 ($\underset{\sim}{L}^T$ arises from the divergence velocity terms in Eqs 12) as:

$$L_a^{AB} = \int_{\Omega^e} N^A N^B_{,a}\, d\Omega \tag{17}$$

is the elemental contribution to node A from node B and direction a; Ω_e = domain of the finite element e. In Eqs 16, a superimposed dot is used to indicate the material derivative (Eq. 13); $\underset{\sim}{M}_w$, $\underset{\sim}{M}_o$ and $\underset{\sim}{C}$ are matrices arising from grouping $\dot{\bar{\underset{\sim}{p}}}_w$ and $\dot{\bar{\underset{\sim}{p}}}_o$ in Eqs. 16; $\underset{\sim}{n}_s$, $\underset{\sim}{n}_w$ and $\underset{\sim}{n}_o$ are nonlinear vector-valued functions of the field-state vectors (solid displacement and fluid pressures); and $\underset{\sim}{f}_s$, $\underset{\sim}{f}_w$ and $\underset{\sim}{f}_o$ are terms contributed from boundary conditions through the weak-form formulation.

The resulting semi-discrete equations in Eqs 16 are a system of coupled ordinary differential equations which are to be integrated in time. This is accomplished by applying a step-by-step integration procedure, and a direct time integration scheme is constructed to advance all field variables (solid displacement and fluid pressures) simultaneously in time. The system being first-order, a simple generalized trapezoidal finite difference time-stepping algorithm (see e.g., Zienkiewicz and Taylor (1991)) is used, viz.,

$$\bar{\underset{\sim}{y}}_{i,n+1} = \bar{\underset{\sim}{y}}_{i,n} + \dot{\bar{\underset{\sim}{y}}}_{i,n+\alpha}\,\Delta t$$

$$\dot{\bar{\underset{\sim}{y}}}_{i,n+\alpha} = \dot{\bar{\underset{\sim}{y}}}_{i,n}\cdot(1-\alpha_i) + \dot{\bar{\underset{\sim}{y}}}_{i,n+1}\cdot\alpha_i \tag{18}$$

where $\Delta t = t_{n+1} - t_n$; and $\bar{\underset{\sim}{y}}_{i,n}$, $\dot{\bar{\underset{\sim}{y}}}_{i,n}$ are the approximations for $\bar{\underset{\sim}{y}}_i(t)$ and $\dot{\bar{\underset{\sim}{y}}}_i(t)$ at time t_n, respectively (note: y=p for i=o,w and y=u^s for i=s). The scalar α_i (i = s,o,w) is an algorithmic parameter which can be chosen to ensure unconditional stability (viz., $0.5 \le \alpha_i \le 1$) and second-order accuracy (viz., α_i=0.5) in the linear case. Maximal high-frequency dissipation is provided by selecting α_i=1.

3.2 *Solution Procedure*

Eqs 16 together with Eqs 18 represent a fully coupled and highly nonlinear system which requires simultaneous solution. Since all the coefficients are dependent on the unknowns, iterations mst be performed within each time step to construct the solution. For that purpose, a fully implicit formulation is used and the resulting system of equations is solved by performing a linearization via a truncated Taylor's series expansion, and result in the following coupled linear system of equations to be solved at each iteration:

$$\begin{bmatrix} \alpha_s \Delta t \, \underset{\sim}{K}_s & -\alpha_w \Delta t \, \underset{\sim}{L} & -\alpha_o \Delta t \, \underset{\sim}{L} \\ \alpha_w \Delta t \, \underset{\sim}{L}^T & \underset{\sim}{M}_w + \alpha_w \Delta t \, \underset{\sim}{K}_w & \underset{\sim}{C} \\ \alpha_o \Delta t \, \underset{\sim}{L}^T & \underset{\sim}{C} & \underset{\sim}{M}_o + \alpha_o \Delta t \, \underset{\sim}{K}_o \end{bmatrix} X \begin{bmatrix} \Delta \underset{\sim}{\bar{v}}^{(i)}_{s,n+1} \\ \Delta \underset{\sim}{\dot{\bar{p}}}^{(i)}_{w,n+1} \\ \Delta \underset{\sim}{\dot{\bar{p}}}^{(i)}_{o,n+1} \end{bmatrix} = \begin{bmatrix} \underset{\sim}{r}^{(i)}_{s,n+1} \\ \underset{\sim}{r}^{(i)}_{w,n+1} \\ \underset{\sim}{r}^{(i)}_{o,n+1} \end{bmatrix} \quad (19)$$

where the superscript i is used to index the nonlinear iterations; $\underset{\sim}{K}_s$, $\underset{\sim}{K}_w$ and $\underset{\sim}{K}_o$ are solid, water and oil stiffness matrices arising from linearization of Eqs 16, respectively; $\underset{\sim}{r}_s$, $\underset{\sim}{r}_w$ and $\underset{\sim}{r}_o$ denote the residuals. All the matrices on the diagonals are (or can be made to be) symmetric but the overall matrix is not. In many other investigations, Eqs 19 were solved simultaneously using a direct solver (see e.g., Sukirman and Lewis (1993)). However, this method is not found suitable in large scale simulations because of the large resulting nonsymmetric matrix problem especially for three-dimensional cases. Further, such an implementation requires that a special software module be developed to combine the field equations. In our implementation, we use an unconditionally stable, robust and computational efficient partitioned procedure for the simultaneous integration of the transient coupled field equations as proposed in Prevost (1996). The procedure does not require that the full matrix system of coupled equations be assembled, and allows use of the existing single-field analysis software modules (viz., solid and reservoir models) to solve the coupled field problem. An iterative partitioned conjugate gradient algorithm is used to avoid having to form and assemble the Schur's complement matrix. Details are presented in Prevost (1996). The coupling matrices thus never need to be formed, resulting in substantial computational savings as illustrated hereafter.

4 APPLICATION

The proposed coupled model formulation and computational procedures, presented in the preceding sections, were developed and implemented into the DYNAFLOW finite-element code (Prevost, 1983). To show the capability of the model for simulating multiphase flow coupled with geomechanics and the computing efficiency of the iterative multistagger procedure for solving the coupled equations, a three-dimensional, five-spot waterflood problem is considered. The section of a quarter, five-spot for a hypothetical homogeneous and isotropic oil reservoir is modeled. The reservoir rock is treated as a linear-elastic material. The central, water injection well and one of the four production wells are placed at the lower, left corner and the upper, right corner of the quarter section, respectively. The block dimensions of the modeled section are 250 m by 250 m with a reservoir thickness of 5 m and the modeled section

Table 1: Reservoir data for the simulation problem studied

Porosity 0.3
Water viscosity 1.0 cp
Oil viscosity 4.0 cp
Water density 1.0 g/cm^3
Oil density 0.8 g/cm^3
Absolute permeability 2.0 Darcy
Young modulus 1.0x10^6 psi
Poisson ratio 0.25
Initial water saturation 0.2
Reservoir oil pressure 2828 psi

Table 2: Oil and water relative permeabilities and capillary pressure as a function of water saturation

Water Saturation	Relative permeability water	Relative permeability oil	Capillary pressure (psi)
0.200	0.00	0.60	1.45
0.225	0.01	0.53	1.20
0.250	0.02	0.47	1.03
0.300	0.04	0.38	0.84
0.350	0.07	0.31	0.71
0.400	0.09	0.25	0.61
0.450	0.13	0.18	0.55
0.500	0.17	0.13	0.48
0.550	0.22	0.09	0.44
0.600	0.28	0.05	0.40
0.650	0.35	0.02	0.36
0.700	0.45	0.00	0.34

Table 3: Simulation results from the iterative multistagger solution procedure

Day	Distance (m)	S_w	p_o (psi)	Vertical Strain ($x10^{-5}$)
10				
	17.7	0.700	3025.0	-1.470
	53.0	0.385	2971.0	-0.404
	88.4	0.227	2934.0	-0.048
	123.7	0.200	2912.0	0.038
	159.1	0.200	2898.0	0.035
	194.5	0.200	2887.0	0.036
	229.8	0.200	2879.0	0.035
	265.2	0.200	2872.0	0.047
	300.5	0.200	2863.0	0.031
	335.9	0.200	2850.0	0.354
30				
	17.7	0.700	3111.0	-1.120
	53.0	0.700	3070.0	-0.516
	88.4	0.515	3039.0	-0.297
	123.7	0.277	3015.0	-0.049
	159.1	0.206	2997.0	0.022
	194.5	0.200	2982.0	0.024
	229.8	0.200	2969.0	0.025
	265.2	0.200	2954.0	0.070
	300.5	0.200	2934.0	0.026
	335.9	0.200	2892.0	1.050
70				
	17.7	0.700	3192.0	-0.869
	53.0	0.700	3157.0	-0.280
	88.4	0.700	3132.0	-0.170
	123.7	0.656	3112.0	-0.174
	159.1	0.436	3093.0	-0.111
	194.5	0.248	3074.0	-0.013
	229.8	0.204	3055.0	0.020
	265.2	0.200	3034.0	0.092
	300.5	0.200	3001.0	0.022
	335.9	0.200	2933.0	1.700

Table 4: Simulation results from the simultaneous solution procedure

Day	Distance (m)	S_w	p_o (psi)	Vertical Strain ($x10^{-5}$)
10				
	17.7	0.700	3025.0	-1.480
	53.0	0.385	2971.0	-0.402
	88.4	0.227	2934.0	-0.047
	123.7	0.200	2912.0	0.038
	159.1	0.200	2898.0	0.036
	194.5	0.200	2887.0	0.036
	229.8	0.200	2879.0	0.035
	265.2	0.200	2872.0	0.047
	300.5	0.200	2863.0	0.031
	335.9	0.200	2850.0	0.354
30				
	17.7	0.700	3111.0	-1.120
	53.0	0.700	3070.0	-0.510
	88.4	0.515	3039.0	-0.297
	123.7	0.277	3015.0	-0.053
	159.1	0.206	2997.0	0.022
	194.5	0.200	2982.0	0.024
	229.8	0.200	2969.0	0.025
	265.2	0.200	2954.0	0.070
	300.5	0.200	2934.0	0.026
	335.9	0.200	2892.0	1.050
70				
	17.7	0.700	3192.0	-0.869
	53.0	0.700	3157.0	-0.280
	88.4	0.700	3132.0	-0.166
	123.7	0.658	3112.0	-0.176
	159.1	0.438	3093.0	-0.113
	194.5	0.248	3074.0	-0.012
	229.8	0.204	3055.0	0.021
	265.2	0.200	3034.0	0.092
	300.5	0.200	3001.0	0.022
	335.9	0.200	2933.0	1.700

is uniformly divided into a 10 by 10 by 5 finite-element mesh. The simulated waterflood consists of 400 m^3/day water injection at the injection well (i.e., 100 m^3/day water injection for each quarter section) for 70 days and the water flow was assumed to be uniformly distributed along the 5-m layer. At the production well, the boundary condition of constant oil pressure was assumed to hold during the 70-day period. At day 70 the injected water volume was less than 8% of the total reservoir pore volume. Reservoir data, water and oil relative permeabilities and capillary pressure, used for the simulation problem, are shown in Tables 1 and 2, respectively.

In Tables 3 and 4 simulation results for the waterflood problem investigated from the multistagger and conventional simultaneous procedures are presented, respectively. Calculated values of water saturation S_w, oil pressure p_o, and vertical strain, at the mid plane of the reservoir layer, are presented as functions of injection time and diagonal distance from the injection well to the production well. Tables 3 and 4 clearly show that the numerical results from both procedures were about the same. However, the multistagger procedure is more computationally efficient than the conventional procedure. The proposed multistagger method was very stable and gave solutions much faster than the conventional method. For a 70 time-step simulation, it took 592 CPU seconds using the multistagger method and 7179 CPU seconds using the conventional method, run on a IBM/R6000 Model 590 workstation. This result shows that the proposed multistagger method is about 12 times faster than the conventional method.

5 CONCLUSIONS

In this study, an effort has been made to develop a computer method for the analysis of coupled geomechanics and multiphase-flow problems that are complicated and computationally intensive. Coupled field equations for isothermal two-phase fluid flow in a deformable porous system and the corresponding finite element equations are developed. The resulting system of coupled equations was numerically solved by both a multistagger iterative method and the conventional simultaneous method. Based on the cases investigated, the numerical results show that the proposed multistagger procedure is robust, accurate, and efficient for analyzing coupled field problems and is more economical in the computational cost compared to the conventional simultaneous procedure. The key features of the proposed multistagger method should be helpful in attempts to analyze large-scale, three-dimensional field problems that geomechanics and multiphase flow are coupled.

ACKNOWLEDGMENTS

The authors acknowledge with thanks permission to publish this paper from Phillips Petroleum Company. The opinions expressed in the paper are those of the authors and do not necessarily represent those of Phillips Petroleum Company.

REFERENCES

Aziz, K. and Settari, A. (1979). Petroleum Reservoir Simulation. Elsevier Applied Science Publishers, London and New York.

Biot, M.A. (1941). General Theory of Three-Dimensional Consolidation. J. Appl. Phys., **12**, 155-164.

Biot, M.A. (1955). Theory of Elasticity and Consolidation for a Porous Anisotropic Solid. J. Appl. Phys., **26**, 182-185.

Biot, M.A. (1956). Theory of Deformation of a Porous Viscoelastic Anisotropic Solid. J. Appl. Phys., **27**, 452-469.

Ertekin, T. and Farouq Ali, S.M. (1979). Numerical Simulation of the Compaction-Subsidence Phenomena in a Reservoir for Two-Phase Non-isothermal Flow Conditions. Proc. 3rd Int. Conf. Numerical Methods in Geomechanics, Aachen.

Gutierrez, M. and Hansteen, H. (1994). Fully Coupled Analysis of Reservoir Compaction and Subsidence. SPE paper 28900, Proc. European Petroleum Conference, Oct. 25-27, London, 339-347.

Lewis, R.W. and Schrefler, B.A. (1987). The Finite Element Method in the Deformation and Consolidation of Porous Media. John Wiley & Sons, New York.

Lewis, R.W. and Sukirman, Y. (1993). Finite Element Modelling of Three-Phase Flow in Deforming Saturated Oil Reservoirs. Int. J. Num. Anal. Meth. Geom., **17**, 577-598.

Lewis, R.W. and Sukirman, Y. (1993). Finite Element Modelling for Simulating the Surface Subsidence above a Compacting Hydrocarbon Reservoir. Int. J. Num. Anal. Meth. Eng., **18**, 619-639.

Norman, R.M. (1990). Wettability and its Effects on Oil Recovery. J. Pet. Tech., 1476-1484.

Prevost, J.H. (1983). DYNAFLOW manual. Princeton University, Princeton, New Jersey.

Prevost, J.H. (1996), Partitioned Solution Procedure for Simultaneous Integration of Coupled-Field Problems. to appear in Comm. Num. Meth. Eng.

Settari, A. and Mourits, F.M. (1994). Coupling of Geomechanics and Reservoir Simulation Models. Proc. 8th Int. Conf. on Computer Methods and Advances in Geomechanics, Morgantown, West Virginia, May 22-28, 1994, 2151-2158.

Sukirman, Y. and Lewis, R.W. (1993). A Finite Element Solution of a Fully Coupled Implicit Formulation for Reservoir Simulation. Int. J. Num. Anal. Meth. Geom., **17**, 677-698.

Terzaghi, U. (1943). Theoretical Soil Mechanics. Wiley, New York.

Tortike, W.S. and Farouq Ali, S.M. (1979). Modelling Thermal Three-Dimensional Three-phase Flow in a Deforming Soil. Proc. 3rd Int. Conf. Numerical Methods in Geomechanics, Aachen.

Tortike, W.S. and Farouq Ali, S.M. (1993). Reservoir Simulation Integrated with Geomechanics. J. of Canadian Petroleum Technology, Vol. 32, No. 5, 28-37.

Verruijt, A. (1969). Elastic Storage in Aquifers. Flow Through Porous Media, R.J.M. De Wiest (ed.), Academic Press, 331-376.

Zienkiewicz, O.C. and Taylor, R.L. (1991). The Finite Element Method. Vol. 1 & 2, McGraw-Hill, London.

Computer Methods and Advances in Geomechanics, Yuan (ed.) © *1997 Balkema, Rotterdam, ISBN 90 5410 904 1*

Uncoupled method of thermo-hydro-mechanical problem in porous media: Accuracy and stability analysis

Zhang Wenfei & Li Xiaojiang
Daqing Petroleum Institute, Qinhuangdao, China

ABSTRACT: An uncoupled numerical method to solve thermo-hydro-mechanical problems is presented. In this method, the linear equations relative to displacement u, pressure P and temperature T can be solved separately while in the conventional coupled method u, P and T are calculated simultaneously. The uncoupled method can reach the same accuracy as the coupled method by using less calculation work. The theoretical formula of the uncoupled method and its stability analysis are given in detail in this paper. Numerical examples verify and demonstrate the uncoupled method.
KEY WORDS: thermo-hydro-mechanics, coupled, uncoupled, numerical method, iteration, stability

1. INTRODUCTION

Since Biot (1941) systematically presented the coupled theory of hydro-mechanics, the numerical method in this field has been paid much attention (Zienkiewicz & Shiomi, 1984, Skoczylas & Shahrour, 1992). In recent years more and more applied problems have been discussed (Carter & Book,1994, Selvadurai & Yue,1994, Reddish et al,1994, Coussy,1991). This problem includes three different variables, displacement u, pressure P and temperature T. Generally we can solve the coupled problem using the finite element method. For some engineering problems the uncoupled method is advantageous. Even though we call this method uncoupled, the physical equations are still coupled. We can only use an uncoupled numerical method to simplify the calculation procedure. In some applications, a physical uncoupling is possible, especially, displacement and pressure variation usually have minor effect on the temperature; the temperature equation can thus be solved independently of the other two equations.

There are many possible approaches to uncouple the thermo-hydro-mechanical problem. Here we consider the most complete, that is, to solve each of the three equations independently. Thus we obtain superior implementation flexibility and the best economic effect. It is evident that this method can be extended to any possible combination of the three equations or to any two equations. We discuss such applications in this paper. Only the elastic model is considered. The accuracy, stability and effectiveness are analyzed in detail. Numerical examples of coupled and uncoupled methods are compared.

2. BASIC CONSTITUTIVE EQUATIONS

Here we introduce the physical equations directly.
Mechanical equation:

$$d\sigma = E d\varepsilon - b I dP - \alpha_b K_b I dT \tag{1}$$

$$\mathrm{div}\sigma + f_b = 0 \tag{2}$$

Hydrodiffusion equation:

$$\frac{1}{\eta}\frac{\partial P}{\partial t} + b\frac{\partial \varepsilon_{kk}}{\partial t} - \frac{L\rho_o}{\eta T_o}\frac{\partial T}{\partial t} = \frac{k}{\mu}\nabla^2 P \tag{3}$$

Thermodiffusion equation:

$$(c_k\Gamma_o + \frac{\rho_o L^2}{\eta T_o})\frac{\partial T}{\partial t} - \frac{\rho_o L}{\eta}\frac{\partial P}{\partial t} + \alpha_b K_b T_o\frac{\partial \varepsilon_{kk}}{\partial t}$$

$$= \lambda\nabla^2 T + \frac{T_o k\rho_o}{\mu}\mathrm{grad}P(\frac{c_p}{T_o}\mathrm{grad}T - \frac{\alpha_f}{\rho_o}\mathrm{grad}P) \tag{4}$$

Boundary conditions:
For the mechanical problem

$$u = u_b \quad \text{on} \quad S_u, \qquad f = f_t \quad \text{on} \quad S_f \tag{5}$$

For the hydraulic problem

$$P = P_b \quad \text{on} \quad S_P, \qquad \frac{k}{\mu}\frac{\partial P}{\partial n} = Q \quad \text{on} \quad S_Q \tag{6}$$

For the temperature problem

$$T = T_b \quad \text{on} \quad S_T, \qquad \lambda\frac{\partial T}{\partial n} = q \quad \text{on} \quad S_q \tag{7}$$

3. DISCRETIZATION OF SPACE AND TIME

We assume

$$u = N\bar{u} \qquad \varepsilon = B\bar{u}$$
$$P = N'\bar{P} \qquad \text{grad}P = B'\bar{P}$$
$$T = N'\bar{T} \qquad \text{grad}T = B'\bar{T} \tag{8}$$

Where N and N' mean that we can use different shape functions to approximate u, P and T. B and B' are matrix differential operators (Zienkiewicz & Shiomi,1984, Jaeger & Cook,1969). For the time variable we assume

$$\bar{P} = \theta\bar{P}_{t+\Delta t} + (1-\theta)\bar{P}_t$$
$$\frac{\partial\bar{P}}{\partial t} = (\bar{P}_{t+\Delta t} - \bar{P}_t)/\Delta t_t \tag{9}$$

where $0 \le \theta \le 1$. After some work we can get

$$R\bar{U}_{t+\Delta t} = S\bar{U}_t + \bar{D}_t \tag{10}$$

Where

$$R = \begin{pmatrix} R_U & R_{UP} & R_{UT} \\ R_{PU} & R_{MP} + \theta\Delta t R_P & R_{PT} \\ R_{TU} & R_{TP} + \theta\Delta t R_{CT} & R_{MT} + \theta\Delta t R_T \end{pmatrix}$$

$$\bar{D}_t = \begin{Bmatrix} d\bar{f} \\ \Delta t(\theta\bar{Q}_{t+\Delta t} + (1-\theta)\bar{Q}_t \\ \Delta t(\theta\bar{q}_{t+\Delta t} + (1-\theta)\bar{q}_t \end{Bmatrix}$$

$$\bar{U}_{t+\Delta t} = \{\bar{u}_{t+\Delta t} \quad \bar{P}_{t+\Delta t} \quad \bar{T}_{t+\Delta t}\}^T$$

$$\bar{U}_t = \{\bar{u}_t \quad \bar{P}_t \quad \bar{T}_t\}^T$$

$$S = \begin{pmatrix} R_U & R_{UP} & R_{UT} \\ R_{PU} & R_{MP} - (1-\theta)\Delta t R_P & R_{PT} \\ R_{TU} & R_{TP} - (1-\theta)\Delta t R_{CT} & R_{MT} - (1-\theta)\Delta t R_T \end{pmatrix}$$

$$R_U = -\int_\Omega B^T E B d\Omega$$

$$R_{UP} = R_{PU}^T = \int_\Omega B^T b I N' d\Omega$$

$$R_{UT} = R_{TU}^T = \int_\Omega B^T \alpha_b K_b I N' d\Omega$$

$$R_P = \int_\Omega B'^T \frac{k}{\mu} B' d\Omega$$

$$R_{MP} = \int_\Omega N'^T \frac{1}{\eta} N' d\Omega$$

$$R_{PT} = R_{TP} = \int_\Omega N'^T (\frac{-L\rho_o}{\eta T_o}) N' d\Omega$$

$$R_T = \int_\Omega B'^T \frac{\lambda}{T_o} B' d\Omega$$

$$R_{MT} = \int_\Omega N'^T (\frac{c_k\Gamma_o}{T_o} + \frac{\rho_o^2 L^2}{\eta T_o^2}) N' d\Omega$$

$$R_{CT} = \int_\Omega N'^T (\frac{\alpha_f k}{\mu}\text{grad}P - \frac{k\rho_o c_p}{\mu T_o}\text{grad}T) B' d\Omega$$

$$d\bar{Q} = \int_{S_Q} N'^T dQ$$

$$d\bar{f} = -\int_\Omega N^T df_b d\Omega - \int_{S_F} N^T df_t dS$$

$$d\bar{q} = \int_{S_q} N'^T \frac{1}{T_o} dq \tag{11}$$

where superscript T means transposition. From (11) we see that because of the term of convection, associated with R_{CT}, the matrix R is not symmetric. Further detail derivation of the above equations are given by (Zienkiewicz & Shiomi,1984, Coussy,1991, Jaeger & Cook,1969).

4. UNCOUPLED METHOD

To solve equation (10) more efficiently we use the iterative formula as

$$\begin{pmatrix} R_U & & \\ & R_{MP} + \theta\Delta t R_P & \\ & & R_{MT} + \theta\Delta t R_T \end{pmatrix} \begin{Bmatrix} \bar{u}_{t+\Delta t} \\ \bar{P}_{t+\Delta t} \\ \bar{T}_{t+\Delta t} \end{Bmatrix} =$$
$$\begin{pmatrix} 0 & -R_{UP} & -R_{UT} \\ R_{PU} & 0 & -R_{PT} \\ -R_{TU} & -R_{PT} - \theta\Delta t R_{CT} & 0 \end{pmatrix} \begin{Bmatrix} \bar{u}_{t+\Delta t} \\ \bar{P}_{t+\Delta t} \\ \bar{T}_{t+\Delta t} \end{Bmatrix} + S\bar{U}_t + \bar{D}_t \tag{12}$$

In the left side of equation (12) the stiffness matrix is already a diagonal block matrix so we can solve $\bar{u}_{t+\Delta t}$, $\bar{P}_{t+\Delta t}$ or $\bar{T}_{t+\Delta t}$ separately in each time step. Because we do not know $\{\bar{u}_{t+\Delta t} \quad \bar{P}_{t+\Delta t} \quad \bar{T}_{t+\Delta t}\}^T$ in the right side of the equality during the calculation we express it as $\{\bar{u}^*_{t+\Delta t} \quad \bar{P}^*_{t+\Delta t} \quad \bar{T}^*_{t+\Delta t}\}^T$. Generally, we have two ways to approximate $\{\bar{u}^*_{t+\Delta t} \quad \bar{P}^*_{t+\Delta t} \quad \bar{T}^*_{t+\Delta t}\}^T$. The first way is to use

$$\{\bar{u}^*_{t+\Delta t} \quad \bar{P}^*_{t+\Delta t} \quad \bar{T}^*_{t+\Delta t}\}^T = \sum_{i=1}^{n}\alpha_i\{\bar{u}_t \quad \bar{P}_t \quad \bar{T}_t\}^T \tag{13}$$

where α_i is a weight factor. This is the well-known and universal extrapolation method. The second way is the iterative method. That is to calculate (12) iteratively using

$$\begin{pmatrix} R_U & & \\ & R_{MP}+\theta\Delta t R_P & \\ & & R_{MT}+\theta\Delta t R_T \end{pmatrix} \begin{Bmatrix} \bar{u}^K_{t+\Delta t} \\ \bar{P}^K_{t+\Delta t} \\ \bar{T}^K_{t+\Delta t} \end{Bmatrix} =$$

$$-\begin{pmatrix} 0 & R_{UP} & R_{UT} \\ R_{PU} & 0 & R_{PT} \\ R_{TU} & R_{PT}+\theta\Delta t R_{CT} & 0 \end{pmatrix} \begin{Bmatrix} \bar{u}^{K-1}_{t+\Delta t} \\ \bar{P}^{K-1}_{t+\Delta t} \\ \bar{T}^{K-1}_{t+\Delta t} \end{Bmatrix} + S\bar{U}_t + \bar{D}_t \tag{14}$$

The convergence condition is

$$\left\| \begin{pmatrix} \bar{u}^K_{t+\Delta t} & \bar{P}^K_{t+\Delta t} & \bar{T}^K_{t+\Delta t} \end{pmatrix}^T - \begin{pmatrix} \bar{u}^{K-1}_{t+\Delta t} & \bar{P}^{K-1}_{t+\Delta t} & \bar{T}^{K-1}_{t+\Delta t} \end{pmatrix}^T \right\| \le \varepsilon_1 \tag{15}$$

Where K is the number of iterations and ε_1 is a given small positive number. Generally, in each time step we first solve the equations of T and then the equations of u and P until condition (15) is satisfied. In the prediction method (13), it is very difficult to choose suitable α_i. The second way is always effective, but we have to do more calculation work. However, for the elastic problem, few iterations are sufficient for convergence, so the iterative method will be used here.

5. STABILITY ANALYSIS

In our problem, there are two different conceptions of stability. One is the stability of equation (10). That is the stability of discretization of the time variable. The sufficient condition of stability of (10) is

$$\rho\left(R^{-1}S\right) < 1 \tag{16}$$

Where $\rho\left(R^{-1}S\right)$ is the largest absolute value of the eigenvalues of $R^{-1}S$. Many authors have studied this problem and there have been many theoretical conclusions so we do not discuss it here. Another is the stability of equation (12). If we want to obtain the same result by using the uncoupled form (12) and the coupled form (10) the stability of (12) is necessary. The sufficient condition of the stability of (12) is

$$\rho\left(\begin{pmatrix} R_U & & \\ & R_{MP}+\theta\Delta t R_P & \\ & & R_{MT}+\theta\Delta t R_T \end{pmatrix}^{-1} \begin{pmatrix} 0 & -R_{UP} & -R_{UT} \\ -R_{PU} & 0 & -R_{PT} \\ -R_{TU} & -R_{PT}-\theta\Delta t R_{CT} & 0 \end{pmatrix}\right) < 1 \tag{17}$$

When condition (17) is satisfied the iterative formula (12) is usable. We calculated (17) numerically and analyzed the influence of some parameters to ρ. The numerical results show that $\|\rho\| < 1$ is always held except for some values of no practical use.

6. UNCOUPLED METHOD APPLIED TO OTHER COUPLED PROBLEMS

We have discussed the uncoupled algorithm of the thermo-hydro-mechanical problem. In fact this algorithm can be easily used to solve other kinds of coupled problems, for instance the hydro-mechanical problem, the thermo-mechanical problem and the thermo-hydraulic problem. As an example, the uncoupled equations of the hydro-mechanical problem are:

$$\begin{pmatrix} R_U & \\ & R_{MP}+\theta\Delta t R_P \end{pmatrix} \begin{Bmatrix} \bar{u}^K_{t+\Delta t} \\ \bar{P}^K_{t+\Delta t} \end{Bmatrix} = -\begin{pmatrix} 0 & R_{UP} \\ R_{PU} & 0 \end{pmatrix} \begin{Bmatrix} \bar{u}^{K-1}_{t+\Delta t} \\ \bar{P}^{K-1}_{t+\Delta t} \end{Bmatrix} +$$

$$\begin{pmatrix} R_U & R_{UP} \\ R_{PU} & R_{MP}-(1-\theta)\Delta t R_P \end{pmatrix} \begin{Bmatrix} \bar{u}_t \\ \bar{P}_t \end{Bmatrix} + \begin{Bmatrix} d\bar{f} \\ \Delta t(\theta\bar{Q}_{t+\Delta t}+(1-\theta)\bar{Q}_t) \end{Bmatrix} \tag{18}$$

In the same way we can obtain the equations of other coupled problems. The stability analysis of these uncoupled algorithms can be stated by using the same method as given in (17).

7. NUMERRICAL EXAMPLES AND COMPARISION

In this part we shall give out some numerical examples of the above algorithm and compare the numerical results of coupled and uncoupled methods. As an example we consider the column model shown in fig. 1. We use the following values of the parameters: Young's modulus = 10000 MPa, Poisson's ratio = 0.2, ρ_o = 1000 kg/m^3, Γ_o = 2000 kg/m^3, η = 4000 MPa, μ = 10^{-6} MPa.s, k = 10^{-16} m^2. α_b = 9X10^{-7}/^{o}C, α_f = 3X10^{-6}/^{o}C, b = 1, L=1.71X10^{4}cal/kg, c_p=1000 cal/kg/^{o}C, λ = 0.2 w/m/^{o}C, C_k = 40 J/m^3 ^{o}C. In all calculations we use 8 node isoparametric elements and the 4 point Gaussian integration scheme. We have also

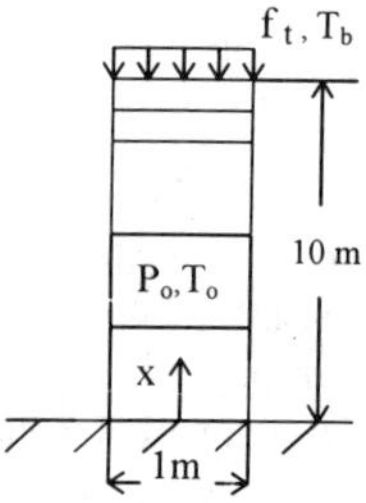

fig.1 column model of thermo-hydro-mechanical problem

calculated some examples with elements of different shape functions for u, P and T, respectively. That is for displacement u we use 8 node isoparametric elements, while for P and T we use 4 nodes linear elements. The numerical results tell us that this kind of combination is reasonable and very good accuracy can be obtained.

7.1 Numerical example of thermo-mechanical problem

For the thermo-mechanical problem the uncoupled equation (12) is simplified to

$$\begin{pmatrix} R_U & \\ & R_{MT}+\theta\Delta t R_T \end{pmatrix}\begin{Bmatrix} \bar{u}^K_{t+\Delta t} \\ \bar{T}^K_{t+\Delta t} \end{Bmatrix} = -\begin{pmatrix} & R_{UT} \\ R_{TU} & \end{pmatrix}\begin{Bmatrix} \bar{u}^{K-1}_{t+\Delta t} \\ \bar{T}^{K-1}_{t+\Delta t} \end{Bmatrix} +$$

$$\begin{pmatrix} R_U & R_{UT} \\ R_{TU} & R_{MT}-(1-\theta)\Delta t R_T \end{pmatrix}\begin{Bmatrix} \bar{u}_t \\ \bar{T}_t \end{Bmatrix} + \begin{Bmatrix} d\bar{f}_t \\ \Delta t(\theta \bar{q}_{t+\Delta t}+(1-\theta)\bar{q}_t) \end{Bmatrix} \tag{19}$$

To verify the program we compare to some analytic solutions. We take $\Delta T=50\ ^oC$ and $f_t=0$. Comparisons of the calculated results using the uncoupled method and the theoretical result are given in fig. 2.

7.2 Numerical example of hydro-mechanical problem

For the example of hydro-mechanical problem, we take Po=10 MPa, and f_t=10 MPa. Because the relative influence between pressure P and displacement u is very important, several iterations are necessary to reach the satisfactory accuracy. The numerical results are shown in fig. 3. After two iterations the results of the coupled and the uncoupled methods converge to the same line.

7.3 Numerical example of thermo-hydro-mechanical problem

In the numerical example we use Po=10 MPa, $\Delta T=50\ ^oC$ and f_t =10 MPa. The result obtained by using the coupled method and the uncoupled method are compared in fig. 4 to fig. 5. Usually only one iteration is sufficient to get very accurate solution for temperature T. For pressure P and displacement u, 3 or 4 iterations are required.

7.4 Numerical result of stability analysis

To study the stability of the algorithm presented here we calculated (17) numerically. The parameters we used are the same as the ones given in the above examples. The result is positive, that is $\rho<1$. But using only one particular group of parameters is not sufficient to verify the stability of the uncoupled algorithm and it is very important to know whether the algorithm is still stable when we use different parameters. For this purpose we studied the influence of each parameter on the stability of the algorithm respectively. We give the numerical results in fig. 6 to fig. 11. From fig. 6 and fig. 11 we see that for all the parameters in the given range ρ is always less than one. So we can conclude that the uncoupled method stable. In fact, only when Young's modulus E is less than 4000 MPa, ρ is great than one, or when thermal expansion coefficient of solid α_b is greater than $5X10^{-4}$, ρ is great than one. Fortunately, in most practical problems these two parameters are

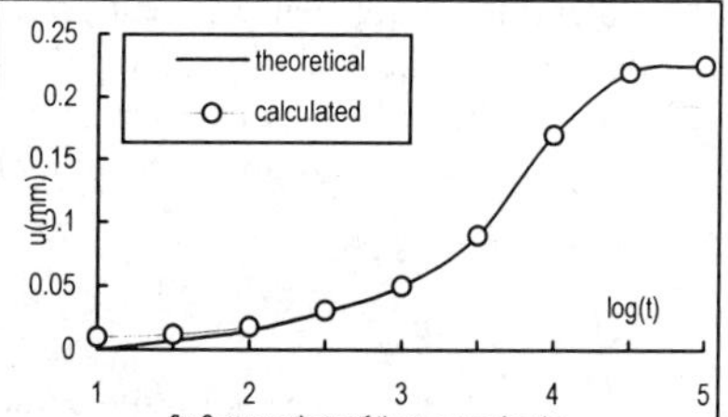

fig.2 comparison of thermo-mechanics

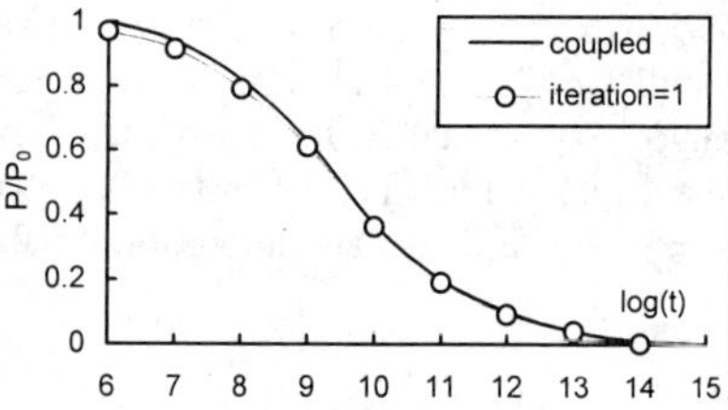

fig.3 comparison of hydro-mechanics, Z=0.85H

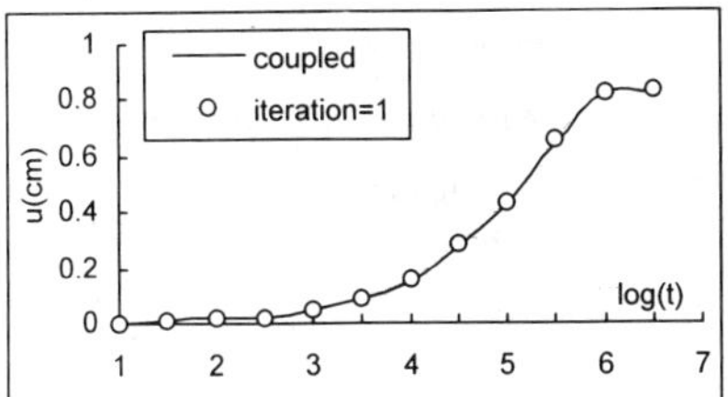

fig.4 comparison of thermo-hydro-mechanics, Z=H

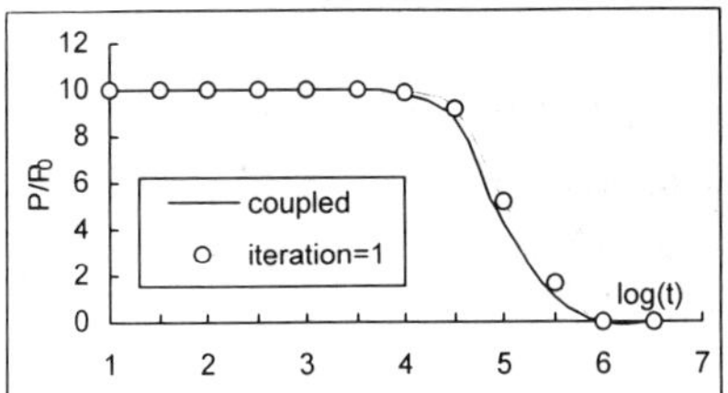

fig.5 comparison of thermo-hydro-mechanics, Z=0

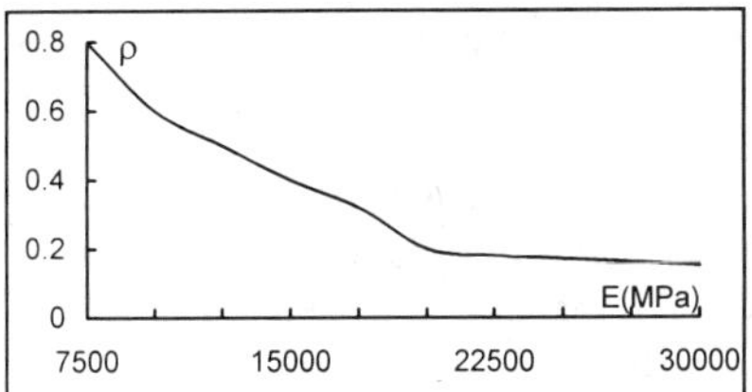

fig.6 influence of Young'S modulus E to ρ

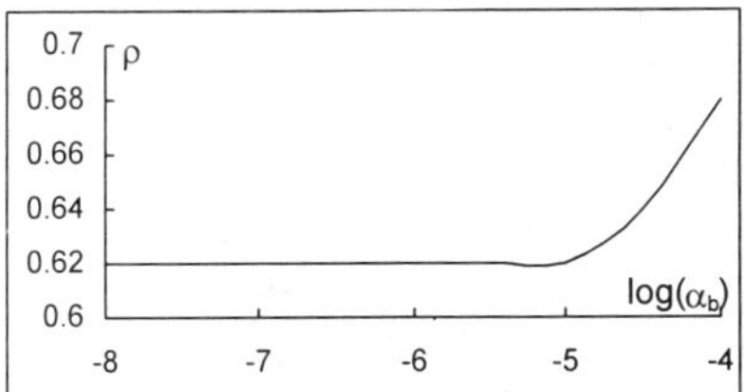

fig.7 influence of α_b to ρ

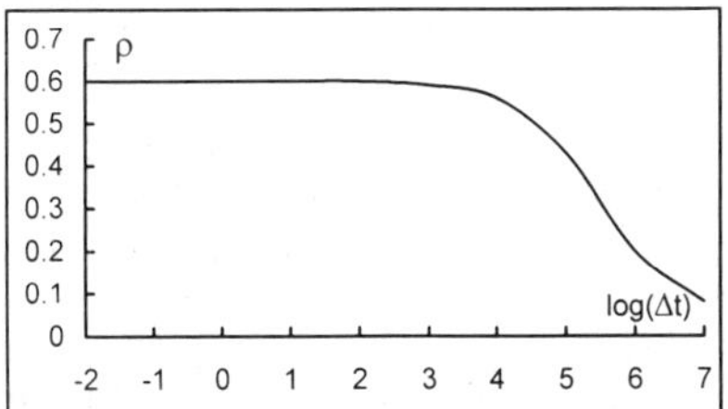

fig.8 influence of Δt to ρ

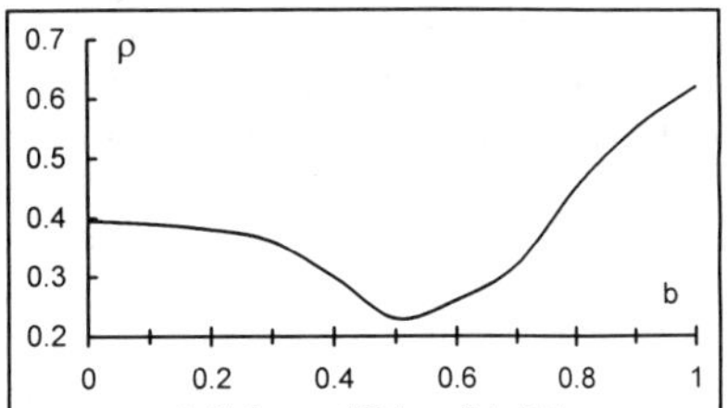

fig.9 influence of Biot's coefficient b to ρ

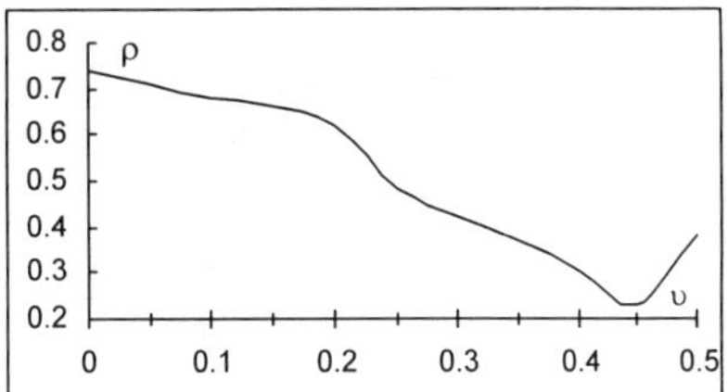

fig.10 influence of Poisson's ratio υ to ρ

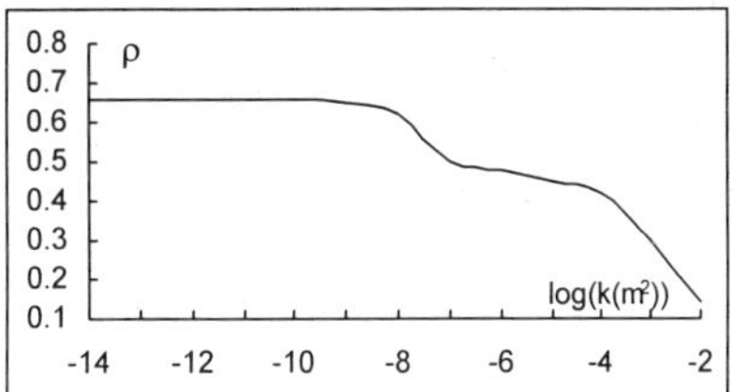

fig.11 influence of permeability k to ρ

always in the stable range. Although the other parameters have some influence on the largest eigenvalue ρ however they do not influence the stability of the algorithm. The results of parameters which have less influence on ρ and which do not influence the stability of the algorithm, for example, the size of space grids, are not given here.

8. CONCLUSION AND SOME REMARKS

In this paper we have developed a computational procedure for solving the coupled thermo-hydro-mechanical problem, analyzed stability, accuracy and effectiveness of the uncoupled algorithm, and calculated many examples. For the approach presented here we conclude that: 1) The uncoupled method is correct. It is a stable method. 2) The uncoupled method is efficient, especially when the number of nodes is large. 3) The uncoupled method is versatile. There are 7 different combinations for the system of u, P and T, and we can solve all of them using only one program. When there is coupled convection the general stiffness matrix is nonlinear. With the uncoupled algorithm we only need to change the matrix associated with convection, hence we can use less calculation and reach better accuracy than with the coupled method, which only uses the linear stiffness matrix. 5) Generally, when we change the value of some physical parameters in the equations, the speed of convergence is also changed a little, but the calculation is still convergent.

In this study, we assumed elasticity and a single phase fluid model. We are sure that the uncoupled method can also be used to solve the multiphase fluid problem. And that will be even more effective than for a single phase fluid. Although the ideas of the uncoupled method can be extended to the plastic model, we can not simply conclude that the uncoupled method for the plastic problem is usable. For the plastic problem the uncoupled procedure may need a little modification and the stability study will be more complicated.

LIST OF SYMBOLS

b coefficient of Biot
c_k specific heat of solid in isochoric state
c_p specific heat of fluid
E matrix of elasticity
f vector function of external force
$\bar{f}$ discrete vector of total external force
f_b vector function of body force
f_t vector function of applied surface force
I unit operator
k permeability
K_b compression modulus of the solid
L latent heat of unit mass of fluid
N matrix of interpolation function for displacement field u
P pressure
N' matrix of interpolation function for P and T
$\bar{P}$ discrete vector of pressure
P_o initial pressure
P_b given boundary pressure
q density of heat flow
Q density of injected fluid
S given boundary
t time variable
T temperature
$\bar{T}$ discrete vector of temperature
T_o initial temperature
T_b given boundary temperature
Δt increment of time
ΔT increment of temperature
u vector function of displacement
$\bar{u}$ discrete vector of displacement
u_b vector function of boundary displacement
α_b thermal expansion coefficient of solid
α_f fluid thermal expansion coefficient
θ calculative coefficient
σ vector function of stress
ε vector function of strain
ε_{kk} volumetric strain
υ Poisson's ratio
μ dynamic viscosity of the fluid
η modulus of Biot
λ thermo conductivity of the fluid
ρ_o fluid density
Γo solid density
Ω considered field

REFERENCES

Biot M.A., *General theory of three-dimensional consolidation*, J. Appl. Phys., 12, 155-164, 1941.

Carter J.P. and Book J.R., *Analysis of fully coupled thermo-mechanical behavior around a rigid cylindrical heat source buried in clay*, Int. J. Numer. and Analy. Meth. Geomech., Vol. 18, 177-203, 1994.

Coussy O., *Mécanique des milieux poreux*, Editions Technip, 1991.

Jaeger J.C. and Cook N.G.W., *Fundamentals of Rock Mechanics*, Chapman and hall Ltd. and Science Paperbacks,1969.

Lewis R.W. and Schrefler B.A., *The finite element method in the deformation and consolidation of porous media*, John Wiley & Sons, 1987.

Reddish D.J., Yao X.L. and Waller M.D., *Computerised prediction of subsidence over oil and gas fields*, Eurok'94, 1994 Balkema, Rotterdam, ISBN 90 5410 502 X.

Selvadurai A.P.S. and Yue Z.Q., *On the indentation of poroelastic layer*, Int. J. for Numer. and Analy. Meth. Geomech., Vol. 18, 161-175, 1994.

Skoczylas F., Shahrour I. and Henry J.P., *Numerical analysis of a thermo-hydro-mechanical coupling in rocks*, Revue de L'institut Français du Pétrole, Vo. 47, No. 1, 1992.

Zienkiewicz O.C. and Shiomi T., *Dynamic behavior of saturated porous media; The generalized Biot formulation and its numerical solution*, Int. J. Numer. & Analy. Meth. Geomech., Vol. 8, 71-96, 1984.

Computer Methods and Advances in Geomechanics, Yuan (ed.) © 1997 Balkema, Rotterdam, ISBN 90 5410 904 1

Numerical model of pressure-based coupled heat and moisture transfer through unsaturated soils using the integrated finite difference method

D.M.Li & A.M.Crawford
Department of Civil Engineering, University of Toronto, Ont., Canada
K.C.Lau
G.E.O., Hong Kong, China

ABSTRACT: Designing a nuclear waste repository requires the ability to model the performance of the buffer barriers with respect to heat generated by the waste. Previously, a moisture-based computer program, TRUCHAM, has been developed to evaluate the performance of the buffer barrier system. For a more realistic evaluation of the buffer performance, a pressure-based model was required. This model has now been developed and incorporated into the computer program TRUCHAMP. The Integrated Finite Difference Method is used to solve for the governing equations of the pressure-based model. The IFD formulation and the results of analyses using TRUCHAMP are presented in this paper.

1 INTRODUCTION

This paper describes numerical modeling for the analysis of the near-field thermal and hydraulic performance of an underground geological disposal vault in which the nuclear waste containers are emplaced and surrounded by a clay-based engineered barrier. Considering a scenario that the nuclear waste emplacement will take place over several decades, the emplaced buffer in the disposal room will undergo an initial phase of thermal drying. The drying phase may affect the integrity of the engineered barrier because of the potential for shrinkage cracks and a decrease in its thermal conductivity which results in increased container temperature. The initial phase of thermal drying will be followed by a wetting phase as the heat generation of the nuclear waste canisters decays and ground water re-establishes itself.

With the existence of temperature gradients in a soil medium, the moisture in the medium will redistribute itself to new equilibrium values to conform with the applied temperature gradient. Subsequently, this new moisture content distribution will affect the temperature distribution in the soil medium. This process indicates the coupling effect between the temperature and moisture distribution in a porous medium. The governing equations for the transient coupled heat and mass transfer are derived from the moisture-based model of Philips and deVries (1957) and later converted to a potential-based model by Thomas and King (1991).

A pressure-based model is more desirable than a moisture-based model in the study of the performance of the buffer barrier. A pressure-based model allows for a more realistic analysis of the thermal and hydraulic regimes since continuity of pressure head and temperature are maintained for different materials in contact. Furthermore, the pressure-based model solves for the pressure head that enables the calculation of the stress-strain behaviour of the medium to give a complete thermal-hydraulic-mechanical analysis of the buffer barrier.

To numerically solve the governing equations of the potential-based model, an Integrated Finite Difference Method (IFDM) is utilized. The technique accommodates complex geometries, wide variation of nonlinear material properties, and Neuman and Dirichlet boundary conditions. Effects of heat and moisture diffusion on the thermal and hydraulic performance of clay-based engineered barrier can be studied by means of this numerical model.

2 INTEGRATED FINITE DIFFERENCE FORMULATION

The governing equations of the pressure-based coupled heat and moisture transfer in unsaturated media

(Thomas and King, 1991) can be expressed in a multi-dimensional form as:

$$C_{\psi\psi}\frac{\partial\psi}{\partial t}+C_{\psi T}\frac{\partial T}{\partial t}=\nabla\left(K_{\psi\psi}\nabla\psi\right)+\nabla\left(K_{\psi T}\nabla T\right)+\frac{\partial k}{\partial z} \quad (1)$$

$$C_{T\psi}\frac{\partial\psi}{\partial t}+C_{TT}\frac{\partial T}{\partial t}=\nabla\left(K_{T\psi}\nabla\psi\right)+\nabla\left(K_{TT}\nabla T\right) \quad (2)$$

The coefficients and variables in equations (1) and (2) are not described here but can be found in Thomas and King (1991).

Moisture and heat source terms can be included in the right-hand side of equation (1) and (2), respectively. For simplicity, these terms are included in the final stages of the following derivation. To solve for the unknown variables pressure head ψ and temperature T, the Integrated Finite Difference Method is used (Narasimham, Witherspoon & Edwards 1978; Radhakrishna, Lau, & Crawford, 1984). The coupled governing equations (1) and (2) can be written into the integral form as:

$$\begin{aligned}&\int_V C_{\psi\psi}\frac{\partial\psi}{\partial t}dV+\int_V C_{\psi T}\frac{\partial T}{\partial t}dV\\&=\int_V\nabla\left(K_{\psi\psi}\nabla\psi\right)dV+\int_V\nabla\left(K_{\psi T}\nabla T\right)dV+\int_V\frac{\partial k}{\partial z}dV\end{aligned} \quad (3)$$

$$\begin{aligned}&\int_V C_{T\psi}\frac{\partial\psi}{\partial t}dV+\int_V C_{TT}\frac{\partial T}{\partial t}dV\\&=\int_V\nabla\left(K_{T\psi}\nabla\psi\right)dV+\int_V\nabla\left(K_{TT}\nabla T\right)dV\end{aligned} \quad (4)$$

By applying the Divergence Theorem to the right-hand side of (3) and (4), the equations can be expressed as:

$$\begin{aligned}&\int_V C_{\psi\psi}\frac{\partial\psi}{\partial t}dV+\int_V C_{\psi T}\frac{\partial T}{\partial t}dV\\&=\int_\Gamma\left(K_{\psi\psi}\nabla\psi\right)\cdot n d\Gamma+\int_\Gamma\left(K_{\psi T}\nabla T\right)\cdot n d\Gamma+\int_\Gamma k\cdot n_z d\Gamma\end{aligned} \quad (5)$$

$$\begin{aligned}&\int_V C_{T\psi}\frac{\partial\psi}{\partial t}dV+\int_V C_{TT}\frac{\partial T}{\partial t}dV\\&=\int_\Gamma\left(K_{T\psi}\nabla\psi\right)\cdot n d\Gamma+\int_\Gamma\left(K_{TT}\nabla T\right)\cdot n d\Gamma\end{aligned} \quad (6)$$

Let the flow domain be discretized into appropriately small volume elements (Figure 1) in which the variation of ψ is not rapid and the average properties of a volume element are associated with a representative nodal point l. Furthermore, let the volume element be so chosen that the lines joining the nodal point l to its neighbours are normal to the interface between the respective elements. By assuming that the average properties between adjacent nodal points can be represented by a linear relation which is independent of time, equation (5) can be written as:

$$\begin{aligned}C_{\psi\psi}V_l\frac{\Delta\psi_l}{\Delta t}+C_{\psi T}V_l\frac{\Delta T_l}{\Delta t}&=\sum_m K1_{l,m}\left(\frac{\psi_m-\psi_l}{d_{l,m}+d_{m,l}}\right)\cdot\Gamma_{l,m}\\&+\sum_m K2_{l,m}\left(\frac{T_m-T_l}{d_{l,m}+d_{m,l}}\right)\cdot\Gamma_{l,m}\\&+\sum_m K5_{l,m}\left(\frac{z_m-z_l}{d_{l,m}+d_{m,l}}\right)\cdot\Gamma_{l,m}\end{aligned} \quad (7)$$

where

$$K1_{l,m}=\frac{\left(K_{\psi\psi}\right)_l\left(K_{\psi\psi}\right)_m}{\left(K_{\psi\psi}\right)_l d_{m,l}+\left(K_{\psi\psi}\right)_m d_{l,m}}\left(d_{l,m}+d_{m,l}\right)$$

$$K2_{l,m}=\frac{\left(K_{\psi T}\right)_l\left(K_{\psi T}\right)_m}{\left(K_{\psi T}\right)_l d_{m,l}+\left(K_{\psi T}\right)_m d_{l,m}}\left(d_{l,m}+d_{m,l}\right)$$

$$K5_{l,m}=\frac{(k)_l(k)_m}{(k)_l d_{m,l}+(k)_m d_{l,m}}\left(d_{l,m}+d_{m,l}\right)$$

Similarly, equation (6) can be written as:

$$\begin{aligned}C_{T\psi}V_l\frac{\Delta\psi_l}{\Delta t}+C_{TT}V_l\frac{\Delta T_l}{\Delta t}&=\sum_m K3_{l,m}\left(\frac{\psi_m-\psi_l}{d_{l,m}+d_{m,l}}\right)\cdot\Gamma_{l,m}\\&+\sum_m K4_{l,m}\left(\frac{T_m-T_l}{d_{l,m}+d_{m,l}}\right)\cdot\Gamma_{l,m}\end{aligned} \quad (8)$$

where

$$K3_{l,m}=\frac{\left(K_{T\psi}\right)_l\left(K_{T\psi}\right)_m}{\left(K_{T\psi}\right)_l d_{m,l}+\left(K_{T\psi}\right)_m d_{l,m}}\left(d_{l,m}+d_{m,l}\right)$$

$$K4_{l,m}=\frac{\left(K_{TT}\right)_l\left(K_{TT}\right)_m}{\left(K_{TT}\right)_l d_{m,l}+\left(K_{TT}\right)_m d_{l,m}}\left(d_{l,m}+d_{m,l}\right)$$

Through mathematical manipulation, the IFD form of the governing equations are simplified such that in equation (7) pressure head change $\Delta\psi$ is the only time dependent variable, and in equation (8) temperature

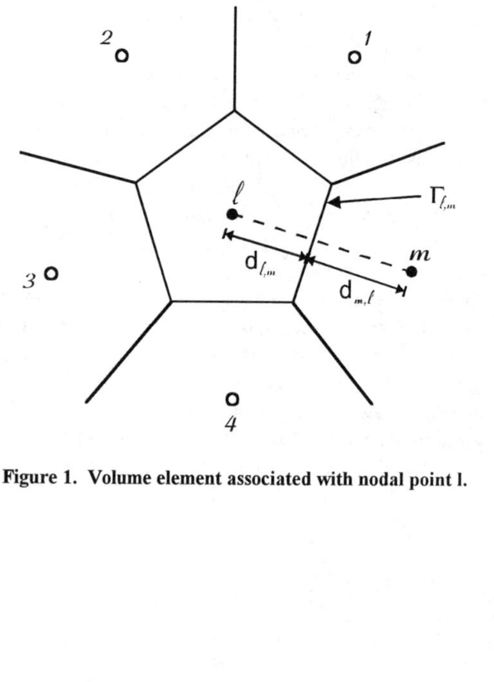

Figure 1. Volume element associated with nodal point l.

change ΔT is the only time dependent variable. Thus equation (7) leads to:

$$\Delta\psi_l = \frac{\Delta t}{A1_l V_l}\left[\sum_m B1_l\left(\frac{\psi_m - \psi_l}{d_{l,m} + d_{m,l}}\right)\cdot\Gamma_{l,m} + \sum_m C1_l\left(\frac{T_m - T_l}{d_{l,m} + d_{m,l}}\right)\cdot\Gamma_{l,m} + \sum_m D1_l\left(\frac{z_m - z_l}{d_{l,m} + d_{m,l}}\right)\cdot\Gamma_{l,m}\right] \quad (9)$$

where $A1_l = \dfrac{C_{T\psi}C_{\psi T} - C_{\psi\psi}C_{TT}}{C_{\psi T}}$

$$B1_l = \frac{C_{TT}}{C_{\psi T}} K1_{l,m} - K3_{l,m}$$

$$C1_l = \frac{C_{TT}}{C_{\psi T}} K2_{l,m} - K4_{l,m}$$

$$D1_l = \frac{C_{TT}}{C_{\psi T}} K5_{l,m}$$

Similarly, simplifying equation (8) leads to:

$$\Delta T_l = \frac{\Delta t}{A2_l V_l}\left[\sum_m B2_l\left(\frac{\psi_m - \psi_l}{d_{l,m} + d_{m,l}}\right)\cdot\Gamma_{l,m} + \sum_m C2_l\left(\frac{T_m - T_l}{d_{l,m} + d_{m,l}}\right)\cdot\Gamma_{l,m} + \sum_m D2_l\left(\frac{z_m - z_l}{d_{l,m} + d_{m,l}}\right)\cdot\Gamma_{l,m}\right] \quad (10)$$

where $A2_l = \dfrac{C_{TT}C_{\psi\psi} - C_{\psi T}C_{T\psi}}{C_{T\psi}}$

$$B2_l = K1_{l,m} - \frac{C_{\psi\psi}}{C_{T\psi}} K3_{l,m}$$

$$C2_l = K2_{l,m} - \frac{C_{\psi\psi}}{C_{T\psi}} K4_{l,m}$$

$$D2_l = K5_{l,m}$$

The summation terms in equation (9) represent the flux of moisture flow across the interface between elements l and m. For simplicity, the coefficients $X1_{l,m}$, $X2_{l,m}$, and $X3_{l,m}$ are introduced to represent the rate of fluid transfer between nodal point l and m per unit difference in pressure head, temperature, or elevation, respectively. With the consideration of boundary conditions and sources or sinks, equation (9) can be generalized as:

$$\Delta\psi_l = \frac{\Delta t}{A1_l V_l}\left[\sum_m X1_{l,m}(\psi_m - \psi_l) + \sum_m X2_{l,m}(T_m - T_l) + \sum_m X3_{l,m}(z_m - z_l) + B_\psi + G_\psi\right] \quad (11)$$

where $X1_{l,m} = B1_l \dfrac{\Gamma_{l,m}}{d_{l,m} + d_{m,l}}$

$$X2_{l,m} = C1_l \frac{\Gamma_{l,m}}{d_{l,m} + d_{m,l}}$$

$$X3_{l,m} = D1_l \frac{\Gamma_{l,m}}{d_{l,m} + d_{m,l}}$$

$$B_\psi = \sum_b X1_{l,b}(\psi_b - \psi_l) + \sum_b X2_{l,b}(T_b - T_l) + \sum_b X3_{l,b}(z_b - z_l)$$

b = nodal number of boundary elements

Similarly equation (10) can be generalized as:

$$\Delta T_l = \frac{\Delta t}{A2_l V_l}\left[\sum_m Y1_{l,m}(\psi_m - \psi_l) + \sum_m Y2_{l,m}(T_m - T_l) + \sum_m Y3_{l,m}(z_m - z_l) + B_T + G_T\right] \quad (12)$$

where $Y1_{l,m} = B2_l \dfrac{\Gamma_{l,m}}{d_{l,m} + d_{m,l}}$

$$Y2_{l,m} = C2_l \frac{\Gamma_{l,m}}{d_{l,m} + d_{m,l}}$$

$$Y3_{l,m} = D2_l \frac{\Gamma_{l,m}}{d_{l,m} + d_{m,l}}$$

$$B_T = \sum_b Y1_{l,b}(\psi_b - \psi_l) + \sum_b Y2_{l,b}(T_b - T_l) + \sum_b Y3_{l,b}(z_b - z_l)$$

2.1 *Explicit Scheme*

Let $\psi_l = \psi_l^P$, $\psi_m = \psi_m^P$, $T_l = T_l^P$, and $T_m = T_m^P$ in equation (11) and (12) to obtain an explicit form for determining the $\Delta\psi$ and ΔT unknowns. If the values in the beginning of the time step at each nodal point are known, the change in pressure head and temperature can be directly calculated from the explicit equations:

$$\Delta\psi_{l,\exp} = \frac{\Delta t}{A1_l V_l}\left[\sum_m X1_{l,m}(\psi_m^P - \psi_l^P) + \sum_m X2_{l,m}(T_m^P - T_l^P) + \sum_m X3_{l,m}(z_m - z_l) + B_\psi + G_\psi\right] \quad (13)$$

where

$$B_\psi = +\sum_b X1_{l,b}(\psi_b - \psi_l^P) + \sum_b X2_{l,b}(T_b - T_l^P) + \sum_b X3_{l,b}(z_b - z_l)$$

$$\Delta T_{l,\exp} = \frac{\Delta t}{A2_l V_l}\Bigg[\sum_m Y1_{l,m}\left(\psi_m^P - \psi_l^P\right) + \sum_m Y2_{l,m}\left(T_m^P - T_l^P\right) + \sum_m Y3_{l,m}\left(z_m - z_l\right) + B_T + G_T\Bigg] \tag{14}$$

where

$$B_T = \sum_b Y1_{l,b}\left(\psi_b - \psi_l^P\right) + \sum_b Y2_{l,b}\left(T_b - T_l^P\right) + \sum_b Y3_{l,b}\left(z_b - z_l\right)$$

The superscript p denotes the known initial values at the beginning of the time interval Δt.

By setting up equations (13) and (14) at each nodal point l in a flow domain and then solving simultaneously, the coupled temperature and pressure head distribution in the flow domain can be determined after each time step using the following forward difference equations:

$$\psi_l^{P+1} = \psi_l^P + \Delta\psi_{l,\exp} \tag{15}$$

$$T_l^{P+1} = T_l^P + \Delta T_{l,\exp} \tag{16}$$

2.2 *Stability Criteria*

Equations (13) or (14) can become unstable with time if the time step, Δt, exceeds a critical value. Following the method of stability analysis by Narasimhan, Witherspoon & Edwards(1978), the time step which is critical to instability of the solution in the vicinity of element l is given by:

$$\Delta t \;\le\; \left[\beta_\psi, \beta_T\right]_{\min} \tag{17}$$

where $\beta_\psi = \dfrac{A1_l V_l}{\sum_m \left(X1_{l,m} + X2_{l,m} + X3_{l,m}\right)}$

$$\beta_T = \frac{A2_l V_l}{\sum_m \left(Y1_{l,m} + Y2_{l,m} + Y3_{l,m}\right)}$$

Physically, β_ψ and β_T represent the approximate time required for element l to react significantly to changes in potential or temperature in the adjacent elements to which l is connected. Obviously, if the time step exceeds the lessor of β_ψ or β_T, the solution becomes unstable. To alleviate the numerical instability, the time step may be reduced. Often, the restrictions on the size of the time step can become so severe that the explicit method of solution may be uneconomical to employ. Therefore, an implicit approach is needed to enable the handling of large time steps. The implicit method used in TRUCHAMP will not be presented due to the limitation of space but can be found in the work of Li (1997).

An Integrated Finite Difference computer code, TRUCHAMP, for numerically modeling the potential-based transient coupled heat and moisture flow in a multi-dimensional unsaturated soil domain subjected to time-dependent heat and moisture generation is presented.

3 PROGRAM VALIDATION

Validation of the program has been made with data from theoretical solutions. Additionally, comparison of results obtained from the numerical model with those generated by independently developed numerical models (Thomas & King, 1991; Kanno, Kato & Yamagata, 1996) yielded excellent correlation.

The analysis of the temperature and volumetric liquid content of a heated horizontal cylinder of unsaturated porous soil is used to demonstrate the utility of TRUCHAMP.

The geometry and material properties of the sample are described in detail by Ewen & Thomas (1989); Thomas et al. (1996). The inner boundary at A comprises a heat source of constant temperature of 333K. The outer boundary at B maintains a constant temperature of 293K. Both boundaries are impermeable to moisture. The soil has an initial volumetric liquid content of 0.05.

Although the pressure-based model solves for the temperature and pressure head, it is easier to interpret variations of moisture content than pressure heads. Therefore the presentation of the results are in terms of temperature and volumetric liquid content.

Steady state temperatures are plotted in Figure 2, and the transient temperature profiles in Figure 3. Figure 2 indicates that the temperature gradient at steady state is not constant along the radius. This is a result of the variation in thermal conductivity of the soil with the moisture content. In Figure 3, a rapid temperature increase occurs radially within 3 hours of heating. As the moisture redistributes due to the temperature gradient, the associated thermal characteristics of the soil change to cause the temperatures to gradually fall toward their equilibrium values.

Figure 4 illustrates the distribution of volumetric liquid content in the cylinder at steady state. It can be noted that the drier region extends a greater distance vertically above the heat source than in other directions. This is a result of ponding due to gravity in the lower regions of the cylinder. The variations of volumetric liquid content with time along the line AB are given in

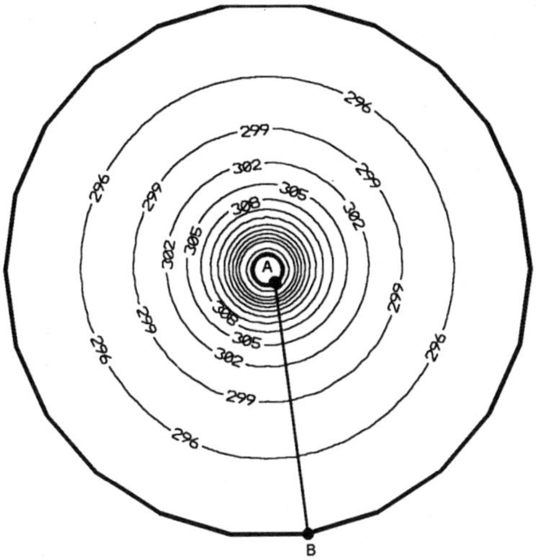

Figure 2. Temperature [K] distribution at steady state.

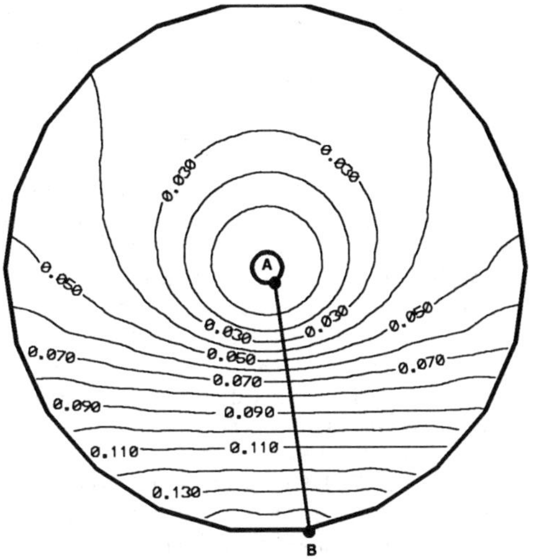

Figure 4. Volumetric liquid content distribution at steady state.

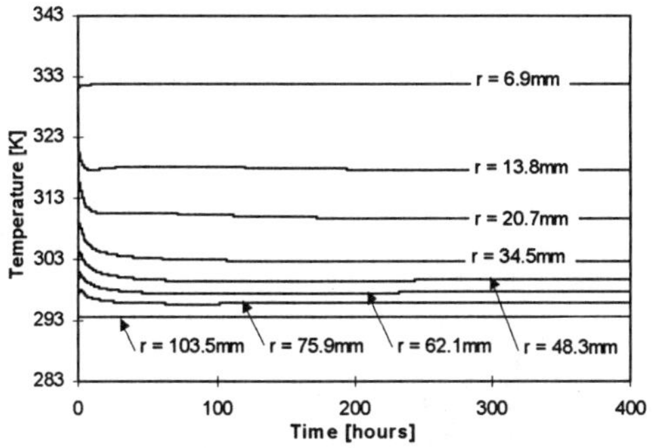

Figure 3. Temperature variation with time along line AB.

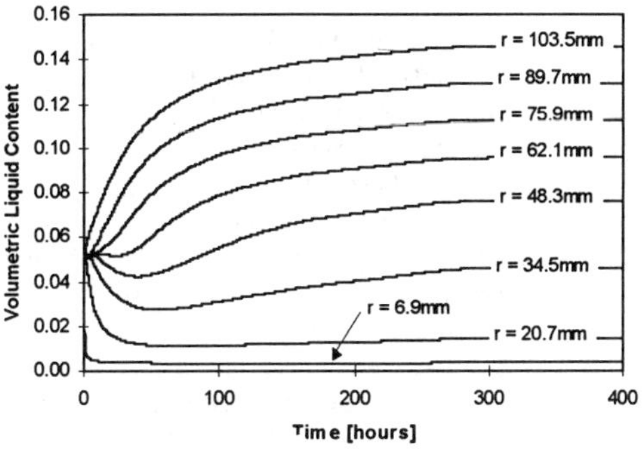

Figure 5. Variation of volumetric liquid content with time along line AB.

Figure 5. The region nearest to the heated boundary immediately experiences rapid drying and subsequently the moisture is forced towards the cooler boundary.

A more detailed description of this work along with the numerical solution of several problems are contained in the work of Li (1997).

4 CONCLUSIONS

The program, TRUCHAMP, has been validated for two-dimensional problems using non-linear material parameters. Results from TRUCHAMP show that it is able to handle the coupled heat and moisture flow in unsaturated materials with large differences in material properties.

With proper characterization of the material properties and boundary conditions, TRUCHAMP can be employed as a useful tool in assessing the thermal and hydraulic response of the prototype disposal vault. This program will also be useful in solving many heat and mass transport problems encountered in the field of heat storage, underground power cables, and nuclear waste disposal programs.

REFERENCES

Ewen, J., & Thomas, H. R. 1989. Heating unsaturated medium sand. *Geotechnique*. 39(3):455-470.

Kanno, T., Kato, K. & Yamagata, J. 1996. Moisture movement under a temperature gradient in highly compacted bentonite. *Engineering Geology*. 41:287-300.

Li, D. M. 1997. Numerical analysis of pressure-based coupled heat and moisture flow in unsaturated soils using the integrated finite difference method. *M.A.Sc. Thesis*. Department of Civil Engineering, University of Toronto.

Narasimhan, T. N., Witherspoon, P. A. & Edwards, A. L. 1978. Numerical model for saturated-unsaturated flow in deformable porous media: 2 - The algorithm. *Water Resources Research*. 14(2):255-261.

Philip, J. R. & de Vries, D. A. 1957. Moisture movement in porous materials under temperature gradients. *Trans. American Geophysical Union*. 38(2):222-232.

Radhakrishna, H. S., Lau, K. C. & Crawford, A. M. 1984. Coupled heat and moisture flow through soils. *Journal of Geotechnical Engineering, ASCE*. 110(12):1766-1784.

Thomas, H. R. & King, S. D. 1991. Coupled temperature/capillary potential variation in unsaturated soil. *Journal of Engineering Mechanics, ASCE*. 117(11):2475-2491.

Thomas, H. R. et al. 1996. On the development of a model of the thermo-mechanical-hydraulic behaviour of unsaturated soils. *Engineering Geology*. 41:197-218.

APPENDIX I. NOTATION

$d_{l,m}$ = perpendicular distance from nodal point l to the interface between element l and m

G_ψ = pressure head source or sink term

G_T = thermal source or sink term

n = unit normal of the surface

n_z = unit vector in the vertical direction

$\Gamma_{l,m}$ = surface area of interface between elements l and m

T_l = temperature of element l

Δt = time step

V_l = volume of element l

$\Delta\psi_l$ = change in pressure head of element l

ΔT_l = change in temperature of element l

ψ_l = pressure head of element l

z_l = elevation of nodal point associated element l

$$C_{L\psi} = \left.\frac{\partial\theta_L}{\partial\psi}\right|_T$$

$$C_{LT} = \left.\frac{\partial\theta_L}{\partial T}\right|_\psi$$

$$K_{L\psi} = k$$

$$C_{V\psi} = \frac{\rho_o h}{\rho_L}\left[\frac{(\eta-\theta_L)g}{R_V T} - \left.\frac{\partial\theta_L}{\partial\psi}\right|_T\right]$$

$$C_{VT} = \frac{(\eta-\theta_L)h}{\rho_L}\left(\beta - \rho_o\frac{\psi g}{R_V T^2}\right) - \frac{\rho_o h}{\rho_L}\left.\frac{\partial\theta}{\partial T}\right|_\psi$$

$$K_{V\psi} = \frac{D_{atm}\upsilon\eta\rho_o}{\rho_L}\frac{g}{R_V T}h$$

$$K_{VT} = \frac{D_{atm}\upsilon\eta}{\rho_L}\frac{(\nabla T)_a}{(\nabla T)}\left(h\beta - \rho_o\frac{\psi g}{R_V T^2}h\right)$$

$$C_{\psi\psi} = C_{L\psi} + C_{V\psi}$$

$$C_{\psi T} = C_{LT} + C_{VT}$$

$$K_{\psi\psi} = K_{L\psi} + K_{V\psi}$$

$$K_{\psi T} = K_{VT}$$

$$C_{T\psi} = L\cdot C_{V\psi}$$

$$C_{TT} = \frac{H}{\rho_L} + L\cdot C_{VT}$$

$$K_{T\psi} = L\cdot K_{V\psi}$$

$$K_{TT} = \frac{\lambda}{\rho_L} + L\cdot K_{TT}$$

Computer Methods and Advances in Geomechanics, Yuan (ed.) © 1997 Balkema, Rotterdam, ISBN 90 5410 904 1

Modeling of stress, deformation and flow in oil sands

Sri T. Srithar
Golder Associates Pty Ltd, Melbourne, Vic., Australia

Peter M. Byrne
Department of Civil Engineering, University of British Columbia, B.C., Canada

ABSTRACT: An analytical model is developed to predict the stresses, deformations and fluid flow in oil sand masses and implemented in a two dimensional finite element program. This program has been applied to predict the responses of a horizontal well pair in the underground test facility of Alberta Oil Sand Technology and Research Authority (AOSTRA). The predictions are discussed and compared with the measured responses wherever possible. This indicate that the analysis gives insights into the likely behaviour during an oil recovery process by steam injection.

1 INTRODUCTION

Oil recovery schemes involve open pit mining in the shallow oil sand formations, and in-situ extraction techniques such as tunnels and well-bores in the deep oil sand formations. In the in-situ extraction procedures some form of heating is often required as the very high viscosity of the bitumen makes conventional recovery by pumping impractical.

In-situ thermal methods, such as steam injection through vertical well-bores, have generally been used and are relatively effective for the recovery of heavy oil from deep seated formations. During steam injection, high pore fluid and stress gradients are created around the well-bore which can lead to the instability and collapse of the well casing.

When there is an increase in temperature, if the volume change of the pore fluid components is greater than that of the voids in the soil skeleton, there will be an increase in pore pressure and consequently a reduction in effective stress. The effective stresses may become zero and liquefaction may occur, if the oil sand is subjected to rapid increase in temperature and if an undrained condition prevails.

Modelling of deformation and flow during an oil recovery process in oil sands is relatively complex because of the nature of the oil sand and the recovery process involved. Oil sand comprises four phases; solid, water, bitumen and gas, whereas, a general soil consists of three phases; solid, water and air. The presence of bitumen and gas makes the oil sand behave differently compared to a general soil. A comprehensive study of the behaviour of oil sands can be found in Sobkowicz and Morgenstern (1984).

In this study, modelling of deformation and flow behaviour of oil sand is divided into two parts; modelling of the behaviour of the sand skeleton and modelling of the behaviour of the pore fluid.

2 MODELLING OF SAND SKELETON BEHAVIOUR

Modelling the behaviour of the oil sand skeleton is essentially a stress-strain modelling of a dense sand. The oil sand skeleton exhibits significant shear induced volume expansion or dilation. Realistic modelling of dilation is important because it will increase the pore space and thereby increase the permeability and reduce the pore pressure. These changes will have significant effect in the overall deformation and flow predictions.

Furthermore, oil recovery methods are cyclic processes which will cause the sand skeleton to undergo loading and unloading sequences resulting in irrecoverable plastic strains. This necessitates the use of an elasto-plastic stress-strain model.

The stress-strain model postulated in this study is a double-hardening type and consists of cone and

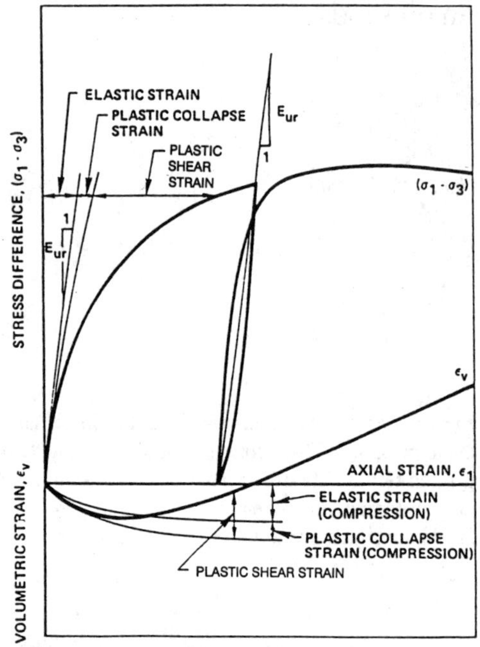

Figure 1. Components of strain increment

cap-type yield surfaces. The cone-type yield surface is based on the model proposed by Nakai and Matsuoka (1983) and the cap type yield surface is based on the model proposed by Lade (1977).

Detailed description of the stress-strain model, development of the constitutive matrix in a general three dimensional Cartesian coordinate system, its implementation in three dimensional, two dimensional plane strain and axisymmetric conditions are not presented herein because of space limitations and can be found in Srithar (1994).

In simple terms, the stress-strain matrix consists of three components; the plastic shear matrix obtained from cone-type yield surface, the plastic collapse matrix obtained from cap-type yield surface and the elastic matrix obtained using Hooke's law. A schematic illustration of the strain components are shown in Figure 1.

The strain increments are expressed as,

$$\{d\varepsilon^s\} = [C^s]\{d\sigma'\}$$
$$\{d\varepsilon^c\} = [C^c]\{d\sigma'\}$$
$$\{d\varepsilon^e\} = [C^e]\{d\sigma'\} \qquad (1)$$

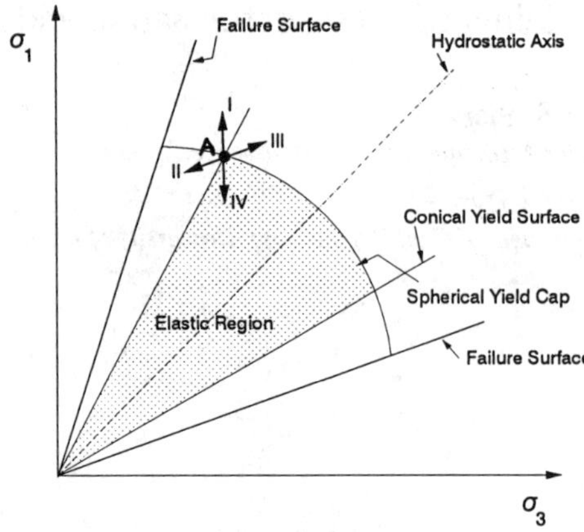

Figure 2. Possible loading conditions

where, $\{d\varepsilon\}$ is the strain increment, $[C]$ is the stress-strain matrix and superscripts s, c and e denote plastic shear, plastic collapse and elastic components.

The total strain increment is obtained by adding the relevant components depending on the loading conditions. A schematic illustration of possible loading conditions are shown on a two dimensional stress space in Figure 2. For example, stress path II from stress condition A will produce only elastic and plastic shear strains.

2.1 *Stress Strain Model Parameters and Verification*

The stress-strain model requires thirteen dimensionless parameters; seven parameters for plastic shear, two for plastic collapse and four for elastic stress-strain matrices.

All of these model parameters can be obtained from standard triaxial compression tests and details of this can be found in Srithar(1994).

The triaxial test results under various stress paths reported by Kosar (1989) on Athabasca McMurray formation interbedded oil sands were considered to verify the applicability of the stress-strain model to oil sands. The triaxial tests were carried out on undisturbed samples obtained from the AOSTRA underground test facility at various depths from 152 m to 161 m.

The predicted and the measured responses of the triaxial tests on oil sand samples for various stress paths are shown in Figure 3. The stress paths of the triaxial tests are shown in the insert in Figure 3.

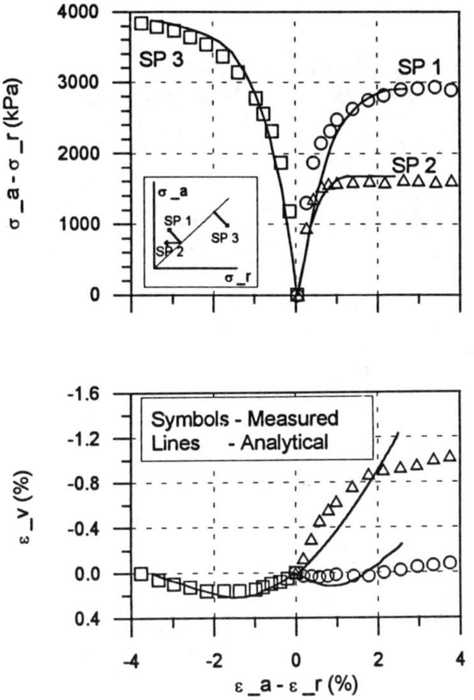

Figure 3. Measured and predicted results of triaxial tests

3 MODELLING OF FLUID FLOW

In petroleum reservoir engineering, the flow in oil sand is often analysed as multi-phase flow, but solely as a flow problem without paying much attention to the porous medium. The most widely used model to analyse the flow in oil sand is called 'β-model' or 'the black-oil model' (Aziz and Settari, 1979) and it makes the following assumptions:

- There are three distinct phases; oil, water and gas.
- Water and oil are immiscible and they do not exchange mass or phases.
- Gas is assumed to be soluble in oil but not in water.
- Gas obeys the universal gas law.
- Gas exsolution occurs instantaneously.

With these assumptions, and considering volume change in the sand skeleton due to the effects of stresses and temperature changes, a flow continuity equation is derived from the general equation of mass conservation. However, the flow equations are not considered separately for individual phases as in petroleum reservoir engineering. The flow continuity equations for the three phases were combined and a single effective equation is formulated. Basically, the derived flow continuity equation is similar to a single phase flow equation in geomechanics but the permeability and compressibility terms have been changed to include the effects of different phase components.

The flow continuity equation is expressed as,

$$k_{EQ}\,\nabla^2 P + \frac{\partial \varepsilon_v}{\partial t} - c_{EQ}\,\frac{\partial P}{\partial t} + \alpha_{EQ}\,\frac{\partial \theta}{\partial t} = 0 \quad (2)$$

where k, c and α with subscript EQ are the equivalent permeability, compressibility and coefficient of thermal expansion respectively. ε_v is the volumetric strain, P is the pore pressure, θ is the temperature and t is the time.

The equivalent terms include the contributions from the individual phase components, their variation with saturation, pressure and temperature. Details of the fluid flow model can be found in Srithar (1994).

4 ANALYTICAL MODEL

The models of oil sand skeleton behaviour and the multi-phase fluid flow are coupled and solved as a consolidation problem. The Analytical equations developed are similar to Biot's consolidation equations and are of incremental form in terms of time, temperature and loads.

The analytical equations are solved using a finite element procedure which includes the following:

- Galerkin's weighted residual scheme
- A fully implicit time marching scheme
- Frontal solution scheme for unsymmetric matrices

The analytical model has been incorporated in a finite element program. Various aspects of the model have been checked by comparing the results with closed form solutions and laboratory test results and are presented in Srithar (1994).

5 APPLICATION TO AN OIL RECOVERY PROBLEM

The finite element program has been applied to predict the response of an oil recovery process by steam injection. The Phase A pilot in the

Underground Test Facility (UTF) of AOSTRA is considered herein for analysis. The UTF uses a steam assisted gravity drainage process with horizontal injection and production wells. A brief description of the UTF and the problem to be analysed are presented here. Further details about the UTF can be found in Scott et al. (1991) and in AOSTRA reports on UTF.

The geological stratification at the UTF comprises a number of soil layers. However, it can be simplified as consisting of three different soil types, in a broad sense. Devonian Waterway formation limestone exists below a depth of 165 m. Overlying the limestone is the McMurray formation oil sand which is about 40 m thick. The top 125 m overburden can be classified as Clearwater formation shale.

A schematic 3-dimensional view and a plan view of the UTF are shown in Figure 4. There are two shafts accessing the tunnels in the limestone beneath the oil sand layer. The tunnels were constructed at a depth of about 178 m with the roof being about 15 m below the limestone-oil sand interface. Three pairs of horizontal injection and production wells were drilled from the tunnels up into the oil sands at about 24 m spacing.

A vertical section of the well pairs was instrumented with thermocouples for measuring temperatures, pneumatic and vibrating wire piezometers for measuring pore pressures and extensometers and inclinometers for measuring horizontal and vertical displacements.

Modelling of all three well pairs with their steaming histories and with the detailed geological stratification would be complex as the steaming of different well pairs started at different times. To illustrate the problem and to demonstrate the applicability of the program in a simple manner, only one well pair is considered here for analysis.

In the analysis, the injection and production wells are modelled by nodes with known pore pressures. The steam injection pressure is assumed to be maintained at 2800 kPa (1300 kPa above the in-situ pore pressure) and the production pressure is assumed to be at 2000 kPa (500 kPa above the in-situ pressure).

The temperature-time histories of the nodes were specified as an input to the program. These were obtained from the field measurements made at the UTF. The temperature contours in the oil sand layer at different times are shown in Figure 5.

The steam chamber which is the region in the oil sand layer where the temperature is the same as the steam temperature, grows with time as shown in the figure. At time t = 30 days, the steam chamber extends to a distance of about 10 m horizontally and vertically from the injection well.

Even though a larger domain is analysed, the results are plotted only for the oil sand layer which is of primary interest. The predicted excess pore pressures in the oil sand layer are shown in Figure 6.

The injection and production wells are also indicated in the figure. With time, the region of higher pore pressure expands as shown indicating the growth of the steam chamber which is also implied by the temperature contours in Figure 5.

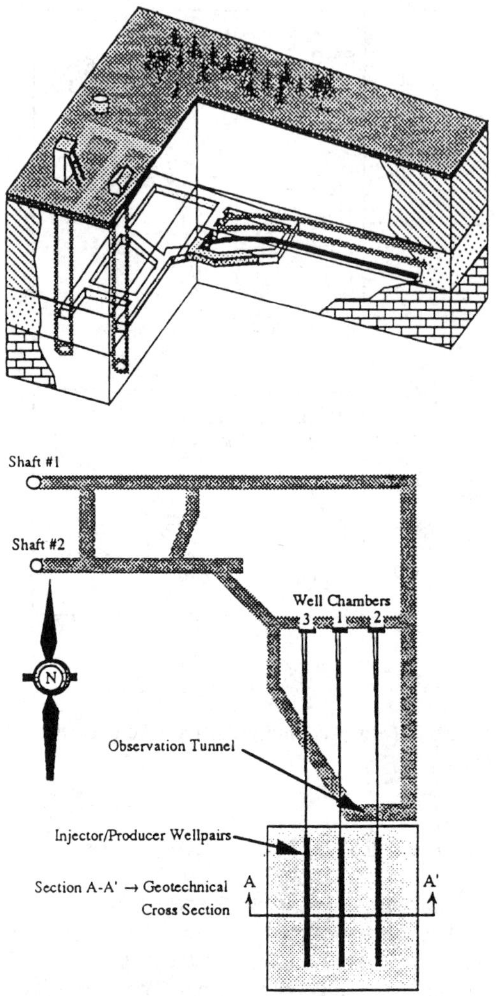

Figure 4. Schematic 3-dimensional and plan view of the UTF (adopted from Scott et al., 1991)

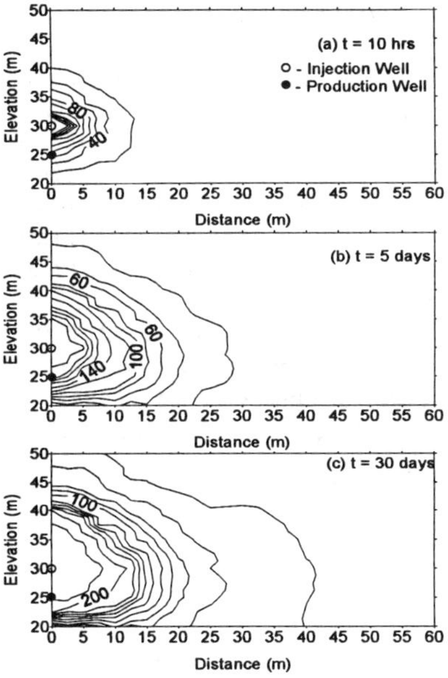

Figure 5. Measured temperature contours in the oil sand layer (°C)

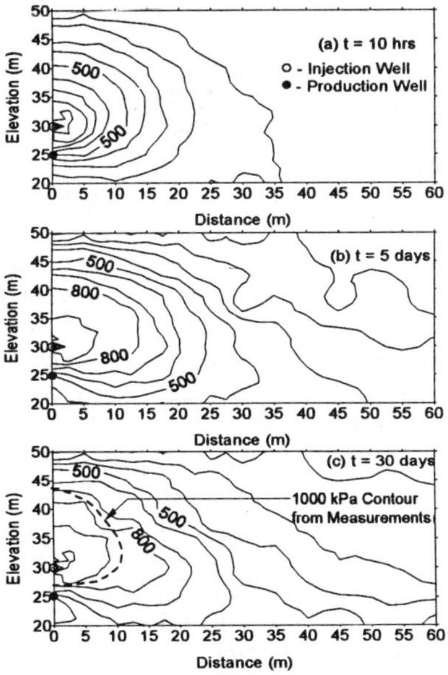

Figure 6. Predicted excess pore pressure contours in the oil sand layer (kPa)

The 1000 kPa excess pore pressure contour from the field measurements is also shown in Figure 6. It

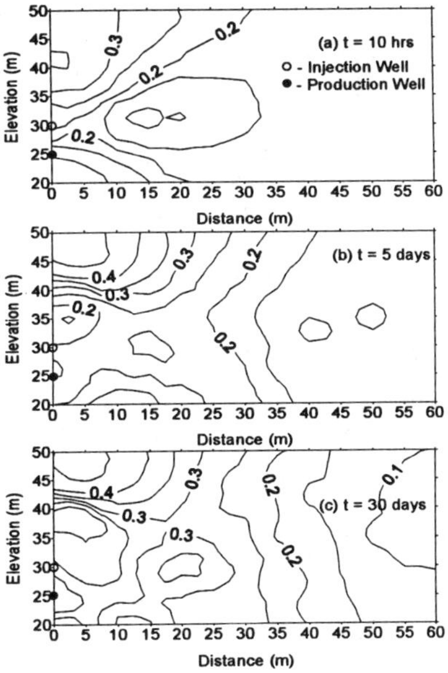

Figure 7. Predicted failure stress ratio contours in the oil sand layer

can be seen from the figure that the measured zone of 1000 kPa is slightly larger than the predicted zone. However, the shapes of the predicted pore pressure contours are similar to the measured ones.

The failure stress ratio which is an index giving the current stress state relative to the failure stress state is shown in Figure 7. It appears that the shape of the steam chamber and the corresponding temperature increases create higher shear stresses in the region above the steam chamber. This is implied by higher failure stress ratios and a maximum failure stress ratio of about 0.45 is predicted in the region about 15 m above the injection well. Since the predicted stress ratios are below unity, there would not be any failure.

Figure 8 shows the predicted horizontal displacement along a vertical line at 7 m from the wells, at time t = 30 days. The field measurements made in a instrumented bore hole at about the same distance away from the wells are also shown in the figure. It can be seen that the field measurements are slightly larger than the predictions at some locations, but in the overall picture, the predictions are in reasonable agreement with the measurements.

To show the importance of this type of analytical study, the same oil recovery problem has been analysed with the permeability of the oil sand reduced by a factor of 10. The highest failure stress

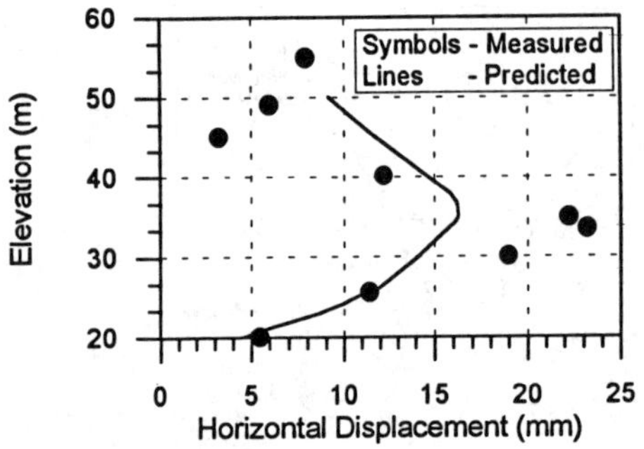

Figure 8. Comparison of predicted and measured horizontal displacements at 7 m from the wells

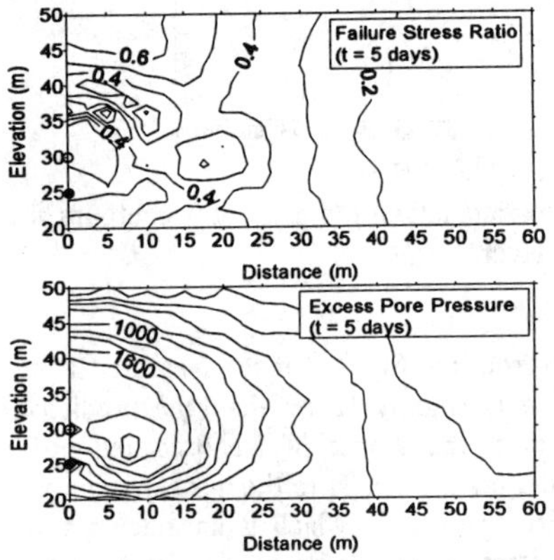

Figure 9. Predicted excess pore pressure and failure stress ratio contours for reduced permeability.

ratio for this case occurs at about t=5 days and a maximum excess pore pressure of 2200 kPa is predicted.

The maximum predicted excess pore pressure is higher than the excess steam injection pressure of 1300 kPa. The reason for this higher excess pore pressure is that the pore fluid expands more than the solids and since the permeability is low, there is not enough time for the expanded pore fluid to escape.

The predicted excess pore pressure contours and failure stress ratio contours at time t=5 days are shown in Figure 9. A maximum failure stress ratio of about 0.7 is predicted in a region just above the steam chamber.

6 SUMMARY AND CONCLUSIONS

A coupled stress-deformation-flow model was developed to analyse the geotechnical aspects in an oil recovery process and implemented in a finite element procedure. The finite element code has been applied to model a horizontal well pair in the underground test facility of AOSTRA. Results have been presented in terms of displacements and pore pressures and failure stress ratios.

The type of analytical study presented in this study could give insights into the likely behaviour in terms of stresses, deformations and flow in oil recovery projects. The results from the analysis suggest that in oil sands with low permeability, higher rates of heating could cause shear failure.

ACKNOWLEDGEMENTS

The authors are grateful to AOSTRA for their financial support. The authors are also grateful to Dr. Marcis Kurzeme for his review of this paper.

REFERENCES

Aziz, K. & Settari, T. 1979. *Petroleum reservoir simulation*. Applied Sci. Publ.

Kosar, K.M. 1989. *Geotechnical properties of oil sands and related strata*. Ph.D. Thesis, Dept. of Civil Engr. Univ. of Alberta, Canada.

Lade, P.V. 1977. Elasto-plastic stress-strain theory for cohesionless soil with curved yield surfaces. *Int. J. of Solids Structures*: Vol. 13.

Nakai, T. & Matsuoka, H. 1983. Constitutive equation for soils based on the extended concept of spatial mobilised plane and its application to finite element analysis. *Soils & Found.*: Vol 23. No.4.

Scott, J.D., Chalaturnyk, R.J., Stokes, A.W. & Collins, P.M. 1991. The geotechnical program at the AOSTRA Underground Test Facility. *44th Can. Geot. Conf.*, Calgary, Canada.

Sobkowicz, J. & Morgenstern, N.R. 1984. The undrained equilibrium behaviour of gassy sediments. *Can. Geot. J*: Vol 21. No.3.

Srithar, T. 1994. *Elasto-plastic deformation and flow analysis in oil sand masses*. Ph.D. Thesis, Univ. of British Columbia, Vancouver, Canada.

Computer Methods and Advances in Geomechanics, Yuan (ed.) © *1997 Balkema, Rotterdam, ISBN 90 5410 904 1*

Research on a model of seepage flow of fracture networks and modelling for the coupled hydro-mechanical process in fractured rock masses

He Shaohui
Tunnel Section of Department of Civil Engineering, Northern Jiaotong University, Beijing, China

ABSTRACT: In this paper the graph theory is introduced into using to describe the properties of fracture networks in fractured rock masses. On the basis of the cubic law of a single fracture, a model of seepage flow of fracture networks in fractured rock masses is established. Meanwhile, algorithms for modelling a coupled hydro-mechanical process in fractured rock masses are given.

1 INTRODUCTION

Fractures have a major influence on rock masses behaviour, and fracture studies have always been a central part of rock mechanics. Ground water flow in rock masses is today an important research area, mainly in conjunction with the assessments of underground disposal of hazardous waste and engineering stability. In consequence, many rock mechanics researchers have focused on the hydromechanical properties of rock fractures and coupled hydro-mechanical model during about the last decade.

The conductivity of rock masses is governed by the geometry of the void space between the two fracture surfaces and the interconnected properties of fractures. This geometry is complex and its detailed nature is not very well known. The aim of this paper is to give some new algorithms that have been used to describe the fracture geomtry and research on a model of seepage flow of fracture networks and modelling a coupled hydro-mechanical process in fractured rock masses.

2 MODEL OF SEEPAGE FLOW OF FRACTURE NETWORKS IN FRACTURED ROCK MASSES

2.1 *Some definitions*

Definition 1:Groundwater flow in rock masses is regarded as seepage flow in fracture networks.

Definition 2:Seepage force produced by seepage flow in fracture networks is regarded as a surface force acting on two fracture surfaces not a body force.

Definition 3:Seepage flow in a single fracture is obeyed the cubic law, namely

$$k_{ij} = {\gamma b_j^2}/{12\mu} \tag{1}$$

Where k_{ij} is the conductivity coefficient of groundwater of fracture, γ is the specific gravity of water, μ is the kinematic viscosity, b_j is the apertures of fractures.

2.2 *Matrix expressions of fracture networks in fractured rock masses*

Fracture networks in fractured rock masses is treated as a graph (See Fig. 1) of the graph theory

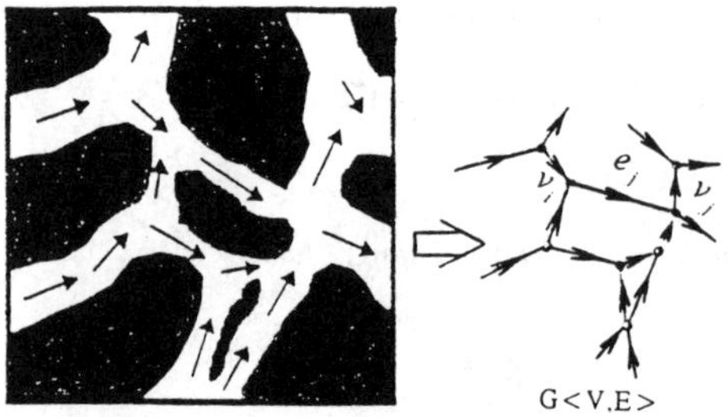

Fig. 1 Seepage flow in fracture networks corresponding with the graph G<V,E>

$$G<V,E> \quad (2)$$

Where V is the set of vertexes of a graph, V={v_1, v_2, ..., v_i}, i=1, 2, ..., n, E is the set of sides of a graph, E={e_1, e_2, ..., e_j}, j=1, 2, ..., m.

The incidence matrix of oriented fracture networks in fractured rock masses, M, is as follows:

$$M=\left[m_{ij}\right]_{n\times m} \quad (3)$$

Where

$$m_{ij}=\begin{cases}1 & v_i \textit{ is the beginning vertex of the side } e_j\\ 0 & v_i \textit{ is not incident with the side } e_j\\ -1 & v_i \textit{ is the end vertex of the side } e_j\end{cases}$$

The loop matrix of fracture networks in fractured rock masses, C, follows:

$$C=\left[c_{ij}\right]_{n\times m} \quad (4)$$

Where

$$c_{ij}=\begin{cases}1 & \textit{if the side } e_j \textit{ is included in the loop } i\\ 0 & \textit{else}\end{cases}$$

It is assumed that T is a spanning tree of a interconnected graph(its rank is n and the number of its sides is m),G<V,E>, that the sides e_1, e_2, ..., e_{m-n+1} are the hypotenuses of the spanning tree T and that the loops C_r (r=1, 2, ..., m-n+1) are born of the spanning tree T plus the hypotenuses e_r (r=1, 2, ..., m-n+1). Therefore, the loop C_r is called the basic loop of the fracture networks, G<V,E>, corresponding to the hypotenuse e_r.

The matrix C_f is the basic loops matrix. According to the graph theory, the incidence matrix M is perpendicular to the basic loop matrix C_f, namely

$$M\cdot C_f^T=0 \quad \text{or} \quad C_f\cdot M^T=0 \quad (5)$$

The adjacency matrix of a assignment weight fracture networks G<V,E> is expressed as follows:

$$A=\left[a_{ij}\right]_{m\times m}=\begin{cases}p_{ij} & \textit{if } (v_i,v_j)\in E \textit{ is possessed of the weight } p_{ij}\\ 0 & \textit{if } (v_i,v_j)\notin E\end{cases}$$

The apertures b_j (j=1,2,...,m) of fractures are regarded as the weights of the fracture networks G<V,E>. So the adjacency matrix A is expressed as another form:

$$A=\begin{bmatrix}b_1 & 0 & \cdots & 0\\ 0 & b_2 & \cdots & 0\\ \cdots & \cdots & \cdots & \cdots\\ 0 & 0 & \cdots & b_m\end{bmatrix} \quad (6)$$

Where b_j (j=1,2, ... ,m) is the aperture of fractures

2.3 *A model of seepage flow of fracture networks in fractured rock masses*

According to the cubic law, the matrix of conductivity coefficient of ground water of fracture networks in fractured rock masses is defined by:

$$K=\left[k_{ij}\right]_{m\times m}=\begin{cases}0 & (v_i,v_j)\notin E\\ k_j & (v_i,v_j)\in E, \textit{fracture filled}\\ \dfrac{\gamma b_j^2}{12\mu} & (v_i,v_j)\in E, \textit{fracture not filled}\end{cases} \quad (7)$$

On the basis of Wittke's model[1], the nodal equations of conservation of flow capacity of fracture networks in fractured rock masses are as follows:

$$\sum_{n_i'} q_j-\sum_{n_i'} P_j+Q_i=0 \qquad i=1,2,\ldots,n \quad (8)$$

Where q_j is the flow rate of fracture elements, P_j is the infiltration capacity of precipitation, Q_i is the source flow rate of intersection nodes of fractures.

The basic loop equation of seepage flow networks in fractured rock masses follows:

$$\sum_{m_c'} \Delta H_j = 0 \qquad c=1,2,\dots,r \tag{9}$$

Where ΔH_j is the water head difference value between one node of fractures and another one.

Combining formula (2) and (3) gives the matrix forms of equation (8) and (9) as follows:

$$M \cdot q - M_d \cdot P + Q = 0 \tag{10}$$

$$C_f \cdot \Delta H = 0 \tag{11}$$

Where M_d is the matrix that elements -1 of the incidence matrix M are replaced by the number 1.

Meanwhile, equation (11) is converted to the following form by equation (5):

$$\Delta H = M^T \cdot H \tag{12}$$

Where H is the column matrix of water head H_i of nodes.

In accordance with the cubic law, the column matrix q of flow rate of fracture elements is as follows:

$$q = A \cdot K \cdot \Delta H \tag{13}$$

To synthesize equation (10), (12) and (13), a model of seepage flow of fracture networks in fractured rock masses is established as follows by matrix operations:

$$M \cdot A \cdot K \cdot M^T \cdot H = M_d \cdot P - Q \tag{14}$$

If the lengths ℓ_j (j=1,2, ... ,m) of fracture elements of fracture networks are treated as the weights of fracture networks G<V,E>, the adjacency matrix A is changed into the following form:

$$D = \begin{bmatrix} \ell_1 & 0 & \cdots & 0 \\ 0 & \ell_2 & \cdots & 0 \\ \cdots & \cdots & \cdots & \cdots \\ 0 & 0 & \cdots & \ell_m \end{bmatrix} \tag{15}$$

Combining equation (2), (7), (12) and (15) defines the column matrix V of seepage flow velocity of fracture elements of fracture networks in fractured rock masses:

$$V = K \cdot D^{-1} \cdot M^T \cdot \Delta H \tag{16}$$

2.4 *Solution of the model of seepage flow of fracture networks in fractured rock masses*

The LU resolution method of linear equations set [2] and the initial flow method [3] are used to solve the model of seepage flow of fracture networks in fractured rock masses.

3 ALGORITHM FOR A COUPLED HYDRO-MECHANICAL PROCESS IN FRACTURED ROCK MASSES

3.1 *Determination of water pressure*

Water pressure in fracture networks is determined by:

$$P_w = \gamma (H - Z) \tag{17}$$

Where Z is the height mark, H is the water head.

3.2 *A coupled hydro-mechanical process in fractured rock masses*

The flow rate (Q) along a fracture (Fig. 2b) is affected by changes in the effective stress normal to the fracture plane (σ_n'). It is a coupled process where a change in effective stress causes mechanical deformation which in turn affects the fracture aperture and fracture transmissivity.

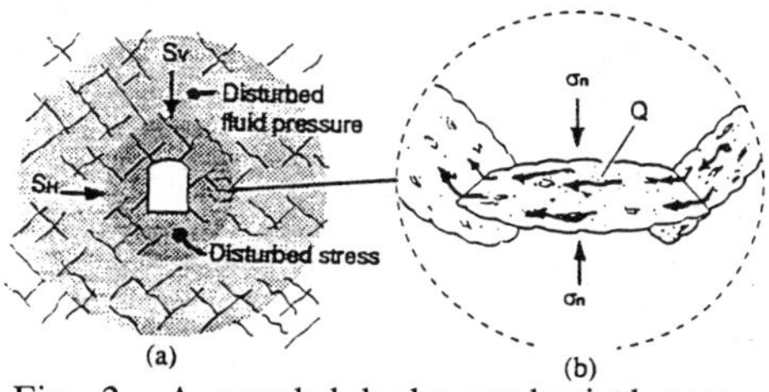

Fig. 2 A coupled hydro-mechanical process in fractured rock masses around a tunnel

The mechanical displacements (u_j, v_j) of a fracture plane are related to the effective stress (σ'_n, τ) according to Goodman's classical model[4]:

$$\begin{cases} u_j = \dfrac{\sigma'_n}{k_n} = \dfrac{\sigma_n - P_w}{k_n} \\ v_j = \dfrac{\tau}{k_s} \end{cases} \tag{18}$$

Where u_j is the possible closure at the effective stress, v_j is the possible sliding, k_n and k_s are respectively the normal and tangental stiffnesses of a fracture.

Because of variations of fracture apertures caused by stress field in fractured rock masses, the conductivity property of fracture networks is correspondingly varied. The adjacency matrix discribing variations of fracture apertures of fracture networks at stress field and the transmissivity coefficient matrix discribing variations of transmissivity property of fracture networks at stress field can be defined by:

$$A' = \left[a'_{ij}\right]_{m\times m} = \begin{bmatrix} \left(b_1 + u_{j1} + v_{j1}tg\alpha_{j1}\right) & \cdot & \cdot & 0 \\ \cdot & & \cdot & \cdot \\ \cdot & & \cdot & \cdot \\ 0 & \cdot & \cdot & \left(b_m + u_{jm} + v_{jm}tg\alpha_{jm}\right) \end{bmatrix} \tag{19}$$

Where b_j (j=1,2,…,m) is the initial aperture of a fracture, α_{jj} is the dilation angle of a fracture.

$$K' = \left[k'_{ij}\right]_{m\times m} = \begin{cases} \dfrac{\gamma}{12\mu}\left(b_j + u_{jj} + v_{jj}tg\alpha_{jj}\right)^2 & i = j \\ 0 & i \neq j \end{cases} \tag{20}$$

The equation (19) together with (20) are used to describe a coupled hydro-mechanical relation for a fracture networks.

3.3 *Numerical modelling for a coupled hydro-mechanical process*

The finite element method is used to analyse a coupled hydro-mechanical process. As stated in definition 2, seepage forces produced by seepage flow in fractured rock masses are regarded as surface forces. Therefore, water pressures in fracture networks can be alloted to nodal loads for FEM analysis (See Fig. 3).The allotment method of water pressures in fracture networks for FEM analysis follows:

$$\{R\}^e = \int_s [N]^T \{P_w\} t ds \tag{21}$$

Where $\{R\}^e$ is the column matrix of nodal loads of elements, [N] is the matrix of shape function of elements, $\{P_w\}$ is the column matrix of water pressures, t is the thickness of elements.

In brief, numerical simulation for a coupled hydro-mechanical process in fractured rock masses follows the following steps:

Step 1 : to solve the model of seepage flow of fracture networks, equation (14), and determine water pressures in fracture networks.

Step 2 : to allot water pressures in fracture networks to nodal loads of joint elements, analyse stress and deformation in fractured rock masses by FEM and determine varitions of fracture apertures.

Step 3: to adjust fracture apertures of fracture networks in turn and solve the model of seepage flow of fracture networks once again and determine the conductivity property of fracture networks that is affected by stress field in fractured rock masses.

The three steps from step 1 to 3 above are repeated until the demands for modelling a coupled hydro-mechanical process are satisfied.

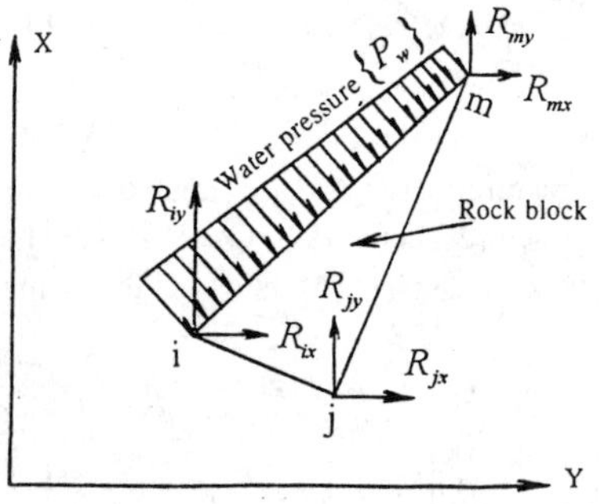

Fig. 3 Allotment water pressure in fractures to nodel loads

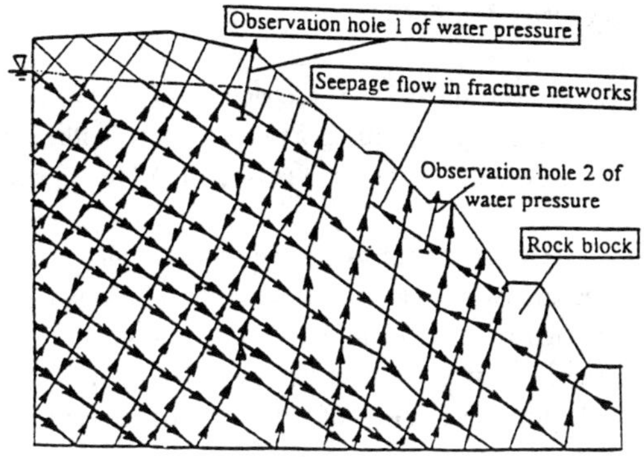

Fig. 4 Seepage flow graph of this slope

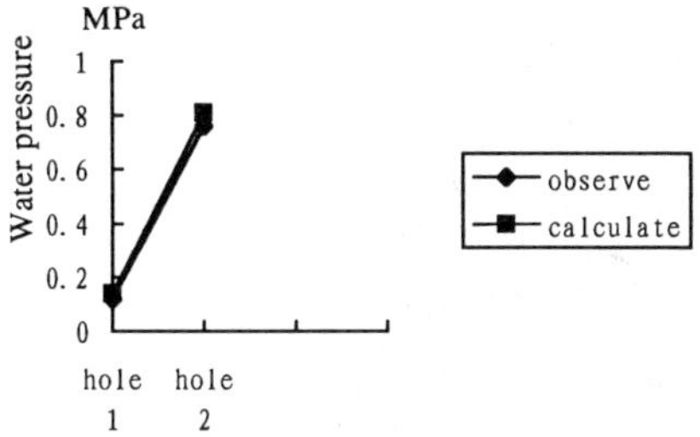

Fig. 5 Comparison observation result of water pressure of two holes with calculation result

4 HISTORY CASE

This model and algorithm for a coupled hydro-mechanical process in fractured rock masses are used to evaluate influence of seepage flow of fracture networks in fractured rock masses on stability of the Southwest side of the open pit slope of Da Gushan Iron Mine.

4.1 *Seepage flow in fracture networks of the Southwest side of this slope*

The fracture networks of this slope are first determined by measurement and statistical breakdown of fractures for this slope, and then water pressures and flow graph (See Fig. 4) in fracture networks of this slope are obtained by solving this model. Comparison observation result of water pressure of two observation holes with calculation result is shown in Fig. 5. The calculation result of water pressures in fracture networks approximates to the observation result. Fig. 5 shows that the model of seepage flow of fracture networks in fractured rock masses is reasonable.

4.2 *FEM analysis for influence of seepage flow of fracture networks in fractured rock mases on stability of the slope*

The FEM model was constituted by 173 nodes, 229 isoparametric elements and 243 joint elements. The mechanical parameters and the material constants are listed in table 1.

Table 1 Mechanical parameters and material constants

Items		Used values
Young's modulus of migmatite	E	1000MPa
Poisson's ratio of migmatite	ν	0.25
Density of migmatite	ρ	$25\,KN/m^3$
Tensile strength of joints	S_t	0.1MPa
Shear stiffness of joints	k_s	$100\,MN/m^3$
Normal stiffness of joints	k_n	$900\,MN/m^3$
Frictional grip of migmatite	C	$500\,KN/m^3$
Angle of internal friction of migmatite	ϕ	35°

As mentioned before, water pressures in fracture networks of this slope are allotted to nodal loads for FEM analysis. The stress and strain of a dry slope and a water-saturated slope are evaluated by FEM analysis. The results of stress and strain of FEM analysis are shown in Fig. 6 and Fig. 7.

As shown in Fig. 6 and Fig. 7, the stress and strain patterns of the top and toe of a dry slope are greatly different from that of the top and toe of a water-saturated slope. The plastic deformations of the top and toe of a water-saturated slope are more serious than that of the top and toe of a dry slope. The results of FEM analysis show that seepage flow of fracture networks in fractured rock masses greatly influences on stability of the Southwest side of the open pit slope of Da Gushan Iron Mine.

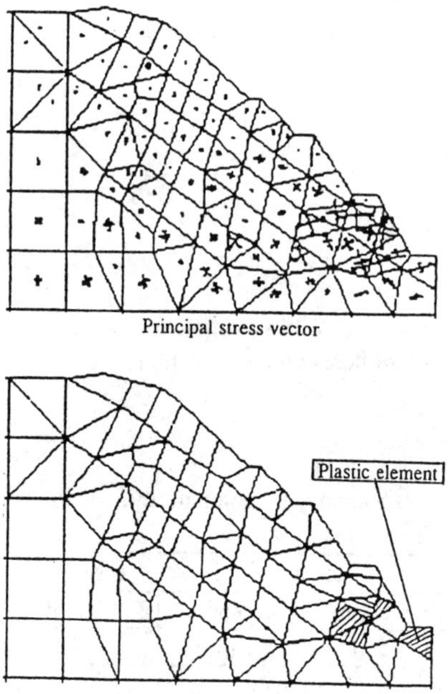

Fig. 6 Principal stress vectors of elements and plastic elements of a dry slope

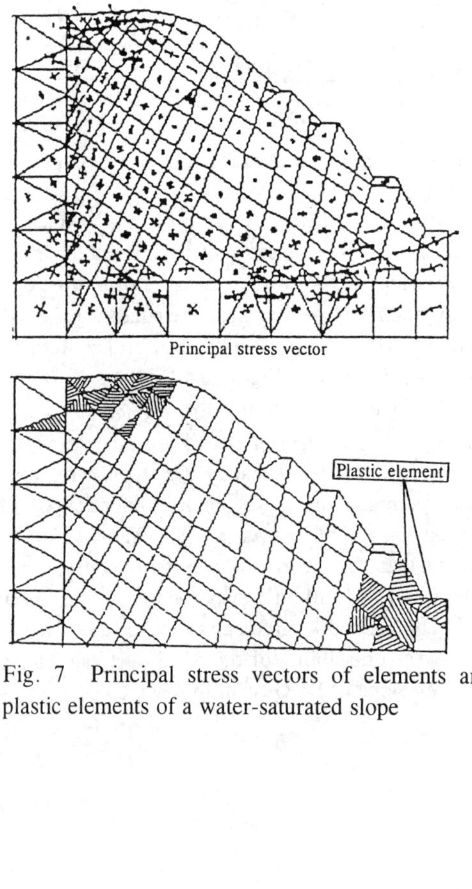

Fig. 7 Principal stress vectors of elements and plastic elements of a water-saturated slope

5 CONCLUSIONS

1. The model of seepage flow of fracture networks in fractured rock masses correctly described the state of seepage flow in fracture networks.

2. The algorithm for modelling a coupled hydro-mechanical process correctly evaluated influence of seepage flow of fracture networks in fractured rock masses on stability of rock engineering and in turn influence of variations of stress field on properties of seepage flow of fracture networks in fractured rock masses.

REFERENCES

Wittke W. 1970. Improvement of the properties of rock masses. Proc. 2nd Cong. ISRM.

Xu Chuiwei. 1986. Methods of calculations. Beijing: Higher Education Press.

Zhang Youtian. 1988. Initial flow method for seepage analysis with free surface. J. Irrigation Works.8:18-26.

Goodman R. E. 1974. The mechanical properties of joints. Proc. 3rd Cong. ISRM. 1A:127-140. Denver.

Computer Methods and Advances in Geomechanics, Yuan (ed.) © *1997 Balkema, Rotterdam, ISBN 90 5410 904 1*

Numerical model for hydro-mechanical behaviour in deformable porous media: A benchmark problem

D.Gawin
Technical University of Lodz, Poland
L.Simoni & B.A.Schrefler
Construction and Transport Department, Padua University, Italy

ABSTRACT: A model for numerical simulation of gas and water flow in deformable porous media is presented. A benchmark problem, based on the desaturation experiment performed by Liakopoulos (1965), is proposed. Different solution techniques, applicable to the transition from fully saturated to partially saturated state are discussed and used for the proposed test problem.

1 MATHEMATICAL MODEL

The mathematical model, used in the present analysis, is an isothermal version of that proposed by Gawin et al. (1995), Gawin & Schrefler (1996a). It consists of three balance equations: mass of dry air, mass of water species (both liquid water and vapour, phase change is taken into account) and linear momentum of the multiphase medium. An appropriate set of constitutive and state equations, as well as some thermodynamic relationships complete the model.

The governing equations of the model, expressed in terms of the chosen state variables: gas pressure p_g, capillary pressure p_c and displacement vector $\mathbf{u}$, are as follows:

Dry air conservation equation:

$$\phi \frac{\partial}{\partial t}\left[(1-S)\rho_{ga}\right] + \alpha(1-S)\rho_{ga}\frac{\partial}{\partial t}(\nabla\cdot\mathbf{u}) + \nabla\cdot\left(\rho_{ga}\mathbf{v}_g\right) - \nabla\cdot\left(\rho_g \mathbf{v}^d{}_{gw}\right) = 0 \tag{1.1}$$

where ϕ is porosity, S - degree of saturation with liquid water, ρ - density, t - time, α - the Biot's constant, $\mathbf{v}$ - velocity, $\mathbf{v}^d$ - diffusion velocity. Subscript *ga* is related to dry air, *gw* to water vapour, and *g* - to their binary mixture, i.e. moist air.

Water species (liquid and vapour) conservation equation:

$$\phi \frac{\partial}{\partial t}\left[(1-S)\rho_{gw}\right] + \alpha(1-S)\rho_{gw}\frac{\partial}{\partial t}(\nabla\cdot\mathbf{u}) + \nabla\cdot\left(\rho_{gw}\mathbf{v}_g\right) + \nabla\cdot\left(\rho_g \mathbf{v}^d{}_{gw}\right) = -\phi\,\rho_l\frac{\partial S}{\partial t} - \alpha S\rho_l\frac{\partial}{\partial t}(\nabla\cdot\mathbf{u}) - \nabla\cdot\left(\rho_l\mathbf{v}_l\right), \tag{1.2}$$

where subscript *l* is related to liquid water.

Linear momentum balance equation for the mixture:

$$\nabla\cdot\sigma + \rho\mathbf{b} = \mathbf{0}, \tag{1.3}$$

where a constant (time independent) specific body force (normally corresponding to the acceleration of gravity) $\mathbf{b}$ is assumed, and ρ is the averaged density of the multiphase medium:

$$\rho = (1-\phi)\,\rho_s + \phi\,S\,\rho_l + \phi\,(1-S)\,\rho_g, \tag{1.4}$$

where subscript *s* is related to the solid matrix.

The constitutive relationship for the solid skeleton is assumed in the form:

$$d\sigma' = \mathbf{D}\,(\,d\varepsilon - d\varepsilon^{o}\,) \tag{1.5}$$

together with the definition of the strain matrix $\mathbf{B}$ relating strain and displacement:

$$\varepsilon = \mathbf{B}\,\mathbf{u} \tag{1.6}$$

where $\mathbf{D}$ is a tangent matrix, and $d\varepsilon^{o}$ represents the autogeneous strain increments and the irreversible part of the thermal strains.

For fluid phases the multiphase *Darcy law* has been applied as constitutive equation, thus for liquid phase we have:

$$\mathbf{v}_l = -\frac{\mathbf{K}K_{rl}}{\mu_l}\left(\nabla p_l - \rho_l \mathbf{b}\right), \qquad p_c < p_b$$

$$\mathbf{v}_l = -\frac{\mathbf{K}K_{rl}}{\mu_l}\left(\nabla p_g - \nabla p_c - \rho_l \mathbf{b}\right), \; p_{ir} > p_c \geq p_b \quad (1.7)$$

$$\mathbf{v}_l = \mathbf{0}, \qquad p_c \geq p_{ir}$$

while for the gas phase holds:

$$\mathbf{v}_g = -\frac{\mathbf{K}K_{rg}}{\mu_g}\nabla p_g, \qquad p_c \geq p_b, \quad (1.8)$$

$$\mathbf{v}_g = \mathbf{0}, \qquad p_c < p_b$$

where $\mathbf{K}$ is intrinsic permeability tensor, $K_{r\pi}$ is relative permeability of the π-phase, $\mathbf{v}_\pi$ - velocity of the π-phase relative to the solid, p_{ir} - the capillary pressure value corresponding to irreducible saturation and p_b - the bubbling pressure.

Fick's law is applied for the description of the diffusion of the binary gas species mixture (dry air and water vapour):

$$\mathbf{v}^d_{ga} = -\frac{M_a M_w}{M^2} D_{eff} \nabla\left(\frac{p_{ga}}{p_g}\right) = -\mathbf{v}^d_{gw} \quad (1.9)$$

where M_a, M_w and M are molar masses of dry air, water and moist air, respectively, and D_{eff} is the effective diffusion coefficient of vapour in air inside the pores of the medium.

Equation of state of perfect gases and *Dalton's law* are assumed for dry air, vapour and their mixture (moist air):

$$p_{ga} = \rho_{ga} T R / M_a, \qquad p_g = p_{ga} + p_{gw},$$
$$p_{gw} = \rho_{gw} T R / M_w, \qquad \rho_g = \rho_{ga} + \rho_{gw}. \quad (1.10)$$

where R is the universal gas constant.

For the solid skeleton the *effective stress principle* is assumed in its *finite* form as

$$\sigma' = \sigma + \left(p_g - S p_c - p_{atm}\right)\mathbf{I}, \quad (1.11)$$

where $\mathbf{I}$ is the second order unit tensor, p_{atm} - the atmospheric pressure.

Finally, for the closure of the model, some thermodynamic relations are used.

The *Kelvin equation* gives the relative humidity R.H. of the moist air inside the pores:

$$\text{R.H.} = \frac{p_{gw}}{p_{gws}} = \exp\left(-\frac{p_c M_w}{\rho_l R T}\right), \quad (1.12)$$

where the water vapour saturation pressure p_{gws} is constant, because we deal with isothermal case.

The model is completed by appropriate initial and boundary conditions.

The *initial conditions* specify the full fields of gas pressure, capillary pressure and displacements:

$$p_g = p_g^o, \qquad p_c = p_c^o, \qquad \mathbf{u} = \mathbf{u}^o \qquad \text{at } t=0. \quad (1.13)$$

The boundary conditions can be of the first kind or *Dirichlet's boundary conditions* on Γ_j^1:

$$p_g = \hat{p}_g \text{ on } \Gamma_g^1, \; p_c = \hat{p}_c \text{ on } \Gamma_c^1, \; \mathbf{u} = \hat{\mathbf{u}} \text{ on } \Gamma_u^1, \quad (1.14)$$

of the second kind or *Neumann's boundary conditions* on Γ_j^2:

$$-\left(\rho_{ga}\mathbf{v}_g - \rho_g \mathbf{v}^d_{gw}\right)\cdot\mathbf{n} = q_{ga} \qquad \text{on } \Gamma_g^2$$
$$-\left(\rho_{gw}\mathbf{v}_g + \rho_l \mathbf{v}_l + \rho_g \mathbf{v}^d_{gw}\right)\cdot\mathbf{n} = q_{gw} + q_l \text{ on } \Gamma_c^2 \quad (1.15)$$
$$\sigma\cdot\mathbf{n} = \mathbf{t} \qquad \text{on } \Gamma_u^2$$

and of the third kind or *Cauchy's (mixed) boundary conditions* on Γ_j^3:

$$\left(\rho_{gw}\mathbf{v}_g + \rho_l \mathbf{v}_l + \rho_g \mathbf{v}^d_{gw}\right)\cdot\mathbf{n} = \beta_c\left(\rho_{gw} - \rho_{gw\infty}\right) \qquad \text{on } \Gamma_c^3 \quad (1.16)$$

where the boundary $\Gamma = \Gamma_j^1 \cup \Gamma_j^2 \cup \Gamma_j^3$, $\mathbf{n}$ is the unit normal vector, pointing toward the surrounding gas, q_{ga}, q_{gw} and q_l are, respectively, the imposed dry air-, vapour- and liquid- flux, $\mathbf{t}$ - the imposed traction, $\rho_{gw\infty}$ - the mass concentration of water vapour in the undisturbed gas phase distant from the interface, while β_c is convective mass transfer coefficient.

The governing equations (1.1)-(1.16) have been discretized in space by use of the finite elements method and in time domain using a backward implicit finite difference scheme. The resulting set of nonlinear equations has been solved by means of a monolithic Newton - Raphson method.

2 DESCRIPTION OF THE BENCHMARK

The proposed benchmark is based on the experiment performed by Liakopoulos (1965) on a column of Perspex, 1 meter high, packed with sand and instrumented to measure the moisture tension at several points along the column during its desaturation due to gravitational effects. Before the start of the experiment ($t<0$) water was continuously added from the top and was allowed to drain freely at the bottom through a filter, until uniform flow conditions were established. At $t=0$ the water supply was ceased and the tensiometer readings were recorded as well as outflow and outflow rate at the bottom. The porosity and hydraulic properties of the sand were measured by Liakopoulos (1965).

The approximated equations of the Liakopoulos' saturation-capillary pressure and relative permeability of water-capillary pressure relationships, valid for saturations $S \geq 0.91$, have the following form:

$$S = 1 - 1.9722 \cdot 10^{-11}\, p_c^{\;2.4279}, \qquad K_{rl} = 1 - 2.207(1 - S)^{1.0121} \tag{2.1}$$

This test problem has been numerically solved previously by Liakopoulos (1965), Narasimhan & Whitherspoon (1978), Schrefler & Simoni (1988), Zienkiewicz et al. (1990), Schrefler & Zhan (1993), Gawin et al. (1995), as well as by Gawin & Schrefler (1996) to check their numerical models.

In the present solution the column is divided into 20 four-, eight- and nine-node isoparametric finite elements of equal size. At the beginning mechanical equilibrium state is assumed.

The *boundary conditions* are the following: for the lateral surface, $u_h = 0$, where u_h means horizontal displacement of soil; for the top surface, $p_g = p_{atm}$, where p_{atm} is atmospheric pressure, for the bottom surface, $p_g = p_{atm}$, $p_c = 0$, $u_h = u_v = 0$, where u_v is the vertical displacement of soil.

Young's modulus of soil, Poisson's ratio and Biot's constant are assumed similarly as by Schrefler & Simoni (1988).

The relative permeability of gas phase is assumed according to the relationship of Brooks & Corey (1966):

$$K_{rg} = (1 - S_e)^2 \left(1 - S_e^{5/3}\right), \qquad S_e = (S - 0.2)/(1 - 0.2) \tag{2.2}$$

Table 1. Material data

Young's modulus	$E = 1.3$ MPa
Poisson's ratio	$\nu = 0.4$
Biot's constant	$\alpha = 1$
Solid grain density	$\rho^s = 2000$ kg m^{-3}
Liquid density	$\rho^w = 1000$ kg m^{-3}
Porosity	$\phi = 0.2975$
Intrinsic permeability	$K = 4.5 \cdot 10^{-13}$ m^2
Water viscosity	$\mu^w = 1 \cdot 10^{-3}$ Pa·s
Air viscosity	$\mu^g = 1.8 \cdot 10^{-5}$ Pa·s
Gravitational acceleration	$g = 9.806$ m s^{-2}
Atmospheric pressure	$p_{atm} = 101325$ Pa

Table 1 shows the complete set of physical data used in calculations. First, the case of initial hydrostatic water pressure distribution, then the case of initial steady flow is considered.

3 NUMERICAL PROPERTIES OF THE DISCRETIZED MODEL

Let **A** be the global matrix for one element in the discretized problem. This matrix presents the following peculiarities:

-the matrix is not symmetric;

-the matrix is not diagonally dominated;

-the order of magnitude of diagonal terms is very different.

When solving the discretized set of non-linear equations, several numerical difficulties were encountered, hence different analyses were performed to investigate this fact.

The first study performed, to assess the different response characteristics of the interacting fields, was the eigenproblem solution for the matrix $\mathbf{A}^T\mathbf{A}$, because the matrix **A** is not symmetric. This analysis was performed at different time steps of the solution.

The square roots of the calculated eigenvalues are in the following ranges:

-displacement field (discarding the rigid body motion): $2 \cdot 10^7$ - $2 \cdot 10^5$;

-water field: $2 \cdot 10^{-6}$ - $2 \cdot 10^{-14}$;

-gaseous field: $2 \cdot 10^{-9}$ - $2 \cdot 10^{-15}$.

Only the displacement field presents zero (or nearly zero) eigenvalues with associated eigenvectors corresponding to constant distribution of the field variable. This happens for the rigid body motions, which are ruled out by the boundary conditions. All other eigenvectors correspond to distributions of the field variable of the same type of the approximating functions. This analysis was performed using the coefficient matrices at the end of the iteration process of the non-linear solution, in different time instants and for different elements in the mesh. Hence the conclusion is, that the system is ill-conditioned due to the response characteristics of the different fields, but no particular problems arise from this situation, except when applying the model close to fully saturated conditions: one equation, in fact, the gas mass balance vanishes so that a numerical technique is necessary to avoid the ensuing numerical problems.

The following three methods have shown high efficiency during calculations close to fully saturated conditions:

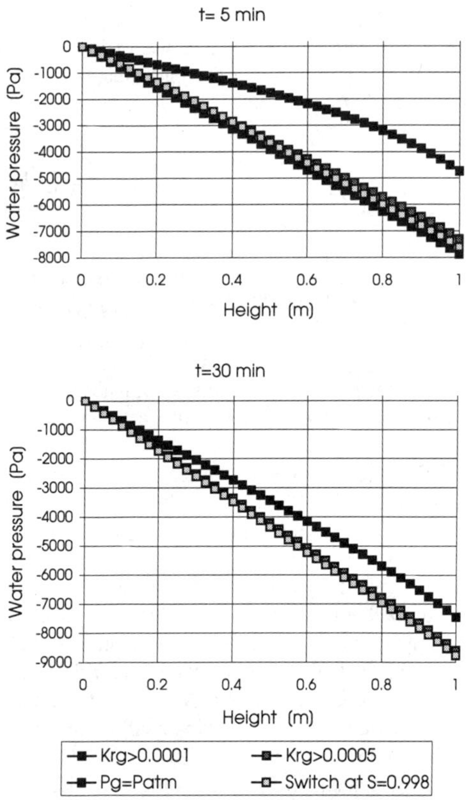

Figure 1. Comparison of the water pressure distributions at two time stations for the undeformable case

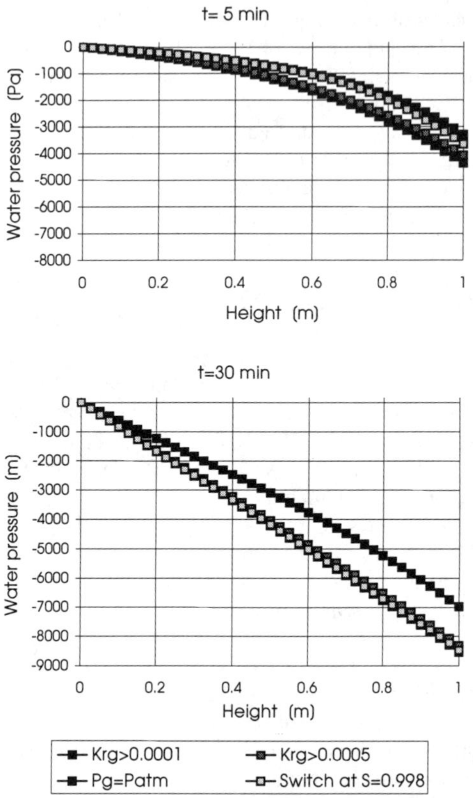

Figure 2. Comparison of the water pressure distributions at two time stations for the deformable case

-fixing gas pressure value in the partially saturated zone equal to atmospheric pressure (1-phase flow approach), Schrefler & Simoni (1988), Schrefler & Zhan (1993), Gawin et al. (1995);

-assumption of a very small, but finite value of gas relative permeability even for fully saturated state. This can be obtained by an according modification of the capillary pressure - saturation relationship Schrefler & Zhan (1993) and/or of the relative permeability curve Gawin & Schrefler (1996);

-application of "switching" between one-phase flow and fully two-phase solutions for a certain value of saturation (usually corresponding to bubbling pressure) Gawin et al (1995), Gawin & Schrefler (1996). Usually also application of an additional lower limit for the gas relative permeability is necessary to avoid oscillations in solution for pressures.

4 RESULTS OF THE CALCULATIONS

The calculations were performed for four cases, corresponding to the first and the third from the above mentioned methods, both for deformable and undeformable (Young's modulus value 10^6 times higher) material:

-one-phase flow - case 1;

-two-phase flow with switching at p_c= 2000 Pa (S=0.998, no gas flow below bubbling pressure) and additional lower limit for the gas relative permeability, $(K_{rg})_{min}$= 0.0001 - case 2;

- two-phase flow with switching at p_c= 0 Pa (S=1, no gas flow for fully saturated state) and additional lower limit for the gas relative permeability, $(K_{rg})_{min}$= 0.0005 - case 3;

- two-phase flow with switching at p_c= 0 Pa (S=1, no gas flow for fully saturated state) and additional lower limit for the gas relative permeability, $(K_{rg})_{min}$= 0.0001 - case 4.

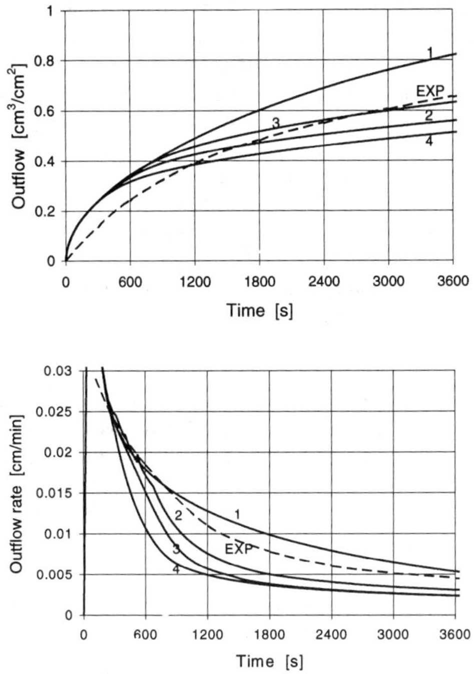

Figure 3. Comparison of the total outflow- and outflow rate-history with experimental data of Liakopoulos (1965) (broken line) for initial hydrostatic water pressure distribution (numbers in figures correspond to the cases analysed)

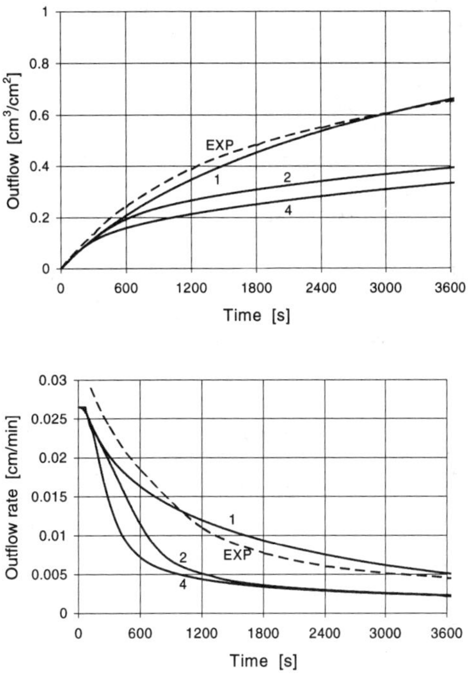

Figure 4. Comparison of the total outflow- and outflow rate-history with experimental data of Liakopoulos (1965) (broken line) for initial steady water flow conditions (numbers in figures correspond to the cases analysed)

Independently from the spatial- (10 or 20 elements) and time- discretization ($\Delta\tau$=10s, 1s or 0.5s) the same results were obtained. Furthermore 3x3 and 2x2 Gauss' points integration schemes were applied without visible differences in the obtained solution.

The water pressure solutions of the analysed cases, for two different time stations: τ=5 and 30 minutes, are compared in Figure 1 for deformable- and in Figure 2 for undeformable material. The results concerning other state variables, not shown here, are presented in Gawin & Schrefler (1996b).

Comparison of outflow and outflow rate at the bottom of the column allows an important check of all the obtained results. In fact outflow is due not only to desaturation, but also to the "squeezing" effect related to soil deformability, see Figure 5. This fact was firstly observed by Narasimhan & Witherspoon (1978), who, only by accounting deformability of soil were able to obtain pressure profiles similar to the experimental ones.

Fig. 3 presents a comparison of the experimental and numerical results, corresponding to the four cases analysed, for total outflow and outflow rate history at the bottom of the sand column.

The initial conditions of the analysed case are inconsistent (a too high outflow rate during the first seconds) with the original results of Liakopulos (1965). In fact, starting computations from the hydrostatic water pressure distribution results in zero outflow rate in first instant, which then increases rapidly (Figure 3) instead of outflow rate decreasing gradually from its initial value.

Thus, for better accordance of the proposed benchmark with the experiment, the ICs have been changed, i.e. p_c= 0 assumed for all the nodes, what corresponds to steady flow of the water through the sand column. As before the *mechanical equilibrium* state was assumed for t=0.

The time history of the total outflow and outflow rate through the bottom of the column for three of the analysed cases with the changed ICs are compared in Figure 4 with the experimental data of Liakopoulos (1965). After ceasing of water supply, at t=0, gradually decreasing outflow rate can be observed, what corresponds to the real physical conditions of the Liakopoulos' experiment. It is worth to

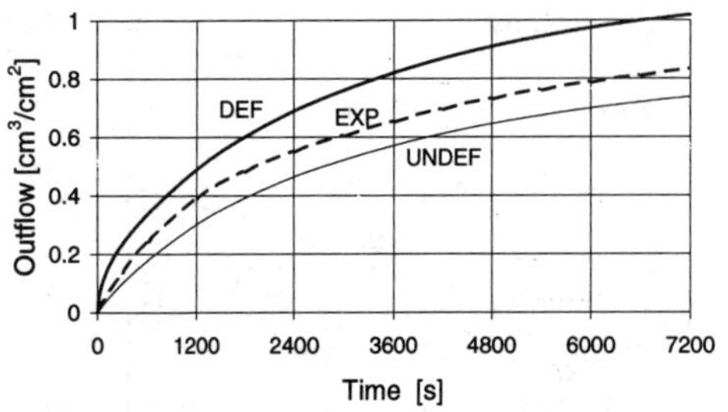

Figure 5. Comparison of the total outflow history (case 1) for initial hydrostatic water pressure for deformable (DEF) and undeformable (UNDEF) material with experimental data (EXP) of Liakopoulos (1965)

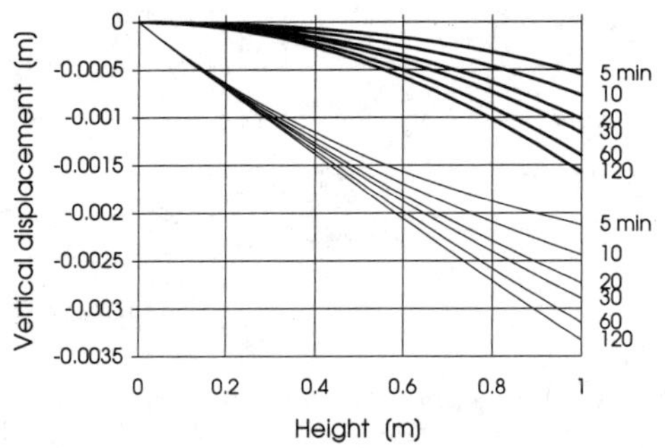

Figure 6. Comparison of the displacement profiles for one-phase flow solutions with two different ICs: initially steady water flow (normal line) and hydrostatic water pressure distribution (fine line)

underline, that the change of the initial conditions results also in a change of the curvature of the displacement diagrams as well as the maximal values of displacement, as shown in Figure 6.

5 CONCLUSIONS

Transition from fully- to partially- saturated state (or vice-versa) results in a delicate numerical problem, depending on the assumed mathematical model:

-no particular problems were experienced under the assumption that air phase remains at atmospheric pressure (one-phase flow). In this case only one mass balance (for water) is used;

-when dealing with two fluids flowing, problems arise from the necessity to switch from one mass balance equation (for water) to two mass balance equations (for water and gas) and vice-versa.

In the second case, three different techniques were presented for solving these problems: the proposed numerical procedures are however sensitive to the applied techniques.

It is possible to conclude that the problem is very demanding both from point of view of mathematical modelling and used numerical techniques. Hence this proposal for a benchmark problem, in which have been clearly specified:

-the material properties, in particular the lower threshold of the relative permeability;

-the numerical discretizations;

-the different switching procedures.

The steady flow initial conditions give a better representation of the outflow during the onset of the experiment without substantially affecting pressure distributions.

The solution for incremental formulations of linear momentum balance equation and effective stress has been presented in Schrefler & Gawin (1996). The ensuing solutions give practically the same results as those presented here.

REFERENCES

Brooks, R.N. & A.T.Corey 1966. Properties of porous media affecting fluid flow, *J. Irrig. Drain. Div. Am. Soc. Civ. Eng.*, 92 (IR2): 61-68

Gawin, D., P.Baggio, B.A.Schrefler 1995. Coupled heat, water and gas flow in deformable porous media. *Int. J. Num. Meth. in Fluids*, 20: 969-987

Gawin, D. & B.A.Schrefler 1996a. Thermo- hydro- mechanical analysis of partially saturated porous materials, *Engng.Comp.* 13(7): 113-143

Gawin, D. & B.A.Schrefler 1996b. Modelling of saturated-unsaturated transient in porous media. *Proc. of Int. Joint Conf. of Italian Group of Computational Mechanics and Ibero-Latin American Association of Computational Methods in Engineering*: 173-176. Padova, Italy

Liakopoulos, A.C. 1965. *Transient Flow through Unsaturated Porous Media*, Ph. D. thesis, Univ. of California, Berkeley .

Narasimhan, T.N & P.A. Witherspoon 1978. Numerical model for saturated-unsaturated flow in deformable porous media. 3. Applications, *Water Resour. Res.*, 14: 1017-1034.

Schrefler, B.A. & L.Simoni 1988. A unified approach to the analysis of saturated-unsaturated elastoplastic porous media. In G. Svoboda (ed), *Numerical Methods in Geomechanics*: 205-212, Rotterdam, Balkema.

Schrefler, B.A. & L.Simoni 1991. Comparison between different finite elements solutions for immiscible two phase flow in deforming porous media, in G.Beer & J.R.Booker (eds.), *Computer Methods and Advances in Geomechanics*, 1215-1220. Rotterdam: Balkema

Schrefler, B.A. & Z.Xiaoyong 1993. A fully coupled model for water flow and airflow in deformable porous media. *Water Resour. Res.*, 29: 155-167

Schrefler, B.A. & D.Gawin 1996. The effective stress principle: incremental or finite form? *Int. J. for Num. and Anal. Meth. in Geomechanics*. 20 (11): 785-815

Zienkiewicz, O.C., Y.M.Xie, B.A.Schrefler, A.Ledesma, N.Bicanic 1990. Static and dynamic behaviour of soils: A rational approach to quantitative solutions, II, Semi-saturated problems, *Proc. R. S. London.* A 429: 311-321

Computer Methods and Advances in Geomechanics, Yuan (ed.) © 1997 Balkema, Rotterdam, ISBN 90 5410 904 1

Research on mechanic hydraulic coupling analysis of bedrock for dams

Huang Jineng
Conghua City Architecture Design Institute, Guangzhou, China
Xu Weizu & Zeng Yingjuan
North China Institute of Water Resources and Hydropower, Beijing, China

ABSTRACT: Aiming at mechanic and hydraulic analyses of bedrock for dams, a method of seepage-stress coupling analysis is presented in the paper, in which a continuous and a discrete mathematical models are separately used. The influences of geometric parameters of the fissures, and especially of the existence of thin weakened layers within the rockmass on the results of the coupling analysis are investigated according to the method, and conclusions are gained.

1. INTRODUCTION

There are normally in big amount of tectonic surfaces such like fissures, layers and faults in the rockmass resulted from a variety of natural processes. The close interactions between mechanic and hydraulic behaviours of the rockmass may be practically observed as that the deformation manner of fissures could be changed due to the change of stress state in the rockmass, which results in a deviation of seepage features of the rockmass on the one hand, and the altered seepage charateristics would induce an associated change of the volumetric seepage forces on the rockmass, which makes changes of the stress state in the rockmass on the other hand. For the rock base of the hydraulic structures, a realistic prediction of seepage behaviours of the rockmass considering the seepage stress coupling would be important to prevent the accidents of dams caused by unpredicted seepage forces on the rockmass(Bellier, J. and Londe, P. 1976). The seepage-stress coupling analysis for fissured rockmass has been developed by the geotechnic engineering researchers since the recent years. L.C.Louis (1974) obtained a correlation between seepage coefficient K and normal stress σ_n as:

$$K=K_0\exp(-\alpha\sigma_n) \qquad (1)$$

where K_0 is a reference seepage coefficient, or surface seepage coefficient; α is an empirical factor. Zhang Youtian (1993), M.Oda (1986), Yang Yanyi and Zhou Weiyuan (1990), and P.J.Erban and K.Bell (1988) got significant results in succession in their seepage-stress coupling researches.

In this paper, a method of seepage-stress coupling analysis by separately employment of a continuous model and a discrete model is introduced. A computer program according to a two dimensional model of the method has been worked out. The sensitivity investigation of the analytical results upon geometrical parameters of the fissures, especially upon the existence of thin weakened layers within the rockmass is performed through a systematical analysis.

2. SEEPAGE FLOW ANALYSIS OF ROCKMASS FOR DAMS

2.1 *Formation of Seepage Fissure Network in Rockmass for Dams and Dissection of Seepage Flow Elements*

According to statistical analysis of the measured geometrical parameters of fissures, the propability models of the fissure parameters as attitude, trace length and width for different sets of fissures are established. The attitude, location and number of the fissures are then simulated by computer on the basis of Monte Carlo principle, and the charts of interconnected fissure networks could be drawn up. The fissure segment between two adjacent intersection nodes in the fissure network can be thereafter regarded as an one dimensional seepage flow element.

For seepage flow in narrow fissures, the Darci's law would be conformed, and the seepage coefficient K can be expressed by the following formula amended by Louis (1974):

$$K = \frac{\beta g a^2}{12 \nu c} \qquad (2)$$

where β is the ratio between interconnected area and total area of the fissures; g is the gravity acceleration; a is the fissure width; υ is the kinematic vicosity factor of water; c is a coefficient for revision of the fissure surface roughness, and can be determined by:

$$c = 1 + 8.8(\frac{\Delta}{d})^{1.5} \qquad (3)$$

Δ is the unevenness of the fissure surface; d is the hydraulic radius.

The seepage flow in fissures would be described by a one dimensional quadratic model in the analysis.

2.2 *Seepage Flow Analysis for Thin Weakened Layers*

The 6-nodes isoparametric elements with curved sides are used for simulating the seepage flow in the weakened layers. The seepage coefficients along the layer surface (K_ξ) and perpendicular to the layer surface (K_η), respectively, can be gained according to the literature (Means, R.E. and Parcher, J.V. 1964) by:

$$K_\xi = A_\xi \frac{e^2}{1+e}$$
$$K_\eta = A_\eta \frac{e^2}{1+e} \qquad (4)$$

where, e —— void ratio of the material filling up the layers; A_ξ, A_η —— coefficients determined from tests.

By employment of the variational prineiple, the seepage flow equations can be obtained as:

$$[\Phi]\{H\}=\{Q\} \qquad (5)$$

where, $[\Phi]$ is the seepage matrix for the being investigated region; $\{H\}$, $\{Q\}$ are the water head and water discharge colunm vectors, respectively.

By introducing the hydraulic boundary conditions, and the equilibrium condition of Σ Q=0 at each node with unknown water head, the seepage flow field in the rockmass can thus be solved.

3. ELASTO PLASTIC STRESS ANALYSIS OF ROCKMASS

The rockmass is regarded as an elasto plastic medium, and the thin weakened layer is simulated by the wellknown model of Goodman. The Drucker Prager yielding criterion with the isoclinic hardening features is accepted in the analysis. The basic finite element equations are then solved by using the increment initial stress method.

4. PERFORMANCE OF SEEPAGE STRESS COUPLING ANALYSIS

By ignoring the effect from shear expansion of the fissures, and consulting the fitting curves provided by Barton, the relationship between normal stress σ_n and normal displacement $\tilde{\delta}_{n,T}$ on the fissure surface can be expressed as:

$$\sigma_n = \frac{K_{n,0} + \tilde{\delta}^*_{n,T}}{\mu}\left[(\frac{\tilde{\delta}^*_{n,T}}{\tilde{\delta}^*_{n,T} - \tilde{\delta}_{n,T}})^{\mu} - 1\right] \qquad (6)$$

where $K_{n,0}$ is the gradient of the fitting curve at the origin point; μ is a positive parameter, the limit value of $\tilde{\delta}^*_{n,T}$ may be taken as the initial average width a of the fissures.

Equation (6) can be used for determining the fissure width(a) under a certain stress state, and then the associated seepage coefficient K_f can be obtained from the equation(2). For the layers, the invariant total volume of the solid particles (V_S) inside the layers (i.e. without any occurence of piping or quick soil in the layers) is assumed. Under this assumption, from the volumetric strain under a certain stress state of the rockmass, the associated void ratio e_1 can be calculated from the initial void ratio e_0 by the following equation:

$$e_1 = (1+e_0)\frac{1-2\mu}{E}(\sigma_1+\sigma_2+\sigma_3)+e_0 \qquad (7)$$

The seepage coefficient corresponding to e_1 can thus be determined from equation (4).

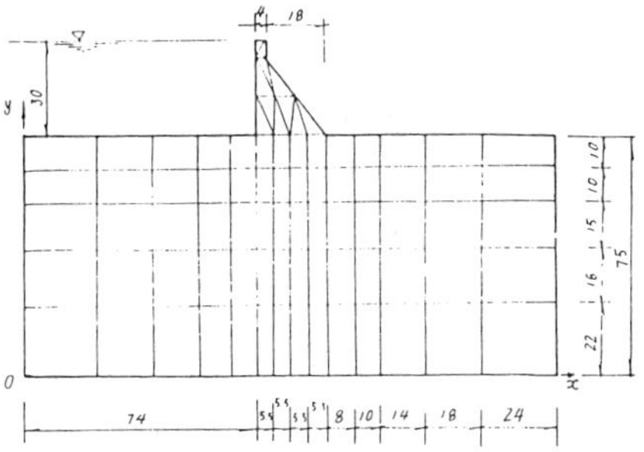

Fig.1 Schematic Diagram of Finite Element Mesh in Dam and Dam Base Rock

5. EXAMPLES

(1), From the first example shown in Fig.1, two sets of fissures are selected, and the effect due to the change of their concentration, initial width, inclination and trace length on the results of coupling analysis are investigated through calculations. The results from coupling and noncoupling analyses are also obtained, respectively, and compared each other.

The finite element mesh used in stress analysis is shown in the figure. The parameters of the rockmass are: Elasticity modulas E=1.67 × 10^4 MPa; Poisson ratio ν =0.25; Density γ =25kN/m^3, cohesional resistance c=9200Pa ; Internal friction angle Φ= 41 ° ; the parameters of fissures are: $K_{n,0}$=10^4MPa; μ =0.5; and $\tilde{\delta}^x_{n,T}$ is the initial width of the fissure element.

From the results of the above calculations, it is indicated that the water heads for the whole seepage flow field are getting increased and more uniform with the increment of the fissure concentration. When the inclination angle of the fissures with the horizontal plane is bigger than 60 ° or less than 30 ° , the seepage pressure head at the dam heel (upstream edge) point will be bigger than that at this point when the inclination angle is 45 ° . The water heads in the whole seepage flow field will be totally increased when the initial fissure width is bigger. The water heads at the upstream portion will be increased, but decreased at the downstream portion when the trace length is increased and inducing the improvement of connectivity. The comparison between the results from coupling and noncoupling analyses is shown in Fig.2.

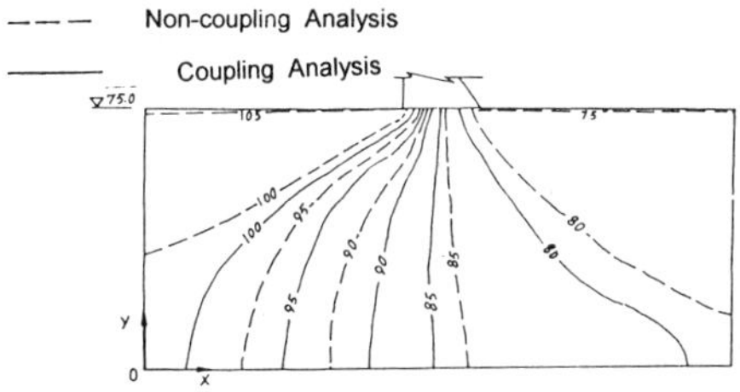

Fig.2 Iso-head Lines in the Seepage Flow Field

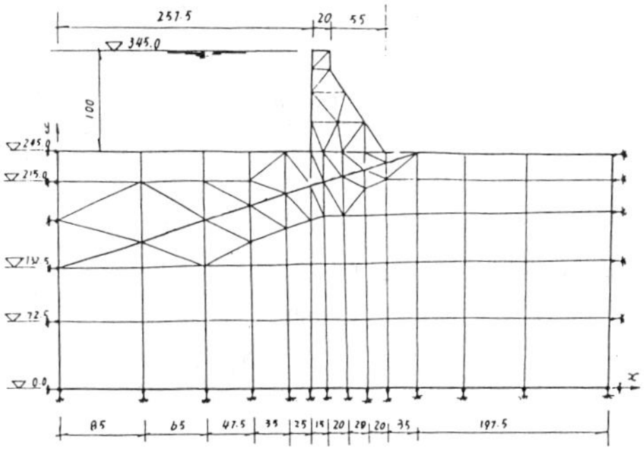

Fig.3 Schematic Diagram of Finite Element Mesh in Dam and Dam Rock in Case of Existence of a Weakened Layer

(2), Another example given in Fig.3 is used to illustrate the effect of changes of weakened layer thickness upon the results of the coupling analysis.

The parameters for the concrete dam:Elasticity modulas E=0.22 × 10^5 MPa; Poisson's ratio ν =0.167; Density γ =24kN/m^3; The parameters for the rockmass: Elasticity modulas E=0.22 × 10^5MPa; Poisson ratio ν =0.167; Internal friction angle Φ=41 ° ; Cohesion resistance c=92kPa; For the

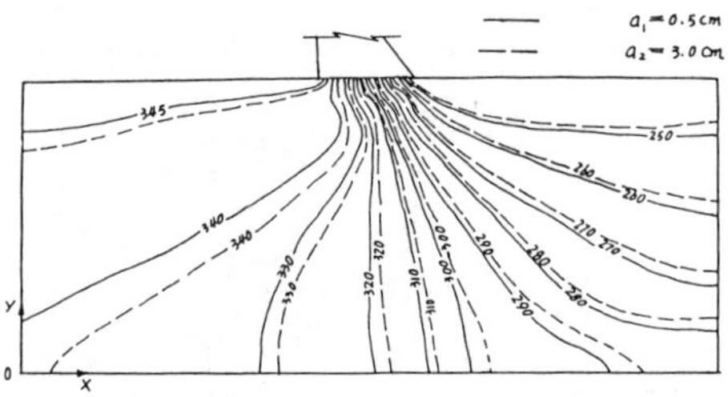

Fig.4 Iso-head Lines in the Seepage Flow Field in Deviation of Layer Thickness

layers:Elasticity modulas E=0.2 × 10^5MPa; Shearing modulas G=0.68 × 10^3MPa; Internal friction angle Φ =30 ° ; Cohesion resistance c=0.1MPa; Void ratio e=0.65.

The results of isolines in the seepage flow field in a deviation of layer thickness are shown in Fig.4.

6. CONCLUSIONS

(1) There are apparent interactions between seepage and stress in the fissured rockmass, and a coupling analysis would be important for insurance of the dam safety.

(2) On the basis of a seepage flow analysis with the discrete model, and a separately performed elasto plastic stress analysis with the continuous model, the seepage-stress coupling analysis of rockmass could be carried out through a statistical approach. This method proposed in the paper is proved to be feasible in prediction of the seepage flow and stress states in the bedrock of dams under the critical conditions.

REFERENCES

Louis, C. 1974. Rock Hydraulics, 《 *Rock mechanies* 》

Zhang Youtian & Liu Zhong 1993. Hydraulic behavior of rock fractare analyzed by using fractals, *International Symposium on Application of Computer Methods in Rock Mechanics and Engineering.*

Oda, M. 1986. An Equivalent Continuum Model for Coupled Stress and Fluid Flow Analysis in Jointed Rock Masses, *Water Resources Research,* Vol.22, No.13.

Yang Yanyi & Zhou Weiyuan 1991. A Coupled Seepage-damage Analysis Model for Jointed Rock Masses and Its Application to *Rock* Engineering (in Chinese). *Shuili Xuebao*, No.5.

Erban, P.J. & Gell, K. 1988. Consideration of Interaction between Dam and Bedrock in Coupled Mechanic Hydraulic FE Program, *Rock Mechanics and Rock Engineering, Vol.21, No.2.*

Bellier, J. & Londe, P. 1976. The Malpasset Dam, Engineering Foundation, *Conference on Evaluation of Dam Satety.*

Means, R.E. & Parcher, J.V. 1964. 《 *Physical Properties of Soils* 》

Computer Methods and Advances in Geomechanics, Yuan (ed.) © 1997 Balkema, Rotterdam, ISBN 90 5410 904 1

Coupled THM behaviour of crushed salt: Analysis of an in situ test simulating a repository

S.Olivella, A.Gens & E.E.Alonso
Departamento de Ingeniería del Terreno, ETSICCPB UPC, Barcelona, Spain

ABSTRACT: A formulation for coupled thermo-hydro-mechanical behaviour of salts (rock and porous backfills) is summarised in this paper. The balance equations are presented together with some aspects of the mechanical constitutive law for crushed salts. An application to simulate the thermo-mechanical analysis of an in situ test is presented. The predictions of the model are in quite good correspondence with the in situ measurements.

1 INTRODUCTION

Salt rocks are regarded, among other geological media, as one of the alternatives for radioactive waste disposal in underground repositories. Considering this alternative has motivated research on the coupled thermo-hydro-mechanical behaviour of such media. A comprehensive approach which includes constitutive modelling of crushed salts based on laboratory creep tests and balance equations for mass (water, air and salt), momentum (equilibrium of stresses, Darcy's and Fick's laws) and energy (equation of heat flow in a multiphase system) was developed. The theoretical approach was transformed into its numerical counterpart leading to the finite element simulator (CODE_BRIGHT). In this paper, this approach is used to model an in situ test which is in progress at the Asse salt mine (Germany).

The test simulates the emplacement of a nuclear canister in an opening excavated in salt rock. Repository conditions are reproduced by means of a heater which is surrounded by crushed salt. Heating started at the end of 1990 and the test is still in progress. From the beginning of the test, temperatures, drift convergences, stresses and porosities have been measured. Initially, a self-settlement of the backfill took place so, a gap developed between the fill and the roof. This gap closed up almost immediately after heating started. Higher temperatures in rock accelerate the convergence of the opening and subsequent compaction of the crushed salt.

In this paper the theoretical approach developed by the authors to analyse saline media behaviour is presented. This approach consists of balance equations and constitutive equations. The formulation contains equations of flow of water and air which are presented for completeness but are not used in the application included in this paper. Since this analysis involves essentially thermo-mechanical behaviour, the stress-strain constitutive model is described in more detail. This formulation has been the basis for the development of a 3-D FEM code called CODE_BRIGHT (Olivella et al, 1996). Using this program, an in situ experiment simulating a repository in a salt rock formation has been analysed.

2 GOVERNING EQUATIONS

This section summarises the coupled approach that includes the balance equations and some constitutive laws.

The medium is composed by three components: salt, water and air; which are present in three phases: solid, liquid and gas (air is not permitted in the solid phase and salt is not permitted in the gas phase). The balance equations are listed below using the following notation:

ϕ: porosity; ρ: density;
ω: mass fraction; $\theta=\omega\rho$: partial density;
S_l: liquid degree of saturation;
S_g: gas degree of saturation;
j: total mass flux; **q**: advective flux;
i: non-advective mass flux;
u: solid displacements; σ: stress tensor;
b: body forces;
$\mathbf{i}_c$: conductive heat flux;
E: specific internal energy;
$\mathbf{j}_E$: energy fluxes due to mass motion;
f: source/sink terms;
superscripts *h*, *w* and *a* refer to salt, water and air, respectively;
subscripts *s*, *l* and *g* refer to solid, liquid and gas phase, respectively.

The governing equations include: stress equilibrium equations (1, 2 or 3 according to the dimensions of the problem), mass balance equations (different species) and energy balance equation for the medium as a whole (thermal equilibrium is assumed).

Mechanical equilibrium equations (1, 2 or 3 dimensions):

$$\nabla \cdot \sigma + \mathbf{b} = \mathbf{0}$$

Water mass balance:

$$\frac{\partial}{\partial t}\left(\theta_s^w(1-\phi)+\theta_l^w S_l\phi+\theta_g^w S_g\phi\right)+ \\ +\nabla\cdot(\mathbf{j}_s^w+\mathbf{j}_l^w+\mathbf{j}_g^w)=f^w$$

Air mass balance:

$$\frac{\partial}{\partial t}\left(\theta_l^a S_l\phi+\theta_g^a S_g\phi\right)+\nabla\cdot(\mathbf{j}_l^a+\mathbf{j}_g^a)=f^a$$

Energy balance:

$$\frac{\partial}{\partial t}\left(E_s\rho_s(1-\phi)+E_l\rho_l S_l\phi+E_g\rho_g S_g\phi\right)+ \\ +\nabla\cdot(\mathbf{i}_c+\mathbf{j}_{Es}+\mathbf{j}_{El}+\mathbf{j}_{Eg})=f^Q$$

Water in inclusions balance:

$$\frac{\partial}{\partial t}\left(\theta_s^w(1-\phi)\right)+\nabla\cdot(\mathbf{j}_s^w)+f_s^w=0$$

Salt mass balance:

$$\frac{\partial}{\partial t}\left(\theta_s^h(1-\phi)+\theta_l^h S_l\phi\right)+\nabla\cdot(\mathbf{j}_s^h+\mathbf{j}_l^h)=f^h$$

The following variables are considered as the unknowns, i.e. the variables considered as independent: displacements, $\mathbf{u}=(u_x,u_y,u_z)$, liquid pressure, P_l (MPa), gas pressure, P_g (MPa) temperature, T (°C), mass fraction of water in incl., ω^w_s(-) and porosity, ϕ(-); respectively for the equations listed above.

The stress equilibrium equations are a simplified form of the balance of momentum for the porous medium. Mass balance of water, salt and air are established. Since the assumption of equilibrium is made, the mass of each species as present in any phase (solid, liquid or gas) is balanced for the porous medium as a whole. In this way, one equation for each species is obtained. The equilibrium assumption implies that partition functions are required to compute the fraction of each species in each phase. Exception is made for the treatment of brine inclusions. In this case, water in inclusions cannot be considered in equilibrium with water in pores. Hence an extra equation to balance water in solid is required.

Each partial differential equation is naturally associated to an unknown. These unknowns can be solved in a coupled way, i.e., allowing all possible cross coupling processes that have been implemented. Alternatively, the code allows one to solve any uncoupled problem to obtain a single unknown. More details of the formulation presented here can be found in Olivella et al (1994).

Associated to this set of balance equations, there is a set of constitutive relationships. These are summarised as follows:

Equilibrium restrictions		
Solubility	Dissolved salt	ω_l^h
Henry's law	Dissolved air	ω_l^h
Psychrome-tric law	Vapour mass fraction	ω_g^w

and:

Constitutive equations		
Darcy's law	liquid and gas advective flux	$\mathbf{q}_l$, $\mathbf{q}_g$
Fick's law	vapour and salt non-advective flux	$\mathbf{i}_g^w$, $\mathbf{i}_l^h$
Inclusion migration law	brine inclusions non-advective flux	$\mathbf{i}_s^w$
Fourier's law	conductive heat flux	$\mathbf{i}_c$
Retention curve	degree of saturation	S_l, S_g
Mechanical constitutive model	Stress tensor	σ
Phase density	liquid density	ρ_l
Gases law	gas density	ρ_g

3 MECHANICAL CONSTITUTIVE MODEL

Among the constitutive laws listed above, one that has received special attention is the mechanical constitutive model to describe the creep behaviour of crushed salts.

The model is based on two mechanisms of deformation:

- Fluid assisted diffusional transfer creep (FADT).
- Dislocation creep (DC).

Using an idealised geometry, it has been shown that strain rates for these two mechanisms could be obtained as:

$$\frac{d\varepsilon}{dt} = \frac{d\,\varepsilon^{e}}{dt} + \frac{d\,\varepsilon^{FADT}}{dt} + \frac{d\,\varepsilon^{DC}}{dt}$$

where the following forms were adopted for each mechanism:

$$\frac{d\,\varepsilon^{e}}{dt} = \mathrm{D}^{-1}\frac{d\sigma'}{dt}$$

$$\frac{d\,\varepsilon^{FADT}}{dt} = \frac{1}{2\eta^{d}_{FADT}}(\sigma' - p'\mathrm{I}) + \frac{1}{3\eta^{v}_{FADT}}p'\mathrm{I}$$

$$\frac{d\,\varepsilon^{DC}}{dt} = \frac{1}{\eta^{d}_{DC}}\Phi(F)\frac{\partial G}{\partial \sigma'}$$

in which D is the elastic stiffness matrix, and the viscous coefficients (η's) for FADT and DC mechanisms are defined below.

For FADT:

$$\frac{1}{\eta^{v}_{FADT}} = \frac{16B(T)\sqrt{S_l}}{d_0^3}g^{v}_{FADT}(e)$$

$$\frac{1}{2\eta^{d}_{FADT}} = \frac{16B(T)\sqrt{S_l}}{d_0^3}g^{d}_{FADT}(e)$$

For DC:

$$F = G = \sqrt{q^2 + \left(\frac{p}{\alpha_p}\right)^2}$$

$$\Phi(F) = F^n$$

$$\alpha_p = \left(\frac{\eta^{v}_{DC}}{\eta^{d}_{DC}}\right)^{\frac{1}{n+1}}$$

where:

$$\frac{1}{\eta^{v}_{DC}} = A(T)g^{v}_{DC}(e)$$

$$\frac{1}{\eta^{d}_{DC}} = A(T)g^{d}_{DC}(e)$$

Both for FADT and DC a function of void ratio appears (g^v and g^d), which come from the adopted geometry of grains and pores. These functions are plotted in Fig. 1.

It can be seen that the FADT contribution has been put into a viscoelastic form (with nonlinear viscosities) while the DC contribution is cast in a viscoplastic form.

For the elastic counterpart (ε^e) we have used a linear form with young modulus E and Posson's ratio υ.

The definition of FADT mechanism requires: $B(T)$: temperature dependent parameter; d_0: grain size ; e: ratio between volume of pores and volume of solid (void ratio); S_l: ratio between volume of brine and volume of pores. In the analysis presented in this paper, dry conditions have been assumed, so S_l=0 and FADT does not operate.

For DC mechanism: $A(T)$: temperature dependent parameter (same value as for rock); e: ratio between volume of pores and volume of solid (void ratio); n: stress power (same value as for rock), are required.

This model was developed in Olivella et al (1993) where more details can be found.

4 ANALYSIS OF TSS IN SITU TEST

An in situ test simulating a repository in rock salt is currently in progress in the Asse salt mine (Germany). The TSS (Thermal Simulation of Drift Emplacement) test consists of two parallel tunnels (length=70 m) separated 14.8 m where a total of 6 heaters (length=5.5 m, diam.=1.5 m) with a thermal power of 6.4 kW each were emplaced (GRS, Droste et al 1996).

A preliminary analysis simulating this in situ test has been performed. The calculations have been carried out using CODE_BRIGHT (Olivella et al, 1996).

A two-dimensional finite element mesh (Fig. 2) with 754 nodes has been used to solve the coupled thermo mechanical problem. The domain modelled includes

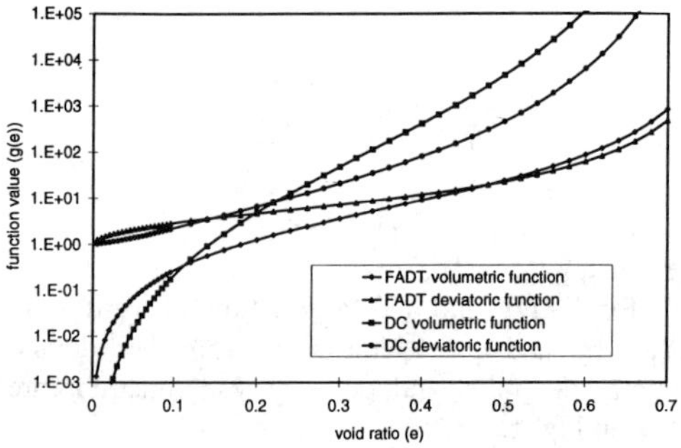

Figure 1. Functions of void ratio.

the cross section of one tunnel where the heaters are situated and part of the confining host rock. The assumption of symmetry along the plane between the tunnels is quite realistic (field measurements confirm it). Initial and boundary conditions have been obtained from published reports by GRS (Droste et al 1996).

A stress of 10 MPa has been used as boundary condition on the top of the simulated domain. It should be mentioned, however, that the depth of the tunnels is about 750 to 800 m which would lead to an overburden stress of 16 to 17 MPa. The lower value used is explained by the stress reduction caused by the different openings existing near the test area. In fact, in situ measurements indicate stresses of the order of 12 MPa.

For this analysis the following creep parameters for DC mechanism have been used:

$$A(T) = 5 \times 10^{-6} \exp(\frac{-59640}{RT}) \text{ s}^{-1}\text{MPa}^{-n}$$

$$n = 5.375$$

which are based on published reports from GRS. The elastic properties used are: elastic modulus of 200 MPa for the fill and 25000 MPa for the rock, and Poisson's ratio 0.3 in both cases. The creep model for the rock salt is obtained using void ratio equal or near to zero in the same law for crushed salt.

On the other hand, for this analysis the backfill is assumed dry as mentioned above, so the FADT counterpart of the constitutive model does not operate.

The initial porosity of the backfill is 0.35 and at the end of the simulated period (5 years: 1991-1995) a value in the order of 0.28 has been achieved according to the calculations.

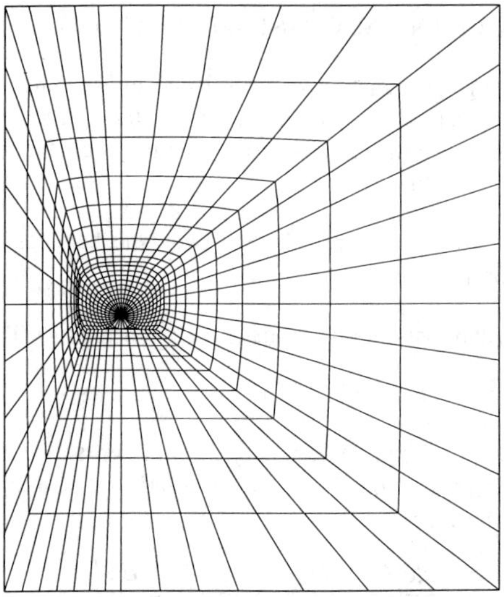

Figure 2: Finite element mesh

During the process of model development the grain sizes of the materials used in laboratory tests were quite uniform. This is justified by the fact that the main interest is the understanding of the deformation processes. However, the materials used to backfill the openings have a wide range of sizes. For this reason, the theoretically derived model is consistent with initial void ratios above 0.75 (porosity>0.42) which are obtained without compaction if the grain size is uniform. When the grain size is not uniform the initial void ratios obtained (without compaction) are lower than 0.75. For instance in this in situ test a value of 0.54 (porosity=0.35) was measured after backfill emplacement (without compaction). Since the model does not contain explicitly a dependence on this initial void ratio, a modification of the model was introduced by means of an effective void ratio.

$$e_{eff} = e + \frac{e^3}{e_{nat}^3}(e_{max} - e_{nat})$$

where e_{max} represents a value that would be obtained using uniform grain size, e_{nat} is the value actually obtained without compaction, and e is void ratio at any time during volumetric creep. In this analysis 0.75 and 0.54 have been used, respectively, for e_{max} and e_{nat}. This modification allows a reduction of the stiffness of the material for high void ratios while keeping a similar behaviour to the predictions of the theoretical law as void ratio reduces.

Thermal conductivity is a key parameter because it changes with porosity, which reduces in the porous fill as convergence progresses. The following linear function has been used:

$$\lambda(\phi) = \lambda_r + (\lambda_0 - \lambda_r) \times \left(\frac{\phi}{\phi_0} \right)$$

where λ_r and λ_0 represent thermal conductivities of rock (5.5 W/mK) and porous reference material (0.01 W/mK), respectively, and ϕ and ϕ_0 (0.37) are the current and reference porosities. This law gives a low thermal conductivity for the backfill which progressively increases towards the value for rock salt as porosity reduces.

The results obtained in the analysis are summarised in Figures 3, 4, and 5, one showing the temperature evolution in the heater and in the tunnel roof, another one showing the convergence evolution, both vertical and horizontal, and the last one showing the stresses in the fill.

This thermal conductivity variation causes the temperature evolution to show a maximum value and afterwards to decrease as thermal conductivity of the porous material increases.

The predicted convergences are in good agreement with the measurements. However, it should be mentioned that the convergences are essentially controlled by the behaviour of the host rock. This is because the stresses in the fill are quite low and hence, the fill does not play a significant supporting role.

Also, at the beginning of the in situ test, a gap developed between the backfill and the tunnel roof. This gap closed almost immediately. This was also incorporated in the simulation and it is the reason for the stress reduction (absolute value) at the beginning of the stress evolution (see Figure 5).

The predicted stresses are below the measured values. Since the computed convergences are in accordance with the measured ones, it seems that the stiffness of the porous material should be higher. This can be achieved by modification of the effective void ratio described above. However, this would lead to slower convergence rates. Therefore, both, the properties of the backfill and the host rock (probably including boundary conditions) should be modified in a new analysis.

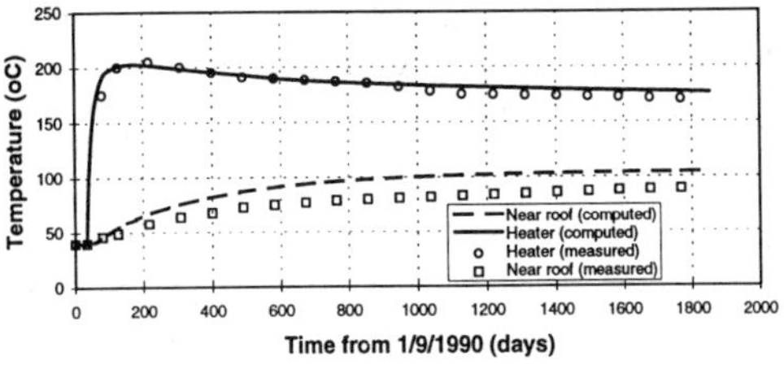

Figure 3. Temperature evolution

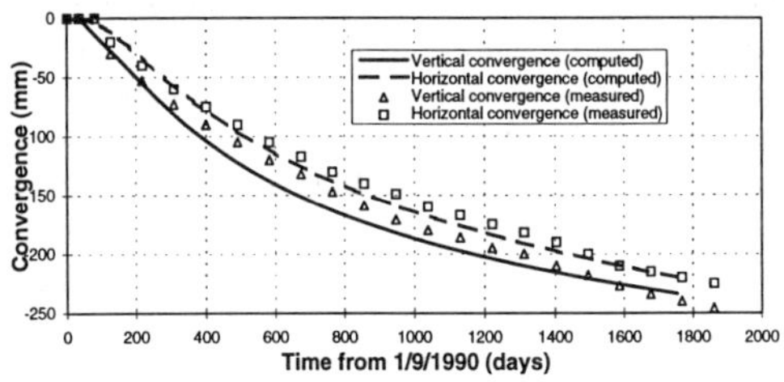

Figure 4. Convergence evolution

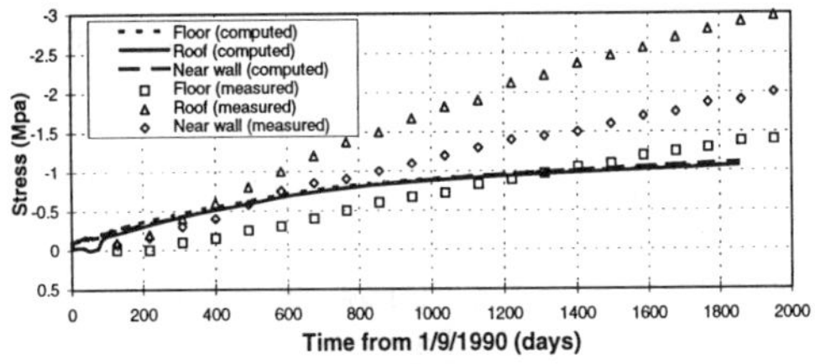

Figure 5. Stresses in the backfill

Since, small amounts of brine have been measured in the field, the possible influence of the FADT mechanism should also be investigated.

Another important factor that has not been mentioned yet is the three-dimensionality of the problem. The simulation has been carried out on a hot section (including the heater) but the measurements are available also in cold sections (due to heaters discontinuity). Therefore, a more appropriate analysis would require a 3-D representation of the domain.

5 CONCLUSIONS

A formulation for the coupled THM analysis of saline media has been summarised in this paper. A thermo-mechanical analysis shows that the behaviour of an in situ test simulating a repository in a salt mine can be properly modelled. Further work is necessary in order to improve the predictions obtained so far. Probably modifications of the parameters used for the creep behaviour of the crushed salt together with modifications of boundary conditions would lead to more accurate calculation of the stresses in the backfill.

On the other hand, the influence on the behaviour of such systems caused by the possible presence of small amounts of moisture is one of the less known aspects. These small quantities of brine may be sufficient for causing dissolution/ precipitation of salt induced by stress concentration at the pore scale level which may cause porosity reductions. Therefore, it should be investigated in further analyses if discrepancies between measurements and predictions obtained in analyses assuming dry conditions could be explained by the presence of humidity.

6 ACKNOWLEDGEMENTS

This work has been carried out on the framework of a project funded by ENRESA (Empresa Nacional de Residuos, Spain) and the European Community Commision.

7 REFERENCES

Olivella, S.; A. Gens, J. Carrera, and E. E. Alonso, 1993: Behaviour of Porous Salt Aggregates. Constitutive and Field Equations for a Coupled Deformation, Brine, Gas and Heat Transport Model. Mechanical Behaviour of Salt III, ISBN 0-87849-100-7.

Olivella, S., J. Carrera, A. Gens, E. E. Alonso,1994. Non-isothermal Multiphase Flow of Brine and Gas through Saline media. Transport in Porous Media, 15, 271:293

Droste, J. H.K. Feddersen, T. Rothfuchs and U. Zimmer, 1996, The TSS Project: Thermal Simulation of Drift Emplacement.(GRS).

Olivella, S., A. Gens, J. Carrera, E. E. Alonso, 1996, 'Numerical Formulation for a Simulator (CODE_BRIGHT) for the Coupled Analysis of Saline Media " Engineering Computations, Vol 13, No 7, pp: 87-112

Computer Methods and Advances in Geomechanics, Yuan (ed.) © 1997 Balkema, Rotterdam, ISBN 90 5410 904 1

Numerical modeling of unsaturated porous media as a two and three phase medium: A comparison

G. Klubertanz, L. Laloui & L. Vulliet
Soil Mechanics Laboratory, EPF Lausanne, Switzerland

ABSTRACT: A numerical simulation of the hydro-mechanical behaviour of an unsaturated porous medium is presented. The coupled hydro-mechanical model used is based on the continuum theory of mixtures and treats the unsaturated soil as a three phase porous medium (solid, liquid and gas). Principal variables are the solid deformation, the liquid pressure and the gas pressure. The two fluid phases are in motion and a non-linear pore pressure - saturation relation is used. The resulting system of equations is discretized in space using the finite element technique and in time by the Θ - method. The comparison of numerical results and experimental test data shows that this hydro-mechanical model is capable to reproduce the principal phenomenon in the case of a hydric solicitation. The three phase formulation is compared to a simplified version considering only a two phase medium with static air phase. Merits and shortcomings of the two approaches are shown.

1. HYDRO-MECHANICAL MODEL

1.1 Introduction

The hydro-mechanical behavior of unsaturated soils can be modeled via a formulation based on the momentum and mass balance equations of its three phases: solid, liquid (water) and gas (air). If one allows for interaction between the phases, flow of the fluids and skeleton deformation (in the following linear elastic), the numerical model achieved covers most of the important features of unsaturated soils.

In this paper, we use the definition of unsaturated soil as a mixture of three phases: solid, water and air. However, according to the mixture theory, such a mixture would be saturated in the sense that the three phases occupy the full volume (no empty space).

In the following, we will report the main equations that can be found using an approach based on work by Hassanizadeh & Gray (1979) and Schrefler *et al.* (1990). For details of the derivation the reader is referred to Klubertanz *et al.*(1997)

1.2 Balance equations

The momentum balance of the three phase mixture, reads

$$\nabla \bullet (\sigma'_{ij} - S^w p^w \delta_{ij} - (1-S^w)\, p^a\, \delta_{ij}) + \rho g_i = 0 \qquad (1)$$

where σ'_{ij} is the effective stress tensor, p^w and p^a the water and the air pressure, g_i the gravity vector and ρ the density of the mixture as $\rho = (1\text{-}n)\, \rho^s + n\, S^w \rho^w + n\, S^a \rho^a$. Note that here ρ^s, ρ^w and ρ^a are the solid, water and air true densities respectively. S^w is the water saturation, defined as the ratio of water volume divided by the total pore volume and S^a the degree of air saturation, defined as the ratio of air volume divided by the total pore volume. Further n denotes the porosity, δ_{ij} the Kronecker delta and $\nabla\bullet$ the divergence operator. Please note that throughout this paper quantities related to the solid, water or air phase are denoted by the superscripts s, w and a respectively.

The mass balance equations read (Klubertanz *et al.*, 1997)

$$\left(\frac{n}{S^w}\frac{\partial S^w}{\partial p^c}+\frac{n}{\rho^w}\frac{\partial \rho^w}{\partial p^w}\right)\frac{\partial p^w}{\partial t}-\frac{n}{S^w}\frac{\partial S^w}{\partial p^c}\frac{\partial p^a}{\partial t}+$$

$$\nabla \bullet v^s+\nabla \bullet v^{ws}+v^w\,\nabla n+\frac{n}{S^w}v^w\nabla S^w+ \qquad (2)$$

$$\frac{n}{\rho^w}v^w\nabla\rho^w-\frac{1}{\rho^s}v^s\nabla n=0$$

for the summed water - solid balance and

$$\left(\frac{n}{1-S^w}\frac{\partial S^w}{\partial p^c}+\frac{n}{\rho^a}\frac{\partial \rho^a}{\partial p^a}\right)\frac{\partial p^a}{\partial t}-\frac{n}{1-S^w}\frac{\partial S^w}{\partial p^c}\frac{\partial p^w}{\partial t}+$$

$$\nabla \bullet v^s+\nabla \bullet v^{as}+v^a\,\nabla n-\frac{n}{1-S^w}v^a\nabla S^w+$$

$$\frac{n}{\rho^a}v^a\nabla\rho^a-\frac{1}{\rho^s}v^s\nabla n=0 \qquad (3)$$

for the summed air - solid balance, both for incompressible grains.

In (2) and (3) v^π means the Lagrangian velocity of the π-th phase (π = s,w,a).

Note in passing that for the derivation of (2) and (3) $S^a = 1-S^w$ and that $v^{\pi s} = n(v^\pi - v^s)$ is used. Additionally, it is assumed that the degree of water saturation S^w is only a function of the capillary pressure defined as $p^c = p^w - p^a$ (see below), as well as ρ^π being a function of the pressure p^π exclusively (exept for the grain, where ρ^s is constant).

1.3 *Simplified mass balances*

The solid is assumed to be linear elastic and the air phase governed by the ideal gas law; furthermore, the relative fluid - solid velocity is governed by Darcy's law (Gray & Hassanizadeh, 1991).

Under those assumptions and neglecting higher order terms one finally finds

$$\left(\frac{n}{S^w}\frac{\partial S^w}{\partial p^c}+\frac{n}{\rho^w}\frac{\partial \rho^w}{\partial p^w}\right)\frac{\partial p^w}{\partial t}-\frac{n}{S^w}\frac{\partial S^w}{\partial p^c}\frac{\partial p^a}{\partial t}+ \qquad (4)$$

$$\nabla \bullet v^s+\nabla \bullet\left(\frac{k_{rw}K}{\mu^w}\left(\nabla p^w-\rho^w\mathbf{g}\right)\right)=0$$

for the water - solid balance and

$$\left(\frac{n}{1-S^w}\frac{\partial S^w}{\partial p^c}+\frac{n}{\rho^a}\frac{\partial \rho^a}{\partial p^a}\right)\frac{\partial p^a}{\partial t}-\frac{n}{1-S^w}\frac{\partial S^w}{\partial p^c}\frac{\partial p^w}{\partial t}+ \qquad (5)$$

$$\nabla \bullet v^s+\nabla \bullet\left(\frac{k_{ra}K}{\mu^a}\left(\nabla p^a-\rho^a\mathbf{g}\right)\right)=0$$

for the air - solid balance.

The relative permeability k_{rw} of water is supposed to be a function of the degree of saturation and of the porosity (Seker, 1983), the relative permeability of air k_{ra} being a function of the porosity only. K is the geometric permeability.

A capillary pressure - saturation relation reported by Seker (1983) which has been obtained by tests on different materials has been used:

$$S^w=\frac{1}{\left(\frac{1}{\Psi_0}\log\left(-\frac{c_d p^c}{g_z\rho^w}\right)\right)^{\frac{1}{\Psi_1}}+1} \qquad (6)$$

where Ψ_0, Ψ_1 and c_d are material parameters and g_z represents the vertical component of the gravity vector.

Equations (1), (4) and (5) represent a set of three equations for the three unknowns solid displacement, u_i, air pressure p^a and water pressure p^w (u_i being, of course, linked to σ'_{ij} in (1) via a constitutive equation, in this case linear elastic, and to v^s via derivation in time). Other choices for the set of unknowns are of course possible.

1.4 *Two phase model*

A more commonly used formulation consists in a two phase model with so called static air phase. This corresponds to an unsaturated mixture in the sense of the mixture theory, i.e. the pore volume is partly empty. Assuming constant air pressure $p^a = p^{atmospheric}$ and setting this reference pressure $p^{atmospheric} \equiv 0$, one can easily derive such a model from (1) and (4), yielding

$$\nabla\bullet(\sigma'_{ij} - S^w p^w\delta_{ij}) + \rho g_i = 0 \qquad (7)$$

for the momentum balance and

$$\left(\frac{n}{S^w}\frac{\partial S^w}{\partial p^c}+\frac{n}{\rho^w}\frac{\partial \rho^w}{\partial p^w}\right)\frac{\partial p^w}{\partial t}+ \qquad (8)$$

$$\nabla \bullet v^s + \nabla \bullet v^{ws} = 0$$

for the solid - water mass balance. The remaining system is of two equations for the two unknowns u_i and p^w.

2. DISCRETIZED SYSTEM

Using a standard Galerkin procedure and the Θ-Method for discretisation (e.g Zienkiewicz, 1977), one finds the following system of equations for the three phase model, equations (1),(4),(5)

$$\mathbf{K}\,\mathbf{u}_{n+1} + \mathbf{L}^w\, p^w_{n+1} + \mathbf{L}^a\, p^a_{n+1} = \mathbf{F}_{n+1}$$

$$\mathbf{L}^{wT}\,\mathbf{u}_{n+1} + (\mathbf{M}^w + \Delta t\,\Theta\,\mathbf{H}^w)\, p^w_{n+1} + \mathbf{N}^{wa}\, p^a_{n+1} = \mathbf{L}^{wT}\,\mathbf{u}_n + (\mathbf{M}^w - (1-\Theta)\,\Delta t\,\mathbf{H}^w)\, p^w_n + \mathbf{N}^{wa}\, p^a_n + \Delta F^w_{n+1}$$

$$\mathbf{L}^{aT}\,\mathbf{u}_{n+1} + (\mathbf{M}^a + \Delta t\,\Theta\,\mathbf{H}^a)\, p^a_{n+1} + \mathbf{N}^{aw}\, p^w_{n+1} = \mathbf{L}^{aT}\,\mathbf{u}_n + (\mathbf{M}^a - (1-\Theta)\,\Delta t\,\mathbf{H}^a)\, p^a_n + \mathbf{N}^{aw}\, p^w_n + \Delta F^a_{n+1}$$

(9 a-c)

In (9 a-c) **K** is the rigidity matrix, $\mathbf{L}^w$ and $\mathbf{L}^a$ the liquid and gas flow matrices receptively and $\mathbf{F}^\pi$ the loading vectors. The other matrices are given in the appendix.

For the case of the two phase model, the discretized system (9) would consist in equations (9 a,b) without the p^a - terms.

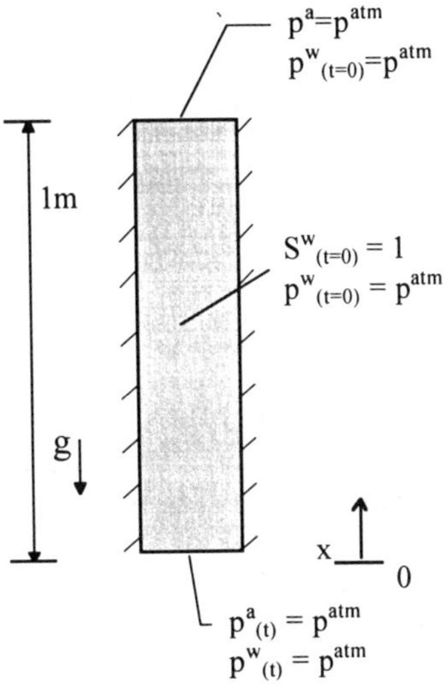

Figure 1: Initial and boundary conditions

Table 1: Material parameters

Parameter	Value
Young's modulus	$E=1.3*10^6$ Pa
Poisson's ratio	$\nu=0.4$
porosity	$n=0.297$
gravity acceleration	$g=9.81$ m/s^2
water viscosity	$\mu^w=10^{-3}$ Pas
air viscosity	$\mu^a=1.8*10^5$Pas
solid (grain) density	$\rho^s=2700$ kg/m^3
water density	$\rho^w=1000$ kg/m^3
air density	$\rho^a=1$ kg/m^3
intrinsic permea-bility	$K=4.5*10^{-13}$ m^2

(see also Gawin *et al.*, 1996)

3. EXPERIMENT

The case studied is the drainage of a sand column. It can be considered representative for a wide range of gravity - governed unsaturated soil behaviour since no external loads are applied and the deformation and desaturation depends only on the soil and fluid parameters. Coupling between the phases is expected to be crucial.

The corresponding laboratory experiments have been carried out by Liakopoulos (1964). In these tests water is allowed to flow through a sand column. The measurement starts when the water disappears from the surface. Both upper and lower end of the column are exposed to atmospheric pressure. The side walls are rigid and impermeable for both water and air. The initial conditions are consequently $S^w=1$ and $p^w=p^{atm}$ (=0) all over the column (see also Fig.1). Material parameters are given in Table 1.

Measurements of pressure at several points of the column and outflow at the bottom are available as discrete functions of time. Liakopoulos (1964) measured pore pressure and water outflow during desaturation.

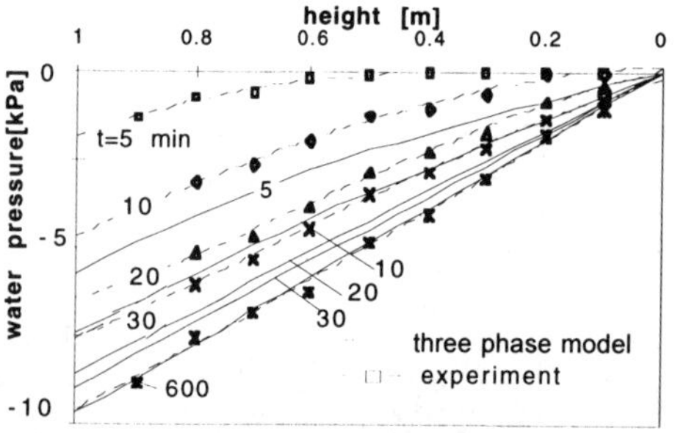

Figure 2: Pore pressure vs. column height; experimental results (dotted lines with symbols) and three phase calculation (solid lines)

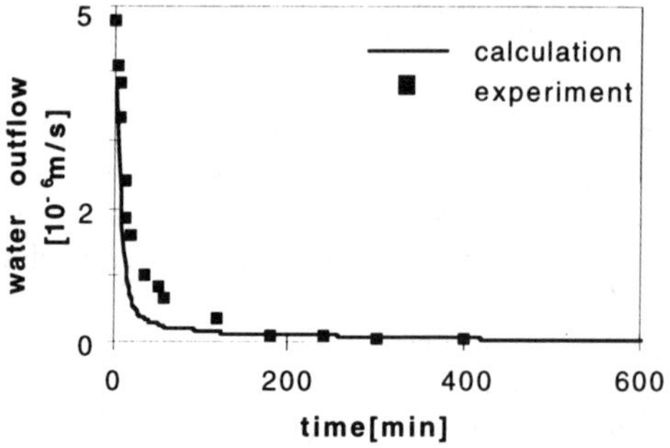

Figure 3: Water outflow vs. time

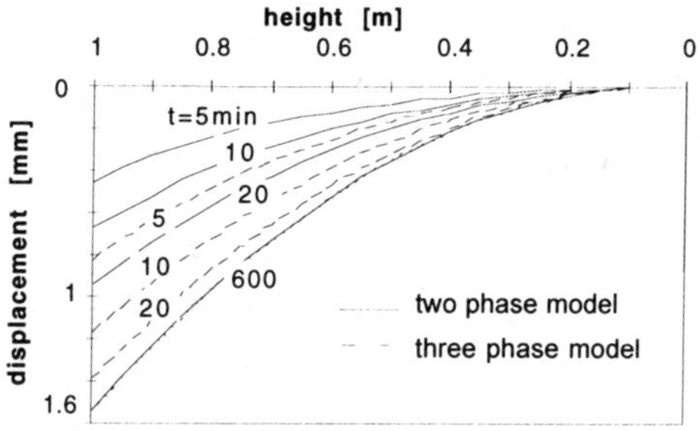

Figure 4: Displacement for two (solid lines) and three (dotted lines) phase calculations (results for t = 600s are identical)

4. NUMERICAL SIMULATION

The above models are applied to Liakopoulos (1964) experiment introduced in the previous chapter. Material parameters in Table 1 are used. With Ψ_0=3.1, Ψ_1=0.18 and the dimensional constant c_d=100 m^{-1} in (6) the capillary pressure - saturation data given by Liakopoulos is reproduced reasonably well. The residual degree of air saturation is set to S^a_r = 1%. All results are calculated for a reference pressure of $p_0 = 0$.

4.1 Comparison with experiments

In Figure 2 the calculated results of the three phase model for the pore pressure vs. column height for several time steps (solid lines) are given and compared with the experimental data (doted lines with symbols).

Already after 20 minutes there is a good agreement between computations and measurements, even if discrepancies in the early part can be seen. Those may partly be due to the simplifications during the derivation of equations (4) and (5) (for a detailed discussion, see Klubertanz *et al.*, 1997). Another possible source of error is the employed capillary pressure - saturation relation (6). Figure 3 shows the computed outflow at the lower end of the column compared with experimental data. The calculated results match the experimental data very well.

4.2 Two and three phase calculation

Figures 4 to 8 show comparisons between the two phase model with static air phase (i.e. $p^a_{(t)}$ = const. = $p^{atmospheric}$, Equations (7),(8)) and the full three phase model ((1),(4),(5)).

In the three phase case equilibrium is reached much faster than the two phase case (see Figures 4 and 5). One can see that in the early stages of the experiment, the two phase model matches the experimental results better than the three phase model (cf. Fig. 5 and 2).

Figure 6 represents the air pressure for selected time steps over the column heights. It can be seen that this pressure reaches important values and is of the same order of magnitude as the porewater pressure. This is important for the suction $p^c = p^w - p^a$ (see Figure 7), which shows significant differences compared with the two phase case, as would have been expected.

The consequences on the degree of saturation, which is, for the chosen relation (6), a nearly linear function of the suction in the range of interest, are important, as can be seen in Figure 8. Note that the maximum degree of saturation is only 99%, which

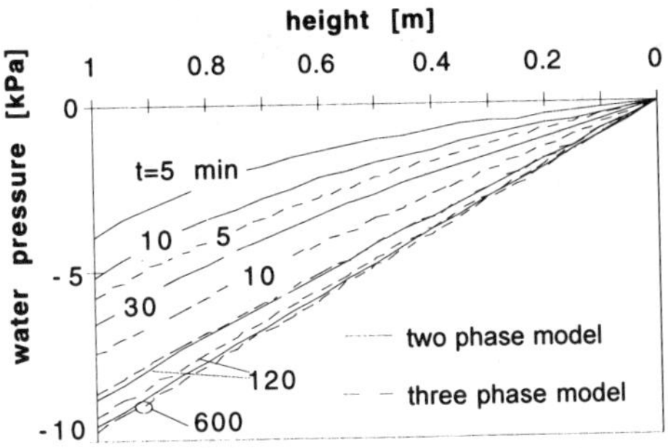

Figure 5: Water pressure for two (solid lines) and three (dotted lines) phase calculations

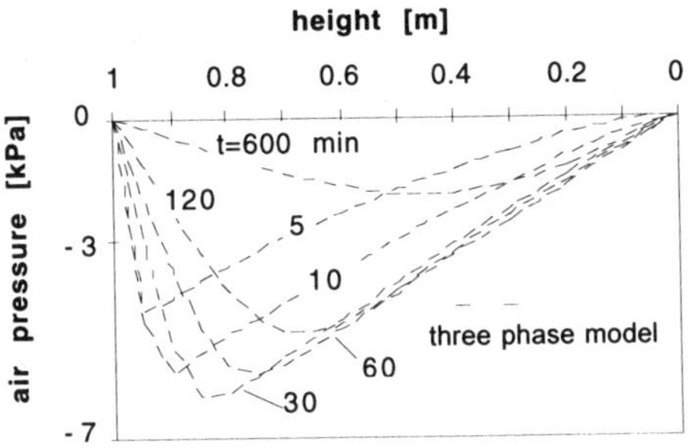

Figure 6: Air pressure for three phase calculations

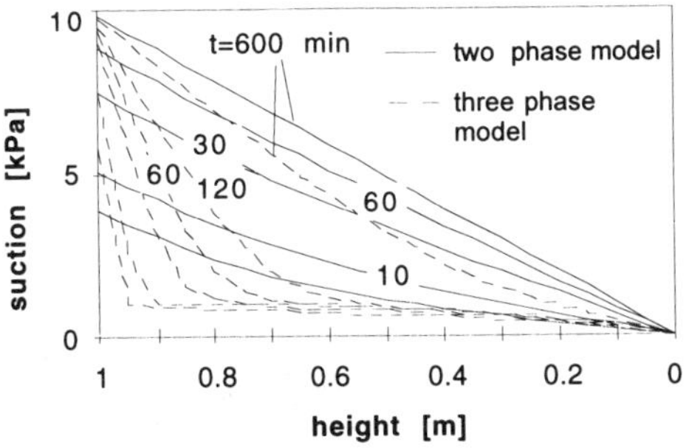

Figure 7: Suction for two (solid lines) and three (dotted lines) phase calculations

is due to computational stability reasons. Other ways of stabilizing the solution for degrees of saturation near 100% are possible and lead to comparable results.

In the above figures differences between two and three phase calculations are significant and underline, at least for this particular case, the

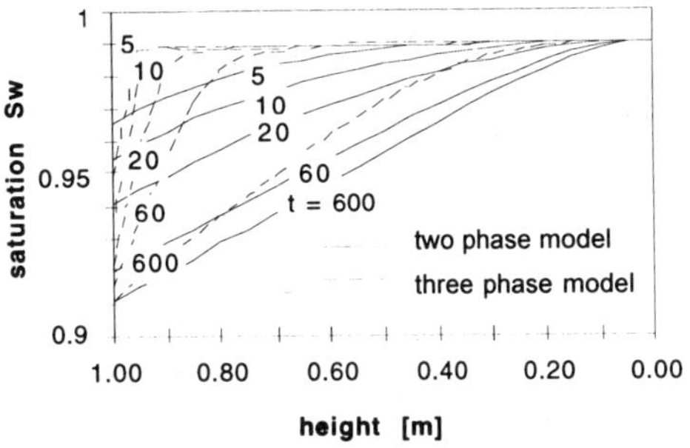

Figure 8: Degree of saturation for two (solid lines) and three (dotted lines) phase calculations

importance of a fully coupled three phase approach to model unsaturated soils. With only two phases, the calculation seems to miss characteristic features of the unsaturated soil behavior which in certain applications may become important, especially for the prediction of the evolution of S^w.

As far as the early part of the experiment is concerned, the two phase model seems to work better. Physically the soil is almost fully saturated at the beginning (degree of saturation near S^w=1) and the air phase is discontinous, violating the assumption in the three phase model. For such a particular situation, the two phase model is consequently more appropriate.

5.CONCLUSION

It has been shown that the proposed models reproduce relatively well most of the experimental results (pore water pressure and outflow). In the early part of the transient process the two phase model shows better results. It can be seen that the differences between a two and a three phase model are important and that it is necessary to consider a fully coupled three phase calculations in order to correctly simulate unsaturated soil behavior, namely to compute the degree of saturation. The model can be extended to two or three dimensional problems and to include elasto-plastic solid behavior.

AKNOWLEDGEMENTS

This research has been supported by the board of

the Swiss Federal Institutes of Technology. Some of the work has been carried out in the context of the ALERT Geomaterial Network.

REFERENCES

A. W. Bishop, The principle of effective stress, *Teknisk Ukeblad*, 39, 1959, 859 - 863

D. Gawin, Simon I, L., Schrefler, B. A., Proposal of a benchmark, *ALERT Geomaterials*, 1996

W. G. Gray and S. M. Hassanizadeh, Paradoxes and realities in unsaturated flow theory, Water res. research, 27/8, 1991, pp. 1874 - 1854

M. Hassanizadeh and W. G. Gray, General conservation equations for multi-phase systems: 3. Constiutive theorie for porous media flow, *Adv. in water resources*, 2, 1979, pp. 25 - 40

A. C. Liakopoulos, Transient flow through unsaturated porous media, *Ph D. thesis*, University of Berkely, California, 1964,

B. A. Schrefler, Simoni, L., Xikui, LI, O. C. Zienkiewicz, Mechanics of unsaturated porous media, *Numerical methods and constitutive modeling in geomaterials*, Springer Verlag, 1990, 169 - 209

E. Seker, Etude de la déformation d'un massif de sol non saturé, *Doctoral thesis 183*, EPFL, Lausanne, 1983

G. Klubertanz, L. Laloui, L. Vulliet, Numerical modeling of the hydro-mecanical behaviour of unsaturated porous media, *NAFEMS world congress*, Stuttgart, 1997 (to appear)

O. C. Zienkiewicz, *The finite element method*, Mc Graw - Hill, London, 1977

APPENDIX

The matrices in equation (9 a-c) read in detail, using standart notation (Zienkiewicz, 1977):

$$M^w = \int_\Omega n\left(\frac{\partial S^w}{\partial p^c} + \frac{S^w}{\rho^w}\frac{\partial \rho^w}{\partial p^w}\right)N^T\ N\,d\Omega$$

$$H^w = \int_\Omega S^w\, K^w\, \nabla N^T\ \nabla N\, d\Omega$$

$$N^{wa} = \int_\Omega n\frac{\partial S^w}{\partial p^c}N^T\ N\,d\Omega$$

$$M^a = \int_\Omega n\left(\frac{\partial S^w}{\partial p^c} + \frac{1-S^w}{\rho^a}\frac{\partial \rho^a}{\partial p^a}\right)N^T\ Nd\Omega$$

$$H^a = \int_\Omega \left(1-S^w\right)K^a\,\nabla N^T\ \nabla N\, d\Omega$$

$$N^{aw} = \int_\Omega n\frac{\partial S^w}{\partial p^c}N^T\ N\,d\Omega$$

Computer Methods and Advances in Geomechanics, Yuan (ed.) © 1997 Balkema, Rotterdam, ISBN 90 5410 904 1

A general and automatic programming set-up for thermo-hydro-mechanical and physico-chemical evolutions in heterogeneous media

P. Jouanna & M.A. Abellan
DTMC, UMR UM2-CNRS, Université Montpellier II, France

ABSTRACT: This paper shows how to derive a general and automatic set-up of constitutive equations and associated variables in any model of heterogeneous media, whatever the number of constituents may be and whatever the complexity of the evolutions may be (thermo-hydro-mechanical and physico-chemical), on the basis of the generalized approach of heterogeneous media as derived previously by the authors.

1 INTRODUCTION

The modern engineering problems in heterogeneous media deal with coupled physico-chemical, mechanical and heat phenomena. These multiple couplings need to put some order at a conceptual level and in the computing process.

The first attempt to put some order at a conceptual level is found in the field theory (Truesdell and Toupin, 1960). This work has been recently extended introducing an arbitrary reference movement to describe the movement of the different constituents and a link between the physical and the phenomenological approach. This generalized approach (Jouanna & Abellan, 1995) leads to an automatic way of producing the constitutive relations, including balance, state and non equilibrium relations, plus complementary relations between phenomenological and physical variables. The generalized fundamental relations for a generic constituent π are recalled in section 2, the associated variables and notations being defined in Appendix.

These constitutive equations, in presence of N constituents (π=1,N) can be derived under their physical, eulerian or lagrangian forms, leading to a total of 3 sets of 13 N relations. Handling such a huge number of coupled differential equations and associated variables requires, before any numerical processing, to generate a strategy for an automatic numerical set-up of variables and equations as proposed in section 3.

Finally, an illustration of this automatic programming set-up is outlined in section 4, in the case study of a non-saturated soil over a seasonal heat storage. Four constituents (solid, liquid water, water vapour and air) and 90 variables (Lagrange, Euler and physical variables) are considered, allowing to follow the mass, momentum and heat transfers and the associated couplings.

2 FUNDAMENTAL RELATIONS IN A GENERALIZED APPROACH OF HETEROGENEOUS MEDIA

In the study of heterogeneous media the basic kinematics difficulty is due to the different velocity $\mathbf{v}_\pi$ followed by each constituent π. To overcome this difficulty, the generalized approach proposes to choose a common and arbitrary reference velocity field $\mathbf{v}^*$ for all the constituents.

On this basis, the fundamental relations are given herebelow in Euler variables ($\mathbf{x}$,t) for any constituent π, including the mass, momentum and entropy balance relations following $\mathbf{v}^*$, the material state relations, non-equilibrium relations and the relations linking phenomenological to physical variables (Jouanna & Abellan, 1995, chap. 1, pp. 1-127).

Some complementary relations, related to definitions or to constraint relations on the total set of constituents, are recalled in Appendix.

• Generalized mass balance relation

$$\frac{\partial}{\partial t}\rho_\pi(\mathbf{x},t)+\mathrm{div}[\rho_\pi(\mathbf{x},t)\mathbf{v}^*(\mathbf{x},t)]=-\,\mathrm{div}[\rho_\pi(\mathbf{x},t)\mathbf{w}_\pi^*(\mathbf{x},t)]$$
$$+\,\rho_\pi(\mathbf{x},t)m_\pi(\mathbf{x},t)+\hat{c}_\pi(\mathbf{x},t) \qquad (1)$$

• Generalized momentum balance relation

$$\frac{\partial}{\partial t}[\rho_\pi(\mathbf{x},t)(\mathbf{w}_\pi^*(\mathbf{x},t)+\mathbf{v}^*(\mathbf{x},t))]$$
$$+\mathrm{div}[\rho_\pi(\mathbf{x},t)(\mathbf{w}_\pi^*(\mathbf{x},t)+\mathbf{v}^*(\mathbf{x},t))\otimes\mathbf{v}^*(\mathbf{x},t)]$$
$$=-\,\mathrm{div}\{\underline{\sigma}_\pi(\mathbf{x},t)+\rho_\pi(\mathbf{x},t)(\mathbf{w}_\pi^*(\mathbf{x},t)+\mathbf{v}^*(\mathbf{x},t))\otimes\mathbf{w}_\pi^*(\mathbf{x},t)\}$$
$$+\rho_\pi(\mathbf{x},t)\mathbf{f}_\pi(\mathbf{x},t)+(\mathbf{w}_\pi^*(\mathbf{x},t)+\mathbf{v}^*(\mathbf{x},t))[\rho_\pi(\mathbf{x},t)m_\pi(\mathbf{x},t)+\hat{c}_\pi(\mathbf{x},t)]$$
$$+\,\hat{\mathbf{p}}_\pi(\mathbf{x},t) \qquad (2)$$

• Generalized entropy balance relation

$$\frac{\partial}{\partial t}[\rho_\pi(\mathbf{x},t)s_\pi(\mathbf{x},t)] + \mathrm{div}[\rho_\pi(\mathbf{x},t)s_\pi(\mathbf{x},t)\mathbf{v}^*(\mathbf{x},t)]$$
$$= -\,\mathrm{div}\frac{\mathbf{q}_\pi(\mathbf{x},t)}{T_\pi(\mathbf{x},t)} - \mathrm{div}[\rho_\pi(\mathbf{x},t)s_\pi(\mathbf{x},t)\mathbf{w}_\pi^*(\mathbf{x},t)]$$
$$+\frac{1}{T_\pi(\mathbf{x},t)}\rho_\pi(\mathbf{x},t)r_\pi(\mathbf{x},t) + \frac{1}{T_\pi(\mathbf{x},t)}\mu_\pi(\mathbf{x},t)\hat{c}_\pi(\mathbf{x},t)$$
$$-\frac{1}{T_\pi(\mathbf{x},t)}[\underline{\sigma}_\pi(\mathbf{x},t):\underline{\mathbf{grad}}\mathbf{v}_\pi(\mathbf{x},t)]^{irrev}$$
$$-\frac{1}{T_\pi(\mathbf{x},t)}\mathbf{v}_\pi(\mathbf{x},t)\,\hat{\mathbf{p}}_\pi(\mathbf{x},t) - \frac{1}{T_\pi(\mathbf{x},t)}\frac{\mathbf{q}_\pi(\mathbf{x},t)}{T_\pi(\mathbf{x},t)}\mathbf{grad}T_\pi(\mathbf{x},t)$$
$$+\frac{1}{T_\pi(\mathbf{x},t)}\hat{e}_\pi(\mathbf{x},t) \qquad (3)$$

• Material state relation relative to mass for π

$$d\mu_\pi = \sum_{\pi'=1}^{N}[A_{\rho\pi'}\,d\rho_{\pi'} + \mathbf{A}_{\sigma\pi'}d\underline{\mathbf{grad}}\mathbf{u}_{\pi'} + A_{T\pi'}dT_{\pi'}] \qquad (4)$$

• Material state relation relative to momentum for π

$$d\underline{\sigma}_\pi = \sum_{\pi'=1}^{N}[B_{\rho\pi'}\,d\rho_{\pi'} + \mathbf{B}_{\sigma\pi'}d\underline{\mathbf{grad}}\mathbf{u}_{\pi'} + B_{T\pi'}dT_{\pi'}] \qquad (5)$$

• Material state relation relative to heat for π

$$d(\rho_\pi s_\pi) = \sum_{\pi'=1}^{N}[C_{\rho\pi'}\,d\rho_{\pi'} + \mathbf{C}_{\sigma\pi'}d\underline{\mathbf{grad}}\mathbf{u}_{\pi'} + C_{T\pi'}dT_{\pi'}] \qquad (6)$$

• Material non-equilibrium relation giving $\hat{c}_\pi$ for π

$$\hat{c}_\pi = M_\pi \sum_{r=1}^{R}\{\nu_\pi{}^r \sum_{r'=1}^{R}[L^{rr'}\sum_{\pi o=1}^{k^{r'}}(\frac{1}{T_{\pi o}{}^r}\tilde{\mu}_{\pi o}\,\nu_{\pi o}{}^{r'})]\} + \sum_{\pi'=1}^{N}L_{\pi\pi'}\frac{1}{T_{\pi'}} \qquad (7)$$

• Material non-equilibrium relation giving $\hat{\mathbf{p}}_\pi$ for π

$$\frac{1}{T_\pi}\hat{\mathbf{p}}_\pi = \sum_{\pi'=1}^{N}[\underline{\mathbf{L}}_{\pi\pi'}\,\mathbf{w}_{\pi'}{}^* + \Delta_{\pi\pi'}\frac{1}{T_\pi}\mathbf{grad}T_{\pi'}] \qquad (8)$$

• Material non-equilibrium relation giving $\hat{e}_\pi$ for π

$$\hat{e}_\pi = M_\pi \sum_{r=1}^{R}\{\nu_\pi{}^r \sum_{r'=1}^{R}[L'^{\,rr'}\sum_{\pi o=1}^{k^{r'}}(\frac{1}{T_{\pi o}{}^r}\tilde{\mu}_{\pi o}\,\nu_{\pi o}{}^{r'})]\} + \sum_{\pi'=1}^{N}L'_{\pi\pi'}\frac{1}{T_{\pi'}} \qquad (9)$$

• Material non-equilibrium relation giving $\mathbf{q}_\pi$ for π

$$\frac{1}{T_\pi}\mathbf{q}_\pi = \sum_{\pi'=1}^{N}[\underline{\mathbf{L}}'_{\pi\pi'}\,\mathbf{w}_{\pi'}{}^* + \Delta'_{\pi\pi'}\frac{1}{T_\pi}\mathbf{grad}T_{\pi'}] \qquad (10)$$

• Material non-equilibrium tensorial relation :

- for a fluid (π=f)

$$\frac{1}{T_f}\delta[\underline{\sigma}_f]^{irr} = -\sum_{f'=1}^{F}\underline{\underline{\mathbf{K}}}_{ff'}{}^{\pm}\,\delta[\underline{\mathbf{grad}}\,\mathbf{v}_{f'}]^{\pm} - \sum_{s'=1}^{S}\underline{\underline{\mathbf{G}}}_{fs'}{}^{\pm}\,\delta\underline{\sigma}_{s'}{}^{\pm}$$

- for a solid (π=s)

$$\frac{1}{T_s}\delta[\underline{\mathbf{grad}}\,\mathbf{v}_s]^{irr} = -\sum_{f'=1}^{F}\underline{\underline{\mathbf{K}}}'_{sf'}{}^{\pm}\,\delta[\underline{\mathbf{grad}}\,\mathbf{v}_{f'}]^{\pm} - \sum_{s'=1}^{S}\underline{\underline{\mathbf{G}}}'_{ss'}{}^{\pm}\,\delta\underline{\sigma}_{s'}{}^{\pm} \qquad (11)$$

• Volume ratio differential relation

$$d_{v^*}n_\pi(\mathbf{x},t) = [d_{v^*}\rho_\pi(\mathbf{x},t) - n_\pi(\mathbf{x},t)\,d_{v^*}(\frac{\rho_\pi(\mathbf{x},t)}{n_\pi(\mathbf{x},t)})]\frac{n_\pi(\mathbf{x},t)}{\rho_\pi(\mathbf{x},t)} \qquad (12)$$

• Stress ratio differential relation

$$d_{v^*}n_{\sigma\pi}(\mathbf{x},t)$$
$$= [\underline{\sigma}_\pi(\mathbf{x},t)]^{-1}[d_{v^*}\underline{\sigma}_\pi(\mathbf{x},t) - n_{\sigma\pi}(\mathbf{x},t)d_{v^*}(\frac{\underline{\sigma}_\pi(\mathbf{x},t)}{n_{\sigma\pi}(\mathbf{x},t)})]n_{\sigma\pi}(\mathbf{x},t) \qquad (13)$$

3 A STRATEGY FOR AN AUTOMATIC NUMERICAL SET-UP OF VARIABLES AND EQUATIONS

3.1 Different steps of the proposed strategy

• The 1st step consists in listing the constituents and phases to be considered and in choosing v* consistent with the boundary conditions.
• The 2nd step consists in listing the variables to be considered under the special assumptions of each case study. This listing is to be extracted from Table 1 which presents an exhaustive listing of all possible variables (physical, eulerian or lagrangian).
• The 3rd step consists in describing where these variables appear in the equations. Such a picture is obtained starting from the most general matrix presented in Table 2. In a given equation refered e_i, a variable refered v_j can be linked to a fluid or a solid (f/s), only to a solid (s) or only to a fluid (f).
• The 4th step consists in choosing the strategy for solving the different equations relative to any constituent π. Indeed different strategies can be considered. However in order to arrive to an automatic numerical-set up, it is convenient to associate only one variable to only one relation, the values of the other variables being supposed to be known in an iterative process at the previous numerical step. Such a bijection between equations and associated variables leads to replace the general matrix of Table 2 by a simpler extracted matrix. Among all the possibilities offered, an example is presented in Table 3 where equations are reordered under a symmetric form.

Table 1. Physical, eulerian and lagrangian variables

		Mass				Momentum					Heat				
Type of variable		V. R.	Pilot var.	Dual rev.	Source terms	Stress ratio	Pilot var.	Dual rev.	Dual irrev.	Source terms	V. R.	Pilot var.	Dual rev.	Dual irrev.	Source terms
Physical	fluid (f)	n_π	μ'_π	ρ'_π	/	$n_{\sigma\pi}$	v'_π	σ'^{rev}_π	σ'^{irr}_π	/	n_π	T'_π	s'_π	q'_π	/
	solid (s)	n_π	μ'_π	ρ'_π	/	$n_{\sigma\pi}$	σ'_π	u'^{rev}_π	u'^{irr}_π	/	n_π	T'_π	s'_π	q'_π	/
Euler	fluid (f)	/	μ_π	ρ_π	$\hat{c}_\pi$	/	w_π^*	σ_π^{rev}	σ_π^{irr}	$\hat{p}_\pi$	/	T_π	s_π	q_π	$\hat{e}_\pi$
	solid (s)	/	μ_π	ρ_π	$\hat{c}_\pi$	/	σ_π	u_π^{rev}	u_π^{irr}	$\hat{p}_\pi$	/	T_π	s_π	q_π	$\hat{e}_\pi$
Lagrange	fluid (f)	/	μ_π^*	ρ_π^*	$\hat{C}_\pi^*$	/	W_π^*	Σ_π^{*rev}	Σ_π^{*irr}	$\hat{P}_\pi^*$	/	$\mathcal{T}_\pi^*$	S_π^*	Q_π^*	$\hat{E}_\pi^*$
	solid (s)	/	μ_π^*	ρ_π^*	$\hat{C}_\pi^*$	/	Σ_π^*	U_π^{rev}	U_π^{irr}	$\hat{P}_\pi^*$	/	$\mathcal{T}_\pi^*$	S_π^*	Q_π^*	$\hat{E}_\pi^*$
Var. reference		v1	v2	v3	v4	v5	v6	v7	v8	v9	v1	v10	v11	v12	v13

• The 5th step consists in elaborating the initial conditions. The field of each considered variable, extracted from table 1, has to be fixed in the domain at the beginning of the computation. This problem is delicate because an arbitrary set of initial values leads generally to incompatibilities in the fundamental relations, i.e. physical inconsistencies, which are finally made obvious by numerical divergences. One method to tackle this problem is described in the illustration.

• The 6th step is of course to fix the boundary conditions. The term boundary refers generally to the boundary of a specified constituent, the movement of which is taken for convenience as the reference velocity. On this boundary variables or fluxes of dual reversible variables are to be fixed :

- One can fix either the following variables
v2 for (f/s), v6 for (f) or v7 for (s), v10 for (f/s)
- Or the following fluxes through this boundary
  the flux of v3 for (f/s)
  the flux of v6 for (s) or v7 for (f)
  the flux of v11 for (f/s)

• The 7th step consists in choosing the integration method which can be monolithic, staggered or individual (Abellan, 1994, t. 1, p. 151). Whatever the method may be, at each computation step it is convenient to consider the physical variables (for initial and boundary conditions), Euler variables (for presenting the results) and Lagrange variables (for the space - time mesh).

3.2 Advantages of the proposed strategy

The above strategy proposes a very clear presentation of the total set of equations and associated variables and leads to the following advantages:
(i) It allows to verify that the number of equations is equal to the number of unknowns, for the considered constituents.
(ii) It leads to an automatic iterative resolution process, one subroutine being specified for one equation, one variable and one constituent due to the bijection between equations and associated variables.
(iii) The discretisation time step and space step can be modulated depending on the considered variable in an individual method, or the considered subset of variables in a staggered method.
(iv) If some modification is related to one equation or one variable, only one subroutine is concerned.
(v) The bijection presented in Table 3 can vary depending on the constituent considered, in view to obtain the simplest bijection taking into account special assumptions particular to each case study.
(vi) It gives a method for fixing the initial and boundary conditions in any complex medium following any complex thermodynamical path, without redundancy or forgetting.
(vii) The results are directly accessible for any variable and any constituent according to Table 3.

Table 2. Matrix of relations and variables

e	v1	v2	v3	v4	v5	v6	v7	v8	v9	v10	v11	v12	v13
e1			f/s	f/s		f	s	s					
e2			f/s	f/s		f/s	f/s	f/s	f/s				
e3			f/s	f/s		f/s	f/s	f/s	f/s	f/s	f/s	f/s	f/s
e4		f/s	f/s			f	s	s			f/s		
e5			f/s			f/s	f/s	f/s			f/s		
e6			f/s			f	s	s		f/s	f/s		
e7		f/s		f/s						f/s			
e8						f	s	s	f/s	f/s			
e9		f/s								f/s			f/s
e10						f	s	s		f/s		f/s	
e11						f/s	f/s	f/s					
e12	f/s		f/s										
e13					f/s	s	f	f					

Table 3. Extracted and reordered matrix

e	v1	v2	v3	v4	v5	v6	v7	v8	v9	v10	v11	v12	v13
e12	f/s												
e4		f/s											
e1			f/s										
e7				f/s									
e13					f/s								
e2						f/s							
e5							f/s						
e11								f/s					
e8									f/s				
e6										f/s			
e3											f/s		
e10												f/s	
e9													f/s

Table 4. Physical, eulerian and lagrangian variables considered in the illustration

		Mass				Momentum					Heat				
Type of variable		V. R.	Pilot var.	Dual rev.	Source terms	Stress ratio	Pilot var.	Dual rev.	Dual irrev.	Source terms	V. R.	Pilot var.	Dual rev.	Dual irrev.	Source terms
Physical	π=a,v,w	n_π	μ'_π	ρ'_π	/	n_π	v'_π	$\underline{\sigma}'^{rev}_\pi$	$\underline{\sigma}'^{irr}_\pi$	/	n_π	T'	s'_π	q'_π	/
	π=s	n_s	♦	ρ'_s	/	n_s	$\underline{\sigma}'_s$	$u'_s{}^{rev}$	$u'_s{}^{irr}$	/	n_s	T'	s'_s	q'_s	/
Euler	π=a,v,w	/	μ_π	ρ_π	$\hat{c}_\pi$	/	w_π*	$\underline{\sigma}_\pi^{rev}$	$\underline{\sigma}_\pi^{irr}$	$\hat{p}_\pi$	/	T	s_π	q_π	0
	π=s	/	♦	ρ_s	0	/	$\underline{\sigma}_s$	$u_s{}^{rev}$	u_s^{irr}	$\hat{p}_s$	/	T	s_s	q_s	0
Lagrange	π=a,v,w	/	μ_π*	ρ_π*	$\hat{C}_\pi$*	/	W_π*	Σ_π*rev	Σ_π*irr	$\hat{P}_\pi$*	/	$\mathcal{T}$*	S_π*	Q_π*	0
	π=s	/	♦	ρ_s*	0	/	Σ_s*	$U_s{}^{rev}$	U_s^{irr}	$\hat{P}_s$*	/	$\mathcal{T}$*	S_s*	Q_s*	0
Var. reference		v1	v2	v3	v4	v5	v6	v7	v8	v9	v1	v10	v11	v12	v13

4 ILLUSTRATION OF THE METHOD

The numerical treatment (Jouanna and Abellan, 1995, pp. 301-355) simulates the phenomena occuring in a non-saturated soil embankment placed above an underground heat storage. The different steps of the automatic programming set-up developped are, for this application, the followings:

- In the 1st step, 3 chemical species, 4 constituents and 3 phases are considered : π = a refers to dry air, π = v to the water vapour, g to the gas phase (air + water vapour), π = w to the liquid water, π = s to the solid. The common and arbitrary reference velocity field v* is chosen for all constituents equal to the velocity field of the solid v_s, on which boundary conditions are fixed.
- In the 2nd step the exhaustive list of the Euler, Lagrange and physical variables to be taken into account is obtained by application of table 1 to each constituent (π = a, v, w, s). The list of the remaining variables is given in Table 4 after considering the following assumptions:

[H0]: Stress ratio $n_{\sigma\pi}$ is equal to the volume ratio n_π
[H1]: Constituents temperatures are identical, $\hat{e}_\pi = 0$.
[H2]: For all constituents, $m_\pi = 0$.
[H3]: Air and solid are inert, $\hat{c}_a = \hat{c}_s = 0$.
[H4]: The external force considered is gravity (- g).
[H5]: External specific heat sources $r_\pi = 0$.
[H6]: Air and water vapour are perfect gases
[H7]: Fluids are ideal.

- In the 3rd step, the general matrices, deduced from Table 2, are given in Table 5 (π=a), Table 6 (π=v), Table 7 (π=w) and Table 8 (π=s).
- In the 4th step, the extracted matrices, deduced from Table 3, are given in Table 9 (π=a), Table 10 (π=v), Table 11 (π=w) and Table 12 (π=s).
- In the 5th step, initial conditions cannot be obtained in a simple manner at the end of construction of the compacted embankment, because the soil is not homogeneous and not in equilibrium. The method used consists in rebuilding numerically the embankment prior any other computation, layer after layer, the only initial conditions to be known being derived from the state of the top layer after its compaction.
- In the 6th step, boundary conditions are to be fixed on the boundary of the solid, its velocity being chosen as v*. The values of variables v2, v6 or v7, and v10 are fixed for all the constituents (π = a, v, w, s) in the top and bottom layers.
- In the 7th step, an individual integration method is applied. In this approach each variable is computed by its associated equation, in one associated subroutine, in which the values of all necessary variables are estimated at the previous iteration step. The resulting flow-chart is presented in Table 13, where appear, under the main, the subroutines linked to the different constituents and the different variables. This table gives by the same way an automatic presentation of the results.

Table 5. Matrix for air

e	v1	v2	v3	v4	v5	v6	v7	v8	v9	v10	v11	v12	v13
e1			a			a							
e2			a			a			a				
e3			a			a	a	a	a	a	a	a	
e5			a				a			a			
e6							a			a	a		
e8						a,v	a,v						
e10										a		a	
e11						a		a					
e12	g		a										

Table 6. Matrix for water vapour

e	v1	v2	v3	v4	v5	v6	v7	v8	v9	v10	v11	v12	v13
e1			v	v		v							
e2			v	v		v	v	v	v				
e3			v	v		v	v	v	v	v	v	v	
e4		v					v			v			
e5			v				v			v			
e6							v			v	v		
e7		v,w		v									
e8			a,v			a,v				v			
e10										v		v	
e11						v		v					
e12	g		v										

Table 7. Matrix for liquid water

e	v1	v2	v3	v4	v5	v6	v7	v8	v9	v10	v11	v12
e1			w	w		w						
e2			w	w		w	w	w	w			
e3			w	w		w	w	w	w	w	w	w
e4		w	w				a,v,w			w		
e5			w				a,v,w			w		
e6			w				w			w	w	
e7				v,w								
e8			w			w	w			w		
e10										w		w
e11						w		w				
e12	w		w									

Table 8. Matrix for solid

e	v1	v3	v6	v7	v8	v9	v10	v11	v12
e1		s		s	s				
e2		s	s	s	s	s			
e3		s	s	s	s	s	s	s	s
e5			s	a,v,w,s					
e6		s	s				s	s	
e8						a,v,w,s			
e10							s		s
e11			s	a,v,w	s				
e12	s	s							

Table 9. Extracted matrix for air

e	v1	v3	v6	v7	v8	v9	v10	v11	v12
e12	g								
e1		a							
e8			a						
e5				a					
e11					a				
e2						a			
e3							a		
e6								a	
e10									a

Table 10. Extracted matrix for water vapour

e	v1	v2	v3	v4	v6	v7	v8	v9	v10	v11	v12
e12	g										
e4		v									
e1			v								
e7				v							
e8					v						
e5						v					
e11							v				
e2								v			
e3									v		
e6										v	
e10											v

Table 11. Extracted matrix for liquid water

e	v1	v2	v3	v4	v6	v7	v8	v9	v10	v11	v12
e12	w										
e4		w									
e1			w								
e7				w							
e8					w						
e5						w					
e11							w				
e2								w			
e3									w		
e6										w	
e10											w

Table 12. Extracted matrix for solid

e	v1	v3	v6	v7	v8	v9	v10	v11	v12
e12	s								
e1		s							
e11			s						
e5				s					
e2					s				
e8						s			
e3							s		
e6								s	
e10									s

Table 13 Flow-chart

MAIN

AIR	WATER VAPOUR	GAZ	LIQUID WATER	BULK	TEMPERATURE
	μ_v		μ_w		T
	$\hat{c}_v$		$\hat{c}_w$		
ρ_a	ρ_v		ρ_w	ρ_s	
$\underline{\sigma}_a^{rev}$	$\underline{\sigma}_v^{rev}$		$\underline{\sigma}_w^{rev}$	$\underline{\sigma}_s$	
$\mathbf{w}_a^*$	$\mathbf{w}_v^*$		$\mathbf{w}_w^*$	$\mathbf{u}_s^*$ rev	
$\underline{\sigma}_a^{irr}$	$\underline{\sigma}_v^{irr}$		$\underline{\sigma}_w^{irr}$	$\mathbf{u}_s^*$ irr	
$\hat{\mathbf{p}}_a$	$\hat{\mathbf{p}}_v$		$\hat{\mathbf{p}}_w$	$\hat{\mathbf{p}}_s$	
q_a	q_v		q_w	q_s	
s_a	s_v		s_w	s_s	
		n_g	n_w	n_s	

5 CONCLUSION

As seen on the above illustration, a problem in heterogeneous media becomes rapidly intricated and the large number of equations and variables cannot be handled without a strong methodogical approach.

The automatic programming set-up proposed hereabove in the most general case is a powerful tool which can be used for deriving a systematic treatment of any application, whatever the number of constituents and phenomena may be.

This automatic set-up leads to a very simple organization of the programme, as illustrated by table 13, which can be extended to any number of constituents and to all possible variables appearing in a thermo-hydro-mechanical and physico-chemical problem in an heterogeneous medium.

REFERENCES

Abellan, M-A. 1994. Approche phénoménologique généralisée et modélisation systématique de milieux hétérogènes. Illustration sur un sol non-saturé en évolution thermo-hydro-mécanique et physico chimique. Tome 1 (Concepts) et Tome 2 (Illustration), *Thèse de Doctorat*, Université Montpellier II, France.

Abellan, M-A., Jouanna, P., Saix, C. 1995. Non saturated consolidation under thermo-hydro mechanical actions. An in situ heat storage facility in a clayey silt, in "*Modern issues in non-saturated soils*", Chapter 7, C.I.S.M. courses and lectures No. 357, edited by A. Gens/ P. Jouanna/ B.A. Schrefler, Springer-Verlag, Wien, New York, 301-355.

Jouanna, P., Abellan, M-A. 1995. Generalized approach to heterogeneous media, in "*Modern issues in non-saturated soils*", Chapter 1, C.I.S.M. courses and lectures No. 357, edited by A. Gens/ P. Jouanna/ B.A. Schrefler, Springer-Verlag, Wien, New York, 1-127.

Truesdell, C. & Toupin, R. 1960. *The classical field theories*, Encyclopedia of Physics, ed. by Flügge, Volume III/1, Principles of Classical Mechanics and Field Theory, Springer Verlag, Berlin, Göttingen, Heidelberg.

APPENDIX:

• Notations non explicitly defined in the text

$\hat{c}_\pi$, $\hat{C}_\pi^*$ mass supply to π due to other constituents per volume and time unit in kg m^{-3} s^{-1}

d differential symbol

d_{v^*} differential following the virtual movement

$\hat{e}_\pi$, $\hat{E}_\pi^*$ energy supply to π due to other constituents per volume and time unit in W m^{-3}

$\mathbf{f}_\pi$ external force acting on the mass unit of π

irr irreversible part of a quantity

M_π molar mass of π in kg mole^{-1}

m_π external mass source of π in s^{-1}

n_π volume ratio of π defined by $\rho_\pi = n_\pi \rho'_\pi$

$n_{\sigma\pi}$ stress ratio of π defined by $\underline{\underline{\sigma}}_\pi = n_{\sigma\pi} \underline{\underline{\sigma}}'_\pi$

$\hat{\mathbf{p}}_\pi$, $\hat{\mathbf{P}}_\pi^*$ momentum supply vector to π due to other constituents in kg m^{-2} s^{-2}

$\mathbf{q}'_\pi$, $\mathbf{q}_\pi$, $\mathbf{Q}_\pi^*$ heat influx vector relative to π

rev reversible part of a quantity

r_π external supply of heat per mass unit of π and time unit in J kg^{-1} °C^{-1}s^{-1}

s'_π, s_π, S_π^* specific entropy of π in J kg^{-1} °C^{-1}

T', T, $\mathcal{T}^*$ temperature in °C or °K

$\mathbf{u}'_\pi$, $\mathbf{u}_\pi$, $\mathbf{U}_\pi$ displacement of π in m

var abbreviation for variable

$\mathbf{v}'_\pi$, $\mathbf{v}_\pi$ velocity vector of π

V.R. abbreviation for volume ratio

$\mathbf{w}'_\pi, \mathbf{w}_\pi^*$, $\mathbf{W}_\pi^*$ relative velocity vector with $\mathbf{w}_\pi^* = \mathbf{v}_\pi - \mathbf{v}^*$

δ variation symbol

μ'_π, μ_π, μ_π^* specific chemical potential of π

π index of any constituent

ρ'_π, ρ_π, ρ_π^* mass density of π in kg m^{-3}

$\underline{\underline{\sigma}}'_\pi$, $\underline{\underline{\sigma}}_\pi$, $\underline{\underline{\Sigma}}_\pi^*$ stress tensor relative to π in its movement

$\tilde{\mu}_\pi$ molar chemical potential of π defined by $\tilde{\mu}_\pi = M_\pi \mu_\pi$

ν_π^r stochiometric coefficient of π in the chemical reaction r

$A_{\rho\pi'}$, $\mathbf{A}_{\sigma\pi'}$, $A_{T\pi'}$, $B_{\rho\pi'}$, $\mathbf{B}_{\sigma\pi'}$, $B_{T\pi'}$, $C_{\rho\pi'}$, $\mathbf{C}_{\sigma\pi'}$, $C_{T\pi'}$, $L^{rr'}$, $L_{\pi\pi'}$, $\mathbf{L}_{\pi\pi'}$, $\Delta_{\pi\pi'}$, $L'^{rr'}$, $L'_{\pi\pi'}$, $\mathbf{L}'_{\pi\pi'}$, $\Delta'_{\pi\pi'}$, $\underline{\underline{\mathbf{K}}}_{ff'}^{\pm}$, $\underline{\underline{\mathbf{G}}}_{fs'}^{\pm}$, $\underline{\underline{\mathbf{K}}}'^{\pm}_{sf'}$, $\underline{\underline{\mathbf{G}}}'^{\pm}_{ss'}$ phenomenological coefficients

• Definition relations

- the Euler displacement vector of π is given by:

$$\mathbf{u}_\pi(\mathbf{x},t) = \mathbf{u}_\pi^{rev}(\mathbf{x},t) + \mathbf{u}_\pi^{irr}(\mathbf{x},t)$$

- the Lagrange velocity vector of π is given by:

$$\mathbf{V}(\mathbf{X}^*,t) \equiv \frac{d}{dt}\,\mathbf{U}(\mathbf{X}^*,t)$$

• Constraint relations on the total set of constituents are given here below in Euler variables $(\mathbf{x},t)$:

$$\sum_{\pi=1}^{\pi=N} \hat{c}_\pi(\mathbf{x},t) = 0$$

$$\sum_{\pi=1}^{\pi=N} \hat{\mathbf{p}}_\pi(\mathbf{x},t) = \mathbf{0}$$

$$\sum_{\pi=1}^{\pi=N} \hat{e}_\pi(\mathbf{x},t) = 0$$

$$\sum_{\pi=1}^{\pi=N} n_\pi(\mathbf{x},t) = 1$$

Computer Methods and Advances in Geomechanics, Yuan (ed.) © 1997 Balkema, Rotterdam, ISBN 90 5410 904 1

Application research on a numerical model of two-phase flows in deformation porous medium

Y. Sun & S. Sakajo
Kiso-Jiban Consultants Co., Ltd, Tokyo, Japan
M. Nishigaki
Department of Environmental Design, Okayama University, Japan

ABSTRACT: Numerical simulation of multiphase flow in porous media is important in many branches of science and engineering. In this paper, a F.E numerical model of two-phase flows in deformation porous medium (Y.Sun,1995) was introduced. For showing the application values and method on a F.E numerical model of two-phase flows in deformation porous medium, several F.E analysis for some typical kinds of multiphase flow problems in porous media have been performed. Through summering and analyzing the calculating results of those examples, it can be concluded that a F.E numerical model of two-phase flows in deformation porous medium presented in this paper can provide well results that more closing practice state than other method.

1. INTRODUCTION

Numerical simulation of multiphase flow in porous media is important in many branches of science and engineering. These include groundwater hydrology, petroleum engineering., soil science, and geomechnical engineering. The simultaneous flow of water and air through unsaturated soils is one example of such a multiphase system.

The influence of the air phase on the motion of the water is one possible reason for studying the unsaturated zone as a two-phase(air-water) system, a second more compelling reason involves modeling transport of volatile organic compounds in unsaturated soils. Such components of the air phase, and so an understanding of the movement of the air phase as well as the water phase is important.

When there is a path for the air to escape in unsaturated zone, it has sufficiently large mobility that it usually equilibrates quickly to (close to) atmospheric pressure. Thus Richards's equation is usually a good approximation for describing the water phase motion under most transient water movement episodes. But, what conditions dose the air phase behavior substantially alter the flow of water ? based on laboratory column experiments and subsequent numerical simulations, the answer was that the air can significantly retard the motion of water only when air is trapped inside the soil column, that is, when(1) the top of the column is saturated or nearly so, (2) the sides of the column does not allow air to escape. Under this combination of conditions the air phase has a significant influence on the infiltrating water. In generally, it is assumed that air phase mobility is much greater than the water phase mobility so that the air phase remains at essentially atmospheric pressure

In other way, a method for remediating an aquifer, which is contaminated by organic(solvents, gasoline) trapped in the saturated zone, is to inject air or oxygen into the aquifer. Injection of air may enhance microbial degradation and volatilization. Direct injection of air in the saturated zone, known as air sparging, together with vapor extraction in the unsaturated zone, has been put forward as an effective in-situ remediation technique. Injection in fine clay soils requires high air entry pressures, which may cause fracturing and deformation of the soil. This may results in the requirement for the coupling analysis method of two-phase flow in deformation porous medium.

A finite-element numerical model for immiscible two-phase fluid flow in a deforming porous media was presented by Li Xikui(1990). The displacements, the pressures and the saturation of wetting phase are taken as primary unknowns of the model as the constraint about the capillary pressure between two fluid phases is applied. The numerical result for some examples also illustrate the performance and capability of the approach. But, because of taking the pressures and the saturation of wetting phase as primary unknowns of the model, there is some difficulty to deal with the air pressure boundary problem and air injection volume problem using this numerical model.

In this paper, a F.E numerical model of two-phase flows in deformation porous medium (Y.Sun,1995) was introduced. The displacements, the pressures p_a and p_w are taken as primary unknowns in the model. For showing the application values and method on a F.E numerical model of two-phase flows in deformation porous medium presented by author, several F.E analysis for some typical kinds of multiphase flow problems in porous media have been performed. Through summering and analyzing the calculating results of those examples, the performance and capability of the F.E numerical model of two-phase flows in deformation porous medium presented in this paper will be illustrated

2. THEROY AND SOLVING PROBLEME

In the last years, author(Y.SUN 1995) has derived a set of coupling governing equations of two-phase fluid flow in a deforming porous media. Based on this, a finite-element numerical model and program(SWACOP-2D) for analyzing the generalized coupling problem in porous medium were also developed. The displacements, the pressures p_a and p_w are taken as primary unknowns of the model. By use of this method, the changing of wetting front, distribution of stress and strain and the changing of pore pressure in porous medium can easily be calculated in any time step. The degree of saturation in the element can be calculated in according to the relationship between the suction $p_c = p_a - p_w$ and volumetric water content θ. Because of taking the air and water pressures air as primary unknowns in the model, it will be easy to deal with the air pressure boundary problem and air injection volume problem using this numerical model.

2.1 FE Analysis Method

By use of the equilibrium equation, the continue equations of mass, and the motion equations of two-phase flow, the governing equations of coupled two-phase flow in deformation porous medium can be established conveniently. If the increments of the displacements du_{ij}, the pore water pressure p_w and pore air pressure p_a are accepted as the independent variables, the finite element formulas can be written as follows

$$\begin{bmatrix} K_{nm}^{ik} & C_{nm}^{i} & \overline{C}_{nm}^{i} \\ A_{nm}^{i} & D_{nm}-\frac{\Delta t}{2}B_{nm} & E_{nm} \\ G_{nm}^{i} & H_{nm} & J_{nm}-\frac{\Delta t}{2}L_{nm} \end{bmatrix} \begin{bmatrix} \delta \overline{U}_{k}^{m} \\ \overline{P}_{w}^{m} \\ \overline{P}_{a}^{m} \end{bmatrix}_{t+\Delta t} = \begin{bmatrix} \Delta \overline{F}_{n}\big|_{t+\Delta t} + C_{nm}^{i}\overline{P}_{w}^{m}\big|_{t} + \overline{C}_{nm}^{i}\overline{P}_{a}^{m}\big|_{t} \\ \frac{\Delta t}{2}\delta Q_{nm}^{w}\big|_{t+\Delta t} + (D_{nm}-\frac{\Delta t}{2}B_{nm})\overline{P}_{w}^{m}\big|_{t} + E_{nm}\overline{P}_{a}^{m}\big|_{t} \\ \frac{\Delta t}{2}\delta Q_{nm}^{w}\big|_{t+\Delta t} + H_{nm}\overline{P}_{w}^{m}\big|_{t} + (J_{nm}-\frac{\Delta t}{2}L_{nm})\overline{P}_{a}^{m}\big|_{t} \end{bmatrix} \tag{1}$$

In here,

$$K_{nm}^{ik} = \sum_{a=1}^{N_o}\int_{v_a} N_{n,j} C_{ijkl} N_{m,l} dv_a$$

$$c_{nm}^{i} = \sum_{a=1}^{N_o}\int_{v_a} N_{n,j}\chi\delta_{ij} N'_{m} dv_a$$

$$\overline{c}_{nm}^{i} = \sum_{a=1}^{N_o}\int_{v_a} N_{n,j}(1-\chi)\delta_{ij} N'_{m} dv_a$$

$$F_{nm}^{i} = \sum_{a=1}^{N_o}\int_{v_a} N_{n,j} N'_{m}\delta \hat{T}_{i}^{m} dv_a + \sum_{a=1}^{N_o}\int_{v_a} N_{n}\delta\gamma'_{s} b_{i} dv$$

$$A_{nm}^{i} = \sum_{a=1}^{N_o}\int_{v_a} N'_{m} a_{1}\delta_{ij} N_{m,i} dv_a$$

$$B_{nm}^{i} = \sum_{a=1}^{N_o}\int_{v_a} N'_{n,1}\frac{k_w}{\gamma_w}\delta_{ij} N'_{m,i} dv_a$$

$$D_{nm} = \sum_{a=1}^{N_o}\int_{v_a} N'_{n} a_{2} N_{m} dv_a$$

$$E_{nm} = \sum_{a=1}^{N_o}\int_{v_a} N'_{n} a_{3} N_{m} dv_a$$

$$Q_{nm} = \sum_{a=1}^{N_o}\int_{v_a} N'_{n} N'_{m}\delta \hat{Q}(x,t) ds_a$$

$$G_{nm} = \sum_{a=1}^{N_o}\int_{v_a} N'_{m} a_{4} N_{m,i} dv_a$$

$$L_{nm}^{i} = \sum_{a=1}^{N_o}\int_{v_a} N'_{n,i}\frac{k_a}{\gamma_a}\delta_{ij} N'_{m,i} dv_a$$

$$H_{nm} = \sum_{a=1}^{N_o}\int_{v_a} N'_{n} a_{2} N_{m} dv_a$$

$$J_{nm} = \sum_{a=1}^{N_o}\int_{v_a} N'_{n} a_{5} N_{m} dv_a \tag{2}$$

where, k_w, k_a are the permeability, which are the functions of saturation. γ_w, γ_a are densities, b_i is body force C_{ijkl} is tensor of material modules, $a_1 \sim a_5$ are the variables that can be calculated by the relationship of water volume content θ and suction $\phi_c = p_a - p_w$. χ is equal to saturation degree.

Resolving eq.1, the initial conditions and boundary conditions will be used that shown as follows.

(1) Initial conditions

An initial state on the field concerned should be determined before the transient process is analyzed. Initial pressures p_{w0}, p_{a0}, and initial deformation u_{i0} may be given or evaluated by a analysis. As initial p_{w0}, p_{a0}, and initial deformation u_{i0} on the domain Ω are prescribed or evaluated, the initial external force.

$$U_i(x_i,t) = U_i(x_i,0)......(3a)$$

$$H(x_i,t) = H(x_i,0)........(3b)$$

$$p_w(x_i,t) = p_w(x_i,0)......(3c)$$

$$p_g(x_i,t) = p_g(x_i,0)......(3d)$$

water head $H = P_w/\gamma_w + z$

(2) Boundary conditions

The natural boundary conditions

$$\sigma_{ij}(x_i,t)n_j(x) = T_i(x_i,t)...(4a)$$

$$H(x_i,t) = H(x_i,t)........(4b)$$

$$p_w(x_i,t) = p_w(x_i,t)......(4c)$$

$$p_g(x_i,t) = p_g(x_i,t)......(4d)$$

have been automatically satisfied.

The force boundary condition should be prescribed. It is noted that the boundary displacements and pressures prescribed with zero value. The nodal displacements, pressures and pore pressure will be ruled out from the global primary unknown vector.

The flow volume boundary condition

$$\left.\begin{aligned}\{(\frac{k_w}{\gamma_w})\nabla^2 H\}n_j &= -\hat{Q}_w(x,t..)\\ \{(\frac{k_a}{\gamma_a})\nabla^2 p_g\}n_j &= -\hat{Q}_a(x,t.)\end{aligned}\right\} \quad (5)$$

have been prescribed.

3. NUMERICAL EXAMPLE

In the last years, the numerical model presented above has been used to simulate a lot of coupling problems of permeability and deformation, for examples, slope stability analysis due to raining, the consolidation analysis of unsaturated soil and the collapse settlement analysis due to water permeability(Y.Sun 1995). Those analysis results show that the numerical model presented by author is one effective method for coupling problems in unsaturated soil. In this section, we will verify the effective of this model for bounded water infiltration, air pressure and air injection boundary condition problems by comparing with the results of experiment and numerical simulating of some examples.

3.1 Water Infiltration into Bounded Column

To examine the effect of air phase for water infiltration, the experiments of Vachaud(1974) are used. The problems reported by Vachaud(1974) were shown in Figure 1. This is one dimensional and transient. The appropriate functional relationships for moisture content and relative permeability are shown in Figure 2. These relationships were derived from laboratory measurements reported by Vachaud(1974).

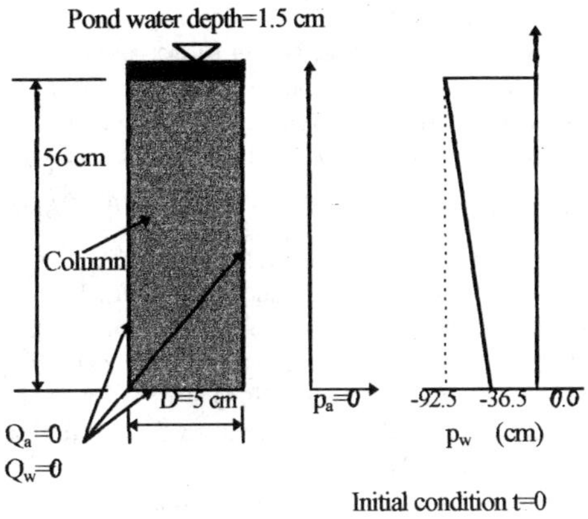

Figure 1. Experiment sketch initial condition reported by Vachaud(1974)

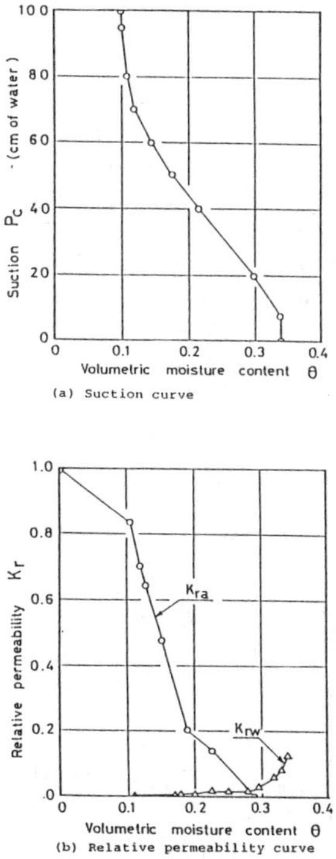

Figure 2. Water content(infiltration) and permeability curves as described by Vachaud(1974)

This problem is a good test of a two-phase simulator as it involves combinations of both fixed head and flux boundary conditions and, as Vachaud(1974) have shown, this is a case where the air phase has a significant retarding effect on the water movement.

This example involves a hydrostatic initial condition with water applied to the top boundary at the 1.5 cm pond water. The column is sealed at the bottom so that neither air nor water can escape, and air is initially set to be at atmospheric pressure at the surface. In this case the flux of water at the top of the column exceeds k_{ws}, so that the water content continues to build up until at the top of the column $\theta = \theta_{ws}$. At this time, air is at residual saturation at the top of the column, so that air cannot escape from the column. The air then builds up pressure until capillary head reaches a critical value, defined by $h_{crit} = h_{ac}$ cm, where h_{ac} is the air entry pressure, or bubbling pressure, estimated by Vachaud(1974) as h_{ac}=8 cm. At this point the air pressure head is defined so that the capillary pressure is equal to 8 cm.

Numerical results of water pressure and moisture content profiles for several different times are shown in Figure 3, 4. Water pressure changes with time increase during water infiltration from top part to bottom part of the column. Saturation degree along the elevation of column also rises with time increase. Comprising with the experiment results of Vachaud(1974), it is found that satisfactory agreement between numerical and experiment results is achieved.

In other way, numerical model of one-phase water flow(Akai.K.,Ohnishi.Y.,&Nishigaki.M. 1977) also was used to calculate the pond water infiltration problem. The calculating results are shown in figure 5. Comparing figure 5, with figure 3, it can be found that water pressure changes calculated by one-phase flow model are faster than that of two-phase flow model . This results verifies that there is obviously influence effect of air pressure for water seepage in unsaturated zone as the boundary condition of unsaturated zone is so closed.

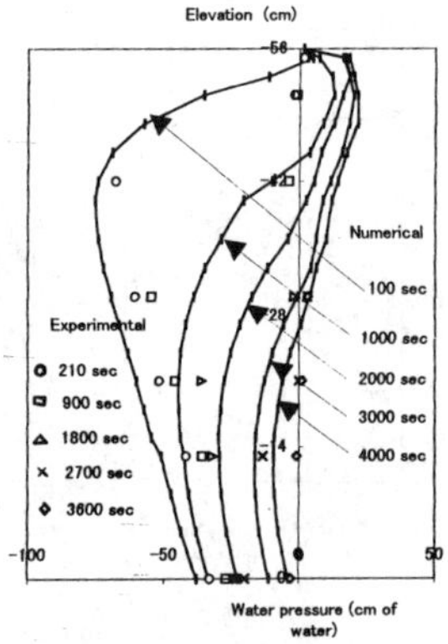

Figure 3. Time evolution of water pressure profile comparing the numerical results

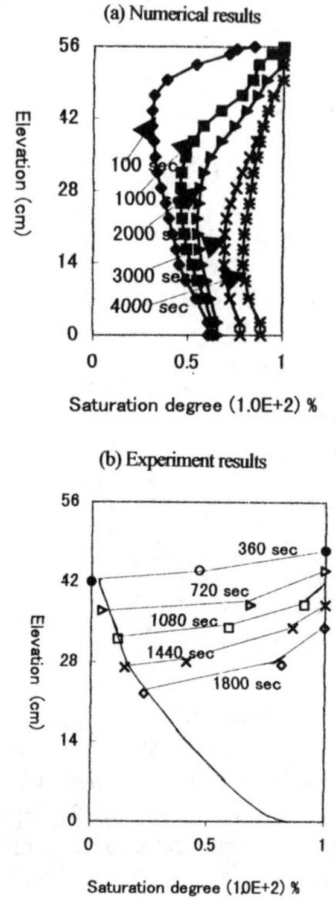

Figure 4. Time evolution of saturation degree profile comparing the numerical results

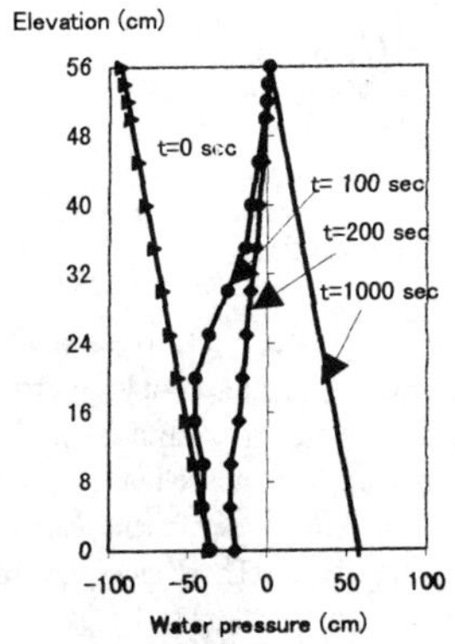

Figure 5. Time evolution of water pressure profile comparing the numerical results (one-phase flow model)

For showing the changes of air pressure in closed boundary condition, 1-D raining permeability experiment results reported by Y.Ishihara &E. Shimojima (1983) was introduced as shown in Figure 6. This result shows that air pressure in closed boundary condition raised continually with raining water infiltration.

From those experiment and numerical analysis results

described above, it can be found that there is obviously influence effect of air pressure for water seepage in unsaturated zone as the boundary condition of unsaturated zone is so closed that air pressure can not escape easily. Therefor, it is necessity to use the numerical model that consider air and water flow in deformation porous medium instead of that of one-phase flow analysis in some geomechnical engineering problems. For example, in slope stability analysis during raining process, due to the air phase has a significant retarding effect on the water movement, the water seepage toward depth of slope will become slow, the shallow layer of slop will immediately become saturated state, raining water flow in shallow part of slope will produce from upper part of slope to foot of slope. Therefore, it is very easy to cause failure of slope in the shallow part. The analysis results by F.E model presented in this paper is very closed practice state of shallow slope failure (Y.Sun ,1995) . In the analysis of one-phase flow model, the raining water will seepage continually toward the depth of slope till bottom parts of slope become saturation state. Before the bottoms of slope become saturated state, the seepage force will increase stability of slope. It can be concluded that the F.E model of one-phase flow is not fit in with practice state.

3.2 AIR PRESSURE PROBLEM

To demonstrate the effective of the numerical model of two-phase flow in deformation porous medium for air pressure and air injection boundary condition problems, a series of numerical experiments were performed. Two of the numerical experiment about air pressure and air injection volume are constant in the top boundary of column are simulated here. The numerical experiments used a column that is bounded at the bottom so that neither air can escape.

(1) Air pressure P_a = constant

As shown in Figure 7, in the top boundary of open column, air pressure pa=2.0, 200 cmAq are given. The top boundary condition for water is set to be 25 cm and allows

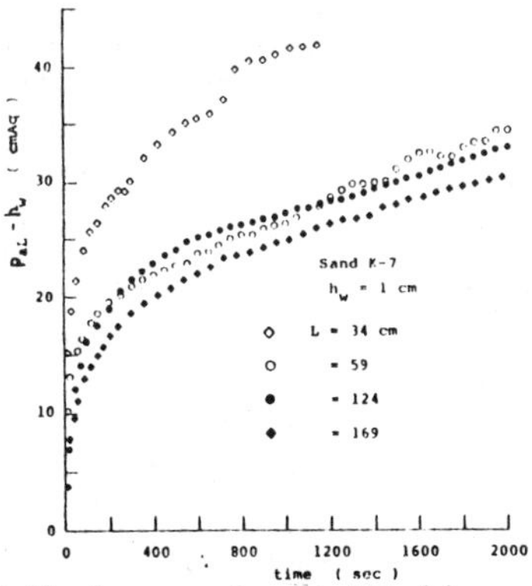

Figure 6. The air pressure change in 1-D raining permeability experiment reported by Y. Ishihara &E. Shimojima (1983)

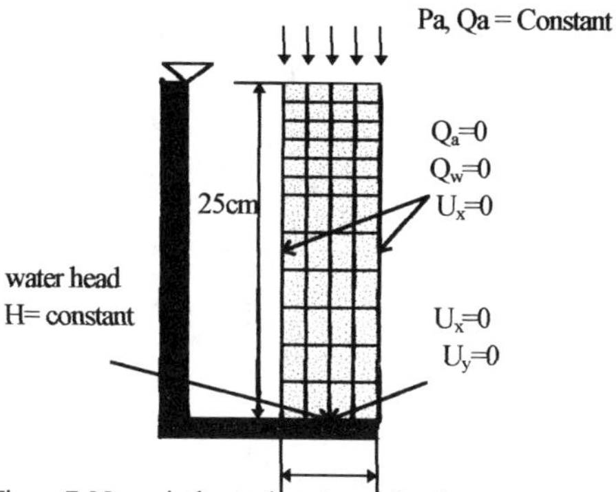

Figure 7. Numerical experiment model and F.E analysis condition

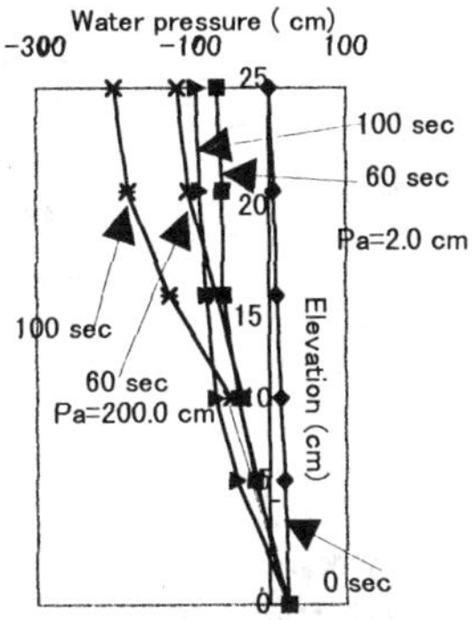

Figure 8(a). Water pressure changes with time in p_a=const

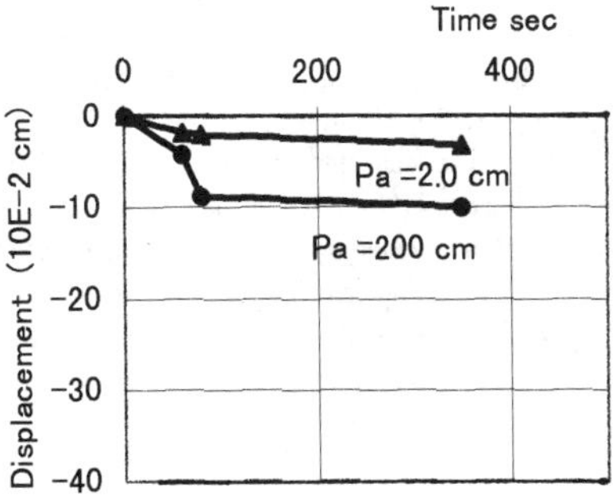

Figure 8(b). Displacement changes of top boundary side with time in p_a=const

water to escape through the bottom. The deformation changes in top boundary and water pressure or water content changes in the column are analyzed. Figure 8(a) shows the change of water pressure, which has became mines value from initial plus value due to the action of air pressure in the up part of column. Because of the influence of water pressure at bottom, after some time, a balance state of water and air pressure in the column could be achieved. Figure 8(b) shows the

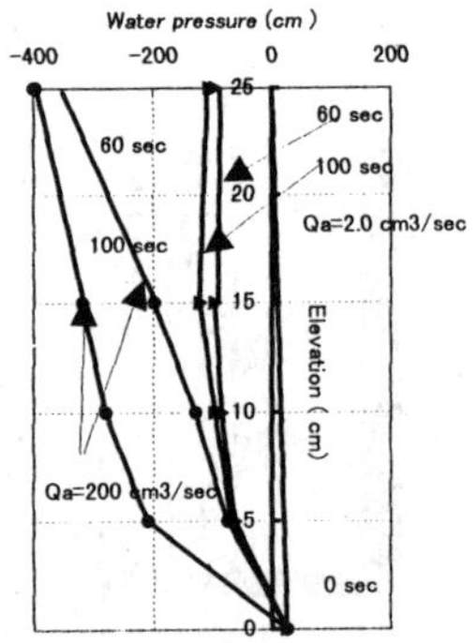

Figure 9(a). Water pressure changes with time in Q_a=const

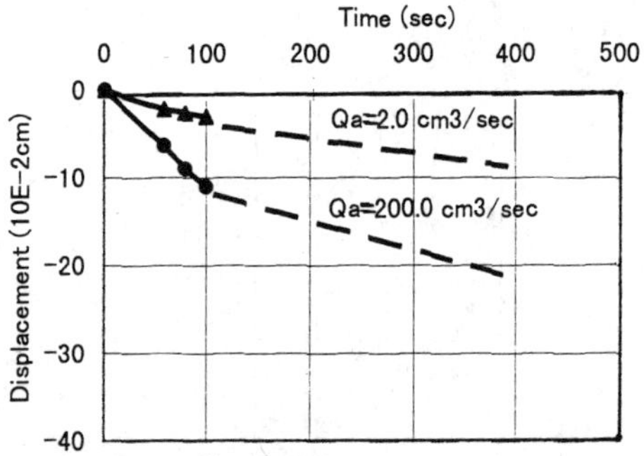

Figure 9(b). Displacement changes of top boundary side with time in Q_a=const

deformation change in the top boundary side caused by the action of air pressure, it can be found that the air pressure used in top boundary side is bigger, the deformation in top boundary is bigger. The deformation caused by air pressure is quickly get to stability state, this deformation characteristics is different with that caused by external force. The reason is that air pressure from top boundary is mainly act to water phase through the gap of porous medium to increase the suction value of porous medium. And the increase of suction in unsaturated zone will cause the shrinking of unsaturated soil frame. When a balance state of water and air pressure in the column is arrived, the suction change of the column is get to small and the deformation of column is also get to stability.

(2) Air injection volume Q_a = constant

In the top boundary of open column, air injection volume Qa =2.0, 200 cm^3/sec are given. Numerical experiment model and F.E analysis condition are shown in Figure 7. The top boundary condition for water is set to be 25 cm and allows water to escape through the bottom. The deformation in top boundary and water pressure or water content in the column are analyzed. Figure9(a),(b) shows the change of water pressure and he deformation change in the top boundary side caused by the action of air injection in the column. Because when air injection volume is big, the air pressure preserving air injection volume constant is also high and with the increase of air injection zone the air pressure used in air injection process is also need to be raised. Comparing with the results of both air pressure constant and air injection volume constant, it can be found that the deformation in the top boundary side caused by the action of air injection increases continue with time increase during air injection process. When air injection volume is big, the change of water pressure in unsaturated zone is also big. Because of the influence of water pressure at bottom, when air injection volume value is small, a balance state of water and air pressure in the column could be arrived after some time, for example,Qa=2.0 cm3/sec case.

4. CONCLUDE

1. A lot of the results of experiment show when air is trapped inside the soil, air can significantly retard the motion of water. This may results in the requirement for the numerical model of two-phase flow instead of one-phase flow model in many branches of science and engineering.

2. Through summering and analyzing the calculating results of those examples, it can be concluded that a F.E numerical model of two-phase flows in deformation porous medium presented in this paper can provide well results that more closing practice state than other method.

3. The numerical results show the air pressure or air injection volume used in top boundary side is bigger, the deformation in top boundary and the changes of water pressure in the unsaturated zone are bigger.

REFERENCE

Akai.K.,Ohnishi.Y.,&Nishigaki.M. (1977) "Finite element analysis of saturated-unsaturated seepage in soil", Jnue. of civil eng. No.264.pp87-97(inJapanise)

Ishihara,Y. &E.Shimojima(1983) "A role of pore air in infiltration process", Bull. Disas. Prev. Res. Inst., Kyoto Univ., Vol. 33, part 4, 1983, pp.163-222.

Li Xikul(1990) "A Numerical Model for Immiscible The Phase Fluid Flow in A Porous Medium and Its Time Domain Solution", International Journal for Numerical Methods in Engineering, Vol.30. pp1195-1212.

Michael A. Celia and Philip Binning(1992) "A Mass Conservative Numerical Solution for Two-Phase Flow in Porous Media With Application to Unsaturated Flow", Water Resources Research, Vol. 28, No.10, pp2819-2828, October 1992.

Sun Yao (1995) " A Study on the Coupling Theory of Two-Phase Flow in Deformation Porous Medium and Geomechanical Behaviors of Unsaturated Soil" Ph.D. thesis, University of Okayama ,1995 (In Japanes).

Computer Methods and Advances in Geomechanics, Yuan (ed.) © 1997 Balkema, Rotterdam, ISBN 90 5410 904 1

Computer simulations for borehole instability under in-situ conditions

Yutian Luo & Maurice Dusseault
Department of Earth Sciences, University of Waterloo, Ont., Canada

Abstract: The computer analyzing package to simulate borehole behaviours under true in place conditions around a borehole is developed. These conditions include, (1) principle stresses with s_1 ³ s_2 ³ s_3, instead of simply with s_2= s_3; (2) effects of fluid, heat, and mass transports; (3) time-dependent behaviour, focusing on borehole continuous deformation and damages. The software is designed to provide in both parametric analysis and case simulation.

1 INTRODUCTION

Borehole stability analyses must focus on the coupled processes of stress redistribution and transport (Darcy, Fourier and Fickian processes). *In situ* stresses are the sum of effective stress ($s¢_{i,j}$) and pore pressure (p_o). Stress redistribution and rock behaviour depends on effective stresses and mechanical response through elastic and plastic deformation parameters (Young' modulus E, Poisson's ratio n, yield criteria, and plasticity parameters), whereas transport behaviour depends on pressure and diffusion parameters (porosity n, heat capacity c, hydraulic conductivity K, thermal conductivity k, ionic diffusivity D, permeability k, compressibility C, and so on). In well-defined cases, transport effects can be functionally combined with stresses by non-coupled approaches.

Determination of effective stress requires study of the transmission and distribution of stresses both vertically and horizontally in the ground. Vertical stresses arise mainly from the weight of the overburden. Horizontal stresses are also affected by the overburden weight through the Poisson's ratio (n) effect (i.e.: $s¢_h$ = $s¢_v$(n/(1-n)) for a perfectly elastic gravitating horizontal system. However, we know that tectonic effects (lateral loading or unloading) override the Poisson effect in natural earth systems, leading to stress fields that can be strongly anisotropic (s_{hmin} ¹ s_{HMAX} ¹ s_v) and governed by the geological history (tectonics, diagenesis, thermal effects, ¼). For example, considering the effects of surface forces only, the horizontal tectonic force acting between rigid tectonic plates is a major factor, and can lead to conditions of thrust, normal, or wrench faulting, where the rock fabric yields under the highly deviatoric loads imposed by crustal movements. These processes can be greatly aided by high pore pressures generated by impeded compaction drainage, mineral transformations (e.g.: smectite→illite, gypsum→anhydrite, calcite→ dolomite), thermal effects, or hydrocarbon fluid generation (kerogen→oil and gas). The magnitude of s_{hmin}, s_{HMAX} may act to or decrease the local fluid transport properties through slow alterations in the rock fabric. This arises mainly through diagenetic alterations, which are enhanced by high effective stresses which compress the formation and accelerate geochemical processes of mass transfer, or by generation of extensional fracture networks normal to s_3 by high pore pressures. The directions of the major far-field stresses with respect to the orientation of a borehole greatly affect transport behaviour because of inherently anisotropic transport properties, and because borehole creation generates oriented damage fields (microfissures) in the surrounding rock as a result of the stress concentrations.

The uniform far-field fluid pressure also acts in the rock; it is generally partially or totally counterbalanced by an elevated borehole pressure (p_w) generated by a drilling fluid (foam, air, drilling mud with high density granular material). The effect of p_w ¹ p_o is to generate a load at the wall to partially balance the far-field total stresses. Imbalance of these forces ($p_w < p_o$) is more likely to lead to borehole instability and damage, particularly if the rocks are weak, but this condition, called "underbalanced" drilling, is also used in areas where the rocks are strong in order to achieve faster drilling rates. If fluids (water, oil, gas) in a porous stratum are encountered when $p_w < p_o$, massive influx can occur, and this is exceedingly dangerous condition which may lead to well loss through a blowout. Therefore, in these conditions, $p_w > p_o$ is enforced. Alternatively, if the condition $p_w > s_3$ is reached, it becomes possible to hydraulically fracture the strata around the borehole, leading to lost circulation and possible loss of well control. Thus, in general, the condition for the drilling fluid pressure throughout the open-hole section of a borehole is kept as s_3 ³ p_w ³ p_o. This may be impossible to achieve in long open-hole sections where the horizontal stress and pore pressure gradients are changing, and installation of steel casings are then used to isolate upper sections of the borehole to permit hole advancement.

During drilling, it may be judged desirable to minimise fluid transport across the borehole wall not only to reduce the chance of instability, but also to minimise invasion of damaging fluids into a reservoir interval. Clearly, in the case of oil and gas production, it is economically desirable to maximise flux into the wellbore, and this leads to substantial hydrodynamic drag forces (an inwardly-directed body force linked to the pressure gradient) applied to the strata, which may lead to spalling and the incorporation of formation solids (usually sand) into the production stream. Whatever the goals, it is clear that the far-field stresses and the intrinsic pore pressures dictate the local stresses and the fluxes, therefore rational calculations should account for these effects. This must be done through coupled analytical or numerical solutions for stresses and pressures, accounting for the presence of the borehole and the mechanical and transport properties of the surrounding strata..

2 EFFECTS OF TECTONIC FORCES

Considering the generally flat-lying stratigraphy and the great lateral extent of most basins compared to the thickness, it is reasonable to assume that the tectonic forces act in plane stress condition; that is, they are specified by s_r and s_q only.

When a basin is subjected to tectonic forces, the stress distribution within the basin is associated with this loading. If a well is drilled, the far-field stress distributions are perturbed around this borehole. Using polar coordinates, the stress distributions before, during, and after drilling must satisfy the general compatibility equation:

$$(\frac{\partial^2}{\partial r^2}+\frac{1}{r}\frac{\partial}{\partial r^2}+\frac{1}{r^2}\frac{\partial^2}{\partial \theta^2})(\frac{\partial^2 \phi}{\partial r^2}+\frac{1}{r}\frac{\partial \phi}{\partial r}+\frac{1}{r^2}\frac{\partial^2 \phi}{\partial \theta^2}) = 0 \quad (1)$$

where f is the stress function which depends on r and/or q.

The solution satisfying the boundary condition of uniform far-field loading is:

$$\sigma_r\Big|_{R\to\infty} = S(\frac{r_0^2}{r^2}-1) \quad (2)$$

$$\sigma_\theta|_{R\to\infty} = -S(\frac{r_0^2}{r^2}+1) \quad (3)$$

And that the stress distribution around the borehole arising under a uniaxial far-field stress is:

$$\sigma_r|_{R\to\infty} = \frac{S}{2}(\frac{r_0^2}{r^2}-1)-\frac{S}{2}(1+\frac{3r_0^4}{r^4}-\frac{4r_0^2}{r^2})\cos 2\theta \quad (4)$$

$$\sigma_\theta|_{R\to\infty} = \frac{S}{2}(\frac{r_0^2}{r^2}+1)+\frac{S}{2}(1+\frac{3r_0^4}{r^4})\cos 2\theta \quad (5)$$

Where R is the radius of a huge ring around the borehole, r_0 is the radius of the borehole. Above solutions are obtained under the condition that effects of pore pressure are ignored.

If no borehole exists, the formation is subjected to an orientated uniform stress; at the borehole wall, f_r=0, and f_q =S+2Scos2q. Note that in the neighbourhood of the borehole both f_r and f_q are q depended, that means, the borehole breakout and other instabilities are occurred orientated. Later in this paper we will discuss all possible cases coupling with different pore pressure regimes, using the software we developed.

3 FLUID FLOW AROUND A BOREHOLE

Suppose the formation fluid exerts an uniform pressure P_0 in a basin. At the area around a borehole, this pressure changes because an opening exists. If no fluid source exists within the basin, the fluid flows around the borehole can be specified by the Laplace equation:

$$\frac{\partial^2\Psi}{\partial r^2}+\frac{1}{r}\frac{\partial\Psi}{\partial r}+\frac{1}{r^2}\frac{\partial^2\Psi}{\partial\theta^2}=0 \quad (6)$$

Where Y is a pressure tendency which depends on r, the diameter of a centralis ring area with a borehole, and q, the orientation angle from a direction we assigned. This pressure function can then be separated into functions R(r) and Q(q):

$$\Psi(r,\theta) = R(r)\Theta(\theta) \quad (7)$$

Then the Laplace eigne solution is obtained:

$$\Psi(r,\theta) = C_0 + D_0 \ln r + \sum_{m}^{\infty}(A_m \cos m\theta + B_m \sin m\theta(C_m r^m + D_m \frac{1}{r^m}) \tag{8}$$

Where m is eigenvalues (m = 0, 1, 2,...).

The diameter of borehole r_0 is so small, compared with the size of the basin, that the pressure changes caused by the borehole cannot affect the area farther away. Thus, the far-field pressure tendency is still:

$$\Psi|_{r\to\infty} = \_P_0 r \cos\theta \tag{9}$$

And the pressure tendency at the borehole wall is defined as "0" to simplify the solution (The pressure tendency is a relative value):

$$\Psi|_{r=r_0} = 0 \tag{10}$$

Using these boundary conditions we finally obtain:

$$\Psi(r,\theta) = -P_0 r \cos\theta + P_0 \frac{r_0}{r} \cos\theta + D_0 \ln \frac{r}{r_0} \tag{11}$$

The physical meaning of the different items on the right-hand side of this solution can explain how the borehole affects below:

1. The first item of the right-hand side represents the original distribution of the pressure tendency with no borehole. The item takes a form the same as the pressure tendency distribution at an area far from the borehole.
2. The second item represents the redistribution of the pressure tendency when an opening exists. That is why this item vanishes at the area far from the borehole ($r \to \mu$). Thus, the different effects of a borehole can be discussed by modifying this item.
3. The last item, which includes an uncertainty constant D_0, reflects the fluid flowing across the borehole wall. D_0 exists because we omitted to define if there is any fluid transport through the borehole wall. In the case of drilling a well and other purposes we attempt to make no fluid transport acrose the borehole wall (neglecting the filtration), that means, D_0=0. So, the solution is more simple and easier to be used to the pressure borehole submitted. For example, the points A and C on the borehole wall as shown in Fig. 1 exerts the maximum and minimum pore pressure respectively, and at points B and D the pressure tendency is largest.

4 FLUID TRANSPORT COUPLED WITH STRESSES

Since s_r = 0 at the borehole wall, we need to analyze the effects of s_q only. On the basis of poro-elastic assumption, pore pressure (p) must be taken into account when analyzing the effective stress. Pore pressure around a borehole can be obtained from the gradient of the pressure tendency equation (19):

$$p = -\frac{d\Psi}{dr} = D_0 \frac{1}{r} + P_0 \cos\theta + P_0 \frac{r_0}{r} \cos\theta \tag{12}$$

In practice an internal pressure p_b is applied from the borehole to keep the borehole stable. Thus, the total stress that the borehole subjected to is:

$$\sigma_\theta = \frac{1}{2}(S+p) + \frac{r_0}{2r_0}(S+p-2p_b) - \frac{1}{2}(S-p)(1-\frac{3r_0^4}{r^4})\cos 2\theta \tag{13}$$

This formula couples the stress and pressure around a borehole. With the varying of diameter r from r_0 (borehole wall) to a distance far enough from the hole, the s and p charge properly with this variation. So, all possible cases of borehole behaviour have been covered.

However, this stress distribution is deduced based on a circular hole, but in practice borehole shapes are not so regular. The borehole may be an ellipse if the far-field stress is of deviation. It is time consuming, and difficult for us to deduce a new set of formulas responding to this shape change, because the change is continuing, and the geometry types are numerous. We introduce the _________ transformation to take advantage of meeting most types of shapes, and of simplifying the formulas by inheriting the provers solutions for circular holes.

Suppose the borehole has already become an ellipse with focal distance ***c***. The acting direction of the far-field stress is parallel to the major axis ***a*** (may be in any angles). On the z-plane (this is a complex plane where _________ function is defined) equations (14), (17) and (18) become:

$$\Delta\Psi = 0 \tag{14}$$

$$\Psi|_{\zeta\to\infty} = P_0 \zeta \tag{15}$$

$$\frac{\partial\Psi}{\partial n}|_{n\perp wall} = 0 \tag{16}$$

Using _________ function:

$$\frac{\zeta(z)}{c} = \frac{1}{2}(z + \frac{1}{z}) \tag{17}$$

the subsidiary equations on an z-plane (a normal complex plane) are:

$$\Delta\Psi = 0 \tag{18}$$

$$\Psi|_{z\to\infty} = P_0\frac{c}{2}(r+\frac{1}{r})\cos\psi|_{z\to\infty} = \frac{c}{2}P_0 r\cos\psi \tag{19}$$

$$\frac{d\Psi}{dr}|_{r=\frac{a+b}{c}} = 0 \tag{20}$$

Now the ellipse with focal distance ***c*** is transformed to a circle with diameter ***(a+b)/c*** on the z-plane. It is possible to inherit equation (19):

$$\Psi = \mathrm{Re}[\frac{1}{2}cP_0 z + \frac{1}{0}P_0\frac{(a+b)^2}{c}\frac{1}{z} + D_0 \ln\frac{cz}{(a+b)^2}] \tag{21}$$

Inverse back to z-plane:

Using ________ transform we easily get the pressure tendency around an elliptical borehole so easy. For other formulas we can proceed in similar manner.

$$\Psi = \mathrm{Re}[\frac{1}{2}P_0(\xi+\sqrt{\xi^2-1}) +$$
$$\frac{1}{2}P_0(a+b)^2(\xi-\sqrt{\xi^2-1}) + D_0\ln\frac{\xi+\sqrt{\xi^2-1}}{(a+b)^2} \tag{22}$$

5 SOFTWARE PACKAGE OF BOREHOLE INSTABILITY EVALUATION

The previous analyses are incorporated into a software system. The effect of individual factor can be analyzed, and then, simulation can be performed based on the data from these analyzing, as well as from tests or field sites. For example, this allows us to do borehole breakout analysis in combining the tectonic and local drilling or injection information. We also expect using this software to analyze the scatter of practical breakout data. All simulation models are coded into a window-based user friendly software package. A multi-tasking graphical screen is also developed through which the different monitoring information windows, or the different expressions of the simulation can be observed at the same time. This is very useful for analyzing, or modelling the cases with complicated in place conditions. Here we briefly introduce these basic functions in the same order as the main menu of the software.

File: The pull-down menu provides the standard file management operations. Users can do things of "Open", "Save", "Print", and so on the same as do in the standard windows software.

Stress: This sub-menu provides the function to do stress analysis under several different conditions like "Isotropic", or "Anistropic" with the specified direction or orientation.

Transport: Here users can select to simulation with coupling different materials which may suffer transportation, including "Fluid", "Heat", "Mass". The different parameters may be supplied for different simulations using pop-up sub-menus.

Application: Here we go further more step to simulate the different applications like "Injection", and "Flooding". All these are done under a in place stress assumption.

Graphics: Three kinds of graphical displays and views for the simulation and calculation are provided at this time. They are "XY_Plot", "Top_View", and "Cross_Section". We will continue to design and develop other displays soon.

6 CONCLUSION

Software to study some of the effects of far-field stresses on local fluid transport around a borehole has been developed. Stress magnitudes and orientation are variable to account for different borehole orientations, and the model can also account for non-circular borehole shapes which may arise through borehole damage (breakouts, differential spalling, sloughing of fissile shale). Borehole instability is best understood as a critical condition, and understanding what leads to this critical condition must be based on case histories. Use of a model which is more amenable to studying the variations in stresses and pressure values and their consequences may aid in understanding the geomechanical behaviour around a borehole, and allow optimisation during drilling or oil and gas production.

7 ACKNOWLEDGMENTS

We wish to thank the Natural Sciences and Engineering Research Council of Canada, a number of private oil companies for their financial assistance. Highly competent technical personnel at the University of Waterloo are responsible for machining and setting up the apparatus, and their involvement is invaluable.

Computer Methods and Advances in Geomechanics, Yuan (ed.) © 1997 Balkema, Rotterdam, ISBN 90 5410 904 1

Finite element analysis of transient deformation and seepage process in unsaturated soils

Xikui Li & Yiqun Fan
Dalian University of Technology, China

ABSTRACT: This paper presents a numerical model for the finite element analysis of transient deformation and seepage problems in unsaturated soils. A definition of the generalized effective stress for unsaturated soils is proposed on the basis of the two stress variables formula. The Alonso constitutive model to simulate the non-linear mechanical behaviour of unsaturated soils is integrated into the model. With introduction of the natural coordinates in the stress space a consistent algorithm for the non-smooth multisurface pressure dependent plasticity is derived.

1 GOVERNING EQUATIONS

The stress state of a local point in unsaturated soils can be described by two stress variables: the net stress σ^* and the suction p_c defined by

$$\sigma^* = \sigma + \mathbf{m}p_g ; \quad p_c(S_w, p_n) = p_g - p_w \tag{1}$$

where σ is the total stress tensor and $\mathbf{m}^T = [1 \ 1 \ 1 \ 0 \ 0 \ 0]$ for 3D case, p_c is assumed as a function of the saturation degree S_w of porous water and the mean net stress p_n, p_g and p_w the pressures in porous air and water. The net stress rate can be written as

$$\dot{\sigma}^* = \dot{\sigma}' + H_e^{-1}\mathbf{D}_e\mathbf{m}\dot{p}_c \tag{2}$$

with $\dot{\sigma}' = \mathbf{D}_e\dot{\varepsilon}^e = \mathbf{D}_{ep}\dot{\varepsilon}$ defined as the generalized effective stress and $\mathbf{D}_e$ and H_e being the material elastic moduli, $\dot{\varepsilon}$ and $\dot{\varepsilon}^e$ total and elastic strain rates, $\mathbf{D}_{ep}$ the tangent elasto-plastic modulus matrix. Substitution of (1) into (2) gives

$$\dot{\sigma}' = \dot{\sigma} + \dot{p}\mathbf{m} \tag{3}$$

$$\dot{p} = \chi_s \dot{p}_w + (1-\chi_s)\dot{p}_g \tag{4}$$

$$\chi_s = 3K_t H_e^{-1}; \quad K_t = \frac{E}{3(1-2\nu)} \tag{5}$$

where p can be regarded as the equivalent pore pressure acting on the solid grains by porous fluid mixture.

Let denote the Darcy's velocities for the porous water and air as $\dot{\mathbf{w}}_w$ and $\dot{\mathbf{w}}_g$, The mass conservation for porous fluids and overall momentum equilibrium equations can be given by

$$\left[\frac{(1-n)}{K_s} + \frac{n}{K_w}\right]S_w\dot{p}_w + +S_w\dot{u}_{i,i} + \dot{w}_{wi,i} + \frac{\dot{w}_{wi}}{\rho_w}\frac{\partial \rho_w}{\partial x_i} + \left[\frac{(1-n)S_w(1-\chi_s)}{K_s}\frac{dp_c}{dS_w} + n\right]\dot{S}_w = 0 \tag{6}$$

$$\left[\frac{(1-n)}{K_s} + \frac{n}{K_g}\right]S_g\dot{p}_w + S_g\dot{u}_{i,i} + \dot{w}_{gi,i} + \frac{\dot{w}_{gi}}{\text{-}\rho_g}\frac{\partial \rho_g}{\partial x_i} + \left[[\frac{(1-n)S_g(1-\chi_s)}{K_s} + \frac{nS_g}{K_g}]\frac{dp_c}{dS_w} - n\right]\dot{S}_w = 0 \tag{7}$$

$$\sigma_{ij,j} + \rho b_i - \rho\ddot{u}_i = 0 \tag{8}$$

where n stands for the porosity of the medium, S_g the saturation degree of air, $\dot{\mathbf{u}}$ and $\ddot{\mathbf{u}}$ the velocity and the acceleration of the solid, ρ_s, ρ_w, ρ_g are the density of the solid, water and air. We note that the average density ρ of the solid-water-air mixture can be written as

$$\rho = \rho_s(1-n) + n(S_w\rho_w + S_g\rho_g) \tag{9}$$

The Darcy's velocities of the porous water and air can be given from the momentum equilibrium equations for the porous water and air as

$$\dot{w}_{wi} = (k_w)_{ij}(-p_{w,j} + \rho_w(b_j - \ddot{u}_j)) \qquad (10)$$

$$\dot{w}_{gi} = (k_g)_{ij}(-p_{g,j} + \rho_g(b_j - \ddot{u}_j)) \qquad (11)$$

where $(k_w)_{ij}$ and $(k_g)_{ij}$ are the permeability coefficients of porous water and air.

2 FINITE ELEMENT DISCRETIZATION AND $\mathbf{u} - p_w - S_w$ FORMULATIONS

Substitution of equations (10) and (11) into mass conservation equations (6) and (7) and the finite element discretization in space for equations (6) - (8) lead to three discrete equations

$$\mathbf{M}_p\ddot{\mathbf{u}} + \mathbf{Q}_p\dot{\mathbf{u}} + \mathbf{C}_{pp}\dot{\mathbf{p}}_w + \mathbf{C}_{ps}\dot{\mathbf{S}}_w + \mathbf{H}_{pp}\mathbf{p}_w + \mathbf{H}_{ps}\mathbf{S}_w = \mathbf{F}_p$$

$$\mathbf{M}_s\ddot{\mathbf{u}} + \mathbf{Q}_s\dot{\mathbf{u}} + \mathbf{C}_{sp}\dot{\mathbf{p}}_w + \mathbf{C}_{ss}\dot{\mathbf{S}}_w + \mathbf{H}_{ps}^T\mathbf{p}_w + \mathbf{H}_{ss}\mathbf{S}_w = \mathbf{F}_s$$

$$\int_\Omega N^u_{k,i}\sigma''_{ij} d\Omega - \mathbf{Q}_p^T\mathbf{p}_w - \mathbf{Q}_s^{*T}\mathbf{S}_w + \mathbf{M}_u\ddot{\mathbf{u}} = \mathbf{F}_u \qquad (12)$$

where $\mathbf{Q}_p^T = \int_\Omega \mathbf{B}^T\alpha_b \mathbf{m}\mathbf{N}_p d\Omega$

$$\mathbf{Q}_s^{*T} = \int_\Omega \mathbf{B}^T\alpha_b(1-\chi)\frac{p_c}{S_w}\mathbf{m}\mathbf{N}_s d\Omega$$

$$\mathbf{B} = \mathbf{L}\mathbf{N}_u ; \mathbf{M}_u = \int_\Omega \mathbf{N}_u^T\rho\mathbf{N}_u d\Omega$$

$$\mathbf{M}_p = \int_\Omega (\nabla\mathbf{N}_p)^T(\mathbf{k}_w\rho_w + \mathbf{k}_g\rho_g)\mathbf{N}_u d\Omega$$

$$\mathbf{F}_u = \int_\Omega \mathbf{N}_u^T\rho\,\mathbf{b}d\Omega + \int_{\Gamma_t}\mathbf{N}_u^T\bar{\mathbf{t}}\,d\Gamma ; \chi = 3K_t\overline{H}_e^{-1}$$

$$\mathbf{C}_{pp} = \int_\Omega \mathbf{N}_p^T(\frac{\alpha_b - n}{K_s} + \frac{nS_w}{K_w} + \frac{nS_g}{K_g})\mathbf{N}_p d\Omega$$

$$\mathbf{C}_{ps} = \int_\Omega \mathbf{N}_p^T(\frac{(\alpha_b - n)(1-\chi_s)}{K_s} + \frac{nS_g}{K_g})\frac{dp_c}{dS_w}\mathbf{N}_s d\Omega$$

$$\mathbf{H}_{pp} = \int_\Omega (\nabla\mathbf{N}_p)^T(\mathbf{k}_w + \mathbf{k}_g)(\nabla\mathbf{N}_p)d\Omega$$

$$\mathbf{H}_{ps} = \int_\Omega (\nabla\mathbf{N}_p)^T\mathbf{k}_g\frac{dp_c}{dS_w}(\nabla\mathbf{N}_s)d\Omega$$

$$\mathbf{Q}_s = \int_\Omega \mathbf{N}_s^T\mathbf{m}^T\alpha_b S_g\frac{dp_c}{dS_w}\mathbf{B}d\Omega$$

$$\mathbf{H}_{ss} = \int_\Omega (\nabla\mathbf{N}_s)^T\mathbf{k}_g(\frac{dp_c}{dS_w})^2(\nabla\mathbf{N}_s)d\Omega$$

$$\mathbf{C}_{sp} = \int_\Omega \mathbf{N}_s^T(\frac{(\alpha_b - n)}{K_s} + \frac{n}{K_g})S_g\frac{dp_c}{dS_w}\mathbf{N}_p d\Omega$$

$$\mathbf{M}_s = \int_\Omega (\nabla\mathbf{N}_s)^T\frac{dp_c}{dS_w}\mathbf{k}_g\rho_g\mathbf{N}_u d\Omega$$

$$\mathbf{C}_{ss} = \int_\Omega \mathbf{N}_s^T[(\frac{(\alpha_b - n)(1-\chi_s)}{K_s} + \frac{n}{K_g})S_g\frac{dp_c}{dS_w} - n]\times \times\frac{dp_c}{dS_w}\mathbf{N}_s d\Omega$$

$$\mathbf{F}_s = \int_\Omega (\nabla\mathbf{N}_s)^T\frac{dp_c}{dS_w}\mathbf{k}_g\rho_g\mathbf{b}d\Omega - \int_{\Gamma_g}\mathbf{N}_s^T\frac{dp_c}{dS_w}\bar{\dot{\mathbf{w}}}_g^T\mathbf{n}d\Gamma$$

$$\mathbf{F}_p = \int_\Omega (\nabla\mathbf{N}_p)^T(\mathbf{k}_w\rho_w + \mathbf{k}_g\rho_g)\mathbf{b}d\Omega - \int_{\Gamma_w}\mathbf{N}_p^T\bar{\dot{\mathbf{w}}}_w^T\mathbf{n}d\Gamma - \int_{\Gamma_g}\mathbf{N}_p^T\bar{\dot{\mathbf{w}}}_g^T\mathbf{n}d\Gamma \qquad (13)$$

3 ELASTO-PLASTIC CONSTITUTIVE MODEL AND CONSISTENT INTEGRATION ALGORITHM

The elasto-plastic constitutive model proposed by Alonso (Alonso, Gens and Josa, 1990) is utilized to describe the general behaviour of unsaturated soils. The boundary of the elastic domain of the model is defined by three yield surfaces interacting in a non-smooth fashion in $p_n - q - p_c$ stress space, where q is the second stress invariant of the net deviatoric stress tensor. In this section, a consistent integration algorithm for the model with non-smooth triplicative yield surfaces , which is composed of the return mapping algorithm for integration of the rate constitutive equations and the consistent elasto-plastic tangent modulus matrix, is developed.

Let $\hat{\sigma}^T = [\sigma_I \; \sigma_{II} \; \sigma_m \; \tau_{xy} \; \tau_{yz} \; \tau_{zx}]$ and $\hat{\varepsilon}^T = [\varepsilon_I \; \varepsilon_{II} \; \varepsilon_m \; \varepsilon_{xy} \; \varepsilon_{yz} \; \varepsilon_{zx}]$ denote the stress and the strain vectors referred to the natural coordinates and they can be transformed from σ and ε by

$$\hat{\sigma} = \mathbf{T}\sigma \; ; \qquad \hat{\varepsilon} = \mathbf{T}\varepsilon \qquad (14)$$

The transformation matrix $\mathbf{T}$ is defined by

$$\mathbf{T} = \begin{bmatrix} 1/\sqrt{2} & 0 & -1/\sqrt{2} & 0 & 0 & 0 \\ -1/\sqrt{6} & \sqrt{2/3} & -1/\sqrt{6} & 0 & 0 & 0 \\ 1/\sqrt{3} & 1/\sqrt{3} & 1/\sqrt{3} & 0 & 0 & 0 \\ 0 & 0 & 0 & 1 & 0 & 0 \\ 0 & 0 & 0 & 0 & 1 & 0 \\ 0 & 0 & 0 & 0 & 0 & 1 \end{bmatrix} \qquad (15)$$

With the definitions for p_n and q we have

$$p_n = \sigma_m/\sqrt{3} + p_g = \sigma'_m/\sqrt{3} + \chi\, p_c \qquad (16)$$

$$q = (\frac{1}{2}\sigma^{*T}\mathbf{P}\sigma^*)^{1/2} = (\frac{1}{2}\hat{\sigma}^{*T}\hat{\mathbf{P}}\hat{\sigma}^*) = (\frac{1}{2}\hat{\sigma}'^T\hat{\mathbf{P}}\hat{\sigma}')^{1/2} \qquad (17)$$

Substitution of (2) into (3) with transformation of stress vectors referred to the original coordinates into those referred to the natural coordinates gives

$$\dot{\hat{\sigma}}' = \dot{\hat{\sigma}}^* - \sqrt{3}\chi_s \mathbf{1}_m \dot{p}_c \qquad (18)$$

where $\hat{\sigma}^*, \hat{\sigma}'$ are the net stress vector and the generalized effective stress vector referred to the natural coordinates and $\hat{\mathbf{P}} = diag(3, 3, 0, 6, 6, 6)$; $\mathbf{1}_m^T = [0 \quad 0 \quad 1 \quad 0 \quad 0 \quad 0]$.

The three component yield surfaces in the model are: the Critical State Lines (CSL); the State Boundary Surface (SBS) and the Suction Increasing (SI) yield surfaces. They can be given as

$$\text{CSL: } F_1 = F_1(q, \sigma'_m, p_c, \bar{\varepsilon}^p) = q + A_c\sigma'_m + B_c = 0 \qquad (19)$$

with $A_c = \dfrac{2\,Sin\phi}{\sqrt{3}(3 - Sin\phi)} = \dfrac{M}{\sqrt{3}}, B_c = -M(p_s - \chi\, p_c)$.

p_s is the cohesion, as a piece-wise linear hardening rule with p_c and the equivalent plastic strain $\bar{\varepsilon}^p$ is considered, $p_s = c_0 + k p_c + h_{ep}\bar{\varepsilon}^p = p_s(\bar{\varepsilon}^p, p_c)$. M is the slope of the CSL in $p_n - q$ plane.

$$\text{SBS: } F_2 = q^2 - M^2(-p_n + p_s)(p_0 + p_n) \qquad (20)$$

Substitution of (15) into (20) gives alternative form of F_2 as

$$F_2 = F_2(q, \sigma'_m, p_c, \bar{\varepsilon}^p, \varepsilon_v^p)$$
$$= q^2 - M^2(-\frac{\sigma'_m}{\sqrt{3}} - \chi\, p_c + p_s)(p_0 + \frac{\sigma'_m}{\sqrt{3}} + \chi\, p_c) \qquad (21)$$

According to the model definitions p_0 is evaluated by

$$p_0 = p^c \left(\frac{p_0^*}{p^c}\right)^{\frac{\lambda(0)-\kappa}{\lambda(p_c)-\kappa}} \qquad (22)$$

$$\lambda(p_c) = \lambda(0)[(1-r)\exp(-\beta_\lambda p_c) + r] \qquad (23)$$

where the saturated preconsolidation stress p_0^* is a hardening parameter depending on the plastic volumetric strain ε_v^p. The evolution of $p_0^*(\varepsilon_v^p)$ with ε_v^p is governed by

$$\dot{p}_0^* = -\frac{p_0^*(1+e_0)}{\lambda(0) - \kappa}\dot{\varepsilon}_v^p \qquad (24)$$

$\lambda(0), p_0^*(0), r, \beta_\lambda, \kappa, p^c$ should be given as the material parameters to determine the current value of $p_0 = p_0(\varepsilon_v^p, p_c)$.

$$\text{SI: } F_3 = F_3(p_c, \varepsilon_v^{ps}) = p_c - p_{c,0}(\varepsilon_v^{ps}) = 0 \qquad (25)$$

where the evolution of $p_{c,0}$ obeys the hardening rule

$$\dot{p}_{c,0} = -\frac{(1+e)(p_{c,0} + p_{at})}{\lambda_s - \kappa_s}\dot{\varepsilon}_v^{ps} \qquad (26)$$

The rate constitutive equation for elasto-plastic materials is generally given by

$$\dot{\sigma}' = \mathbf{D}_e\dot{\varepsilon}^e = \mathbf{D}_e(\dot{\varepsilon} - \dot{\varepsilon}^p) \qquad (27)$$

where $\dot{\varepsilon}^p$ is the plastic strain rate. Pre-multiplying the transformation matrix $\mathbf{T}$ to the both sides of (27) gives

$$\dot{\hat{\sigma}}'' = \hat{\mathbf{D}}_e\dot{\hat{\varepsilon}}^e = \hat{\mathbf{D}}_e(\dot{\hat{\varepsilon}} - \dot{\hat{\varepsilon}}^p) \qquad (28)$$

where $\hat{\mathbf{D}}_e = diag(2G, 2G, 3K_t, G, G, G)$ for the elastic modulus matrix of the isotropic materials.

For simplicity in discussion, the evolution of the plastic strain vector $\dot{\varepsilon}^p$ for the multi- surface plasticity is assumedly given by the associate plasticity rule as follows

$$\dot{\hat{\varepsilon}}^p = \sum_{i=1}^{m}\dot{\hat{\varepsilon}}^{pi} = \sum_{i=1}^{m}\dot{\lambda}_i\frac{\partial F_i}{\partial\hat{\sigma}''} \qquad (29)$$

In contrast with single surface plasticity, violation of a yield condition by the trial elastic stress does not ensure that the yield condition is active. A backward Euler integration procedure with determination of the active yield conditions on the basis of the Kuhn-Tucker conditions is to be developed as an extension of the classical radial return mapping algorithm. To calculate the plastic strain vectors defined by (29) we consider that particular forms of the flow vectors for CSL, BSB and SI component models and have

$$\Delta\hat{\varepsilon}^{p1} = \lambda_1(\frac{1}{2q}\hat{\mathbf{P}}\hat{\sigma}' + A_c\mathbf{1}_m);\ \Delta\hat{\varepsilon}^{p2} = \lambda_2(\hat{\mathbf{P}}\hat{\sigma}' + A_b\mathbf{1}_m)$$

$$\Delta\hat{\varepsilon}^{p3} = \frac{1}{\sqrt{3}}\Delta\varepsilon_v^{ps}\mathbf{1}_m = -H_{mp}^{-1}\Delta p_{c,0}\mathbf{1}_m$$

$$H_{mp}^{-1} = \frac{1}{\sqrt{3}} \frac{(\lambda_s - \kappa_s) h(\Delta p_{c,0})}{(1+e)(p_{c,0} + p_{at})} \quad (30)$$

with $\Delta p_{c,0} = p_c - p_{c,0} \geq 0,\ A_b = \frac{M^2}{\sqrt{3}}(2p_n + p_0 - p_s)$ and $h(*)$ is the Heaviside function.

It has been shown that there are two internal state variables introduced in the above three yield functions, i.e. $\bar{\varepsilon}^p$ and ε_v^p, which control the evolution of the stress vector $\hat{\sigma}'$ with given strain and suction p_c. The incremental equivalent plastic strain can be given (Duxbury and Li, 1996) by

$$\begin{aligned} \Delta\bar{\varepsilon}^p &= \lambda_1 + 2q\lambda_2 + sign(\sigma'_m)\frac{\lambda_1 A_c + \lambda_2 A_b}{\sqrt{3}} \\ &= \Delta\bar{\varepsilon}^p(\lambda_1, \lambda_2, A_b, \alpha, q^E) \end{aligned} \quad (31)$$

The constitutive equation for an incremental load step is given by

$$\hat{\sigma}' = (\alpha \mathbf{1}_d + \beta \mathbf{1}_h)\hat{\sigma}'^E \ ; q = \alpha q^E ; \ \sigma'_m = \beta \sigma'^E_m \quad (32)$$

where the trial elastic stress vector is defined as

$$\hat{\sigma}'^E = \mathbf{D}_e(\hat{\varepsilon}_{t+\Delta t} - \hat{\varepsilon}_t^p) ; q^E = (\frac{1}{2}\hat{\sigma}''^{ET}\hat{\mathbf{P}}\hat{\sigma}''^E)^{1/2} \quad (33)$$

with

$$\alpha = \frac{q^E - 3G\lambda_1}{q^E(1+6G\lambda_2)} = \alpha(\lambda_1, \lambda_2, q^E) \quad (34)$$

$$\begin{aligned} \beta &= 1 - \frac{3K_t(\lambda_1 A_c + \lambda_2 A_b - \Delta p_{c,0} H_{mp}^{-1})}{\sigma''^E_m} \\ &= \beta(\lambda_1, \lambda_2, A_b, \sigma'^E_m, p_c) \end{aligned} \quad (35)$$

$$\mathbf{1}_d = diag(1, 1, 0, 1, 1, 1); \mathbf{1}_h = diag(0, 0, 1, 0, 0, 0) \quad (36)$$

The plastic volumetric strain is taken as the internal state variable to control the evolution of SBS. The incremental plastic volumetric strain can be computed by

$$\begin{aligned} \Delta\varepsilon_v^p &= \sqrt{3}\Delta\varepsilon_m^p = \sqrt{3}(\lambda_2 A_b - \Delta p_{c,0} H_{mp}^{-1}) \\ &= \Delta\varepsilon_v^p(\lambda_2, A_b, p_c) \end{aligned} \quad (37)$$

Substitution of (31) for q and σ'_m into the yield function (18) and (20) and evaluation of variations $\dot{\alpha}, \dot{\beta}, \dot{q}^E, \dot{\sigma}''^E_m, \dot{\bar{\varepsilon}}^p, \dot{\varepsilon}_v^p$ lead to

$$(\partial_{\lambda_1} F_1)\dot{\lambda}_1 + (\partial_{\lambda_2} F_1)\dot{\lambda}_2 + (\partial_\varepsilon F_1)^T \dot{\hat{\varepsilon}} + (\partial_{p_c} F_1)\dot{p}_c = 0 \quad (38)$$

$$(\partial_{\lambda_1} F_2)\dot{\lambda}_1 + (\partial_{\lambda_2} F_2)\dot{\lambda}_2 + (\partial_\varepsilon F_2)^T \dot{\hat{\varepsilon}} + (\partial_{p_c} F_2)\dot{p}_c = 0 \quad (39)$$

For a typical integration point in plastic yielding the backward-Euler algorithm returning to the yield surfaces at the current iteration k can be written in an iterative manner as

$$\begin{bmatrix} \partial_{\lambda_1} F_1 & \partial_{\lambda_2} F_1 \\ \partial_{\lambda_1} F_2 & \partial_{\lambda_2} F_2 \end{bmatrix}^{k-1} \begin{Bmatrix} \Delta\lambda_1^k \\ \Delta\lambda_2^k \end{Bmatrix} = \begin{Bmatrix} -F_1^{(k-1)} - (\partial_{p_c} F_1)^{k-1} \Delta p_{c,0} h(\Delta p_{c,0}) \\ -F_2^{(k-1)} - (\partial_{p_c} F_2)^{k-1} \Delta p_{c,0} h(\Delta p_{c,0}) \end{Bmatrix} \quad (40)$$

if both CSL and SBS yield surfaces are active. It is remarked that by application of the above implicit backward Euler scheme the return mapping algorithm is devised as a constrained optimization problem governed by discrete Kuhn - Tucker conditions.

With the consistency conditions (38) and (39) the variations $\dot{\lambda}_1$ and $\dot{\lambda}_2$ can be expressed in terms of the variations $\dot{\hat{\varepsilon}}$ and $\dot{p}_c$, then both of $\dot{\alpha}$ and $\dot{\beta}$ can be expressed in terms of $\dot{\hat{\varepsilon}}$ and $\dot{p}_c$, finally it is given from (32) that

$$\dot{\hat{\sigma}}'' = \hat{\mathbf{D}}_{ep}\dot{\hat{\varepsilon}} + \mathbf{Q}_{ep}\dot{p}_c \quad (41)$$

where the consistent tangent modulus matrix $\hat{\mathbf{D}}_{ep}$ and the consistent tangent modulus vector $\hat{\mathbf{Q}}_{ep}$ are

$$\begin{aligned} \hat{\mathbf{D}}_{ep} &= (\alpha \mathbf{1}_d + \beta \mathbf{1}_h)\hat{\mathbf{D}}_e + \mathbf{1}_d \hat{\sigma}''^E \mathbf{C}^T_{\alpha\varepsilon} + \mathbf{1}_h \hat{\sigma}''^E \mathbf{C}^T_{\beta\varepsilon} \\ \hat{\mathbf{Q}}_{ep} &= (C_{\alpha p}\mathbf{1}_d + C_{\beta p}\mathbf{1}_h)\hat{\sigma}''^E \end{aligned} \quad (42)$$

4 NUMERICAL EXAMPLES

The example considers a flexible footing resting on a strip of unsaturated soil, which is 12m deep, 46m wide and of infinite length in the horizontal direction. An uniformly distributed pressure loading is applied to the central 8m section which increases to 1200 kN within 1.5 seconds. The example is modeled as a transient plane strain problem for the deformation of solid skeleton of the unsaturated soil coupled with the seepage process of immiscible porous water and air flow. The mesh along with boundary conditions is shown in Figure 1 in which, by symmetry, only one half with 23m wide is taken and discretized.

As the initial condition suction on the top surface outside the footing area is set to zero to simulate a permanent wetting of the surface. In addition, both the pore air pressure and the pore water pressure on

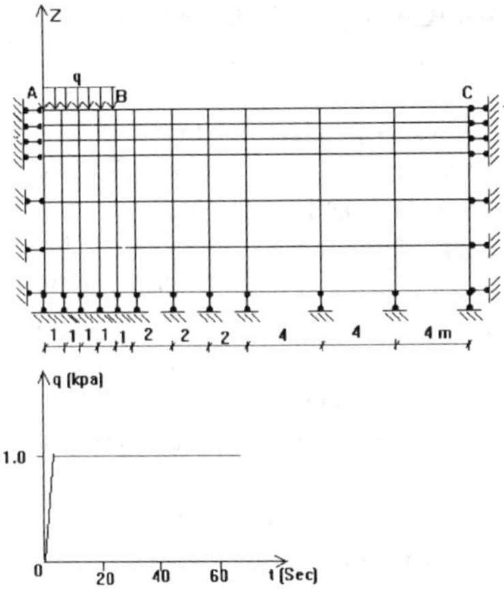

Figure 1 Mesh of centrally loaded soil strip and the load history

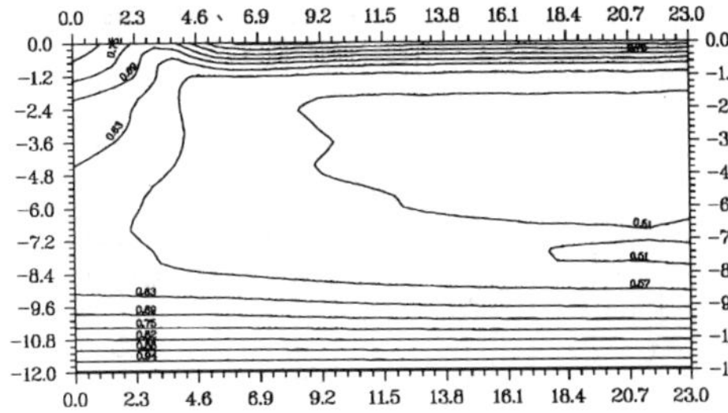

Figure 2 The saturation contours

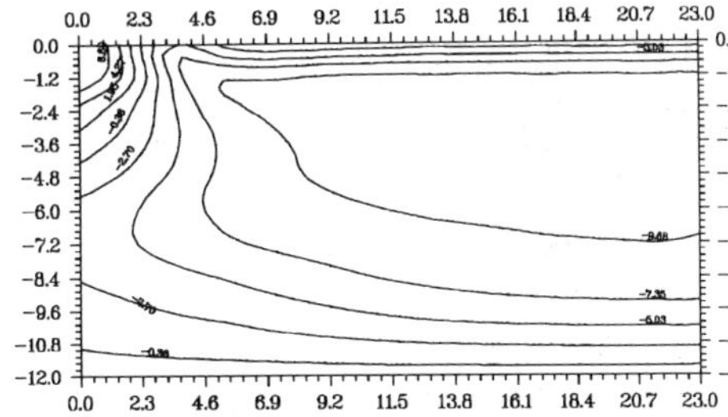

Figure 3 The contours of the pore water pressure

the surface are fixed to zero due to the passive air assumption. To simulate the saturated zone beneath the water table the saturation is kept equal to one at the bottom of the domain throughout the analysis. The initial saturation at time t = 0 is set to 0.5 over the domain. The initial effective stresses acting on the soil solid skeleton are assumed as $\sigma_z' = 25.44z(kpa)$ with z being the coordinate value in z axis and $\sigma_x' = \sigma_y' = 0.65\sigma_z'$.

The suction is assumed to relate the saturation degree and the void ratio as

$$p_c = (a_s \ln e)^{-1} \ln(1 - (S_{r0} - S_w)/b_s) \tag{43}$$

where the parameters $a_s = 0.1 m^2/kN$, $b_s = 0.85$ and $S_{r0} = 0.999$ are used. The void ratio is updated according to the current values of the mean net stress and the suction as (Lloret and Ledesma, 1993)

$$e = e_0 + a_e p_n + b_e p_c + c_e p_n p_c \tag{44}$$

where $a_e = -1.5 \times 10^{-4}\ 1/kpa$, $b_e = -1.0 \times 10^{-4}\ 1/kpa$ and $c_e = 1.0 \times 10^{-6}\ 1/kpa$ are used with the initial value e_0 of the void ratio.

The material properties for solid phase of the unsaturated soil are: $E = 3.6 \times 10^3 kpa$, $\upsilon = 0.3$, the initial uniform porosity $n = 0.322$. The bulk moduli for soil grains and water are $K_s = 6.146 \times 10^7 kpa$ and $K_w = 1.724 \times 10^6 kpa$. The density for the solid phase and water are: $\rho_s = 2.647\, kNSec^2/m^4$ and $\rho_w = 1 kNSec^2/m^4$. The nonlinear material parameters for the Alonso model are: (1) the initial cohesion $c_0 = 10kpa$ and the hardening parameters $k = 0.8$ and $h_{ep} = 0$; the internal frictional angle $\phi = 45^0$; (2) the parameters used to determine $\lambda(p_c)$ in equation (23) are: $\lambda(0) = 0.2$, $r = 0.75$, $\beta_\lambda = 0.0125$ and $\kappa = 0.02$ in (24). (3) the initial value of $p_0^* = 300kpa$ and the reference stress $p^c = 150kpa$ used in equation (22); (4) the parameters for SI surface used in (26) are: $\lambda_s = 0.08$ and $\kappa_s = 1.667 \times 10^{-4}$ and the initial suction at each integration point is taken as the threshold of SI yielding $p_{c,0}$.

To model the porous fluid flows within the unsaturated soil the following formulae (Lloret and Ledesma, 1993) for evaluation of isotropic water and air permeabilities, which take into account the effects of the saturation S_w and the void ratio e, are utilized

$$k_w = b_w \left(\frac{S_w - S_r}{1 - S_r} \right)^3 a_w^e \tag{45}$$

$$k_g = b_g [\beta_g (p_g + p_{at}) \frac{g}{\mu_g}] [e S_g]^{a_g} \tag{46}$$

where the material parameters used are: $a_w = 0.1 \times 10^6$; $b_w = 3 \times 10^{-9} m/s$; $S_r = 0.25$; $a_g = 3.98$; $b_g = 0.97 \times 10^{-9} m/s$; $\beta_g = 0.1189 \times 10^{-4} s^2/m^2$; $\mu_g = 0.185 \times 10^{-7} kNs/m^2$; $p_{at} = 10\, kN/m^2$, $g = 9.81 m/s^2$.

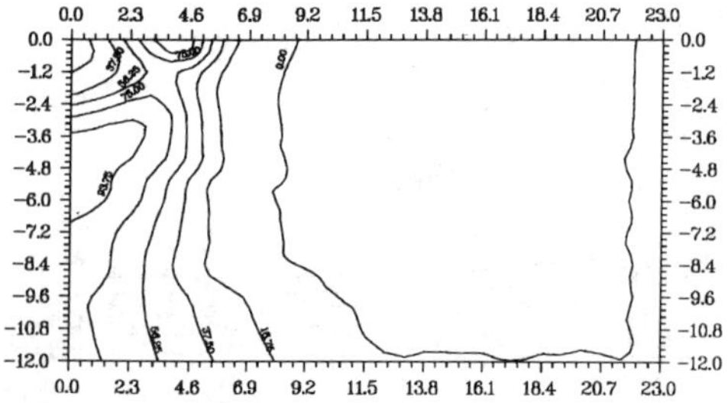

Figure 4 The contours of the equivalant plastic strain

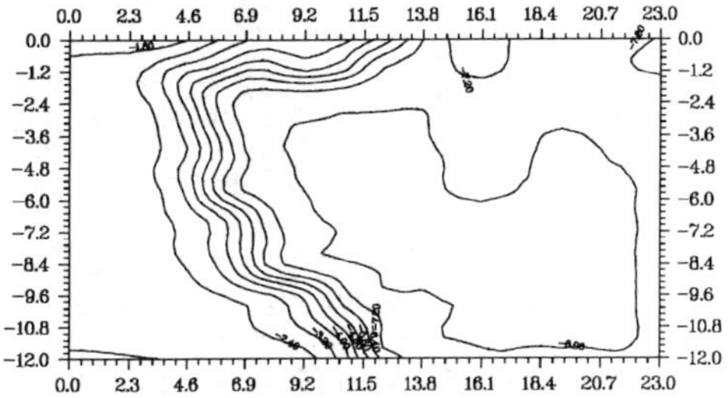

Figure 5 The contours of the volumetric plastic strain ($\varepsilon_v^p = -10^s$, where s stands for the value indicated on the contours; s = -8 implies no volumetric plastic strain)

Figure 2 - Figure 5 present the distributions of the saturations, the pore water pressure, the equivalent plastic strains and the volumetric plastic strains throughout the domain at t = 5 minutes, when the severe collapse under the footing occurs.

ACKNOWLEDGMENT

This work is sponsored by the State Scientific and Technological Commission of China through the National key project on basic research and applied research: Applied research on safety and durability of major civil and hydraulic engineering projects. This support is greatly acknowledged.

REFERENCES

Bear J. and Verruijt A. 1987. Modeling Groundwater Flow and Pollution, D.Reidel Publishing Company (Reprinted 1992).

Terzaghi K. 1943. Theoretical Soil Mechanics, Wiley, New York.

Fredlund D.G.and Rahardjo H. 1993. Soil Mechanics forUnsaturated Soils, John Wiley & Sons, Inc., New York.

Li Xikui, Zienkiewicz O.C. and Xie Yimin 1990. A numerical model for immiscible two-phase fluid flow flow in a porous medium and its time domain solution, Int. J. Numer. Methods Eng., **30**, 1195-1212.

Li Xikui and Zienkiewicz O.C.1992. Multiphase flow in deforming porous media and finite element solutions, Computers & Structures, **45**, 211-227.

Alonso E.E., Gens A. and Josa A. 1990. A Constitutive model for partially saturated soils, Geotechnique, **40**, 405-430.

Duxbury P.G. and Li Xikui 1996. Development of elasto- plastic material models in a natural coordinate system, Comp. Meth. Appl. Mech. Eng., **135**, 283-306.

Zienkiewicz O.C. and Shiomi T. 1984. Dynamic behaviour of saturated porous media: the generalized Biot formulation and its numerical solution, Int.J. Numer.Anal. Methods Geomech., **8**, 71-96.

Leiws R.W. and Schrefler B.A. 1987.,The Finite Element Method in the Deformation and Consolidation of Porous Media, John Wiley, New York.

Lloret A. and Ledesma A. 1993. Finite element analysis of deformations of unsaturated soils, Civil Eng. European Courses, Barcelona.

Zienkiewicz O.C. and Taylor R.L. 1991.The Finite Element Methods, 4th Edn.,Vol.2, MrGraw-Hill.

Computer Methods and Advances in Geomechanics, Yuan (ed.) © 1997 Balkema, Rotterdam, ISBN 90 5410 904 1

Progress in coupled analysis of a thermo-hydro-mechanical experiment in fractured rock

Chin-Fu Tsang & Jonny Rutqvist
Earth Sciences Division, Ernest Orlando Lawrence Berkeley National Laboratory, Calif., USA

ABSTRACT: A field experiment to study coupled thermo-hydro-mechanical effects in fractured rock is currently being performed at Kamaishi Mine in Japan (Fujita *et al.*, 1996a, b, c). The experiment is being conducted in an alcove of 5x8 meters drilled off an existing tunnel about 250 meters below ground surface. A test pit of 1.7 meters in diameter and 5 meters depth was drilled, in which a heater was placed and surrounded by bentonite. The rock is highly fractured. However, there are three major sub-vertical fractures intersecting the test pit. A large number of boreholes were drilled to monitor rock hydraulic and mechanical responses prior, during and post heater turn-on over the next two years. We use the code ROCMAS (Noorsihad *et al.*, 1992) to simulate the Kamaishi experiment and analyze rock behavior at the site. In the first stage of the work reported in this paper, the hydromechanical effects are studied and compared with available data. The results show the importance of the three fractures and certain parameter data sets. This provides the insight for a fully 3D coupled modeling analysis of the experiment currently underway.

1 INTRODUCTION

The study of coupled thermo-hydro-mechanical (T-H-M) processes is an important part of the near field performance assessment of high level nuclear waste repositories in rocks. These processes include effects of heat flow from the nuclear waste canister and mechanical effects of excavation. Both the elevated temperature and excavation affect the hydraulic permeability of the rock mass by closure and or opening of existing fractures. An *in situ* T-H-M experiment is being conducted at Kamaishi Mine in Japan from 1995 to 1998. The experiment is conducted at a depth of 250 meters and simulates the conditions at a potential future nuclear waste repository in fractured hard rocks (Figure 1). The experiment is a major Test Case in the international cooperative project DECOVALEX (Stephansson *et al.*, 1996) for the development of coupled T-H-M models and their validation against field experiments. One objective of the test is to observe near-field coupled T-H-M phenomena *in situ* and to build up confidence in coupled mathematical models of these processes. The experiment is divided in several phases, starting with a hydromechanical evaluation of the effects of excavation and ending with a fully coupled T-H-M analysis of the heater test.

In this paper the main interest is to evaluate the influence of fractures in the hydromechanical response in the near field area as a first step towards fully coupled T-H-M analysis. We are modeling the hydromechanical responses due to the excavation of the test pit. These calculations are then compared to the actual field measurements that are now available. The experience from this first hydromechanical phase will provide insight and can be regarded as a preparation for the later fully coupled T-H-M analysis of the heater test.

2 KAMAISHI MINE TEST CASE

2.1 Geometry

The Kamaishi mine heater test (Fujita *et al*, 1996a) is being conducted in an 5 by 7 meters alcove excavated off an existing drift (Figure 1). A test pit of 1.7 meters in diameter and 5 meters in depth is drilled in the floor of the alcove. The test pit simulates a deposition hole of a nuclear waste repository. The heat output from the waste package in the test pit will be simulated by installing a heater which is embedded in a buffer clay material. All measurements take place within a radius of about two meters from the center of the test pit where most of the T-H-M effects are expected to occur. For the first hydromechanical phase the measurements are focused on mechanical displacements of the test pit and along main fractures, and on the fluid pressure distribution and water inflow to the test pit.

2.2 Far field properties and boundary conditions

The far field properties and geometry will affect near field behavior of fluid pressure and stress. In the rock mass at the level of the test drift many fractures are striking NE with a steep dip. The minimum principal stress is sub-vertical and the maximum principal stress is oriented almost perpendicular to the axis of the test drift. The maximum hydraulic conductivity is oriented parallel to the dominating fracture orientation and perpendicular to the maximum principal stress orientation. This indicates that the hydraulic conductivity is correlated to the fracture orientation rather than to the *in situ* stress.

The fluid pressure distribution as a function of depth from the ground surface at Kamaishi does not correspond to the hydrostatic pressure gradient with depth due to continuous drainage in the existing drift system. At the level of the T-H-M test the fluid pressure in the nearby 100 meters area varies between 0.1 and 0.4 MPa with an average pressure of about 0.3 MPa.

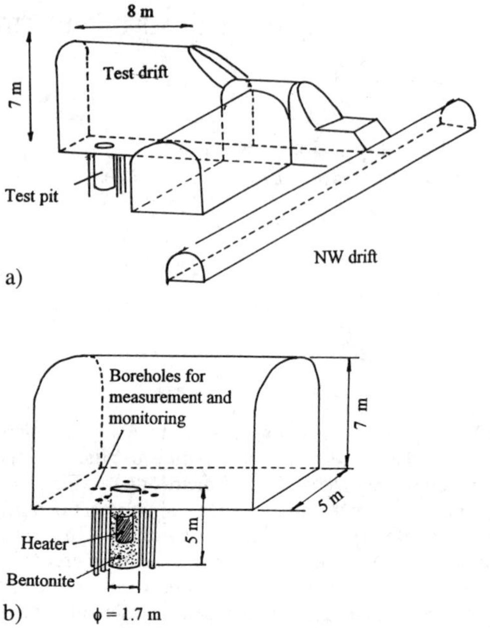

Figure 1. Layout of the T-H-M experiment at Kamaishi Mine (reproduced from Fujita *et al*, 1995): a) Test drift and nearby drift system. b) Test drift and test pit.

2.3 Near field properties

Figure 2 presents the fractures mapped on the floor of the drift. These fractures are striking preferentially NE with a steep dip. In Figure 2a, there are too many fractures to include discretely in a finite element model. On the other hand it may be too few fractures to be treated as an equivalent continuum near the test pit. However, there are three large shear fractures adjacent to the test pit which are striking approximately EW (Fracture 1, 2 and 3 in Figure 2b). These fractures have been sheared and reported to be the most open fractures on the floor and contain soft mineral filling of up to 20 mm in thickness. They may therefore be considerably softer and more conductive than other fractures in the near field. Similar fractures with the same strike and dip can be found on the roof which are also dominating the inflow into the drift.

The mechanical and hydraulic properties of the intact rock material have been determined in laboratory experiments. A number of tests have also been conducted on fractures in the laboratory as well as in the field. Laboratory test on clean joints indicated a normal stiffness of 200-800 GPa/m (at normal stress of 0-10 MPa) and a shear stiffness of 1-10 GPa/m. However, since Fracture 1, 2 and 3 are sheared fractures partly filled with mineral filling, they maybe softer than these values.

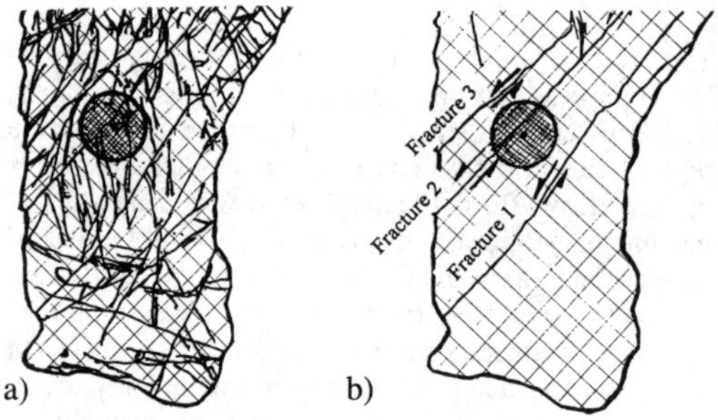

Figure 2. Sketch of fractures and the test pit on the floor of the test drift: (a) Fracture traces longer than 0.5 meters. (b) Fractures with observed displacement (Fujita *et al*, 1996c).

The hydraulic conductivity and fluid pressure were measured in seven boreholes located within a radius of two meters from the center of the unexcavated test pit. The results of the hydraulic tests indicate that the hydraulic conductivity as a whole is small but dominated by a few open fractures. These fractures have hydraulic transmissivities ranging from 10^{-7} to 10^{-6} m^2/s which represent an equivalent hydraulic aperture of about 50 to 100 microns. It is very difficult to correlate hydraulic connections between different boreholes with intersecting fracture planes. This may imply that the flow takes place in channels between closed or mineral filled parts of the fractures.

During excavation of the test pit, the rock will strain to allow the stresses to be redistributed. Most of it occurs within a distance of three diameters from the test pit. In response to the changed stress, fractures will close or open with corresponding changes in the rock permeability. The fluid pressure at the site is only 0.3 MPa and it will not be the cause of any considerable changes in fracture apertures. Thus, the hydromechanical coupling can be considered as a one way coupling and is due to the disturbed stress field around the test pit excavation.

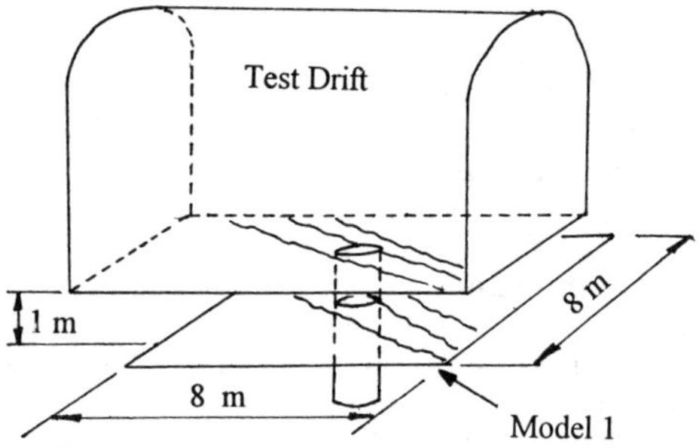

Figure 3. Sketch showing a horizontal section of the model below the test drift.

3 MODELLING APPROACH

In the modeling of the Kamaishi heater test, we intend to use the three-dimensional coupled thermo-hydromechanical finite element program ROCMAS (Noorishad *et al*, 1992). We may use one of the following alternative modeling approaches:

1) Anisotropic equivalent properties of a rock mass with ubiquitous fractures.
2) Porous isotropic matrix with discrete fractures.
3) Combination of alternatives 1 and 2.

The first approach is most appealing from a practical point of view since the discretization of the finite element mesh is much simpler without the discrete fractures. However, we may have to include some major fractures (third approach) to accurately model the near field behavior. The question is then which fractures should or can be included in a three-dimensional model? To answer this question we start up with some simple two-dimensional calculations. Since there is only a one-way coupling, we start with a mechanical calculation which then serves as an input to a hydraulic calculation.

We make a study with parametric variation over a range of reasonable values of the material properties. We will try to predict the general hydraulic and mechanical behavior around the test pit within a range of maximum and minimum values. We do also focus on the near field consequences of including Fracture 1, 2 and 3 in the model. This first modeling effort can therefore be viewed as a preparation for the more intensive three-dimensional fully coupled T-H-M modeling of the heater test.

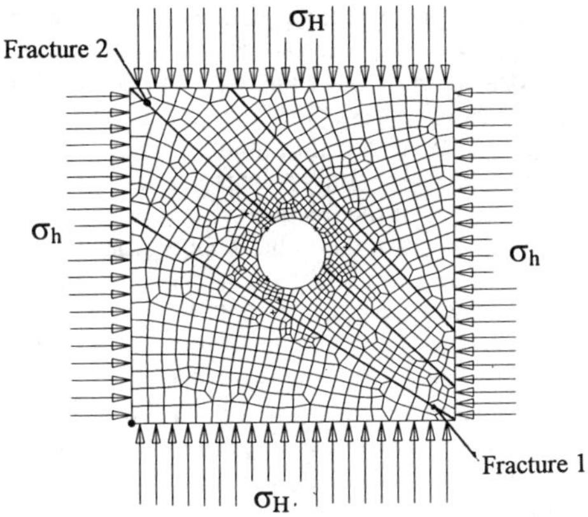

Figure 4. Finite element disctretization of an 8x8 meters horizontal section and boundary stress of Model 1.

Table 1. Four alternative property sets used for parametric variation

Property set	Initial normal stiffness at a normal stress of 12 MPa (GPa/m)	Shear stiffness (GPa/m)
0 joint	-	-
1 joint	500	10
3 joints	500	10
3 joints (low stiffness)	144	1

4 MODELING OF MECHANICAL EFFECTS IN 2D SECTIONS

A two-dimensional model of 8 by 8 meters was constructed for horizontal planes below the drift floor (Figure 3). The model contains Fracture 1, 2 and 3, but one or several of the fractures may be omitted by changing them to solid material elements. The model was discretized into 1016 elements of which 96 were joints elements (Figure 4). The boundary stresses on the model were estimated using results from a separate two-dimensional simulation of the drift before excavation of the test pit.

A parametric variation was conducted with four sets of mechanical fracture properties (Table 1). The first property set is a linear elastic material with no fractures. The second and the third set consist of one and three fractures, respectively, with the same shear stiffness and normal stiffness. These fracture properties are taken as average values of the laboratory measurements conducted on the clean fractures at Kamaishi. The fourth set represents fractures with considerable lower stiffness.

Figure 5 compares the modeling results of maximum principal stress for the first and fourth property sets corresponding to no fractures and three soft fractures cases respectively. The result using set 1 (Figure 5a), which can be considered as a verification exercise, compares well to classical solutions of a circular opening in a linear elastic medium of infinite extension. The fractures in the fourth case influence the maximum principal compressive stress magnitude in areas close to each fracture (Figure 5b). Furthermore, the stresses tend to be relocated from the right wall of the test pit giving increased stress on the left wall. The modeling results of the minimum principal stress (Figure 6) shows a similar behavior but with a stronger influence of the fractures.

Figure 7 compares the radial convergence of the test pit calculated for sets 1 and 4. These results show that the convergence pattern is strongly influenced by the fracture intersecting the test pit. The results also indicate considerable shear displacement in Fracture 2 when the shear stiffness is low. However, this shear displacement has little effect on the convergence of the measurement points (MC1 to MC4) on the test pit surface (Figure 7a). The maximum diametrical expansion, on the other hand, is very sensitive to the properties of Fracture 2.

The displacements of Fractures 1 and 2 were calculated between anchor points of a joint deformeter. Without any joint, displacements are in the orders of tens of microns. When the soft joints are included the shear displacement increases by the order of a millimeter. For Fracture 1, the direction of shear and fracture opening are sensitive to the properties of the joints.

5 MODELING OF INFLOW IN A 2D SECTION

The inflow into the test pit was modeled with a two-dimensional model in the plane of Fracture 2 which was extended to a boundary 25 meters from the test pit (Figure 8). The pressure at the outer boundary was set to 0.3 MPa and to zero at the boundary of the excavation. After a long time we obtain a steady flow, and a hydraulic potential gradient (Figures 8 and 9) which are essentially directed downwards and towards the drift and the test pit. At the floor of the test pit, water is flowing out from the test pit because the floor is flooded and the fluid pressure is kept constant.

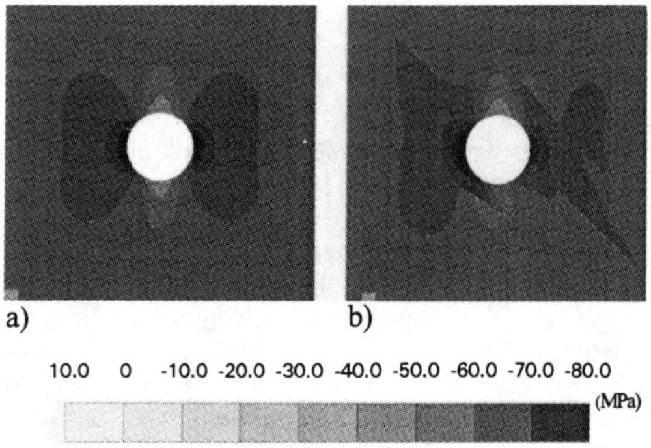

Figure 5. Magnitude of maximum principal stresses for a model a) without and b) with soft fractures.

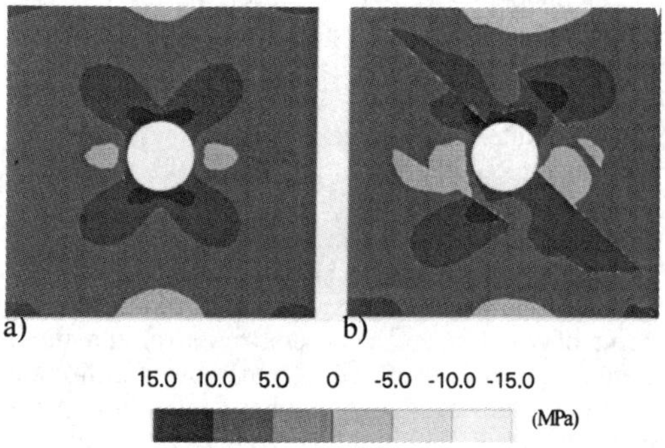

Figure 6. Magnitude of minimum principal stresses for model a) without and b) with soft fractures.

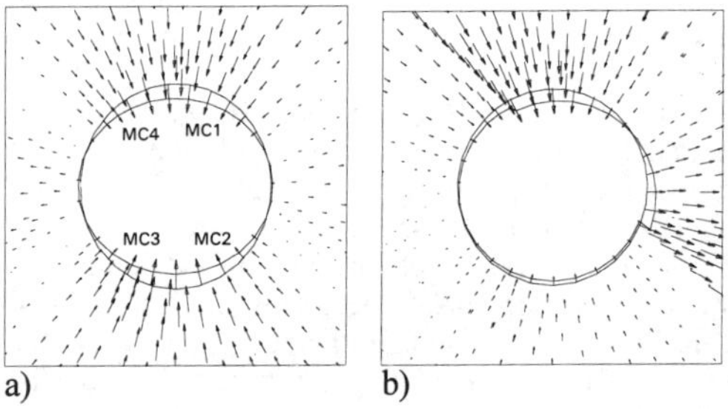

Figure 7. Radial convergence of the rock for a model a) without and b) with soft fractures.

The resulting inflow is presented in Figure 10 for a hydromechanical response. This simulation is compared to a pure hydraulic calculation with a constant hydraulic aperture equal to 50 microns. The aperture in the hydromechanical case was obtained from the mechanical calculations of the horizontal 2D-sections for property set 3.

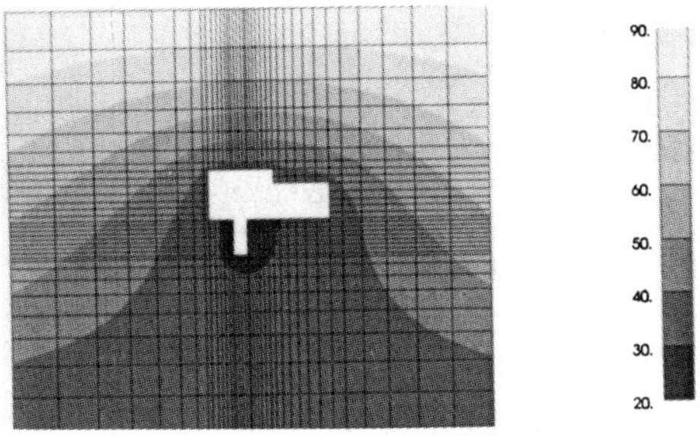

Figure 8. Finite element model in the plane of Fracture 2 for flow calculation and hydraulic potential in meters at steady flow.

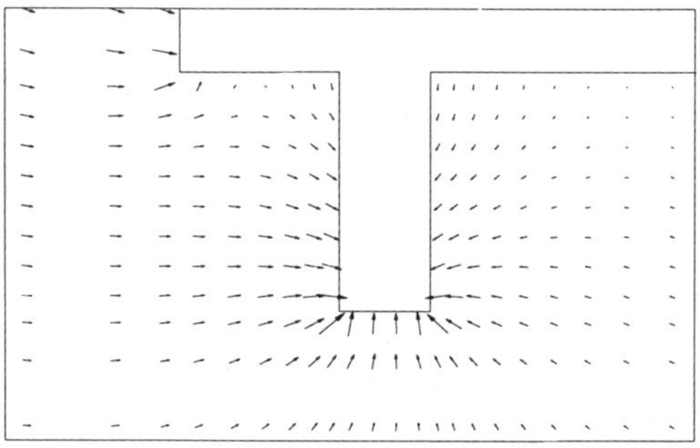

Figure 9. Detail of the flow pattern near the test pit.

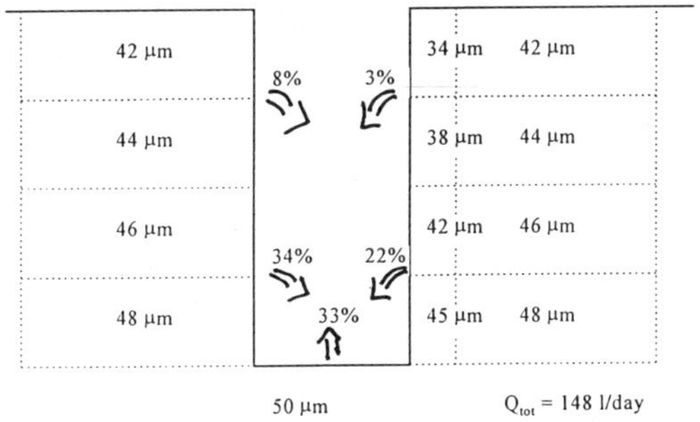

Figure 10. Inflow into the test pit for a hydromechanical calculation

The inflow is 163 l/day for the purely hydraulic case and 148 l/day for the hydromechanical case. Most of the fracture closure takes place at the upper part of the test pit where the inflow is relatively low. In the lower part and at the bottom, where most of the inflow takes place, the hydromechanical effects are weaker. A third calculation was performed with a constant hydraulic aperture of 100 microns and resulted in an inflow of 1324 l/day. This indicates that the initial aperture is the most important parameter and that the effects of the

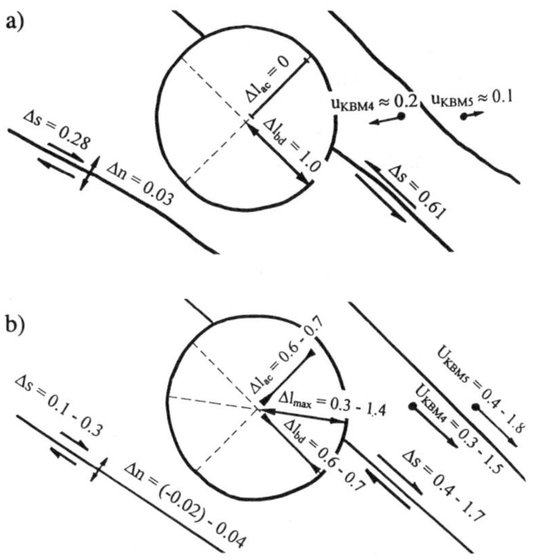

Figure 11. Comparison of displacements in millimeters: a) Measurements and b) range of modeling with properties sets 3 and 4.

hydromechanical response on the inflow is small compared to the variation due to the uncertainty of the initial fracture permeability.

6 COMPARISON OF PREDICTION TO MEASURED DATA

The convergence measurements indicated that the test pit is expanding in one direction. The modeling shows that it is possible to have expansion of the test pit in certain directions (see Figure 7) because of shear displacement in the intersecting fracture. The maximum diametrical expansion was calculated to be in the range from 0.3 to 1.4 mm depending on the properties of Fracture 2 (Figure 11b). This implies that the magnitude of the measured expansion of 1 mm is within the range of the calculated maximum expansion. However, the direction of the calculated expansion does not agree with the direction of the measured expansion.

The shear displacement and opening of Fracture 1 are within the range of the predictions both concerning magnitude and direction. The results indicate that property set 4 with a lower fracture stiffness gives a better agreement. In Fracture 2 the magnitude of shear displacement is within the prediction range but the shear direction is reversed. Also the displacement in KBM4 is reversed. One possible explanation is that the boundary stresses in Model 1 are incorrect. A full three-dimensional model may give slightly different stress field. However, it is not expected to change direction completely to produce a reversed shear displacement. Furthermore, a reversed boundary stress would also reverse the displacement in Fracture 1 which would disagree with measurements. The inward displacement at KBM4 and at the same time the outward displacement at KBM5 seems contradictory.

The measured inflow into the drift was 280 liters/day when the drift floor was flooded. The modeling gave an inflow of between 163 to 1324 liters/day for an aperture of 50 and 100 microns, respectively. Thus, the actual inflow was within the wide range of prediction. The measurements also showed that the inflow was dominated by flow through Fracture 2.

7 CONCLUDING REMARKS

In a numerical modeling we will never be able to predict the exact response in every point in the rock mass because the local geometry and properties of fractures are essentially unknown before excavation. However, we may be able to study the general hydromechanical response of the rock around the deposition hole and give approximate values of quantities such as total inflow and maximum displacement. A detailed comparison with data however, would require much more measurements, in order to obtain a more accurate prediction of the overall displacement field.

From our two-dimensional modeling we may conclude that it is essential to include Fracture 2 as a discrete feature because it dominates the inflow into the test pit. This fracture has a rather well defined geometry since it can be followed on both the tunnel floor and the wall of the test pit. However, its properties may vary widely over its plane and therefore cannot be properly modeled with a perfectly planar joint element of homogenous properties. In the field, the fracture is found to be partly filled with minerals, curved and sometimes branching into several parallel fractures. Furthermore, the mechanical monitoring during the experiment is not detailed enough to provide data for a very detailed model with thin joint elements. For instance the displacement measurement over Fracture 2 is conducted between anchors which 80 cm apart. According to the fracture log, this 80 cm of the borehole contains 14 fractures of which 7 are described as open. A simpler and more practical approach for modeling may be to simulate the fracture with solid elements of highly anisotropic properties. This may be accurate enough to define high and low flow zones in the test pit for the future thermo-hydro-mechanical modeling of the heater test.

ACKNOWLEDGEMENTS

Financial support was provided by the Swedish Nuclear Power Inspectorate. A grant for post-doctoral studies from the Wenner-Gren Center Foundation in Sweden to the second author is gratefully acknowledged. Work is also partially supported by Office of Energy Research, Office of Basic Energy Sciences, Engineering and Geosciences Division the Department of Energy, of under contract No. DE-AC03-76SF00098.

REFERENCES

Chijumatsu, M., T. Fujita, Y. Sugita, H. Ishikawa & Y. Moro 1996a. Coupled thermo-hydro-mechanical experiment at kamaishi mine. Instrumentation. Power reactor and nuclear fuel development corporation (PNC), Japan TN8410.

Chijumatsu, M., T. Fujita, Y. Sugita & H. Ishikawa 1996b. Coupled thermo-hydro-mechanical experiment at kamaishi mine. Initial data around the THM experiment area. PNC TN8410.

Chijumatsu, M., T. Fujita, Y. Sugita, H. Ishikawa & A. Kobayashi 1996c. Coupled thermo-hydro-mechanical experiment at kamaishi mine. Hydraulic tests. PNC TN8410.

Chijumatsu, M., T. Fujita, Y. Sugita & H. Ishikawa 1996d. Coupled thermo-hydro-mechanical experiment at kamaishi mine. Laboratory rock property tests. PNC TN8410.

Fujita, T., Y. Sugita, T. Sato, H. Ishikawa & T. Mano 1996a. Coupled thermo-hydro-mechanical experiment at kamaishi mine. Plan. PNC TN8020.

Fujita, T., Y. Sugita, H. Chijimatsu & H. Ishikawa 1996b. Coupled thermo-hydro-mechanical experiment at kamaishi mine. Mechanical properties of fracture. PNC TN8410.

Fujita, T., Y. Sugita, H. Chijimatsu & H. Ishikawa 1996c. Coupled thermo-hydro-mechanical experiment at kamaishi mine. Fracture characteristics. PNC TN8410.

Noorishad, J., C.-F. Tsang & P.A. Witherspoon 1992. Theoretical and field studies of coupled behavior of fractured rocks - 1. Development and verification of a numerical simulator. *Int. J. Rock. Mech. Min. Sci. & geomech. Abstr.* 29:401-409.

Stephansson, O., L. Jing & C.-F. Tsang. 1996. *Coupled thermo-hydro-mechanical processes of fractured media. Mathematical and experimental studies.* Amsterdam: Elsevier.

10 Disaster and geo-environmental engineering

Computer Methods and Advances in Geomechanics, Yuan (ed.) © 1997 Balkema, Rotterdam, ISBN 90 5410 904 1

A boundary integral equation formulation of contaminant transport in fractured and non-fractured porous media

J.R.Booker
Centre for Geotechnical Research, University of Sydney, N.S.W., Australia
C.J.Leo
Faculty of Engineering, University of Western Sydney, N.S.W., Australia

ABSTRACT: This paper presents a boundary integral equation method for contaminant transport of a contaminant species in porous media which consists of both fractured and non-fractured material. The method developed in this paper utilises a double porosity model to deal with contaminant transport in the presence of the discontinuities and solid blocks. When there are no discontinuities, the material reduces to a single continuum with one porosity and so the formulation is able to deal with deposits which consist of both fissured and non-fissured material. This approach is used to develop a boundary integral equation formulation which solves a wide range of problems of practical interest.

1. INTRODUCTION

It is not uncommon to find in landfills that the integrity of clay layers is adversely affected by the presence of desiccation and shrinkage cracks. In many instances, it has also been reported that in clayey tills where many landfill or waste sources are situated structural defects have been found to occur widely (e.g. Cherry, 1989; Rudolph et al, 1991; Jorgensen and Fredericia, 1992). The fact that material non-homogeneity and discontinuities are present quite regularly in the porous media and that they can have significant impact on the contaminant transport process suggest that it is both useful and necessary to develop appropriate models which take account of these effects.

In an earlier paper, Leo and Booker (1993) have presented a boundary integral equation method for analysing contaminant transport in fractured porous media in which the sets of fracture planes are assumed to be orthogonal to each other. Leo and Booker (1996a, b) have also presented a boundary integral equation method for the analysis of non-homogeneous non-fractured porous media. In this paper, it will be shown that both techniques can be combined to develop a more general boundary integral equation technique which may be used to analyse sites consisting of zones of a number of different materials which may be either fractured or non-fractured.

2. FRACTURED AND NON-FRACTURED POROUS MEDIA

The theories relating to a double porosity model of contaminant transport in fractured material have been discussed in detail elsewhere previously (e.g. Bibbly, 1981; Barker, 1985; Rowe and Booker, 1989, 1990; Dykhuizen, 1990; Sudicky, 1990; Leo and Booker, 1993; Zimmerman et al. 1993).

In the model used by Rowe and Booker (1989), Leo and Booker (1993) the fracture planes are assumed to be orthogonal to each other. Figure 1 shows a schematic diagram of the fracture system where the fracture spacings are denoted by $2H_1, 2H_2, 2H_3$ and the effective apertures are shown as $2h_1, 2h_2, 2h_3$. Set 1 of the fractures is parallel to the $\tilde{y}\tilde{z}$ plane, set 2 to the $\tilde{z}\tilde{x}$ plane and set 3 to the $\tilde{x}\tilde{y}$ plane and $\tilde{x}, \tilde{y}, \tilde{z}$ are a set of local axes.

For a homogeneous fractured porous media where the groundwater advection is temporally steady and spatially uniform, the equation of contaminant transport in the local co-ordinate $(\tilde{x}, \tilde{y}, \tilde{z})$ axes is given by,

$$\nabla \cdot (\mathbf{D_a} \nabla c) - \mathbf{V_a} \cdot \nabla c = (n_f + \beta_f K_f)(\frac{\partial c}{\partial t} + \gamma c) + q \tag{1}$$

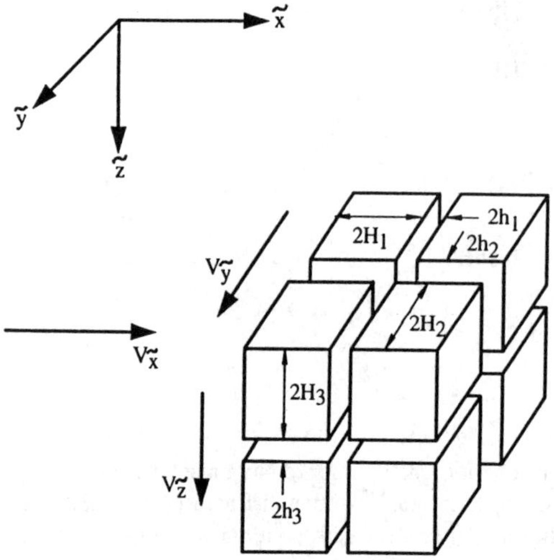

Figure 1: Fracture system in porous media

where $\mathbf{D}_a$ is the 'effective' tensor of hydrodynamic dispersion, the components of which are given by are given by (Leo and Booker, 1993):

$$D_{a\tilde{x}\tilde{x}} = (D_0 + \alpha_L V_{\tilde{x}})(\frac{h_2}{H_2} + \frac{h_3}{H_3})$$

$$D_{a\tilde{y}\tilde{y}} = (D_0 + \alpha_L V_{\tilde{y}})(\frac{h_3}{H_3} + \frac{h_1}{H_1})$$

$$D_{a\tilde{z}\tilde{z}} = (D_0 + \alpha_L V_{\tilde{z}})(\frac{h_1}{H_1} + \frac{h_2}{H_2})$$

$D_{aij} = 0$ when $i \neq j$ and i, j spans the index set $(\tilde{x}, \tilde{y}, \tilde{z})$, D_0 is coefficient of molecular diffusion, α_L longitudinal dispersivity in the fractures, V_i is the component of the seepage velocity in the direction of ith co-ordinate axis, $\nabla = (\partial/\partial\tilde{x}, \partial/\partial\tilde{y}, \partial/\partial\tilde{z})^T$, $\mathbf{V}_a$ is the vector of the components of the Darcy velocity, $n_f = h_1/H_1 + h_2/H_2 + h_3/H_3$ is the fracture porosity, $\beta_f = 1/H_1 + 1/H_2 + 1/H_3$ represents the surface area per unit volume, K_f is the linear distribution coefficient defined as the mass of contaminant sorbed per unit area of surface divided by the concentration of the contaminant, γ is the sum of the first order radioactive and biodegradation constants.

The quantity q is the rate at which the contaminant migrates into the matrix per unit volume of the fracture-matrix system, which has to be determined experimentally or derived theoretically.

Taking the Laplace transform:

$$\bar{c} = \int_0^\infty e^{-st} c(x, y, z, t) dt \tag{2}$$

of equation (1), it is found that,

$$\nabla \cdot (\mathbf{D}_a \nabla \bar{c}) - \mathbf{V}_a \cdot \nabla \bar{c} = (n_f + \beta_f K_f)(s+\gamma)(\bar{c} - \frac{c_0}{s+\gamma}) + \bar{q} \tag{3}$$

and as shown by Rowe and Booker (1989, 1990), the Laplace transform of q is found to be,

$$\bar{q} = (s+\gamma)\bar{\eta}\left(\bar{c} - {c_0}/{s+\gamma}\right) \tag{4}$$

For the fracture system shown in Fig 1, the theoretical value of $\bar{\eta}$ is given by:

$$\bar{\eta} = n_m R_m \left[1 - 8 \sum_{j,k,l} \frac{s+\gamma_m}{\left[s+\gamma_m + {D_m}/{R_m}\left(\alpha_j^2 + \beta_k^2 + \omega_l^2\right)\right]} \cdot \frac{1}{(\alpha_j H_1)^2} \cdot \frac{1}{(\beta_k H_2)^2} \cdot \frac{1}{(\omega_l H_3)^2} \right] \tag{5}$$

where $\alpha_j = (j - 1/2)\pi/H_1$, $\beta_k = (k - 1/2)\pi/H_2$, $\omega_l = (l - 1/2)\pi/H_3$, D_m is the coefficient of diffusion in the matrix blocks, $R_m = 1 + \rho_m K_m / n_m$ is the retardation coefficient of the matrix, n_m is the porosity of the solid matrix, K_m is the distribution coefficient of the solid matrix.

It is perhaps worth pointing out at this stage that the behaviour of a non-fractured material can be recovered from that of a fractured material merely by setting q to zero and then replacing the parameters for the fractured material by the corresponding parameters for the non-fractured material, this leads to the equation (Leo and Booker, 1996a, b):

$$\nabla \cdot (\mathbf{D}_a \nabla \bar{c}) - \mathbf{V}_a \cdot \nabla \bar{c} = (n + \rho K_d)(s+\gamma)(\bar{c} - \frac{c_0}{s+\gamma}) \tag{6}$$

where $\mathbf{D_a}$ is the 'effective' hydrodynamic dispersion tensor of the contaminant in the non-fractured material. The coefficients of the 'effective' tensor for an isotropic material are defined as: $D_{ak\ell} = n\left[\left(D_0 + \alpha_T V\right)\delta_{kl} + \left(\alpha_L - \alpha_T\right)\frac{V_k V_\ell}{V}\right]$, where the terms in square brackets are the values of hydrodynamic dispersion tensor (Bear, 1979). V is the magnitude of the seepage velocity, α_T is the coefficient of tranversal dispersivity, n is the porosity of the non-fractured material, V_i is the component of the Darcy velocity in the *ith* co-ordinate axis of the non-fractured material, ρ is the dry density of the non-fractured material, K_d is the linear distribution coefficient of the of the sorbed contaminant onto the solid grains of the non-fractured material

It can be observed that that the general form of equation (3) and (6) remains identical thus enabling a boundary integral equation method to be developed which can accommodate both fractured and non-fractured materials.

3. GOVERNING EQUATIONS FOR ZONED POROUS MEDIA

For prismatic landfills which are commonly found in practice, it will suffice to consider only plane conditions. The attention in this paper will therefore be restricted to developing a method for plane analysis only. The porous media will usually consists of zones of several different materials, some of which may be fractured while others are not. Each zone will be assumed to consist of fairly homogeneous material. In each zone j, the Laplace transform equation of contaminant transport for a single species in local $(\tilde{x},\tilde{z})$ space is given by,

$$\nabla(\mathbf{D}_{aj}\cdot\nabla\bar{c}) - \mathbf{V}_{aj}\cdot\nabla\bar{c} = \Theta_j(\bar{c} - \frac{c_{0j}}{\Lambda_j}) \quad (7)$$

where for definiteness, a subscript j has been added to each of the quantities defining the properties in zone j. For a non-fractured zone it follows from equation (6) that,

$$\Theta_j = \left(n_j + \rho_j K_{dj}\right)\left(s + \gamma_j\right)$$
$$\Lambda_j = s + \gamma_j$$

and for a fractured zone it follows from equation (3) and (4) that,

$$\Theta_j = \left(n_{fj}R_{fj} + \overline{\eta}_j\right)\left(s + \gamma_j\right)$$
$$R_{fj} = 1 + \frac{\beta_{fj}K_{fj}}{n_{fj}}$$
$$\Lambda_j = s + \gamma_j$$

The local $(\tilde{x},\tilde{z})$ co-ordinates are related to the global (x,z) co-ordinates as follows,

$$\begin{aligned} \tilde{x} &= x\cos\theta_j + z\sin\theta_j \\ \tilde{y} &= -x\sin\theta_j + z\cos\theta_j \end{aligned} \quad (8)$$

where θ_j is the angle between the local and global co-ordinate system. Thus equation (8) can be used to perform co-ordinate transform between the global and local co-ordinate systems.

If the local co-ordinate system is chosen so that the local axes are parallel to the principal directions of $\mathbf{D_{aj}}$, then equation (7) reduces to,

$$D_{a\tilde{x}\tilde{x}j}\frac{\partial^2\bar{c}}{\partial\tilde{x}^2} + D_{a\tilde{z}\tilde{z}j}\frac{\partial^2\bar{c}}{\partial\tilde{z}^2} - V_{a\tilde{x}j}\frac{\partial\bar{c}}{\partial\tilde{x}} - V_{a\tilde{z}j}\frac{\partial\bar{c}}{\partial\tilde{z}} = \Theta_j(\bar{c} - \frac{c_{0j}}{\Lambda_j}) \quad (9)$$

and the normal component of the mass flux on the boundary is given by,

$$\bar{f}_{\tilde{n}} = V_{a\tilde{n}j}\bar{c} - \left(\mathbf{D}_{aj}\nabla\bar{c}\right)\cdot\mathbf{1}_j$$

where $V_{a\tilde{n}j}$ is the component of Darcy velocity normal to the boundary of zone j, $\mathbf{1}_j$ is the unit normal vector to the boundary of zone j. Now suppose that, $\bar{c}_p = \sigma_j$ is a particular solution of equation (9). The simplest and commonest case will be when the initial concentration in the zone c_{0j} is spatially constant, then it is easily found that $\bar{c}_p = \frac{c_{0j}}{\Lambda_j}$. It follows that,

$$\Delta\bar{c} = \bar{c} - \sigma_j \quad (11)$$

satisfies the equation,

$$D_{a\tilde{x}\tilde{x}j}\frac{\partial^2\Delta\bar{c}}{\partial\tilde{x}^2}+D_{a\tilde{z}\tilde{z}j}\frac{\partial^2\Delta\bar{c}}{\partial\tilde{z}^2}-V_{a\tilde{x}j}\frac{\partial\Delta\bar{c}}{\partial\tilde{x}}-V_{a\tilde{z}j}\frac{\partial\Delta\bar{c}}{\partial\tilde{z}}=\Theta_j\Delta\bar{c} \quad (12)$$

and the normal component of the mass flux, $\bar{f}_{\tilde{n}\sigma j}$ on the boundary, due to σ_j is,

$$\bar{f}_{\tilde{n}\sigma j}=V_{a\tilde{n}j}\sigma_j-\left(\mathbf{D}_{aj}\nabla\sigma_j\right)\cdot\mathbf{l}_j \quad (13)$$

so that,

$$\Delta\vec{f}_{\tilde{n}}=\bar{f}_{\tilde{n}}-\bar{f}_{\tilde{n}\sigma j} \quad (14)$$

It is possible to reduce equation (12) to the mathematically convenient modified Helmholtz equation by using a second series of co-ordinate transform as follows:

$$\tilde{x}=u_jX \quad (15a)$$
$$\tilde{z}=w_jZ \quad (15b)$$
$$V_{aXj}=w_jV_{a\tilde{x}j} \quad (15c)$$
$$V_{aZj}=u_jV_{a\tilde{z}j} \quad (15d)$$
$$\Delta\bar{f}_N=\left(w_jl_{\tilde{x}j}L_{Xj}+u_jl_{\tilde{z}j}L_{Zj}\right)\Delta\bar{f}_{\tilde{n}} \quad (15e)$$

where $l_{\tilde{x}j},l_{\tilde{z}j}$ and L_{Xj},L_{Zj} are the direction cosines of the normals on the boudary of zone j in the $(\tilde{x},\tilde{z})$ and (X,Z) spaces respectively, and

$$u_j=\left(\frac{D_{a\tilde{x}\tilde{x}j}}{D_{aj}}\right)^{\frac{1}{2}}$$

$$w_j=\left(\frac{D_{a\tilde{z}\tilde{z}j}}{D_{aj}}\right)^{\frac{1}{2}}$$

$$D_{aj}=\left(D_{a\tilde{x}\tilde{x}j}D_{a\tilde{z}\tilde{z}j}\right)^{\frac{1}{2}}$$

Then, introducing the change of variable:

$$\Delta\bar{c}=\Delta\bar{c}^*e^{(\varpi_jX+\lambda_jZ)}$$
$$\Delta\bar{f}=\Delta\bar{f}^*e^{(\varpi_jX+\lambda_jZ)}$$

where,

$$\varpi_j=\frac{V_{aXj}}{2D_{aj}}$$
$$\lambda_j=\frac{V_{aZj}}{2D_{aj}}$$

equation (12) is reduced to the familiar modified Helmholtz equation,

$$D_{aj}\nabla^2\Delta\bar{c}^*=S_j\Delta\bar{c}^* \quad (16)$$

where,

$$S_j=\Theta_j+D_{aj}\left(\varpi_j^2+\lambda_j^2\right)$$

Thus proceeding formally in the manner outlined in Booker and Leo (1996a,b), equation (16) can be used to formulate the boundary integral equation,

$$\varepsilon(\mathbf{r}_0)\Delta\bar{c}^*(\mathbf{r}_0)=\int_{\Gamma^j}\left(\Delta\bar{c}^*\Delta\bar{f}_N^{\#}-\Delta\bar{c}^{\#}\Delta\bar{f}_N^*\right)d\Gamma \quad (17)$$

where,

$$\varepsilon(\mathbf{r}_0)=\begin{cases}1 & \text{if } \mathbf{r}_0 \text{ is within the domain of zone } j\\ 0 & \text{if } \mathbf{r}_0 \text{ is outside the domain of zone } j\\ \frac{1}{2} & \text{if } \mathbf{r}_0 \text{ lies on a smooth boundary of zone } j \text{ or value of the subtended angle} \div 2\pi \text{ if the boundary is not smooth}\end{cases}$$

$\mathbf{r}_0$ is the position vector of the point of disturbance, Γ^j is the transformed boundary of zone j in the (X,Z) space, $\Delta\bar{c}^{\#}=\frac{1}{2\pi D_{aj}}K_0\left(\sqrt{S_j/D_{aj}}R^{\#}\right)$, the fundamental solution of equation (16) , K_0 is the modified Bessel function of the second kind of order zero. $\Delta\bar{f}_N^{\#}=\frac{V_{aNj}}{2}\Delta\bar{c}^{\#}-D_{aj}\frac{\partial\Delta\bar{c}^{\#}}{\partial N}$, V_{aNj} is the normal component of the Darcy velocity on the boundary of zone j in (X,Z) space, $R^{\#}=\left[(X-X_0)^2+(Z-Z_0)^2\right]^{\frac{1}{2}}$, X_0,Z_0 are the co-ordinates of the point of disturbance.

An approximation of the boundary integral equation (17) can be formulated using conventional techniques widely described in literature (e.g. Brebbia and Walker, 1980; Barnejee (1981), Leo and Booker, 1996a,b) and takes the form of,

$$\mathbf{H}_j^*\Delta\mathbf{c}^*=\mathbf{G}_j^*\Delta\bar{\mathbf{f}}_N^* \quad (18)$$

where $\mathbf{H}_j^*$, $\mathbf{G}_j^*$ are the fully populated influence matrices, $\Delta\bar{\mathbf{c}}^*$, $\Delta\bar{\mathbf{f}}_N^*$ are the vectors of nodal values on the boundary elements of zone j. If the co-ordinate transformations and the change in variables were reversed, it is found that equation (18) can be expressed in natural co-ordinates and in terms of the nodal values of the variables $\bar{c}$, $\bar{f}_n$ as,

$$\mathbf{H}_j\bar{\mathbf{c}} = \mathbf{G}_j\bar{\mathbf{f}}_n + \mathbf{B}_j \qquad (19)$$

The vector $\mathbf{B}_j$ arose due to the presence of the non-zero initial concentration in the zone. Using equation (19) the contributions from all the zones are then assembled to form a global system,

$$\mathbf{H}\bar{\mathbf{c}} = \mathbf{G}\bar{\mathbf{f}}_n + \mathbf{B} \qquad (20)$$

where $\mathbf{H}$, $\mathbf{G}$ and $\mathbf{B}$ are the global influence matrices and vector. It is noted that the influence matrices in the global system are not fully populated. Equation (20) is used to solve for the unknown concentration and normal mass flux on the external boundaries and on the interfaces between the zones. Once these values are known the value of the concentration can be solved at any internal point in the domain.

It should however be observed that the solution found by applying these equations will be in the Laplace transform domain and thus has to be inverted to obtain the solution in the time domain. This is done by performing a numerical inversion of the Laplace transform using the algorithm by Talbot (1979).

4. CONCLUSION

A boundary integral equation formulation is presented for solving contaminant transport problems in which the porous media may or may not be fractured. To deal with the fractured material, a double porosity model has been used. Laplace transform and a series of co-ordinate transforms are then used to formulate a boundary integral equation from which the system of approximating algebraic equations can be obtained.

5. REFERENCES

1. Barker, J.A. 'Block-geometry functions characterizing transport in densely fissured media', *Journal of Hydrology*, Vol. 77, pp263-279, 1985.
2. Barnerjee, P.K. and Butterfield, R., 'Boundary element methods in engineering science', *McGraw-Hill, London,* 1981.
3. Bear, J., 'Hydraulics of Groundwater', *McGraw-Hill, New York,* 1979.
4. Bibbly, R. 'Mass transport of solutes in dual-porosity media', *Water Resources Research,* Vol.17(4), pp.1075-1081, 1981.
5. Brebbia, C.A. and Walker, S., 'Boundary element techniques in engineering', *Newnes-Butterworths, London,* 1980.
6. Cherry, J.A., 'Hydrogeologic contaminant behaviour in fractured and unfractured clayey deposits in Canada' *Contaminant Transport in Groundwater, edited by H.E. Kobus and Kinzelbach, W., A.Balkema, Rotterdam, Netherdams*, pp11-20, 1989.
7. Dykhuizen, R.C., 'A new coupling term for dual-porosity models', *Water Resources Research*, Vol.26, pp.351-356, 1990.
8. Jorgensen, P.R. and Fredericia, J., 'Migration of nutrients, pesticides and heavy metals in fractured clayey tills', *Geotechnique,* V42, pp67-77, 1992.
9. Leo, C.J. and Booker, J.R., 'Boundary element analysis of contaminant transport in fractured porous media', *International Journal for Numerical and Analytical Methods in Geomechanics*, Vol.17(7), pp.471-492, 1993.
10. Leo, C J. and Booker, J.R., 'A boundary element method for analysis of contaminant transport in porous media I: Homogeneous porous media', Submitted to the *International Journal for Numerical and Analytical Methods in Geomechanics*, 1996a.
11. Leo, C J. and Booker, J.R., 'A boundary element method for analysis of contaminant transport in porous media II: Non homogeneous porous media', Submitted to the *International Journal for Numerical and Analytical Methods in Geomechanics*, 1996b.
12. Rowe, R.K. and Booker, J.R., 'A semi-analytic model for contaminant migration in a regular two- or three-dimensional fractured network: conservative contaminants', *International Journal for Numerical and Analytical Methods in Geomechanics*, Vol.13, pp.531-550, 1989.
13. Rowe, R.K. and Booker, J.R., 'Contaminant migration in a regular two- or three-dimensional fractured network: reactive contaminants', *International Journal for Numerical and*

Analytical Methods in Geomechanics, Vol.14, pp.401-425, 1990.
14. Rudolph, D.L., Cherry, J. And Farvolden, R., 'Groundwater flow and solute transport in fractured lacustrine clay near Mexico City, *Water Resouces Res.,* V9(5), pp1332-1356, 1991.
15. Sudicky, E.A., 'The Laplace transform Galerkin technique for efficient time-continuous solution of solute transport in double-porosity media, *Geoderma*, 46, pp.209-232, 1990.
16. Talbot, A., 'The accurate numerical integration of Laplace transforms', *Journal Inst. Mathematical Applications*, Vol.23, pp.97-120, 1979.
17. Zimmerman, R.W., Chen, G., Hadgu, T. And Bodvarson, G.S., 'A numerical dual-porosity model with semianalytical treatment of fracture/matrix', *Water Resources Research*, Vol.29(7), pp.2127-2137, 1993.

Computer Methods and Advances in Geomechanics, Yuan (ed.) © *1997 Balkema, Rotterdam, ISBN 90 5410 904 1*

Numerical analysis of lateral movement of quay walls during the Hyogo-Ken Nanbu Earthquake, 1995

Jafril Tanjung & Abdul Hakam
Andalas University, Japan
Makoto Kawamura
Department of Regional Planning, Toyohashi University of Technology, Japan
Tatsuo Uwabe
Ministry of Transportation, Japan

ABSTRACT: In this study, two dimensional nonlinear FEM analysis for gravity type walls was carried out taking into account interaction between structure, soils and sea water. One of the damage quay walls in the Port of Kobe was selected as a model of the numerical analysis assuming the slip surfaces in the subsoil. The horizontal and vertical strong motion record in the deep layer of the Port of Kobe was used as the input motion. The estimated lateral movement of the damaged gravity quay wall was about one meter and it is relatively small compared with those observed in the Hyogo-ken Nanbu Earthquake. It is considered that horizontal strong motion caused a large lateral displacement.

1 INTRODUCTION

The Hyogo-Ken Nanbu earthquake of January 17, 1995 suffered major damage of harbour in Hansin region. Most of gravity walls in the Port of Kobe were damaged, moved laterally more than a few meters and settled. These large displacement of gravity walls are presumed to be caused by the strong ground motion and/or liquefaction of subsoil. In the Port of Kobe, large displacements of quay walls due to the Hyogo-Ken Nanbu earthquake were observed as shown in Figure 1. Unfortunately, the type of more than 90 percent of quay walls in the Port of Kobe was same gravity caisson type. Most of gravity type quay walls were severely damaged in the same manner so that they do not facilitate. On the other hand damage of the dolphin type of quay wall was relatively small.

Nonlinear FEM analysis for gravity type wall were carried out to find out the characteristics of lateral movement of the wall. In the analysis motion of pore water is coupled with the motion of the soil taking into account the soil-structure and water interaction. This FEM code, which is called BEAD, was developed by Uwabe.

In the analysis, ground motion records of the Hyogo-Ken Nanbu Earthquake at the Port of Kobe were used as input motion data to compare the lateral displacements of the quay walls with those of damaged quay walls.

In the series of calculation several factors such as magnitude, frequency and number of cycles of horizontal sinusoidal input motion were changed to see how much those factors affect the movement of quay walls which are similar type with the model mentioned above.

To find out a counter measure for a large lateral displacement, different shape and dimension of quay walls were analyzed.

2 CHARACTERISTICS OF LATERAL MOVEMENT OF THE QUAY WALLS FOR DIFFERENT INPUT MOTIONS

A model of quay wall shown in Figure 2 was selected as a typical example in the Port of Kobe. The height and width of the concrete caisson are 15 meter and 10 meter, respectively. A back fill zone behind the quay wall and gravel layer under the wall were considered. Joint elements were introduced between different type of materials such as between the concrete caisson and back fill soils and original soils and back fill soils. Water elements were considered for sea water. Interaction elements were employed at interface between sea water and the concrete caisson and between sea water and soils. Stress-strain relations of soils, used in the analysis, are non-linear elastic while stress-strain relation of concrete is assumed linear elastic.

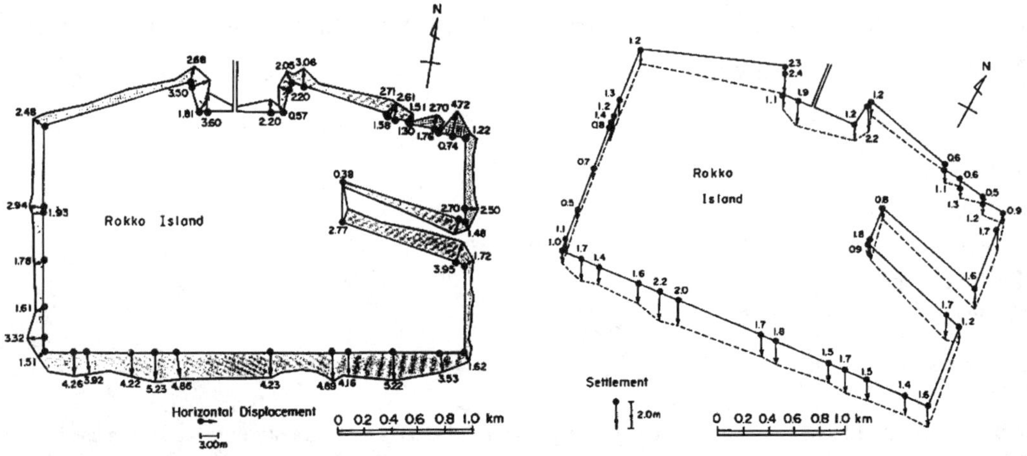

Figure 1. Displacement of quay wall at Rokko Island, horizontal and vertical displacements, respectively.

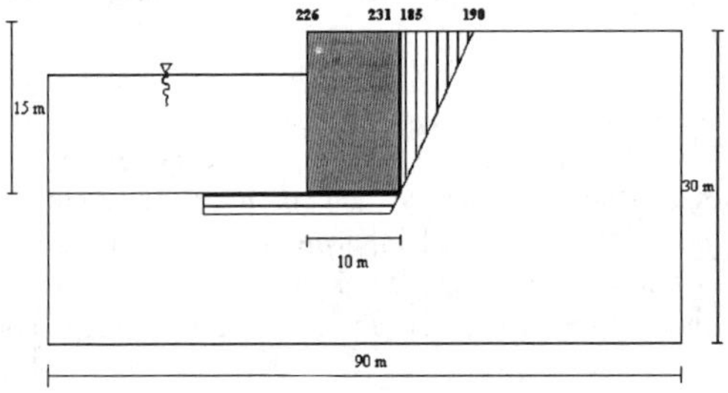

Figure 2. A model of quay wall.

Material properties that are used in analysis are listed in Table 1, Table 2, and Table 3, for concrete caisson, soils and joint elements, respectively.

Table 1. Material properties of concrete caisson elements.

No.	Density (t/m^3)	Poisson's Ratio	Shears Mod. (t/m^2)	Frictional Angle (^{0})
1.	2.100	0.167	1.285 E+6	0

Table 2. Material properties of soils elements.

No.	Density (t/m^3)	Poisson's Ratio	Shears Mod. (t/m^2)	Frictional Angle (^{0})
1.	2.000	0.330	2.000 E+4	30
2.	1.500	0.330	1.800 E+4	30
3.	1.500	0.330	1.000 E+4	30
4.	1.960	0.380	3.500 E+3	30
5.	2.000	0.330	1.750 E+4	30
6.	2.000	0.330	9.000 E+3	30
7.	1.960	0.380	1.000 E+4	30

Table 3. Material properties of joint elements.

No.	Tangential Stiffness (t/m^3)	Normal Stiffness (t/m^3)	Angle of Residual Curve (^{0})
1.	1.000 E+3	1.000 E+6	30
2.	1.000 E+3	1.000 E+6	10

The observed input motions at 28 meter below the surface of the Port of Kobe are shown in Figure 3.

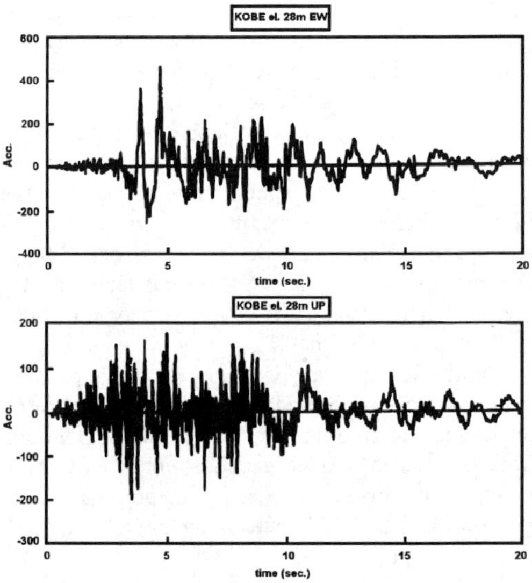

Figure 3. Observed record of acceleration at 28 meter below the surface of Port of Kobe.

The displacements of the wall which were calculated using these observed records as input motions, are shown in Figure 4.

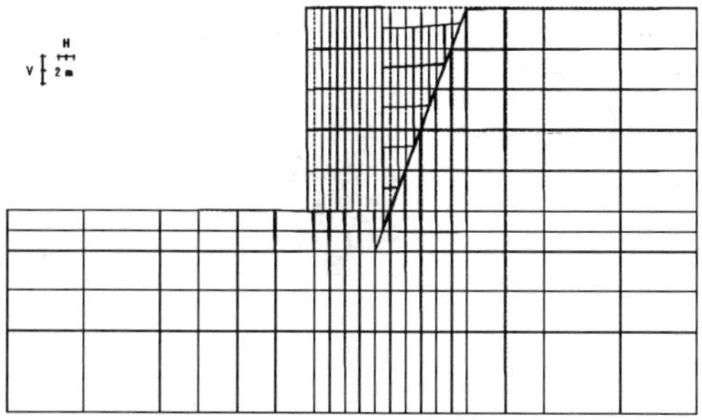

Figure 4. Displacement of quay wall in the case where the observed horizontal and vertical record of acceleration were used as input motion.

The displacement of top of the wall and its settlement are 1.07 meter and 0.07 meter, respectively. These results are relatively small compared with the observed values because the failure of the foundation soil layer is not considered in this analysis. The displacement of the wall in the case where only the vertical record is used as the input motion, are shown in Figure 5. The effect of the vertical input motion is small.

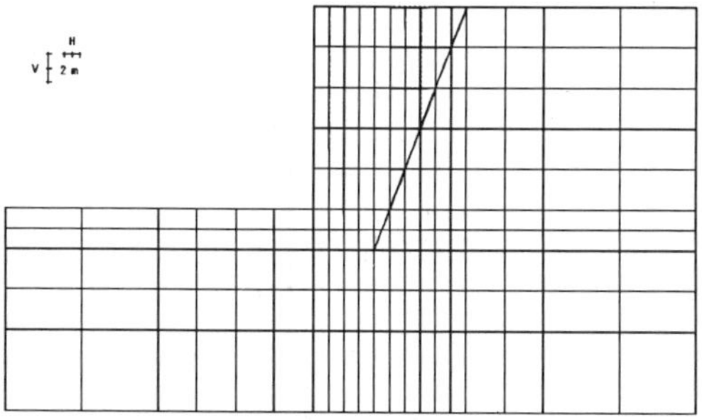

Figure 5. Displacement of quay wall in the case where only vertical observed was applied.

Displacements of the surface of back fill which is adjacent to the wall are shown in Figure 6 for different type of sinusoidal horizontal input motion. Negative value of a and y direction correspond with the displacement toward sea and settlement, respectively.

According to these figures, more than one meter horizontal displacement toward sea and settlement were obtained for the case that the frequency of the sinusoidal input motion and the number of cyclic wave are 1 Hz and 3, respectively. The horizontal acceleration which is larger than 200 gals and has the period longer than around 0.3 second, easily cause larger than one meter movement of the gravity type quay wall.

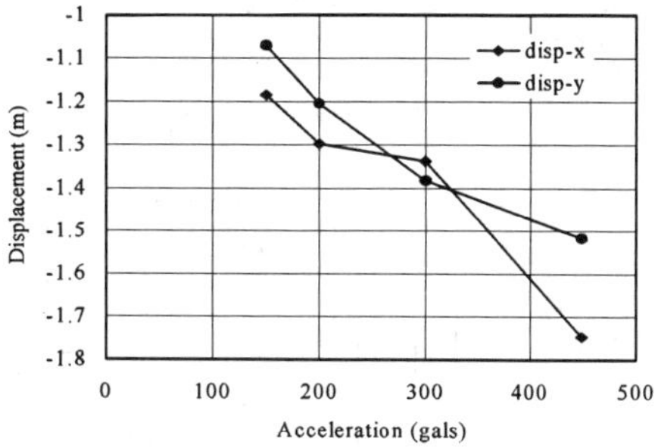

(a) Frequency = 1 Hz, Number of cycles = 3

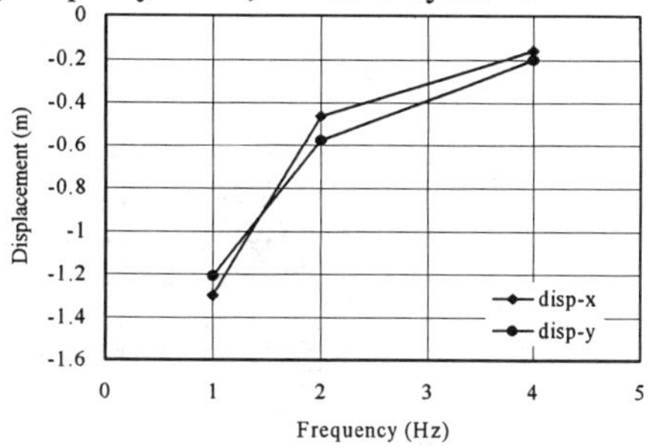

(b) Acceleration = 200 gals, Number of cycles = 3

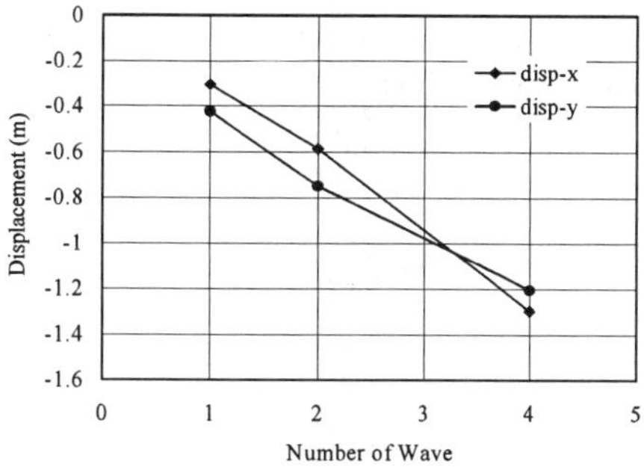

(c) Acceleration = 200 gals, Frequency = 1 Hz

Figure 6. Displacement of the surface back fill adjacent to the wall for acceleration, frequency and number of cycle of sinusoidal input motion data.

3 DISPLACEMENT OF THE WALL FOR THE DIFFERENT SHAPE

To investigate the relation between the width of the caisson wall and lateral movement of the quay walls, three type of model were analyzed using the observed earthquake motion shown in Figure 3. The results are shown in Figure 7. The lateral displacement of the quay walls in case 3 is about 0.4

times as large as those in case 2, while the width of the wall in case 3 is 1.5 times as large as the one in case 2. Increase of the width of the wall is effective to reduce the lateral displacement of the wall although it requires increase of the cost for the construction.

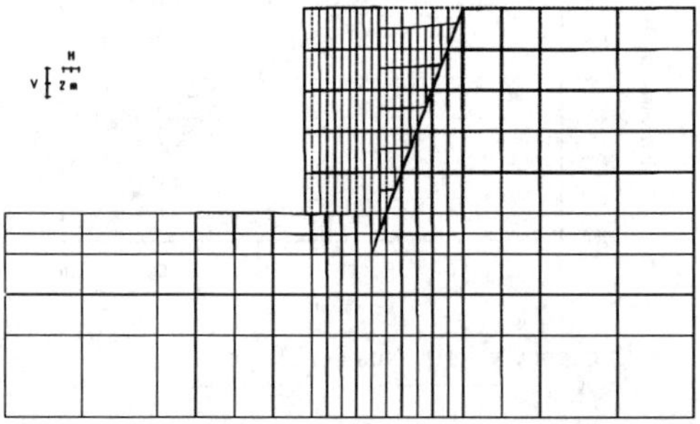

Width of the wall = 10 m (a) Case 1

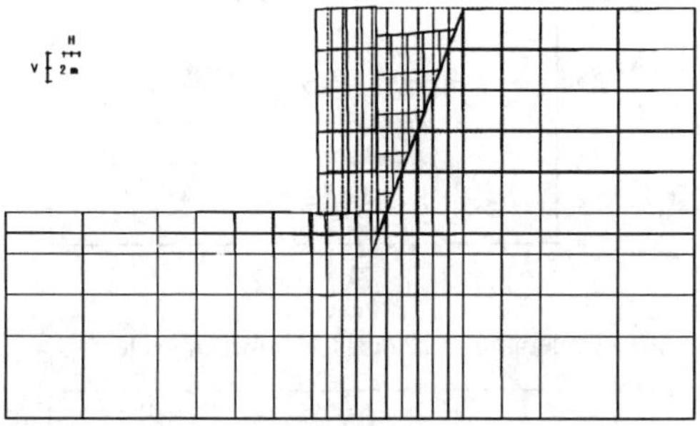

Width of the wall = 8 m (b) Case 2

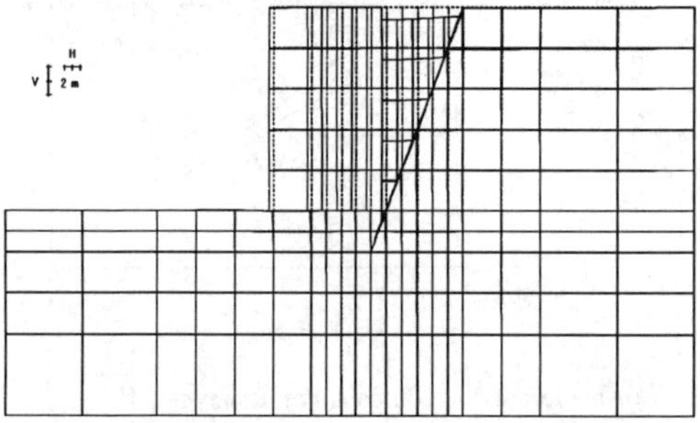

Width of the wall = 15 m (c) Case 3

Figure 7. Displacement of quay wall which have different dimensions

To find out the new type of the wall which is more effective to reduce the lateral movement and the cost, the displacements of the wall were calculated for the two types of the wall which are shown in Figure 8. These new types of the wall here a small projection at the base of the caisson to prevent slip at the base of the wall. Comparing the results shown in Figure 9. The small projection of the wall base at the front side, which corresponds to Type B, is more effective to reduce the horizontal displacement of the wall. The soils around the small projection in Type B is more stiff than those in Type A. It is presumed that the more stiff soils are effective to prevent the slip at the wall base.

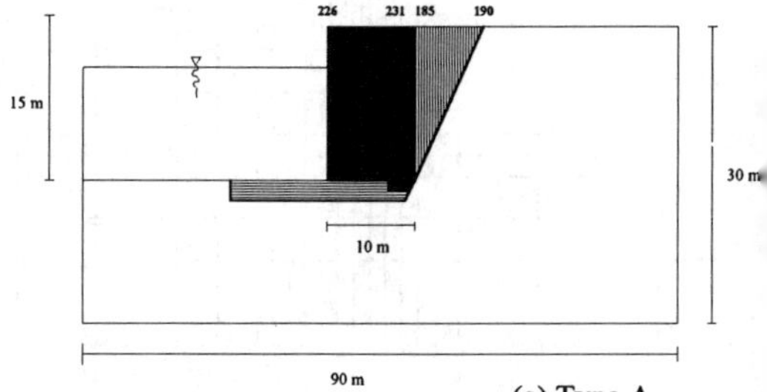

(a) Type A

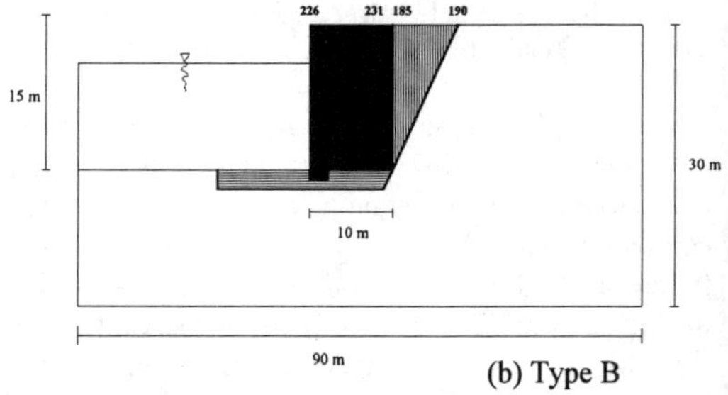

(b) Type B

Figure 8. Models which have a small projection of the walls

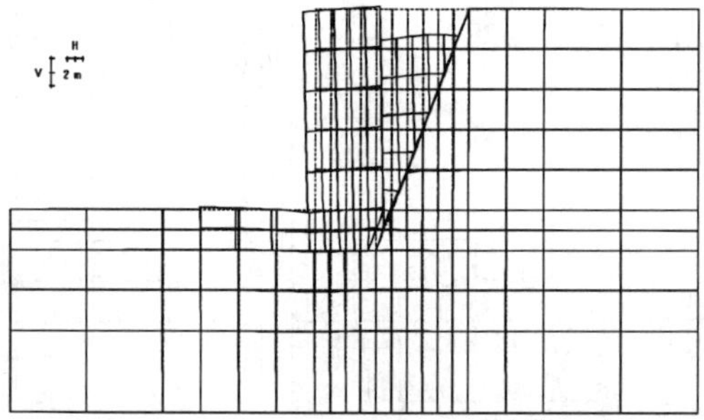

(a) Type A

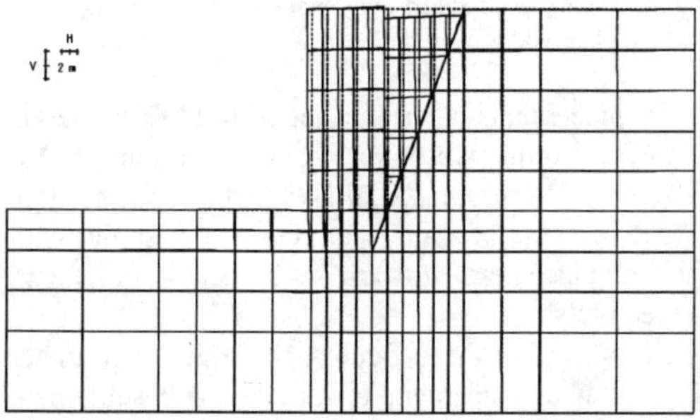

(b) Type B

Figure 9. Displacement of the walls which is shown in Figure 8.

4 CONCLUSION

As the results, the followings are made clear.

(1) The estimated lateral movement of the damaged gravity quay wall was about one meter and it is relatively small compared with those observed in the Hyogo-ken Nanbu earthquake because the failure of the foundation soil layer is not considered in the analysis. It is considered that horizontal strong motion caused a large lateral displacement.

(2) As the result of the parameter study it is found that the horizontal acceleration which is larger than 200 gals and has the period longer than around 0.3 second will easily cause a large lateral movement of the gravity type quay wall larger than one meter.

5 REFERENCES

Das, Braja M.1983. *Fundamental of soil dynamics.* New York: Elsevier.

Das, Braja M. et. al. 1995. Permanent settlement of a square shallow foundation sand due to cycle load. *Proceeding of the First International Conference on Earthquake Geotechnical Engineering*: 779-783. Rotterdam: Balkema.

Goodman, Richard E. et al. 1968. A model for the mechanics of jointed rock. *Journal of Soil Mechanics and Foundation Division, Proceeding of the American Society of Civil Engineers.* 94(SM1): 637-659. Michigan: ASCE.

Uwabe, T. et al. 1982. Coupled hydrodynamic response characteristics and water pressures of large composite breakwaters. *Fourteenth Joint Meeting US-Japan on Wind and Seismic Effects.* UJNR. Washington.

Uwabe, T. et al. 1985. Hydrodynamic response analysis based on strong-motion earthquake record of fill type breakwater. *Proceedings of the International Symposium on Ocean Space Utilization'85, Tokyo*: 361-368. Tokyo: Springer-Verlag.

Computer Methods and Advances in Geomechanics, Yuan (ed.) © 1997 Balkema, Rotterdam, ISBN 90 5410 904 1

3D DEM study of thermo-mechanical responses of a nuclear waste repository in fractured rocks – Far- and near-field problems

Lanru Jing, Håkan Hansson & Ove Stephansson
Royal Institute of Technology, Stockholm, Sweden

Baotang Shen
Division of Exploration and Mining, CSIRO, Kenmore, Qld, Australia

ABSTRACT: Three-dimensional Distinct Element Method (DEM) models were used to study the coupled thermo-mechanical processes in a fractured rock mass surrounding a repository for spent nuclear fuel. The effects of glaciation, *in-situ* stresses, heat released from the spent fuel, swelling pressure from the buffer materials and the disturbances caused by the excavations was studied by using one far-field and three near-field models. Impacts on the global and local stability of rocks, canister safety and local hydraulic permeability were evaluated with safety and performance assessment measures.

1 INTRODUCTION

Coupled thermo-mechanical impact on the fractured rock mass surrounding a nuclear waste repository is an important aspect of the performance and safety assessment of nuclear waste repository. Consideration must be given to the ultimate thermal loads, changes of strain and stress fields, repository stability, changes of fracture apertures and system response to possible future loading mechanisms such as glaciation and deglaciation over time. The presence of fractures of different scales in the rock mass also requires that the responses of rock fractures be closely studied since fractures form the pathways which conduct water, and therefore, are pathways of nuclides transport. The excavation of the storage tunnel system will change the stress state in the surrounding rocks which may also have an impact on the hydraulic properties of the near-field rocks.

Mathematical models and numerical methods are applied to study responses of nuclear waste repositories to the various loading mechanisms because that closed form solutions to the problems concerned do not exist in general, and that full scale experimental studies are impractical (since the spatial scale should be large and extremely long time span of the repository service life cannot be duplicated). The discrete element methods (DEM) with explicit representation of fractures and rock matrix (blocks) have advantages over the continuum approaches such as FEM and BEM. DEM offers more realistic representation of the fracture system geometry and direct formulation of interactions between contacting rock blocks for both small and large deformations. These advantages become more important when water flow through fracture networks are considered.

This contribution presents a DEM study of the coupled thermo-mechanical responses of a potential repository under loading of *in-situ* stresses, heating from the waste (spent nuclear fuel), a glaciation-deglaciation cycle, swelling pressure of the buffer material and excavation of a tunnel-deposition hole system, based on the KBS III concept of repository (SKB, 1992). A three-dimensional Discrete Element Method (DEM) code, 3DEC, was used for the work (ITASCA, 1994) which treats the fractured rocks as an assemblage of polyhedral blocks defined by completely connected fractures of polygonal shapes. The primary purpose of the studies is to investigate: i) temperature distribution and thermal gradients; ii) stress distribution and fracture deformation due to thermal loading and glaciation; and iii) stability conditions of both the global rock formations and the local tunnel-deposition hole system. To achieve these aims, four DEM computational models were used. One large scale far-field model without excavations of tunnels and deposition holes and three small scale near-field models based on different fracture network realizations at Äspö area with consideration of excavation of tunnels and deposition holes and the swelling pressure from the buffer material

The problem studied is a part of project SITE' 94 initiated by the Swedish Nuclear Power Inspectorate (SKI, 1996; Hansson et al., 1995; Shen and Stephansson, 1995). In the following sections, the geological background, computational models, material parameters, loading sequences, results and interpretations and conclusions are presented.

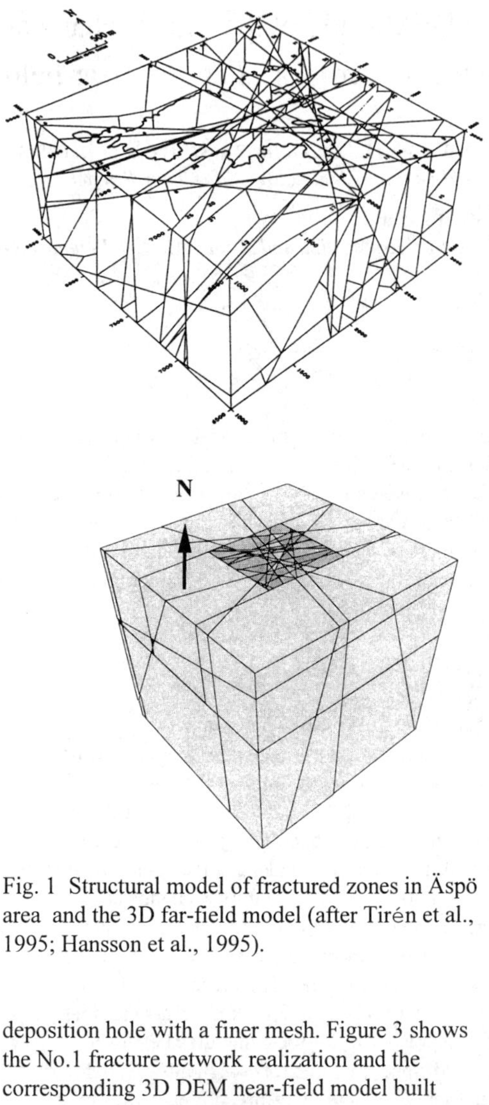

Fig. 1 Structural model of fractured zones in Äspö area and the 3D far-field model (after Tirén et al., 1995; Hansson et al., 1995).

2 GEOLOGICAL AND COMPUTATIONAL MODELS

2.1 *Model Geometry, Size and Boundary Conditions*

The hypothetical nuclear waste repository under study is located at about 500 m below the ground surface in Äspö area, Southern Sweden. The site contains 52 major faults and fracture zones and numerous small scale discontinuities (Tirén et al., 1995). Figure 1 illustrates the structural model of the Äspö area and the established 3D DEM far-filed model of size 4 x 4 x 4 km. It contains 23 fracture zones, 1024 blocks and 84574 finite difference elements. The isolated and inpersistent fracture zones were ignored and closely located fractures were merged to simplify the model set-up. An inner region which contains the potential repository and is of 1.5 x 1.5 x 1.5 km in size was specified with finer mesh. The potential repository is assumed to have 4000 canisters in total and they are arranged in 10 parallel tunnels with a 25 meter spacing. Along each tunnel 40 deposition holes are drilled with a 6 meter spacing.

The near-field model is based on a tunnel-deposition-hole system adopted for a potential nuclear waste repository (Fig. 2). Three near-field DEM models were generated based on this concept and three realizations from a statistical fracture distribution in the Äspö area of 139, 202 and 126 fractures, respectively (Geier, 1995). The dimensions of the three near-field DEM models are 25×25×18 m, containing an inner region of 7.5 ×7.5×7.5 m which surrounds a tunnel and a deposition hole with a finer mesh. Figure 3 shows the No.1 fracture network realization and the corresponding 3D DEM near-field model built according to this network with some simplifications.

The orientations of the models are such that the axes of the two principal horizontal *in-situ* stress components are normal to the vertical boundaries of the models and parallel to the z- and x-axes of the coordinate system, respectively. The y-axis points upwards. For the global far-field model, the bottom surface was fixed and the top surface is free, representing the ground surface. The top surface was assumed to have a +6° C temperature, and all other vertical boundaries have natural thermal conditions. The initial temperature field in the model was specified with a gradient of

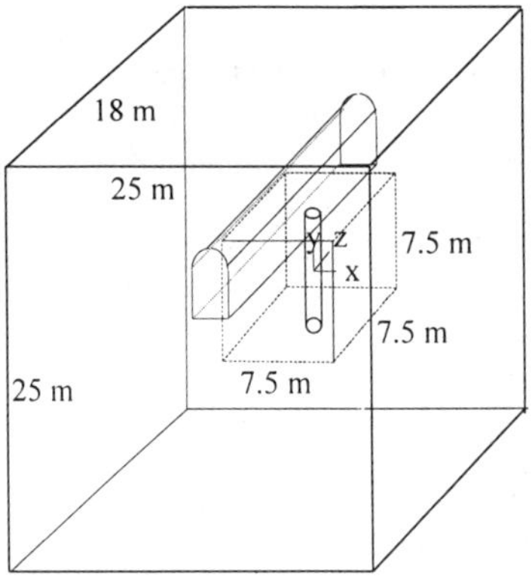

Fig. 2 Tunnel-deposition hole system by the KBS-III concept

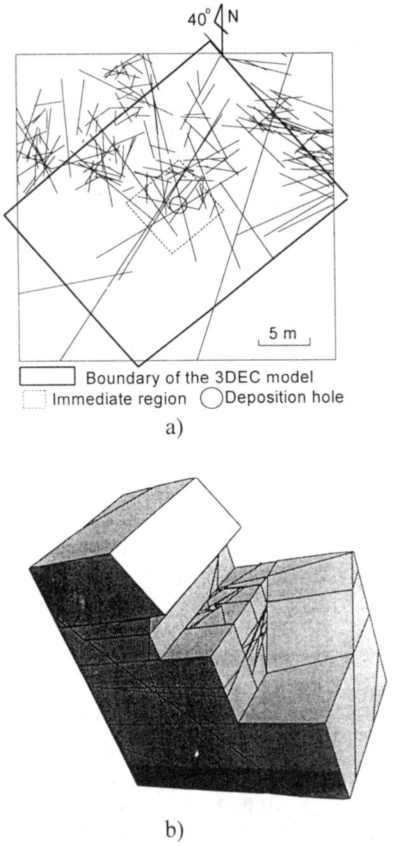

Fig. 3 Fracture pattern of the No. 1 fracture network realization (Geier, 1995) (a) and the corresponding DEM model after simplification (b) (after Shen and Stephansson, 1995).

16° C / km in the vertical y-direction. The *in-situ* and boundary stresses on the vertical boundaries are specified (unit: MPa), by (Bjarnasson et al. 1989)

$$\begin{cases} \sigma_{zz} = 7.44 + 0.068y \\ \sigma_{xx} = 3.74 + 0.037y \end{cases} \quad (y \geq -515\text{m}) \qquad (1)$$

$$\begin{cases} \sigma_{zz} = 0.021 + 0.042y \\ \sigma_{xx} = -0.10 + 0.023y \end{cases} \quad (y < -515\text{m}) \qquad (2)$$

Because of their small size of the near-field models the *in-situ* and vertical boundary stresses were assumed to be constant. At the depth of 500 meters below the ground surface, the stress components are given by (unit: MPa)

$$\sigma_{zz} = -26.56,\ \sigma_{xx} = -14.76,\ \sigma_{yy} = -13.50 \qquad (3)$$

The top surfaces of the near-field models was given a constant vertical stress σ_{yy} according to equation (3) and the bottom surfaces were fixed all the time. Adiabatic thermal boundary conditions were used for all thermal calculations.

2.2 Loading Sequences

In this study the impacts of four events were studied: (i) *in-situ* stresses; ii) weight of an future ice sheet of 2,200 m thickness due to glaciation and its removal (deglaciation) with a remaining 200 m thick ice sheet; (iii) thermal loading due to heat generated by emplaced waste; iv) excavation of the tunnel and deposition hole, (v) swelling pressure of 10 MPa on the wall of the deposition hole due to expansion of compacted bentonite backfill. The far-field model considered only impacts from event i), ii) and iii). The effect of glaciation was considered by a vertical boundary stress, equivalent to the intrinsic weight of the ice sheet at the top surfaces of the models. The swelling pressure was applied as boundary stress on the wall of the deposition hole for the near-field models. The loading conditions for glaciation follows the SITE'94 central scenario (King-Clayton et al., 1995).

For the far-field model, the heat release intensity at the time of emplacement is 6.1 MW per square kilometer. The heat source function (unit: Watt) is given by Thunvik and Braester (1991) as

$$Q = 1066\left[0.75312e^{-0.021761t} + 0.24688e^{-0.001278t}\right] \quad (4)$$

for the far field model and

$$Q = 803e^{-0.2176t} + 263e^{-0.1277t} \quad (5)$$

for the near-field model. The parameter t is time in years. The thermal response of the model was calculated for 200, 400, 100, and 60,000 years after the emplacement of the waste canisters. Thermal response during glaciation/deglaciation and the friction between the ice sheet and ground surface were ignored. The glaciation/deglaciation cycle was simulated with the thickness of ice sheet in the following sequence: 0 m (free surface) - 1000 m - 2200 m - 1000 m - 200 m - 0 m (free surface).

2.3 *Material Models and Properties*

The rockmasses were assumed to be linearly elastic material for far-field model and Mohr-Coulomb plasticity material model for near-field models. A Coulomb friction law with constant normal and shear stiffness was assumed for the fractures. For the far-field model, due to the lack of experimental data, the normal (k_n) and shear (k_s) stiffness of the large scale fracture zones was calculated as

$$k_n = E/d, \quad k_s = G/d \quad (6)$$

where d is the width of the fracture zones, E and G are the Young's modulus and shear moldulus of the material inside the fracture zone. They are assumed as E=10 GPa and G = 5 GPa, respectively. The material properties for intact rock used in the study are given as: density = 2700kg/m^3, Young's modulus= 37.5 GPa (far-field model) and 65 GPa (near-field model), shear modulus = 24.5 GPa (far-field model) and 24.59 GPa (near-field model), cohesion = 10 MPa, tensile strength = 10 MPa, internal friction angle = 40°, dilation angle = 15°, specific heat = 2 MJ/(m^3 °C), linear thermal expansion coefficient = $8.5 \times 10^{-6} / ^{o}C$, thermal conductivity = 3 W/m°C. The fracture properties are given as: normal stiffness = shear stiffness = 50 GPa/m (near-field model), friction angle = 20° (far-field model) and 30° (near-field model), and cohesion = 5 MPa.

3 MODELLING RESULTS

3.1 *Far-field Model*

Results gave a peak temperature about 48° C at the center of the repository after 200 years of the emplacement of the waste canisters. This peak temperature will then remain for another 200 years, followed by a gradual decrease until returning to the original temperature field at 60,000 years (Fig. 4). Figures 5 and 6 show the variation of shear and normal displacement of one of the fracture zones (No. 43) and the comparison of the calculated and measured stress components as functions of depth along one of the boreholes (KAS02) at the Äspö site. The maximum shear and normal (closure) displacement of fractures zones are 82 mm and 65 mm, respectively, and the maximum fracture opening is only 6 mm. The maximum net increase of stress magnitudes due to heating and glaciation from the far-field model calculation are:

σ_{zz} = 10 (MPa) (due to heating)
= 16 - 22 (MPa) (due to glaciation)
σ_{xx} = 10 (MPa) (due to heating)
= 10 - 18 MPa (due to glaciation)
σ_{yy} = 3 (MPa) (due to heating)
= 10 - 30 (MPa) (due to glaciation).

3.2 *Near-field Model*

Results from the DEM models show a major change of stress state in the vicinity of the tunnel and deposition hole due to excavation, swelling pressure, thermal loading and glaciation (Fig. 7). The maximum shear and normal displacement of the fractures are presented in Fig. 8. Thermal loading after 200 years from waste emplacement has the most severe effects on both near-field stress state and fracture deformation, but the swelling pressure has only a minor effect. Plastic deformation or yielding appeared in some local areas of intact rocks near the wall of the tunnel and deposition hole due to excavation, indicating

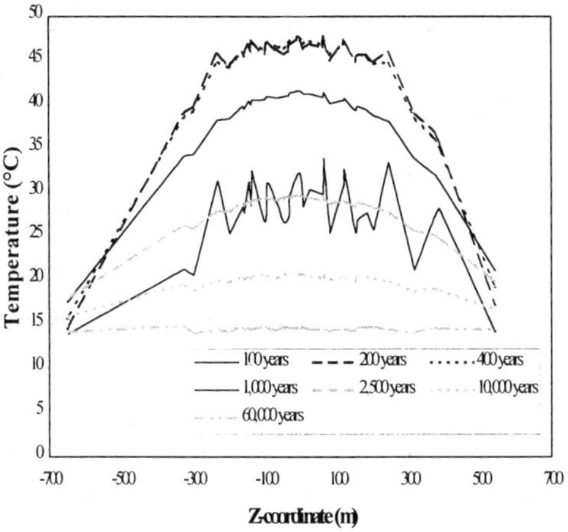

Fig. 4 Temperature distribution along a central profile of the repository during heating (after Hansson et al. 1995)

potential intact rock failure. This yielding becomes more widely spread with the heating and glaciation, especially around intersections of fractures (Fig. 9). The maximum temperature is 25.31 ° C, less than that calculated by the far-field model since the contributions from nearby canisters were not considered. The swelling pressure can reduce the degree of stress concentration around the deposition holes by 12%, therefore is a positive supporting measure to the deposition holes. Table 1 lists the temperature, deformation and material failure caused by different loading mechanisms for both far and near-field models. Effect of swelling pressure was excluded because it has not very significant effect on the overall on these measures.

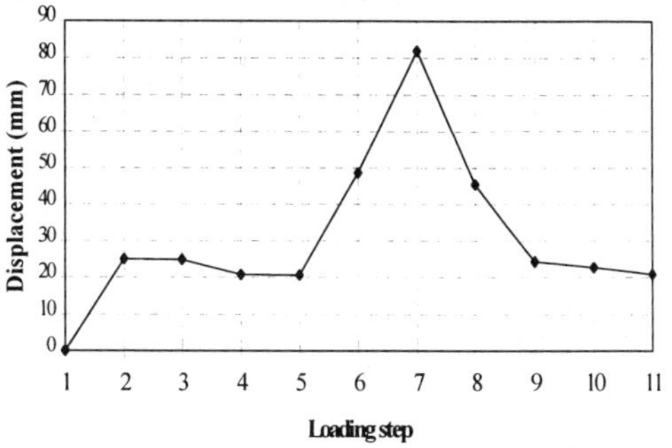

a)

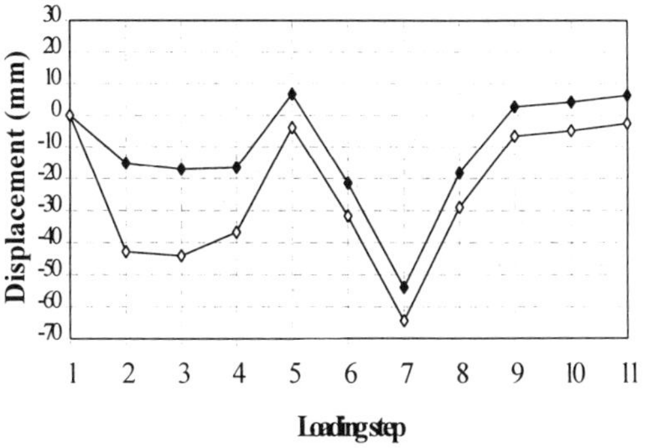

b)

Fig. 5 Maximum shear displacement (a) and maximum and minimum normal displacement (b) for fracture zone 43. Opening of fracture zone is positive. Loading steps: 1) present conditions; 2) 200 years heating; 3) 400 years heating; 4) 1,000 years heating. 5) 60,000 years heating; 6) 1,000 m ice thickness; 7) 2,200 m ice thickness; 8) 1,000 m ice thickness; 9) 200 m ice cliff; 10) 100 m water; 11) end of glaciation (after Hansson et al, 1995).

4 IMPACTS ON PERFORMANCE AND SAFETY OF REPOSITORY ISSUES

The aim of this study is to investigate the impact of excavation, heating, swelling pressure of buffer materials and glaciation on: (1) global rock mass stability surrounding the tunnel system of the repository , (2) local rock stability around tunnels and deposition holes; (3) canister safety and (4) hydraulic properties of near-field rocks. The modelling results are interpreted according to these concerns. Table 2 summarizes the impacts of different loading mechanisms on the these measures.

5 CONCLUSIONS

The three-dimensional distinct element code 3DEC was used to simulate the thermo-mechanical responses of fractures rocks hosting a hypothetical nuclear waste repository. Both the far and the near-field responses were analyzed. The study considers the thermal loading from canisters, swelling

Table 1 Summary of mechanical responses due to different loading mechanisms

Response	Excavation (near-field)	Heating	Glaciation
Maximum Temperature		25 - 31 °C (near-field model) 48 °C (far-field model)	
Hole convergence	0.8 mm		
Maximum shear displacement of fractures	0.4 mm	1.0 mm (near-field model) 25 mm (far-field model)	0.85 mm (near-field model) 82 mm (far-field model)
Maximum normal displacement of fractures	0.14 mm (Opening)	0.4 mm closure (near-field model) 42 mm closure (far-field model) 6 mm opening (far-field model)	0.5 mm (near-field model) 65 mm (far-field model)
Material failure (yielding)	Close to the tunnel and hole	wide spread around fracture intersections	wide spread around fracture intersections

Table 2 Impacts on Safety and Performance Assessment measures by different loading mechanisms

Loading Mechanism	Far-field rock stability	Near-field rock stability	Canister safety	Near-field hydraulic property
Excavation	Not studied	Concentration of stresses and minor to moderate failure of intact rocks.	Effect on canister safety by nearby tunnel not studied.	Fracture closure near walls of excavations, reducing near-field permeability
Swelling pressure	Not studied	Improvement of stability of deposition holes.	Positive protection to canisters	Not studied
Heating	No significant effect on global rock stability	Significant stress increase, material failure, and fracture deformation	Minor influence on canister safety	Significant reduction of rock permeability due to significant fracture closure.
Cooling	No significant effect on global rock stability	Significant stress relief, material failure, and fracture deformation	Minor effects	Fracture reopening and increase of global permeability.
Glaciation	Significant deformation of existing fracture zones and possible creation of new faults by great increase of vertical stresses	Major increase of stresses, failure zones and fracture deformation, possible new fracture generation and propagation	Possible canister damage by large shear deformation of fractures across the deposition holes.	Significant closure of existing fractures, decreasing near-field permeability and possible new flow pathways by newly created fractures
Deglaciation	Large cyclic fracture zone deformation and vertical stress relief; ground surface rebound, permanent fracture deformation remains	Cyclic stress relief and irreversible fracture deformation.	Minor to moderate effect	Fracture re-opening and permeability increase, but not reversible to initial state due to permanent fracture deformation.

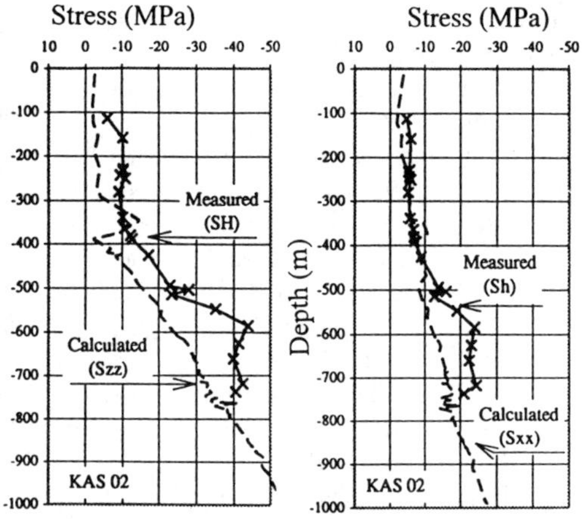

Fig. 6 Comparison of measured and calculated stress components along borehole KAS02.

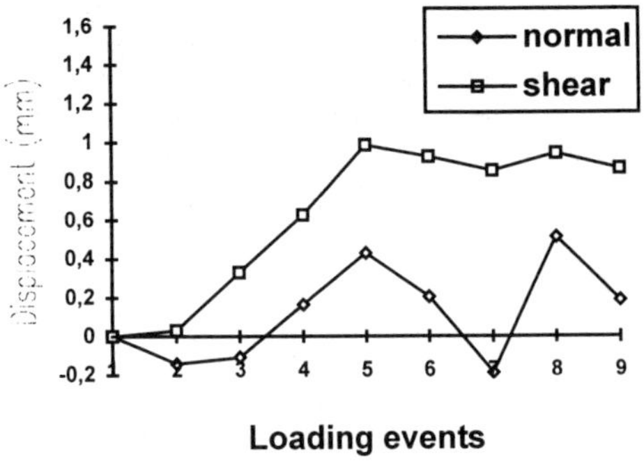

Fig. 8 Variation of shear and normal displacements of fractures as functions of loading events (same as in Fig. 7). Fracture closure is taken as positive (after Hansson et al., 1995).

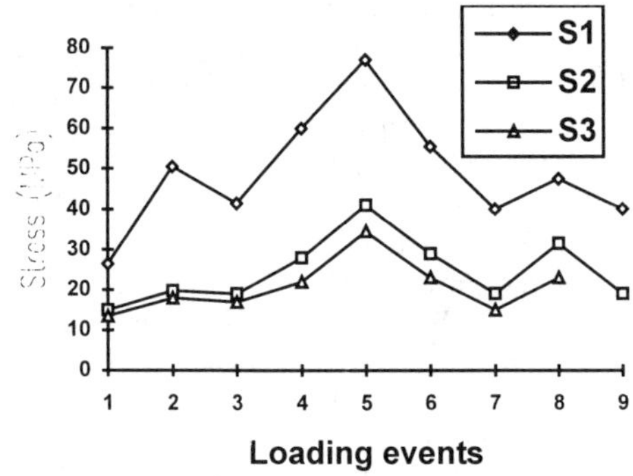

Fig.7 Variation of magnitudes of principal stresses at a point on the wall of the deposition hole as a function of loading events: 1) in-situ stress; 2) excavation; 3) swelling pressure from buffer; 4) heating for 1 year; 5) heating for 200 years; 6) heating for 1000 years; 7) heating for 10000 years; 8) glaciation; 9) deglaciation (after Hansson et al., 1995).

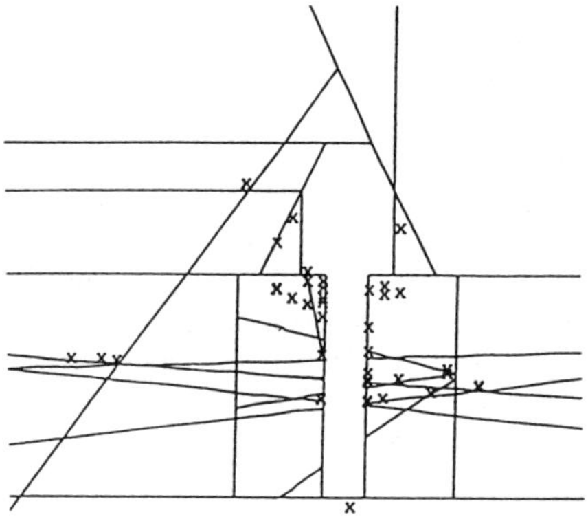

Fig. 9 Distribution of yielding zones (marked by the x signs) around the tunnel and deposition holes (after Shen and Stephansson, 1995).

pressure from buffer materials, stress concentration due to excavation and the impacts of glaciation. The impact of these different loading mechanisms on four measures of safety and performance assessments for rock stability, canister safety and local hydraulic property were evaluated. From the evaluation of the numerical results, we draw the following conclusions.

During the life time of a waste repository, the cyclic heating-cooling and glaciation-deglaciation were found to have the most significant impacts on the stability and safety of the repository. The modelling shows that it is possible for these events to create instability of both far-field and near-field rocks, cause fracture propagation and coalescence at some locations near major faults or fracture zones, and thereby change pathways for water flow. The swelling pressure in the deposition hole is found to have only minor effects on the rock stability and fracture deformation. The shear deformation of the fractures are size-dependent.

For near-field studies, the fractures used were of small sizes (about 1 to 10 m of mean lengths), resulting in a shear displacement of about 1 mm. For the far-field model, it reached to about 82 mm, which may pause a significant threat to the canister safety if the deposition hole is crossed by a larger fracture zone.

Closure is the major mode of fracture deformation during heating, which reduces the water inflow into the tunnel, and increase the fracture shear strength, but may also increase the stress concentration in the intact rocks around the tunnels and deposition holes, which may, in turn, have a negative impact on tunnel stability. On the other hand, during cooling of the repository and deglaciation, rock volume contraction will cause some fractures to re-open, thereby changing the water flow pathways, reducing the shear strength of the fractures, and also reducing the stress concentration of the intact rock. The water flow in the vicinity of the repository is expected to be affected by thermal loading and glaciation, due mainly to changes of fracture aperture and to changes in the connectivity of the fracture network by fracture propagation and coalescence. The analysis of such a coupled thermo-hydro-mechanical process in fractured rocks is only possible with a more sophisticated computer code which can also handle initiation and propagation of fractures in three-dimensional space. Such a code has not been developed and validated up to date.

ACKNOWLEDGEMENTS

This work was initiated and financially supported by the Swedish Nuclear Power Inspectorate (SKI) during 1992-1994 as a part of project SITE'94. The authors wish to express their gratitude to Johan Andersson, Fritz Kautsky and Öivind Toverud at the Radioactive Waste Division of SKI, for their assistance during the project.

REFERENCES

Bjarnasson, B., Klasson, H., Leijon, B., Trindell, L. and Öhman, T. Rock stress measurements in boreholes KAS02, KAS03 and KAS05 on Äspö. Progress Report 25-89-17, Swedish Nuclear Fuel and Waste Management Co., 1989.

Geier, J., SITE-94: Development of discrete-fracture models for the near-field, SKI report, Swedish Nuclear Power Inspectorate, Stockholm, 1995.

Hansson, H., Shen, B., Stephansson, O. and Jing, L., Rock mechanics modelling for the stability and safety of a nuclear waste repository: executive summary. Technical Report to Swedish Nuclear Power Inspectorate, Division of Engineering Geology, Royal Institute of Technology, Stockholm, 1995.

ITASCA, 3-D Distinct Element Code, Version 1.50 User's Manual. ITASCA Consulting Group Inc., Minneapolis, 1994.

King-Clayton, L. M., Chapman, N. A., Kautsky, F., Svensson, N.-O., de Marsily, G. and Ledoux, E., The central scenario for SITE-94. SKI report, Swedish Nuclear Power Inspectorate, Stockholm, 1995.

Raiko, H. and Salo, J.-P., The design analysis of ACP-canister for nuclear waste disposal, Report YJT-92-05, Nuclear Waste Commission of Finnish Power Companies, Helsinki, 1992.

Shen, B. and Stephansson, O., Near-field rock mechanics modelling for nuclear waste disposal, SKI report, Swedish Nuclear Power Inspectorate, Stockholm, 1995.

Swedish Nuclear Waste and Fuel Management Co. (SKB), Project on alternative systems study (PASS), final report. SKB Technical Report 93-04, Stockholm, Sweden, 1992.

Swedish Nuclear Power Inspectorate (SKI), SKI SITE'94, deep repository performance assessment project. SKI Report 96:36, Swedish Nuclear Power Inspectorate, 1996.

Thunvik, R. and Braester, C. Heat propagation from a radioactive waste repository. SKB 91 Reference canisters. SKB TR 91-61, Stockholm, 1991.

Tirén, S., Askling P., Beckholmen, M. and Carlsten, S. Development of SITE'94 geologic structural model of the Äspö site, Southeastern Sweden. SKI report (in press), 1995.

Computer Methods and Advances in Geomechanics, Yuan (ed.) © 1997 Balkema, Rotterdam, ISBN 90 5410 904 1

Possibility prognosis landslide on the open pit mine

Lj. Ilić
Kolubara-Projekt 14220, Lazarevac, Yugoslavia
S. Krstović
Department of Mining Faculty, University of Belgrade, Yugoslavia

ABSTRACT: In this research study We have presented the results of monitoring and measuring of the instability appearance on the large open-pit coalfield mines in Serbia like: KOLUBARA, KOSOVO, KOSTOLAC, as well as the possibility of caving time forecast and/or caving preventing.

1. INTRODUCTION

In Serbia three large coalfields have been developed: the ones of KOLUBARA, KOSOVOMETOHIYA, and KOSTOLAC. KOLUBARA coalfield is situated in the central part of Serbia, 50 km. away, to the south, from the city of Beograd, near by the town of Lazarevac.The coalfield have been investigated by deep boreholes and underground mining as well as by the existing open-pit mining. The coal belongs to the Tertiary respectively to Pliocene formation, and represents the younger lignites. In Kolubara coalfield the geological reserves of 3.8 billion tons of lignite have been found, and all of them can be exploited by open-pit mining. In Kolubara coalfield three open-pimines have been opened, and they are: "Polje B", "Polje D", and "Tamnava -Istok field". The forurth one - "Tamnava-Zapad field" is in the phase of opening. The open-pit mine "Polje D" on which certain investigations have been made, gives the production of 15 million tons of lignite per year, and after the reconstruction, which is under way, it should reach capacity of 18.8 million tons of lignite per year. Excavation of overburden and coal is being done by the ECL (Excavator, belt Conveyor, Loader) and ECS (Excavator, belt Conveyor, Stacker) systems. The excavated coal is transported to the separation plant where it is crushed and separated. The fine fractions are used by termal power plants, and the coarse ones by the other consumers.

2. THEORETICAL THESES

The results of many year research of the instability appearance on the coalfields KOLUBARA, KOSOVO and KOSTOLAC, that is to say on their open-pit mines: Polje D, Tamnava-Istok, Belacevac, and Drmno, have been presented in this text. The research consisted of the field observation and measuring as well as of the laboratory testing of the respective samples obtained by drilling.The results of the monitoring of the former many year instability apperances (2,3,5) show that it is the type of progressive fracture with the caving machanism shown on Fig. 1. Knowledge of the type and mechanism of caving enables the respective measuring during present and future research.

The progressive fracture appears in overconsolidated clays which contain micro cracks and have big heights and slopes of the working floors. When conditions for the progressive fracture are fulfilled local fractures take place, and that causes decrease of pressure of overconsolidation, and by that decrease of shear strength. In case of elastic behaviour of the overconsolidated clay, decrease of the inner stress will as a result have relaxation and lifting of clay, and formation of sliding plane. The appearance of sliding causes reverse energy and deformation of overconsolidated clay, which reduces strength of sliding from top to residual values and causes appearance of the progressive fracture, and it can
be expressed by decrease of rigidity index I_b or increase of residual factor R, as in the following equation:

$I_b < 0.50;\ R>0.85 \rightarrow \quad \gamma\ \ h\ \ \sin\alpha \geq c + \sigma\ tg\varphi$ 1

Mathematical procedure of the progressive fracture solving is based on the previously described mechanism, that is to say that the limit state of shear does not become at the same time

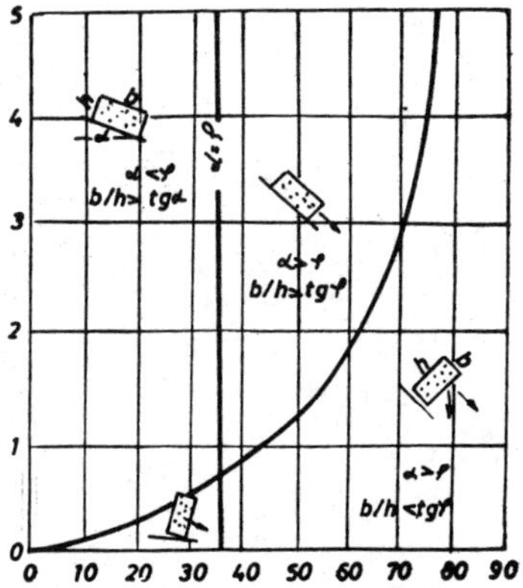

Fig. 1. Landslide mechanism

on all the points, but progressively, depending and geometry of the block and its position in the ground. Mathematical model shown on Fig. 2. is based on certain hypotheses and they are:
- Blocks are taken as firm objects in relation to the crack filling which is subject to deformity.
- There is a deformability of the filling in the direction of shearing, and deformities in the direction vertical onto the crack are considered to be negligibly small.

Movement of the blocks in a row occurs by sliding along the cracks, and fracture occurs when the shear strength in the cracks is transgressed.
In order to form static system it is necessary to add some new conditions to the above mentioned initial hypotheses. They are: 1. Balance conditions; 2. Movement plan; 3. Connection between movement and reactive forces. During movement of the blocks along the cracks or interlayers by shearing, balance conditions are: $\Sigma x = 0$ and $\Sigma y = 0$ which, applied to the system of blocks on Fig. 2., amounts:

B3: $T_3 \cos \alpha_3 - N_3 \sin \alpha_3 - T_{23} \cos \alpha_{23} + N_{23} \sin \alpha_{23} = H + Z_3$ 2

$T_3 \sin \alpha_3 + N_3 \cos \alpha_3 + T_{23} \sin \alpha_{23} + N_{23} \cos \alpha_{23} = G + V_3$ 3

B2: $T_{23} \cos\alpha_{23} - N_{23} \sin \alpha_{23} - T_2 \cos \alpha_2 - N_2 \sin \alpha_2 - T_1 \cos\alpha_{12} + N_{12} \sin\alpha_{12} = Z_2$ 4

$T_{23}\cos\alpha_{23} - H_{23}\cos\alpha_{23} + T_2\sin\alpha_2 + T_{12}\sin\alpha_{12} + N_{12}\cos\alpha_{12} = G_2$ 5

B1: $T_{12}\cos\alpha_{12} - N_{12}\sin\alpha_{12} + T_1\cos\alpha_1 - N_1\sin\alpha_1 = Z_1$ 6

$T_{12} \sin \alpha_{12} - N_{12} \cos\alpha_{12} + T_1 \sin\alpha_1 + N_1 \cos\alpha_1 = G_1$ 7

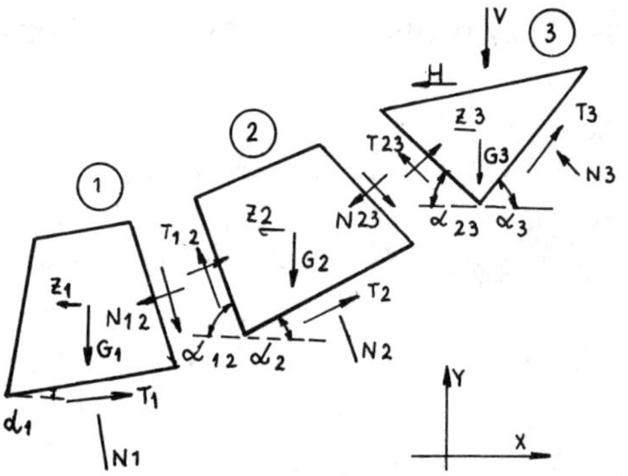

Fig. 2. Mathematical model of the sliding body

Having in mind that the conditions of balance were given in 6 equations, and that there are 10 unknowns, the system is indefinite, and should be supplemented by the additional equations, which will be obtained from the movement plan, that is, from connections between shearing and reactive
forces in the cracks, which can be seen from the kinematics chain , Fig.3.

$a_n u = u_n \rightarrow u_2 = a_2 u;\ u_{12} = a_{12} u$ 8

By using the last equationion, it is possible, adequate to Fig.2., to establish kinematic chain of movement in a row of blocks, from which the remaining four equations can be obtained:

$S_3 = (m_3N_3 + n_3T_3) - S_{23}(-m_{23}N_{23} + n_{23}T_{23}) = \varnothing$ 9

$S_{23} = (-m_{23}N_{23} + n_{23}T_{23}) - S_2(-m_2N_2 + n_2T_2) = \varnothing$ 10

$S_2 = (-m_2N_2 + n_2T_2) - S_{12}(-m_{12}N_{12} + n_{12}T_{12}) = \varnothing$ 11

$S_{12} = (-m_{12}N_{12} + n_{12}T_{12}) - S_1(-m_1T_1 + n_1T_1) = \varnothing$ 12

By solving these 10 equations we get the values of the reactive forces in the cracks (Ti,Ni), and with them values of the shearing movement "u". Just these "u" movements should be registered at the beginning.
Registration of the movement is possible by slide measuring device installed in the respective borehole on the respective depths. On the base of this registration it is possible to calculate the deformations and the time of caving of the slope, which is given by the equation No. 13 and Fig. 4.

$t_R - t_1 = 0.5\ (t_2 - t_1)^2 / (t_2 - t_1) - 0.5\ (t_3 - t_1)$ 13

Parallelly with monitoring on the field, the respective geomechanical and laboratory testings are necessary to be done with the aim to define

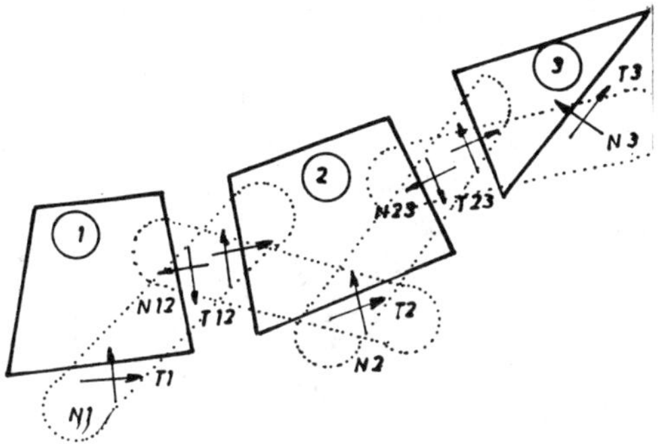

Fig. 3. Kinematic chain of slope caving.

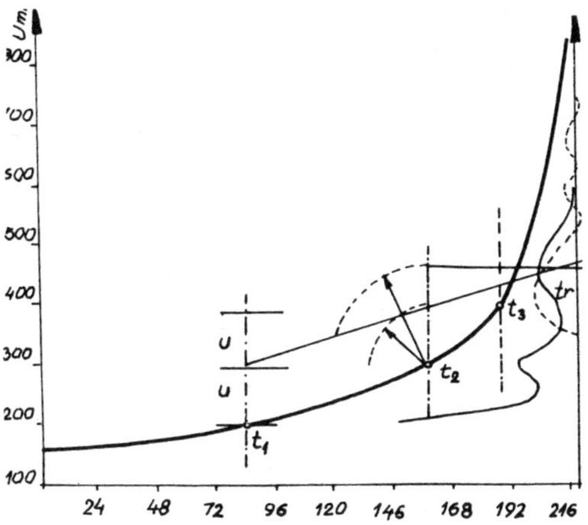

Fig. 4. Slope deformation diagram

the overconsolidation degree, which depend upon the loading history and technology, that is to say dynamics of excavation, as given by the following equations:

$\sigma_p/\sigma_0 > 4 \rightarrow \sigma_3 > \sigma_1 \rightarrow A > \varnothing \quad B \geq \varnothing \rightarrow \Delta u > \varnothing$ 14

$\Delta u = B[\Delta\sigma_3 + A(\Delta\sigma_1 - \Delta\sigma_3)] \rightarrow A = \Delta u_d/(\Delta\sigma_1 - \Delta\sigma_3)$..15

With the stated relations the necessary observations and calculations have been made on the occasion of monitoring and rehabilitation of the endangered slopes on the above mentioned mines.

3. PRACTICAL APPLICATION

The first monitoring and measuring of instability phenomenon were done in 1975, when a big landslide appeared on the Kosovo coalfield open-pit mine Belacevac (2), the results of which, as well as the results obtained by monitoring of the later landslides on the other mines will be described chronologically:

Later appearance of instability in Kolubara coalfield on the open-pit mine žPolje D' and then on the open-pit mine žTamnava-Istok' enabeled further monitoring of sliding and deformation mechanism. So, on the Polje D' on the north slope, detailed geotecnical measurements, consisting of geomechanical drilling, laboratory geomechanical testings, field surveying observations (3) were done. The results obtained by "in situ" testing on large samples (3) which have been shown on Fig. 8. point out the decisive influence of the interlayers and micro-cracks on the appearance of deformations and instability.

Based on the obtained measuring results (3), the respective technical documentation with the tecnological operations was made, and those operations acted preventively on rehabilitation of the initial instability.

The second case of instability was in Kolubara coalfield, on the western terminal slope on "Tamnava -Istok' open-pit mine.

The first signs of instability were registered the day before caving of the slope, that is on September 13, 1993. at 6 o'clock A.M. Namely, on that day a crack was noticed on the local asphalt road shown on Fig. 6 and 9.

Some time later, that is about 2 o'clock P.M. a new crack appeared westward of the existing one, and that one was 850 m. long, 0.5 m. wide, and with the supposed depth of 70 m.

About 5 o'clock P.M. it was noticed that the the whole slope with all coal and overburden working floors was moving over the interlayer between coal layers. Continuous movement of the whole coal layer as well as of the roof sediments was continued during the next 24 hours, and could be monitored and measured in the interlayer on the lowest coal working floor. The next day, that is on September 14, 1993. at

Fig. 5."In situ" testing on" Polje D",

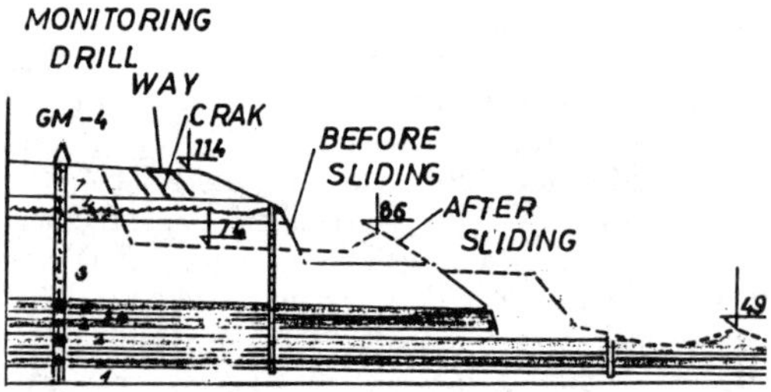

Fig. 6. Geotecnical profile of the slide

about 2:30 o'clock A.M., an abrupt movement, that is an "explosion" of the slope occured. That mechanism will be explained in the following text.

The western terminal slope of Tamnava-Istok open-pit mine, on which the slope fracture occured, was composed of the following lithological members (as shown on the geotechnical profile, Fig. 9): The highest parts of the soil were made of a complex of Kvaternary clays (7) with alluvial sands (6) and a complex of alluvial gravel (4) bellow them. The thickest lithological member is a complex of Pontian clays (7) whose average thickness is about 50 m., and which lies on the coal layer. The average thickness of the coal layer is 30 m., and it is devided out by clay interlayers (2b), from 0.1 to 2 m. thick,- and with very bad geomechanical characteristics. Below the coal layer there are saturated sands (1). From the hydrogeological point of view, two outcrops are existing in this deposit: the upper one in gravel, and the lower one in the floor sands, which are subartesian. In the geomechanical sense the western terminal slope on the slided location represented an instable area due to the interlayers in coal which have very bad geomechanical characteristics. ($C=15$ kn/m^2; $P=8^0$; $\gamma=16.40$ kn/m^3), so that it could not bear its own load, because it was excavated under the general inclination of $\alpha = 22^0$, and with the hight of $h = 80$ m. The described slided mass amounting 4.5 million of cubic meters of coal and overburden, was 850 m. long, 300 m. broad, and 80 m. high, with the half-moon shape.

At the moment of fracture the slope changed the shape, as can be seen from the geotechnical profile on Fig. 6. The highest parts lowered as much as 35 m. under inclination of 65 degrees, with the simultaneous breaking of the coal working floor and the plateau, and raising of the broken coal mass for 10 m. on the bench mark 40, and for 12 m. on the bench mark 74. Kinematics and dynamics of the landslide occured as shown on the Fig. 1 and 2., and was

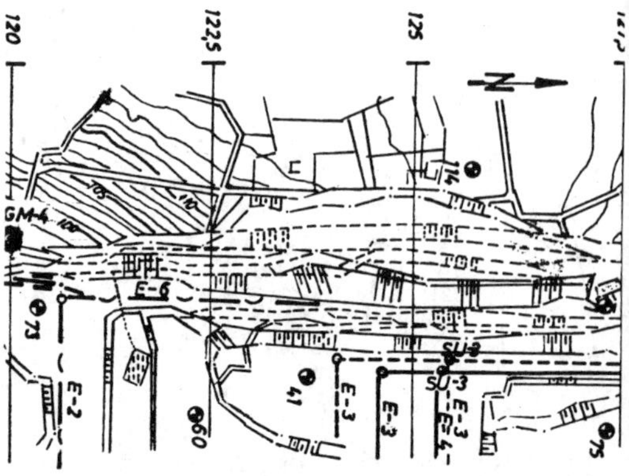

Fig. 7. Layout of the landslide on Tamnava-Istok open-pit mine.

very similar to the described appearance which had prevously occured on Kosovo open-pit mines. When the slided mass calmed down, the additional relaxation of movements were occuring during the next 90 days, on which occasion the further raising and movenent of the landslide foot was occuring. So, after 90 days when the slided mass, thank to the reclamation works, calmed down, its foot passed the distance of 120 m. and stopped.

On the base of the former landslide dynamics monitoring results it has been concluded that registering of the initial movement of the sliding mass was in no case successful, but it was noticed by chance in a progressed phase. With the aim to register promptly the initial movement of the instable mass, it is necessary to install the adequate slide measuring devices into the already drilled boreholes. In this case the sliding measuring devices were installed in the firm part of the terminal western slope, on the location where the mining excavation was to come. Namely, the installed measuring device, shown on Fig. 10. and 11., was on geological profile 120 in the geological borehole GM4.

The measuring device installed in the borehole GM4 was on the depth of 74 m. and anchored in the coal floor itself with four measuring bodies installed in the critical interlayers. The measuring results information conveying was foreseen to be done by monitoring system in order to provide data reliability.The first element of the installed measuring device was on the depth of 42.70 m. from the soil on the contact between roof and coal.The second element was installed on the depth of 52.80 m. from the soil in the first overburden interlayer in coal. The third element was installed od the depth of 64.50 m. in the second coal interlayer. The fourth element was

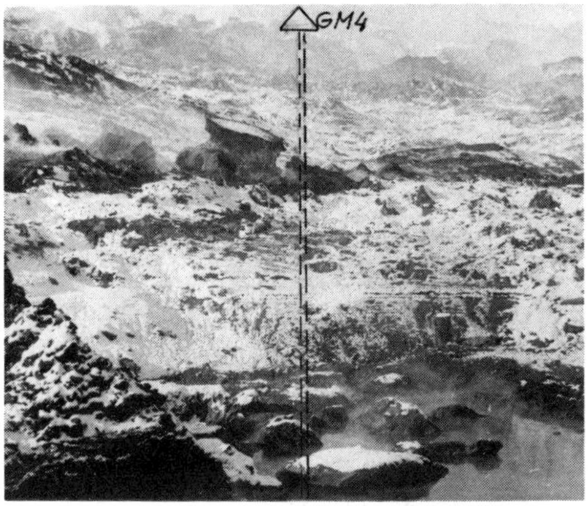

Fig. 8. Layout of the installed sliding measuring device.

on the depth of 73 m. in the coal and floor contact interlayer. Conveying of information to the measuring board was made by the elements - indicators, and the system of levers and cables, as shown on Fig.9. In the second phase, monitoring system was foreseen to be completed by transformation of the mechanical movements into electrical impulses, and their introduction into the distribution centre monitor. Monitoring of the cracks is continued with the aim to obtain the necessary data and undertake the necessary preventive mining, geological, and geotechnical works, which, as a result, provide safe performance of the terminal western slope on Tamnava-Istok. open-pit mine. Rehabilitation of the described landslide was performed on the base of the prepared Design of rehabilitation, made by "Kolubara-Projekt " company from Lazarevac, and "Geosonda" and "Soiltest" companies from Beograd. Rehabilitation according to the above Design was performed so as to decrea se tangential stresses by

Fig. 9. Mechanical slide measuring device

excavation of the upper parts of working floor, and transportation of the excavated material onto the waste dumps at the foot of the working floor. With the described technological way safety factor increased from F1 = 1.0 to F2 = 1.4 on the rehabilitated landslid, and F3 = 1.5 on the firm terminal slope southward of the landslide, which is being formed by the existing mining works, and in which a slide measuring device was installed, and a monitoring system formed with the aim to monitor behaviour of the massif durin excavation.

CONCLUSION

In this scientific research study the results of exploration, concerning slope stability monitoring on large open-pit coal mines in Serbia, were presented. On large open-pit mines in Serbia, like: Kolubara, Kosovo, Kostolac, instabilities during excavation appear from time to time, and it is necessary to register them in time with the aim of successful preventive actions and/or rehabilitation. Registration of the initial appearances of instability, as described in this study, is the basic condition for the slope breakage forecast, that is to say, for its rehabilitation and prevention.

ACKNOWLEDGMENT

I take the opportunity to thank to my collegues and to the managers on the open-pit mines of Kolubara coalfield, as well as to the Ministry of Science of Serbia, without whose help this research could not be made.

REFERENCES

1. Ilić Lj. 1974. Der Abau von braunkohle aus einem Begrutsch in tagebau Kosovo, Montreux

2. Ilić Lj.1989. Studija izbora realnih geomehaničkih parametara na polu D,Lazarevac.

3. Ilić Lj. 1991. Geomehanika na površinskim kopovima, Beograd

AUTHORS ADDRESSES
Dr. Ljubimko Ilić
"Kolubara-Projekt"
Kolubarskgi trg 8
14220 Lazarevac
Yugoslavia
Phone No.:(381 11) 8123-590

COMPANY ADDRESSES
Miss Maja Ilić, Manager
"SOILTEST"
Petrarkina 10/11
11000 Belgrade
Yugoslavia
Phone No.:(381 11) 407-443

Computer Methods and Advances in Geomechanics, Yuan (ed.) © 1997 Balkema, Rotterdam, ISBN 90 5410 904 1

A new numerical method for contaminant transport in unsaturated soils

Zhao Weibing, Gao Junhe, Wei Yingqi & Shi Jianyong
Geotechnical Engineering Institute, College of Civil Engineering, Hohai University, Nanjing, China

ABSTRACT: Contaminant transport in unsaturated subsoils is a subject which is encountered frequently in environmental geotechnical engineering problems . At present , all of the proposed numerical methods have defects . In this paper , a new numerical method--Streamline Upwind / Petrov-Galerkin (SUPG) finite element method is derived. Calculated results are compared with field measured values and those of other numerical methods . It shows that this method is right , has good accurancy and is better than other numerical methods . It can be applied to contaminant transport problems in unsaturated foundation soils.

1 INTRODUCTION

Contaminant transport in unsaturated subsoils is a subject which is encountered frequently in many environmental geotechnical engineering problems.

Two numerical methods are proposed to simulated the subject at present,that are, finite difference method and characteristic finite element method (Huang Kangle1988). But both of them have defects: the former has numerical dispersion and numerical oscillation difficulties; the latter can avoid such difficulties, however, its formulation is got through a mobilized coordinate system , the computer program is so complicate that it can't be accepted by practicing engineers. For this reason, an new numerical method SUPG (Streamline Upwind /Petrov-Galerkin) finite element method is proposed in this paper.

2 NUMERICAL APPROACH

If the complex physical and chemical actions among pollutant and soils are ignored, The basic contaminant transport in unsaturated soils is described as :

$$\frac{\partial(\theta c)}{\partial t}=\frac{\partial}{\partial z}\left(\theta D\frac{\partial c}{\partial z}\right)-\frac{\partial(q c)}{\partial z} \quad (1)$$

where t is time ; z denotes the vertical coordinate assumed positive downward; θ is volumetric moisture content ; D is unsaturated seepage velocity ; and c is concentration of pollutant .

By means of weighted residual technique ,Equation (1) becomes:

$$\int_\sigma \frac{\partial(\theta c)}{\partial t}U_i dz=\int_\sigma\left\{\frac{\partial}{\partial z}\left(\theta D\frac{\partial c}{\partial z}\right)-\frac{\partial(q c)}{\partial z}\right\}U_i dz \quad (2)$$

in which $\sigma=[z_{i-1},z_{i+1}]$, i is inner point and the weight function U_i is defined as:

$$U_i=N_i+P_i \quad (3)$$

where N_i is interpolation function and is constitutive on element boundaries ; P_i is the disturbance of weight function and is disconstitutive on the boundaries, but its limitation exist and is limit on the both sides of the boundaries . The basic water flow equation for one-demension is taken as :

$$\frac{\partial\theta}{\partial t}=-\frac{\partial q}{\partial z} \quad (4)$$

Substituting Equation (4) into Equation (2) one obtains :

$$\int_{z_{i-1}}^{z_{i+1}}\left(\theta\frac{\partial c}{\partial t}+q\frac{\partial c}{\partial z}\right)(U_i+N_i)dz=$$

$$\int_{z_{i\_1}}^{z_{i+1}}\left[D\frac{\partial\theta}{\partial z}\frac{\partial c}{\partial z}+\theta D\frac{\partial c}{\partial z}+\theta D\frac{\partial^2 c}{\partial^2 z}\right](N_i+U_i)dz \quad (5)$$

When $|z_{i-1}-z_i|$ and $|z_{i+1}-z_i|$ are very small , some hypotheses are presented as :

1 in $[z_{i-1},z_i]$, θ, c adop linear interpolation functions :

$$\theta=\frac{z_i-z}{\Delta z}\theta_{i-1}+\frac{z-z_{i-1}}{\Delta z}\theta_i \quad (6)$$

$$c=\frac{z_i-z}{\Delta z}c_{i-1}+\frac{z-z_{i-1}}{\Delta z}c_i \qquad (7)$$

$$N_i=\frac{z-z_{i-1}}{\Delta z} \qquad (8)$$

2 in $[z_i - z_{i+1}]$, θ,c adopt the similar functions :

$$\theta=\frac{z_{i+1}-z}{\Delta z}\theta_i+\frac{z-z_i}{\Delta z}\theta_{i+1} \qquad (9)$$

$$c=\frac{z_{i+1}-z}{\Delta z}c_i+\frac{z-z_i}{\Delta z}c_{i+1} \qquad (10)$$

$$N_i=\frac{z_{i+1}-z}{\Delta z} \qquad (11)$$

3 $P=\alpha\Delta t\frac{\partial N_i}{\partial z}$ (12)

where α is a coefficent

4 some functions of θ such as D, q etc. are not linear functions of θ, however when $|z_{i-1}-z_i|$ and $|z_{i+1}-z_i|$ are very small,we can linearlized D, D' and q as herein:

$$D_{i-\frac{1}{2}}\approx\frac{1}{2}(D_{i-1}+D_i) \qquad (13)$$

$$D_{i+\frac{1}{2}}\approx\frac{1}{2}(D_i+D_{i+1}) \qquad (14)$$

$$D'_{i-\frac{1}{2}}\approx\frac{1}{2}(D'_{i-1}+D'_i) \qquad (15)$$

$$D'_{i+\frac{1}{2}}\approx\frac{1}{2}(D'_i+D'_{i+1}) \qquad (16)$$

$$q_{i-\frac{1}{2}}\approx\frac{1}{2}(q_{i-1}+q_i) \qquad (17)$$

$$q_{i+\frac{1}{2}}\approx\frac{1}{2}(q_i+q_{i+1}) \qquad (18)$$

Each part of Equation (5) is developed as follow :

The right side of Equation (5) :

$$D'=\frac{\partial D}{\partial z} \qquad (19)$$

$$\int_{z_{i-1}}^{z_{i+1}}D\frac{\partial\theta}{\partial z}\frac{\partial c}{\partial z}N_i dz=\int_{z_{i-1}}^{z_i}D\frac{\theta_i-\theta_{i-1}}{\Delta z}\frac{c_i-c_{i-1}}{\Delta z}\cdot\frac{z-z_{i-1}}{\Delta z}dz+\int_{z_i}^{z_{i+1}}D\frac{\theta_{i+1}-\theta_i}{\Delta z}\frac{c_{i+1}-c_i}{\Delta z}\frac{z_{i+1}-z}{\Delta z}dz$$
$$=\frac{1}{2}D_{i-\frac{1}{2}}\frac{\theta_i-\theta_{i-1}}{\Delta z}(c_i-c_{i-1})+\frac{1}{2}D_{i+\frac{1}{2}}\frac{\theta_{i+1}-\theta_i}{\Delta z}\cdot(c_{i+1}-c_i) \qquad (20)$$

$$\int_{z_{i-1}}^{z_{i+1}}\theta D'\frac{\partial c}{\partial z}N_i dz=\int_{z_{i-1}}^{z_i}D'\left(\frac{z_i-z}{\Delta z}\theta_{i-1}+\frac{z-z_{i-1}}{\Delta z}\theta_i\right)\cdot\frac{c_i-c_{i-1}}{\Delta z}\frac{z-z_{i-1}}{\Delta z}dz+\int_{z_i}^{z_{i+1}}D'\left(\frac{z_{i+1}-z}{\Delta z}\theta_i+\frac{z-z_i}{\Delta z}\theta_{i+1}\right)\cdot\frac{c_{i+1}-c_i}{\Delta z}\frac{z_{i+1}-z}{\Delta z}dz$$
$$=D'_{i-\frac{1}{2}}\left(\frac{1}{6}\theta_{i-1}+\frac{1}{3}\theta_i\right)(c_i-c_{i-1})+D'_{i+\frac{1}{2}}\left(\frac{1}{3}\theta_i+\frac{1}{6}\theta_{i+1}\right)(c_{i+1}-c_i) \qquad (21)$$

$$\int_{z_{i-1}}^{z_{i+1}}\theta D\frac{\partial^2 c}{\partial^2 z}N_i dz=\cdot\theta_{i-\frac{1}{2}}D_{i-\frac{1}{2}}\int_{z_{i-1}}^{z_i}\left[\frac{\partial}{\partial z}\left(\frac{\partial c}{\partial z}N_i\right)-\frac{\partial N_i}{\partial z}\frac{\partial c}{\partial z}\right]dz+\cdot\theta_{i+\frac{1}{2}}D_{i+\frac{1}{2}}\int_{z_i}^{z_{i+1}}\left[\frac{\partial}{\partial z}\left(\frac{\partial c}{\partial z}N_i\right)-\frac{\partial N_i}{\partial z}\frac{\partial c}{\partial z}\right]dz=0 \qquad (22)$$

$$\int_{z_{i-1}}^{z_{i+1}}D\frac{\partial\theta}{\partial z}\frac{\partial c}{\partial z}P_i dz=\frac{1}{2}D_{i-\frac{1}{2}}\frac{\theta_i-\theta_{i-1}}{\Delta z}\frac{c_i-c_{i-1}}{\Delta z}\int_{z_{i-1}}^{z_i}\alpha\Delta t\frac{1}{\Delta z}dz+\frac{1}{2}D_{i+\frac{1}{2}}\cdot\frac{\theta_{i+1}-\theta_i}{\Delta z}\frac{c_{i+1}-c_i}{\Delta z}\int_{z_i}^{z_{i+1}}\alpha\Delta t\frac{-1}{\Delta z}dz=\frac{1}{2}\alpha\Delta tD_{i-\frac{1}{2}}\frac{\theta_i-\theta_{i-1}}{(\Delta z)^2}\cdot(c_i-c_{i-1})-\frac{1}{2}\alpha\Delta tD_{i+\frac{1}{2}}\frac{\theta_{i+1}-\theta_i}{(\Delta z)^2}(c_{i+1}-c_i) \qquad (23)$$

$$\int_{z_{i-1}}^{z_{i+1}}\theta D'\frac{\partial c}{\partial z}P_i dz=\int_{Z_{i-1}}^{z_i}D'\left(\frac{z_i-z}{\Delta z}\theta_{i-1}+\frac{z-z_{i-1}}{\Delta z}\theta_i\right)\frac{c_i-c_{i-1}}{\Delta z}\alpha\Delta t\frac{\partial N_i}{\partial z}dz+\int_{z_i}^{z_{i+1}}D'\left(\frac{z_{i+1}-z}{\Delta z}\theta_i+\frac{z-z_i}{\Delta z}\cdot\theta_{i+1}\right)\frac{c_{i+1}-c_i}{\Delta z}\alpha\Delta t\frac{\partial N_i}{\partial z}dz=\frac{1}{2}D'_{i-\frac{1}{2}}\alpha\Delta t\frac{\theta_{i-1}+\theta_i}{\Delta z}\cdot(c_i-c_{i-1})-\frac{1}{2}D_{i+\frac{1}{2}}\alpha\Delta t\frac{\theta_i+\theta_{i+1}}{2\Delta z}(c_{i+1}-c_i) \qquad (24)$$

$$\int_{z_{i-1}}^{z_{i+1}}\theta D'\frac{\partial^2 c}{\partial^2 z}P_i dz=\frac{\alpha\Delta t}{\Delta z}\theta_{i-\frac{1}{2}}D_{i-\frac{1}{2}}\frac{\partial c}{\partial z}\Big|_{z_{i-1}}^{z_i}-\frac{\alpha\Delta t}{\Delta z}\theta_{i+\frac{1}{2}}D_{i+\frac{1}{2}}\frac{\partial c}{\partial z}\Big|_{z_i}^{z_{i+1}}$$
$$=\frac{\alpha\Delta t}{\Delta z}\left[\theta_{i-\frac{1}{2}}D_{i-\frac{1}{2}}(c_i-c_{i-1})-\theta_{i+\frac{1}{2}}D_{i+\frac{1}{2}}(c_{i+1}-c_i)\right] \qquad (25)$$

The left side of Equation (5) :

$\int_{z_{i-1}}^{z_{i+1}}\theta\frac{\partial c}{\partial t}N_i dz$ Neuman's idea[Ne-zheng Sun,1989]

$$\left(\frac{\partial c}{\partial t}\right)_i\int_{z_i}^{z_{i+1}}\theta N_i dz=\left(\frac{\partial c}{\partial t}\right)_i\left[\int_{z_{i-1}}^{z_i}\left(\frac{z_i-z}{\Delta z}\theta_{i-1}+\frac{z-z_{i-1}}{\Delta z}\theta_i\right)\frac{z-z_{i-1}}{\Delta z}dz+\int_{z_i}^{z_{i+1}}\left(\frac{z_{i+1}-z}{\Delta z}\theta_i+\frac{z-z_i}{\Delta z}\theta_{i+1}\right)\cdot\frac{z_{i+1}-z}{\Delta z}dz\right]=\left(\frac{\partial c}{\partial t}\right)_i\left(\frac{1}{6}\theta_{i-1}+\frac{1}{6}\theta_{i+1}+\frac{2}{3}\theta_i\right) \qquad (26)$$

$$\int_{z_{i-1}}^{z_{i+1}} q\frac{\partial c}{\partial z}N_i\,dz =$$

$$\frac{1}{2}q_{i-\frac{1}{2}}\left(c_i - c_{i-1}\right)+\frac{1}{2}q_{i+\frac{1}{2}}\left(c_{i+1}-c_i\right) \tag{27}$$

$$\int_{z_{i-1}}^{z_{i+1}}\theta\frac{\partial c}{\partial t}P_i dz = \left(\frac{\partial c}{\partial t}\right)_i \int_{z_{i-1}}^{z_i}\left(\frac{z_i - z}{\Delta z}\theta_{i-1}+\frac{z - z_{i-1}}{\Delta z}\theta_i\right)$$

$$\cdot\frac{a\Delta t}{\Delta z}dz + \left(\frac{\partial c}{\partial t}\right)_i \int_{z_1}^{z_{i+1}}\left(\frac{z_{i+1}-z}{\Delta z}q_i + \frac{z - z_i}{\Delta z}q_{i+1}\right)$$

$$\cdot\frac{-\alpha\Delta t}{\Delta z}dz = \frac{1}{2}\alpha\Delta t\left(\frac{\partial c}{\partial t}\right)_i\left(\theta_{i-1}-\theta_{i+1}\right) \tag{28}$$

$$\int_{z_{i-1}}^{z_{i+1}} q\frac{\partial c}{\partial z}P_i dz = \int_{z_{i-1}}^{z_i} q\frac{c_i - c_{i-1}}{\Delta z}\alpha\Delta t\frac{1}{\Delta z}dz +$$

$$\int_{z_i}^{z_{i+1}} q\frac{c_{i+1}-c_i}{\Delta z}\alpha\Delta t\frac{-1}{\Delta z}dz =$$

$$q_{i-\frac{1}{2}}\frac{\alpha\Delta t}{\Delta z}\left(c_i - c_{i-1}\right) - q_{i+\frac{1}{2}}\frac{\alpha\Delta t}{\Delta z}\left(c_{i+1}-c_i\right) \tag{29}$$

Substituting Equation(19)-(29) into Equation (5) and defining $\left(\frac{\partial c}{\partial \mathrm{t}}\right)_i$ as :

$$\left(\frac{\partial c}{\partial \mathrm{t}}\right)_i = \frac{c_i^{k+1}-c_i^k}{\Delta t} \tag{30}$$

one obtains :

$$\left[q_{i-\frac{1}{2}}\frac{\Delta z + 2\alpha\Delta t}{2\Delta z} - D_{i-\frac{1}{2}}\left(\frac{1}{2\Delta z}+\frac{\alpha\Delta t}{2(\Delta z)^2}\right)\left(\theta_i - \theta_{i-1}\right)\right.$$

$$-D'_{i-\frac{1}{2}}\left(\frac{1}{6}\theta_{i-1}+\frac{1}{3}\theta_i\right) - \frac{\alpha\Delta t}{2\Delta z}D_{i-\frac{1}{2}}\left(\theta_{i-1}-\theta_i\right) - \frac{\alpha\Delta t}{\Delta z}\cdot$$

$$\left.D_{i-\frac{1}{2}}\theta_{i-\frac{1}{2}}\right]c_{i-1}^{k+1} + \left[\frac{\Delta z - 2\alpha\Delta t}{2\Delta z}q_{i+\frac{1}{2}} - \frac{\Delta z + 2\alpha\Delta t}{2\Delta z}q_{i-\frac{1}{2}} +\right.$$

$$D_{i-\frac{1}{2}}\frac{\Delta z + \alpha\Delta t}{2\Delta z}\frac{\theta_i - \theta_{i-1}}{\Delta z} + D_{i+\frac{1}{2}}\frac{\alpha\Delta t - \Delta z}{2\Delta z}\frac{\theta_{i+1}-\theta_i}{\Delta z} +$$

$$\cdot D'_{i-\frac{1}{2}}\left(\frac{1}{6}\theta_{i-1}+\frac{1}{3}\theta_i\right) - D'_{i+\frac{1}{2}}\left(\frac{1}{3}\theta_i + \frac{1}{6}\theta_{i+1}\right) + \frac{\alpha\Delta t}{2\Delta z}\cdot$$

$$D'_{i-\frac{1}{2}}\left(\theta_{i-1}+\theta_i\right) + \frac{\alpha\Delta t}{2\Delta z}D'_{i+\frac{1}{2}}\left(\theta_i + \theta_{i+1}\right) + \frac{\alpha\Delta t}{\Delta z}\left(\theta_{i-\frac{1}{2}}\right.$$

$$\left.\cdot D_{i-\frac{1}{2}} + \theta_{i+\frac{1}{2}}D_{i+\frac{1}{2}}\right) - \frac{1}{\Delta t}\left(\frac{1}{6}\theta_{i-1}+\frac{1}{6}\theta_{i+1}+\frac{2}{3}\theta_i\right) +$$

$$\left.\frac{\alpha}{2}\left(\theta_{i-1}-\theta_{i+1}\right)\right]c_i^{k+1} + \left[q_{i+\frac{1}{2}}\left(\frac{\alpha\Delta t}{\Delta z}-\frac{1}{2}\right)+\frac{1}{2\Delta z}D_{i+\frac{1}{2}}\right.$$

$$\cdot\left(\theta_{i+1}-\theta_i\right) - \frac{\alpha\Delta t}{2(\Delta z)^2}D_{i+\frac{1}{2}}\theta_{i+\frac{1}{2}} - \frac{\alpha\Delta t}{2(\Delta z)^2}D_{i+\frac{1}{2}}\left(\theta_{i+1}-\right.$$

$$\left.\left.\theta_i\right) + D'_{i+\frac{1}{2}}\left(\frac{1}{3}\theta_i + \frac{1}{6}\theta_{i+1}\right) - \frac{\alpha\Delta t}{2\Delta z}D'_{i+\frac{1}{2}}\left(\theta_i + \theta_{i+1}\right)\right]c_{i+1}^{k+1}$$

$$= \left[-\frac{1}{6}\theta_{i-1} - \frac{1}{6}\theta_{i+1} - \frac{2}{3}\theta_i - \frac{1}{2}a\left(\theta_i - \theta_{i+1}\right)\Delta t\right]\frac{1}{\Delta t}c_i^k \tag{31}$$

Similar equation to Equation(31) for every node 'i' can be derived and a linear algebraic set of equations obtained . Giveing the α a proper value, then one can makes the coefficent matrix of the set of equations become diagonal. Combining the equations with water flow problem (Zhu Xueyu1994),yeilds the value of c.

3 CASE ANALYSIS

3.1 Case survey

The case is a field experiment and was conducted by Wrrick et al . 1971 . The soil is calssified as Panoche clay loam, a deep alluvial loam soil of the Cenral valley of California , which initialy had an average moisture of $0.20\,cm^3/cm^3$, was wetted with 7.62 cm of 0.2N $CaCl_2$, followed immediately by 22.9 cm of solute--free water . The total infiltration occurred in17.5hours. Tensiometers and solute samplers had previously been installed (Figure 1).

3.2 Mathematics models and parameters

The case can be simplifilied as one-dimensional problem .The mathematics models are:

The water flow model :

$$\frac{\partial\theta}{\partial t} = \frac{\partial}{\partial z}\left[D(\theta)\frac{\partial\theta}{\partial z}\right] - \frac{dK(\theta)}{dz}\frac{\partial\theta}{\partial z}$$

$$\theta(t,0) = 0.38 \qquad\qquad z>0$$

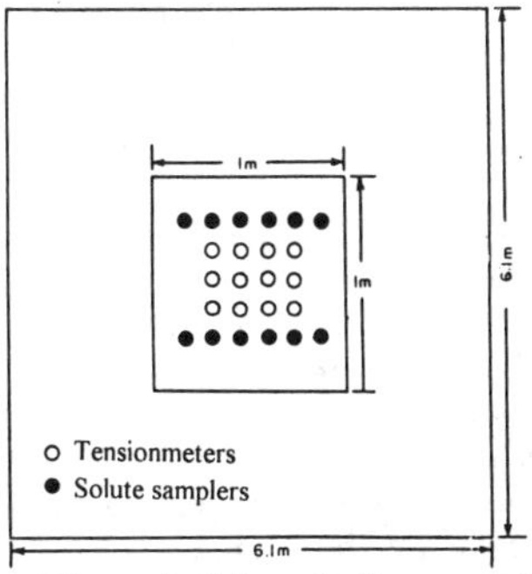

Figure 1. Schmatic diagram of field site showing relative position of tensionmeters and samplers

Table 1. Parameters

parameter	value
$K(\theta)$	$4.67\times10^{-5}e^{35.8\theta}$
$D(\theta)$	$4.42\times10^{-4}e^{25.3\theta}$ $\theta\le0.35$
$D(\theta)$	$8.662e^{0.5928\theta}$ $\theta>0.35$
D	$0.8\|v\|+0.6$

$$\theta(0,z)=\begin{cases}0.15+0.05z/60, & z\le 60cm\\ 0.2 & z>60cm\end{cases}$$

$$\theta(t,d)=0 \qquad d=200cm$$

where $D(\theta)$ is disffusivity in unsaturated flow, and $K(\theta)$ unsaturated hydraulic conductivity.

The $CaCl_2$ transport model :

$$\frac{\partial(\theta c)}{\partial t}=\frac{\partial}{\partial z}\left(\theta D\frac{\partial c}{\partial z}\right)-\frac{\partial(qc)}{\partial z}$$

$$c(0,z)=0 \qquad z>0$$

$$c(t,0)=\begin{cases}209meq/l & t\le168\min\\ 0 & t>168\min\end{cases}$$

$$c(t,d)=0 \qquad d=200cm$$

Some necessary parameters are listed in table1.

3.3 Simulation procedures

1. Discretization of elements

In this case,the scope from ground to the depth of 200cm are chosen as calculated area . Space step Δz=2cm,100 elements and 101 nodes are discreted, consequently.

2. Discretization of time

The discretization of time step has two choices:

1 equial step at constant,

2 changeable step.

The former choice and Δt=2 min are adopted in this case.

3. Solution of water flow problem

As we know , before contaminant transport problem simulated , water flow problem must be solved to get the values of θ, q and v ,which the former needs.The procedure has been conducted by Zhu Xueyu(1994).

4. Solution of pollutant transport problem

By means of the forgoing SUPG formulation , the authors make a computer program in FORTRAN--77 and get the 'c'value at different time (Figure 2).

5 Calculated results and discussion

The SUPG F.E.M. calculated results,the"precise solution"(Van Genuchten,1982), the Characterrstic--F.E.M. results(Huang Kangle,1988) and the field measured values are compared in Figure2.

From the figure we can find that the calculated results have a good agreement with the field measured values,and approach to those of other methods.

4 CONCLUSIONS

In this paper, the finite element formulation of contamiant transport in unsaturated soils is developed by means of SUPG method. A case is analysed and calculed results are compared with those of other methods.It demonstrates that:

1 The new method can be applied effectively to simulate the pollutant transport in unsaturated soils;

2 It has high accurancy, and can avoid numerical difficulities;

3 During its formulation development, a fix coordinate system is adopted , that makes the program very simple.

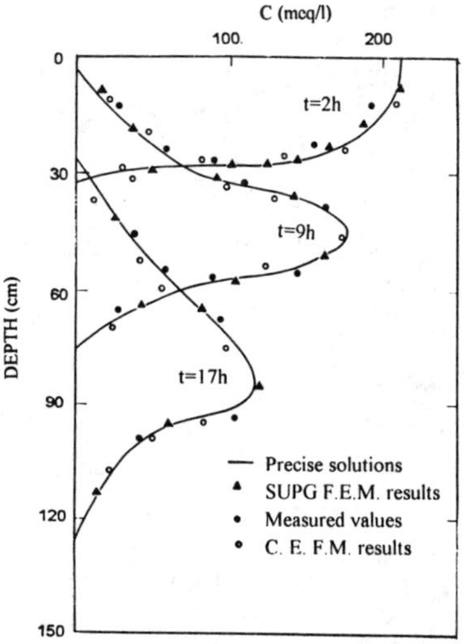

Figure 2. Distributions of chloride of 2,9and17 hours

REFERENCES

Huang Kangle 1988.The characteristic method combined with finite element scheme for solution of solute transport in unsaturated soils. China: *Journal of Wuhan Water and Electric Power University.* 1:88-96.

Ne-zheng Sun 1989 . *Mathematical modeling of groundwater pollution.* Beijing:Geology Press.

Zhu Xueyu & Xie Chunhong1994.Numerical solution of SUPG finite element method for unsaturated flow . *Chinese Journal of Hdraulic Engineeering .* 6:37-42.

Van Genuchten, M.T.1982 . A comparision of numerical solutions of the one-dimensional unsaturate flow and mass transport equations. *Advance in Water Resources.*12:47-55

Warrick,A.W.,Biggar,J.W.,and Nielsen D.R.1971. Simultaneous solute and water transfer for unsaturated soil. *Water Resources Research.* 7(5) : 1216-1225.

Computer Methods and Advances in Geomechanics, Yuan (ed.) © 1997 Balkema, Rotterdam, ISBN 90 5410 904 1

Predicted and observed build-up of water pressure following the installation of piezometers in a deep clay formation

V. Labiouse
Belgian Nuclear Research Centre (CEN·SCK), Mol, Belgium

C. Grégoire
Department of Civil Engineering of the Katholieke Universiteit Leuven (KUL), Belgium

ABSTRACT: The hydro-mechanical response of a deep clay formation during and after the installation of a piezometer is studied. A special attention is paid to the build-up of water pressure in the piezometer. One- and two-dimensional axisymmetric calculations are carried out with the FLAC code and take into account the several steps of the installation procedure. They lead to numerical predictions in good agreement with the in situ measurements and provide very useful information for the elaboration of future instrumentation programmes.

1. INTRODUCTION

As part of the Belgian R&D programme on the disposal of radioactive waste, the Belgian Nuclear Research Centre (CEN•SCK) has initiated studies to demonstrate the technical feasibility and the long-term safety of geological disposal of reprocessed High-Level radioactive Wastes (HLW). As an acceptable repository site is the Tertiary Boom Clay formation below the Mol/Dessel nuclear site, SCK•CEN has built an underground research facility (URF) at 223 m depth where in-situ experiments can be performed in representative disposal conditions (Bonne & al, 1992).

To get a better understanding of the hydro-mechanical behaviour of the clay around experiments performed from this laboratory, several borehole were drilled and instrumented. Among them, 10 single- and multi-piezometers were installed.

The paper is concerned with the numerical analysis of the clay mass response during and after the installation of a piezometer. It takes a special interest in the prediction of the build-up of water pressure in the piezometer and in its comparison with in situ measurements.

2. STATEMENT OF THE PROBLEM

To keep trace of the hydraulic disturbances induced by the construction of a gallery (called Test Drift), a multi-piezometer was installed in the Boom Clay formation. A borehole, 85 mm diameter and 21.5 m length, was first drilled from the underground laboratory. Then a 60 mm diameter stainless steel casing, equipped with five filters of 116 mm length, was introduced in the drill-hole.

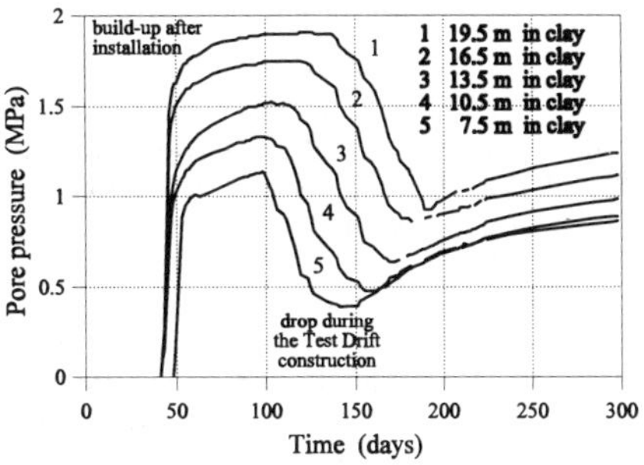

Fig. 1: Evolution of water pressure in the five filters of a multi-piezometer around the Test Drift.

Fig. 1. shows the evolution of water pressure in the five filters. They were first allowed to reach equilibrium (the difference that can be observed between the equilibrium values is due to the presence of already-existing underground openings). Then the excavation of the Test Drift was started and, as its front progresses, a significant drop in water pressures was recorded. This phenomenon is related to the strong hydro-mechanical coupling that exists in Boom Clay.

In this paper, we focus on the rise of water pressure that occurs in the piezometers after their installation. Predictions got from calculations with the FLAC finite difference code (1995) are compared with the observed build-up of pressure. For that purpose, the measurements taken by the farthermost filter in the medium (filter n°1) are chosen, because of the smallest influence of the existing excavations.

3. MODELLING APPROACH

The calculations are performed in the framework of the porous media theory (Coussy 1991), i.e. making due allowance for the deformabilities of the soil skeleton and of the pore fluid, and accounting for hydro-mechanical coupling. The set of data that has been chosen for modelling is summarized in table 1.

Table 1. Characteristics taken for the calculations.

Initial total stress	p_0	4400	kPa
Initial pore pressure	p_{p0}	2200	kPa
Borehole radius	r_i	4.25	cm
Young's modulus	E'	$3\ 10^5$	kPa
Poisson's coefficient	ν'	0.1	-
Cohesion	c'	0	kPa
Friction angle	φ'	21.8	°
Dilation angle	ψ	0	°
Permeability coefficient	k	$4\ 10^{-12}$	m/s
Porosity	n	0.39	-
Water bulk modulus	K_w	$2\ 10^6$	kPa

With regard to the in situ conditions of the Boom Clay formation at -223 m depth, realistic assumptions are made to simplify the problem: homogeneity and isotropy of the soil mass, as well as isotropic in situ stress field : total stress of 4400 kPa (overburden) and pore water pressure of 2200 kPa (water column).

The behaviour of the soil matrix (drained properties) is taken as linear-elastic perfectly-plastic with a Mohr-Coulomb strength criterion. The values of cohesion, friction angle and dilation angle correspond to critical state values. Indeed, the overconsolidation ratio of Boom Clay at -223 m is a little bit higher than 2, and the mean effective stress remains constant in the elastic zone around tunnels driven in axisymmetric conditions. Consequently, based on a critical state soil mechanics reasoning, the clay will plastify at a stress state close to the critical state.

To be representative of the complex reality, the modelling has to account for the several steps of the installation procedure (Grégoire 1996) : drilling of the borehole, installation of the piezometer, closure of the hole around the casing due to consolidation, saturation of the filter with pore water flowing out of the medium (36 days determined by in situ measurements) and finally increase of water pressure in the piezometer up to equilibrium.

Owing to the in situ conditions of the Boom Clay formation (continuous medium and nearly isotropic stress field), the analyses are mainly based on two methods: new analytical solutions in the framework of the "convergence-confinement" method (Labiouse & Bernier 1997), and numerical calculations with the FLAC code. The former one is only available for the immediate stability of the borehole and under small strain condition. On the other hand, the later is appropriate for short and long term stabilities (making allowances for the pore water pressure dissipation), and allows for both small and large strain analyses.

The calculations were first performed with an one-dimensional mesh and a small strain assumption. Several other runs were then carried out to assess the influence of some modelling characteristics like the size of the grid and the large strain mode. Finally, a two-dimensional configuration has been tried to assess whether it could lead to different predictions.

4. ONE-DIMENSIONAL CALCULATIONS

In this first configuration, the influence of the front of the borehole is neglected. Consequently, the problem can be studied in any cross section perpendicular to the axis of the borehole and a plane strain axisymmetric configuration is thus appropriate (Fig. 2).

The next subheadings describe the main modelling steps that were used to account for the successive stages of the piezometer installation.

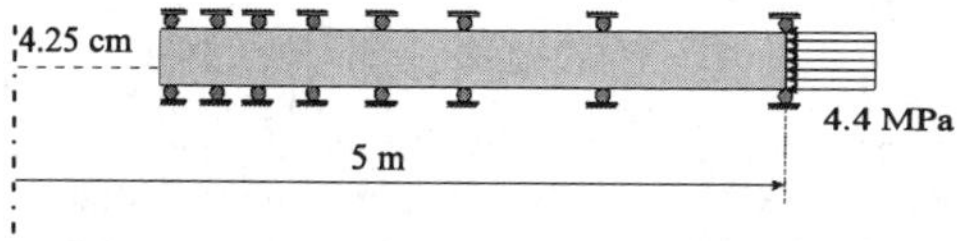

Fig. 2: One-dimensional configuration.

4.1 *Drilling of the borehole*

For the analysis of this first stage, a fast drilling rate is assumed, and accordingly the situation after the execution of the borehole can be idealized to an undrained one (no flow of water). The drilling is then simply simulated by removing the total pressure applied at the internal boundary of the mesh. Figure 3 represents the distribution of total stresses (radial σ_r, tangential σ_t and longitudinal σ_l), pore water pressure p_p and displacement u_r in the medium. It clearly illustrates the disturbed zone that develops in the surroundings of the borehole and the drop of pore water pressure in this zone.

The main results of this short term equilibrium state are summarized in table 2 : convergence u_{ri} and pore water pressure p_{pi} at the borehole wall as well as plastic zone extent R_{pe}. The values calculated with the FLAC code are in excellent agreement with the predictions got from the closed-form solutions.

Table 2. Main results of the short term equilibrium.

	u_{ri} [cm]	p_{pi} [kPa]	R_{pe} [cm]
FLAC	1.18	-1276	40
Analytic	1.18	-1309	40.44

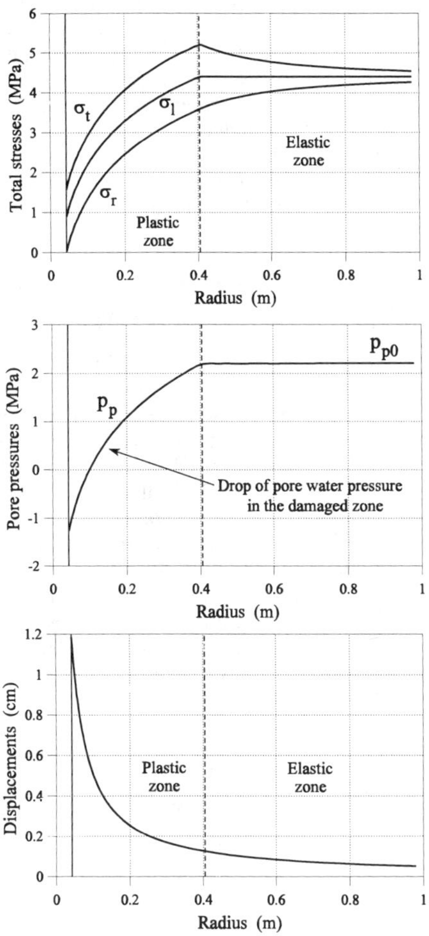

Fig. 3: State around the borehole after drilling.

4.2 *Saturation of the piezometer*

As the pore pressure distribution after the completion of the borehole is not uniform in the clay mass (see Fig. 3), fluid flow will occur, causing time-dependent displacements and the progressive closure of the hole around the piezometer. To account for the presence of the casing, the convergence undergone by the wall is fixed when it reaches the overgap value of 1.25 cm (difference between the radius of the drilling head and the radius of the piezometer tube).

A drained hydraulic boundary condition is imposed during the 36 days that follow the installation of the piezometer. Indeed, this duration corresponds to the time (measured in situ) that has been necessary to saturate the filter with pore water flowing out of the medium.

As the dimènsions of the filters are known, the mean in situ flow into the piezometer is estimated at 12.7 10^{-13} m^3/s. On the other hand, the mean flow of water calculated by FLAC during the 36 days of filter saturation is equal to 17.5 10^{-13} m^3/s. Considering the uncertainties of the geotechnical data and the simplifications made in the modelling, the agreement between both values is reasonably good.

The total pressure acting on the casing stabilizes at 829 kPa and induces a small recompression of the clay surrounding the borehole.

4.3 *Rise in water pressure*

When the saturation is completed, the filter is closed and the water pressure is allowed to increase up to equilibrium. To model this last stage of the "installation" procedure, we impose an impervious boundary condition at the inner radius of the mesh and we keep trace of the rise in water pressure at that location.

When performing the calculations with a 2 GPa pore water bulk modulus, two difficulties arose: a very long computation time and the development of numerical instabilities (oscillations in the results). In order to overcome these problems, the fluid bulk modulus was reduced by a factor of 20 (≡0.1 GPa). This solution has clearly the advantage to improve the speed and the stability of the calculations, but it has the drawback to slow down the water flux and the soil response during consolidation.

To account for this difference in consolidation speed, a time correction has to be applied. It can be estimated either by comparing the evolution of pore water pressure in both calculations, or by means of a simplified tool (Labiouse & Bernier 1996) based on Poisson equation (flow in a poro-elastic medium under axisymmetric condition). In the present calculations, the pore pressure changes computed for pure water (K_w = 2 GPa) were 1.8 as fast as their evolution with a 20 times more compressible fluid; and consequently the time correction factor was taken as 1.8.

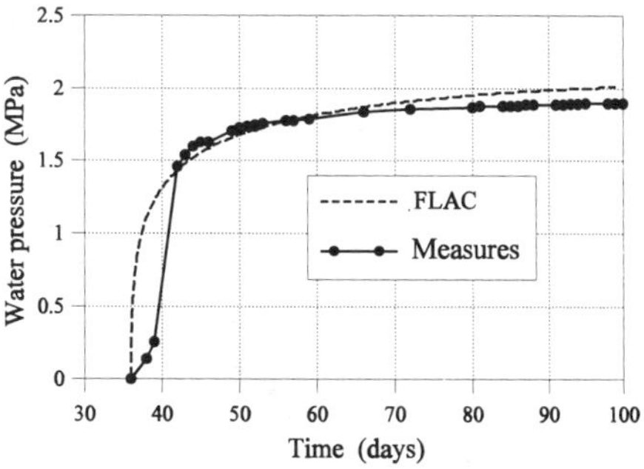

Fig. 4: Predicted and observed build-up of water pressure in the piezometer.

Figure 4 reports the build-up of water pressure that occurs in the filter after the 36 days saturation; the markers ● on the solid line represent the in situ measurements, while the dashed line is related to the numerical predictions. Their agreement is quite good, except at the beginning of the response. The slower increase in the measurements is probably due to the presence of some air remaining in the filter after its closure (≡ non-fully saturation).

On the other hand, a small discrepancy can also be noticed for the equilibrium water pressure: the results got from the numerical calculations are increasing continuously up to 2200 kPa (i.e. the imposed in situ pore water pressure), while the measures never reach 2000 kPa. This lower value could be explained, either by the hydraulic disturbance that exists in the vicinity of the existing excavations, or by an in situ pore water pressure lower than the water column pressure.

4.4 *Long term equilibrium*

The long term hydro-mechanical equilibrium is reached after about 200 days. Fig. 5 represents the distributions of total stresses around the borehole at that moment (dashed lines) as well as, for comparing purposes, the total stresses after drilling (solid lines).

During consolidation, the total pressure exerted by the clay mass on the piezometer tube is increasing up to 2350 kPa (among which 2200 kPa of pore water pressure). This significant rise in pressure on the casing induces a recompression of the surrounding clay. As a consequence, the disturbed zone that arise during the drilling of the borehole undergoes an elastic recompression and its extent (40 cm) doesn't grow during the consolidation stage.

4.5 *Complementary modelling works*

In Boom Clay, as in many soils, pore water contains some dissolved air and gas, which could significantly reduce its bulk modulus K_w (one to two orders of magnitude!). Consequently, to assess the influence of this factor on the response of the clay around the borehole, a parametrical study has been carried out. It has been established that :

1. For the short term equilibrium, a reduction of water bulk modulus induces a larger extent of the plastic zone, a smaller variation of pore pressures in this zone and larger convergences. However, this influence becomes only significant for values lower than 0.1 GPa.
2. As already mentioned, the higher the pore water bulk modulus, the faster the rate of pore pressure dissipation and the rise in water pressure in the filter.

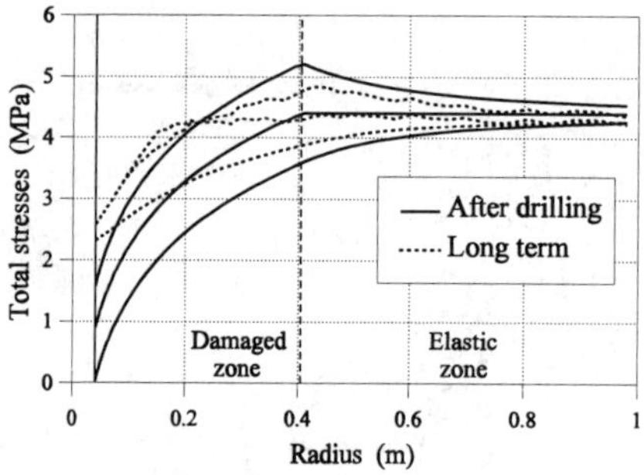

Fig. 5: Short and long term distributions of total stresses around the borehole.

To assess the sensitivity of the numerical results to the size and the shape of the zones taken in the mesh, two complementary runs were performed; the former with very thin elements near the borehole (grid n°2) and the later with an optimized aspect ratio (height to width) of the zones (grid n°3). With regard to the short term equilibrium state, the results are summarized in table 3, together with the already-presented results of the basic mesh (grid n°1) and of the closed-form solutions. As it could be expected, the more refined mesh (n°2) improves a little bit the accuracy near the borehole; but on the other hand, the best-shaped grid (n°3) leads to more stable and more accurate predictions during the consolidation stage.

Table 3. Influence of the grid size and shape.

	u_{ri} [cm]	p_{pi} [kPa]	R_{pe} [cm]
Grid n°1	1.18	-1276	40
Grid n°2	1.18	-1281	40
Grid n°3	1.16	-1215	39.3
Analytic	1.18	-1309	40.44

All the above-presented numerical results were taken from small strain analyses. To appraise the influence of this modelling assumption, a large strain run has been performed with the same set of data.

The short term equilibrium state that develops in the medium after the drilling of the borehole is reported in Fig. 6 for both small and large strain modes (respectively represented by the solid and dashed curves). One clearly observes that the disturbances predicted by the large strain analysis are less important: smaller extent of the plastic zone and smaller convergences. The main numerical results are given in table 4.

Table 4. Small and large strain calculations.

	u_{ri} [cm]	p_{pi} [kPa]	R_{pe} [cm]
Small strain	1.18	-1276	40
Large strain	0.85	-1269	31.3

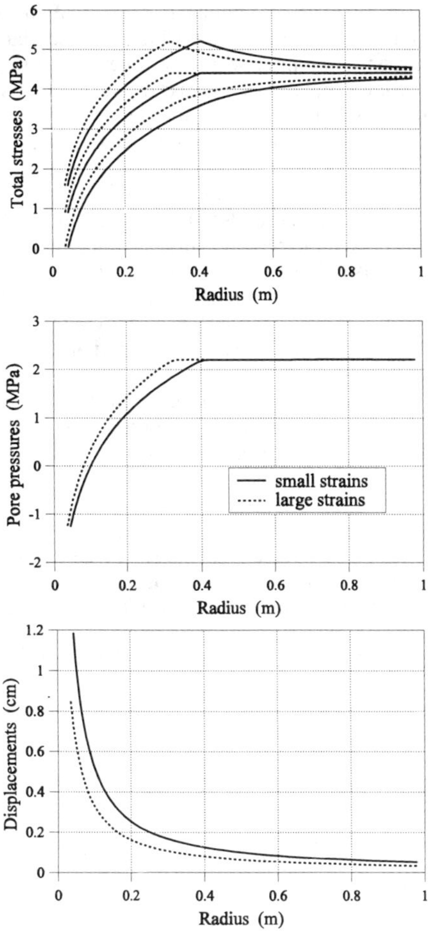

Fig. 6: Short term equilibrium state for small and large strain calculation modes.

During the consolidation stage that follows the installation of the piezometer, the differences between small and large strain results become less significant. In particular, it is established that the calculation mode is of small account on the predicted build-up of water pressure in the filter.

5. TWO-DIMENSIONAL CALCULATIONS

The one-dimensional calculations presented in the first part of the paper are based on the assumption that the three-dimensional problem can be replaced by the plane strain and axisymmetric study of a borehole section. In order to account for the real 3D-response of the clay, especially near the front of the hole, a two-dimensional axisymmetric calculations is performed.

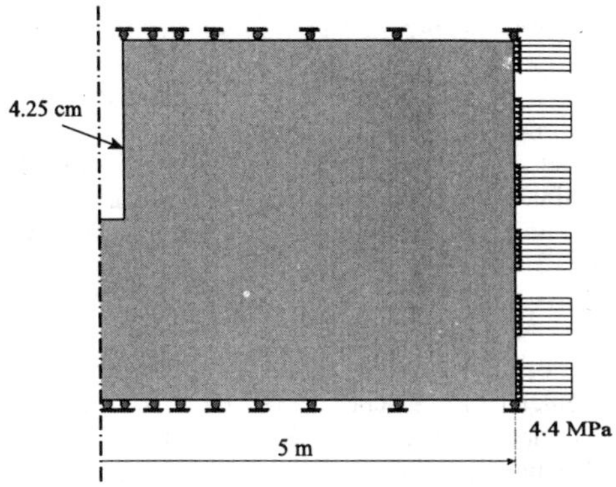

Fig. 7: Two-dimensional configuration.

The geometry (at the end of the run) and boundary conditions are schematically represented in Fig. 7.

The modelling steps account for the same sequence of piezometer installation as before: drilling of the borehole, installation of the instrument, closure of the hole around the casing, saturation of the filter and finally build-up of water pressure up to equilibrium.

5.1 *Drilling of the borehole*

The progressive drilling of the borehole is simulated by removing successively the elements located just ahead of the front. The equilibrium is achieved after each modelling step.

The filled contour plot of Fig. 8 represents the drop of pore water pressure arising from the mechanical deformations undergone by the clay mass during drilling. This hydraulic disturbance mainly occurs in the plastic zone that develops around the borehole (Bernier & al 1996). However, in the elastic zone ahead of the face, due to non-isochoric deformations, a small increase of pore water pressure can be noticed.

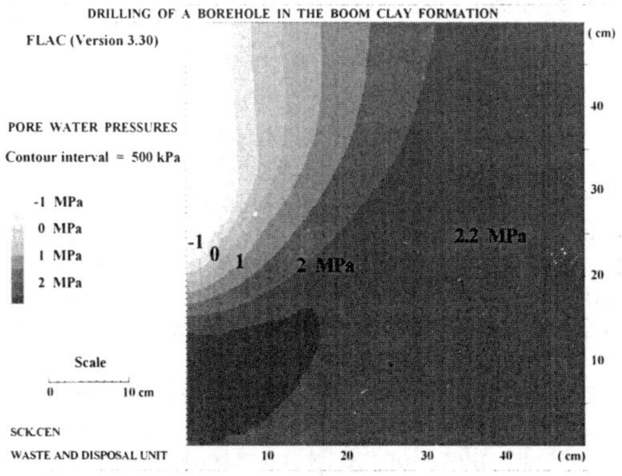

Fig. 8: Filled contours of pore water pressures.

5.2 *Pore water pressure dissipation*

The rise in pore water pressure calculated with the two-dimensional configuration can hardly be distinguished from the predictions made with the simplified one-dimensional mesh; and so it is for other results :

1. During the saturation of the filter, a mean flow of water of 16 10^{-13} m^3/s is calculated, i.e. within less than 10 % of the 17.5 10^{-13} m^3/s flow evaluated by the one-dimensional mesh.

2. The total pressure exerted by the clay mass on the piezometer tube reaches 2350 kPa at the end of the consolidation stage, as for the simplified configuration.

3. The distributions of total stresses and displacement at long term equilibrium are very similar to those found with the plane-strain axisymmetric grid.

From all these establishments, we come to the conclusion that the choice of a two-dimensional axisymmetric configuration is not relevant to analyse the response of the clay mass during and after the installation of a piezometer (except if a special attention has to be paid to some results near the front of the borehole). Plane strain axisymmetric calculations are sufficient and much less time consuming.

6. USEFUL IMPLICATIONS

From the performed numerical analyses, useful information can be inferred for the elaboration of future instrumentation programmes :

1. The extent of the disturbed zone that develops around the borehole should be taken into account for the location of the instruments, especially with regard to the interference that could occur between different measuring devices.

2. The time necessary to reach the long term hydro-mechanical equilibrium around the borehole provides an idea of how long in advance the instruments have to be installed before starting some experiment.

3. The total pressure acting on the casing at the end of the consolidation could be useful to improve the design of the measuring devices.

7. CONCLUSIONS

The present paper is concerned with the numerical analysis of the Boom Clay mass response during and after the installation of a piezometer. It takes a special interest in the prediction of the build-up of water pressure in the piezometer and in its comparison with in situ measurements.

The calculations are mainly carried out with the axisymmetric feature of the FLAC finite difference code. They allow for the hydro-mechanical coupling and take into account the several steps of the installation procedure: drilling of the borehole, installation of the piezometer, closure of the hole around the piezometer tube (due to consolidation), saturation of the filter with pore water flowing out of the medium, and finally increase of the water pressure in the piezometer up to equilibrium.

Both two-dimensional and one-dimensional (plane strain) axisymmetric configurations are used and their results are found to fit well with the in situ measured rise in water pressure. Moreover, it is established that the 2D axisymmetric calculations do not improve the predictions of the 1D axisymmetric model.

Complementary calculations have also been performed to assess the influence of some modelling characteristics like the size of the grid and the large strain mode.

Several information deduced from the performed study should lead to improvements in the design of future instrumentation programmes

Acknowledgements

The results presented in the paper are taken from an undergraduate thesis of the "Faculté Polytechnique de Mons" mainly carried out at the Waste and Disposal department of the Belgian Nuclear Research Centre. The second Author is grateful to both institutes.

REFERENCES

Bernier F., Labiouse V., Verstricht J. 1996. *Praclay project : Geotechnical modelling and instrumentation programme of the connecting gallery.* Report to NIRAS/ONDRAF, CEN•SCK, Mol, Belgium. In print.

Bonne A., Beckers H., Beaufays R., Buyens M., Coursier J., De Bruyn D., Fonteyne A., Genicot J., Lamy D., Meynendonckx P., Monsecour M., Neerdael B., Noynaert L., Voet M., Volckaert G. 1992. *The HADES demonstration and pilot project on radioactive waste disposal in a clay formation.* Final report to EC, EUR 13851.

Coussy O. 1991. *Mécanique des milieux poreux.* Editions Technip, Paris.

FLAC 1995. Manuals of the FLAC code. Itasca Consulting Group, Minneapolis, USA.

Grégoire C. 1996. *Comportement hydro-mécanique de l'argile de Boom autour d'un forage piézométrique.* Undergraduate thesis. Faculté Polytechnique de Mons, Belgium.

Labiouse V., Bernier F. 1996. *Influence of pore water compressibility on the short and long term stability of nuclear waste disposal galleries in clay.* Geomechanics'96, Roznov, Czech Republic. 6 pp.

Labiouse V., Bernier F. 1997. *Hydro-mechanical disturbances around excavations.* Feasibility and Acceptability of Nuclear Waste Disposal in the Boom Clay Formation, CEN•SCK, Mol, Belgium. pp. 15-26

Computer Methods and Advances in Geomechanics, Yuan (ed.) © 1997 Balkema, Rotterdam, ISBN 90 5410 904 1

Influence of compaction stresses on cracking of clayey liners

Jayantha Kodikara & Md. Rahman Fashiur
Department of Civil and Building Engineering, Victoria University of Technology, Melbourne, Vic., Australia

ABSTRACT: In engineered waste containment systems, compacted clayey liners are commonly used as hydraulic barriers which are expected to maintain a low hydraulic conductivity during their design life. Under some field conditions, the compacted clay structure can display forms of cracking which lead to large increases in hydraulic conductivity. Cracking can occur due to desiccation when the clayey liners undergo moisture loss due to drying. The clay structure can also flocculate, which has similar effects to cracking, when exposed to certain chemical liquids and leachates. The current paper presents a simplified theoretical analysis for the initiation and depth of cracking under these conditions. It is found that in addition to overburden stresses, the residual compaction stresses play a significant role in inhibiting crack initiation and growth.

1 INTRODUCTION

Compacted clayey liners are commonly used in engineered waste containment systems as hydraulic barriers to minimise the potential contamination of the geo-environment. The most important requirement for an effective clayey liner is to maintain a low hydraulic conductivity during its design life. However, previous research has indicated that the changes in clay structure which may take place after the compaction of the liner can lead to large increases in the hydraulic conductivity. One such situation is when the liner undergoes desiccation due to moisture loss. This could happen if the liner is left uncovered during construction or the cover material is not sufficiently thick to prevent moisture evaporation. Liners used as landfill caps are more susceptible to evaporative moisture loss because they generally have a relatively small amount of cover above them. Base liners could suffer desiccation if the temperature at the top of the liner rises above that of the bottom owing to high temperatures that could develop within some waste fills (Jessberger, 1995). The hydraulic conductivity of compacted clayey soils could also increase significantly due to changes in clay structure when exposed to some chemical liquids and leachates. A notable review on this topic has been published by Mitchell and Madsen (1987).

Currently, experimental methods are predominantly used to examine the likelihood of above clay structure changes. Reliable theoretical methods are not readily available mainly because of the complexity of these mechanisms and their dependence on many interacting variables. However, moisture losses during desiccation can be examined relatively accurately using unsaturated flow theory, provided reliable input parameters are available. But relating these moisture losses to the initiation and extent of cracking is not straightforward. Jessberger (1995) has examined some of the influencing variables on the basis of a theoretical investigation of desiccation cracking in clayey soils given by Morris et al. (1992). In the model presented by Morris et al, cracking occurs when the tensile horizontal stresses created by matric suction exceed the compressive stresses due to overburden pressures in excess of the tensile strength of the clayey soils. While this approach appears to be reasonable for cracking of natural or loosely dumped clayey soil deposits (e.g. mine tailings), for compacted clayey soils the residual stresses generated during the compaction process should also have some inhibitive effect on the crack formation. Previous investigations have shown that significant residual horizontal stresses can develop when soils are compacted adjacent to rigid retaining walls (e.g. Clayton and Symons, 1992; Filz and Duncan, 1992). Although direct experimental evidence is not readily available, it can be expected that substantial residual stresses would also develop during clayey liner compaction and these need to be accounted for in theoretical stress analyses.

In the current paper, simplified theoretical models for structural changes due to desiccation and chemical effects are developed. Residual compaction stresses are modelled using the semi-analytical method developed by Duncan and Seed (1986) and subsequently modified by Filz and Duncan (1992). The model presented by Morris et al (1992) is extended to account for residual compaction stresses and is then used to examine desiccation cracking. An attempt was also made to use the same model to predict the likelihood of structural changes in clay due to chemicals using the double layer theories of clay chemistry.

2 MODELLING OF RESIDUAL COMPACTION STRESSES

2.1 *Modelling with continuum mechanics*

An attempt was made to model the compaction process using continuum mechanics. FLAC (Fast Lagrangian Algorithm of Continuua, Itasca, 1993) an explicit finite difference code, was used for the modelling. A one dimensional soil column with lateral restraints (i.e. uniaxial *or* K_0 conditions) was subjected to monotonically increasing strains applied vertically downwards at the top surface, representing the application of external compaction loading. These boundary conditions are generally considered applicable to soil compaction in the laboratory and are assumed to approximate the field condition, too. Subsequently, the direction of vertical strains was reversed and the soil column was relieved from external compaction loading. The stress paths followed by vertical and horizontal stresses were monitored during this simulation.

Firstly, a conventional elasto-plastic model with Mohr-Coulomb shear failure criterion (i.e. $c-\phi$ soil) was used. The initial principal stress system included gravity vertical stresses (σ_v) and horizontal stresses (σ_h) given by $K_0\sigma_v$ where K_0 is given by well known expression of $(1-\sin\phi)$. The model showed that initially, the horizontal and vertical stresses increase elastically so that their incremental ratio ($\Delta\sigma_h/\Delta\sigma_x$) is given by: $K_e=\nu/(1-\nu)$ where ν is Poisson's ratio of the soil. If the coefficient of active earth pressure K_a (=$(1-\sin\phi)/(1+\sin\phi)$) is less than K_e, then shear failure would occur and the incremental ratio of horizontal to vertical stresses would be maintained at K_a. If K_a is greater than K_e, then the soil will behave elastically indefinitely. The unloading of the model would take place with the stress gradient given by K_e. The residual stresses resulting from these stress paths do not conform to experimental evidence of K_o loading, and this clearly indicates the inadequacy of the Mohr-Coulomb elasto-plastic model for simulation of the compaction behaviour. In particular, firstly, it cannot maintain the loading along the K_0 line and, secondly, active shear failure develops even without any lateral straining.

Subsequently, a double-yield (cap) model allowing for plastic volumetric strains under isotropic stresses (or irreversible compaction strains) was utilised in the modelling. This also failed to establish K_0 loading and the loading path changed from elastic to volumetric failure and then to shear failure. Based on this simplistic study, which ignores the pore pressure development in unsaturated soils, it is apparent that these continuum models are not capable of modelling the compaction process effectively. After all, the original Jaky expression for K_0 was developed using particulate mechanics and, therefore, it is not surprising that continuum models were unable to capture that behaviour. It would be interesting to test newly emerged particle flow codes to examine uniaxial loading and unloading behaviour without lateral straining.

2.2 *Modelling with semi -analytical model of Filz and Duncan (1992).*

Based on the experimental evidence of K_0 loading, Duncan and Seed (1986) developed a model to compute residual horizontal stresses adjacent to a rigid wall subsequent to compaction of granular soils in multiple lifts. This model assumes that a transient vertical stress due to compaction loading (e.g. roller weight) is applied on the soil and removed as the roller moves away from the wall. The model simulates this by K_0 (= $1-\sin\phi$) loading of the transient vertical stress and subsequent unloading based on a non-linear path for removal of the transient stress. This results in a residual horizontal stress. Placement and compaction of subsequent lifts of soils are also modelled using hysteretic loading paths. This model deals with cohesionless soils and no pore pressure generation is taken into account. Filz and Duncan (1992) have extended this model to partially saturated cohesive soils where pore pressure generation is considered by an application of Skempton's *A* and *B* parameters. The model has been implemented in a computer program EPCOMPAC, and has been validated against several field and laboratory tests. However, Ishihara and Duncan (1993) pointed out that the model needs further refinement and validation, particularly in the area of pore pressure generation. Despite that, this model appears to be the only currently available for modelling compaction stresses in cohesive soils, and hence is used in the current study.

The EPCOMPAC program was used to simulate the compaction of a 1.5m thick clayey liner in 10 equal lifts. It was assumed that the compaction stresses generated would be equivalent to those adjacent to a rigid retaining wall, and that K_0 conditions apply. This seems reasonable because the stresses were computed at the centre of the layers along the centre line of the rollers. Parametric conditions used in these simulations are given in Table 1. A medium size roller and a heavy roller, representing standard and modified compaction efforts respectively, were used in the simulation. The total stresses computed for these two conditions are shown in Table 2. It can be seen that substantial residual horizontal stresses can develop within the liner depending on the compaction effort used. Clayton and Symons (1992) have noted that residual horizontal stresses in compacted cohesive backfill adjacent to rigid walls have been measured to be in the range of 0.4 c_u to 0.8c_u for highly plastic soils where c_u is the undrained shear strength of the soil. If the undrained shear strength of compacted clay with moisture content a few percentage points wetter than optimum is assumed to be about 100 kPa, the above range will give residual stresses in the range 40 to 80 kPa. On this basis, it appears that the computed residual stresses are in the right order.

3 INFLUENCE OF RESIDUAL COMPACTION STRESSES ON CRACKING

In order to appreciate the influence of residual compaction stresses on cracking of clay liners, a simplified numerical study was undertaken using FLAC program. A 1.5 m thick compacted clayey liner with parametric conditions and residual stresses as in Section 2 were used in the study. It was assumed that cracks would develop at 1 m spacing. within the clayey liner Hence, as shown in Figure 1, a block of soil 0.5 m long and 1.5 m high was considered with the boundary conditions shown and with the initial stress conditions and parameters given in Section 2.

Table 1. Parameters used in compaction modelling

Parameter	Roller 1	Roller 2
Roller load (kg)	20,000	32,400
Liner thickness (m)	1.5	1.5
Lift thickness (mm)	150	150
K_0	0.5	0.5
ϕ' (degrees)	30	30
c'(kPa)	57	57
ϕ (degrees)	22	22
c (kPa)	65	65
A	0.5	0.5
*E (MPa)	35	35
Unit weight, γ (kN/m^3)	20	20

**E is used in FLAC modelling only.*

Table 2. Residual total stresses after compaction of a clayey liner of 1.5 m thickness

Depth below surface (mm)	Roller 1		Roller 2	
	Total Stress (kPa)	Pore pressure (kPa)	Total Stress (kPa)	Pore pressure (kPa)
75	47.3	-15.6	56.9	-14.4
225	51.1	-14.8	61.3	-13.5
375	54.1	-14.0	64.9	-12.6
525	56.7	-13.3	67.9	-11.9
675	58.9	-12.6	70.6	-11.2
825	60.8	-12.0	72.8	-10.5
975	62.6	-11.4	74.8	-9.9
1125	64.2	-10.9	76.7	-9.3
1275	65.7	-10.3	78.3	-8.7
1425	67.2	-9.7	79.9	-8.1

An ideal elasto-plastic model with Mohr-Coulomb criterion was assumed for the soil constitutive behaviour. Then, the lateral restraining on the right hand vertical face was removed to simulate the formation of a crack. The results indicated that the block deformed towards the right side with the release of residual horizontal stresses. Several computations were conducted with different levels of overburden pressures applied on the liner surface. Figure 1 shows the lateral deformation of the right hand top corner of the block plotted against the overburden pressure. It is clear that the block behaves elastically up to about 160 kPa, and then shear failure takes place causing large deformations. Similar behaviour was observed for the situation with residual stresses generated by using the heavy roller, although the deformations in the elastic range were higher with proportional increases to residual stresses.

This study indicates that the residual horizontal stresses generated by the compaction process have the capacity to inhibit the formation of cracks in the

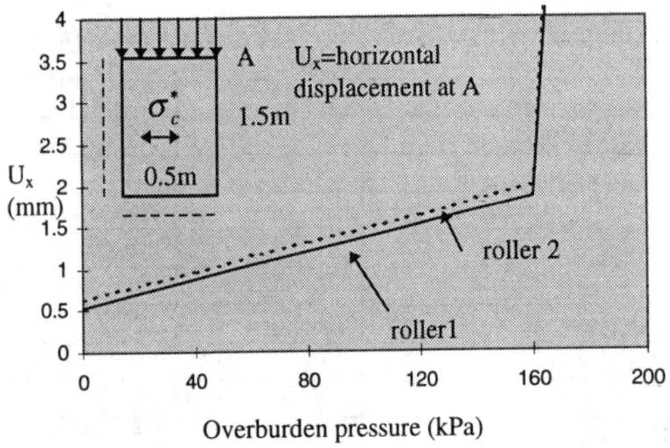

Figure 1. Results of FLAC deformation analysis of a compacted liner section

clayey liner, and in these examples, they can accommodate displacements up to about 1.5 mm before a crack is formed. As expected, the overburden stresses also have the same effect, and if they are high enough, shear failure can occur in the liner closing the cracks. Furthermore, the study indicates that it is possible to approximate the liner behaviour as purely elastic for low overburden pressures and purely plastic for higher pressures. It follows then that these factors play significant roles in clayey liner cracking.

4 PREDICTION OF CRACK FORMATION FOR CLAYEY LINERS

4.1 *Desiccation cracking*

Morris et al. (1992) have shown that, in an elastic soil with matrix suction $(u_a - u_w)$, the horizontal strain ε_x resulting from internal stresses σ_x, σ_y and σ_z is given by:

$$\varepsilon_x = \frac{\sigma_x - u_a}{E} - \frac{\nu}{E}\left(\sigma_y + \sigma_z - 2u_a\right) + \frac{\left(u_a - u_w\right)}{H} \qquad (1)$$

where u_a and u_w are the air and water pressures in the soil and E is a compression modulus for changes in net total stress relative to the pore air pressure $(\sigma - u_a)$. H is an elastic modulus for changes in matrix suction $(u_a - u_w)$, and ν is Poisson's ratio. Here, x and y directions are horizontal directions and z is the direction vertically down. Because the stress changes take place in the horizontal plane, cracks are considered to form in the vertical direction, and the x direction is assumed to be perpendicular to the cracks, whereas the y direction is parallel to the cracks.

Since lateral straining is not allowed before cracking, and assuming $\sigma_y = \sigma_x$, Equation (1) can be re-arranged as:

$$\sigma_x - u_a = \frac{\nu}{(1-\nu)}\left(\sigma_z - u_a\right) - \frac{E}{H}\frac{\left(u_a - u_w\right)}{(1-\nu)} \qquad (2)$$

For a compacted clayey liner subjected to a residual horizontal compaction stress of σ_c^*, and a vertical overburden pressure of p_O , Equation (2) may be modified as:

$$\sigma_x - u_a = \sigma_c^* + \frac{\nu}{(1-\nu)}p_0 + \frac{\nu}{(1-\nu)}\left(\sigma_z - u_a\right) - \frac{E}{H}\frac{\left(u_a - u_w\right)}{(1-\nu)} \qquad (3)$$

Morris et al. (1992) have shown that it is reasonable to use $E/H = (1-2\nu)$. It is assumed that cracking initiates when the net horizontal stress $(\sigma_x - u_a)$ falls below the tensile strength of the soil. Substituting these conditions and using $\sigma_z = \gamma z$, where γ is the unit weight of liner soil and z is the depth measured from top of the liner surface and $u_a = 0$, an expression for depth of cracking (z_c) can be established as:

$$z_c = \frac{(1-2\nu)}{\nu\gamma}S_0 + \left[t - \sigma_c^* - \frac{\nu}{(1-\nu)}p_o\right]\frac{(1-\nu)}{\nu\gamma} \qquad (4)$$

where S_O is the average matric suction within the clayey liner. It should be noted that here a constant suction value was assumed to exist within the liner. However, this will produce conservative results with greater crack depths if a significant suction gradient exists, for example due to a water table in the ground below. As demonstrated by Morris et al., Equation 4 can be easily modified for such a situation. The tensile strength of clayey soils can increase with suction. In the current paper, as derived by Morris et al, tensile strength t is assumed to be given by $-(u_a - u_w)\tan(\phi' - 5)$ where $\phi'$is in degrees, ignoring the tensile strength at saturation.

Equation 4 is applied to a parametric evaluation of cracking of clayey liners, and the results are presented in Table 3. Cracking depths were determined for two Poisson's ratio values (0.3 and 0.4), two overburden pressures (0 and 20 kPa) and three values of residual compaction stresses (0,50,70 kPa) under a range of suction values (20 to 500 kPa). S^* gives the suction values required to initiate cracking.

Table 3. Results of desiccation cracking calculations

Suction (kPa) (pF)	$\nu = 0.3$						$\nu = 0.4$					
	$p_0 = 0$			$p_0 = 20$			$p_0 = 0$			$p_0 = 20$		
	$\sigma_c^* =$ 0	$\sigma_c^* =$ 50	$\sigma_c^* =$ 70	$\sigma_c^* =$ 0	$\sigma_c^* =$ 50	$\sigma_c^* =$ 70	$\sigma_c^* =$ 0	$\sigma_c^* =$ 50	$\sigma_c^* =$ 70	$\sigma_c^* =$ 0	$\sigma_c^* =$ 50	$\sigma_c^* =$ 70
20 (2.30)	0.8	nc	nc	nc	nc	nc	0.15	nc	nc	nc	nc	nc
40 (2.60)	1.6	nc	nc	0.6	nc	nc	0.3	nc	nc	nc	nc	nc
60 (2.78)	2.4	nc	nc	1.4	nc	nc	0.45	nc	nc	nc	nc	nc
80 (2.90)	3.2	nc	nc	2.2	nc	nc	0.6	nc	nc	nc	nc	nc
100 (3.00)	3.9	nc	nc	2.9	nc	nc	0.75	nc	nc	nc	nc	nc
200 (3.30)	7.9	2.1	nc	6.9	1.1	nc	1.5	nc	nc	0.5	nc	nc
500 (3.70)	19.7	13.9	11.6	18.7	12.9	10.6	3.8	0.01	nc	2.8	nc	nc
S^* (kPa)	>0	147	207	>0	173	232	>0	500	700	135	632	830

Notes:
nc - no cracking
S^* *- the suction required to initiate cracking*
all the pressures and stresses are in kPa.

The results indicate that, for compacted clay under zero overburden pressure, cracking would develop even at very low suctions (high saturations) if the residual compaction stresses are ignored (see $\sigma_c^* = 0$ and p_0=0 results). This result is contrary to the generally observed behaviour of clayey soils where cracking is not normally observed at these suctions or saturation levels. The suction necessary to create cracking, however, increases markedly when residual compaction stresses are taken into account. For example, when Poisson's ratio is 0.4, and under zero overburden stress, cracking would not initiate until suction reaches 500 kPa and 700 kPa respectively for residual compaction stresses of 50 and 70 kPa. For Poisson's ratio of 0.3, cracking would initiate at lower suction levels. These above calculations show that higher compaction effort tends to produce more resistance to cracking. It is also clear that by increasing overburden stress to 20 kPa (approx. 1m of soil), cracking can be minimised. It should be noted that covering with additional soil cover will be of more advantage in minimising the moisture loss and, therefore, the generation of high suction levels.

The preceding analysis is a simplified account of the influence of residual compaction stresses on desiccation cracking, but nevertheless highlights their importance. The depths of cracks in Table 3 were computed directly from Equation 4, and may not be applicable to a clayey liner of limited thickness. Furthermore, if ground water is present near the surface, then that would also control the suction profile and the cracking depth. The cracking depths may also be influenced by stress concentrations at crack tips as well as shear failure of soils (Morris et al. 1992).

4.2 *Clay structure changes due to chemicals*

It is believed that some chemical liquids reduce the clay double layer thickness and this causes flocculation and increases hydraulic conductivity. The diffuse double layer theory based on Gouy-Chapman equation is commonly used to explain these effects. Based on this equation, the physico-chemical properties of leachate which can lead to these changes include high electrolyte concentration, low dielectric constant and high cation valence. The effect of these properties is to bring the clay particles closer together and, therefore, it can be argued that this effect is similar to that of shrinkage of soils due to desiccation. On this basis, it may be possible to establish a relationship between the stress system within the compacted clay and the physico-chemical properties of the leachate at the onset of these clay structural changes.

In the current paper, only the effect of high electrolyte concentration of the leachate is considered in some detail. In this situation, the liner is most likely to be flooded with leachate, and, it is reasonable to assume that the clay is saturated. Therefore, the effective stresses are used to compute the lateral strain which can be given as:

$$\varepsilon_x = \frac{\sigma'_x}{E}(1-\nu) - \frac{\sigma'^*_c}{E}(1-\nu) - \frac{\nu}{E}\sigma'_z + \frac{S}{H} - \frac{u}{H} \qquad (5)$$

S is solute suction generated by electrolytes in the leachate, and u is the gravity water pressure. Here, the effective residual compaction stress σ'^*_c is conveniently approximated by total stress σ^*_c. This conservative assumption is not unreasonable because the pore pressure generated after compaction is not apparently very significant (see Table 2). Imposing these conditions and using $\varepsilon_x = 0$ as well as the $\sigma'_x = t$ condition, Equation 5 can be rearranged to give the solute suction S at the onset of clay structural changes as:

$$S = \frac{\nu}{(1-2\nu)}(p'_0 + \gamma' z) + \frac{(1-\nu)}{(1-2\nu)}(\sigma^*_c - t) + u \qquad (6)$$

where z is the depth measured from the top of the liner and γ' is the submerged unit weight of soil.

Table 4 shows a parametric study undertaken using Equation 6, in which the suctions required to initiate structural changes at the top (*z=0*) of a clayey liner were computed for a range of conditions assuming that $t = -10$ kPa and *u*=0. It was deemed interesting to assess the likely electrolyte concentrations which would cause these suctions. For this purpose, an equation presented by Mitchell (1976) is used: namely $S = RTC$, where R is the gas constant (8.314 J/K/mol); T is absolute temperature (K); and C is the difference in the cation concentration between leachate and the mid planes of the double layers of clay particles. The results shown in Table 4 were computed for NaCl leachates at 25^0 C. They again highlights the importance of compaction stresses in providing resistance to the effects of structural changes. The results also indicate that liners seem to be unaffected by leachate unless NaCl concentrations exceed about 4200 mg/L for the conditions assumed.

5. CONCLUDING REMARKS

It is important to account for residual compaction stresses in the assessment of likely structural changes in clayey liners due to either desiccation cracking or

Table 4. Parametric results for chemical effects

p'_0	$\sigma^*_c = 0, t = -10$ ν = 0.4		$\sigma^*_c = 50, t = -10$ ν = 0.4		$\sigma^*_c = 70, t = -10$ ν = 0.4	
	S	C	S	C	S	C
0	30	708	180	4245	240	5660
20	70	1651	220	5189	280	6604
40	110	2594	260	6132	320	7547
60	150	3538	300	7075	360	8491
80	190	4481	340	8019	400	9434
100	430	5425	380	8962	440	10377

Notes:
All the pressures and stresses are in kPa. Electrolyte concentration is in mg/L.

chemical effects. Equations were derived to predict the conditions required for these changes to occur, and some parametric results were presented. Although these equations are considered to represent the likely mechanisms of liner behaviour, they are based on a number of simplifying assumptions, specially in relation to chemical effects. Therefore, while these equation may be useful in examining the relative effects of various influencing variables, further research is need to examine the validity of these assumptions in more detail.

ACKNOWLEDGMENT

The financial support provided to the second author by EPA Victoria and the Victorian Education Foundation is gratefully acknowledged. The authors wish to thank Prof. J.M. Duncan for providing the EPCOMPAC program. The comments made by Mr. Don Jordan were also helpful.

REFERENCES

Clayton, C.R.I. and Symons, I.F. (1992). The pressure of compacted fill on retaining walls, Geotechnique, Vol. 42, pp.127-130.

Duncan, J.M. and Seed, R.B. (1986). Compaction-induced earth pressures under K_0 -conditions, J. of Geotech. Engg. Division, ASCE, 117(2), pp. 1833-1847.

Filz, G.M. and Duncan, J.M. (1992). An experimental and analytical study of earth loads on rigid retaining walls, Research Report, Virginia Polytechnic Institute and State University, USA.

Ishihara, K. and Duncan, J.M. (1993). At rest and compaction-induced earth pressures of moist soils, Research Report, Virginia Polytechnic Institute and State University, USA.

Jessberger, H.L. (1995). Waste containment with compacted clay liners, Geoenvironment 2000 edited by Acar, Y.B. and Daniel, D.E., Geotech. Special Publication No. 46, ASCE, Vol. 1, pp. 463-483.

Mitchell, J. K., (1976). Fundamentals of Soil Behaviour, John Wiley & Sons, p. 422.

Mitchell, J. K., & Madsen, F. T., (1987). Chemical effects on clay hydraulic conductivity, Geotech. Practice for Waste Disposal ' 87, GT Div., ASCE, Geotech. Special Publication 13 , pp. 87-116.

Morris, P.H., Graham J. and Williams, D.J. (1992). Cracking in drying soils. Canadian Geotech. Journal, Vol. 29, pp. 263-277.

Computer Methods and Advances in Geomechanics, Yuan (ed.) © 1997 Balkema, Rotterdam, ISBN 90 5410 904 1

Modelling of coal ash leaching

Gavin Mudd & Jayantha Kodikara
Department of Civil and Building Engineering, Victoria University of Technology, Melbourne, Vic., Australia

ABSTRACT: The safe disposal of coal ash produced from power stations involves the assessment of rates of production and quality of ash leachate. Laboratory column experiments have been used to obtain data on the ash leachate. The extrapolation of this data to the field conditions requires theoretical modelling of the underlying chemical leaching and transport mechanisms. In the current paper, a model was developed to simulate the leaching of ash which exists in saturated or near-saturated conditions. The model was applied to simulate the experimental results, and good agreement was obtained between the experimental and model results.

1 INTRODUCTION

In the Latrobe Valley region of Victoria Australia, large quantities of brown coal ash are produced annually in the power station complexes which supply the majority of the electricity requirement for Victoria. One of the principal environmental concerns associated with the management of these facilities is the hydrogeological impact of the disposal of coal ash. Currently, all coal ash produced by the power stations is slurried to on-site disposal ponds. Since the capacity of the ash ponds is limited, options for the future management of the ash ponds are considered, and the feasibility of depositing of aged ash in a nearby overburden dump is investigated. In the environmental impact assessments of the ash disposal strategies, assessment of the leaching of some chemicals from the ash is essential. The current paper relates to chemical leaching of coal ash which exists under saturated or near-saturated conditions as would be found in disposal ponds.

There has been considerable research undertaken on Latrobe Valley brown coal ash, and the early research was predominantly concentrated on solving problems in transport pipes such as scaling, or on predicting ash quality based on coal quality. Deed (1981) reported early research on ash leaching on the basis of field experiments, but the interpretation of results was difficult owing to poor experimental design and lack of detailed chemical analyses of ash and leachate. Subsequently, Black (1990a, 1990b) undertook a study comprising a series of column and sequential batch leaching tests under controlled laboratory conditions. This study was targeted to investigate the chemical leachate quality simulating more closely those expected in the ash ponds.

Laboratory experiments using small columns have generated data on leachate quality and production rates in small columns. A basis to extrapolate this data to the field conditions is then required. For example, one may want to assess the quality and production rates of leachate at the base of an ash pond on the basis of experimental findings. For this purpose, Black (1990a) has used the linear transformation of the leaching rates relative to the height of the leaching mass. This approach is elementary and ignores the influence of all the dominant variables involved. In the current paper, a theoretical model for the coal ash leaching is developed, and a numerical scheme to solve the governing equations under given boundary conditions is presented. Numerical simulations were carried out to compare the model and experimental results.

2 THEORETICAL MODEL FOR ASH LEACHING

2.1 *Leaching mechanisms.*

Various mechanisms have been identified, by which chemicals can leach out from solid material. These include bulk diffusion, advection, chemical reactions, and surface transfer phenomena such as matrix dissolution (Côté, 1986a; Côté *et al,* 1986b). Leaching of chemicals by the way of bulk diffusion and advection are more important for solid wastes such as cement based solidified wastes. Nevertheless, for particulate wastes such as coal ash, these mechanisms will play a significant role in the migration of solutes through the porous medium, once they are leached out from the solids. Chemical reactions and surface phenomena are more important for coal ash, and are described on the basis of kinetic formulations for a particular species of interest or using mass transfer coefficients. For coal ash, initial wash-off of chemicals is also considered to be a significant mechanism in exchange of chemicals from surface and aqueous solutions.

Recent advances in chemical analysis and speciation techniques have led to the use of geochemical equilibrium models combined with detailed ash speciation and mineral analysis to predict the chemistry of ash leachates (e.g. Eighmy *et al.*, 1995). However, these were for batch tests only and required intensive analysis of the ashes to obtain detailed quantitative mineral data for input into the geochemical model. While this approach is considered in a limited sense in the overall research project, a more simplified approach is adopted to model the ash leaching as would be applicable to ongoing management of ash disposal.

2.2 *Theoretical Leaching Model*

The model developed for coal ash leaching is based on the work of Straub & Lynch (1982) and subsequently used by others (e.g. Demetracopoulos *et al,* 1984; Lu, 1996) on landfill leaching. It postulates that the leaching rate for a certain chemical is proportional to the ratio of current soluble mass S (per unit volume) to the initial soluble mass S_0 (per unit volume) in the ash particles as well to a concentration deficit from a maximum value (C_{max}) to the current concentration (C) in the aqueous solution. This can be expressed mathematically as:

$$L_R = \alpha\left(\frac{S}{S_o}\right)^{\beta}\left(C_{\max} - C\right) \qquad (1)$$

where α is a mass transfer coefficient applicable to a particular chemical species and β is an exponent signifying the effect of decaying chemical mass. C_{max} represents the maximum concentration of a chemical species, for example the concentration at saturation in aqueous solution. At a particular instant, the concentration deficit ($C_{\max} - C$) provides the driving force for leaching by surface phenomena. When this deficit is zero, it can be assumed that the system is in equilibrium. The model is considered to entail matrix dissolution of chemicals from the surface to the aqueous phase under reducing chemical mass. The initial wash-off is modelled by defining the initial concentration (C_0) in the aqueous solution. It is interesting to note that if β is assumed to be unity and $C_{\max} - C \approx C_{\max}$, the leaching rate given by Equation (1) signifies a situation where the soluble mass decays exponentially with time.

Based on the principle of conservation of mass, the migration of solutes through saturated ash incorporating leaching can be expressed by the advection-dispersion equation (ADE) as:

$$\frac{\partial C}{\partial t} = D_z\frac{\partial^2 C}{\partial z^2} - v_z\frac{\partial C}{\partial z} + L_R \qquad (2)$$

where D_z is the longitudinal dispersion/diffusion coefficient, v_z is the linearised advective velocity, and z is the depth measured from top of the ash surface. This equation assumes one-dimensional uniform flow in a non-deforming saturated medium.

The boundary and initial conditions for the problem are expressed as:

$$\begin{aligned} z &= 0 \qquad C(0,t) = g(t) \\ z &= H \qquad \frac{\partial C}{\partial z} = 0 \quad \text{and,} \\ t &= 0 \qquad C(z,0) = C_0 \end{aligned} \qquad (3)$$

where $g(t)$ is the concentration of the influent water, H is the height of the column or the ash deposit, and C_0 is the initial concentration of the of the pore fluid in the ash depicting initial wash-off of the chemicals.

3 NUMERICAL SOLUTION

3.1 *Finite difference approximation*

The equations, subject the boundary and initial conditions given in Section 2, are solved using a block-centred explicit, upwind finite difference scheme (Peaceman, 1977; Zheng & Bennett, 1995) subject to the following stability criteria

$$\Delta t \leq \frac{\Delta z^2}{2D_z + v\Delta z} \quad (4)$$

The change in soluble mass incorporating the leaching rate can be expressed in difference form as:

$$S_{k+1} = S_k - \left(L_R\right)_k \Delta tn \quad (5)$$

where *k* is the time step and *n* is the porosity.

3.2 Control of numerical dispersion

One of the main difficulties with solving the ADE lies in the fact that the equation has both a parabolic component (dispersion) and a hyperbolic (advection) component to the partial differential equation (PDE). Solution techniques such as finite difference (and other methods), work well for one form of PDE, but not for both combined (Press *et al.*, 1992). Approximating the ADE using finite difference equations gives rise to a phenomenon known as numerical dispersion, which leads to artificial smearing of the advective solute front. Detailed accounts of the source and control of numerical dispersion are given in Zheng & Bennett (1995), Noorishad *et al.* (1992) and Peaceman (1977).

There are two dimensionless parameters which can be used to help minimise the effects of numerical dispersion. They are the Courant number, Cr, (controlling advective flow) and the Peclet number, Pe, (controlling hydrodynamic dispersion). By maintaining low Courant and Peclet numbers, that is, a high spatial and temporal discretization, it is possible to reduce the effects of numerical dispersion. For the one-dimensional case of explicit upwind scheme, the numerical dispersion can be calculated by the following relation (Peaceman, 1977; Zheng & Bennett, 1995).

$$D_{num} = \frac{1}{2} v_z \Delta z (1 - Cr) \quad (6)$$

For advective flow only, where hydrodynamic dispersion (D_z) can be taken as zero, the ADE equation represents predominantly a plug flow with a sharp concentration front, except for the smearing due to numerical dispersion. Nevertheless, if the values of Δz and Δt are chosen so that $Cr = 1$, $D_{num} = 0$ and a step concentration can be modelled exactly without numerical dispersion.

4 EXPERIMENTAL RESULTS

As noted previously, Black (1990b) has conducted a series of column leaching tests on a range of ash samples including Hazelwood precipitator ash (HPA) and Hazelwood ash pond sediment (HAPS). He used a 50 mm diameter by 300 mm long perspex columns packed with dry ash for leaching tests. Glass wool was used at the influent and effluent end to introduce uniform flow conditions Pure water was fed to the top of the columns with a pump at a constant rate of 0.015 mL/minute, and leachate was collected from the bottom at various intervals and subjected to various chemical analyses. The salient data applicable to column tests are summarised in Table 1. Because of the relatively high flow rate used, it is reasonable to expect that the ash was reasonably well saturated when the leachate was collected. The results of the above tests were used to compare the numerical model against experimental results in the following section.

Table 1. Salient data for column tests by Black (1990b)

Ash type	Ash height (mm)	Pore Volume (ml)	Total volume (ml)	Porosity
HPA	161.3	217.3	316.6	0.69
HAPS	192.5	268.3	377.9	0.71

5 NUMERICAL SIMULATIONS

5.1 Parameters and sensitivity analyses

The main parameters needed for the numerical model are the linearised advective velocity (v_z), maximum (saturation) concentration (C_{max}), mass transfer coefficient (α), initial concentration (C_0), hydrodynamic dispersion (D_z) and the initial soluble mass (S_0). Both the linearised velocity and initial mass are derived from the results of Black (1990b). The value for hydrodynamic dispersion was approximated from the general range of values given in Fetter (1993), and was calculated as

$$D_z = \lambda\, v_z + D^* \quad (7)$$

where λ is longitudinal dispersivity, and D^* is the effective molecular diffusion coefficient. Based on Fetter (1992), λ is taken as *H/10*. However, since the flow system in the columns were dominated by advection, the value of λ was decreased to *H/20*.

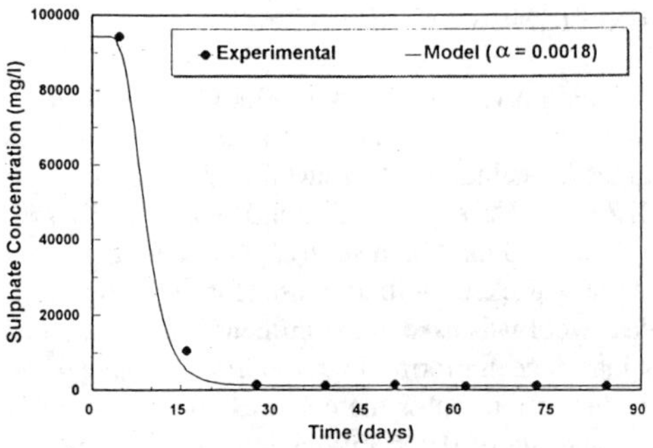

(a) Leaching of SO_4

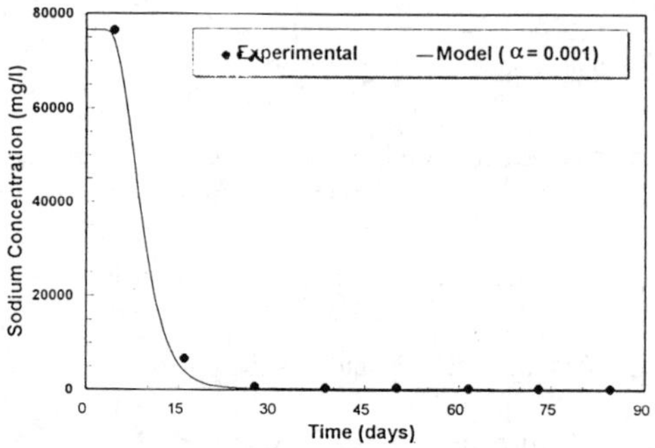

(c) Leaching of Na

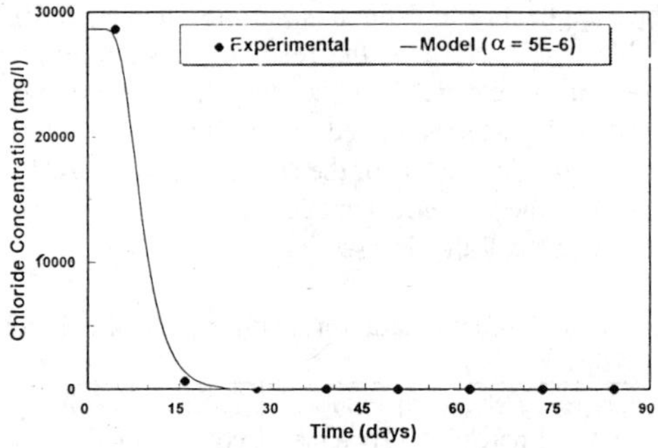

(b) Leaching of Cl

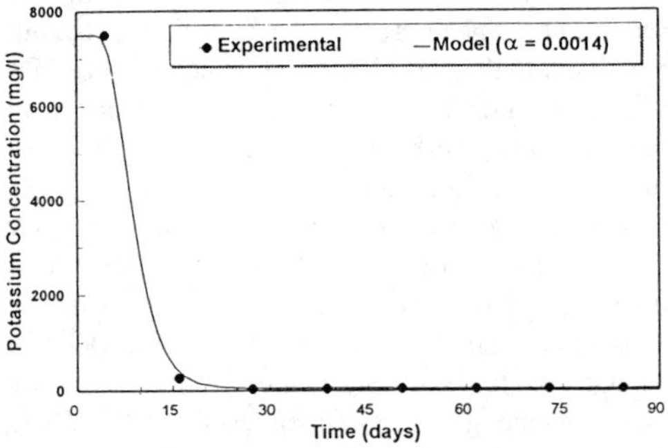

(d) Leaching of K

Figure 1. Comparison of experimental and model results of Hazelwood Precipitator ash.

Assuming a coefficient of molecular diffusion of D^* of 4.32 x 10^{-5} m^2/day, Equation (7) gave D_z of 1.72 x 10^{-4} and 1.97 x 10^{-4} m^2/day for HPA and HPAS respectively. Numerical simulations indicated that the value of λ chosen in this range does not significantly influence leachate concentration curves. Furthermore, it is not possible to use experimental data to examine the influence of this parameter in detail because only a few data points on the initial leaching phase are available.

The maximum concentration C_{max} was assumed to be the first available leachate concentration from the experiments. Since it is assumed that ash became significantly saturated prior to leaching, the initial concentration C_0 was approximated to be equal to C_{max}.. The mass transfer coefficient α was determined by fitting the numerical results to the experimental data. The value of α principally controls the residual concentration values after the initial wash-off, and has a relatively low influence on the initial drop in concentration. The effect of β was also examined, and it was found that β does not have a significant impact primarily because the leaching is dominated by initial wash-off. Hence, the value of unity was adopted for β. In these numerical simulations, the goodness of fit was determined by the eye.

5.2 *Comparison with experimental results*

Comparisons of simulated and experimental results for SO_4, Cl, Na and K are shown in Figures 1 and 2 for Hazelwood precipitator ash (HPA) and Hazelwood ash pond sediment (HAPS) respectively. These chemical species were chosen to cover a range of chemical leaching characteristics. Parameters used for these simulations are shown in Table 2. Owing to the high initial concentrations present in the ash, it is difficult to see clearly the residual leaching profiles. Therefore, the model and experimental residual concentrations at about 90 days of leaching are separately compared in Table 3.

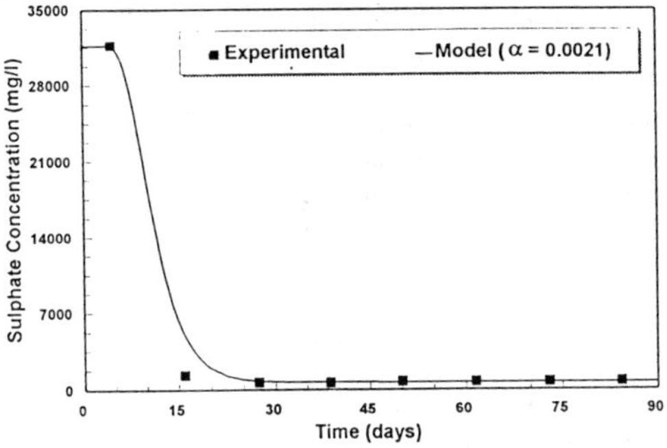

(a) Leaching of SO_4

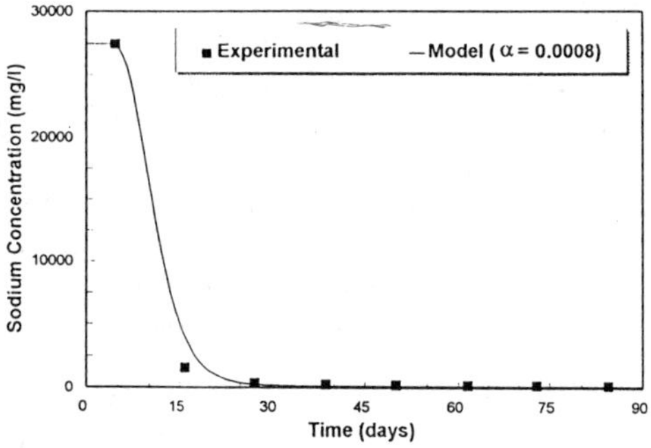

(c) Leaching of Na

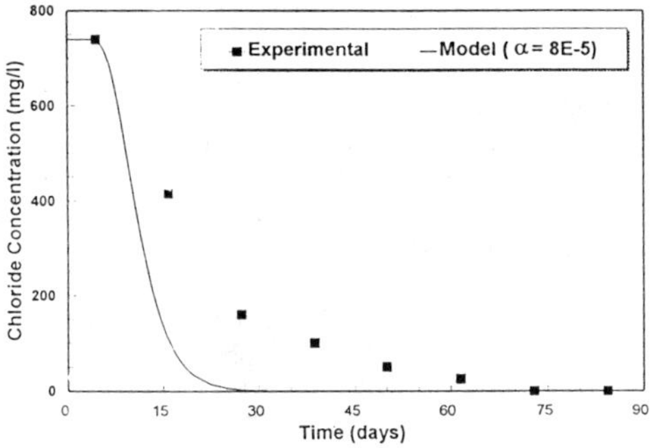

(b) Leaching of Cl

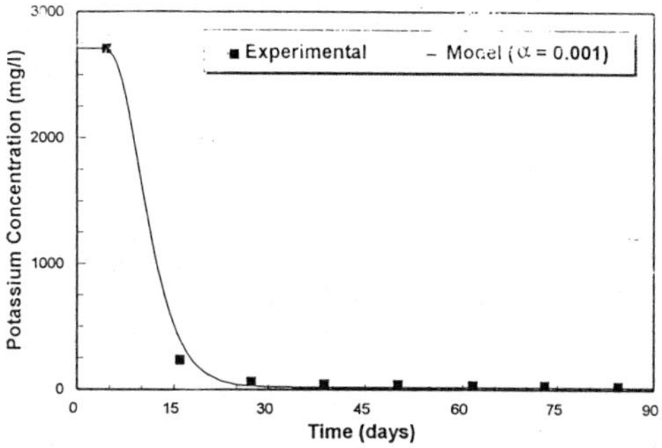

(d) Leaching of K

Figure 2. Comparison of experimental and model results of Hazelwood pond sediment.

Table 2. Parameters used for numerical simulations

Species	α x 10^3 (day^{-1})	C_{max} mg/L	S_0 (mg/L)
HPA - SO_4	1.8	94315	125063
Cl	0.005	28636	3790
Na	1.0	76570	49273
K	1.4	7494	4737
HPAS - SO_4	2.1	31690	86679
Cl[1]	0.08	740	1032
Na	0.8	27440	17464
K	1.0	2709	2382

[1] The original value was below the detection limit (<0.1%), hence a minimum value of 0.13% was assumed to avoid a negative soluble mass in the computations.

Table 3. Residual leachate concentrations (90 days)

Species	Residual concentrations (mg/L)			
	HPA		HPAS	
	model	experiment	model	experiment
SO_4	1100	1034	650	645
Cl	<1	<1	<1	<1
Na	330	358	13.8	160
K	44	40	17	31

5.3 Discussion of results

Good agreement was obtained between the experimental and model results. It can be seen that all the chemical species considered leach out primarily by initial wash-off and subsequent advection dominated flow. The species SO_4 and Na have some residual leaching fuelled by high initial soluble masses. The species Cl and K have not shown significant residual concentrations partly because of relatively small initial soluble masses in the ash. In general, HAPS ash contained lower initial soluble masses because of the slurrying process removing some of the chemicals. Impact of this is evident for Cl leaching given in Figure 2(b) where it leaches out at a slower rate than for fresh ash shown in Figure 1(b). This effect is also evident where the significant difference in α values was obtained for Cl for the two types of ashes.

6 CONCLUDING REMARKS

The theoretical model developed appears to capture reasonably well the major mechanisms of leaching of coal ash which exist in saturated or near-saturated conditions. It appears that the leaching under these conditions are dominated by initial wash-off and advective flow. While this form of model represents the field condition in ash ponds, other factors such as lateral flows, ash burial process, densification of ash due to overburden stresses and perhaps, the subsequent reductions in permeability should still be considered. In addition, the modelling of field conditions such as aged ash deposits in dry land would require the consideration of unsaturated flow and non-regular infiltration. Hence, the model will need to be extended to cover unsaturated flow in order to apply for these situations.

ACKNOWLEDGMENT

This research is conducted under a joint Australian Postgraduate Award (Industry)/Geo-Eng Australia scholarship. The authors wish to thank Mr. Chris Black (HRL Technology), Mr. Terry McKinley (Geo-Eng Australia), Dr. Dominic Caridi and Jason Beard (Dept. Chemistry & Biology, VUT) for their contributions and ongoing support.

REFERENCES

Black, C. J. (1990a); "Yallourn W Ash Leaching"; State Electricity Commission of Victoria; Report SC/90/174.

Black, C. J., (1990b); "Leachability of Morwell Brown Coal Ash"; *M. App. Sci. (Env. Eng.) Thesis,* Dept. of Chemical Engineering, University of Melbourne.

Côté, P. (1986a), "Contaminant Leaching From Cement-based Waste Forms Under Acidic Conditions", *PhD Thesis*, McMaster University.

Côté, P. (1986b), "An Approach for Evaluating Long-Term Leachability From Measurement of Intrinsic Waste Properties", in *Hazardous and Industrial Solid Waste Testing and Disposal - Vol. 6.*, ASTM, Editor D. Lorenzen *et al.*

Deed, R. (1981), "Behaviour of Some Latrobe Valley Ashes on Their Internment in Overburden Dumps", State Electricity Commission of Victoria, Research & Development Department., Report No. LO/81/139.

Demetracopoulos, A. C., Sehayek, L. & Erdogan, H. (1986), "Modeling Leachate Production From Municipal Landfills", *Journal of Environmental Engineering Division, ASCE* **112**, 849-866.

Eighmy, T. T., Eusden, J. D. (Jr.), Krzanowksi, J. E., Domingo, D. S., Stampeli, D., Martin, J. R., Erickson, P. M., (1995) "Comprehensive Approach Toward Understanding Element Speciation and Leaching Behavior in Municipal Solid Waste Incineration Electrostatic Precipitator Ash", *Environmental Science & Technology* **29**, 629-646.

Fetter, C. W., (1993) "Contaminant Hydrogeology", Macmillan Publishing, New York.

Lu, C. (1996), "A Model of Leaching Behaviour From MSW Incinerator Residue Landfills", *Waste Management & Research* **14**, 51-70.

Noorishad, J., Tsang, C. F., Perrochet, P. & Musy, A. (1992) "A Perspective on the Numerical Solution of COnvection-Dominated Transport Problems: A Price to Pay for the Easy Way Out", *Water Resources Research* **28**, 551-561.

Peaceman, (1977) "Fundamentals of Numerical Reservoir Simulation", Developments in Petroleum Science , 6, Elsevier Scientific Publishing Company, Amsterdam.

Press, W. H., Teukolsky, S. A., Vetterling, W. T. & Flannery, B. P., (1992) "Numerical Recipes" 2nd Edition, Cambridge University Press, New York, 963 pages.

Straub, W. A. & Lynch, D. R. (1982); "Models of Landfill Leaching : Moisture Flow and Inorganic Strength", *Journal of Environmental Engineering Division, ASCE* **108**, 231-250.

Zheng, C. & Bennett, G. D. (1995) "Applied Contaminant Transport Modeling : Theory and Practice", Van Nostrand Reinhold, New York, 440 pages.

Computer Methods and Advances in Geomechanics, Yuan (ed.) © 1997 Balkema, Rotterdam, ISBN 90 5410 904 1

Studies on the interaction between municipal solid waste and biogas extraction systems in sanitary landfills

O. Del Greco, A. M. Ferrero & C. Oggeri
Departamento Georisorse e Territorio, Politecnico di Turino, Italy

ABSTRACT: In the M.S.W. landfills it is fundamental the extraction of the biogas which is originated from the decomposition of the organic fraction of the wastes. The gas can induce high pressures inside the waste pile and, in particular, in both lining systems placed at the top and at the bottom of the waste body, causing ruptures of the isolation envelope. In order to reduce the pressure, biogas extraction systems are constructed in every landfill being active during its life time. The biogas extraction systems are made of vertical columns.
Due to the phenomenon of consolidation of the wastes, a sort of negative friction arises between the wastes and the biogas dredging columns, involving more critical stress conditions between the columns and the isolation layer at the base of the landfill.
The aim of this note is the description of the mechanical features of the consolidation phenomenon of wastes, the study of the interaction between waste and the biogas dredging columns and of the interaction between the dredging columns and the base layer. A numerical model has been set up with the purpose to simulate the described aspects and to analyze with more details the interaction between the various elements that are part of the complex structure of a landfill.

1 INTRODUCTION

Municipal and industrial waste landfills are a particular engineering domain. A more rational approach to this problem is relatively new (no more than twenty years) and other rational waste disposal methods or processing are under study. The preminent idea is devoted to study the reduction of the amount of waste to be arranged in landfills, by means of preventive selections of recyclable materials, by means of combustion or special biological treatment. Nevertheless, it seems that the need of landfill disposal will not be eliminated at all, due to wastes which do not have any other way after processing that the arrangement in landfills. Scientific researches, aimed to increase the safety conditions in landfills, have obtained during last years considerable, and therefore usefulness, success. Some problems are still to be studied in order to ameliorate the environmental and technical performances in landfills.

The main requirement of a landfill is that of ensuring a containment of waste as a tightness envelope, without permitting any spontaneus outlet of fluid components: biogas and leachate. This fluid parts cannot nevertheless remain inside the landfill body, but they must be extracted by means of prepared systems, all along the lifetime of the landfill.

Lanfills are waste bodies which can assume different shapes, but in general they can be divided in two main types (Figure 1): a) filling of excavations with the possibility of raising an embankment; b) construction of an artificial enbankment in the shelter of a slope, with a bottom and lateral containment.

In every case, anyway, the waste bodies must be contoured by a top and a bottom isolation liners. These systems have the role of avoiding contacts between waste (in solid or fluid forms) and the

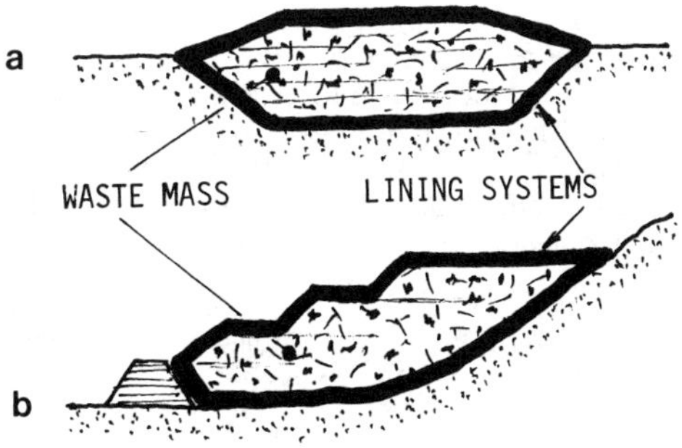

Figure 1. Landfill geometries: a) filling of an excavation; b) accumulating along a natural slope.

natural environment (geologic site and atmosphere). In order to fit this role, isolation liners are made of different materials, both natural and geosinthetics. Therefore they have a multylayered structure, in which every constituent presents particular features that allow the material to perform their specific role. The overall continuity of the liners must also be encountered, in order to avoid singular diffusion ways of contaminants.

2 DESCRIPTION OF ISOLATION AND DRAINAGE SYSTEMS

The bottom isolation of a landfill can be guaranted by a low permeability geologic formation, as that the geologic formation can act as a barrier. Unfotunately such geological condition is not often encountered and, in any case, it is rather difficult to prove the perfect vertical and lateral continuity of the formation. For this reason quite all national laws foresee the construction of a multilayered structure.

The most used system is made of a bed of compacted natural clay. Over this layer there is a high density poliethilene membrane (HDPE). The base soil is compacted each 20-25 cm layer, reaching at the end the desired thickness (in Italy 1 m for municipal waste and 2 m for industrial hazardous waste). The permeability coefficient of this layer must not overtake the value of 10^{-9} m/s.

The impermeabilization of the lateral slope of the landfill can also be made of a geosynthetic composed of a thin coating of sodic benthonite inside two geotextiles instead of the clay layer (permeability < 10^{-11} m/s). This geocomposite layer can also be used to ameliorate the tightness of the bottom multilayer liner. The HDPE geomembrane lies just over the compacted clay has a thickness of 1.5 - 2 mm and its lateral continuity is ensured by a thermic process of welding of long stripes 8-10 m large.

The leachate produced in the landfill drain through the waste body and accumulate over the bottom isolation liner. Here the liquid phase moves through a draining stratus of sand and gravel (about 30 cm thick) and reach the lower zone of the basin where there is an extraction system.

At the end of the landfill lifetime a covering liner has made. This structure must provide a streaming and drainage of meteoric water, must allow differential settlements and isolate the waste from the environment.

3 DRAINAGE OF BIOGAS AND GEOTECHNICAL PROBLEMS

The organic fraction of waste decomposes and after about 6 months it starts producing biogas; then it continues the biochemical reactions for at least ten years, with a maximum of biogas production 2-5 years after the landfill closure. The decomposition gas is mainly constitued by CO_2 (40-45 %) and hydrocarbons like CH_4 (52-55 %). The amount of biogas production is about 150-200 m^3 each 1 m^3 (700-800 kg) of municipal waste. The CO_2 combines with leachate and they drain together, while the CH_4, ligther than the atmospheric air, moves up to the top covering system and it must be extracted in order to avoid eccessive pressure against the liners and to prevent ruptures and pollution.

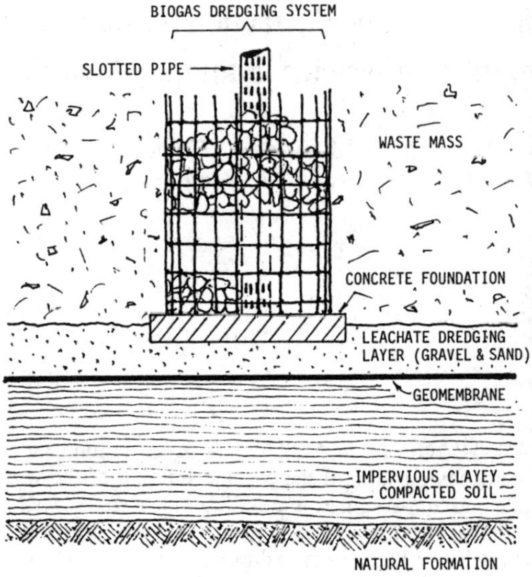

Figure 2. Biogas dredging system founded on leachate collecting layer and bottom lining system.

The drainage and the extraction of biogas need suitable way out able to allow the quick outlet of the biogas. The extracted gas can be combusted immediately or used for energy recovery. The most used drainage system in Italy is made of a vertical columnar cilinder made of gravel and pebbles contained in a metallic grid. Inside this element there is a slotted HDPE pipe. This well has a concrete foundation slab, generally posed on the drainade stratum at the base of the landfill (Figure 2). The biogas well is elevated together with the waste body just up the covering system. The wells are posed with a spacing grid of 20-30 m of radius.

The biogas wells can have two types of mechanical interaction with the elements of the landfill. The first is the friction phenomena which develope all along the wells and the waste body, as a consequence of different stiffness and consequent differential settlements of waste body. The second is the interaction, at the base of the well, between the concrete ring and the bottom isolation liner, which act in this sense as a foundation formation. In this last situation a problem of bearing capacity is posed, in the sense that it is necessary to determine the maximum load that may be applied by the well foundation to the underlaying strata without causing failure. As far as the geosynthetic liners is concerned, a failure mode associated with punching shear is probable. A particular attention on the shear characteristics at the interface between compacted clay and geosynthetics is emphasized, because they act as discontinuities, characterized by shear resistance parameters.

4 NUMERICAL MODELLING

The geometry of the model represents a typical well for the biogas extraction in a axisymmetric scheme.

The development of the model represents the construction of a waste disposal by considering the superimposition of several strata of material over the in situ formation. The first strata represent the impermeabilization system formed by tick clay layer (1m) and by a HDPE geomembrane.

On the isolation system a sand and gravel stratum is then set in place and, subsequently, the biogas well, composed by: the slotted pipe contoured by pebbles, a reinforced concrete ring at the pipe base and a concrete platform foundation.

The waste disposal foresees an alternating superimposition of strata of waste and gravel up to 1.5 m from the top of the landfill. Here the waste is covered and isolated by a stratum of clay and by a HDPE geomembrane. Finally the vegetal soil is modelled too.

The model has been carried out by means of the finite difference methods and the code used is FLAC (ITASCA, 1993).

The model is composed by 4000 axisymmetric elements and the well axis corresponds with the axis of symmetry. Between the quadrilateral elements reproducing the biogas well and the waste material, special joint elements have been introduced to modelled the presence of a discontinuity where large sliding movements can occurred.

The material stress strain behaviour is elastic ideally plastic and the strength criterion is the Mohr-Coulomb law. The pipe is considered simply elastic.

Within the same described geometrical scheme four different models have been set up by changing the waste strength features, while two more models have been set up respectively with a reduced plate foundation and without the plate foundation. Table 1 shows the different model features, whilst table 2 shows the deformability and strength characteristics of all involved materials in models A, B, C, D, E described in table 1.

In model F material features have been changed during the landfill construction to simulate the variation of the waste behaviour during its life time due to organically decomposition phenomena. In this case, elastic modulus changes linearly with principal

TABLE 1 - Model description

MODEL TYPE	OBJECTIVE
• (A) Geometrical scheme as fig. 1 friction angle waste - well = 20°	Evaluate the influence of friction angle on biogas system
• (B) Geometrical scheme as fig. 1 friction angle waste - well = 25°	Evaluate the influence of friction angle on biogas system
• (C) Geometrical scheme as fig. 1 friction angle waste - well = 29°	Evaluate the influence of friction angle on biogas system
• (D) Different geometrical scheme, no foundation plate, friction angle waste - well = 25°	Evaluate the influence of the presence of the foundation plate.
• (E) Different geometrical scheme, reduced foundation plate, friction angle waste - well =25°	Evaluate the influence of the foundation geometry.
• (F) Geometrical scheme as fig. 1 friction angle waste - well = 25°, variation of strength and deformability features	Study the effect of organically decomposition.

TABLE 2 - Mechanical features of modelled materials

	Bulk Modulus K (MPa)	Shear Modulus G (MPa)	Cohesion c (MPa)	Friction angle φ (°)	Specific weight γ (kg/m^3)
foundation soil	1670	556	1	35	2100
clay	139	36.2	0.1	30	1800
gravel	125	57.7	0	36	2000
geomembrane	1800	1390	4	40	940
waste	0.741	0.303	0.02	25	476
pipe	1800	1390	-	-	960
vegetal soil	1580	567	0.1	35	1900

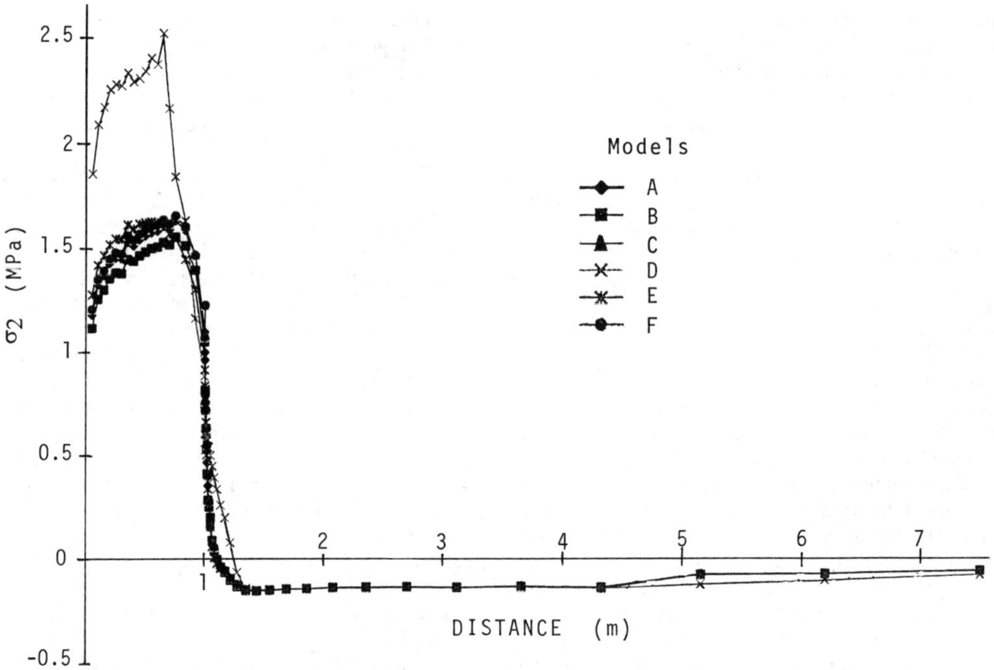

Figure 3. Calculated minimum principal stress in the HDPE geomembrane of the bottom lining (tensile stress are positive) vs the distance from the well axis, at the end of the filling time.

stress ($E=f(\sigma_1)$), friction angle changes according to experimental data and waste specific weight increases with waste compactation.

Models A, B, C have the same geometry but different friction angles (20°, 25°, 29°) between the well and the waste since reliable value of this data seams very difficult to be obtained by bibliography.

The analysis of the results obtained by numerical models were mainly aimed to evaluate:

1. the stress state in the HPDE geomembranes place on the base and on the top of the waste disposal;
2. the static function of the plate foundation;
3. the state of stress in the isolation clay strata.

The first point has been studied by analysing the tensile and compressive stresses in the HDPE geomembrane, placed at the waste disposal base. The graphs of figure 3 report the minimum principal stress (σ_2) at increasing distances from the model axis for all different described models and they show as traction (positive in the graph) is present for all models up to 1.2 m from the well axis. Considering that the HDPE geomembrane has a tensile strength of 26 MPa, the geomembrane factor of safety has been computed as the ratio between the strength and maximum tensile value computed. Table 3 reports the factor of safety for all considered models. Factor of safety are always higher than 10 as recommended by EPA (Environmental Protection Agency) due to the important role played by the isolation system.

Maximum tensile values, and consequently minimum factor of safety, have been computed for model (D) (without foundation plate). Factor of safety increases for increasing dimension of the foundation plate. Maximum values of deformations and plasticized areas of the clay stratum have been determined for case (D) too.

The same kind of analysis have been performed

TABLE 3 - Minimum principal stress and factor of safety of the base HDPE geomembrane

MODEL	σ_2 (MPa)	Factor of safety
(A)	1.63	15.9
(B)	1.55	16.8
(C)	1.65	15.8
(D)	2.52	10.3
(E)	1.62	16
(F)	1.65	15.8

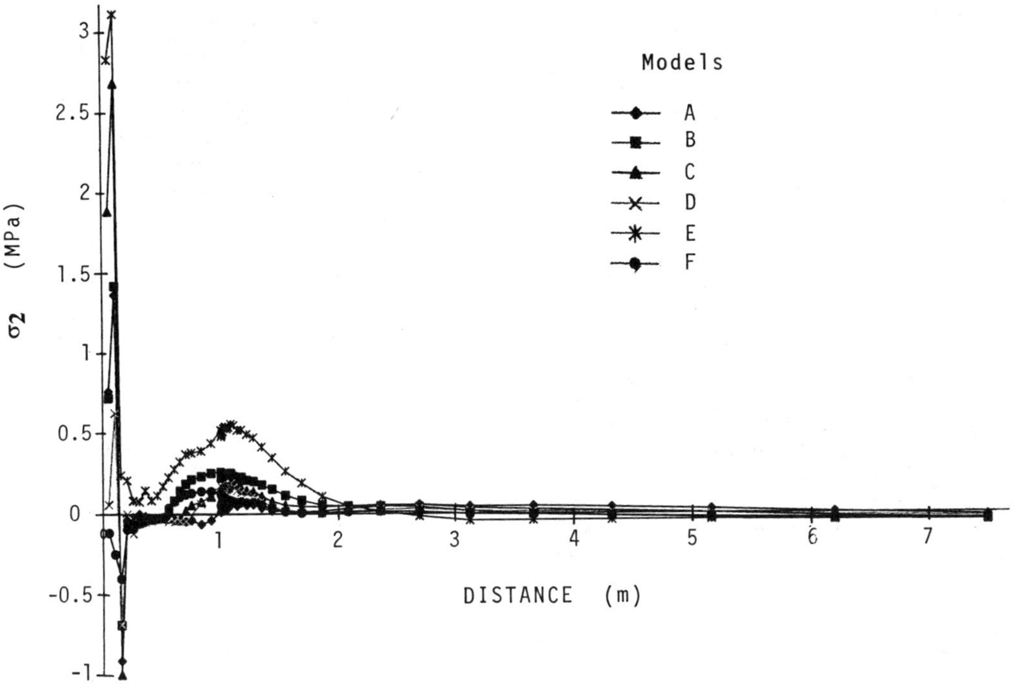

Figure 4. Calculated minimum principal stress in the HDPE geomembrane of the top lining (tensile stress are positive) vs the distance from the well axis.

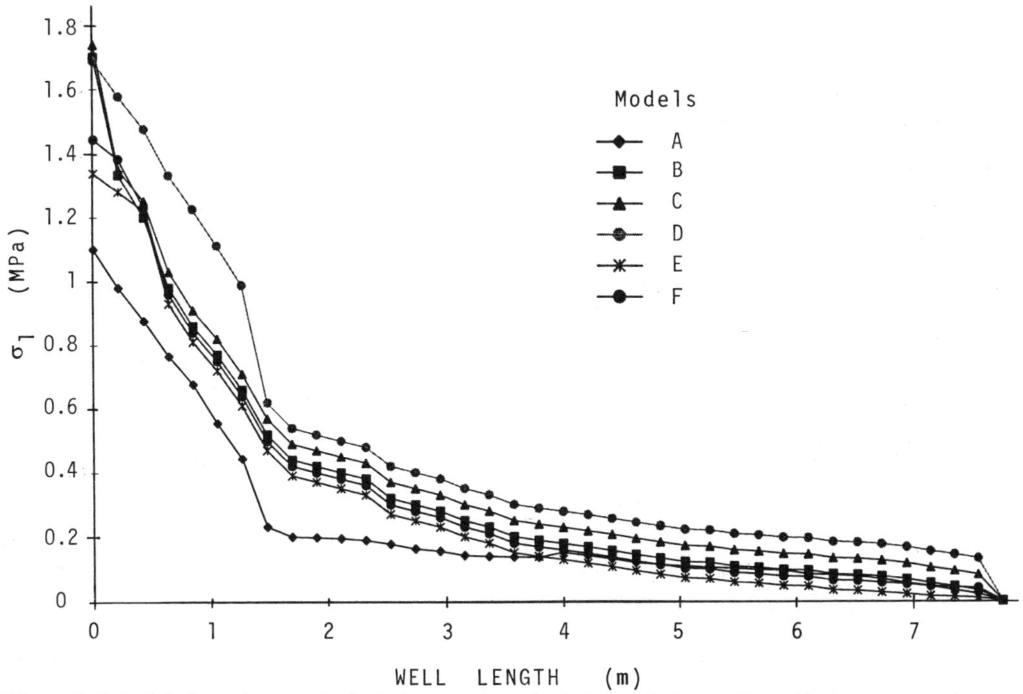

Figure 5. Calculated maximum principal stresses along the interface between the well structure and the waste mass.

TABLE 4 - Minimum principal stress and factor of safety of the top HDPE geomembrane

MODEL	σ_2 (MPa)	Factor of safety
(A)	1.36	19.1
(B)	1.42	18.3
(C)	1.32	19.7
(D)	1.02	25.5
(E)	3.11	8.4
(F)	1.32	19.7

for the isolation system at the top of the waste (Figure 4 and Table 4). In this case higher tensile values correspond to a reduced foundation area (Model E).

Finally the interaction between the well and the waste have been analysed. Negative friction effects have been observed in all cases due to high deformability of waste. Stress values in the foundation system increases for higher values of the friction angle between the well and the waste with a consequent decrease of global well stability (Figure 5).

5 CONCLUSIONS

On the base of the described results some remarks on the biogas well construction system can been obtained: in order to reduce the traction values in the base isolation system flexible foundation of large dimension are suggested to avoid both geomembrane fail for traction and plasticization of the clay stratum; a low friction angle between the well and waste reduces the negative friction, and consequently the state of stress in the foundation, this aim could be achieved by taking a special care in the construction of the external part of the well structure in order to avoid protrusion of pebbles.

Computer Methods and Advances in Geomechanics, Yuan (ed.) © 1997 Balkema, Rotterdam, ISBN 90 5410 904 1

Numerical modeling of groundwater contamination with volatile material by multi-component flow formulation

Kazumasa Itoh, Sei-ichiro Mori & Yasunori Otsuka
Kawamoto Research Institute, OYO Corporation, Omiya, Japan

Hiroyuki Tosaka
The University of Tokyo, Japan

ABSTRACT: In numerical prediction of groundwater contamination by volatile organic compounds, it is necessary to consider dissolution and volatilization, that has scarcely introduced into numerical simulation. In this study, multi-phase flow formulation has been modified and applied to groundwater contamination problem. And transient diffusion modeling has been introduced to express dissolution and volatilization. Through numerical experiment and field application, validity and usefulness of this formulation has been presented.

1 INTRODUCTION

For construction of waste disposal site, it is necessary to predict contaminant flow from landfill in siting and site investigation stage for the risk assessment. In waste disposal site, many kinds of contaminant materials would be target for prediction.

In order to predict flow and transport of contaminant in groundwater pollution problem, or evaluate effect of remediation, numerical simulation with convection-diffusion model has been utilized in many cases. However, in the case of groundwater contamination with materials such as VOCs, which is slightly soluble to water and volatile in air, conventional numerical simulation with convection-diffusion model can not be applied directly.

In the case of groundwater contamination with volatile material, contaminated area would widely spread because of volatilization in unsaturated zone and diffusion in air and dissolution into groundwater in unsaturated and saturated zone according to rainfall.

2 FORMULATION OF CONTAMINAITON

2.1 Basic Formulation

For numerical simulation of contaminant transport in such cases, the authors has been developing numerical modeling with multi-component flow formulation. In this method, water-gas-contaminant three phase, and five components, as pure contaminant, water, air, contaminant solute, evaporated contaminant flow has been introduced as numerical model of contaminant transport. And, volatilization, dissolution from pure material to water, dissolution from vapor to water has been modeled as phase transfer. In this model, transient volatilization and dissolution has been modeled as one dimensional diffusion. And, diffusion of solute and evaporated material has been introduced in formulation.

The basic formula of multi-component simulation is as follows;

(1) Material balance of water

$$\nabla \frac{kk_{rcw}}{\mu_{cw} B_{cw}} \nabla \psi_{cw} + \nabla \frac{kk_{rcc}}{\mu_{cc} B_{cc}} \nabla \psi_{cc} - q_{ws}^{cw} - q_{ws}^{cc} = \frac{\partial}{\partial t}\left(\frac{\phi S_{cw}}{B_{cw}} + \frac{\phi S_{cc}}{B_{cc}} \right) \quad (1)$$

(2) Material balance of air

$$\nabla \frac{kk_{rca}}{\mu_{ca} B_{ca}} \nabla \psi_{ca} - q_{as}^{ca} = \frac{\partial}{\partial t}\left(\frac{\phi S_{ca}}{B_{ca}} \right) \quad (2)$$

(3) Material balance of pure contaminant

$$\nabla \frac{kk_{rcc} R_{cc}}{\mu_{cc} B_{cc}} \nabla \psi_{cc} - f_{cs}^{cc-cw} - f_{cs}^{cc-ca} - f_{cs}^{cc-r} - q_{cs}^{cc} = \frac{\partial}{\partial t}\left(\frac{\phi S_{cc} R_{cc}}{B_{cc}} \right) \quad (3)$$

(4) Material balance of contaminant solute

$$\nabla \frac{kk_{rcw}R_{cw}}{\mu_{cw}B_{cw}} \nabla \psi_{cw} + \nabla D_{cw} \nabla \left(\frac{R_{cw}}{\alpha_{cw}} \right) + f_{cs}^{cc-cw} + f_{cs}^{cw-ca}$$

$$- f_{cs}^{cc-r} - q_{cs}^{cw} = \frac{\partial}{\partial t} \left(\frac{\phi S_{cw} R_{cw}}{B_{cw}} \right) \qquad (4)$$

(5) Material balance of evaporated contaminant

$$\nabla \frac{kk_{rca}R_{ca}}{\mu_{ca}B_{ca}} \nabla \psi_{ca} + \nabla D_{ca} \nabla \left(\frac{R_{ca}}{\alpha_{ca}} \right) + f_{cs}^{ca-cc} - f_{cs}^{ca-cw}$$

$$- f_{cs}^{ca-r} - q_{cs}^{ca} = \frac{\partial}{\partial t} \left(\frac{\phi S_{ca} R_{ca}}{B_{ca}} \right) \qquad (5)$$

where, k : permeability of ground $[L^2]$

k_r : relative permeability of each phase[-]

μ : viscosity of each phase $[ML^{-1}T^{-1}]$

B : formation volume factor [-]

R : volumetric dissolution/volatilization ratio

D : diffusion coefficient $[LT^{-2}]$

q : sink/source term $[T^{-1}]$

p : porosity [-]

S : saturation of each phase [-]

ψ : pressure of each phase $[MLT^{-2}]$

And, each subscription cw, cc, ca denotes contaminated water phase, chemical contaminant phase and contaminated air phase.

In these equations, unknown parameters are ψ_{cw}, y_{cc}, y_{ca}, S_{cw}, S_{cc}, S_{ca}, R_{cw}, R_{ca}, R_{cc}. And, supplemental equations are added to basic equation in order to reduce unknown parameters.

$$\psi_{cw} = P_{cw} - \rho_{cwR} gZ = P_{cc} - P_{c,cw} - \rho_{cwR} gZ \qquad (6)$$

$$\psi_{cc} = P_{cc} - \rho_{ccR} gZ \qquad (7)$$

$$\psi_{ca} = P_{ca} - \rho_{caR} gZ = P_{cc} + P_{c,ca} - \rho_{caR} gZ \qquad (8)$$

$$S_{cw} + S_{cc} + S_{ca} = 1 \qquad (9)$$

where, $P_{c,cw}$: capillary pressure of water

$P_{c,ca}$: capillary pressure of air

ρ_{cwR} : density of water under pressure

ρ_{ccR} : density of chemical under pressure

ρ_{caR} : density of air under pressure

Using equations (6) - (9) and, considering R_{cc} as constant, unknown parameters have been reduced to P_{cc}, S_{cw}, S_{ca}, R_{cw}, R_{ca}.

In this formulation, volumetric dissolution / volatilization ratio R and volumetric concentration α has been defined as shown in Figure.1.

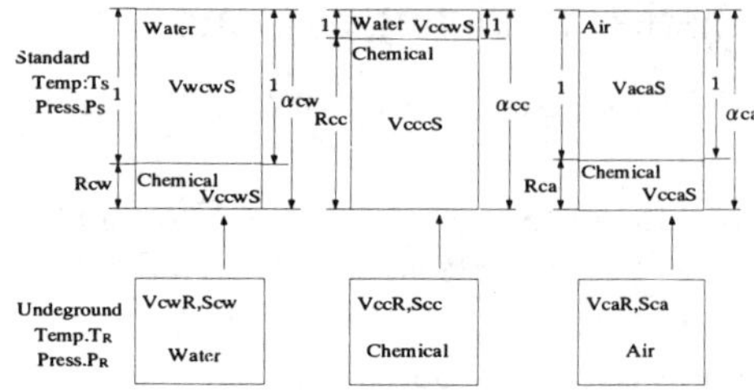

Figure.1 Definition of concentration

According to this definition, Volumetric ratio and concentration has been formulated as follows.

$$R_{cw} = \frac{V_{ccws}}{V_{wcws}} \qquad (10)$$

$$R_{cc} = \frac{V_{cccs}}{V_{wccs}} \qquad (11)$$

$$R_{ca} = \frac{V_{ccas}}{V_{acas}} \qquad (12)$$

$$\alpha_{cw} = \frac{(V_{wcws} + V_{ccws})}{V_{wcws}} \qquad (13)$$

$$\alpha_{cc} = \frac{(V_{wccs} + V_{cccs})}{V_{wccs}} \qquad (14)$$

$$\alpha_{ca} = \frac{(V_{acas} + V_{ccas})}{V_{acas}} \qquad (15)$$

Here, relationship between R and a has been calculated as next equation.

$$\alpha = 1 + R \qquad (16)$$

In this method, formation volume factor has been defined from apparent formation volume factor B', defined as follows,

$$B'_{cw} = \frac{V_{cwR}}{(V_{wcwS} + V_{ccwS})} \qquad (17)$$

$$B'_{cc} = \frac{V_{ccR}}{(V_{wccS} + V_{cccS})} \qquad (18)$$

$$B'_{ca} = \frac{V_{cca}}{(V_{acaS} + V_{ccaS})} = \frac{P_{ref}}{P_{ca}} \qquad (19)$$

$$B_{cw} = \frac{V_{cwR}}{V_{wcwS}} = \frac{(V_{wcwS} + V_{ccwS})}{V_{wcwS}} \cdot \frac{V_{cwR}}{(V_{wcwS} + V_{ccwS})}$$

$$= \alpha_{cw} B'_{cw} = (1 + R_{cw}) B'_{cw} \qquad (20)$$

$$B_{cc} = \alpha_{cc} B'_{cc} = (1 + R_{cc}) B'_{cc} \qquad (21)$$

$$B_{ca} = \alpha_{ca} B'_{ca} = (1 + R_{ca}) \frac{P_{ref}}{P_{ca}} \qquad (22)$$

where, P_{ref} : reference pressure

P_{ca} : total pressure of air phase

In these formula, phenomena of dissolution and volatilization has been considered as material exchange between each phase shown as f in basic equations.

From these formula, pressure of contaminant phase, saturation of water phase and gas phase, volumetric content of solute and gaseous contaminant has been calculated with Finite Difference Method.

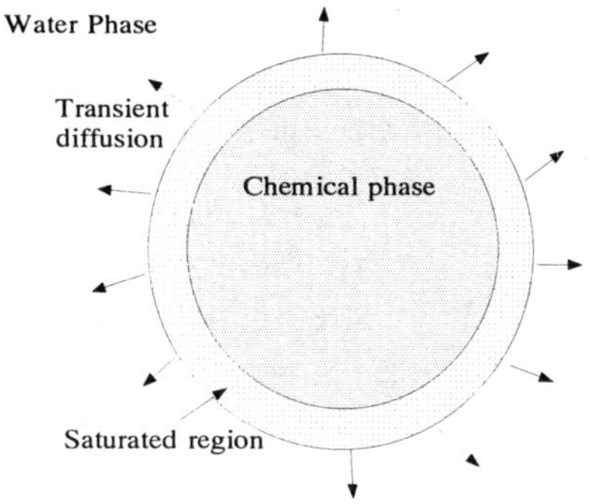

Figure 2 Schematic diagram of dissolution

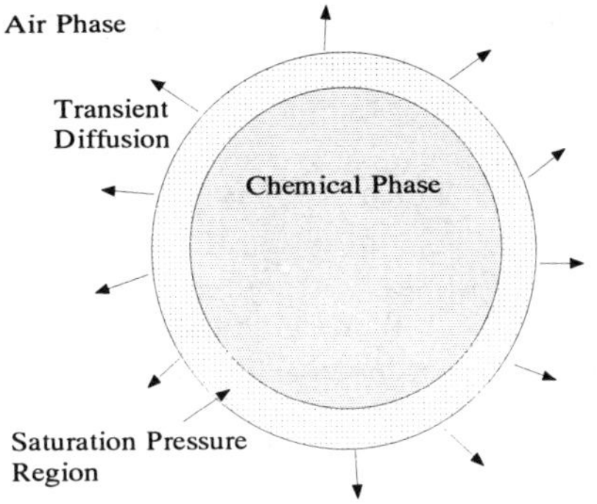

Figure 3 Schematic diagram of volatilization

2.2 Modeling of dissolution and volatilization

In order to construct mathematical model of dissolution and volatilization, the authors has introduced three kind of material exchange between each phase.

1) Dissolution of chemical from chemical phase to water phase
2) Volatilization of chemical from chemical phase to air phase
3) Dissolution of chemical from air phase to water phase

And, for the modeling of each material exchange, transient diffusion by Higbie has been introduced.

1) Numerical modeling of dissolution from chemical to water

Schematic diagram of dissolution is shown in Figure 2.

Considering dissolution from chemical to water as one dimensional diffusion, concentration of solute chemical in water is calculated from Fick's second law.

$$\frac{\partial C}{\partial t} = D_{cw} \frac{\partial^2 C}{\partial Z^2} \tag{23}$$

From analytical solution of (23) with initial condition $C = C_{sw}$, and boundary condition $C = C_{sat}$ (at Z=0), $C = C_{sw}$ (at Z=∞), distribution of concentration in water phase has been calculated as (24).

$$\frac{C - C_{sw}}{C_{sat} - C_{sw}} = \mathrm{erfc}\left(\frac{Z}{2\sqrt{D_{cw} t}}\right) \tag{24}$$

From this equation, dissolution rate from saturated region through unit area is shown as equation (25).

$$N_\ell = -D_{cw} \left.\frac{\partial C}{\partial Z}\right|_{Z=0} = \sqrt{\frac{D_{cw}}{\pi t}}(C_{sat} - C_{sw}) \tag{25}$$

And, average dissolution rate during time Δt through unit area is shown as equation (26).

$$u_{cs} = \frac{1}{t}\int_t^{t+\Delta t} N_\ell dt = 2\sqrt{\frac{D_{cw}}{\pi \Delta t}}(C_{sat} - C_{sw}) \tag{26}$$

As a result, average dissolution rate has been calculated by multiplying contact area (A).

$$f_{cs}^{cc-cw} = 2A\sqrt{\frac{D_{cw}}{\pi \Delta t}}\left(\frac{R_{cc}}{1+R_{cc}} - \frac{R_{cw}^{(n+1)}}{1+R_{cw}^{(n+1)}}\right) \tag{27}$$

And, if contact area is assumed to be proportional to m1 power of the volume of chemical phase, and volumetric ratio of dissolution to water and chemical phase, (27) has been transformed to next equation.

$$f_{cs}^{cc-cw} = \lambda(V(1 - S_{cw}^{(n+1)} - S_{ca}^{(n+1)}))^{m_1} \frac{S_{cw}^{(n+1)}}{S_{cw}^{(n+1)} + S_{ca}^{(n+1)}} \cdot \sqrt{\frac{D_{cw}}{\pi \Delta t}}\left(\frac{R_{cc}}{1+R_{cc}} - \frac{R_{cw}^{(n+1)}}{1+R_{cw}^{(n+1)}}\right) \tag{28}$$

2) Numerical modeling of volatilization

For numerical modeling of volatilization of chemical component to air phase, transient diffusion formulation has been also applied. Schematic diagram of volatilization is shown in Figure 3.

After the same process as dissolution, volumetric ratio of volatilization from chemical phase is expressed as next equation.

$$f_{cc}^{ca-cc} = \lambda(V(1 - S_{cw}^{(n+1)} - S_{ca}^{(n+1)}))^{m_1} \frac{S_{ca}^{(n+1)}}{S_{cw}^{(n+1)} + S_{ca}^{(n+1)}} \cdot \sqrt{\frac{D_{ca}}{\pi \Delta t}}\left(\frac{R_{ca(sat)}}{1+R_{ca(sat)}} - \frac{R_{ca}^{(n+1)}}{1+R_{ca}^{(n+1)}}\right) \tag{29}$$

Volumetric concentration in air phase in equation (29) has been expressed with total pressure and saturation pressure.

$$f_{cc}^{ca-cc} = \lambda(V(1-S_{cw}^{(n+1)}-S_{ca}^{(n+1)}))^{m_1}\frac{S_{ca}^{(n+1)}}{S_{cw}^{(n+1)}+S_{ca}^{(n+1)}}$$
$$\cdot\sqrt{\frac{D_{ca}}{\pi\Delta t}}\left(\frac{P_{cs}}{P_{cc}^{(n+1)}+P_{c,ca}}-\frac{R_{ca}^{(n+1)}}{1+R_{ca}^{(n+1)}}\right) \qquad (30)$$

On the other hand, material loss in chemical phase has been calculated with conditional equation of gas. In this condition, pressure of evaporated chemical has been considered as saturation pressure in saturated region. Mol number of evaporated chemical has been calculated as next equation.

$$n = \frac{P_{cs}f_{cc}^{ca-cc}}{RT} \qquad (31)$$

where, R : gas constant
T : temperature

And, volumetric volatilization rate in liquid chemical phase has been calculated.

$$f_{cc}^{cc-ca}$$
$$= \frac{xP_{cs}f_{cc}^{ca-cc}}{\rho_c RT}$$
$$= \lambda(V(1-S_{cw}^{(n+1)}-S_{ca}^{(n+1)}))^{m_1}\frac{S_{ca}^{(n+1)}}{S_{cw}^{(n+1)}+S_{ca}^{(n+1)}}$$
$$\cdot\frac{xP_{cs}}{\rho_c RT}\sqrt{\frac{D_{ca}}{\pi\Delta t}}\left(\frac{P_{cs}}{P_{cc}+P_{c,ca}}-\frac{R_{ca}^{(n+1)}}{1+R_{ca}^{(n+1)}}\right) \qquad (32)$$

3) Numerical modeling of dissolution from gas phase to water phase

Dissolution of evaporated chemical to water has been modeled with Henry's law.

$$c_w = HP_c \qquad (33)$$

where, c_w : weight concentration of chemical in water phase
H : Henry's constant

In this model, material exchange between gas phase and water phase is caused only by dissolution into water, and concentration of chemical in water calculated by Henry's law is the upper limit of dissolution.

And, transient diffusion model has been imported to construct numerical model of transient dissolution, as is shown in former chapters.

With Henry's law and transient diffusion model, material exchange rate in gas phase has been calculated by next equation.

$$f_{cs}^{cw-ca} = \lambda(VS_{ca}^{(n+1)})^{m_1}\frac{S_{cw}}{1-S_{ca}}$$
$$\cdot\sqrt{\frac{D_{cw}}{\pi\Delta t}}\left(\frac{HR_{ca}^{(n+1)}(P_{cc}+P_{c,ca})}{\rho_c(1+R_{ca}^{(n+1)})}-\frac{R_{cw}^{(n+1)}}{1+R_{cw}^{(n+1)}}\right) \qquad (34)$$

And, volumetric exchange rate in water phase has been calculated by similar method to volatilization modeling as is shown in next equation.

$$f_{cs}^{ca-cw} = \lambda(VS_{ca}^{(n+1)})^{m_1}\frac{\rho_c RTS_{cw}^{(n+1)}}{x(1-S_{ca}^{(n+1)})}\frac{(1+R_{ca}^{(n+1)})}{R_{ca}^{(n+1)}(P_{cc}+P_{c,ca})}$$
$$\cdot\sqrt{\frac{D}{\pi\Delta t}}\left(\frac{HR_{ca}^{(n+1)}(P_{cc}+P_{c,ca})}{\rho_c(1+R_{ca}^{(n+1)})}-\frac{R_{cw}^{(n+1)}}{1+R_{cw}^{(n+1)}}\right) \qquad (35)$$

3. NUMERICAL EXPERIMENT OF MATERIAL EXCHANGE

In order to validate numerical modeling of material exchange phenomena, the authors have carried out some numerical experiment of material exchange, in which convection term cased by flow has not been considered.

At first, simulation of dissolution has been carried out. In this calculation, initial volumetric ratio of water phase and chemical phase has been set to 1:1. And, diffusion coefficient about dissolution has been changed. Result of simulation is shown in Figure 4a.

From this result, it is shown that numerical modeling of dissolution with transient diffusion model would be valid, and diffusion coefficient shows significant influence to dissolution rate.

And, for the next step, simulation of volatilization has been carried out. In this case, initial volumetric ratio of chemical phase and gas phase has been set to 1:1. And, initial pressure in gas phase has been set to 1.033 kgf/cm^2. Chemical material has been assumed to be PCE, of which, saturation pressure is 20 mmHg. In this simulation, transient total pressure in gas phase has been calculated.

Result of volatilization simulation is shown in Figure 4b.

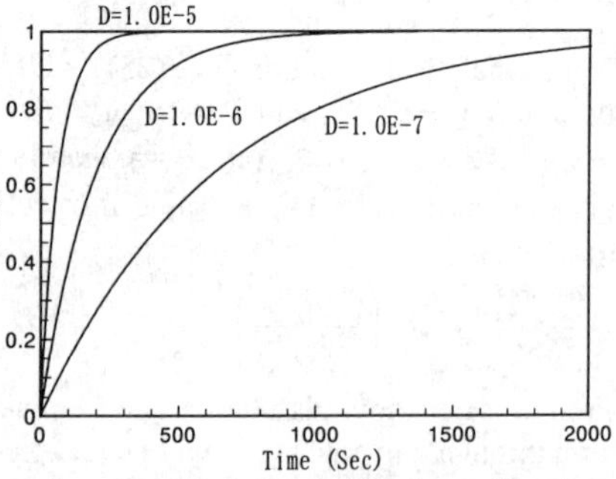

Figure 4a Result of simulation about dissolution

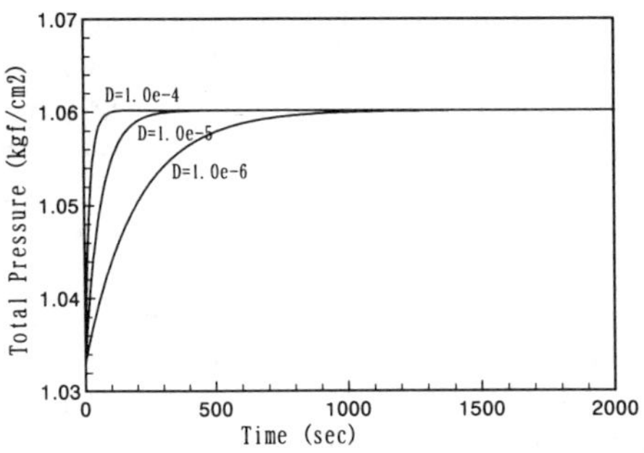

Figure 4b Result of simulation about volatilization

In this case, partial pressure of air in gas phase has not changed, although, partial pressure of evaporated chemical in gas phase has changed according to volatilization. And, transient change of pressure has been similar to concentration change in dissolution simulation.

Although, these material exchange phenomena has been able to modeled and calculated, the assumed parameters, such as diffusion coefficient in transient diffusion model must be identified by laboratory experiment.

4. FIELD APPLICATION

The authors has applied numerical simulation to actual field application. In this case, groundwater contamination with PCE caused by small laundry factory has been calculated.

Grid model for numerical simulation is shown in Figure 5.

In this area, multiple aquifer has been formed with impermeable clay layers. However, in this study, top aquifer has been imported in numerical model. The horizontal extent of calculated model has been 350m x 250m, and for vertical direction, the extent between surface and impermeable layer has been divided to finite difference grid.

Boundary condition has been set as fixed head, that has been determined from calculation with wider model.

Contaminant source has been set as is shown in Figure 5, with seepage of pure PCE during 20 years. Flow rate of PCE seepage has been set to 5l/day in former 10 years, and 10l/day in latter 10 years.

As the result of calculation, transient change of concentration has been picked up at the center of grid that includes monitoring borehole No.1 and 2.

The result of calculation with measured value is shown in Figure 6. At the bottom of No.2, measured

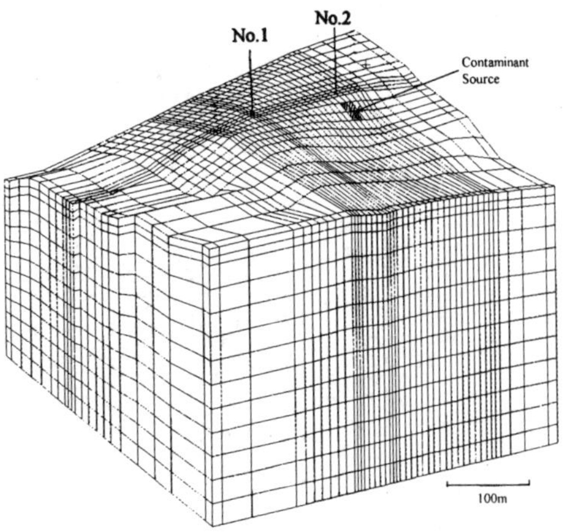

Figure 5 Grid model for field application

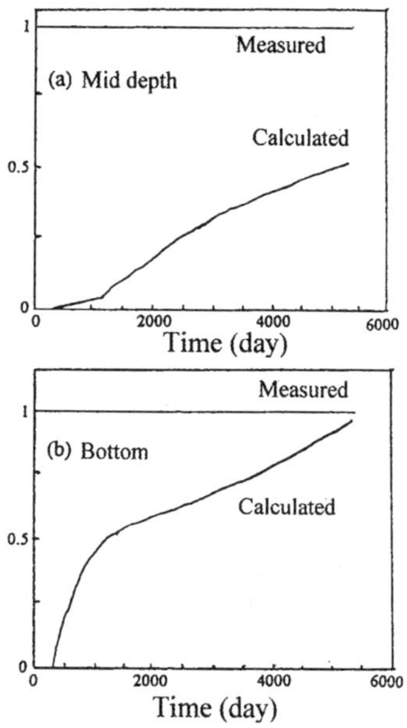

Figure 6 Results of calculation in comparison with measured concentration

and calculated concentration shows good accordance with each other. However, at the mid depth, measured value is about twice of calculated value. It would be according to low solubility of PCE, and relatively high downward velocity of pure PCE.

And, horizontal distribution of concentration at mid depth is shown in Figure 7.

According to water table that inclines from upper right corner to lower left corner in this figure, concentration distribution spreads from contaminant

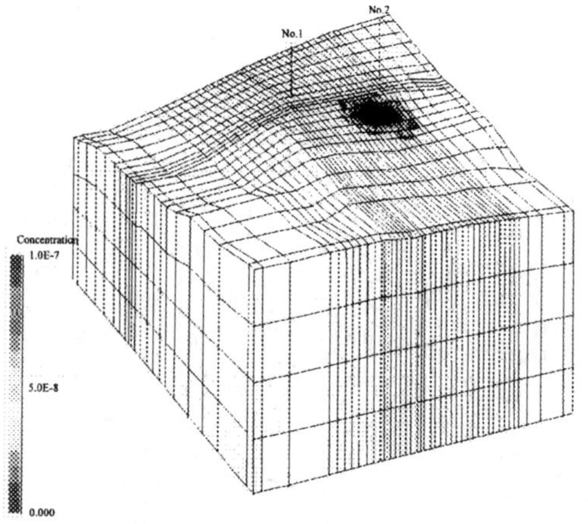

Figure 7 Horizontal distribution of concentration from numerical simulation

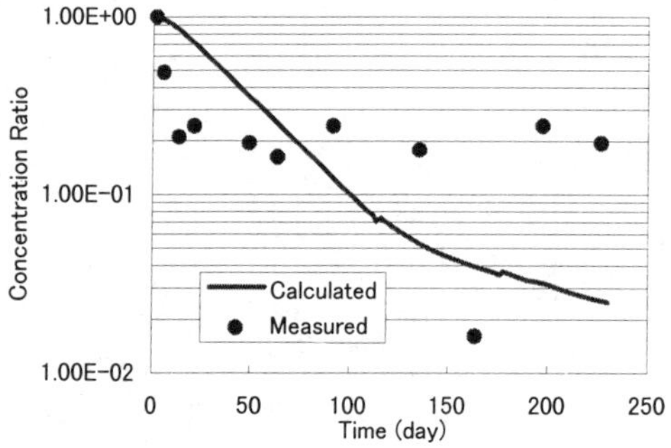

Figure 8 Result of numerical simulation of pumping

source to the direction of No.1. And, this distribution shows relatively good correspondence to real distribution obtained from field survey.

In this area, at present situation, artificial remediation by pumping is now being operated. For the second step, numerical simulation of pumping has been carried out. Figure 8 shows the result of calculated and measured concentration that are normalized by initial concentration.

From this result, comparing with the measured concentration, calculated value shows the rapid drop. However, because of the fluctuation of measured value, these two series of data shows relatively good accordance with each other.

5. CONCLUSION

In this study, the authors have presented the multi-phase flow formula for groundwater contamination problems that imply dissolution or volatilization, and some results of numerical experiments and actual field application.

Followings are the result of this study.

1) Multi-phase flow formulation is appropriate method to express flow of contaminants with dissolution and volatilization.
2) Transient diffusion modeling and numerical simulation for dissolution and volatilization has been validated.
3) Multi-phase simulation has been applied to actual contaminated field, and this method has shown relatively good correspondence with measured value.

For further application of this method, some parameters in transient diffusion model must be identified through laboratory experiments.

REFERENCES

Tosaka, H., ITOH, K., EBIHARA, M., INABA, K., ITOH, A. & KOJIMA, K. 1996. Comprehensive Treatment of Groundwater Pollution by Multi-component, Multi-phase convection / Diffusion Modeling. Journal of Groundwater Hydrology. Vol.20 No.3 pp.167-180 (in Japanese with English abstract).

SLEEP, B.E. & SYKES J.F. 1993 Compositional Simulation of Groundwater Contamination by Organic Compounds. 1. Model Development and Verification. WATER RESOURCES RESEARCH. Vol. 29 No.6 pp.1697-1708

Computer Methods and Advances in Geomechanics, Yuan (ed.) © 1997 Balkema, Rotterdam, ISBN 90 5410 904 1

Numerical modeling of miscible pollutant transport by ground water in unsaturated zones

Xikui Li
Dalian University of Technology, China

J.P. Radu & R.Charlier
University of Liège, Belgium

ABSTRACT: A numerical model to simulate miscible pollutant transport through unsaturated soils is presented. A modified version of the characteristic Galerkin method and a fully explicit algorithm are developed for discretization of the equations governing the pollutant transport phenomena and the numerical solutions. The numerical examples validate the performance and the capability of the model and algorithms.

1 INTRODUCTION

The numerical model to simulate pollutant transport through unsaturated soils is composed of the two coupled phases : the hydro-mechanical analysis of the two immiscible fluids through the porous medium to determine the velocity field of the porous water and air flows and the saturation degrees; the pollutant transport analysis through the porous medium. The emphasis of the present paper for the numerical modelling is specially given to the second phase.

The main governing phenomena of the miscible pollutant transport modelled in the present work include: convection; mechanical dispersion and molecular diffusion. In addition, three other phenomena responsible for temporary pollutant storage (or release) and inducing retardation effects, that is adsorption, degradation and immobile water effect, are also integrated into the model (Bear and Verruijt, 1987).

Solute transport in porous media can be governed by transient advection-diffusion equations with primary unknowns c_m, where c_m is the pollutant volumetric concentration in porous mobile water, in a generalized sense as certain retardation terms appear in the equations due to such phenomena as adsorption, degradation and immobile water effect integrated into the model. It is remarked that the pollutant concentration in the immobile water and in the solid particles are treated as state variables at the element integration points.

The difficulties in numerical solution of the advection - diffusion equations have long been recognized. Introduction of the characteristic Galerkin method gives great impetus to the extensive development of the finite element procedure for the solution of the acvection - diffusion equations. The success of the method in solving the advection - diffusion equations is achieved by a suitable operator splitting procedure. The key to such a split lies in a fractional step method derived originally by (Chorin,1967) and subsequently developed by others for incompressible flows. Zienkiewicz and Taylor(1991) and Zienkieiwcz and Codina (1995) applied the characterized Galerkin method to the advection - diffusion problems and developed an explicit algorithm for the problems. To take into account the retardation effects due to such phenomena as adsorption, degradation and immobile water effect integrated into the model, a modified version of the characteristic Galerkin method and corresponding explicit algorithm are developed in the present paper for the discretization of the equations governing the pollutant transport and the numerical solutions.

The numerical examples validate and demonstrate the performance and the capability of the presented numerical model.

2 GOVERNING PHENOMENA AND CONSTITUTIVE FORMULAE FOR POLLUTANT TRANSPORT IN UNSATURATED SOILS

The constitutive formulae to describe each of the phenomena governing the pollutant transport in unsaturated soils are described as follows.

2.1 Advection

The advective flow of the pollutant carried in the porous mobile water at c_m is defined by

$$\mathbf{J}_c = c_m \dot{\mathbf{U}}_w \tag{1}$$

where the intrinsic phase velocity of the mobile water

$$\dot{\mathbf{U}}_w = \dot{\mathbf{u}} + \frac{\dot{\mathbf{w}}}{\theta_m} \approx \frac{\dot{\mathbf{w}}}{\theta_m} \tag{2}$$

with the volumetric portion held by the phase of porous mobile water

$$\theta_m = n(S_w - S_{w0}) \tag{3}$$

in the above $\dot{\mathbf{w}}$ is the Darcy's (average) velocity of the porous water and $\dot{\mathbf{u}}$ the velocity of the solid skeleton of the soil, which may be approximately omitted, n is the porosity, S_w the saturation degree of the porous water, S_{w0} the portion of the immobile water in S_w, nS_{w0} is the volumetric proportion held by the immobile porous water symbolized by

$$\theta_{im} = nS_{w0} \tag{4}$$

2.2 Molecular diffusion

The molecular diffusive mass flux produced by the random movement of molecules in the mobile porous water due to concentration gradient is defined by the Fick's law as

$$\mathbf{J}_m = -D_m \nabla c_m \tag{5}$$

where D_m is the molecular diffusion coefficient in the porous medium, which is related to the porosity and the saturation degree.

2.3 Mechanical and hydrodynamic dispersion

The mechanical dispersive flux due to the variation of seepage velocity flow on the cross section during pollutant migration has been proposed to obey a Fick's type law

$$\mathbf{J}_d = -\mathbf{D}_d \nabla c_m \tag{6}$$

where $\mathbf{D}_d$ is a second rank symmetric tensor called mechanical dispersion tensor. For an isotropic porous medium, it can be expressed by

$$\mathbf{D}_d = a_T \left\| \dot{\mathbf{U}}_w \right\| \mathbf{I} + (a_L - a_T) \frac{\dot{\mathbf{U}}_w \dot{\mathbf{U}}_w^T}{\left\| \dot{\mathbf{U}}_w \right\|} \tag{7}$$

where $\mathbf{I}$ is the unit matrix, a_L is the longitudinal dispersity, i.e. the dispersity along the streamline direction and a_T is the transversal dispersity, the dispersity along the normal to the streamline.

Therefore, the dispersive flux and the diffusive flux can be added up to obtain the so-called hydrodynamic dispersive flux as

$$\mathbf{J}_h = -\mathbf{D}_h \nabla c_m \tag{8}$$

where $\mathbf{D}_h$ is the hydrodynamic dispersive tensor.

2.4 Immobile water effect

Immobile, or stagnant water may be due to the water occupying dead-end pores, or local zones with very low permeability, or may also occur, in unsaturated flow, in pendular rings of drained pores. With c_{im} denoting the pollutant concentration in the immobile water, the net rate of pollutant exchange, f_{im}^m, is often expressed by

$$f_{im}^m = \alpha_d^* (c_{im} - c_m) \tag{9}$$

where α_d^* is a transfer coefficient from the fluid in motion to the immobile water ($\sec^{-1}$) that depends on the molecular diffusion coefficient and the geometry of the contact area between immobile and mobile water.

2.5 Adsorption

Adsorption is the phenomenon of increase in the pollutant mass on the solid at a fluid-solid interface in order to minimize the surface tension. The main factors affecting the adsorption and desorption of pollutants to or from the solid are the physical and chemical characteristics of the considered pollutant and of the surface of the solid.

Adsorption is considered instantaneous and reversible. The Freundich type isotherm law of the form is employed in the present model and extended to both mobile and immobile water. The adsorbed quantities F_m and F_{im} of the pollutants by means of mobile water and immobile water can be respectively written by

$$F_m = (1-p) K_d^m c_m^{N1}; \quad F_{im} = p K_d^{im} c_{im}^{N2} \tag{10}$$

where K_d^m and K_d^{im} are called the division or distribution coefficients (m^3/kg), and exponents $N1$ and $N2$ are parameters, which are, in general, specific to the particular soil material and pollutant species under consideration. Generally

$K_d^m = K_d^{im} = K_d$ and $N1 = N2 = N$ are assumed. The formulae (12) can then be written as

$$F_m = (1-p)K_d c_m^N ; \quad F_{im} = pK_d c_{im}^N \tag{11}$$

If $N = 1$ is assumed the model is degraded to a linear adsorption one. p is the portion of the fluid-solid contact surface which concerns immobile fluids.

The adsorped mass F of pollutant species per unit mass of the solid composed of two parts: one towards mobile water and the other immobile water

$$F = F_m + F_{im} = (1-p)K_d c_m^N + pK_d c_{im}^N \tag{12}$$

2.6 Degradation

Several processes, such as chemical degradation, radioactive disintegration and etc. are responsible for the degradation phenomenon. As a linear degradation model is considered the pollutant mass degradation per second and per unit mass at each phase can be expressed by

$$\Gamma_s = -k_s F = -k_s(F_m + F_{im}) \tag{13}$$

$$\Gamma_m = -k_m \, c_m / \rho_w \tag{14}$$

$$\Gamma_{im} = -k_{im} \, c_{im} / \rho_w \tag{15}$$

where k_s, k_m and k_{im} are degradation coefficients ($\sec^{-1}$) in the solid, the mobile and immobile water.

3 GOVERNING EQUATIONS OF MISCIBLE POLLUTANT TRANSPORT

The component's mass balance equations for the pollutant in the mobile and immobile water and in the solid can be given in the forms

$$\frac{\partial(\theta_m c_m)}{\partial t} + div[\theta_m(c_m \dot{\mathbf{U}}_w - \mathbf{D}_h \nabla c_m)] = \theta_m \rho_w \Gamma_m + f_{im}^m - f_m^s + Q^* c^* \tag{16}$$

$$\frac{\partial(\theta_{im} c_{im})}{\partial t} = \theta_{im} \rho_w \Gamma_{im} - f_{im}^m - f_{im}^s \tag{17}$$

$$\frac{\partial(\theta_s \rho_s F)}{\partial t} = \theta_s \rho_s \Gamma_s + f_m^s + f_{im}^s \tag{18}$$

where $\theta_s = 1 - n$ is the volumetric proportion held by the solid phase, ρ_s is the bulk density of the solid phase and $\rho_s F$ can be regarded as the concentration of the adsorbed pollutant in the solid, Q^* is the water source flux with the concentration c^* of the pollutant species, f_x^y is the quantity of the pollutant migrated from the phase x to the phase y per unit time in a representative element volume.

Substitution of the formulae (12) and (13) into equation (18) gives the adsorbed pollutant from the mobile and immobile water

$$f_m^s = \frac{\partial}{\partial t}[(\theta_s \rho_s (1-p)K_d c_m^N] + k_s \theta_s \rho_s (1-p) K_d c_m^N \tag{19}$$

$$f_{im}^s = \frac{\partial}{\partial t}[(\theta_s \rho_s pK_d c_{im}^N] + k_s \theta_s \rho_s pK_d c_{im}^N \tag{20}$$

Substitution of equation (19) and (9) into (16) gives

$$R_m \frac{\partial(\theta_m c_m)}{\partial t} + A_m(\theta_m c_m) + \dot{\mathbf{U}}_w^T \nabla(\theta_m c_m) - div(\theta_m \mathbf{D}_h \nabla c_m) - \alpha_d^* c_{im} - Q^* c^* = 0 \tag{21}$$

where

$$R_m = 1 + \frac{\theta_s}{\theta_m} \rho_s (1-p) K_d N c_m^{N-1} \tag{22}$$

$$A_m = div\dot{\mathbf{U}}_w + k_m + \alpha_m + \frac{\partial}{\partial t}[\frac{\theta_s}{\theta_m} \rho_s (1-p) K_d] c_m^{N-1} + \frac{\theta_s}{\theta_m} \rho_s (1-p) K_d k_s c_m^{N-1} \tag{23}$$

$$\alpha_m = \alpha_d^* / \theta_m \tag{24}$$

Now consider the governing equation (17) for the immobile water. Substitution of equation (20) into (17) gives

$$R_{im} \frac{\partial c_{im}}{\partial t} + A_{im} c_{im} = \alpha_{im} c_m \tag{25}$$

where

$$R_{im} = 1 + \frac{\theta_s}{\theta_{im}} \rho_s pK_d N c_{im}^{N-1} \tag{26}$$

$$A_{im} = k_{im} + \alpha_{im} + \frac{1}{\theta_{im}} \frac{\partial \theta_{im}}{\partial t} + \frac{\partial}{\partial t}[\frac{\theta_s}{\theta_{im}} \rho_s pK_d] c_{im}^{N-1} + \frac{\theta_s}{\theta_{im}} \rho_s pK_d k_s c_{im}^{N-1} \tag{27}$$

$$\alpha_{im} = \alpha_d^* / \theta_{im} \tag{28}$$

It is noted that equation (25) is linear if N=1 is assumed and we may still analytically solve equation (25) in the time interval for c_{im} at time $t \in [t_n, t_{n+1}]$ by means of the values of the quantities at time n as follows

$$c_{im}(t) = c_{im}(t_n)\exp(-\frac{A_{im}}{R_{im}}(t-t_n)) + \frac{\alpha_{im}}{A_{im}}[(t-t_n)\dot{\bar{c}}_m$$
$$+(c_m(t_n) - \frac{R_{im}}{A_{im}}\dot{\bar{c}}_m)[1-\exp(-\frac{A_{im}}{R_{im}}(t-t_n))] \quad (29)$$

where a linear function c_m in the time interval Δt is assumed and the average rate $\dot{\bar{c}}_m$ is defined as

$$\dot{\bar{c}}_m = \frac{c_m(t_{n+1}) - c_m(t_n)}{\Delta t} \quad (30)$$

Substitution of the solution (29) for c_{im} into equation (21) gives

$$\frac{\partial(\theta_m c_m)}{\partial t} + \dot{\mathbf{U}}_w^{*\,T}\nabla(\theta_m c_m) + A\theta_m c_m$$
$$-\frac{1}{R_m^*} div(\theta_m \mathbf{D}_h \nabla c_m) + Q = 0 \quad (31)$$

where

$$A = \frac{A_m^*}{R_m^*}; \dot{\mathbf{U}}_w^* = \frac{\dot{\mathbf{U}}_w}{R_m^*}; Q = (Q_{im}(t) - Q^* c^*)/R_m^* \quad (32)$$

$$R_m^* = R_m - \frac{\alpha_{im}\alpha_m}{A_{im}}(t-t_n) +$$
$$\frac{\alpha_{im}\alpha_m R_{im}}{A_{im}^2}[1-\exp(-\frac{A_{im}}{R_{im}}(t-t_n))] \quad (33)$$

$$A_m^* = A_m + \frac{\alpha_{im}\alpha_m}{A_{im}\theta_m}\frac{\partial\theta_m}{\partial t}\times$$
$$\times\left[(t-t_n) + \frac{R_{im}}{A_{im}}[1-\exp(-\frac{A_{im}}{R_{im}}(t-t_n))]\right] \quad (34)$$

$$Q_{im}(t) = -\alpha_{im}[(\theta_{im}c_{im})_{t=t_n}\exp(-\frac{A_{im}}{R_{im}}(t-t_n))$$
$$+\frac{\alpha_m}{A_{im}}(\theta_m c_m)_{t=t_n}[1-\exp(-\frac{A_{im}}{R_{im}}(t-t_n))]] \quad (35)$$

4 FINITE ELEMENT FORMULATIONS USING CHARATERISTIC GALERKIN METHOD

According to the characteristic Galerkin method, we regard $\dot{\mathbf{U}}_w^*$ as the generalized convective velocity in equation (31) and, correspondingly, the generalized convective operator can be split from the diffusive operators. It is remarked that the real convective velosity $\dot{\mathbf{U}}_w$, instead of the generalized convective velocity $\dot{\mathbf{U}}_w^*$, is used for the diffusive operator in equation (31).

Consider time t_{n+1} as reference time. The discretization in time domain using characteristic Galerkin method and the values of the relevant quantities at time level t_n leads to a fully explicit scheme for the advection - diffusion equation (31) as

$$\Delta(\theta_m c_m) = (\theta_m c_m)^{n+1} - (\theta_m c_m)^n$$
$$= \Delta t[-\dot{U}_{w,j}^{*\ n}\frac{\partial(\theta_m c_m)^n}{\partial x_j}$$
$$+[\frac{1}{R_m^*}div(\theta_m \mathbf{D}_h \nabla c_m)]^n - (A\theta_m c_m)^n - Q^n] +$$
$$\frac{\Delta t^2}{2}\dot{U}_{w,k}^{*\ n}\frac{\partial}{\partial x_k}[\dot{U}_{w,i}^{*\ n}\frac{\partial(\theta_m c_m)^n}{\partial x_i}] +$$
$$\frac{\Delta t^2}{2}\dot{U}_{w,k}^{\ n}\frac{\partial}{\partial x_k}[-[\frac{1}{R_m^*}div(\theta_m \mathbf{D}_h \nabla c_m)]^n +$$
$$(A\theta_m c_m)^n + Q^n] = 0 \quad (36)$$

The discretization in space for $(\theta_m c_m)$ and c_m in equation (36) is carried out by the standard Galerkin method as the equation is derived from a self-adjoint problem along the characteristic and the spatial discretization by Galerkin method is optimal.

With the finite element interpolation approximation

$$(\theta_m c_m) = \sum N_j(\bar{\theta}_m c_m)_j \quad (37)$$
$$c_m = \sum N_j(c_m)_j \quad (38)$$

where $(*)_j$ is the value of function $*$ at nodal point j, N_j is the shape function at nodal point j for the interpolation of the function $*$ within the element, and denoting the vectors $(\bar{\theta}_m \mathbf{c}_m)$ and $\mathbf{c}_m$ for nodal values of $(\theta_m c_m)$ and c_m, the finite element equation for the explicit algorithm can be given by

$$\mathbf{M}[(\bar{\theta}_m \mathbf{c}_m)^{n+1} - (\bar{\theta}_m \mathbf{c}_m)^n] =$$
$$-\Delta t[(\mathbf{C}_1 + \mathbf{C}_2)(\bar{\theta}_m \mathbf{c}_m)^n + \mathbf{K}(\mathbf{c}_m)^n + \mathbf{f}^n +$$
$$\Delta t[(\mathbf{K}_{u1} + \mathbf{K}_{u2})(\bar{\theta}_m \mathbf{c}_m)^n + \mathbf{f}_s^n + \mathbf{D}(\mathbf{c}_m)^n]]$$
$$+\frac{\Delta t^2}{2}[\int_\Gamma \mathbf{N}^T[(\dot{\mathbf{U}}_w^*)^{n^T}\mathbf{n}](\dot{U}_{w,j}^*)^n\frac{\partial(\theta_m c_m)^n}{\partial x_j}d\Gamma +$$
$$\int_\Gamma \mathbf{N}^T[(\dot{\mathbf{U}}_w)^{n^T}\mathbf{n}](A\theta_m c_m)^n d\Gamma]$$
$$+\frac{\Delta t^2}{2}[\int_\Gamma \mathbf{N}^T[(\dot{\mathbf{U}}_w)^{n^T}\mathbf{n}]Q^n d\Gamma -$$
$$\int_\Gamma \mathbf{N}^T[(\dot{\mathbf{U}}_w)^{n^T}\mathbf{n}][\frac{1}{R_m^*}div(\theta_m \mathbf{D}_h \nabla c_m)]^n d\Gamma] \quad (39)$$

where $\mathbf{n}$ is the unit normal vector to the boundary Γ, $\mathbf{N} = [N_1\ N_2 \ldots\ldots N_{NNODE}]$ is the row vector, and

$$\mathbf{M} = \int_{\Omega} \mathbf{N}^T \mathbf{N} d\Omega \; ; \; \mathbf{C}_1 = \int_{\Omega} \mathbf{N}^T \dot{U}^{*\,n}_{w,k} \frac{\partial \mathbf{N}}{\partial x_k} d\Omega$$

$$\mathbf{C}_2 = \int_{\Omega} \mathbf{N}^T A \mathbf{N} d\Omega \; ; \; \mathbf{f}^n = \int_{\Omega} \mathbf{N}^T Q d\Omega$$

$$\mathbf{K} = \int_{\Omega} \frac{1}{R_m^*} \frac{\partial \mathbf{N}^T}{\partial x_k} \theta_m D_{h,kl} \frac{\partial \mathbf{N}}{\partial x_l} d\Omega$$

$$\mathbf{K}_{u1} = \frac{1}{2} \int_{\Omega} \frac{\partial (\mathbf{N}^T \dot{U}^*_{w,k})}{\partial x_k} \dot{U}^*_{w,l} \frac{\partial \mathbf{N}}{\partial x_l} d\Omega$$

$$\mathbf{K}_{u2} = \frac{1}{2} \int_{\Omega} \frac{\partial (\mathbf{N}^T \dot{U}_{w,k})}{\partial x_k} A \mathbf{N} d\Omega$$

$$\mathbf{f}_s^n = \frac{1}{2} \int_{\Omega} \frac{\partial (\mathbf{N}^T \dot{U}_{w,k})}{\partial x_k} Q d\Omega$$

$$\mathbf{D} = \frac{1}{2} \int_{\Omega} \frac{\partial (\mathbf{N}^T \dot{U}_{w,k})}{\partial x_k} \frac{\theta_m}{R_m^*} \frac{\partial}{\partial x_l} \left(D_{h,lm} \frac{\partial \mathbf{N}}{\partial x_m} \right) d\Omega \qquad (40)$$

5 NUMERICAL RESULTS

As the first example an one dimensional problem is considered. The bar is 30 meters long and 1 meter high meshed by 30 $\times 1$ uniform four noded isoparametric elements. A Gaussian concentration distribution of unit initial amplitude centred at co-ordinate x = 5.0, shown in Figure 1, is applied. The steady and uniform intrinsic velocity filed with $\dot{U}_{w,x} \equiv 1$ over the bar is assumed. In the first case, the transport of pollutants by pure advective flow, i.e. with the infinite value of Peclet number, is only involved. The Courant number used is $C_r = 0.25$. Figure 1 illustrates the concentration distributions at time t = 10 sec.. It is observed that the results obtained by using the characteristic Galerkin method agree very well with the exact solutions and perform better than those given by SUPG method. Figure 2 gives the decay curve of the concentration amplitude through the nodal points along axis X, which also shows the better performance of the characteristic Galerkin method than SUPG method in this example.

To illustrate the effects of various mechanisms on the pollutant transport, we consider five test cases for the bar subjected to the same initial concentration distribution and the steady advective flow as the above. The pollutant transport for the five test cases are governed respectively by (1) the convection velocity in the porous fluid at the concentration. (C); (2) the convection and hydrodynamic dispersion with $D_m = 0$ and $a_L = 0.5$ (C & DI); (3) the convection, the hydrodynamic dispersion and immobile water effect with the parameters $\alpha_d^* = 8.0448 \times 10^{-3}$ and the immobile water portion $S_{w0} = 0.25$ in $S_{w0} = 0.5$ (C & DI & IM); (4) the convection, the hydrodynamic dispersion, immobile water effect and degradation with the parameters $k_m = k_{im} = 0.01$ (C & DI & IM & D); (5) the convection, the hydrodynamic dispersion, immobile water effect, the degradation and adsorption with the parameters $k_s = 0.5$, $K_d = 0.01$ and $p = 0.5$ (C & DI & IM & D & A). The curves shown in Figure 3 illustrate the effects of the various phenomena governing the pollutant transport on the pollutant distributions.

The second example, as a 2d problem is taken from (Brooks and Hughes 1982). The transient solutions for the example can be regarded as a device to obtain the steady state solution. The domain is 10×10 meters square and a 10×10 square mesh in space is

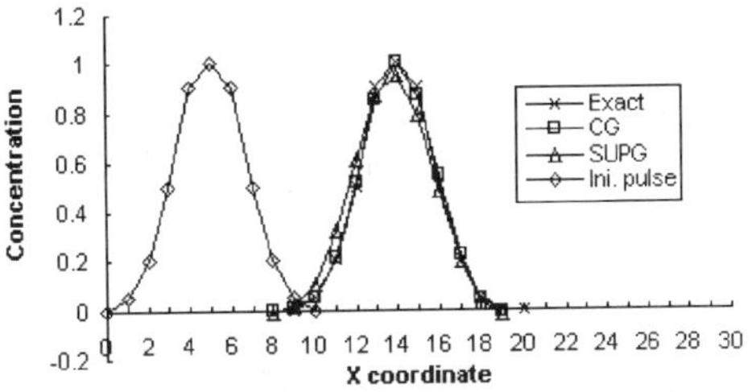

Figure 1 Concentration distribution at t=0 (initial pulse)

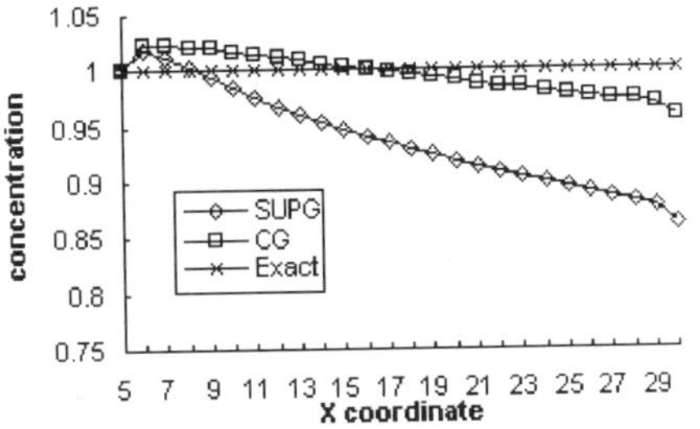

Figure 2 Concentration amplitute curves for pulse problem

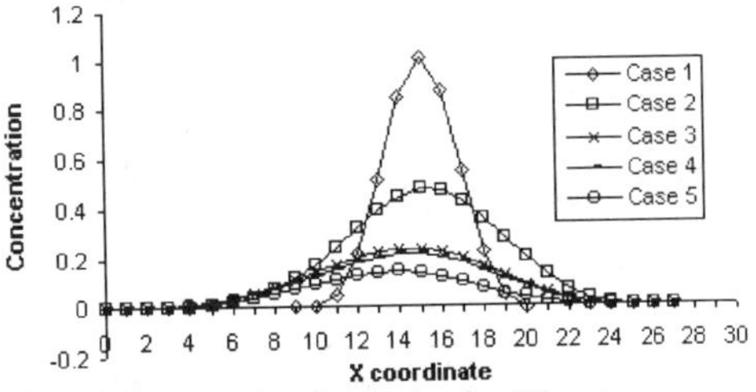

Figure 3 Concentration distribution for different cases

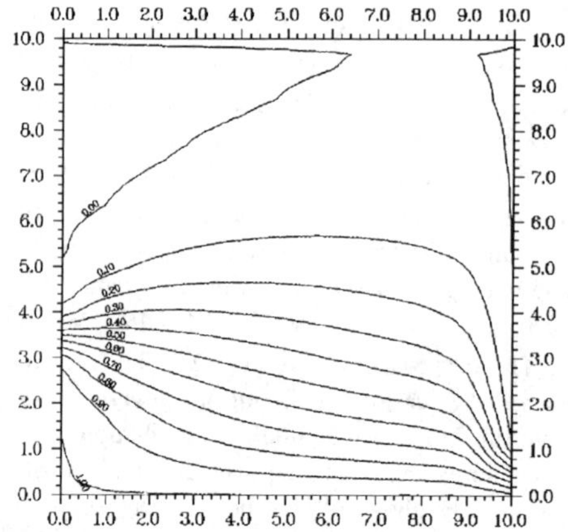

Figure 4 Concentration contour at t = 10 sec.

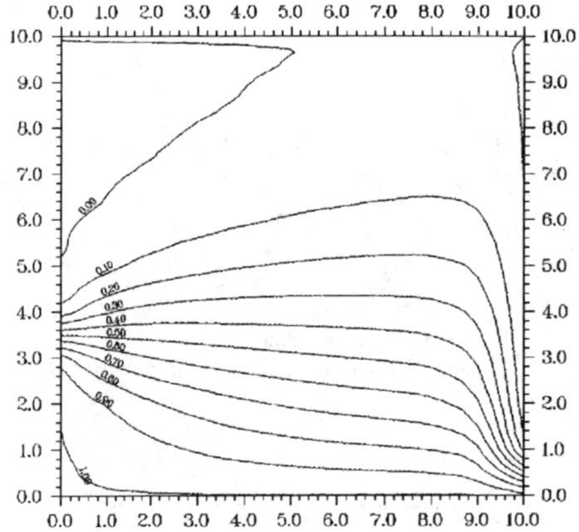

Figure 5 Concentration contour at t = 30 sec.

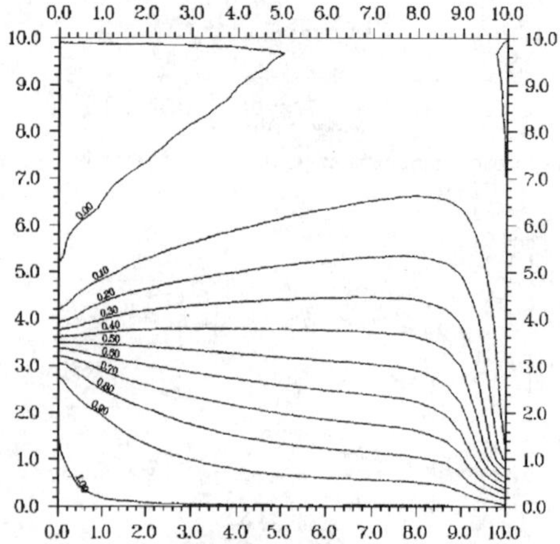

Figure 6 Concentration contour at t = 40 sec.

used. The concentration $c_w = 1$ is imposed along the bottom boundary edge and the lower 3 meters of the left boundary edge and $c_w = 0$ along the rest of the boundaries. The convective velocity $\|\dot{\mathbf{U}}_w\| = 1$ and is oriented with an angle of 30° with respect to the bottom boundary edge. In the example, all the governing phenomena, i.e. the convection, the dispersion($a_L = a_T = 0.5$ and $D_m = 0$),the immobile water effect ($\alpha_d^* = 8.0448 \times 10^{-3}$, $S_{w0} = 0.25$ in $S_w = 0.5$), the degradation($k_s = k_m = k_{im} = 0.01$) and the adsorption($K_d = 0.01$ and $p = 0.5$) are involved.

The contours given in Figure 4-6 illustrate the concentration distributions of the pollutants for the test case at the three time instants, i.e. t = 10 sec., 30 sec. and 40 sec. respectively. It is observed that the resulting concentration distributions well converge to to the steady solutions.

ACKNOWLEDGMENT

This work is sponsored by the European Commission Research Contract CI1*CT94-0014(DG12 HSMU) and the National Natural Science Foundation of China (Research Project No. 59478014). This support is greatly acknowledged.

REFERENCES

Bear J. and Verruijt A. 1987. Modelling Groundwater Flow and Pollution, D.Reidel Publishing Company, Holland.

Radu J.P., Biver P., Charlier R. and Cescotto S. 1994. 2D and 3D finite element modelling of miscible pollutant transport in groundwater,below the unsaturated zone, Int.Conf. on Hydrodynamics, Wuxi, China.

Chorin A.J. 1967. A numerical method for solving incompressible viscous problems, J.Comput.Phys., 2:12-26.

Zienkiewicz O.C. and Taylor R.L. 1991. The Finite Element Methods, Chaper 12, 4th Edn.,Vol.2, MrGraw-Hill.

Zienkiewicz O.C. and Codina R. 1995. A general algorithm for compressible and incompressible flow - Part I. The split, characteristic-based scheme, Int. J.Numer. Methods Fluids, 20: 869-885.

Brooks A.N. and Hughes T.J.R. 1982. Streamline upwind Petrov Galerkin formulation for convection dominated flows with particular emphasis on the incompressible Navier Stokes equation, Computer Methods in Appl. Mech. and Engng., **32**,199-259.

Computer Methods and Advances in Geomechanics, Yuan (ed.) © 1997 Balkema, Rotterdam, ISBN 90 5410 904 1

Effects of non-linear response of soil deposits on earthquake motion

H. Modaressi & A. Mellal
BRGM, Direction de la Recherche, Orléans, France

ABSTRACT: The influence of non-linear behaviour of soils on the seismic ground response is investigated using a numerical model. Cyclic and dynamic analyses are carried out and the role of coupling between acceleration components is studied. It is shown that such a coupling may increase the risk of liquefaction in loose sands. The well-known equivalent-linear approach is compared to the numerical model which is based on an elastoplastic behaviour assumption for the soil.

1 INTRODUCTION

Strong-motion records during recent earthquakes such as *Loma Prieta* (1989), *Manjil* (1990), *Northridge (1994)* and *Hyogoken-Nambu* (*Kobe*-1995) and observations in *SMART1* strong motion array (*Taiwan*) have confirmed that on soft soils the seismic motion on the ground surface is largely affected by non-linear behaviour of soils. "Evidence of a significant shift in the site's fundamental period and of low-frequency transfer of seismic energy is provided by data from strong earthquake recordings" as reported by Mohammadioun (1997). This fact has been largely admitted by geotechnical engineers and has been much debated since very recently by seismologists. In the seismological community the non-linear ground response resulting in the reduction of shear wave velocity and the increase of damping have been ignored as, in general, simultaneous recordings of weak and substantially strong motions recorded in the near field were not available.

The variation of shear modulus and damping ratio with distortion, known as *G/D-γ* curves, has been known as a significant feature of the soil behaviour submitted to cyclic loading since the pioneer works by Seed and Idriss (1968, 1970). This observation resulted in equivalent-linear approach which has been extensively used since then. Albeit its shortcomings have been repeatedly enumerated in the past, due to its versatility, it has become the major tool in practical engineering applications. On the other hand, the emergence of cyclic elastoplastic constitutive models for soils in the late 70s and early 80s has open a new horizon for soil dynamics studies (see for example Prévost and Höeg 1975, Ghaboussi and Dikmen 1978, Aubry *et al.* 1982).

The information concerning the capability of these models in representing the variation of the shear modulus and the damping ratio in a wide range of distortion, namely from 10^{-6} to 10^{-2} is scarce. Pande and Pietruszczak (1986) compared several constitutive models and reported that almost all of them, except those based directly on this property, were unsuccessful in reproducing this feature of the soil behaviour. In section 2 we will give some numerical results showing that it is possible to simulate realistic *G/D-γ* curves using elastoplastic models.

The use of such models is more suitable than equivalent-linear approach as they represent a rational mechanical process. The results obtained using the elastoplastic model for a soil column are compared to those computed by a linear-equivalent approach in section 3.

Obviously, the non-linearity associated to the elastoplasticity highlights the flaw of computations assuming body-wave decomposition hypothesis; i.e. SH-P/SV decomposition. In section 4 we will investigate such coupling effects.

The computations and output displays presented in this paper have been performed by *CyberQuake* software developed by authors.

2 FUNCTIONING OF THE MODEL

The constitutive model used in this study may be considered as a generalization of the Coulomb Friction law in which some aspects such as the dependency on the compactness state and the evolving plasticity, similar to those proposed by Aubry *et al.* (1982) and Hujeux (1985), are included. It represents a two-dimensional plane shearing and incorporates dilatancy-contractance in the direction normal to the plane.

The dynamic model is developed with a one-dimensional geometry and three-dimensional kinematics assumption; i.e. laterally infinite soil layers with three-dimensional displacement (velocity, acceleration) field. In fact, the simple geometry consideration is usually an inherent feature of simplified methods. The principal step for simplified models is to introduce important aspects of the soil behaviour such as strain shearing and/or pore-pressure build-up induced by the cyclic loading during earthquakes. Moreover, the seismic motion in the multilayer system is adequately represented considering a plane plastic shear mechanism.

Figures 1, 2, 3 and 4 show the capability of the constitutive model in simulating the monotonous and cyclic behaviour of soils. Figures 5 and 6 exhibit its efficiency in reproducing G/D-γ curves.

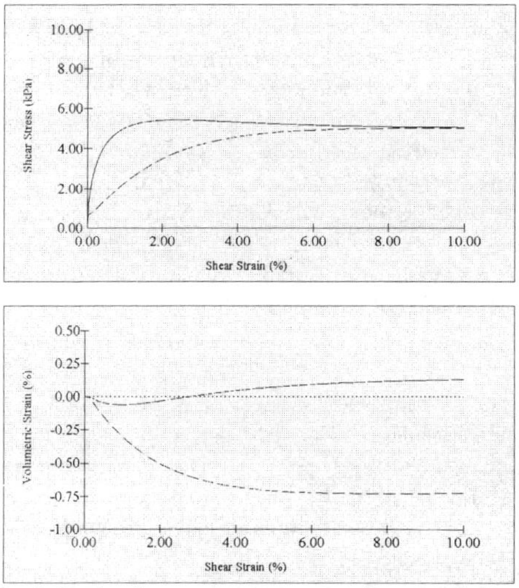

Fig. 1. Drained monotonous test on loose and dense sands

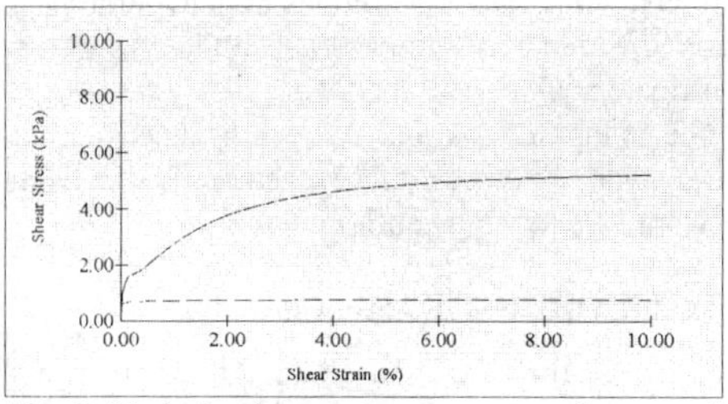

Fig. 2. Undrained monotonous test on loose and dense sands

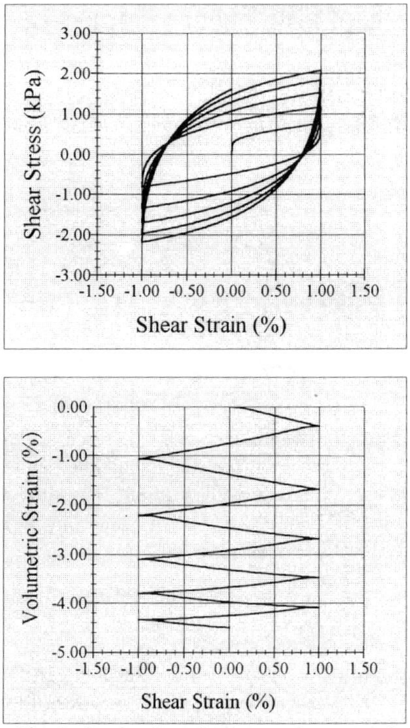

Fig. 3. Drained cyclic test on loose sand

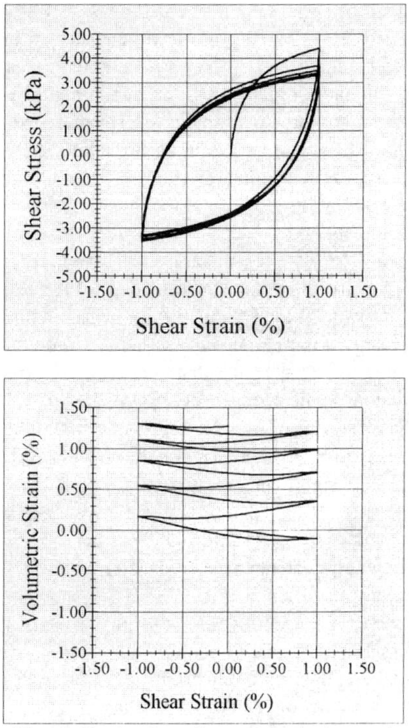

Fig. 4. Drained cyclic test on dense sand

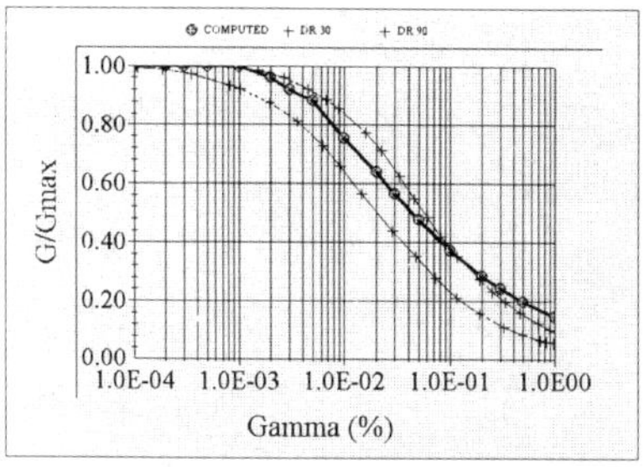

Fig. 5. Computed G-γ curve *vs.* experimental curves for loose and dense sands

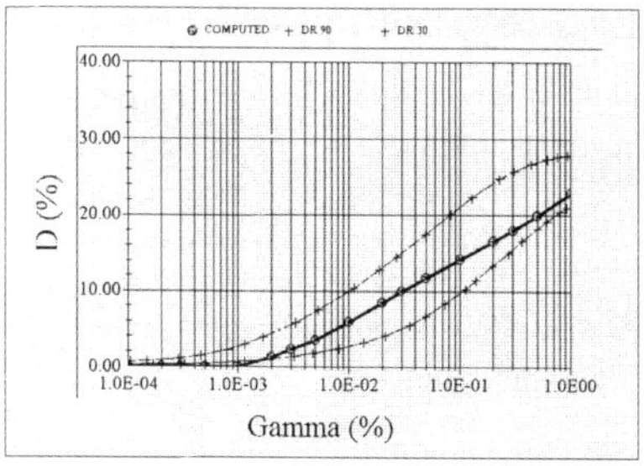

Fig. 6. Computed D-γ curve *vs.* experimental curves for loose and dense sands

3 LINEAR-EQUIVALENT/ELASTOPLASTICITY

In this section we compare results obtained assuming elastoplastic behaviour for the soil with those obtained by equivalent-linear approach for three input amplitudes. The soil column is 10 m high and totally drained. The shear and dilational wave velocities are respectively 100 and 142 m/s. The G/D-γ curves used in equivalent-linear computations are those obtained by elastoplastic constitutive model with a confining stress corresponding to the dead weight of the soil at the mid-height. Obviously, in the equivalent-linear approach the soil is supposed to be initially homogeneous. In the elastoplastic model, the plastic behaviour depends on the mean effective stress. Thus, the soil may not be considered as homogeneous. This fact should be always considered in any comparison between these methods.

The results show that for input motions resulting in small distortion ($<10^{-4}$%) identical results are obtained by equivalent-linear and elastoplastic models and correspond to an elastic behaviour of sediments (Fig. 7). For moderate distortion (10^{-4}% to 10^{-2}%) shown in figure 8 almost the same acceleration is obtained at the ground surface. For larger input motion the results differ and in elastoplastic computation irreversible vertical and horizontal displacements occur (Fig. 9). The equivalent-linear approach is obviously unable to predict such a phenomenon.

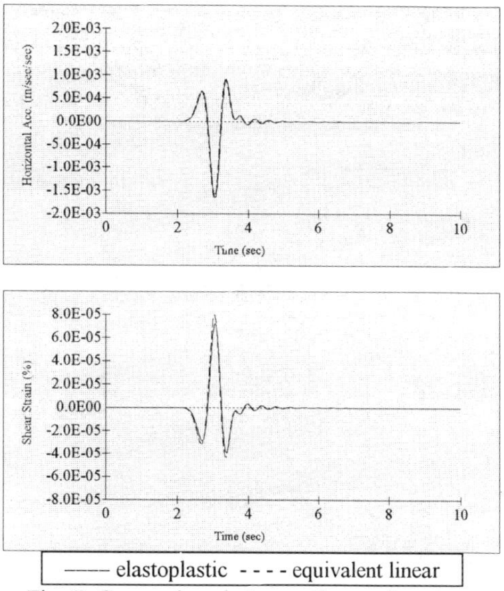

Fig. 7. Comparison between the accelerations on the ground surface and the shear strains at the middle of the soil column (small distortion)

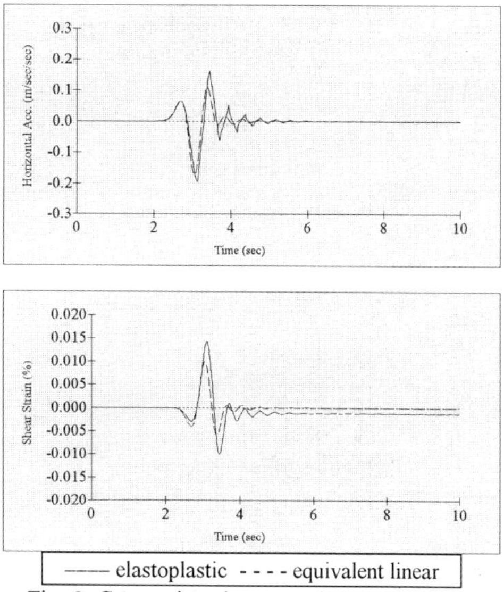

Fig. 8. Comparison between the accelerations on the ground surface and the shear strains at the middle of the soil column (moderate strain)

The frequency content of the accelerations computed on the ground surface shows that for small strain induced by the low amplitude input motion no difference is observed between equivalent-linear and elastoplastic models. For moderate strain only a slight difference is obtained especially associated to high frequency generation. For large strain two phenomena take place. With the elastoplastic model the main peak shifts toward lower frequencies and higher frequencies (absent in the input motion) are generated. The former is conform to seismic recordings on soft soil deposits and the latter is a well-known process of sub-harmonic generation.

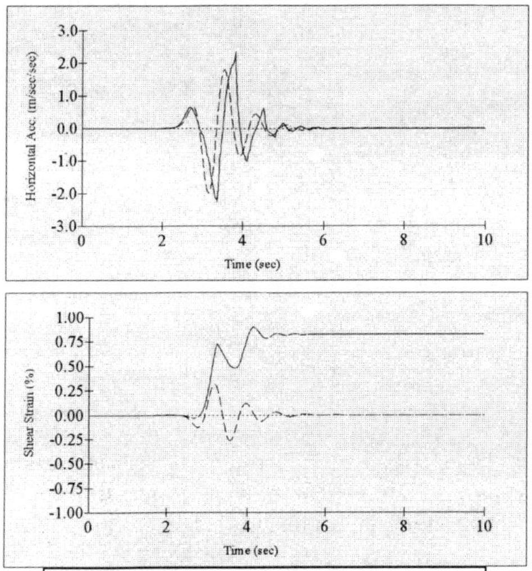

—— elastoplastic - - - - equivalent linear

Fig. 9. Comparison between the accelerations on the ground surface and the shear strains at the middle of the soil column (large strain)

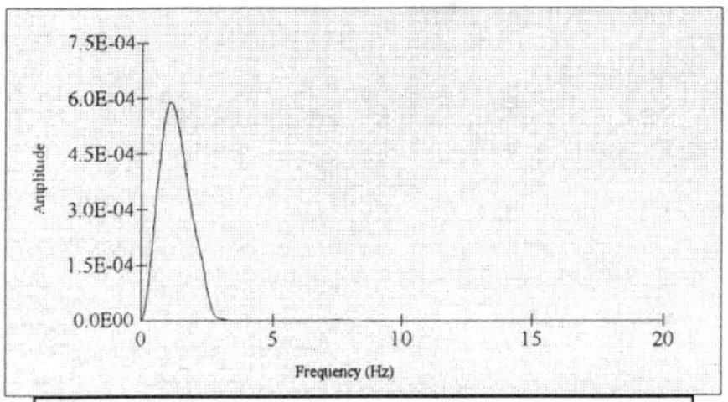

—— elastoplastic - - - - equivalent linear

Fig. 10. Fourier Transform of the acceleration on the ground surface (small strain)

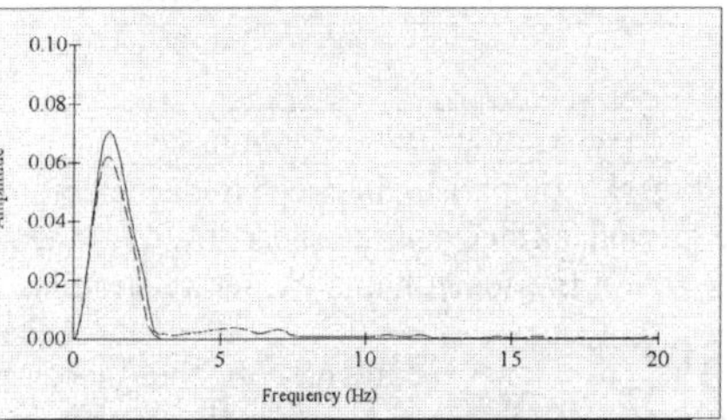

—— elastoplastic - - - - equivalent linear

Fig. 11. Fourier Transform of the acceleration on the ground surface (moderate strain)

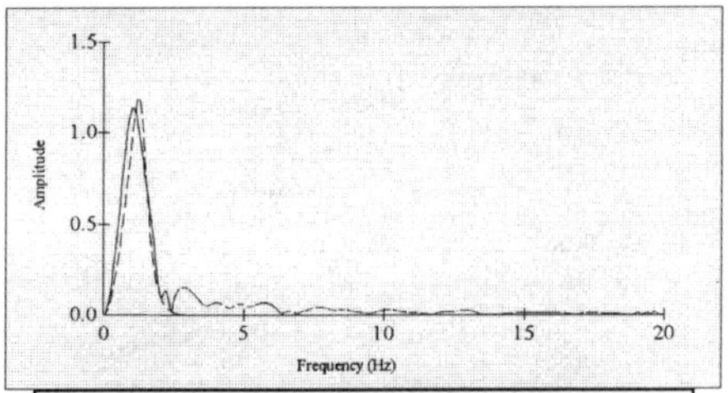

—— elastoplastic - - - - equivalent linear

Fig. 12. Fourier Transform of the acceleration on the ground surface (large strain)

4 COUPLING DUE TO PLASTICITY

In modelling seismic events near the ground surface, the incident motion is often considered as a combination of body waves. When the behaviour of the material may be considered as linear and isotropic, it is possible to decompose the input motion into SH and P-SV body waves. Hence, for computational efficiency, in-plane and out-of-plane analyses may be carried out separately. In soils exhibiting non-linear behaviour (distortion approximately more than 10^{-4}%) this decomposition is not theoretically admissible and the three-dimensional kinematics has to be considered. We have examined this aspect through cyclic tests in homogeneous conditions as well as initial-boundary value computations.

4.1. *Numerical Simulation of Cyclic Tests*

In undrained cyclic shear tests presented in the figures 13 to 18 the shear strain is applied respectively in one direction, in two orthogonal directions or consecutively in the first and the second direction. The result is always displayed in the same direction in order to make the comparison possible. As expected, the results are different due to the presence of coupling between shear deformations themselves and shear and volumetric strains.

For the loose saturated material considered here, the simultaneous shearing results in the liquefaction of the soil (Fig. 14 and 17) and the consecutive shearing brings the soil in smaller mean effective stress state (Fig. 15 and 18) compared to the shearing in only one direction (Fig. 13 and 16). The preliminary loading in the y direction results in an effective mean pressure decrease before the loading in the x direction starts. Henceforth, the deviatoric resistance of the material decreases but still an almost hysteretic response is observed and the liquefaction is delayed.

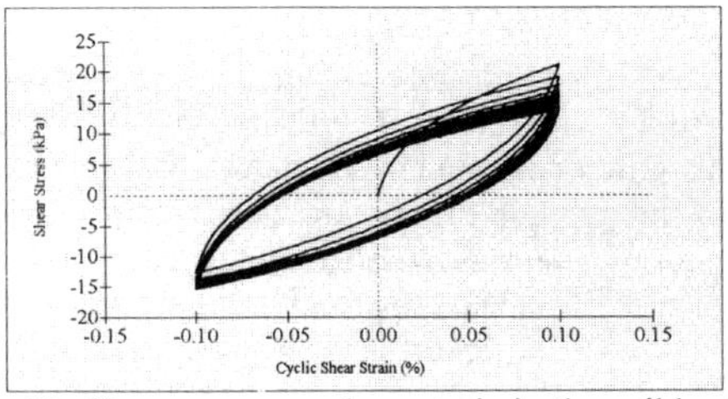

Fig. 13. Shear stress *vs.* shear strain in the soil loaded in the x direction

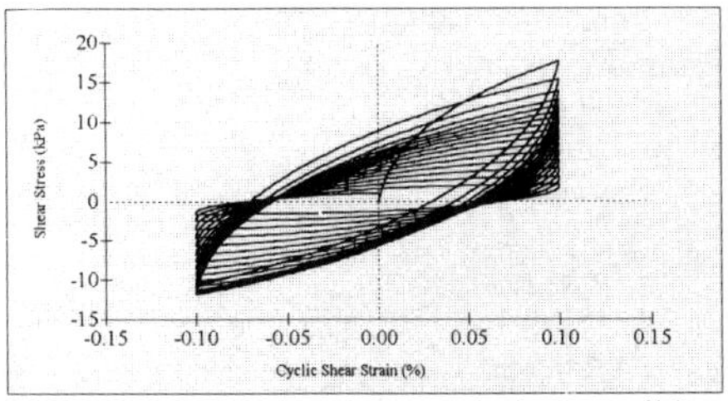

Fig. 14. Shear stress *vs.* shear strain in the soil loaded in the x and y directions simultaneously

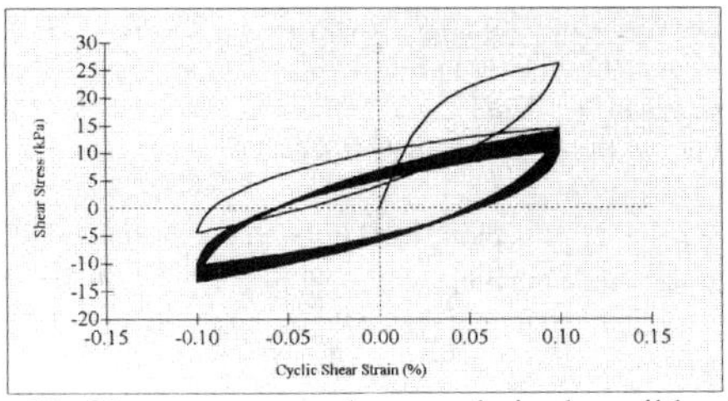

Fig. 15. Shear stress *vs.* shear strain in the soil loaded in the y then in the x direction

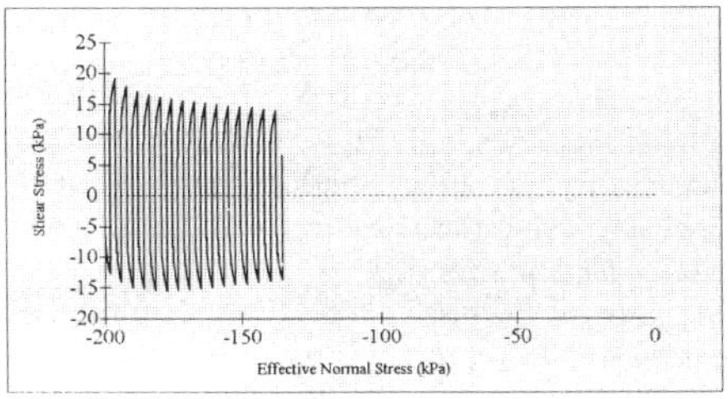

Fig. 16. Shear *vs.* effective normal stress in the soil loaded in the x direction

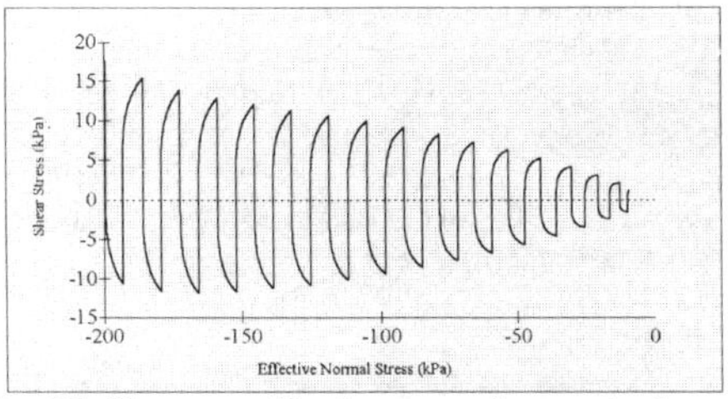

Fig. 17. Shear *vs.* effective normal stress in the soil loaded in the x and y directions simultaneously

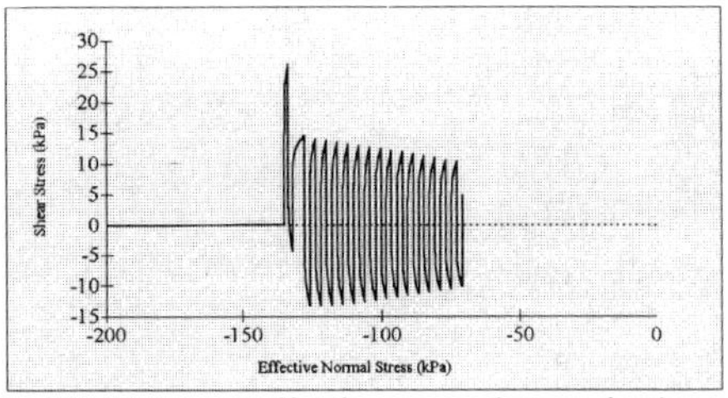

Fig. 18. Shear *vs.* effective normal stress in the soil loaded in the y then in the x direction

4.2. *Numerical Simulation of Wave Propagation*

In this paragraph we consider a 10 m high soil profile subjected either to harmonic or real accelerograms given at the bedrock outcrop. The input motion is applied either in one or two horizontal directions simultaneously and results are always displayed following the same direction (x axis). In the case of harmonic signal both horizontal components are assumed to be in phase. The water table is located at the ground surface. The results obtained in the case of harmonic input motion show that the horizontal acceleration at the ground surface is dramatically smaller (Fig. 20 compared to Fig. 19) when both horizontal components of the input motion are taken into account.

The time history of pore-pressures (Fig. 21 and 22) explain such a difference. In fact, in the case of simultaneous loading the pore-pressure increase results in the liquefaction of the soil layer. Consequently, the horizontal acceleration decreases while the horizontal displacement increases.

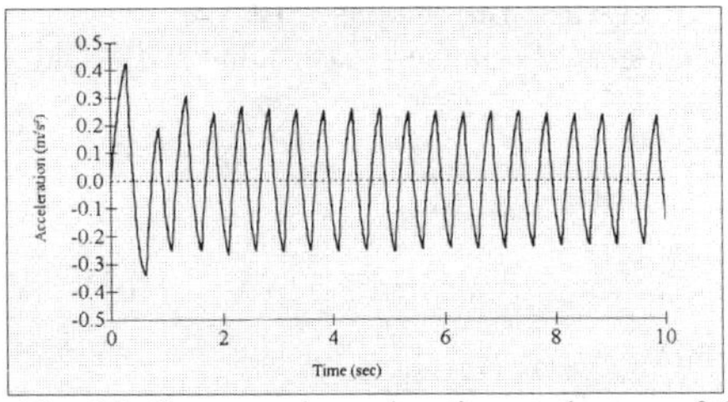

Fig. 19. Computed acceleration at the top of the soil column loaded only in the x direction

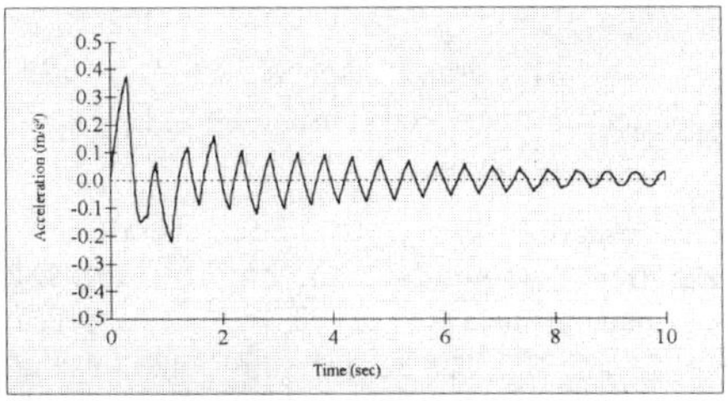

Fig. 20. Computed acceleration at the top of the soil column loaded in the x and y directions simultaneously

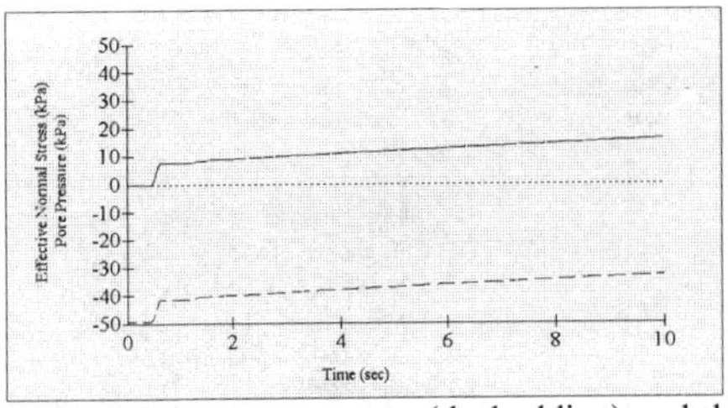

Fig. 21. Pore water pressure (dashed line) and the effective normal stress (solid line) for the loading in the x direction only

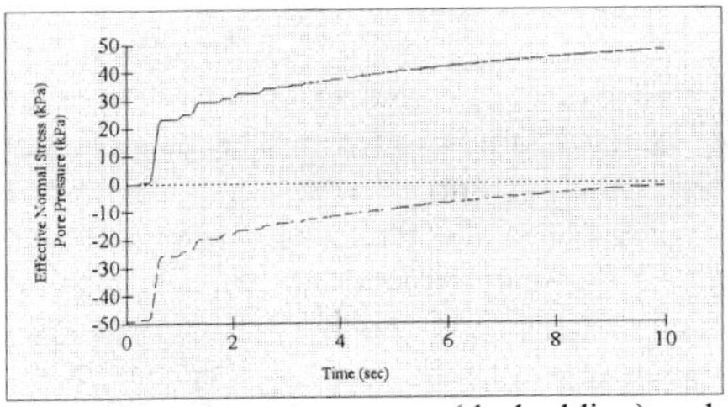

Fig. 22. Pore water pressure (dashed line) and the effective normal stress (solid line) for the loading in the x and y directions simultaneously

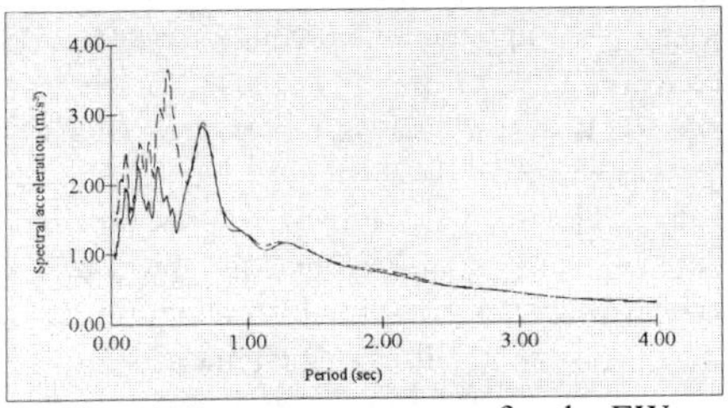

Fig. 23. Response spectrum for the EW component: EW applied (dashed line), EW and NS applied (solid line)

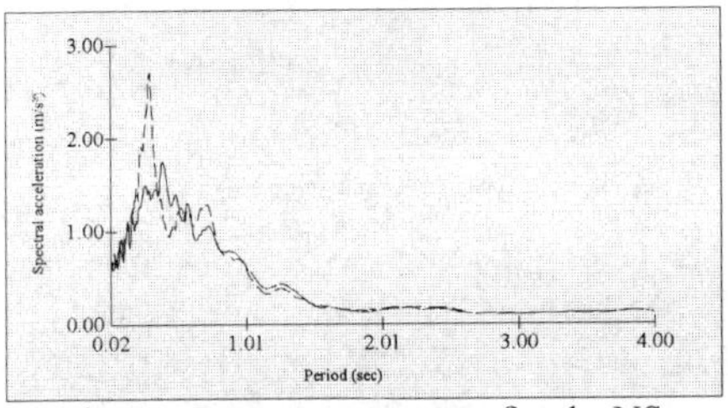

Fig. 24. Response spectrum for the NS component: EW applied (dashed line), EW and NS applied (solid line)

The pseudo-velocity response spectra of the acceleration computed at the top of soil column when subjected to a real accelerogram are presented in Fig. 23 and 24. The results show the considerable role of coupling on the frequency and amplitude modulation of the final result.

5 CONCLUSION

The need to consider the coupling aspects in seismic analysis of soft soils is investigated by a numerical model. Often neglected in seismic analysis this result is not surprising in the soil mechanics context. More specifically, we showed that in the case of loose sediments the standard approach considering in-plane and out-of-plane motions separately may not be conservative. The model is also compared to the equivalent-linear approach. We show that for input motions inducing small to moderate strains and for totally drained conditions both techniques give comparable results. However, the standard equivalent-linear approach may not be recommended as the above-mentioned coupling is not be included.

REFERENCES

Aubry, D., Hujeux, J.C., Lassoudière, F. & Meimon, Y. 1982, A double memory model with multiple mechanisms for cyclic soil behaviour, *Int. Symp. Num. Mod. Geomech.*, Zurich.

Ghaboussi, J. & Dikmen, S.U. 1978, Liquefaction analysis of horizontally layered sands, *J. Geotech. Eng. Div.*, ASCE, 104 (GT3): 341-356.

Idriss, I.M. & Seed, H.B. 1968, Seismic response of horizontal soil layers, *J. Soil Mech. Found. Div.* ASCE, 94:1003-1031.

Modaressi, H., Faccioli, E., Aubry, D. & Noret, C. 1995a, Numerical modelling approaches for the analysis of earthquake triggered landslides, *3rd Int. Conf. Rec. Advan. Geotech. Earthquake Engineering and Soil Dynamics* , St. Louis, Ed. S. Prakash, vol. 2, 833-843.

Modaressi, H., Foerster, E., Aubry, D. & Modaressi, A. 1995b, Research *vs.* professional computer-aided dynamic analysis of soils, *1st Int. Conf. Earthquake Geotech. Eng. IS-TOKYO 95,* Earthquake Geotechnical Engineering, (Ed. K. Ishihara), Balkema, Rotterdam, ISBN 90 5410 578 X, 1171-1176.

Mohammadioun, B. 1997, Non-linear response of soils to horizontal and vertical bedrock earthquake motion, *J. Earthquake Eng.*, vol. 1.

Pande, G.N. & Pietruszczak, S. 1986, A critical look at constitutive models for soils, Geomechanical Modelling in Engineering Practice, (Ed. R. Dungar and J. Astuder), 369-395.

Prévost, J.H. & Höeg, K. 1975, Analysis of pressuremeter in strain-softening soil, *J. Geotech. Eng. Div.*, ASCE, GT8: 717-733.

Seed, H.B. & Idriss, I.M. 1970, Soil moduli and damping factors for dynamic response analysis of horizontally layered sites, *Earthquake Engineering Research Center*, Univ. of California, Berkeley, CA, Report No. UCB/EERC 70-10.

11 Tunnel engineering

Computer Methods and Advances in Geomechanics, Yuan (ed.) © 1997 Balkema, Rotterdam, ISBN 90 5410 904 1

Finite element analysis of Biot's consolidation in tunnel excavation based on a constitutive model with strain softening

T. Adachi
Civil Engineering Department, Kyoto University, Japan

Liu Jun
Daiho Consulting Engineers, Japan

ABSTRACT: In this paper, based on the Biot's consolidation theory and Adachi-Oka's elastoplastic constitutive model with strain-hardening and strain-softening, a FEM coupling analysis is carried out for the mechanical behavior of surrounding ground due to the excavation.

1. INTRODUCTION

When a tunnel is excavated in fully-saturated soft sedimentary rocks, the following two types of time dependent behavior occur in the surrounding ground. The first is due to the intrinsic rate dependent characteristics of the materials, such as creep deformation or stress relaxation. The second is caused by the movement of pore water due to the change of pore water pressure distributing in the surrounding ground. The attention of this paper is focused on investigating the second type of time dependent behavior of the surrounding ground. Generally speaking, for over-consolidated clay and soft rock, the mechanical behavior usually shows not only strain-hardening and strain- softening but also time dependency. In this paper, based on the Biot's consolidation theory and Adachi-Oka's elastoplastic constitutive model with strain-hardening and strain-softening, a FEM coupling analysis is carried out for the mechanical behavior of surrounding ground due to the excavation.

2. CONSTITUTIVE MODEL WITH STRAIN SOFTENING

By introducing the stress history tensor, Oka-Adachi(1985) developed an elasto-plastic constitutive model for geologic materials in order to express both strain-hardening and strain-softening characteristics. The stress history tensor is a function of the stress history with respect to the strain measure and is given by:

$$\sigma^*_{ij} = \frac{1}{\tau}\int_0^z \exp\left(-\frac{z - z'}{\tau}\right)\sigma_{ij}(z')dz' \qquad (1)$$

$$dz = (de_{ij}de_{ij})^{1/2} \qquad (2)$$

where τ is the material parameter which controls the strain-hardening and strain-softening phenomena, z is the strain measure, and e_{ij} is the deviation strain tensor.

In the derivation of the model, the incremental plastic strain is obtained by the following non-associated flow rule:

$$d\varepsilon^p_{ij} = \Lambda\frac{\partial f_p}{\partial \sigma_{ij}}df_y \qquad (3)$$

where f_p is the plastic potential function, f_y is the yield function. We assume that the material behaves elastically in the over consolidated region only if the stress history tensor ratio η^* is kept constant. Namely, the plastic deformation takes place whenever the stress history tensor ratio changes. Therefore, we define the plastic yield function as follows:

$$f_y = \eta^* - \kappa \qquad (4)$$

where η^* is stress history tensor ratio defined by:

$$\eta^* = \left(s^*_{ij}s^*_{ij} / \sigma^{*2}_m\right)^{1/2} \qquad (5)$$

and κ is hardening-softening parameter. The evolution equation of the strain hardening-softening parameter is given by:

$$\kappa = \frac{(M^*_f - \kappa)^2 G'\gamma^p}{M^{*2}_f} \qquad (6)$$

In the case of proportional loading, it can be integrated as:

$$\kappa = \frac{M^*_f G'\gamma^p}{M^*_f + G'\gamma^p} \qquad (7)$$

where M^*_f is the value of stress history tensor ratio η^* at the residual strength state, and G' is the initial tangential modulus of the hyperbolic relation.

The plastic potential function f_p is assumed to be a function of the actual stress and is defined as

$$f_p = \bar{\eta} + \bar{M}\ln\left[(\sigma_m + b)/(\sigma_{mb} + b)\right] = 0 \quad (8)$$

where

$$\bar{\eta} = \left(s_{ij}s_{ij}/(\sigma_m + b)^2\right)^{1/2} \quad (9)$$

Furthermore, we introduce an overconsolidated boundary surface f_b which defines the boundaries of normally consolidated region(f_b>b) and over consolidated region(f_b<0). The function of boundary surface is also given by:

$$f_b = \bar{\eta} + \bar{M}_m\ln\left[(\sigma_m + b)/(\sigma_{mb} + b)\right] = 0 \quad (10)$$

In the function, σ_{mb}(corresponding to the pre-consolidated pressure for over-consolidated clay) and b are material parameters, η is the stress ratio invariant defined by the equation(9) and $\bar{M}_m$ is the value of η when the maximum compression taking place. By using the yield function and the plastic potential function in the non-associated flow rule together with the Prager's consistency condition, we obtain the plastic strain increment as follows:

$$d\varepsilon_{ij}^p = \Lambda\left[\frac{\bar{\eta}_{ij}}{\bar{\eta}} + \left(\bar{M} - \bar{\eta}\right)\frac{\delta_{ij}}{3}\right]\left[\frac{\eta_{kl}^*}{\eta^*} - \eta^*\frac{\delta_{kl}^*}{3}\right]\frac{d\sigma_{kl}^*}{\sigma_m^*} \quad (11)$$

$$\Lambda = \frac{M_f^{*2}}{G'\left(M_f^* - \eta^*\right)^2}\frac{\gamma^p\bar{\eta}}{e_{mn}^p\bar{\eta}_{mn}} \quad (12)$$

where

$$\bar{\eta}_{ij} = \frac{S_{ij}}{(\sigma_m + b)}, \quad \eta_{ij}^* = \frac{S_{kl}^*}{\sigma_m^*} \quad (13)$$

There are 8 constitutive parameters, that is, Young's modulus E, Poisson Ratio v, stress history tensor parameter τ, strain hardening-softening parameters M^*_f and G', plastic potential parameters b and σ_{mb}, and boundary surface parameters $\bar{M}_m$. They can be determined by a series of drained triaxial compression tests on the material.

3. THE COMBINATION OF BIOT'S THEORY AND FEM

Biot's consolidation equations are expressed as follows:

$$\frac{\partial\sigma'_{ij}}{\partial x_j} + \frac{\partial u_e}{\partial x_i} = 0 \quad (14)$$

$$\frac{\partial\theta}{\partial t} = -\frac{k}{\gamma_w}\nabla^2 u_e \quad (15)$$

where u_e is the pore water pressure, θ is the volumetric strain, k is the coefficient of permeability, γ_m is the unit weight of water.

Eq.(14) shows the equilibrium condition of differential element and eq.(15) shows the relation between the volume change and the drain amount respectively. In general, the body force is considered in eq.(14). But in the analysis we think that the compaction procedure due to self-weight has finished, so the influence of self-weight compaction is not considered in this study and the pore water pressure is regarded as a unknown.

On the other hand, the equation which describes the relation between volumetric strain and time adopts the form of backward difference which is more stable. The backward difference equation expressed as follows:

$$\frac{\theta|_{t+\Delta t} - \theta|_t}{\Delta t} = -\frac{k}{\gamma_w}\nabla^2 u_e|_{t+\Delta t} \quad (16)$$

We suppose that the pore water pressure is a unknown and it is a constant in a element, therefore, according to the energy principle and the related boundary conditions we can derive the expression as follows:

$$\begin{bmatrix} F \\ V \end{bmatrix} = \begin{bmatrix} K & L \\ L^T & 0 \end{bmatrix}\begin{bmatrix} \bar{u} \\ u_e \end{bmatrix} \quad (17)$$

where u is the displacement of node , F is the equivatent node force vector, K is the rigidity matrix of the usual element, L^T is the transformation vector from the node displacement into the volumetric strain. The parameters above can be expressed by suitable shape functions and their differential.

Refer to Fig.1 and consider the continuous eq.(15), the following expression can be obtained:

$$V|_{t+\Delta t} = V|_t - \{\alpha u_e - \sum \alpha_i u_{ei}\} \quad (18)$$

Let s_i be equal to the distance from point O shown in Fig.1 to the centre of element i and bi be equal to the width of drainage surface.

Then the coefficients α, α_i can be expressed as follows:

$$\alpha = \frac{k\Delta t}{\gamma_w}\sum\frac{b_i}{s_i} \quad (19)$$

$$\alpha_i = \frac{k\Delta t}{\gamma_w}\frac{b_i}{s_i} \quad (20)$$

Arrange eq.(17) in order, the following expression can be derived :

$$\begin{bmatrix} F|_{t+\Delta t} \\ V|_t \end{bmatrix} = \begin{bmatrix} K & L \\ L^T & \alpha \end{bmatrix}\begin{bmatrix} \bar{u} \\ u_e \end{bmatrix}_{t+\Delta t} - \sum \alpha_i u_{ei}|_{t+\Delta t} \quad (21)$$

Eq.(21) is true at the time of $t+\Delta t$. In order to solve nonlinar probem using increment method, eq.(21) is transformed into the following form:

$$\begin{bmatrix} \Delta F|_{t+\Delta t} + Lu_e|_t \\ 0 \end{bmatrix} = \begin{bmatrix} K & L \\ L^T & \alpha \end{bmatrix}\begin{bmatrix} \Delta\bar{u} \\ u_e|_{t+\Delta t} \end{bmatrix} - \sum \alpha_i u_{ei}|_{t+\Delta t} \quad (22)$$

where ΔF is the increment of node force during time Δt, Δu is the increment of displacement during time Δt.

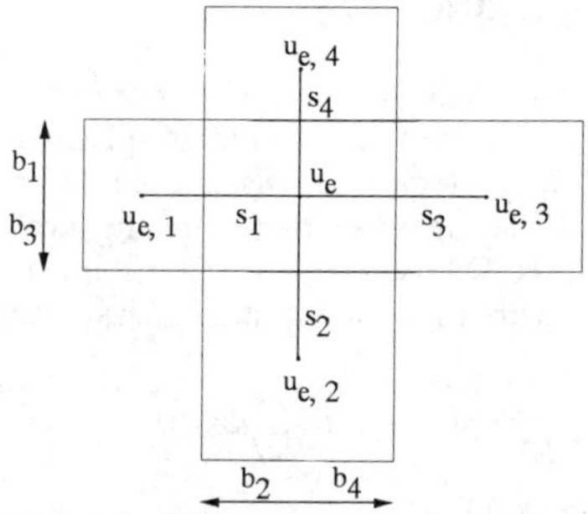

Fig.1 Model of the pore water dissipation

Table 1 Material parameters

G'	454.0	τ	0.06
σ_{mb}(kpa)	24.0	M_f^*	0.84
b_v (kpa)	1.91	ν	0.40
b_z (kpa)	1.0×10^{-5}	E (kgf/cm^2)	9.8×10^{3}
γ_s (kgf/cm^3)	0.00232	k (cm/sec)	1.0×10^{-6}

4. BOUNDARY CONDITION OF NUMERICAL ANALYSIS AND MATERIAL PARAMETERS

The material parameters used in the analysis are given in Table 1.

Fig.2 shows the finite element mesh used in numerical analysis. The mechanical behavior for the element A and element B are investigated respectively on condition that the gravitational stress field is first evaluated with an initial stress, and the lateral pressure coefficient k_0 is equal to 0.67. In addition, the following supposed conditions are also used:

1). The boundary condition is that there are only horizontal restrictions at the left side and the right side, and there are both horizontal and vertical restrictions at the bottom side.

2).In the calculation procedure, we assume that pore water pressure suction or dissipation occurs on the whole boundary of the analysis region.

3). The release method of the initial stress of the surrounding ground takes the following two cases:

Case-1 The initial stress of the surrounding ground is released stage by stage, and 2% of the stress is released once with the time increment of 100 seconds.

Case-2 The total amount of the initial stress is released immediately after the tunnel excavation is completed.

5. CONCLUSIONS AND DISCUSSION

5.1 Case-1

Fig.3 shows the variance of the softening layer with the release of initial stress, in which it is known that the softening layer lies near the horizontal parts of the inside wall of the tunnel. Form Fig.3 it is clear that the softening occurrence is only related to the position, and it can't be reputed that the softening occurrence considering the pore water pressure is earlier than that without the consideration of the pore water pressure.

Fig.4 shows the relation between the stress, pore water pressure and the release rate of initial stress for element A and B. From Fig.4(a) the stress variance of element A is known with the release of initial stress, that is, the stress decreases at first, then changes to increase. Therefore, considering the mechanical behavior of earth crust fully, we can say that it is very important to take a ideal costruction time. For element B shown in Fig.4(b), considering the pore water pressure, when the release rate of

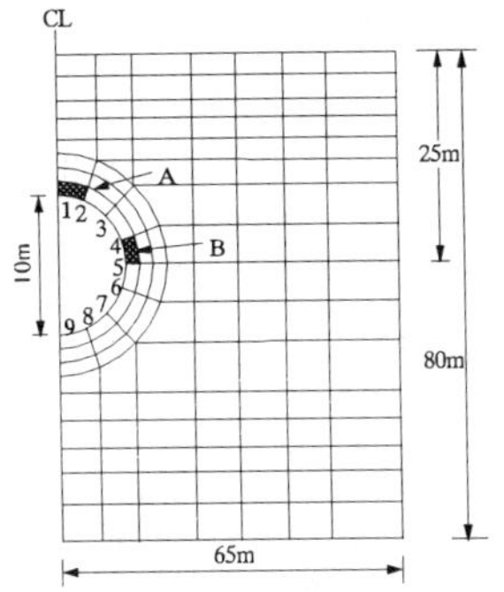

Fig.2 Finite element mesh

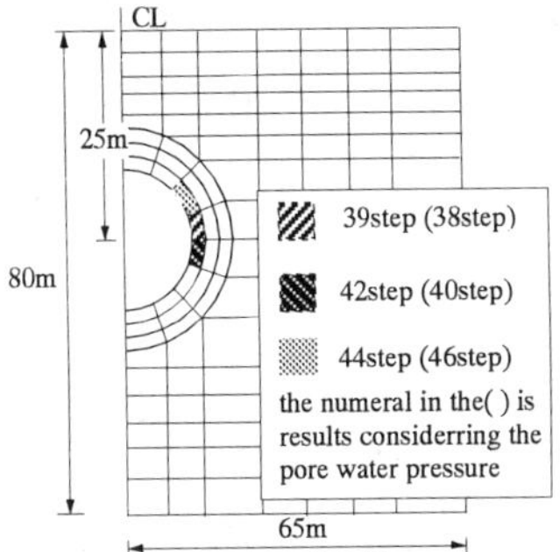

Fig.3 Development of failure zones

initial stress reaches a certain value, the element B will soften. Furthermore, the residual stress strength considering the pore water pressure is higher than that without the pore water pressure. Therefore , considering the pore water pressure is useful to the stability of the earth crust.

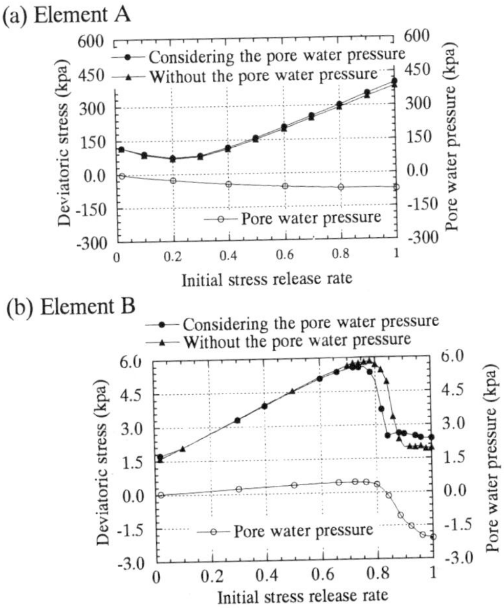

Fig.4 Relation between the stress, pore water pressure and the release rate of initial stress

Fig.5 shows the relation between the stress path and the variance of pore water pressure for elements A and B. From Fig.5(a) it is shown that the mean stress always decreaes without the consideration of the pore water pressure, but considering the pore water pressure the mean stress increases a little at first then decreases. As viewed from the decrease amount of the mean stress, it can be said that considering the pore water pressure is useful to the stability of the earth crust. From Fig.5(b), it is shown that the stress and mean stress increase at first, then decrease without the consideration of pore water pressure. But considering the pore water pressure, the mean stress decreases, this is because the element B is considerd to be in compressive state and the positive pore water pressure occures. When the element B softens, the pore water pressure decreases and becomes a negative value. Due to the influence of the negative pore water pressure, the residual strength increaes slightly.

5.2 Case-2

Fig.6 shows the region of positive pore water pressure at the moment when the all initial stress released. According to that region we can approximately know the compressive domain around the tunnel.

Fig.7 shows the relations between the stress and the mean stress for element A. In order to investigate the influence to the top of the tunnel for the pore water pressure, the stress and the mean stress for both before the pore water pressure dissipation and after the dissipation are calculated respectively with the

(a) Element A

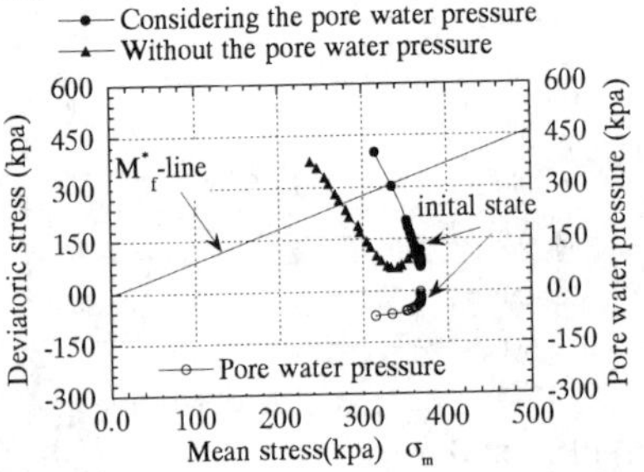

(b) Element B

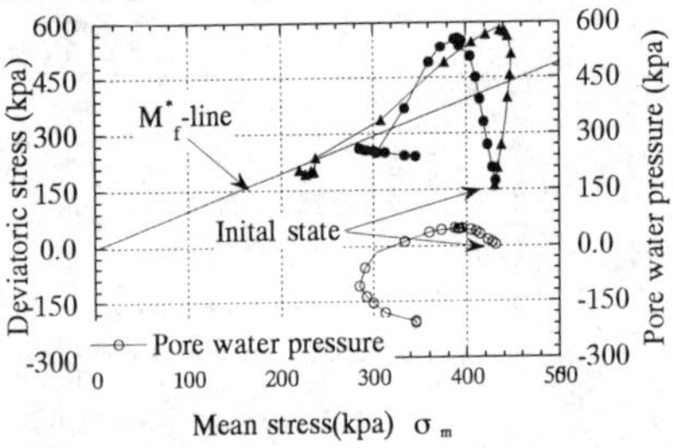

Fig.5 Stress path and the variance of pore water pressure

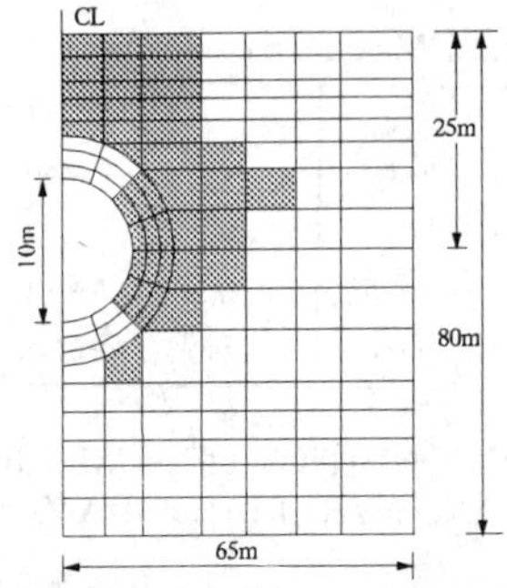

Fig.6 Region of positive pore water pressure

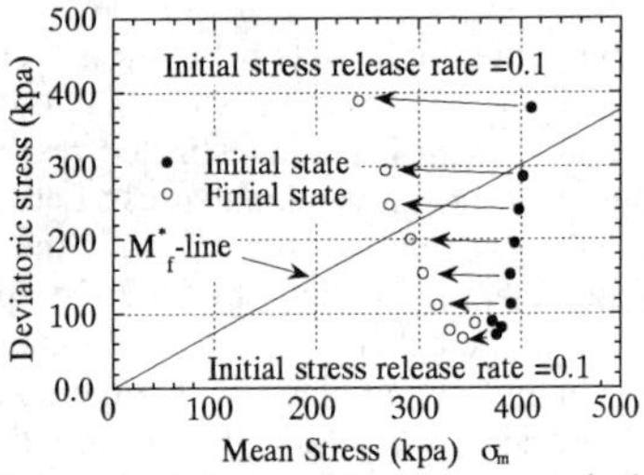

Fig.7 Relations between the stress and the mean stress

different initial stress release rate. From Fig.7 it is known that the mean stress in the finial state is smaller than that in the initial state and the stress in the finial state increases. Therefore, it can be said that the pore water pressure dissipation is not useful to the stability of the earth crust.

6. CONCLUSIONS

From above analysis, the following conclusions are obtained.

1. The softening region around the tunnel can be determined by using Adachi-Oka model
2. From the analysis results of case(2) it is shown that the tunnel is steady after excavation immediately due to the appearance of the negative pore water pressure and then the tunnel turns to unsteady with the disappearance of the negative pore water pressure.
3. The ground stress at a certain place decreases first and then increases with the release of inital stress.

REFERENCES

Oka, Adachi & Okano 1986. Two-dimensional consolidation analysis using an elasto-viscoplastic constitutive equation. International journal for numerical and analytical methods in geomechanics, vol. 10: 1-16.

Adachi & Liu.J 1996. Finite element analysis of Biot's consolidation in slope excavationbased on a constitutive model with strain softening. Balkema, Rotterdam.ISBN 9054108185 vol. 2: 1131-1136.

Computer Methods and Advances in Geomechanics, Yuan (ed.) © 1997 Balkema, Rotterdam, ISBN 90 5410 904 1

The effect on underground tunnel's stability caused by permeating in fractured rockmass

Zhu Zhende & Sun Jun
Tongji University, Shanghai, China

Wang Chongge
Shandong Institute of Mining and Technology, Taian, China

ABSTRACT: On the basis of discussing the fractured rockmass's permeability tensor, the relation of fractured rockmass's permeability tensor and its normal stress was obtained by making use of deformating curve of structure surface of concrete block tested. Softening of rock strength caused by water was analyzed by means of damage mechanics theory. The dynamical seepage force in fissure of wall rock can bring about opening of fissure, aggravating of damage, raising of seepage coefficient, and thus results in disadvantageous effect on wall rock's stability of underground tunnel.

1 INTRODUCTION

Permeability in fractured rockmass can't be neglected in rock engineerings, such as hydro-electric, oil and mines. Water exists as a seepage force which can imfluence rock's strength and deformation characteristics through changing its mechanical, physical and chemical properities. Fissure water make rock's strength decrease while its deformation increase. Sometimes, the fissure water has decisive effect on the stability of rock engineerings. Statistics shows that more than 90 percent of rock slopes' failure are related to underground water, more than 60 percent shaft construction accidents are related to pour water, more than 30~40 percent dam failure accidents are caused by seepage flow. So, studying the permeating effect on rock's strength and deformation become a very important subject.

Louis (1974) obtained experiential formula between seepage coefficient and normal stress after analyzed the experiment of drilling hole pump at one dam. Erichson (1987) set up the coupled relationship between stress and seepage flow by the idea of analyzing the deformation caused by shear or compression stress. Oda (1986) expressed the formula for seepage flow and deformation according to geometry tensor of fissures. Nolte (1989) suggested an exponent formula to describle the relationship of seepage flow and stress with the crack closing displacement. Zhou Ruiguang and Xiang Jiannan (1985) studied the variation of rock strength with the moisture content W. Kilsall (1984) studied the variation of seepage coefficient of surrounding rocks after the excavation of underground caverns. Snow (1984) obtained the formula for seepage coefficient of one group of parallel fractures by experiments.

On the basis of above results, the paper would discussing the relationship between the fractured rockmass's permeability coefficient and its normal stress from the experiments of structure surface of concrete block. Further more, the seepage coefficient of surrounding rocks after excavation is also analyzed, and the variation of permeability would affect the stability of rockmass.

2 DAMAGED MECHANICAL PROPERITIES OF ROCKS WITH WATER

The weakness extent of rock's strength caused by water is related to its physical property, initial stress state, moisture content, density and so on. A lot of experiments showed that Young's modulus of rocks are assumed to have linear relationship with mositure content in mines and to be expressed as:

$$E = E_0 - A(w - w_0) \quad (1)$$

Where: E_0 = initial Young's modulus; w_0 = initial moisture content; w = moisture content; E = instantaneous Young's modulus; A = constant.

Density of rocks varies with stress can be expressed as:

$$r_d = r_{do}/(1 - \varepsilon_v) \quad (2)$$

Where: r_{do} = density of rocks at natural state; ε_v = volume strain; r_d = instantanous density of rocks.

and $w = k_s r_w (1/r_d - 1/r_s)$ (3)

Where: k_s = coefficient of saturation water; r_w = density of water; r_s = density of granular in rocks.

Substituting (2) to (3), then substituting (3) to (1), the weakness equation of Young's modulus of rocks with no dilatancy is:

$$E = E_0 - A\{k_s r_w[\frac{1}{r_{do}} - \frac{1-2\mu}{E_0 r_{do}}(\sigma_x + \sigma_y + \sigma_z) - \frac{1}{r_s}] - w_0\} \quad (4)$$

According to continuum damage mechanics, damage tensor D of rock with water is:

$$D = 1 - \frac{E}{E_0} \quad (5)$$

Substituting (4) to (5), then the damage evolution equation is:

$$D = \frac{A}{E_0}\{k_s r_w[\frac{1}{r_{do}} - \frac{1-2\mu}{E_0 r_{do}}(\sigma_x + \sigma_y + \sigma_z) - \frac{1}{r_s}] - w_0\} \quad (6)$$

3 THE RELATION OF PERMEATING TENSOR OF FRACTURED ROCKMASS AND STRESS

Assuming rockmasses are continuum and with the principle of equal flow, equivalent continuum model can be used to analyze the permeability in fractured rockmass. Assume there are S groups of dominent fractures, snow D. T.[3] and Pomm E. C expressed the seepage tensor as:

$$[k] = \sum_{r=1}^{s} \frac{g b_r^3 m_r}{12 r}(\delta_{ij} - n_i^{(r)} n_j^{(r)}) \quad (7)$$

Where: $n_j^{(r)}$ = normal cosine of rth group fractures; b_r = opening displacement of rth group of fractures; m_r = density of rth group of fractures; r = dynamically visco coefficient of water.

In order to describle the relationship between stress and seepage tensor, deformation characteristics of concrete samples with one crack under uniaxial compression state is obtained. Figure 1 shows the relationship between deformation and axial stress. It can be concluded that with the increasing of axial stress, the opening displacement would decreasing until it is thoroughly closed.

Assuming axial normal stress to be σ_n, the maximum opening displacement of fractures to be U_{f_0}, then,

$$U = U_{f_0}(1 - e^{-\frac{\sigma_n}{k_n}}) \quad (8)$$

Where: U = displacement of fractures; k_n = equivalent closing stiffness of structure surface.

From equation (8), the formula for opening displacement structure surface as follows:

$$b_r = U_{f_0} \cdot e^{-\frac{\sigma_n}{k_n}} \quad (9)$$

Substituting (9) to (7), then

$$[k] = \sum_{r=1}^{s} \frac{g m_r}{12\gamma} U_{f_0}^3 \cdot e^{-\frac{3\sigma_n}{k_n}}(S_{ij} - n_i^{(r)} n_j^{(r)}) \quad (10)$$

Considering of permeating pressure in fractured rockmass, σ_n in equation (10) should be effective normal stress and defined as follows:

$$\sigma_n = \sigma - p = \sigma - r_w(H - Z) \quad (11)$$

Where: σ = normal stress on RVE, and $\sigma = \sigma_{ij} n_i n_j$, for plane state, $\sigma = l^2\sigma_x + m^2\sigma_y + 2lm\tau_{xy}$; P = permeating pressure acted on fractures; Z = position of RVE; H = total head of water.

Substituting (11) to (9), the relationship between the opening displacement of fractures and permeating pressure in rockmass can be expressed as:

$$b_r = U_{f_0} e^{-\sigma_{ij} n_i n_j / k_n} \cdot e^{r_w (H-Z)/k_n} \quad (12)$$

Substituting (11) to (10), then

$$[k] = \sum_{r=1}^{s} \frac{g m_r}{12\gamma} U_{f_0}^3 \{exp[-3(\sigma_{ij}^n n_i^{(r)} n_j^{(r)} - P)/k_n]\} \cdot (\delta_{ij} - n_i^{(r)} \cdot n_j^{(r)})) \quad (13)$$

Assuming surrounding rocks to be isotropic, and only the softening damage caused by water is considered, it is clear that this kind of damage should be isotropic, then equation (11) would be:

$$\sigma_n = \tilde{\sigma} - P \quad (14)$$

$$\tilde{\sigma} = \sigma/(1 - D) \quad (15)$$

Substituting (6) to (15) and (14), then

$$\sigma_n = \sigma_{ij} n_i n_j \{1 - \frac{A}{E_0}[k_s \gamma_w (\frac{1}{r_{do}} - \frac{1-2\mu}{E_0} I_1 - \frac{1}{\gamma_s})] - w_0\}^{-1} - P \quad (16)$$

Where: σ_n = normal stress acted on softening damaged structure surface of rock caused by water.

Substituting (16) to (12) and (13), the relationship between opening displacement and pemeating tensor is as follows:

$$b_r = U_{f_0} \cdot e^{-\sigma_{ij} n_i n_j / k_n (1-D)} \cdot e^{r_w (H-Z)/k_n} \quad (17)$$

$$[k] = \sum_{r=1}^{s} \frac{g m_r}{12\gamma} U_{f_0}^3 \{exp[-3(\sigma_{ij}^{(r)} n_i^{(r)} n_j^{(r)} (1 - D)^{-1} - P)/k_n]\} \cdot (\delta_{ij} - n_i^{(r)} n_j^{(r)}) \quad (18)$$

4 APPLICATION

Figure 2 shows circle underground tunnel with one structure surface AB in surrounding rocks. With

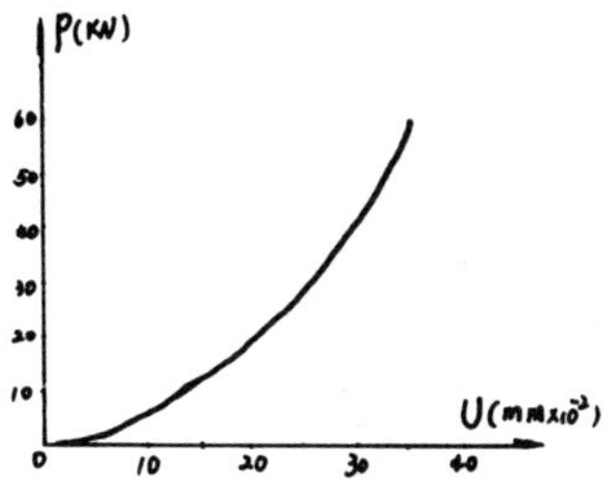

Fig1. deformation curve of fractures in concrete samples

the excavation of caverns, stress in surrounding rocks would redistribute, the fracture would open and fissure water would permeat into surrounding rocks. Here, seepage coefficent in fracure AB would be discussed.

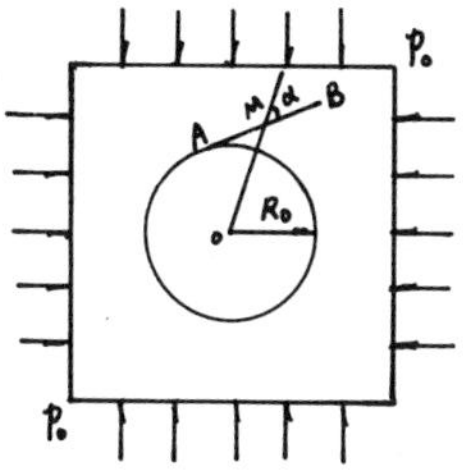

Fig2. Stress state in surrounding rocks

Stress in surrounding rocks is as follows:

$$\begin{cases} \sigma_r = P_0(1 - \frac{R_0^2}{r^2}) \\ \sigma_v = P_0(1 + \frac{R_0^2}{r^2}) \end{cases} \quad (19)$$

Normal stress of point M at fracture AB is:

$$\sigma_{n_1} = P_0(1 + \frac{R_0^2}{r^2}\cos2\alpha) \quad (20)$$

Normal stress of point M before excavation is:

$$\sigma_{n_2} = P_0 \quad (21)$$

With the acting of normal stress σ on a group of fractures, and water in fractures is laminar flow, then,

$$k_f = \frac{gb^2}{12\gamma} = \frac{g}{12\gamma}U_{f_0}^2 e^{-2\sigma_n/k_n} \quad (22)$$

Where: k_f = seepage coefficient; b = opening displacement of fractures.

Subsuiting (20), (21) to (22), then

$$k_{f_1} = \frac{g}{12\gamma}U_{f_0}^2 e^{-2P_0(1+\frac{R_0^2}{r^2}\cos2\alpha)/k_n} \quad (23)$$

$$k_{f_2} = \frac{g}{12\gamma}U_{f_0}^2 e^{-2P_0/k_n} \quad (24)$$

Let $n_1 = k_{f_1}/k_{f_2} = e^{-2P_0\cos2\alpha(r^2k_n)^{-1}}$ (25)

Where: n_1 = ratio of seepage coefficient

Seepage coefficient on every point of fracture is different because of the stress redistribution in surrounding rocks, thus results in the variation of hydraulic properites of rock.

If drenched phenomena is considered, then equation (23) should be:

$$k_{f_3} = \frac{g}{12\gamma}U_{f_0}^2 e^{-2[P_0(1+\frac{R_0^2}{r^2}\cos2\alpha)-P]/k_n} \quad (26)$$

$$n_2 = k_{f_3}/k_{f_1} = e^{2P/k_n} = e^{2r_w(H-Z)/k_n} > 1 \quad (27)$$

It is very clear that permeability water makes fissure seepage flow a violent change.

Furthermore, if softening of surrounding rocks is considered, then equation (23) should be:

$$k_{f_4} = \frac{g}{12\gamma}U_{f_0}^2 e^{-2[P_0(1+\frac{R_0^2}{\gamma^2}\cos2\alpha)(1-D)^{-1}-P]/k_n} \quad (28)$$

$$n_3 = k_{f_4}/k_{f_1} = e^{-2P_0(1+\frac{R_0^2}{\gamma^2}\cos2\alpha)D(1-D)^{-1}/k_n} e^{2P/k_n} \quad (29)$$

Equation (29) shows that softening damage could make seepage coeffieient different on every point of fracture AB.

Considering the softening damage of surrounding rocks, stress in surrounding rocks should be:

$$\begin{cases} \sigma_{rw} = P_0(1 - \frac{R_0^2}{\gamma^2})(1 - D)^{-1} \\ \sigma_{Qw} = P_0(1 + \frac{R_0^2}{\gamma^2})(1 - D)^{-1} \end{cases} \quad (30)$$

Considering of symmetry of tunnel, $\tau_{r\theta}$ in surrounding rocks is zero, the criterion of Mohr-coulomb is:

$$\sigma_{\theta w} = \frac{1 + \sin\varphi_w}{1 - \sin\varphi_w}\sigma_{rw} + \frac{2C_w \cdot \cos\varphi_w}{1 - \sin\varphi_w} \quad (31)$$

Where: C_w = cohension of soaked rock; φ_w = friction angle of soaked rock.

The plastic failure criterion of rockmass around tunnel wall is:

$$\sigma_{\theta w} = 2P_0(1 - D)^{-1} \leqslant \frac{2C_w \cdot \cos\varphi_w}{1 - \sin\varphi_w} \quad (32)$$

or $2P_0 \leqslant \frac{2C_w \cdot \cos\varphi_w}{1 - \sin\varphi_w} \cdot (1 - D)$

For dry tunnels, the equation (32) would be:

$$\sigma_\theta = 2P_0 \leqslant \frac{2C\cos\varphi}{1 - \sin\varphi} \quad (33)$$

Comparing equation (32) with (33), it shows that cohension and friction angle of surrounding. rocks decrease because of the softening damage. So, $C/C_w < 1$, $\varphi/\varphi_w < 1$.

If gravitational force water exists in

Table 1

n_i \ 2α	0	30°	45°	60°	90°	120°	135°	150°	180°
n_1	0.9444	0.9517	0.9604	0.9718	1.0	1.029	1.0412	1.0507	1.0588
n_3	0.9504	0.9538	0.9579	0.9633	0.9763	0.9895	0.9950	0.9993	1.0
n_2	1.0029								

rockmass, especially bearing pressure water, the shear strength of rock would decrease because fissure pressure can offset part of normal stress.

So, $\tau_w = (\sigma_n - P)\text{tg}\varphi_w + C_w$ (34

$$\Delta\tau = \tau - \tau_w = C + \sigma_n \cdot \text{tg}\varphi - (\sigma_n - P)\text{tg}\varphi_w - C_w = (C - C_w) + \sigma_n(\text{tg}\varphi - \text{tg}\varphi_w) + P\text{tg}\varphi_w \quad (35)$$

Equation (35) shows the mulitiple mechanical properites of rock caused by water.

The parameters of one circle tunnel in Xingwen mining bureau is as follows:

R_0 = 2.5m; r_{do} = 23.2KN/m^3; r_s = 27.4KN/m^3; w_0 = 0.7%; P_0 = 10MPa; P = 0.5MPa; C = 7MPa; φ = 38°; C_w = 6.7MPa; φ_w = 36°; k_s = 0.8; A = 54254MPa; k_n = 350MPa

The result is as follows:

Damage tensor, D = 0.32; shear incrementalism, $\Delta\tau$ = 0.67 MPa.

If not considering the effect of water; tangential stress. σ_θ = 28.7 MPa, the surrounding rocks still in elasticity.

If considering the effect of water, tangential stress: $\sigma_{\theta w} = 29.4 > \dfrac{2C_w \cdot \cos\varphi_w}{1 - \sin\varphi_w} = 26.3$ MPa, some surrounding rocks are in plasticity be cause of the rocks, softening damage, bolting or grouting measures should be taken.

If fracture AB in Fig2 is on the tunnel wall, then ratio of seepage oefficient is shown in table 1 let $r = R_0$.

According to above results, it can be concluded that stress in surrounding rocks increase, shear strength decrease, and aggravate seepage of fissures because of softening of rocks with water.

5 CONCLUSIONS

1. The relationship between permeating tensor and normal stress of fractures is obtained by considering the softening damage of rock because of water.

2. The variation of stress field would result in the variation of permeating tensor, and the softening damage of surrounding rocks would make the shear strength decrease.

3. The effect of permeating flow on the mechanical properties of rocks should be consided while designing and supporting underground tunnels.

REFERENCES

Seedsman. The behaviour of clayshales in water. *Can. Geotech. J.*, 1986, 23, 18－22.

Y.P. Chugh. Effects of moisture on strata control in coal mines. *Engineering Geology*, 1981, 17, 241－255.

Snow. D. T. Anisotropic permeability of fractured media, *Water Resource Research*, 1967, 5, *NO*.6.

M. Kachanov, Continuum model of medium with cracks, *J. of Eng. Mech. Division, ASCE*, 1980, Vol.106.

Kilsall P. C. et al.. Evaluation of excavation induced changes in rock permeability. *Int. J. of Rock Mech. and Min. Sci*. 1984, NO.3.

Shun Guangzhong. *Foundation of Rock Mechanics*, *Science Press*, 1983

Xiang Jiannan. Discussing the imfluence of mechanical properites of weak beddings caused by water. *Hydrogeology Engineering*. 1986, NO.1

Liu Jishan. Coupled relationship between mechanical paramters of structural surface and hydropower parameters and its application, *Hydrogeology Engineering*. 1988, NO.2

Computer Methods and Advances in Geomechanics, Yuan (ed.) © 1997 Balkema, Rotterdam, ISBN 90 5410 904 1

The Pinglin EB TBM penetration through Chingyin Fault

Y.Y.Tseng, H.C.Tsai, C.T.Tseng & Bennie Chu
Ret-Ser Engineering Agency, Taiwan, China

ABSTRACT : From the project reports of Construction Industry Research and Information Association (London), the primary reason for poor performance of TBM is variable ground with few exceptions. The Pinglin main tunnel is a world wide difficult TBM project in terms of adverse geologies tunnel size and length. RSEA would like to devote ourselves to the TBM tunnelling in the variable ground and share our experience with international construction industry. This paper presents the prior planning and actual performance of Pinglin Eastbound TBM penetration through Chingyin Fault.

I INTRODUCTION

The Pinglin tunnels of Taipei-Ilan Expressway consist of a pilot tunnel (ϕ = 4.8m)and two main tunnels (ϕ = 11.74m). All three tunnels have the length about 12.9 km. The full-face TBM(Tunnel Boring Machine) will be applied for tunnel excavation ahead from east to west.

The Pinglin tunnels with the maximum overburden of 700m are situated within the fold-and-thrust belt structural region in Taiwan. The identified faults are shittsao, northern and southern branches of shihpai, paling, Tachingmen, Shanghsin and Chingyin, ranging from 10 m to 60 m wide. The underground water is abundant, especially in the immediate vicinity of faults. The hardest rock area about 4 km long near the east portal is the quartzite of Szeleng with uniaxial compressive strength 1670 kg/cm^2 and total hardness 150. The large size TBM penetrating through the wide water-bearing fault zone is a worldwide known challenge.

The Contract stipulated that the first 720 m of the excavation of Eastbound tunnel have been done by DB(drill and blast) method and there after by TBM, starting from 39K+512. Since

1.overseas experts have serious reservations about TBM viability in the fault,

2.the heavy support deformation of 5.8 cm/day at Westbound sta. 39K+387 continues in the zone of Chingyin Fault excavated conventionally,

3.the No. 6 and No.7 stoppages of pilot TBM occur within the zone of Chingyin Fault,

4.the boreholes of drilling investigation from pilot tunnel are spontaneously collapsing and squeezing throughout most of its length, then the drill string get stuck in ground exerting extremely high squeezing pressure,

the top heading excavated through Chingyin Fault by DB method with strong and heavy supports is proposed to carry over 120m . The inrush water behind the Chingyin Fault and the problematic ground will be treated beforehand. The TBM will then be used to mine out the bench of tunnel.

II GEOLOGY

2.1 Zone 1 : EB sta.39^K+512 to ± 39^K+461 \ 39^K+451 as shown in Fig.1.

Rock type : argillite, minor clay gouge.

Rock mass class : very poor becoming extremely poor towards Chingyin Fault.

Water : of no major influence.

Possible tunnelling hazards :

(I) mild squeezing pressure.

(II)small scale loose blocky ground throughout.

(III)possibility of large-scale fault-dimensioned collapse.

(IV)tendency for over excavation and ravelling.

2.2 Zone 2 : EB sta. 39^K+461 to 39^K+421 \ 39^K+451 to 39^K+412.

Rock type : argillite, sheared and crushed with significant proportion of clay.

Rock mass class : extremely poor throughout.

Water : minor localised inflow possible.

Possible tunnelling hazards :

(I) mild to high squeezing pressure.

(II)tendency for ravelling failure leading to

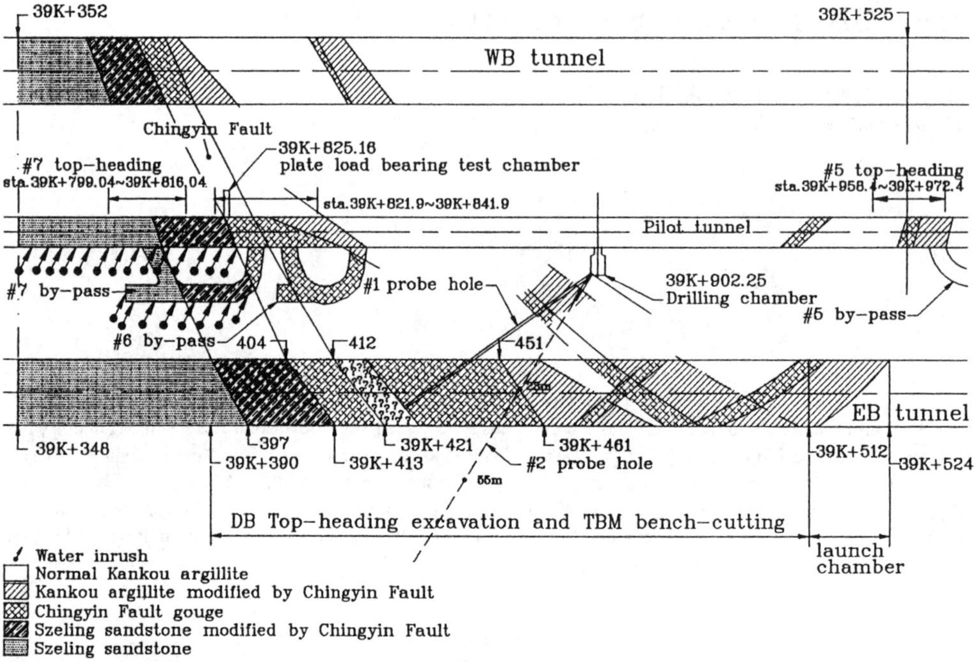

Fig. 1 The geology of Chingyin Fault

progressive major collapse.

(III)lining design probably inadequate, so that segments may break down.

2.3 Zone 3 : EB sta. 39^{K}+421 to 39^{K}+413 \ 39^{K}+412 to 39^{K}+404.

Rock type : clay, firm to stiff with micro and macro slick sides; some fragments of argillite or crushed tuff may be present.

Rock mass class : not applicable as this is more of a soil than a rock.

Water : unlikely.

Possible tunnelling hazards :

(I) heavy squeezing pressure.

(II)ravelling collapse inevitable.

(III)lining design inadequate, so that concrete segments may collapse under squeezing pressure.

2.4 Zone 4 : EB sta. 39^{K}+413 to 39^{K}+397 \ 39^{K}+404 to 39^{K}+390.

Rock type : sheared highly fractured sandy argillite with some gouge on fault zones, also intrusions of distinctive pale green/grey tuff.

Rock mass class : extremely poor.

Water : maybe found in distinct zones or conduits, i.e. localised in fault/shear zones and pressure 15 bars maximum.

Possible tunnelling hazards :

(I)possibility of run-of-ground due to water under high pressure localised in zone of weak material having low internal shear strength.

(II)small scale loose blocky ground throughout.

2.5 Zone 5 : EB sta. 39^{K}+397 onwards \ 39^{K}+390 onwards.

Rock type : closely bedded sandstone, siltstone and sandy argillite.

Rock mass class : fair or poor.

Water : very minor but gradually increasing up to the No. 8 aquifer stoppage area.

Possible tunnelling hazards :

(I)seamy rockmass, tendency for slabs of rock to fall out of the crown of tunnel.

(II) the face should be stable.

Ⅲ METHODOLOGY

3-1 DB top-heading excavation :

1. For the purpose of investigation and drainage, the probe holes at least 3 m shall be drilled.
2. Due to the large support deformation of WB tunnel excavated conventionally in the zone of Chingyin Fault, the top-heading shall be over excavated at least 20 cm to keep enough clearance for TBM boring, as shown in Fig 2.
3. Forepoling and lagging are support measures

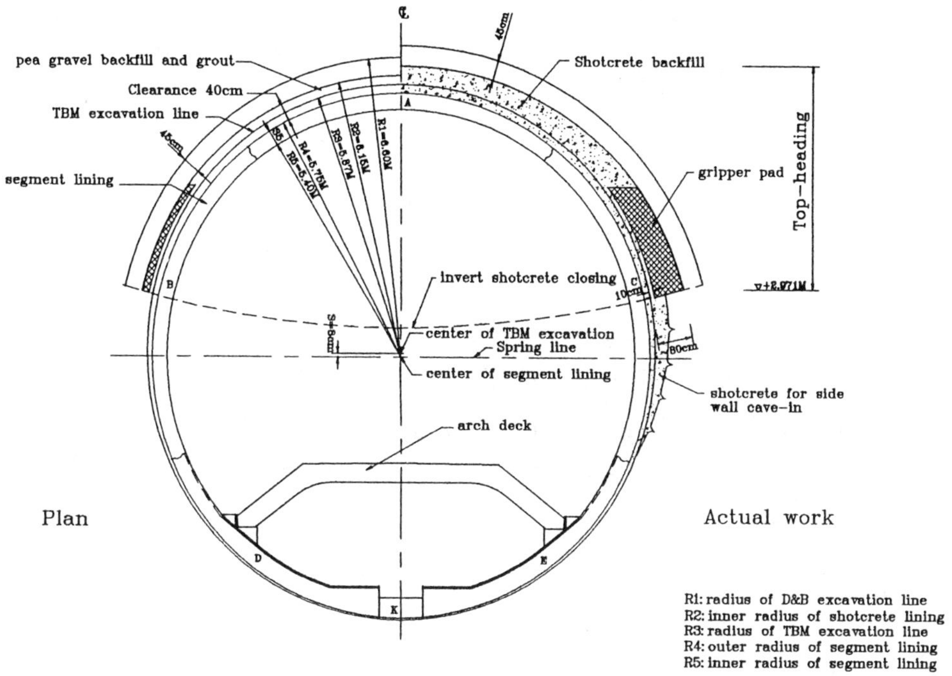

Fig.2 The plan and actual work through Chingyin Fault

installed in the tunnel longitudinal direction prior to the excavation. H100 × 100 steel beam will be used as forepoling in case of caving-in.

4. To prevent the face from frequent collapses, the central section of the face will not be taken out and only a peripheral slot is gouged out. Shotcrete may be applied to avoid a loosening of the surrounding rock mass. A H200 × 200 @ 1m(or double if necessary) steel arch rib with side wings will then be inserted. Three layers of mesh-reinforced shotcrete and 6m @ 1.5m systematic rockbolts as shown in Fig 3 will be installed. The invert consisting of two layers of geogrid-reinforced shotcrete is required to construct in the unstable ground. Each side wing of steel rib is underpinned with three 6m rockbolts and cement grouting as shown in Fig 3.
5. The steel ribs are linked together with anchor bolts and mesh-reinforced shotcrete sidewalls. The sidewalls play the roles of stabilizer pad and gripper pad, which are necessary for TBM boring.

3-2 TBM launching and bench-cutting

1. A launch chamber will be constructed 12m long from 39^K+512 ± to 39^K+524 ± approximately as shown in Fig 4.
2. The reaction structure will be erected in position. The structure shall be designed to take the load required to hold the first ring in position when TBM pushing forward.
3. After TBM boring (or bench-cutting), the ring will be erected. The temporary support of H100 × 100 steel arch will be fitted as the ring is built. The invert will be filled with pea gravel and grouted later.
4. After 4 to 5 rings are built, the gap between the segment and the shotcrete lining (or rock) will be filled with shotcrete, as shown in Fig 2 and Fig 5. The temporary support will be removed after the shotcrete has set and must be strong enough to hold the weight of segment in position.
5. The TBM will advance with grippers reacting partly on the concrete side wall and partly on the side of cut. Special care must be taken to control line and level. The front of TBM shall be checked for face stability and clearance.

IV PERFORMANCE

A new Wirth TB 1172 H/TS as shown in Table I was erected and ready to walk into the tunnel in Nov. 1995. Although a baptism of fire, i.e. TBM full-face boring through the Chingyin Fault, was justified because there would be worse Faults and hazards further into the mountain ,the plan of DB top-heading excavation through the Chingyin Fault and TBM bench-cutting was finally

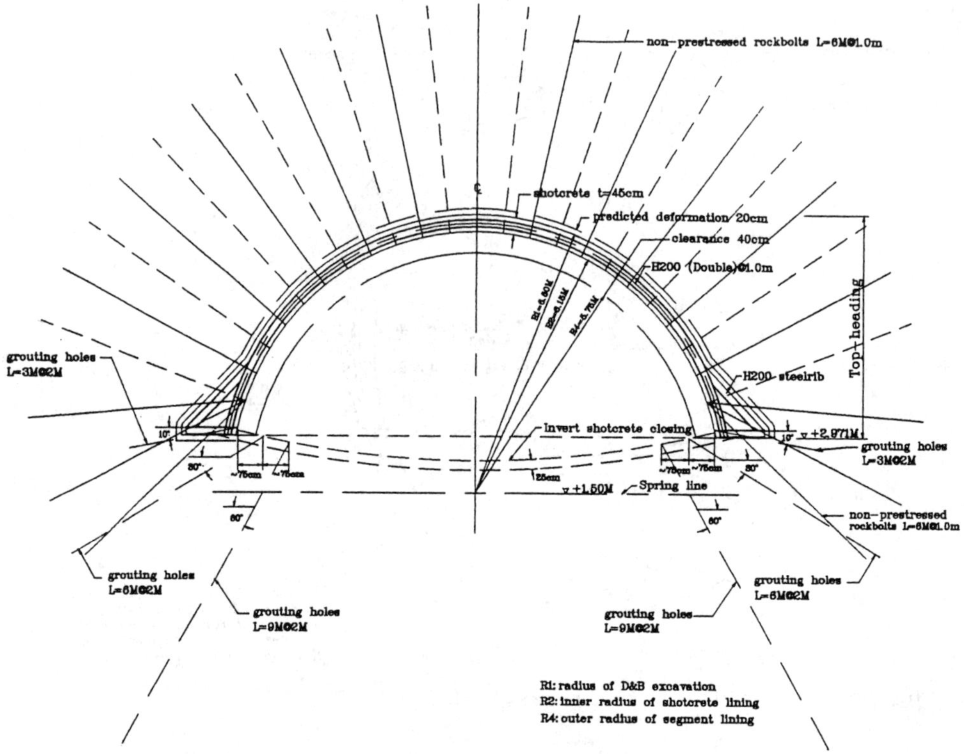

Fig.3 The support of DB top–heading excavation

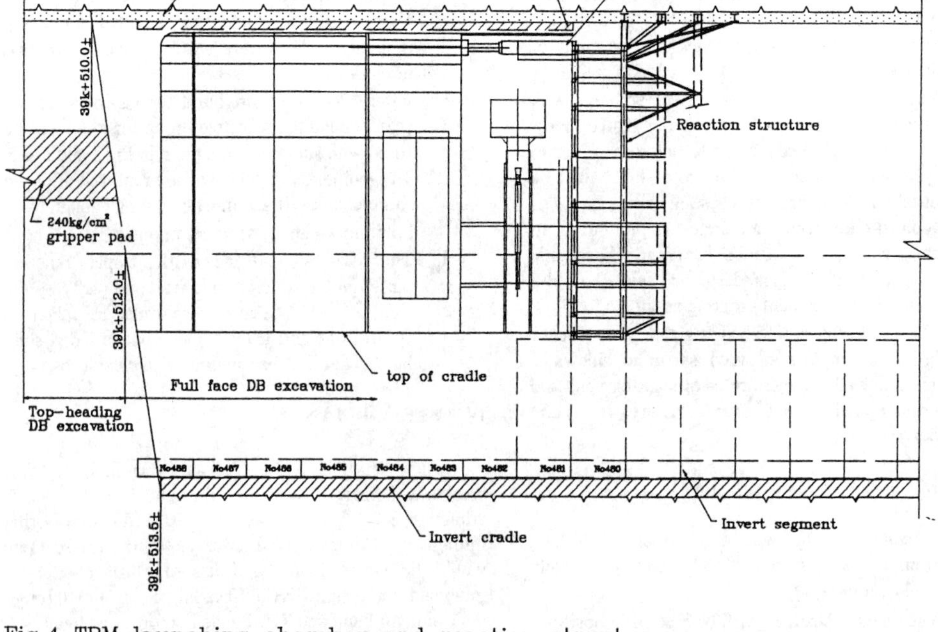

Fig.4 TBM launching chamber and reaction structure

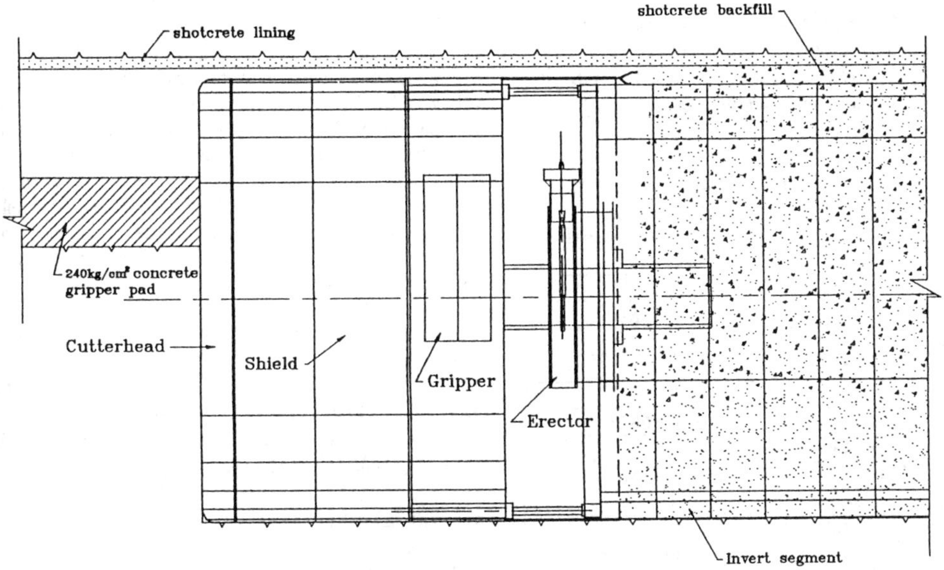

Fig.5 TBM bench-cutting

Table I specification of WIRTH TB 1172 H/TS

Bore diameter	11,740 mm
Cutterhead	
-power	4,000 kw
-speed	0-4.0 rpm
-torque at 4.0 rpm and 75% efficiency	7,200 kNm
-breakout torque, at 0.95 rpm	30,000 kNm
-thrust of 18 telescope cylinders	50,600 kN
-thrust of 28 push jacks	78,700 kN
-stroke	1.5 m
-motors	18
Gripper clamping force	65,000 kN
Gripper configuration	T (3-point anchoring)
Hydraulic system pressure(bar)	max.405
Total machine power	7,550 kw
Muck handling capacity	1150 m^3/n
Machine length	250 m
Estimated weight (TBM+BU)	3000 t
Diameter of cutter disc	432 mm
Number of cutting discs	
-face	71
-gauge	3
-center	6
-overcutter	3
Transformers installed	2 × 3150 kva(690v) 1 × 1250 kva(440v)

approved by the Client. It took two weeks to remobilize including 4000 M^3 ramp backfill. The excavation of top-heading by the conventional DB method was advanced with the average rate of 0.8 m/day, and interrupted by the holidays of Chinese New year.

The jobs of TBM walk-in and boring were mainly driven by two shifts in ten hours each, working 6 days per week. The dirt clearance, grouting, curing, maintenance and repairs were carried out by using the 4-hour left and Sunday.

During the period of TBM walk-in, logistic installations, railing system, ventilation, etc for TBM boring were constructed. The best record of TBM walk-in from the portal to the launching chamber was 57 m in two 10-hour shifts (i.e. 20 hrs). And the best record of TBM penetration through Chingyin Fault was 7.5 m in one 10-hour shift, as shown in Table II. The most critical reason which delayed the advance was shotcreting for the face stability, side-wall cave-in and backfill. The average amount of shotcrete was 18 M^3 per ring. In addition, the tunnelling was also delayed by the erection of reaction structure and machine problems such as :

1.main bearing oil leakage,
2.conveyor belt tear,
3.segment crane breakdown ,
4.power shutdown,
5.oil and water hose break .

The micro cracks were found at two side concrete segments B and C from sta. 39^K+432.9 to 39^K+452.4 in the zone of Chingyin Fault. Concrete segments may collapse if cracks propagate under squeezing pressure. Therefore the concrete segments A 、 B and C from ring 528 to ring 541 were reinforced by installing one H150 × 150 steel rib per ring.

Table II Performance date

Date	Rings	Shotcrete backfill
8.14 (W)	No480	
8.19 (M)	No481	
8.24 (Sa)		No480 ($3M^3$)
8.28 (W)	No482	
9.03 (T)	No483	
9.06 (F)		No480 ($16M^3$)
9.07 (Sa)		No480 ($6M^3$)
9.09 (M)	No484~485	
9.11 (W)		No485 ($15M^3$)
9.12 (Th)	No486	
9.13 (F)	No487~488	No486~488 ($7M^3$)
9.14 (Sa)	No489~490	No489~490 ($8M^3$)
9.16 (M)	No491~492	No488~490 ($15M^3$)
9.17 (T)	No493~494	face ($8M^3$)
9.18 (W)		No491~493 ($38M^3$)
9.19 (Th)	No495	No491~493 ($26M^3$)
9.20 (F)	No496~499	No494~497 ($16M^3$)
9.21 (Sa)		No494~498 ($28M^3$)
9.23 (M)	No500~502	No494~498 ($8M^3$)
9.24 (T)	No503	No499~502 ($44M^3$)
9.25 (W)	No504	No499~502 ($28M^3$)
9.26 (Th)	No505~508	
9.27 (F)		No503~507 ($37M^3$)
9.28 (Sa)		No503~507 ($35M^3$)
9.30 (M)	No509~510	No503~507 ($22M^3$)
10.01 (T)	No512~513	No509~512 ($8M^3$)
10.02 (W)		No508~512 ($52M^3$)
10.03 (Th)	No514~516	No508~512 ($20M^3$)
10.04 (F)	No517~518	No513~517 ($16M^3$)
10.05 (Sa)		No513~517 ($50M^3$)
10.07 (M)	No519~522	No518~521 ($8M^3$)
10.08 (T)		No518~521 ($55M^3$)
10.09 (W)	No523~527	No522~526 ($14M^3$)
10.10 (Th)		No522~526 ($40M^3$)
10.11 (F)	No528~529	
10.12 (Sa)	No530~532	No527~531 ($48M^3$)
10.14 (M)	No533~537	No527~531 ($14M^3$)
10.15 (T)		No532~536 ($60M^3$)
10.16 (W)	No538~542	No537~541 ($10M^3$)
10.17 (Th)		No537~541 ($54M^3$)
10.18 (F)	No543~547	No542~546 ($32M^3$)
10.19 (Sa)		No542~546($40M^3$)
10.20 (S)		No547~551 ($24M^3$)
10.21 (M)	No548~552	
10.22 (T)		No547~551 ($39M^3$)
10.23 (W)	No553~556	No552~556 ($28M^3$)
10.24 (Th)	No557	No552~556 ($54M^3$)
Totals	78 rings	1026 M^3

V REMARKS

Many people believed that the method of top-heading followed by the TBM was impractical and not accepted for the long distance (122M) and large dimension (11.74m) of tunnel . The cave-in of top-heading as TBM proceeded through the Chingyin Fault and the huge side chimmney were expected and would be extremely difficult to backfill. However, RSEA completed the difficult task successfully and obtained much more valuable experience.

Pinglin tunnels are not located in the ground which offers the best TBM tunnelling prospect. More serious problems and incidents are expected further into the mountain. The versatility, creative thinking and cooperation shall be encouraged to overcome all those difficulties together.

VI REFERENCE

1. Parkes D.B, The performance of Tunnel Boring Machine in rock , CIRIA Special Publication 62 .
2. Dolcini G. , Grandori R. and Marconi M. , " Water supply revamp for Bogota " , Tunnels & tunnelling , vol. 22, No. 9, PP. 33-38(1990) 。
3. Grandori C. , "Development and current experiences with double TBM in ITALY" , Tunnelling and Underground Space Technology , vol. 4 , No. 3 , PP. 333-336 。
4. Hester J. C. , Criigton G. S. Curist D. J. , Channel Tunnel UK-FRANCE : The UK TBM Drives , RETC proceeding , PP. 751-762 (1991) 。
5. Hunter P. W. and Aust M. I. E. , "Excavation of a major tunnel by double shield TBM through mixed ground of basalt and clayey soils" , RETC proceeding , vol. 1 , PP. 527-582 (1987) 。
6. Kelley M. N. , Denmark's Great Belt Fixed Link Bored Tunnel Project , RETC proceeding , PP. 881-892 (1991) 。
7. Robbins R. J. , " Tunnel boring machine for high water inflows " , International Tunnelling Association Congress " Tunnel and water " (1982) 。
8. Robbins R. J. , " Tunnel machines in hard rock " USA-ROC and ROC-USA Economic Council's 14th Joint Business Conference Taipei (1990) 。
9. Sinotech Engineering Consultants, Inc. "Taiwan Area National Expressway Engineering Bureau Ministry of Communications final report on geological exploration for detail design of Taipei-Ilan Expressway project" (June 1993) 。

Computer Methods and Advances in Geomechanics, Yuan (ed.) © 1997 Balkema, Rotterdam, ISBN 90 5410 904 1

An introduction to the construction of Headrace Tunnel by TBM in Shihlin Hydropower Project

Chung-chun Fu & Tsang-hsien Chen
Liyutan Project Office, Taiwan, China Power Company, Taiwan, China
Te-ho Hsiao
Sinotech Engineering Consultants, Ltd, Taiwan, China

ABSTRACT: The geological investigation conducted for the Shihlin Hydropower Project indicated that the rock condition is feasible for using TBM to excavate the headrace tunnel. Therefore, in order to expedite the construction work, TBM was adopted for the headrace tunneling work, and the NGI-CSIR rock mass classification method was adopted for design of tunnel excavation support. Due to the terrain limitation at the portal area of the construction adit, the TBM had to be erected in segments on the portal platform and then assembled inside the tunnel. During construction, the TBM confronted some fractured rocks and invasion of groundwater. As a result, the boring machine was damaged and the cutterhead was stuck. However, these difficulties were eventually overcome by implementation of appropriate measures. To date, the TBM has completed approximately 40% of designated tunnel excavation.

1 INTRODUCTION

The Shihlin Hydropower Project, located in central Taiwan (see Figure 1), is the second phase of the Liyutan Reservoir Project. When completed, the project will divert water from the Taan River into the Liyutan Reservoir and generate power by taking advantage of the available head. In the first phase of the Liyutan Reservoir Project, construction of a dam on a tributary of the Taan River, the Chin San Creek, had already been completed for water impounding since November 1992.

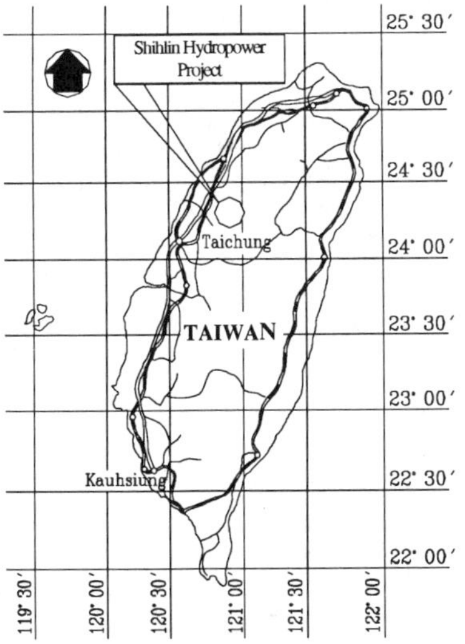

Figure 1. Project Site

The Shihlin Hydropower Project consists of a dam on the Taan River at an elevation of 594m, a headrace tunnel about 5,525m in length, an underground powerhouse and a tailrace tunnel leading to the Chin San Creek. One important target of the project is to divert water from the Taan River through the headrace tunnel and the discharging device into the Liyutan Reservoir to meet the demand of 900,000 tons of water per day for agricultural irrigation and public water supply in the Taichung area before generating power.

2 ENGINEERING LAYOUT

The main items of the project are as follows:

1. Concrete Gravity Dam: 21.0m (H) × 253.5m (L)
2. Bell Mouth Shaped Intake:
   41.0m (W) × 19.0m (H) × 32.0m (L)
3. Pressure Headrace Tunnel:
   3.5m (I.D.) × 5,525m (L)

4. Restricted Orifice Surge Tank:
   10.0m (I.D.) × 55.5m (H)
5. Underground Declining Penstock:
   3.5m ~ 1.5m (I.D.) × 490m (L)
6. Underground U-shaped Power Cavern:
   Dimensions =14.0m (W) × 36.9m (H) × 49.2m (L)
   Generator Capacity = 2 @ 40,000 kW each
   Yearly Yield = 266,000,000 kWh
7. Gravity Tailrace Tunnel:
   Horseshoe shape cross section - 4.3m (I.D.) × 1,070m
8. Discharging Device:
   Steel Pipe Section - 1.6m (I.D.) × 175m (L)
   Concrete Channel Section - 8.0m (W) × 2.0m (H) × 243m (L)

The general layout of the project is shown in Figure 2. Since construction of the headrace tunnel is one of the critical paths of this project, expedition of this construction work is the most important task in this project in order to enable water diversion as early as possible.

3 GEOLOGY

Rock formations exposed along the headrace tunnel alignment are mainly sedimentary rocks of Miocene age, and consists mainly of sandstone, siltstone, shale and their alternations. Stratum distribution from east to west as well as from older to younger is the Nankang Formation, Nanchung Formation and Kuechulin Formation (see Figure 3), and the uniaxial compressive strength of the intact rock decreases gradually from 1000 kg/cm^2 to 100 kg/cm^2 accordingly. Those rocks are generally composed of 70% of quartz, 26 ~ 30% of matrix and a small amount of feldspars.

Major geological structures expected to be encountered along the tunnel alignment include Shimasheam Fault, shear zones S1 ~ S6 and Shihlin Anticline. The Shimashan Fault and shear zones S1 ~ S4 have been encountered and treated successfully in the section excavated by the D&B method while the other two shear zones are expected to be encountered in the section excavated by the TBM method. The geological investigation also indicates that a natural gas reservoir is confined underneath the Shihlin Anticline and coal seams are preserved within the Nanchung Formation. Combustible gas could invade the tunnel due to pressure relief during excavation. In addition, a great amount of water might flow into the tunnel at the region near Shihlinshan with elevated groundwater table. Therefore, preventive measures against combustible gas and groundwater must be implemented to ensure construction safety.

4 EXCAVATION METHOD

The geological investigation indicates that 4 shear zones and one fault would be encountered in the upstream 1,300m, and only 1 ~ 2 shear zones in the remaining 4,225m. In addition to geological conditions, several other factors were also assessed for the selection of appropriate excavation methods. Those factors include construction duration, environmental impacts, access to the construction site, availability of reliable skillful labors on site, and the plan to promote tunneling techniques in Taiwan.

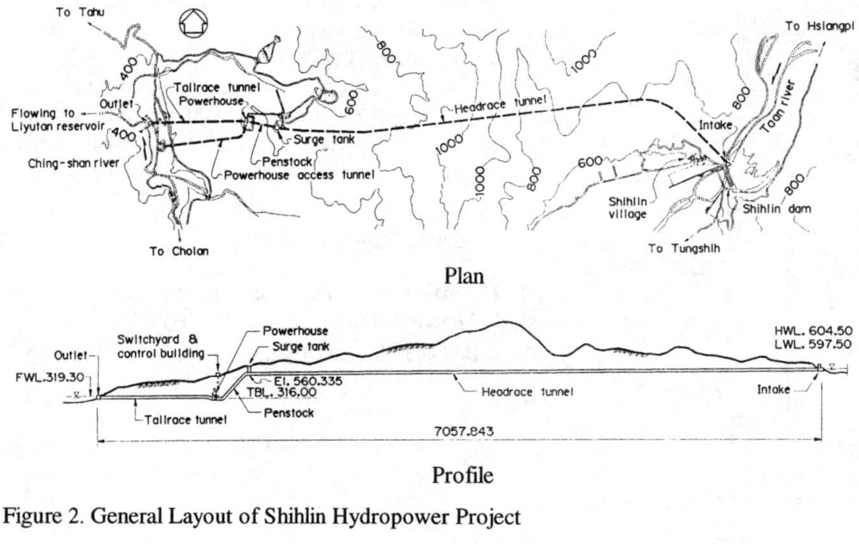

Figure 2. General Layout of Shihlin Hydropower Project

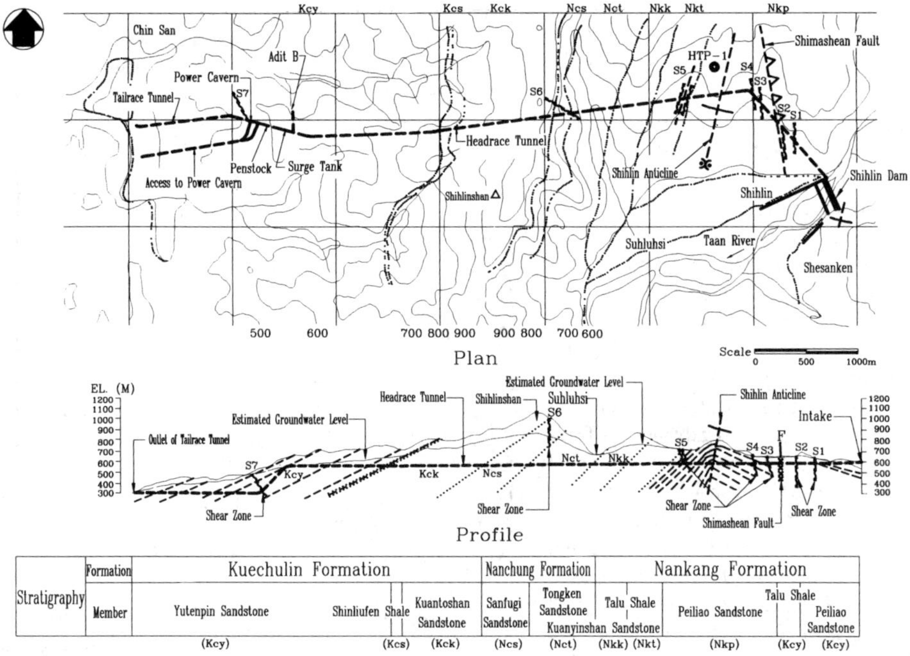

Figure 3. Project Geology

After extensive assessment of all the factors, the D&B method was adopted for the upstream 1,300m of the tunnel, and the remaining 4,225m of the tunnel would be driven by using TBM (Tunnel Boring Machine).

5 TUNNEL SUPPORTS

The better known and commonly used rock mass classifications of NGI/CSIR systems are adopted for the design of headrace tunnel support in the TBM excavation process. Five types of support systems were designed as a part of the permanent structure to protect the tunnel. The relationship between rock mass classification and tunnel support types is shown in Table 1. Details of the support types are shown in Table 2.

Table1 Support Types and Rock Mass Classification

Support Types	I	II	III	IV	V
Rock Mass Quality	Very Good	Good	Fair	Poor	Very Poor
NGI-Q Values	>40	10~40	4~10	1~4	<1
CSIR/RMR	>77	65~77	56~65	44~56	<44

Table 2. Tunnel Supports

Type I	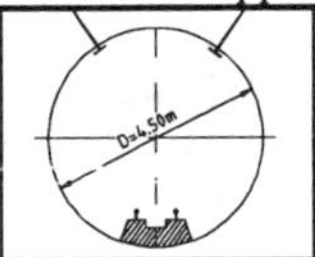	2 Rock Anchors L= 2m. as required
Type II	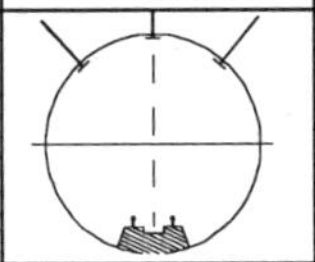	3 or 2 Rock Anchors, L = 2m, @ 2.0m × 2.0m U-shaped Crown Ring
Type III	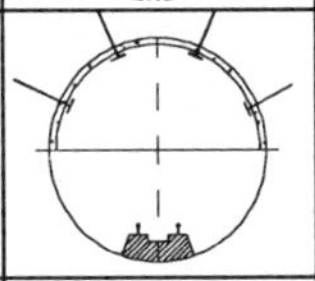	4 or 3 Rock Anchors, L = 2m, @ 2.0m × 1.5m 5cm Shotcrete in a range of 180°;U-shaped Crown Ring
Type IV	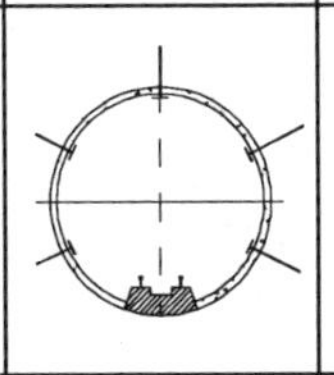	5 or 4 Rock Anchors, L = 2.5m, @ 2.0m × 1.5m 10cm Shotcrete and ⌀ 3.2mm, 7.5cm × 7.5cm wiremesh in a range of 320° U-shaped Crown Ring
Type V	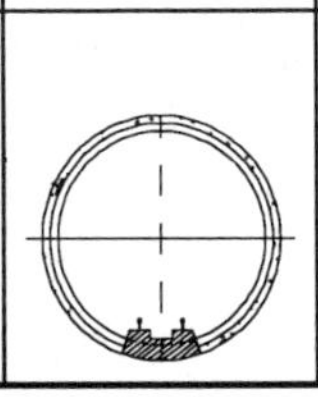	10cm Shotcrete and ⌀ 3.2mm, 7.5cm × 7.5cm wiremesh in a range of 320° H Shape(100mm×100mm) Steel Ribs, 360° ,@ 0.75m ~ 1.25m

6 TYPE OF TBM

Since the headrace tunnel of this project is of the pressure type, water tightness of the tunnel is essential. Therefore, the cast-in-place concreting method was selected for tunnel lining and precast invert segments, used as foundation for rail and installed concurrently with TBM excavation, should be a part of the final lining. Considering the specified tunnel lining method, low uniaxial compressive strength of the rocks and the shear zones that would be encountered, the contractor has chosen a refurbished Open Main-Beam Machine manufactured by ROBBINS Company of the US for the project. This machine was once used in the Evinos-Marnos Water Transfer Tunnel Project in Greece. The greatest advantage of this type of machine is that it allows the installation of rock supporting elements immediately behind the cutterhead support, only 3m from the face. Therefore, rockfall during excavation in unfeasible geology could be minimized. Typical rock supporting elements used are rock anchors, steel ribs and shotcrete. The total length of the TBM is 210m (see Table 3). Of the total length, the primary section is only 18m long and the back-up system take the remaining 192m.

The primary section includes cutterhead, cutter discs, cutterhead support, steel rib erector, TBM-Band, grippers, main beam, TBM operator room, rear section. crown ring erector, roof drill, guiding device, probe drill and gas monitors with sirens (see Figure 4). The back-up system includes working platform, equipment platform, invert segment handling unit, muck conveyor system, dust scrubber system, ventilation system, etc.

Furthermore, additional subsidiary facilities are available in the portal area of the construction adit for TBM maintenance, materials storage and muck transportation purposes.

Table 3. General Specification of TBM (Open Main-Beam Type, ROBBINS)

1. Cutterhead:	
Bore Diameter	4.53m
Cutters & Number	ROBBINS 17inch disc V mounted, 31
Max. Available Cutterhead Thrust	7200 kN at 330 bar
Cutterhead Horsepower & Torque	1072 Hp(800kW), 779kNm
Rotation Speed	10.2 rpm
Boring Stroke	61 inch (1.55m)
2. Hydraulic System:	
Max. System Pressure	330 bar
Electric Power	100 Hp (74KW) (2 motors)
3. Electric System:	
Motor Circuits	380V, 3-phase, 50Hz
Incoming Power	10,000V, 3-phase, 50Hz
Transformer (1) (2)	650kVA×2 (for TBM) 250kVA×1(for back up)

7 CONSTRUCTION WITH TBM

7.1 *TBM Assembly*

According tothe original construction schedule, 210m long of the headrace tunnel at the downstream end and an access adit should be excavated by the D&B

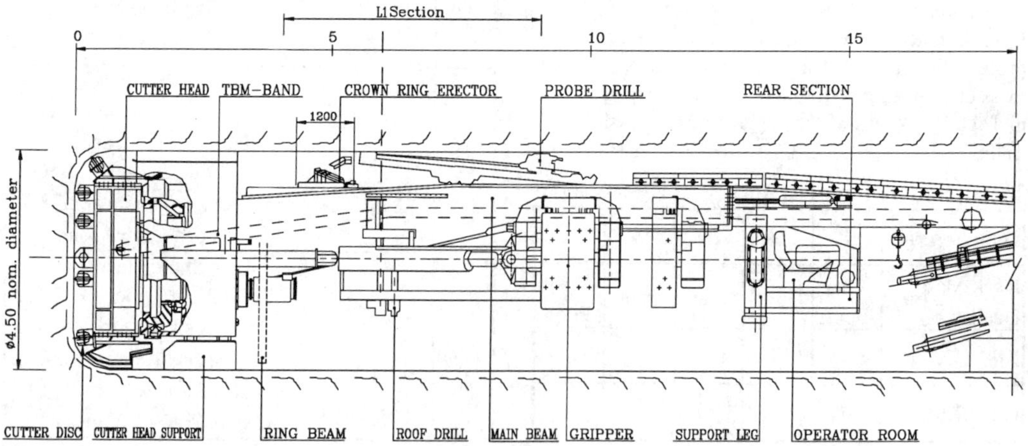

Figure 4. TBM Primary Section Layout

method for TBM assembly prior to the arrival of the TBM. However, the excavation progress was retarded by rockfall accidents during the adit excavation. When the TBM was delivered to the construction site, only the adit and 130m of the tunnel excavation were completed. To avoid delay to the start of TBM boring, TBM assembly proceeded in two stages. The first stage included the assembly from the cutterhead to the conveyor system totaling 110m in length, and the second stage included the assembly of muck loading and trail systems in a length of 100m. The first stage of assembly was completed on May 2, 1996. After driving through about 144m on May 26, 1996, the second stage of assembly commenced and the TBM assembly was completed on June 3, 1996. Because of the terrain limitation in the portal area of the adit, the TBM had to be erected in segments in the portal area in advance and then delivered inside the tunnel for final assembly.

7.2 *Geological Investigation Prior to Excavation*

Performance of TBM excavation is closely related to knowledge obtained on geological conditions ahead of the excavation face. To acquire core samples for geological investigation on site before TBM launching, an advance horizontal exploratory hole in a length of 700m was driven. For excavation beyond 700m deep, HSP (Horizontal Seismic Profiling) tests or 30m long probe drills would be performed. Additional horizontal exploratory holes should be driven if necessary.

7.3 *Construction Process*

After excavation, installation of rock supporting systems is performed in two phases. In the first phase, protection works for the crown area include installation of crown rock anchors, wire mesh, U-shaped crown rings (for Type II to IV) and steel ribs (for Type V). In the second phase, the remaining supporting elements including rock anchors and shotcrete are then installed. Supporting work for the first phase is carried out utilizing the space between the cutterhead and the grippers located in a distance of 3 ~ 8m from the excavation face. This procedure enables immediate installation of the supporting elements as close to the excavation face as possible, and hence is very important for tunneling through fractured rocks. The second phase of tunnel support installation, performed on a working platform 32 ~ 44m away from the face, proceeds concurrently with the excavation work to promote construction efficiency.

A snaking conveyor belt transports the muck from the back of cutterhead to the loading point and loads it to muck cars. A diesel locomotive pulls the cars to a rotation tipping device in the portal area. Between May 2 and May 26, 1996, TBM was in the preliminary operation stage, and bored 144m (sta. 5k+398 ~ 5k+254). Beginning on June 4, the TBM was fully operational, and it had advanced 1,658m (sta. 5k+254 ~ 3k+596) by November 30. The average daily advancement was about 14.2m during these 117 working days (refer to Table 4 for construction rate from June through November).

Table 4. TBM Excavation Rate

Descriptions		Units	June	July	Aug.	Sep.	Oct.	Nov.	Monthly Average
(1)Excavated Length		m	338	459	395.7	184	252.6	28.3	276.26
(2)Actual Boring Time		hr	202.5	238	169	113.5	139	35	
(3)Total working Time		hr	530	638	432	452	468	180	
(4)Boring Rate; **(1)÷(2)**		m/hr	1.669	1.928	2.341	1.621	1.817	0.808	
(5)Usage; **(2)÷(3)×100%**		%	38.21	37.30	39.12	25.11	29.7	19.44	31.48
(6)TBM Excavation Rate; **(1)÷(3)**		m/hr	0.638	0.720	0.916	0.407	0.540	0.157	
(7)Working Days		day	23	27	19	19	20	9	19.5
(8)Average Construction Rate		m/day	14.69	17.00	20.82	9.68	12.63	3.14	14.2
Daily Average Construction Rate for Various Types of Support	Type I	m/day	0	0	0	0	0	0	0
	Type II		29.54	30.3	27.1	0	0	0	28.98
	Type III		19.36	21.2	18.8	22.94	20.51	0	20.56
	Type IV		12.0	0	15.0	11.61	9.36	10.2	11.63
	Type V		7.58	13.48	10.32	8.47	7.07	3.63	8.47

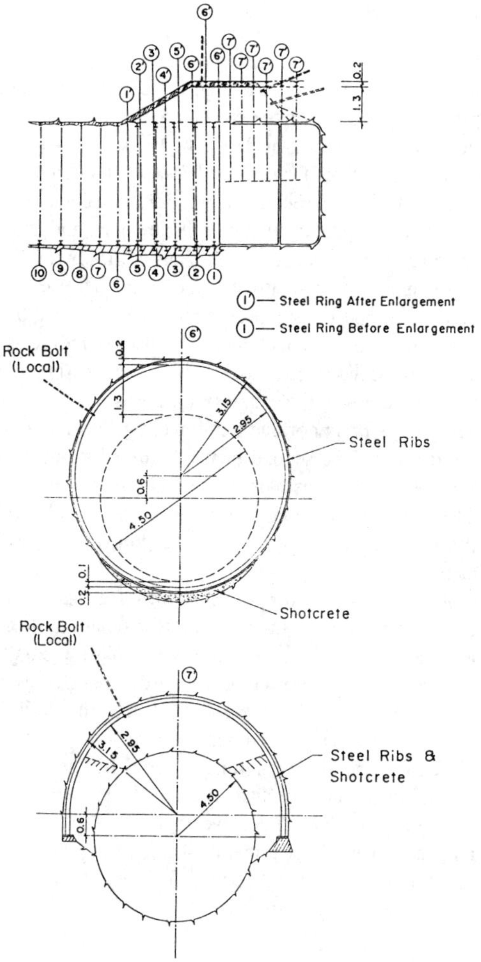

Figure 5. Treatment Process on Sta. 3k+600 ~ 610

7.4 *Construction Hazard*

In the early period, tunnel excavation proceeded rather smoothly because rock conditions were favorable and there was little groundwater. Monthly advancing rate could even reach 400m. However, as the excavation entered the region of high mountains, groundwater increased enormously and geological variation became more complex. TBM construction began to slacken consequently.

Major hazardous incidents that have occurred during excavation are as follows:

1. At sta. 3k+ 877±, a large amount of groundwater was encountered during excavation. Muddy mixture of excavated materials and water seeped into the TBM and damaged the grease seals on the cutter disc shaft. Therefore, excavation work was temporarily stopped and repair of the TBM took about 7 days.

2. Between sta. 3k+610 and sta. 3k+600, rock became fractured and groundwater entered the tunnel from the crown area at a rate of 36 L/sec. The already installed invert segments were uplifted by water and the cutterhead was stuck due to the deformation of the surrounding rocks. Consequently, the invert segments and a certain depth of soft rocks had to be removed and backfilled by shotcrete first, then the invert segments were reinstalled. To free the TBM cutterhead, the cross-section area around the cutterhead was enlarged from crown to springline. Finally, steel ribs and shotcrete were used to protect the enlarged face. The treatment took about 25 days to complete (see Figure 5).

8 CONCLUSIONS

The TBM excavation method has already been adopted in various tunnel projects all over the world. However, this method is still quite new in Taiwan. To answer the question of whether TBM's can work satisfactorily under the complex geological conditions in Taiwan, more trials as well as the devotion of the engineers are needed. In the Shihlin Hydropower Project, the TBM advanced 1,802m from May to November 1996. This construction rate, which may not be much for TBM's in general, has certainly set a new record for tunnel excavation in Taiwan. Hopefully, the experience obtained from this project will contribute to the further development of the tunneling industry in this country.

Computer Methods and Advances in Geomechanics, Yuan (ed.) © 1997 Balkema, Rotterdam, ISBN 90 5410 904 1

Tunnel face stability of hydroshield tunnelling

Mona Mansour
Faculty of Engineering, Helwan University, El-Mataria, Cairo, Egypt

G. Swoboda
Innsbruck University, Austria

ABSTRACT: Prediction of settlements above shield-driven tunnels is one of the major designer questions, predominantly in urban tunnelling. This paper reports on a three-dimensional parametric study based on the Finite Element Method, which was conducted to investigate the impact of various parameters on stability of the tunnel face. Considering homogeneous soil, the study involves the effect of various soil parameters, such as shear strength parameters and coefficient of earth pressure at rest, and the cover depth of the ground above the tunnel crown. For tunnelling techniques applying positive pressures to support the tunnel face, such as: compressed air, earth pressure-balanced (EPB) and slurry methods, the results of this analysis have a direct practical significance for evaluating the pressure required to maintain tunnel face stability. A design criteria based on the yielding concept is suggested to define the critical stability state of the tunnel face.

1 INTRODUCTION

Ground control using shield methods has succeeded, to a great extent, in preventing soil deformations at the circumferential boundary of the tunnel. Results of the analyses performed in previous studies show that a rigid shield followed by an adequate lining and a perfect grouting system can efficiently restrict radial soil deformations (Mansour 96), (Swoboda & Mansour 95), (Swoboda & Mansour 96). Face stability of tunnels driven in soft ground, however, remains a major concern for excavation design. The most important requirement is to prevent any excessive deformations at the tunnel face, which might lead to face failure and consequent undesirable subsidence at the ground surface. In shield tunnelling, various face support techniques are used to minimize face deformations to an acceptable limit and to control groundwater, if it exists. Positive ground control methods such as compressed air, earth pressure-balanced shields and bentonite slurry shields are mostly used.

For the purpose of understanding tunnel face behaviour under various parametric and geometrical configurations, studies based on several concepts have been carried out. These studies are mainly based on experimental tests, limit state analysis and finite element analysis.

Experimental tests using the centrifugal technique were successfully performed to study the behaviour of the tunnel face (Chambon et al. 1991), (Chambon & Corté 1994), (Kimura & Mair 1981). Failure mechanisms, which developed from gradual reduction of face support pressure, could be defined for various soil conditions. The influence of different parameters on tunnel face stability was investigated. The experimental studies require sophisticated models to be built, which are processed using special machines. These facilities are not always available.

The limit state analysis makes it possible to estimate stability conditions at the tunnel face using lower bound and/or upper bound solutions (Davis et al. 1980), (Leca & Dormieux 1990), (Anagnostou & Kovari 1994). Relations to assess the face support pressure were given . An estimate for slurry pressure, necessary to support the tunnel face under various soil conditions and tunnel geometries, was introduced. The proposed methods using this technique are applicable only for homogeneous soils.

Finite element models have been employed to study the behaviour of the tunnel face and to estimate the critical support pressure under various conditions (Romo & Diaz 1981), (Chaffois et al.1988), (Lee & Rowe 1990), (Eisenstein & Ezzeldine 1994). The advantage of these models is that they can be adapted for various geometric configurations, loading conditions and material properties. Complicated geological formations, such as stratified soil, can be simulated. Models based on axisymmetric analysis have been used to predict the critical yield pressure and to study the impact of various parameters on this pressure. Axisymmetric models can not be applied well for shallow tunnels. Few studies based on three-dimensional analysis have been carried out.

2 APPLIED NUMERICAL MODELS

Both experimental studies and theoretical analyses have shown that stability of the tunnel face represents a real three-dimensional problem. Because face stability is mainly a failure problem, nonlinear soil behaviour is essentially considered, where an iterative technique is followed to redistribute the unbalanced stresses. An elasto-plastic model for soil, based on the Mohr-Coulomb failure criterion is adopted in the analysis.

The shield, the lining and the grouting systems are presumed to provide the circumference of the tunnel with the full support needed to prevent soil deformation except through the face. Therefore, elements having very high stiffness parameters are used to represent the rigid supporting system in this region.

3 DESIGN CRITERIA OF FACE SUPPORT PRESSURE

Estimation of support pressure required to maintain stability of the tunnel face, must be based on a certain design criterion that meets the site requirements. This criteria can be based on two different concepts:

(a) yielding concept,

(b) deformation concept.

Criterion based on the yielding concept, represents the most critical situation, where the determined support pressure can be defined as: "the pressure below which, deformation at the tunnel face begins to rapidly increase due to yielding". This pressure should also ensure that the plastic region at the tunnel face does not extend to the ground surface. Because the yield criterion provides a critical stability condition, a safety factor must be introduced.

The other criterion based on the deformation concept is applied when the surface deformation resulting from the yielding concept exceeds the acceptable limits imposed by the site conditions. This case is particularly likely to occur in soils having a low elastic modulus. The face pressure that achieves this criterion can be defined as: "the pressure that attains ground surface deformation within prerequisite values". A factor of safety is not needed in this concept.

In the present analysis, the yield criterion is considered because it is a general face stability concern and it must be realized for any tunnel, regardless of the site conditions. To find the critical face pressure at yield, a simple technique is followed similar to that used by (Eisenstein & Ezzeldin 1994), (Lee & Rowe 1990), (Chaffois et al.1988) and (Romo & Diaz 1981) to find the relation between face pressure and the corresponding face displacement. A circular tunnel is presumed to be excavated in a certain soil condition. The circumference of the tunnel is supported by a rigid liner and its face is under initial ground pressure. The initial face pressure is gradually reduced, resulting in an axial displacement at the face towards the gallery. The critical state of face stability is reached when the magnitude of face pressure is reduced such that the rate of increase of the face displacement associated with face pressure reduction follows an accelerating trend. At this point, yield pressure at the face is recorded. A check of whether the plastic zone at this pressure is extended to the ground surface or not, is carried out. The following definitions are used:

- initial face pressure: the following two conditions are considered, as shown in Fig. 1:
  (a) face pressure equals the initial ground pressure at the tunnel face and it varies linearly between a minimum value of $C\gamma K_o$ at the crown, to a maximum value $(C + D)\gamma K_o$ at the invert, where γ is the unit weight of soil, K_o is the coefficient of earth pressure at rest and C & D as defined in Fig. 2,
  (b) uniform face pressure equals the initial soil pressure at the tunnel axis and is given by, $(C + \frac{D}{2})\gamma K_o$
- face pressure ratio (PF): the ratio between active face pressure, after a certain reduction in the initial value, and the initial ground pressure at the tunnel face
- yield pressure ratio (P_y): the ratio between face pressure at yield, and the initial ground pressure at the tunnel face
- relative face displacement (δ_x/D): the ratio between horizontal face displacement δ_x at the point of maximum deformation at yield, and the tunnel diameter D
- cover depth ratio (C/D): the ratio between soil depth above the crown level C, and the tunnel diameter D

The first distribution considered for the initial face pressure, condition (a), results in a face support pressure having the same distribution as the soil pressure. This condition represents the earth pressure-balanced method and, to some extent, the slurry method when excavation is performed above the groundwater table. However, uniform face pressure, represented by condition (b), simulates the face pressure distribution when the slurry method is applied below groundwater, and when compressed air is applied above the groundwater table, which is not a common practice.

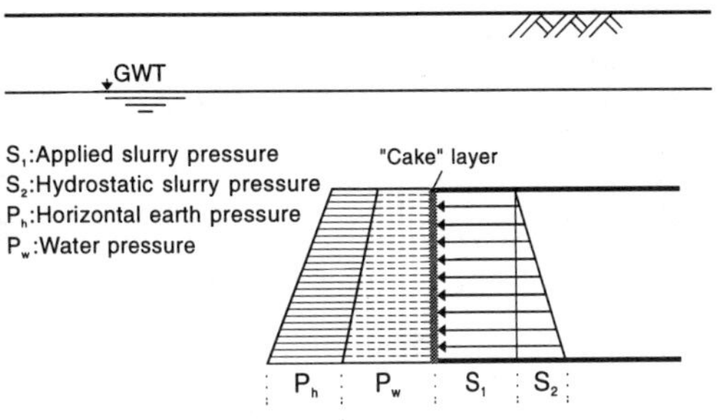

(a) Slurry pressure during excavation

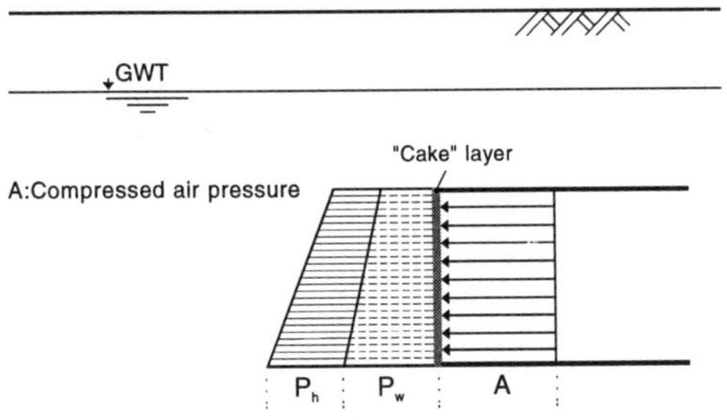

(b) Compressed air during stops

Figure 1: Face pressure distributions

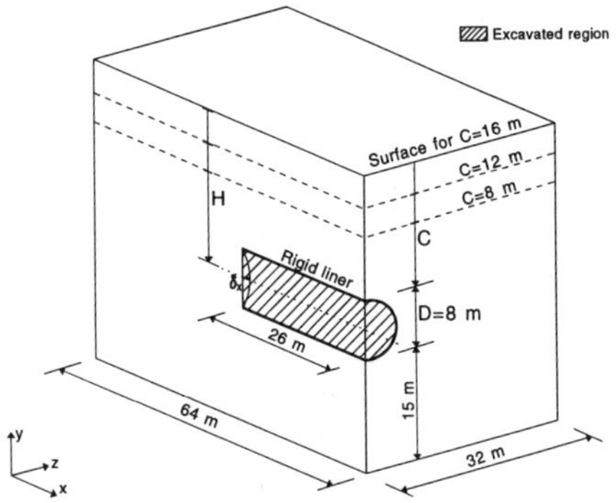

Figure 2: Dimensions of the finite element model

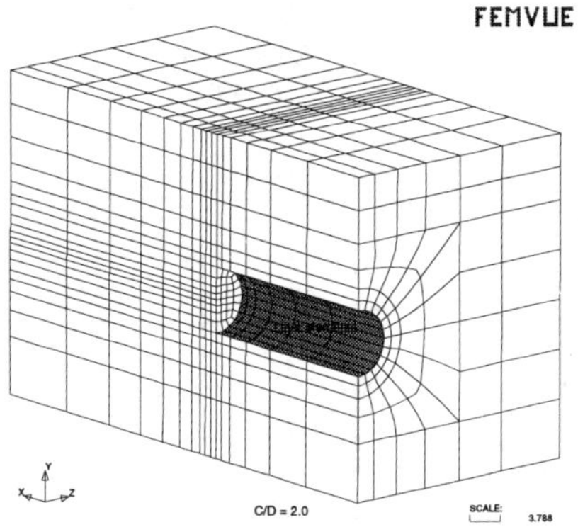

Figure 3: Finite element mesh for C/D=2.0

4 THREE-DIMENSIONAL FINITE ELEMENT ANALYSIS

Figure 2 shows the geometrical configuration of the tunnel and the soil mass used in the finite element analysis. Three cover depth ratios (C/D=1.0, 1.5, 2.0) are considered. The soil mass is modelled using 20-node brick elements (IPQS). SHELL elements, each of eight nodes and a relatively large thickness of 80 cm are used to simulate the rigid tunnel liner. The finite element mesh corresponding to the cover depth ratio $C/D = 2.0$ is shown in Fig. 3 after tunnel excavation to illustrate the shell elements representing the tunnel liner.

Face stability for tunnels excavated in homogeneous soils is examined for both cohesive and frictional soils. Table 1 and Table 2 show different parameters investigated for cohesive soil and cohesionless soil, where :

c' : coefficient of cohesive resistance of soil.
ϕ : angle of internal friction of soil.
K_o : coefficient of earth pressure at rest.

The two alternative conditions of initial face pressure are described as follows:

(a) : non-uniform initial face pressure equals initial soil pressure at the face.

(b) : uniform initial face pressure equals initial soil pressure at the tunnel axis.

In all cases, Young's modulus for the elastic state and Poisson's ratio are chosen to be 30.0 MPa and 0.3, respectively.

5 FACE STABILITY IN COHESIVE SOILS

For the three cover depth ratios, stability of the tunnel face in clayey soils, having different cohesive resistance values, c', and a constant small angle of internal friction, $\phi = 10^o$, was studied.

Relationships between face pressure ratio, PF, and relative face displacement, δ_x/D, at the center point are plotted for various values of the coefficient of cohesive resistance, c'. Figures 4 and 5 show these relations for cover depth ratios, C/D, equal 1.0 and 2.0, respectively. As a general trend, the face displacement gradually increases as the face pressure is released (from $PF = 100\%$ to 0%). The yield point is not well defined and a graphic construction is needed to identify the yield pressure ratio, P_y. The relationship

Table 1: Parameters for cohesive soils

No.	C/D	ϕ (^{o})	c' (kPa)	K_o
C1		10.0	10.0	0.8
C2	1.0	10.0	30.0	0.8
C3		10.0	50.0	0.8
C9		10.0	10.0	0.8
C10	2.0	10.0	30.0	0.8
C11		10.0	50.0	0.8

Table 2: Parameters for frictional soils

No.	C/D	ϕ (^{o})	c' (kPa)	K_o	Initial face pressure (a)	(b)
F1		20.0	0.0	0.6	*	*
F2		25.0	0.0	0.6	*	
F3		30.0	0.0	0.6	*	*
F4		35.0	0.0	0.6	*	
F5	1.0	40.0	0.0	0.6	*	*
F6		20.0	5.0	0.6	*	
F7		30.0	5.0	0.6	*	
F8		40.0	5.0	0.6	*	
F9		20.0	0.0	0.6	*	*
F10		25.0	0.0	0.6	*	
F11		30.0	0.0	0.6	*	*
F12		35.0	0.0	0.6	*	
F13		40.0	0.0	0.6	*	*
F14		20.0	5.0	0.6	*	
F15	1.5	25.0	5.0	0.6	*	
F16		30.0	5.0	0.6	*	
F17		35.0	5.0	0.6	*	
F18		40.0	5.0	0.6	*	
F19		30.0	5.0	0.4	*	
F20		30.0	5.0	0.5	*	
F21		30.0	5.0	0.7	*	
F22		30.0	5.0	1.0	*	
F23		20.0	0.0	0.6	*	*
F24		25.0	0.0	0.6	*	
F25		30.0	0.0	0.6	*	*
F26		35.0	0.0	0.6	*	
F27	2.0	40.0	0.0	0.6		*
F28		20.0	5.0	0.6	*	
F29		30.0	5.0	0.6	*	
F30		40.0	5.0	0.6	*	

* : condition considered in the analysis

between the yield pressure ratio required to achieve the critical face stability and the cohesive resistance parameter is plotted for various cover depth ratios as

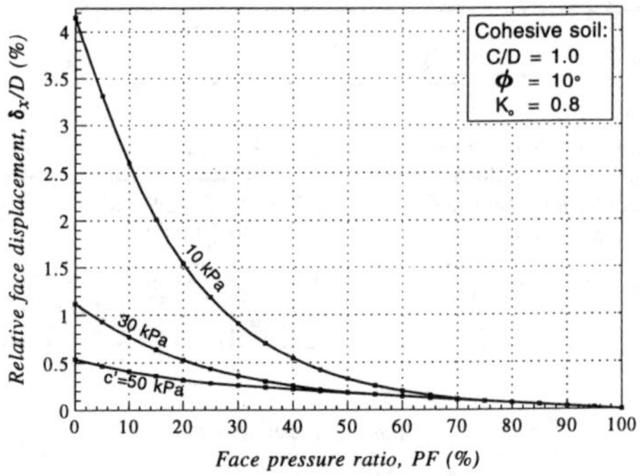

Figure 4: Face displacement for cohesive soils, C/D=1.0

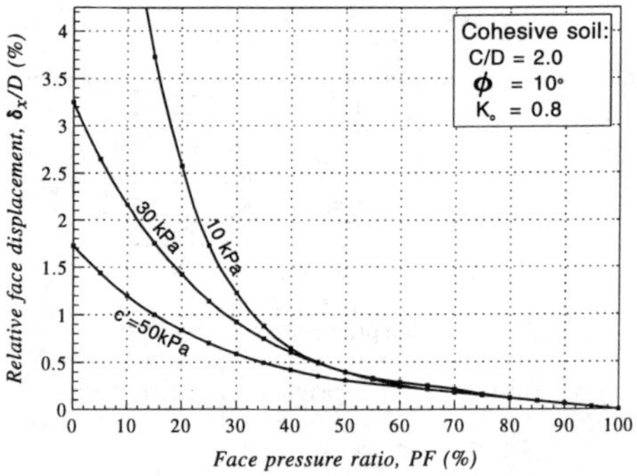

Figure 5: Face displacement for cohesive soils, C/D=2.0

shown in Fig 6. These relations represent a summary for all the studied cases. The following findings were obtained:

- *Effect of cohesive resistance:*
  for all cover depths, yield pressure ratio P_y decreases as the factor of cohesive resistance of soil c' increases. The reduction of P_y begins with a low rate, which becomes slightly higher for $c' > 30kPa$. The difference between P_y required for very soft clay (c' = 10 kPa) and that required for stiff clay (c' = 50 kPa), does not exceed 14% for each cover depth ratio.

- *Effect of cover depth:*
  the yield pressure ratio P_y increases slightly with the increase in cover depth ratio. Duplication of cover depth ratio, however, results in an about 6% increase in the required P_y at any value of cohesive resistance parameter.

- A yield pressure ratio of 40% is sufficient to maintain the critical stability state of the tunnel face, for the range of soil conditions and tunnel geometries considered in the present study.

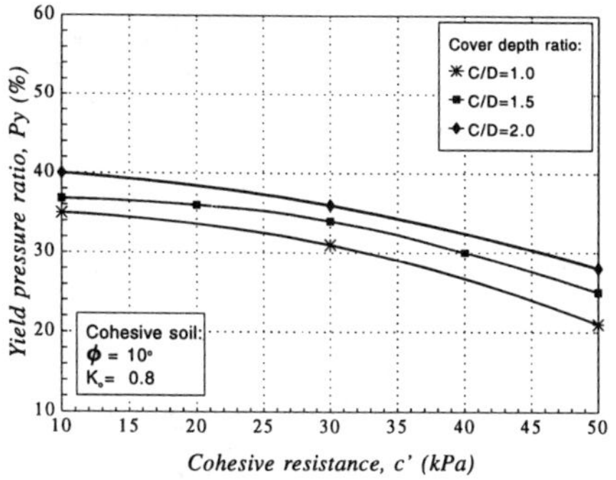

Figure 6: Yield pressure ratio for cohesive soils

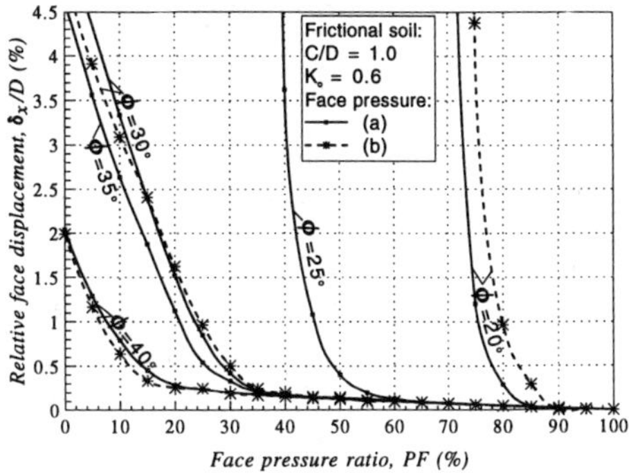

Figure 7: Face displacement for ϕ soils, C/D=1.0

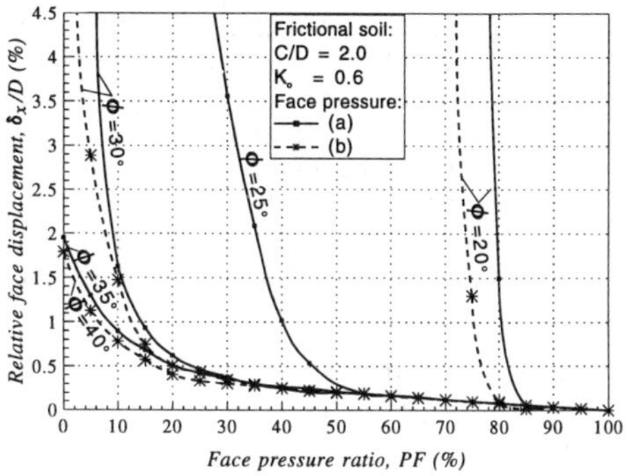

Figure 8: Face displacement for ϕ soils, C/D=2.0

5.1 *Failure mechanism in cohesive soil*

For all the studied cases, the plastic zone at full pressure release at the face is found to occur in a local zone limited to the vicinity of the tunnel face, except for the cohesive resistance parameter, $c' = 10\ kPa$, where the plastic zone extends to the ground surface. As the cohesive resistance parameter increases, the developed plastic zone decreases.

Propagation of the failure mechanism to the ground surface is associated with undesirably high surface subsidence, and an external face pressure is essential to control it from the beginning. For low values of c', the applied yield pressure reduces the plastic zone and prevents its propagation to the surface. Whereas for higher values of c', it reduces the extension of the developed plastic zone. Consequently, the ground settlement is reduced dramatically in cases of low cohesive resistance values and reasonably in cases of higher values.

It can be concluded that the failure mechanism is an important parameter when deciding the factor of safety used to calculate the working face pressure.

6 FACE STABILITY IN FRICTIONAL SOIL

For the three cover depth ratios, the stability of the tunnel face in frictional soils with various angles of internal friction was investigated. Two different conditions were examined:
(i) pure frictional soils, $c' = 0.0$, and
(ii) frictional soils having a small cohesive resistance $c' = 5.0\ kPa$.

Also, the influence of the coefficient of earth pressure at rest was studied. Two conditions of initial face pressure distribution, representing the EPB and slurry techniques, are considered according to Table 2. The relationships between face pressure ratio. PF, and relative face displacement, δ_x/D, at the point of maximum deformation at yield, are plotted for several values of the angle of internal friction, ϕ. These relations are illustrated in Figures 7 and 8 for the considered cover depth ratios C/D, 1.0 and 2.0, respectively. The solid lines represent case (a) where the initial face pressure equals the initial soil pressure at the face, and the dashed lines are for case (b) where the initial face pressure is uniform and equals the initial soil pressure at the center point of the face. In the elastic state, the face displacement increases slightly as the face pressure is released; when yield starts, face displacement rapidly increases at an accelerating rate. Contrary to cohesive soil, the yield point for frictional soil is clear and can be defined easily. It is seen that there is no considerable difference in the results obtained for the two cases of face pressure distribution.

A summary of yield pressure ratios P_y obtained for various angles of internal friction ϕ and various cover depth ratios, is given in Fig 9. For each cover depth ratio, a curve is fitted for the ϕ–P_y relation. From these relations the following findings can be obtained:

- *Effect of angle of internal friction:*
  The angle of internal friction ϕ has a considerable effect on the required yield pressure ratio P_y. As the value of ϕ increases, the yield pressure ratio P_y decreases, namely at a high rate for small angles of internal friction ϕ and then at a lower

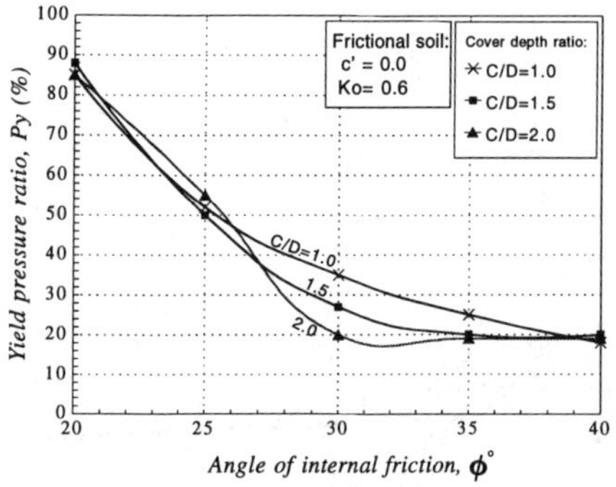

Figure 9: Yield pressure ratio for ϕ soils

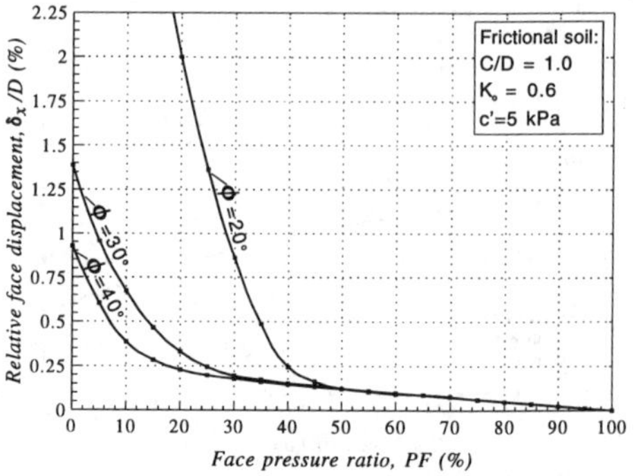

Figure 10: Face displacement for $\phi - c'$ soils, C/D=1.0

rate. It is seen that the effect of ϕ on the yield pressure ratio P_y becomes negligible when its value reaches 30°, for $C/D = 1.5$, and 35° for $C/D = 2.0$. For the shallowest tunnel, $C/D = 1.0$, the angle of internal friction ϕ continues to affect P_y up to $\phi = 40^o$. It can be concluded that as the tunnel becomes deeper the range of the angle of internal friction ϕ, that affects the tunnel face stability, becomes smaller.

- *Effect of cover depth:*
  The cover depth ratio C/D has no significant effect on the yield pressure ratio at low values for the angle of internal friction ϕ and also at the highest value, $\phi = 40^o$. For intermediate values of ϕ a small influence is found where the yield pressure ratio P_y increases as C/D decreases. The same conclusion was obtained from experimental tests performed on tunnels in sand , where the cover depth slightly affected the face pressure at collapse.

6.1 *Failure mechanism in frictional soil*

For all the cases studied in pure frictional soil, the plastic zone developed due to full pressure release at the tunnel face was found to extend up to the ground surface. At the predetermined yield pressure, propagation of the plastic zone is restricted such that it is limited to the vicinity of the face. As an exception, for the shallowest tunnel, $C/D = 1.0$, a pressure higher than the calculated yield pressure is required to prevent the plastic zone from extending to the surface.

Experimental tests (Chambon and Corté 1994) have shown that for the same soil and tunnel diameter the failure mechanism at yield pressure has almost the same pattern for various cover depths. However, for very shallow tunnels, $C/D < 1.0$, the failure envelope is cut by the ground surface, causing secondary failure at the surface, whereas for deeper tunnels the failure envelope closes before reaching the ground surface.

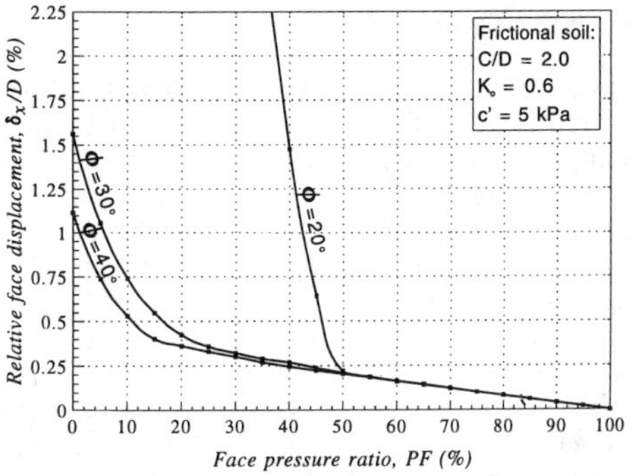

Figure 11: Face displacement for $\phi - c'$ soils, C/D=2.0

From the results of numerical analysis, which are supported by experimental results, it can be concluded that for very shallow tunnels the failure mechanism is the most critical criterion in designing the face support pressure in cohesionless soils.

6.2 *Effect of cohesion*

The existence of a small percentage of fine particles, i.e. clay particles, mixed with the frictional soil may influence the face stability. For the three cover depth ratios, face stability in frictional soil having various angles of internal friction ϕ and a small cohesive resistance of $c' = 5\ kPa$ was examined. The deformation pattern of the tunnel face is found to have the same trend as in pure frictional soils.

The relationship between face pressure ratio and relative face displacement is plotted for the three cover depths, as shown in Figs 10 and 11. It is seen that this small cohesive resistance provided the face in frictional soil with reasonable stability, which is greater than that in pure frictional soil. As shown in Fig 12, this effect is more evident in soils having small values of ϕ, where the added cohesive resistance represents a

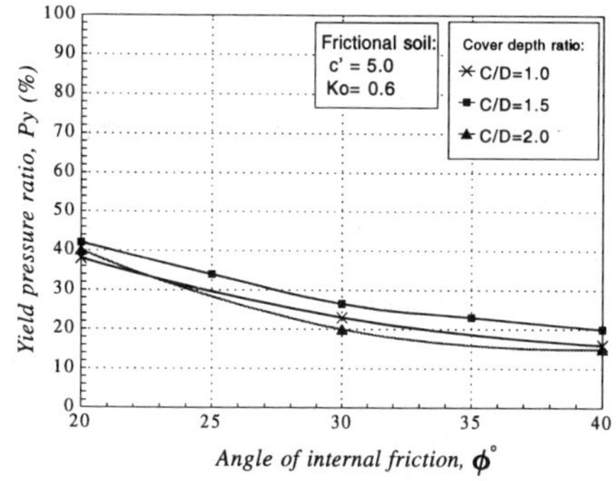

Figure 12: Yield pressure ratio for ϕ-c' soils

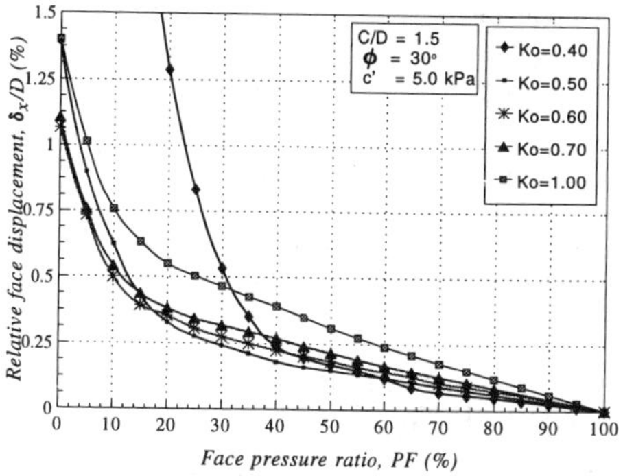

Figure 13: Face displacement at different K_o

considerable percentage of the soil's shear strength. On the other hand, for high values of ϕ the added cohesion does not significantly affect the yield pressure.

At full release of face pressure, the plastic zone extends to the ground surface only at low values of the angle of internal friction ϕ. In general, the small cohesive resistance used can reduce the plastic zone developed in the case of pure frictional soils such that at high values of ϕ the plastic zone is limited to the vicinity of the tunnel face.

6.3 *Effect of coefficient of earth pressure at rest*

The coefficient of earth pressure at rest, K_o, represents the ratio between horizontal and vertical soil stresses at any point. As K_o increases, the confining stress increases and hence yielding occurs at higher stresses. For frictional soil having an angle of internal friction of $\phi = 30^o$ and a small cohesive resistance of $c' = 5\,kPa$, the effect of the coefficient of earth pressure at rest K_o on face behaviour was studied.

Figure 13 shows $PF - \delta_x/D$ relations for several K_o values. When face pressure ratio reduces from $PF = 100\%$ until the yield starts, the soil behaves elastically. In this range of elastic behaviour the relative face displacement increases slightly as K_o increases

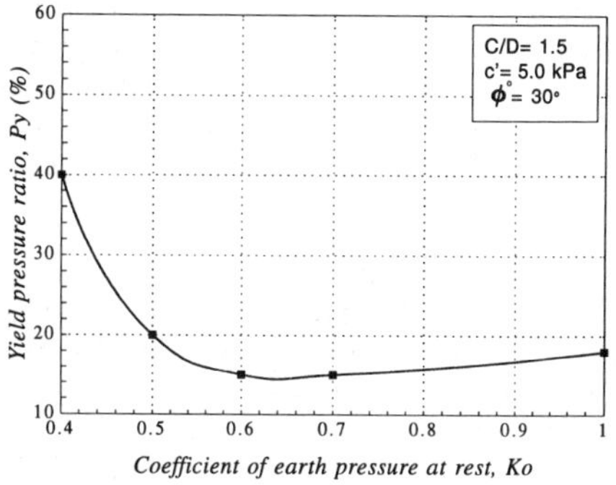

Figure 14: Yield pressure ratio variation with K_o

due to the increase of initial horizontal face pressure. At the same time, the confining pressure increases as K_o increases, resulting in a delay in yield onset. The relationship between K_o and P_y given in Fig 14 shows that at $K_o = 0.4$ a yield pressure ratio of 40% is obtained. For $K_o \geq 0.5$, the yield pressure ratio is reduced to about 20% and is not considerably influenced by the change in K_o.

The plastic zone at full pressure release extends to the ground surface at low values of $K_o \leq 0.5$. For higher values, the plastic zone ends below the ground surface.

7 CONCLUSIONS

Three-dimensional numerical analysis is absolutely important in order to obtain reliable results when studying the stability of the tunnel face. From the above parametric study based on the Finite Element Method, the following conclusions can be obtained:

- in cohesive soils, the yield at the tunnel face occurs gradually with a reduction of face pressure, while in frictional soils it occurs suddenly.

- no significant differences are obtained when various face pressure distributions simulating various tunnelling techniques, such as EPB and the slurry method, are considered in the analysis.

- tunnel face stability in cohesive soils is affected by the value of the cohesive resistance parameter c'. However, a pressure equal to 40% of the initial face pressure is found to be sufficient to reach the critical face stability in all the analyzed cases.

- the angle of internal friction, ϕ, is seen to influence considerably the tunnel face stability in frictional soils.

- for cohesive soils, the cover depth above the crown level has a minor effect on tunnel face stability.

- for frictional soils, the cover depth ratio C/D exerts no significant effect on the yield pressure ratio at low values for the angle of internal friction ϕ or at high values. For intermediate ϕ values a small influence is found where the yield pressure ratio P_y increases as C/D decreases.

- for all the cases studied in cohesive soils, the plastic zone at full pressure release at the face is found to occur in a local zone limited to the vicinity of the tunnel face, except for very soft clay, $c' = 10\ kPa$, where the plastic zone extends to the ground surface. The obtained yield pressure could reduce the plastic zone and prevent its propagation to the surface.

- for all the studied cases in pure frictional soil, plastic zone developed due to full pressure release at the tunnel face is found to extend up to the ground surface. At the predetermined yield pressure, propagation of the plastic zone is restricted such that it is limited to the vicinity of the face, except in very shallow tunnels.

- addition of a small cohesive resistance to the pure frictional soils improves their performance for tunnel face stability. This phenomenon is more evident at low values for the angle of internal friction.

- a coefficient of earth pressure at rest greater than 0.5 does not affect the required yield pressure. At lower values, the yield pressure increases considerably. With regard to failure mechanism, at $K_o < 0.5$ a pressure slightly higher than the yield pressure may be required to stop the plastic zone before it reaches the ground surface.

ACKNOWLEDGMENT

The work reported here was partially supported by the research project P10077-ÖTE "Saftey of Tunnels" of the Austrian National Science Foundation.

REFERENCES

Anagnostou, G. & K. Kovari 1994. The face stability of slurry-shield-driven tunnels. *Tunnelling and Underground Space Technology* 9(2):165–174.

Broms, B.B. & H. Bennermark 1967. Stability of clay at vertical opening. *J. Soil Mech. Found. Div., ASCE* 93:71–94.

Chaffois, S., P. Laréal, J. Monnet & C. Chapeau 1988. Study of tunnel face in a gravel site. Swoboda (ed.), *Numerical Methods in Geomechanics, Innsbruck*:1493–1498. Rotterdam:Balkema.

Chambon, P., J-F. Corté, J. Garnier & D. König 1991. Face stability of shallow tunnels in granular soils. Hon-Yim Ko & Francis G. Mclean (eds.), *Centrifuge91* :99–105. Rotterdam:Balkema.

Chambon, P. & J-F. Corté 1994. Shallow tunnels in cohesionless soil: stability of tunnel face. *Journal of Geotechnical Engineering, ASCE* 120(7):1148–1165.

Davis, E.H., M.J. Gunn, R.J. Mair & H.N. Seneviratne 1980. The stability of shallow tunnels and underground openings in cohesive material. *Geotechnique* 30(4):397–416.

Eisenstein, Z. & O. Ezzeldine 1994. The role of face pressure for shields with positive ground control. Abdel Salam (ed.), *Proc. of the International Congress on Tunnelling and Ground Conditions, Cairo, Egypt* :557–571. Rotterdam:Balkema.

Kimura, T. & R.J. Mair 1981. Centrifugal testing of model tunnels in soft clay. *Proc. 10th International Conference on Soil Mechanics and Foundation Engineering, Stockholm* 1:319–322.

Leca, E. & L. Dormieux 1990. Upper and lower bound solutions for the face stability of shallow circular tunnels in frictional material. *Geotechnique* 40(4):581–606.

Lee, K.M. & R.K. Rowe 1990. Finite element modelling of the three-dimensional ground deformations due to tunnelling in soft cohesive soils: Part 2 -Results. *Computers and Geotechnics* 10:111–138.

Mansour, Mona 1996. Three-dimensional Numerical Modelling of Hydroshield Tunnelling. *Ph.D. Thesis, Innsbruck University, Innsbruck, Austria.*

Romo, M.P. & M.C. Diaz 1981. Face stability and ground settlement in shield tunneling. *Proc. 10th International Conference on Soil Mechanics and Foundation Engineering ,Stockholm* 1:357–360.

Swoboda, G. & M. Mansour 1995. Three-dimensional numerical modelling of slurry shield tunnelling. In Wagner, H. and Shulter, A. (eds), *Proc. of the Inter. Lecture Series TBM Tunnelling Trends, Hagenberg, Austria* :27–41. Rotterdam:Balkema.

Swoboda, G. & M. Mansour 1996. Three-dimensional numerical modelling of slurry shield tunnelling. In Chang-Koon Choi, Chung-Bang Yun and Dong-Guen Lee (eds), *Proc. of the Third Asian-Pacific Conference on Computational Mechanics, Seoul, 16–18 Sept. 1996* 4:2659–2665. Korea:Techno-Press.

Computer Methods and Advances in Geomechanics, Yuan (ed.) © 1997 *Balkema, Rotterdam, ISBN 90 5410 904 1*

A rock-lining interaction formula for concrete lining designs of pressure tunnels

Wern-ping Chen
Sverdrup Civil, Inc., Boston, Mass., USA

ABSTRACT: This paper presents a rock-lining interaction formula for concrete lining designs of pressure tunnels. This formula is derived from the formulation by Kruse (1970). Geotechnical parameters in this derivation include the deformation modulus of the disturbed rock adjacent to tunnel excavation, as well as that of the intact rock out of the tunnel excavation influence zone. Comparisons and discussions are made between the presented and the U.S. Army Corps of Engineering's (1978) formulations. Practical design considerations and a numerical algorithm, based on the derived formulation, are given.

1 INTRODUCTION

It is customary to design the reinforced concrete linings of pressure tunnels by steel lining design formulations. In this type of approach which is an approximation, the amount of reinforcing steel for a concrete lined tunnel is translated directly from that of a steel lining analysis, i.e., translating the steel lining thickness to reinforcing steel area directly. Brekke (1987) has a detailed review for steel lining design formulations. Since a steel lined tunnel is a plane strain problem and a reinforced concrete lined tunnel is not, a specific design formation for concrete lined tunnels is needed.

The Corps of Engineers (COE) Design Manual (1978) provides a formulation for concrete lining designs of pressure tunnels; however, it does not consider the change of rock deformation modulus caused by the impact of tunnel excavation. This paper proposes a formulation which includes the change of rock deformation modulus from tunnel excavation.

Though the mechanical and pore water pressure coupling effect is an important phenomenon for pressure tunnel designs, it is not addressed in this paper in order to simply the derivation. Schleiss (1986, 1988) and Fernández (1994) have insightful investigations for this effect.

2 DERIVATION

The proposed derivation is based on Kruse's (1970) formulation for steel lining designs of pressure tunnels. It is reviewed below.

2.1 *Kruse's formulation*

The Kruse's formulation was derived for steel lining designs of pressure tunnels. In this formulation, the radial steel lining displacement is equal to the sum of the radial displacements of the backfill concrete, disturbed rock, intact rock, and the annular gap between the steel lining and the rock. Equation (1), reformulated by Brekke (1987), describes this formulation. Equations (2) to (5) provide detailed expressions of each displacement component accordingly.

$$\Delta_S = \Delta_C + \Delta_D + \Delta_I + \Delta_G \tag{1}$$

$$\Delta_S = \frac{P_s R_s^2}{t_s E_s}(1-\nu_s^2) \tag{2}$$

$$\Delta_C = \frac{P_1 R_1}{E_c} \ell_n(\frac{R_s}{R_1}) \tag{3}$$

$$\Delta_D = \frac{P_1 R_1}{E_{RD}} \ell_n\left(\frac{R_x}{R_2}\right) \quad (4)$$

$$\Delta_I = \frac{P_1 R_1}{E_R} (1+\nu_R) \quad (5)$$

Δ_G is approximately 0.0003 times R_s for a 46^{0}F temperature differential.

where,
Δ_S: the radial displacement of the steel lining
Δ_C: the radial displacement of the backfill concrete
Δ_D: the radial displacement of the disturbed rock
Δ_I: the radial displacement of the intact rock
Δ_G: the annular gap between the steel lining and the backfill concrete
R_s: the radius of the steel lining
R_1: the radius of the inner concrete surface
R_2: the radius of the outer concrete surface
R_x: the radius of the inner intact rock surface
P_i: the tunnel internal hydraulic pressure
P_1: the radial pressure at R_1 = the pressure shared by the rock
P_s: the tunnel internal pressure shared by the steel lining = P_i - P_1
E_s:Young's modulus of steel, 29000 ksi
E_c:Young's modulus of concrete
E_r: the deformation modulus of intact rock
E_{rd}: the deformation modulus of disturbed rock, approximately equal to 1/3 of E_r
ν_s: Poisson's ratio of steel, 0.3
t_s : thickness of the tunnel lining

To derive the formulation for reinforced concrete lining designs of pressure tunnels, the only term which needs to be modified in Equation (1) is the term $\Delta_{S,}$ as described in Equation (2).

2.2 *Proposed concrete lining design formulation*

Figure 1 shows the geometry of a concrete lined pressure tunnel. The radial displacement of a reinforcing steel can be derived from the relationship between the thrust, P_s times R_s, and the reb radius, R_s, as shown in Figure 2. Equations (6) to (9) present these derivations.

$$2\pi\, R_s + \Delta_H = 2\pi\, R_s' \quad (6)$$

$$R_s' = R_s + \frac{\Delta_H}{2\pi} = R_s + \Delta_S \quad (7)$$

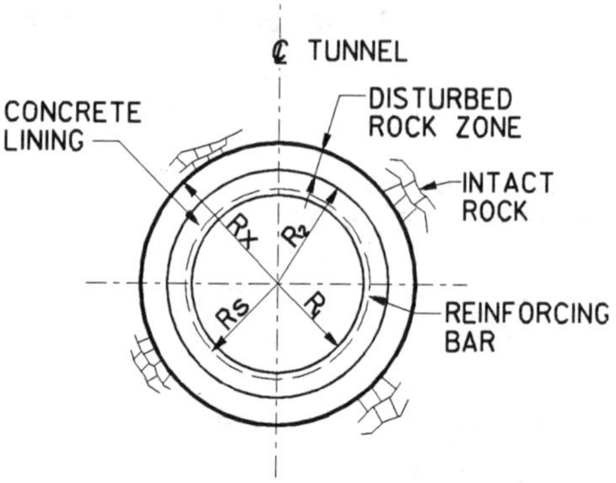

Figure 1 Geometry of a concrete lined pressure tunnel.

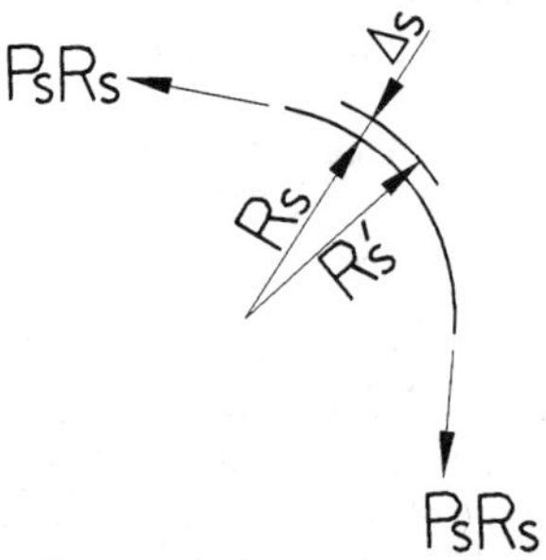

Figure 2 Radial displacement, Δ_S, of the reinforcing steel bar.

$$\Delta_H = \frac{P_s R_s L}{E_s A_s}$$

$$= \frac{P_s R_s (2\pi R_s)}{E_s A_s} = \frac{2\pi\, P_s R_s^2}{E_s A_s} \quad (8)$$

$$\therefore, \quad \Delta_S = \frac{P_s R_s^2}{E_s A_s} \quad (9)$$

where,
Δ_S: the radial displacement of the reinforcing steel bar
Δ_C: the average radial displacement of the concrete lining
Δ_H: the total elongation of the reinforcing steel bar
R_s: the radius of the steel reinforcing bar
R_s': the radius of the steel reinforcing bar after elongation
P_s: the tunnel internal pressure exerting on the reinforcing bar = P_i - P_1

L: the reinforcing bar perimeter = $2\pi R_s$
A_s: the area of the reinforcing steel bar
For other definitions: see Section 2.1

The two unknowns, P_l and P_s, are determined from the compatibility and the equilibrium conditions. Equation (1), the compatibility condition, can be consolidated into two unknowns, P_l and P_s, by substituting Equations (3), (4), (5), and (9) for Equation (1). The equilibrium condition requires that the total internal tunnel pressure be equal to the sum of the pressure exerting on the reinforcing bar and the pressure resisted by the surrounding rock, i.e., $P_i = P_s + P_l$. Once P_s and P_l are determined, the amount of steel reinforcement for the tunnel can be designed based on the allowable stress and permissible deformation of the reinforcing steel.

2.3 *Numerical comparisons*

Numerical comparisons were made between the proposed formulation and that of the COE's. The basis of the COE's formulation is the one listed as "the steel lining, concrete reinforcement and rock combined" in the Corps' manual. Since the COE's formulation includes both the steel lining and the reinforcing steel, the steel lining thickness is assumed to be null in order to compare with the proposed formulation. The numerical parameters are: a 16-ft excavated diameter tunnel; 1-ft concrete lining; $A_s = 0.2$ in^2/ft; $P_i = 200$ psi; $E_c = 3605$ ksi; $E_s = 29000$ ksi; $E_r = 500$ ksi; $E_{rd} = 0.33\ E_r$; $\nu_s = 0.3$; $\nu_r = 0.25$ (Poisson's ratio of rock); $\nu_c = 0.17$ (Poisson's ratio of concrete); the concrete has a 28-day compressive strength of 4000 psi; and the disturbed rock zone is assumed to be 1-ft for a TBM excavated tunnel (Nishida, 1982). The subgrade reaction constant of the rock around the tunnel of the COE's formulation is taken as $E_r/(R_1*(1+\nu_r))$. The results are shown in Table 1.

Table 1 demonstrates that the pressure exerted on reinforcing steel by the presented formulation is higher than that of the COE's. This phenomenon is also reflected on the lower rock sharing pressure and the higher radial reinforcing steel displacement of the presented formulation. This is consistent with our experience, since the presented formulation considers the effect of disturbed rock, while the COE's treats the rock as a homogeneous intact medium.

Table 1. Numerical comparisons between the COE's and the proposed formulations.

Formulation	COE's	Proposed
P_s (psi)	29.3	32.7
P_l (psi)	170.7	167.3
Δ_S (inches)	0.0410	0.0456

3 DESIGN CONSIDERATIONS

Pressure tunnels in highly fractured rock, having a low RQD, require a steel lining to mitigate water exfiltration. Pressure tunnels in slightly fractured rock, having high RQD, are usually more economical when designed as concrete lined tunnels. Transient pressures from hydraulic hammer effects are usually neglected in concrete lined tunnels because the concrete lining is permeable in the sense that cracks are inevitable.

The amount of steel reinforcing for a pressure tunnel depends highly upon the permeability of the in situ rock. For a low permeability ground, with $k < 10^{-5}$ cm/sec, the requirement to satisfy both the allowable steel stress and a less stringent lining crack width controls the lining design, since water exfiltration is of less concern. For a high permeability ground, with $k > 10^{-5}$ cm/sec, the requirement to satisfy a stringent lining crack width controls the lining design, since water exfiltration needs to be minimized.

In the following discussions, the longitudinal crack is defined as a crack in the tunnel axis direction, and the circumferential crack is defined as a crack in the direction transverse to the tunnel axis. The purpose of hoop reinforcement is to confined longitudinal cracks, and the longitudinal reinforcement to confined circumferential cracks.

3.1 *Hoop reinforcing - low permeability ground*

In low permeability ground, the hoop reinforcing is determined from allowable reinforcing stress and permissible lining crack width. The allowable reinforcing stress is about 40 percent of the yield stress of a reinforcing steel. From a lining crack width, the corresponding reinforcing steel stress level can be derived. This stress level must be less than the allowable reinforcing steel stress.

The total radial reinforcing displacement, Equation (10), is equal to the sum of the elastic radial displacement of the concrete lining, Equation (11), and the radial displacement translating from a permissible crack width, the second term in Equation (10). Its corresponding reinforcing steel stress level is derived from Equation (12), and is shown in Equation (13).

$$\Delta_S = \Delta_h + \frac{N\,W}{2\pi} \qquad (10)$$

$$\Delta_h = \frac{f_t\,R_c}{E_c} \qquad (11)$$

$$\epsilon_s = \frac{\Delta_S}{R_s} \qquad (12)$$

$$f_s = \epsilon_s\,E_s = \frac{\Delta_S\,E_s}{R_s} \qquad (13)$$

Δs: the radial displacement of the reinforcing bar
Δh: the average radial concrete lining displacement
N: total number of cracks in concrete lining
W: the allowable crack width
R_c: the average concrete lining radius
R_s: the radius of the reinforcing steel
E_c: Young's modulus of the concrete
E_s: Young's modulus of the reinforcing steel
ε_s: the allowable reinforcing steel strain
f_s: the allowable reinforcing steel stress which shall be smaller than 40% of f_y; f_y is the reinforcing steel yield stress
f_t: concrete tensile capacity = 40% of the modulus of rupture = $0.4 * 7.5\ f_c'^{1/2}$

The permissible crack width, W, is limited to 0.1" to prevent tunnel instability from washing out of joint fillings (Schleiss, 1987, 1988). In a unreinforced concrete lining, two longitudinal cracks are usually developed (Fernández, 1994). For a reinforced concrete lining, the number of longitudinal cracks, N, can be determined from the crack spacing, S_c, evaluated by Rizkalla and Hwang (1984). Hendron et al. (1987) have simplified S_c to: $S_c = d / (10\ \rho_g)$, where the d is the diameter of the reinforcing bar and ρ_g the ratio of the steel area to the concrete area. Conservatively, two cracks can be assumed for reinforced concrete linings.

3.2 *Hoop reinforcing - high permeability ground*

In high permeability ground, the hoop reinforcing is determined solely by the permissible crack width. ACI 224R-90 provides a relationship between the crack width and the reinforcing steel stress. The allowable reinforcing stress level, Equation (15), is derived from Equation (14).

$$\omega = 0.1\ .\ f_s\ (d_c\ A)^{(1/3)}\ .\ 10^{(-3)} \qquad (14)$$

$$f_{sa} = \frac{10{,}000\ .\ \omega}{(d_c\ A)^{(1/3)}} \qquad (15)$$

where,
ω: the crack width
f_s: the reinforcing steel stress level corresponding to an ω value
d_c: distance from the concrete tension surface to the center of a reinforcing bar; 2" is the maximum value in Equation (15) (Rice, 1984)
A: effective tension area of concrete divided by the number of reinforcing bars = $2\ d_c\ S_b$; S_b is the spacing between reinforcing bars in the tunnel axis direction
f_{sa}: allowable reinforcing steel stress based on the permissible crack width, ω_a

The crack width, W, is recommended to be 0.004" in ACI 224R-90 Table 4.1 for water retaining structures; however, this value is 0.008" in ACI 224.2R-86. The author believes that the 0.008" permissible crack width is a reasonable number for most concrete lined pressure tunnels, and that the 0.004" criterion is too stringent

3.3 *Longitudinal reinforcing*

The amount of longitudinal reinforcing steel required in a concrete lined pressure tunnel depends upon the permeability of the surrounding ground and the code of practice. Table 2 provides a comparison of reinforcing steel ratios from different sources. The reinforcing ratio, ρ_g, is the ratio of steel area to the gross concrete area.

For concrete lined tunnels, the spacing between circumferential cracks is one to two tunnel diameters. On this basis and ACI 350R-89, ρ_g is about 0.0033 for a 20-ft diameter tunnel. The British Code is the most stringent code for this issue. Its ρ_g depends upon the crack width, the spacing between reinforcing bars, the temperature differential, the reinforcing steel size, and the concrete compressive strength. The Australian Code adopts a single value of 0.004.

For a practical design purpose, ρ_g is recommended to be 0.0045 for tunnels in high permeability ground, and 0.0018 for a tunnel in low permeability ground.

4 THE NUMERICAL ALGORITHM

A numerical algorithm, based on the presented formulation, for concrete lined pressure tunnels, is outlined here.

1. Consolidate Equation (1) by substituting it from Equations (3), (4), (5), and (9).
2. Try a reinforcing steel area of A_s.
3. Solve P_s and P_l from the consolidated Equation (1), in step 1, and the relationship $P_i = P_l + P_s$. Find the reinforcing steel stress, f_s, by dividing P_s by A_s.
4. Hoop reinforcing steel: For low permeability ground, check Δ_s and f_s against the limits from Equations (10) and (13). For high permeability ground, check f_s against the limit from Equation (15).
5. Repeat steps 2 to 4 until both Δ_s and f_s satisfy the permissible limits.
6. Longitudinal reinforcing steel: Use ρ_g = 0.0045 for high permeability ground and 0.0018 for low permeability ground. The area of longitudinal steel is equal to the product of A_g and ρ_g.

For permissible limits of Δ_s and f_s, see Section 3.

Table 2. Comparisons of longitudinal reinforcing steel ratios.

Source	ρ_g
ACI 318-95 for structural slabs	0.0018
ACI 318-95 for walls	0.002 to 0.0025
ACI 350 R-89 (Fig. 2.5) s = the spacing between circumferential cracks or the length between shrinkage dissipating joints	0.0028, for s=25 ft. 0.0033, for s=30 ft. 0.0041, for s=40 ft
ASCE (1993)	0.0028
Australian Code for water retaining structures	0.004
British Code of Practice	0.0033 to 0.008

5 CONCLUSIONS

1. A rock-lining interaction formulation for concrete lined pressure tunnels is developed. Its behavior is within the expectations, as demonstrated by comparisons to COE's formulations.
2. It is reasonable to assume that two circumferential cracks will develop in a concrete lined pressure tunnel. The maximum permissible crack width to prevent tunnel instability is about 0.1". For high permeability ground, the maximum crack width is limited to between 0.004" and 0.008". The later is a reasonable number for design.
3. Reinforcing ratios of 0.0045 and 0.0018 are recommended for longitudinal reinforcing steel for high and low permeability grounds, respectively.

REFERENCES

ASCE 1993. Strength design of reinforced concrete hydraulic structures. Technical engineering and design guides as adapted from the U.S. Army Corps of Engineers, No.2.

ACI 224.2R-86. Cracking of concrete members in direct tension.

ACI 224R-90. Control of cracking in concrete structures.

ACI 318-95/318R-95. Building code requirements for structural concrete and commentary.

ACI 350R-89. Environmental engineering concrete structures.

Brekke, T.L. & Ripley, B.D. 1987. Design guidelines for pressure tunnels and shafts. *EPRI Document*, RP-1745-17.

Corps of Engineering, Department of the Army. 1978. Engineering and design tunnels and shafts in rock, EM 1110-2-2901.

Fernández, G. 1994. Behavior of pressure tunnels and guidelines for liner design. *J. Geotech. Eng.*, ASCE, vol.120, no.10: 1768-1791.

Hendron, A.J., Fernández, G., Lenzini, P., and Hendron, M.A. 1987. Design of pressure tunnels. *The art and science of geotechnical engineering at the dawn of the twenty first century,* Prentice-Hall, Englewood Cliffs, N.J.: 161-192.

Kruse, G.H. 1970. Rock properties and steel tunnel liners. *J. of the Power Division*, ASCE, vol.96, no. Po3: 415-435.

Nishida, T., Matsumura, Y., Miyanaga, Y. & Hori, M. 1982. Rock mechanical viewpoint on excavation of pressure tunnel by tunnel boring

machine. *Proceedings of the International Symposium on Rock Mechanics*: 812-826.
Rice, P.F. 1984. Structural design of concrete sanitary structures. *Concrete International,* October: 339-341.
Rizkalla, S.H., Hwang, L.S. 1984. Crack prediction for members in uniaxial tension. ACI Struct. J., (Nov.-Dec.): 572-579.
Schleiss, A.J. 1987. Design Criteria for pervious and unlined pressure tunnels. *Proceeding of the international conference on Hydropower*, Oslo, Vol. 2: 667-678.
Schleiss, A.J. 1988-89. Design of reinforced concrete-lined pressure tunnels. *Proceedings of the international conference on tunnels and waters*, Madrid: 1127-1133.

Computer Methods and Advances in Geomechanics, Yuan (ed.) © 1997 Balkema, Rotterdam, ISBN 90 5410 904 1

Some remarks on 2-D-models for numerical simulation of underground constructions with complex cross sections

H. F. Schweiger & H. Schuller
Institute for Soil Mechanics and Foundation Engineering, Technical University Graz, Austria
R. Pöttler
Kling-Consult, Germany

ABSTRACT: The majority of analyses performed for tunnelling in practical engineering are still 2-D-calculations. The most commonly used methods for taking into account 3-D effects in 2-D analyses are the load reduction and the stiffness reduction method, at least as far as simulation of tunnels constructed according to the principles of the NATM are concerned. In both methods the deformations occuring ahead of the tunnel face have to be estimated. In this paper it is shown that the stress relief factors for both methods cannot be easily related by considering equivalent spring models and a new relationship is proposed. However it follows from the numerical study presented that this correlation only holds for linear elastic material behaviour and simple geometric conditions. Furthermore it is shown that the load reduction methods has advantages if nonlinear material behaviour and complex construction sequences with temporary linings have to be considered.

1 INTRODUCTION

Numerical methods have been applied to solve problems in tunnelling in various geological conditions ranging from deep tunnels in jointed rock to shallow tunnels in soft soil. It is obvious that tunnel construction involves threedimensional stress redistributions in the ground and a number of 3-D analyses have been performed and published in the literature (e.g. Semprich 1980, Kielbassa & Duddeck 1991, Ostermeier & Schikora 1992, Augarde et al. 1995). However these analyses have been made mainly for research purposes and involved simplified assumptions either with respect to the material behaviour, the discretization or the modelling of the excavation stages. Full 3-D models for complex multistage excavations as they are common e.g. for large underground openings constructed according to the principles of the New Austrian Tunnelling Method (NATM) are very rare in practical engineering and justified only for outstanding constructions. Methods have been proposed therefore in the literature to account for 3-D effects in 2-D-analyses in an approximate manner (e.g. Schikora & Fink 1982, Laabmayr & Swoboda 1986) and the most commonly used ones are the load reduction method and the stiffness reduction method although other approaches have been developed and successfully applied in practice. The problem with all these approximations is twofold: firstly one has to guess a so-called pre-relaxation or stress relief factor which is not known a priori if only limited experience for a given construction method in given ground conditions is available. These pre-relaxation factors can be determined by in situ measurements but are then available only for adjusting the numerical calculations during construction but not at the design stage. Secondly one has to recall that this concept of pre-relaxation is reasonably well understood for modelling tunnel construction involving only a few excavation steps in a cross section. For more complex excavation sequences commonly encountered in the construction of e.g. underground stations this concept may not be applied with the same degree of confidence and thus the results may strongly depend on the method used.

In this paper the two methods (load reduction and stiffness reduction method) are briefly recalled for continuity. The commonly used relationship between the two stress relief factors which is based on a simple spring analogue is then discussed and it is shown that this relationship does not account for the geometry of the excavation which turns out to be an important factor. An extended relationship is suggested which allows an easier correlation of the two methods. However the finite element study presented clearly indicates that results obtained from both methods differ significantly when complex excavation stages and nonlinear behaviour have to be modelled and that the load reduction method has some advantages over the stiffness reduction method because the assumptions made are much clearer from a computational point of view.

2 2-D MODELLING OF TUNNEL EXCAVATION

The basic idea behind all 2-D models for simulating tunnel excavation is to capture the deformation which occurs ahead of the tunnel face by some means without performing a full 3-D analysis. This is especially important when the design of the temporary lining is of concern. It has been mentioned that a number of approaches have been proposed ranging from design charts developed from 3-D analyses (Kielbassa & Duddeck 1991) to simply assigning fictitious stiffness properties to the tunnel lining (e.g. Schikora & Fink 1982). The two most commonly adopted approaches for modelling tunnel excavation based on the principles of the NATM are the so called load reduction method and stiffness reduction method (Schikora & Fink 1982, Laabmayr & Swoboda 1986).

2.1 *Stiffness reduction method*

Figure 1 illustrates the required computational steps. Starting from an initial state of stress σ_0 equivalent nodal forces are calculated on the boundary of the tunnel cross section and applied in a first computational step whereas the Youngs's modulus of the "core" E_c inside the cross section of the tunnel is modified, so that $E_c = \alpha.E_0$, E_0 being the Youngs's modulus of the ground. This calculation step obviously leads to stress redistributions around the tunnel and the resulting stress state σ^* is then taken as initial stress for the second computational step where the full excavation is modelled with the lining in place (Figure 1). For multistage excavations this procedure is repeated for each construction stage.

The factor α is chosen in such a way that the deformations calculated from computational step 1, which is a "fictitious" load case, matches the deformations measured before the shotcrete lining is in place. At the design stage α has to be estimated from previous experience under similar conditions.

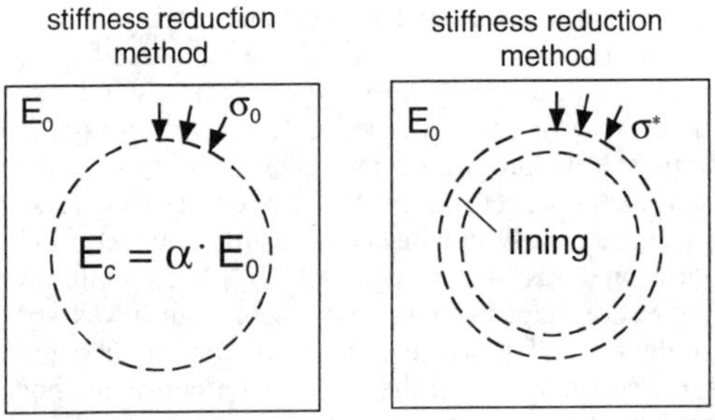

Figure 1. Illustration of computational steps for stiffness reduction method

For shallow tunnels the assumption made is usually verified directly by in situ measurements during construction using e.g. extensometers, for deep tunnels however a direct check is usually not possible.

2.2 *Load reduction method*

A similar approach is adopted in the load reduction method. Again equivalent nodal forces according to the initial state of stress σ_0 are calculated. A supporting pressure $\sigma_s = \beta.\sigma_0$ is then applied at the inside of the tunnel for the first computational step, ie. only $(1-\beta).\sigma_0$ is acting as "load" on the ground (Figure 2).

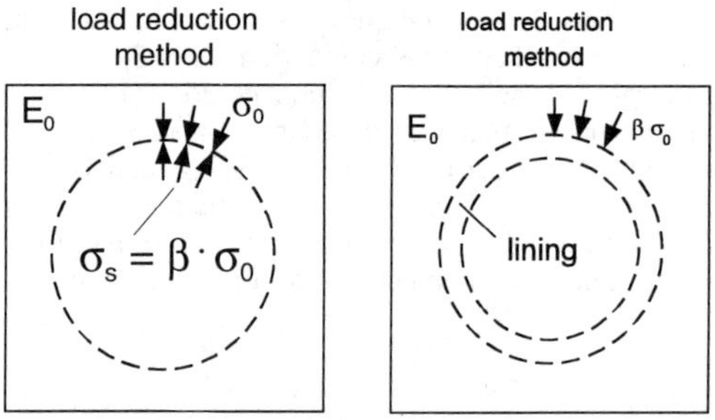

Figure 2. Illustration of computational steps for load reduction method

In the second computational step $\beta.\sigma_0$ is applied on the system including the lining (Figure 2).

Again the parameter β is chosen in such a way that deformations occuring before placement of the lining are matched with field data and the same uncertainties exist as for the parameter α. Similar to the stiffness reduction method, for multistage excavations the procedure is repeated and varying parameters β may be used.

2.3 *Comparison of both methods*

It has always been a matter of personal preference or dependent on the capabilities of the employed software whatever method was chosen and both have been successfully applied in practice. However in some cases it is at one hand desirable to relate the parameters α and β and on the other hand to study the performance of both methods for multistage excavation sequences and nonlinear material behaviour. In the latter case the load reduction method seems to have some advantages because the assumptions made on what part of the "load" is applied to the ground and what part to the system including the lining is much better defined. In the stiffness reduction method the nonlinear behaviour of the "core" may result in unwanted stress distributions especially if a high number of excavation steps with temporary linings has to be modelled. This problem is addressed briefly in section 4.

The question whether it is possible to achieve the same result for a given situation with either method by correlating α and β in an appropriate way is frequently discussed. Laabmayr & Swoboda 1986 suggested a relationship

$$\alpha = \frac{\beta}{1-\beta} \tag{1}$$

which is based on an equivalent spring model. It can be easily shown that this relationship only holds if the actual geometry of the problem, e.g. the tunnel diameter compared the discretized domain, is ignored. If one considers a (still) onedimensional model but taking into account the relation of the diameter of the tunnel and the discretized area one ends up with a relationship given by

$$\alpha = \frac{\beta}{1-\beta} \cdot \frac{l_c}{l} \tag{2}$$

with l_c being the length representing the supporting core and l representing the discretized domain. Equation (2) is exact for a linear elastic material and a constant initial stress state and cannot be readily transfered to plane strain conditions with initial stresses increasing with depth and arbitrary k_0 conditions. However one could base a relationship between α and β on the well known elastic solution of a hole in an infinite medium with constant stress σ_0. If one recalls the distribution of the radial stress for this case one can see that from a distance of about 2.5r (r = radius of hole) the radial stress is hardly influenced by the opening thus the area contributing to the deformation is defined by approximately 2.5r. Applying the same reasoning as above for the equivalent spring model the following empirical relationship can be established

$$\alpha = \frac{\beta}{1-\beta} \cdot \frac{1}{2.5} \tag{3}$$

As will be shown in the next section this correlation allows an almost exact comparison of both methods under certain conditions.

3 PARAMETRIC STUDY FOR SIMPLIFIED EXCAVATION PROBLEM

The main aim of this section is twofold. First it is demonstrated that it is possible to achieve almost identical results applying the load reduction and stiffness reduction method for certain conditions by using the correlation given in Equation (3). Secondly it is shown that a correlation between these two methods is not possible for nonlinear material behaviour.

The geometry and set up of the example chosen follows from Figure 3 and the properties used are given in Table 1.

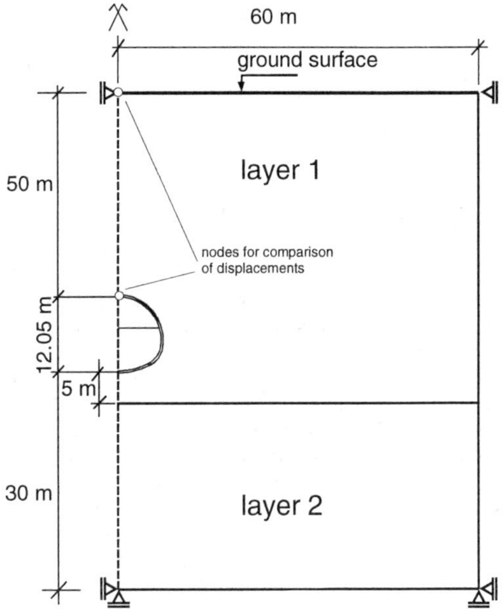

Figure 3. Set up for test example

Table 1. Properties used for layers 1 and 2

	Layer 1	Layer 2
E	50 000 kN/m^2	200 000 kN/m^2
ν	0.3	0.25
φ	28°	40°
c	20 kN/m^2	50 kN/m^2
γ	21 kN/m^3	23 kN/m^3
K_o	0.5	0.6

Shotcrete (d = 25 cm): linear elastic
$E_1 = 5\,000 \times 10^3$ kN/m^2
$E_2 = 15\,000 \times 10^3$ kN/m^2
$\nu = 0.15$

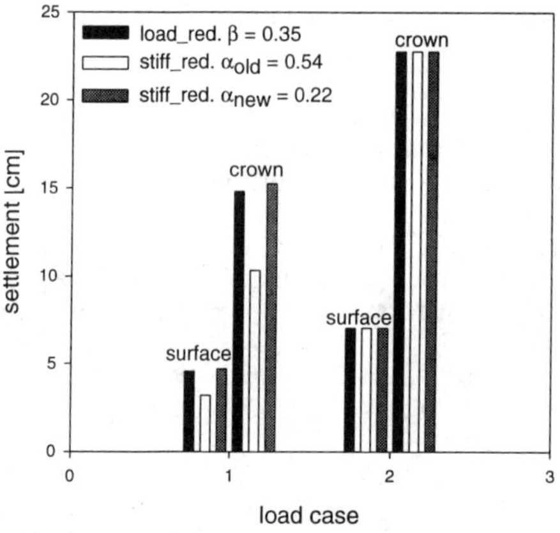

Figure 4. Comparison of crown and surface settlements - elastic solution / no shotcrete lining

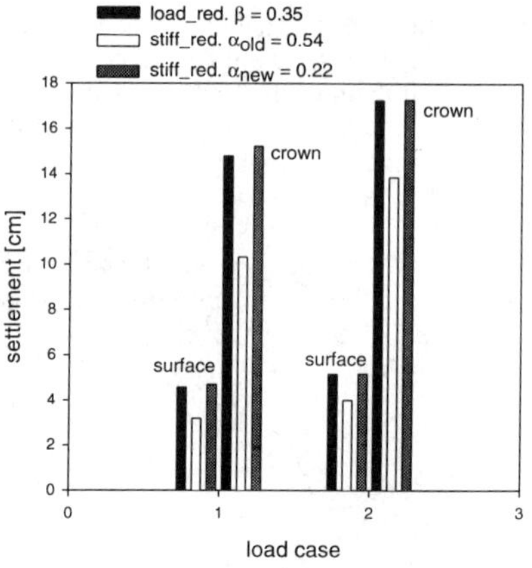

Figure 5. Comparison of crown and surface settlements - elastic solution / with shotcrete lining

3.1 *Circular tunnel with constant stress field*

The first example serves as verification for Equation (3) and is a further simplification of the set up shown in Figure 3. Only one layer and a single excavation step is considered and a constant initial stress of $\sigma_{0,v}$ = 1 000 kPa with K_0 = 0.5 is assumed. E_2 is assumed for the shotcrete.

Three pre-relaxation factors for the load reduction method have been investigated, namely β = 0.35, 0.5 and 0.65. The respective factors for the stiffness reduction method are with respect to Equation (3) α = 0.22, 0.4 and 0.74. This values are denoted with the subscript 'new' in the following to establish the difference to Equation (1) which yield values of α = 0.54, 1.0 and 1.86, and these are denoted with the subscript 'old'. Only the results for β = 0.35 are presented here because the general trends hold for all values of β.

Figure 4 can be regarded as check because it shows results from an elastic analysis involving no shotcrete lining and therefore for load case 2 (= full excavation) all results are identical. However for load case 1 (= pre-relaxation) significant differences are observed if Equation (1) is used whereas the use of Equation (3) allows comparison of the two methods.

Figure 5 shows the same for the case with shotcrete lining and it follows that now the application of Equation (1) yields different results also for load case 2. It is obvious that stresses calculated in the lining differ also significantly.

Figure 6 plots surface and crown settlements for the nonlinear analysis using a Mohr-Coulomb criterion and it becomes clear that a direct comparison of the two approaches is no longer possible but Equation (3) still compares better with the load reduction method than Equation (1).

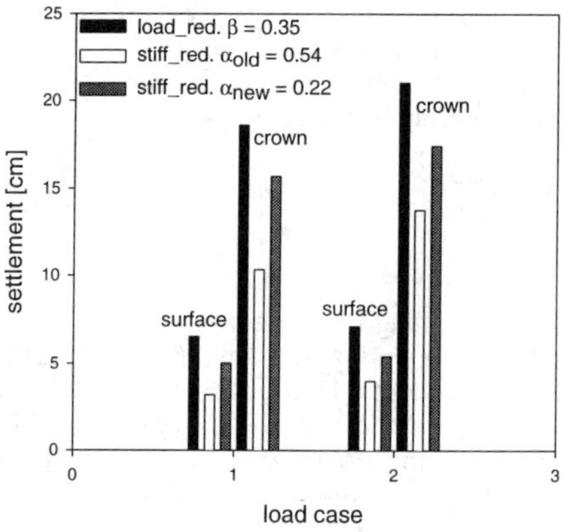

Figure 6. Comparison of crown and surface settlements - plastic solution / with shotcrete lining

3.2 *Tunnel excavation with initial stress increasing with depth*

In this section results are presented by considering initial stresses as given in Figure 3 and Table 1.

The pre-relaxation factor assumed for the load reduction method is again β = 0.35 and the respective factors for the stiffness reduction method are α_{new} = 0.22 and α_{old} = 0.54. As in the previous example in the case of a single step excavation load case 1 corresponds to computational step 1 and load case 2 to computational step 2 referring to the description given in section 2.

Figure 7 corresponds to Figure 5 and it can be seen that Equation (3) still holds as long as the material behaviour is elastic despite the fact that the initial

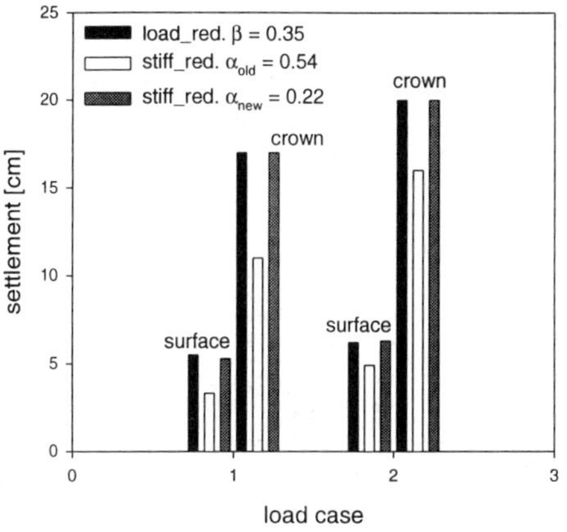

Figure 7. Comparison of crown and surface settlements - elastic solution

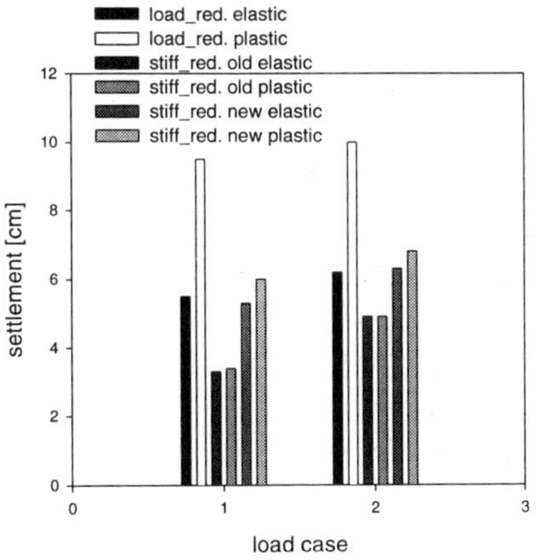

Figure 9. Comparison of elastic and plastic solution for settlements

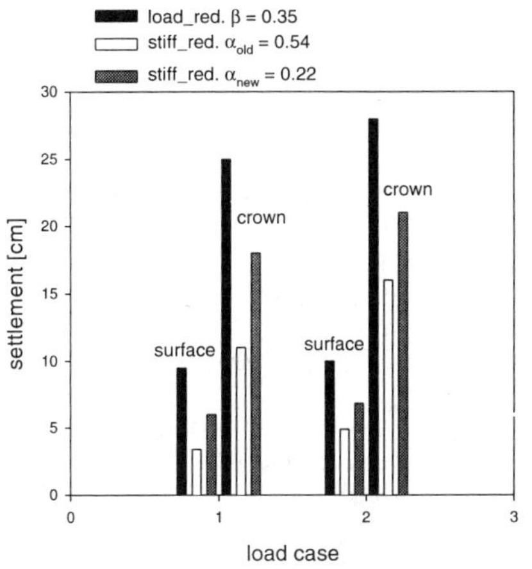

Figure 8. Comparison of crown and surface settlements - plastic solution

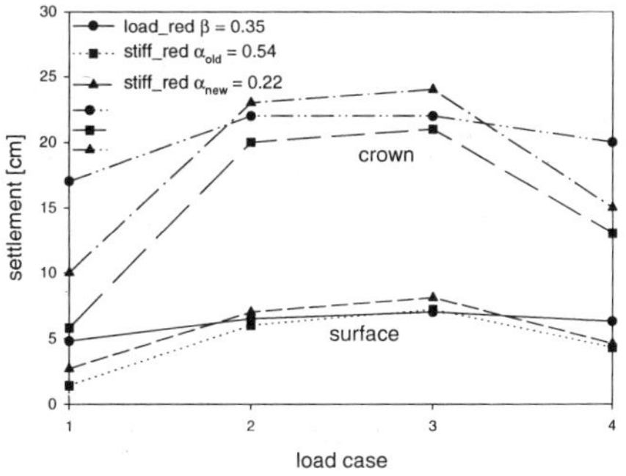

Figure 10. Crown and surface settlements for excavation in 2 steps

stress is no longer uniform. Figure 8 shows the displacements obtained for the nonlinear analysis and again both methods give now significantly different results. Figure 9 is very important from a practical point of view because it depicts calculated surface settlements from the elastic and plastic analysis and it is obvious that the stiffness reduction method yields significantly less plastic deformations as compared to the load reduction method. From a design point of view this has important consequences because it influences the stresses calculated in the lining.

In Figure 10 the development of displacements for the surface and the crown are shown for the case where two excavation stages have been modelled. As mentioned before the same strategy has now to be adopted twice. In this case the lining of the top heading had been assigned E_1 for "load case" 2 and E_2 for "load cases" 3 and 4 (see Table 1) in order to account roughly for the time dependent increase of stiffness of the shotcrete. The lining for the bench and invert in "load case" 4 was given E_1. It is now interesting to note that the application of the stiffness reduction method leads to significant heave in the system for later excavation stages when the full shotcrete lining is in place. This deficieny is well known and is sometimes overcome by introducing varying α-parameters or by combining both methods with varying β and α (Laabmayr & Swoboda 1986). However results then become even more difficult to assess.

4 PRACTICAL EXAMPLE

Inorder to emphasize the findings presented in the

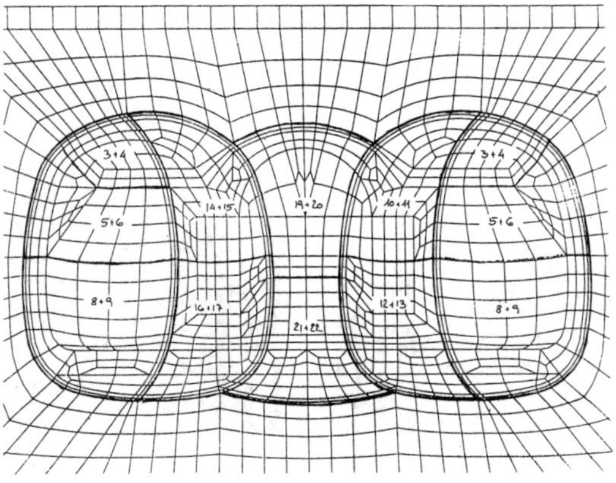

Figure 11. Example for complex excavation sequence

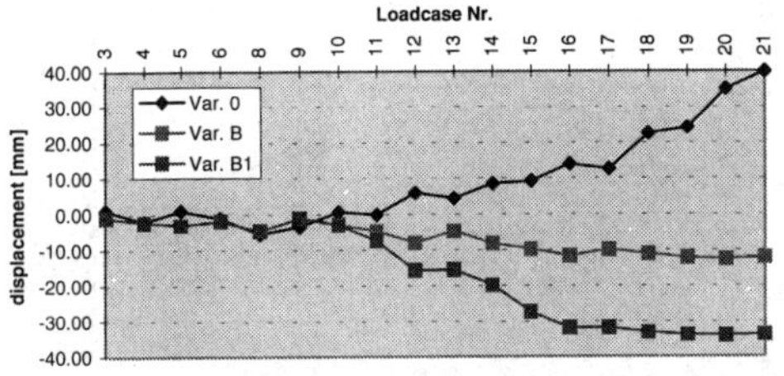

Figure 12. Comparison of load reduction and stiffness reduction method for complex excavation sequence

previous section a result from a practical problem is shown. A detail of a finite element mesh of a typical cross section for an underground station is given in Figure 11. Figure 12 presents the surface settlements for all excavation stages in the centre of the station for the load reduction method (Var. B and B1) and the stiffness reduction method (Var. 0). The unfavourable and not very realistic heave obtained from application of the stiffness reduction method is obvious whereas settlements in the range to be expected are calculated by applying the load reduction method (Var. B and Var. B1 consider slightly different excavation steps as far as the temporary supports are concerned).

5 CONCLUSION

Threedimensional finite element modelling of tunnel excavation is still not state of the art in practical engineering. It is therefore essential to use reasonable methods to take into account deformations occuring ahead of the tunnel face at least in an approximate way. Two very common methods, namely the load reduction method and the stiffness reduction method, have been compared in this paper.

It is shown that previously used relationships to correlate both methods can be improved and allow comparison of both approaches for certain conditions. However once nonlinear material behaviour is considered the methods are very difficult to compare.

It could be shown that for a Mohr-Coulomb material less plastic deformations are developed when using the stiffness reduction method. If complex excavation sequences and nonlinear material behaviour are to be modelled the load reduction method tend to give results which are easier to be interpreted and is thus preferable for practical engineering. The main reason being that the load applied to the ground and the system ground/lining is well defined in the load reduction method. Applying the stiffness reduction method the anticipated distribution of the load is partially lost due to nonlinear stress redistributions inside the core which also influence the stress state around the tunnel.

REFERENCES

Augarde, C.E., H.J. Burd & G.T. Houlsby 1995. A three-dimensional finite element model of tunnelling. In G.N. Pande & S. Pietruszczak (eds.), *Proc. 5th Int. Symp. Numerical Models in Geomechanics, Davos, 6-8 Sept. 1995*: 457-462. Rotterdam: Balkema.

Kielbassa, S. & H. Duddeck 1991. Stress-Strain Fields at the Tunnelling Face - Three-dimensional Analysis for Two-dimensional Technical Approach. *Rock Mechanics and Rock Engineering* 24: 115-132.

Laabmayr, F. & G. Swoboda 1986. Grundlagen und Entwicklungen bei Entwurf und Berechnung im seichtliegenden Tunnel - Teil1. *Felsbau* 4: 138-143.

Ostermeier, B. & K. Schikora 1992. Ergebnisse räumlicher Berechnungen für Tunnelvortriebe im Lockergestein nach der Spritzbetonbauweise. *Bauingenieur* 67: 19-25.

Schikora, K. & T. Fink 1982. Berechnungsmethoden moderner bergmännischer Bauweisen beim U-Bahn-Bau. *Bauingenieur* 57: 193-198.

Semprich, S. 1980. Berechnung der Spannungen und Verformungen im Bereich der Ortsbrust von Tunnelbauwerken im Fels. *Wittke (ed.), Veröffentlichungen des Institutes für Grundbau, Bodenmechanik, Felsmechanik und Verkehrswasserbau, RWTH Aachen*, Heft 8.

Computer Methods and Advances in Geomechanics, Yuan (ed.) © 1997 Balkema, Rotterdam, ISBN 90 5410 904 1

Application of large strain analysis for prediction of behavior of tunnels in soft rock

Mitsuo Nakagawa
CRC Research Institute, Inc., Osaka, Japan

Yujing Jiang & Tetsuro Esaki
Faculty of Engineering, Kyushu University, Fukuoka, Japan

ABSTRACT: To estimate the stability of tunnels excavated in soft rock masses which are governed by the strain-softening and dilatancy behaviors during the post-failure, general numerical methods based on the infinitesimal strain theory are difficult to give a appropriate assessment, especially for the large deformation phenomena with plastic flow. In this paper, the large strain analysis with explicit finite difference code FLAC(Fast Lagragian Analysis of Continua) is carried out, and its applicability to rock engineering is also investigated. Simulations of uniaxial compression test of soft rock material and tunnel excavations in soft rock masses with shallow overburden are tried to verify the large strain analysis method.

1 INTRODUCTION

It is an important subject to evaluate exactly the mechanical behavior of ground in consideration of the safety and economical maintenance for tunnel construction. The cases of excavating tunnels in the deeper soft ground seem to be more and more increased. Deformation and stability of tunnels depend greatly on the mechanical features of soft rock mass. It has been pointed out, according to laboratory test, that soft rock has the features of the transition from strain softening to residual state and large dilatancy during the post-failure. However, it is difficult to represent the large deformation phenomena and plastic flow by using a general numerical approach, such as the FEM(finite element method), based on the infinitesimal strain theory. The explicit finite difference code, FLAC (Cundall,1988), is being considered as an efficient simulation technique for the cases.

This paper is to explain the basic principles of the large deformation method in FLAC code and applications concerned with tunnel behavior in soft rock masses. First, the feature and mathematical constitution of the strain-softening and dilatancy behaviors of soft rock masses based on the laboratory test are illustrated, and then realization of large strain analysis in FLAC code is clarified. Finally, the applicability of the proposed method to rock engineering is verified by simulations of uniaxial compression test and tunnel excavations in soft rock mass.

2 FEATURE OF SOFT ROCK MASS AND EXPRESSION OF LARGE STRAIN ANALYSIS

2.1 *Empirical feature of Soft Rock Masses*

It has been made clear that soft rock represents the strain-softening and dilatancy under low confining pressure(Jiang, 1993). Dilatancy means that volumetric increases with shear deformation during plastic flow under constant compression stress. The transition condition from brittle failure (i.e. strain-softening) to ductile failure(perfect-plasticity) is summered in Figure 1, by means of the relation between confining pressure(σ_3) and principal stress difference($\sigma_1 - \sigma_3$) (Shigeki, 1978). Strain-softening appears clearly if confining pressure is below the brittle-to-ductile transition pressure. The empirical relation between confining pressure and dilatancy in sedimentary soft rock mass is shown in Figure 2. It is also can be understood that dilatancy appears remarkably in the low confining pressure condition and the increasing extent with the principle strain in the strain-softening phase differs from that in the plastic flow phase.

2.2 *Realization of Large Strain Analysis in FLAC*

It is necessary to introduce large strain analysis when the behavior of tunnels in soft rock masses with strain-softening and dilatancy mentioned above is needed to be estimated appropriately. In general, large strain analysis means to solve problems with large deformation. The form of control equation depends on reference coordinate system measured in

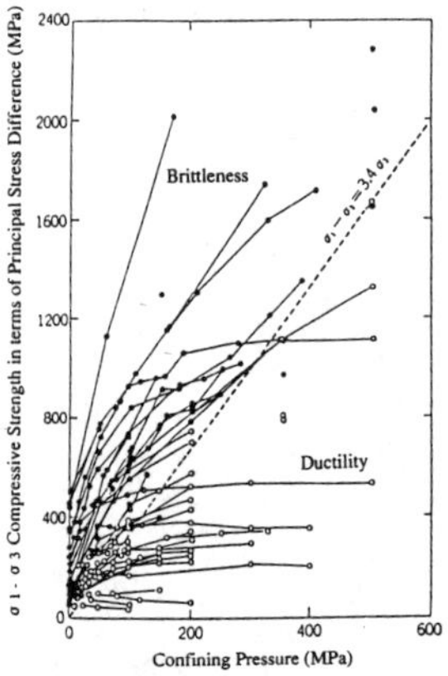

Figure 1. Difference between brittle failure and ductile failure in rock mass.

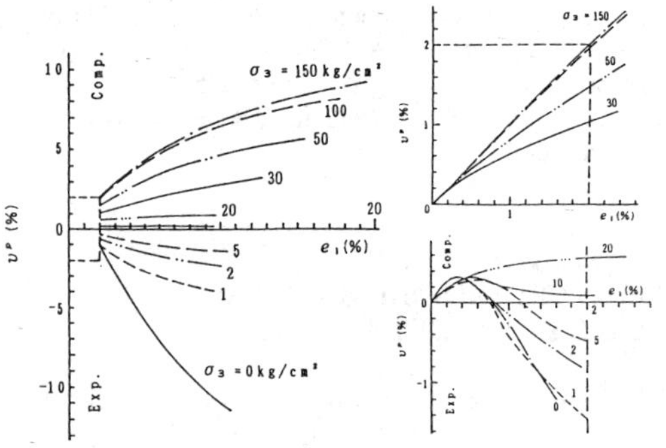

Figure 2. Relation between deviatoric strain and volumetric strain in sedimentary rock mass.

case of large deformation and large rigid rotation. Then control equation may associate with geometrical variable. This nonlinearity of control equation is called geometrical nonlinearity. This must be considered in large strain problem.

FLAC is to solve these problems in explicit time-marching procedure with finite difference method and discretization is implemented by finite difference for both space and time. This method involves the full dynamic equation of motion, even with modeling system that is essentially static so as to ensure that numerical scheme is stable when the physical system being model is unstable with both material and geometrical nonlinearity. In this paper, some factors which realize large strain analysis will be made up(Cundall, 1995) and explanation of formulation of FLAC itself is omitted.

(a) Treatment of reference coordinate system

The grid coordinates are modified at each time step by means of the Updated Lagragian formulation, for incremental deformation according to Equation (1).

$$x_i^{(t+\Delta t)} = x_i^{(t-\Delta t)} + \dot{u}_i^{(t+\Delta t/2)}\Delta t \quad (1)$$

$$\Delta t \le 2\sqrt{\frac{m}{k}} \quad (2)$$

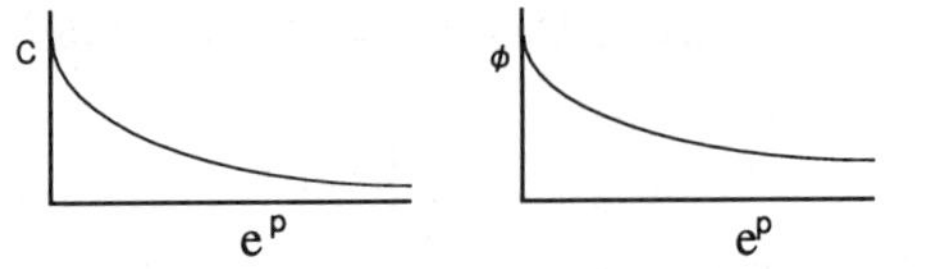

Figure 3. Modeling of reduction of strength with increase of plastic shear strain.

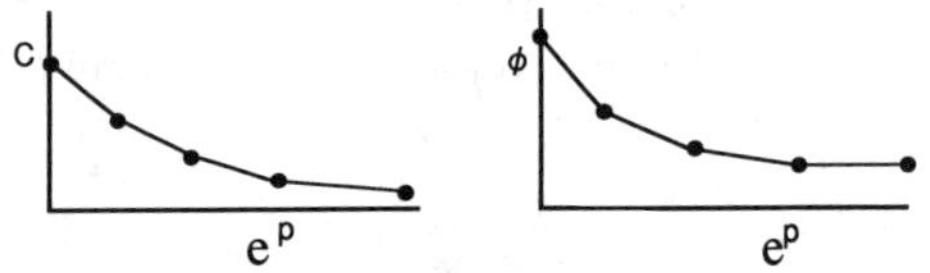

Figure 4. Approximation of reduction of strength by means of assembly of line segment.

The time step Δt, computed by Equation (2), is very small, so that deformation in one time step can be treated as a small strain problem.

(b)Definition of strain rate

Stresses are not related directly to strains, but an incremental formulation involving Euler's strain rate is adopted when invoking the constitutive relation. Euler's strain rate is the linear definition used in small strain analysis.

(c)Definition of stress tensor

The various definitions of stress tensor are attempt to represent the stress in an objective way when a body experiences increments of large strain and rotation. All stress measures are identical for small strain and rotation. The stress formulations incorporate terms that refer to the original, undeformed state of the body. In the updated Lagrangian formulation used here, each increment is treated as a small strain increment involving Cauchy stresses, since a very small time step is taken.

(d)Objectivity of stress tensor

The stress tensor is updated at each time step to account for finite rotation, as required by the condition of frame indifference. The rotation correction terms are identical with those contained in the definition of the Jaumann stress tensor, and are given in the following Equation.

$$\sigma_{ij} := \sigma_{ij} + (\omega_{ik}\sigma_{kj} - \sigma_{ik}\omega_{kj})\Delta t \quad (3a)$$

$$\omega_{ij} = \frac{1}{2}\left(\frac{\partial \dot{u}_i}{\partial x_j} - \frac{\partial \dot{u}_j}{\partial x_i}\right) \quad (3b)$$

(e)Constitutive relation

The usual stress-strain relations, formulated in terms of small strain theory, are used in this approach because very small time-steps are taken.

Consequently the operations to realize geometrical non-linearity lie in update coordinate in equation (1) and correction of stress tensor in equation (3) . Represent of large deformation of system is realized by allowance for large deformation of each element.

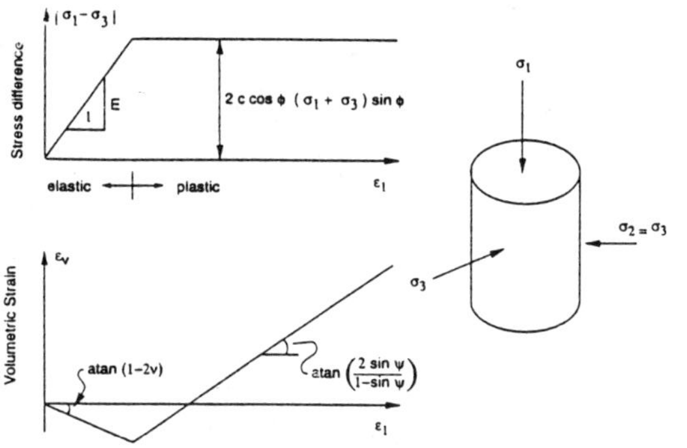

Figure 5. Idealized relation of dilatancy from triaxial test results (Vermeer and de Borst, 1984).

2.3 *Modeling of strain-softening and dilatancy*

(a)Modeling of strain-softening

In this paper, the concept shown in Figure 3 is introduced to model strain-softening behavior, which cohesion and/or friction drop gradually with increase of plastic shear strain e^p after yielding. In real implementation, this concept is approximated as assembly of line segment as shown in Figure 4. When a set of principal plastic strain is given, plastic shear strain used here is derived from Equation (4).

$$e^p = \left\{\frac{1}{2}(e_1^p - e_m^p)^2 + \frac{1}{2}(e_m^p)^2 + \frac{1}{2}(e_3^p - e_m^p)^2\right\}^{\frac{1}{2}} \quad (4a)$$

$$e_m^p = \frac{1}{3}(e_1^p + e_3^p) \quad (4b)$$

(b)Modeling of dilatancy

The extent of dilatancy is represented by dilatancy angle Ψ. In this paper, the idealized relation of volumetric strain-axial strain based on Mohr-Coulomb yield criterion, shown in Figure 5 is used for the decision of dilatancy angle Ψ. The relation of volumetric strain-axial strain is shown in Equation (5).

$$\frac{\dot{\varepsilon}_v^p}{\dot{\varepsilon}_a^p} = \frac{-2\sin\Psi}{1 - \sin\Psi} \quad (5)$$

3 APPLICATION TO COMPRESSION TEST

In this section, the simulation of compression test is carried out based on the result of uniaxial compression test to verify the applicability of large strain analysis.

3.1 *Summary of Test and Modeling*

The material properties of the specimen are shown in Table 1. Axial stress and axial strain mean the macroscopic representation as a whole of specimen, and compression is treated as positive below.

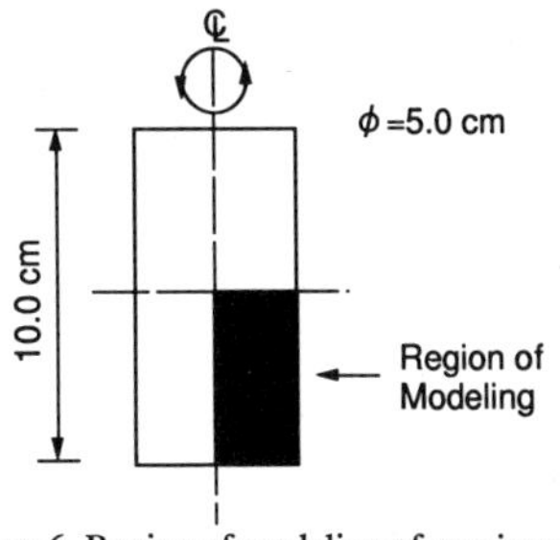

Figure 6. Region of modeling of specimen.

Table 1. Material properties of specimen.

Young's modulus	1200.0 (MPa)
Poisson's rate	0.0875
Density	1860.0(kg/m^3)
Uniaxial strength	4.80(MPa)
Friction angle	30.0(deg)
Cohesion	1.38(MPa)
Dilation angle	25.5(deg)

Table 2. Parameters of strain-softening.

Plastic shear strain e^p	Cohesion C (MPa)
0.0	1.38
0.0839	0.173
1.0	0.173

As shown in Fig. 6, the specimen is axi-symmetrical and is modeled as the half part below the middle height of specimen in consideration of symmetry between the upper and lower parts. The Mohr-Coulomb yield criterion and non-associated flow rule are assumed, the plastic potential function has the same type with the yield function while the angle of friction ϕ in yield function is replaced by the dilatancy angle Ψ. The model in simulation is subjected to the constant velocity at the top of the surface up to 10% of axial strain on the condition that the system equilibrium is maintained. The stress-strain relation from the result of the test is approximated as three line segments (i.e. elastic phase, strain-softening phase, plastic flow phase) which are represented by the relation of plastic shear strain e^p shown in Table 2 verse the corresponding cohesion C. The internal friction angle ϕ with e^p is assumed constant.

3.2 *Simulation and Discussion*

(a) Deformation and Propagation of Plastic Failure

Deformation, displacement vectors and development of plastic zone in the model are shown in Figure 7. The sign of ∗ in the element denotes that the element is currently at yield and the sign of × denotes that the element is not currently at yield but has been in the past. The strain-softening appears in the entire

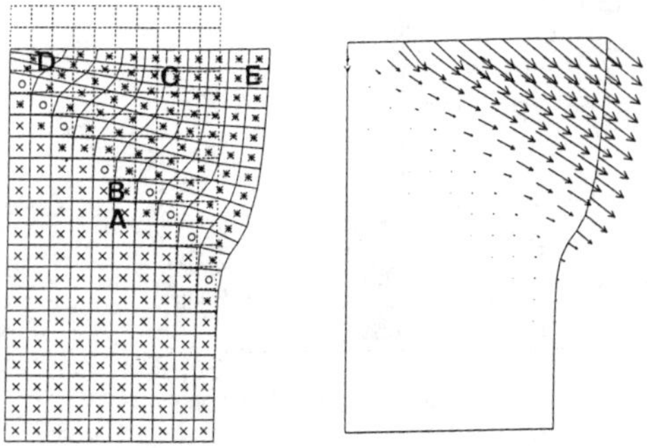

Figure 7. Deformation and plastic failure zone.

(a)

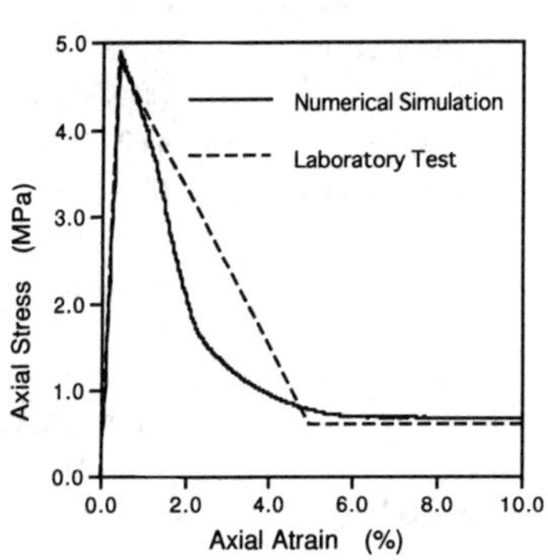

(b)

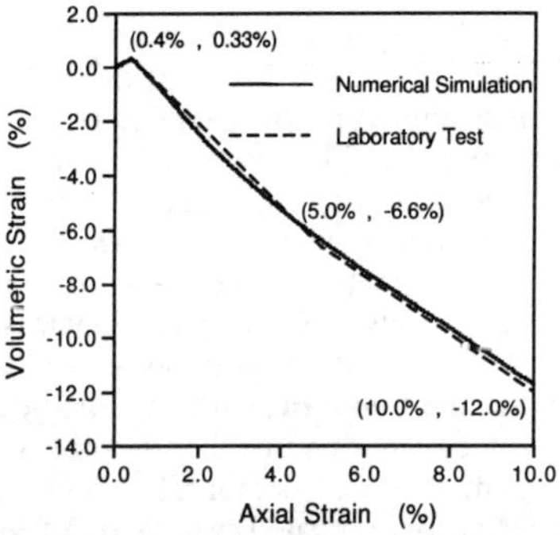

Figure 8. Comparison between the test and simulation: (a) Axial stress-axial strain, (b) Volumetric strain-axial strain.

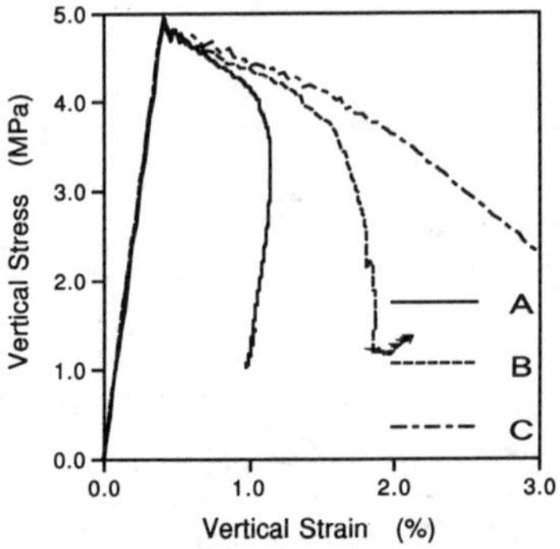

Figure 9. Responses of vertical stress-vertical strain in elements A, B and C.

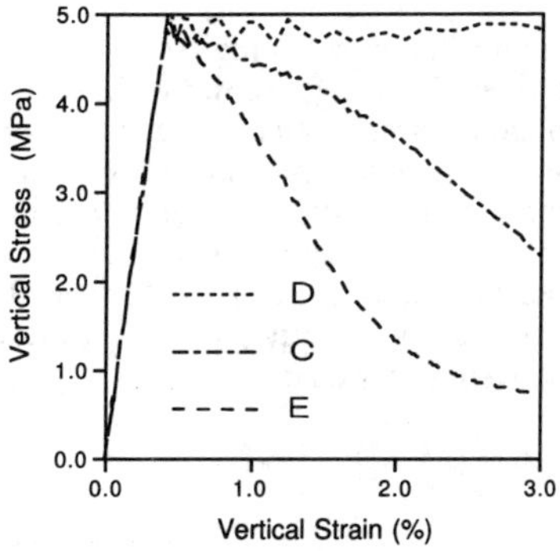

Figure 10. Responses of vertical stress-vertical strain in elements D, C and E.

region above the hypothetical plane, which is inclined about 45 deg through the center of the specimen, and elastic region appears in the entire region below the plane. Thus, a shear band is formed around this hypothetical plane. The maximum lateral displacement at the center of lateral outside surface is 6.14 mm.

(b) Comparison with Test

The comparison between the approximated linear stress-strain relation segments based on the test and the numerical simulation is shown in Figure 8. It is found that the simulation is close to the approximated results of the test. However the axial stress-axial strain relation in the simulation differs from the test and exhibits a curved line in the strain-softening phase. The reason of the difference is considered as that the inputted material properties are defined for each element locally, while the relation obtained from the test has the macroscopic meaning as a whole of specimen. As a result of above mentioned, it is clear that the faithful reproduction of strain-softening and dilation behaviors of soft rock is possible by using the approach.

(c) Local Behavior in the Model after Yield

The relations of vertical stress-vertical strain of the elements A, B, C in Figure 7, which are arranged lengthways, are shown in Figure 9. It is found that the C-element, which lay above the slip plane, represents almost completely the strain-softening behavior. On the other hand, the A-element below the slip plane yielded in the past and then has returned to elastic state due to unloading of inner stress as strain-softening has been formed locally. The B-element shows the intermediate behavior between A and B elements.

Based on above results, it is cleared that amount of

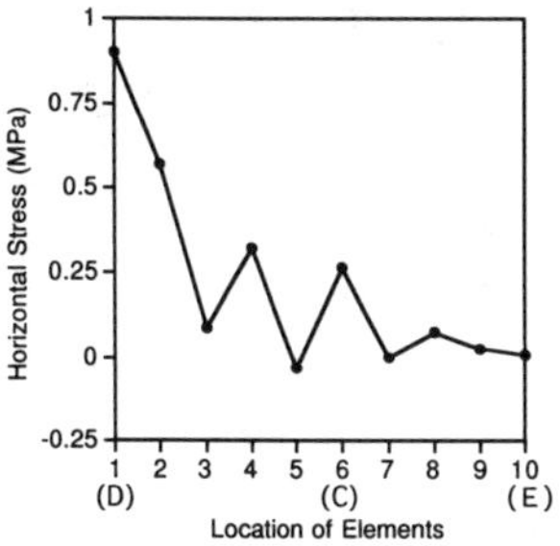

Figure 11. Distribution of local lateral stress in the horizontally arranged elements.

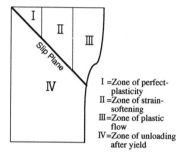

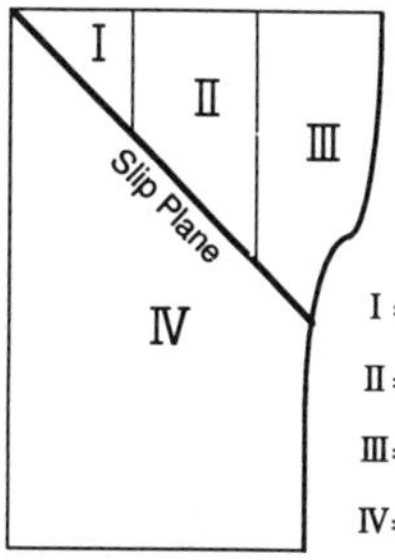

Figure 12. Zoning of failure in the model of specimen.

axial strain of the model during strain-softening agrees with the difference between amount of compression in strain-softening region and one of restitution due to unloading.

Next, the relations of vertical stress-vertical strain in the elements D, C, E in Figure 7, which are arranged horizontally, are shown in Figure 10. The reduction of strength can not be seen in D-element, which lay at the nearest of center axis of specimen and its behavior seems to be as perfect-plasticity. On the other hand, reduction of strength is remarkably in E-element, which lay near the lateral surface of specimen, compared with C-element. This is considered to be caused by the confining effect from the surrounding elements, as shown in Figure 11.

Summarily, the local behaviors in the inside of specimen can be considered to be divided into four zones as shown in Figure 12. Zone I denotes perfect-plasticity, zone II denotes strain-softening prior to plastic flow, zone III denotes plastic flow after strain-softening, and zone IV denotes unloading after yield. In this simulation, lateral pressure is limited to zero. However, change of four zones will be followed by the increment of lateral pressure. In this case, zone I will be expanded laterally and zone III will be reduced and then vanish due to the effect of confinement . Apparent behavior of specimen is represented by the total of above four zones.

Table 3. Material properties of specimen.

Young's Modulus	1000.0 (MPa)
Poisson's rate	0.30
Density	2500.0(kg/m³)
Uniaxial strength	0.05(MPa)
Friction angle	30.0(deg)
Cohesion	0.0173(MPa)
Dilation angle	10.0(deg)

Table 4. Parameters of strain-softening.

Plastic shear strain e^p	Cohesion C (MPa)
0.0	0.05
0.01	0.01
1.0	0.01

4 APPLICATION TO STABILITY ANALYSIS OF TUNNELS IN SOFT ROCK

4 .1 *Summary of Simulation*

A model of a unsupported tunnel of 10 m diameter is prepared to simulate the large deformation phenomenon due to excavation. The overburden thickness over the tunnel is 5 m. Both the developing process of large deformation around the tunnel and the collapse caused when plastic regions reach at ground surface are simulated here. Mohr-Coulomb yield criterion, strain-softening and dilatancy are assumed as the previous section. Region of modeling is half of cross section area based on the symmetry condition. Material properties used here are shown in Table 3. The initial in-situ stress field with the lateral pressure coefficient 1.0 is achieved. Parameters concerned with strain-softening are shown in Table 4, where cohesion C is reduced as increment of e^p but ϕ remains constant.

4 .2 *Result and Discussion*

Collapse of ground over the tunnel is induced when plastic region due to excavation reaches at ground surface in case of shallow overburden. Progress of large deformation and plastic region just after excavation is shown in Figure 13. The signs of $*$ and $\times$ in the elements denote the same meaning with the previous section, and the signs of $+$ and # denote that the elements are currently at tension yield. The phenomena that development of failure around tunnel crown induces shear failure at the upper region of lateral wall is well reproduced. This approach(i.e. explicit finite difference method) makes the track of failure behavior and progress of plastic region possible. This is characterized in Stage 4~Stage 5. It can be considered that the traditional numerical approaches based on the infinitesimal strain theory, such as FEM(Finite Element Method), may represent the behavior up to Stage

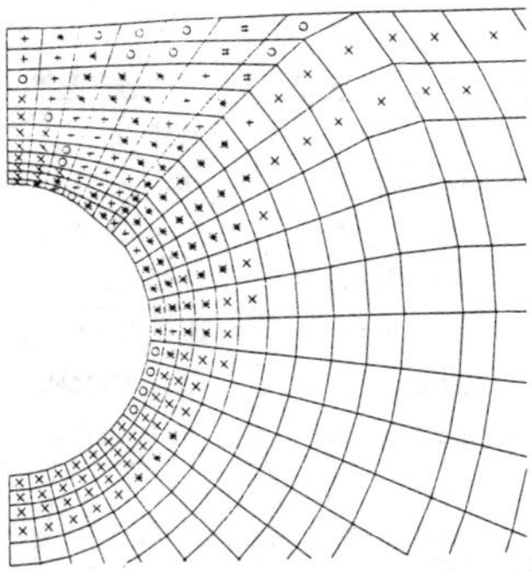

(Stage 1)

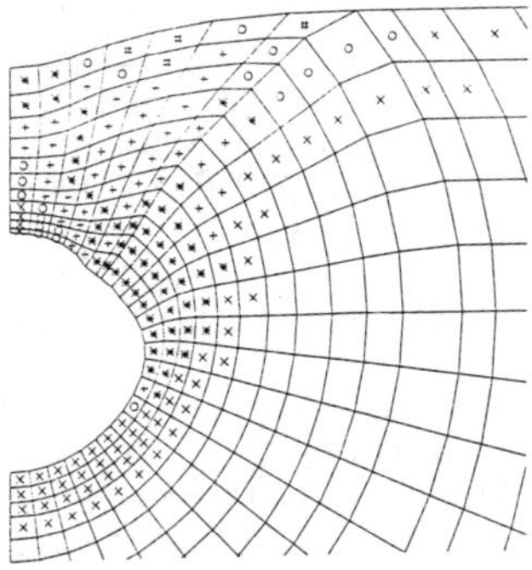

(Stage 2)

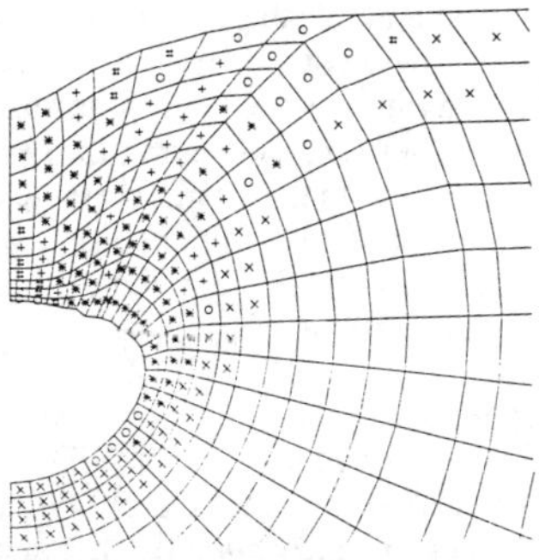

(Stage 3)

1~Stage 2, but impossible to track progress of large deformation and plastic region during post-failure.

5 CONCLUSION

In this paper, a numerical approach for the analysis of large deformation behavior with FLAC code is proposed. Simulations of compression test of rock material and tunnel excavation in soft ground at shallow depth are implemented as examples to verify its appli-

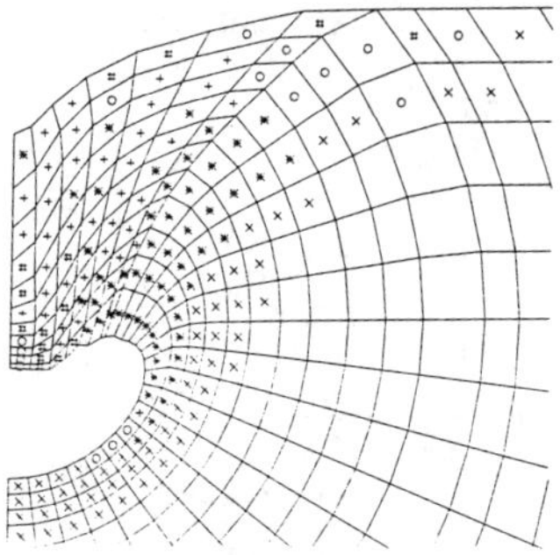

(Stage 4)

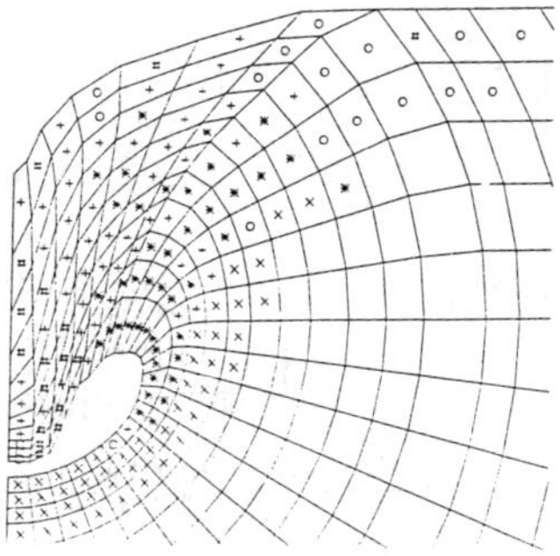

(Stage 5)

* :Shear yield
#／+ :Tension yield
× :Elastic due to unloading after yield

Figure 13. Progress of collapse of tunnel crown at shallow overburden.

cability to rock engineering. As a result, it is clarified that the proposed approach can be used to faithfully reproduce large deformation with plastic flow and progress of plastic region during the post-failure.

REFERENCES

1. Cundall, P.A., Board, M. : A microcomputer program for modeling large-strain plasticity problem, *Proc. 6th Int. Conf. on Numerical Methods Geomechanics*, Innsbruck, Austria,1988.
2. Yujing Jiang : Theoretical and experimental study on the stability of deep underground opening, PhD. thesis, Kyushu University, Japan, 1993.
3. Cundall, P.A. : Personal Communication, 1995-1997.
4. Vermeer,P.A., and R. de Borst. : *Non-Associated Plasticity for Soils, Concrete and Rock*, Heron, 29(3), 1984.

Computer Methods and Advances in Geomechanics, Yuan (ed.) © 1997 Balkema, Rotterdam, ISBN 90 5410 904 1

Designing tunnel linings in tectonic regions with the rock technological heterogeneity to be taken into account

N.N. Fotieva, N.S. Bulychev & A.S. Sammal
Tula State University, Russia

ABSTRACT: Method of designing tunnel linings upon the action of tectonic forces in rock mass whose deformation modulus decreases near the opening due to drilling and blasting operations is proposed in the paper presented. The presence of the rock zone strengthened by grouting around the tunnel also may be taken into account.

1 INTRODUCTION

As it is known the main peculiarity of the rock mass subjected to the action of forces of the tectonic origin is a substantial difference of the initial stress field on the depth from the one caused by the rocks' own weight. So, the horizontal components of the initial stresses may be much more than the vertical ones and the main axes of the initial stresses may be inclined to the horizontal and the vertical under some angle. It has been confirmed by the results of full-scale measurements fulfilled for example by Turchaninov & Markov (1966).

The method of designing tunnel linings of an arbitrary cross-section shape described in the paper presented allows to take into account both the properties of the initial stress field and the influence of the technological heterogeneity of the rock mass i.e. continuous changing of the deformation modulus of the rock with moving off from the opening surface caused by drilling and blasting operations or the presence of the rock layer strengthened by grouting. In the first case for describing continuous technological heterogeneity of the rock the results of experimental data or formula proposed by Rukin & Ruppeneit (1968) may be applied.

2 THE DESIGN METHOD

For determining rock mass stress state around an opening the new analytical solution of the elasticity theory plane contact problem for a multi-layer ring of an arbitrary shape (with one axis of symmetry) supporting a hole in a linearly deformable medium is being applied. The design scheme is shown in Figure 1.

The S_j ($j = 1,...,N-1$) ring layers the materials of which have E_j ($j = 1,...,N-1$) deformation modules and ν_j ($j = 1,...,N-1$) Poisson ratios simulate the zone where the rock technological heterogeneity takes place (here continuous changing deformation modulus is being substituted for a discrete one and a big enough number of layers of small thickness is being considered to reach the admitting accuracy). The S_N layer simulates the tunnel lining.

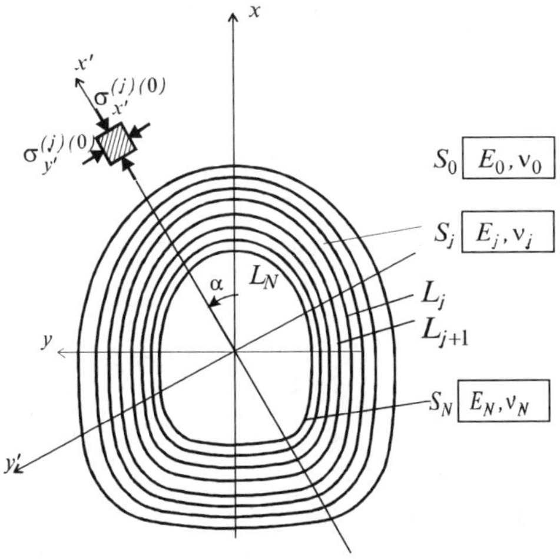

Figure 1. Design scheme.

The S_0 linearly deformable medium simulates the rock mass in a natural condition. There are initial stresses in the S_0 medium and in the S_j ($j = 1,...,N-1$) layers caused by the action of tectonic forces:

$$\sigma_{x'}^{(j)(0)} = -N_1\alpha^*, \quad \sigma_{y'}^{(j)(0)} = -N_2\alpha^*,$$
$$\tau_{x'y'}^{(j)(0)} = 0, \qquad (1)$$

where N_1, N_2 are the principal stresses acting under an arbitrary α angle to the vertical and horizontal, α^* = the correcting multiplier introduced for the approximate registration of an influence of the l_0 distance between the lining being constructed and the tunnel face. That multiplier may be determined by the empirical formula (Manual 1983):

$$\alpha^* = \exp(-1.3\frac{l_0}{R_{N-1}}) \qquad (2)$$

where R_{N-1} = the average radius of the opening.

The conditions of displacements and complete stresses vectors continuity are satisfied on the L_j ($j = 0,...,N-1$) contact lines. The L_N internal outline is free from loads.

The solution of the problem formulated above has been obtained with the application of the complex variable analytic functions theory using the apparatus of conform mapping and complex series. That solution is the generalisation of the solution by Fotieva & Sammal (1988) for the absence of forces symmetry to be taken into account.

After representing complete stresses in the rock mass in the form of sums of initial tectonic stresses determined by formulae

$$\sigma_x^{(j)(0)} = -N_1\alpha^*\left[\frac{1+\xi}{2} + \frac{1-\xi}{2}\cos 2\alpha\right],$$
$$\sigma_y^{(j)(0)} = -N_1\alpha^*\left[\frac{1+\xi}{2} - \frac{1-\xi}{2}\cos 2\alpha\right], \qquad (3)$$
$$\tau_{xy}^{(i)(0)} = -N_1\alpha^*\frac{1-\xi}{2}\sin 2\alpha, \qquad (j=0,...,N-1)$$

where

$$\xi = \frac{N_2}{N_1} \qquad (4)$$

and the $\sigma_x^{(j)}, \sigma_y^{(j)}, \tau_{xy}^{(j)}$ additional stresses caused by the presence of the opening and introducing the $\varphi_j(z)$, $\psi_j(z)$ ($j = 0,...,N$) complex potentials regular in corresponding S_j ($j = 0,...,N$) areas and connected with stresses by known formulae (Muskhelishvili, 1966) the contact problem for the determination of the additional stresses and displacements is being transformed into the boundary problem of the complex variable analytic functions theory at the following boundary conditions:

$$\varphi_{j+1}(t) + t\overline{\varphi'_{j+1}(t)} + \overline{\psi_{j+1}(t)} = \varphi_j(t) + t\overline{\varphi'_j(t)} +$$
$$+\overline{\psi_j(t)} - \lambda_{j,N-1} N_1\alpha^*\left[\frac{1+\xi}{2}t - \frac{1-\xi}{2}\bar{t}e^{2i\alpha}\right]$$

upon L_j ($j=0,...,N-1$)

$$æ_{j+1}\varphi_{j+1}(t) - t\overline{\varphi'_{j+1}(t)} - \overline{\psi_{j+1}(t)} =$$
$$= \frac{\mu_{j+1}}{\mu_j}\left[æ_j\varphi_j(t) - t\overline{\varphi'_j(t)} - \overline{\psi_j(t)}\right] \qquad (5)$$
$$\varphi_N(t) + t\overline{\varphi'_N(t)} + \overline{\psi_N(t)} = 0 \qquad \text{upon } L_N.$$

where $æ_j = 3 - 4\nu_j$, $\mu_j = \frac{E_j}{2(1+\nu_j)}$, ($j=0,...,N$),

t = affix of the corresponding outline point.

The solution of the problem is being obtained in the same way as in the work by Fotieva & Sammal (1988).

With the aim of $z = \omega(\zeta)$ rational function the conform mapping of the exterior of the circle having the $R_N < 1$ radius in the plane of ζ variable on the exterior of the L_N outline in the z plane so that the circumference of the $R_0 = 1$ radius turns into the L_0 outline is being fulfilled. Then the circles of the R_j ($j = 1,...,N-1$) radii turn into the corresponding L_j ($j = 1,...,N-1$) outlines.

The boundary conditions (5) in the transformed area after some operations become the following:

$$\overline{\varphi}_{j+1}(\frac{R_j}{\sigma}) + \frac{\overline{\omega}(\frac{R_j}{\sigma})}{\omega'(R_j\sigma)}\varphi'_{j+1}(R_j\sigma) + \psi_{j+1}(R_j\sigma) =$$

$$= \overline{\varphi}_j(\frac{R_j}{\sigma}) + \frac{\overline{\omega}(\frac{R_j}{\sigma})}{\omega'(R_j\sigma)}\varphi'_j(R_j\sigma) + \psi_j(R_j\sigma) - \qquad (6)$$

$$-\lambda_{j,N-1}N_1\alpha * \left\{ \frac{1+\zeta}{2}\overline{\omega}(\frac{R_j}{\sigma}) - \frac{1-\zeta}{2}\omega(R_j\sigma)e^{-2i\alpha} \right\},$$

$(j=0,...,N-1)$

$$æ_{j+1}\overline{\varphi}_{j+1}(\frac{R_j}{\sigma}) - \frac{\overline{\omega}(\frac{R_j}{\sigma})}{\omega'(R_j\sigma)}\varphi'_{j+1}(R_j\sigma) - \psi_{j+1}(R_j\sigma) =$$

(7)

$$= \frac{\mu_{j+1}}{\mu_j}\left[æ_j\overline{\varphi}_j(\frac{R_j}{\sigma}) - \frac{\overline{\omega}(\frac{R_j}{\sigma})}{\omega'(R_j\sigma)}\varphi'_j(R_j\sigma) - \psi_j(R_j\sigma) \right]$$

$$\overline{\varphi}_N(\frac{R_N}{\sigma}) + \frac{\overline{\omega}(\frac{R_N}{\sigma})}{\omega'(R_N\sigma)}\varphi'_N(R_N\sigma) + \psi_N(R_N\sigma) = 0, \quad (8)$$

where $\sigma = e^{j\theta}$.

The $\varphi_0(\zeta)$, $\psi_0(\zeta)$ complex potentials are regular in the S_0 area and turn up to zero on the infinity. They are represented in the transformed area as complex series on negative degrees of ζ variable.

The $\varphi_j(\zeta)$, $\psi_j(\zeta)$ (j=1,...,N) complex potentials regular in corresponding ring layers are represented in the form of Loran series. Due to the absence of the forces symmetry all unknown coefficients of the series mentioned above are the complex values. It results in substantial complication of the problem solution because it is necessary to obtain two systems of algebraic equations for determining real and imagine parts of unknown coefficients. However in the case considered also it is possible to obtain the recurrent relationships connecting the real and imagine parts of the coefficients of the series characterising the complex potentials in S_{j+1} and S_j (j = 0,...,N) contacting areas from the (6),(7) boundary conditions as it has been made in the work by Fotieva & Sammal (1988) concerning the similar problem for the unsupported opening. It offers the opportunity of fixing many layers in the rock mass for the approximation with the necessary accuracy of any law of deformation modulus changing depending on the distance from the opening surface. The computer program developed allows up to 20 layers to be under consideration. It is evident that the same model may be applied with the aim of the presence of the rock zone including several weakened layers being preliminary strengthened by grouting to be taken into account.

If the rock mass is being grouted through the bore-holes buried in the lining on the some l_1 distance from the tunnel face the stresses in the lining are being summed from the ones appearing during the face advance till distance of l_1 (when the layer of the strengthened rock is till absent) and taking place already on the strengthening works being finished. The detail description of the design procedure for tunnel linings constructed in the homogeneous rock mass including the case when the rock is subjected to the linear hereditary creep has been given in the paper by Fotieva, Sammal & Chetirkin (1988). Here that technique is the same but it is possible to take into account the simultaneous presence of both the strengthened and the weakened rock layers.

3 EXAMPLES OF THE DESIGN

The example of designing tunnel lining taking into account the rock mass weakening near the opening caused by drilling and blasting operations is given below. The form and sizes of the lining cross-section are shown in Figure 2. The average radius of the opening is R= 3.86 m.

The change of the rock deformation modulus with moving off from the opening surface is determined by formula (Rukin & Ruppeneit, 1968)

$$E(r) = E_0\left(1 - \frac{0.8R^2}{r^2}\right) \quad (9)$$

where r = the distance from the centre of the opening.

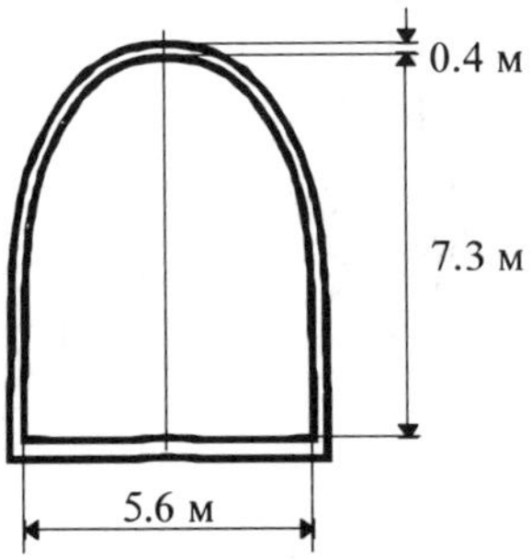

Figure 2. The lining cross-section.

For calculation 10 layers have been assumed in the rock mass around the opening. The relative Δ_j / R_{10} (j = 1,...,10) thickness of the layers and corresponding E_j/E_0 (j = 1,...,10) dimensionless relations being assumed as constants within each layer are given in the Table 1 (the numbers of layers are counted off from the outside).

The deformation characteristics of the lining material are E_{11}/E_0 = 2.0, ν_{11} = 0.2. The relation between principal initial stresses in the rock mass is ξ = 0.3; the angle of their bowing is α=40 degrees.

Table 1. The relative Δ_j / R_{10} thickness of layers being assumed in weakened rock around the opening and the E_j/E_0 (j = 1,...,10) values.

Characteristics	Layers									
	1	2	3	4	5	6	7	8	9	10
Δ_j / R_{10}	0.60	0.60	0.50	0.50	0.20	0.10	0.10	0.10	0.10	0.09
E_j/E_0	0.90	0.85	0.78	0.70	0.61	0.55	0.49	0.43	0.34	0.25

The results of the design determined in the parts of the $N_1\alpha^*$ value namely the $\sigma_\theta^{(in)}$, $\sigma_\theta^{(ex)}$ normal tangential stresses on the internal and external outlines of the lining cross-section are given in Figure 3,*a,b* by solid lines; for comparison the same stresses appearing in the lining located in homogeneous rock mass having the E_0 deformation modulus are given in Figure 3,*a,b* by dotted lines.

The example of designing the same lining constructed in previously grouted rock subjected later to weakening by drilling and blasting operations is given below. The thickness of grouted rock zone is 4.6 m ($1.19 \cdot R_{10}$); the change of the rock deformation modulus is given in the Table 2.
The stresses in the lining are shown in Figure 4.

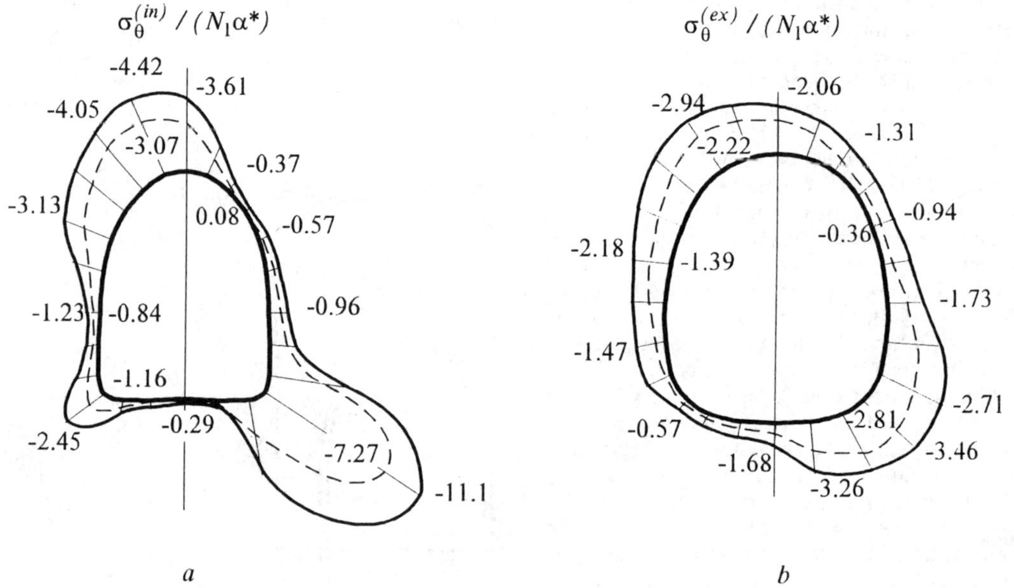

Figure 3. Distributions of stresses in the lining.

Table 2. The Δ_j/R_{10} and E_j/E_0 ($j = 1,...,10$) values assumed in previously grouted rock.

Characteristics	Layers									
	1	2	3	4	5	6	7	8	9	10
Δ_j/R_{10}	0.60	0.60	0.50	0.50	0.20	0.10	0.10	0.10	0.10	0.09
E_j/E_0	0.90	0.85	0.78	1.54	1.34	1.21	1.08	0.94	0.75	0.55

Table 3. The change of E_j/E_0 value depending on the distance from the tunnel surface.

Characteristics	Layers									
	1	2	3	4	5	6	7	8	9	10
Δ_j/R_{10}	0.60	0.60	0.50	0.50	0.20	0.10	0.10	0.10	0.10	0.09
E_j/E_0	0.90	0.85	0.78	0.70	0.70	0.70	0.70	0.70	0.70	0.70

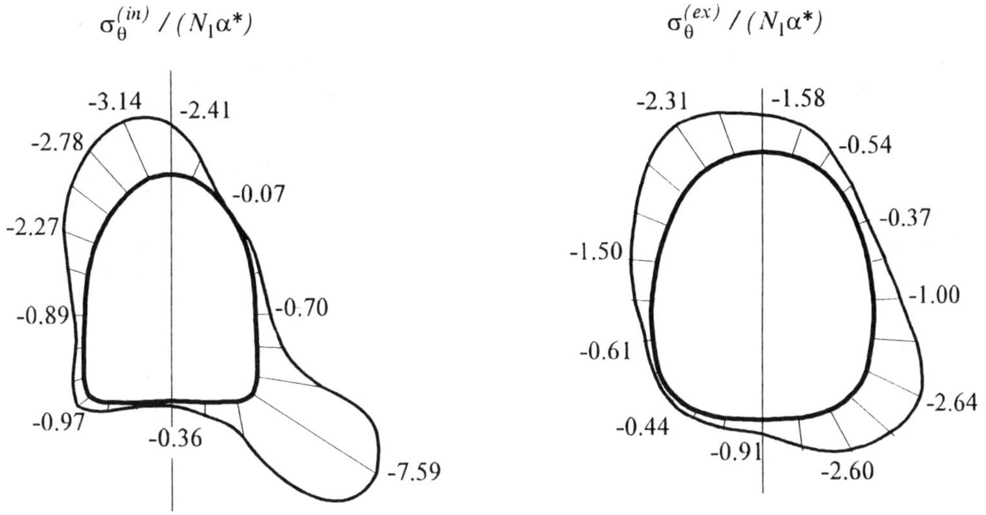

Figure 4. Distributions of stresses in the lining constructed in previously grouted rock.

Further we consider the same lining constructed on the relative distance $l_0/R_{10} = 0.3$ from the tunnel face in the weakened rock, deformation modulus of which is being changed by the law showed in the Table 1, with the application of grouting being fulfilled through the lining
in the relative distance $l_1/R_{10} = 0.6$ from the tunnel face.

The grouted rock zone embraced 6 layers of the weakened zone. The change of the rock deformation modulus moving off from the opening surface is shown in the Table 3. The stresses (in the parts of the N_1 value) in the lining constructed with the application of the rock grouting through the structure are shown in Figure 5.

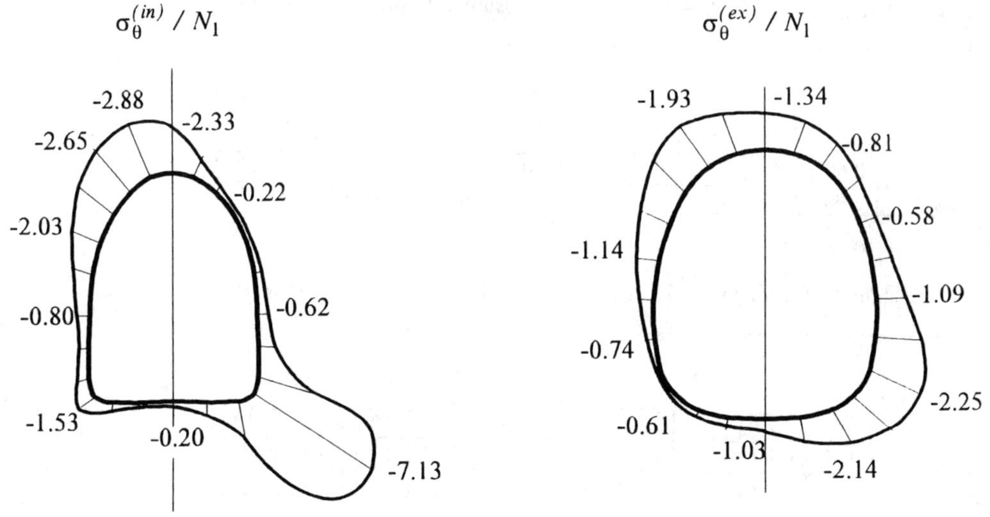

Figure 5. Distributions of stresses in the lining constructed in the rock grouted after lining erection.

In conclusion we can mark that the technological heterogeneity of rocks has a substantial influence on the tunnel linings stress state and must be taken into account at their design and construction.

REFERENCES

Bulychev,N.S. & Nina N. Fotieva 1977. The estimation of the stability of rocks surrounding the mining openings. *Mining construction.* 3: 16 22.

Fotieva, N.N., Sammal, A.S. & N.S.Chetyrkin 1988. Influence of soil creep on stress state of the underground structures being built with the strengthening massif. *Proc. of the Int. Conf. on Reology and Soil Mech.* Coventry UK, London.

Fotieva, N.N. & A.S. Sammal 1988. Determining stresses aroùnd workings taking rock technological heterogeneity into account. *Proc. of Int.Symp. on Modern Mining Technology*: Tayan, Shandong, P.R.C.: 286–293.

Fotieva, N.N. 1980. *The design of non–circular tunnels linings.* Moscow: Stroyizdat.

Muskhelishvili, N.I. 1966. *Some basic problems of the mathematical elasticity theory.* Moscow: Nauka.

Manual in designing underground mining workings and support computation, 1983. *VNIMI, VNIIOMShS, Stroyizdat,* Moscow.

Rukin, V.V. & K.V. Ruppeneit 1968. *Mechanism of the interaction of the lining of the tunnel under a head with the rock's mass.* Moscow: Nauka.

Turchaninov, I.A. & G.A. Markov 1966. The influence of a new tectonic on the stress-state of rocks in Xibin apatite minings. *Izvestiya AN SSSR* 8: 83-89.

Computer Methods and Advances in Geomechanics, Yuan (ed.) © 1997 Balkema, Rotterdam, ISBN 90 5410 904 1

Tunnel excavation in Bangkok: Analytical and field measurement

S.Teachavorasinskun
Sirindhorn International Institute of Technology, Thammasat University, Thailand

ABSTRACT: The main objective of the present study is to establish a system for prediction of the behavior a soil mass. The system should include at least two main components; namely, data determination and analytical tool. To accomplish the former, the borehole log in Bangkok area and her vicinities was collected and used for preparation of the contour maps. Simulation of the tunnel excavation in Bangkok soft clay was, then, carried out at several locations; e.g., Hua-Lumpong (Bangkok station), using the corresponding soil properties extrapolated from the proposed contour maps. The actual ground subsidence measured in the field, though limited in the amount of data, was used for the comparison and calibration purposes. The calculation results, though very crude, can be used as an indicator for selection of the further elaborate analysis and/or preventive measures against damages.

1 INTRODUCTION

As the complication of the newly developed constitutive models increases, the number of material properties required as input parameters also considerably increases. For example, the elasto-plastic cap model using the non-associated flow with kinematic unloading surface proposed by Teachavorasinskun, et al.(1995) requires the total of 11 parameters to complete the analysis. Moreover, the determination of those parameters is so complicated and needs the assistance of the sophisticated testing equipment such as the stress path controlled triaxial testing apparatus and etc. It seems, however, that as the complexity of the constitutive model increases, the chance that the model being utilized in the actual application is being lost. This is because 1) it costs too much to carry out such complicated tests in order to obtain a reliable set of model parameters and 2) practical engineers do not well understand the details of the models, and as a result, they are not confident in using them.

It is believed that, in the very near future, the advanced analytical techniques, such as the finite element analysis (FEM) and the discrete element analysis (DEM), will take the important role in verification of the design which is generally followed the conventional method. Therefore, in order to promote the usage of such analytical techniques among practical engineers, the development of a system composing of the following components is necessary,

1. Procedure for determination of necessary soil information.
2. A simple constitutive model having a certain degree of reliability when uses with the information obtained from (1).
3. Computer software capable of carrying out the pre-processing, processing and post-processing analysis

In the actual earth work, the material properties obtained from the laboratory and in-situ tests are usually limited to the SPT *N*-value, undrained shear strength, S_u, and plasticity index. It is, therefore, the aim of the present study to establish a preliminary route which effectively utilizes the above mentioned three properties. This paper presents the very preliminary results obtained from the first few calibration analyses.

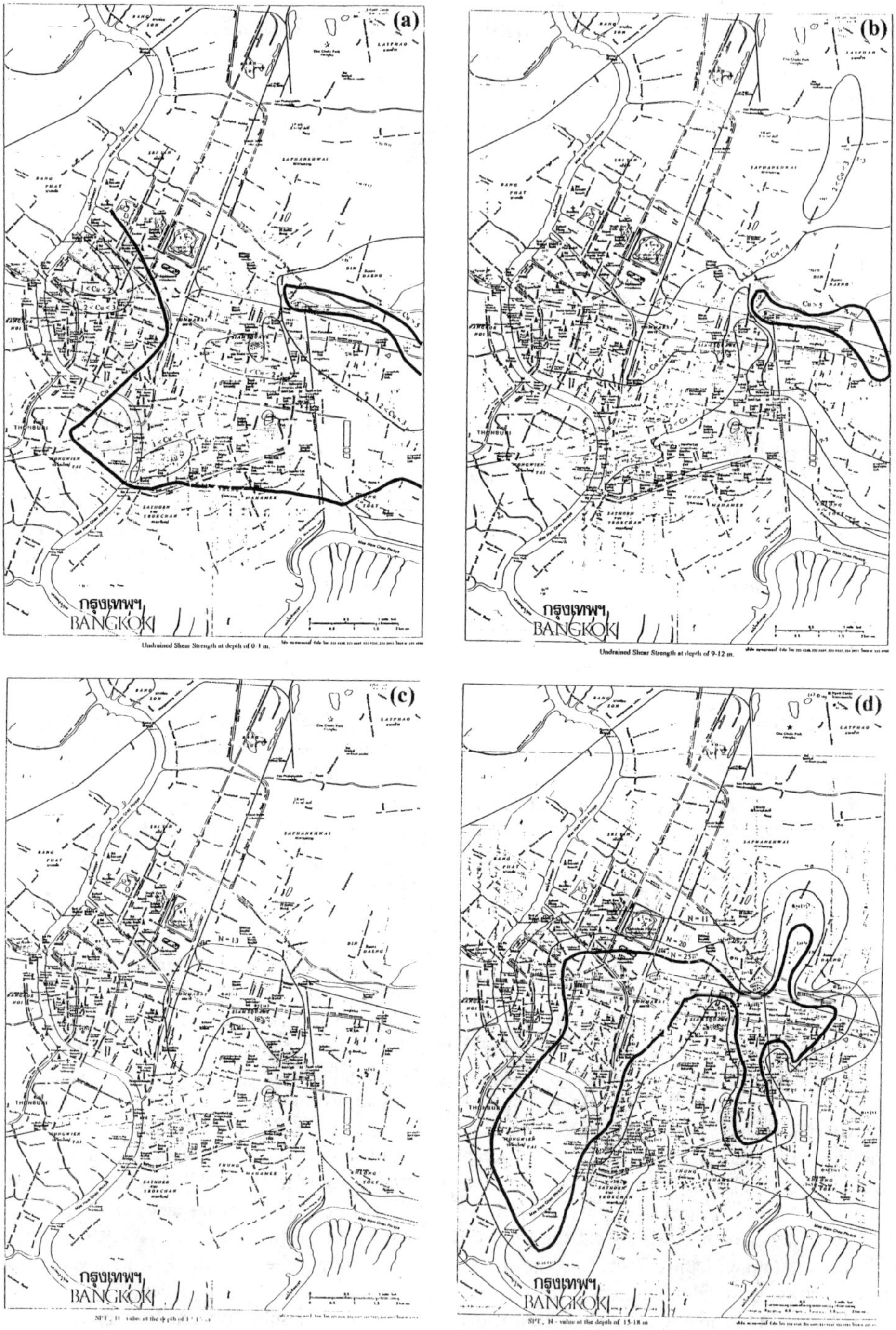

Fig.1 Examples of the proposed contour maps (a) - (b) Undrained shear strength and (c) - (d) SPT N-value

2 SOIL PROFILE IN BANGKOK AND HER VICINITIES

More than 150 borehole data had been collected around Bangkok Metropolitan and her vicinities. For each borehole log, the properties; i.e., the SPT *N*-value and the undrained shear strength, S_u, were averaged for every 3 meters depth. These averaged values were used in preparation of the contour maps. The locations of all boreholes were, first, identified on the map, and the values were plotted onto the corresponding locations. The isoline of each value was drawn to form a contour of the property. The examples of such contour map are shown in Fig.1(a) to (d).

From the contour map (Fig.1c), the first sand layer is found at the depth of about 12-15 meters in the central area of Bangkok (Siam Square). As the depth increases, sand layer enlarges to cover the overall area of Bangkok Metropolitan (Fig.1d). The expansion of sand layer seems to have its center around the 'Siam Square' area, and expands outward like a flat dome. Note that the depth shown in the map are the actual depth from the ground level of those locations. The values were not adjusted to be referenced to any standard level (e.g., mean sea level). The information obtained or extrapolated from the maps is very preliminary and it should be used with great care. Table 1 shows the examples of the soil properties of the sites 'Hua-Lumpong (Bangkok station)' and 'Victory Monument' extrapolating from the proposed contour maps. The information shown in the table is the typical form of data generally obtained from the simple site investigation, therefore, it will be very encouraging if these two values can be effectively used and yields a reasonable analytical result.

Table 1 Examples of soil properties extrapolated from the contour maps

Depth (m)	Hua-Lumpong station		Victory Monument	
	N-value	S_u (t/m^2)	N-value	S_u (t/m^2)
0-3	-	3.5	-	3.0
3-6	-	1.5	-	1.5
6-9	-	2.5	-	2.0
9-12	-	3.5	-	3.5
12-15	-	6.0	13	-
15-18	25	-	20	-
18-21	20	-	20	-
21-24	25	-	31	-
24-27	33	-	33	-

3 DRUCKER-PRAGER MODEL

The selection of the suitable constitutive model will decide the success of the analysis. It is considered that there is no a constitutive model being superior to the others, since, the reliability of any model depends mostly on interpretation of the result based on a number of calibration analyses and experience of the engineer. In the present study, the Drucker-Prager (D-P) model was selected to represent the behavior of the soil during yielding. The D-P model resumes an open cone shape in the three dimensional stress space as shown in Fig.2a and becomes a straight line when it is viewed on the $\sqrt{J_2} - I_1$ plane (Fig.2b). The yield surface, *f*, can be expressed as:

$$f = \alpha I_1 + \sqrt{J_2} - k = 0$$

where $I_1 = \sigma_{kk}$ is the first invariant tensor, $J_2 = \frac{1}{2} s_{ij} s_{ij}$ is the second deviatoric invariant tensor, and σ_{ij}, s_{ij} and δ_{ij} are the stress tensor, the deviatoric stress tensor and the Knocnecker delta respectively. The constants α and *k* in the above equation can be obtained by fitting the D-P yield surface to the Coulomb criterion. Since the Coulomb, in three dimensional stress space, have the hexagonal cone shape, the parameter α and *k* are very dependent on the edge where the D-P is fitted to. In the present study, the inner cone of the Coulomb criterion was used as a reference (tensile meridian), since it results in smaller strength and, therefore, yields more conservative analytical results (Chen and Mizuno, 1990). The expressions of α and *k* in terms of friction angle, ϕ, and cohesion, *c*, are:

$$\alpha = \frac{2 \sin\phi}{\sqrt{3}\,(3 + \sin\phi)}$$

$$k = \frac{6 c \cos\phi}{\sqrt{3}\,(3 + \sin\phi)}$$

Note that in order to obtain both friction angle, ϕ, and cohesion, *c*, for a soil, at least three full triaxial tests are needed to be carried out on the undisturbed samples. With the limited information obtained from the in-situ test (borehole data), it is very difficult to have the full *c*-ϕ information of a soil. This seems be fine for sandy soil, since the apparent cohesion in sand is known to be very small and can be neglected ($c = k = 0$). However, when dealing with clayey soil as always be the case in the construction of shallow

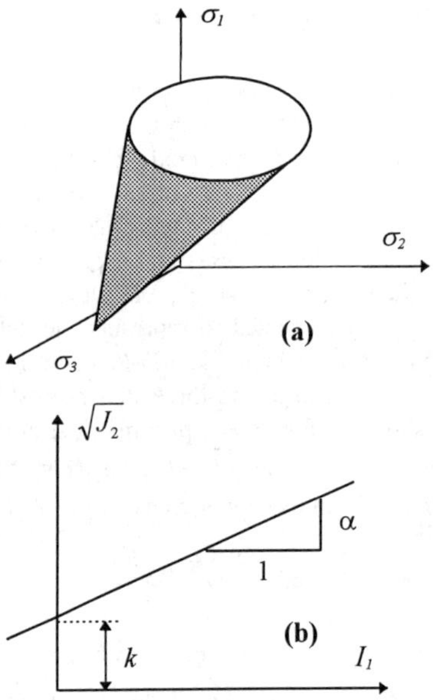

Fig.2 Schematic view of the Drucker Prager surface (a) in stress space and (b) in two dimensional stress plane.

tunnel, the conditions when clay is assumed to be $\phi = 0$ and c-ϕ material can result in quite different behaviors.

In summary, for clayey soil, due to the limited in soil information, the D-P model was reduced to be the von-Mises surface ($\phi = 0$) and, for sandy soil, the parameter k was always kept to be zero.

4 DETERMINATION OF ELASTIC YOUNG'S MODULUS

The methods for determination of the elastic Young's modulus, E, are still very ambiguous. There are several methods quoted in the literature to be the better ways in estimation of E. In the past, these could be divided into two main groups; namely, the dynamic and static Young's modulus. It was not until the early of this decade that researchers become realizing that the difference is, in fact, due to the strain level dependency characteristic of the soil. Namely, the conventional static E is measured at much larger strain levels than that encountered in the dynamic measurements (Tatsuoka and Shibuya, 1992). At very small strain levels; i.e., $\varepsilon < 10^{-4}\%$, a unique value of E is confirmed from either method. Although the fact has been realized, the use of the elastic Young's modulus seems unpopular in the general engineering practice, except in some dynamic problem. This is because:

- Complicated tests; e.g., resonant column test, in-situ wave propagation velocity measurement, etc., are required in order to obtain the elastic Young's modulus.
- A new non-linear constitutive model which can cope with the highly non-linear behavior of soil is necessary.

Therefore, in the present study, conventionally proposed empirical equation between the Young's modulus and the SPT N-value and the undrained shear strength, S_u, were adopted.

For cohesive soil, the modulus of elasticity, E, could be obtained form the simple empirical correlation between E and undrained shear strength, S_u, as (Bowles, 1996):

Normally consolidated to lightly overconsolidated clay:

$$E = (750 \text{ to } 1200)\ S_u$$

Heavily overconsolidated clay:

$$E = (1500 \text{ to } 2000)\ S_u$$

For Sandy soil, several empirical correlation have been proposed by several organizations and researchers. For example, the Japanese Highway Standard proposed that:

$$E = (26 \text{ to } 29)\ N \quad (kg/cm^2)$$

The further step following this study should be to calibrate such empirical correlation using the local information. The most suitable correlation for Bangkok clay should be established in order to obtain a more reliable computation result. By the time being, the above correlations were adopted in the calculation.

5 DETERMINATION OF INTERNAL FRICTION ANGLE

The internal friction angle of sandy soil can also be estimated from the simple empirical correlation with the SPT N-value. The example of such empirical equations are:

Japanese Highway Department :

$$\phi = 15 + \sqrt{15N} \leq 45$$

$$\phi = 1.85\left(\frac{N}{\sigma_v + 0.7}\right)^{0.8} + 26$$

when σ_v is the vertical stress (kg/cm^2). Meyerhof (1957) proposed the correlation between the SPT *N*-value and the relative density, D_r, as:

$$\frac{N}{D_r^{\,2}} = A + B p_0$$

where p_0 is the overburden stress (kPa) and A and B are site dependent parameters. Skempton (1957) proposed that A and B should be about 15 to 54 and 0.306 to 0.204, respectively. After the relative density is determined, the internal friction angle could be obtained by:

$$\phi = 28 + 15D_r \ (\pm 2^\circ)$$

It is to be mentioned that although such empirical correlations are very rough, the results could provide some indications which may help in selection of a more proper analysis and/or preventive measures against damages.

6 TUNNEL EXCAVATION ANALYSIS

The soil properties at two locations; Hua-Lumpong (Bangkok station) and Victory Monument, were used as examples for tunnel excavation analysis (Table 2). The commercially available program called 'ANSYS' was used for carrying out the analysis. The finite element model representing the soil profiles was, first, generated. Note that, since the contour maps providing the soil properties at every 3 meters, the generated FEM mesh can be used for any location by simply changing the values of the material properties as long as the position of the tunnel was not changed which was the case in the present study.

The analytical flow-chart adopted for the present study are shown in Fig.3. The tunnel was assumed to have a diameter of about 3.2 m and center of the tunnel was located at the depth of about 16 m from the ground surface. The soil was represented by an 4-node plane strain solid element, while the tunnel lining (reinforced concrete) was represented by a 2-node beam element. From Fig.3, the initial stress analysis (self weight) was, first, carried out in order to assign the initial stresses to all elements. Then, the excavation was simulated by the deactivation of the prescribed elements, and at the same time, the beam elements representing the lining of the tunnel were activated. The stresses carrying by the deactivated elements were transferred to the adjacent existing elements as the external nodal forces, hence, settlement was induced. The stiffness and section properties of the beam element were calculated based on a reinforced concrete segment having a thickness of 20 cm.

It should be noted that the initial stress analysis was carried out using the linear elastic assumption and the settlement induced during this step was ignored, only the stresses were kept as the initial stresses. This is to avoid some numerical instabilities occurred at very low stress levels due to the non-linearity algorithm. In the simulation of excavation, the D-P model was, then, utilized.

7 ANALYTICAL RESULTS

The example of the computed deformed shape of the ground after excavation at Hua-Lumpong (Bangkok station) is shown in Fig.4. Note that the initial settlement due to initial stress analysis is also included in the figure. The settlement was larger above the center of the tunnel and became smaller at the point away from the tunnel. It was at the distance of about three times of the diameter of the tunnel that the settlement became negligible. The summary of the settlement of the ground surface and at the crown of the tunnel is shown in Table 2.

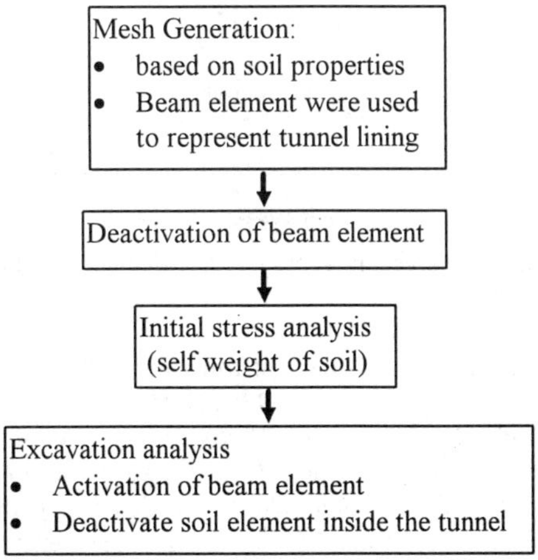

Fig.3 Analytical flow adopted for tunnel excavation analysis

Table 2 Summary of the settlement obtained from the analysis

Location	Settlement (mm)*	Location	Settlement (mm)*
Hua-Lumpong		Victory Monument	
• Ground surface	11.77	• Ground surface	15.64
• Tunnel crown	53.95	• Tunnel crown	70.70

*excluding settlement due to initial stress analysis

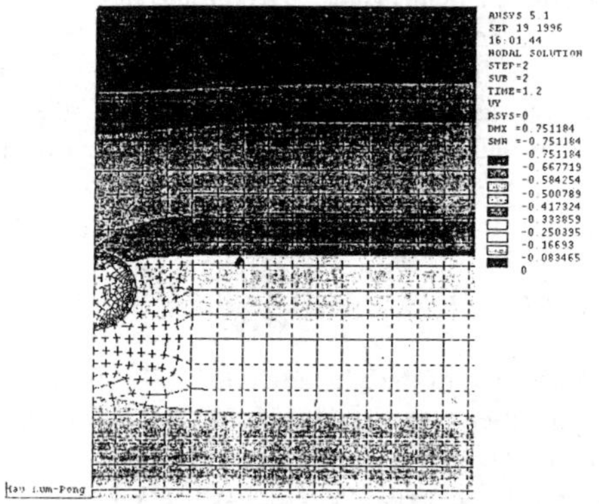

Fig.4 Deformed shape after tunnel excavation

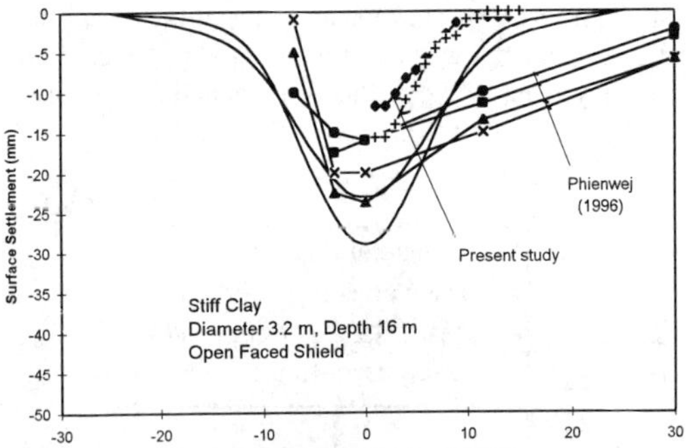

Fig.5 Observed and computed settlement troughs

Fig.5 shows the settlement trough actually observed in the small open shield tunnel constructed in Bangkok area (Phienwej, 1996). The configuration of the tunnel is very similar to that used in the present analysis. The maximum observed settlement varies from about 15 to 25 mm. When the settlement troughs obtained from the analysis are plotted into the figure, it can be seen that the amount of predicted ground surface subsidence is fairly well similar the average measured values. However, the width of the settlement troughs is much narrower that the observed ones. This might due to that the observed data is obtained from the open shield tunneling method which always results in larger settlement than other shield tunnel methods due to the lack of constrained at the face of the tunnel.

8 CONCLUSIONS

The role of the FEM analysis, in the future, is to verify the design done by the conventional method. Therefore, a system which assist the user with 1) parameter determination, 2) material modeling and 3) finite element analysis program should be established. Since in the actual practice, only limit number of soil parameters will be obtained, the less complicated model providing a reasonable accuracy will be the most suitable model in fulfilling the task. To be confident, calibration analyses should have be carefully done.

REFERENCES

Bowles, J. E. (1996):"Foundation analysis and design," 5th Edition, McGraw Hill.

Chen, W. E. and Mizuno, E.(1990):" Nonlinear analysis in soil mechanics: Theory and Implementation," Developments in Geotechnical Engineering 53, Elsevier.

Meyerhof, G. G.(1957):"Discussion on sand density by spoon penetration,"Proc. of the 4th ICSMFE, Vol3., pp.110.

Phienwej, N.(1996):"Tunneling and cut-and cover excavation in Bangkok soils," Similar on Urban and Traffic Engineering and Geotechnical Engineering in Delta Areas, Thammasat University, Bangkok, Thailand.

Skempton, A. W. (1986):"Standard penetration test procedure," Geotechnique, Vol 36, No.3, pp.425-447.

Tatsuoka, F. and Shibuya, S.(1991):"Deformation characteristics of soils and laboratory tests," State-of-the-Art, Proc.9th Asian Regional Conference on Soil Mechanics and Foundation Engineering, Vol.2, pp.101-170

Teachavorasinskun, S., Tanizawa, F. and Sueoka, T.(1995):"Modified cap model with kinematic hardening for dynamic behavior of sand," Proc.of the 7th Canadian Conference on Earthquake Engineering, Montreal, Canada, Vol.1, pp.1210-1214

Computer Methods and Advances in Geomechanics, Yuan (ed.) © *1997 Balkema, Rotterdam, ISBN 90 5410 904 1*

The influences of in-situ stresses on the deformations of tunnels

Yao-Chung Chen & Yao-Ming Lin
National Taiwan Institute of Technology, Taipei, Taiwan, China

ABSTRACT: To understand the influences of the axial stresses on the deformations of tunnels, three-dimensional analyses are performed by using a three-dimensional numerical code, FLAC-3D. The tunnel is excavated in three stages -- top heading, bench I and bench II. Different sets of lateral stress coefficients are assumed. Numerical results showed that axial stresses have great influences on the deformations of tunnels. Increasing axial stresses will increase the deformations and plastic zones around the tunnels. Increasing axial stresses will decrease the plastic zones ahead of the excavation face. It is better excavated along the maximum lateral stress direction than along the minimum lateral stress direction.

1 INTRODUCTION

Tunnel excavation is actually a three-dimensional problem, but it is usually analyzed by two-dimensional program.. For homogeneous rock condition, isotropic stresses and circular tunnel shape, it can be analyzed by two-dimensional axial-symmetric model. Three-dimensional analysis usually takes a lot of computing time and is more difficult to generate the mesh as compared with two-dimensional analysis. So, engineers are reluctant to use three-dimensional analysis. However, two-dimensional analyses can not consider the stress in the direction of tunnel axis, that is the axial stress. The effect of axial stress on the deformations of tunnel and the support pressures might be important. Also, the pre-support deformations should be properly considered in two-dimensional analysis to obtain reasonable simulation of three-dimensional conditions.

The deformations that occur right after tunnel excavation and before the installation of support are called pre-support deformations. The deformations at excavation face are about 30% of the final deformations for numerical analysis of elastic rock conditions, circular tunnel shape, and full face excavation. The percentages of pre-support deformation are about 25 to 80% for numerical analysis of elastic-plastic rock conditions, circular tunnel shape, and full face excavation (Chen & Pong 1994). The percentages are influenced by the properties of rock mass and the stiffness of supports. When tunnels are excavated in multiple stages, the pre-support deformations in each stage can only be obtained by three-dimensional analysis. Chen & Lu (1996) investigated this problem by using three-dimensional program FLAC3D. The deformation behavior of tunnel, excavated in three stages, was investigated for different rock mass conditions and lateral stress coefficients. They found that most of the crown deformations occur after the excavation of top heading and excavation of bench I and bench II has little effects on crown deformations but has obvious effects on wall deformations. The ratios between crown and wall deformations and the distribution of plastic zones are greatly influenced by the lateral stress coefficients. Two-dimensional numerical analysis could obtain good results if the pre-support deformations due to top heading excavation are properly considered.

The purpose of this research is to investigate the effects of in-situ stresses on the deformation behavior of tunnels excavated in multiple stages by three-dimensional numerical analysis. The deformations of tunnel, the plastic zones around tunnel, and the stresses in the support will be studied. In this study, vertical stress is σ_z, horizontal stress perpendicular to tunnel axis is σ_x, and horizontal stress along tunnel axis (the axial stress) is σ_y. The lateral stress coefficient K_x is the ratio of σ_x to σ_z, and the lateral stress coefficient K_y is the ratio of σ_y to σ_z. Two kinds of stress fields are considered in this paper. The first one assumes that the vertical stress σ_z is equal to the horizontal stress σ_x and the axial stress σ_y is changing (K_y =0.7, 1, 1.5). This is to understand the effects of axial stress. The second one assumes that the vertical stress is constant while the two horizontal stresses are changing, either σ_y is larger or smaller than σ_x. This is to study the effects of tunnel driving direction in respect with the maximum horizontal stress direction.

2 NUMERICAL ANALYSIS

2.1 *Numerical code*

The numerical code used for three-dimensional analyses is FLAC3D (Itasca 1994). It is an explicit finite difference code that simulates the behavior of structures built of soil, rock or similar materials that may undergo plastic flow when their yield limit is reached. Materials are represented by elements, or zones, which form a grid that is adjusted by the user to fit the shape of the object to be modeled. Each element behaves according to a prescribed linear or non-linear stress-strain law in response to the applied forces or boundary restraints.

2.2 *Procedures of numerical analysis*

1. Set up mesh and boundary conditions: The geometry of tunnel is shown in Figure 1. From the results of preliminary studies, it is found that to give reasonable accuracy and calculation time the boundary should be 80 m away from the tunnel center. The boundary conditions are assumed to be hinges or rollers.

2. Decide properties and apply in-situ stresses: Decide the properties and parameters of rock mass; apply in-situ stresses and make the program reach equilibrium.

3. Excavate and support: Excavate the tunnel in three stages that are top heading, bench I and bench II. Each run is 2 m in length. Assume a support delay of 2 m. After the first run of top heading excavation, let the tunnel reach equilibrium without supports; excavate the second run; apply support on the first run; then allow the tunnel to reach equilibrium. This process is repeated until 20 runs of excavation of top heading is completed. The deformations at the beginning several runs will be influenced by the boundaries and their results are disregarded. After excavation of top heading, the excavation of bench I and bench II will follow the same procedures until 20 runs are completed. The deformations are recorded at five points as shown in Figure 1.

2.3 *Assumptions and parameters*

1. Properties of rock mass: The rock mass are assumed to be homogeneous and isotropic. They are modeled as an elastic-plastic material with a Mohr-Coulomb yield criterion and non-associated flow rule (dilation angle ψ=0°). The elastic modulus of rock mass (E_m) is calculated by the following empirical equations (Bieniawski 1978, Serafim and Pereira 1983).

For RMR>50, E_m=2RMR-100 (GPa) (1)

For RMR<50, $E_m=10^{[(RMR-10)/40]}$ (GPa) (2)

In which, RMR is the rock mass rating of the geomechanics classification system as proposed by Bieniawski (1978). The RMR used in this study is 30. The E_m is calculated as 3.162 GPa. The Poisson's ratio of rock mass is 0.25. The strength parameters, cohesion c and friction angle ϕ for the Mohr-Coulomb criterion are calculated according to Hoek (1987). The cohesion is 0.821 MPa and friction angle is 8.63°.

2. In-situ stresses: The center of tunnel is assumed to be 200 m below ground level. The in-situ vertical stress is calculated as 5.4 MPa, based on a unit weight of 27 kN/m^3. The in-situ lateral stress is calculated by multiplying the vertical stress with lateral stress coefficient (K_x or K_y). Three different values of K were used in this study (0.7, 1.0 and 1.5).

3. Support: A layer of 15 cm thick shotcrete is applied on the surface of tunnel as the supporting system. The compressive strength of shotcrete (f'_c) is 210 kg/cm^2 and Poisson's ratio is 0.17. The elastic modulus of shotcrete (E_c) is calculated as follows.

$E_c=15100(f'_c)^{0.5}$ (kg/cm^2) (3)

4. Tunnel shape and excavation stages: As shown in Figure 1, a kind of horse-shoe shape tunnel is assumed for numerical analysis. The width and height of the tunnel are 12.6 m. Three stages of excavation are used for construction, which are top heading, bench I and bench II.

3 RESULTS AND DISCUSSIONS

3.1 *Effects of tunnel axial stress*

When the lateral stress coefficient K_x is equal to one and the lateral stress coefficient K_y is set to be

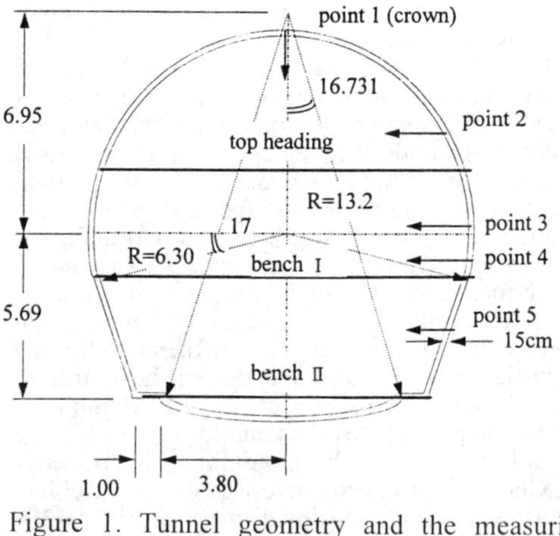

Figure 1. Tunnel geometry and the measuring points of tunnel deformations

0.7, 1, or 1.5, the effects of axial stress can be observed. Figure 2(a) shows the results of tangential stresses of the lining with normalized distance S/B , where S is the distance from excavation face and B is the tunnel width. From the figure, it is shown that lining tangential stresses increase with increasing distance and reach to the final values at distance about 2B from excavation face. Increasing axial stress will increase the lining stresses. The stresses at crown are larger than the stresses at wall. Figure 2(b) shows the results of deformations at crown and wall with normalized distance. From this figure, it is shown that increasing axial stress will increase the deformations, and the deformations at wall are larger than the deformation at crown. Because the crown is an arch and the support dose not form a closed ring, it is expected to have more deformations at wall and more stresses at crown.

Table 1 summarizes the deformations at the measuring points for each construction stage. It is shown that most of the crown deformations occur during the top heading excavation stage, later bench excavation has little influence. The wall deformations, however, are greatly influenced by the bench excavation. Table 2 summarizes the percentage of pre-support deformations at the measuring points for each construction stage. It is shown that the range of percentage is about 32% to 57%. In the top heading excavation stage, increasing axial stress will decrease the percentage of pre-support deformations. In the bench excavation stages, the percentage is not much influenced by the axial stress.

Figures 3 & 4 show the distribution of plastic zones around tunnel for lateral stress coefficients of 0.7 & 1.5. It is clearly observed from these figures that increasing axial stress will increase the extent of plastic zones around the tunnel, but will decrease the extent of plastic zones ahead of the tunnel.

3.2 *Effects of in-situ stresses*

In this study, the vertical stress is kept constant; one of the lateral stresses is equal to the vertical stress; and the other stress is either larger or smaller than the vertical stress. The tunnel is excavated along the maximum horizontal stress direction for one case and along the minimum horizontal stress for another case. Two sets of lateral stress coefficients are used, they are (0.7, 1) and (1, 1.5). The purpose of this study is to understand the influences of tunnel excavation direction in respect to the in-situ stresses.

Figure 5(a) shows the results of tangential stresses of the lining with normalized distance S/B for the stress set of (1, 1.5). It is shown that tangential stresses of the lining support increase with increasing normalized distance. When tunnel is excavated along the maximum horizontal stress direction the lining stress is lower. When tunnel is

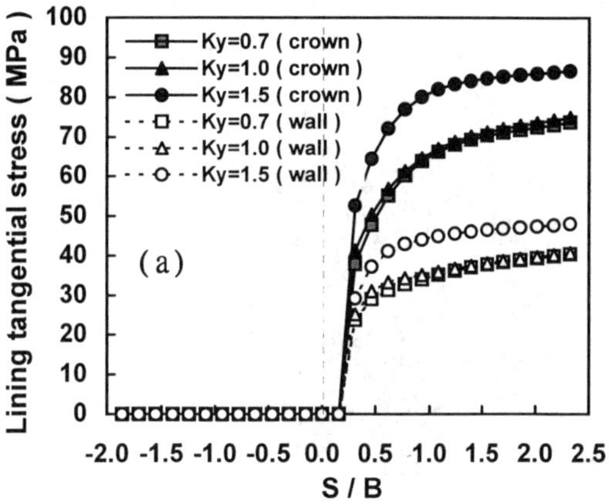

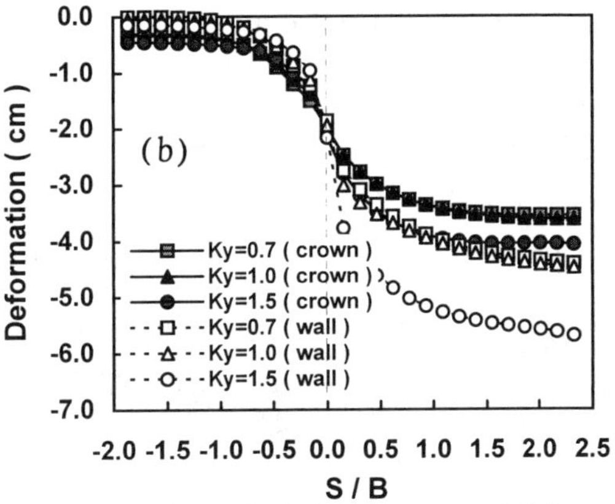

Figure 2. Effects of axial stress on tunnel deformations and lining stresses.

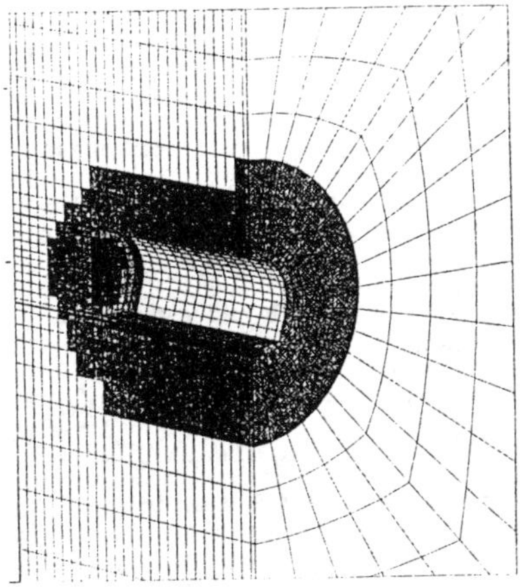

Figure 3. Distribution of Plastic zones for K_y=0.7, K_x=1.0.

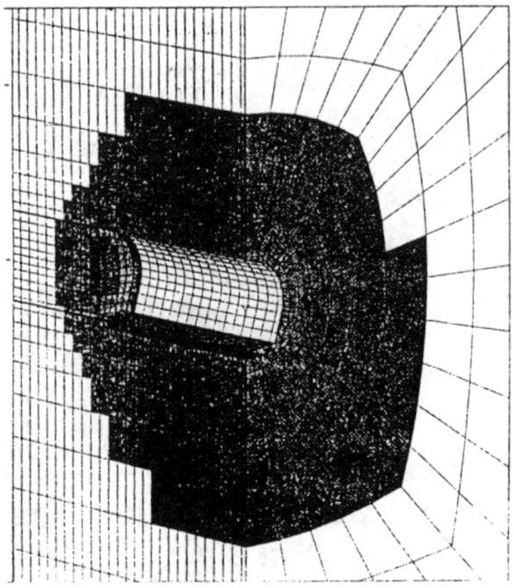

Figure 4. Distribution of Plastic zones for K_y=1.5, K_x=1.0.

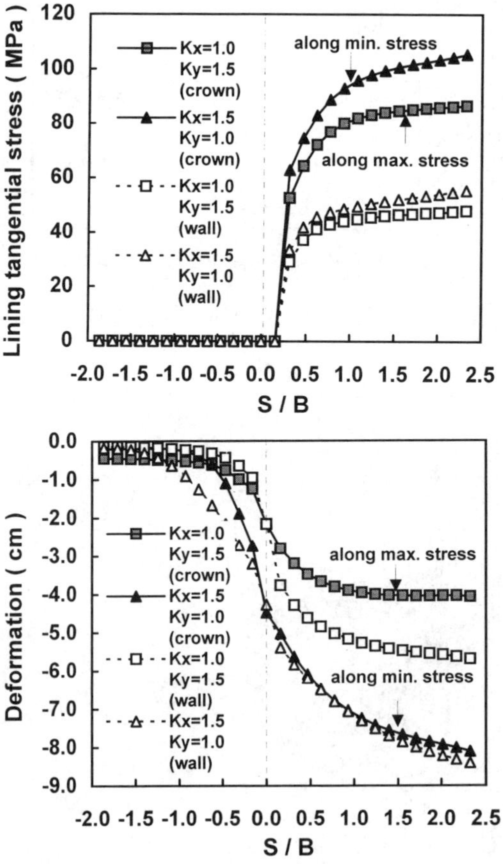

Figure 5. Effects of in-situ stresses on tunnel deformations and lining stresses (1.0 & 1.5)

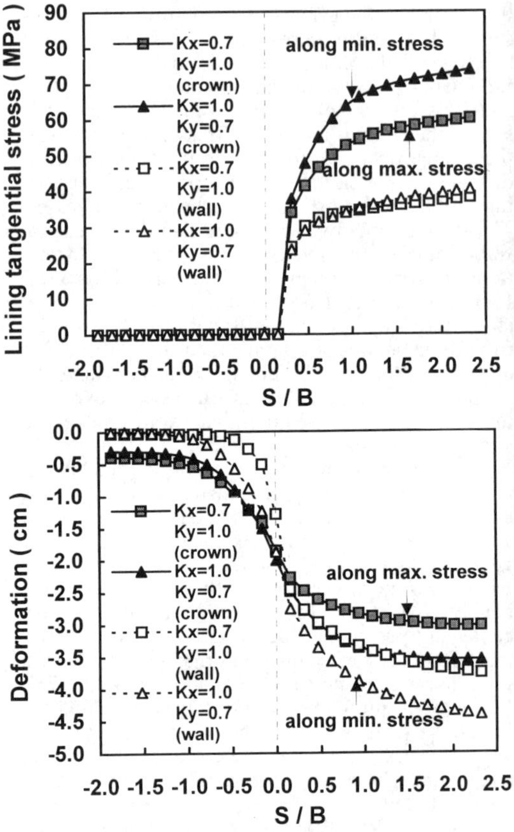

Figure 6. Effects of in-situ stresses on tunnel deformations and lining stresses (1.0 & 0.7)

excavated along the minimum horizontal stress direction the lining stress is higher. The stresses at crown are larger than the stresses at wall. Figure 5(b) shows the results of deformations at crown and wall with normalized distance. From this figure, it is shown that excavated along the minimum horizontal stress direction will get larger deformations, and the deformations at wall are larger than the deformation at crown. Figures 6(a) & (b) show the results for the stress set of (0.7, 1). Their results are the same as the results of Figure 5.

Table 3 & 4 summarize the deformations at measuring points for each construction stage for the two sets of stress coefficients (0.7, 1) and (1, 1.5), respectively. It is shown in Table 3 that excavated along the two different stress directions will have different deformations and the ratios between them after bench II excavation are about 1.2. This is close to the ratio 1.4 between the two horizontal stresses (1/0.7). In Table 4, the ratios of deformations between the two different excavation directions are about 1.3 to 2 for different tunnel locations. The difference is larger at crown and smaller at wall. The ratio of the two horizontal

Table1. Tunnel deformations for different axial stress conditions.

deformation / stage (cm) Ky	top heading			bench- I			bench- II		
	point 1 crown	point 3 wall	point 5	point 1 crown	point 3 wall	point 5	point 1 crown	point 3 wall	point 5
Ky=0.7	2.93	1.45	1.51	3.31	3.07	2.37	3.55	4.40	4.59
Ky=1.0	2.98	1.46	1.51	3.37	3.12	2.40	3.61	4.47	4.65
Ky=1.5	3.37	1.93	1.89	3.80	3.97	3.06	4.06	5.69	5.91

Table 2. Percentages of pre-support deformations for different axial stress conditions.

deformation / stage (cm) Ky	top heading			bench- I			bench- II		
	point 1 crown	point 3 wall	point 5	point 1 crown	point 3 wall	point 5	point 1 crown	point 3 wall	point 5
Ky=0.7	55.6	40.6	30.3	56.6	43.4	37.8	41.7	42.5	35.9
Ky=1.0	55.0	37.6	27.0	54.5	44.3	37.4	34.3	42.9	36.1
Ky=1.5	51.2	29.1	20.7	56.4	41.1	32.1	43.3	41.9	36.3

Table3. Tunnel deformations for excavation along different stress directions.

deformation / stage (cm) K_x , K_y	top heading			bench- I			bench- II		
	point 1 crown	point 3 wall	point 5	point 1 crown	point 3 wall	point 5	point 1 crown	point 3 wall	point 5
K_x=0.7, K_y=1.0	2.44	0.67	0.57	2.76	2.35	1.36	3.02	3.76	3.60
K_x=1.0, K_y=0.7	2.93	1.45	1.51	3.31	3.07	2.37	3.55	4.40	4.59
Ratio	1.2	2.2	2.6	1.2	1.3	1.7	1.2	1.2	1.3

Table 4. Tunnel deformations for excavation along different stress directions.

deformation / stage (cm) K_x , K_y	top heading			bench- I			bench- II		
	point 1 crown	point 3 wall	point 5	point 1 crown	point 3 wall	point 5	point 1 crown	point 3 wall	point 5
K_x=1.0, K_y=1.5	3.37	1.93	1.89	3.80	3.97	3.06	4.06	5.69	5.91
K_x=1.5, K_y=1.0	6.47	4.47	4.32	7.57	6.87	5.77	8.11	8.41	8.83
Ratio	1.9	2.3	2.3	2.0	1.7	1.9	2.0	1.5	1.5

stresses is 1.5 and it is close to the ratio of deformations at wall. From these comparisons, it is clearly shown that the differences in the deformations are closely related to the differences in the horizontal stresses, but the geometry of tunnel and the distribution of plastic zones will affect the ratios of deformations. When applying two-dimensional analysis in tunnel design the effects of axial stress on tunnel deformations may be approximately considered by multiplying the deformations with the stress ratios.

4 CONCLUSIONS

1. Increasing axial stress will increase the tunnel deformations, the lining stresses, and the plastic zones around the tunnel. But the percentage of pre-support deformations after top heading excavation and the plastic zones ahead of excavation face will both decrease with increasing axial stress.

2. The differences in the deformations when tunnel is excavated along different stress directions are closely related to the differences in the stresses. The ratios of the deformations are about the same as the stress ratios.

3. When tunnel is excavated along the maximum horizontal stress direction the tunnel deformations, the lining stresses, and the plastic zones will be smaller. So, it is better to design the tunnel parallel to the direction of maximum stress.

REFERENCES

Bieniawski, Z. T. 1978. Determining Rock Mass Deformability : Experience from Case Histories. *Int. J. Rock Mech. Min. Sci.* 15: 237-248

Chen Y. C. & Y.R. Pong 1994. Effects of support delay on the deformation of tunnels. *J. of the Chinese Institute of Civil & Hydraulic Engineering*. 6(1): 11-118 (in Chinese).

Chen Y.C. & M.S. Lu 1996. Numerical analysis of multi-stage excavation of tunnel. *Proc. 3rd Asian-Pacific Conf. Computational Mech., Seoul, 16-18 Sept. 1996*: 2257-2262. Korea: Techno-Press.

Hoek, E. 1987. Estimating Mohr-Coulomb Friction and Cohesion Values from the Hoek-Brown Failure Criterion. *Int. J. Rock Mech. Min. Sci. & Geomech. Abstr*. 27(3): 227-229.

Itasca Consulting Group, Inc. 1994. *FLAC3D Version 1.0*. Minneapolis, Minnesota.

Serafim, J. L. & J.P. Pereira 1983. Considerations of the Geomechanics Classification of Bieniawski. *Proc. Int. Symp. on Engng. Geol. and Under-ground Constr.*, LNEC, Lisbon, Portugal.

Computer Methods and Advances in Geomechanics, Yuan (ed.) © 1997 Balkema, Rotterdam, ISBN 90 5410 904 1

Numerical analysis on a three-tube highway tunnel – A case study

Tze-Pin Lin
Department of Mineral and Petroleum Engineering, National Cheng Kung University, Taiwan, China

ABSTRACT: A three-tube highway tunnel in the northeastern Taiwan is under construction. A numerical model is employed in this study to simulate and evaluate the excavation and performance of the tunnels. The measured data from the monitoring of the pilot tunnel are used to back analyze the rock mass properties of the site. The timing of instrumentation and support placement is modeled by decreasing the virtual support pressure step by step accordingly. The results show that the performance of the tunnels could be predicted satisfactorily and the interaction between tunnels in multiple tube excavation is significant.

1 INTRODUCTION

The Taipei-Ilan Expressway is under construction in the northeastern Taiwan. Along the highway, the 12.9 km long Pinglin Tunnel is most challenging to engineers. The tunnel consists of one pilot and two main tubes of the sizes of 5.2 m and 12.5 m, respectively. Except the first few hundred meters, the tunnels are being drilled by full-face tunnel boring machines (TBMs). The configuration and excavation sequence of the tunnels in the drill and blast section are shown in Figure 1. The pilot tunnel had advanced for around 600 m before the main tunnels commenced. In some sections of the pilot tunnel, there were cracks and damages developing on the shotcrete linings and steel set supports, and the development seems to follow the progress of the main tunnels. Although the stress disturbance due to excavation is negligible when the distance is one diameter away from the excavation (Hoek & Brown 1980), it is suspected that these damages are caused by the stresses induced by excavation of the main tunnels. A numerical model is therefore employed to evaluate the performance of the tunnels, especially the tunnel safety during the benching operation of the main tunnels.

2 GENERAL GEOLOGY

The Pinglin Tunnel is driven through the northern end of the Hsueshan Range which was initially uplifted and folded by the orogenies in Pliocene, resulting in large scale folding and reverse faulting structures. The orogenies were succeeded by the splitting of the Okinawa Trough (a back arc basin), causing intensive high angle faulting. These folds and faults have strikes of east-northeast in general and dip southeastward. The alignment of the tunnels is roughly perpendicular to these major geologic structures.

The tunnels are mainly driven through slightly metamorphosed Tertiary sedimentary rocks. The lithostratigraphic units in this area are:

Miocene	Fangchiao Formation	Alternations of sandstone and shale, intercalated with coal seams
	Makang Formation	Thick sandstone and shale
Oligocene	Tatungshan Formation	Argillite with sandstone interbeds
	Tsuku Formation	Alternations of argillite and sandstone
	Kankou Formation	Massive argillite and slate
Eocene	Szeleng Sandstone	Quartzite intercalated with argillite

In general, these rocks distribute, in ascending order, from the southeast to the northwest along the tunnels. The drill and blast section of the tunnels is mainly in the Kankou Formation, i.e. massive but fractured and jointed argillite and slate. The rock mass rating (RMR) based on Bieniawski's classification at the sections that the measured data were used at this study is around 15, and the NGI-Q value is around 0.02 (Bieniawski 1989; Barton et al 1974).

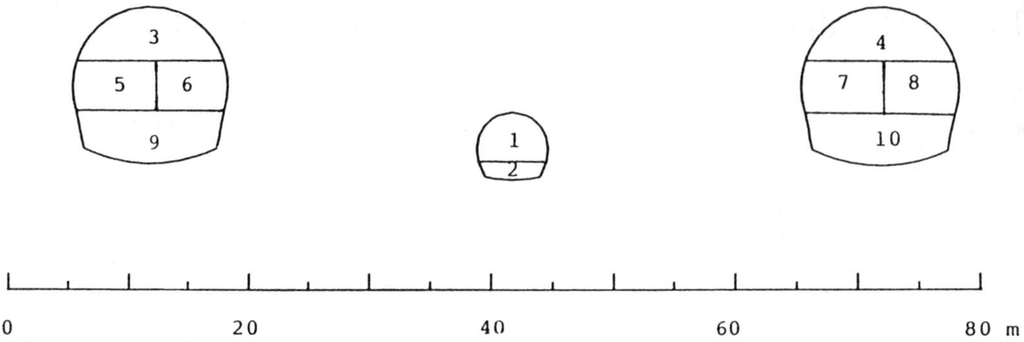

Figure 1. The configuration and excavation sequence of the Pinglin Tunnel.

3 NUMERICAL MODELING

The induced stress and strain changes in rock during tunnel driving are very complicated in a three-dimensional manner. It is usually treated, however, as a two-dimensional plain strain problem for simplicity. Since the deformation of the tunnel wall is constrained by the unexcavated portion, the convergence of the tunnel wall, instead of completed as excavated, is increasing with the tunnel advancing. In order to simulating this effect with a two-dimensional model, an axisymmetric finite difference mesh was used to calculate the deformation of a circular tunnel. A two-dimensional model was then employed to calculate the virtual support pressure needed to maintain that same deformation at certain sections along the tunnel's alignment. The timing of instrumentation and support placement could then be estimated from the relationship between the virtual support pressure, the tunnel deformation and the position of the interested sections (Lin 1997).

Since there is no feasible way to estimate the virtual support pressure in benching operation, it is decided to model the timing of support placement with two extreme cases. One is to simulate the extremely early supported situation by applying the support right after excavation. The other is to simulate the extremely late supported condition by applying the support after the convergence has occurred. The real situations should fall in-between the two extremes.

In the modeling, the support and instrumentation data are obtained from the field records. The rock mass properties are determined from back analyzing the measured deformations of the pilot tunnel. The performance of the main tunnels is predicted and compared with the observations after the excavation of main tunnels.

4 RESULTS AND DISCUSSION

The measured deformation data at the completion of the first stage benching operation of the main tunnels are compared with the calculated results in Table 1. It is shown that the values of the measured roof settlement well fall in the predicted ranges, while the horizontal convergence is overestimated. This may be caused by underestimating the measured convergence because of delay in instrumentation, since the increases in measured convergence due to benching operation alone match the predicted values well.

Table 1. Comparison of calculated and measured tunnel deformations.

Monitoring Section	Measured Data (mm)		Calculated Values (mm)	
	Roof Settlement	Horizontal Convergence	Roof Settlement	Horizontal Convergence
Pilot Tunnel	10	123	12	125
Main Tunnel 1	133	166	80^{*} ~ $317^{\#}$	230^{*} ~ $553^{\#}$
Main Tunnel 2	173	66		
Main Tunnel 3	146	150		
Main Tunnel 4	198	89		
Main Tunnel 5	181	107		

Note: * early supported; # late supported.

Axial stresses on supports of the pilot tunnel are shown in Figure 2. It is evident that the stresses increase dramatically when the main tunnels are excavated, especially when the support placement of the main tunnels are delayed (Figure 3). The way to reduce the stresses on the supports of the pilot tunnel is to apply the supports on the main tunnels as soon as possible. The linings of the pilot tunnel can sustain the stresses induced by the excavation of the main tunnels only when the supports are placed on the main tunnels extremely early. It was predicted based on Figure 3 that the linings of the pilot tunnel would be over-stressed during the benching operation of the main tunnels. This is confirmed by field observations in which the shotcrete linings of the pilot tunnel began to crack when the benching operation of the main tunnels was in progress (Sinotech Engineering Consultants, Inc. 1995).

It is interesting to look at the stress distribution on the supports of the main tunnels. The axial stresses on the linings of early supported tunnels are higher but more uniformly distributed, and the tensions on the rock bolts are less (Figures 4a and 4b). The bending moments of the linings are larger at the lower part of the side walls of tunnels in early supported situation, while they are larger at the shoulder for late supported tunnels (Figures 4c and 4d). This could be confirmed by observations in the field. Some of the shotcrete cracked and steel sets buckled at the shoulder of the tunnels while some of the lower parts of the side walls squeezed in up to half a meter.

The numerical model could be validated with the distribution of the calculated failure zones around the tunnels too. The depth of the failure zones ranges from 2 to 8 meters for main tunnels depending on the timing of support placement, that could be compared with the borehole extensometer measurements of 4 to 9 meters.

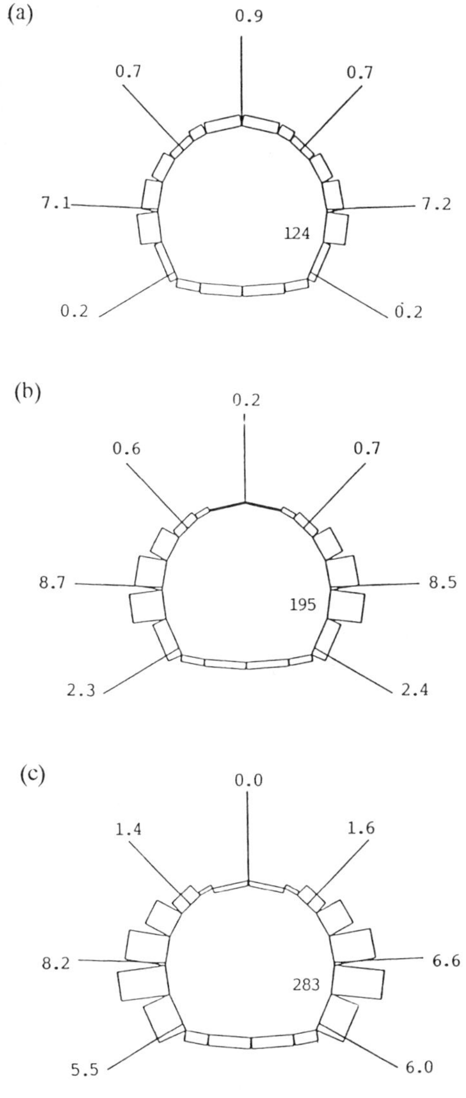

Figure 2. Axial stresses on supports of the pilot tunnel; the unit is in kg/cm^2 for shotcrete and steel set linings, and in tons for rock bolts. (a) at completion of the pilot tunnel; (b) at completion of the early supported main tunnels; (c) at completion of the late supported main tunnels.

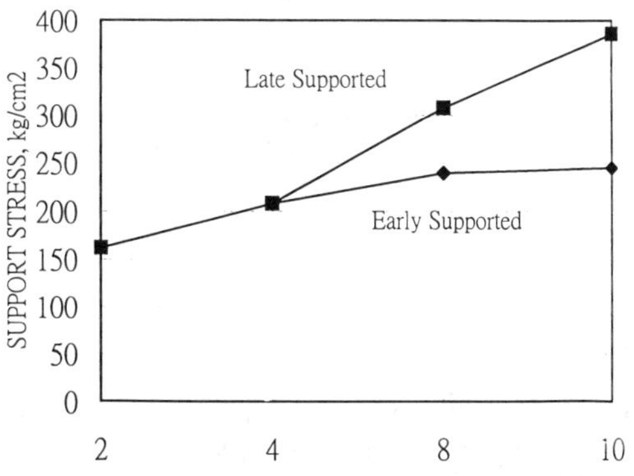

Figure 3. Axial and moment stresses on supports at the side walls of the pilot tunnel increase as excavation progressing. The excavation sequence is indicated in Figure 1

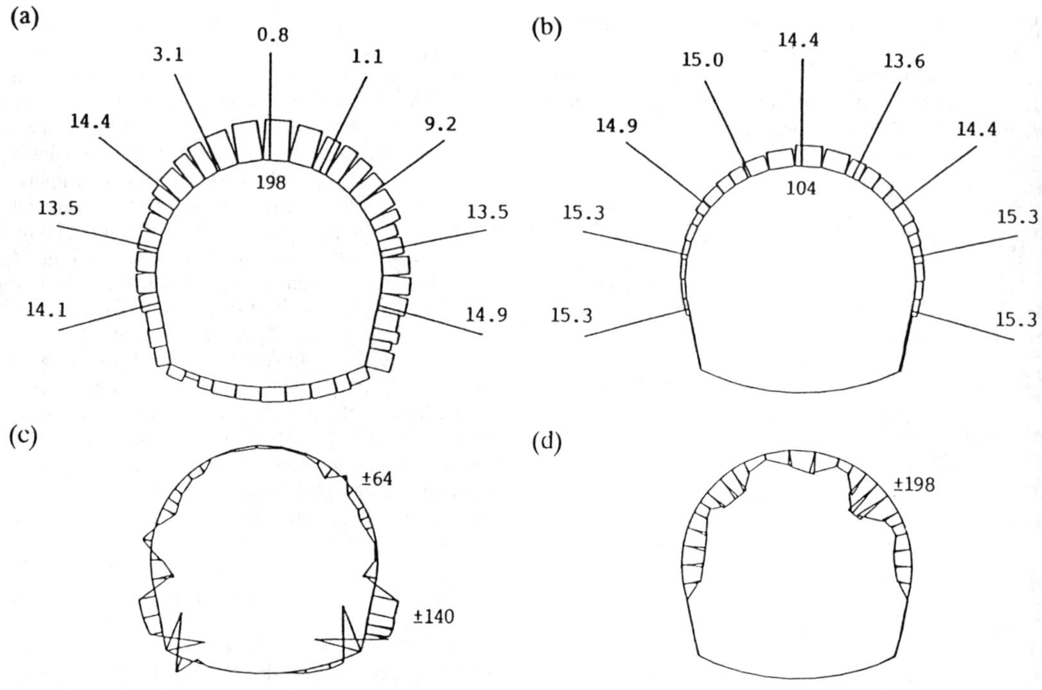

Figure 4. Stresses on supports of the main tunnels; the unit is in kg/cm^2 for shotcrete and steel set linings, and in tons for rock bolts. (a) axial stresses, early supported; (b) axial stresses, late supported; (c) moment stresses, early supported; (d) moment stresses, late supported.

5 CONCLUSIONS

1. The virtual support pressure could be used to simulate the three-dimensional effect in tunnel driving with a two-dimensional model.

2. The performance of the main tunnels could be predicted satisfactorily using parameters derived from back analyzing the measured data of the pilot tunnel.

3. The interaction between tunnels in multiple tube excavation is significant. The way to reduce the stresses on supports of the pilot tunnel is to place the supports on main tunnels as soon as possible.

REFERENCES

Barton, N. R., Lien, R. and Lunde, J. 1974. Engineering Classification of rock masses for the design of tunnel support. *Rock Mech.* 6:189-239.

Bieniawski, Z. T. 1989. *Engineering Rock Mass Classifications.* New York: Wiley.

Hoek, E. & Brown, E. T. 1980. *Underground Excavations in Rock.* London: Institution of Mining and Metallurgy.

Lin, T.-P. 1997. Simulation of support timing with virtual support pressure. *Proceedings of the 7th Geotechnical Engineering Conference, Taiwan, 28-29 August 1997.*

Sinotech Engineering Consultants, Inc. 1995. Damage inspection on the linings of the pilot tube of the Pinglin Tunnel. Internal report, in Chinese, not published.

Computer Methods and Advances in Geomechanics, Yuan (ed.) © 1997 Balkema, Rotterdam, ISBN 90 5410 904 1

Shield tunnelling of the Chungho Line of Taipei MRT

Ling-San Lin
Dorts, Taipei Municipal Government, Taiwan, China

Jau-Ling Chang & Daniel C.P.Chu
SDPO, Dorts, Taipei Municipal Government, Taiwan, China,

ABSTRACT: Bored tunnelling of the Chungho Line of Taipei MRT was challenged by the conditions of soft ground, high water table, presence of cobbles, drift woods and methane, stacked tunnel configruations and existing buildings lying above the alignment. The problems were overcome by cost effective counter-measures and the performance of tunnelling has been satisfactory.

1. INTRODUCTION

The Initial Network of the Taipei Rapid Transit Systems (TRTS) serving the Metropolitan Taipei Area is totally 88 km in length. The Network comprises five heavy-capacity lines (i.e., the Tamshui Line, Hsintien Line, Nankang Line, Panchiao Line and Chungho Line) and one medium-capacity line (i.e., the Mucha Line).

Extending from the southern portion of the Taipei Basin to further south, the Chungho Line is 5.4 km in length and consists of 4 underground stations, (O16 to O19) and 4.3 km length of twin bored tunnels of 6.1 m in outer diameter. The line joins the Hsintien Line after crossing the Hsintien River. Figure 1 shows the layout of the Initial Network of the TRTS System.

The tunnelling for the Chungho Line has been difficult. The tunnels were driven under many existing buildings with small clearances from the foundations. In one stretch, between Stations O16 and O17, the alignment passes below many buildings over a length of about 300 m. Due to the constraints of the narrow roads, Stations O17 and O18 are stacked-platform stations. The twin tunnels between these two stations are also in a stacked position. The smallest clearance between the two stacked tunnels is about 4 m. The most problematic section of tunnelling for the Chungho Line is the 250 m stretch between Stations O18 and O19. On the approach to Station O19, the alignment is shallow, with a tight curve of 200 m in radius, in soft clay and passes directly below a large number of shabby buildings with poor foundations.

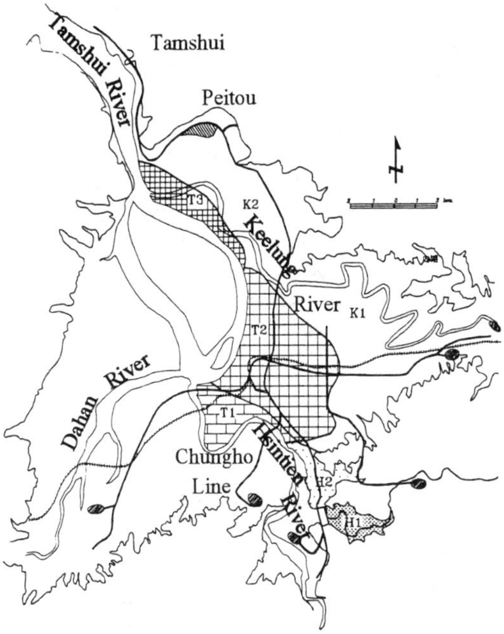

Fig. 1 Network of the TRTS system and the subzoning of the Taipei basin

This paper describes the construction of the tunnels with emphasis on experience dealing with obstructions of cobbles, drift woods and methane. The performance of bored tunnelling is assessed by the magnitude of ground settlement and is compared with tunnelling in other areas in Taipei Basin.

2. SUBSOIL CONDITIONS

2.1 Subsoil profile

The subsoil in Taipei Basin composes of sedimentary deposits of the Sungshan Formation of thickness varying from 40 m to 70 m underlain by gravelly deposits of the Chingmei Formation and bedrock of sandstone or shale. According to Woo and Moh (1990), the Sungshan Formation consists of sub-layers of silty sand and silty clay which were deposited alternately. In the central area of the Taipei City, typically six sub-layers classification for the Sungshan Formation is applicable. Beginning from bottom, sub-layers 1, 3 and 5 are silty sand layers and sub-layers 2, 4 and 6 are silty clay layers.

Based on the geological origin and sedimentary environment, the Sungshan Formation has been divided into 7 subzones. They are T1 to T3 of the Tamshui River Zone, K1 and K2 of the Keelung River Zone and H1 and H2 of the Hsintien River Zone.

Figure 2 presents the subsoil profiles along the Chungho Line, indicating the 4 to 6 alternatively deposited sandy or clayey layers, which are comparable to the typical 6 sub-layers in the central city area. Tentatively the subsoil could be divided into 2 subzones. The northern subzone, Y1, is characterized by the presence of the upper gravel layer, sub-layer 5a, at the depth from 8 m to 18 m. The southern subzone, Y2, is dominant by thick clayey deposits. It appears that the Y1 and the Y2 subzones are similar to the T2 and the K1 subzones both in stratigraphy and in soil properties, respectively.

2.2 Upper gravel and drift woods

The upper gravel layer, sub-layer 5a, at the depth from 8 m to 18 m would be an obstruction to the driving of bored tunnels. Based on observations made during the excavation for the cut-and-cover stations, the maximum size of cobbles in this layer is around 400 mm. The cobbles are in-filled by matrix of sand and gravel.

Drift woods are present in the upper gravel layer. One large tree trunk of about 3 m in length was recovered from a basement excavation site in vicinity of Station O16 and was reported in the geotechnical report prepared at the design stage. During the construction stage, a dozen of drift woods of 2 m to 5m in length were encountered in the upper gravel layer within the 5000 m^2 excavation area for Station

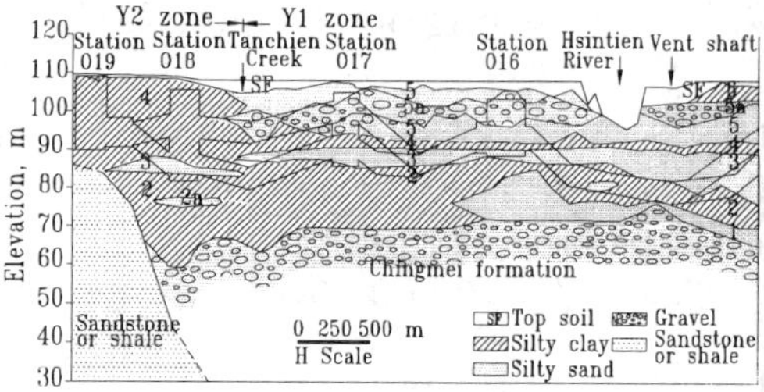

Fig.2 Subsoil profile along the Chungho line

O17. Driving of the up-track tunnel between Stations O17 and O18 was also obstructed by drift woods at the depth of 10 m. As shown in Fig.3, the presence of drift woods in the Taipei Basin was summarized by Hwang (1996).

2.3 Gas problems

The presence of gas in the Taipei Basin was reported by Woo and Moh (1990). As shown in Fig. 3, two of the gas emission sites are located along the Chungho Line. The gas emission site next to the Hsintien River was confirmed by supplementary drillholes conducted during the construction stage. A pocket of gas was penetrated by drillholes BA8 and BA9 at depth of 28.7 m to 27.5 m. The gas was under a high pressure which sent the water and debris in the drillhole to a height of as much as 15 m. Results of gas analysis on samples collected from drillhole BA8 indicated a concentration of 17.8% of methane. Since methane is highly explosive at the concentration of 2.5% to 15%, this gas hazard poses significant concerns on the safety for the driving of the bored tunnels and for the operation of the system.

2.4 Groundwater conditions

From north to south, the ground elevation along the Chungho Line varies from RL 107 m to RL 109 m. In 1991, prior to the commencement of the construction, the piezometric elevation for sub-layer 5 varied from RL 102 m at the north to RL 107 at the south end of the Chungho Line. Piezometric elevation for the Chimgmei Formation varied from RL 99 m to RL 96 m throughout the Line. It should be noted that RL 100 m corresponds to the mean sea level at Keelung Harbour.

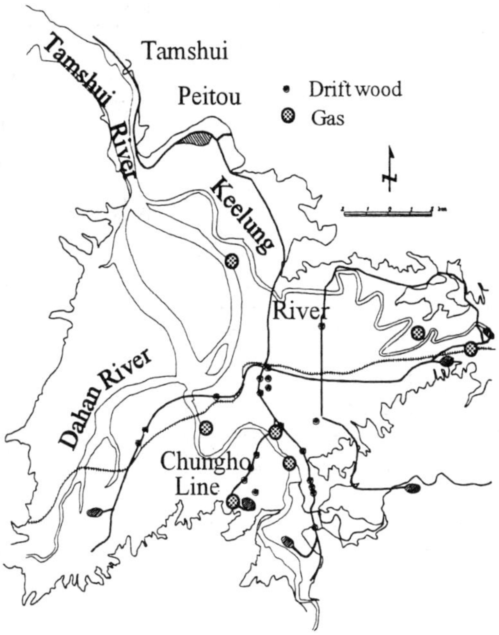

Fig.3 Location of sites with gas emissions and drift woods (after Hwang 1996)

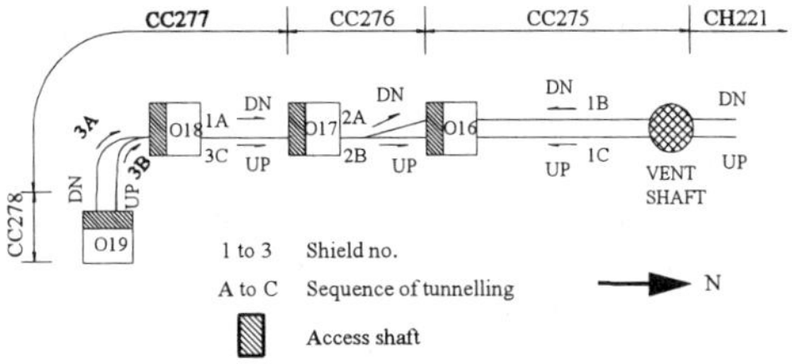

Fig.4 Sequence of bored tunnelling

Table 1. Progress of bored tunnelling for Chungho Line

Shield No.	Tunnel No.	Section of Tunnel	Down/Up Track	Launching Date	Arrival Date	No. of Rings	Average Rate m/day
1	1A	O18-O17	Down	8 Feb 95	16 Oct 95	877	4
	1B	VS-O16	Down	26 Feb 96	30 May 96	1107	12
	1C	VS-O16	Up	13 Jul 96	29 Oct 96	1087	10
2	2A	O17-O16	Down	13 Jun 95	2 Oct 95	984	9
	2B	O17-O16	Up	4 May 96	3 Dec 96	988	5
3	3A	O19-O18	Down	6 Nov 95	28 Nov 95	242	11
	3B	O19-O18	Up	26 Jan 96	8 Feb 96	232	17
	3C	O18-O17	Up	8 Apr 96	29 Nov 96	871	4

Note: VS stands for ventilation shaft located at the interface with Contract CH221.

3. BORED TUNNELLING

3.1 Tunnelling machines

Three earth pressure balancing shield machines of identical specifications were used for driving the bored tunnels. The shields are 6270 mm in outer diameter, 6190 mm in length and weighs 600 kN each. Although over-cutter with an outer-diameter of 6280 mm was available, the shields were equipped with articulate jacks to negotiate with the tight curve of 200 m in radius.

Since it was anticipated that obstructions such as drift woods and boulders would be encountered, the shields were equipped with 2 pressure chambers behind the bulkhead for compressed air. The 2 pressure chambers have a total volume of 2.32 m^3 space. The maximum design pressure for the chambers was 300 kPa. Up to 3 people can stay in the pressure chambers. Bentonite slurry was provided to the soil chamber through 4 outlets on the bulkhead to improve the workability of the soil slurry. The maximum injection pressure for bentonite slurry was 600 kPa.

The screw conveyor was 700 mm in diameter. The maximum capacity was 125 m^3/hr at a maximum rate of 10 rpm. It had a maximum torque of 195 kN-m. The maximum size of cobble which could be conveyed was 200 mm.

The shield machines were provided with the facility for foam injection so that the soil chamber pressure could be maintained when driving through gravelly soils. The foam was produced by injecting polymeric agent and compressed air into the soil chamber. Mixed with the soil, the moist foam provides higher viscosity and better workability of the gravelly soil can be achieved. The sequence and the progress of tunnelling are shown in Fig. 4 and in Table 1, respectively.

3.2 Mitigation measures for gas hazards

Gas mitigation measures for driving bored tunnels include the following:

1. the shield machine and auxiliary equipment, which include power generating system, muck conveying system, lighting system and ventilation system, were equipped with spark-proof devices,
2. gas detection devices installed on the shield machines and portable gas detectors were provided.

During tunnelling under the Hsintien River where

gas was found to be a problem, the following particular measures were enforced:

1. two gas detectors were installed on the shield machine. One of them was installed above the pressure chamber while the other was installed at the end of the screw conveyer. These detectors were connected to an automatic power interrupting system. When a concentration of methane of 1.25% was detected, power supply for the shield machine will be automatically shut down, the outlet gate of the screw conveyor will be closed and the gap at the tail and at the articulate joint of the shield will be sealed. The shield machine will only be re-activated as the concentration of methane reduced to 1% or lower.

2. a close-type slurry pipe system was used for discharging the excavated material. Compared with the muck cart system, which were used at other bored tunnel sections, this mitigation measure reduces the risk of releasing gas into the tunnel.

3. the capacity of the ventilation system has been increased from 780 m^3/min to 1560 m^3/min,

4. a safety manual for underground construction was provided to every tunnelling worker,

5. a safety officer responsible for all safety aspects on gas hazards was assigned to the tunneling crew.

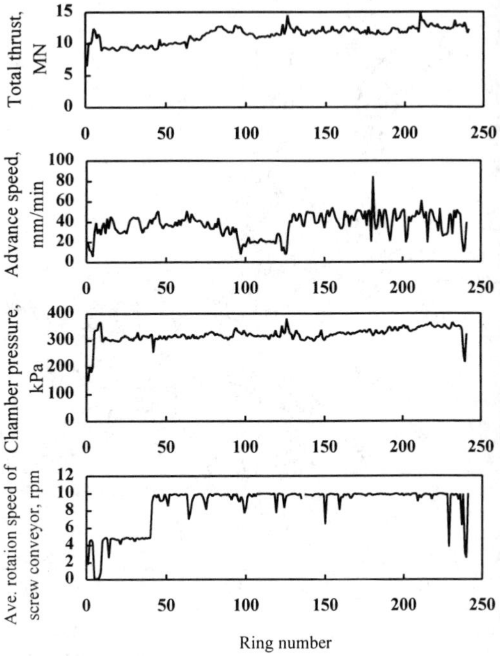

Fig.5 Record of tunnelling operations for down track tunnel between Stations O19 and O18

3.3 Ground treatment at tunnel portals

In order to prevent collapsing and piping failure after the opening of the tunnel portal, ground treatment was conducted at the launching and the arrival shafts behind diaphragm walls. Jet grouting using double tube technique such as RODINJET-2 and JSG was applied. The methods of jet grouting are similar to those reported by Kauschinger and Welsh (1989). Typically the treated soil achieved unconfined compressive strengths of 4.0 to 6.6 MPa and coefficients of permeability of 0.03 to 1 $\times 10^{-7}$ m/sec.

3.4 Control of tunnelling

The shield machines were operated by semi-automatic control. The operating parameters include chamber pressure, rotation speed of screw conveyor, jack speed, jack thrust, cutter torque, amount and pressure of backfill grouting, attitude of the shield, amount of foam injection and other relevant data. These information were transmitted electronically or manually to the operator so that he could adjust the operation to achieve safe and stable tunnel driving.

One of the important factor for operation of earth-pressure balancing shield is the controlling the earth chamber pressure and a criteria of 300 kPa was adopted. Fig.5 presents the record of driving the down track tunnel between Stations O19 to O18.

3.5 Backfill grouting

Ground settlements due to tunnelling could generally be controlled by backfill grouting. The tail void, i.e., the annular gap between the outer diameter of the shield skin and that of the tunnel lining, was backfilled with sand/cement grout simultaneously with the driving of the shield. The sand/cement grout has an additive to retard the setting time to about 5 hours so that workability could be maintained during transportation. Backfill grouting was applied to 4 groutholes one by one on the third lining behind that one which had been erected. The maximum grouting pressure was limited to 500 kPa. Injection volume ranged from 60% to 230% of the theoretical tail void.

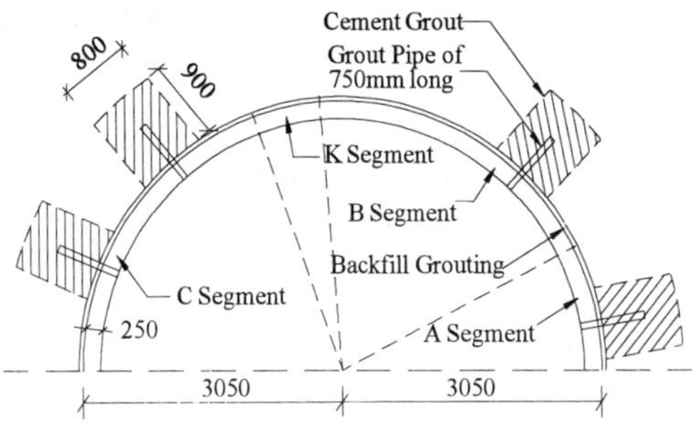

Fig.6 Arrangement of the secondary backfill grouting

3.6 *Secondary backfill grouting*

For the bored tunnel section between Stations O18 and O19 in which the tunnels are directly beneath existing buildings, secondary backfill grouting was applied to further minimise the ground and foundation settlements. Secondary grouting was applied at the ninth or the tenth lining behind the shield. With an average driving rate of 11 m per day, secondary grouting was applied on the same day and was implemented in the evening shift. The water/cement ratio in weight was 0.5 to 0.6. Cement grout was injected through 3 or 4 groutholes on the lining at a pressure of 200 to 500 kPa and the average injected volume was 94 litre per ring. As shown in Fig. 6, the injection pipe was 34.2 mm in diameter and 750 mm in length. There were 6 holes provided along the 250 mm outer end of the injection pipe.

3.7 *Obstructions during tunnelling*

While the screw conveyor had been designed to handle cobbles less than 200 mm in size, there were no occasions which required clearing of cobbles inside the earth chamber during driving. On the other hand, fragments of drift woods were randomly observed in the spoil although the rate of tunnel driving had not been slowed down significantly.

There was an occasion that the presence of drift wood caused the ground to collapse when tunnelling proceeded from Station O18 to O17. A sinkhole of 5 m in diameter occurred right above the head of the shield. Workers entered the soil chamber through the pressure chamber recovered two pieces of drift wood which were 500 mm and 400 mm in length. The tunnel crown was at a depth of 10 m. The sinkhole could be attributed to over-excavation and unbalanced soil chamber pressure. Remedial measures included the following:

1. concrete was poured into the ground to backfill the sinkhole,
2. cement bentonite grout was injected in stages through grout pipes extending from the grout injection ports at the bulkhead,
3. secondary backfill grouting was carried out for a section of 25 m behind the tail of the shield,
4. a zone of 4 m ahead of the shield was treated by using jet grouting.

In order to prevent collapsing from happening again, the remaining 200 m length of the bored tunnel was treated with sodium silicate base grout using the tube-a-manchette technique.

The merits of the provision of the pressure chamber with the shields are demonstrated. The chamber enabled obstructions encountered at the head of the shield to be cleared in a quick and effective manner.

3.8 *Methane*

Gas with concentration of methane ranged from 20% to 57% was encountered during the driving of a section of 60 m long at the down-track tunnel of the river crossing. After the alarm system was activated, the shield machine stopped automatically and the rate of ventilation was increased. Workers inside the tunnel were immediately evacuated. Tunnel driving resumed after the concentration of methane dropped to 1.0%. However, the actual progress of tunnelling was not much affected by the evacuation since the average driving rate of 20 m/day was still maintained.

4. SETTLEMENT DUE TO TUNNELLING

The performance of bored tunnelling could be assessed by the magnitude of ground settlement. Generally settlement over tunnels could be represented by an error function expressed by the following (Peck 1969):

$$\delta = \frac{VA}{2.5i}\exp[\frac{-x^2}{2i^2}] \tag{1}$$

where δ =settlement, V=ground loss, A=sectional area of tunnel, x=horizontal distance to the tunnel centre and i=trough width parameter and is the transverse distance to the point of inflection.

Clough and Schmidt (1981) proposed the following relationship for soft ground:

Table 2. Ground losses and indices of consolidation settlements for single drive

Zone	Major Soil Layers	Contract	Depth Z_0, m	Surface Settlement δ_{10}, mm	Ground Loss V_{10}, %	Index of Consolidation α, mm
Y1	Sandy	CC275	15-28	7-21	0.46-1.90	1-4
		CC276	18-25	4-32	0.71-2.33	6-18
		CC277	19-24	4-5	0.25-0.97	3-16
Y2	Clayey	CC277	19-25	13-21	0.98-1.63	3-10
		CC277(a)	19-25	9-11	0.61-0.79	8-13
K1	Clayey	CN256(b)	13-19	2-32	0.19-1.49	5-14
T1	Sandy	CN262(b)	13-14	18-40	0.85-2.03	6-12
T2	Sandy	CH218(b)	17-21	16-24	1.11-1.64	5-10

(a) With secondary backfill grouting
(b) After Hwang, Sun and Ju (1996)

$$i = \frac{D}{Z}\left(\frac{Z_o}{D}\right)^{0.8} \tag{2}$$

where D=tunnel diameter and Z_o=vertical distance from surface to centre of tunnel. The ground loss, V, can generally be back-calculated from case histories where the ground settlements were actually monitored.

According to Hwang, Sun and Ju (1996), the settlement occurred in 10 days after passing of the head, δ_{10}, and the index of consolidation, α, the slope of the settlement curve in the semi-log plot, shall be calculated to assess the performance of bored tunnelling. Ground loss values, V_{10}, for bored tunnelling in the Y1 zone and in the Y2 zone are presented in Table 2. In the Y1 zone where majority of the bored tunnels were driven in sandy layers, ground loss ranged from 0.23% to 2.33%. For tunneling in the Y2 zone where the clay layer is predominant, ground loss ranged from 0.25% to 1.63%.

The effects of secondary grouting on minimising settlement can be inferred from Table 2. In Y2 zone ground loss was reduced to 0.6% to 0.8% where secondary grouting was applied. Table 2 shows that ground loss values are comparable with those observed at bored tunnelling in the K1, T1 and T2 zones which were reported by Hwang, Sun and Ju (1996). It appears that the performance of tunnelling for the Chungho Line has been as good as that for other Lines in the TRTS System.

5. CONCLUSIONS

Due to the conditions of soft ground, high water table, presence of cobbles, drift woods and methane, stacked tunnel configurations, existing buildings lying directly above the alignment and sub-standard conditions of the existing buildings, the Chungho Line tunnels were difficult to construct. However, the problems were overcome by adopting meticulous counter-measures which included the following:

1. extensive site investigation conducted in both the design stage and the construction stage enabled proper understanding of the subsoil and groundwater conditions.
2. the purposely designed shield machines had the capability of handling obstructions caused by cobbles, drift woods, methane and tight radius of curvature of the alignment. It proves to be cost effective measures during the driving of the tunnels.

Notwithstanding the difficulties encountered, observations of surface settlement along the tunnel alignment indicate that the performance of bored tunnelling for Chungho Line is as good as tunnelling of other Lines of the TRTS. A limited number of observations indicate that secondary backfill grouting further reduced ground settlement by one half.

ACKNOWLEDGEMENTS

The authors wish to express their gratitude to Bilfinger+Burger/Eastern Co. for providing field data which were presented in this paper and are grateful to Mr. L. W. Wong, Mr. H. S. Kao and Dr. R. Hwang of Moh and Associates, Inc. for reviewing the manuscript.

REFERENCES

Clough, W. and Schmidt, B. (1981), Design and performance of excavations and tunnel in soft clay, *Soft Clay Engineering*, Elsevier Amsterdam.

Hwang, N.R. (1996), *Personal Communication.*

Hwang, N. R., Sun, R. L. and Ju, D. H. (1996), Settlement Over Tunnels - TRTS Experience, *Twelfth Southeast Asian Geotechnical Conference,* May 1996, Kuala Lumpur, Malaysia.

Kauschinger, J. L. and Welsh, J. P. (1989), Jet Grouting for Urban Construction, *Proceedings of the 1989 Seminar, Design, Construction and Performance of Deep Excavations in Urban Areas*, MIT, 1989.

Peck, R. B. (1969), Deep Excavations and Tunnelling in Soft Ground, *Proc., 7th, ICSMFE, State-of-the-art Volume,* Mexico City, Mexico:225-290.

Woo, S. M. and Moh, Z. C. (1990), Geotechnical Characteristics of Soils in the Taipei Basin, *Tenth Southeast Asian Geotechnical Conference*, April 1990, Taipei, Taiwan, CHINA

Computer Methods and Advances in Geomechanics, Yuan (ed.) © *1997 Balkema, Rotterdam, ISBN 90 5410 904 1*

'Sense as you advance' automation in trenchless technology

Pradeep U. Kurup & Mehmet T. Tumay
Louisiana Transportation Research Center, Louisiana State University, La., USA
Beyongseock Lim
Louisiana State University, La., USA

ABSTRACT: Existing methods for determining engineering soil properties relevant to soft ground tunneling cannot obtain continuous soil profiles along the center line of the tunnel. This paper describes a real time site characterization and evaluation software using a new concept to "Sense As You Advance" developed for remotely controlled tunnel boring operations. The soil properties and ground water conditions ahead of the machine are obtained by a specially designed tunnel boring machine instrumented with an automatically advancing and retracting miniature piezocone penetrometer in front of it. The acquired data is processed in real time to evaluate the engineering properties of the soil that are also displayed on a computer screen in graphic form. Such real time data acquisition gives continuously updated information of the soil ahead which can be used to accurately control the tunnel boring operation using an intelligent closed-loop computer-controlled feedback process.

1 INTRODUCTION

The overcrowding and congestion of large cities due to population and industrial growth has necessitated an increasing demand in exploiting the vast resource of underground space. Underground space are being exploited for gas, electricity, telecommunication, water and sewage utility pipelines, roads and railways for transportation, shelter from natural and man-made disasters, and for storage and waste disposal. Trenchless technology has become an effective and economical alternative to open-cut construction. It is ideal for installations beneath buildings, highways, railways, rivers and oceans. Compared to open-cut construction, trenchless technology methods are less disruptive to the public and environment, less hazardous, causes minimal damage to property and existing utilities, allows work to be performed in a wide range of ground and hydrostatic conditions, highly productive and is often cost-effective. Pipe jacking and tunneling are two widely used trenchless technology methods. Both methods essentially consists of a Tunnel Boring Machine (TBM) usually cylindrical in shape within which the excavation takes place. In pipe jacking the boring head is advanced by the main cylinder inside the access shaft or by intermediate cylinders pushing the pipe behind it. In tunneling the TBM propels itself by jacking against the in-place liner segments using cylinders positioned around the barrel of the machine. After the shield (the front part of the machine) is pushed forward by the full stroke of the propulsion jacks, the jacks are retracted, and the next ring of liner segment is erected within the tail of the shield. Soil excavation is performed using a rotating cutter head with provisions to confine the spoil or slurry behind the cutter head and to control the ground water. Typical examples of cutter head machines with provisions of spoil control include the earth pressure balance, water pressure balance, slurry, and mud or slime shields. The pressure applied to the tunnel face counter-balances, in theory, the existing earth and hydrostatic pressures. It needs to be properly balanced, because if it is too large, it can produces an up heave at the surface or even failure. If the pressure applied to the tunnel face is too small, it may lead to excessive settlements at the surface. The magnitude of the face pressure also affects the performance of the TBM. Large pressures may require the use of excessive power that can slow down the rate of advance of the excavation with the potential of stacking the machine. Prior knowledge of detailed stratigraphy, soil type and its engineering properties are imperative for successful application of trenchless technology.

2 SUBSURFACE PROFILING

Soft ground tunneling using trenchless technologies poses a unique challenge to geotechnical engineers since a wide range of soils (silt, sand, clay and gravel) are encountered during the boring operations. Modern Tunnel Boring Machines (TBM) are capable of extending the range of soil types they can operate in by changing the cutter head in situ. However this requires assessment of the soil properties ahead of the TBM. A detailed site investigation and evaluation is essential well in advance before the actual design and tunnel boring operation. Existing methods for determining engineering soil properties relevant to soft ground tunneling consists of laboratory tests performed on samples obtained from vertical bore holes or in situ tests performed vertically at certain discontinuous intervals along the proposed path of the tunnel. Borings are typically spaced 80 to 100 meter intervals depending on soil variability. Depending on the tunnel diameter, borings are located 3 meters or greater from the center line of the tunnel (Figure 1) to avoid slurry losses through the boreholes and also to avoid tunneling through grouted boreholes or through well casings. Hence it is not possible to obtain continuous, detailed and precise information of soil properties along the tunnel path using conventional soil investigation methods. Unexpected soil conditions in between boreholes and in situ test locations could cause improper TBM operation. This could lead to surface heave, settlement or even stacking of the TBM.

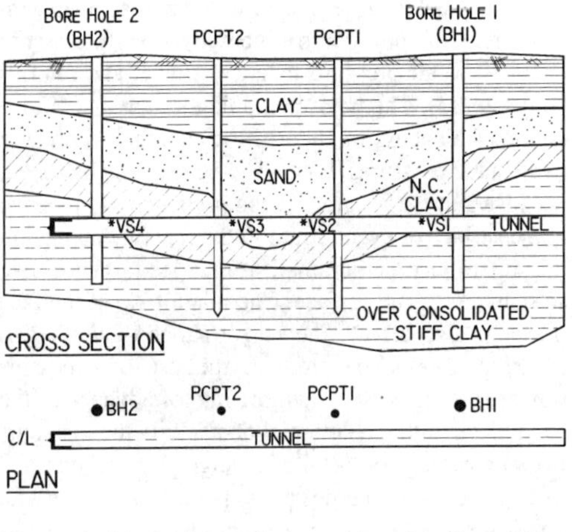

Figure 1. Limitations of conventional soil profiling methods for trenchless excavation

As illustrated in figure 1 conventional soil profiling methods are unable to detect the sand layer between sections VS2 and VS3. The vertical bore holes and in situ tests are also not capable of exactly delineating the location of the interface between different soil types at sections VS1, VS2, VS3 and VS4. It is very important to locate these soil interfaces because tunneling in mixed phase soil conditions can create serious problems for the TBM operation.

2.1 *Laboratory tests*

Borehole sampling and laboratory testing of disturbed and undisturbed samples is similar to that conducted for many geotechnical projects. It is recommended that boreholes be taken down to a depth of 2 m or twice the tunnel diameter which ever is greater, below the tunnel invert. Sampling is usually continuous from one diameter above and one diameter below the proposed tunnel boundary. The type of laboratory tests that are usually performed in cohesive soils are hydrometer analyses, atterberg limits, moisture content and unit weight, shear strength, consolidation and permeability. In cohesionless soils, gradation and particle shape characteristics, moisture content and unit weight, flow characteristics, elastic constants and shear strength may be determined. In rock samples unconfined compressive strength and point load tests are performed to estimate strength, and petrographic analyses conducted to determine mineralogy.

2.2 *In situ tests*

Laboratory testing of borehole samples are slow and expensive. The soil samples are disturbed even in the so called undisturbed samples. In contrast in situ tests are performed under existing in situ state and environmental conditions. They are much faster and economical compared to laboratory testing. In situ tests may be performed inside or in between boreholes. Some of the in situ tests used for soil profiling before tunneling are the pressuremeter, vane shear, dilatometer, standard penetration test (SPT), cone penetration test (CPT) and piezocone penetration tests (PCPT). PCPT essentially consists of pushing into the soil at a rate of 2 cm/s an electronic device known as the piezocone penetrometer. The data recorded during a PCPT are the cone resistance, sleeve friction and pore water pressures. Coarse grained soils are characterized by high cone resistance and low friction ratio which is the ratio of the sleeve friction to the cone resistance expressed as a percentage. Fine grained soils are characterized by low cone resistance and high friction ratio. The measured pore water pressure is the sum of the hydrostatic pore pressure and the excess pore pressure due to cone intrusion. In coarse grained soils the excess pore pressure dissipates almost instantaneously due to the large void size whereas in fine grained soils the excess pore pressure dissipation is slow. This feature is utilized in estimating the consolidation characteristics and hydraulic conductivity of fine grained soils. CPT and PCPT give continuous soil profiles and have become very popular for soil characterization and estimation of engineering soil properties. However as mentioned earlier current methods of geotechnical investigations do not provide a continuous profile of soil characteristics along the proposed tunnel path as they are performed vertically at intermittent locations.

3 TBM OPERATION VARIABLES

The jacking forces or thrust, cutter head rotational speed and rate of spoil removal are the main variables that have to be appropriately controlled for effective TBM operation. These variables have to be continuously varied depending on the type of soil encountered during the boring process. Any ground settlement or heave may be practically eliminated by controlling the earth or slurry pressure in front of the TBM during advance.

In the earth pressure balance method the earth pressure at the tunnel face is controlled by keeping the chamber between the cutter head and bulk head filled with spoil and adjusting the rotation of the screw conveyor, which controls the volume of spoil being discharged out of the screw conveyor. In slurry shield method the soil excavated by the cutter head is mixed with the charging water or bentonite and converted to slurry which is then pumped out. The slurry pressure at the TBM face is controlled by adjusting the inlet slurry charging pump rate and the outlet slurry discharge pump rate. The pressure in the cylinders of the thrust jacks are controlled to adjust the rate of advance. This also affects the TBM face pressure. Subsurface investigations to identify and quantify soil properties and to determine ground water conditions are fundamental for controlling various TBM design and working variables for its safe and efficient operation.

4 SENSE AS YOU ADVANCE

A prototype real time site characterization and evaluation software using a new concept to “Sense As You Advance” developed for remotely controlled tunnel boring operations in soft soils is described here. A schematic diagram of the "Sense As You Advance" technology is shown in figure 2. The soil type and properties ahead of the machine (to a length greater than the TBM or installation pipes whichever is greater) is obtained by a specially designed TBM instrumented with an automatically advancing and retracting miniature piezocone penetrometer in front of it. Before the TBM is advanced into the ground from the jacking pit the soil properties ahead of the TBM is obtained by performing a Horizontal Piezocone Penetration Test (HPCPT). The acquired data is processed in real time to evaluate the engineering properties of the soil that are also displayed on a computer screen in graphic form. The penetrometer is retracted back into the machine and tunnel boring operations starts. During tunnel boring the rotational speed (rpm) of the cutter head, earth pressure and pore pressures at the face of the TBM and the rate of slurry charge and discharge are continuously monitored. The soil data and TBM operation variables are input into a computational analysis software. This stage could involve fuzzy logic control and even rigorous finite element analysis. The results from the computational analysis stage is used by a feed back control process to adjust the jacking pressures behind the TBM, the rotational speed of the cutter head, and the rate of slurry charge and discharge or spoil removal. Thus the TBM operation variables are continuously monitored and varied depending on the soil properties and earth pressures in front of the TBM. Ground movements are virtually eliminated by maintaining a consistent counter balance pressure on the tunnel face always between the active and passive earth pressures. This counterbalance pressure may be applied by either earth pressure or slurry pressure depending on the type of boring technique used and these pressures are controlled by the rate of spoil or slurry removal and the rate of advance of the TBM. The thrust of the jack and the speed at which they are extended is carefully synchronized with the rate of excavation. After the machine has fully penetrated the ground, excavation is halted, the jacks are retracted and the first section of pipe is lowered into the jacking pit. The pipe is then connected to the back of the TBM to form a water tight fit. During this time interval the piezocone penetrometer is advanced to sense the soil type and

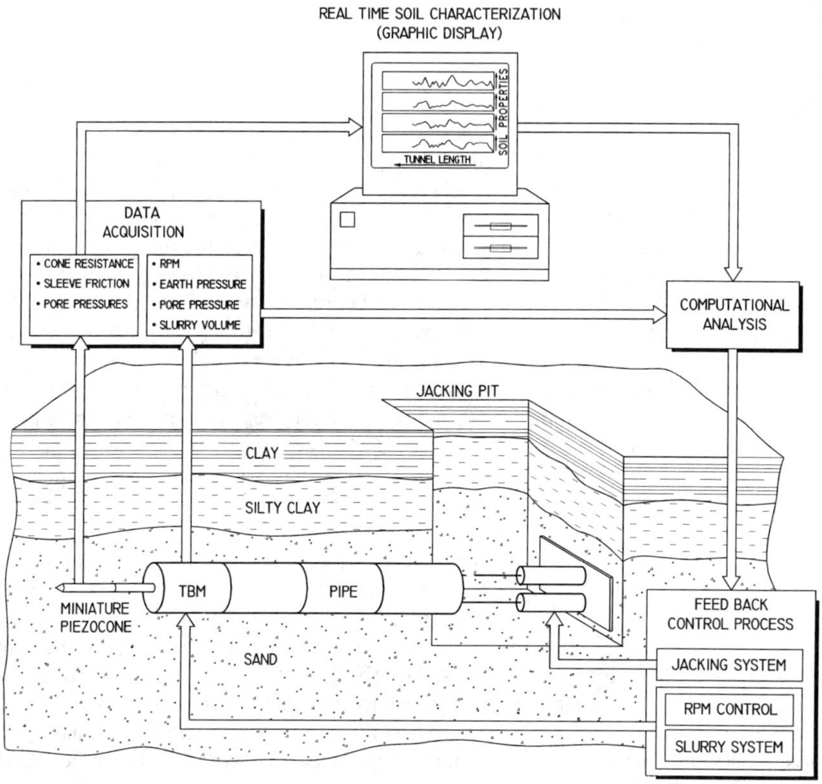

Figure 2. Schematics of "Sense As You Advance" technology for trenchless excavation

properties ahead of the TBM. A short term dissipation test may also be performed to estimate the flow characteristics. In coarse grained soils the hydrostatic pressure can be determined. The piezocone is then retracted and the pipe and TBM are jacked together as excavation at the face of the machine continuous. This process is continued until all pipes are installed and the tunnel is complete.

The innovative feature of this new concept is the incorporation of a novel "Seeing Ahead" technology during the trenchless excavation process. Such real time data acquisition gives continuously updated information of the soil and ground water conditions in advance, which can then be used to accurately control the tunnel boring operation using an intelligent closed-loop computer-controlled feedback process.

4.1 *Horizontal piezocone penetration test (HPCPT)*

HPCPTs give more detailed soil profiles along the center line of the proposed tunnel path. They are able to capture variations in soil profile between vertical in situ tests and bore hole locations. They can also detect the exact location of interfaces between different strata along the excavation alignment. HPCPTs do not entirely eliminate conventional methods for obtaining geotechnical properties but complements it to obtain a more accurate control of TBM operation as it bores through various soil types. HPCPTs performed during tunneling can however reduce the number of vertical bore holes and vertical in situ tests performed by increasing the spacing between them and thereby result in cost savings.

4.2 *Real time data acquisition, processing and display*

The authors have developed a software in Turbo-Pascal to acquire cone resistance (q_h), sleeve friction (f_{sh}), and pore pressures measured at the tip (u_{1h}) and at the shoulder (u_{2h}) during HPCPT's performed ahead of the tunnel boring operation. The data is analyzed to

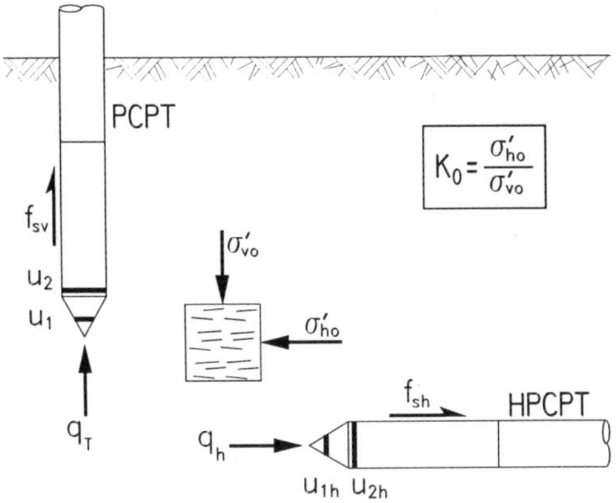

Figure 3. Stress and fabric anisotropy

estimate various soil and in situ state properties such as the undrained shear strength, deformation modulus, lateral stress coefficient, and overconsolidation ratio. In coarse grained soils the friction angle and relative density may also be obtained. The main difference between conventional PCPT and HPCPT is that of anisotropy of in situ stress and soil fabric (Figure 3).

In conventional PCPT the effective overburden vertical stress (σ'_{vo}) increases with penetration depth and is parallel to the direction of cone intrusion (ie: it acts parallel to the friction sleeve). The effective in situ lateral stress (σ'_{ho}) acts normal to the friction sleeve. In HPCPT however the in situ lateral stress acts parallel to the direction of cone intrusion. The overburden stress acts normal to the friction sleeve and does not vary appreciably during the test. It changes as the density of the overburden soil varies and with changes in TBM elevation. In fact even from the stress and fabric anisotropy point of view one might presume it might be more appropriate to base tunnel design upon HPCPT data because the actual tunneling process takes place along the horizontal direction. The expressions adopted in this paper for estimating soil properties from HPCPT data are essentially similar to those used to interpret vertical PCPTs. However new calibration studies will be required to obtain empirical factors pertaining to HPCPT in light of the stress and fabric anisotropy described earlier. Calibration may be performed in the field by conducting in situ HPCPTs from inside vertical pits. Laboratory calibration chamber studies under controlled stress and boundary conditions in high quality homogeneous soil samples subjected to known stress histories may also be performed. The software developed by the authors display (Figure 4) the following soil properties during tunneling in cohesive soil deposits:

Undrained Shear Strength:
The undrained shear strength is estimated using the following empirical equation:

$$s_u = \frac{q_h - \sigma_{vo}}{N_{kh}} \tag{1}$$

where N_{kh} = empirical cone factor pertaining to HPCPT, and σ_{vo} = vertical stress. Several factors such as plasticity, stress history, stiffness, sensitivity and fabric are known to influence N_{kh}. Calibration chamber studies on normally consolidated soils (Kurup, 1993, 1994) have indicated that N_{kh} is greater than N_{kT} (pertaining to PCPT) because of the influence of lateral stress coefficient on the empirical cone factors.

Deformation Modulus:
The soil one dimensional deformation modulus M is estimated from the cone resistance using a simple relationship:

$$M = \alpha \, q_h \tag{2}$$

where α is an empirical factor that depends on plasticity and grain size.

Lateral Stress Coefficient:
Based on an empirical relationship suggested by Sully and Campanella (1990), the following equation is used to estimate the in situ lateral stress coefficient (K_o):

$$K_o = a + b\left[\frac{u_{1h} - u_{2h}}{\sigma'_{vo}}\right] \tag{3}$$

where a and b are constants. New calibration is required for values of 'a' and 'b' pertaining to HPCPT.

Overconsolidation Ratio:
The following expression proposed by Kurup (1993) and Tumay et al. (1995) is used to estimate OCR:

$$OCR = 2\left[\frac{3}{(1.95M+1)}\left(\frac{q_h - u_{2h}}{\sigma'_{vo}(1+2a)+2b(u_{1h}-u_{2h})}\right)\right]^{1.33} \tag{4}$$

where M is the slope of the critical state line.

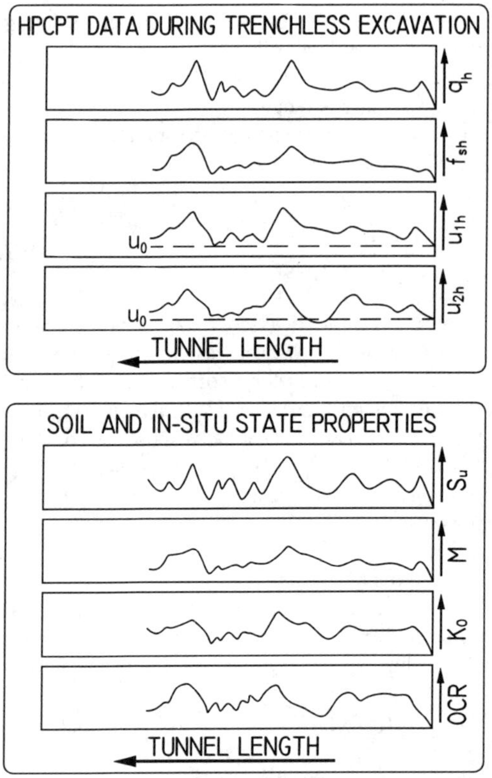

Figure 4. Graphic display of soil and in situ properties

5 CONCLUSIONS

Prior knowledge of soil properties and ground water conditions along the excavation alignment are imperative for successful application of trenchless technology. A prototype real time site characterization and evaluation software using a new concept to "Sense As You Advance" is developed for remotely controlled tunnel boring operations in soft soils. The soil type and properties ahead of the machine is obtained by a specially designed TBM instrumented with an automatically advancing and retracting miniature piezocone penetrometer in front of it. The acquired data is processed in real time to evaluate the engineering properties of the soil that are also displayed on a computer screen in graphic form. Such real time data acquisition gives continuously updated information of the soil ahead which can be used to accurately control the tunnel boring operation to maximize system efficiency and productivity.

6 ACKNOWLEDGMENTS

The financial support from the National Science Foundation under Grant CMS 9531782 is gratefully acknowledged. The support of Louisiana Transportation Research Center is also appreciated. The authors would like to thank Yigal Bynoe for his help in preparing the figures.

7 REFERENCES

Kurup, P.U. 1993. *Calibration Chamber Studies of Miniature Piezocone Penetration Tests in Cohesive Soil Specimen, Ph.D. Dissertation.* Louisiana State University, Baton Rouge, LA.

Kurup, P.U., Voyiadjis, G.Z. and Tumay, M.T. 1994. Calibration Chamber Studies of Piezocone Tests in Cohesive Soils. *ASCE, Journal of Geotechnical Engineering Division*, 120(1): 81-107.

Sully, J. P. and Campanella, R. G. 1991. Effect of Lateral Stress on CPT Penetration Pore Pressures. *ASCE, Journal of Geotechnical Engineering*, 117(7): 1082-1088.

Tumay, M.T., Kurup, P.U., and Voyiadjis, G.Z. 1995. Profiling Lateral Stress Coefficient and Overconsolidation Ratio from Piezocone Penetration Tests. *Proceedings, International Symposium on Cone Penetration Testing, CPT'95, Linkoping, Sweden, October 1995*: 337-342.

Computer Methods and Advances in Geomechanics, Yuan (ed.) © *1997 Balkema, Rotterdam, ISBN 90 5410 904 1*

The research on the construction mechanics for building double-tube parallel tunnels

Zeng Xiao-qing
Department of Geotechnical Engineering of Tongji University, Shanghai, China

ABSTRACT: This paper is designed for applying the construction mechanical analytical method to the two parallel tunnels construction case in which both shields are driving in the same direction. The author applies the three-dimension time & space dynamics simulation to the construction process of double-tube tunnels with a computer. This paper put forward construction mechanical analytic method which embodies properly the effects of time and space for the tunnel construction.

1 INTRODUCTION

Since communication nets in urban district are quite complicated, and the restrictions of geologic conditions and construction technique, different kinds urban subways appear not always in the form of a single tunnel. For example, the four-tube express subway tunnels in high speed public communication system in Singapore. Most of the subways horizontal double-tube parallel tunnels have been adopted in Shanghai, Guangzhou and other cites in China. For the layout of subway stations, two tunnels going along the street are close to each other. It is often that a single shield excavates one tunnel and then another parallel one, for the sake of reducing the effects produced from between the two tunnels. But sometimes in order to speed up the construction schedule, two shields should be operated simultaneously in the same or opposite direction. In this case if no measures were adopted, considerable effects would be produced in practice. Since excavation orders , shield driving methods, machines and tools for construction are various, and earth digout volumes are different. Characters of mutual actions are different, too. Therefore, seeking the laws of interactions between tunnels has obviously much practical significance .

The process of tunneling is a problem of three-dimensions which varies constantly with time and space. As the excavation goes on, the original balance of physics and mechanics in the underground rock and soil in the urban district will vary. Hence, it is certain that changes and variations in mechanic state and deformations of the surrounding rock and soil medium. It will take place as well between the rock and soil medium and the tunnel, including other building. In order to research the mutual effects being produced in the process of two shields driving, an analytic method of construction mechanics is introduced into this paper. In the adopted numerical simulation of construction mechanics, the loading system and the geometry-physical parameters are quite related with time. Taking the object in this research, the author uses the theoretical research method and tactics that conform with the construction reality and in this way to analyze the mutual effects in the process of the shield tunnel construction.

2 THEORETICAL MODEL

This author researches the mutual effects between the two shields both driving in the same direction in the horizontal double-tube tunnels, taking the shield construction at Shanghai Subway as the practical background.

The load choice : the machine load and the excavation load. The former includes the shield thrust and the friction force of the shield produced around the surrounding soil when driving forward. The latter is the excavating release force.

In order to reflect the problem of the three-dimension time & space in tunnel construction, a semi-analytic numerical method is adopted in this paper. That is to reduce the problem of the mutual actions between the three dimension medium and the structural system into the one-dimension finite numerical problem.

Regarding to the geometry model of calculation elements , the semi-analytical infinite ring element is adopted. A plane which is perpendicular to the main axis of the tunnel separates the structure and the finite media of soil into a certain amount of infinite layer element. There are three types of elements: single-tube element(I), double-tube element(II), and non-tube element(III), as figure 1 shows.

As the tunnel construction goes on, double tube elements and single tube elements are increasing, and non-tube elements decreasing. As the geometrical shape changes with time goes by, the general rigidity matrix also changes in various stages of time.

When composing the semi-analytical displacement function, a column coordinate system(as shown in Figure 1) is adopted, analyzing along the circular θ and in the radial r direction while discreting along the axial X. The formula of semi-analytical function is as follows:

$$u(x,r,\theta,t)=\sum_{m=1}^{P}\sum_{n=1}^{Q} H_{mn}(r,\theta) \sum_{k=1}^{S} N_k(x)\, u_{mn}^{(k)}(t)$$

Where $N_k(x)$ stands for polynomial interpolation function, $u_{mn}^{(k)}(t)$ stands for generalized valuable , $H_{mn}(r,\theta)$ stands for analytical function in the direction r and direction θ . The region of one of the analytical directions is infinite. Further more, the integration region of variation functional infinite. It makes up the semi-analytical infinite element. In this way, the limitation of the finite element method in solving infinite problems has been somewhat gotten rid of . Though the provided analytical function would not be able to reflect the exact solution in the direction of the problem perfectly. The approaching to the exact solution of the analytical function family could be achieved properly owing to combining with the generalized variable of the discrete directions and giving satisfaction to the boundary condition as well as to the variation equation.

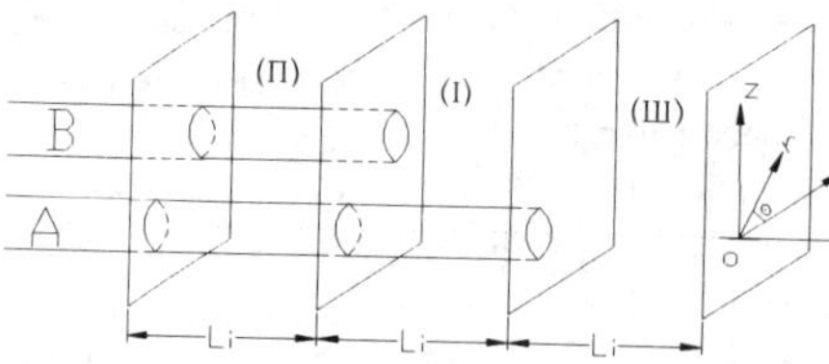

Fig.1 three kinds of calculation element

This paper takes the earth surrounding as the medium of the same quality in different directions with linear and viscoelastical properties. The constitutive mode is considered as a Kelvin-Voigt mode. The lining of tunnel is taken as a linear elastical body.

3. THE NUMERICAL SIMULATION OF CONSTRUCTION MECHANICS

3.1 A brief on calculation method of construction mechanics

A construction of building double tube parallel tunnels with two shields driving in the same direction is taken as a sample for studying the mutual effects of the two tunnels produced from the shield construction. In order to explain the analytical method of construction mechanics in a concise way, we take a concrete and simplified case here. We suppose the distance between first and second shields Diz is 20 m. Each time pace takes one day. And for each day one shield drives 10 m. The region is taken as 100 m. The region is divided into 10 elements with the plane which is perpendicular to the axis. Thus, the whole time to build two tunnels in this entire region with double front and back shields driving in the same direction is 12 days. Then, the calculating progress of the construction process within the region can be obtained according to Fig.2 and Table 1 . In Table 1, Pi(t) stands for the mechanic state of time effect caused by excavation at the day i.

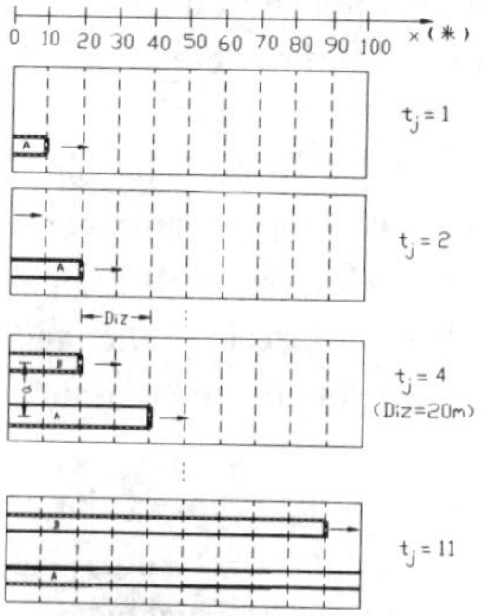

Fig.2 Construction progress of simulation calculation

Table 1

Time(t_j) (day)	Variations of elements	Mechanics state		
		Moment	Time effect	Mechanical state at that time
t_0=0	10 elements of non-tube			
t_1=1	9 elements of non-tube 1 elements of single-tube	$P_1(t_1)$		
t_2=2	8 elements of non-tube 2 elements of single-tube	$P_2(t_2)$	$P_1(t_2)$	$P_2(t_2)+P_1(t_2)$
t_3=3	7 elements of non-tube 2 elements of single-tube 1 elements of double-tube	$P_3(t_3)$	$P_2(t_3)+P_1(t_3)$	$P_3(t_3)+P_2(t_3)+P_1(t_3)$
$3 \le t_j \le 10$	10-j elements of non-tube 2 elements of single-tube j-2 elements of double-tube	$P_j(tj)$	$\sum_{i=1}^{j-1} P_{i(t_j)}$	$\sum_{i=1}^{j} P_{i(t_j)}$
t_{11}=11	1 elements of single-tube 9 elements of double-tube	$P_{11}(t_{11})$	$\sum_{i=1}^{10} P_{i(t_{11})}$	$\sum_{i=1}^{11} P_{i(t_{11})}$
t_{12}=12	10 elements of double-tube (finish construction)	$P_{12}(t_{12})$	$\sum_{i=1}^{11} P_{i(t_{12})}$	$\sum_{i=1}^{12} P_{i(t_{12})}$

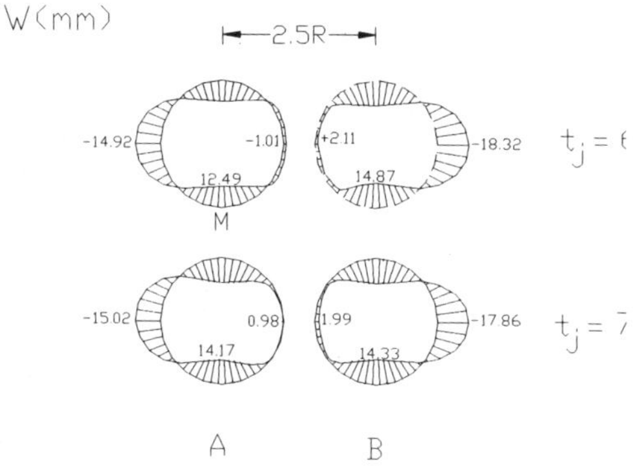

Fig.3 Chart of displacement when a=2.5R

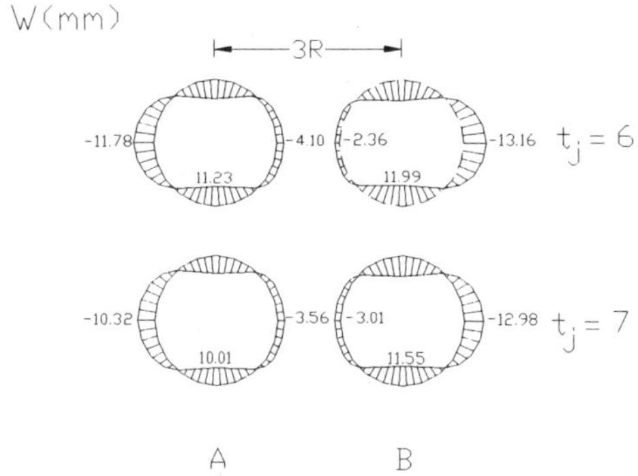

Fig.4 Chart of displacement when a=3R

3.2 Calculation parameters

Considering the limitations of the capacity of computer and the time of calculation etc., the

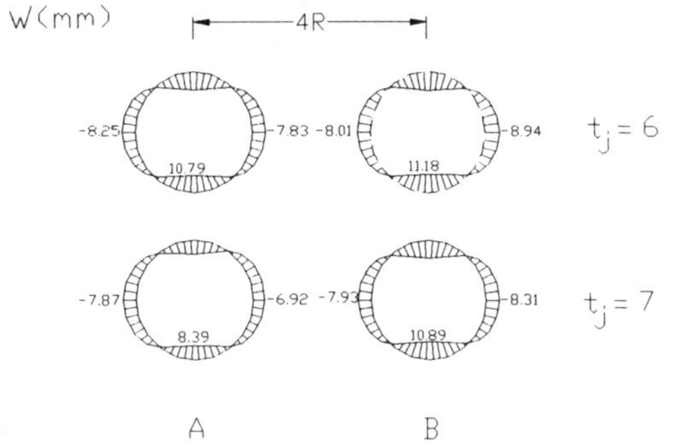

Fig.5 Chart of displacement when a=4R

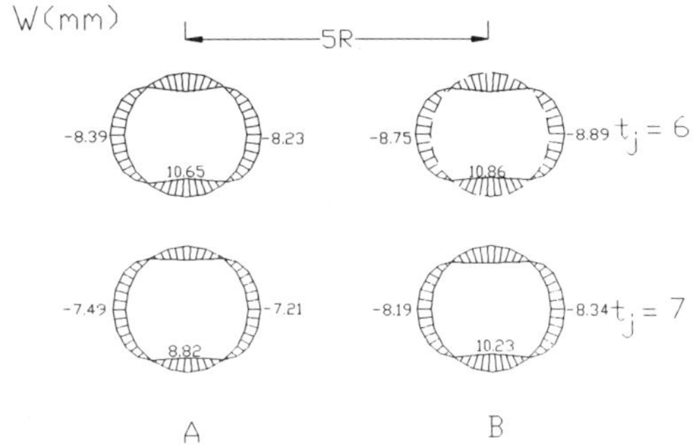

Fig.6 chart of displacement when a=5R

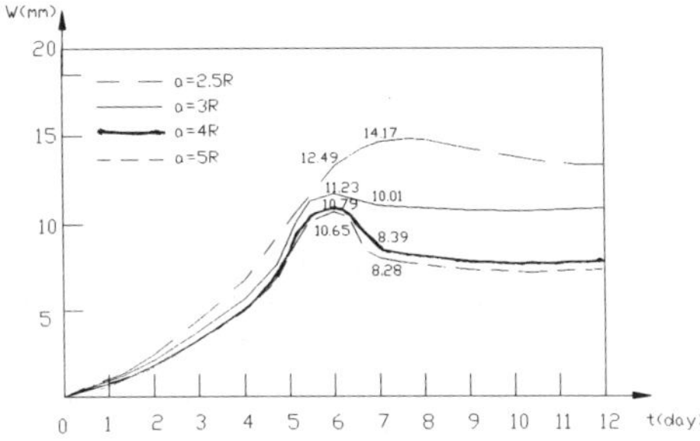

Fig.7 The displacement curve of M point

following hypotheses are adopted in this paper.

(1) Calculation region x $\in$ [0 , 100], displacement = 0 on the boundary line x= 0m and x= 100m.

(2) Take outside diameter of tunnel as 6.2m, inside diameter of tunnel as 5.5m, length of element as 1m, and center to center distance between two tunnels as 9 m,12 m, 15 m, 18 m separately; distance between tube center and ground surface H_o = 10 m.

Table 2

Distance between centers a	Max. lateral displacement (mm)	Max. longitudinal displacement (mm)
2.5R	18	15
3R	13	12
4R	9	11
5R	9	10

(3) Suppose adhesive of rock and soil media C=13.6 kPa, inner friction angle φ =10°, unit weight γ = 17.6 k N/ m^3, elastic modules E_R= 4.2 × 10^3 K p a, Poisson ratio μ_r=0.25, lateral pressure coefficient λ =0.75, Kelvin-Voigt model parameter E_H = 69 × 10^3 kPa, E_k = 6.0 × 10^3 kPa , η_k=1.2, elastic modules of lining structure Es=3.5 × 10^7kPa, Poisson ratio μ_s=0.17.

3.3 The Results of calculation

In this article the output of calculation program is the displacement values, which are on various points of both tunnels at the cross section where x=50m, in different time. Fig.3-Fig.6 show the displacements in different center distance. Fig.7 shows the displacement curve of M point at the bottom of tunnel A at different moments in the entire construction program(M point shown in Fig.3).

Table 2 which contrasts the various displacements can be obtain from Fig.3-Fig.6.

4. CONCLUSION

It is clear from the results of the above simulation calculation that the effects of mutual displacement will not striking when the center to center distance of the two tunnels exceeds or equals 4R. Since the parallel tunnel construction with both shields have a center to center distance exceeding 4R, the effects produced between the two tunnels, when construction goes on, will not to be significant enough.

Calculating with the construction mechanical method, the soil displacement values change with time. It inflect the effects of construction process of the state of soil mechanics and the time. Hence, it can better simulate the changes in the structure of construction process with shields and the mechanical state of the media. It can also be seen that under the shields driving on, the displacement of same point in the sail develops from small to noticeable and then again tend to stable.

REFERENCE

1, E. soliman, H. Duddeck and H. Ahrens, Two-and Three-dimensional Analyses of Closely Spaced Double-tube Tunnels, Tunneling and Underground space Technology, 1993, Vol. 8, No. 1, 13-18

2, Z.Eisenstein & L.Samarasekera, Stability of unsupported tunnels in clay, Can.Geotech.J.29, 609-613,(1992)

3, R. K. Rome &K. M. LEE, An evaluation of simplified techniques for estimating three-dimensional undrained ground movements due to tunneling in soft soils, Can. Geotech. J. , 20, 11-22(1983)

Computer Methods and Advances in Geomechanics, Yuan (ed.) © 1997 Balkema, Rotterdam, ISBN 90 5410 904 1

Role of temporary restraints during installation of sewer linings

Salek M. Seraj
Bangladesh University of Engineering and Technology, Dhaka, Bangladesh
Uday K. Roy
University of Tokyo, Japan

ABSTRACT: The paper summarises some of the results of a numerical parametric study aimed at understanding the structural response of various circular and non-circular (viz. egg-, inverted egg-, elliptical- and horseshoe-shaped) sewer linings. The effects of various restraint conditions, which simulate different temporary support systems that may be used by the contractors during installation of the lining and different loading configurations which may arise at different stages of grouting the annulus gap between the lining and the sewer, have been thoroughly investigated. It has been shown that, by introducing additional temporary restraints before grouting around circular or non-circular sewer linings, considerably higher grouting pressures, leading to a more reliable grouting operation, can usually be attained. A comparison between the various types of restraints on a specific lining has led to enhancement factors for the permissible grouting pressure or, alternatively, to reduction factors in terms of the lining thickness that could be used in designing lining systems. Use of appropriate temporary restraint during installation of sewer linings is expected to lead to a better lining-sewer-soil system.

1 INTRODUCTION

Although preventive maintenance and renovation in one form or another have taken place from the earliest days of sewer construction, it is only relatively recently that sewer rehabilitation has become a subject of increasing interest to the engineering community. Of particular importance is lining the existing sewers of different shapes with glass reinforced plastic, glass reinforced cement, etc., which, besides improving hydraulic characteristics, leads to the enhancement of the structural capacity of the sewer-soil system. Linings also prevents the sewage and waste water from going to the surrounding soil and thereby arrest contamination.

The shape of the lining follows that of the sewer after allowing for an annulus gap so that the sewer lining fits within the existing sewer with a roughly uniform gap between the lining and the sewer walls. Figure 1 shows details of the geometry of the egg-shaped (ES), inverted egg-shaped (IES), horseshoe-shaped (HSS) and semielliptical-shaped (SES) linings that will be studied, in addition to circular linings, under various boundary and installation conditions in this paper. The annulus gap is filled with a cementitious grout which, when set, creates a composite sewer-lining-soil system. Whereas comprehensive design curves pertaining to various criteria may be found in Arnaout, et al. (1988), Pavlovic, et al. (1997) and Seraj, et al. (1997a, 1997b, 1997c), the present paper focuses primarily on the beneficial effect of temporary restraints during installation of linings.

2 GROUTING TECHNOLOGY

2.1 *Method of grouting*

In sewer lining, staged or partial grouting and full grouting are usually adopted. Grout is usually injected through the bottom of the lining. In case of staged grouting, grouting is performed in two stages. The first stage involves grouting the annulus up to the springings, and this is followed by a second stage carried out after the grout of first stage has set. On the other hand, full grouting is performed in a single stage. This technique is more practical than staged grouting.

2.2 *Restraint conditions*

Since the performance of linings of different shapes is particularly sensitive to the type of support provided during grouting, the structural analysis of the sewer linings have been carried out for three different support systems that may be used during installation. These consist of hardwood wedges packed at different locations around the cross-section of the lining on the outside, together with internal struts positioned at the same locations. It is assumed that the packing between the sewer and the lining is closely spaced so that the structure can be studied by means of a two-dimensional finite element model. The three possible support systems considered in the present study are shown in Figure 2 with reference to circular lining.

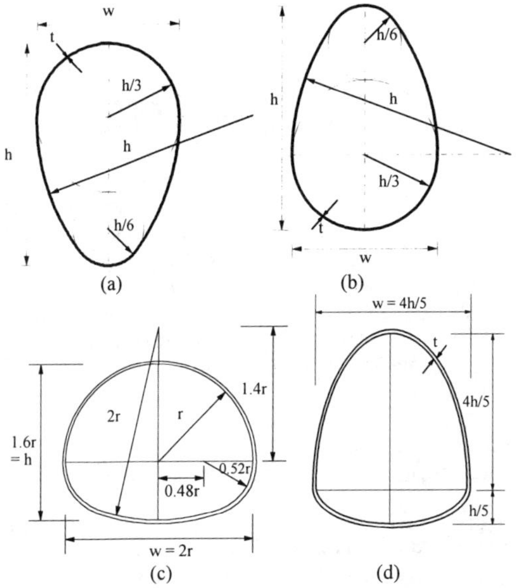

Fig. 1 (a) Egg-, (b) inverted egg-, (c) horseshoe- and (d) semielliptical-shaped sewer linings

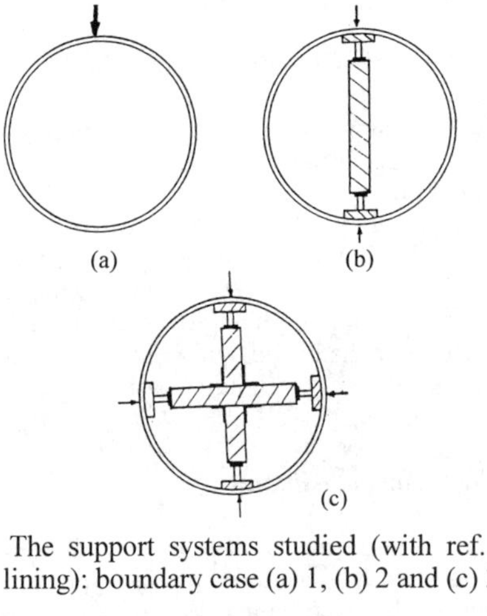

Fig. 2 The support systems studied (with ref. to circular lining): boundary case (a) 1, (b) 2 and (c) 3

2.3 *Load during installation of linings*

Three loading configurations, namely staged grouting pressure, flotation pressure and uniform pressure are adopted throughout the analysis unless otherwise specified. The loading conditions are shown in Figure 3 with reference to HSS lining.

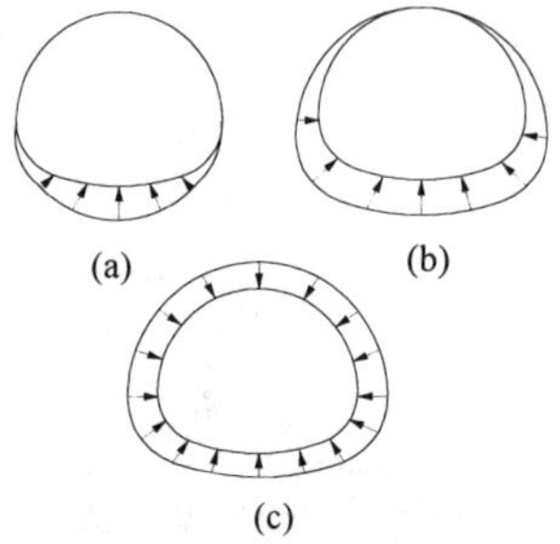

Fig. 3 The loading conditions studied (with reference to horseshoe-shaped lining): (a) staged grouting, (b) pressure up to crown only and (c) uniform pressure

2.4 *Basis of design*

All properly designed sewer linings must satisfy both stress-, deflection-limit and, if applicable, buckling criteria. Here, the stress-limit criteria is so defined that the maximum bending stress developed during grouting must not exceed the allowable bending stress of the lining material. For deflection-limit criteria, a maximum allowable deflection in the lining not exceeding 3 percent of the width of the lining, as advocated by the Water Research Centre (1983), has been adopted for all the linings.

3 ANALYSIS TECHNIQUE

3.1 *Parameters used*

Geometrical parameters include width, height and thickness of lining (w, h and t). Material parameters include allowable short-term bending stress (S_S), short-term modulus of elasticity (E_S) and Poisson's ratio (ν) of lining. Load parameters include unit weight of grout mix (G) and excess head of grout measured from crown of lining conforming to uniform pressure load (H).

3.2 *Equations involved*

The dimensionless equations corresponding to the bending stress S and the deflection δ at any point on the lining can be written for the three load cases as follows:

(a) Staged Grouting (Figure 3a)

$$S / Gw = A(w / t)^2 \quad (1)$$

$$\delta / w = (B_x^2 + B_y^2)^{1/2} K \quad (2)$$

(b) Flotation (Figure 3b)

$$S / Gw = C(w / t)^2 \quad (3)$$

$$\delta / w = (D_x^2 + D_y^2)^{1/2} K \quad (4)$$

(c) Uniform Pressure (excess head *H*) (Figure 3c)

$$S / Gw = E(H / w)(w / t)^2 \quad (5)$$

$$\delta / w = (F_x^2 + F_y^2)^{1/2} (H / w) K \quad (6)$$

where $K = (Gw / E_s)(w / t)^3$ (7)

In these equations, *S/Gw* can be regarded as a non-dimensional stress while δ/w is the deflection related to the size of the lining and *K* is a measure of lining flexibility. Here, A, C, E, B_x, B_y, D_x, D_y, F_x *and* F_y are all constants which depend on the boundary set-up adopted during the grouting of the annulus and loading configurations used in the analysis.

The total bending stress S_t and the total deflection δ_t at any point in a lining subjected to a head of grout which is greater than the lining height *h* (i.e. full flotation) can be divided into values of bending stress and deflection resulting from the two loading cases of pressure up to the crown (i.e. flotation) and uniform pressure. This implies that, by adding Equations 3 and 5, and Equations 4 and 6, the following dimensionless equations for the total bending stress and the total deflection can be found.

$$S_t / Gw = \left|(C + E(H / w))(w / t)^2\right| \quad (8)$$

$$\delta_t / w = (M_x^2 + M_y^2)^{1/2} K \quad (9)$$

where,

$$M_x = D_x + F_x (H / w) \quad (10a)$$

$$M_y = D_y + F_y (H / w) \quad (10b)$$

Since the maximum bending stress and the maximum defection in a lining must not exceed the respective values of S_s and $0.03\,w$, the values of S_t and δ_t in Equations 8 and 9 can be replaced by S_s and $0.03w$, respectively. As the point of injection of the grout is usually located at the invert of the lining, it is convenient to replace the value of *H* in Equations 8 and 9 by the equivalent expression $(p/G) - h$, where *p* is measured from the invert of the lining. As a result, Equations 8 and 9 can be rewritten to produce the following design equations.

$$R = \left|C + E(p / Gw - h / w)\right| \quad (11)$$

where,

$$R = (S_s / Gw)(t / w)^2 \quad (12)$$

and

$$0.03 / K = (N_x^2 + N_y^2)^{1/2} \quad (13)$$

where,

$$N_x = D_x + F_x (p / Gw - h / w) \quad (14a)$$

$$N_y = D_y + F_y (p / Gw - h / w) \quad (14b)$$

In cases where buckling is to be considered, dimensionless Equations 15 and 16 apply for flotation and uniform pressure cases, respectively (Pavlovic, et al. 1995). Here, *M* corresponds to the membrane stress at any point in the lining.

$$M / Gw = \alpha(w / t) \quad (15)$$

$$M / Gw = \beta(w / t)(H / w) \quad (16)$$

This leads to the following dimensionless equation for the total membrane stress (M_m) at any point in the lining under full grouting:

$$(M_m / Gw)(t / w) = (\alpha + \beta(H / w)) \quad (17)$$

Equating M_m with the critical buckling stress of a hinged arch of equivalent radius and unrestrained length (Timoshenko and Gere 1961), as reported by Pavlovic, et al. (1995) for circular lining, the stiffness of lining (S_F) is approximately given by the following dimensionless equation:

$$(S_F / Gw) = (1 / 4Q)(\alpha - \beta + \beta(p / Gw)) \quad (18)$$

where, $S_F = (1/12)(E_s / (1 - \nu^2))(t / w)^3$ and Q is equal to 3 for boundary condition 2 and 15 for boundary condition 3.

For any particular lining geometry and material properties, the above equations must be satisfied at the locations of maximum bending stress, deflection and axial stress in the lining. The maximum allowable grouting pressure *p* which can be applied on the lining during grouting is the minimum of the *p* values as determined by all the criteria.

3.3 *Numerical simulation*

A linear two-dimensional finite-element (FE) model is used in order to simulate the behaviour of sewer linings of different shapes under various probable loads during installation. The thickness of the lining is assumed to be constant all around the cross-section. Due to symmetry of the lining geometry, loading and boundary conditions about the vertical axis, only half of the cross-section is analysed. The elements used in the analysis are two-noded beam elements each having three degrees of freedom (horizontal and vertical displacement, and rotation) at each node.

The restraints due to the support system shown in Figure 2 are simulated numerically in the analysis by fixing the horizontal and vertical components of displacement at the corresponding nodal points. This involves a small approximation in that the deformation in the restraining struts is ignored, the strut being very stiff compared with the lining.

The various loading configurations shown in Figure 3 have been simulated by applying equivalent point loads at appropriate nodes. A typical two-dimensional finite element mesh, adopted in the analysis of inverted-egg shaped sewer linings, is given in Figure 4. Mesh for other linings are similar.

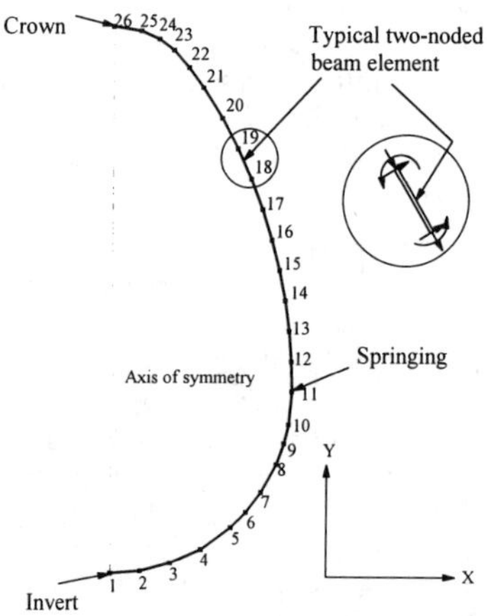

Fig. 4 IES lining: Two-dimensional finite element mesh adopted in the analysis

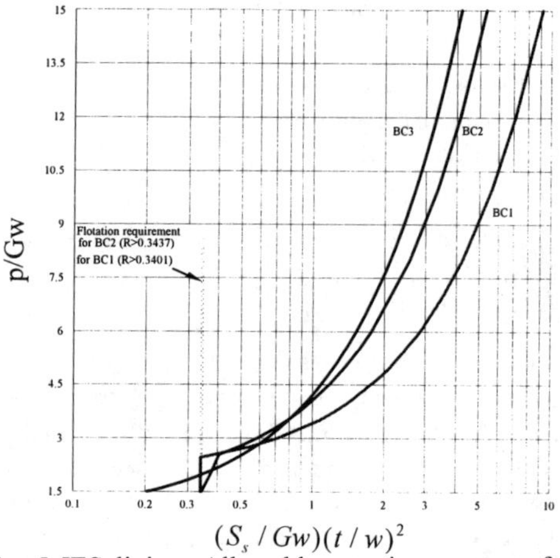

Fig. 5 IES lining: Allowable grouting pressure for different boundary conditions, based on stress-limit criteria

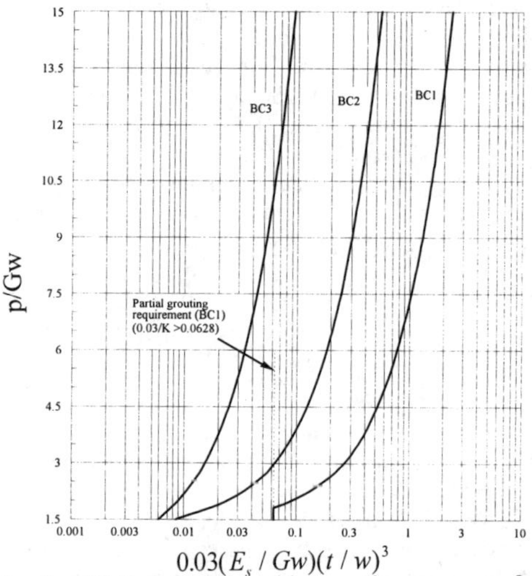

Fig. 6 IES lining: Allowable grouting pressure for various boundary cases, based on deflection-limit criteria

3.4 *Design curves*

For each load and boundary case, the parametric analysis is carried out by varying one parameter at a time. The results (bending stresses, deflections and axial stresses) are given in terms of dimensionless equations linking all the independent parameters together. The non-dimensional bending stress (S/Gw) and deflection (δ/w) are plotted against $(w/t)^2$ and lining flexibility K, respectively for staged grouting and flotation load, and against $(H/w)(w/t)^2$ and $(H/w)K$ for uniform pressure. Similarly, the non-dimensional membrane stress (M/Gw) is plotted against (w/t) and $(w/t)(H/w)$, respectively for flotation and uniform pressure cases. From these plots, constants for the maximum bending stress, maximum deflection and maximum membrane stress in the lining are computed for different boundary cases and different loading configurations. The value of these constants are employed in getting relationship of p with R, K and M. Once a boundary case is selected and the geometric and material parameters are chosen, a value of allowable grouting pressure based on the stress-limit, deflection-limit and, if applicable, buckling-criteria can be determined using appropriate curves. For circular linings, buckling criterion alone has been considered as membrane stress dominates in such linings. While comprehensive design curves pertaining to various criteria may be found in Arnaout, et al. (1988), Pavlovic, et al. (1997) and Seraj, et al. (1997a, 1997b, 1997c), Figures 5 and 6 give design curves for inverted-egg shaped sewer linings for different boundary conditions, based on stress- and deflection-limit criteria. The least of the two grouting pressures is the allowable grouting pressure for this particular lining. It is clear that as the amount of restraints increases, the allowable grouting pressure increases.

3.5 *Enhancement factors*

Both the maximum bending stress and the maximum deflection in a lining that arise from grouting pressure can be reduced by introducing additional restraints during installation. Similarly, additional restraints also result in an increase in resistance against buckling of the lining. This implies that an enhancement in the value of the grouting pressure can be achieved, thus ensuring adequate grouting of the annulus. This gives rise to the introduction of what can be termed an enhancement factor (EF). Here, for stress-limit and deflection-limit criteria, the enhancement factor is defined as the ratio of the allowable grouting pressure which could be applied on any particular lining using boundary case 2 or 3 to the one corresponding to boundary case 1, i.e.

$$EF_i = p_i / p_1 \tag{19}$$

Here i corresponds to boundary cases 2 or 3. Since, in the present study, only boundary cases 2 and 3 have been considered in the buckling analysis, the corresponding enhancement factor for this third criterion is given by p_3/p_2.

Values of EF are determined for each of the stress limit, deflection-limit and, if applicable, buckling criteria. It has been observed that the EFs for deflection-limit criteria are much higher than their stress-limit and buckling counterparts, and, thus, *will not govern* the design.

Enhancement factors has been calculated for boundary conditions 2 and 3. Such factors for horseshoe-shaped linings based on stress-limit and buckling criteria are plotted in Figure 7. Similar curves for other linings types may also be found.

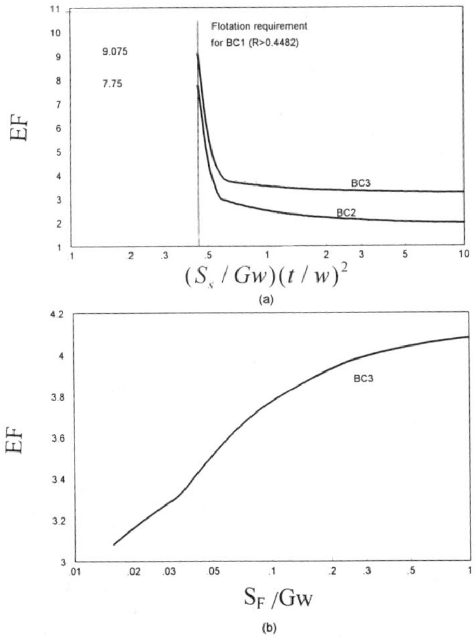

Fig. 7 HSS lining: Enhancement factors for allowable grouting pressure based on (a) stress-limit criteria and (b) buckling criteria

From the figure, it can be deduced that the highest possible enhancement in the allowable grouting pressure, for stress-limit criteria, that can be achieved for adopting boundary conditions 2 and 3, instead of boundary condition 1, are 7.75 and 9.075, respectively. As the value of R increases gradually from 0.4482 to 0.5820, the value of enhancement factor sharply decreases from 7.75 to 2.99 for boundary case 2 and from 9.075 to 3.770 for boundary case 3. Beyond this range of R, the EF gradually decreases, attaining virtually constant values for both boundary conditions. The enhancement factors corresponding to the buckling criterion are also above 3, for using boundary condition 3 instead of 2 during installation of HSS linings. Similar enhancements have also been found for other linings.

3.6 *Reduction factors*

Once a value of allowable grouting pressure is determined for any particular lining using a certain restraint set-up, a considerable reduction in the allowable thickness of the lining can usually be achieved if additional restraints are used instead. This gives rise to the introduction of another factor, called the reduction factor (RF), which is defined below.

For stress-limit criteria, the reduction factor is defined as the ratio of the lining thickness resulting from the use of boundary case 2 or 3 to the one corresponding to boundary case 1. The equation used to calculate the values of RF is as follows:

$$RF_i = t_i / t_1 \tag{20a}$$

where

$$t_i = [C_i + (p / Gw - 0.8)E_i]^{1/2} \, [Gw^3 / S_s]^{1/2} \tag{20b}$$

and

$$t_1 = [C_1 + (p / Gw - 0.8)E_1]^{1/2} \, [Gw^3 / S_s]^{1/2} \tag{20c}$$

with i corresponding to boundary cases 2 or 3, and other variables being defined earlier.

The reduction factors corresponding to buckling criteria are given by the ratio t_3/t_2 since, as for the enhancement factors considered earlier, only boundary cases 2 and 3 are considered. Here, the RF is found as follows:

$$RF_3 = t_3/t_2 \tag{21}$$

The reduction factors for semielliptical-shaped linings based on stress-limit and buckling criteria are plotted in Figure 8. Once again, similar curves for other linings types were also studied. It can be seen from Figure 8(a) that considerably larger reduction factors are achieved, under stress-limit criteria, by using boundary case 3 when compared with boundary case 2. Again, the figure shows that, for boundary case 2, a minimum value of p_1/Gw equal to 6.7 is needed in order to achieve reduction factors less than one. Thus no beneficial effect can result from the use of boundary case 2 if the value of p_1 is less than 6.7 *Gw*.

The reduction factors based on the buckling criterion, as shown in Figure 8(b), reveals that, whereas under boundary condition 2, either stress or buckling criteria may govern the determination of reduction factors, under boundary condition 3, stress-limit criteria will invariably turn out to be the most critical consideration.

Again, stress and buckling limitations proved to be more critical for the determination of RFs, and, hence reduction factors under deflection-limit criteria are not reported here. Similar conclusions are valid for other linings, as well.

It has been found that restraints in circular linings do not reduce appreciably the membrane stresses. A recent study (Pavlovic, et al. 1997) has reported some test results showing that the restraint systems provided only a minor contribution to the reduction of stresses/strains in the lining since, as a result of these restraints, the values of membrane strains and displacements were usually only slightly reduced. Nevertheless, additional restraints lead to a stiffer structure with higher critical buckling pressures.

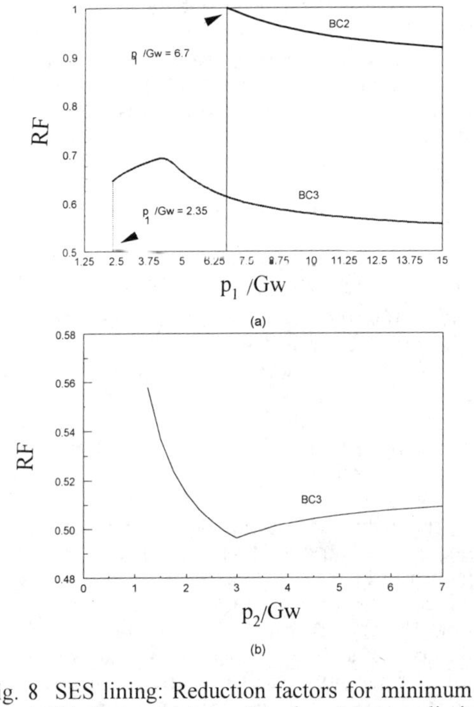

Fig. 8 SES lining: Reduction factors for minimum permissible lining thickness based on (a) stress-limit criteria, and (b) buckling criteria

4 CONCLUSIONS

It has been shown that, by introducing additional temporary restraints before grouting around sewer linings, usually considerably higher grouting pressures, leading to a more reliable grouting operation, can be attained. In the study, it has been assumed that all restraints are fully effective, so that the restrained points of the lining are prevented from moving in any direction. Such ideal conditions will very nearly be realised if internal supports coupled with external packing are effectively provided.

In the case of the approximate buckling analysis, introduction of additional restraints reduced the effective length of the arch between the restraints, thus leading to a stiffer structure with higher critical buckling pressure. The presently considered buckling criterion assumes an uniform pressure intensity on the arch; such a criterion may easily be achieved if linings of adequate thickness and acceptable physical properties are employed in design.

In general, boundary case 3 having both horizontal and vertical restraints has been found to lead to most satisfactory grouting operation leading to most efficient lining-sewer-soil system

REFERENCES

Pavlovic, M. N., Arnaout, S., & Seraj, S. M. 1997. Some Aspects of Composite Circular Sewer Linings Under Installation Conditions. Submitted for Publication.

Arnaout, S., Pavlovic, M. N. & Dougill, J. W. 1988. Structural Behaviour of Closely Packed Egg-Shaped Sewer Linings During Installation and under Various Restraint Conditions. *Proc. ICE (Part 2)*. 85: 49-65.

Seraj, S. M., Roy, U. K. & Pavlovic, M. N. 1997a. Structural Behaviour of Closely Packed Inverted Egg-Shaped Sewer Linings During Installation and under Various Restraint Conditions. *Thin-Walled Structures*. In Press.

Seraj, S. M., Roy, U. K. & Pavlovic, M. N. 1997b. Structural Design of Closely Packed Horseshoe-Shaped Sewer Linings During Installation. Submitted for Publication.

Seraj, S. M., Roy, U. K. & Pavlovic, M. N. 1997c. Semielliptical-Shaped Sewer Linings Under Installation Conditions. In Preparation.

Timoshenko, S. P. & Gere, J. M. 1961. *Theory of Elastic Stability.* NY: McGraw-Hill.

Water Research Centre. 1983. *Sewerage Rehabilitation Manual.* Swindon.

Computer Methods and Advances in Geomechanics, Yuan (ed.) © *1997 Balkema, Rotterdam, ISBN 90 5410 904 1*

Analysis of three-dimensional rheologic behavior for soft rock tunnel

Xiao Ming, Yu Yutai & Li Jianping
Department of Hydroelectric Engineering, Wuhan University of Hydraulic and Electric Engineering, China

ABSTRACT : In this paper, the rheologic behavior of soft rock tunnel is analyzed by a three-dimensional visco-elastoplastic mechanical model. The paper presents an iterative method of variable visco-plastic stiffness and gives an accelerating iterative formula. The method both has faster computation rate and can keep convergence stability. The analysis results of an engineering show that the computation method can fully describe the rheologic behavior of soft rock tunnel.

1 INTRODUCTION

As the development of high-speed road and underground railroad in cities, people have paid great attention to the support pattern, structural style and construction procedure for soft rock tunnel. Because of the obvious rheologic behavior in soft rock, all factors will be changed during construction as the time. To keep the stability of surrounding rock, it is extremely important to choose reasonable supporting pattern and supporting time for the construction of soft rock tunnel. In order to study the stability behavior and supporting pattern, it is necessary to analyze the rheologic behavior of soft rock tunnel with three-dimensional visco-elastoplastic FEM.

Nowadays the general methods of rheologic analysis are initial stress or initial strain method. Those methods not only are of slower convergence rate but also can not keep the plastic stress flow along the yield surface. This paper presents a method of variable visco-plastic stiffness according to three-dimensional rheologic model. This method combines the merits of initial stress and variable stiffness method. It not only accelerates the calculation rate but also can keep the plastic stress flow along yield surface during iterative process. By analyzing for an engineering of high speed road tunnel, it shows that computation method can fully describe the rheologic behavior of soft rock tunnel.

2 VISCO-ELASTOPLASTIC MODEL

After the excavation of soft rock tunnel, in generally, the total strain of rock masses can be divided into three parts which are shown in figure 1. It is:

$$\{\varepsilon\}=\{\varepsilon_e\}+\{\varepsilon_v\}+\{\varepsilon_p\} \qquad (1)$$

where $\{\varepsilon_e\}$, $\{\varepsilon_v\}$ and $\{\Delta\varepsilon_p\}$ stands respectively for elastic strain, visco-elastic and visco-plastic strain. In above all strains can be represented by constitutive equation:

$$\{\Delta\varepsilon_e\} = [H_e]^{-1}\{\sigma\} \qquad (2)$$

$$\{\varepsilon_v\}_{t+\Delta t} = e^{-b\Delta t}\{\varepsilon_v\}_t + \frac{a}{b}[A]\{\sigma\}(1-e^{-b\Delta t}) \qquad (3)$$

$$\{\varepsilon_p\}_{t+\Delta t} = \{\varepsilon_p\}_t + \frac{\Delta t}{\eta_3}<\frac{F}{F_0}>\{\frac{\partial F}{\partial \sigma}\} \qquad (4)$$

in which $a=\eta_2$ and $b=E_2/\eta_2$, $[H_e]$ is a symmetric elasticity matrix($[H_e]^{-1}$ is a compliance matrix), η_2 、 η_3 are the materials coefficient of visco-elasticity and visco-plasticity. F is a yield function. F_0 denotes here any convenient reference value of F to render the non-dimensional expressions. [A] is matrix of Poisson's ratio.

When time internal Δt is small enough, we may make $F_0=1$. To ensure no visco-plastic flow below the yield limit we define:

$$<\frac{F}{F_0}>=\begin{cases} 0 & \text{if} \quad F<0 \\ F/F_0 & \text{if} \quad F\geq 0 \end{cases} \qquad (5)$$

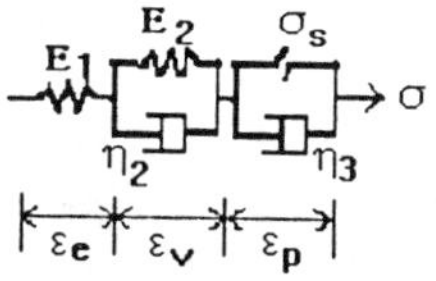

Fig. 1 rheologic model

To substitute formula (2), (3) and (4) into equation (1), we can get the stresses of surrounding rock after excavation of soft rock tunnel.

$$\{\sigma\}=[H_e]\{\varepsilon_e\}=[H_e](\{\varepsilon\}-\{\varepsilon_v\}-\{\varepsilon_p\} \quad (6)$$

or $\{d\sigma\}=[H_e](\{d\varepsilon\}-\{d\varepsilon_v\}-\{d\varepsilon_p\})$ (7)

3 ACCELERATING ITERATIVE METHOD

In order to accelerate computation rate and keep convergence stability, a iterative method of incremental load with variable visco-plastic stiffness is applies in this paper. The method changes the visco-plastic stiffness of element which have produced plastic flow according to the stress level of the forward time interval. It can keep visco-plastic stiffness constant during iterative process in every time interval. According to the formula of rheologic stress, we can obtain the fundamental balance equation of *ith* time interval after integrating for the equation (7):

$$[K_e]\{\Delta\delta\}_i=\{\Delta R\}_i+\{\Delta R_v\}_i+\{\Delta R_p\}_i \quad (8)$$

where $\{\Delta\delta\}_i=\{\delta\}_i-\{\delta\}_{i-1}$, $\{\Delta R\}_i$ is excavation load, $\{\Delta R_v\}_i$ and $\{\Delta R_p\}_i$ are incremental visco-elastic and visco-plastic load during *ith* time interval:

$$\{\Delta R_v\}_i=\int_v [B]^T[H_e]\{\Delta\varepsilon_v\}_i dv \quad (9)$$

$$\{\Delta R_p\}_i=\int_v [B]^T[H_e]\{\Delta\varepsilon_p\}_i dv \quad (10)$$

The incremental strain of visco-elastic can be calculated as following formula based on the fundamental relation of visco-elastic strain (3):

$$\{\Delta\varepsilon_v\}_i=e^{-b\Delta t}\{\Delta\varepsilon_v\}_{i-1}+\frac{a}{b}[A]\{\Delta\sigma\}_{i-1}(1-e^{-b\Delta t}) \quad (11)$$

The incremental strain of visco-plastic can be obtained by the value of yield function F_i based on the equation (4):

$$\{\Delta\varepsilon_p\}_i=\{\varepsilon_p\}_i-\{\varepsilon_p\}_{i-1}=\frac{\Delta t}{\eta_3}\cdot\left\{\frac{\partial F}{\partial\sigma}\right\}\cdot F_i \quad (12)$$

In order to make state of stress flow along yield surface, it is necessary to bring the stresses outside yield surface $\{\Delta\sigma_p\}_i$ back along the normal to the yield surface after every iteration is finished. When the calculating time interval Δt is enough small and the value of yield function F_i is not too large, the change of the yield function can be expressed as:

$$F_i-F_s=\left\{\frac{\partial F}{\partial\sigma}\right\}^{-T}\cdot\{\Delta\sigma_p\}_i \quad (13)$$

Where F_s is the value of yield function on yield surface, so $F_s=0$. $\{\Delta\sigma_p\}_i$ are the change of stress required to bring the state of stress back along the normal to the yield surface. To substitute equation (13) into equation (12) we have:

$$\{\Delta\varepsilon_p\}_i=\frac{\Delta t}{\eta_3}\left\{\frac{\partial F}{\partial\sigma}\right\}\left\{\frac{\partial F}{\partial\sigma}\right\}^T[H_e][B](1-s)\{\Delta\delta\}_i \quad (14)$$

Where $\{\Delta\delta\}_i$ is the incremental displacement of *ith* time interval, *s* is element coefficient of elasticity. When state of stress goes to elasticity from elasticity *s*=1 and to plasticity from plasticity *s*=0. If state of stress goes to plasticity from elasticity, *s* can be computed according to initial stress $\{\Delta\sigma_0\}$ and incremental stress $\{\Delta\sigma\}_i$:

$$F(\{\Delta\sigma_0\}+s\{\Delta\sigma\}_i)=0 \quad (15)$$

To substitute formula (14) into equation (10), we can obtain:

$$\{\Delta R_p\}_i=\frac{\Delta t}{\eta_3}\int_v C[B]^T[H_p][B](1-s)\{\Delta\delta\}_i dv=[K_\eta]\{\Delta\delta\}_i \quad (16)$$

Where $C=A_k+\left\{\frac{\partial F}{\partial\sigma}\right\}^T[H_p]\left\{\frac{\partial F}{\partial\sigma}\right\}$, $[K_\eta]$ denotes visco-plastic stiffness. $[H_p]$ is a plasticity matrix. A_k is a hardening parameter. So equation (8) may be written

$$[K_e]\{\Delta\delta\}_i=\{\Delta R\}_i+\{\Delta R_v\}_i+[K_\eta]\{\Delta\delta\}_i \quad (17)$$

Obviously above equation must use iterative method to compute $\{\Delta\delta\}_i$. In order to accelerate convergence of iterative process, we present the fundamental iterative formula according to J. N. Thomas' idea of accelerating iterative computation. Iterative formula of *nth* step are:

$$\{\Delta\delta_n\}_i=[K_e]^{-1}[K_\eta](\{\Delta\delta_{n-1}\}_i-\{\Delta h_{n-1}\}_i) \quad (18a)$$

$$\{\Delta h_n\}_i=[K_e]^{-1}[K_\eta]\alpha_{n-1}\{\Delta\delta_n\}_i \quad (18b)$$

$$\alpha_n=\{\Delta\delta_n\}_i^T(\{\Delta\delta_n\}_i+\{\Delta h_n\}_i)/(\{\Delta\delta_n\}_i^T\{\Delta\delta_n\}_i) \quad (18c)$$

$$\{\Delta\delta\}_i=\sum_{j=0}^{n}\{\Delta\delta_j\}_i+\sum_{j=0}^{n}\{\Delta h_j\}_i \quad (18d)$$

Where $\{\Delta\delta_n\}_i$ is incremental displacement and $\{\Delta h_n\}_i$ is modified displacement. α_n is an accelerated coefficient and $\{\Delta\delta\}_i$ is whole displacement. When n=0, let $\{\Delta h_0\}_i=0$, $\alpha_0=1$, $[K_\eta]=0$, as equation (17) we can obtain:

$$\{\Delta\delta_0\}_i=[K_e]^{-1}(\{\Delta R\}_i+\{\Delta R_v\}_i) \quad (19)$$

When relative value of *nth* step satisfy:

$$\left|(\Delta\delta_n-\Delta\delta_{n-1})_i/\Delta\delta_{ni}\right|\le\varepsilon \quad (20)$$

we consider that iterative computation conform to convergence criteria .

4 ENGINEERING EXAMPLE

This paper analyses a soft rock tunnel on the line of high-speed road. The tunnel located at the south of China. Its shape of section is one circular arc with three driveway and the size of section is 16 × 12m. The buried depth of tunnel are 8--30m. The physical parameters of rock mass are shown in Table 1. Computation network of three-dimensional are divided into 4040 space elements and 6452 nodes. The supporting patterns of tunnel include the first shotcrete and second reinforced lining, steal arched frame and anchor bar. The construction procedure of tunnel is divided into four step. First step is to excavate the pilot of tunnel(time=30 days). Second step is to excavate the arch of tunnel and to exert the second lining of pilot (time=20 days). The third step is to excavate the inverted arch and to exert the second lining at arch of tunnel(time=30 days). The fourth step is only to exert the second lining of the inverted arch(time=50 days). First lining all is exerted after excavating of every step.

4.1 The time-dependent behavior of fracture

When time=24 hours, the plastic zone locate only at local region of arch and lateral wall (Fig.2). As the development of time, the plastic zones are spread incessantly toward the periphery of tunnel. In the end of first excavation(time=30 days), the plastic zones have distributed all periphery of tunnel and the first shotcrete support of lateral wall all enter the plastic state(Fig.3). After the arch crown is excavated (the second stage time=40); there are greater plastic zones at arch abutment and bottom of arch. When the inverted arch is excavated(the third stage), spread of the plastic zone at arch abutment is not much developed, but at the bottom of arch is very rapid (Fig.4). The analysis result in above show that the rheologic behavior and construction procedure of soft rock tunnel is very much affected for the distribution and development of the plastic zone of soft rock tunnel.

Table 1 The physical parameters of rock mass

rock stratum	E (GP_a)	C (MP_a)	Φ	μ	R (KN/M^3)	η_2 ($10^4 P_a h$)	η_3
completely weathered	0.12	0.04	20.0	0.4	20.0	5.8	58.0
intensely weathered	0.65	0.07	26.0	0.4	21.0	5.8	58.0
slightly weathered	1.2	0.1	28.0	0.4	27.0	5.8	58.0
relaxed zone	0.3	0.04	25.0	0.4	27.0	5.8	58.0

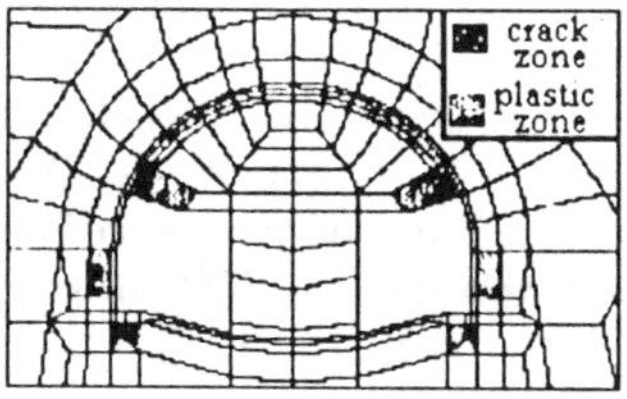

Fig. 2 plastic zone when time=24 hours

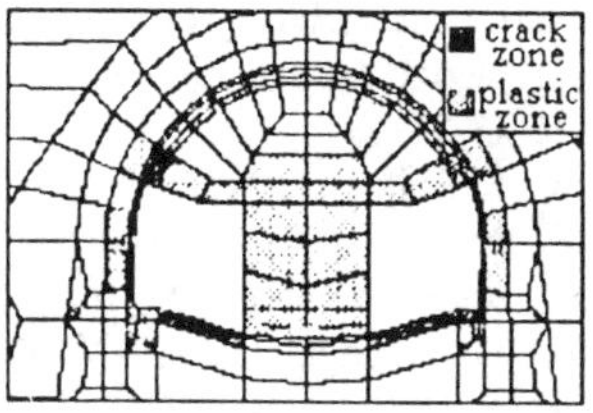

Fig. 3 plastic zone when time=30 days

4.2 The time-dependent characteristics of stress

After excavation of tunnel, the distribution of first principal stress vector are basically bearing of tangential along tunnel boundary and stress concentration at the bottom of arch rather obvious(Fig. 5). The stress history development is shows in figure 6. Elastic radial stress of surrounding rock is immediately released and shear stress is abruptly increased when the tunnel is excavated. As soon as the adjust of elastic stress is finished, the stress of surrounding rock enters the visco-elastoplastic state. As the time development, the radial stress of arch crow is still released and shear stress of arch crown keeps to increase. And yet radial and shear visco-plastic stress of arch abutment all are increased during the second stage. When the time last two month, stress change

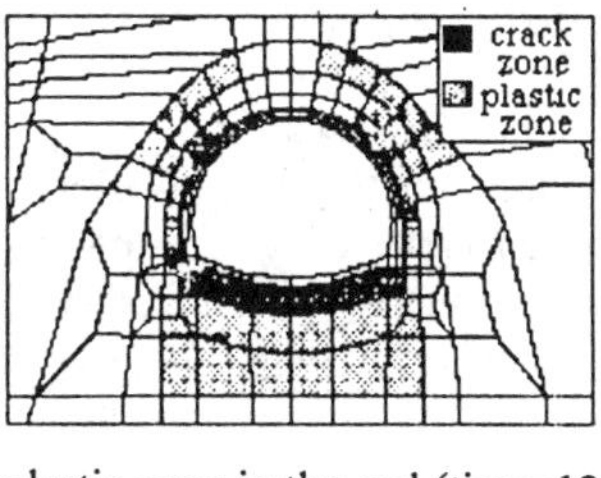

Fig. 4 plastic zone in the end (time=120 days)

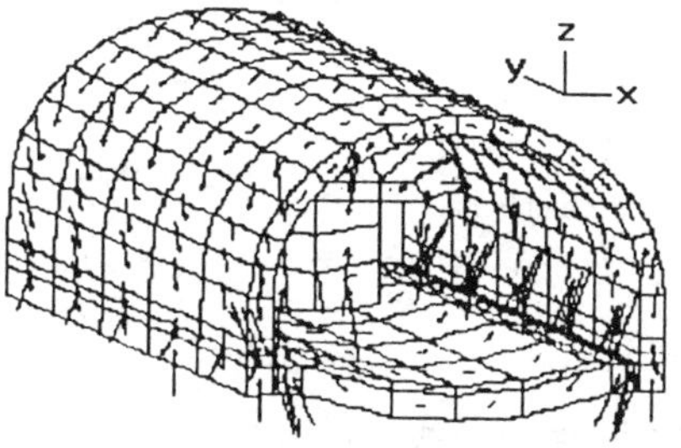

Fig. 5 distribution of first principal stress

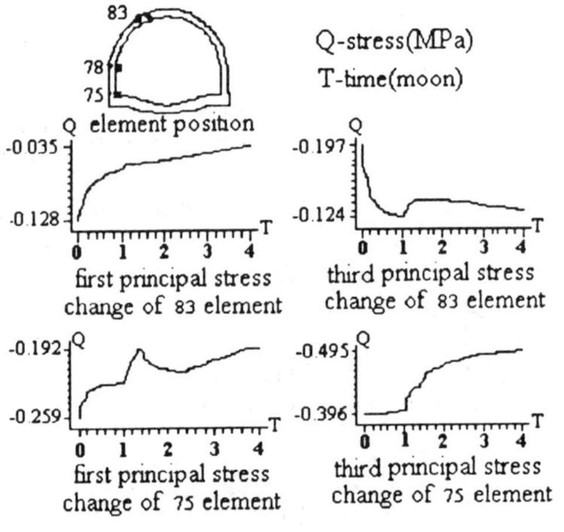

Fig. 6 stress change of time-dependent

stress of lateral wall and tunnel bottom is much affected by the rheologic behavior of soft rock. Because of the excavation of the inverted arch in end, the change of visco-plastic stress in the lateral wall and inverted arch is not stability until time last for four months. These results serve to illustrate that excavation sequence are of greater influence for the distribution regularity of visco-plastic stress.

4.3 The regulation of deformation of tunnel

After soft rock tunnel is excavated and before the second reinforced concrete lining is exerted, development of the tunnel boundary displacement is very swift (Fig. 7). As soon as the second reinforced concrete lining is exerted, the boundary displacement is limited at once. The time-displacement curve shows that the change of displacement basically keep stability after time lasts for two months. Because support of reinforced concrete lining at lateral wall is early exerted, the deformation of lateral wall is effectively limited. This shows that supporting time of the reinforced concrete is of rather large influence for the deformation of tunnel.

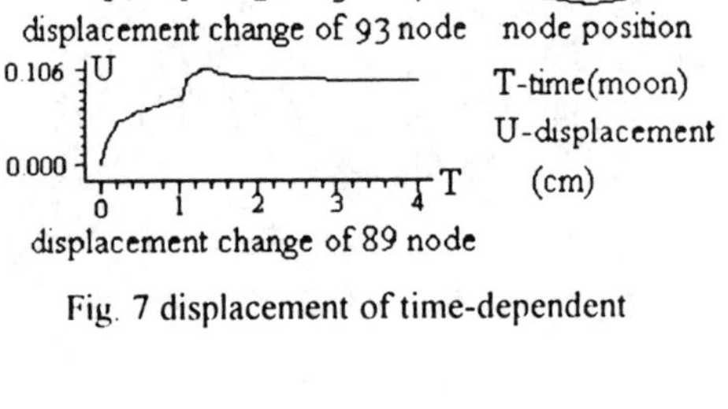

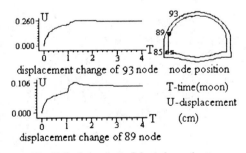

Fig. 7 displacement of time-dependent

4.4 Influence of construction method and supporting time for the stability

When we exert the second reinforced concrete lining of the inverted arch at once after the pilot is excavated, the plastic zones at arch abutment will be obviously enlarged. So the stiff support of the inverted arch can not exert too early for soft rock tunnel. The suitable supporting time of reinforced concrete lining for inverted arch is after soft rock tunnel all is excavated. This example shows fully that reasonable supporting pattern and suitable supporting time are very important to the surrounding rock stability of soft rock tunnel.

5 CONCLUSION

Iterative method of variable visco-plastic stiffness is a powerful efficient computation method for soft rock tunnel. It not only has better convergence rate and can accelerate computation speed, but also shows fully the influence of rheologic behavior for surrounding rock stability of soft rock tunnel.

Rheologic behavior and construction procedure of soft rock tunnel is of greater influence for the distribution of plastic zone and surrounding rock stress. Reinforced concrete lining can efficiently limit the rock deformation and the spread of plastic zones of soft rock tunnel. However, the choice of supporting time is not reasonable, it may obtain opposite results. So it is very important to chose reasonable supporting pattern and supporting time for surrounding rock stability of soft rock tunnel.

REFERENCES

H. J. Siriwardane and C. S. Desai, Computational procedures for non-linear three-dimensional analysis with some advanced constitutive laws. Int. J. for Numerical and Analytical Methods in Geomechanics. Vol. 7. 143-171, 1983.

J. N. Thomas, An improved accelerated initial stress procedure for elastoplastic FEM analysis, Int. J. Numerical and Analysis Methods in Geomechanics. Vol. 8. 56-69, 1984.

Yu Yutai and Xaio Ming, Three-dimensional elasto-plastic FEM analysis for the surrounding rock stability of large-scale underground opening. Chinese Journal Rock Mechanics and Engineering. Vol. 6. No.1 45-56.

G. Meschke, C. Kropik and H. A. Mang, Numerical analyses of tunnel lining by means of a viscoplastic material model for shotcrete, Int. J. for Numerical Methods in Engineering, Vol. 39, No. 18 3145-3162(1996).

Computer Methods and Advances in Geomechanics, Yuan (ed.) © 1997 Balkema, Rotterdam, ISBN 90 5410 904 1

Coupling algorithm of extended Kalman Filter-FEM and its application in tunnel engineering

Shuping Jiang
Chongqing Highway Research Institute, Ministry of Communication, China

ABSTRACT: The nondeterminative dynamics of surrounding rock along with tunnel excavation are analysed using the stochastic process and optimal estimation. Combining the functions of filtering modification and innovation production of the Kalman filter with the iterative calculation and the zone analysis in FEM, the author has set up an extended Kalman Filter – FEM algorithmic model that can reflect the dynamic nonlinear stochastic process in surrounding rock, and worked out a practical algorithmic program.

Using the field measurement values as the set of measurements, and the initial stress and plastic modules as the process state variables, the author computed and analysed nondeterministically the dynamic state of the surrounding rock, thus revealing its stochastic nature and providing a foundation for design and construction engineering.

1. Preface

In order to solve problems such as fuzziness and unpredictablity in tunnel engineering, the study of deterministic back analysis, as well as its applications, have been greatly developed since the 1980s. However, engineering practices show that considerable errors exist between the solutions of this deterministic algorithm and the results of site surveys. This is mainly caused by the undeterministic feature of the dynamic state of the surrounding rocks, not including the instability in the computing models and its errors.

When determining the rock parameters at a certain moment t_c, the conventional determinative back analysis method adopts the rock deformation variables at the moment t_c as the input data. What it considers is only a determinative value. It leaves out absolutely those deformation variates at the previous moments $t_{c-n}(n = 1,2,\cdots)$ that would have influence on the data from the present moment t_c. It also leaves out the dynamic process that reflects the relationship between rock parameters and its deformation variables.

On the contrary, a nondeterministic back analysis method regards the relationship between deformation variates (receiving signals) and rock parameters (outgoing signals) as a stochastic system, . This is a stochastic process along the way of excavation, i. e, to take full consideration of the impact of the variates from the previous moment t_{c-n} upon the present moment t_c. and thus to modify the parameters (called "filtering"), and in the end get the optimal state estimation to the unknown parameters. In order to ascertain the state variables of the surrounding rock parameters in its stochastic process, the author, combining with FEM, introduced the Kalman Filter, which originally belonged to the domain of optimal control theory and information theory, and worked out this new coupling algorithm – the extended Kalman filter – FEM algorithmic program.

2. Back Analysis Model Account for the Function of Volumetric Strain

The units where a larger difference exists between field measurement displacement values and computing values, can be regarded as non – plastic ranges. (These units are usually around the tunnel peripheries.) In these ranges, not only the initial stress, but also the influence of volumetric strains should be considered. Volumetric strain is caused by the cohesion function after excavation, where as initial stress already exsits within the surrounding rocks so long before excavation that we cannot get to know it. We use the concept "initial strain" only to set off the influence of volumetric strains on the stability of surrounding rocks after excavation. Here the initial strains are regarded as the volumetric stress and further converted as equivalent loadings of peripherals (hence "strain equivalent loading"):

$$\{F_s\} = -\int_{V_s} [B]^T[D]\{\varepsilon_0\}dV_s \quad (1)$$

where V_s is ranges influenced by volumetric strain, $\{\varepsilon_0\}$ is parameters of volumetric strain. then, with the exception of the three common unknown parameters (σ_{x0}/E, σ_{y0}/E, τ_{xy0}/E) that need inversion, the volumetric strain parameter $\{\varepsilon_0\}$, also, needs to join the inversion. $\{\varepsilon_0\}$ is originally formed by $[\varepsilon_{x0}, \varepsilon_{y0}, \gamma_{xy0}]^T$, but here we can give consideration to volumetric strain and suppose that ε_{x0} equals ε_{y0} and γ_{xy0} equals 0. Thus, the equivalent loadings that exert on the peripheral nodes turn to be the sum of initial stress equivalent loadings plus the volumetric loadings in non - plastic ranges, as in the equation:

$$\{F\} = \int_V [B]^T\{\sigma_0\}dx - \int_{V_s}[B]^T[D]\{\varepsilon_0\}dV_s \quad (2)$$

where $\{\sigma_0\} = [\sigma_{x0}\ \sigma_{y0}\ \tau_{xy0}]^T$ is initial stress, V is excavation units.

According to the algorithm raised by S. Sakurai, the item $\{u\}$ in the stiffness equation ($\{F\} = [K]\{u\}$) can be divided into a measure node displacement $\{u_1\}$, unknown node displacement $\{u_2\}$ and boundary node displacement equals zero, $\{u_3\} = 0$; liminting the unknown node displacement $\{u_3\}$ with an algebraic expression and a method of least squares, we can get the following:

$$\{C\} = [[A]^T[A]^{-1}][A]^T\{u_1\} \quad (3)$$

Here, matrix $[A]$ is a physical solution concerning stiffness matrix and equivalent loading vectors.

$$\{C\} = \{\sigma_{x0}/E\ \sigma_{y0}/E\ \tau_{xy0}/E\ \varepsilon_{01}\ \varepsilon_{02} \cdots \varepsilon_{0n}\}^T \quad (4)$$

Obviously, unknown parameters of $\{C\}$ can be obtained so long as the node displacement measuring values are given. But the quantities of measuring lines must be more than that of the unknown parameters.

3. Extended Kalman Filter – FEM Algorithm

The basic of Kalman Filter is to replace stochastic process with stochastic time series, and to approximate to the stochastic differential equation with the stochastic diffrence equation. After this procession and then computation, the Kalamn Filter computing formula of the non - linear stochastic system can be written as the following:

Filter equation

$$\hat{x}_{t/t} = \hat{x}_{t/t-1} + K_t[z_t - h_t(\hat{x}_{t/t-1})] \quad (5)$$

$$\hat{x}_{t+1/t} = f_t(\hat{x}_{t/t})$$

Kalman gain matrix (n × m)

$$K_t = P_{t/t-1}H_t^T[H_tP_{t/t-1}H_t^T + R_t]^{-1} \quad (6)$$

Estimated error covariant matrix (n × n)

$$P_{t/t} = P_{t/t-1} - K_tH_tP_{t/t-1} \quad (7)$$

$$P_{t+1/t} = F_tP_{t/t}F_t^T + G_tQ_tG_t^T \quad (8)$$

Initial conditions

$$\hat{x}_{0/-1} = \bar{x}_0, \qquad P_{0/-1} = \sum\nolimits_0 \quad (9)$$

Where, H, F are defined correspondently as:

$$H = \left.\frac{\partial h_m(x,t)}{\partial x_n}\right|_{x=\hat{x}_{t/t-1}} \quad (10)$$

$$F = \left.\frac{\partial f_n(x,t)}{\partial x_n}\right|_{x=\hat{x}_{t/t}} \quad (11)$$

In this paper, the sensitive coefficient method is applied to compute the partial differential determinant H. It means to compute each element of the determinant by the following formula:

$$\frac{\partial h_m(\hat{x})}{\partial x_n} = \frac{h_m(\hat{x}_n + \Delta\hat{x}_n e_n) - h_m(\hat{x}_n)}{\Delta\hat{x}_n}$$

Where $\Delta\hat{x}_n$ is a slight increment of the unsolved parameter state variable, e_n is an unit matrix corresponding to the n - th state variable.

In the aforementioned formulas, $\hat{x}_{t/t}$ is a state variable estimate (observation renewing) at moment t_i under the condion of z_t; $\hat{x}_{t+1/t}$ is a state variable forecasting estimate at moment t_{i+1} (state transition); $P_{t/t}$ is an estimated error covariant matrix at moment t_i (n × n); K_t is a gian matrix at moment t_i(n × m); Z_t is a m - dimensional observation vector at moment t_i (Sample realization of $Z(t_i)$); H_t is an observation matrix at moment t_i (m × n); F_t is a state transition matrix (n × n); G_t is a state noise coefficent matrix; R_t is an observation noise; Q_t is a system noise covariant matrix.

where, the state variables are the unsolved rock mass parameters as σ_{x0}/E, σ_{y0}/E, τ_{xy0}/E, ε_{01}, ε_{02},...... Considering that $\{u_1\}$ showed in equation (3) is a vector function to be estimated, it could be written by the following formula(u is used to replace u_1).

$$\{u\} = h(\sigma_{x0}/E,\ \sigma_{y0}/E,\ \tau_{xy0}/E,\ \varepsilon_{01}, \varepsilon_{02}, \cdots\cdots) \tag{12}$$

In a filtering equation, $h_t(\hat{x}_{t/t-1})$ and an observation matrix $H_t(\hat{x}_{t/t-1})$ are given individually by equation (12) and (10), and equation (12) is transformed by (3).

The contents aforementioned have formed an algorithmic model of the extended Kalman Filter – FEM, however, a question of how to solve inverse matrix $[H_t P_{t/t-1} H^T + R]^{-1}$ would be met in computing, thus the work of calculating becomes great, and even appears as a divergence. In this paper we use the sequential process method to avoid those.

In addition, the radical root propagation is used to compute the estimated error covariant matrix to guarantee its nonnegativity and avoid divergence.

In a nondeterminative back analysis, the initial value (initial condition) of its filter is most important to the precision of the filtering computing. Even if the filter estimation ultimately could be solved by setting a certain value as the initial value of parameter to be estimated, however, the amount of computing would be several times greater, and also the divergence would appear. Therefore, in this article, the computing result of determinative back analysis as an initial value of parameter to be estimated will reduce the numbers of filtering and avoid unnecessary computing, and at the same time guarantee the essential precision of filtering.

4. An Example of Engineering Application

A road tunnel is 1336m in length, 10m in net width for each tube, and 7.2m in net height.

The formation is limestone, and there is some degree of compresso – crushed zone in the rock matrix. In order to make supporting and lining reasonably, we work on surrounding rock monitoring and information feedback.

A numerical section is buried 140m in depth. The density of rock is 2.74g/cm^3, the poisson's ratio is 0.3, the friction angle is 30°, and the critical strain is 0.6%. The investigated route times IRT = 5 (see Figure 1), investigating times ITT = 13, continue for 37 days.

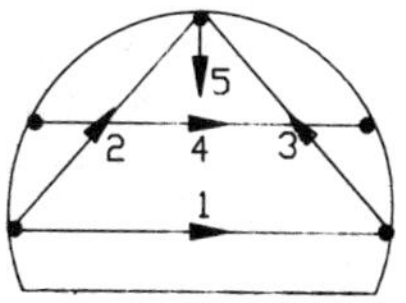

Figure 1. Arrangement of measuring lines

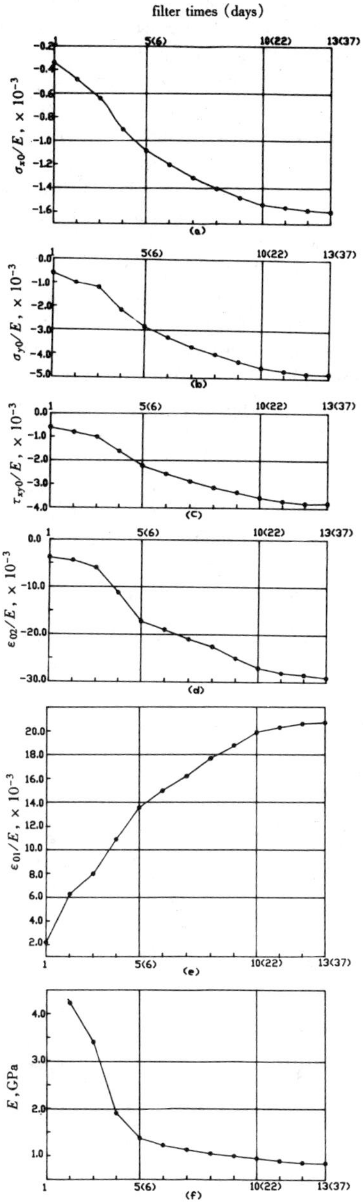

Figure 2. Stochastic time series filtering estimate of surrounding rock parameters

For initial condition, the determinative back – analysis counting results are used as initial value of the state vector filtering, and then the numerical trial is done, taking $P_0 = 1.0 \times 10^3$, $R_0 = 1.0 \times 10^{-4}$ (accounting for white noise, and the value R unchangeable in each step). In addition, in order to accelerate filtering process, the numerical trial shows that it is appropriate to increase 4 times gain matrix K. The observation matrix H is calculated by using sensitivity vector algorithm, in which, the incremental parameters are 15% of the parameters themself. The finite element mesh is 104 elements (including 8 excation elements) and 329 nodal points (including 9 excavating nodal points). Taking filtering steps ITT = 12 and responding to time series, the estimated parameters results are shown in Figure 2.

From Figure 2, the initial increments for each parameter are larger, and tending to weaken in the end. This indicates that the filter is in a stable condition. The estimated error covariant P changes along with filtering (P is a logarithm in Figure 3), fast in the beginning and then steady as shown in Figure 3, this is consistent with the variable tendency for the estimated parameters.

After the time series filtering estimate of surrounding rock parameters being computed, the shear strain distribution and filtering estimate in plastic area are immediately obtained for the relevant period. (shown in Figure 4.)

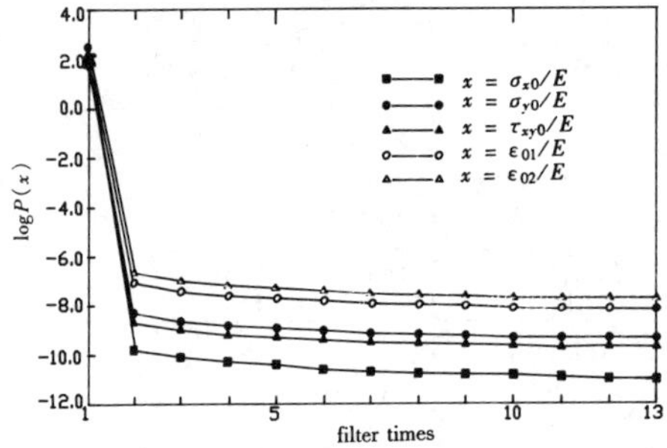

Figure 3. Filtering changes in parameter states estimated error covariant P

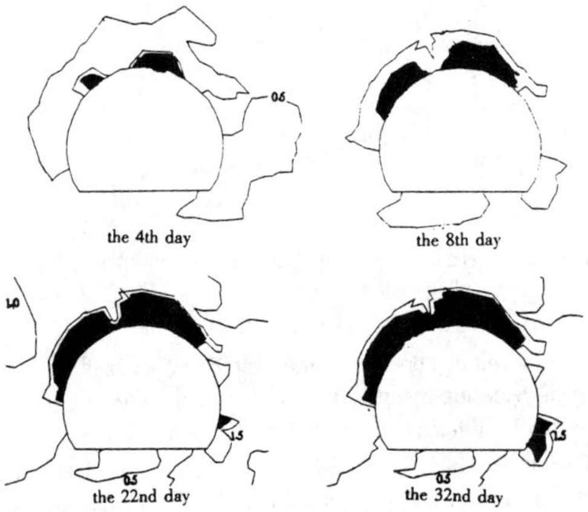

Figure 4. Maximal shear strain distribution and plastic area time seires changes upon a nondeterministic back analysis method

5. Conclusions

(1). Kalman Filters have getting rid of the old and taking in the new, and innovation production of parameter filtering, Combining this function with the finite element method, we can analyze the objective dynamic state in surrounding rock. In this paper, a nondeterminative back – analysis method is presented.

(2). The estimated error covariant matrix, gain matrix and observation matrix are functions of the above parameters, and are dependent on the above observation displacement.

They must be calculated on-line instead of setting up priorly.

(3). The initial conditons influence the estimated filtering value and its accuracy. By using the determinative back analysis results as initial value of the nondeterminative solution and proper increasing the incremental value properly, the filtering times can be lowered and thus the task can be achieved quickly.

(4). At the initial period of tunnel excavation, the elastic modulus of surrounding rock decrease suddenly, with the passage of time, the incremental displacement in surrounding rock decreases, and the parameter of surrounding rock changes steadily. This observation is very important in reasonable excavating or supporting.

6. References

S. Sakurai etc., Back Analysis of Displacement Measurements in Tunnelling, Proceedings of the Japan society of civil engineers, 1987. 337: 137 – 145.

Sun Jun etc., An optimization method for the elasto-plastic inversion of parameters in rock mechanics, Chinese Journal of rock mechanics and engineering, 1992, 11(3): 221 – 229.

Kalman R E, A New Approach to Linear Filtering and Prediction Problems. Trans ASME, J Basic Eng, 1960, 82D(1).

Kalman R E, Bucy R S. New Results in Linear Filtering and Prediction Theory. Trans ASME, J Basic Eng, 1961, 83D(1).

Makoto Suzuki, Estimation of Spacial Variation of Soil Properties Using Extended Kalman Filter Algorithm, Proceedings of the Japan society of civil engineers, 1989. 406: 71 – 77.

Jiang Shuping, Research on Back – analyzing the Nondeterminative Dynamic of Underground Openings Rock Surrounding by Extended Kalman Filter-FEM (a doctor – degree paper), Tongji University, 1993.

T. Katayama, Applied Kalman Filter. Tokyo Asakura Bookstore, Japan, 1989.

Computer Methods and Advances in Geomechanics, Yuan (ed.) © 1997 Balkema, Rotterdam, ISBN 90 5410 904 1

Effect of pore water behaviour on soft rock tunneling

He Lingfeng
Global Environment Engineers, Kyoto University, Japan

Liu Jun
Daiho Consulting Engineers, Japan

ABSTRACT: In this paper, based on Adachi-Oka's elastoplastic constitutive model, the tunnel excavation is simulated, and the influence of the pore water pressure on the mechanical behavior of ground is discussed in the plane strain condition.

1. INTRODUCTION

It is known that the failure of a excavated tunnel often occurs immediately or during several decades after excavation. The reasons of the tunnel instability can be regarded as follows: the dissipation of pore water pressure, the deterioration of materials and the progressive failure. When a tunnel is excavated in fully-saturated soft sedimentary rock, two types of time dependent behaviors occur in the surrounding ground. The first comes from the intrinsic rate dependent characteristic of the materials such as creep deformation or stress relaxation. The second is caused by the loss of apparent strength of materials due to the dissipation of the excessive pore water pressure in the surrounding ground. The purpose of this paper is to investigate the mechanical behavior of a excavated tunnel due to the dissipation of the pore water pressure. The ground materials generally are of the dilatancy, especially for dense sand, over-consolidated soil and soft rock. Therefore, the two aspects of strain-hardening and strain-softening should be considered in investigation.

In this paper, based on Adachi-Oka's elastoplastic constitutive model, the tunnel excavation is simulated, and the influence of the pore water pressure on the mechanical behavior of the ground is discussed in the plane strain condition.

2. NUMERICAL PROCEDURE DEALING WITH STRAIN SOFTENING

For nonlinear problem, the whole strain increment can be divided into two parts, elastic and plastic strain incremenet:

$$d\varepsilon_{ij} = d\varepsilon^{e}_{ij} + d\varepsilon^{p}_{ij} \tag{1}$$

According to the definition:

$$d\sigma_{ij} = [D]\, d\varepsilon^{e}_{ij} = [D]\, d\varepsilon_{ij} - [D]\, d\varepsilon^{p}_{ij} \tag{2}$$

here [D] is the elastic stiffness matrix. In plane strain condition, only the whole strain increment $\triangle\varepsilon_z$ is equal to zero, the plastic strain increment $\triangle\varepsilon_z^{p}$ is not necessary to be zero. Therefor Eq.(2) in plane strain condaition will take the form as:

$$\Delta\sigma_z = G\beta(\Delta\varepsilon_x + \Delta\varepsilon_y) - G(\beta(\Delta\varepsilon^{p}_x + \Delta\varepsilon^{p}_y) + \alpha\Delta\varepsilon^{p}_z) \tag{3a}$$

$$(\Delta\sigma) = (C)\{\Delta\varepsilon\} - (\Delta\sigma') \tag{3b}$$

in which:

$$[C] = G\begin{bmatrix} \alpha & \beta & 0 \\ \beta & \alpha & 0 \\ 0 & 0 & 1 \end{bmatrix} \quad , \quad \{\sigma\} = \begin{Bmatrix} \sigma_x \\ \sigma_y \\ \tau_{xy} \end{Bmatrix}$$

$$\alpha = \frac{K}{G} + \frac{4}{3}, \quad \beta = \frac{K}{G} - \frac{2}{3}$$

here $\{\Delta\sigma^{r}\} = G\{\Delta\varepsilon^{p}\}$

is so called relaxation stress tensor. The equilibrium equation and geometric equation for finite element method can also be expressed as follow in incremental way,

$$\int [B]^{T}\{\Delta\sigma\}dv = \{\Delta R\} \tag{4}$$

$$\{\Delta\varepsilon\} = [B]\{\Delta u\} \tag{5}$$

Substituting Eq.3 into Eq.4 will result in following equation:

$$[K]\{\Delta u\} = \{\Delta R\} + \{\Delta R^{r}\} \tag{6}$$

among the equation:

$$[K] = \int_{v} [B]^{T}[C][B]dv \tag{7}$$

$$\{\Delta R'\} = \int_{v} [B]^{T}\{\Delta\sigma'\}dv \tag{8}$$

In Adachi-Oka model with strain softening, a stress history tensor defined in Eq.(9) is introduced:

$$\sigma_{ij} = \frac{1}{\tau}\int_0^z \exp\left(-\frac{z-z'}{\tau}\right)\sigma_{ij}(z')dz' \tag{9}$$

It is easy to show that there exists a relation such that:

$$\left\{\Delta\sigma'\right\} = \left\{ f\left(\sigma_{ij}, \overset{*}{\sigma}_{ij}, \varepsilon^p_{ij}\right) \cdot \frac{\Delta z}{\tau} \right\} \tag{10}$$

(10) can also be expressed as:

$$\left\{\frac{d\sigma^r}{dz}\right\} = \left\{ f\left(\sigma_{ij}(z), \overset{*}{\sigma}_{ij}(z), \varepsilon_{ij}(z)\right) \cdot \frac{1}{\tau} \right\} = \left\{ F\left(\left\{\sigma^r(z), z\right\}\right)\right\}$$

(if $\triangle \Rightarrow 0$) (11)

By denoting the symbol as $\{\sigma^r\}=\boldsymbol{\sigma}^r$, Eq. (11) can be rewritten as:

$$\frac{d\sigma^r}{dz} = \boldsymbol{F}\left(\boldsymbol{\sigma}^r(z), z\right) \tag{12}$$

For such kind of ordinary differential differential equations of one degree, Runge-Kutta Method seems very suitable in obtaining the numerical solution. The Method dose not need any specific condition that the function $\mathbf{F}(\mathbf{y},t)$ in Eq.14 must satisfied for convergence, if and only if the $\mathbf{F}(\mathbf{y},t)$ has the continuous forth order differential and satisfy the Lipschize condition, that is:

for $\left\{\begin{matrix} a \le t \le b \\ -\infty \le y \le \infty \end{matrix}\right\}$ there exist a postitive L which is independent of $(\mathbf{y},t)$ and enable the following relation be valid:

$$\left|F(y_1,t) - F(y_2,t)\right| \le L\left|y_1 - y_2\right| \tag{13}$$

It is easy to show that the Adachi-Oka Model satisfies this condition. The formulas of the Runge-Kutta Method is expressed in the Eq.(14)

$$\begin{aligned} &\frac{dy}{dt} = F(y,t) \\ &\boldsymbol{K}_1 = \Delta t\, \boldsymbol{F}\left(y_n, t_n\right) \\ &\boldsymbol{K}_2 = \Delta t\, \boldsymbol{F}\left(y_n + \frac{\boldsymbol{K}_1}{2}, t_n + \frac{\Delta t}{2}\right) \\ &\boldsymbol{K}_3 = \Delta t\, \boldsymbol{F}\left(y_n + \frac{\boldsymbol{K}_2}{2}, t_n + \frac{\Delta t}{2}\right) \\ &\boldsymbol{K}_4 = \Delta t\, \boldsymbol{F}\left(y_n + \boldsymbol{K}_3, t_n + \Delta t\right) \\ &y_{n+1} = y_n + \left(\boldsymbol{K}_1 + 2\boldsymbol{K}_2 + 2\boldsymbol{K}_3 + \boldsymbol{K}_4\right)/6 \end{aligned} \tag{14}$$

3. THE COMBINATION OF BIOT'S THEORY AND FEM

Biot's consolidation equations are expressed as follows:

$$\frac{\partial \sigma'_{ij}}{\partial x_j} + \frac{\partial u_e}{\partial x_i} = 0 \tag{15}$$

$$\frac{\partial \theta}{\partial t} = -\frac{k}{\gamma_w}\nabla^2 u_e \tag{16}$$

where u_e is the pore water pressure, θ is the volumetric strain, k is the coefficient of permeability, γ_m is the unit weight of water.

Eq.(15) shows the equilibrium condition of differential element and eq.(16) shows the relation between the volume change and the drain amount respectively. In general, the body force is considered in eq.(15). But in the analysis we think that the compaction procedure due to self-weight has finished, so the influence of self-weight compaction is not considered in this study and the pore water pressure is regarded as a unknown.

On the other hand, the equation which describes the relation between volumetric strain and time adopts the form of backward difference which is more stable. The backward difference equation expressed as follows:

$$\frac{\theta|_{t+\Delta t} - \theta|_t}{\Delta t} = -\frac{k}{\gamma_w}\nabla^2 u_e|_{t+\Delta t} \tag{17}$$

We suppose that the pore water pressure is a unknown and it is a constant in a element, therefore, according to the energy principle and the related boundary conditions we can derive the expression as follws:

$$\begin{bmatrix} F \\ V \end{bmatrix} = \begin{bmatrix} K & L \\ L^T & 0 \end{bmatrix}\begin{bmatrix} \bar{u} \\ u_e \end{bmatrix} \tag{18}$$

where u is the displacement of node , F is the equivatent node force vector, K is the rigidity matrix of the usual element, L^T is the transformation vector from the node displacement into the volumetric strain. The parameters above can be expressed by suitable shape functions and their differential.

Refer to Fig.1 and consider the continuous Eq.16, the following expression can be obtained:

$$V|_{t+\Delta t} = V|_t - \left\{\alpha u_e - \sum \alpha_i u_{ei}\right\} \tag{19}$$

Let s_i be equal to the distance from point O shown in Fig.1 to the centre of element i and bi be equal to the width of drainage surface.

Then the coefficients α, α_i can be expressed as follows:

$$\alpha = \frac{k\Delta t}{\gamma_w}\sum\frac{b_i}{s_i} \tag{20}$$

$$\alpha_i = \frac{k\Delta t}{\gamma_w}\frac{b_i}{s_i} \tag{21}$$

Arrange eq.(18) in order, the following expression can be derived :

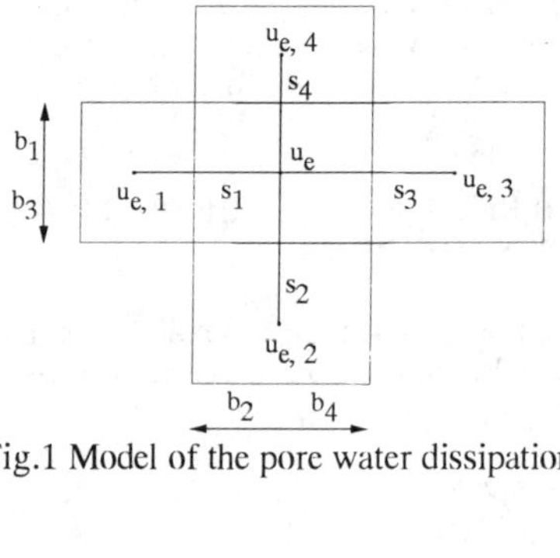

Fig.1 Model of the pore water dissipation

Table 1 Material parameters

G'	452.0	τ	0.006
σ_{mb}(kpa)	4.0×10^{3}	M_f^{*}	2.0
b(kpa)	2.0×10^{2}	ν	0.40
M	2.0	E (kpa)	1.5×10^{5}
γ_s(kgf/cm^3)	0.00262	k (cm/sec)	1.0×10^{-6}

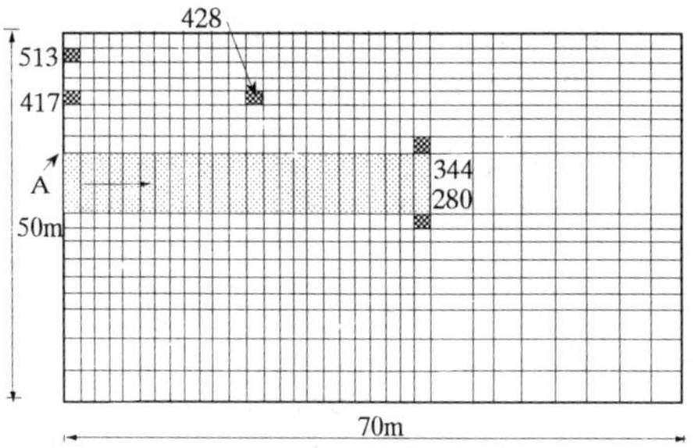

Fig.2 Finite element mesh

$$\begin{bmatrix} F|_{t+\Delta t} \\ V|_t \end{bmatrix} = \begin{bmatrix} K & L \\ L^T & \alpha \end{bmatrix} \begin{bmatrix} \bar{u} \\ u_e \end{bmatrix}_{t+\Delta t} - \sum \alpha_i u_{ei}|_{t+\Delta t} \quad (22)$$

Eq.(22) is true at the time of $t+\Delta t$. In order to solve nonlinar probem using increment method, eq.(22) is transformed into the following form:

$$\begin{bmatrix} \Delta F|_{t+\Delta t} + Lu_e|_t \\ 0 \end{bmatrix} = \begin{bmatrix} K & L \\ L^T & \alpha \end{bmatrix} \begin{bmatrix} \Delta\bar{u} \\ u_e|_{t+\Delta t} \end{bmatrix} - \sum \alpha_i u_{ei}|_{t+\Delta t} \quad (23)$$

where ΔF is the increment of node force during time Δt, Δu is increment of displacement during time Δt.

4. BOUNDARY CONDITION OF NUMERICAL ANALYSIS AND MATERIAL PARAMETERS

The material parameters used in the analysis are given in Table 1.

The other parameters are as follows: the tunnel's diameter is 10 meters, the lateral coefficient K_0 is equal to 0.67, and the gravity is taken as the initial stress field. The boudary conditions used in the analysis are as follows: the boundary condition is that there are only horizontal restrictions at the left side and the right side, and there are both horizontal and vertical restrictions at the bottom side. The pore water pressure disspates round the tunnel and the earth surface. Considering the actual construction circumstames, we suppose that the release rate of the initial stress is 20%, and the initial stress of 20% release completely after 100 seconds. On the above conditions, the both cases, one is considering the pore water pressure, the other is not considering the pore water pressure, are analysed respectively. The finite element meshes used in the analysis is shown in Fig.2.

5. THE RESULTS AND ANALYSIS

5.1 The relation among A point's displacement, the vertical stress of element 417 and 513, and the pore water pressure

From Fig.3(a) it is known that when the pore water pressure is not considered, the vertical stress of element 417 decreases fast at first with the increasing of the tunnel top's displacement, but after the tunnel top's displacement reaches a certain value the vertical stress is almost invariable. From above fact we can infer that the relaxtion region where element 417 lies does not extend further. Considering the pore water pressure, the vertical stress also increases while the tunnel top's displacement increases due to the influence of the negative pore water pressure. But with the dissipation of the pore water pressure, the vertical stress decreases. Making a comparison between the two vertical stress values, we can know that the finial vertical stress value considering the pore water pressure is larger than that without the consideration of the pore water pressure.

From Fig.3(b) it is shown that the vertical stress of element 513 decreases at first while the tunnel top's displacement increases, but after the tunnel top's displacement reaches a certain value the vertical stress becomes to increase reardless of the consideration of the pore water pressure or not. Therefore, it can be reputed that the region where the element 513 located at becomes to compaction region from relaxation region. Especially, considering the pore water pressure, the finial vertical stress exceeds the initial vertical stress with the dissipation of pore water

(a) Element 417

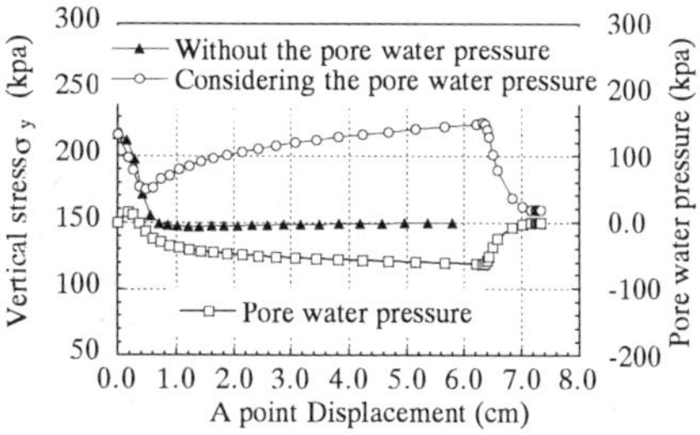

(a) Element 513

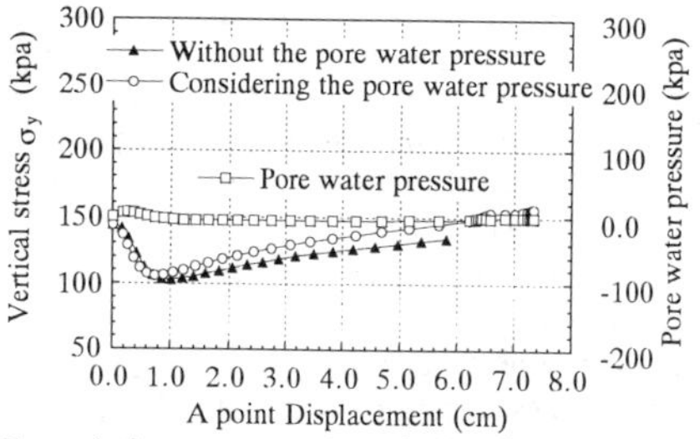

Fig.3 The relation among A point's displacement, the vertical stress and the pore water pressure

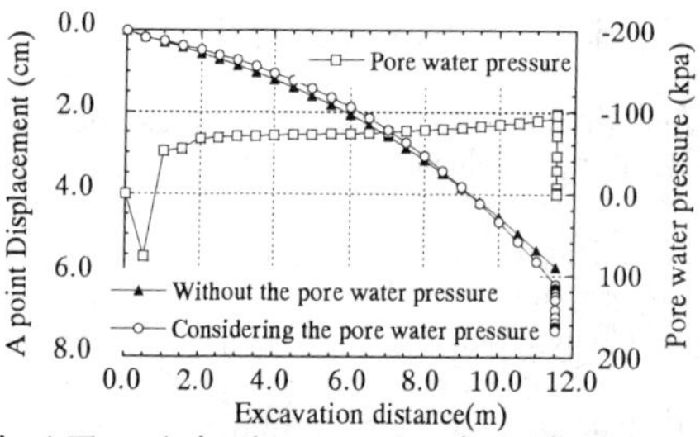

Fig.4 The relation between A point's displacement and the excavation distance

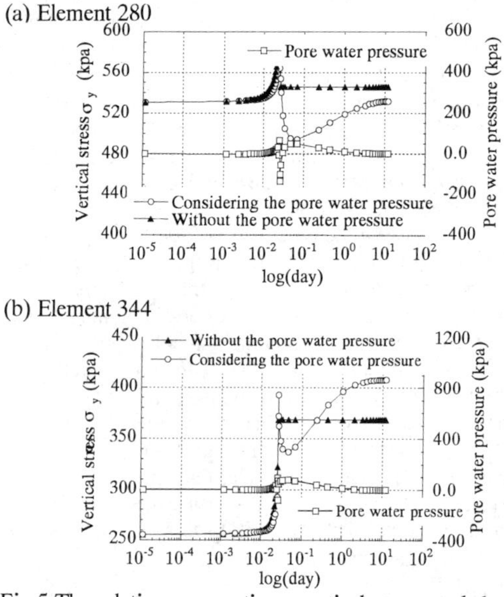

Fig 5 The relation among time, vertical stress and the pore water pressure

pressure, so it is reputed that the element 513 is in compaction region. Comparison between Fig.3(a) and Fig.3(b) it is indicated that the finical vertical stress considering the pore water pressure is larger than that without the pore water pressure.

5.2 The relation between A point's displacement and the excavation distance

From Fig.4 we know that the displacement of point A has not much relations withthe pore water pressure while the tunnul is excavated. But after the pore water pressur dissipates, the finial settlement of point A is larger than that without the pore watre pressure.

5.3 The relation among time, vertical stress and the pore water pressure

When the pore water pressure is not considered, the vertical stress for element 344 and element 280 increas with the excavation. But considering the pore

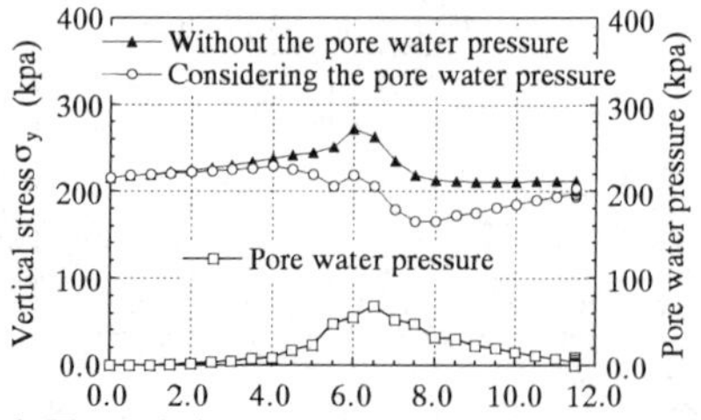

Fig.6 The relation among the excavation distance, vertical stress and pore water pressure

water pressure, the vertical stress for element 344 increases and the finial vertical stress for element 280 decreases slightly.

5.4 Relation among the excavation distance, vertical stress and pore water pressure for element 428

Form Fig.6 it is known that the pore water pressure for the element 428 increases gradually with the increasing of the excavation distance, and it reaches a maximum when the excavation distance extends to the bottom of element 428, then it decreases due to the influence of expansion. Comparing the vertical stresses between considering the pore water pressure and without the pore water pressure, it is shown that the fomer is smaller due to the influence of the positive pore water pressure.

6 CONCLUSIONS

In this paper the simulation calculation of the tunnel excavation is analysed. The following conclusions are obtained.

(1) With the increasing of the tunnel top's displacement, the vertical stress for the region far from the top point decreases at first, then becomes to increase. The vertical stress considering the pore water pressur is larger than that without the pore water pressure.

(2) With the increasing of the tunnel top's displacement, the vertical stress for the region near the top point decreases at first, then it is almost invariable. We can say that the relaxation region near the top point is not related to the top's displacement.

(3) The tunnel top's displacement considering the pore water pressure is larger than that without the pore water pressure with the increasing of the excavation distance.

(4)The element located at the tunnel top of the finial excavation surface, its vertical stress considering the pore water pressure is larger than that without the pore water pressure. On the contrary, the element located at the tunnel bottom of the finial excavtion surface, its vertical stress considering the pore water pressure is slightly smaller than that without the pore water pressure.

REFERENCES

Adachi, Oka, Yashima & Zhang 1994. Analysis of earth tunnel strain softening constitutive model:879-882. XIII CISMFE, New Delhi, India.

Oka, Adachi & Okano 1986. Two-dimensional consolidation analysis using an elasto-viscoplastic constitutive equation. International journal for numerical and analytical methods in geomechanics, vol. 10: 1-16.

Computer Methods and Advances in Geomechanics, Yuan (ed.) © *1997 Balkema, Rotterdam, ISBN 90 5410 904 1*

The conception and imitation calculation of tunneling displacements limit

Zhu Yongquan, Liu Yong & Zhang Sumin
Shijiazhuang Railway Institute, China

ABSTRACT:This paper proposed the conception of stability and displacements limit of tunnel. The stability criterion and design method are established by the measured displacements and limit displacements of tunnel. By this way, the characteristics of " feedback " design and construction, and information feedback in NATM can be really expressed.The essential obstacle is the determination of limit displacements.This paper proposed the limit displacements from imitation calculation based on the measured displacements and inverse calculation , and additional physical model tests and field investigation data.

The limit displacements of standard railway tunnel are imitated in this paper.

1 THE CONCEPTION OF TUNNELING STABILITY

The tunnel should be considered a integrated body which conclude surrounding rock and supported structure, or supported structure and surrounding rock are looked as supported system. In order to take measures before hand and make reasonable decision on design and construction ,it is necessary to analyze the stability of the system.The tunneling stability refer to stability degree of the system. In this state, supported structure does not occur abnormal crack and failure,the last result of displacement developing is asymptotic convergence, and the last displacement does not access limited boundary.

2 THE STABILITY OF TUNNEL CAN BE EXPRESSED BY THE DISPLACEMENT

Corresponding to the mechanics of materials, rock mechanics emphasizes stress and strength, and uses their relation to evaluate the state of rock in traditional method, so is the design of supported structure in tunneling. Based on the supposition of initial geostress and the dimension and distribution of loading , both method use the relation of stress and strength to evaluate the state of supported structure .Because tunneling has so many uncertainties and complex variablities , the stress system method has been proved difficult to use by a host of fact .

The developing of displacement reflects dynamic mechanical action of supported structure . Tunnel is a underground engineering ,we can survey crack of inner surface in certain distance, but observe hole picture difficultly . Displacement of inner surface can be measured by special equipment . Although the mechanism is complex ,the displacement can express the reaction of different action . Using displacement to uncover mechanics dynamic state is a direct and easy method, the stability of tunnel should be express by the displacement .

The stability criterion is established by the limit displacement of control points in certain limit state of tunnel.

With the developing of NATM ,we can judge the stability of tunneling by measured displacements of surrounding rock and supported structure in time. Based on the displacement of tunneling, we can establish the stability criterion and link the stability criterion with the field measured data. By this means,the characteristics of " feedback " design and construction,and information feedback in NATM can be really expressed.

3 THE APPROACH TO DETERMINE THE LIMIT DISPLACEMENT

The essential obstacle is the determination of limit displacement. When the property of rock and supported structure and construction condition do not content certain function, the displacement of tunnel express limit displacement, it can be deter-mined by theoretical analysis 、 field measure-ment and physical model test.

By field survey, we not only statistics displacement of tunnel in normal state, but also collect really displacement before failure ,which is difficult to collect . In order to obtain limit displacement in different state of rock and supported structure and different construction, the price of field survey and physical model test is very high. Based on the field geotechnical parameters obtained by measured displacements and inverse calculation,the limit displacements of tunnel are gained,addition physical models test and case studies are carried out for validation.

4 THE LIMIT STATE OF TUNNELING DISPLACEMENT

The initial geostress starts to release, and the stress of surrounding rock redistributes with tunneling excavation, the process and character are related to the index of physical mechanics and construction space-time effect. During the process, the displacement of tunnel occurs. After excavation, if we do not support the rock , a plastically zone can be find and move to the tunnel, and the rock will relax and supported structure doesn't occur abnommal crack and failure collapse because of concentrated stress. According to the character of rock 、 the dimension and direction of initial geostress and the size and shape of tunnel, the process and conclusion are different. It may be self-stability in integral hard rock, it may collapse sharply in soft rock. The surrounding rock do the process in common condition.

In the integral and hard rock, tensile stress exceeds its limit stength, and causes partial collapse or layer splitting of tunneling because of the releasing of geostress and gravity action. The first condition relate to geological structure, it can be determined by

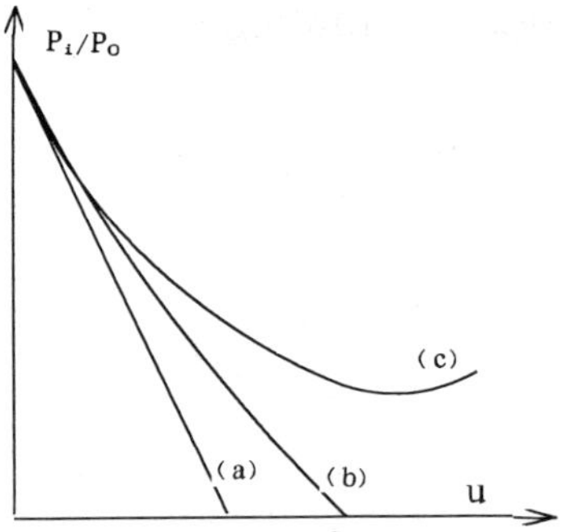

Fig.1 The forms of the curve of rock convergence

geological survey and drawing and mechanics analysis ; the second one relate to high geostress commonly, it can be defined by the theory of rock burst.

In order to utilize the self-carrying capacity of surrounding rock,we should make primary supporting in time, which let the rock occur a certain plastic deformation , but the harmful relax must be controlled. So we consider the state as limit state, the cor-responding displacement as limit displacement of tunneling.

The relation between deformation and resistance can be expressed by the curve with abscissa of deformation U and ordinate of resistance Pi,it can be called the curve of rock convergence or the characteristic curve. The dis-placement increases with releasing of geostress. After elastic stage, the plastic zone and plastic displacement occur and develop ,so the shape is curve . The resistance Pi increases with the deformation, when looseness pressure of surrounding rock increases as figure 1(c). According to the property of rock 、 condition of tunnel and constructing technique, the shape of curve is different. For example , in figure 1, curve (a) is a line elastic deformation; curve (b) is a elasticoplastic convergence deformation, which ex-presses the tunneling stability ; curve (c) is a divergent one ,in this state ,because the effect of loosen stress is very large ,the supported structure should be made to keep tunnel stability.

When the displacement is little then limit one ,in order to utilize the self-carrying capacity ,we allow occurrence of plastic deformation ,but must control harmful relax. Therefore, this displacement can be defined as limit displacement of tunneling.

5 THE IMITATION OF LIMIT DISPLACEMENT

In the condition of ideal elasticoplastic rock 、 Mohr-coulomnb yield criterion and releasing load divided into 20 equal parts, as the depth of tunnel(H)、 elastic modules(E)、 Poisson's ratio (μ)、 angle of internal friction(φ) and cohesive force(C) equal 300m 、 0.2×10^4 MPa 、 0.32 、 32.5° and 0.2MPa respectively, this paper imitates III classification standard railway tunnel ,the conclusion is expressed in figure 2. From figure 2 ,following releasing of initial geostress, the foot of wall yields firstly , along the wall,the yielded elements increase and develop to deep parts of surrounding rock, so do the elements of arch part and invert lastly, this expresses the increasing of plastic radius.

The U-Pi curve of arch part is expressed in figure 3, shape from line to curve expresses the increasing of plastic range, it render harmful loosening eventually. Following plastic deformation developing, mechanics property of yielded rock decreases, because of the action of gravity ,loosen stress taking place. we can simply consider the gravity of loosen zone as the loosen press.

Form Figure 3 , when the ratio of Pi to Po (Pi is resistance and Po value of initial geostress) equals 0.1, the curve will inflect. Corresponding figure 2 ,the yielded range and depth of arch part and bottom are large,so we must support the rock to restrain the harmful deformation and protect the rock loosening . The limit displacement is expressed in table 1. In order to compare with the measured data in case ,the displacement in table 1 is the relative deformation between the measuring points (as Figure 4).

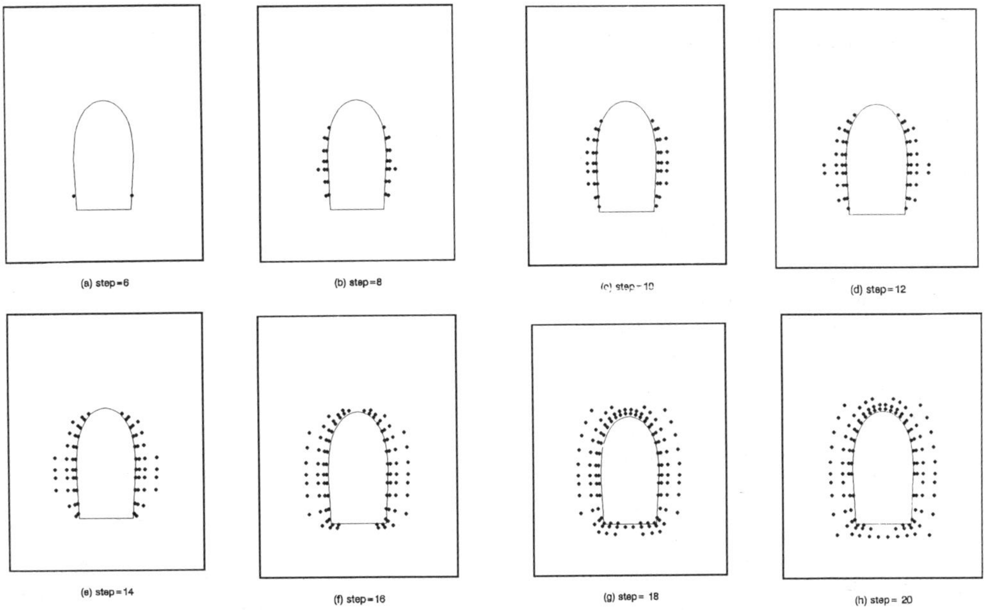

Fig.2 The process rule of yielded surrounding rock

Table 1 The limit relative deformation between the measure points(mm)

measured line	1	2	3	4	5	6	7	8	9	10
relative deformation	21.38	50.86	63.69	49.34	12.69	9.79	23.55	30.30	30.62	29.71

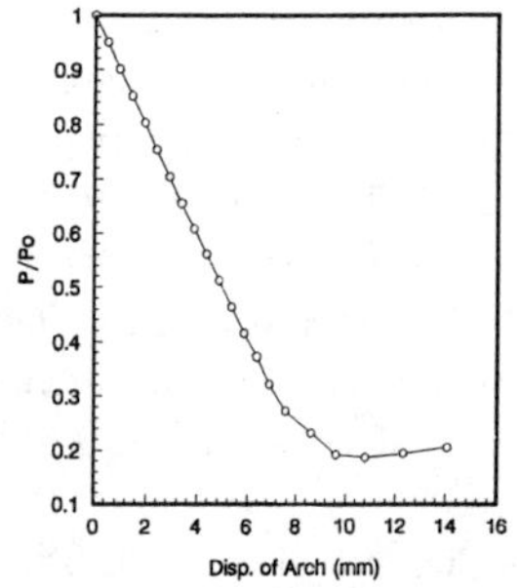

Fig.3 The curve of rock convergence

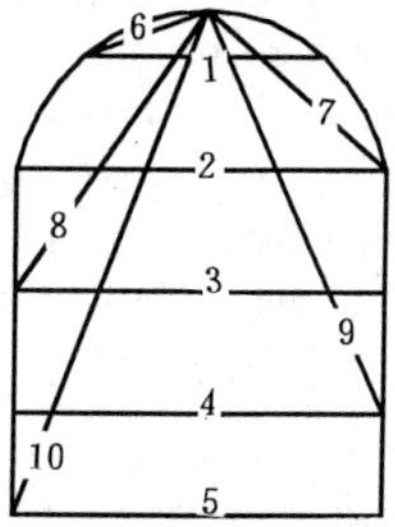

Fig.4 The measure lines of tunnel outline for relative deformation

REFERENCES

National standard of china , The criterion of anchor and jet-concrete for support. GBJ86-85,1985.

Zhu Yongquan etc, A Monte-carlo element method for back analysis of field geotenical parameters. *Rock and Soil Mechanics,* Vol.16, No.3.

Zhu Yongquan etc, Stochastic back analysis of lode on tunnel lining structure. *Rock and Soil Mechanics,*Vol.17,No.2.

J.J.Scott,Underground mining practical rock mechanics, *Mining Engnieering*, June 1991:641-646.

K.Hurt, New development in rock bolting , *Colliery Guardian*, July 1994:133-143.

M.Panter, A.Guenot, Analysis of convergence behind the face of a tunnel. *Laboratoire Central des ponts et chaussees*, Paris Frans.

T.Kitagawa, Application of convergence confinement analysis to the study of preceeding displacement of a aqueezing rock tunnel. *Rock Mechanics and Rock Engineering* 24,1991:31-51.

J.P.Barlow&P.K.Kaiser,Interpretation of tunnel ocnvergence measurements.*Pro.8th Int.Confer. on Rock Mech.*,Canada,1987:787-792.

E.T.Brown,etal, Ground response curves for rock tunnel. *Geotech. Engng.*,Vol.109,No.1,1983:15-39.

Chen Zhiyin ,Time series analysis applied to displacement prediction in NATM, *Proc.Int.Symp.on Mondern Mining Technology*, oct 1988,taian:315-320.

X.Nie and Q.zhang, A system of monitoring and dimensioning tunnel support.*Rock Mech.Rock Engng.*(1994)27(1):23-36.

Computer Methods and Advances in Geomechanics, Yuan (ed.) © 1997 Balkema, Rotterdam, ISBN 90 5410 904 1

Modeling of soil-structure interaction during tunnelling in soft soil

Sylviane Bernat, Bernard Cambou & Purwaningsih Santosa
LTDS-UMR, Ecole Centrale de Lyon, France

ABSTRACT: This paper discusses the modelling of loading induced by tunnelling in soft soil. Only two dimensional analyses are considered. Two kinds of modelling are analysed and compared : a simple decrease of the stress vector applied at the periphery of the excavation (described by a scalar parameter λ named the "deconfinement" rate), which is supposed to take into account in a very simple way the complex interaction between the lining and the surrounding soil, and a more complex modelling taking into account four different phases simulating the different kinds of interaction between the tunnel and the soil (deconfinement, weight change of the excavated zone, pore pressure applied on the lining, grout injection and consolidation of the grout). This comparison is achieved using two constitutive models : an elastic-perfectly plastic model (the Mohr Coulomb ("MC") model) and a more complex elastic-plastic model considering hardening mechanisms (the "CJS" model). The results obtained show that, if the analysis is achieved only to evaluate the surface settlements, the first simple modelling can be used with a good approximation, but only if the CJS model is considered. If more complete results are needed, in particular the loading applied on the lining and the displacement of the lining, it is necessary to consider the second more complex modelling. The two kinds of modelling are compared on a simple case of shallow tunnel with homogeneous soil conditions, and then, on a real site where numerous data have been obtained during the construction of the metro of Lyon. This site benefitted from coordinated research in which our Laboratory was in charge of the numerical modelling.

1 WORKSITE PRESENTATION

The urban conditions in the LYON-VAISE area required the underground excavation of the metro. The infrastructure consists of two shallow tunnels excavated between June 1993 and March 1995. The poor mechanical properties of the ground encountered required the use of a slurry shield. The tunnel is lined with precast concrete segments installed under the protection of the tail. The annular space remaining between the external side of the tunnel lining segments and the ground is filled with an inert grout injected during the excavation process. The instrumentation along the layout is divided into two monitoring areas, "zone 1" (composed of 2 sections S1 and S2) and "zone 2".

This instrumentation has been explained in detail in previous papers [B1]. The measurements conducted during the excavation of the tunnels concerned the vertical movements (multipoint extensometers), the surface settlements (topographic measurements), the horizontal movements (inclinometers), the extrados segment injection pressure and the tunnel displacements.

2 PHYSICAL ANALYSIS OF THE TUNNELLING PROCESS

An analysis of the ground displacements during the tunnelling process (Fig. 1) permits the distinguishing of different phases :

- *During the excavation (Ph0)* :

The stability of the working face is ensured by the slurry pressure, and the problem considered is essentially 3D, but the technology used is so reliable that the movements in front of the TBM (tunnelling boring machine) are negligible.

- *Passing the TBM (Ph1) :*

A displacement of the soil around the TBM may be due to the possible overcut of the working face, to the conical shape of the tail, or to the guiding problem of the TBM : it corresponds to a beginning of the deconfinement process. It can be noted that the variation of weight of the tunnel (initially constituted of soil and water, and then replaced by the TBM) was negligible in the analysed case.

- *Escaping the tail (Ph2) :*

The filling of the annular void between the lining and the soil requires a grouting process which can induce a temporary outward displacement of the soil at the periphery of the tail. The tunnel, initially created by the TBM, is replaced by the concrete segment and the grout, and then, consequently its weight changes. If located under the water table, the tunnel could have an upward movement due to the water pressure. At the end of this phase, the concrete segments are pressurised by the soil and the grout.

- *Consolidation of the grout (Ph3) :*

The grout was injected in a fluid state with a high value for the water content. For several weeks after the digging of the tunnel, the consolidation of the grout occurs, inducing displacements in the surrounding soil. It can be noted that the soil located around the tunnel, which has been disturbed by the tunnelling process, can at the same time also be submitted to a consolidation process.

3 COMPUTATION STRATEGY

The proposed computation strategy is based on:

1) a constitutive model of the soil adapted to the stress paths encountered during tunnelling (elastic-plastic CJS law [CA] developed at the LTDS Laboratory of the Ecole Centrale de Lyon),

2) a methodology to identify the model constants based on laboratory tests (triaxial and oedometer tests) and in situ tests (pressuremeters),

3) a computation code (the CESAR finite elements software developed at LCPC),

4) an excavation modelling using a two-dimensional approach, in a plane which is orthogonal to the tunnel, and using a stress release at the periphery of the excavation called the deconfinement process.

Refer to [B1] for more details on these four points.

4 EXCAVATION MODELLING

The stress generated by tunnel excavation is essentially three-dimensional, since it depends on the distance to the working face. A significant (but conventional) simplification was applied considering a plane strain calculation in the plane which is orthogonal to the shield advance. This 2D approach appeared to be justified since the analysed measurements show that the settlements recorded ahead of the working face were always very low (less than 2 mm).

Two procedures are possible for modelling the tunnelling in 2D (see Fig. 2) :

."Procedure 1" : Tunnelling is simulated by a stress boundary condition at the periphery of the tunnel, fitted on a local displacement measure

The principle is to delete the soil inside the excavation and to replace it by fictitious stress supporting vectors which equilibrate the initial state. Tunnelling modelling then consists in progressively releasing all the components of stress vectors acting at the excavation periphery by reducing them using the same factor with respect to their initial value. This reducing factor is called the "deconfinement rate" written λ. In such a simulation with no support in the tunnel, the stress boundary condition can go until the total cancelling of the vectors for a rate λ=100%. But in the case of lined tunnel, λ never reaches 100% because the stabilisation occurs at a lower value corresponding to the interaction between soil and structure. The problem is to calculate this final rate λf taken to be representative of the whole tunnelling process. It is usually difficult to estimate this parameter *a priori* but it can be determined *a posteriori* by comparison between simulations and measurements. This work was done on the Vaise worksite [B1] [B2].

."Procedure 2" : The ground-lining interaction is modelized

This more complex approach deals with the modelling of the interaction between soil and structure. A first stress release until a deconfinement rate $\lambda 1$ is considered. The lining is then modelized and λ is increased to 100%. The final stress state will result from the interaction between the lining and the ground.

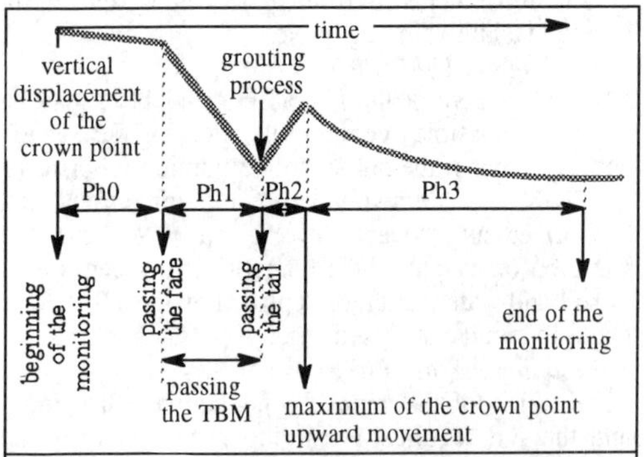

Fig. 1 : 4 different phases of the crown vertical displacement

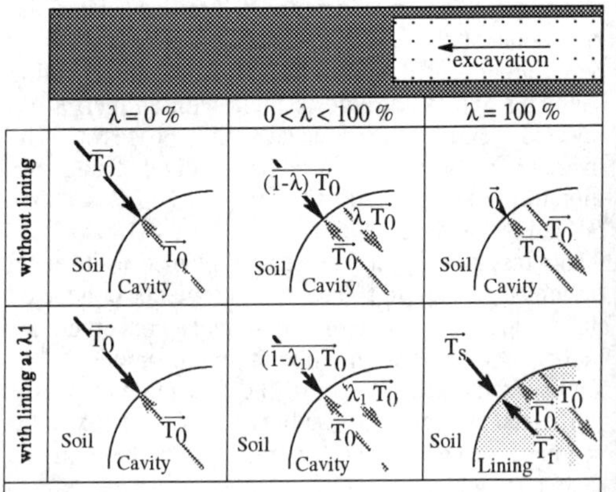

Fig. 2 : The 2 ways of tunneling's modelling

These two kinds of modelling were analysed, and a confrontation was achieved between them in order to value the degree of validity of the former. Indeed, it has the advantage of taking into account the soil-structure interaction in a global way. Therefore it is not necessary to identify the mechanical characteristics of the grout and of the concrete segments.

5 TUNNELLING SIMULATION USING THE DECONFINEMENT RATE AS A STRESS BOUNDARY CONDITION

5.1 *Analysis in the simple case of a shallow tunnel*

To begin, an analysis was achieved on a simple case of a shallow tunnel to obtain a better understanding of the phenomenon generated by this type of loading. The soil was taken to be homogeneous and composed of a silty sand. Constants of the two models (MC & CJS) were defined using triaxial test results. To account for the dependence of the Young modulus with the depth, fictitious 3 m thick layers were considered.

. *Effective stress paths around an unlined tunnel :*
The stress paths of various points (located at the periphery of the excavation) obtained by a finite element analysis were drawn, for the two models considered, in the principal stress plan (Fig. 3a - 3b).
- All effective stress paths start from the K0-line.
- At the crown and invert, excavation produces a decrease in the vertical stress, following an elastic unloading path until the isotropic state where the principal directions are inverted (for λ near 50%). Then, the vertical stress continues to decrease and the path reaches the failure criterion.
- At the point located at the haunches, excavation induces an horizontal unloading which leads, more quickly in MC than in CJS, to the plastic state (for λ near 30%). Then, the paths follow the plastic criterion, which is not exactly the same for the 2 models because of the strain plane paths analysed.
- The points located at 45° over and below the haunches follow the same kind of paths, but more progressively. The rotation of the principal directions occurred here so that the initial one (which was vertical and horizontal) becomes radial and orthoradial. It can be noted that for values of λ less than 30%, the MC modelling leads to stress states satisfying elastic conditions. Even for high values of λ, the calculation shows that the plastic area remains located near the haunches of the tunnel, in a very small area.

. *Horizontal displacements :*
MC and CJS models predict inward horizontal displacements near the tunnel periphery. Only the CJS model predicts horizontal outward movements of the ground, beginning at one radius from the haunches. This phenomena has been observed on the inclinometric measurements.

5.2 *Application to the work of the Lyon metro*

. *Modelling using λ fitting on final displacements values for Zone 1 :*
For each instrumented area, the final stress release rate λf was estimated using the final vertical

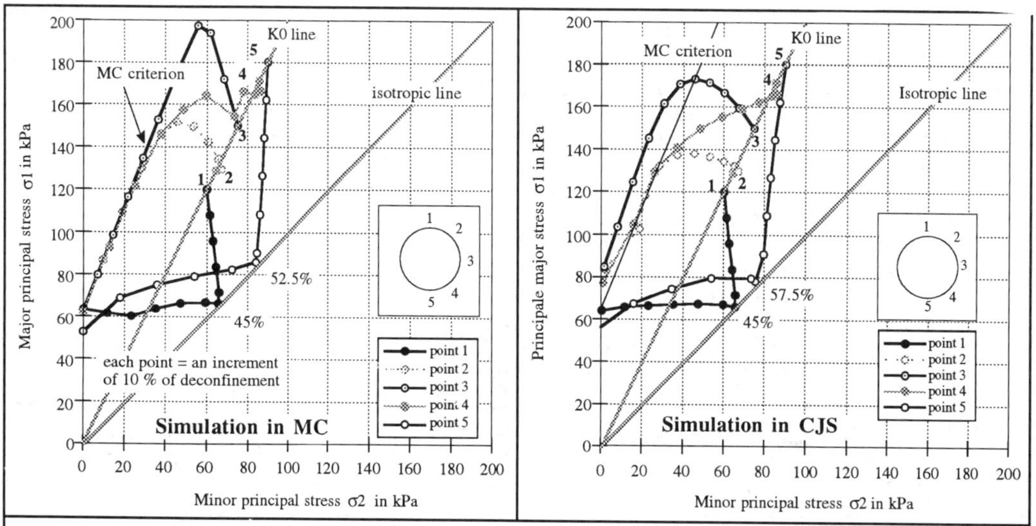

Fig. 3a et 3b : Effective stress path around an unlined tunnel : MC (3a) and CJS (3b) simulations until λ = 100 %

displacement of the measurement point the closest to the excavation, located at the crown of the tunnel. This adjustment of λf proved that the final deconfinement rate depends on the technological process and on the layout conditions.

. *Modelling using cycle of loading-unloading in λ for Zone 2 :*
The simulation in one phase of deconfinement is incompatible with the experimental results in Zone 2 where the settlements due to Ph1 were so small that the upward movement in Ph2 was predominant : the final settlements were in an inverse order in relation to the 3 other sections. The surface settlement was greater than the crown settlement, and Phase Ph2 can no longer be neglected. Simulations were performed by imposing an evolution of λ adjusted onto the settlement changes with time at the crown points. The phase by phase simulation give results which are in good agreement with all the measurements [B2].

. *Increasing of λt versus λr :*
Some variants of this method have been analysed, such as increasing the tangential stress release λt at the excavation periphery, in order to take into account the fluid behaviour of the grout injected behind the segment. These simulations [B2] demonstrated that the displacements measured are compatible with significant tangential stress release rates, and that this increase of λt furthered the lateral outward movement.

. *Conclusion :*
The various comparisons between experimental and simulation curves demonstrated that the proposed procedure obtained simulations in good agreement with the phenomena observed under tunnelling conditions. The deconfinement used as a boundary condition, until a final rate experimentally determined, or eventually with an intermediate "reconfinement" cycle, permits the prediction of the displacements measured in the ground and in particular in the surface settlements. But the limitation of this kind of modelling is that λ is only a mathematical stratagem but has no physical reality. It can be noted particularly that, in the case of a shallow tunnel, the weight equilibrium is not satisfied by such a modelling, and the loads at the periphery of the excavation are surely not representative of the real values applied on the concrete segments. Therefore, if precise information is needed for the segment design, it is better to attempt to take into account the real interaction between soil and structure.

6 MODELLING OF THE SOIL-STRUCTURE INTERACTION

6.1 *Difference between the 2 kinds of modelling : Change of the soil stiffness inside the excavation*

Some simulations, achieved imposing deconfinement on tunnels of different stiffness, showed that 2 types of mechanisms could appear :
- on an empty or a flexible tunnel, deconfinement generates an horizontal ovalization (upward movement of the bottom, crown settlement), and in a lower order, settlements of the surface points,
- under a sufficiently rigid tunnel (such as concrete segments), the deconfinement generates an upward movement of the tunnel (with a little vertical ovalization), and in a lower value, an upward movement of the surface points.

6.2 *Comparison of the 2 kinds of modelling in the simple case of a shallow tunnel*

In the case of an underwater tunnel, the final rate of 100% of "procedure 2" (cf. § 4) applied in effective stresses is only equivalent to a loss of 100% of the effective weight of the tunnel area. In order to obtain a tunnel which is really in equilibrium the water pressure must be applied to the external face of the segment, and the weight of the concrete segment must also be taken into account (Fig. 4a). The settlements scheme for each step of calculation is presented in Fig. 5a. This type of modelling in 4 steps has been applied to the simple case explained in 5.1, for the MC and the CJS model, and for different values of $\lambda 1$ (Fig. 5b). Obviously, the greater $\lambda 1$ is, the greater the settlements (step a) will be. The earlier the lining is placed, the more effective it will be. It is then interesting to note that the curves obtained are parallel straight lines in each of the following steps. The

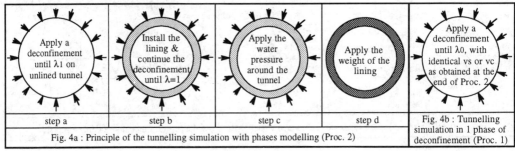

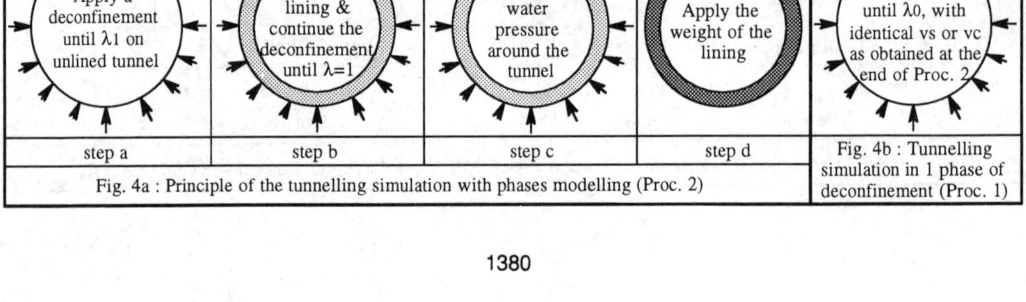

Fig. 4a : Principle of the tunnelling simulation with phases modelling (Proc. 2)

Fig. 4b : Tunnelling simulation in 1 phase of deconfinement (Proc. 1)

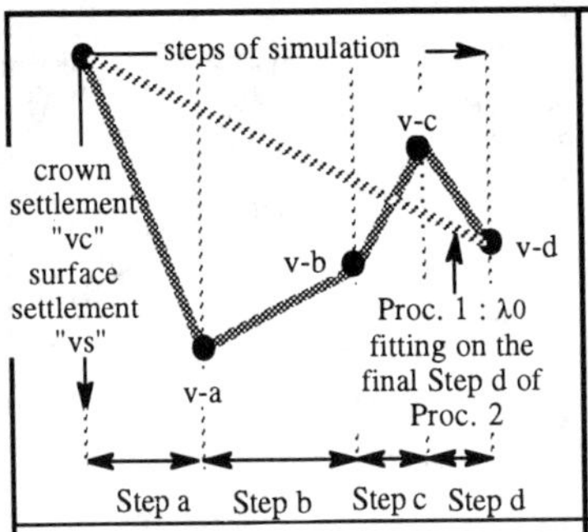

Fig. 5a : Settlement for each calculation step

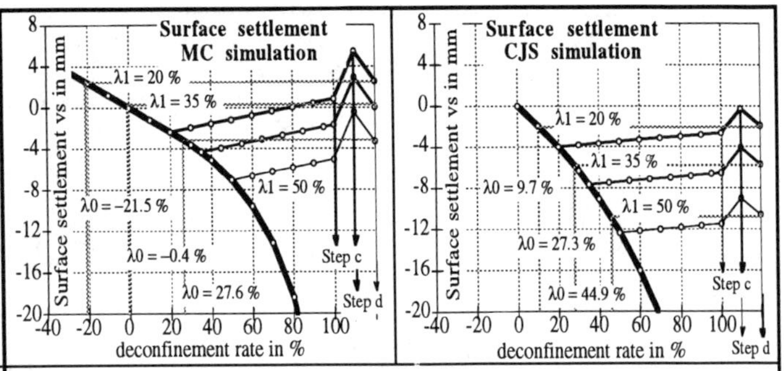

Fig. 5b : Fitting of λ0 on the final surface settlement of Proc. 2 depending on λ1

effect of lining (step b) is linear in reducing the settlement. Applying the water pressure (step c) will force an upward movement of the tunnel. On the contrary, applying the weight of the lining (step d) will cause a settlement. Note that steps c and d do not depend on λ1.

This type of simulation in 4 phases (only depending on λ1) was compared with a "procedure 1" modelling (using a simple deconfinement) conducted until λ0 which is directly adjusted onto the surface "vs" or the crown "vc" settlements of the 4 steps simulation final state (Fig. 4b & 5a).

A linear relationship between λ1 and λ0 can be observed (Fig. 6), but it can be noticed that the CJS and the MC calculations give very different results. In the CJS, this relation is almost the same for "vs" and "vc". Therefore, field data for surface settlement can be used to define λ with the same level of accuracy as the measurements near the crown. This information can not be obtained using the MC model. For the MC, negative values of λ are obtained for positive settlements that resulted with the final step. The settlement trough is not realistic, as it shows a heave of the soil in surface if λ1 < 50%. The CJS analysis gives very similar final surface settlement trough for the 2 kinds of modelling (phases and deconfinement modelling).

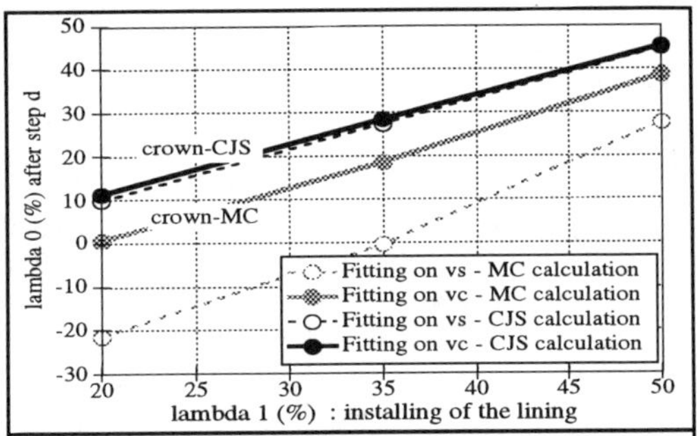

Fig. 6 : relationship between λ0 and λ1 in CJS & MC

6.3 *Application to the Lyon metro worksite*

Such a phases simulation was applied for the Lyon metro worksite. The concrete segment is modelized and the final deconfinement rate is increased until 100%. The proposed simulation takes into account the interaction between the soil, the grouting, the water and the concrete segments. The measurements achieved at the crown of the tunnel allow 3 phases to be defined for the simulation (Fig. 7) :

Ph0+1) A first stress release,
until a deconfinement rate λ1 experimentally determined from in situ measurements, simulating the excavation and the interaction with the tail.

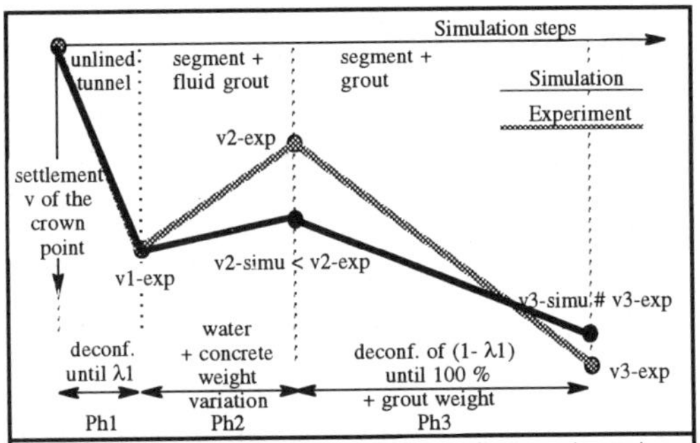

Fig. 7 : Principle of the simulation modelled soil-stucture interaction

Ph2) The concrete segments and the grout in its fluid state are now considered.

The water pressure exerted on the watertight lining is modelized. Under this loading the calculations allow both the global uplift of the tunnel and of the points in the ground located near the tunnel to be put in evidence. These movements depend greatly on the characteristics considered for the grout in its fluid state. For the sake of simplicity an elastic perfectly plastic model (MC) was chosen to represent the behaviour of the pasty grout which is considered in undrained conditions when it is injected. The chosen constants correspond to an incompressible (ν=0.49), frictionless (φ=0°) solid with very poor elastic

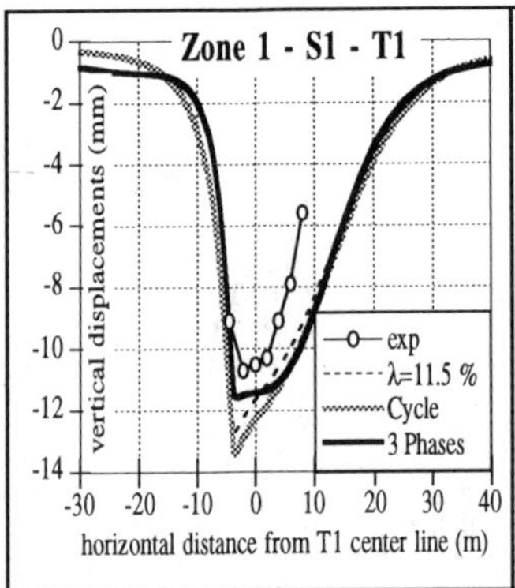

Fig. 8a : Settlement trough

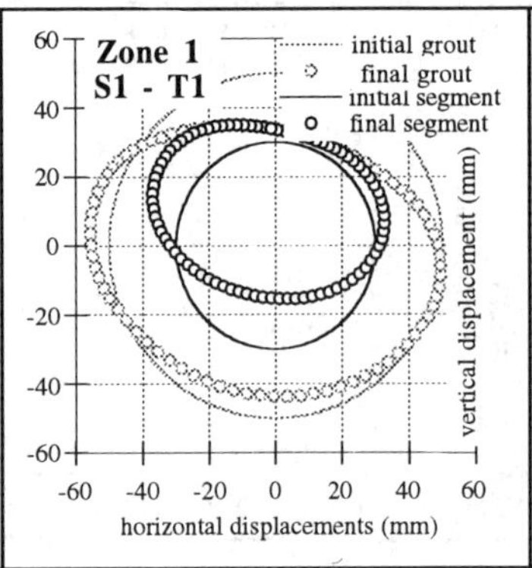

Fig. 8b : Tunnel upward movement

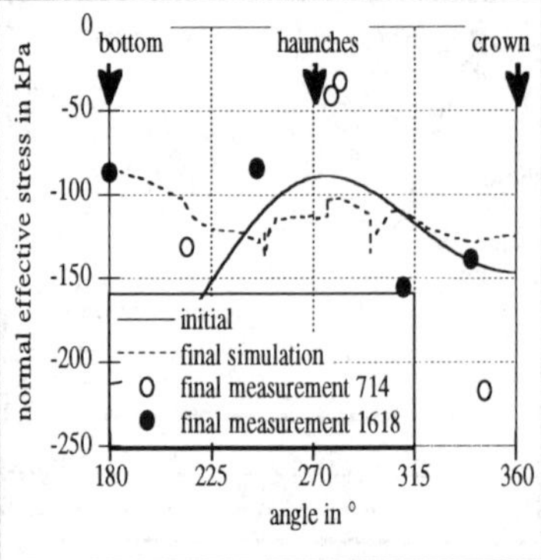

Fig. 8c : Grout pressure

characteristics (E=10 kPa) and shearing strength (Cu=10 kPa). The values of E and Cu have been determined by a fitting technique considering for this phase an upward movement close to 50 mm for the tunnel and 5 mm for the extensometric crown point (given by the in situ measurements). This phase Ph2 is physically related to the water pressure applied on the watertight lining and to the grouting effect.

Ph3) Complete soil-grout-lining interaction :

At this stage the tunnelling machine is no longer located in the tunnel, and then the effective weight of the excavated area will decrease and the interaction between the lining, the gout and the soil will reach its equilibrium state. The loading is an evolution of the deconfinement rate until $\lambda=100\%$ (complementary stress release of Ph1) which allows the evolution of effective weight of the excavated zone to be taken into account. Furthermore, in this phase the grout will consolidate so its characteristics will be changed for the drained ones which were obtained from laboratory tests (oedometer and shearing box) and take the following values : $E' = 10$ MPa, $\nu = 0.33$, $c' = 15$ kPa, $\varphi' = 35°$.

It can be noted that the grouting is here only supposed to backfill the annular void, and the upward movement is only linked with the water pressure. This simulation uses only one fitted parameter $\lambda 1$ and needs the determination of the mechanical properties of the grout in its fluid and solid states. Then, the calculated settlements obtained at the end of Ph. 2 and 3 are no longer fitted on experimental results but are calculated from the interactions between the lining, the water, the grout and the soil. All the observed phenomena are correctly described by this simulation [B3] (Fig. 8a). Particularly, it is able to simulate the upward movement of the concrete segment inside the grout (Fig. 8b) and to evaluate the grout pressure (Fig. 8c) and the loading applied to the lining.

7 CONCLUSION

The set of measurements, obtained on the experimental sites of the Lyon metro, was used to qualify a tunnelling simulation procedure using the finite element method. This site enabled the highlighting of an important phenomenon : if the initial phase of settlements Ph1 is small, the analysis of final conditions is not sufficient and the various tunnelling phases must be taken into account. Modelling using λ, eventually with an intermediary "loading-unloading" cycle, seems to give realistic results concerning the surface displacements. However, if the aim is to estimate the displacements and the loads in the lining, the complete soil structure interaction during the tunnelling process has to be modelized. 2 types of parameters must be determined:

- the initial deconfinement rate $\lambda 1$ (which depends on the process of the TBM, on the layout conditions, on the speed of the TBM) : it is difficult to be estimated *a priori* but can be estimated on measurements at the beginning of the work,
- the mechanical properties of the grout (under undrained conditions when it is injected, and under drained conditions for the consolidation phase).

This modelling has the advantage of ensuring the weight equilibrium of the tunnel and of explaining all the experimental phenomena, while giving a realistic estimation of the loads in the lining.

REFERENCES

[CA] : Cambou, B. Jaffari K. 1988. Modèle de comportement des sols non cohérents. *Revue française de géotechnique* n°44: 43-45.

[B1] : Bernat, S. Cambou, B. & al 1995. Modelling of tunnel excavation in soft soil. *Proc. 5th Int. Symp. Numerical models in Geomechanics, Numog V*: 471-476. Davos, Suisse.

[B2] : Bernat, S. Cambou, B. & al 1996. Numerical modelling of tunneling in soft soil. *Proc. Int. Symp. Underground Construction in Soft Ground, TC 28*. Londres.

[B3] : Bernat, S. Cambou, B. & al 1997. Tunnel à faible profondeur pour le métro de Lyon. Instrumentation sur site et modélisation. *Proc. XIV Int. conf. Soil mechanics & fondation engineering*. Hamburg.

Computer Methods and Advances in Geomechanics, Yuan (ed.) © 1997 Balkema, Rotterdam, ISBN 90 5410 904 1

Shotcrete-realistic modeling for the purpose of economical tunnel design

R.Galler
Institute of Geomechanics, Tunneling and Heavy Construction Engineering, Mining University Leoben, Austria

ABSTRACT: From 1990 till 1995 the Institute of Geomechanics, Tunneling and Heavy Construction Engineering of Mining University Leoben, Austria, executioned long-term tests in order to show shotcrete's realistic material behaviour. On the one hand the essential part of creep, on the other hand the in theory well known part of relaxation had to be tested in laboratory. The obtained material parameters were used as input parameters in a numerical 3D-simulation taking the Finite Element Program ABAQUS. For the analyse the modified rate of flow method was used. By using this method programming of subroutine UMAT was necessary. The subroutine worked well for a 3D-single element. Good conformity between laboratory tests and calculation was achieved. By applying the routine UMAT to a ralistic tunnel model numerical problems appeared. With the routine creep, which is fix incorporated in ABAQUS, it was possible to run large 3D-problems including realistic long-term-behaviour of shotcrete. The results of calculation show significant relaxation of normal forces and moments in the shotcrete shell with increasing age. So the conclusion by using shotcrete in tunneling is that time works for an increasing factor of safety.

1 INTRODUCTION

In order to simulate the shotcrete shell in numerical studies often an elastic material law is in use. So in order to consider creep, relaxation and shrinkage of shotcrete a material law had to be developed. Such a development needs a large number of laboratory tests and parameter studies, which all were carried out at the Institute of Geomechanics, Tunneling and Heavy Construction Engineering, Mining University Leoben, Austria.

2 LABORATORY TESTS AND RESULTS

The samples were manufactured in one go with the shotcrete lining at four different building sites in Austria (tunnel „Kaponig", tunnel „Sonnberg", tunnel „Galgenberg", tunnel „Semmering").

The laboratory tests contained 329 uniaxial compression tests, thereby 200 short-term- and 129 long-term tests.

Table 1:Number and kind of laboratory short-term tests

short-term tests	number of tests
shotcrete age: 6 hours	40
shotcrete age: 28 days	110
shotcrete age: 7,14,... days	50

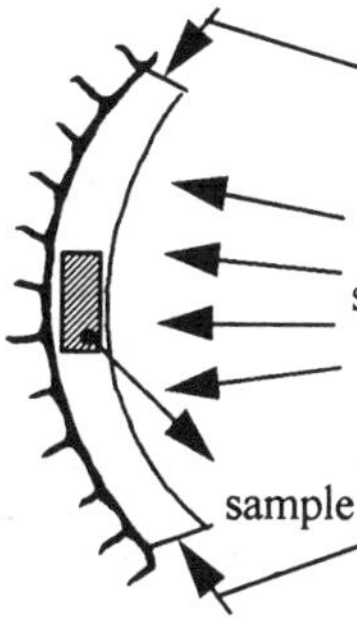

Figure 1: sampling

For the long-term testing the first series of samples was forced to a longitudinal strain of 3000µm/m within the first 100 hours. For the following samples the strain-rate was raised step by step (fig.2). With an longitudinal shortening less than 1% within the first 100 hours no cracks were observed. Axial shortening between 1% and 1,7% caused fine cracks, nevertheless the samples didn't break down.

The stress developement curves (fig.3) show that maximal stress in the tested shotcrete samples is nearly independent from the axial shortening, there are values between 8 and 10 MPa after 100 hours. Constant axial shortening for a period of another 300 hours shows a relaxation of stress to values near 4 MPa.

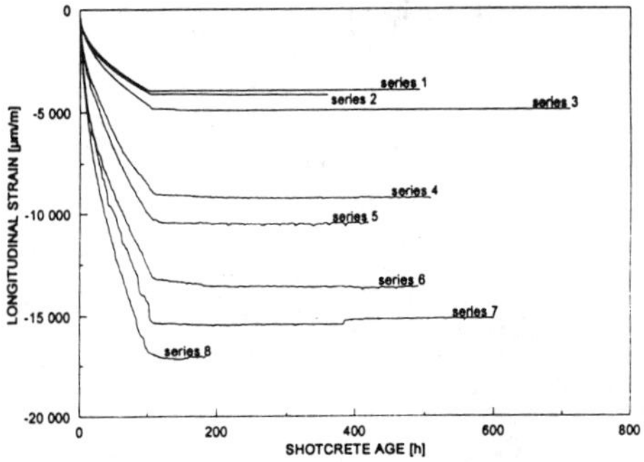

Figure 2: strain controled tests

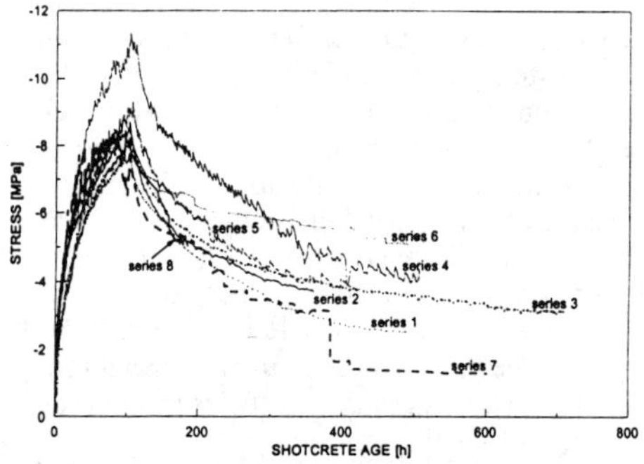

Figure 3: stress developement curve

3 IMPLICATION OF THE MODIFIED RATE OF FLOW METHOD INTO THE SUBROUTINE UMAT

3.1 The rate of flow method after Schubert

The rate of flow methode was worked out by England and Illston [5], in order to simulate creep strain of hardened concrete. Schubert [1] adapted this methode to shotcrete.

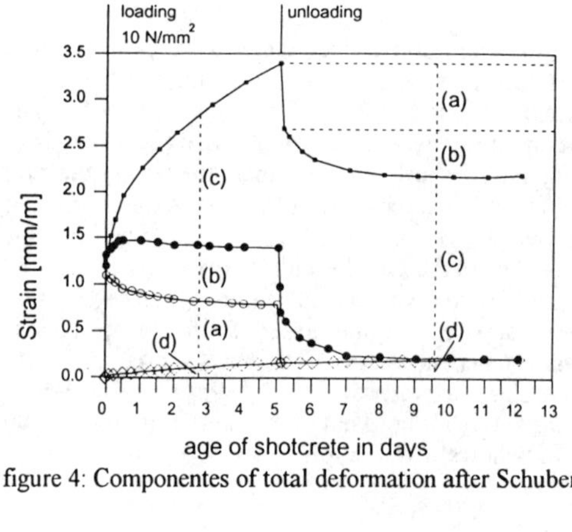

figure 4: Componentes of total deformation after Schubert

After Schubert there are following parts of deformation:

a) spontaneous elastic deformation, proportional to the time dependent Young's Modulus
b) reversible (delayed elastic) creep deformation
c) plastic time dependent creep deformation
d) deformation because of shrinkage and temperature

$$\Delta\varepsilon(t) = \frac{\sigma_1 - \sigma_2}{E(t)} + \sigma_2 \Delta C(t) + \Delta_{\varepsilon_d} + \Delta_{\varepsilon_{sh}}$$

$\Delta\varepsilon(t)$ = change of deformation during an interval
σ_1 = stress at the beginning of interval
σ_2 = stress at the end of interval
$\Delta\varepsilon_d$ = change of delayed elastic deformation
$\Delta\varepsilon_{sh}$ = change of shrinkage and temperature deformation
$E(t)$ = time dependent Young's Modulus
$C(t)$ = irreversible creep deformation

3.2 The subroutine UMAT

With the subroutine UMAT ABAQUS provides an interface whereby any mechanical constitutive model can be added to the library. The following flow chart shows the programming of the above mentioned model:

handing over parameters from the programm into routine Umat
stress, statev, ddsdde

feed back constants into the Common-block
A, Cd, Q, ST, ESCHW

feed back results from the foregoing step of calculation
(get automatically 0 at the first increment)
eddeltaold, etempold, cold,
viscoelxxold, viskoelyyold, viscoelzzold

calculation of the strain incrementes, because of shrinkage and temperature

calculation of principal stresses, [S],
by using the subroutine SPRINC

calculation of the Young's Modulus
$E(t) = E_{28}\sqrt{t / (4.2 + 0.85 * t}$

calculation of the principal strains, [E],
by using the subroutine SPRINC

stresses calculated by using the rate of flow methode

connection of the calculated stresses in 3 dimensions
$Stress(1) = \sigma_{xx} + \nu(\sigma_{yy} + \sigma_{zz})$
$Stress(2) = \sigma_{yy} + \nu(\sigma_{xx} + \sigma_{zz})$
$Stress(3) = \sigma_{zz} + \nu(\sigma_{xx} + \sigma_{yy})$

calculation of eigenvectors of the rotation matrix
by using the subroutine SPRIND
result: matrix AN(i,j)

$$AN = \begin{bmatrix} l_1 & m_1 & n_1 \\ l_2 & m_2 & n_2 \\ l_3 & m_3 & n_3 \end{bmatrix}$$

calculation of the rotation matrix
in order to rotate stresses

calculation of the general stresses

$$\{\sigma'\} = [T_E]^{-T}\{\sigma\} \qquad [T_E]^{-T} = \begin{bmatrix} AN & 2AN \\ 1/2AN & AN \end{bmatrix}$$

$\{\sigma'\}$... general stresses $\{\sigma\}$... principal stress
$[T_E]^{-T}$... rotation-invariant-transpose matrix

feed back of the stress vector

determinant of Jacobian gets 0
DDSDDE (i,j) = 0.0

$$\begin{bmatrix} \frac{\partial\Delta\sigma_{xx}}{\partial\Delta\varepsilon_{xx}} & \frac{\partial\Delta\sigma_{xx}}{\partial\Delta\varepsilon_{yy}} & \frac{\partial\Delta\sigma_{xx}}{\partial\Delta\varepsilon_{zz}} & 0 & 0 & 0 \\ \frac{\partial\Delta\sigma_{yy}}{\partial\Delta\varepsilon_{xx}} & \frac{\partial\Delta\sigma_{yy}}{\partial\Delta\varepsilon_{yy}} & \frac{\partial\Delta\sigma_{yy}}{\partial\Delta\varepsilon_{zz}} & 0 & 0 & 0 \\ \frac{\partial\Delta\sigma_{zz}}{\partial\Delta\varepsilon_{xx}} & \frac{\partial\Delta\sigma_{zz}}{\partial\Delta\varepsilon_{yy}} & \frac{\partial\Delta\sigma_{zz}}{\partial\Delta\varepsilon_{zz}} & 0 & 0 & 0 \\ 0 & 0 & 0 & G & 0 & 0 \\ 0 & 0 & 0 & 0 & G & 0 \\ 0 & 0 & 0 & 0 & 0 & G \end{bmatrix}$$

Jacobian matrix:
calculation of principal diagonal

$$\frac{\partial\Delta\sigma_{xx}}{\partial\Delta\varepsilon_{xx}} = \frac{\sigma_{xx(\varepsilon_{ii}+d)} - \sigma_{xx(\varepsilon_{ii})}}{d}$$

Jacobian matrix:
calculation of the elements above and below
the principal diagonale

$$\frac{\partial\Delta\sigma_{yy}}{\partial\Delta\varepsilon_{xx}} = \frac{P(\sigma_{xx(\varepsilon_{ii}+d)} - \sigma_{xx})}{d}$$

calculation of the rotation matrix [T], in order to determine
the general Jacobianmatrix.

$$[T] = \begin{bmatrix} AN & 0 \\ 0 & AN \end{bmatrix}$$

transpose of rotation matrix

calculation of the general Jacobianmatrix
jacdreh (i,j)

$$[K'] = [T][K][T]^{-1}$$

[K']... general Jakobianmatrix [K]... principal Jakobianmatrix
[T]... orthogonal matrix

storage of values for the next calculation step
epsschwin, etemp, viskoelxx, viscoelyy, viscoelzz

return

Flow chart of programming subroutine UMAT

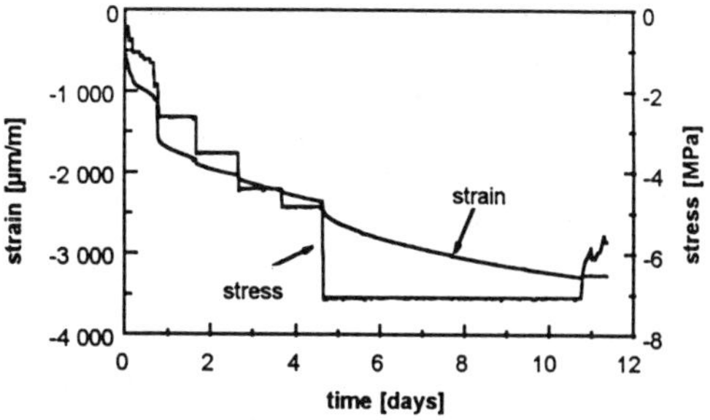

figure 5: stress controled laboratory test

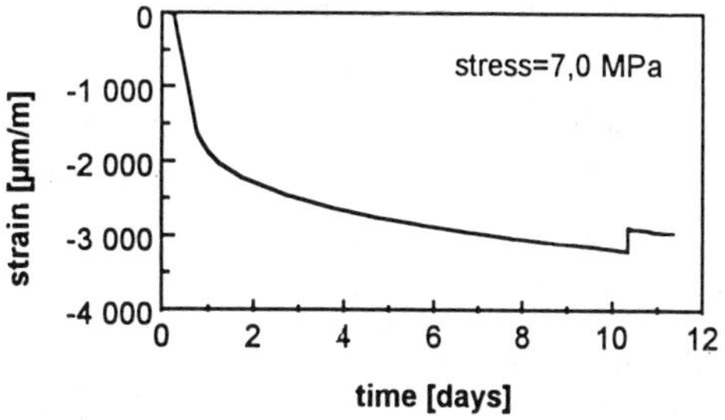

figure 6: calculated longitudinal strain

3.3 Results by using the programmed UMAT

Following figures show the calculated results of a single-element-test in comparison with the measurements:

Following problems occured:

1. With a constant ν it isn't possible to simulate time dependent progress of lateral strain.
2. By using the subroutine SPRINC sometimes stresses in zz-direction get out in yy-direction.

As above mentioned problems occured, the fix incorporated routine CREEP was used for running large 3D-problems in order to simulate long-term behaviour of shotcrete.

$$\varepsilon(t) = \frac{A}{m+1}\sigma^n t^{m+1}$$

A, m, n ... material constants

A,m,n have to be determined by long-term tests. They have no physical meaning.

Of course it was realized that in this way it isn't possible to calculate a realistic short-term behaviour of shotcrete. Nevertheless calculation of long-term behaviour can be done in a satisfactoring way (fig.7).

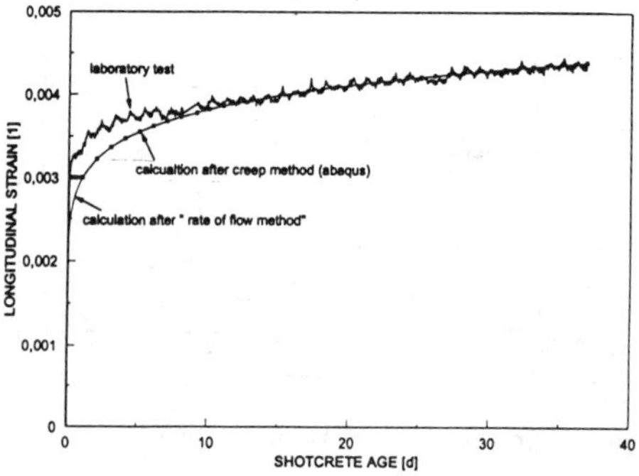

figure 7: comparison between measured and calculated longitudinal strain by using the routine CREEP

4 NUMERICAL SIMULATIONS OF „REAL" 3D-PROBLEMS BY USING THE ROUTINE CREEP

The Finite Element Model which was used for these simulations had a length of 60 m and a width and heigth of 40 m. The diameter of the tunnel was 8 m, the thickness of the shotcrete shell measured 20 cm. The model consisted of about 8000 solid elements and 200 shell elements. In a first step the primary stress was realized, in a second step the young's modulus of the elements in the face of heading had to be decreased, these elements were removed in a third step. After that the stressless activation of the shotcrete shell elements was carried out. For the advancing rock face the described process had to be repeated.

After ending the advance of the rock face the long-term behaviour of the shotcrete shell was tested by carrying out a step time of about 100 days. As mentioned in chapter 3 subroutine CREEP was used for this purpose.

Normal force in the shotcrete shell: At a shotcrete age of 0.5 days the trend of normal force shows an increase from roof to side wall.

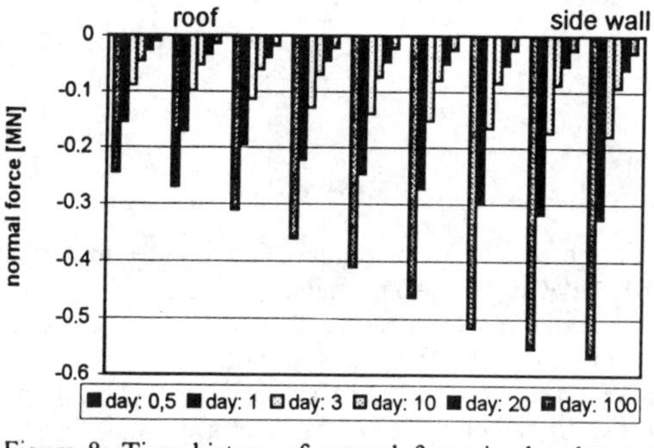

Figure 8: Time history of normal force in the shotcrete shell

Table 3: Values of normal force in the shotcrete shell [MN]

shotcrete age	0.5 days	10 days	100 days
roof	-0.25	-0.05	-0.013
side wall	-0.57	-0.09	-0.03

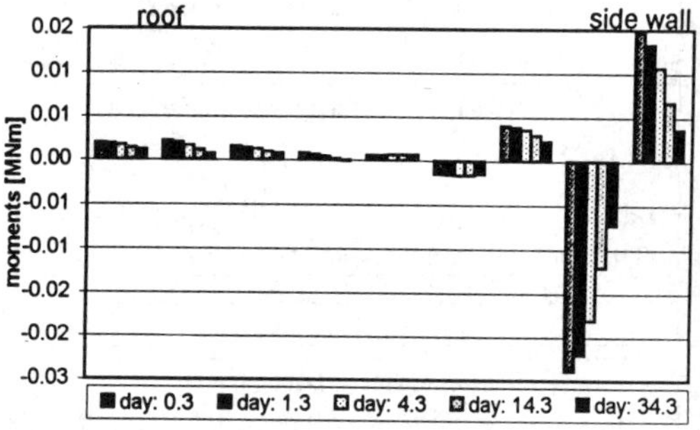

Figure 9: Time history of moments in the shotcrete shell [MNm]

With proceeding time significant reduction of the normal force in the shotcrete shell can be realized. Besides this fact the difference of the normal force between roof and side wall decreases (fig.8)

Moments in the shotcrete shell: As figure 9 shows there is a significant reduction of the moments in the shotcrete shell with increasing time, too.

So the question is, where are the forces going to? The numerical simulations show that the maximum of the stress state is moving from the side wall deeper into the rock with increasing time. During the observed time period its value increases in a small extent.

5 CONCLUSION

Economical tunnel design needs a realistic material law for shotcrete. The modified rate of flow method after Schubert [1] works quite well for single elements but refuses in large three dimensional problems. Therefore subroutine CREEP was used to simulate the long-term behaviour of shotcrete in large problems. These simulations show that both normal forces and moments in the shotcrete shell significantly decrease with time. So the safety factor of the shotcrete shell gets higher with increasing time.

REFERENCES

[1] Schubert: „ Beitrag zum rheologischen Verhalten des Spritzbeton", Felsbau 6 (1988); Nr.3

[2] Aldrian: „Beitrag zum Materialverhalten von früh belastetem Spritzbeton“, Dissertation am Institut für Geomechanik, Tunnelbau und Konstruktiven Tiefbau der Montanuniversität Leoben, Mai 1991

[3] Golser: „Kontrolle der Spritzbetonbeanspruchung im Tunnelbau“, BHM,135. Jahrgang (1990), Heft 10

[4] Golser: Praxisorientiertes Rechenmodell für Tunnel und Kavernen“, Gemeinsamer Forschungsauftrag des Bundesministeriums für wirtschaftliche Angelegenheiten, 2. Zwischenbericht

[5] Golser : „Materialgesetz für Spritzbeton“, Forschungsauftrag des Bundesministeriums für wirtschaftliche Angelegenheiten, Institut für Geomechanik, Tunnelbau und Konstruktiven Tiefbau der Montanuniversität Leoben

[6] England, Illston: „Methodes of computing stress in concrete from a history of measured strain“, Civil Engineering and Public Works Review, 60, Parts 1,2 and 3, April, May and June1965, pp. 513-17, 692-4, 846-7

[7] Mang: „Tunnelabzweigungen in kriechaktiven Böden“, Forschungsauftrag des Bundesministeriums für wirtschaftliche Angelegenheiten, Institut für Festigkeitslehre der Technischen Universität Wien

[8] Mosser: „Numerische Implementierung eines zeitabhängigen Materialgesetzes für jungen Spritzbeton in ABAQUS“, Diplomarbeit am Institut für Geomechanik, Tunnelbau und Konstruktiven Tiefbau der Montanuniversität Leoben, November 1993

[9] Pichler: Untersuchungen zum Materialverhalten und Überprüfung von Rechenmodellen für die Simulation des Spritzbetons in Finite Elemente Berechnungen“, Diplomarbeit am Institut für Geomechanik, Tunnelbau und Konstruktiven Tiefbau an der Montanuniversität Leoben, Dezember 1993

Computer Methods and Advances in Geomechanics, Yuan (ed.) © 1997 Balkema, Rotterdam, ISBN 90 5410 904 1

Numerical analysis of tunnels in strain softening soil

Donatella Sterpi & Annamaria Cividini
Department of Structural Engineering, Politecnico di Milano, Italy

ABSTRACT: Some aspects are discussed of the finite element analysis of underground excavations in soil and rock masses characterized by strain softening behaviour. First the basic features are recalled of a procedure for detecting the bifurcation condition and of a stress-strain relationship taking into account the loss of the mechanical resistance with increasing shear deformation. Then the results of a small scale test on a 3D tunnel model, aimed at investigating the stability at the tunnel face, are compared with those obtained through 3D and 2D finite element analyses, accounting also for softening material behaviour. Finally some comments are presented on the effectiveness of the adopted numerical models.

1. INTRODUCTION

A reliable evaluation of the stability of tunnels would permit a prediction of the induced displacements and, in the case of shallow excavations, of the area interested by the surface settlements. However the stability analysis presents non negligible problems when dealing with stiff soil and rock masses characterized by strain softening behavior. Usually this effect is associated with localization of strains in narrow shear bands [e.g. Vardoulakis, 1988].

The attention is focused here on the numerical modelling of softening effects as observed during laboratory tests on the small scale model of a shallow tunnel. In fact, the comparison between experimental and numerical results allows to check the effectiveness of the numerical analysis of strain softening effects.

In the following, the main aspects of the adopted softening material model and of the laboratory model tests are illustrated. Then the experimental data are compared with those obtained by elastoplastic, three-dimensional finite element analyses, showing that this calculation can reproduce only the initial stages of the test.

Subsequently, in order to model also the behaviour observed when approaching the collapse of the tunnel face, a softening material model is adopted in the framework of a two dimensional finite element scheme. Finally some comments are presented concerning, in particular, the influence of the size of the finite elements on the numerical results.

2. SOFTENING MODEL

In general terms, a strain softening material model should contain two main ingredients. They consist of a suitable criterion for detecting the onset of localization and of a procedure for taking into account the consequent loss of shear strength. Both can be related to different possible hypotheses on the physical nature of the phenomenon [e.g. Sterpi et al., 1995].

Two assumptions have been considered in this study and implemented in the finite element program SOSIA2, for SOil-Structure Interaction Analysis, [Cividini & Gioda, 1992]. In both cases the loss of strength and stiffness of the material is accounted for, but the initiation of this process is associated to different causes [Sterpi, 1997].

In the first case, referred to in the following as case A, the onset of localization, and the formation of a plane of strain discontinuity is strictly related to the non associated nature of the plastic flow rule.

In the second case (B) the loss of mechanical resistance takes place in those zones where a limit of the permanent deformation is reached, regardless of the characteristics of the plastic flow rule.

In the following the main aspects of the localization conditions and of the post-peak behavior are briefly recalled.

2.1 *Localization Condition*

The approach adopted in case A for detecting the onset of localization and the formation of a discontinuity plane follows the procedure proposed by Ortiz et al. [1987].

In plane strain regime, consider a discontinuity plane having an unknown inclination in a reference system $\underline{x}$ and denote with $\underline{n}$ the unit vector normal to it. In the following, infinitesimal increments are denoted by a superposed dot while the symbol Δ indicates the difference between the values of the same variable evaluated on the two sides of the discontinuity plane.

In matrix form, the equilibrium conditions for the stress rate $\dot{\underline{\sigma}}$ across the plane read

$$\underline{\underline{N}}^T \Delta\dot{\underline{\sigma}} = \underline{0} \quad , \tag{1}$$

where the entries of $\underline{\underline{N}}$ are the direction cosines of $\underline{n}$.

The difference between the displacement rates $\Delta\dot{\underline{u}}$ on the two sides of the plane can be expressed as

$$\Delta\dot{\underline{u}} = \left\{\dot{\underline{u}} - \dot{\underline{u}}\right\} = \dot{g}\,\underline{m}\left(\underline{n}^T\underline{x}\right) \; . \tag{2}$$

In the above equation ($\underline{n}^T\underline{x}$) represents the distance from the plane; $\dot{g}$ is an amplitude factor and $\underline{m}$ is the unknown unit vector that defines the direction of the difference $\Delta\dot{\underline{u}}$ of the displacement rates. Eq.(2) shows that the displacement field remains continuous after the onset of localization and that the difference between the displacement increments on the two sides of the plane varies linearly with the distance from it.

The gradient of $\Delta\dot{\underline{u}}$ can be expressed by the Maxwell's form of the compatibility conditions [Thomas, 1961] and this leads to the following matrix form of the "jumps" of the strain rates across the discontinuity

$$\Delta\dot{\underline{\varepsilon}} = \dot{g}\,\underline{\underline{N}}\,\underline{m} \quad . \tag{3}$$

Considering now an elasto-plastic material, it is assumed that the constitutive relationship holds also for the jumps in the stress and strain rates across the discontinuity plane,

$$\Delta\dot{\underline{\sigma}} = \underline{\underline{D}}^{ep}\Delta\dot{\underline{\varepsilon}} \tag{4}$$

where $\underline{\underline{D}}^{ep}$ is the tangent elasto-plastic constitutive matrix. This implies that the material behaviour on both sides of the plane is governed by the same tangent constitutive matrix or, in other words, that when bifurcation occurs the material on both sides of the plane remains on the loading branch of the elasto-plastic relationship and that no elastic unloading takes place.

By substituting eq. (3) into eq. (4), and taking into account eq. (1), the following homogeneous expression is arrived at

$$\left[\underline{\underline{N}}^T\underline{\underline{D}}^{ep}\underline{\underline{N}}\right]\underline{m}\,\dot{g} = \underline{0} \quad . \tag{5}$$

According to eq. (5) localization can take place only if the determinant of the matrix within square brackets vanishes.

In the presence of a non associated plastic flow rule, this condition can lead to real solutions also in the case of perfectly plastic or positive hardening behaviour [Rudnicki & Rice, 1975]. Observing that in the former case the localization condition depends only on the stress state and on the material parameters, it is possible to evaluate the stress states in which the above condition is fulfilled, as shown for the Drucker-Prager yield criterion by Sterpi [1997].

In the approach adopted in case B, the initiation of softening is simply related to the attainment of a predefined limit of the plastic strains [Cividini & Gioda, 1992].

2.2 *Post-peak Behaviour*

Once the criterion discussed in the previous section, depending on case A or B, is fulfilled, a loss of strength (and stiffness) takes place with increasing permanent deformation. In particular, it is assumed that the shear strength parameters (e.g. cohesion and friction angle) are linear functions of a measure of the irreversible deformation represented by the square root of the second invariant of the deviatoric plastic strains. The parameters remain unchanged, and equal to their peak values, until the previously discussed condition is fulfilled. Then, they are reduced with increasing plastic deformation until their residual values are reached.

This material scheme has been implemented in a finite element code and details of the step-by-step iterative solution technique for the post-peak analysis have been presented in [Cividini & Gioda, 1992].

3. LABORATORY MODEL TESTS

The experimental investigation on a small scale tunnel model has been carried out at the Takasago Laboratory of Mitsubishi Heavy Industries, Ltd [Sterpi et al., 1996]. Here, for sake of briefness, only the main features of this investigation are recalled.

The experimental setup consists of a 3.85x3x3.5m steel container within which a horse-shaped metallic pipe can be placed which represents the tunnel lining. A vinyl bag is inserted within the pipe as a support of the tunnel face. The test preparation initiates by spreading out a first layer of uniformly grained sand from a hopper, and by placing the tunnel model on it. To ensure the repeatability of tests, and the homogeneity of the soil mass, the deposition continues in layers until a level over the tunnel crown is reached, equal about to the tunnel diameter.

The vinyl bag is pressurized by air to ensure the stability of the tunnel face during the sand deposition. The air pressure was evaluated following a procedure used in actual tunnel excavations for ensuring the stability of the face [e.g. Ward & Pender, 1981]. This pressure is then reduced during the test by small steps until failure occurs.

For monitoring the test progress, various slope transducers and some load cells are placed in the sand during the preparation of the tests. In addition nine displacement transducers are installed to record the surface settlements (cf. position 'h...' in fig.1).

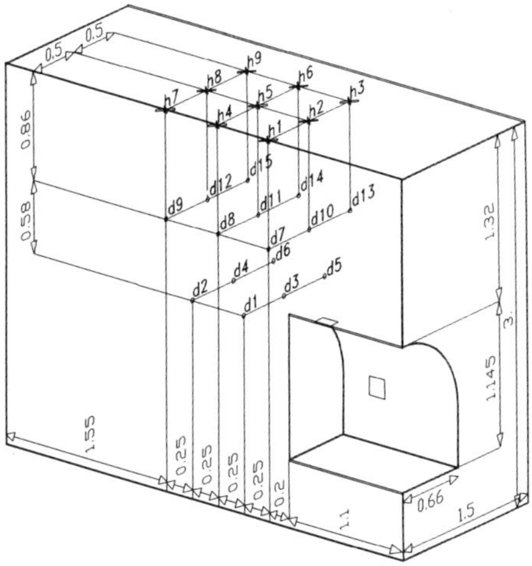

Fig.1 - Axonometric view of one-half of the laboratory model and transducer location.

The described apparatus has been used in a series of tests, adopting three different tunnel models, two of which include a pre-lining. The surface settlements recorded in the tests without pre-lining are used in the next Section for a quantitative comparison between experimental and numerical results.

4. NUMERICAL INVESTIGATION

The numerical part of the study includes 3D and 2D elastic perfectly plastic finite element analyses and some 2D calculations in which the mentioned softening models A and B were introduced.

For the elastic-perfectly plastic analysis in 3D regime, the soil was modelled by 8-node brick elements, while 4-node elastic shell elements were used for the tunnel liner. Starting from the initial geostatic condition, the "excavation" forces were gradually applied to the nodes of the tunnel face, until reaching the collapse of the face.

The increase of the surface settlements during the test and the corresponding numerical results are shown in fig.2. Only the initial portion of the recorded data is reported in these diagrams, since in the subsequent part of the test the soil mass was no longer homogeneous due to the formation of a sliding surface. This strain localization represents a condition that cannot be modelled through the adopted perfectly plastic behavior.

Considering a simplified 2D geometry and the softening approaches recalled in Sec.2, an attempt was made to simulate the tunnel test up to the collapse of its face. Even if the plane strain regime of these analyses does not permit a quantitative comparison with the experimental results, these analyses may lead to some qualitative conclusions on the effectiveness of the adopted material models.

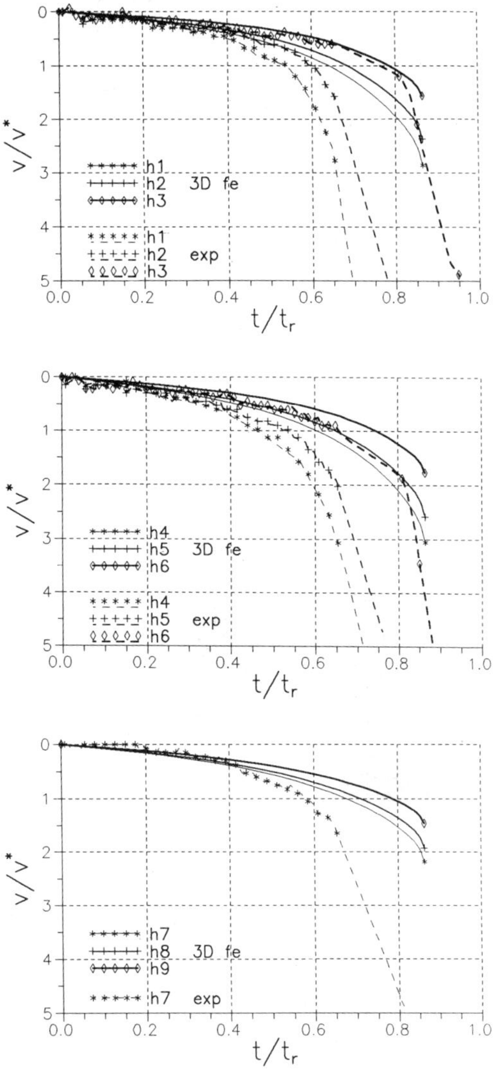

Fig.2 - Variation of recorded surface settlements during the test and comparisons with the results of the 3D elasto-plastic analysis.

The results of calculations are summarized in figs.3 and 4. The first one shows the deformed meshes at collapse for perfectly plastic and softening cases, while fig.4 shows the contour lines of the square root of the second invariant of deviatoric plastic strains for the same cases.

The marked influence of the softening behaviour is clearly shown by these results. In fact, in the case of softening the plastic zone spreads up to the ground surface, with a consequent large increase of the surface settlements, while in the perfectly plastic case the plastic zone develops only in the vicinity of the tunnel face.

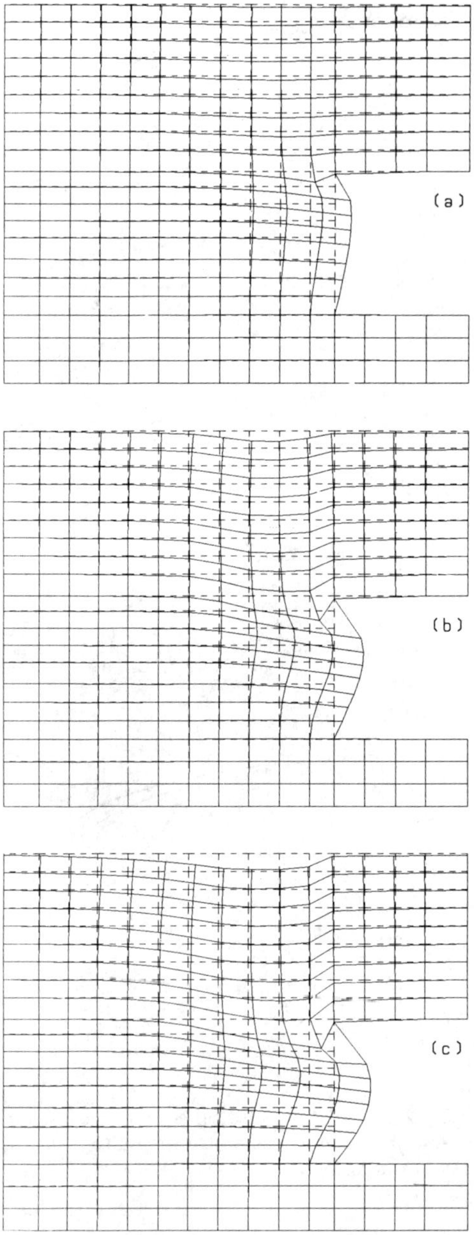

Fig.3 - Deformed finite element coarse grids at the tunnel face adopting (a) elastic-perfectly plastic and (b,c) softening material models A and B (displacement magnification factor = 7).

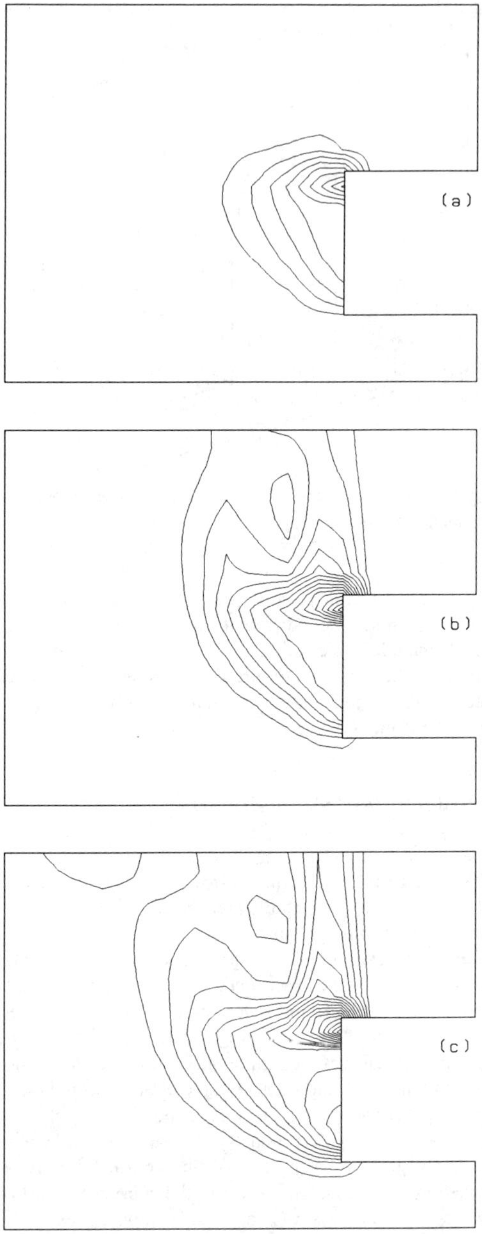

Fig.4 - Contour lines of the second invariant of the deviatoric plastic strains for the cases in fig.3 (minimum contour value 0.005, contour increment 0.5).

It is known that the results of finite element softening analyses are influenced by the size of the adopted elements. To reach some insight into this problem, the calculations were repeated using a refined grid. Figs. 5 and 6 report, respectively, the deformed meshes and the plastic strain contour lines at collapse, obtained with the fine grid. Again, as a consequence of softening, two zones or "bands" develop in which the shear strains tend to concentrate. The first one, characterized by a large strain gradient, is located along the vertical line through

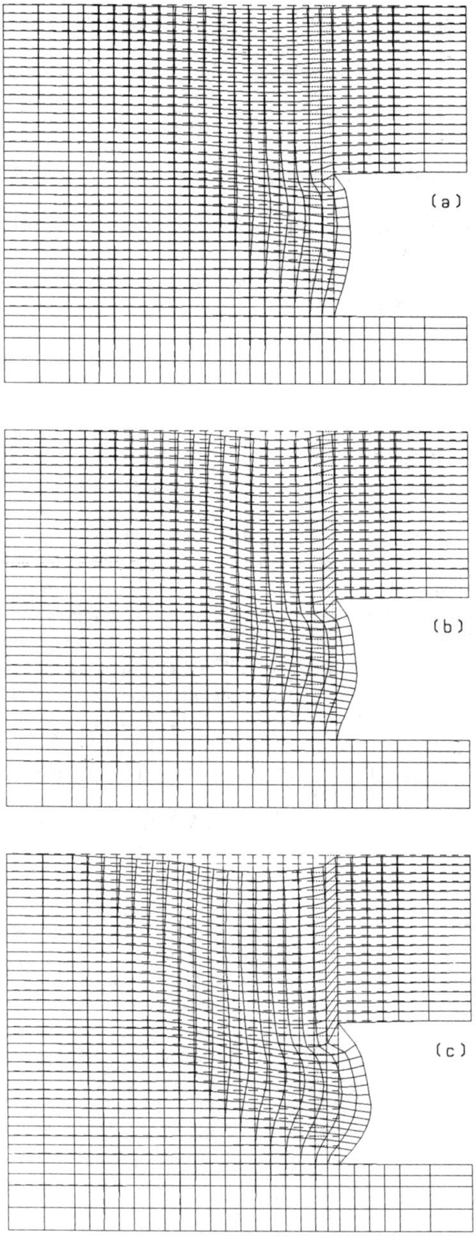

Fig.5 - Deformed finite element fine grids at the tunnel face adopting (a) elastic-perfectly plastic and (b,c) softening material models A and B (displacement magnification factor = 7).

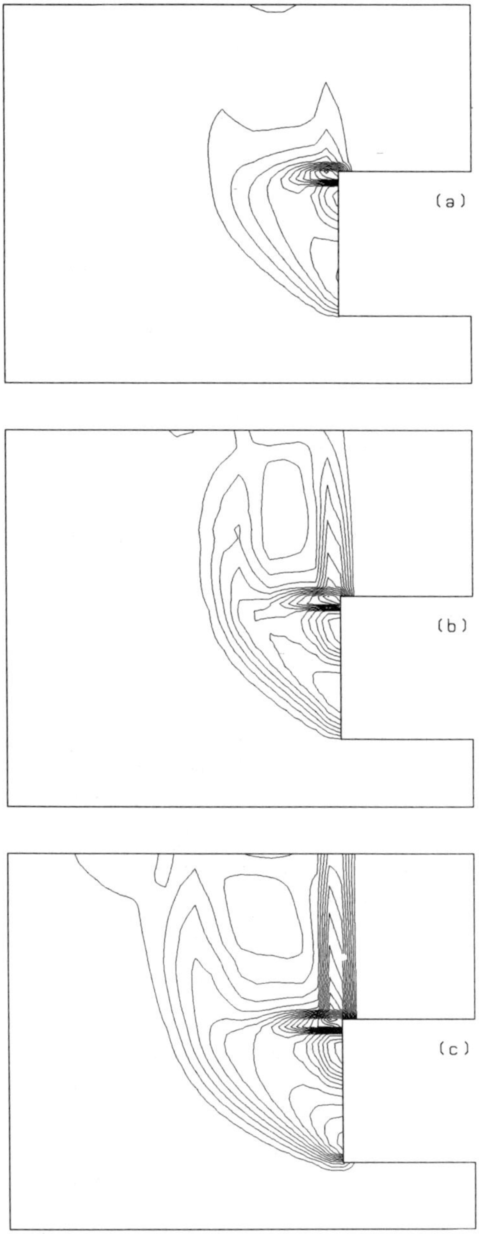

Fig.6 - Contour lines of the second invariant of the deviatoric plastic strains for the cases in fig.5 (minimum contour value .005, contour increment .5).

the crown, while the second one has a curved shape and starts at the excavation bottom.

To limit the influence of discretization, the rate of reduction of shear strength with increasing shear deformations was related to the "average" size of the elements as suggested by Pietruszczak & Mroz [1981].

This turned out to be an effective provision, in fact the influence of the element size in the case of softening was quite similar to that observed in standard perfectly plastic calculations as shown in fig.7.

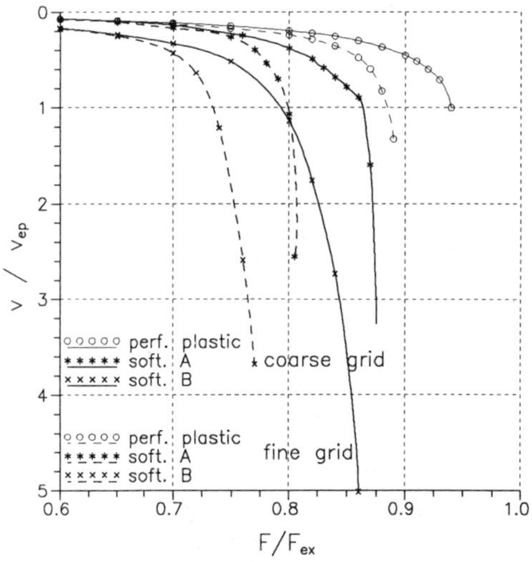

Fig.7 - Variation of vertical displacement at location 'h4' with increasing excavation forces assuming elastic-perfectly plastic and softening material behaviour and adopting coarse (solid lines) and fine (dashed lines) grids.

On the basis of the presented results, it can be concluded that both softening approaches are able to model the progressive changes toward the collapse of the tunnel face and, in addition, they are able to predict the surface displacements. They led to slightly different results, e.g. in the softening case A the plastic zone has extension slightly smaller than that obtained in case B, however this is due to different way in which the onset of localization is evaluated.

5. CONCLUSIONS

Two strain softening models have been adopted in 2D finite element analyses aimed at the interpretation of the behavior recorded in laboratory tests on a small scale tunnel.

In the first case the onset of localization is seen as the effect of a structural instability of the material related to a non associated plastic flow law. Once localization has occurred, the analysis is continued considering the reduction of shear strength parameters with increasing irreversible deformation. In the second case the loss of shear strength takes places in the zones where a particular limit in the permanent deformation is reached.

From a qualitative point of view, it was observed that the softening models are able to simulate the deformation that gradually develops from the tunnel face towards the ground surface, which is missed by standard perfectly plastic analyses.

Both the approaches have a potential for application to engineering problems. However, a quantitative assessment of the advantages of such models can be obtained solely by carrying out 3D softening analysis and by comparing their results with laboratory and field measurements.

ACKNOWLEDGMENTS

The financial support of the Ministry of the University and Research of the Italian Government and of the Construction Engineering Research Institute Foundation (Kobe, Japan) are gratefully acknowledged.

REFERENCES

Cividini A. & Gioda G. (1992). A finite element analysis of direct shear tests on stiff clays. Int. J. Numer. Anal. Methods in Geomechanics, 16:869-886.

Ortiz M., Leroy Y. & Needleman A. (1987). A finite element method for localized failure analysis. Computer Meth. Applied Mech. & Eng., 61: 189-214.

Pietruszczak S. & Mroz Z. (1981). Finite element analysis of deformation of strain softening materials. Int. J. Num. Methods in Eng., 17:327-334.

Rudnicki J.W. & Rice J.R. (1975). Conditions for the localization of deformation in pressure-sensitive dilatant materials. J. Mech. Phys. Sol., 23:371-394.

Sterpi D., Cividini A. & Donelli M. (1995). Numerical analysis of a shallow excavation. Proc. 8th ICRM, Tokyo, 2: 545-549.

Sterpi D., Cividini A., Sakurai S. & Nishitake S. (1996). Laboratory model test and numerical analysis of shallow tunnels. Proc. ISRM Int. Symposium Eurock '96, Torino, Italy.

Sterpi D. (1997) Influence of the strain localization phenomenon on the stability of underground openings (in italian). Ph.D.Thesis, Politecnico di Milano, Italy.

Thomas T.Y. (1961). Plastic flow and fracture in solids. Academic Press, New York.

Vardoulakis I. (1988). Stability and bifurcation in geomechanics. Proc. 6th ICONMIG, Innsbruck, 155-168

Ward W.H. & Pender M.J. 1981. Tunneling in soft ground. Proc. 10th ICSMFE, Stockholm, 4: 261-275.

Computer Methods and Advances in Geomechanics, Yuan (ed.) © 1997 Balkema, Rotterdam, ISBN 90 5410 904 1

Tunneling in gravel formations – Investigation and numerical modeling

M.H.Wang
Civil Engineering Division, Provisional Engineering Office of High Speed Rail, Taipei, Taiwan, China
C.T.Chang, T.S.Yen & Y.K.Wang
Sinotech Engineering Consultants Ltd, Taipei, Taiwan, China

ABSTRACT: Gravel formations are widely distributed along the western foothills geologic province. Tunneling in rural areas of this formation with shallow to medium overburden requires a different design philosophy than the urban tunneling to justify the safety and economy of the tunnel support design. An investigation program is conducted and a numerical modeling and verification scheme is developed to meet the objective.

1 INTRODUCTION

Gravel Formations, generally in the middle to late Pleistocene age, are widely distributed along the piedmont area west of the western foothills geologic province as shown in Figure 1. The gravel formation, from a few decades to hundreds meters thick, is mainly composed of gravel layers of well graded and loosely cemented gravel and intercalation of sand and clay beds or lenses.

With the rapid infrastructure development, more and more underground construction projects are concentrated in these areas. Among these projects, the high speed rail tunnel project, cross section of 160 m^2, underpassing the terrace gravel formation with overburden thickness of a few meters to over one hundred especially make formidable challenges to the geotechnical and tunnel engineers.

The gravel formation can be categorized as competent ground with respect to foundation engineering, but as weak ground with respect to tunneling. Unlike the urban tunneling in weak ground with shallow overburden where both stability and ground settlement are important, the shallow cover sections of tunnel in the terrace gravel formation are located in rural areas where considerable less ground settlement restriction is required to the design and construction of the tunnel. Therefore, the analysis and design of shallow overburden tunnel which emphasizes the stability, justifies the tunnel safety and economy and extends to medium and high overburden are attempted.

Since little tunneling experience associated with such tunnel dimension and overburden in the gravel formation is available, a geotechnical investigation program to look into the strength and compressibility characteristics of the gravel formation is conducted and the numerical modeling to simulate tunnel excavation and support, which serves as parametric studies and design verifications, is carried out to justify the primary support design of the tunnels.

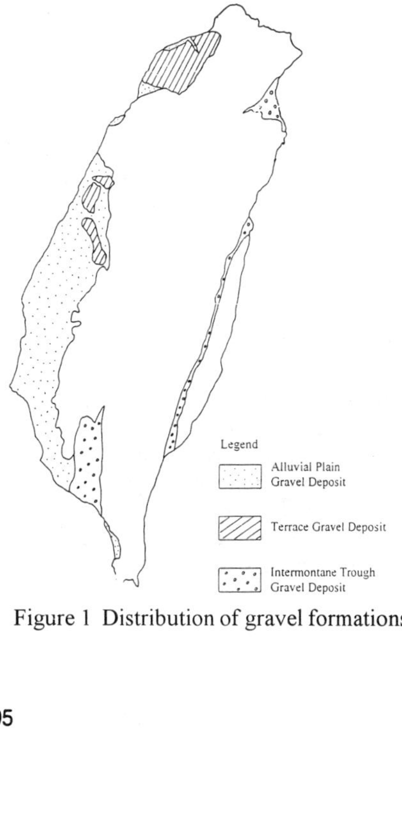

Figure 1 Distribution of gravel formations in Taiwan

2 GEOTECHNICAL INVESTIGATION

The rockfill materials have been extensively studied regarding the characteristics of strength and compressibility using remoulded samples in large-scale triaxial compression tests (Marsal 1967 and Marachi et al. 1972). The remoulding process can be justified for the rockfill materials; however, the process can not reconstitute the in-situ particle structure and the bonding or cementation contributed by the matrix fines of the gravel formations which may lead to underestimate of strength and compressibility of the formations. In the sampling aspect, although undisturbed sampling by ground freezing method for gravel materials shows some promising results (Goto et al. 1992), the undisturbed sampling in gravel formation is still very difficult or practically impossible, if large diameter of gravel particles are present. Therefore, the geotechnical investigation of the gravel formation can mainly resort to the in-situ tests.

A series of investigation programs including percussion drilling with resistivity logging, test pit, field sieve analysis, field unit weight, field direct shear, and field plate loading tests are conducted to look into the engineering properties of the ground.

Percussion drilling using jumbo drilling rig with resistivity logging verification is conducted to assess the vertical distribution of the gravel formation. Field unit weight of the gravel formation is measured in a volume of 1 cubic meter within the test pit. To obtain the field characteristics of the grain size distribution, the field sieve analysis is performed on air dry samples with quantity approximately equal to 100 times of the particle weight of D_{90} (grain size corresponding to 90% passing by weight).

Field direct shear tests following the ISRM recommended procedures (Lama & Vutukuri 1978) on sample of 80cm × 80cm × 40cm in dimension with a pre-set shear gap of about 5 cm are conducted to assess the shear strength of the material. Although the shear gap no less than the maximum grain size of the sample is recommended (Head 1982) to eliminate the possible interference of particle rolling confined within the RC cap which may induce additional friction during shearing, the recommended set-up is practically impossible for this gravel material because bearing type of failure may occur in the shear gap due to normal force prior to the shearing application. The plate loading test following the ASTM D1194-72 recommended procedure is performed with a steel plate of 87.4 cm in diameter to evaluate the deformability of the formation.

3 ENGINEERING PROPERTIES OF THE GRAVEL FORMATION

The unit weight of the gravel formation is found to be around 19 to 23 kN/m^3. The typical grain size distribution of the gravel material is shown in Figure 2 which indicates the maximum nominal diameter of the particle of about 30 cm, however boulders up to 100 cm are occasionally found in the field. The gravel content (USCS) of the formation is generally high, ranges from 50% to as high as 90%.

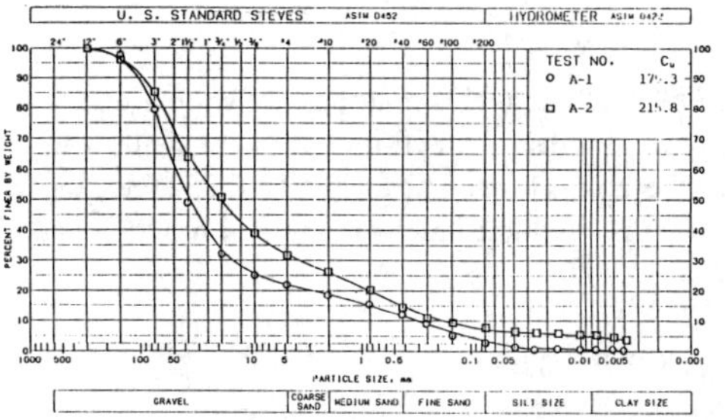

Figure 2. Typical grain size distribution

The geological and geotechnical investigation of the terrace gravel formation indicates that the formation can be categorized into clast-supported and matrix-supported gravel depending on the amount of gravel content. Studies have shown that the gradation has a significant influence on the engineering properties of the rockfill materials (Matheson 1986). Although some direct shear test results show significant difference between clast-supported and matrix-supported gravel, quantification of the shear strength with respect to gravel content based on 16 sets of field direct shear tests on this formation is somehow inconclusive (Tsai et al. 1995) due to scattering of the data. However, shear strength parameters are found to be: apparent cohesion 15 ~ 45 kN/m^2 and angle of shearing resistance 32 ~ 45 degrees.

Shearing deformation is generally accompanied by dilatancy in gravel materials. The angle of dilation ψ defined by

$$\psi = \text{Arctan}(dY/dX)$$

where dY and dX are the displacement increment in the normal and shearing directions respectively. From the field direct shear test results, the angle is found to be 8 ~ 12 degrees within the range of normal stresses applied.

The elastic modulus of the ground is estimated using elastic solution for circular plate loading on a half infinitive space based on the unloading-reloading curve of the field plate loading test to exclude the possible ground disturbance during set-up. The values are between 200,000 and 1,000,000 kN/m^2 depending on the gravel content and packing of the particle structure.

The effects of confining pressure to the engineering properties of rockfill materials such as shear strength, deformation modulus, and volumetric and axial strains at failure have been proved to be quite significant (Marachi et al. 1972 and Charles and Watts 1989), but due to limitation of the in-situ test program, the effects to this gravel formation is not investigated. Nevertheless, the influence is supposed to hold true for this gravel formation.

4 TUNNEL DESIGN CONCEPT AND NUMERICAL SIMULATION

The mountain tunnel with high overburden in competent rock is generally designed based on the engineering experience and the performance of existing tunnels in similar ground. The design concept is to utilize the ground surrounding the opening to support itself as much as possible to reach an economic design. While the urban tunnel design, generally of shallow overburden and in weak ground, is concerned more with the ground settlement and the consequence to the existing structures. Therefore tunnel design and construction to limit the ground deformation has become the first priority and results in quite heavy tunnel support requirement.

For tunnels in rural area with shallow to medium overburden and in weak ground, a different design thought other than that developed for the urban tunneling is required. For rural tunnels with shallow overburden in weak ground, less ground settlement restriction is required, thus the ground surrounding the tunnel opening should be utilized to some extent and stability be maintained to justify a safe and economic tunnel support design.

Little tunneling experience with such a tunnel dimension in this terrace gravel formation is available. Hence, a numerical modeling to assess the tunnel excavation and support to meet the objectives set forth is conducted. In this study, a two dimensional plain strain model using a commercial program FLAC by ITASCA consulting group, Inc. which implements finite difference solving scheme is used to simulate the interaction of ground and tunnel excavation and support.

In a numerical modeling of the tunnel support and ground interaction assuming the ground as a continuum, the confining pressure required from tunnel support will approach an asymptotic limit if the deformation of ground is allowed to develop large enough which is generally termed as the ground reaction curve. The ground reaction curve and ultimate load on tunnel is studied using a plain strain trap door numerical modeling by Koutsabeloulis and Griffiths (1989). The trap door simulation is quite straight forward numerically. However, a two dimensional plain strain model to simulate the three dimensional tunnel excavation and support needs some knowledge of the pre-deformation which is the ground deformation before tunnel drive face is reached. Unfortunately, the pre-deformation is generally not known, because most of the tunnel monitoring system only measures the ground deformation after support installation (except for shallow cover tunnel monitoring). Nevertheless, the effect of pre-deformation in numerical modeling is generally simulated in practice by stress release of the initial stress of the ground after tunnel excavation. The percentage of stress release is somehow subjected to engineering judgment and becomes quite controversial. In this study a more objective method to justify the percentage of stress release is attempted for the case which concerns mainly the tunnel stability or in other words the ground load at failure.

Localized deformations or the displacement discontinuities are generally found within soil mass at failure. The shear strain gradient is expected to be the greatest along the discontinuities, although the discontinuities do not coincide with those planes which are associated with the maximum shear strain (Roscoe 1970). Nevertheless, the development and propagation of localizations within the soil mass is still best illustrated using contours of the maximum shear strain (Stone and Muir Wood 1992).

Stability, in other words the ground load at failure, is the major interest of the tunnel design here. The maximum shear strain developed within the soil mass in the numerical analysis can be utilized for the judgment of impending ground failure surrounding the tunnel opening as long as the maximum shear strain at failure of the ground material can be estimated. And hence, the adequacy of tunnel primary support can be verified.

The numerical modeling to simulate the sequential tunnel excavation and support procedures and the verification of tunnel support is briefly described as follows:

(1) the primary stress condition prior to tunnel excavation which is a function of overburden is calculated;

(2) the ground in the top heading is removed and internal pressure is applied by say 50% of the equivalent primary stresses to simulate the percentage

of stress release before the application of the primary support (shotcrete and rockbolts); then the shotcrete and rockbolts in the top heading is installed and simulated until the equilibrium of the entire system is reached;
(3) repeat step (2) for the bench and invert excavation and support;
(4) the load carrying members of the primary support are checked against the ultimate strength with defined safety factors;
(5) the ground surrounding the opening is checked against the impending failure for verification as outlined in the above concept.

The maximum shear strain pattern of a case of the numerical analysis is shown in Figure 3 (a). The tunnel dimensions of the simulated case are width of 14 m and height of 12 m. The overburden is 15 m above the crown. The elastic modulus is 300,000 kN/m^2 and ν of 0.35. The gravel material is assumed to follow the Mohr-Coulomb plasticity with an apparent cohesion of 20 kN/m^2 and angle of shearing resistance of 34 degrees. The dilation angle is assumed to be zero. The unit weight of the ground is 22 kN/m^2. The coefficient of lateral earth pressure at rest K_0 is 0.4. The stress release in the top heading excavation is set to be 40% of the initial stresses.

It is found that the dilation angle makes the greatest influence to the collapse load using trap door modeling (Koutsabeloulis and Griffiths 1989). A parametric study to look into this aspect is conducted using this modeling scheme. The stress release ratio is maintained to be 40% and the dilation angle is changed from 0 to 20 degrees using a non-associated flow rule. The results reflecting on the concentration factor, Cv, which is defined as

$$Cv = N/(\gamma t \times H \times R) \times 100\%$$

where N: axial load on the tunnel support at spring line level; γt: total unit weight of the ground; H: the overburden above tunnel crown and R: the radius of tunnel cross section, is found to be less significant, all cases Cv values around 45%. However the effect is rather significant to the maximum shear strain pattern developed around the tunnel as shown in Figure 3.

In Figure 3 (a) where the dilation angle is 0, the maximum shear strain contours are well developed and extended to the ground surface, however the 3 (b) where the dilation angle is 20 degrees, the general failure pattern of the ground is still not significant. For the design purpose, the dilation angle of the ground can be taken as zero for conservatism. Because, for materials with higher angle of dilation to develop the similar pattern as in the case of zero dilation angle, a higher percentage of stress release is required and results in less ground load at failure and lighter tunnel support.

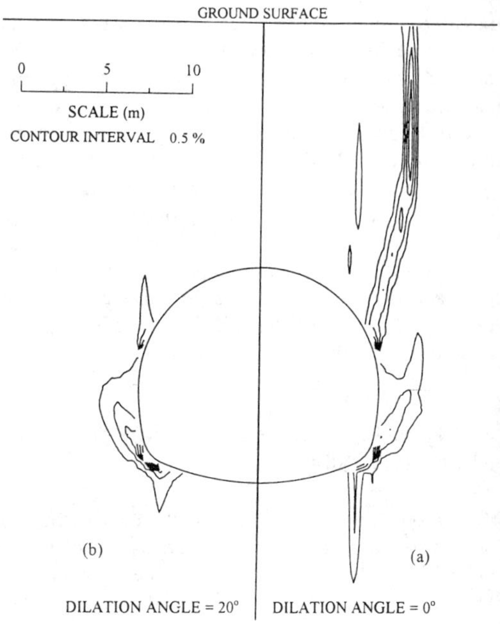

Figure 3. Shear strain pattern around tunnel with dilation angles of 0 and 20

In the invesigation of this terrace gravel formation, no strains are measured in the in-situ tests. The volumetric and axial strains at failure of the large consolidated drained triaxial compression tests on Oroville dam material (Marachi et al. 1972), which has the best similarity in particle shape and gradation to the studied terrace gravel material, are used to estimated the maximum shear strain at failure under different confining pressure. The maximum shear strains at failure under different confining pressures of 132 and 792 kN/m^2 are, thus, estimated to be 1.9% and 2.9% respectively for the gravel material.

Since the maximum shear strain at failure at higher confining pressure is larger than that at lower ones, the percentage of stress release of the numerical simulation at higher overburden can be higher than that of shallow overburden. Hence, a rational tunnel support requirement of medium to high overburden can be justified with the same numerical modeling scheme.

5 CONCLUSIONS

A numerical modeling scheme with the verification procedures for the primary tunnel support design where stability is of the concern is proposed for shallow overburden tunnel and extents to medium and high overburden. However, It has to be emphasized that it is not the intention of the authors to replace the tunnel engineering experience with such a modeling and verification scheme of the tunnel support design. It is only to provide as a preliminary analysis of the tunnel support where little engineering experience is available.

In the investigation program of the terrace gravel material, the particle size effects on the strength and compressibility is excluded due to the practical difficulties and budget limit to the dimensions of the test samples. More versatile in-situ tests should be developed to investigate the engineering properties of materials with large grain particles.

The estimation of the maximum shear strain at failure within the soil mass in the verification procedures using triaxial testing apparatus has to notice that the volumetric and axial strains at failure show different magnitude between active and passive failure modes. The ground failure surrounding the tunnel opening is generally the active one. The test set-up to measure the volumetric and axial strains should be properly arranged if triaxial testing apparatus is to be used.

The tunnel support design of the high speed rail tunnel project using the above procedure is completed in the design stage. The construction is about to initiate in the near future. Verification from the construction performance of the tunnel is needed to justify the proposed tunnel support design procedures in this study.

REFERENCES

Charles, J.A. and K.S. Watts, 1980. The influence of confining pressure on the shear strength of compacted rockfill. *Geotechnique* 30 (4): 353-367.

Goto, S., Y. Suzuki, S. Nishio & H. Oh-oka 1992. Mechanical properties of undisturbed tone-river gravel obtained by in-situ freezing method. *Soil and Foundations.* Vol. 32 (3): 15-25.

Head, K.H. 1982. *Manual of soil laboratory testing,* Vol. II:554-559.

Koutsabeloulis, N.C. & D.V. Griffiths 1989. Numerical modeling of the trap door problem. *Geotechnique* 39 (1): 77-89.

Lama, R.D. & V.S. Vutukuri 1978. *Handbook on mechanical properties of rocks* Vol. III.

Marachi, N.D., C.K. Chan & H.B. Bolton 1972. Evaluation of properties of rockfill materials. *J. Soil Mechanics and Foundations, ASCE.* Vol. 98 (SM1):95-114.

Marsal, R.J. 1967. Large scale testing of rockfill materials. *J. Soil Mechanics and Foundations, ASCE.* Vol. 93 (SM2):27-43.

Matheson, G.M. 1986. Relationship between compacted rockfill density and gradation, *J. Geotechnical Engineering, ASCE.* Vol. 112 (12):1119-1124.

Roscoe, K.H. 1970. The influence of strains in soil mechanics. The Tenth Rankine Lecture.

Stone, K.J.L. & D. Muir Wood 1992. Effects of dilatancy and particle size observed in model tests on sand. *Soils and Foundations.* Vol. 32 (4):43-57.

Tsai, M.H., J.C. Chern & M.D. Wang 1995. The study of shear strength of gravel formations in western Taiwan. *Proc. Int. Symp. Underground Construction in Gravel Formations.* March 23-24, 1995, Taipei, Taiwan, CHINA

Computer Methods and Advances in Geomechanics, Yuan (ed.) © 1997 Balkema, Rotterdam, ISBN 90 5410 904 1

Research on computation methods for structural reliability index of tunnel lining

Gaobo, Tan Zhongsheng & Guan Baoshu
Department of Underground and Geomechanics Engineering, Southwest Jiaotong University, Chengdu, China

ABSTRACT: This paper applies the method that combined the finite element response surface method with Monte-Carlo simulation to Calculate the characteristic parameters of force in tunnel lining structure.On these grounds,the validity of these reliability analysis methods is verified in calculating the structural reliability index of cut tunnel lining and cover tunnel lining.Comparing and analogizing the process and results,this paper makes a conclusion that TRYM method has advantage over than other methods.

Keywords: tunnel, lining structure, reliability index

1 INTRODUCTION

According to existing Tunnel Design Code,the single safety coefficient in the design method of certainty value has these shortcomings as follows (1)Because the safety coefficient is obtained from experiential method,it is inconsistent with accurate computation at present.(2)The single safety coefficient based on certainty of loads and resistance can not comprehensively reflect the actual capacity of structures.(3)Structural reliability may not be increased in proportion which structural safety coefficient is increased.Hence,it is essential that the limit state design based on reliability theory is applied in tunnel and underground engineering.The material uncertainty,geometric uncertainty,loading uncertainty and model uncertainty are considered in this paper,best method to calculate structural reliability of tunnel lining is researched by means of theory analysis and actual engineering exemples.

2 ESTABLISH LIMIT STATE EQUATION

To determine the failure state of tunnel lining structure,limit state equation is established by using strength test formula for existing tunnel design code.When the distance which axial force deviates from centre line $e_0 \leqslant 0.2d$,the lining section is controlled by compression strength,and when $e_0 \leqslant 0.2d$, the section is controlled by tension strength.Therefore, limit state equations of lining structure are derived as follows:[1]

$$g=\phi\ \alpha\ d\sigma_c b-N=0 \qquad e_0 \leqslant 0.2d \tag{1}$$

$$g=\frac{1}{6}(1.75\ \phi\ \sigma_t bd_2+Nd)-M=0 \qquad e_0 < 0.2d \tag{2}$$

in which b is width of section,generally b=1m; d is thick of section; σ_c is limit compressive strength of concrete; σ_t is limit tensile strength of concrete; M is moment; N is axial force; α is effect coefficient α $=1-1.5e_0/d$; ϕ is axial wanding coefficient, generally $\phi=1$.

3 ANALYSIS METHOD OF STRUCTURAL RELIABILITY

There are many methods to calculate structural reliability index β , in this paper, five methods are applied to calculate structural reliability of tunnel lining. They include improved Monte-Carlo Method, JC method, improved design point method. (TRYM),RBDF method and a new iterative method (New-ZWQ).

3.1 Improved Monte-Carlo method

At pressent, the Monte-Carlo method in structural reliability analysis usually requests a great deal of simulation, it is about $100/p_f$ times,and it has not enough high precision. Therefore, this paper applies improved monte-carlo method to calculate structural fuilure probability, basic theory is as follows.[2]

If structural limit state equation is

$$z=g(x_1,x_2,\cdots,x_n)=0 \tag{3}$$

in which $x_1,x_2,\cdots,x_n$ are independent basic random variables,structural failure probability is given by

$$P_f = \int\int\cdots\int_{\Omega} f(x_1,x_2,\cdots,x_n)dx_1 dx_2\cdots dx_n$$

$$= \int_{\Omega_1} f_{x_1}(x_1)dx_1 \int_{\Omega_2} f_{x_2}(x_2)dx_2\cdots\int_{\Omega_n} f_{x_n}(x_n)dx_n \tag{4}$$

in which Ω is failure domain; $f(x_1,x_2,\cdots,x_n)$is union probability density function; Ω_i is intergral domain of random variable x_i; $f_{x_i}(x_i)$ is probability density function of x_i.

If assmuing a domain $[x_{i1},x_{i2}]$ in random variable x_i, let $\int_{x_{i1}}^{x_{i2}} f_{x_i}(x_i)dx_i \approx 1$, domain $D=[x_{11},x_{12}]\times[x_{21},x_{22}]\times\cdots\times[x_{n1},x_{n2}]$ can be obtained, moreover $\Omega\subset D$.

If M is the failure times in N times of simulation, failure probability can be obtained by

$$P_f \approx \frac{|\Omega|}{M}\sum_{\Omega}\prod_{i=1}^{n} f_{x_i}(x_i) \tag{5}$$

If samples are fairly well-distributed in domain D, the following equations can be obtained

$$\frac{|\Omega|}{M} = \frac{|D|}{N} \tag{6}$$

$$\frac{|D|}{N}\sum_{D}\prod_{i=1}^{n} f_{x_i}(x_i) \approx 1 \tag{7}$$

Hence, from eqs.(5)、(6) and (7), failure probability is given by

$$P_f \approx \frac{\sum_{\Omega}\prod_{i=1}^{n} f_{x_i}(x_i)}{\sum_{D}\prod_{i=1}^{n} f_{x_i}(x_i)} \tag{8}$$

Eqs. (8) is formula of improved Monte-Carlo method to calculate failure probability, it has higher precision, and fewer times of simulation.

3.2 JC method

According to relative document, the iterative formulations of JC method are given by

$$\begin{cases} x_i^* = \mu_{x_i} - \sigma_{x_i}\beta\alpha_i \\ \alpha_i = \left.\dfrac{\partial g}{\partial x_i}\right|_* \sigma_{x_i} \Big/ \sqrt{\sum_{i=1}^{n}(\sigma_{x_i}\dfrac{\partial g}{\partial x_i}|_*)^2} \\ g(x_1^*,x_2^*,\cdots,x_n^*)=0 \end{cases} \tag{9}$$

in which * is design checking point; x_i^* is the value of random variable x_i in checking point; μ_{x_i} and σ_{x_i} are mean and standard variance.

3.3 TRYM method

In this method, a linear approximation to limit state equation is obtained by Taylor's series expanding at design checking point as follows

$$z = g(x_1^*,x_2^*,\cdots,x_n^*) + \sum_{i=1}^{n}(x_i - x_i^*)\frac{\partial g}{\partial x_i}|_* \tag{10}$$

Since checking point is on limit state surface, so $g(x_1^*,x_2^*,\cdots,x_n^*)=0$, and the mean and standard variance of z are given by

$$\mu_z = \sum_{i=1}^{n}(\mu_{x_i} - x_i^*)\frac{\partial g}{\partial x_i}|_* \tag{11}$$

$$\sigma_z = \sum_{i=1}^{n}\alpha_i\sigma_{x_i}\frac{\partial g}{\partial x_i}|_* \tag{12}$$

$$\alpha_i = \sigma_{x_i}\frac{\partial g}{\partial x_i}|_* \Big/ \sqrt{\sum_{i=1}^{n}(\sigma_{x_i}\frac{\partial g}{\partial x_i}|_*)^2} \tag{13}$$

According reliability index $\beta = \frac{\mu_z}{\sigma_z}$, checking point can be obtained by

$$x_i^* = \mu_{x_i} - \sigma_{x_i}\beta\alpha_i \tag{14}$$

Eqs.(11) ~ Eqs.(14) make up iterative formulae, reliability index and checking point can be obtained at the same time by using this method.

3.4 RBDF method

In limit state equations of tunnel lining Eqs.(1) and Eqs.(2), let

$$\begin{cases} R = \phi\alpha d\sigma_c b \\ C = \frac{1}{6}(1.75\phi\sigma_t bd^2 + Nd) \end{cases} \tag{15}$$

Therefore, reliability design expressions are shown as follows[4]

$$\begin{cases} \lambda_R\mu_R \geq \lambda_N\mu_N & e_0 \leq 0.2d \\ \lambda_c\mu_c \geq \lambda_M\mu_M & e_0 > 0.2d \end{cases} \tag{16}$$

in which λ_R、 λ_C are reliability coefficicent of resistance; λ_M、 λ_N are reliability coefficient of load effects.

In theory, statistic parameters of R and C should be obtained from random variables σ_t, σ_c, d, M,and N by using Monte-Carlo method to simulate,thus, to simple, after neglecting some variables uncertainty, statistic parameters of R and C can be obtained by

$$\begin{cases} \mu_R = \alpha d\mu_{\sigma_c} \\ \delta_R = \delta_{\sigma_c} \end{cases} \tag{20}$$

$$\begin{cases} \mu_c = \frac{1}{6}(1.75d^2)\mu_{\sigma_t} + d\mu_N) \\ \delta_c = \frac{\sqrt{1.75^2\sigma_{\sigma_t}^2 d^4 + d^2\sigma_N^2}}{1.75\mu_{\sigma_t}d^2 + \mu_N d} \end{cases} \tag{20'}$$

3.5 New-ZWQ method

Let $x=\{x_1, x_2, \cdots, x_n\}$ be random space, and $x_1, x_2, \cdots, x_n$ are independent random variables, $Y=\{y_1, y_2, \cdots, y_n\}$ is standard normal random space. According to the transformation theorm of equal probability, x_i can be transformed to y_i by[3]

$$F_{X_i}(x_i) = \Phi(y_i) \tag{21}$$

$$x_i = F_{X_i}^{-1}(\Phi(y_i)) \tag{21'}$$

where F_{X_i} is probability distribution function of random variable x_i.

In space Y, limit state equation (3) becomes

$$z = g(F_{X_1}^{-1}[\Phi(y_1)], F_{X_2}^{-1}[\Phi(y_2)], \cdots, F_{X_n}^{-1}[\Phi(y_n)]) \\ = G(y_1, y_2, \cdots, y_n) \tag{22}$$

Let $Y^{(k)} = \{y_1^{(k)}, y_2^{(k)}, \cdots, y_n^{(k)}\}$ be iterative point on limit state surface. Gradient of limit state surface at this point $\nabla G = \left\{\frac{\partial G}{\partial y_1}, \frac{\partial G}{\partial y_2}, \cdots, \frac{\partial G}{\partial y_n}\right\}$. After moving distance λ along negtive gradient from point $Y^{(k)}$, point $Y_\lambda^{(k)}$ is obtained by

$$Y_\lambda^{(k)} = Y^{(k)} - \lambda\nabla G \tag{23}$$

Because the negtive gradient direction at any point on limit state surface points to the failure domain,it is assured that point $Y_\lambda^{(k)}$ is in failure domain if only giving appropriate λ .Hence,

$$Y^{(k+1)} = \beta^{(k+1)} \cdot \alpha^{(k+1)} \tag{24}$$

in which $\beta^{(k+1)}$ is reliability index that iterative times is k+1, $\alpha^{(k+1)}$ is sensibility , $\alpha^{(k+1)} = \{\alpha_1^{(k+1)}, \alpha_2^{(k+1)}, \cdots, \alpha_n^{(k+1)}\}$, and

$$\alpha_i^{(k+1)} = \frac{y_i^{(k)} - \lambda\frac{\partial G}{\partial y_i^{(k)}}}{\left\|Y^{(k)} - \lambda\nabla G\right\|} \tag{25}$$

$$\beta^{(k+1)} = \frac{\nabla G \cdot [Y^{(k)}]^T - G(Y^{(k)})}{\frac{\partial G}{\partial Y^{(k)}} \cdot [\alpha^{(k+1)}]^T} \tag{26}$$

3.6 Transformation of dependent random variables

In all methods above, basic random variables in limit state equations are assumed to be independent. However,some basic random variables in a lot of engineering are dependent,therefore,to apply directly these method above,dependent random variables are transformed to independent random variables at first. Covariance matrix of random variables $x_1, x_2, \cdots, x_n$ is given by

$$[C_X] = \begin{bmatrix} \sigma_{X_1}^2 & \mathrm{cov}(x_1, x_2)\cdots\mathrm{cov}(x_1, x_n) \\ \vdots & \\ \mathrm{cov}(x_n, x_1)\mathrm{cov}(x_n, x_2) & \cdots \quad \sigma_{X_n}^2 \end{bmatrix} \tag{27}$$

where $\mathrm{cov}(x_i, x_j) = \rho_{x_i x_j}\sigma_{x_i x_j}$ is covariance of x_i and x_j, $\rho_{x_i x_j}$ is dependent coefficicent.

If $x_1, x_2, \cdots, x_n$ are linear independent, matrix $[C_X]$ is only diagonal matrix. Hence,matrix $[C_X]$ can be transformed to diagonal matrix by using linear transformation method, and diagonal elements are only variance,so

$$[C_Y]=[A]^T[C_x] \tag{28}$$

in which [A] is transformed matrix, it consists of charateric vector of matrix $[C_x]$.

Therefore, dependent random variables x= $\{x_1, x_2, \cdots, x_n\}$ can be transformed to independent random variables y=$\{y_1, y_2, \cdots, y_n\}$, following formula is obtained by means of this method

$$\{Y\}=[A]^T\{X\} \tag{29}$$

where staistics of {Y} is obtained from the expression of linear combination of {X}. And Eqs. (29) may also be written as follows

$$\{x\}=([A]^T)^{-1}\{Y\} \tag{30}$$

Hence,the limit state equation for dependent basic random variables {x} can be transformed to a new limit state equation for independent basic random variables {Y}, in this way,reliability index can be solved by using directly these methods above.

4 EXAMPLES AND RESULT ANALYSIS

4.1 Calculating reliability of cover tunnel lining

According standard design for railway tunnel in China,straigth wall lining for IV catagory rock is selected to calculate its reliability,loading-structure model is adopted,in addition,elastic resistance of rock is also considered.

According to references[1][4][5], the statistic parameters and the probability distribution of basic random variables are obtained,and they are shown in table 1.

Table1 statistics of basic random variables

name	Q	γ	Er	Ec	d	σ_t	σ_c	μ
mean	73.47 (kpa)	26.27 (KN/m^3)	39922 (MPa)	24869 (MPa)	0.44 (m)	1573 (kpa)	23680 (kpa)	0.2
coeff. var.	0.65	0.06	0.79	0.06	0.29	0.03	0.11	0.05
distri-bution	log-normal	normal	log-normal	normal	normal	normal	log-normal	normal

Lining internal force is calculated by using finite elment-response surface method,it is known that lining internal force (including axial force N and moment M) is relative with loosed load Q. lining concrete elastic modulus Ec,surroding rock elastic modulus Er,etc.After variability of other random variables are neglected,internal force M and N are given by

$$\begin{cases} N = N(Q, Ec, Er) \\ M = M(Q, Ec, Er) \end{cases} \tag{31}$$

Let Q=x_1,Ec=x_2,Er=x_3, and y be lining internal force M,N,hence,Eqs.(31) becomes

$$y=y(x_1,x_2,x_3) \tag{32}$$

in which x_1,x_2,x_3 are transformed to standard normal random variables ξ_1,ξ_2,ξ_3 by using Eqs.(23),and the relativity of ξ_1,ξ_2,ξ_3 is equal to the relativity of x_1,x_2,x_3 ($\rho_{\xi_i\xi_j} \approx \rho_{x_ix_j}$). Since there is not statistic data for ralativity of Q,Er and Ec,assume that they are independent here.

Let y be given by

$$\begin{aligned} y = a_0 + a_1\xi_1 + a_2\xi_2 + a_3\xi_3 + a_4\xi_1\xi_2 \\ + a_5\xi_1\xi_3 + a_6\xi_2\xi_3 + a_7\xi_1\xi_2\xi_3 \end{aligned} \tag{33}$$

therefore, eight equations about coefficient a_i(i=0,1, $\cdots$,7) are obtained by taking $\xi_i = \pm 1$(i=1,2,3). According to the values of ξ_i,the values of x_1,x_2,x_3 are solved, and then internal force y is solved by using finite element method, finally, a_i is obtained from Eqs.(33). Therefore, the samples of M,N are obtatried by using Monte-Carlo method, after taking statisitic analysis for samples, their statistics are obtained.

In this example, the statistics of internal force of lining arch top are μ_M=0.64KNm, μ_N=101.6KN, δ_M=0.61, δ_N=0.6,and M,N are lognormal distribution.

Since $e_0 \leqslant 0.2d$ in the section of lining arch top,Eqs.(1) is selected as limit state equation to calculate reliability index. In eqs.(1), basic random variables σ_c,d,N are obviouslly independent, hence, reliability index can be directly calculated by using these methods above, calculating results are shown in table 2.

Table 2 reliability index β

method	times	d^*(m)	σ_c^*(kpa)	N^*(KN)	β
improved M-C	35000				4.06
JC	divergent				
Trym	4	0.081	23590.6	111.22	4.09
RBDF		0.44	23680	101.63	9.11
New-ZWQ	5	0.069	23460.9	94.54	4.10

4.2 Calculating reliability of cut tunnel lining

This example takes from reference [4], the statistics of basic random variables is shown in table 3.

Since $e_0 \leqslant 0.2d$ in calculating section,eqs.(2) is adopted as limit state equation, in which M,N are obviouslly dependent. When using improved Monte-Carlo method to calculate reliability index in this

Table 3 statistics of basic random variables

name	σ_c	σ_t	d	M	N
mean	20 (MPa)	4 (MPa)	0.55 (m)	40 (KNm)	260 (KN)
coeff.var.	0.18	0.21	0.09	0.25	0.22
distribution	log-normal	log-normal	log-normal	log-normal	log-normal

paper, two cases are considered: $\rho_{MN}=0$ and $\rho_{MN}=0.8$, when usingother four methods, the case of $\rho_{MN}=0$ is only considered. Calculating results are shown in table 4.

Table 4 reliability index β

method	times	σ_t^* (kpa)	N^* (KN)	M^* (KNm)	d^* (m)	β
improved M-C	35000 ($\rho_{MN}=0$)					4.06
	60000 ($\rho_{MN}=0.8$)					3.96
Trym	4	309.21	241.49	46.43	0.53	4.47
RBDF		4000	260	40	0.55	4.49
New-ZWQ	8	1994.96	226.32	115.94	0.42	6.32
JC	divergent					

4.3 Calculating results analysis

It illustries from two examples that all of methods but for JC method can be applied to calculate reliability index of tunnel lining.Improved Monte-carlo method is of higher precision and fewer times of simulation.The result of TRYM method is close to the result of improved Monte-Carlo method, and other methods have large error.

5 ACKNOWLEDGEMENTS

(1) When calculating reliability index of tunnel lining structure, improved Monte-carlo method can be considered as comparing standard of other methods.

(2) Because JC method is easily affected by nonlinear limit state equation, it is usually divergent. In design of tunnel lining, it is suggestied that the TRYM method is better method to calculate structural reliability index.

REFERENCES

1. Design Code of Railway Tunnels (TBJ-85), China Railway Publishing House, Beijing,1986
2. Wu shiwei, the papers of structural safety and reliability analysis, press of He-Hai university, 1988
3. Gong jingxin,a new iterative method to calculate structural reliability index, structural mechanics of computation and application,1995.8
4. Zhang mi,The research of loading statistics for railway cut tunnel, North Jiaotong university, 1995.9
5. Xie jinchang,Tan Zhongsheng,Optimum fit for the probability distribution function of loosened rock load on railway tunnel lining in China,the ITA Annual meeting, China, 1990

Computer Methods and Advances in Geomechanics, Yuan (ed.) © 1997 Balkema, Rotterdam, ISBN 90 5410 904 1

Research on the prediction of fractured zone of the tunnel in soft and weak rock mass

Chunsheng Qiao
The Northern Jiaotong University, Beijing, China

ABSTRACT: Through the in situ test and the numerical simulation, it has been known that there is no relation between the extent of the fractured zone around a tunnel and the location of the peak axial force induced in a full-grouted rock bolt. It is dangerous to note simply the peak axial force of rock bolt or their location. It may be possible to estimate the size of the fractured zone from radial displacement distribution of the tunnel measured by use of multiple-point borehole extensometers.

1 INTRODUCTION

Tunnel in soft and weak rock mass conditions is usually characterized by large ground displacement or convergence. These results from the development of a fractured zone (plastically deformed region) produced around the tunnel by stress changes duo to excavation. The stability of the tunnel is largely affected by the fractured zone, it is necessary to estimate the size of the fractured zone by theoretical analyses and in situ measurements.

In general, there are two ways to quantitatively determine the extent of the fractured zone: the first is by numerical analysis in which reasonable adoption of rock mechanical parameters is a time consuming duties; the second uses different instruments monitoring rock performance in situ such as recording wave propagation velocities or extensometers etc. In this paper another way to analyze the fractured zone was discussed

The fully-grouted rock bolts with resin or cement are becoming increasingly popular for tunnel and mining support. The difference between point anchored bolt and fully-grouting bolt is that in the later case the bolt is fully anchored with the rock by mortar or resin and deformed together with the rock around it. Several published results of in situ measurements (Okubo et al 1982 & Freeman 1978) shows that axial force induced in fully-grouted rock bolt has distinctive distribution with peak. A very interested question is put forward: does this peak value have some relation to the extent of the fractured zone? If the boundary of the fractured zone coincides with the peak location of induced axial force, the estimation of the extent of these zone will be easily performed.

For seeking after this relation mentioned above some experiments in situ and numerical simulations were carried out. The primary result was introduced in this paper.

2 THE IN SITU EXPERIMENTS

The experimental site is located at the 38# test tunnel of Matsumine mine in Japan. The rock is a kind of typical soft rock, brecciated tuff, containing gypsum component which will expand when touch with water. The tunnel has a half circular section, wide of 4.2m , height of 3.6m and under overburden of 300m, being driven by full face drill and blast methods with steel set supports being installed after each mucking circle. There are three testing sections disposed with a space of 1.7m apart each others (shown in Fig.1 and Fig.2).

Three bolts were penetrated on section 1 and five bolts on section 3 respectively, with the same 3m of length and 1in. of bolt diameter. These bolts are fully-grouting by mortar and are used as a monitoring anchor (Ax) through 6 strain gages stuck on the bolt at different positions to record their axial strain distribution. Before the performance of the test in situ the bolt with strain gages was calibrated at the laboratory so as to get the relation between the strain and axial force inside the bolt.

At the section 2 there are three 3-points extensometers (Ex) for recording the relative displacements between anchor points inside the rock mass and the tunnel boundary. The anchors is located at 1m, 2m and 3m of depth from the tunnel wall respectively. Besides, anchors (Cv) for measuring the convergence of the side wall and crown were also installed.

The monitoring duration lasted for half year, starting from the excavation of the section 1 and stopping when a distance of 12.0m apart from the section 1 reached.

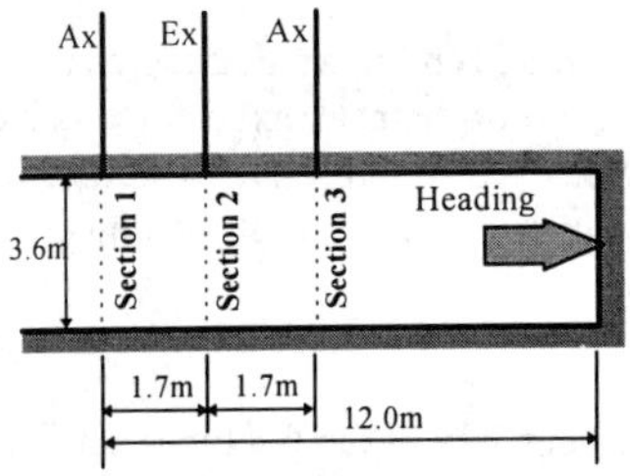

Fig.1 Diagram of the test tunnel.

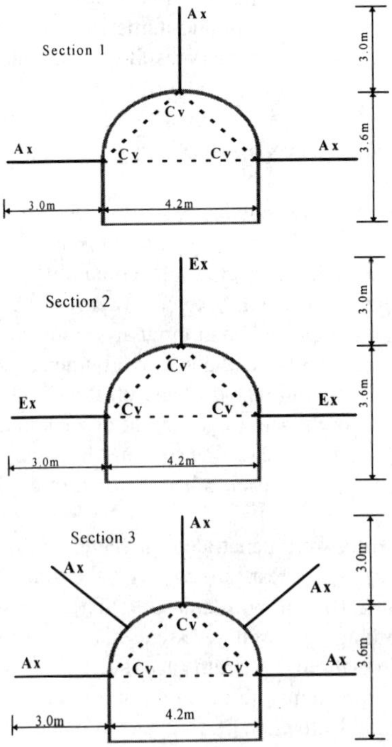

Fig.2 Setting pattern of the instruments installed.

3 MONITORING RESULTS

The deformation results of these three sections measured in this test are shown in Fig.3~5. It is found that there are transparent differences in deformations of these three sections, except section 3 has a well-distributed convergence, the section configurations of the other two have very strong distortions in which a symmetrical triangle changes into a oblique one with same directions of distortions. The displacement toward right side wall in section 2 is much more than those toward the left side and around the crown.

For the most of rock bolts, except individuals, appearances of the axial forces induced were observed after their erections and the axial force increased rapidly during the heading advance, but further enhancement of the force ceased after the tunnel face advance terminated.

Most of the peak values are located at the 0.8m~1.6m along the bolt length. Although the magnitude of the peak value varied in aforementioned manner and the convergence of the tunnel section still increased linearly due to the rock creep, but their peak points keep their positions without changes with the tunnel face advance and the time elapse. Therefore, from this result there is no definite relationships to

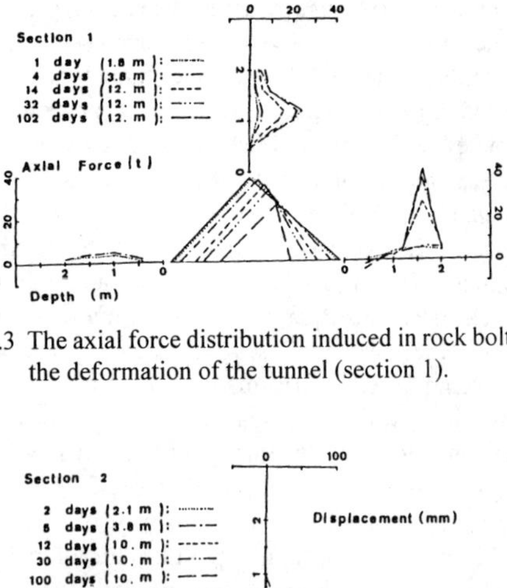

Fig.3 The axial force distribution induced in rock bolt and the deformation of the tunnel (section 1).

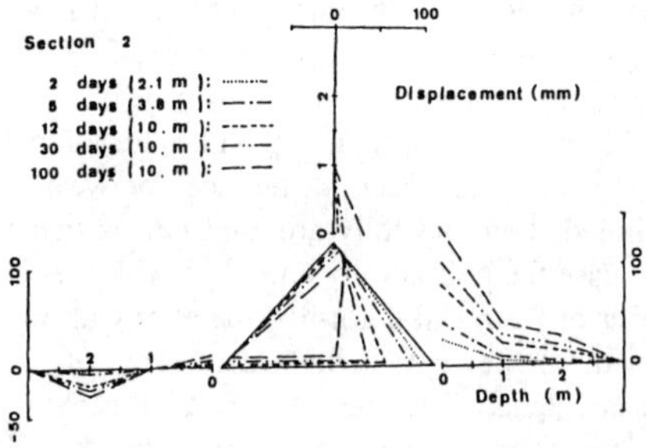

Fig.4 The radial displacement of the surrounding rock and the deformation of the tunnel (section 2).

express the extent of the fractured zone connected with the position of peak point of axial force distribution alone the rock bolt.

There are some things special at the section 1 and 3. On the section 1 the axial force of the bolt on the right side wall is much more big beyond the others and on the section 3 big axial forces with opposite signs again each other appeared at the positions of 0.8m and 1.0m respectively alone the bolt length. It is interested to find that the times taking place these distinctive axial forces at section 1 and 3 are the same as that happening to start the deformation of the rock 1m behind of right side wall. Kaitowski et al (1980). showed from experiments that when a bolt is passed through a discontinuous plane very high tensile stress and compressive stress happen in the bolt on the two sides of this plane. Saito and Amano (1982) pointed out that a peak axial force also was found when the bolt met a discontinuity. From their statements, it is possible that there is a new discontinuous plane existed on the right side alone the direction parallel with the center line of this tunnel. Therefore, it is seemed that a distinctive distribution of the axial force often corresponds a new formation of a discontinuity in rock mass.

4 A NUMERICAL SIMULATION OF THE TUNNEL DEFORMATIONS AND AXIAL FORCE OF ROCK BOLT

Since the relationship between the extent of the fractured zone and the bolt axial force can not precisely understand from these in situ tests, a numerical simulation have been carried out. In this analysis the following points are concerned: feature of non-linear deformation of soft rock with consideration of strain softening, the bolt-rock

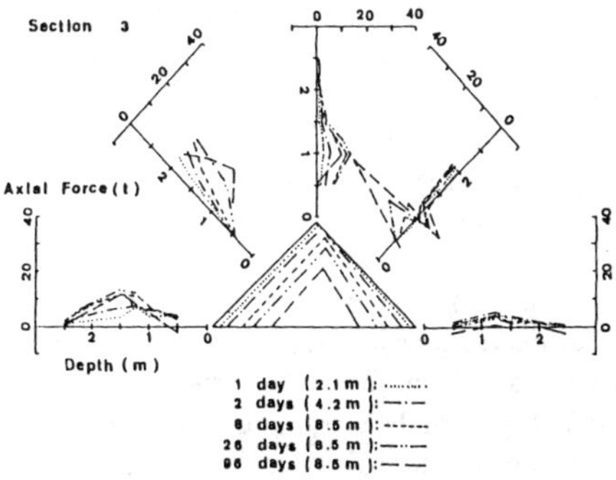

Fig.5 The axial force distribution induced in the rock bolt and the deformation of the tunnel (section 3).

interactions and the effects of the movement of the cutting face when the excavation is carried on. Actually, this is a 3-dimensional non-linear problem. The following treatments are adopted for a simplification.

4.1 *A simulation of excavation of the tunnel*

Consider a simple case which a circular tunnel being driven by full face drill method. The horizontal and vertical in situ stresses are assumed to be equal and to have a magnitude p_0 (Yamatomi 1984). The operation of excavations regards as a process of relaxation, step by step, of initial stress p_0 subjected on the periphery part of rock. As the p_0 approaches zero, that means the tunnel is driven out, then the force p_i subject on the periphery part of rock excavated after some time exists the following relation with p_0:

$$p_i = (1 - r_{ex})p_0$$

where r_{ex} is a coefficient of excavation, which equals to 0 and 1 before and after driving respectively. The expression of r_{ex} versus driven distance of cutting face d was given by Yamatomi (1984). As the Poisson's ratio equal to 0.19, the relation of d versus r_{ex} is shown in table 1. A simulation of cutting face moving forward may be performed by using these data, which means that stresses and displacements in the rock surrounding the tunnel, according to the conditions of plane strain, can be regarded as the stresses and displacements induced by the moving cutting plane.

Table 1 Relationship between coefficient of excavation and distance from the tunnel face.

Distance from the tunnel face (m)	Coefficient of Excavation
0.5	0.543
1.5	0.793
2.5	0.893
3.5	0.941
4.5	0.967
5.5	0.982
6.5	0.991
7.5	0.996
8.5	1.0

4.2 *A elasto-plastic analysis of tunnel deformation by FEM*

Suppose that rock is a homogeneous, isotropic medium, the extended Von Mises criterion for yield (Yamatomi 1984) may be adopted as the following:

$$F = (\alpha + 1)\sqrt{3J_2} - \alpha I_1 - \sigma_Y(\bar{\varepsilon}^P) = 0 \qquad (1)$$

where I_1 and J_2 are the first invariant of the stress tensor and the second invariant of the deviatoric stress tensor respectively, $\sigma_Y(\bar{\varepsilon}^P)$ is the yield stress under axial compression test, which is a function of equivalent plastic strain $\bar{\varepsilon}^P$. From the results of tests, most of the rocks fulfill the following relation:

$$\sigma_Y = H(\bar{\varepsilon}^P) = H_1 e^{-H_2 \bar{\varepsilon}^P} - H_3 e^{-H_4 \bar{\varepsilon}^P} + H_5 \bar{\varepsilon}^P \quad (2)$$

For considering the behavior of strain-softening of rock and effect of confining pressure, α in (1) regards as a function of equivalent plastic strain :

$$\alpha = \alpha_1[1 + \alpha_2(1 - e^{-\alpha_3 \bar{\varepsilon}^P})] \quad (3)$$

among the (2) and (3) H_1, H_2, H_3, H_4, H_5, α_1, α_2, α_3. are material constants, which are obtained by regression of testing results.

4.3 *Method for calculation of the axial force in a rock bolt*

As for the fully-grouting bolt installed, if there is no deformation of the rock, no axial force exists in the bolt. Therefore, an interaction between bolt and rock must be put into consideration. Based on Saito's model (1983), considering the equilibrium of the element in the bolt as shown in Fig.5, the following equation is obtained,

$$d\sigma/dx = -2\tau/r \quad (4)$$

where r is the radium of a bolt, τ is the shear stress subjected on the surface of the bolt and σ is the axial stress in the bolt. If the axial stress is

$$\sigma = -E_r\, dv/dx \quad (5)$$

where E_r is the modulus of bolt material, from equation (4) and (5)

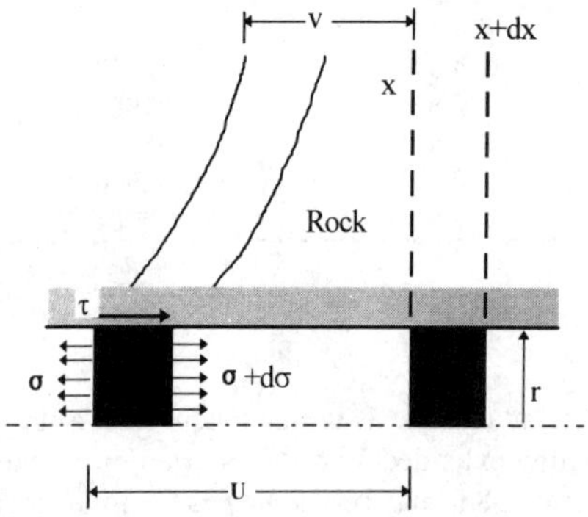

Fig.6 Equilibrium in the grouted rock bolt.

$$\frac{d^2v}{dx^2} = -\frac{2\tau}{E_r r} \quad (6)$$

Suppose shear stress is induced by the relative displacements, then

$$\tau = c(v - u) \quad (7)$$

where c is a shear coefficient of bolts, determined from the cement material and from the pull-out test of rock bolt. Replace (7) into (6) the differential equation is obtained

$$\frac{d^2v}{dx^2} - \alpha^2 v + \alpha^2 u = 0 \quad (8)$$

The axial force of bolt may be calculated from (8), since the distribution of rock displacements u is given.

4.4 *Parameters used in computation*

Computing parameters used in this analysis were obtained from tests both in laboratory and in situ shown in Table 2

Table 2 Mechanical parameters adopted in calculation

Property	Value
Young's modulus of rock	0.21GPa
Poisson's ratio of rock	0.19
H_1	41.01MPa
H_2	706.8
H_3	38.94MPa
H_4	807.9
H_5	0.195MPa
α_1	0.191
α_2	1.520
α_3	458.6
Radius of bolt	17.7mm
Young's modulus of bolt	136.0GPa

Shear coefficient c is $50 kg/cm^3$ based on the in situ pull out test of bolt in a same bolt conditions of the test tunnel.

5 RESULTS OF COMPUTATION AND ITS ANALYSIS

The variations of the distribution of radial displacements of rock around the tunnel, responding the process driven forward of the cutting face is shown in Fig.7. At the beginning of the excavation the deformations of the rock around the tunnel were very small and belonged to a elastic range. As the cutting face departing some distances from this testing

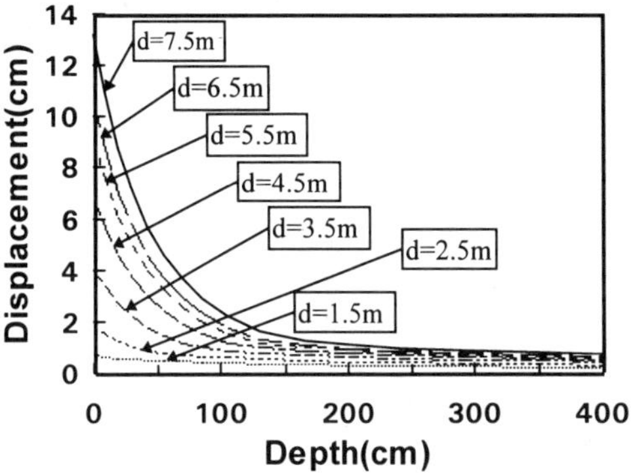

Fig.7 Radial displacement distribution in the surrounding rock.

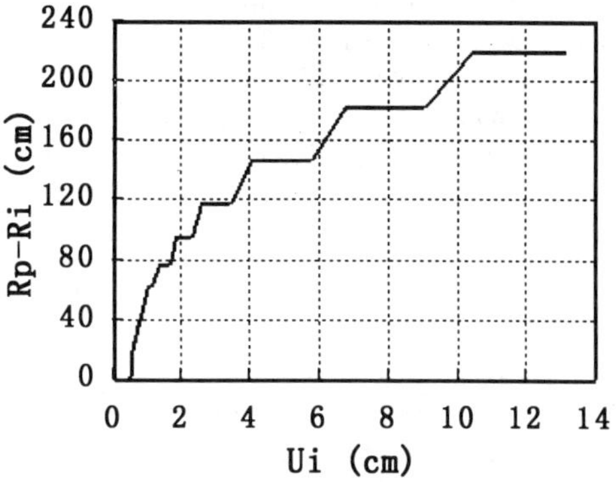

Fig.8 Extent of the fractured zone versus radial displacement of the tunnel wall.

section the displacements increased rapidly in the rock around this section because of a fractured zone formed in the surrounding rock. Rock in this zone is in a plastic state with very large plastic deformations. Rock should have enough strength at the elasto-plastic boundary since it just enters a yield state, therefore, the position happening a strong deformation should correspond to the inside of the fractured zone instead of at the boundary.

The extent of fractured zone (R_P-R_i) versus the radial displacements of periphery of the tunnel (U_i) is shown in Fig.8. At the beginning of the excavation there is no fractured zone happened since all rock is in a elastic state. The fractured zone is formed only if the cutting face departed this section from some distances and is extended its extent while the cutting face is driven more forward apart from this test section. From that it is said that the fractured zone is experienced a process from zero volume gradually to the extent induced by the process of excavation.

It has the same obvious peak points in the distributions of axial forces in Fig.9 both from the calculations and from the monitoring in situ. All axial forces are divided by the peak value in Fig.9 in order to observe the regularity of the variation in peak positions. The movement of the peak point of axial force during the process of excavation is shown in Fig.9. The peak point is moved gradually with a inside direction toward to the periphery of the tunnel instead of that outside toward the rock, which shows clearly a disagreement with the extent enlarged in fractured zone. Therefore, there is also no any relation between the peak point position and the extent of the fractured zone based on the calculations. It is concluded from both of the computation and

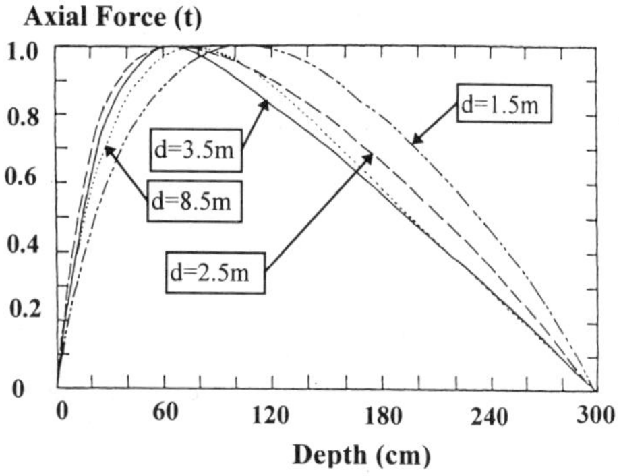

Fig.9 Axial force distribution of rock bolt calculated.

monitoring in situ that a determination of a fractured zone relys only on the position of peak point in axial force is not reasonable.

A total process of the formation and its expansion of the fractured zone clearly realized from Fig.7. If mutiple-point borehole extensometers with enough points and precision are used to monitoring the deformation of surrounding rock in the duration of excavation, an estimation of the extent of fractured zone from the distribution of the displacements can be completed.

6 CONCLUSION

For searching a practical method to determine the extent of fractured zone around a tunnel in a soft and weak rock a study of application of axial force of a fully-grouting rock bolt to solve this issue was carried on. From this research the extent of fractured zone developed from zero to a volume closely connected with the process of excavation. The peak points of the axial force in a fully-grouting bolt moves toward a direction opposite from that of the extension of the fractured zone as the cutting face drives apart from this bolt, that means there is no explicit relation between the peak point and the extent of the fractured zone.

There is one point to which attention should be paid that if a discontinuity formed in the range of the bolt length, the axial force of the bolt will show some distinctive distribution, which may be a kind of addition of reference even this phenomena can not predict the extent of fractured zone.

It is concluded that a stability evaluation of tunnels bases only on monitoring the axial force of the rock bolts is not applicable. An application of mutiple-point borehole extensometers with enough precision to monitor the extent of the fractured zone of the tunnel, at the same time to refer to the variation of the axial forces is recommended.

REFERENCES

Okubo, S., Amano, K., Koizumi, S. & Nishimatsu, Y. 1984. In-situ measurement on the effect of roadway support. Journal of the Metallurgical Institute of Japan 100: 11～16.

Tanimoto, C., Hata, S. & Kariya, K. 1981. Interaction between fully bounded bolts and strain softening rock in tunneling. Proc. of 22$^{nd.}$ Symp. On Rock Mechanics. 347～352.

Kwitowski, A. & Wade, L. V. 1980. Reinforcement mechanics of untensioned full-column resin bolts. Report of Investigations (USBM), RI 8439.

Saito, T. & Amano, S. 1982. Fundamental Study on effects and Design of rock bolting. Proc. of 14th Sympo. on Rock Mechanics of JSCE. 76~80.

Yamatomi, J. 1984. Rock mechanics studies on the underground excavation in soft and weak rockmass. Doctor's Thesis, University of Tokyo.

Saito, T. 1983. Rock bolting and cable bolting in Hiraki Mine. MMIJ/AusIMM Joint Symposium. 119～133.

Computer Methods and Advances in Geomechanics, Yuan (ed.) © 1997 Balkema, Rotterdam, ISBN 90 5410 904 1

The construction method of starting tunnelling in special geological condition

Feng Weixing, Wang Keli & Liu Yong
Shijiazhuang Railway Institute, China

ABSTRACT: One of the biggest difficulties of tunnelling in most seriously weathering granite is that the starting tunnelling can not be carried out, because the landslide of the mountain slope happens as soon as the hillside at tunnel portal is excavated. Taking Qiling Tunnel on Beijing - Jiulong railway line for example,this paper introduces the construction method of starting tunnelling in most seriously weathering granite, including the option of scheme of starting tunnelling, construction method, technical measures, primary support etc. By this method, remarkable technical and economical benefits are achieved in construction practice.

1 INTRODUCTION

Qiling tunnel, 2536 meters in length, located in the division from Jian to Dingnan on Beijing-Jiulong railway line, is a double track electrification railway tunnel. The inlet of the tunnel is on a straight section with 4.8 ‰ up gradient. The length of open cut tunnel at inlet is 33 meters. The inlet portal is designed at the foot of hillside where there are flourishing bushes and lower topgraphy. About 380m tunnel near to inlet is in most seriously weathering black mica granite which is equivalent to Scale I of Chinese Railway Tunnel Surrounding Rock Scale.

During the excavation of cutting in April of 1993, it is the rainy season. Because of taking in water, the strength of the weathering rock is quickly dropped. Under the action of both surface water and ground water, the drawing landslide of hillside happens, and cracks and benches appear on the groundsurface. The workiong face for tunnelling can not be formed. Therefore, the normal tunnel excavation is difficult to carry out.

2 GENERAL TECHNICAL MEASURES FOR DESIGN AND CONSTRUCTION

Considering the big difficulty of starting tunnelling for Qiling tunnel and ensuring the time limit and safe construction,we take following synthetic technical measures:

2.1 Synthetic water treatment measures

1. Strengthen cutting dewatering, enlarge side drainage ditches, improve tunnel dewatering system.
2. Ensure the dewatering ditch over tunnel portal being in good conditions. Maintenance work will be timely carried out if there are cracks and water leakage.
3. Tamp the groundsurface pits above the portal to reduce water accumulation and to prevent from water penetration.
4. Take the measure of small pumping shafts to lower ground water table during the construction of open cut tunnel.

2.2 Deep hole pregrouting to strengthen strata

Deep hole - double liquid slurry - simple injection - high pressure - splitting grouting is adopted to strengthen the strata at portal section with the strengthening scope of 46m in longitudinal direction,30m in transverse direction and the depth from groundsurface to 5m below the invert bottom (see Fig.1).

1. Grouting material

Cement-sodium silicate double liquid slurry is used as the main grouting material in which the cement may be 525# or 425# Portland cement, and the modulus of the sodium silicate is about 2. 4 to 3. 2 and its concentration is about 30 to 40 B'e. Slow condensating agent of sodium phosphate type is used in the grouting.

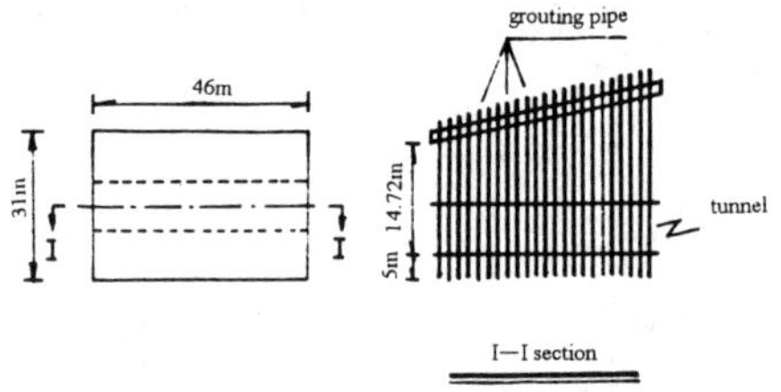

Fig.1 Grouting scope

2. Mixing ratio
W:C 0.5:1 to 1:1
Volumn ratio between cement and sodium silicate 1:0.6
Slow condensating agent : 3 % of cement amount

3. Grouting hole arrangement
The central distance between holes both in longitudinal direction and in transverse direction is 2m and the holes are positioned interleavingly. The grouting pipe consists of two parts, the upper and the lower. The upper part is a seamless steel pipe with 2.8m in length (0.8m above groundsurface and 2.0m below groundsurface). and its corresponding borehole diameter is 80mm. The lower part is a general steel pipe with side holes and with the length from 2m below groundsurface to 5m below tunnel invert (The general steel pipe may be replaced by hard plastic pipe in the excavation area of the tunnel).

4. Grouting pressure
The final grouting pressure is 6 MPa.

5. Diffusion radius $R_k = 1.2m$

6. Grouting amount of single hole

$$Q_l = \pi R_k^2 n l \alpha$$

where,l — hole depth (m);
n — strata porosity, n = 0.4 according to geological exploration;
α— slurry filling ratio, α = 0.5.

7. Grouting machine
2TGZ-60/120 double liquid slurry pump: 2 , or BW-250 mud pump: 4.

2.3 Lengthen open cut tunnel and stabilize the foot of hillside.

2.4 Adopt reasonable construction programs such as constructing the open cut tunnel from outer to inner, and using the excavation method of piperoof partial excavation or side heading method for normal tunnel according to real conditions.

3 SCHEME DESIGN OF STARTING TUNNELLING FOR THE TUNNEL

The first 30m construction of the tunnel is the key of safe starting tunnelling and the prerequisit for smooth constructing the whole tunnel. Based on the adopted assistant construction mesaures (mainly the groundsurface deep hole grouting to strengthen the strata) and former construction experiences, we adopt the following construction methods for starting tunnelling at the inlet of the tunnel.

3.1 Piperoof partial excavation

Piperoof is a kind of preadvace supporting mainly consisting of steel pipes and steel arch (or reinforcement lattice arch). It has been widely used in soft-weak rock tunnel which has bad self-stability and is a effective supporting method for constructing tunnel in soft-weak strata.

1. Construction procedure
The construction procedure of piperoof partial excavation in this project is illustrated in Fig.2.

(a) Piperoofing (step Ⅰ)
A row of ringlike piperoof is set at the outer closely along the tunnel excavation line. Insert reinforcement bars and grout sand-cement slurry into the pipe to enhance the stiffness of the pipes and the cohension between the pipes and strata.

(b) Ringlike excavating of upper bench (step ②) and shotcreting 3 ~ 5cm in thickness mixing with quick-harding agent.

(c) Arch shaped primary supporting (step Ⅲ)
Insert reinforcement lattice arch into the crown with 0.5m spacing in longitudinal direction after the completion of step ②. Formed steels are matted under the arch feet. Longitudinal reinforcement bars are set between two reinforcement lattice arches with 1.0m spacing along the arch. Shot 30cm thick concrete to form arch shaped primary support (The shotcreting is carried out in two times and it is necessary to bury the the lattice reinforcement with concrete). A 3.0m long steel pipe bolt with side holes is set at each foot of the arch shaped primary support and grouting is done for the pipe bolt to strengthen the cohension between the foot and surrounding rock.

(d) Excavating the central soil of upper bench (step ④)
The central soil of upper bench is excavated by small machines or workmen behind step ② 2.0 to 3.0m.

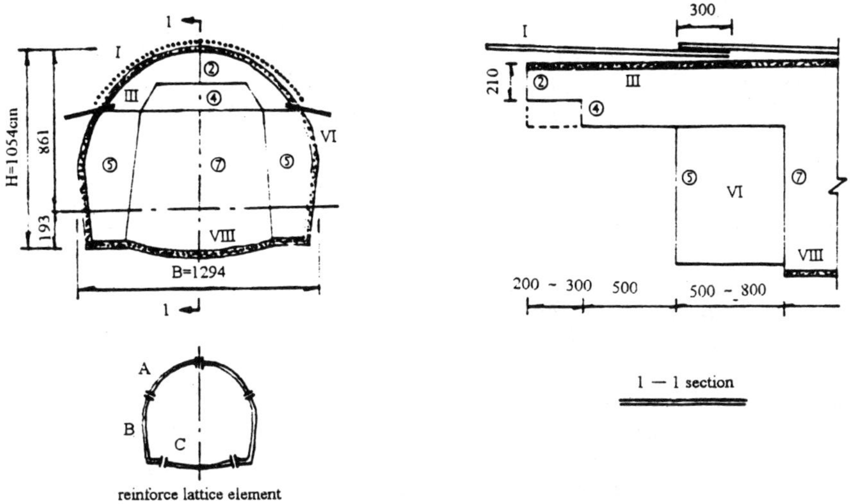

Fig.2 Construction procedure of piperoof partial excavation

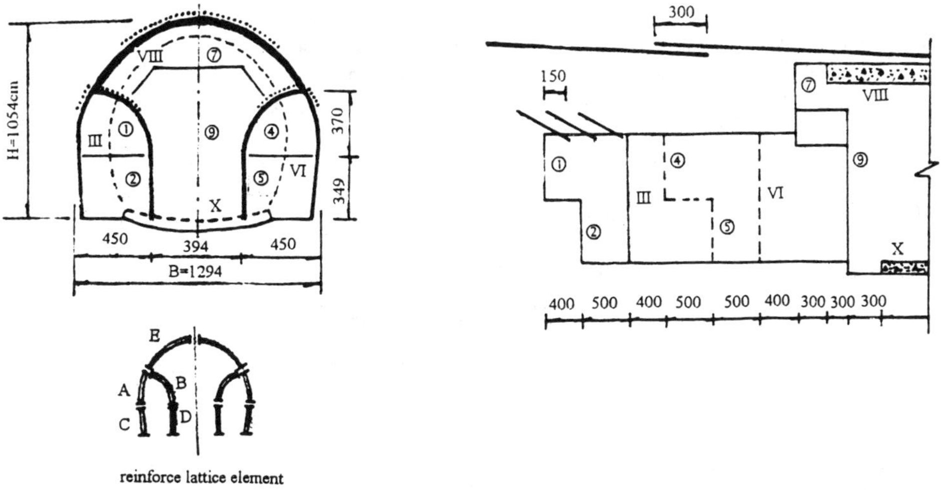

Fig.3 Side heading method

(e) Side walls (step ⑤) are excavated at intervals behind step ④ about 5.0m and 2.5m thick shotcreting with quick-harding agent is set on time.

(f) Put the side wall lattice steel together with the arch lattice steel as soon as the step ⑤ is completed. Steel plate is matted under each foot of the side wall lattice steel and shotcreting is commenced to form primary support (step VI).

(g) The work of excavating central soil (step ⑦) is carried out by back-shovel excavator coordinated with tip trucks behind step ⑤ 5.0 to8.0m. The longitudinal length of the excavation of central soil is varied from 3.0 to 4.0m each time.

(h) Commence the work of invert primaary support (step VIII) to form enclosed loop of primary support.

(i) Erect workforms to pour permanent lining (step IX).

2 Piperoof design

(a) Piperoof design

The scope of crown piperoofing: 65 ° × 2 = 130 ° , and its corresponding arc length is 1521.06 cm.

The distance between the centre of steel pipe and lining outer line: 7.0 cm.

The central distance between two pipes: 33.0 cm.

Steel pipe: Φ 108 × 9 hot-rolling seamless steel pipe.

The number of each row of piperoof: 45.

The length of each piperoof is 18.0m (consisting of 3 × 6.0m steel pipe connected by 10 to 15 cm screwing).

Insert reinforcement cage (consisting of 4 × Φ 16) and grouting C30 sand- cement slurry into the pipe.

Machines used for erecting piperoof: Tuxing 881 piperoof driller made in China.

Horizontal projection overelap length of two adjacent rows of the piperoof in longitudinal direction 3.0m.

(b) Reinforcement lattice design

The reinforcement lattice is composed of three parts: the arch (element A), the side wall (element B) and the invert (element C). They have same section of 25 × 25 cm consisting of 4 × Φ 22 reinforcement. The shape of the lattice is equivalent to that of the outer line of tunnel lining. Angle steels are welded at the end of the lattice element for the purpose of connection. The lattice elements are connected by screw to form integral structure. Longitudinal Φ 22 reinforcement bars are welded between two lattice structures.

3.2 Side heading method

1 Construction procedure

The side heading method adopted in this project is demenstrated in Fig.3.

(a) Excavating ①

Small pipe grouting is carried out above the area of excavation before excavating ①. The small pipe is Φ 42 steel pipe in 5. 0m length with side holes. After the completion of small pipe grouting, excavate the ① soil and erect reinforcement lattice, and shot concrete to form primary support. The spacing between two reinforcement lattices in longitudinal direction is 0.5m.

(b) Excavating ②

Excavate ② soil behind step ① about 4.0m and timely put the lattice together with that of corresponding to step ① and shot concrete to form primary support.

(c) Workform concrete lining (step III)

Check the excavation demensions, and erect workforms to pour C20 concrete lining III.

(d) Excavating ④

Excavate ④ soil behind step ① about 13.0m with the same construction procedure of step ①.

(e) Excavating ⑤

Excavate ⑤ soil behind step ④ about 4.0m with the same construction procedure of step ②.

(f) Workform concrete lining (step VI)

Check the excavation demensions and set workforms to pour C20 concrete lining VI.

(g) Excavating ⑦

First, the piperoof is set outer of the excavation line. The pipe is Φ 108 × 9 hot-rolling seamless steel pipe with 8. 0m length. Then, excavate ⑦ soil behind step VI about 4. 0m and timely erect arch reinforcement lattice and connect it together with the side wall lattices, and shot concrete in accordance with design requirements.

(h) Workform concrete lining (step VIII)

Check the excavation demensions and set workforms to pour C20 concrete lining VIII.

(i) Excavating ⑨

The central soil ⑨ is excavated by back-shovel excavator coordinated with tip trucks behind step VIII about 3.0m.

(j) Invert pavement (step X)

Pave C15 concrete to form tunnel invert behind step ⑨ about 3.0m.

4 CONCLUDING REMARKS

The special construction scheme design of starting tunnelling for Qiling tunnel on Beijing-Jiulong railway line is, in this paper, introduced. It had been put into real construction after the scrutiny done by the departments of design, construction and investment. During the early time, the piperoof partial excavation method is adopted. When the settlement of arch feet of primary support appears comparatively big values, the piperoof partial excavation method is replaced by side heading method which can quickly form side walls and enable the arch primary support to have hard and stable base (the top of side wall) . The difficulties of mud inflow and big surrounding rockpressure are overcome by using the side heading method. It is the side heading method that plays very important role in starting tunnelling and time limit guarantee,and that gives out preliminary experiences and design parameters for the study of shallow buried satured soft plastic soil tunnel construction.

REFERENCES

Xu Fuqiang, The Application of Small Pipe Advance Grouting in Qingshui River Tunnel, *Proceedings of 7th Conference of China Tunnel & Underground Engineering Society*, Beijing: 1992, 276-278

Chief Bureau of Capital Construction of Railway Ministry, *Railway Tunnel NATM Guideline*, Beijing: China Railway Press, 1988

Computer Methods and Advances in Geomechanics, Yuan (ed.) © 1997 Balkema, Rotterdam, ISBN 90 5410 904 1

Effect of construction speed on the behavior of NATM tunnels

Ahmed Moussa
Faculty of Engineering, El-Mataria, Cairo, Egypt

Harald Wagner
D2 Consult, Linz, Austria

ABSTRACT: Normally shotcrete tunnels are constructed by sequential excavations and shotcreting. In most cases the design of the supporting means for the excavations, ignores the effect of excavation rate and the delay in the erection of the lining. In this research, three dimensional finite element analysis has been used to study the influence of the rate of face advance and time of support application on the behavior of the tunnel. The excavations are simulated by stress elimination on steps. The simulation of the hardening process of shotcrete is based on the experimental data for the development of shotcrete strength with time. Based on parametric studies, critical values for support capacity and installation time can be predicted. The study can help to insure the tunnel safety, satisfy the deformation limits, and optimize the construction costs.

1 INTRODUCTION

According to the New Austrian Tunnelling Method, tunnels are constructed by sequential excavations and shotcreting. The method takes, as much as possible, the self supporting advantage of the surrounding soil, to allow some deformations before the shotcrete ring gets reasonable stiffness to support the ground movement.

The available mathematical methods of analyses are much more refined than are the properties that constitute the structural model. Hence, in most cases it is more appropriate to investigate alternative possible properties of the model or even different models, than to aim for more refined model For most cases, it is preferable that the structural model employed and the parameters chosen for the analyses are lower-limit cases. This may prove that even for unfavorable assumptions, the tunnelling process and the final tunnel are sufficiently safe. In general, the structural design model does not try to represent exactly the actual condition in the tunnel, although it covers these conditions (Duddek 1992).

The supporting pressure introduced by the shotcrete lining depends on many factors, such as; the excavation rate or face advance, the delay in support application in terms of the length of unsupported part, and the rate of shotcrete hardening in addition to geological and geometrical conditions. Successful tunnelling should optimize the lining pressure to insure tunnel safety and economy, and satisfy the deformation limits.

2 STRUCTURAL DESIGN OF TUNNELS

The engineering approach for designing a tunnel structure differs very much from structures above the ground in which the loads and strength are mostly well defined. For tunnels, the stress release in the ground and the soil strength are the main features for design consideration.

To design a tunnel we have to predict the magnitude of the expected straining actions on the lining and the distribution of the displacements around the tunnel. Different methods are currently used to design tunnels. El-Nahhas (1992) classified these methods according to the approach and assumptions on which the problem was formulated as follows:

1. Methods based on subgrade reactions
2. Relative stiffness approach
3. Convergence-confinement approach
4. Finite element analysis
5. Observational approach.

The first two methods fail to include the effect of the construction procedure on the soil-lining interaction. many design approaches incorporate inconsistent and even conflicting assumptions which simultaneously lead to inaccurate results (Eisenstein 1985). Generally, all analytical methods have many limitations in the application, such as; geometry of the tunnel section, stratification of surrounding soil, and loading conditions.

Two-dimensional finite element analysis has been widely used employing different methods simulating the three-dimensional effect. Although the use of three dimensional method to study the ground movement caused by tunnelling has the appeal of reducing modeling errors, it suffers from the important disadvantages of complexity and time consuming (Augarde 1995).

3 PROPOSED METHOD

In this study three-dimensional finite element analysis has been employed to predict the behavior of a tunnel under different excavation and construction conditions. The great software advance achieved by Innsbruck university in finite element programs, mesh generator, and post-processing programs (Swoboda 1996), helps very much to handle complicated systems and to make full analysis of the results. Rapid developments in hardware reduces the calculating time.

3.1 *Finite element model*

An Isoparametric Quadratic Solid element (IPQS) has been used to model soil inside and around the tunnel. The element has 20 nodes as shown in Fig. 1. The shape function in terms of the natural coordinates ζ, η, ξ can be written as follows:

For corner nodes [i=1,2,3,4,13,14,15,16]

$$N_i = \frac{1}{8}(1+\zeta_o)(1+\eta_o)(1+\xi_o)(\zeta_o+\eta_o+\xi_o-2) \quad (1)$$

For mid side nodes [i=5,7,17,19]

$$N_i = \frac{1}{4}(1-\zeta^2)(1+\eta_o)(1+\xi_o) \quad (2)$$

For mid side nodes [i=6,8,18,20]

$$N_i = \frac{1}{4}(1-\eta^2)(1+\zeta_o)(1+\xi_o) \quad (3)$$

For mid side nodes [i= 9,10,11,12]

$$N_i = \frac{1}{4}(1-\xi^2)(1+\eta_o)(1+\zeta_o) \quad (4)$$

where $\zeta_o = \zeta\zeta_i$, $\eta_o = \eta\eta_i$ and $\xi_o = \xi\xi_i$.

The tunnel lining is modelled by the Isoparametric Sandwich Shell Quadratic (ISSQ) element shown in Fig. 2. The multilayer system of the element permits very good simulation of shotcrete application. The quadratic shape function of the element is based on Ahmad's concept for degenerated isoparametric elements, whereby the geometry is described by the mid-surface. The coordinates of shell nodes can be determined by means of the mid-surface's coordinates, the shell thickness, and a direction vector normal to the mid-surface (Swoboda 1990). The displacement function used belongs to the family of Lagrange polynomes and is of $C1$ continuity. The shape function for the element nodes can be written as:

$$N_i = \zeta\zeta_i(1+\zeta\zeta_i)\,\eta\eta_i(1+\eta\eta_i)/4 \qquad i=1,2,3,4 \quad (5)$$

$$N_i = \zeta\zeta_i(1+\zeta\zeta_i)(1-\eta^2)/2.0 \qquad i=6,8 \quad (6)$$

$$N_i = \eta\eta_i(1+\eta\eta_i)(1-\zeta^2)/2.0 \qquad i=5,7 \quad (7)$$

$$N_i = (1-\zeta^2)(1-\eta^2) \qquad i=9 \quad (8)$$

3.2 *Simulation of excavations*

The tunnel construction is simulated by loading steps. Three-dimensional finite element mesh is generated specially to enable step by step excavations. The excavations are simulated by eliminating the elements. This means that if the elements concerned in one round are from n to m, their modulus of elasticity is reduced to zero, and their stresses, σ, are integrated and applied as external loads $\{F_{ex}\}$

$$\{F_{ex}\} = \sum_{i=n}^{m}\int_V [B]^T\{\sigma_o\}dV \quad (9)$$

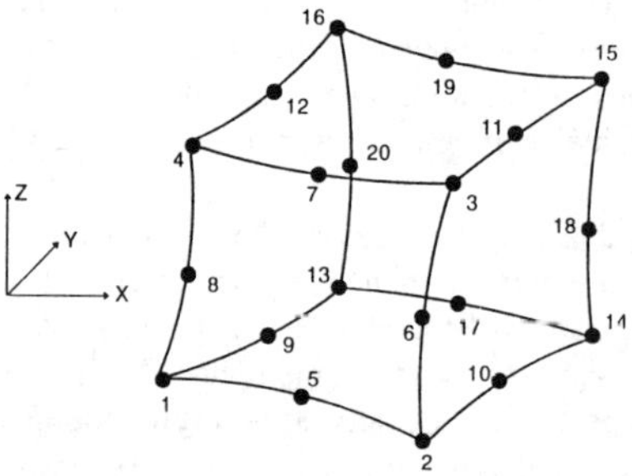

Figure 1. IPQS element for modelling soil.

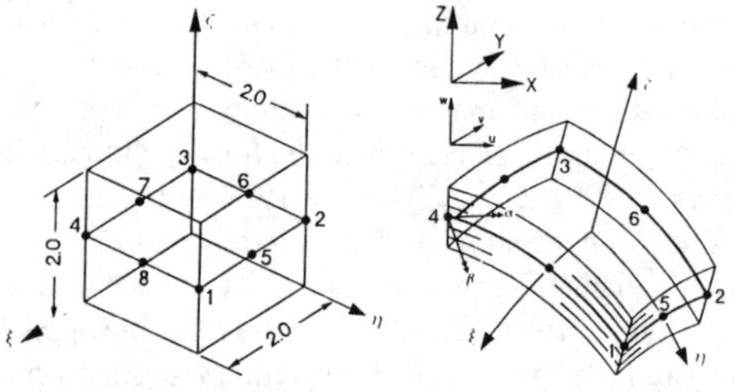

Local coordinates Cartesian map

Figure 2. Quadratic sandwich shell element for modelling of shotcrete lining.

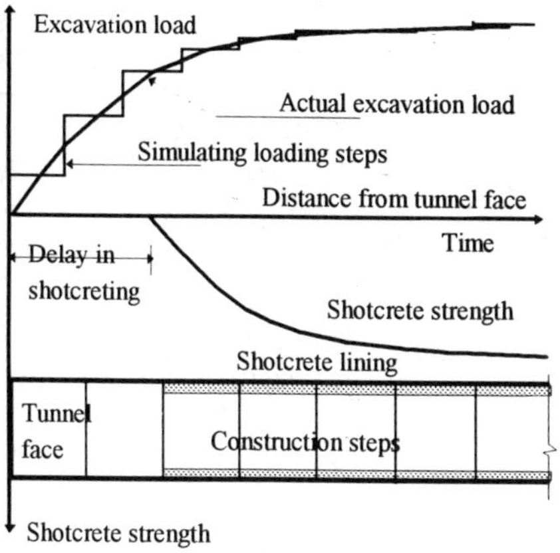

Figure 3. Simulation of shotcrete hardening with the face advance

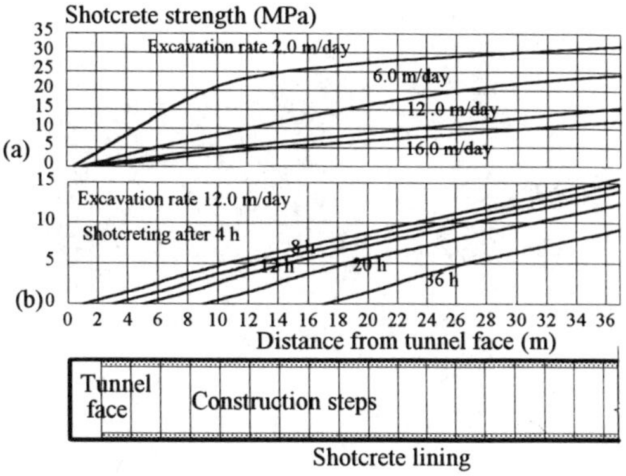

Figure 4. Shotcrete strength along the tunnel for: (a) different excavation rates, (b) application time

The elapsed time between the excavation, and the installation and setting of shotcrete lining causes loss of support in the radial direction of the tunnel sections. The soil stresses have to be redistributed by arching in the longitudinal direction of the tunnel.

The whole behavior of the tunnel near the face is expected to be function of the unsupported length of the tunnel which depends on:

1. The excavation rate
2. The delay in shotcreting
3. The shotcrete hardening and quality

The relation between the face advance, and excavation load and shotcrete hardening can be illustrated by Figure 3.

4 APPLICATIONS

4.1 *Tunnel geometry*

The three-dimensional model has been applied to a shallow subway tunnel of 9.4 meters diameter, and 12 meters overburden. The finite element mesh used is shown in Figure 6. Making use of the tunnel symmetry, the dimensions of the studied part have been reduced to 50 m. width, 56 m. depth, and 70 m. length. The mesh has been generated on strips to enable the simulation of excavations on steps, the length of each is 2.0 meters.

4.2 *Simulation of shotcrete hardening*

Based on the experimental data of the development of shotcrete strength with time (Kusterle 1985), the strength of shotcrete lining along the tunnel has been evaluated for different excavation rates and starting times as shown in Fig. 4.

5 GENERAL BEHAVIOR OF THE TUNNEL

The general behavior of the tunnel can be described by the distribution of deformations and stresses resulting from three-dimensional finite element analysis. The typical contour lines for vertical deformation of the soil surrounding the tunnel are given in Figure 5. The horizontal deformations in the direction of the tunnel axis are as shown in Figure 8. Both Figures present the deformations after 18 loading steps, each simulates 2 m. excavation. The contour lines for radial bending moments on the lining are given in Figure 6. The axial forces in the lining in the longitudinal direction of the tunnel are described by contour lines in Figure 9. Tension forces on the lining are resulted towards the tunnel face on the crown and the inverts, while compression forces acts on the tunnel bench. These forces are mostly due to the tendency of the surrounding soil to move towards the unsupported tunnel face.

6 EFFECT OF EXCAVATION RATE

The effect of excavation rate or, speed of face advance on the tunnel behavior has been studied by considering different excavation rates of 2, 6, 12, and 16 m/day. The strength of shotcrete lining at different sections is as shown in Figure 4-a. The predicted distributions of the vertical crown deformations along the tunnel axis for the different cases are plotted in Figure 13. The relationship

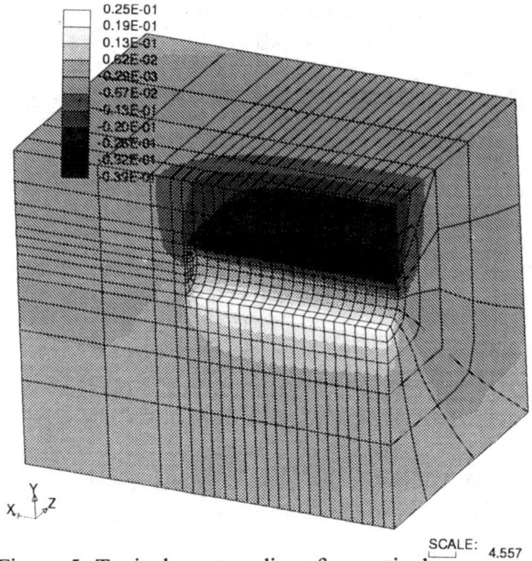

Figure 5. Typical contour lines for vertical deformations.

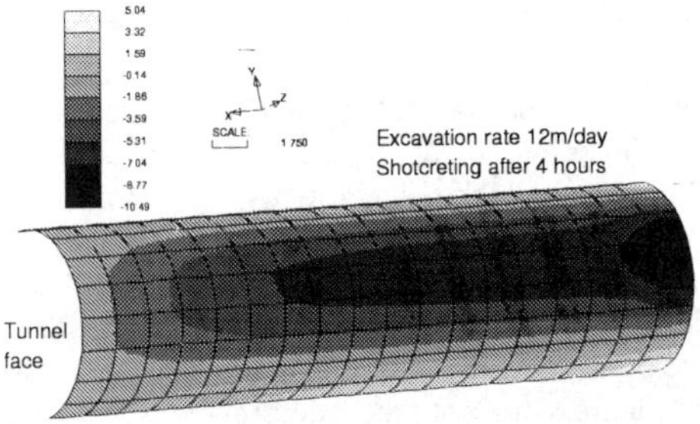

Figure 6. Contour lines for bending moments on the lining in the radial direction (M_y).

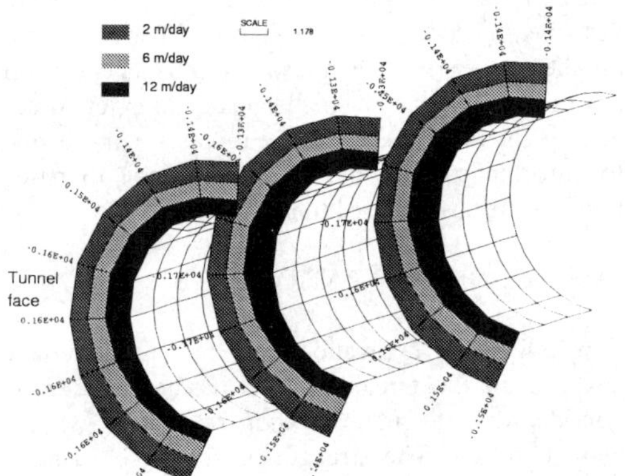

Figure 7. Effect of the excavation rate on the axial forces diagrams for the lining (N_y)..

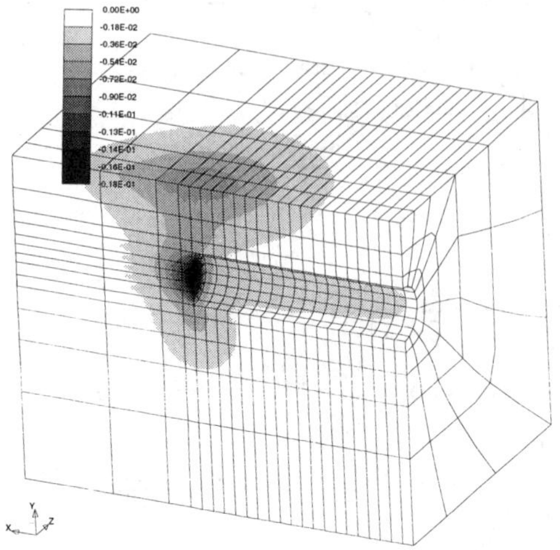

Figure 8. Typical contour lines for vertical deformations

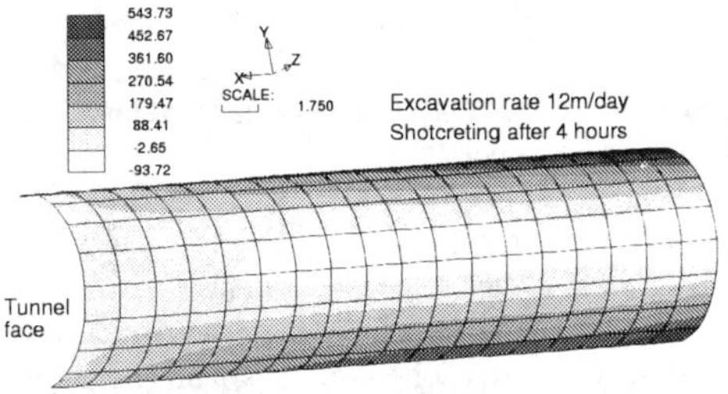

Figure 9. Contour lines for axial forces on the lining in the longitudinal radial direction (N_x).

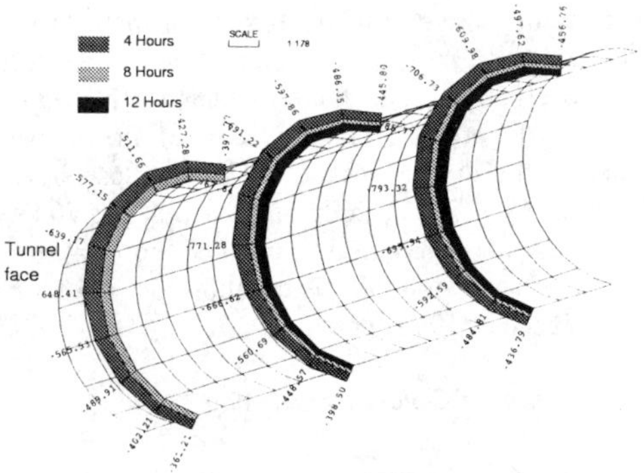

Figure 10. Effect of the delay in shotcreting on the axial forces diagrams for the lining (N_y)..

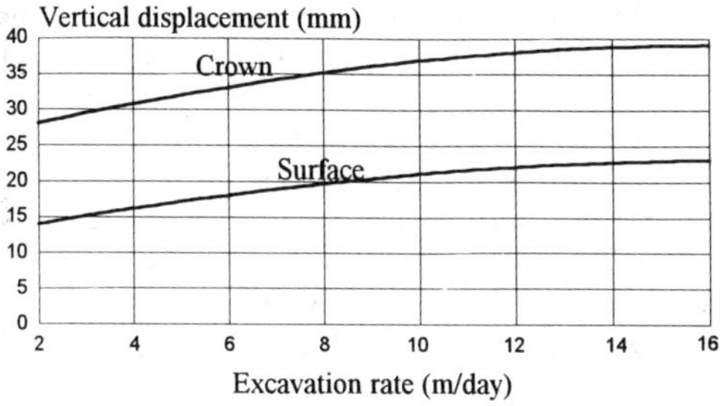

Figure 11. Effect of excavation rate on the vertical ground displacement

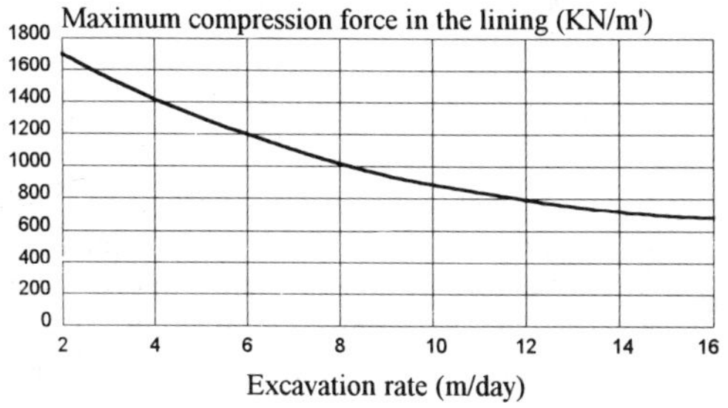

Figure 12. Effect of excavation rate on the axial forces in the lining.

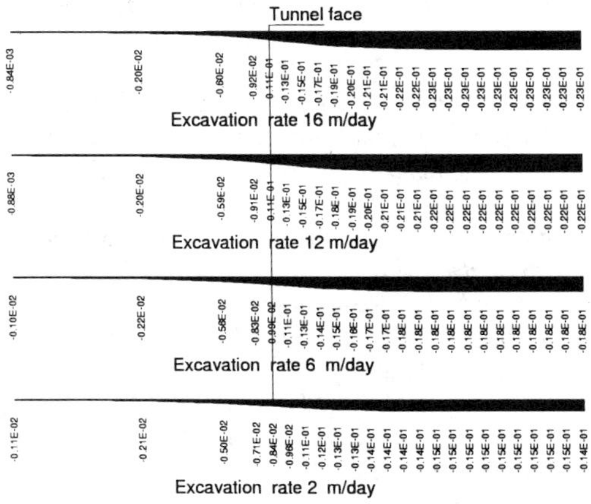

Figure 13. Effect of the excavation rate on the settlement of the tunnel crown.

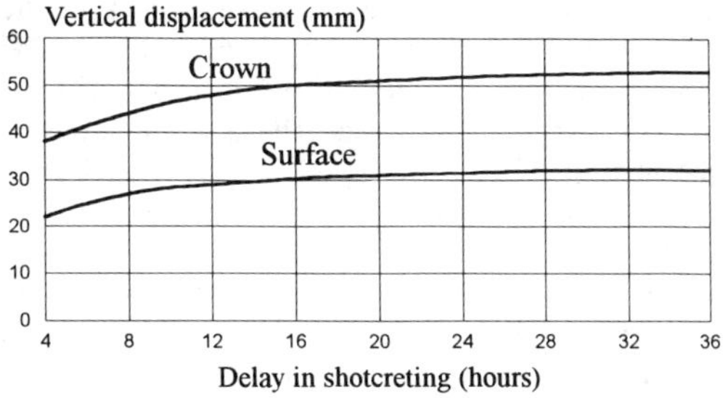

Figure 14. Effect of delay in shotcreting on the vertical ground displacement.

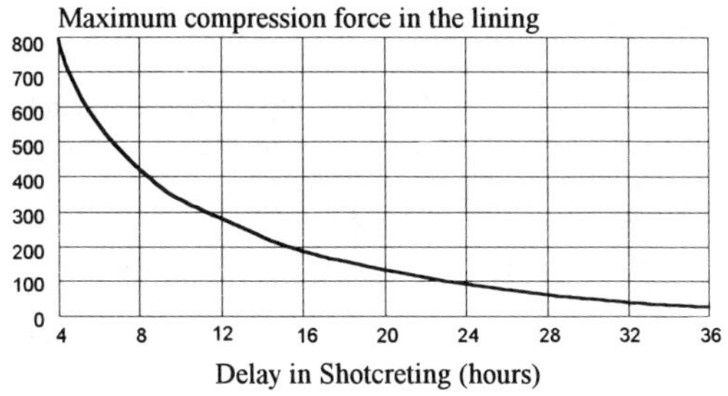

Figure 15. Effect of delay in shotcreting on the axial forces in the lining.

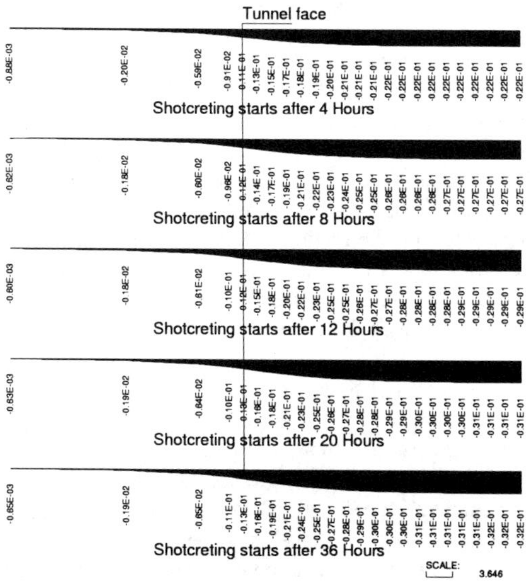

Figure 16. Effect of delay in shotcreting on the settlement of the tunnel crown.

between the final vertical deformation and the excavation rate has been plotted in Figure 11. The figures show that final deformation increases with the increase of excavation speed. The axial ring forces in the lining are plotted in Figure 7 for tunnel sections at distances 6, 18, and 32 m from the tunnel face for different excavation rates. The relationship between the excavation rate and the maximum axial force in the lining at distance of 32 m from tunnel face is as shown in Figure 12. It is clear from the figure that the lining forces decrease with the increase in the excavation rate. These effects are due to the fact that most of the redistribution of soil stresses (excavation load) happens before the complete hardening of shotcrete.

7 EFFECT OF DELAY IN SHOTCRETING

The effect of delay in shotcreting has been studied by analyzing the tunnel for different starting times of shotcrete construction after excavation. The considered cases are 4, 8,12, 20 and 36 hours after excavation. The excavation rate is constant (12 m/day) for all cases. The strength of shotcrete lining at different sections is as shown in Figure 4-b. The predicted ground settlements at crown along the tunnel axis for the different cases are plotted in Figure 16. The predicted relationship between the final ground settlement and the delay in shotcreting has been plotted in Figure 14. The figures show that final deformation increases with the shotcreting delay. The axial forces in the lining are plotted in Figure 10 for tunnel sections at distances 6, 18, and 32 m from the tunnel face for different shocreting times. The relationship between the delay in shotcreting and the maximum axial force in the lining at distance of 32 m from tunnel face is as shown in Figure 15. It is clear from the figure that the lining forces decrease with the delay in shotcreting. These results are mainly because big part of the redistribution of soil stresses (excavation load) happens before the construction and hardening of shotcrete lining.

8 CONCLUSIONS

Three-dimensional finite element analysis has been successfully employed to predict the influence of the construction speed on the behavior of shotcrete tunnels. A subway tunnel has been analyzed under different cases of excavation rates and different times of shotcrete application. The shotcrete strength for different sections are assumed based on experimental data. The results show an increase in the tunnel deformations and a decrease in the lining forces as the excavation rate and delay in shotcreting increase. This is mainly because big part of the redistribution of soil stresses happens before the complete hardening of shotcrete lining.

Through similar studies on any tunnel, the face advance and shotcrete application time can be optimized to get safe and economic tunnel which satisfy the deformation limits.

ACKNOWLEDGMENT

The authors express their gratitude to the Austrian national science foundation, for the promotion of the research project (No. P10077-OGE) with the title 'Safety of tunnels'.

REFERENCES

Augarde, C.E., H.J. Burd & G.T. Houlsby. 1995. A three-dimensional finite element model of tunnelling. Proc. of the Fifth Int. Symp. on Num. Models in Geomech. NUMOG V, Davos, Switzerland, 6-8 Sept. 1995: 457-462. Rotterdam: Balkema.

Duddeck, H. 1992. Tunnelling in rock masses. in Bell, F.B. *Engineering in rock masses.* Butterworth, Heinemann:423-439.

El-Nahhas, F., F. El-Kadi & A. Ahmed. 1992. Interaction of tunnel linings and soft ground. *Tunnelling and Ground Space Technology*. Vol. 7, No. 1: 33-43.

Eisenstein, Z. & A. Negro. 1985. Comprehensive design method for shallow tunnels. Proc. *Underground Structures in Urban Areas Conf.*, Prague. Vol 1:375-391.

Kusterle, W. 1985. Fruhfestigkeiten des Spritzbeton. *Berichtsband der Inter. Fachtagung Spritzbeton-technologie.* 35-38. Innsbruck.

Swoboda G. 1990. Numerical Modeling of Tunnels, in C. S. Desai & G. Gioda. *Numerical Methods and Constitutive Modeling in Geomechanics, Courses and Lectures*, No.311:277-318 Wien: Springer-Verlag.

Swoboda, G., 1996 *Programmsystem FINAL - finite element analysis program for linear and nonlinear structures*, Version 7.0, Austria: University of Innsbruck

12 Underground opening and mining engineering

Computer Methods and Advances in Geomechanics, Yuan (ed.) © 1997 Balkema, Rotterdam, ISBN 90 5410 904 1

Constitutive modelling of rock at an underground power house, India

A. Varadarajan & K.G. Sharma
Indian Institute of Technology, Delhi, India
C.S. Desai
University of Arizona, Tucson, Ariz., USA

ABSTRACT: A power house cavern is being constructed as a part of a hydroelectric project across Satluj river in Himachal Pradesh state in India. A research project is in progress to use a finite element procedure with HISS model for various rocks and joints and analyse the stability and field performance of the power house cavern. In the first stage, limited triaxial tests have been conducted on the quartz mica schist samples collected from the cavern site. δ_1 version of HISS model has been used to characterise the behaviour of the rock. The overall prediction by this model is satisfactory.

1 INTRODUCTION

A hydroelectric power project is under construction in the middle reaches of the Satluj river in Himachal Pradesh state in the northern part of the Indian subcontinent. The layout of the project is shown in Fig.1. The project envisages construction of a 60.5m high concrete dam across Satluj river at Nathpa Village, a 27.78 km long and 10.15 m dia head race tunnel on the left bank of the river and a large 211 m long, 20 m wide and 49 m high underground power house at Jhakri village with an installed capacity of 1500 MW (6 x 250 MW).

A research project is in progress with an objective to use a finite element procedure with HISS model for various rocks and joints and anlayse the stability and field performance of the power house cavern. Herein the testing and constitutive modelling of a rock type in the cavern site is presented.

2 ROCK

The rock type present in the project area comprises of various metamorphic rocks like augen gneiss, granite gneiss, biotite schist, chlorite muscovite schist with amphibolites and pegmatite. Quartz veins are also present in abundance. These rocks belong to Jeori-Wangtu Gneissic complex of Pre-cambrien Age and form the basement of Rampur group of rocks lying at higher stratigrafic position. The rocks are openly folded and a number of anticlines and synclines have been identified in the project area.

At the underground power house site, the rocks are essentially quartz mica schist and biotite schist with thin bands of sericite schist. For the present study, quartz mica schist samples have been used. This rock contains quartzitic bands of varying thickness of 1 to 1.5 cm with uneven distribution and micaceous bands of 1 to 2 mm thickness and at times showing pygmatic folding. The rock is coarse grained having granular texture and light to medium greyish white colour. An x-ray diffractogram on similar rock samples showed presence of 31.4% quartz, 26.5% of chlorite and 22.3% of mica in abundance followed by clay minerals such as kaolinite, sepiolite and illite forming the rest of its constituents (Nasseri, 1992). A scanning electron micrograph study shows the coarse nature of grains in this rock along with strong preferred orientation of mica.

The physical properties of the rock as determined by conducting tests as per ISRM/IS specifications are as follows:
Specific Gravity : 2.83
Density : Dry=26 kN/m^3, Saturated=27.4 kN/m^3
Water Content at Saturation : 0.28-0.64%
Void ratio : 0.09-1.54%

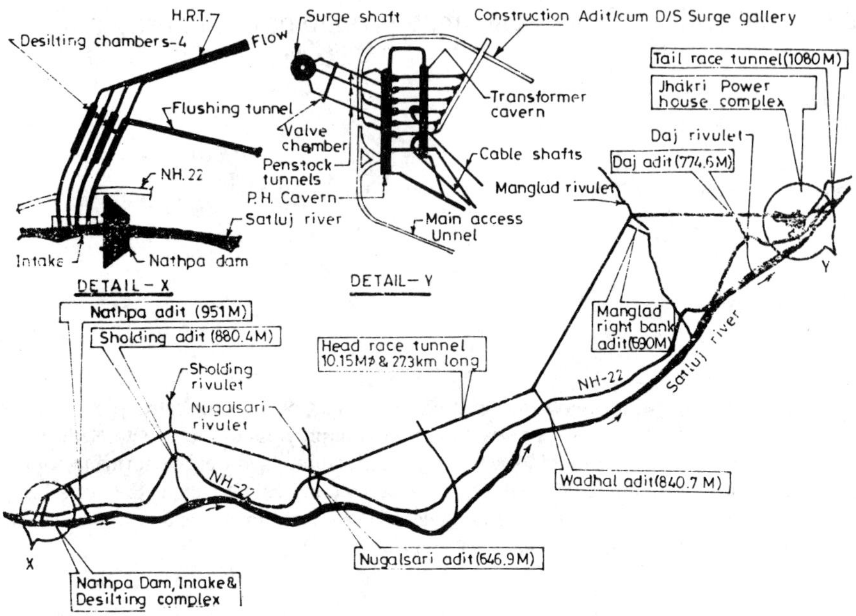

Fig.1 Layout plan of Nathpa - Jhakri project

3 EXPERIMENTAL STUDY

Specimens of size 3.8 cm dia and 7.60 cm long were cored from rock blocks collected from the project site and lapped to obtain specimens which met tolerance limits specified by ISRM. The specimens were ovendried at 105°C for 24 hours and then cooled in desiccator before testing.

Triaxial tests were conducted using a high pressure triaxial cell, as shown in Fig. 2 (Sharma and Bagde, 1994). The triaxial cell had the following features: ram capacity of 2MN, maximum cell pressure of 140 MPa, balanced ram and provision for electrical wires for the measurement of strains through electrical resistance strain gauges. Axial loading was applied using a 5MN loading frame and a hydraulic cylinder unit. The axial load was measured with a load cell of 1 MN capacity. For applying cell pressure a hydraulic pressure pump was used to maintain and control cell pressure upto 140 MPa. A 10-channel auto scanning digital strain indicator was used to measure axial and diametral strains in rock specimens through bakelite based electrical resistance strain gauges.

Conventional triaxial tests on the rock specimens were conducted with the confining pressures of 0, 5, 15, 35 and 50 MPa.

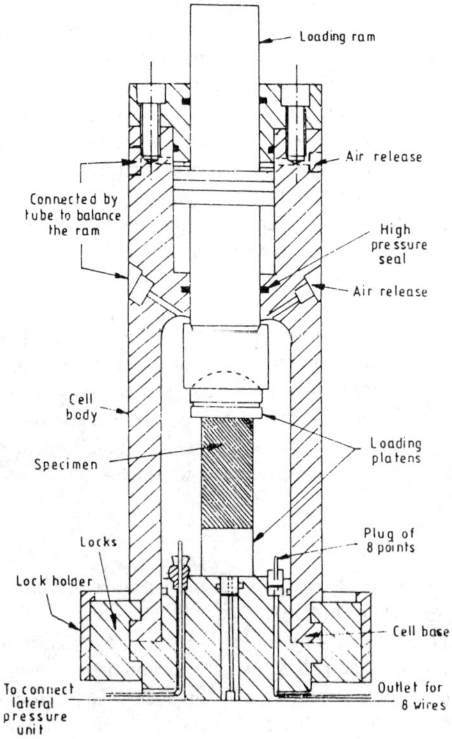

Fig.2 High pressure triaxial cell

4 CONSTITUTIVE MODEL

As a first step towards developing a suitable constitutive model, the hierarchical single surface (HISS) model developed by Desai and his coworkers has been adopted herein. The details of the model are contained in a number of publications (for example, Desai et al. 1986). The model is briefly described in the following.

The δ_1 model which depicts the behaviour of initially insotropic material hardening with nonassociative plasticty has been used.

The yield function F is written as

$$F = \frac{J_{2D}}{P_a^2} - [-\alpha(\frac{J_1}{P_a})^n + \gamma(\frac{J_1}{P_a})^2](1-\beta S_r)^m \quad \text{(1a)}$$

$$= \frac{J_{2D}}{P_a^2} - F_b F_s \quad \text{(1b)}$$

where, J_{2D} is the second invariant of the deviatoric stress tensor; J_1 is the first invariant of the stress tensor; P_a is the atmospheric pressure; γ,β and m are the material response functions associated with the ultimate behaviour, m=-0.5, α is a hardening function; n is the phase change parameter and S_r is the stress ratio given by

$$S_r = \frac{\sqrt{27}}{2} J_{3D} \cdot J_{2D}^{-1.5} \quad (2)$$

where, J_{3D} is the third invariant of deviatoric stress tensor.

F_b is the basic function describing the shape of the yield function in the $J_1 - \sqrt{J_{2D}}$ space and F_s is the shape function which describes the shape in the octahedral plane.

The hardening function, α is given by

$$\alpha = \alpha(\xi, \xi_v, \xi_D, r_v, r_D) \quad (3)$$

where, $\xi = \int (d\epsilon_{ij}^P \cdot d\epsilon_{ij}^P)^{1/2}$ is the trajectory of the plastic strains part of ξ respectively, and r_v and $r_D = \xi_v/\xi$ and ξ_D/ξ respectively.

Table 1 Material Constants for Quartz Mica Schist

Young's Modulus, E =	$KP_a(\frac{\sigma_3}{P_a})^n$
	K = 26,000, n = 0.18, P_a = 0.1014 MPa.
Poisson's Ratio, ν = 0.25	
Tensile Strength, R = 12.35 MPa	
Ultimate Parameters	
	m = -0.50
	γ = 0.0257
	β = 0.75
Phase Parameter, n = 3.25	
Hardening Parameters	
	a_1 = 0.34302E-8
	η_1 = 0.85
Nonassociative Parameter, κ = 0.6	

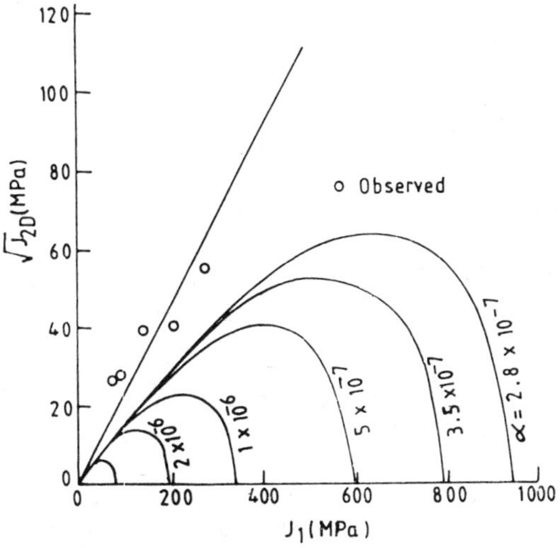

FIG. 3 PLOT OF F_b IN $J_1 - \sqrt{J_{2D}}$ SPACE (S_r = 1)

A simple form of α is

$$\alpha = \frac{a_1}{\xi^{\eta_1}} \quad (4)$$

where a_1 and η_1 are the material constants for the hardening behaviour.

For the nonassociative model, δ_1, the plastic potential function, Q is defined as

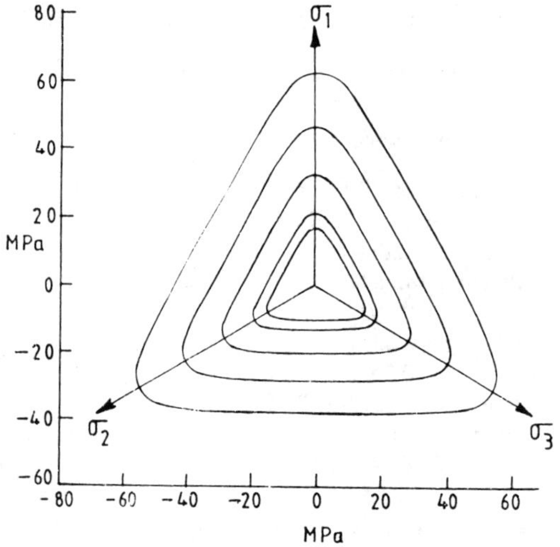

FIG. 4 PLOT OF F_s IN OCTAHEDRAL PLANE

$$\frac{J_{2D}}{P_a^2} - [-\alpha_Q(\frac{J_1}{P_a})^n + (\frac{J_1}{P_a})^2]\, F_s \qquad (5)$$

where,

$$\alpha_Q = \alpha + \kappa\,(\alpha_o - \alpha)\,(1-r_v) \qquad (6)$$

in which α_o is the value of α at the end of initial hydrostatic loading and κ is a nonassociative parameter.

5 DETERMINATION OF MATERIAL CONSTANTS

Six material constants, γ, β, n, a_1, η_1 and κ are required for the δ_1 model in addition to the two elastic constants, Young's modulus, E and Poisson's Ratio, ν.

The material parameters have been determined following the procedure outlined in various references (for example, Desai and Varadarajan, 1987). In this study, the experimental results of all the tests have been used to determine the average material constants. The constants obtained are given in Table 1.

6 VERIFICATION OF THE MODEL

The material constants obtained have been used to backpredict the observed behaviour of the rock. The incremental constitutive equation is used on the basis of the normality rule of plasticity and consistency condition dF=0 as

$$\{d\sigma\} = \{C^{ep}\}\,\{d\epsilon\} \qquad (7)$$

where, $\{d\sigma\}$, $\{d\epsilon\}$ are vectors of incremental stress and strain components respectively and $\{C^{ep}\}$ is the constitutive matrix containing the material constants. To predict the behaviour under a given stress-path, Eq.(7) is integrated along that path starting from the initial hydrostatic state.

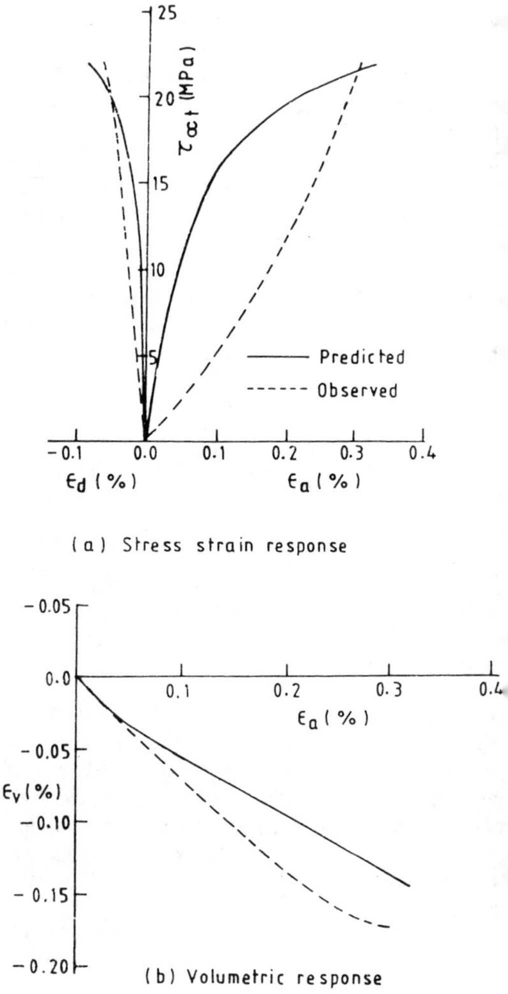

FIG. 5 COMPARISON BETWEEN PREDICTIONS AND OBSERVATIONS FOR CTC PATH (σ_0 = 15MPa)

Figure 3 shows the plot of the basic function, F_b

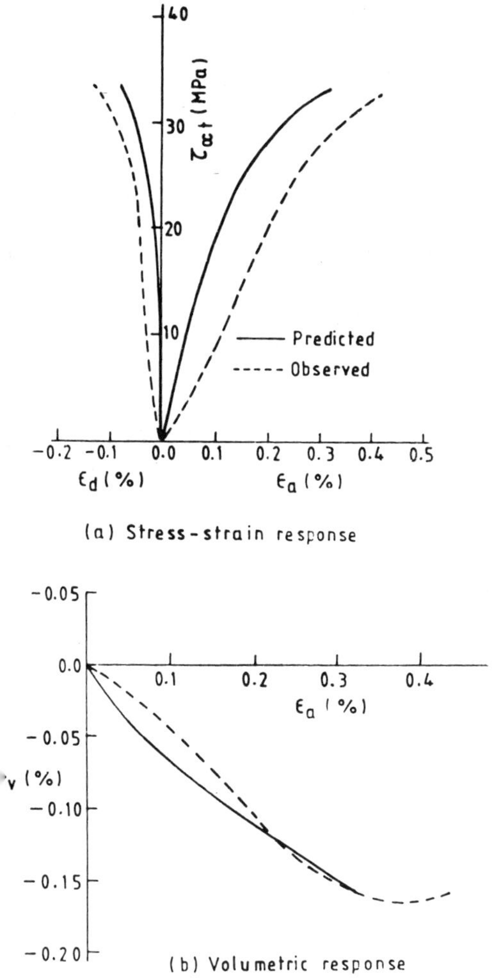

FIG. 6 COMPARISON BETWEEN PREDICTIONS AND OBSERVATIONS FOR CTC PATH (σ_0 = 35 MPa)

in $J_1 - \sqrt{J_{2D}}$ stress-space. The plot at ultimate condition (α=0) is also shown along with the observed results. In Fig. 4 is shown the plot of the shape function, F_s in octohedral plane for various J_1 values. The model provides satisfactory predictions.

Figures 5-7 show comparison of predicted and observed stress-strain-volume change response of CTC test at various confining pressures.

The prediction of stress-strain response by δ_1 model improves with confining pressure. The volume change behaviour shows volumetric compression for all the tests. The predicted behaviour is generally satisfactory, but the magnitudes are different.

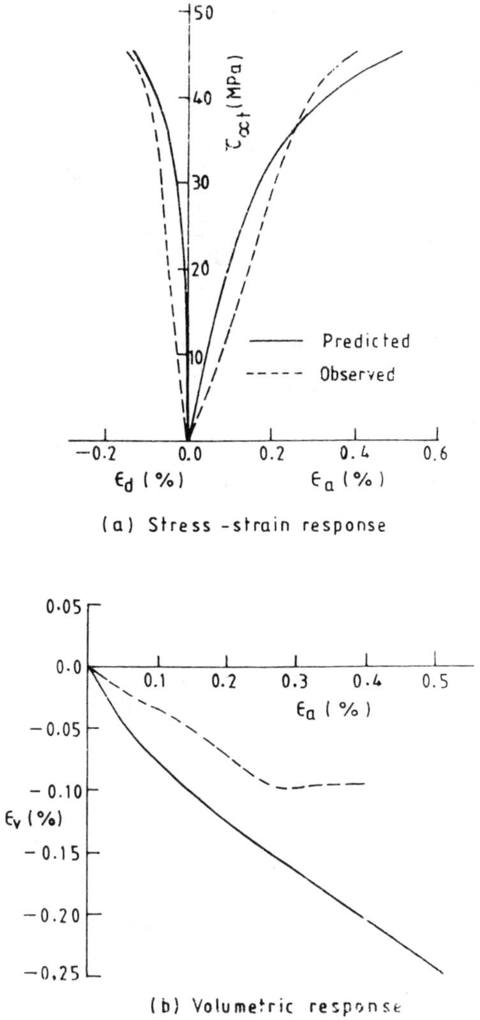

FIG. 7 COMPARISON BETWEEN PREDICTIONS AND OBSERVATIONS FOR CTC PATH (σ_0 = 50 MPa)

7 CONCLUSIONS

A limited quartz mica schist rock samples obtained from the power house cavern area of the Nathpa-Jhakri hydroelectric project have been subjected to triaxial tests in the laboratory using CTC path. As a first step, the δ_1 version of HISS model has been adopted to characterize the behaviour of the rock. The model provides, in general, satisfactory predictions. Further work is in progress in testing more number of samples using a stiff testing machine under various stress-paths. Improved version of HISS model is proposed to be used to capture the complex behaviour of the rock including strain-softening.

ACKNOWLEDGEMENTS

The research reported herein was supported by Grant Nos. Int. 9213752 and 9300828 from International Program, National Science Foundation, Washington, D.C., USA (Dr. Marjorie Lueck, Program Director), under U.S.-India Cooperative Project between Indian Institute of Technology, New Delhi, and The University of Arizona, Tucson, U.S.A. The assistance and support provided by the Nathpa-Jhakri Project is gratefully acknowledged.

REFERENCES

Comprehensive Geotechnical Report, No.2, (1991), On the studies carried out for Nathpa Jhakri Hydroelectric Project, Shimla and Kinnaur Districts (H.P), March 88-Jan. 91, *Geological Survey of India.*

Desai, C.S., Somasundaram, S. and Frantziskunis, G, (1986), A hierarchical approach for constitutive modelling of geologic materials, *Int. J. Num. Analyt. Meth. in Geomech.,* Vol.10, No.3.

Desai, C.S. and Varadarajan, A., (1987), A constitutive model for short term behaviour of rock salt, *J. of Geophysical Research,* Oct.

Nasseri, M.H.N.B., (1992), Strength and deformation responses of schistose rocks, *Ph.D. Thesis,* I.I.T. Delhi.

Sharma, M.K. and Bagde, M.N., (1994), Evaluation of rock properties and analysis of underground cavern by FEM: Nathpa Jhakri Project, *M.Tech. Thesis,* I.I.T. Delhi.

Computer Methods and Advances in Geomechanics, Yuan (ed.) © *1997 Balkema, Rotterdam, ISBN 90 5410 904 1*

In-situ stress determination by stress perturbation

K.J.Wang
Institute of Rock and Soil Mechanics, The Chinese Academy of Sciences, China

C.F.Lee
Department of Civil and Structural Engineering, The University of Hong Kong, China

ABSTRACT: The normal borehole deformation meter is used to measure the radial deformations of borehole induced by stress perturbation, as which is taken here the normal stress-relief overcoring technique. For the cases with and without knowledge of material parameters, the induced deformations of borehole are computed with different methods and compared with measured ones. In this way, the three dimensional in-situ stresses can be determined uniquely for actual complicated geological conditions and only one borehole is needed.

1 INTRODUCTION

The important difference between geomaterial and usual engineering material is that there are initial stresses, usually called in-situ stresses, in geomaterial before engineering perturbation. The magnitude and direction of in-situ stresses are affected with many factors, which is an unsolved problem to be explored (Lee & Wang 1995).

It is difficult, even impossible, to determine the in-situ stresses quantitatively in theory. Instead, the most practical and available way is to make measurement of in-situ stresses in field.

The common measurement of in-situ stresses is made with instrument installed in borehole to measure the response of stress perturbation nearby. In principle, any kind of stress perturbation near a borehole can lead to its deformation related to the in-situ stresses. The simpler the operation and the more sensitive is the borehole deformation to the operation, the more reliable is this method of in-situ stress determination by stress perturbation. Pending the availability of stress perturbation techniques of a more advanced nature, the normal stress-relief overcoring technique is still used presently in conjuction with the normal borehole deformation meter to measure the radial deformations of borehole. The overcoring depth will, however, be the mid-depth of the borehole deformation meter only, such that the deformations measured at the various measuring points on the deformation meter are non-uniform, which is different with normal overcoring technique where the overcoring depth is over the depth of deformation meter to make the measured deformations uniform.

Usually, the material parameters can be determined with the test of core sample. In this case, the so-called deformation factors are computed, which are the induced radial deformations of measuring points on the deformation meter under the action of a unit stress component and given stress perturbation. With the computed deformation factors and measured deformations of borehole, the in-situ stresses can be determined with the superposition principle and least square method. In the case without knowledge of material parameters, which occurs especially when the geological condition is very complicated, the induced deformations of the measuring points are computed repeatedly with variable material parameters and in-situ stresses to make the computed deformations approach measured ones. In this way, the in-situ stresses are also determined.

With either of the above methods, the three dimensional in-situ stresses can be determined uniquely for actual complicated geological conditions and only one borehole is needed.

Along the train of thought of this paper, it may be very simple and easy to make measurement of in-situ stresses in field if there is a new kind of stress perturbation to occur, whose operation is much easier than the existing overcoring technique.

2 PRINCIPLE AND FORMULA

As mentioned above, the stress-relief overcoring technique is used in this paper as the way of stress perturbation to a borehole.

The normal borehole deformation meter is installed in the borehole to obtain the radial deformations of borehole periphery. The overcoring depth will, however, be the mid-depth of the borehole deformation meter only, such that the deformations obtained at various measurement points are non-uniform (Figure 1).

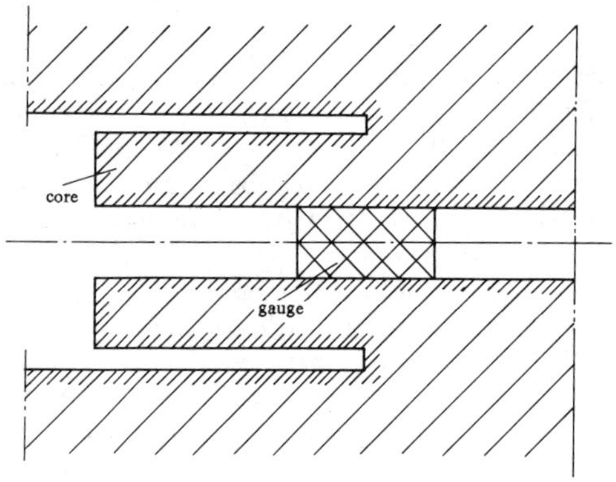

Figure 1. Stress perturbation to a borehole

For the case with knowledge of material parameters, a three dimensional elastic FEM program is used to determine the deformation factors, which are the induced radial deformations of measuring points on the deformation meter under the action of a unit stress component and the given stress perturbation, as followes.

Under the action of a unit stress component

$$\sigma_x = 1$$
$$\sigma_y = \sigma_z = \tau_{xy} = \tau_{yz} = \tau_{zx} = 0$$

the deformations of every measurement point in different directions, called as deformation factors, are δ_{1j}, $j = 1, 2, \ldots, n$, where n is the total number of measurement points on the borehole deformation meter and larger than 6. So do δ_{2j}, δ_{3j}, δ_{4j}, δ_{5j}, and δ_{6j}, respectively. Besed on the superposition principle for small deformation, under the action of actual in-situ stresses (σ_x, σ_y, σ_z, τ_{xy}, τ_{yz}, τ_{zx}), the deformation of measurement point j on the borehole deformation meter due to stress perturbation is

$$D_j = \sigma_x\delta_{1j} + \sigma_y\delta_{2j} + \sigma_z\delta_{3j} + \tau_{xy}\delta_{4j} + \tau_{yz}\delta_{5j} + \tau_{zx}\delta_{6j} \qquad (1)$$

For n measurement points, there are n equations as equation (1), which compose a system of overdetermined linear equations. The system of equations can be solved using least square method with Chebeshov transformation (Hornbeck 1975) to eliminate the ill condition of the normal equations, and the three dimensional in-situ stresses (σ_x, σ_y, σ_z, τ_{xy}, τ_{yz}, τ_{zx}) are uniquely determined.

For the case without knowledge of material parameters, the same FEM program is used repeatedly to compute the induced deformations of the same measuring points under the actions of variable in-situ stresses with variable material parameters to make the object fundtion

$$S = \sum_{j=1}^{n}(D_j^c - D_j^m)^2 \qquad (2)$$

approach a minimum value, where D_j^c and D_j^m are the computed and measured induced deformations, respectively, of measuring point j. Obviously, the object fundtion S is correlated with material parameters and in-situ stresses, and it is necessary to use an optimization routine in conjuction with the FEM program. The material parameters and in-situ stresses are considered as real ones as soon as they make the object fundtion S smaller than a certain value.

After the in-situ stress components (σ_x, σ_y, σ_z, τ_{xy}, τ_{yz}, τ_{zx}) are determined with the measured deformations of borehole and either of the above algorithms, the principal stresses σ_i and the direction cosines of principal directions l_i, m_i and n_i, i = 1, 2, 3 can be determined with the formula shown in the textbooks on the theory of elasticity and plastisity (Westergaard 1952). Then with coordinate transformation

$$\begin{cases} l_i = \cos\beta_i \cos(\alpha_i - \theta) \\ m_i = \cos\beta_i \sin(\alpha_i - \theta) \\ n_i = \sin\beta_i \end{cases} \qquad (3)$$

the dip and azimuth angles β_i and α_i of the principal stress σ_i can be given to meet the requirement for

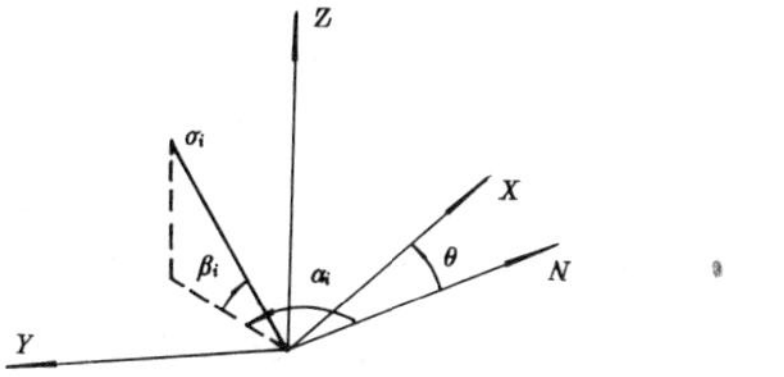

Figure 2. Coordinate transformation

engineer (Figure 2). In Figure 2, (x , y , z) is the Cartesion coordinate system corresponding stress components, in which z is vertical ordinate, θ is the angle of the due north to x axis.

3 EXCUTION PROCESSES

The deformation factors and deformations of measurement points for the cases with and without knowledge of material parameters can be obtained using either analytical methods for idealized cases or numerical methods for real complicated conditions. For practical application, a simple three dimensional elastic FEM program is used to determine them. Figure 3 shows a quarter of finete element mesh.

The base of the mesh, with the coordinate $z = 0$, is fixed, and the outer surfaces of the mesh

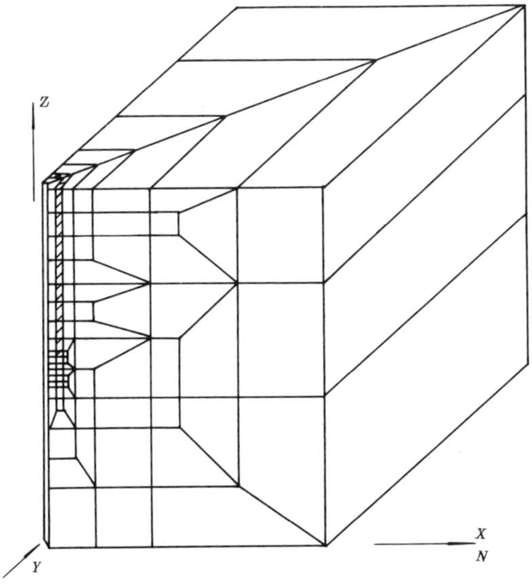

Figure. 3 Finite element mesh

are 5 or more times of the diameter of overcoring slot from the borhole deformation meter to eliminate the boundary effects.

It should be noted that the deformation factors and deformations of measuring points for the two cases are the induced deformations with stress perturbation under the action of certain stresses. So they are the differences between two kinds of deformations with and without stress perturbation, which is the stress-relief overcoring ring slot here and modelled with dashed elements in Figure 3.

This is similar to that of back analysis condsidering the unobserved displacements before installation of measuring instruments in surrounding rock masses of tunnels and the same computing method, called as method of parameter covariation, can be used here (Wang 1993).

In order to reduce the computation scale and time, so called relaed-degree-of-freedom solver (Wang 1992) is used. The point numbers near measurement points are put in the last part and only the equations corresponding to the related freedoms of this part are kept in core after the elimination loop for the smaller point numbers. In this way, a very large system of equations can be reduced to a very small group of equations, which can be solved easily and shortly on the microcomputers.

For the case with knowledge of material parameters, the number of loading cases is 6 for $\sigma_x = 1$, $\sigma_y = 1$, $\sigma_z = 1$, $\tau_{xy} = 1$, $\tau_{yz} = 1$ *and* $\tau_{zx} = 1$, respectively. Fortunately, the coefficients of left-hand terms of the equations keep the same values for the 6 loading cases, and the arithmetic operation for the right-hand terms can be made separately after the elimination process of the equation coefficients. In this way, the solutions for the 6 loading cases can be given once.

For the homogeneous isotropic media with nearly the same Poisson ratio, the computation of deformation factors is made only for the first project. The deformation factors for the other projects are equal to the existing deformation factors multiplied by the ratio of elastic modula E/E_0 , where E_o is the elastic modulus for the first project, and E for the new project.

With the deformation factors, δ_{1j}, δ_{2j}, δ_{3j}, δ_{4j}, δ_{5j} and δ_{6j} , solved using the above method and the observed deformations of measuring point, D_j , the overdetermined equations (1) are made up. The usual method to solve overdetermined equations (1) is to solve their determined normal equations and the solution is equivalent to that of overdetermined equations with the principle of least square. Unfortunately, the normal equations of overdetermined equations are ill-conditioned and the solution is sensitive to the computation accuracy. In order to overcome the difficulty, the Chebeshev transition is made and a stable solution can be obtained in the area [−1,1] (Hornbeck 1975).

As long as the equations (1) are solved, the is-situ stress components (σ_x, σ_y, σ_z, τ_{xy}, τ_{yz}, τ_{zx}) are obtained.

For the case without knowledge of material parameters, the same FEM program and same FEM mesh are used. In addition, an independent optimization module is used for comparative analysis of FEM results and choice of variable material parameters and in-situ stresses. The computer will exercise operation command on the repeated execution of the optimization routine and FEM program (Wang & Lee 1995) until the object function (2) approaches a minimum value (Figure 4). As a result, the three dimensional stress are determined.

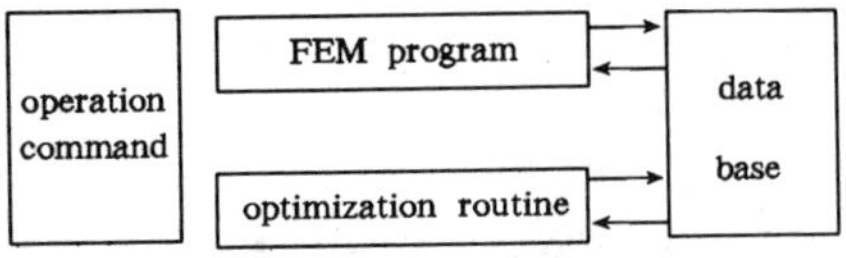

Figure 4. Operation of independent modules

After the stress components are determined with either of the above algorithms, the principal stresses (σ_1, σ_2, σ_3) and their dip and azimuth angles (β_i, α_i, $i = 1,2,3$) can be given from the related formula.

The above process can be executed for all the sampling points of in-situ stress measurement. The number of the sampling points and the measurement area are usually very limited because of the limitation of cost and condition. Therefore, it is necessary to construct a stress field for engineering interest with a limited amount of measured data (Wang & Lee 1997).

The whole execution flow chart is shown in Figure 5.

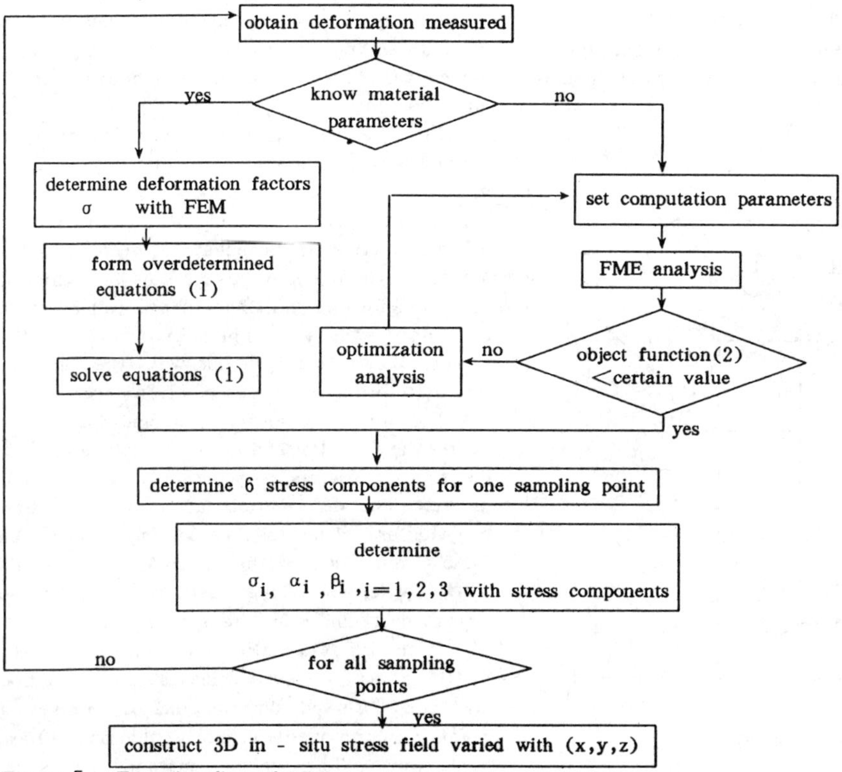

Figure 5. Excution flow chart

4 EXAMPLE

A practical measurement result of in-situ stresses through 3 boreholes with different orientations is taken here as an example to compare with the result obtained with the method discussed in this paper.

The location of measurement borehole and coordinate system are shown in Figure 6.

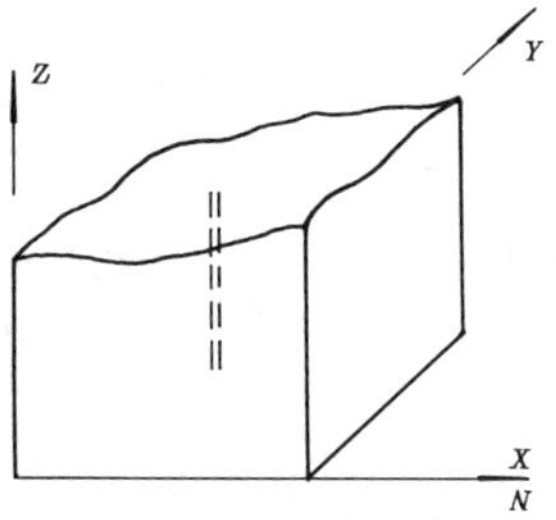

Figure 6. Location of borehole and coordinate system

The sampling point taken here is 307 m far from the ground surface. The elastic modulus $E = 20000\text{MPa}$ and Poisson ratio $\nu = 0.2$. The measured data for in-situ stresses of the point are

$$\sigma_1 = 14.12\text{MPa} \quad \alpha_1 = -6° \quad \beta_1 = 82°$$
$$\sigma_2 = 10.89\text{MPa} \quad \alpha_2 = -140° \quad \beta_2 = 5°$$
$$\sigma_3 = 8.52\text{MPa} \quad \alpha_3 = 130° \quad \beta_3 = 6° \quad (4)$$

where the stress is compressive if its value is positive.

With coordinate transformation, the corresponding stress components are

$$\sigma_x = 9.87\text{MPa}$$
$$\sigma_y = 9.63\text{MPa}$$
$$\sigma_z = 14.03\text{MPa}$$
$$\tau_{xy} = 1.60\text{MPa}$$
$$\tau_{yz} = -0.30\text{MPa}$$
$$\tau_{zx} = 1.10\text{MPa} \quad (5)$$

The stresses are acted on the outer surfaces of the model shown is Figure 3, and the overcoring-slot-induced deformations of borehole are computed with FEM. Then the induced deformations are taken here as 'measured' deformations of the borehole.

As for the deformation factors, they can be determined with FEM for each cases, or equal to the last deformation factors multiplied by the ratio of elastic modula for two cases if the Poisson ratios are the same for the two cases and so are the meshes and geometry conditions.

As mentioned above, with the computed deformation factors and 'measured' deformations of borehole, the overdetermined equations (1) are formed. Following the excution flow shown in Figure 5, the principal stresses and principal directions are determined for the sampling point as follows

$$\sigma_1 = 14.09\text{MPa} \quad \alpha_1 = -6.10° \quad \beta_1 = 81.80°$$
$$\sigma_2 = 10.91\text{MPa} \quad \alpha_2 = -139.61° \quad \beta_2 = 4.91°$$
$$\sigma_3 = 8.48\text{MPa} \quad \alpha_3 = 130.07° \quad \beta_3 = 6.11° \quad (6)$$

If the elastic modulus E and Poisson ratio ν were not known previonsly, the FEM program has to be run repeatedly in conjunction with an optimization routine (Figure 4), until the object function S (eq. 2) is smaller than a given small value. The results with this method are $E = 20100\text{MPa}$, $\nu = 0.20$.

$$\sigma_1 = 14.20\text{MPa} \quad \alpha_1 = -5.80° \quad \beta_1 = 80.67°$$
$$\sigma_2 = 10.91\text{MPa} \quad \alpha_2 = -139.01° \quad \beta_2 = 5.15°$$
$$\sigma_3 = 8.48\text{MPa} \quad \alpha_3 = 133.03° \quad \beta_3 = 6.22° \quad (7)$$

The slight differences among equations (4), (6) and (7) result from the computation errors and can be allowed for engineering requirement.

In order to show the capacity of the method proposed in this paper for practical geological conditions, a little change is made for the FEM mesh shown in Figure 3. There is a fault through the bottom of deformation meter with the strike of S—N and a dip of 25° to the horizontal plane (Figure 7). The normal and shear stiffnesses of the fault are 1000MPa/cm ane 200MPa/cm, respectively. The elastic modulus and Poisson ratio for the continuous media, and the 'measured' deformations of borehole all keep the same values in the case without fault. But, the deformation factors are quite different with FEM computatuon, and the results of in-situ stresses are

$$\sigma_1 = 12.07\text{MPa} \quad \alpha_1 = -5.87° \quad \beta_1 = 74.51°$$
$$\sigma_2 = 11.15\text{MPa} \quad \alpha_2 = -139.50° \quad \beta_2 = 9.23°$$
$$\sigma_3 = 10.31\text{MPa} \quad \alpha_3 = 131.32° \quad \beta_3 = 10.01° \quad (8)$$

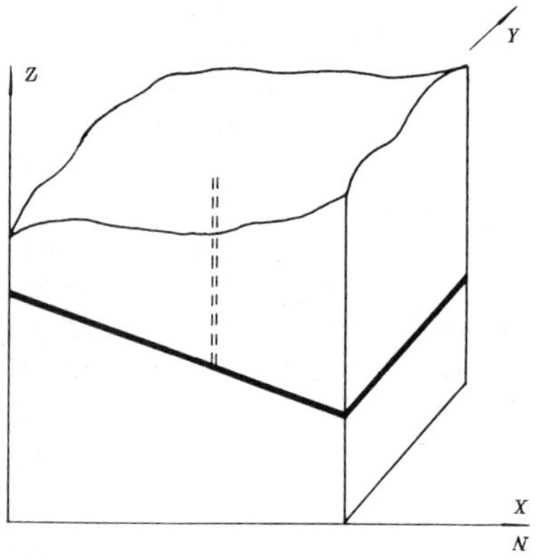

Figure 7. The media with a fault

For more complicated conditions, such as anisotropic, unhomogeneous, and discontinuous properties, the methods proposed in this paper are still available. The measurement process is not changed, only the computation time is longer than idealized case.

After the in-situ stresses are determined for all the sampling points, a three dimensional stress field can be constructed, which is varied with the coordinates (x, y, z). Figure 8 shows the stress field constructed with the measurement data of 6 sampling points in the borehole of Figure 6(Wang & Lee 1997).

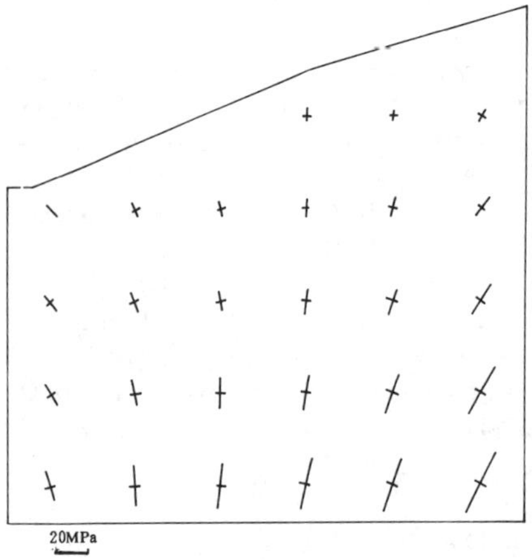

Figure. 8 Stress projection to *XOZ* plane through measurement hole

5 CONCLUSION

With existing normal borehole deformation meter and stress-relief overcoring technique in only one borehole, the three dimensional stresses of a sampling point can be determined for the two cases with and without knowledge of material parameters.

The methods proposed in this paper are available to practical complicated geological conditions because the FEM used here can take the complicated material properties in account.

There is no limitation for the kind of stress perturbation, so that it is possible to make the measurement of in-situ stresses simpler and easier than existing methods if threr is some new advanced stress perturbation to occur.

REFERENCES

Hornbeck, R. W. 1975. *Numerical Methods*. New York: Quantum Publisher Inc.

Lee, C. F. & K. J. Wang 1995. Analysis on occurrence and origin of high horizontal in-situ stresses. Chinese Journal of Rock Mechanics and Engineering. 14(3): 193—200

Wang, K. J. 1992. A related-degree-of-freedom solver for numerical analysis on microcomputer. *Proc. 4th Int. Conf. EPMESC*. Dalian, 30 July—2 August 1992: 130—135. Dalian: The Dalian University of Technology Press.

Wang, K. J. 1993. Back analysis of displacements directly approaching to the observation values. *Proc. Int. Conf. APCOM'93*, Sydney, 1—4 August 1993: 949—954. Rotterdam: Balkema.

Wang, K. J., C. F. Lee & Y. K. Cheung 1995. The program structure of independent modeles applied to complicated problems. *Proc. 5th Int. Conf. EPMESC*, Macao, 1—4 August 1995: 633—638. Korea: Techno-Press.

Wang, K. J. & C. F. Lee 1997. Random approximation of measured in-situ stress values by three-dimensional stress function fitting. *Proc. Int. Symp. Rock Mechanics and Enviromental Geotechnical Engineering*, Congqing, 1—4 April 1997

Westergaard, H. M. 1952. *Theory of Elasticity and Plasticity*. New York: John Willey & Sons, Inc.

Computer Methods and Advances in Geomechanics, Yuan (ed.) © 1997 *Balkema, Rotterdam, ISBN 90 5410 904 1*

Design and effect of anchoring supporting for weak rockmass surrounding galleries

Quansheng Liu, Duowen Ding & Weishen Zhu
Institute of Rock and Soil Mechanics, The Chinese Academy of Sciences, Wuhan, China

ABSTRACT: Based upon rock mechanics experiments and numerical analyses, anchoring support design for galleries is carried out and rational modes and parameters of anchoring are put forward according to the mechanical property, geostress, plastic zone and deformability of the surrounding rockmass. A self-developed extension-type compound bolt is used to reinforce the underground gallaries in combination with flexible shotecrete support. In site monitoring shows that the deformation and the stability of the gallery anchored using this method meet the specified requirments.

1 INTRODUCTION

A gallery has an overburden depth of 640m and a constructing azimuth of N60°E. The gallory lies in the strata of sedimentary rock mainly consisting of sandstone, shale and clay rock. The strata have seriously fractured in structure. The excavation region of the gallery is located in the affecting zone of faults F_{10-9} to F_{10-5}, between the feather branches of F_{10-5} (10) and F_{10-5} (11) derived from F_{10-5}. F_{10-5} (10) has a strike of 240° and a dip of 80°; F_{10-5} has a strike of 230° and a dip of 65° (Fig. 1). Owing to the action of fracture structure, most of the original structures of the rockmass have been fractured beyond recognition. The lithology around the gallery is complicated and the strata have fractured so seriously that they lost their original strength and stability. In the fracture zone, some rocks have become pelinite and can even be moulded with hands into any shape; after hydration, they look like paste. Seams can be rolled into powders with hands. Sandy shale is too fractured to be prepared for specimens; clay rock will be siltized when encountering water, and the siltized clay rock, while stoving, it will be crumbled into pieces or become mud-paste like if encountering water again. Because of great overburden above the gallery and of serious effects of structures, both vertical and horizontal stresses are very high. The measured vertical and two horizontal components of the original geostress field are respectively σ_2 = 19.5MPa (1.27 times the self-weight stress), σ_x = 15.7MPa (3.07 times the horizontal stress, $\frac{\nu}{1-\nu}\gamma H$, caused by the gravitational stress field) and σ_y = 12.8MPa (2.5 times the horizontal stress of $\frac{\nu}{1-\nu}\gamma$). All these data have shown that the rookmass surrounding the gallery bears not only the action of the gravitational field but also an intensive geological structural movement causing in a high structural stress, which creates unfavourable effects on both constructing and supporting of the gallery.

The structural movement strongly affects the surrounding rockmass, making the rockmass blocky in structure and developed in structural plane. The cementation between layers is poor and the media are easily softened characterized by occasional swelling and crumbling. The structural planes have cut the rockmass into a number of unstable blocks having a uniaxial saturated strength of $R_b < 300$MPa. The P wave of

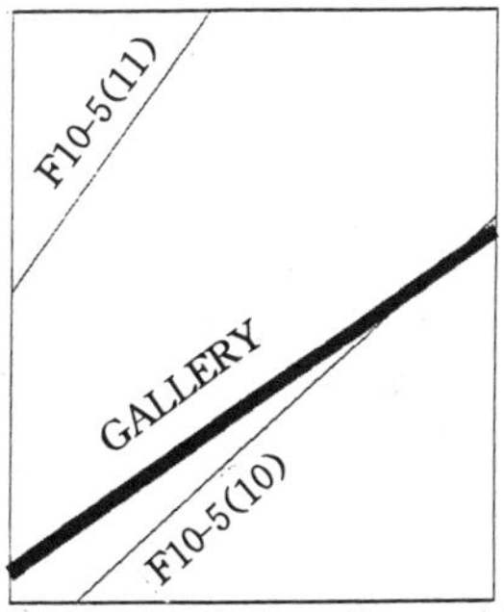

Figure 1. Plan of gallery and faults distribution

the rockmass has a velocity less than 2.0×10^3m/s. The discrete blocks and high geostresses make cave-in or side-fall of the gallory take place easily and no doubt it is very difficult to support the gallery.

2 NUMERICAL SIMULATION OF GALLERY'S PLASTIC ZONE AND SELECTION OF ANCHORING SCHEMES

The cross sectional shape of the gallery is shown in Fig. 2. The measured geostress is $\sigma_y = 19.5$MPa and $\sigma_x = 12.8$MPa; the physic-mechanical parameters of the rockmass are: unit weigh, $\gamma = 2.4$ ton/m^3; initial elastic module, $E = 0.4 \times 10^4$MPa; residual elastic module, $E_r = 0.18 \times 10^4$MPa; initial cohesion, $c = 0.9$MPa; residual cohesion, $c_r = 0.4$MPa; initial frictional angle, $\varphi = 22.5°$; visco-plastic rehological coefficient, 8.3×10^4MPa/hr. Considering the elastic-visco-plastic constitutive relation of the rockmass, we select a rheological mechanical model as shown in Fig. 3 for FEM analysis under plane strain presumption. The structure is axially symetrical, so only half a region is taken for computation. The size of the rockmass to be calculated is 37×64m^2 in cross section. Altogether, there are 48 four-node isotropic parametric elements in computing. The calculated convergence curve of the gallery is shown in Fig. 4, which shows, that the maximum displacement takes place at the vault with a max settlement of 92mm or so, the floor has a small displacement and max convergence of the two sidewalls is as high as 160mm. Fig. 5 gives the development of the plastic zone around the periphery obtained from calculation, which shows that the plastic zone of the gallery mainly occur at the sidewalls and more often than not instability failure firstly takes place in these areas. Accordingly, priority should be given to supporting these areas or otherwise the overall instability will be led to. It is advisable to introduce more dense and special installation of bolts and to use some other supporting measure in these areas. The calculated thickness of relaxation zone is about 1.5m. The above analysing results suggest the following design scheme of anchoring support. For roof and sidewall, the bolt length is 1.7m, the diamond-shaped arrangement is used with a spacing of 600mm × 600mm (a more dense spacing is used in dangerous zones). For floor, the bolt length is 1.2m with a installing spacing of 600mm×600mm. A compound extension-type bolt developed by our Institute is used, which consists of two layers of bolt body, a washer, outer welding rod, a cap

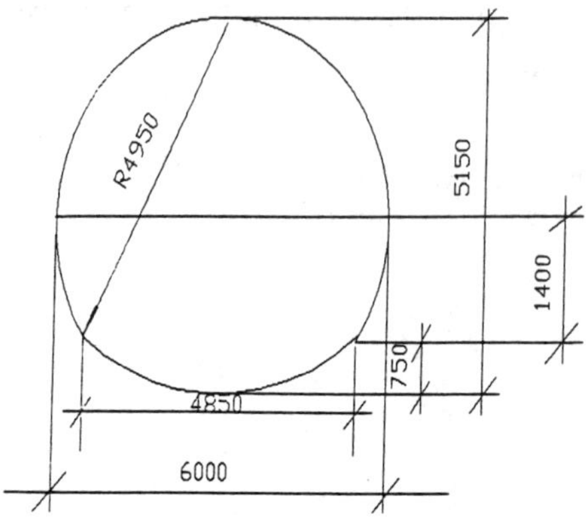

Figure 2. Designed shape and size of the gallery cross section

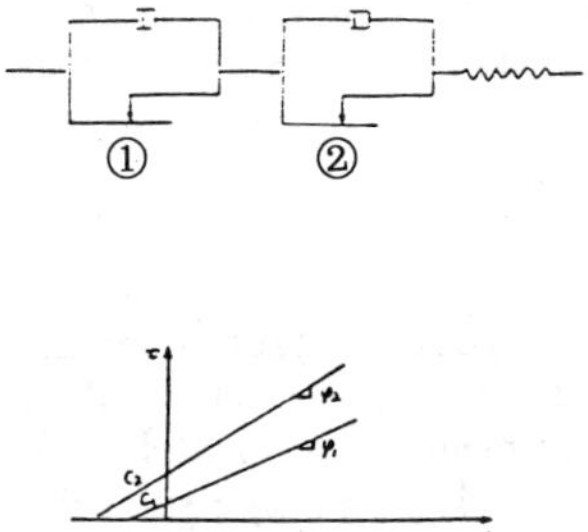

Figure 3. Rheological mechanical model and strength criterion

plate and nuts. The outer body is fully bonded mortar bolt and the inner body is end-anchored bolt with a lower deformation module and a higher tensile strength. The outer body has a bending resistance and a shearing resistance, both matching with the rockmass, which helps the outer lining to bear a portion of tension. This compond bolt makes use of the deformability of the lining itself to cause a continuous deformation along the full bolt length, so it overcomes the advantage of common extension-type of bolts — local extension and contraction creat a sudden deformation leading to laryer seperation inside the rockmass. Polyvinyl alcohol aldehyde is used to strengthen the grouted layer, which makes the compressive strength of shotcrete layer 30% higher than the ordinary concrete and deformation 5—10 times as large as the ordinary. The arranging scheme of bolts is as follows: an indined cross installation of bolts is introduced with an installing angle (with the normal of gallery's periphery) of ±22.5°. A series of biaxial compressive modelling tests have been conducted (Zhu Weishen, Liu Quansheng 1988) to compare the anchoring effects of different anchoring parameters. The testing results

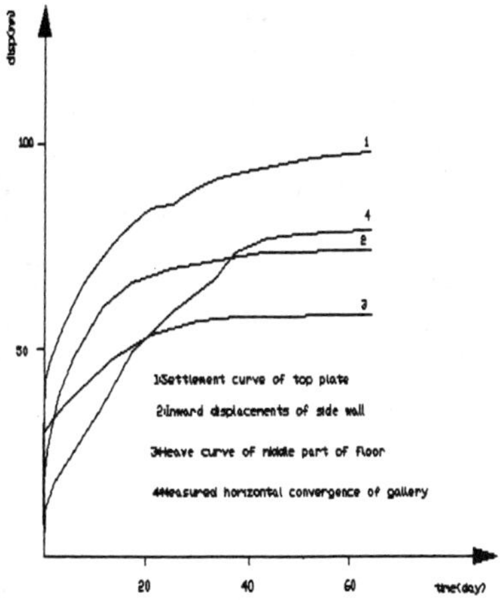

Figure 4. Variation law of convergence curve and comparison between measured displacement curve

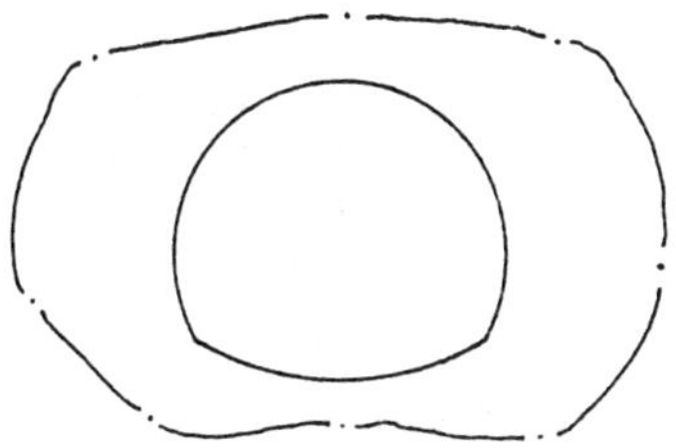

Distribution of plastic zone when $t=0$

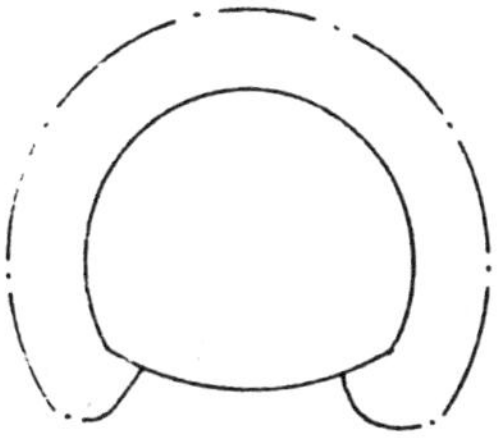

Distribution of plastic zone under stable deformations

Figure 5. Development of plastic zone around gallery's periphery

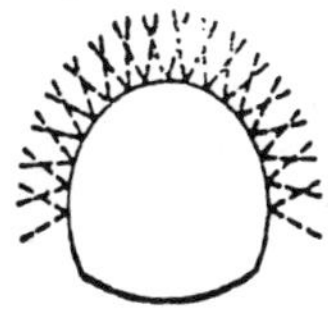

Figure 6. Layout of cross installation of bolts

indicate the above angle is the most rational. Fig. 6 shows the layout of cross arrangement of bolts.

3 ANALYSIS ON ANCHORING EFFECT OF GALLERY

Based upon the in site measurement performed in the course of construction and reinforcement of the gallery, the analysis on anchoring effects has been carried out, which leads to the following measuring items:

(1) surface displacement (periphery convergence) of the rockmass

(2) deep displacement

(3) deformation and bearing force of bolts

(4) intactness of rockmass and acoustic wave measuring in relaxation fracture zones.

The in site monitoring results of the anchored gallery are described below.

(1) The convergence deformation of the gallery is shown as in Fig. 7b. Because an inclined cross installation of bolts with a rational installing angle was used, the deformation of the surrounding rockmass has been restrained considerably and the convergence of the gallery controlled within a very small range.

(2) The deep displacement measuring results are shown in Fig. 7c. The displacement obtained using a 3-point extensometer shows that the displacement curves, U_{A1} and U_{A2}, measured in

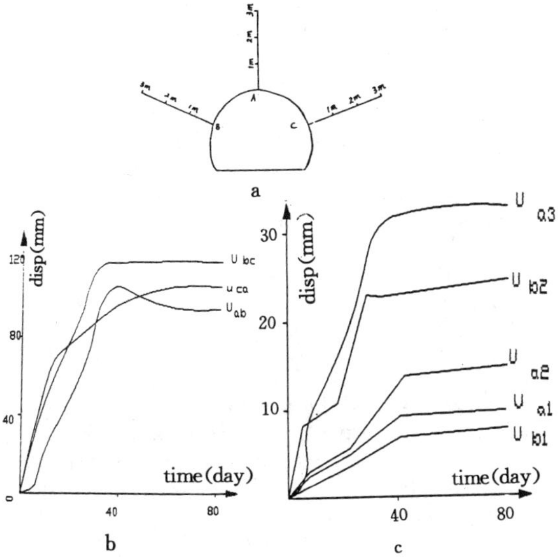

Figure 7. Arrangement of displacement measuring points and monitoring results

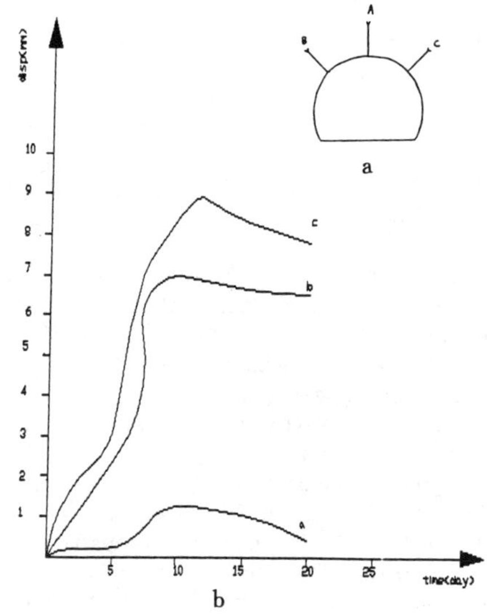

(a) arrangement of measuring bolts
(b) deformation — time relationship curve
Figure 8. Deformation measurement of bolts

borehole A at the depths of 1m and 2m are coincident with each other, having a small defference between them, which indicates an overall settlement tendancy of the roof.

(3) Shown in Fig. 8 are the monitoring results of bolt deformation and its bearing force. In the obliquely upper part of the two sides of the vaults, the measured deformation and bearing force of bolts are larger whereas those measured in the vault are smaller, which shows that the rockmass near the vault tends to moving downwards in the whole.

(4) Measuring results of the intactness of the surrounding rockmass are as follows. Acoustic wave method was used in the later excavation-supporting period to monitor the rockmass intactness. Within the area 500mm thick around the gallery surface there exists a low velocity zone. Although the wave velocity is low, its variation with depths is not obvious, the wave velocity curve is gentle, which shows that the stress and the rock structure tend to being stable after adjustment. Beyond this area, say 500 ~1500mm beyond the hole mouth, the wave velocity increases with the increment of depth and the curve becomes steeple, which shows that the structure of the rockmass in this area is in an unstable state, the stress and the structure of the rock mass are still under adjustment. Beyond the area of stress adjustment, the wave velocity again becomes gentle and the stress state and the rockmass structure gradually tend their original state respectively.

(5) A long-term monitoring plan has been made on the gavllery supported by anchoring and grouted concrete for many years, the monitoring results indicate that the gallery is stable, meeting the engineering requirements.

The above monitoring results have shown that the gallery reinforced by anchoring has a small convergence displacement to guarantee the gallery's stability and the anchoring effect is satisfying.

4 CONCLUSIONS

(1) This paper, based upon the theory of rock mechanics, puts forward a rational method for anchoring design accoridng to the mechanical property, geostress, plastic zone and deformability of the surrounding rockmass.

(2) An inclined cross installation of bolts has been introduced with an installing angle, α, (to the normal of the gallery's peripher) of $\pm 22.5°$. In the relatively dangerous zones, it is advisible to introduce more dense installation of bolts. This design scheme can better restrain the rockmass deformation thus to increase the stability of the gallery considerbly.

(3) In site monitoring results have shown that although the overall settlement of the roof of the anchored gallery occurs, the rockmass finally became stable with a small displacement. The rockmass surrounding the anchored gallery can be divided into three zones, i. e., adjustment zone of stress and rockmass structure, transition zone of stress and zone of original stress.

REFERENCES

Zhu Wenshen, Liu Quanshen, Wang Ping. 1988. Determination of optimum anchoring parameters for reinforcing soft rockmass using full length bonded bolts, *Proc. of Int. Sym. on Modern Mining Techniques*

Liu Quanshen et al. 1990. Extension-type bolt and grouted laryer supporting using flexible shotcrete, research report, *Inst. of Rock & Soil Mech.*, Wuhan, Academia Sinica.

Computer Methods and Advances in Geomechanics, Yuan (ed.) © 1997 Balkema, Rotterdam, ISBN 90 5410 904 1

Numerical analysis on geomechanical problems in underground pipe jacking works

Feng Dingxiang, Feng Shuren & Ge Xiurun
Institute of Rock and Soil Mechanics, The Chinese Academy of Sciences, Wuhan, China

ABSTRACT: Some aspects concerning geomechanics in the design of the wall thickness of underground pipe jacking works are analysed and discussed using the numerical analysis method, including study of earth pressures on the pipe and back calculation of actual pressure distribution on the outer wall of the pipe according to the measured variation of the pipe in its inner diameter. In combination with *in-situ* measured deformation data in Yangluo Power Plant, the main force-bearing state of the jacked pipe is analysed and a new way is suggested for designing pipe's wall thickness taking into account construction controlling conditions.

1 INTRODUCTION

Pipe jacking method is a constructing method to jack underground pipes into soil layers directly. There is no need to carry out surface earth works when using this method, so construction can effectively keep away from surface buildings and structures and in many cases the method can considerably speed up construction. There has been no ripe or well-considered method yet for the wall thickness design of steel pipes due to various affecting factors and complicated force-bearing conditions encountered in the construction of pipe jacking works. The wall thickness is empirically selected as 1% of the pipe diameter generally at present. In order to find out a more rational designing method for the pipe wall thickness, we put forward a new designing way based upon practical experiences concerning pipe jacking works, viz., estimate earth pressures on the pipe using numerical analysis and design the wall thickness of a steel pipe in combination with the controlling condition of construction. This paper will give a brief description about numerical analyses concerning geomechanical aspects of pipe jacking works. The analyses are combined with *in-situ* measured deformations and stresses of the admitting pipe of Yangluo Power Plant.

Yangluo Power Plant is located in Hubei Province, its steel admitting pipe was laid in clayed soils and silt sandstones with an overburden thickness of 16~19m. The pipe measures 3m in diameter, 3cm in wall thickness and 89.3m in length. Three measuring items were involved in the study, viz., radial deformation of the pipe: diametrically arranged three or four pairs of measuring points on the inner pipe wall surface to monitor the diameter variation during jacking pipe and after completion of construction using micrometers (0.01mm); measurement of inner wall's strain: arranged five 8-point (equally spaced) measuring sections, each point having a trefoil strain rosette stick on it; and measurement of earth pressures on the outer pipe wall: arranged two sections and burried six load cells altogether.

The measuring results show that in the whole period of observation, most parts of the pipe bore a stress less than 50MPa and the rest bore a stress beyond 100MPa with the exception of some rare locations (beyond 160MPa); two principle stresses point towards the axis and the circumference of the pipe respectively; the pipe's radial deformation (with a maximum value of up to 27mm) is mainly caused by the pressure of the surrounding soilmass, seriously affected by the drifting of the pipe direction and less related with the axial jacking force.

Described below will be, in combination with the *in-situ* measured data in pipe jacking works in Yangluo Power Plant, the numerical analysis on geomechanical aspects in the study of pipe wall thickness.

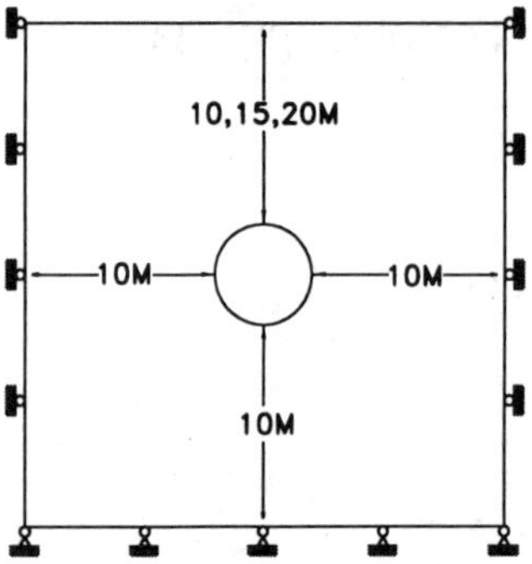

Figure 1. FEM calculation model for force-bearing state of jacked pipe

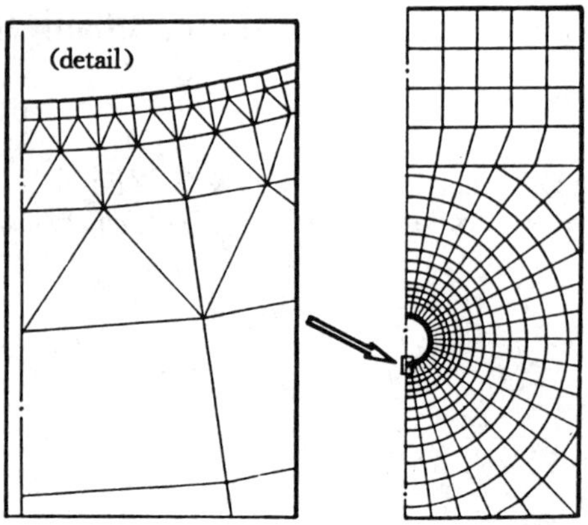

Figure 2. Schematic diagram of computation mesh

2 FEM COMPUTATION OF EARTH PRESSURES ON OUTER PIPE WALL

The magnitude of the earth pressure on the outer pipe wall is of paramount importance to the design of pipe wall thickness. In the past, the vertical pressure of clayey soils on a burried pipe was generally calculated according to effective overburden pressures. If the pipe is laid in soil layers with a high shearing strength and an overburth thickness beyond three times the outer pipe diameter, the theory of pressure arch may be used. Although by using this theory, corresponding formulations can be established and computing parameters can be adjusted according to measured data, it is difficult to completely describe the interaction between soilmass and pressure pipe and to reveal the force-bearing mechanism of the pipe because the theory analyses the soilmass and the pipe independently. In addition, the empirical paramters involved, such as lateral pressure coefficient, lateral reaction coefficient of soil layers and effective coefficient of pipe's stiffness vary too considerably to be determined easily and their effects on computing results are so decisive that practical application of the formulation often encounters great difficulty. FEM analysing, however, possesses an obvious property capable of better reflecting the interaction of soilmass and pressure pipe.

To counter the general situation of pipe jacking method, FEM analysing has been carried out, mainly aimed at the distribution law of the earth pressure on the outer pipe wall. The model and the mesh for computation are respectively shown in Fig. s 1 and 2. The pressure pipe, measuring 3m in inner diameter and 3cm in wall-thickness, is 10m away from the left, right and bottom boundaries and three different overburden depths of 10m, 15m and 20m (Fig. 1) are considered. Because of symmetry, only half a computing domain is considered. In the case of 20m thick overburden, altogether 1028 quadralateral and trangular elements of soilmass, 160 joint elements between soil and pipe and 929 nodes are involved. Prelimilary computation is performed using a 2—D elasticity programme. Three moduli of deformation (3MPa, 5MPa and 7MPa) are introduced according to the general range of clay soil's and sandsoil's physical indexes, the corresponding Poisson's ratios are 0. 3, 0. 35 and 0. 4, the unit weight of the soilmass is 18. 5kN/m^3, the tangential and normal stiffnesses of the joint element are respectively 5. 0MN/m^3 and 50. 0MN/m^3.

Shown in Fig. 3 are partial computing results of normal pressures, σ_n, on the outer pipe wall under the above conditions. The abscissa stands for joint element number. An anticlockwise coding system for elements is used, joint no. 1 is located in the lowest and joint no. 160 the upmost. In the Figure, each curve is marked with its deformation module (E), Poission's ratio (ν) and overburden depth (H) for computing. The Figure shows a remarkable effect of Poisson's ratio on the earth pressure's distributing law. Based upon the computing results, we can summarize a list in which an engineer can find out the rough distributing law and approximate value of the earth pressure according to the mechanical indexes of soil layers for his design of pipe wall thickness.

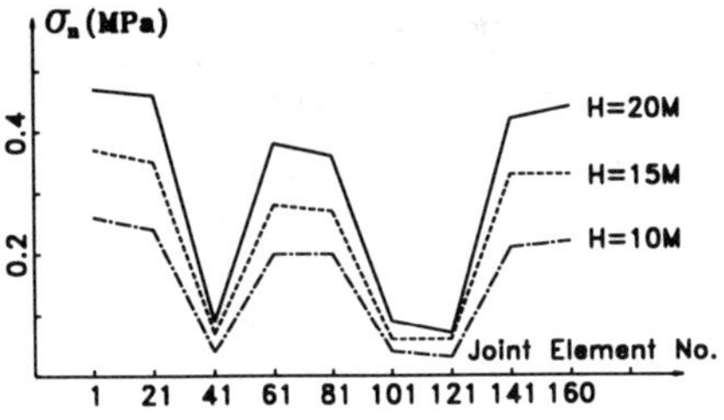

(a) E=3.0MPa,ν= 0.3

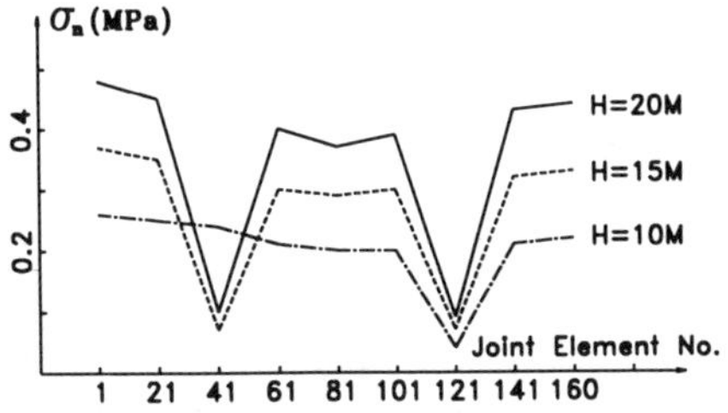

(b) E=3.0MPa,ν=0.35

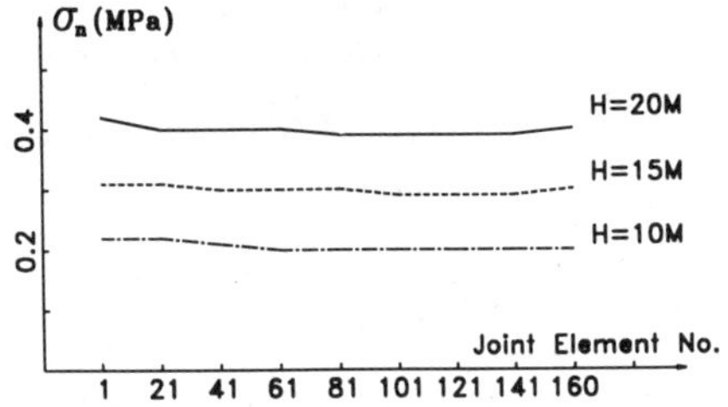

(c) E=7.0MPa,ν= 0.4

Figure 3. Distribution curve of earth pressures on outer pipe wall

3 BACK CALCULATION OF NORMAL PRESSURE ON OUTER PIPE WALL FROM RADIAL DEFORMATION

The measuring results obtained from the previously stated three measuring items (section 1) in Yangluo are all directly related to geological environments and regime of the pipe jacking works and they can be mutually verified.

Based upon the measured data of pipe's radial deformations, the earth pressure on the pipe can be back-calculated. The steel pipe itself behaves like an ideal elastic medium before yielding, the back calculation result is no doubt rather reliable. The back-calculated earth pressure is different from that defined in terms of general concept, the former including the resisting force of the soilmass caused by pipe deformations. In back calculating by FEM method, it is assumed that the pressures on the pipe's outer wall point

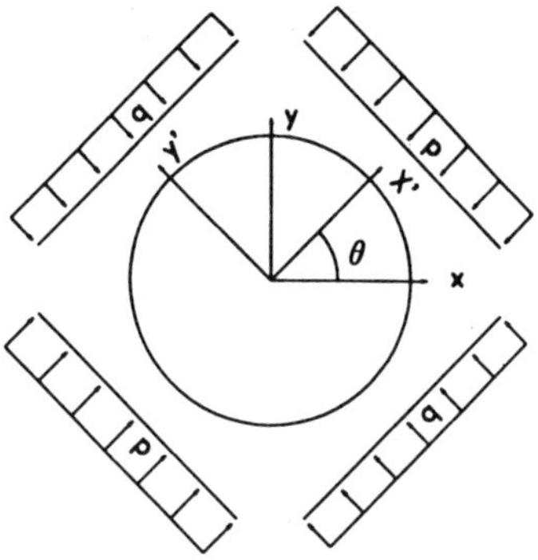

Figure 4. Schematic diagram of back-calculating condition for earth pressure from radial deformation

towards two orthogonal principle directions, the maximum uniform compressive stress value is p at an angle of θ with horizontal orientation and the uniform compressive stress perpendicular to p is q (Fig. 4). In calculating, the elastic module and Poisson's ratio of the steel pipe are $E = 2.1 \times 10^5$MPa and $\nu = 0.3$ respectively, the FEM mesh contains 2884 nodes and 2160 quadralateral elements. The trial and error method is used for computation: firstly, given a set of values of p, q, θ according to the measured value in pipe's diametral variation, calculate the deformation of the pipe diameter caused by this set of p, q, θ and then adjust p, q, θ to make the calculated diametral variation as fully coincident with the measured one as possible. Table 1 lists the comparison between measured values and some typical caclulating results in the 9th of September, 1992.

The solution from such a back analysis is not unique. For example, in the case of a set of pipe's diametral variations measured along three orientations, a set of p, q, θ can be generally determined. Under the action of this external loading, the calculated variation in pipe diameter lengths is basically coincident with the measured one. However, in the case of the measured results containing four orientations, they are less coincident with the calculated values. The reason is that the force-bearing state of pipe's outer wall is more complicated than that from the previous assumption.

The above computing results based on the measured data show that the variation of pipe diameter considerably depends upon the difference between the two principle compressive stresses. As a matter of fact, the combination of a set of p, q can be decomposed into two components, viz., uniform compressive load, q, on the outer

Table 1. Comparison of computing and measuring results of pipe's diametral deformations

location no. of pipe	computing parameters	calcutated deformation (mm)		measured deformation (mm)	
		measuring line	diametral	measuring line	diametral
3	$p=300$kPa $q=284$kPa $\theta=5°$	14 25 36	14.55 5.38 −22.42	14 25 36	14.33 5.75 −21.62
8	$p=300$kPa $q=290$kPa $\theta=94°$	14 25 36	0.32 −12.33 11.68	14 25 36	1.53 −12.37 12.33
10	$p=300$kPa $q=284$kPa $\theta=92°$	15 26 37 48	−16.99 22.99 15.85 −22.11	15 26 37 48	−16.15 18.69 12.44 −25.87

wall and superposition of uniorientational pressures of p and q along the direction of maximum compression. The elastic module of the steel pipe is very high, resulting in a very low diametral deformation under a uniform compressive load. For instance, for a pipe with a diamter of 3m and a wall thickness of 3cm, the variation of the pipe inner diameter and the circumferential compressive stress are 0.0736mm and 5.15MPa respectively when a uniform load of 0.1MPa is acted on the outer wall, whereas the above two values are changed into high up to 14.3mm and 22MPa when a uniorientational load of 0.01MPa is acted.

4 AXIAL STRESS CAUSED BY ECCENTRIC JACKING FORCES

When an eccentric moment of jacking forces occur, the resultant maximum and minimum axial stresses (σ_{max}, σ_{min}) acting on the pipe section are expressed by the following equation:

$$\left.\begin{array}{l}\sigma_{max}=P(1+x/r_0)/F\\ \sigma_{min}=P(1-x/r_0)/F\end{array}\right\} \tag{1}$$

where P — jacking force (kN)
F — cross section area of pipe (cm^2)
x — eccentric distance (cm)
r_0 — jacking core radius of cross section

$$r_0=D[1+(D_n/D)^2]/8 \tag{2}$$

where D_n — pipe inner diameter (cm)
D — pipe outer diameter (cm)

It can be obtained from the above equations that for a pipe with a diamter of 3m and a wall thickness of 3cm, the maximum compressive stress is 44.36MPa when a jacking force of 10000kN is exerted with an eccentric distence of 20cm, whereas the maximum compressive stresses are 49.02MPa and 53.69MPa respectively for eccentric distences of 30cm and 40cm.

5 STRESS IN PIPE WALL CAUSED BY PIPE BENDING IN PIPE JACKING WORKS

The real pipe central line in construction will undoubtedly deviate from the designed one owing to the complicated and changeable geological conditions, possible variation of the resultant force direction of the back jacking seat from pipe axis, effect of nonuniform head-on resisting force of the pipe, and so on. The stress in the pipe wall caused by the bending of the central line can be calculated from eq. (3)

$$\sigma=ER_0/\rho \tag{3}$$

where E — elastic module of pipe
R_0 — outer diameter of pipe
ρ — bending radius of pipe

$$\rho=(b^2+4bh^2)/8h \tag{4}$$

where b — hypotenuse length of bended central line
h — hyptenuse height of bended central line

For a pipe of 3m in diameter and 3cm in wall thickness and 2.1×10^5MPa in elastic module, the calculation from eq.s (3) and (4) gives the results of maximum bending stresses in the pipe as 80. 3MPa, 107. 1 MPa and 160. 7MPa respectively for bending radii of 4000m, 3000m, 2000m and 1000m, which shows that the bending stress in the pipe is very high, so it becomes a governing factor in the pipe wall thickness design.

6 CHECKING COMPUTATION OF PIPE STRENGTH

Checking computation of pipe stability should be carried out in the design of pipe wall thickness. Through the above analysis, it can be seen that the checking calculation for both stability and stiffness of the pipe can be easily realized, whereas the checking calculation for pipe strength is the crux in the wall thickness design. The composition of the stress in the pipe contains mainly: (a) axial stress caused by jacking forces including that caused by eccentric jacking; (b) axial force caused by bending and (c) circumferential stress caused by surrounding soil pressures. The second stress in the above is related to the eccentricity during jacking, but the real reason causing it is the pressure of surrounding soils. The axial stress caused by the earth pressure and that caused by the jacking force should be superposed together accordingly.

The stress in the pipe wall mainly point towards circumference, approximately composed of two parts, viz., uniform normal stress around pipe's outer wall and unioriemtational uniform compression. Owing to the flxibility of the steel pipe, the normal stress around the outer wall tends to being uniform under the mutual coupling action of soilmass and pipe.

The above described combinations of stresses are very complicated. The maximum compressive stresses caused by various factors such as eccentric jacking, bending of pipe and earth pressure are not generally same located and they often occur in different time. In the case of hard soilmass, not very long pipe and short constructing period, for example, providing that the pressure caused by the soilmass self-weight does not take place, then the checking calculation of the strength of a pipe with a diameter of 3m can meet requirements under the conditions of a pipe wall thickness of 3cm and of a jacking force of 10000kN when the following limitations are made: eccentric distance is not greater than 40cm and pipe bending radius is not less than 3000m anywhere or the former is not greater than 30cm and the latter is not less than 2500m.

The above discussions show that the design of pipe wall thickness must be linked with the technical requirements in construction. Further analyses indicate that the designed value of the admitting pipe wall thickness of 28mm for a pipe of 3m in diameter meets the engineering requirements in Yangluo Power Plant for the local geological environments if the constructing requirements of eccentric distance not greater than 20cm and pipe bending radius not less than 4000m can be guaranteed.

7 AKNOWLEDGMENTS

The authors would like to express their thanks to Sr. engineer Xu Shenyang and Sr. engineer Chen Huiming of Zhongnan Design Institution of Electric Power, Chinese Ministry of Energy Sources for their kind supporting and helping in research work and to Prof. Li Guangyu of author's Institute for his offering measured data.

REFERENCES

Gao Naixi, Zhang Xiaozhu. 1994. Book of *Pipe Jacking Techniques*. Publishing House of Chinese Building Industry

Computer Methods and Advances in Geomechanics, Yuan (ed.) © 1997 Balkema, Rotterdam, ISBN 90 5410 904 1

Countermeasures for some problems related to underground storage of low temperature materials

Yoshinori Inada & Naoki Kinoshita
Ehime University, Matsuyama, Japan
Tetsu Nishioka & Kenzo Ochi
Tokyu Construction Co., Ltd, Sagamihara, Japan

ABSTRACT: The authors propose a "resin lining system" as a countermeasure for preventing leakage of gas and liquid in case of low temperature materials storage in openings excavated in rock mountain. On the other hand, to reduce thermal stress around the openings, the authors propose to use adiabatical materials. Then, a "combination lining system of resin and adiabatical material" are also proposed as a countermeasure for preventing leakage of gas and liquid from openings and reduction of thermal stress at the same time.

1 INTRODUCTION

From the view points of multiple-utilization of land in Japan (which has little available land), reduction of heat loss and maintenance costs and environmental safeguards, utilization of underground openings excavated in mountains should be considered.

The authors have shown by theoretical analysis and experiment that in the case of low temperature materials such as LNG, LPG and frozen food etc., during storage in openings excavated in rock mountain, the openings will shrink toward the mountain and cracks will occur around the openings and develop with time (Inada et al. 1995,1996).

Therefore, countermeasures for preventing leakage of gas and liquid from the opening become important. Also, countermeasures for the reduction of thermal stress become important for maintaining the stability of the openings. As countermeasures for preventing leakage, the authors have been proposing a "water curtain system" and an "ice lining system"(Inada et al., 1991,1992).

In this study, as another method the authors proposed a "resin lining system". The results of the tests of strength, deformation characteristics and thermal properties of resin at low temperature are described. Using these values, the temperature distribution was analyzed by using FDEM (Inada et al., 1983). Then, cracks around openings caused by thermal stress were analyzed by CAM (Inada et al., 1987). From the results of the analysis mentioned above, the effects of resin lining for preventing leakage are discussed. Then, the authors propose a "combination lining system of resin and adiabatical material" as a countermeasure for preventing leakage and reduction of thermal stress around openings at the same time.

2 CHARACTERISTICS OF STRENGTH AND DEFORMATION OF RESIN AT LOW TEMPERATURES

2.1 Resin used for experiment

In this study, Urethane resin was used for lining material to prevent leakage of liquid and gas in case of storage of low temperature materials in openings. Three main types of composition of urethane resin which were different in viscosity were chosen; 0.45Pa・s(sample 1),0.60Pa・s(sample 2),1.16Pa・s (sample 3) at 25℃. After these main compositions were hardened by a hardening agent, they were used for experiment.

2.2 Compressive and tensile strength

The specimen size was ϕ 3×6cm for the uniaxial compression test. The cooling rate of the specimen was set at -1℃/min. After reaching the required

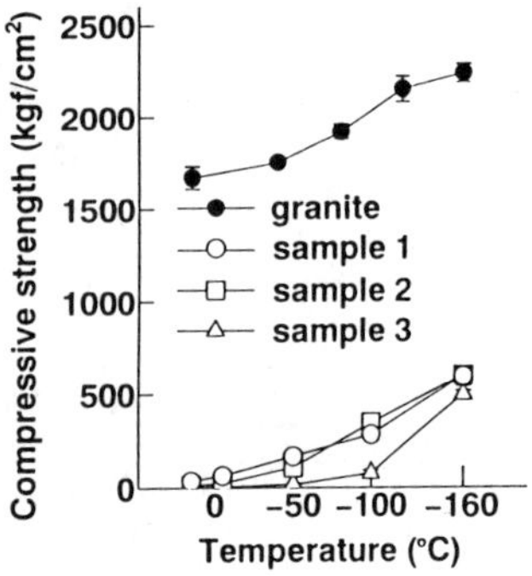

Fig.1 Compressive strength of resin.

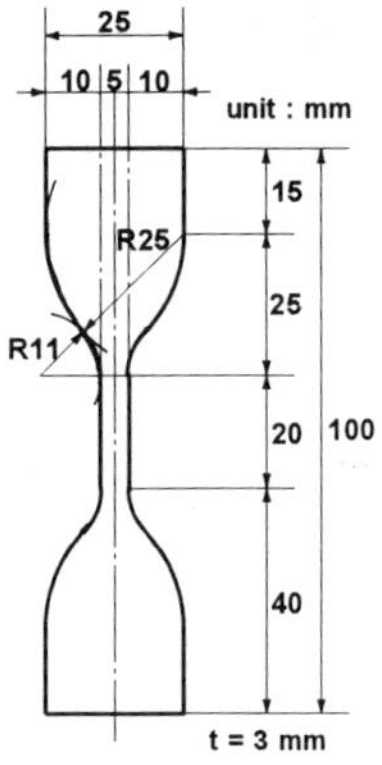

Fig.2 Schematic diagram of specimen.

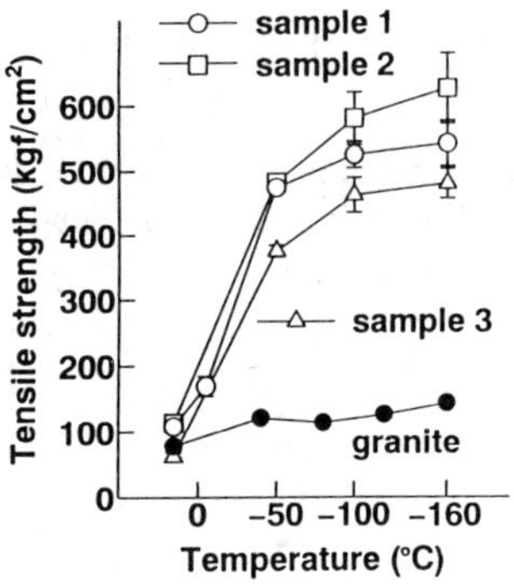

Fig.3 Tensile strength of resin.

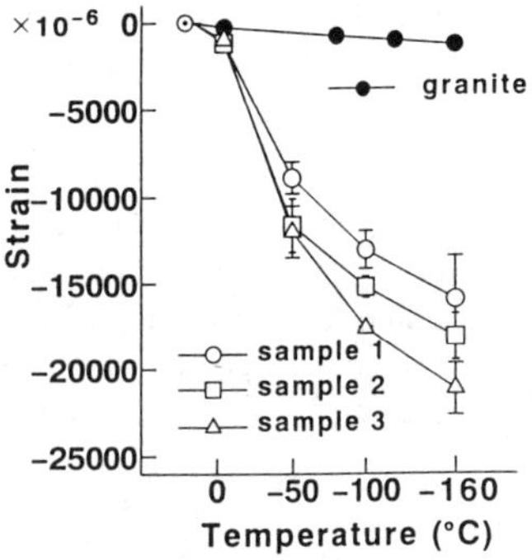

Fig.4 Strain of resin.

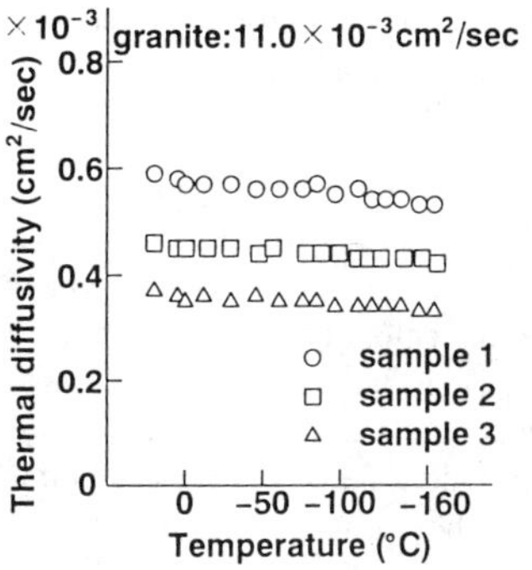

Fig.5 Thermal diffusivity of resin.

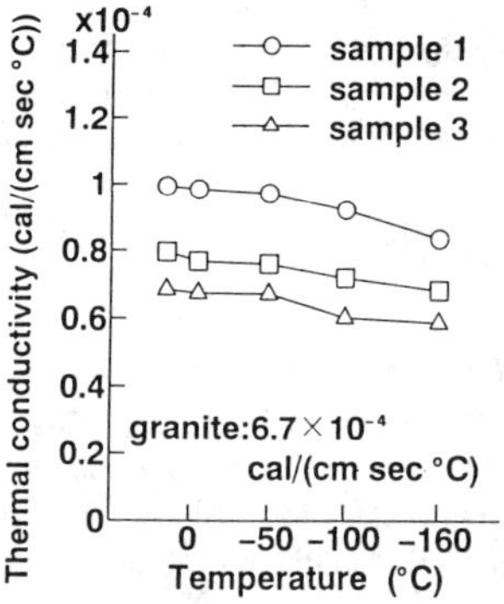

Fig.6 Thermal conductivity of resin.

temperature, the specimen was kept at that temperature for one hour. The results of the compression test are shown in Fig.1 From this figure, it is found that compressive strength rises with falling temperatures and the value of resin is smaller than that of granite at all temperatures, however the ratio of the rising strength is greater than that of granite.

The specimens for uniaxial tension test of resin were formed as shown in Fig.2. The size of granite specimen was $\phi\, 3 \times 3$cm for the radial compression test (Brazilian test). Fig.3 shows tensile strength. It is found the value rises with falling temperatures and the value of all resins is larger than that of granite at very low temperatures, and the value of sample 2 is the largest.

Tangential Young's modulus and Poisson's ratio were also obtained from 30% of failure stress in stress-strain curves which were obtained when compression tests were carried out.

2.3 Strain

The thermal expansion of resins at low temperatures was obtained by using the comparison method with quartz glass rod (Inada et al., 1971). Results are shown in Fig.4. It is estimated that though all specimens shrink with falling temperatures, the amount of shrinkage of sample 1 is smaller than that of sample 2 and sample 3. The shrinkage of resin is extremely large compared with that of granite.

3 THERMAL PROPERTIES AT LOW TEMPERATURES

3.1 Thermal properties of resin

The change of thermal diffusivity at low temperatures was obtained by adopting an experimental value to FDEM. The specimen which was prepared

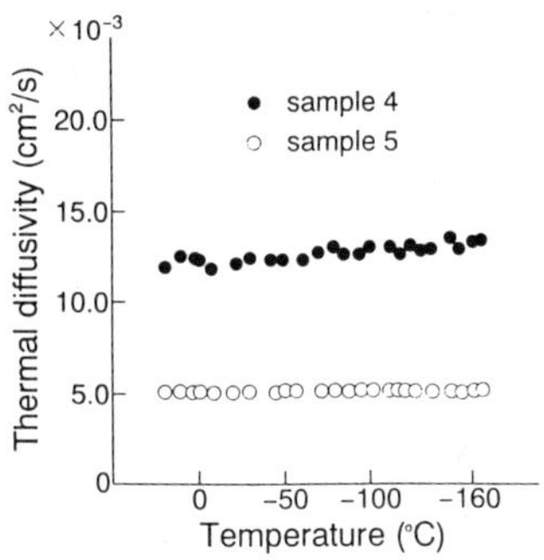

Fig.7 Thermal diffusivity of adiabatical material.

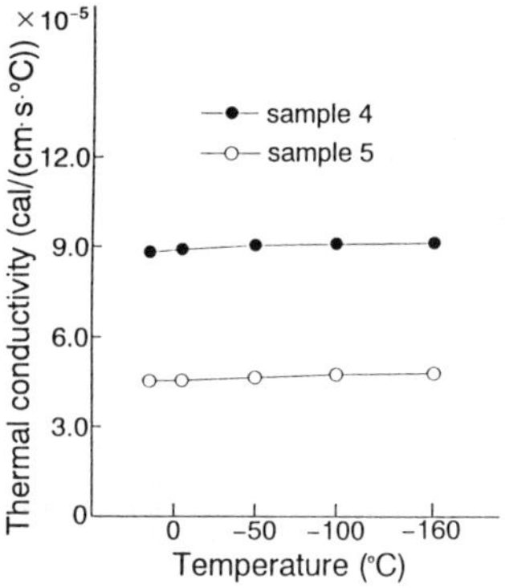

Fig.8 Thermal conductivity of adiabatical material.

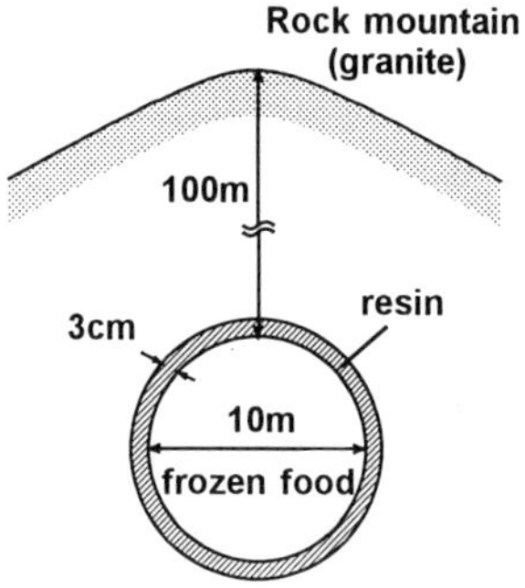

Fig.9 Schematic diagram of model used for analysis.

as 15×10×25cm was covered with adiabatical material except one surface (10×15cm) to obtain one-dimensional heat conduction. Liquid nitrogen as a source was touched with this surface and temperature changes were measured with thermometers which were arranged at regular intervals of 5mm from the surface to 25mm distance. Thermal diffusivity was obtained by adopting temperature-time curves to FDEM. The results are shown in Fig.5. It can be seen that the value of all samples are constant in practical application. From these results, it is found that the thermal diffusivity of resin at low temperatures is about 1/20-1/30 of that of granite.

Fig.6 shows thermal conductivity which was calculated from thermal diffusivity and heat capacity. It can be seen that the value is constant in practical application in range of 15℃ to -160℃, and about 1/70-1/100 of that of granite.

From the results for strength, deformation characteristics and thermal properties mentioned above, it was found that tensile strength of sample 2 was the largest. Considering this fact, it was judged that sample 2 was the most effective material for lining.

3.2 *Thermal properties of adiabatical material*

The adiabatical materials used for experiment were polystyrene foam and rigid urethane foam. We call the former sample 4 and the latter sample 5. Both specimens are adiabatical materials widely used for construction.

Thermal diffusivity of adiabatical material at low temperatures is shown in Fig.7. It can be seen that the value of both materials are constant in practical application in the range of 15℃ to -160℃. From these results, it is found that the value of the adiabatical materials at low temperatures is about same in sample 4 and 1/2 in sample 5 of that of granite.

Fig.8 shows thermal conductivity of adiabatical materials. As these values are 1/100 of that of granite, it is assumed that adiabatical materials conduct a heat extremely slowly.

4 PREVENTION OF LEAKAGE OF LIQUID AND GAS AND REDUCTION OF THERMAL STRESS

4.1 *Analysis method*

It is supposed that an opening is excavated in granite rock mountain to a depth of 100m beneath the ground surface with a diameter of 10m, and LNG(-162℃), frozen food(-60℃) and LPG(-43℃) were stored within.

Temperature distribution around an opening was analyzed by using FDEM which was developed for adoption of composite materials. Cracks around openings caused by thermal stress were analyzed by CAM using temperature distribution.

4.2 *Prevention of leakage of liquid and gas*

In the case of low temperature materials storage, as the authors have made clear by theoretical analysis, cracks will occur radially due to shrinkage of openings toward the mountain and will develop with time. Therefore, prevention of leakage of liquid and gas along the cracks becomes an important problem.

In this study, a "resin lining system" is proposed to

Table 1 Physical properties of granite.

Temperature (℃)	Expansion coefficient (1/℃)×10⁻⁴	Young's modulus (kgf/cm²)×10⁶	Poisson's ratio	Compressive strength (kgf/cm²)	Tensile strength (kgf/cm²)
20~ 10	0.0000	0.494	0.250	-1670.0	79.0
10~ 0	0.1510	0.494	0.250	-1670.0	79.0
0~ -10	0.1490	0.495	0.250	-1678.0	95.0
-10~ -20	0.1450	0.495	0.250	-1689.0	104.0
-20~ -30	0.1400	0.496	0.250	-1711.0	111.0
-30~ -40	0.1360	0.496	0.250	-1741.0	116.0
-40~ -50	0.1320	0.497	0.240	-1778.0	121.0
-50~ -60	0.1270	0.497	0.240	-1819.0	124.0
-60~ -70	0.1230	0.498	0.240	-1859.0	127.0
-70~ -80	0.1190	0.498	0.240	-1899.0	129.0
-80~ -90	0.1140	0.499	0.240	-1938.0	132.0
-90~-100	0.1100	0.499	0.230	-1978.0	133.0
-100~-110	0.1060	0.500	0.230	-2017.0	135.0
-110~-120	0.1010	0.500	0.230	-2057.0	137.0
-120~-130	0.0970	0.501	0.230	-2097.0	138.0
-130~-140	0.0930	0.501	0.230	-2137.0	140.0
-140~-150	0.0880	0.501	0.220	-2176.0	142.0
-150~-160	0.0840	0.502	0.220	-2216.0	143.0
-160~-170	0.0800	0.502	0.220	-2256.0	145.0

Table 2 Physical properties of sample 2.

Temperature (℃)	Expansion coefficient (1/℃)×10⁻⁴	Young's modulus (kgf/cm²)×10²	Poisson's ratio	Compressive strength (kgf/cm²)	Tensile strength (kgf/cm²)
20~ 10	0.0000	1.759	0.433	-13.8	114.4
10~ 0	0.2815	3.325	0.439	-28.2	152.0
0~ -10	0.5630	4.892	0.445	-41.9	188.5
-10~ -20	0.7468	10.510	0.442	-55.7	255.0
-20~ -30	0.9306	16.140	0.437	-69.4	325.6
-30~ -40	1.1144	21.760	0.435	-83.1	395.7
-40~ -50	1.2982	27.380	0.432	-96.9	460.8
-50~ -60	1.4820	33.010	0.429	-110.6	503.1
-60~ -70	1.4554	34.190	0.424	-157.6	523.0
-70~ -80	1.4288	48.750	0.419	-204.5	537.5
-80~ -90	1.4022	62.590	0.414	-251.5	554.0
-90~-100	1.3756	76.430	0.409	-298.4	570.3
-100~-110	1.3490	93.180	0.406	-345.4	583.9
-110~-120	1.3175	109.900	0.406	-387.8	590.6
-120~-130	1.2860	126.700	0.407	-430.1	600.2
-130~-140	1.2545	143.400	0.408	-472.5	610.8
-140~-150	1.2230	160.200	0.409	-514.8	612.5
-150~-160	1.1915	176.900	0.409	-557.2	618.1
-160~-170	1.1600	193.600	0.410	-599.5	629.7

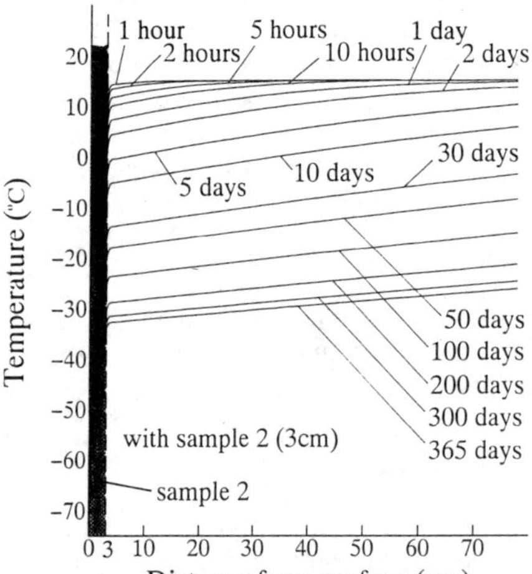

Fig.10 Temperature distribution around the opening for the case of setting up sample 2 as a lining.

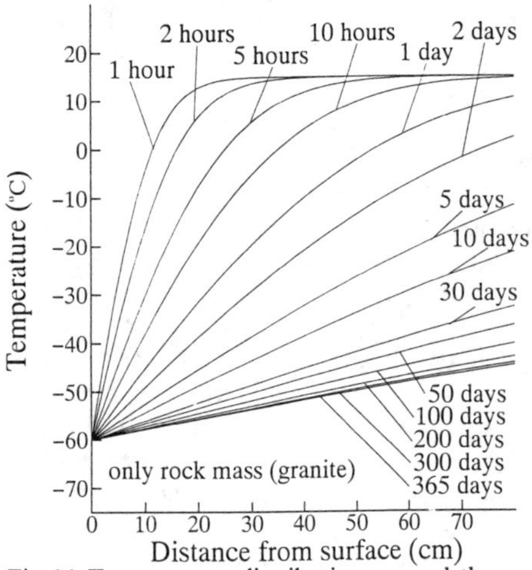

Fig.11 Temperature distribution around the opening for the case of granite rock mass only.

prevent leakage of liquid and gas from openings. As mentioned above, from the results of experiments, it was found that sample 2 was the most suitable resin for the lining in the case of low temperature material storage. In this analysis, it is assumed that 3cm of sample 2 was set up on the surface of the opening as a lining when frozen food was stored in the opening. Fig.9 is a schematic diagram of model used for analysis. Physical properties of granite and sample 2 at low temperatures used for analysis are shown in Table 1 and Table 2.

Fig.10 shows the change of temperature distribution around the opening with time. For comparison, Fig.11 shows the results for the case of granite rock mass only. It is found that the temperature gradient is extremely sharp at the beginning of storage, but becomes gentler with time. The temperature enters a semi-steady state after 1 year. Comparing both figures, it is found that heat conduction is slow for resin. It is supposed that this was due to the fact that the thermal conductivity of resin is much smaller than that of granite.

The results of stress analysis are shown in Fig.12(a). It was found that cracks occurred due to shrinkage of rock mass around openings and developed with time. Also, cracks occur in the lining after 30 days. Fig.12(b) shows the results of analysis of the case of setting up a waterproof sheet between the rock mass and the resin, so that thermal behavior of the rock mass and the resin are independent of each other. It was found that cracks occurred due to shrinkage of rock mass around openings, however

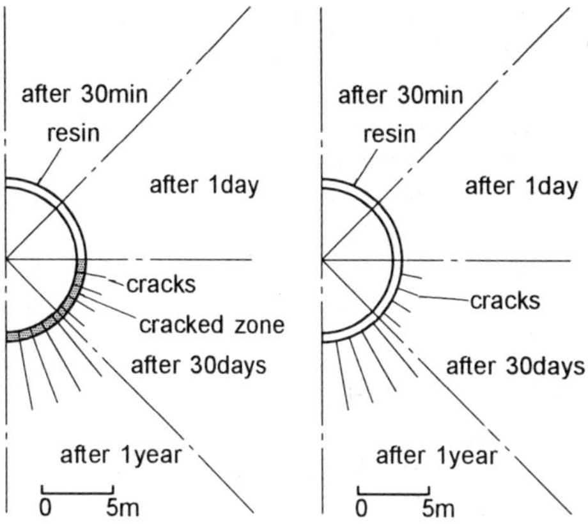

(a)direct (b)indirect

Fig.12 Cracks and cracked zone around the opening.

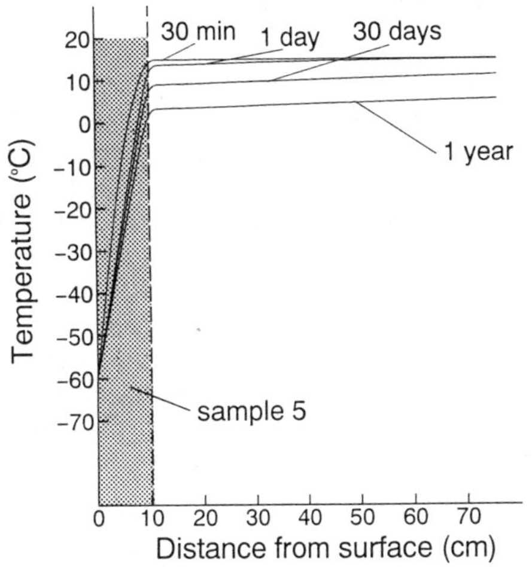

Fig.13 Temperature distribution around the opening for the case of setting up sample 5.

the layer of resin is stable after 1 year.

From these facts, it was found that as it makes the thermal behavior of rock mass and the lining independent, the "resin lining system" will become an effective way to prevent leakage of liquid and gas from the opening.

4.3 Reduction of thermal stress

Considering the effect of the repeated heat cycle and latent cracks in rock mass, in case of low temperature material storage, as much reduction of thermal stress around openings as possible becomes

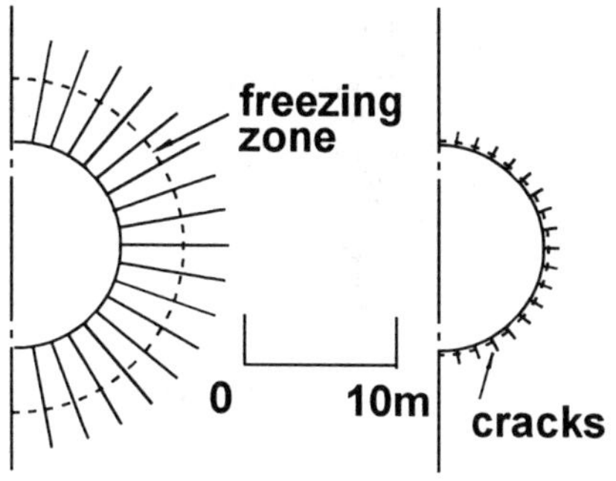

(a)rock mass only (b)with sample 5

Fig.14 Cracks and cracked zone around the opening.

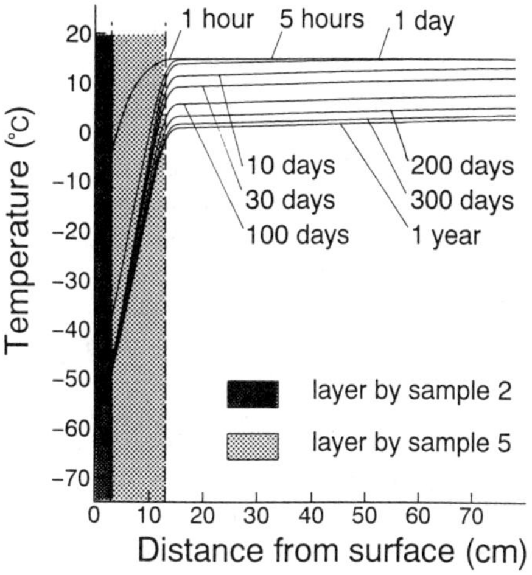

Fig.15 Temperature distribution around the opening for the case of the combination lining system.

important.

In this study, it is found that setting up a 10cm layer of adiabatical material (sample 5) on the surface of the openings is one of the useful techniques for reducing thermal stress. Fig.13 shows the change of temperature distribution around openings. In comparison with the case of rock mass only (Fig.11), it is found that heat is conducted slowly for adiabatical materials, and the temperature gradient of adiabatical material has an extreme shape. It is supposed that this was due to the fact that the thermal conductivity of adiabatical materials is much smaller than that of granite.

Fig.14(a),(b) shows cracks around openings after 1 year. It is found that thermal stress was reduced and development of cracks was controlled by the use of adiabatical materials.

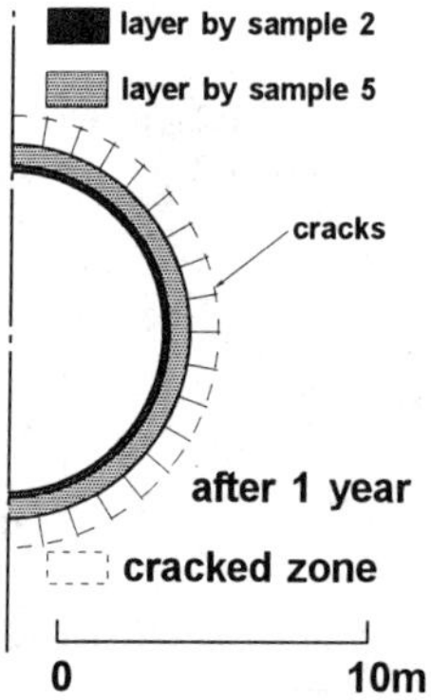

Fig.16 Cracks and cracked zone around the opening.

4.4 Consideration of a "combination lining system of resin and adiabatical material"

A "combination lining system of resin and adiabatical material" as a countermeasure for preventing leakage of liquid and gas from openings and reduction of thermal stress at the same time is proposed.

Temperature distributions were analyzed in the case of a sample 5 layer of 10cm on the surface of the opening and a sample 2 layer of 3cm on the surface of sample 5. The change of temperature distribution with time in the case of storage of frozen food is shown in Fig.15. It is found that the temperature gradient of sample 5 is extreme compared to the others. Fig.16 shows cracks around the opening after 1 year. In this case, it is found that the cracks were developing around the opening radially. But, development of the cracks was controlled by setting up adiabatical materials. On the other hand, in the layer of sample 2, the theoretical fracture stress does not occur and the layer is stable after 1 year.

5 CONCLUSION

The main results obtained in this study are as follows :

1. Compressive and tensile strength of urethane resin at low temperatures become stronger with falling temperatures. Tensile strength of urethane resin is larger than that of granite.

2. The thermal conductivity of adiabatical material at low temperatures is about 1/100 of that of granite.

3. To make the thermal behavior of rock mass and the lining of resin independent, the "resin lining system" will be an effective way of preventing leakage of liquid and gas from the opening.

4. The "combination lining system of resin and adiabatical material" will be an effective way of reducing thermal stress and preventing leakage of liquid and gas at the same time.

ACKNOWLEDGEMENTS

The authors would like to express gratitude to T. Matsuo, a graduate student of Ehime University at that time.

REFERENCES

Inada, Y., Kinoshita, N. & Seki, S. ; Thermal behavior of rock mass around openings affected by low temperature, *Proc. 8th Int. Cong. on Rock Mech. (ISRM)*, 721-724, 1995.

Inada, Y., Kinoshita, N. et al. ; A few remarks on thermal behavior of rock mass around openings affected by low temperature, *J. of Japan Soc. of Civil Eng.*, 547(Ⅲ-36), 211-220, 1996.

Inada, Y. & Kohmura, Y. ; Effect of water curtain system on storage of low temperature materials in underground opening, *Proc. 7th Int. Conf. on Computer Meth. and Adv. in Geomecha.*, 1565-1570, 1991.

Inada, Y. et al. ; 1992. Basic research on preventive measures against leakage of liquid and cold air from cracks due to storage of low temperature materials in opening excavated in mountain, *J. of Japan Soc. of Civil Eng.*, 445(Ⅲ-18), 65-73, 1992.

Inada, Y. & Shigenobu, J. ; Temperature distribution around underground openings excavated in rock mass due to storage of liquefied natural gas, *J. Min. Meta. Inst. Japan*, 99(1141), 179-185, 1983.

Inada, Y. & Taniguchi, K. ; Plastic zone around underground openings excavated in rock mass due to storage of liquefied natural gas, *J. Min. Meta. Inst. Japan*, 103(1192), 365-372, 1987.

Inada, Y. et al. ; On the coefficient of thermal expansion of rocks, *Suiyokwai-si*, 17(5), 200-203, 1971.

Computer Methods and Advances in Geomechanics, Yuan (ed.) © 1997 Balkema, Rotterdam, ISBN 90 5410 904 1

Interrelated CAD/FEM/BEM/DEM system for education in civil engineering and underground structure design

O.K.Postolskaya, D.V.Ustinov & I.N.Voznesensky
Moscow State University of Civil Engineering, Russia

ABSTRACT: The set of professional programs is being used for the educational purposes. Both gains and drawbacks of such applications are discussed with practical recommendations.

1 INTRODUCTION

"Frische Köpfe braucht das Land" - "The country needs fresh heads" - declared Hans Baumann in the German newspaper "Die Welt" on 8 March, 1997. This notice is absolutely true not only for Germany, but for every other country of the world too. However, a specific situation exists if we consider the problems of computer-based or computer-oriented education in Russia and some other now independent countries of the former USSR.

In the country of tremendous natural resources and human potential, where fundamental science was always at the level or higher if compared to that in the world, in the very beginning of the fifties cybernetics and computer science were politically outlawed. Of course, obviously, computers existed in the military-oriented industries and effort was directed towards local production of such machines, but time was gone, - and as it was said by the high-ranking representative of a major US computer manufacturer in the mid-seventies at the computer exhibition in Moscow - forever.

Let us be more optimistic in this respect, but we still can not state that all consequences of the situation in the previous years are overcome. Civil engineers were the last in the line to get a computer for a university. First implementations of the FEM analyses in the civil engineering projects appeared in the USSR almost ten years later, than in the West, not only due to the tedious work of the in-code programming, because a FEM program written in one of the high-level languages such as FORTRAN, would have been too slow to expect practical results on the available computers, but mostly due to unavailability of any computers.

For our purposes we should outline two most important factors. First and positive is the fact, that experience of programming for slow and unreliable computers, where not lost, gives the ability to produce extremely effective codes for large and complex problems run on PCs. Second and negative fact results from the situation, when even in the end of seventies an average student in civil engineering could have used or just seen a computer once or twice in all five years of his university life. Many at the moment active Professors of civil engineering where educated at that or earlier time and are not always able to overcome the feeling of uneasiness when business leads to this damned machine. On the contrary, recent students (and in some cases kids of those same Professors), have PCs at home and are able to use any given software generally without understanding how it works, which brings us to the core of this paper.

2 GENERAL CONSIDERATIONS

What may be required from a newly graduated civil engineer by his/her first employer - a construction company, a design or research institution or consulting bureau? In Russia at the moment, with exception of cases when young engineers start their careers at the construction site, these are good command of computers and foreign languages and only after - basic engineering education. It is sad and difficult to admit, but computers and foreign languages are at the moment the weakest points in most Russian technical universities teaching civil engineering for one and the same reason - the education is state budget financed and this budget can not pay for good computers and good language instructors. But let us leave the financial problems beyond the scope of this paper.

Civil engineering education in Russian universities of technology takes 4 years to get the BSc degree, 5 - to get the engineer's diploma and 6 - for the MSc degree. Very few stop at the BSc level, not many apply for the MSc degree, therefore we shall discuss the basic, i.e. engineer's education.

Five years may be conditionally divided in four levels or periods of uneven duration. The first two years are devoted to the general basic education, which is important for understanding of most of subjects to be learned later. The third year, with inclusion of some subjects started in the second year and some ending in the forth, provides basic engineering education. The fourth and first half of the fifth year ends the educational process with subjects of selected specialty. During the second part of the fifth year the engineer-to-be has to pass state exams and to produce diploma work or diploma project. The "work" is more theoretical as compared to the project and is done mostly by scientifically oriented students.

Since our greatest interest is the role of computer methods in the educational process the question arise: when, at what level do we start to teach numerical methods? This question appears to be of great importance, and many wrong steps were made and hot heads cooled before we came to the procedure below.

3 PERSONAL APPROACH

Center of Underground and Special Engineering of the Moscow State University of Civil Engineering teaches subjects of specialty to students, graduating with diplomas in underground hydraulic engineering (i.e. tunnels, caverns and pressure shafts of underground hydraulic power stations, irrigation projects, etc.) and underground construction for public and industrial use (i.e. everything underground except of hydraulic engineering projects).

At the same time the staff works in research, design and consulting spheres for underground construction projects. Close relation with the industry demand somewhat personal approach to students - i.e. people who will be working for our current and potential clients.

However the quality of education related to knowledge of the most frequently used computer methods remained law for quite a long period of time, until we changed the sequence of teached subjects, methodology and most of the lecturers as well.

4 AS SIMPLE AS POSSIBLE - BUT NOT SIMPLER!

This advice, often related to Einstein, provided the approach to solution of the problem.

As a normal practice, numerical methods are introduced to students along with related chapters of high mathematics and computer programming. It looks like the most reasonable procedure with one serious drawback - the students at this stage do not know why and what for they need to know such things like finite, boundary, distinct and other elements. Later, when it comes to practical implementation of these methods, most of the previously acquired knowledge is already forgotten being not on demand for a year or more.

Additional effort is needed to bring all up to date, and this is exactly the effort we want to avoid without loosing in quality of education. It was decided, that one introductory course is needed in order to avoid problems and, if not the sequence, then the contents of several subjects should be changed.

The lecture course tentatively named "Introduction to the specialty" starts at the very first day of study. Prepared and given as illustrative and as simple as possible, lectures describe the most famous projects in underground engineering along with design methods. Impressive graphical presentation of results of numerical modelling for these and current projects develops interest to study those methods, providing a better audience to lecturers of numerical methods.

It is understood, that virtually everything given at this course should be repeated at the more comprehensive level almost three years later, but such a small loss of time is compensated by a better understanding of what happens and what is going to happen in the university.

Later we arrange close cooperation of lecturers from different chairs and departments of the university in order to coordinate material presented to students.

5 HOW IT WORKS

For the Figure 1 we selected only the most related to our problems lecture courses and the most frequently used numerical methods and software packages. For all activities described herein PCs are provided by the Center. So, how the whole idea works?

Starting to understand the power of the computer methods and their importance for analysis and design of structures and rock/soil masses from the introductory course, students get practical command of CAD packages. At the same time they get basical knowledge of the theoretical background of the most frequently used numerical methods - FEM, BEM & DEM. Theory is supplemented with illustration of solutions of different practical problems.

It is always stressed, that geometry of the problem can be prepared with the use of a CAD package and later a mesh generator of, say, a FEM package may produce the mesh from a CAD file. The students should make several practical examples themselves. Parallel to this basic knowledge of Geographic Information Systems (GIS) is provided so that courses in Geology and hydrogeology may refer to practical implementation of relatively new information technologies. Certain ideas of coupling GIS with numerical models are also explained.

For the courses tentatively grouped under "Level 2" on Figure 1 numerical modelling packages serve as practical tool to illustrate the explained phenomenae, also comparisons of numerical and closed-form solutions are provided. If the study at levels 1 and 2 was successful, the student easily uses all needed software packages while studying the subjects of specialty, preparing the graduation work and, hopefully, the rest of his/her professional career.

The question arises what kind of software is constantly used in the described process? It would have

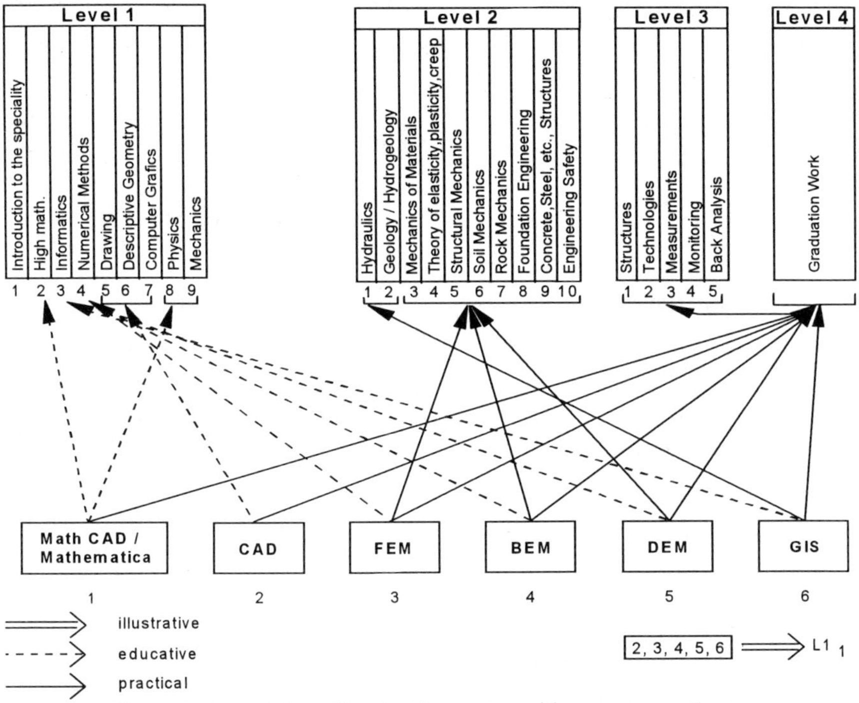

Figure 1. Interrelation of levels of education with computer methods

been quite logical to teach future engineers structure analyses and design using the same software, which will be used in practical work after the graduation. With small exceptions it is impossible because too many different packages are in practical use.

As it already could have been noticed, the FEM appears to be more frequently used as compared to BEM and DEM. Here we use STATAS FEM/BEM code (Yufin 1979, Yufin et al, 1993) developed in the Center and used for real practical work since 1972. ITASCA's UDEC DEM code is used to teach how discrete systems can be dealed with. Some lecturers use other codes, which may be more problem-oriented and popular commercial codes are described too.

The described software runs under the IBM OS/2 warp 4 operation system and all new programming at the moment is done under WATCOM Fortran/C/C++ v.10.6 shell. This gives additional possibilities of a better graphic presentation, etc., using IBM system programs, such as Presentation Manager, for one's own applications.

Authors consider presented system of continuous teaching of numerical methods for all period of the university educational cycle worth to be followed.

ACKNOWLEDGMENTS

The authors express their appreciation to the ITASCA Consulting Group and personally to Dr. Roger D.Hart for given possibility to use the educational version of the DEM UDEC code in this educational project.

The appreciation is also expressed to Dr. C.Katz and Dr. B.Protopsaltis for provision of descriptive material and illustrative microdisks for the SOFiSTiK package.

REFERENCES

IBM Corp., 1994, OS/2 Warp, Version 3. Technical Library. G25H-7116.

Yufin, S.A. 1979, Numerical analysis of rock structures considering material nonlinearities. Proc. 20th US Symposium on Rock Mechanics, Austin, TX, pp.265-272.

Yufin, S.A., Postolskaya, O.K. & Rechitsky, V.I., 1993. Stability of rock caverns as viewed from the back analysis data. Proc.of the International Symposium Eurock'93, Luis Ribeiro e Sousa & N.F. Grossmann (eds), Balkema, pp.751-758.

Computer Methods and Advances in Geomechanics, Yuan (ed.) © *1997 Balkema, Rotterdam, ISBN 90 5410 904 1*

Non-linear analysis of rock masses in underground openings and determination of support pressure

A. Fahimifar
Department of Civil Engineering, Amirkabir University of Technology, Tehran, Iran

ABSTRACT :In attempting to develop a comprehensive non-linear analytical solution, applicable to various excavation shapes under different loading conditions, an elasto-viscoplastic model has been adopted . The advantage of this model is that it reflects the rock mass behaviour much more better than the elastoplastic models and it is capable of evaluating the time-dependent characteristics of the rock mass around underground openings . The Hoek-Brown and Mohr-Coulomb yield criteria were used to define the yield condition for the rock, and a " no-tension criterion " was used to determine the likely tension areas around the excavation .

A computer program was prepared for determining elasto-viscoplastic stresses and displacements around the underground opening and the finite element method was used as the numerical solution .

A horse-shoe tunnel and a circular tunnel were analyzed and the results obtained were compared with the elastic solution . The results reveal the significance of the elasto-viscoplastic model as a very relevant method for these type of problems .

1 INTRODUCTION

Analysis of rock mass behaviour surrounding underground openings is a complicated theoretical problem due to the large number of factors which must be taken into account in order to derive a comprehensive solution . Closed-form solutions such as those proposed by Ladanyi (1974) and Brown et al (1982) based on several simplifying assumptions , particularly they are only applicable to circular tunnels and shafts in a hydrostatic stress field .

In this paper on the basis of an elasto-viscoplastic model with using Mohr-Coulomb and Hoek-Brow yield criteria and with the aid of finite element method a relatively comprehensive solution for determination of stress and strain fields in the rock mass around underground spaces are proposed and examined .

2 ELASTIC-VISCOPLASTIC MODEL

Figure 1 shows an elasto-viscoplastic model consists of a spring in series with a system of slider and viscous elements . With applying a force P to the system , the spring deforms elastically and after unloading the deformation recovers entirely . As P increases to a magnitude overcoming the frictional constraints in slider , irrecoverable deformation will occur in the system . But, due to the existence of the element D (Dashpot) in the system the irrecoverable deformation will occur with time . In this model the element S represents yielding of the material and element D represents time dependent behaviour of the system .

The following assumptions may be made for this model :

a) It is assumed that at the beginning of loading the material behaves perfectly elastic .

b) In unstable states the stress field in the material can be considered out of the yield surface . Stable

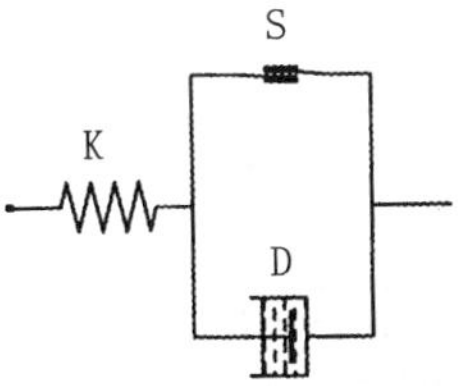

Figure 1 : The elastic-cviscoplastic model

state in the system is the state in which the stress and strain fields do not change with time.

c) For the stresses out of the yield surface there is a rate of increase in the time-dependent viscoplastic strain which is obtained from the viscoplastic flow equation. The increase in residual viscoplastic strain will continue until the stress field reaches to the yield surface (i.e. stable state). Also, for stress fields within the yield surface (elastic range), the rate of increase in viscoplastic strain is zero.

d) Other concepts of theory of plasticity such as flow rule, yield function, and hardening rule are valid in this model and have similar applications.

The following equation has been proposed for rate of change of viscoplastic strain (Naylor, et al, 1981):

$$\dot{\underset{\sim}{\varepsilon}}^{VP} = \eta . < \phi(F) > . \frac{\partial Q}{\partial \underset{\sim}{\sigma}} \qquad (1)$$

$Q, \dot{\varepsilon}^{VP}, \eta, \phi(F)$, and σ represent potential surface, rate of change of viscoplastic strain with time, fluidity parameter, flow function and stress field respectively.

As mentioned, in stable state, stress fields will be given by yield surface. In fact, in stable state, the elastoplastic analysis of material is the same as its viscoplastic analysis. For this reason, it is possible to analyze the elastoplastic problems using viscoplastic theory. For associated flow rule equation (1) is given by:

$$\frac{d\underset{\sim}{\varepsilon}^{VP}}{dt} = \eta . < \phi(F) > . \frac{\partial F}{\partial \underset{\sim}{\sigma}} \qquad (2)$$

The corresponding algorism is illustrated graphically as in figure 2. By approximating dt=Δt, the magnitude of assumed visco-plastic strain, for the first iteration, is determined as follows:

$$\Delta \underset{\sim}{\varepsilon}^{VP} = \eta . < F > . \frac{\partial F}{\partial \underset{\sim}{\sigma}} . \Delta t \qquad (3)$$

Then, the vector for correction of the nodal forces on each element is determined as:

$$r_{ie} = \int_{vol} B^T . D . \Delta \varepsilon_1^{VP} . dv \qquad (4)$$

In next iteration the relation K.U = R (the basic equation in finite element analysis) is corrected as follows.

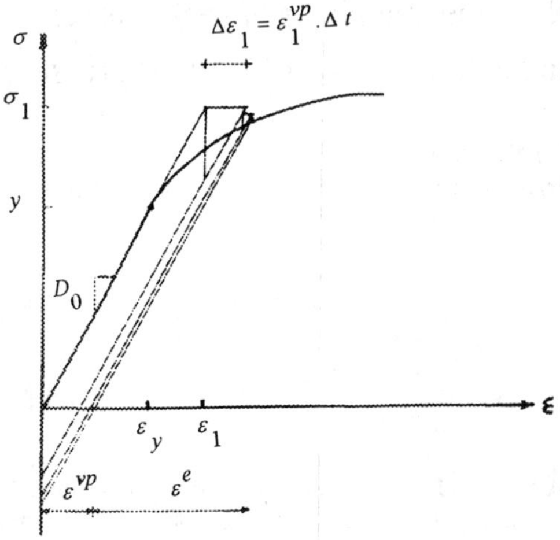

Figure 2 : Graphical presentation of viscoplastic method.

$$\text{K.U} = \text{R} + \sum_{n=1}^{nel} r_{ie} \qquad (5)$$

The parameter nel represents number of elements. This procedure is continued until a pre-determined tolerance is obtained.

The most significant advantages of the elasto-viscoplastic model may be stated as:

a) relatively simple mathematical relations are required for analysis. In other models the most important factor is the determination of mathematical relations for the elastoplastic matrix (D^{ep}) which is not needed in this model.

b) The elasto-viscoplastic model reflects the rock mass behaviour much more better than the elastoplastic models. Using the computer program the model is able to evaluate time-dependent characteristics of the rock mass around underground openings.

3 YIELD CRITERIA

In this work Mohr-Coulomb and Hoek-Brown yield criteria and a no-tension criterion were used to define the rock mass behaviour. The following assumptions were made in the analysis:

a) The rock mass around the excavation is homogeneous and isotropic.

b) The yield criteria are perfect plastic.

c) The associated flow rule is used for determining plastic deformation and strain.

On the basis of these assumptions the required mathematical relations were derived for viscoplastic analysis . In this paper the no-tension criterion is only described and its corresponding relations for using in the analysis are determined .

Identification of possible tension areas around the underground excavation before the elastoplastic analysis is of paramount importance. A no-tension analysis shows the active cracked regions due to tension stresses . For this purpose the yield function may be given by :

$$F = D_{ten} - \sigma_3 \tag{6}$$

σ_3 is the minor principal stress and D_{ten} is a small constant which is practically selected very near to zero . When F < 0 there is no tension areas on the excavation surface , and for F > 0 there will be tension areas . This is a useful criterion for qualitative determination of possible tensile areas on the excavation surface , but not a quantitative one .

The necessary relationships for no-tension criterion are determined as follows :

Using the stress invarients , the equation (6) is given by :

$$F = D_{ten} - (\sigma_s - \frac{\sigma_d}{2}) \tag{7}$$

Regarding the relation (2) , in stable state , $\phi(F) = F$ and rate of change of viscoplastic strain is given by :

$$\frac{d\underset{\sim}{\varepsilon}^{VP}}{dt} = \eta . F \frac{\partial F}{\partial \underset{\sim}{\sigma}} \tag{8}$$

Therefore , by derivation of equation (7) rate of viscoplastic strain is determined :

$$\frac{\partial F}{\partial \sigma_x} = -\frac{\partial \sigma_s}{\partial \sigma_x} + \frac{1}{2}\frac{\partial \sigma_d}{\partial \sigma_x} = -\frac{1}{2} + \frac{(\sigma_s - \sigma_y)}{\sigma_d} \tag{9}$$

$$\frac{\partial F}{\partial \sigma_y} = -\frac{1}{2} + \frac{(\sigma_s - \sigma_x)}{\sigma_d} \tag{10}$$

$$\frac{\partial F}{\partial \tau_{xy}} = \frac{1}{2}\frac{\partial \sigma_d}{\partial \tau_{xy}} = \frac{2\tau_{xy}}{\sigma_d} \tag{11}$$

Thus , rate of viscoplastic strains for no tension criterion are given as follows :

$$\frac{d\varepsilon_x^{VP}}{dt} = \eta . F . \left[\frac{(\sigma_s - \sigma_y)}{\sigma_d} - 0.5\right] \tag{12}$$

$$\frac{d\varepsilon_y^{VP}}{dt} = \eta . F . \left[\frac{(\sigma_s - \sigma_x)}{\sigma_d} - 0.5\right] \tag{13}$$

$$\frac{d\gamma_{xy}^{VP}}{dt} = 2.\eta . F . \tau_{xy} / \sigma_d \tag{14}$$

The same procedure was used for determining the required relationships for Mohr-Coulomb and Hoek-Borwn criteria .

4 COMPUTER PROGRAM

A computer program was prepared for determining stresses and displacements around underground excavations , and finite element method was used as the numerical solution .

The program is able to analyze the elastic and/or elastoplastic stress and strain fields in a two dimensional continuum statically , it is also possible to determine time dependent stress and strain in a two dimensional continuum .

As the main objective of this investigation is to determine the support pressure and to assess rock mass deformation around underground spaces , the program is capable of calculating yield functions and rate of viscoplastic strain on the basis of Mohr-Coulmb and Hoek-Brown criteria .

The program is able to calculate the elastoplastic deformation in the rock mass around the tunnel for different stages of excavation . It is also capable of comparing the multi-stage to one stage excavations . This program is capable of processing upto five excavation stages .

5 DISCUSSION OF RESULTS

In order to examine the integrity of the method and the performance of the program, two different underground spaces including a circular tunnel and a horse-shoe tunnel , under various loading conditions were analyzed . Figures 3 and 4 show the two cases .

The mechanical parameters for circular tunnel have presented below the figure.

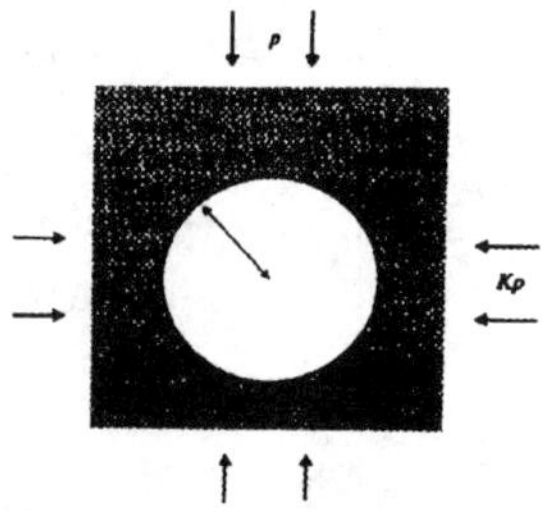

$\Phi = 32^{\circ}$ $C = 1.5\,MPa$

$p = 7\ MPa$ $p_i = 0$

Figure 3 : Circular tunnel section and assumed parameters.

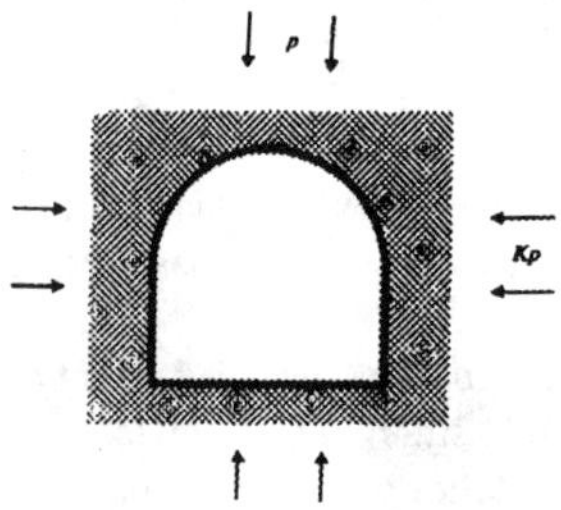

P =14 Mpa , Ko = 1.5 , $\nu = 0.2$

Figure 4 : Horse-shoe tunnel section and assumed parameters.

The analysis was carried out for various magnitudes of Ko , C , ϕ (Mohr-Coulomb criterion) m , and s (Hoek-Brown criterion). Idealization of geometric boundary conditions , after examining several finite element meshes , have been illustrated in figure 5 . The results obtained have good agreement with the previous works (Varadrajan et al , 1985) .

The analysis of circular tunnel for two cases of elastic and elastoplastic , and for a range of Ko from 0.75 to 2 has been presented in figures 6,7 and 8. The plastic zone around circular tunnel for different magnitudes of Ko has been plotted in figure 6 . With increasing Ko the plastic zone at the tunnel crown increases .

Figure 7 llustrates dislacement of the side walls for two cases of elastic and elastoliastic analysis for various magnitudes of Ko.With increasing Ko,displacements increas in two cases. In elastic analysis there is a lineara relation, however, in elastoplastic analysis the relation is non-linear with an increaing rate.

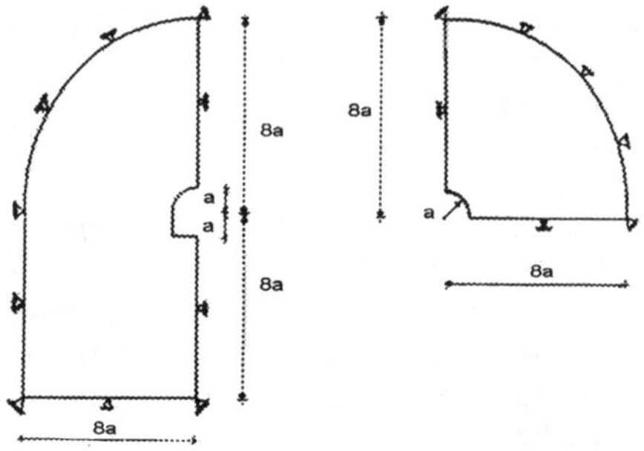

Figure 5 : Idealization of geometric boundary conditions for two tunnels.

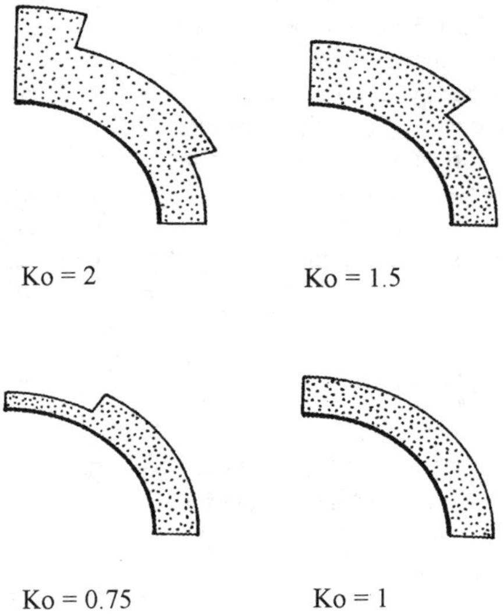

Figure 6 : Plastic zones around circular tunnel for different magnitudes of Ko .

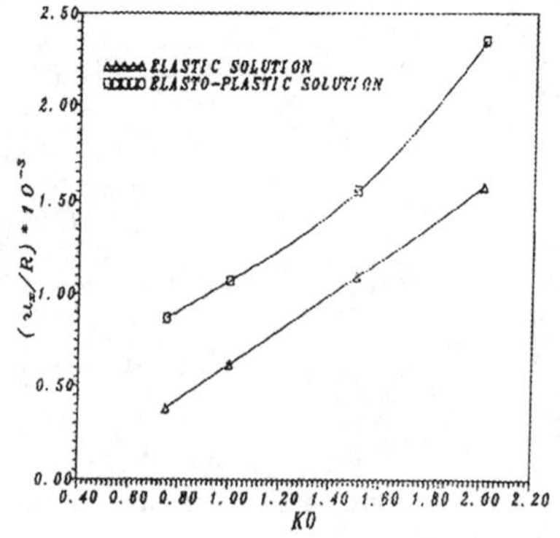

Figure 7 : Displacement of side walls for various magnitudes of Ko.

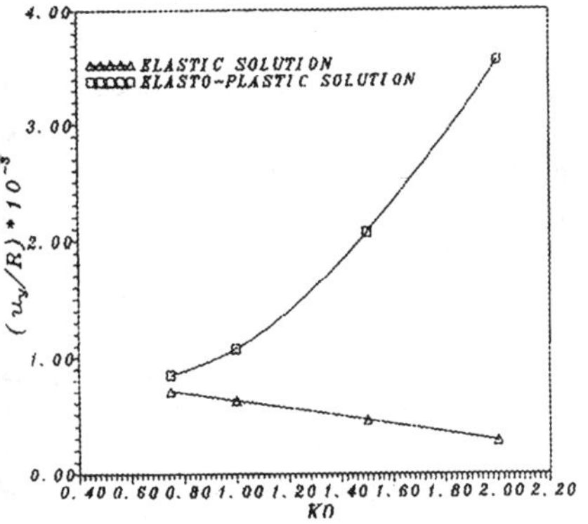

Figure 8 : Displacement at the circular tunnel crown for different magnitudes of Ko.

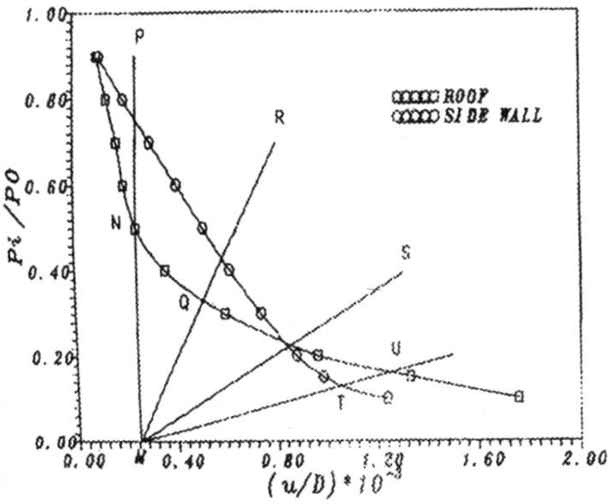

Figure 9: Ground characteristics curves for horse-shoe tunnel on the basis of Mohr- Coulomb criterion, and support lines for various stiffnesses.

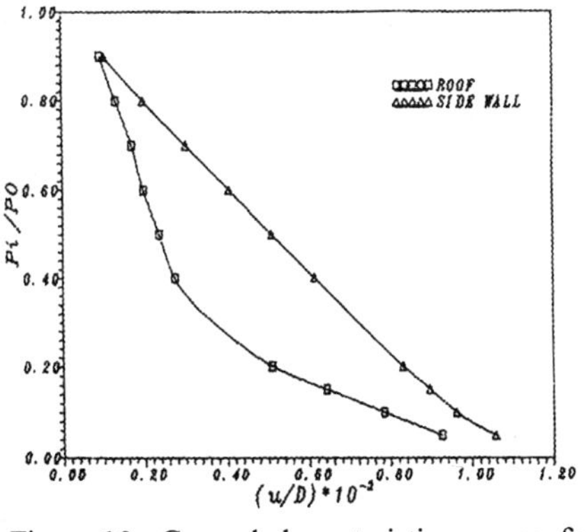

Figure 10 : Ground characteristics curves for horse-shoe tunnel on the basis of Hoek-Brown criterion .

Figure 8 shows deformation at the tunnel crown for various magnitudes of Ko , for two cases of elastic and elastoplastic analyses . In elastic analysis as Ko increases , deformation at the tunnel crown decreases linearly , however , in elastoplastic analysis, with increasing Ko, displacement at the tunnel crown increases non-linearly with an increasing rate . That is, the two cases show entirely different results . The results of elastoplastic analysis are in agreement with the results of figure 6 . Thus , it may be implied that elastic analysis dose not reflect the real behaviour of rock around underground structures .

The horse-shoe tunnel was analyzed for various amount of support pressures , and the characteristics curves for roof and sidewalls were plotted . Figures 9 and 10 illustrate the support characteristics curves for roof and side walls for Mohr-Coulomb and Hoek-Brown yield criteria respectively . By plotting appropriate support reaction curve , the required support pressure is obtained . In figure 8 four types of support reaction line ranging from very stiff to very soft support systems , have been plotted .

Figures 11 and 12 represent deformation of the tunnel surface and the plastic zone around the horse-shoe tunnel respectively for one and two stage excavations . For running the program , C=1.5 MPa , $\phi = 32^{\circ}$ and $p_i / p_o = 0.2$ (support pressure

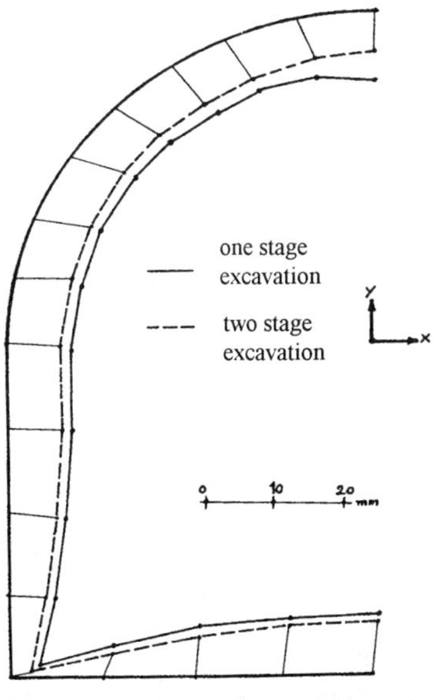

Figure 11 : Tunnel surface deformation for one and two stage excavations .

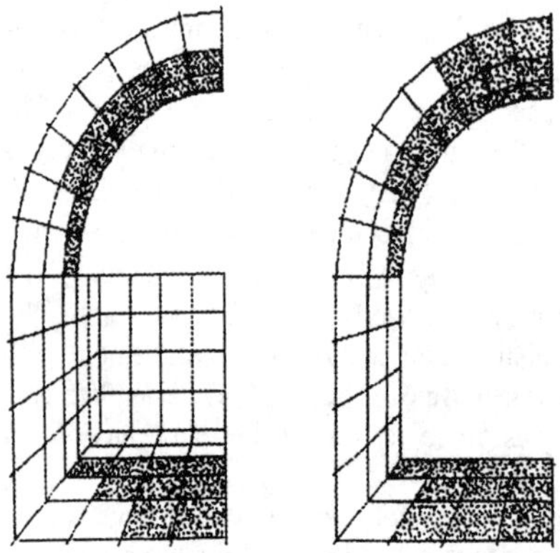

Figure 12 : Comparison of the plastic zones for one and two stage excavations.

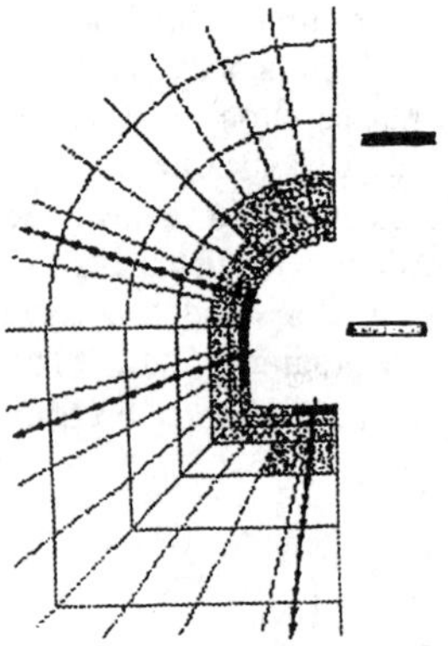

Figure 13 : The results of no-tension analysis and the required locations for supporting .

factor=0.2) were selected . As figure 11 shows , deformation of the tunnel surface for two stage excavation is less than that of the one stage .
From figure 12 it is also clear that the plastic zone for two stage excavation is smaller than that of one stage .
Figure 13 shows the plastic zone around the horse - shoe tunnel for elastoplastic and no-tension analyses , with no support system . In fact , no-tension analysis is able to predict the zones around the tunnel which are capable of cracking qualitatively . The quantity of the required support for such locations may be determined by elastoplastic analysis . In figure 13 the tension zones have been supported by appropriate rock bolts .

6 SUMMARY AND CONCLUSIONS

In this paper application of the elasto-viscoplastic model in underground openings was examined . The results obtained for two tunnels with different geometries were discussed . Regarding various conditions in different sites , such as variations in material properties , Ko factor , shape of section for the underground space and other parameters , it becomes clear that closed-form solutions on the basis of hydrostatic stress field and circular sections and assuming linear elastic behaviour , are very simplified solutions and far from reality .

REFERENCES

Ladanyi , B 1974 . Use of the long-term strength concept in the determination of ground pressure on tunnel linings . Advances in rock mechanics , proc 3rd congr ., Int . Soc . Rock Mech ., Denver , 2B , 1150-56 . Washington .
Brown , E . T . and J . W . Bray 1982 . Rock-support interaction calculations for pressure shafts and tunnels . Proc . ISRM sym . , Aachen .
Naylor , D. J., G. N. Pande , and B. Simpson 1981 . Finite elements in Geotechnical Engineering ., Pineridge press, Swansea , U. K .
Sharma , K. G., A. V. Varadarajan and R. K. Srivastava , 1985 . Elasto-viscoplastic finite element analysis of tunnels , 5th Int . Con . Num . Meth . Geomech ., Nagoya .

Computer Methods and Advances in Geomechanics, Yuan (ed.) © 1997 Balkema, Rotterdam, ISBN 90 5410 904 1

Finite deformation elastoplastic analysis of underground structures and retaining walls

T. Tanaka, H. Mori & M. Kikuchi
Department of Agriculture, Meiji University, Kawasaki, Japan

ABSTRACT: Nonlinear geotechnical problems are often solved by using an incremental Newton-Raphson algorithm. A major part of the computational effort is spent in computing displacements by solving a linear matrix equation. This paper describes how a dynamic relaxation method can be used to effectively solve the finite deformation elastoplastic problems. The constitutive model for the nonassociated strain hardening-softening material was used. A yield function of Mohr-Coulomb type and a plastic potential function of Drucker-Prager type are employed. Rowe's stress-dilatancy relation is used for the evaluation of dilatancy. A generalized return mapping algorithm including a shear band effect is implemented. An integration of spatial rate constitutive equations is applied. In order to demonstrate the performance of the method, the collapse of underground arch structures and passive retaining wall problems are analyzed. The calculated results are compared with the results of model experiments using Toyoura sand.

1. INTRODUCTION

Many elastoplastic geotechnical problems with material nonlinearity also involve geometric nonlinearity. If the nonlinearity due to large deformation is neglected in this class of problems, the results of analysis do not agree with those of experiment. The collapse of underground arch structures, including buckling, and the passive retaining wall problem with rotation about the toe are considered to belong to this class of problems.

In this study, we attempted to estimate the feasibility of finite element analysis, by comparing the results of experiments (a underground arch structure and retaining wall) with those of finite element analyses (infinitesimal deformation analysis and finite deformation analysis). The constitutive model for nonassociated strain hardening-softening elastoplastic material was employed [Tanaka,1991]. In order to avoid numerical instability due to the singularity of the nonassociated Mohr-Coulomb model, the constitutive model based on the yield function of Mohr-Coulomb type and the plastic function of Drucker-Prager type were employed.

2. FINITE ELEMENT ANALYSIS

The failure of a sand mass is generally progressive and depends on the development of a shear band of localized deformation. Nonlinear geotechnical problems are often solved using an incremental Newton-Raphson algorithm with constituting the stiffness matrix. Recently, the dynamic relaxation method has been shown to be more effective than the modified Newton-Raphson method [Tanaka,1992]. In this analysis, the dynamic relaxation method combined with the generalized return mapping algorithm instead of constituting the stiffness matrix was applied to the integration algorithms of elastoplastic constitutive relations including the shear band effect.

In the return mapping algorithm [Simo and Ortiz,1986], the elastically predicted stresses (σ_A) are relaxed onto a suitably updated yield surface (σ_B). A change in stresses can cause an associated change in the elastic strains given by;

$$\varepsilon^e = [D]^{e-1}(\sigma_B - \sigma_A) \tag{1}$$

where $[D]^e$ is the elastic matrix. As the total strain does not change during the relaxation process, the change in plastic strain is balanced by an equal and opposite change in the elastic strain;

$$S\varepsilon^p = -\varepsilon^e = -[D]^{e-1}(\sigma_B - \sigma_A) \tag{2}$$

where S is the ratio of the shear band area to the finite element area. The plastic strain increments are proportional to the gradient of the plastic potential (Φ);

$$\varepsilon^p = \lambda(\frac{\partial \Phi}{\partial \sigma}) \quad (3)$$

where λ is a positive scalar multiplier to be determined with the aid of the loading-unloading criterion. Combining Eq.2 and Eq.3 gives;

$$\sigma_B = \sigma_A - S\lambda[D]^e(\frac{\partial \Phi}{\partial \sigma}) \quad (4)$$

As the plastic strains change, there is a change in the hardening-softening parameter (κ);

$$\varepsilon_B^p = \varepsilon_A^p + \lambda(\frac{\partial \Phi}{\partial \sigma}) \quad (5)$$

$$\kappa_B = \kappa_A + d\kappa \quad (6)$$

The yield function and the plastic potential function are given by;

$$f = \frac{\sqrt{J_2}}{g(\theta)} - 3\alpha(\kappa)\sigma_m = 0 \quad (7)$$

$$\Phi = \sqrt{J_2} - 3\alpha'(\kappa)\sigma_m = 0 \quad (8)$$

where

$$\alpha = \frac{2\sin\phi}{\sqrt{3}(3-\sin\phi)} \quad (9)$$

$$\alpha' = \frac{2\sin\varphi}{\sqrt{3}(3-\sin\varphi)} \quad (10)$$

where J_2 is the second invariant of deviatoric stresses, σ_m is mean stress, θ is the Lode angle, ϕ is the mobilized internal friction angle and φ is the mobilized dilatancy angle. In the case of the Mohr-Coulomb model, $g(\theta)$ in Eq.11 takes the form;

$$g(\theta) = \frac{3-\sin\phi}{2\sqrt{3}\cos\theta - 2\sin\theta\sin\phi} \quad (11)$$

The frictional hardening-softening functions are expressed as;

$$\alpha(\kappa) = \left(\frac{2\sqrt{\kappa\varepsilon_f}}{\kappa+\varepsilon_f}\right)^m \alpha_p$$

$$(\kappa \le \varepsilon_f) \text{ :hardening-regime} \quad (12)$$

$$\alpha(\kappa) = \alpha_r + (\alpha_p - \alpha_r)\exp\{-(\frac{\kappa-\varepsilon_f}{\varepsilon_r})^2\}$$

$$(\kappa \ge \varepsilon_f) \text{ :softening-regime} \quad (13)$$

where m, ε_f and ε_r are material constants, and α_p and α_r are given by Eq.9. The peak friction angle ϕ_p can be estimated from the empirical relations proposed by Bolton [1986,1987], and the mobilized dilatancy angle φ can be estimated from the modified Rowe's stress-dilatancy relationships. The elastic moduli are estimated using the equations;

$$G = G_o \frac{(2.17-e)^2}{1+e}\sqrt{\sigma_m} \quad (14)$$

$$K = \frac{2(1+\nu)}{3(1-2\nu)}G \quad (15)$$

where e is the void ratio, ν is Poisson's ratio and G_o is a material constant.

The dry density (γ_d), residual internal friction angle (ϕ_r), Poisson's ratio (ν), initial shear modulus (G_o) and shear band thickness ($S.B.$) were assumed to have the following values based on the data from a test of air-pluviated dense Toyoura sand [Tatsuoka et al.,1993].

$$\gamma_d = 1.60g/cm^3, \quad \phi_r = 34^o, \quad \nu = 0.3,$$

$$G_o = 19620KN/m^2, \quad S.B. = 0.3cm$$

Material constants m, ε_f and ε_r were assumed to have the following values.

$$m = 0.3, \quad \varepsilon_f = 0.1, \quad \varepsilon_r = 0.6$$

Material constants m and ε_f which are related to hardening, were determined by a plane strain compression test (PSC) at low stress levels. Material constant ε_r which is related to softening, should be determined from the stress-strain relationship in the shear band. Tatsuoka et al. [1986] reported that the shear band thickness was about 20~30 times the mean grain diameter (D_{50}). Tanaka [1992] discussed the influence of shear band thickness on collapse loading of strip footings and showed that the strain softening becomes less marked with the shear band thickness.

The finite deformation analysis based on the updated Lagrangian formulation proposed by Nagtegaal and Reblo (N-R formulation) [1988] is applied. We can directly integrate the constitutive rate equation to yield the corotational stress.

$$\bar{\sigma} = d\bar{\sigma} + \sigma \quad (16)$$

$$d\bar{\sigma} = L^{ep} d\varepsilon^{RN} \quad (17)$$

where L^{ep} is the elastoplastic modulus for the small strain analysis and $d\varepsilon^{RN}$ is the rotation neutralized strain increment. The Cauchy stress at the increment is obtained by the transformation.

$$\sigma = R\bar{\sigma}R^T \quad (18)$$

where R is the rotation tensor calculated from polar decomposition of the deformation increment. The rotation neutralized strain increment is given by the following equation with good approximation.

$$d\varepsilon^{RN} = R^T d\varepsilon R \quad (19)$$

We used the total Lagrangian formulation given by Hibbitt, Marcal and Rice [1970] to examine the buckling problem of underground arch structures. Also, for passive retaining wall problems, we carried out the calculations which were based on the updated Lagrangian formulation proposed by Nagtegaal and De Jong (N-D formulation) [1981].

3. TEST APPARATUS

The test apparatus of the underground arch structure shown in Fig.1 consisted of a glass-walled box, an air-bag, an arch structure (polyester sheet) and a hopper. The air pressure was overburdened by the air-bag until the arch structure had broken. The test apparatus of the retaining wall shown in Fig.2 consisted of a glass-walled box, a rigid wall (steel plate) and a hopper. A rough wall face was prepared by attaching sandpaper to the steel plate face. Experimental results were measured using three earth pressure cells and a load cell. The retaining wall was rotated about the toe (in passive mode) by means of a screw

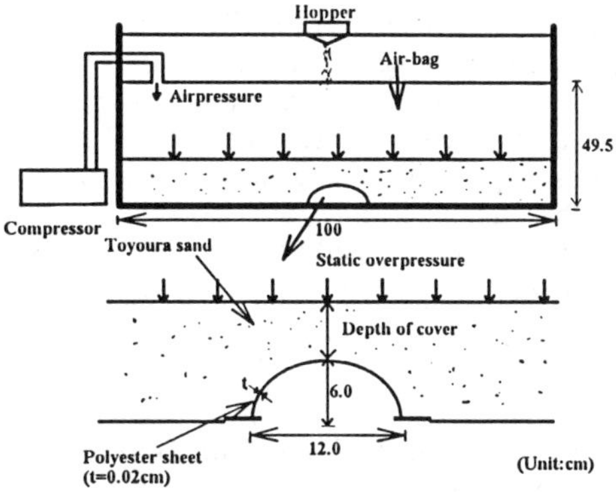

Fig.1 Test apparatus (underground arch structure)

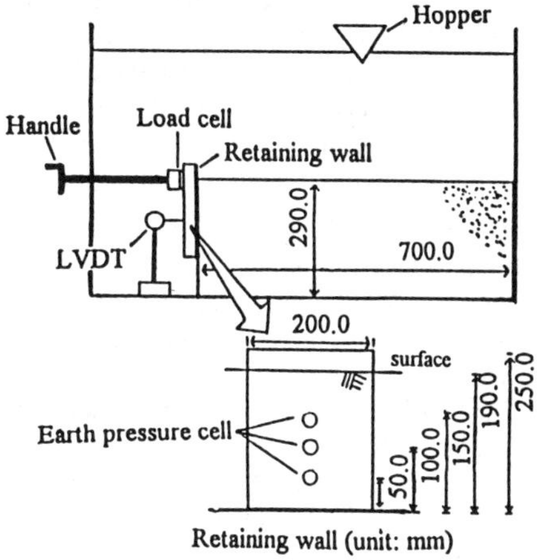

Fig.2 Test apparatus (retaining wall)

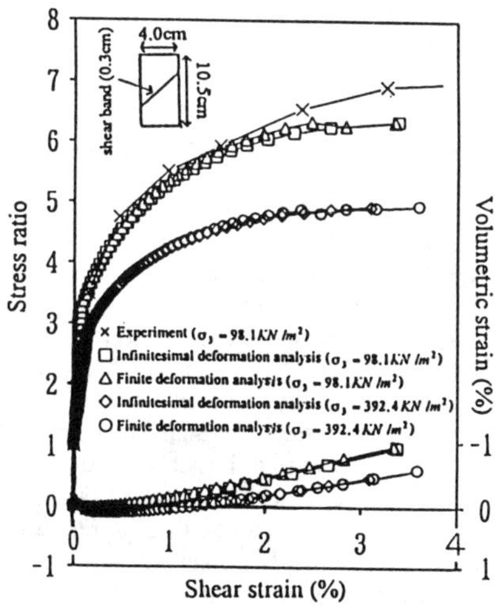

Fig.3 Stress-shear-volume changes by analysis

handle until the complete primary failure surface appeared in the sand mass. For lubrication to reduce the side-wall friction, a latex rubber membrane and a thin layer of silicone grease were used. The sand mass was prepared by pouring air-dried sand into the testing box through the hopper. The sand used in these tests was Toyoura sand ($G_s = 2.64$, $e_{max} = 0.977$, $e_{min} = 0.605$, $D_{50} = 0.16mm$). The dry density was about $1.60g/cm^3$.

4. NUMERICAL AND EXPERIMENTAL RESULTS

The back-prediction of PSC by finite element methods (the infinitesimal deformation analysis and finite deformation analysis) using one element was carried out employing the above mentioned material properties using load control. The calculated stress ratio -shear strain-volumetric strain change relations under $\sigma_3 = 98.1KN/m^2$ or $\sigma_3 = 392.4KN/m^2$ are shown in Fig.3. The experimental stress-shear strain relationship under $\sigma_3 = 98.1KN/m^2$ obtained by Tatsuoka et al. [1986] is also shown. The results of analysis exhibits good agreement with those of the experiment.

The finite element mesh used for underground arch structure analyses is shown in Fig.4. The calculations were carried out using load control (the equivalent distributed load increment was $1.962KN/m^2$ and maximum number of iterations 5000). The finite element mesh used for retaining wall analyses is shown in Fig.5. The calculations were carried out using load control (the load increment was 4.9N and maximum number of iterations 100000).

For an underground structure, Fig.6 shows relations between the limit pressure and depth of cover obtained analytically and experimentally. The limit pressure tended to increase with increments of depth of cover. Fig.7 shows relations between overpressures and vertical displacement of the crown obtained by infinitesimal deformation analysis, finite deformation analysis and experiment (the depth of cover is 2.5cm). Although the limit pressure by infinitesimal deformation analysis is not attained and that given by finite deformation analysis (the updated Lagrangian formulation) is larger than experiments, that given by finite deformation analysis (the total Lagrangian formulation) tended to agree with the experiment (glass-wall). Fig.8 shows contour lines of calculated maximum shear strain with mesh deformations. The zone of concentrated shear strain extended from the middle third of the arch structure and the

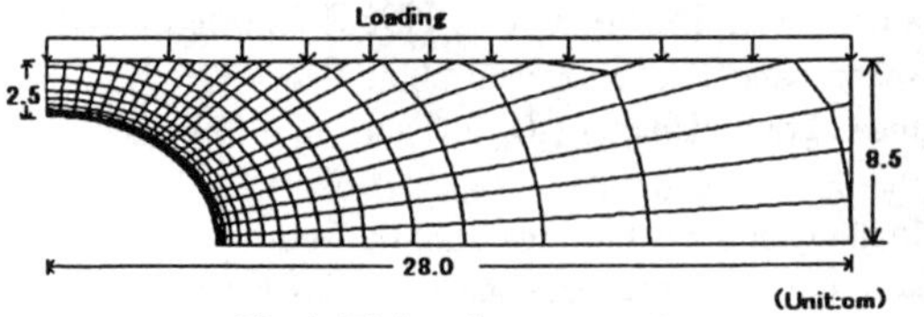

Fig.4 Finite element mesh (underground arch structure)

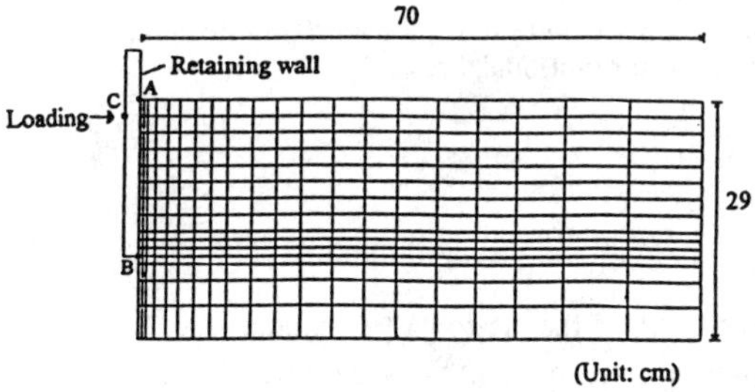

Fig.5 Finite element mesh (retaining wall)

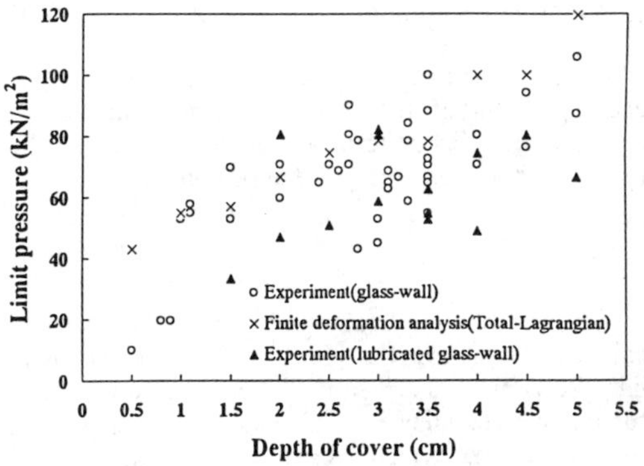

Fig.6 Limit pressures and depths of cover by analyses and experiments

direction of the localized zone found analytically is approximately identical with those in the mode experiment.

Next, results for a retaining wall are given. The angle of wall friction employed in this analysis was assumed to be equal to the internal friction angle of sand ($\delta = \phi$). Fig.9 shows relations between loads and rotation angle obtained by infinitesimal deformation analysis, finite deformation analysis and experiment. The infinitesimal deformation analysis gave the residual state too early, whereas the results of finite deformation analysis (the updated Lagrangian formulation) tended to be identical to those of the experiment. The N-R formulation gave approximately the

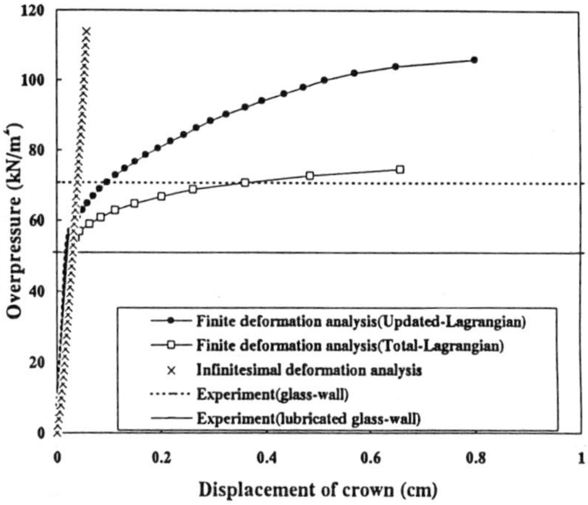

Fig.7 Overpressures and vertical displacements of crown by analyses and experiment

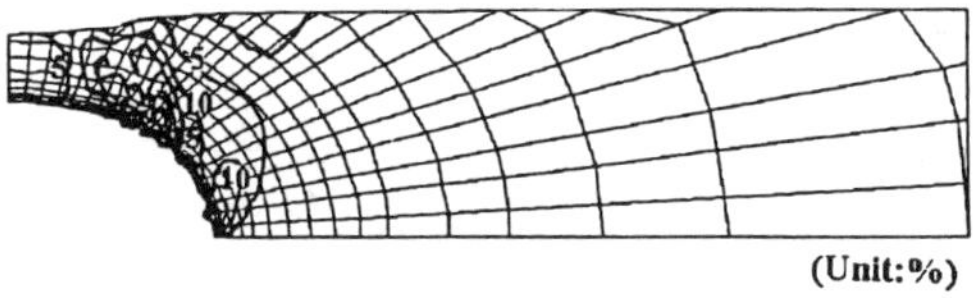

Fig.8 Contour lines of calculated maximum shear strain with mesh deformations (underground arch structure)

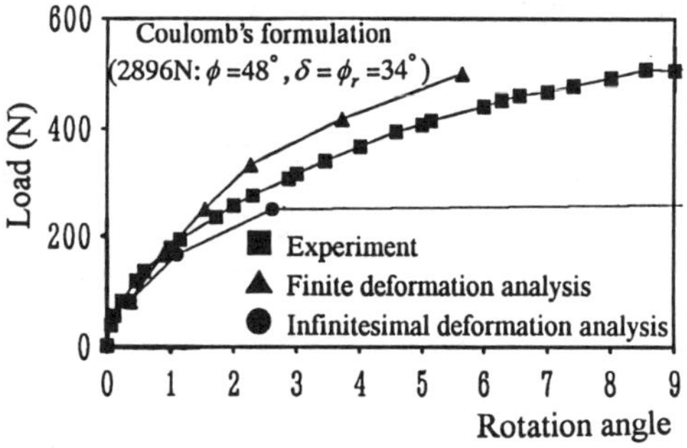

Fig.9 Loads obtained by analyses and experiment

same results as the N-D formulation. Fig.10 shows contour lines of calculated maximum shear strain with mesh deformations. The concentrated zone of shear strain extended from the top of the wall and the direction of the localized zone by the analysis was approximately identical to the direction of outermost shear band development observed experimentally.

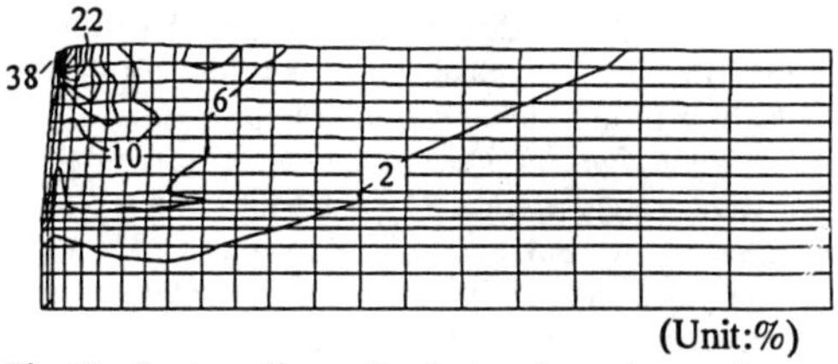

Fig.10 Contour lines of calculated maximum shear strain with mesh deformations (retaining wall)

5. CONCLUSION

An underground arch structure and retaining wall were analyzed, and the results were compared with those obtained in model experiments. In the case of an underground arch structure, although the limit pressure determined by infinitesimal deformation analysis was not attained, finite deformation analysis gave good results. In the case of a retaining wall, finite deformation analysis gave better results than infinitesimal deformation analysis when the wall friction was assumed to be equal to the internal friction of sand. It was shown that finite deformation analysis would be useful in solving buckling problems or passive earth pressure problems.

REFERENCES

M. D. Bolton (1986): "The strength and dilatancy of sands", *Geotechnique*, **36**, pp.56-78

M. D. Bolton (1987): "Discussion on the strength and dilatancy of sands by Tatsuoka", *Geotechnique*, **37**, pp.219- 226

H. D. Hibbitt, P. V. Marcal and J. R. Rice (1970): "A finite element formulation for problems of large strain and large displacement", *Int. J. Solids Struct.* **6**, pp.1069-1086

J. C. Nagtegall and J. E. De Jong (1981): "Some Computational Aspects of Elastic-Plastic Large Strain Analysis", *International Journal For Numerical Methods in Engineering*, **17**, pp.15-41

J. C. Nagtegall and N. Reblo (1988): "On the development of a general purpose finite element program for analysis of forming processes", *Int. J. Nume. Meth. Engng*. **25**, pp.113-131

J. C. Simo and M. Ortiz (1986): "An analysis of a new class of integration algorithms for elasto-plastic constitutive relations", *Int. J. Nume. Meth. Engng*. **23**, pp.353-366

T. Tanaka (1992): "Deformation and stability analysis by finite element method; In principal of soil mechanics [1st revision]", *The Japanese Society of Soil Mechanics and Foundation Engineering*, pp.109-154

T. Tanaka (1991): "Elastoplastic constitutive model for softening with localization and finite element analysis of footing", *Transactions Japanese Society Irrigation, Drainage and Reclamation Engineering, Japan*, **154**, pp.83-88

F. Tatsuoka, M. Sakamoto, T. Kawamura and S. Fukushima (1986): "Strength and deformation characteristics of sand in plane strain compression at extremely low pressures", *Soils and Foundations*, **26** (1), pp.65-84

F. Tatsuoka, M. S. A. Siddiquee, C. S. Park, M. Sakamoto and F. Abe (1993): "Modelling stress-strain relations of sand", *Soils and Foundations*, **33** (2), pp.66-81

Computer Methods and Advances in Geomechanics, Yuan (ed.) © 1997 Balkema, Rotterdam, ISBN 90 5410 904 1

Interaction between the vertical shaft supporting structure and rock mass

Z.Tomanović
Faculty of Civil Engineering, University of Montenegro, Yugoslavia

P.Anagnosti
Faculty of Civil Engineering, University of Belgrade, Yugoslavia

ABSTRACT: In this paper the "step by step" shaft construction method is described by using the FEM formulation of the problem in 3D conditions. This procedure enables the analysis of successive stages of construction and the resulting interaction of the supporting structure and rock-mass. The concept of the method and its physical interpretation are presented. The paper also proposes a new technique of treating the materials with inelastic properties. The interpreted results of the stress and strain analysis of the rock mass and the vertical circular shaft lining are obtained by using the sequential approach.

1. INTRODUCTION

Development of construction methods and use of new materials in excavation and construction of underground structures is associated with new approaches in structural analyses, particularly in the field of numerical models that can reflect impacts caused by applied construction methods. It is in the nature of civil engineering structures including underground structures that they are dependent on the factor of time. However, the direct introduction of time in structural analyses as a separate variable creates considerable problems from the definition of the proper relationship to the resolution of the interrelations of time with stress-strain state in actual structural problems. To enable a rational solution for the cited problem the continuous construction process can be substituted by a number of construction steps with instantaneous changes in the structure within selected time intervals.

Basing on the cited principles, the continuous construction procedure can be treated, within the frames of FEM, in pseudostatic conditions, as a single step construction with the lining formed just after a full completion of excavations i.e. after the equilibrium is attained in the rock mass itself. This approach overestimates the stresses in rock mass and the lining is not loaded by rock mass. The other approach based on the assumption of the existence of a lining in advance of the application of gravity forces to rock mass configuration (the most frequent current method) leads to an overestimation of stresses in the lining and underestimation of the strains in rock mass. The correct solution lays between the cited methods and can be analysed by applying a step by step simulation of the construction procedure in a three dimensional field of stresses and strains, as it is shown in this paper. By the described procedure one can simulate the underground construction of any shape and size, but its principle can be also applied for other construction procedures like fill dam construction etc.

The cited numerical method concerned with the problem of interaction includes an approximative representation of rock mass as a pseudo-elastic medium with determination stresses and strains, permanently taking into account the impact of rock mass on the shaft support. Therefor it can be used for analysing the above mentioned problem of excavation and underground structure lining in as close accordance with the construction process as possible. The partial stress relaxation in the rock mass of the yet unlined part of excavation which is displaced forward with the successive process of concrete lining is thus taken into consideration.

It should be noted that the method described above is of general character as regards geometry, the number of construction sequences and the relationship between stresses and strain. This technique also enables the simulation of the construction of an opening of any shape and size, such as caverns or tunnels, as well as fill dam and ather staged construction procedures.

2. DESCRIPTION OF THE SEQUENTIAL APPROACH

The finite element method is a widely used numerical method of analysing the underground structures and the surrounding rock mass stress states. Nevertheless, the original concept of the FEM does not easily allow for the analysis of systems characterised by the following: exclusion or inclusion of certain elements from or into the system, changes in boundary conditions, or the "freezing" of the stress-strain states in some elements of the system, in order to carry out structural analysis in certain construction phases, i.e. time sequences. It is for this reason that solutions for time-dependent problems are usually obtained by the "step-by-step" methods.

By using FEM, the sequential approach, which as a rule alternates the excavation lining sequences, is usually carried out on a series of models whose geometry corresponds with characteristic construction phases, instantenously applying gravity forces on the system (most frequent current method). However, this method does not take into account previous stress states,

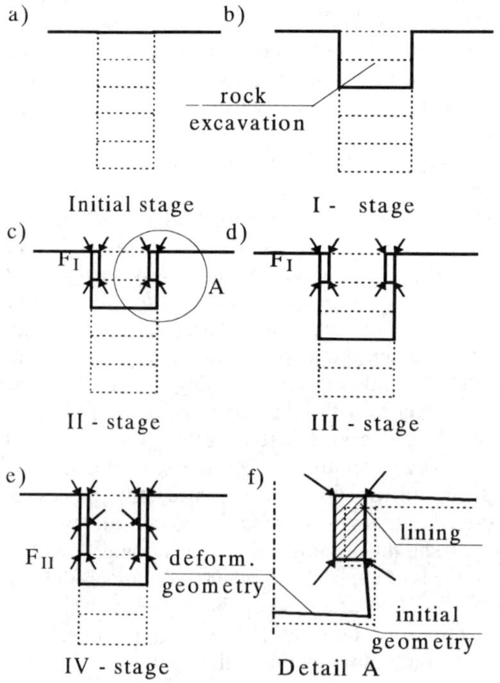

Figure 1. The simulation scheme of shaft construction

i.e. the deformed geometry of the system, which results in serious errors in the structural analysis of the stress and strain states. This paper will present the procedure which enables the introduction of the stress states and the deformed geometry of the previous model in the next model of the system.

The construction of a vertical circular shaft and its lining that was selected for application of the cited numerical procedure is presented in Fig.1 (P.Anagnosti & Z.Tomanović, 1996). The excavation of two lifts is chosen to be the first construction sequence (Fig.1.b). Due to this excavation, the previous state of stresses in the surrounding rock mass is modified and the new state of stresses takes place followed by respective displacements, and the excavation surface is gets a new, deformed shape.

The next step consists in the introduction of lining. In case one introduces the lining element into the model by simply adding one or more elements with appropriate geometry and material properties, the impact of the lining stiffness and weight will change the response of the surrounding rock. However, this impact can not be transferred to the rock (excluding concrete shrinkage and similar effects) by a lining placed on the already deformed excavation surface.

In order to avoid this impact the "equivalent - fictive" forces are to be introduced at the nodal points (F_I) which deform the lining element in such a way that the resulting displacement are those existent without lining. (Fig. 1. f) (W.Witke, 1990). In addition to the definition of these "equivalent - fictive" forces, the own weight of the lining element is to be assigned to the nodal points.

The next construction step is the excavation of another rock mass bench, as shown in Fig.1.d. The model for this step is obtained by a transformation of the step II model i.e. by excavating and removing elements from the third bench of elements within the excavation boundary. From this step on, the first support ring takes over part of the forces which have resulted from further excavation and the weight of the next support rings. It should be noted that, due to the "equivalent-fictive forces" impact, there is some initial stress state in the support elements. Determination of the real stress state in the next construction steps requires the substraction of the initial-fictive stress state from the one existing in the given construction sequence. It is in this way only that a proper structural analysis of each support element can be carried out.

Step IV consists in the construction of the next

support ring (Fig.1.e.). The newly constructed ring adheres to the nodal points of the rock mass elements and support structure, and a new system of fictive forces F_{II} develops in them. As with the first support ring, the fictive forces should also be determined. The following, alternating construction steps - excavation and lining construction, are modelled according to the described procedure.

3. PHYSICAL INTERPRETATION OF THE APPLICATION AND DETERMINATION OF THE EQUIVALENT FICTIVE FORCES

The procedure used in the application of the "equivalent - fictive" forces has a sound physical meaning, as shown in Fig. 2 (Z.Tomanović, 1996) where zone D_I (the rock mass) which has already been deformed and the associated zone D_{II} (the lining) are given. The geometry of the boundaries S_I and S_{II} are to be the same, which requires additional "equivalent - fictive " nodal forces to bring the boundary of these two zones with different stiffness to the same geometry. In FEM formulation this requirement is fulfilled when the nodal displacements of the boundary S_{II} of the zone D_{II} under the "equivalent-fictive" forces are equal to the displacements of boundary S_I of the zone D_I under gravity forces. These "equivalent-fictive" forces can be computed in FEM formulation by expression (1).

The equivalent - fictive forces can therefore be also determined before connecting the system S_{II} to the system S_I, provided that boundary nodal displacements of the former are equal to the already determined nodal displacements of the latter. Obviously, there has to be an equilibrium of the equivalent - fictive forces. This method of displacement "equivalent - fictive" forces computation is considerably shorter because the

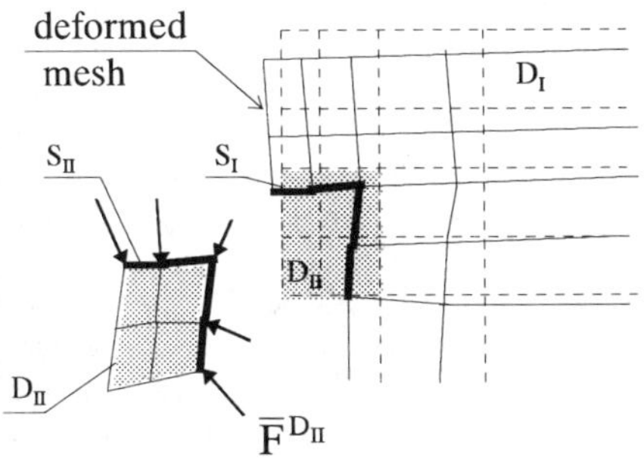

Figure 2. Scheme of the system and the "equivalent - fictive" forces

number of equations, i.e. unknowns is to a great extent smaller than the number of displacement unknowns in both systems.

Using FEM, the "equivalent-fictive" forces can be theoretically determined by applying expression (1) (W.Witke, 1990):

$$\{F\} = [K_b]\{\delta\} \qquad (1)$$

where: $\{F\}$ = vector of the external nodal fictive forces; $[K_b]$ = element stiffness matrix; and $\{\delta\}$ = vector of nodal displacements.

However, in performing the cited computations on a real case the application of "an available software" may lead to different procedures, particularly when it comes to determination of the "equivalent-fictive" forces. The software package SAP90 which is used for the analysis of the vertical shaft construction can be applied in a manner that by introducing as an input the nodal displacements one can obtain as the output nodal forces. Following this procedure, the "equivalent - fictive forces" are obtainable by using as an input the known displacements of the previous sequence, but for the next sequence of excavation the nodal points are made free to follow the changes in excavation boundary. (P.Anagnosti & Z.Tomanović 1996[b]).

These forces being determined, the lining element weight is introduced so that the rock mass stress state is influenced by the weight of the newly constructed lining rather than its stiffness, which is the ultimate aim of this method.

4. APPLICATION OF THE SEQUENTIAL APPROACH TO MATERIALS WITH NON-LINEAR BEHAVIOUR

The described sequential procedure has been applied to a linear elastic rock-mass and concrete lining for a vertical shaft construction. The basic aim of the analysis is determination of concrete lining stresses. The real state of stresses is to be computed by substraction of the stresses caused by the introduced "equivalent - fictive forces" from the resultant stress state (W.Witke 1990). This approach is in accordance with linear elasticity used for rock-mass and concrete lining elements.

The substraction of the cited initial stresses or strains may be considered as the transformation of the co-ordinate system $\sigma - \varepsilon$ into the system $\overline{\sigma} - \overline{\varepsilon}$, as it is shown in Fig. 3 (Z.Tomanović, 1996). The relationship between these two systems is : $\sigma = \sigma_0 + \overline{\sigma}$ and $\varepsilon = \varepsilon_0 + \overline{\varepsilon}$. Thus, the real state of stresses can be followed in the system $\overline{\sigma} - \overline{\varepsilon}$. The

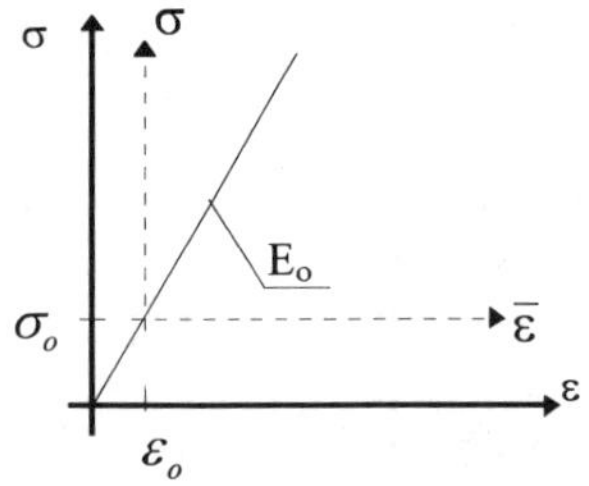

Figure 3. $\sigma - \varepsilon$ The linear elastic model diagram

elements in t he rock mass in all stress - strain states are described in the system $\sigma - \varepsilon$, due to the lack of "equivalent - fictive" forces in the rock mass elements. It may be also noted that linear elasticity is not a restrictive assumption for representation of rock mass, provided that the available FEM package is capable to perform the computations beyond the linear elastic formulation of the strains.

In case of the analysis of non-linear elastic behavior of the concrete lining the described procedure needs to be modified, and in this paper the procedure for bilinear elastic model will be outlined as it is shown in Fig. 4 (Z. Tomanović, 1996). The first step in determining the "equivalent - fictive" nodal forces has to be based on the initial elasticity modulus (Eo). Every next step in the analysis can be performed using the assumed non-linear stress - strain relationship and the transform of the co-ordinate system whose origin is to follow the stress strain curve as it is shown in Fig. 4 in accordance with the transform relationship: $\hat{\sigma}^i = \sigma + \sigma^i_o$ and $\hat{\varepsilon}^i = \varepsilon + \varepsilon^i_o$, where (i) is the indication of the element of the concrete lining.

The real state of stresses or strains is observed within the $\sigma - \varepsilon$ system, which implies that the real stress state in each element can be obtained by subtracting the initial fictive stress σ^i_o or strain ε^i_o from the stresses and strains characteristic of the current phase of the analysis.

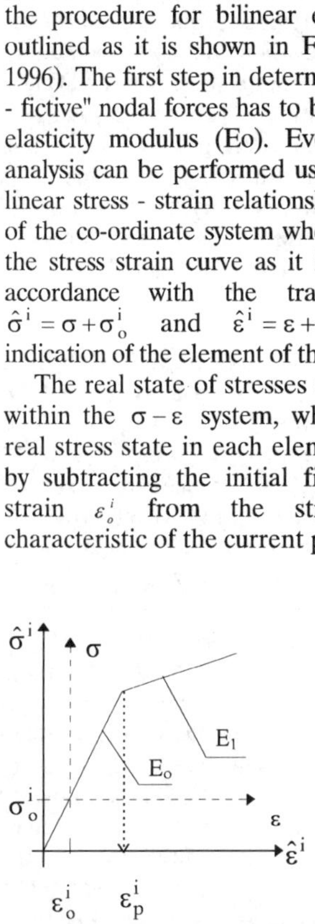

Figure 4. $\sigma - \varepsilon$ The bilinear concrete behaviour diagram

Yielding, i.e. transfer to the other side of the σ–ε diagram, takes place as soon as the strains reach value $\hat{\varepsilon}^i_p$. In a general case, each element has a different yield value if it is observed in the co-ordinate system $\hat{\sigma} - \hat{\varepsilon}$, despite the fact that all elements describe the same material. However, in the σ–ε co-ordinate system this value is the same for all the elements with same mechanical features and stress states.

The procedure described above enables the analysis of models with non-linear stress-strain function. In this case, an initial fictive modules of elasticity $\hat{E}^i_o$ is assumed and the analysis follows the steps of the procedure for a bilinear model.

5. INTERPRETATION OF THE STRUCTURAL ANALYSIS OF A VERTICAL SHAFT BY USING THE SEQUENTIAL APPROACH

Construction of an underground open space causes significant changes in the initial state of stress in the surrounding ground. This change requires a proper structural analysis and appropriate decisions regarding its shape and size, dimensioning of the lining thickness and stiffness etc. In the following part of this paper typical parts of the performed analysis are presented, and particular emphasis is placed on the comparison of the stress states obtained by "current methods" and those determined by the sequential procedure described above. In the analyses we have used the SAP90 software package for rock masses with a wide range of elastic properties (Es) and a variable size of the shaft diameter ranging from 5 to 15 m.

5.1. Stress state in the rock mass.

Excavation and successive lining construction cause changes in the primary stress state in the rock mass. The secondary state of stresses in rock mass computed by using sequential approach and "current methods" is presented Fig. 5. The charts show the changes in vertical, tangential and radial stresses at different depths of a shaft with a constant diameter. As the tangential stress increases with the depth, the radial stress decreases, and the stress difference also increases with depth. The state of stresses shows a large deviation from the primary state at the level of shaft bottom, and then the stresses gradually tend to the initial-"primary" state at the depth of approximately two diameters distance below the shaft bottom. The radial stress at

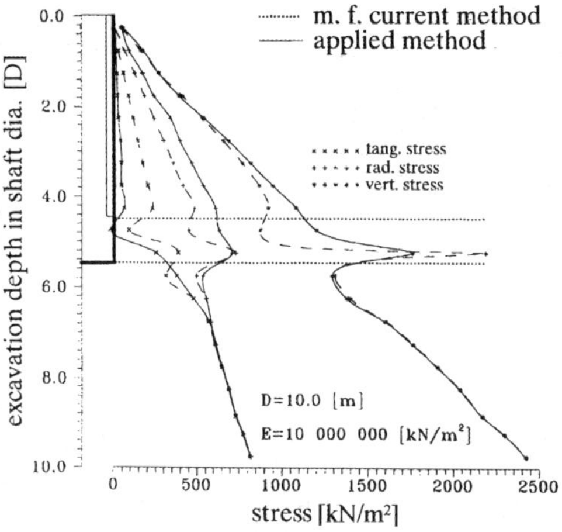

Figure 5. A diagram of rock mass stress

the unsupported part of the shaft is always equal to zero as it is the requirement for a free surface.

5.2 Stress state in the lining

Lining which is gradually constructed becomes loaded by a part of the radial stresses during each further excavation. These stresses are usually defined as "the loading of rock mass on the lining". In case of the vertical shaft under consideration, the lining is under tangential pressure along the whole depth, due to which tangential stresses in the lining govern its dimensioning. It is for this reason that stress analysis of the lining deals primarely with tangential stresses and with the factors influencing their values and distribution.

The values of tangential stresses in concrete lining along the shaft depth obtained by the "current method" and the sequential procedure are shown in Fig. 6, and the extent to which these stresses are overestimated in case of neglection of the elastic deformations of the unlined parts of the rock at each construction sequence may be noted easily.

Besides the difference in maximal values of the cited stresses (computed by the current versus sequential procedure) there is a significant difference in distribution along the shaft depth. The tangential stress does not reach its maximum value at the bottom "ring" of the lining but at a distance equal to 3 - 4 "rings" above the bottom of the shaft., as it is presented in Fig. 6. The very bottom "ring" of the lining is placed on the already deformed rock surface and the tangential stresses do not develop in this ring. By using the "current procedure" one would obtain the largest tangential stresses in the last "ring" of the lining, which clearly demonstrates the principal advantage of the sequential procedure in the stress analysis of concrete lining.

The first upper ring of the lining also does not have the minimum loading value, in contrast to the second and the third "ring", as it can be determined by the current method, as in Fig.6 . An explanation of this loading can be found in the fact that there is no structure and rock mass

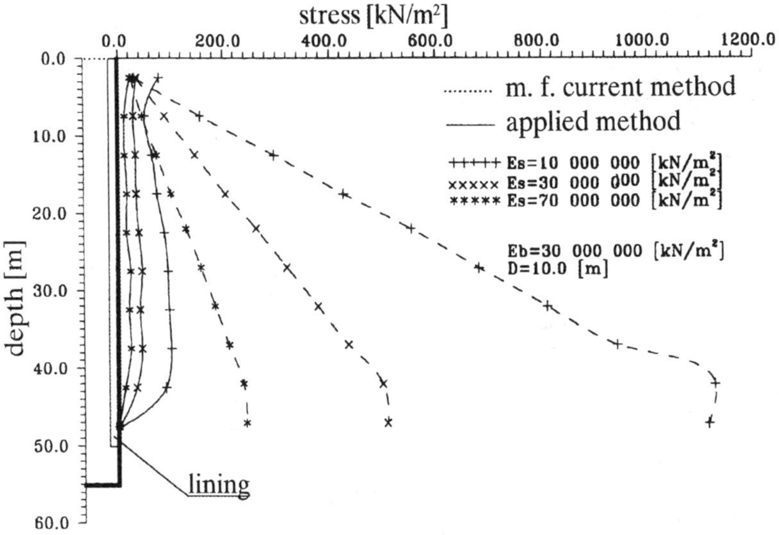

Figure 6. Tangential stresses in the lining dependent on depth

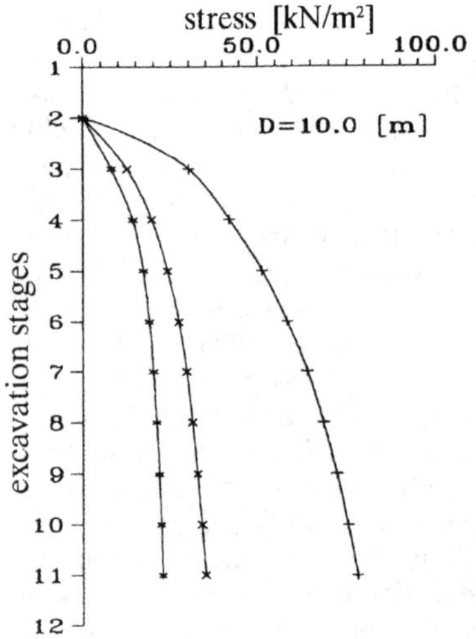

Figure 7. The stress increment of the stress in various construction sequences

above the first "ring" so that all the loadings resulting form further excavation are supported at this level. All lower ring, excavations are loading the ring supports already placed above the lowest ring excavation. Due to this the stress values in the second and the third ring are slightly lower than the ones in the first ring. This effect disappears with distance and the increase of stress becomes constant.

One of the most significant advantages of the sequential approach is the possibility to follow the stress-strain states development in the lining and rock mass at various construction phases. The diagram showing the increase of the first "ring" tangential stresses in different construction sequences is given in Fig.7

6. CONCLUSION

The performed analysis of the stress-strain state in concrete lining and the surrounding rock mass for a vertical shaft construction clearly indicates that the numerical analyses which are not based on the sequential determination of stresses can not properly describe the actual stress states either in the lining or in the surrounding rock mass. The differences that are taking place are quite beyond the "academic level" of discussion and the overestimation of stresses stemming from the "most frequent current" methods is quite high.

The sequential method enables a considerably better insight into the stress states existing during underground construction. Further development of this method with inclusion of more complex relationships between stresses and strains in rock mass and, first of all, time-dependent solutions may lead to the creation of a powerful numerical means for a precise structural analysis of underground construction.

REFERENCES

Anagnosti, P. & Z. Tomanović 1996. *Numerical Simulation of Vertical Shaft Construction.* Yugoslav Society for Materials and Structures Investigation and Research. Cetinje, volume 2, pp.1-6 (in Serbian).

Anagnosti, P. &Z. Tomanović 1996. *FEM Analysis of Sequential Tunnel Lining Impact on Stress-stain State.* The International Conference - Trends in the Development of geotehnics, Belgrade, pp.57-66.

Grabinsky, M. W. F., J.H. Curran 1993. *Efficient Mesh Generation Procedures for Finite Element Analysis of Underground Structures. Int.* J. Rock Mech. Min. Sci. & Geomech. Abstr., Vol.30. No.6, pp.591-600.

SAP90 1990. *Computer Programs for the Finite Elements of Structures,* Verification Manual, University Avenue, Berkeley,California 94704, USA.

Tomanović, Z., 1995. *Analysis of Stress Conditions around Vertical Shaft with Circular Cross-section by using the Finite Element Method.* Researches, Faculty of Civil engineering of the University Montenegro, pp.401-417

Tomanović, Z. 1996. *Analysis of interaction between vertical shaft supporting structure and rock mass.* M.Sc. Thesis, Faculty of Civil engineering of the University of Belgrade (in Serbian).

Wittke, W., 1990. *Rock Mechanics - Theory and Applications with Case Histories.* Springer-Verlag, Berlin Heidelberg, pp.296-299

Computer Methods and Advances in Geomechanics, Yuan (ed.) © *1997 Balkema, Rotterdam, ISBN 90 5410 904 1*

Rockburst synthetic criterions of high geostress area for Laxiwa powerstation underground caverns

Liu Xiaoming
Department of Hydropower Engineering, University of Hydraulic and Electrical Engineering, Wuhan, China
C.F. Lee
Department of Civil and Structural Engineering, University of Hong Kong, China

ABSTRACT: A rockburst is the sudden and violent failure of rock. During the failure process, excess energy is liberated which cause the surrounding rock mass to vibrate. The level of the elastic strain energy gathering rock increase during the excavation. The level of the elastic strain energy gathering not only depends on surrounding rock properties but also surrounding rock stress conditions. Therefore we could summarize two main factors from many causing factors. One is rock properties which reflect the ability of rock storing energy. It is internal cause of rockburst. The other is surrounding rock stress condition. It is external cause of rockburst. Putting the two sides together, a synthetic criterion of rockburst is proposed in this paper.

1 INTRODUCTION

Rockburst is a sudden and violent failure of rock. During the failure process, excess energy is liberated which cause injury to persons or damage to underground workings. And it is one of the major problem in mining industry and in powerstation underground plant caverns.

Based on the understanding and modeling the mechanism of such a phenomenon, prediction of rockburst and safe rock engineering design are very importance for underground opening. The general and essential feature of rockburst is their sudden, violent nature. According to their amount of release energy or the violent level, the rockburst can be delimited to several classes (Ortlepp 1992). The slight rockburst cause rock fragment peal off and has no injury to persons or damage. But intense rockburst can cause injury to persons and damage working and cause the surrounding rock mass to strong vibrate which are felt both underground and on surface. In the early 1960's, Cook ((1966) postulated that to control bursting, it was necasary to relieve stored energy in small enough amounts so that it would be dissipated nonviolently. It is important to understand the theoretical principles underlying energy release, which might contribute towards the evaluation of the amount of energy that is released when a rock structure failure. In this regard, monitoring of rock movements and seismic wave energy concentration can give certain approximate values. If one considers an overall energy balance, a feel for the amount of released energy can be obtained.

Although, it is now widely accepted that rockburst failure process excess energy is liberated, which cause the surrounding rock to vibrate. But why and how could the surrounding rock store the sufficient strain energy ? And which are the main factors for rockburst occur? The rockburst phenomenon is associated with the rapid discharge of energy which had been stored in a large volume of rock. It is not always necasary that rockbursting occurs where large accumulations of strains and stresses are in evidence, since this phenomenon depends on the strength properties and behavior of the type of rock mass in question.

Blake (1972) uses a finite-element digital model in which he incorporated the results of microseismic monitoring to show that a pillar in the Galena Mine suffered rockburst due to the average stress exceeding the strength of the pillar and to the mine stiffness being lower than the pillar stiffness. Using the stiffness mechanism, it is possible to explain the presence of violent failure in hard and brittle rock structure. Accordingly, rockburst control could be accomplished by reducing the stiffness of rockburst-prone structures Crouch, S.L.(1974).

The level of the strain energy gathering not only depends on surrounding rock property but also surrounding rock stress condition. Rockburst does not occur in any surrounding rock condition, although the surrounding rock has the property of rockburst. The occurring must be in a certain stress condition or storing enough energy. Therefore we could summarize two main factors from many causing factors. One is rock property which reflect the ability of rock storing strain energy and it is internal cause of rockburst. The other is surrounding rock stress condition. It is external cause of rockburst.

In this paper, a synthetic criterion of rockburst is proposed. The synthetically criterion has two sides. By using finite element method analysis the surrounding stress redistribute regularity during underground caverns opening. According to the rockburst strength criterion, we can obtain one side of the synthetically criterion. By analyzing the surrounding rock property of energy storing and transmission, we can study what kinds of rocks which possess excess liberated energy during its failure process. The other side of the synthetic criterion of rockburst is obtained from rock special rock properties. We used the synthetic criterion of rockburst analyzing the LAXIWA powerstation underground caverns and the slight rockburst will occur. The results are basically agreement on the appearance of scene rockburst failure.

2 STRESS CONDITION OF ROCKBURST

The stress condition of surrounding rock is the external cause of rockburst. According to a large number rockburst on the spot investigation and studies, it is found high geostress area surrounding rock stress increased by a wide main during underground caverns excavation which is the important reason. The higher of surrounding rock stress is, the larger of rockburst possibility is and the violent of rockburst intensity is. Oppositely, rockburst usually does not occur in low geostress area. Therefore we are concerned with using which surrounding rock stress paramagnets to judge rockburst and the critical failure value. The surrounding rock stress condition study must include not only indicating rockburst occurring stress state but also as simple as rock failure strength criterion. But it is not equal to the strength criterion which is obtained in lab tests. Because the surrounding rock stress is dynamic state with the process of caverns excavation and we must consider the influence of rock fault, crack, water and so on.

For this reason, many rockburst strength criterions have been proposed. They are usually empirical criterions. Some of them are as follows:

<1> Rockburst criterion was proposed by Barton (Sweden, 1974)

$\sigma_1 = (0.2 \sim 0.4)\sigma_c$ (moderate rockburst)

$\sigma_1 > 0.4\sigma_c$ (intense rockburst) (1)

where σ_1: Surrounding rock maximum principal stress, σ_c: Uniaxial compressive strength

<2> Rockburst criterion was proposed by Russenes (Norway 1985)

According to maximum tangential stress σ_θ and rock point load strength I_s the rockburst intensity had been built. We can get:

$\sigma_\theta < 0.2\sigma_c$ (no rockburst)

$0.2\sigma_c \le \sigma_\theta < 0.3\sigma_c$ (slight rockburst)

$0.3\sigma_c \le \sigma_\theta \le 0.55\sigma_c$ (moderate rockburst)

$\sigma_\theta > 0.55\sigma_c$ (intense rockburst) (2)

<3> The combinations of surrounding rock stress σ_1, σ_2 are considered in the criterion by (Hou 1990). It is proposed that if the combinations are difference then the condition of rockburst are difference. Assume the combination is $\sigma_2/\sigma_1 = 0.5$ we can get:

$\sigma_1 < 0.3\sigma_c$ (no rockburst)

$0.3\sigma_c \le \sigma_1 < 0.37\sigma_c$ (slight rockburst)

$0.37\sigma_c \le \sigma_1 < 0.62\sigma_c$ (moderate rockburst)

$\sigma_1 > 0.62\sigma_c$ (intense rockburst) (3)

3 ROCK PROPERTIES OF ROCKBURST

Of course, surrounding rock stress condition is a main factor for rockburst. Considering the same stress condition, it may be or not occurring rockburst with different surrounding rock mechanical properties. It is also due to rock deformation and some specific storing energy properties which is verified by engineering. Even if there is in high geostress area and the cavern radius convergence with several centimeters, soft surrounding rock could not occur rockburst. Because the external work is mainly dissipated by plastic deformation and there is little elastic strain energy storing in surrounding rock. it could not have enough elastic strain energy for rockburst. This indicates there are difference of ability for storing energy properties between various rocks. According to above reason, some reachers proposed rock special property index which can be aimed rockburst.

Poland mining science institute proposed rock elastic energy index (W_{ET}).
They have given the rockburst criterion:

$$
\begin{aligned}
&W_{ET} \geq 5 && \text{(intense rockburst)} \\
&W_{ET} = 2.0 \sim 4.9 && \text{(moderate rockburst)} \\
&W_{ET} < 2.0 && \text{(no rockburst)} \qquad (4)
\end{aligned}
$$

Hou Faling (1990) proposed rockburst tendency index (W_{qx}).

$$W_{qx} = \Omega_Z / \Omega_H \qquad (5)$$

where, Ω_Z is the storing elastic energy which is the left below part of the rock complete stress-strain curve. Ω_H is the right below part. He gave the rockburst criterion according to a great number of engineering information.

$$W_{qx} > 1.5 \qquad (6)$$

4 ROCKBURST SYNTHETIC CRITERION

Because surrounding rock stress condition and surrounding rock mechanical properties are two main factors which cause rockburst occurring. So it is not overall for only considering surrounding rock stress condition as rockburst criterion or only considering rock mechanical properties. It is rational which put them together.

Building a rational rockburst criterion must be considered several basic points below:

(a) Considering the stress condition of surrounding rock during opening stage.

(b) Influence of the surrounding rock stress components combinations.

(c) Reflect rock properties which are aimed to the ability of rock storing elastic strain energy.

Due to above principle, equations (1),(2), are not suitable stress criterions for not considering the surrounding rock stress components combinations. As rock strength failure criterion must include lateral compression or the effect of middle principal stress under multiaxial stress state. So we suggest using the tangential stress σ_θ and longitudinal stress σ_l of underground cavern to build rockburst stress criterion. This kind of stress criterion is rational not only in reflecting the stress state of surrounding rock side wall well but also in considering the combinations of the surrounding rock stress components. The criterion is shown as (Hou 1986):

$$\sigma_\theta \leq K_J \sigma_c \qquad (7)$$

where, K_J is called stress combinative factor.
Table 1 shows K_J value with different combinative case between σ_θ and σ_l .

Table 1. The relationship of combinative cases and K_J

σ_l / σ_θ	0.00	0.25	0.50	0.75	1.00
K_J	0.19	0.29	0.38	0.40	0.42

On the other side, we suggest using the rockburst damage energy index (W_D) as the rockburst criterion for surrounding rock property (Liu 1995). Based on a great number of test and theories analysis, the rock elastic-brittle damage mode was built. The rockburst damage energy index was obtained by analyzing rock energy storing, releasing and changing (see Figure 1). W_D is the specific value of the elastic strain energy release quantity and damage dissipative energy during its failure stage (segment BC). If W_D=1 ,then the elastic strain energy in rock transfer completely to damage dissipative energy. There is no remained energy which transfer to external. So rockburst could not occur. If W_D>1 ,part of elastic strain energy transfer from rock internal to rock external (including kinetic energy,acoustic energy,thermoenergy et.) rockburst may occur. The more the W_D is , the more the possibility of rockburst is. And the rockburst is more intense. For this reason, we can obtain the criterion of rockburst :

$W_D>1$ (8)

W_D can be given as:

$$W_D = \frac{(1-D_B)Y_B - (1-D_C)Y_C}{\int_{D_B}^{D_C} Y dD} \quad (9)$$

where, D is a function of ε, Y is a function of σ and D. W_D can be obtained according to rock uniaxial compressure stress-strain curve. D_B, D_C are the damage value of points B and C. Y_B, Y_C are the damage energy release rate of points B,C. Due to the process of W_D calculating, we can find the W_D value reflect the material damage character, strength and material constitutive property.
Table 2 gives some kinds of rock W_D value for engineering and actual rockburst.

Table 2 The test results of various rocks mechanical properties

Hydropower station Engineering	σ_c (MPa)	ε_s 10^{-3}	n	Y_c (MPa)	W_D	Remark column
ReTan (orthoclase)	220.0	6.2	32	0.56	1.32	occur rockburst
TianShenQiao (dolomite)	88.7	4.8	12	0.18	1.15	occur rockburst
LaXiWa (granite)	176.0	5.5	25	0.32	1.27	occur rockburst

* σ_c is rock uniaxial compressive strength; ε_s, n, Y_c are rock damage property paramagnets ;
W_D is rock damage energy index

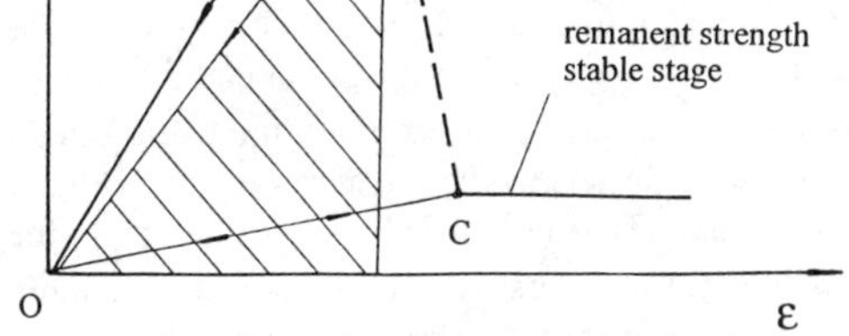

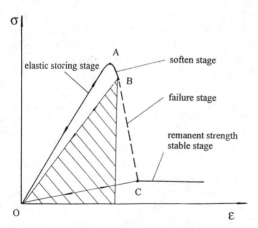

Figure 1 Brittle rock uniaxial compressure stress-strain curve

After above analysis, we propose a synthetic criterion for rockburst:

$$\begin{cases} \sigma_\theta \geq K_J \sigma_c \\ W_D > 1 \end{cases} \quad (10)$$

This synthetic criterion is overall in not only reflecting surrounding rock stress condition for rockburst but also reflecting rock storing, dissipating, releasing energy mechanical properties. According to analyzing surrounding rock cavern stress by using finite element method or other methods and the results of rock lab testing, we can obtain the stresses of the most dangerous part of cavern and the rock special mechanical property W_D. Then we can get the basic judgment whether the rockburst would be occurring by using the synthetic criterion (10).

5 ROCKBURST ANALYSIS ON LAXIWA POWERSTATION UNDERGROUND CAVERN

LAXIWA powerstation will be situated at the upper reaches of the Yellow River Northeast QingHai province in China. The rock mass of dam area is intrusive granite. The test cavern is located at the right bank of the river 250m below surface. The cavern's length, width, high are 27.5m, 2.5m, 6.7m and its shape is arch culvert. It is approximate intact granite rock in the region of test cavern.

There were nearly tens times failure occurrence during the opening stage of the testing cavern. On 8 July, 1991, there is a greater blast of failure sound and shake with several rock pieces flying from side wall to the other side about several meters away. The weight of the ejected rock pieces are about 0.2Kg~1.0Kg. We can look there are many slices which are like the scales of a fish on the fresh rockburst failure surface and the surface is felt warm.

According to the parameters of LAXIWA engineering, we have calculated the stresses near the advanced face of the opening cavern using 3-D finite element method. We have obtained the face distributive law of surrounding rock tangential stress σ_θ and longitudinal stress σ_l (see table 3 figure 2)
From the calculate results, stress combinative factor of the most dangerous position is
$K_J = 0.32$, $K_J \sigma_c = 50.24 MPa$

Therefore, the surrounding rock tangential stress σ_θ on the most dangerous points at the most dangerous section is just exceed the limit value of

Table 3 Surrounding rock tangential stress σ_θ (MPa) and longitudinal stress σ_l (MPa)

	0+3.92m		0+7.84m		0+11.76m		0+15.68m		0+19.60m		0+23.52m	
	σ_θ	σ_l	σ_θ	σ_l	σ_θ	σ_l	σ_θ	σ_l	σ_θ	σ_l	σ_θ	σ_l
1*	-16.96	-17.19	-28.93	-16.61	-31.48	-11.74	-23.84	-14.23	-18.12	-14.75	-5.63	-16.39
2	-28.64	-22.46	-50.24	-22.94	-53.57	-22.70	-47.35	-21.97	-17.85	-17.97	-5.28	-17.26
3	-13.00	-17.58	-25.27	-12.51	-28.86	-11.75	-28.22	-8.97	-18.64	-19.25	-6.67	-17.76
4	-1.51	-16.33	-3.17	-8.79	-4.87	-6.97	-4.48	-3.61	-13.09	-17.56	-6.705	-18.91
5	-2.07	-16.08	-5.34	-8.83	-7.34	-7.45	-6.81	-3.96	-13.18	-17.95	-6.78	-19.00
6	-5.77	-16.67	-12.82	-10.85	-15.36	-10.14	-14.21	-7.20	-15.07	-18.10	-5.95	-19.09
7	-27.09	-20.77	-48.05	-18.39	-52.82	-19.72	-45.96	-17.01	-19.17	-19.62	-5.30	-17.13
8	-29.68	-21.04	-49.95	-18.92	-55.43	-20.01	-46.71	-17.63	-19.04	-19.49	-4.61	-16.67

* 1~8 are the key points on cavern surface (see figure 2)

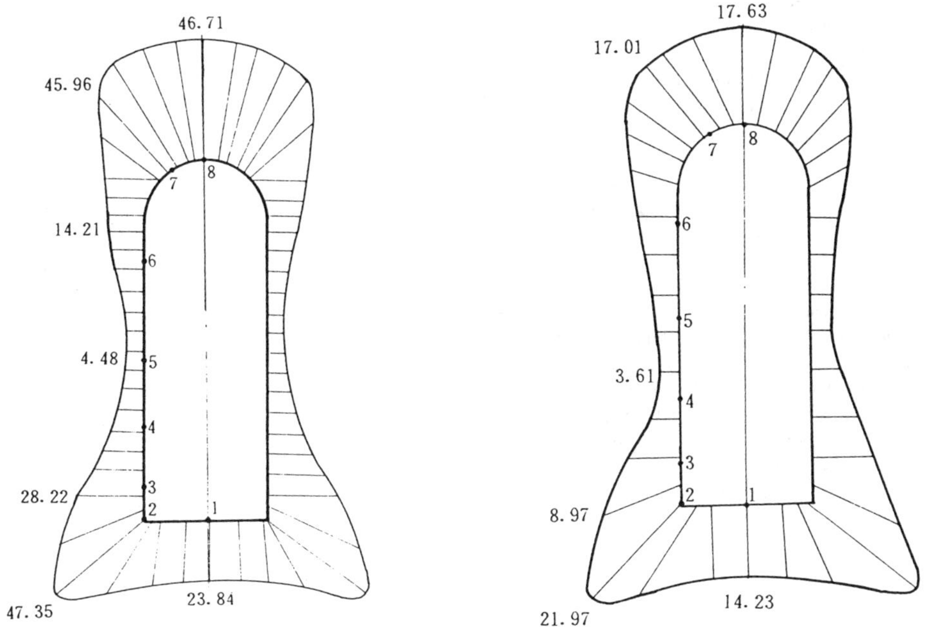

(a) Tangential stress σ_θ

(b) Longitudinal stress σ_l

Fig 2. Stress distribution law on the surrounding rock surface(0+15.68m)

rockburst stress criterion. Due to the synthetic criterion (9), we can obtain the judgment results: there would be occurring rockburst during opening cavern at LAXIWA underground engineering. The graduation of rockburst is slight and the position may be at the top center and bottom corner of cavern.

6 CONCLUSIONS

It is not suitable for rockburst criterion only considering surrounding rock strength criterion or only considering surrounding rock mechanical properties. The synthetic criterion of rockburst is rational with putting surrounding rock stress condition and surrounding rock mechanical properties together. It is necessary for considering the combination of the surrounding criterion.

According to surrounding rock stress analyzing during the opening of testing cavern and the lab testing results of rock properties, LAXIWA powerstation will occur slight rockburst by using the synthetic criterion.

REFERENCES

W.D.Ortlepp The design of support for the containment of rockburst damage in tunnels-An engineering approach *Int. symposium rock support in mining and underground construction,* 1992 pp 593-609 Rottdam:Balkema

Cook, N.G.W. et al. 1966 Rock mechanics applied to the study of rockburst. *J. of the S. Afr. Inst. of min. & Met.* 66:435-528

Blake, W. 1972. Rock burst mechanics. *Q. of the colorado school of mines* 67(1).

Crouch, S.L. 1974 Analysis of rock burst in cut-and-fill stopes. *Min.Engng.Aime 256(december):298-302*

N.Bartan, R. Liem, and J. Lunde, Engineering classification of rock masses for the design of, *Tapir Publishers, 1985*

Faling Hou, Yile Song, The complete stress-strain response on the rock and the analysis of rockburst tendency index, *CSME Mechanical Engineering Forum 1990, University of Toronto Campus*, 1990. pp177-180

tunnel support, 1974

Gu Zhaoqi Experiences in Norway Hydropower Engineering Hou Faling, Jia Yuru The relations between rockburst and surrounding rock stress in underground chambers, *Int. symposium on Engineering in complex rock formation science press* Beijing, China, 1986

Liu Xiaoming, Failure mechanism test study for brittle rock and LAXIWA hydropower station caverns rockburst analysis *Ph.D thesis Wuhan university of hydraulic and electrical engineering*, 1995

Computer Methods and Advances in Geomechanics, Yuan (ed.) © 1997 Balkema, Rotterdam, ISBN 90 5410 904 1

Dynamic characteristic and earthquake response of underground structures of a hydroelectric project

Wei Mincai, Kang Shilei & Wang Mingshan
Kunming University of Science and Technology, Yunnan, China

ABSTRACT: In this paper, the model with 8-node inconcordent isoparametric element of three dimensions is successfully applied to the dynamic analysis of underground structures, and the wave front method for solving static finite element is also used to the subspace itration method. The results show: The choice of calculation boundary conditions affect dynamic characteristic of underground cavern considerably, the dynamic responses of underground structure reach a peak in the direction of earthquake action, the deep underground cavern is safer in the case of earthquake since it's dynamic response is not very large.

1 MAIN CONDITIONS

1.1 The hydropower project is located on the Lancang River in the south-west of China. It is a underground powerhouse with six generators. The section of powerhouse has a width of 29.5m and a high of 65.5m. The main transformer chamber is 22m × 31.5m. The tailrace gate chamber is 19m × 68m. The stresses of computed stress field around the power house area were 6.71 ~ 22.5 Mpa before earthquake. The caverns are located in slight weathering gneissic granite, their depth is about 330 ~ 630m under the ground.

1.2 The computed model and rage ane as Figure 1 and 2. The upper boundary is the surface of the earth. It includes three zones of 4#、 5#、 6# genertor. Two faults is through th area of the cavens.
The mechanical characteristics of rocks is showed on the table 1.

Table 1. The Characteristics of Rock

Kind of rock	E(Mpa)	μ	ρ(kg/m^3)	C(Mpa)	φ(°)
gneissic granite	4 × 10^4	0.25	2500	3.5	55
fault zone	0.37 × 10^4	0.30	2200	0.7	5

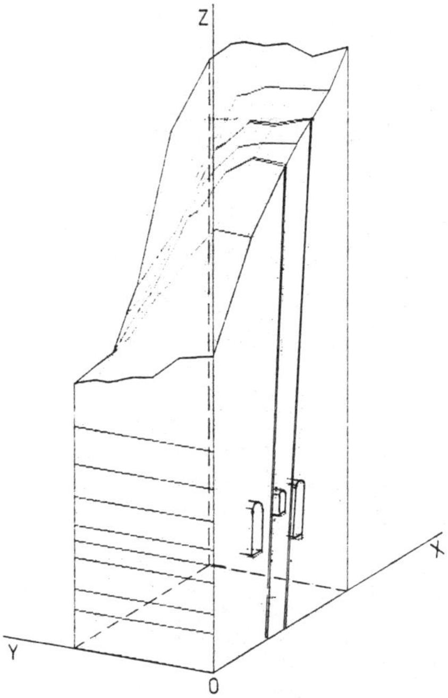

Fig.1 The model of calculation

1.3 The wave of the major aftershock of the Lancang-Gengma earthquakes of China in 1988 is adopted in the dynamic calculation. The wave form is as Fig.3.

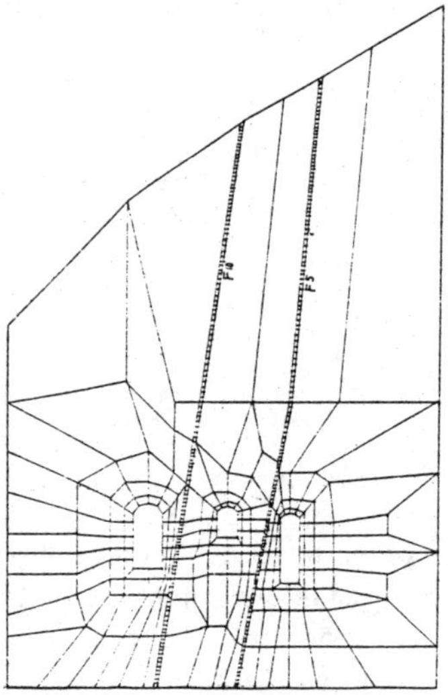

Fig.2 The Horizontal Section

On November 30, when the major affershock with the magnitute of 6.7 attached Lancang District again, a strong accelerogram with the peak values of 0.5g in the horizontal direction and o.39g in the vertical direction was obtained by the Zhutang mobile station located 3.8km far from the epicenter.

2 THE PRINCIPAL AND METHOD OF CALCULATION

2.1 The 8-node in conordand isoparametric elements are adopted in the program of dynamic calculation, and the wave front method for solving static finite element is applied to subspace itration method. First of all the natural frequency and vibration mode of undergrond stricture are calculated, and then the dynamic response under dynamic loads is also computed with the aid of vibration mode superposition method. The O.C.Ziekiewcz-G.n.pende hyperbolicyield criterion is used to judge the yield condition of rock element.

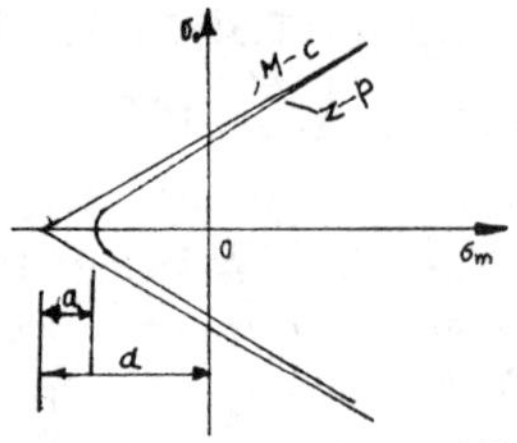

Fig.4 the yield curve in the meridian plane

2.2 Rayleigh Damping matrix

The dynamic equation of the discrete bodies is

$$[M]+\{\ddot{\delta}_s\}+[C]\{\dot{\delta}_s\}+[K]\{\delta_s\}=\{R_s\} \qquad (1)$$

where

$[M]$ -mass matrix

$[K]$-rigidity matrix

$\{\delta_s\}$-displacement matrix

$[C]$-damping matrix

$\{R_s\}$-loading matrix

The damping matrix of element is

$$[C]^e = a[m]^e + b[K]^e \qquad (2)$$

where

$$a=\frac{2\omega_i\omega_j(\omega_i\lambda_j-\omega_j\lambda_i)}{\omega_i^2-\omega_j^2}$$

$$b=\frac{2(\omega_i\lambda_j-\omega_j\lambda_i)}{\omega_i^2-\omega_j^2}$$

In the equation above, the ω_i and ω_j are the natural

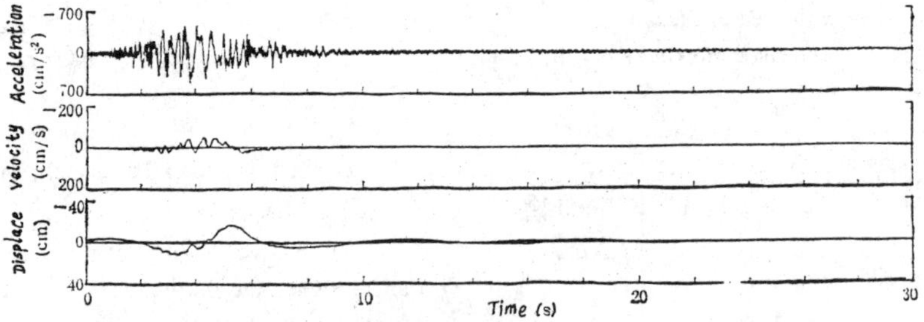

Fig.3 The wave form of the earth surface of Lancang earthquake aftershock (The magnitude of 6.7)

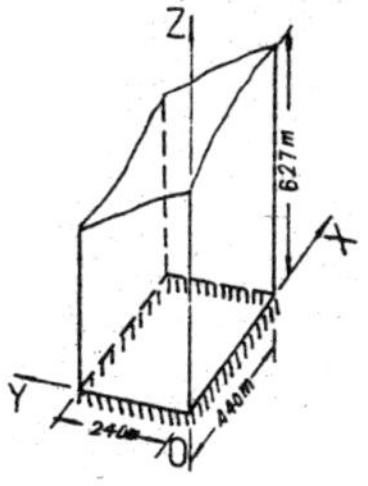

Model I

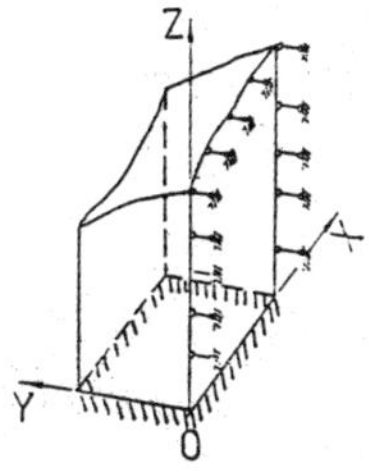

Model Ⅱ

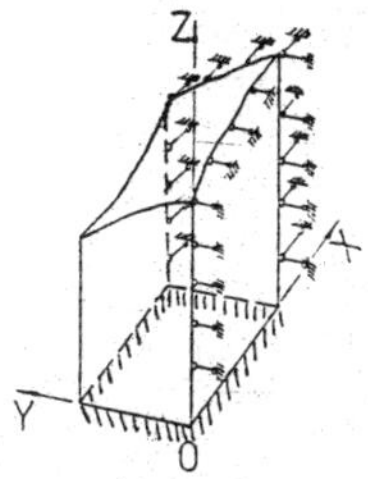

Model Ⅲ

Fig.5 three Models for calculation

circular frequency of two neighbor, the damping ratio λ_i, λ_j equal 0.02 ~ 0.24 in general. In the computation, λ equaled 0.05.

2.3 Yield criterion

Zienkiewcz-Pende hyperbolic yield criterion is as the figure 4. not only it abolishes noncontinvity in the corners of Mohr-Coulomb criterion, but also corresponds with the condition of the rocks. It's yield function is

$$F = \sqrt{-\xi}\left(\sigma_m + \frac{\eta}{2\xi}\right) + \sqrt{\sigma_o + \varsigma - \frac{\eta^2}{4\xi}} \qquad (3)$$

where

$\sigma_m = \left(\sigma_x + \sigma_y + \sigma_z\right)/3$

$\sigma_o = \sqrt{J_2}/g(\theta_\sigma)$

$J_2 = \left(S_x^2 + S_y^2 + S_z^2\right)/2 + \tau_{xy}^2 + \tau_{yz}^2 + \tau_{zx}^2$

$S_i = \sigma_i - \sigma_m \quad (i = x, y, z)$

$g(\theta_\sigma) = \dfrac{2K}{(1+K) - (1-K)\sin(3\theta_\sigma)}$

$K = \dfrac{3 - \sin\varphi}{3 + \sin\varphi}, \quad \beta = \dfrac{12}{(3 - \sin\varphi)^2}$

$\xi = -\beta\sin^2\varphi, \quad \eta = 2\beta C\sin\varphi\cos\varphi$

$\varsigma = \beta\left(\alpha^2\sin^2\varphi - C^2\cos^2\varphi\right)$

θ_σ—Lode angle

α—Shape coefficient

3 THE BOUNDARY CONDITIONS INFLUENCE ON THE DYNAMIC CHARACTERISTIC OF UNDERGROUND STRUCTURE

The models of three different boundary conditions are applied during computation, it is shown in Figure 5.

Model I: The foot of this madel is fixed only. The continuous rockmass is separated simply to a alone hill. It's natural frequency is smaller, and the period

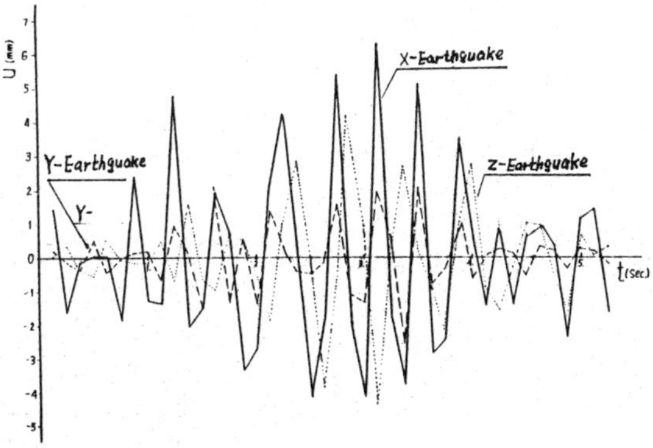

Figure 6. The U~t curve of a node on the side wall of the powerhouse cavern

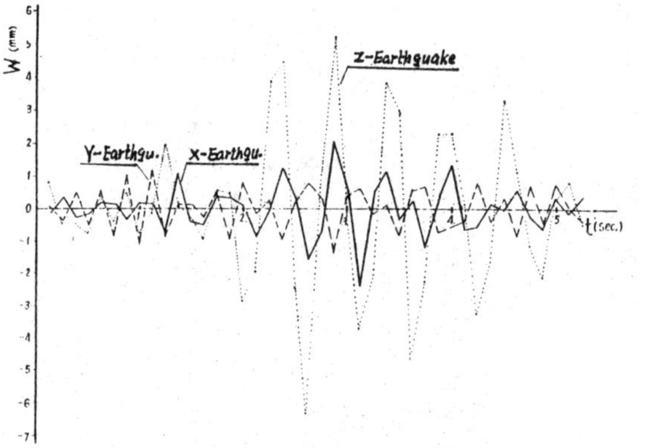

Figure 7. The W~t curve of a node on th crown of the powerhouse cavern

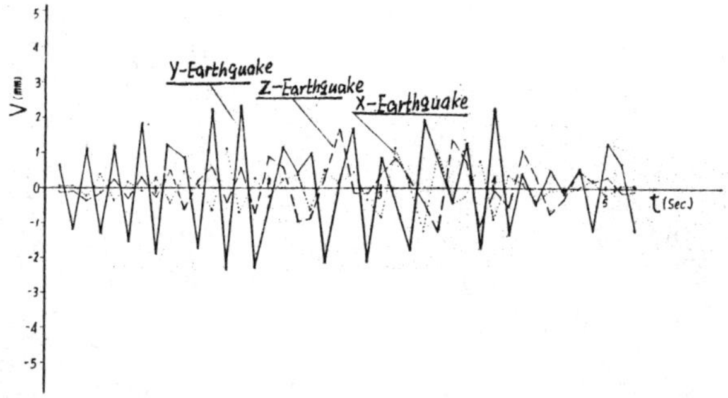

Figur 8. The V~t curve of a node on the head wall of the powerhouse cavern

of vibration is longer. It is not adopted because the all elements of rock are broken in the dynamic calculation and it is not correspond with topography of the project area.

Model Ⅱ: Its foot is fixed, and a side of it is bound by links in horizontal direction as a hill connected with the mountain , the period is reduced. f=0.69 ~4.80Hz, T=1.45 ~ 0.21sec. The rockmass is still broken serionsly in the eartnquake calculation.

Model Ⅲ: This model is bound by links in the horizontal direction of parallel to the cavern longitudinal axis on the basis of model Ⅱ. Calculation shows: f=1.83 ~ 5.42Hz, T=0.55 ~ 0.81sec. This model has only three free surfaces, it reflects objective condition that the east of topography of the powerhouse area is higher than the west. It is the calculation model of dynamic anylysis of the project.

4 THE EARTHQUAKE RESPONSE OF THE UNDERGROUND CAVERS

In the earthquake proof calculation on the model Ⅲ, the three different earthquake waves are imported that is as horizontal and perpendicular to the longitudinal axis of caver, or as horizontal and parallel to the Longitudinal axis of caver, or as vertical up or down.

4.1 The breaking status of the surrounding rock

In earthquake dynamic calculation the values of

Table2. The Maximum Displacements(mm)

earthquake	powerhouse		transformer chamber		gate chamber	
direction	wall	roof	wall	roof	wall	roof
x	6.59	2.50	5.14	2.87	3.28	2.42
y	2.16	1.36	1.66	2.21	1.19	2.20
z	4.56	6.31	3.64	7.85	2.36	7.50

Table 3. The seismic stress maximum (Mpa)

earthquake direction		powerhouse roof	powerhouse wall	powerhouse foot	transformer chamber roof	transformer chamber wall	transformer chamber foot	gate chamber roof	gate chamber wall	gate chamber foot
x	σ_t	1.0	0.5	0.8	1.1	1.0	0.9	1.3	1.0	0.8
	σ_c	0.4	0.1	0.2	0.5	0.1	0.4	0.5	0.0	0.2
y	σ_t	0.8	0.6	0.7	1.0	0.7	0.7	1.3	1.1	0.6
	σ_c	0.2	0.1	0.2	0.1	0.2	0.1	0.0	0.2	0.0
z	σ_t	1.9	3.2	1.1	2.2	3.4	1.2	1.4	2.8	1.1
	σ_c	0.7	0.3	0.7	0.9	0.7	0.9	0.8	0.3	0.7

principal stress have been 10~25Mpa, and all of they have been compressed stresses. All of rock elements have not been broke except for the fault elements. Therefore, the underground caverns have been generally safer in the case of earthquake.

4.2 The displacements round cavern

In earthquake the maximum displacements of node round cavern is shown in table2.

The displacement~time curve of a node round the cavern is as Figure 6,7,8.

The table 2 and Fig.6~8 show that the larger horizontal displacement on the side wall of cavern appears when the horizontal earthquake of vertical longitudinal axis of cavern is action, the larger horizontal displacement appears on the head wall of cavern when the horizontal earthquake paralleled with the longitudinal axis is in action, the down displacement of crown of cavern is largest when the up and down earthquake is in action.

4.3 The seismic stress of earthquake

The maximum of seismic stress of the elements round the cavern is shown in the table 3.

The table 3 show:

(1) When the earthquake force is in action, the tensile stresses appear round the cavern. The maximums are on the up left angle near the slope and on the down right angle of the mountain. The earthquake response is smaller.

(2) when the up-down earthquake is in action, the response of the underground cavern is the most intense. The wall of the cavern is extended in the vertical direction. The dynamic tensile stress is much big as 3.4 Mpa. If there were no the initial stresses of the rockmass before earthquake, the surrounding rock might damage because of pulling.

(3) In earthquake the dynamic compression stresses weren't big, they weren't more than 0.9 Mpa.

(4)The proportion of the seismic stress to the total stress (superposed the seismic stress with initial stress) is smaller, when the earthquake in the x direction it is 4.2~17.1%, is 11.5~19.2% in the y direction, is 4.9~11.5% in the z direction. Therefore the underground structure wouldn't be broken in earthquake becausethe initial stresses of rockmass before earthquake play a controlling part in the earthquake.

Computer Methods and Advances in Geomechanics, Yuan (ed.) © 1997 Balkema, Rotterdam, ISBN 90 5410 904 1

Study of large volume gas storage in rock salt caverns by FEM

M. Lu
SINTEF, Rock and Mineral Engineering, Trondheim, Norway
E. Broch
Norwegian University of Science and Technology, Trondheim, Norway

ABSTRACT: Finite Element analyses are performed to study the feasibility of large volume gas storage caverns in rock salt created by leaching. A number of operation phases have been studied including leaching, gas storage and an accidental gas blow-out. Interesting results are gained and numerical problems are revealed as well.

1 INTRODUCTION

A comprehensive research program to develop a technology of storing natural gas in large volume caverns in rock salt has been carried out since 1994. As a part of the program, a research project, including laboratory tests, field instrumentation and numerical analyses, has been performed by a team of researchers from a number of universities and research organisations in both Norway and Germany. The goal of the project is to establish a general numerical model which handles the mechanical responses of the caverns under various loading conditions properly. Special attention is paid to cavern convergence resulting from the creep property of the rock salt.

Laboratory triaxial creep tests simulating the cavern real loading situations are underway. Further work on in-situ rock stress state in rock salt deposit and adjacent rock formations has been planned. Special instrumentation techniques, which are needed in monitoring cavern convergence, are under consideration. Numerical simulations are performed with both axisymmetric and three dimensional models. The axisymmetric simulations are for parametric study, whilst the 3-D models are used in simulating the real engineering projects, which are to be located at Bernburg, Germany. In addition to the behaviour of a single storage cavern, a cavern system consisting a group of caverns will also be studied. But this paper will concentrate on single caverns.

ABAQUS, a general purpose commercial FE code, is selected for the simulations, considering its general applicability in non-linear computations and flexibility in user subroutines.

2 NUMERICAL MODELS

2.1 Geometric models

The cavern geometry is simplified as a 170 m high cylinder with elliptic horizontal cross sections as shown in Figure 1. The major/minor axes at the cavern center is 200/100 m, which reduce to 80/40 and 130/65 m at roof and floor, respectively, making a total volume about 2.3 million m^3. As shown in Figure 2. the FE model includes only a quarter of the cavern, considering the geometric symmetry. The model consists of 16276 elements and 18709 nodes.

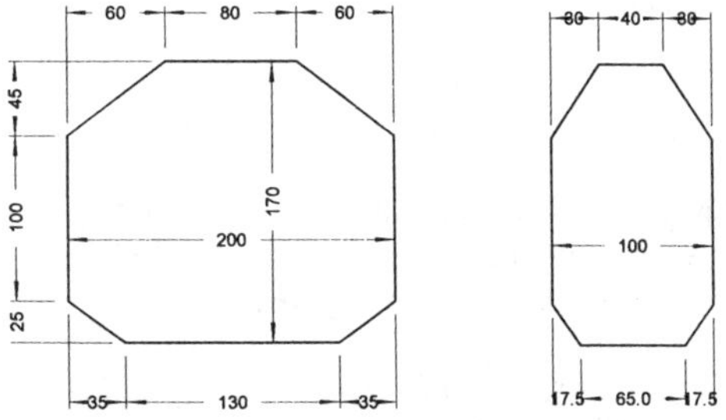

Figure 1. Cavern dimensions.

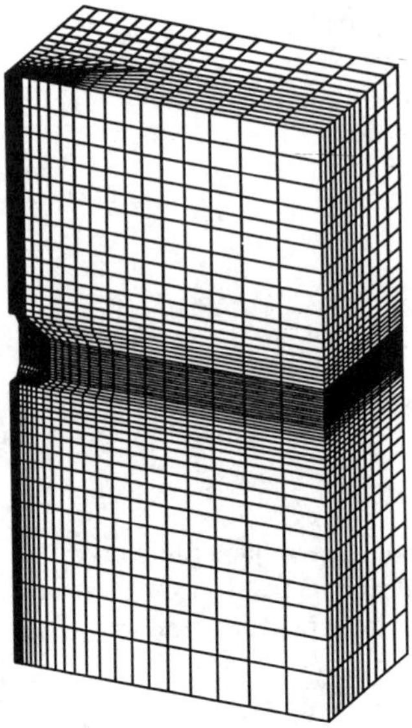

Figure 2. 3-D Finite Element mesh.

2.2 Material model and in-situ rock stress

The rock mass in the cavern site consists of a 490 m thick rock salt deposit, under and above which there are sandstone and other rock formations. The cavern is situated entirely within the rock salt leaving only 40 m margin to the adjacent sandstone underneath. The rock stress is measured in nearby salt mines and a hydrostatic in-situ stress state is suggested for the rock salt formation. The elastic constants and in-situ stresses of all rock masses are given in Table 1.

Table 1. Elastic constants and in-situ stresses.

Stratum	Height [m]	E [GPa]	ν	σ_v	σ_h/σ_v
Anhydride	90	35	0.30	σ_g	0.7
Rock salt	510	25	0.25	σ_g	1.0
Sandstone	190	35	0.30	σ_g	0.5
Basal anhydride	640	35	0.30	σ_g	0.7

* σ_g is gravitational stress.

A hyperbolic sine law with strain hardening and a power law with time hardening are proposed for the primary and secondary creep, respectively:

$$\dot{\varepsilon}_{cr} = ASinh\frac{\sigma}{\sigma_0}\exp\left(\frac{-\varepsilon_{cr}}{\varepsilon_0}\right)\ [1/h] \tag{1}$$

$$\dot{\varepsilon}_{cr} = M\sigma^N\ [1/h] \tag{2}$$

where ε_{cr} and $\dot{\varepsilon}_{cr}$ are creep strain and creep strain rate. Constants A=9.0E-7, σ_0 =2.5MPa, ε_0=3.0E-3, M=4.0E-13 and N=4.8 for the rock salt at the Bernburg site. An empirical shear strength criterion, as shown in Figure 3, is also proposed based on laboratory triaxial tests:

$$\sigma_1 = \sigma_D\left(1+\frac{k\sigma_3}{\sigma_Z}\right)^{\frac{1}{k}} \tag{3}$$

where σ_D=27MPa is the uniaxial compressive strength, constants k=3.1 and σ_z=2.3MPa.

2.3 Computation scheme

Computations initiate with the explicit integration in time domain, then shift to implicit integration when the code feels necessary, i.e. the time increments are too small in comparison to the step time period, restricted by the stability requirements. The implicit integration is unconditionally stable, so large time increments can be used. The computation accuracy is controlled by the maximum difference in the creep strain increment in two consecutive time increments. In most cases of the study, the automatic time incrementing provided by ABAQUS is also adopted, which ensures an economic computing. The geometric non-linearity is ignored.

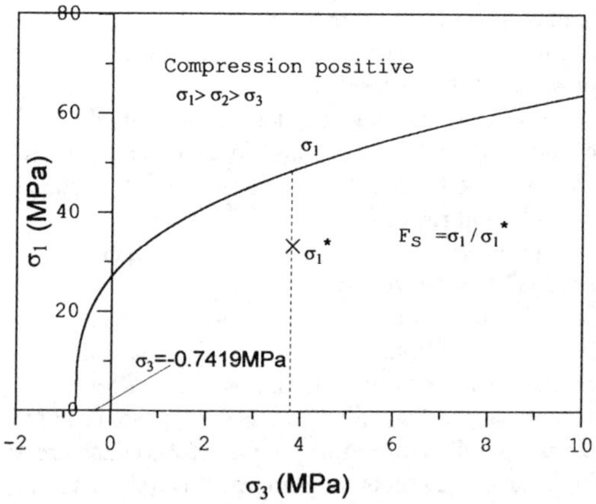

Figure 3. Empirical shear strength criterion and definition of factor of safety against shearing.

3 CONTENTS OF 3-D SIMULATIONS

The main engineering concerns in mechanical point of view are cavern stability, both in normal operating conditions and during an accidental gas blow-out, as well as the long term convergence. In accordance, following simulations are carried out:

1. Initial stress field generation.

2. Leaching in five stages, in which the leached elements are removed and hydrostatic pressure is applied on the cavern surface. The secondary creep law is adopted.

3. First three years of gas storage. In these computations the hydrostatic pressure is replaced by the gas pressure, which follows the real annual variation cycle recorded at the Bernburg site, as shown in Figure 4. The maximum and minimum gas pressure is 9.5 and 2.5 MPa, respectively. The primary creep law is adopted considering the varying loading.

4. Next 27 years of gas storage. Considering the long term effect, the annually mean gas pressure of 6.43 MPa is adopted in these computations instead of using the varying annual pressure cycle. The secondary creep law is used again.

5. An accidental gas blow-out, during which the gas pressure drops linearly to zero within 12 days and the pressure-free state remains for three months for repairing work. The primary and the secondary creep laws are adopted for the first 12 days and the following 3 months, respectively.

4 3-D COMPUTATION RESULTS

Displacements, Mises stress, creep strain and factor of safety against shear yielding at representative locations including, cavern floor center, cavern roof center, cavern mid-wall at major and minor axes, are shown in Figures 5-8.

The major part of deformation occurs in the first 3 years gas storage, whilst during the leaching and the last 27 years operation only a small deformation takes place. The maximum displacement of about 2 m is predicted in the mid-wall at the minor axis of the cavern boundary. The total volume convergence in 30 years is estimated as 5 %, which is less than expected. It can also be seen that the deformation develops smoothly, meaning that there is no much influence from the fluctuations of the gas pressure applied in the first 3 years.

Figure 6 shows that the stresses are disturbed by the gas pressure fluctuations. However, a uniformly distributed Mises stress about 6 MPa is gradually approached in long term on the cavern boundary. The maximum effective creep strain predicted is about 3 %, as shown in Figure 7. However, on the

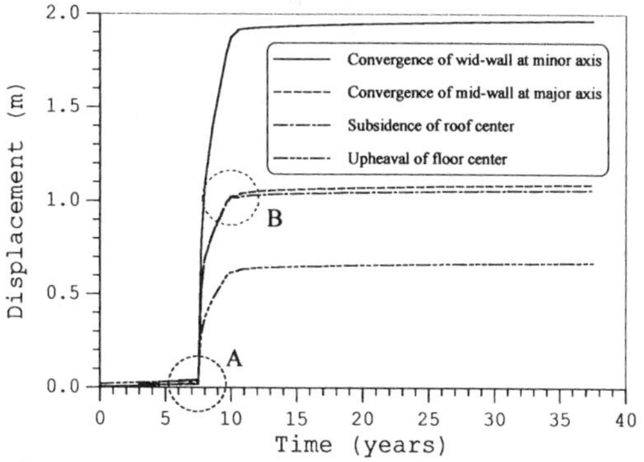

Figure 5. Representative displacements.

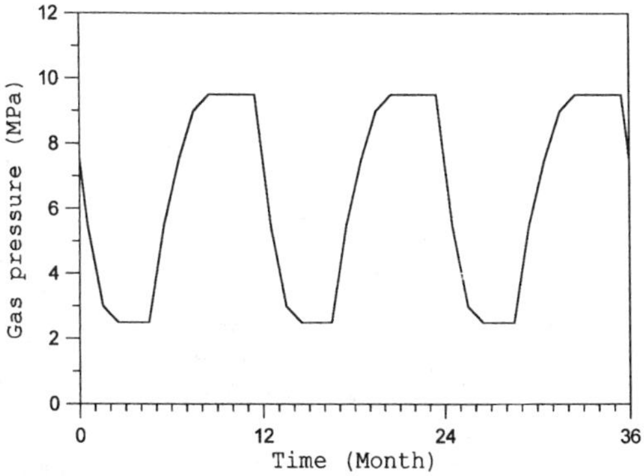

Figure 4. Annual variation of gas pressure.

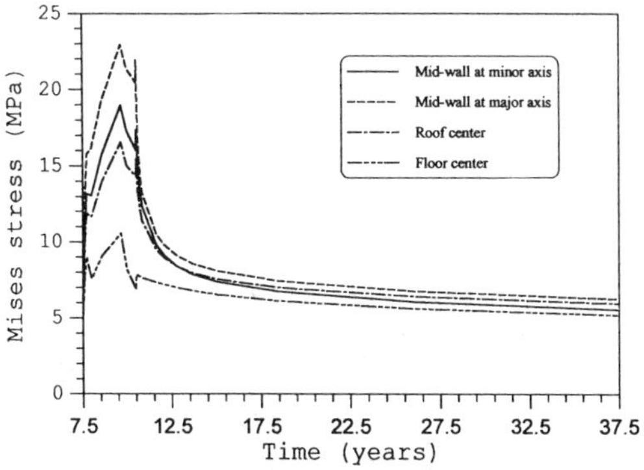

Figure 6. Representative Mises stresses.

contrary to the displacement, it takes place on the major axis of the cavern boundary.

Figure 8 shows the factor of safety against shear yielding at some locations around the cavern boundary. The factor of safety is defined in Figure 3. The figure indicates that yielding does not occur even when the gas pressure drops to zero and the undesirable stress state recovers with time due to the creep nature of the rock salt.

Detailed descriptions of the 3-D simulations are given in Lu 1997 c.

5 IS THE ANALYSIS JUSTIFIED?

The computation result presented above gives a pleasant picture: General stability is guaranteed and convergence is not significant. However, a closer study of the result leads to suspicions. In Figure 5 sharp variations in displacements appear at the beginning and end of the first three years gas storage, as marked with dashed circles. The same is for the creep strains shown in Figure 7. It is believed that such sharp turnings in displacements and strains could not happen in reality and they must have been caused by numerical bias. At the beginning of the gas storage, the leaching water pressure is replaced by the gas pressure. The difference in value of the pressures is not so great. But, the creep law is changed from secondary to primary creep. At the end of the third year gas storage, the varying gas pressure is replaced by an annually average gas pressure and the creep law is changed back from primary to secondary creep.

Numerical analyses with single element cube models and axisymmetric models are then carried out to study three potential factors: Applicability of the creep laws; Using an equivalent constant gas pressure as well as ignoring the geometric non-linearity (Lu 1997 a, b).

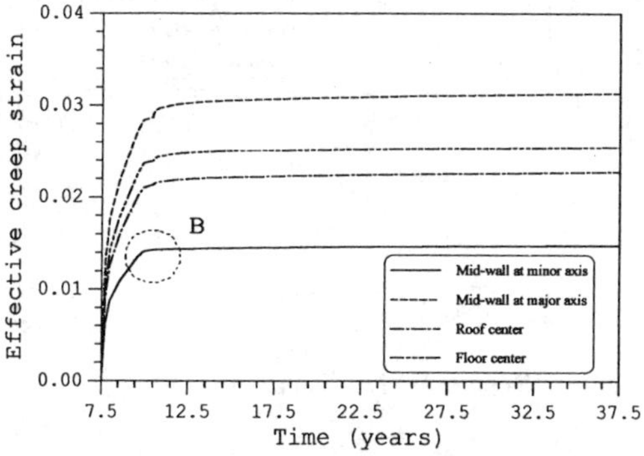

Figure 7. Representative effective creep strains.

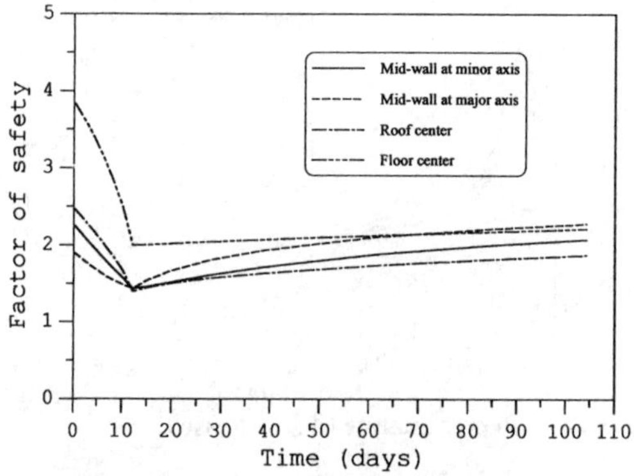

Figure 8. Factor of safety during gas blow-out.

6 CREEP LAWS

It has been mentioned previously that two creep laws are used: One for the primary creep and the other for the secondary creep. The primary creep law has a varying creep rate under a constant stress, whilst that for the secondary creep is constant. Therefore, a logic way of combining the two laws is: (1) Initiate with the primary creep law when the stress state starts shifts from the hydrostatic stress state; (2) The primary creep is applied until the creep rate equals that of the secondary creep under the same stress and (3) Then the secondary creep law takes over, as shown in Figure 9 with the solid line.

It is also interesting to note from Figure 9 that if an unsuitable creep law is adopted the creep strain will always be underestimated, no matter the secondary creep law is used for the primary creep or vice versa. Considering the stress is dependent on

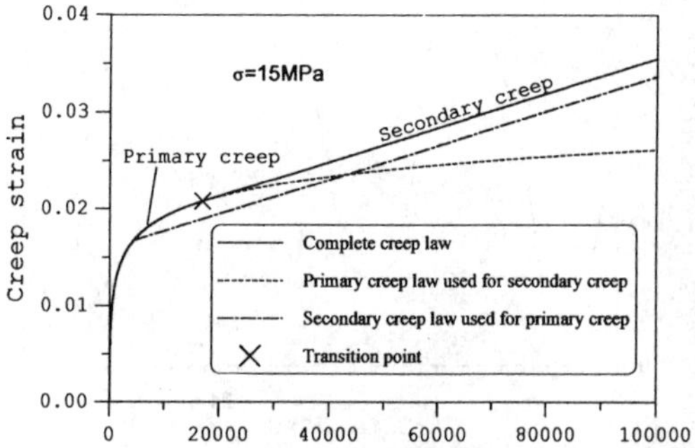

Figure 9. Complete creep law and underestimation of creep strain.

both time and locations, the applicability of the creep law is also a spatio-temporal function. A numerical technique has been developed to detect the applicable creep law automatically based on current creep strain rates. Figures 10 (a) and (b) demonstrates the distributions of the applied creep law at the end of the leaching process and after 30 years gas storage. In the figures the dark and bright areas refer to the regions in which the secondary and primary creep law is active, respectively. It is clear that the primary creep is dominating even after 30 years, with only a limited area around the cavern being switched to secondary creep.

Since the secondary creep is used in the complete leaching period for the entire mode in the 3-D simulations, there is no doubt that creep in this period is underestimated significantly. Figure 11 demonstrates a comparison of the upheaval of the cavern floor center during the entire 37.5 years including leaching and gas storage. The dashed line is from the computations with specified creep laws defined exactly the same as that used in the 3-D simulations, while the solid line is from the computations with the auto-detected creep laws.

(a) (b)

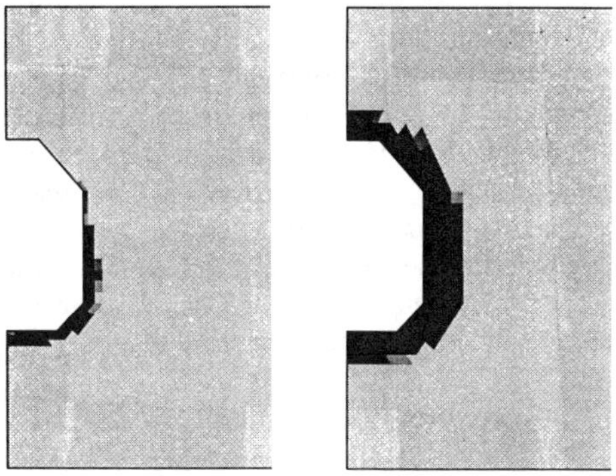

Figure 10. Distributions of applied creep laws: (a) At the leaching end. (b) After 30 years gas storage.

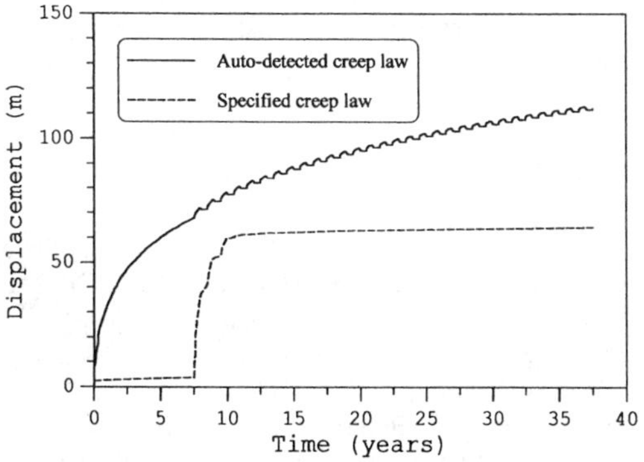

Figure 11. Comparison of computations with specified and auto-detected creep laws.

7 GEOMETRIC NON-LINEARITY

The maximum displacement and creep strain obtained from 3-D simulations are about 2 m and 0.03, respectively, without taking into account the geometric non-linearity resulting from large deformation. In order to investigate the effect of ignoring the geometric non-linearity parallel computations are performed with the axisymmetric models, of which one considers the large deformation, the other not, with all other conditions being exactly the same. Figure 12 shows the comparison of the upwards displacement of the cavern floor center. The difference is marginal. More detailed study indicates the maximum difference is 1.5 and 3.0 % in displacement and volume convergence, respectively. This seems within the limit of acceptance from engineering point of view.

8 FLUCTUATION OF GAS PRESSURE

Investigation has also been made on whether adopting the real varying gas pressure or using an equivalent constant pressure, e.g. the annually average one. Figures 13 and 14 show the development of displacement and Mises stress at the cavern floor center. It can be seen that the stress reflects essentially the pressure variation while the

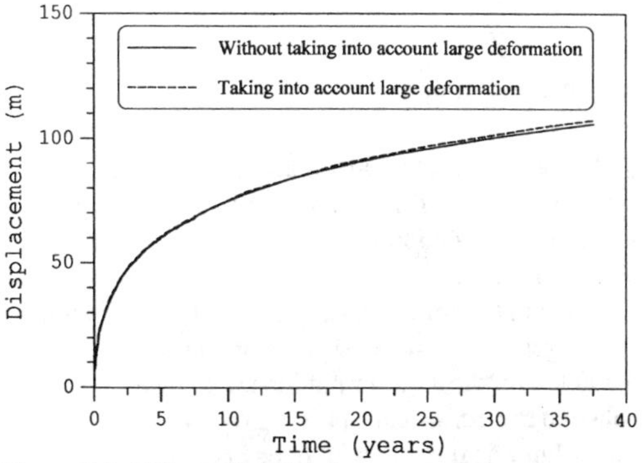

Figure 12. Effect of taking into account large geometric non-linearity.

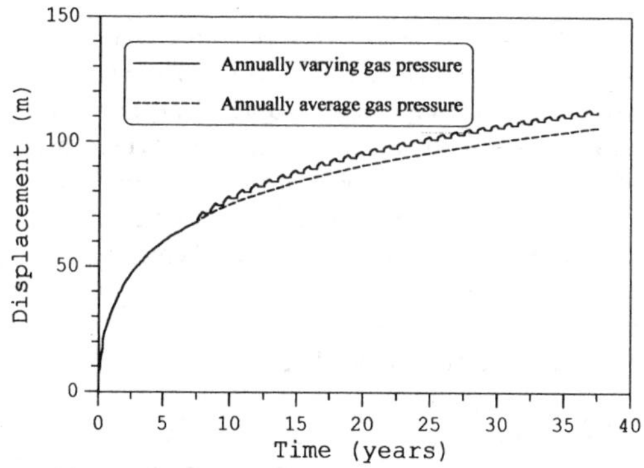

Figure 13. Comparison of constant and varying gas pressure in displacement.

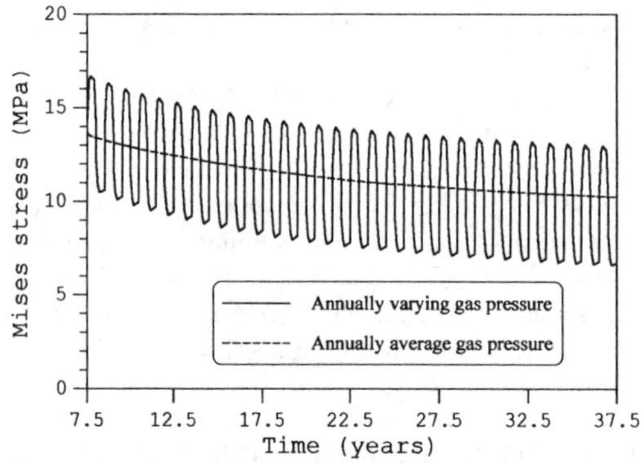

Figure 14. Comparison of constant and varying gas pressure in Mises stress.

difference in deformation is negligible. If the rock strength is high and the main concern is the cavern convergence, it is well acceptable to adopt a constant gas pressure. However, a way has to be worked out to estimate properly the equivalent pressure, so that the deformation is computed correctly.

9. CONCLUSIONS

1. The 3-D numerical simulations indicate a general stability of the storage cavern, even in the severe loading conditions during an accidental gas blow-out.

2. A maximum displacement of 2 m and a volume convergence of 5 % in 30 years operation is predicted by the 3-D simulations. However, it is believed the deformation is underestimated.

3. Numerical study with cube and axisymmetric models confirms that the suitable creep law is a function of both time and space. It is essential to adopt the correct creep law in computations. The creep strains will always be underestimated if a 'wrong' creep law is used.

4. A creep law auto-detection technique is developed, which ensures a correct creep law to be used.

5. For such gas storage projects, it is found that creep initiates following the primary creep law, then switches to the secondary creep some time later. The secondary creep region starts from around the cavern and develops with time. However, even after a long period of time, say 30 years, secondary creep takes place only within a limited region around the cavern, whilst the primary creep is still dominating outside the region.

6. It is not necessary to take into account the geometric non-linearity from engineering point of view, considering the limited discrepancy of less than 3 % in stress and deformation.

7. The fluctuations of gas pressure affect deformation only marginally. However, the stress development reflects the pressure variations. Therefore, if the rock strength is high and the main engineering concern is the cavern convergence, it is well acceptable to adopt an constant gas pressure. However, a way has to be worked out to estimate the equivalent pressure properly.

8. It is extremely essential to develop measuring techniques to monitor the cavern deformation, so that the numerical models can be validated.

ACKNOWLEDGEMENTS

The authors of this paper would like to thank the sponsors of this project, Statoil in Norway and VNG in Germany, for their generous support. Thanks are extended to the project partners in TU Bergakademie, Freiberg and in IfG, Leipzig, Germany, for providing input data and useful discussions.

REFERENCES

Lu, M. 1997 a. Numerical tests of creep simulations with cube models. SINTEF report STF22 F97021

Lu, M. 1997 b. Numerical simulations of gas storage in rock salt with axisymmetric models. SINTEF report STF22 F97022

Lu, M. 1997 c. Three dimensional Finite Element analysis of gas storage in rock salt: Single cavern. SINTEF report (to be published)

Computer Methods and Advances in Geomechanics, Yuan (ed.) © 1997 Balkema, Rotterdam, ISBN 90 5410 904 1

An analytical and experimental investigation on stability of underground openings

S. Akutagawa & S. Sakurai
Department of Civil Engineering, Kobe University, Japan
K. Ogawa
Kyowa Design Co., Ltd, Osaka, Japan

ABSTRACT : The paper discusses about the results of a series of model experiments in which tunnel excavation is simulated by reducing pressure of an air bag installed in an virtual ground made by alminium bars. The results of these experments are compared with prediction made by a rigid-plastic finite element analysis. The findings from these comparative and prametric studies show clearly relative differences in the level of stability for underground openings of different sizes constructed at different depths.

1 INTRODUCTION

An effective use of underground space is one of the important strategies to be taken in areas such as, for example, Japan where it is uneasy to allocate available spaces to civil structures of considerable dimension on ground surface not only from shortage of space but also from economical reasons in heavily developed urban areas. The effective usage of underground spaces requires optimally designed underground openings of certain dimension to be constructed with minimum cost while not jeopardizing stability both during construction and after completion. This problem has been addressed by many researchers both from analytical and experimental viewpoints (for example, Murayama and Matsuoka, 1971, Adachi et al, 1985). However, relatively little efforts have been made to investigate explicitly stability of underground openings at their limit states. Of several important technical aspects to be studied in this regard, this paper deals with two key parameters shown in **Fig.1** for assessment of stability, or factor of safety, of an unsupported openings with varying dimension of opening and overburden height.

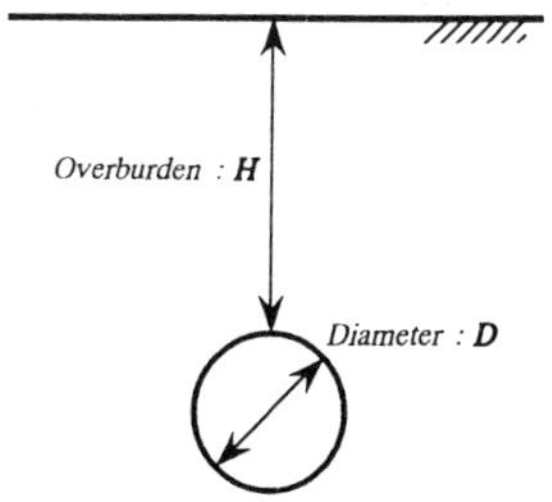

Fig.1 Two key parameters

Firstly, the paper discusses about the results of a series of model experiments in which tunnel excavation is simulated by reducing pressure of an air bag installed in an virtual ground made by aluminium bars (Sakurai et al, 1994). Secondly, a modification made to the original formulation of the rigid-plastic finite element method, namely *RPFEM*, (Tamura, 1990 and Tamura et al, 1984) is briefly presented in which a loading system is divided into two groups to enable simulation of the model tests. Thirdly, the results of these experiments are compared with prediction of numerical simulation from three different viewpoints. The findings from these comparative and parametric studies show relative differences in the level of stability for underground openings of different sizes constructed at different depths.

2 EXPERIMENTAL SIMULATION

2.1 *Apparatus and experimental procedure*

Fig.2 shows an experimental apparatus with which a tunnel excavation is simulated by reducing air pressure of a vinyl bag installed within aluminium bars. Aluminium bars are 5 cm long and consist of two kinds with diameters of 1.6mm and 3mm, which are mixed with 60%-40% weight ratio. A model specimen is set up initially by equating the air bag pressure to the overburden pressure calculated at tunnel crown. The pressure is then lowered gradually by manually controlling a pressure regulator, while the applied pressure is monitored by a manometer. When the air pressure is greater than a threshold value, the simulated ground around tunnel reaches an equilibrium state. When it reaches the threshold value, which we call *collapse pressure*, deformation of sudden nature occurs, in the order of few millimeters around tunnel surface, which soon develops continuously and endlessly before terminating experiments. With certain intervals, photographs are taken of specimen surface, which are later processed by digitizer to obtain displacement and strain fields. This configuration can be adjusted to model tunnels of different diameters for varying overburden heights.

2.2 *Results*

Fig.3 shows distribution of maximum shear strain for those cases where overburden *H* was varied from 0.5*D* to 2*D*, while *D* (tunnel diameter) was set to 10cm. These strain values were calculated by differentiating displacement fields obtained at or immediately after collapse states were reached during experiments. In most cases, the maximum displacements in those states were around 10mm.

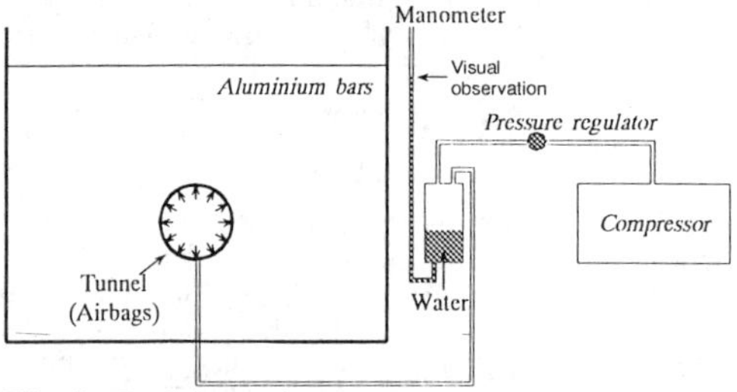

Fig.2 Experimental apparatus

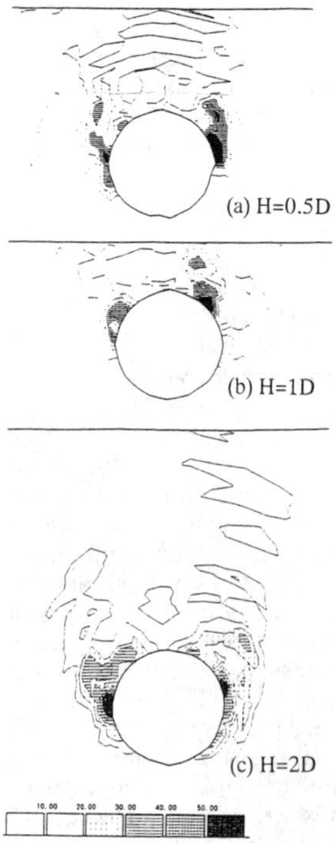

Fig.3 Maximum shear strain distribution

In all three cases, zones of high shear strain develop from tunnel wall, rising up, especially for cases with small overburden heights, to the ground surface. A closer look at these results may suggest that the zones of high shear strain develop from around tunnel SL (spring line) for a deep tunnel (*H=2D*), whereas it starts from locations closer to tunnel crown for a shallow tunnel (*H=0.5D*). Displacement vectors obtained for *H=2D* case is also shown in **Fig.4**. These results will be compared with analytical results in the forthcoming sections.

3 RIGID PLASTIC FINITE ELEMENT FORMULATION

In this section, a mathematical definition of the rigid-plastic finite element simulation is briefly introduced. The procedure employed in this study is a modified version of the procedure given by Tamura (1990), in which a loading system is divided into two groups as shown in **Fig.5**. The first group, *G*, representing the gravity forces is kept constant throughout an iterative analysis procedure. The second group, *F*, representing air pressure is varied with a load factor μ, seeking for a collapse pressure at a limit state. The following gives a brief description of the mathematical procedures based on this concept of two loading systems.

A system of governing equations is defined by the following three equations.

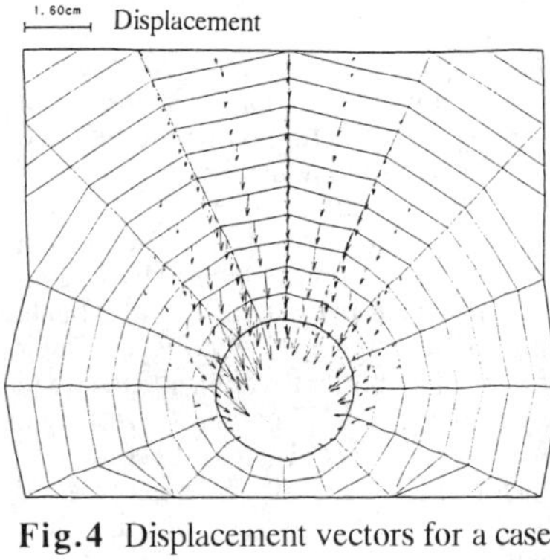

Fig.4 Displacement vectors for a case of H=2D

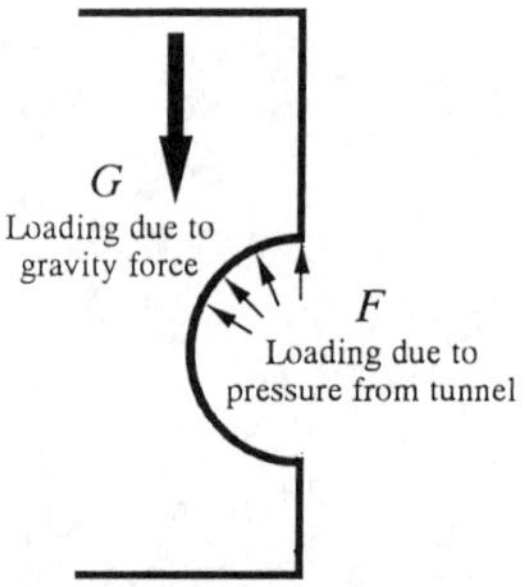

Fig.5 Basic concepts : Two loading systems

$$M\dot{u}_l^e + N\bar{\lambda}_l = \mu F_l + G_l \tag{1}$$

$$\int_{V_l} (L\dot{u}_l^e - \frac{3\sqrt{2}\alpha}{\sqrt{1+6\bar{\alpha}^2}}\bar{\dot{e}}_l)dV_l = 0 \tag{2}$$

$$F^T\dot{u}^e = 1 \tag{3}$$

Note that M and N are given as follows;

$$M = \int_{V_l} \frac{\sqrt{2(1+6\bar{\alpha}^2)}k}{\sqrt{1+6\alpha\bar{\alpha}}} \frac{B_l^T Q B_l}{\bar{\dot{e}}_l} dV_l \tag{4}$$

$$N = \int_{V_l} (L^T - \frac{3\sqrt{2(1+6\bar{\alpha}^2)}\alpha}{1+6\alpha\bar{\alpha}} \frac{B_l^T Q B_l}{\bar{\dot{e}}_l}\dot{u}_l^e)dV_l \tag{5}$$

Notations used for above equations are given as below. Note that a subscript l denotes association with element l.

- α : material constant associated with yield function
- $\bar{\alpha}$: material constant associated with potential function
- k : material constant associated with yield function
- B_l : displacement velocity - strain matrix
- Q : operator to translate engineering strain to tensor
- $\bar{\dot{e}}_l$: effective strain rate
- V_l : element volume
- $\dot{u}_l^e$: nodal displacement velocity
- L : volume strain rate - strain rate matrix
- $\bar{\lambda}_l$: hydrostatic stress
- μ : load factor
- F_l : nodal forces acting on excavation surface
- G_l : body forces generated by gravity

In plane strain problems, α and k can be described in terms of c and ϕ by comparing yield functions of Drucker-Prager and Mohr-Coulomb criteria (Drucker and Prager, 1952) as follows;

$$\alpha = \sqrt{\frac{(3+2r\sin^2\phi) - \sqrt{9+12r(1-r)\sin^2\phi}}{6r^2(3+\sin^2\phi)}} \quad (for\ r \neq 0) \tag{6}$$

$$\alpha = \frac{\sin\phi}{3} \quad (for\ r = 0) \tag{7}$$

$$k = \frac{3\alpha c}{\tan\phi} \tag{8}$$

where $r = \bar{\alpha}/\alpha$ ranges between 0 and 1. When r is 1, a flow rule is an associated one and dilatancy is present. If r is 0, a flow rule is non-associated and dilatancy is non-existent.

For a finite element implementation, the equations (1), (2), and (3) are expanded around $\dot{u}$ and solved simultaneously to obtain $d\dot{u}$ to define a new state of displacement velocity field. Procedure is iterative and convergence is achieved when $d\dot{u}$ becomes sufficiently small.

4 COMPARISON OF DEFORMATIONAL CHARACTERISTICS

First, some parametric studies were conducted to observe relationships between deformational characteristics predicted by the RPFEM simulation and those obtained from the experiments for the case of a tunnel diameter 10cm. The input parameters used for the analyses are given in **Table 1**.

Table 1 Input parameters used for RPFEM simulation

Parameter		*Value*	*Unit*
Unit weight	γ	2.18	gf/cm^3
Cohesion	c	2.6	gf/cm^2
Friction angle	ϕ	29	degree
Dilatancy parameter	r	1, 0.5, 0	

For each case of varying overburden heights, namely 0.5*D*, 1*D*, 2*D*, analysis was conducted for three cases having different dilatancy parameters of 1, 0.5, and 0. For most cases, zones of high shear strain developed from tunnel face up to the ground surface, although their shapes or extent of higher displacements varies slightly from case to case. Of these results, the maximum shear strain distribution and displacement vectors are shown in **Fig.6** and **Fig.7**, respectively, for the case of dilatancy parameter of 0.5, which simulated the most favourably the results of the experiments. These results suggest that the deformational mechanisms obtained from RPFEM simulation resemble satisfactorily those obtained from the experiments, assuring the significance of comparison of collapse pressure and state of stresses to be discussed in the next step.

5 COMPARISON OF COLLAPSE PRESSURE AND FAILURE MECHANISMS

5.1 *Preliminary*

Having confirmed that the RPFEM simulations give realistic deformational behaviour at collapse, it is now important to examine if collapse pressure predicted by analysis agrees with those obtained from the experiments, in order to make complete an understanding of collapse mechanism of underground openings. In addition, there exist

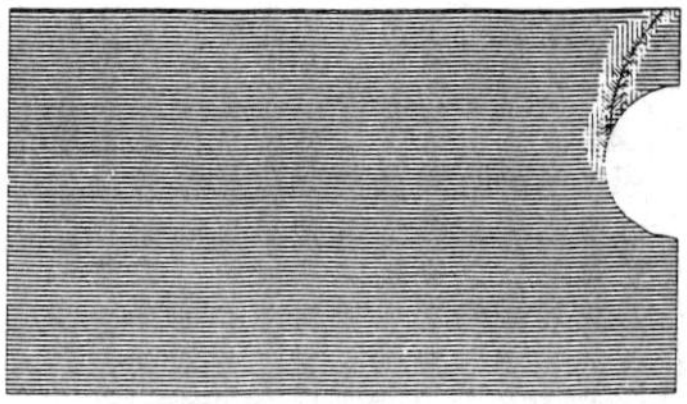

(a) H=0.5D

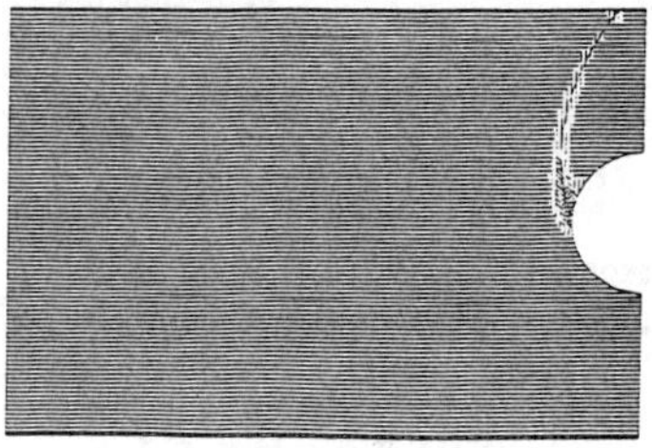

(b) H=1D

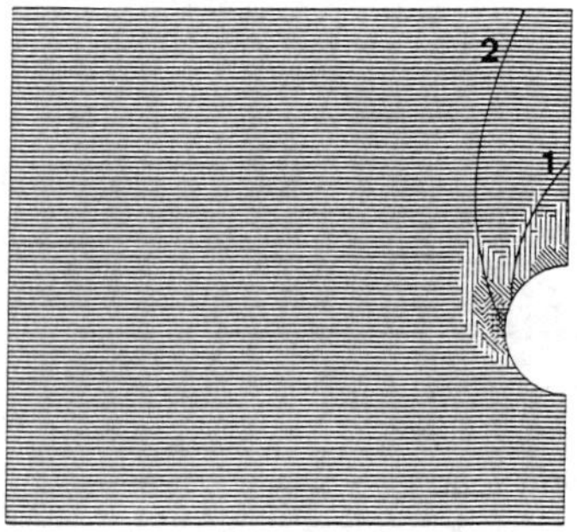

(c) H=2D

Fig.6 Results of RPFEM simulation for r=0.5

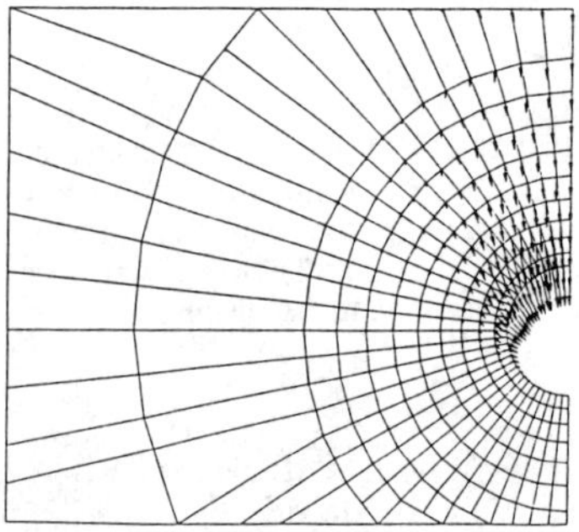

Fig.7 Displacement vectors for r=0.5 (for case of H=2D only)

fundamental, yet important problems to be addressed with regard to safe construction of underground openings from the following three viewpoints.

Viewpoint 1 : How does safety of an opening of a given size vary with varying overburden height ?

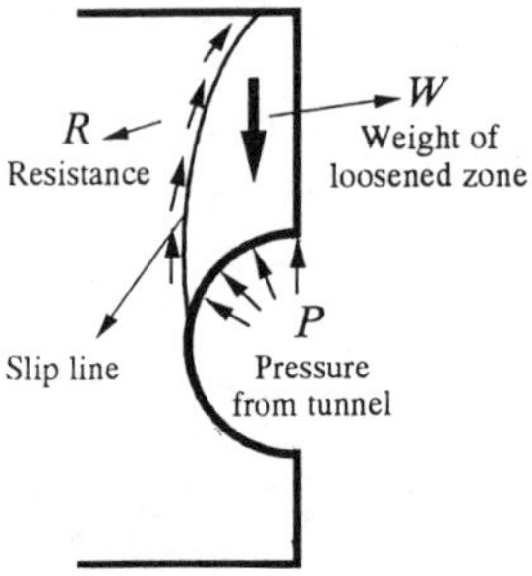

Fig.8 Forces in equilibrium at state of collapse

Viewpoint 2 : How does safety of an opening vary if D is varied while keeping H/D constant ?

Viewpoint 3 : How does safety of an opening vary if D is varied while H is kept constant ?

In presenting results of analyses for discussion with these viewpoints in mind, the three forces shown in **Fig.8** are defined according to the findings made this far as to the failure modes of an underground opening. W represents weight of a *unstable zone* above tunnel which is divided from the surrounding by a slip line, as indicated in **Fig.6** for the case of D=10cm, running in the middle of highly sheared zone. W is then counteracted by resisting force R acting along the slip line and supporting force P given by air pressure. Naturally, these force vectors are balanced so that a relation

$$W+R+P=0 \tag{9}$$

holds. Note that R is computed from results of analysis by integrating both shear and normal stresses acting on slip surface. In the following discussion on force balance, only the vertical components of force vectors are cosidered.

5.2 *Viewpoint 1*

Firstly, stability of a tunnel of a given diameter is investigated for varying overburden. D is set to

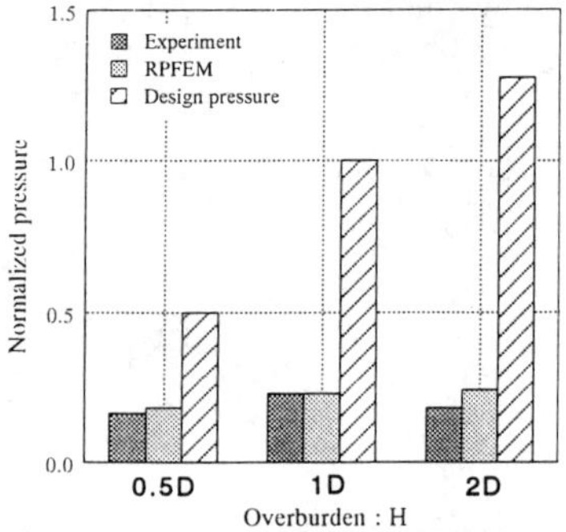

Fig.9 Collapse pressures for D=10cm

10cm, while H is taken to be $0.5D$, $1D$, and $2D$. The results are plotted in **Fig.9** in which collapse pressures are normalized by overburden stress γH in which γ is unit weight. Also plotted in the figure is values of tunnel pressures computed from Terzaghi's formula commonly used in design practices.

The collapse pressures obtained from the RPFEM analyses show good agreement with those of the experiments, indicating reliability of the analytical procedure not only for simulation of deformational behaviour, but also for simulating force equilibrium at limit state. The collapse pressures of both analysis and experiment remain more or less constant regardless of overburden height, whereas the tunnel pressure computed from Terzaghi's formula, which are in general much greater than collapse pressures, increases as a tunnel gets deeper. Based on the slip lines indicated in **Fig.6** for the three cases, W, R, and P are calculated, for unit thickness of model specimen, and plotted in **Fig.10**. Note that for the case of $H=2D$, two slip lines were identified and plotted in the figures.

The shape of slip lines is case-dependent. Therefore, W varies for each value of overburden. However for all cases, the force required to balance the weight of unstable zone can be very small and of the same order because of the resisting force supplied from the surrounding ground material. Namely, when a tunnel is located at different depth, deformational mechanism and shape of unstable zone will be different accordingly, though the minimum pressure required to sustain tunnel stability remains almost unchanged.

5.3 *Viewpoint 2*

Secondly, stability of a tunnel is investigated with an emphasis on scale effect. The cases compared here are $0.5D$, $1D$, $2D$ for D=10cm versus $0.5D$, $1D$, $2D$ for D=20cm. The results for the D=10cm case has already been given in **Fig.9**. Those for D=20cm are shown in **Figs.11** and **12**. Comparing normalized collapse pressures between **Figs.9** and **11**, it is recognized that the values for D=20cm are generally greater than those for D=10cm, indicating that a tunnel of the same geometrical configuration in terms of H/D ratio becomes more unstable, in a relative sense, as the tunnel size increases.

The deformational characteristics, judged from the distribution of maximum shear strain distribution for both cases of D=10 and 20cm, were found to be almost identical in a proportional sense. That is if you have a maximum shear strain distribution diagram for D=10cm, you can simply enlarge it by a factor of 2 to obtain a corresponding version for D=20cm. This makes that ratio of W between D=10 and 20cm will be nearly 4. Accordingly, if values of stresses along slip line become exactly double those for D=10cm, collapse pressures will also be doubled, resulting in a perfectly proportionate case having, in a relative sense, an identical level of stability. The detailed investigation of the stress distribution along the slip lines (Ogawa, 1996), however, revealed that the shear stresses along the slip lines for D=20cm tend to be less than the doubled version of stresses obtained for D=10cm. This difference of shear stress distribution is believed to make larger tunnels more unstable at limit states.

5.4 *Viewpoint 3*

Thirdly, stability of a tunnel of varying sizes is investigated for a constant value of H. Situation of this nature is encountered if one seeks for the

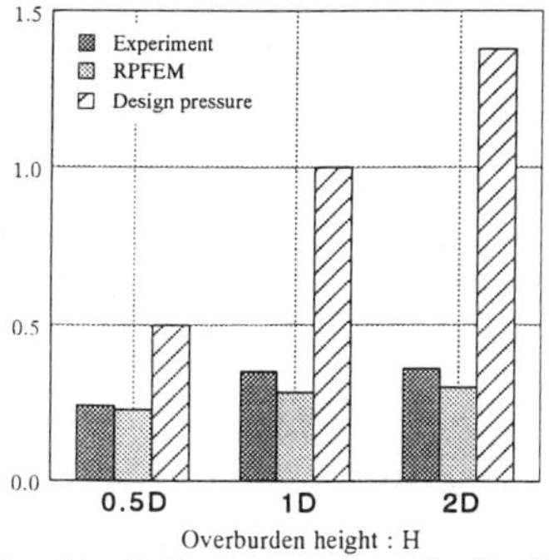

Fig.11 Collapse pressures for D=20cm

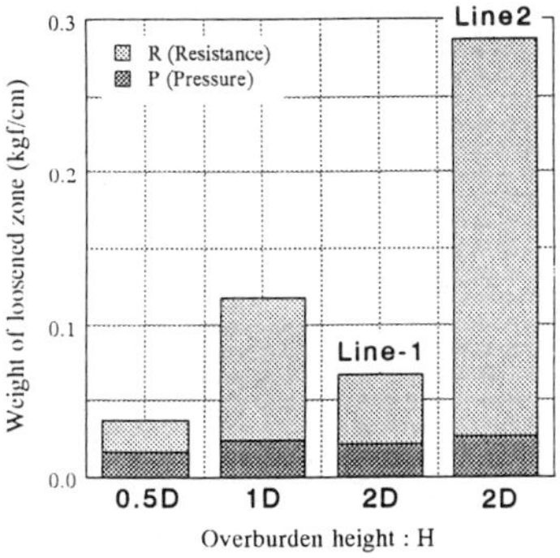

Fig.10 Force equilibrium for D=10cm

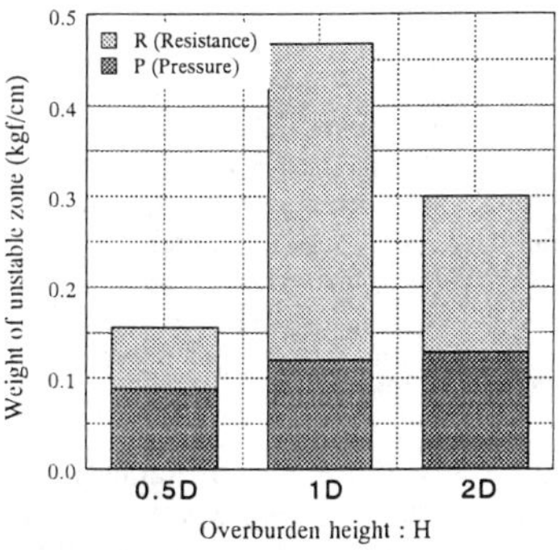

Fig.12 Force equilibrium for D=20cm

Table 2 Normalized collapse pressures for H=10 and 40cm

(a) H=10cm

	D=10cm	*D=15cm*	*D=20cm*
Experiment	0.23	0.37	0.48
RPFEM	0.23	0.35	0.45

(b) H=40cm

	D=10cm	*D=15cm*	*D=20cm*
Experiment	0.04	0.11	0.18
RPFEM	0.07	0.11	0.15

Table 3 Forces at state of collapse for H=10 and 40cm

(a) H=10cm (Unit : kgf/cm)

	P	*R*	*W*
D=10cm	0.024	0.094	0.118
D=15cm	0.047	0.079	0.126
D=20cm	0.087	0.070	0.157

(a) H=40cm (Unit : kgf/cm)

	P	*R*	*W*
D=10cm	0.028	0.072	0.100
D=15cm	0.070	0.130	0.200
D=20cm	0.129	0.171	0.300

maximum size of a stable tunnel at certain overburden height. Cases considered here are tunnels of diameter 10, 15, and 20cm at depths of 10 and 40cm. Collapse pressures and forces at collapse are given in **Tables 2** and **3**. For both cases of H=10 and 40cm, tunnel stability decreases as diameter increases. From a viewpoint of collapse pressure, the rate of stability deterioration is almost identical for both cases H=10 and 40cm. However, the rate of changes in tunnel pressure P is greater for the case of D=10cm.

The investigation (Ogawa, 1996) identified that this difference may be attributed to the geometric characteristics of high strain zone for the three cases for H=10cm. As diameter increased from 10 to 20cm, the location of slip line changed gradually in such a way that resisting force integrated over the line was reduced. On the contrary, for the case of H=40cm, geometry of high strain zone remained almost constant. Deterioration of stability for this case therefore may be attributed simply to an increase in the size of unstable zone and its weight. The results given here suggests that stability of underground openings decreases as their diameter increases regardless of where they are situated. However, the mechanism of stability deterioration is different for each value of overburden height.

6 CONCLUSION

The fundamental aspects on stability of underground openings were studied through lab experiments and rigid-plastic finite element analyses. It was noticed firstly that the proposed RPFEM procedure, in which external forces are treated in two different groups, is readily applicable to estimate deformational characteristics, state of stress, and collapse pressure at a limit state of equilibrium. The following comparative and parametric studies revealed relative differences in the level of stability for underground openings of different sizes constructed at different depths. These findings are made with respect to limit states of failure, however, the insights given by the results presented here are expected to help establish more reliable and effective design procedure of underground openings constructed with clearly defined safety factors.

REFERENCES

Adachi, T., T. Tamura, A. Yashima & H. Ueno 1985. Behaviour and simulation of sandy ground tunnel, *Journal of Geotechnical Engineering*, JSCE, **III-3**(358) 129-136.

Asaoka, A., T. Kodaka, G. Pokharel & T. Miyagawa 1993. *Proc. 28th National Symp. on Soil Mechanics and Foundation Engineering,* 2811-2814.

Drucker, D.C. & W. Prager 1952. Soil mechanics and plastic analysis or limit design, *Quarterly Journal of Applied Mathematics*, **10**:157-165.

Murayama, S. & H. Matsuoka 1971. Earth pressure on tunnels in sandy ground, *Proc. of JSCE*, (187):95-107.

Ogawa, K. 1996. An investigation on stability of tunnel by means of rigid-plastic finite element analysis, *Master's thesis*, Kobe University.

Sakurai, S., I. Kawashima, Y. Kawabata & A. Saragai 1994. Model tests on deformation and loosening pressure of shallow tunnel, *Journal of Geotechnical Engineering*, JSCE, **III-26**(487) 271-274.

Tamura, T. 1990. Rigid-plastic finite element method in geotechnical engineering, Current Japanese materials research, *Computational Plasticity*, **7**:135-164, Elsevier Applied Science.

Tamura, T., S. Kobayashi & T. Sumi 1984. Limit analysis of soil structure by rigid plastic finite element method, *Soils and Foundations*, **24**(1): 34-42.

Computer Methods and Advances in Geomechanics, Yuan (ed.) © 1997 Balkema, Rotterdam, ISBN 90 5410 904 1

Influence of rock temperature on stress in the spiral casing structure of the underground hydropower station

Wang Linyu
Department of Civil Engineering, Zhejiang University, Hangzhou, China

Zhong Bingzhang & Lin Zhongxiang
Department of Mechanics, Zhejiang University, Hangzhou, China

ABSTRACT: The temperature field and thermal stress on the spiral casing structure of an underground hydropower station are analyzed by the finite element method in this paper. With the first and third temperature boundary conditions, namely temperature and heat flow boundary conditions, the axisymmetric steady temperature fields of the structure are obtained and two kinds of the thermal stresses are compared. According to the calculating results, it is known that the thermal stress calculated with the third boundary condition is smaller than that with the first boundary condition and that the spiral casing structure bears considerable thermal stress. Furthermore, the elastic resistance of the surrounding rock is considered.

1 INTRODUCTION

To the concrete structure of the large volume, temperature loads not only cause notable changes on the stress state of the structure, but also maybe make the structure cracking. So the integrity and durability of the structure could be influenced. It is important for the safety of a reinforced concrete structure to analyze the temperature field and thermal stress of the structure.

The spiral casing with its surrounding concrete structure is one of the important structures for a hydropower station. Generally, the influence of temperature loads on the structure is not considered in the past. The reinforcing bars are designed in the surrounding concrete only according to the stresses caused by the internal water pressure and external loads. Some engineers consider the influence of temperature, but they simply consider the influence of thermal stresses by extending the cross sectional area of reinforcing bars. So the thermal stress of the structure can not be considered effectively. In fact, with the continuous increase capacity of the single-unit in the hydropower station, the spiral casing structure becomes larger and larger. The influence of the thermal stress also increases gradually. So it is inappropriate not to consider thermal stress.

In this paper the temperature field and thermal stress on the spiral casing structure of an underground hydropower station are calculated with the first and third temperature boundary conditions. The thermal stresses of the concrete in the tangential direction are arranged. The calculated results are quite satisfactory.

2 PARTIAL DIFFERENTIAL EQUATION OF HEAT CONDUCTION

2.1 *Partial differential equation of heat conduction*

To study the theory of heat conduction, it is important to define temperature distribution, namely to define the temperature field. For the three-dimensional problems, we can obtain the partial differential equation of heat conduction

$$\nabla^2 t + \frac{q_v}{\lambda} = \frac{1}{a}\frac{\partial T}{\partial t} \qquad (2\text{-}1)$$

where q_v is the density of the internal heat sources, λ is the heat conduction coefficient of the material, $a = \lambda / (\rho C_p)$ is the temperature conduction coefficient of material, ρ is the density, C_p is the specific heat, T is the temperature of the body, and

t is the time.

Many problems of stress analysis can be represented by a two-dimensional model to an adequate degree of approximation. A special case of three-dimensional problems, namely axisymmetric solids, can be considered as two-dimensional. For the axisymmetric problems, we can conclude the equation

$$\frac{\partial^2 T}{\partial r^2}+\frac{1}{r}\frac{\partial T}{\partial r}+\frac{\partial^2 T}{\partial z^2}+\frac{q_v}{\lambda}=\frac{1}{a}\frac{\partial T}{\partial t} \quad (2\text{-}2)$$

where r is the radius, and z is the symmetric axis.

For an axisymmetric steady temperature field and in the absence of heat sources we get the equation

$$\frac{\partial^2 T}{\partial r^2}+\frac{1}{r}\frac{\partial T}{\partial r}+\frac{\partial^2 T}{\partial z^2}=0 \quad (2\text{-}3)$$

2.2 *Temperature boundary conditions*

There are four types of temperature bounding conditions. Two types of them which are used in this paper are introduced.

1) The first boundary condition (Temperature boundary condition)

Temperature prescribed at the surface of the body

$$T_w = f(s,t) \quad (2\text{-}4)$$

where $f(s,t)$ is a preassigned function of time t and position s on the surface.

2) The third boundary condition (Heat flow boundary condition)

Prescribed heat flux across the surface of the body

$$q_n = g(s,t) \quad (2\text{-}5)$$

Here again $g(s,t)$ is a preassigned function and the subscript n denotes the normal to the surface. If the heat flux q_n is induced by convection, Equation (2-5) may be put into the form

$$-\lambda\left(\frac{\partial T}{\partial n}\right)_w = \alpha(T_f - T_w) \quad (2\text{-}6)$$

where α is the convective heat transfer coefficient. T_f is the temperature of the surroundings, T_w is the temperature on the surface of the body, and n is an outward normal.

3. ELASTIC RESISTANCE COEFFICIENT OF THE ROCK

In hydropower projects some constructions are built on the rock, such as the dam, and some of them are built in the rock, such as the plant house of the underground hydropower station. So the problem of the rock resistance need be considered.

The spiral case is endured by the internal water pressure. And it makes the surrounding concrete yield the radial deformation so as to press the surrounding rock. So the rock resistance is produced. The rock can be looked as the elastic medium when the deformation is small. The rock resistance coefficient k can be expressed as the expression on the stress p and the corresponding displacement u of an arbitrary point, namely

$$p = ku \quad \text{or} \quad k = \frac{p}{u} \quad (3\text{-}1)$$

Experiments indicate that the coefficient k of the same rock is not a constant, but has an inverse ratio with the extracting radius r of the rock. In order to unify the elastic resistance coefficient for the same rock, the unit elastic resistance coefficient k_0 can be looked as the coefficient k of the extracting radius $r = 100\text{cm}$ when Young's modules E and Poisson's ratio μ are determined, the unit elastic resistance coefficient k_0 of the rock can be calculated by the following formula

$$k_0 = \frac{E}{100(1+\mu)} \quad (3\text{-}2)$$

The resistance coefficient of the surrounding rock is

$$k = \frac{100k_0}{r} \quad (3\text{-}3)$$

From equation (3-2) and (3-3), we can obtain

$$k = \frac{E}{r(1+\mu)} \quad (3\text{-}4)$$

For the surrounding concrete structure of the spiral case, the elastic resistance coefficient of any corner node of the finite element is

$$k_1^{'} = k \times \frac{A}{n} \quad (3\text{-}5)$$

the elastic resistance coefficient of other node is

$k_2' = k_1'/2$ (3-6)

where A is the surface area of the body for the whole spiral casing structure, and n is the elemental numbers of the surface. k_1' and k_2' of the axisymmetric structure are constants as that of the whole structure.

4 COMPUTING MODEL AND ANALYSES OF THE SPIRAL CASING STRUCTURE

4.1 *Introduction*

The spiral casing structure of an underground hydropower station lies in the rock. Its upstream and downstream directions and the bottom get in touch with the rock. The right and left side of the structure are temperature joints due to settlement of the unit. Its volume is very large, and the shape is complicate. There are temperature differences between the inner and outer surfaces. The structure is an axisymmetric structure approximately. It can be computed according to the axisymmetric problem under the premise of meeting the engineering request. Considering the thermal stress of the cross section in the downstream direction is bigger than other cross sections, we calculate the axisymmetric temperature field and thermal stress of a cross section on the position. The mesh divisions of the structure are adopted by the axisymmetric elements of the triangle and quadrangle. The computing model and element division are shown in figure 1. Temperatures which are shown in figure 1 are ones in summer, so the temperature of the inner surface is lower than that of the outer surface on the structure. The temperature field and thermal stress in winter are not arranged owing to the limitation of space on the article.

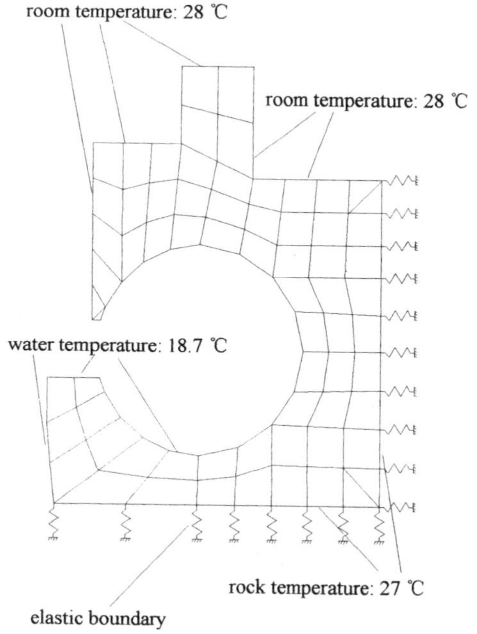

Figure 1. Computing model of thermal stresses

4.2 *The form and computation on the axisymmetric temperature field*

In this paper the axisymmetric steady temperature fields of the computing model are calculated by the finite element method with the first and third temperature boundary conditions respectively. To the first boundary condition, the surface temperature T_w of the body is the inner and outer surface temperature of the concrete which is shown in figure 1. To the third boundary condition, the temperature of figure 1 is the surrounding temperature T_f. The surface temperature T_w of the body must be calculated from equation (2-6).

In the process of the computation on the temperature field, the heat conduction coefficient and the convective heat transfer coefficient can be used. According to the request of the engineering, we adopt the following values

$$\lambda = 1.249 \text{Kcal} / (\text{m} \cdot \text{h} \cdot {}^\circ\text{C})$$

$$\alpha = 16 \text{ Kcal} / (\text{m}^2 \cdot \text{h} \cdot {}^\circ\text{C})$$

The final temperature values of all nodes can be obtained by the finite element computation on heat conduction of the steady state. The steady temperature is shown in figure 2 and 3.

4.3 *Computing results and analyses of the thermal stresses*

Firstly, we should exert the nodal temperatures which are shown in figure 2 and 3 on each node in the former disposal of the temperature computation. Then we can analyze the axisymmetric thermal stress of the computing model according to the heat expansion coefficient of the concrete. To adapt for the concrete reinforcement, the stress of the tangential direction is arranged on the basis of

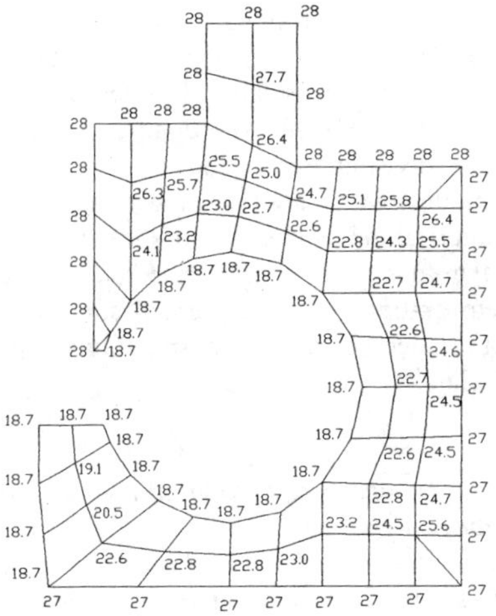

Figure 2. Temperature field of the first boundary condition （Unit: ℃）

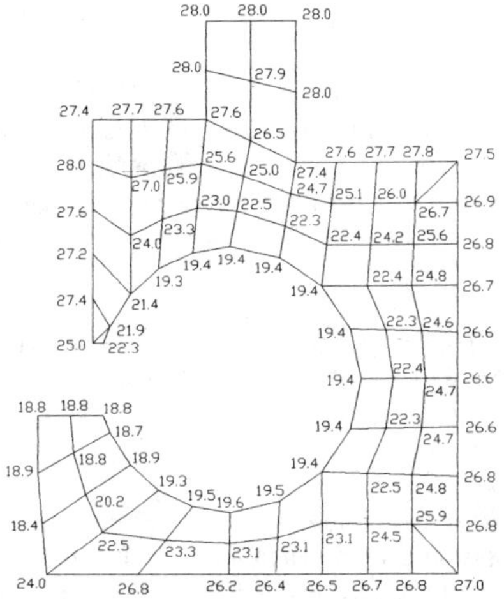

Figure 3. Temperature field of the third boundary condition （Unit: ℃）

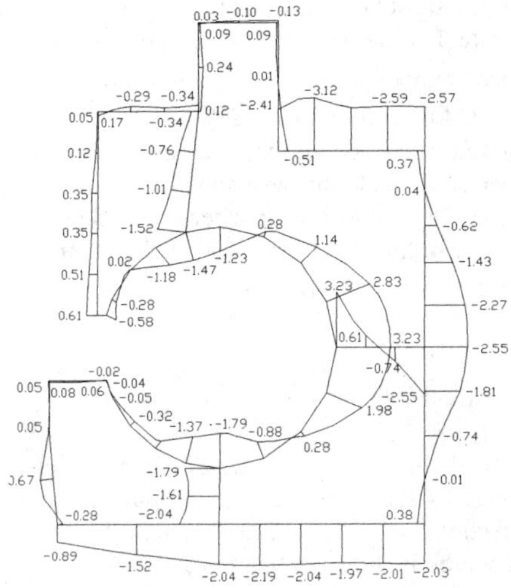

Figure 4. Thermal stresses of the tangential direction with the first boundary condition （Unit: MPa）

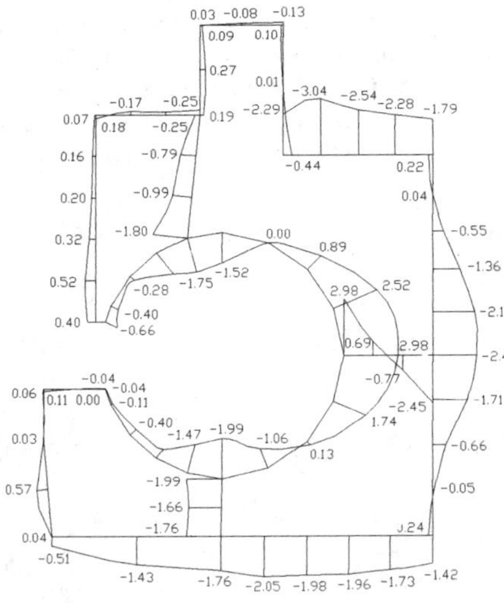

Figure 5. Thermal stresses of the tangential direction with the third boundary condition （Unit: MPa）

the computing results. The thermal stresses are shown in figure 4 ~ 5 （The positive value is tensile stress） and analyzed as follows:

（1） The thermal stresses in the tangential direction under the first temperature boundary condition are shown in figure 4. From the figure we can find that the stress distributions which are along with the concrete surface are very unequal. The largest tensile stress is 3.23MPa, it exceeds the tensile resistance strength of the concrete greatly. The tensile stress area mainly lies in the inner surface of the concrete on the right, and its influence on the structure is disadvantageous. Why this phenomenon took place is that the concrete deformation is restricted by the elastic boundary of the surrounding rock. Because the temperature differences of the place are 8 ℃ ~ 9 ℃, they must cause a large temperature gradient in a small region and cause the results of the quite bigger tensile stress in the inner surface and the big compressive stress in the outer surface.

(2) The thermal stresses under the third temperature boundary condition are shown in figure 5. The stress distributions of figure 5 and 4 are approximately same. But the stresses have a certain degree of descents. To the first boundary condition, temperatures are directly exerted on the structural boundary, and to the third boundary condition, temperatures are transferred to the structure through heat convection of the structure and its surroundings. So the temperature difference, temperature gradient and thermal stress of the latter are smaller than that of the former for the spiral casing structure. For the third temperature boundary condition the stresses are reliable because the practical heat conduction is considered.

5 CONCLUSIONS

1. To the spiral casing structure, the thermal stress is very large under certain circumstances. It must be considered when the concrete structure is designed.

2. The thermal stress is closely related to the temperature boundary condition. The thermal stress calculated with the third boundary condition is smaller than that of the first boundary condition. So the third temperature boundary condition which is more reasonable should be used.

3. Generally the elastic resistance of the surrounding rock is advantageous to structural stresses, but it is disadvantageous under temperature loads for the spiral casing structure of the underground hydropower station.

4. In the practical engineering the temperature loads are very complicated, and the related factors are varied.

REFERENCES

O.C.Ziankiewicy 1977. *The finite element method.* McGraw-Hill.

Hans Ziegler. 1983. *An introduction to thermomechanics.* North-Holland publish- ing company-Amsterdam. New York. Oxford.

Ivar Holand and Kolbein Bell. 1972. *Finite element methods in stress analysis.*Published by TaPIR, the technical university of Norway. Third printing.

J.L.Nowinski 1978. *Theory of thermoelasticity with supplications.* Printed in the Netherlands. Ireland.

H.P.Yagoda 1995. An extraction technique for accurately computing steady state temperatures in high gradient regions — theory. *Mechanics research communications,* Vol.22, No.2, pp.137 ~ 150.

Computer Methods and Advances in Geomechanics, Yuan (ed.) © 1997 Balkema, Rotterdam, ISBN 90 5410 904 1

The study of reliability analysis method for shallow-buried subway tunnel constructing with excavation method

Wu Kangbao, Song Yuxiang & Jing Shiting
Shijiazhuang Railway Institute, China

ABSTRACT : The paper describes the method of calculating the effects of the lining structure of shallow-buried subway tunnel by the stochastic finite element method (SFEM) of first — order perturbation, based on the " Action — Reaction " model. The limit state equations are established, according to the ultimate limit states and serviceability limit states. According to the probability statistical principle, the data, coming from our in situ investigation, of the main random variables influencing on the reliability of the structure of shallow-buried subway tunnel are studied. The reliability of the lining structure of subway tunnel is calculated. The study method and the results of this paper will be provided for reference in the reliability analysis of subway tunnel and in the modification of " Design Specification of Railway Tunnel " .

1 INTRODUCTION

The environmental condition of underground structure are extremely complicated. The interaction effects between rock mass and supports hasn't been fully realized. As a result, the design is still performed on a semi-empirical, semi-theoretical and deterministic basis. However, the loads acting on the shallow-buried underground structure are relatively obvious, it is reasonable to analyze this kind of structure with " Action — Reaction " model and easy to use the methods and experience of reliability analysis for ground structure. This paper takes the shallow-buried subway tunnel as an example, and study the method of reliability analysis of the structure. Therefore, the load effects and reliability index of the structure can be obtained. Finally, the variation of some of the main random variables, which influence the reliability of the underground structure, has been nalyzed.

2 THE METHOD OF LOAD EFFECT ANALYSIS FOR SHALLOW-BURIED SUBWAY TUNNEL

Nowadays the internal forces of lining structure are most calculated with finite element method for underground structures. when the randomness of calculated parameters are considered, the stochastic finite element method is often used. The perturbation SFEM has been widely used in stochastic analysis for ground structure. Therefore, the SFEM of first — order perturbation has been used in load effects analysis in this paper.

In regard of the supporting structure of shallow-buried subway tunnel, as the loads are obvious, the " Action — Reaction " model is often used in structural analysis, in that case the supporting structure is usually divided into beam-column elements. The local deformation theory assumed by Winkler is used for simulating the interaction effects between rock mass and supports, which are reflected by the elastic resistance. The column element having relatively high rigidity (only compressive, not tensile and bending) are used for simulating the interaction effects between primary support and secondary lining. Each rock column as an elastic rod, one point of which is sustained on the nodal point of which is sustained on the nodal point of the beam-column element and another is sustained on the rigid support. The calculating model is shown in figure 1.

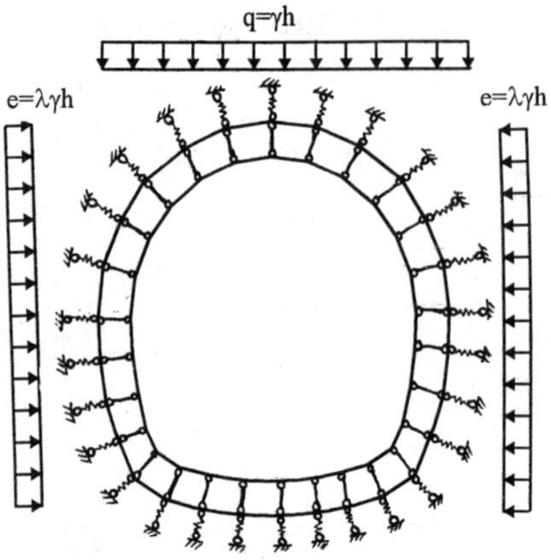

Fig.1 The calculating model

2.1 The randomness of material property and geometric size of lining structure

1 The stiffness matrix of beam column element

Consider the beam-column element with six degrees of freedom, in certain analysis, the element stiffness matrix in the global coordinate system is established as :

$$K_0^1 = T^T K_0^0 T \qquad (1)$$

where T is the coordinate exchange matrix and is also deterministic, and K_0^0 is the element stiffness matrix in the local coordinate system, which is related to the elastic modules of the supporting structure (E), supporting structure thickness (d) and the length of the element (l). According to the practical work of the supporting structure of the tunnel, E and d are considered as random variables. then, they are represented by the sum of a deterministic value and a small random fluctuation, therefore :

$$\begin{cases} E_i = E_i^0 + E_i' \\ d_i = d_i^0 + d_i' \end{cases} \qquad (2)$$

Where:

$$\begin{cases} E(E_i') = E(d_i') = 0 \\ D(E_i') = \sigma_{E_i'}^2 \\ D(d_i') = \sigma_{d_i'}^2 \end{cases} \qquad (i=1,2,\cdots,N_1)$$

N_1 is the total number of elements with randomness of material property.

Now substitute equation (2) into equation (1) ,and consider the first-order approximation of the Taylor expansion of matrix K_1 around the mean values of random variables E and d ,the beam-column element stiffness matrix in the global coordinate system is established as follows :

$$K_1 = K_0^1 + K_0^{1E} E_i' + K_0^{1d} d_i' \qquad (3)$$

Certainly, the first deterministic term of the above expression agree with the element stiffness matrix used in the conventional finite element method, $K_0^{1E}E_i'$ and $K_0^{1d}d_i'$ are the terms under the influence stochastic parts of random variables E and d, respectively. The matrices K_0^{1E} and K_0^{1d} are the first-order partial derivatives of matrix K_0^1 in regard to random variables E and d, respectively.

2. The stiffness matrix of elastic rod element

The i-th elastic rod element is shown in figure 2.

The element stiffness matrix is derived from the elastic resistance K_r, and the displacements u_{3i-1} and u_{3i-2}, according to the assumption of Winker. Now consider the beam-column element length l as determinate, and the coefficient of elastic resistance of rock mass (K_r) as random variable, which is represented by the sum of a deterministic value and a small random fluctuation, that is :

$$K_{ri} = K_{ri}^0 + K_{ri}' \qquad (4)$$

where:

$$\begin{cases} E(K_{ri}') - 0 \\ D(K_{ri}') = \sigma_{K_{ri}'}^2 \end{cases} \qquad (i=1,2,\cdots,N_2)$$

N_2 is the total number of elastic rod elements.

The elastic rod element stiffness matrix in the global coordinate system is established as follows :

$$K_2 = K_0^2 + K_0^{2Kr} K_{ri}' \qquad (5)$$

where K_0^2 is the deterministic part.

$$K_0^2 = K_r L_i \begin{bmatrix} \sin^2\varphi & -\sin\varphi_i\cos\varphi_i & 0 \\ -\sin\varphi_i\cos\varphi_i & \cos^2\varphi_i & 0 \\ 0 & 0 & 0 \end{bmatrix}$$

$$L_i = \frac{1}{2}(l_i + l_{i+1})$$

$K_0^{2K_r}K_{ri}'$ is the term under the influence of the random

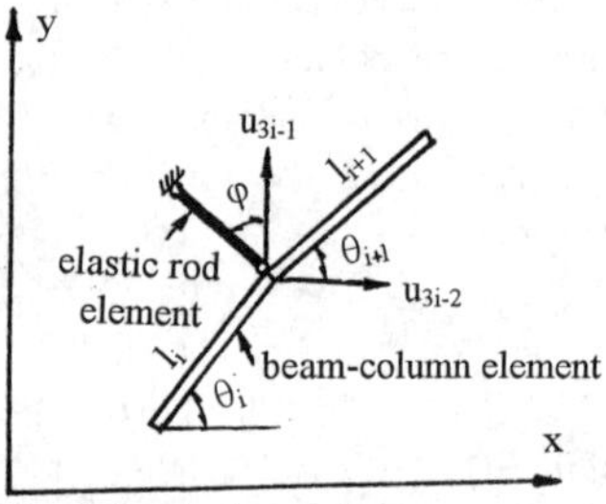

Fig.2 The elastic rod element

variable K_r, and $K_0^{2K_r}$ is the first derivative of K_0^2 in regard to the random variable K_r.

2.2 The randomness of displacement boundary conditions

Displacement can be divided into two parts, that is : $U=[U_1 \quad U_2]^T$. Where U_2 is the unknown nodal displacement which is to be solved, while U_1 is the given displacement, and is uncertain. The displacement U_1 is represented by the sum of a deterministic value and a small random fluctuation, that is :

$$U_{1i} = U_i^0 + U_i' \tag{6}$$

where :

$$\begin{cases} E(U_i') = 0 \\ D(U_i') = \sigma_{U_i'}^2 \end{cases} \qquad (i=1,2,\cdots,N_3)$$

N_3 is the total number of nodal points with random displacement boundary conditions.

2.3 The randomness of the load

The load can also be divided into two parts, $P=[P_1 \quad P_2]^T$. Where P_1 is the unknown boundary resistance, which is related to the parameter of the property of material, and the boundary conditions of displacement. P_2 is the applied load, which has no connection with the structure itself, and is uncertain. The load vector P_2 is represented by the sum of a deterministic value and small random fluctuation, that is :

$$P_{2i} = P_i^0 + P_i' \tag{7}$$

where :

$$\begin{cases} E(P_i') = 0 \\ D(P_i') = \sigma_{P_i'}^2 \end{cases} \qquad (i=1,2,\cdots,N_4)$$

N_4 is the total number of nodal point with the effects of random applied loads.

2.4 The establishment and solution of the stochastic finite element equilibrium equation of the supporting structure

Since the global stiffness matrix of the supporting structure has a connection with the parameter of the property of material, the geometric size and the coefficient of elastic resistance of rock mass, all of which are stochastic. The stiffness matrix K is expanded into the parametric equation in terms of random variables E , d, and K_r. Thus, we have :

$$K = K_0 + \sum_{i=1}^{N_1} K_{0i}^{1e} E_i' + \sum_{i=1}^{N_2} K_{0i}^{1d} d_i' + \sum_{i=1}^{N_2} K_{0i}^{2K_r} K_{ri}' \tag{8}$$

where K_0 is the deterministic term of the global stiffness matrix in the global coordinate system, and is calculated by the conventional finite element method.

The nodal displacement vector U in the global co-ordinate system is in connection with the parameter of material property, the geometric size, the boundary conditions and the types of the lodes, they are all considered as random variables. The nodal displacement vector U is expanded into the parametric equation in terms of random variables E, d, Kr, U_1 and P_2. as a result, we obtained:

$$U = U_0 + \sum_{i=1}^{N_1} U_i^E E_i' + \sum_{i=1}^{N_1} U_i^d d_i' + \sum_{i=1}^{N_2} U_i^{K_r} K_{ri}' + \sum_{i=1}^{N_3} U_i^U U_i' + \sum_{i=1}^{N_4} U_i^P P_i' \tag{9}$$

where U_0 is the nodal displacement in the global coordinate system, and is solved by the conventional finite element method.

The nodal load vector P consists of the unknown boundary resistance and applied loads. As the boundary resistance has connection with the nodal displacements, so that the nodal load vector P is also related to the parameter of material property, the geometric size, the boundary conditions and the types of the lodes. The nodal load vector P is expanded into the parametric equation in terms of random variables E, d, Kr, U_1 and P_2. Thus, we have:

$$P = P_0 + \sum_{i=1}^{N_1} P_i^E E_i' + \sum_{i=1}^{N_1} P_i^d d_i' + \sum_{i=1}^{N_2} P_i^{K_r} K_{ri}' + \sum_{i=1}^{N_3} P_i^U U_i' + \sum_{i=1}^{N_4} P_i^P P_i' \tag{10}$$

where P_0 is the load vector which is calculated by the conventional finite element method.

Substituting now Eqs.(8), (9) and (10) into the standard finite method equation (11).

$$K\,U = P \tag{11}$$

The equation (12) can be obtained as:

$$[K_0 + \sum_{i=1}^{N_1} K_{0i}^{1E} E_i' + \sum_{i=1}^{N_2} K_{0i}^{1d} d_i' + \sum_{i=1}^{N_2} K_{0i}^{2K_r} K_{ri}']$$
$$\cdot\, [U_0 + \sum_{i=1}^{N_1} U_i^E E_i' + \sum_{i=1}^{N_1} U_i^d d_i' + \sum_{i=1}^{N_2} U_i^{K_r} K_{ri}' + \sum_{i=1}^{N_3} U_i^U U_i' + \sum_{i=1}^{N_4} U_i^P P_i']$$
$$= [P_0 + \sum_{i=1}^{N_1} P_i^E E_i' + \sum_{i=1}^{N_1} P_i^d d_i' + \sum_{i=1}^{N_2} P_i^{K_r} K_{ri}' + \sum_{i=1}^{N_3} P_i^U U_i' + \sum_{i=1}^{N_4} P_i^P P_i'] \tag{12}$$

We can obtain the following equation group

(Eq.(13)) of the SFEM of the first-order perturbation under the influence of al random variables, by comparing the same coefficients at the two side of equation (12).

$K_0U_0=P_0$

$K_0U_i^E+K_{0i}^{1E}U_0=P_i^E$

$K_0U_i^d+K_{0i}^{1d}U_0=P_i^d \qquad (i=1,2,\cdots,N_1)$

$K_0U_i^{K_r}+K_{0i}^{2K_r}U_0=P_i^{K_r} \qquad (i=1,2,\cdots,N_2)$

$K_0U_i^U=P_i^U \qquad (i=1,2,\cdots,N_3)$

$K_0U_i^P=P_i^P \qquad (i=1,2,\cdots,N_4) \qquad (13)$

First, the vector U_0 is solved according to first equation of Eq.(13), then we may bring U_0 into other equations of Eq.(13), therefore, the vector U_i are solved, respectively.

2.5 The expectations and variances of internal forces of section

In order to work out the expectations and variances of the stresses on section, the expectations and variances of the internal forces of section should be solved first. Then according to the relationship between internal forces and stresses on section of the structure, therefore, the expectations and variances of stresses are obtained.

In conventional finite element analysis, the internal forces of section are written as follows:

$$S=TK_1U \qquad (14)$$

However, in stochastic finite element analysis methodology, now substituting Eqs.(3) and Eqs.(9) into equation (14), and ignoring the terms up to the second order products of random fluctuations, thus, we have:

$$E(S)=TK_0^1U_0 \qquad (15)$$

assume: $A=TK_0^{1E}$, $B=TK_0^{1d}$, $C=TK_0^1$

Now if we consider A_i as the i-th row vector of matrix A, and S_i as the i-th component of he internal forces of beam-column element, the variances of internal forces on section can be derived as follows:

$$D(S)=(A_iU_0)^2\sigma_{E_i}^2+(B_iU_0)^2\sigma_{di}^2 + \sum_{j=1}^{N_1}\sum_{k=1}^{N_1}(C_iU_j^E)(C_iU_k^E)Cov(E_j',E_k') + \sum_{j=1}^{N_1}\sum_{k=1}^{N_1}(C_iU_j^d)(C_iU_k^d)Cov(d_j',d_k') + \sum_{j=1}^{N_2}\sum_{k=1}^{N_2}(C_iU_j^{k_r})(C_iU_k^{k_r})Cov(K_{rj}',K_{rk}') + \sum_{j=1}^{N_3}\sum_{k=1}^{N_3}(C_iU_j^U)(C_iU_k^U)Cov(U_j',U_k') + \sum_{j=1}^{N_4}\sum_{k=1}^{N_4}(C_iU_j^P)(C_iU_k^P)Cov(P_j',P_k') \qquad (16)$$

3 THE ESTABLISHMENT OF LIMIT STATE EQUATIONS OF STRUCTURAL RELIABILITY ANALYSIS FOR SHALLOW-BURIED SUBWAY TUNNEL

3.1 Ultimate limit state

In the structural reliability analysis all preliminary functions are expressed with limit state, which usually consists of the ultimate limit state and the service ability limit state. The two kinds of limit state equations can be established based on the " Design Specification of Railway Tunnel " and the " Design Specification of Concrete Structure " .

To shallow-buried metro, the limit state equation of compressive bearing capacity of inner lining is:

$$z=\varphi R_a(h-1.5e_0)-N=0 \qquad (17)$$

where: R_a is the ultimate compressive strength of concrete, φ is the longitudinal bending coefficient of member, $\varphi=1$ for subway tunnel, e_0 is the eccentricity on section, N is the axial pressure of section, h is the thickness of section.

The ultimate limit state equations of outer lining made of grid and shotcrete are as follows:

$$Z=f_{cm}bx+f_y'A_s'-\sigma_sA_s-N=0 \text{ (axial force control) } (18)$$

$$Z=(f_{cm}bx+f_y'A_s'-\sigma_sA_s)e_0-M=0 \text{ (moment control)} (19)$$

3.2 The equation of serviceability limit state

In regard of the supporting structure of shallow-buried subway tunnel, the limit state mainly refers to excessive deformation, crack formation (crack limit) and development (width limit of crack). To restrict crack from taking place, current tunnel specifications have provided formula for calculating crack (tensile strength of eccentric compressive member of rectangular section of concrete). From it we can get the limit state equation in normal using state as follows :

$$Z=\left[1.75R_lbh/(6e_0/h-1)\right]-N=0 \qquad (20)$$

where: R_l is the ultimate tensile strength of concrete, and the symbols mean the same as formula (17).

4 A NUMERICAL EXAMPLE OF THE STRUCTURAL RELIABILITY ANALYSIS OF SHALLOW-BURIED SUBWAY TUNNEL

For probability analysis of above limit state, first all random variables must be analyzed by means of probability theory, then the statistical characteristic of load effects are evaluated, and last reliability index is obtained. The probabiliticic characteristic random variables are shown in table 1. For some shallow-buried subway tunnel in Beijing, the lining structure consists of primary support which is made of grid and shotcrete (30cm thick), and the secondary lining, which is made of concrete(25cm thick). The shape of the cross-section of the support structure of shallow-buried subway tunnel is shown in Fig.3. The calculating model of lining structure is shown in Fig. 4.

In this paper, the load effects are calculated by SFEM of first-order center perturbation. The two conditions of construction stage with only primary support, ordinary using stage after completion are calculated, according to different load combinations of load respectively. The statistical characteristics of load effects in each of the two stages are shown in table 2 and table 3. We can see that the load effects approximately obey lognormal distribution.

After calculating the statistical characteristics of the load effects of the lining structure under two stages, the reliability index of control section of lining structure in each stage are obtained with the centerpoint method of first-order second-moment and point method of first-order second-moment respectively, according to the limit state equation. The results are shows in table 2 and table 3.

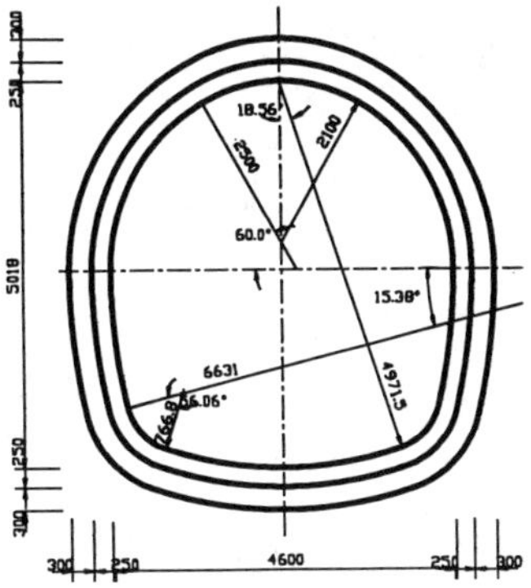

Fig.3 The cross-section of support of shallow-buried tunnel

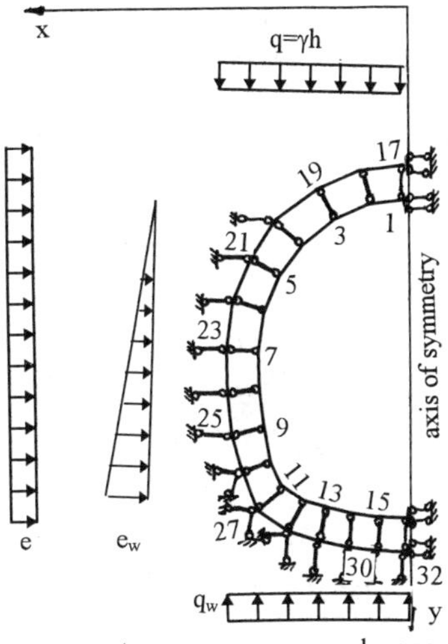

e_w,q_w — water pressure; e,q — rock pressure

Fig.4 The calculating model of lining structure

Table 1 Probabilistic characteristics of random variables

random variables		Mean-value	Standard deviation	robability distribution
volume-weight of ground (γ)		19.5KN/m^3	0.599KN/ m^3	Normality
Overburden thickness (h)		11.27m	1.17m	Normality
Coefficient of horizontal (λ)		0.3675	0.0671	Lognormal
Elastic modules of	I*	2.1 × 10^{5}KN/m^3	3.78 × 10^{4}KN/m^3	Lognormal
lining structure (E)	O*	3.0 × 10^{5}KN/m^3	4.20 × 10^{4}KN/m^3	Lognormal
Thickness of	I*	25cm	3.75cm	Normality
lining structure (d)	O*	30cm	4.50cm	Normality
Coefficient of elastic	H*	44978KN/m^3	7701KN/m^3	Lognormal
resistance of ground (K)	V*	50114KN/m^3	9539KN/m^3	Lognormal

I*— Inner lining, O*— Outer lining, H*— Horizontal, V*— Vertical

Table 2 Expectations and standard deviations of load effects and reliability index of lining structure in construction stage

section number	expectation of axial force (KN)	expectations of moment (KN-M)	standard deviation of moment (KN)	standard deviations of moment (KN-M)	reliability index	
					center point method	checking point method
17	-375.415	52.373	103.612	14.014	4.51	4.58
22	-526.020	-23.671	142.045	6.343	5.50	5.63
26	-649.827	-104.739	181.952	28.279	2.58	2.67
27	-594.305	-101.346	165.482	27.923	2.27	2.35
32	-501.559	85.937	141.943	24.163	2.79	2.86

Tab.3 Expectations and standard deviations of load effects and reliability index of lining structure in normal using stage

section number	expectation of axial force (KN)	expectations of moment (KN-M)	standard deviation of moment (KN)	standard deviations of moment (KN-M)	reliability index	
					center point method	checking point method
1	-189.057	14.235	46.318	3.524	5.17	5.29
5	-193.882	-9.581	50.603	2.443	7.52	7.68
9	-182.063	-7.312	45.697	1.810	5.89	6.01
10	-167.809	-12.665	41.955	3.178	6.93	7.20
16	-183.414	17.135	46.037	4.204	4.47	4.58
17	-295.126	34.221	73.485	8.501	4.70	4.79
22	-376.230	-21.941	95.186	5.583	6.37	6.49
26	-491.291	-62.223	122.336	15.395	4.35	4.42
27	-472.496	-64.874	118.127	15.634	3.94	4.02
32	-443.283	68.975	111.708	17.247	3.63	3.72

5 CONCLUSIONS

From above analysis we can see that the stochastic calculation for the shallow-buried metro by applying the probabilistic design method on the basis of reliability theory is practicable, more reasonable, more scientific and better reflecting the structural reliability than traditional design method.

The load effects of shallow-buried metro structure are calculated by using the SFEM of first-order center perturbation and the reliability index of the structure is got according to the probability limit state equations established to meet the demand of different functions and the demand of different functions and based on the " Action — Reaction " model. That will be improved the design theory of lining structures. Therefore, it's reasonable to a shallow-buried subway tunnel structure whose load condition is relatively obvious. The study method in this paper will be an important reference to the modification of " Design Specification of Railway Tunnel " .

The method studied in this paper can be used not only in the reliability analysis of shallow-buried subway tunnel structures, but also in that of other underground structures.

REFERENCES

Hisada, T. and Nakagiri, S. Stochastic finite element method developed for structural safety and reliability. *Proc. 3rd int. Conf. on structural Safety and Reliability*, 1981, Norway.

Deodatis,G. Bounds on response variability of stochastic finite element system: effect of stochastic dependence. *Journal of Probabilistic Engineering Mechanics*, Vol. 5, No.2, 1990, USA.

Design Specification of Railway Tunnel (TBJ3-85),*China Railway Publishing Company*

Design Specification of Concrete Structure (GBJ10-89), *Chain Planing Publishing Company*

Computer Methods and Advances in Geomechanics, Yuan (ed.) © 1997 Balkema, Rotterdam, ISBN 90 5410 904 1

An intelligence analysis system of the underground engineering supporting decision

Jianxi Ren
Institute of Rock and Soil Mechanics, The Chinese Academy of Sciences, Wuhan, China

ABSTRACT: The main idea and general structure of a neural network intelligence analysis software of rock underground engineering stablity appraised and supporting decision which is combined with the neural network technology, expert system technology, the pattern recognition technology, and the finite element method analysis in the course of researching the software, is outlined in the paper. The results show that this system is practical in underground engineering.

1 INTRODUCTION

There are a large of undetermined knowledge problem and fuzzy information in the underground engineering supporting decision. The research of expert system about underground engineering supporting is one of forward position research task in the intelligence rock mechanics. The combination artificial neural network (CANN) model is combined with the expert system technology, the finite element method, and the pattern recognition technology in the system. It is the study goal that the system provides an effective intelligence assistant tool with a rock underground engineering researcher who processes the problem of rock stablity appraised and supporting decision.

2 GENGRAL STRUCTURE

There are knowledge base , inference machine, explanation machine , study machine, data base, and user-system interface in the system. The main train of thought is showed in the Fig. 1.

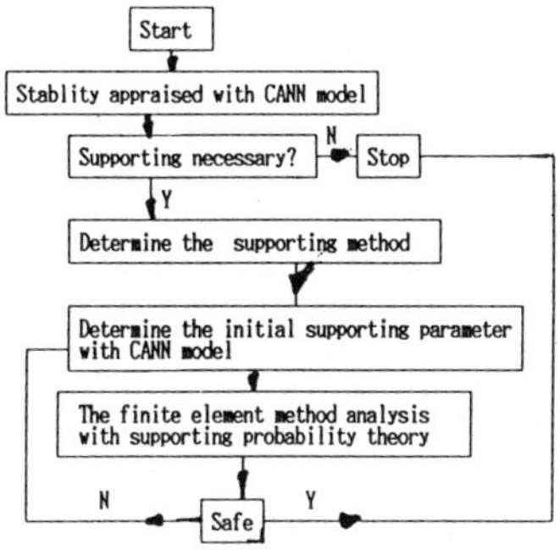

Fig. 1 General structure

3 CANN MODEL BASE THEORY

3.1 Structure

Multi-layered neural network based on BP method is a kind of neural network model which is used most extensively and effectively . In order to solve the rock engineering problem, the CANN model which is developed from the BP model is used in this system. Fig. 2 is the main structure.

3.2 Base theory

Suppose a having N nodal point network, the

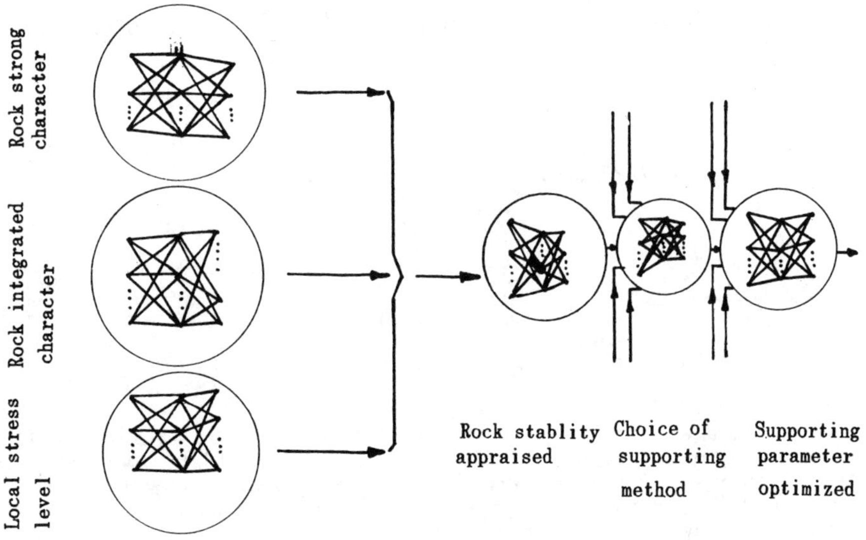

Fig. 2 CANN model

function of all nodal point is Sigmoid function, in order to describe easily, suppose the network has only one output y, the output of nodal point i is O_i, and suppose there are n samples (x_k, y_k) $(K=1, 2, \ldots n)$, to a input x_k, the output of network is y_k, the output of nodal point i is O_{ik}, the input of nodal point j is

$$net_{jk}=\sum_i W_{ij}\, O_{jk} \qquad (1)$$

The error function is

$$E=1/2 \sum_{k=1}^{N} (y_k-\hat{y}_k)^2 \qquad (2)$$

here, y_k is the real output,

Definition

$E_K=(y_k-\hat{y}_k)^2$

$net_{jk}=\sum_i W_{ij} O_{jk}$

$\delta_{jk}=\partial E_k/\partial net_{jk}$

here,

$O_{jk}=f(net_{jk})$

So

$$\frac{\partial E_k}{\partial W_{ij}}=\frac{\partial E_k}{\partial net_{jk}}\cdot\frac{\partial net_{jk}}{\partial W_{ij}}=\frac{\partial E_k}{\partial net_{jk}}\cdot O_{ik}$$

$$=\delta_{jk}O_{ik} \qquad (3)$$

1) when j is the output nodal point, $O_{jk}=\hat{y}_k$

$$\delta_{jk}=\frac{\partial E_k}{\partial y_k}\cdot\frac{\partial y_k}{\partial net_{jk}}=-(y_k-\hat{y}_k)\,f'(net_{jk}) \qquad (4)$$

2) when j is not the output nodal point, there is

$$\delta_{jk}=\frac{\partial E_k}{\partial net_{jk}}=\frac{\partial E_k}{\partial O_{jk}}\cdot\frac{\partial O_{jk}}{\partial net_{jk}}=$$

$$\frac{\partial E_k}{\partial O_{jk}}\cdot f'(net_{jk}) \qquad (5)$$

$$\frac{\partial E_k}{\partial O_{jk}}=\sum\frac{\partial E_k}{\partial net_{mk}}\cdot\frac{\partial net_{mk}}{\partial O_{jk}}$$

$$=\sum\frac{\partial E_k}{\partial net_{mk}}\cdot\frac{\partial}{\partial O_{jk}}\cdot\sum_i W_{mi}O_{ik}=\sum_m \delta_{mk}W_{mj}$$

So

$$\begin{cases} \delta_{jk}=f'(net_{jk}) \cdot \sum_m \delta_{mk} W_{mj} \\ \dfrac{\partial E_k}{\partial W_{ij}} = \delta_{jk} \cdot O_{ik} \end{cases} \quad (6)$$

If the network has M layers, and the M layer has only output nodal point, the first layer is input nodal point, then the BP theory is

1) Choice the initial connected strength W_{ij}

2) Repeating follow processes until weaken

a. k=1 to n

(a) compute $O_{jk} \cdot net_{jk}$ and $\hat{y}_k$

(b) back compute all layer from M to 2

b. from formula 4, 5, and 6 compute δ_{jk} about the same layer nodal point $\nabla_j \in M$

3) revise connected strengthes

$$W_{ij}=W_{ij} - \mu \frac{\partial E}{\partial W_{ij}} \quad , \mu >0 \quad (7)$$

where,

$$\frac{\partial E}{\partial W_{ij}} = \sum_{k=1}^{n} \partial E_k / \partial W_{ij} \quad (8)$$

4 ROCK STABLITY APPRAISED CANN MODEL SUBSYSTEM

4.1CANN base

The affects of rock stablity have many differences because of difference rock classifiction and width. The decisive effect of rock stablity is decided by the rock strong character and rock integrated character. So, CANN base has been built according to the difference rock classifiction and difference width in the system. For example, there are 5 CANNs according to I grade rock (Tab. 1)

4.2 Affect factors and rock stablity appraised grades

The effect of underground water, joint grown level, opening way (all section opened and multi-step opened), opening method (opened by hand, opened by machines, and opened by blasting method etc.), the ratio between height and width are considered in the rock stablity appraised. The final appraised grades are very stablity (VS), stablity (S), relatively stablity (RS), stablity poor (SP) and not stablity (NS).

Tab. 1 CANN base about I grade rock

width (m)	<5	5-10	10-20	20-30	>30
Kind	ICANN1	ICANN2	ICANN3	ICANN4	ICANN5

Tab. 2 Introduce rules of joint grown level

Command degree				Input
NR	RR	R	VR	data
1	0	0	0	1
0	1	0	0	2
0	0.2	0.2	0.6	3.4

NOTE: NR=Not rich, RR=Relatively rich
R=Rich, VR=Very rich

4.3 Introduce rules about input nodal point

Suppose a character has I kind data , the L_i of every data is the command degree ((1<i< I) , then,

$$V=\begin{cases} j & \text{If } MAXL_j=1 \text{ and } L_i=0 (i\neq j) \quad 1<j<I \\ \sum_{j=1}^{I} j \cdot L_j / \sum_{j=1}^{I} L_j & \text{others} \end{cases} \quad (9)$$

Tab. 2 is a example.

4.4 Choice of the hide layer nodal point number

After computing the CANN model with a group engineering samples (The hide layer nodal point is 3, 4, 5, 6, 7, 8, 9, 20, and 25) , we find that the repeatedly compute number is reduced when the hide layer nodal point number is increased. When the nodal point number is some values, the repeatedly compute number is not change, then the repeatedly compute number is

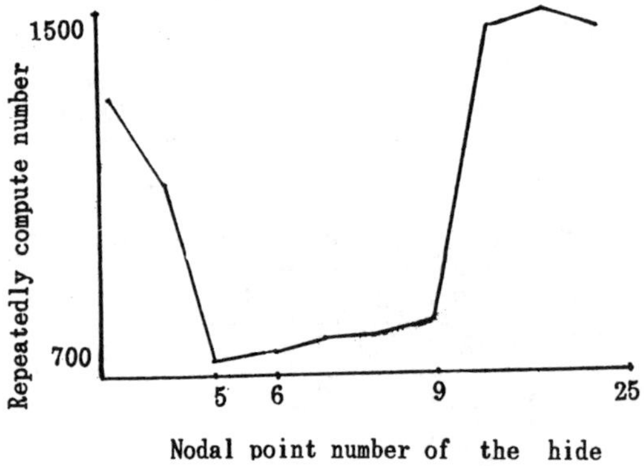

Fig. 3 Choice of hide layer nodal point

increased. The hide layer nodal point number is smallness or largeness, the repeatedly compute number is increased, it causes that the compute time increased. In fact, the function of the hide layer nodal point number is a middle-measure. If middle- measures are very small, the burden of others nodal point will be increased, and the compute time will also be incredsed. On the other hand, if the middle-measure number is largeness, the surplus nodal points are occured. It means that the output is a constant about every one smples , and the study time is increased. For a example, the hide layer nodal point number of the CANN of anchor jet supporting method is 6 (Fig. 3)

4.5 Explantion method

One hope output vector of CANN is (1, 0, 0, 0, 0), then the grade of rock stablity appraised is VS. In fact, this vector is difficult computed by BP model. The follow method is used in this system.

$$\text{IF } MAXO_i > h \text{ and } (MAXO_i) - O_j > 2\delta, \text{ THEN } S_t = i. \quad (10)$$

where, the O_i (i=1, 2, 3, 4, 5) is output vector , δ is error, S_t is the grade of rock stablity, and S_t =1 means VS, S_t=2 means S, S_t=3 means RS, S_t = 4 means SP, S_t=5 means NS. Suppose the region

Tab. 3 CANN base about anchor jet supporting method

Depth (m)	Rock classifiction				
	I	II	III	IV	V
<300	ANNM11	ANNM12	ANNM13	ANNM14	ANNM15
300-500	ANNM21	ANNM22	ANNM23	ANNM24	ANNM25
500-800	ANNM31	ANNM32	ANNM33	ANNM34	ANNM35
800-1000	ANNM41	ANNM42	ANNM43	ANNM44	ANNM45
>1000	ANNM51	ANNM52	ANNM53	ANNM54	ANNM55

data is h, $MAXO_i$ is the largest data of output vectors, O_j $(i \neq j)$ is the others.

After studied, it shows that when h=0. 8, the explanation of output is conform to reality, the error is 0. 01.

5 SUPPORTING DECISION

5.1 Supporting method and supporting CANN base

There are a large of affect factors in the rock engineering supporting decision. Only between similar engineering geology condition and supporting parameters have a group of stablity weight rights . According to the difference of supporting method and rock classifiction, difference CANN models are built (Tab. 3). It's possibly that there are many supporting methods (e. g. anchor jet supporting, whole casting concrete and etc.) which can be used in practice. The supporting method can be chosen by user with the human-computer interface .

5.2 Supporting parameter

The output pattern of CANN is a numerical vector which data rnge is [0, 1] . There are different explanation method in difference application range. According to the feature of underground engineering supporting decision problem , an explanation method based on "practice engineering " is used. If the input pattern of CANN has n features, the output

Tab. 4 A example of supporting parameter

Input data						Output data										Note
I1	I2	I3	I4	I5	I6	01	02	03	04	05	06	07	08	09	010	
						10	200	10	60	22	275	60	60	280	A	R
110	X0	6.9	6.85	9800	2.7	10	200	12	20	22	300	80	80	280	A	C

where,
I1: depth (m)
I2: section form
I3: general span (m)
I4: total height (m)
I5: elastic modules (Mpa)
I6: rock density (T/m^3)
01: jet material thickness (m)
02: concrete standard
03: diameter of reinforcing bar net (cm)
04: reinforcing bar net mould distance (cm)
05: anchor diameter (m)
06: anchor length (cm)
07: interval distance of two anchors (cm)
08: interval distance of two anchors (cm)
09: aonchor tensil strength (Mpa)
010: supporting area
R: the result used in the engineering
C: the result computed by this system

pattern of CANN has k features, it can be understood as: a n dimension Euclidean space (IV) is maped to a k dimension Euclidean space (OV). $S=\{S_i\}$ is a group of samples (i=1, 2, ... q), I is a N dimension characteristic vector corresponding samples S_i.
0 is a K dimension characteristic vector corresponding sample S_i, then

$$S=\{(I_i, 0_i)\}, i=1, 2, \ldots q \quad (11)$$

Hpothesis PE is a practice engineering which will be sloved. Input pattern vector is NSI, after computed by CANN, output pattern vector is NSO. {0}is a group of output in S. $\{S_{min}\} \in S$ is a group of samples which distance is the shortest between NSO and $0_j \in \{0\}$ in Euclidean space.

$$D_{ov}\{0_i\}=\min||NSO-0_j||^{0.5} \quad (12)$$

there are many 0_j which can be matched with formula (12).

Hypothesis $\{I_{min}\}$ is a group of input vector of $\{S_{min}\}$. The sample of the shortest distance between NSI and $I_j \in \{I_{min}\}$ in the IV space is

$$D_{iv}\{I^*\}=\min\in \{I^*\}||NSI-I_j||^{0.5} \quad (13)$$

where, I^* is input vector of $\{S_{min}\}$,
that is to say
$(I^*, 0^*) \in \{S_{min}\}$
then, $(I^*, 0^*)$ corresponding sample is an explanation of practice engineering PE. The supporting parameter of samples $(I^*, 0^*)$ can explained a initial supporting decison parameter of engineering PE.

5.3 Supporting probability analysis

After determined the initial supporting parameters, the supporting probability with finite element method is used in this system. Following rules is the relation between rock supporting stablity and probability. Suppose P means probability.

Rule1, IF $P > 90\%$, THEN rock supported stablity grade is VS.

Rule2, IF $90\% > P > 80\%$, THEN rock supported stablity grade is S.

Rule3, IF $80\% > P > 70\%$, THEN rock supported stablity grade is RS.

Rule4, IF $70\% > P > 60\%$, THEN rock supported stablity grade is SP.

Rule5, IF $P < 60\%$, THEN rock supported stablity grade is NS.

6 EXAMPLES

6.1 Example 1

The output vector of network about a engineering is (0.83, 0.31, 0.22, 0.03, 0.15), according to formula (3), the grade of rock stablity appraised is VS.

6.2 Example 2

Tab. 4 is a example which is used by this system. After computed the probability of this tunnel engineering with finite element method, the rock supporting probability of this tunnel is over 80%, so the supporting parameter computed from this system is reasonable.

7 CONCLUSIONS

1) A profitable attempt that the "compute intelligence"has been used instead of the "symbol intelligence" in this softwre. The difficulty of the "knowledge blasting"and "match contradiction " is avoided.

2) Building a rock underground engineering supporting decision expert system based on the CANN model is significance. It provided a new method to increase the intelligence level of geomechanics expert system.

3) There is not a good method with the choice of good layer numbers of CANN model and hide nodal point in BP model at present. The part minimum problem and the slowness of convergence velocoity is existnce in BP model. It's needed to study in future.

4) CANN expert system can be finished on a large scale integrated circuit technology. It estimate that it will be translated into cheap commodity, and used widespreadly in future. It make a good prospect to the development of expert system and its appliction in rock engineerings.

REFERENCES

Jiao licheng, 1990, 12. Neural network system theory, Xian Univ. of Elec. & Tech. Press, CHINA

Jianxi Ren, 1993, 5. Research on an expert system of supporting decision in rock underground engineering based on the numerical analysis. (Master dissertation). Xian CHINA.

Computer Methods and Advances in Geomechanics, Yuan (ed.) © 1997 Balkema, Rotterdam, ISBN 90 5410 904 1

A new method to control floor heave in heterogeneous strata

Chaojiong Hou, Yanan He & Xiao Li
China University of Mining and Technology, Xuzhou, China

ABSTRACT: By means of FEM numerical calculations, the occurrence, development and the end form of the plastic zone surrounding gateway and the relation between floor heave of gateway and sinking of two walls are studied. A new principle and method are proposed to control floor heave by means of reinforcing the two walls and corners in soft and weak rocks surrounding gateway. Numeric analysis using computer shows that reinforcement of different positions of the gateway will help to control the floor heave, but the results of reinforcement of two walls and corners are best. Underground commercial test is made in Huantanglin coal mine, remarkable result is obtained in control of floor heave.

1 INTRODUCTION

Various measures to control floor heave in gateway so far are generally focused on the floor strata and the roles which they have played can be classified into following: 1) increase the deformation resistance of floor strata; 2) heighten the strength of floor strata; 3) lower the stress in shallow portion of floor strata and transfer the excavation-induced high stress to the deep portion of surrounding strata; 4) low the stress in floor strata and heighten its strength simultaneously (Oldengott, 1981; Callis and Newson, 1987; Afrouz, 1990; Hou and Zhang, 1987; Черняк, 1978; Кошелев и Петренко и Новиков, 1990; Пирский и Стовпник 1990; Литвинский, 1986).

The authors have studied the fundamental principles of controlling the floor heave in heterogeneous and stratified ore bodies and raised the new method to control floor heave by means of reinforcing the two walls and corners (chiefly the bottom corners) in soft and weak rocks surrounding gateway. Both the FEM calculations and the field tests have shown that the best effects could be achieved to control floor heave by reinforcing the two walls and corners in soft and weak rocks. Moreover, the new method is simple in construction and could be popularized and employed easily.

2 FUNDAMENTAL PRINCIPLES TO CONTROL FLOOR HEAVE IN GATEWAY

Gateways in coal seams are generally excavated in heterogeneous and stratified strata. The two walls of gateway are normally coal with lower strength in comparison with roof and floor strata. The non-linear FEM numerical simulating analysis was conducted with program ADINA.

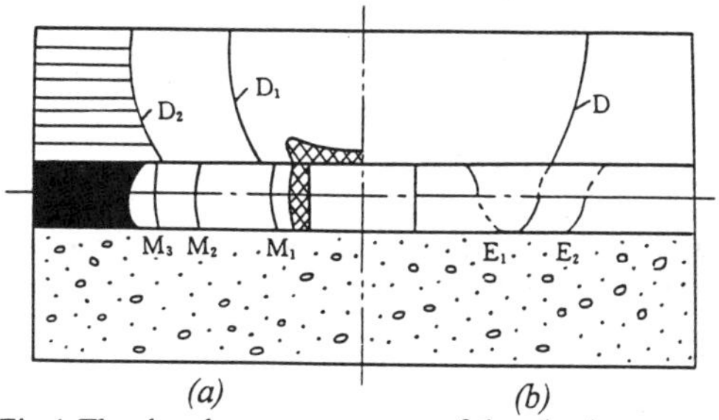

Fig.1 The development process of the plastic zone

The analysis reveals that the plastic zone of the surrounding strata of the gateway develops gradually. It is closely related with the mining activity. The development process of the plastic zone will be illustrated by the analysis result from model of the solid coal way (with the nether roof, but without the immediate bottom).

When starting heading the way, a plastic zone, which is not very large, develops in the surrounding strata under the stress concentration (see the double line shade of Fig.1(a)). When under the moving abutment pressure in front of the working face, the stress on the upper roof begins to increase by 1.5, 2.0 or 2.5 times, accordingly, the plastic zone in the two side walls extends drastically to M1, M2 and M3 position. The zone in the roof extends upward to the upper stable roof rock, and its width develops to line D1 and D2. As the zone develops gradually, the plastic boundary lines like M1, M2, M3 and D1, D2 will go through the whole plastic zone. The plastic zone lines can be taken to be a componet of a group of smooth curves (E1 and E2). The existance of the free face in side walls makes the upper part of E1 and the lower part of E2 fully develop, which causes the coal body to deform along them and the two sides come out into the free space (see Fig.2(b)).

If there is immediate bottom in model, the process of the plastic break of bottom is very like the roof.

The research conducted by FEM numerical calculation indicates that under the action of secondary stress after excavation, the plastic zone in surrounding rock starts from the two walls of the lowest strength and the corners of high stress concentration, shown in Fig.2. With the development of plastic zone in two walls and corners, the plastic zone in other parts also gradually extends. However, the plastic zone in two walls and corners are the largest finally (Ma and Hou, 1990). The bigger the plastic zone in two walls and corners develops, the larger the floor heave of gateway will be that resulted from plastic deformation, visco-plastic flow and volume bulking of broken strata.

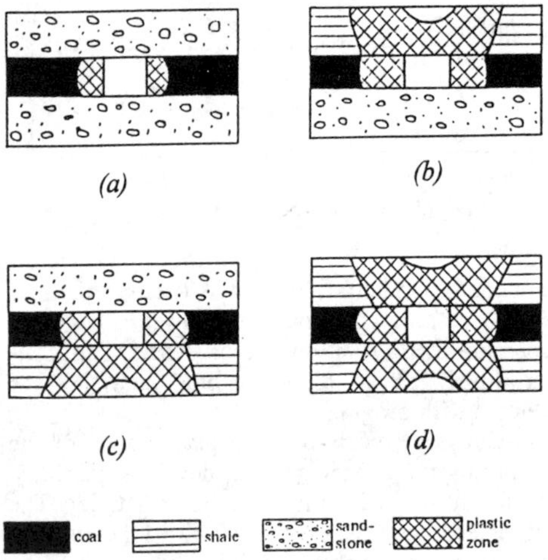

Fig.2 Development of plastic zone in surrounding rock of gateway

(a) rocks around two walls with low strength
(b) rocks around two walls and roof with low strength
(c) rocks around two walls and floor with low strength
(d) rocks around gateway with low strength

Besides, mining-induced high stress may compress each layer of strata around gateway and cause them sinking (Hou and Ma, 1993). As soon as the two walls sink, they'll accelerate the rupture and sliding in floor strata so that the floor heave will be more drastic. The softer the two walls, the bigger the sinking induced by compression and the larger the floor heave. The FEM calculated results based on the strata condition around gateway in Huangtanglin coal mine illustrate that with the increase in overburden depth, there has the increase in wall sinking and floor heave. The relation between the amount of floor heave in floor center and the amount of sinking in wall center is shown in Fig.3.

In order to control the floor heave, the two walls and corners (mainly the bottom corners) in soft and weak strata should be reinforced as early as possible after being excavated. Its functions are:

1) Weaken the stress concentration around corners and form the bearing arch with high self-bearing capacity in two walls and corners so as to control the development of plastic zone in two walls and corners.

2) Heighten the self-bearing capacity of strata around two walls and corners (especially the bottom corners) and lower the sliding floor heave caused by compression and sinking of loose and broken strata in two walls.

3 TEST STUDIES IN HUANGTANGLIN COAL MINE

3.1 Numerical Simulation Study

The non-linear FEM numerical simulation calculation was conducted based on the condition of field test gateway in Huangtanglin coal mine.

The coal being mined out in Huangtanglin coal mine belongs to lignite in Tertiary of Cainozoic group. The strata dip is 7 to 8 degrees. The roof and floor of coal seam are thick mudstone. The clay minerals are dominated by kaolinite and montmoril onite. The test gateway is driven along the roof in coal-body with a elevation of -198m. The uniaxial compressional strength of roof, floor and coal seam are 19.6 MPa, 13.4 MPa and 12.1 MPa respectively.

To reflect the practical loading state of gateway, we utilize the calculation method of "loading first and excavation followed". To study the amount of floor heave due to different portion of strata being reinforced, we adopted the method of decreasing the rockmass strength in plastic zone to simulate the loose zone of surrounding rock (Song and Sun, 1991). Plastic zone is considered to be the loose range, the strength of rockmass in this range is lowered 40 to 50 percent and the strength of

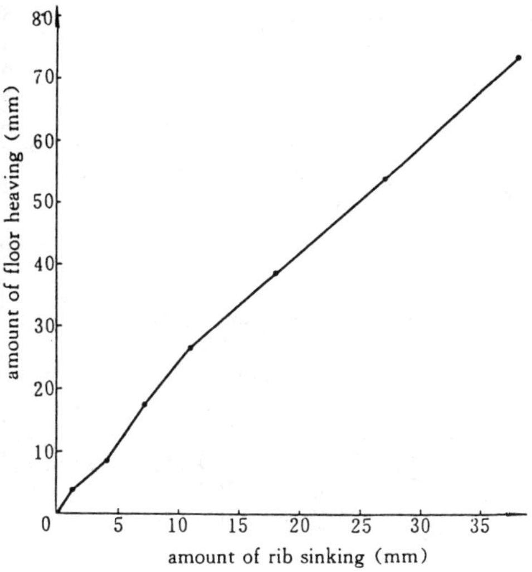

Fig.3 Relation between floor heave of roadway and sinking of two walls in Huangtanglin coal mine

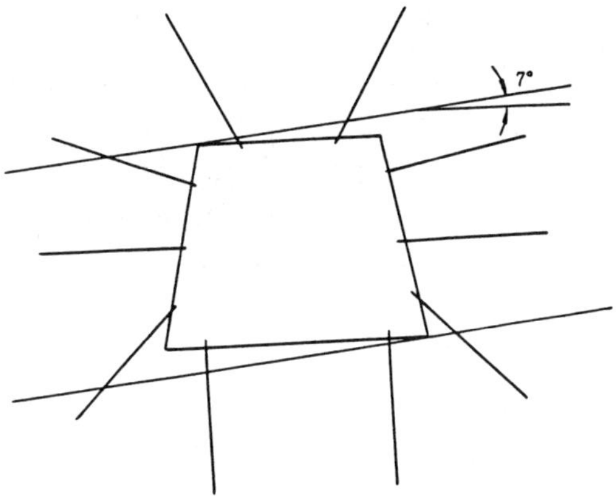

Fig.4 The arrangement of bolt in the test gateway, Huangtanglin coal mine

rockmass in the reinforced zone is decreased 8 to 10 percent. For the comparison of plans with the same conditions, the method can reflect the general regularity.

Five plans were considered to reinforce the surrounding rock. The amount of floor heave and the relative decreasing ratio of floor heave in floor center of gateway for various plans are listed in Table 1. It shows that under the production and geological conditions of Huangtanglin coal mine, the plastic zone of surrounding rock after excavation occurs and develops at two walls and corners of gateway, and the effects of controlling floor heave are the best by way of reinforcing the two walls and corners in five plans.

3.2 Field test study

The combined support of extendable bolt (∮ 14 × 1700 mm) and bidirectional yieldable support with No.11 mine-used I-steel is employed in the test gateway. To control the floor heave, the bolt is arranged in the pattern of reinforcing the walls and corners shown in Fig.4, where the bolt is anchored by two swelling-cement-cartridges of ∮ 38 × 200 mm and the anchorage force reaches more than 70 kN.

The monitoring on the floor elevation of test gateway and headentry that is not reinforced at elevation of -163m was carried out by means of leveling. The results are listed in Table 2. For the test gateway, its service period is 291 days. The convergence of roof-to-floor and wall-to-wall are 409 mm and 260 mm, respectively. The monitored data show that under the condition of soft and weak strata, the amount of compressional sinking in various strata caused by the secondary stress is relatively large. Since the development of plastic zone in walls and corners and the sinking of surrounding rock around two-wall are all restrained after the soft and weak walls and corners around gateway are reinforced so that the floor heave is controlled. The variation of floor elevation by leveling displays the minus values representing the compressional sinking of various strata.

Since the numerical simulation only studied the effects of reinforcing the different portion of loose zone and didn't consider the compressional sinking effect of strata, therefore the calculated results are only the comparative values of relative effects to control floor heave in different reinforcing plans and not the real values of floor deformation.

Table 1 Comparison of effects to control floor heave in different reinforcing plans

calculation plans	1	2	3	4	5
portion to be reinforced	No	Roof	Walls	Floor	Walls & Corners
amount of floor heave, mm	23.3	21.8	17.7	16.2	15.6
decreassing ratio of floor heave	0	6%	24%	31%	33%

Table 2. The variation of elevation for measuring points in gateway floor of -198 m and -163 m

gateway location	reinforcement state of surrounding rock	start & end date for measurement	time interval	variation of elevation in floor	note
-163 m	No	Sep. 6 - Dec.1, 89	86 days	+285 mm	Positive value shows floor going up; Netative value shows floor going down.
-198 m	Walls & corners are reinforced	Jul. 26 - Oct.5, 90	71 days	-83 mm	
		Jul. 26, 90 - Jan.3, 91	161days	-397mm	

4 CONCLUSION

1. When the coal gateway is driven in heterogeneous and stratified ore body, the plastic zone in surrounding rock usually occurs and develops in the two walls and corners around gateway where has the weakest strength and stress concentration due to excavation-induced secondary stress. Meanwhile, the secondary stress can also lead to the compressional sinking in various strata around gateway. Therefore, the soft and weak walls and corners of gateway should be reinforced as early as possible after being excavated so as to restrain the development of plastic zone and the sinking of two walls and finally to control the floor heave.

2. Both the FEM numerical simulation and the field test underground indicated that the reinforcement of walls and corners (mainly the bottom corners) in soft and weak surrounding rock is a new and effective method to control the floor heave.

3. The reinforcement of walls and corners in soft and weak surrounding rock could be realized by grouting and bolting. Compared with the method of controlling floor heave by reinforcing floor heave completely, the proposed method not only has the best effect to control floor heave, but also is simple and convenient in construction and easy to be popularized and applied.

REFERENCES

M. Oldengott, Massnahmen zur verringerung der sohlenhebung, Verlag Glückauf GmbH. ESSEN, 1981.

A. V. Callis and S. R. Newson, Progress into Roadway Reinforcement Techniques in the UK, The Mining Engineer, 1987, No.11, 233-242.

A. Afrouz, Methods to reduce floor heave and sides closure along the arched gate roads, Mining Science and Technology, 1990, No.10, 253-263.

Hou Chaojiong & Zhang Shudong, Control of Gateroad Floor Heaves by a New Type of Ring Support, Mining Science & Technology (Proceedings of the Intenational Symposium on Mining Technology and Science, September 1985), China Coal Industry Publishing House & Trans Tech. Publications, 1987, 147-157.

И.Л. Черняк, Предотврашение пучения почвы горных выработок, Москва "Недра", 1978.

К.В. Кошелев, Ю.А. Петренко, А.О. Новиков, Охрана и ремонт горных выработок, Москва "Недра", 1990.

А.А. Пирский, С.Н. Стовпник, Опытно-промышленные испытания способа упрочнения пород ддя борьбы с пучением почвы, Угопь Украины, 1990, No. 4, 9—11.

Г.Г. Литвиский, Новый способ сооружения обратного свода крепи, Шахтное строитедьство, 1986, No. 2, 24—26.

Nianjie Ma & Chaojiong Hou, A research into plastic zone of surrounding strata of gateway affected by mining abutrment stress, Proceedings of the 31st U. S. Symposium 211-217, A. A. Balkma / Rotterdam / Brookfield / 1990.

Chaojiong Hou & Nianjie Ma, The integral sinking of rocks surrounding actual mining roadway and its mechanic analysis, Proceedings of the 2nd International Symposium on Mining Technology & Science 509-517, China University of mining & Technology, Xuzhou, Jiangsu, P. R. C. 1991.

Song Dezhang & Sun Jun, Mechanism of study on the Mechanical shotcrecte-rock bolt support system, Chinese Journal of Rock Machanics and Engineering, Vol. 10, No.2, 197-204, 1991.

Yanan He & Yongnian He, Control of roadway deformation due to dynamic pressure and control of heaving in soft rock in Maoming by using combined yieldable supports, Journal of China Coal Society, Vol. 17, No.4, 17-23, 1992.

Computer Methods and Advances in Geomechanics, Yuan (ed.) © 1997 Balkema, Rotterdam, ISBN 90 5410 904 1

Ground movement and control of UCG in elevated temperature

Wang Zaiquan & Hua Anzeng
China University of Mining and Technology, Xuzhou, China

ABSTRACT: This paper firstly introduces Underground Coal Gasification (UCG) Technique and the related strata fall, movement and subsidence. It secondly analyzes the physical and mechanical characteristics of coal and surrounding rock in elevated temperature. Taking the semi-industry test of UCG in Xinhe Mine for example, the authors study cavity stability, ground movement and subsidence law associated with UCG by numerical simulation. Then the stability of cavity, subsidence, state of post-burn cavity is surveyed in the field. Finally, discuss the problems to be solved in the future.

1 INTRODUCTION

Underground Coal Gasification (UCG) is a chemical mining method which changes in situ coal into burning gas through the controlled burning of coal underground (Fig. 1). UCG can increase the ratio of utilization of coal source, decrease the labour intensity of workers, improve the safty condition, greatly diminish the environment destruction and air pollution, and obtain a better economic results with the modification of UCG technique. Thus most coal producing countries payed attention to this technique.

Obviously, the gasification cavity will be gradually developed with the coal gasification, and the mechanical properties and stability state of coal and rock strata will be changed greatly. When a cavity developed, to some extent, the stability of cavity will be decreased, and the coal and adjoining rock will fall. The falling of coal and rock strata not only disturbs the normal process of gasification (such as the choke of channel, blanking of burning flame), but also causes the possibility of surface water imbibing into gasifier, seeping of gas from gasifier, and the destruction of surface building.

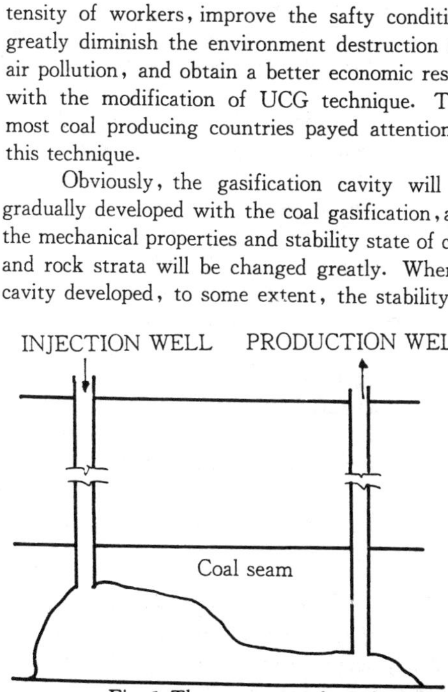

Fig. 1 The structure of gasifier

So for the realization of continuous, stable, safe production of coal gas and efficient control of the gasification process, the stability state of gasifier and the related ground movement, subsidence must be analyzed and controlled. Taking the semi-industry test of UCG in Xinhe Mine as an example, this paper firstly analyzes the physical and mechanical characteristics of coal and surrounding rock in elevated temperature, then studies cavity stability, ground movement and subsidence law associated with UCG by numerical simulation. The stability of cavity, subsidence, state of post-burn cavity is surveyed in situ. Finally, discusses the advantages of UCG in diminishing of ground movement and subsidence and the problems to be solved in the future.

2 THEORETICAL BACKGROUND

The general equations governing the coupled thermal-mechanical-moisture diffusion response of in situ geological materials are:

(i) An energy equation for conductive-convective heat transfer through a porous medium with moisture flux, heat source/sink terms, and dilatational coupling

$$\rho_s \cdot C_s \frac{\partial T}{\partial t} = (k_{ij}T_{,j})_i - \varphi\rho_g \cdot Cg \cdot V_i T_{,i}$$

$$+ H(T) + \alpha_1 \frac{\partial M}{\partial t} + \alpha_2 \frac{\partial \varepsilon}{\partial t} \quad (1)$$

where ρ is the mass density, C is the specific heat, k_{ij} is the thermal conductivity , H is the heat source/sink term, T is the temperature, φ is the porsity , M is the moisture function (pore pressure function), V_i is the convective flux coefficient, $\varepsilon = U_{k,k}$, is the dilatation and subscipts s and g are the solid and gas designations, respectively.

(ii) The momentum equations with inertia, body force, thermal gradient, and moisture gradient terms. These equations derived from variational considerations are

$$(D_{ijkl}\varepsilon_{kl})_{,j} + F_i = \rho_s \ddot{U}_i + \alpha_{ij}T_{,j} + \beta_{ij}M_{,j} \quad (2)$$

where D_{ijkl} is the temperature dependent material property tensor, ε_{kl} is the strain tensor, F_i is the body force vector, U_i is the displacement vector, α_{ij}, β_{ij} are thermal and moisture diffusion coefficients respectively.

(iii) The diffusion equation with temperature coupling

$$\rho_s \cdot C_m \frac{\partial M}{\partial t} = (m_{ij}M_{,j})_{,i} + (t_{ij}T_{,j})_{,i} \quad (3)$$

where C_m is the moisture capacity, m_{ij} is the moisture conductivity, and t_{ij} is the coupling thermo-gradient coefficient. In the above equations and throughout the paper indices i, j, k and l assume values 1 to 3 (2 in 2-dimension) and the summation convention is implied whenever the indices are repeated. Partial differenation with respect to the x_i coordinate is denoted by " ,i".

For above governing equations, combined with the characteristics of UCG, They can be simplified the following equations:

$$\rho_s C_s \frac{\partial T}{\partial t} = [K_{ij}^M M_{,j} + K_{ij}^T T_{,j}]_{,i} + H_{(T)} \quad (4)$$

$$\sigma_{ij}, j + F_i = 0 \quad (5)$$

$$\sigma_{ij} = D_{ijkl} \cdot \varepsilon_{kl} + \alpha_{ij}\Delta T + \beta_{ij}\Delta M \quad (6)$$

Where K_{ij}^M, K_{ij}^T is the coefficient of conductivity of temperature and moisture. For Xinhe mine $K_{ij}^M = 0$, $\beta_{ij} = 0$. From above equations, the physical properties are the thermal heat C_s, conductivity k_{ij} and temperature expansion coefficient α_{ij}. Obviously, these properties are dependent on temperature. Fig2. to Fig. 4 showed the relation between physical properties and temperature (RE. Glass). The elastic modulus of mechanical properties is also the function of temperature (Fig. 5) (RE. Glass). From consideration of FEM simulation, there are conclusions concerning the effects of properties on modeling effort: the thermal properties can be treated as istropic, the expansion coefficients are such that tensile stresses will dominate in the vicinity of the hot cavity. The value of above properties lies in their use in a cavity growth model for UCG, The model consists primarily of twostages:

The finite element simulation of temperature distribution by ANINAT.

The finite element simulation of stresses, displacement and the determination of failure zone by ANIDA, and is used to gererate the probable shape of cavity formed during UCG.

3 ANALYSIS OF CAVITY STABILITY AND GROUND MOVEMENT IN UCG

3.1 *Cavity characteristics in process of UCG*

When gasification begans, according to the difference of gasification reaction and temperature, the gasification zone may be divided into three zones. It is known that the temperature field in the coal seam and adjoining rock strata is different in different reacton zones. Thus the stability of cavity is different. From consideration of the gasification technique, the stability of oxdination cavity is the must important for normal gasification of coal seam. In this zone, burning cavity stability and roof collapse may be forecast by means of numerical simulation. In the simulation of cavity stability, we must be consider the thermal soften of coal and rock strata and the parameter changes with temperature, moisture.

When a gasification process ended, the temperature in the coal seam and adjoining rocks will gradually decrease and finally cool to normal temperature. The drying distillation action of coal seam will make the coal carbonize, the roof and floor rock strata (such as shale, clay) vitrify. The carbonizing and vitrifing will make the bearing capacity of coal and rock strata increase, thus stability of post-burn cavity will be improved. On the other hand, the ash (production of coal burning) and collapsed coal and rock will stow the gasification cavity. Howeer if the post-burn cavity is too large to support the overburden load, strata movement and subsidence will also take place. In this circumstances, some measures must be adopted to improve and controll the stability of the cavity.

3.2 *Stability analysis of cavity associated with UCG*

The stability analysis of pre-gasification and post-burn are similar to that of normal mining, thus this paper only simply introduces the dynamic stability method and conclusion of gasification. The soften-

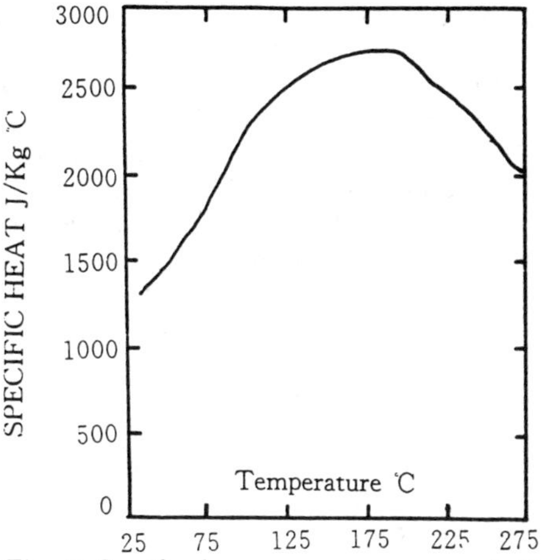

Fig. 2 Specific heat-temperature curve (RE. Glass)

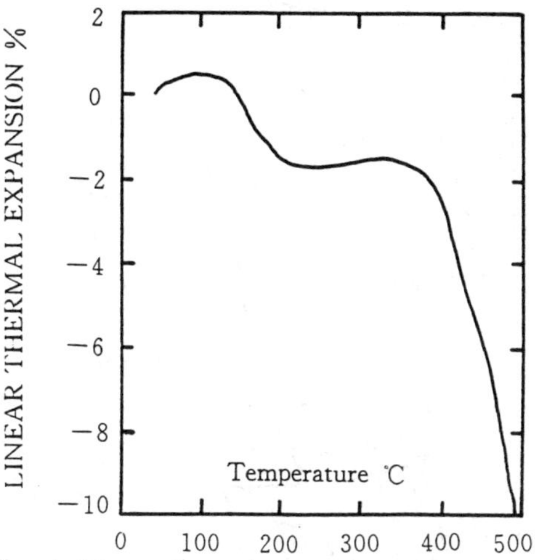

Fig. 4 Linear thermal expansion-temperature curve(RE. Glass)

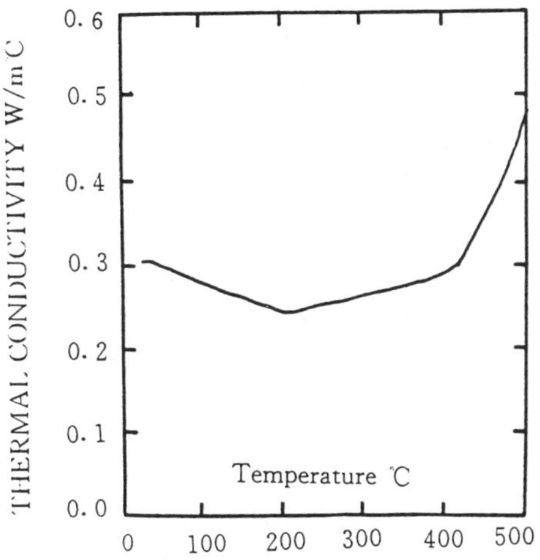

Fig. 3 Thermal conductivity-temperature curve (RE. Glass)

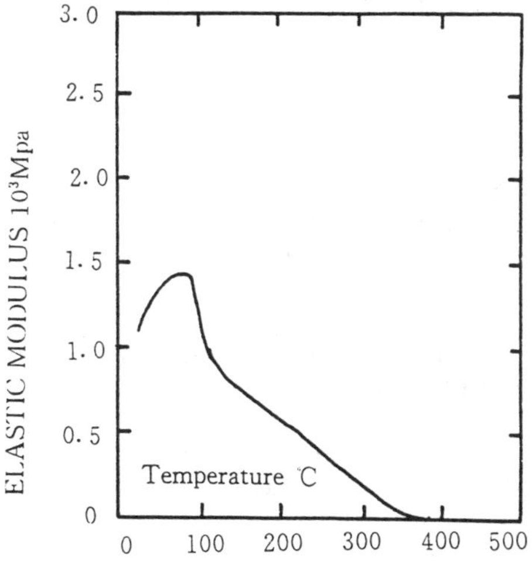

Fig. 5 Elastic modulus-temperature curve (YANG, 1978)

ing of coal and overburden and thermal stress are the most important factors, so the paper mainly considers above two factors in analyzing cavity stability.

Firstly, according to the temperature distribution of coal and shale overburden, different physical parameters are adopted for different temperature elements. For example, the elastic moduli in different temperatures (softening properties of coal and overburden) are used in FEM simulation, and the thermalelastic model is adopted to solve the stresses, displacements and the stability state. If the temperature is greater than 316℃, the elastic modulus value of related element is taken as 1/50 of initial value (normal temperature codition), because the elastic modulus will be decreased to zero when temperature is greater than about 316℃

The follow conclusions can be made from the prelimiary analysis and external similar results associated with UCG:

In the condition of Xinhe Mine (the final combustion height was about 10m), only local loss of stability happened in the upper coal and upper roof of the caity (Fig. 6), the entire cavity stability was assured, and had no subsidence tendency. The strata movement was limited in the scope of shale overburden, thus gas leakage from gasifier and

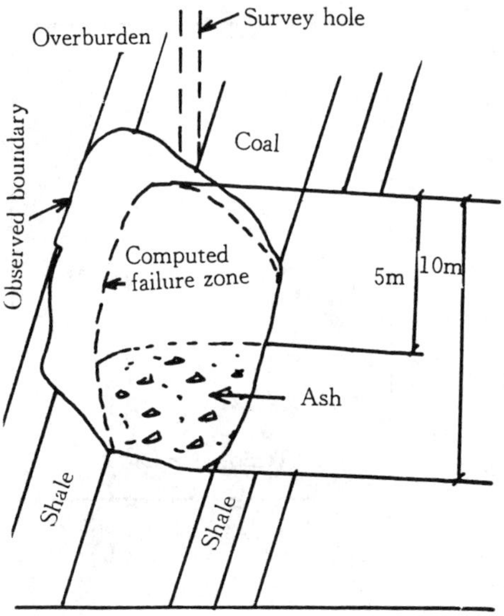

Fig. 6 Failure zones induced by UCG

water intaking into gasifier were avoided, this has also been verfied by the field survey.

The softening of coal and shale makes the thermal stress decrease greatly because thermal stress is mainly determined by the elastic modulus. Thermal stress is specially small in the softening zone of cavity, but displacement is large.

4 FIELD SURVEY OF ROOF COLLAPSE, SUBSIDENCE AND CAVITY STATE

The surface subsidence was monitored. The monitored results show that there has not been obvious subsidence to date.

The geophysical prospecting of post-burn cavity was done to determine the scale of cavity. From the abnomal information of electric resistence ratio surveyed in the field, we can conclude the post-burn height of the cavity is about 10m. (Fig. 6) From the results of borehole television survey of post-burn cavity, we know that the post-burn cavity is mainly filled with ash and roof coal and rock collapsed, the height of remaining cavity is only about 5m (Fig. 6), which is beneficial to the prevention of extra strata movement and subsidence.

From above research, we can come to the conclusion that the UCG process does not cause obvious subsidence, and roof collapse does not seriously affect the gasification process because the scale of collapse is limited in the allowence. The field survey results are also in accordance with the simulation results.

5 CONCLUSION

The elevated temperature associated with UCG makes the mechanical properties of coal and overburden change greatly, thus the cavity stability analysis must consider this serious changes.

In Xinhe's condition of gasification, the scope of roof collapse is limited and had not induced subsidence. The post-burn stability state of cavity is also improved. Further study is needed on complicated temperature and moisture field, and the transformation process of heat, and gas, and the influences each other, for the simulation of dynamic stability of cavity associated with UCG.

REFERENCES

Advani, S. Y. T. Lin and L. Z. Shock, 1983, Thermal and structure response, evaluation for UCG, Society of Petrolem Engineers Journal, Vol, 17, No. 6: 413—422.

Yang -Tai, Thomas lin, 1978, Structural mechanics simulation associated with UCG, Ph. D. Dissertation, West Virginia University, Morgantown, West Virginia: 1—40

Advani, J. K. Lee. etal, 1983, Stresses mediated response associated with UCG cavity and subsidence predition modeling, Proceedings of the Ninth Annual UCG Symposium, August: 282—287.

R. E. Glass. The thermal and structural properties of a Hanna basin coal, Journal of Energy Resources Technology, Transactions of the ASME, Vol. 106, June, 1984: 266—271

H. H. Mortazari, etal. The effect of moisture on the structural stability of a coal cavity. Journal of Energy Resources Technology, Transactions of the ASME, Vol. 108. Sept: 246—253

R. E. Glass, etal. The thermal and structural properties of the coal in the big seam. In situ, 8 (2) 1984: 193—205

Rlbert. J. Mae. Kinon. Moving grid finite lement imulation of avity growth and rock response associated with UCC. 27th US Symposium on Rock Mechanics: Key to Energy Production: 716—723

Computer Methods and Advances in Geomechanics, Yuan (ed.) © 1997 Balkema, Rotterdam, ISBN 90 5410 904 1

Calculating the location parameters of roadways by using numerical method

Xizheng Zhou & Jianbiao Bai
China University of Mining and Technology, Xuzhou, China

ABSTRACT: Under the condition of the strata composed of bedded sedimentary rocks simplified to macroscopic isotropy and according to the abutment pressure redistribution originated from mining and the stability coefficient of roadways, this paper discusses numerical method to calculating the location parameters of mine roadways.

1 SYNOPSIS

The activity of underground mining will produce the area of increased stress and the area of decreased stress. The stability of mine roadways is affected not only by original stress but also by the stress situation redistributing around working faces. So it is the location of roadways to working faces that is an important content in roadway design.

Generally, the main location parameters of roadways are as follows: the safe height between roadways and above extraction space, the rational horizontal distance between roadways and extraction boundary, and the stable width of protecting pillars. The main purpose of this paper is to calculate above parameters and the corresponding stability coefficient of roadways under the conditions of different mining depth and different strength of surroundings by the means of numerical analyses.

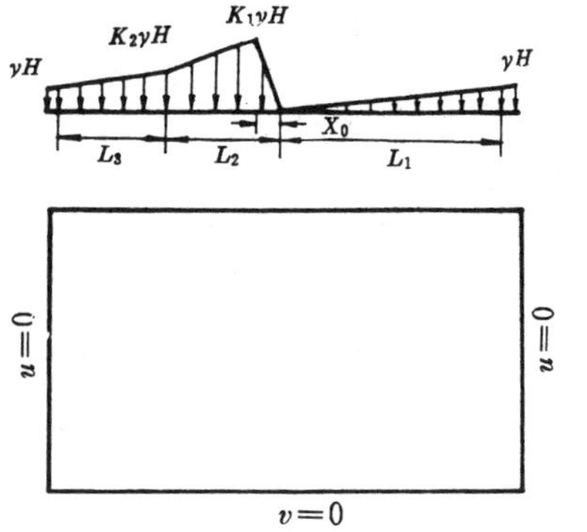

Figure 1. Calculating pattern

2 CALCULATING PATTERN

In coal mining, ore body in sedimentary environment generally lies in bedded strata whose thickness and mechanical property stochastically vary.Under the condition of certain hypotheses, floor strata can be substituted with equivalent macroscopic isotropy (Tan XueShu 1994). The calculating pattern is indicated as Figure 1 and Figure 2. In the figures,H indicates mining depth (m); γ indicates the average unit weight of the superincumbent strata above coal face (kNm^{-3}); H_1 indicates the thickness of the superincumbent strata above induced fracture zone (m). The range of abutment pressure distribution and the coefficient of stress increase are showed in Table 1 (Jang JinQuan 1993). The stress distribution in

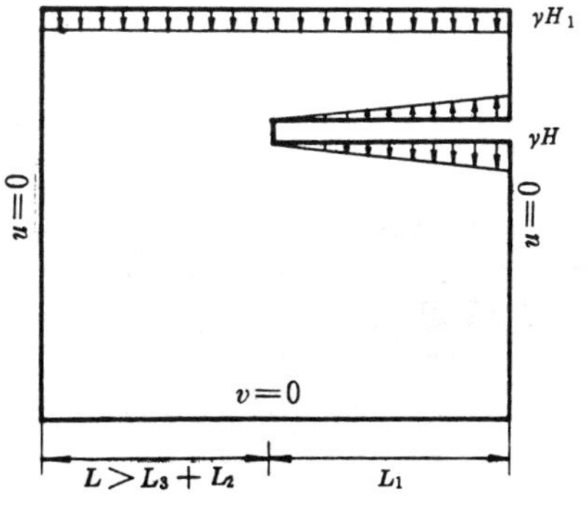

Figure 2. Calculating pattern

Table 1 Some parameters of abutment pressure

Mining depth	Range of abutment pressure (m)				Coefficient of stress increase	
H (m)	L_1	L_2	L_3	X_0	K_1	K_2
<200	95	20	40	5		
300	125	25	50	5		
400	150	30	60	6		
500	165	30	60	6	3	1.8
600	180	30	60	6		
800	205	35	70	7		
>900	215	40	80	7.5		

floor can be calculated with the analysis of FEM.

In the calculating model, the floor strata of coal seams is regarded as semi-infinite. It is dealt with in the light of the linear-elasticity problems of plane strain with a longitudinal section or along with the advancing direction of working faces or along with inclined direction, the floor is part of sandshale. In the calculating model 2 the roof strata is remained proper thickness and the load of covering strata is even load.

3 STABILITY COEFFICIENT OF ROADWAYS

The most important elements of numerous factors influencing the stability of roadways are the stress of surrounding rocks, the strength of surrounding rocks and the relation between them. It is difficult to accurately measure the stress and strength of surrounding rocks. In practice, the ratio of the maximum principal stress of the location ahead of the excavation of roadways to the uniaxial compressive strength of the rocks around roadways is defined as the stability coefficient of roadways. Based on numerous observed data, the relation between the stability coefficient of roadways and the stability degree of roadways can be determined as Table 2 (Li ShiPing 1986).

The stability of roadways is reflected by the attainment of safety and minimum necessary section during the working period of roadways. In theory, when the surrounding stress exceeds or is equal to the strength of surroundings, roadways start being destroyed and are in the state of instability. The stability coefficient of roadways and the displacement of surroundings shown in Table 2 are the bases of selecting the pattern of supporting and determining the parameters of supporting.

Table 2 The stability coefficients of roadways and displacement of surrounding rock

Stability of roadways	Stability coefficient	Displacement of surrounding rocks (mm)
Stability	< 0.25	< 50
Medium-stability	0.25 ~ 0.4	50 ~ 200
Instability	0.4 ~ 0.65	> 200

4 CALCULATING THE LOCATION PARAMETERS OF ROADWAYS

Based on the required stability of roadways and the strength of the surrounding rocks of roadways, the scope of the permitted maximum principal stress around roadways need be determined. Then based on the stress distribution in floor which has been worked out, the location parameters of roadways can be obtained under different extraction depth.

When roadways are affected by above extraction activity, they at least remain medium-stability.The range of the maximum principal stress is determined in accordance with formula (1) and then the safe

Table 3 The height between roadways and above extraction space h (m)

Mining depth	Uniaxial compressive strength of surrounding rocks R (MPa)							
H (m)	<20	30	40	50	60	80	100	120
<200	50(0.37)	50(0.25)	50(0.18)	45(0.15)	40(0.12)	40(0.09)	35(0.07)	30(0.06)
300		65(0.38)	60(0.29)	55(0.23)	50(0.20)	45(0.15)	40(0.13)	35(0.11)
400			70(0.39)	65(0.31)	60(0.27)	55(0.20)	45(0.17)	40(0.15)
500				75(0.37)	65(0.32)	55(0.25)	50(0.20)	40(0.18)
600					70(0.37)	60(0.29)	50(0.23)	45(0.21)
800						65(0.39)	60(0.32)	50(0.28)
>900							65(0.37)	55(0.32)

height between roadways and above extraction space is calculated as Table 3.The numbers among brackets in Table 3 are the relative stability coefficient of roadways.

$$\sigma_{max} / R \leqslant 0.40 \quad (1)$$

Where σ_{max}= the maximum principal stress around roadways,MPa; R = the uniaxial compressive strength of surrounding rocks, MPa.

The rational horizontal distance between roadways and extraction boundary can keep the roadways in the area of decreased stress. The maximum principal stress around roadways is calculated in accordance with formula (2); the rational horizontal distance has been worked out as Table 4; in Table 4, Z means the distance between roadways and the bedding plane of seams.

$$\sigma_{max}/R \leqslant 0.25 \quad (2)$$

The distance between roadways and coal seams should exceed the width of direct destroyed area in floor owing to mining. According to practical cases of mining engineering, the distance between roadways and coal seams should not be shorter than the distance shown in Table 5, but generally not be longer than 50 meters.

The stable width of the pillars protecting roadways is the width influenced by the abutment pressure of working space. When the width of pillars protecting roadways is wide enough, the stress in pillars presents saddle shape and the stress in the centre of pillars is near to original stress.

The stable width of protecting pillars should make the maximum principal stress in the centre of pillars (that is the location of roadways) fit for formula (1). The stable width of the protecting pillars in one side of roadways is shown in Table 6.

The location parameters that are calculated in the paper are adapt to the conditions as follows: coal seams are inclined or gently inclined; the roof of working space is the type of moderate caving; rigid support is used for roadways; and there is not thick sand stone in the space between roadways and coal seams. If the conditions vary, the parameters should

Table 4 The rational distance between roadways and extraction boundary X (m)

Mining depth	Compressive strength	Distance between roadways and the bedding plane of seams Z (m)					
H(m)	R(MPa)	10	15	20	30	40	50
300	<30			25(0.12)	30(0.15)	35(0.16)	40(0.19)
	30~60	15(0.07)	15(0.11)	20(0.10)	25(0.11)	30(0.12)	35(0.13)
	>60	10(0.08)	10(0.12)	12(0.12)	15(0.11)	17(0.12)	20(0.12)
600	30~60			25(0.13)	30(0.16)	35(0.17)	40(0.17)
	>60		17(0.13)	20(0.14)	25(0.16)	30(0.15)	35(0.15)
900	>60			25(0.12)	30(0.17)	35(0.18)	40(0.17)
1000	>60			25(0.14)	30(0.18)	35(0.19)	45(0.19)

Table 5 The distance between roadways and the bedding plane of coal seams Z(m)

Mining depth	Uniaxial compressive strength of surrounding rocks R (MPa)		
H (m)	<30	30~60	>60
300	20	10	10
600		20	15
900			20

Table 6 The stable width of the protecting pillars in one side of roadways B (m)

Mining depth	Uniaxial compressive strength of surrounding rocks R (MPa)							
H(m)	<20	30	40	50	60	80	100	120
< 200	75(0.28)	55(0.21)	45 (0.18)	40(0.15)	35(0.13)	30(0.11)	30(0.09)	30(0.07)
300	75(0.40)	70(0.30)	60(0.25)	45(0.24)	40(0.21)	35(0.18)	35(0.14)	35(0.12)
400		75(0.40)	70(0.30)	65(0.25)	50(0.25)	40(0.22)	40(0.17)	40(0.14)
500			75(0.40)	65(0.33)	55(0.29)	45(0.25)	45(0.20)	45(0.17)
600				75(0.35)	70(0.31)	65(0.26)	50(0.25)	50(0.21)
800					75(0.39)	70(0.31)	65(0.27)	60(0.25)
> 1000						75(0.37)	70(0.31)	65(0.28)

be timed by a coefficient.

Because there are a few hypotheses in the paper and some data come from actual measurement, the calculated location parameters and the corresponding stability coefficient can be expressed in common.

5 CONCLUSION

The stability of roadways is influenced not only by original stress field but also by redistributed stress field in strata during mining activity. Based on the model of macroscopic isotropy that is simplified from coal-bearing sedimentary strata with the nature of the random variation of its stratified thickness and mechanical character, the distribution of stress in floor strata can be calculated under the condition of the abutment pressure as a result of mining activity. According to the stability coefficient of roadways. It has practical sense to the rational layout of roadways to calculating rational location parameters of roadways to mining space under different conditions.

REFERENCES

Tan XueShu et al., 1994. Compound rock mass mechanices theory and its applications, *Coal Industry Publishing House*, PP119 ~ 120.

Jiang JinQuan et al., 1993. Stress and activity of rock surrounding working face, *Coal Industry Publishing House*, PP116 ~ 117.

Li ShiPing, 1986. Brief study course of rock mechanices, *China University of Mining and Technology Publishing House*, PP164 ~ 165.

Computer Methods and Advances in Geomechanics, Yuan (ed.) © 1997 Balkema, Rotterdam, ISBN 90 5410 904 1

Calibration and validation of a FLAC model for numerical modelling of backfilling at Jinchuan No. 2 mine

Q.L.Cao, B.Stillborg & C.Li
Division of Rock Mechanics, Luleå University of Technology, Sweden

Abstract: The present study is a part of a numerical modelling exercise carried out of the backfill at Jinchuan No.2 mine, China. Based on laboratory tests of rocks of the mine, a set of rock mass classification results obtained by using the Q-system is used to evaluate the mechanical properties of the rock mass. A FLAC model of undercut-and-fill mining at the No.2 mine is calibrated through a comparison between the vertical displacements calculated by the FLAC model and the one measured at the 1300 m level of the mine. The FLAC model is validated by the comparisons between the development of stress in the backfill and the extent of yielding zones predicted by the model and those measured or observed in the mine. Finally, the double-yield model of the backfill is replaced by the Mohr-Coulomb model to further validate the FLAC model.

1 INTRODUCTION

Mining with backfill has long been recognised as a potential method for ground control. The backfill plays many important roles in the cut-and-fill mining, providing the mining industry with the benefits of increased extraction and decrease in the rate of accidents. Nevertheless, one serious drawback inherent in the hydraulic backfill is its low stiffness and low compressive strength.

In the recent decades, a number of rock mechanics investigations into the support performance of backfill have been carried out through a variety of research approaches, which include laboratory testing, in-situ instrumentation and numerical modelling. Numerous papers and several symposia (e.g. Granholm 1983, Hassani 1989, Glen 1993) associated with mining using backfill have documented the development of practice and research in cut-and-fill mining.

Based on findings by past researches, it has been commonly accepted that the backfill can effectively support the surrounding rock and prevent it from ravelling. The following behaviour of backfill have been recognised:

1. The behaviour of backfill material is highly non-linear and characterised by the time-dependent consolidation and a portion of irrecoverable strain. The double-yield model is appropriate to simulate backfill materials.
2. Backfill has a significant influence on the residual strength of the surrounding rock.
3. The support effect of backfill depends on the stiffness ratio of the backfill to the surrounding rock.
4. Gaps in backfill have significantly negative effect on the support performance of the fill.

The support performance of backfill and the stability of the surrounding rocks at Jinchuan No.2 mine were studied by numerical modelling. The No.2 mine consists of two orebodies, which overlap and are offsett by a fault. The No.1 orebody is approximately 1,800 m long and 900 m high. Undercut-and-fill mining method is adopted for the mining of the No.1 orebody. The mining is conducted downwards in slice by slice, which is cut and filled in every third drift with 4×4 m in size.

The No.2 mine was brought into production in 1982. The designed capacity for the first mining phase is 8000 t/d. The mining is conducted on a single level 50 m in height. According to its second phase design, the production capacity should be increased to 17000 t/d. To accommodate for the increased production capacity, mining with the current method must be simultaneously conducted on 2 ~ 3 levels. During mining in the second production phase of the No.2 mine, the rock mechanics concerns are: (1) the stability of the backfill and its support performance to sustain itself and the surrounding rocks, (2) the downward movement and failure of the low grade orebody in the overlying strata and hanging-wall, (3) the stability of the sill pillar between two adjacent levels on which mining operations are conducted simultaneously.

Before the numerical analyses of the above rock mechanics problems were conducted, an appropriate model was calibrated and validated by using the two-dimensional finite difference code FLAC (ITASCA 1992). Based on the FLAC simulations for the calibration and validation of the FLAC model, the following specific aspects were addressed:

1. Development of the failure zone in the low grade ore in the single-level mining.
2. Downward movement of the low grade ore and the backfill.
3. Effect of the constitutive model of the backfill on the support performance of backfill.

2 CALIBRATION OF THE FLAC MODEL

To evaluate the ground stability problems in the No.2 mine, two categories of numerical models are employed. 2-D mine-scale models were established to study: (1) the support performance of cemented backfill as a global support element, (2) the ability of the backfill to resist the movement of the surrounding rock mass and the stope roof, and (3) the development of a yielding zone in the low grade orebody in different mining depths and mining sequences. The effect of the poor contact and gap between two backfilled slices was studied by using a stope model.

A reasonable numerical model should be calibrated and validated before it can be used to predict the behaviour of the rock and the backfill.

Table 1 Rock mass moduli and strength parameters

Parameter	Low grade ore	High grade ore	Two-pyroxene peridotite
m_i	9.05~26.0	3.0~20	12.3~20.6
Q-value	0.06~1.4	0.69~2.5	0.14~2.83
RMR	18~47	40~52	26~53.4
E_m (GPa)	1.58~8.41	5.62~11.22	2.5~12.13
m	0.48~3.92	0.36~3.6	0.87~3.71
s ($\times 10^{-4}$)	1.10~ 27	12.7~48	2.686~48

The present calibration is mainly conducted through one comparison, or back-analysis, by using the displacement measurements carried out during mining of the first three slices. The calibration was carried out by adjusting the rock properties in the FLAC model until the stress and displacement responses at typical points calculated by the FLAC model fit to those measured in the field. The calibrated properties of the rock mass were finally used as the input data to the other FLAC models.

2.1 Properties of the rock mass and backfill in the No.2 mine

In the No.2 mine, there are three major rock types in the region of concern: high grade ore, low grade ore, and two-pyroxene peridotite. The low grade ore in the foot-wall is mined together with the high grade ore, but a part in the hanging-wall and overlying strata is not mined at present. The orebody and rocks are so highly fractured that the entire rock mass can be approximately treated as a homogeneous medium. The failure criterion used in the FLAC model is the Mohr-Coulomb criterion, in which the equivalent values of the cohesion and friction angle were determined from the tangent to the envelop to the principal stress circles defined by the Hoek-Brown criterion (Hoek 1990, Hoek & Brown 1980).

The determination of the mechanical properties of the in-situ rock mass remains one of the most difficult problems in the field of rock mechanics. The rock mass classification of the No.2 mine was conducted by using the Q-system in 1986. It shows that the rock mass quality is very poor to poor (Ludvig, 1986). Therefore, the deformation moduli of the rock masses in the No.2 mine can be estimated from the empirical equations proposed by Bieniawski (1976) and Sarafim and Pereira (1983). Based on the laboratory experimental data of rocks and the rock mass classification results, the moduli and strength parameters of rock masses were estimated and are listed in Table 1, where, m_i, m and s are material constants in the Hoek-Brown failure criterion for intact rock and rock mass respectively, Q is the Tunnelling Quality Index, RMR is the rock mass ratings, and E_m is the young's modulus of the rock mass.

The backfill in the No.2 mine is used for a number of purposes. One of the most important roles is to provide a support to the low grade orebody in the hanging-wall and overlying strata. The backfill is cemented milled Gobi sands. A backfill with a cement to sand ratio of 1:4 is placed in a 2m layer for the lower part of the stope and 1:8 for the upper part. Its initial void volume ratio is 47.1%. The measurements of stope closure, pressure and strain in backfill were conducted in-situ and showed a strain hardening behaviour of the backfill. Thus the double-yield model was used to simulate the backfill. The parameters in the double-yield model were estimated by making a best-fit to the experimental results of a uniaxial strain test through a FLAC model developed by ITASCA (1992).

2.2 In-situ measurements

When the undercut and fill mining was tested in the No.2 mine, a number of instruments were installed to determine the response of the rock and the backfill to mining during the time of August 1985 to July 1986 (Ludvig 1986). Levelling was conducted in order to measure the vertical movement occurring in the stopes and the drifts due to mining. Initial measurements were taken in February 1986 when the first mining slice was almost finished.

The measuring points No.23, 5 and 6 were installed 1 m above the 1300 m level at line 17 of the mine. Point No.5 is generally located at the centre of the orebody in the crosscut direction. Its downward displacement was 92 mm when the mining of the third slice began. The levelling points in the hanging-wall were points No.23 and No.6, which were located closely, their downward displacements were 30 mm and 28 mm respectively (Ludvig, 1986). The stress monitoring has shown that the stress in the backfill at line 17 was 2.94~3.13 MPa, and approached a stable value of 2.35 MPa during the mining of the second slice (Chang et al., 1994).

2.3 FLAC model formulation and boundary conditions

A mine-scale model was developed along line 17 with a large span near the centre of the orebody to simulate the present mining operations and calibrated using the displacements observed in-situ. A close-up of the FLAC grid is shown in Figure 1.
Since FLAC provides a function to attach two sub-grids with different zoning density together, an extended coarse grid can be wrapped around a finely discretized, inner grid. In the present case, a 190×11 strip of zones with a progressive decreasing size towards the top edge is wrapped around the finely-zoned central region of interest by attaching the top edge of the coarse grid strip to the central block at all points on the inner surface of the strip. The "tail" of

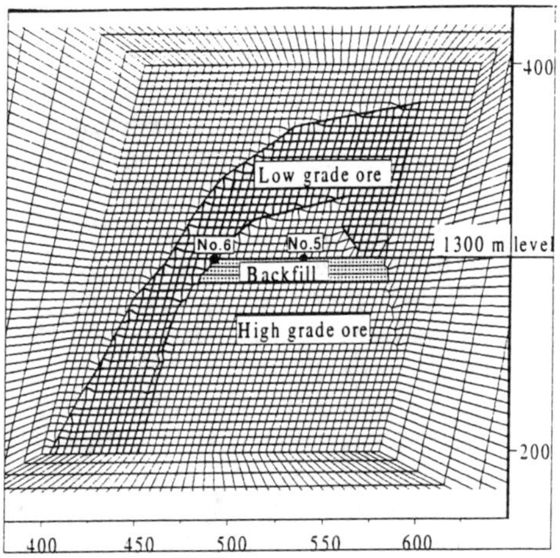

Figure 1 A close-up of the FLAC grid.

the coarse strip is then attached to the "head".
As seen in Figure 1, the mined region is enclosed by the low grade ore, 20~30 m thick in the hanging-wall and 40~60 m in height in the overlying strata. In the roof of 1300 m level, there is a block of high grade ore left to stabilise the roof.

The model is horizontally fixed along the left and right boundaries and vertically fixed in the bottom. The surface of the model is free but loaded with a horizontal stress of 3 MPa. The model is loaded by initiating the in-situ stresses described by following equation (Tian 1995):

$$\begin{aligned}\sigma_1 &= 3.0+0.0425h \quad \text{(MPa)}\\ \sigma_2 &= 3.0+0.027h \quad \text{(MPa)}\\ \sigma_3 &= \gamma h \quad \text{(MPa)}\end{aligned} \qquad (1)$$

where, h is depth from the surface and γ is the mean unit weight of the rock. Output from the first excavation provides the starting stress state for the second step and so on.

2.4 Model calibration

Model calibration is the process of adjusting model parameters so that the model approximates field measurements and observations. Figure 1 shows that the mining of 3 slices has been completed starting from 1300 m level. The high grade and low grade ore in the foot-wall are excavated by declaring the appropriate zones to be null, and the backfilling is modelled by assigning the properties of the backfill to the nullified zones. In order to compare with actual measurements, displacements or stresses are "monitored" at a few gridpoints in the FLAC model. As seen in Figure 1, two grid points, located in the roof and hanging-wall respectively, are selected to monitor the vertical displacements corresponding to the levelling points No.5 and No.6 in-situ. The development of the stress in the backfill was "monitored" by using a zone in the backfill region.

In the calibration of the FLAC model, the input data of the mechanical parameters are selected from the range of parameters listed in Table 1 for each type of rock mass. The outputs of the FLAC model, specifically the vertical displacements at points No.5 and No.6, were used to compare with corresponding measurements in-situ. The comparisons of simulated vertical displacements to the in-situ measurements are only conducted during mining of slices 2 and 3, when the in-situ measurements were carried out.

The predicted values shown in Figure 2 are the best result after a number of adjustments to the mechanical properties of the rock masses. During the calibration, the properties of the backfill are kept constant because the behaviour of the backfill material varies relatively little no matter in the laboratory or in-situ. The relative errors between the vertical displacements predicted by the FLAC model and the one measured in-situ were: 4.76% in the roof, 9~16.8% in the hanging-wall.

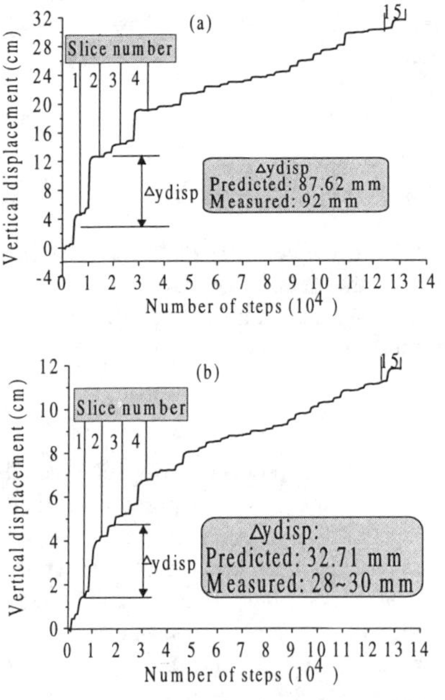

Figure 2 Comparison of vertical displacements at (a) point No.5, (b) points No.23 and No.6 measured in-situ and predicted by the FLAC model.

3 VALIDATION OF THE FLAC MODEL

3.1 Yielding zones

The validation was first carried out by comparing the extent of yielding zones predicted by the model and observed in the field. Yielding in the rock mass and backfill was examined during the simulation of each mining step. The results indicate that the yielding zones mainly extend to both the overlying rock mass and underlying high grade ore in the form of an arch.

When the excavation on the fourth slice was completed, the yielding zone extends 72 m in height in the roof and spread to the low grade ore in the hanging-wall. In the subsequent mining, the yielding zones develop gradually at a decreasing rate. Comparing the yielding indicator plots in the cases of mining of slices 4 and 12, it indicates that a little difference exists in the development of the yielding zones in the overlying rock mass between the two cases. This means that the yielding of the rock mass in the roof is not closely related to the mining depth or the number of slices mined.

During mining operations, the observations of the failure phenomenon in-situ indicated that (Chang & Zhang, 1994):

Observation 1: When a drift was excavated in the rock mass and orebody at the 1350 m level, the surrounding rock mass around the drift was shown to be intact without failure.

Observation 2: A new haulage drift was developed in the foot-wall 6 m away from backfilled region and this was also stable.

When and where the above observations were conducted is not clear to the authors. However, they could be assumed to be conducted after the mining of slice 12 because the number of mining slices does not have much influence on the distribution of yielding zones as mentioned above. Figure 3 shows the distribution of the yielding zone after 12 slices of mining. The development of the yielding zone is concentrated to the hanging-wall and the underlying high grade ore. A portion of the low grade ore in the roof and the rock mass in the foot-wall between the levels 1300 and 1350 are in elastic state. Thus the non-failure area on the 1350 m level predicted by FLAC model might be considered to be consistent with the observation 1. It is observed from Figure 3 that one zone was at yield in past in the foot-wall on 1300 m level. Thus the yielding zone is only extended 4 m into the foot-wall away from backfill on the 1300 m level, this is in good agreement with the observation 2.

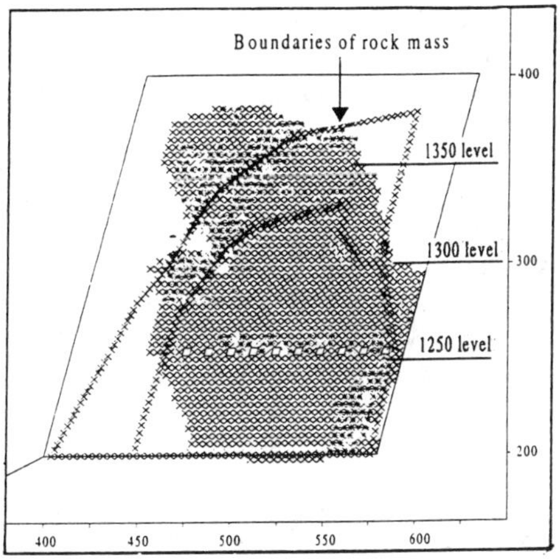

Figure 3 Plastic zones after mining of 12 slices.

3.2 Stresses in the backfill

The stresses were "monitored" in the model and compared with the in-situ measurements. Figure 4 shows the major principal stress in two contiguous zones within the roof and the backfill respectively. The stress in the backfill is gradually built up to a level of 12.27 MPa during mining of the first slice. When the second slice is mined, the stress in the backfill drops to 2.37 MPa. This stress decreases very slightly during mining of the subsequent slices. These results agree very well with the in-situ measurements, which showed that the stress in the backfill was 2.35 MPa during the mining of slice 2 and was maintained at this level in the following mining steps (Chang & Zhang 1994).
On the other hand, it is clearly seen that the stress in the roof starts to increase gradually during mining of the second slice due to the increase in the shear stresses along the contacts of the backfill with hanging-wall and foot-wall. This effect is clearly seen in Figure 5. The effect of the backfill support to the surrounding rock mass is mobilised by these shear stresses. The backfill itself is mainly supported by transferring its weight to the hanging-wall and foot-wall through the shear stress and arch effect. Owing to the transfer of the load from the backfill to the surrounding rocks, the vertical pressure on the mining structure underlying the backfill will not increase linearly with depth. Consequently, the mining is conducted under a stable backfill roof. This has been conformed by mining operations.

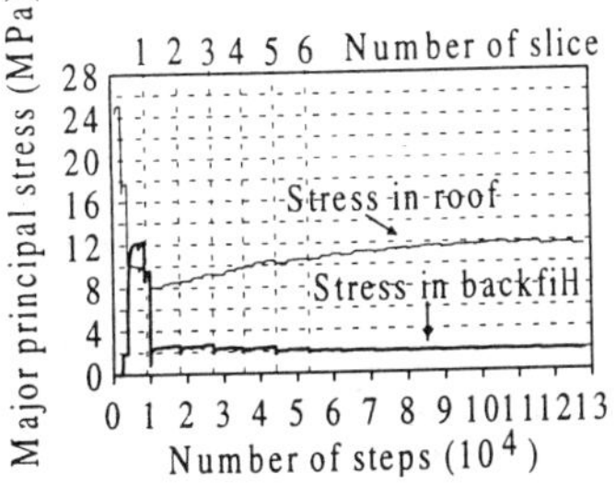

Figure 4 The major principal stresses in the roof and the backfill.

3.3 Modelling of the backfill by the Mohr-Coulomb model

Some researches (e.g. Fourie, 1993) indicate that the commonly used Mohr-Coulomb model always under-estimated the displacement response. In turn, the potential failure parameters is underestimated. Instead of the double-yield model, if the Mohr-Coulomb model is used to simulate the backfill material in the FLAC model, the simulated results could be used to further validate the calibrated FLAC model by comparing the displacement responses to the double-yield model and the Mohr-Coulomb model.

Figure 6 shows the deflection of the immediate roof along the length of the slice at 1300 m level. Although the Mohr-Coulomb model was assigned essentially the same modulus as the double-yield model, it showed significantly smaller vertical displacement, especially in the central portion. This is because the backfill with the double-yield model starts to undergo yielding at much earlier stage of calculation than in the Mohr-Coulomb model. The extent of the yielding zone in the Mohr-Coulomb model is smaller than in the double-yield model, resulting in much smaller total strains and displacements.

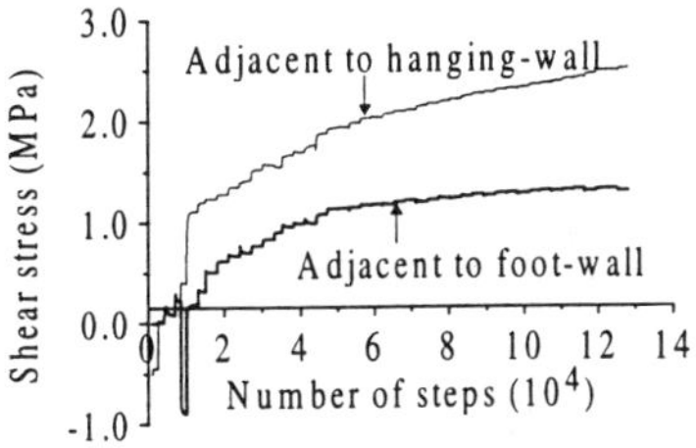

Figure 5 Shear stresses within the backfill adjacent to the hanging- and foot-wall.

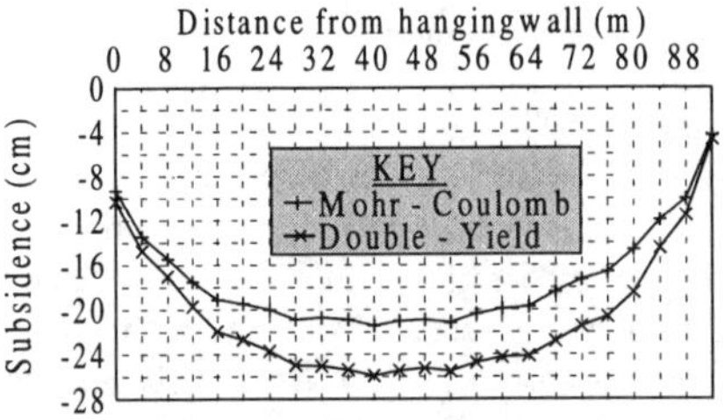

Figure 6 Deflection of the ore roof of 1300 m level.

The under-estimation of displacements by the Mohr-Coulomb model is further illustrated in Figure 7. Figure 7 shows the convergence between the hanging-wall and foot-wall at 1300 m level. The difference in the convergence by the two models becomes clear after mining of the second slice, when the yielding zone extends 56 m into the roof for the double-yield model. Whereas the extent of yielding zone is only 24 m for the Mohr-Coulomb model. Owing to the smaller extension of yielding zone, a smaller convergence and higher stresses are predicted by the FLAC model with Mohr-Coulomb backfill material.

4 CONCLUSIONS

1. The in-situ vertical displacements of surrounding rocks measured during mining of slices 2 and 3 are used to calibrate a FLAC model for stability analysis of Jinchuan No.2 mine. The relative errors of the vertical displacements predicted by the FLAC and measured in-situ are 4.76% for the roof and 9~16.8% for the hanging-wall.

2. The shear stress along the contacts between the backfill and the low grade in the hanging- and foot-wall increases with mining depth, resulting in a support effect of the backfill to sustain movement of the surrounding rock mass and the backfill itself. The development of stress in the backfill validates the FLAC model.

3. The FLAC model is further validated by the in-situ observations of the yielding zones, which is in good agreement with the extent of yielding zones as modelled by the FLAC model.

4. The use of Mohr-Coulomb model for the backfill under estimates the displacements compared to the use of the double-yield model.

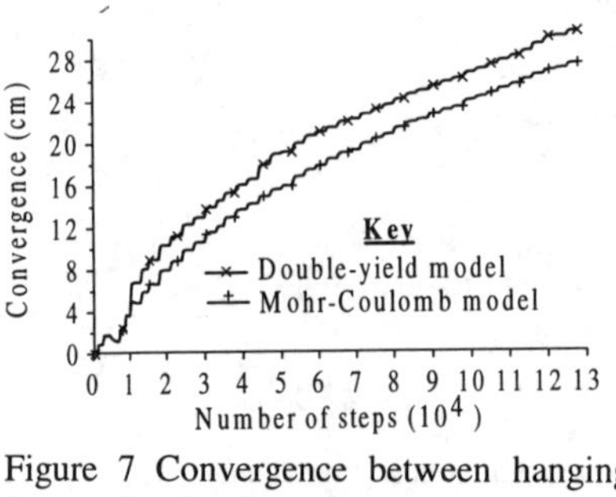

Figure 7 Convergence between hanging-wall and foot-wall in the first slice.

REFERENCES

Bieniawski Z.T. 1976. Rock mass classification in rock engineering. *Exploration for rock engineering*, Vol. 1, 97-106.

Chang Zhongyi et al., 1994. The Ground control of undercut and fill mining without pillar in large area. *Nonferrous Metals Mining (in Chinese).* No.3~4, 28-34.

Fourie A. B., et al., 1993. An evaluation of four constitutive models for the simulation of backfill behaviour. *Minefill 93*. 33-38.

Glen H.W., 1993. Minefill 93. *The proc. 5th Int. Symp. on Mining with backfill.* SAIMM.

Granholm S.,1983. Mining with backfill. *Proc. Int. Symp. on Mining with backfill.* Luleå. Balkema.

Hassani F. P. & M. J. Scoble & T.R. Yu, 1989. Innovations in mining backfill technology. *Proc. 4th Int. Symp. on Mining with backfill.* Montreal, Balkema.

Hoek E., and E. T. Brown, 1980. Underground excavations in rock. London: *Instn Min. Metall.*

Hoek E., 1990. Estimating Mohr-Coulomb friction and cohesion values from the Hoek-Brown failure criterion. *Int. J. Rock Mech. Sci. & Geomech. Abstr.*, 27(3), 227-229

ITASCA, 1992. FLAC, Version 3.2. *User's Manual.* Itasca Consulting Group Inc.

Ludvig B., et al., 1986. The Jinchuan mine project--Sino-Swedish Technical Corporation for No.2 Mine Area. *Final Report.*

Sarafim J.L. et al., 1983. Consideration of the geo-mechanical classification of Bieniawski. *Proc. int. Symp. On engineering geology and underground construction.* Lisbon 1(2). 33-34.

Tian Y.S., 1995. Engineering characteristics and rheological property of rock mass in Jinchuan Nickel mine. *Proc. 8th Int. Cong. on Rock Mech.*, Tokyo, Vol. 1, 9-11.

Computer Methods and Advances in Geomechanics, Yuan (ed.) © 1997 Balkema, Rotterdam, ISBN 90 5410 904 1

The landslide process of Yanchihe phosphorus mine simulated by discrete element method

Zhou Xiaoqing & Wang Yuanhan
Department of Civil Engineering, Huazhong University of Science and Technology, Wuhan, China

ABSTRACT: The discrete element method is a numerical method suitable for solving problems associated with uncontinuous medium, and it is especially suitable for analyzing the stability problem of jointed rock mass. In this paper, the method is used to simulate the dynamic landslide process of phosphorus mine in Yanchihe. A corresponding result based on geological analysis is obtained by numerical simulation. Both the geological reality and the numerical simulation can be firmed each other.

1. INTRODUCTION

On June third, 1980, a large-scale and disastrous landslide occurred at Yanchihe phosphorus mine in Yichang region, which destroyed all the buildings of the mine, and none of the workers in the site was survival. It caused a serious loss.

In order to analyze the process of motivation and the reasons for destruction of the mountaintop, as well the decisive factors for landslide, we have used the discrete element method to simulate the dynamic process of landslide based on geological investigation. The result is very similar to the real moving process of the sliding mountaintop.

2. BRIEF INTRODUCTION ON YANCHIHE LANDSLIDE

2.1 *Geological condition*

Yanchihe phosphorus mine was administered by Yuan'an county, which is located 87 kilometers north to Yichang city. Its geography coordinate is : east longitude 118°13′, north latitude 31°31′. The mine was located on north west flank of the mountain, which was steep cliff. And the sliding mountaintop was a cliff up in the air on three directions. The relative difference in elevation between the mountain valley and the mountain top was about 400 meters.

The geological section of the sliding mountaintop is in Figure 1. From Figure 1 we can find : the mountain is mainly composed of dolomite and sandy shale. The dolomite rock mass is very hard and complete, while the sandy shale has many joints, and it may be soften by water. So the mountain was jointed with a hard rock layer and a soft rock layer. What's more, there is a layer of shale about 10 meter deep at the bottom of the sliding mountaintop, which may become a sliding plane for mountaintop when it soften by rain water.

2.2 *Situation about mining*

Yanchihe phosphorus mine was established in 1969, and was reformed in 1975. The enlarged preliminary design was producing 150000 tons of phosphorous ore per year, while the real production was 80000 tons of phosphorus ore. The mining method was general mining and regular mineral columns was remained, and the mining scheme of combining flat holes and a central slanting shaft was used. The back mining was from top to bottom, while in each section the mining sequence was from inside to outside or from south to north. Because the sandy shale on the bottom was very soft, the deformation of the central slanting shaft was serious during the mining.

2.3 *Surface crack and the process of landslide*

Ten main cracks was found in the sliding

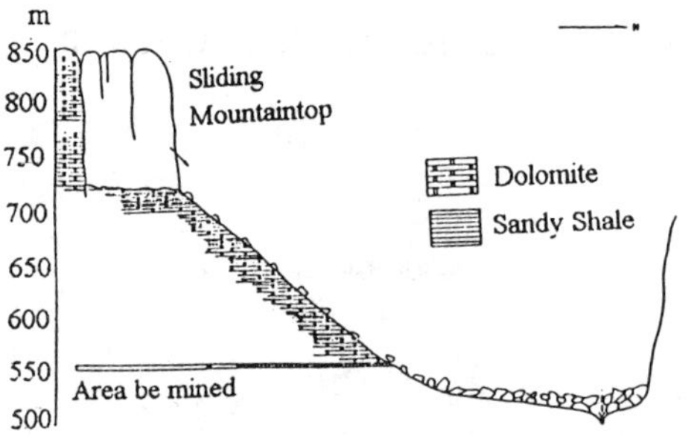

Figure 1. The geological section of the sliding mountaintop

mountaintop and on the surface of the part what was mined since 1978. And there were six cracks cutting the mountain.

It was raining on May 30th and 31st, the rainfall reached 74mm. The speed and the count of crack's deformation growth in a large degree. Small landslides occurred continuously after the rain, and the volume of landslide became more and more large, until the disastrous landslide occurred.

3 LANDSLIDE SIMULATED BY DISCRETE ELEMENT METHOD

3.1 *Discrete element method*

Discrete element method is a numerical method suitable for analyzing problems associated with uncontinuous medium. This method divide the considered rock mass into a series of rock blocks cut by jointed plane, and each block can move independently. By examining the contact and the overlap between every two blocks, we can calculate the force subject to each block , then the Second Newtonian Motion Law may be used to analyze the force-deformation relationship of the rock mass.

1. Equation of motivation

If the mass of a block is m, and the force subject to it is $\sum F(t)$, according to the Second Newtonian Motion Law, we can get the acceleration of the block on time t :

$$\ddot{u}(t) = \frac{\sum F(t)}{m} \tag{1}$$

2. Equation of physics

The deformation of blocks is the overlap between blocks. Assuming the normal force between blocks is in direct proportion to the normal overlap between blocks, that is

$$F_n = k_n u_n \tag{2}$$

where k_n is the normal stiffness.

While the shearing force has relationship with the history of force, so the shearing force is given by

$$\Delta F_s = k_s \Delta u_s \tag{3}$$

where k_s is the tangent stiffness, and Δu_s is the tangential relative displacement.

3. Damping

Because the elastic system will oscillate forever, we must give damping to reach stable equilibrium. Considering the stiffness and the damping of each block, equation (1) becomes the fundamental equation for discrete element method:

$$m\ddot{u}(t) + c\dot{u}(t) + ku(t) = f(t) \tag{4}$$

where c is the damping coefficient, k is the stiffness and f is the load.

3.2 *Selection of calculating simulate*

Based on the real structure of the rock mass in Yanchihe phosphorus mine area, we can summarize the calculating simulate (Figure 2.) from the geological section figure (Figure 1.). This figure has some features such as:

(a) Considering that the mountain has a rock mass medium structure with a hard layer and a soft layer, and the soft layer is very thin, we neglect the thickness of the sandy shale layer and then look it as a joint plane.

(b) Null zone is used to simulated the mined phosphorus ore.

(c) The four main cracks is achieved by reducing its cohesion and its friction coefficient.

(d) A block sticking with the ground is used to simulate the building on the toe of the slope.

(e) The area far away from the sliding mountaintop , the ground 50 meter lower from the phosphorus pit and the mountain on the opposite side are supposed to be fixed.

3.3 *Selection of mechanical parameters*

Because we can not get the accurate mechanical parameters of the rock in Yanchihe , the parameters used in the numerical simulating of the landslide are achieved by consulting some references. The mechanical parameters of dolomite is : modules of elasticity E = 3.72GPa, Poisson's ratio $\nu = 0.25$, unit weight of rock mass $\gamma = 2.65 \, t/m^3$, cohesion c=100MPa, angle of internal friction $\phi = 28°$.

The chosen parameters may be not in agreement with the real parameters of the rock mass, but the mechanical parameters of the same rock would not vary in a large degree, and our calculation is only qualitative analysis, so the chosen parameters is feasible.

3.4 *Calculating process*

The landslide is simulated in two steps:

(a) Null zones are used to simulate the mined phosphorus ore, then the upper rock blocks fall under the action of gravity.

(b) When the calculation is stable, static water pressure caused by rainfall is applied in the cracks of the mountaintop. At the same time, the contact parameters of some joints are changed because the sandy shale was soften by rain water. Then, the rock blocks in the sliding mountaintop has been falling under the action of static water pressure , gravity and friction.

3.5 *Results*

Because our purpose is to simulate the dynamic process of the landslide, the results are showed by figures. Figure 3 is the block distribution and the stress distribution of the whole mountain when the mountain is stable after the phosphorus ore be mined. From the figure we can get: the rock blocks up the area where phosphorus ore was mined has downward displacement under the action of gravity. The block distribution and velocity distribution of the whole mountain with different steps under the action of static water pressure are showed in Figure 4 to Figure 9. During this process, firstly parts of the sliding mountaintop rock blocks fell down, and the velocity of the blocks went to maximum when they got to the toe of the slope; then the building was pushed by the falling blocks; lastly the building was pushed to the toe of the opposite mountain, and the

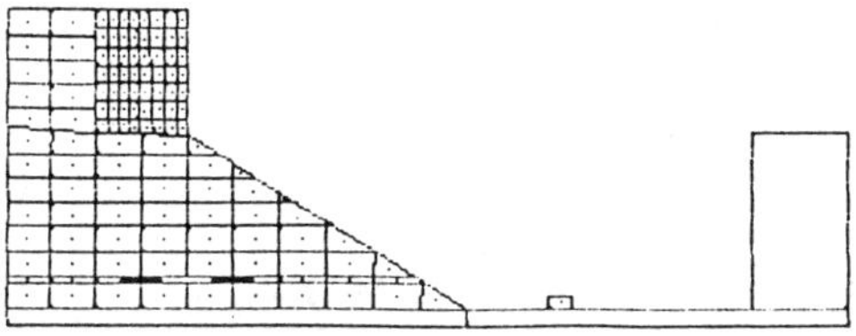

Figure 2 The calculating model

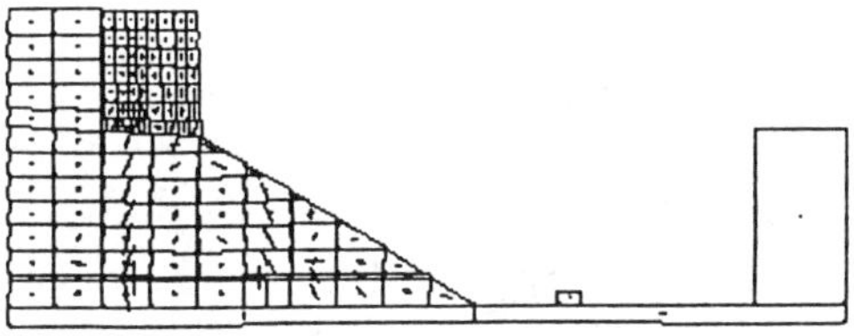

Figure 3. The stress distribution of the whole mountain when the mountain is stable after the phosphorus ore be mined.

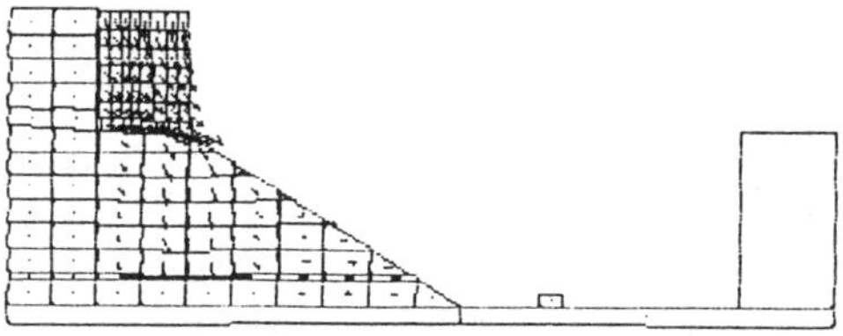

Figure 4. The block distribution and velocity distribution at t = 2.581

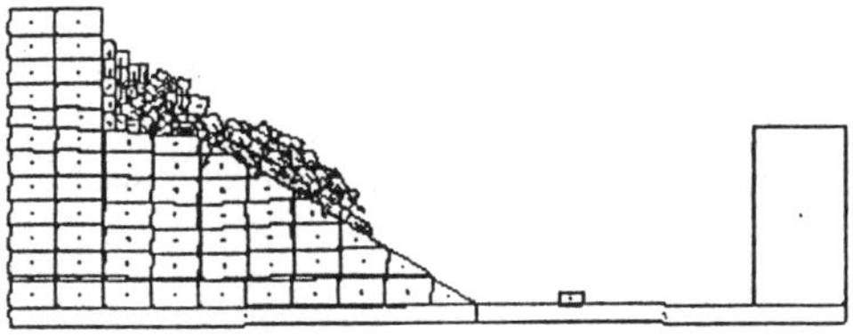

Figure 5. The block distribution and velocity distribution at t = 34.29

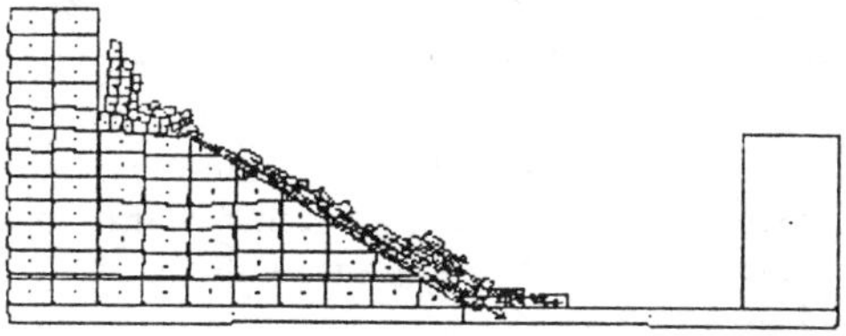

Figure 6. The block distribution and velocity distribution at t = 69.21

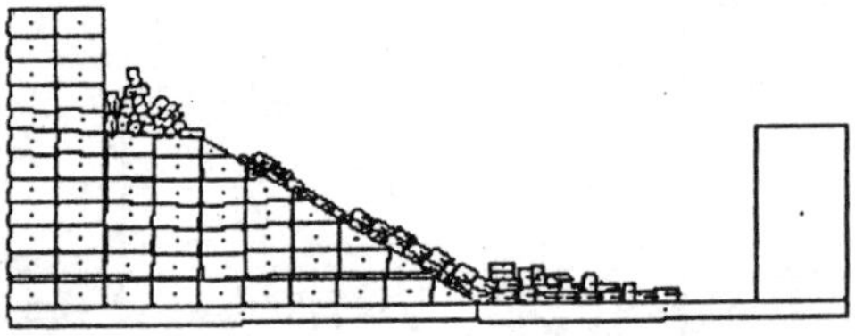

Figure 7. The block distribution and velocity distribution at t = 104.5

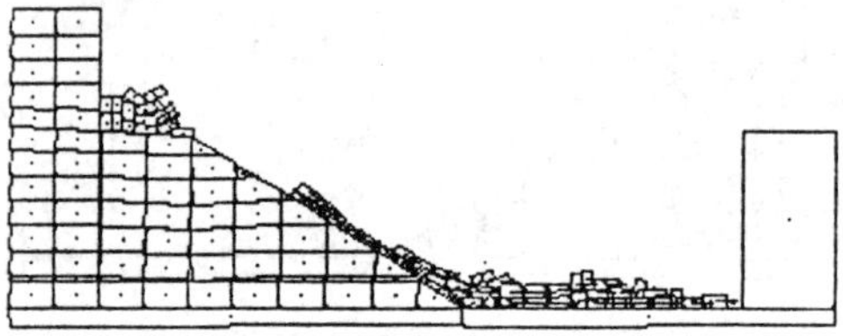

Figure 8. The block distribution and velocity distribution at t = 140

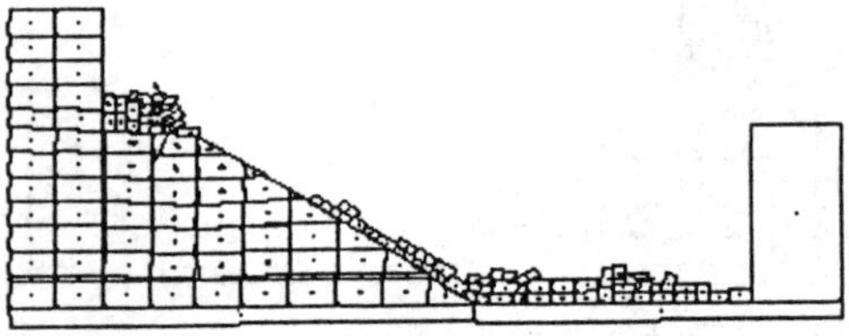

Figure 9. The block distribution and velocity distribution at t = 211.7

whole mountain was up to a new stable condition with blocks anywhere.

The result matched the real condition of landslide chiefly.

4 CONCLUSION

The results of the process of Yanchihe landslide simulated by discrete element method in this paper indicate: as a numerical method for analyzing uncontinuous medium, discrete element method can simulate the process of rock deformation and motivation effectively. So, we can say that this method is very suitable for analyzing the stability of rock mass.

ACKNOWLEDGMENTS

The work is supported by the Ph.D. Program of State Education Commission, China and the Research Grants Council, Hong Kong.

REFERENCES

Wang Yongjia , Xing Jibo Discrete Element Method and its Use in Rock and Soil Mechanics. Published by Press in N.E. Institute of Technology, Shenyang, Liaoning, 1991.

Wei Qun The Fundamental Principle , Numerical Method and Program of Discrete Element Method. Published by Science Publishing House, Beijing, 1991.

Cundall P. A. A Computer Simulate for Simulating Progressive Large Scale Movements in Blocky Systems. Proceedings of the Symposium of the International Society of Rock Mechanics , Nancy , France , 1971 ; Vol . 1, 2–8 .

Cundall P. A. UDEC — A Generalized Distinct Element Program for Simulating Jointed Rock. Report PCAR-1-80, Peter Cundall Associates, European Research Office , U. S. Army , March 1980.

Qi Xiaojun Engineering Geology and Hydrogeology. Published by Water Conservancy and Electrify Power Publishing House , 1985.

The Teaching and Researching Section of Geology, Department of Hydraulics, Gezhouba Water Conservancy and Electrify Power Institute of Technology , Introduction about the Landslide of Yanchihe Phosphorus Mine in Yuan'an county , Hubei Province.

Computer Methods and Advances in Geomechanics, Yuan (ed.) © 1997 Balkema, Rotterdam, ISBN 90 5410 904 1

Fracture features and bolting effect on surrounding rocks of actual mining roadways

Tianhe Kang
Institute of Rock and Soil Mechanics, The Chinese Academy of Sciences, Wuhan, China

Yadong Xue & Zhongming Jin
Shanxi Mining Institute, Taiyuan, China

ABSTRACT: The investigation results of 20 seams of the roadway surrounding conditions in ten coal mines show that the surrounding rocks are generally soft, thin, bedded strata cut by many sets of fractures. The roadways are sandwiched between upper and lower hard rock strata. Using similar model test and finite element simulation, the fracture features of the surrounding rocks and the bolting effect are analyzed. Several aspects to which attention should be paid in bolting are put forward.

1 INTRODUCTION

Countless facts have proved that bolting is an effective method in rock and soil engineering. It is still difficult, however, to popularize and apply this method in the actual mining roadways of coal mines due to complex geological conditions, poor surrounding strata, and periodic action of long duration and repeated mining dynamic load. The safety reliability of the bolting system has not yet been developed. This paper tries to indicate the surrounding strata structure and fracture features of the roadways and explore the control effect of bolting on the surrounding rocks.

2 SURROUNDING STRATA STRUCTURES AND MECHANICAL PROPERTIES

2.1 *Roof rock strata structures and rock properties*

The investigation results of 20 seams of the roadway surrounding rocks in ten mines of Xishan, Fenxi, and Huozhou mine districts in China show that the roof structures of the roadways have multi-thin bedded strata, double or compound strata, monolithic-thick stratum and coal seam, etc., four types, in which the multi-thin bedded strata and double or compound strata roofs constitute the majority (Kang and Xue 1996). The test results on the mechanical properties show that the roof rocks have a low strength index and a high deformation index.

2.2 *Rock properties of roadway sides*

The roadways are driven along the coal seams. The sides are soft and fractured coal. The uniaxial compressive strengths of coals are 5~35 MPa. So the roadways are sandwiched in soft stratum between upper and lower hard strata.

2.3 *Rock physical and mechanical properties of floors*

The test results of 13 roadway floors in three mining districts show that the ultimate floor specific pressures drop after the rocks are soaked in water. The average ultimate specific pressure of the siltstones drops 53% and the stiffness drops 72%. The average ultimate specific pressure of the mudstones drops 71% and the stiffness 96%. The average ultimate specific pressure of the shales drops 47% and the stiffness 87%.

3 FRACTURE FEATURES OF SURROUNDING ROCKS

3.1 *Basic features of fracture distribution*

The measurement results about the basic parameters of the surrounding rock fracture, such as sets, azimuth, and denseness, show that the basic features are as follows:

1. The fractures have many sets, heavy denseness, and small intervals.
2. The fracture degree and the distribution in the different strata in the same mining district are different.
3. The fracture sets and denseness in the soft strata are more and heavier than that in the hard strata.
4. The fractures in the multi-thin bedded strata are

more than that in the monolithic-thick strata.

Table 1 shows the statistics of the fracture sets and the amounts. Generally there are 3~4 sets of fractures and the fracture amounts are 5~15 in 1 m^2 area.

Table 1. Fracture statistics in surrounding strata

Seam	Rock	Fracture sets	Fractures in 1m^2
Tuanbai 1#	Sandstone	4	5.0
Xinzhi 2#	Sandstone	2	7.5
Bailong 1-2#	Sandstone	4	7.0
Tuanbai 2#	Sandy mud	3	5.4
Tuanbai 1#	Mudstone	4	15.7
Xinzhi 11#	Mudstone	4	7.9
Caocun 11#	Mudstone	3	13.18
Zhendi 3#	Mudstone	4	14.5
Cao 9-10#	Limestone	3	4.72
Zhendi 8#	Limestone	3	5.4
Average		3.4	8.6

3.2 *Fractal study on fracture distribution*

The fracture distribution in rock mass is very complicated. The basic parameters describing it, such as fracture sets, azimuth, and denseness, etc., cannot quantitatively describe the fracture scale or connectivity. So the fractal method of fracture scale distribution is used. The research results (Kang and Zhao 1995) have verified that the relation between fracture scales and the amounts in rock mass obeys the law

$$n(L) = a_0 L^{-d} \qquad (1)$$

where $n(L)$ is the fracture amount of the length $\geqslant L$ in measured area; a_0 is the fitting coefficient; d is the fractal dimension on the fracture scale distribution.

Table 2 shows the fractal measurement and calculation results on 10 roadways of roof fracture scale distribution. The initial fractal scale used in the research is $L_0 = 500$ mm. $n_1(1000)$ is the fracture amount, calculated by formula $n_i(L) = m_0 a_0 L^{-d}$ (Kang and Zhao 1995), of length $\geqslant$ 1000mm in 1 m^2 area, in which $m_0 = (L_E / L_0)^2$. L_E is the characteristic scale of the area calculated, here L_E = 1000mm. The data in Table 2 are the average value for five measurement areas. Table 3 shows the fractal measurement and calculation results on two sides of coal mass fractures of 6 roadways. It can be seen from the data in table 2 and table 3 that the connective fracture amount in 1 m^2 area of the roof rocks is more than that of two sides of coal mass, but the fractal dimension d of the roof rocks is less than that of two sides of coal mass. So small scale fractures in two sides of coal mass are more than that in roof rocks.

3.3 *Expression on fracture scale distribution*

The mechanical properties of rock mass are decided by not only lithological characters, fracture sets, and denseness, but also by fracture scale. The fractal dimension d can express the distribution feature of fracture scales. d is a parameter reflecting the relative variation of fracture amounts versus breaking scales. If a large scale fracture amount is relatively less and a small scale breaking amount is relatively more, the fractual dimension value d is greater; conversely, d is less. So the product $N_i\, d$ of the fracture amount N_i under a certain scale and the fractal dimension d can be considered as a comprehensive parameter to reflect the fracture distribution feature. The parameter value decides the mechanical properties of rock mass. The parameter can be called fracture index K_i (K_i=$N_i d$).

Table 2. Fractal measurement and calculation results on roof fracture

Roadway	Rock	$n(L_0)$	$n(L_0/2)$	$n(L_0/4)$	$n(L) = a_0 L^{-d}$	$n_1(1000)$
Gaoyang 6106	Sandstone	3.0	14.5	41.0	$n(L) = 411748L^{-1.8880}$	3.57
Gaoyang 6103	Sandstone	1.7	5.7	17.3	$n(L) = 57645L^{-1.6761}$	2.16
Gaoyang 4302	Sandy mud	3.7	10.3	44.0	$n(L) = 232244L^{-1.7886}$	4.00
Gaoyang 7204	Sandstone	5.0	14.7	43.0	$n(L) = 77945L^{-1.5535}$	6.81
Guandi 26402	Mudstone	2.7	12.0	40.0	$n(L) = 507201L^{-1.9461}$	2.94
Caocun 10308	Limestone	2.0	6.3	16.3	$n(L) = 25307L^{-1.5146}$	2.88
Zhendi 12410	Sandstone	5.5	15.0	54.5	$n(L) = 155589L^{-1.6570}$	6.65
Zhendi 18107	Sandy mud	4.2	14.3	43.7	$n(L) = 156671L^{-1.6911}$	5.29
Xinzhi 2-402	Mudstone	3.5	11.3	29.3	$n(L) = 49637L^{-1.5321}$	5.03
Xinzhi 11-101	Mudstone	2.6	7.4	22.6	$n(L) = 42224L^{-1.5622}$	3.48
Average		3.4	11.2	35.2	$n(L) = 171621L^{-1.6809}$	4.28

Table 3. Fractal measurement and calculation results on two sides of coal mass fractures

Roadway	$n(L_0)$	$n(L_0/2)$	$n(L_0/4)$	$n(L)=a_0L^{-d}$	$n_1(1000)$
Xinyao 8920	2.15	8.91	30.23	$n(L)=318003L^{-1.9104}$	2.36
Fengshan 2314	3.48	12.26	46.20	$n(L)=367836L^{-1.8630}$	3.80
Yangquan 8701	2.35	10.32	31.47	$n(L)=285613L^{-1.8738}$	2.73
Wang 4309	3.39	10.36	34.79	$n(L)=115783L^{-1.6825}$	4.15
Shuiyu 1061	5.14	16.61	60.58	$n(L)=325021L^{-1.7823}$	5.85
Xinzhuang 7301	3.00	9.25	28.00	$n(L)=70748L^{-1.6137}$	4.08
Average	3.25	11.29	38.55	$n(L)=247167L^{-1.7876}$	3.82

3.4 *Effect of fracture index K_i on rock strength*

Figure 1 shows the uniaxial compressive strength of some roof rocks versus the breaking index K_0 ($K_0=N_0d$). In the research the test blocks are 50 × 50 × 100mm cuboid. The N_0 is the connective fracture amount on the top surface of the tested blocks. The initial fractal scale is L_0=50mm. The rocks tested are as follows:

1. medium-grained sandstone of Bailong seam roof;
2. fine-grained sandstone of Xinzhuang seam roof;
3. medium-grained sandstone of Tuanbai seam roof;
4. medium-grained sandstone of Wangtai seam roof;
5. limestone of Dalong seam roof.

It can be seen that the fracture index K_0 affects prominently rock strength at linear law.

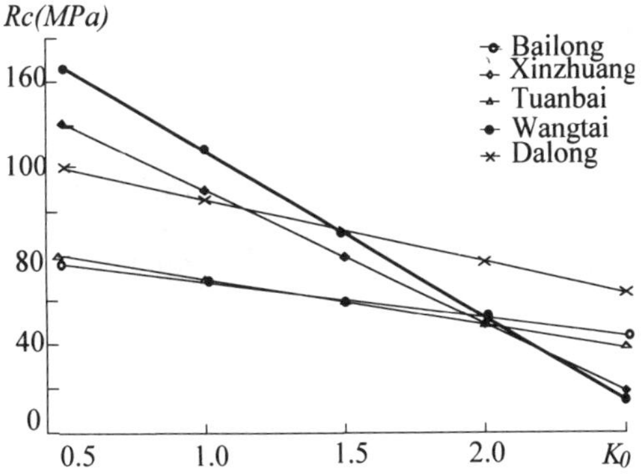

Figure 1. Uniaxial compressive strength R_c versus fracture index K_0

3.5 *Effect of mining dynamic load on surrounding rock fracture*

With the mining working face advancing, the roadway surrounding rocks being ahead of the face are acted on periodically by the mining dynamic load produced by the periodic fracture of the main roof. The action results are that the two sides of coal mass of the roadway are heavily fractured and the roof convergence largely increases, but the change of the roof rock breaking degree is not obvious. Figure 2 shows some observation results on the breaking fractal dimension value d of two sides of coal mass of the roadways versus the distance L to the mining working face walls. With the mining working faces approaching, the fractal dimension values decrease. It indicates that the amounts of connective fracture increase. Therefore, under the action of the mining dynamic load, the fracture evolution process of the wall rocks is that the original fractures first expand, then connect, and then the new fractures emerge.

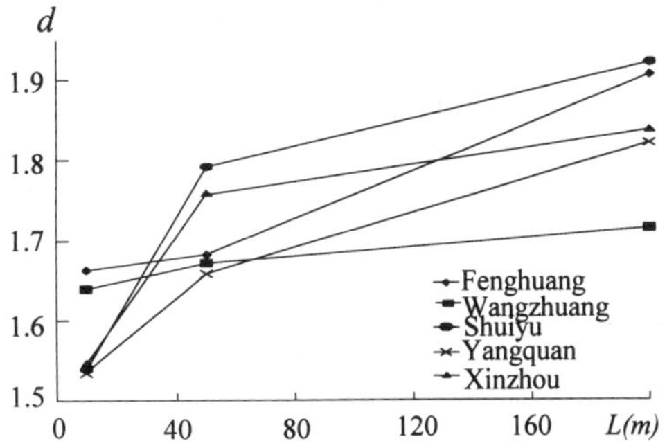

Figure 2. Fractal dimension d of wall rock versus the distance to the mining face wall

4 SIMILAR MODEL TESTS ON FRACTURE FEATURE OF SURROUNDING ROCKS

4.1 *Basic hypotheses*

Plane strain hypothesis: Underground roadways can be regarded as a plane strain problem. Boundary condition hypothesis: The limit of the stress concentration zone around roadways is about 3~5 times roadway radium. The periphery of the rock mass is considered an in-situ rock stress state.

4.2 Test equipment

The sizes of the model frame are length × width × height=770 × 230 × 480mm.The vertical load is placed using two hydraulic jacks. The sides are bound by a channel iron frame and polymethyl methacrylate plates.

4.3 Test program

The test models can be divided into two parts. The first part is two no-bolting models. Their rock strata structures and rock uniaxial compressive strengths are shown in the figure 3a,b. The roadway cross sections are squares of 3.0 × 3.0m. The main purpose is to know the breaking feature of the surrounding rocks under a no-bolting condition. The second part is two bolting models. The bolt arrangement is shown in Figure 4. The other conditions are the same as the models in the first part. The main purpose is to know the effect of bolting on surrounding rock breaking.

Ordinal number	Columnar section	Thickness (m)	Rock	Strength (MPa)
4		4.5	Sandstone	70.0
3		1.35	Mudstone	25.0
2		3.0	Coal	20.0
1		3.0	Sandstone	70.0

a

Ordinal number	Columnar section	Thickness (m)	Rock	Strength (MPa)
7		1.45	Limestone	45.0
6		1.25	Mudstone	25.0
5		0.57	Coal	20.0
4		1.45	Limestone	45.0
3		1.25	Mudstone	25.0
2		3.0	Coal	20.0
1		3.0	Sandstone	70.0

b

Figure 3. Rock strata and mechanical properties of test models

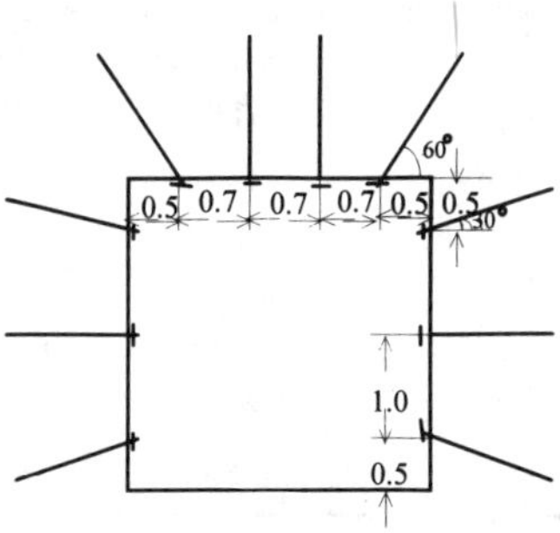

Figure 4. Bolt layout of the second part models

4.4 Model design

The geometric similar constant is α_l=25. The volume weight similar constant is α_γ =1.5. The stress similar constant is α_σ=37.5. The aggregates of the model materials are sand and 3% mica powder. The main cementing materials are cement and plaster.

4.5 Fracture features of surrounding rocks

The general fracture feature of the no-bolting models is that the fractures start at the top or bottom parts and expand to the middle parts of two sides of coal mass and form arched bedded breaking. The fractures develop gradually to the coal mass depths until the roof limiting span is reached and the breaking arches in the roof rocks are formed. Figure 5 shows the breaking evolution photographs of the no-bolting model one.

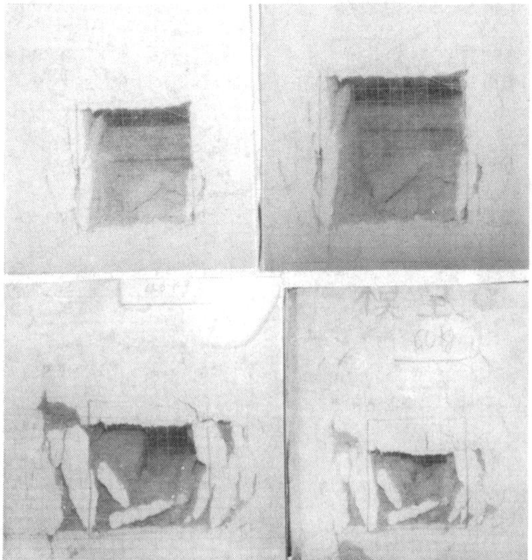

Figure 5. Breaking evolution of no-bolting model one

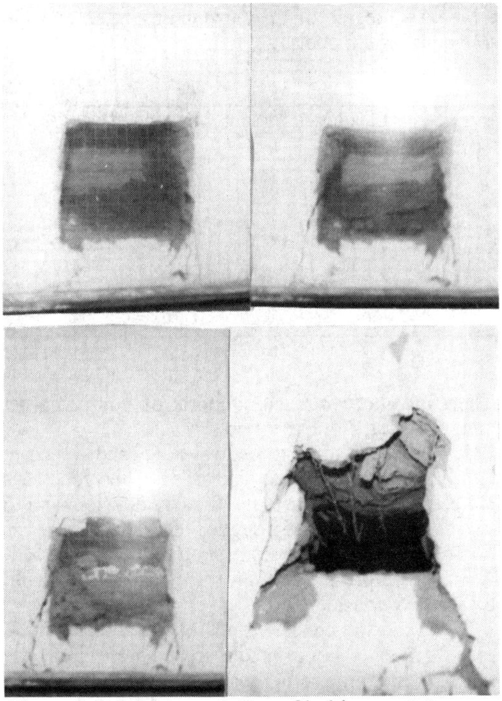

Figure 6. Breaking evolution of bolting model one

The main distinctions of the bolting-models are that their wall breaking loads are greater than that of the no-bolting models, and their fractures are mainly around the top or the bottom corners of the roadways and develop gradually to rock depths. Figure 6 shows the breaking evolution photographs of the bolting model one.

5 NUMERICAL SIMULATION ON FRACTURE FEATURE OF SURROUNDING ROCKS

5.1 Numerical simulation program

Two kinds of models with different surrounding rock structures are simulated. The first is the model with a thick mudstone roof. Its strata structure is sandstone floor 10.0m, medium hard coal 3.0m, mudstone roof 6.0m, and limestone roof 11.0m. The second is the model with the double or compound strata roof. Its strata structure is sandstone floor 10.0m, medium hard coal 3.0m , mudstone roof 1.35m, sandstone roof 4.5m, and limestone roof 11.5m. Four programs — no bolting, roof bolting, sides bolting, both roof and sides bolting — are considered for each model. The basic vertical load is the weight of 120m rock strata. The stress concentration coefficients considered are 1.0, 1.7, 2.2, 2.5, 2.8, 3.0, 3.2, 3.3. The rock mechanical parameters are the same as the similar material models.

5.2 Numerical simulation technique

The external loading model is used when the in-situ rock stress is calculated. The internal loading model is used when the excavation process is simulated. The excavation is considered step by step (Yu 1991).The directional failure element is used to simulate rock stratifications. The rod element is used to simulate the rock bolts (Yu 1983). Drueker-Prager criterion is used to define the plastic failure coefficient *f*. The *f* is used to evaluate the breaking degree of the surrounding rocks.

$$f = \frac{\beta I_1 + J_2^{1/2}}{K} \qquad (2)$$

where $\beta = \frac{\sin\phi}{(9+3\sin^2\phi)^{1/2}}$; $K = \frac{3C\cdot\cos\phi}{(9+3\sin^2\phi)^{1/2}}$; C is rock cohesive force ; ϕ is rock coefficient of internal friction ; I_1 is the first stress invariant ; J_2 is the second stress invariant.

5.3 Simulation results

Figure 7 shows the plastic element evolution of the foure programs in the first model as load increases. The checks in the figure are the elements divided in the surrounding rocks. The numbers in the elements indicate the loading steps when the elements reach plastic failure.

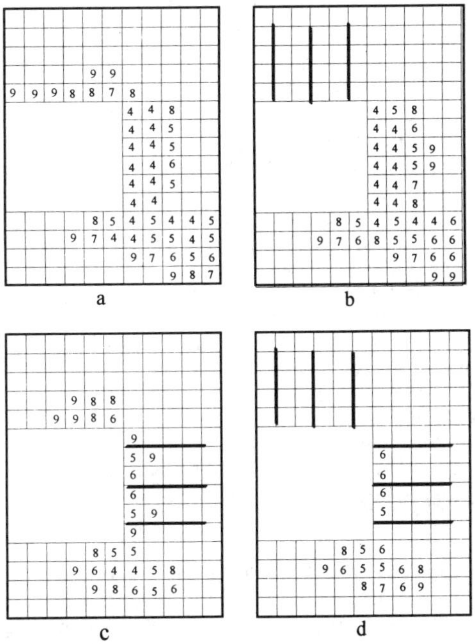

Figure 7. Plastic element evolution of the first model as load increases (foure programs)

Table 4 shows the distribution of the plastic zones, the breaking degree of the surrounding rock, and the bolting effect on the surrounding rocks of the eight programs in the two models.

Table 4. Numrical simulation results

Program	Plastic zone	Plastic coefficient	Bolting effect (%)
Model 1, no-bolting	Sides, top and bottom corners	20.64	
Model 1, roof bolting	Sides, bottom corners	18.16	20.22
Model 1, sides bolting	Top and bottom corners	6.24	46.20
Model 1, roof and sides bolting	Bottom corners	6.09	54.54
Model 2, no-bolting	Sides, top and bottom corners	19.32	
Model 2, roof bolting	Sides, bottom corners	17.32	16.68
Model 2, sides bolting	Top and bottom corners	5.57	42.16
Model 2, roof and sides bolting	Bottom corners	5.96	54.95

6 EFFECT OF BOLTING ON SURROUNDING ROCK FRACTURE

The breaking of the roadway surrounding rocks is very developed. The roadways are sandwiched in the soft strata. Under the action of mining dynamic abutment pressure, the sides of the roadways are easily broken. Bolting can change the breaking state of the surrounding rocks. Three aspects of the effect of bolting on the surrounding rocks can be seen by comprehensively analyzing the research results. These are as follows:

1. Bolting can change the stress state of the surrounding rocks. The two-dimensional stress state is turned into a three-dimensional stress state by bolting. According to Mohr's theory of strength, the tangential breaking limit stress of the peripheral rocks will be increased from the uniaxial compressive strength σ_c to $\sigma_c' = \sigma_c + P/ab \cdot tg^2(\varphi/2 + 45^\circ)$, where P is the bolting force and a and b are the cross and line intervals of the bolts.

2. The bolting can increase the internal friction angle of the surrounding rocks. The similar model test results show that the average internal friction angle of the bolting-models is 15° greater than that of the no-bolting models. Hence, the bolting can increase the internal friction angle and the cohesive force of the surrounding rocks.

3. The fractured surrounding rocks when bolted still have enough strength and load-carrying capacity. The breaking state of the no-bolting models is arched bedded spalling from the sides to the surrounding rock depths, while the breaking state of the bolting models is large wedge-shaped blocks. The blocks squeeze and articulate mutually; hence, the fractured surrounding rocks bolted still have enough strength and load carrying capacity.

7 CONCLUSIONS AND SUGGESTIONS

1. The surrounding rocks of the actual mining roadways have the characteristics of being soft, bedded, and cut by many sets of fractures, etc. Therefore the combined supports of bolt-beam-net and bolt-truss should be used.

2. The actual mining roadways are sandwiched in soft and fractured strata. The sides of the roadways are easily broken. Due to the fracture development to two sides of coal mass depths, the final destabilization results in the total destruction of the whole roadways. Therefore the bolting of the road sides should be strongly considered.

3. Due to the cross-sections of the actual mining roads generally being rectangle and trapezium, the top and bottom corners are stress concentration zones. The breaking of the surrounding rocks starts generally around the corners and then develops to the surrounding depths; therefore, the boltings near the corners should be strengthened.

REFERENCES

Kang Tianhe, Xue Yadong & Jin Zhongming 1996. Bolting design criteria of actual mining roadways based on surrounding rock condition and dynamic load action. Chinese Journal of Rock Mechanics and Engineering. 15(supp): 571-576.

Kang Tianhe, Zhao Yangsheng & Jin Zhongming 1995. Fractal study on crack scale distribution in coal mass. Journal of China Coal Society. 20(4): 391-398.

Yu Xuefu 1991. Rock and soilmechanics in information times. Beijing: Science Publishing House.

Yu Xuefu 1983. Surrounding stability of underground engineering. Beijing: Coal Publishing House.

Computer Methods and Advances in Geomechanics, Yuan (ed.) © 1997 Balkema, Rotterdam, ISBN 90 5410 904 1

Research of preventing water-inrush and rock-burst disasters in strong seismic coal fields in China

Li Baiying & Wen Xinglin
Shandong Institute of Mining and Technology, Taian, China

ABSTRACT: Observations have been made for a long time and in a wide range on deformation and damage of floor strata in deep tunnels. Floor strata movement, advance bearing pressure, floor rock osmosis and water-resisting ability were observed. Based on the measured data and the data of the equivalent material and computative modeling, the regularity of damage and strata movement has been obtained. According to the data of the hydraulic fracturing test in situ and in lab, the quantitative evaluation of water-resistance of the rock mass has been presented. Finally, the Zhao colliery is taken as an example to discuss the possibility to predict the rockburst and the water-irruption occuring on the floor as well as the safety evaluation of deep mining.

Tangshan city is a famous coal city in China, which suffered from a strong earthquake in 1976. After this seism, some collieries occurred various phenomena of hydrogeological engineering, i.e. rockbrusts waterbrusts. The largest colliery waterbrust happened in this mine area is worth mentioning. In order to predict the seism and secure the mine safety, the comprehensive and highly accurate observations have been held in this coal district, for example, the Zhaogezhuang colliery in the Tangshan seism coal district.

The Zhaogezhuang colliery has a mining history over one hundred years, the mining depth being over 1000m, beneath the rock strata of seam floor, the pressure of confined water in Ordovician limestone up to 10MPa, at 1975, the large floor waterburst happened and resulted in flood partial colliery. A few years ago, a rockburst occurred at the mining face. In order to protect the seism and flooding induced by mining, the systematical observative research has been carried out on the characters of the underground pressure, primary ground stress and rock mass hydraulics in the deep mining area, at the same time, the various modeling experiments and comprehensive analyses have been performed in the laboratory, their major content is as follows :

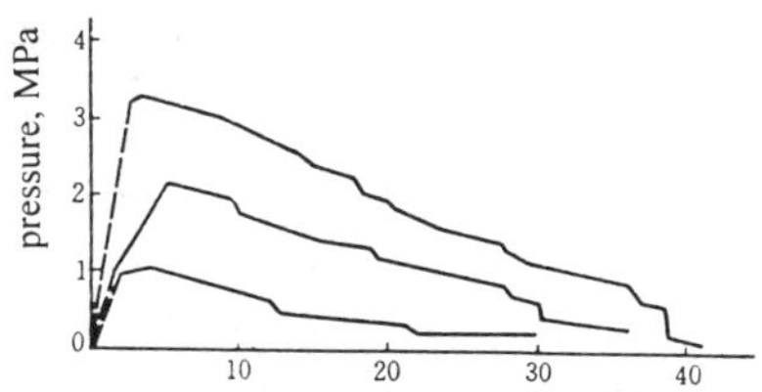

Fig. 1 Measured bearing pressure distribution

1 REGULARITY OF THE EFFECT BY UNDERGROUND PRESSUR ON THE FLOOR AND LAW OF DAMAGE

The field site was in the working face 1237 of Zhao colliery with an average mining depth about 900m, an average thickness of coal seams is 10 m and a dip angle 26°. The mining methods used were the inclined slicing longwall method along strike. The number of slicing was 5 and the lift was 2 m . The average strike length of this working face was 185m . The method of roof control was the bulk caving. The structure was simple and the strata were of uniclinal structure.

1.1 Observations of the advance bearing pressure

In the haulage road way in front of the working face the horizontal boreholes were drilled to put in the compression gauge. The borehole depth was 2m. Fig. 1 shows the measured supporting pressure distribution.

1.2 Observations of the borehole water-injection experiment

The water-injection boreholes were drilled in the floor at different distances from the coal seam (shown in Fig.2). The 2m strata in the deepest orehole were tbhe water-injection segment and the rest was the pipe liner and the space outside the pipe was grouting. The two slices were measured to study the effect and damage in the floor. Fig. 3 are the curve of water-injection capacity

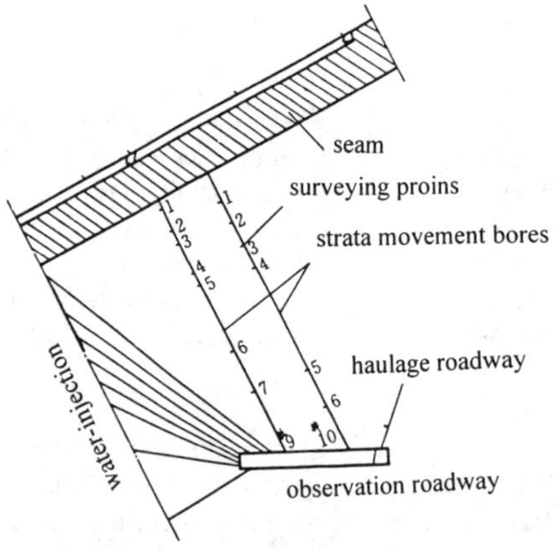

Fig. 2 Layout of the observations in-situ

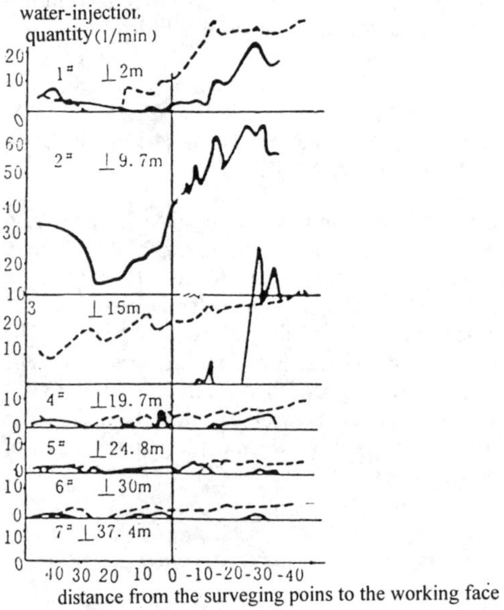

—— the first slice

······ the second slice

Fig. 3 The curves of water-injection quantity

The curves of water-injection capacity in Fig.3 are very typical. All their processes are the down-up-down type which is corresponding to the stress state in the floor, that is, compression-expansion-compression.

1.2.1 The depth of the damaged zone

The borehole whose capacity of water-injection changed most greatly was the third borehole whose distance from the coal seam was 15m. The capacity was 0.11 /min before mining but it increased to 47 l/min after mining. The second slicing condition was similar. The water-injection curves of the borehole 4. 5.6 and 7 were relatively even. So we can find that he damaged zone is situated between the third borehole and the fourth borehole.

The statistical analysis was made on the former data of the water-injection tests, so the criteria to determine the damaged depth were obtained as follows:

1) $k_2 / k_1 = 2$

2) $\Delta k=0.2$ (m/d)

where k_1---- the permeability coefficient before mining (m/d)

k_2----the permeability coefficient before mining (m/d)

$\Delta k=k_2 - k_1$ (m/d)

According to the above criteria and the interpolation calculation, the damaged depth of the first and second slices were 19m and 22m respectively. But to add the thickness of unmined coal seam, the depth increased only one meter.

1.2.2The mining influence depth

The changing law of the water-injection curve of the seventh borehole were the same as that of the other boreholes and it coincided basically with the changing law of the advance bearing pressure. This shows that the mining influence depth was bigger than the distance of the seventh borehole from the coal seam, i.e., > 45.5m.

According to the analysis above, we can obtain that the permeability of the floor beds increased obviously, but the damaged depth increased a small quantity on the conditions of the subslicing of thick coal seam.

1.3 The model experiments of similar material in the lab

According to the laws of similarity, the models

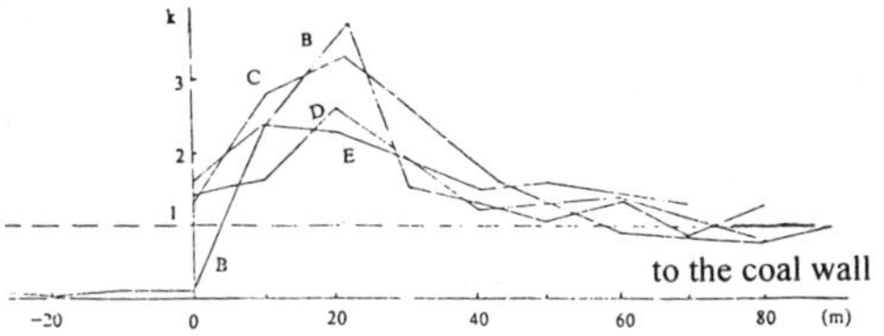

The adance is 110m

Fig. 4 Stress distribution in floor of first slicing

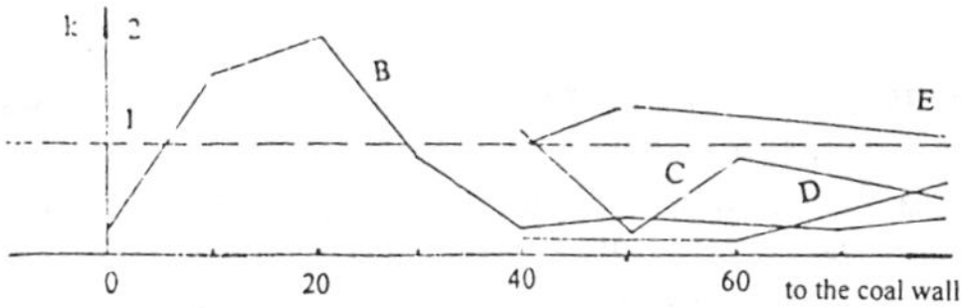

The adance is 51m

Fig. 5 Stress distribution in floor of second slicing

were established to simulate the 1237 working face in the deep at Zhao colliery. Fig. 4 is the pressure distributions in the floor of the first slicing.

The figure shows: the mining influence limits in the horizontal direction is 70m. The obvious influence limit is 40m. In the mined-out area, the floor stress approached the original stress where the distance from the coal wall is 75m. The stress attenuates greatly in the shallow above the 21.2m floor depths, but it attenuates slowly in the deep.

The second slice began to be mined in no time after the first slice was mined out. The advance concentrated stress zone was not formed completely in the floor. Even if the concentrated stress occurs, the values is smaller than that of the first slice (shown in Fig. 5). This shows that the damaged degree of repeated mining will decrease greatly.

The strata movements of the floor proved the above conclusion

1.4 Simulation by the finite element

Seventeen models were established to simulate the 1237 working face. Fig. 6 is the vertical stress distribution.

The peak value of the vertical stress attenuates by the similar negative exponent with the floor depth increasing.

The stretch and compression occurred in the floor beds along with the rise and decline of the strata. The tensile stress occurred in the limits of 3m floor under the goaf with very small values.

The damage of the floor beds developed below the coal wall of the working and along the diagonal.

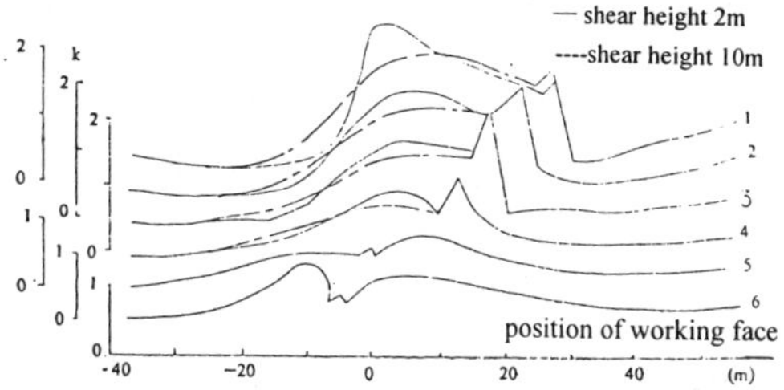

Fig. 6 Vertical stress in floor

floor depth: 1 — 1.5m, 2 — 7.5m, 3 — 11.5m
4 — 18.0m, 5 — 28.0m, 6 — 36.0m

This occured in the same directions as the shearing stress.

The structure planes (the bedding plane, the tender interlayer, the joint etc.) had a low strength. So the damage was easy to develop along these structure planes.

In a word, the changing and the time-space laws of the damaged depth were obtained in the deep on the conditions of subslicing of the thick coal seam. The damaged depth increased 1m after each subslicing on the conditions of Zhao colliery.

2. THE REGULARITY OF STRATA MOVEMENT EXISTING IN THE MINING PROCESS

The borehole 9 and 10 (as shown in Fig. 2) were used to observe the strata movement in bores. The surveying points in bores were fixed by the compressed wood and were linked by the steel ropes. The observation instrument was the mechanical displacement instrument. Fig. 7 shows the measured displacement curves in bore 9.

The following strata movement laws can be seen from Fig. 7:

2.1 Relationships between movement and advance

The general process is divided into 4 stages. Which are the decline before mining, rise, decline and the stable after mining in order .

The first two stages were observed in this experiment.

2.2 Relationships between the movement and the distance from the seam

The displacement decreased with the distance increasing. The mining influence zone was up to 67.28m.

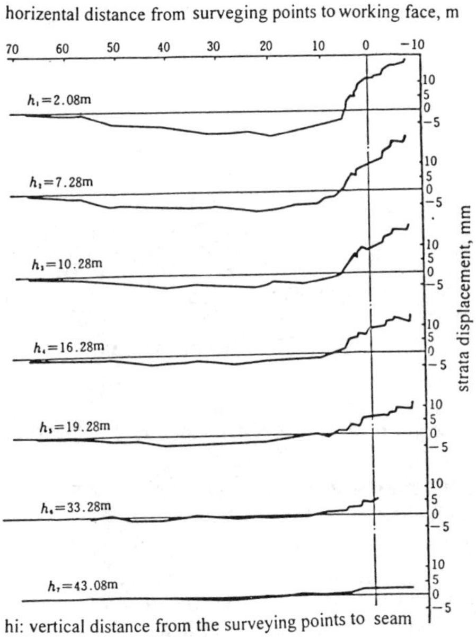

Fig. 7 Measured strata displacement curves in bore 9

2.3 *Relationships between movement and underground pressure*

Strata movement is controlled by the underground pressure. when the floor is in the zone of the advanced bearing pressure, the floor beds are in the down-going conditions.

2.4 *Irregular strata movements*

When the surveying points are at the down-going stage, the expansion will occur in the strata between two surveying points. This is the main reason for creating the separation beds and the bedding cracks.

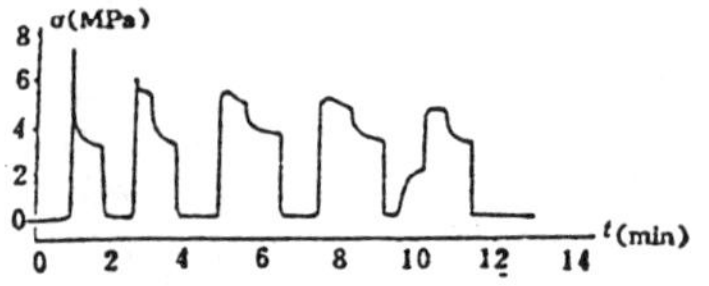

Fig. 8 stress — time recording curve in bore 2

3. THE QUANTITATIVE EVALUATION OF THE CHARACTERS OF ROCK MASS HYDRAULICS AND WATER-RESISTANCE

In order to obtain the principal stress in the deep and the fracturing pressure of the rock mass in-situ, the three bores were drilled to do the site hydraulic fracturing test in the crosscut roadways at level 5 .The seven strata were fractured in this test.

The table 1 includes the measured data of the site hydraulic fracturing. Fig. 8 is the stress-time curves in the borehole 2.

According to the data of the site hydraulic fracturing test and the principal stress data measured in the shallow in this region, the relationships between the principal stress and the depth were obtained as follows:

$$\sigma_v = 0.0235\,H \quad \text{(MPa)}$$
$$\sigma_H = 1.63 + 0.0194\,H \quad \text{(MPa)} \qquad (1)$$
$$\sigma_h = 1.88 + 0.0077H \quad \text{(MPa)}$$

The hydraulic fracturing tests of the 3-dimension surrounding pressure in the lab had been done to study the internal relationships of the hydraulic fracturing between the site and the lab and the action of the principal stress. The surrounding pressure was given according to the formula (1) (the depth is 1100m).

According to the hydraulic fracturing data obtained in the lab, the major factors which affect the fracturing pressure of rock mass are as follows:

1)The structure, the joints and the cracks of the sample,

2)The surrounding pressure, i.e., the principal stress,

3)The characters and the compressive strength, etc..

Table 1

No.	Depth (m)	lithologicol characters	P_b (Mpa)	P_s (Mpa)	P_r (Mpa)	P_0 (Mpa)	σ_H (Mpa)	σ_h (Mpa)	σ_v (Mpa)	T_c (Mpa)	τ_{Hmax} (Mpa)	σ_H direction
1	446	mudstone	12.62	5.12	5.12	0.12	10.18	5.12	10.48	7.50	2.53	
	453	sandstone	10.44	4.07	3.94	0.19	8.08	4.07	10.64	6.50	2.01	
	460	bauxitic rock	4.89	4.26	2.26	0.26	10.26	4.26	10.81	2.63	3.00	
2	453	siltstone	7.69	4.31	4.56	0.19	8.18	4.31	10.64	3.13	1.94	N87° E
3	453	sandstone	13.44	4.56	4.56	0.19	8.93	4.56	10.64	8.88	2.19	N87° E
	471	sandstone		6.00	3.12	0.37	14.50	6.00	11.07		4.25	N73° E
	476	siltstone		8.17	9.17	0.42	14.90	8.17	11.19		3.37	

Among them, the primary cracks affect the fracturing pressure most greatly. For example, for the sandstone, the P_b is 2.2 MPa when the primary cracks exist in the sample, but it is 14.0 MPa when the sample is intact.

In order to obtain the quantitative evaluation of the water-resistance of rock mass, the new concept of the water-resistant coefficient has been presented in this paper.

$$Z_f = P_b / R \quad (2)$$

where:

Z_f ---- The water-resistant coefficient (MPa / m)

P_b---- The fracturing pressure of rock mass (MPa)

R ---- The fracturing radius (m)

P_b and R can be obtained by the site measurement or computation. For example, the measured P_b of siltsand was 10.44 MPa, the fracturing radius were gained according to the anology and the value was 43m. So Zf is 0.242 MPa/m.

In view of the above, the water-resistant coefficients of different lithological characters can be obtained respectively and the total water-resistant ability of the strata can as well.

4. DISCUSSION AND EVALUATION

Rockburst is the particular phenomenon in the underground engineering. A condition to create the rockburst is the stress concentration of the surrounding rock. Especially, the rockburst is easy to occur in the fragile rock and the structure zone. For the circular roadway, the maximum stress condition in the circle is $\sigma = 3\ \sigma_1 - \sigma_3$. If the value of the ($\sigma_1 - \sigma_3$) is bigger, it is more possible to create the rockburst.

According to Formula 1, the principal stress at the 1200m depth in Zhao colliery is as follows:

$\sigma_v = 28.2$ MPa $\sigma_H = 24.9$MPa $\sigma_h = 10.12$

Because the $\sigma_v / \sigma_h > 2.5$, rockburst is easy to occur in this stress conditions.

Although the deep mining at Zhao colliery had not created big rockburst, the disasters of the coal burst happened in 1992. In the mean time, the pressure bump in the deep is bigger and bigger, accidents of roof-fall and others. happen frequently. Therefore, in order to predict the rockburst, the stress measurement in the deep should be strengthened and it can predict the occured of the earthquake. The important work is to understand the conditions of the stress distribution. When the stress concentration occurs, artificial caving can be used to avoid rockburst.

Rockburst is in close relationship with water-irruption, for example, water-irruption can occur along with rockburst. No profound study has been made of rockburst prodiction until recently, but the theories of the water-irruption prediction are opportune. The main theories are the "down-three-zone" theories.

The "down-three-zone" indicates that the floor is divided into three zones after mining, i.e., 1) damaged zone, 2) effective aquiclude and 3) water conducting zone. The thickness and the water-resisting ability of effective aquiclude control the occurance of water-irruption.

The Zhao colliery is the example to show the general steps to predict water-irruption by use of the "down-three-zone" theories.

1) To gain the basic data.

For example, the thickness of the floor (M=95~135m) and the water pressure (Pw = 11.98MPa).

2) To obtain the damaged depth (h_1):

The methods are measurement, anology and computation. For 1237 working face, h1 will be up to 25m after slicing 5.

3) To determine the conducting height of the confined water (h_3)

According to statistical analysis on the boreholes which get into the Ordovician limestone, we can gain the following height.

For the normal area, $h_3 \leq 5$m

For the domain, $h_3 = 5\sim10$m

4) To determine the thickness of the effective aquiclude (h_2)

$h_2 = M - h_1 - h_3$

For the normal area, the minimum thickness is 70m.

5) Safety evaluation

a) Evaluation by use of the total water-resisting ability.

If the total water-resisting ability is bigger than the water pressure, the mining is safe. Otherwise it is dangerous.

For normal area at Zhao colliery, it is safe.

b) Evaluation by the fracturing conditions

According to the hydraulic fracturing laws $P_b = 3\ \sigma_3 - \sigma_1 - P_o + T$, if $P_b > P_w$, the confined water can not fracture the rock, it is safe. For Zhao colliery, P_b is 13.3 MPa < P_w, so it is safe.

5. CONCLUSION

Zhao colliery is the representative one in the strong seismic coal field. It is applicable to each strong seismic coal field. Water-irruption and rockburst are the major accidents in this colliery. Prevention and safety evaluation should be made based on the "down-three-zone" theories and the laws of stress distribution.

REFERENCES

Gao Yanfa & Xiao Hongtian 1992. Observation of Strata Movement and Back Analysis of Displacements in the Coal floor of a Deep Mine *China Taian*.

Zhang Wenquan & Wen Xinglin 1993. Research on the relationship between the mining parameters and the breaken depth of the floor of gently-inclined seam *China Taian.*

Computer Methods and Advances in Geomechanics, Yuan (ed.) © 1997 Balkema, Rotterdam, ISBN 90 5410 904 1

Mine-wide ground stability assessment and mine planning using the finite element technique

Y.S.Yu, N.A.Toews, S.Vongpaisal, R.Boyle & B.Wang
CANMET, Natural Resources Canada, Ottawa, Ont., Canada

ABSTRACT: As mines get older and deeper, the dilution and mine stability problems get worse. These problems could be further complicated by future automation of underground mining. Regardless of the degree of mining automation underground, a good mine design is still considered one of the most important mining components in reducing mining costs and increasing safety and productivity.

A rational mine design and optimization of mining sequences could be greatly improved with the application of state-of-the-art modelling techniques. Modelling technology will help the mine designers to effectively identify potential stability problems in advance and take preventive measures accordingly.

In this paper, a powerful FE modelling tool is discussed and three case studies are presented. This work has demonstrated that modelling technology can be applied successfully to mine-wide stability assessment and for long term ground control and mine planning.

1 INTRODUCTION

Underground mine structures tend to be very large and complex as well as difficult to model numerically. With most current three-dimensional (3-D) finite element (FE) models it is unavoidable to significantly simplify the geometry and geology of the mine.

Numerical modelling has gained popularity due to its great potential application in the design of large underground mine openings, ground control and mine planning, as well as the trend towards the application of bulk mining at depth. A number of 3-D boundary element (BE) programs were developed in Canada in the early 1990's for modelling complex mine geometries [Yu et al, 1991, Curran et al, 1991, Wiles, 1991]. However, the FE method is more versatile and can easily simulate heterogeneity of rock masses. But most of the current FE programs usually generate a large number of data and require a large disk space, thus restricting the size of models, and preventing their routine use as a mine planning tool.

A mine-wide stability analysis of relatively simple mining operations can easily consist of $10x10^4$ brick elements with the solution of approximately $30x10^4$ equations. Even with such a model, a lot of details in mining geometry and geology could not be considered. Until recently, the computer time required to run such a model would take several days for an elastic solution on a workstation. This lengthy turn-around time definitely prevents its use as a routine mine planning tool. An effort has been made to improve the efficiency of a FE model by re-structuring the data handling procedure and by taking advantage of the recent advances in computer technology. By implementing a preconditioned conjugate gradient iterative solution procedure, a new FE program has been developed and the capabilities of modelling large-sized mine structures has been greatly enhanced. It is now possible to model rock (mine) structures with up to 10^6 elements on a workstation within an acceptable turnaround time. Therefore, the new program can now be used as an engineering tool for ground control in minimizing mine-wide ground instability and in maximizing ore recovery, as well as in evaluating mining sequences and alternate mining methods for long term mine planning.

2 COMPUTER PROGRAM

ROCK3D^{FE} is a static linear elastic, 3-D FE program, specifically developed for handling large-sized models efficiently in terms of data preparation and computer time. It has been used for a number of in-house cooperative research projects [Yu et al, 1996]. A linear elastic analysis for hard rock mining applications was fully validated, and the nonlinear analysis based on a localized non-linearity scheme and Drucker-Prager failure criterion is being implemented. To speed up the analytical process, an advanced pre- and post-processing module, based on OpenGL graphics, has been incorporated.

2.1 *Solution Procedure*

The equations that have to be solved in linear finite element analysis are:

$$Ax = b \qquad (1)$$

where A, b, and x are, respectively, the stiffness matrix, load vector and displacement vector. For two-dimensional (2-D) structures Eq. 1 has been very successfully solved using a banded Gaussian solver. However for 3-D structures, this solution procedure has been less efficient.

To better appreciate the size and numbers involved for large-sized FE models, consider a square 2-D 100x100 elements structure. The statistics can be summarized as: number of nodes = 10^4, number of equations (2 per node) = $2*10^4$, size of A (taking into account symmetry) $\approx 2*10^8$, half band size ≈ 200, and banded representation of A $\approx 4*10^6$ = 16 Mbytes (single precision).

Note that for a regular structure (each node associated with at most 4 elements) the maximum number of nonzeros in a half band is less than or equal to 10. The number of nonzeros in A is therefore $\approx 20*10^4$. For today's typical microcomputer this problem is easily solvable.

Now consider a cube 3-D 100x100x100 elements structure. The statistics are: number of nodes $= 10^6$, number of equations (3 per node) $= 3*10^6$, size of A (taking into account symmetry) $\approx 4.5*10^{12}$, half band size ≈ 30000, and banded representation of A $\approx$ $9*10^{10} \approx 360$ gigabytes (single precision).

For a regular structure (each node associated with at most 8 elements) the maximum number of nonzeros in a half band is less than or equal to 42. The number of nonzeroes in A is therefore $\approx 126*10^6 \approx 504$ Mbytes (single precision). Given the capacity and speed of the most advanced microcomputers currently available this problem is not solvable using a banded solver but is possible using sophisticated solvers and data structures involving only nonzeros.

The basic solution technique adopted in ROCK3D^{FE} is the preconditioned conjugate gradient iterative procedure [Angeleri et al, 1989 and Tan et al 1991]. The preconditioning is an incomplete Cholesky LDL^T decomposition of the global stiffness matrix. The decomposition is incomplete because all matrix fill-in is ignored during decomposition. The resulting approximate LDL^T decomposition has a nonzero structure that is identical to the global stiffness matrix.

The global stiffness matrix (A) and the approximate decomposition (LDL^T) are stored as column compressed vectors. This needs one additional vector, *KA*, a vector of length *neq* (no.of equations) which defines the location of the diagonal elements in the compressed stiffness matrix.

The iterative procedure used closely follows that outlined in the reference [Tan and Bathe, 1991]. The stiffness matrix A and force vector b are assembled. A is stored in a compressed column data structure (no zero values are stored). Then an approximate LDL^T decomposition of A is carried out. L is a unit lower triangular matrix, L^T is the transpose of L, and D is a diagonal matrix. No fill-in is allowed. This LDL^T decomposition is called K in the description below.

$$r_0 = b - Ax_0$$
$$z_0 = K^{-1}r_0 \qquad (2)$$
$$p_0 = z_0$$

The initial guess x_0 is taken to be a zero vector. The procedure then iterates as follows:

$k = 0, 1, 2, \ldots$

if $\|r_k\| < \varepsilon$ procedure terminates.

Else

$$\alpha_k = \frac{(z_k, r_k)}{(p_k, Ap_k)}.$$
$$x_{k+1} = x_k + \alpha_k p_k$$
$$r_{k+1} = r_k - \alpha_k Ap_k$$
$$z_{k+1} = K^{-1}r_{k+1}$$
$$\beta_k = \frac{(z_{k+1}, r_{k+1})}{(z_k, r_k)}$$
$$p_{k+1} = z_{k+1} + \beta_k p_k$$

There are two major operations in each iteration, a matrix-vector multiplication Ap_k and a solution of equations $Kz_{k+1} = r_{k+1}$. The solution involves only a forward and backward substitution.

The procedure converges fairly quickly for well conditioned problems such as elastic rock behavior. Figure 1, as shown above, is a plot of the solution time versus the number of equations, for a typical mining problem. The lower and upper limits of the number of

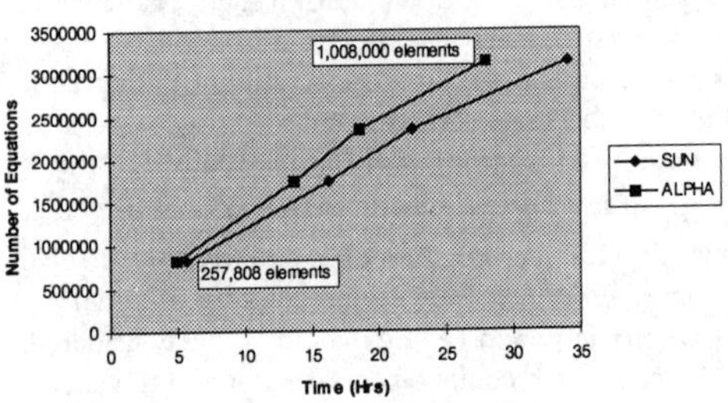

Figure 1 Solution speed versus no. of equations

equations correspond to approximately 257,808 and 1,008,000 brick elements. The testing was conducted on both a Sun SparcStation 10 and a Digital AlphaStation 200 4/233. The results indicated that the solution speed was slightly faster on the AlphaStation 200-4/233. For example, the time required to run a model of 553,358 elements were 13.7 and 16.3 hours, respectively, on the Alpha and Sun workstations.

3 CASE STUDIES

The ROCK3D^{FE} technology has been successfully applied to Canadian hard rock mines on a cooperative research and cost recovery basis. Three examples are presented here as case studies. Two of the mines are located in Quebec and the third in Northern Ontario. The 3-D model-size ranges from $2x10^5$ to $6x10^5$ elements depending on the requirements of the analyses. 15-20 models were constructed and analyzed for each of the three mines. The project objectives, in general, were to assess the overall mine stability and to provide design guidelines to the operators for making ground control decisions with respect to mining sequences and mine stability; specific issues for each mine were also dealt with.

The mechanical properties of the mine rocks were determined by laboratory testing. To estimate in-situ deformation modulus, E_m, the empirical relationship based on the Geomechanics Rock Mass Rating (RMR) was adopted:

$$E_m = 10^{((RMR-10)/40)}$$

The pre-mining stress conditions were established, based on in-situ stress determinations, as: σ_v = 0.029 MN/m; σ_{EW} = $K_{EW}\sigma_v$ and σ_{NS} = $K_{NS}\sigma_v$. Where σ_v is the vertical stress; σ_{EW} and σ_{NS} are the horizontal stresses parallel and perpendicular to the strike directions (EW and NS) of the ore bodies, respectively. K_{EW} and K_{NS} vary, respectively, from 1.5-2.13 and 2.0-2.5 depending on the mine sites.

In addition to the conventional shear and tensile failures, stability analyses were also based on a strain energy density criterion. Through model calibration and back analyses, for each mine, an allowable limit of strain energy density was established for identifying potential ground failure.

Some typical longitudinal and transverse sections, and stresses and strain energy density contours from selected areas of interest are presented as examples for the Mines A, B and C, at the end of the paper.

3.1 *Mine A*

The mine, has been operating since 1937 and has produced over four (4) million ounces of gold since then. Shrinkage, room-and-pillar, and cut-and-fill mining are employed , with a production rate of about 525,000 tones per year.

The gold deposit occurs within a band of andersite volcanics rock units intruded by a porphyritic diorite and bounded to the north and south by two east-trending sub-vertical main shear zones. The east-west oriented, steeply dipping ore zones form a mineralization block of about 200 m wide extending to a depth of 1500 m. For the study, three ore zones have been identified. These are M-Zone, M-Branch and M-North. The principal rock types include diorite porphyry, porphyry dykes and diabase dykes.

The main objectives of this project were: (a) assessing mine-wide ground stability based on the mine plans, and (b) providing essential design guidelines for making ground control decisions with respect to optimizing pillar and sill pillar dimensions. The following conclusions were drawn:

- A thickness of 9.1 m for the sills was considered adequate. The average strain energy density of the sills identified in the study ranged from 0.25 to 0.42 MJ/m^3
- To minimize instability of the permanent pillars and the access areas under the permanent pillars, mining should proceed from east to west between levels 2500 and 2600. The 27.4 m stope pillar dimension, east of 2602EA stope, should be adequate for temporary support for the area.
- Based on the stresses and strain energy distribution, the shaft, the cross-cuts and the ore passes between levels 2400 and 2800 should evoke no major concerns in terms of stability.
- The model study indicated that 2704E pillar in M-Zone North was subjected to high strain energy. This might cause rock burst problems. Therefore, proper ground support in this area should be planned.
- CANMET strain rings were already installed at two strategic locations in the M-Zone. The stress changes at the strategic locations should be monitored regularly and recorded properly according to the mining events. This information would be useful to determine intact resistance or residual strength, for ground control as well as for model calibrations for future studies.

3.2 *Mine B*

The gold-bearing deposit occurs within complex rock units of gabbro, rhyolite and basalt. The mine employs a shrinkage stoping method, producing approximately 1400 tons/day.

The objectives of this project were to assess the mine-wide ground stability and to provide guidelines for mine design and ground control to minimize ground failure problems.

In addition, specific ground stability concerns were also examined: (a) the optimum size of the sill pillars, stope pillars and stope spans; (b) the effect of a partial extraction of sill pillars on mine stability; (c) the stability of the draw points at lower levels; (d) a strategic study of mining ore blocks for the years 2000 - 2003, below the 2350 level; and (e) The influence of

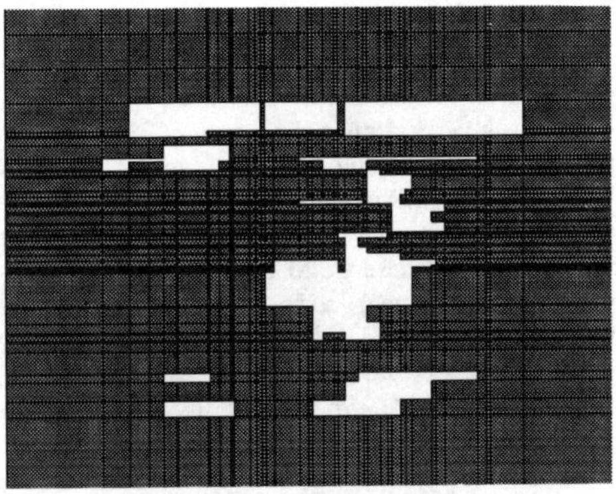

Fig. 2 A typical longitudinal view for one of the ore zones - Mine A.

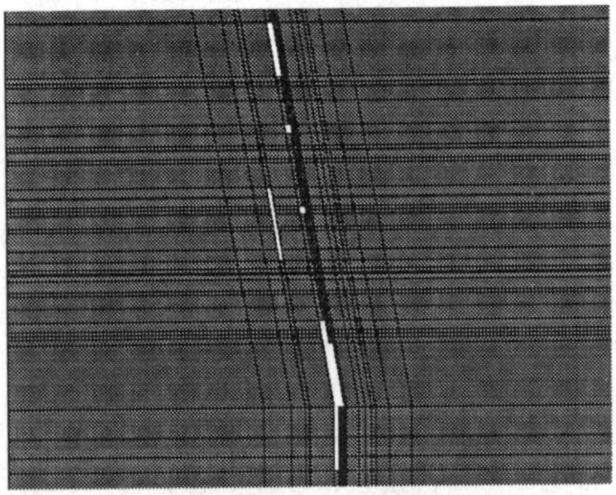

Figure 3 A simplified transverse view of the mine A

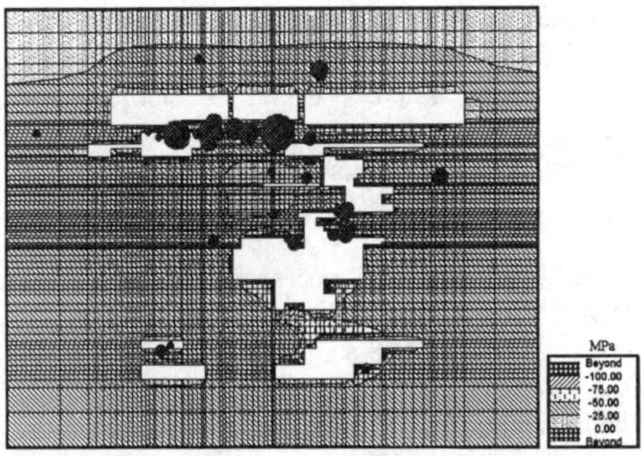

Fig. 4 σ_1 stress contours (comp. ' - ') for a longitudinal section, 1 m from the hanging wall (spheres indicate seismic events), Mine A.

mining-induced stresses of the 18-0-5N stope in the South zone on the stability of the adjacent stopes in the Main zone. The study indicated that:

- Based on previous studies at other mines and back analyses at the Mine A, a strain energy density level

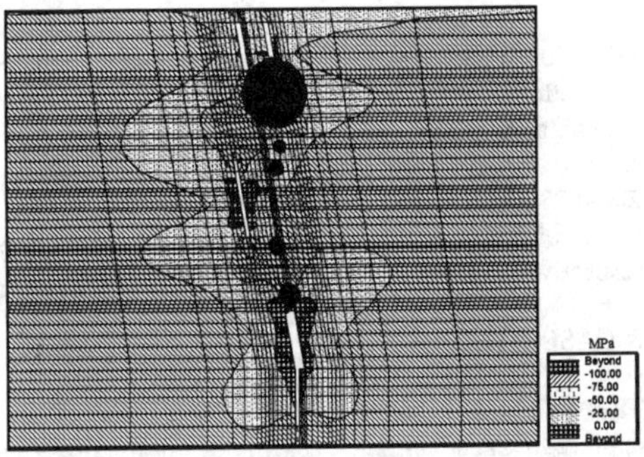

Fig.5 σ_3 stress contours (comp. ' - ') for a transverse section (zoomed, spheres indicate seismic events), Mine A.

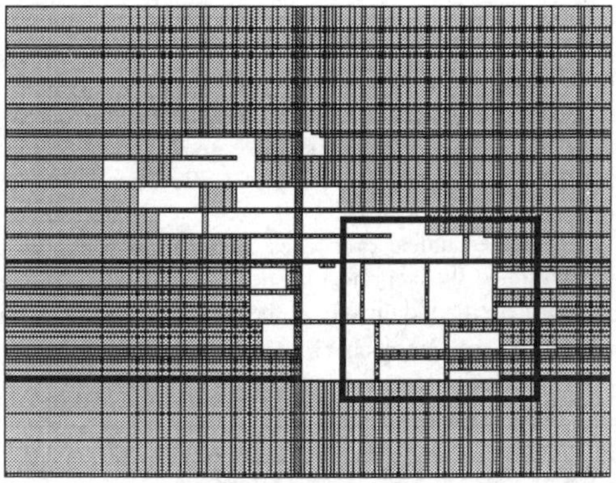

Fig. 6 A longitudinal view from one of the ore zones, 1 m from the hanging wall (box indicates the area to be zoomed for viewing stresses, Fig. 7), Mines B.

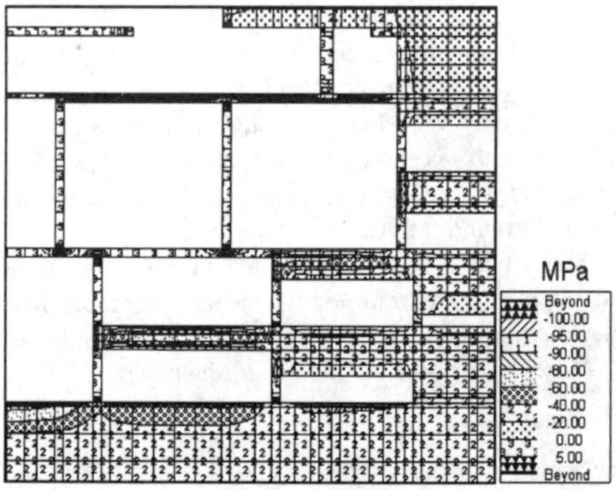

Fig. 7 σ_1 stress contours (comp. ' - ') for a zoomed area as shown in Fig. 6, Mine B.

of 0.6 MJ/m^3 could be used as a guide or allowable design limit for identifying potential ground

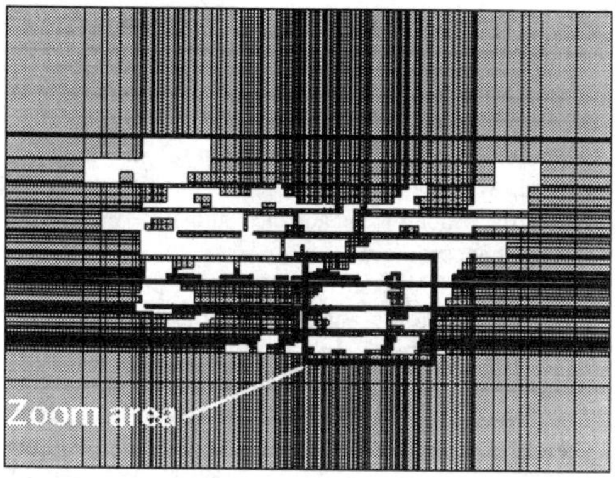

Fig. 8 A longitudinal view from one of the ore zones, 0.5 m from the footwall (box indicates the area to be zoomed for viewing stresses, Figs. 9 and 10), Mines C.

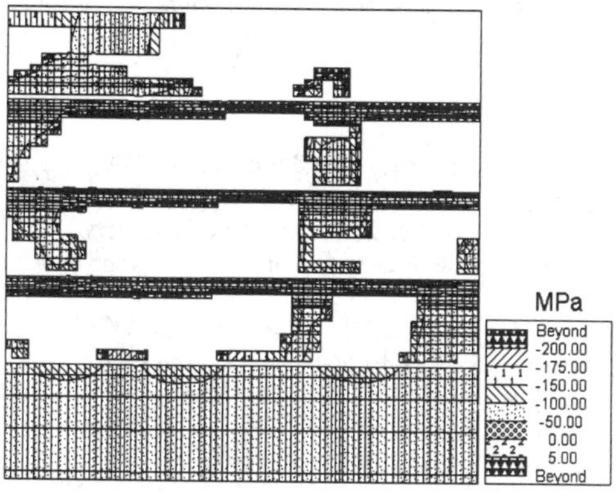

Fig. 9 σ_3 stress contours (comp. ' - ') for a zoomed area as shown in Fig. 8, 0.5 m from the footwall, Mines C.

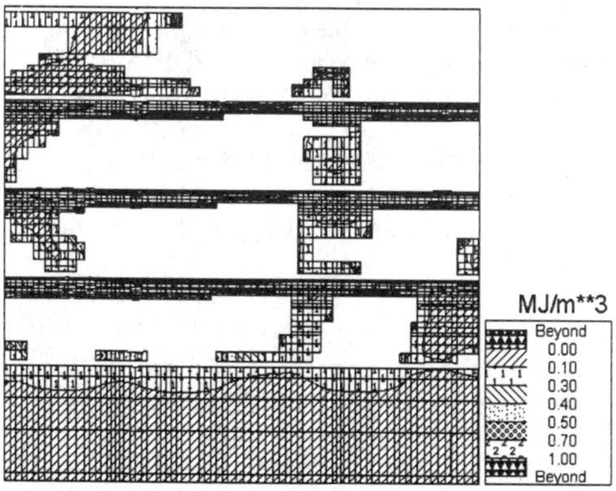

Fig. 10 Strain energy distribution around the openings for a zoomed area as shown in Fig. 8, 0.5 m from the footwall, Mines C.

instability problems or for potential rockburst prediction.

- Higher stresses and strain energy density would be expected in the sill on the 1650 level and subsequently, the ground conditions around the 18-A-5 stope would deteriorate as mining progressed deeper at the lower levels. Tensile zones, with a magnitude of about 1-5 MPa, would penetrate into the hangingwall and footwall to a distance of about 3-5 m in most of the stopes above level 2175. The extent of the tensile zones would increase as mining progressed deeper. The tensile stresses combined with unfavorable geological weakness could cause structural instability, and cause more dilution.
- The stresses resulting from a partially extracted sill pillar were very high (>300 MPa) and far exceeded the strength of the mine rocks. The strain energy density level in one of the partial sill pillars was reaching 2.9 MJ/m^3 which could cause serious rockbursting problems. Hence, from a ground control point of view, the mine plan calling for a partial extraction of sill pillars was not a favorable option and was not recommended for implementation. Longhole mining was recommended for extracting such sill pillars.
- Draw points with dimensions of 3 m wide, 9.1 m long and 9 m centre to centre were considered to be stable. The average strain energy density around the draw point was about 0.33 MJ/m^3 .
- With regard to sill pillar extraction, a minimum of 15 m of sill pillar in thickness should be left before any sill extraction was attempted for the alternate levels. With a stope width of 1.4 m, the stope span should be kept at 90 m wide or less and mining should proceed with caution. A longhole mining method was recommended for extracting the remaining sill. To provide a warning system against sudden sill failure, strain rings or other types of instrumentation should be installed at strategic locations at the sill.
- The influence of mining-induced stresses from stope 18-0-5N in the South zone on the stability of the adjacent stopes in the Main zone was quite significant. Shrinkage stoping beyond the 1700 level would require ground support of the hangingwall in order to minimize dilution.

3.3 *Mine C*

The mine has operated since 1948, and produced more than 3,000,000 ounces of gold. Over 80 km of drifts have been driven and approximately 200 stopes mined. Mining has progressed to about 1700 m below surface, employing conventional cut-and-fill and open stoping methods. The current production was reduced from 320,000 to 234,000 tons/year due to depleting ore reserves. In order to compensate for reduced reserves, sill pillars between levels 19 (844 m below surface) and 22 (981 m) were identified for extraction, These pillars was highly stressed and there was no methodology to address the safe mining of highly stressed sill pillars.

The study was carried out in terms of strategic mine planning to maximize sill pillar extraction while minimizing the risk of potential rockburst damage. Based on the results of this investigation and on engineering judgment, the following observations and conclusions are made:

- From the strain energy distribution and ground control point of view, the optimized mining sequence was to mine sills from the lowest level followed by mining the upper sill pillars on the upper levels. The lower stope should be filled before extracting the sill pillar immediately above.
- A waste pillar would be useful for localized ground control, but it would cause rockburst problems if it had a solid core of less than 6 m. Allowing a 2 m skin effect of cracking at both ends, a minimum 10 m long waste pillar should be left. The empirical equation, $ED = 1.5 * (P_L)^{-0.5}$, was established for determining the minimum length of waste pillars at this mine, where ED is stored strain energy density (MJ/m^3) and P_L is the length of waste pillar (m).
- High strain energy density levels, $>1.0\ MJ/m^3$, were concentrated in the skin of the pillar and in the brow areas. This high energy density zone extended to a depth of approximately 1 m from the pillar skin surfaces, and to a depth of approximately 2 m in the brow areas. This energy magnitude exceeded the allowable design limit, and therefore, cracks or localized ground failures were anticipated.
- Major ground failures were not anticipated during the sill pillar recovery trial. However, when the stope length was greater than 64 m, attention should be given to minimize ground failure problems. This required careful observations and monitoring of ground conditions; and the mining panels should be planned as large as possible for each blast, as the high strain energy zone would increasingly penetrate towards the sill pillar core. Observation and recorded data should be fed back for computer model verification and design modification.
- All thin portions of sill pillars on the west side were liable to sustain higher stress concentration than the remaining portions of the sill pillars. Therefore, special attention should be given to drilling and blasting, as well as to supporting the ground. The recovery of sill pillars on the east side should have less ground failure problems because they were less liable to sustain high strain energy concentrations than those on the west side.

4 CONCLUSIONS

A good mine design is considered one of the most important mining components in reducing mining costs, increasing productivity and minimizing ground hazards. A rational mine design and optimization of mining sequences for mine planning could be greatly improved with the application of modelling techniques. Modelling technology, if properly applied, will help the miners to identify potential instability problems in advance and take preventive measures accordingly

The ROCK3D^{FE} software has been proven to be a powerful tool in the analysis of large complex mine geometries for hard rock mining. Models can be built, solved and analyzed in a reasonable turn-around time with relatively inexpensive equipment. Therefore, it can be used as a routine mine planning tool.

5 ACKNOWLEDGMENTS

The close cooperation received from management and the technical personnel at the mines is greatly appreciated.

REFERENCES

Angeleri, F., Sonnad, V. and Bathe, K.J. 1989. Studies of finite element procedures - an evaluation of preconditioned iterative solvers. *Computers & Structures* Vol. 32, No. 3/4, pp 671-677.

Bathe, K.J. 1993. ADINA - A finite element program for automatic dynamic incremental nonlinear analysis. *Military Institute of Technology, Massachusetts, Report No. 82448-1*. pp. 435. (Sept. 1975, rev. 1993)

Corkum, B.T., Curran, J.H. and Grabinsky, M.W. 1991. EXAMINE^{3D}: A three-dimensional visualization tool for mine datasets. *Proc. 2nd Can. Conf. on Computer Applications in the Mineral Industry, Vol.2, 679-688,* Vanc., Canada.

Tan, L. H. and Bathe, K. J. 1991. Studies of finite element procedures - the conjugate gradient and GMRES methods in ADINA and ADINA-F. *Computers and Structures* Vol 40, No. 2, pp441-449.

Wiles, T.D. 1991. MAP3D - Mining analysis program in 3-dimensions, user manual. *Mining Modeling Limited Report*, Copper Cliff, Ontario, Canada.

Yu, Y.S., Closset, L. and Toews, N.A. 1991. A Practical mining application using BEAP3D Software Package, *Proc. 2nd Can. Conf. on Computer Applications in the Mineral Industry, Vol.2,713-722,*Vancouver,Canada.

Yu, Y.S., Toews, N.A., Boyle, R., and Vongpaisal, S. 1996. Very large-scale 3-D modelling of mine structures using the finite element technique. *Proc. 2nd North American Rock Mech. Symp.*, June 16-19 1996. pp 1871-1878, Montreal, Quebec.

Computer Methods and Advances in Geomechanics, Yuan (ed.) © 1997 Balkema, Rotterdam, ISBN 90 5410 904 1

Borehole stability in shales – A scientific and practical approach to improving water-base mud performance

Fersheed K. Mody
Baroid Drilling Fluids, Inc. (A Dresser Industries, Inc. Company), Houston, Tex., USA

ABSTRACT: This paper summarizes the complex phenomenon of the physic-chemical interaction between drilling fluids and shale, and suggests strategies to apply this understanding to design of water-based muds (WBM) to combat borehole instabilities problems in shales while drilling.

1 INTRODUCTION

Shale makes up over 75 percent of drilled formations and causes over 90 percent of wellbore instability problems. The drilling of shale can result in a variety of problems ranging from washout to complete collapse of the hole. More typically, drilling problems in shales are experienced as bit balling, sloughing, or creep. The problem is severe and, for the industry, has been estimated to be a conservative $500 million/year problem[1-4].

As oil reserves deplete and the cost of drilling increases, the need to drill extended-reach wells with long open hole intervals will also increase. In the past, oil-based muds (OBM) have been the system of choice for difficult drilling. Their application has been typically justified on the basis of borehole stability, fluid loss, filtercake quality, lubricity, and temperature stability. As the environmental concerns restricted the use of oil-based muds, the industry provided innovative means to obtain OBM performance without negatively impacting the environment. One such example is the successful introduction and application of ester-based biodegradable invert emulsion drilling fluids in the past decade. These systems provided attractive alternatives to traditional OBM in accessing hydrocarbon reserves located at horizons difficult to reach and also in environmentally sensitive regions. However, there is a general consensus that a new breed of WBMs that mimics the performance of OBM will serve as a very attractive alternative. But, conventional WBM systems have failed to meet key performance criteria obtained with OBM in terms of rate of penetration (ROP), bit and stabilizer balling, lubricity, filtercake quality and thermal stability. More importantly, severe borehole (in)stability problems are encountered when drilling shale formations with conventional WBM, leading to significant increases in the overall well cost.

Past efforts to develop improved WBM for shale drilling have been hampered by a limited understanding of the drilling fluid/shale interaction phenomenon. This limited understanding has resulted in drilling fluids designed with nonoptimum properties required to prevent the onset of borehole instability. Historically, problems have been approached on a trial-and-error basis, going through a costly multi-well learning curve before arriving at reasonable solutions for optimized operations and systems. Recent studies[1-5] of fluid-shale interactions, however, have produced fresh insights into the underlying causes of borehole (in)stability, and these studies suggest new and innovative approaches as to the design of shale drilling fluids.

This paper summarizes the complex phenomenon of the physio-chemical interaction between drilling fluids and shale, and suggests strategies to apply this understanding to the design of WBM to combat borehole instability problems in shales.

2 PROBLEM BACKGROUND AND DISCUSSION

In many cases, borehole instability arises from insufficient mud pressure support on the borehole wall from an inadequate mud weight or a time-dependent increase in near-wellbore pore pressure. This results in higher water contents in the near-wellbore region which can be translated into lower shale strengths[2,3]. The movement of the water (fluid) in or out of the shale is governed by several mechanisms[1,2,4]; the two most relevant are the hydraulic pressure difference (ΔP) between the wellbore pressure (mud weight) and the shale pore pressure, and the chemical potential differences (Δu) between the drilling fluid and the shale pore fluid. Mody et al (1993) presented data based on the two fundamental driving forces and transport mechanisms (hydraulic and osmotic) to explain how an OBM could be effective at stabilizing the wellbore.

The interaction process and the mechanisms of transport for WBM/shale system can be quite different and more complex. The molar free energies of all of the constituents within the shale and the water-based drilling fluid provide the driving forces that result in the transfer of water, cations, anions, etc. The sum of all these potentials (hydraulic, chemical, electrical, and thermal) can result in a net flow. The equilibrium conditions will be dictated by the sum of these potentials. The difficulty is in the mathematical treatment of this coupled physio-chemical interaction between the WBM and shale. The knowledge of driving forces with the associated transport mechanisms are required in modeling this phenomenon. Several investigators have used the nonequilibrium thermodynamic approach in the treatment of the transport process in shales[1,2,4,6]. Nonequilibrium thermodynamics allow the incorporation of cross effects between different phenomenon, such as flux of a solution with different ionic species caused by the hydraulic gradients and/or chemical potential gradient of that species, as well as thermal and electrical gradients.

This complex nonequilibrium thermodynamic problem can be simplified by restricting the discussions in WBM/shale systems to the transfer of water alone, and thus, an osmotic mechanism. This assumption may have limited utility, because it requires that the shale itself under in situ conditions must possess characteristics similar to that of a semipermeable membrane. The contention that shales can have characteristics resembling a semipermeable membrane under in situ conditions has not been totally resolved by the scientific community.

If one accepts the concept that under in situ conditions low permeability, nonfractured, clay rich shales provide a leaky semi-permeable membrane, then the relationship between hydraulic flow and flow caused by the differences in chemical potential will define the net flow of fluid for WBM/shale systems. The alteration in the near wellbore pore pressure and the effective stress state will be influenced by this fluid flow. Experimental evidence reported by Mody et al (1993), and van Oort et al (1994) show that the chemical potential driving force, created from the activity difference between the drilling fluid and shale, can counteract the hydraulic driving force (the difference between the mud weight and pore pressure). The presence of this phenomenon in the WBM/shale system has great implications in optimizing the salinity of WBMs and in the modeling of borehole (in)stability in shales. As stated earlier, one of the root causes of shale (in)stability is the time-dependent increase in near-wellbore pore pressure caused by the invasion of drilling fluid into the shale, driven by the hydraulic pressure gradient across the wellbore interface from mud to the shale pore fluid. Drilling fluids that can eliminate, counteract, or minimize the invasion of this pressure front (flow) into the shales would be expected to reduce the effects that lead to borehole instability.

Conventional WBMs are not specifically designed to inhibit the flow of aqueous filtrate into shales. The progressive near-wellbore pore pressure build up with these drilling fluids can eventually induce borehole destabilization when the stress around the borehole exceeds the strength of the shale. The fluid invasion not only alters the stress state, but it can also impact the near wellbore strength of the shale[1-4]. Given this scenario, the ability to drill trouble-free holes in shales with conventional WBMs may sometimes be a matter of good luck as well as a matter of good judgment.

3 RELATIONSHIP BETWEEN HYDRAULIC AND OSMOTIC PRESSURE GRADIENTS AND ITS EFFECT ON THE DIFFUSION RATE OF WATER

The conceptual basis of the coupled mechanical/chemical potential borehole stability model was presented by Mody et al (1993). The fundamental basis of this model was the assumption that an osmotic barrier (i.e. semipermeable membrane) exists between the shale and the drilling fluid. For OBMs this assumption was validated[3]. In the case of WBMs, experiments conducted under in situ conditions demonstrate that the shales can have properties that resemble a range of perm-selective characteristics, ranging from semi-permeable membrane (i.e. complete ion restrictive) to a leaky-membrane (selective ion restriction)[1,2, and 4]. The concept of "reflection coefficient" was first introduced by Staverman in 1951, to quantify the membrane efficiency (i.e. leakiness). The fundamental basis of this concept was applied to understanding drilling fluid/shale interaction and its impact on borehole (in)stability[1]. In this section a theoretical equation based upon the thermodynamics of the system is presented relating the hydraulic and osmotic pressure gradients to the rate of movement of water. A relationship is obtained in terms of flow rates, osmotic pressures and hydraulic pressures and can be expressed as follows:

$$Q = -KA\overline{V}_o\left[\frac{dp}{dx} - \frac{d\pi}{dx}\right] \qquad \ldots\ldots\ldots (1)$$

Where: Q = volume flow rate
K = transmission constant
V_o = partial molar volume
A cross-section area of the diffusive path
p = pressure (hydraulic)
π = osmotic pressure

Thus, based upon the thermodynamic principles, it is shown (Equation #1) that osmotic pressures acts as a negative hydraulic pressure in regulating the rate of water movement. Equation #1 can be applied to represent the diffusion of water in any system which is heterogeneous with respect to dissolved solutes, especially if the dissolved solutes are restricted in their movement by physical forces and/or mechanical barriers. It can be applied to the movement of water through shales where the net ionic movement could be restricted when shales are subjected to in situ stresses. It can also be applied to the diffusion of water through the semipermeable membrane provided by the OBM.

4 BRIEF OVERVIEW OF FLOW AND DRIVING FORCES

The state of a mass of liquid water has been described as a somewhat diffused or incomplete crystal. Because of the incomplete nature of its crystal form, the water "crystal" is easily deformed by forces such as pressure. This deformation and the consequent translatory movement of mass of water in response to a pressure are known as viscous flow. These are the dominant form of water movement in large shale pores.

In addition to this viscous movement, which involves motion of the whole semicrystalline mass of water, diffusion takes place. The theories of viscous and diffusive movements in bulk solution are well advanced. However, there has been a reticence toward applying these theories directly to the movement of water in finely porous media like shale. The fact that clay-water-ion interaction occurs has raised doubts concerning the applicability of these theories to such a system. Kemper (1961) presented a method of adapting these classic equations to porous media, with the understanding that much is not known about the micro-mechanisms involved in the clay-water-ion system, and that secondary interactions probably exist between diffusive and viscous types of movements. Direct superposition of the classic equations for diffusive and viscous flow was used to develop a theory to handle systems where both types of flow occur. In context with the objectives of this paper, it is appropriate to skip the theoretical intricacies, and instead, focus the discussion on the practical implications of the theory to gain the following important insights:

1. Viscous flow takes place in response to the pressure (hydraulic) gradient in the water, while diffusive flow takes place in response to the gradient of the molar-free energy of water.
2. The influence of film thickness (i.e., average pore diameter) is significant on the relative effectiveness of the molar-free energy gradients compared to the pressure gradients as water moves, and in general, the smaller the average pore diameter, the more effective is the influence of the molar free energy to cause movement of water. For smaller dimensions of the film (@ 20Å) the influence of molar-free energy is comparable to that of equivalent pressure (hydraulic) gradients.
3. For larger pore diameters the free energy gradient equivalent to an osmotic stress gradient plays a minor role in determining the direction and magnitude of flow compared to an equivalent pressure (hydraulic) gradient. It is important to note that the above observation was based upon the assumption that clay-water interaction does not enter the picture. When clay-water interactions are known to exist, adequate adaption of the theory may be necessary to minimize the error between theoretical and observed values.

The concepts presented above were extended to include restrictions of solute by membranes and consequent effects on the movement of water. It was found that the osmotic pressure and hydraulic pressure differences are equally effective in moving water through the membrane when the solutes are completely restricted by the membrane. One of the key conclusions of this study was that the viscous flow is the major mechanism by which water moves through membranes (solute completely restricted) in response to osmotic and hydraulic pressure gradients and that the flow caused by diffusive mechanism is negligible. When solutes are not completely restricted by the membrane (e.g., WBM/shale system), it was found that the osmotic pressure differences (as anticipated) are less effective than hydraulic pressure differences in moving water through membranes. For partial restriction of solute, an equation was derived including the sizes of the solute molecule and the pores to compare the flow rates due to partial restriction of solutes in response to osmotic and hydraulic pressure gradients of equal size. The equation is expressed as follows:

$$\left[\frac{(\Delta Q/\Delta t)_o}{(\Delta Q/\Delta t)_H}\right]_{\Delta P = -\Delta\pi} = \left[1 - \frac{(a-r_s)^2}{(a-r_w)^2}\right] \qquad (2)$$

where $(\Delta Q/\Delta t)_o$ = flow rate due to osmotic pressure gradient
where $(\Delta Q/\Delta t)_H$ = flow rate due to hydraulic pressure gradient
ΔP = hydraulic pressure gradient
$\Delta\pi$ = osmotic pressure gradient
a = pore radius
r_s = solute radius
r_w = radius of water molecule

It should be noted that the above equation (#2) is valid only when the radius of the solute is equal to or less than the radius of the pore (i.e. $r_s \leq a$). When $r_s > a$ the solute is completely restricted and the osmotic and hydraulic pressures are equally effective in causing flow through a membrane.

For electrically neutral systems (i.e., uncharged membranes), the Staverman reflection coefficient σ was shown to be equal to the right hand side of the above equation.

$$\sigma = \left[1 - \frac{(a-r_s)^2}{(a-r_w)^2}\right] \quad \ldots\ldots\ldots\ldots \quad (3)$$

It is very important to note that when osmosis occurs through a charged membrane (i.e., a shale), the above theoretical formulation may not be entirely valid. The proportionality relationship of the flow rate to the concentration difference of the solute is no longer obeyed.

5 DISCUSSIONS

Based on Equation 3, it is clear that one of the key parameters that governs the magnitude of the reflection coefficient σ (i.e., membrane ideality) of a "leaky system" is the relative size of the solute radii "r_s" compared to the average radii of the pore throat "a." When $r_s > a$, the solute is completely restricted and the system behaves as a perfect semipermeable membrane. It has been shown[3] that oil films in invert oil emulsion muds (IOEMs) constitute highly selective membranes ($\sigma = 1$). Thus, by optimizing the water phase activity of the IOEM using solutes, an osmotic flow from the formation to the drilling fluid can be induced. This flow can compensate or over take the effective hydraulic flow. This process can lead to increase in formation strength (due to reduction in near wellbore water content) and alteration of near wellbore effective stresses (as a result of change in near wellbore pore pressure). Both processes (when optimized) are shown to benefit borehole stability[1].

In WBMs, the issue of capillary entry pressure is irrelevant and no semipermeable "film" promoting high effective osmotic flow exists. However, for a shale/WBM system to acquire rate/perm-selectivity properties comparable to that of IOEM (i.e., $\sigma = 1$), the effective radius (hydrated) of the solute molecule that imparts activity to the drilling fluid has to be, in theory at least, equal to or greater than the average effective pore throat radii of the shale in situ.

One important general observation is that even when the size of the solute molecule is several factors smaller than the pore size, relatively high values of σ are estimated. It is critical to understand that the magnitude of the osmotic flow is not governed by the magnitude of σ alone but is governed by the product of σ and the theoretical osmotic pressure difference $\Delta\pi$ (between the drilling fluid and the shale pore fluid). It is this product (σ x Δ π) that needs to be optimized (either by manipulating the activity of the WBM or by manipulating the magnitude of σ) to fully exploit the potential of osmotic flow in shales with WBM for shale stabilization. For example, the osmotic flow in shales with small (but non zero) reflection coefficients may be increased by lowering the activity of the drilling fluid which generates high value of $\Delta\pi$.

6 GUIDELINES FOR FUTURE WORK

Future research and development should focus on the development of cost-effective, superior performance, environmentally friendly water-based shale drilling fluids. The conceptual framework outlined below can serve as a useful blueprint to achieve this objective:

- Conduct controlled experiments in the laboratory under realistic downhole conditions on actual downhole preserved shale core with known WBM systems to calibrate and validate the model.
- Develop test procedures and/or downhole tools to more accurately characterize the in situ pore structure in shales.
- Research efforts in the design of new WBM systems need to be directed at developing high molecular weight low water activity novel solutes.

REFERENCES

Mody, F. K. and Hale, A. H.: "A Borehole Stability Model to Couple the Mechanics and Chemistry of Drilling Fluid/Shale Interaction", paper IADC/SPE 25728, presented at the IADC/SPE Drilling Conference, Amsterdam, February 23-25, 1993.

van Oort, E., Hale, A. H., Mody, F. K., and Roy, S.: "Critical Parameters in Modelling The Chemical Aspects of Borehole Stability in Shales and in Designing Improved Water-Based Shale Drilling Fluids", paper SPE 28309, presented at the 69th Annual Technical Conference and Exhibition of SPE, New Orleans, LA, September 25-28, 1994.

Hale, A. H. and Mody, F. K., and Salisbury, D. F.: "Experimental Investigation of the Influence of Chemical Potential on Wellbore Stability", paper IADC/SPE 23885, presented at the IADC/SPE Drilling Conference, New Orleans, LA, February 18-21, 1992.

van Oort, E., Hale, A. H., and Mody, F. K.: "Manipulation of Coupled Osmotic Flows for Stabilization of Shales Exposed to Water-Based Drilling Fluids", paper SPE 30499, presented at the 70th Annual Technical

Conference and Exhibition of SPE, Dallas, TX, October 22-23, 1995.

van Oort, E.: "A Novel Technique for the Investigation of Drilling Fluid Induced Borehole Instability in Shales", paper SPE/ISRM 28064, presented at the SPE/ISRM Conference, Delft, August 29-31, 1994.

Wong, S. W., and Heidug, W. K.: "Borehole Stability in Shales: A Constitutive Model for the Mechanical and Chemical Effects of Drilling Fluid Invasion", paper SPE/ISRM 28059, presented at the SPE/ISRM Conference, Delft, August 29-31, 1994.

Kemper, W. D.: "Movement of Water as Effected by Free Energy And Pressure Gradients: I. Application of Classical Equations for Viscous and Diffusive Movements to the Liquid Phase in Finely Porous Media", Proc. Soil Sci. Soc. Am., v.25, pp 255-260, (1961).

Staverman, A. J.: "The Theory of Measurement of Osmotic Pressures", Recueil des Travaux Chimiques des Pays-Bas, v.70, pp 344-352, (1951).

Computer Methods and Advances in Geomechanics, Yuan (ed.) © 1997 Balkema, Rotterdam, ISBN 90 5410 904 1

Study on application of principle of dynamic stress detour to increase resistance of protective structure

Zheng Quanping
Luoyang Hydraulic Engineering Technology Institute, Luoyang, China
Zhou Zaosheng
Nanjing Engineering Institute, China
Zheng Hongtai
Southwest Jiaotong University, Chengdu, China

ABSTRACT: According to the principle of dynamic stress detour and characteristic of protective structure, a method is presented in this paper which increase considerably the shock-resistant capacity of the structure indirectly. As a result of functions of guiding of stress detour created by the construction described in this paper, the travelling direction of wave is changed, the most stress kinetic flow has made the principal part of the enormous kinetic energy carried by the wave not to flow over the opening and structure, so the wave load on the structure is decreased effectively. A great deal of calculation results given in this paper verify that the strength magnitude of protective structure can be considerably increased when the principle mentioned above is introduced.

1 INTRODUCTION

When waves propagate in the media of rock and soil, and meets with a hole or a condition that the material property of medium has a sudden change, stress streamlines alter, which causes stress concentration and the alteration of deformation pace. In the action zone of explosion wave, underground structure certainly produces the consequence of the effect of this concentration and alteration, too. In the structure and the medium which transmits wave load, if the wave load is so sufficient that the effect of the deformation pace shows that the asynchronism of deformation pace goes beyond the enduring limit of material, the damage phenomenon takes place. People used to reinforce the structure and surrounding rock in order to increase the protective capacity in research field and engineering practice before. This is a method which restrains directly a relative quantity of asynchronization of deformation. The method has no change in direction and route of transmission of stress wave, and even increases stress concentration. Therefore, it is certain that the enormous kinetic energy carried by the wave takes an effect on the structure and surrounding rock reinforced to cause the possibility of destruction.

According to the possibility mentioned above that the wave produces the phenomenon of stress detour flow during propagation, a method is presented in this paper. That is, after underground opening excavated and surrounding rock reinforced, and before the structure constructed, a blast cavity layer is formed by drilling, blasting or other engineering measure in the deep rock outside the range reinforced. Consequently, the rock pillars between every two cavities in this layer bear gravity and other statical load in normal times. But when a strong wave front arrives at the layer, serious stress concentration occurs in near zone of the rock pillars. For this reason, as the wave fronts move forward, the destruction of the rock pillars takes place successively, which results in that the cavity layer becomes a loose crush belt. This loose crush belt prevents most of the later wave from passing to the structure inside, and causes the phenomenon of dynamic stress detour and the direction of wave propagation is forced to change in the outer rock. Consequently, most of the enormous kinetic energy carried by the later wave is outside the belt and far away from the opening and structure, which assures the reduction of the possibility of destruction greatly and the increment of resistance of the structure substantially correspondingly. A plenty of calculation results given in this paper verifies the effectiveness and rational significance of the increasing correspondingly of the capacity of the anti-explosion of structure.

2 INCREASING THE CAPACITY OF PROTECTIVE STRUCTURE BASED ON DYNAMIC STRESS DETOUR PRINCIPLE

2.1 Mechanical effects of contained explosion

Contained explosion means that it has not enough energy to get a exit in the direction of the least burden under conditions of the strength and overlying thickness of rock. As shown in Fig.1, in such conditions and after exploding, a cavity R_1 is formed as the intensive detonating explosive pressure(up to 10 ~ 100MPa) acts on the wall of charge room and compresses the rock; a dense crush zone R_2 is formed as the surrounding rocks are crushed; the stress waves continue to reduce and a fissured zone R_3 is formed later because of the action of reflection at the fissures. Finally, the energy of the stress wave reduces further and the media in zone R_4 only vibrate and don't change the states, and form elastic wave to propagate outwards.

It is shown from test data that dimensions of above various zones are closely related with characters of rock. For limestone, if R_0 is assumed as the radius of the charge room,

$$R_1 = 3R_0, \quad R_2 = 6R_0, \quad R_3 = 10R_0 \tag{1}$$

in this paper, R_2 is defined as the effective cavity radius of a contained explosion, i.e.

$$r_e = R_2 = 6R_0 \tag{2}$$

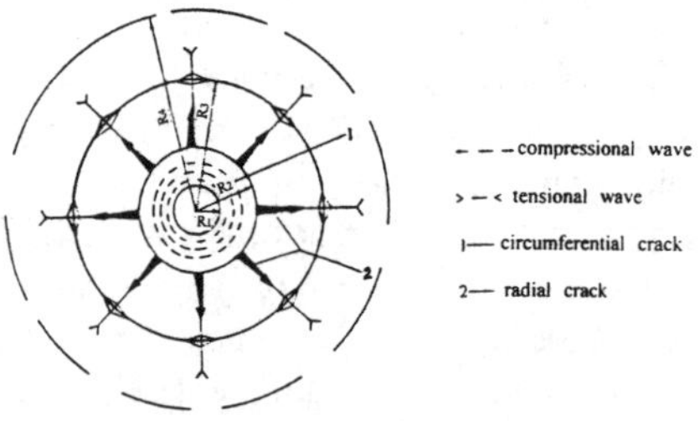

Fig.1 Mechanical effects of contained explosion

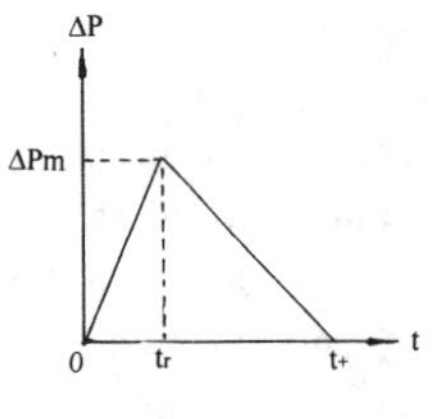

Fig.2 Load curve of stress wave

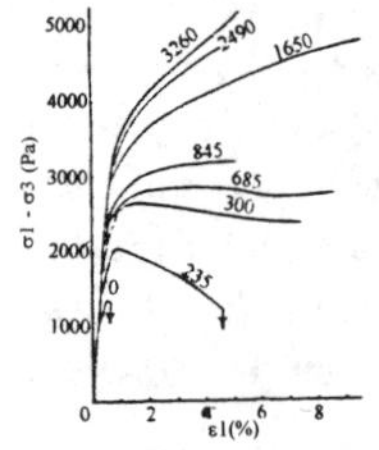

Fig.3 Effect of confining pressure on rock strenghth

2.2 Technique of diverging dynamic stress

As you know, stress wave loads are of the characteristic of time process. This time process is divided into increasing and reducing parts which shows in Fig.2. This is one of the characteristics based on in this paper. The other is the characteristic of rock failure.

Rock failure are closely related with its stress conditions. It is shown in Fig.3 that confining pressures have a great influence on rock strengths. In uniaxial or near unianxial conditions, rock strengths and its bearing capacities are the smallest, and when the stress of rock exceeds its uniaxial strength, rock fails and loses its bearing capacities. With the increase of confining pressure rock strengths increase obviously. Especially when the confining stress exceeds a certain value, rock will not be a brittle material and become a ductile one. At this time, it may get larger deformations and don't lose its bearing capacities. Therefore, man-made free or close free surface is able to reduce rock strength, and to have a tendency to brittle failure. Man-made cavity may produces stress detour, makes a change of the direction on its circumference, that is, it changes the transmitting routes of stress and strain, and simultaneously has a stress concentration so that the stress increases correspondingly near the zone of its circumference, and leads to have a change of the compatibility of deformation nearer places and to disrupt the synchronism of deformation. Based on the above reasons, in order to have a small difference of deformation and safety and no destruction in the structure and surrounding rock, a loop of sphere cavity is formed by engineering measures in rock at a suitable distance from the opening. When the wave fronts arrive at this loop, the zones between two cavities become a collecting corridors of stress flow to pass through, and because the stress state in this zone is close to uniaxial, the rock in this zone will take a destruction very quickly under the action of later wave and lose its bearing capacities. In the same way, the above similar destructive phenomenon will happen in the zone between the other two cavities. Because wave load is of characteristic of time process, after the rock pillars between every two cavities take a destruction in succession by the part of the wave fronts, the later waves pass through ones and the cavities near ones produces a loss of stress. So most of wave loads can't pass through ones, and takes a detour in the rock outside the loop to causes the changing of its transmission direction, and propagates and flows along the outer zone of the loop. This innovation above and the utilization of the time process will attain the effect of divergence and decrement on the wave load. The rock pillars between every two cavities play a role in the transmission of dead loads before shocked by the wave. When a explosion happens and its strong waves intrude, the rock pillars take a destruction, the loop of obstructing wave is formed, diverging stress and consuming energy come into action. At this time, the strong stress wave and its enormous kinetic energy are changed in their passing route to keep away far from the underground structure, and transmit and propagate into the deep rock because of the existence of the cavity loop obstructing wave and the action of consuming energy and diverging stress. In the result, only a few loads of the stress wave transmit into the underground structure and surrounding rock reinforced, so that it is ensured that the underground structure and surrounding rock are used in safety. The method described above is named as the technique of diverging dynamic stress (TDDS).

2.3 Design parameters

As shown in Fig.4, diameter of the opening is assumed to be R_0. Radius R_d of the cavity layer, thickness S of rock pillar between two cavities and radius r_d of the cavity can be expressed as the following equations

$$R_d = R_0 + l + 10,\ S = r_d \tag{3}$$

All variations in the equation are in meter. l is the length of anchors or depth of the reinforcement. If the structure span inside the opening is d_0 and the thickness of the soft filling between the structure and the opening is δ (seen in Fig.4), their relations can be expressed as follows:

$$R_0 = \frac{d_0}{2} + h + \delta \tag{4}$$

r_d in equation (3) is the effective radius of the cavity of a contained explosion in equation (2). In engineering practice, r_d is required to be as large as possible. Its limit is that there

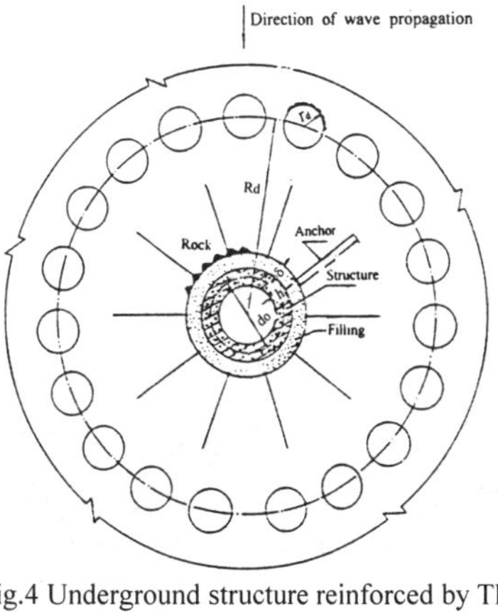

Fig.4 Underground structure reinforced by TDDS

are no falling rock blocks in the reinforced opening when explosion is used to form the cavities. So r_d is related with the embedment depth $(l+10)$ of the cavities and the characteristics of the surrounding rock. Its value must be determined in site or by model test according to engineering site conditions.

After R_d and r_d are determined, S can be obtained from equation(4). Up to now, all the required parameters are determined. As to non-circle opening, equivalent radius R_0 can be used.

2.4 Construction method

After opening is excavated and reinforced, a suitable drilling machine, its diameter can reach 20 cm and above at present, can be selected according to the requirements. The pilot hole can be drilled to a depth of $(l+10)$, then a explosive charge and its detonator can be placed at the designed depth by a steel bar. The holes are filled by high strength cement, and when the cement reaches its strength, the explosives are detonated and the required cavity is formed.

It should be noted in engineering practice that the explosives to form the cavities must be detonated simultaneously. In this way there are two advantages. First, it is avoided that the later detonating compresses and destroys the cavities formed previously. Second, as pressing simultaneously in circumferential and longitudinal directions, its intense explosion pressures are simultaneously forced to propagate along the radial direction.

3 CHIEF RESULTS OF NUMERICAL ANALYSIS

3.1 Analytical model

For research object in underground engineering, since its longitudinal dimensions are far larger than its cross sections, it can be simplified as a plane strain problem, and separated from its real zone according to its influence range. A triangular load of the stress wave, which has the same pattern and magnitude as the pressure wave of the corresponding place of free field (see Fig.2), exerts uniformly on the upper boundary of this segment.

In the selection of the segment, the kinetic influence of inertia and the characteristics of the boundary reflection of waves are considering. Its horizontal width is taken as 100m, and the distance between the top of the opening and its upper boundary 85m, and one between the bottom of the opening and its lower boundary 195m according to the propagation theory of wave and the features of the wave load as shown in Fig.2.

In order to make a comparison of the difference of the structure respondence between TDDS is adopted and is not done, two calculations are respectively carried on. One is that the conventional reinforced method is adopted as shown in Fig.4 when there are no cavities. The other is that the conventional reinforced method and TDDS are adopted as also shown in Fig.4 when there are cavities. Two kinds of nets are the same except the cavity loop and its limited small zone. the net figures in and around the structure are shown in Fig.5.

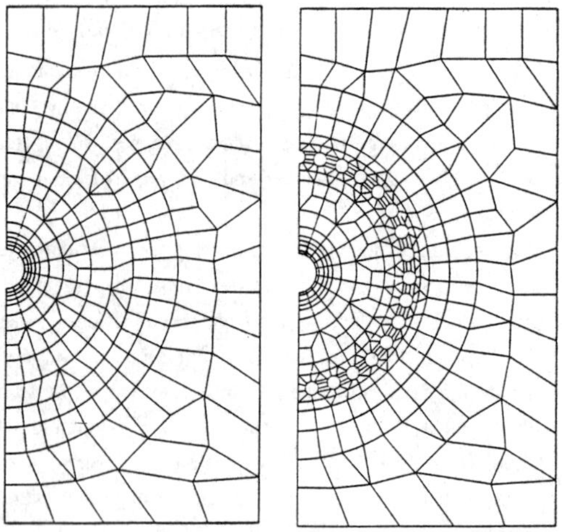

a. TDDS is not adopted　　　b. TDDS is adopted

Fig.5 Finite element net near the structure

Tab.1 Mechanical parameters of materials

Parameters / Materials	Ed (GPa)	ρ (t/m^3)	μ	C (MPa)	φ (degree)	Rc (MPa)
Reinforced concrete structure	52.5	2.35	0.23	5.80	54.0	/
Soft backfilling of foam concrete	1.25	0.88	0.11	0.44	36.0	/
Rock	82.5	2.65	0.28	4.00	47.0	/
Pillars between cavities and neighboring	82.5	2.65	0.23	/	/	50.0
Anchors *	206.0	7.80	0.25	/	/	142.0

* Areas of anchors are taken as $0.00071m^2$ under plane strain condition

The rise time t_r and the positive duration t_+ of the triangular wave load are taken respectively as 0.0065s and 0.045s in the numerical analyses of this paper. The positive phase is only considered. In order to compare the results of influence caused by TDDS, five peak values of impulse load, 30MPa, 50MPa, 80MPa, 100MPa and 120MPa, are considered.

Drucker-Prager material model is selected for structure, filling and rock mass, and Von. Misses for anchors, pillars between cavities and neighboring rock. Their mechanical parameters are taken as listed in Tab.1.

3.2 Principal stress fields and their streamlines

Fig.6 is the principal stress fields of the structure, filling and surrounding rock when ΔP_m = 80MPa. By comparison between a, b two situations in the figure it can be seen that, because TDDS has been adopted the magnitudes and the directions of principal stress have a great change in the deep rock near the cavity layer, the stress transmitting routes take a change during the action of the wave, and the stress in rock transmits downwards along the circumferential direction of the outer zone of the cavity layer. Thus the magnitude of stresses in structure and surrounding rock is decreased considerably. It can be seen more obviously from stress streamline in Fig.7 that the man-made cavity layer in the rock around the structure changes effectively the streamlines and routes of stresses, and debars the stress

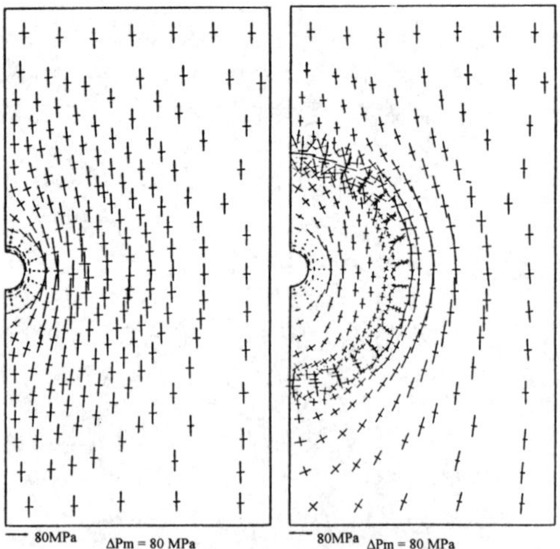

a. TDDS is not adopted　　　b. TDDS is adopted

Fig.6 Principal stress fields near the structure

wave from transmitting into the opening and structure, and creates that the stresses are mostly guided to travel along the outer rock of the cavity layer and then into the lower and deeper rock. The consistence between TDDS theory and its realization has been presented sufficiently in Fig.7

It is shown further in Fig.7 that, the part of the stress wave, which is not obstructed, still intrudes into the structure and surrounding rock, and travels downwards into the rock under the opening. But its magnitude has been decreased considerably. This shows that it is not necessary to construct the cavities in the surrounding rock under structure.

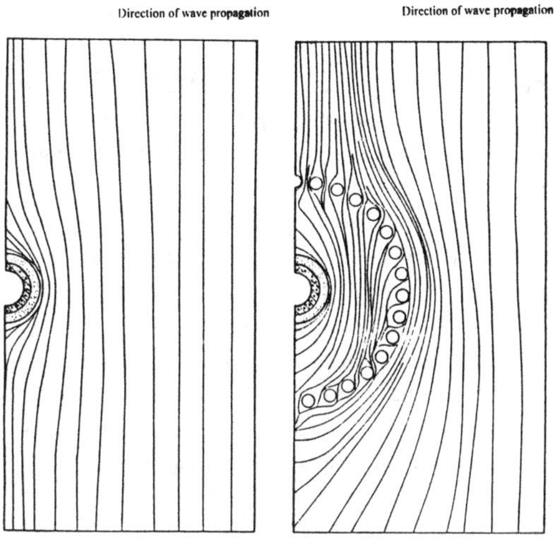

a. TDDS is not adopted b. TDDS is adopted

Fig.7 Stress streamline near the structure

3.3 Structure deformation energy

Fig.8 is the curves of the structure deformation energy of unit length along longitudinal direction. The unit of energy is taken as joule. Those curves show that, when TDDS is adopted, the deformation energies of the underground structure decrease obviously, those curves have a platform and then descend slowly. The time of the platform is 4 ~ 5ms. The platform is one of the main reasons which decreases the deformation energy of underground structure. Appearance of the platform is due to that, when the peak values of stress wave are relatively large, the front part of stress waves arrives at the cavity layer to cause enough serious stress concentration so that the pillars between cavities are overexerted to damage until destroying. Thus the later stress waves are more effectively obstructed to travel into and to press onto the structure and surrounding rock, and to be compelled to transmit downwards along the outer rock of the cavity layer. It is the time difference of the process of wave transmission that is utilized by TDDS in order to produce coupling and self-locking of process. This thinking has made the action of TDDS more obvious when the wave loads are considerably large(seen Tab.2)

Due to the limitation of paper length required, there are some other important results of calculations, such as structure yielding rate, process curves of principal stress at the middle of structure, plastic yielding zone of surrounding rock, process curves of principal stress at the middle of opening and so on, which can not be presented here and will have to do in another paper. All the results got show that, TDDS can decrease very considerably the wave loads on the underground structure and surrounding rock, and range of plastic yielding zone.

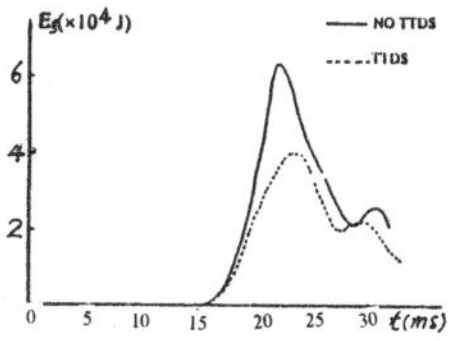

a. ΔPm = 30MPa

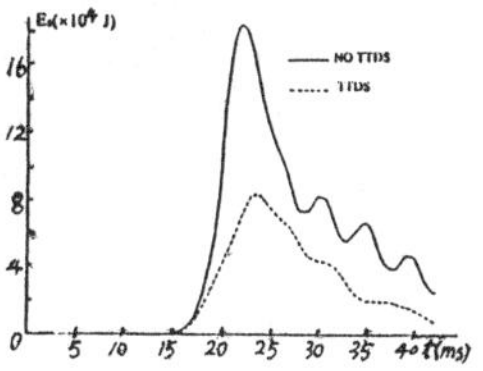

b. ΔPm = 50MPa

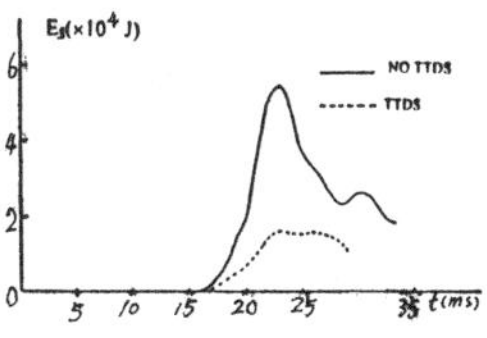

c. ΔPm = 80MPa

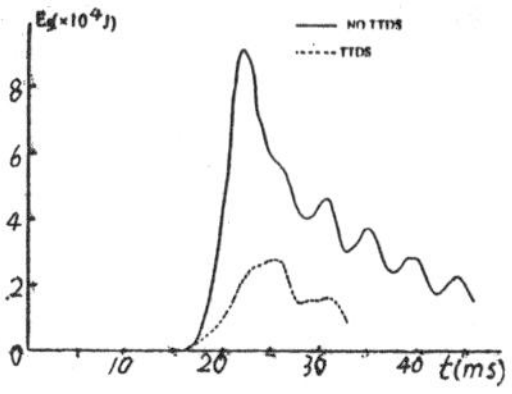

d. ΔPm = 100MPa

Fig.8 Deformational energy of the structure

Tab.2 The deformational energy of structure of unite length along longitudinal direction (KJ/m)

ΔPm(MPa) / Energy	30.0	50.0	80.0	100.0	120.0
Es (TDDS isn't adopted)	63.0	184.3	544.5	905.2	/
E's (TDDS is adopted)	39.5	82.6	160.5	273.6	334.1
$\frac{Es - E's}{Es} \times 100\,\%$	37.3	55.2	70.5	69.8	/

4 CONCLUSIONS

Previous results show that, adopting TDDS makes full use of the failure characteristics of rock and the process feature of time, and can effectively debar the stress wave from transmitting into and pressing onto the underground structure and surrounding rock. The cavity layer plays the role of diverging dynamic stress indeed that can be summarized as follows:

1 It changes effectively the transmitting route of principal stress in the surrounding rock around underground structure buried depth, and decreases very considerably the ratios of maximum values of principal stress in the structure and surrounding rock to the peak values of stress wave.

2 A platform appears during the responding of the deformational energy of the underground structure. The times of the platforms are 4 ~ 5ms. The peak values of the deformational energy have been reduced by 40 ~ 70%.

3 The common reinforced method used ordinarily can not change stress transmitting route and wave travelling direction. Therefore, the resistance of structure increased by one is very limitational. Adopting TDDS may considerably increases the realistic resistance of underground structure. Actually, the larger load of stress wave, the more striking the effect of its action appears.

REFERENCES

1 Zhu Ruigeng 1979. Explosion in soil and rock and its mechanical effect (in Chinese). *Rock and soil mechanics.* Vol.2,1979.

2 J.C.Jaeger and N.G.W.Cook 1979. *Fundamentals of Rock Mechanics (Third Edition)*. London: Chapman and Hall.

3 Zheng Hongtai and Guo Peijun 1996. The Present status and prospects of rock and soil mechanics (in Chinese). *Symp. of the Century Anniversary of the Fouding of Southwest(Tangshan) Jiaotong University.* Southwest Jiaotong University Press.

Computer Methods and Advances in Geomechanics, Yuan (ed.) © 1997 Balkema, Rotterdam, ISBN 90 5410 904 1

Prediction and measurement of hangingwall movements of Detour Lake Mine SLR stope

B.Wang & S.Vongpaisal
Mining Laboratories, CANMET, Natural Resources Canada, Ottawa, Ont., Canada
K.Dunne
Detour Lake Mine, Placer Dome Canada Limited, Timmins, Ont., Canada
R.Pakalnis
University of British Columbia, Vancouver, B.C., Canada

ABSTRACT: This paper presents measurements and a successful prediction of the hangingwall movement of the Detour Lake Mine SLR (Sub-Level Retreat) stope. Detour Lake Mine is an underground narrow vein gold mine operating at 3750 tons per day. The ore body extends from surface down to the 745 metre level striking east-west and dipping about 70° to 80° to the north. The SLR stope block extends from 560 to 660 m level and varies in width from 5 to 10 m and is approximately 300 m along strike. The hangingwall movements were measured using extensometers at various sections. Numerical modelling studies were carried out to back analyze two opened areas and to predict the stability of an area to be mined. A Block-Spring Model (BSM) was used in this study to simulate the jointed rocks. The back analyses of two mined sections agreed well with the monitoring data that allowed the prediction of the new stope to be made with confidence. The measurements of the hangingwall displacements at the analyzed cross-sections are presented. A comparison indicated that the displacements of the hangingwall reached the magnitude predicted when the stope was mined. This study indicated that the back analysis of an open mine stope and numerical model calibration can be very useful for prediction of stope behaviour.

1 INTRODUCTION

Detour Lake Mine (DLM) is located 300 km northeast of Timmins, Ontario, Canada. It is an underground gold mine operating at 3750 tons per day. The ore body extends from surface down to the 745 meter level striking east-west and dipping about 70° to 80° to the north (Fig. 1.a). The Sub-Level Retreat (SLR) stope block extends from 560 to 660 m level and varies in width from 5 m to 10 m and is approximately 300 m along strike (Fig.1.b). There are two predominant sets of joints: east-west and north-south, both vertically dipping. The hangingwall rocks located to the north of the ore are competent mafic volcanics with potassic alteration. The footwall rocks are less competent chloritic greenstone and talc chlorite schist.

The hangingwall displacements of the stope between 560 m and 660 m levels were monitored with extensometers installed before blasting. Each extensometer registered a sudden displacement increase when the section was blasted.

Back analyses of two cross-sections were conducted using the Block-Spring Model (BSM), a numerical modelling software for stress and deformation analysis of discontinuous rocks. The displacements from the numerical models were compared with the extensometer data. A prediction of the wall movement at a third cross-section was made before blasting took place. The back analysis results and the prediction have been reported by Wang et al. (1995). The back analysis results were verified in the paper but the prediction could not be confirmed since the data were not available when the manuscript was completed. A verification of the BSM results with the measurements after blasting may provide useful information for the future mine modelling studies.

2 BACK ANALYSIS

Two cross-sections of the SLR stope were analyzed. The geometries of the stopes are as shown in Figs. 2 and 3. The locations of the two

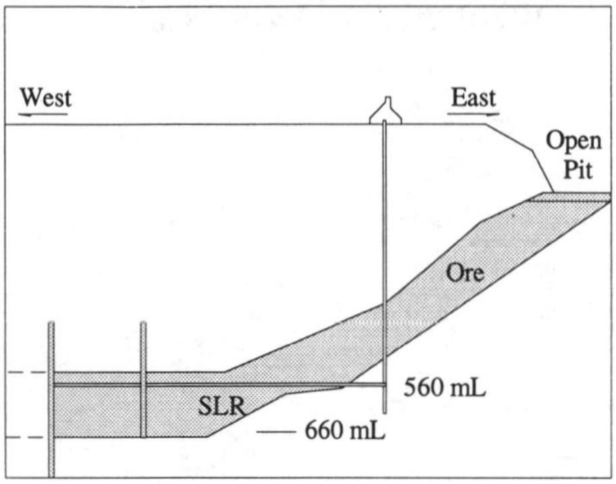

Fig. 1.(a) Longitudinal section of the mine

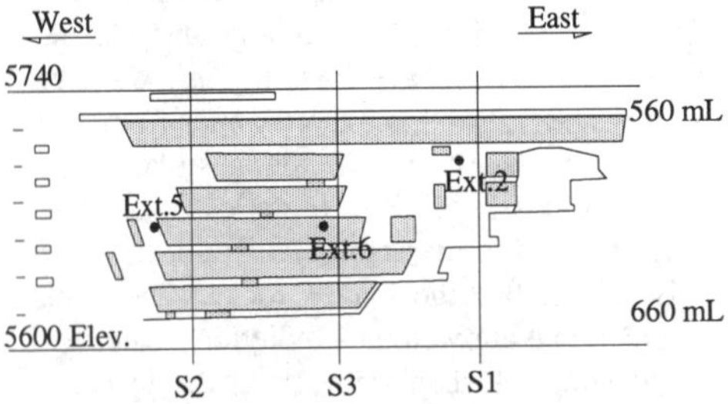

Fig. 1.(b) Longitudinal section of the SLR stope

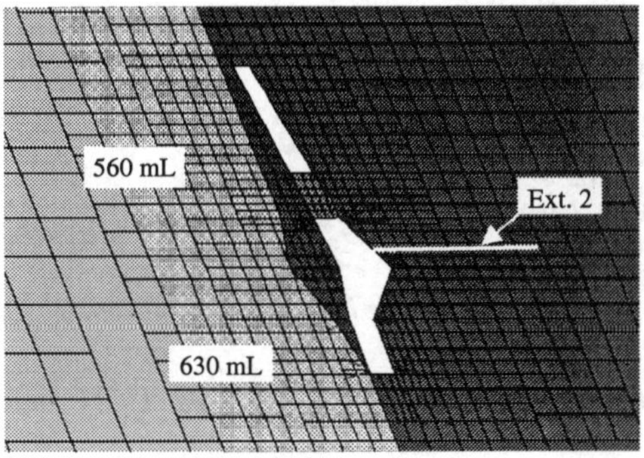

Fig. 2 Geometry of cross-section 1 (Location: S1 in Fig. 1.b)

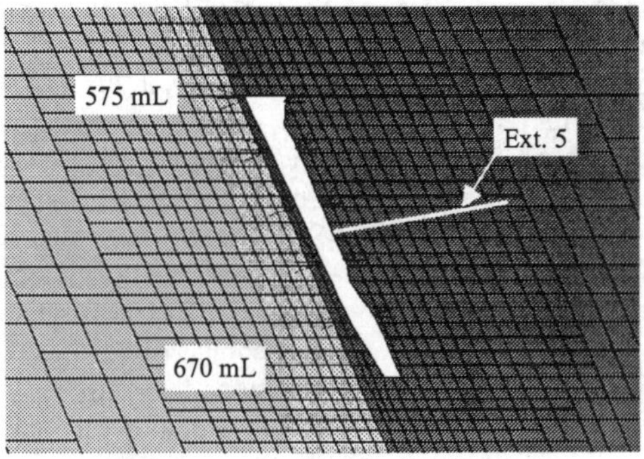

Fig. 3 Geometry of cross-section 2 (Location: S2 in Fig. 1.b)

sections are shown in the longitudinal section as S1 and S2 in Fig. 1.(b). The stopes are approximately 560 m to 660 m below surface. There was a horizontal sill pillar at the 560 m level separating this stope from the one at an upper level.

An extensometer was installed close to each of the two sections as shown in Fig. 1.(b) as Ext. 2 and Ext. 5. The extensometers were projected to the section planes as indicated in Figs. 2 and 3. The instruments measured the hangingwall displacements before and after the stope was mined as shown in Figs. 4 and 5. A sudden increase of the displacements was indicated at both sections after the stopes were opened. The displacements at the hangingwall face were recorded approximately 0.2 cm and 0.5 cm respectively at the two sections. It should be noted that the displacement figures given throughout this paper are relative to the far end of the corresponding extensometer.

Back analyses of the two sections of the open stope were conducted using the BSM model. The rock blocks were generated based on two predominant joint families intersecting with the section plane: one parallel to the stope and the other approximately horizontal. The blocks are about 1 to 3 m in size near the stope face. The BSM model assumes that the blocks are isolated and may slide and detach. The initial ground stress was estimated by the gravity of the overlaying rocks for the vertical components and 1.3 times higher for the horizontal (North-South) components.

The BSM results of the rock displacements along the extensometer lines are plotted in Figs. 6 and 7 and compared with the maximum displacements recorded by the extensometers #2 and #5 respectively. The BSM results shown in the charts are the relative values referred to the far end of the corresponding extensometer. Same convention applies to the rest of the text.

It is noted from Figs. 6 and 7 that the BSM results are within the range of the instrumentation data. The back analyses of these two cross-sections therefore indicated that the BSM model can be used to predict the behaviour of other stopes.

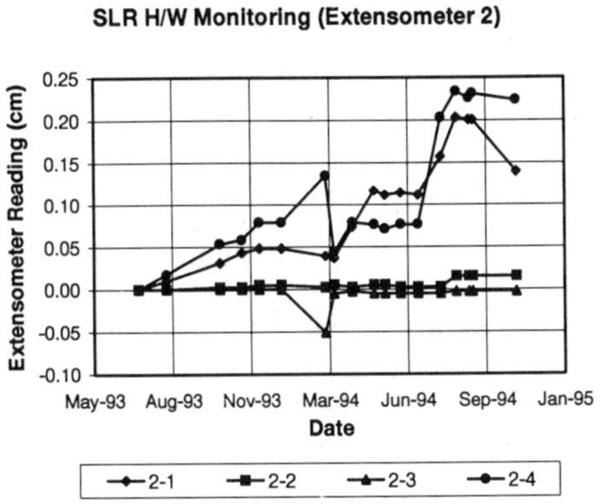

Fig. 4 Measurements of extensometer #2 (Anchor 2-1 close to stope)

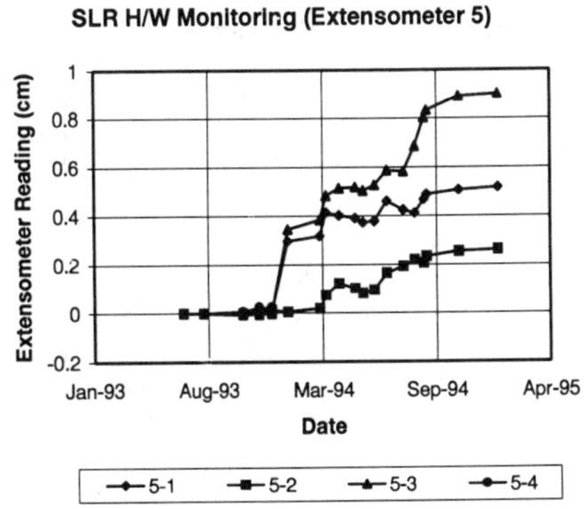

Fig. 5 Measurements of extensometer #5 (Anchor 5-1 close to stope)

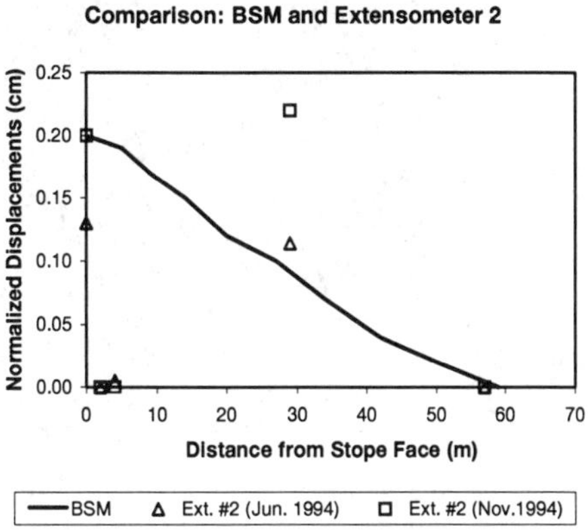

Fig. 6 Section 1: Hangingwall displacements from BSM and extensometer #2.

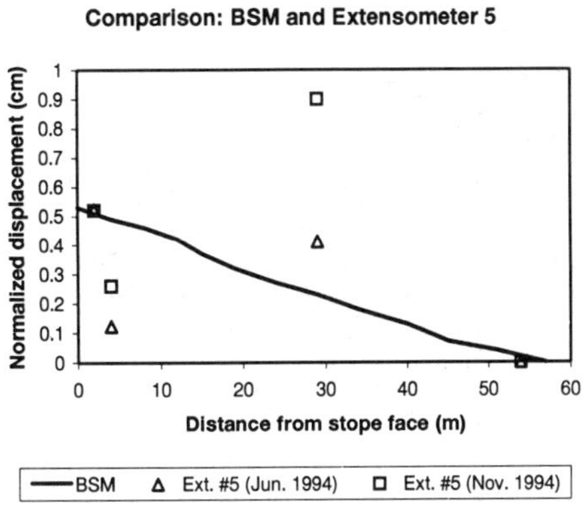

Fig. 7 Section 2: Hangingwall displacements from BSM and extensometer #5

3 PREDICTION AND VERIFICATION

Prediction of section 3 of the SLR stope (S3 in Fig. 1.b) was conducted using the BSM model. Fig. 8 shows the geometry of this section and part of the BSM mesh around the stope. A projection of the extensometer (Ext. 6) near this section is also shown in the figure. The rocks were classified into four groups:

HW - hangingwall mafic / potassic mafic flow;
MZ - main zone (ore);
TC - footwall talc chlorite;
VC - footwall volcanoclastic biotite and metasediments beyond the talc chlorite zone.

There is a footwall fault of about 0.2 to 1 m thick which is approximately parallel to the stope and is about 3 to 6 m away from the stope face at this section. The fault zone is mainly chlorite or clay gouge.

The footwall talc chlorite zone (TC) extends 20 to 50 m from the stope. The ground condition at this section is similar to the previous two sections analyzed that provided a base for the prediction. The material properties used in the prediction of this cross-section are listed in Table 1. The intact rock properties are based on Placer Dome Canada Ltd. (1992). In the BSM model, the friction angles of all the rock types were reduced by 5°, and the cohesions were reduced by a factor of 0.1, taking into account the rock mass conditions. The parameters have been verified with the BSM models in the back analyses of the previous two cross-sections.

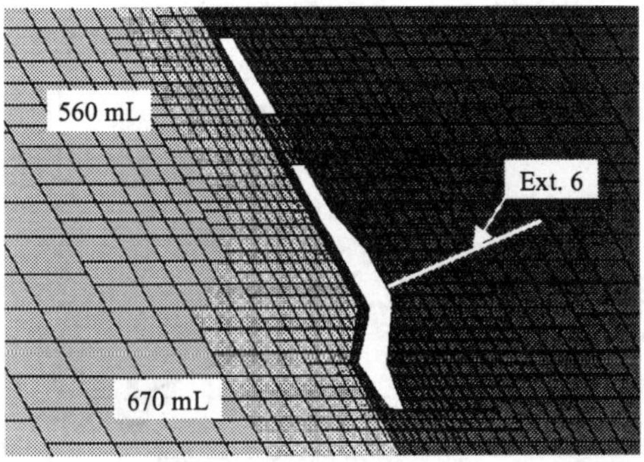

Fig. 8 Geometry of cross-section 3 and the BSM mesh around the stope (Location: S3 in Fig. 1.b)

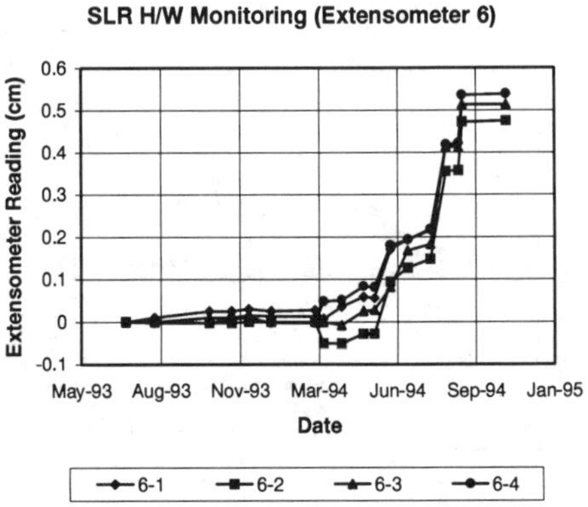

Fig. 9 Measurements of extensometer #6

Table 1: Material properties

Material	HW	MZ	TC	VC	Fault
E (MPa)	88000	93000	32000	50000	10000
μ	0.25	0.26	0.30	0.25	0.30
γ (MN/m^3)	0.029	0.030	0.030	0.030	
ϕ (°)	50 (45†)	51 (46†)	42 (37†)	50(45†)	32
C (MPa)	60(6.0†)	44(4.4†)	20(2.0†)	40(4.0†)	2
K_n (MPa/m)					10000
K_s (MPa/m)					10000

Note: E = Young's modulus;
μ = Poisson's Ratio;
γ = Unit weight;
ϕ = Friction angle;
C = Cohesion;
† Used in the BSM model.

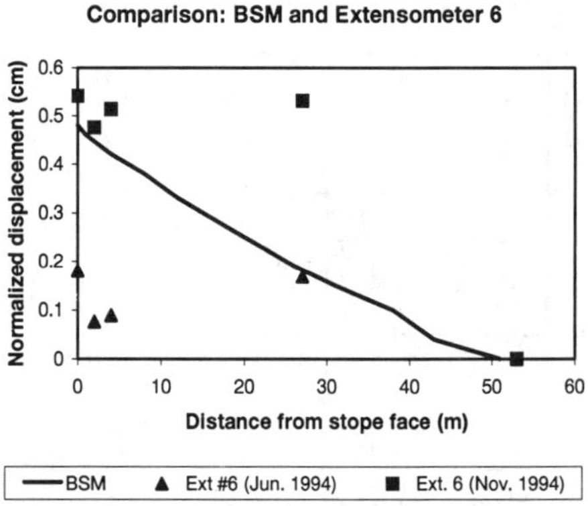

Fig. 10 Section 3: Displacements predicted by BSM and comparison with the measurements from extensometer #6

Cables were installed in the hangingwall and the footwall at the sub-levels of 20 meters apart (Placer Dome Canada Ltd., 1993). The lengths of the cables were 7 to 15 m in the hangingwall and 7 to 9 m in the footwall. At each sub-level, three cables were installed in the hangingwall with an inclination angle of 0°, 20° and 40° from the horizontal line and three in the footwall at -20°, 0° and +20°.

The extensometer recorded the hangingwall displacements before and after the stope was mined as shown in Fig. 9. At the time when the BSM prediction was made, the extensometer data were up to June 1994 when the stope was not fully opened yet.

A comparison of the BSM results with the extensometer data is given in Fig. 10. The pre-mining measurements from the extensometer were around 0.1 cm while the BSM predicted a maximum displacement of 0.5 cm, obviously higher than the measurements as seen in Fig. 10. However, when the stope was mined in September 1994 the extensometer registered a hangingwall displacement of 0.5 cm (Fig. 9), the same magnitude as predicted by the BSM model. For comparison purpose, the post-mining measurements from the extensometer are also given in Fig. 10. The measurements confirmed the prediction of the BSM model.

This study has shown that the back analysis of the existing underground opening may provide reliable data for the prediction of the new openings. It may therefore increase the confidence on the design of the mine stopes or other underground openings. This case study also shows that the BSM model can be a useful tool for analysis of jointed rocks.

4 CONCLUSIONS

A prediction of the hangingwall movements of the Detour Lake Mine SLR stope was conducted using the Block-Spring Model. Back analyses of two cross-sections were also conducted to provide information for the prediction. The hangingwall movements before and after the stope blasting were measured using extensometers. The extensometer data confirmed the prediction to be correct. The study illustrated that back analysis may provide valuable information for the new stope analysis. It also indicated that the BSM model may be a useful tool for conducting stability analysis on jointed rocks.

5 REFERENCES

Wang, B., Yu, Y.S. and Vongpaisal, S., (1995); A Case Study of Sub-Level Retreat Mining at Detour Lake Mine Using BSM Models. The 2nd International Conference on Mechanics of Jointed and Faulted Rocks, Vienna, Austria, pp.927-932.

Placer Dome Canada Ltd. - Detour Lake Mine, R. Pakalnis, 1992. *Ground Stability Guidelines for Cut and Fill Mining of Wide Ore Bodies, Final Report to CANMET.* CANMET Project Number: 9-9145.

Placer Dome Canada Ltd. - Detour Lake Mine, R. Pakalnis, 1993. *Design Guidelines for Sub-Level Retreat Mining Method, Milestone #2, Interim Report.* CANMET Project Number: 9-9145.

Computer Methods and Advances in Geomechanics, Yuan (ed.) © *1997 Balkema, Rotterdam, ISBN 90 5410 904 1*

Some computational aspects of basic inverse problems in land subsidence theory

I.V. Dimov & V.I. Dimova
University of Mining and Geology, Sofia, Bulgaria

ABSTRACT: When coal seams are mined under build up surfaces the following problem arises: how to plan the mining sequence in order to achieve specific subsidence pattern of the surface, which is determined by the respective building codes for preserving the surface items (protection problem)?

The answer to the above problem is achieved as a result of the solution of an operator equation of I kind. The new problem is incorrect in J. Hadamard's sense.

The solution is achieved by using A. Tikhonov's reguralization method. The cases when the right sight of the equation and operator itself are defined approximately are discussed. The subject of determining the parameter of reguralization is also discussed.

1 INTRODUCTION

We suggest in the land subsidence mechanics the following two disparate problems to be distinguished (Dimov & Dimova 1985) (Dimov & Dimova 1994) (Figure 1):

- Direct problem (prediction problem) - approximately given the subsidence of the immediate roof $w^0 = w^0(\xi, \eta)$, determine the subsidence equation on the Earth's surface $w = w(x, y)$;
- Inverse problem (protection problem) - approximately given the subsidence equation $w = w(x, y)$, determine the mining order, i.e. determine the subsidence on the immediate roof $w^0 = w^0(\xi, \eta)$.

In the direct problem the subsidence of the immediate roof is set approximately, because is determined either on the base of the geodesic measurements or on the base of expert estimate.

In the inverse problem the subsidence equation is approximately given (from...to...), being determined

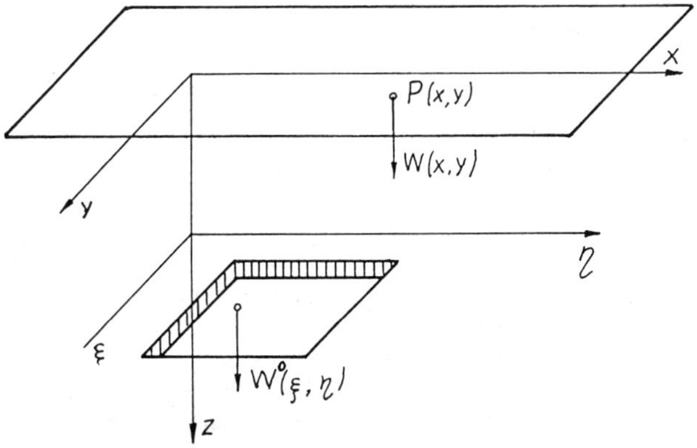

Figure 1. Basic scheme.

on the base of the building standards for the preservation of the surface sites.

Here we will consider the inverse problem.

2 SOLUTION OF THE INVERSE PROBLEM

It is known, that if: 1) the mined seam is horizontal; 2) the mined area is known; 3) the superposition principle is valid, then the subsidence of a surface point with coordinates (x,y) can be determined according to the formula (Whittaker & Reddish 1989)

$$w(x,y)=\iint_{(F)}\varphi(x-\xi,y-\eta)w^0(\xi,\eta)d\xi d\eta \qquad (1)$$

where $w^0(\xi,\eta)$ is the subsidence of the immediate roof, F is the mined area, φ is the influence function, which according to different authors and for different regions, takes different forms. For instance, according to King $\varphi=(1/Htg\beta)\,\mathrm{sech}(x/Htg\beta)$, where β is the influence angle and H is the bedding depth of the mined seam.

In the inverse problem are given approximately both $w=w(x,y)$ and $\varphi=\varphi(x-\xi,y-\eta)$. It is sought $w^0=w^0(\xi,\eta)$, which is equivalent to seeking the mining order. For a better formulation of the problem, we descretize equation (1) and the received result we rewrite in the form of system of linear algebraic equations:

$$Aw^0=w \;,\quad w^0\in R^n \;,\quad w\in R^m \qquad (2)$$

where w is the given vector, w^0 is the sought vector, $A=\{a_{ij}\}$, $i=1,\dots m;\ j=1,\dots n$.

Now we will introduce some concepts. A standard solution of (2) means the solution, which is less distant from the origin of the coordinate system (with respect to $(0\dots0)^t$).

It is possible the system to have no solution, but to have a pseudosolution. According to the least squares method (Gauss), the pseudosolution is achieved by minimization of the quadratic functional $\Phi(w^0)=\|Aw^0-w\|^2$. For obtaining the pseudosolution of the system (2) some methods for minimization of the quadratic functional $\Phi(w^0)$ are applied. For example, we could write

$$\Phi'(w^0)=2(A^*Aw^0-A^*w)=0 \qquad (3)$$

(A^* is a matrix, conjugate with the matrix A). In that way we receive the so called system of normal equations, which satisfy the pseudosolution of (2) (or which satisfy the solution, in case it exists):

$$A^*Aw^0=A^*w \qquad (4)$$

How we can find the standard pseudosolution? This problem, we will set in the form: to be found

$$\min\|w^0\| \qquad (5)$$

provided that $\|Aw^0-w\|=\mu=\min\|Aw^0-w\|$ (or $A^*Aw^0=A^*w$).

In that way the posed problem can be solved by means of the pseudoinversion method of Moore-Penrose. Here we will, however, use an equivalent approach (A. Tikhonov), by formulating the following problem (Tikhonov & Arsenin 1977): to be found

$$\min M^\alpha(w^0)=\|Aw^0-w\|^2+\alpha\|w\|^2 \qquad (6)$$

where $\alpha>0$ is a parameter. The solution of (6) will be received by determining the gradient of $M^\alpha(w^0)$, i.e. $[M^\alpha(w^0)]'=2(A^*Aw^0+\alpha w^0-A^*w)$ and setting it equal to zero. Thus we receive the system of linear algebraic equations

$$(A^*A+\alpha E)w^0=A^*w \qquad (7)$$

where E is the single matrix

$$E=\begin{pmatrix}1\dots\dots\dots0\\ \dots..1\dots.0\\ \dots\dots\dots..1\end{pmatrix}$$

The system (7) has a unique solution for each $\alpha>0$:

$$w^0=(A^*A+\alpha E)^{-1}A^*w \qquad (8)$$

where $(A^*A+\alpha E)^{-1}$ is a matrix, inverse to the matrix $A^*A+\alpha E$.

With this, however, the problem is still not solved. The reason is that the above applied method is stable in respect to the errors in setting the vector w, but is unstable in respect to the errors in setting the matrix. So the problem is incorrectly posed.

Let us remind that in our case the function φ is approximately given (the angle β is given on the base of measurements), hence the matrix A is also approximately given.

So the pseudoinversion methods (as fare as the least squares method) doesn't allow to receive a stable approximation to the standard pseudosolution.

The fundament for creating stable methods for solving incorrect problems, is making the sense of the fact, that generally the ill-posed problems are not predetermined. In solving them, additional information should be used. It should be defined criteria for selection the approximate solutions. If these criteria are formulated, it is possible successfully to be constructed a stable method for solving unstable problems. It means to be constructed the so-called regularising algorithm.

Now, let us suppose, that the approximations of the matrix A_h and the vector w_δ are known from an experiment. They are such, that $\|A - A_h \le h\|$, $\|w - w_\delta \le \delta\|$. The errors hold $h \ge 0$, $\delta \ge 0$. We introduce the notation $\eta = (h,\delta)$. We want to construct a stable method for solving the system of linear algebraic equations (2), given the approximate data A_h, w_δ, η. This will permit to receive an approximate solution w_η^0, such, that $w_\eta^0 \to \overline{w}_H^0$ for $\eta \to 0$ ($\overline{w}_H^0$ is the exact standard pseudosolution of (2)).

In seeking a solution of the posed problem, Tikhonov's reguralization method will be used (Tikhonov & Arsenin 1977) (Dimova 1994). We will call Tikhonov's functional the following functional

$$M^\alpha\left[w^0\right] = \left\|A_h w^0 - w_\delta\right\|^2 + \alpha\left\|w^0\right\|^2 \qquad (9)$$

where $\alpha > 0$ is reguralization parameter.

The first member in $M^\alpha\left[w^0\right]$ is the discrepancy, which shows how well the vector w^0 satisfies the equation $A_h w^0 = w_\delta$. The second member is the "fine" for the vector's deviation from the zero. We pose the extreme problem: to be found

$$\min_{w^0} M^\alpha\left(w^0\right) \qquad (10)$$

The solution of this problem, which exists and is unique for any $\alpha > 0$, we will denote as $w_\eta^{0\alpha}$.

For determining the extremal $w_\eta^{0\alpha}$, we calculate the gradient of $M^\alpha\left[w^0\right]$:

$$\left(M^\alpha\left[w^0\right]\right)' = 2\left(A_h^* A_h w^0 - \alpha w^0 - A_h^* w_\delta\right)$$

After equation the above expression to zero, we receive the equation

$$\left(A_h^* A_h + \alpha E\right) w^0 = A_h^* w_\delta \qquad (11)$$

which solution has the form

$$w_\eta^{0\alpha} = \left(A_h^* A_h + \alpha E\right)^{-1} A_h^* w_\delta \qquad (12)$$

For the inversion of the matrix $\left(A_h^* A_h + \alpha E\right)$ (symmetrical and positively determined) the well-known effective methods are used (Tikhonov & Arsenin 1977). Now it is necessary to adjust the values of the reguralization parameter α with the errors in setting the data $\eta = (h,\delta) : \alpha = \alpha(\eta)$, such that $w_\eta^{0\alpha(\eta)} \to \overline{w}_H^0$. This can be done in the following way:

Over the extremals of the functional (9), we will determine the function

$$\rho_\eta(\alpha) = \left\|A_h w_\eta^{0\alpha} - w_\delta\right\| - \hat{\mu}_\eta - \left(\delta + h\left\|w_\eta^{0\alpha}\right\|\right),$$

where

$$\hat{\mu}_\eta = \min\left(\delta + h\left\|w^0\right\| + \left\|A_h w^0 + w_\delta\right\|\right)$$

This function is called generalized discrepancy. Let

$$\left\|w_\delta\right\| > \hat{\mu}_\eta + \delta \qquad (13)$$

It can be seen, that $\rho_\eta(\alpha)$ is differentiable for $\alpha > 0$, increases strictly monotonically and it's values exhaust the interval $\left(-\delta\ ,\ \left\|w_\delta\right\| - \hat{\mu}_\eta - \delta\right)$. It means, that it exists a unique root of the equation

$$\rho_\eta(\alpha) = 0 \qquad (14)$$

which we will denote with $\alpha(\eta)$. For obtaining the root, the known methods for solving transcendent equations can be applied (Kichikov & Kuramshina 1989).

It can be proved, that $w_\eta^{0\alpha(\eta)} \underset{\eta \to 0}{\to} \overline{w}_H^0$, i.e. the constructed algorithm is regiralizing by A.

Tikhonov. There exist well-developed methods for finding the extremal of the functional $M^{\alpha}\left[w^{0}\right]$ (Kichikov & Kuramshina 1989) (Tikhonov & Arsenin 1977).

Our experience in the solution of the inverse problems in geomechanics allows us to recommend in the applied inverse problems to be used the above described numerical method. Our results are stated in the works (Dimov & Dimova 1944) (Dimova 1994) (Dimova 1993).

REFERENCES

Dimov, I.V. & V.I. Dimova 1985. Nekorektni zadachi v geomehanikata (Incorrect problems in the gomechanics)(In Bulgarian). *NIPRORUDA*, no. 4, Sofia, Bulgaria.

Dimov, I.V. & V.I. Dimova 1994. Some 2D inverse problems arising in strata mechanics. *World Mining Congress*, (Proc. of the 16th World Mining Cong.,12-16 Sep. 1994,vol.5, Sofia, Bulgaria).

Dimova, V.I. 1994. Numerical solution of the 3-D basic inverse and incorrect problem in the theory of Earth's surface protection in mining areas. *Proc. of the Eight Intern. Conference on Computer Methods and Advances in Geomechanics, 22-28 May 1994, Morgantown, USA.*

Dimova, V.I. 1993. Subsidence as a modern theory of environmental protection. In: *Safety and Environmental Issues in Rock Engineering, EUROCK'93*, (Proc. of the ISRM Int. Symp., 21-24 June 1993, Lisbon, Portugal).

Kichikov, I.V. & G.M. Kurrmashina 1989. *Chislennye metody i kolebatelnoi spektroskopii (Numerical methods and oscillation spectroscopy)* (in Russian), Matematika i kibernetika, 1, 1989, Znanie, Moscow.

Tikhonov, A.N. & Arsenin, V.Y. 1977. *Solution of ill-posed problems*. V.H. Whinston & Sons, Washington, D.C., USA.

Whittaker, B.N. & D.J. Reddish 1989. *Subsidence.* Amsterdam-Oxford-N.Y.-Tokyo: Elsevier.

Computer Methods and Advances in Geomechanics, Yuan (ed.) © 1997 Balkema, Rotterdam, ISBN 90 5410 904 1

The rock hydraulics model and FEM of blocked medium of mining above confined aquifer

Y.S. Zhao, D. Yang & S. H. Zheng
Shanxi Mining Institute, Taiyuan, China

ABSTRACT: Mining above Confined aquifer is a vital problem on facing in China and other countries all over the world. On the basis of studying law of deformation, of seepage and of coupled of fracture and rock block, this paper presents the rock hydraulic model of blocked medium or coupled mathematical model of solid-fluid, and introduces the numerical method and programmable design in detail, and introduce the research achievement about the Xiyu Coal Mine in Shanxi Province of China, including the design of protecting coal pillars of fault, law of stress, deformation and permeability of fracture in the process of mining coal.

1. Introduction

Mining above confined aquifer is a crucial productive problem which the mining industry of China as well as many other countries is facing. About 60% of coal mine are related to Ordovician limestone in China and there are 285 ones affected by Ordovician water among 601 nationally managed coal-mines. The water outburst has taken place 1600 times, which submerges the whole coal-shaft about 200 times. This caused economical damage RMB 4 billion. Several dozens of billion tons of coal can't be cut owing to the water. Thus the solution to the problem is very important to China and many other countries in the world.

Mining above confined aquifer or water outburst of bottom is a event that the deformation or damage of impermeable stratum causes Ordovician water to pour into mining spaces, or even the whole coal-mine under the circumstance of natural stress, digging stress and water pressure. It involves Rock Mechanics,Fluid Mechanics, Mining Engineering and Geology of Water and Engineering, etc. But the essence of the problem is the solid-liquid coupling or rock fluid mechanics problem.

Many scholars have studied the problem for many years. But the theoretical model and the used method are almost experience, such as the water-outburst parameter brought up by Chinese scholars and the pure seepage mechanics method. The results obtained from these methods are very different from the real ones. It's imperative to present the theory of the block medium of rock fluid mechanics to correctly solve the problem.

2. The rock hydraulics Model of blocked medium

Blocked medium of rock mass is cut by structural plane which are equal to the engineering scale. This kind of rock mass has not REV, so only the rock structure mechanics method can be used. That is to say, the rock mass is considered as the combination of rock block and their structural plane. According to the equilibrium of their interaction, the engineering character of rock mass can be resolved.

When the model is built, the rock block is considered as quasi-continum media model and fracture as fracture media model. The seepage, deformation and their coupled equations of rock block and fracture are constructed respectively. Then the rock hydraulics model of blocked medium can be set up.

2.1 *Physic Basis*

The following basic assumptions are introduced to facilitate constructing the model and for the completeness of the theory.

1. On a small stress gradient, seepage equation complies with linear Darcy's law:

$$q = K(\Theta, p)\frac{\partial p}{\partial x} \quad (2.1)$$

where $K(\Theta,p)$ is the coefficient of permeability, is a function of volumetric Θ stress and porous pressure p.

2. Deformation and stress of rock block complies with Hooke's law:

$$\sigma_{ij} = \lambda\delta_{ij}e + 2\mu\varepsilon_{ij} \quad (2.2)$$

3. Effective stress of the solid framework complies with revised Terzaghi's effective stress principle:

$$\sigma_{ij} = \sigma'_{ij} + \alpha p\delta_{ij} \quad (2.3)$$

4. The media are saturated only by water.

5. The fracture seepage complies with Darcy's law:

$$q = k_f\frac{\partial p}{\partial s} \quad (2.4)$$

where $k_f = \frac{b^2}{12\mu}$, b, the width of fracture, is a function of porous pressure and stress.

6. Deformation of fracture complies with the model of Goodman's joint element.

7. The effective stress law of fracture element complies with Terzaghi's effective stress principle:

$$\sigma' = \sigma - p_f \quad (2.5)$$

where q is the seepage rate; p is the porous pressure , $\sigma_{ij}, \sigma'_{ij}$ are the stress tensor and the effective stress tensor respectively; α is the equal coefficient of porous pressure; σ and σ' are the normal and effective stress of fracture respectively; δ_{ij} is kronecker's symbol; λ, μ are Lame's constant; s is tangential coordination of fracture.

2.2 *Seepage equation of fracture*

We consider the conservation of mass of water in any control element, Hence we obtain:

$$div(\rho q) = \frac{\partial(n\rho)}{\partial t} + w \quad (2.6)$$

$$\frac{\partial}{\partial s}(k_f\frac{\partial p}{\partial s}) = \rho\frac{\partial\phi}{\partial t} + \beta\phi\frac{\partial p}{\partial t} + w \quad (2.7)$$

where $\rho\frac{\partial\phi}{\partial t}$ is the change of fracture porosity; $\beta\phi\frac{\partial p}{\partial t}$ mainly indicates the change of water in fracture element due to the compressibility of water; w is the item of source and sink.

2.3 *Seepage equation of rock block*

Rock block can be considered as the equivalent continuum medium and it is homogeneous on a large scale. Suppose the volumetric deformation of rock block includes two parts: deformation of the rock framework and the pore. $\alpha = (1-n)\alpha_s + n\alpha_p$, Where α_s and α_p are deformation of solid framework and pore respectively; n is porosity; $(1-n)\alpha_s \ll \alpha_p$, hence the volumetric deformation of the equivalent continuum media is equal to the pore deformation: $\alpha = n\alpha_p$;

we consider the conversation of mass of water in any control element

$$div(\rho q) = \frac{\partial(n\rho)}{\partial t} + w \quad (2.8)$$

$$div(\rho q) = \rho\frac{\partial n}{\partial t} + n\frac{\partial\rho}{\partial t} + w$$

when the compressibility of water and the deformation of framework are considered, we can obtain:

$$\frac{\partial n}{\partial t} = \frac{\partial e}{\partial t}, \frac{\partial\rho}{\partial t} = \beta\rho\frac{\partial p}{\partial t}$$

we take the above formulas into (2.8),so

$$\frac{\partial}{\partial x}(k_x\frac{\partial p}{\partial x}) + \frac{\partial}{\partial y}(k_y\frac{\partial p}{\partial y}) = \frac{\partial e}{\partial t} + \beta\rho\frac{\partial p}{\partial t} + w \quad (2.9)$$

where ρ is water density;β is coefficient of water compressibility; w is the item of source and sink; e is volumetric deformation; equation (2.7) and (2.9) constitute the seepage equation of rock hydraulics model of blocked medium.

2.4 deformation *equation of rock block*

Because rock block are considered as the equivalent continuum medium, the equilibrium equation can be expressed by the total stress as

$$\sigma_{ij,j} + F_i + (\alpha p)_{,i} = 0 \quad (2.10)$$

we take equation (2.2) into (2.10)

$$(\lambda+\mu)U_{j,ji} + \mu U_{i,jj} + F_i + (\alpha p)_{,i} = 0 \quad (2.11)$$

The difference between this equation and the general elastic one is that this equation are added the item of equal porous pressure $(\alpha p)_{,i}$, where α is an equal porous pressure coefficient, which is a function of volumetric stress and porous pressure.

2.5 *deformation equation of fracture*

Fracture deformation uses the model of Goodman's equal width joint element. According to the effective stress law of fracture, the stress of fracture element can be expressed as

$$\sigma_n' = \sigma_n - p \qquad \sigma_s' = \sigma_s \tag{2.12}$$

Deformations equation is

$$\begin{cases} \sigma_n' = D_n \varepsilon_n \\ \sigma_s' = D_s \varepsilon_s \end{cases}, \begin{cases} \varepsilon_n = \delta_n / b \\ \varepsilon_s = \delta_s / b \end{cases}, \begin{cases} \delta_n = \dfrac{b}{D_n}(\sigma_n - p) \\ \delta_s = \dfrac{b}{D_s}\sigma_s \end{cases} \tag{2.13}$$

To no width joint element, D_s and D_n are tangential and normal resistance of a unit length of a fracture element respectively.
To equal width joint element, they can be expressed by the general elastic constant.

2.6 *rock hydraulics Model of blocked medium*

To sum up, we can obtain a complete rock hydraulics model of blocked medium.

Seepage equation of rock block is:

$$\frac{\partial}{\partial x}(k_x \frac{\partial p}{\partial x}) + \frac{\partial}{\partial y}(k_y \frac{\partial p}{\partial y}) = \frac{\partial e}{\partial t} + \beta p \frac{\partial p}{\partial t} + w$$

deformation equation of rock block is

$$(\lambda+\mu)U_{j,ji} + \mu U_{i,jj} + F_i + (\alpha p)_{,i} = 0$$

Seepage equation of fracture is:

$$\frac{\partial \phi}{\partial t} + \frac{\partial}{\partial s}(k_f \frac{\partial p}{\partial s}) + w = 0$$

deformation equation of fracture is

$$\delta_n = (\sigma_n - p) b / D_n$$
$$\delta_s = \sigma_s b / D_s$$

Coupled equation is

$$K_{fi} = f(\Theta, p)$$
$$\alpha = \alpha(\Theta, p)$$
$$k_f = f(\sigma_n)$$
$$\varphi = \varphi(\sigma_n)$$
$$\lambda, \mu, D_n, D_s = f(p, \eta)$$

It's a typical coupled mathematical model of blocked medium of solid deformation and water seepage. Compared with the general equation, it differs in the following respects: (a) the influence of water pressure on the solid framework deformation is considered in the equilibrium equation of solid deformation; (b) $\frac{\partial e}{\partial t}$ is added to the seepage equation, this indicates that the effective stress of the solid framework has a marked influence on water flow.

3. The FEM of rock hydraulics model in blocked medium

To such complex mathematical model, the only effective way is to use numerical methods. This article mainly discuss our work used by FEM.

3.1 *Discretization of the fracture seepage equation*

Fracture seepage equation is:

$$\beta\phi \frac{\partial p}{\partial t} + \frac{\partial \phi}{\partial t} + \frac{\partial}{\partial s}(k_f \frac{\partial p}{\partial s}) + w = 0 \tag{3.1}$$

we assume that the water pressure in fracture element is linear (Fig.1)

$$P(s) = a_1 + a_2 s \tag{3.2}$$

it can be expressed by basic interpolation function N_1, N_2 as:

$$p(s) = [N_i, N_j]\{p_i, p_j\}^T \tag{3.3}$$

where $N_i = (l-s)/L$; $N_j = s/L$; s is tangential coordination; p_i, p_j are water pressure of joint i,j respectively.
Using Galerkin method, we can obtain discrete fracture seepage equation.

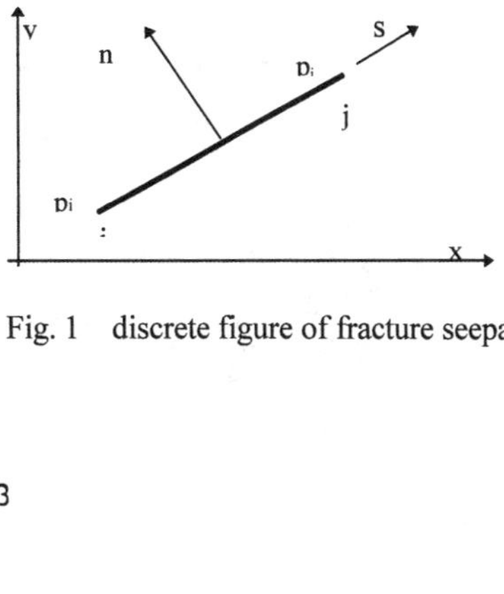

Fig. 1 discrete figure of fracture seepage element

$$\int_\Omega N_i[\beta\phi \frac{\partial p}{\partial t} + \frac{\partial \phi}{\partial t} + \frac{\partial}{\partial s}(k_f \frac{\partial p}{\partial s}) + w]d\Omega = 0 \qquad (3.4)$$

$$\int_\Omega N_i \frac{\partial}{\partial s}(k_f \frac{\partial p}{\partial s})d\Omega + \int_\Omega N_i[\beta\phi \frac{\partial p}{\partial t} + \frac{\partial \phi}{\partial t} + w]d\Omega = 0 \qquad (3.5)$$

using integration by parts on the first item and the second boundary condition, we have

$$-k_f\begin{bmatrix} \frac{b}{L} & \frac{-b}{L} \end{bmatrix}\begin{Bmatrix} p_i \\ p_j \end{Bmatrix} + \frac{L}{2}\bar{q} \qquad (3.6)$$

In the second item, the fracture porosity φ is a function of the length and width of fracture.

$$\phi = b' \cdot s, b' = b - \delta_n$$

$$\frac{\partial \phi}{\partial t} = \frac{\partial b'}{\partial t} = -\frac{\partial \delta_n}{\partial t}, \delta_n = \frac{\sigma_n - p_f}{k_n} \cdot b \qquad (3.7)$$

$$\frac{\partial \phi}{\partial t} = \frac{b}{k_n}(\frac{\partial p_f}{\partial t} - \frac{\partial \sigma_n}{\partial t})$$

$$\beta\phi\frac{\partial p_f}{\partial t} = \beta(b - \delta_n)\frac{\partial p_f}{\partial t} = \beta b(1 - \frac{\sigma_n - k_f}{k_n})\frac{\partial p_f}{\partial t} \qquad (3.8)$$

We combine the above two formulas into one.

$$\frac{\partial \phi}{\partial t} + \beta\phi\frac{\partial p_f}{\partial t}$$
$$= \frac{b}{k_n}\frac{\partial p_f}{\partial t} - \frac{b}{k_n}\frac{\partial \sigma_n}{\partial t} + \beta b\frac{\partial p_f}{\partial t} - \beta b\frac{\sigma_n - p_f}{k_n}\frac{\partial p_f}{\partial t}$$
$$= \frac{b}{k_n}[\frac{b}{k_n} + \beta k_n - \beta\frac{\sigma_n - p_f}{k_n}]\frac{\partial p_f}{\partial t} - \frac{b}{k_n}\frac{\partial \sigma_n}{\partial t}$$
$$= a_1\frac{\partial p_f}{\partial t} - a_2\frac{\partial \sigma_n}{\partial t} \qquad (3.9)$$

Then the second item of Galerkin integrated equation can be expressed as :

$$\int_\Omega N_i a_1 \frac{\partial p_f}{\partial t}d\Omega - \int_\Omega N_i a_2 \frac{\partial \sigma_n}{\partial t}d\Omega + \int_\Omega N_i \overline{w}_i d\Omega \qquad (3.10)$$

Integrating the above formula according to basic interpolation function, we have:

$$a_1\begin{bmatrix} \frac{bL}{3} & \frac{bL}{6} \end{bmatrix}\begin{Bmatrix} \frac{\partial p_i}{\partial t} \\ \frac{\partial p_j}{\partial t} \end{Bmatrix} - \frac{bLa_2}{2p}\frac{\partial \sigma_n}{\partial t} + \frac{bL}{2}w_i \qquad (3.11)$$

Combining the first item, we obtain the Discretization equation of fracture element seepage:

$$-k_f\begin{bmatrix} \frac{b}{L} & -\frac{b}{L} \end{bmatrix}\begin{Bmatrix} p_i \\ p_j \end{Bmatrix} + \frac{L}{2}q_0 + a_1\begin{bmatrix} \frac{bL}{3} & \frac{bL}{6} \end{bmatrix}\begin{Bmatrix} \frac{\partial p_i}{\partial t} \\ \frac{\partial p_j}{\partial t} \end{Bmatrix} \qquad (3.12)$$
$$-\frac{bLa_2}{2}\frac{\partial \sigma_n}{\partial t} + \frac{bL}{2}w_i$$

3.2. *The Discrete Equation of Rock block Seepage*

Used Galerkin method to discrete, the detail analysis of such Discretization can refer to other reference. We can acquire the discrete equation of blocked medium rock mass by combining the discrete equation of fracture with that of rock block by related nodal:

$$[T]\{p\} + [S]\{\frac{\partial p}{\partial t}\} + \{H\} = 0 \qquad (3.13)$$

where [T] is permeable arrays, [S] is liquid storage arrays, [H] is column vector.

Suppose the rock block is impermeable, then the permeable passage of blocked medium rock mass and liquid storage space is completely made up of fracture, and its seepage discrete equation is as follows:

$$\sum_{k=1}^{m}\{a_{1k}\begin{bmatrix} \frac{b_k L_k}{3} & \frac{b_k L_k}{6} \end{bmatrix}\begin{Bmatrix} \frac{\partial p_i}{\partial t} \\ \frac{\partial p_j}{\partial t} \end{Bmatrix} - \frac{b_k L_k a_{2k}}{2}\frac{\partial \sigma_n}{\partial t} + \frac{b_k L_k}{2}w_{ik}$$
$$-k_f\begin{bmatrix} \frac{b_k}{L_k} & -\frac{b_k}{L_k} \end{bmatrix}\begin{Bmatrix} p_i \\ p_j \end{Bmatrix} + \frac{L_k}{2}q_0\} = 0 \qquad (3.14)$$

3.3. *Discrete Equation of Deformation of Rock block*

We used quadrilateral and four point isoparametric element and triangle element in Discretization rock block . Fractures are treated as Goodman joint element. What is different from general Discretization equation of solid deformation is, in our equation, added an item which is related to porous pressure. Here, by using initial stress method, we give an detail explanation about this item. To general element, its equal loading is :

$$\{f_i\} = \int_v [B]^T\{\alpha p_i\}dv \qquad (3.15)$$

But to fracture element, we can use [B] array of fracture element because water pressure in fracture element is supposed to comply with lineal distribution. So the equal loading of fracture element is as follows:

$$\{F\} = \sum_{k=1}^{m}\{f_{ek}\}$$
$$f_{ek} = \int [B]^T\{p\}dv = b_k L_k [B]^T\{p_{ek}\}$$
$$= L_k\begin{bmatrix} -0.5 & 0 & -0.5 & 0 & 0.5 & 0 & 0.5 & 0 \\ 0 & -0.5 & 0 & -0.5 & 0 & 0.5 & 0 & 0.5 \end{bmatrix}\begin{Bmatrix} p_{kx} \\ p_{ky} \end{Bmatrix}_e \qquad (3.16)$$

The two formulas above show how to calculate equal additional loading of porous pressure. When we calculate all the element in the area, we get equal loading of each nodal. Then the discrete equation of rock mass deformation can be simplified as:

$$[K]\{u\}=\{F\} \quad (3.17)$$

3.4. *Program Design of rock Hydraulics Model of Blocked Medium*

Obviously, the system above is a complex one, so we work out following plans to solve: treat solid deformation system and water seepage system as two different system, coupled and alternated to solve according to time sequence. The detail steps are: according to time sequence, first calculate solid deformation at $t=t_0$, and combine it into seepage equation to get water pressure $p(t_1)$ of each nodal in the area at $t_1=t_0+\Delta t$, then combine it into solid deformation equation to get solid deformation at t_1, combine it into seepage equation to get $p(t_2)$ at $t_2=t_1+\Delta t$, and so on. At last, obtain the deformation and stress of each element and the pressure of each nodal in the area. We can take two means to assure the precision is satisfactory, one is dividing time fine, another is alternated two systems more times in same time step before calculating next time step.

4. Numerical Simulation of Mining above Confined Aquifer

The main factors of water outburst of mining above confined aquifer are: a. water-bearing of aquifer and water pressure; b. fracture distribution; c. mining method and its effects on bottom stability. In the numerical simulation, we consider the change of mining space. The program frame is showed Fig. 2. the numerical simulating area is Taiyuan Xiyu Coal Mine. Xiyu coal Mine locates in the east edge of Xishan Coal Filed. Here geologic tectonics are complex, and there are many faults and caved-in-pillar. Mining depth of this mine is +600m, the lowest level is about +450m, but the static water level of Ordovician period FengFeng group limestone is about +830m, and the highest confined aquifer

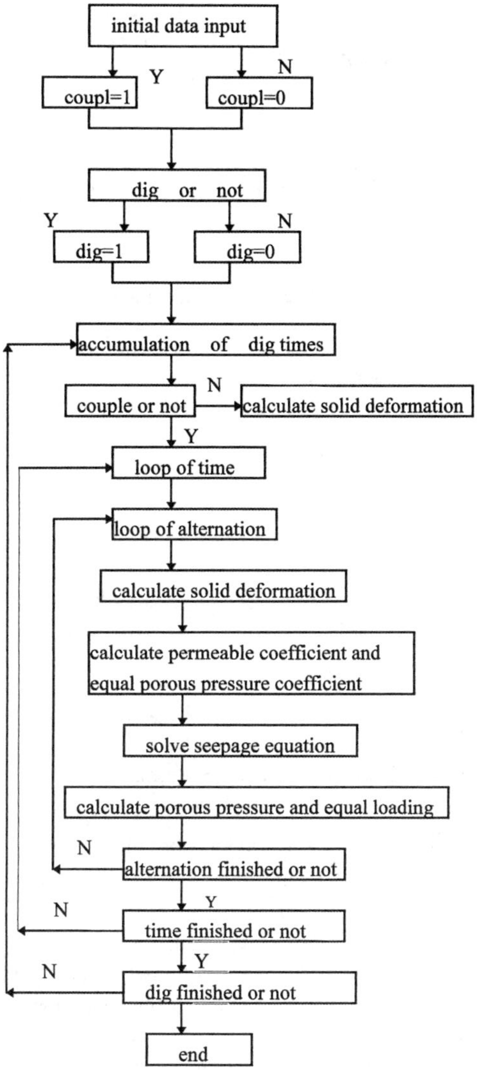

Fig. 2 frame figure of computer program

pressure is 3.8Mpa, so having a delicately research of movement of confined aquifer, water outburst of mine, bottom stability of coal seam and rational mining design become necessary. Our institute has cooperated with Xiyu Coal Mine, doing such work between 1995 and 1996, and analyzing some typical sections, one is showed in figure 3, the numerical experiment plans are showed in table 1. These plans have research bottom deformation, water seepage and water outburst under two case: different width of

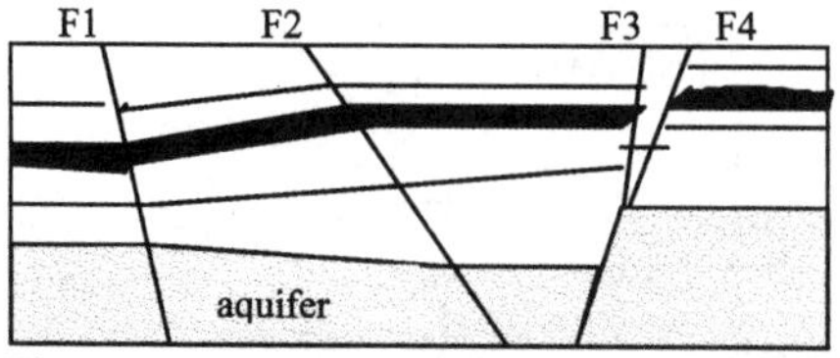

Figure 3. Simplify figure of numerical calculating model

Table 1 Numerical experiment plan

plan	advanced rate(m/times)	width of coal pillars(m)
I	100	25
II	200	25
III	100	75
IV	200	75

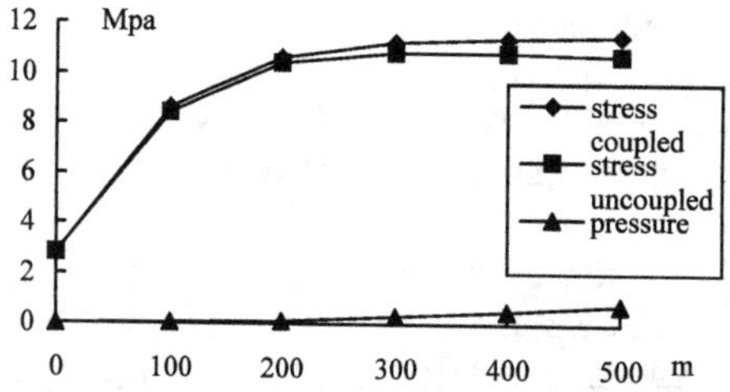

Fig. 4 stress and water pressure of pillar in the process of mining

protecting pillar and different rate of mining(m/times). The result of this research can give advice to the rational mining method decision on confined aquifer.

From our work, we get following conclusion:

1. To bottom deformation of coal seam, the coupled is greater than the uncoupled. Among four plans, the highest difference is 11%. To stress, this difference is about 18%.

2. Effect of protecting pillar. The experiment show the rational width of fault protecting pillar can resistant the permeability of confined aquifer along fault effectively. The relations between water pressure in protecting pillar and advance distance of working face are summed up in figure 4. When working face is advanced to 500m, if the protecting pillar width is 25m, the water pressure in protecting pillar is 0.55Mpa. If the width is 75m, the pressure is 0.44Mpa, the range is about 25%.

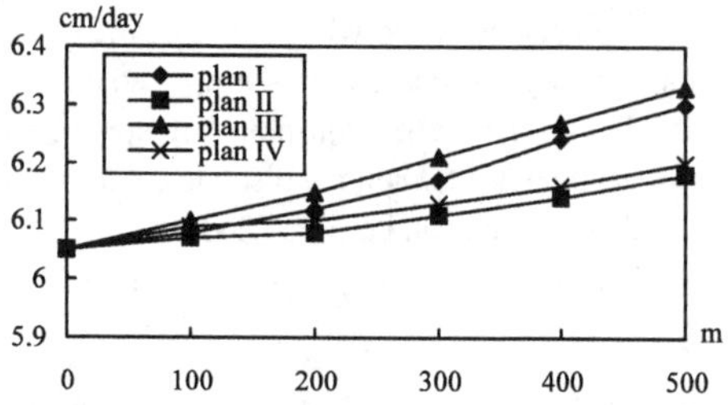

Fig.5 permeability of fracture in the process of mining

3. Analysis of fault deformation and permeability in the process of mining, Figure 5. shows the change of permeability and width of fault with different width of fault protecting pillar and different advanced distance of working face. From the figure we can see, to pillar width is 25m, when working face advances from 50m to 550m, permeable coefficient rise from 6.05 cm/day to 6.35 cm/day. But to pillar width is 75m, permeable coefficient rise form 6.05 cm/day to 6.30 cm/day.

5. Conclusions

From the research above, we can see, rock hydraulics model of block medium and its FEM is a effective method to solve mining problem above confined aquifer, and it can provide effective guidance to mining method decision and mining design.

But because of the ignorance of fracture distribution feature and the difficult of net dissection of numerical calculation and calculation itself, this paper selected different sections to simulate the space feature of the area, so there must be error. To solve this problem, the cubic Rock hydraulics model of blocked medium is the imperative work to do.

[Reference]

[1] Bear J. 1972. Dynamics of Fluids in Porous Media, American Elsevier, Publishing Company Inc.

[2] Yangsheng Zhao 1994.12. Rock Fluid Mechanics, The Coal Industry Publishing of China

[3] Louis C. 1974. Rock Hydraulics in Rock Mechanics, 1974.

[4] Cook N.G.W. 1992. Natural Joints in Rock, Int. J. Rock Mech. Min. Sci.& Geo. Abstr. Vol.29.No.3, p198-223.

Computer Methods and Advances in Geomechanics, Yuan (ed.) © 1997 Balkema, Rotterdam, ISBN 90 5410 904 1

A laminated span failure model for highwall mining span stability assessment

B. Shen & M. E. Duncan Fama
CSIRO Division of Exploration and Mining, Brisbane, Australia

ABSTRACT: Unsupported span stability is a major issue for the successful use of highwall mining methods in coal mines. To predict the stability of unsupported spans in closely bedded and laminated coal roof, a simple analytical model has been developed in which the immediate roof is treated in 2D as a cracked beam. The analytical model provides a very simple solution, and agrees well with numerical results using UDEC.

1 INTRODUCTION

Highwall mining is a method used in surface coal mines to extract coal by cutting rectangular drives into seams exposed at the base of open pits which have reached their economic depth limit. This method is reliant upon the self supporting capacity of the rock mass for success. Consequently, unsupported span stability is one of the primary considerations for highwall mining.

For many years, perhaps ever since the establishment of rock mechanics, span stability has been one major focus for underground rock engineering. Consequently, many methods and techniques have been developed to predict span stability for different purposes and different geological conditions. Bieniawski (1993) and Barton (1974) suggested the use of rock classification methods and developed the RMR system and Q-system, respectively. The rock classification method requires a considerable amount of data from previous rock engineering practice in similar geological conditions. Considerable experience has been gained for hard rocks, but little experience for soft and bedded rocks in coal measures. Goodman and Shi (1985) developed a key block theory to predict the roof failure caused by block fall; Beer and Meek (1982) used the voussoir beam model to study the stability of bedding roof with vertical joints. The latter method has been discussed and extended by many researchers (Brady and Brown, 1993, Hoek et al, 1995, Sofianos, 1996) and has often been applied in mining design for thick bedded roofs. The disadvantage of this model is that it does not provide a closed form analytical solution but the solution is obtained by numerical iterations. There are also many numerical codes developed in recent years which hold the potential for span stability assessment, such as UDEC (Itasca, 1993) and 3DEC (1994) for blocky rock masses and FLAC (Itasca, 1991) and ABAQUS (Hibbitt et al., 1994) for intact and soft rock masses. Most of the numerical methods require a comprehensive rock and joint properties and are often not easy to use by mining engineers.

The typical roof geology of mining sites in central Queensland is characterised by thinly laminated and interbedded sedimentary rocks. This makes direct use of the existing methods difficult. In this paper, a new but very simple span model: the laminated span failure model (LSFM) is described.

2 THE LAMINATED SPAN FAILURE MODEL (LSFM)

The analysis was treated in two dimensions, because the entries or drives were parallel and extended a considerable distance into the highwall. Out of plane jointing was ignored. During highwall mining operations, span failure was often observed to follow three steps (Figure 1):

a. The immediate roof delaminated into several thin layers, each behaving as an individual plate, ie. in 2D, a beam (Figure 1b);

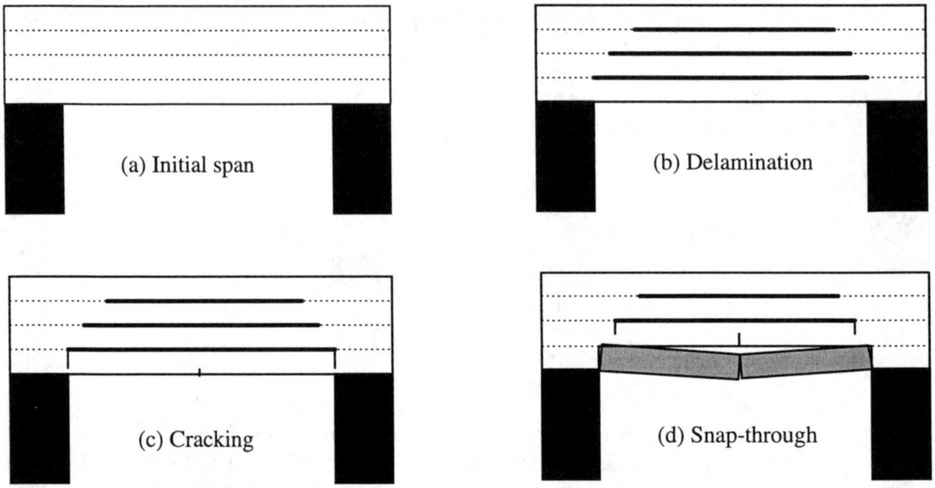

Figure 1. Span failure process observed from the highwall mining entry.

b. Cracks initiated in the roof beam near the side walls of the entry as well as at the center of span (Figure 1c);

c. The cracked beam snapped through at the middle of the span (Figure 1d).

Stability or otherwise of the span was often controlled by whether the immediate roof beam could snap through. Here we study the critical conditions of snap-through. In order to get an analytical solution, we simplify the cracked beam as a rod-roller system (Figure 2). The system is loaded by the weight of the rods (distributed load p) and supported by the abutments. The load will compress and deform the rods, and in certain conditions, will cause the rods to snap through.

2.1 Simple solution for elastic beam

For the system to reach a state of equilibrium, the axial force of the rods (F) will be

$$F=\frac{ps}{4\sin\theta} \tag{1}$$

where the notation is as shown in Figure 2.

At the position of equilibrium (θ), the relation between the axial force in the rods and their angle is:

$$F=\Delta l/l\cdot Et=Et\left(1-\frac{\cos\theta_0}{\cos\theta}\right) \tag{2}$$

where t is beam thickness, E is the Young's modulus.

Combining Eqs.(1) and (2), we obtain,

$$\frac{ps}{4\sin\theta}=Et\left(1-\frac{\cos\theta_0}{\cos\theta}\right) \tag{3}$$

Using the relation $\tan\theta_0=t/s$. Eq.(3) becomes

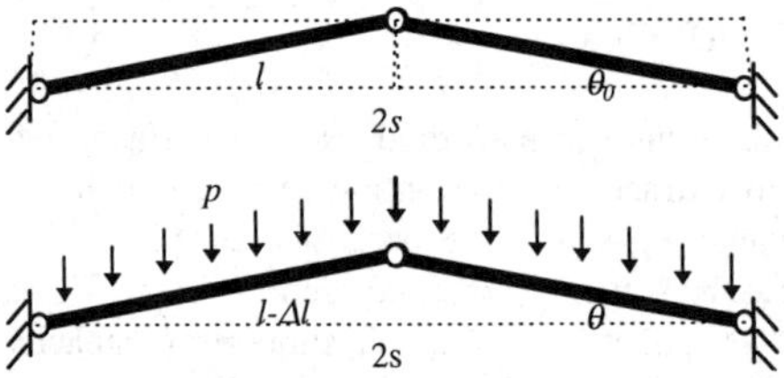

Figure 2. Mathematical model of a cracked roof beam.

$$\frac{p}{4E}=\left(1-\frac{\cos\theta_0}{\cos\theta}\right)\tan\theta_0\sin\theta \tag{4}$$

or in another form

$$\frac{p}{4E}=\tan\theta_0\sin\theta-\sin\theta_0\tan\theta \tag{5}$$

Because the beam is thin and long ($\theta_0<<1$, $\theta<<1$), we can use the following approximations for both θ_0 and θ:

$$\sin\theta=\theta-\frac{\theta^3}{6}+O(\theta^5)$$

$$\tan\theta=\theta+\frac{\theta^3}{3}+O(\theta^5)$$

After simplification, Eq.(5) becomes:

$$\frac{p}{4E}=\frac{1}{2}(\theta_0^3\theta-\theta_0\theta^3)+O(\theta_0^5)+O(\theta^5) \tag{6}$$

Neglecting the small terms $O(\theta_0^5)+O(\theta^5)$, eq.(6) becomes:

$$\frac{p}{2E}=\theta_0^3\theta-\theta_0\theta^3 \tag{7}$$

or in the standard form of a cubic equation for θ:

$$\theta^3 - \theta_0^2\theta + \frac{p}{2E\theta_0} = 0 \quad (8)$$

It is not difficult to prove that the critical condition for stability is the condition that Eq.(8) have only one positive solution. This is $q^3 + r^2 = 0$ (Shen and Duncan Fama, 1996), where:

$$q = -\frac{\theta_0^2}{3}, r = -\frac{p}{4E\theta_0}$$

or

$$-\frac{\theta_0^6}{27} + \frac{p^2}{16E^2\theta_0^2} = 0 \quad (9)$$

The solution of Eq.(9) is

$$\theta_0 = \left(\frac{3\sqrt{3}}{4} \cdot \frac{p}{E}\right)^{1/4} \quad (10)$$

The critical stable span (S=2s) can then be obtained by using the relation $\theta_0 \approx t/s$:

$$S = 2\left(\frac{4}{3\sqrt{3}}\right)^{1/4} \cdot t \cdot \left(\frac{E}{p}\right)^{1/4} \quad (11)$$

The above solution shows that the critical stable span is controlled by the roof beam thickness (t), the roof rock Young's modulus (E) and the load on the cracked beam (p).

3 COMPARISON BETWEEN THE ANALYTICAL MODEL AND UDEC

3.1 A brief description of UDEC

UDEC, Universal Distinct Element Code, is a two dimensional code developed by Itasca Consulting Group (Itasca, 1993) to model the mechanical behaviour of a block assembly such as a jointed rock mass. It considers joints as block boundaries along which elastic contact, sliding or separation is allowed. Block contact is simulated by introducing an artificial "overlap" of joint surfaces. UDEC is an explicit finite difference code which uses time step iteration to reach an equilibrium solution or the collapse of the system. This makes it possible to monitor the process of rock movement and joint deformation before a final equilibrium state is reached. UDEC does not simulate failure caused by cracking in the context of the theory of fracture mechanics. However, there is an alternative way to estimate the cracking process with UDEC, by using the concept of an "artificial joint". The artificial joint should have the same strength as the intact rock so that it can represent the failure of intact rock. If the artificial joint slides or separates it means that cracking of intact rock occurs due to shear or tension, respectively.

3.2 UDEC models and modelling results for span stability

As part of the numerical investigation for unsupported span stability, a large number of UDEC simulations were conducted. The UDEC model was composed of one rectangular entry of width 3.5 metres and two half pillars. Each pillar had a width of 3 metres. The mesh was bounded by two vertical roller boundaries, which represented the symmetric lines down the centre of the pillars. Closely bedded roof rock, typical in central Queensland, was simulated and the bedding planes were modelled as very weak and easily separated. Vertical artificial joints were introduced to simulate intact rock cracking. The maximum stable span was obtained by gradually increasing the span width until a roof fall occurred.

Based on a previous study (Shen et al, 1996), the following parameters were identified as most influential upon unsupported span stability in a weakly bedded roof: (1) *In situ* stress ratio and magnitude: σ_x (horizontal) and σ_y (vertical); (2) Intact rock strength and modulus: E, ν, σ_t (tensile strength) ϕ (friction angle) and c (cohesion); (3) Spacing of bedding planes: t.

Three values of each of the three groups of parameters were considered. They represented low,

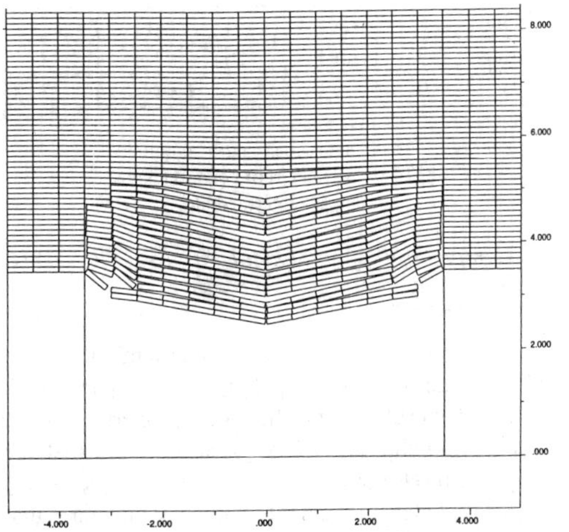

Figure 3. Typical roof failure in closely bedded rock, UDEC simulation (combination: s3r1t1, see Table 2).

intermediate, and high values of the group of parameters, see Table 1. Each UDEC simulation used one of the combinations, resulting in 27 UDEC simulations.

Table 1. Parameters used in UDEC modelling

	Low	Intermediate	High
In situ stress ratio and magnitude	$\sigma_x/\sigma_y = 0.5$ $\sigma_y = 2.4MPa$	$\sigma_x/\sigma_y = 1.0$ $\sigma_y = 2.4MPa$	$\sigma_x/\sigma_y = 2.0$ $\sigma_y = 2.4MPa$
Intact rock strength and modulus	$E=1GPa$ $\nu=0.24$ $\sigma_t = 0.5MPa$ $\phi=30°$ $c=1.0MPa$	$E=3GPa$ $\nu=0.24$ $\sigma_t = 1.0MPa$ $\phi=40°$ $c=2.0MPa$	$E=5GPa$ $\nu=0.24$ $\sigma_t = 2.0MPa$ $\phi=50°$ $c=3.0MPa$
Immediate roof beam thickness	$t=0.1m$	$t=0.25m$	$t=0.5m$

The results showed a wide variation in the maximum stable span ranging from 5 metres to 23 metres for different property combinations, see Table 2. A typical roof fall predicted by UDEC is shown in Figure 3.

Table 2. Maximum stable span predicted by UDEC for different property combinations.

Comb*	Span (m)	Comb	Span (m)	Comb	Span (m)
s1r1t1	5	s1r1t2	9	s1r1t3	15
s2r1t1	5.5	s2r1t2	10.5	s2r1t3	15
s3r1t1	6.5	s3r1t2	11.5	s3r1t3	16.5
s1r2t1	5.5	s1r2t2	11	s1r2t3	17
s2r2t1	6	s2r2t2	13	s2r2t3	19
s3r2t1	7.5	s3r2t2	15	s3r2t3	20
s1r3t1	7	s1r3t2	14	s1r3t3	21
s2r3t1	9	s2r3t2	16	s2r3t3	22
s3r3t1	9	s3r3t2	17	s3r3t3	23

Combination: s - *In situ* stress ratio
r - Intact rock strength and modulus
t - Immediate roof beam thickness
1 - Low; 2 - Intermediate; 3 - High

3.3 *Comparison of analytical results with UDEC results*

The simple solution of the analytical laminated span failure model (LSFM), Eq.(11), was compared with the UDEC results using the same properties as for the corresponding UDEC simulation. Note that the analytical model only uses a few of the mechanical properties listed in Table 1, and does not take into account the effect of *in situ* stresses. The analytical and UDEC results are plotted against a factor F: $F=t(E/p)^{1/4}$ in Figure 4. In general, three different UDEC span values are obtained for each F value, one for each of the different *in situ* stress ratio values in Table 1. On average, the analytical results agree well with the UDEC results.

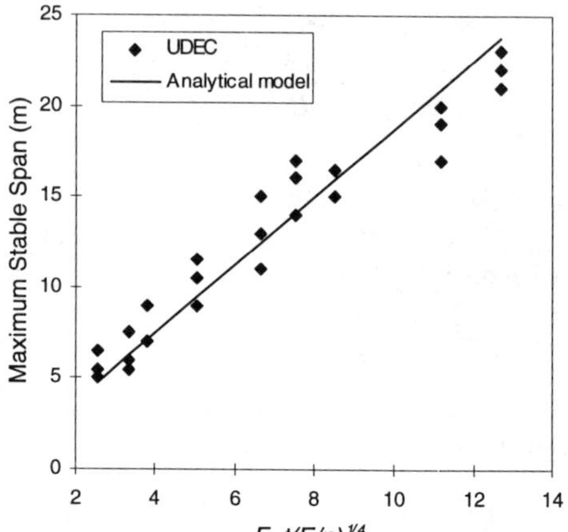

Figure 4. Comparison of the predicted analytical maximum stable unsupported span with the UDEC predictions.

4 DISCUSSION

The laminated span failure model (LSFM) has been developed for the purpose of unsupported span stability assessment in closely bedded highwall mining roof conditions. This model provides a simple and easy-to-use equation for calculating the maximum stable span. Even though the LSFM has used some simplifications and assumptions for what is a complicated roof failure process, it captures the main factors which control the stability of a unsupported span. It agrees well with UDEC results for a wide range of roof geological parameters. The good agreement suggests that the analytical model can be used instead of the more time-consuming numerical model, UDEC, for unsupported span stability assessment and prediction.

The LSFM may be combined with a probabilistic approach to take into account the variability of the roof rocks over the large area of highwall mining reserve. The current model has been extended to consider local yielding of the roof beam. The extended LSFM with the probabilistic approach has been found to give reasonable predictions for some highwall mining pits in central Queensland (Follington et al. 1996). Details of the extended

LSFM and its applications can be found in Shen and Duncan Fama (1996).

ACKNOWLEDGMENTS

This research was funded by the Australian Coal Association Research Program (ACARP), BHP Coal Pty. Ltd., MIM Holdings Limited, Thiess Contractors Pty. Ltd., Mining Technologies Australia Pty. Ltd. and CSIRO. The contribution of I. Follington and B. Leisemann to this project and helpful comments and input from H. Asche, S. Craig and J. Shi, are gratefully acknowledged.

REFERENCES

Barton, N., R. Lien and J. Lunde 1974. Engineering classification of rock masses for design of tunnel support. *Rock Mech.* 6:183-236

Beer, G. and J.L. Meek 1982. Design curves for roofs and hanging walls in bedded rock based on voussoir beam and plate solutions. *Trans Instn Min. Metall.* 91, A18-22.

Bieniawski, Z.T. 1993. Classification of rock masses for engineering: the RMR system and future trends. In: *Comprehensive rock engineering - Vol 3*, *Rock testing and site characterisation.* 553-573

Brady B.H.G. and E.T. Brown 1993. Rock mechanics for underground mining. *George Allen & Unwin (Publisher).* 2nd Ed.

Follington, I.L., M.E. Duncan Fama, B. Shen, B.E. Leisemann, R. McPhee and T.P. Medhurst 1996. Program 3 Report - Monitoring of highwall mining operations in Pit 17DU South, Moura Mine, Qld. CSIRO Exploration and Mining Report No. 264C (Confidential).

Goodman R. and G.H. Shi 1985. *Block theory and its application to rock engineering* (London: Prentice-Hall, International)

Hibbitt, Karlsson & Sorensen, Inc. 1994. ABAQUS - Version 5.4. User's manuals, 1080 Main Stress, Pawtucket, RI 02860-4847.

Hoek, E., P.K. Kaiser and W.F. Bawden 1995. *Support of underground excavations in hard rock.* Rotterdam: A.A. Balkema.

Itasca 1991. FLAC - Fast Lagrangian Analysis of Continua, Version 3.1, User's Manual. Itasca Consulting Group, Inc., Minneapolis, Minnesota, USA.

Itasca 1993. UDEC - Universal Distinct Element Code, Version 2.1., User's Manual. Itasca Consulting Group, Inc., Minneapolis, Minnesota, USA.

Itasca 1994. 3DEC - Three-Dimensional Distinct Element Code, Version 1.5, User's Manual. Itasca Consulting Group, Inc., Minneapolis, Minnesota, USA.

Shen, B., M.E. Duncan Fama and I.L. Follington 1996. A preliminary study of roof failure mechanisms for highwall mining. CSIRO Division of Exploration and Mining Report No.229F.

Shen B. and M.E. Duncan Fama 1996. Span stability prediction for highwall mining — Analytical and numerical studies. CSIRO Division of Exploration and Mining Report No.316F.

Sofianos, A.I. 1996. Analysis and design of an underground hard rock voussoir beam roof. *Int. J. Rock Mech. Min. Sci. & Geomech. Abstr.* 33(2), 153-166.

Computer Methods and Advances in Geomechanics, Yuan (ed.) © 1997 Balkema, Rotterdam, ISBN 90 5410 904 1

Subsidence analyses over leached salt gas storage caverns

D. Nguyen Minh & E. Quintanilha de Menezes
LMS, Ecole Polytechnique, France

G. Durup
Gaz de France, France

ABSTRACT: A numerical simulation of natural gas underground storage field in bedded rock salt is proposed, and compared to available in situ measurements in two sites, Tersanne and Etrez (France). Evolution of surface subsidence and cavity volume loss is of main concern. Although the problem is three dimensional, an axisymmetric model for a single cavity is used together with a superposition method previously validated. A good determination of surface subsidence is obtained by a hybrid BEM-FEM model. There is a reasonable agreement between in situ data and numerical values. A long term evalaution of subsidence is then proposed for the next hundred years.

1 INTRODUCTION

Underground gas storage in large leached cavities in salt layers or salt domes is a worldwide practise.

Subsidence over such storage fields results from closure of cavities, due to creep of salt. The phenomenon, of rather limited amplitude, compared to mining areas, is however delayed over a long time period, and extends over a large surface. On an environmental point of view, a careful survey of subsidence is needed, with a typical observation zone extending over about 4000 m diameter, using a network of survey benchmark over the storage field (Figure 1). By the same time, observations on cavities allow to have a good knowledge of the current delayed properties of rock salt. Contrarily, the elastic constants of the overburden are rather unknown, and eventually, the behavior of joints between layers.

Numerical analyses of subsidence in Tersanne (France) have been already attempted (Nguyen Minh et al, 1991, 1993). This paper deals with a comparative analysis of subsidence on Tersanne and and an other site Etrez, which have been observed by Gaz De France over tens of years. This allows to confirm some general features of the subsidence phenomenon, by detailed analyses with taking into acount of both similarities and differences between these two sites. A long term prediction is then proposed, with adequate choice of constitutive model for rocksalt.

The three dimensional problem is approximated by a superposition method of single cavity solution in an infinite medium, validated in the previous study (Nguyen Minh et al, 1993).

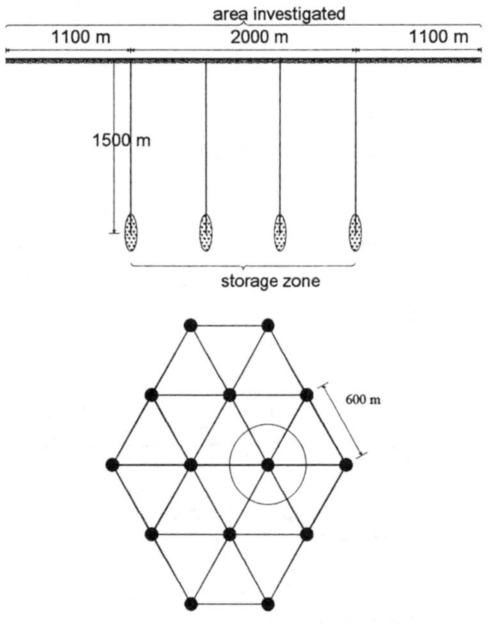

Figure 1 - Theoretical display of the 14-Cavity gas storage field and subsidence observation zone

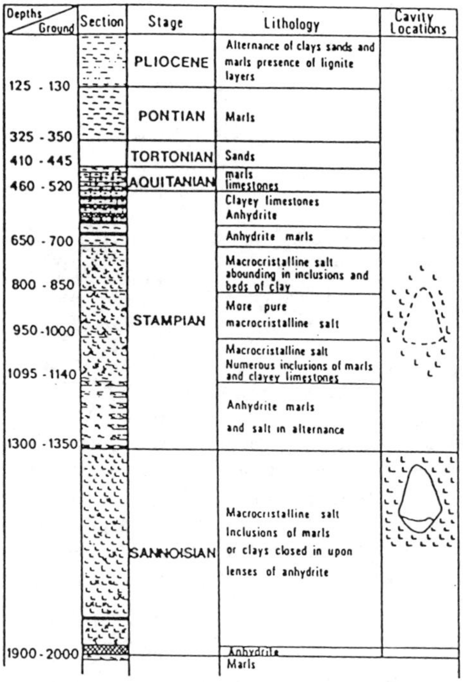

Figure 2a Geological section of Etrez site

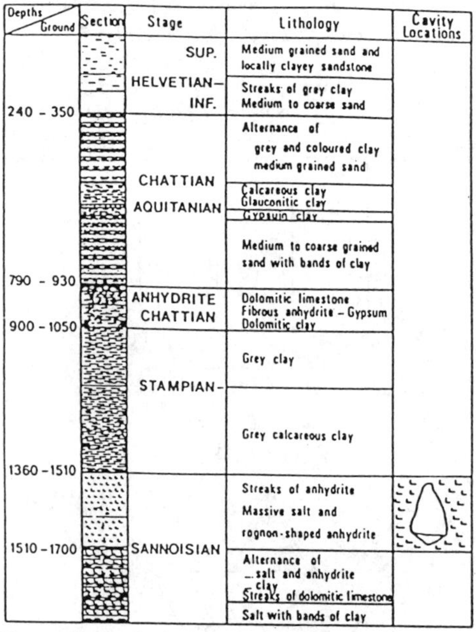

Figure 2b-Geological section of Tersanne site

Although the rockmass medium was non linear, with viscous rocksalt layer and linear elastic cover, such method has proved to give a fair good accuracy.

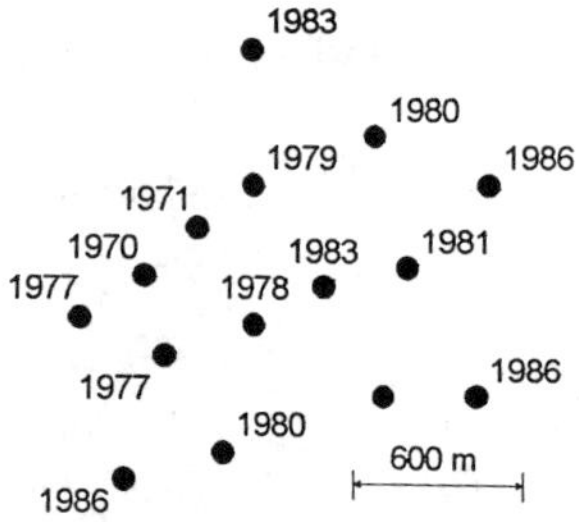

Figure 3. The Tersanne field develppment

For this axisymmetrical analysis of the single cavity, we use a newly developped hybrid FEM-BEM software which takes into account the specificity of this problem, and allows to exempt doing a two step FEM calculation as proposed formerly

2. DESCRIPTION OF THE TWO SITES

The geological structure of the two sites, shows a horizontally bedded rockmass, with sedimentary layers overlying a deep hard rockmass. There is one thick salt layer in Tersanne, and two ones in Etrez, whose cavities are located in the lower layer (Figures 2a and 2b).

Both sites have developped over a period of some ten years since leaching of the first cavity to the last one. The Tersanne field has extended in a definite direction, while the Etrez site, which was created some ten years latern, extended more regularly, almost according to the theoretical hexagonal pattern (Figures 3 & 4).

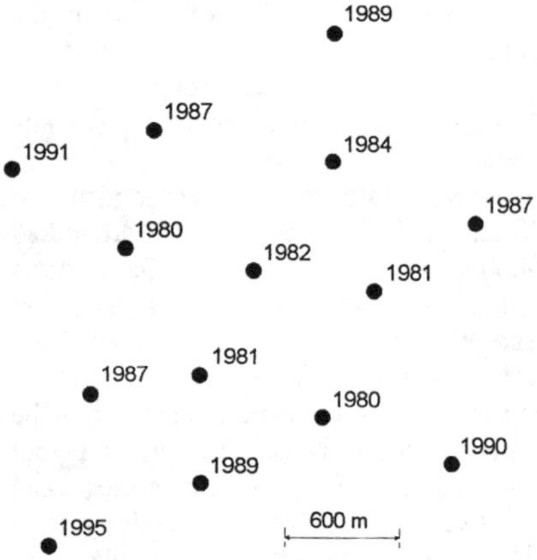

Figure 4- Development of Etrez site

Cavities are also variable in size in each site, the mean cavity volume in Tersane is about 200 000 m^3, while this volume is 300 000 m^3 in Etrez. Conversely, Tersanne cavities are located deeper than in Etrez. Mean depth is about 1550 m in the first site, and 1450 m for the second one.

Rocksalt rheology is also different for both sites, the Tersanne salt is more creeping than Etrez salt.

These features lead to a different subsidence evolution between the two sites. Subsidence in Etrez is about ten times smaller than in Tersanne for analogous dates.

3 NUMERICAL MODEL DESCRIPTION

3.1 The hybrid BEM-FEM model

For this particular problem, with an unbounded field, the application of a hybrid BEM-FEM method appears to be quite appropriate. Boundary elements need only a discretisation of the border of the elastic domain, while finite elements deal with the non linear viscous behavior in rocksalt layer. Moreover, infinite BE elements and infinite elements in each domain to allow for infinite conditions (Bettess, 1992).

The boundary element region is treated as a larger finite element. For the Finite Element region the finite element matrices are $\boldsymbol{KU=F}$, where $\boldsymbol{K}$ is the stiffness matrix and $\boldsymbol{F}$ the equivalent nodal forces. The boundary elements matrices are different: $\boldsymbol{HU=GP}$, where $\boldsymbol{U}$ represents the displacements and $\boldsymbol{P}$ the surface tractions. To combine the two equations one must reduce the BE expression to a finite element form: $\boldsymbol{G^{-1}HU=P}$. Then the tractions values must be converted by means of the distribution matrix $\boldsymbol{M}$, i.e., $\boldsymbol{F=MP}$. Finally the BE equation will be $\boldsymbol{M(G^{-1}H)U=MP=F}$ or $\boldsymbol{K'U=F}$. The two matrices in BE and FE regions can now be assembled to form the global stiffness matrix (Brebbia, 1984; Pande, 1990).

A special procedure is also used to speed up numerical calculations (Menezes & Nguyen-Minh, 1993).

3.2 Constitutive laws for the rockmass model

Two alternative models for the rocksalt layer are used herein : the Norton-Hoff (NH), and Lemaitre-Menzel-Schreiner (LMS) laws. Gaz De France uses L-M-S law, which supposes a continuously strain rate decreasing creep, which has proved to be suitable for middle term observation time periods (Vouille et al, 1983). The parametres of this law are identified on cavity closure observations, and will be considered herein as saltrock current time reference parameters. N-H law is a steady state creep (Langer et al, 1984), its parameters will be determined here so that a cavity at 1500 m depth, and with 16.5 MPa constant internal pressure (equivalent to the actual cyclic pressure variation) will suffer a same volumetric loss than the cavity with L-M-S model after ten years. This evaluation of the model will be used for pessimistic long term prediction.

If we consider a cavity sustaining stationary creep in small strain formulation (steady stress state), then its volumetric rate dV/Vdt will express in a same formulation for both laws, as follows :

$$\frac{\mathbf{dV}}{\mathbf{V\,dt}} = -\mathbf{10^{-6}}\alpha\left(\frac{\mathbf{P_{\infty}-P_i}}{\mathbf{A}}\right)^{\beta}\mathbf{t^{\alpha-1}}$$

$$\mathbf{A} = \left(\frac{\mathbf{2}}{\mathbf{3}}\right)^{\frac{\beta+1}{\beta}}\mathbf{K\beta} \qquad (3.1)$$

where $\alpha=1$ pour N-H, and $\alpha<1$ for L-M-S.

The parameters are the following :

- Tersanne (geothermal temperature 70 °C)

$\alpha = 0.5$ $\beta = 3.6$ $K = 0.85\,MPa$ L-M-S (given by GDF)

$\alpha = 1.0$ $\beta = 3.0$ $K = 2.60\,MPa$ N-H (equivalent to L-M-S)

- Etrez upper salt layer, (50 °C geothermal temperature)

$\alpha = 0.36$ $\beta = 2.98$ $K = 1.1\,MPa$ L-M-S (given by GDF)

$\alpha = 1.0$ $\beta = 3.00$ $K = 6.34\,MPa$ N-H (Equivalent to L-M-S)

- Etrez lower layer (55°C geothermal temperature)

$\alpha = 0.44$ $\beta = 3.93$ $K = 1.95\,MPa$ L-M-S (given by GDF)

The rocksalt elastic parameters are :

$\nu = 0.25$ $E = 25000\,MPa$

$\nu = 0.25$ $E = 25000\,MPa$

The other sedimentary layers are supposed to be linear elastic. Their small influence on the subsidence phenomenon is secondary, as shown by parametric analyses, so standard elastic parameters are choosen for all the sedimentary layers ::

$\nu = 0.4$ $E = 6000\,MPa$

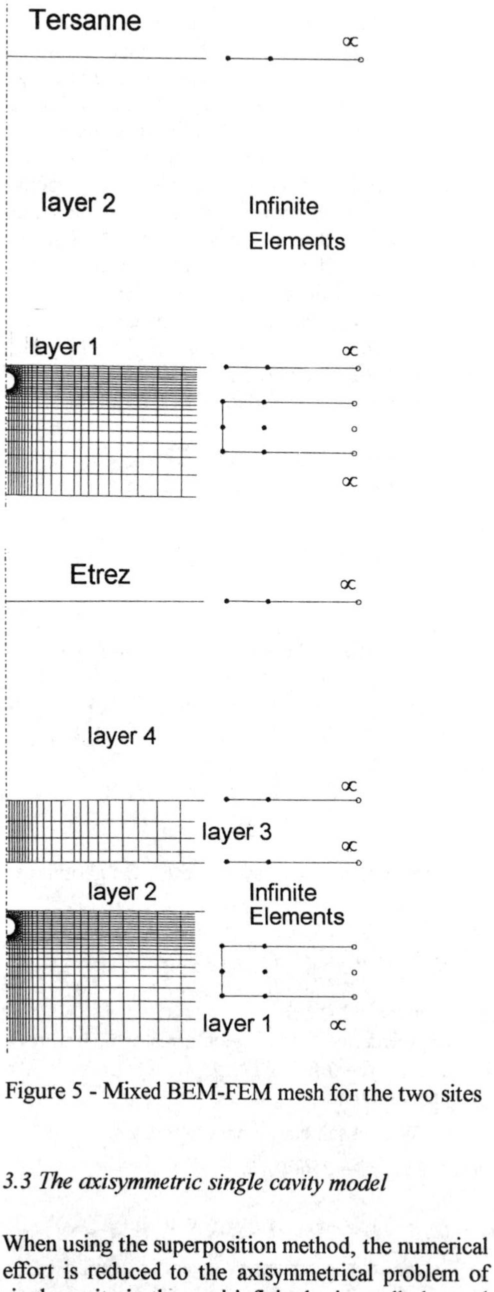

Figure 5 - Mixed BEM-FEM mesh for the two sites

3.3 The axisymmetric single cavity model

When using the superposition method, the numerical effort is reduced to the axisymmetrical problem of single cavity in the semi-infinite horizontally layered rockmass.

For each site, cavity has a typical 40 m radius and is 1500 m deep. The rockmass modellisation takes account of the geological differences of each site (Figure 5). The extension radius of the dicretized domain, from the symmetry axis , is about 3 000 m at the salt layer and

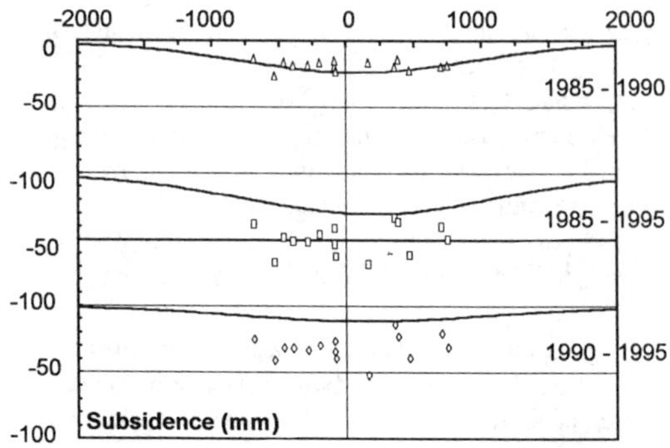

Figure 6 - Comparison of numerical predictions (solid lines) with Tersanne subsidence data (points).

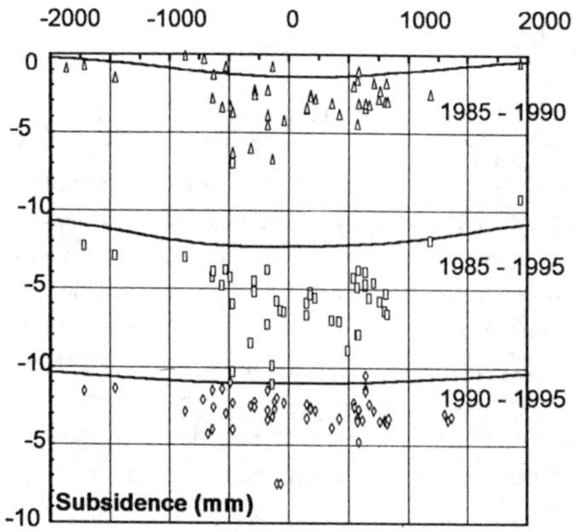

Figure 7 - Comparison of numerical predictions (solid lines) with Etrez subsidence data (points).

increases up to 6 000 m at the surface, to obtain a good description of the surface subsidence. The mixed mesh thus develops to a larger distance from the rotational symmetry axis when arriving up to the surface. A fine mesh of the large surface boundary by BE does not much empede the memory size necessary for the calculations.

4 COMPARISON WITH IN SITU DATA

In former studies, the comparison with Tersanne data was based on the subsidence rate and satisfactory results were obtained. Comparison here is concerned with differential subsidence for given periods of time, say for 10 years intervalls from1985 to 1995. The numerical simulations presented herein take advantage of the superposition method, by taking into account the operation date of each cavity, its location, and its volume.

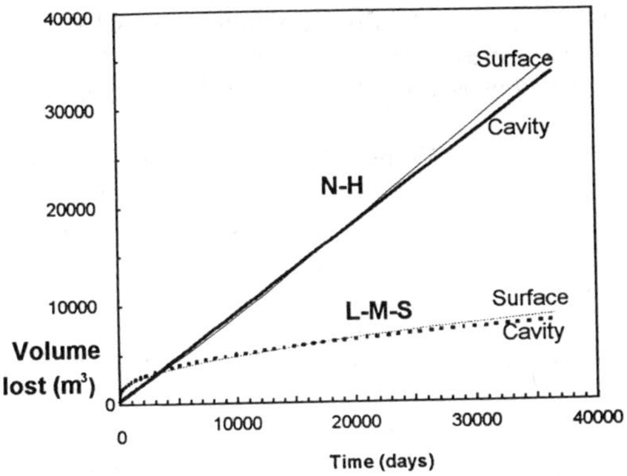

Figure 8 - Cavity volume losses (Etrez).

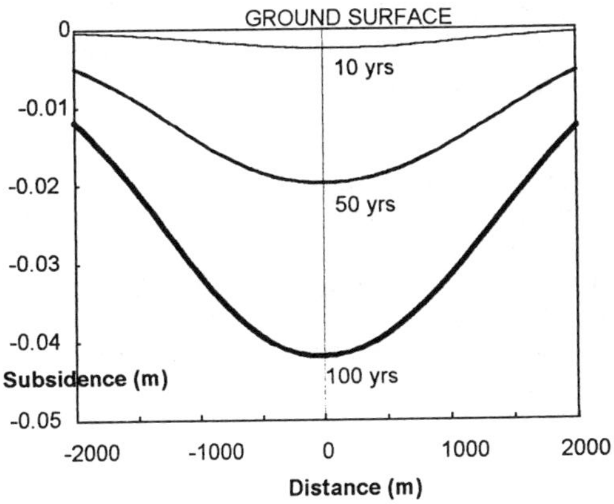

Figure 9 - Subsidence prediction for Etrez (N-H law)

A fairly satisfactory interpretation of Tersanne subsidence is obtained with L-M-S model, and still better agreement is obtained with the N-H model.

This confirms the previous interpretations based on the subsidence rate (Nguyen Minh et al, 1993).

Contrarily, although using the same procedure, it can be seen that there is no good agreement for Etrez. Indeed, the subsidence data are very low here, and are of the order of the accuracy of measurements. So we can say that theoretical predictions are correct as they reveal a small subsidence (of the order of 5 mm in ten years).

5 LONG TERM SUBSIDENCE PREDICTION

Evolution of cavity volume loss, and subsequent subsidence volume loss is shown for the Etrez case, for the two alternative constitutive laws for rocksalt.

It can be seen that :

- cavity volume loss and subsidence volume loss are practically identical.
- Volume loss evolution law is practically linear for N-H law.

These results are due to the small loading level in Etrez, and indicates that the structure quickly turns to a steady state behavior, which was not the case in Tersanne, where the ratio between subsidence volume loss and cavity volume loss was 60% (at least in the transient regime).

For superposition, the long term prediction does not need taking into account the different operation time for each cavity. Indeed, all the cavities are supposed to be created instantly, with the same initial volume, and are arranged according to the theoretical hexagonal pattern.

Figure 8 shows the subsidence prediction for Etrez, which remains limited (maximum 4.5 cm after 100 years), and is about one tenth of predicted subsidence for Tersanne (42 cm).

If the apparent under-evaluation of subsidence for the current observation time in Etrez was effective, and not due to measurement uncertainties, (see Figure 7) then, N-H parameters could be easily readjusted by changing parameter K in (3.1), in order to increase by two or three times the maximum predicted subsidence. Consequently, the maximum predicted subsidence for long term will be increased by the same amount, as subsidence vs time response with N-H law is quasi linear. But the final subsidenec prediction will remain anyway limited.

Horizontal displacements vs radial distance r to the centre of the storage field are also of interest. They allow to determine the border of the compressive horizontal stress zone in the central part of subsidence through, near from the extremum of the horizontal displacement vs radius r (Figure 10). That compressive zone in the central part of the subsidence through could be less pervious than the surrounding tensile zone, and contribute, with the depth of subsidence to retain surface water flow.

Moreover, that radius which is about 1500 m for Etrez, is also near from the zone where the maximal horizontal subsidence gradient arises, i.e., in the range 1000 m-1500 m (compare Figures 9 and 10). This maximum gradient is also an important parameter to check for surface structures.

Finally, for Etrez, that point is practically stationary, and is situated at a radius of 1500 m, while the maximal horizontal gradient is limited to 2.10^{-5} at a radius 1300 m. Similar results are obtained for Tersanne, but maximum gradient and maximum subsidence are about ten times higher.

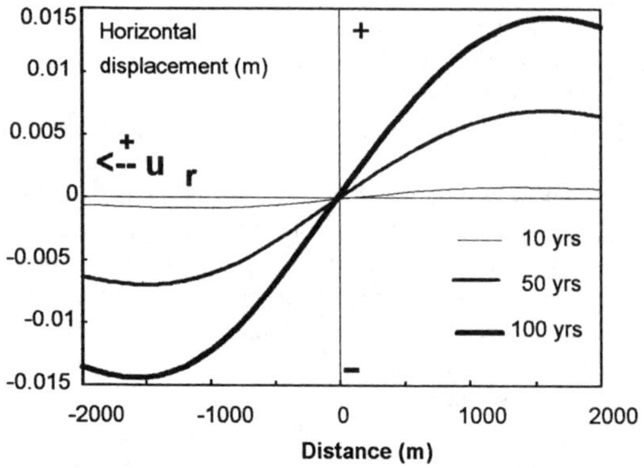

Figure 10 - Horizontal displacement prediction for Etrez (N-H law)

For both sites, subsidence through is about 3000 m radius.

6 CONCLUSION

Superposition method allows for an efficient simulation of the subsidence, with taking into account of the parameters of the specific site under study. The mixed BEM-FEM calculation allows for an accurate determination of the far-field displacements, i.e. subsidence.

The comparative analysis of the two sites of Tersanne and Etrez allows to confirm that the rocksalt properties do control the subsidence phenomenon. For long term prediction, we have proposed to bracket the subsidence between two different laws the N-H one, and L-M-S, which equivalence is based on the identical loss of volume of a single cavity 1500 m deep, after 10 years.

Finally, the predicted subsidence for 100 years, show that Tersanne site is creeping about ten times the one in Etrez, but it remains limited, and does not endager the surface building nor hydrogeology.

REFERENCES

Bettess, P., 1992. *Infinite Elements*. Penshaw Press.Brebbia, C.A., Telles, J.C.F., and Wrobel, L.C., 1984. *Boundary Element Techniques*. Springer-Verlag.

Durup, J.G., 1990. Surface subsidence measurements on Tersanne cavern field. *Solution Mining Research Inst.*, Paris, October 14-19.

Durup, J.G., 1991. Relationship between subsidence and cavern convergence at Tersanne (France). *Solution Mining Research Inst.*, Atlanta, April 28-30.

Langer, M., Wallner, M., and Wassmann Th.H., 1984. Gebirgsmechanische Bearbeitung von Stabilitatsfragen bei Deponiekavernen. *Kali u. Steinsalz* 9, 2, pp.66-76, Essen.

Menezes, J.E.Q., and Nguyen-Minh, D., 1993. Incompressible numerical modelling for long term convergence evaluation of underground works in salt. *3rd Conf. Mechanical Behavior of Salt*, Palaiseau, France.

Nguyen-Minh, D.; Braham, S., and Durup, J.G., 1993. Modelling Subsidence of the Tersanne underground gas storage field. *3rd Conf. Mechanical Behavior of Salt,* Palaiseau, France.

Nguyen-Minh, D.; Maitournam, H.; Braham, S., and Durup, J.G., 1993. A tentative interpretation of long-term surface subsidence measurements over a solution mined cavern field. *7th Symposium on Salt*, Vol.1 441-449, Elsevier.

Pande, G.N., Beer, G., and Williams, J.R., 1990. *Numerical Methods in Rock Mechanics*. J.Wiley & Sons.

Vouille, G., S.M.Tijani, S.M., and Hugout, B., 1983. Le sel gemme en tant que liquide visqueux. Proceedings *5th Int.Cong.Rock Mechanics*, pp.D241-D246, Melbourne, Australia.

Computer Methods and Advances in Geomechanics, Yuan (ed.) © 1997 Balkema, Rotterdam, ISBN 90 5410 904 1

BEM computation of bolt supporting structures*

Yan-Chang Wang & De-Cheng Zhang
Ningxia University, Yinchuan, China

ABSTRACT: In this paper, a new computational method of BEM for the interaction between frictional bolts and the rock mass is developed by formulationg the surface frictional forces of the bolts in displacements of the rock. The corresponding formulation for the underground openings is derived. The computer code for the approach has been compiled and the results computed demonstrate the validity of the method.

1 INTRODUCTION

Owing to the advantages that the discretization of elements is carried out only in the boundary of the body under consideration and , thus, the less input data are required, The boundary element method (BEM) is playing an important part in the analysis of the rock mechanics and the rock stability. However, there are still many difficulties in the applications to the multi-medium problem and the underground supporting structure for BEM. In view of the fact that the bolt supporting technology is extensively being appied to underground excavation engineering and rock stabilization engineering, it has become an important research subject how to effectivelly consider the behaviors of the bolt. In this paper, a new Computation method of BEM for the interaction between the frictional bolts and the rock mass is developed by formulating the surface frictional forces of the bolts in displacements of the rock. The BEM formulation for the underground excavation involving frictional bolts is derived. Based on the formulation, a computer code has been compiled and the illustrative results for a circle cave demonstrate the potential of the method.

2 THE BEM FORMULATION FOR THE FRICTIONAL BOLTS.

Let us consider a small bolt element $\triangle r$ subjected to an axial surface frictional force and denote the shear stress of unit bolt's length by fr , the cross section area of the bolt by A and the axial tension stress by σ_r (Fing. 1). In terms of the equilibrium condition of the small element, we have.

$$\triangle r fr + A(\sigma_r + \triangle\sigma_r) = A\sigma_r$$

When the $\triangle r$ is enough small, we can derive form above equation that

$$fr = -A\frac{\partial\sigma_r}{\partial r} \quad (1)$$

Further, the equation(1) can be expressed as

$$fr = -AE\frac{\partial^2 u_r}{\partial r^2} \quad (2)$$

where, E is the young's modulus of the bolts and u_r is the axial aisplacment.

The BEM formulation can be expressed, in the absence of volume force, as

$$cu + \int_\Gamma p^* u d\Gamma = \int_\Gamma u^* p d\Gamma \quad (3)$$

in which, u^* and p^* are the Kelvin's fundamental solutions of the displacment and traction. The u and p are displacment and traction vectors. Γ is the boundary of the rock mass under consideration. Let's divided the Γ into two part, Γ_1 and Γ_2 , with Γ_1 being the ordinary boundary and Γ_2 being the common boundary of the rock mass and

$$fr = -A\frac{\triangle\sigma_r}{\triangle r}$$

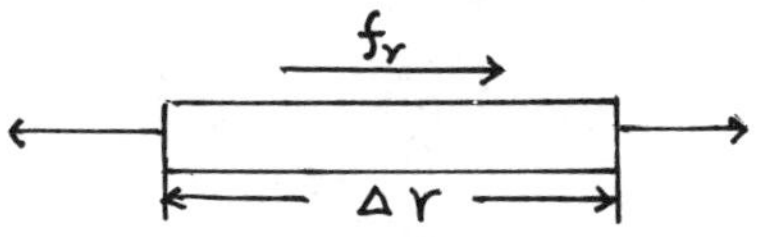

Fig. 1.

* The subjets is financially aided by the National Natural science Funds of P.R.China

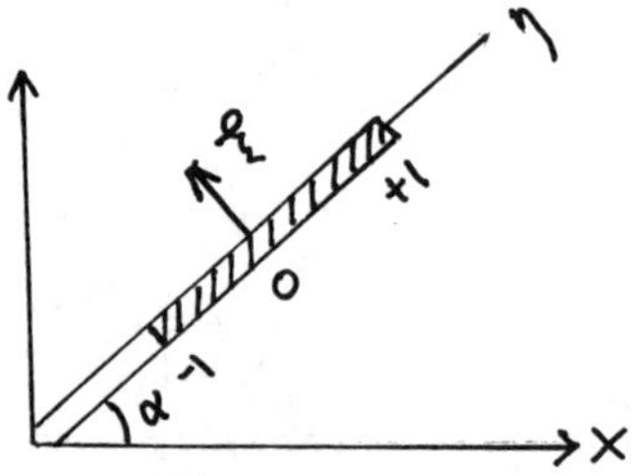

Fig.2.

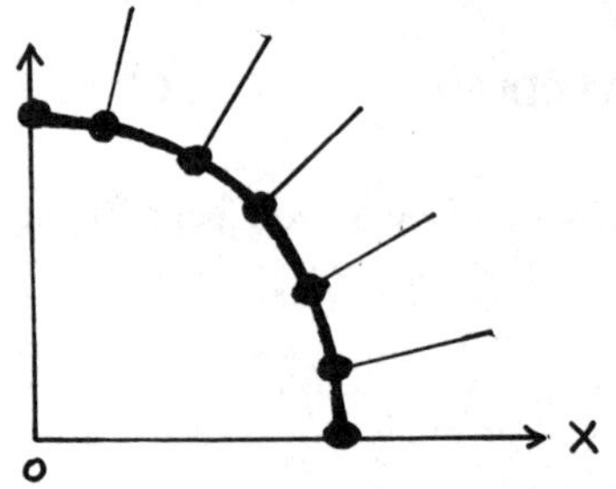

Fig.3.

bolts. Thus, the equation (3) becomes

$$cu + \int_{\Gamma} p^* u d\Gamma = \int_{\Gamma_1} u^* p d\Gamma + \int_{\Gamma_2} u^* p' d\Gamma \qquad (4)$$

in the above formulation, the left integral and the righ first integral can be calculated with ordinary BEM procedure. The right second integral can be calculated by the method described in the following.

Let's consider an typical three node element from Γ_2 (Fig.2.)

The interpolation functions are.

$$\varphi_1 = -\frac{1}{2}(1-\eta)\eta \quad \varphi_2 = (1-\eta)(1+\eta) \quad \varphi_1 = \frac{1}{2}(1+\eta)\eta \quad (5)$$

The shear force of bolt can be expressed by x- and y- direction forces;

$$p_1' = cos\alpha fr = -EA\frac{\partial^2 u_x}{\partial r^2}$$

$$p_2' = sin\alpha fr = -EA\frac{\partial^2 u_y}{\partial r^2} \qquad (6)$$

in which displacments u_x and u_y can be calculated through interpolation

$$u_x = \varphi_1 u_{1x} + \varphi_2 u_{2x} + \varphi_3 u_{3x}$$

$$u_y = \varphi_1 u_{1y} + \varphi_2 u_{2y} + \varphi_3 u_{3y} \qquad (7)$$

where u_1, u_2, u_3 are the nodal displacments of the bolt element. The integral of the jth element over Γ_2 can be calculated by

$$\int_{\Gamma_j} u_{kl}^* p_l' d\Gamma \qquad k = 1,2; \quad l = 1,2 \qquad (8)$$

where u_{kl}^* is the Kelvin's displacment solutions of the 2 - D elastic problem.

$$u_{kl}^* = \frac{1}{8\pi G(1-\nu)}[(3-4\nu)ln(\frac{1}{r})\delta_{kl} + r,_k r,_l] \qquad (9)$$

taking $k = 1$

$$\int_{\Gamma_j} u_{kl}^* p_l' d\Gamma = \int_{\Gamma_j} u_{11}^* p_1' d\Gamma + \int_{\Gamma_j} u_{12}^* p_2' d\Gamma$$

$$= -EA\{\int_{-1}^{+1}[u_{11}^*\frac{\partial^2\varphi_1}{\partial\eta^2}, u_{12}^*\frac{\partial^2\varphi_1}{\partial\eta^2}, u_{11}^*\frac{\partial^2\varphi_2}{\partial\eta^2}, u_{12}^*\frac{\partial^2\varphi_2}{\partial\eta^2}, u_{11}^*\frac{\partial^2\varphi_3}{\partial\eta^2}, u_{12}^*\frac{\partial^2\varphi_3}{\partial\eta^2}]\frac{1}{|J|}d\eta\}$$

$$\times [u_{1x}, u_{1y}, u_{2x}, u_{2y}, u_{3x}, u_{3y}]^T \qquad (10)$$

and taking $k = 2$

$$\int_{\Gamma_j} u_{kl}^* p_l' d\Gamma = \int_{\Gamma_j} u_{21}^* p_1' d\Gamma + \int_{\Gamma_j} u_{22}^* p_2' d\Gamma$$

$$= -EA\{\int_{-1}^{+1}[u_{21}^*\frac{\partial^2\varphi_1}{\partial\eta^2}, u_{22}^*\frac{\partial^2\varphi_1}{\partial\eta^2}, u_{21}^*\frac{\partial^2\varphi_2}{\partial\eta^2}, u_{22}^*\frac{\partial^2\varphi_2}{\partial\eta^2}, u_{21}^*\frac{\partial^2\varphi_3}{\partial\eta^2}, u_{22}^*\frac{\partial^2\varphi_3}{\partial\eta^2}]\frac{1}{|J|}d\eta\}$$

$$\times [u_{1x}, u_{1y}, u_{2x}, u_{2y}, u_{3x}, u_{3y}]^T \qquad (11)$$

From equation (5), we have

$$\frac{\partial^2\varphi_1}{\partial\eta^2} = 1 \quad \frac{\partial^2\varphi_2}{\partial\eta^2} = -2 \quad \frac{\partial^2\varphi_3}{\partial\eta^2} = 1 \quad \frac{1}{|J|} = \frac{2}{lj} \qquad (12)$$

$$[g]_j = \frac{2}{lj}\begin{bmatrix} \int_{-1}^{+1}u_{11}^* d\eta, \int_{-1}^{+1}u_{12}^* d\eta, \int_{-1}^{+1}-2u_{11}^* d\eta, \int_{-1}^{+1}-2u_{12}^* d\eta, \int_{-1}^{+1}u_{11}^* d\eta, \int_{-1}^{+1}u_{12}^* d\eta \\ \int_{-1}^{+1}u_{21}^* d\eta, \int_{-1}^{+1}u_{22}^* d\eta, \int_{-1}^{+1}-2u_{21}^* d\eta, \int_{-1}^{+1}-2u_{22}^* d\eta, \int_{-1}^{+1}u_{21}^* d\eta, \int_{-1}^{+1}u_{22}^* d\eta \end{bmatrix} \qquad (13)$$

$$[u]_j = [u_{1x}, u_{1y}, u_{2x}, u_{2y}, u_{3x}, u_{3y}]^T \qquad (14)$$

Hence:

$$\int_{\Gamma_j} u^* p' d\Gamma = [g]_j [u]_j^T \qquad (15)$$

Useing the above equations, the surface forces of the bolts may be expressed by displacment of the rock, which can be calculated by the routine BEM, and so do stress of the rock mass

3 EXAMPLE

We take a circle underground cavity as an computational example. The radius of the cavity is 10 metres. There is a radial pressure p = 15MPa action on the interior surface. The young's modulus of the rock is 21000 MPa, and the Posson coefficient ν is 0.3. The young's modulus of the bolt is 200000 MPa, and the cross-section area is 3×10^{-4} square metre. Considering symmetry, Fig.3 shows schematically the conputational model.

The interior surface of the cavity was discretized into 6 linear boundary elements with 7 nodes. For the purpos of comparison, the displacment radius is calculated in the two case. In the first case, there are not bolts involved in the cavity and the computation was carried out by routine elastic BEM. The computed radial displacment is 7.738×10^{-3} metres. In the second case, there are 7 bolts exerte in the cavity and the radial displacmement computed by the method introduced in this paper is 7.621×10^{-3} metres. These results are in the agreement with those in reference [3], in where the FEM was used and the cavity is divided into about 30 elements. The discretization of boundary is easy and the input date is small, the compulational result demonstrate that the method developped in this paper has the distinct advantage over the FEM.

REFRENCES

Brebbia, C. A. and Tells, J. C. F. (1984), BoundaryElement Techniques: Theory and Applications in engineering, Springer-Verlag, Rio de Janeiro, 229 – 303.

Brebbia, C. A. and Dominguez, J. C. (1988), Boundary Elements, An Introductory course, McGraw – Hill, New York, 327 – 408.

Xiao wei, Gao. (1993), "Practibale Computation of Elasicity FEM for Frictional Bolts", Journal of Ningxia Institute of Technology, Vol. 5(4), 64 – 69.

Computer Methods and Advances in Geomechanics, Yuan (ed.) © 1997 Balkema, Rotterdam, ISBN 90 5410 904 1

Spline infinite element-QR methods for analysis of underground engineering

Rong Qin
Department of Civil Engineering, GuangXi University, Nanning, China

ABSTRACT: In this paper, the spline infinite element-QR methods for analysis of interactions of underground structures and soil/rock medium is presented. It has both the advantages of spline infinite element method and the features of spline finite element method. Therefore, this method has great advantages over the conventional boundary element method and finite element method. It can easily be carried out on a micro-computer. This is the project supported by natural science foundation of GuangXi and nation of China.

1 INTRODUCTION

The applications of underground structures are very wide, they play an important part in the engineerings. The analysis of interactions of underground structures and rock (or soil) medium is an important problem, but the accurate analysis of this problem is very difficult. Therefore, the interaction problems are solved by the numerical methods . If the finite element method is used for the analysis of interaction problems of underground structures and soil/rock medium, then the number of degree of freedom are so large that a large computer is needed and the cost of computation is high. In this paper, a new computational method for analysis of interaction problems of underground structures and soil/rock medium is presented. The principle of this method as follows: (1) Establish the control equations of the underground structures and its medium in near field by QR — method. (2) Establish the control equations of infinite soil/rock medium by spline infinite element methods. (3) Establish the control equations of the coupling problems of underground structares and soil/rock medium. (4) Determine the displacements and the stresses of couplig system. This method is called spline infinite element-QR method (SIE-QRM). It has both the advatages of spline infinite element method and the features of spline finite element method . The main features of this method are its higher accuracy, smallest amount of data input and more economy in computer storage and time requirements. This method can easily be carried out on a micro-computer. Therefore, this method has great advantages over the conventional boundary element method and finite element method. It is a powerfal and economical method.

This paper aims to introduce briefly the fundamentals and the applications of the SIE-QRM for the analysis of the interaction problems of underground structures and soil/rock medium (Figure 1).

2 QR—METHOD

2.1 *Displacement functions*

Figure 1 is an underground structure in semi-infinite soil/rock medium. If we take it as a plane strain problem, then the displacement functions of underground structure and soil/rock medium in near field may be expressed by following forms

$$u_i = \sum_{m=1}^{r} u_{im} X_{im}(y)$$

$$v_i = \sum_{m=1}^{r} v_{im} Y_{im}(y) \qquad (1)$$

where u_i and v_i are the displacement functions on $x = x_i$, X_{im} and Y_{im} are orthogonal functions/orthogonal polynomials. For example, if underground structure and its medium all around is symmetrical, then

$$X_{im} = \sum_{n=1}^{m} (-1)^{n-1} \frac{n(m+n)!}{(m-n)!(n-1)!(n+1)!} \left(\frac{y}{b}\right)^{n-1}$$

$$Y_{im} = \sum_{n=1}^{m} (-1)^{n-1} \frac{(m+n)!}{(m-n)!(n-1)!(n+1)!} \left(\frac{y}{b}\right)^{n} \qquad (2)$$

From eq. (1) may obtain

$$\{V\}_i = [N]_i \{\delta\} \qquad (3)$$

where $\{V\}_i = [u_i \quad v_i]^T$

$$[N]_i = \mathrm{diag}(0, \cdots, [N^*]_i, 0, \cdots) \qquad (4)$$

$$\{\delta\} = [\{\delta\}_0^T \quad \{\delta\}_1^T \quad \cdots \quad \{\delta\}_N^T]^T$$

$$[N^*]_i = [[N]_1 \; [N]_2 \cdots [N]_r] \qquad (5)$$

$$\{\delta\}_i = [\{\delta\}_1^T \quad \{\delta\}_2^T \quad \cdots \quad \{\delta\}_r^T]^T$$

$$[N]_m = \mathrm{diag}(X_{im}, Y_{im}) \quad \{\delta\}_m = [u_{im} \quad v_{im}]^T$$

2.2 *Basic equation*

If we make the underground structure and its medium in near field into rectangular mesh or triangular mesh, as shown in figure 2, then the functional of total potential energy of the element is

$$\Pi_e = \frac{1}{2}\{V\}_e^T [k]_e \{V\}_e - \{V\}_e^T \{f\}_e \qquad (6)$$

where $[k]_e$ and $\{f\}_e$ are respectively the stiffness matrix and the load vector of element, $\{V\}_e$ is the nodal displacement vector of the element, or

$$\{V\}_e = [N]_e \{\delta\} \qquad (7)$$

where

$$[N]_e = [[N]_A^T \quad [N]_B^T \quad [N]_C^T \quad [N]_D^T]^T \qquad (8)$$

for the rectangular element with 4-nodes.

$$[N]_e = [[N]_A^T \quad [N]_B^T \quad [N]_C^T]^T \qquad (9)$$

for the triangular element with 3-nodes.
we substituting eq. (7) into eq. (6) may obtaim

$$\Pi_e = \frac{1}{2}\{\delta\}^T [G]_e \{\delta\} - \{\delta\}^T \{F\}_e \qquad (10)$$

where

$$[G]_e = [N]_e^T [k]_e [N]_e \quad \{F\}_e = [N]_e^T \{f\}_e \qquad (11)$$

Since the functional of total potential energy of underground structure and its medium in near field is

$$\Pi = \sum_{e=1}^{M} \Pi_e \qquad (12)$$

We substituting eq. (10) into eq. (12) may obtain

$$\Pi = \frac{1}{2}\{\delta\}^T [G] \{\delta\} - \{\delta\}^T \{f\} \qquad (13)$$

where

$$[G] = \sum_{e=1}^{M} [G]_e \qquad \{f\} = \sum_{e=1}^{M} \{F\}_e \qquad (14)$$

Using the variation principle, we obtain

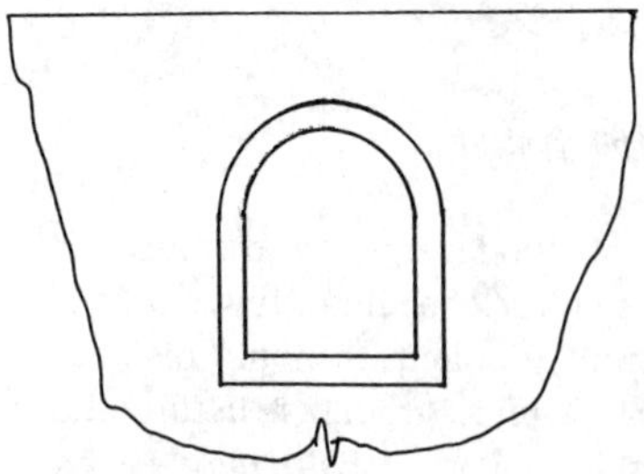

Figure 1. Underground structure

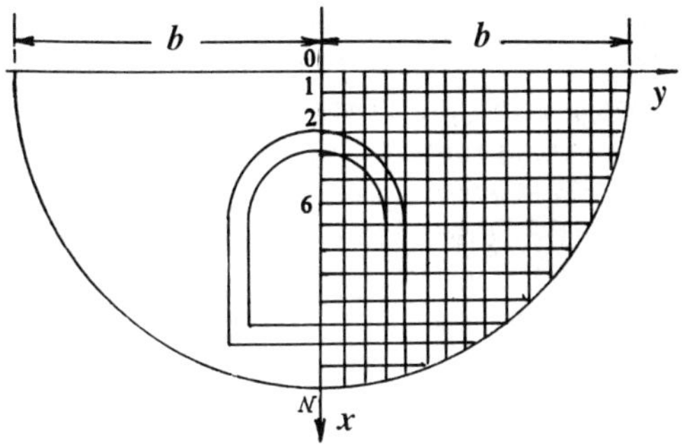

Figure 2. Plane mesh

$$[G]\{\delta\} = \{F\} \tag{15}$$

This is the stiffness equation of underground structures and its medium in near field.

From the above mentioned we may know that eq. (15) is built up from the elements of the underground structure and the medium, nevertheless its number of unknowns has nothing to do with the number of nodes of mesh of structure and medium, it is only relevant to the collocation points and r (number of items). Thus , the number of unknown of eq. (15) is very small, Eq. (14) can directly superpose, which no extension. The above method is called QR-method (Qin Rong, 1995).

3 SPLINE INFINITE ELEMENT METHODS

3. 1 *Displacement functions*

For soil/ rock medium in for field (Figure 3), its displacement functions may be expressed by following forms:

$$u_r = \sum_{j=0}^{P} u_{rj}\Phi_j(\theta)f(r)$$
$$u_\theta = \sum_{j=0}^{P} u_{rj}\Phi_j(\theta)f(r) \tag{16}$$

where u_r and u_θ are respectively radial and tangential displacements, $\Phi_j(\theta)$ and $f(r)$ are respectively basis function and attenuate function ,or

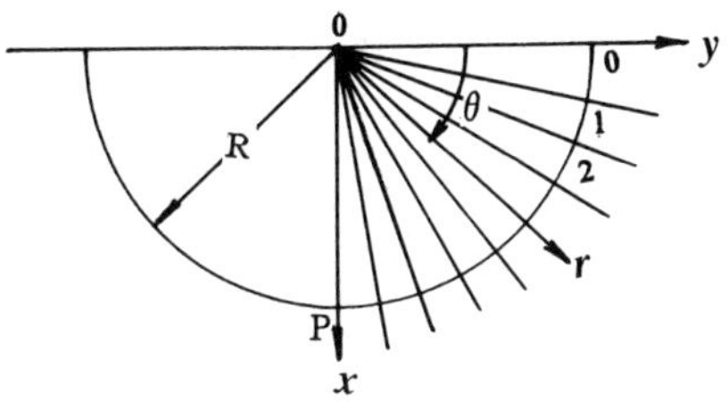

Figure 3. Spline partition of medium far field.

$$\begin{aligned}\Phi_j(\theta) = {} & \frac{10}{3}\varphi_3(\frac{\theta}{h} - j) \\ & - \frac{4}{3}\varphi_3(\frac{\theta}{h} - j + \frac{1}{2}) \\ & - \frac{4}{3}\varphi_3(\frac{\theta}{h} - j - \frac{1}{2}) \\ & + \frac{1}{6}\varphi_3(\frac{\theta}{h} - j + 1) \\ & + \frac{1}{6}\varphi_3(\frac{\theta}{h} - j - 1) \\ & j = 0,1,2\cdots,P \end{aligned} \tag{17}$$

$$f(r) = (\frac{r}{R})^{-n} \quad r \geqq R \tag{18}$$

in which $\varphi_3(\theta)$ is cubic B-spline function (Qin Rong ,1985).

From eq. (16) may obtain

$$\{V\} = [N]\{a\} \tag{19}$$

where

$$\{V\} = [u_r \quad u_\theta]^T$$

$$[N] = [[N]_0 \quad [N]_1 \quad \cdots \quad [N]_P] \tag{20}$$

$$\{a\} = [\{a\}_0^T \quad \{a\}_1^T \quad \cdots \quad \{a\}_P^T]^T$$

in which

$$[N]_j = \mathrm{diag}(\Phi_j, \Phi_j)f(r) \tag{21}$$

$$\{a\}_j = [u_{rj} \quad u_{\theta j}]^T$$

3. 2 *Basic equations*

If the medium in far field is symmetrical, then its functional of total potential energy is

$$\Pi = \frac{1}{2}\int_R^{\infty} 2\int_0^{\frac{\pi}{2}} (\varepsilon^T\sigma - 2\{V\}^T q) r d\theta dr$$
$$- 2(\int_R^{\infty} \{V\}^T p dr)_{\theta=0}$$
$$- 2(\int_0^{\frac{\pi}{2}} \{V\}^T p r d\theta)_{r=R} \qquad (22)$$

where

$$\varepsilon = [\varepsilon_r \ \varepsilon_\theta \ \varepsilon_{r\theta}]^T \qquad \sigma = [\sigma_r \ \sigma_\theta \ \sigma_{r\theta}]^T$$

$$q = [q_r \ q_\theta]^T \qquad p = [p_r \ p_\theta]^T$$

in which

$$\varepsilon_r = \frac{\partial u_r}{\partial r} \qquad \varepsilon_\theta = \frac{u_r}{r} + \frac{1}{r}\frac{\partial u_\theta}{\partial \theta}$$
$$\varepsilon_{\theta\partial} = \frac{1}{r}\frac{\partial u}{\partial \theta} + \frac{\partial u_\theta}{\partial r} - \frac{u_\theta}{r} \qquad (23)$$

We substituting eq. (16) into eq. (23) may obtain

$$\varepsilon = [A]\{a\} \qquad \sigma = [D]\varepsilon \qquad (24)$$

$$[A] = [[A]_0 \ [A]_1 \ \cdots \ [A]_p] \qquad (25)$$

$$[A]_j = \begin{bmatrix} \Phi_j(\theta) f'(r) & 0 \\ \frac{1}{r}\Phi_j(\theta) f(r) & \Phi_j'(\theta) f(r) \\ \frac{1}{r}\Phi_j'(\theta) f(r) & \Phi_j(\theta)[f'(r) - f(r)] \end{bmatrix} \qquad (26)$$

in which $\Phi_j'(\theta)$ and $f'(r)$ are respectively the derivatives of $\Phi_j(\theta)$ and $f(r)$.

We substituting eq. (24) into eq. (22) may obtain

$$\Pi = \frac{1}{2}\{a\}^T[G]_s\{a\} - \{a\}^T\{F\}_s \qquad (27)$$

where

$$[G]_s = 2\int_R^{\infty}\int_0^{\frac{\pi}{2}} [A]^T[D][A] r d\theta dr \qquad (28)$$

$$\{F\}_s = \{F_1\}_s + \{F_2\}_s \qquad (29)$$

$$\{F_1\}_s = 2\int_R^{\infty}\int_0^{\frac{\pi}{2}} [N]^T q r d\theta dr$$
$$+ 2(\int_R^{\infty} [N]^T p dr)_{\theta=0} \qquad (30)$$

$$\{F_2\}_s = 2(\int_0^{\frac{\pi}{2}} [N]^T p r d\theta)_{r=R} \qquad (31)$$

in which q and p are respectively the body force and face force, [D] is the elastic matrix of the plane strain problems of soil/ rock medium. Using the variation principle, we obtain

$$[G]_s\{a\} = \{F\}_s \qquad (32)$$

This is the stiffness equation of soil/ rock medium in far field. The above method is called spline infinite element methods (Qin Rong, 1995).

4 SIE-QRM

4.1 *Consistent conditions*

From the consistent conditions on r = R may obtain

$$\{a\} = [H]\{\delta\} \qquad (33)$$

where

$$\{a\}_j = [T]_j[N]_j\{\delta\} \qquad (34)$$

$$[T] = \begin{bmatrix} \sin\theta & \cos\theta \\ \cos\theta & -\sin\theta \end{bmatrix} \qquad (35)$$

in which

$$[T]^{-1} = [T]^T = [T] \qquad (36)$$

$$[N]_j = [N(y_j)] \qquad (37)$$

Eq. (37) is determined by eq. (4). we substituting eq. (33)into eq. (27) may obtain

$$\Pi_1 = \frac{1}{2}\{\delta\}^T[G^*]\{\delta\} - \{\delta\}^T(\{F_1^*\} + \{F_2^*\}) \qquad (38)$$

where

$$[G^*] = [H]^T[G][H] \qquad (39)$$

$$\{F_1^*\} = [H]^T[F_1]_s \quad \{F_2^*\} = [H]^T\{F_2\}_s \qquad (40)$$

From eq. (13) may obtain

$$\Pi_2 = \frac{1}{2}\{\delta\}^T[G]\{\delta\} - \{\delta\}^T(\{F_1\} + \{F_2\}) \quad (41)$$

From consistent conditions on r = R, we may obtain

$$\{F_2\} = - \{F_2^*\} \quad (42)$$

Thus, from $\Pi = \Pi_1 + \Pi_2$, we can obtain

$$\Pi = \frac{1}{2}\{\delta\}^T[K]\{\delta\} - \{\delta\}^T\{f\} \quad (43)$$

where

$$[K] = [G] + [G^*]$$

$$\{f\} = \{F_1\} + \{F_1^*\} \quad (44)$$

From variation principle, $\delta\Pi = 0$, we can obtain

$$[K]\{\delta\} = \{f\} \quad (45)$$

This is the stiffness equation of the interaction of underground structures and soil/rock medium, The above method is called spline infinite element-QR methods (SIE-QRM).

5 NONLINEAR PROBLEMS

This paper aims to introduce briefly spline infinite element-QR methods for elasto-plastic analysis of underground structures and soil/rock medium.

5.1 *QR-method*

For underground structures and soil/rock medium in near field (Figure 2), its elasto-plastic stiffness equation may be established by QR-method. From QR-method, we can obtain

$$[G]\{\delta\} = \{F_1\} + \{F_2\} + \{F_1^P\} \quad (46)$$

where $[G]$, $\{F_1\}$ and $\{F_2\}$ are all same with the stiffness matrix and the load vector of elastic problems, they are determined by eq. (14). $\{F_1^P\}$ is the additional load vector arises from the viscoplastic stress, or

$$\{F_1^P\} = \sum_{e=1}^{M}[N]_e^T\{f_1^p\}_e \quad (47)$$

where

$$\{f_1^p\}_e = \int_e [B]_e^T \sigma_0 t dx dy \quad (48)$$

in which $[B]_e$ is the strain matrix of elastic element, t is the thickness of element, σ_0 is the viscoplastic stress vector (Qin Rong, 1995).

5.2 *Spline infinite element methods*

For soil/rock medium in far field (Figure 3), its elasto-plastic stiffness equation may be established by spline infinite element methods. From this method we can obtain

$$[G^*]\{\delta\} = \{F_1^*\} + \{F_2^*\} + \{F_2^P\} \quad (49)$$

where $[G^*]$, $\{F_1^*\}$ and $\{F_2^*\}$ are all same with the elastic problems, they are determined by eq. (39) and eq. (40). $\{F_2^P\}$ is the additional load vector arises from the viscoplastic stress, or

$$\{F_2^P\} = [H]^T\{F_2^P\}_s \quad (50)$$

where

$$\{F_2^P\}_s = 2\int_R^\infty \int_0^{\frac{\pi}{2}} [A]^T \sigma_0 r d\theta dr \quad (51)$$

5.3 *Coupling methods*

From eq. (46), eq. (49) and eq. (42), we may obtain

$$[K]\{\delta\} = \{f\} + \{f^P\} \quad (52)$$

where

$$\{f^P\} = [F_1^P] + \{F_2^P\} \quad (53)$$

Eq. (52) is the elasto-plastic stiffness equation of interaction of underground structures and soil/rock medium. The above method is called spline infinite element-QR methods.

6 DYNAMIC RPOBLEMS

The dynamic equations of interaction of underground structures and soil/rock medium may be established by spline infinite element-QR methods . From this method, we can obtain

$$[M]\{\ddot{U}\} + [C]\{\dot{U}\} + [K]\{U\} = \{P\} \qquad (54)$$

This is the dynamic equation of interactions of underground engineering. where $[M]$, $[C]$and $[K]$are respectively the mass matrix ,damping matrix and stiffness matrix of the underground engineering. $\{\ddot{U}\}$, $\{\dot{U}\}$, $\{U\}$ and $\{P\}$are respectively the acceleration vector, velocity vector, displacement vector and disturbed force vector of the underground engineering, and they are all function of time.

7 NUMERICAL EXAMPLES

This example aims to study the dynamic interaction problems of underground structures with the section of shape of a horse hoof (Figure 4) in semi-infinite elastic medium. This is a spatial problem.

The underground structures: $E_s = 2.1 \times 10^{11} N/m^2$, $\rho_s = 7840 kg/m^3$, $\mu_s = 0.25$, $t = 0.22m$, where t is the thickness of shells.

The medium is the granite: $E_0 = 6.2 \times 10^{10} N/m^2$, $\rho_0 = 2679 kg/m^3$, $\mu_0 = 0.25$. The distance of the point A (Figure 4) to the ground is $2a$.

Figure 5 represents the response of radial displacement at point A. The results are analysed by spline infinite element-QR methods (SIE-QRM). In figure 5, t is time, c_l and G are respectively the velocity of longitudinal wave and shear modulus of medium. From here we know that SIE-QRM is a powerful and economical method.

8 CONCLUSIONS

Spline infinite element-QR methods is a powerful and economical method for analysis of

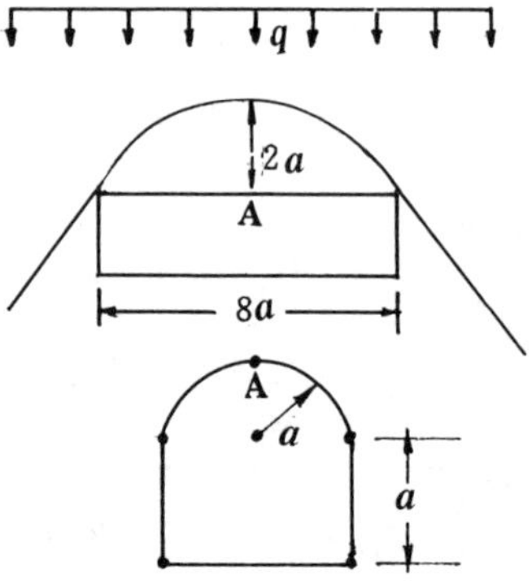

Figure 4 Dynamic interactions of underground engineering

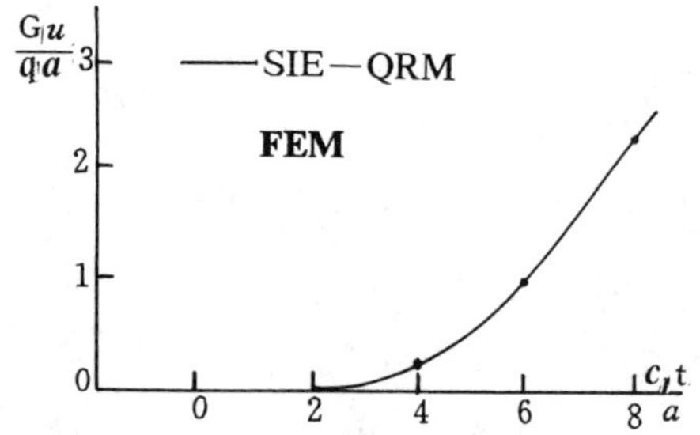

Figure 5 Response of radial displacement at point A.

the underground engineering. This method has wide applicabitity. It can easily be carried out with a micro-computer to analyse the nonlinear problems of underground engineering, such as three-dimensional preblems , plane strain preblems and axisymmetric problems.

REFERENCES

Qin Rong (1985), Spline function methods in structural mechanics, GuangXi press of China.

Qin Rong (1988), Spline boundary element methods, GuangXi press of China, ISBN 7-8065-044-6/O. 4.

Qin Rong (1995), New method for nonlinear analysis of reinforced concrete structures, 《Building For The 21st Century》, Vol. 1, ISBN 0-86857-671-9.

Qin Rong (1994), Spline function methods for elasto-plastic analysis of layered foundations, 《Engineering Mechanics》, ISSN 1000-4750, China.

Computer Methods and Advances in Geomechanics, Yuan (ed.) © 1997 Balkema, Rotterdam, ISBN 90 5410 904 1

A theoretical study on determination of rock pressure acting on shaft-lining by back analysis method

Chou Wanxi & Cheng Hua
Huainan Mining Institute, China

ABSTRACT: On the basis of the analysis of shaft-lining's stress state and displacement state, the author ot this paper proposed the back analysis method for determination of the elliptic load acting on shaft-lining. In this paper described are shaft-lining's stress state and displacement state, presented are some formulas as well as process of calculation of elliptic load and discussed are the application of proposed back analysis method.

1 IN GENERAL

The vertical shaft in the collery has long been known as a troat of the coal mine and it is an important underground structure and access for mining transportation and ventilation. In addition in-spit of frequent shaft-lining failure at home and abroad the shaft-lining problems are very noticeable. A good example is that the in-situ measurements of ground pressure acting on the shaft-lining are carried out by the pre-Soviet Union Mining Construction Research Institute in decade of 60's and chinese higher learning institutions and research and designing institutes in East-China during decade of 70's. Since that time the NATAM method has been accepted by many contries and then on the basis of their works a back analysis method is proposed.

At the end of decade of 80's a comprehensive study on analysis of shaft-lining failure cause is rised in China. Part of those works is the in-situ measurement of shaft-lining's stress which has been carried out under author's direcion, and it is the first example to measure the shaft-lining's stress with sleeve fracturing technique. The work of this paper is a continuation of those works mentioned above and its purpose is to seek a simple, fast and low cost in-situ measurement method of ground pressure acting on shaft-lining.

Th author's back analysis method is proposed on the basis of stress and displacement analysis of shaft-lining. The author hopes that the propsed back analysis method will play an improtant role not only in shaft-lining's research but also in roadway concrete support research.

2 THE STRESSES IN SHAFT-LINING SUBJECTED TO UNEQUAL LOAD

2.1 *Unequal load with elliptic distribution*

It is assum that a shaft-lining subjected to a unequal load, has an inner radius a and an outer radius b, the normal load can be expressed as a formulas used widely:

$$P_\theta = P_0 + P_2 \cos 2\theta \quad (1)$$

where $P_0 = \dfrac{P_{max} + P_{min}}{2}$, $P_2 = \dfrac{P_{max} - P_{min}}{2}$

these formalas show that the normal load can be expressed with maximum normal load component P_{max} and minimum normal load component P_{min}, as shown in Fig. 1. In addition, when the tangential displecement at shaft-lining's outer surface is restrained by the rock wall of the shaft, a shear stress which is directly proportional to the tangential displacement, can be caused. The maximum shear stress is existed at outer periphery which makes an a angle of 45°to the maximum normal load or minimum normal load. The minimum displacement is existed at axes of maximum normal load and minimum normal load. The shear stress at the outer periphery ot shaft-lining can be expressed as

$$\tau = t \sin 2\theta \quad (2)$$

whene τ— maximum shear stress at the outer periphery of the shaft-lining. Analysis ot shaft-lining's stresses shows that the existence of the

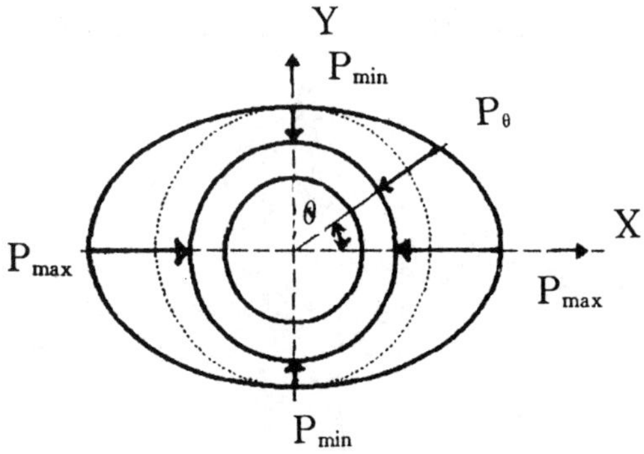

Fig. 1 Elliptic normal load

shear stress τ is favorable for the shaft-lining subjected to the forces. In order to guard , the role of the shear stress can be regleted.

2. 2 *The stress and displacement in shaft-lining*

Based on the theory of elasticity, the stress components in polar coordinate axes can be expressed with Airy stress founction as:

$$\sigma_r=\frac{1}{r}\frac{\partial F}{\partial r}+\frac{1}{r^2}\frac{\partial^2 F}{\partial\theta^2}$$
$$\sigma_\theta=\frac{\partial^2 F}{\partial r^2} \qquad (3)$$
$$\tau_{r\theta}=\frac{\partial^2}{\partial r\partial\theta}\left(\frac{F}{r}\right)$$

where σ_r — radial stress; σ_θ — tangentialstress; $\tau_{r\theta}$ — shear stress; f — Airy stress founction, and

$$F=A_0r^2+B_0\ln r+(A_2r^2+B_2r^4+C_2r^{-2}+D_2)\cos 2\theta \qquad (4)$$

The stresses components existed in shaft-lining are obtained from formulas (3) and (4):

$$\sigma_r=2A_0+B_0r^{-2}-2\cos 2\theta(A_2+3C_2r^{-4}+2D_2r^{-2})$$
$$\sigma_\theta=2A_0-B_0r^{-2}+2\cos 2\theta(A_2+6B_2r^2+3C_2r^{-4})$$
$$\tau_{r\theta}=2\sin 2\theta(A_2+3B_2r^2-3C_2r^{-4}-D_2r^{-2}) \qquad (5)$$

From the boundary condition of the inner surface of shaft-lining $\sigma_r=\tau_{r\theta}=0$ for $r=a$ and $\sigma_r=-P_0-P_2\cos 2\theta$ for $r=b$, the constants in Eq. (5) can be expressed as

$$A_0=\frac{-P_0b^2}{2(b^2-a^2)}$$
$$B_0=\frac{P_0a^2b^2}{b^2-a^2}$$
$$A_2=\frac{-P_2b^2(b^4+a^2b^2+2a^4)}{2(b^2-a^2)^3}$$
$$B_2=\frac{-P_2b^2(3a^2+b^2)}{6(b^2-a^2)^3}$$
$$C_2=\frac{P_2a^4b^4(a^2+3b^2)}{6(b^2-a^2)^3} \qquad (6)$$
$$D_2=\frac{-P_2a^2b^2(a^4+a^2b^2+2a^4)}{2(b^2-a^2)^3}$$

The radial displacement and tangential displacement are determined by the retained condition at the ends of the shaft-lining. It can be considered as plan strain problem when the ends are retained. if there is longituginal stress existet at the ends it can be considered as a plan stress problem.

To determine the above displacements the following expressions are required:

$$\varepsilon_\theta=\frac{u}{r}+\frac{1}{r}\frac{\partial v}{\partial\theta}$$
$$\varepsilon_r=\frac{\partial u}{\partial r}$$
$$r_{r\theta}=\frac{1}{r}\frac{\partial u}{\partial\theta}+\frac{\partial v}{\partial r}-\frac{v}{r} \qquad (7)$$
$$\varepsilon_r=\frac{1}{E}(\sigma_r-\mu\sigma_\theta)$$
$$\varepsilon_\theta=\frac{1}{E}(\sigma_\theta-\mu\sigma_\theta)$$

where ε_r — radial strain; ε_θ — tangential strain; μ — Poisson ratio. From Eq. (5) and (7) the displacements for plan stpess problem can be obtained:

$$U=\frac{r}{E}\{2A_0(1-\mu)-B_0r^{-2}(1+\mu)$$
$$-2\cos 2\theta[A_2(1+\mu)+2B_2r^2\mu-C_2r^{-4}(1+\mu)$$
$$-2D_2r^{-2}]\} \qquad (8)$$
$$V=\frac{2r\sin 2\theta}{E}[A_2(1+\mu)+B_2r^2(3+\mu)$$
$$+C_2r^{-4}(1+\mu)+D_2r^{-2}(\mu-1)]$$

For plan strain problem, the displacements in shaft-lining are obtained:

$$U=\frac{r}{E}\{2A_0(1-\mu-2\mu^2)-B_0r^2(1+\mu)$$
$$-2\cos 2\theta[A_2(1+\mu)+2B_2r^2\mu(1+\mu)$$
$$-C_2r^{-4}(1+\mu)-2D_2r^{-2}]\} \qquad (9)$$
$$V=\frac{2r(1+\mu)}{E}\sin 2\theta[A_2+B_2r^2(3-2\mu)$$
$$+C_2r^{-4}-D_2r^{-2}]$$

It is difficult to use Eg. (5) directly for strength checking computation, since the selection of dangerous poimt in shaft-lining is not specified and the expression is too enormous to use. The strength checking computation will be very simple. if the size is chosed. Substituting the constants expressed with Eg. (6) into eq. (5) and let $r=b$ the tangential stress in shaft-ling can be obtained:

$$\sigma_r=-P_0-P_2\cos 2\theta$$

$$\sigma_\theta=\frac{1}{m-1}[\cos 2\theta\times \times\frac{-P_(m^2+6m+1)+2t(2m-m^2+1)}{m-1} -P_0(m+1)] \quad (10)$$

$$\tau_{r\theta}=-t\sin 2\theta$$

If let $r=a$ the stresses at inner periphery of shaft-lining can be expressed as

$$\sigma_r=0$$

$$\sigma_0=\frac{2m}{m-n}[2\cos 2\theta\frac{P_2(m+1)-t}{m-1}-P_0] \quad (11)$$

$$\tau_{r\theta}=0$$

where $m=(\frac{b}{a})^2$

The tangential stress σ_θ has extreme value at points of θ equal 0°, 90°, 180° and 270°. For θ equal 0° and 180° at inner surface of shaft-lining the maximum value is apeared; for θ equal 90° and 270°- the minimum value.

3 BACK ANALYSIS METHOD FOR DETERMINATION OF ELLIPTIC LOAD

3.1 *Stress back analysis*

Proposed back analysis method for determination of elliptic load acting on shaft-lining can be divided in two kinds: stress back analysis method and displacement back amalysis method, those methods are suitable for elliptic load distrubution. What is called stress back analysis method is means that the stress back analysis using the stress exprssion of shaft-lining's inner surface is based on measured tangential stress at 3 points with angular separation of 45°.

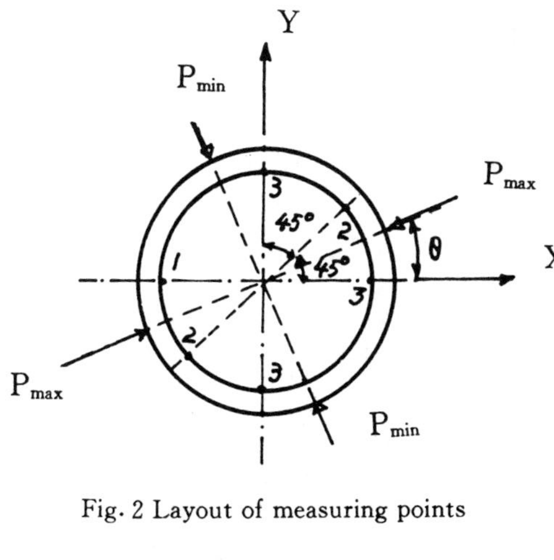

Fig. 2 Layout of measuring points

Hence, it is require that Eq. (11) is simplified as

$$\sigma_\theta=\alpha P_0+\beta P_2\cos 2\theta \quad (12)$$

where $\alpha=-\frac{2m}{m-1}$, $\beta=\frac{4(m+1)m}{(m-1)^2}$

$m=(\frac{b}{a})^2$

If the tangential stresses σ'_θ, $\sigma''_\theta$and σ'''_θ at 3 respectively points dispersed with angular separation of 45° along inner periphery of shaft-lining bave been measured, 3 simaltaneous equations may be given as

$$\sigma'_\theta=\alpha P_0+\beta P_2\cos 2\theta$$
$$\sigma''_\theta=\alpha P_0+\beta P_2\cos 2(\theta+45°) \quad (12)$$
$$\sigma'''_\theta=\alpha P_0+\beta P_2\cos 2(\theta+90°)$$

Solving the equation set (12) leads to

$$P_0=\frac{\sigma'_\theta+\sigma'''_\theta}{2\alpha}$$

$$P_2=\frac{1}{2\beta}\sqrt{[\sigma'_\theta-(\sigma'_\theta+\sigma'''_\theta)]^2+[2\sigma''_\theta-(\sigma'_\theta+\sigma'''_\theta)]^2} \quad (13)$$

$$\tan\theta=\frac{2\sigma''-(\sigma'_\theta+\sigma'''_\theta)}{2\sigma'_\theta+(\sigma'_\theta+\sigma'''_\theta)}$$

Above formalas show that each component of elliptic load and maximum load direction may be found.

3.2 *Displacement back analysis method*

The suitable condition of the displacement back anlysis method is the same as stress analysis method. its difference is that the back analysis is based on test value of radial displacement of shaft-lining. Similarly, substituting Eq. (6) into first equation of Eq. (8) and let $r=a$, the radial displacement of inner periphery of shaft-lining can be expressed as

$$U=\alpha P_0-\beta P_2\cos 2\theta \quad (14)$$

where $\alpha=\frac{a}{E}[2F(1+\mu)-Ga^{-2}(1+\mu)]$

$$\beta=\frac{a}{E}[2H(1+\mu)+2I\mu-Ja^{-1}(1+\mu) -2Ka^{-2}]$$

where $F=\frac{-b^2}{2(b^2-a^2)}$

$$G=\frac{a^2b^2}{b^2-a^2}$$

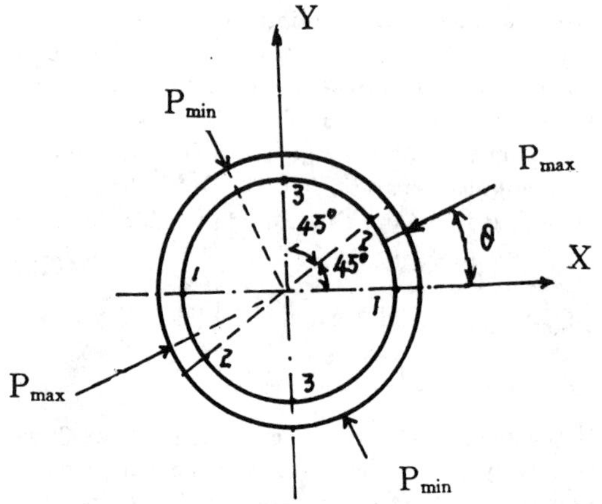

Fig. 3 Layout of displacement measuring points

$$
\begin{aligned}
H&=\frac{b^2(b^4+a^2b^2+2a^4)}{2(b^2-a^2)^3}\\
J&=\frac{-b^2(3a^2+b^2)}{6(b^2-a^2)^3}\\
J&=\frac{a^2+3b^2}{6(b^2-a^2)^3}\\
K&=-\frac{a^2b^2(a^4+a^2b^2+2b^4)}{2(b^2-a^2)^3}
\end{aligned}
\tag{15}
$$

In case the displacements U_1, U_2 and U_3 along 3 radial directions respectively at inner periphery of shaft-lining are determined, the radial displacement expression may be found as (see Fig. 3).

$$
\begin{aligned}
U_1&=\alpha P_{\partial}-\beta P_2\cos 2\theta\\
U_2&=\alpha P_0-\beta P_2\cos 2(\theta+45°)\\
U_3&=\alpha P_0-\beta P_2\cos 2(\theta+90°)
\end{aligned}
\tag{16}
$$

Solving the Eq. (16) leads to

$$P_0=\frac{U_1+U_2}{2\alpha}$$

$$P_2=\frac{1}{2\beta}\times$$

$$\times\sqrt{[2U_1-(U_1+U_3)]^2+[2U_2-(U_1+U_3)]^2}$$

$$\tan\theta=\frac{2U_2-(U_1+U_3)}{2U_1-(U_1+U_3)}$$

4 DUSCUSSION

The elliptic load distrubution adopted in this paper as computational load is based on such cause: the elliptic load is most harmful in comparison with other, hence the strength reserve of shaft-lining can be increased; comparatively speaking the elliptical load is more approximate to practic and is widely used. It is suitable that the tangetial stress value of shaft-lining's inner surface is measured with sleeve fracturing technique. Since obtained ralue from these technique is comparatively reriable. Author's back analysis method is saitable for circular roadway concrete support except shaft-lining.

REFERENCE

Chou wanxi, Determination of ground pressure in a vertical shaft by back analysis. Journal of China Coal Society. 1989, No. 3. 37—44.

Chou wanxi, Wei Shanbin, Zang Desheng, Measurement and analysis of shaft-lining stress, Chinese Journal of Geotechnical Engineering 1995. No. 1, 61—65.

W. Flügge. Handbood of Engineering Mechanics, McGRAW-HILL BOOK COMPANY INC, 1962. 37—11—38—4.

G. A. Klupenikov et al. Interact of Rockmass with shaft-lining, Moscow, Nedla, 1966, 268—281.

Author index